Advances in Intelligent Systems and Computing

Volume 868

Series editor

Janusz Kacprzyk, Polish Academy of Sciences, Warsaw, Poland
e-mail: kacprzyk@ibspan.waw.pl

The series "Advances in Intelligent Systems and Computing" contains publications on theory, applications, and design methods of Intelligent Systems and Intelligent Computing. Virtually all disciplines such as engineering, natural sciences, computer and information science, ICT, economics, business, e-commerce, environment, healthcare, life science are covered. The list of topics spans all the areas of modern intelligent systems and computing such as: computational intelligence, soft computing including neural networks, fuzzy systems, evolutionary computing and the fusion of these paradigms, social intelligence, ambient intelligence, computational neuroscience, artificial life, virtual worlds and society, cognitive science and systems, Perception and Vision, DNA and immune based systems, self-organizing and adaptive systems, e-Learning and teaching, human-centered and human-centric computing, recommender systems, intelligent control, robotics and mechatronics including human-machine teaming, knowledge-based paradigms, learning paradigms, machine ethics, intelligent data analysis, knowledge management, intelligent agents, intelligent decision making and support, intelligent network security, trust management, interactive entertainment, Web intelligence and multimedia.

The publications within "Advances in Intelligent Systems and Computing" are primarily proceedings of important conferences, symposia and congresses. They cover significant recent developments in the field, both of a foundational and applicable character. An important characteristic feature of the series is the short publication time and world-wide distribution. This permits a rapid and broad dissemination of research results.

More information about this series at http://www.springer.com/series/11156

Kohei Arai · Supriya Kapoor
Rahul Bhatia
Editors

Intelligent Systems and Applications

Proceedings of the 2018 Intelligent Systems Conference (IntelliSys) Volume 1

Set 1

Springer

Editors
Kohei Arai
Faculty of Science and Engineering
Saga University
Saga, Japan

Rahul Bhatia
The Science and Information
(SAI) Organization
Bradford, UK

Supriya Kapoor
The Science and Information
(SAI) Organization
Bradford, UK

ISSN 2194-5357 ISSN 2194-5365 (electronic)
Advances in Intelligent Systems and Computing
ISBN 978-3-030-01053-9 ISBN 978-3-030-01054-6 (eBook)
https://doi.org/10.1007/978-3-030-01054-6

Library of Congress Control Number: 2018955283

This Springer imprint is published by the registered company Springer Nature Switzerland AG
The registered company address is: Gewerbestrasse 11, 6330 Cham, Switzerland

Editor's Preface

Welcome to the Intelligent Systems Conference (IntelliSys) 2018 which is held on September 6–7, 2018, in London, UK.

The technology nowadays has reached the point where intelligent systems are replacing human intelligence in aiding in the solution of very complex problems as well as in decision-making processes. In many cases, intelligent systems have already outperformed human activities. Several directions took place in this field of computational intelligence. Massive access and use of intelligent systems in everyday applications have created the need for such an international conference which serves as a venue to report on up-to-the-minute innovations and developments.

IntelliSys 2018 provided a setting for discussing a wide variety of topics including deep learning, neural networks, image/video processing, intelligent transportation, artificial intelligence, robotics, data mining, smart health care, natural language processing, ambient intelligence, machine vision, and the Internet of things. This two-day conference program covers four keynote talks, contributed papers, special sessions, poster presentations, workshops, and tutorials on theory and practice, technologies, and systems.

IntelliSys 2018 has attracted 568 submissions from 50+ countries. After the double-blind review process, we finally selected 194 full papers including 13 poster papers to publish. The conference, IntelliSys 2018, has a wide range of featured talks including keynotes which provide visions and insights into future research directions and trends.

We would like to express our deep appreciation to the support of many people: authors, presenters, participants, keynote speakers, session chairs, volunteers, program committee members, steering committee members, and people in other various roles. We would also like to express our sincere gratitude to all their valuable suggestions, advice, dedicated commitment, and hard work.

We are looking forward to our upcoming Intelligent Systems Conference that will be held on 2019 at the same location. We hope that it will be as interesting and enjoyable as it has been in all of its three predecessors.

Hope you find IntelliSys 2018 in London both enjoyable and valuable experience!

Kind Regards,

Kohei Arai
Conference Chair

Contents

ViZDoom: DRQN with Prioritized Experience Replay, Double-Q Learning and Snapshot Ensembling

Christopher Schulze(✉) and Marcus Schulze

Austin, USA
cschulze7@gmail.com, mschulze765@gmail.com

Abstract. ViZDoom is a robust, first-person shooter reinforcement learning environment, characterized by a significant degree of latent state information. In this paper, double-Q learning and prioritized experience replay methods are tested under a certain ViZDoom combat scenario using a competitive deep recurrent Q-network (DRQN) architecture. In addition, an ensembling technique known as snapshot ensembling is employed using a specific annealed learning rate to observe differences in ensembling efficacy under these two methods. Annealed learning rates are important in general to the training of deep neural network models, as they shake up the status-quo and counter a model's tending towards local optima. While both variants show performance exceeding those of built-in AI agents of the game, the known stabilizing effects of double-Q learning are illustrated, and priority experience replay is again validated in its usefulness by showing immediate results early on in agent development, with the caveat that value overestimation is accelerated in this case. In addition, some unique behaviors are observed to develop for priority experience replay (PER) and double-Q (DDQ) variants, and snapshot ensembling of both PER and DDQ proves a valuable method for improving performance of the ViZDoom Marine.

Keywords: Reinforcement learning · Priority experience replay
Double-q learning · Deep learning · Recurrent neural networks
Snapshot ensembling

1 Introduction

Increasingly, deep reinforcement learning (DRL) is the topic of great discussion in the artificial intelligence community. For reinforcement learning (RL) experts, RL has always shown great promise as a robust construct for solving task oriented problems. Recently, the DRL variant has gained significant public attention. But to be clear, much remains to be done at large in the field concerning high state-action dimensional problems that approximate well some real world problems

This work was not supported by any organization.

K. Arai et al. (Eds.): IntelliSys 2018, AISC 868, pp. 1–17, 2019.
https://doi.org/10.1007/978-3-030-01054-6_1

of interest. From board games like Go in recent engagements with agents from Deep Mind to very high state-action dimensional games like those from the real time strategy (RTS) and first person shooter (FPS) video game genres, DRL has established itself, presently, as a prime construct for solving highly complex problems without direct supervision for most, if not all, of its training.

Given the success, much work has been done recently to improve upon models of DRL, including that of the Deep Q-Network (DQN). Priority experience replay (PER) and Double-Q learning (DDQ) are two such methods that can be used to improve a DQN agent's rate of improvement or degree of learning stability, respectively, during training. However, to the authors knowledge, these two methods have yet to be tested under the ViZDoom [8] environment - a setting characterized by a higher degree of latent information relative to Atari and other popular RL environments - in the context of a specific, effective deep recurrent learning architecture. Replicating the DRQN structure in the paper from Lample et al. [5], the authors in this paper test the benefits offered by PER and DDQ methods under an efficient ensembling method (Fig. 1).

Fig. 1. Defend the center scenario: melee enemies converge towards the center.

2 Deep Reinforcement Learning Overview

2.1 Reinforcement Learning, Q-Learning, and Deep Q-Networks

The field of reinforcement learning frames a learning problems using two separate entities: an agent and an environment. As Sutton discusses [10], it is useful to conceptualize this framework as follows: an agent is defined as an actor attempting to learn a given task. The agent is delineated from the environment by defining its characteristics and action set as items encompassed by its realm of influence and under its complete control; all other aspect of the learning problem are then attributed to the environment. Generally, the environment can be described as the setting in which the learning problem takes place. The agent receives its current state s_t from the environment, and it proceeds to interact and affect the environment through action a_t, which is derived from a policy π_t. The environment takes in action a_t and updates the current state of the agent to s_{t+1} along with sending a reward signal r_t to the agent, indicating the value of s_t to s_{t+1} transition via action a_t. Given the RL framework, we can define discounted rewards at time $i = 0$ as:

$$R = \sum_{i=0}^{T} \gamma^i r_i, \tag{1}$$

where $\gamma \in [0,1)$. Rewards are discounted to simulate the concept of delayed rewards to the agent, the importance of the discounted being directed by the value of γ. γ values close to 0 indicate immediate rewards are more important; whereas γ values close to 1 indicate to the agent that longer of sequences of actions are important to consider to achieve high rewards. The goal of the agent is to maximize its rewards as seen by the environment over the course of a single experience or series of experiences (i.e. games) by developing a policy. This policy dictates what action an agent performs given its current state.

To obtain an approximation of optimal values corresponding to state-action pairs, the temporal difference method [11] Q-learning, first developed by Watkins [3] in 1989, can be used. The Q-function for a given state-action pair with policy π is as follows:

$$Q^\pi(s, a) = E[R_i | s_i = s, a_i = a] \tag{2}$$

The task is then to find the optimal policy, giving:

$$Q^*(s, a) = max_\pi(Q^\pi) = max_\pi(E[R_i | s_i = s, a_i = a]) \tag{3}$$

Thus, Bellman optimality is then satisfied by the optimal Q-function, as the above can be rewritten as:

$$Q^*(s, a) = max_\pi(Q^\pi) = E[r + \gamma * max_{a'}(Q^*(s', a'))|s, a] \tag{4}$$

For most problems - with large state-action spaces - direct, recursive methods, including dynamic programming methods, are impractical. Rather, the optimal Q-function is approximated by a parametrized method. In particular, deep learning architectures have proven very successful at approximating the optimal value of high-dimensional target spaces. Let Q_w be defined as a Q-function with parameters w. For updates to the Q_w function, the loss can be defined as:

$$L_t(w_t) = E_{s,a,r,s'}[l_t(y_t - Q_w(s, a))], \tag{5}$$

where l_t is any reasonable transformation of the temporal difference error for training deep neural networks, including L1 (proportional to mean absolute value error) and L2 (proportional to mean squared error) losses, among others, and

$$y_t = r + \gamma * max_a(Q_w(s', a))$$

Given successes with stochastic gradient descent updates, we can instead lose the expectation and give stochastic updates to Q_w using the following as loss for backpropagation:

$$L_t(w_t) = l_t(y_t - Q_w(s, a)) \tag{6}$$

To gain experiences in the environment, ϵ-greedy training can be used, wherein the agent randomly acts at with probability ϵ or choses what it deems its best action with probability $1 - \epsilon$. Using this strategy, ϵ is decayed over the course of training, usually starting with a value of 1 and having a minimum value of 0.1.

To stabilize Q-values during learning, a replay memory is used to remember s, a, r, s' experiences as the agent interacts with the environment; the agent then learns by sampling from this replay memory uniformly after a set number of games played and fitting against the selected experiences. The replay memory along with a target Q-function were introduced to counter the algorithms strong affinity to local optima given these greedy updates.

Enter the modern framework for Deep Q-Networks in full. Deep Q-Networks (DQNs) are a parametrized, model-free, off-policy method; specifically, it uses Q-learning for value estimation. Two deep neural network architectures are used to learn the Q-values for each experience sampled from replay memory. An online network Q_w learns in a greedy fashion, and the target network $\tilde{Q}_w$ acts as a tether point to reduce the likelihood of the online network from falling into a local optima due to its greedy updates.

Let $Q(s, a)$ be defined as the online network, and $\tilde{Q}(s, a)$ be defined as the target network. During training using replay memory after a number of games have been played, the online network is updated using the following target:

$$Q_{target}(s, a) = r + \gamma max_a(\tilde{Q}(s', a)) \tag{7}$$

2.2 Deep Recurrent Q-Networks

A variant of DQN, deep recurrent Q-Networks (DRQNs) have shown exceptional performance in a variety of RL problems. In environments characterized by significant latent information, recurrency offers a way to fill in the gaps of missing knowledge for a given state [7]. The RL environment of ViZDoom is no exception; as a first person shooter, the agent is bound by a first-person, 90-degrees view of objects in front of it. There is no radar - a HUD display of enemies around the player in a 360° arc is present in many other FPS games - or other indicators of what could be in the other 270°.

As such, rather than receiving states s_t as in the RL framework, it is more aptly put that the agent in ViZDoom receives partial observations o_t of its current state s_t [7]. Thus, using a DQN, the objective is to estimate not $Q(s_t, a_t)$ but instead $Q(o_t, a_t)$, putting the agent at a distinct disadvantage concerning state level information. One way to counter this is to include state information from the previous sequence of states, thereby allowing the agent to instead estimate $Q(o_t, h_{t-1}, a_t)$ where h_{t-1} represents information gathered at state $t-1$ and passed to state t. The long-term, short-term memory cell, LSTM, is one recurrent construct capable of doing this; at a given time t, the LSTM cell takes in o_t and h_{t-1} and outputs h_t. Instead of $Q(o_t, a_t)$, the network then estimates $Q(h_t, a_t)$, increasing the level of information available to the agent for a given state-action pair.

2.3 Double-Q Learning

Double-Q learning advances upon DQN in a simple, yet remarkably effective way: let the loss used in learning by the agent be defined by the value-maximizing actions of the online network with the Q-values of the target network associated with those maximizing actions. With the decoupling of the maximizing action from its value, it is possible to eliminate the maximization bias present in Q-Learning [6]. Specifically, the loss is defined as follows:

$$L_t(w_t) = l_t(y_t - Q_w(s, a)), \tag{8}$$

where l_t is a transformation of the temporal difference (TD) error and

$$y_t = r + \gamma \tilde{Q}_w(s, a_{online}))$$

$$a_{online} = argmax_a(Q(s, a))$$

This innovation was spurred by the observation that Q-learning tends to overestimate the values of state-actions pairs. Through experimentation of deep double-Q learning using myriad, diverse Atari game environments, Hasselt et al. find that deep double Q-networks (DDQNs) show marked improvement in the estimation of values for state-action pairs [14]. Furthermore, Double-Q learning introduces a level of stability in Q-learning updates, which allows for more complex behavior to be learned.

2.4 Prioritized Experience Replay

During training, the DQN agent samples uniformly from its replay memory to populate a batch training set and subsequently learn on this set. This is done many times over the course of the experiment, allowing the agent to continue to learn from previous experiences. The construct of replay memory was created to simulate learning from experiences that are sampled i.i.d (independently and identically distributed). Without approximating i.i.d sampling, the agent can quickly overfit to recent state-action pairs, inhibiting learning and adaptation. The method of prioritized experience replay innovates on this front by biasing the sampling of experiences [12]. Specifically, experiences are weighted and sampled with higher probability according to the TD error observed for that sample - the larger the TD error, the higher the probability with which a given experience will be sampled. The intuition behind this is that experiences that significantly differ from the agents expectation will have more didactic potential. Concerning empirical validation of the method, Schaul et al. created PER and showed that it affords significant performance improvement across many Atari-based RL environments. The authors here apply PER to a DRQN model and train and test it in the defend-the-center ViZDoom scenario (see Sect. 4.1. Scenario section).

2.5 Ensembles of Deep Q-Networks: Snapshot Ensembles

Ensembling of models has proven useful in many situations, countering, to an extent, the tendency of nonlinear models to over-fit to a given training distribution. However, the generation of ensembles for deep neural networks can prove onerous, requiring the training of multiple networks in parallel or - even worse, in terms of total training time - sequentially. Recently, methods have been proposed to gain the advantages of ensembling while reducing the time to generate such a set of models. Snapshot ensembling is one such example, employing a cosine annealing learning rate to generate M models over T total epochs from the training of a single model over those T epochs [4]. This is done by using the cosine annealing learning rate to train the single model and take snapshots of its current weights every T/M epochs. Thus, only a single model is trained while, at the same time, providing a diverse model population for the ensemble through use of the cosine annealing learning rate.

The authors here use the snapshot ensemble method to analyze performance improvement of the aforementioned model (DRQN) in the context of the ViZDoom Reinforcement Learning Environment, utilizing the learning enhancement methods of PER and DDQ.

2.6 Review of Modern RL Game Environments

For Reinforcement Learning, Why Video Games?

A burgeoning field of work has been created using the VizDoom environment due to the unique problems that the 3-D first-person shooter (FPS) engine can provide. Navigation, recognition of objects in the space, and decision making when other objects or actors are encountered are all obstacles that an FPS environment presents. Another important obstacle is that the agent will never have a complete view of the given state space. Recent deep reinforcement learning methods applied to this environment have focused primarily on the free-for-all death match setting to train their models. Instead, the authors here choose to train and test models using the Defend the Center mode.

Including ViZDoom, there are a number of diverse game-based RL frameworks in current use. In general, video games have increasingly become an important tool for AI research and development. It is easy to see why, as video games allow for an environment rich with parameters, feedback, and end goals for agents to gauge their success. Originally, the frameworks tested were simple 2-D environments from games originally on the Atari 2600 that allowed for simple movement and near fully observable states. On the other hand, the engines for FPS environments provide a wealth of data and multiple obstacles and objectives for the AI agents to learn from. One of the most important obstacles provided by the FPS setup is the lack of complete information for a given state. This obstacle mimics real world problems for autonomous agents, as they must be able to act with a limited view of the world around them.

Arcade Learning Environments

The classic 2D video games of the past have been used to a great extent for deep learning. The main platforms for these games are the Atari 2600, Nintendo NES, Commodore 64 and ZX Spectrum. One of the most used emulator environments is Stella, which uses the Atari 2600 and has 50 games available. Methods previously used in these environments include Double DQN, Bootstrapped DQN, Dueling DQN, and Prioritized DQN. Montezuma's Revenge is a notable game from this genre as it requires memorization of other rooms that aren't immediately available to the agent. Methods used in this space are DQN-PixelCNN, DQN-CTS, and H-DQN [2,9].

Real Time Strategy

StarCraft: Brood War is a popular real time strategy (RTS) game in which the objective is to destroy all of the enemy structures and claim victory. Players move towards victory by gathering resources to create attacking units that can then be directed in various actions. The obstacles for the agent are many as the state space is complex and not fully observable by the agent at any one time. There are three factions available in this environment that all have their own unique characters and abilities. Even if the agent is limited to learning to play only a single faction, there are still 3 different match ups that could be encountered. Each of these match ups will have their own strategies and units that will be needed to counter different compositions of units built from different structures.

Another important skill that the agent needs to learn is intelligence gathering - understanding of both map layout and composition of the opponent's army. At the start of the game, the map is under what is called "fog of war", which blacks out any area that doesn't have a controlled unit to provide vision. In summary, the agent must navigate the environment and gather intelligence on the opposing player, build an appropriate base from which to train units to defend and attack the opponent, maneuver these units into advantageous positions in the environment, engage opponent units and manage the abilities available to units, and manage resources. All of these tasks must also be performed without complete knowledge of the state space. The combination of the many problems experienced in the course of one game has led deep learning researchers to focus on specific problems in the game as the sparsity of rewards makes the training of highly non-linear functions, such as neural networks, difficult. The main problem that has been the focus of many researchers has been the micromanagement of units in combat scenarios. Methods such as IQL, COMA, Zero Order, and BiCNet have been performed with promising results [9].

Open World Games

Open world games are another avenue of research, as the nature of open world games positions issues of exploration and objective setting as the main challenges to the agent. Project Malmo is an overlay built onto the Minecraft engine that can be used to define the environment and allow for objectives to be imposed on an agent in a normally free task environment. These large and open problems are commonly used to test reinforcement learning methods. Methods such as H-DLRN and variations of NTMs have been successful in preforming varying tasks in the space [9].

Racing Games

Racing games are another popular genre for AI research as there are many challenges here as well. Depending on the game chosen, the inputs for control can be as complex as having a gear stick, clutch and handbrakes, while others are much more simplified. Challenges for this environment include positioning of the vehicle on the course for optimal distance traveled, adversarial actions to block or impede other drivers when other vehicles are present, and sometimes the management of resources. This genre is also useful because the entire state is not available to the agent in the first or third person view. Methods that have been used in this genre include Direct Perception, Deep DPG and A3C, with a popular environment for this genre being the simulator TORCS [9].

First-Person Shooter

VizDoom is based off of the popular video game Doom and is a first person shooter. In this environment there are many challenges that make it a useful tool for training AI. There are many modes of play within the VizDoom environment, two of which are mentioned here: Defend the Center and Deathmatch. In Defend the Center, the agent is spawned in the center of a circular room and is limited to turning left or right to find and eliminate the enemies that spawn. At most, only 5 melee enemies are present on the map at any one time. These are spawned against the wall; they then make their way towards the agent in the center. As enemies are eliminated, others are spawned to take their place. The objective is to hold out for as long as possible with a limited amount of ammunition. Having limited ammo helps to constrain episodes to a limited time limit, as when ammo is depleted, defeat is inevitable. In Deathmatch the objective is to reach a certain amount of frags (i.e. kills) before other actors on the map or to have the most frags when time runs out. The view of the agent is determined by the resolution of the screen chosen and varies between 90° and 110°, meaning that the agent doesn't have complete knowledge of its full state at any time. Other obstacles include the recognition of enemies in the space, aiming weapons, and navigation

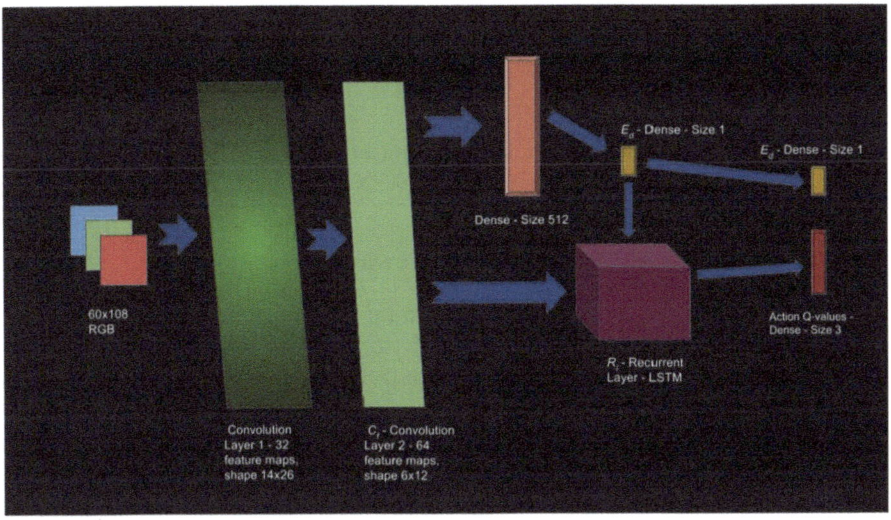

Fig. 2. DRQN architecture.

of space. Methods that have been used in this environment include DQN+SLAM, DFP, DRQN+Auxiliary Learning, and A3C+Curriculum Learning [8,9].

3 Model Architecture and Agent Training

3.1 Deep Neural Network Architecture

The authors here use the DRQN model architecture (Fig. 2) specified by Lample et al. [5]. This model architecture was used in all experiments, as it was noteworthy in its performance without the additions of PER and DDQ; thus the authors here endeavored to use PER and DDQ to observe how such variants can aid a deep learning architecture that is well suited to this RL problem. A general description of the architecture is as follows:

- Input to the network is a single frame, i.e. the 3 channels (RGB) of a frame, with each frame being resized to 60×108 pixels.
- Input is sent through two convolutional layers.
- The output of the final convolutional layer C_f is flattened and sent to two destinations.
- Output of C_f is sent to a dense layer and then to a subsequent dense layer E_d with sigmoid activation of size 1. This is sent to a recurrent layer R_l and is also used for loss.
- Output of C_f is flattened and then sent directly to the recurrent layer R_l.
- R_l then outputs to a dense layer with ReLu activation and then to a subsequent dense layer with linear activation of size three, as there are three actions that can be performed at a given time (turn left, turn right, and shoot) in the Defend-the-Center scenario.

The size-1 dense layer is a boolean indicator for enemy detection and is fitted by querying the game engine for enemies in the agents vision; if there are any enemies in the agents view, the true value is 1, otherwise it is 0. The feeding of predicted enemy detection information into the recurrent layer was found by Lample et al. to be very beneficial in training in a related ViZDoom RL scenario known as Deathmatch, where agents engage in a competitive, FPS match.

3.2 Frame Skipping, Fitting, and Reward Structure

Frame skipping is quite crucial in the training of RL agents that use video data with a reasonably fast frames-per-second (fps) value [10]. For example, using a frame-skip value of 2, the agent will make an action on frame 0, f_0. That same action will be performed for the skipped frames, f_1 and f_2. There are many reasons for using frame-skipping, one of the main benefits being that it prevents the agent from populating the replay memory with experiences that differ almost imperceptibly from one another. Without frame-skipping, this can pose a serious issue for training DRQNs, as the elements of sequences that are sampled for training will be practically the same, often causing the agent to learn degenerate, loop-like behavior where it performs a single action over and over. For all experiments, the ViZDoom engine was run at 35 fps. The authors here experimented initially with various frame skip values, opting for a frame-skip value of approximately 10 for reported results.

For training of a DRQN, sequential updates are important in order to take advantage of the recurrent layer. Here, the authors sampled sequences of length 7, used the first four experiences as primer for training - passing LSTM cell state information sequentially from the previous experience to the next - and trained on the final three experiences.

Agents were trained for a total of 11500 games in the defend-the-center scenario, using ϵ-greedy training. Here, ϵ was decayed from 1.0 to 0.1 over the course of training. For the three experiments without snapshot ensembling, a linearly decaying, cosine annealed (small amplitude) learning rate was used. For the other three using snapshot ensembling, a cosine annealed learning rate was used. Instead of using Adam, which combines classical momentum with the norm-based benefits of RMSProp, all models were optimized using the Nesterov Adam optimizer, as this swaps out classical momentum for an improved momentum formulation, Nesterov's accelerated gradient (NAG) [13]. In terms of reward structure, the agent was given a reward during training of $+1$ for frags of enemy units, and a penalty of -0.5 for deaths. A penalty proportional to the amount of health lost was also included.

3.3 Hyperparameters

Hyperparameter values for all experiments are as follows:

- γ value of 0.9.

- For PER, a β_0 value of 0.5 was used, and β was increased to 1.0 linearly over the course of each experiment.
- For PER, α value of 0.7.
- For PER, ϵ offset of 0.05 for non-singular priority values.
- A batch size of 20 with sequence length 7 was used for replay memory training.

4 Experiments

4.1 Scenario

Previous literature concerning ViZDoom has focused on using the Deathmatch scenario to train their agents in navigation, combat, and enemy recognition. The authors here focus on the problem of spatial recognition as opposed to navigation. In order to emphasize this, the game mode Defend the Center available in the Vizdoom engine was chosen. This goal of this game type is to frag as many adversaries as possible before you are overrun. The agent is allowed no movement in the center of this circular arena other than the adjustment of its angle of view. With this limitation, the authors here aim to have agents learn spatial awareness and to prioritize the targets based on distance from the agent. As a review of Sect. 2.6, in the Defend the Center scenario, the agent is spawned in the center of a circular room and is limited to altering the degree of its view to find and eliminate the enemies that spawn. At most only 5 melee enemies are spawned against the wall (Fig. 3) that will then make their way towards the agent in the center. The objective is to hold out for as long as possible with a limited amount of ammunition.

Fig. 3. Defend the Center Scenario: enemies spawn at a distance from the agent.

4.2 Software and Hardware

Six experiments in total were performed, two sets of 3 experiments each. The first set was to test the base DRQN and PER and DDQ variants without snapshot ensembling, while the second set employed the use of snapshot ensembling. In both cases, the authors here use the proportional PER version, rather than rank PER. In all experiments, agents were trained for approximately 12 h for a total of 11500 games of the ViZDoom defend-the-center scenario on an NVIDIA GeForce Titan X GPU using Python 3.7 and the neural network library Keras, with Tensorflow backend and the NVIDIA CUDA Deep Neural Network library (cuDNN).

4.3 Results

The aim of these experiments is to test the learning benefits offered by DDQ and PER - relative to the baseline DRQN architecture that was formulated by Lample et al. and reproduced here - in the context of ViZDoom, an environment where a great deal of state information is hidden from the agent at each time step. Specifically, the authors here studied effects of DDQ and PER on the early phases of learning, using a sizeable frame-skip value. As stated above, a frame-skip value of approximately 10 was used for training of all agents. Given that most research using ViZDoom has studied RL agents using much smaller values of frame-skip (around 5), this work allows a look at RL agent learning rate improvements at higher frame-skip values in the context of this FPS environment. As noted by Braylan et al., using larger frame-skip values greatly accelerates learning [1]. Such behavior was also noted by authors here; larger frame-skips led to significant accelerations in learning rate, albeit at the cost of the agents precision. With larger frame-skip values, the agent can overshoot an enemy when turning toward the target, decreasing the ability of the agent to center on target before shooting. At the other end of the spectrum, too few frame-skips can cause the agent to learn degenerate, repetitive behavior - as noted earlier in this paper - such as assigning higher value, irrespective of input, to the action with the highest variance in value.

For testing, snapshots of each agent - base DRQN, DRQN with DDQ, and DRQN with PER - were taken at 100 game intervals over the course of the full 11500 games of training. These snapshots were then tested in defend-the-center games and, as with training, tasked with gaining as many frags as possible. Specifically, each snapshot was given 100 games to accumulate frags. In addition, three sets of five models each were used in creating snapshot ensembles - one ensemble each for the base DRQN, DRQN with DDQ, and DRQN with PER agent types.

Given the higher frame-skip used, the authors, unsurprisingly, observed that in all cases the performance of agents plateaued after roughly 6000 out of the 11500 games. Given that the subject of interest here is the rate of learning in early development of the agent, this is not an issue. However, it does serve as a basis for further research. In particular, one might ask what the upper limit of performance would be at smaller frame-skip values using the aforementioned DRQN architecture with the addition of either PER or DDQ. This is one of a set of future targets for further investigation.

Concerning metrics of performance, cross entropy loss and K/D ratio are used to measure agents success at various time points in development. Cross entropy loss was used to measure the agent's recognition of an enemy or enemies

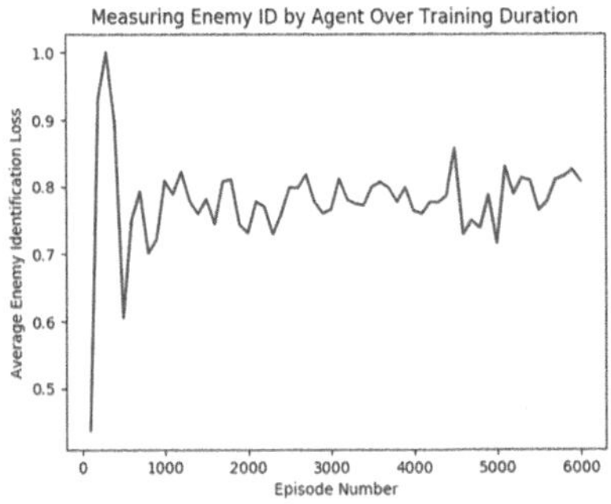

(a) DRQN

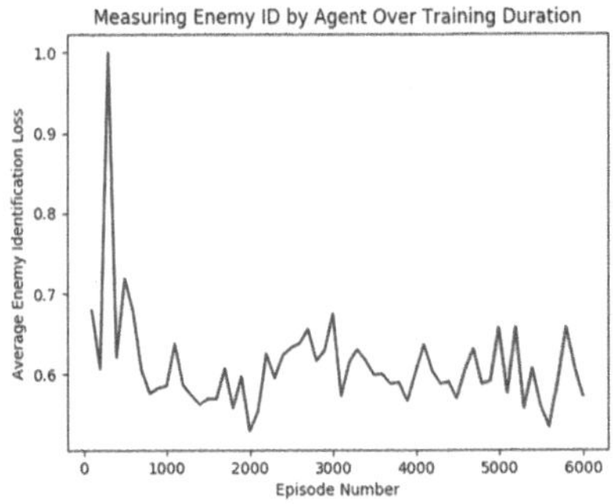

(b) DDQ DRQN

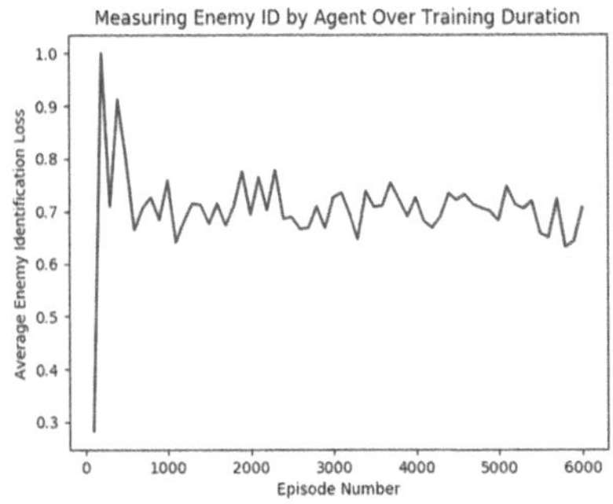

(c) PER DRQN

Fig. 4. Average enemy ID cross entropy loss - normalized by max loss.

(a) DRQN

(b) DDQ DRQN

(c) PER DRQN

Fig. 5. Average K/D over 100 defend the center games.

in the current input frame, calculated using the output of the size-1 dense layer (see Fig. 2). K/D ratio is the frags-to-deaths ratio, i.e. the number of total frags by the agent over its total number of lives. Roughly speaking, this translates to frags-to-games-played ratio.

In the case of DRQN with PER, the agent quickly achieved impressive levels of performance, reaching an average K/D of 5.62 after 400 games trained (see Fig. 5). The maximum average value (averaged over 100 games for each model and then finding max over all models) for enemy identification cross entropy loss was 0.2196 (see Fig. 4 - all loss was normalized by the max loss value observed for each graph), with a mean average K/D of 4.82 and an average K/D standard deviation of 0.614. The snapshot ensemble method improved upon the mean average K/D of DRQN with PER by over a standard deviation, achieving an average K/D of 5.51. Likely, the snapshot ensemble of PER was so successful due to PER's aggressive tendency to fit against challenging samples mixed with the snapshot ensemble method's ability to create similar and yet competitively diverse committees using its cosine annealing learning rate (Table 1).

Table 1. K/D and enemy ID loss

	DRQN	DDQ DRQN	PER DRQN
Mean average K/D	4.65	3.90	4.82
Average K/D standard Dev	0.620	0.393	0.614
Max enemy ID loss	0.269	0.298	0.2196

For the DRQN with DDQ, the authors report a set of agents characterized by consistency. Cumulatively, these agents had a mean average K/D of 3.90 and an average K/D standard deviation of 0.393, and the maximum average value for enemy identification cross entropy loss was 0.298. The snapshot ensemble method improved upon DRQN with DDQ by over a standard deviation as well, achieving an average K/D of 4.37.

Using the DRQN architecture mentioned here (without PER or DDQ), the agent achieved a max average K/D of 5.25, never reaching an average K/D to rival the max average K/D observed by the agent with PER. Furthermore, the base DRQN agent had an average K/D standard deviation of nearly double that of the agent with DDQ, illustrating its lack of consistency relative to the DDQ version of itself. To elaborate, the base DRQN agent had a mean average K/D of 4.65, an average K/D standard deviation of 0.620, and the maximum average value for enemy identification cross entropy loss was 0.269. No improvement was observed using the snapshot ensemble of base DRQNs, indicating potentially that increased value overestimation in such an environment led to development of a committee of models that were too dissimilar to allow for advantages offered by ensembling. Note that the DRQN architecture used here is efficient and very competitive, allowing for quick learning of complex tasks. Thus, PER and DDQ should be looked on as boosts to a well tuned learning architecture.

In all experiments, successful enemy detection was achieved by 1000 games by all agent types. In terms of qualitative notes on agent behavior, DDQ DRQN agents were better at developing and maintaining an ammo conservation strategy. Specifically, it was noted that DDQ agents would hold fire longer than other agents, allowing enemies to come closer. This in turn, reduced the possibility of missing a shot. DRQN PER agents, on the other hand, were observed to more quickly respond to enemies attacking from behind. When the agent is damaged, its screen will flash red, and the amount of health remaining is indicated in the lower left of the screen. PER agents were seen to more quickly turn to address and neutralize attackers from behind as well as from the flanks.

5 Conclusion

In this paper, the authors tested double-Q learning and prioritized experience replay in the context of the ViZDoom reinforcement learning environment. In addition, these methods were coupled with the use of the efficient deep neural network ensemble creation method known as snapshot ensembling. This breaks new ground in the ViZDoom environment, as DDQ and PER had not been tested using this clever, ensembling method - populating members of an ensemble by effectively training a single model, in this case a deep recurrent neural network, with the use of a cosine annealing learning rate. Furthermore, snapshot ensembling yielded significant results, improving DDQ and PER agent performance by over a standard deviation above the mean in both cases.

Following this, there are a number of avenues for further research. One of these consists of examining other DRL models at higher frame-skip values using ViZDoom. To what extent are other DRL model structures (dueling network architectures, actor-critic models, asynchronous variants, etc.) affected? Not only does frame-skipping accelerate training, it can also be looked on as modeling a potential scenario for autonomous agents in physical 3D space. If on a mission for example, a search-and-rescue agent's visual sensors are damaged and only every 11th frame on average of its video feed is a valid image, a reasonable question that arises is: will the agent, trained on a frame-skip distribution with a mean of x be able to cope with the new, shifted frame-skip distribution with mean $x+k$, where k is the additional number of frames, on average, that must be skipped before a valid frame is observed? At what values of k does performance begin to degrade significantly. This is also assuming the distribution shape remains unchanged; what are the limitations of learning from a certain type of frame-skip distribution (uniform, Gaussian, Landau, etc.)? Given its emerging research community and solid RL environment, the authors here note ViZDoom as a promising setting to test further research questions such as these.

Acknowledgment. The authors here would like to thank: (1) the ViZDoom development team for their continued maintenance and extension of this remarkable RL framework, and (2) IEEE CIG for their support of the annual ViZDoom Limited and Full Deathmatch Competitions.

References

1. Braylan, A., Hollenbeck, M., Meyerson, E., Miikkulainen, R.: Frame skip is a powerful parameter for learning to play Atari. Space **1600**, 1800 (2005)
2. Bellemare, M.G., Naddaf, Y., Veness, J., Bowling, M.: The arcade learning environment: an evaluation platform for general agents. J. Artif. Intell. Res. (2012)
3. Watkins, C.J.C.H.: Learning from delayed rewards. Ph.D. thesis, University of Cambridge England (1989)
4. Huang, G., Li, Y., Pleiss, G., Liu, Z., Hopcroft, J.E., Weinberger, K.Q.: Snapshot ensembles: Train 1, get M for free. ICLR submission (2017)
5. Lample, G., Chaplot, D.S.: Playing FPS games with deep reinforcement learning. arXiv preprint arXiv:1609.05521 (2016)
6. Hasselt, H.V.: Double q-learning. In: Advances in Neural Information Processing Systems, pp. 2613–2621 (2010)
7. Hausknecht, M., Stone, P.: Deep recurrent q-learning for partially observable MDPS. arXiv preprint arXiv:1507.06527 (2015)
8. Kempka, M., Wydmuch, M., Runc, G., Toczek, J., Jakowski, W.: Vizdoom: a doom-based AI research platform for visual reinforcement learning. In: IEEE Conference on Computational Intelligence and Games (2016)
9. Justensen, N., Bontrager, P., Togelius, J., Risi, S.: Deep Learning for Video Game Playing. ArXiv:1708.07902
10. Sutton, R.S., Barto, A.G.: Introduction to Reinforcement Learning. MIT Press, Cambridge (1998)
11. Sutton, R.S.: Learning to predict by the methods of temporal differences. Mach. Learn. **3**(1), 9–44 (1988)
12. Schaul, T., Quan, J., Antonoglou, I., Silver, D.: Prioritized experience replay. In: International Conference on Learning Representations (ICLR). http://arxiv.org/abs/1511.05952 (2016)
13. Dozat, T.: Incorporating Nesterov momentum into Adam. Technical Report, Stanford University (2015). http://cs229.stanford.edu/proj2015/054_report.pdf
14. Van Hasselt, H., Guez, A., Silver, D.: Deep reinforcement learning with double q-learning. arXiv preprint arXiv:1509.06461 (2015)

Ship Classification from SAR Images Based on Deep Learning

Shintaro Hashimoto[1(✉)], Yohei Sugimoto[2], Ko Hamamoto[3], and Naoki Ishihama[1]

[1] Research Unit III, Research and Development Directorate, Japan Aerospace Exploration Agency, JAXA, Tsukuba, Japan
{hashimoto.shintaro,ishihama.naoki}@jaxa.jp
[2] System Technology Unit, Research and Development Directorate, Japan Aerospace Exploration Agency, JAXA, Tsukuba, Japan
sugimoto.yohei@jaxa.jp
[3] Satellite Applications and Operations Center, Space Technology Directorate I, Japan Aerospace Exploration Agency, JAXA, Tsukuba, Japan
hamamoto.ko@jaxa.jp

Abstract. Ship classification based on remote sensing data is an important task of maritime/sea border security and surveillance applications. Therefore, in this research, ship classification was performed by using deep learning aiming for improving the accuracy of ship classification with respect to conventional approaches. The research focuses on not only ship or non-ship classification but also ship classification by its type and length, which is difficult using such filtering methods as CFAR and a standard deviation filter. The ship type classification resulted in promising accuracy for specific ship types with distinguishable features from SAR images. The ship length classification error resulted in 26% on average. Moreover, this research could classify ship or not with equivalent accuracy as GPU by using FPGA.

Keywords: Ship classification · SAR image · Deep learning Convolutional neural network · FPGA

1 Introduction

For the purpose of ensuring the safety of human life on the seas, specific ships are obliged to be equipped with the Automatic Identification System (AIS). This AIS realizes the safe navigation of a ship, monitoring of checking in and out of ports, etc. by exchanging navigation information mutually between AISs using radio waves [1]. AIS provides such information as unique identification, ship name, position, course, and speed. Although this equipment provides measures against terrorism, many ships are actually shutting down the AIS signal even if equipped with AIS because AIS information may be abused by pirates, sea jacks, etc. and since the obligation to install AIS is limited to large ships that satisfy certain conditions, and given the high cost of installing AIS, small ships are rarely equipped with AIS. Although pirates and sea jacks do not transmit the AIS signal, they may camouflage AIS information in some cases.

K. Arai et al. (Eds.): IntelliSys 2018, AISC 868, pp. 18–34, 2019.
https://doi.org/10.1007/978-3-030-01054-6_2

Thus, the ship needs a navigator to visually confirm safety. However, there are cases where "a navigator did not notice other ships" or "collusion cannot be avoided when a navigator finally notices other ships", as seen in such recent maritime accidents as the collision involving a U.S. Aegis warship [2]. Hence, monitoring of ships that do not rely on AIS is in great demand. Airborne and spaceborne radar systems such as Synthetic Aperture Radar (SAR) are known to be a useful tool to detect moving objects like ships in the ocean because they can see through the clouds during the day and night times. For the purpose of contributing to "the monitoring of ships that do not rely on AIS", this research conducts ships classification from SAR images based on deep learning. Ship classification of this research classifies the presence/position, type, and length of a ship. These types of classifications are adopted because they offer very important information for maritime security in order to judge "the dangers of the ships/the dangers facing the ships" and "conduct an additional investigation". In this research, deep learning is processed by using a GPU on the ground. However, assuming implementing deep learning/neural network model onboard the satellite (FPGA-based onboard processor more specifically) should also be considered in this research.

2 Related Research

Already on the ground, services to detect ships from sea scenes observed by SAR are beginning to appear [3]. In such services, threshold filtering, such as Constant False Alarm Rate (CFAR) and a relative standard deviation filter, is adopted for ship detection [4, 5]. Although It seldom uses the Haar Cascades for ship detection, this research used Haar Cascades [6]. Figures 1, 2 and 3 shows the results of ship detection using CFAR, standard deviation filter, and Haar Cascades. Table 1 shows accuracy and speed of ship detection using existing methods. In the test data, 3 scenes (32 ships included) were not used as training data. The size of one scene has about 13,000 pixels by 13,000 pixels.

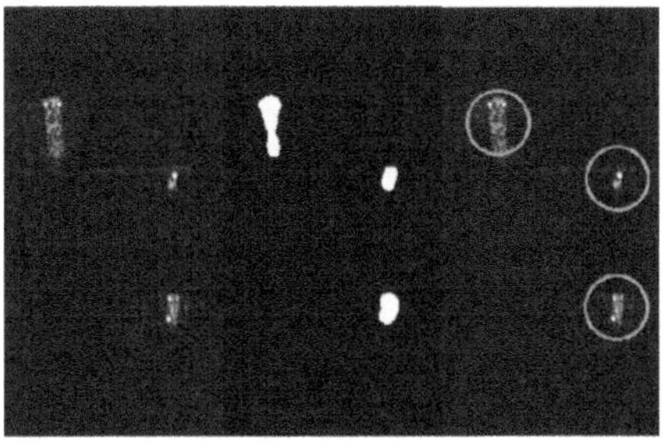

Fig. 1. Ship detection using CFAR (Left: SAR image; Middle: CFAR filtering; Right: Estimated position from CFAR).

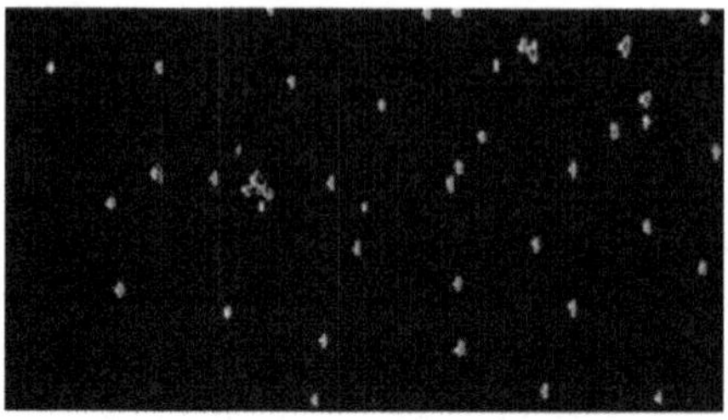

Fig. 2. Ship detection using standard deviation filter.

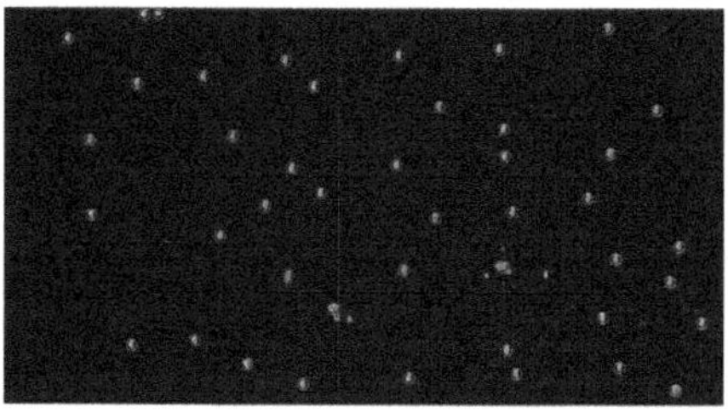

Fig. 3. Ship detection using Haar Cascades.

Table 1. Accuracy and speed of ship detection using existing methods

Name	Accuracy	Speed/image
CFAR	100%	30 s
Standard deviation filter	100%	30 s
Haar Cascades	59%	8 s

As these methods of ship detection by threshold filtering do not consider the shape of a ship, it is only possible to detect objects blighter than the background sea to a certain degree above the threshold in SAR images (i.e., classify presence/position of ship candidates). Consequently, classification based on the type or size of a ship requires a combination of other methods by computer or judgment by experts.

In addition to the presence/position of a ship, this research also classifies types and length of ships. As a classification method in this research, convolutional neural network (CNN) of deep learning was used.

There are quite works of literature studying ship classification in airborne/spaceborne optical and remote sensing images based on deep learning. For example, Bousetouane and Morris have proposed a pipeline processing build upon the GPU implementation of Fast-R-CNN for a collection of ground-based optical images/videos captured in high resolution (2032 $\times$ 1072 pixels) [7]. Zhang et al. have proposed ship detection method based on CNN, namely S-CNN for optical images [8]. The S-CNN has a CNN that can classify ship and non-ship classes. One of the remarkable features of their study is that it uses two ship models (V-shape model for ship head and ||-shape model for ship body) to localize ship proposals, for example, in an optical image of a busy harbor full of ships. Bentes et al. have studied ship classification in TerraSAR-X images with CNN, in which they proposed a multi-input resolution CNN model in order to improve the ability to derive features from SAR images [9]. Their target classes are cargo, tanker, windmill, harbor, and oil platform. Their algorithm uses the CFAR detector to identify a region of interest, which is widely used in maritime SAR image processing. The classification algorithm proposed by Makedonas et al. also used the threshold filtering method as pre-processing because it reduces the overall processing time for "ship classification by using NN" [10]. CNN/NN requires enormous calculations, but detailed classification is possible. The amount of calculation is proportional to the size of the input (in this research, the size of an SAR image). For this reason, features are extracted using the threshold filtering method of low processing before CNN. Thus, the amount of information to be input to CNN can be reduced.

In contrast, this research adopts deep learning/CNN with regard to pre-processing and achieves high speed processing. By adopting deep learning, several benefits are obtained, such as:

- There is no longer a need to change the threshold according to nonuniform sea-level clutter in the sea scene observed by SAR; and
- It is also possible to consider the shape of a ship and the environment around the ship, such as the wake if it is cruising.ran

3 Overview of Ship Classification

This section describes the scope of objects to be classified and the data to be used in ship classification of this research.

3.1 Ship Classification Policy

The purpose of this research is to classify ships in maritime/sea-only scenes observed non-uniformly with waves, sea clutter, convection in the air, etc. Classification the lands and man-made structures (e.g., wind power stations, etc.) offshore is out of the research scope because they can be preliminary eliminated by coastline detection and masking through geo referencing or, if referencing is not available, using a supervised/unsupervised technique [11, 12].

3.2 Usage Data

The data used for ship classification uses SAR images of level 1.5 (a.k.a. L-1.5) intensity map with spatial resolution of 10 m as observed by the Phased Array type L-band Synthetic Aperture Radar (PALSAR-2) mounted on the Advanced Land Observing Satellite (ALOS-2). PALSAR-2 is an L-band Synthetic Aperture Radar (SAR) sensor, a microwave sensor that emits L-band radio waves and receives their reflection from the ground to acquire information. Since PALSAR-2 is L-band, the synthetic aperture time is long, and even when the resolution (10 m) is high, the blur appears much stronger compared with X-band. The polarization of SAR adopted HV polarization. This is because, in HV images, sea areas appear darker than they do in HH images regardless of sea conditions.

3.3 Types of Ship Classification

From SAR images, this research classifies:

- presence/position of ships
- type of ships
- length of ships

3.4 Development/Evaluation Environment

Table 2 shows the development/evaluation environment.

Table 2. Development/evaluation environment

Name	Detail
OS	Ubuntu 14.04
CPU	Intel$^{(R)}$ Core$^{(TM)}$ i7-5930 K CPU
GPU	GeForce GTX TITAN X
Memory	32 GB
Type of Storage	SSD

4 Method of Ship Classification by Using Deep Learning

This section describes ship classification using deep learning—the proposed method of this research. When trying to build a model of deep learning that classifies the items listed in Sect. 3.3, it is inefficient to express such a three-in-one model due to the mixture of classification and regression problems. Therefore, ship classification was conducted by combining multiple models of deep learning. In this section, Subsect. 4.1 explains the overall process, and Subsect. 4.2 to 4.5 explain the specific contents of each process (model of deep learning).

4.1 Sequence of Ship Classification

Figure 4 shows the sequence of ship classification. The alphabet (e.g. "(a)") in Fig. 4 correspond to the alphabet enumerated in this section. The role of processing (inference machine) in each sequence is described below.

(a) *Classification of presence/position (Global):* This is a classification machine that classifies ships at high speed from SAR images and extracts candidate ships. The feature of this classification machine is that it classifies global noise such as sea clutter and waves with high frequency and it also omissions of ship classification are little. Generally, threshold-based filtering, such as CFAR or a relative standard deviation filter, is used in this phase. However, this research adopted deep

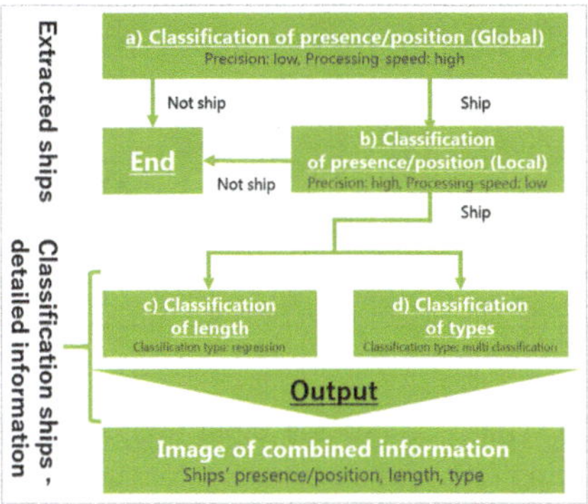

Fig. 4. Sequence of ship classification

learning. Hence, it eliminates the need to change the threshold according to the nonuniformity of the sea scene.

(b) *Classification of presence/position (Local):* This is a classification machine that classifies ships with high precision regarding whether the ship extracted by the classification machine described in (a) above is actually a ship. This classification machine has high classification precision. However, the processing speed is very slow. This classification machine and the classification machine of (a) have a trade-off relationship. Specific precision and speed will be explained in detail in Sect. 5. This classification machine eliminates local noise such as sea clutter mistakenly extracted by the classification machine described in (a) above. Moreover, it adjusts the size of the area where the ships are located (presented). As a result of this processing, a single ship is extracted and input into the classification machines described in (c) and (d) below.
(c) *Classification of length:* This is a regression machine that classifies ship length from the ship extracted by the classification machine described in *(b)* above.
(d) *Classification of type:* This is a multi-classification machine that classifies ship type from the ship extracted by the classification machine described in *(b)* above.

Combining the classification machines described in *(a)* to *(d)* above enables the classification of a ship's position, its length, and its type. As a final result, an image combining these classification results is output.

4.2 Classification of Presence/Position (Global)

The input of this classification machine is each image consisting of 1,000 pixels by 1,000 pixels. It can interpret this as a big array of numbers. In other words, it is to be a 1 M-dimensional (1000 pixels by 1000 pixels) vector. The output of this classification machine is a 100-dimensional (10 grids by 10 grids) vector per image (Fig. 5). Since the SAR image is the size of about 13,000 pixels by 13,000 pixels, an input image of 1,000 pixels by 1,000 pixels is generated by a raster scan of the SAR image. Each value of output represents the presence of a ship by a numerical value: 0 to 1. By matching the input image with the output, the position of the ship can be classified in an area of

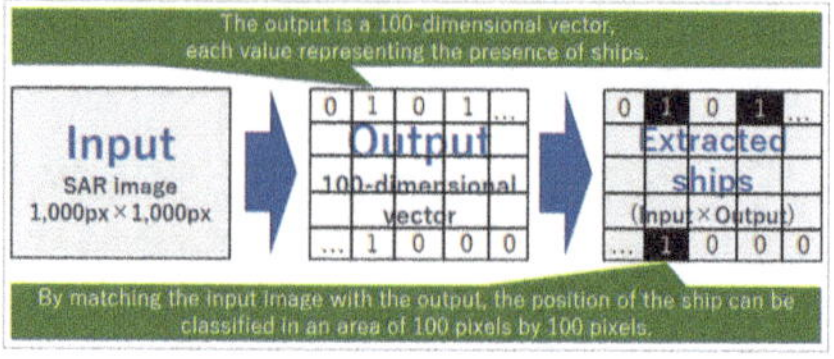

Fig. 5. Processing of classification of ships' presence/position (global).

100 pixels by 100 pixels. That is, when the input image is split by 100 pixels by 100 pixels, 10 grids by 10 grids (100-dimensional vector) can be assumed.

Then, this classification machine will explain the loss function. The loss function for learning used (6), which is composed of (1) and (2). The error with respect to the output node is calculated from teacher data t_k and output value y_k of each class k by the square error of (1). Output value y_k does not apply the softmax function. c of the variable is the number of classes.

Equation (2) gives a larger loss value as a sum of penalties when the output value is more distant from the teacher data (e.g., when the classification output indicates that the target is not-ship although it is a ship according to the teaching data). Equations (4) and (5) are the conditional expressions of (2). In (4), the minimum value is set to 0 so that the loss value does not become negative. Equation (5) must also be a numeric value equal to or greater than 0 so that it does not become zero. Also, n of the variable in (5) adjusts the value by the scale of the penalties. The smaller the value of n, the greater the penalty.

$$Loss_1 = \frac{1}{2}\sum_{k=1}^{c}(y_k - t_k)^2 \tag{1}$$

$$Loss_2 = \sum_{k=1}^{c} p_k \tag{2}$$

$$p_k = \frac{t_k}{y_k} - 1 \tag{3}$$

$$p_k = \begin{cases} p_k = 0 & if\ p_k \leqq 0, \\ p_k & if\ p_k > 0. \end{cases} \tag{4}$$

$$y_k = \begin{cases} y_k = 1e^{-n} & if\ y_k \leqq 0, \\ y_k & if\ y_k > 0. \end{cases} \tag{5}$$

$$E = Loss_1 + Loss_2 \tag{6}$$

The training data is generated in the following fashion. First, 10 sea images (5,000 pixels by 5,000 pixels) where a ship is not displayed are extracted from SAR images. Also, 3,104 images of ships were extracted manually, referring to the AIS information.

Next, it synthesizes sea images, images of ships, and generates learning data. Specifically, it randomly crops 1,000 pixels by 1,000 pixels from sea images (5,000 pixels by 5,000 pixels) and randomly adds 0, 90, 180, and 270-degree rotation. Then, ships of 0 to 10 are added randomly to the cropping sea image, with rotation being added randomly to each ship. Approximately 60,000 items of training data were prepared. Figure 6 shows an example of the generated learning data.

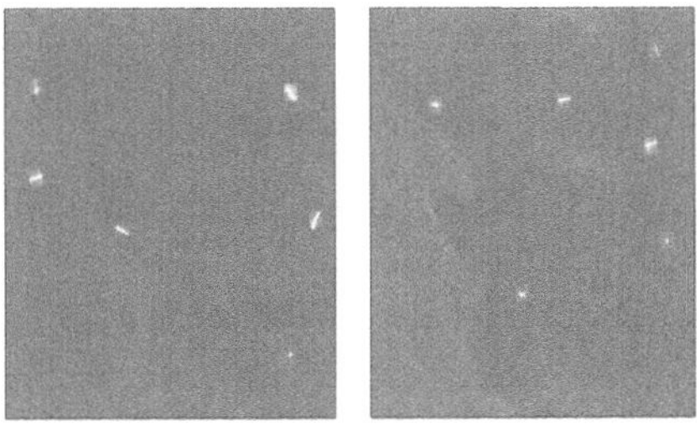

Fig. 6. Generated training data.

The model of CNN in this subsection is shown below.

Model:
Conv (10, 10, 1, 16) → Pool → Conv (7, 7, 16, 16) → Pool → Conv (5, 5, 16, 16) → Pool → Conv (5, 5, 16, 16) → Pool → Conv (3, 3, 16, 32) → Pool → Conv (3, 3, 32, 64) → Pool → Fc ((16 * 16 * 64), 1024) → Fc (1024, 100)

Legend:
Conv: Convolution (Kernel_x, Kernel_y, Input, Output)
Pool: Max Pooling (Kernel size: 2x2; Stride size: 2x2)
FC: Fully Connected (Input, Output)

Since the input image is large, Batch normalization was not used from the viewpoint of processing speed and memory [13]. If pooling and convolution layer is increased, feature amount disappears. For this reason, output is large even in near the final layer. Depending on the initial value, it often does not converge. This also occurs when Batch normalization and Adam optimizer are used [14]. Thus, eliminating bias and initializing method of weight by He was used to converge loss in training [15].

Figure 7 shows the result of classification with SAR images using this classification machine. In Fig. 7, the left side is the non-processed data; the right side is the adjusted image, such as the contrast. The white rectangles in Fig. 7 indicate the presence/position of ships. An area of white rectangles is input as a candidate for a ship to the classification machine described in Subsect. 4.3 below. However, multiple ships on one grid pose the problem of not being able to divide the area for each ship. The classification machine described in Subsect. 4.3 below will deal with this problem.

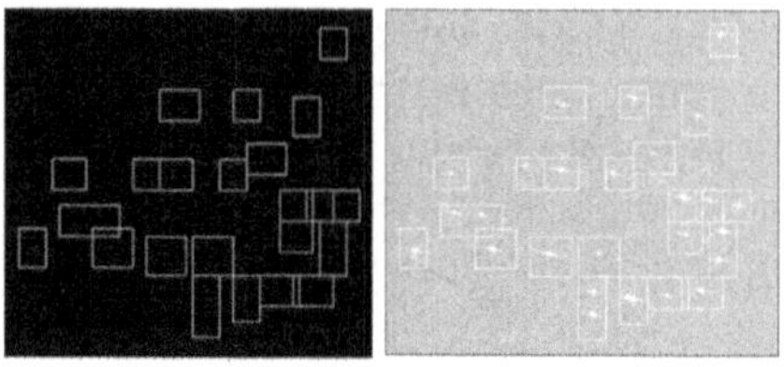

Fig. 7. Processing result of classification of ships' presence/position (global).

4.3 Classification of Presence/Position (Local)

In the classification machine described in this subsection, the ship candidate extracted in Subsect. 4.2 above is input and classified into two classes: the ship class and the non-ship class (i.e., sea clutter and wave). Since the input of this classification machine is fixed length (3,600-dimensional vector), the area of the ship candidate extracted in Subsect. 4.2 above is raster scanned, and the fixed length is cropped as input data of this classification machine.

There are about 80,000 total items of learning data. As a breakdown, there are approximately 40,000 images of ships obtained by adding rotation to 3,104 images of ships, and about 40,000 images of noise such as sea clutter.

The model of CNN in this subsection is shown below.

```
Model:
Conv (3, 3, 1, 64) → Conv (3, 3, 64, 64) → Pool → Conv
(3, 3, 128, 128) → Conv (3, 3, 128, 128) → Pool →
Conv(3, 3, 256, 256) → Conv(3, 3, 256, 256) → Pool →
Conv(3, 3, 256, 256) → Conv(3, 3, 256, 256) → Pool →
Conv(3, 3, 512, 512) → Conv(3, 3, 512, 512) -> Pool →
Fc((1*1*512), 2048)-> FC(2048, 1024) -> FC(1024, 512) -
> FC(512, 2)
```

In training of the model, initializing method of weight by He and Adam optimizer was used.

Figure 8 shows an image obtained by this classification machine. This figure shows the same place as shown in Fig. 7. In Subsect. 4.2 above, it extracted multiple ships on one grid. However, this classification machine can separate each ship. By combining the two classification machines described in Subsects. 4.2 and 4.3 above, it becomes strong against noise. That was because such global noise as waves can be eliminated by the classification machine described in Subsect. 4.2 above, and such local noise as sea clutter can be eliminated by this classification machine. Also, the processing speed can be increased by combining the two classification machines described in Subsects. 4.2 and 4.3 above.

Each ship extracted by this classification machine will be the input of the classification machines described in Subsects. 4.4 and 4.5 below.

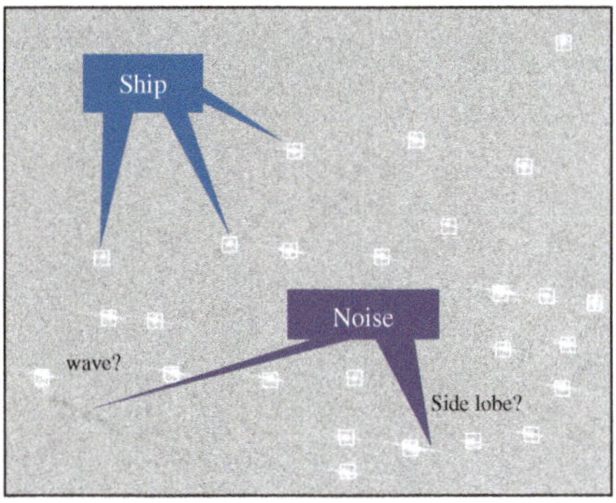

Fig. 8. Processing result of classification of ships' presence/position (local).

4.4 Classification of Length

In the classification machine described in this subsection, with the ship extracted in Subsect. 4.2 above as an input, the length of ships is classified by a regression. Training data consist of 2,987 images of ships. Augmentation processing of training data such as rotation was not done due to the difficult geometric transformation of the ships. A label (answers) in training data is represented by a real value of 0.0 to 1.0, with the maximum value of ship length being 1.0.

The model of CNN in this subsection has changed only output number in Subsect. 4.3 to 1.

4.5 Classification of Types

In the classification machine described in this subsection, with the ship extracted in Subsect. 4.2 as an input, the types of ships are classified. There are four classes of output according to the main types of ships: Fishing, Passenger, Cargo, and Tanker. Approximately 30,000 images obtained by randomly rotating the 1,922 images of ships are used as training data.

The model of CNN in this subsection has changed only output number in Subsect. 4.3 to 4.

Figure 9 shows the images overlaid with the ship's length and type information classified. This figure is an image of the final output obtained through Subsects. 4.2 to 4.3 above.

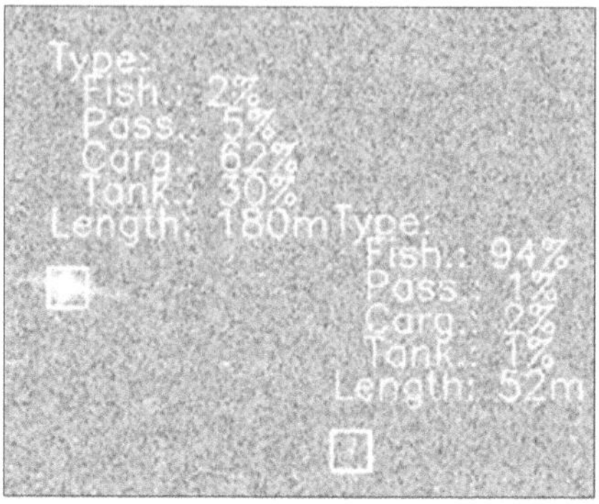

Fig. 9. Final result.

5 Evalution of Ship Classification

This section describes the evaluation of the classification machines described in Sect. 4. The test data used for each evaluation in this Section was not used as training data. Moreover, the test data did not perform any special image processing.

5.1 Classification of Presence/Position (Global)

In order to test the classification accuracy, test data is prepared, which includes a total of 513 ships which are intentionally excluded from the training data generation. Input images of test data were extracted by raster scan from SAR images according to the input size in the same fashion as described in Subsect. 5.2, Sect. 4. The extracted image is real data that has not been processed except for raster scan (extracting). As a result of ship classification based on test data (not used as training data), 512 ships (99.8%) out of 513 ships were classified (Table 3). A total of 36 SAR images included test data showing different noise for each image. An error of up to 30% is allowed against a ship's likelihood. Increasing the value of this allowed ship's likelihood can reduce the missing extraction (detection) of ships. However, given the high possibility of such noise as waves being extracted as a candidate for a ship, a trade-off exists whereby the classification machine described in Subsect. 5.3, Sect. 4 requires a longer processing time.

Table 3. Classification result of ships' presence/position (global)

	Number of ships	Accuracy rate
Ships of AIS	513	99.8%
Ships of Classification	512	

Table 3 shows how many ships can be classified by this classification machine when "there is ships' location information by AIS" and "there are ships at the location of AIS in SAR images". As shown in Fig. 10, the number of overlapping AIS (white circles) and classification results (white rectangles) was counted. Even if this classification machine classified a ship without AIS, due to not knowing whether it is really a ship, it is intentionally removed from statistics. For example, the white circles in Fig. 11 (030, 032 of the number) are the position information of AIS. However, the object that seems to be a ship does not exist. It may interpret in the following ways: "AIS is disguised", "garbled by communication error", "not captured by SAR because the ship cruising speed is too fast", or "the sea current state was extremely unstable". For this reason, this classification machine excluded such a state from the statistical data. The white rectangles in Fig. 11 are the ships classified by this classification machine. The lack of AIS excludes the ships from statistical data. Figure 12 shows the SAR image in the vicinity of an offshore wind power station. However, this is also excluded from the statistics.

Fig. 10. Superposition of AIS (white circles) and classification result (white rectangles).

As a reason, in addition to the fact that SAR images including terrestrial and maritime structures are outside the scope of this research, there are cases where ship types are not registered in AIS near the offshore wind power plant. Thus, it cannot be decided whether it is an AIS indicating an offshore wind power plant, a ship, or a buoy.

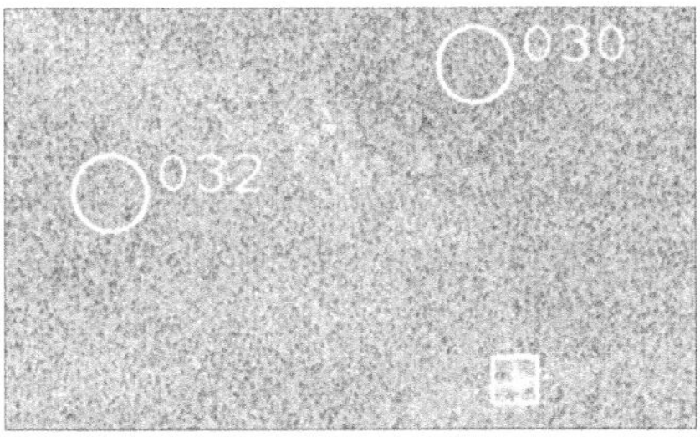

Fig. 11. Example of ships not present at the location of AIS.

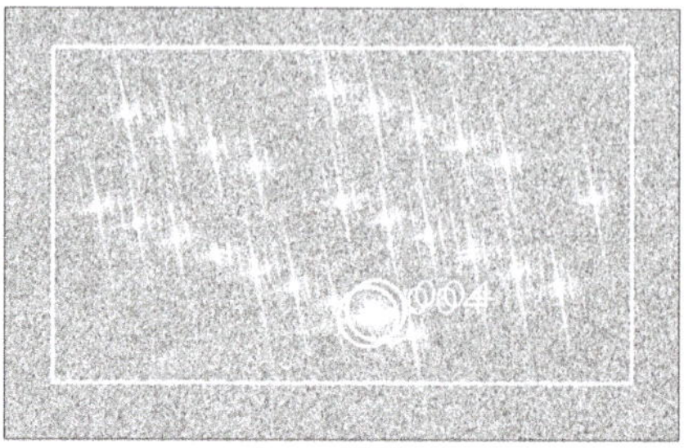

Fig. 12. Example of proximity to an offshore wind power plant.

Figure 13 shows a ship that could not be classified. Compared with other ships, this ship was very small in size and could not be separated from noise. It can be interpreted that the backscattering coefficient is considered to be very low from the small luminance value and the side lobe of this ship. Therefore, it is considered to be a small ship made of wooden or FRP material. In order to classify such as a ship, it is necessary to increase the learning data of a similar ship. Such ships were not included in the training data.

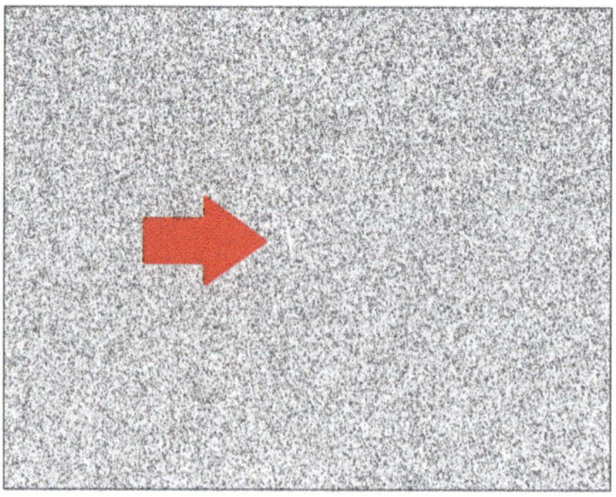

Fig. 13. Example of cannot classify examples of difficult ship classification.

5.2 Classification of Presence/Position (Local)

As a result of ship classification based on test data (not used as training data), 513 (100%) out of 513 ships were classified (Table 4). As a result of noise classification such as sea clutter based on test data (not used as training data), 511 (99.6%) out of 513 noises were classified (Table 4). Table 5 shows the F-measure of this classification result.

Table 4. Classification result of ships' presence/position (local)

		True value	
		Ships	Noises
Classification result	Ships	513	2
	Noises	0	511

Table 5. F-measure of ships' presence/position classification (local)

Accuracy	99.8%
Precision	99.6%
Recall	100%
F-measure	99.8%

And when using this classification machine after extracting a ship candidate by using the classification machine described in Subsect. 5.2, Sect. 4, the processing speed was confirmed as being about 120 times faster than the raster scan of the SAR image by only using this classification machine. The processing speed in the GPU was 3, 500 ms per image.

5.3 Classification of Length

As a result of ship length classification for 299 ships based on test data (not used as training data), the dispersion was 13%. The median error was 17%, and the average was 26%. Figure 14 shows a scatter diagram of the error. This scatter diagram was scaled excluding outliers due to its visibility.

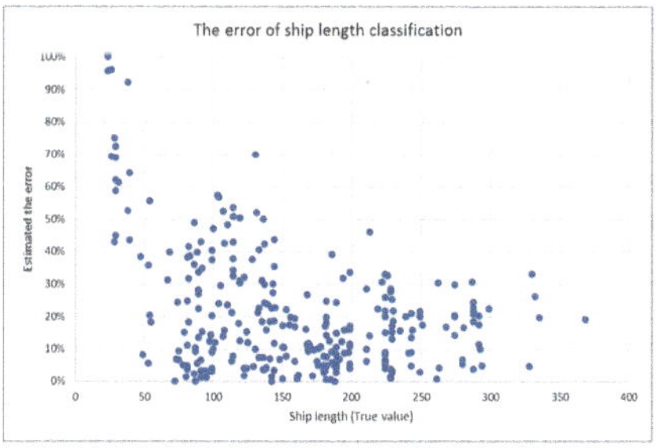

Fig. 14. Error dispersion of ship length classification.

5.4 Classification of Types

As a result of ship type classification based on test data (not used as training data), 224 (65.7%) out of 341 ships was the correct answer rate. Figure 15 shows the confusion matrix as a detailed classification result. From Fig. 15, it was found that although the classification rate of Fishing and Cargo was high, Passenger and Tanker were hardly classified. One of the reasons, in addition to the limited amount of training data, is that distinct differences between the two classes cannot be found in their SAR images. As a possible solution, it is conceivable to increase the amount of training data, use multiple polarizations and optical images, and employ other measures.

		Class				
		Fishing	Passenger	Cargo	Tanker	
Classification result	Fishing	25	0	9	0	73.5%
		7.3%	0.0%	2.6%	0.0%	26.5%
	Passenger	1	1	13	1	6.3%
		0.3%	0.3%	3.8%	0.3%	93.8%
	Cargo	8	1	176	12	89.3%
		2.3%	0.3%	51.6%	3.5%	10.7%
	Tanker	4	2	66	22	23.4%
		1.2%	0.6%	19.4%	6.5%	76.6%
		65.8%	25.0%	66.7%	62.9%	65.7%
		34.2%	75.0%	33.3%	37.1%	34.3%

Fig. 15. Confusion matrix of ship type classification.

6 Installation on FPGA

This research also considered installing ship classification on FPGA. The algorithm of ship classification considered is Subsect. 5.3, Sect. 4 only. For quick implementation, this research used the PYNQ-Z1 and PYNQ-Z1's standard library. For this reason, there are some restrictions on the implementation, and this research implemented it under the following conditions.

- Model: CNV & Binarized Neural Network [16].
- Network architecture: Three convolutional layers (each of the layer consist of two convolutional layers & pooling layer) and two fully connected (512 units) layers.
- Input size: 28pixels by 28 pixels.
- Output size: 2 classes (ship or not).

Other conditions such as training data are the same as Subsect. 5.3, Sect. 4.

Deep Learning mainly has the processes of training and inference. Since the training process requires matrix operation with a large batch size, large memory, high throughput, and high precision are required against hardware. However, since inference process has a small batch size, low memory, low throughput, and low precision are sufficient. Rather, low power consumption and low latency are required. In other words, GPU is suitable for training processes, FPGA is suitable for inference processes. For this reason, in this research, training process was performed with GPU and only inference process was performed with FPGA.

Table 6. Classification result of ships' presence/position (local) on FPGA

		True value	
		Ships	Noises
Classification result	Ships	513	0
	Noises	0	513

As a result of ship classification on FPGA, 513 (100%) out of 513 ships were classified (Table 6). As a result of noise classification such as sea clutter based on test data (not used as training data), 513 (100%) out of 513 noises were classified (Table 6). This result shows that the accuracy is better than Subsect. 5.2, Sect. 5. The main reason is considered to have become easier to classify ships and noises by compressed images. It is also confirmed by the similarity algorithm of Subsect. 5.3, Sect. 4 that the accuracy is improved by compressing the image. Although this research did not happen, feature quantities may be reduced by image compression, which may increase misclassification of ships. From this perspective, compressing images is not preferable.

The processing speed on FPGA was about 330 ms per image. In the case of GPU, it was about 1,165 ms per image. In the case of CPU, it was about 7, 381 ms per image. This value excluded overhead due to such as reading data and initializing. In order to reduce the measurement error, it inferred 513 images at a time. The number of microseconds per image is divided by the inferred number of images.

Logic Elements usage rate was 30, 605 out of 53, 200 (58%). PYNQ-Z1 is installed FPGA of Z-7020 (XC7Z020). This FPGA has 53, 200 LUTs (6-input).

7 Conclusion

In this research, the presence/position, length, and type of ships were classified from SAR images by using deep learning. By combining the extraction of ship candidates from a global area using grid division and ship classification by sub regions, it was possible to classify the presence/position of ships at high speed and with high accuracy. Moreover, this research could classify ship or not with equivalent accuracy as GPU by using FPGA. The length of ships could be classified at 13% dispersion error using image regression. Although the correct answer rate was 66% in ship type classification, there remains a problem whereby some ship types that look similar in SAR images such as Passenger and Tanker are difficult to classify. In the future, in addition to improving the accuracy of each classification machine, heterogeneous learning using not only luminance values but also other information like phase values of complex images should be incorporated.

References

1. International Maritime Organization: SOLAS, Chap. 5 (2002)
2. Anna Fifield: There wasn't a lot of time as water flooded U.S. destroyer below decks. The Washington Post (2017)
3. Ai, J., Yang, X., Zhou, F., Dong, Z., Jia, L., Yan, H.: A correlation-based joint CFAR detector using adaptively-truncated statistics in SAR imagery. Sensors (Basel) (2017)
4. El-Darymli, K., McGuire, P., Power, D., Moloney, C.R.: Target detection in synthetic aperture radar imagery: a state-of-the-art survey. J. Appl. Remote Sens. **7**(1), 071598 (2013)
5. Arii, M.: Improvement of ship-sea clutter ratio of SAR imagery using standard deviation filter. In: 2011 IEEE International Geoscience and Remote Sensing Symposium (IGARSS) (2011)
6. Viola, P., Jones, M.: Rapid object detection using a boosted cascade of simple features. In: Computer Vision and Pattern Recognition (2001)
7. Bousetouane, F., Morris, B.: Fast CNN surveillance pipeline for fine-grained vessel classification and detection in maritime scenarios. In: 2016 13th IEEE International Conference on Advanced Video and Signal Based Surveillance (AVSS), pp. 242–248, Colorado Springs, CO (2016)
8. Zhang, R., Yao, J., Zhang, K., Feng, C., Zhang, J.: S-CNN-based ship detection from high-resolution remote sensing images. In: Proceedings of the ISPRS-International Archives of the Photogrammetry Remote Sensing Spatial Information Sciences, pp. 423–430 (2016)
9. Bentes, C., Velotto, D., Tings, B.: Ship classification in TerraSAR-X images with convolutional neural networks. IEEE J. Ocean. Eng. **PP**(99), 1–9 (2017)
10. Makedonas, A., Theoharatos, C., Tsagaris, V., Costicoglou, S.: A multilevel approach to ship classification on sentinel-1 sar images using artificial neural networks. In: LPS16 ESA Symposium (2016)
11. Li, R., et al.: DeepUNet: a deep fully convolutional network for pixel-level sea-land segmentation. arXiv:1709.00201 (2017)
12. Garzelli, A., Zoppetti, C., Pinelli, G.: Computational efficient unsupervised coastline detection from single-polarization 1-look SAR images of complex coastal environments. In: Proceedings of the SPIE 10427, Image and Signal Processing for Remote Sensing XXIII, 1042714 (2017)
13. Loffe, S., Szegedy, C.: Batch normalization: accelerating deep network training by reducing internal covariate shift. arXiv:1502.03167 (2015)
14. Kingma, D.P., Ba, J.: Adam: a method for stochastic optimization. arXiv:1412.6980 (2014)
15. He, K., Zhang, X., Ren, S., Sun, J.: Delving Deep into Rectifiers: Surpassing Human-Level Performance on ImageNet Classification. arXiv:1502.01852 (2015)
16. Umuroglu, Y., et al.: FINN: a framework for fast, scalable Binarized neural network inference. In: 25th International Symposium on Field-Programmable Gate Arrays. arXiv: 1612.07119 (2017)

HIMALIA: Recovering Compiler Optimization Levels from Binaries by Deep Learning

Yu Chen[1,2(✉)], Zhiqiang Shi[1,2], Hong Li[1,2], Weiwei Zhao[3], Yiliang Liu[4], and Yuansong Qiao[5]

[1] Institute of Information Engineering, Chinese Academy of Sciences, Beijing, China
{chenyu9043,shizhiqiang,lihong}@iie.ac.cn
[2] School of Cyber Security, University of Chinese Academy of Sciences, Beijing, China
[3] School of Information Science and Engineering, Lanzhou University, Lanzhou, China
zhaoww15@lzu.edu.cn
[4] Arab Academy, Beijing International Studies University, Beijing, China
2016220312@stu.bisu.edu.cn
[5] Software Research Institute, Athlone Institute of Technology, Athlone, Ireland
ysqiao@research.ait.ie

Abstract. Compiler optimization levels are important for binary analysis, but they are not available in COTS binaries. In this paper, we present the first end-to-end system called HIMALIA which recovers compiler optimization levels from disassembled binary code without any knowledge of the target instruction set semantics. We achieve this by formulating the problem as a deep learning task and training a two layer recurrent neural network. Besides the recurrent neural network, HIMALIA is also powered by two other techniques: instruction embedding and a new function representation method. We implement HIMALIA and carry out comprehensive experiments on our dataset consisting of 378,695 different functions from 5828 binaries compiled by GCC. The results show that HIMALIA exhibits accuracy of around 89%. Moreover, we find that HIMALIA's learnt model is explicable: it can auto-learn common compiler conventions and idioms that match our prior knowledge.

Keywords: Binary analysis · Reverse engineering · RNN
Feature embedding · Model explicable

1 Introduction

Modern compilers such as GCC, Clang and ICC implement a large number of optimization flags which have a different impact on code quality, compilation time, code size, energy consumption, etc. For ease of use, compilers typically provide a limited number of standard optimization levels that combine various

K. Arai et al. (Eds.): IntelliSys 2018, AISC 868, pp. 35–47, 2019.
https://doi.org/10.1007/978-3-030-01054-6_3

optimization flags providing a number of trade-offs between multiple objective functions such as code quality, compilation time and code size [1]. For example, there are 5 commonly used optimization levels in GCC, Clang and ICC, namely `-O0`, `-O1`, `-O2`, `-O3` and `-Os`. The same piece of source code could generate binaries with varied instructions, data flows and control flows when different compiler optimization levels are used.

There are a lot of important applications of compiler level detection in the security and software engineering, such as vulnerability discovering [2], binary similarity detection [3] and authorship identification [4]. For example, in C/C++ languages, "undefined behavior" is a serious problem. That is certain behaviors are not defined in the C/C++ standard, hence the compilers are free to decide how to handle such cases during optimization. This arbitrary behavior in some cases leads to security vulnerabilities. For example, CVE-2016-9843 and CVE-2016-9840 are recently reported critical vulnerabilities in zlib that arise due to the arbitrary handling of undefined behavior during compiler optimization. However, compiler optimization levels are not available in binaries. We also can not get such information by iteratively compiling the source code with different compiler optimization levels and comparing the binaries with the target one, since source code is often unavailable for COTS binaries. It is worth noting that binaries are made up of functions which are typically linked from different compilation units,[1] so each function in the binary code may have a different compilation level. Thus, the goal of this paper is to recognize the compiler optimization level of each function of binaries without explicitly encoding any semantics specific to the instruction set being analyzed or the conventions of the compiler used to produce the binary. Because of the interdependencies among different parts of a function introduced by compilers, the feature is hard to extracted from several independent instructions. Besides, there are more than one hundred compilers and each of them may have totally different optimization strategies or rules. Therefore, the direct algorithmic approach which rely on human or rule-base algorithm are very time consuming.

In this work, we present the first end-to-end system called HIMALIA which trains a Recurrent Neural Network (RNN) to recover compiler optimization levels of binaries. The design of such a system undergoes a long term of trial-and-error, sifting through the choice of network architectures, trying different instruction represent methods, etc. For example, we find that the compilation level `-O2` can be easily confused with `-O3`, which also has been reported by Egele *et al.* [5]. Thus, HIMALIA trains two model: one for classifying `-O0`, `-O1`, `-O2/-O3` and `-Os`, the other for classifying `-O2` and `-O3`. In summary, HIMALIA consists of three modules: input preprocess module, 4-classification model and 2-classification model. The input preprocess module transforms the input binary file into a list of functions and represents each function in a well-designed way which could retain the semantic information of the instructions as much as possible. The 4-classification model classifies functions into four categories: `-O0`, `-O1`, `-O2/-O3` and `-Os`.

[1] Different compilation units that make up a single executable can be optimized as a single module only if Link Time Optimization (LTO) is enabled by the compiler.

The functions with label -O2/-O3 are further classified into -O2 and -O3 by the 2-classification model. We restrict our study to Linux x64 applications compiled by GCC, though the techniques presented can be easily extended to other compilers, OS platforms and instruction sets.

The contributions of this paper are summarized as follows:

- To the best of our knowledge, we are the first to present an end-to-end system called HIMALIA which employs deep learning to recover compiler optimization levels from disassembled binary codes.
- We propose a new function representation method and employ an embedding layer in the network architecture which is trained together with recurrent layers. These methods can help to retain and capture the semantic information of the instructions as much as possible.
- We build a dataset consisting of 378,695 different functions from 5828 binaries and perform comprehensive experiments. The results show that HIMALIA exhibits high accuracy and the results are explicable.

2 Related Work

Traditional machine learning algorithms have been extensively used in the context of reverse engineering such as malware detection and function identification. Abou-Assaleh *et al.* [6] applied the N-Gram analysis (CNG) method based on byte n-gram analysis in the detection of malicious code. Rieck *et al.* [7] proposed a learning-based approach to automatically classify malware behaviors. Rosenblum *et al.* [8] formulated the function identification as a structured classification problem which incorporated content and structure features of binary code to identity the starting byte of each function in stripped binaries. Bao *et al.* [9] proposed an automatic function identification algorithm which utilize weighted prefix trees to improve the efficiency of function identification.

In recent years, deep learning has also been used in the fields of reverse engineering. Shin *et al.* [10] applied RNNs to the task of function boundary identification. They showed that deep learning based method could dramatically reduce the computational time while achieving a higher accuracy. Chua *et al.* [11] employed a RNN with 3 layers to identify function types from the x86/x64 machine code of a given function. Their method could achieve a high accuracy by automatically learning relationships among instructions, compiler conventions, stack frame setup instructions, use-before-write patterns, and operations relevant.

Our work is motivated by these studies, but different in three aspects: (1) to the best of our knowledge, we are the first to study the problem of recovering the compiler's optimization level from the binary code of a given function, and propose the first end-to-end solution based on deep learning; (2) we apply a sophisticated operands abstraction mechanism to retain the instruction's information as much as possible; (3) we employ instruction embedding techniques that have not been used in studying the learnt models for binary analysis tasks.

3 Problem Definition

Given a binary code, we assume to have the following knowledge: (a) the boundaries of a function; (b) the boundaries of instructions in a function; (c) the `mnemonic` and `operands` of an instruction.

The input to our final model Ψ is a target function for which we are recovering the compiler optimization level. Functions are represented in disassembled form, such that each function is a sequence of instructions. Let T_f denote the disassembled code of a target function f consisting of p instructions. Then T_f can be defined as:

$$T_f := \langle I_f[1], I_f[2], \cdots, I_f[t], ..., I_f[p] \rangle \tag{1}$$

where $I_f[t]$ is the tth instruction of the target function f. Each instruction $I_f[t]$ consists of one `mnemonic` and a list of `operands` which are denoted by op^t and $\langle d_1^t, d_2^t, \cdots, d_m^t \rangle$ respectively. In this work, we use the operand type instead of the operand as the input, thus $I_f[t]$ is defined as:

$$I_f[t] := \langle op^t, \Phi(d_1^t), \Phi(d_2^t), \cdots, \Phi(d_m^t) \rangle \tag{2}$$

where $\Phi(d)$ is the operand type of operand d.

With the above definitions, we are now ready to state our problem definition. Our goal is to learn a model Ψ, which is used to classify the compiler's optimization level for a target function f. The optimization level τ is defined as:

$$\tau ::= -O0 | -O1 | -O2 | -O3 | -Os \tag{3}$$

4 Detailed Design

The overall architecture of HIMALIA is shown in Fig. 1. HIMALIA has three modules: preprocess module, 4-classification model and 2-classification model. Given a binary file, the input preprocess module converts it to a list of functions with each function consisting of a set of instructions which are represented in a well-designed way. For each function, HIMALIA recovers its optimization level in two steps. The processed data first go to the 4-classification model which outputs one of the following four labels: `-O0`, `-O1`, `-O2/-O3` and `-Os`. If the output label is `-O2/-O3`, the data are then fed to the 2-classification model which outputs the final label, namely `-O2` or `-O3`.

4.1 Input Preprocess

In this step, we transform the binary file into a list of functions which will be fed to our machine learning model. The input preprocess consists of three steps. First, we disassemble the binary file. Then, we employ the standard Linux `objdump` utility to locate the boundaries of the functions. Since disassembling the binaries and locating the boundaries of functions are not our contribution, we do not discuss them in more details.

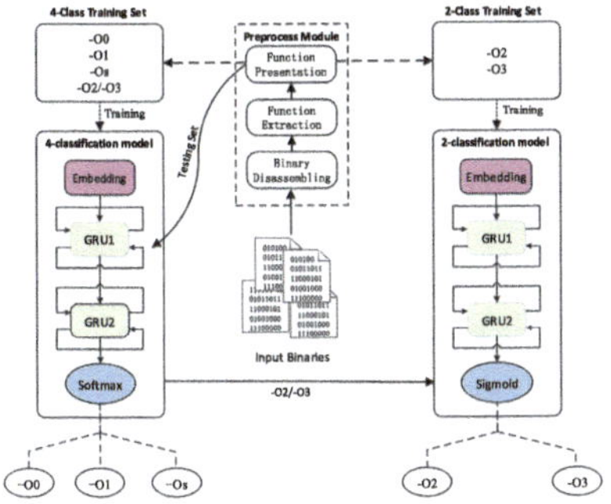

Fig. 1. Architecture of HIMALIA.

In the third step, we represent the functions. Bao *et al.* [9] define the functions as a set of bytes. They treat the code itself as a sequence of bytes $C[0], C[1], \cdots, C[l]$, where $C[i] \in Z_{256}$ is the ith byte in the sequence. However, the above method does not support the explicability of the results of deep learning models and does not capture the semantic information of the instructions. In this work, a function is represented in a new way which can retain the semantic information of the instructions as much as possible. Given the disassembled form of a function T_f, the function is represented as a list of instructions $\langle I_f[1], I_f[2], \cdots, I_f[t], ..., I_f[p]\rangle$ where $I_f[t]$ is the tth instruction of the function. For each instruction $I_f[t] : \langle op^t, d_1^t, d_2^t, \cdots, d_m^t\rangle$, the mnemonic op^t remains unchanged, but the operands $d_1^t, d_2^t, \cdots, d_m^t$ are turned into the operand types, denoted as $\Phi(d_1^t), \Phi(d_2^t), \cdots, \Phi(d_m^t)$ where $\Phi(\cdot)$ is a function that maps an operand to the operand type [12]. All the operand types are listed in Table 1. For example, the instruction `movzx ecx, byte ptr [rdi]` will be turned into `movzx R, B`.

Table 1. Operand types

Operand type	Description
R	The operand is a general register
S	The operand is a control register
I	The operand is an immediate
N	The operand is a near address
F	The operand is a far address
M	The operand is a memory reference
B	The operand has a "base + index" addressing mode
D	The operand has a "base + index + displacement" addressing mode

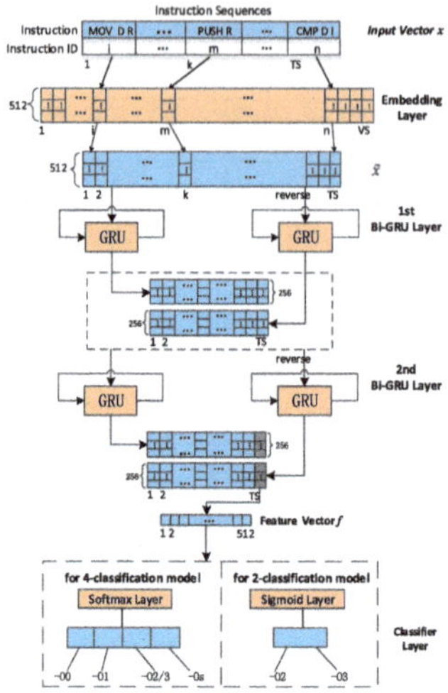

Fig. 2. Network architecture of our classification model.

4.2 Network Architecture

As shown in Fig. 2, Both 4-classification model and 2-classification model have a similar network architecture except for the classifier layer. The input to the network is a function represented by a set of processed instruction. The output of the network is the optimization level of the input function. To capture semantics of the instructions, we employ an embedding layer which will be trained together with recurrent layers. The detail methods are introduced as follows.

4.2.1 Function Vectorization

We convert function to vector by the following step. First, we create a dictionary D which contains all possible instructions.[2] Second, given a function $T_f = \langle I_f[1], I_f[2], \cdots, I_f[t], \cdots, I_f[p] \rangle$, we convert it to $\boldsymbol{x} = \langle i_1, i_2, \cdots, i_t, \cdots, i_p \rangle$ where i_t is the index of the instruction $I_f[t]$ in the dictionary D. Third, since the functions may have different lengths, so each vector is padded or reduced to a fixed length L which is computed as $L = \mu + SD$, where μ is the average length of all function and SD is the standard deviation length of all functions. There are plenty of important information at the beginning and ending of the function, so we will reserve these two parts. For each vector $\boldsymbol{x}$, if $|\boldsymbol{x}|$ is smaller than L, $L - |\boldsymbol{x}|$ zeros are padded in the middle of $\boldsymbol{x}$; otherwise, $|\boldsymbol{x}| - L$ elements in the middle of the vector are removed from $\boldsymbol{x}$.

[2] All the instructions are processed in the way described in the input preprocess section.

4.2.2 Instruction Embedding

After function vectorization, each instruction is represented by its index in the dictionary. However, this representation can not capture the semantics of instructions from their contextual use in the binary. One general approach to extract contextual relationships is to employ a technique called word embedding which converts each instruction to an m-dimensional vector. For example, Chua *et al.* used skip-gram negative sampling for word embedding and trained it independently. However, their method is an unsupervised-learning-based method, and can not learn the semantics of the instructions as a whole. In this work, we randomly initialize an embedding layer and train it together with other layers of the network for instructions embedding. Given a function $\boldsymbol{x} = \langle i_1, i_2, \cdots, i_t, \cdots, i_L \rangle$ as the input, the embedding layer converts it to a 2D tensor $\tilde{X} = \langle \boldsymbol{v}_1, \boldsymbol{v}_2, \cdots, \boldsymbol{v}_t, \cdots, \boldsymbol{v}_L \rangle$ where $\boldsymbol{v}_t$ is a m-dimensional vector representing the tth instruction of the function. The embedding layer has a trainable weight matrix $\tilde{W}$ with size $(|D| + 1) * m$ where $|D|$ is the size of dictionary D (in this work we set m to be 512). One of the advantages of using embedding layer is that during the training phase, the values of the parameters in the layer can be updated by their gradients to the loss function for the specific task.

4.2.3 RNNs for Semantic Capturing

We finally used two bidirectional GRU (BIGRU) [13] layers for semantic capturing. To design a module for capturing the semantics of an instruction sequence, we have considered and tested various architectures, like CNN, RNN, and their variations such as LSTM and GRU, etc. We find that the RNN model is a suitable choice because it has a notion of "memory" and can capture contextual information as far as possible when learning instruction representations.

4.2.4 Classifier Layer

We employ a simple full connected layer as our classifier layer which takes the feature vector $\boldsymbol{f}$ as input. For the 4-classification model, the layer is activated by a softmax function and has 4 output units which represent the probability distribution over `-O0`, `-O1`, `-O2`/`-O3` and `-O4`. We take the index of the max value as the predicted optimization level. For 2-classification model, the layer is activated by a sigmoid function and has only one output unit. If the output is larger than 0.5, the predicted label is `-O2`; otherwise, is `-O3`.

4.3 Training

The 4-classification model and the 2-classification model are trained separately using the RMSprop optimizer for 30 epochs. We used the default values of Keras v2.0.6: $lr = 0.001$, $rho = 0.9$ and $epsilon = 1e - 06$. The mini-batch size is set to 256. We initialize all weights of the network by uniform distribution random value and use the cross-entropy as our loss function. We randomly sample 80% of our dataset and use them as the training set and the remaining are used as the testing set.

5 Evaluation

In this section, we describe the experimental results on our own dataset and report our findings about the explicability of HIMALIA. We ran our experiments on a 48 cores Intel Xeon E5-2650 v4 machine with 256 GB of RAM equipped with 2 Tesla P100 cards. The NVIDIA driver version is 384.90 and the CUDA SDK version is 8.0. We implemented our models using Keras v2.0.6 with Tensorflow v1.2.1. All training and testing models in this work spent a total of a few days.

5.1 Dataset and Ground Truth

Because we are the first to employ machine learning to recover compiler information from binaries, we build our own dataset and make it publicly available to seed future improvements. Our dataset consists of 378, 695 different functions from 5828 binaries compiled by gcc-4.8.5, gcc-5.3.1 and gcc-6.2.1, as shown in Table 2. Our dataset is generated as follows. We first download the source code of 399 popular open-source projects such as bash, openssh, putty, sqlite, ntp and gzip, and then compile the source code with five optimization levels, namely `-O0`, `-O1`, `-O2`, `-O3`, `-Os`. We obtain the ground truth for the function optimization levels from the compiler setting when we compiling the source code. In order to simplify our dataset, we only keep the functions whose binary form are all different when 5 different optimization levels are used. Meanwhile, we filter out the functions that contain less than 20 instructions, because such functions are mostly stub functions [3] which almost have no value for binary analysis.

Table 2. Our dataset

Compiler	Version	Number of functions	Number of functions for each optimization level
GCC	4.8.5	131425	26285
	5.3.1	129890	25978
	6.2.1	117380	23476

5.2 Performance of Our System

We employ two BIGRU layers for semantic capturing in HIMALIA. We first evaluate the performance of HIMALIA with two BIGRU layers, and then compare performance with other 6 different models: `1-layer LSTM`, `2-layer LSTM`, `2-layer CONV`, `3-layer CONV`, `2-layer CONV` + `1-layer LSTM` and `1-layer LSTM + 2-layer CONV`. The experimental results are shown in Table 3. We can have the following observations from the results.

[3] Stub functions are typically functions which have been defined but have no real code in them. For examples, most run-time library functions are stub functions.

Table 3. Classification results

Model Layers	Parameters	Preision					Recall					Accuracy
		O0	O1	O2	O3	Os	O0	O1	O2	O3	Os	
CONV*2	15,821,828	0.9995	0.9846	0.6674	0.5275	0.9334	0.9917	0.9251	0.5880	0.6434	0.9588	0.8219
CONV*3	15,887,620	0.9997	0.9557	0.5850	0.5721	0.9576	0.9476	0.9288	0.6017	0.6287	0.9257	0.8135
CONV*2+LSTM	16,873,988	0.9994	0.9977	0.7174	0.7133	0.9931	0.9998	0.9962	0.7137	0.7202	0.9898	0.8828
LSTM+CONV*2	17,397,252	1.0000	0.9971	0.6569	0.5773	0.9924	0.9996	0.9967	0.6064	0.6304	0.9920	0.8442
BIGRU*2	**17,987,588**	**0.9997**	**0.9977**	**0.7356**	**0.7405**	**0.9953**	**0.9999**	**0.9976**	**0.7385**	**0.7408**	**0.9907**	**0.8934**
LSTM	17,200,132	0.9998	0.9976	0.7263	0.7067	0.9926	0.9997	0.9975	0.7100	0.7235	0.9925	0.8842
LSTM*2	18,775,044	0.9998	0.9967	0.7515	0.7048	0.9943	0.9997	0.9973	0.7162	0.7436	0.9908	0.8890

- HIMALIA exhibits accuracy of around 89.3%. The results show that it can recover compiler optimization levels of binaries with a high accuracy.
- BIGRU model has a higher accuracy than other network architectures, particularly with regards to `-O2` and `-O3`. For `-O0`, `-O1` and `-O3`, it also outperforms other network architectures, but the gap is not very significant.
- The experimental results also show that the accuracy of recovering `-O0`, `-O1` and `-Os` is much higher than the accuracy of recovering `-O2` and `-O3`. The results are in agreement with artificial experience of reverse engineering. We hypothesize that the optimization strategy of `-O2` is very similar with that of `-O3`. If we treat `-O2` and `-O3` as the same class, the model would have a very high accuracy (exceeding 99%).

5.3 Effectiveness of Our Instruction Preprocessing

During the data preprocess, we propose a new instruction representation method which represents an instruction by its mnemonic and the types of its operands. In the evaluation, we compare the performance of our method (mnemonic + operand type) with Bao's method (raw byte) [9]. Figure 3 shows the F1 scores for each compiler level under above two methods. The results show that our method outperforms Bao's method. In particular, the F1 scores of `-O2` and `O3` when using our method are much larger than Bao's method. The reason is that our method can effectively retain the semantic information of the instructions.

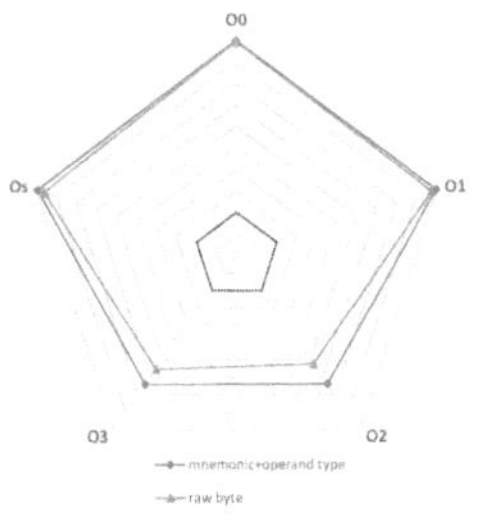

Fig. 3. F1 Scores for each compiler optimization level under different instruction representation methods.

5.4 Explicability of Our Model

Neural networks have long been known as "black boxes", because it is difficult to understand exactly how it works due to the complex structure and the large number of parameters. To explore how HIMALIA "understands" the input binary file, we employ saliency map technique to visualize the importance of each instruction in a prediction.

5.4.1 Saliency Map

Saliency map [14] is commonly obtained by computing the gradient of the network's output with respect to the input. Let $\boldsymbol{g}_t$ denote the gradients of the feature vector $\boldsymbol{f}$ with respect to $\boldsymbol{v}_t$ which is the representation of the tth instruction. The saliency of tth instruction is a scalar s_t and is calculated by the following equation:

$$s_t = \sum_{i=1}^{m} g_t[i] \times |g_t[i]| \tag{4}$$

where $g_t[i]$ is the ith element of g_t and m is the length of $g_t[i]$. So for each function, we get a vector $\boldsymbol{S} = \langle s_1, s_2, \cdots, s_t, \cdots, s_L \rangle$. For visualize our saliency map, all the elements of $\boldsymbol{S}$ are scaled to $0 \sim 100$ by z-score and min-max methods.

5.4.2 Findings

We take the function `log_SB_buf` (shown in Fig. 4) as an example to demonstrate the explicability of HIMALIA. We can observe that HIMALIA considers instructions in `line 1,2,5,6,11,17` in Fig. 4`(a)` as the 6 most important instructions for the compiler optimization level `-O0`. On the other hand, instructions in `line 1,9,16` in Fig. 4`(b)` are considered as the 3 most important instructions for the compiler optimization level `-O2`. After analyzing the above results, we hypothesize that HIMALIA learns to use the following compiling conventions in the classification:

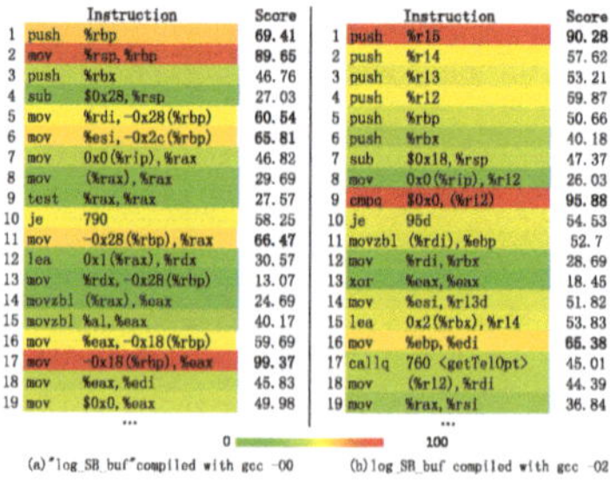

Fig. 4. Relative score of importance generated by saliency map for each instruction.

- *Function Prologue Convention:* When the optimization level is set to `-O0`, the compiler usually uses the function prologue `push %rbp` and `mov %rsp,%rbp` (line 1 and line 2 in Fig. 4(a)) to save the callers stack frame. But when the optimization level is set to `O2`, the compiler does not maintain the stack frame. Instead, it pushes the registers that will be used at the beginning of the function (line 1 in Fig. 4(b)), which can help to improve the efficiency. We can observe that all the three instructions have very high scores, which indicate that HIMALIA has learnt the function prologue convention.
- *Function Parameter Passing Convention:* When the optimization level is set to `-O0`, the compiler usually uses register `%rdi`, `%esi`, `%rsi`, `%rdx`, `%rcx`, etc. to pass parameters to sub-functions. The function typically saves the parameters in the general registers to the stack frame before they are used (line 5 and line 6 in Fig. 4(a)). When the parameters are used, they are then moved from the stack frame to the general register (line 11 in the Fig. 4(a)). Besides, the compiler typically uses the register *%edi* to pass the parameter for `-O2` (line 16 in Fig. 4(b)). We can observe that `mov %rdi,-0x28(%rbp)`, `mov %esi,-0x2c(%rbp)`, `mov -0x28(%rbp),%rax` (line 5, line 6 and line 11 in Fig. 4(a)) and `mov %epb,%edi` (line 16 in Fig. 4(b)) are all given high scores, which agrees with the above conventions.

Beside, it is more exciting to find that HIMALIA seems to have learned semantic information from contexts of instructions:

- *Redundant Instruction:* `-O0` is typically the compiler's default optimization level which turns off optimization entirely to reduce compilation time, and therefore produces several redundant instructions. We observe that `mov -0x18(%rbp),%eax` (line 17 in Fig. 4(a)) and `mov %eax,-0x18(%rbp)` (line 16 in Fig. 4(a)) locate next to each other in Fig. 4(a). The semantic meaning of these two instructions is to move the value in the stack `-0x18(%rbp)` to the register `%eax`, and then move the value in the register `%eax` back to the stack `-0x18(%rbp)`, which logically means the second instructions are redundant. In Fig. 4(a), HIMALIA assigns the instruction `mov -0x18(%rbp),%eax` (line 17 in Fig. 4(a)) a high score for `-O0`. This case shows that HIMALIA can identify redundant instructions through their contexts, and use them as clues to the recover `-O0`.
- *Multi-Functional Instruction:* `-O2` is an advanced compiler option which tries to produce faster code, and therefore usually uses multi-functional instructions. We hypothesize this is why HIMALIA gives a high score to the instruction `cmpq %0x0,(%r12)` (line 9 in Fig. 4(b)) which locates in a classic load-compare-jump sequence: `mov 0x0(%rip),%r12`, `cmpq %0x0,(%r12)`, `je 95d` (line 8, line 9 and line 10 in Fig. 4(b)).

6 Conclusion and Future Work

In this paper, we propose a neural-network-based system called HIMALIA to recover compiler optimization levels of binaries. We implement and test its performance on our dataset. The experimental results show that HIMALIA can achieve high accuracy and the results are also explicable.

Although we have tested many models to distinguish `-O2` and `-O3`, but the results can be still further improved. Besides, the saliency map in this work is obtained by a simple way. In the future, we will compute a more fine-grained saliency map which is expected to reveal more explicability of our model.

Acknowledgment. This work was supported by National Key Research and Development Program of China (2016YFB0800202); National Natural Science Foundation of China under Grants No. U1636120; Fundamental Theory and Cutting Edge Technology Research Program of Institute of Information Engineering, CAS; SKLOIS (No. Y7Z0361104 and No. Y7Z0311104) and Science Foundation Ireland under Grant Number 13/SIRG/2178.

References

1. Hoste, K., Eeckhout, L.: Cole: compiler optimization level exploration. In: IEEE/ACM International Symposium on Code Generation and Optimization, pp. 165–174 (2008)
2. Wang, X., Zeldovich, N., Kaashoek, M.F., Solar-Lezama, A.: Towards optimization-safe systems: analyzing the impact of undefined behavior. In: Twenty-Fourth ACM Symposium on Operating Systems Principles, pp. 260–275 (2013)
3. David, Y., Partush, N., Yahav, E., David, Y., Partush, N., Yahav, E., David, Y., Partush, N., Yahav, E.: Similarity of binaries through re-optimization. ACM Sigplan Not. **52**(6), 79–94 (2017)
4. Caliskanislam, A., Yamaguchi, F., Dauber, E., Harang, R., Rieck, K., Greenstadt, R., Narayanan, A.: When coding style survives compilation: de-anonymizing programmers from executable binaries (2016)
5. Egele, M., Woo, M., Chapman, P., Brumley, D.: Blanket execution: dynamic similarity testing for program binaries and components. In: Proceedings of the 23rd USENIX Conference on Security Symposium (2014)
6. Abou-Assaleh, T., Cercone, N., Keselj, V., Sweidan, R.: N-gram-based detection of new malicious code. In: International Computer Software and Applications Conference - Workshops and FAST Abstracts, pp. 41–42 (2004)
7. Rieck, K., Holz, T., Willems, C., Dssel, P., Laskov, P.: Learning and classification of malware behavior. In: International Conference on Detection of Intrusions and Malware, pp. 108–125 (2008)
8. Rosenblum, N., Zhu, X., Hunt, K., Hunt, K.: Learning to analyze binary computer code. In: National Conference on Artificial Intelligence, pp. 798–804 (2008)
9. Bao, T., Burket, J., Woo, M., Turner, R., Brumley, D.: Byteweight: learning to recognize functions in binary code. In: Usenix Security Symposium (2014)
10. Shin, E., Song, D., Moazzezi, R.: Recognizing functions in binaries with neural networks. In: USENIX Security Symposium, pp. 611–626 (2015)
11. Chua, Z.L., Shen, S., Saxena, P., Liang, Z.: Neural nets can learn function type signatures from binaries. In: 26th USENIX Security Symposium (USENIX Security 17), pp. 99–116. USENIX Association, Vancouver, BC (2017). https://www.usenix.org/conference/usenixsecurity17/technical-sessions/presentation/chua

12. Intel, I.: Intel 64 and ia-32 architectures software developers manual. Volume 3A: System Programming Guide, Part, 1(64) (2016)
13. Cho, K., Van Merriënboer, B., Gulcehre, C., Bahdanau, D., Bougares, F., Schwenk, H., Bengio, Y.: Learning phrase representations using RNN encoder-decoder for statistical machine translation. arXiv preprint arXiv:1406.1078 (2014)
14. Simonyan, K., Vedaldi, A., Zisserman, A.: Deep inside convolutional networks: Visualising image classification models and saliency maps. Comput. Sci. (2013)

Architecture of Management Game for Reinforced Deep Learning

Marko Kesti(✉)

Faculty of Social Sciences, University of Lapland, Rovaniemi, Finland
marko.kesti@ulapland.fi

Abstract. This article proposes that bona-fide theory for human resource (HR) management connection to performance should include a scientifically approved architecture with explaining power and game theoretical approach that address management behavior tendencies and workplace problems countertendencies to human performance. Management practices have tendencies to improve workers' human performance. Workplace problems have tendencies to reduce human performance. Game theory is useful because management practices are situation-sensitive, with causal effect on business performance. Deep reinforcement learning with artificial intelligence provides emerging new possibilities, which may revolutionize organizations' HR-management. This article presents human capital theories with a game theoretical approach. The stochastic Bayesian game seems to be suitable for describing leaders' behavior meaning to staff performance and annual profit. Using Bayesian management game, it is possible to simulate the management learning outcome where both well-being and business performance flourish. In this case, the managers (players) succeed in achieving the Nash equilibrium between staff quality of working life and sustainable profitability.

Keywords: Management game · Human resource management · Leadership Game architecture · Q-learning · AI

1 Introduction

Critical scientists Fleetwood and Hesketh [6] and Ehrhart et al. [5] say that science connecting HR-management to performance is broken. This means that the connection between staff human qualities and performance is more complex than previous research anticipated. Correlation is not necessarily telling that causality occurs, and causality is not enough without a proper explanation of a phenomenon behind causality.

As long as scientists are not able to explain how human qualities affect business performance, leadership programs cannot solve the problem of the performance at operational level. Explanatory power is necessary for making the development of organization performance effective. Furthermore, management performance may include psychological issues. Solving work-related problems requires time, which reduces the working capacity to make revenue. A decrease in monthly revenue results in a decrease

K. Arai et al. (Eds.): IntelliSys 2018, AISC 868, pp. 48–61, 2019.
https://doi.org/10.1007/978-3-030-01054-6_4

in profit. However, solving problems will reduce fuss in the future. Therefore, if executives require monthly revenue and profit, the leader may have a social dilemma-whether improve staff's well-being or maximize the monthly profit.

The best practices of HR management seem to have tendencies to improve workers' performance. Furthermore, workplace problems have tendencies to decrease human performance. If these tendencies are identified, there can be created tendential predictions for each management activity [6].

HR development may improve the profit in two ways: by cost savings and by business profit efficiency. Increasing the effective working time may allow making more revenue. More revenue with the same costs increases the profit, thus it improves the efficiency in making profit. Cost savings on staff absence and turnover are easy to measure and calculate. Better work efficiency is more difficult. It needs the measurement and analysis of staff's quality of working life (QWL). An improvement of the QWL may reduce staff costs and increase work efficiency. The effects of fiscal performance can be analyzed by the function of human capital productivity.

The game of human capital management is non-symmetric because each player has different roles and reward functions. Workers get reward from their improved self-esteem, which is measured by QWL index. Supervisors get reward from the team profit. The article management game is signaling game where workers give opinions (signals) on problems that threaten their QWL. Supervisors may enhance the problem solving because a leader knows that unsolved problems cause fuss and therefore eat the long-term profit.

Game theory is useful when, due to the complexity of the domain, the solution is difficult to analyze with preprogrammed agents' behaviors [3]. In multi-agent reinforcement learning, the agents must discover a solution on their own, using experience-based rational learning [3, 4]. The Bayesian game is a strategic game with imperfect information [21]. The player has a certain base strategy that she/he updates according to the information. As the information is imperfect, a stochastic Bayesian learning phenomenon occurs.

2 Architecture of a Management Game

The aim of performance management is to measure and develop an organization's performance at all levels, and align it with strategic targets. Nevertheless, how can the effect of employee performance on business performance be reliably analyzed? This question is important when seeking competitive advantages through human capital productivity. The new science provides a theoretical architecture for the management game.

First, it is necessary to understand how employees produce economic value. Staff is not only a cost. Staff work motivation and innovativeness may provide the source for competitive advantage. In addition, successful investment in technology or product development may form a competitive advantage. The main value of HR consists in the organization's capabilities to utilize human intangible assets. Thus, effective HR management and development are keys to success. Staff QWL is a new index that defines employee intangible performance. It is one of the most important scorecards of HR management and a production parameter in the theory of Human Capital Production Function [11, 16, 18].

The function of human capital production is

$$R = K * L * TWh * (1 - Ax) * QWL \tag{1}$$

where

R = Revenue [$]
K = Coefficient for effective working time revenue relation,
HR business ratio [$/h]
L = Labor capacity in full-time equivalent [pcs]
TWh = Theoretical yearly working time [h]
QWL =Quality of working life, indicating utilization of human capital intangible asset (0–100%)
Ax = The auxiliary working time of the total theoretical working time (vacation, absence, family leave, orientation, training, HR practices, and HRD) (%)
(1 − Ax) = (100% − Ax) = Time available for actual work (time spent at work)
(1 − Ax) * QWL = Effective working time from the theoretical working time

The production monetary volume is the revenue, and the operating profit (EBITDA) is the revenue minus operative costs.

Figure 1 illustrates an example of the use of the function of human capital production at team level. By improving the QWL, the company makes €4,000 more EBITDA per employee. In practical case studies, these levels of profit increases have been achieved [14].

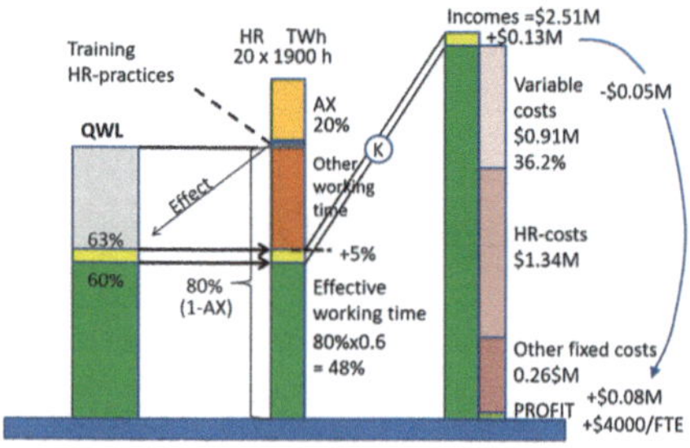

Fig. 1. The phenomenon of the improvement in team level productivity.

The QWL determines the effective working time. The staff effective working time produces the revenue. The coefficient K describes the business area, tangible investments, and business logic. The improvement of QWL requires that HR development time go to auxiliary working time (Ax). Thus, it reduces the time for work. When efficient HR development is conducted, the effective working time can be increased, despite the time that is invested in management practices. In addition, if absence and turnover are high, HR development may reduce them. As a result, the annual Ax-time may not increase, and a positive effect on profit may occur.

Staff well-being consists in self-esteem categories that affect human performance in different phenomena. Anxiety will only affect the performance, if it is present. Creativity will boost the performance, but only if negative feelings are not taking the staff's focus. Thus, it seems obvious that well-being connection to performance is not simple statistics. A new scientific method solves this problem: the QWL index [17].

In Herzberg's [8] motivation theory, human performance is hygiene factors multiplied by motivation factors. Hygiene factors cause distress, while motivation factors tend to increase the performance. The theory of the QWL index includes three self-esteem categories; each has a unique effect on performance. The self-esteem categories are (Fig. 2):

(1) Physical and emotional safety (PE),
(2) Collaboration and identity (CI),
(3) Objectives and creativity (OC).

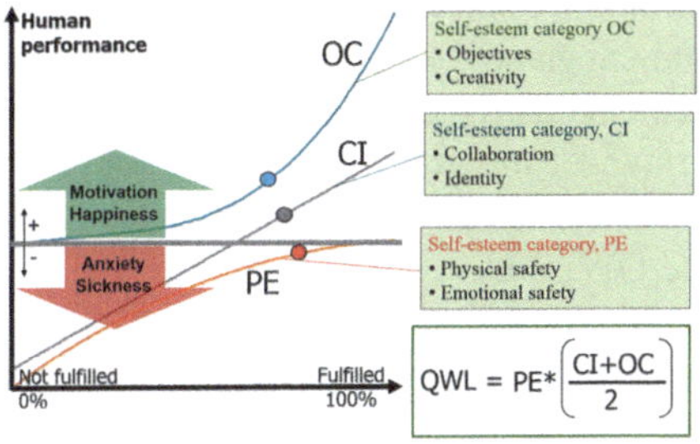

Fig. 2. Self-esteem categories of the QWL-index.

Chosen categories and their effect on performance form the theory of the QWL index. It is also important to know that the QWL index has logical connection to customer satisfaction, in accordance with Kano et al.'s model [10].

The QWL index is calculated using the following equation [17]:

$$QWL = PE(x1) * ((CI(x2) + OC(x3))/2) \tag{2}$$

Where,

QWL is the quality of working life index (0 … 1) PE(x1) is the function of physical and emotional safety CI(x2) is the function of collaboration and identity OC(x3) is the function of objectives and creativity.

The functions of the self-esteem categories are adjusted so that the result is always between 0 and 100% (0 … 1). Thus, the QWL index is the scorecard for employee performance and for the utilization of human intangible assets [17].

3 Game Theoretical Approach to Management

It seems that the Bayesian theorem is useful in management simulation with dynamic analysis of sequences. The player's (leader) strategy hypothesis guides the actions at different events. Management practices and workplace problems have both certain predicted tendencies affecting workers' self-esteem. As the tendencies consequence data update the state after each sequence, the player may update his/her leadership strategy, which further guides the next actions. Bayesian probability is related to the probability of a player's subjective behavior that rational thinking will lead to optimal result as the new information comes available [24].

In a management game, the player should learn the optimal leadership strategy without knowing the exact reward function or state transition function. This approach is called model-free reinforcement learning and can be defined with the Q-learning approach [7]. The leader has prior belief about the state of the nature of an organization's business. The state of the nature comes from the architecture of the management game that has predictive capabilities.

The game is non-symmetric since the game stakeholders have unique roles in the organization. A management game has three stakeholder groups: workers, team leaders, and executives (management board members). Executives are responsible for the organization's overall performance, investments in technology, and business logic. These decisions may improve K-coefficient, but also the tendency to cause fuss in the organization and, thus, reduce QWL index.

Workers are responsible for their work performance. Therefore, they are interested in the QWL and give signals, if problems that threaten their self-esteem occur. Workers preference strategy is to give their leader opinions about the problems. In a simplified digital team of leaders' learning game, the worker's strategy may be stationary, meaning that workers' behavior may be chosen in advance when the events scenario is known.

A team leader is responsible for the team economic performance, that can be, for example, EBITDA. The leader registers workers, and signals and makes own prior belief for own leadership strategy, $\rho_{Li}(\tau_{wi}(t_{wi})) > 0$ for all $t_{wi} \in T_{wi}$. The leader also registers signals from business outcomes, in form of monthly and cumulative profit, $\rho_{Li}(\Omega_{Profit}, \tau_{wi}(t_{wi})) > 0$. The leader has a prior belief strategy on how to act on these signals. The leader is rewarded by the profit at the end of the year and by positive monthly feedback, if the QWL is improved. The leader knows that the yearly profit is the sum of cumulative monthly profit. After acting, the leader will get outcome signals (Ω_{Profit}) from the state change and from the worker's response signal $\tau_{wi}:\Omega_{\Delta QWL} \rightarrow T_{\Delta wi,}$ where $\Omega_{\Delta QWL}$ is the state change at workers' QWL. Outcome signals and reward results may cause changes in the leader's preference strategy for the next sequence.

The leader reward function is $\gamma_L(\omega_{Li}, \pi_{Li})$, where ω_{Li} is the combination of monthly profit situation and QWL change. π_{Li} is the leader's strategy in the current month. It seems that, at the beginning, the leader strategy is weighted at the monthly profit, but that later the QWL change starts to be more interesting in adjusting the strategy for optimizing cumulative yearly profit. In this paper, the stochastic nature of the leadership game is a key to learning the Nash [20] general sum equilibrium between the QWL and profit.

$$\gamma_L() = \text{Revenue} - \text{Costs} = \text{Profit} \tag{3}$$

The QWL is improved by leadership actions that reduce the monthly working time for making the revenue. Thus, improving the QWL reduces monthly revenue and profit, but may increase effective working time in the future and so increase the future profit. On a monthly basis, this phenomenon may be contradictory and confusing, but, by practice, the best reward is achieved where both workers' and leader's payoff functions flourish. This means the Nash equilibrium [20], where yearly QWL is improved with high cumulative profit. In Nash equilibrium, the leader's choices are the best response to the workers' signals and business cumulative outcome at the end of the year.

The function of human capital production with the QWL index forms the architecture for the management game. This architecture is supplemented by several empirical evidence-based rules [11–14, 22]. These rules make the changes of the simulation state more realistic. So far, the following rules are included:

- The QWL can be improved by effective management practices (tendential improvement predictions of HR practices).
- Implementing management practices requires working time.
- Work-related problems decrease the QWL (tendential counter predictions).
- The effect of HR practices is dependable on the leaders' skills on a chosen practice.
- In some HR practices, the effect of improvement decreases if they are used too often.
- As self-esteem increases, its further improvement becomes more difficult (self-esteem factors have productivity improvement boundary).

All these rules act simultaneously and thus make the game domain rather complicated to understand deeply. However, with simulation rational learning, the player may find an optimal solution, according to reinforcement learning methodology [3]. The player makes decisions in the management game simulation as follows:

- The player makes decisions based on the chosen strategy and HR-practice skills.
- The player may use workers' opinions as signals to choose effective situational HR practices.
- The player makes decisions based on signals from fiscal results and team QWL measurement.

In a team management game, the players (workers and team leader) reward functions are related since the QWL affects long-term profit. The equilibrium of the dominant strategy is achieved when the workers and the leader know that the best outcome is the situation where problems are solved effectively. This situation needs cooperation where workers give helpful opinions (signals) for identifying and solving the problems. Many times the equilibrium is difficult to find because workplace problems tend to be fussy and problem-solving skills may be inadequate. In a good organization culture, the parties know that a non-cooperative state of mind forms barriers to problem identification and solving, and unsolved problems will eventually reduce both the QWL and profit.

In some cases, the executive management may cause harmful social dilemma at supervisors and team level. Short-term monthly or quarterly profit requirement may force the leader to choose a strategy where HR activities are minimized to maximize the time for work. The leadership's social dilemma is the situation where the leader neglects the workers' signals to be able to meet the executives' profit target. However, the leader may know this strategy will eat the work well-being and reduce the long-term profit. Neglecting workers opinions may eventually change the game from cooperative to non-cooperative mode, where workers do not give verbal signals from work-related problems. The equilibrium of competitive advantage is very difficult to achieve at non-cooperative mode. Without a right signal interpretation, the leader cannot choose optimal HR actions to prevent a decrease in the QWL. When a negative spiral occurs, the leader will have to make decisions solely by the fiscal data from the past.

Bayesian stochastic strategic non-symmetric signaling learning game follows Markov's decision process [1, 19, 23]:

$$\langle S, A, \gamma, \rho \rangle \quad (4)$$

Where,

S is discrete state space (scenario)
A is discrete action space
γ: $S \times AA \rightarrow R$ is reward function
ρ: $S \times A \rightarrow \Delta$ is the transition function, where Δ is the set of probability distributions over state space S

Information is incomplete, but perfect. The agents (workers and leader) do not know other agents' payoff functions in detail, but they can observe other agents' immediate payoffs and actions from past months. A leader does not know exactly which actions would be the best, but he/she can choose actions that should be good enough. The leader will get workers' emotional feedback immediately and stochastic information from profit change. After several game rounds, the player (leader) will learn the optimal actions to improve both the QWL and annual profit. Thus, the player will achieve the Nash equilibrium of stochastic Markov learning game [23]. Leadership game learning function is:

$$Q_{t+1}(s, a) = (1 - \alpha_t)Q_t(s, a) + \alpha_t\left[\gamma_t + \beta\pi'(s_{t+1})Q_t(s_{t+1})\right] \quad (5)$$

Where,
$\beta \in [0,1]$ is the discounted reward factor
$\alpha_t \in [0,1]$ is the learning rate $(1 - \alpha_t)$
$\gamma_t = \gamma_{Profit} + \gamma_{QWL\epsilon}$ (where γ_{Profit} is the monthly profit and $\gamma_{QWL\epsilon}$ is the QWL change effect at a future profit).

The algorithm of the stochastic leadership game can be formed according to the multiagent reinforcement learning process [9, 19, 23]. The following algorithm is used:

(a) Initialize: Let $t = 0$ (January)
(b) For all s in S and a^1 in A^1 and a^2 in A^2 let $Q^1_{t+1}(s, a^1, a^2) = 1$
(c) Initialize first signal S_0 from workers
Loop

*Choose HR action a_t^1 based on strategy $\Omega^1(s_{t,})$ which is players's attempt behavior of Nash equilibrium solution in a strategy learning game $Q_t^1(s_t)$.
*Observe workers response to the change of QWL and reward γ_t^2 indicating the change at the QWL index.
*Observe rewardγ1t indicating the monthly and cumulative EBITDA.
*Observe S_{t+1} workers signals of the next month state (problem). State transition function can be predefined, for example according to the market situation (i.e., cash cow, recession, and growth).
*Update Q^1 such that
$Q_{t+1}^1(s, a^1, a^2) = (1 - \alpha t)\, Q^1 t(s, a^1, a^2) + \alpha_t\, [\gamma^1 t + \beta\pi^1(s_{t+1})\, Q^1 t(s_{t+1})\pi^2]$ where $\pi^1(s_{t+1})$ is mixed strategy Nash solution of mixed behavior strategy game $Q_t^1(S_t)$. Where, reward γ_t is $\gamma_{Profit} + \gamma_{QWL\epsilon}$, where agent 2 is workers' behavior according to stationary strategy π^2.

(d) Let t: = t + 1
When a player practices several simulation sequences, the optimal strategy π^* should emerge. In this learning, the player starts with arbitrary actions and observes the rewards.
Then, the player updates the strategy assumptions to more optimal assumptions for maximizing the operating profit. Therefore, the player updates the learning based on the following equation:

$$Q_{t+1}^1(s, a) = (1 - \alpha_t) Q_t(s, a) + \alpha_t \left[\gamma_t + \beta \max_b Q_t(s_1, b)\right] \tag{6}$$

Watkins and Dayan [25] proved that the sequence in this equation converges to the optimal $Q_t(s, a)$. Our practical studies seem to confirm Watkins and Dayan's findings that the Q-learning sequence converges to the optimal $Q_t(s, a)$.

4 Results

Our practical solution for management game is digital Unity-based learning game that allows studying the incremental Q-learning function (see Fig. 3). In the digital simulation, the players can learn essential leadership practices and strategy without the noise factors of a real-life working environment. In the simulation, the workers give opinions (signals) on possible problems. Three different state change scenarios occur, according to the market situation (i.e., cash cow, recession, and growth). Each has preliminarily adjusted problems that threaten the workers QWL. Workers are imaginary real persons with behaviors that are related to fixed state scenario. Workers give five different emotional expressions and verbal comments, according to the state change of QWL. Among workers' emotions, the leader can observe team QWL index and fiscal data of profit and revenue, both monthly and cumulative. The player has approximately 30 different management practices as action states to choose each month. These actions form the discrete action space for each month state space. One simulation year will take approximately 20 min at the learning game.

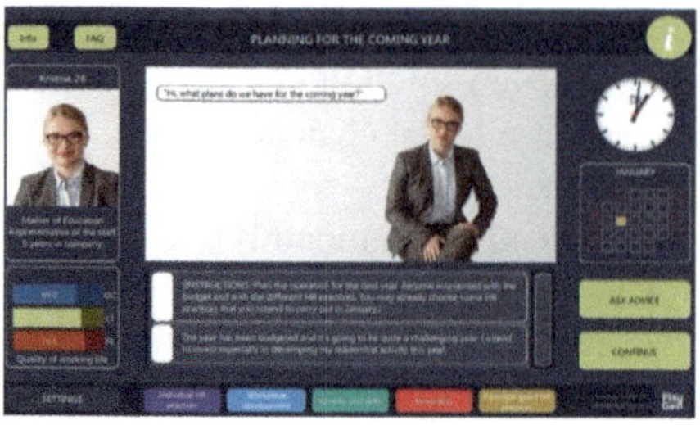

Fig. 3. User interface for game-based learning [15].

Practical studies indicate that management game theory with new digital applications provide completely new possibilities for management research, teaching and business analytics. In fact, a management game provides the methodology to make the most precise analytics for an organization's human performance.

Game simulation meets the principle of contingency theory, according to which an optimal leadership strategy is situation-sensitive and therefore may be different for each team. Companies can input their own company's data to the simulation and then analyze the best roadmap to success. Management can find optimal strategies to achieve a sustainable competitive advantage at different kinds of situations, scenarios, and leadership skills.

The focus of the management game Q-learning is on finding the best leadership strategy and behavior that fosters the team performance. Problems should be solved in time to maintain a basic performance level. Higher performance requires also HR practices that nurture employee motivation and innovativeness. In this way, the company can achieve competitive advantage where both well-being and business performance flourish.

Practical studies indicate that learning an optimal strategy is not so easy. Even though the game follows simplified reality with clear signals and without confusing noise factors, many students needed additional advice to learn the optimal strategic behavior. The learning game should not include too complicated social behavioral and strategic phenomena–the simulation of the whole reality would make it too complicated for learning purposes. The management game theory reveals the fact that achieving organizational competitive advantage is difficult. It seems evident that the management game theory with digital learning tools will speed up the management learning process.

The following paragraph provides a sample of a typical player's Q-learning history, which is divided into four learning phases (Figs. 4, 5, 6 and 7):

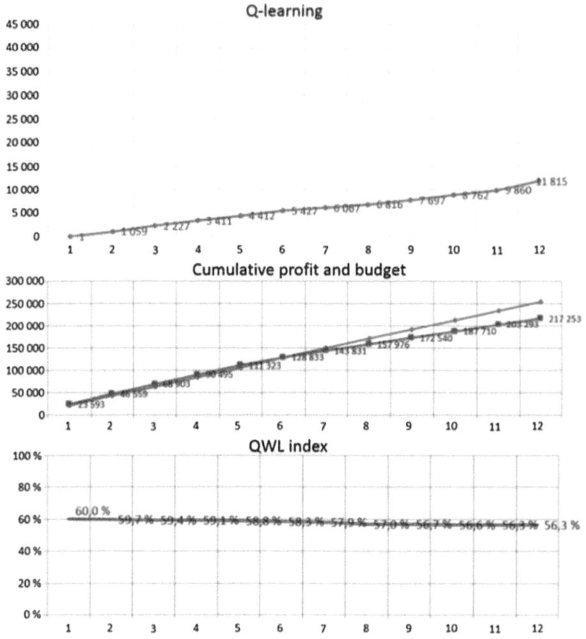

Fig. 4. Q-learning results when the player is not doing any management practices.

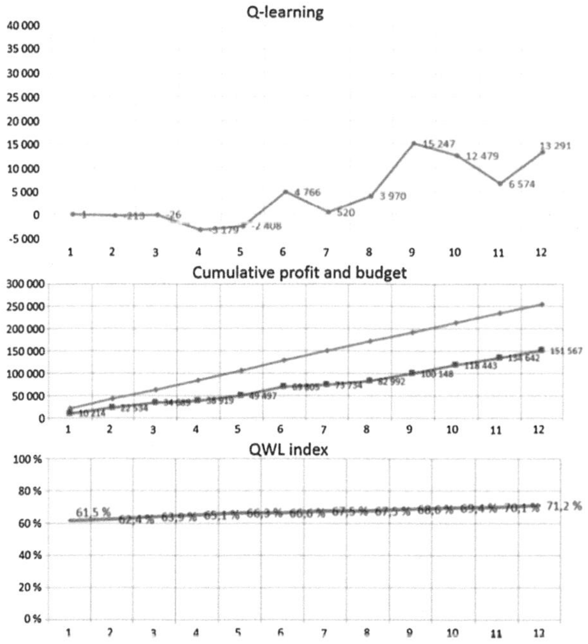

Fig. 5. Q-learning results when the player tests random management practices.

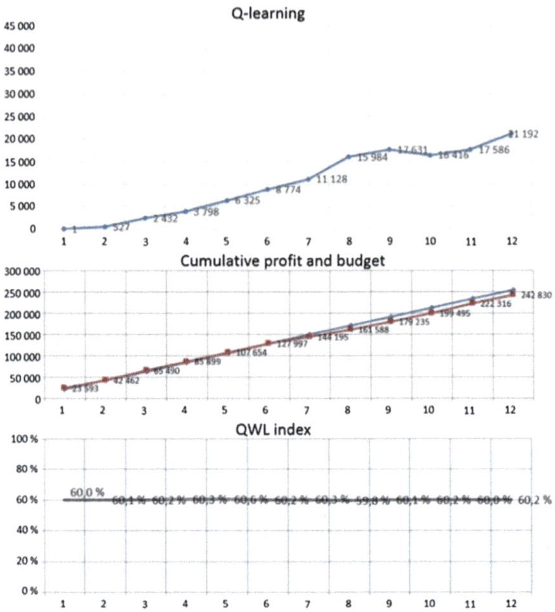

Fig. 6. Q-learning results when a player tries to solve workers' problems.

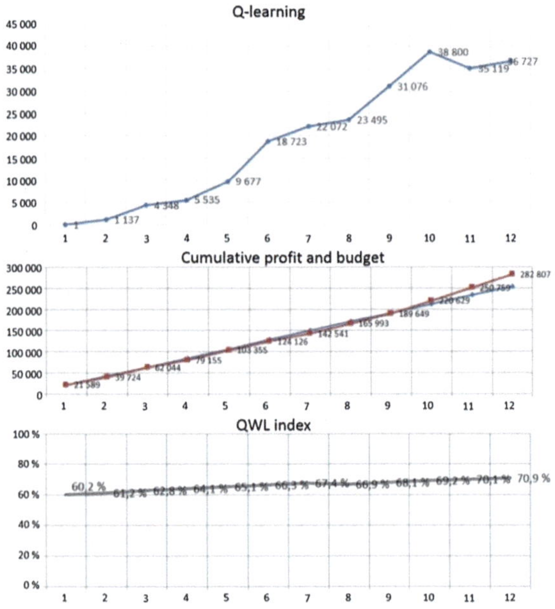

Fig. 7. Q-learning results when the player follows the yearly management practices plan and tries to solve the workers' problems.

Phase 1: A player tests the game without doing any HR practices, thus simulating the team without a leader. The QWL reduces and the profit goes under the budget. In the figures, the blue line is the cumulative budget.
Phase 2: A player tests the game by choosing HR actions randomly, just to see what happens. Even if the QWL improves, the economy collapses. Therefore, this strategy does not seem appropriate.
Phase 3: The player uses a strategy where problems are solved according to workers' signals. The QWL improves slightly and the profit reaches close to the budget.
Phase 4: The player does systematic HR practices (e.g., development discussions, target setting, and well-being inquiry). The QWL increases and the profit exceed the budget.

We analyzed 57 students' learning results, which supports the argument that rational players' Q-learning converges to the optimal. However, this was not so obvious because 23% of the students needed additional help and advice to reach good results. The learning results can be divided into three categories:

- 46% reached excellent results without help.
- 32% achieved good results without help.
- 23% needed additional help to reach good results.

Practical studies seem to meet Watkins and Dayan's (1992) argument that Q-learning sequence converges to the optimal Qt(s, a). However in some cases it needs additional supervised help.

5 Discussion for the Future

The architecture for a management game seems to work at the simulation learning game with the Q-learning approach. The game seems to generate curiosity about the management effect on team performance. Despite the learning game is simplified reality, some students had problems in learning the optimal solution. This may be due to the management social dilemma: it is necessary to sacrifice some profit to gain better profit later. The more a company tries to maximize the short-term profit, the more likely it will not maximize it in a longer period. HR development follows the phenomenon of investment.

Next, we will implement AI in the learning game. We believe that learning curiosity and efficiency can increased by an AI-powered educational game. When a player's learning is jammed, AI will help choosing the most optimal management actions. This would also add the element of interactivity, which enhances investment-based thinking when choosing the actions for different events. AI can give timely suggestions and so create a guided learning experience. In creating AI assistance, we will utilize Bellman advantage function.

In the future, the management game theory can be added with executive level strategic behavior, which contains investments and organization changes. This would add one layer more, since now the game focus is on team leaders learning and analytics. Research indicates that a management game theoretical approach provides a methodology for much more valuable management simulations in the future. Reinforced learning algorithms can be connected with AI neural network; which will create organization specific deductive rules for the simulation.

Bersin [2] suggests that the leadership development paradigm that many companies around the world follow is simply not delivering what is expected and necessary. It seems that AI powered management game theory cab foster organization productivity and may even revolutionize HR management. In a management learning game, real-time data can be utilized from different teams. Each team leader will simulate his/her own team's performance and learn to make optimal management activities in different coming situations. This approach will promote real-time situational awareness to maintain or improve the performance in different challenges. AI will assist line-managers in improving team performance and preventing possible problems.

References

1. Bellman, R.: A Markovian decision process. J. Math. Mech. **6**(5), 679–684 (1957)
2. Bersin, J.: Global Human Capital Trends 2014: Engaging the 21st-century workforce. Deloitte University Press (2014)
3. Busoniu, L., Babuska, R., De Schutter, B.: A comprehensive survey of multiagent ˇ reinforcement learning. IEEE Trans. Syst. Man Cybern. Part C Appl. Rev. **38**(2), 156–172 (2008)
4. Duryea, E., Ganger, M., Hu, W.: Exploring deep Reinforcement learning with multi Q-learning. Intell. Control. Autom. **7**, 129–144 (2016)
5. Ehrhart, M.G., Schneider, B., Macey, W.H.: Organizational Climate and Culture, An Introduction to Theory, Research and Practice. Routledge, New York (2014)
6. Fleetwood, S., Hesketh, A.: Explaining the Performance of Human Resource Management. Cambridge University Press, Cambridge (2010)
7. Harsanyi, J.C.: Games with Incomplete Information Played by Bayesian players, I-III. Manage. Sci. **14**(3), 159–183 (1967)
8. Herzberg, F., Mausner, B., Snyderman, B.: The Motivation to Work, 2nd edn. Wiley, New York (1959)
9. Hu, J., Wellman, M.P.: Multiagent reinforcement learning: theoretical framework and an algorithm. In: Proceedings of the 15th International Conference on Machine Learning, San Francisco, CA, pp. 242–25 (1998)
10. Kano, N., Seraku, N., Takahashi, F., Tsuji, S.: Attractive quality and must-be quality. Hinshitsu: J. Jpn. Soc. Qual. Control. **14**(2), 39–48 (1984)
11. Kesti, M., Syväjärvi, A.: Human resource intangible assets connected to the organizational performance and productivity. In: Ravindran, A., Shirazi, F. (eds.) Business Review: Advanced Applications, pp. 136–173. Cambridge scholars publishing, Cambridge (2013)
12. Kesti, M.: Strateginen henkilöstötuottavuuden johtaminen, Strategic human capital management, Talentum, Helsinki (2010)
13. Kesti, M.: The tacit signal method at human competence based organization performance development. University of Lapland (2012)
14. Kesti, M.: Human capital production function. GSTF J. Bus. Rev. **3**(1), 22–32 (2013)
15. Kesti, M.: AI Software Practical Test Using Game Theory Q-Learning Approach. PlayGain Inc (2017)
16. Kesti, M., Syväjärvi, A.: Human capital production function in strategic management. Technology and Investment **6**, 12–21 (2015)

17. Kesti, M., Leinonen, J., Syväjärvi, A.: A multidisciplinary critical approach to measure and analyze human capital productivity. In: Russ, M. (ed.) Quantitative Multidisciplinary Approaches in Human Capital and Asset Management, pp. 1–317. IGI Global, Hershey (2016). (1-22)
18. Kesti, M., Leinonen, J., Kesti, T.: The productive leadership game: from theory to game-based learning. In: Public Sector Entrepreneurship and the Integration of Innovative Business Models. IGI Global 2017, pp. 238–260 (2017)
19. Littman, M.L.: Markov games as a framework for multi-agent reinforcement learning. In: Proceedings of the 11th International Conference on Machine Learning, New Brunswick, pp. 157–163 (1994)
20. Nash, J.F.: Non-cooperative games. Ann. Math. **54**, 286–295 (1951)
21. Osborne, M.J., Rubinstein, A.: A Course in game theory, MIT Press (1994)
22. Pietiläinen, V., Kesti, M.: Johtamisen tilanneherkistyminen ja asiantuntijuus. In: Perttula, J., Syväjärvi, A. (eds.) kirjassa Johtamisen psykologia (2012)
23. Puterman, M.L.: Markov Decision Processes: Discrete Stochastic Dynamic Programming. Wiley, New York (1994)
24. Sutton, R.S., Barto, A.G.: Reinforcement Learning: An Introduction. MIT Press/Bradford Books (1998)
25. Watkins, C., Dayan, P.: Q-learning. Mach. Learn. **3**, 279–292 (1992)

The Cognitive Packet Network with QoS and Cybersecurity Deep Learning Clusters

Will Serrano(✉)

Intelligent Systems and Networks Group, Imperial College London, London, UK
G.Serrano11@imperial.ac.uk

Abstract. This paper proposes the insertion of Deep Learning (DL) Clusters to the Cognitive Packet Network (CPN). Packet routing and traffic management in the CPN are based on the Random Neural Network (RNN) Reinforcement Learning (RL) algorithm. The RNN represents the transmission of information between neurons firing excitatory and inhibitory impulsive spikes; the additional Deep Learning clusters reproduce the technique, the human brain applies when learning, memorizing and taking decisions. This paper proposes the combination of both learning algorithms, RNN and DL as the complete brain model with the addition of a DL cluster structure for Quality of Service parameters; Cybersecurity certificates and finally DL Management clusters for final routing decisions. The presented model has been tested in several simulation settings and network sizes; it has been validated with the CPN itself without DL clusters. The obtained results are encouraging; the presented CPN with DL clusters as a method to transmit, information, learn the environment and take decisions successfully emulates the operation of the human brain.

Keywords: Random neural network · Deep learning clusters
Cognitive packet network · Quality of service (QoS)
Cybersecurity · Routing

1 Introduction

The human brain consists of clusters of neurons [1] that specialise in learning from the senses and transmit information sending positive and negative impulses or signals. It operates with two distinct memories [2]; quick decisions and task related actions use short term memory whereas identity and security is preserved by the long term memory. In addition, the brain performs its functions with two operation modes [3]; usual actions are conscious and decisions taken under emergency situations such as an external attack are unconscious. The human brain performs simultaneously different functions, it learns the environment properties from its five senses, it stores memories to retain its identity, it makes decisions on diverse scenarios and finally, it defends itself against external attacks or dangers.

This paper proposes the connection between the most intricate biological organisms; the human brain with the most elaborated artificial structure developed for large data networks: the Internet; the information infrastructure of the World Wide Web and the Big Data. The association between both the biological and the artificial model is the

K. Arai et al. (Eds.): IntelliSys 2018, AISC 868, pp. 62–85, 2019.
https://doi.org/10.1007/978-3-030-01054-6_5

Random Neural Network [16–18]. Data networks receive information from its users and transmit it to different physical positions; therefore they have to make packet routing decisions based on several Quality of Service (QoS) parameters and store the routing table in memory, actions performed while at risk of Cybersecurity attacks. This paper proposes the Cognitive Packet Network (CPN) [11–15] with additional Deep Learning (DL) clusters [31, 32] that reproduces the brain operation. Deep Learning (DL) clusters emulate the long term memory acquiring the data network identity: QoS values and Cybersecurity certificates. In addition, this proposed DL structure includes a layer of DL Management Clusters to make the final routing decision. The CPN-RL routing algorithm is used in routine or conscious operations as short term memory because its quick and adaptable routing learning whereas DL clusters are used to take routing decisions in emergency situations, such external cyber-attacks, or unconscious operation using the long term memory as a robust and safe and although inflexible and inefficient routing algorithm.

Section 2 presents the RNN, CPN, DL clusters and Cybersecurity concept and literature review. Section 3 describes the mathematical model of CPN with DL clusters. Section 4 presents the validation of the proposed DL structure with several QoS and Cyber situations in small 9 node network with 1 decision layer and medium 16 node network with 2 decision layers is described in Sect. 5. Sections 6 and 7 draw conclusions and related bibliography, respectively. The appendix shows the CPN-DL neural schematic.

2 Related Work

2.1 Cybersecurity

New technological, industrial, and social services and applications are possible due to the increment of network connectivity enabled by the Ethernet and Internet protocols however applications and users are progressively under new cybersecurity risks and attacks. Ericsson [4] presents Cybersecurity threats and concerns in a smart grid infrastructure where network vulnerabilities and information security domains are analysed in Power Communications Systems. Ten et al. [5] carry out a survey on the Cybersecurity aspects of critical infrastructure; in addition they present an SCADA framework based on four techniques: real time monitoring, anomaly discovery, impact assessment and final mitigation plan. An attack tree evaluation is modelled with an algorithm for cybersecurity assessment that includes the examination of ports and password strategies. Cruz et al. [6] describe a distributed intrusion detection system for SCADA systems that incorporates several security agents configured for three specific areas: the development of the device, process level and network security capabilities; the integration of anomaly and signature based methods against identified and rogue threats and finally the incorporation of a distributed multi layered message design to transmit predefined events between elements. Wang et al. [7] present a framework that enables the development of rival resilient Deep Neural Networks (DNN) by intro ducing a data conversion module between the rival and the DNN that identifies attacker samples with a reduced impact on the precision on the classifier. Tuor et al. [8] propose an unsupervised Deep Learning method to identify irregular network activity in real time. Network events are extracted from system logs as features and the DNN learns users' normal behaviour to detect potential malicious activity. Wu et al. [9] describe a classification of risks and cyber-attacks in manufacturing systems with promising mitigation actions such as unsupervised machine learning for anomaly recognition and supervised machine learning for threat classification. Kim [10] presents a new computer control system architecture with cyber defensive features based on system hardware diversification and unidirectional data transmission with the principle that the discovery and avoidance of cyber attacks will never be exhaustive.

2.2 Deep Learning

Deep Learning uses a cascade of l-layers of non-linear processing modules for feature extraction and transformation where each input of a level corresponds to the output from the previous one. Deep Learning obtains several levels of data representations that relate to the different levels of abstractions creating a concept hierarchy where the higher the layer, the more abstract concepts are learned. Schmidhuber [24] surveys Deep Learning in Artificial Neural Networks. Bengio et al. [25] analyse recent research in the field of Deep Learning and unsupervised feature learning as well as improvements in probabilistic methods. They present an innovative probabilistic model that considers likelihood based probabilistic methods; reconstruction based methods, including auto encoders, and geometrically based learning methods. Jie et al. [26] present an advanced model to optimize the deep learning of neural networks. They blend the capacity of Deep Learning methods to learn complex and abstract internal representations with the stability of linear methods. They insert a linear loss layer between the input and the first hidden non-linear layers of the conventional deep model. Le et al. [27] research the benefits and weaknesses of optimization methods with regards to the simplification and the learning speed of pre-training the unsupervised deep learning feature. Ngiam et al. [28] present an application of deep networks to learn features over several modalities to demonstrate that cross modality feature learning performs better than single modality learning. Sutskever et al. [29] propose a sequence learning method that takes marginal assumptions on the structure of the sequence applying a multi-layered Long Short Term Memory (LSTM) to map the input sequence to a fixed dimensional vector. Bekker et al. [30] present a Deep Learning intra cluster training strategy with an implementation to language identification where linguistic clusters define a cost function to train the neural network.

2.3 Cognitive Packet Network

The Cognitive Packet Network (CPN) was presented by Gelenbe [11–15], it assigns routing and traffic management functions to the packets instead of the nodes. QoS targets are allocated to Cognitive Packets (CP) within the CPN, which they pursue when taking routing decisions with small reliance on the network nodes. CPs learn from experience of other CP packets with whom they interchange network information using n Mailboxes (MB) and their own inspection about the network. Cognitive Packets store network information in their private Cognitive Map (CM).

Certain Goal G is based on Quality of Service parameters that the CP requires to meet, $G = \alpha Delay + \beta Loss + \gamma Bandwidth$, with its related reward R which is $R = 1/G$. Consecutive measured R figures are represented by R_l, $l = 1, 2\ldots$ which are then used to calculate a decision threshold $T_l = \alpha T_{l-1} + (1-\alpha)R_l$. The CP takes a routing decision based on this value; if the calculated reward is larger than the related node threshold; the CPN recompenses the previous decision taken; otherwise; it punishes it. The CPN routing algorithm is executed by the RNN with Reinforcement Learning; a recurrent RNN takes routing decisions and stores the CM which weights are updated so that decisions are reinforced or weakened depending on how they have meet the QoS goal. The CPN has been tested in large complex networks up to 100 nodes with best and worst case performance scenarios (Fig. 1).

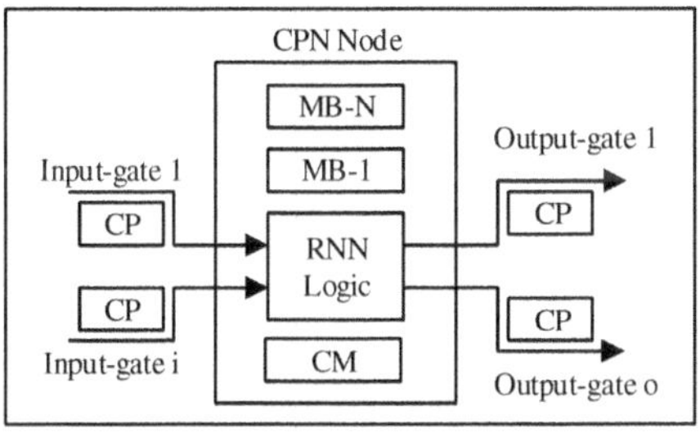

Fig. 1. Cognitive packet network.

2.4 Random Neural Network

The RNN [16–18] models more accurately the transmission of signals in many biological neural networks where they propagate as spikes or impulses, rather than as analogue signal levels. The RNN is a recurrent spiking stochastic model for neural networks; its main analytical properties are the "product form" and the existence of the unique network steady state solution. It has been applied in different applications including network routing with Cognitive Packet Networks based on reinforcement learning algorithm, which requires the search for routes that meet specific Quality of Service requirements [11–15], exit route optimization for evacuees in emergency scenarios [19, 20], search for specific objects based on patterns [21], video compression [22], and image texture learning and generation [23].

The RNN is formed of M neurons where each one receives excitatory (positive) and inhibitory (negative) spike signals from external sensory sources or internal neurons. These spike signals happen following independent Poisson processes of rates $\lambda^+(m)$ for the excitatory spike signal and $\lambda^-(m)$ for the inhibitory spike signal respectively to neuron $m \in \{1, \ldots M\}$. Each neuron is represented at time $t \geq 0$ by its internal state $k_m(t)$ which is a non-negative integer. The arrival of a negative spike to neuron m at time t results in the decrease of the internal state by one unit: $k_m(t^+) = k_m(t) - 1$ if $k_m(t) \geq 0$, or has no effect if $k_m(t) = 0$. On the other hand, the arrival of an excitatory spike to neuron m always increases the neuron's internal state by 1; $k_m(t^+) = k_m(t) + 1$ (Fig. 2).

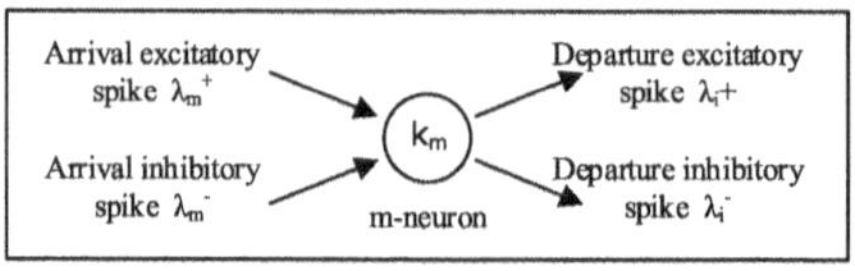

Fig. 2. Random neural network.

2.5 Clusters of Neurons

Deep Learning with Random Neural Networks is presented by Gelenbe and Yin [31, 32]. This model is based on the generalized queuing networks with triggered customer movement (G-networks) where customers are either "positive" or "negative" which can be transferred from queues or leave the network. G-Networks were defined by Gelenbe [33, 34]; where an extension to this model was also developed by Gelenbe et al. [35] in which synchronised interactions of two queues could add a customer in a third queue.

The model defines a specific network M(n) composed of n identically connected neurons, each which has a firing rate r and external inhibitory and excitatory signals λ^- and λ^+ respectively. The state of each neuron is denoted by q which receives an inhibitory input from the state of an external neuron u which does not belong to M(n). For any neuron k Є M(n) there is an inhibitory weight $w^-(u) \equiv w^-(u, k) > 0$ from u to k (Fig. 3).

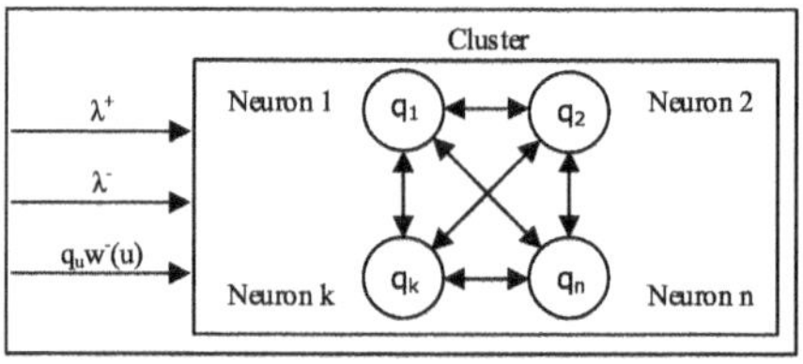

Fig. 3. Clusters of neurons.

The Deep Learning Architecture is formed of several C clusters, each of which is composed of a M(n) cluster with n hidden neurons. For the c cluster, c = 1, …, C, the state of each of its identical neurons is denoted by q_c. In addition, there are U input neurons which do not belong to these C clusters with the state of the u neuron u = 1,…, U denoted by $\overline{q_u}$. The cluster network has U input neurons and C clusters (Fig. 4).

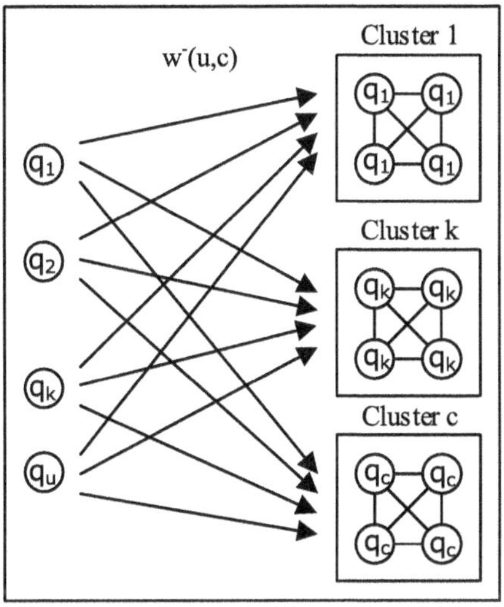

Fig. 4. Deep learning clusters.

2.6 Deep Learning Cluster Structure

The Deep Learning cluster structure considers:

- $I = (i_1, i_2, \ldots, i_u)$, a U-dimensional vector $I \in [0,1]^U$ that assigs the input state $\overline{q_u}$ to the neuron u;
- $w^-(u, c)$ is the U x C matrix of weights from the U input neurons to the neurons in each of the C clusters;
- $Y = (y_1, y_2, \ldots, y_c)$, a C-dimensional vector $Y \in [0,1]^C$ that assigns the neuron state q_c for the cluster c.

The cluster network learns iteratively the U x C weight matrix $w^-(u, c)$ for the input I and output Y pair (i_u, y_c) using Gradient Descent learning algorithm.

2.7 Deep Learning Management Cluster Structure

The Deep Learning management cluster structure was presented by Serrano et al. [36]. It makes management decisions based on the inputs obtained from different Deep Learning clusters.

The Deep Learning management cluster structure defines:

- $I_{mc} = (i_1^{mc}, i_2^{mc}, \ldots, i_u^{mc})$, a C-dimensional vector $I_{mc} \in [0,1]^C$ that assigns the input state $\overline{q_c}$ for the cluster c;
- $w^-(c)$ is the C-dimensional vector of weights from the C input clusters to the neurons in the Management Cluster mc;
- Y_{mc}, a scalar $Y_{mc} \in [0,1]$, the neuron state q_{mc} for the Management Cluster mc that represents its final decision (Fig. 5).

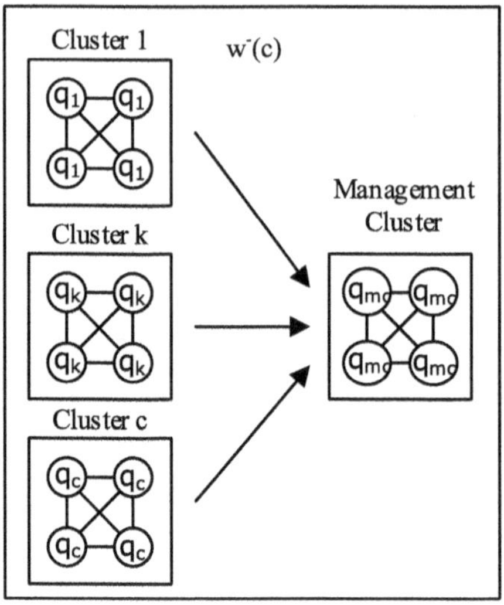

Fig. 5. Random neural network with a management cluster.

3 Deep Learning Cluster Structure in the Cognitive Packet Network

The Cognitive Packet Network updates its network weights instantaneously after the direct observations of the network QoS parameters enabling its routing algorithm to make fast decisions highly adjustable to QoS variations. The CPN emulates the conscious human brain using short term memory when making quick decisions in normal operation based on the direct data collected from the senses.

This paper presents the addition of Deep Learning clusters to the Cognitive Packet Network in which each DL cluster structure specialises in several Cybersecurity certificates (User, Packet and Node), QoS values (Delay, Loss and Bandwidth) and the optimum packet route for each QoS value. In addition, the DL structure includes a hierarchical layer composed of DL management clusters (QoS, Cyber and CEO) that make the final routing decision based on the inputs CPN-RL algorithm and Deep Learning QoS clusters. Deep Learning algorithm adjust slowly to QoS variations, the proposed structure uses the DL structure as a safe and robust packet routing when the CPN is suffering a Cyber attack. This represents the subconscious human brain using long term memory; where it makes minimum decisions only for protection or subsistence (Fig. 6).

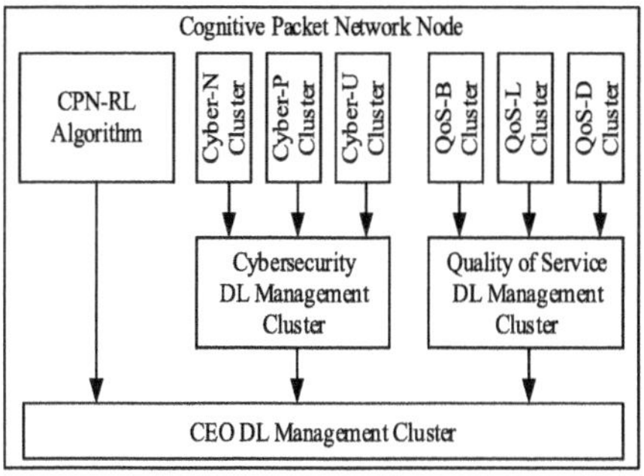

Fig. 6. RL algorithm and DL clusters CPN structure.

3.1 Cybersecurity DL Cluster

A Deep Learning cluster is allocated per Cybersecurity certificates: Node, Packet and User. The node cybersecurity DL cluster validates the CPN nodes to prevent the CPN against impostor nodes. The packet cybersecurity DL cluster confirms the transmitted packets are legitimate to protect the CPN against Denial of Service attacks. The user cybersecurity DL Cluster validate the user or application that is transmitting packets. The proposed cybersecurity network weights can be learned by the CPN nodes in initialization mode or directly configured by the CPN network administrator.

Each Cyber DL cluster identifies its relevant parameters when a CPN node receives a Cognitive Packet (CP) and uses them as its input and output values. The CPN node decides the certificate as invalid, or equivalently, the CPN is under a Cyber attack if the quadratic error between the Cyber DL cluster output vector and the input vector is greater than a set threshold.

The Cybersecurity Node DL cluster is defined as:

- $I_{C\text{-}N} = (i_1^{C\text{-}n}, i_2^{C\text{-}n}, \ldots, i_u^{C\text{-}n})$ a U-dimensional vector where $i_1^{Cyber\text{-}n}, i_2^{Cyber\text{-}n}, \ldots, i_u^{Cyber\text{-}n}$ represent the Cybersecurity Node certificate from the CP;
- $w_{C\text{-}N}(u,c)$ is the U x C matrix of weights of the Cybersecurity Node DL Cluster;
- $Y_{C\text{-}N} = (y_1^{C\text{-}n}, y_2^{C\text{-}n}, \ldots, y_c^{C\text{-}n})$ a C-dimensional vector where $y_1^{Cyber\text{-}n}, y_2^{Cyber\text{-}n}, \ldots, y_c^{Cyber\text{-}n}$ represent the Cybersecurity Node certificate from the DL cluster.

Cybersecurity Packet DL cluster:

- $I_{C\text{-}P} = (i_1^{C\text{-}p}, i_2^{C\text{-}p}, \ldots, i_u^{C\text{-}p})$ a U-dimensional vector where $i_1^{C\text{-}p}, i_2^{C\text{-}p}, \ldots, i_u^{C\text{-}p}$ represent the Cybersecurity Packet parameters from the CP;
- $w_{C\text{-}P}(u, c)$ is the U x C matrix of weights of the Cybersecurity Packet DL Cluster;
- $Y_{C\text{-}P} = (y_1^{C\text{-}p}, y_2^{C\text{-}p}, \ldots, y_c^{C\text{-}p})$ a C-dimensional vector where $y_1^{C\text{-}p}, y_2^{C\text{-}p}, \ldots, y_c^{C\text{-}p}$ represent the Cybersecurity Packet certificate from the DL cluster.

Cybersecurity User DL cluster:

- $I_{C\text{-}U} = (i_1^{C\text{-}u}, i_2^{C\text{-}u}, \ldots, i_u^{C\text{-}u})$ a U-dimensional vector where $i_1^{C\text{-}u}, i_2^{C\text{-}u}, \ldots, i_u^{C\text{-}u}$ represent the Cybersecurity User parameters from the CP;
- $w_{C\text{-}U}(u,c)$ is the U x C matrix of weights of the Cybersecurity User DL cluster;

- $Y_{C\text{-}U} = (y_1^{C\text{-}u}, y_2^{C\text{-}u}, \ldots, y_c^{C\text{-}u})$ a C-dimensional vector where $y_1^{C\text{-}u}, y_2^{C\text{-}u}, \ldots, y_c^{C\text{-}u}$ represent the Cyber User certificates from the DL cluster.

3.2 Quality of Service DL Cluster

A Deep Learning cluster is allocated to each QoS network parameter: Bandwidth, Packet Loss and Delay. The QoS DL cluster learns its best assigned QoS figure with its best CPN node gates to deliver it. When a CPN node detects a route with a better QoS parameter; it learns its value and inserts the gate on the first position of its QoS DL routing table.

The QoS Bandwidth DL cluster is defined as:

- $I_{QoS\text{-}B} = (i_1^{QoS\text{-}b}, i_2^{QoS\text{-}b}, \ldots, i_u^{QoS\text{-}b})$ a U-dimensional vector where $i_1^{QoS\text{-}b}, i_2^{QoS\text{-}b}$, and $i_u^{QoS\text{-}b}$ represents the same QoS Bandwidth parameter;
- $w_{QoS\text{-}B}(u, c)$ is the U x C matrix of weights of the QoS Bandwidth DL Cluster;
- $Y_{QoS\text{-}B} = (y_1^{QoS\text{-}b}, y_2^{QoS\text{-}b}, \ldots, y_c^{QoS\text{-}b})$ a C-dimensional vector where $y_1^{QoS\text{-}b}$ is the Bandwidth QoS figure and $y_2^{QoS\text{-}b}, \ldots, y_c^{QoS\text{-}b}$ represents the CPN node's Bandwidth best routing gates.

QoS Loss DL cluster:

- $I_{QoS\text{-}L} = (i_1^{QoS\text{-}l}, i_2^{QoS\text{-}l}, \ldots, i_u^{QoS\text{-}l})$ a U-dimensional vector where $i_1^{QoS\text{-}l}, i_2^{QoS\text{-}l}$, and $i_u^{QoS\text{-}l}$ represents the same QoS Loss parameter;
- $w_{QoS\text{-}L}(u,c)$ is the U x C matrix of weights of the QoS Loss DL Cluster;
- $Y_{QoS\text{-}L} = (y_1^{QoS\text{-}l}, y_2^{QoS\text{-}l}, \ldots, y_c^{QoS\text{-}l})$ a C-dimensional vector where $y_1^{QoS\text{-}l}$ represents the Loss QoS figure and $y_2^{QoS\text{-}l}, \ldots, y_c^{QoS\text{-}l}$ represent the node's Loss best routing gates.

QoS Delay DL cluster:

- $I_{QoS\text{-}D} = (i_1^{QoS\text{-}d}, i_2^{QoS\text{-}d}, \ldots, i_u^{QoS\text{-}d})$ a U-dimensional vector where $i_1^{QoS\text{-}d}, i_2^{QoS\text{-}d}$, and $i_u^{QoS\text{-}d}$ represents the same QoS Delay parameter;
- $w_{QoS\text{-}D}(u,c)$ is the U x C matrix of weights of the QoS Delay DL Cluster;
- $Y_{QoS\text{-}D} = (y_1^{QoS\text{-}d}, y_2^{QoS\text{-}d}, \ldots, y_c^{QoS\text{-}d})$ a C-dimensional vector where $y_1^{QoS\text{-}d}$ represents the Delay QoS figure and $y_2^{QoS\text{-}d}, \ldots, y_c^{QoS\text{-}d}$ are the node's Delay best routing gates.

3.3 DL Management Cluster

The final routing decision is made by the Deep Learning management clusters which are configured in a hierarchal structure. The Cybersecurity and Quality of Service DL management clusters consider the output from their associated Cyber and QoS clusters respectively and report cluster information to the CEO DL management cluster. If the Cybersecurity DL management cluster confirms the network certificates the CEO DL management cluster choses the route provided by the CPN-RL routing algorithm as normal operation. However, if the Cybersecurity DL management cluster identifies an error in the network certificates, the CEO DL management cluster routes packets as safe operation using the QoS DL clusters.

The Cybersecurity DL Management cluster is defined as:

- I_{Cmc}, a C-dimensional vector $I_{Cmc} \in [0,1]^C$ with the values of the certificate errors for each Cybersecurity cluster;
- $W^{-}_{Cmc}(c)$ is the C-dimensional vector of weights that represents the priority of each Cybersecurity cluster;
- Y_{Cmc}, a scalar $Y_{Cmc} \in [0,1]$ that represents if the packet has been successfully Cyber validated.

The Quality of Service DL management cluster:

- I_{Qmc}, a C-dimensional vector $I_{Qmc} \in [0,1]^C$ with the values of the QoS figures for each QoS cluster;
- $w^{-}_{Qmc}(c)$ is the C-dimensional vector of weights that defines the Goal, $G = (\alpha_{Delay}, \beta_{Loss}, \gamma_{Bandwidth})$;
- Y_{Qmc}, a scalar $Y_{Qmc} \in [0,1]$ that defines the best QoS routing decision to be taken by the CEO cluster.

CEO management cluster as:

- I_{CEOmc}, a scalar $I_{CEOmc} \in [0,1]$ with the figure of the Quality of Service DL management cluster;
- w^{-}_{CEOmc} a scalar $w^{-}_{CEOmc} \in [0,1]$ that defines the validation error by the Cyber DL management cluster;
- Y_{CEOmc}, a scalar $Y_{CEOmc} \in [0,1]$ that defines the final routing decision (Fig. 7).

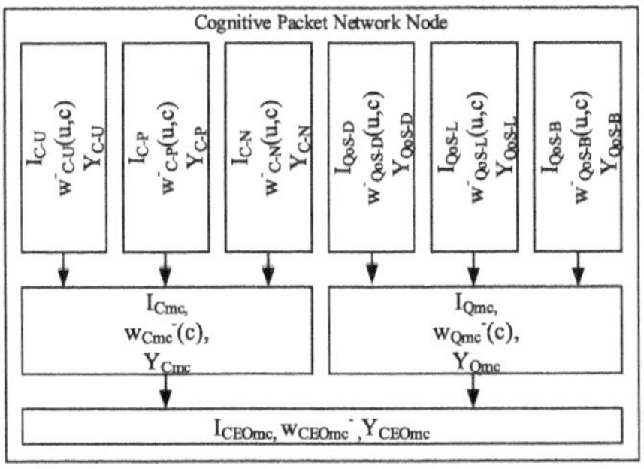

Fig. 7. Cognitive packet netowork node with DL clusters model.

4 Implementation

The Cognitive Packet Network with Deep Learning clusters is implemented in the Network Simulator Omnet 5.0. The simulation comprises various n × n node square networks in which the nodes in the same and adjacent layers are interconnected between each other. The first node (Node 1) is the only transmitter and the last node (Node n) is the only receiver for the simulations where the other nodes route the Cognitive Packets. A model of a 4 × 4 node network is represented in the figure below (Fig. 8).

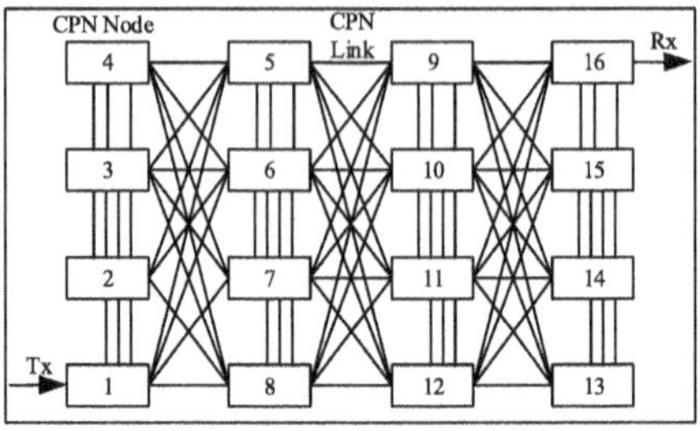

Fig. 8. 4 × 4 Node CPN-DL network.

Every node has assigned a normalized Quality of Service Delay, Loss and Bandwidth parameters in relation to their number; node i has an associated of Delay: i*10; Loss: (n−i)*5 and Bandwidth: 5 + (i*10) respectively in a n × n network. This allocation is shown in the table below for a 4 × 4 node network (Table 1).

Table 1. Initial QoS values – 4 × 4 network

Node 4 Delay: 40 Loss: 65 Bandwidth: 45	Node 5 Delay: 50 Loss: 60 Bandwidth: 55	Node 9 Delay: 90 Loss: 40 Bandwidth: 95	Node 16 Delay: 160 Loss: 05 Bandwidth: 165
Node 3 Delay: 30 Loss: 70 Bandwidth: 35	Node 6 Delay: 60 Loss: 55 Bandwidth: 65	Node 10 Delay: 100 Loss: 35 Bandwidth: 105	Node 15 Delay: 150 Loss: 10 Bandwidth: 155
Node 2 Delay: 20 Loss: 75 Bandwidth: 25	Node 7 Delay: 70 Loss: 50 Bandwidth: 75	Node 11 Delay: 110 Loss: 30 Bandwidth: 115	Node 14 Delay: 140 Loss: 15 Bandwidth: 145
Node 1 Delay: 10 Loss: 80 Bandwidth: 15	Node 8 Delay: 80 Loss: 45 Bandwidth: 85	Node 12 Delay: 120 Loss: 25 Bandwidth: 125	Node 13 Delay: 130 Loss: 20 Bandwidth: 135

The QoS figures switches between each internal node assigned within the same column after two Cognitive Packets are sent by Node 1; as shown in the below table for a 4 × 4 network (Table 2).

Table 2. Final QoS values 4 × 4 network

Node 4 Delay: 40 Loss: 65 Bandwidth: 45	Node 5 Delay: 80 Loss: 45 Bandwidth: 85	Node 9 Delay: 120 Loss: 25 Bandwidth: 125	Node 16 Delay: 160 Loss: 05 Bandwidth: 165
Node 3 Delay: 30 Loss: 70 Bandwidth: 35	Node 6 Delay: 70 Loss: 50 Bandwidth: 75	Node 10 Delay: 110 Loss: 30 Bandwidth: 115	Node 15 Delay: 150 Loss: 10 Bandwidth: 155
Node 2 Delay: 20 Loss: 75 Bandwidth: 25	Node 7 Delay: 60 Loss: 55 Bandwidth: 65	Node 11 Delay: 100 Loss: 35 Bandwidth: 105	Node 14 Delay: 140 Loss: 15 Bandwidth: 145
Node 1 Delay: 10 Loss: 80 Bandwidth: 15	Node 8 Delay: 50 Loss: 60 Bandwidth: 55	Node 12 Delay: 90 Loss: 40 Bandwidth: 95	Node 13 Delay: 130 Loss: 20 Bandwidth: 135

This model proposes to set the initial CPN-RL network weight values with initialization packets sent at random gates.

4.1 Cybesecurity Deep Learning Clusters

The Cybersecurity DL clusters are composed of ten input neurons (u = 10) and ten output clusters (c = 10) where the certificate is a 10 dimension vector. $i_u^{C\text{-}u}$, $i_u^{C\text{-}p}$, $i_u^{C\text{-}n}$ are configured with a value between 0.1 and 0.9 with a distance of 0.1Δ. The Cybersecurity DL clusters learns the network weights where the value of the input cluster I is the same as the output cluster Y (Table 3).

Table 3. Cybersecurity DL cluster implementation

Cluster	Input-Output	Value (Input = Output)
Cyber User	$i_1^{C\text{-}u} \ldots i_{10}^{C\text{-}u}$ $y_1^{C\text{-}u} \ldots y_{10}^{C\text{-}u}$	(0.9,0.8,0.9,0.8,0.9, 0.8,0.9,0.8,0.9,0.8)
Cyber Packet	$i_1^{C\text{-}p} \ldots i_{10}^{C\text{-}p}$ $y_1^{C\text{-}p} \ldots y_{10}^{C\text{-}p}$	(0.7,0.6,0.7,0.6,0.7, 0.6,0.7,0.6,0.7,0.6)
Cyber Node	$i_1^{C\text{-}n} \ldots i_{10}^{C\text{-}n}$ $y_1^{C\text{-}n} \ldots y_{10}^{C\text{-}n}$	(0.5,0.4,0.5,0.4,0.5, 0.4,0.5,0.4,0.5,0.4)

4.2 Quality of Service Deep Learning Clusters

The QoS DL clusters are composed of three input neurons (u = 3) and three output clusters (c = 3). The QoS Delay DL cluster is set to $i_1^{QoS\text{-}d} = 0.5$; $i_2^{QoS\text{-}d} = 0.5$ and $i_3^{QoS\text{-}d} = 0.5$; $y_1^{QoS\text{-}d}$ represent the best QoS Delay value, $y_2^{QoS\text{-}d}$ the gate to best QoS Delay route and $y_3^{QoS\text{-}d}$ the gate to second best Delay route. A similar configuration is followed for the Quality of Service Loss and Bandwidth DL clusters respectively (Table 4).

Table 4. QoS DL cluster implemenatation

Cluster	Input	Value	Output	Value
QoS Delay	$i_1^{QoS\text{-}d}$	0.5	$y_1^{QoS\text{-}d}$	Best QoS Delay Parameter
QoS Delay	$i_2^{QoS\text{-}d}$	0.5	$y_2^{QoS\text{-}d}$	Best QoS Delay Gate
QoS Delay	$i_3^{QoS\text{-}d}$	0.5	$y_3^{QoS\text{-}d}$	Second Best QoS Delay Gate
QoS Loss	$i_1^{QoS\text{-}l}$	0.6	$y_1^{QoS\text{-}l}$	Best QoS Loss Parameter
QoS Loss	$i_2^{QoS\text{-}l}$	0.6	$y_2^{QoS\text{-}l}$	Best QoS Loss Gate
QoS Loss	$i_3^{QoS\text{-}l}$	0.6	$y_3^{QoS\text{-}l}$	Second Best QoS Loss Gate
QoS Bandwidth	$i_1^{QoS\text{-}b}$	0.7	$y_1^{QoS\text{-}b}$	Best QoS Bandwidth Parameter
QoS Bandwidth	$i_2^{QoS\text{-}b}$	0.7	$y_2^{QoS\text{-}b}$	Best QoS Bandwidth Gate
QoS Bandwidth	$i_3^{QoS\text{-}b}$	0.7	$y_3^{QoS\text{-}b}$	Second Best QoS Bandwidth Gate

The model normalizes the outputs of the DL clusters to (0.5 + QoS figure/1000) and (0.5 + Best Gate/100), respectively.

4.3 Deep Learning Management Clusters

The inputs of the Cybersecurity DL management cluster correspond to the certificate errors measured by each Cyber DL cluster; its network weights are set with same Figure (0.1) to assign the same priority to different cyber DL clusters. The output Y_{Cmc} is the cybersecurity network status.

The inputs of the Quality of Service DL management cluster are the best QoS figures from each QoS DL cluster; its networks weights matches the Goal G = (α_{Delay}, β_{Loss}, $\gamma_{Bandwidth}$). The output Y_{Qmc} is quantified best QoS route decision.

The input of the CEO DL management cluster is the figure given by the QoS DL management cluster; its network weight is the figure given by the Cybersecurity DL management cluster. The output corresponds to the final routing decision between the gates provided by the RNN algorithm and the Quality of Service DL Clusters (Table 5).

Table 5. DL Management Cluster Implementation

Cluster	Input	Network	Output
Cyber	(User Error, Packet Error, Node Error)	(0.1, 0.1, 0.1)	0.0 if $Y_{Cmc} > 0.999$ 0.999 if $Y_{Cmc} \leq 0.999$
QoS	(Delay Value, Loss Value, Bandwidth Value)	(α_{Delay}, β_{Loss}, $\gamma_{Bandwidth}$)	0.1 if $Y_{Delay\text{-}qmc} > Y_{Loss\text{-}qmc}$ and $Y_{Delay\text{-}qmc} > Y_{Bandwidth\text{-}qmc}$ 0.5 if $Y_{Loss\text{-}qmc} > Y_{Delay\text{-}qmc}$ and $Y_{Loss\text{-}qmc} > Y_{Bandwidth\text{-}qmc}$ 0.9 if $Y_{Bandwidth\text{-}qmc} > Y_{Delay\text{-}qmc}$ and $Y_{Bandwidth\text{-}qmc} > Y_{Loss\text{-}qmc}$
CEO	(0.1, 0.5 or 0.9)	(0.0 or 0.999)	CPN-RNN if $0.6 < Y_{CEOmc} < 1$ DL-Delay if $0.4 < Y_{CEOmc} < 0.6$ DL-Loss if $0.2 < Y_{CEOmc} < 0.4$ DL-Bandwidth if $0.1 < Y_{CEOmc} < 0.2$

5 Experimental Results

Two n $\times$ n networks are simulated, 3 $\times$ 3 and 4 $\times$ 4, with different Cybersecurity certificates, Quality of Service parameters and variable Goals to assess the adaptability and performance of our proposed solution.

5.1 Cybersecurity DL Cluster Validation - 3 $\times$ 3 Network

The three Cyber DL clusters have been validated where the security certificates are modified at node 1 and validated at node 4 once the CPs decide the optimum route. The certificates are increasingly changed from the correct value with 0.1Δ increments applied to the ten certificate dimensions. Table below shows the cybersecurity validation error (Tables 6, 7 and 8).

Table 6. Cybersecurity user validation

Dimension	$\Delta = 0.0$	$\Delta = 0.1$	$\Delta = 0.2$	$\Delta = 0.3$	$\Delta = 0.4$
1	9.75E-11	0.0102	0.0409	0.0921	0.1638
2	9.7537E-11	0.0213	0.0851	0.1915	0.3406
3	9.7537E-11	0.0326	0.1305	0.2938	0.5226
4	9.7537E-11	0.0451	0.1806	0.4067	0.7238
5	9.7537E-11	0.0576	0.2306	0.5195	0.9249
6	9.7537E-11	0.0715	0.2867	0.6465	1.1519
7	9.7537E-11	0.0851	0.3414	0.7703	1.3732
8	9.7537E-11	0.1006	0.4038	0.9119	1.6273
9	9.7537E-11	0.1153	0.4633	1.0470	1.8698
10	9.7537E-11	0.1323	0.5321	1.2038	2.1526

Table 7. Cybersecurity packet validation

Dimension	Δ = 0.0	Δ = 0.1	Δ = 0.2	Δ = 0.3	Δ = 0.4
1	4.72E-10	0.0108	0.0431	0.0970	0.1725
2	4.7238E-10	0.0233	0.0933	0.2104	0.3747
3	4.7238E-10	0.0373	0.1497	0.3382	0.6036
4	4.7238E-10	0.0533	0.2147	0.4864	0.8709
5	4.7238E-10	0.0707	0.2855	0.6488	1.1659
6	4.7238E-10	0.0904	0.3664	0.8363	1.5101
7	4.7238E-10	0.1112	0.4527	1.0379	1.8831
8	4.7238E-10	0.1347	0.5509	1.2701	2.3192
9	4.7238E-10	0.1592	0.6541	1.5160	2.7853
10	4.7238E-10	0.1866	0.7711	1.7993	3.3325

Table 8. Cybersecurity node validation

Dimension	Δ = 0.0	Δ = 0.1	Δ = 0.2	Δ = 0.3	Δ = 0.4
1	9.19E-10	0.0114	0.0458	0.1032	0.1838
2	9.1918E-10	0.0200	0.0800	0.1800	0.3200
3	9.1918E-10	0.0258	0.1259	0.2835	0.5044
4	9.1918E-10	0.0400	0.1600	0.3600	0.6400
5	9.1918E-10	0.0515	0.2061	0.4639	0.8250
6	9.1918E-10	0.0600	0.2400	0.5400	0.9600
7	9.1918E-10	0.0715	0.2863	0.6442	1.1455
8	9.1918E-10	0.0800	0.3200	0.7200	1.2800
9	9.1918E-10	0.0916	0.3664	0.8246	1.4661
10	9.1918E-10	0.1000	0.4000	0.9000	1.6000

The Cybersecurity DL cluster certification error greatly increments even with only one 0.1Δ increase. The validation error results are consistent between the three Cybersecurity DL clusters. Certificate changes generate a greater error if the increments are combined in the same dimension rather than divided into other different dimensions.

5.2 Quality of Service DL Cluster Validation - 3 × 3 Network

The Cognitive Packet Network is simulated with a constant flow of 160 Cognitive Packet where the first 20 packets initialize the network. Goal varies after 20 packets whereas QoS parameters update 2 packets after the new Goal is assigned (Table 9).

Table 9. Qos cluster validation simulation parameters

Packet	Goal	QoS
1-20	Network Initialization Packets	
21-22	1*Delay	Initial Values
23-40	1*Delay	Final Values
41-42	1*Loss	Initial Values
43-60	1*Loss	Final Values
61-62	1*Bandwidth	Initial Values
63-80	1* Bandwidth	Final Values
81-82	0*5Delay + 0.5*Loss	Initial Values
83-100	0*5Delay + 0.5*Loss	Final Values
101-102	0*5Delay + 0.5*Bandwidth	Initial Values
103-120	0*5Delay + 0.5* Bandwidth	Final Values
121-122	0*5Loss + 0.5*Bandwidth	Initial Values
123-140	0*5Loss + 0.5* Bandwidth	Final Values
141-142	0*3Delay + 0*3Loss + 0.3* Bandwidth	Initial Values
143-160	0*3Delay + 0*3Loss + 0.3* Bandwidth	Final Values

Table 10. Goal: 1*Delay – 3 × 3 Network

Packet	CPN Route	DL Route	Best Route	Goal 1/Reward	1/Threshold
21	1-4-9	1-4-9	1-4-9	130.00	130.00
22	1-4-9	1-4-9	1-4-9	130.00	130.00
23	1-4-9	1-4-9	1-6-9	150.00	130.00
24	1-4-9	1-4-9	1-6-9	150.00	131.76
25	1-4-9	1-4-9	1-6-9	150.00	133.38
26	1-4-9	1-4-9	1-6-9	150.00	134.87
27	1-5-9	1-4-9	1-6-9	140.00	136.25
28	1-4-9	1-4-9	1-6-9	150.00	136.61
29	1-2-6-9	1-4-9	1-6-9	150.00	137.84
30	1-6-9	1-4-9	1-6-9	130.00	138.97
31	1-6-9	1-6-9	1-6-9	130.00	138.02
40	1-6-9	1-6-9	1-6-9	130.00	132.99

The Quality of Service DL clusters have been validated with seven flexible Goals for the same Cognitive Packet flow. The Cognitive Packet route decision taken by the CEO DL Management Cluster under two scenarios is shown in table below. The CEO selects the DL route obtained from the QoS DL management cluster when the Cybersecurity DL management cluster confirms the packet security certificate has not pass its validation; on the other hand the CEP choses the CPN-RL route when the Cybersecurity management cluster has confirmed the packet certifications (Table 10).

The first two Cognitive Packets chose the optimum route (130 Delay) whereas the third CP acknowledges the QoS figure has updated to 150 Delay. The Threshold value adjusts gradually following the Goal deterioration. Cognitive Packets change route after the CPN-RL weights are updated with the updated QoS figures finding the optimum route after seven CPs. The route provided by the QoS DL clusters remains unaltered because of its slow learning algorithm until the CPN-RL discovers the new updated optimum route where DL provides the best route after 8 CPs. The CPN Threshold adjusts increasingly to its initial figure after the new optimum route is learnt (Table 11).

Table 11. Goal: 0.5*Delay + 0.5*Loss – 3 × 3 Network

Packet	CPN Route	DL Route	Best Route	Goal 1/Reward	1/Threshold
81	1-4-9	1-4-9	1-4-9	82.50	82.50
82	1-4-9	1-4-9	1-4-9	82.50	82.50
83	1-4-9	1-4-9	1-6-9	87.50	82.50
84	1-5-9	1-4-9	1-6-9	85.00	82.97
85	1-6-9	1-4-9	1-6-9	82.50	83.17
86	1-6-9	1-6-9	1-6-9	82.50	83.10
87	1-6-9	1-6-9	1-6-9	82.50	83.04
88	1-6-9	1-6-9	1-6-9	82.50	82.99
89	1-6-9	1-6-9	1-6-9	82.50	82.94
90	1-6-9	1-6-9	1-6-9	82.50	82.90
91	1-6-9	1-6-9	1-6-9	82.50	82.86
100	1-6-9	1-6-9	1-6-9	82.50	82.64

The CPN-RL adjusts immediately to a combined QoS Goal learning the route after two CPs. The route offered by the QoS DL clusters updates a CP later (Table 12).

Table 12. Goal: 0.3*Delay + 0.3*Loss + 0.3*Bandwidth

Packet	CPN Route	DL Route	Best Route	Goal 1/Reward	1/Threshold
141	1-4-9	1-4-9	1-4-9	101.66	101.66
142	1-4-9	1-4-9	1-4-9	101.66	101.66
143	1-4-9	1-4-9	1-6-9	111.66	101.66
144	1-4-9	1-4-9	1-6-9	111.66	102.58
145	1-4-9	1-4-9	1-6-9	111.66	103.42
146	1-4-9	1-4-9	1-6-9	111.66	104.18
147	1-5-9	1-4-9	1-6-9	106.66	104.89
148	1-6-9	1-4-9	1-6-9	101.66	105.06
149	1-6-9	1-6-9	1-6-9	101.66	104.71
150	1-6-9	1-6-9	1-6-9	101.66	104.40
151	1-6-9	1-6-9	1-6-9	101.66	104.12
160	1-6-9	1-6-9	1-6-9	101.66	102.60

The results for a combined Goal that includes the three different QoS parameters are also consistent with previous validation results. CPN-RL adjusts instantaneously to the Goal change finding the new optimum route after five Cognitive Packets.

5.3 DL Management Cluster Validation - 3 × 3 Network

The Deep Learning management clusters (Cyber, QoS and CEO) are validated on this section with two opposite Cybersecurity situations: Δ = 0.0 for normal status and Δ = 0.1 for Cyber attack (Table 13).

Table 13. DL management cluster validation - 3 × 3 Network

Variable	Packet 30		Packet 85		Packet 148	
	Δ = 0.0	Δ = 0.1	Δ = 0.0	Δ = 0.1	Δ = 0.0	Δ = 0.1
	G 1*D	G 1*D	G 0.5*D + 0.5*L	G 0.5*D + 0.5*L	G 0.3*D + 0.3*L + 0.3*B	G 0.3*D + 0.3*L + 0.3*B
Cyber I_{cmc}	5E−11	3.4E−4	5E−11	3.4E−4	5E−11	3.4E−4
Cyber Y_{cmc}	1.0E+00	1.0E+00	1.0E+00	1.0E+00	1.0E+00	1.0E+00
QoS-D I_{qmc}	6.3E−01	6.3E−01	3.2E−01	3.2E−01	2.1E−01	2.1E−01
QoS-L I_{qmc}	0.0E+00	0.0E+00	2.6E−01	2.6E−01	1.8E−01	1.8E−01
QoS-B I_{qmc}	0.0E+00	0.0E+00	0.0E+00	0.0E+00	2.1E−01	2.1E−01
QoS-D Y_{qmc}	1.8E−01	1.8E−01	3.0E−01	3.0E−01	3.9E−01	3.9E−01
QoS-L Y_{qmc}	1.0E+00	1.0E+00	3.4E−01	3.4E−01	4.4E−01	4.4E−01
QoS-B Y_{qmc}	1.0E+00	1.0E+00	1.0E+00	1.0E+00	3.9E−01	3.9E−01
CEO I_{CEOmc}	1.0E−01	1.0E−01	1.0E−01	1.0E−01	9.0E−01	9.0E−01
CEO $w_{CEO}(c)$	0.0E+00	1.0E+00	0.0E+00	1.0E+00	0.0E+00	1.0E+00
CEO Y_{CEOmc}	1.0E+00	5.7E−01	1.0E+00	5.7E−01	1.0E+00	1.3E−01
Routing Decision	CPN Gate 4 Node 6	DL-D Gate 2 Node 4	CPN Gate 4 Node 6	DL-D Gate 2 Node 4	CPN Gate 4 Node 6	DL-B Gate 2 Node 4

Note- D: Delay; L: Loss and B: Bandwidth.

Three different strategic Cognitive Packets with different Goals are selected (CP 30, CP 85 and CP 148). The figures provided by the DL management cluster confirm the proposed DL structure. The precise quantification of the DL management cluster neuron potential and the configuration of relevant thresholds to take optimum routing decisions.

5.4 Quality of Service DL Cluster Validation - 4 × 4 Network

The CPN Deep Learning structure has been validated with seven variable Goals for the same packet flow although the QoS = 1*Delay is only shown on the table below for simplicity as the results for the other QoS values are similar with the previous validation.

The figures obtained by the 4 × 4 node network are consistent to the 3 × 3 node network. The first two CPs follow the optimum route (300 Delay) whereas the third CP confirms the Quality of Service figures have adapted to 360 Delay. CPN-RL discov-erers the optimum route after 24 CPs followed by DL that acquired the updated route a CP after (Table 14).

Table 14. Goal: 1*Delay – 4 × 4 Network

P	CPN Route	DL Route	Best Route	Goal	1/T
1	1-5-9-16	1-5-9-16	1-5-9-16	300.00	300.00
2	1-5-9-16	1-5-9-16	1-5-9-16	300.00	300.00
3	1-5-9-16	1-5-9-16	1-8-12-16	360.00	300.00
4	1-5-9-16	1-5-9-16	1-8-12-16	360.00	300.50
5	1-5-9-16	1-5-9-16	1-8-12-16	360.00	301.00
6	1-5-9-16	1-5-9-16	1-8-12-16	360.00	301.49
7	1-5-9-16	1-5-9-16	1-8-12-16	360.00	301.98
8	1-6-9-16	1-5-9-16	1-8-12-16	350.00	302.47
9	1-7-9-16	1-5-9-16	1-8-12-16	340.00	302.88
10	1-2-6-10-16	1-5-9-16	1-8-12-16	360.00	303.21
11	1-8-9-16	1-5-9-16	1-8-12-16	330.00	303.69
12	1-4-5-10-16	1-5-9-16	1-8-12-16	390.00	303.93
13	1-3-5-11-16	1-5-9-16	1-8-12-16	370.00	304.61
14	1-5-11-16	1-5-9-16	1-8-12-16	340.00	305.15
15	1-6-11-16	1-5-9-16	1-8-12-16	330.00	305.46
16	1-7-10-16	1-5-9-16	1-8-12-16	330.00	305.69
17	1-2-7-11-16	1-5-9-16	1-8-12-16	340.00	305.91
18	1-4-6-12-16	1-5-9-16	1-8-12-16	360.00	306.22
19	1-8-10-16	1-5-9-16	1-8-12-16	320.00	306.68
20	1-3-6-12-16	1-5-9-16	1-8-12-16	350.00	306.80
21	1-5-11-16	1-5-9-16	1-8-12-16	340.00	307.18
22	1-4-3-7-12-16	1-5-9-16	1-8-12-16	380.00	307.48
23	1-2-8-11-16	1-5-9-16	1-8-12-16	330.00	308.07
24	1-6-12-15	1-5-9-16	1-8-12-16	320.00	308.27
25	1-7-12-15	1-5-9-16	1-8-12-16	310.00	308.39
26	1-3-4-8-12-16	1-5-9-16	1-8-12-16	370.00	308.40
27	1-8-12-16	1-5-9-16	1-8-12-16	300.00	308.92
28	1-8-12-16	1-8-12-16	1-8-12-16	300.00	308.82
29	1-8-12-16	1-8-12-16	1-8-12-16	300.00	308.73
30	1-8-12-16	1-8-12-16	1-8-12-16	300.00	308.64
40	1-8-12-16	1-8-12-16	1-8-12-16	300.00	307.80

5.5 DL Management Cluster Validation - 4 × 4 Network

Similar as the previous 3 × 3 node validation, the DL Management Clusters (Cyber, QoS and CEO) are validated with two opposite Cybersecurity situations: Δ = 0.0 for normal status and Δ = 0.1 for Cyber attack. Three different strategic Cognitive Packets with different Goals are selected (CP 107, CP 228 and CP 341) (Table 15).

Table 15. Dl management cluster validation 4 × 4 network

Variable	Packet 107		Packet 228		Packet 341	
	Δ = 0.0	Δ = 0.1	Δ = 0.0	Δ = 0.1	Δ = 0.0	Δ = 0.1
	G 1*D	**G 1*D**	**G 0.5*D + 0.5*L**	**G 0.5*D + 0.5*L**	**G 0.3*D + 0.3*L + 0.3*B**	**G 0.3*D + 0.3*L + 0.3*B**
Cyber I_{cmc}	5E−11	3.4E−4	5E−11	3.4E−4	5E−11	3.4E−4
Cyber Y_{cmc}	1.0E+00	1.0E+00	1.0E+00	1.0E+00	1.0E+00	1.0E+00
QoS-D I_{qmc}	8.0E−01	8.0E−01	4.0E−01	4.0E−01	2.7E−01	2.7E−01
QoS-L I_{qmc}	0.0E+00	0.0E+00	2.9E−01	2.9E−01	1.9E−01	1.9E−01
QoS-B I_{qmc}	0.0E+00	0.0E+00	0.0E+00	0.0E+00	2.7E−01	2.7E−01
QoS-D Y_{qmc}	1.4E−01	1.4E−01	2.5E−01	2.5E−01	3.4E−01	3.4E−01
QoS-L Y_{qmc}	1.0E+00	1.0E+00	3.2E−01	3.2E−01	4.1E−01	4.1E−01
QoS-B Y_{qmc}	1.0E+00	1.0E+00	1.0E+00	1.0E+00	3.3E−01	3.3E−01
CEO I_{CEOmc}	1.0E−01	1.0E−01	1.0E−01	1.0E−01	9.0E−01	9.0E−01
CEO $w_{CEO}^{-}(c)$	0.0E+00	1.0E+00	0.0E+00	1.0E+00	0.0E+00	1.0E+00
CEO Y_{CEOmc}	1.0E+00	5.7E−01	1.0E+00	5.7E−01	1.0E+00	1.3E−01
Routing Decision	CPN Gate 6 Node 8	DL-D Gate 3 Node 5	CPN Gate 6 Node 8	DL-D Gate 6 Node 8	CPN Gate 6 Node 8	DL-B Gate 3 Node 5

Note- D: Delay; L: Loss and B: Bandwidth.

The figures provided by the 4 × 4 network DL management cluster validation are consistent with the 3 × 3 network.

5.6 Quality of Service DL Cluster Validation – Overall Results

The overall results for the DL cluster structure are shown on the below table. P represents the number of CPs the QoS DL clusters requires to be send before it adjusts to the optimum route and G is the final network Goal (Table 16).

Table 16. QoS DL cluster validation – overall results

QoS	Network	
	3 × 3	4 × 4
1*D	P:8 G:130	P:25 G:300
1*L	P:4 G:25	P:26 G:75
1*B	P:7 G:140	P:25 G:315
0.5*D + 0.5*L	P:3 G:82.5	P:26 G:202.5
0.5*D + 0.5*B	P:7 G:135	P:25 G:307.5
0.5*L + 0.5*B	P:6 G:87.5	P:26 G:210
0.3*D + 0.3*L + 0.3*B	P:6 G:101.65	P:19 G:240

6 Conclusions

This paper has proposed the Random Neural Network with a Deep Learning cluster structure, a biological inspired learning algorithm that models the human brain. Deep Learning algorithm stores routing information in long term memory adjusting slowly to Quality of Service variations as it learns from the CPN-RL algorithm. On the other hand the CPN-RL algorithm adapts very quickly to flexible QoS variations with fast decisions based in short term memory. In addition a layer of DL management clusters is included to take final routing decisions. The CEO DL management cluster makes the correct routing decisions based on the information from the Cybersecurity and QoS DL management clusters respectively. This enables the CPN to select an optimum route under normal operation or a safe route in case of Cyber attack. The DL structure has been validated it against two different n × n networks, 3 × 3 small size with one decision layer and 4 × 4 medium size with two decision layers.

The inclusion of DL clusters specialised in distinct functionalities (Cybersecurity, Quality of Service, and Management) enables an adjustable configuration analogous to the human brain operation; DL clusters are capable to adapt and being assigned where more resources are necessary such as memory, computing or routing. Similarly to human behaviour, an unbalanced adjustment of the CPN to variations of QoS may result in CPN "anxiety" because of CPN node rewards and different optimum routes provided by DL and CPN-RL algorithms may trigger CPN "depression" in the long term due CPN node thresholds.

Future research will expand gradually the network to large scale 7 × 7 (49 nodes, 5 decision layers) and very large scale 10 × 10 networks (100 nodes, 8 decision layers) in which the CPN "anxiety" and "depression" properties will be further analysed.

Appendix: Cognitive Packet Network with Deep Learning Clusters - Neural Schematic

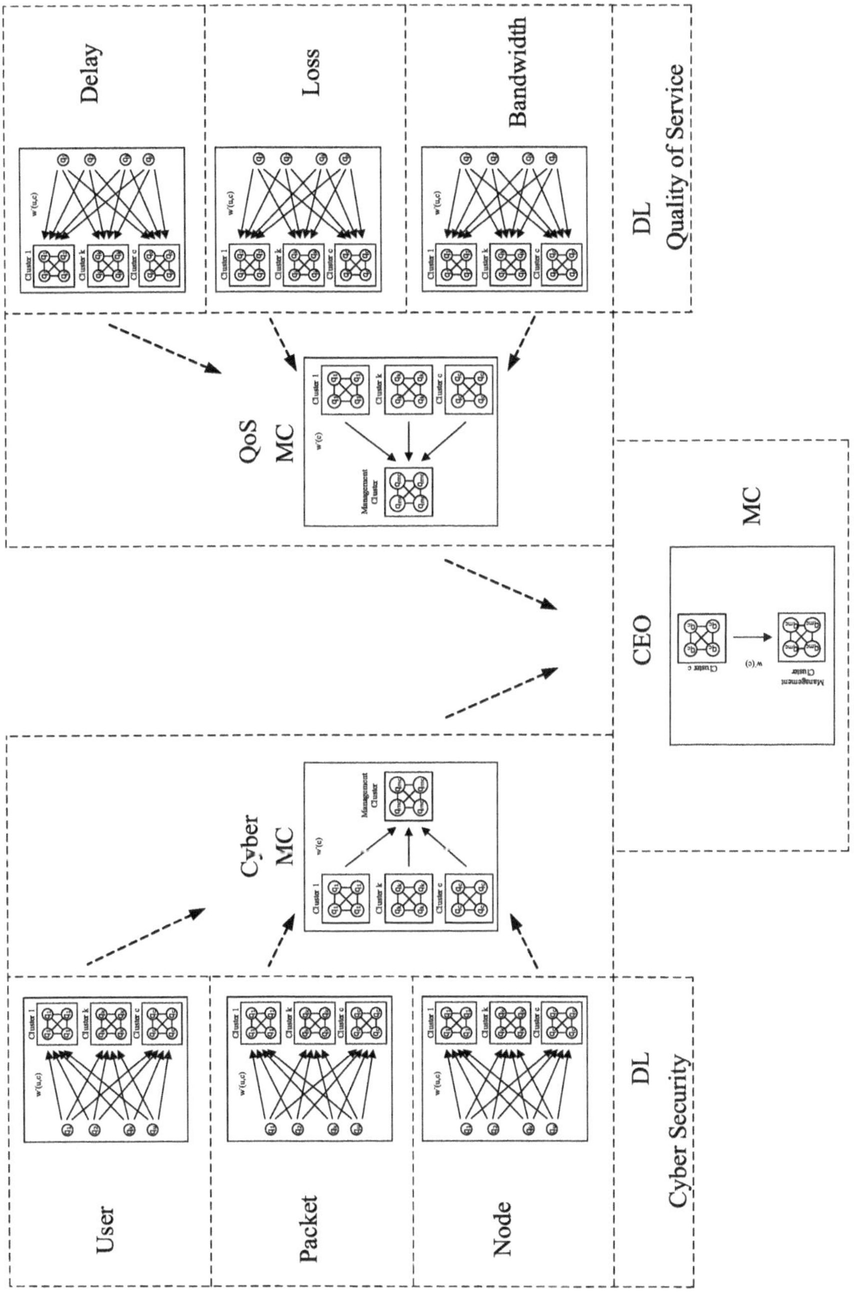

References

1. Danielle, S.B., Bullmore, E.: Small-world brain networks. Neurosci. Rev. J. Bringing Neurobiol. Neurol. Psychiatry **12**, 512–523 (2007)
2. Squire, L.R.: Declarative and nondeclarative memory: multiple brain systems supporting learning and memory. J. Cogn. Neurosci. **4**(3), 232–243 (1992)
3. Grossberg, S.: The link between brain learning, attention, and consciousness. Conscious. Cogn. **8**(1), 1–44 (1999)
4. Ericsson, G.N.: Cyber security and power system communication, essential parts of a smart grid infrastructure. IEEE Trans. Power Delivery **25**(3), 1501–1507 (2010)
5. Ten, C.-W., Manimaran, G., Liu, C.-C.: Cybersecurity for critical infrastructures: attack and defense modeling. IEEE Trans. Syst. Man Cybernet. Part A **40**(4), 853–865 (2010)
6. Cruz, T., et al.: A cybersecurity detection framework for supervisory control and data acquisition systems. IEEE Trans. Ind. Inform. **12**(6), 2236–2246 (2016)
7. Wang, Q., et al.: Learning Adversary-Resistant Deep Neural Networks. CoRR abs/1612.01401 (2016)
8. Tuor, A., et al.: Deep learning for unsupervised insider threat detection in structured cybersecurity data streams. In: AAAI, pp. 4993–4994 (2017)
9. Wu, M., Song, Z., Moon, Y.B: Detecting cyber-physical attacks in CyberManufacturing systems with machine learning methods. J. Intell. Manufact. (2017)
10. Kim, C.: Cyber –defensive architecture for networked industrial control systems. Int. J. Eng. Appl. Comput. Sci. (2017)
11. Gelenbe, E.: Cognitive packet network. Patent US 6804201 B1 (2004)
12. Gelenbe, E., Xu, Z., E.: Seref: cognitive packet networks. In: ICTAI 1999, pp. 47–54 (1999)
13. Gelenbe, E., Lent, R., Xu, Z.: Networks with cognitive packets. In: MASCOTS 2000, pp. 3–10 (2000)
14. Gelenbe, E., Lent, R., Zhiguang, X.: Measurement and performance of a cognitive packet network. Comput. Netw. **37**(6), 691–701 (2001)
15. Gelenbe, E., Lent, R., Montuori, A., Xu, Z.: Cognitive packet networks: QoS and performance. In: MASCOTS 2002, p. 3 (2002)
16. Gelenbe, E.: Random neural networks with negative and positive signals and product form solution. Neural Comput. **1**(4), 502–510 (1989)
17. Gelenbe, E.: Stability of the random neural network model. Neural Comput. **2**(2), 239–247 (1990)
18. Gelenbe, E.: Learning with the recurrent random neural network. In: IFIP Congress (1), pp. 343–349 (1992)
19. Gelenbe, E., Fang-Jing, W.: Large scale simulation for human evacuation and rescue. Comput. Math Appl. **64**(12), 3869–3880 (2012)
20. Filippoupolitis, A., Hey, L.A., Loukas, G., Gelenbe, E., Timotheou, S.: Emergency response simulation using wireless sensor networks. In: AMBI-SYS, p. 21 (2008)
21. Gelenbe, E., Koçak, T.: Area-based results for mine detection. IEEE Trans. Geosci. Remote Sens. **38**(1), 12–24 (2000)
22. Gelenbe, E., Sungur, M., Cramer, C., Gelenbe, P.: Traffic and video quality with adaptive neural compression. Multimedia Syst. **4**(6), 357–369 (1996)
23. Atalay, V., Gelenbe, E., Yalabik, N.: The random neural network model for texture generation. IJPRAI **6**(1), 131–141 (1992)
24. Schmidhuber, J.: Deep learning in neural networks: an overview. Neural Netw. **61**, 85–117 (2015)

25. Bengio, Y., Courville, A.C.: Pascal Vincent: Unsupervised Feature Learning and Deep Learning: A Review and New Perspectives. CoRR abs/1206.5538 (2012)
26. Jiea, S., Zhichenga, Z., Feia, S., Annia, C.: Progressive framework for deep neural networks: from linear to non-linear. J. China Univ. Posts Telecommun. **23**(6), 1–7 (2016)
27. Le, Q.V., Ngiam, J., Coates, A., Lahiri, A., Prochnow, B., Ng, A.Y.: On optimization methods for deep learning. In: ICML, pp. 265–272 (2011)
28. Ngiam, J., Khosla, A., Kim, M., Nam, J., Lee, H., Ng, A.Y.: Multimodal Deep Learning. In: ICML, pp. 689–696 (2011)
29. Sutskever, I., Vinyals, O., Le, Q.V.: Sequence to sequence learning with neural networks. In: NIPS, pp. 3104–3112 (2014)
30. Bekker, A.J., Opher, I., Lapidot, I., Goldberger, J.: Intra-cluster training strategy for Deep Learning with applications to language identification. In: MLSP, pp. 1–6 (2016)
31. Gelenbe, E., Yin, Y.: Deep Learning with random neural networks. In: IJCNN, pp. 1633–1638 (2016)
32. Yin, Y., Gelenbe, E.: Deep Learning in Multi-Layer Architectures of Dense Nuclei. CoRR abs/1609.07160 (2016)
33. Gelenbe, E.: G-networks: a unifying model for neural nets and queueing networks. In: MASCOTS, pp. 3–8 (1993)
34. Fourneau, J.-M., Gelenbe, E., Suros, R.: G-networks with multiple class negative and positive customers. In: MASCOTS, pp. 30–34 (1994)
35. Gelenbe, E., Timotheou, S.: Random neural networks with synchronized interactions. Neural Comput. **20**(9), 2308–2324 (2008)
36. Serrano, W., Gelenbe, E.: The deep learning random neural network with a management cluster. In: International Conference on Intelligent Decision Technologies, pp. 185–195 (2017)

Convolution Neural Network Application for Road Asset Detection and Classification in LiDAR Point Cloud

George E. Sakr(✉), Lara Eido, and Charles Maarawi

Faculty of Engineering (ESIB), St. Joseph University of Beirut, Beirut, Lebanon
georges.sakr@usj.edu.lb

Abstract. Self-driving cars (or autonomous cars) can sense and navigate through an environment without any driver intervention. To achieve this task, they rely on vision sensors working in tandem with accurate algorithms to detect movable and non-movable objects around them. These vision sensors typically include cameras to identify static and non-static objects, Radio Detection and Ranging (RADAR) to detect the speed of the moving objects using Doppler effect and Light Detection and Ranging (LiDAR) to detect the distance to objects. In this paper, we explore a new usage of LiDAR data to classify static objects on the road. We present a pipeline to classify point cloud data grouped in volumetric pixels (voxels). We introduce a novel approach to point cloud data representation for processing within Convolution Neural Networks (CNN). Results show an accuracy exceeding 90% in the detection and classification of road edges, solid and broken lane markings, bike lanes, and lane center lines. Our data pipeline is capable of processing up to 20,000 points per 900ms on a server equipped with 2 Intel Xeon processors 8-core CPU with HyperThreading for a total of 32 threads and 2 NVIDIA Tesla K40 GPUs. Our model outperforms by 2% ResNet applied to camera images for the same road.

Keywords: Self driving car · Convolution neural networks
LiDAR point cloud

1 Introduction

As the industrial age continues to unfold, self-driving cars are seen as the next logical step in automation [1]. It is estimated that self-driving cars will lead to many environmental benefits with regard to fuel economy [2,3] through the optimization of highways traffic flow [4,5], the reduction of the entire vehicle fleet to only 15% of the current amount by leveraging the sharing economy [6], and enabling platoon driving, which is projected to save to 20–30% fuel consumption [7]. In addition, self-driving cars are expected to cause a decline in accident rates, which constitute the eight highest death cause worldwide by the World Health Organization [8]. Another consequence of handing the reins over to autonomous

K. Arai et al. (Eds.): IntelliSys 2018, AISC 868, pp. 86–100, 2019.
https://doi.org/10.1007/978-3-030-01054-6_6

systems is stress reduction [9] and increased availability of parking space by 75% of the current capacity [10]. Leading technology and automotive companies in this space are targeting the end of the next decade as the self-driving car era in which the number of self-driving cars will surpass the number of regular cars.

For cars to drive themselves safely, they need algorithms which are very accurate in detecting and tracking the position of both moving and non-moving objects. Generally, in machine vision, this can be achieved using two different approaches: using the more traditional image processing techniques with hand-coded features and labels, or using a more automated machine learning approach, where Artificial Neural Networks *learn* features within the data. Both approaches are well established in this domain but there is no agreed upon answer on which approach is better as both are common in the industry. In this study, we evaluate the latter by employing CNN for the detection and classification of pseudo-static roadway features in LiDAR point cloud data. The novelty in this research is summarized in three points:

(1) Using LiDAR point cloud to detect and classify static road assets, where the usual usage of LiDAR was to calculate the distance between the car and other objects.
(2) A new data pipeline that transforms 3D point cloud into a 2D hyper-image that is compatible with CNN.
(3) A new CNN-based deep architecture for high accuracy in detection and classification of static road assets.

The rest of the paper is organized as follows: Sect. 2 discusses previous techniques of road asset detection. Sect. 3 describes the experimental setup, data collection procedure and the LiDAR data format. In Sect. 4 we introduce the network of choice in our study (CNN) and discuss its advantages. Section 5 introduces a novel 3D data representation method allowing it to fit in the CNN model, and Sect. 6 describes the CNN architecture. The experimental results and the comparison with CNN on camera images are presented in Sect. 7 and finally the execution time of the pipeline is discussed in Sect. 8.

2 Related Work

In this section we present some of the previous research for road asset detection. Chan Yee Low et al. combine the use of Canny edge detection and Hough transform for lane marker detection. The system captures images from a front viewing vision sensor placed facing the road behind the windscreen as input. Canny edge detection performs feature recognition which is followed by a Hough transform for lane generation [11]. This method requires empirical tuning of multiple parameters and manual selection of the features to detect. Aly presented a real-time approach to lane marker detection in urban streets by generating top view images and using selective oriented Gaussian filters and fast RANSAC algorithm for fitting Bezier splines through the lane markings. This method achieved comparable results to previous techniques [12]. Weng et al. uses LiDAR point cloud for traffic signs recognition. Their algorithm starts by using the intensity of the received point cloud to detect traffic signs which are known to be painted with highly reflective materials. Then, they make use of the geometric shape and the pairwise 3D shape context to make the final classification. The results show that the proposed method is efficient in detecting and classifying traffic signs from LiDAR point clouds [13]. The main issue with the aforementioned research was the need to manually handcraft the features that the algorithm should monitor in order to perform detection. In the deep learning approach, the first layers of the neural network will have the ability to automatically detect important features and feed them to the deeper part of the network for detection. Luca et al. [14] took advantage of CNN for road detection using only LiDAR data. The LiDAR point cloud was used to create top-down images encoding several basic statistics like elevation and density. Those top view images are 2D images with the basic statistics included. They can be fed to a CNN using the traditional way. The shortcoming of this method is that it removes an entire dimension from the data, which could contain valuable information. Maturana et al. [15] coupled a volumetric occupancy map with a 3D CNN to detect small and potentially obscured obstacles in vegetated terrain. However, their work did not cover road assets such as lines and road edges instead they were interested in differentiating low vegetation that are safe to land on from solid objects that might be hazardous for landing.

This paper specifically aims to present a pipeline that takes incoming LiDAR pulses to detect and classify different static road assets. This research focuses on the detection and classification of road edges, yellow single solid lines, yellow single broken lines, bike lanes and lane center line using deep neural networks. To achieve this aim, a new method is introduced at the start of the pipeline to transform the 3D LiDAR point cloud into hyper-images compatible with CNN. The pipeline terminates with a new CNN architecture that will detect and classify static road assets from the transformed hyper-images. The realization of this pipeline needs an annotated dataset of static road assets. The data collection and annotation is described in the next section.

3 Data Collection and Formatting

Training a classifier to detect and classify road assets requires the collection of annotated structured data. This section describes the data collection phase as well as the format of the acquired data.

3.1 Data Collection

In this study we evaluate our extraction and classification methodology on LiDAR point cloud data collected with a mobile mapping setup. A survey-grade Inertial Navigation System (INS) was utilized in combination with a laser scanner. The model for the INS is Novatel IGM-S1 STIM. The LiDAR that was used is the Velodyne HDL-32 that was mounted tilted backwards at a 45° pitch angle.

This mounting configuration is appropriate in surveying and mapping applications. The sensors were time synchronized using a pulse per second (PPS) signal originating from the GPS receiver. The INS was used to measure the vehicle trajectory through the environment, and LiDAR was used to extract points in the vicinity of the vehicle. Data was collected from a single trip on Marin Ave in Berkeley, CA. Data was registered to the UTM (Zone 10N) frame of reference to create a point cloud. Relative fit and loop closure methods were employed to ensure a point cloud free of artifacts. To create a ground truth dataset from the input data, a map of 3D vectors of splines was generated with manual annotation using custom made tools.

3.2 Data Format and Annotation

The LiDAR outputs a point cloud which is at its most basic, a collection of points in three dimensions, each point having an x, y and z coordinate defined in a coordinate frame of reference. In addition to geometry, each point also has an intensity value, which corresponds to the amount of light reflected back to the laser scanner. However, the reflected points are not annotated and it is not possible to identify the reflecting object. Hence the need to use the ground truth map that was annotated manually.

In the simplest of forms the ground truth map is a JSON file that contains a summary of the static objects found in the point cloud. A triangular road sign is represented by its three vertices. Any point from the point cloud that falls withing these three vertices is annotated as road sign. Similarly, a road edge is represented by a spline which is a sequence of continuous lines. If a point falls on any road-edge spline it is annotated as road edge. This process can be also applied to bike lanes, solid and broken lines. In general we examined the raw points which fall within the shape defined in the map, and applied the appropriate annotation. For those points which do not fall within any predefined shape, they get the $NULL$ label. The $NULL$ label is extremely important as it will be used to test if the final model can detect an asset or will it be missed and labeled as $NULL$. The program responsible for labeling the points was written in C++ and 32 threads were created in order to accelerate it. The choice of 32 threads was due to the server's capability of running up to 32 threads simultaneously.

The described method generates annotated point cloud. This annotated point cloud will be used as an input to our classifier. Hence the need to define which classifier to use and how can the data be formatted to be compatible with this classifier. Next section will answer those needs.

4 CNN

Pattern recognition and image captioning were greatly impacted by Convolution Neural Networks [16], [17]. The challenging part before the CNN era was handcrafting the features and tuning the classifier. This previous challenge was revolutionary solved by CNN which gave us the ability to learn the features directly from the training data. CNN also has an architecture that makes it specifically efficient in image recognition by implementing the convolution operation that captures the 2D nature of an image. CNN have been making small progress in commercial use for the past 20 years [18], however, the adoption of CNN has grown exponentially in the last seven years because of two main reasons. First, the publication of large and labeled data sets such as the Large Scale Visual Recognition Challenge (ILSVRC) [19] and second, the development of the massively parallel graphics processing unit (GPU) which is used to accelerate the weights optimization process for CNN.

One drawback in regular neural network is that it treats all pixels on equal terms. This is a result of the fact that every neuron is connected to all the pixels of the image, hence treating far away pixels the same way it treats the close pixels which destroys, from the neuron perspective, the spacial structure of the image. In contrast CNN takes advantage of spatial structure in an image by using their special architecture which is greatly suitable to image recognition. A deep neural network is usually made up of a succession of convolution neural networks, pooling layers and dense layers. There are of course other types of deep networks but in this research we used the traditional structure of a deep network. The convolution layer is based on a small local receptive field also called a kernel. The size of this small receptive field is a hyper parameters that must be tuned using validation data. For example if the kernel size is 3×3 then every neuron in the convolution network would be connected to 9 pixels. The kernel slides horizontally and vertically by a certain amount of pixels called stride. The stride is also a hyper parameter that should be tuned. Whenever the kernel slides it covers a new part of the image which by itself will be connected to a different neuron. So every receptive field is connected to a different neuron. However, the second important principle in a convolution network is weight sharing. All the neurons in the same layer share the same weights. This gives the network the ability to learn the same pattern across the image. Hence if the layer learns to detect horizontal edges, in on part of an image it will also be able to detect it anywhere in the image. This technique makes the convolution layer shift invariant. But this also restricts the layer to learn one thing in a image. This is why a full convolution layer is made by stacking many layers on top of each others. So every layer will learn to detect a different thing across the whole image. To reduce overfitting a dropout layer can be used after the convolution stage. A dropout layer simply drops some of the neurons with a probability ϵ that is also tuned on the validation set. Finally the output of the convolution layers is flattened into 1 vector which forms the input of the dense layer. The dense layer is just a regular fully connected layer that will bring together all the learnt features from the convolution part and used a softmax to output the decision.

5 Data Representation and Labeling

As described in the previous section, CNN expects a list of 2D images to be inputed for training. However, the point cloud is just a list of annotated quadruple (x, y, z, r). In this section we present the algorithm that creates hyper-images from the point cloud. Hyper-images are designed to be compatible with CNN. This section also presents the labeling method used to give every created hyper-image a corresponding label.

5.1 Transformation Algorithm

The algorithm for transforming the point cloud into a compatible hyper-image is defined as follows:

(1) find the minimum and maximum values of x, y and z in the point cloud.
(2) create a 1 m^3 voxel starting at x_{min}, y_{min} and z_{min}.
(3) split the voxel into 1000 sub-voxels of $10 \times 10 \times 10\,cm^3$ (Fig. 1).
(4) find all the points from the point cloud that fall within every sub-voxel.
(5) give a value to every sub-voxel equal to the average reflectivity of the total points that fall inside it.
(6) slide the voxel by a stride of 25 cm in all directions, respectively.
(7) repeat from Step 3 until you reach x_{max}, y_{max} and z_{max}.

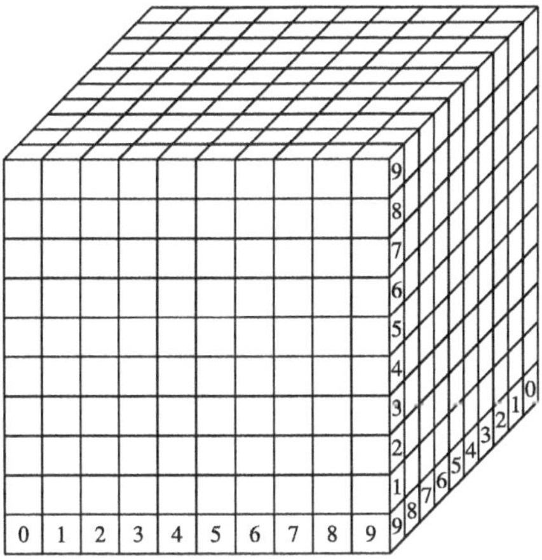

Fig. 1. 1 m^3 divided into 1000 sub-voxels of $10 \times 10 \times 10\,cm^3$

The above steps will transform the point cloud into a group of 3D-like images. The final step is to create the hyper-images that are 2D-like images. Normally a 2D image is formed by a group of pixels each having R, G, B components. We decided to take every sub-voxel from the x-y surface and give it 10 components instead of 3 (RGB) components. Those 10 components are the values of the sub-voxels lying along the z-axis of the surface sub-voxel. Figure 2a shows the bottom right vector along the z-axis before the flattening procedure and Fig. 2b shows the 1 pixel with its 10 components that replaces the original vector. Hence a big voxel

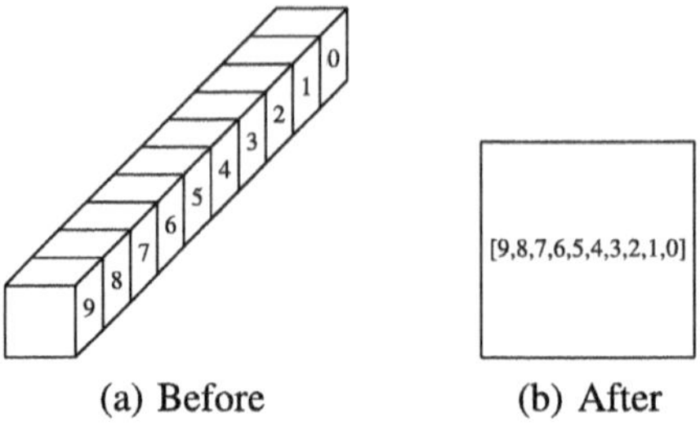

Fig. 2. Flattening of the bottom right vector of the 3D voxel

is a now a hyper-image having a width and height of 10 pixels. Each pixel has 10 components instead of the regular 3 (RGB) components. As a summary we say that every hyper-image is (10,10,10). Finally all N hyper-images are generated and grouped in one array making the training set a 4D array (N,10,10,10) which is perfectly compatible with the input of a CNN. This transformation allows also the localization of a detected asset to within $1\,m^3$ because it scans the input point cloud and creates $1\,m^3$ hyper-images and classifies them. Finally this transformation results in a set of unlabeled hyper-images.

5.2 Labeling the Hyper-Images

The hyper-images are formed by flattened sub-voxels. Every sub-voxel contains a number of annotated points generated from the JSON file. We define the dominant sub-label as the most frequent asset from all the points that fell inside this sub-voxel.

(1) Every sub-voxel is sub-labeled by its dominant sub-label.
(2) The dominant label in a hyper-image is the most frequent sub-label among all of its sub-voxels.
(3) Label the hyper-image by its dominant label.
(4) If all sub-voxels are sub-labeled null then the hyper-image receives the null label.
(5) In case the dominant label was null but other sub-voxels are not sub-labeled as null, then the second most frequent sub-label will be given to the hyper-image.

The output of the above algorithms is a dataset of labeled hyper-images. The hyper-images are fed into a CNN-based deep architecture introduced in the next section.

6 Classifier Architecture

The classifier is designed to minimize the categorical cross entropy error defined by:

$$L = -\sum_{i=1}^{n}\sum_{c=1}^{C} t_{ic} log(y_{ic})$$

where N is the total number of samples in the batch, C is the total number of classes, $t_{ic} = 1$ if and only if sample i belongs to class c and y_{ic} is the probability that sample i belongs to class c. The training set of hyper-images is fed into the classifier which then computes a proposed class for each image. The proposed class is compared to the desired class for that hyper-image and the weights are adjusted to minimize the error. In minimizing the error, the network brings its output closer to the desired output. The weight adjustment is accomplished using back propagation as implemented in the KERAS machine learning package. The update of the weights occurs after the batch is sent into the network, the error on that batch propagates backwards and every weight is updated by a small amount proportional to the error. The proportionality factor is called the learning rate α. An epoch is completed when all the batches have passed through the network. The weights are saved after every epoch. Early stopping was used with a delay of three epochs. This allows the network to stop the optimization process after three consecutive epochs in which the validation error does not decrease. The weights that yielded the smallest error are saved. A block diagram of our training system is shown in Fig. 3. Once trained the network can classify hyper-images as shown in Fig. 4.

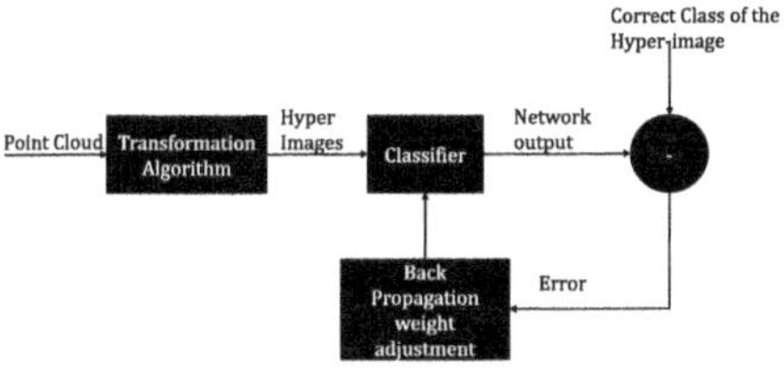

Fig. 3. Training phase

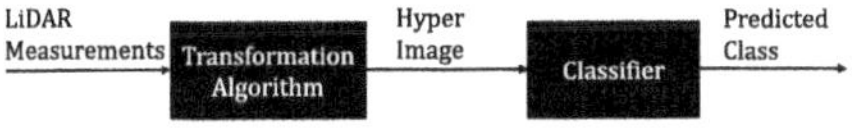

Fig. 4. Testing phase

The classifier block is a deep neural network made up of convolution layers, dropout layers and fully connected dense layers. The number of convolution layers, dense layers as well as the dropout probability are varied to obtain a better accuracy. The kernel size of every convolution layer is also a parameter that can be varied and the stride at which it slides on the image is also another parameter to tune. Finally the batch size is usually set based on the amount of memory available. All of the above mentioned parameters and their values are discussed in next section.

7 Results and Discussion

This section describes the architectures used to detect and classify 2, 4 and 6 different road assets. It also discusses the tuning of the different network parameters used to achieve this accuracy. The accuracy of our model is compared to ResNet's [20] accuracy on camera images. The ResNet was tuned on images corresponding to the same LiDAR point cloud.

7.1 Detecting Bike Lanes

The first models were evaluated by classifying two classes: Bike Lane and Null (images that do not correspond to any road asset). The data set was balanced between each class as well as between training and testing: 51,566 images for training from which 5156 were used for validation and 58,014 images for testing which were never used during the training phase. The testing and training data each belongs to a different part of the road. This allows us to demonstrate the ability of our classifier to generalize to unseen scenarios. Many different architectures were used and the one that yielded the highest validation accuracy was used on the testing set. Table 1 shows the results for the different architectures on the validation set. Note that the kernel size of the convolution layer and the dropout layer was also varied and the reported accuracy is the for the architecture that yielded the highest accuracy. Finally the dropout layer (if used) is placed after all the convolution layers and before the dense layer. We also tried to put it between the convolution layers but we did not obtain a better accuracy. The reference ResNet accuracy on camera images for this model is 96%.

Table 1. 2-Class validation accuracy

Model (convolutional, dropout, dense)	Accuracy (%)
7 Convolutional, 1 Dropout, 4 Dense	97
7 Convolutional, 1 Dropout, 3 Dense	98.3
7 Convolutional, 1 Dropout, 2 Dense	98.4
7 Convolutional, 0 Dropout, 3 Dense	97.8
6 Convolutional, 1 Dropout, 2 Dense	97.7
5 Convolutional, 1 Dropout, 2 Dense	97.6
4 Convolutional, 1 Dropout, 2 Dense	98
4 Convolutional, 0 Dropout, 2 Dense	96.8
3 Convolutional, 1 Dropout, 2 Dense	97.3

As it can be seen in Table 1, the architecture that yields the highest accuracy on the validation set is the one with seven convolution layers, one dropout layer and two dense layers. All convolution layers used the ReLu activation function and 0 padding. The first convolution layer used 16 kernels each of size 3×3, the second consists of 32 kernels of size 3×3, the third used 64 kernels of size 3×3, and the last four used 128 kernels of size 3×3. The convolution layers were followed by one dropout layer with dropout probability of 0.2. Two dense layers were used after the dropout later, the first had a size of 512 with a ReLu activation function and followed by another dense layer with size 2 and a softmax activation function. The batch size for training was 128. The Adam optimizer was used with adaptive learning rate to minimize the categorical cross-entropy loss. This model gave an accuracy of 98.1% over the testing set. A high accuracy given that the testing set represents a part of the road that is unseen before by the network. This shows that our classifier is able to generalize to unseen scenarios. It also shows that LiDAR-CNN outperforms RestNet by more than 2%.

7.2 Four Assets

In this part we present models evaluated to classify: Bike Lane, Lane Centerline, Road Edge and Null. The data set was balanced between all four classes (21,234 image of each class). The images were shuffled and then 61,153 images were used for training, 6,795 for validation and 16,988 for testing (20% of the total number of images). Many different architectures were used and the one that yielded the highest validation accuracy was used on the testing set. Table 2 shows the results for the different architectures on the validation set. Note that the kernel size of the convolution layer and the dropout layer was also varied and the reported accuracy is the for the architecture that yielded the highest accuracy. The reference ResNet accuracy for this experiment is 92%. The accuracy was obtained using the same procedure used for the two class experiment.

Table 2. 4-class validation accuracy

Model (convolutional, dropout, dense)	Accuracy (%)
7 Convolutional, 1 Dropout, 2 Dense	93.4
6 Convolutional, 1 Dropout, 3 Dense	93.2
6 Convolutional, 1 Dropout, 2 Dense	93.6
5 Convolutional, 1 Dropout, 2 Dense	94.5
4 Convolutional, 1 Dropout, 2 Dense	94.6
4 Convolutional, 0 Dropout, 2 Dense	94.1
3 Convolutional, 1 Dropout, 2 Dense	94

As it can be seen in Table 2, the architecture that yields the highest accuracy on the validation set is the one with four convolution layers, one dropout layer and two dense layers. All convolution layers used the ReLu activation function and 0 padding. The first convolution layer used 16 kernels each of size 3×3, the second consists of 32 kernels of size 3×3, the third used 64 kernels of size

3×3, and the last one used 128 kernels of size 3×3. The convolution layers were followed by one dropout layer with dropout probability of 0.2. Two dense layers were used after the dropout later, the first had a size of 512 with a ReLu activation function and followed by another dense layer with size 4 and a softmax activation function. The batch size for training was 128. The Adam optimizer was used with adaptive learning rate to minimize the categorical cross-entropy loss. This model gave an accuracy of 94.3% over the testing set. A very good accuracy given that the testing set represents a part of the road that is unseen before by the network. This shows the ability of training the model on a small part of the road and extrapolate to unseen territories. This shows that our classifier is able to generalize to unseen scenarios. We also notice that LiDAR-CNN outperforms ResNet by around 2.6% for the best model obtained.

7.3 Six Assets

In this part we present models evaluated to classify: Bike Lane, Lane Centerline, Road Edge, Yellow Single Solid Line, Yellow Single Broken Line and Null. The data set was balanced between all six classes (9,246 image of each class). The images were shuffled and then 55,473 images were used for training, 5,547 for validation and 13,869 for testing (25% of the total number of images). Many different architectures were used and the one that yielded the highest validation accuracy was used on the testing set. Table 3 shows the results for the different architectures on the validation set. Note that the kernel size of the convolution layer and the dropout layer was also varied and the reported accuracy is the for the architecture that yielded the highest accuracy. The reference accuracy that was given by ResNet for this experiment is 87.5%.

Table 3. 6-class validation accuracy

Model (convolutional, dropout, dense)	Accuracy (%)
7 Convolutional, 1 Dropout, 2 Dense	89.5
6 Convolutional, 1 Dropout, 2 Dense	90.2
5 Convolutional, 1 Dropout, 2 Dense	87.6
5 Convolutional, 0 Dropout, 2 Dense	89.7
4 Convolutional, 2 Dropout, 2 Dense	88.1
4 Convolutional, 1 Dropout, 3 Dense	89.5
4 Convolutional, 1 Dropout, 2 Dense	89.5
4 Convolutional, 0 Dropout, 3 Dense	90
4 Convolutional, 0 Dropout, 2 Dense	90.6
3 Convolutional, 1 Dropout, 2 Dense	89.4
3 Convolutional, 0 Dropout, 2 Dense	88.7

As it can be seen in Table 3, the architecture that yields the highest accuracy on the validation set is the one with four convolution layers, zero dropout layer and two dense layers. All convolution layers used the ReLu activation function and zero padding. The first convolution layer used 16 kernels each of size 3×3, the second consists of 32 kernels of size 3×3, the third used 64 kernels of size 3×3, and the last one used 128 kernels of size 3×3. Two dense layers were then used, the first had a size of 512 with a ReLu activation function and followed by another dense layer with size 6 and a softmax activation function. The batch size for training was 128. The Adam optimizer was used with adaptive learning rate to minimize the categorical cross-entropy loss. This model gave an accuracy of 90.3% over the testing set. A good accuracy given that the testing set represents a part of the road that is unseen before by the network. This shows that our classifier is able to generalize to unseen scenarios. This proves the ability of training the model on a small part of the road and extrapolate to unseen territories. LiDAR-CNN also outperforms RestNet by around 3%.

8 Execution Time

The latency introduced in every part of the pipeline is introduced in this section. Since the latency depends on the hardware and software used, they are introduced first then the execution time of every part of the pipeline is presented.

8.1 Hardware

The training of a deep network like the ones presented above with a large training set is challenging on a regular computer due to memory and processing speed limitations. It has to be done offline on a high performance computer and the resulting model is implemented on a regular computer. To train the classifier we used a server equipped with 2 Intel Xeon processors 8-core CPU with Hyper-Threading for a total of 32 threadsand, 2 NVIDIA Tesla K40 GPU with 12 GB of memory each dedicated to the GPU. 128 GB of RAM were available for the Xeon CPUs and 2TB of SSD hard drive were used to store the data locally. The presence of the 2GPUs was of absolute importance to train the model in a reasonable amount of time. For instance the training phase took between 5 and 7 min per model.

8.2 Software

Keras API was installed on Ubuntu server to easily operate with CNN libraries. Keras is an open source API capable of running on top of Tensorflow [21]. Tensorflow is an open source library for machine learning developed by Google. For data labeling C++ was used with Json libraries to access the Json file.

8.3 Pipeline Execution Time

The execution time of all phases of the pipeline was evaluated:

- Generating 3D voxels from points received by the LiDAR.
- Formatting the images into Hyper-Images.
- Analyzing the hyper-images by the CNN model.

In order to simulate the reception of data points by the LiDAR, a file containing 20,000 points in a $3 \times 3 \times 3$ m space was created. The 35 images were generated in the process and fed to the CNN model. The execution time of each stage of the pipeline was computed several times and the average value was calculated. The results are presented in Table 4.

Table 4. Execution time

Stage	Mean time (ms)	Std (ms)
Read File Into RAM	21.36	2.14
Calculate Space Boundaries	0.05	0.004
Generate Voxels	390.86	51.96
Generate Hyper-images	20.975	5.57
Write Images to Disk (Pickle)	1.92	0.19
Load Images from Disk (unpickle)	3.26	0.04
Load CNN Model from Disk	7872.9	53.53
Classify 35 Hyper-Images	411.22	30.41

Table 4 shows the execution time for the different stages of the pipeline. It is noted that during runtime, the CNN model will be loaded once so the 8 seconds should not be counted, as well as writing and reading from disk. So in total 20,000 LiDAR points in a $3 \times 3 \times 3$ environment takes about 825ms to classify. We presume that the execution time could be further reduced in case CNN was implemented in C++.

9 Conclusion and Future Work

In this research we presented a new application for LiDAR combined with deep learning for road asset detection. We presented also a new method for LiDAR data representation to create hyper-images that fits the input of a convolution neural network. The network was able to differentiate between six different assets with high accuracy $> 90\%$. The limitation of our model is its inability to locate the detected asset within the 1 m cube voxel. Another limitation is the execution time in real time which takes around 825 ms to process 20,000 LiDAR points. This time must be reduced to less than 100 ms to be able to accommodate the amount of data generated by the LiDAR. As a future work we will consider locating assets inside the voxel, we will implement the model in C++ to further accelerate the processing time.

Acknowledgment. This project has been funded with the joint support from the National Council for Scientific Research in Lebanon and the St. Joseph University of Beirut. The authors would like to thank the dean Fadi Geara of the faculty of engineering at St Joseph university for providing us with the material needed for this study, namely, the new deep learning server with multiple NVIDIA Tesla GPUs in collaboration with Murex. We would also like to thank Civil Maps for providing us with the LiDAR data and labels used in this research and Dr. Fabien Chraim from civil maps for his innovative ideas and reviewing the paper.

References

1. Rosenzweig, J., Bartl, M.: A review and analysis of literature on autonomous driving. E-J. Mak.-Of Innov. (2015)
2. Payre, W., Cestac, J., Delhomme, P.: Intention to use a fully automated car: attitudes and a priori acceptability. Transp. Res. Part F: Traffic Psychol. Behav. **27**, 252–263 (2014)
3. Luettel, T., Himmelsbach, M., Wuensche, H.-J.: Autonomous ground vehiclesconcepts and a path to the future. In: Proceedings of the IEEE, vol. 100, no. Special Centennial Issue, pp. 1831–1839 (2012)
4. Le Vine, S., Zolfaghari, A., Polak, J.: Autonomous cars: the tension between occupant experience and intersection capacity. Transp. Res. Part C: Emerg. Technol. **52**, 1–14 (2015)
5. Jamson, A.H., Merat, N., Carsten, O.M., Lai, F.C.: Behavioural changes in drivers experiencing highly-automated vehicle control in varying traffic conditions. Transp. Res. Part C: Emerg. Technol. **30**, 116–125 (2013)
6. Ross, P.E.: Robot, you can drive my car. IEEE Spectr. **51**(6), 60–90 (2014)
7. Weyer, J., Fink, R.D., Adelt, F.: Human-machine cooperation in smart cars. An empirical investigation of the loss-of-control thesis. Saf. Sci. **72**, 199–208 (2015)
8. Violence and Injury Prevention - World Health Organization: Global status report on road safety 2013: supporting a decade of action. World Health Organization (2013)
9. Rudin-Brown, C.M., Parker, H.A., Malisia, A.R.: Behavioral adaptation to adaptive cruise control. In: Proceedings of the Human Factors and Ergonomics Society Annual Meeting, vol. 47, no. 16, pp. 1850–1854. SAGE Publications Sage CA, Los Angeles (2003)
10. Alessandrini, A., Campagna, A., Delle Site, P., Filippi, F., Persia, L.: Automated vehicles and the rethinking of mobility and cities. Transp. Res. Procedia **5**, 145–160 (2015)
11. Low, C.Y., Zamzuri, H., Mazlan, S.A.: Simple robust road lane detection algorithm. In: 2014 5th International Conference on Intelligent and Advanced Systems (ICIAS), pp. 1–4. IEEE (2014)
12. Aly, M.: Real time detection of lane markers in urban streets. In: Intelligent Vehicles Symposium, 2008 IEEE, pp. 7–12. IEEE (2008)

13. Weng, S., Li, J., Chen, Y., Wang, C.: Road traffic sign detection and classification from mobile lidar point clouds. In: 2015 ISPRS International Conference on Computer Vision in Remote Sensing, p. 99 010A. International Society for Optics and Photonics (2016)
14. Caltagirone, L., Scheidegger, S., Svensson, L., Wahde, M.: Fast lidar-based road detection using convolutional neural networks. arXiv preprint arXiv:1703.03613 (2017)
15. Maturana, D., Scherer, S.: 3D convolutional neural networks for landing zone detection from lidar. In: 2015 IEEE International Conference on Robotics and Automation (ICRA), pp. 3471–3478. IEEE (2015)
16. LeCun, Y., Boser, B., Denker, J.S., Henderson, D., Howard, R.E., Hubbard, W., Jackel, L.D.: Backpropagation applied to handwritten zip code recognition. Neural Comput. **1**(4), 541–551 (1989)
17. Krizhevsky, A., Sutskever, I., Hinton, G.E.: Imagenet classification with deep convolutional neural networks. In: Advances in Neural Information Processing Systems, pp. 1097–1105 (2012)
18. Jackel, L.D., Sharman, D., Stenard, C.E., Strom, B.I., Zuckert, D.: Optical character recognition for self-service banking. AT& T Tech. J. **74**(4), 16–24 (1995)
19. Berg, A., Deng, J., Fei-Fei, L.: Large scale visual recognition challenge (ILSVRC) (2010). http://www.image-net.org/challenges/LSVRC
20. He, K., Zhang, X., Ren, S., Sun, J.: Deep residual learning for image recognition. arXiv preprint arXiv:1512.03385 (2015)
21. Abadi, M., Agarwal, A., Barham, P., Brevdo, E., Chen, Z., Citro, C., Corrado, G.S., Davis, A., Dean, J., Devin, M., et al.: Tensorflow: large-scale machine learning on heterogeneous distributed systems. arXiv preprint arXiv:1603.04467 (2016)

PerceptionNet: A Deep Convolutional Neural Network for Late Sensor Fusion

Panagiotis Kasnesis[1(✉)], Charalampos Z. Patrikakis[1], and Iakovos S. Venieris[2]

[1] Department of Electrical and Electronics Engineering, University of West Attica, Athens, Greece
pkasnesis@yahoo.gr, bpatr@puas.gr

[2] School of Electrical and Computer Engineering, National Technical University of Athens, Athens, Greece
venieris@cs.ntua.gr

Abstract. Human Activity Recognition (HAR) based on motion sensors has drawn a lot of attention over the last few years, since perceiving the human status enables context-aware applications to adapt their services on users' needs. However, motion sensor fusion and feature extraction have not reached their full potentials, remaining still an open issue. In this paper, we introduce PerceptionNet, a deep Convolutional Neural Network (CNN) that applies a late 2D convolution to multimodal time-series sensor data, in order to extract automatically efficient features for HAR. We evaluate our approach on two public available HAR datasets to demonstrate that the proposed model fuses effectively multimodal sensors and improves the performance of HAR. In particular, PerceptionNet surpasses the performance of state-of-the-art HAR methods based on: (1) features extracted from humans, (2) deep CNNs exploiting early fusion approaches, and (3) Long Short-Term Memory (LSTM), by an average accuracy of more than 3%.

Keywords: Convolutional neural network · Deep learning · Feature learning Human activity recognition · Sensor fusion

1 Introduction

The proliferation of the Internet of Things (IoT) over the last few years has contributed to the collection of huge amounts of time-series data. An IoT device with high sampling rates, such as a wearable, produces hundreds of data every second, resulting to a data explosion, considering the vast number of such devices connected over the internet. Through real-time or batch data processing, meaningful information is extracted, revealing daily patterns of individual owners or social groups. This information can be exploited by context-aware applications in order to enhance wellbeing [1], health status [2], facilitate smart environments [3] and improve security [4, 5].

Context-aware applications are capable of discovering and reacting to changes in the environment, their user is situated in [6, 7]. Their effectiveness depends on four entities [8]: identity, location, time, and status (activity). The first three entities can be

K. Arai et al. (Eds.): IntelliSys 2018, AISC 868, pp. 101–119, 2019.
https://doi.org/10.1007/978-3-030-01054-6_7

easily tracked without the need of deploying sophisticated algorithms. Extracting knowledge from sensors to define the human activity however, is a complex task, where the use of signal processing is mandatory.

Conventional signal processing techniques in HAR, apply mathematical, statistical or heuristic functions over raw motion data, in order to extract valuable features, identified as handcrafted features (HCFs). Concretely, D. Figo et al. [9] categorize the feature extraction techniques in three domains: the *time* domain (e.g., mean, max, mean values), the *frequency* domain (e.g., Fast Fourier Transformation) and *discrete representation* domain (e.g., Euclidean-based distances). These hand-crafted features, feed a classification algorithm, which after a training phase, is able to recognize a human activity (e.g., walking, cooking etc.). The accuracy of the classification algorithm depends heavily on the extracted features, while the feature extraction process is time consuming.

Deep Learning (DL) [10] can provide a solution to this problem. DL is a branch of Machine Learning (ML), in particular, Artificial Neural Networks (ANNs), and has the ability to automatically extract features. What is more, previous implementations of DL approaches to computer tasks such as computer vision [11], speech recognition [12] and natural language processing [13], have outperformed past techniques based on HCFs. As a result, state-of-the-art HAR methods (e.g., [14, 15]) are based on DL. In this paper, we adopt a DL approach, to process and analyze multimodal sensor data produced by mobile motion sensors. In our proposal, we apply late sensor fusion (2D convolution), to recognize the patterns of human activities. More specifically, this is the first 2D convolution on multimodal raw motion sensor data (to the best of our knowledge). The contributions and innovations of our proposal can be summarized in the following:

- Proves that late sensor fusion is more effective in Deep Convolutional Neural Networks.
- Applies a 2D convolution to vertically stacked motion sensor data.
- Utilizes global average pooling over feature maps in the classification layer, instead of traditional fully connected layers.
- Outperforms other state-of-the-art deep learning techniques in HAR.

The rest of the paper is organized as follows: In Sect. 2, an overview of the state-of-the-art in deep learning approaches to HAR is presented. In next Sect. 3, the proposed architecture of the deep convolutional neural network is explained. Section 4 describes the experimental set up, while Sect. 5 presents the results of the proposed method using two public datasets. Finally, Sect. 6 concludes the paper and proposes future work directions.

2 State-of-the-Art

After the raw data collection, a typical human activity recognition system deploys tools and techniques for preprocessing, segmentation, feature extraction and classification [16]. However, DL approaches to HAR have revealed that the step of feature extraction is included in the DL algorithm [17]. One of the greatest advantages of Deep Neural Networks (DNNs) is their ability to extract their own features, which manage to express complex relations/patterns between the data [18]. Thus, DL approaches to HAR are considered to be the state-of-the-art, and are categorized as follows: (a) *Autoencoders*, (b) *Convolutional Neural Networks*, (c) *Deep learning on spectrograms* and (d) *Convolutional Recurrent Neural Networks*. In the following paragraphs, a summary of existing DL work towards HAR, in all four categories will be presented.

2.1 Autoencoders

Autoencoders are a specific field of ANNs, based on unsupervised learning (i.e., machine learning based on unlabeled data), where an ANN consisting of one hidden layer, tries to produce as output the input values. Many stacked Autoencoders form a DNN. In [19] an Autoencoder technique was used to extract key features for HAR. In particular, the authors used Restricted Bozltman Machines (RBMs) [20], which are a particular form of log-linear Markov Random Fields (i.e., a non-directed probabilistic graphical model) and has been applied successfully for dimensionality reduction in computer vision [21]. Another Autoencoder approach was proposed in [22], where C. Vollmer et al. use a Sparse Autoencoder [23] for extracting features. Sparse Autoencoders have been successfully applied to computer vision problems, such as medical image analysis [24, 25]. However, Autoencoders are fully connected DNN models, and as a result, do not manage to capture the local dependencies of the time-series sensor data [26, 27].

2.2 Convolutional Neural Networks

Yann LeCun et al. introduced LeNet, a Convolutional Neural Network (CNN) in [28]. Based on the mathematical operation of convolution (i.e., the combination of two functions to form a third one), LeNet managed to outperform the other classification algorithms in recognizing hand-written digits. However, ConvNets (Convolutional Networks) drew public attention almost 15 years later, where the deep ConvNet of Krizhevsky et al. [11], called AlexNet, surpassed the performance of the runner-up algorithm by almost 5%, using the ImageNet dataset [29]. Since then, DL has become the state-of-the-art method in various computer tasks (e.g., natural language processing).

The first CCN approach to HAR was introduced in [30]. The authors used as input a 1D array representation of the motion signals, unlike image analysis that uses a 2D array of pixels. In this way, the signals are stacked into channels (channel-based stacking). For example, a tri-axial accelerometer sensor produces 3 channels (X, Y, Z axis), similarly to colored images (Red, Blue, Green channels). As a result, the convolution operation is applied to each signal individually. After two successive *Convolutional* and *Pooling* (*ConvPool*) operations [31], they concatenated the channels into a 1D array and applied a Multilayer Perceptron (MLP), otherwise a Dense layer, to do the classification.

Similarly, in [27] they propose the same architecture, but they use a more shallow CNN (only one *ConvPool* operation), having as input 3D acceleration time series. The results they acquired showed that CNN approach outperforms the Autoencoders. In addition to this, Ronao et al. [14, 32, 33] and J. B. Yang et al. [34] propose a deeper CNN (3 *ConvPool* layers) approach to HAR, relying on a dataset with tri-axial gyroscope and accelerometer data. Moreover, introducing additional information as input, by adding the Fourier transformation, acquired, almost, a 1% higher accuracy [14]. It should be noted that all the aforementioned CNN approaches fuse the motion data in the first hidden layer and use as final hidden layer an MLP (i.e., similarly to Autoencoders some local time-dependent patterns are not discovered). Thus, we will refer to them for the rest of the paper using the abbreviation CNN-EF (Early Fusion).

2.3 Deep Learning on Spectrogram

An interesting approach which is applied in audio signals is that of converting the input sensor signal to spectrogram, during feature extraction step, providing a representation of the signal as a function of frequency and time. Afterwards, the spectrogram image feeds a DNN, similarly to the image analysis process.

Alsheikh et al. [35] adopt a hybrid approach of Deep Learning and Hidden Markov Models (DL-HMM) for sequential activity recognition. Specifically, the tri-axial accelerometer signal is translated into spectrogram and afterwards a RBM is applied. Furthermore, a non-mandatory HMM step, which has as input the emission probabilities out of the DNN, is used for modeling temporal patterns in activities. A more sophisticated technique is adopted in [36], where the CNN has as input an *activity image*. According to the authors, the raw tri-axial accelerometer and gyroscope signals are stacked row-by-row into a signal image, based on an algorithm. Subsequently a 2D Discrete Fourier Transform (DFT) is applied to the *signal image* and its magnitude is represents the *activity image*. This way, signal sequences are adjacent to other sequences, enabling the DNN to extract hidden correlations between neighboring signals. However, it should be noted that, conversion of time-series data into the frequency domain is not as effective in HAR as in audio classification, and time statistical features have proven to be more essential [9].

2.4 Convolutional Recurrent Neural Networks

Recurrent Neural Networks (RNNs) [37] are a family of neural networks for processing a sequence of values. As a result, RNNs are applied broadly to time-series data. Moreover, because in case of sequential data, a value xi depends on a set of previous n values $\{x_{i-1}, x_{i-2}, \ldots, x_{i-n}\}$ and on a set of next n values, $\{x_{i+1}, x_{i+2}, \ldots, x_{i+n}\}$ a mechanism named LSTM (Long Short-Term Memory) [38] is applied to enhance the memory of the network.

A hybrid approach called Convolutional LSTM is presented in [15]. This network consists of four consecutive 1D convolution operations, which have as input vertically stacked motion signals, where the output of the last feeds a LSTM layer. Afterwards, a final LSTM layer is used to predict the class of the activity performed. As a result, the signals are not fused during the convolutions, but they are convolved using the same filter. This RNN approach was applied on two HAR datasets and managed to model temporal dependencies more effectively than a conventional ConvNet.

Furthermore, the same authors, using the same network architecture, studied the possibilities of transfer learning (e.g., transferring trained filters) in CNNs layer by layer for activity recognition based on wearable sensors [39]. The experimental results show that the performance of the model for the same application domain was not affected by transferring features of the first layer, while there was an improvement in training time (~17% reduction). However, the accuracy of the algorithm was significantly reduced after applying transfer learning between different applications domains, between sensor locations and between sensor modalities.

Following the heuristic that the motion signals are correlated with each other, which is denoted in [36], in this paper we propose a different representation of time-series sensor data (vertical stacking) that applies a late sensor fusion and allows a 2D convolution operation over them.

3 PerceptionNet

In this paper, we introduce the concept of applying a late 2D convolution on HAR data in an attempt to avoid overfitting and discover more general activity patterns, emanating from the cross-correlation between high-level features of the motion signals. We developed a deep CNN model, named PerceptionNet, having as input vertically stacked motion signals in order to exploit the semantics and the grid-like topology of the input data, in contrast with conventional ANNs. The intuition for applying late sensor fusion, the components/layers of PerceptionNet and the selected optimizer are described below.

3.1 Convolutional Layer

The convolution operation manages to obtain a less noisy estimate of a sensor's measurements, by averaging them. Because of the fact that some measurements should contribute more in the average, the sensors measurements are convolved with a weighting function w [31]. Consequently, in our case the input, which is the preprocessed and segmented motion signal, is combined with the filters (weights), which are trained in order to discover the most suitable patterns (e.g., peaks in the signal). Moreover, each filter is replicated across the entire signal. As a result, the replicated units share the same weight vector and bias and form a *feature map* (or *activation map*), which is the product of several convolutions in parallel of the signal and the *filters*.

In other words, all the signal values in a convolutional layer respond to the same feature within a specific receptive field [40]. This iteration over all the units allows for motion features to be detected regardless of their position in the sensor signal (translation invariance property). In particular, the ith product element of a discrete 1D convolution between input array x and a 1D filter w equals:

$$c_i^{l,q} = b^{l,q} + \sum_{d=1}^{D} w_d^{l,q} x_{i+d-1}^{l-1,q} \quad (1)$$

where l is the layer index, q is the activation map index, D is the total width of the filter w, and b is the bias term. However, in case there are more than one channels, (i.e., the sensor signals are stacked by the channel axis), the *i*th product elements (c_i) of the sensor signals are added, producing a new element ($c_{i,j}$):

$$c_{i,j}^{l,q} = b^{l,q} + \sum_{h=1}^{H} \sum_{d=1}^{D} w_{d,h}^{l,q} x_{i+d-1}^{l-1,q} \tag{2}$$

where *h* is the channel index. This way, the translation invariance property is lost, since the specific receptive fields of the signals may not be correlated. In addition to this, motion signals produced by low-cost, not well-calibrated sensors, suffer from sampling rate instability (regularity of the timespan between successive measurements) [41].

In order to understand this issue better, imagine a picture showing a cat, where the R, G, B channels are not stacked correctly (e.g., the nose in the red channel matches the eye in the blue channel, and the mouth in the green channel). If a 2D convolution was applied to this picture, it would detect in each channel different edges, and, as result, by adding them it would extract features only with respect to this particular R, G, B topology, resulting to overfitting. However, if the CNN model identified the low-level features (e.g., edges), and the mid-level features (e.g., nose) for each channel separately, it would have the capability to “see the whole picture” (i.e., perceive the high-level features) and generalize what it learned.

3.2 Architecture

The architecture of PerceptionNet, illustrated in Fig. 1, consists of the following layers:

Layer 1: 48 1D convolutional filters with a size of (1, 15), i.e., W^1 has the shape (1, 15, 1, 48). This is followed by a ReLU [11] activation function, a (1, 2) strided 1D max-pooling operation and a dropout [42] probability equal to 0.4.

Layer 2: 96 1D convolutional filters with a size of (1, 15), i.e., W^2 has the shape (1, 15, 48, 96). This is followed by a ReLU activation function, a (1,2) strided 1D max-pooling operation and a dropout probability equal to 0.4.

Layer 3: 96 2D convolutional filters with a size of (3,15) and a stride of (3,1), i.e., W^3 has the shape (3, 15, 96, 96). This is followed by a ReLU activation function, a global average-pooling operation [43] and a dropout probability equal to 0.4.

Layer 4: 10 output units, i.e., W^4 has the shape (96, 6), followed by a softmax activation function.

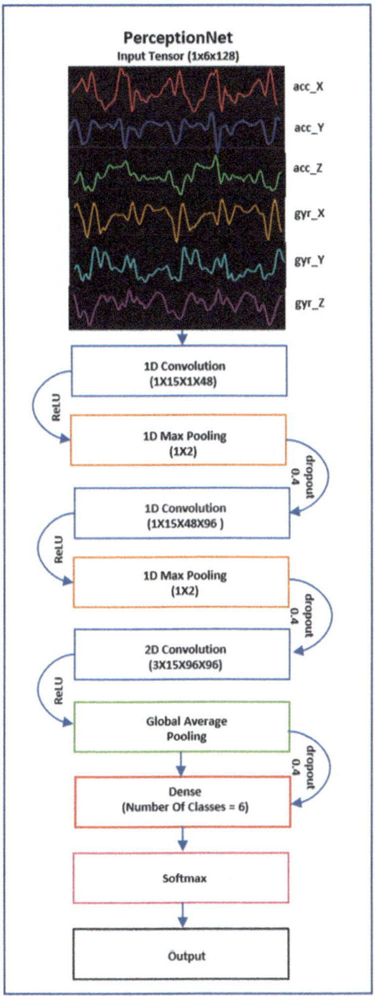

Fig. 1. PerceptionNet architecture.

3.3 Adadelta Optimizer

We selected the *adadelta* [44] as optimizer in our network. The reason for this was because *adadelta* adapts dynamically over iterations, does not need manual tuning of the learning rate and appears robustness to noisy gradient information, different model architecture choices, various data modalities and selection of hyper-parameters. According to [44], let $L'(\theta_t)$ be the first derivative of the loss function f with respect to the parameters θ at time step t. Here g_t is called the second order moment of $L(\theta_t)^2$. Given a decay term ρ and an offset ε we perform the following updates:

$$g_t = (1 - \rho)L'(\theta_t)^2 + \rho g_{t-1} \tag{3}$$

where $g_0 = 0$ and $s_0 = 0$. The term s_t denotes the 2nd moment of $\Delta\theta_t^2$ for updating the parameters:

$$\Delta\theta_t = -\frac{\sqrt{s_{t-1} + \varepsilon}}{\sqrt{g_t + \varepsilon}} L'(\theta_t) \tag{4}$$

$$s_t = (1 - \rho)\Delta\theta_t^2 + \rho s_{t-1} \tag{5}$$

$$\theta_{t+1} = \theta_t + \Delta\theta_t \tag{6}$$

4 Experimental Set Up

The experiments were executed on a computer workstation equipped with an NVIDIA GTX Titan X GPU, featuring 12 gigabytes RAM, 3072 CUDA cores, and a bandwidth of 336.5 GB/s. We used Python as programming language, and specifically the Numpy library for matrix multiplications, data preprocessing and segmentation, the scikit-learn library for implementing the t-SNE algorithm, and the Keras high-level neural networks library using as backend the Theano library. In order to accelerate the tensor multiplications, we used CUDA Toolkit in support with the cuDNN, which is the NVIDIA GPU-accelerated library for deep neural networks. The software is installed on a 16.04 Ubuntu Linux operating system.

4.1 Datasets

We evaluated PerceptionNet on two public available HAR datasets, UCL [45] and PAMAP2 [46]. The first one was used to tune the hyper-parameters, and we compared, afterwards, our CNN's performance against the state-of-the-art approaches: (a) CNN-EF [14], (b) Convolutional LSTM [15], (c) CNN on spectrograms [35], and (d) SVM method based on HCF [45]. Finally, in order to test the general applicability of our approach we used the same CNN architecture on the PAMAP2 dataset and compared it with the conventional CNN-EF [14] and the Convolutional LSTM [15] methods.

(1) *UCL*

The UCL HAR dataset consists of tri-axial accelerometer and of tri-axial gyroscope sensor data, collected by a waist-mounted smartphone (Samsung Galaxy S II smartphone). A group of 30 volunteers, with ages ranging from 19 to 48 years, executed six daily activities (standing, sitting, laying down, walking, walking downstairs and upstairs). The mobile sensors produced 3-axial linear acceleration and 3-axial angular velocity data with a sampling rate of 50 Hz and were segmented into time windows of 128 values (2.56 s), having a 50% overlap. Furthermore, the dataset is separated into train data.

The obtained dataset contains 10,299 samples, which are partitioned into two sets, where 70% of the volunteers (21 volunteers) was selected for generating the training data (7,352 samples) and 30% (9 volunteers) the test data (2,947 samples). Moreover, following the results in [47] where it was shown that subject independent validation techniques should be applied for the evaluation of activity monitoring systems to tune the hyper-parameters, during the validation phase we followed a *Leave-3-Subject-Out* approach. Concretely, the samples of 3 volunteers (27, 29 and 30) were used as validation set, which are equal to 15% of the training set.

Finally, we normalized each sensor's values (x_i) by subtracting the mean and dividing by the standard deviation:

$$z_i = \frac{x_i - \mu_i}{\sigma_i} \tag{7}$$

(2) *PAMAP2*

The PAMAP2 HAR dataset contains 12 lifestyle activities (such as walking, cycling, ironing, etc.) from 9 participants wearing 3 Colibri wireless inertial measurement units (IMU) and a heart rate monitor. The 3 IMUs had a sampling frequency of 100 Hz, were placed on the dominant arm, on the chest and on the dominant side's ankle, and produced tri-axial accelerometer, gyroscope and magnetometer data. In order to obtain the same sampling rate with the UCL dataset and the same sensor signals, we downsampled the PAMAP2 dataset to 50 Hz and selected only the accelerometer and gyroscope data. The resulting dataset had 18 dimensions, with the same time window (2.56 s) and overlap (50%) as the UCL dataset.

Since, the data were collected by only 9 participants and for the reason that only 4 subjects (1, 2, 5, and 8) have enough samples of all the activities, we selected a *Leave-1-Subject-Out* approach, for the test and the validation set. More specific, the samples of subject 1 were used for the test set and the samples of subject 5 for the validation set. The training set contained 13,980, the test set 2,453 and the validation set 2,688 samples. The PAMAP2 samples were, also, normalized using (7).

4.2 Performance Metrics

We used precision, recall, weighted (w) F1-score (otherwise F-measure), and accuracy as performance measures. It should be noted that accuracy, in contrast with the other 3 metrics, takes only into account the total number of samples and not class imbalance. On the other hand, precision, recall, and wF1-score consider total number of samples for each class separately. The above metrics are described as:

$$accuracy = \frac{1}{N}\sum_{i=1}^{N}\left(TP_i + TN_i\right) \tag{8}$$

$$precision = \frac{1}{N}\sum_{i-1}^{N}\left(\frac{TPi}{TPi + FPi}\right) \tag{9}$$

$$recall = \frac{1}{N}\sum_{i=1}^{N}\left(\frac{TPi}{TPi + FNi}\right) \quad (10)$$

$$F_1 = \sum_{i} 2 * w_i \frac{precision_i * recall_i}{precision_i + recall_i} \quad (11)$$

where TP, TN, FP, FN represent the true positive, true negative, false positive and false negative predictions, respectively. It should be noted that in a multiclassification problem, the recall, precision and wF1-score metrics iterate over all the classes by selecting the samples belonging to one of them as positive (class of interest), while they consider the rest samples as negative (rest of the classes).

Moreover, we selected the confusion matrix as a visualization of the classification performance of PerceptionNet. The confusion matrix is easy to be interpreted; it shows where the classification algorithm "confused" a class with another one (i.e., it predicted lying activity instead of standing activity). In mathematical terms, the confusion matrix can be described by M_{ij}, with i denoting the "actual" classes and j the "predicted" classes. By summing all entries in the row i of the matrix it shows the total number of the samples annotated as activity i, while by summing all entries in the column j of the matrix it shows the total number of the samples predicted as activity j.

5 Results

5.1 Validation Phase

Before, testing our approach, we used the validation set of UCL to tune its hyper-parameters. Table 1 contains all the hyper-parameters and their possible values. Since it is time consuming to train a deep CNN model, and because the UCL dataset was thoroughly examined in [14], we did not evaluate exhaustively all potential combinations of the hyper-parameters. Thus, we based the selection of the most promising model on the most significant factors (filter shape, dropout probability, layer of 2D convolution, and use of dense layer or global average pooling) that differentiate our method from that described in [14].

Since the convergence and the prediction accuracy of a Deep Neural Network depend a lot on the weight initialization [48], in order to obtain more representative results for each hyper-parameter, we ran the experiments 10 times. The mean value of each different hyper-parameters model was used as selection criterion. It should be noted that we selected a variation (i.e., the random numbers were sampled from a uniform instead of a normal distribution) of the weight initialization introduced by K. He et al. [49], whose upper and lower thresholds are given by:

$$w_i = \pm\sqrt{\frac{2}{N_{in}}} \quad (12)$$

Table 1. Experimental set-up for tuning hyper-parameters

Symbol	Parameter	Values
–	Batch size	64
α	Learning rate	1.0
ρ	Rho	0.95
ε	Epsilon	1e-08
–	Number of channels	1
–	Input height	6
–	Input width	128
–	Number of convolutional layers	3
–	1D convolution size	1X5-1X17
–	2D convolution size	3X15
–	2D strides size	3X1
–	1D max pooling size	1X2
–	1D max pooling stride	1X2
–	Dropout	0-0.7
–	Activation map channels	32-192
–	Dense layer size	0-1500
–	2D convolutional layer	1-3
–	Maximum epochs	2000
–	Early stopping criterion epochs	100

where N_{in} represents the total numbers of neurons that outputs the $(i-1)$*-th* layer, which are the input regarding the next layer. The adadelta optimizer had the following hyper-parameters: learning rate equal to 1, ρ equal to 0.95, and ε equal to 1e-08. Moreover, we set the batch size equal to 64 and the minimum number of epochs to 2,000, but the training procedure was automatically terminated if the best training accuracy had not improved after 100 epochs. The model that achieved the lowest error rate on the validation set was saved, and its filters were used to obtain the accuracy of the model on the test set.

Before, testing our approach, we used the validation set of UCL not only to tune its hyper-parameters (Table 1), but to show also that fusing sensor signal in the latest convolutional layer is more effective. Thus, we applied the 2D convolutions on three different convolutional layers, each of them having filter size 3×15 and stride equal to (3,1), for the vertical and the horizontal axis respectively. Figure 2 shows that applying the 2D convolution on the last convolutional layer increases the accuracy of the model. The increased performance of applying a late 2D convolution is, also, illustrated in Fig. 3, which shows that by applying the t-SNE algorithm [50] after the last convolutional operation most of the instances of the six activity classes are easily categorized.

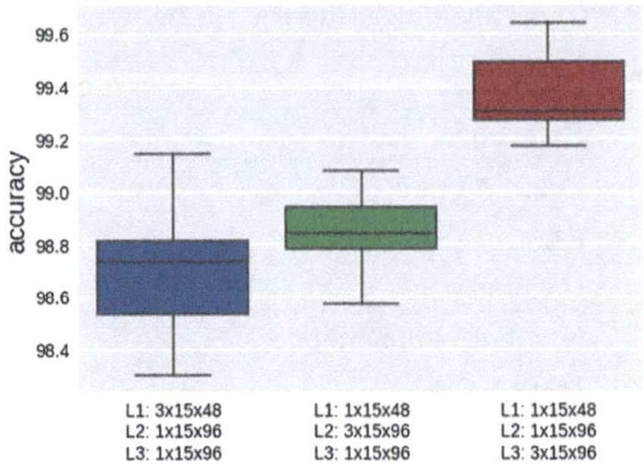

Fig. 2. Accuracy results of 2D convolutions on the 1st, 2nd and 3rd convolutional layer of the UCL validation set.

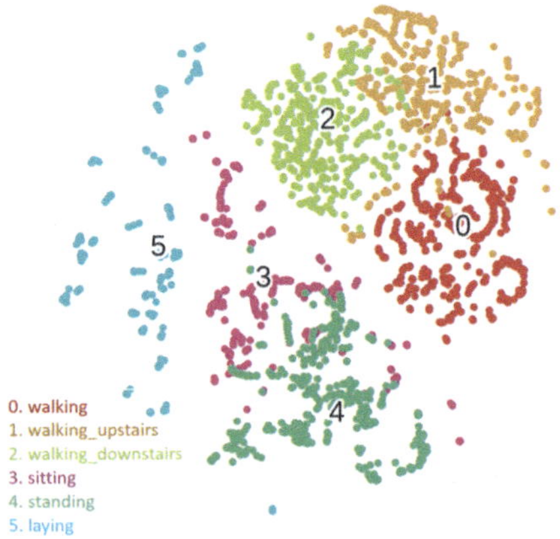

Fig. 3. t-SNE visualization of the test set's last hidden layer representations in PerceptionNet for six activity classes.

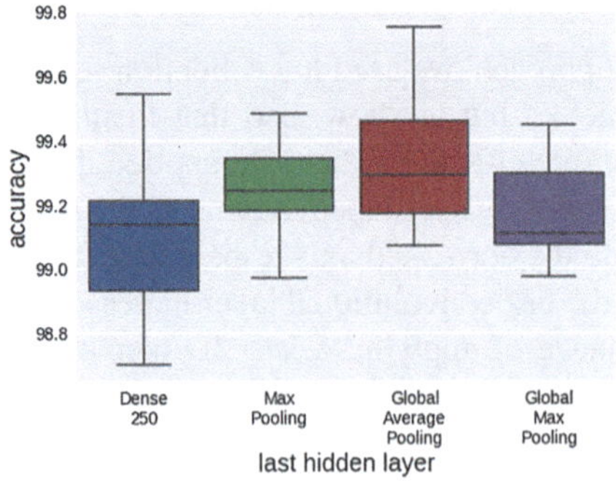

Fig. 4. Accuracy results based on different last hidden layers of the UCL validation set.

Moreover, we examined the concept of having a dense layer, as last hidden layer and that of using a global average pooling or a max average pooling operation instead of max pooling operation. As it is shown in Fig. 4, we obtained the highest mean accuracy with a global average pooling operation (99.38%).

5.2 UCL Test Phase

The results we obtained using PerceptionNet model on the test set are presented in Fig. 5, and have a range from 0.9620 to 0.9752. After obtaining the results, we developed a CNN ensemble, based on probability voting (e.g., the probability of the 10 runs of our model were added and divided afterwards by 10). Figure 6 presents the CNN ensemble. The best epoch on the validation data achieved 0.9725 accuracy on the test data, which is about 2.5% higher than the ConvNet described in [14]. Table 2 compares the accuracy obtained from the PerceptionNet against those obtained by state-of-the-art models, whose results are reported in the literature, and our implementation of the Convolutional LSTM [15].

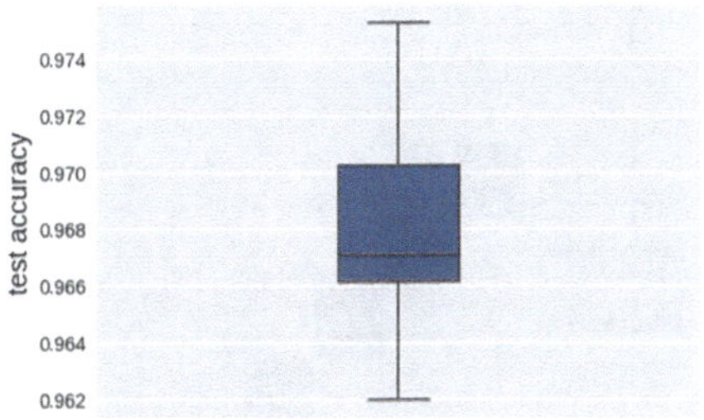

Fig. 5. Test accuracies of PerceptionNet on the UCL set.

```
function probability_ensemble(trained_cnn_models, num_classes, test_data):
        foreach i belonging in test_data
                foreach j belonging in num_classes
                        foreach k belonging in trained_cnn_models
                                total_preds(i, j, k) ← predict(i, j, k)
        mean_total_preds ← mean_trained_cnn_models(total_preds)
        ensemble_preds ← argmax_trained_cnn_models(mean_total_preds)
        return ensemble_preds
```

Fig. 6. Pseudocode for the average probability ensemble.

Table 2. Comparison of perceptionnet to other state-of-the-art methods

Method	Accuracy on test data
CNN-EF [14]	94.79%
CNN-EF + FFT features [14]	95.75%
SVM on HCF [45]	96.00%
CNN on spectrogram [35]	95.18%
Convolutional LSTM [15]	92.59%
PerceptionNet	**97.25%**

The accuracy per subject of our 2D ConvNet is presented in Table 3, while Fig. 7 presents the confusion matrix (precision: 0.9731, recall: 0.9725, and wF1-score: 0.9724), which reveals the difficulty of distinguishing standing from sitting and the reverse.

Table 3. Accuracy of perceptionnet on each subject of the test set

Subject	Accuracy on test data
2	98.01%
4	98.10%
9	90.95%
10	92.49%
12	96.24%
13	100.0%
18	99.45%
20	98.01%
24	100.0%

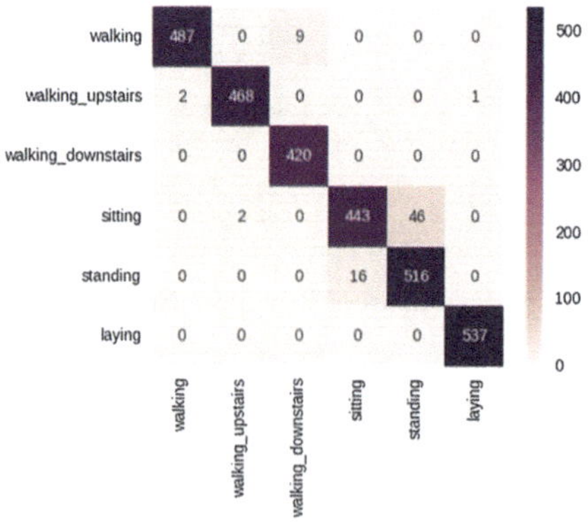

Fig. 7. Confusion matrix of PerceptionNet on the UCL test data.

5.3 PAMAP2 Test Phase

PAMAP2 dataset was selected as a second dataset, in order to examine the general applicability of our approach. Table 4 presents the precision, the recall, the wF1-score and the accuracy of PerceptionNet ensemble on the test data, compared with the ensembles of the conventional CNN (channel-based stacking) and the Convolutional LSTM methods. PerceptionNet achieved almost 2% higher accuracy form the Convolutional LSTM and 4% higher than the CNN-EF approach.

Table 4. Comparison of perceptionnet to other state-of-the-art methods on the PAMAP2 test data

Method	Precision	Recall	wF1-score	Accuracy
CNN-EF [14]	85.51%	84.53%	84.57%	84.53%
Convolutional LSTM [15]	87.75%	86.78%	86.83%	86.78%
PerceptionNet	**89.76%**	**88.57%**	**88.74%**	**88.56%**

Figure 8 shows the confusion matrix of our model on the PAMAP2 test data. Not surprisingly, the model again struggles to distinguish the sitting activity from the standing activity. The authors of [51, 52] argue that this misclassification is a common problem, and an extra IMU on the thigh would be a solution. Finally, it should be noted that the ironing class had very high recall (0.9913), but very low precision (0.6930) indicating a large number of False Positives.

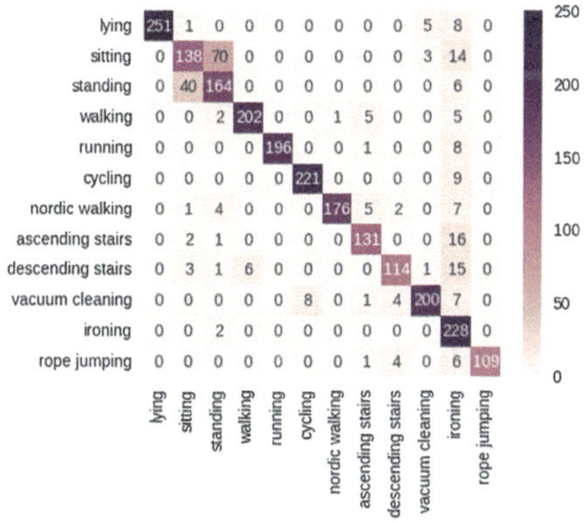

Fig. 8. Confusion matrix of PerceptionNet on the PAMAP2 test data.

6 Conclusion

In this paper, we propose a deep Convolutional Neural Network (CNN) for human activity recognition (HAR) that performs a 2D convolution on the last convolutional layer. CNNs have proven to be capable of automatically extracting the temporal local dependencies of time-series 1D signals, but state-of-the-art approaches fuse motion signals without extracting feature from them separately and combine the high-level extracted features using as last hidden layer a dense layer. We argue that motion signals should be treated discretely, and a late 2D convolution operation on them discovers more efficient activity patterns. The experiments performed during the validation phase of our method (PerceptionNet) justified our intuition, since we managed to reduce the overfitting. Moreover, applying a global average pooling layer to our model, instead of a dense layer, improved significantly the accuracy. Our approach was evaluated on two public available HAR datasets and outperformed the other deep learning state-of-the-art methods. However, despite the fact that PerceptionNet achieved high accuracy, it struggled to distinguish standing from sitting activity, and misclassified a lot of samples as ironing.

Future steps towards improving the model's performance include the use of Capsule Networks and the Dynamic Routing [53] mechanism to achieve a more efficient sensor fusion. Moreover, in order to reduce the training time without affecting negatively the PerceptionNet's performance, transfer learning techniques should be studied. Finally, the features that are extracted after the global average pooling layer can be used as embedding and one-shot learning techniques should be investigated in the future.

Acknowledgment. This work is funded by the European Commission under project TRILLION, grant number H2020-FCT-2014, REA grant agreement no [653256]. Moreover, we gratefully acknowledge the support of NVIDIA Corporation with the donation of the Titan-X GPU used for this research.

References

1. Bosems, S., van Sinderen, M.: Model-driven development for user-centric well-being support from dynamic well-being domain models to context-aware applications. In: 3rd International Conference on Model-Driven Engineering and Software Development (MODELSWARD), pp. 425–432 (2015)
2. Ongenae, F., Claeys, M., Dupont, T., Kerckhove, W., Verhoeve, P., Dhaene, T., Turck, F.D.: A probabilistic ontology-based platform for self-learning context-aware healthcare applications. Expert Syst. Appl. **40**(18), 7629–7646 (2013)
3. Seo, D.W., Kim, H., Kim, J.S., Lee, J.Y.: Hybrid reality-based user experience and evaluation of a context-aware smart home. Comput. Ind. **76**, 11–23 (2016)
4. Li, W., Joshi, A., Finin, T.: SVM-CASE: An SVM-based context aware security framework for vehicular ad-hoc networks. In: IEEE 82nd Vehicular Technology Conference (VTC2015-Fall), pp. 1–5 (2015)
5. Patrikakis, Ch.Z., Kogias, D.G., Loukas, G., Filippoupolitis, A., Oliff, W., Rahman, S.S., Sorace, S., La Mattina, E., Quercia, E.: On the successful deployment of community policing services the TRILLION project case. In: IEEE International Conference on Consumer Electronics (ICCE 2018) (2018)
6. Abolfazli, S., Sanaei, Z., Gani, A., Xia, F., Yang, L.T.: Rich mobile applications: genesis, taxonomy, and open issues. J. Netw. Comput. Appl. **40**, 345–362 (2014)
7. Schilit, B.N., Theimer, M.M.: Disseminating active map information to mobile hosts. IEEE Netw. **8**(5), 22–32 (1994)
8. Dey, A.K., Abowd, G.D., Salber, D.: A conceptual framework and a toolkit for supporting the rapid prototyping of context-aware applications. Hum.-Comput. Interact. **16**, 97–166 (2001)
9. Figo, D., Diniz, P.C., Ferreira, D.R., Cardoso, J.M.P.: Preprocessing techniques for context recognition from accelerometer data. Pers. Ubiquitous Comput. **14**, 645–662 (2010)
10. LeCun, Y., Bengio, Y., Hinton, G.: Deep learning. Nature **521**, 436–444 (2015)
11. Krizhevsky, A., Sutskever, I., Hinton, G.E.: Imagenet classification with deep convolutional neural networks. In: Advances in Neural Information Processing Systems, pp. 1097–1105 (2012)
12. Graves, A., Mohamed, A., Hinton, G.: Speech recognition with deep recurrent neural networks. In: Proceedings of ICASSP 2013, Vancouver, Canada, May 2013

13. Socher, R., Perelygin, A., Wu, J., Chuang, J., Manning, C.D., Ng, A.Y., Potts, C.: Recursive deep models for semantic compositionality over a sentiment treebank. In: Proceedings of the 2013 Conference on Empirical Methods in Natural Language Processing, pp. 1631–1642 (2013)
14. Ronao, C.A., Cho, S.-B.: Human activity recognition with smartphone sensors using deep learning neural networks. Expert Syst. Appl. **59**, 235–244 (2016)
15. Ordóñez, F.J., Roggen, D.: Deep convolutional and LSTM recurrent neural networks for multimodal wearable activity recognition. Sensors **16**(1), 1–25 (2016)
16. Reiss, A., Stricker, D.: Creating and benchmarking a new dataset for physical activity monitoring. In: The 5th Workshop on Affect and Behaviour Related Assistance (ABRA) (2012)
17. Kasnesis, P., Patrikakis, ChZ, Venieris, I.S.: Changing the game of mobile data analysis with deep learning. IEEE ITPro Mag. **19**(3), 17–23 (2017)
18. Vincent, P., Larochelle, H., Bengio, Y., Manzagol, P.A.: Extracting and composing robust features with denoising autoencoders. In: Proceedings of the Twenty-Fifth International Conference on Machine Learning (ICML 2008), pp. 1096–1103. ACM (2008)
19. Plötz, T., Hammerla, N.Y., Olivier, P.: Feature learning for activity recognition in ubiquitous computing. In: Proceedings of the Twenty-Second IJCAI, vol. 2, pp. 1729–1734. AAAI Press (2011)
20. Hinton, G., Osindero, S., Teh, Y.-W.: A fast learning algorithm for deep belief nets. Neural Comput. **18**(7), 1527–1554 (2006)
21. Hinton, G.E., Salakhutdinov, R.R.: Reducing the dimensionality of data with neural networks. Science **313**, 504–507 (2006)
22. Vollmer, C., Gross, H.-M., Eggert, J.P.: Learning features for activity recognition with shift-invariant sparse coding. In: Artificial Neural Networks and Machine Learning–ICANN, pp. 367–374 (2013)
23. Lee, H., Ekanadham, C., Ng, A.Y.: Sparse deep belief net model for visual area V2. In: Advances in Neural Information Processing Systems (NIPS), vol. 20 (2008)
24. Zhang, Y.-D., Zhang, Y., Hou, X.-X., Chen, H., Wang, S.H.: Seven-layer deep neural network based on sparse autoencoder for voxelwise detection of cerebral microbleed. In: Multimedia Tools and Applications, pp. 1–18 (2017)
25. Jia, W., Yang, M.: Wang: Three-category classification of magnetic resonance hearing loss images based on deep autoencoder. J. Med. Syst. **41**(10), 165 (2017)
26. Abdel-Hamid, O., Mohamed, A.-R., Jiang, H., Penn, G.: Applying convolutional neural networks concepts to hybrid nn-hmm model for speech recognition. In: Acoustics, Speech and Signal Processing (ICASSP), pp. 4277–4280 (2012)
27. Zeng, M., Nguyen, L.T., Yu, B., Mengshoel, O.J., Zhu, J., Wu, P., Zhang, J.: Convolutional neural networks for human activity recognition using mobile sensors. In: MobiCASE, pp. 197–205. IEEE (2014)
28. LeCun, Y., Bottou, L., Bengio, Y., Haffner, P.: Gradient-based learning applied to document recognition. In: Procedings of the IEEE, pp. 2278–2324, November 1998
29. Deng, J., Dong, W., Socher, R., Li, L.-J., Li, K., Fei-Fei, L.: ImageNet: a large-scale hierarchical image database. In: Computer Vision and Pattern Recognition (CVPR09) (2009)
30. Zheng, Y., Liu, Q., Chen, E., Ge, Y., Zhao, J.L.: Time series classification using multi-channels deep convolutional neural networks. In: Proceedings of International Conference on Web-Age Information Management, pp. 298–310 (2014)
31. Goodfellow, I., Bengio, Y., Courville, A.: Deep Learning (Adaptive Computation and Machine Learning series). MIT Press, Chap. 10, pp. 330–372 (2016)

32. Ronao, C.A., Cho, S.-B.: Deep convolutional neural networks for human activity recognition with smartphone sensors. In: Neural Information Processing, pp. 46–53. Springer (2015)
33. Ronao, C.A., Cho, S.-B.: Evaluation of deep convolutional neural network architectures for human activity recognition with smartphone sensors. In: Proceedings of the KIISE Korea Computer Congress, pp. 858–860 (2015)
34. Yang, J.B., Nguyen, M.N., San, P., Li, X., Krishnaswamy, S.: Deep convolutional neural networks on multichannel time series for human activity recognition. In: IJCAI 2015 Proceedings of the 24th International Conference on Artificial Intelligence, pp. 3995–4001 (2015)
35. Alsheikh, M.A., Selim, A., Niyato, D., Doyle, L., Lin, S., Tan, H.-P.: Deep Activity Recognition Models with Triaxial Accelerometers. In: The Workshops of the Thirtieth AAAI Conference on Artificial Intelligence (2015)
36. Jiang, W., Yin, Z.: Human activity recognition using wearable sensors by deep convolutional neural networks. In: Proceedings of the 23rd ACM international conference on Multimedia, pp. 1307–1310 (2015)
37. Rumelhart, D.E., Hinton, G.E., Williams, R.J.: Learning internal representations by error propagation. In: Parallel Distributed Processing, vol. 1, Chap. 8, pp. 318–362. MIT Press, Cambridge (1986)
38. Hochreiter, S., Schmidhuber, J.: Long short-term memory. Neural Comput. **9**(8), 1735–1780 (1997)
39. Ordóñez, F.J., Roggen, D.: Deep convolutional feature transfer across mobile activity recognition domains, sensor modalities and locations. In: Proceedings of IEEE 20th International Symposium on Wearable Computers (ISWC), pp. 92–99 (2016)
40. Luo, W., Li, Y., Urtasun, R., Zemel, R.: Understanding the effective receptive field in deep convolutional neural networks. In: 29th Conference on Neural Information Processing Systems (NIPS) (2016)
41. Stisen, A., Blunck, H., Bhattacharya, S., Prentow, T.S., Kjærgaard, M.B., Dey, A., Sonne, T., Jensen, M.M.: Smart devices are different: assessing and mitigating mobile sensing heterogeneities for activity recognition. In: Proceedings of the 13th ACM Conference on Embedded Networked Sensor Systems, pp. 127–140, November 2015
42. Srivastava, N., Hinton, G., Krizhevsky, A., Sutskever, I., Salakhutdinov, R.: Dropout: a simple way to prevent neural networks from overfitting. J. Mach. Learn. Res. **15**, 1929–1958 (2014)
43. Lin, M., Chen, Q., Yan, S.: Network in network. In: International Conference on Learning Representations (2014)
44. Zeiler, M.D.: ADADELTA: An Adaptive Learning Rate Method. Technical report, arXiv 1212.5701
45. Anguita, D., Ghio, A., Oneto, L., Parra, X., Reyes-Ortiz, J.L.: A public domain dataset for human activity recognition using smartphones. In: 21th European Symposium on Artificial Neural Networks, Computational Intelligence and Machine Learning, ESANN 2013, April 2013
46. Reiss, A., Stricker, D.: Introducing a new benchmarked dataset for activity monitoring. In: The 16th IEEE International Symposium on Wearable Computers (ISWC) (2012)
47. Reiss, A., Weber, M., Stricker, D.: Exploring and extending the boundaries of physical activity recognition. In: IEEE SMC Workshop on Robust Machine Learning Techniques for Human Activity Recognition, pp. 46–50 (2011)
48. Glorot, X., Bengio, Y.: Understanding the difficulty of training deep feedforward neural networks. In: Proceedings of the 13th International Conference on Artificial Intelligence and Statistics (AISTATS), pp. 249–256 (2010)

49. He, K., Zhang, X., Ren, S., Sun, J.: Delving deep into rectifiers: surpassing human-level performance on ImageNet classification. In: IEEE International Conference on Computer Vision (ICCV) (2015)
50. van der Maaten, L.J.P., Hinton, G.E.: Visualizing high-dimensional data using t-SNE. J. Mach. Learn. Res. **9**, 2579–2605 (2008)
51. Ermes, M., Pärkkä, J., Mäntyjärvi, J., Korhonen, I.: Detection of daily activities and sports with wearable sensors in controlled and uncontrolled conditions. IEEE Trans. Inf. Technol. Biomed. **12**(1), 20–26 (2008)
52. Reiss, A., Stricker, D.: Creating and benchmarking a new dataset for physical activity monitoring. In: Proceedings of the 5th International Conference on PErvasive Technologies Related to Assistive Environments, Article no. 40, June 2012
53. Sabour, S., Frosst, N., Hinton, G.E.: Dynamic routing between capsules. In: 31st Conference on Neural Information Processing Systems (NIPS) (2017)

Reinforcement Learning for Fair Dynamic Pricing

Roberto Maestre(✉), Juan Duque, Alberto Rubio, and Juan Arevalo

BBVA Data and Analytics, Av. de Burgos, 16D, 28036 Madrid, Spain
{roberto.maestre,juanramon.duque,
alberto.rubio.munoz,juanmaria.arevalo}@bbvadata.com

Abstract. Unfair pricing policies have been shown to be one of the most negative perceptions customers can have concerning pricing, and may result in long-term losses for a company. Despite the fact that dynamic pricing models help companies maximize revenue, fairness and equality should be taken into account in order to avoid unfair price differences between groups of customers. This paper shows how to solve dynamic pricing by using Reinforcement Learning (RL) techniques so that prices are maximized while keeping a balance between revenue and fairness. We demonstrate that RL provides two main features to support fairness in dynamic pricing: on the one hand, RL is able to learn from recent experience, adapting the pricing policy to complex market environments; on the other hand, it provides a trade-off between short and long-term objectives, hence integrating fairness into the model's core. Considering these two features, we propose the application of RL for revenue optimization, with the additional integration of fairness as part of the learning procedure by using Jain's index as a metric. Results in a simulated environment show a significant improvement in fairness, while at the same time maintaining optimisation of revenue.

Keywords: Reinforcement learning · Dynamic pricing · Fairness
Jain's index

1 Introduction

Determining the right price of a product or service for a particular customer is a necessary, yet complex endeavour; it requires knowledge of the customer's willingness to pay, estimation of future demands, ability to adjust strategies to competition pricing [1], etc. Dynamic pricing [2,3] represents a promising solution for this challenge due to its intrinsic adjustment to customer expectations. Indeed, with the advent and establishment of digital channels, unique opportunities for the application of dynamic pricing are arising, thus enhancing research in the field [4,5]. Equality is a critical aspect of dynamic pricing as it influences customers' perceptions of fairness [6–11]. As long as perceptions of fairness are used by people as a heuristic for trust, inequality may lead to a destruction of

K. Arai et al. (Eds.): IntelliSys 2018, AISC 868, pp. 120–135, 2019.
https://doi.org/10.1007/978-3-030-01054-6_8

that trust. For instance, large price differences between groups of customers, or a discrepancy between the distribution of cost and profit are known to affect the relationship between customers and sellers [12]. Should this relationship be damaged, it could eventually generate substantial financial losses for a company in the medium to long-term [13]. Although the definition of fairness depends on the domain context and, consequently, has a diffuse definition [14], we propose clear design principles for fairness in dynamic pricing with a focus on elements such as group or individual schemes (in which policy decisions are applied), and equity or definitions of equality. This clear design allows us to specify a well-defined metric and to include it in the overall model in order to ensure fairness.

Different approaches have been proposed to address the problem of maximizing revenue [2]. Among these is a promising technique consisting of optimizing pricing policies with Reinforcement Learning (RL) applied to different market scenarios (uni or multi-agent) [15–18]. Despite the extensive application of RL models to dynamic pricing, concepts such as equality and fairness are rarely incorporated into the learned policies. Moreover, the *black–box* nature of Machine Learning models [19–22] means that it is of utmost importance to ensure a good trade-off between revenue (goal for companies) and equality (principle of fairness and a goal for customers) [23]. There are also ethical issues in artificial intelligence such as algorithm bias, in which if the input data (or market environment itself) reflects unfair biases of the broader society [24–27], the output model can potentially capture such biases, hence perpetuating unfair policies. These biases can have a significant impact on dynamic pricing outcomes, as the efficiency of policies (revenue) has to be balanced against the need to achieve equality in the treatment of customers [28].

Many metrics have been proposed in order to measure fairness from the *resource allocation distribution* point of view [29]. Such a perspective has been widely studied [30–32] in order to avoid unfair resource allocation for nodes in wireless networks. In the present paper we apply concepts from resource allocation distribution to dynamic pricing so as to interpret fairness as *similarly distributed prices among groups of customers* (equality between groups of customers). Identification of different groups of customers is required for an effective price discrimination. In each group, certain assumptions about price sensitivity have to be made; groups could be defined, for instance, by customer's income level, gender, location, communication channel, etc. Our learning procedure provides homogeneous prices among such groups, but at the same time, takes into account price sensitives within each group to maximize revenue. As we will show, maintaining a balanced price distribution among groups of customers will boost perceptions of fairness and thus increase levels of trust. This is because the price for different offers is chosen by considering equality.

The present study is organized as follows. Section 2 provides some useful definitions from the fairness design principles, the resource allocation distribution literature, and reviews the main concepts of RL. The experimental methodology is then presented in Sect. 3, where synthetic groups of customers are introduced, and RL is applied to the dynamic pricing problem. Next, we show our experimental results in Sect. 4. Finally, we draw some conclusions in Sect. 5.

2 Background

In this section we start by first reviewing the main concepts of Reinforcement Learning (RL), more specifically Q-Learning as a variation of RL. We also define the approximation of Q-values by means of Neural Networks (NN) minimizing a loss function. Then, we propose our design principles for fairness. Finally, we introduce the main Jain's index properties, which will allow us to measure fairness in dynamic pricing policies.

2.1 Reinforcement Learning Concepts

Reinforcement Learning is an area of machine learning that studies how an agent takes actions in an environment to achieve a given goal [33]. RL differs from supervised learning as the agent learns by trial and error while interacting with the environment, as opposed to learning with labeled data. As such, RL is best suited for learning as part of sequential decision problems.

Q-learning is a variant of RL introduced by Watkins [34], whereby the heuristics of the model are directly related to the rewards provided by the environment in each iteration. Q-Learning can be applied to dynamic pricing as each action will modify the state of the environment (related to fairness), thus providing a reward (the bid itself). Equation (1) represents the generic way in which Q-learning is expressed:

$$Q(s,a) \leftarrow (1-\alpha)Q(s,a) + \alpha\Big(r' + \gamma \arg\max_{a^*} Q(s',a^*)\Big|s,a\Big). \tag{1}$$

Here, α is the learning rate, γ the discount factor and r' is the reward obtained after performing action a in state s. Thus, the goal of RL is to find a policy $\pi_* : s \rightarrow a$ that maximizes the expected discounted utility with some exploration methods. We use ϵ-greedy strategy that follows a random action with ϵ probability and action with $1-\epsilon$ probability.

Q-Learning in its simplest form uses a table (Q-table) to represent state-action values, i.e. Q-Values. The problem, however, becomes intractable as the number of states and actions increases [15,35]. In general terms, the Q-value function in (1) can be estimated by function approximation, i.e. $Q(s,a) \approx Q(s,a;\theta)$ [36]. Since NN act as universal function approximators [37], they provide an excellent framework for Q-value estimation. Indeed, recent advances in NN applied to Q-learning have shown them to greatly outperform traditional RL methods [35,38–40]. The process of learning Q-value estimations is as follows: at each iteration step i, approximated Q-values $Q(s,a;\theta_i)$ are trained by Stochastic Gradient Descent, by minimizing the loss $L(\theta_i)$:

$$L(\theta_i) = \left[\Big(y - Q(s,a;\theta_i)\Big)^2\right], \tag{2}$$

where $y = \left[r' + \gamma \arg\max_{a^*} Q(s', a^*; \theta_{i-1}) \middle| s, a \right]$.

Reinforcement learning algorithms constitute a suitable method for learning pricing policies, whenever the expected revenue for taking a pricing action is unknown in the absence of complete information of the environment [41].

2.2 Fairness Design Principles

We describe a policy in dynamic pricing as fair when the policy provides *similarly distributed prices among groups of customers.* From this definition, we can establish the following principles: (A) *equality*, which drives homogeneous prices, and (B) *groups* as atomic *entities* for applying prices, rather than individuals. Taking into account these two principles, we further propose to use Jain's Index [42] as a fairness metric because: (I) it works for multiple groups, (II) it provides an indicator of how homogeneous the policy is as a clear indicator of fairness in percentage terms (from 0 to 100% fair). Other metrics for measuring equality and fairness suited to be applied to dynamic pricing, such as entropy, min-max ratio, etc. are out of the scope of this paper. A full and comprehensive study on several fairness metrics and their properties can be found in the reference list [29].

2.3 Metrics for Fairness: Jain's Index

Let $\mathcal{C}$ be a portfolio of customers. Each customer c_i belongs to one group $g_\alpha \in \mathcal{G}$, where $\mathcal{G}$ is a collection of groups covering $\mathcal{C}$. We use $\bar{g}_\alpha$ to represent the average price allocated in group g_α.

The Jain's index is defined as

$$J = \frac{\left(\sum_{g \in \mathcal{G}} \bar{g}\right)^2}{|\mathcal{G}| \sum_{y \in \mathcal{G}} \bar{g}^2}.$$

It provides a measurement of how fair the average price allocation is in $\mathcal{G}$ (here, $|\mathcal{G}|$ is the number of groups in $\mathcal{G}$). Jain's index poses two convenient characteristics when used to define the heuristics in a RL model: (I) values are continuous between $[0, 1]$ and (II) the index applies to any number of groups. First feature fits perfectly with our definitions of states in Sect. 3.1 while the second feature is valuable when several groups of customers are defined.

As mentioned earlier, Jain's index offers an excellent way of knowing the fairness of our price allocation among groups in $\mathcal{G}$. For instance, if there are only two groups and either $\bar{g}_1 \ll \bar{g}_2$ or $\bar{g}_2 \ll \bar{g}_1$, Jain's index is 0.5, reflecting an unfair situation. Therefore, the model should learn to increase prices when possible in group g_1 and/or decrease the prices in g_2. This two-group case is represented in Fig. 1. As observed, values on the diagonal are 1.0, because both groups have the same average price, i.e.: $\bar{g}_1 = \bar{g}_2$. Thus, the policy is fair with an homogeneous price between groups. However, when $\bar{g}_1 = 50, \bar{g}_2 = 0$ or $\bar{g}_1 = 0, \bar{g}_2 = 50$ the

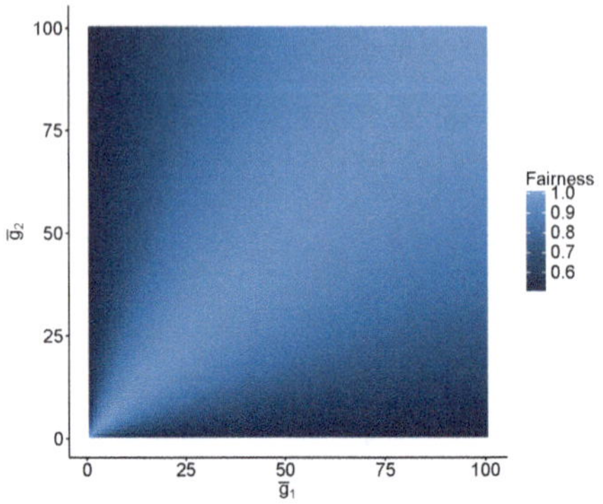

Fig. 1. Jain's index for two groups of customers. Please, notice that the index equals here the percentage of people for which a pricing policy is fair.

policy is fair only with the 50% of the groups (heterogeneous case). Thus, by means of this metric, we can constrain the differences between groups and provide fairer prices.

3 Methodology

We define four synthetic groups $\mathcal{G} = \{g_1, g_2, g_3, g_4\}$ to experiment with fairness and revenue. We simulate customers $c \in \mathcal{C}$ belonging to any group in $\mathcal{G}$. Each customer responds to each price bid with the following logistic function $\phi \in [0, 1]$:

$$\phi(a, g \mid a \in \mathcal{A}, g \in \mathcal{G}) = \left[1 + e^{-(b+w\cdot a)}\right]^{-1} \tag{3}$$

where a is the action chosen by the agent (the bid), and b_g and w_g are parameters defining the sensitivity of each group, see Table 1.

Table 1. Weights for the logistic function ϕ

Group	b	w
1	18.229	−2.369
2	4.4757	−1.1526
3	−1.09195	0.34000
4	0	0

This simple environment provides a scenario in which four different behaviours are simulated. Customers belonging to g_1 will accept much higher prices than customers in g_2. On the other hand, customers from g_3 exhibit an inverse behaviour: the larger the price, the higher the acceptance probability.

This group can be seen as customers whose perception of the product quality increases with price. Finally, g_4 customers are not sensitive to any changes in price. Figure 2 represents the probability of acceptance of a given price for each group (agent action or bid value).

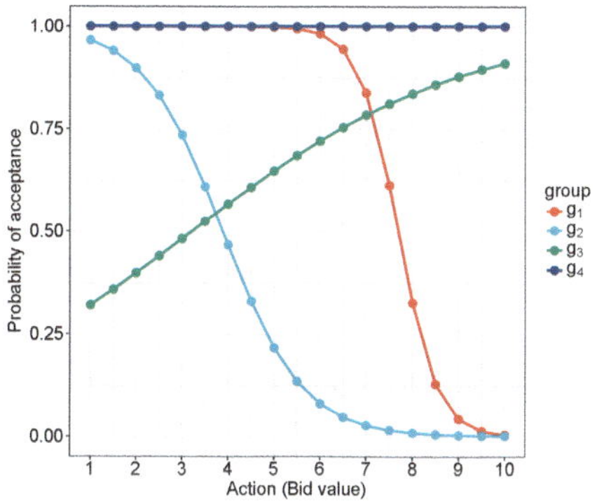

Fig. 2. Prices acceptance probabilities for the 4 groups defined in (3) and Table 1.

In order to achieve both the maximum revenue and fairness, the model should learn to increase the price within each group (maximizing the revenue), while reducing the differences between groups when possible, so that fairness is also maximized. Note that maximizing revenue entails finding the maximum price for each group in which the probability of acceptance is the highest [43].

The statistics in the experiments are computed by running 350 epochs. In each epoch 1000 bids are simulated with a 100 customers randomly selected from $\mathcal{C}$. In order to achieve a good balance between exploration and exploitation, for every epoch t we decrease ϵ according to the next ratio $\epsilon = 1/e^{\frac{t}{20}}$ (starting by $\epsilon = 1$).

The following subsections define the main elements of RL proposed in the present paper for a fair dynamic pricing. To this end, we define a state that allows us to identify both the customer group, $g \in \mathcal{G}$, and the average allocated price, $\bar{g}$; this way, both revenue and fairness can be maximized. As will be shown, revenue maximization is included in the reward through the value of each bid (see (5)).

3.1 States

We define the state as $\mathcal{S} = \{s_c, s_f\}$, where s_c represents the customer group, and s_f represents the global fairness. We propose to code s_c and s_f as independent one-hot vectors. For the latter, s_f, we define a partition $\mathcal{I}$ of mesh p in the interval $[0, 1]$, $\mathcal{I} \equiv 0 < p < 2p < ... < 1 - p < 1$, so that there is a k such that $\mathcal{I}_k \leq s_f \leq \mathcal{I}_{k+1} \forall s_f$. Given this partition, we define the one-hot encoding of s_f as a vector with a 1 at position k, and zeros otherwise. Finally, $\mathcal{S}$ is a vector resulting from the concatenation of s_c and s_f one-hot encoding, and has the final dimension $n = |\mathcal{G}| + 1/p + 1$. As an example, if we define a partition with mesh $p = 1/3$, one state with customer $c_\alpha \in g_2$ and a global Jain's index with value 0.89, the vector $\mathcal{S}$ will be represented as follows:

$$\mathcal{S}_\alpha = \begin{array}{ccccccc} \multicolumn{4}{c}{\overbrace{\hspace{4cm}}^{\mathcal{G}}} & \multicolumn{3}{c}{\overbrace{\hspace{4cm}}^{J}} \\ g_1 & g_2 & g_3 & g_4 & [0,\frac{1}{3}) & [\frac{1}{3},\frac{2}{3}) & [\frac{2}{3},1] \\ \downarrow & \downarrow & \downarrow & \downarrow & \downarrow & \downarrow & \downarrow \\ [\,0 & 1 & 0 & 0 & 0 & 0 & 1\,] \end{array}$$

3.2 Actions

We define the action space, $\mathcal{A}$, as a partition of mesh q in the interval $[\min, \max]$, i.e.

$$\mathcal{A} \equiv \min < \min + q < \min + 2q < ... < \max - q < \max\,.$$

The action space has dimension $m = 1/q + 1$. Note that as long as $\mathcal{A}$ is constrained between $\{\min, \max\}$ prices, the model will never bid excessively unfair prices –i.e., a non-sense price.

Extensive experimentation show that best values of p, q for discretizing the state and action vectors lie at around 0.01. With this selection we provide a good balance between expansion and shrinkage over the state/action space. On the one hand, a fine grid mesh (large p and q) make certain over–the–space defined functions (such as customer sensitivity in (3)) more expressive. On the other hand, small discretizations negatively affect learning convergences. Thus, the selection of parameters p and q is critical.

3.3 Q-Value Approximation

In accordance with our definitions of state and action, we propose a simple lineal approximation of the policy:

$$\begin{aligned} T_0(w_0, b_0) :& \mathbb{R}^n \rightarrow \mathbb{R}^m \\ & s \rightarrow a = w_0 \cdot X + b_0, \end{aligned} \tag{4}$$

where $w_0 \in \mathbb{R}^{m \times n}$ and $b_0 \in \mathbb{R}^m$ are learnable parameters. Thus, we are approximating Q-values with a linear, one-hidden layer network. Depending on the problem complexity, the NN can be expanded with more layers and different activation functions. With this compact method only two vectors (weights and bias for every layer) are needed to represent all Q-values.

3.4 Reward

Reward definition in dynamic pricing modeling with RL is generally challenging, as it represents the heuristics of the model itself. In the present paper, a multi-objective heuristic is required to account for both revenue and fairness optimization. Thus, hyper-parameters are needed to balance and be established as specific goals for each objective. For instance, one goal might be learning a policy with a fairness value equal to 0.85 while maximizing the bidding price, or vice versa. Bearing this in mind, we define the reward as:

$$r = \beta_p \exp\left(-\frac{(p - p_t)^2}{\sigma_p}\right) + \beta_f \exp\left(-\frac{(f - f_t)^2}{\sigma_f}\right). \tag{5}$$

The first term in (5) stands for price optimization, while the second intends to balance it with fairness. Here β_p, σ_p, p_t and β_f, σ_f, f_t are hyper-parameters that balance and establish a specific goal for p and f, respectively. Variable $p \in \mathbb{R}[0, 1]$ relates to the price given in the ith bidding, with p_t the target price. Similarly, $f \in \mathbb{R}[0, 1]$ is the fairness index, and f_t the desired (target) fairness. In order to model the reward, we have chosen a normal distribution of prices and fairness as it exhibits two convenient properties [44]: it points in the direction of the target price p_t and target fairness f_t; it also provides a uniform decay for states distant from the main goal. The following Fig. 3 shows several examples of hyper-parameter choices. NB: The reward is scaled between $[0, 1]$.

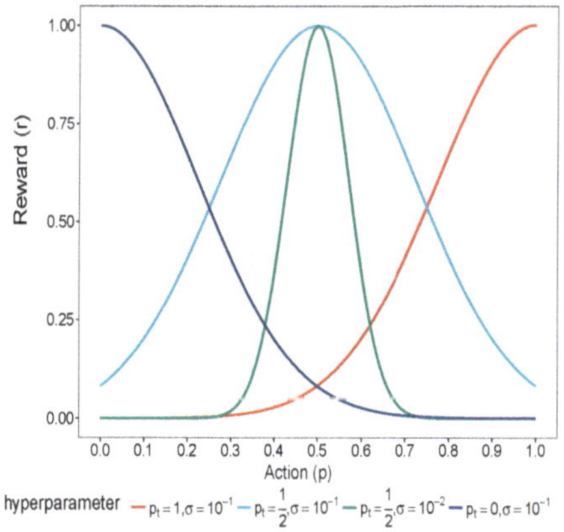

Fig. 3. Gaussian reward function.

In a real environment, customers will accept or reject each bid, i.e., the bid will be the value itself or 0. Thus, we define p as follows:

$$p = \begin{cases} \nu(a) & \text{if } \mathcal{B}\big(\phi(a, g)\big) = 0 \\ a & \text{if } \mathcal{B}\big(\phi(a, g)\big) = 1 \end{cases} \tag{6}$$

Here, $\mathcal{B}$ is the binomial distribution, and the function $\phi(a, g \mid a \in \mathcal{A}, g \in \mathcal{G})$ is given by (3). The $\nu(a)$ function represents the penalization when a bid is rejected. It could be modeled with a zero value, a constant negative value to penalize the rejection, or even a value proportional to the bid itself. However, in dynamic pricing it is important to keep the rejection rate low because the expected revenue of a policy is calculated as the ratio between the average bid value and the rejection ratio. Hence, it is convenient to take ν =constant. From an industrial perspective, we use a large penalization ($\nu = -0.5$) in order to avoid customer rejection.

Next, we define the variable f, that accounts for fairness optimization in (5). From a resource allocation point of view, given a fixed number of groups of customers ($|\mathcal{G}|$ fixed), a large and homogeneous value of the resources ($\bar{g}$ large) will increment Jain's index, and thus fairness. However, in the dynamic pricing context, increasing the average bid price will certainly not be perceived as fair. Therefore, we propose to distribute low and homogeneous prices among groups ($\bar{g}$ low), in contrast to increasing prices where possible to maximize revenue. These ideas can be expressed mathematically as follows:

$$f = \frac{\left(\sum_{g \in G} \max\{\mathcal{A}\} - \bar{g}\right)^2}{|\mathcal{G}| \sum_{g \in G} \left(\max\{\mathcal{A}\} - \bar{g}\right)^2} \in [0, 1], \tag{7}$$

which can be interpreted as a rotated Jain's index after comparing Figs. 2 and 4. Nevertheless, our transformation keeps the main properties of Jain's index invariant (i.e. population size, scale and metric independence, boundedness or continuity). Despite the simplicity of this approach, this rotation effectively allows for the balance between revenue and fairness.

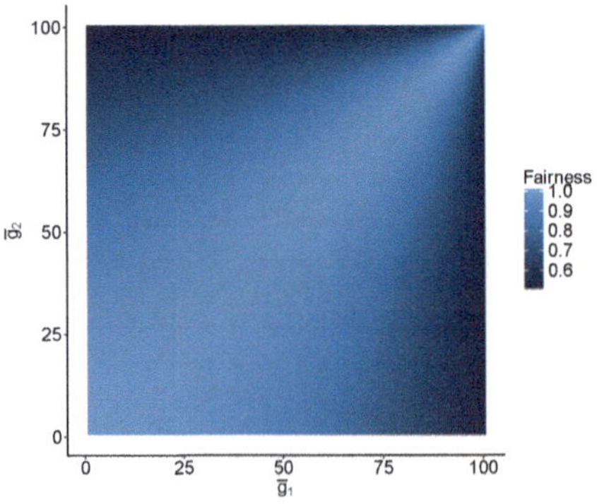

Fig. 4. Rotated Jain's index

Finally, we should point out that the selection of reward's hyper-parameters in (5), along with the discretization of the state/action space, must be carried out with care as it has a dramatic impact on the learning process.

3.5 Reinforcement Learning with NN

The following pseudo-code defines a standard way for training a NN to approximate Q-values. As illustrated in Algorithm 1, the agent takes each bid a to evaluate the fairness using averages prices in $\mathcal{G}$. The fairness value is calculated with the latest information given by $\mathcal{G}$. With both the bid (p) and fairness information (f), the reward r is obtained. Finally, a gradient descent step is executed to reduce the loss produced between the current and the target Q-value, see (2). In the experiments, we use the ADAM gradient descent algorithm [45] with a learning rate equal to 0.01.

Algorithm 1. Q-learning with NN approximation

Initialize Neural Network with Random weights
for $e = 1, E$ **do** ▷ E total epochs
  $\forall \bar{g} \in \mathcal{G}, \bar{g} = 0$ ▷ Initialize revenue distribution
  for $i = 1, I$ **do** ▷ I total iterations
    $a_i = \begin{cases} \arg\max_{a^*} Q(s_i, a^*; \theta_i) & \text{w.p. } 1 - \epsilon \\ \text{Random action} & \text{w.p. } \epsilon \end{cases}$
    bid a_i on the environment ▷ One one customer
    Update $\mathcal{G}$
    Calculate fairness f_i based on Eq. (7) ▷ On $\mathcal{G}$
    Get reward r_i based on Eq. (5)
    Get state $s^{'} = s_{i+1}$
    $y_i = \begin{cases} \nu & \text{if } \mathrm{r}_i = 0 \\ \mathrm{r}_i + \gamma \arg\max_{a^*} Q(s^{'}, a^*; \theta_{i-1}) & \text{o/w} \end{cases}$
    $\Big(y_i - Q(s, a; \theta_i)\Big)^2$ ▷ Gradient descent step
  end for
end for

4 Results and Discussion

The evaluation of the agent's learning procedure is challenging, as the binomial distribution in (6) introduces statistical fluctuations that lead to changes in the states/actions distribution. In order to alleviate these instabilities, we use the average cumulative bid ($\overline{F_p}$) and average fairness ($\overline{f}$). Although the bid p can take negative values in (6), we constrain $p \in [0, +\infty)$ for a better representation of the actual revenue. Regarding the reward r in (5), we scale the cumulative reward for each experiment between $[0, 1]$. This normalization is applied for a better comparison due to the fact that different weights are used in each experiment (β).

We conduct a number of experiments in order to understand the effect of the different competing objectives in (5) (*i.e.*, revenue and fairness). Experiments *I* and *II* are designed to test both objectives separately (revenue and fairness).

Experiments *III* and *IV* to find a policy with a desired target fairness. Experiment *V* is proposed as a null model which has no learning mechanism, thus all actions are randomly selected ($\epsilon = 1.0$) with zero reward value. Table 2 provides the main results by comparing the average cumulative bid, the averaged fairness, the reject ratio and the expected cumulative revenue for the previously defined experiments. The expected revenue $\overline{r}$ is calculated as result of the product between $\overline{F_p}$ and rejects.

Table 2. Results for the Proposed Experiments

Exp.	Hyperparameters				Results			
	β_p	β_f	p_t	f_t	$\overline{F_p}$	$\overline{f}$	Rejects	$\overline{r}$
I	1	0	1	-	**2761.3**	0.62	0.10	**2485.17**
II	0	1	-	1	1326.2	**0.99**	0.25	994.65
III	1	1	1	0.90	2282.1	0.88	0.11	2031.21
IV	1	1	1	0.75	2527.4	0.76	0.11	2249.36
V	0	0	-	-	1708.9	0.97	**0.31**	1179.12

For each experiment, Table 3 shows the average bid allocated in each group. This table provides valuable[1] information related to the fairness achieved for each policy. Related to the fairness principles design defined in Sect. 2.2, we can observe that the more target fairness given by f_t, the more homogeneous the prices among groups. In this sense, the policy incorporates the proposed fairness principles definition.

Table 3. Average bid on segments $g_\alpha \in \mathcal{G}$

Exp	$\overline{g_1}$	$\overline{g_2}$	$\overline{g_3}$	$\overline{g_4}$
I	5.8	2.0	9.5	9.8
II	2.7	2.9	2.6	2.7
III	5.6	2.0	6.6	6.8
IV	5.5	2.1	8.5	6.6
V	5.4	5.5	5.4	5.5

Figures 5, 6 and 7 show the evolution of the average cumulative bid, fairness and reward, respectively. In experiment *I* the agent learns a policy that minimizes fairness average and maximizes bid average in contrast to experiment *II* in which fairness is maximized obtaining the minimum bid average. Note that the profit in experiment *III* is the highest because the algorithm learns to increase

[1] $\epsilon = 1.0$. In null model all actions are randomly selected.

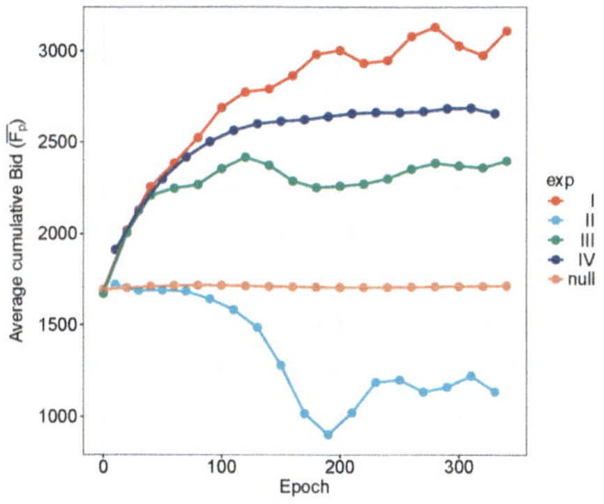

Fig. 5. Averaged cumulative bid. In exp. *I* the model learns to maximize revenue in contrast to exp. *II* in which learns to be fair. Exp. *III* and *IV* are proposed to learn a model balancing revenue and fairness with a specific targets. In the exp. *V* all actions are taken randomly

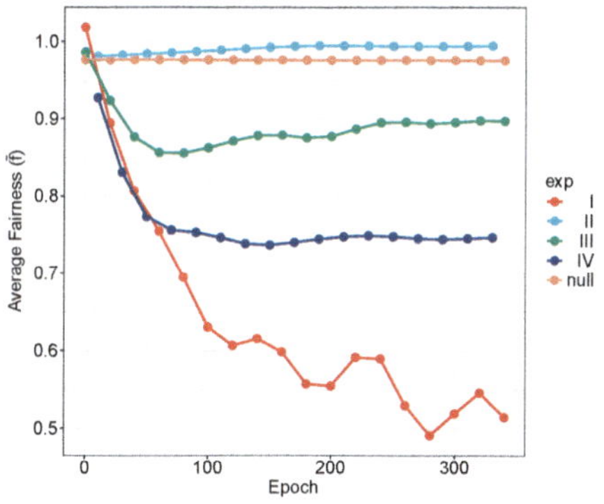

Fig. 6. Averaged fairness (see description of experiments in Fig. 5)

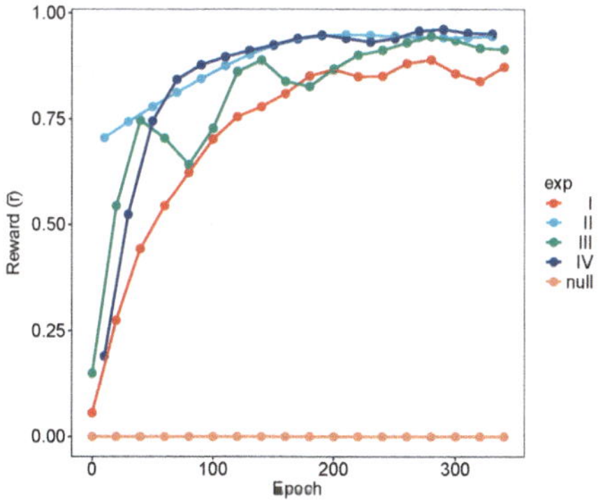

Fig. 7. Average reward

the bid value in groups g_3, g_4 and decrease the bid value in groups g_1 and g_2 (see Table 3 of $\bar{g}_\alpha$ values). In this policy, the bid values are adjusted to minimize the rejection ratio, defined as the price acceptance probability, see (3) and Fig. 2).

Experiments *III* and *IV* are proposed with the aim of achieving a target fairness f_t. Indeed, the averaged values map to the selected targets given by f_t ($f_t = 0.90, \overline{f} = 0.88$ and $f_t = 0.75, \overline{f} = 0.76$ for experiments *III* and *IV*, respectively). In experiment *III* the policy keeps the average price low on g_3 to reach a reasonable fairness ($\overline{f} = 88\%$). Conversely in experiment *IV* the calculated policy slightly increases the bid value in g_3 obtaining a less fair policy ($\overline{f} = 76\%$) with more revenue. As a heavy penalization for bid rejections is used in the experiments, given by $\nu = -0.5$, the bid rejected ratio remains low.

An interesting result arises from the comparison between fairness in experiments *II* and *V* (null model). The average prices in every group are ~ 5.5 in the null model because they are randomly and uniformly selected from $\mathcal{A}$. However, due to our definition of the rotated Jain's index in (7), the average fairness in *II* is slightly higher than that in *V*.

Related to the learning convergence, a descent in the average learning rate is a good indicator of learning convergence [34]. Let p a pair with the form $(s, a \mid s \in \mathcal{S}, a \in \mathcal{A})$ and $\mathcal{P}^n$ the set of all pairs p visited by the agent. Then, the learning rate is defined as:

$$l = \mathbb{E}\big[\{C(x) \mid \forall x \in \mathcal{P}\}\big], \text{ where } C(x) = \sum_{p \in \mathcal{P}} \delta_{xp}. \tag{8}$$

Figure 8 shows the convergence rate of learning rates for the experiments proposed. As observed, the learning rate exhibits a descent to 0, guaranteeing the convergence of the algorithm.

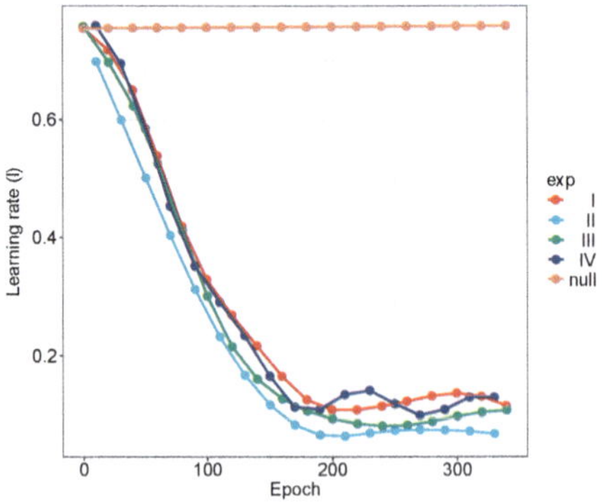

Fig. 8. Learning rate convergence

5 Conclusion

Fairness, as an ethical definition, is sometimes seen to be negative by constraining development [46]. However, the present paper represents an example in which both customers and companies can benefit from transparency. Even if there is a disagreement with regards the exact nature of a fair policy in dynamic pricing, the fairness design principles proposed in Sect. 2.2 provide an initial definition based on equality, that constitutes a promising way to open a rich dialogue. In this paper we have demonstrated that an unfair scenario can, to a certain degree, be *unbiased* (the fairness targets given by f_t), by integrating fairness metrics as part of the model optimization. We cover the integration of fairness from a *resource allocation point of view*, and include it as a design principle.

In terms of future work, there are possible extensions to this study. Firstly, there are many parameters within the learning process $(\lambda, \beta_p, \beta_f, \nu, \sigma_p, \sigma_r...)$ or the states-actions space (p, q) that need some more tuning if we consider the challenging and time- consuming task required to evaluate each combination of parameters. Secondly, the comparison with other state-of-the-art models, such as evolutionary algorithms [47], is a valuable asset for an effective comparison with the Q-learning method proposed. Finally, more complex environments can be defined including more variables as time or external factors.

We demonstrate empirically that an effective balance between revenue and fairness maximization can be achieved in real time models such as RL, in which the model learns by performing actions in a certain environment. We proposed a fairness metrics based on a rotated Jain's index. We have developed a synthetic environment to experiment with four different customer sensitivities. We have also tested several parametrized experiments, demonstrating that a given balance between revenue and fairness can be an option.

References

1. Deksnyte, I., Zigmas Lydeka, P.: Dynamic pricing and its forming factors. Int. J. Bus. Soc. Sci. **3**(23) (2012)
2. Narahari, Y., Raju, C.V.L., Ravikumar, K., Shah, S.: Dynamic pricing models for electronic business. Sadhana **30**(2), 231–256 (2005)
3. den Boer, A.V.: Dynamic pricing and learning: Historical origins, current research, and new directions. Surv. Oper. Res. Manag. Sci. **20**(1), 1–18 (2015)
4. Adamy, J.: E-tailer price tailoring may be wave of future (1999). http://articles.chicagotribune.com/2000-09-25/business/0009250017_1_prices-amazon-spokesman-bill-curry-don-harter
5. Reinartz, W.: Customizing prices in online markets. Symphonya. Emerging Issues in Management, no. 1 Market-Space Management, 5 (2002)
6. Garbarino, E., Lee, O.F.: Dynamic pricing in internet retail: effects on consumer trust. Psychol. Mark. **20**(6), 495–513 (2003)
7. Lee, S., Illia, A., LawsonBody, A.: Perceived price fairness of dynamic pricing. Ind. Manag. Data Syst. **111**(4), 531–550 (2011)
8. Xia, L., Monroe, K.B., Cox, J.L.: The price is unfair! a conceptual framework of price fairness perceptions. J. Mark. **68**(4), 1–15 (2004)

9. Weisstein, F.L., Monroe, K.B., Kukar-Kinney, M.: Effects of price framing on consumers' perceptions of online dynamic pricing practices. J. Acad. Mark. Sci. **41**(5), 501–514 (2013)
10. Haws, K.L., Bearden, W.O.: Dynamic pricing and consumer fairness perceptions. J. Consum. Res. **33**(3), 304–311 (2006). D. I. served as editor, and E. A. served as associate editor for this article
11. Odlyzko, A.: Privacy, economics, and price discrimination on the internet. In: Proceedings of the 5th International Conference on Electronic Commerce, ICEC 2003, pp. 355–366. ACM (2003)
12. Kimes, S.E.: A retrospective commentary on discounting in the hotel industry: a new approach. Cornell Hotel. Restaur. Adm. Q. **43**(4), 92–93 (2002)
13. Kahneman, D., Knetsch, J., Thaler, R.: Fairness as a constraint on profit seeking: entitlements in the market. Am. Econ. Rev. **76**(4), 728–41 (1986)
14. Finkel, N.J.: Not Fair! The Typology of Commonsense Unfairness, 1st ed. American Psychological Association (APA) (2001)
15. Kutschinski, E., Uthmann, T., Polani, D.: Learning competitive pricing strategies by multi-agent reinforcement learning. J. Econ. Dyn. Control. **27**(11), 2207–2218 (2003)
16. Knnen, V.: Dynamic pricing based on asymmetric multiagent reinforcement learning. Int. J. Intell. Syst. **21**(1), 73–98 (2006)
17. Gupta, M., Ravikumar, K., Kumar, M.: Adaptive strategies for price markdown in a multi-unit descending price auction: a comparative study. In: IEEE International Conference on Systems, Man and Cybernetics, vol. 1, pp. 373–378 (2002)
18. Menon, R.B., Menon, S.B., Srinivasan, D., Jain, L.: Online reinforcement learning in multi-agent systems for distributed energy systems. In IEEE Innovative Smart Grid Technologies - Asia (ISGT ASIA), pp. 791–796 (2014)
19. Skirpan, M., Gorelick, M.: The authority of "fair" in machine learning. CoRR vol. abs/1706.09976 (2017)
20. Burrell, J.: How the machine thinks: understanding opacity in machine learning algorithms. Big Data Soc. **3**(1) (2016)
21. Bostrom, N.: Superintelligence: Paths, Dangers, Strategies, 1st ed. Oxford University Press (2014)
22. Cerquitelli, T., Quercia, D., Pasquale, F.: Transparent Data Mining for Big and Small Data, 1st edn. Springer Publishing Company, Incorporated (2017)
23. Mikians, J., Gyarmati, L., Erramilli, V., Laoutaris, N.: Detecting price and search discrimination on the internet. In: Proceedings of the 11th ACM Workshop on Hot Topics in Networks, HotNets-XI, pp. 79–84. ACM (2012)
24. Bolukbasi, T., Chang, K.-W., Zou, J.Y. Saligrama, V., Kalai, A.T.: Man is to computer programmer as woman is to homemaker? debiasing word embeddings. In: Advances in Neural Information Processing Systems 29, pp. 4349–4357. Curran Associates, Inc. (2016)
25. Caliskan, A., Bryson, J.J., Narayanan, A.: Semantics derived automatically from language corpora contain human-like biases. Science **356**(6334), 183–186 (2017)
26. Hardt, M., Price, E., Srebro, N.: Equality of opportunity in supervised learning. CoRR, vol. abs/1610.02413 (2016)
27. Ryu, H.J., Mitchell, M., Adam, H.: Improving smiling detection with race and gender diversity. ArXiv e-prints (2017)
28. Weiss, R., Mehrotra, A.: Online dynamic pricing: efficiency, equity and the future of E-commerce. Va. J. Law Technol. **6**(2) (2001)
29. Lan, T., Kao, D.T.H., Chiang, M., Sabharwal, A.: An axiomatic theory of fairness. CoRR, vol. abs/0906.0557 (2009)

30. Arianpoo, N., Leung, V.C.: How network monitoring and reinforcement learning can improve tcp fairness in wireless multi-hop networks. EURASIP J. Wirel. Commun. Netw. **2016**(1), 278 (2016)
31. Sirajuddin, M., Rupa, C., Prasad, A.: Techniques for enhancing the performance of tcp in wireless networks. In: Suresh, L.P., Dash, S.S., Panigrahi, B.K. (eds.) Artificial Intelligence and Evolutionary Algorithms in Engineering Systems, pp. 159–167 (2015)
32. Zhang, X.M., Zhu, W.B., Li, N.N., Sung, D.K.: Tcp congestion window adaptation through contention detection in ad hoc networks. IEEE Trans. Veh. Technol. **59**(9), 4578–4588 (2010)
33. Sutton, R.S., Barto, A.G.: Introduction to Reinforcement Learning, 1st ed. MIT Press (1998)
34. Watkins, C.J., Dayan, P.: Technical note: Q-learning. Mach. Learn. **8**(3), 279–292 (1992)
35. Dini, S., Serrano, M.: Combining q-learning with artificial neural networks in an adaptive light seeking robot (2012)
36. van Hasselt, H., Wiering, M.A.: Reinforcement learning in continuous action spaces. In: IEEE International Symposium on Approximate Dynamic Programming and Reinforcement Learning, pp. 272–279 (2007)
37. Hornik, K., Stinchcombe, M., White, H.: Multilayer feedforward networks are universal approximators. Neural Netw. **2**(5), 359–366 (1989)
38. Riedmiller, M.: Neural fitted q iteration - first experiences with a data efficient neural reinforcement learning method. In: Gama, J., Camacho, R., Brazdil, P.B., Jorge, A.M., Torgo, L. (eds.) Machine Learning: ECML 2005, pp. 317–328. Springer, Heidelberg (2005)
39. Lin, L.-J.: Reinforcement learning for robots using neural networks. Ph.D. dissertation, School of Computer Science, uMI Order No. GAX93-22750 (1992)
40. Mnih, V., Kavukcuoglu, K., Silver, D., Graves, A., Antonoglou, I., Wierstra, D., Riedmiller, M.: Playing atari with deep reinforcement learning. In: NIPS Deep Learning Workshop 2013 (2013). arxiv:1312.5602Comment
41. Rana, R., Oliveira, F.S.: Dynamic pricing policies for interdependent perishable products or services using reinforcement learning. Expert. Syst. Appl. **42**(1), 426–436 (2015)
42. Jain, R., Durresi, A., Babic, G.: Throughput fairness index: an explanation. The Ohio State University, Technical report, February 2010
43. Phillips, R.: Pricing and Revenue Optimization, Stanford Business Books. Stanford University Press (2005)
44. Matignon, L., Laurent, G.J., Fort-piat, N.L.: Improving reinforcement learning speed for robot control. In: IEEE/RSJ International Conference on Intelligent Robots and Systems, pp. 3172–3177 (2006)
45. Kingma, D.P., Ba, J.: Adam: a method for stochastic optimization. CoRR, vol. abs/1412.6980 (2014)
46. Bowie, N.E.: Organizational Integrity and Moral Climates, pp. 183–205. Springer (2013)
47. Salimans, T., Ho, J., Chen, X., Sutskever, I.: Evolution strategies as a scalable alternative to reinforcement learning. CoRR, vol. abs/1703.03864 (2017)

A Classification-Regression Deep Learning Model for People Counting

Bolei Xu[1(✉)], Wenbin Zou[1], Jonathan Garibaldi[2], and Guoping Qiu[1,2]

[1] College of Information Engineering, Shenzhen University, Shenzhen, China
xubolei@gmail.com, {wzou,qiu}@szu.edu.cn
[2] School of Computer Scinece, University of Nottingham, Nottingham, UK
jon.garibaldi@nottingham.ac.uk

Abstract. In this paper, we construct a multi-task deep learning model to simultaneously predict people number and the level of crowd density. Motivated by the success of applying "ambiguous labelling" to age estimation problem, we also manage to employ this strategy to the people counting problem. We show that it is a reasonable strategy since people counting problem is similar to the age estimation problem. Also, by applying "ambiguous labelling", we are able to augment the size of training dataset, which is a desirable property when applying to deep learning model. In a series of experiment, we show that the "ambiguous labelling" strategy can not only improve the performance of deep learning but also enhance the prediction ability of traditional computer vision methods such as Random Projection Forest with hand-crafted features.

Keywords: People counting · Deep learning · Ambiguous labelling

1 Introduction

In many application scenarios, there is a need to count the number of people at a scene. For example, in public spaces such as airports and railway stations, knowing the number of people present at the scene can help better manage the space and to ensure public security. With the wide spread installation of visual surveillance cameras almost everywhere in such public space, it is possible to perform automatic people counting through analyzing surveillance videos. Using computer vision and machine learning techniques for people counting has therefore attracted a lot of interest in the literature. However, like many computer vision applications, people counting in video is also a very challenging problem.

In recent years, deep learning neural networks have emerged as a powerful technique for many computer vision problems. In this paper, we are inspired by the significant performance of deep learning on various vision tasks [1–3] and apply the deep learning method to extract deep feature for the crowd counting problem. In previous work, several kinds of deep learning models have been proposed to address the people counting problem. Zhang *et al.*, [4] and Wang *et al.* [5] construct deep networks to directly output people number. Some later

K. Arai et al. (Eds.): IntelliSys 2018, AISC 868, pp. 136–149, 2019.
https://doi.org/10.1007/978-3-030-01054-6_9

works [6, 7] apply deep learning network to produce density map instead of people number to achieve better performance. The density map presents the position of human heads and thus is able to provide people number. However, such method requires to label human position when constructing training datasets, which limits their scalability to the real world application. On the other hand, occlusion is a severe problem in crowd counting. In the case of high density crowd, it is difficult for human to label accurate head positions and provide reliable people numbers for the training datasets.

Inspired by the success of multi-task deep learning method [8], we propose a classification-regression deep learning model which treats the whole surveillance image as the input image, and the deep learning model not only outputs the person number but also estimates the level of crowd density. We show that such multi-task network structure is able to learn more discriminative feature representation than a network solely outputs people number, because the task of estimating the density level could provide a coarse counting number which is less affected by the variation of image scale. In the work of [8], they simultaneously produce density map and 10-way crowd count classification. We differ from their method by predicting people number instead of producing density map. Apart from the aforementioned reasons, directly predicting people number requires less computational resources, since producing density map is usually based on a convolutional layer with filter size of 1×1 to map feature map to the density map. In contrast, the performance of our method is comparable to that of [8] through using only one fully-connected layer after the base network.

In order to address occlusion problem, we also adopt a strategy called "ambiguous labelling" method. The "ambiguous labelling" was first applied to solve the age estimation problem [9–11], since the faces of neighbouring ages usually present similar image features. Thus, in the previous work of age estimation, authors could assign ambiguous labels to input face images and take the problem as a classification task. We reason that it is also possible to apply "ambiguous labelling" strategy to the crowd counting problem. One reason is that people counting problem is similar to the problem of age estimation, for instance, the image of 500 people is similar to the image of 510 people. On the other hand, the size of people counting dataset is usually small, which is not sufficient to train a deep learning model with large number of parameters. To solve this problem in the deep learning context, "ambiguous labelling" method enables us to create various people number labels for the input image that can augment training dataset for the deep learning model. We provide detail analysis in Sect. 3. In the experiment, we show that this method is effective not only for the deep learning model, but also for the traditional computer vision methods such as random projection forest model [12].

2 Related Work

The crowd counting task was initially solved by the detection method. Different kinds of features are used to detect the body of pedestrians including motion features [13], histogram-of-gradients [14] or Bayesian model-based segmentation [15]. However, occlusion becomes a serious problem when applying to estimate high density crowd. Then the part-based detection methods are developed to solve this problem [16,17]. These methods usually take a long time to count people since they have to exhaustively scan each frame of the video with the trained detector. Another approach is to cluster the trajectories which have coherent motion and then the number of clusters is used to estimate the moving pedestrians [18,19]. One problem of the clustering method is that it can only provide accurate result when reliable trajectories can be extracted. Thus, this approach is not able to handle the occlusion problem and low video frame rates due to the broken feature tracks. Foroughi *et al.* [20] take the people counting task as a classification problem. They apply sparse representation to capture the hidden structure and semantic information in the image data, and the feature dimension is further reduced by random projection. However, one serious problem with the classification method is if any label information (i.e. the number of people) in the testing set is not included by the training set, this method cannot achieve high accuracy result, which means their algorithm requires large training set to cover almost all the possible situation in the testing set.

A more suitable approach to solving the aforementioned problems is to count by regression. Low-level features are firstly extracted and then mapped to the people number by the regression model. As this kind of approach does not require to detect and track individual person, it has relatively low computational cost and demonstrates promising results on solving the occlusion problem. A variety of features have been used by previous works to estimate the crowd density, such as total area [21,22], edge count [23,24] and texture features [25]. Chan *et al.* [26] take the perspective distortion into account and experiment with additional features such as Minkowski fractal dimension to estimate the irregularity of edges.

The traditional approaches are suffering from two main problems. Firstly, they heavily rely on the background segmentation techniques to remove noise. Secondly, an unavoidable step in the traditional approaches is to extract hand-crafted features. However, designing hand-crafted features is not an easy step and it is usually difficult to find out optimal hand-crafted feature representation. The deep learning approach can well-solve both problems. It does not have to apply background segmentation method to pre-process images and it is able to count people number from different perspectives [6]. Another advantage is that deep learning can be constructed as an end-to-end model, which takes whole image as input and outputs people number or the head position. It means feature designing is not a necessary step when applying deep learning.

Some previous work apply deep learning method to address the problem of people counting. At the initial stage, the deep learning framework is usually employed to directly output people number. Zhang *et al.* [4] propose a Convolutional Neural Network (CNN) based framework to extract deep features of crowd scene and use a data-driven method to fine-tune the CNN model to the target scene. Wang *et al.* [5] also construct a deep network in order to estimate extremely dense crowds. Marsden *et al.* [27] apply a scale aware deep learning model with a single column fully convolutional network that takes multiple scales of image as the input in the prediction stage. Each scale of image produces a people number and the final counting number is to take the average of these estimates.

Apart from directly predicting people number, another way to apply deep learning is to generate density map and then count people number from density map. Zhang *et al.* [6] first develop this method to count people number from density map. They use a Gaussian kernel to convolve a labelled image and then compute people number by summarizing pixel value. There are also some following work to produce density map based on deep learning approach. Boominathan *et al.* [28] combine one deep network and one shallow network to predict a density map for a given crowd image. Sindagi *et al.* [8] propose a cascaded deep network structure to simultaneously classify crowd into different levels and produce density map. However, the approach based on density map has to label the head positions for the whole dataset, which is a time-consuming process when applying to the high density crowd or the large scale datasets.

3 Application of Ambiguous Labels to People Counting

We here illustrate the rationales that we apply "ambiguous labelling" strategy for the people counting problem.

Firstly, we show that people counting problem is similar to the age estimation problem. Figure 1 presents a typical case in the people counting problem. The ground-truth number for Fig. 1(a) is 26 persons while the person number in Fig. 1(b) is 31. Although the people numbers are totally different, the major contents of both images are very close. It is confirmed by the traditional features extracted from both images. Two main features (segment area and perimeter-area) employed by the previous work [26] are almost the same. If we look into the details of both images, there are three minor differences leading to different person numbers: (1) In the red bounding box, a woman is pushing a stroller for a baby but the size of baby body is small in the image. (2) In the green bounding box, a walking woman's body is occluded by an obstruction while only part of woman body is shown in image. (3) In the yellow bounding box, three persons' heads appear on the image. However, only piece of their heads can be seen in the image. Thus, we can see that similar image features do not always refer to the same person number. It is the same as the age estimation problem that neighbouring age might present similar image features. This is the main reason that we could assign various labels to each input image as done in the previous age estimation work.

Secondly, "ambiguous labelling" strategy enables us to create augmented training dataset for deep learning model. As insufficient training data could lead to over-fitting problem, a desirable training dataset should have multiple images for each image label. However, the mainstream people counting datasets (UCSD and Mall datasets) usually contain limited number of images for each

(a) GT=26; SA=0.59 ;PAR=0.56.

(b) GT=31; SA=0.57; PAR=0.57.

Fig. 1. Both images are captured from Mall dataset. 'GT' refers to ground-truth, 'SA' refers to segment area and 'PAR' refers to perimeter-area ratio. All values are provided by the original dataset author

people number. Consequently, we could improve the predicting ability of model by enlarging the size of training dataset. By assigning various labels to the images in the training dataset, we can obtain a much larger size of training datasets than that of the original one. It means for each specific people number (training label), we can find a variety of crowd scenes (training image) in the training dataset. The deep learning model can thus learn more discriminative features with sufficient number of training images.

4 Label Ambiguity Construction

In this section, we introduce our method to model the randomness of people number and thus to create ambiguous labels for each input surveillance image. For each scalar-valued people number label $l \in \mathbb{R}$ of the input image, we seek a label distribution that should satisfy two criteria: (1) the ground truth value should have the highest possibility of being assigned to the image; and (2) when the labels are farther from the ground truth, they should be assigned to the image with lower probabilities. In this paper, we adopt the Gaussian distribution in the experiment to model the ambiguous labels for each surveillance image as shown in Fig. 2, whose mean value μ is equal to the ground-truth value. The corresponding standard deviation σ for the Gaussian distribution is usually an unknown factor but can work well when it is carefully chosen [11]. We thus empirically set σ to 2 in the experiment. By constructing a Gaussian distribution, we can randomly sample M labels for each input image. As the problem of occlusion usually appears in the relatively high density crowd, we only apply the "ambiguous labelling" strategy to the images of people number over 15.

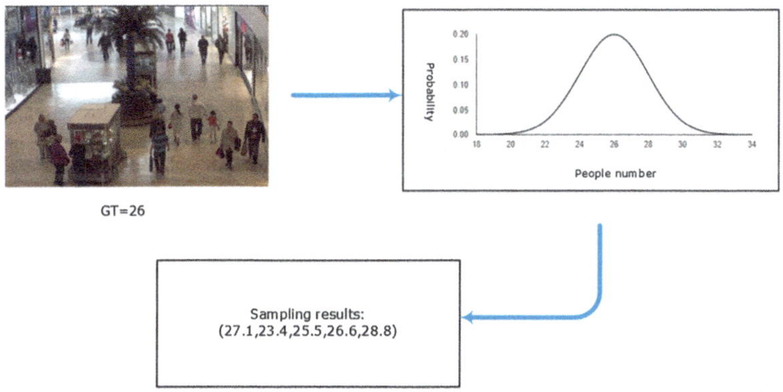

Fig. 2. The process of how to assign ambiguous labels to an input image. The ground-truth value of input image is regarded as the mean value μ for the Gaussian distribution. We randomly sample $M = 5$ labels for each image

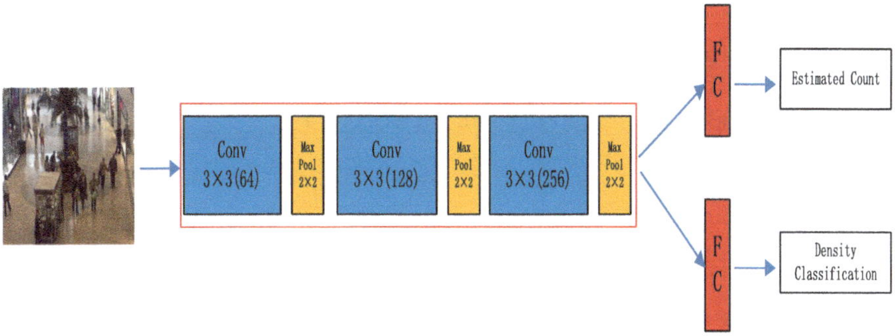

Fig. 3. The network architecture of our classification-regression deep learning model. The regression branch outputs the accurate crowd density, while the classification branch predicts the coarse people number

5 Deep Classification-Regression Learning Model

In this paper, we do not apply Resnet [1] or VGG deep learning model [2] as the base convolutional network to address the problem. The reason is that the size of crowd counting datasets is relatively small (usually around 2000 images), which is not sufficient to train the Resnet or VGG network with large number of parameters. For this crowd counting problem, we construct the convolutional network based on a custom network structure as shown in Fig. 3. We construct the multi-task deep learning model by connecting two parallel sub-networks to the base convolutional network. One sub-network is used to predict people number and another sub-network is used to estimate the crowd density level.

The people counting network is consisting of one fully-connected layer with 256 neurons and Rectified Linear Unit (ReLU) is taken as the activation function. This branch finally produces people number $\hat{l_k}$ for the input image x_k with label l_k, and we use Mean Squared Error (MSE) as the objective function for this branch:

$$\mathcal{L}_{MSE} = \frac{1}{K}\sum_{k=1}^{K}\frac{1}{2}\|l_k - \hat{l_k}\|^2. \tag{1}$$

The classification layer aims to classify input image to one of the density levels. We create classification labels for each dataset with an interval of 10 people. For instance, if the maximum people number in the training dataset is 100, then we can create 11 labels for the dataset. The level-1 density refers to the people number of 0 to 10, and level-2 refers to people number of 11 to 20. The rest can be done in the same manner where level-11 refers to the people number above 100. The classification layer also contains a fully-connected layer that has 256 neurons with ReLU activation function. We use softmax function as classifier and use the cross-entropy error as the loss function:

$$\mathcal{L}_{level}(p, q) = -\sum_{x} p(x)\log q(x), \tag{2}$$

where p is the ground-truth distribution of density level, and q is the estimated class probabilities produced by the softmax classifier. Then the total loss for the whole deep learning model can be written as:

$$\mathcal{L}_{total} = \lambda\mathcal{L}_{MSE} + \mathcal{L}_{level}, \tag{3}$$

where λ is a weighting factor.

6 Experiment

6.1 Experiment Setup

For the parameter settings, we initialize the whole deep network with Gaussian distribution of zero mean and set its standard deviation to 0.01, and bias to zeros. We empirically set $\lambda = 2$ in (3). We then optimize the network by Stochastic

Gradient Descent (SGD) with a learning rate of 0.01 and the size of mini-batches is 128. In the experiment, the network usually convergences around 30 epochs. We conduct all the experiments over the UCSD pedestrian dataset and Mall dataset. When creating ambiguous labels for each dataset, we randomly sample $M = 5$ labels from Gaussian distribution for each image. The input image is resized to 256×256 for the deep learning model.

We test our proposed algorithm on the UCSD pedestrian database [26] and Mall dataset [29], which are two well-known datasets on the evaluation of people counting algorithms. Both datasets contain 2000 frames that are captured by a stationary camcorder from outdoor and indoor scene respectively. The example images from two datasets are shown in Fig. 4.

(a) UCSD dataset

(b) Mall dataset

Fig. 4. Crowd scenes of UCSD dataset and Mall dataset. The UCSD dataset captures outside scene while the Mall dataset captures indoor scene

We separate the datasets as previous work: in UCSD dataset, frames 601-1400 are employed for training; in Mall dataset, the first 800 frames are used. The rest frames in each dataset are applied for testing. Two evaluation metrics are applied for numerical testing and comparison with the state-of-the-art algorithms. The first one is called *mean absolute error* (MAE) to estimate the average absolute error of each testing frames:

$$\epsilon_{mae} = \frac{1}{N} \sum_{i=1}^{N} |m_i - \widetilde{m}_i|, \tag{4}$$

where N is the total number of test images, m_i is the ground truth for ith test image, and $\widetilde{m}_i$ is the corresponding prediction result. The second one is *mean squared error* (MSE) which assesses the average mean squared error:

$$\epsilon_{mse} = \frac{1}{N}\sum_{i=1}^{N}(m_i - \widetilde{m}_i)^2. \tag{5}$$

Table 1. Performance Comparison with the Traditional Computer Vision Methods on Two Datasets

Model	UCSD		Mall	
	MAE	MSE	MAE	MSE
GPR [26]	2.24	7.97	3.72	20.10
MORR [29]	2.25	7.82	3.59	19.00
CA-RR [30]	2.07	6.86	3.43	17.70
RPF (hf)	1.90	6.01	3.22	15.52
RPF (fc1)	1.78	5.46	3.02	13.80
RPF (fc2)	1.62	4.84	2.86	11.44
Ours	**1.48**	**3.24**	**2.73**	**10.20**

6.2 Comparing with Hand-Crafted Features

In the first experiment, we compare our deep learning method with the traditional computer vision methods including our random projection forest that employ hand-crafted features. Table 1 presents the results of this experiment. It can be seen that our deep learning method significantly outperforms other traditional methods. We also conducted an experiment on the Random Projection Forest (RPF) [12], which employs different kinds of feature. One is the same hand-crafted features (hf) as [26], and another one is the deep feature from the FC layer in the regression branch (fc1), and the FC layer in the classification branch (fc2). It can be seen that the deep features from deep learning model are more discriminative than the hand-crafted features, and the features from fc1 is better than that from fc2, which is caused by the regression branch is able to predict more detail people density scenario than the classification branch.

6.3 Comparing with CNN-based Approaches

We also compare our method with the CNN-based approaches. These approaches include Zhang *et al.* [4], Kumagai *et al.* [31], Sam *et al.* [32], and Sheng *et al.* [33].

Table 2. Performance Comparison with the CNN-based Methods on Two Datasets, where '-' Indicates no Result Reported

Model	UCSD		Mall	
	MAE	MSE	MAE	MSE
Zhang *et al.* [4]	1.60	3.31	-	-
Kumagai *et al.* [31]	-	-	2.75	13.40
Sam *et al.* [32]	1.62	**2.10**	-	-
Sheng *et al.* [33]	2.86	13.0	**2.41**	**9.12**
Ours	**1.48**	3.24	2.73	10.20

Form Table 2 we can see that our CNN method achieves the best performance on the UCSD dataset with MAE as the evaluation criteria, and slightly worse performance than Sam *et al.* [32] on the MSE evaluation. On the Mall dataset, our deep learning approach provides comparable performance when comparing to other CNN methods. Comparing with other approaches, the classification branch in our model can provide a coarse estimation to the people density, which is less influenced by the variation of perspectives and image scale.

Table 3. Evaluation of ambiguous label on random projection forest and deep learning model

Model	UCSD		Mall	
	MAE	MSE	MAE	MSE
RPF without ambiguous labels	1.90	6.01	3.22	15.52
RPF with ambiguous labels	1.78	5.10	3.04	13.82
Deep learning without ambiguous labels	1.62	4.82	2.92	12.61
Deep learning with ambiguous labels	1.48	3.24	2.73	10.20

6.4 Evaluation of Ambiguous Labelling

Then we conduct an experiment to evaluate the effectiveness of ambiguous labelling strategy. We apply ambiguous labelling method to both deep learning model and also the random projection forest model. From Table 3 we can seen that by employing ambiguous labelling method can increase the performances of both deep learning model and the random projection forest model with larger size of training dataset. It confirms the effectiveness of ambiguous labelling method that it is not only effective on age estimation problem in previous work but also helpful on the crowd estimation problem.

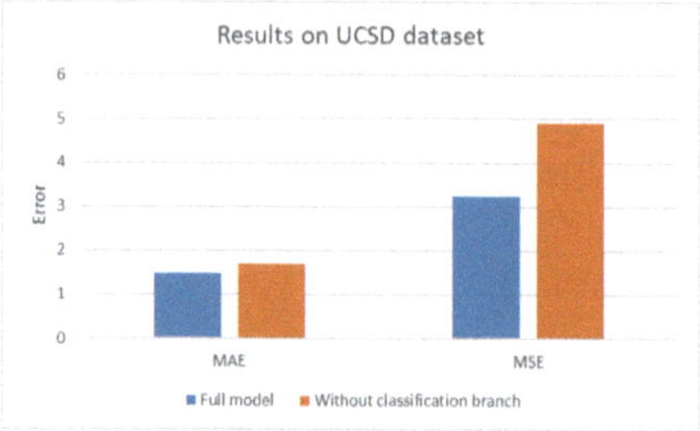

(a)

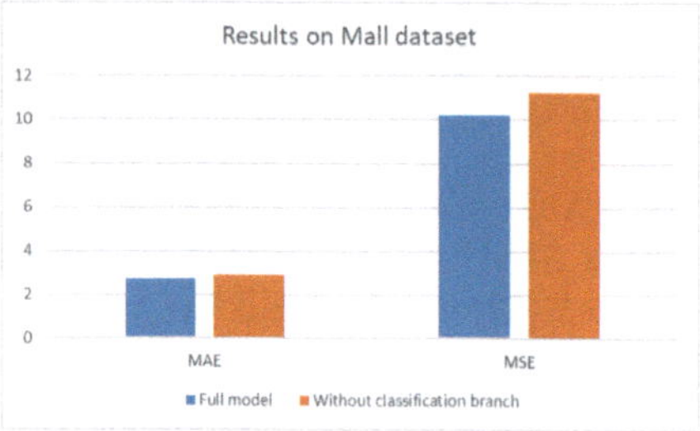

(b)

Fig. 5. Comparison results on UCSD and Mall datasets with and without classification branch. Figure (a) is the results on UCSD dataet, and Fig. (b) is the results on Mall dataset

6.5 Evaluation of Necessity of Classification

As we propose a multi-task deep learning model, it is also necessary to evaluate the necessity of the classification branch in the deep learning model. We compare two models: one is the full model and another one is the model without the classification branch. From Fig. 5 we can see that it is necessary to include the classification branch to the model. The classification branch provides a coarse counting number that is less influenced by the image scale and the variation of perspectives. Thus, from the experiment result, we can see that the full model with two branches shows much better performance than the model without the classification branch on both datasets.

Table 4. We Construct Four Different Models to Evaluate how the Size of Training Dataset Affect the Prediction Performance

Model	MAE	MSE
Model 1	1.32	2.89
Model 2	1.48	3.24
Model 3	2.52	9.00
Model 4	2.73	10.20

6.6 Evaluation of the Influence of Dataset Size

One inevitable problem when applying deep learning model is the size of dataset. Insufficient training dataset size would lead to the over-fitting problem and reduce the generalization ability of the model.

In this experiment, we modify the training dataset to evaluate the influence of dataset size. When testing on the UCSD dataset, we also include the whole Mall dataset into the training dataset. When testing on the Mall dataset, we add the whole UCSD dataset to the training dataset. It results in four kinds of model:

(1) Model 1: Training on the training dataset of UCSD and whole dataset of Mall, and testing on the testing dataset of UCSD.
(2) Model 2: Training on the training dataset of UCSD, and testing on the testing dataset of UCSD.
(3) Model 3: Training on the training dataset of Mall and whole dataset of UCSD, and testing on the testing dataset of Mall.
(4) Model 4: Training on the training dataset of Mall, and testing on the testing dataset of Mall.

From Table 4 we can see that when the training data grows, the performance produced by the deep learning increases as well. It verifies the assumption that the larger dataset would lead to the better performance when applying deep learning model.

7 Conclusion

In this paper, we have constructed a multi-task deep learning model for the crowd estimation problem. We show that the deep learning method is able to outperform previous computer vision methods based on hand-crafted features. Apart from employing deep feature, we propose an ambiguous labelling method to create various label for each input image. The experiment result confirms the effectiveness of the ambiguous labelling method, which is able increase the performance of both deep learning method and also our previous random projection forest method.

References

1. He, K., Zhang, X., Ren, S., Sun, J.: Deep residual learning for image recognition. In: Proceedings of the IEEE Conference on Computer Vision and Pattern Recognition, pp. 770–778 (2016)
2. Simonyan, K., Zisserman, A.: Very deep convolutional networks for large-scale image recognition. arXiv preprint arXiv:1409.1556 (2014)
3. Ghifary, M., Kleijn, W.B., Zhang, M., Balduzzi, D., Li, W.: Deep reconstruction-classification networks for unsupervised domain adaptation. In: European Conference on Computer Vision, pp. 597–613. Springer (2016)

4. Zhang, C., Li, H., Wang, X., Yang, X.: Cross-scene crowd counting via deep convolutional neural networks. In: Proceedings CVPR (2015)
5. Wang, C., Zhang, H., Yang, L., Liu, S., Cao, X.: Deep people counting in extremely dense crowds. In: Proceedings of the 23rd ACM international conference on Multimedia, pp. 1299–1302. ACM (2015)
6. Zhang, Y., Zhou, D., Chen, S., Gao, S., Ma, Y.: Single-image crowd counting via multi-column convolutional neural network. In: Proceedings of the IEEE Conference on Computer Vision and Pattern Recognition, pp. 589–597 (2016)
7. Onoro-Rubio, D., López-Sastre, R.J.: Towards perspective-free object counting with deep learning. In: European Conference on Computer Vision, pp. 615–629. Springer (2016)
8. Sindagi, V.A., Patel, V.M.: CNN-based cascaded multi-task learning of high-level prior and density estimation for crowd counting. arXiv preprint arXiv:1707.09605 (2017)
9. Chen, K., Kämäräinen, J.-K.: Learning with ambiguous label distribution for apparent age estimation. In: Asian Conference on Computer Vision, pp. 330–343. Springer (2016)
10. Geng, X., Wang, Q., Xia, Y.: Facial age estimation by adaptive label distribution learning. In: 2014 22nd International Conference on Pattern Recognition (ICPR), pp. 4465–4470. IEEE (2014)
11. Gao, B.-B., Xing, C., Xie, C.-W., Wu, J., Geng, X.: Deep label distribution learning with label ambiguity. IEEE Trans. Image Process. **26**(6), 2825–2838 (2017)
12. Xu, B., Qiu, G.: Crowd density estimation based on rich features and random projection forest. In: 2016 IEEE Winter Conference on Applications of Computer Vision (WACV), pp. 1–8. IEEE (2016)
13. Viola, P., Jones, M.J., Snow, D.: Detecting pedestrians using patterns of motion and appearance. In: Ninth IEEE International Conference on Computer Vision, Proceedings, pp. 734–741. IEEE (2003)
14. Dalal, N., Triggs, B.: Histograms of oriented gradients for human detection. In: IEEE Computer Society Conference on Computer Vision and Pattern Recognition, CVPR 2005, vol. 1, pp. 886–893. IEEE (2005)
15. Zhao, T., Nevatia, R.: Bayesian human segmentation in crowded situations. In: 2003 IEEE Computer Society Conference on Computer Vision and Pattern Recognition, Proceedings, vol. 2, pp. II–459. IEEE (2003)
16. Wu, B., Nevatia, R.: Detection of multiple, partially occluded humans in a single image by bayesian combination of edgelet part detectors. In: Tenth IEEE International Conference on Computer Vision, ICCV 2005, vol. 1, pp. 90–97. IEEE (2005)
17. Lin, S.-F., Chen, J.-Y., Chao, H.-X.: Estimation of number of people in crowded scenes using perspective transformation. IEEE Trans. Syst., Man Cybern., Part A: Syst. Hum.S **31**(6), 645–654 (2001)
18. Brostow, G.J., Cipolla, R.: Unsupervised bayesian detection of independent motion in crowds. In: 2006 IEEE Computer Society Conference on Computer Vision and Pattern Recognition, vol. 1, pp. 594–601. IEEE (2006)
19. Rabaud, V., Belongie, S.: Counting crowded moving objects. In: 2006 IEEE Computer Society Conference on Computer Vision and Pattern Recognition, vol. 1, pp. 705–711. IEEE (2006)
20. Foroughi, H., Ray, N., Zhang, H.: Robust people counting using sparse representation and random projection. Pattern Recogn. (2015)

21. Paragios, N., Ramesh, V.: A MRF-based approach for real-time subway monitoring. In: Proceedings of the 2001 IEEE Computer Society Conference on Computer Vision and Pattern Recognition, CVPR 2001, vol. 1, pp. I–1034. IEEE (2001)
22. Cho, S.-Y., Chow, T.W., Leung, C.-T.: A neural-based crowd estimation by hybrid global learning algorithm. IEEE Trans. Syst., Man, Cybern., Part B: Cybern. **29**(4), 535–541 (1999)
23. Davies, A.C., Yin, J.H., Velastin, S.A.: Crowd monitoring using image processing. Electron. Commun. Eng. J. **7**(1), 37–47 (1995)
24. Regazzoni, C.S., Tesei, A.: Distributed data fusion for real-time crowding estimation. Signal Process. **53**(1), 47–63 (1996)
25. Marana, A., da Costa, L., Lotufo, R., Velastin, S.: On the efficacy of texture analysis for crowd monitoring. In: International Symposium on Computer Graphics, Image Processing, and Vision, Proceedings, SIBGRAPI 1998, pp. 354–361. IEEE (1998)
26. Chan, A.B., Liang, Z.-S., Vasconcelos, N.: Privacy preserving crowd monitoring: Counting people without people models or tracking. In: IEEE Conference on Computer Vision and Pattern Recognition, CVPR 2008, pp. 1–7. IEEE (2008)
27. Marsden, M., McGuiness, K., Little, S., O'Connor, N.E.: Fully convolutional crowd counting on highly congested scenes. arXiv preprint arXiv:1612.00220 (2016)
28. Boominathan, L., Kruthiventi, S.S., Babu, R.V.: Crowdnet: a deep convolutional network for dense crowd counting. In: Proceedings of the 2016 ACM on Multimedia Conference, pp. 640–644. ACM (2016)
29. Chen, K., Loy, C.C., Gong, S., Xiang, T.: Feature mining for localised crowd counting. In: BMVC, vol. 1, no. 2, p. 3 (2012)
30. Chen, K., Gong, S., Xiang, T., Loy, C.C.: Cumulative attribute space for age and crowd density estimation. In: 2013 IEEE Conference on Computer Vision and Pattern Recognition (CVPR), pp. 2467–2474. IEEE (2013)
31. Kumagai, S., Hotta, K., Kurita, T.: Mixture of counting cnns: Adaptive integration of cnns specialized to specific appearance for crowd counting. arXiv preprint arXiv:1703.09393 (2017)
32. Sam, D.B., Surya, S., Babu, R.V.: Switching convolutional neural network for crowd counting. arXiv preprint arXiv:1708.00199 (2017)
33. Sheng, B., Shen, C., Lin, G., Li, J., Yang, W., Sun, C.: Crowd counting via weighted VLAD on dense attribute feature maps. IEEE Trans. Circuits Syst. Video Technol. (2016)

The Impact of Replacing Complex Hand-Crafted Features with Standard Features for Melanoma Classification Using Both Hand-Crafted and Deep Features

Binu Melit Devassy[1], Sule Yildirim-Yayilgan[2(✉)], and Jon Yngve Hardeberg[1]

[1] Department of Computer Science, Norwegian University of Science and Technology, Gjøvik, Norway
binum@stud.ntnu.no, jon.hardeberg@ntnu.no
[2] Department of Information Security and Communication Technology, Norwegian University of Science and Technology, Gjøvik, Norway
sule.yildirim@ntnu.no

Abstract. Melanoma is the deadliest form of skin cancer and it is the most rapidly spreading cancer in the world. An earlier detection of this kind of cancer is curable; hence, earlier detection of melanoma is pre-eminent. Because of this fact, a lot of research is being done in this area especially in automatic detection of melanoma. In this paper, we are proposing an automatic melanoma detection system which utilizes a combination of deep and hand-crafted features. We analyzed the impact of using a simpler and standard hand-crafted feature, in place of complex usual hand-crafted features e.g. shape, texture, diameter, or some custom features. We used a convolutional neural network (CNN) known as deep residual network (ResNet) to extract the deep features and utilized the scale invariant feature descriptor (SIFT) as the hand-crafted feature. The experiments revealed that combining SIFT did not improve the accuracy of the system however, we obtained higher accuracy than state-of-the-art methods with our deep only solution.

Keywords: Melanoma detection · ResNet · SIFT

1 Introduction

Melanoma, the treacherous skin cancer is primarily caused by UV radiation from sun or other sources. Around 10,130 people died annually in US because of melanoma [1]. In Norway, incidence rate (number of cases per 100,000 per year) has dramatically increased from 1.9 and 2.2, for women and men respectively, to 19.6 and 19.0 over a period of 57 years from 1953 to 2010 [2]. The major cause for skin cancer is determined as the exposure to intense radiations from sun. Therefore, the primary prevention should be avoiding long term exposure to solar radiation or use some protection against sun exposure. Skin cancer could have been broadly classified as melanoma (Fig. 1(a)) and non-melanoma (Fig. 1(b)). Non-melanoma skin cancers are more common and have a higher chance of recovery than melanoma, because they are less capable of spreading

K. Arai et al. (Eds.): IntelliSys 2018, AISC 868, pp. 150–159, 2019.
https://doi.org/10.1007/978-3-030-01054-6_10

from one part to the other parts of the body [3]. An early detection of melanoma cancer is almost curable, if not it can be fatal. The most common diagnosis method is visual inspection with dermoscopy followed by histopathological examination as required [4]. The ABCD(E) rule is the widely accepted technique for clinical inspection of skin lesions [5].

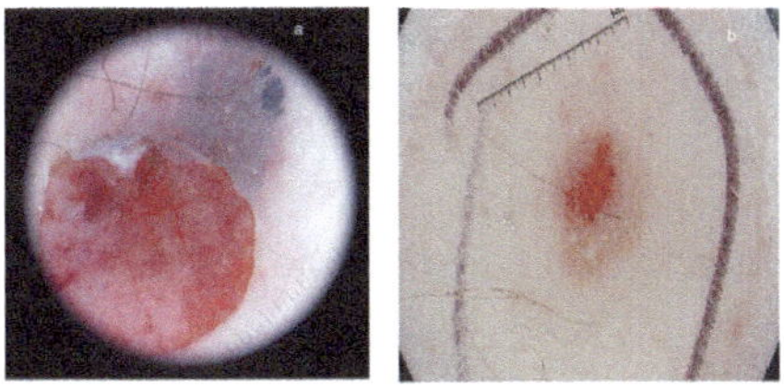

Fig. 1. Examples of skin lesions (a) melanoma and (b) seborrheic keratosis (images taken from ISIC challenge 2017) [6]

The detection of melanoma still continues to be a difficult problem. The manual inspection is considered as the effective solution, but it is also difficult and time consuming because most people have many non-cancerous lesions on their skin. Due to this fact and its importance, many researchers are working on automatic and semi-automatic detection of melanoma. The recent trend in this field is to use a hybrid classifier that employs both deep and hand-crafted features for detecting melanoma. Most of these hybrid classifiers are using computationally expensive complex features as hand-crafted features [7].

In this paper, we are comparing the degradation or improvement in accuracy while replacing the complex features with simple well-known features in hybrid melanoma classifiers. We compared our result with the result from the ISIC 2017 Melanoma Detection challenge [6]. Here we are interested in detecting melanoma and not interested in other classes of non-melanoma skin lesions.

This paper is organized as follows. In Sect. 2, we will discuss about the related work. In Sect. 3, the proposed method will be explained. In Sect. 4, we will discuss the details of the experiments. In Sect. 5, we will analyse the results obtained. Section 6 will discuss about the contributions of this paper and Sect. 7 has the conclusions and future work.

2 Related Work

Using deep learning algorithms for skin lesion classification have already shown improvement in the classification performance over the algorithms that extract traditional and quantitative handcrafted features and training them for classification [7–10]. Additionally, some pre and post processing is necessary while extracting hand-crafted features. However, could a combined use of traditional hand-crafted and deep learning features improve the overall classification performance?

We can find many examples of work related to combining deep features with hand crafted features particularly for skin image classification [7], and generally for medical image classification. Consider the work done by Rahul Paul et al. [11]. The authors obtained an accuracy of 90% in predicting lung cancer by combining deep and hand-crafted features. Initially they had 77.5% accuracy from hand crafted features and 77.5% from deep features. [12] is another example of hybrid implementation for chest pathology detection. Bram van Ginneken et al. [13] also used a combination of CNN and hand-crafted features to detect Pulmonary Nodule and they obtained a significantly better result. These works used sophisticated hand-crafted features or a mixture of them. Our work is different from the predecessor because of the simplicity of the hand-crafted feature that we have used. As mentioned earlier we have used a combination of hand-crafted and deep features. We used SIFT (scale invariant feature transform) [14] a well-known feature descriptor as hand crafted feature and used ResNet [15] for extracting deep features.

We have used the data set provided by the ISIC challenge 2017 for training and evaluation to make a reasonable comparison with state of the art results. The data set contains total 2000 skin lesion images, 374 melanoma images and 254 seborrheic keratosis images. We have done data augmentation to increase the robustness of the training. We will discuss the process flow of our approach and the experiments in the upcoming sections.

3 Proposed Method

Image pre-processing, segmentation, feature extraction and classification are the basic building blocks of an algorithm to detect skin lesions [3]. In our proposed method (Fig. 2), we introduced data augmentation to increase the robustness of the training. The input image will be passed through the hand-crafted and deep feature extraction methods and both features will be combined in the form of a single vector and passed into the classifier. We will describe each block in detail in the following sections.

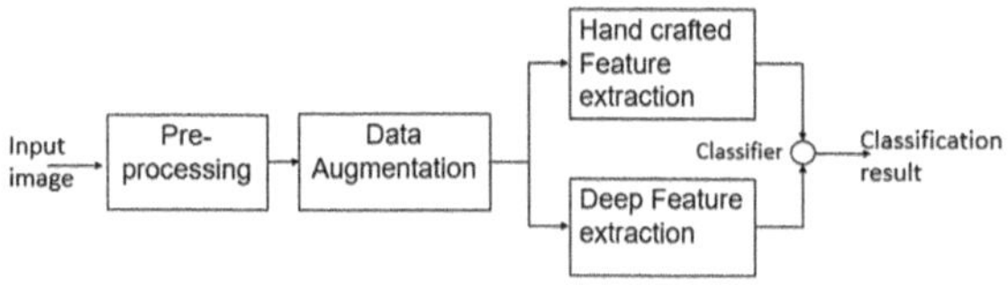

Fig. 2. High level diagram of the proposed framework.

3.1 Pre-processing

Currently there is no standard protocol available for capturing and transmitting skin lesion images. Skin lesion images are coming from entirely different set of conditions like varying illuminance, different capturing devices and different angles of capturing. Here we are more concerned about the changes in the illuminant. It will affect the color appearance of the lesion images and thus the accuracy of the system. To overcome this, we are applying color constancy algorithms on the input images. There are several approaches that are available in the literature like Gray world, Max_RGB, Shades of Gray [16] etc., and we obtained promising results with the Gray world method.

3.2 Data Augmentation

It is a good practice to do data augmentation to improve the performance of deep neural networks especially when we have a small amount of training data [17]. Data augmentation can improve the robustness of the deep neural network. The common methods for data augmentation include various geometric transformations, cropping and flipping. In our approach, we are more interested in orientation changes and cropping because, as we mentioned earlier, there is no protocol available for capturing and transmitting the skin lesion images, hence different clinicians will use different angles for capturing the image. We have used a combination of two crops and four rotations (45, 90, 135 and 180°), hence for each input image we have generated eight augmented versions. The cropping rectangles are the largest inner rectangles ensuring that all pixels belong to the original image. Finally, we performed a scaling to obtain square of size 225×225 since our CNN requires a square image as input.

3.3 Feature Extraction

Feature extraction is the core part of any classification problem. As we mentioned earlier most of the hybrid-methods are using a complex hand-crafted feature with deep network. In our approach, we are using SIFT descriptor as our hand-crafted feature and ResNet for extracting deep features. We have selected SIFT because it is a proven and well known descriptor in computer vision applications, which is invariant to most of the image transformations [18]. As we mentioned earlier in section A, there is no standard protocol for skin lesion image acquisition. Hence we will receive skin lesion images with different orientation and scaling. That is why we decided to extract the dominant features using SIFT, which is also invariant to orientation and scaling.

Another fact behind choosing SIFT is the computation cost. SIFT algorithm is already used in many real-time applications [19, 20]. However, we do not have any computational time matrix related to the other hand-crafted features for a comparison. The most important fact is that the hybrid approach with SIFT and deep features is unique in melanoma detection.

While using SIFT as a feature set, we faced the problem of feature set reduction. It was a challenging problem because each SIFT descriptor has 128 features and for each image we got an average of 125 SIFT descriptors. We approached this problem with two different methods for feature set reduction. The first approach was using a SIFT match to obtain the best features and in the second approach we have used the bag of words method. In the first method, we used a SIFT matching [21] of the original image with its corresponding augmented images to obtain the matching descriptor. From the matching descriptors, we have chosen top twenty features based on their scores. In the second approach, we have used the standard bag of words using k-means [22]. We grouped the descriptors into different number bags (10, 20, 50 and 100) and we found that twenty bags gives the best performance while combined with deep features.

In deep learning, more network depth is a desirable feature, but deeper networks are difficult to train. The major problem with a deeper network is the accuracy saturation and then a rapid degradation, and adding more layers will end up with a higher training error [15].

The deep residual network also known as ResNet is made up of building blocks known as residual blocks as shown in Fig. 3 and each block contains several convolutional layers, batch normalization layer and ReLU layers. The residual block will allow to bypass few convolution layers at a time [23]. Therefore, ResNet is capable of overcoming the limitation of other deep networks by adding the shortcut connections, hence, we decided to choose the ResNet for extracting deep features. We have used a pre-trained ResNet model to perform the classification, which was generated using the ImageNet ILSVRC challenge data [23]. The last layer of this network has 2048 features extracted from the input image and we tap these features for doing the classification.

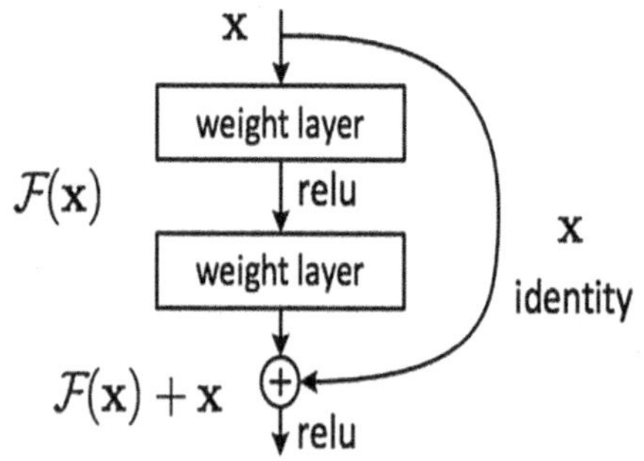

Fig. 3. Fundamental building block of a residual network [23].

3.4 Classification

Since we have the features set from both the deep network and the SIFT descriptors, here we will combine both of the features and feed into the classifier. Our case is a binary classification, melanoma or not. We explored SVM and RusBoost [24] algorithms to generate the classifier. We will discuss the performance difference between these two in the results section.

3.5 Comparison with State of the Art

In this section, we are doing a comparison of state of the art method [25] with ours. In general, our approach is a canonical approach with all necessary modules while the other method has some extra modules for additional processing. The major differences are:

- We are using a standalone CNN while the other method is based on an ensemble of CNNs.

- The authors are using age and sex information with the features extracted from the input image for prediction but we are using only the input image.
- They are using more complex pre-processing, we are using relatively simple pre-processing.

4 Experiments

We have used MatConvNet tool box [26] for extracting the deep features and used vlfeet SIFT library [27] for extracting the SIFT descriptors. We performed a number of experiments to understand the impact of combining SIFT features with deep features. For that, first, we performed training and evaluation only using the deep features, then only using the SIFT features and finally we have done the hybrid evaluation using both the deep and the SIFT descriptors. We have also done all the above experiments with both SVM and RusBoost classifiers. Initially we had 2000 images that are augmented into 16000 images using the combinations of two crops and four orientations. We divided the entire data set into ten groups and performed a cross validation by using nine groups for the training and one for the evaluation.

5 Results and Evaluation

The classification results are validated based on the accuracy values. Table 1 summarizes the results for deep features plus SIFT features, deep features only, SIFT features only and state of the art result. We can observe that the result from deep only features with the SVM classifier is giving the highest accuracy and SIFT only features with the

Table 1. Average Results After Cross-Validation using 20 BOW SIFT

	Deep + SIFT	Deep Only	SIFT Only	State of the art (K. Matsunaga et al.)
SVM	Sn: 0.3435	Sn: 0.3063	Sn: 0.1029	NA
	Sp: 0.9113	Sp: 0.9568	Sp: 0.9434	Sp: 0.851
	Acc: 0.8041	Acc: 0.8324	Acc: 0.7849	Acc: 0.828
	Pr: 0.4592	Pr: 0.6022	Pr: 0.2989	NA
	Rc: 0.3435	Rc: 0.3063	Rc: 0.1029	NA
	MCC: 0.2811	MCC: 0.3442	MCC: 0.0738	NA
	Fs: 0.3812	Fs: 0.3972	Fs: 0.1450	NA
RUSBoost	Sn: 0.4536	Sn: 0.4408	Sn: 0.6364	NA
	Sp: 0.9080	Sp: 0.9182	Sp: 0.5981	NA
	Acc: 0.8217	Acc: 0.8266	Acc: 0.6044	NA
	Pr: 0.5137	Pr: 0.5321	Pr: 0.2383	NA
	Rc: 0.4536	Rc: 0.4408	Rc: 0.6364	NA
	MCC: 0.3723	MCC: 0.3784	MCC: 0.1753	NA
	Fs: 0.4681	Fs: 0.4678	Fs: 0.3467	NA

RusBoost classifier is giving the lowest among all. From the result it is clear that, combining SIFT with deep features is decreasing the accuracy around 2% in the SVM classifier and remains unchanged with the RusBoost classifier. We can derive the reason for this behavior from the SIFT only and deep only results where the accuracies are 0.7849 and 0.8324 respectively as follows: while combining both, we are getting an accuracy approximately equal to the average of both values 0.8041. In the case of RusBoost classifier, we are getting the result (0.821) almost same as the highest (0.8266) in the combination, which is the deep feature only, instead of the average value. The important fact is that with the deep only features we achieved an accuracy (0.8324) greater than that of the state of the art method (0.828) (Fig. 4).

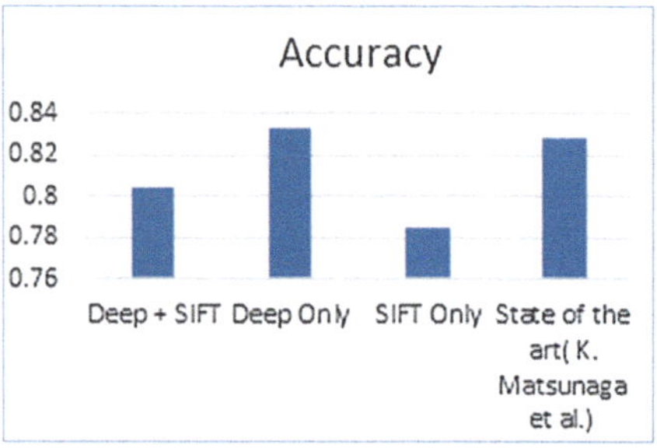

Fig. 4. The comparison result of SVM based. We can see that our deep only method has higher accuracy than state of the art.

We also measured the speed of our processing for a single image, and the deep only approach is taking 0.372 s average processing time for evaluating and with the SIFT it is taking 0.7945 s in the average. These are the measurements taken using Matlab R2017_a, on a system with a 12 core Intel Xeon processor with 64 GB RAM without using GPU acceleration.

6 Contribution

Through this experiment we want to analyze that, whether combining the SIFT features with the deep features will make any improvement in the final result. We are also interested in comparing our results with state of the art results. We achieved this using the limited number of data from the ISIC challenge 2017 using data augmentation, ResNet, and SIFT with SVM and RUSBoost classifiers.

7 Conclusion and Future Work

From the results discussed above, we can come to the main conclusion that combining SIFT with deep feature did not improve the results. For the case of the SVM, the accuracy reduced around 2% after combining the deep features with the SIFT features. We are considering this work as an initial mile stone in this area where a combination of the

SIFT features and deep features are applicable. However still we can see some unexplored area where future researchers can work: the first area is the feature reduction method. We tried only two, still there exists few other methods to explore. The other opportunity is to explore the various feature selection methods and their combinations [28]. Also, we can explore other classifiers like Decision Trees [29], Random Forest [30], Symmetric Uncertainty Feature Selector [31], etc.

We can observe the processing time is very small and we can improve it by using low level programming languages (C++) with GPU acceleration. We could not make a comparison of processing time because of unavailability of a standard processing time. We believe that this could be another opportunity as future work, to improve the speed along with accuracy to achieve a real-time scanning.

The next important finding is that, the performance of our deep only feature with SVM classifier out performs state of the art. We achieved this by using rather simpler canonical approach than a complex approach. They also used age and sex information along with the images that is another future work. Moreover, we need a larger data set to do a perfect training and evaluation because all the available deep networks are trained with approximately one million images from ImageNet [32].

We are not doing any segmentation of the skin lesion as a part of the pre-processing step. We believe that our results will improve if we can integrate some reliable segmentation method as stated in [33]. Some studies show that pattern recognition yields better results over other approaches [34], hence the researches in future can emphasize on some other standard and simpler hand-crafted features to uncover the pattern from the skin lesion images instead of the SIFT descriptor.

References

1. Melanoma http://www.skincancer.org/skin-cancer-information/melanoma. Accessed 16 Oct 2017
2. Larsen, K.: Cancer in Norway (2010)
3. Oliveira, R.B., Filho, M.E., Ma, Z., Papa, J.P., Pereira, A.S., Tavares, J.M.R.S.: Computational methods for the image segmentation of pigmented skin lesions: a review. Comput. Methods Programs Biomed. **131**, 127–141 (2016)
4. Wurm, E.M., Peter Soyer, H.: Scanning for melanoma. Aust. Prescr. **33**(5), 150–155 (2010)
5. Shaw, H.M., Rigel, D.S., Friedman, R.J., Mccarthy, W.H., Kopf, A.W.: Early diagnosis of cutaneous melanoma revisiting the ABCD criteria. JAMA **292**(22), 2771–2776 (2004)
6. ISIC 2017: Skin Lesion Analysis Towards Melanoma Detection. https://challenge.kitware.com/#challenge/n/ISIC_2017%3A_Skin_Lesion_Analysis_Towards_Melanoma_Detection. Accessed: 28 Nov 2017
7. Majtner, T., Yildirim-Yayilgan, S., Hardeberg, J.Y.: Combining deep learning and hand-crafted features for skin lesion classification. In: 2016 6th International Conference on Image Processing Theory, Tools Application IPTA 2016 (2017)
8. Yuan, Y., Chao, M., Lo, Y.: Automatic skin lesion segmentation using deep fully convolutional networks with jaccard distance **62**(c), 1–11 (2017)
9. Menegola, M.F., Pires, R., Bittencourt, F.V., Avila, S., Valle, E.: Knowledge transfer for melanoma screening with deep learning. In: Proceedings of the - International Symposium Biomedcal Imaging no October, pp. 297–300 (2017)

10. Esteva, et al.: Dermatologist-level classification of skin cancer with deep neural networks. Nature **542**(7639), 115–118 (2017)
11. Paul, R., et al.: Deep feature transfer learning in combination with traditional features predicts survival among patients with lung adenocarcinoma. Tomogr. J. Imaging Res. **2**(4), 388–395 (2016)
12. Bar, Y., Diamant, I., Wolf, L., Lieberman, S., Konen, E., Greenspan, H.: Chest pathology detection using deep learning with non-medical training. In: 2015 IEEE 12th International Symposium on Biomedical Imaging (ISBI), pp. 294–297 (2015)
13. van Ginneken, A.A.A.S., Jacobs, C., Ciompi, F.: Off-the-shelf convolutional neural network features for pulmonary nodule detection in computed tomography scans. In: 2015 IEEE 12th International Symposium on Biomedical Imaging (ISBI), pp. 286–289 (2015)
14. Lowe, G.: Object recognition from local scale-invariant features. In: Proceedings of the Seventh IEEE International Conference on Computer Vision, vol. 2, pp. 1150–1157 (1999)
15. Wu, S., Zhong, S., Liu, Y.: Deep residual learning for image steganalysis. Multimed. Tools Appl. 1–17 (2017)
16. Celebi, M.E., Mendonca, T., Marques, J.S.: Dermoscopy Image Analysis [Book review]. IEEE Trans. Med. Imaging **35**(4), 1147–1148 (2016)
17. Kumar, J.K., Lyndon, D., Fulham, M., Feng, D.: An ensemble of fine-tuned convolutional neural networks for medical image classification. IEEE J. Biomed. Heal. Informatics **21**(1), 31–40 (2017)
18. Ke, Y., Sukthankar, R.: PCA-SIFT: a more distinctive representation for local image descriptors. In: Proceedings of the 2004 IEEE Computer Society Conference on Computer Vision and Pattern Recognition. CVPR 2004, vol. 2, p. II-506-II-513 Vol.2 (2004)
19. Dardas, N.H., Georganas, N.D.: Real-time hand gesture detection and recognition using bag-of-features and support vector machine techniques. IEEE Trans. Instrum. Meas. **60**(11), 3592–3607 (2011)
20. Lalonde, M., Byrns, D., Gagnon, L., Teasdale, N., Laurendeau, D.: Real-time eye blink detection with GPU-based SIFT tracking. In: Fourth Canadian Conference on Computer and Robot Vision. CRV 2007, pp. 481–487 (2007)
21. Lowe, G.: Distinctive image features from. Int. J. Comput. Vis. **60**(2), 91–110 (2004)
22. Venegas-Barrera, C.S., Manjarrez, J.: Visual categorization with bags of keypoints. Rev. Mex. Biodivers. **82**(1), 179–191 (2011)
23. He, K., Zhang, X., Ren, S., Sun, J.: Deep residual learning for image recognition. In: 2016 IEEE Conference on Computer Vision and Pattern Recognition (CVPR), pp. 770–778 (2016)
24. Seiffert, C., Khoshgoftaar, T.M., Van Hulse, J., Napolitano, A.: RUSBoost: a hybrid approach to alleviating class imbalance. IEEE Trans. Syst. Man, Cybern. - Part A Syst. Humans 40(1), 185–197 (2010)
25. Matsunaga, K., Hamada, A., Minagawa, A., Koga, H.: Image classification of melanoma, nevus and seborrheic keratosis by deep neural network ensemble. 1 Isic 2017, pp. 2–5 (2017)
26. Vedaldi, A., Lenc, K.: MatConvNet - convolutional neural networks for MATLAB. In: Proceeding of the ACM International Conference on Multimedia (2015)
27. SIFT detector and descriptor. http://www.vlfeat.org/overview/sift.html. Accessed 29 Nov 2017
28. Chen, Y., Lin, C.: Combining SVMs with various feature selection strategies. Featur. Extr. **324**(1), 315–324 (2006)
29. Quinlan, J.R.: Induction of decision trees. Mach. Learn. **1**(1), 81–106 (1986)
30. Ho, T.K.: The random subspace method for constructing decision forests. IEEE Trans. Pattern Anal. Mach. Intell. **20**(8), 832–844 (1998)

31. Yu, L., Liu, H.: Feature selection for high-dimensional data: a fast correlation-based filter solution. In: International Conference on Machine Learning, pp. 1–8 (2003)
32. Deng, J., Dong, W., Socher, R., Li, L.-J., Li, K., Fei-Fei, L.: ImageNet: a large-scale hierarchical image database. In: 2009 IEEE Conference on Computer Vision and Pattern Recognition, pp. 248–255 (2009)
33. Codella, N., et al.: Deep learning ensembles for melanoma recognition in dermoscopy images. **61**(4), 1–28 (2016)
34. Carli, P., et al.: Pattern analysis, not simplified algorithms, is the most reliable method for teaching dermoscopy for melanoma diagnosis to residents in dermatology. Br. J. Dermatol. **148**(5), 981–984 (2003)

Deep Learning in Classifying Depth of Anesthesia (DoA)

Mohamed H. AlMeer[1(✉)] and Maysam F. Abbod[2]

[1] Computer Science and Engineering Department, College of Engineering, Qatar University, POB 2713 Doha, Qatar
almeer@qu.edu.qa

[2] Department of Electronic and Computer Engineering, Brunel University London, Uxbridge, UB8 3PH, UK
Maysam.Abbod@brunel.ac.uk

Abstract. This present study is what we think is one of the first studies to apply Deep Learning to learn depth of anesthesia (DoA) levels based solely on the raw EEG signal from a single channel (electrode) originated from many subjects under full anesthesia. The application of Deep Neural Networks to detect levels of Anesthesia from Electroencephalogram (EEG) is relatively new field and has not been addressed extensively in current researches as done with other fields. The peculiarities of the study emerges from not using any type of pre-processing at all which is usually done to the EEG signal in order to filter it or have it in better shape, but rather accept the signal in its raw nature. This could make the study a peculiar, especially with using new development tool that seldom has been used in deep learning which is the DeepLEarning4J (DL4J), the java programming environment platform made easy and tailored for deep neural network learning purposes. Results up to 97% in detecting two levels of Anesthesia have been reported successfully.

Keywords: Deep learning · DeepLearning4J · Depth of anesthesia · DoA
Neural networks

1 Introduction

Recently electrical brain signals have been researched extensively to serve different applications and especially the application in biomedical engineering. The Brain Computer Interface (BCI) has been designed around brain signals and it connects computers to human mind and translates his intention to commands that is used to communicate with other devices. A major benefit of BCI is the assistance it provides for disabled people to easily communicate with other humans. Besides that, biomedical applications such as the diagnosis of brain related disease such as Alzheimer, Epilepsy, and the severely affected brains which resulted from traumas or those leading to Nicoma. In addition to all that, BCIs has succeeded recently to effectively assist as one of the biometric Identification methods.

K. Arai et al. (Eds.): IntelliSys 2018, AISC 868, pp. 160–169, 2019.
https://doi.org/10.1007/978-3-030-01054-6_11

Electroencephalography (EEG) is a recording of brain low voltage signals emerging from currents flowing within the brain neurons and gives the unique reference for brain electrical activities which can be measured and recorded. That EEG signals contains considerable amount of information related to time and frequency classified into four different bands; Beta (13 to 30 Hz), Alpha (8 to 13 Hz), Theta (4 to 8 Hz), and Delta (0.5 to 4 Hz). The human physical and conditions can be identified from each band giving a unique feature for such conditions, as each band reflects differently to the type of physical stimulus.

Nevertheless, a major challenge in detecting (EEG) signal level arises due to the fact that brain signals come in very low amplitude. Other activities done by the patients such as eye blinks, muscular movements, teeth movements, and even heart beats could interfere with the EEG signal and introduces considerable distortion. The EEG signals needs to be processed in order to obtain appropriate features, hence it is normally analyzed by time domain algorithms, or frequency domain algorithms, beside time-frequency processing algorithms. But the frequency content of those bands is more informative and hence more often used in EEG analysis relying on the Fast Fourier Transform (FFT) or other similar types of transform [1].

Many techniques has been proposed by researchers to clean EEG signal which diverges from temporal filtering [2, 3], spatial filtering [4–6], feature extraction [7, 25], and selection [8–11], besides dimensionality reduction [12–14]. Power Spectral Density (PSD) is one of the dominant feature extraction technique currently and extensively used in EEG classification. Self-organizing Maps (SOM), as well as correlation [16] and entropy [17], beside Support Vector Machines [15, 18, 19], are considered some of the statistical feature extraction methods that were successfully been used in EEG preprocessing.

This paper presents a novel EEG-Anesthesia Level recognition algorithm to overcome this constraint by fully utilizing a large Deep Neural Networks (DNN) when its hyperparameters carefully tailored, to give the best performance. In this work, a supervised training method was adopted to preliminary training of each layer, then use supervised training method to fine-tune of the whole network. Finally, pattern classification is implemented by SoftMax classifier. The mentioned model accepts the raw EEG as input with complete disregard to any feature engineering solutions. When fully trained, the DNN is tested on 7 out of 23 different subjects collected from operation rooms EEG recorder collecting. The results showed that its speed while training reached up to 15 min with a round 10K training epochs has finally attained an accuracy of 97% in identifying 100,000 testing data samples covering the 22 patients but with special focus on 7 of them.

After discussing the Deep Learning concepts and approaches to EEG classification and deep neural networks in Sect. 2, we present some related works in Sect. 3, although there were very few which combines DoA level prediction and Deep Learning approach. We present the methodology of Feed Forward Deep Neural Networks which operates on different activation functions in Sect. 4. In Sect. 5, we present the material used for this study; while Sect. 6 shows the results. And finally, we conclude the work with discussion on future directions.

2 Deep Learning

Deep Machine Learning is based on the concept of computational models used to represent information with similar characteristics close to that of the human brain. It tries to model the high-level abstractions contained in data. The "Deep Learning" topic, the hottest in the artificial intelligence and machine learning techniques, is one of the recently gained the most of the researcher's attentions and currently and widely used in solving many engineering problems. Those solutions diversify from those related to computer vision [20], speech recognition [22], and even in natural language processing [21]. Basically, is known that the Deep Learning is considered a hierarchical structural that has the capability of extracting advanced level features from lower ones constructed within a multi-layer network and hence overcomes the traditional problem faced by shallow neural networks.

Based on neural networks, the Deep Learning (DL) is a machine learning network topology that tries to model high level abstractions contained in data. Different from older learning algorithms or so called shallow learning, but now deep learning can process ultimately huge numbers of data using many layers containing many neurons. In detail the DNN comprises of layers of nonlinear processing units or nodes called neurons. These neurons change their parameters during learning process to reach best fit. A deep neural network consists of an input layer, output layer and multiple hidden layers in between. Now each layer processes the output from the previous layer and deliver to the next layer. The layers memorize low-level features up to high-level features embedded in data as layers go deeper and deeper. So, when further we dive into the network, the more complex features the network can represent and memorize. Nevertheless, in order to find the best network to do such a good work, many variations of hidden layer configurations beside other learning parameters need to be tested.

Many types of activation function for the neuron are employed. This makes it another axis of freedom to be considered during the creation of a deep neural network. Examples of activation functions are *ReLU* (Rectified Linear Unit), the Logistic, the *TanH*, and the *SoftExponential*. The output layer for classification of multiple classes normally uses the *softmax* function for the activation. The function that learns the weight vector is called the optimizer function. A popular optimizer is *SGD* (Stochastic Gradient Descent) is typically used in training. The training dataset is fed into the network in batches. The process of passing of each data in the training dataset is called an epoch. In training a deep neural network, the optimizer, number of epochs, and batch size are parameters to be considered and makes the great difference in the final layer (*softmax* layer).

It has been noted that Deep Learning not yet been widely used in detecting Anesthetic levels although of its early applications covered most of bioengineering applications. Only few studies pinpointed the deep learning importance in EEG-based Anesthetic level detection.

3 Related Works

Underneath the deep learning model, we can find deep neural networks, convolutional neural networks, beside recurrent neural networks. For our classification of Anesthesia level using EEG signal the deep neural network (DNN) is successfully applied. The first researches combining EEG signals and DL, has started with classifying brain's motor activity [23], then brain-computer interface [24, 26], in addition to BCI using motor imagery [25], and more. Although, Anesthetic level detection in EEG is studied extensively, little research has been done to implement end-to-end detection and classification networks. In this thesis, Deep Neural Networks (DNNs) are used to detect and classify the EEG data for possible Anesthetic levels for patients under surgery, and their further classification into two classes (Wake and Anesthetized).

Artificial neural network (ANN) based detection of anesthetic levels have been researched by several researchers. Watt [27] uses a three-layered feed forward neural network for detection of the spectral signatures within EEG recording giving three distinct levels of anesthesia. He claimed the overall accuracy rate obtained is 77%. Krikic [28] suggested a new method which uses spectral entropy and embedded eigen-spectrum features. He used a type of neural network namely radial basis pattern classifier to reach overall accuracy rate of 98%.

4 Methodology

The general form of computation done in neurons in any neural network is following the next two operations: to combine linearly inputs

$$y = w0 + \sum_{i=1}^{n} wi \ . \ xi \tag{1}$$

And a non-linear transformation

$$z = f(x) \tag{2}$$

Using some of the non-linear activation functions, usually two activation functions are widely used in classical ANN framework and they are the sigmoidal function beside the tanh function as shown next, respectively:

$$f(x) = \frac{1}{1 + e^{-x}}. \tag{3}$$

$$f(x) = \frac{e^{x} - e^{-x}}{e^{x} + e^{-x}} \tag{4}$$

We can have fully connected layers of those neurons following previous paradigm, but conventional ANN paradigm shows limit to the approximations and conversion. Based on that researchers have proposed some alteration on the number of hidden layers as to be deeper beside some presentation of newer activation functions. Loss function is also another newer concept in Deep Learning with its Deep Neural Networks.

Activation functions such as *Rectified Linear Units (RLU)* and *SoftMax* are newly introduced with deep Neural networks. The next shows their formulas, respectively:

$$f(x) = max(0; x) \tag{5}$$

$$f(x) = \frac{e^{xi}}{\sum_{j=1}^{n} e^{xj}} \tag{6}$$

The weights in the ANN is adjusted during the training phase until the correct output vector is generated from a given input. This operation continues until the global error is minimized to an acceptable value. It is worth noting that we use the Feedforward Neural Network (FNN) to refer to the a more basic ANN architecture used. The neurons are connected forward in series way and its activation propagates in unidirectional from the input layer to the output layer [1].

5 Materials

It has been known that there is a strong relation between Bispectral Index and the EEG graphic features and found that the anesthetic features are linearly correlated with the BIS reading for all levels of anesthesia. As many literatures has approached the classification of anesthetic levels, only the next levels illustrated in Table 1 are of concern to us. Human experts have assessed the anesthetic levels and reached 5 classifications labels; Awake state, light anesthesia, moderate anesthesia, deep anesthesia, and near suppression level. As noted earlier, we are implementing an artificial deep neural model to extract most dominant features from EEG signal in supervised manner that were classified by human experts in prior. The resulted DNN would then be used to classify test sets of data. In this work, the raw EEG data were not being processed by any preprocessing algorithms or filter intentionally done, therefore bandpass filters were not used beside down or up sampling or even baseline removals. The reason behind that is two folded, one for leaving the processing light and fast to work noticeably fast for real time applications, while the other is to show the effectiveness of the deep learning when used in processing raw EEG data.

Table 1. Test results conducted on all subject's data but distributed based on 65% for training and 35% for testing. Batch size are different, but the data epochs is 3610 X 128

Batch size	Accuracy	Precision	Recall	F1 score
<1000	**100%**	100%	100%	100%
1000	**92%**	85%	73%	77%
3610	**90%**	87%	86%	87%

The EEG data generated from the 22 subjects were taken under full Anesthesia for different surgery time spanning from ½ h to 2 h. Although noise and deformation of the EEG were quite high with the beginning of the surgery giving unclear EEG signal patterns, but some were in good shape and were successfully extracted for training the DNN model. Finally, the resulted good records that span equally between the two anesthetic levels were only witnessed in only 7 subjects in which we based our training sets on.

For each experiment only one second epoch was extracted (128 samples). And to investigate performance of the proposed DNN model for DoA prediction, we created 7 test data that were showing quite clear Awaken epochs and plenty of Anesthetized epochs. For each subject (patient) 2 sets of datasets create, i.e., one for Awaken state and the other for Anesthetized state. Two schemes to organize training datasets were also followed; one to construct data set for a particular subject for sole training; while the others work as test sets; and in another scheme, all datasets except one are used for training leaving the one set for testing.

Totally 121,000 epochs were obtained from 7 subjects and later the other 15 subjects were taken some of their epochs for testing. We have to bear in mind that, the datasets were measuring different brain activities for different stages of hypnosis stages.

Training sets were intentionally extracted to fill 90% of the total available sets while the remaining 10% epochs were selected for testing. Anyway, we further re-distributed trainig_to_testing datasets to balance different configuration that we thought would be more helpful in determining the best distribution raises to the best DNN performance.

We have used a so-called comma-separated values (CSV) data, the recorded data organized to, in this work (series of EEG data till 128 samples per epoch). It was exported from MATLAB program responsible to open the long EEG file and cut to small epochs. Then the result is de-normalized on multiple sets of data then joined together resulting in the CSV data that the DeepLearning4J the JAVA programming uses for all neural network computation.

6 Results

The complete data of five subjects have been used to train and test the network with. In a total of 3610 epochs each with 128 readings giving a total of 462,080 data readings. Then data has been arranged such that the one which gives a specific class are to be grouped together then comes the other class group. In a total of two classes cover the two anesthetic levels, Waken and Anesthetized.

Three different configuration of data training has been implemented: the *train_and_test_all* in which 65% used for training and remaining 35% is kept for testing. Another configuration is the train one and test all. Also, *train_all_except_one* for test. And finally, *train_one_and_test_on_each* other alone. Next table shows results.

Table 1 shows the accuracy of the five subjects validation results emerged from DeepLearning4J package when all are used for train and subsequent test. The percentage of the taken epochs for train to the test is 65% to 35%. Accuracy up to 100% reached when chosen a batch size of less than 1000. Table 2 shows validation results for the configuration of taking one subject for training and leaving the others for testing. This one has resulted in an accuracy up to 93%. The reason behind other lower accuracies is the EEG subject difference and the abundance of clean signals prior to and during anesthesia.

Table 2. Test results on the 5 subjects data but taken as train one and test one. The training data size is different according to subject trained

Epochs	Trained	Tested	Accuracy
235 Class 0	P1	P2	85%
235 Class 1		P6	71%
		P7	**93%**
		P19	86%
141 Class 0	P2	P1	68%
558 Class 1		P6	44%
		P7	**74%**
		P19	69%
157 Class 0	P6	P1	40%
1032 Class 1		P2	61%
		P7	68%
		P19	**75%**
290 Class 0	P7	P1	**85%**
290 Class 1		P2	84%
		P6	72%
		P19	79%
60 Class 0	P19	P1	**35%**
623 Class 1		P2	38%
		P6	38%
		P7	38%

Table 3. Test results on the same subjects data but taken as train one and test all. Readings are given for different batch sizes. The training data size is different according to subject trained

Trained On	Tested	Accuracy	Precision	Recall	F1 score
P1	P (2, 6, 7, 19)	78%	76%	76%	76%
		71%	67%	76%	66%
		82%	50%	82%	90%
		79%	50%	79%	88%
P2	P (1, 6, 7, 19)	**90%**	85%	83%	84%
		87%	87%	83%	85%
		82%	83%	86%	82%
		69%	50%	69%	82%
P6	P (1, 2, 7, 19)	**82%**	81%	74%	76%
		79%	80%	74%	75%
		67%	71%	76%	66%
		48%	50%	48%	65%
P7	P (1, 2, 6, 19)	**87%**	84%	73%	77%
		85%	84%	73%	76%
		83%	84%	73%	76%
		78%	82%	73%	74%
P19	P (1, 2, 6, 7)	**80%**	80%	67%	69%
		77%	80%	67%	68%
		72%	78%	67%	66%
		64%	73%	66%	62%

Tables 3 and 4 depict the other two training configurations where one subject is taken for training and the others are grouped in one set for testing and vice versa. An accuracy of 90% and 99% for the consequent two methods has been reached. The result indicators in the following tables are based on the next interpretations:

(1) Accuracy: is defined as the percentage of epochs that were correctly identified by our model.
(2) Precision: is the number of true positives divided by the number of true positives and false positives.
(3) Recall: The number of true positives divided by the number of true positives and the number of false negatives.
(4) F1 Score - Weighted average of precision and recall.

Table 4. Test results on the same subject's data but taken as train all except one then test that one. Readings given for different batch sizes. The training data size is different according to subject trained

Trained On	Tested	Accuracy	Precision	Recall	F1 score
P (2, 6, 7, 19)	P1	**82%**	86%	82%	81%
		79%	82%	82%	79%
		73%	72%	82%	70%
P (1, 6, 7, 19)	P2	**99%**	99%	99%	99%
		99%	99%	99%	99%
		99%	99%	99%	99%
P (1, 2, 7, 19)	P6	**85%**	6*%	67%	67%
		82%	73%	67%	69%
		75%	73%	66%	67%
P (1, 2, 6, 19)	P7	89%	89%	89%	89%
		90%	87%	88%	88%
		92%	62%	86%	67%
		91%	50%	91%	95%
P (1, 2, 6, 7)	P19	**96%**	89%	91%	90%
		96%	92%	91%	93%
		96%	94%	91%	92%
		96%	94%	91%	93%

Here the Accuracy measures the model overall performance, while the Precision, Recall and F1 measure a model's relevance.

For Comparison purposes another four references has been used for benchmarking. In [29] studied the use of neural networks in the classification of anesthesia depth level using recurrent neural networks (shallow neural networks) and reached an accuracy of 99.6%. Another work [30] has used Elman Neural Networks and Multilayer Perceptron MLP to reach an accuracy of 95% and 99%, respectively. Both of those works have used three Anesthetic levels of low, medium, and high. A combined wavelet and neural network classifiers has been studied in [31] with an accuracy of classification reaching 96.6%. All those works have reported results based on three level of anesthesia and inclusively using features extracted from the EEG signal for trained data, beside using shallow neural networks. Nevertheless, our works have used a deep neural network based on deep learning operating exclusively on raw EEG signals, classifying only two anesthetic levels and reported a maximum accuracy of 97%.

7 Conclusion

This experimented study demonstrated that the DL training with EEG data set is able to detect two levels of anesthesia in an untrained (unseen) data set corresponding to new subjects with high accuracy up to 97%. Experiments have shown that multiple factors increase the training and the recognition accuracies, such as increasing training data, having more subject's diversities contributing those data, beside choosing cleaner training epochs, in addition to tuning the hyperparameters of the network and in particularly the training batch size.

References

1. Yuan, L., Cao, J.: Patients' EEG data analysis via spectrogram image with a convolution neural network. In: International Conference on Intelligent Decision Technologies, pp. 13–21. Springer, Cham (2017)
2. Thomas, K.P., Cuntai, G., Lau, C.T., Vinod, A.P., Kai Keng, A.: A new discriminative common spatial pattern method for motor imagery brain computer interfaces. IEEE Trans. Biomed. Eng. **56**, 2730–2733 (2009)
3. Wei, Q., Wei, Z.: Binary particle swarm optimization for frequency band selection in motor imagery based brain-computer interfaces. Bio-Med. Mater. Eng. **26**, S1523–S1532 (2015)
4. Aghaei, A.S., Mahanta, M.S., Plataniotis, K.N.: Separable common spatio-spectral patterns for motor imagery BCI systems. IEEE Trans. Biomed. Eng. **63**, 15–29 (2016)
5. Li, X., Lu, X., Wang, H.: Robust common spatial patterns with sparsity. Biomed. Sig. Process. Control **26**, 52–57 (2016)
6. Wu, W., Chen, Z., Gao, X., Li, Y., Brown, E.N., Gao, S.: Probabilistic common spatial patterns for multichannel EEG analysis. IEEE Trans. Pattern Anal. Mach. Intell. **37**, 639–653 (2015)
7. Zhang, A., Yang, B., Huang, L.: Feature extraction of EEG signals using power spectral entropy. In: International Conference on BioMedical Engineering and Informatics, Sanya, pp. 435–439 (2008)
8. Sharma, A., Imoto, S., Miyano, S.: A top-r feature selection algorithm for microarray gene expression data. IEEE/ACM Trans. Comput. Biol. Bioinf. **9**, 754–764 (2012)
9. Luo, J., Feng, Z., Zhang, J., Lu, N.: Dynamic frequency feature selection based approach for classification of motor imageries. Comput. Biol. Med. **75**, 45–53 (2016)
10. Sharma, A., Paliwal, K.K., Imoto, S., Miyano, S.: A feature selection method using improved regularized linear discriminant analysis. Mach. Vis. Appl. **23**, 775–786 (2014)
11. Sharma, A., Paliwal, K.K., Imoto, S., Miyano, S.: A feature selection method using improved regularized linear discriminant analysis. Mach. Vis. Appl. **25**, 775–786 (2014)
12. Sharma, A., Paliwal, K.K.: A deterministic approach to regularized linear discriminant analysis. Neurocomputing **151**(Part 1), 207–214 (2015)
13. Sharma, A., Paliwal, K.K.: Rotational linear discriminant analysis technique for dimensionality reduction. IEEE Trans. Knowl. Data Eng. **20**, 1336–1347 (2008)

14. Sharma, A., Paliwal, K.K.: A gradient linear discriminant analysis for small sample sized problem. Neural Process. Lett. **27**, 17–24 (2008)
15. Li, X., Chen, X., Yan, Y., Wei, W., Wang, Z.J.: Classification of EEG signals using a multiple kernel learning support vector machine. Sensors **14**, 12784–12802 (2014)
16. Rocca, D.L., Campisi, P., Vegso, B., Cserti, P., Kozmann, G., Babiloni, F., et al.: Human brain distinctiveness based on EEG spectral coherence connectivity. IEEE Trans. Biomed. Eng. **61**, 2406–2412 (2014)
17. Sabeti, M., Katebi, S., Boostani, R.: Entropy and complexity measures for EEG signal classification of schizophrenic and control participants. Artif. Intell. Med. **47**, 263–274 (2009)
18. Liu, N.-H., Chiang, C.-Y., Chu, H.-C.: Recognizing the degree of human attention using EEG signals from mobile sensors. Sensors **13**, 10273–10286 (2013)
19. Brown, L., Grundlehner, B., Penders, J.: Towards wireless emotional valence detection from EEG. In: Annual International Conference of the IEEE Engineering in Medicine and Biology Society (EMBC), pp. 2188–2191 (2011)
20. Krizhevsky, A., Sutskever, I., Hinton, G.E.: ImageNet classification with deep convolutional neural networks. In: Advances in Neural Information Processing Systems, pp. 1–9 (2012)
21. Graves, A., Mohamed, A., Hinton, G.: Speech recognition with deep recurrent neural networks. In: IEEE International Conference on Acoustics, Speech and Signal Processing, (2013). https://doi.org/10.1109/icassp.2013.6638947
22. Collobert, R., Weston, J.: A unified architecture for natural language processing: deep neural networks with multitask learning. In: Proceedings of the Twenty-Fifth International Conference on Machine Learning, vol. 20, 160–167 (2008)
23. Morimoto, T., Sketch, S.: Classifying brain's motor activity via deep learning. Stanford University, CS 229 Machine Learning (2014)
24. Nurse, E., Mashford, B.S., Yepes, A.J., Kiral-Kornek, I., Harrer, S., Freestone, D.R.: Decoding EEG and LFP signals using deep learning. In: CF 2016 Proceedings of the ACM International Conference on Computing Frontiers, pp. 259–266 (2016)
25. Walker, I.: Deep convolutional neural networks for brain computer interface using motor imagery. Thesis, Department of Computing, Imperial College London (2015)
26. Zaitcev, A., Cook, G., Wei, L., Paley, M., Milne, E.: Feature extraction for BCIs based on electromagnetic source localization and multiclass Filter Bank Common Spatial Patterns. In: 2015 37th Annual International Conference of the IEEE Engineering in Medicine and Biology Society (EMBC), pp. 1773 1776 (2015)
27. Watt, R.C., Navabi, M.J., Scipione, P.J., Hameroff, S.R., Maslana, E.S.: Neural network estimation of anesthetic level using EEG spectral signatures. In: Annual International Conference of the IEEE Engineering in Medicine and Biology Society, November 1990
28. M. Krkic. S. J. Roberts, I. A. Rezek, and C. Jordan: EEG-based assessment of anesthetic-depth using neural network. In: Proceedings of the Institute of Electronics Engineering. AI Methods in Biosignal Analysis, vol. 10, pp. 1–6, April 1996
29. Srinivasan, V., Eswaran, C., Sriraam, N.: EEG based automated detection of anesthetic levels using a recurrent artificial neural network. Int. J. Bioelectromagnetism **7**, 267–270 (2005)
30. Lalitha, V., Eswaran, C.: Automated detection of anesthetic depth levels using chaotic features with artificial neural networks. J. Med. Syst. **31**(6), 445–452 (2007)
31. Benzy, V.K., Jasmin, E.A.: A combined wavelet and neural network based model for classifying depth of anaesthesia. Procedia Comput. Sci. **46**, 1610–1617 (2015)

Content Based Video Retrieval Using Convolutional Neural Network

Saeed Iqbal(✉), Adnan N Qureshi, and Awais M. Lodhi

Faculty of Information Technology, University of Central Punjab, Lahore, Pakistan
saeediqbalkhattak@gmail.com, {dr.qureshi,awais.lodhi}@ucp.edu.pk

Abstract. We present a content-based retrieval method for unconstrained video. To achieve the goal content based video retrieval we segmented the videos, detecting objects that match the human-defined interest. Object detection and classification is basic task in video analysis. For object detection we separate the foreground object from background and perform localization on each extracted frames from videos, and measure intensity histogram of 8-oriented frames and then perform haar-cascade, Gabor filter, active appearance model (AAM) and Convolutional Neural Network (CNN) algorithms get the authentic result. We used two datasets: Youtube and SegTek containing more than 2000 videos and using cluster computing to get the state-of-the-arts result of object detection and segmentations.

Keywords: Content-based image retrieval
Convolutional neural network · Active appearance model
Support vector machine · KMeans · KNN · Image classification

1 Introduction

According to the paper [1], currently the digital data has reached 281 exa-bytes. Furthermore, according to the IDC report the digital data is to be 44 times larger than in 2020 than in 2009. Currently most of the data is unstructured such as images, video, text, and musics. In information retrieval, content wise video retrieval is the basic problem. Generally user access the information from a web to use the Query-Process model. User submits a query in query-process model to access the desired information. Currently the diverse content of queries and result have been annotated in advance, and saved with time and locations. Currently video and images based retrieval has the prominent research area due to social media and interaction among different cultures (digital universe). Although retrieval of accuracy of multimedia data is increasing day by day. To retrieve multimedia, annotation is a problem to annotate multimedia content with keywords. Generally, it is difficult to accurately annotate the all type of multimedia content.

In order to resolve the mentioned problem, we proposed our framework to retrieve the multimedia content according to the user query. In query-process

K. Arai et al. (Eds.): IntelliSys 2018, AISC 868, pp. 170–186, 2019.
https://doi.org/10.1007/978-3-030-01054-6_12

model, user submit a(n) image/video using query interface, our proposed framework will returned result based on the content of the query. For video processing and analyzing several techniques and models/algorithms has been proposed. Object detection and classification are the main task in video analytics and become the starting point for other applications. These all are processed via machine learning algorithms and computationally extensive machines. To processing the huge datasets of a videos [2,3] we used machine learning algorithms which includes Neural Network, Convolutional Neural Network (CNN), Deep Neural Network (DNN) and Computer vision for object detection.

Neural Network implement the functionality and structure of biological neural networks in the form of mathematical model. Neuran is the basic unit of every Neural Network which is a simple mathematical model and NN can be seen in its mathematically form mentioned below (1):

$$y(t) = F(\sum_{i=0}^{l} w_i(t).x_i(t) + bias) \tag{1}$$

All the Inputs are multiplied with their corresponding weights. Weighted inputs and bias are added and pass to transfer function F(x). Where `y(t)` is the final value at discrete time `t`, `F` is transfer function, $x_i t$ is the initial value at discrete time `t` and `i` goes to `0.,.,.,.,m` and $w_i t$ is the weight value at discrete time. Figure 1 depicted the structure of Neural Network and Mathematical model of NN.

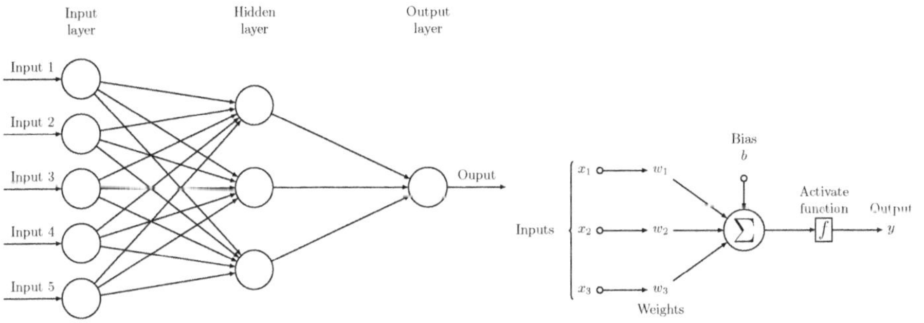

Fig. 1. Neural network architecture.

Machine learning is a type of artificial intelligence and it has Convolutional Neural Network (CNN). Basically CNN is use for the visual recognition. It is highly successful in image classification and object recognition tasks. Convolutional Neural Network is consist of neurons which have learn-able biases and weights. Initially every neuron accept some initial values (inputs), calculate a function (dot product) and then spontaneously follows it with a non-linearity. The complete network still declare a single differentiable score function from the basic image pixels on one end to class scores at the other [4].

CNN architecture build an accurate assumption on the basis of initial inputs (images), that invoke the system to encode some properties into the architecture. After that they make the forward function more efficient to implement, largely reduce amount of parameters in the network [5].

Computer Vision is another field for understanding images and gets information from them. The use of computer vision is Image search, Audio-visual recognition and object detection.

2 Background

The larger part of earlier work in visual recognition has focused on classifying images with fixed set of visual sections and progress was achieved in these struggles.

It has been noted in visual cognition literature that humans does not mainly focus their attention on whole scene by just looking them at once [6]. Rather, they focus step by step on various parts of the scenes to get suitable information.

Mechanisms are indifferent to different parts of the image/video. With the latest surge of interest in deep neural networks, attention based models have been shown to obtain promising results on many challenging tasks, along with caption generation [7], machine translation [8], game-playing and tracking [9], as well as image recognition (e.g. Street View House Numbers dataset [10]). Many of these models have employed long short-term memory (LSTM) [11] based RNNs and have shown good results in learning sequences.

In 2011, Google started a project with the name of The Google Brain to dig the theme of deep neural networks both for research and for extensive analysis of Google's products. Early work of the Google, the DistBelief, is scalable distributed training and inference system [12], they have done an extensive work and research on DistBelief including work on classification/regression (supervised learning) [13], language representation [14,15], models for image classification and object detection [16,17], video classification [18], speech recognition [19–21], on several computational device/gadgets and shared parameters or states with in all collaborating devices.

Parallelism is achieved and allow a wide variety of several approaches via modest changes in computation [22–24]. In larger deployment, TensorFlow consistently updating the parameters and take advantage of synchronization environments. Currently the achievement of TensorFlow is better and their programming model is more extensible as compare to others and it supports a broader range of hardware platform and training/testing modules. Our internal users of DistBelieft have been switched to TensorFlow [12]. Users rely on research and production of TensorFlow and the diversification of deep neural network in image classification specially on face recognition using Convolutional Neural Network.

In this paper a method to of matching statistical models of appearance to images is given. This method is used to generate synthetic images from input images by generating a model of the original image which is named as the Active Appearance Model (AAM) which is a type of shape model. This model in one way tracks the deformation in every new frame to the previous one. A model is trained on a set of given images where the main features are marked. In this paper a large amount of different post face alignment discussed and result is generated via Convolutional aggregation of local evidence (CALE) which based on a Convolutional Neural Network (CNN) architecture.

For object detection the most common dataset is COCO, the COCO (Common Objects in Context) dataset is a Multipath Network [25] for object detection which contains objects at multiple scales in its database in which the objects clutter each other hence resulting in heavy OCCLUSION. The algorithm uses a metric which results in the most precise object localization and classification. The three key modifications that the network uses are:

(1) Skip connections.
(2) Providing Context through foveal regions.
(3) Integral Loss Function to maximize accuracy.

The algorithm works in such a way that the object proposal algorithm generates candidate regions that contain the target object and then a CNN classifier classifies those proposal regions. The Multipath network, coupled with DeepMask object proposals, achieves major gains on COCO detection.

Object is vary size by size, color to color and also in multi shapes. Hassan Sadrnia et al. [26] classified the long type watermelon depending on the fruit shapes. Physical characteristics of watermelon such as mass, volume, dimensions, density, spherical coefficient and geometric mean diameter were measured. Arivazhagan et al. [27] proposed an efficient fusion of color and texture features for fruit recognition. The recognition was done by the minimum distance classifier. Objects have multi shapes and single image contains multiple objects, this study presents an algorithm for video base monitor, to detect and also maintain the count of total detected objects (vehicles). The unique features of this algorithm is to calculate approximate speed value of speed, Gaussian Mixture Model (GMM) background modeling [28], object histogram and pyramidal Lucas Kanade method [29]. According to research paper, it was possible to obtain such high accuracy of 94.5 percent, 92.4 percent and 97.4 percent even in the afternoon when we were under the influence of the shadow. It was confirmed that the accuracy was reduced at a crowding driving or low-speed driving [30].

Real time object detection and to analyses those result in real time is also a complex problem. In this paper, the researcher present a comprehensive reviews of collect data and manage traffic lights system in real time, which include manages and monitoring and giving high priority public vehicles and also government vehicles and also guide other people by giving instruction signal via electronic traffic boards. Instructions like helping the people to park their vehicles in free parking spaces with variable contents [31].

3 Proposed Framework

In our system, we used Query-Process model to retrieve the suitable result against query. User input an image/video that they want to search from the datasets. Our proposed method search that image (object or face) into dataset (video) and retrieve the time slot where that image is found and also the list of result (videos). For a particular sentence or word to search in datasets (video) user input a word or sentence and it will retrieve the time slot where it is appear in datasets (video) and list of videos (Fig. 2).

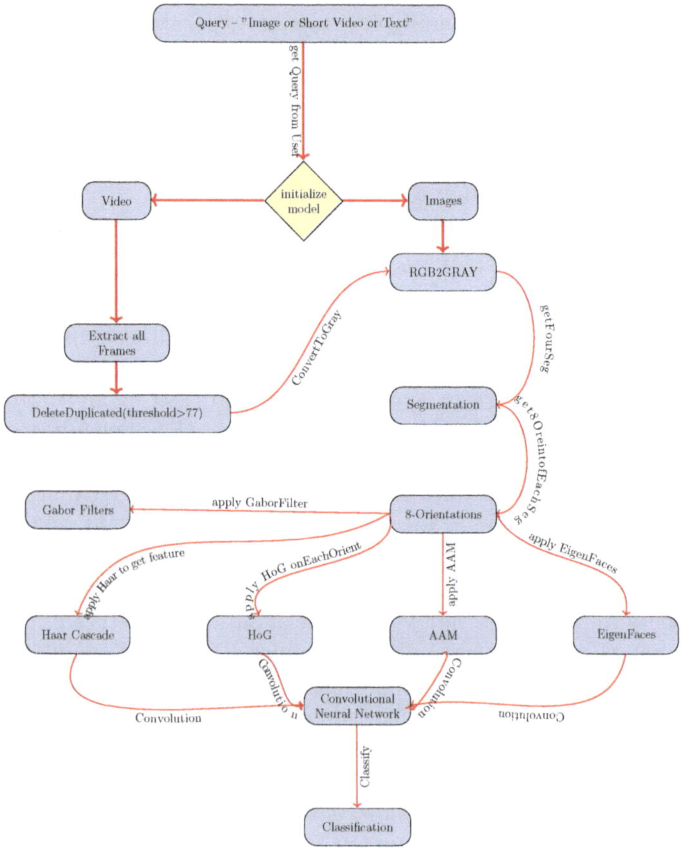

Fig. 2. Proposed architecture.

3.1 Video Processing

In video processing section, user submit a frame/short clip of video (maximum 5 s video) via query-process model. We have limit the scope (retrieve video based on faces and objects) of our paper. An image or extracted frames will be converted to gray and remove noise using filters and then segmented into four coordinates. Each segmented image is converted into eight orientation (i.e. 45, 90, 135, 180,

225, 270, 315, and 360). In each oriented image, we apply different filters, feature extraction methods and techniques, classification/clustering algorithm and Convolutional Neural Network to speed up and accurate the classification/grouping process. The details are discussed in Sect. 4.

4 Implementation Detail

We applied the same techniques as presented by other authors in literature reviews. We used two datasets: Youtube [3] and SegTek [2,32] containing more than 2000 videos and using cluster computing to get the state-of-the-arts result of object detection, segmentations and video indexing. For video indexing, we extract frames from videos and delete duplicated frames with threshold 77 or greater. In the resulted frames, events and object are the main features to indexed the videos in a datasets. In literature there are several features (color, texture, motion, edges and shapes) to retrieve videos from a datasets.

We have used HAAR Cascading, histogram of oriented gradients (HoG), active appearance model (AAM) and EigenFaces for face detection and object detection from extracted frames and retrieve video from a datasets. Figure 3 depicted the comparison of face/object detection techniques. Haar Cascading detected 13 object (face) from 30 frames, Histogram of Oriented detect 21 objects from 30, Acative Appearnace Model detect 26 objects from 30 frames, Eigenface 18 objects from 30 and gabor filter detect 21 from 30 frames. Active Appearance model give us a plausible result to achieve high accuracy. The technique that we currently using is Haar Cascading Classifier is fully trained with hundreds views/images of a particular object (in our case lips and eyes). In this technique we are using haar cascade for segmentation. We have segmented the faces using haar cascade from the image. All these features are stored for further processing/training. Same procedure is done for queries images. As discussed earlier the face image is segmented into further facial parts, i.e. left eye, right eye and mouth. In time and space complexity EigenFaces and gabor cascading give us best result as compare to other techniques.

For classification and grouping we have used different classification/clustering algorithm on a clustered environment. First we have choose a very simple classification algorithm Naive Bayes, to get the relevant video against query.

Naive Bayes algorithm classify the object on bases of feature presence and absence. For instance a human have two hands, two legs and one head, all these feature have contribution to decide regarding a human. Naive Bayes is totally based on probability model. During training and testing phase, it worked very well in real world problems. It need very less amount of training data to calculate the features.

For Naive Bayes classification, we have used a portion of a datasets (200 videos) due to limited resources (low processing power). Figure 4 depicted the progress of K-Means, KNN and SVM algorithms. We randomly increased the datasets (100 to 125 videos contains none of the object that we are searching through query) so the number of retrieved result(videos) going down. After 41

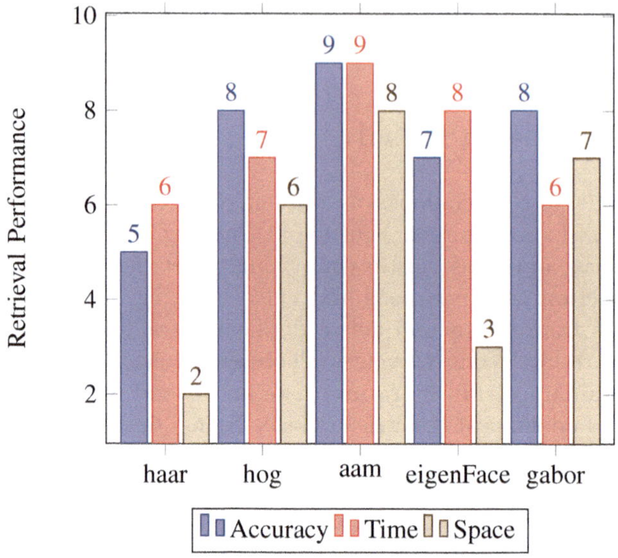

Fig. 3. Comparison of algorithms.

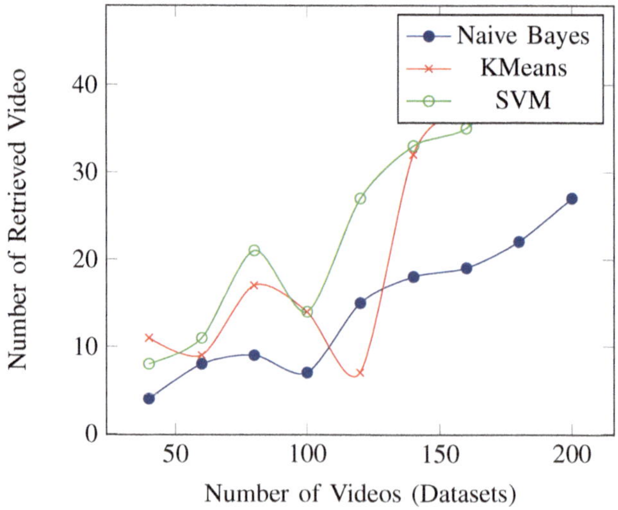

Fig. 4. Line Chart of Naive Bayes, SVM, and KMeans Algorithms.

hours (due to low processing power), we get the following confusion matrix result from Naive Bayes algorithm. In query, we upload two image (256*256) that contains car and human body respectively and upload one short video (5 sec) that contains bicycle and other objects. It returned 60 video (38 correct videos and 22 incorrect videos) from a dataset against the first image. The Second result contains 57 correct videos and 8 incorrect videos. The third result which against video, it returned 75 videos (43 correct videos and 32 incorrect) (Fig. 5).

Confusion Matrix of Naive-Bayes		Predicted			
		CAR	BODY	BICYCLE	Total
Actual	CAR	38	7	15	60
	BODY	3	57	5	65
	BICYCLE	13	19	43	75
	Total	54	83	63	200

Fig. 5. Confusion matrix of Naive Bayes Algorithm – Training.

During testing we get the following confusion matrix Fig. 6 using Naive Bayes algorithm, the result is good, but taking too much time because of low processing machinery. We get 32 car's video from the dataset out of 49 retrieved videos, 63 humans video from random dataset out of 74 videos retrieve by system and bicycle video is 54 out of 75 videos in a retrieved list.

After Naive Bayes algorithm, we move to another classification algorithm K-Nearest Neighbor, the main usage of KNN is to find the similar object in a datasets. During training it finds similar groups of K objects that are closed to the testing objects. It has weighting scheme, the nearer object is contributing more to the average than the more distant ones, such as common weighting scheme consists in giving each neighbor a weight of 1/d, where d is the distance to the neighbor. Neighbors are taken from the known class. There are three main elements:

- Labeled Object sets.
- K Value.
- Nearest neighbor Numbers.

KNN have also some issues regarding the performance of K-Nearest Neighbor (KNN). The result will be sensitive to noisy points if the value of K is too small

Confusion Matrix of Naive-Bayes – Testing		Predicted			
		CAR	BODY	BICYCLE	Total
Actual	CAR	32	5	12	49
	BODY	4	63	7	74
	BICYCLE	9	12	54	75
	Total	45	80	73	198

Fig. 6. Confusion matrix of Naive Bayes Algorithm – Testing.

and other side it will include too many other point from other classes if the K values is large. It is very easy to understand and implement.

For KNN classification, we collect some portion of datasets that contains 731 videos of different objects (human,car, bi-cycle, tractor, dog, cats and horse etc.) In query, we upload three image (256*256) that contains car, human body and bi-cycle, respectively. It returned 223 video (211 correct videos and 12 incorrect videos) from a dataset against the first image. The Second image get result which contains 223 correct videos and 23 incorrect videos. The third image result are 243 correctly classified videos and 19 wrongly classifies as depicted in Fig. 7.

Confusion Matrix of KNN		Predicted			
		CAR	BODY	BICYCLE	Total
Actual	CAR	211	4	8	223
	BODY	11	223	12	246
	BICYCLE	7	12	243	262
	Total	229	239	263	731

Fig. 7. Confusion matrix of KNN Algorithm – Training.

During testing we achieve the depicted confusion matrix Fig. 8 using KNN testing algorithm.

Confusion Matrix of KNN Testing		Predicted			
		CAR	BODY	BICYCLE	Total
Actual	CAR	229	13	19	261
	BODY	16	136	8	160
	BICYCLE	13	7	234	254
	Total	258	156	261	675

Fig. 8. Confusion matrix of KNN Algorithm – Testing.

One of the most popular unsupervised algorithm in machine learning is K-Means clustering. It clusters the data on the bases of non-similarity and similarity. Euclidean, Manhattan Distance is used for similarity measure. It divide the datasets into several cluster on the bases of similar content or features. The members of resulting groups bear statistical similarity but the identification of type of similarity is not possible through K-Means clustering. Detailed review of K-Means clustering is available in [33,34].

- **Initialize:** Randomly select the middle point of every cluster (K).
- **Repeat:** Allocate each point to the nearest center of a group.

For each repetition, we calculate N x K comparison, which determines the time complexity for one repetition cycle. In KMeans classification, we used the same dataset (705 videos) that we have used in KNN with small changes (delete duplicated videos). In query, we upload three image (256*256) that contains car, human body and bi-cycle, respectively. It returned 221 video (181 correct videos and 30 incorrect videos) from a dataset against the first image. The Second image return 257 correct videos and 30 incorrect videos. The third image result are 176 correctly classified videos and 31 wrongly classifies as shown in Fig. 9.

Following Fig. 10 shown the result of K-Means testing algorithm. During testing the system correctly classify car data to 213 out of 225 videos, human body 261 out of 285 result and 239 bicycle out of 270 results.

Support vector machine is a powerful tool for binary classification, capable of generating very fast classifier functions following a training period. There are two types of methods in machine learning, one is supervised and another is unsupervised. Supervised learning based on learn by result and unsupervised

Confusion Matrix of KMeans - Training		Predicted			
		CAR	BODY	BICYCLE	Total
Actual	CAR	181	19	11	221
	BODY	21	257	9	287
	BICYCLE	17	14	176	207
	Total	219	290	196	705

Fig. 9. Confusion matrix of KMeans – Training.

Confusion Matrix of KMeans Testing		Predicted			
		CAR	BODY	BICYCLE	Total
Actual	CAR	213	7	5	225
	BODY	11	261	13	285
	BICYCLE	14	17	239	270
	Total	238	285	257	780

Fig. 10. Confusion matrix of KMeans – Testing.

based on learn by example. Supervised learning takes input as set of training data. Support vector machine is a supervised learning technique that analyze data and identify pattern used for classification. It takes a set of input, read it and for each input desired output form [35,36]. This type of method is known as classification. The main feature of SVM is to build a hyper planes or a set of hyper planes with the help of support vectors in a higher dimension space which are mainly used for classification. It divides the space into two half space. A 'good separation' is attain by hyper planes that has the largest distance to the nearest data points [35].

For Support Vector Machines classification, we used the same dataset (705 videos) that we have used in KMeans. In query, we upload three image (256 * 256) that contains car, human body and bi-cycle respectively. It returned 221 video

(203 correct videos and 18 incorrect videos) from a dataset against the first image. The Second image return result which contains 277 correct videos and 10 incorrect videos. The third image result are 181 correctly classified videos and 16 wrongly classifies as shown in Fig. 11.

Confusion Matrix of SVM – Training		Predicted			
		CAR	BODY	BICYCLE	Total
Actual	CAR	203	7	11	221
	BODY	6	277	4	287
	BICYCLE	10	6	181	197
	Total	219	290	196	705

Fig. 11. Confusion matrix of SVM – Training.

Figure 12 depicted the result of SVM testing algorithm. During testing the system correctly classify car data to 321 out of 343 videos, human body 296 out of 330 result and 269 bicycle out of 287 results.

Confusion Matrix of SVM – Testing		Predicted			
		CAR	BODY	BICYCLE	Total
Actual	CAR	321	9	13	343
	BODY	16	296	18	330
	BICYCLE	7	11	269	287
	Total	344	316	300	960

Fig. 12. Confusion matrix of SVM – Testing.

The above classification/cluster algorithm are taking too much time to retrieve the result against queries. To achieve a better result in a short time, we trained our model using Convolutional Neural Network (CNN) with the help of theano [37] and TensorFlow [38]. In traditional neural network and deep learning the major difference are the number of hidden layers and their connections. Traditionally, Neural Network has three layers they are trained for classification or supervised data. They haven't the capability of generalizable optimization, they are optimized only for single specific task [39].

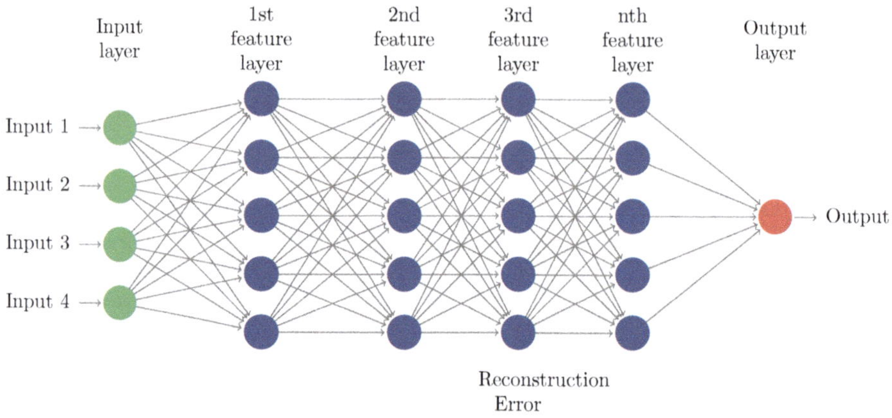

Fig. 13. Convolutional neural network.eps

Deep learning observed the pattern based upon the initial data that they received in input layers. The features of deep learning layers are not designed or extracted by human, but they are learned from learning procedures [40]. It process the initialized values in a non-linear manner and the subsequent hidden layer are learn by generalization. These learning are then put into supervised layer to tune the complete network via back propagation, max-pooling and dropout methods. Figure 13 depicted the whole structure of Convolutional Neual Network.

Input layer contains the values of the image pixel (256×256). Convolution Layer calculate the output of the neurons using dot product between their value and weights. For element wise activation function we use ReLu Layer with thresholding is zero. For down sampling we use pooling layer to reduce the dimensions, it also control over-fitting problem.

We used a very small convolutional neural network which consist of input layer, twenty six hidden layers and one input. To increase the performance of our architecture, we use ReLu to rectified the gradient problem. For training and testing, sampling dataset get randomly from YouTube [3] and SegTrack [2] dataset. Figure 14 depicted the confusion matrix obtained from our training and testing. After getting outstanding result during training in a very short time, we

Confusion Matrix of Conv. Neural Network - Training		Predicted			
		CAR	BODY	BICYCLE	Total
Actual	CAR	476	11	21	508
	BODY	6	359	7	372
	BICYCLE	7	13	405	425
	Total	489	383	433	1305

Fig. 14. Confusion matrix of Convolutional Neural Network – Training.

test our convolutional neural network on a sample of dataset with the same procedure (Query-Process Model). Query contains three images (Car, human body, and Bicycle), Fig. 15 represent the confusion matrix of Convolutional Neural Network testing. It return 128 car result out of 137, 96 human body out of 108 and 134 bicycle result out of 154.

Sensitivity and Specificity is calculate by complete retrieved result (list). Relevant and irrelevant result are calculated from the retrieved list and actual videos in a datasets.

Confusion Matrix of Conv. Neural Network - Testing		Predicted			
		CAR	BODY	BICYCLE	Total
Actual	CAR	128	4	5	137
	BODY	3	96	9	108
	BICYCLE	7	13	134	154
	Total	138	113	148	399

Fig. 15. Confusion matrix of convolutional neural network – testing.

5 Conclusion

In this paper, we presented a Query-Process Model that retrieved a list of videos from a datasets using image processing techniques (haar, histogram of gradient, active appearance model and eigen-faces), and classification/clustering algorithms (K-Nearest Neighbor, KMeans, Support Vector Machine and Naive Bayes) and result (training and testing) are discussed with their confusion matrix (Sensitivity and Specificity).

Convolutional Neural Network results are outstanding as compare to other traditional classification algorithms. Due to limited hardware resources (only CPU), we cannot achieve as much good result that we expected from our system.

Future research effort should be allocated to general object detection using complex Convolutional Neural Network (CNN), Deep Belief Network (DBN) with multiple hidden layer using ReLu, and dropout architecture using unlimited hardware resources (GPU or TPU).

References

1. Chute, C., Manfrediz, A., Minton, S., Reinsel, D., Schlichting, W., Toncheva, A.: The diverse and exploding digital universe. In: IDC White Paper (2008)
2. Tsai, D.: Georgia Tech Segmentation and Tracking Dataset (GT-SegTrack) (2017). http://cpl.cc.gatech.edu/projects/SegTrack/
3. Prest, A., Leistner, C., Civera, J., Schmid, C., Ferrari, V.: Learning object class detectors from weakly annotated video. In: 2012 IEEE Conference on Computer Vision and Pattern Recognition (CVPR), pp. 3282–3289. IEEE (2012)
4. Mehmood, Z., Mahmood, T., Javid, M.A.: Content-based image retrieval and semantic automatic image annotation based on the weighted average of triangular histograms using support vector machine. Appl. Intell. **48**(1), 166–181 (2018)
5. Johnson, J., Karpathy, A., Fei-Fei, L.: DenseCap: fully convolutional localization networks for dense captioning. In: Proceedings of the IEEE Conference on Computer Vision and Pattern Recognition, pp. 4565–4574 (2016)
6. Kevin O'Regan, J., Deubel, H., Clark, J.J., Rensink, R.A.: Picture changes during blinks: Looking without seeing and seeing without looking. Vis. Cogn. **7**(1–3), 191–211 (2000)
7. Xu, K., Ba, J., Kiros, R., Cho, K., Courville, A., Salakhudinov, R., Zemel, R., Bengio, Y.: Show, attend and tell: neural image caption generation with visual attention. In: International Conference on Machine Learning, , pp. 2048–2057 (2015)
8. Bahdanau, D., Cho, K., Bengio, Y.: Neural machine translation by jointly learning to align and translate. arXiv preprint arXiv:1409.0473 (2014)
9. Mnih, V., Heess, N., Graves, A., et al.: Recurrent models of visual attention. In: Advances in Neural Information Processing Systems, pp. 2204–2212 (2014)
10. Ba, J., Mnih, V., Kavukcuoglu, K.: Multiple object recognition with visual attention. arXiv preprint arXiv:1412.7755 (2014)
11. Hochreiter, S., Schmidhuber, J.: Long short-term memory. Neural Comput. **9**(8), 1735–1780 (1997)

12. Dean, J., Corrado, G., Monga, R., Chen, K., Devin, M., Mao, M., Senior, A., Tucker, P., Yang, K., Le, Q.V., et al.: Large scale distributed deep networks. In: Advances in neural information processing systems, pp. 1223–1231 (2012)
13. Le, Q.V., MarcAurelio Ranzato, R.M., Devin, M., Chen, K., Corrado, G.S., Dean, J., Ng, A.Y.: Building high-level features using large scale unsupervised learning. arxiv. org (2011)
14. Mikolov, T., Chen, K., Corrado, G., Dean, J.: Efficient estimation of word representations in vector space. arXiv preprint arXiv:1301.3781 (2013)
15. Vinyals, O., Kaiser, Ł., Koo, T., Petrov, S., Sutskever, I., Hinton, G.: Grammar as a foreign language. In: Advances in Neural Information Processing Systems, pp. 2773–2781 (2015)
16. Frome, A., Corrado, G.S., Shlens, J., Bengio, S., Dean, J., Mikolov, T., et al.: DeVISE: a deep visual-semantic embedding model. In: Advances in Neural Information Processing Systems, pp. 2121–2129 (2013)
17. Szegedy, C., Liu, W., Jia, Y., Sermanet, P., Reed, S., Anguelov, D., Erhan, D., Vanhoucke, V., Rabinovich, A.: Going deeper with convolutions. corr abs/1409.4842 (2014)
18. Karpathy, A., Toderici, G., Shetty, S., Leung, T., Sukthankar, R., Fei-Fei, L.: Large-scale video classification with convolutional neural networks. In: Proceedings of the IEEE Conference on Computer Vision and Pattern Recognition, pp. 1725–1732 (2014)
19. Chen, G., Parada, C., Heigold, G.: Small-footprint keyword spotting using deep neural networks. In: 2014 IEEE International Conference on Acoustics, Speech and Signal Processing (ICASSP), pp. 4087–4091. IEEE (2014)
20. Hinton, G., Deng, L., Yu, D., Dahl, G.E., Mohamed, A.-R., Jaitly, N., Senior, A., Vanhoucke, V., Nguyen, P., Sainath, T.N.: Deep neural networks for acoustic modeling in speech recognition: the shared views of four research groups. IEEE Signal Process. Mag. **29**(6), 82–97 (2012)
21. Heigold, G., Vanhoucke, V., Senior, A., Nguyen, P., Ranzato, M., Devin, M., Dean, J.: Multilingual acoustic models using distributed deep neural networks. In: 2013 IEEE International Conference on Acoustics, Speech and Signal Processing (ICASSP), pp. 8619–8623. IEEE (2013)
22. Al Rfou, R., Alain, G., Almahairi, A., Angermueller, C., Bahdanau, D., Ballas, N., Bastien, F., Bayer, J., Belikov, A., Belopolsky, A., et al.: Theano: a python framework for fast computation of mathematical expressions. arXiv preprint (2016)
23. Goodfellow, I.J., Bulatov, Y., Ibarz, J., Arnoud, S., Shet, V.: Multi-digit number recognition from street view imagery using deep convolutional neural networks. arXiv preprint arXiv:1312.6082 (2013)
24. Recht, B., Re, C., Wright, S., Niu, F.: HOGWILD: a lock-free approach to parallelizing stochastic gradient descent. In: Advances in Neural Information Processing Systems, pp. 693–701 (2011)
25. Zagoruyko, S., Lerer, A., Lin, T.-Y., Pinheiro, P.O., Gross, S., Chintala, S., Dollár, P.: A multipath network for object detection. arXiv preprint arXiv:1604.02135 (2016)
26. Sadrnia, H., Rajabipour, A., Jafary, A., Javadi, A., Mostofi, Y.: Classification and analysis of fruit shapes in long type watermelon using image processing. Int. J. Agric. Biol. **1**, 68–70 (2007)
27. Arivazhagan, S., Shebiah, R.N., Nidhyanandhan, S.S., Ganesan, L.: Fruit recognition using color and texture features. J. Emerg. Trends Comput. Inf. Sci. **1**(2), 90–94 (2010)

28. Insuasti-Ceballos, D., Bouwmans, T., Castellanos-Dominguez, G.: GMM background modeling using divergence-based weight updating. In: Progress in Pattern Recognition, Image Analysis, Computer Vision, and Applications: 21st Iberoamerican Congress, CIARP 2016, Lima, Peru, 8–11 November 2016, Proceedings, vol. 10125, p. 282. Springer (2017)
29. Bruhn, A., Weickert, J., Schnörr, C.: Lucas/kanade meets horn/schunck: combining local and global optic flow methods. Int. J. Comput. Vis. **61**(3), 211–231 (2005)
30. Jang, H., Won, I.-S., Jeong, D.-S.: Automatic vehicle detection and counting algorithm. Int. J. Comput. Sci. Netw. Secur. (IJCSNS) **14**(9), 99 (2014)
31. Nield, D.: Denmark just installed environmentally friendly traffic lights that give priority to bikes and buses (2017). https://www.sciencealert.com/copenhagen-just-installed-environmentally-friendly-traffic-lights-that-give-priority-to-buses-and-bikes
32. Tsai, D., Flagg, M., Nakazawa, A., Rehg, J.M.: Motion coherent tracking using multi-label mrf optimization. Int. J. Comput. Vis. **100**(2), 190–202 (2012)
33. Iqbal, S., Shaheen, M., et al.: A machine learning based method for optimal journal classification. In: 8th International Conference for Internet Technology and Secured Transactions (ICITST), pp. 259–264. IEEE (2013)
34. Hartigan, J.A., Wong, M.A.: Algorithm as 136: a k-means clustering algorithm. J. R. Stat. Soc. Series C (Appl. Stat.) **28**(1), 100–108 (1979)
35. Jain, S.: A machine learning approach: SVM for image classification in CBIR. Int. J. Appl. Annovation Eng. Manag. (IJAIEM) **2**(4) (2013)
36. Kotsiantis, S.B., Zaharakis, I., Pintelas, P.: Supervised machine learning: a review of classification techniques (2007)
37. Bergstra, J., Breuleux, O., Lamblin, P., Pascanu, R., Delalleau, O., Desjardins, G., Goodfellow, I., Bergeron, A., Bengio, Y., Kaelbling, P.: Theano: deep learning on GPUs with python (2011)
38. Abadi, M., Barham, P., Chen, J., Chen, Z., Davis, A., Dean, J., Devin, M., Ghemawat, S., Irving, G., Isard, M., et al.: Tensorflow: a system for large-scale machine learning. In: OSDI, vol. 16, pp. 265–283 (2016)
39. Williams, J.M.: Deep learning and transfer learning in the classification of EEG signals (2017)
40. Voulodimos, A., Doulamis, N., Doulamis, A., Protopapadakis, E.: Deep learning for computer vision: a brief review. Comput. Intell. Neurosci. 2018 (2018)

Proposal and Evaluation of an Indirect Reward Assignment Method for Reinforcement Learning by Profit Sharing Method

Kazuteru Miyazaki[1(✉)], Naoki Kodama[2], and Hiroaki Kobayashi[3]

[1] National Institution for Academic Degrees and Quality Enhancement of Higher Education, Kodaira, Tokyo, Japan
teru@niad.ac.jp
[2] Tokyo University of Science, Noda, Chiba, Japan
7418701@ed.tus.ac.jp
[3] Meiji University, Kawasaki, Kanagawa, Japan
kobayasi@meiji.ac.jp

Abstract. We know Profit Sharing method that can guarantee a rationality in the case of acquiring a reward in reinforcement learning. This paper proposes a method that generates indirect rewards by Profit Sharing method in order to assure the rationality. The proposed method is applied to Deep Q-Network and the method is named DQNbyPS. It is shown that DQNbyPS can reduce the number of trial and error searches than the original Deep Q-Network in Pong that is one of Atari 2600 games.

Keywords: Reinforcement learning · Profit sharing · Deep learning Deep Q-network

1 Introduction

Among machine-learning approaches, reinforcement learning (RL) focuses on goal-oriented learning from interaction with an environment [25]. RL generally uses rewards and penalties as teacher signals and the learning is driven by those signals. It is attractive that the Dynamic Programming (DP) can be used to analyze the behavior of RL. RL methods based on DP can optimize the behavior in the Markov Decision Processes (MDPs) if RL user defines rewards and penalties appropriately.

Despite important applications [7–9,13,24,26,27,29], the real world application is rather difficult because, first, it requires many number of trial and error searches and, second, there is no guideline how to design values of rewards and penalties. Though these are essentially neglected in theoretical researches [1,4,22], they are serious problems in application to real world. If we assign inappropriate values of rewards and penalties, unexpected results would be arised.

K. Arai et al. (Eds.): IntelliSys 2018, AISC 868, pp. 187–200, 2019.
https://doi.org/10.1007/978-3-030-01054-6_13

There is another approach known as exploitation-oriented learning (XoL) [15], where reward and penalty signals are treated independently and adjustment of values among them is not necessary. Furthermore it can reduce the number of trial and error searches significantly by enhancing successful experiences strongly. XoL, first, aims to guarantee the rationality of the policy with as small number of trial and error searches as possible rather than to pursue the optimality, and second, XoL dose not require to assign values to rewards and penalties. XoL requires however to set the order of priority among rewards and penalties.

XoL is a new approach for learning based on trial and error searches and it has some certain aspects. Table 1 summarizes four features among XoL and RL. (1) XoL can learn with fewer number of trial and error searches than RL by strongly following a success experience. (2) Xol does not require sophisticated design of values of rewards and penalties since they are treated as independent signals. This allows us to handle these signals intuitively and easily. (3) XoL does not pursue the optimality but the rationality. The optimality can be guaranteed by the multi-start method [12]. (4) XoL is efficient and effective for the class that exceeds MDPs since XoL is a Bellman free method.

Table 1. Approaches for learning based on trial and error searches

	XoL	RL
The number of trial and error searches	Less	More
Design for a reward and a penalty	Priority among them	Sophisticated values
The optimality in MDPs	Multi-start method	Assured
Beyond MDPs	Strong	Weak

Examples of XoL with a reward include Profit Sharing (PS) method [10], Rational Policy Making (RPM) algorithm [12], PS-r# [15], and so on. Furthermore, the following methods can treat a reward and a penalty at the same time: Penalty Avoiding Rational Policy Making (PARP) algorithm [13], Improved PARP [27] and Expected Failure Probability (EFP) algorithm [16].

In this paper, we introduce an indirect reward to RL to accelerate the learning. In this case, it is important not to destroy the rationality of the policy by the indirect reward. PS can guarantee the rationality in the case of acquiring a reward. Therefore, after relevant terms are explained in Sect. 2, a method that generates a indirect reward by PS is proposed in Sect. 3. Then, the method is applied to Deep Q-Network (DQN) [19,20], and the method is named *DQNbyPS* Finally, it is shown that DQNbyPS can learn with fewer trial and error searches than the original DQN in Pong that is one of Atari 2600 games in Sect. 4. In the last Sect. 5, this paper is summarized and future works are addressed.

2 The Domain

2.1 Definition of Terms

Consider an agent placed in an unknown environment. After the agent perceives a sensory input from the environment, it selects an action from a set of discrete types of actions. Time is discretized by one sensory input and action selection cycle. The input from the environment is named *a state*. The number of the discrete type actions is named *Type of actions*.

A pair SA that consists of a state S and an action A selected in the state is named *a rule* (it may be written as $rule(S, A)$ more explicitly). Rewards or penalties are provided from the environment as a result of action sequence. If our purpose has been achieved, a reward is given to a state that was transited or an action selected at that time. On the other hand, if our purpose has been failed, a penalty is given to the state or the action.

A rule sequence that begins from a reward, a penalty, or an initial state and ends with the next reward or penalty state is named *an episode*. For example, when the agent selects rule S_0A_1, S_0A_0, S_1A_0, S_2A_0, S_1A_1, S_0A_0, S_2A_0 and S_1A_1 in Fig. 1(a), there exist two episodes $(S_0A_1S_0A_0S_1A_0S_2A_0S_1A_1)$ and $(S_0A_0S_2A_0S_1A_1)$ as shown in Fig. 1(b). If an episode includes rules with the same state but paired with different actions, the partial rule sequence from one of the rules until the next one is named *a detour*. For example, the episode $(S_0A_1S_0A_0S_1A_0S_2A_0S_1A_1)$ includes two detours (S_0A_1) and $(S_1A_0S_2A_0)$ as shown in Fig. 1(b).

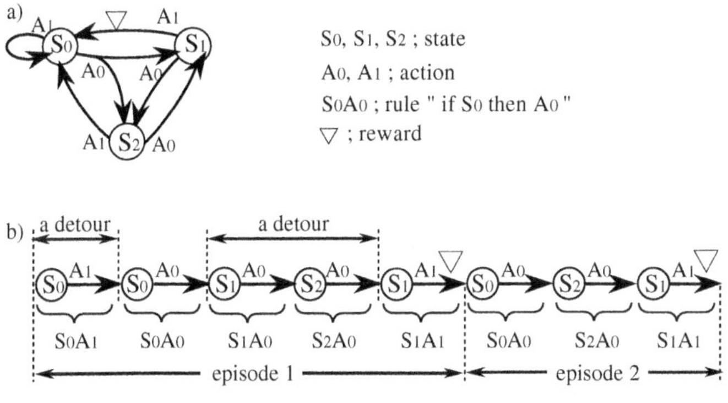

Fig. 1. (a) An example of an environment. (b) Episodes and detours included in (a).

A rule always included in a detour is named *an ineffective rule*, and otherwise named *an effective rule*. After obtaining the episode 1 in Fig. 1(b), rules S_0A_1, S_1A_0 and S_2A_0 are ineffective rules and rules S_0A_0 and S_1A_1 are effective rules. Then if the episode 2 is experienced, the rule S_2A_0 changes to an effective rule. If a rule was previously considered as an effective rule and, after some time, always included in a detour, then this situation is named *type 2 confusion* [12]. Every time the agent gets a reward, the effectiveness of the rule on the episode experienced at that time is judged.

A rule that directly contributes to receive a penalty is named *a penalty rule.* If all rules that can select in a state are penalty or ineffective rules, the state is named *a penalty state.* If a destination state after using a rule is a penalty state, the rule is also named *a penalty rule.*

A function that maps states to actions is named *a policy.* The policy where an amount of reward acquisition is larger than zero is named *a rational policy.* Especially, a rational policy receiving no penalty is named *a penalty avoiding rational policy. The optimal policy* is a policy that can maximize the amount of a reward.

2.2 Q-Learning

We know Q-learning (QL) [28] that is a representative RL method. QL uses the *Q-value* to memorize the value of rules. The learning of QL proceeds by updating the *Q-value* each time the agent selects an action. The following equation is used to update $Q(s_t, a_t)$, the Q-value of rule(s_t, a_t), when the agent selects the action a_t in the state s_t, acquires a reward r_t, and transits to the next state s_{t+1};

$$Q(s_t, a_t) \leftarrow (1-\alpha)Q(s_t, a_t) + \alpha(r_t + \gamma \max_{a'} Q(s_{t+1}, a')), \tag{1}$$

where α and γ are the learning and the discount rates, respectively.

The previously proposed theorem in the paper [28] guarantees that QL can obtain the optimal policy in MDPs. That is, the optimal policy is given when the agent selects an action with the maximum Q-value at the state in condition that the Q-value has converged through reducing the learning rate. However, no rationality is guaranteed by this theorem before the convergence of Q-values. Furthermore, as mentioned in the paper [11], in general, an enormous number of trial and error searches are required to converge Q-values since interaction with an environment is required not only to update the Q-values but also to identify the unknown environment.

2.3 Profit Sharing

When a reward is given, PS propagates a reward backward in the episode in order to acquire a rational policy. Consider an action selection at time t, that is, the action a_t was used at the state s_t in the episode and let f_t and $Q(s_t, a_t)$ be a reward shared to $rule(s_t, a_t)$ and the total amount of rewards of the rule at time t, respectively. The following geometric decreasing function is used to propagate the reward backward:

$$f_t = \lambda^{N-t} R, \tag{2}$$

$$Q(s_t, a_t) \leftarrow Q(s_t, a_t) + f_t, \tag{3}$$

$$t = N, N-1, ..., 1, 0 < \lambda < 1,$$

where R, N and λ are the reward value, the episode length and the *discount rate*, respectively. The function to propagate a reward is named a *reinforcement function.*

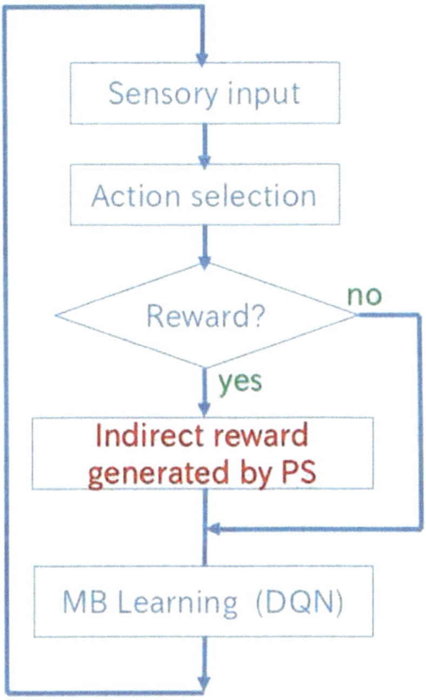

Fig. 2. DQNbyPS algorithm.

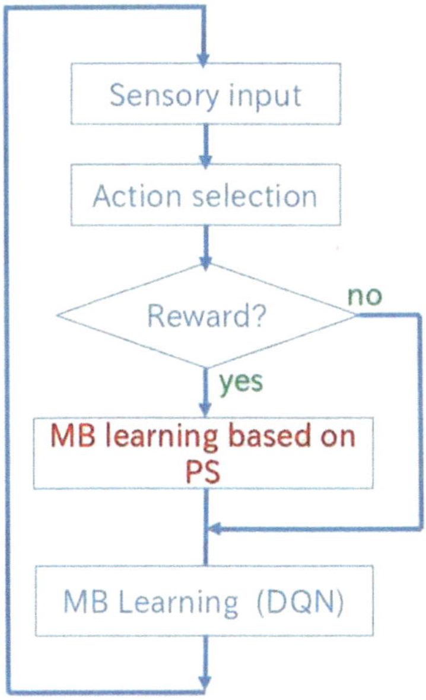

Fig. 3. DQNwithPS algorithm.

The rationality theorem of PS [10] gives a necessary and sufficient condition to acquire a rational policy in the class where types of reward is one and there is no type 2 confusion. It can be shown that (2) with $\lambda = \frac{1}{Types\ of\ actions}$ satisfies the condition. Moreover, the *rationality theorem of PS in multi-agent learning* [14] gives a necessary and sufficient condition in order to distribute a reward to agents that do not receive a reward directly.

PS is one of the XoL methods. PS generally can acquire a rational policy with fewer trial and error searches than QL. Furthermore, it can guarantee to acquire a rational policy in the class where types of reward is one and without type 2 confusion. In contrast, the rationality of RL methods based on DP such as QL is unknown for a class that exceeds MDPs.

2.4 Deep Q-Network

We know Deep Q-Network (DQN) that successfully combines reinforcement learning and deep learning. In DQN, the Q-value of QL is approximated by convolutional neural networks (CNN). DQN was applied to 49 Atari 2600 games [20]. It has out-performed existing methods and the game scores are higher than human experts at some games.

In DQN, the game screen is inputted directly to the CNN. Then the perceptual aliasing problem [2] typically occurs in general. In order to reduce the effects of the problem, DQN inputs the last four frames to the CNN at each cycle. The perceptual aliasing problem in DQN has been investigated in the paper [6].

The network configuration in our experiments at Sect. 5, is the same as the previously reported one [20]. DQN uses *a replay memory* that stores *the experiences* comprised of the four-tuple (s_t, a_t, r_t, s_{t+1}) in order to update the network parameters. First of all, the replay memory continues to store four-tuples in the order that the agent experienced. If the storage amount reaches to the available memory capacity, the contents of the memory is discarded from the oldest experience. Next, Minibatch (MB) learning is performed in order to update CNN parameters as follows: First, from the replay memory, certain number of experiences are randomly sampled. Next, the gradient is calculated for each experience and the target value sets to $r_j + \gamma \max_{a'} Q(s_{j+1}, a')$. Finally, according to the calculated gradients, CNN parameters are updated.

3 Proposal of an Indirect Reward Assignment Method

3.1 Proposed Method

RL uses a reward to learn but the speed is rather slow. If we want to accelerate the learning speed, one means is to use an indirect reward. In this case, it is important to assure some kind of rationality for the indirect reward. For PS, the rationality theorem described above can guarantee the rationality for distributed rewards. Therefore, when a reward was given, if we distributes the reward to relevant rules according to (2) keeping the rationality and we use the distributed value as an indirect reward for QL, then the indirect reward will have the same rationality and the learning will speed up.

In this paper, based on the idea, we propose a method where Q-values are updated with direct and indirect rewards. That is, when a reward R is given, we calculate indirect rewards and update relevant Q-values as follows:

$$r_t = \lambda^{N-t} R \tag{4}$$

$$Q(s_t, a_t) \leftarrow (1-\alpha)Q(s_t, a_t) + \alpha(r_t + \gamma \max_{a'} Q(s_{t+1}, a')) \tag{5}$$

$$t = N, N-1, ..., 1, \ 0 < \alpha, \gamma, \lambda < 1.$$

Note that the reward is direct for $t = N$ but indirect for the others. Next we show an example of the method named *DQNbyPS*.

3.2 An Example: DQNbyPS

Typically, DQN requires many trial and error searches in order to update CNN parameters of DQN. In this section, we aim to reduce the number of trial and error searches with the algorithm in Fig. 2. We refer to the method as *DQNbyPS*.

Recently, another variation of XoL named *DQNwithPS* [17] was proposed. DQNwithPS executes MB learning based on PS as shown in Fig. 3. On the other hand, DQNbyPS does not execute MB learning based on PS but it just generates indirect rewards by PS. We compare performance of both algorithms as well as the original DQN.

3.3 Related Works

There is a method to improve the score of DQN [21]. Some studies have attempted to increase the speed of learning using a model [5] and concerning the discount rate γ of QL [3]. These previous studies except for DQNwithPS are based on QL. On the other hand, we focus on PS, that is our original learning approach, in terms of improving the learning speed by introducing PS to DQN.

4 Application to Atari 2600 Game

4.1 Experimental Setting

DQNbyPS is evaluated using Pong in one of Atari 2600 games in this section. Consider two agents playing Pong game. One is a learning agent and the other does not learn. When the learning agent hits a ball and the opposite agent cannot hit it back, the learning agent gets a point of Pong game and acquires a reward 1.0. In contrast, when the learning agent fails to hit a ball, it loses the point and receives a penalty −1.0. The game is ended when the point of either agent has reached to 21, The game score is calculated from the difference of scores of two agents. The score of Pong therefore is an integer value between −21 and 21.

The performance of DQNbyPS is confirmed by changing λ. The learning result is evaluated based on the score. We gather scores of 10 non-learning games after every 10 learning games, and compare their average values to evaluate the performance.

The sizes of MB is 32. The replay memory has 1,000,000 frames. The discount rate for QL, γ, is 0.99. We use (2) that is one of a geometric decreasing reinforcement functions of PS. The discount rate λ is determined later.

The action is selected using an ϵ-greedy strategy. We set $\epsilon = 0.01$ in the evaluation phase. It means that the action with the maximum Q-value and a random action are selected with probability of 0.99 and 0.01, respectively. For the learning phase we fix $\epsilon = 0.01$ in DQNbyPS taking advantage of the experience-oriented aspects of PS, while for DQN, ϵ is changed according the following strategy [19]:

- $\epsilon = 0.1$ (if the number of actions exceeds one million)
- $\epsilon = 1.0 - 0.9 * \frac{The\ number\ of\ actions}{1000000}$ (otherwise)

We use a chainer deep learning framework on the 64-bit version of Ubuntu 16.04.

4.2 Results and Discussion

The results are shown in Figs. 4, 5, 6, 7, 8 and 9. These figures show scores of the learning agent plotted against the number of actions. Here, the score is the average of 10 experiments. λ was changed to 0.2, $0.\dot{3}$, 0.5, 0.7, and 0.9. The value of $0.\dot{3}$ ($= \frac{1}{Types\ of\ actions} = \frac{1}{3}$) satisfies the rationality theorem of PS. These figures include the results for DQN.

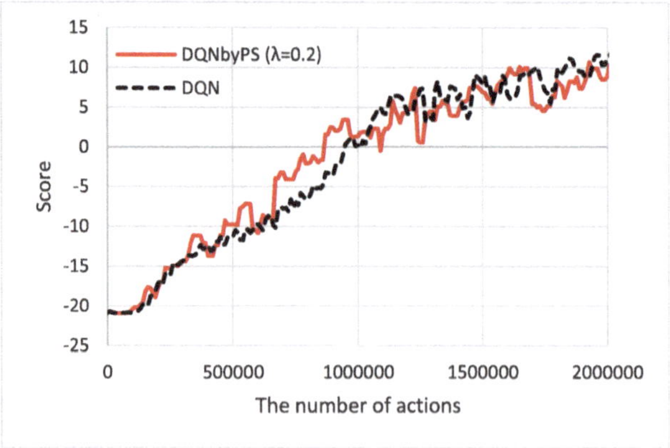

Fig. 4. Comparison with DQN and DQNbyPS ($\lambda = 0.2$).

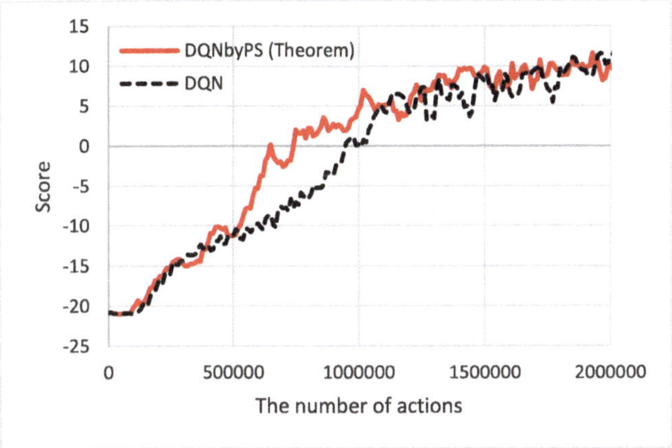

Fig. 5. Comparison with DQN and DQNbyPS (Theorem) where "Theorem" means that $\lambda = \frac{1}{Types\ of\ actions} = \frac{1}{3} = 0.\dot{3}$.

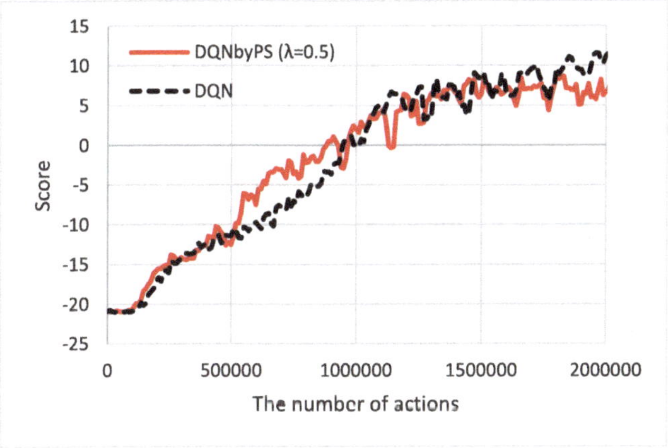

Fig. 6. Comparison with DQN and DQNbyPS ($\lambda = 0.5$).

As shown in Figs. 4, 5, 6, 7 and 8, DQNbyPS can learn with the fewest actions when $\lambda = 0.\dot{3}$ that satisfies the rationality theorem of PS. It means that the rationality theorem of PS is effective for our proposed indirect reward assignment method.

Figure 9 shows comparison between DQN with scheduled ϵ and DQN with fixed $\epsilon = 0.01$. From Fig. 9 we cannot confirm that the difference between them. Through numerical experiments, we can confirm the effectiveness of DQNbyPS.

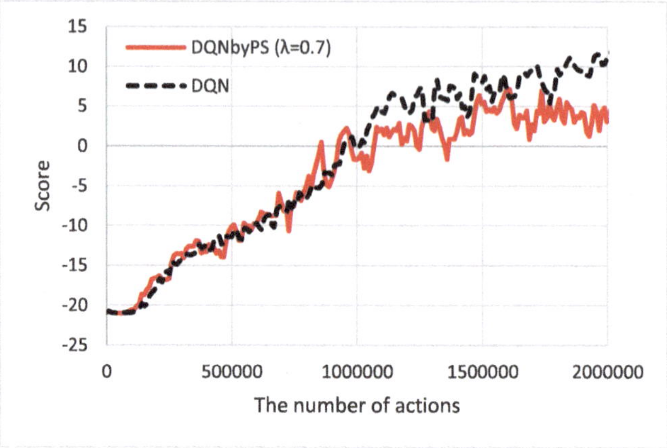

Fig. 7. Comparison with DQN and DQNbyPS ($\lambda = 0.7$).

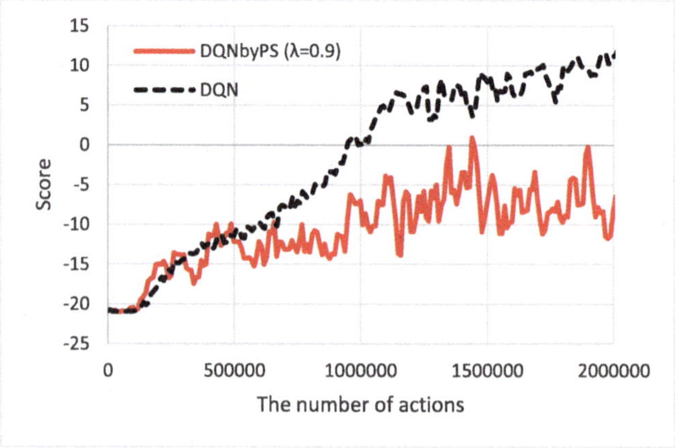

Fig. 8. Comparison with DQN and DQNbyPS ($\lambda = 0.9$).

Figure 10 shows results of DQNbyPS (Theorem) and DQNwithPS ($\lambda = 0.99$). Though both methods show almost same performance, DQNwithPS with λ for the rationality theorem is inferior in the performance to the one for $\lambda = 0.99$ as shown in Fig. 11. It shows the difficulty to design an appropriate λ in DQNwithPS.

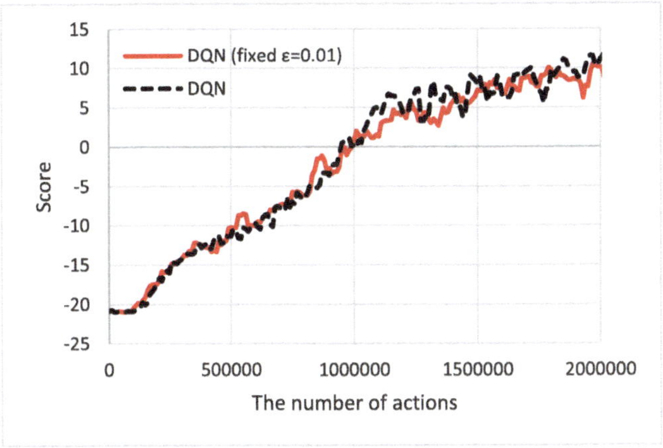

Fig. 9. Comparison with DQN and DQN with fixed $\epsilon = 0.01$.

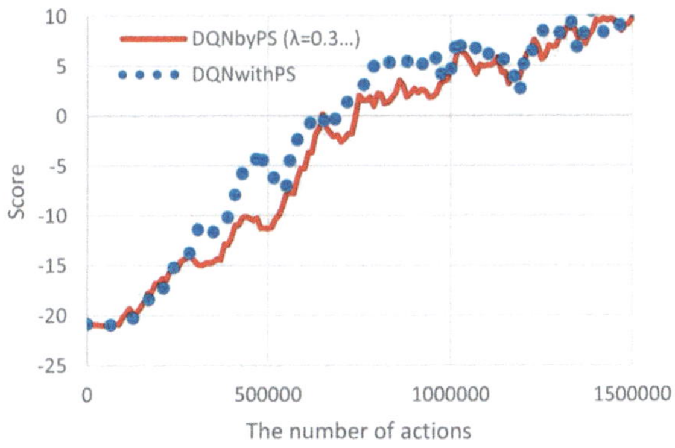

Fig. 10. Comparison with DQNbyPS (Theorem) and DQNwithPS ($\lambda = 0.99$).

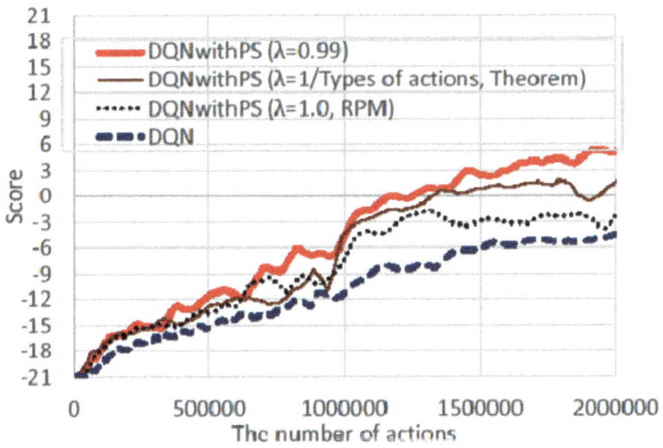

Fig. 11. Results with varying the λ in DQNwithPS.

5 Conclusion

We proposed a method where indirect rewards for QL are generated by Profit Sharing. The proposed method was applied to DQN and the method was called DQNbyPS. It was shown that DQNbyPS can learn with fewer trial and error searches than the original DQN in Pong that is one of Atari 2600 games.

There are many methods to improve DQN [3,5,6,21,23]. The proposed DQNbyPS method can be easily combined with these previous methods since it can simply introduce PS to DQN and does not need to change the essence of the original one.

Currently, we aim to combine EFP to PS in DQNbyPS. EFP is an XoL method that aims to avoid a penalty through estimating an expected failure probability of each rule. If EFP can be incorporated in DQNbyPS, QL can be eliminated and more efficient learning can be expected. In order to realize the combination, it is necessary to prepare independent CNNs for EFP and PS, respectively. It is an important future work. Furthermore, we aim to combine DQNbyPS to some variations of DQN.

In addition to the above improvements of our method, we plan to apply DQNbyPS to the other issues such as Keepaway task [26,27], a biped walking robot [7], a consciousness system [18].

Acknowledgment. This work was supported by JSPS KAKENHI Grant Number 17K00327.

References

1. Abbeel, P., Ng, A.Y.: Exploration and apprenticeship learning in reinforcement learning. In: Proceedings of 22nd International Conference on Machine Learning, pp. 1–8 (2005)
2. Chrisman, L.: Reinforcement learning with perceptual aliasing: the perceptual distinctions approach. In: Proceedings of the 10th National Conference on Artificial Intelligence, pp. 183–188 (1992)
3. Francois-Lavet, V., Fonteneau, R., Emst, D.: How to discount deep reinforcement learning: towards new dynamic strategies. In: NIPS 2015 Deep Reinforcement Learning Workshop (2015)
4. Gosavi, A.: A reinforcement learning algorithm based on policy iteration for average reward: empirical results with yield management and convergence analysis. Mach. Learn. **55**(1), 5–29 (2004)
5. Gu, S., Lillicrap, T., Sutskever, I., Levine, S.: Continuous deep Q-learning with model-based acceleration. arXiv:1603 (2016)
6. Hausknecht, M., Stone, P.: Deep recurrent Q-learning for partially observable MDPs. arXiv:1507 (2015)
7. Kuroda, S., Miyazaki, K., Kobayashi, H.: Introduction of fixed mode states into online reinforcement learning with penalties and rewards and its application to biped robot waist trajectory generation. J. Adv. Comput. Intell. Intell. Inf. **16**(6), 758–768 (2013)

8. Matsui, T., Goto, T., Izumi, K.: Acquiring a government bond trading strategy using reinforcement learning. J. Adv. Comput. Intell. Intell. Inf. **13**(6), 691–696 (2009)
9. Merrick, K., Maher, M.L.: Motivated reinforcement learning for adaptive characters in open-ended simulation games. In: Proceedings of the International Conference on Advanced in Computer Entertainment Technology, pp. 127–134 (2007)
10. Miyazaki, K., Yamamura, M., Kobayashi, S: On the rationality of profit sharing in reinforcement learning. In: Proceedings of the 3rd International Conference on Fuzzy Logic, Neural Nets and Soft Computing, pp. 285–288 (1994)
11. Miyazaki, K., Yamamura, M., Kobayashi, S.: k-certainty exploration method: an action selector to identify the environment in reinforcement learning. Artif. Intell. **91**(1), 155–171 (1997)
12. Miyazaki, K., Kobayashi, S.: Learning deterministic policies in partially observable Markov decision processes. In: Proceedings of the 5th International Conference on Intelligent Autonomous System, pp. 250–257 (1998)
13. Miyazaki, K., Tsuboi, S., Kobayashi, S.: Reinforcement learning for penalty avoiding policy making and its extensions and an application to the Othello game. In: 7th International Conference on Information Systems Analysis and Synthesis (ISAS 2000), vol. 3, pp. 40–44 (2001)
14. Miyazaki, K., Kobayashi, S.: Rationality of reward sharing in multi-agent reinforcement learning. New Gener. Comput. **19**(2), 157–172 (2001)
15. Miyazaki, K., Kobayashi, S.: Exploitation-oriented learning PS-r$^{\#}$. J. Adv. Comput. Intell. Intell. Inf. **13**(6), 624–630 (2009)
16. Miyazaki, K., Muraoka, H., Kobayashi, H.: Proposal of a propagation algorithm of the expected failure probability and the effectiveness on multi-agent environments. In: SICE Annual Conference 2013, pp. 1067–1072 (2013)
17. Miyazaki, K.: Exploitation-oriented learning with deep learning - introducing profit sharing to a deep Q-network. J. Adv. Comput. Intell. Intell. Inf. **21**(5), 849–855 (2017)
18. Miyazaki, K., Takeno, J.: The necessity of a secondary system in machine consciousness. Procedia Comput. Sci. **41**, 15–22 (2014)
19. Mnih, V., Kavukcuoglu, K., Silver, D., Graves, A., Antonoglou, I., Wierstra, D., Riedmiller, M.: Playing Atari with deep reinforcement learning. In: NIPS Deep Learning Workshop 2013 (2013)
20. Mnih, V., Kavukcuoglu, K., Silver, D., Rusu, A.A., Veness, J., Bellemare, M.G., Graves, A., Riedmiller, M., Fidjeland, A.K., Ostrovski, G., Petersen, S., Beattie, C., Sadik, A., Antonoglou, I., King, H., Kumaran, D., Wierstra, D., Legg, S., Hassabis, D.: Human-level control through deep reinforcement learning. Nature **518**, 529–533 (2015)
21. Nair, A., Srinivasan, P., Blackwell, S., Alcicek, C., Fearon, R., Maria, A.D., Suleyman, M., Beattie, C., Petersen, S., Legg, S., Mnih, V., Silver, D.: Massively parallel methods for deep reinforcement learning. In: ICML Deep Learning Workshop (2015)
22. Ng, A.Y., Harada, D., Russell, S.J.: Policy invariance under reward transformations: theory and application to reward shaping. In: Proceedings of 16th International Conference on Machine Learning, pp. 278–287 (1999)
23. Osband, I., Blundell, C., Pritzel, A., Roy, B.V.: Deep exploration via bootstrapped DQN. arXiv:1602 (2016)
24. Randlϕv, J., Alstrϕm, P.: Learning to drive a bicycle using reinforcement learning and shaping. In: Proceedings of the 15th International Conference on Machine Learning, pp. 463–471 (1998)

25. Sutton, R.S., Barto, A.G.: Reinforcement Learning: An Introduction. A Bradford Book. MIT Press, Cambridge (1998)
26. Stone, P., Sutton, R.S., Kuhlamann, G.: Reinforcement learning toward RoboCup soccer keepaway. Adapt. Behav. **13**(3), 165–188 (2005)
27. Watanabe, T., Miyazaki, K., Kobayashi, H.: A new improved penalty avoiding rational policy making algorithm for keepaway with continuous state spaces. J. Adv. Comput. Intell. Intell. Inf. **13**(6), 675–682 (2009)
28. Watkins, C.J.H., Dayan, P.: Technical note: Q-learning. Mach. Learn. **8**, 55–68 (1992)
29. Yoshimoto, J., Nishimura, M., Tokita, Y., Ishii, S.: Acrobot control by learning the switching of multiple controllers. J. Artif. Life Rob. **9**(2), 67–71 (2005)

Eye-Tracking to Enhance Usability: A Race Game

A. Ezgi İlhan(✉)

Faculty of Fine Arts, Design and Architecture, Department of Industrial Design, Atılım University, Ankara, Turkey
ezgi.ilhan@atilim.edu.tr

Abstract. An important field of research in human-computer interaction studies is the usability of computer games. This paper provides brief definitions of human-computer interactions and usability, and also describes the relevance of these interactions to computer games. Design decisions concerning game elements such as graphical user interface, feedback messages, position and the colour of functional buttons located on the game screen play an important role in identifying the usability and playability of computer games. This study uses eye-tracking technology in order to record eye movements to focus the action of "seeing", which reflects the inner world of humans. A managerial racing game was chosen as an example to analyse its usability. In this context, the design of the social race game was reviewed by recording eye movement data of the participants. The results of eye-tracking data were supported by user comments, which were finally used to improve the design and usability features of the game.

Keywords: Human-computer interaction · Usability · Eye-tracking
Social race game

1 Introduction

Human-computer interaction studies become more and more popular in the modern world due to the high number of computer users. Human-computer interaction is an interdisciplinary study field that covers the design, application and improvement of technologies. It mainly focuses on the relationship between humans and the computer. Thus, this field is related to human-centred disciplines such as psychology, sociology, cognitive sciences, ergonomics and industrial design and computer-focused disciplines, such as computer science and software engineering. Human-computer interaction studies can be conducted with four constituents, i.e. user, task, product and the context. Users are observed while performing some of the given tasks by using the products of related technology in the focal context. According to these observations, necessary data are obtained and used in order to improve the systems. Human-computer interaction is a strong study field for designers to ensure the success of a technological product. This product can be an application, web site or a computer game. The visual designs like colours, images, page layout and graphical user interface can be analysed within the scope of human-computer interaction studies [1, 2].

K. Arai et al. (Eds.): IntelliSys 2018, AISC 868, pp. 201–214, 2019.
https://doi.org/10.1007/978-3-030-01054-6_14

Usability, on the other hand, can be defined as an easy, efficient and effective usage of a product. The technological product can be considered as usable if the target user group can accomplish determined tasks after receiving the necessary training and technical support related to product use. Usability also identifies the user satisfaction showing the success percentage on the tasks that are required to be done [2].

One of the significant fields of human-computer interaction studies is the issue of usability. Use of the technology varies depending on the personality and needs of different people. This change in the interaction with technical systems requires more research on usability. Hence, designing more usable technologies can be achieved by human-computer interaction studies [3].

In recent years, usability of computer games gained a strong research potential under human-computer interaction field [4]. Usability, playability, graphical user interface, and user friendliness are important terms of games that necessitate deeper analysis from players/users' point of view. Game controls should be enjoyable, clear and easy to be controlled by the players. Usability of the game is as important as the success of game mechanics and dynamics [5]. Furthermore, in order to increase the number of players by creating positive user experience, usability problems of the game should be eliminated. Thus, usability and playability tests should be conducted with the target player group before the game release.

Testing of the game is an important process to understand the player behaviours and expectations. For that reason, some studies related to game and usability adapts a set of heuristics of playability with usability research during the development cycle [6]. In order to detect usability problems of the game, eye-tracking technology can be used. Eye-tracking technology introduces location, duration and the path of eye movements of participants related to a product. It is used in human-computer interaction studies in various fields like human-related academic research, cognitive science, marketing and user-experience design. According to the sequence and path of eye movement and duration of gaze time, it creates heat, attention and interest maps [7].

Eyes reflect emotions of people and give clues about how they perceive the world. People get more than 90% of information from the outside world through their eyes [8]. However, eye movements are deleted quickly from the short-term memory. 47% of people forget or ignore where they looked [9]. For this reason, eye-tracking studies are beneficial for designers to record eye movement and gaze points, to understand visual interest, to reveal excitement and to analyse the usage scenarios of the product. Eye tracking studies answer usability questions like what participants see/cannot see, where they look at, what they think about and how they behave/feel related to the focal product. They evaluate the effect of the design and show the interesting points that are appealing to users.

This paper will analyse the usability of a social race game with the help of the data provided by eye-tracking technology. This section provides the background of the study giving a general overview of the topics concerning human-computer interaction, usability of games and eye-tracking technology. The information gathered from the literature reviews helped to design the methodology of the study. Therefore, Sect. 2 of the paper will explain the methodology of the study including the selection of game tasks, choice of the user group and place of the study, and collection of data. It will also mention the purpose and design of the study. Section 3 will present qualitative and quantitative analysis of collected data. It will discuss eye-tracking data in terms of the usability and playability of the game. Section 4 will cover general outcomes and conclusions of the study. Final Sect. 5 will indicate limitations of the study and make suggestions for the future work.

2 Methodology

2.1 Selection of Game Tasks

Main purpose of the study is to analyse the usability of the game via recording eye movements of the participants. A car race management game belonging to simulation and strategy genres was selected for the study. In the game, the player, who acts as the manager, races in different paths. After developing the cars, pilots and the pit team in the city, the player registers to the new races. The priority of game tasks was determined firstly in order to be prepared for game analysis:

(1) *Main goal of the game*: To race-several paths
(2) *Assistive goals of the game*:
- To modify the car - the garage
- To heal the car - the garage
- To improve the pilot by tactics and skills - the academy
- To improve the pit team by skills - the academy
- To rest the pilot – the tavern
- To entertain the pilot – the tavern
- To make the pilot lose weight– the tavern.

These tasks were grouped according to the different places in the city. Since it is a managerial race game, the only task assigned to the players was to participate in the race at least once. Other than that, participants were not assigned any special tasks in the game. They were totally free to play the game as they desire in order to define the usability and playability problems of the game. Their actions in the garage, academy and the tavern were collected to analyse the success of the assistive purposes of the game.

2.2 Selection of Participants

The game is considered under mid-core category. It is going to be played on a social platform and the potential player group is approximately 20 year-old boys. Hence, the aim is to find young participants who can give beneficial feedback to improve the usability and playability of the game.

For the validity of usability tests, the minimum number of participants should be 5. Usability problems can be assessed with 75% success, when there are at least 5 participants. The gestures and mimics, comments and voices and eye movements of users can be collected to test the usability of the product [10]. At the beginning of the study, a pilot study was conducted in order to check if technical tools supported the study, the process was working as initially intended and the observation time was enough to collect necessary data. Upon ensuring these in the pilot study, final studies were started with the focus group of participants. The participants were selected according to their interest in computers and computer games. 10 players, who have experience in these subjects, participated in the study. 10 players were considered to give beneficial and sufficient feedback so as to ascertain the usability problems of the game. The study took nearly half an hour for one participant.

2.3 Venue and Equipment

Usability tests can be conducted in human-computer interaction research laboratories. The participant and the observer are placed in separate rooms in these laboratories. The observer has the control to watch, hear and record the reactions of the participant [2]. That study was held in the Human-Computer Interaction Research and Application Laboratory of Computer Centre in Middle East Technical University in Ankara, Turkey. Eye-tracking analyses were made by Tobii Studio software, since Tobii 1750 eye tracker was used to record eye movements.

2.4 Data Collection Tools

The study collects both eye-tracking data and questionnaire results of the participants. Eye-tracking analysis will be explained in detail within the scope of this paper.

(1) *Eye-Tracking:* In design and development processes, designers have some priorities. They focus on some aspects of the game more than the others. Yet, the usability and playability features of the game are determined as a result of the usage path and the action choice of players. How players respond to the game with their playing habits, emotions and attitudes contributes to the game development process. How players interact with the game during eye-tracking reveals different usability and playability aspects of the game.

For performing data collection through eye-tracking, the observer takes place in the laboratory to welcome participants. With already prepared questions, it is time for the observer to get more detailed information during the game - player interaction period.

While recording eye movements, the goal is to collect detailed information from players' perspectives to interpret game features. The questions that need answers related to the race, garage, academy and the tavern are listed below:

The Race

- **Registration:** What is the process of registration for the race?
- **Race screen:** Is it easy to realize tactic buttons in the race? How often are they used (see Fig. 1)?
- Is it easy to realize the pit stop button? How many of them cannot see this button and were eliminated from the race? Is it easy to take actions and control the game in the pit stop? Can players use all of the actions?
- Do players control health bar displays on Race GUI?

Fig. 1. Race screen.

- How do players use the interfaces and information display screens during the race? Can they understand the given information easily?
- **Race type:** How do players feel in different types of races (ranking or elimination)?
- **Race ending:** What kind of actions do players take after the race?

The Garage

- Can players understand how to heal the car (see Fig. 2)?

Fig. 2. The garage.

- Can players understand how to modify the parts of the car? Do players compare the properties of new parts during the modify process?
- Can players understand how to change the car from the action panel?

The Academy

- Can players understand how to improve the pilot and the pit team (see Fig. 3)?

Fig. 3. The academy.

- Can players easily register their pilots to new classes by scheduling?
- Can players understand how to change the pilot from the action panel?

The Tavern

- Can players understand how to improve the pilot (see Fig. 4)?

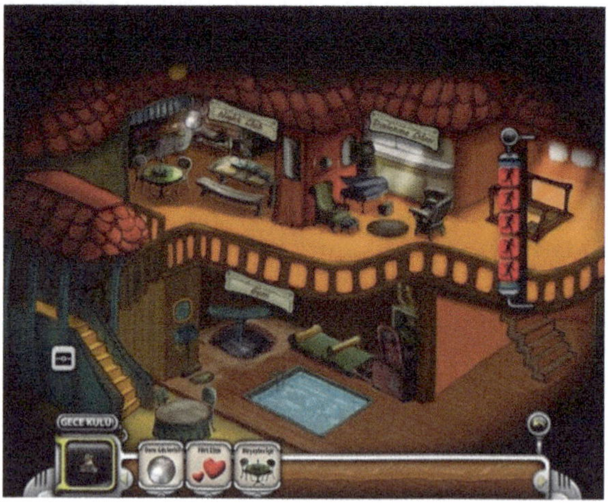

Fig. 4. The tavern.

- Which rooms do players use mostly?
- Can players understand how to change the pilot from the action panel?

(2) *Questionnaire:* In order to understand the players' feelings about the game, a questionnaire was used in addition to recording eye movements. Since the priority is the race, questions and answers related to the race are far more critical than the city buildings (the garage, the academy and the tavern). The scope of the study is comprised of analysing the eye-tracking data of participants. Hence, technically useful comments, collected with the questionnaire to support eye movements, are given in detail in this paper. However, the results of other questionnaire questions will not be explained in this paper.

3 Analysis and Discussion

According to eye-tracking data of 10 participants, the heat maps of different scenes were created. Both these maps and the answers of the questionnaires were analysed together sensitively to interpret game features. In order to improve the usability of the game, some assertions are done concerning the race, the garage, the academy and the tavern screens.

The Race

- **Registration:** During the registration, the players choose the time of the race they play. The system presents the races in each minute but does not let them register to the races that will start in less than two minutes. This makes players confused in that they cannot register to races for that period of time and they look for other actions to be occupied with. Since the race is the core part of the game, lack of feedback is a serious problem causing race preventions.
- After the registration, the time remaining for the race is shown on the map screen. However, it is placed under the panel, which made it impossible to control when the registration panel is open. The time for the race on the city screen also has a problem of not being realized due to its shape and colour.
- The race notification appears on the screen giving the information that there is less than five minutes before the race starts. However, the player misunderstands that the race would start after five minutes, which necessitates a change in this wording.
- The type of the race is provided on the registration panel. Although it is only text, it was designed like a button, which confused the players.
- **Race screen:** Race GUI and the buttons caused some confusion for the players. While they were trying to watch their car on the track, tactic buttons and the GUI drew attention, which disrupted the players while following their cars.
- Eye movements of participants focused on the tactic buttons located on the lower left part of the screen. Although the focus is the race, players give tactics from that part and the eye movements were generally on this part of the race screen. Thus, there is a need to change the place of controls (see Fig. 5).

Fig. 5. Heat map of eye movements during the race.

- Players do not understand if they actually drive the car or just give managerial instructions as giving tactics. It seems like they prefer driving.
- Once players realize that they need to give tactics during the race, they start to feel the autonomy to choose the needed tactics. The problem is the hesitation to use tactics because they consume pilot energy. For the first several levels, energy loss should be in the minimum level to encourage users to race.
- Tactic costs (how much they decrease the pilot energy) and benefits (what they provide for the race) cannot be easily understood due to the lack information while they are being used.
- Even if players find the tactic panel they are not keen to use it because tactic names are not appealing.
- Some of the tactic animations make players misunderstand that they do totally different actions rather than their own benefits.
- Players do not understand the effect area of the handbrake tactics. This is one of the things that will be learnt by playing.
- Some of the players have no idea about the word "pit", so the "P" letter on the button is meaningless for them. It may be written as "enter the pit stop" on the button to give a clue about that action.
- Due to the distance of pit stop button from other controls, it was impossible to see the red warning light to click and use the pit. The control displays and the pit button should be closely related to each other on the screen.
- When players clicked the pit button accidentally during the race, they could not cancel that action. As a result, they entered the pit and lost time in the race. Therefore, it was indicated as a boring experience to lose time due to the lack of the cancelation action.

- In the race, the initial position of each car is different than the others. There are three different locations to start. In order to start at the forefront, players wanted to have some actions to do until the race.
- **Race type:** Once the players start to control the black display that has the ranking, best lap time and best pit time, they want to get better values and race to be the top ranking player (see Fig. 5).
- Players generally don't understand how they are eliminated in the race because they generally miss the moment of elimination. Therefore, there is a need for an instant feedback.
- When player's car is captured by the destructor car, it is excluded from the race. However, players don't understand the function of the black destructor car, which follows other cars behind them. An animation may be added in order to show that the black destructor car beats other cars.
- Players have difficulty when they play on the second path in the type of destructor elimination races. The playability problem should be addressed by redesigning to improve the balance within levels.
- **Race ending:** When players are out of race due the exhaustion, they want to know the main reason of their elimination from the race. When their curiosity is clearly answered by a race finish panel, they could have a solution to play better in the following race.
- If pilot energy and car health drops down to zero, the player is out of the race. When he does not heal them in the city, these values are saved and the next race starts with these values. In that case, the player is eliminated in the first lap of the race, which is a meaningless loss of time and money for the registration. If the pilot or the car has no health, the race can start with a health of 20%. Lack of information about the car and the pilot after the race make players feel uncomfortable. The final values of the car and pilot should be shown to the players in the city. This would make players heal them in the city in order to prepare for the following races.

The Garage

- Players cannot see the attributes of the car easily, so they do not know which actions to choose in the garage at the first look, as illustrated by the heat map of eye movements in Fig. 6. They try to focus the car; however they are not informed what to do exactly for that purpose. In order to understand the need of either healing the car or modifying the parts, the car attributes should be shown clearly on the action panel.
- Tapping area of the car parts cannot be clicked easily. Car parts should be distributed better on the screen and the click area overlay should be fine-tuned for the parts.
- Players want more functional options for the cars to be able to have competitive advantage in the races over other players. There should be more customization options like different engines with different fuel consumption.
- Players want to distinguish themselves from others with different cars. Thus, they want to change the body of the car. There may be more alternatives in terms of different forms and colours even if they do not functionally change anything in the game.

Fig. 6. Heat map of eye movements in the garage.

The Academy

- The heat map of the eye movements in the academy is shown in Fig. 7. It is clear that eye movements focus on the action panel on the lower part of the screen. Through this action panel, players register their pilots to academy classes expecting that the

Fig. 7. Heat map of eye movements in the academy.

classes would start right away. 77% of them do not understand that there is a schedule and pilots take classes automatically according to the schedule. Feedback message should be clear and users should receive a notification when a class starts.

- Tactics' costs and benefits are confusing for the players. Hence, there should be a tooltip for every tactic and skill class in the academy showing them each detail.
- Names of the rooms and classes are not clear. Wording must be lucid and tell the player what it is directly related to.
- Players want to control previously learnt tactics in the academy. This can be seen in the inventory, but a skill tree should be adapted into the academy action panel, as well.

The Tavern

- Players want to see the attributes of pilots in detail, therefore the heat map of eye movements in the tavern focused on the action panel, as seen in the Fig. 8. However, they cannot see this information easily so they do not know which actions to be taken in the tavern. In order to minimize this confusion, pilot attributes should be shown on the action panel located on the lower side of the screen.

Fig. 8. Heat map of eye movements in the tavern.

- Players cannot recognize the selected room in the tavern due to lack of visuals. The rooms should be noticeable when clicked. It should be achieved by applying an overlay to the selected room.
- Players want to see a meaningful animation of the actions taking place in the rooms. It may not be feasible to create animations for all actions but at least there may be

some motion, movement and some kind of visual feedback telling that there is something happening there.

- Since the name 'tavern' does not cover all the actions available in the building, this name of the building should be changed.
- Weight loss of the pilots does not result in a noticeable change in the game play as it is so small compared to the car weight. It can be applied with a multiplier and can have a higher effect on the race, which can improve the playability feature of the game.

For the open-ended question of the questionnaire, which asks for the first thing that players remember from the game, most of the players commented that they mostly remembered the city buildings and their functions. The buildings can be considered helpful for the playability of the game in order to support the race part. The usability of city buildings was evaluated as being easy, clear and enjoyable.

4 Conclusion

This study was focused on the triangular relationship of human-computer interaction, usability and computer games. It provided tangible data collected from eye movement recordings of the participants. It focused on a managerial race game to question and to analyse usability problems.

For the game example, playability and usability factors play as significant roles as the fun factor, game mechanics and dynamics. As mentioned in the literature, playability and usability properties support each other mutually in the game. If one is lacking, the success of the game is debatable. Thus, in this study participants played the game in front of the observer and generated useful data in order to enhance the usability of the game. At this point, one answer given by players to one of the questionnaire questions should be remembered. They need to have instant and detailed feedback in every step and they do not want to wait unnecessarily during gameplay.

As a result of conducting eye-tracking studies, usability related problems such as unclear waiting, insufficient feedback messages, confusing graphical user interface, crowded game screen elements, inefficient controls, and unobtrusive images were detected.

The elements of graphical user interface and visual components like icons, buttons and display controls were redesigned to support gameplay. Their arrangements on the screen were organized again to let players take actions easily. In order to eliminate usage-related problems, the game was enriched by clear feedback messages.

This study helped to understand that eye-tracking technology could beneficially be used to improve a game in the design and developing process. Besides games, in order to improve the features of all digital products and interfaces, human-computer interactions and eye-tracking technology can be used effectively to determine the shortcomings during development steps. Taking the advantage of human-computer interaction research, necessary changes can be made before the final product release.

5 Limitations and Related Work

Since a game is a voluntary choice of players to have fun, satisfaction of players defines the success of the game. Having further ideas concerning a game design by the help of feedbacks from its users means further usability and experience development. This research shows that the more test groups are analysed with a higher number of participants, the more the game can be improved to achieve usability and playability success and the fun factor.

For the time being, the study was conducted with the user group involving young players. Nonetheless, since it is a mid-core social racing game, it can be preferred by different participants having distinct sex, interest and age properties. Several changes in these properties of the test group may lead to more exciting and marginal results for the development of the game. For the future work, the target group can be altered and the number of participants may be increased to compare and ensure the results with current study. Related work may be applied to hard-core games played on different platforms or mobile games to test their success in terms of usability and playability, as well. The laboratory circumstances can be adapted to collect eye-tracking data from different types of platforms. Moreover, interaction between the user and the computer to improve user experience related to separate products can be interpreted usefully via using eye-tracking technology.

References

1. Shneiderman, B.: Designing the User Interface: Strategies for Effective Human-Computer Interactions. Addison-Wesley, Boston (1997)
2. Acartürk, C., Çağıltay, K.: İnsan bilgisayar etkileşimi ve ODTÜ'de yürütülen çalışmalar [Human computer interaction and research at the Middle East Technical University]. 8. Akademik Bilişim Konferansı. Pamukkale University, Denizli, 9–11 February 2006
3. Booth, P.: An Introduction to Human-Computer Interaction. Lawrence Erlbaum Associates, Hove (1989)
4. Jorgensen, A.H.: Marrying HCI/Usability and computer games: a preliminary look. In: Proceedings of the 3rd Nordic Conference on Human-Computer Interaction (NordiCHI 2004), pp. 393–396. ACM Press, New York (2004)
5. Rollings, A., Adams, E.: On Game Design. New Riders Publishing, Thousand Oaks (2003)
6. Desurvire, H., Wiberg, C.: Game usability heuristics (PLAY) for evaluating and designing better games: the next iteration. In: Ozok, A.A., Zaphiris, P. (eds.) Proceedings of the 3rd International Conference on Online Communities and Social Computing: Held as Part of HCI International (OCSC 2009), pp. 557–566. Springer, Berlin (2009)
7. Bergstrom, J.R., Schall, A.J.: EyeTracking in User Experience Design. Morgan KaufmannPublishersInc, San Francisco (2014)
8. Guiping, C., Axu, H., Yonghong, L., Hongzhi, Y.: Review of linguistics understanding based on eye tracking system. J. Northwest Univ. Natl. (Nat. Sci.) **32**(2), 49–55 (2011)

9. Guan, Z., Lee, S., Cuddihy, E., Ramey, J.: The validity of the stimulated retrospective think-aloud method as measured by eye tracking. In: Grinter, R., Rodden, T., Aoki, P., Cutrell, E., Jeffries, R., Olson, G. (eds.) Proceedings of the SIGCHI Conference on Human Factors in Computing Systems, pp. 1253–1262. ACM Press, New York (2006)
10. Nielsen, J.: Usability Engineering. AP Professional, Boston (1993)

A Survey of Customer Review Helpfulness Prediction Techniques

Madeha Arif(✉), Usman Qamar, Farhan Hassan Khan, and Saba Bashir

College of Electrical and Mechanical Engineering, National University of Sciences and Technology (NUST), Islamabad, Pakistan
madeha.arif15@ce.ceme.edu.pk

Abstract. Online user generated reviews are now a vital source of product evaluation to both consumer and retailer. There is a need of knowing the factors, generally affecting the helpfulness of reviews and how to identify them. Various studies and researches have been conducted for finding the helpfulness value of online reviews in past recent years. In this paper we have summarized and then analyzed the past work of review helpfulness prediction in a systematic way. The paper provides brief of methodology of each study and how it is contributing towards this domain. It also emphasizes on the pros of the methods used in past and how they are lacking in determining few other aspects of the review helpfulness. The survey discovers that the most popular techniques used for helpfulness prediction are supervised ones and most frequently used are Regression Models and SVM.

Keywords: WOM (Word of Mouth) · Helpfulness · Product reviews Helpfulness ratio · Reviewer history

1 Introduction

Online shopping places allow users to purchase items and help to make purchase decisions as well [1]. Customer reviews on the commerce websites has led to proliferation of these platforms. Customer reviews are very important for both buyer and the business managers. Buyers can make purchase decisions based on the experiences of the users while the investors of the business can get help in making investment decisions in the future stocks. Online reviews are word of mouth these days. Online reviews are like unstructured big data [2]. Consumers can get peer suggestions out of them. Practically, it is not possible for consumer to undergo reading all the reviews and make buying decisions. So, it is a problem to organize all the reviews in a structured way and get the useful information out of them. There are almost thousands of reviews for a single product and one cannot simply read all of them. Online reviews have both positive and negative impact on the product [3].

There are multiple popular sites like Amazon, IMDB, Yelp.com etc. that are hosting reviews and managing multiple data of reviews and reviewers. Such large volume of data is very challenging for both consumers and businesses [4]. Only

K. Arai et al. (Eds.): IntelliSys 2018, AISC 868, pp. 215–226, 2019.
https://doi.org/10.1007/978-3-030-01054-6_15

volume and valence of Word of Mouth (WOM) is not enough to determine the helpfulness of a review. Past most researches have focused on this aspect only, few have identified various other factors present in the review text and reviewer attributes. Ghose et al. [5] have focused on lexical, semantic and grammatical features of reviews. There are few past researches which have provided a complete survey of the contributed work in this domain and all are not from the recent years [20, 21]. So, there is a need to do the updated survey of the most recent work. This paper presents a survey of previous researches which contributed towards finding the helpfulness of online reviews. We summarize the old researches from the recent past years to provide helpful content to the researchers of this domain for further improvement and innovation of ideas.

The research is organized as: Sect. 2 presents the research methodology adopted to carry out the survey. Section 3 shows the major findings and explanation of previous studies. It also includes analysis of the past work based on parameters. Discussion in Sect. 4 represents the major findings of the survey and recommendations based on it. Section 5 includes conclusion and future work of the study.

2 Research Methodology

This research is carried out in a systematic manner. Data collection is done by a search process that includes finding researches related to the area from various databases. The selected researches and then analyzed for the quality content and the data is extracted from those researches. We have selected researches from three databases i.e. IEEE, SPRINGER and ELSEVIER as shown in Fig. 1.

Fig. 1. Papers selected from scientific databases.

Following are the certain quality factors that were devised before carrying the research.

2.1 Effective Technique Proposed

It is main concern of the selected paper that it should propose any technique or model regarding finding the helpfulness of the online reviews.

2.2 Results Validation

All those papers which do not provide any assessment of the results based on the validation by some dataset are excluded from the very start of the research.

2.3 Repetition

Only papers with new and unique research are considered. Those representing same methodology or models are excluded.

2.4 Recent Researches

Most of the papers from the recent 3 years i.e. 2014–2016 and the current year i.e. 2017 are collected for analysis as they are the most updated researches. Recent research counts per year are shown in Fig. 2.

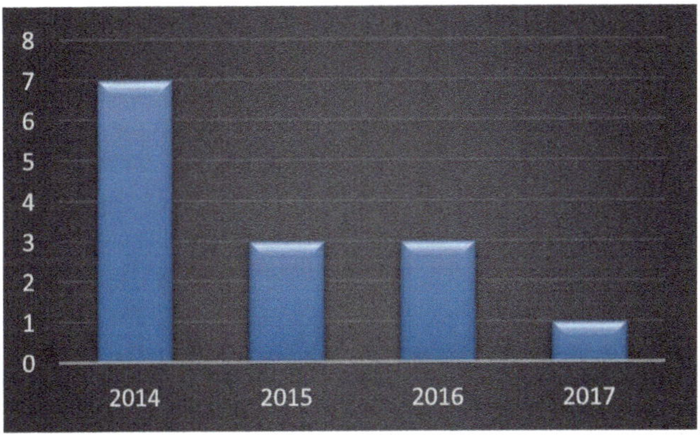

Fig. 2. Papers selected per year 2014–17.

3 Analysis and Results

This portion presents the major findings from the studies selected for the analysis. It provides brief of the methodologies adopted in each past research. Researches are categorized based on the classification techniques.

3.1 Support Vector Machine (SVM)

Zhang et al. [6] presented a technique that uses a machine learning approach to predict helpfulness of online reviews. Data collection was done from Amazon.com. Scrapy, an open source web scraping framework was used to scrap data from Amazon. Predicting helpfulness was considered a classification problem. Helpfulness was made label class with values in percentage obtained from helpful votes and total votes. This study used SVM as their predicting model. Tool used is WEKA tool. Sequential minimal optimization algorithm (SMO) proposed by John Platt is used for training purpose. Best accuracy produced by the model is at 73.3%. Average precision of the model appears to be 68.7%.

Krishnamoorthy et al. [12] have used three types of classifiers for determining the helpfulness of online reviews: SVM, Random Forest and Naïve Bayes. Two datasets were used, and random Forest outperformed the remaining two classifiers. For Dataset1 accuracy was recorded 81.33% and for Dataset2 accuracy was 77.02%.

3.2 Regression Models

Zhang [7] authors have used correlation based approach to determine the MAE and RMSE to determine the difference of predicted helpfulness and observation of product designers. PMCC, Pearson Product Moment Correlation Coefficient, has been used for finding association. Dataset is crawled from Jingdong.com. Linear Regression, SMOG, MLP and REPTree *are used as classifiers*. Best features are selected from each subset by PCA and Relief. Relief provided best results of feature selection.

Chen et al. in [9] have proposed that online review helpfulness can only be predicted based on its word embedding information. It is a word context of semantic representation of word. A word dictionary is prepared from N-Grams by removing all stop words and those words which have frequency lower than 10. TF-IDF value is considered as weight. Genism tool is used for word embedding features. Word embedding size is set to 100. Three regression models were used for results computation: LR-Linear Regression, SVR-Support Vector Regression and LSVR-Linear Support Vector Regression. Performance is evaluated using Root Mean Square Error.

Chua et al. [11] have applied multiple regression analysis to identify the review helpfulness. Dataset is taken from Amazon.com. Total vote count by the user was considered as control variable. Authors used ANOVA Tool to check the availability of helpfulness in a review.

Qazi et al. [14] have applied Tobit regression model to identify the helpful reviews from the feature set obtained from TripAdvisor.com. Regression model was selected based on the type of dependent variable i.e. helpfulness. For results verification authors have used Efron's pseudo R2 value. Its value was 0.167 and was found to be the most achievable value.

Ngo-Ye et al. [18] authors have used text regression model to predict online review helpfulness. Bag of Words (BOW) is made from the review text. RFM value of customer is obtained by RFM analysis. RFM can be referred to Frequency, Recency and Financial Value of a customer's transactional history. Additionally, review text and its helpfulness are also calculated. After preprocessing and feature selection, BOW is generated. Then Support Vector Regression Model is applied and it is evaluated by RMSE, RRSE, MSE and RAE.

3.3 Ensemble Learning Technique

Singh et al. [10] used Ensemble learning technique called Gradient Boosting Algorithm. Large datastream is divided into small chunks of data. Classifiers are trained on each of these small data chunks then developed a heuristic rule. Gradient boosting algorithm develops models based on an ensemble of tree, trains all the trees of the ensemble for unlike labels and combines the trees. It was found that review polarity is the most effective variable in deciding the helpfulness of a review. Dataset was obtained from Amazon.com.

3.4 Naïve Bayes

Krishnamoorthy et al. [12] used Naïve Bayes in addition to SVM. It is already explained in Section A.

3.5 Neural Networks

Lee et al. [17] presented the application of effect of reviews on economic transactions and decisions. Evaluation method is based on multi layered network topology with feed forward mechanism. Activation function used is:

$$\mathrm{OUT} = \frac{1}{1 + e^{-NET}}$$

OUT is based on binary values of (0,1) and final output is the weighted average of all the outputs. Initial weight was 0.3 and momentum was 0.1. HPNN was adopted with back propagation algorithm because of its good performance and simplicity. For prediction of accuracy of regression and HPNN, v-fold cross validation was used.

Zhang et al. [7] uses Neural Network approach and it has already been discussed in Sect. 3.2.

3.6 K-Nearest Neighbor Technique

Thuan et al. [16] have used Hybrid model to find similarity between users by using RFS preference and RFS helpfulness. User similarity matrix was constructed. Top N-nearest neighbor method used to find nearest neighbors of the target user. Mean Absolute Error has been used to evaluate the efficiency.

3.7 Statistical Hypothesis Tests

Chen et al. [8] proposed that helpfulness of a review as perceived by a customer is directly related to their purchase decisions. Authors presented three hypotheses.

Empirical verification of the hypotheses is performed by using dataset from Amazon.com. Top 100 reviewers were selected. Then review rating provided by each reviewer is found and compared that what is the most rating value given by a reviewer. For proving hypothesis 1, helpfulness percent is calculated by using helpful votes and total votes. For testing hypothesis 2, average helpfulness of most purchased products category of a consumer is calculated. For proving hypothesis 3, two review helpfulness vectors are calculated. First vector is average of review helpfulness of 100 reviewers. Second one is the average helpfulness of reviews that are voted as helpful by at least

100 times. 67 consumers fulfilled this condition. Average ratio of all 67 consumers is 33.5%.

Ullah et al. [13] used Natural Language Processing (NLP) technique to find the review helpfulness from emotional content. A word list was constructed from the reviews of Stanford morpheme analyzer; adjectives, nouns, verbs, special characters and other morphemes. Positive and Negative bag of Words (BOW) is created and dependency between polarity of a word and category by using Coefficient Correlation (CC). Dataset was taken from IMDB.

Hypothesis of emotional content effect on helpfulness was proved by checking the deviation of positive or negative emotional content from the helpfulness percentage.

Zhang et al. [15] proposed SENTRAL algorithm for sentiment analysis from review. A dependencies list was generated, and a tree was created. The DAL score is based on the level of pleasantness the 200,000 words list invokes in human's mind. On 1 to 3 scales, 1 is for the most unpleasant and 3 is for the utmost pleasant one. They normalized the score on a scale of 0–1. Error rate difference was 0.01 only.

Wan et al. [19] investigates about reliability of online review helpfulness based on the votes given by the review readers. Reviews are ranked based on helpfulness ratio value and then Pearson Chi-square test is conducted to identify the difference between review helpfulness based on features of reviewer.

3.8 Analysis Parameters

For analysis, parameters are defined in order to compare the proposed techniques in previous researches. Following are the parameters:

(1) *Research Problem*

Problem domain of the research, either it is economic review, sales analysis, determining the customer's product selection trend, finding review or reviewer characteristics or to prove some other hypotheses.

(2) *Proposed Approach*

Methodology proposed to solve the research problem. It can be any process, or a framework devised to classify the reviews as helpful or unhelpful.

(3) *Dataset Source*

Any e-commerce website from which authors have prepared the feature set. It can include Amazon, Yelp, IMDB, Google Play, etc.

(4) *Feature Count*

Total number of features used in dataset.

(5) *Type of Features*

Type of features indicates either only review or reviewer features are used, or both are considered.

Table 1. Techniques using dataset from Amazon

Source/Category	Research problem	Proposed approach	Feature count	Type of features		Validation results
				Review	Reviewer	
[6] A	A general model for automatic prediction of the online review helpfulness	Predictive model generated by SVM on WEKA Tool	22	✓	✓	Average accuracy: 68.7%
[8] G	Determine helpfulness of online reviews from consumer's perspective	Statistical verification of hypotheses	3	✓		Empirical tests proved all the three proposed hypotheses
[9] B	Use of Word Embedding information to identify helpfulness of reviews	LR, LSVR, SVR	7	✓		RMSE of the proposed feature set: 0.248
[10] C	Determine helpfulness of online user reviews	Ensemble learning technique to minimize the MSE	26	✓	✓	MSE decreases for 100 trees and then remains constant
[11] B	Review helpfulness to determine review sentiment, review quality and type of product	ANOVA tool multiple regression analysis	8	✓		All the six regression analysis results were more than acceptable threshold i-e 0.2
[12] A, D	Effect of linguistic features on review helpfulness	Naïve Bayes, SVM, Random Forest	10	✓		D1: 81.33% D2: 77.02% Random Forest performs best
[15] G	Helpfulness of online reviews using Confidence and probabilistic distribution	Prior beta distribution and Confidence Interval	2	✓		AbsoluteError: 0.014 RMSE: 0.03984
[18] B	Reviewer engagement characteristics and their impact on online review helpfulness prediction	Text regression model. BOW full model constructed for all the words	10	✓		Hybrid model of BOW and RFM performed best in all of the models
[19] G	Check the reliability of online reviews helpfulness based on customer voting	Chi-square tests to identify difference of helpfulness ratio	5		✓	Pearson Chi-Square results = gender: 1 age: 3 ethnicity: 3 Income: 3 Mobile use: 4

Table 2. Techniques using dataset from other websites

Source/Category	Research problem	Proposed approach	Dataset source	Feature count	Type of features		Validation results
					Review	Reviewer	
[7] B, E	Relationship between review valency, votes and its helpfulness	PCA and RELIEF for features selection. Used algorithms: MLPNN, LR, SVR, FDT	Jingdong	26	✓		MAE and RMSE are used for error Std Error was: 0.03 to 0.04
[13] G	Impact of emotions on movies reviews helpfulness	Coefficient Correlation (CC) used for measurement of dependency between word and its polarity category	IMDB	6	✓		Hypothesis was proved based on deviation of positive/Negative content from helpfulness percentage
[14] B	Factors contributing towards online review helpfulness	TOBIT Regression Model	TripAdvisor	3	✓		Efron's pseudo R2 value of
[16] F	A recommender system with review helpfulness features	Top N-nearest neighbor method to find nearest neighbors of the target user	Epinions Dataset	12	✓		MAE = 0.77
[17] E	Using Neural Network approach to determine online review helpfulness	HPNN and Regression. HPNN outperformed Regression	TripAdvisor	19	✓	✓	v-fold cross-validation. Wilcoxon test: z = 3.059 and significance = 0.002
[18]	Influence of reviewer engagement characteristics on review helpfulness	Text regression model. BOW full model constructed for all the words	Amazon, Yelp	10	✓	✓	Hybrid model of BOW and RFM performed best in all of the models

(6) *Validation Results*

Results of the validation process of the proposed technique like accuracy, Root Mean Square Error or Absolute Error etc.

Table 1 shows results of Amazon datasets while Table 2 shows analysis results of techniques using dataset from other websites like IMDB yelp, etc.

4 Discussion

Volume of online reviews is increasing, and it is becoming hard to analyze the opinions provided by users in their reviews. There are several techniques proposed for predicting online reviews helpfulness. In this paper we have analyzed techniques of prediction of online review helpfulness. All the techniques are efficient in some way and are proving their proposed hypotheses through empirical data.

Methodology of [6, 12] have used SVM classification to identify the reviews as helpful. SVM performs better classification but in [12] authors have statistically proved that Random Forest has outperformed in case of the data they have used. Authors in [7, 9, 11, 18] have used regression models. RMSE and MAE are evaluation parameters in regression models. Tobit regression is also used as a predictor of review helpfulness in [14]. Most of the datasets are prepared from Amazon.com by crawlers and features are different for each research except few overlapping ones. Figure 3 shows distribution of algorithms in existing researches. Table 3 shows future works and limitations of past researches.

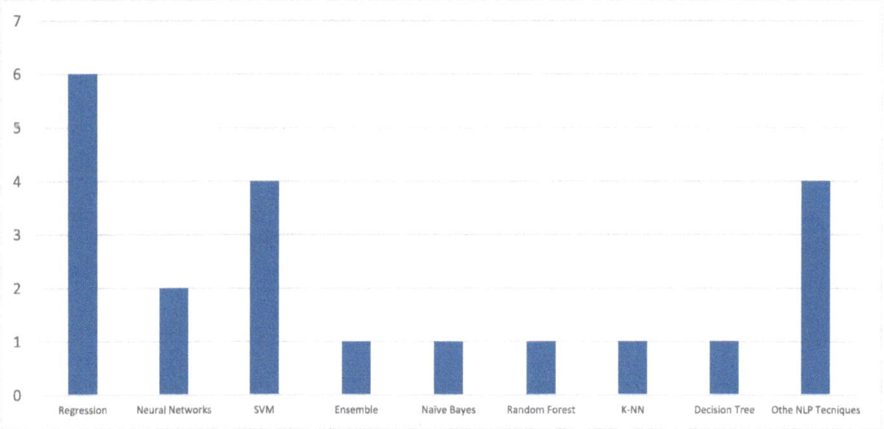

Fig. 3. Distribution of algorithms in previous researches.

Table 3. Future work and limitations of past researches

Source	Future work/limitations
[6]	To consider more factors, greater product categories for classification, more hypotheses, and parameters to construct a quantitative helpfulness model
[7]	There are significant differences between online reviews and traditional customer requirement, so there exist some problems in verifying the obtained customer requirement
[8]	Paper explores relationship between review and helpfulness factors, but quantitative model is not constructed. Review data is collected from Amazon.com only and divided into very few categories. Data from other websites is not verified
[9]	Some of the variables like in the data are proxies for the actual measure that one would need for more advanced empirical modeling. It only focused on sales data. While in future work can look at real demand data
[10]	One of the major limitations of this work is that the non-English words are counted as wrong words in this study, but several non-English words contain polarity information. These non-English words may be considered for polarity calculation in a future study Only e-commerce website reviews were considered
[11]	–
[12]	Only e = commerce retailer websites are considered for the study
[13]	NLP technique is not as efficient as human nature Future research should also examine the objective and subjective nature of reviews and product-type reviews, which can provide more insights regarding customer reviews
[14]	Results are strictly generalizable only to hotel reviews. For future work, researchers are encouraged to sample from a different domain
[15]	Developing new ways for both review helpfulness prediction and review ranking mechanism, as well as in applications with large online data for e-business services
[16]	–
[17]	–
[18]	Use datasets from other domains to verify the results and make the methodology generic for all type of products
[19]	Improve ranking process and further expanding the review samples and testing the extent of our findings in different product categories and domains

5 Conclusion and Future Work

Research is conducted by keeping in view the need of exploring the summary of recent researches for assistance in further researchers. This research is structured and systematic, so it can help readers catch the previous knowledge of this domain. Limitations and future work can help to get the need for research information or research gap to be filled in by further studies.

For future work, we will propose an efficient model for predicting online review helpfulness that will be independent of type of dataset and will be more accurate than the existing techniques.

References

1. Casalo, L.V., Flavián, C., Guinalíu, M.: Understanding the intention to follow the advice obtained in an online travel community. Comput. Hum. Behav. **27**(2), 622–633 (2011)
2. Zhu, F., Zhang, X.: Impact of online consumer reviews on sales: the moderating role of product and consumer characteristics. J. Mark. **74**(2), 133–148 (2010)
3. Lee, J., Park, D.H., Han, I.: The effect of negative online consumer reviews on product attitude: an information processing view. Electron. Commer. Res. Appl. **7**(3), 341–352 (2008)
4. Cao, Q., Duan, W., Gan, Q.: Exploring determinants of voting for the "helpfulness" of online user reviews: a text mining approach. Decis. Support Syst. **50**, 511–521 (2011)
5. Ghose, A., Ipeirotis, P.G., Li, B.: Designing ranking systems for hotels on travel search engines by mining user-generated and crowdsourced content. Mark. Sci. **31**, 493–520 (2012)
6. Zhang, Y., Zhang, D.: Automatically predicting the helpfulness of online reviews. In: 2014 IEEE 15th International Conference on Information Reuse and Integration (IRI), pp. 662–668, August 2014
7. Zhang, Z., Qi, J., Zhu, G.: Mining customer requirement from helpful online reviews. In: Enterprise Systems Conference (ES), pp. 249–254, August 2014
8. Chen, Y., Chai, Y., Liu, Y., Xu, Y.: Analysis of review helpfulness based on consumer perspective. Tsinghua Sci. Technol. **20**, 293–305 (2015)
9. Chen, J., Zhang, C., Niu, Z.: Identifying helpful online reviews with word embedding features. In: International Conference on Knowledge Science, Engineering and Management, pp. 123–133, October 2016
10. Singh, J.P., Irani, S., Rana, N.P., Dwivedi, Y.K., Saumya, S., Roy, P.K.: Predicting the "helpfulness" of online consumer reviews. J. Bus. Res. **70**, 346–355 (2017)
11. Chua, A.Y., Banerjee, S.: Helpfulness of user-generated reviews as a function of review sentiment, product type and information quality. Comput. Hum. Behav. **54**, 547–554 (2016)
12. Krishnamoorthy, S.: Linguistic features for review helpfulness prediction. Expert Syst. Appl. **42**, 3751–3759 (2015)
13. Ullah, R., Zeb, A., Kim, W.: The impact of emotions on the helpfulness of movie reviews. J. Appl. Res. Technol. **13**, 359–363 (2015)
14. Qazi, A., Syed, K.B.S., Raj, R.G., Cambria, E., Tahir, M., Alghazzawi, D.: A concept-level approach to the analysis of online review helpfulness. Comput. Hum. Behav. **58**, 75–81 (2016)
15. Zhang, Z., Wei, Q., Chen, G.: Estimating online review helpfulness with probabilistic distribution and confidence. In: Foundations and Applications of Intelligent Systems, pp. 411–420 (2014)
16. Thuan, T.T., Puntheeranurak, S.: Hybrid recommender system with review helpfulness features. In: TENCON 2014-2014 IEEE Region 10 Conference, pp. 1–5, October 2014
17. Lee, S., Choeh, J.Y.: Predicting the helpfulness of online reviews using multilayer perceptron neural networks. Expert Syst. Appl. **41**, 3041–3046 (2014)
18. Ngo-Ye, T.L., Sinha, A.P.: The influence of reviewer engagement characteristics on online review helpfulness: a text regression model. Decis. Support Syst. **61**, 47–58 (2014)
19. Wan, Y., Nakayama, M.: The reliability of online review helpfulness. J. Electron. Commer. Res. **15**, 179 (2014)

20. Karimi, S., Wang, F.: Online review helpfulness: impact of reviewer profile image. Decis. Support Syst. **96**, 39–48 (2014)
21. Ghose, A., Ipeirotis, P.G.: Estimating the helpfulness and economic impact of product reviews: mining text and reviewer characteristics. IEEE Trans. Knowl. Data Eng. **23**, 1498–1512 (2011)

Automatized Approach to Assessment of Degree of Delamination Around a Scribe

Petr Dolezel[1(✉)], Pavel Rozsival[1], Veronika Rozsivalova[2], and Jiri Tvrdik[1]

[1] Faculty of Electrical Engineering and Informatics, University of Pardubice, Pardubice, Czech Republic
{petr.dolezel,pavel.rozsival}@upce.cz, jiri.tvrdik@student.upce.cz
[2] Metal Trade Comax, Velvary, Czech Republic
veronika.jaskova@mtcomax.cz

Abstract. The aim of this paper is to select a suitable methodology and a sequence of procedures for assessment of degree of delamination around a scribe. This should be achieved with as high level of automation as possible. To keep the universality of application, regulation ISO 4628-8:2007 was followed. The procedure proposed in the paper implements an optical device for data acquisition and two approaches of image processing. It also provides a possibility of manual correction of the results to maintain the robustness of the solution. The preliminary evaluations of the results indicate, that more than 90% of the samples are interpreted correctly without any human operator interactions.

Keywords: Delamination · Optical system · Image processing ISO 4628

1 Introduction

All paint coatings degrade over time, no matter what they are exposed to. The resulting defects can be very diverse; for example, chalking, blistering, flaking or rusting of the painted metal. The evaluation of these defects must be defined in the most universal way possible, so that various involved entities, such as product vendors, can communicate efficiently with other interested parties. ISO 4628 "Paints and varnishes. Evaluation of degradation of coatings. Designation of the quantity and size of the defects and intensity of uniform changes to appearance" was proposed for the purpose of the assessment and quantification of the main defects that can occur in coatings [1].

ISO 4628 is a comprehensive regulation, which provides more than necessary number of tools for every possible type of coating defects. Nevertheless, manual evaluation of samples is implied [2], and, surprisingly, not many methods for automatized application of ISO 4628 have been presented, so far. Whitfield et al. presented an approach to quantify corrosion susceptibility based upon electrical resistance and colorimetric outputs [3]. A new concept of corrosion surface damage analysis by using the digital image processing is proposed in [4]. Other

K. Arai et al. (Eds.): IntelliSys 2018, AISC 868, pp. 227–236, 2019.
https://doi.org/10.1007/978-3-030-01054-6_16

approaches, which use optical systems, are presented by Kapsalas et al. [5] or by Cringasu et al. [6]. The issue of cited works is the focus only for a limited range of materials or surfaces. Conversely, industrial entities ask for robust systems with a wide range of scope. Some techniques applied to different phenomena, which could be promising to use to the discussed issue have been published too [7–9].

Based on these facts, the aim of this work is to develop and present an approach for application of one part of ISO 4628 to a wide range of coatings, regardless of color, asperity or reflectivity.

In the following paragraphs, the aim of the our project is properly defined, the methodology of the assessment method is described, the solution is proposed and, eventually, several examples are presented.

2 Aim of the Project

Producers of coating compositions need to test their products in various corrosive environments. One of the regulations, which defines the degree of coating degradation, is called "Assessment of degree of delamination and corrosion around a scribe" (ISO 4628-8:2007). This standard defines the criteria for evaluating both intensity and quantity of the delamination of paint coating or corrosion of the covered metal. A numerical scale of 0 to 5 is adopted for evaluation, where 0 signifies the absence of changes, while 5 means defects so notable that further discrimination is not reasonable [10].

The testing procedure starts with sample preparation. A set of metal plates is covered with coating composition and, after some time, a series of testings is performed; an assessment of degree of delamination around a scribe is one of them. Thus, the plates are scribed with a sharp edge and exposed to a corrosive environment for a defined period of time (120 h, 240 h, 480 h, 720 h, ...). After exposure, the plates are rinsed with tap water and the residues of water are then removed using compressed air. A loose coating is cut off with a knife blade held at an angle. A boundary should be situated where the coating becomes tightly adhered to the plate, see Figs. 1, 2 and 3 for some examples of coated metals prepared by this procedure.

The area of delamination could be then determined by measurement and calculation or by comparison to pictorial standards. The issue is, that the degree of delamination is determined either manually, or by the autonomous system, which is, however, prepared only for one type of coating. Therefore, the project, which should provide a general autonomous tool for the degree of delamination assessment, has been set. The resulting device is presented in this paper.

3 Methodology

According to ISO 4628-8:2012 [10], two possibilities of the assessment of degree of delamination around a scribe can be implemented.

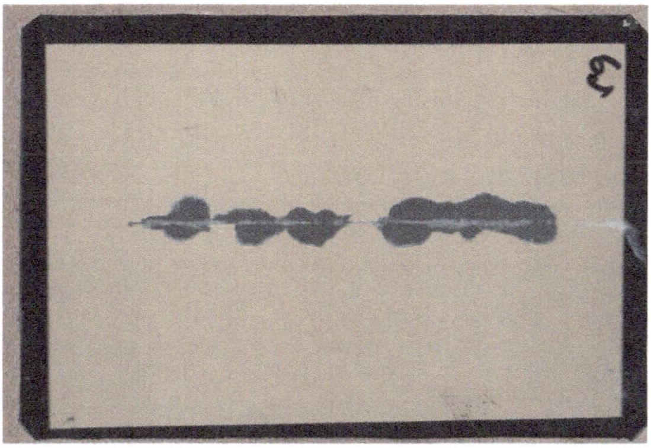

Fig. 1. Example 1.

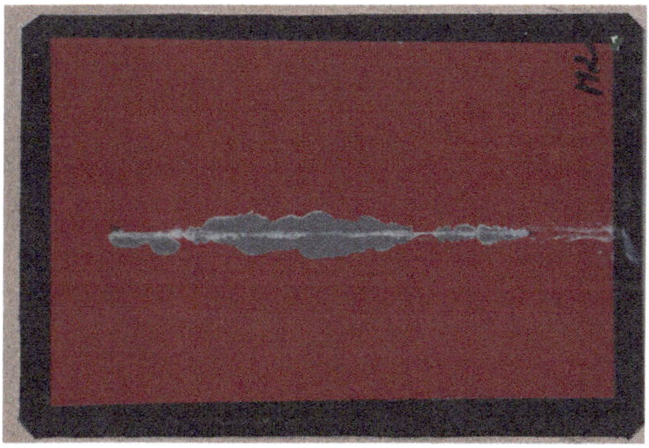

Fig. 2. Example 2.

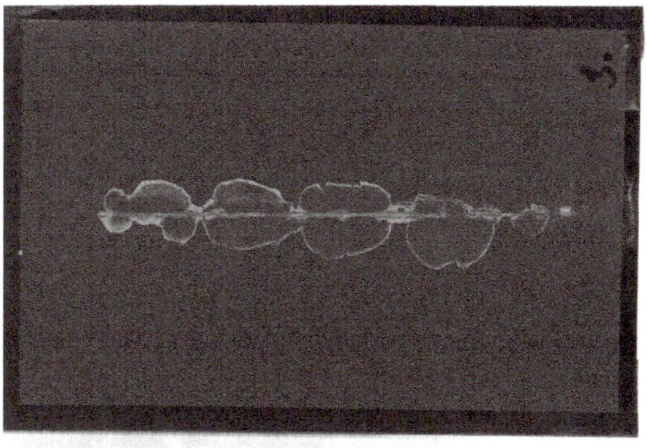

Fig. 3. Example 3.

3.1 First Option

The width of the area of delamination has to be measured at a minimum of six points uniformly distributed along the scribe. With that, the arithmetic mean is determined and the resulting value is designated as the mean overall width of the zone of delamination, d_1, in millimeters.

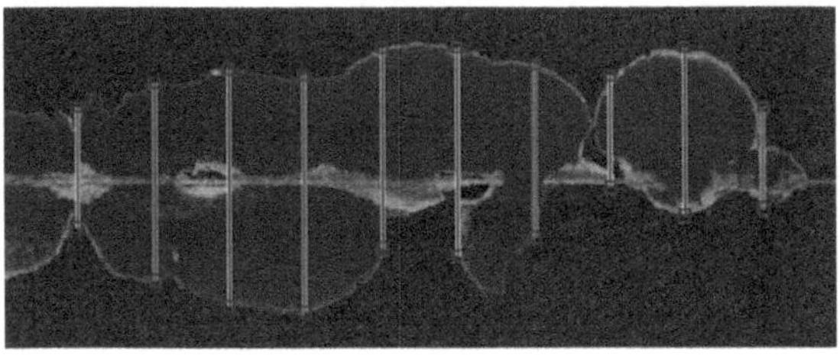

Fig. 4. Mean overall width of the zone of delamination is, in this case, defined as a sum of the lengths shown in the figure divided by 10.

Then, the degree of delamination d, in millimeters, can be calculated using the equation:

$$d = \frac{d_1 - w}{2}, \tag{1}$$

where w is the width of the original scribe, in millimeters. See Fig. 4 for an example of application.

3.2 Second Option

The area of delamination should be determinable for this approach. The norm proposes laying transparent millimeter-grid paper over the plate and counting the number of squares corresponding to the area. Then, the degree of delamination d, in millimeters, can be calculated using

$$d = \frac{A_d - A_l}{2l}, \tag{2}$$

where A_d is the area of delamination, including the scribe area, in square millimeters, A_l is the area of the scribe in the area evaluated, in square millimeters, and l is the length of the scribe in the area evaluated, in millimeters. See Fig. 5 for an example of application.

4 Implementation

Apparently, the second approach of the degree of delamination assessment described in Sect. 3.2 is more accurate and the procedure for acquiring the black and white image shown in Fig. 5 seems to be feasible. Hence, implementation

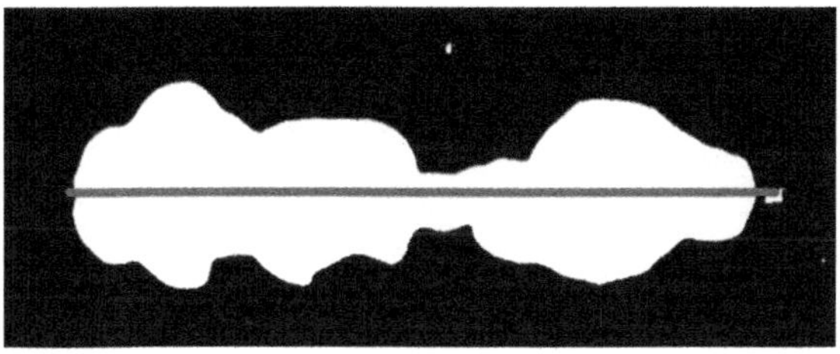

Fig. 5. White area is the area of delamination, gray is the area of the scribe.

of the second approach is tempting. However, after extensive literature research as well as after many experiments, we were not able to find a robust enough image processing technique, which can provide a correct black and white image regardless of color, asperity or reflectivity of coating. Therefore, the intended device should provide a functionality for possible corrections performed by a human operator. And, a considering user friendly interface for these corrections, it would be much more convenient for the human operator to operate with a finite number of abscissae (Fig. 4) rather than with continuous areas (Fig. 5).

In consequence, the first approach (Sect. 3.1) is selected for implementation. A block diagram of the intended device is shown in Fig. 6.

A generic color scanner is used for image acquisition in a first prototype of the device. After that, the acquired image has to be transformed into a black and white form, where white color represents the area of delamination. Clearly, this is the most important step of the whole procedure. After many experiments and using a trial and error method, two possibilities of this transformation were implemented into a control software.

The first approach implements an active contours method without edges proposed by Chan and Vese in 2001 [11]. This complex approach was initially designed for object detection, and it is based on techniques of curve evolution, Mumford-Shah functional for segmentation, and level sets. Adversely, the other approach is based more on "brutal force" and it implements a set of very basic image processing operations:

- Subtracting the image from the image acquired at the beginning of the process. The result should roughly represent the area of delamination.
- Transformation of the image into a grayscale image. To be specific, an intensity image is determined, where grayscale values I are computed by forming a weighted sum of the R, G, and B components

$$\mathrm{I} = 0.2989\mathrm{R} + 0.5870\mathrm{G} + 0.1140\mathrm{B}. \tag{3}$$

- Filtration of the image using 2D median filtering [12], where each output pixel contains the median value of neighboring pixels in the input image. The size of neighborhood corresponds with the size of the original image; 0.01% of the image is used for the neighborhood.

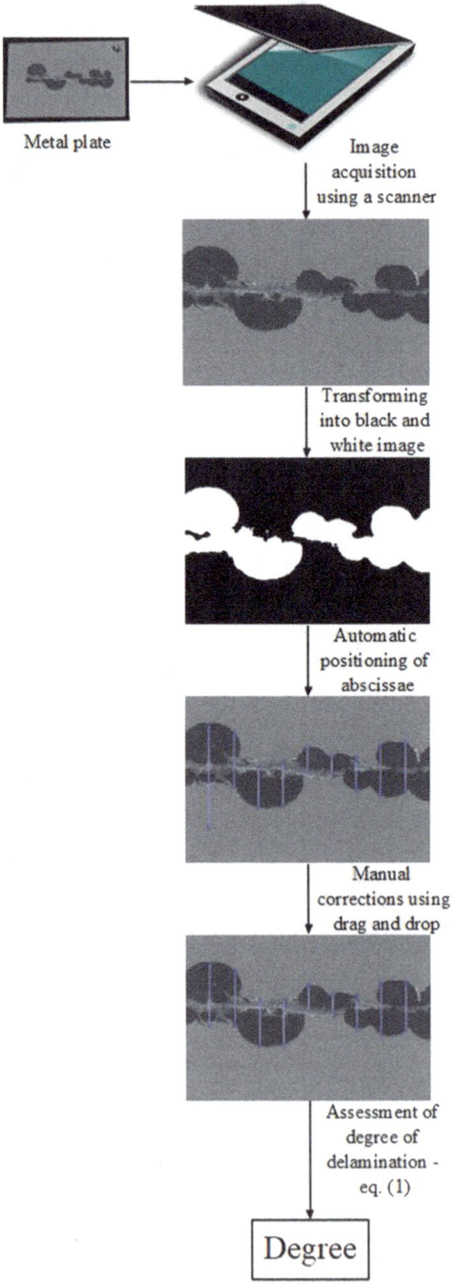

Fig. 6. Block diagram of intended device.

- Transformation of the image into a black and white image using a threshold function after adjusting the image intensity.

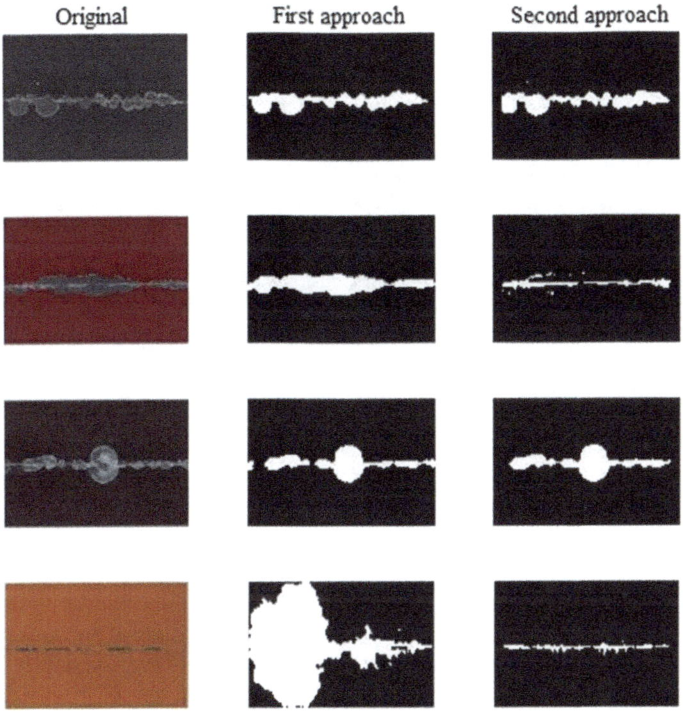

Fig. 7. Transformation into black and white image. Most of the samples were transformed sufficiently (see first and third row). In several cases, only one approach provided good results (see second and fourth row).

For the purposes of the project, 112 samples of metal plates were provided by the contracting authority. At least one of the previously described approaches worked sufficiently with the majority of the provided samples. More detailed results are summarized at the end of the paper. Some examples are shown in Fig. 7.

As a next step of the processing, the positioning of the abscissae is automatically performed using a black and white image gained by one of the previously mentioned approaches. The correct positions of the abscissae are necessary for the application of (1). Column coordinates of the abscissae are known since the number of them is defined by the user (should be equal or bigger than 6 according to ISO 4628-8) and their positions should be distributed uniformly. The row coordinates are found simply as the transitions from black to white, starting from the top and bottom of the image.

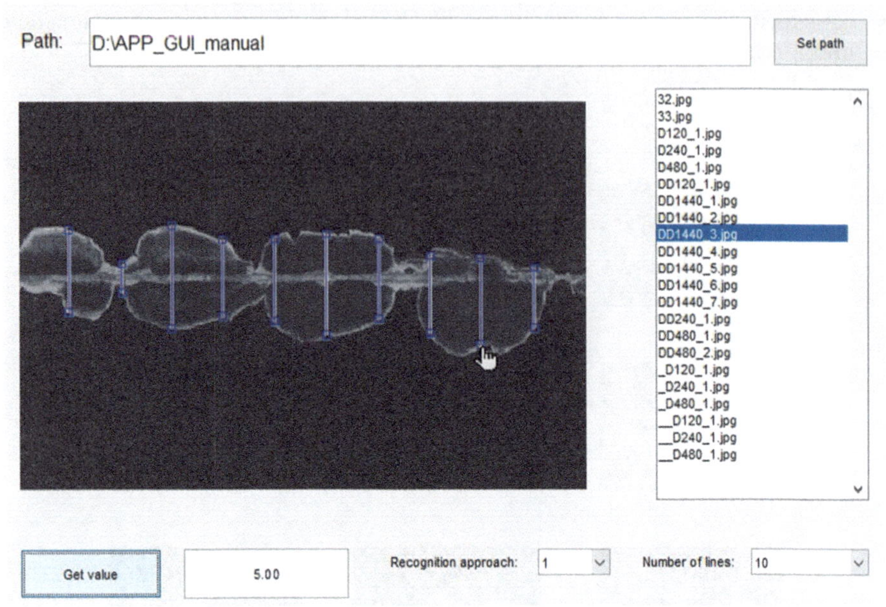

Fig. 8. Detail of human-computer interface.

The last step of the processing consists of possible manual manipulation with the abscissae to correct their positions. This manipulation is sometimes necessary to get reliable results and it definitely rises the robustness of the device. A human operator can manipulate the abscissae using a drag and drop approach with a computer mouse. The main human-computer interface of the device is shown in Fig. 8.

5 Case Study

A prototype of a device proposed in previous sections was constructed and tested with an available dataset. A color scanner, which is used for image acquisition, provides 24bit RGB 3600×2400px images. Image processing is performed by a standard personal computer. The results obtained with 112 metal plates provided by the contracting authority are summarized in the table below. Three types of results are defined there:

- *Sufficient evaluation with both approaches* means that the sample is automatically evaluated correctly regardless of the approach selected for transformation into a black and white image.
- *Sufficient evaluation with at least one approach* means that the sample is automatically evaluated correctly with at least one approach selected for transformation into a black and white image (includes previous subset).

- *Human operator intervention required* means that neither of approaches provided a correct solution and it was necessary to correct the positions of abscissae manually.

Note, that the dataset contains various kinds of coatings including different colors, surfaces of diverse granularity, or paintings with scattered reflectance. On the other hand, the number of samples is not entirely predicative. Thus, consider these results as an initial insight into the issue rather than a comprehensive statistical analysis. From the contracting authority perspective, the device is being applied in an operating laboratory with very positive feedback. Summary of the results is shown in Table 1.

Table 1. Summary of the results

Sufficient evaluation with both approaches	68%
Sufficient evaluation with at least one approach	91%
Human operator intervention required	9%

6 Conclusion

The device, which implements the autonomous assessment of the degree of delamination around a scribe according to ISO 4628-8, is proposed and tested in this contribution. The device was designed on demand and its first functional prototype is already used by the contracting authority.

The proposed procedure implements the optical device for data acquisition, two ways of image processing and provides a possibility of manual correction of the results. The preliminary evaluations of the results indicate, that more than 90% of the samples are interpreted using full automation, while the rest of the samples required operator intervention. Apparently, the prototype is fully open to further improvements, which will be definitely implemented with an extension of the dataset.

Acknowledgment. The work has been supported by the Funds of University of Pardubice, Czech Republic. This support is very gratefully acknowledged. We would also like to show our gratitude to METAL TRADE COMAX, one of the first representatives in continuous coil coating in Europe. METAL TRADE COMAX made the initial request to create the device for the autonomous assessment of the degree of delamination around a scribe and also provided the dataset of metal plates.

References

1. Paints and varnishes – Evaluation of degradation of coatings – Designation of quantity and size of defects, and of intensity of uniform changes in appearance – Part 1: General introduction and designation system. International Organization for Standardization, Geneva, CH, Standard, January 2016
2. Hanus, J.: Selection and evaluation of singlelayer coating compositions in corrosive environments. Acta Universitatis Agriculturae et Silviculturae Mendelianae Brunensis **59**(5), 53–64 (2011)
3. Whitfield, M., Bono, D., Wei, L., Van Vliet, K.: High-throughput corrosion quantification in varied microenvironments. Corros. Sci. **88**, 481–486 (2014)
4. Choi, K., Kim, S.: Morphological analysis and classification of types of surface corrosion damage by digital image processing. Corros. Sci. **47**(1), 1–15 (2005)
5. Kapsalas, P., Zervakis, M., Maravelaki-Kalaitzaki, P.: Evaluation of image segmentation approaches for non-destructive detection and quantification of corrosion damage on stonework. Corros. Sci. **49**(12), 4415–4442 (2007)
6. Cringasu, E.C., Dragomirescu, A., Safta, C.A.: Image processing approach for estimating the degree of surface degradation by corrosion. In: 2017 International Conference on Energy and Environment (CIEM), pp. 275–278, October 2017
7. Dobrovolny, M., Bezousek, P., Hajek, M.: Application of a cumulative method for car borders specification in image. Radioengineering **17**(4), 2.75–2.79 (2008)
8. Skrabanek, P., Majerik, F.: Detection of grapes in natural environment using hog features in low resolution images. J. Phys. Conf. Ser. **870**(1), 012004 (2017)
9. Rahul, S., Honc, D., Dusek, F., Gireesh, K.: Frontier based multi robot area exploration using prioritized routing. In: 30th European Conference on Modelling and Simulation, ECMS 2016, pp. 25–30, June 2016
10. Paints and varnishes – Evaluation of degradation of coatings – Designation of quantity and size of defects, and of intensity of uniform changes in appearance – Part 8: Assessment of degree of delamination and corrosion around a scribe or other artificial defect. International Organization for Standardization, Geneva, CH, Standard, November 2012
11. Chan, T.F., Vese, L.A.: Active contours without edges. IEEE Trans. Image Process. **10**(2), 266–277 (2001)
12. Lim, J.S.: Two-Dimensional Signal and Image Processing. Prentice Hall, Englewood Cliffs (1990)

Face Detection and Recognition for Automatic Attendance System

Onur Sanli and Bahar Ilgen(✉)

Department of Computer Engineering, Istanbul Kültür University, Istanbul, Turkey
onursanli03@gmail.com, b.ilgen@iku.edu.tr

Abstract. Human face recognition is an important part of biometric verification. The methods for utilizing physical properties, such as human face have seen a great change since the emergence of image processing techniques. Human face recognition is widely used for verification purposes, especially if individuals attend to lectures. There is a lot of time lost in classical attendance confirmations. In order to solve this time loss, an Attendance System with Face Recognition has been developed which automatically tracks the attendance status of the students. The Attendance System with Face Recognition performs daily activities of the attendance analysis which is an important aspect of face recognition task. By doing this in an automated manner, it saves time and effort in classrooms and meetings. In the scope of the proposed system, a camera attached to the front of the classroom continuously captures the images of the students, detects the faces in the images, compares them with the database, and thus the participation of the student is determined. Haar filtered AdaBoost is used to detect the real-time human face. Principal Component Analysis (PCA) and Local Binary Pattern Histograms (LBPH) algorithms have been used to identify the faces detected. The paired face is then used to mark course attendance. By using the Attendance System with Facial Recognition, the efficiency of lecture times' utilization will be improved. Additionally, it will be possible to eliminate mistakes on attendance sheets.

Keywords: Face detection · Face recognition
Principal Component Analysis (PCA) · Local Binary Pattern Histogram (LBPH)
Viola-Jones

1 Introduction

Automatic facial analysis, including facial recognition and facial expression recognition, has become a very active topic in computer vision area. Face authentication/verification is utilized in several tasks including civil and security applications, electronic transactions. It is also useful in the areas such as interactive games and movies, image searching and wearable systems.

Biometric face recognition is basically applied in three main areas: (1) continuous participation systems and employee management, (2) visitor management systems, and (3) minimal authorization systems and access control systems. In the scope of the

K. Arai et al. (Eds.): IntelliSys 2018, AISC 868, pp. 237–245, 2019.
https://doi.org/10.1007/978-3-030-01054-6_17

research area, holistic methods, such as *Principal Component Analysis* (PCA) and *Local Binary Pattern Histograms* (LBPH) have been utilized frequently.

Face recognition task in classrooms or meeting rooms is one of the most popular applications in the research area. Traditionally, the attendance information of the students is supplied manually using the attendance sheet issued by the members of the class, which is a time-consuming activity. It is also very difficult to verify the attendance information of a single student without considering his/her respond in a large classroom environment with distributed branches. The facial recognition shows how to use an effective attendance system to automatically record the presence of a registered person in a designated place. The proposed system also maintains a daily record for each individual with a universal system time.

Although it is straightforward for human perception, automatic facial recognition requires considering several challenges. These difficulties result from variations between the appearances of the same face and similar appearances of the different faces. The former is known as intrapersonal variations. This includes the variations of the same face caused by the issues such as head pose, lighting and face expression. There may be additional difficulties such as hair styling and aging which can also reduce the accuracy of the system. The latter difficulty is also known as interclass similarity. It usually results from similar faces of relatives, twins or any people.

An image can be expressed in mathematics as a multidimensional matrix and a matrix value. It can be treated as a vector with image size and direction. It is known as an image vector. If x represents a p x q image vector, then x is the matrix of the image vector. Thus, the image matrix can be represented as $x = \{x_1, x_2, \ldots, x_n\}^T$, where T is the transpose of the matrix x. It is also known that the certain cases such as definition of glass in an image matrix might be very difficult and it requires additional approaches that can handle these limitations. The algorithm proposed in this paper successfully overcomes these limitations. The rest of the paper is organized as follows. Section 2 summarizes previous work on the research area. In Sect. 3, the proposed methodology is explained. Sections 4 and 5 present our experimental results, conclusion and the future work respectively.

2 Related Work

There are several efforts in the face recognition area which provides support to numerous applications. The automated attendance system in [1] involves automating attendance follow-up with face recognition and perception techniques. The purpose of the system is to keep track of the students' attendance in the classes and to ensure that the faculties have access to the information of the students who are kept in these registers. In the study, Deep Reinforcement Learning approach with *Convolutional Neural Networks* (CNNs) was used to compare user identities for mobile payments, focusing on a face recognition process under uncertain facial features. The OpenCV [2] library was used in system development. PCA algorithm is also utilized for learning and training phase of face images. The images of the users can be taken with the phone camera. The central database can be accessed in the proposed mobile-based system.

The results showed that the facial image of CNN schemas could increase sensitivity to recognition compared to average sensitivity.

In the scope of the research carried out by Symbiosis International University [3], Viola & Jones algorithm is used for face detection while PCA and LDA algorithms are used for face recognition. Face detection and recognition were performed with a camera module connected to Raspberry Pi. The result of the study indicates that the face recognition-based system was more efficient than the biometric and *Radio Frequency Identification* (RFID) based systems. In [4], OKAO vision library provided by OMRON Corporation for face recognition and detection is utilized. Two types of cameras have been used in the system. One is a sensor camcorder on the ceiling to acquire the views of seats which students are sitting on. The other is a camera in front of the seats to capture images of the faces of the students. In the scope of the study, it has been observed that the continuous observation has increased the estimation performance for attendance.

Face images can be considered as the concatenated form of micro images. In [5] face is divided into small cells to extract LBP histogram for efficient and concatenated face representation. Both shape and texture information are utilized for the representation. Nearest neighbor classifier is used in the experiments. The results indicate that the proposed simple and efficient method provides fast feature extraction. Furthermore, there is a modified LBPH algorithm [6] as an alternative to the LBPH algorithm used for face recognition. The LBPH algorithm has low recognition rate under illumination variance, expression variation, and attitude deviation. In order to solve this problem, a modified neighborhood gray-median (MLBPH) based LBPH algorithm has been proposed. The results have shown that the MLBPH algorithm has a higher recognition rate than the LBPH algorithm.

Another work is the face recognition system [7], which is made up of smart glasses that emerge with the advancement in today's technology. The aim of the study is to develop an effective face recognition system on smart glasses. The achieved results of the proposed efficient system were accurate and satisfactory.

In a face recognition system, the traditional method of data size reduction is the rearrangement of image vectors in the face. This causes the loss of structural properties by the data itself. As a result, the sensitivity of recognition cannot be high. The efforts in the area [8] aim to develop a data reduction method based on *Multilinear Discriminant Subspace Projection* (MDSP). As a result of the experiments, it is concluded that MDSP has better results than conventional size reduction algorithms in terms of classification accuracy.

Furthermore, there is a *Neural Aggregation Network* (NAN) study [9] for video face recognition. This work was carried out using a set of video or images with more than one hundred instances of a person. To achieve this, a fixed-size feature representation for face recognition was created. Experiments on IJB-A, YouTube Face, and Celebrity-1000 video face recognition tests have shown in the end that it performs better than pure aggregation methods and provides the best correctness. An automated attendance management system has been also developed to overcome the difficulties [10] such as pose, illumination, rotation and other factors. The system integrates several techniques such as integral images, AdaBoost and Haar-like features.

3 Methodology

In the scope of this study, we use a two-stage mechanism in the system we have developed. It consists of an initial face detection that is followed by face recognition. Haar classifier and Viola-Jones algorithm are utilized for the face detection. PCA and LBPH algorithms have been used in the face recognition phase.

3.1 Viola-Jones Algorithm

The Haar cascade classifier can be found in the OpenCV library [11]. Paul Viola and Micheal Jones [12] used Haar-like features shown in Fig. 1 for object detection. It is also known as the Viola and Jones object detection framework (Viola and Jones object detection structure). In the most basic sense, objects that are desired to be found by a certain algorithm are first introduced to the computer. Then the images or video frames in which similar shapes exist are scanned to find that object. There is a need for positive pictures that contain the object sought for training as well as the negative pictures that the object does not exist. In classifier training, objects in positive pictures are scanned as frame sets in certain sizes. The sum of the pixel values of the black region in the frame and the darkness values in the white region are summed to generate specific target values. Edge, line and center surround features are shown in Fig. 1.

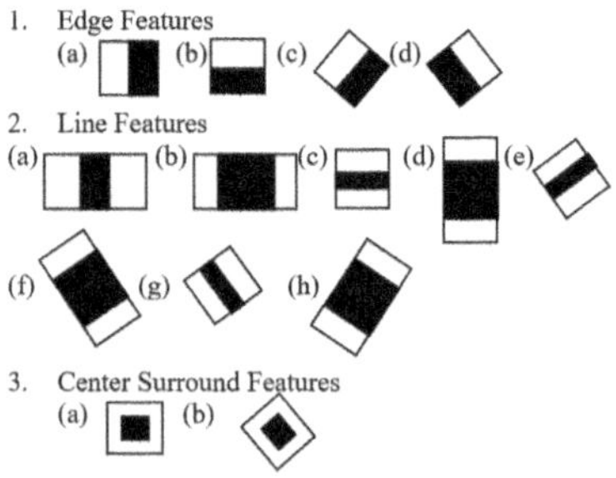

Fig. 1. Edge, line and center surround features.

These classifiers including features, are called weak classifiers. An object will have many of these weak classifiers, and there is an object sought with high accuracy in the point where these weak classifiers are gathered.

3.2 Principal Component Analysis (PCA)

PCA is one of the most frequently used method for data compression, dimensionality reduction and data decorrelation. The objective of the PCA [13] is to obtain a small number of mutually independent data from a large number of data with linear relationships. When Principal Component Analysis is applied in face recognition problem the steps below are followed.

- Centre and Standardize calculations

Let $\{a_1, a_2, a_3, \ldots a_n\}$ be our data set for training. Data set average

$$Avg = \frac{1}{n}\sum\nolimits_{i=1}^{n} a_n \tag{1}$$

- Calculate the covariance matrix

Each item in the training data set is different from Yi = ai-Avg vector. Covariance matrix Cov:

$$Cov = \frac{1}{n}\sum\nolimits_{i=1}^{n} Yi.Yi^T \tag{2}$$

- Find the eigenvectors of covariance matrix and transform the data to be in terms of the components

Pick n important eigenvectors of Cov and calculate the k differences of weight vectors W_{ik} for each element in the data set from 1 to n.

$$w_{ik} = \sum\nolimits_{k=1}^{n} \sum\nolimits_{i=1}^{n} e_k^T.(a_i - Avg) \tag{3}$$

3.3 Local Binary Pattern Histograms (LBPH)

The LBP [14] is a simple and efficient texture measurement operator which utilizes thresholding the neighborhood of pixel for pixel labelling. The LBP operator creates a tag for each pixel of the view. It consists of one or more zeros. These labels are formed by comparing the pixels adjacent to the center pixel N × N. In general, $LBP_{P,R}$ can be defined by three different circular neighborhoods. P represents the neighboring number and R represents the sampling Radius as it is shown "Fig. 2" [15, 16].

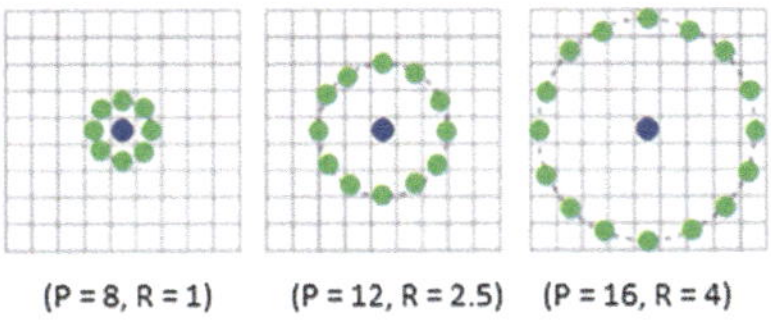

Fig. 2. Various circular $LBP_{P,R}$ operators

The operator LBP8.1 is used in this study. That is, neighborhood analysis is done using 3 × 3 matrices.

$$\mathrm{LBP}_{8,1}(\mathrm{x_c}) = \sum_{p=0}^{p-1} u(x_p - x_c)2^p \tag{4}$$

$$\mathrm{u(y)} = \begin{cases} 1, y \geq 0 \\ 0, y < 0 \end{cases} \tag{5}$$

In (5), y is the difference between the center pixel and the neighboring pixel. x_c is the center pixel produced by the LBP tag, x_p is the neighbor of center pixel, and u(y) is the bit produced by the LBP operator. An example is shown in "Fig. 3" [15].

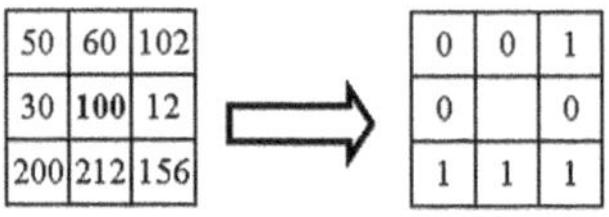

Fig. 3. LBP8,1 operator application

4 Experimental Results

In the scope of this study, all system requirements were applied and successfully completed. Depending on the test results applied to the system, face recognition and detection can be used to keep an entry log. Additionally, a person who is not registered to the system can also be added to the system by introducing his/her own face. As a first attempt, a certain number of gray scale face samples have been gathered. Once the registration has been made, the attendance log can be kept for the person. We used our own face database on the system and using these faces we have realized training processes with LBPH and PCA algorithms. While comparing the efficiencies of the PCA and LBPH algorithms, a threshold value for each algorithm was determined in experiments with examples. Based on these thresholds, experiments were conducted according to the number of face samples found. Reliability rates based on the number of these facial specimens were examined. We evaluated certain number of samples and noticed that the reliability ratio that emerged after the identification of one person's face did not exceed the threshold values we have seen. While these ratios show less fluctuations in the EigenFace algorithm that we applied with PCA, we found more deviations in the LBPH algorithm. Finally, the EigenFace algorithm we implemented with PCA was a little bit more efficient than LBPH, but the LBPH algorithm gave the opposite result when the PCA algorithm caused the program to visibly slow down during the processing of the data and polling acquisitions. As a result of the comparison, PCA gave more reliable ratios with lower performance. The performance was better for the LBPH algorithm when the PCA was well below the reliability ratios.

A sample interface of our automatic attendance system is shown in Fig. 4. When the face matching is performed, the face of the student is circled in the green box. Then the student ID and name are displayed by the system. The course attendance of the

Fig. 4. Sample screen of automatic face detection and recognition system.

matching student is marked in the database and information about the day and time is saved immediately. During the face recognition process, if the student does not match with a sample in the face database, an additional option of "adding as new student" is selected. Then the face samples are taken as soon as the add button is pressed. After receiving face samples, the system starts to train these face samples and records the information in the training set file.

Table 1 shows the test cases that have been successfully completed in the scope of this study.

Table 1. Automatic attendance system test cases

Objective	Description	Result
Verify login details	– Login to the system using username and password with test data	– User should login to the system
Verify the lectures after login	– Login to system – Be redirected to course selection page	– User logins to the system and displays lectures
Verify the attendance lists after lecture selection	– Login to system – Select course from the table	– User logins to the system and displays attendance list
Verify the detecting and recognizing face	– Login to system using password and test data – Start course selection – Select classifier – Start video capture – Select new student and enter related info	– User logins to the system – User displays lectures – Redirecting attendance page – User opens camera – Detecting and recognizing face – Adding new student to the system

5 Conclusions and Future Work

This study shows that the proposed approach provides better solution compared to traditional attendance tracking methods, biometric fingerprint and RFID systems. Algorithms such as PCA and Viola-Jones for biometric face recognition and face detection were used efficiently in the system. Our experiments have been conducted using face images of 10 students. The reliability results of face recognition system were obtained between the ranges 75% and the 95% during the experiments. Although the developed system has been designed for schools and classrooms, it can also be adapted and used in work places and security systems. It is also important to bear in mind that better face recognition results may be yielded in terms of LBPH performance on such systems after some improvements on the utilized algorithms. As a future work, the system can be further improved by introducing additional features in recognition phase such as body, object and motion detection.

References

1. Wang, P., Lin, W.H., Chao, K.M., Lo, C.C.: A face-recognition approach using deep reinforcement learning approach for user authentication. In 2017 IEEE 14th International Conference on e-Business Engineering (ICEBE), pp. 183–188. IEEE, November 2017
2. Bradski, G., Kaehler, A.: OpenCV. Dr. Dobb's J. Softw. Tools **3** (2000)
3. Patil, A., Shukla, M.: Implementation of classroom attendance system based on face recognition in class. Int. J. Adv. Eng. Technol. **7**(3), 974 (2014)
4. Kawaguchi, Y., Shoji, T., Weijane, L.I.N., Kakusho, K., Minoh, M.: Face recognition-based lecture attendance system. In: The 3rd AEARU Workshop on Network Education, pp. 70–75 (2005)
5. Ahonen, T., Hadid, A., Pietikainen, M.: Face description with local binary patterns: application to face recognition. IEEE Trans. Pattern Anal. Mach. Intell. **28**(12), 2037–2041 (2006)
6. Zhao, X., Wei, C.: A real-time face recognition system based on the improved LBPH algorithm. In: 2017 IEEE 2nd International Conference on Signal and Image Processing (ICSIP), pp. 72–76. IEEE, August 2017
7. Xu, W., Shen, Y., Bergmann, N., Hu, W.: Sensor-assisted multi-view face recognition system on smart glass. IEEE Trans. Mob. Comput. **17**(1), 197–210 (2018)
8. Mei, M., Huang, J., Xiong, W.: A discriminant subspace learning based face recognition method. IEEE Access (2017)
9. Yang, J., Ren, P., Chen, D., Wen, F., Li, H., Hua, G.: Neural aggregation network for video face recognition. arXiv preprint (2017)
10. Shirodkar, M., Sinha, V., Jain, U., Nemade, B.: Automated attendance management system using face recognition. Int. J. Comput. Appl. (2015). (0975–8887) International Conference and Workshop on Emerging Trends in Technology (ICWET 2015)
11. Bradski, G., Kaehler, A.: Learning OpenCV: Computer Vision with the OpenCV Library. O'Reilly Media, Inc., Sebastopol (2008)
12. Soni, L.N., Datar, A., Datar, S.: Implementation of Viola-Jones Algorithm based approach for human face detection. Int. J. Curr. Eng. Technol. **7**, 1819–1823 (2017)

13. Wimmer, H., Powell, L.: Principle component analysis for feature reduction and data preprocessing in data science. In: Proceedings of the Conference on Information Systems Applied Research ISSN, vol. 2167, p. 1508 (2016)
14. Shan, C.: Learning local binary patterns for gender classification on real-world face images. Pattern Recogn. Lett. **33**(4), 431–437 (2012)
15. Tuncer, T., Engin, A.V.C.I.: Yerel İkili Örüntü Tabanli Veri Gizleme Algoritmasi: LBP-LSB. Türkiye Bilişim Vakfı Bilgisayar Bilimleri ve Mühendisliği Dergisi **10**(1), 48–53
16. Ojala, T., Pietikainen, M., Maenpaa, T.: Multiresolution gray-scale and rotation invariant texture classification with local binary patterns. IEEE Trans. Pattern Anal. Mach. Intell. **24**(7), 971–987 (2002)

Fine Localization of Complex Components for Bin Picking

Jiri Tvrdik(✉) and Petr Dolezel

Faculty of Electrical Engineering and Informatics, University of Pardubice, Pardubice, Czech Republic
jiri.tvrdik@student.upce.cz, petr.dolezel@upce.cz

Abstract. The aim of this paper is to present verified parameters of a particular approach to one sub-procedure of the bin picking problem. To successfully implement an automatic bin picking application using a robotic arm, it is necessary, among others, to detect the precise position and rotation angle of a selected object. In this approach, the procedure is considered as a two step operation. The first step provides an initial guess of both position and rotation angle, while the second one should specify the pose as exactly as required for following operations. The goal of the paper is to determine a correct relation between those two mentioned steps, i.e. to specify the maximal possible degree of uncertainty provided by the first step so that the second step works correctly. The proposed problem is dealt with as an industrial contract and the results are clearly dependent on specific conditions. However, it can be plainly used as a first insight into the problematics of bin picking, and provide a good starting point for deeper investigations.

Keywords: Bin picking · Point clouds · Pose estimation · Robotic arm

1 Introduction

'Bin Picking Problem', a generalization of 'Pick and Place Problem', has been a very attractive field of research for many years. Although being a very popularized phenomenon among the public nowadays, it is still able to cause great interest to scientists and engineers for its complexity and variability. In simple words, a solution of the 'Bin Picking Problem' should allow the robotic arm (or group of them) to grab a particularly shaped object, located randomly in a box with other items, and place it in a defined position.

The industrial bin picking applications often include sets of two-and three-dimensional cameras, conveyor belt with containers and a group of robotic manipulators, see Fig. 1.

From the engineering point of view, two main challenges are set. On the one hand, the object has to be precisely positioned (its position and rotation angle has to be estimated), the grab coordinates have to be computed considering various constraints and, eventually, the object has to be robustly grasped. On

K. Arai et al. (Eds.): IntelliSys 2018, AISC 868, pp. 246–255, 2019.
https://doi.org/10.1007/978-3-030-01054-6_18

Fig. 1. Bin picking in industry [1].

the other hand, motion planning has to be performed, considering both robotic manipulator limitations and the surroundings of the station. Then, the object can be placed in its desired position, see [2] for a nice review of the whole process.

In this contribution, the first mentioned challenge is discussed. In order to securely implement it, three comprehensive technical issues have to be dealt with [3]:

- The nimble robot grippers that are able to firmly handle a variety of parts in boxes.
- The real-time 6 degree-of-freedom positioning system that can locate a particular object in its container among others.
- The real-time grip planner that is able to estimate a firm and robust grasp.

Within those three points, let us focus on the second one. Considering object detection/identification and pose estimation, many approaches have been proposed during the last decades [4–6]. Almost every proposed approach uses computer vision of some sort.

Recently, due to an increasing availability of advanced laser 3D scanners, point cloud processing has often been implemented for pose estimation [7]. With point clouds, a new family of pose estimation algorithms becomes available to implement, and an Iterative Closest Point (ICP) algorithm is one of them [8]. An ICP algorithm can be implemented very efficiently and it has been used for bin picking applications many times [9,10].

One of the key features of an ICP algorithm is its sensitivity to an initial guess of a position and rotation angle of the searched object [11]. Clearly, the sensitivity rate depends on a very complex set of features, including size and complexity of the object or the resolution of the point cloud [12]. Therefore, a sensitivity rate relevant to the particular conditions of the application should be estimated in order to develop a robust bin picking application.

In the following paragraphs, the procedure of the sensitivity rate for one bin picking application is presented. The aim of the contribution is properly defined in Sect. 2. Then, the experiments with the results are presented and, finally, the paper is concluded with some discussion.

2 Problem Formulation

A long-term contractor of our local University ordered the improvement for an existing bin picking system. A UR3 type robotic arm by Universal Robots (see Fig. 2) is used to pick objects and place them to a defined position for future processing. PhoXi 3D scanner M by Photoneo is used for point cloud acquisition. This scanner provides point clouds with absolute accuracy of less than 100 µm.

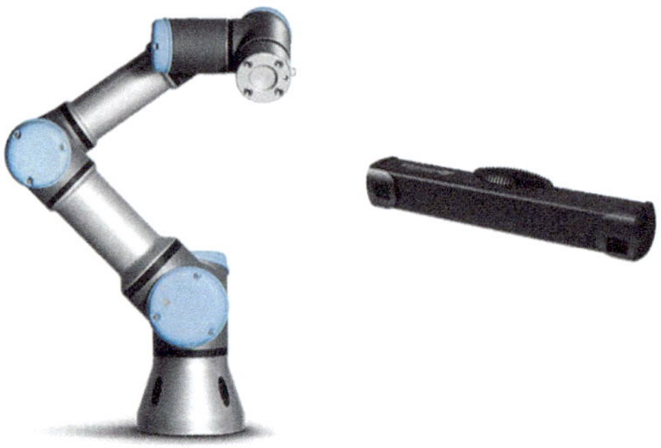

Fig. 2. UR3 and PhoXi 3D scanner M.

Fig. 3. Object of interest.

The aim of the setting is to find a position of an optimally located object (see Fig. 3) in a bin with randomly loaded components (see Fig. 4), grab it and move to a desired position. The pose estimation, as a part of the bin picking procedure, is used as follows.

- Get the 3D point cloud of the object and the 3D point cloud of the scene using PhoXi 3D scanner as the inputs to the procedure.
- Get the initial guess of the optimal position and rotation angle of the searched object using an in-house approach - this approach is proposed by the contracting authority and we are not allowed to publish it.

Fig. 4. Bin with components.

- Use the ICP algorithm to fine localization of the object.

An example of the point cloud of the searched object is in Fig. 5 and the point cloud of the scene is in Fig. 6.

The ICP algorithm is a straightforward method to align two shapes. Hence, if the position and orientation of one shape is set (in this case, the shape of the scene in Fig. 6) it iteratively tries to find a particular operation composed of translation and rotation, which transforms the other shape (in this case, the object in Fig. 5) to the pose which minimizes differences between each couple of corresponding points found in both shapes.

As the iterative algorithm, it requires an initial guess to be set. The issue of the algorithm is, that it diverges with an inappropriately selected initial guess. Thus, it is necessary to determine a sensitivity rate relevant to the conditions defined by the situation. Therefore, a set of experiments is necessary to perform

Fig. 5. Point cloud of the object.

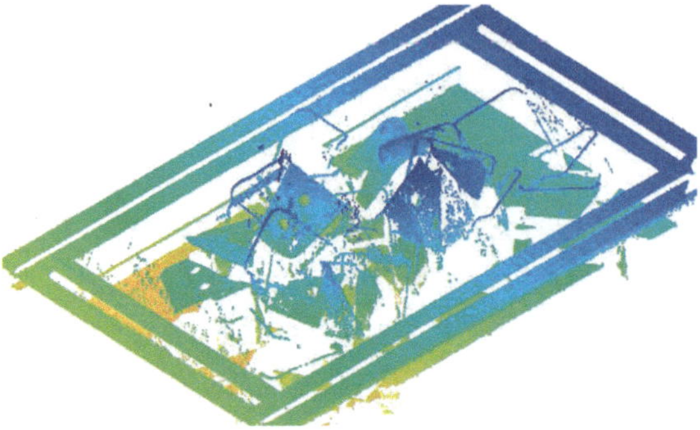

Fig. 6. Point cloud of the scene.

in order to detect the maximal difference in both position and rotation angle suitable enough for the ICP algorithm to converge. In other words, the aim is to detect the acceptable rate of error provided by the in-house approach for initial guess estimation.

3 Experiment Setting

An experimental stand was prepared to perform a testing procedure, see Fig. 7. It was designed to emulate the conditions in an industrial environment; at least in terms of geometric dimensions. Using this device, a set of experiments was performed. A blind search approach was implemented as a search algorithm. To be more specific, a set of initial guesses was prepared using translation and rotation of the object, see Table 1 for detailed information. The ICP algorithm was

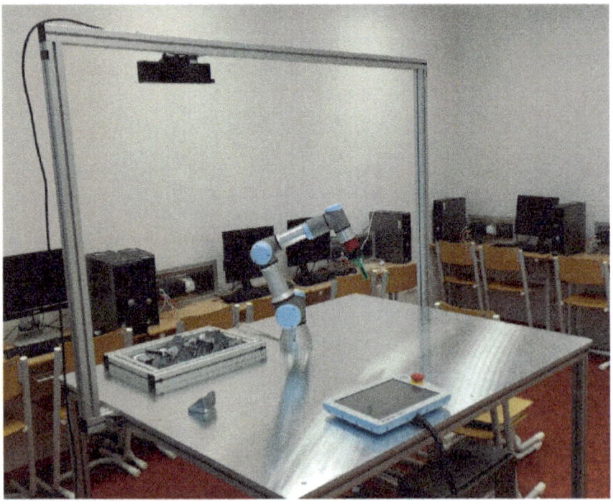

Fig. 7. Experimental device.

Table 1. Parameters of testing set

Number of samples	100000
Translation in X-axis	±100 mm
Translation in Y-axis	±100 mm
Translation in Z-axis	±100 mm
Rotation in X-axis	±30°
Rotation in Y-axis	±30°
Rotation in Z-axis	±30°

then used for fine localization. Three different initial guesses are shown in Fig. 8 as an example. Red clouds represent the situation, where the ICP algorithm did not converge, while the green cloud indicates convergence.

The procedure of the ICP algorithm is summarized in Fig. 9, see [8] for a detailed description of each step. In our implementation of the algorithm, the maximum number of iterations is set to 1 000 epochs and the stopping criterion is defined as the 2-element vector $D = [\delta_1, \delta_2] = [0.001, 0.001]$, that represents the absolute difference in translation and rotation estimated in two consecutive iterations. δ_1 measures the Euclidean distance between two translation vectors, δ_2 represents the angular difference in radians.

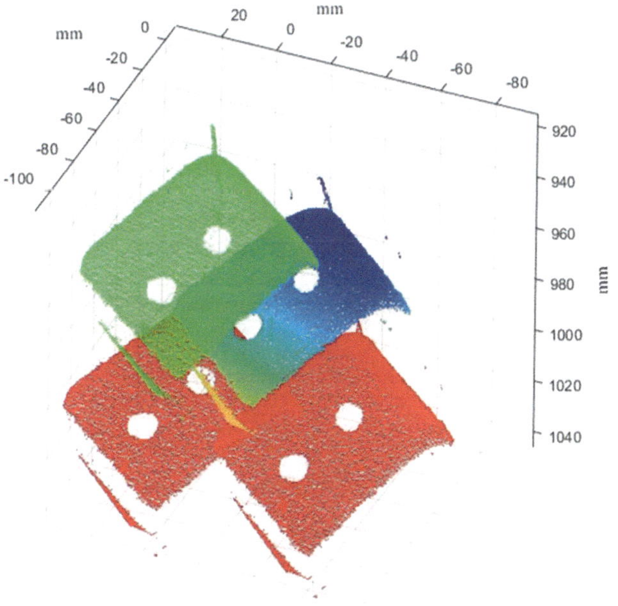

Fig. 8. Example of initial guesses; green objects converge, while red objects do not. Blue object is the desired position.

4 Results

According to the performed testings, the safe tolerance of the initial guess of a pose is summarized in Table 2. Clearly those results are significantly dependent on the experiment conditions, especially the shape and complexity of the object and the density of the point cloud. The visualization of the perimeter of positions, which still indicate convergence to the correct pose, is shown in Fig. 10.

Apparently, related to specific conditions of the experiment, the initial guess setting significantly affects the overall hit rate, and it is recommended to keep the initial guess uncertainty within a few centimeters and degrees of rotation angle.

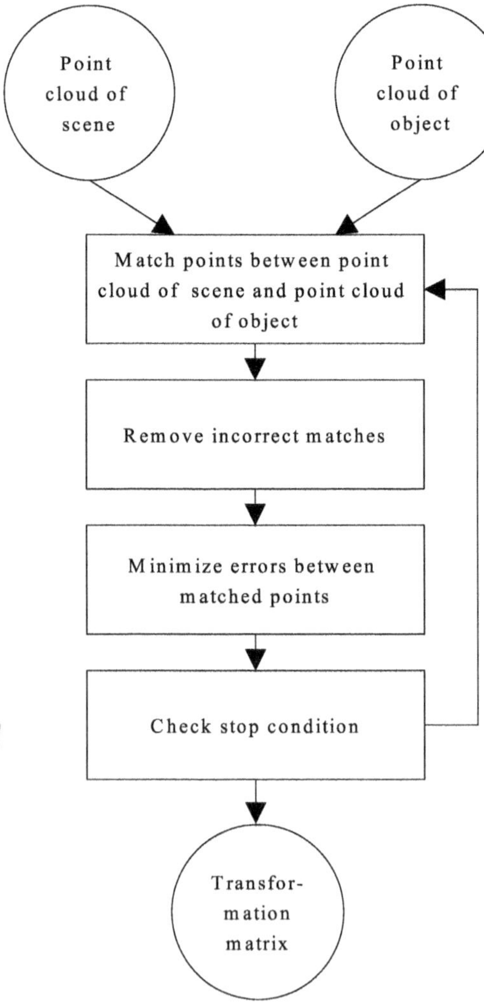

Fig. 9. ICP algorithm.

Table 2. Limiting values of positions, which still converge to correct pose

Translation in X-axis	$[-30; 30]$ mm
Translation in Y-axis	$[-30; 30]$ mm
Translation in Z-axis	$[-30; 30]$ mm
Rotation in X-axis	$[-15; 15]^{\circ}$
Rotation in Y-axis	$[-15; 15]^{\circ}$
Rotation in Z-axis	$[-15; 15]^{\circ}$

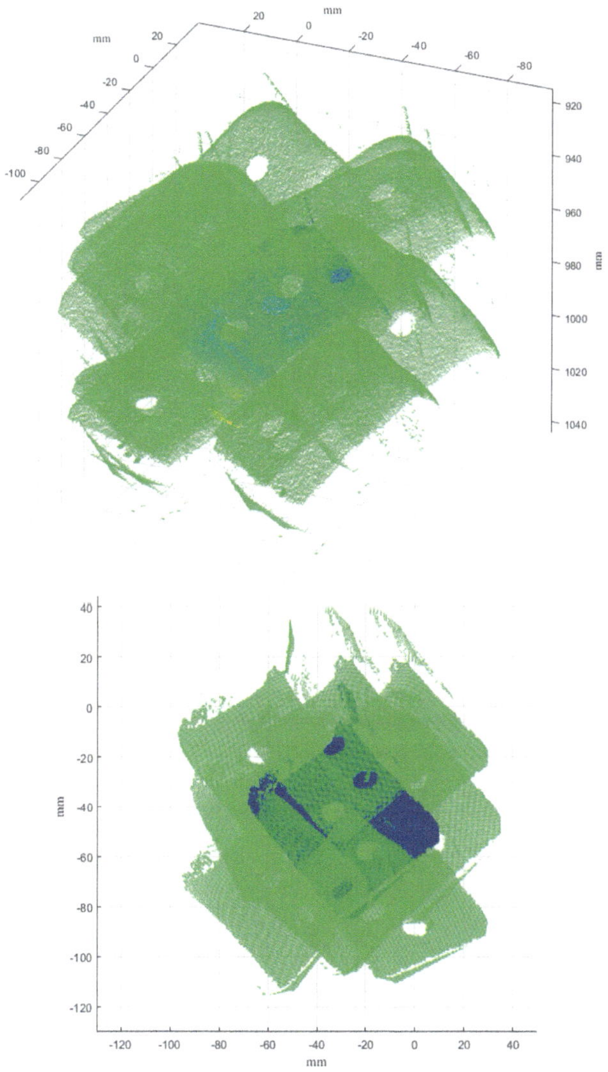

Fig. 10. Perimeter of positions, which still converge to correct pose.

5 Conclusion

An accurate localization of complex components depending on the initial guess of the position and the rotation angle is discussed in this paper. Although the issue is related to a specific demand from a contracting authority, the results can be roughly generalized into a group of similar situations. To be more specific, dealing with the positioning of complex components using point clouds, the ICP algorithm is significantly sensitive to an initial guess. Thus, special care must be taken to select the appropriate method for choosing the initial guess. In our specific case, the dominant dimension of the object is about 60 mm, while the tolerance of initial guess uncertainty is about 30 mm. The knowledge of the absolute tolerance in an initial guess procedure can significantly help with the overall stability of the solution.

Acknowledgment. The work has been supported by the Funds of University of Pardubice, Czech Republic. This support is very gratefully acknowledged.

References

1. Sopalski, M., Schyja, A., Miegel, V.: Flexible bin-picking (2017). https://www.vdi-wissensforum.de/news/flexible-bin-picking/
2. Pochyly, A., Kubela, T., Singule, V., Cihak, P.: 3D vision systems for industrial bin-picking applications. In: Proceedings of 15th International Conference MECHATRONIKA, pp. 1–6, December 2012
3. Shi, J., Koonjul, G.S.: Real-time grasping planning for robotic bin-picking and kitting applications. IEEE Trans. Autom. Sci. Eng. **14**(2), 809–819 (2017)
4. Kuo, H.Y., Su, H.R., Lai, S.H., Wu, C.C.: 3D object detection and pose estimation from depth image for robotic bin picking. In: 2014 IEEE International Conference on Automation Science and Engineering (CASE), pp. 1264–1269, August 2014
5. Ulrich, M., Wiedemann, C., Steger, C.: Combining scale-space and similarity-based aspect graphs for fast 3D object recognition. IEEE Trans. Pattern Anal. Mach. Intell. **34**(10), 1902–1914 (2012)
6. Rahardja, K., Kosaka, A.: Vision-based bin-picking: recognition and localization of multiple complex objects using simple visual cues. In: Proceedings of the 1996 IEEE/RSJ International Conference on Intelligent Robots and Systems 1996, IROS 1996, vol. 3, pp. 1448–1457, November 1996
7. Gong, X., Chen, M., Yang, X.: Point cloud segmentation of 3D scattered parts sampled by realsense. In: 2017 IEEE International Conference on Information and Automation (ICIA), pp. 1–6, July 2017
8. Besl, P.J., McKay, N.D.: A method for registration of 3-D shapes. IEEE Trans. Pattern Anal. Mach. Intell. **14**(2), 239–256 (1992)
9. Tian, X., Hua, F., Wang, T.: An innovative localization system of transformer substation automation robots based on pattern recognition and ICP algorithm. In: 2016 4th International Conference on Applied Robotics for the Power Industry (CARPI), pp. 1–5, October 2016
10. Boehnke, K., Otesteanu, M.: Progressive mesh object registration. In: 2008 IEEE/SICE International Symposium on System Integration, pp. 6–11, December 2008

11. Oh, J.K., Lee, C.H., Lee, S.H., Jung, S.H., Kim, D., Lee, S.: Development of a structured-light sensor based bin-picking system using ICP algorithm. In: ICCAS 2010, pp. 1673–1677, October 2010
12. Attia, M., Slama, Y.: Efficient initial guess determination based on 3D point cloud projection for ICP algorithms. In: 2017 International Conference on High Performance Computing Simulation (HPCS), pp. 807–814, July 2017

Intrusion Detection in Computer Networks Based on KNN, K-Means++ and J48

Mauricio Mendes Faria(✉) and Ana Maria Monteiro

Faculty of Campo Limpo Paulista (FACCAMP),
Rua Guatemala, 167, Campo Limpo Paulista, SP 13231-230, Brazil
mauriciofaria@gmail.com, anammont.alu@gmail.com

Abstract. The diversification of web, desktop and mobile applications has made information security an issue in public and private organizations. Large amounts of data are generated by applications leading to many network requests that generate considerable volumes of traffic data that must be analyzed quickly and effectively to avoid unauthorized access. Analyzing these network data, it is possible to extract knowledge to detect if applications are experiencing instability on behalf of malicious users. Tools called IDS (Intrusion Detection System) are used to detect malicious accesses. An IDS can use different techniques to classify a network connection as intrusion or normal. This work analyses data mining algorithms that can be integrated into an IDS to detect intrusions. Experiments were conducted using the WEKA environment, the NSL-KDD dataset, the supervised algorithms KNN (K Nearest Neighbours) and J48, and the unsupervised algorithm K-means++.

Keywords: Intrusion detection · Data mining · K-means++ · KNN J48

1 Introduction

Applications in different computational architectures produce data in exponential quantities and require efficient processes of knowledge discovery so that they can be used for the benefit of the organization that holds them. Much of this data comes from application access logs. Information, such as user logins, hosts, IP (Internet Protocol), access ports, access protocol types, access date and time must be stored, in large amounts, in log files and can be considered as an Achilles' heel, as they provide malicious individuals with opportunities to obtain valuable information from data that represents the daily routine of system users. But also, through the logs it is possible to detect if the system is going through some instability due to attempts of invasion. To investigate these weaknesses, organizations point out the need to invest in Intrusion Detection System (IDS) for taking preventive actions.

An IDS monitors network traffic using multiple sensors to detect external and internal network intrusions. An IDS analyzes the information collected by the sensors and returns a summary to the system administrator or to the intrusion prevention system. Different approaches can be used to detect these intrusions or anomalies. One of these approaches is based on data mining algorithms. In this work, KNN, K-Means++ and J48

K. Arai et al. (Eds.): IntelliSys 2018, AISC 868, pp. 256–271, 2019.
https://doi.org/10.1007/978-3-030-01054-6_19

algorithms were chosen for future integration into an IDS. These algorithms were chosen because they have the advantage of their simplicity when compared to other more complex models, such as Support Vector Machine algorithms or Neural Networks, for example.

The rest of this article is organized as follows. Section 2 overviews the problem of intrusion detection and analysis, Sect. 3 presents the algorithms used for intrusion detection. Section 4, in turn, describes the configuration of the experiments, the data and the formulas used to evaluate the performance of the algorithms. Section 5 presents the results obtained, Sect. 6 analyzes them and finally Sect. 7 presents the conclusion.

2 Problem of Intrusion Analysis and Detection

Intrusion detection is the process of determining a system invasion by observing available information about the state of that system and monitoring its users' activities [11]. Intruders can be outside entities or users from within the system trying to access unauthorized information [5]. Based on these observations the intruders can be divided into external intruders, which are those who do not have authorized access to the system they are dealing with, and the internal intruders who are those who have authorized access to only part of the system and surpass their access rights. But regardless of type, the purpose of these intruders is to do something that transgresses the security policy of the system, so intruders must be detected, and measures must be taken to minimize the consequences of the intrusions.

2.1 Intrusion Detection

When a system connection is made, a sequence of TCP (Transfer Control Protocol) packets is initiated and finalized in some well-defined time with data flowing from the IP source to the IP destination under some well-defined protocol. Each connection can be assigned a normal or an attack label. Attacks, in turn, can be labeled in categories: (a) DoS (Denied of Service), such as Syn Flood. (b) R2L (Remote to Local Attack), an unauthorized access from a remote machine, for example, guessing password. (c) U2R (User to Root), an unauthorized access with super user privileges, for example, buffer overflow. (d) Probing, monitoring and trial-and-error method, for example Port Scanning. Recalling the four categories mentioned above are not necessarily intrusions but may evidence a server's fragility in a sequence for a later intrusion attempt.

Intrusion detection is the second line of a network defense and it is done through IDS. Several techniques are used to detect intrusion in computer networks. Based on [4], these techniques can be grouped into the following categories: based on knowledge, machine learning techniques, statistical analysis and signal analysis. Data from user connections are rich in hidden information and can be analyzed using data mining techniques for extracting models that describe normal or abnormal connections and these models can be used to make decisions [6].

The objective of this work is to analyze the behavior of KNN, K-Means++ and J48 algorithms with the purpose of being used in an IDS to detect anomalies (intrusions) in computer networks. Tests were made to address the performance (accuracy) of these algorithms.

3 Algorithms for Intrusion Detection

The following is a description of the algorithms used for intrusion detection.

3.1 K-Nearest Neighbor (KNN)

The KNN (k-Nearest Neighbor) algorithm belongs to the family of algorithms IBL (Instance-based Learning) [7]. The algorithms of this family store all the training data and when a new instance (data) is presented to the algorithm to be classified, a set of similar data (next) to the new instance is retrieved from the training set and used to classify the new data.

To classify a new instance, the KNN algorithm retrieves the k nearest neighbors and the class assigned to the new instance is the most frequent class among those k neighbors. To determine the closest or more similar neighbors, the concept of distance between the instance to be classified and the data in the training set is used. The most commonly used measure to determine similarity is the Euclidean one, but Manhattan and Chebyshev measures can also be used [10].

3.2 J48

The J48 algorithm is a Java implementation of the C4.5 algorithm in the WEKA environment. This algorithm, developed by Quinlan [9], uses a training data set to construct a decision tree to classify new instances. It belongs to the class of supervised learning algorithms.

A decision tree has decision nodes and leaves. Each decision node contains a test for one of the data attributes and each descendant branch corresponds to a possible value of this attribute, the leaves in turn are associated with a class and, each tree path, from the root of the tree until a leaf corresponds to a classification according to the values of the attributes present in the new instance to classify [10]. The attribute chosen as a test for a node is the one that best separates the data set associated with that node, and the criterion is based on entropy and information gain [10].

3.3 K-Means++

Clustering brings together a set of unsupervised machine learning algorithms to discover patterns in data. A clustering algorithm deals with unclassified data that partitions into clusters (groups) based on the similarity between these data.

Among the clustering algorithms, the K-Means algorithm is one of the most used. A parameter value k is supplied by the user. In this algorithm, initially, k points are randomly chosen (the initial centroids of k groups). Then, each point in the data set is associated with the nearest centroid, thus forming k groups. Then the centroid of each group is updated to reflect the average of the points belonging to the group. The process repeats until no changes in the groups.

K-Means is a simple and efficient algorithm, but it does not necessarily find the optimal group configuration and is also very sensitive to the set of centroids initially chosen [2]. To avoid degradation due to random initial centroids, an improvement of the K-Means algorithm, named K-Means++, was proposed. In the K-Means++ algorithm the initial centroids are not chosen according to a uniform random distribution, but according to a weighted probability distribution in which a point x is chosen with probability proportional to the square of its distance from the nearest centroid [1]. The idea of modifying the K-Means algorithm is to select a better set of centroid values that translate into an improvement in execution time and quality of results as the number of clusters increases [1].

4 Experiments

Two experiments were carried out. The first one called "Anomaly Detection" (AD) aims to determine which of the algorithms used can better classify the data set instances as normal access or anomaly. The second one called "Anomaly Type Detection" (ATD) aims to determine which algorithm can better classify the data set instances as a normal access or as an access belonging to one of the four categories of anomalies: DoS, R2L, U2R and Probing. Both experiments used the KNN, K-Means++ and J48 algorithms available in the WEKA[1] environment. The details of the configuration of each algorithm used in the experiments are presented in *A*, the dataset used for the tests is described in B, experiment settings in C and the attribute selection in D.

4.1 KNN, K-Means++ and J48 Algorithms Setup in the WEKA Environment

For the best configuration of each algorithm, it was established the combination of the basic parameters of each algorithm.

The KNN algorithm has as parameters the value of k, which corresponds to the number of nearest neighbors, the distance measure used and the type of normalization. The combinations between these parameters, in the AD and ATD experiments, are described in Table 1.

For the NNSearch = covertree option, only the Euclidean distance setting is allowed.

The K-Means++ algorithm has as parameters the value of k, which corresponds to the number of centroids (or groups), the distance measure used and, as a pre-processing condition, normalization. The combinations between these parameters in the AD and ATD experiments are described in Table 2.

[1] The WEKA (Waikato Environment for Knowledge Analysis) environment began to be written in 1993, using Java, at the University of Waikato in New Zealand, being acquired later by a company in late 2006. This environment aims to aggregate algorithms from different approaches area of artificial intelligence dedicated to the study of machine learning.

Table 1. KNN algorithm parameter combinations for AD and ATD experiments

k	Normalization	Type	NNSearch	Distance
1, 3, 5, 7	No	–	Covertree	Euclidean
	Yes	Attribute		
	Yes	Instance		

Table 2. Algorithm K-Means++ parameter combinations for AD and ATD experiments

k		Norm.	Type normalization	Distance
AD	ATD			
2	5	Yes	Instance	Euclidian
		No	–	
		Yes	Instance	Manhattan
		No	–	
		Yes	Attribute	Euclidian
		Yes	Attribute	Manhattan

The value of k for the AD experiment was set to 2, since the purpose is to classify each instance of the test data set as "normal" or "anomaly". For the ATD experiment, the configuration of the value of k is 5, since the intention is to classify each instance of the test data set as "normal", "dos", "r2l", "u2r" or "probing".

The J48 algorithm has as parameters the option of pruning or not the induced tree, and the normalization. The combinations between these parameters are described in Table 3.

4.2 Data Set NSL-KDD

The NSL-KDD data set is derived from the KDDCup99 dataset that was designed through a TCP-Dump process logging the traffic of a network. NSL-KDD is a refinement of KDDCup99 [8] dataset in which redundant records have been eliminated while maintaining the essence records of the original dataset. The choice of the NSL-KDD data set is because it was designed in an analytical way with more homogeneous records so as not to influence the result of the techniques used [11].

This data set consists of files in two formats: text and Attribute-Relation File Format (ARFF), which allows them to be used in tools such as WEKA or in applications developed for specific purposes. Like the predecessor, NSL-KDD has data that represents users' access over a network. Data such as IP, type of service, duration of connection, protocol, type of attack, etc. provide relevant information that can be used by data mining algorithms or other algorithms to identify intrusions in computer networks. Each record file consists of 42 attributes with information ranging from basic characteristics of the connection, knowledge associated with TCP connections and traffic in a 2 s window. These attributes are described in Table 4.

Table 3. Algorithm J48 parameter combinations for AD and ATD experiments

Pruning	Normalization	Type normalization
Yes	Yes	Instance
Yes	Yes	Attribute
No	Yes	Instance
No	Yes	Attribute
Yes	No	–
No	No	–

Table 4. Information and attributes of the archives comparing the NSL-KDD data set

Attribute information	Attribute
Basic features of an individual connection	Duration, protocol_type, service, src_byte, dst_byte, Flag, Land, wrong_fragment, urgent
Knowledge of a TCP connection	Hot, num_failed_logins, logged_in, num_comprissed, root_shell, su_attempted, num_root, num_file_creations, num_shells, num_access_files, num_outbund_cmds, is_hot_login, is_guest_login
Traffic computed using a 2-s window	Count, serror_rate, rerror_rate4, same_srv_rate4, diff_srv_rate4, srv_count4, srv_serror_rate, srv_rerror_rate5, srv_diffe_host_rate5
Undocumented attributes	dst_host_count, dst_host_srv_count, dst_host_same_srv_rate, dst_host_diff_srv_rate, dst_host_same_src_port_rate, dst_host_srv_diff_host_rate, dst_host_serror_rate, dst_host_srv_serror_rate, dst_host_rerror_rate, dst_host_srv_rerror_rate

4.3 Experiments Settings

For the AD and ATD experiments, the subsets described in Table 5 were selected from the NSL-KDD data set. The data subsets used for the ATD experiment were changed from the original KDDTrain + .txt and KDDTest + .txt contained in the NSL-KDD data set, so that they have the type of attack (dos, r2l, u2r, probing) instead of the "anomaly" or "normal" content.

In Fig. 1, samples and the layout of the subset of training and test data are presented for the AD and ATD experiments, the latter with the appropriate modifications "unpublished" [3].

In the AD experiment, the algorithms consider only the class attribute, which indicates whether the instance is normal or anomal access, when it is in training. This attribute is ignored in test execution. In the ATD experiment the algorithm will ignore the classatack attribute and difficult, either for training or test, and will only consider the class category attribute, which indicates whether the instance is a "normal" access or belongs to the "anomaly" categories "dos", "R2l", "u2r" and probing.

Table 5. Dataset selected for experiments AD and ATD

Exp.	Algorithms	Type	Data set training	Data set test
AD	KNN	Supervised	KDDTrain + .arff	KDDTest + .arff
	J48			
	K-Means++	Not supervised	–	KDDTest + .arff
ATD	KNN	Supervised	por_tipo_de_ataque-KDDTrain + .arff	por_tipo_de_ataque-KDDTest + .arff
	J48			
	K-Means++	Not supervised	–	por_tipo_de_ataque-KDDTest + .arff

Samples of the NSL-KDD KDDTrain + .arff training dataset
0,tcp,ftp_data,SF,491,0,0,0,0,0,0,0,0,0,0,0,0,0,0,0,0,0,2,2,0.00,0.00,0.00,0.00,1.00,0.00,0.00,1
50,25,0.17,0.03,0.17,0.00,0.00,0.00,0.05,0.00,**normal**
Samples of the NSL-KDD KDDTest + .arff Test dataset
0,tcp,private,REJ,0,0,0,0,0,0,0,0,0,0,0,0,0,0,0,0,0,0,229,10,0.00,0.00,1.00,1.00,0.04,0.06,0.00
,255,10,0.04,0.06,0.00,0.00,0.00,0.00,1.00,1.00,**anomaly**
Samples of the NSL-KDD por_tipo_ataque_KDDTrain + .arff training dataset
0,tcp,private,S0,0,0,0,0,0,0,0,0,0,0,0,0,0,0,0,0,0,0,199,3,1.00,1.00,0.00,0.00,0.02,0.06,0.00,2
55,13,0.05,0.07,0.00,0.00,1.00,1.00,0.00,0.00,neptune,**dos**,21
Samples of the NSL-KDD por_tipo_ataque_KDDTest + .arff Test dataset
1,tcp,telnet,RSTO,0,15,0,0,0,0,0,0,0,0,0,0,0,0,0,0,0,0,1,8,0.00,0.12,1.00,0.50,1.00,0.00,0.75,
29,86,0.31,0.17,0.03,0.02,0.00,0.00,0.83,0.71,mscan,**probing**,11

Fig. 1. Samples of training subsets of NSL-KDD data set for AD and ATD experiments.

4.4 Attribute Selection

Having established the choice of data sets and algorithms, the next step was to choose the attributes to be used in the tests. The selected attributes are shown in Table 6.

5 Performance Evaluation

Once the classifiers and clusters corresponding to the execution of the algorithms have been obtained, the performance of each algorithm must be evaluated. There are different metrics to evaluate the results obtained. Among these metrics, the most frequently used are the accuracy and error rate.

Four concepts are required to calculate these metrics: true positive (TP), true negative (TN), false positive (FP) and false negative (FN). These concepts will be defined according to the positive instances, which are the instances of the main interest class and the negative instances, which are all the remaining instances. True positives: positive instances correctly classified as positive. True negatives: negative instances correctly classified as negative. False positives: Negative instances misclassified classified as positive. False negatives: positive instances misclassified as negative. A common way of presenting the above concepts is by a cross-tabulation between the class predicted by the model and the actual (real) class of the instances. This tabulation is called confusion matrix. In the WEKA environment the results of the AD and ATD

Table 6. Common attributes for experiments with supervised and not supervised algorithm for the tests AD and ATD

Attributes
Duration, protocol_type, servisse, Flag, src_bytes, dst_bytes, Land, wrong_fragment, urgente, Hot, num_failed_logins, logged_in, num_compromised, root_shell, su_attempted, num_root, num_file_creations, num_shells, num_access_files, num_outbound_cmds, is_host_login, is_guest_login, Count, srv_count, serror_rate, srv_serror_rate, rerror_rate, srv_rerror_rate, same_srv_rate, diff_srv_rate, srv_diff_host_rate, dst_host_count, dst_host_srv_count, dst_host_same_srv_rate, dst_host_diff_srv_rate, dst_host_same_src_port_rate, dst_host_srv_diff_host_rate, dst_host_serror_rate, dst_host_srv_serror_rate, dst_host_rerror_rate, dst_host_srv_rerror_rate

experiments are made available through matrices of confusion with specific layouts that are presented in Figs. 2 and 3, respectively.

In Fig. 2, TP represents the number of instances correctly classified as "normal". TN represents the number of instances belonging to the "normal" class that were incorrectly classified as "anomaly". FP represents the number of instances that are incorrectly classified as "normal" and that belong to the class "anomaly". TN represents the number of instances correctly classified as "anomaly".

In Fig. 3, TP (NO | NO) represents the number of instances correctly classified as "normal". TP (DO | DO) represents the number of instances correctly classified as "dos". TP (R2 | R2) represents the number of instances correctly classified as "r2l". TP (U2 | U2) represents the number of instances correctly classified as "u2r". TP (PR | PR) represents the number of instances correctly classified as "probing".

FP (NO | DO) represents the number of instances incorrectly classified as "normal" and belonging to the "dos" class. FP (NO | R2) represents the number of instances classified as "normal" and belonging to class "r2l". FP (NO | U2) represents the number of instances classified as "normal" and belonging to class "u2r". FP (NO | PR) represents the number of instances classified as "normal" and belong to the "probing" class.

FP (DO | NO) represents the number of instances incorrectly classified as "dos" and belonging to the "normal" class. FP (DO | R2) represents the number of instances classified as "dos" and belonging to class "r2l". FP (DO | U2) represents the number of instances classified as "dos" and belonging to class "u2r". FP (DO | PR) represents the number of instances classified as "dos" and belonging to the "probing" class.

FP (R2 | NO) represents the number of instances incorrectly classified as "r2l" and belong to the "normal" class. FP (R2 | DO) represents the number of instances classified as "r2l" and belong to class "dos". FP (R2 | U2) represents the number of instances classified as "r2l" and belonging to class "u2r". FP (R2 | PR) represents the number of instances classified as "r2l" and belong to the class "probing".

FP (U2 | NO) represents the number of instances incorrectly classified as "u2r" and belong to the "normal" class. FP (U2 | DO) represents the number of instances classified as "u2r" and belong to the "dos" class. FP (U2 | R2) represents the number of instances classified as "u2r" and belonging to class "r2l". FP (U2 | PR) represents the number of instances classified as "u2r" and belong to the class "probing".

		PREDICTED	
		NORMAL	ANOMALY
TRUE	NORMAL	TP	FN
	ANOMALY	FP	TN

Fig. 2. Layout of the confusion matrix provided in the WEKA environment for the AD experiments.

		PREDICTED				
		NORMAL	DOS	R2L	U2R	PROBING
TRUE	NORMAL	**TP (NO|NO)**	FP (DO| NO)	FP (R2|NO)	FP (U2|NO)	FP (PR|NO)
	DOS	FP (NO|DO)	**TP (DO|DO)**	FP (R2|DO)	FP (U2|DO)	FP (PR|DO)
	R2L	FP (NO|R2)	FP (DO|R2)	**TP (R2|R2)**	FP (U2|R2)	FP (PR|R2)
	U2R	FP (NO|U2)	FP (DO|U2)	FP (R2|U2)	**TP (U2|U2)**	FP (PR|U2)
	PROBING	FP (NO| PR)	FP (DO|PR)	FP (R2|PR)	FP (U2|PR)	**TP (PR| PR)**

Fig. 3. Layout of the confusion matrix provided in the WEKA environment for the ATD experiments.

Finally, FP (PR | DO) represents the number of instances classified as "probing" and belong to class "dos". FP (PR | R2) represents the number of instances classified as probing and belonging to class "r2l". FP (PR | U2) represents the number of instances classified as probing and belonging to class "u2r".

In the interpretation of this confusion matrix, when the analysis of one of the quadrants, for example that of TP (NO | NO), the other quadrants TP (DO | DO), TP (R2 | R2), TP (U2 | U2) and TP (PR | PR) become TN.

5.1 Equations for the Calculation of Performance

To evaluate the results obtained in the AD experiments, the following equations were used: Accuracy rate ($T_{ACCURACY1}$ (1)), which indicates the proportion of hits that occurred in the classification. False positive error rate for the "normal" class (TE_{NORMAL} (2)), which is the ratio of errors occurring in the normal class classification. Error rate for false positives for class "anomaly" ($TE_{ANOMALY}$ (3)), which is the proportion of errors occurred in the classification of class anomaly. Total Error Rate (TE_{TOTAL1} (4)), which indicates the total proportion of errors occurred in the classification.

$$T_{ACCURACY1} = \frac{TP + TN}{TP + TN + FP + FN} \tag{1}$$

$$TE_{NORMAL} = \frac{FP}{FP + TN} \tag{2}$$

$$TE_{ANOMALY} = \frac{FN}{TP + FN} \tag{3}$$

$$TE_{TOTAL1} = \frac{FP + FN}{TP + TN + FP + FN} \quad (4)$$

To evaluate the results obtained in the ATD experiments, the following equations were used: Accuracy rate ($T_{ACCURACY2}$ (5)), which indicates the general proportion of correctness occurred in the classification. Total error rate (TE_{TOTAL2} (6)), which indicates the total proportion of errors occurred in the classification. False positive error rate for the "dos" class (TE_{DOS} (7)), which is the ratio of errors occurring in the class classification "dos". False positive error rate for class "r2l" (TE_{R2L} (8)), which is the ratio of errors occurring in the classification of class "r2l". False positive error rate for class "u2r" (TE_{U2R} (9)), which is the proportion of errors occurred in the classification of the class "u2r". False positive error rate for the "probing" class ($TE_{PROBING}$ (10)), which is the proportion of errors occurred in the classification of the class "probing".

$$T_{ACCURACY2} = \frac{TP_{NO|NO} + TP_{DO|DO} + TP_{R2|R2} + TP_{U2|U2} + TP_{PR|PR}}{Total\, Instances} \quad (5)$$

$$TE_{TOTAL2} = 1 - T_{ACCURACY2} \quad (6)$$

$$TE_{NORMAL} = \frac{Total\, FP\, Normal}{Total\, FP\, Normal + Total\, TN_{DO,R2,U2\,e\,PR}} \quad (7)$$

$$TE_{R2l} = \frac{Total\, FP\, R2l}{Total\, FP\, R2l + Total\, TN_{NO,DO,U2\,e\,PR}} \quad (8)$$

$$TE_{U2r} = \frac{Total\, FP\, U2r}{Total\, FP\, U2r + Total\, TN_{NO,DO,R2\,e\,PR}} \quad (9)$$

$$TE_{Probing} = \frac{Total\, FP\, Probing}{Total\, de\, FP\, Probing + Total\, de\, TN_{NO,DO,R2\,e\,U2}} \quad (10)$$

Where,

$$Total\, FP\, Normal = FP_{NO|DO} + FP_{NO|R2} + FP_{NO|U2} + FP_{NO|PR}$$

$$Total\, TN_{DO,R2,U2\,e\,PR} = (TN_{DO|DO} + TN_{R2|R2} + TN_{U2|U2} + TN_{PR|PR})$$

$$Total\, FP\, Dos = FP_{DO|NO} + FP_{DO|R2} + FP_{DO|U2} + FP_{DO|PR}$$

$$Total\, TN_{NO,R2,U2\,e\,PR} = TN_{NO|NO} + TN_{R2|R2} + TN_{U2|U2} + TN_{PR|PR}$$

$$Total\, FP\, R2l = FP_{R2|NO} + FP_{R2|DO} + FP_{R2|U2} + FP_{R2|PR}$$

$$Total\, TN_{NO,DO,U2\,e\,PR} = TN_{NO|NO} + TN_{DO|DO} + TN_{U2|U2} + TN_{PR|PR}$$

$$Total\,FP\,U2r = FP_{U2|NO} + FP_{U2|DO} + FP_{U2|R2} + FP_{U2|PR}$$

$$Total\,TN_{NO,DO,R2\,e\,PR} = TN_{NO|NO} + TN_{DO|DO} + TN_{R2|R2} + TN_{PR|PR}$$

$$Total\,FP\,Probing = FP_{PR|NO} + FP_{PR|DO} + FP_{PR|R2} + FP_{PR|U2}$$

$$Total\,TN_{NO,DO,R2\,e\,U2} = TN_{NO|NO} + TN_{DO|DO} + TN_{R2|R2} + TN_{U2|U2}$$

5.2 KNN Algorithm in AD and ATD Experiments

The parameters configuration to execute the KNN algorithm in the WEKA environment established the values Cross Validate: false, Debug: false, Distance Weighting: Mean Square: false and Nearest Neighbor Search Algorithm: CoverTree (Euclidian Distance). These settings are valid for both the training step and the AD and ATD experiments. The value of k is the only variable parameter value of this configuration, which is chosen for each experiment using the KNN.

The application of normalization filter and the k values were established for each experiment, according to Table 1. The five best results obtained using the KNN algorithm in the AD experiment are shown in Fig. 4. The result of the code 8 experiments, highlighted in bold and italic, shows the best configuration of the KNN algorithm, in the AD experiment, regarding classification quality. For the ATD experiment, the results are presented in Fig. 5. The 1000 code experiment, presents the best configuration of the KNN algorithm.

5.3 K-Means++ Algorithm in AD and ATD Experiments

For the configuration of the K-Means++ algorithm in the WEKA environment, we used the Display Std Devs: False, Dont Replace Missing Values: False, Max Iterations: 500, Preserve Instances Order: False and Sedd: 10 parameters for the testing step. In addition to the configuration described above the configuration of the number of clusters (NumCluster) was 2 for the AD experiment and 5 for the ATD, as presented in Section A. The Distance Function option that allows to select the type of similarity measure has been configured for the Euclidean or Manhattan distances depending on the experiment.

The application of the normalization filter and the k values were established, for each experiment, according to Table 2. The five best results obtained using the K-Means++ algorithm in the AD and ATD experiments are shown respectively in Figs. 6 and 7. The result of the code 38 experiment, highlighted in bold and italic, shows the best configuration of the K-Means++ algorithm in the AD experiment, and the 1018 code experiment, the best result for the ATD experiment, regarding classification quality.

KNN (Lazy IBK)												
			CORRECT INSTANCES		INCORRECT INSTANCES							
COD. EXP.	NORM.	k	CLASS NORMAL (TP)	CLASS ANOMALY (TN)	CLASS NORMAL (FN)	CLASS ANOMALY (FP)	TE_{NORMAL}	$TE_{ANOMALY}$	$T_{ACCURACY}$	TE_{TOTAL}	DISTANCE
8	***NÃO***	***1***	***9342***	***8548***	***369***	***4285***	***0,334***	***0,038***	***0,794***	***0,206***	***EUCLID.***
11	SIM	1	9096	8646	615	4187	0,326	0,063	0,787	0,213	*EUCLID.*
10	NÃO	7	9346	8375	365	4458	0,347	0,038	0,786	0,214	*EUCLID.*
7	NÃO	3	9345	8372	366	4461	0,348	0,038	0,786	0,214	*EUCLID.*
31	SIM	5	9032	8545	679	4288	0,334	0,070	0,780	0,220	*EUCLID.*

Fig. 4. The 5 best results for the AD experiment for the KNN algorithm.

KNN (Lazy IBK)											
COD. EXP.	NORM.	k	QTD. INST.	TE_{NORMAL}	TE_{DOS}	TE_{R2L}	TE_{U2R}	$TE_{PROBING}$	$T_{ACCURACY2}$	TE_{TOTAL2}	DISTANCE
1000	***NÃO***	***1***	***22544***	***0,348***	***0,020***	***0,012***	***0,003***	***0,024***	***0,771***	***0,229***	***EUCLID.***
1003	NÃO	3	22544	0,357	0,021	0,002	0,002	0,024	0,771	0,229	EUCLID.
1006	NÃO	5	22544	0,363	0,021	0,003	0,000	0,023	0,769	0,231	EUCLID.
1009	NÃO	7	22544	0,359	0,022	0,003	0,000	0,027	0,769	0,231	EUCLID.
1007	SIM	5	22544	0,343	0,026	0,003	0,000	0,040	0,765	0,235	EUCLID.

Fig. 5. The 5 best results for the ATD experiments for the KNN Algorithm.

5.4 Algorithm J48 in the AD and ATD Experiments

For the configuration of the J48 algorithm in the WEKA environment, we used the BynarySlits values: False, ConfidenceFactor: 0.25, Debug: False, MinNumObj: 2, NumFolds: 3, ReducedErrorPruning: False, SaveInstanceData: False, Seed: 1, SubtreeRaising: True and UseLaplace False. The pruning parameters and the application of the normalization filter were established, for each experiment, according to Table 3. The five best results obtained using the J48 algorithm in the AD and ATD experiments are shown respectively in Figs. 8 and 9. The result of the code 2 experiment, highlighted in bold and italic, shows the best configuration of the J48 algorithm in the AD experiment, and the best 1015 code experiment for the ATD experiment, regarding classification quality.

6 Analysis of Results

After the end of the experiments AD and ATD, the results were analysed.

6.1 AD Experiment

For the KNN algorithm, in the AD code 8 experiments, values of 0.334 for TE_{NORMAL} and 0.038 for $TE_{ANOMALY}$ were obtained. In the chosen configuration, it is possible to observe that the KNN algorithm can classify in a significant way the instances that belong to class "anomaly", as can be verified by the lower error rate. For the K-Means++ algorithm, in the code 38 AD experiment, values of 0.185 for TE_{NORMAL} and 0.64 for

K-Means++ (Simple K-Means)												
		CORRECT INSTANCES		INCORRECT INSTANCES						NORMALIZATION		
COD. EXP.	k	CLASSE NORMAL (TP)	CLASS ANOMALIA (TN)	CLASS NORMAL (FN)	CLASS ANOMALY (FP)	TE_{NORMAL}	$TE_{ANOMALY}$	$T_{ACCURACY1}$	TE_{TOTAL}	INSTANCE	ATTRIBUTE	DISTANCE
38	*2*	*7401*	*10175*	*2658*	*2310*	*0,185*	*0,264*	*0,780*	*0,220*	*YES*	*NO*	*MANHAT.*
37	2	7375	10183	2650	2336	0,187	0,264	0.779	0,221	YES	NO	EUCLID.
34	2	9477	7018	5815	7018	0,500	0,380	0,562	0,438	NO	NO	EUCLID.
35	2	9477	7018	5815	7018	0,500	0,380	0,562	0,438	NO	YES	EUCLID.
36	2	9616	5729	7104	5729	0,500	0,425	0,545	0,455	NO	NO	EUCLID.

Fig. 6. The 5 best results for the AD experiments for the K-Means++ Algorithm.

K-Means++ (Simple K-Means)											
COD. EXP.	k	NORMALIZATION		TE_{NORMAL}	TE_{DOS}	TE_{R2L}	TE_{U2R}	$TE_{PROBING}$	$T_{ACCURACY2}$	TE_{TOTAL2}	DISTANCE
		ATRIB.	INST.								
1018	*5*	*NÃO*	*SIM*	*0,244*	*0,441*	*0,664*	*1,000*	*0,734*	*0,580*	*0,420*	*EUCLIDEAN*
1020	5	NÃO	SIM	0,271	0,421	0,710	1,000	0,694	0,574	0,426	MANHATTAN
1021	5	NÃO	NÃO	0,208	0,582	0,678	1,000	0,679	0,553	0,447	MANHATTAN
1023	5	SIM	NÃO	0,208	0,582	0,678	1,000	0,679	0,553	0,447	MANHATTAN
1019	5	NÃO	NÃO	0,243	0,588	0,597	1,000	0,665	0,548	0,452	EUCLIDEAN

Fig. 7. The 5 best results for the ATD experiments for the K-Means++.

J48										
	CORRECT INSTANCES		INCORRECT INSTANCES						NORMALIZATION	
COD. EXP.	CLASS NORMAL (TP)	CLASS ANOMALY (TN)	CLASS NORMAL (FN)	CLASS ANOMALY (FP)	TE_{NORMAL}	$TE_{ANOMALY}$	$T_{ACCURÁCIA1}$	TE_{TOTAL}	INSTANCE	ATTRIBUTE
2	*9448*	*8933*	*263*	*3900*	*0,304*	*0,027*	*0,815*	*0,185*	*NO*	*NO*
3	9443	8903	268	3930	0,306	0,028	0,814	0,186	NO	NO
6	1725	11744	7986	1089	0,085	0,822	0,597	0,403	NO	YES
4	722	11335	8989	1498	0,117	0,926	0,535	0,465	NO	YES
14	2858	5437	6853	7396	0,576	0,706	0,368	0,632	YES	NO

Fig. 8. The 5 best results for the AD test for the J48 algorithm.

J48										
COD. EXP.	NORMALIZATION		UNPRUNED	TE_{NORMAL}	TE_{DOS}	TE_{R2L}	TE_{U2R}	$TE_{PROBING}$	$T_{ACURACY2}$	TE_{TOTAL2}
	ATRIB.	INST.								
1015	*SIM*	*NÃO*	*NÃO*	*0,270*	*0,115*	*0,009*	*0,000*	*0,011*	*0,783*	*0,217*
1017	NÃO	NÃO	NÃO	0,383	0,026	0,000	0,000	0,037	0,753	0,247
1016	NÃO	SIM	SIM	0,392	0,025	0,001	0,000	0,048	0,741	0,259
1013	SIM	SIM	SIM	0,063	0,816	0,007	0,000	0,013	0,446	0,554
1012	NÃO	SIM	SIM	0,585	0,600	0,065	0,002	0,505	0,272	0,728

Fig. 9. The 5 best results for the ATD test for the J48 algorithm.

$TE_{ANOMALY}$ were obtained. In the chosen configuration, it is possible to observe that the K-Means++ algorithm was able to classify the instances that belong to the "normal" class as evidenced by the lower error rate. For the algorithm J48, in the code 2 AD experiment, values of 0.264 were obtained for TE_{NORMAL} and 0.027 for TE_{NORMAL}. With this configuration, the J48 algorithm can better classify the instances that belong to the class "anomaly" as evidenced by the lower error rate.

By using the TE_{NORMAL} and $TE_{ANOMALY}$ indicators shown in Fig. 10, the K-Means++ algorithm can better classify the instances belonging to the "normal" class. Already the algorithm J48, can classify better the instances belonging to the class "anomaly". This is evidenced by the lower error rate. Based on the values of the comparative indicators ($T_{ACCURACY1}$ and TE_{TOTAL1}), the algorithm that obtained the best indicators was J48, the second-best result was obtained by the KNN algorithm and the third best indicator was the K-Means++ algorithm, supervised, which was not far from the first two algorithms, being able to discriminate adequately the two classes "normal" and "anomaly".

6.2 ATD Experiments

The TE_{NORMAL}, TE_{DOS}, TE_{R2L}, TE_{U2R} and $TE_{PROBING}$ were used to verify the performance of the classification of instances as "normal", "R2l", "u2r" and "probing". The second point of the analysis is related to the better performance of TE_{NORMAL}, TE_{DOS}, TE_{R2L}, TE_{U2R}, $TE_{PROBING}$, among the KNN, K-Means++ and J48 algorithms, to point out the strengths of each algorithm in relation to FP. The results of $T_{ACCURACY2}$ and TETOTAL2 were used to point out the algorithm that had a better overall performance in the detection of "normal", "dos", "r2l", "u2r" and "probing" classes.

For the KNN algorithm in the 1000 code ATD experiment, values of 0.348 for TE_{NORMAL}, 0.020 for TE_{DOS}, 0.012 for TE_{R2L}, 0.003 for TE_{U2R} and 0.024 for $TE_{PROBING}$ were obtained. In the chosen configuration, it is possible to verify that the KNN algorithm can classify in a significant way the instances that belong to class "u2r" as it can be observed as a function of the error rate. With respect to the K-Means++ algorithm, in the ATD experiment of code 1018, values obtained were 0.244 for TE_{NORMAL}, 0.41 for TE_{DOS}, 0.64 for TE_{R2L}, 1.00 for TE_{U2R} and 0.734 for $TE_{PROBING}$. In the chosen configuration, it is possible to verify that the K-Means++ algorithm can classify the instances belonging to the "normal" class considerably, as can be inferred from the lower error rate. For the J48 algorithm in the 1015 code ATD experiment, values of 0.270 for TE_{NORMAL}, 0.115 for TE_{DOS}, 0.009 for TE_{R2L}, 0.000 for TE_{U2R} and 0.011 for $TE_{PROBING}$ were obtained. In the chosen configuration it is observed that the algorithm J48 succeeds to classify the instances that belong to class "u2r". This evidence is evidenced by the lower error rate.

It is possible to observe through the TE_{NORMAL}, TE_{DOS}, TE_{R2L}, TE_{U2R} and $TE_{PROBING}$ indicators, shown in Fig. 11, that the K-Means++ algorithm was able to classify the instances belonging to the "normal" class considerably. The KNN algorithm was able to classify in a significant way the instances belonging to class "dos". Already the algorithm J48 managed to classify in a significant way the instances belonging to class "r2l", "u2r" and "probing". Based on the values of the $T_{ACCURACY2}$ and TE_{TOTAL2} values, the algorithm that obtained the best indicators was J48, the second-best result was obtained through the KNN algorithm and the third best result was the K-Means ++ algorithm. It is important to note that the performance analysis, through the general indicators, does not reflect the best values of the indicators found by categories of anomalies.

COMPARATIVE BETWEEN THE KNN, K-Means ++ and J48 ALGORITHMS						
ALG.	*k*	TE_{NORMAL}	$TE_{ANOMALY}$	$T_{ACCURACY1}$	TE_{TOTAL1}	DISTANCE
J48	-	***0,304***	***0,027***	***0,815***	***0,185***	--
KNN	1	0,334	0,038	0,794	0,206	EUCLIDEAN
KMeans++	2	0,185	0,264	0,780	0,220	MANHATTAN

Fig. 10. Consolidation of the comparative results between the KNN, K-Means ++ and J48 algorithms.

Algoritmo	TE_{NORMAL}	TE_{DOS}	TE_{R2L}	TE_{U2R}	$TE_{PROBING}$	$T_{ACCURACY2}$	TE_{TOTAL2}
KNN	0,348	0,020	0,012	0,003	0,024	0,771	0,229
K-Means++	0,244	0,441	0,664	1,000	0,734	0,580	0,420
J48	0,270	0,115	0,009	0,000	0,011	0,783	0,217

Fig. 11. Comparison of TE_{NORMAL}, TE_{DOS} TE_{R2L}, TE_{U2R} and $TE_{PROBING}$, $TE_{ACCURACY2}$ and TE_{TOTAL2}, for the KNN, K-Means++ and J48 algorithms.

7 Conclusion

The results of the AD experiments suggest some conclusions: (1) The supervised algorithms (KNN and J48) tend to classify the data more appropriately, but the results of the non-supervised K-Means++ algorithm are promising. (2) The J48 algorithm was able to obtain the best performance with respect to accuracy and general error rate, KNN the second-best performance and K-Means++ the third best. (3) With respect to TE_{NORMAL}, the K-Means++ algorithm obtained the lowest error rate for FP, making it feasible to use it in the IDS sensing stage, being able to properly detect normal accesses in detriment of anomalies. (4) Regarding $TE_{ANOMALY}$, the J48 algorithm obtained the lowest error rate per FP, making it possible to use it in the sensing step of an IDS, being able to correctly detect the anomalous accesses, in detriment of normal ones. The results of the ATD experiments suggest some conclusions: (5) The supervised algorithms (KNN and J48) also managed to classify the data more appropriately. In this case, the K-Means++ algorithm performed poorly in the AD experiment. (6) The J48 algorithm was able to obtain the best performance with respect to the accuracy and general error rate, KNN presented the second-best performance and K-Means++ the third. The latter was far from the top two. (7) In the TE_{NORMAL} aspect, the K-Means++ algorithm obtained the lowest error rate per FP, being able to detect, in an appropriate way, the normal accesses. About TE_{DOS}, the KNN algorithm obtained the lowest error rate by FP, being able to properly detect the anomalies of the "dos" category. (8) In the TE_{R2L}, TE_{U2R} and $TE_{PROBING}$ aspects, the J48 algorithm obtained the lowest error rate by FP, being able to properly detect the anomalous access belonging to the categories of "r2l", "u2r" and "probing". Although the general performance of the K-Means++ algorithm is lower than the others, in the analysis of the FP indicators by category, it was promising for the classification of "normal" instances. (9) Through the AD and ATD experiments, it is possible to state that the use of data mining techniques, from the algorithms studied, is feasible for intrusions detection, identifying normal accesses from anomalies.

Acknowledgment. The authors are grateful to Faccamp Faculty (Faculty Campo Limpo Paulista) for supporting the development and publication of this work.

References

1. Arthur, D., Vassilvitskii, S.: k-means++: the advantages of careful seeding. In: SODA 2007 Proceedings of the Eighteenth Annual ACM-SIAM Symposium on Discrete Algorithms, pp. 1027–1035 (2007)
2. Bottou, L., Bengio, Y.: Convergence properties of the K-Means algorithms. In: Advances in Neural Information Processing Systems, vol. 7, pp. 585–592 (1995)
3. Faria, M.M.: Detecção de Intrusões em Redes de Computadores com base nos Algoritmos KNN, K-Means++ e J48. Dissertação (Mestrado) - Curso de Ciência da Comuputação, Faculdade Campo Limpo Paulista, Campo Limpo Paulista, (2016). Cap. 5. http://www.cc.faccamp.br/Dissertacoes/MauricioMendesFaria.pdf. Accessed 01 Apr 2017
4. García-Teodoro, P., Díaz-Verdejo, J., Maciá-Fernández, G., Vázquez, E.: Anomaly-based network intrusion detection: techniques, systems and challenges. Comput. Secur. **28**(2009), 18–28 (2009)
5. Jones, A.K., Sielken, R.S.: Computer system intrusion detection: a survey. Technical report, Charlottesville: s.n (2000)
6. Han, L., Kamber, M.: Data Mining Concepts And Techniques, 2nd edn. Morgan Kaufmann & Elsevier, São Francisco (2006)
7. Hart, P.E., Cover, T.M.: Nearest neighbor pattern classification. IEEE Trans. Inf. Theory 13, 21–27 (1967)
8. Lincoln Laboratory Massachusets Institute of Technology, n.d. Cyber Systems and Technology. http://www.ll.mit.edu/ideval/data/. Accessed 27 July 2015
9. Quinlan, J.R.: C4.5: Programs for Machine Learning. Morgan Kaufmann Publisher, San Mateo (1993)
10. Silva, L.M.O.d.: Uma Aplicação de Árvores de Decisão, Redes Neurais e Knn para a Identificação de Modelos Arma Não-Sazonais e Sazonais [dissertação]. Rio de Janeiro (RJ): Pontifícia Universidade Católica do Rio de Janeiro - Puc-Rio (2005)
11. Tavallaee, M., Bagheri, E., Lu, W., Ghorbani, A.A.: A detailed analysis of the KDD CUP 99 data set. In: 2009 Second IEEE Symposium on Computational Intelligence for Security and Defence Applications, Ottawa, pp. 53–58 (2009)

Cooperating with Avatars Through Gesture, Language and Action

Pradyumna Narayana[1(✉)], Nikhil Krishnaswamy[2], Isaac Wang[3], Rahul Bangar[1], Dhruva Patil[1], Gururaj Mulay[1], Kyeongmin Rim[2], Ross Beveridge[1], Jaime Ruiz[3], James Pustejovsky[2], and Bruce Draper[1]

[1] Department of Computer Science, Colorado State University, Fort Collins, CO, USA
prady@cs.colostate.edu

[2] Department of Computer Science, Brandeis University, Waltham, MA, USA

[3] Department of Computer and Information Science and Engineering, University of Florida, Gainesville, FL, USA

Abstract. Advances in artificial intelligence are fundamentally changing how we relate to machines. We used to treat computers as tools, but now we expect them to be agents, and increasingly our instinct is to treat them like peers. This paper is an exploration of peer-to-peer communication between people and machines. Two ideas are central to the approach explored here: *shared perception*, in which people work together in a shared environment, and much of the information that passes between them is contextual and derived from perception; and *visually grounded reasoning*, in which actions are considered feasible if they can be visualized and/or simulated in 3D. We explore shared perception and visually grounded reasoning in the context of blocks world, which serves as a surrogate for cooperative tasks where the partners share a workspace. We begin with elicitation studies observing pairs of people working together in blocks world and noting the gestures they use. These gestures are grouped into three categories: social, deictic, and iconic gestures. We then build a prototype system in which people are paired with avatars in a simulated blocks world. We find that when participants can see but not hear each other, all three gesture types are necessary, but that when the participants can speak to each other the social and deictic gestures remain important while the iconic gestures become less so. We also find that ambiguities flip the conversational lead, in that the partner previously receiving information takes the lead in order to resolve the ambiguity.

Keywords: Gesture recognition · Human computer interfaces
Artificial intelligence

K. Arai et al. (Eds.): IntelliSys 2018, AISC 868, pp. 272–293, 2019.
https://doi.org/10.1007/978-3-030-01054-6_20

1 Introduction

Advances in artificial intelligence are fundamentally changing how we relate to machines. We used to treat computers as tools, but now we expect them to be agents, and increasingly our instinct is to treat them like peers [1]. For example, we talk to them and give them personal names (e.g. Alexa, Siri, Cortana). Unfortunately, the more familiar we become with artificial agents, the more frustrated we become with their limitations. We expect them to see and hear and reason like people. *No, no, Alexa, can't you see that I...*

This paper is an exploration of ideas in peer-to-peer communication between people and machines. It considers what capabilities a machine might need, and presents a prototype system with a limited form of peer-to-peer communication. Two ideas are central to the approach explored here. The first is *shared perception*. When people work together, much of the information that passes between them is contextual and derived from perception. Imagine, for example, two people cleaning a room. They might discuss high-level strategy ("you start here, I'll start over there"), but they don't describe every action they take to each other. They can just look at the room to see what the other person has or has not done. Only the high-level discussion is verbal, and even that is grounded in perception: the definitions of "here" and "over there" depend on knowing where the other person is. In general, when one person changes the state of the world, the other person can see it, and the goal of conversation is to provide additional information beyond what is provided by perception.

The second idea is that reasoning about physical objects is grounded in visualization. Imagine a simple command like "put the book on the table". Is this command feasible? Yes, if there exists both a book and a table, and there is a clear spot on the table at least the size of the book, and if there is a clear path from the book's position to the table. Most of the information needed to evaluate this command comes from perception, and in general the command can be understood if it can be visually simulated.

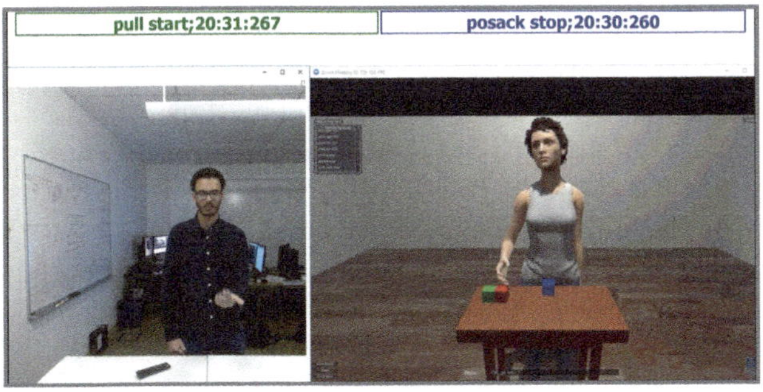

Fig. 1. Prototype peer-to-peer interface. The signaler on the left can communicate through gestures and words. The avatar on the right can communicate through gestures, words and actions

We explore these ideas in blocks world. In particular, we consider a scenario in which one person (the *builder*) has a table with blocks on it, and another person (the *signaler*) is given a target pattern of blocks. Only the builder can move the blocks, so the signaler has to tell the builder what to do. While blocks world is obviously not a real-world application, it serves as a surrogate for any cooperative task with a shared workspace.

We begin our exploration with elicitation studies similar to Wobbrock *et al.* [2], but with differences in how gestures are elicited. In the original elicitation study format, people are given specific actions (called referents), and asked to create a user-defined gesture (called signs) for the action. In our study, we take a more natural approach and simply present a pair of people with a task to complete and observe the actions and gestures that naturally occur. In our elicitation studies, the signaler and builder are both people. They are in separate rooms, connected by a video link. We vary the communication between them across three conditions: (1) the signaler and builder can both see and hear each other; (2) the signaler and builder can see but not hear each other; and (3) the signaler and builder can only hear each other (the signaler can see the builder's table and knows where the blocks are, but cannot see the builder).

Using gestures observed in the elicitation studies, we develop a prototype system in which the signaler is a person but the builder is an avatar with a virtual table and virtual blocks. The signaler can see a graphical projection of the virtual world, and communicate to the avatar through speech and gesture. The avatar communicates back through speech, gesture, and action, where an action is to move a block. Figure 1 shows the set up, with the signaler on the left and the avatar on the right in her virtual world.

Experience derived from using the prototype reveals important features of human-computer cooperation on shared physical tasks. For example, we learned that complex, gesture-based peer-to-peer conversations can be constructed from relatively few gestures, as long as the gesture set includes: (1) social gestures, for example acknowledgement and disagreement; (2) deictic gestures, such as pointing; and (3) iconic gestures mimicking specific actions, such as pushing or picking up a block. When the builder and avatar are allowed to speak, words can replace the iconic gestures, but the social and deictic gestures remain important. We learned that ambiguities arise in the context of conversations not just from questions of reference, i.e. which block to pick up, but also from options among actions, for example whether to put a block down on top of another block or next to it. Fortunately, these ambiguities are easily resolved if the conversational lead is allowed to switch from the signaler to the builder (in this case, the avatar). We also came to appreciate the importance of making two or more gestures at the same time, for example nodding (a social gesture) while signaling for the builder to pick up an object. Finally, we learned how important it was for the avatar to gesture back to the signaler, even when the avatar can speak and move blocks.

From an engineering perspective, we also confirmed that the combination of depth images from inexpensive sensors (Microsoft Kinect v2s) and deep convolutional neural networks is sufficient to recognize 35 common hand poses, and that with GPUs these hand poses can be recognized in real time. The directions of arm movements are also easily detected and are needed for deixis and for supplying directions to representational actions such as push or carry.

2 Related Work

This paper explores multi-modal peer-to-peer communication between people and avatars in shared perceptual domains. As such, it touches on multiple topics that have been studied before, including human/avatar interaction, multi-modal interfaces, and simulation semantics for reasoning, although we know of no previous system that integrates all of these components.

We have long known that people respond differently to avatars than to non-embodied interfaces. Users generally have a more positive attitude toward avatars, and often try to make themselves appear better to the avatar. They also tend to assign personality to avatars [3]. Although usually good, this can backfire: if the avatar is unable to meet a user's goals, the user is more likely to get angry [4]. People respond better to avatars and virtual robots than to non-embodied interfaces, but they respond better still to physically present robots [5]. Although the work here extends primarily to human/avatar interactions, it should extend to interactions between humans and humanoid robots as well.

Multimodal interfaces combining language and gesture have been around since at least 1980, when Bolt introduced "Put-that-there" [6]. Bolt's work anticipated the use of deixis to disambiguate references. More importantly, it inspired a community of researchers to work on multimodal communication, as surveyed in [7,8].

Roughly speaking, there are two major motivations for multimodal interfaces. The psychological motivation, as epitomized by Quek *et al.* [9], holds that speech and gesture are coexpressive, and therefore compliment each other. People are able to process speech and gesture partially independently, so using both modalities to express information increases human working memory and decreases the cognitive load [7]. People therefore retain more information and learn faster when communicating multimodally.

Visual information has been shown to be particularly useful in establishing common ground [10–12], which is important in shared perception scenarios. Other research emphasizes the importance of video and shared visual workspaces in computer-mediated communication [13–16], and highlights the usefulness of non-verbal communication to support coordination between humans. Thus, multimodal interfaces that leverage these human factors have the potential to be more effective collaborators.

The second motivation for multimodal interfaces is practical, as epitomized by Reeves *et al.* [17]. They argue that multimodal interfaces increase the range of users and contexts. For example, a device that can be accessed by either voice or gesture command can be used both in the dark and in noisy environments. They also argue that multimodal interfaces improve security and privacy since, depending on the situation, voice commands might be overheard or gestures might be observed. In addition, Veinott *et al.* draw the implication that the inclusion of video (and gestural information) may be increasingly useful for communication in the presence of language barriers [18].

This paper concentrates on shared physical tasks. When people work together, their conversation consists of more than just words. They gesture and share a common workspace [19–21]. Their shared perception of this workspace supports simulation semantics, and it is this shared space that gives many gestures such as pointing their meaning [22].

If two beings communicate to complete a shared task, they can be considered "agents", who are not only co-situated and co-perceiving but also act, together or individually, in response to communication. To coordinate action there must be agreement of a common goal between the agents, which can be called "co-intent". Together, co-situatedness, co-perception, and co-intent are the first aspects of "common ground". There is a rich and diverse literature on grounded communication [10,23–26]. However, in joint tasks, agents share an additional anchoring strategy—the ability to "co-attend". This ability emerges as central to determining the denotations of participants in shared events. Experienced events differ from events as expressed in language, as language allow us to package, quantify, measure, and order our experiences, creating rich conceptual reifications and semantic differentiations. The surface realization of this ability is mostly manifest through linguistic utterances, but is also witnessed in gestures.

Simulation can play a crucial role in human computer communication by creating a shared epistemic model of the environment. Simulation also creates an environment where two parties may be co-situated and co-attend by giving the agent an explicit *embodiment* [27], and allows the agent to publicly demonstrate its knowledge, providing an additional modality to communicate shared understanding within object and situation-based tasks, such as those investigated by [28–30].

The simulation environment provided includes the perceptual domain of objects, properties, and events. In addition, propositional content in the model is accessible to the discourse, allowing them to be grounded in event logic (a la [31]), and to be distinguished by the agents to act and communicate appropriately. This provides the non-linguistic visual and action modalities, which are augmented by the inherently non-linguistic gestural modality enacted within the visual context.

3 Case Study: Communicating with Gesture, Language and Action

To explore the role of gesture in peer-to-peer communication with shared perception, we first conducted human subject studies. The goal of these studies is to elicit common gestures and their semantic intents for the blocks world task in order to gain insight about how they might be used by people. We then built a prototype human/avatar system to explore communicating with computers.

3.1 Elicitation Studies

We begin with the human subject study design depicted in Fig. 2. As mentioned before, our goal is to elicit a set of gestures and possible actions in blocks world by observing people as they naturally communicate and collaborate with each other to complete a task. Each trial has two subjects, a signaler and a builder. Both subjects stand at the base of a table with a monitor on the other end. A two-way video feed is set up to allow both people to interact as if they were at opposite ends of a long table. The builder is given a set of wooden blocks, while the signaler is given a block layout/pattern. The task is for the signaler to tell the builder how to recreate the pattern of blocks without showing the pattern to the builder. The use of a computer-mediated setup allows us to control the communication based on condition.

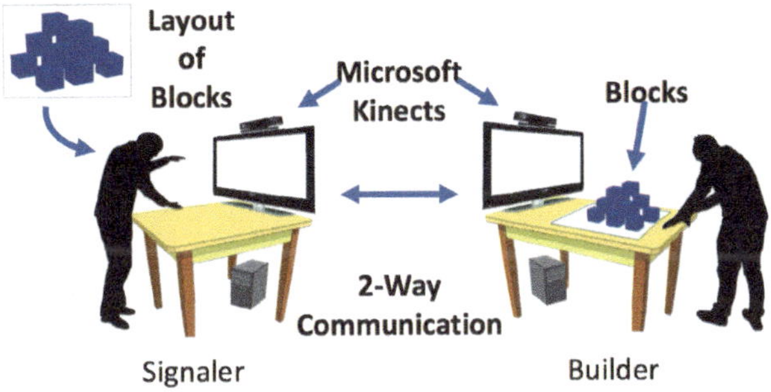

Fig. 2. Human subject study designed to elicit gestures for the blocks world domain and provide insight about how those gestures are used

A total of 439 trials across 60 participants were conducted under the following three conditions:

- Audio + Video: Participants can both see and hear each other through the monitors.
- Video-only: Participants can see but not hear each other, requiring the use of non-verbal communication only.
- Audio-only: Participants can only hear each other (the shared workspace is maintained: the signaler can still see the builder's table and know where the blocks are, but only cannot see the builder).

In all three conditions, RGB-D video is captured of both the signaler and builder using a Microsoft Kinect v2; the Kinect also estimates the 3D coordinates of 17 visible body joints[1] in each frame.

The data set collected and some initial observations have been described elsewhere [32]. Not previously reported, however, are the results below including the overall impact of the visual gestures. As shown in Table 1, participants were able to finish the task in 1:06 (min:sec) on average in the audio + video condition. In the audio-only condition, however, where the signaler was only able to talk to the builder, the average time increased to 1:32.9. This is very similar to

[1] The Kinect v2 estimates the positions of 25 joints, but the 8 lower-body joints are consistently obscured by the table.

the average time to completion for the video-only case, which was 1:35.2. This suggests that gestures are almost as communicative in blocks world as words, and more importantly that words and gestures are not redundant. Their combination is better than either words or gestures alone, in alignment with [9,18].

Table 1. Time to completion (min:sec) for signaler/builder blocks world tasks under three conditions: Audio + Video, Video only, and Audio only

Condition	Trials	Min	Median	Mean	Max
Video + audio	188	0:26.2	0:48.1	1:06.0	6:07.1
Video only	181	0:31.7	1:03.6	1:32.9	8:27.4
Audio only	170	0:09.8	1:06.3	1:35.2	13:36.1

Overall, we collected about 12.5 h of data. The audio + video and video-only trials were hand labeled at the level of left and right hand poses, left and right arm motions, and head motions. Summarizing these labels, we discovered 110 combinations of poses and motions that occurred at least 20 times and were performed by at least 4 different subjects. Of these, 29 were determined to have no semantic intent, as when a subject drops their arms to their side. Of the 81 remaining gestures, many were either minor variations or enantiomorphs of each other. For example, a participant might make the "thumbs up" sign while raising their forearm or pushing it forward. Similarly, the "thumbs up" sign might be made with the right hand, the left hand, or both. We also grouped physically different but semantically similar poses, such as the "thumbs up" and "OK" signs. After grouping similar motions and poses, we were left with 22 unique semantic gestures, as shown in Table 2.

Table 2. Semantic gestures performed at least 20 times in total by at least four different human subjects

Numeral	Representational	Deictic	Social
One	Grab	Point (that/there)	Start
Two	Carry	Tap (this/here)	Done
Three	Push	This group	Positive ack
Four	Push (servo)	Column	Negative ack
Five	Push together	Row	Wait for
	Rotate		Wait (pause)
			Emphasis

The 22 semantic gestures fall into four categories. Deictic gestures, such as pointing or tapping the table, serve to denote objects or locations. Iconic gestures, such as grab, push or carry, mimic actions. Social gestures, such as head nods or thumbs up, address the state of the dialog. Numerals are a form of abstract plural reference. We note that there are broader and more inclusive schemes for categorizing gestures (see [33], Chap. 6), but none are universally accepted and the simple categories above work well for describing the gestures we observed in blocks world.

3.2 Blocks World Prototype

The elicitation studies provide insights about gestures people use in blocks world. Our goal, however, is to explore peer-to-peer communication between people and computers using visually-grounded reasoning in the context of shared perception. To this end, we created a prototype system that replicates the experimental set-up in Fig. 2, except that the builder is now an avatar and the blocks and table are virtual. The system operates in real time, and allows the human signaler to gesture and speak to the avatar. The avatar can gesture and speak in return, as well as move blocks in the virtual world. In some tests we turn off the audio channel, thereby eliminating words and limiting communication to gestures and observation.

This system gives us a laboratory for exploring peer-to-peer communication between people and computers. Using it, we have spent hours building and tearing down simple blocks world structures, and the lessons learned from this experience are summarized in the next section. The rest of this section describes the human/avatar blocks world (HAB) system itself.

HAB has many components. For the purposes of this paper, however, we concentrate on the perceptual module that implements gesture recognition, the grounded semantics module (VoxSim) which determines the avatar's behavior, and the interplay between perception and reasoning. The perceptual module is described in Subsect. 3.2.1, VoxSim is described in Subsect. 3.2.2, and the interactions between perception and VoxSim are described in Subsect. 3.2.3.

3.2.1 Perception

We hypothesize that shared perception is the basis for peer-to-peer communication, particularly when working in a common workspace. To implement shared perception, the human signaler and avatar builder need to be able to see each other as well as the virtual table and its blocks. Perceiving the virtual table and blocks is relatively easy. The avatar has direct access to the virtual world, and can directly query the positions of the blocks relative to the table. The human builder sees the rendering of the virtual world, and therefore knows where the blocks are as well.

More challenging is the requirement for the human and avatar to see each other and interpret each other's gestures. Based on our elicitation studies, we have a lexicon of commonly occurring gestures. We animate the avatar so that she can perform all of these gestures, and rely on the builder's eyes to recognize them. We also have an RGB-D video stream of the human captured by the Microsoft

Kinect v2. The rest of this section describes the real-time vision system used to recognize the builder's gestures in real time.

Gesture recognition is implemented by independently labeling five body parts. The left and right hands are labeled according to their pose. The system is trained to recognize 34 distinct hand gestures in depth images, plus a 35th label ("other") that is used for hands at rest or in unknown poses. The hand poses are directional, in the sense that pointing down is considered a different pose than pointing to the right. Head motions are classified as either nod, shake or other based on a time window of depth difference images. Finally, the left and right arms are labeled according to their direction of motion, based on the pose estimates generated by the Microsoft Kinect [34].

To recognize gestures in real time, the computation is spread across 6 processors, as shown in Fig. 3. The processor shown on the left is the host for the Microsoft Kinect, whose sensor is mounted on top of the signaler's monitor. It uses the Kinect's pose data to locate and segment the signaler's hands and head, producing three streams of depth images. The pose data also becomes a data stream that is used to label arm directions. The hand and head streams are classified by a ResNet-style deep convolutional neural network (DCNN) [35]. Each net is hosted on its own processor, with its own NVIDIA Titan X GPU. The arm labeling process has its own (non-GPU) processor. Finally, a sixth processor collects the hand, arm and head labels and fuses them using finite state machines to detect gestures.

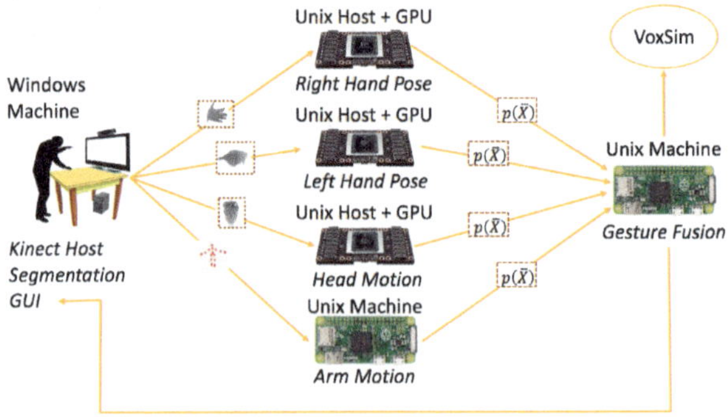

Fig. 3. The architecture of the real-time gesture recognition module

3.2.2 VoxSim

The avatar's reasoning system is built on the VoxSim platform [22,36]. VoxSim is an open-source, semantically-informed 3D visual event simulator implemented in Unity [37] that leverages Unity's graphics processing, UI, and physics subsystems.

VoxSim maps natural language event semantics through a dynamic interval temporal logic (DITL) [38] and the visualization modeling language VoxML [39]. VoxML describes and encodes qualitative and geometrical knowledge about objects and events that is presupposed in linguistic utterances but not made explicit in a visual modality. This includes information about symmetry or concavity in an object's physical structure, the relations entailed by the occurrence of an event in a narrative, the qualitative relations described by a positional adjunct, or behaviors *afforded* by an object's *habitat* [40,41] associated with the situational context that enables or disables certain actions that may be undertaken using the object. Such information is a natural extension of the lexical semantic typing provided within Generative Lexicon Theory [42], towards a semantics of embodiment. This allows our avatar to determine which regions, objects, or parts of objects may be indicated by deictic gestures, and the natural language interface allows for explicit disambiguation in human-understandable terms. The movement of objects and the movement of agents are compositional in the VoxML framework, allowing VoxSim to easily separate them in the virtual world, which means that the gesture used to refer to an action (or program) can be directly mapped to the action itself, establishing a shared context in which disambiguation can be grounded from the perspective of both the human and the computer program.

3.2.3 Perception and VoxSim

To create a single, integrated system we connect the recognition module (and by extension, the human signaler) with VoxSim and its simulated world. VoxSim receives "words" from the gesture recognizer over a socket connection, and interprets them at a contextually-wrapped compositional semantic level. The words may be either spoken or gestured by the (human) builder. For the moment, we have seven multi-modal "words":

(1) *Engage.* Begins when the signaler steps up to the table or says "hello", and ends when they step back or say "goodbye". Indicates that the signaler is engaged with the avatar.
(2) *Positive acknowledge.* Indicated by the word "yes", a head nod, or a thumbs up pose with either or both hands. Used to signal agreement with a choice by the avatar or affirmative response to a question.
(3) *Negative acknowledge.* Indicated by the word "no", a head shake, a thumbs down pose with any combination of hands, or a stop sign gestured with the hand closed, palm forward, and fingertips up. Signals disagreement with a choice by the avatar or negative response to a question.
(4) *Point.* Gestured by extending a single finger, with the optional spoken words "this" or "that". The information given to VoxSim about pointing gestures includes the spot on the tabletop being pointed to. Signals either a block to be used for a future action, or an empty space.
(5) *Grab.* Indicated by a claw-like pose of the hand that mimics grabbing a block or the word "grab". Tells the avatar to grab a block that was previously pointed to.

(6) *Carry.* Indicated by moving the arm while the hand is in the grab position, with the optional spoken word "carry". The information given to VoxSim includes a direction, one of left, right, forward, back, up or down. A "carry up" can be thought of as pick up, and a "carry down" is equivalent to put down.
(7) *Push.* Gestured with a flat, closed hand moving in the direction of the palm, with the optional spoken word "push". Similar to carry, it includes a direction, although *up* and *down* are not allowed. As a special case, a beckoning gesture signals the avatar to push a block toward the signaler.

In addition to the multi-modal "words", there are words that can only be spoken, not gestured. These words correspond to the block colors: black, red, green, blue, yellow and purple. Speech recognition is implemented by simple word spotting, so for example the phrase "the red one" would be interpreted as the single word "red".

The flow of information from the avatar/builder back to human/signaler is similar. The avatar can say the words and perform the gestures mentioned above, with the additional gesture of reaching out and touching a virtual block (a gesture not available to the human builder). One important difference, however, is that the avatar can also communicate through action. Because the human signaler can see the avatar and the virtual blocks world, they can see when the avatar picks up or moves a block.

Any time the recognition module determines that one of the known "words" begins or ends, VoxSim receives a message. VoxSim responds by parsing the meaning of the gesture in context. For example, if the gesture points to a spot on the right side of the table and the avatar is currently holding a block, then the gesture is a request to move the block to that spot. Alternatively, if the avatar is not holding a block, the same gesture selects the block nearest to the point as the subject of the next action.

Gestural ambiguities are common. If two blocks are near each other, a pointing gesture in their direction is ambiguous. Which block did the signaler point to? Similarly, if the user says "the red one" when there are two red blocks on the table, the reference is ambiguous. Actions may also be ambiguous. If the user signals the avatar to put down one block near another, should it stack the blocks or put them side by side?

When presented with ambiguities, VoxSim assumes the initiative in the conversation and asks the user to choose among possible interpretations. In the case of pointing, for example, VoxSim might ask "do you mean the red block?". If the answer is negative, it might then try "do you mean the green block?". VoxSim orders the options according to a set of heuristics that favor interesting interpretations over less interesting ones. For example, if the options are to stack a red block on top of a blue block or put them next to each other, VoxSim favors the stacking option, because stacks are interesting.

4 Lessons Learned

The motivation for the human studies and prototype system described in this paper is to gain first-hand experience with multi-modal peer-to-peer interfaces. We believe the prototype is unique, not because it recognizes natural gestures (although see [43]) but because it interprets those gestures using simulation-based reasoning in the context of a shared perceptual task.

Our experience with HAB has taught us many lessons, and led us to quickly modify and improve it. The next subsection walks the reader through an example of a person and an avatar working together to build a block pattern. The remaining subsections capture and share some of the lessons we have learned through experience, with one important caveat: so far, the only users of the system are its design team. Usability studies with naive users will come later, when the system is more mature.

4.1 Example

We illustrate HAB with an example in which the audio has been turned off, so all communication happens through gestures and actions. The example begins with three blocks on the table: a green block and a red block to the right of the signaler, and a blue one on the left. The signaler's goal is to arrange the blocks in a staircase. The conversation begins when the signaler steps up to the table, causing an *engage* gesture to be recognized and sent to VoxSim.

The signaler points to the left, as shown in Frame A of Fig. 4. VoxSim interprets this gesture as selecting the blue block for the next action. The avatar moves its hand toward the blue block in anticipation; this is a gesture that serves as a form of positive acknowledgment, since it lets the signaler know what the avatar understood. The signaler then beckons, and the avatar pushes the block away from itself and toward the signaler.

Next the signaler points to his right where the red and green blocks are (Frame B of Fig. 4). This is an ambiguous reference, so the avatar reaches toward the red block as a way of asking whether the signaler means the red block. The signaler shakes his head, sending a negative acknowledgment, so the avatar motions toward the green block. This time the signaler nods, resolving the ambiguity. The signaler then beckons again, and the avatar pushes the green block toward the signaler.

Continuing with the example, the signaler points toward the blue block and gestures to slide it to the right (Frame C). The slide gesture is ambiguous, however. Should the avatar slide the block a little ways to the right, or slide it all the way to the green block? Sliding it to the green block is the more interesting option, so this is the one the avatar suggests, and since it is what the signaler wants, he gives a thumbs up (Frame D) and the avatar slides the block.

This style of interaction continues. The signaler selects the red block by pointing and then mimics a grabbing motion. Both the reference and action are unambiguous, so the avatar complies. Next the signaler raises his arm while keeping his hand in the grabbing pose, and brings his arm forward. The avatar

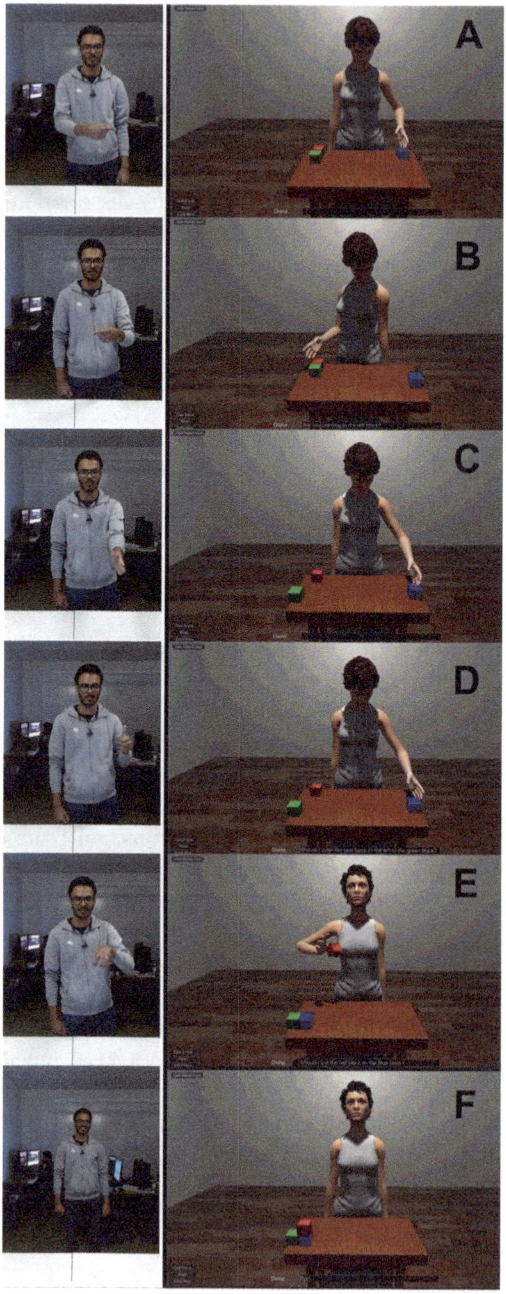

Fig. 4. An example of building a staircase in HAB using only gestures and actions (no spoken language)

responds as shown in Frame E. The signaler then lowers his arm and releases his grip, asking the avatar to put the block down. The placement is ambiguous – should the red block go on the blue block, the green block, or the table top? – but some gestural back and forth quickly clear this up, and the staircase is completed, as shown in Frame F.

4.2 The Uses of Gesture Types (Qualitative Observations)

What did we learn from many interactions like these? We learned qualitative lessons about the roles of different gesture types in dialogs with shared perceptual domains. We were then able to make comparisons between these qualitative conclusions and quantitative results from our human subject studies, suggesting similarities between our human/avatar prototype and true human/human interactions. Finally, we were able to measure the accuracy with which our system recognizes natural human gestures.

The gestures elicited from our human subjects are divided into four categories: numeric, deictic, iconic, and social, as listed in Table 2. HAB currently recognizes 1 deictic gesture (pointing), 3 iconic gestures (grab, carry, and push), and 3 social gestures (engage/disengage, positive acknowledge, and negative acknowledge). The gesture recognition component also recognizes the numbers one through five, although these are not currently supported by the reasoning module (VoxSim).

The iconic gestures for grab, push and carry were among the first gestures we thought to integrate into the system, and more iconic gestures will be added in future work, e.g. stack and rotate. After all, in a physical domain like blocks world, iconic gestures tend to directly represent the underlying actions. Not all action gestures are representational: the beckoning motion used to draw a block toward the user is a social convention, and doesn't mimic a human action. Nonetheless, iconic gestures tend to correspond to actions and be representational.

Interestingly, users feel less comfortable with iconic gestures than deictic ones. The prototype can be run in two modes, with and without the audio channels (i.e. words). We do not have quantitative measures because the prototype is not yet robust enough for naive user studies, but when the audio channel is on, users tend to say the words *grab, push* or *carry* rather than make the gestures (sometimes they do both). This is true even though the words push and carry have to be expanded with directional phrases (e.g. "to the left").

Pointing, on the other hand, seems completely natural. Users do it almost without thinking, and do it whether or not the audio channel is available. This may be because the alternative often requires a description ("the red block" or "the red block on the left" if there are two), although we noted above that actions may also require descriptions in form of directions.

Social gestures turn out to be critically important. They maintain the structure of the dialog. In an early version of HAB, the human users could only gesture and the avatar could only act and ask disambiguating questions. The system was uncomfortable to use, because the user would point to a block and then not be sure whether the avatar had seen the gesture or if they should point again. Ironically, it was almost better when the pointing gesture was ambiguous, because then the avatar would ask a clarifying question. We then gave the avatar her first gesture, reaching toward a block when it was referenced as a form of positive acknowledgment. Immediately the users became more comfortable and tasks were completed more quickly.

The timing of seeking and giving acknowledgment is important. We tested a version of the prototype in which the avatar always waited for confirmation before taking actions. This was slow and frustrated the users. More importantly, because acknowledgments were so common the conversation would often hit an impasse when the avatar was waiting for an acknowledgment that the signaler thought they had already given.

For comfort, peer-to-peer dialogs require that when one partner provides information, the other acknowledges receiving it. This is sometimes called *backchannel* communication. If the information is a request for an action, such as grab, then performing the action is sufficient acknowledgment. If the information is ambiguous, asking to clarify it also serves as acknowledgment. In all cases, however, some form of positive acknowledgment is required from whichever partner, human or avatar, receives the information.

Because acknowledgments are so common, it is important that they be unintrusive. Too many verbal acknowledgments quickly become annoying. The ability to positively acknowledge through head nods is important because it can be done without interrupting the audio stream and without interrupting other gestures by the hands.

Acknowledgments (positive or negative) in response to ambiguity are important, as they allow the system to engage in the conversational act of repair [44], or the use of clarifying and re-referring to correct misunderstandings in discourse. The recognition of the social gestures mentioned before are key to the system's ability to tackle ambiguity. These gestures enable natural and rapid feedback in order to complete tasks and allow the system to function as a human-like conversational agent.

4.3 Human/Human vs Human/Avatar Gesture Usage

The observations above were qualitative, based on our own experience. Studies with naive users are being planned, but further development is required to make sure that system artifacts don't distract naive users and invalidate the data. What we can do now, however, is test if our qualitative predictions match quantitative data from the human/human interaction studies. If so, this supports our predictions and suggests that there are at least similarities between human/human interaction and our prototype human/avatar system.

As shown in Sect. 3.1, we have labeled data from over 180 trials of human/human blocks world interactions in both the audio + video and video-only conditions. Based on our qualitative predictions, social gestures – particularly positive and negative acknowledgments – should outnumber iconic and deictic gestures in both conditions. Iconic gestures should appear more often in the video-only condition than the audio + video condition. Deictic gestures should appear with similar frequencies in both conditions.

Table 3. Frequencies of gestures in human data in the video only and Video + Audio conditions, organized by gesture category. The ratio is frequencies between the two conditions is a measure of how much more likely was a gesture to be made when the audio was turned off, versus when words are available

Gesture	Video only	Video + Audio	Total	Ratio
Iconic gestures				
Translate	234	73	307	3.2
Rotate	225	67	292	3.4
Separate	126	125	251	1.0
Servo Translate	206	44	250	4.7
Bring Together	83	38	121	2.2
Servo Toward	35	20	55	1.7
Iconic total	**909**	**367**	**1,276**	**2.5**
Deictic gestures				
This Block	240	94	334	2.6
That/There	150	106	256	1.4
Here/This	111	42	153	2.6
This Group	86	28	114	3.1
This Column	52	13	65	4.0
This Stack	38	16	54	2.4
These Blocks	24	20	44	1.2
Deictic total	**701**	**319**	**1,020**	**2.2**
Social gestures				
Pos. Acknowledge	693	225	918	3.1
Wait (Pause)	400	246	646	1.6
Start	100	51	151	2.0
Done	81	47	128	1.7
Neg. Acknowledge	109	11	120	9.9
Emphasis	17	27	44	0.6
Social total	**1,371**	**607**	**1,978**	**2.3**
Numeric gestures				
One	85	18	103	4.7
Two	64	9	73	7.1
Three	26	4	30	6.5
Four	21	5	26	4.2
Numeric total	**196**	**36**	**232**	**5.4**

Table 3 shows the frequencies of gestures in the human/human studies, organized by condition and gesture type. The gesture labels do not match up with the gestures recognized by HAB. There are many more gestures in the human data, and the gesture labels are more semantic. All gestures that are intended to make a partner wait, for example, are grouped into one category. *Servo* gestures are the continual gestures as in "a little more... a little more...". For a complete explanation of gestures, see [32].

Our first prediction based on HAB was that social gestures should be more common than iconic or deictic gestures. According to Table 3 this is true, although not by a large margin: 1,371 social gestures versus 1,276 iconic. Social gestures sometimes went unnoticed by the labelers, however, so the true disparity may be larger.

Our second prediction was that iconic gestures should appear more often in the video-only condition than the video + audio condition. It turns out that *all* gestures are more common in the video-only condition, because when people can't hear each other they gesture more. This is why we included the ratio of frequencies between the two conditions in Table 3. While the ratio is positive for all four gesture types, it is higher for iconic (2.5) than for social (2.3). Deictic gestures have the lowest ratio at 2.2. Although not implemented in HAB, numbers have by far the highest ratio at 5.4. Apparently people rarely gesture a number if they can simply say it.

4.4 Coordinate Systems

To ground deixis, the signaler and builder must agree on a coordinate system. In some dimensions the coordinate system is obvious; up is always against gravity, and down is always with it. The other dimensions aren't as clear, however.

Our initial mental model was that the signaler and builder were at different ends of a shared table; think of Fig. 2 without the gap between tables. In this model, the signaler's left is the builder's right, and vice-versa. This seems natural in practice, and re-examining the human subjects data it is indeed the coordinate system that our test subjects used.

This same model would predict that the edge of the table closest to the signaler is farthest from the builder and vice-versa. Thus if a signaler gesturally pushes a block away, the avatar should pull the block toward herself. Conversely, if a signaler pulls a block toward themself, the avatar should push the block away. But somehow, this doesn't seem as natural in practice.

Re-examining the human subjects data, human builders are inconsistent when a human signaler pushes a block away. About half the time, builders pull the block toward themselves. The other half of the time they push the block away. In the video-only trials, this was a source of confusion. In the video + audio trials, signalers and builders resolved the coordinate system verbally, using phrases like "toward you" or "toward me". This suggests an interesting interaction between speech and gesture in the context of deixis.

4.5 Recognition Accuracy

One concern that arose while designing HAB was whether gesture recognition would be accurate enough to support peer-to-peer communication. While there have been many previous 3D gesture-based interfaces, most have been designed to detect large-scale gestures, for example gaming motions or scuba diving signals, as in the ChaLearn challenge [45]. We needed to recognize natural gestures elicited from naive users in the context of blocks world, and were counting on new sensors and recognition techniques to make this possible.

In particular, we were counting on the Kinect sensor to extract accurate depth maps, and the Kinect pose data (a.k.a. skeleton) to reliably identify the locations of the hands and head. We were also relying on ResNet-style deep convolutional neural networks (DCNNs) [35] to recognize hand poses in segmented depth images, and to recognize head motions given a time window of differences of depth images. A significant risk factor with regard to DCNNs was whether we had enough training data to train reliable networks.

Training samples were extracted from the human subjects studies described in Sect. 3.1, which we hand labeled [32]. For 25 hand poses, this yielded a significant number of training samples. There were other hand poses, such as thumbs down, that appeared less often but that we still wanted to include in the system. We therefore supplemented the training samples for 10 more hand poses by having volunteers perform the poses in front of the Kinect. Unfortunately, the data collected in this way turned out to be exaggerated compared to naturally occurring poses.

When the prototype was completed, signalers reported satisfaction with the gesture recognition. In practice, the system rarely missed gestures or inserted false gestures. There are many possible explanations for this, however, including that the signalers were system designers with an interest in gesturing clearly. Furthermore, although the gesture recognition system detects 35 hand poses, only a subset are fused by the finite state machines into the 7 multi-modal gestures integrated with VoxSim. The underlying performance of hand pose recognition was therefore still unknown.

To evaluate the accuracy of hand pose recognition, we collected new data from 14 naive human subjects, using the same experimental setup and protocol as in Sect. 3.1. We didn't have the resources to hand label every frame, so instead we adopted a sampling methodology. The new videos were processed through the DCNN hand pose classifier. To estimate precision, we randomly sampled 60 instances of each hand pose as identified by the DCNN. We then brought in naive raters and asked them whether the detected gestures actually occurred where the DCNN said they did.

The precision results are shown in Fig. 5. The units of the horizontal axis are seconds, so that the leftmost data point is the precision for gesture detections that lasted for $\frac{1}{5}$ of a second or less. The second data point represents detections with a duration between $\frac{1}{5}$ and $\frac{2}{5}$ of a second, and so on. The orange line represents the precision across all 35 poses, while the blue line shows the precision for the 25 poses trained on data from the human subjects studies. The plot shows that even for gestures that last for $\frac{1}{5}$ of a second or less, over 55% of the DCNNs detections are correct. Long duration gestures (one second or more) have a precision of 80%. When we limit the evaluation to the 25 naturally trained poses, these numbers go up to 60% for short gestures and 87% for long ones.

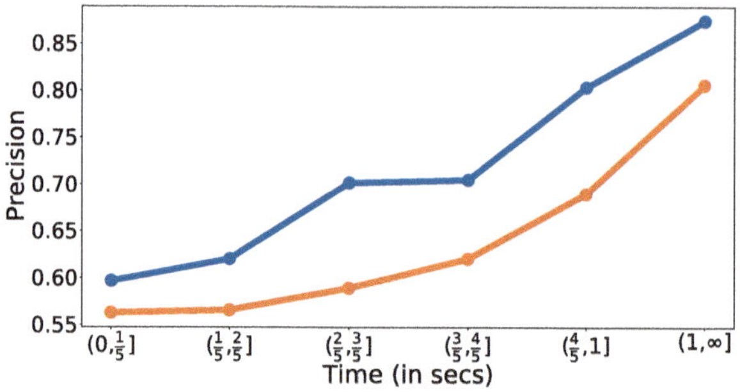

Fig. 5. Precision of hand pose detection. The horizontal axis represents durations of detected hand poses. The vertical axis is the percent of true detections as judged by naive raters. The blue line shows the precision of the 25 hand poses trained on the human subjects data. The orange line adds 10 more poses for which additional, exaggerated training data was collected

Recall was estimated using a similar procedure. In this case, naive raters were given portions of videos, and asked to label any hand poses they saw. For each pose, they selected one frame. (Although not instructed to do so, they usually selected the first frame in which the pose appeared.) We then measured how often the DCNN detected the hand pose at that frame, within a fifth of a second of that frame, within two fifths of a second, and so on, up to a second. The resulting plot is shown in Fig. 6, with the same color scheme as in Fig. 5 in terms of recall for 25 or 35 poses.

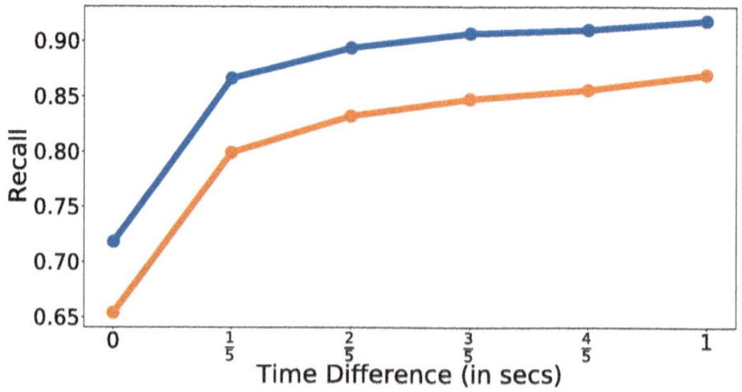

Fig. 6. Recall of hand pose detection. The horizontal axis is the time difference between the frame selected by a naive rater and the automatic detection. The vertical axis is the percent of recall. The blue and orange lines indicate the same distinction between 25 and 35 poses as in Fig. 5

Interestingly, almost half the recall omissions are the result not of mistakes by the DCNN, but of segmentation failures. The depth images passed to the DCNN are windows of the full image centered on the hand, as identified by the Microsoft skeleton. When the hand position is incorrect, failure is inevitable.

5 Conclusion

This paper investigates peer-to-peer communication between people and avatars in the context of a shared perceptual task. In our experiments, people communicate with avatars using gestures and words, and avatars communicate back through gestures, words, and actions. Together, they complete tasks through mixed initiative conversations. The human signaler has the initial goal and tells the avatar what to do, but when ambiguities arise the initiative shifts and the avatar asks the human for clarification. Social cues make this process flow naturally. Overall, we demonstrate an example of peer-to-peer cooperation between people and machines, through shared perception and perceptually-grounded reasoning.

Acknowledgments. This work was supported by the US Defense Advanced Research Projects Agency (DARPA) and the Army Research Office (ARO) under contract #W911NF-15-1-0459 at Colorado State University and the University of Florida and contract #W911NF-15-C-0238 at Brandeis University.

References

1. Küster, D., Krumhuber, E., Kappas, A.: Nonverbal behavior online: a focus on interactions with and via artificial agents and avatars. In: The Social Psychology of Nonverbal Communication, pp. 272–302. Springer (2015)
2. Wobbrock, J.O., Morris, M.R., Wilson, A.D.: User-defined gestures for surface computing. In: CHI 2009, pp. 1083–1092. ACM, New York (2009). http://doi.acm.org/10.1145/1518701.1518866
3. Sproull, L., Subramani, M., Kiesler, S., Walker, J.H., Waters, K.: When the interface is a face. Hum. Comput. Interact. **11**(2), 97–124 (1996)
4. Dastani, M., Lorini, E., Meyer, J.-J., Pankov, A.: Other-condemning anger = blaming accountable agents for unattainable desires. In: Proceedings of the 16th Conference on Autonomous Agents and MultiAgent Systems, pp. 1520–1522. International Foundation for Autonomous Agents and Multiagent Systems (2017)
5. Li, J.: The benefit of being physically present: a survey of experimental works comparing copresent robots, telepresent robots and virtual agents. Int. J. Hum. Comput. Stud. **77**, 23–37 (2015)
6. Bolt, R.A.: "Put-that-there": voice and gesture at the graphics interface. ACM **14**(3), 262–270 (1980)
7. Dumas, B., Lalanne, D., Oviatt, S.: Multimodal interfaces: a survey of principles, models and frameworks. In: Human Machine Interaction, pp. 3–26 (2009)
8. Turk, M.: Multimodal interaction: a review. Pattern Recogn. Lett. **36**, 189–195 (2014)

9. Quek, F., McNeill, D., Bryll, R., Duncan, S., Ma, X.-F., Kirbas, C., McCullough, K.E., Ansari, R.: Multimodal human discourse: gesture and speech. ACM Trans. Comput. Hum. Interact. (TOCHI) **9**(3), 171–193 (2002)
10. Clark, H.H., Brennan, S.E.: Grounding in communication. In: Resnick, L.B., Levine, J.M., Teasley, S.D. (eds.) Perspectives on Socially Shared Cognition, vol. 13, pp. 127–149. American Psychological Association (1991)
11. Clark, H.H., Wilkes-Gibbs, D.: Referring as a collaborative process. Cognition **22**(1), 1–39 (1986). http://www.sciencedirect.com/science/article/pii/0010027786900107
12. Dillenbourg, P., Traum, D.: Sharing solutions: persistence and grounding in multimodal collaborative problem solving. J. Learn. Sci. **15**(1), 121–151 (2006)
13. Fussell, S.R., Kraut, R.E., Siegel, J.: Coordination of communication: effects of shared visual context on collaborative work. In: Proceedings of the 2000 ACM Conference on Computer Supported Cooperative Work, CSCW 2000, pp. 21–30. ACM, New York (2000). http://doi.acm.org/10.1145/358916.358947
14. Fussell, S.R., Setlock, L.D., Yang, J., Ou, J., Mauer, E., Kramer, A.D.I.: Gestures over video streams to support remote collaboration on physical tasks. Hum. Comput. Interact. **19**(3), 273–309 (2004)
15. Kraut, R.E., Fussell, S.R., Siegel, J.: Visual information as a conversational resource in collaborative physical tasks. Hum. Comput. Interact. **18**(1), 13–49 (2003)
16. Gergle, D., Kraut, R.E., Fussell, S.R.: Action as language in a shared visual space. In: Proceedings of the 2004 ACM Conference on Computer Supported Cooperative Work, CSCW 2004, pp. 487–496. ACM, New York (2004). http://doi.acm.org/10.1145/1031607.1031687
17. Reeves, L.M., Lai, J., Larson, J.A., Oviatt, S., Balaji, T., Buisine, S., Collings, P., Cohen, P., Kraal, B., Martin, J.-C.: Guidelines for multimodal user interface design. Commun. ACM **47**(1), 57–59 (2004)
18. Veinott, E.S., Olson, J., Olson, G.M., Fu, X.: Video helps remote work: speakers who need to negotiate common ground benefit from seeing each other. In: Proceedings of the SIGCHI Conference on Human Factors in Computing Systems, CHI 1999, pp. 302–309. ACM, New York (1999). http://doi.acm.org/10.1145/302979.303067
19. Lascarides, A., Stone, M.: Formal semantics for iconic gesture. In: Proceedings of the 10th Workshop on the Semantics and Pragmatics of Dialogue (BRANDIAL), pp. 64–71 (2006)
20. Clair, A.S., Mead, R., Matarić, M.J., et al.: Monitoring and guiding user attention and intention in human-robot interaction. In: ICRA-ICAIR Workshop, Anchorage, AK, USA, vol. 1025 (2010)
21. Matuszek, C., Bo, L., Zettlemoyer, L., Fox, D.: Learning from unscripted deictic gesture and language for human-robot interactions. In: AAAI, pp. 2556–2563 (2014)
22. Krishnaswamy, N., Pustejovsky, J.: Multimodal semantic simulations of linguistically underspecified motion events. In: Spatial Cognition X: International Conference on Spatial Cognition. Springer (2016)
23. Gilbert, M.: On Social Facts. Princeton University Press, Princeton (1992)
24. Stalnaker, R.: Common ground. Linguist. Philos. **25**(5), 701–721 (2002)
25. Asher, N., Gillies, A.: Common ground, corrections, and coordination. Argumentation **17**(4), 481–512 (2003)
26. Tomasello, M., Carpenter, M.: Shared intentionality. Dev. Sci. **10**(1), 121–125 (2007)

27. Bergen, B.K.: Louder than words: the new science of how the mind makes meaning. In: Basic Books (AZ) (2012)
28. Hsiao, K.-Y., Tellex, S., Vosoughi, S., Kubat, R., Roy, D.: Object schemas for grounding language in a responsive robot. Connection Sci. **20**(4), 253–276 (2008)
29. Dzifcak, J., Scheutz, M., Baral, C., Schermerhorn, P.: What to do and how to do it: translating natural language directives into temporal and dynamic logic representation for goal management and action execution. In: IEEE International Conference on Robotics and Automation, ICRA 2009, pp. 4163–4168. IEEE (2009)
30. Cangelosi, A.: Grounding language in action and perception: from cognitive agents to humanoid robots. Phys. Life Rev. **7**(2), 139–151 (2010)
31. Siskind, J.M.: Grounding the lexical semantics of verbs in visual perception using force dynamics and event logic. J. Artif. Intell. Res. (JAIR) **15**, 31–90 (2001)
32. Wang, I., Narayana, P., Patil, D., Mulay, G., Bangar, R., Draper, B., Beveridge, R., Ruiz, J.: Eggnog: a continuous, multi-modal data set of naturally occurring gestures with ground truth labels. In: 12th IEEE International Conference on Automatic Face and Gesture Recognition (2017)
33. Kendon, A.: Gesture: Visible Action as Utterance. Cambridge University Press, New York (2004)
34. Zhang, Z.: Microsoft kinect sensor and its effect. IEEE MultiMedia **19**, 4–10 (2012)
35. He, K., Zhang, X., Ren, S., Sun, J.: Deep residual learning for image recognition. In: Proceedings of the IEEE Conference on Computer Vision and Pattern Recognition, pp. 770–778 (2016)
36. Krishnaswamy, N., Pustejovsky, J.: VoxSim: a visual platform for modeling motion language. In: Proceedings of COLING 2016, the 26th International Conference on Computational Linguistics: Technical Papers. ACL (2016)
37. Goldstone, W.: Unity Game Development Essentials. Packt Publishing Ltd., Birmingham (2009)
38. Pustejovsky, J., Moszkowicz, J.: The qualitative spatial dynamics of motion. J. Spat. Cogn. Comput. **11**, 15–44 (2011)
39. Pustejovsky, J., Krishnaswamy, N.: VoxML: a visualization modeling language. In: Chair, N.C.C., Choukri, K., Declerck, T., Goggi, S., Grobelnik, M., Maegaard, B., Mariani, J., Mazo, H., Moreno, A., Odijk, J., Piperidis, S. (eds.) Proceedings of the Tenth International Conference on Language Resources and Evaluation (LREC 2016). European Language Resources Association (ELRA), Paris, May 2016
40. Pustejovsky, J.: Dynamic event structure and habitat theory. In: Proceedings of the 6th International Conference on Generative Approaches to the Lexicon (GL2013), pp. 1–10. ACL (2013)
41. McDonald, D., Pustejovsky, J.: On the representation of inferences and their lexicalization. In: Advances in Cognitive Systems, vol. 3 (2014)
42. Pustejovsky, J.: The Generative Lexicon (1995)
43. Narayana, P., Beveridge, R., Draper, B.: Gesture recognition: focus on the hands. In: IEEE Conference on Computer Vision and Pattern Recognition (2018)
44. Hirst, G., McRoy, S., Heeman, P., Edmonds, P., Horton, D.: Repairing conversational misunderstandings and non-understandings. Speech Commun. **15**(3), 213–229 (1994). http://www.sciencedirect.com/science/article/pii/0167639394900736
45. Ponce-López, V., Chen, B., Oliu, M., Corneanu, C., Clapés, A., Guyon, I., Baró, X., Escalante, H.J., Escalera, S.: Chalearn lap 2016: first round challenge on first impressions - dataset and results. In: ECCV, pp. 400–418 (2016)

A Safer YouTube Kids: An Extra Layer of Content Filtering Using Automated Multimodal Analysis

Sharifa Alghowinem(✉)

College of Computer and Information Sciences, Prince Sultan University, Riyadh, Saudi Arabia
sghowinem@psu.edu.sa

Abstract. Acknowledging the advantages as well as the dangers of the internet content on kids education and entertainment, YouTube Kids was created. Based on regulations for child-friendly programs, several violations are identified and restricted from viewable content. When a child surfs the Internet, the same regulations could be automatically detected and filtered. However, current YouTube Kids content filtering relies on meta-data attributes, where inappropriate content could pass the filtering mechanism. This research, propose an advanced real-time content filtering approach using automated video and audio analysis as an extra layer for kids safety. The proposed method utilizes the thin-slicing theory, where several one second slices are selected randomly from the clip and extracted. The use of a one-second slice will assure a temporal analysis of the clip content, and ensures a real-time content analysis. For each slice, audio is automatically transcribed using automatic speech recognition techniques to be further analysed for its linguistic content. Furthermore, the audio signal is analysed to detect event and scenes (e.g. explosion). The image frames extracted from the slices are also inspected for its content to avoid inappropriate scenes, such as violence. Upon the success of this approach on YouTube Kids application, investigation of its generalizability to other video applications, and other languages could be performed.

Keywords: Content analyses · Video analyses · Audio analyses
Child safety

1 Introduction

The Internet has become the main source of information for people of all ages, especially young ones. The wide resources from basic websites to interactive web applications and social media have increased the Internet potentials, and therefore, its utilization in educating and entertaining young learners. With the emergence of smartphones and tablets, the demand has further increased for providing diverse content. Children now are few clicks away from their independent knowledge acquiring.

K. Arai et al. (Eds.): IntelliSys 2018, AISC 868, pp. 294–308, 2019.
https://doi.org/10.1007/978-3-030-01054-6_21

Regardless of the benefits of the wide resources of the Internet, their increase also imposes a danger for children who use it. Unlike television, Internet content that could be presented for kids has no regulation [1]. Television regulations for children state certain child-safe periods where advertisement are limited and the show content has several constraints [2]. Due to the nature of the Internet, these regulations cannot be directly applied. Parental control and content filtering mechanisms are utilized to automatically simulate such regulations [3,4]. These mechanisms are mitigation approach to protect kids from viewing inappropriate content.

One example is YouTube Kids created by Google in 2015 as an initiative for a child-safe application. Videos from YouTube are filtered based on advanced algorithms to decide their appropriateness for the little viewers. For the decision, this algorithm analyses the video meta-data, such as title, description, tag, number of views, and ratings, as well as community flagging and comments. Nevertheless, YouTube Kids application content has been a controversial topic because of its sensitive impact on children if inappropriate video passes the filtering mechanism [1,5,6]. Hence, relying on meta-data and community flagging for filtering video content is not enough to ensure child safety [7–9].

Therefore, in this paper, we propose an advanced automated method to filter YouTube videos for the young viewers. This is performed using video content analysis, where images, audio and transcripts are extracted from random one-second slices to be analysed for their content appropriateness. The contribution of this work is as follows:

- We investigate for the first time media appropriateness for children by fusing multimodal aspects of a media. We integrate visual, audio and linguistic features.
- To reduce computational time, we aim for real-time appropriateness analysis by applying the thin slicing theory, where several random one-second slices of a video are analysed.
- We utilize and collect real on-line videos from YouTube Kids that pass the filtering mechanism in order to provide an extra layer of safety for kids.

The paper is structured as follows: Related Work section reviews the literature, where aspects of the proposed study such as video, audio and transcription analysis, as well as fusion of multi-modal analysis are covered. The method section illustrates the detailed process of the proposed investigation, including data collection and slicing, classification approaches and their evaluation from the audio, video and transcript analysis, as well as fusion approaches. A pilot study results are presented in the followed section. Finally, the conclusion section summarizes the proposed study and its limitations.

2 Related Work

Several organizations and commissions in different countries have regulated rules and guidelines for appropriate children television programs and advertisements, such as the U.S. [2], Australia [10], and the UK [11]. The common regulations between these countries are: (1) define specific periods for family-friendly programs, where the content should be appropriate to be viewed by children, and (2) restrict the amount and content of advertisement in these periods. The content of children programs has several criteria, where they should be made specifically for children, well produced using sufficient resources, use suitable level of language understood by children and have educational and entertaining goals. Materials that strictly prohibited are: demeaning any person or group, frightening, unsafe, sexual and violent scenes, as well as drug and alcohol content.

While non-compliance with these regulations could lead the channel to lose their broadcasting license, it cannot be applied to Internet content. With the advancement of technology, an automatic content analysis could detect the presence of offending content. In efforts to help parents, content filtering software has been developed [12], including advanced ones that use artificial intelligence techniques [13]. This content filtering software relies on a defined set of rules and text analysis using natural language processing (NLP). However, such methods do not apply to multimedia content such as videos.

As mentioned earlier, YouTube Kids application uses algorithms to filter the child-friendly videos based on meta-data and community flagging attributes. Even though such method is successful to some extent, some inappropriate videos could pass the algorithm. Enhancing the algorithm with extra layers of filtering by utilizing audio, video, and transcript could result in more reliable results. To the best of our knowledge, there are no studies exist that investigate fusing the three modalities (i.e. audio, video, and transcript) to analyse video content for its appropriateness for children. Nonetheless, several studies investigated different aspects that could be helpful in formulating our proposed method. Such studies include studies that inspected video content using image frames, audio content using speech signal, and natural language processing using the transcription from audio content (e.g. automatic speech recognition).

2.1 Video Content Analysis

Video content analysis has gained a great interest in the past decade, especially with the era of big data analysis [14]. Moreover, several studies explored specific video content classification that could be utilized for classifying content that is suitable for kids.

For instance, violent scenes could be automatically identified through the existence of gunshots, explosions, aggressive human actions and screams in audio or video signals [15], as well as blood and sudden changes in motion in video frames [16]. A dataset of violent scenes was collected for a challenge aim at automatically detecting violent scenes [17,18]. Several studies have used this dataset as a benchmark for accuracy and performance comparison. For example, an accuracy of 72% was obtained in a real-time detection method [19].

Automatic classification of sexual content has been investigated from video frames using advance artificial intelligent techniques. Sexual content was categorised into ten levels in COPINE Scale that range from indicative to sadistic [20]. While most studies focused on the first few levels, few have investigated higher levels of COPINE Scale [21]), where an accuracy of 88% was obtained from video frames. A specific Pornography dataset was collected that contain nearly 80 h of 400 pornographic and 400 non-pornographic videos [22]. This dataset was used to benchmark and compare automatic detection of pornographic content. For instance, [23] used a novel deep neural network that outperformed previous studies, where the accuracy was 95%.

Alcohol and drug contents are one of the prohibited content for children. Automatic detection of such content has been explored in [24], where text, images and video posts were analysed. In their work, the classification results from the video posts performed an accuracy of 79%.

2.2 Audio Content Analysis

On the other hand, the audio content analysis could play a significant part in complementing the video content to enhance the decision of appropriateness of a video to children. The audio signal from a video contains both uttered words (linguistic) and acoustic cues (para-linguistic). Linguistic component of an audio signal is obtained through automatic speech recognition techniques [25], while the para-linguistic component is obtained by processing the audio signal itself [26].

Based on acoustic features, not only gender, age [27], emotion [28], mood [29], etc. of the speakers could be detected, but also music [30] and noise [31]. Moreover, studies investigated scene and event detection from audio signals in a "machine listening" initiative [32]. Events such as: door knock, clearing throat, phone ringing, etc. and scene types such as: bus, busy street, restaurant, supermarket, etc. are classified based on the acoustic features. Specific audio event discovery from YouTube videos was explored, where results were promising [33]. Objectionable sounds such as sexual screams were investigated for automatic detection in [34], which results in a very high classification rate. Detection of speaker characteristics, events that could represent violent scenes such as gunshots, screams, and explosions, as well as events that could represent sexual scenes such as breaths and moans, could prove useful in providing the machine an understanding of the audio content.

2.3 Transcribed Content Analysis

The transcribed content of the media could be analysed using natural language processing to indicate the appropriateness of the used language to the little viewers. Most automatic website filtering uses the textual content analysis to measure the suitability of the website to children [35,36]. Analysis of text complexity and comparisons with child-friendly language models could give an indicator of the focus towards children.

For example, topical and non-topical features were extracted to classify a web page content suitability for children in [37]. Their features included text complexity measures such as readability, language models specifically for children, and children reference analysis. Children reference includes analysing if the child been referred to directly (e.g. kids like you), or been talk about (e.g. kids like yours) could be used to identify the content suitability for kids [37].

Language that demeans a person or a group based on some aspect such as race, or gender (also called hate speech) is restricted especially for children content. Studies have focused on automatically detecting such language from social media posts and comments [38]. Different word features could be extracted such as word- and character-based features, bag-of-words and clustering features, as well as sentiment and lexical analysis for classification. Such features and further prepossessing and classification techniques are surveyed in [38]. Similar techniques could be applied to the transcribed text extracted from the audio in a media content to detect the language suitability for children listener.

2.4 Multimodal Fusion Analysis

Fusing all three parts (audio transcripts, audio acoustic, and video frames) could further enhance the classification results of media for their appropriateness for kids. Multimodal sentiment analysis has been carried out lately, where images, audio and textual information from a video is analysed and fused to detect positive, negative or neutral sentiments [39–41]. Moreover, multimodal emotion recognition has also been investigated, where audio and video signals are fused to recognise expressed emotions [42]. Regardless of the fusion technique used (e.g. feature-level, decision-level), fusing different modalities not only can enhance classification results, but also increase the confidence level of the final results.

Even though there are no studies that aimed directly to automatically analyse the appropriateness of a video to children, studies investigating similar aspect could direct such aim. Approaches that have been used in studies that investigated image, audio and/or textual content of a video could be investigated for classifying child-friendly media. The next section elaborates on our proposed method, as well as steps in collecting video data, extracting image frames, speech signal, and transcription from video content.

3 Method

The general framework for our study is presented in Fig. 1. Generally, videos that pass YouTube Kids filtering mechanism will be collected. The manual annotation will be performed into two classes (i.e. appropriate and not appropriate for kids). To reduce computational time, only one-second slices will be extracted for analysis. From each slice, image frames, audio signal, and transcribed text will be analysed. Fusion techniques will be applied to investigate the enhancement in the classification performance. Detailed techniques of each step are listed in the following subsections.

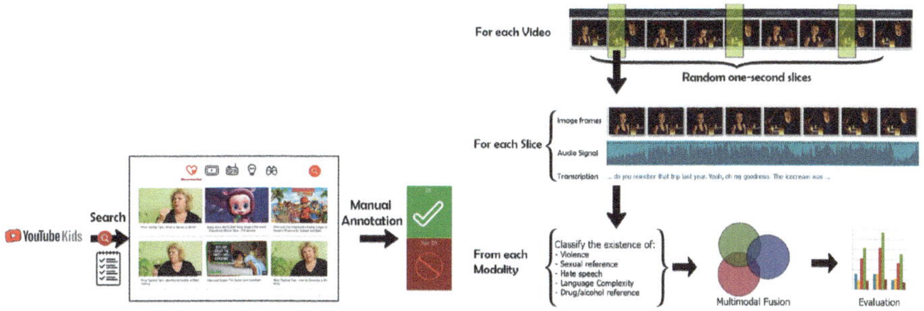

Part I: Manual data collection and annotation. Part II: Automatic analysis of video appropriates for kids.

Fig. 1. The general framework for detecting appropriateness of a media to children

3.1 Data Accusation

In order to validate our proposed method, a dataset of suitable and unsuitable clips that passes YouTube Kids algorithm should be collected. Even though YouTube Kids algorithm blocks list of inappropriate keywords from the search function, some normal keywords could retrieve unsuitable video clips for children. To avoid this, keywords that could potentially lead to such retrieval should be collected first and then used to search for unsuitable videos. To achieve this, feedback from mothers could be collected about keywords that their children have used and resulted in unsuitable videos. A wide spread online survey will be created for this purpose, and distributed through social media and invitations.

Once the list of keywords is collected, a manual search for each keyword will be conducted. The direct and indirect search results from a keyword that are not suitable for children will be collected. Indirect search results include recommendation videos from a viewed video from a keyword. When collecting the video clips a variety of violations will be considered. That is, the collected dataset will include videos that contain: violence, danger and unsafe events, sexual references, drug and alcohol references, hate speech, and language complexity. A large number of videos will be collected in order to give valid results. An equal number of suitable videos (i.e. videos that do not violate any of the child-friendly regulations) will be also collected.

The collected videos will be annotated manually. The videos will be manually labelled into two classes (i.e. suitable and unsuitable). For unsuitable videos, violating aspects will be recorded for further analysis (e.g. violent action). Moreover, the violation source modality will be also recorded. That is, if a video has fire matching scene and language complexity violations, the modalities will be recorded as from video and transcription.

3.2 Media Slicing

To reduce computational time and resources, several random one-second slices will be extracted for automatic analysis and classification. We employ the thin-slicing theory [43] for this purpose. The thin-slicing theory is a theory used in psychology that explains that a behavioural judgment could be derived from a short time of observation as accurate, or even more accurate, as judgment based on long observation.

We propose to select several one-second slices as follows: (1) random one-second slice for each minute of the first three minutes, and (2) random one-second slice every five minutes after the first three minutes. The focus of the first three minutes is based on the concept that most viewers decide to continue watching a video after the first few minutes [44]. Therefore, increasing the analysis in the first parts could potentially protect kids from offensive content. Moreover, the first portion of a video is more likely to represent the remaining of the video.

An experiment will be conducted to compare full analysis and slices analysis. The comparison will be in regard to the classification of the appropriateness of a video and in regrade to computational time. Moreover, the number of slices as well as position of the slices (e.g. every five minutes) will be investigated to optimize the best number and position of slices.

3.3 Classification Approaches

Regardless of the modality (e.g. video, audio, transcription), classification process is similar between each one of them. For all classification approaches, and after the dataset is collected and annotated, and in our cases also sliced, the next step is to perform pre-processing. Preprocessing step is the process of preparing the data for feature extraction stage. This includes cleaning the data from noise (e.g. removing audio noises, removing symbols from a text), locating a particular point of interest (e.g. face in an image frame, an explosion sound event), etc. Preprocessing methods differ not only based on the modality in question, but also for the classification task. The same applies to feature extraction, where each modality (e.g. video) could have different approaches and types of features to be extracted based on the classification problem at hand.

Once the features are extracted, classification is performed, either in supervised or unsupervised classification manner. In our case, even though the final classification should be either suitable or unsuitable video for children, sub classification of the violating aspects of the video should be detected. Sub classifications include: violence, sexual references, hate speech, complex language, drug and alcohol references. Upon the detection of any of these violations in any modality, a video is classified as unsuitable for children. Moreover, depending on the classification method used in each modality to detect each violation, a confidence level of the detection could be used and then used in a combination with other modalities for the final decision. Further elaboration of the classification process for each modality is presented in the following subsections.

3.4 Video Analysis

Technical methods for processing the video for retrieval, indexing, classification and content understanding are surveyed in [45]. Videos are segmented to image frames, where several features are extracted from each frame, including static features such as colour, texture and shapes, object features, and motion features [45]. In our work, the extracted slices from the video will be segmented to image frames and treated as key-frames as a preprocessing step for feature extraction. State of the art features that proved effective in detecting specific violations will be extracted. Motion features have been successfully used in combination with static features to classify key-frames with pornographic content [46]. A Convolutional Neural Networks as a deep learning approach was used which results in outperforming existing methods. In [19] spatio-temporal features including Bag-of-Visual Words (BoVW) and Temporal Robust Features (TRoF) were utilized to detect violent scenes. A linear Support Vector Machine classifier using these features outperformed the previous research on violent scene detection.

This research will follow the proposed approach in [19,46] to detect sexual and violence scenes respectively. Worth noting that these researches used public datasets for their investigation: Pornography-2k dataset [47] and MediaEval Violent Scenes Detection (VSD) dataset [18]. The processes will be utilized and applied to the slices extracted from the video clips to detect such violations.

3.5 Audio Analysis

Preprocessing of the audio signal will serve two aspects: preparing the signal for acoustic feature extraction and preparing the signal for transcription. These approaches are different because the formal should also include the sound events (e.g. explosion sounds), while the latter should only preserve human voices.

For preparing the audio signal for automatic speech recognition, noise and music should be removed. Several methods for noise and music removal for automatic speech recognition have been reviewed in [48]. One of the effective methods is the use of denoising autoencoders to learn features that discriminate between music and speech [49] and between environmental noises and speech [50]. These methods utilize deep neural network (DNN) trained in different music and noise characteristics, along with the use of language model, acoustic model, phonetic dictionary and/or lexicons to identify the spoken words [48]. Post-editing error correction could be performed to decrease word-error-rate, and therefore, enhance the transcription [51]. Further processing and analysis of the transcription modality are presented in the following section.

Automatic recognition of environments, events and scenes from audio signal is reviewed, where several novel methods have been proposed [52]. Acoustic features are extracted such as MFCC, spectrogram, energies, etc. and used as input for the classifier such as deep neural network, Gaussian Mixture Model, ensemble classifiers, etc. Such event detection could be indicators to detect violence, and sexual scenes' events, as mentioned previously in Sect. 2.2

In this research, the best performing methods for both noise and music removal and event and scene detection will be utilized and applied to the slices extracted from the video clips. Such methods are trained on public datasets such as the CHiME-3 for real-life noisy speech dataset [53], and TUT database for acoustic scene classification and sound event detection [54].

3.6 Transcription Analysis

The transcribed text from the audio slices is useful for detecting violent, sexual, drug and alcohol language, as well as hate speech and measuring language complexity. All of which, are indicators for unsuitability of a video for children. The automatic transcription post-processing (e.g. contextual spell checking) would enhance the quality of the produced text. The produced text is then preprocessed before the features could be extracted for the analysis. Preprocessing of text includes: tokenization, stemming, stop words removal, part-of-speech tagging, etc. Textual features are divided into several categories as follows: simple features such as character n-gram, word generalization features such as bag-of-words, sentiment analysis features such as number of negative words, lexical features such as presence of words in a lexicon, linguistic features such as part-of-speech, knowledge-based features such as stereotype assertion [38].

For detecting each of the violations, and to extract the complex features, large datasets are needed. For example, a dataset of annotated hate speech and offensive Language was collected and publicly available for the task of automatic detection of hate speech [55]. A dataset was collected for adult content filtering [56], and another dataset for detecting child grooming (establishing an emotional connection with a child with the objective of sexual abuse) [57]. Such datasets will be utilized in our work to detect violations of violent, sexual, drug and alcohol language, as well as hate speech. For language complexity, the complexity and readability formula will be used to calculate the complexity level of the transcribed language. Text complexity includes syntactic and lexical measures, which could be measured by characters per word, syllables per word, words per sentence, etc. The higher values of these measures, the higher complexity is the text for children.

3.7 Fusion Approaches

As mentioned earlier, the fusion of the different modalities not only could enhance the classification results, but also could increase the confidence level of the final decision. Once the results are obtained from individual modalities for each violation, fusing these results will finalize decision. This could be accomplished in several approaches depending on the classification method used for each modality and for each violation.

Table 1 gives an overview of how each video could be classified as suitable or unsuitable. Child-friendly rule violations are presented along with the modalities that could detect them. Some classification approach gives a confidence level or probability, while other approaches gives a binary classification (e.g. 0 or 1). For each violation, the average confidence level or the number of votes from the used modalities could be used for the fusion results. A threshold should be defined for each violation to get the final result for the violation in question. For example, a threshold of 1 from the number of votes in violence could be set to classify a video as containing violence violation. The final decision of the suitability of the

Table 1. Fusion approach of the violations in each of the three modalities

Modality/ Violation	Violence	Sexual	Hate speech	Complex language	Drug and alcohol	
Video	√	√			√	
Audio	√	√				
Transcription	√	√	√	√	√	
Fusion	ACL/NVC	ACL/NVC	ACL/NVC	ACL/NVC	ACL/NVC	
Decision	Threshold	Threshold	Threshold	Threshold	Threshold	Final decision

ACL: Average Confidence Level; NVC: Number of Voting of the Classification

video will be based on the number of violations detected in the video. Such final decision could be strict (e.g. if one violation exists the video will be labelled as unsuitable), or could use weights for each violation.

3.8 Performance Evaluation

The evaluation of the proposed method will be measured in regard to accuracy and computational time. Accuracy evaluation is measured in three different levels: modality-violation level, fusion-violation level and final decision evaluation. To accomplish these evaluations, a confusion metric is utilized. From which, several measures could be derived such as average accuracy, precision, and F-Measure. Moreover, Receiver Operating Characteristic (ROC) Curve will be used to calculate Area under the ROC curve (AUC) for the evaluation. ROC could be used to select the optimal modal (threshold) for classification problem. Measuring the evaluation of these three levels will give an insight to locate the weakness sources in order to improve them.

Since the ultimate goal is to create a real-time method to filter unsuitable videos for kids, measuring computational time is crucial. Computational time differences in the number of slices, preprocessing approach, extracted feature types, etc. will be compared to identify the best optimization.

4 Pilot Study Results and Discussion

We have informally interviewed four mothers, who their children use YouTube Kids. All mothers have shared their concerns where their children have accidentally watched unsuitable videos through YouTube Kids. For example, a mother shared that a "Superman" keyword led her child to inappropriate dubbed video, that has sexual references. Moreover, "Romantic" keyword led to videos that could be unsuitable for children, as mentioned by one mother. Another example was given about "Depression" and"Disorder" keywords leading to unsuitable videos, as well as depression tests videos. A mother stated that even normal keywords could lead to videos that not meant for children, such as "DIY" or"Pranks".

We browsed YouTube Kids for the mentioned keywords, where the keywords "Disorder", "Romantic" and "Pranks" showed potential videos that are not meant for the little viewers. Moreover, sometimes the keywords do not directly present unsuitable videos, but the recommended/related videos would. We select three videos based on the three keywords, we extract them from YouTube as shown in Table 2. We analyse these videos manually with the view of computer vision, audio signal and transcription described in the methodology. Worth mentioning that the search results also presented several videos that contain no restricted content. However, they are not meant for children because the level of language and level of instruction require a higher level of comprehension (e.g. personal development talks, programming and mathematical tutorials).

Table 2. Selected videos for analysis

Keyword	YouTube ID	Content	Modality
Disorder	giEpegtMIxg	Alcohol reference	Transcription
Romantic	ByCkGKOIEY4	Suicide, Sexual paintings	Video
Pranks	3J6o7hcm8bE	Dangerous tools and actions	Video and Audio

The first video shows a women explaining the differences in diagnosing Major Depression and Bipolar. She explains her experience with alcohol and medications in a duration of 27 min. Even though the video frames and audio have no content violations (e.g. no violence, sexual events), the language from the transcription refers to alcohol. With the thin-slicing theory, the random slices could miss the alcohol reference. However, the remaining of the transcription would show the complexity of the used language for children.

The second video is a collection of classic music in 2.5 h, which also shows paintings. One painting shows a women hanging herself, and another painting shows a couple passionately kissing. The audio part is child-friendly, however the still image paintings could be classified as inappropriate. Given that the paintings in question span for over 2 min each and they are immediately after each other, there is a good chance that the random slices will include them.

The last video shows Fidget Spinner tricks and challenges in 6 min, where some of the spinners are sharp (e.g. Ninja spinner). Shooting spinners, aiming spinners to breaking glass and matches are presented, where it could be dangerous for children if they attempt to imitate. Even though the language in the transcription of the video is suitable, image analysis will detect fire, throwing, and shooting, while audio analysis will detect events such as collisions, glass breaking and screams. This video is short and unsuitable content spans the whole duration of the video. Therefore, it is likely that the classification for violent scenes and event would detect them.

5 Conclusion

In this information era, children having access to all kind of information on the internet is inevitable. Therefore, saving them from accessing inappropriate content has become a critical issue. Several initiatives exist to avoid this issue, including web content filtering and parent control tools. YouTube Kids application is also one of the latest and most popular initiatives that aim to protect kids from unsuitable video content. YouTube Kids relies on algorithms that utilize meta-data and community feedback to filter the videos. However, this method is not enough in protecting kids from inappropriate videos that could pass the filtering mechanism. In order to provide an extra layer of safety for YouTube Kids, we propose a real-time method that utilizes automatic multimodal analysis of a video to judge its appropriateness for little viewers. Videos that passes the filtering mechanism, go through several steps to appropriateness analysis, including: slicing, and image, audio and transcription analysis. Fusion of the three modalities is performed to increase the confidence level of the decision. A pilot experiment was performed, where the results were promising. We believe that the proposed approach could be generalized to other video platforms. Moreover, future work should extend to applying the proposed analysis to different languages. Limitation of the machine vision and machine listening techniques would affect the final results of the proposed method. Therefore, future work should report the identified limitations and methods of overcoming such weaknesses.

References

1. Craig, D., Cunningham, S.: Toy unboxing: living in a (n unregulated) material world. Media Int. Aust. **163**(1), 77–86 (2017)
2. Federal Communications Commission: Policies and rules concerning childrens television programming: revision of programming policies for television broadcast stations. MM Docket, pp. 93–48 (1996)
3. Valcke, M., Bonte, S., De Wever, B., Rots, I.: Internet parenting styles and the impact on internet use of primary school children. Comput. Educ. **55**(2), 454–464 (2010)
4. Livingstone, S., Helsper, E.J.: Parental mediation of children's internet use. J. Broadcast. Electron. Media **52**(4), 581–599 (2008)
5. Burroughs, B.: YouTube kids: the app economy and mobile parenting. Soc. Media+Society **3**(2), 2056305117707189 (2017)
6. Elias, N., Sulkin, I.: YouTube viewers in diapers: an exploration of factors associated with amount of toddlers online viewing. Cyberpsychol. J. Psychosoc. Res. Cyberspace **11**(3) (2017)
7. Eickhoff, C., de Vries, A.P.: Identifying suitable YouTube videos for children. In: 3rd Networked and Electronic Media Summit (NEM) (2010)
8. Aggarwal, N., Agrawal, S., Sureka, A.: Mining YouTube metadata for detecting privacy invading harassment and misdemeanor videos. In: 2014 Twelfth Annual International Conference on Privacy, Security and Trust, pp. 84–93, July 2014

9. Kaushal, R., Saha, S., Bajaj, P., Kumaraguru, P.: KidsTube: detection, characterization and analysis of child unsafe content & promoters on YouTube. In: 2016 14th Annual Conference on Privacy, Security and Trust (PST), pp. 157–164. IEEE (2016)
10. Free, T.: Australia. Commercial Television Industry Code of Practice (2014)
11. Blumenau, J.: Children's media regulations: a report into state provisions for the protection and promotion of home-grown children's media. A report for save kids' tv, April 2011
12. Duerager, A., Livingstone, S.: How Can Parents Support Childrens Internet Safety? EU Kids Online, London, UK (2012)
13. Fuertes, W., Quimbiulco, K., Galárraga, F., García-Dorado, J.L.: On the development of advanced parental control tools. In: International Conference on Software Security and Assurance (ICSSA), pp. 1–6. IEEE (2015)
14. Gandomi, A., Haider, M.: Beyond the hype: big data concepts, methods, and analytics. Int. J. Inf. Manag. **35**(2), 137–144 (2015)
15. Giannakopoulos, T., Makris, A., Kosmopoulos, D., Perantonis, S., Theodoridis, S.: Audio-visual fusion for detecting violent scenes in videos. In: Hellenic Conference on Artificial Intelligence, pp. 91–100. Springer, Heidelberg (2010)
16. Hassner, T., Itcher, Y., Kliper-Gross, O.: Violent flows: real-time detection of violent crowd behavior. In: 2012 IEEE Computer Society Conference on Computer Vision and Pattern Recognition Workshops (CVPRW), pp. 1–6. IEEE (2012)
17. Demarty, C.-H., Penet, C., Gravier, G., Soleymani, M.: The mediaeval 2012 affect task: violent scenes detection. In: Working Notes Proceedings of the MediaEval 2012 Workshop (2012)
18. Sjöberg, M., Ionescu, B., Jiang, Y.-G., Quang, V.L., Schedl, M., Demarty, C.-H.: The mediaeval 2014 affect task: violent scenes detection. In: MediaEval (2014)
19. Moreira, D., Avila, S., Perez, M., Moraes, D., Testoni, V., Valle, E., Goldenstein, S., Rocha, A.: Temporal robust features for violence detection. In: 2017 IEEE Winter Conference on Applications of Computer Vision (WACV), pp. 391–399. IEEE (2017)
20. Taylor, M., Quayle, E., Holland, G.: Child pornography, the internet and offending. Can. J. Policy Res. **2**(2), 94–100 (2001)
21. Vitorino, P., Avila, S., Perez, M., Rocha, A.: Leveraging deep neural networks to fight child pornography in the age of social media. J. Vis. Commun. Image Represent. **50**, 303–313 (2018)
22. Avila, S., Thome, N., Cord, M., Valle, E., AraúJo, A.D.A.: Pooling in image representation: the visual codeword point of view. Comput. Vis. Image Underst. **117**(5), 453–465 (2013)
23. Wehrmann, J., Simes, G.S., Barros, R.C., Cavalcante, V.F.: Adult content detection in videos with convolutional and recurrent neural networks. Neurocomputing **272**, 432–438 (2018). http://www.sciencedirect.com/science/article/pii/S0925231217312493
24. ElTayeby, O., Eaglin, T., Abdullah, M., Burlinson, D., Dou, W., Yao, L.: Detecting drinking-related contents on social media by classifying heterogeneous data types. In: International Conference on Industrial, Engineering and Other Applications of Applied Intelligent Systems, pp. 364–373. Springer (2017)
25. Hannun, A., Case, C., Casper, J., Catanzaro, B., Diamos, G., Elsen, E., Prenger, R., Satheesh, S., Sengupta, S., Coates, A., et al.: Deep speech: scaling up end-to-end speech recognition. arXiv preprint arXiv:1412.5567 (2014)
26. Boashash, B.: Time-Frequency Signal Analysis and Processing: A Comprehensive Reference. Academic Press, Amsterdam (2015)

27. Qawaqneh, Z., Mallouh, A.A., Barkana, B.D.: Deep neural network framework and transformed mfccs for speaker's age and gender classification. Knowl.-Based Syst. **115**, 5–14 (2017)
28. Sudhakar, R.S., Anil, M.C.: Analysis of speech features for emotion detection: a review. In: 2015 International Conference on Computing Communication Control and Automation (ICCUBEA), pp. 661–664. IEEE (2015)
29. Alghowinem, S., Goecke, R., Wagner, M., Epps, J., Parker, G., Breakspear, M., et al.: Characterising depressed speech for classification. In: Interspeech, pp. 2534–2538 (2013)
30. Choi, K., Fazekas, G., Sandler, M., Cho, K.: Convolutional recurrent neural networks for music classification. In: 2017 IEEE International Conference on Acoustics, Speech and Signal Processing (ICASSP), pp. 2392–2396. IEEE (2017)
31. Müller, M.: Fundamentals of Music Processing: Audio, Analysis, Algorithms, Applications. Springer, Cham (2015)
32. Stowell, D., Giannoulis, D., Benetos, E., Lagrange, M., Plumbley, M.D.: Detection and classification of acoustic scenes and events. IEEE Trans. Multimed. **17**(10), 1733–1746 (2015)
33. Jansen, A., Gemmeke, J.F., Ellis, D.P., Liu, X., Lawrence, W., Freedman, D.: Large-scale audio event discovery in one million YouTube videos. In: 2017 IEEE International Conference on Acoustics, Speech and Signal Processing (ICASSP), pp. 786–790. IEEE (2017)
34. Lim, J., Choi, B., Han, S., Lee, C., Chung, B.: Classification and detection of objectionable sounds using repeated curve-like spectrum feature. In: 2011 International Conference on Information Science and Applications (ICISA), pp. 1–5. IEEE (2011)
35. Eickhoff, C., Serdyukov, P., De Vries, A.P.: A combined topical/non-topical approach to identifying web sites for children. In: Proceedings of the Fourth ACM International Conference on Web Search and Data Mining, pp. 505–514. ACM (2011)
36. Gossen, T., NüRnberger, A.: Specifics of information retrieval for young users: a survey. Inf. Process. Manag. **49**(4), 739–756 (2013)
37. Eickhoff, C., Serdyukov, P., de Vries, A.P.: Web page classification on child suitability. In: Proceedings of the 19th ACM International Conference on Information and Knowledge Management, pp. 1425–1428. ACM (2010)
38. Schmidt, A., Wiegand, M.: A survey on hate speech detection using natural language processing. In: Proceedings of the Fifth International Workshop on Natural Language Processing for Social Media, pp. 1–10 (2017)
39. Morency, L.-P., Mihalcea, R., Doshi, P.: Towards multimodal sentiment analysis: harvesting opinions from the web. In: Proceedings of the 13th International Conference on Multimodal Interfaces, pp. 169–176. ACM (2011)
40. Cambria, E., Howard, N., Hsu, J., Hussain, A.: Sentic blending: scalable multimodal fusion for the continuous interpretation of semantics and sentics. In: IEEE Symposium on Computational Intelligence for Human-like Intelligence (CIHLI), pp. 108–117. IEEE (2013)
41. Poria, S., Cambria, E., Howard, N., Huang, G.-B., Hussain, A.: Fusing audio, visual and textual clues for sentiment analysis from multimodal content. Neurocomputing **174**, 50–59 (2016)
42. Poria, S., Cambria, E., Bajpai, R., Hussain, A.: A review of affective computing: from unimodal analysis to multimodal fusion. Inf. Fusion **37**, 98–125 (2017)
43. Slepian, M.L., Bogart, K.R., Ambady, N.: Thin-slice judgments in the clinical context. Annu. Rev. Clin. Psychol. **10**, 131–153 (2014)

44. Kim, J., Guo, P.J., Seaton, D.T., Mitros, P., Gajos, K.Z., Miller, R.C.: Understanding in-video dropouts and interaction peaks in online lecture videos. In: Proceedings of the First ACM Conference on Learning@ Scale Conference, pp. 31–40. ACM (2014)
45. Hu, W., Xie, N., Li, L., Zeng, X., Maybank, S.: A survey on visual content-based video indexing and retrieval. IEEE Trans. Syst., Man, Cybern. Part C (Appl. Rev.) **41**(6), 797–819 (2011)
46. Perez, M., Avila, S., Moreira, D., Moraes, D., Testoni, V., Valle, E., Goldenstein, S., Rocha, A.: Video pornography detection through deep learning techniques and motion information. Neurocomputing **230**, 279–293 (2017)
47. Moreira, D., Avila, S., Perez, M., Moraes, D., Testoni, V., Valle, E., Goldenstein, S., Rocha, A.: Pornography classification: the hidden clues in video space-time. Forensic Sci. Int. **268**, 46–61 (2016)
48. Li, J., Deng, L., Gong, Y., Haeb-Umbach, R.: An overview of noise-robust automatic speech recognition. IEEE/ACM Trans. Audio, Speech, Lang. Process. **22**(4), 745–777 (2014)
49. Malek, J., Zdansky, J., Cerva, P.: Robust automatic recognition of speech with background music. In: 2017 IEEE International Conference on Acoustics, Speech and Signal Processing (ICASSP), pp. 5210–5214. IEEE (2017)
50. Xu, Y., Du, J., Dai, L.-R., Lee, C.-H.: An experimental study on speech enhancement based on deep neural networks. IEEE Signal Process. Lett. **21**(1), 65–68 (2014)
51. Bassil, Y., Alwani, M.: Post-editing error correction algorithm for speech recognition using Bing spelling suggestion. arXiv preprint arXiv:1203.5255 (2012)
52. Virtanen, T., Plumbley, M.D., Ellis, D.: Introduction to sound scene and event analysis. In: Computational Analysis of Sound Scenes and Events, pp. 3–12. Springer, Heidelberg (2018)
53. Barker, J., Marxer, R., Vincent, E., Watanabe, S.: The third chime- speech separation and recognition challenge: dataset, task and baselines. In: IEEE Workshop on Automatic Speech Recognition and Understanding (ASRU), pp. 504–511. IEEE (2015)
54. Mesaros, A., Heittola, T., Virtanen, T.: TUT database for acoustic scene classification and sound event detection. In: 24th European Signal Processing Conference (EUSIPCO), pp. 1128–1132. IEEE (2016)
55. Davidson, T., Warmsley, D., Macy, M., Weber, I.: Automated hate speech detection and the problem of offensive language. arXiv preprint arXiv:1703.04009 (2017)
56. Hammami, M., Chahir, Y., Chen, L.: WebGuard: web based adult content detection and filtering system. In: IEEE/WIC International Conference on Web Intelligence, WI 2003, Proceedings, pp. 574–578. IEEE (2003)
57. Kontostathis, A., Edwards, L., Bayzick, J., Leatherman, A., Moore, K.: Comparison of rule-based to human analysis of chat logs. Commun. Theor. **8**(2) (2009)

Designing an Augmented Reality Multimodal Interface for 6DOF Manipulation Techniques

Multimodal Fusion Using Gesture and Speech Input for AR

Ajune Wanis Ismail[1(✉)], Mark Billinghurst[2], Mohd Shahrizal Sunar[3], and Cik Suhaimi Yusof[1]

[1] School of Computing, Faculty of Engineering, Universiti Teknologi Malaysia, 81310 Skudai, Johor, Malaysia
{ajune,suhaimi}@utm.my

[2] Empathic Computing Laboratory, University of South Australia, Mawson Lakes, SA 5095, Australia
mark.billinghurst@unisa.edu.au

[3] UTM-IRDA Digital Media Centre, MaGICX (Media and Game Innovation Centre of Excellence), Institute of Human Centred Engineering, Universiti Teknologi Malaysia, 81310 Skudai, Johor, Malaysia
shahrizal@utm.my

Abstract. Augmented Reality (AR) supports natural interaction in physical and virtual worlds, so it has recently given rise to a number of novel interaction modalities. This paper presents a method for using hand-gestures with speech input for multimodal interaction in AR. It focuses on providing an intuitive AR environment which supports natural interaction with virtual objects while sustaining accessible real tasks and interaction mechanisms. The paper reviews previous multimodal interfaces and describes recent studies in AR that employ gesture and speech inputs for multimodal input. It describes an implementation of gesture interaction with speech input in AR for virtual object manipulation. Finally, the paper presents a user evaluation of the technique, showing that it can be used to improve the interaction between virtual and physical elements in an AR environment.

Keywords: Augmented reality · Multimodal interface · Hand gesture
Speech input · Human computer interaction

1 Introduction

Augmented Reality (AR) is a technology that allows real and virtual objects to be viewed together in real time on a single display [1]. AR interfaces can enable a person to interact with the real world in ways never before possible [2], such as using physical objects to manipulate virtual content. Interaction is a crucial topic in AR research [3]. In traditional desktop interfaces, a keyboard and mouse is commonly used; however in an AR interface these can produce a very unnatural interaction [4]. So there is a need for research on new AR interaction techniques. One drawback of some of the current input methods for

K. Arai et al. (Eds.): IntelliSys 2018, AISC 868, pp. 309–322, 2019.
https://doi.org/10.1007/978-3-030-01054-6_22

AR systems is that they lack robustness and accuracy. For example, with gesture input, whether people use a stylus or glove or a vision based system, they are often constrained to the recognition of a few predefined hand movements and can be burdened by cables or strict requirements on background color and camera placement [5]. However, concurrent use of two or more interaction modalities can provide more robust interaction with the individual modes. For instance, spoken words can affirm gestural commands, and gestures can disambiguate noisy speech. Gestures that complement speech carry a complete communication message if they are interpreted together with the speech input [6]. So the use of multimodal messages can help reduce the complexity and increase the naturalness of the multimodal interface.

One of the most important research areas in AR is creating appropriate interaction techniques for AR applications to allow users to seamlessly interact with virtual content [2]. Many different interaction methods have been explored including gesture input and other single mode input methods, for example tangible user interface methods such as the Magic-Lens [7], Cubical Interface (CUI) [8], and the Magic-Cup [9]. The Magic Lens (as shown in Fig. 1(a)) uses physical handles to show a virtual lens that can be zoomed or scaling up by moving both hands. The Magic-Cup (shown in Fig. 1(b)) uses the interaction method of "covering" virtual objects with a real object, which employs a novel "shape that can hold an object" technique. The physical motions of the real object can be mapped to virtual object manipulations. Lee et al. [7] created a cube-based tangible AR interface, the CUI (as in Fig. 1(c)) that allowed a user to compose virtual objects using both of their hands to manipulate the real blocks. In each of these cases the user's gestures with real tangible objects are used for virtual object manipulation.

(a) (b) (c)

Fig. 1. (a) CUI, (b) MagicLens, and (c) MagicCup, are the examples of single input methods using unimodal interaction.

More recently, different interaction methods have been explored [10], including multimodal interfaces where users use gesture and speech input for manipulation tasks in the real world, either directly or indirectly. Thus, in recent years, there has been a tremendous interest in introducing various methods for gesture and speech input into AR that could help overcome user interaction limitations in an AR environment. This paper provides an analysis of combined gesture and speech multimodal input methods and different fusion techniques for AR applications. As agreed in [11], AR multimodal interaction is one of the most promising solutions for improving the interaction between the virtual and physical entities since AR supports real-time interaction between real and virtual worlds. The paper reviews several related works on multimodal approaches for AR, multimodal fusion, and different existing tools and techniques. The paper also provides guidelines on multimodal fusion in AR, and how to integrate gesture and speech input in AR environments.

2 Defining AR Coordinate System with Hand Tracking Technique

2.1 3D Hand Gesture Recognition Technique

For gesture recognition we used the Leap Motion hand tracking system [12]. The interaction space is also known as the space in the Leap Motion's field of view. It is the operational space where the Leap Motion sensor can detect the hands and perform the recognition process. It is a rectangular prism region that was determined by the field of view with the height value set for 25 cm. This was the maximum value for the Leap Motion to track the real hands in physical world. We used a process of mapping the coordinates system within the specific interaction space to a normalized range from 0 to 1. The interaction space as shown in Fig. 2, and is determined by the Leap Motion field of view and the user's interaction height setting.

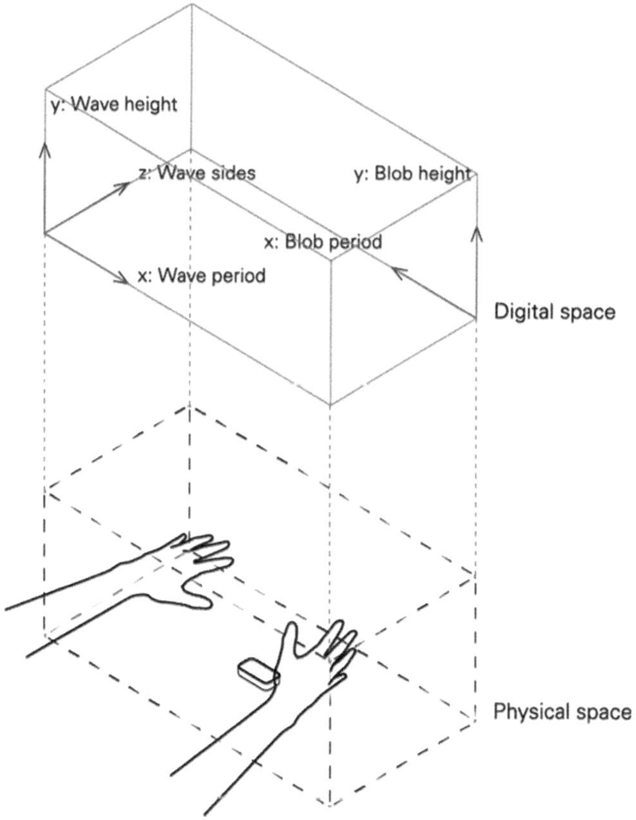

Fig. 2. Mapping coordinate systems between digital space and physical space.

The process of mapping the device's coordinate system (physical space) to the prototype's coordinate system (digital space) can be performed by modifying the interaction space. Within the Leap Motion frame of reference, fingertip position is in unit millimetres. Therefore, let say it was given as (x, y, z) = [100, 100, −100], in millimetres they are, x = +10 cm, y = 10 cm, z = −10 cm. The Leap Controller origin was placed at the top of the hardware itself. So if the middle of the Leap Motion sensor was touched then the coordinates of fingertip would be [0, 0, 0]. By changing the mapping of coordinates from start to end, it covers a larger number of people in the application.

Hand tracking techniques using Leap Motion can return the fingertip position. However, the relative position between hands and AR system is different from the tracking between hands and Leap Motion, which determines that the transformation matrices from two tracking systems cannot be mapped directly. The prototype application needs to find the relationship between two tracking systems and transforms one to the other correctly. In addition to this transformation, the error in both systems should be taken into account. A fingertip position in one system should fully match the position in the other system after transformation as illustrated in Fig. 3. The application collects the data only when Leap Motion detects 5 fingertips for each hand and AR system detects AR tracking markers.

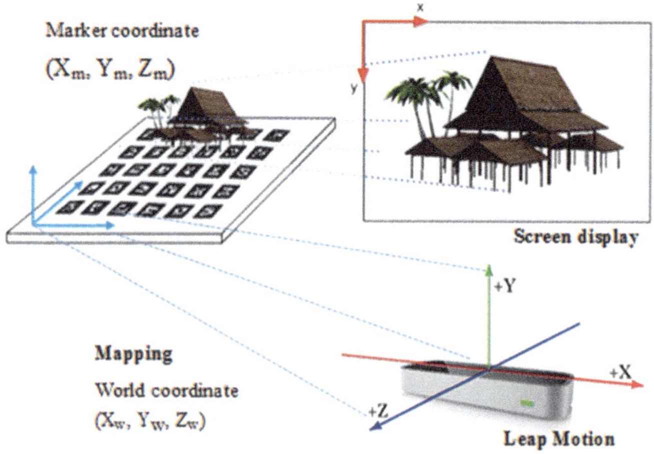

Fig. 3. Configuration leap motion mapping to coordinate system for AR.

We use the ARToolKit [13] tracking library for tracking the position of the AR camera relative to real printed markers. Using this, most of the matrix calculation can be implemented easily. A method we developed to use the image target as the world center and combine with the Leap Motion's coordinate system by multiplying the inverse pose of the main target (image of marker) by the pose of the Leap Motion's target. This creates an offset matrix that can be used to map points from the Leap Motion coordinate frame into ARToolKit image target coordinate system.

We assumed that the AR marker and Leap Motion sensor are both placed at the same origin, where it matches both coordinate systems [14]. This is done for both programmatically, through the rendering process and physically where they are adjusted in the same origin from a camera view perspective. In order to implement a gesture recognition method based on the Leap Motion tracking, the application needs to collect data from both systems. Therefore, users are required to hold their single hand or both hands above the Leap Motion and be facing the AR camera. However, in this prototype, it is presumed that the AR marker (X_m, Y_m, Z_m) and sensor are placed at the same point of origin and in a single world coordinate systems, (X_w, Y_w, Z_w). The recognition process is a process of recognizing the hand-gesture performed by the user. The features extraction process collects data only when Leap Motion detects the user's hands, in which case it finds five fingertips for each hand and some hand pose features. The data acquired by the Leap device as seen in Fig. 4 that will be used in the proposed gesture recognition system, is:

- Position of the fingertips Fi, (i = 1 … N) represent the 3D positions of the detected fingertips (N is the number of recognized fingers). Note that the device is not able to associate each 3D position to a particular finger.
- Palm center, C roughly corresponds to the center of the palm region in the 3D space.
- Hand orientation based on two unit vectors, h is pointing from the palm center to the fingers, while n is perpendicular to the hand (palm) plane pointing downward from the palm center. However, their estimation is not very accurate and depends on the fingers arrangement.

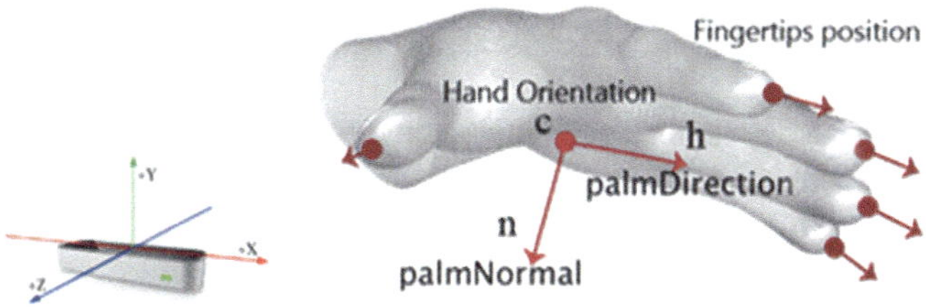

Fig. 4. Mapping coordinate systems between digital space and physical space.

During rendering process, the computer must recognize the position of the camera viewpoint relative to the marker position. When the ARToolKit markers are detected by the camera, the system will render the 3D virtual objects by tracking their position and orientation. The virtual objects are rendered over the live video view, generating an AR view of the real world. Marker detection occurs to allow the camera to show the virtual object registered on the marker in the real world. A viewpoint calculation process is required to obtain the correct transformation matrix. The perspective viewpoints for the transformation matrix will transform the camera viewpoint. This perspective transformation matrix can be found from the default initialization of the calibration process. The calibration process is necessary to pre-define the matrix of the world coordinate for the camera.

2.2 Acquiring Gestures Data Sensing

Using the Leap Motion there are several gestures that can be recognized including tapping, pointing, pinching, grabbing, and stretching (when the user uses both hands). The gestures were chosen based on copying real hand natural movements. The gestures could be mapped onto several actions such as performing selection, shrinking or stretching along different axes, and rotation, either using one hand or both. A tapping gesture is used when an index finger touches tan object one at a time. An action can be triggered either when the user's finger first collides the object, or when user can take their finger off the object. Triggering the action on release is the most common response, and is the approach we use during the tap gesture detection. Implementing tap detection in either the "Select" or "Release" would interpret a tap.

The pinch is a command gesture posture when the distance between the fingertips is calculated to be close together. In our interface the pinch gesture was used to interact with the virtual world instead of pointing, which is frequently used in 2D interfaces. For this work, a pinch recognition algorithm was implemented to move and release 3D objects by triggering a grabbing method. The objects should move relative to the hand coordinates when the user hover their hands over the top of the Leap Motion.

A grabbing function is performed when the pinch is activated, which initializes a vector GrabDistance and generates a bounding sphere around the pinch position of the thumb. This will store the distance from the pinch position to the collided finger. Afterward, it updates the GrabDistance vector with the difference of the pinch position and the collided object right after it identifies each object that collides with the bounding sphere. If the updated distance is less than the GrabDistance of the previous collided object, the object is grabbed and if the PinchDistance is higher, the grabbing and pinching are disabled. The distance between the two nearest base points (such as the thumb and index finger) is used to determine the distance, d, between these two points of a finger. If the 3D coordinates of the two points are (x1, y1, z1) and (x2, y2, z2), the distance is determined by the following formula:

$$d = \sqrt{(x_2 - x_1)^2 + (y_2 - y_1)^2 + (z_2 - z_1)^2} \tag{1}$$

The Algorithm 1 below shows the list of steps used to get the hand's tracked information from the device. Once a hand is detected, a list of finger data is obtained.

Algorithm 1 is implemented in to get pinch method in order to trigger the grab method:

```
Algorithm 1
  assign ThumbPosition = Get thumb fingertip
  calculate threshold based on thumb length
  for each remaining fingers
          get finger position
          PinchDistance = (thumb fingertip, current
          fingertip);
          if (PinchDistance < threshold) Pinching = true
  repeat For Loop in Step 3
  check If Is Pinching? Pinching = false;
  initialize GrabbedObj = null;
  get collided objects on bounding sphere
  initialize GrabDistance;
  for each collided object
          NewGrabDistance = ThumbPosition,
          CollidedObject Position
  repeat For Loop in Line 12
  if NewGrabDistance < GrabDistance, true
  initialize GrabDistance = NewGrabDistance;
  initialize GabbedObj = collidedObject;
  check (GrabbedObj?)
  add Force Vector;
  update
```

Distance points have been marked to appoint a finger base point a reversed normalized direction vector has to be multiplied by the length of a finger. Then, the beginning of this vector is placed in the fingertip position. The end of the vector indicates the finger base point. In our implementation, the fingertip is defined where the finger is initialized to an array of integers from 0 to 4, to represent five fingers. We use a fixed threshold value with 0.7 times proportion to the thumb's size, and obtain the fingertip position of each of the remaining fingers from the Fingers list. Using (1), the distance, is calculated from the thumb to the current finger positions, and if the subsequent PinchDistance is lower than a threshold value, it will trigger a pinch to be executed. The threshold distance is defined for the hands to trigger the desired object.

3 Acquiring Speech Input with Gesture for Multimodal Interface

In our proposed prototype speech recognition is used to give commands directly to the system and is a complementary input modality for gesture. In our work the .NET Microsoft framework was used for speech recognition. The final command recognized by the system depends on both the user's speech and gesture input. The Leap Motion will deliver information about gesture input, which is combined with the spoken command from the grammar log output. The user can then modify a virtual object in the AR scene with a smooth fusion of two modalities input, gesture and speech to complete the task. The commands are stored into a list of grammars as such "delete", "scale", "rotate" and "translate". Speech recognition can be trained for a specific voice or pronunciation, so there is a way to train it for specific words. A common solution to increase the diction accuracy depends on the dictionary, a list of grammar. If user pronounces words that actually exist, the engine will process it by using a dictionary of words.

Figure 5 shows how this works when recognizing the single word "red" spoken by the user. The quality of speech recognition is influenced by many factors such as the microphone hardware, environment, and speech grammar.

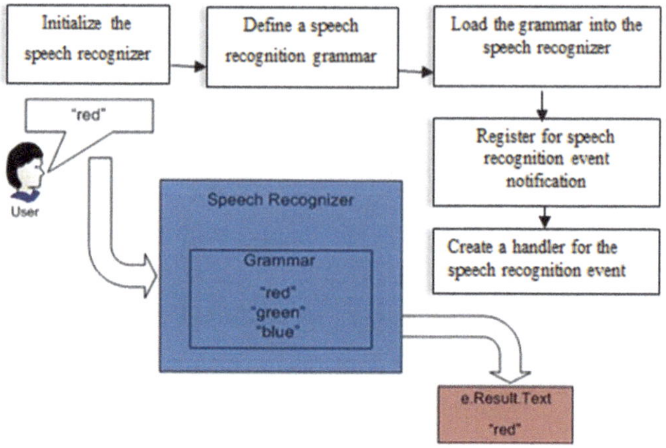

Fig. 5. The speech recognition framework for user interaction technique.

Table 1 presents the list of the commands using multimodal input for virtual object manipulation and the task explanations.

Table 1. User interaction using gesture and speech input

Speech Input	Task Description	Gesture Input
Add [*object_ name*]	**Select an object :** To insert the object into the scene. Objects are stored up to 8 objects in the list. - Gesture and speech work separately to select the object.	Tap on an option 
Translate	**Translation :** To perform translation to the selected object or multiple objects. - Speech input is required for gesture to translate the object	Pinching, Grabbing 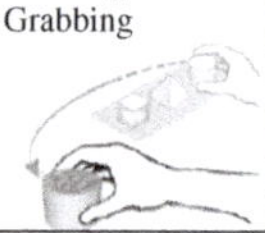
Rotate	**Rotation :** To perform rotation to the objects. - Speech input is required for gesture to rotate the object. - Turning the wrist along x-axis (roll), y-axis (pitch), and z-axis (yaw)	Pinching, Grabbing 
Scale	**Scaling :** To enlarge/shrink an object. - Speech input is required for gesture to resize the objects - Two hands move apart/together along X-axis to enlarge/shrink	Two hands grabbing 
Freeze / Unfreeze	**Freeze :** To freeze object at the current position **Unfreeze :** To unfreeze and enable gesture to manipulate	Freeze, Unfreeze 

The use of two or more modalities may reduce the needed for complex interaction techniques that are used with only a single input modality [15]. For example, the spoken words can inform gestural commands, and gestures can disambiguate speech noise. Gestures that complement speech, on the other hand, carry a complete communicational message only if they are interpreted together with speech. The use of such multimodal inputs can help reduce the complexity and increase the naturalness of the interface for AR.

The user's hands are recognized and processed as natural inputs by the system. In dealing with the virtual object manipulation, there is a need to support six degree of freedom (6DOF) tracking that consists of the ability to move forward or backward, up or down, and right or left as well. In the AR environment the gesture and speech input complement each other, so the interaction performed depends on the gesture interaction together with the speech commands that have been provided. The tracking approach [16] used in this research is to provide 6DOF skeleton-based tracking into the AR prototype.

4 Proposed AR Multimodal Interface Prototype

4.1 Visual Feedback

The proposed prototype provides visual feedback to user about the selection region around each object by showing a virtual wireframe bounding box (as in Fig. 6(a)). The user uses fingertips to pinch the virtual house and if the object is grabbed inside the bounding box then the virtual wireframe will change its color from green to purple (as in Fig. 6(b)). The user can manipulate the virtual house using gestures to perform single object relocation in 6DOF. While the hand still holding the virtual object, the user can freeze and unfreeze the object by issuing the speech commands, "Freeze" and "Unfreeze". When the speech command "Freeze" is executed on the selected object, the user's hand is unable to interact with the virtual object until the speech input "Unfreeze" is issued. The virtual wireframe is in green color which shows that the virtual object is unable to collide (as in Fig. 6(c)). Once the speech command "Unfreeze" is performed, the hand is able to perform tapping, pinching and grabbing actions (as in Fig. 6(d)).

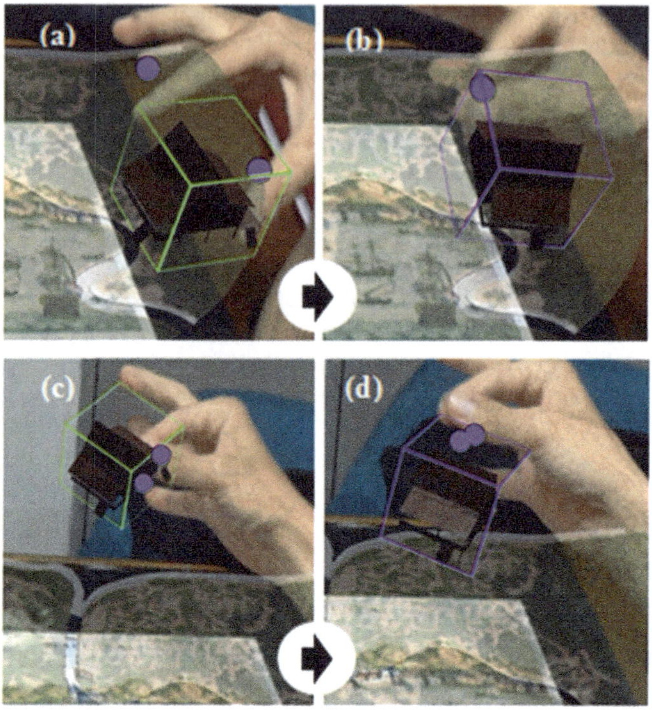

Fig. 6. (a) A virtual wireframe color is changed (b) once the virtual object is grabbed. (c) "Freeze" and (d) "Unfreeze" by adding speech commands.

4.2 Occlusion-Free 3D Object Manipulation

Users are able to manipulate the 3D objects in a natural manner by detecting gesture behaviours and speech input. The list of 3D objects is stored in an array of objects. The Leap Motion controller provides fingertip data to track the real hand of the user, enabling them to interact with the 3D object data. The collisions between the fingertips and the 3D virtual objects are calculated. The geometrical relationship between the 3D object and fingertips is defined in order to detect translation ("translate"), rotation ("rotate"), scaling ("scale") and selection ("select", "select all") actions. With speech inputs, the gesture commands should work for the "delete", "release", and "duplicate" actions while the virtual object remains attached to the gesture. In this condition where the gesture works together with the speech input, the multimodal interaction was executed.

As demonstrated in Fig. 7(a), the virtual menu appears after user executes speech input "Show Menu" and provides a user with several options to choose. The virtual menu pop-ups in the scene attached to the user's left hand while the right hand can be used to tap menu options. Figure 7(b) shows the user using their finger to tap on the virtual menu. The 3D object (Tree) is loaded into the scene while the virtual menu disappears

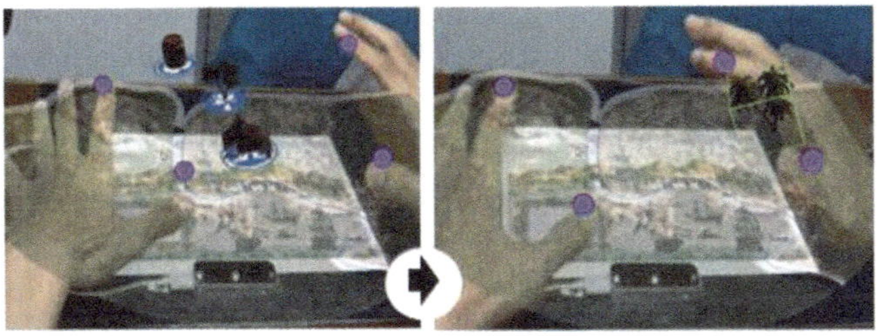

(a) (b)

Fig. 7. (a) Virtual menu appears after user says "Show Menu", (b) User taps on the virtual menu and it disappears once the object is loaded into the scene.

During the sensor-based hand tracking, the motion data is presented in a virtual hand skeleton that is created every time the Leap Motion controller senses a real hand in the motion space. The hand skeleton is being drawn in the AR camera view consistent with the position of the image target. The key to high precision natural hand gesture interaction is high accuracy and high DOF hand pose estimation. Our gesture interaction supports 6DOF manipulation. It provides 6DOF natural hand tracking; tracking of the wrist, 15 joints, and 5 fingertips for both hands every finger at 30 frames per seconds (fps). Virtual objects are correctly overlaid on the AR tracking marker even when the marker is partly covered by user's hand movement, this condition called occlusion-free. The occlusion problem occurs when real hands move around in the AR scene and cover the marker or target image, and the 3D virtual objects are overlaid incorrectly. Usually, occlusion problems happen when the AR application involves 3D object manipulation [17]. The 3D objects should correctly appear on the top of the marker and visible while the real object or user's hands are performing the 3D object manipulation. The user's hands are able to move freely to perform manipulation with multiple objects on the ground. The objects are still correctly registered on the marker even the user's real hands cover half of the marker. This condition enhances the user experience in AR and provides natural user interaction during 6DOF manipulation [18].

In case of scaling, the object size using the distance between the two hands when the object is grabbed by the thumb and index finger of one hand (as in Fig. 8). Once the selected object has been moved, the position is updated at the moment that the user touches the selected object to perform the translation. The intersection between two points connecting each other allows the user to drag the selected objects to update the translation matrices. Multimodal interface may need to consider these conditions:

- Each task required the user to interact with the 3D virtual objects using gestures. However, when working together with speech input, user cannot change the order of the conditions. User need to take into account order effects.
- Speech signal cues are time-sensitive, when working with gesture input for multimodal fusion, both needs to involve simultaneous signals.
- Multiples object manipulation, for example rotation, it can simultaneously execute after scaling. The amount of rotation is indicated by a change in the orientation of the user's hand.

- Multiples object manipulation, for example scaling, when gesture palm moves along the x-axis to enlarge or shrink multiple objects at the same time.
- Unlike using the finger on a pinching method to scale single object, multiple objects are resized using the palm movements. When two or more objects are grabbed after trigger the speech input "Scale All", they likewise respond to palm movement translation.

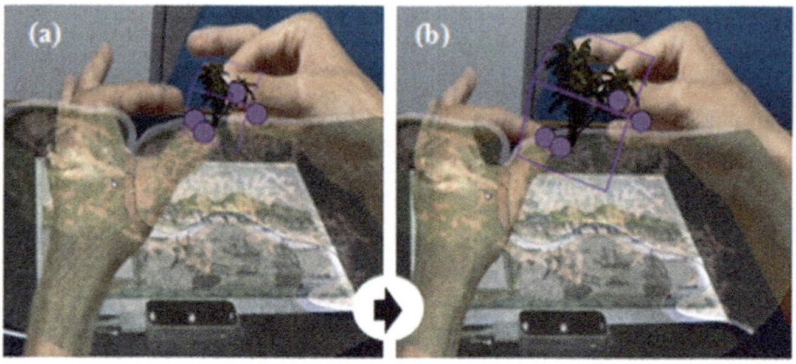

Fig. 8. (a) Using both hands to select and scale a grabbed object, (b) the object resizes according to the pinch gesture motion.

5 Conclusion

In this paper, we describe a designing multimodal interface and an advanced interaction method with AR coordinate system mapping that uses the combination of gestures and speech to interact. Many different interaction methods have been explored to produce appropriate interaction techniques for AR and researchers have started to explore multimodal interaction methods. In our proposed interaction technique, users use their real hand gestures and speech input to imitate real manipulation tasks in the real world. Hence, this paper expands on an appropriate interface by using gesture and speech input for interaction, and multimodal fusion between these input modalities. This proposed design method describes a multimodal interaction using gesture and speech input for AR, and presents a multimodal condition of the method.

The pinch gesture method was implemented using two fingertips of both hands to perform interaction for virtual object manipulation. This occlusion-free interaction technique with gestures was combined with speech input for AR multimodal interaction is developed. The depth data was captured by a Leap Motion through the sensor-based tracking and integrated into a simple prototype to implement virtual object manipulation in 6DOF manipulation using the user bare hands as a tool. Several interactions methods such as translation, rotation and scaling for virtual object manipulation are discussed in this paper. The user's ability to add and remove virtual objects intuitively is implemented with speech input, respectively. The pinch gesture was used to perform grabbing and fingertips used to perform a tapping to select from a virtual menu. These were smooth when the hand tracking runs without delay and the delay must be unnoticed by user.

The research future direction for multimodal interaction in AR is to achieve as intuitive as possible to satisfy user natural interaction. There were two remaining issues for interaction, (1) the occlusion problem; and (2) robust tracking. A number of limitations in tracking but two tracking performances need to be addressed, gesture resolution and gesture recognition accuracy. The recognition issue, a robust gesture recognition method is required to capture the real hand features single or multiple input streams to represent some description of the hand pose. The data produced by recent advanced technology such as Leapmotion can be easily processed to extract the articulated motion of the hand and to derive higher level features, the fingertip locations, pointing direction or force generated by a finger. However, all these need to be interpreted by the application. In future, we expect to produce a multimodal recognition engine that gesture recognition can operate without estimating any 3D features of the hand motion and speech input can naturally detect human accent. The hand poses are computed without a full reconstruction of the hand state. Thus, it can avoid critical occlusions and keep the appearance of the hand in a reasonable range.

Acknowledgment. We would to express our appreciation to Universiti Teknologi Malaysia (UTM) for the funding and supports. We also thank the Human Interface Technology Laboratory New Zealand (HITLabNZ) at University of Canterbury. This work was funded by UTM GUP Funding Scheme.

References

1. Azuma, R., Baillot, Y., Behringer, R., Feiner, S., Julier, S., Blair, M.: Recent advances in augmented reality. IEEE Comput. Graph. Appl., 20–38 (2001)
2. Zhou, F., Duh, H.B.-L., Billinghurst, M.: Trends in augmented reality tracking, interaction and display: a review of ten years of ISMAR. In: Proceedings of the 7th IEEE/ACM International Symposium on Mixed and Augmented Reality. IEEE Computer Society (2008)
3. Kaiser, E., Olwal, A., McGee, D., Benko, H., Corradini, A., Li, X., Cohen, P., Feiner, S.: Mutual disambiguation of 3D multimodal interaction in augmented and virtual reality. In: ICMI 2003: International Conference on Multimodal Interfaces, pp. 12–19, August 2003
4. Jewitt, C.: Technology, Literacy and Learning: A Multimodal Approach. Psychology Press, London (2006)
5. Lim, C.J., Pan, Y., Lee, J.: Human factors and design issues in multimodal (speech/gesture) interface. JDCTA **2**(1), 67–77 (2008)
6. Corradini, A., Cohen, P.: On the relationships among speech, gestures, and object manipulation in virtual environments: initial evidence. In: Proceedings of the International CLASS Workshop on Natural, Intelligent and Effective Interaction in Multimodal Dialogue Systems, pp. 52–61 (2002)
7. Lee, H., Billinghurst, M., Woo, W.: Two-handed tangible interaction techniques for composing augmented blocks. Virtual Reality **15**(2–3), 133–146 (2011)
8. Looser, J., Billinghurst, M., Cockburn, A.: Through the looking glass: the use of lenses as an interface tool for augmented reality interfaces. In: Proceedings of the 2nd International Conference on Computer Graphics and Interactive Techniques in Australasia and South East Asia, pp. 204–211. ACM, June 2004

9. Kato, H., Tachibana, K., Tanabe, M., Nakajima, T., Fukuda, Y.: MagicCup: a tangible interface for virtual objects manipulation in table-top augmented reality. In: 2003 IEEE International Augmented Reality Toolkit Workshop, pp. 75–76. IEEE, October 2003
10. Haller, M., Billinghurst, M., Thomas, B.H. (eds.): Emerging Technologies of Augmented Reality: Interfaces and Design. IGI Global, Hershey (2007)
11. Lee, M.: Multimodal Speech-Gesture Interaction with 3D Objects in Augmented Reality Environments (2010)
12. Leap Motion: Leap Motion Controller (2017). https://developer.leapmotion.com/. Accessed Jan 2017
13. ARToolKit (2017). www.hitl.washington.edu/artoolkit
14. Ismail, A.W., Sunar, M.S.: Multimodal fusion: gesture and speech input in augmented reality environment. In: Computational Intelligence in Information Systems: Proceedings of the Fourth INNS Symposia Series on Computational Intelligence in Information Systems (INNS-CIIS 2014), vol. 331, p. 245. Springer, November 2014
15. Piumsomboon, T., Altimira, D., Kim, H., Clark, A., Lee, G., Billinghurst, M.: Grasp-Shell vs gesture-speech: a comparison of direct and indirect natural interaction techniques in augmented reality. In: 2014 IEEE International Symposium on Mixed and Augmented Reality (ISMAR), pp. 73–82. IEEE (2014)
16. Ismail, A.W., Sunar, M.S.: Intuitiveness 3D objects interaction in augmented reality using S-PI algorithm. TELKOMNIKA Indonesian J. Electr. Eng. **11**(7), 3561–3567 (2013)
17. Olwal, A., Benko, H., Feiner, S.: SenseShapes: using statistical geometry for object selection in a multimodal augmented reality system. In: Proceedings of the Second IEEE and ACM International Symposium on Mixed and Augmented Reality (ISMAR 2003), Tokyo, Japan, 7–10 October 2003, pp. 300–301 (2003)
18. Heidemann, G., Bax, I., Bekel, H.: Multimodal interaction in an augmented reality scenario. In: ICMI 2004: Proceedings of International Conference on Multimodal Interfaces, pp. 53–60 (2004)

InstaSent: A Novel Framework for Sentiment Analysis Based on Instagram Selfies

Rabia Noureen(✉), Usman Qamar, Farhan Hassan Khan, and Iqra Muhammad

Department of Computer Engineering, College of EME, National University of Sciences and Technology (NUST), H-12, Islamabad, Pakistan
{rabia.noureen15, iqra.muhammad15}@ce.ceme.edu.pk, usmanq@ceme.nust.edu.pk, mrfarhankhan@gmail.com

Abstract. Sentiment analysis and opinion mining is the field of study to analyse opinions, attitudes, and emotions of people. It is the most studied research field. With the evolution of social media, many new terms have been evolved including selfies. A selfie has provided a way for to the users of social media to record their personal memories. People are sharing their images on social media in a massive amount. Keeping in mind the big data nature, it is very difficult to analyse selfies manually. In this research study, we have proposed a framework called InstaSent for sentiment analysis based on Instagram selfies. This framework incorporates both text mining and image mining techniques for sentiment prediction. Support vector machine is used for sentiment classification based on the text associated with selfies like captions, hashtags, comments, and emoticons, while the deep learning method Convolutional neural network is used for processing image data for sentiment analyses. By combining the practices for text mining and image mining we believe that this technique will outperform all other techniques presented in this domain. In a nutshell, we believe that our research study has provided novel opportunities for researchers to explore the use of selfies in other domains.

Keywords: Deep learning · Sentiment analysis · SVM
Convolutional neural network · Instagram

1 Introduction

In the past few years, people have become more social on social media sites and this gave rise to a new trend called "selfie". A selfie is a photograph that is taken by a person who is also in that photograph which is taken by using a smartphone or digital camera. A selfie is taken specifically for posting on social media. Selfie became popular after Sony Ericsson Z1010 mobile phone was introduced in late 2003 [1]. This mobile phone introduced a new feature front-facing camera which had a senor for selfies and video calls [1]. They became popular initially among young people but with the passage of time, they gained popularity among other age groups [2, 3]. Time magazine declared selfie in "top 10 buzzwords" of 2012 [4]. The surveys conducted in 2013 proved that two-thirds of Australian women age 18–35 take selfies specifically for

K. Arai et al. (Eds.): IntelliSys 2018, AISC 868, pp. 323–336, 2019.
https://doi.org/10.1007/978-3-030-01054-6_23

posting on social media sites e.g. facebook [3]. Another poll that was conducted by Samsung proved that among all the people who take photos, the 18–24 age group takes 30% selfie in all the other photos [5]. Oxford English Dictionary announced the word "selfie" "word of the year" in November 2013. We can see the popularity of selfies by their existence on social media [6]. According to Wikipedia [1], Instagram has over 53 million photos tagged with the hashtag #selfie. The word "selfie" was mentioned in Facebook status updates over 368,000 times during a one-week period in October 2013 [1, 7]. During the same period on Twitter, the hashtag #selfie was used in more than 150,000 tweets [1, 7].

Sentiment analysis also called as opinion mining is a field that is used to analyze opinions, attitudes, and emotions of people using natural language [8]. The term sentiment analysis was first discussed in [9]. As more and more people are using social media to express their sentiments and for communicating with other people, we have a tremendous amount of data available online in the form of text and multimedia. So research in this field is not only useful in NLP but it is also useful in other domains [10]. Social media sites such as twitter facebook, Instagram and web blogs like 9gag are used for the exchange of data in the form of text, images, videos, etc. which makes it possible for us to perceive the sentiments and opinions of users. A lot of work has been done for sentiment analysis on the basis of text but visual data like images and videos still need to be explored in this field as the people are now more likely to express their feelings by using images. And the multimedia content that is shared by users expresses their sentiments in a most efficient way as compared to text alone [11]. Thousands of selfies that are clicked and posted by users on social media are a major source of sharing opinions and sentiments with their friends and family as users share each and every moment of their life with other in the form of "selfies". These selfies often contain captions in the form of text and hashtags explaining the photo. When users share their selfies usually people that are in their circle comment on them. So these selfies are today the major interest for researchers for sentiment analysis.

The rest of the paper is organized as follows: Sect. 2 describes the problem; Sect. 3 discusses the related work that has been done in image sentiment analysis; Sect. 4 describes our methodology in detail; Sect. 5 discusses the detailed comparative analysis; Sect. 6 proposes the applications of our framework; Sect. 7 proposes the future work of our framework; finally, Sect. 8 concludes the paper.

2 Problem Description

As the evolution of social media gave rise to a concept called selfie. People are sharing their images on social media in a massive amount which depict certain sentiments of the users. Keeping in mind the big data nature it is very difficult to analyse selfies manually. To cope up with this challenge we have proposed a framework for sentiment analysis using Instagram selfies.

3 Background

In this section, we will discuss the major work that has been done in the field of sentiment analysis using image data.

One of the methods is based on a Sentribute algorithm that is discussed in this paper [10]. This algorithm starts by introducing a dataset then low-level features are extracted from this dataset. Then the next step involves building mid-level attributes classifiers from the low-level features extracted in the previous step. Now asymmetric bagging is applied. Eigenface model is used for facial expression detection. And the last step involves the fusion of mid-level attributes and facial expression detection for sentiment prediction [10].

Another method that is discussed in this [12] paper is based on an unsupervised sentiment analysis framework. This framework is used for detecting sentiments from social media images by taking both the textual and visual information and making a unifying model out of this data. This model is then used for sentiment analysis.

A famous technique is presented recently by Borth et al. [13]. He presented comprehensive visual sentiment ontology along with set of detectors (SentiBank [13]) that are comprised of 3,244 noun-adjective pairs. These noun-adjective pairs are created on the basis of Plutchik's Wheel of Emotions [14]. Flickr images are used to train these detectors, and a combination of global and local features is used to represent them.

Baecchi, Claudio, et al. presented a novel unified model for sentiment analysis for social network multimedia. This model was comprised of both text and image data. Textual data were treated by creating a model called CBOW-LR which learns both vector representation and polarity classifier. This model was then extended for visual data using Denoising Autoencoder creating an extended model called CBOW-DA-LR. This novel model showed higher accuracy than SentiBank [13] discussed previously.

You et al. [16] introduced a model for multi-modality sentiment analysis. They used Convolutional Neural Network (CNN) [17] for visual feature extraction and then they fine-tuned CNN. Then visual features were extracted from second to the layers of neural networks by using the fine-tuned model. The textual data (descriptions and titles of images) was trained on a model created by Le and Mikolov [18] they then applied fusion of both visual and textual features for sentiment analysis.

Katsurai and Satoh [19] presented a novel image sentiment analysis method which uses training images and they use feature extraction method from visual, textual and sentimental views. They have done the mapping of these features by using the multi-view CCA framework. The last step is using these projected features to train sentiment polarity classifier.

Ji et al. [20] presented a multimodal correlational model which is based on feature extraction from each modality. These features included words in text view, ANP in image view and symbols are extracted as features in emoticon view. And finally, a hypergraph learning model is developed from all these views by using the bag of words.

Kalayeh et al. recently published a research study on "How to take a good selfie?" [21]. They took Instagram selfie dataset from selfeed.com, which included all the photos from Instagram with the #selfie. They have declared an attribute prediction baseline and then extracted attributes like SIFT descriptors, HOG descriptors. They used deep convolutional neural networks [17] that were trained on ImageNet [22] dataset and extracted 4096-D feature vectors from the selfie photos. And then using the SentiBank [13] dataset they generated a 2089-D vector from selfie photos. As the last step, they used SVM for training the attribute detectors.

Seo et al. [23] proposed a technique in order to acquire training dataset for classifying and predicting sentiments from Images. Their model was able to detect the category of an image and later sentiments can be predicted based on the predictor against that category.

We are going to improve their technique for predicting sentiments from Instagram selfies. As they have only used image data for their research study and completely ignored the textual data in the form of selfie captions, # tags associated with a selfie and no comments posted by users on that particular selfie.

4 Proposed Methodology

Our proposed technique is based on collecting Instagram selfie data which includes textual data in the form of (captions, hashtags and no of comments) and image data (selfie images) and then using different techniques for processing both kinds of data as shown in Fig. 1. We will explain in this section how this process is performed.

4.1 Textual Data

Textual data that is obtained from Instagram is first divided into three subtypes: Caption text, hashtags and no of comments as shown in Fig. 2.

(a) *Caption Text*

It is the text that is written by the users to explain about the selfie while posting on Instagram.

(b) *HashTags*

A hashtag is a keyword or label that is used with # symbol by the users of social media sites. By using hashtags it becomes easy for the people to find messages that are related to a specific topic.

(c) *Comments*

A comment is a statement that is posed by the users of social media to express their opinions about anything posted. So by using captions, hashtags and no. the comments a Meta-model is created. Based on this Meta model the polarity of the data is detected which is then used for sentiment analysis.

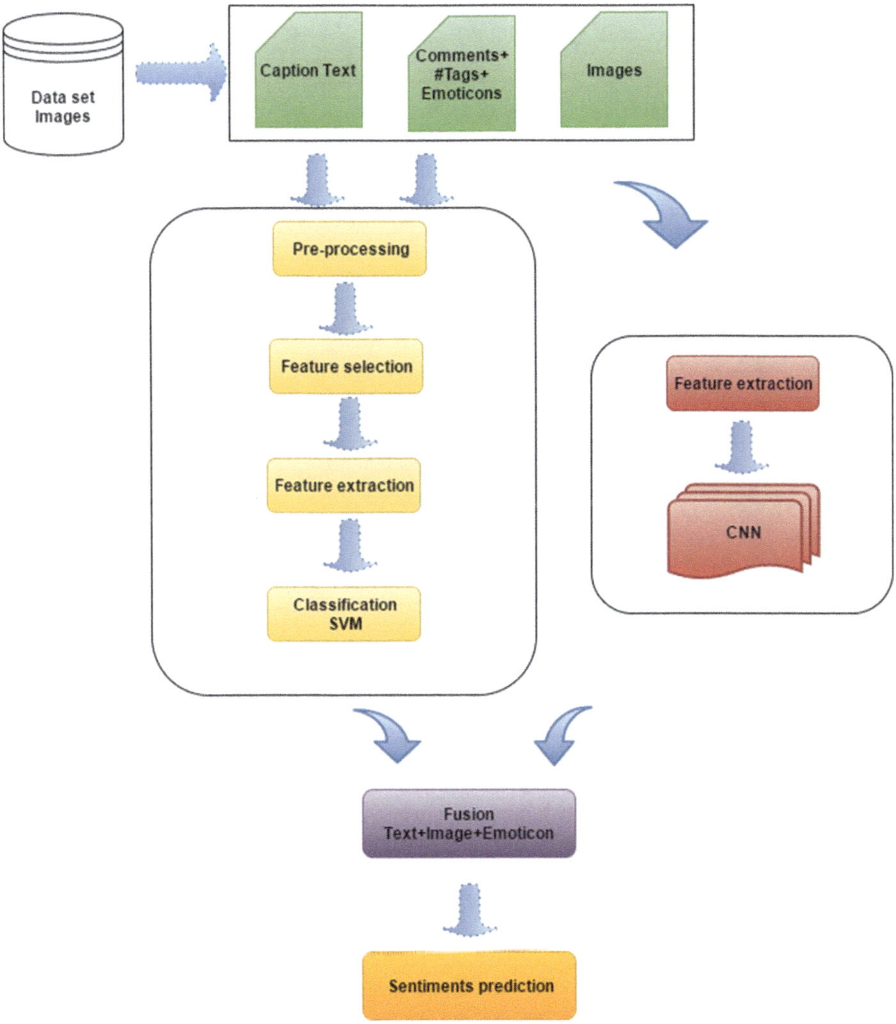

Fig. 1. InstaSent model.

4.2 Preprocessing of Textual Data

The preprocessing step involves some processes before the classification is done. It is the process of preparing and cleaning of the dataset. Mullen, Tony, and Nigel Collier and Agarwal, Apoorv, et al. described some pre-processing tasks in the research studies, we have followed some of them [29, 30]. Pre-processing is done to achieve the following tasks.

(1) Noise reduction
(2) Performance improvement of the classifier
(3) To speed up the process of classification

Fig. 2. Textual data in Instagram selfies.

So this step helps us to achieve real-time sentiment analysis.

(a) *Stop word removal*

Stop words are a list of stop words that are specific to a particular language. These words do not have specific meaning also called “functional words”. Some examples of stop words taken from the English language are “of”, “an”, “it”, “I”, etc. [29]. When we are analysing Natural language we observe that these words should be ignored because they don’t convey any meaning. So we should remove those repeating words that don’t convey any meaning for increasing the performance of our system. So remove these stop words by using this resource[1].

(b) *Stemming*

In this process all the derived words are reduced to their stem or a common representation like all the variants forming, formed; formation can be reduced to a one-word form that is their stem. We can use table look-up approach for this purpose which is a fast method. In this method, all the index terms and their stems are stored in hash tables [29].

(c) *Emoticon replacement*

Emoticon replacement can be done by creating an emoticon dictionary [30]. Emoticon dictionary can be prepared by using the following steps (also see Table 1):

[1] http://www.webconfs.com/stop-words.php

Table 1. Emoticon translation

Emoticon	Translation
:-) :) :o) :] :3 :c) :D C:	Positive
:-(:(:c :[D8 D; D = DX v.v	Negative
: |	Neutral

(1) Label 170 emoticons that are obtained from Wikipedia[2] as positive, negative and neutral.
(2) Then replace all the emoticons in the dataset with their sentiment polarity with the help of emoticon dictionary.

(d) *Acronym replacement*

Create an acronym dictionary from this resource[3] which contains 5,184 acronym translations. So replace all the acronyms in the dataset with their respective translations with the help of a dictionary (Table 2).

(e) *Hashtag removal*

In this process, hashtag symbols # are removed from the words that are associated with the hash tags to make them more understandable for further processing.

(f) *Remove duplicate comments*

When a user posts a selfie image there may be some comments that are duplicate like awesome, nice etc. so we need to eliminate these Duplicate comments to save our processing effort.

4.3 Feature Selection

This process is done for reducing the data that will be used for classification. All the relevant features are identified in this process and irrelevant features will be ignored. In this process, we will select features like:

(1) Captions
(2) Hashtags
(3) Comments
(4) Emoticons

We will ignore features like:

(1) URLs
(2) Targets (@John)

[2] http://en.wikipedia.org/wiki/Listofemoticons
[3] http://www.noslang.com/

Table 2. Acronym translation

Acronym	Translation
LOL	Laugh out loud
ASAP	As soon as possible
BTW	By the way

4.4 Feature Extraction and Polarity Classification

Feature extraction reduces the dimensionality of data by selecting a subset of data. For feature extraction process we will use the process proposed by Baecchi, Claudio, et al. in a modified form. We will create a sliding window for Metadata of each selfie like captions, hashtags, comments, etc. During the entire process, the image is static/fixed while by using the sliding window mechanism text is represented as one window at a time. Local polarity score is computed by processing each window separately. To get the overall polarity of a selfie we will get a total of all windows by adding the polarity of each window and polarity of the emoticon, hence getting a total polarity score.

We will use SVM for classification of textual data while convolutional neural networks [17] will be used for image data. The entire process for polarity classification is shown in Fig. 3.

(a) *SVM*

SVM has showed very good results as compared to other classifiers. Qamar et al. [26] stated support vector machine is a supervised machine learning technique for pattern recognition in data [26]. Support vector machine is used mostly for classification of data and regression analysis. A training instance is assigned to one of two classes [26]. SVM algorithm is used to classify an unknown instance. Wang, Hongbing, et al. declared SVM as no probabilistic binary linear classifier [27].

The data arrangement is done in a way that a hyper-plane is created for the separation of data of different kinds by the application of kernel equation [26].

Kernel function is a function used for the transformation of data that is linearly no separable across domains. Kernel equations are of different types such as linear, quadratic, Gaussian, etc. After category division of data the data is divided into two types of instances for the determination of best hyperplane.

SVM follows the following classification rule (Varguez-Moo, Martha, Francisco Moo-Mena, and Victor Uc-Cetina. [28]):

$$\mathrm{Sgn}\,(\mathrm{f}(\mathrm{x},\mathrm{w},\mathrm{b})) \tag{1}$$

$$\mathrm{f}(\mathrm{x},\mathrm{w},\mathrm{b}) < \mathrm{w}.\mathrm{x} > +\mathrm{b} \tag{2}$$

where x is the example to be classified and the maximum margin hyper plane (w,b) represents a complex problem with a unique solution. The ultimate solution is to minimize $||\mathrm{w}||$ within the specified constraints.

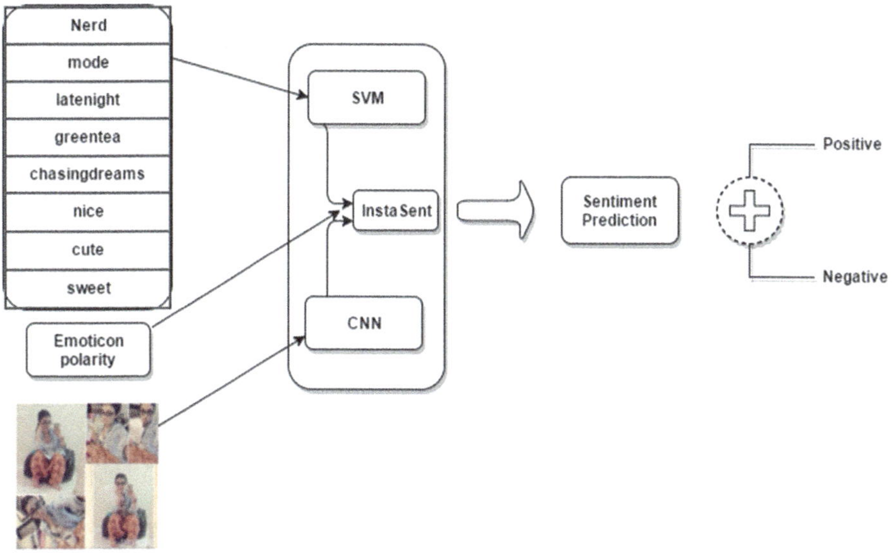

Fig. 3. Polarity classification process.

$$Y_i((w.x_i) + b) \geq 1 \tag{3}$$

In basic SVM framework, the two classes are classified based on a hyper plane, written as:

$$(w.x) + b = 0\, w \in R^n,\ b \in R \tag{4}$$

The linear SVM correctly classifies all training data using following formula:

$$wx_i + b \geq 1 \text{ if } y_i = +1 \tag{5}$$

$$wx_i + b \leq 1 \text{ if } y_i = -1 \tag{6}$$

Maximize the margin by:

$$M = \frac{2}{|w|} \tag{7}$$

(b) *Image data*

Image data in the form of selfies will be processed by using the technique presented by Kalayeh et al. [21].

We will densely extract SIFT, HOG features from images. The next step will be to extract feature vectors by using CNN [17] that is trained on ImageNet [22] dataset.

Another feature will be represented by using 1000 D classification layer. A large network of Overfeat [24] will be used for the implementation of CNN. As a last step 2089-D vector for each image will be generated by using SentiBank [13] ANP. SVM will be used for training attribute detectors [21].

(c) *Convolutional neural networks*

Convolutional neural networks CNN [17] are similar to the ordinary neural networks to some extent. They are comprised of one or more convolutional layers which are then followed by one or more fully connected layers like in traditional multilayer neural networks [25]. A CNN works to transform input 3D volume into an output 3D volume which has a differentiable function that may be having some parameters. The main features of a CNN are:

(1) It takes an assumption that the input data are in the form of images.
(2) They make our activation function fast.
(3) They reduce the no of parameters used in our network.
(4) CNN repeatedly performs one operation on a raw image.

The architecture of CNN is explained in Fig. 4.

So CNN are very efficient and it is a widely used recent technique for image processing. Convolutional neural networks are great as they are able to recognize places, people and things from images.

4.5 Fusion

In this step the total polarity scores of the selfie which will be calculated by adding the polarity of text, images and emoticons. The formula for total polarity is:

$$P_{\text{InstaSent}} = P_t + P_i + P_e \tag{8}$$

Table 3 explains the detailed InstaSent Algorithm.

5 Comparative Analysis

We have compared the state of art techniques that have produced the best results in the field of Text and Image mining. In Table 4, detailed comparative analysis has been performed using the matrices like classifier, accuracy, dataset, and technique used (Either Image mining or Text mining). It can be seen that the 5th approach [17] has used CNN as classifier, has achieved the highest accuracy of 84.7% in Image mining. Hence our InstaSent model that is a fusion of CNN and SVM is expected to increase the accuracy for both image and textual data.

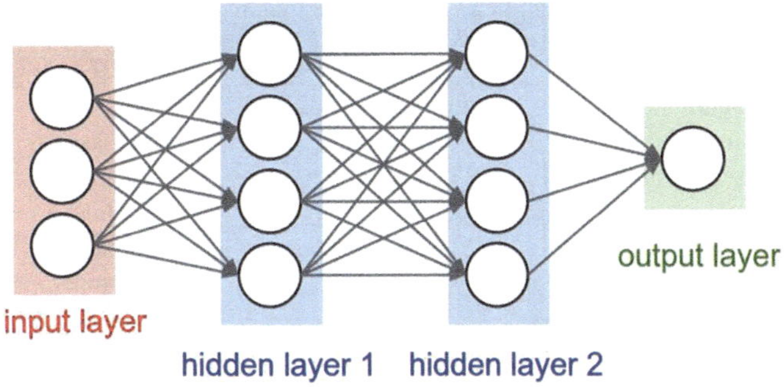

Fig. 4. CNN architecture.

Table 3. Instasent algorithm

TABLE III. INSTASENT ALGORITHM

Algorithm
Input: A set of n selfie images and their data $\{I_i ,.., I_n\}$
Output: k sentiment label for each image I_n
Methods:
Begin
For each image I_i
Split I_i into Caption C_a, Comments C_o, Hashtags H_t, Emoticons E_i & images I_i
endfor
Perform preprocessing for training textual data d_i
Remove stop words
Perform stemming
Replace emoticons
Replace acronyms
Remove hashtags
Remove duplicate comment
Perform feature selection from training data d_i
Perform feature extraction from training data d_i
Compute polarity of textual data using SVM classifier
Predict polarity P_i
endfor
Use images In to extract features Fi
Now extract feature vectors Vi by using CNN
Predict polarity Pi
endfor
Add polarity of emoticons
Use fusion to detect the final polarity

6 Applications

This framework is expected to provide very promising results as it incorporates both textual and visual information. The applications of this framework include:

(1) Sentiment analysis of Instagram user selfies.

Table 4. Comparative analysis

References	Approaches	Classifier	Accuracy	Dataset	Technqiue
[10]	Sentiment analysis using social multimedia	Sentribute	82.35	SUN Database, Twitter, Karolinska Directed Emotional Faces dataset	Image mining
[13]	Large-scale visual sentiment ontology and detectors using adjective noun pairs	SentiBank Logistic Regr	72	Twitter	Text mining & Image mining
[15]	A multimodal feature learning approach for sentiment analysis of social network multimedia	CBOW-DA-LR	83.84	Sanders Corpus, Sentiment140, SemEval-2013, SentiBank, Twitter Dataset	Text mining & Image mining
[16]	Joint visual-textual sentiment analysis with deep neural networks	CNN, Distributed Paragraph Vector Model	77.5	Getty Images, Twitter	Text mining & Image mining
[17]	Imagenet classification with deep convolutional neural networks	CNN	84.7	ImageNet	Image mining

(2) To analyze the opinions of customers about a particular product by making them post their selfies with the product.
(3) To detect the sentiments of employees in a particular organization by using their photos.

7 Future Work

Future work will focus on creating a huge dataset of Instagram selfies from the internet and to train our model; since no standard Instagram selfie dataset is available that can be used for training. Another aspect of our future work will be the implementation of the proposed framework and after performing a series of experiments evaluating our approach through comparative analysis based on similar work that has been done in this field. Furthermore, we will enhance the preprocessing technique to expect extraordinary results.

8 Conclusions

In this paper, we have presented a framework "InstaSent" for sentiment analysis of the users of Instagram based on their posted selfies. To the best of our knowledge, this is the first extensive work done for sentiment analysis on the basis of selfies. InstaSent framework is expected to provide promising results as it uses the best techniques for both text and image mining. SVM has shown very good results in the field of text mining while convolutional neural networks (CNN) based on deep learning concept is a recent advancement in the field of machine learning and computer vision that are extremely popular for recognizing things, faces, etc. in images. We believe that our research study has provided novel opportunities for researchers to explore the use of selfies in other domains.

References

1. https://en.wikipedia.org/wiki/Selfie
2. Bim, A.: The rise and rise of the selfie. The Guardian (London). Accessed 6 April 2013
3. McHugh, J.: Selfies' just as much for the insecure as show-offs. Bunbury Mail, 3 April 2013. Accessed 6 Apr 2013
4. Steinmetz, K.: Top 10 Buzzwords – 9 Selfie, Time, 4 December 2012
5. Melanie, H.: Family albums fade as the young put only themselves in picture telegraph, 13 June 2013
6. The Oxford Dictionaries Word of the Year 2013 is OxfordWords blog, 18 November 2013. Blog.oxforddictionaries.com. Accessed 29 Nov 2013
7. https://en.wikipedia.org/wiki/Selfie#cite_note-McHugh-21
8. Liu, B.: Sentiment analysis and opinion mining. Synth. Lect. Hum. Lang. Technol. **5**(1), 1–167 (2012)
9. Nasukawa, T., Yi, J.: Sentiment analysis: capturing favorability using natural language processing. In: Proceedings of the 2nd International Conference on Knowledge Capture. ACM, pp 70–77 (2003)
10. Yuan, J., You, Q., Luo, J.: Sentiment analysis using social multimedia. In: Multimedia Data Mining and Analytics, pp. 31–59. Springer (2015)
11. You, Q., Luo, J.: Towards social imagematics: sentiment analysis in social multimedia. In: Proceedings of the Thirteenth International Workshop on Multimedia Data Mining, MDMKDD 2013, pp. 3:1–3:8. ACM, New York (2013)
12. Wang, Y., et al.: Unsupervised sentiment analysis for social media images. In: Proceedings of the 24th International Conference on Artificial Intelligence. AAAI Press (2015)
13. Borth, D., et al.: Large-scale visual sentiment ontology and detectors using adjective noun pairs. In: Proceedings of the 21st ACM International Conference on Multimedia. ACM (2013)
14. Plutchik, R.: The nature of emotions. Am. Sci. **89**(4), 344–350 (2001)

15. Baecchi, C., et al.: A multimodal feature learning approach for sentiment analysis of social network multimedia. Multimedia Tools Appl. 1–19 (2015)
16. You, Q., et al.: Joint visual-textual sentiment analysis with deep neural networks. In: Proceedings of the 23rd Annual ACM Conference on Multimedia Conference. ACM (2015)
17. Krizhevsky, A., Sutskever, I., Hinton, G.E.: Imagenet classification with deep convolutional neural networks. In: Advances in Neural Information Processing Systems (2012)
18. Le, Q.V., Mikolov, T.: Distributed representations of sentences and documents. arXiv preprint arXiv:1405.4053 (2014)
19. Katsurai, M., Satoh, S.I.: Image sentiment analysis using latent correlations among visual, textual, and sentiment views (2016)
20. Ji, R., Cao, D., Lin, D.: Cross-modality sentiment analysis for social multimedia. In: 2015 IEEE International Conference on Multimedia Big Data (BigMM). IEEE (2015)
21. Kalayeh, M.M., et al.: How to take a good selfie? In: Proceedings of the 23rd Annual ACM Conference on Multimedia Conference. ACM (2015)
22. Russakovsky, O., et al.: Imagenet large scale visual recognition challenge. Int. J. Comput. Vis. **115**(3), 211–252 (2015)
23. Seo, S., Kang, D.: Study on predicting sentiment from images using categorical and sentimental keyword-based image retrieval. J. Supercomput. 1–11 (2015)
24. Sermanet, P., et al.: Overfeat: integrated recognition, localization and detection using convolutional networks. arXiv preprint arXiv:1312.6229 (2013)
25. http://ufldl.stanford.edu/tutorial/supervised/ConvolutionalNeuralNetwork/
26. Qamar, U., et al.: A majority vote based classifier ensemble for web service classification. Bus. Inf. Syst. Eng. 1–11
27. Wang, H., et al.: Web service classification using support vector machine. In: 2010 22nd IEEE International Conference on Tools with Artificial Intelligence (ICTAI), vol. 1. IEEE (2010)
28. Varguez-Moo, M., Moo-Mena, F., Uc-Cetina, V.: Use of classification algorithms for semantic web services discovery. J. Comput. **8**(7), 1810–1814 (2013)
29. Mullen, T., Nigel, C.: Sentiment analysis using support vector machines with diverse information sources. In: EMNLP, vol. 4 (2004)
30. Apoorv, A., et al.: Sentiment analysis of twitter data. In: Proceedings of the Workshop on Languages in Social Media. Association for Computational Linguistics (2011)

Segmentation of Heart Sound by Clustering Using Spectral and Temporal Features

Shah Khalid(✉), Ali Hassan, Sana Ullah, and Farhan Riaz

Department of Computer Engineering, College of Electrical and Mechanical Engineering, National University of Science and Technology, Islamabad, Pakistan
{shah.khalid85,alihassan,sana.ullah85, farhan.riaz}@ce.ceme.edu.pk

Abstract. Cardiac auscultation is a method used to listen heart sound. Condition of the heart can be predicted with cardiac auscultation because heart generates a specific rhythm of sound and any changes in the rhythm of the heart sound may be due to abnormalities of heart. Auscultation is an easy way to diagnose heart abnormalities; however, it needs training and years of physician's experience to diagnose heart and identify any heart abnormalities. With years of experience, it is still difficult to analyze heart sound. The ability to automatically identify abnormalities or at least support physician decision is relevant to ease the reach of medical diagnosis using mobile or Digi-scope. This paper presents a novel approach for segmentation of S1 and S2 heart sounds by using some of the heart sounds temporal and spectral features. Our method differentiates between S1 and S2 heart sounds and also improves the results as compared to the three finalists.

Keywords: Heart sound segmentation · PCG · Spectral centroid Variation coefficient

1 Introduction

According to WHO (World Health Organization) reports, 97.5 million people died of Cardio-Vascular Diseases (CVD) around the globe in 2015, which is 32.1% of the total human death in the same year [1]. On the basis of these statistics, CVDs are the leading causes of death as no other cause has such a huge margin of death. It is also important to be mentioned that more than three quarters of CVD deaths take place in low and middle income countries.

Any technique that helps in diagnosing or treating CVDs can significantly influence world health. Techniques for diagnosis i.e. echocardiography, electrocardiography (ECG), magnetic resonance imaging (MRI) and ultrasound have provided a much more accurate and direct evidence of heart disease but these are operationally very complex, colossal and expensive. Cardiac auscultation is a more basic, non-invasive, and a very cost effective method for diagnosing CVDs. It is a technique of listening to heart sound, traditionally a stethoscope is used for this process which empowers the doctor to diagnose CVDs. Heart sound is due to blood turbulence and blood turbulence is due opening and closing of heart valves and the fast flow of blood in the chambers. S1 and S2 are the

K. Arai et al. (Eds.): IntelliSys 2018, AISC 868, pp. 337–346, 2019.
https://doi.org/10.1007/978-3-030-01054-6_24

fundamental heart sounds. S1 sound are generated due to closing of atrioventricular valves and S2 is due to semilunar valve. The development of a system or any approach that helps doctors in evaluating these heart sounds might decrease the expense of identification of CVDs and enhance their accuracies.

Until now different heart sound segmentation approaches are reported. Wavelet transformation is used to decompose sound signal into its high and low frequency characteristics. For segmentation of heart sound, some suggested discrete wavelet transformation decomposition and reconstruction of a signal. Feature like peak value and peak duration were extracted for reconstructed signals and fed to a classifier [9]. Calculating Shannon energy of the heart sound signal is also widely applied method [5, 10]. This approach is generally used to help in detection of peaks in sound signal. A comparison study was performed between methods of envelops extraction like Hilbert hung transform and Shannon energy. A probabilistic approach is also proposed for localization of S1 and S1 heart sound by applying Hidden Markov Model [11, 12]. Fuzzy logic was also used for the detection of fundamental heart sounds S1 and S2 [13]. Another approach is used involve decomposition of heart sound signals into atoms and cluster it by dynamic clustering. The dynamic clustering based on density function [14].

Previous work on heart sound segmentation has been reported on the basis of their time domain characteristics. As we know that heart sound is organic signals therefore a constant factor is difficult to find in them. The length or time in between lub and dub or dub and lub often deviates. The amplitude of S1 or S2 changes at different circumstances. Segmentation algorithm based on frequency domain was also proposed by tracking spectrum of heart sound [8] but we know that fundamental heart sounds S1 and S2 don't have any fixed frequencies.

Due to these limitations of heart sounds a unique approach is needed for segmentation of heart sound. In this paper, we used a novel heart sound segmentation approach which involve features extraction from temporal and spectral domain of a phonocardiogram (PCG) signals and a clustering algorithm is used on the basis of those features, for localization of S1 (lub) and S2 (dub) heart sounds. This approach is tested on a publically available dataset that is PASCAL (Pattern Analysis, Statistical Modelling and Computational Learning) heart sound dataset [2]. The rest of the paper is organized as follows. In Sect. 2, the dataset are explained which is used in this paper. Section 3 illustrates our new methodology for segmentation of S1 and S2 heart sounds. In the next section, Sect. 4, we present our results and the results of the three finalists. In the last section, we conclude our paper.

2 Datasets

In this paper PASCAL heart sound dataset are used for analysis and segmentation of heart sound [2]. These data consist of two major subsets, the heart sounds in this dataset are acquired from two different devices. Sub dataset A is acquired from an iPhone app iStethoscope Pro while dataset B is acquired with a Digi-scope in a little smoother and silent clinical environment. Dataset A contains 176 audio files of PCG. It comprises four subsets i.e. Normal (31 files), Murmur (34 files), Extra Heart sound (19 files) and Artifact

(40 files). Dataset B contains total 656 audio files of heart sounds. It is divided into three categories, i.e. Normal (319), Murmur (93), Extrasystol (46).

Normal subset comprises healthy heart sounds and might have background noise i.e. lungs sound due to inhalation and exhalation, sound due to rubbing microphone against skin or cloths. Normal subset are in both dataset A and dataset B. Murmur consists of heart sounds with roaring, whooshing or rumbling sounds in between S1 and S2 or between S2 and S1. As murmur sound is due to problems in heart valves so they are sign to abnormal heart. Both datasets, A and B have murmur set. In Extra heart sound category, there are heart sounds having additional fundamental heart sound, it rhythm is different form normal heart sound, i.e. S1, S1, S2 or S1, S2, S2 and so no. Extra heart sound may or may not be a cause of any abnormalities in heart, it is not usually consider risky. This category is only in dataset A. Artifact contains ranges of different sound, i.e. music, noise, speech. Heart sound is difficult to figure out in here. Such sounds don't help in heart diagnosis; therefore someone collecting the data is advised to try again. This category is only in dataset A. Extrasystol is out of rhythm; it skips or adds extra heart sound. This category is in dataset B only.

These audio files are of different lengths which range from one second to thirty seconds. Dataset A is in .aif format while Dataset B is in .wav format. An annotated data for Normal category of Dataset A and Dataset B are given for training segmentation algorithm. However, an evaluation or test dataset are also given which comprise files selected from Normal category of both Dataset A and Dataset B for testing the proposed segmenting algorithm.

3 Method

In this section, we are going to develop a approach for localization of S1 and S2 heart sounds within the PCG audio files. Our main goal about which we are concern is methods related to reduction or cancelation of noise from the heart sounds, identification of heart sound and segmentation of heart sound into fundamental (S1 and S2) heart sound. Figure 1 shows the flow of our method.

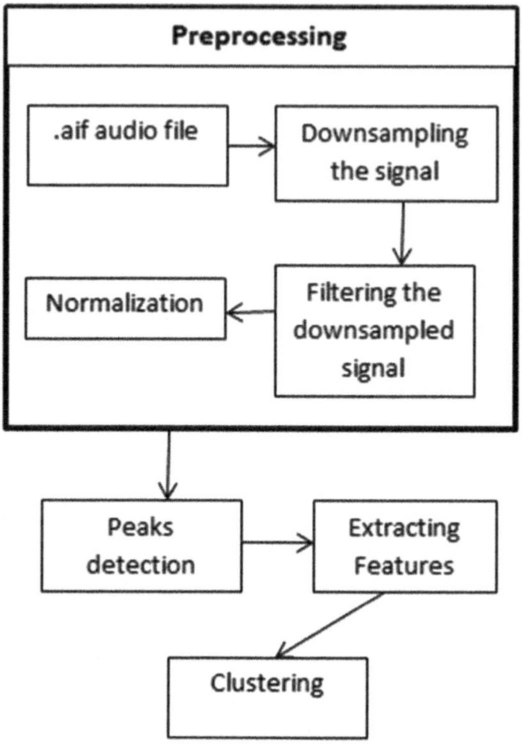

Fig. 1. Flow chart of segmentation methodology.

3.1 Pre-processing

Heart sounds with duration of three seconds or less than three seconds are eliminated. These sounds filed in the targeted dataset are recording of heart sounds contaminated with noise. It is better to clean or reduce the noise from the audio sound.

PCG audio files are initially pre-processed before localizing or segmenting S1 and S2 heart sounds [3]. Down sampling, filtering and normalization are three steps involved in preprocessing. To avoid some computation these audio files are down sampled with the factor of 2, using decimate function of Mat lab (MATLAB, 2015b). Information of heart sound is mostly present in low frequencies while noise on the other hand can be located in the higher frequencies. Therefore, the down sampled signals are filtered with Butterworth bandpass filter of order 6 with cut-off frequency from 25 Hz to 900 Hz to reduce noise which is shown in Fig. 2. Then, the signal is normalized with absolute maximum normalization as shown in Fig. 2. This brings all the signals to a common range of –1 to 1. Normalization is done in two steps, finding extreme absolute value is first step and in second step the signal is divided by maximum value [3].

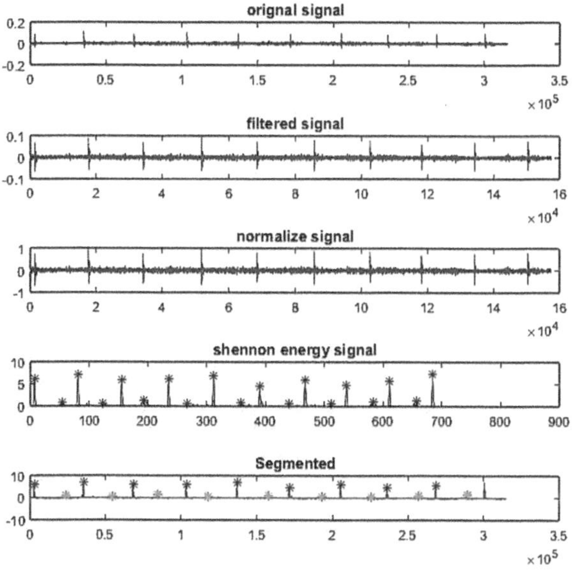

Fig. 2. Heart sound segmentation.

3.2 Peak Detection

Shannon Energy is a powerful technique for extraction of envelope from heart sounds. It helps in split and tooted peaks. It works best in localization of various components in PCG signals. This method converts nonlinear combination of signal into some linear combination of signals. Therefore, Shannon energy is calculated in order to smoothen the preprocessed signal which will help next in finding peaks. Shannon energy has been depicted in the following equation [5]:

$$Shannon\,Energy = -x^2(t)\,log\,x^2(t)$$

Shannon energy for each file is calculated as average Shannon energy. The average Shannon energy is calculated in continuous 0.02 samples per second window with 0.01 samples per second overlap. The average Shannon energy can be represented as follows:

$$E = -\frac{1}{N}\sum_{n=1}^{\infty} x^2(i) log x^2(i)$$

Where x(t) is the processed signal and N is length of the windows which is in this case 0.02 samples per second. In the last average Shannon energy is calculated of every window in the processed sample is then normalized. Normalized Shannon energy of the processed signal can be calculated as follows:

$$P(t) = \frac{E(t) - ME(t)}{SE(t)}$$

Where E(t) is average Shannon energy calculated of window number “t” of size 0.02 sample per second while ME (t) and SE (t) are mean energy and standard deviation of E(t), respectively. P (t) is the normalized average Shannon energy which is also called Shannon Envelope.

Shannon envelope as shown in Fig. 2 smoothen the sound signal and make peaks prominent. Fundamental heart sound peaks are selected on the basis of amplitude thresholding and peaks gap thresholding after calculating Shannon envelope by using a predefined and open source function. Extra peaks with amplitude smaller than threshold are eliminated and those peaks are also rejected whose in-between gaps are smaller than the defined threshold.

After identifying or selecting fundamental heart sound peaks, now it needs to distinguish in between these peaks to analysis the rhythm of heart sound. From that rhythm of heart sounds abnormalities of heart can be identified. The peak selected may be S1 or S2. For this purpose a novel approach are used to distinguish S1 (lub) and S2 (Dub) heart sounds within selected peaks.

3.3 Feature Extraction

Among these peaks, S1 (lab) and S2 (dub) peaks are to be segmented out. For this purpose features are extracted from both time and frequency domains of the Shannon envelope, which is our main contribution in this paper. We extract four features, half of these features are from time domain and the other half is from frequency domain. Temporal features are as follows:

(1) ***Peak value:*** Amplitude value of the peak.
(2) ***Peak Gap:*** Gap between two successive peaks [6].

Whereas Spectral features are:

(1) Spectral centroid
(2) Variation coefficient

The two temporal features, Peak values and peak gaps are calculated from the peaks of the processed signal while spectral centroid and variation coefficients are calculated with our special method.

Some samples are selected before peak sample and the same number of sample after the peak position are selected from the signal obtained after preprocessing phase. Also including sample at peak position. Power Spectral Density (PSD) is calculated for the selected section of the signal for spectral feature extraction [7]. PSD can be calculated with the help of the following equation:

$$P(w) = \sum_{-\infty}^{\infty} r_y[n] e^{-jwn}$$

Where $r_y[n]$ is autocorrelation of the selected region of the signal which can be define as E(y[m]y[m]*), let y[m] is the selected region of the preprocessed signal. The two spectral features are extracted from here.

The first spectral feature Spectral feature extracted is spectral centroid. Spectral centroid can be calculated as follows:

$$C = \frac{\sum wP(w)}{\sum P(w)}$$

P (w) is the amplitude of wth frequency bin in the spectrum. During our studies we analyze that variation coefficient of S1 and S2 was different from one another. Variation coefficient can be calculated as:

$$\sigma^2 = \frac{\sum (w - C)^2 P(w)}{\sum P(w)}$$

Whereas C in spectral centroid which we calculated in the previous equation.

3.4 Clustering

We have extracted four features of each peak, that helps us in identification of S1 (lub) sound peak and S2 (dub) sound peak. There are different machine learning techniques for classification and segmentation that can be used to segment out S1 heart sound and S2 heart sound.

The best approach we find in our case in kmean clustering algorithm. Kmean algorithm gives good result as compere to some other clustering and classification algorithms. It almost successfully identifies S1 and S2 heart sound peaks with the help of features extracted as shown in Fig. 2.

4 Results

We evaluate the results of our approach with the provided training data of Dataset A and Dataset B. This set contains annotated data of the normal heart sound category from Dataset A and B. The annotated data help improve our approach by comparing our method's results and required results.

We also evaluates our results for test dataset and the calculated error of test set A and test set B are shown in Tables 1 and 2, respectively.

Table 1. Results of dataset A

File name	Total of heartbeat	Average error
201101070538.aif	11.5	15719.26087
201101151127.aif	7.5	117933.4
201102081152.aif	9	171266.1111
201102201230.aif	11	584.2727273
201102270940.aif	9.5	189061.1053
201103101140.aif	9.5	20782.73684
201103140135.aif	7.5	73441.6
201103170121.aif	9.5	38229.10526
201104122156.aif	11	182462.1818
201106151236.aif	9.5	43708.52632

Table 2. Results of dataset B

File name	Heartbeat	Average error
103_1305031931979_B.aiff	12.5	88.48
103_1305031931979_D2.aiff	10	38.75
106_1306776721273_B1.aiff	4	46.875
106_1306776721273_C2.aiff	3	60
106_1306776721273_D1.aiff	3.5	117.2857143
106_1306776721273_D2.aiff	7.5	2770.866667
107_1305654946865_C1.aiff	7.5	1623.866667
126_1306777102824_B.aiff	8.5	3239.823529
126_1306777102824_C.aiff	5	1464.6
133_1306759619127_A.aiff	4	61.625
134_1306428161797_C2.aiff	2.5	55.8
137_1306764999211_C.aiff	15	40.13333333
140_1306519735121_B.aiff	11.5	2325.956522
146_1306778707532_B.aiff	18	4246.305556
146_1306778707532_D3.aiff	3	14
147_1306523973811_A.aiff	4	246
148_1306768801551_D2.aiff	8	56.9375
151_1306779785624_D.aiff	4.5	89.66666667
154_1306935608852_B1.aiff	4.5	47.22222222
159_1307018640315_B1.aiff	6	27.75
159_1307018640315_B2.aiff	3	48.16666667
167_1307111318050_A.aiff	13	62.03846154
167_1307111318050_C.aiff	3	249.6666667
172_1307971284351_B1.aiff	3.5	66.71428571
175_1307987962616_B1.aiff	2.5	30
175_1307987962616_D.aiff	7	2521.142857
179_1307990076841_B.aiff	16	496.375
181_1308052613891_D.aiff	3	77.33333333
184_1308073010307_D.aiff	26.5	65.09433962
190_1308076920011_D.aiff	3.5	67.57142857

In Fig. 2, we can see the results of our applied methods for detection of peaks and localization of S1 and S2 heart sounds within the PCG audio file 201102081152.aif. The file 201102081152.aif is selected from the normal category of dataset A. In the first plot, we have the original heart sound signal, whereas, second plot shows signal filtered with Butterworth filter. The third is plot of normalized filtered signal, normalized by absolute maximum normalization while in the fourth plot we can see Shannon energy signal of the processed heart sound signal and in the last we have plot of segmented S1 and S2 heart sound peaks. Green peaks in the last plot represent S1 heart sound while red peaks indicate S2 heart sound.

In Table 1, we depict the segmentation results for the audio files from the category of normal in dataset A. We can see the total numbers of heartbeats in the second column of the table and the third column shows average error for each sound sample which is measure in sample for precision, while the total error of Dataset A is 853188.3002.

Table 2 presents the segmentation results for audio files, which is from the normal category of Data set B. Column 2 in the table shows total heart beats identified while the third column shows average error. The total error of dataset B is 20346.0474.

We can see that the error in dataset B is much better than Dataset A. This might be due to the reason that the Data of dataset B are collected with a digital statoscope in a quite environment of a hospital by expert physician however Dataset A are collected by person with less or no experience in rough condition on a smart phone.

Table 3 shows results of the three finalists and the results of our methodology for both of the datasets A and B.

Table 3. Total error found by three finalists and by our methodology

	Dataset A	Dataset B
ISEP/IPP Portugal	4219736.5	72242.8
CS UCL	3394378.8	75569.8
SLAC Stanford	1243640	76444.4
Our Methodology	835188.3002	20346.4742

5 Conclusion

This paper presents a novel technique for segmentation of S1 and S2 heart sounds, which is the first challenge of "PASCAL Classify Heart Sound Challenge" [2]. In this work, we created an improved segmentation approach by selecting two features in time domain and two features in frequency domain. Feature selection for clustering or segmentation involves these steps. First the sound signals are de noised and normalized. Shannon envelop of the processed signal are calculated. Extra peaks are rejected by peak amplitude thresholding and time in between two successive peak thresholding. In the last four features are extracted from every peak. Our proposed algorithm successfully differentiates between S1 and S2 heart sounds to a great extent while at this stage the two finalists could not achieve substantial success [3, 4]. We also reduced the total error of the Dataset A and Dataset B. We still look forward to completely differentiate between the two heart sounds and to reduce the error. Our next aim is to agree the second challenge of

"PASCAL Classify Heart Sound Challenge" [2] and develop an approach for classification of heart sounds. Major application of this approach is in the area of telemedicine and affordable as it is easy, cost efficient and non-colossal health care.

References

1. Cardiovascular diseases (CVDs), World Health Organization (2017). http://www.who.int/mediacentre/factsheets/fs317/en/
2. Bentley, P., Nordehn, G., Coimbra, M., Mannor, S.: The PASCAL Classifying Heart Sounds Challenge (2011). http://www.peterjbentley.com/heartchallenge/index.htmlK
3. Gomes, E.F., Pereira, E.: Classifying heart sounds using peak location for segmentation and feature construction. In: Workshop Classifying Heart Sounds, La Palma, Canary Islands (2012)
4. Deng, Y., Bentley, P.J.: A robust heart sound segmentation and classification algorithm using wavelet decomposition and spectrogram. In: Workshop Classifying Heart Sounds, La Palma, Canary Islands, (2012)
5. Liang, H., Lukkarinen, S., Hartimo, I.: Heart sound segmentation algorithm based on heart sound envelogram. Comput. Cardiol. **24**, 105–108 (1997)
6. Kumar, D., Carvalho, P., Antunes, M., Gil, P., Henriques, J., Eug, L.: A new algorithm for detection of s1 and s2 heart sound. In: IEEE International Conference on ICASSP 2006, pp. 1180–1183 (2006)
7. Riaz, F., Hassan, A., Rehman, S., Niazi, I.M., Dremstrup, K.: EMD based temporal and spectral features for the classification of EEG signals using supervised learning. IEEE Trans. Neural Syst. Rehabil. Eng. **24**(1), 28–35 (2016)
8. Iwata, A., Ishii, N., Suzamura, N.: Algorithm for detecting the first and the second heart sounds by spectral tracking. Med. Biol. Eng. Comput. **1,** 19–23 (1980)
9. Huiyang, L., Sakari, L., Liro, H.: A heart sound segmentation algorithm using wavelet decomposition and reconstruction. In: Proceeding of the 19th Annual International Conference of the IEEE Engineering in Medicine and Biology Society, pp. 1630–1633, November 1997
10. Ari, S., Kumar, P., Saha, G.: A robust heart sound segmentation algorithm for commonly occurring heart valve diseases. Int. J. Med. Eng. Inform. **32**(6), 456–465 (2008)
11. Gamero, L., Watrous, R.: Detection of the first and second heart sound using probabilistic models. In: Proceedings of IEEE 25th Annual International Conference on Engineering Medicine Biology Society, vol. 3, pp. 2277–2880 (2003)
12. Gill, D., Gavrielli, N., Intrator, N.: Detection and identification of heart sounds using homomorphic envelogram and self–organizing probabilistic model. Comput. Cardiol. **32**, 957–960 (2005)
13. Fanfulla, M., Malcangi, M., Riva, M., Giustina, D.D., Belloni, F.: Cardiac sound segmentation algorithm for arrhythmias detection by fuzzy logic. Int. J. Circuit Syst. Signal Process. **5**(2), 192–200 (2011)
14. Tang, H., Li, T., Qui, T., Park, Y.: Segmentation of heart sounds based on dynamic clustering. Biomed. Signal process. Control **7**, 509–516 (2012)

Evaluation of Classifiers for Emotion Detection While Performing Physical and Visual Tasks: Tower of Hanoi and IAPS

Shahnawaz Qureshi[1](✉), Johan Hagelbäck[2], Syed Muhammad Zeeshan Iqbal[3](✉), Hamad Javaid[4], and Craig A. Lindley[5]

[1] Department of Computer Science, Prince of Songkla University, Hatyai, Thailand
shahnawazqureshi86@gmail.com

[2] Department of Computer Science, Linnaeus University, Växjö, Sweden
johan.hagelback@lnu.se

[3] BrightWare, Riyadh, Saudi Arabia
mzeeshan01@gmail.com

[4] Jinnah International Hospital, Abbottabad, Pakistan
hummad_qazi@yahoo.com

[5] Intelligent Sensing Laboratory, CSIRO, Canberra, Australia
craig.lindley@csiro.au

Abstract. With the advancement in robot technology, smart human-robot interaction is of increasing importance for allowing the more excellent use of robots integrated into human environments and activities. If a robot can identify emotions and intentions of a human interacting with it, interactions with humans can potentially become more natural and effective. However, mechanisms of perception and empathy used by humans to achieve this understanding may not be suitable or adequate for use within robots. Electroencephalography (EEG) can be used for recording signals revealing emotions and motivations from a human brain. This study aimed to evaluate different machine learning techniques to classify EEG data associated with specific affective/emotional states. For experimental purposes, we used visual (IAPS) and physical (Tower of Hanoi) tasks to record human emotional states in the form of EEG data. The obtained EEG data processed, formatted and evaluated using various machine learning techniques to find out which method can most accurately classify EEG data according to associated affective/emotional states. The experiment confirms the choice of a method for improving the accuracy of results. According to the results, Support Vector Machine was the first, and Regression Tree was the second best method for classifying EEG data associated with specific affective/emotional states with accuracies up to 70.00% and 60.00%, respectively. In both tasks, SVM was better in performance than RT.

Keywords: K-Nearest Neighbor (KNN) · Regression Tree (RT) Bayesian Network (BNT) · Support Vector Machine (SVM)

Weka 3: www.cs.waikato.ac.nz/ml/weka/.
The European Clearing House for Open Robotics Development: www.echord.info.

K. Arai et al. (Eds.): IntelliSys 2018, AISC 868, pp. 347–363, 2019.
https://doi.org/10.1007/978-3-030-01054-6_25

Artificial Neural Networks (ANN) · Tower of Hanoi (ToH)
Cognitive psychology; Human Computer Interaction (HCI)
Electroencephalography (EEG)

1 Introduction

As robotic technologies continue to develop at an exponentially increasing rate, interaction among humans and machines is growing. Human-robot interaction is of particular importance in areas where humans and robots work directly together, e.g., clinical, in industrial, military and gaming applications. When it comes to designing real-time systems, it is necessary to study this interaction because robots need to adapt to human guidance and emotions. HRI (Human-Robot Interaction) is concerned with achieving robust cooperation among humans and robots. Achieving good results during collaborative working can be facilitated by providing robots with methods for interpreting human needs and emotions during task performance. One way to study human affective states is by using Electroencephalography (EEG), where EEG spectra can be correlated with states of consciousness and emotions. It demonstrated for first time in almost 80 years ago. However, we are not at a stage where it is feasible to use EEG data as a direct input to technical systems, allowing automated reactions to human states of emotion and consciousness that do not require conscious human contributions into the system. The work reported in this paper conducted in the project PsyIntEC (Psychophysiological Interaction and Empathic Cognition for Human-Robot Cooperative Work); it was part of EU funded ECHORD project. The project aimed to explore practical and cognitive modeling of humans in HRI as a basis for behavioral adaptation. Further, by using psycho-physiological measurements, its purpose was to understand the affective perception of relevant modalities in human-human and human-robot interaction on a collaborative problem-solving task. Furthermore, it explored which type of sensors are more suitable to record human emotions.

2 Related Work

It is hard to evaluate and interpret the interaction between human and robots due to the different human emotions under stressful working situations, especially in collaborative work where uncertainty is higher [1]. The choice of sensors in creating a human-robot workstation is critical for using bio-feedback to adapt to human emotions as measured by the sensors [2].

Considerable previous work has been done by researchers to recognize emotions through text, speech, facial expressions, and gestures [3]. Emotion recognition has been of scientific interest since the discovery of the Electroencephalogram (EEG). Further, there is an increasing amount of interest in controlling physical devices such as home appliances and robots through brain signals derived from EEG [4–9]. Furthermore, Brain-Computer Interface (BCI) Technology translates signals into commands to control the external devices. This technology is helpful for patients having disabilities, e.g., loss of limb movements or gaze control, etc. The BCI based system introduced in [10] and it is efficient and robust after the first phase training to patients. Robotic training partners used for rehabilitative and preventive healthcare [11]. Investigations of the use of physiological signals in human-machine interaction have used linear classifiers to recognize sufficient signatures in signals due to their simplicity, speed and interpretive ability [12–14]. According to work done in [15,16], non-linear classifiers are considered to be most appropriate for feature extraction and deriving cognitive states from signals. A summary of classification methods, along with accuracies when they applied on data sets, is given in Table 1.

Table 1. Overview of classification methods used in studies, as well as, accuracy achieved when they were applied

Reference	Method(s) used	Results
[17]	Sequential Floating Forward Search and Fisher Projection methods	Classify eight emotions with 81% accuracy
[18]	Marquardt Back propagation, Discriminant Function Analysis and K-Nearest Neighbour	Distinguish between six emotions with an accuracy of 83%, 74%, and 71%, respectively
[19]	Probabilistic models may be helpful	To model body expressions, personality of a user in context of interaction
[20]	Artificial Neural Networks	Evaluate the mental workload with mean classification accuracies of 85%, 82% and 86% for the baseline, low and high task difficulty states, respectively
[12]	Support Vector Machine	Embedded in emotion recognizer and provide accuracies of 78.4%, 61.8%, and 41.7% for recognition of three, four and five emotions categories, respectively
[21]	Bayesian Network	Provide accuracy of 74.03%

To human-machine interaction study by using physiological data, two tasks are used most commonly: visual or physical. As a physical task, the Towers of Hanoi (ToH) is a well-known method in the field of clinical and experimental neuropsychology [22]. It helps to comprehend planning concerning information processing. Some test versions are available, but not limited to: a standalone test known as Tower of London, a child/adolescent version, a computerized variant, known as the Stockings of Cambridge test, etc. [23–25]. The correlation between Towers of Hanoi Task performance and EEG studied in [26]. Here, the purpose is to investigate the relationship between psycho-physiological data and the performance of the subjects, when they are performing each task, separately. This study evaluates the performance of classifiers for the data collected while subjects were performing the physical task. The results related to the visual task are present in the study [27]. Further, to the author's best of knowledge, no study compared the performance of classification techniques by using the data of physical and visual tasks.

The primary purpose of this study is to evaluate different machine learning techniques to classify EEG data which is associated with affective/emotional states while performing the physical task. It was involved:

(1) How do psycho-physiological data correlate with the performance of the subjects?
(2) Grouping based on earlier tasks: do they reveal anything about performance in ToH? E.g., In IAPS test the subjects with the lowest arousal are different from the one with the highest arousal during ToH test.

3 Machine Learning Techniques Used to Classify EEG Data

There exist many different machine learning techniques that have successfully used for a wide range of tasks. For classification of EEG data, the most common are Artificial Neural Network and Support Vector Machine [21,28].

3.1 Support Vector Machine (SVM)

SVM is a supervised learning model which is used to analyze data for classification and regression analysis with the help of a kernel function. It was developed by Vapnik [29]. Due to its empirically excellent performance, this technique has successfully applied in many fields, e.g., bioinformatics, image recognition, etc. It may show good performance in a situation where a training data set is small. For the more detailed tutorial on SVM, see [30].

3.2 Artificial Neural Network (ANN)

ANN is a computational model which consist of a set of connected neurons in a weighted manner [31]. An ANN model consists of three layers: (i) an input layer that receives the data; (ii) the hidden layer performs a mathematical function which defines the activation of neurons; and (iii) the third one is an output layer to compute the results. The principle behind ANN model is the backpropagation algorithm for learning the correct weights.

3.3 Regression Tree (RT)

It takes a set of continuous values (i.e., real numbers) as an input to be categorized to many belongings and provides a result [32]. For classification purposes, it used in speech recognition, heart attack, and cancer diagnosis [33,34].

3.4 Bayesian Network (BNT)

Bayesian network model belongs to the family of probabilistic graph model. It is used to characterize the knowledge around an undefined domain [35], and it is useful for the prediction about how the world will behave. It is considered to be an efficient classifier concerning predictive performance and classification accuracy [36]. For more details, see [35].

3.5 K-Nearest Neighbour (KNN)

KNN is a basic and simple classification and regression techniques [37]. It is a non-parametric model which avoids the problem of probability densities [38]. It identifies the K training samples nearest to the test sample and returns the average of K points as a result. KNN used in different fields such as geographic information system (GIS) [39] where points connected with geographical position. For more details, see [38].

4 Experiment Preparation and Execution

To classify the various emotional states in subjects, a ToH game used because it may induce strong emotions. Here, we include states such as attention, concentration, tiredness, etc. in the term 'emotions', while recognizing that these are not emotions as such and have different underlying cognitive and neurological systems and mechanisms associated with them.

For classification of emotions, various systems exist; usually, it observed from two aspects such as Discrete and Dimensional [40]. According to Plutchik, eight primary states of emotions are acceptance, anger, anticipation, disgust, fear, joy, sadness and surprise [41]. The commonly used classification system is the bipolar model, proposed by Russells [42], which considers arousal and valence dimension. In this case, valance dimensions are from negative to positive whereas arousal is from not aroused to excited. The dimensional model is advantageous for emotion recognition because it can determine discrete emotions in its space and most commonly used model for classification of emotions [42,43]. A two dimensional emotional model with **valence** (pleasure to displeasure: the emotion is negative, positive, or neutral) and **arousal** (excitement: physiological activation state of the body ranging from low to high) is used during experiments.

A total of 20 subjects (15 men and five women) participated in the experiment. All subjects were students of Blekinge Institute of Technology, Sweden, and aged from 21 to 35 years. The subjects came from different cultural backgrounds, nationalities, and field of studies. The EEG signals captured from left and right frontal, central, anterior temporal and parietal regions (F3, F4, C3, C4, T3, T4, P3, P4 positions according to the 10–20 system and referenced to Cz [44]).

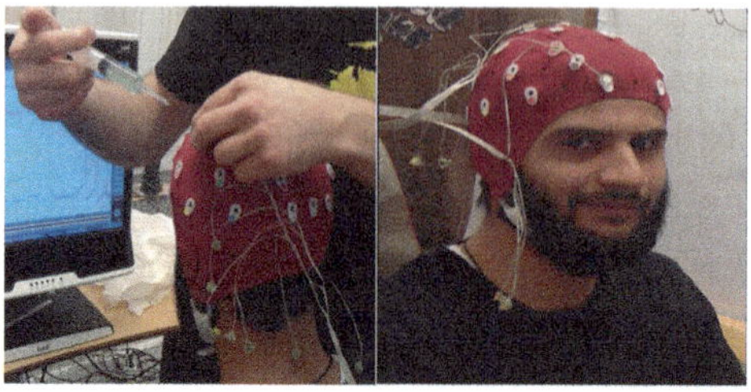

Fig. 1. Biosemi ActiveTwo system. A subject wearing a EEG Biosemi head cap.

Each subject was instructed to solve the puzzle according to the following rules:

- The start and end configuration for ToH puzzle is shown in Fig. 2. The problem has three vertical pegs and discs of growing size (A, B, C, D, and E).
- Initially, all of the discs are at the most left peg with the largest disc on the bottom and the smallest one at the top.
- To solve the puzzle, all the discs are required to move from the most left peg to the most right peg keeping the same order as on the most left peg during the starting configuration. During this, only one disc at a time can be moved, and a larger disc cannot be on the top of smaller one.

To say something potentially meaningful about the task itself, two different tasks tested; one with four discs (4-discs) and one with five discs (5-discs).

- Half of the subjects played the 4-discs first, and then the 5-discs. The other half of the subjects performed the 5-discs first, and then the 4-discs. The purpose of this arrangement was to avoid any possible sequence effects.
- Each subject had 5 min to complete the task.
- The questionnaire was filled by each subject to rate the task and used as a Self-Assessment Manikin (SAM) during the experiment. The questionnaire is available in Appendix A.
- While solving the puzzle the following reactions from subjects observed, e.g., hit the table or shouted as task finishing time limit approaches.

5 EEG Data Analysis and Interpretation

5.1 Data Collection

During the experiments, EEG data for each subject was recorded using BioSemi ActiveTwo System with a sampling rate of 2048 Hz and stored in BioSemi Data Format (BDF) using an ActiView BioSemi acquisition software. Each subject took approximately 20 min individually to complete the experiment (Fig. 1).

Base upon the 10–20 system, Fp1, Fp2, C3, C4, F3, and F4 positions were used to attain the EEG signals. All of the electrodes referenced to Cz.

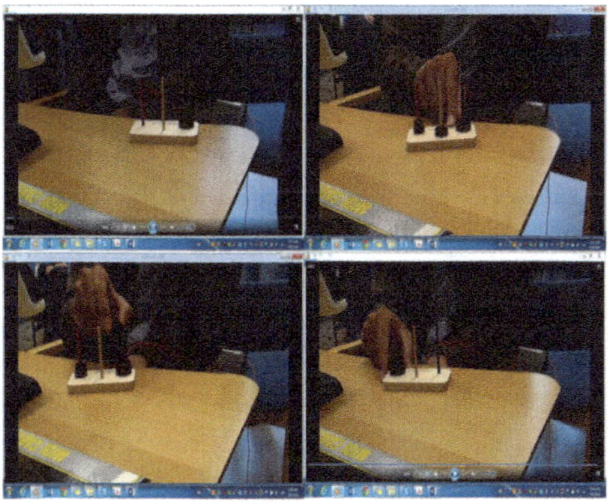

Fig. 2. A subject is at different stages during playing a Tower of Hanoi game and finished it within 5 min.

5.2 Data Screening

The subjects were screened to select EEG data for analysis and processing. The screening was based on SAM to choose the most valuable data to get reliable results. It left 15 subjects out of 20 because the data file(s) corrupted. It further applied to EEG data of the 15 subjects to select the signal ratio to provide sufficient data for the interpretation of emotions based on SAM. The idea behind this was to screen out and separate the data for each emotion. We take the data of two emotional states. Firstly, we divide the data the subjects belongs to positive arousal. Secondly, we separate the data from negative arousal.

EDF browser (a tool for reading and processing of sensor data) was used to reduce the signals individually for the required duration. Each ToH task's duration time is five minutes; hence, two tasks lasted for ten minutes. Due to the baseline inference, the first and last 5 seconds of total 5 min of each task eliminated. It was too narrow down to the exact required data. This process completed for ToH signals with positive, negative and neutral arousal and valence.

5.3 Data Preprocessing

The screened data was preprocessed using EEGLAB MATLAB Toolbox to extract Epoch and Event information, and then Independent Component Analysis (ICA) was performed on the data [45]. The data preprocessing with these various techniques helps to remove the artifacts such as eye blinking, an impedance of the system, the sampling frequency (50–60 HZ due to ground loop), etc. It also made it easier to extract features from the signals.

5.4 Data Processing or Feature Selection

Feature selection is one of the key challenges in affective computing due to the phenomena of person stereotype [46], i.e., different individuals express the same emotion with response patterns having different characteristics for same situations. Each subject involved in the experiment had various physiological indices that showed high correlations with each affective state. The same finding has been observed by [47] and explained by [33]. According to Rani et al., a feature can be considered significant and selected as an input to a classifier if the absolute correlation is more notable for physiological features among subjects [33]. Based on these findings, it observed that the accuracy improved for some techniques (i.e., KNN, BNT, and ANN) when highly correlated features used, while it degraded for the others (i.e., RT and SVM).

The selection of highly correlated features helps to exclude the less essential features for affective state, hence, improve the results [47]. The preprocessed data were further processed to get the real values for the signals using EEGLAB MATLAB Toolbox. The following features were extracted from the real values of each signal to process the data further [48]: Minimum, Maximum, and Mean Value; Standard Deviation (Table 2).

Table 2. A set of features extracted from EEG data

FNo.	Feature
1	$Min\ [A_i]$
2	$Max\ [A_i]$
3	$\mu = \frac{1}{N}\sum_{i=1}^{N} A_i$
4	$\sigma = \sqrt{\frac{1}{N-1}\sum_{i=1}^{N} \lVert A_i - \mu \rVert^2}$

Where vector A is made up of N scalar values.

5.5 Data Formatting

The processed data formatted in Attribute-Relation File Format (ARFF), which is an acceptable file format for the data mining tool WEKA. The values obtained are used as instances in the ARFF file with a binary class value as either negative or positive arousal/valance. Each feature value (Min, Max, Mean and Standard Deviation) for each electrode is a separate attribute in each instance in the file. Six electrodes were used making the total number of characteristics 24 (plus the class value). An independent dataset created for each subject, as well as a combined dataset with data from all subjects.

5.6 Data Classification

Each dataset was classified using machine learning techniques available in WEKA. During the classification, the classifier was trained to classify negative or positive arousal/valence values. The methods used had all the default parameter values as implemented in WEKA. In all experiments, 10-fold cross-validation used.

The selected techniques KNN, RT, BNT, SVM and ANN used for classification in WEKA.

The results from classifying the EEG data for all 15 subjects are presented in Table 3. The highest accuracy obtained with SVM (70.00%) while RT is the second best (60%) accuracy and KNN is third (55%); however, BNT and ANN are almost at the same level (45%) accuracy.

Table 3. Classification accuracies by selected techniques for dataset TB

Techniques	Accuracy
Support Vector Machine (SVM)	70.00%
Regression Tree (RT)	60.00%
K-Nearest Neighbour (KNN)	55.00%
Bayesian Network (BNT)	45.00%
Artificial Neural Networks (ANN)	45.00%

To derive further findings, the Dataset B was further divided into three datasets as given in Table 4: The primary aim of distributing the dataset in this way was to analyze the EEG data for different numbers of subjects to find possible differences among them. The classification accuracies by selected techniques for Dataset 1B, 2B and 3B is shown in Table 5 and Fig. 3.

Table 4. Division of dataset B

Dataset	Description
Dataset 1B	5
Dataset 2B	5
Dataset 3B	5

Table 5. Classification accuracies by selected techniques for dataset 1B, 2B and 3B

Technique	Dataset 1B	Dataset 2B	Dataset 3B
K-Nearest Neighbor	75.55%	68.67%	63.35%
Regression Tree	62.96%	54.44%	49.95%
Bayesian Network	61.26%	53.40%	46.65%
Support Vector Machine	78.78%	74.27%	65.35%
Artificial Neural Networks	72.37%	65.11%	55.24%

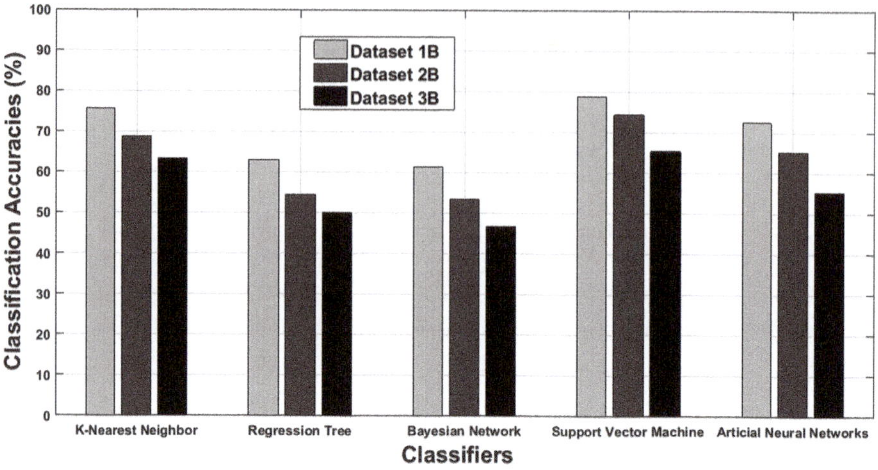

Fig. 3. Classification accuracies by selected techniques for dataset 1B, 2B and 3B.

According to Table 5, there are considerable differences among the accuracies provided by each technique for each dataset. In the last experiment, we used datasets consisting of only a single subject. The same considerable differences observed when the selected techniques were applied to datasets of each subject individually. This was done for the first three subjects. The results are shown in Table 6. In this experiment, KNN was the most accurate classifier with 95.33% accuracy for Subject 3. It is interesting to see that SVM was only able to get 65.00% accuracy on the same subject. BN showed considerable differences in 80.72% accuracy for Subject 2 and only 55.36% for Subject 1 (Fig. 4).

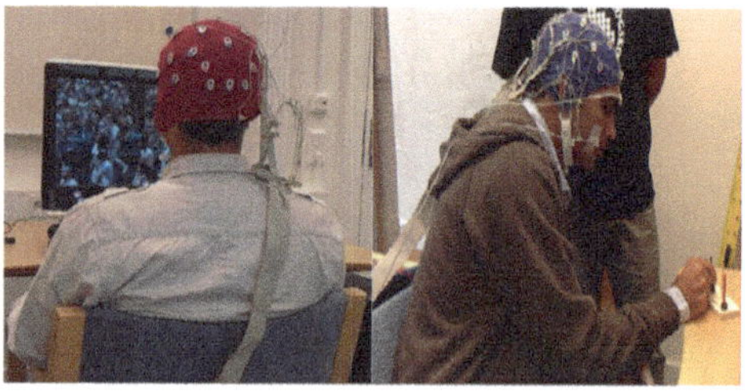

Fig. 4. A subject performing IAPS and ToH tasks.

6 Comparison Between Study Based on IAPS and TOH

Human beings behave differently when in a very similar situation. It observed in the experiment the subjects showed different emotions while performing

Table 6. Classification accuracies and standard deviation by selected techniques for subject 1, 2 and 3

Techniques	Subject 1		Subject 2		Subject 3	
	Accuracy	StDev	Accuracy	StDev	Accuracy	StDev
K-Nearest Neighbor	66.54%	6346	72.72%	313	95.33%	1026
Regression Tree	36.36%	6346	54.54%	313	50.00%	1026
Bayesian Network	55.36%	6346	80.72%	313	66.66%	1026
Support Vector Machine	50.45%	6346	45.45%	313	65.00%	1026
Artificial Neural Networks	52.45%	6346	45.45%	313	50.00%	1026

physical tasks. The subjects having exposure to scenes present in pictures were not stressed; e.g., the subjects from war areas or politically disturb countries elicited calm responses. Compared with the visual task [27], the subjects become aggressive (e.g., hit the desk, etc.) or lose their temper (e.g., shouting, etc.) while solving the Tower of Hanoi puzzle.

According to Table 7, SVM provides the best accuracy among all techniques considered. RT is the second best, and KNN is third; however, BNT and ANN are at the same level. A comparison of these results (based on the physical task performed by subjects) has shown with the results obtained from the experiment, conducted for the study (based on the visual task performed by subjects) [27], to see the differences among both.

The classification accuracies of the selected techniques for Dataset TB formatted for each study (based on the visual task performed by subjects and based on the physical task performed by subjects) individually are shown in Table 7 and Fig. 5:

Table 7. Comparison of classification accuracies by selected techniques for dataset TB.

Techniques	Visual task (IAPS) [27]	TOH
K-Nearest Neighbor (KNN)	52.44%	55.00%
Regression Tree (RT)	52.44%	60.00%
Bayesian Network (BNT)	52.44%	45.00%
Support Vector Machine (SVM)	56.10%	70.00%
Artificial Neural Networks (ANN)	48.78%	45.00%

As shown in Table 7, there are clear differences among results obtained from both studies. The SVM showed a rather large difference of 13.90% compared to the other classifiers and the rest with considerable differences ranging from 2% to 8% approximately. One of the possible reasons for these differences is the difference in tasks performed by the subjects in each study. Performing a task such

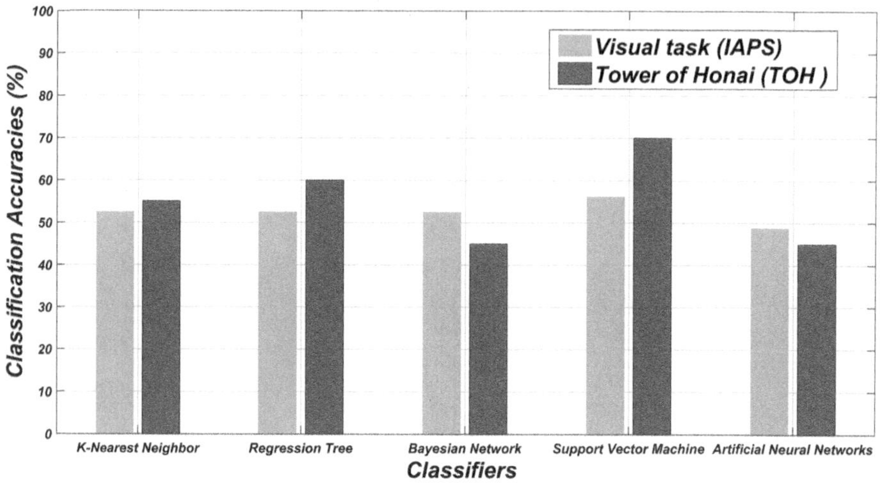

Fig. 5. Comparison of classification accuracies by selected techniques for dataset TB.

as TOH is considered to be more suitable for cognitive psychology to understand planning concerning information processing [32] or cognitive neuroscience [49]. Psychologists in [32,50,51] have come to the same conclusion.

While subjects were experimenting, they made a sequence of moves, visualized the movement of discs over the pegs in their mind, memorized the consequence of each move and evaluated their planned move. All of these focused activities to achieve the future goal lead to generate related brain activity. Hence the EEG data obtained holds highly correlated features which help to make better accuracies in this study as compared to the other.

7 Summary and Conclusion

The TOH puzzle is famous for understanding different aspects of cognitive psychology and is widely used in various research. It is a tool which is used by psychologists for studying information-processing especially.

There are possible differences in brain activities when humans perform a visual task and a physical task. To investigate these differences, an experiment for TOH conducted to test out the psycho-physiological equipment on an interactive, physical task. The primary purpose of this research study was to evaluate different machine learning techniques to classify EEG data obtained while a task performed. To achieve this, five machine learning techniques were selected based on a literature study, their use in empirical studies and accuracy results reported by different authors. The chosen methods such as K-Nearest Neighbour (KNN), Regression Tree (RT), Bayesian Network (BNT), Support Vector Machine (SVM) and Artificial Neural Networks (ANN) evaluated. For validation, the selected techniques were analyzed using EEG data collected from 20 subjects through an experiment in a controlled environment. The data obtained was processed to remove the artifacts (e.g., eye blinking, the impedance of the system) and extract features, and were then formatted. The data was formatted using a model in a form that was acceptable by the classification software to analyze the classification accuracies of the selected techniques.

According to the results, Support Vector Machine (SVM) is the best to classify EEG data associated with specific affective/emotional states with a reported accuracy of 70.00%. Regression Tree (RT) was second best with an accuracy of 60.00%. An obvious observation is: focused brain activities for a single goal can lead to better accuracy results as compared to diverse brain activities for diverse goals. While looking at the findings from the questionnaires, the difference in opinions and data found among the subjects. Table 8 summarizes the conclusions derived from the questionnaire.

Table 8. Findings from questionnaire

Questions	Strongly agree	Strongly disagree
4-Disc Problem was Hard?	30%	70%
5-Disc Problem was Hard?	80%	20%

Different machine learning techniques had difference accuracies, and this can be due to differences of the respective algorithms used. Discovering the differences arising from the algorithms was the primary purpose of the whole point of the study.

8 Future Work

In future, the following dimensions will be explored:

- The EEG data acquisition can be enhanced by using Emotive epoch device.
- The representation of the selected emotions may be improved by recording more than six number of signals.
- To analyze the effect on accuracy by increasing the number of subjects.
- Other EEG data processing techniques such as Fourier or wavelet transform, etc. can be used to seek improvement in data processing.
- To evaluate different machine learning techniques.

Appendix A: Questionnaire

The questionnaire below was used as Self-Assessment Manikin (SAM) during the study for each subject. Subject Number was an anonymous number.

Subject Number:

Strategy Used to Solve Tower of Hanoi (Be Explicit):

The 4-Disc Problem was Hard? (1=Strongly Agree 7=Strongly Disagree)
1 2 3 4 5 6 7

The 5-Disc Problem was Hard? (1=Strongly Agree 7=Strongly Disagree)
1 2 3 4 5 6 7

References

1. Bethel, C.L., Salomon, K., Murphy, R.R., Burke, J.L.: Survey of psychophysiology measurements applied to human-robot interaction. In: The 16th IEEE International Symposium on Robot and Human interactive Communication, RO-MAN 2007, pp. 732–737. IEEE (2007)
2. Hagelbäck, J., Hilborn, O., Jerčić, P., Johansson, S.J., Lindley, C.A., Svensson, J., Wen, W.: Psychophysiological Interaction and Empathic Cognition for Human-Robot Cooperative Work (PsyIntEC), pp. 283–299. Springer, Cham (2014)
3. Liu, Y., Sourina, O., Nguyen, M.K.: Real-time EEG-based human emotion recognition and visualization. In: International Conference on Cyberworlds (CW), pp. 262–269, October 2010
4. Lin, Y.-P., Wang, C.-H., Wu, T.-L., Jeng, S.-K., Chen, J.-H.: EEG-based emotion recognition in music listening: a comparison of schemes for multiclass support vector machine. In: International Conference on Acoustics, Speech and Signal Processing, ICASSP 2009, pp. 489–492. IEEE (2009)
5. Bos, D.O.: EEG-based emotion recognition. Influ. Vis. Audit. Stimuli **56**(3), 1–17 (2006)
6. Horlings, R., Datcu, D., Rothkrantz, L.J.: Emotion recognition using brain activity. In: Proceedings of the 9th International Conference on Computer Systems and Technologies and Workshop for Ph.D. Students in Computing, p. 6. ACM (2008)
7. Murugappan, M., Rizon, M., Nagarajan, R., Yaacob, S., Zunaidi, I., Hazry, D.: Lifting scheme for human emotion recognition using EEG. In: International Symposium on Information Technology, ITSim 2008, vol. 2, pp. 1–7. IEEE (2008)
8. Schaaff, K.: EEG-based emotion recognition. Ph.D. dissertation, Ph.D. thesis, Universitat Karlsruhe (TH) (2008)

9. Li, M., Chai, Q., Kaixiang, T., Wahab, A., Abut, H.: EEG emotion recognition system. In: In-vehicle Corpus and Signal Processing for Driver Behavior, pp. 125–135. Springer, Boston (2009)
10. Fedele, P., Gioia, M., Giannini, F., Rufa, A.: Results of a 3 year study of a BCI-based communicator for patients with severe disabilities (2016)
11. Sørensen, A.S., Nielsen, J., Maagaard, J., Nielsen, J.L., Rasmussen, G., Day, D.: Natural kinesthetic interaction and social relations between training-robots and their users. In: Workshop on Advances and Challenges on the Development, Testing and Assessment of Assistive and Rehabilitation Robots: Experiences from Engineering and Human Science Research, ICRA 2017, vol. 1, p. 44 (2017)
12. Fisher, R.A.: The use of multiple measurements in taxonomic problems. Ann. Eugen. **7**(2), 179–188 (1936)
13. Efron, B., Hastie, T., Johnstone, I., Tibshirani, R.: Least angle regression. Ann. Stat. **32**(2), 407–499 (2004)
14. Jaakkola, T., Jordan, M.: A variational approach to bayesian logistic regression models and their extensions. In: Sixth International Workshop on Artificial Intelligence and Statistics. Citeseer (1997)
15. Wilson, G.F., Russell, C., Monnin, J., Estepp, J., Christensen, J.: How does day-to-day variability in psychophysiological data affect classifier accuracy? In: Proceedings of the Human Factors and Ergonomics Society Annual Meeting, vol. 54(3), pp. 264–268 (2010). SAGE Publications
16. Millan, J.R., Renkens, F., Mouriño, J., Gerstner, W.: Noninvasive brain-actuated control of a mobile robot by human EEG. IEEE Trans. Biomed. Eng. **51**(6), 1026–1033 (2004)
17. Picard, R.W., Vyzas, E., Healey, J.: Toward machine emotional intelligence: analysis of affective physiological state. IEEE Trans. Pattern Anal. Mach. Intell. **23**(10), 1175–1191 (2001)
18. Nasoz, F., Alvarez, K., Lisetti, C.L., Finkelstein, N.: Emotion recognition from physiological signals using wireless sensors for presence technologies. Cogn., Technol. Work. **6**(1), 4–14 (2004)
19. Conati, C.: Probabilistic assessment of user's emotions in educational games. Appl. Artif. Intell. **16**(7–8), 555–575 (2002)
20. Wilson, G.F., Russell, C.A.: Real-time assessment of mental workload using psychophysiological measures and artificial neural networks. Hum. Factors: J. Hum. Factors Ergon. Soc. **45**(4), 635–644 (2003)
21. Rani, P., Liu, C., Sarkar, N., Vanman, E.: An empirical study of machine learning techniques for affect recognition in human-robot interaction. Pattern Anal. Appl. **9**(1), 58–69 (2006)
22. Lourens, S., Zhang, Y., Long, J.D., Paulsen, J.S.: Analysis of longitudinal censored semicontinuous data with application to the study of executive dysfunction: the towers task. Stat. Methods Med. Res. (2014). https://doi.org/10.1177/0962280214560187
23. Zillmer, E., Culbertson, W.C.: Tower of London, Drexel University(TOLDX). Multi-Health System, Chicago, IL (2001)
24. Kirk, U.K.M., Kemp, S.: NEPSY a Development Neuropsychological Assessment Subtest Administration. The Psychological Corporation (1998)
25. Sahakian, B.J., Morris, R.G., Evenden, J.L., Heald, A., Levy, R., Philpot, M., Robbins, T.W.: A comparative study of visuospatial memory and learning in Alzheimer-type dementia and Parkinson's disease. Brain **111**(3), 695–718 (1988)

26. Ruiz-Díaz, M., Hernández-González, M., Guevara, M.A., Amezcua, C., Ågmo, A.: Prefrontal EEG correlation during tower of Hanoi and WCST performance: effect of emotional visual stimuli. J. Sex. Med. **9**(10), 2631–2640 (2012)
27. Sohaib, A.T., Qureshi, S., Hagelbäck, J., Hilborn, O., Jerčić, P.: Evaluating classifiers for emotion recognition using EEG. In: Foundations of Augmented Cognition, pp. 492–501. Springer, Heidelberg (2013)
28. Chen, G., Hou, R.: A new machine double-layer learning method and its application in non-linear time series forecasting. In: International Conference on Mechatronics and Automation, pp. 795–799. IEEE (2007)
29. Vapnik, V.N.: An overview of statistical learning theory. IEEE Trans. Neural Netw. **10**(5), 988–999 (1999)
30. Burges, C.J.: A tutorial on support vector machines for pattern recognition. Data Min. Knowl. Discov. **2**(2), 121–167 (1998)
31. Sasaki, M., et al.: EEG data classification with several mental tasks. In: IEEE International Conference on Systems, Man and Cybernetics, vol. 6, p. 4. IEEE (2002)
32. Breiman, L.: Classification and regression trees (1984)
33. Downey, S., Russell, M.: A decision tree approach to task-independent speech recognition. In: Proceedings-Institute of Acoustics. vol. 14, p. 181 (1992)
34. Brown, L.E., Tsamardinos, I., Aliferis, C.F.: A novel algorithm for scalable and accurate Bayesian network learning. Med. Info. **11**(Pt 1), 711–715 (2004)
35. Ben-Gal, I.: Bayesian networks. Encyclopedia of Statistics in Quality and Reliability (2007)
36. Friedman, N., Geiger, D., Goldszmidt, M.: Bayesian network classifiers. Mach. Learn. **29**(2–3), 131–163 (1997)
37. Parvin, H., Alizadeh, H., Minaei-Bidgoli, B.: MKNN: Modified K-Nearest Neighbor. In: Proceedings of the World Congress on Engineering and Computer Science, pp. 831–834. Citeseer (2008)
38. Dasarathy, B.V.: Nearest neighbor {NN} norms: {NN} pattern classification techniques (1991)
39. Beyer, K., Goldstein, J., Ramakrishnan, R., Shaft, U.: When is nearest neighbor meaningful? In: International Conference on Database Theory, pp. 217–235. Springer, Heidelberg (1999)
40. Mauss, I.B., Robinson, M.D.: Measures of emotion: a review. Cogn. Emot. **23**(2), 209–237 (2009)
41. Plutchik, R.: Emotions and Life: Perspectives from Psychology, Biology, and Evolution. American Psychological Association (2003)
42. Russell, J.A.: Affective space is bipolar. J. Pers. Soc. Psychol. **37**(3), 345 (1979)
43. Chanel, G.: Emotion assessment for affective computing based on brain and peripheral signals. Ph.D. dissertation, University of Geneva (2009)
44. Wu, Y., Ianakiev, K., Govindaraju, V.: Improved k-nearest neighbor classification. Pattern Recognit. **35**(10), 2311–2318 (2002)
45. Speybroeck, N.: Classification and regression trees. Int. J. Public Health **57**(1), 243–246 (2012)
46. Nykopp, T.: Statistical modelling issues for the adaptive brain interface (2001)
47. Posner, J., Russell, J.A., Peterson, B.S.: The circumplex model of affect: an integrative approach to affective neuroscience, cognitive development, and psychopathology. Dev. Psychopathol. **17**(03), 715–734 (2005)
48. Simon, H.A.: The functional equivalence of problem solving skills. Cogn. Psychol. **7**(2), 268–288 (1975)

49. Goel, V., Grafman, J.: Are the frontal lobes implicated in planning functions? Interpreting data from the Tower of Hanoi. Neuropsychology **33**(5), 623–642 (1995)
50. Egan, D.E., Greeno, J.G.: Theory of rule induction: knowledge acquired in concept learning, serial pattern learning, and problem solving (1974)
51. Hayes, J.R., Simon, H.A.: Understanding written problem instructions (1974)

Investigating Input Protocols, Image Analysis, and Machine Learning Methods for an Intelligent Identification System of *Fusarium Oxysporum Sp.* in Soil Samples

Andrei D. Coronel(✉), Maria Regina E. Estuar, and Marlene M. De Leon

Department of Information Systems and Computer Science, Ateneo de Manila University, Quezon City, Philippines
acoronel@ateneo.edu

Abstract. The export of Cavendish cultivars making up one third of the Philippine Banana Export Industry is threatened by the rising concern over the increasing presence of Fusarium wilt in plantations. As Cavendish is susceptible to the infection of the fungus *Fusarium oxysporum sp. cubense*, Tropical Race 4 (Foc TR4), there is a need to develop an early detection mechanism whereby farmers can determine whether the soil is susceptible to the fungi and whether planting Cavendish cultivars in the area should be avoided. Developed in the Philippines, CITAS is a cloud based intelligent total analysis system that uses wireless soil sensor networks and mobile microscopy in determining the presence or absence of Foc TR4. The study implemented a two step approach in the development of an intelligent detection system, specifically using image analysis in magnified soil samples for identification and a machine learning approach for intelligent classification and modeling. The results of the study also served to guide the development of the appropriate soil sampling protocol that would be used for the mobile microscope, as well as the design of the mobile microscope itself. Experiments involve shape detection methods on variable-sized image inputs alongside machine learning techniques such as Artificial Neural Networks (ANN), Convolutional Neural Networks (CNN, a specialized type of ANN), and Support Vector Machines (SVM). The best results of the study with regards to classification accuracy is an 82.23% average after cross-fold validation, produced by ANN on 32×32 pixel image inputs prepared under phase contrast microscopy. However, considering the subsequent implementation of the system on a mobile platform, favored metric is therefore the fastest processing time, yielded by SVM at 8.09 s under similar specimen parameters, with an acceptable accuracy of 81.12%. It is conclusive that a stained soil preparation protocol paired with 100x magnification produces a viable input for a shape-recognition based image analysis technique. The study is an important step towards a multi-parameter approach in the early detection of Foc TR4 infection, thus potentially changing farmer behavior from reactive practices to preventive measures in the context of Fusarium Wilt.

K. Arai et al. (Eds.): IntelliSys 2018, AISC 868, pp. 364–381, 2019.
https://doi.org/10.1007/978-3-030-01054-6_26

Keywords: Image analysis · Machine learning
Smart farming · Wireless sensor networks · Mobile microscopy

1 Introduction

The Philippine export economy continues to depend on the production of Cavendish bananas because it has been generating revenues, as well as a large number of jobs for the Philippine market. Cavendish growers contributed 37.3% of this, equivalent to approximately USD1 Billion to the total amount generated by the agricultural, forestry and fishing sectors. Cavendish bananas are one of the highest valued products as the country remains to be the 2nd largest producers of bananas worldwide, second to Ecuador, providing for 95% of the demand from the Asian market [12]. With much of the economy and livelihood dependent on this industry, there is a rising concern over the spread of Panama disease, also known as Fusarium Wilt, with which if not mitigated may eventually wipe away the Cavendish industry [13].

1.1 CITAS Project

Fusarium Wilt, when detected by farmers through visual inspection of the leaf, indicates that fungi has spread in most areas of plant. In these instances, the plant and surrounding plants are eradicated leaving the soil implantable for the next 10 to 15 years. In developing countries like the Philippines, the objective is to provide a simple, accessible, and affordable solution, such as providing farmers with a smart farm kit. The kit should be designed such that it makes use of a mobile phone to collect image data directly from plants or from soil samples automatically magnified in a mobile microscope. Software should be developed that will allow for the early detection of the disease at the level of the soil, hopefully before it spreads to the plant. CITAS or Cloud-based Intelligent Total Analysis System, is a web and mobile based system that is used to monitor soil quality and soil characteristics of banana farms. CITAS displays raw and analyzed data from soil sensors as well as image data taken from the mobile microscope. The soil sensors transmit near real-time status of the following parameters: soil temperature, pH, conductivity, amount of light, soil moisture and air temperature. The plantscope is an additional tool that can be used to capture microscopic images of soil samples for further image analysis. The end goal of CITAS is to provide a low cost smart farm monitoring toolkit as that will be able to capture, model and forecast the effects of Fusarium Wilt or *Fusarium oxysporum cubense*, TR4 (Foc TR4) on Cavendish bananas cultivated in the Philippines through image analysis and modeling techniques.

1.2 *Fusarium Oxysporum* Fungus

The *cubense* variant belongs to the fungal species *Fusarium oxysporum* (Fo). The morphological features of the fungal forms are recognizable when scrutinized through trained microscopy. The fungal forms are less than 25 μm long, however, the detection of their presence is key in the classification of soil samples with regards to Foc infection. Figures 1, 2, and 3 exhibit the morphological features to be recognized by the image analysis methods, and subsequent machine learning steps [17].

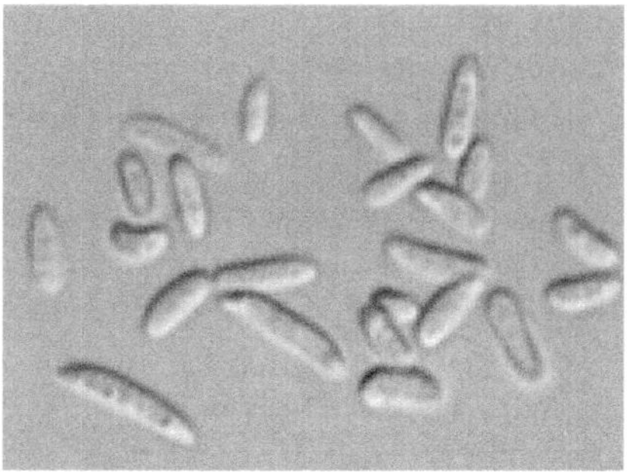

Fig. 1. Foc: Microconidia, from the Fusarium Laboratory Manual, [17]

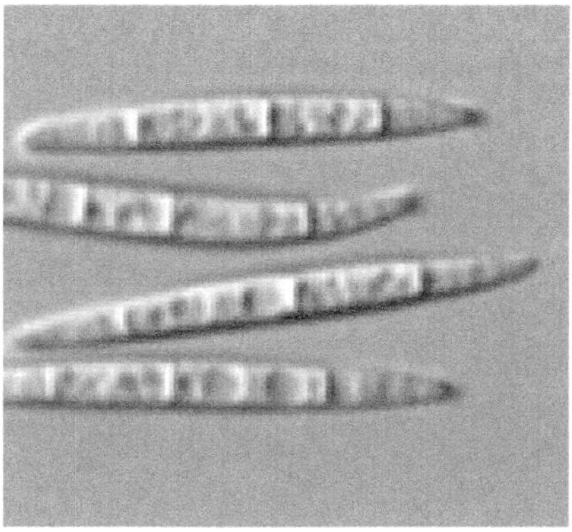

Fig. 2. Foc: Macroconidia, from the Fusarium Laboratory Manual, [17]

1.3 Image Analysis and Machine Learning Within the CITAS Project

Though there are many approaches in the development of an early detection system, CITAS maximizes on image analysis to detect unique characteristics of Foc in magnified soil samples and trains a model for intelligent detection within a mobile application. The prepared soil samples are magnified by a mobile

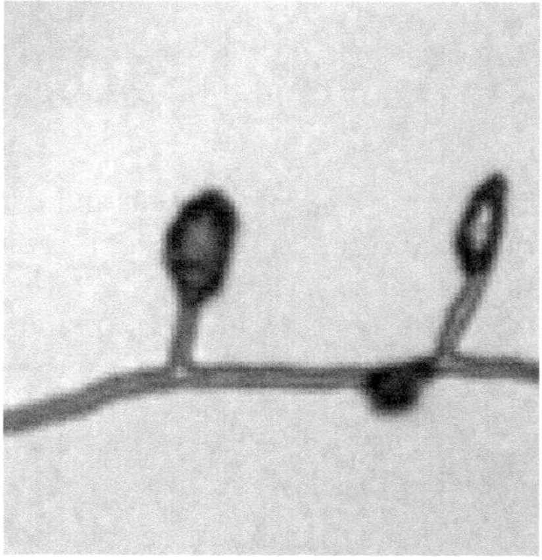

Fig. 3. Foc: Chlamydospores, from the Fusarium Laboratory Manual, [17]

plantscope designed and developed in partnership with the Bioengineering Laboratory of the University of California, Berkeley.

The automated detection is facilitated by image analysis techniques paired with an effective machine learning method to approximate high classification accuracy in categorizing whether or not the soil sample in question is infected with Foc. Given that *Fusarium oxysporum cubense* has distinct morphological features in its multiple fungal forms (i.e. microconidia form, and chlamydospore form, both less than 25 μm), a shape recognition approach to image analysis is ideal, extracting these values alongside intrinsic image features. The extracted values are subjected to classification by machine learning, referencing any new instances with a built classification model trained on a dataset of images acquired and prepared from the field.

1.4 Objectives

The main objective of this study is to be able to recommended the most appropriate machine learning method that can be paired with a shape-detection based image analysis that would yield an acceptable classification accuracy in the context of identifying Foc infection in soil samples. At the same time, the identified methods should be feasibly implementable on a mobile platform, thus resulting in an application-based automated Foc detection module within the CITAS Project. Corollary objectives include fine-tuning soil preparation protocols and directing the design of the mobile microscope that serves to magnify the soil samples prior to image capturing.

2 Related Literature

2.1 Role of Information and Communications Technology in Agriculture

The past decade witnessed the use of information and communications technology (ICT) in the field of agriculture. The application is more commonly referred to as precision agriculture or smart farming. Several studies have demonstrated the use of different ICT components to facilitate gathering, monitoring, and analysis of agricultural data. There are studies that have recognized the importance of communication and interoperability among various components that makes up precision agriculture systems [1–5]. Precision agriculture systems empower farmers by providing tools that will detect anomalies or emergencies concerning produce of crops [5]. CITAS aims to provide the same tool to ordinary farmers, specifically in developing countries, by using data collected from sensors and microscope to create a dashboard that will allows for the monitoring of soil and crops. The system report anomalies such as when soil parameters go beyond normal range or when image detects shapes that can be classified possibly infected.

2.2 Smartphone Applications for Collection of Agriculture Data

There are smartphone applications that exploit the capabilities of embedded sensors to facilitate automatic collection, transmission, processing, and storage of data for precision agriculture. An example of these applications include an energy efficient mobile vision system that capture and process images of suspected diseased plant leaves. Processed plant images that only shows the diseased parts of the plant are sent to laboratory experts for disease identification. This application facilitates a less expensive process for disease detection in plants [6]. Another example, the BaiKhaoNK smartphone application uses mobile vision to estimate chlorophyll in rice leaves. Depending on the color of rice leaves, the smartphone application recommends intervention activities such as the application of a specific amount of nitrogen fertilizers to the rice fields to ensure the crops maintain its optimum health [7]. Accuracy and precision of solutions like these highly depend on a large volume of data set that is fed into a learning model. Moreover, efficiency is a necessary component especially when the solution is deployed in a mobile device.

2.3 Image Analysis and Classification Modeling Methods

Image analysis and machine learning techniques have been used in a multitude of applications for the life sciences. Recommended machine learning methods such as Convolutional Neural Networks (CNN) have been used in intricate investigations involving images, such as the segmentation of mitochondria in electron microscopy [14], and fiber image classification [15]. These methods have also found their way in the domain of agriculture, such as artificial neural network-based image analysis for the evaluation of quality attributes of agricultural produce [16]. Recent breakthroughs in the use of computer and mobile vision to remotely view agricultural phenomena and capture its images for analysis and monitoring using machine learning has been gaining popularity [8]. To demonstrate, accurate and novel solutions to long-standing challenges in agricultural land planning [9], crop yield estimation [11], and crop yield gap analysis [10] were developed using Artificial Neural Networks (ANN). In line with the aforementioned studies, this research is an investigation on the innovative use of ICT for agriculture, leveraging technologies to implement a mobile solution for the detection of an agricultural disease that is anchored on image analysis and machine learning.

3 Methodology

3.1 Methodology in the Context of CITAS

The CITAS Project involves multiple components for both web and mobile platforms, with the goal of providing a smart farm monitoring toolkit for the prevention of Fusarium Wilt on cavendish bananas cultivated in the Philippines. Both web and mobile components of the CITAS system connect to a cloud-based server, that not only serves as the database backend for the system, but also as the facility for the disease modeling of Fusarium Wilt, as seen in Fig. 4 which illustrates the goal of the CITAS project in general, involving the incorporation of a mobile application with automated Foc detection features (through image analysis and machine learning) within its cloud-based mobile and web framework.

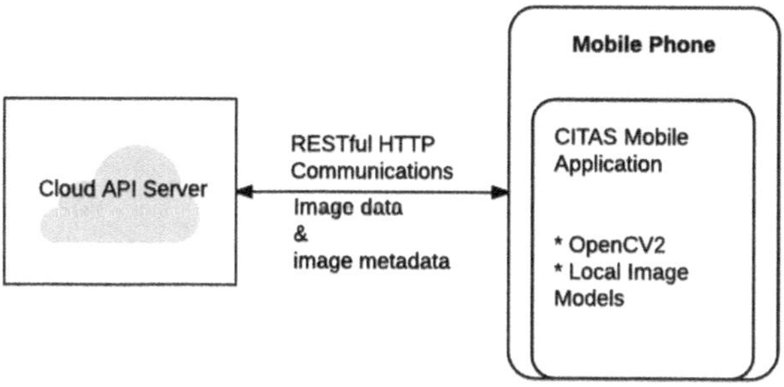

Fig. 4. CITAS overview 7

The modeling of the fungal disease of banana plants is based on the different parameters acquired through the soil sensors of the CITAS system, as well as images magnified by the plantscope, which is a mobile microscope that is paired with a mobile application that is able to classify the infection status of the prepared soil samples. The classification of soil samples implemented at the level of the mobile application of the plantscope was developed through several experiments involving image analysis and machine learning, which is the specific focus of this particular study.

3.2 Methodology for Image Analysis and Machine Learning

(1) Non-mobile and Mobile Platforms of Implementation. The software development methodology involves two major steps with regards to platform implementation. Figure 5 illustrates the initial focus with experiments using a non-mobile platform, the results of which serve to narrow down the technology and algorithm options that can be implemented in a mobile platform (i.e., smartphone). Image analysis and machine learning experiments are done on a non-mobile platform, prior to any mobile implementation. The non-mobile environment allows the identification of a working methodology that can generate results to serve as a reference performance benchmark for the subsequent mobile device implementation.

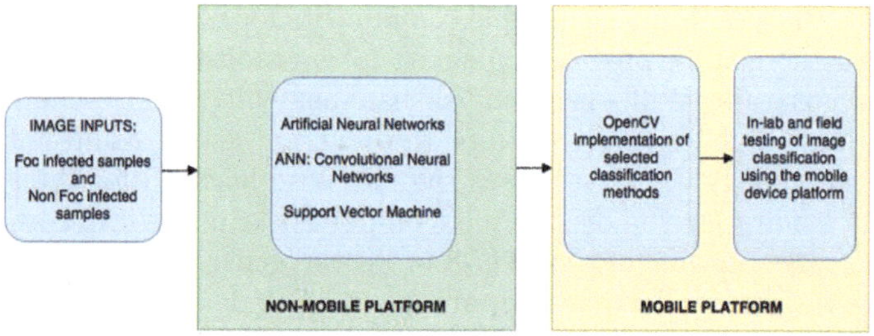

Fig. 5. Non-mobile and mobile platforms involved in the CITAS methodology

Non-mobile platform hardware resources allow for the use of a variety of machine learning and image analysis software technologies, as well as provide a high internal data width and processing throughput. The specifications of the platform consists of processing power up to 2.7 GHz, working on 8 gigabytes of internal memory. This allows the identification of a working methodology that can generate results to serve as a reference performance benchmark for the subsequent mobile device implementation.

(2) Sample Preparation and Dataset Building. The first step prior to executing any implementation level machine learning classification is to build the model for Foc infection based on image analysis parameters. To achieve this, a dataset of images consisting of infected and non-infected soil samples was acquired. Five hundred images with a 50% split between Foc presence and absence was acquired by capturing images of sampled infected and non-infected soil. The initial images were captured using a confocal microscope using both brightfield and phase contrast modalities. Table 1 provides a partial list of the specimens involved in the dataset.

Images are prepared in batches. Two contrast modalities where used, namely: brightfield and phase contrast. Four methods of sample preparation were used, namely: CW stain, CW stain with KOH, Giemsa stain, Write stain and the

Table 1. Partial listing of the dataset

Image batch	Contrast modality	Stain
1	Brightfield	No stain
2	Phase contrast	No stain
3	Brightfield	CW stain
4	Brightfield	CW stain + KOH
5	Phase contrast	CW stain
6	Phase contrast	CW stain + KOH
7	Brightfield	Giemsa stain
8	Phase contrast	Giemsa stain
9	Brightfield	Wright stain
10	Phase contrast	Wright stain

absence of staining. These variations are important during the initial stages of data modeling to determine the best sampling protocol that will be recommended to the farmers during actual implementation.

Though there are a number of histological/biological dyes and stains for microscopy, the main criteria for choosing the staining methods for this study is cost, ease of the staining procedure, and applicability to fungi. The staining method must not only be applicable to Foc, but performing the staining procedure on the field with minimal training is a strong consideration. The end-users of the CITAS toolkit may not necessarily be adept with biological staining, inclusive of the proper handling of the equipment related to the staining procedure. In line with this, staining procedures that require multiple steps were no longer considered. Finally, since one of the secondary goals of the project is to produce a cost-effective toolkit, staining procedures that would require expensive dyes and paraphernalia were also no longer considered.

(3) Image Analysis and Machine Learning Experiments. Prior to machine learning classification, the initial steps involve the manual acquisition of the sample images, and then the extraction of image attributes for image analysis. To achieve this, the steps are as follows:

- Sample preparation using a proper slide preparation procedure.
- Application of staining protocol.
- Mounting the specimen on the microscope for proper magnification.
- Capturing the field of view, resulting in a 512×512 pixel image.
- Extracting intrinsic image attributes and shape-recognition data with Python and R.
- Organizing the extracted data as input for machine learning experiments.

The items on Table 1 were magnified at 100x using a confocal laboratory microscope to match the magnification constraints of the mobile microscope's (i.e. plantscope) physical design. The dimensions of each captured image is 512×512 pixels, and each captured image was treated as a single instance of input which was consequently subjected to image-feature extraction, inclusive of shape and intrinsic image features. Thus, the dataset for machine learning experiments consisted of image attribute values per instance, where an instance represents a captured image (Fig. 6). As seen in the figure, the specific methodology for image analysis and machine learning involves input images of magnified soil samples (infected with Foc and samples without Foc infection) that were subjected to image attribute extraction (Fig. 7), where the values were subjected to classification methods, (i.e. CNN, ANN, and SVM).

There was also a need to normalize the intensity levels of the acquire image input so as to avoid errors caused by classifications that were based on image brightness levels. Subsequent trials progressed by altering the size of image inputs and reverting to the use of microregions, as opposed to using the entire 512×512 region as input. From the original 512×512 input region, the dimensions of the microregions in the progressive experiments were 64×64 pixel dimensions per image input, and 32×32 pixel dimensions per input. Each microregion approximates the area covered by Foc-specific artifacts within the specimen.

The methodology considers the positively identified presence of Foc from the input images as the indicator for categorizing the specific banana plantation area within the farm as infected, regardless of the microregion input size. This therefore means that there is no need to reconsolidate the microregions (i.e. 32×32 pixel dimensions per input image) back to the original 512×512 image input size after subjecting the microregion-based dataset to image analysis and machine learning. Once Foc is identified in any of the input images, the source of the image (i.e. soil site) is considered infected. Further studies could consider the reconsolidation of the inputs for computing different levels of Foc infection based on specimen concentration, however this current study follows the biological protocol of considering an area as infected based on fungal presence.

There is a need to compare the performance of machine learning methods based on classification accuracy. The methods in question are Convolutional Neural Networks (CNN), regular Artificial Neural Networks (ANN), and Support Vector Machine (SVM) classification. Though all methods are proven to be robust classifiers, this study investigated which method is ideal for mobile platform implementation. In so doing, the considerations include, aside from classification accuracy, the processing time, and the computational demands of the inherent algorithms involved.

Convolutional Neural Networks, regular Artificial Neural Networks, and Support Vector Machine classification experiments were implemented across a progressively reduced image input size (Figs. 8 and 9), taking note of the average classification accuracy in the context of cross-validation, as well as the turnaround time of computational processing.

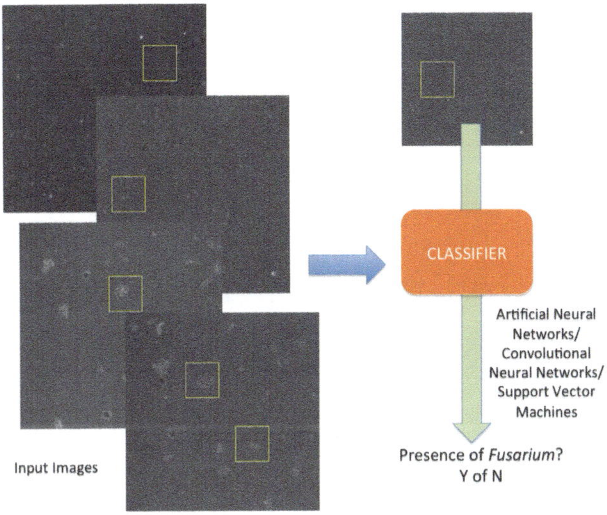

Fig. 6. Machine learning classification applied to a dataset of extracted image feature values per input image, for all images

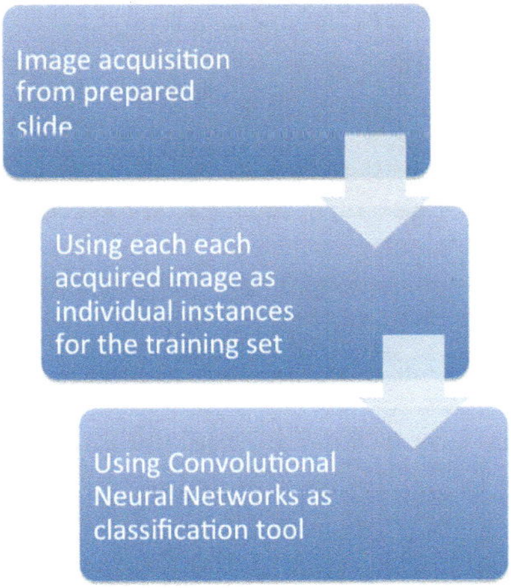

Fig. 7. Initial experiments treat each 512×512 confocal captured image as a single image input in the context of dataset building for machine learning methods

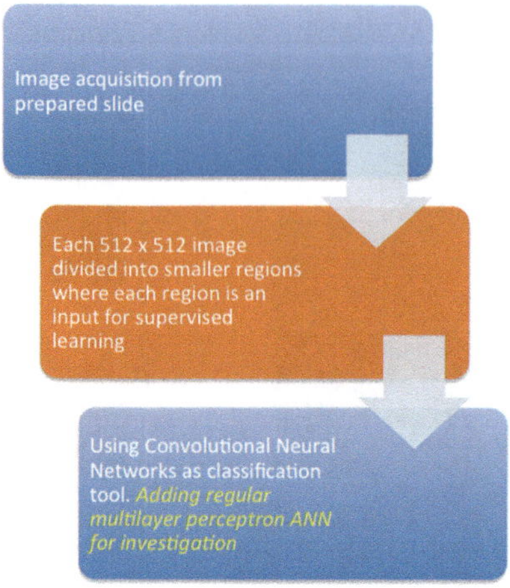

Fig. 8. Subsequent experiments were performed with progressively smaller image input samples (i.e. 64×64, and 32×32)

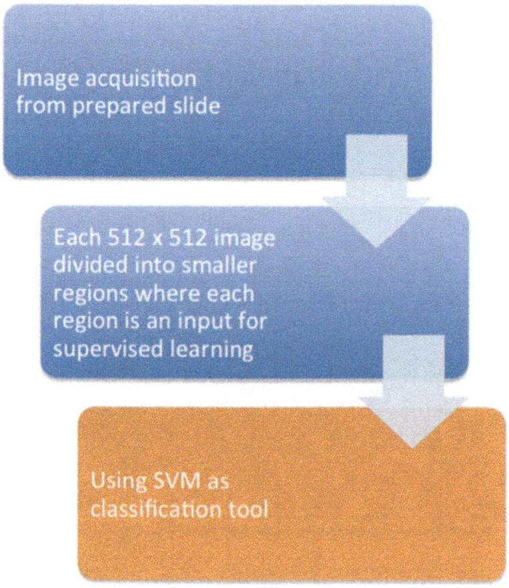

Fig. 9. All experiments performed with CNN and ANN were repeated but with SVM as the classification method

Given the above discussion regarding image analysis and machine learning procedures, the following parameters were considered in the experimental design:

- Size of input image.
- Magnification.
- Contrast modality used in microscopy.
- Sample preparation.
- Image attributes.
- Machine learning method.

3.3 Validation of Laboratory Results

There is a need to validate laboratory-based results with actual field sampling implementation. Figure 10 illustrates parallels between the controlled implementation and the field procedures. The figure shows that the development of the mobile implementation of image analysis and machine learning methods for the automated classification of soil infection regarding Foc is primarily dependent on the source of the image and its preparation prior to analysis. Before any field testing, laboratory prepared infected soil samples (i.e. spiked with Foc) were used to represent field specimens that are positive with Foc. These prepared samples were initially magnified with a confocal microscope. Validation of these procedures occur when the classification results at the level of the mobile application are compared with procedures that employ actual field acquired samples, as opposed to the laboratory simulated samples. Likewise with magnification, validation occurs in the comparison of images produced when magnification is done by the actual plantscope, as opposed to a confocal microscope, using the classification results as benchmark as well.

Classification results for both implementations (lab and field) are done with the same plantscope mobile application. However, it is expected that field testing will reveal variances in the results due to the natural reduction in the constraints

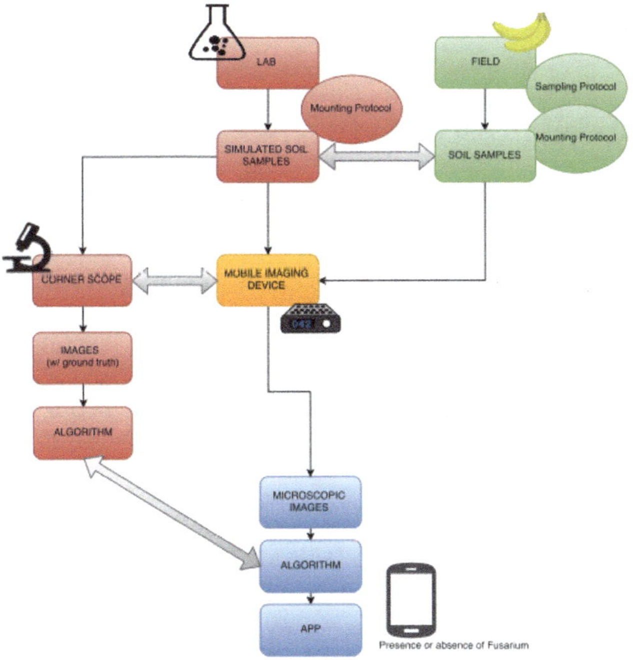

Fig. 10. Laboratory procedures and field implementation: variable in data source and magnification method

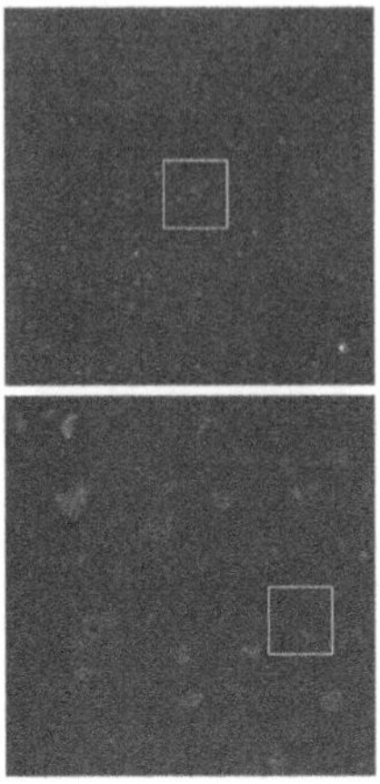

Fig. 11. Examples of 512 × 512 input images taken with a confocal microscope in phase contrast.The top example exhibits the *microconidia* form of Foc, while the bottom example exhibits the *chlamydospore* form of Foc. Both these images classify as infected with Foc

regarding controlled environments. The fine-tuning of results is based on the compensatory adjustments made between the differences garnered by the image source (lab vs. field), and magnification device (confocal microcscope vs. actual plantscope).

4 Results and Discussion

4.1 Results Affected by Image Input Size and Machine Learning Method

The original hypothesis with regards to image input specifications was that a single captured image (512 × 512 pixel dimensions) would correspond to a single instance for image feature extraction. Hence, 100 images inputs would refer to 100 instances of 512 × 512 pixel images (Fig. 11).

The results of machine learning experiments after averaging a cross-fold validation for this set of images are shown in Table 2.

Table 2. Classification accuracy for 512 × 512 pixel image inputs

Machine learning method	Processing time	Average accuracy
Artificial neural networks	201.20 min	45.22%
ANN: convolutional neural networks	243.55 min	42.65%
Support vector machines	70.04 min	48.10%

As seen in the table, the classification results for ANN, CNN, and SVM were 45.22%, 42.65%, and 48.10% respectively, with processing times of 3 h and 21 min (201.20 s), 4 h and 3 min (243.55 s), and 1 h and 10 min (70.04 s). This processing time involves both training and testing time in the context of tenfold cross-validation.

The unsatisfactory results in the previous table could be accounted for by the image input size of 512×512 pixel dimensions. Given that all three forms of Foc are less than 25 µm in length, the resulting feature values in a dataset based on image inputs of this size are not ideal for a training-set in the context of Foc detection. A consistent distribution of Foc specimens across the field of view could improve the results, however, controlling the sample preparation protocol to guarantee an even Foc distribution is more difficult than adjusting the machine learning experiment parameters.

The highest result was rendered by SVM classification at 48.10%, however results below 50% are not ideal for any binary classification procedure. The lead that SVM classification has rendered in this experiment may be accounted for by the training constraints of both neural networks (ANN, and CNN), especially with input images that capture multiple Foc artifacts within a single field of view.

Given the relatively large size of the image inputs, image analysis and machine learning processing with a 2.7 GHz processor supported by 8GB RAM reached a ceiling of 243.55 min (4.06 h) for ANN, and the fastest implementation with SVM at 70.04 min (1.17 h). Both accuracy results and processing time are not ideal for a mobile implementation aiming to approximate real time detection of Foc in field samples.

The above results prompted to pursue the workflow with progressively smaller images, approximating the isolation of the Foc artifacts (i.e. microconidia or chlamydospore form) within a single image input or field of vision.

Tables 3 and 4 show the results of both processing time and classification accuracy of experiments involving 64×64 pixel image inputs and 32×32 pixel image inputs, respectively.

Table 3. Classification accuracy for 64×64 pixel image inputs

Machine learning method	Processing time	Average accuracy
Artificial neural networks	45.46 min	68.83%
ANN: convolutional neural networks	45.46 min	65.25%
Support vector machines	0.258 min (8.09 s)	58.16%

The results of the experiments using the 64×64 pixel sized image inputs resulted in 68.8%, 62.25%, and 58.16% classification accuracy for ANN, CNN, and SVM as well. Likewise, the processing times were observed as 33.43 min, 45.46 min, and 15 s.

Table 4. Classification accuracy for 32 × 32 pixel image inputs

Machine learning method	Processing time	Average accuracy
Artificial neural networks	12.15 min	82.23%
ANN: convolutional neural networks	22.42 min	72.26%
Support vector machines	0.258 min (8.09 s)	81.12%

The experiments with the smallest image inputs at 32 × 32 pixel dimensions rendered results of 82.23%, 72.76%, and 81.12% respectively, with processing times of 12.15 min, 22.42 min, and 8.09 s in the same order.

The highest average classification accuracy after cross-fold validation was rendered by ANN at 82.23%, with a processing time of 12.15 min using the same controlled hardware environment as the rest of the experiments. This a remarkable increase in classification performance relative results in Tables 2 and 3. However, the length of processing is still not acceptable for approximate real-time Foc identification, given a 22.42 min turnaround time on non-mobile platform. The move towards a mobile platform for the deployable version of this application will definitely suffer from a reduction of computational resources and internal memory, thus potentially increasing the processing time.

Given this, the 81.12% classification accuracy rendered by SVM classification is more appropriate for incorporation into the mobile platform, with a processing time of 0.258 min (8.09 s).

4.2 Incorporating Field Acquired Images

Based on the validation methodology illustrated in Fig. 10, the next steps to be taken leading to the progressive improvement of the classification model and mobile platform implementation is to incorporate field-acquired samples on-site (i.e. soil samples from banana plantations in the Philippines), as illustrated in Fig. 12.

The results of the field testing will serve as feedback to the adjustments of several or all of the following parameters:

- Input image size (i.e. potential resizing.)
- Machine learning parameters.
- Sampling protocol (i.e. fine-tuning current methods.)
- Prioritization of image attributes.

As a summary regarding this investigation of protocols and methods, the results prove that smaller regions (denoted by smaller pixel dimensions) that approximately isolate the specimen artifacts in question result in an increased classification accuracy. It was intuitive to expect that CNN would perform as the relative best classifier, however regular ANN was able garner a better performance. Considering subsequent mobile platform implementation, these Neural Network based classifiers would not be ideal due to the length of processing.

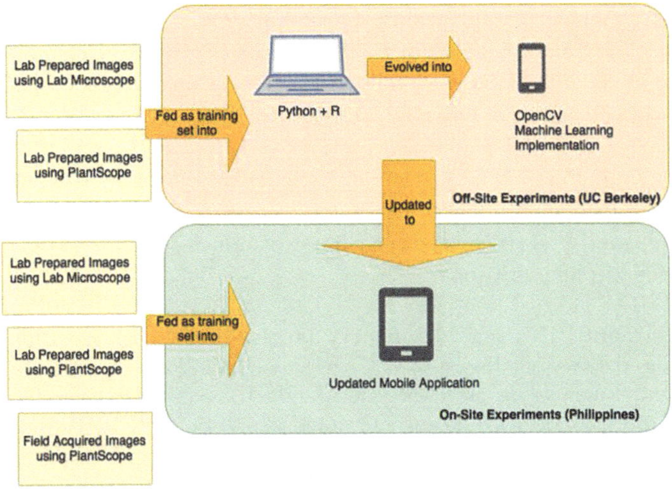

Fig. 12. Field validation for feedback and fine-tuning

The results suggest that in the context of this investigation, support vector machine classification is appropriate for mobile implementation, provided that the image sample preparation protocols as well as the image input size specifications discussed are retained.

5 Conclusion

The CITAS Project aims to provide a smart toolkit for banana farm owners and farmers alike with regards to the detection, and consequently, the prevention of the spread of *Fusarium oxysporum cubense*, or Foc, Tropical Race 4, the fungus responsible for the Fusarium wilt disease. One of the main goals of this project is to model this disease. The system therefore is reliant in its data sources, such as soil sensors and captured images. This study was able to investigate the image aspect, particularly, the microscopic component of the project, where soil samples were subjected to a methodically designed preparation protocol prior to the magnification with a mobile microscope, referred to as the plantscope. The mobile application that accompanies the plantscope is responsible for the automated detection of Foc in the samples. The implementation of the classification system within the mobile app was based on the results of this study, where a shape-recognition based image analysis method was paired with Artificial Neural Networks, Convolutional Neural Networks (a subset of ANN), and Support Vector Machine classification. The results presented consequences that affected the soil preparation protocol, the image capturing and magnification method, and the machine learning algorithm of choice. The results have shown that the relative best results are rendered by using input images of smaller pixel dimensions, in this case, 32×32 pixels, approximating the isolation of the artifact in question. The highest classification accuracy after cross-fold validation was rendered by an ANN classification of these images, captured using phase contrast microscopy, at 82.23%. However, due to the processing time of 12.15 min on a non-mobile platform, it is more appropriate to implement SVM classification, since it returns a classification accuracy of 81.12%, processed within 8.09 s. It is always recommended that further investigations be performed with different technological options (i.e. alternate image libraries, TensorFlow, etc.), but the results of this study serves as a reference performance benchmark for subsequent mobile device implementations.

Acknowledgment. This multidisciplinary project, funded by the Philippine California Advanced Research Institute (PCARI), is a collaboration between the School of Science and Engineering of Ateneo de Manila University and the Bioengineering Department of UC Berkeley. The authors would like to thank PCARI, the Commission on Higher Education (CHED), and the Fletcher Laboratory of UC Berkeley for making this project possible.

References

1. Korduan, P., Bill, R., Blling, S.: An interoperable geodata infrastructure for precision agriculture. In: Proceedings of 7th AGILE Conference on Geographic Information Science, pp. 747–751 (2004)
2. Murakami, E., Saraiva, A.M., Ribeiro, L.C.M., Cugnasca, C.E., Hirakawa, A.R., Correa, P.L.P.: An infrastructure for the development of distributed service-oriented information systems for precision agriculture. Comput. Electron. Agric. **58**(1), 3748 (2007)
3. Nash, E., Korduan, P., Bill, R.: Applications of open geospatial web services in precision agriculture: a review. Precision Agric. **10**(6), 546560 (2009)
4. Nikkila, R., Seilonen, I., Koskinen, K.: Software architecture for farm management information systems in precision agriculture. Comput. Electron. Agric. **70**(2), 328336 (2010)
5. Chen, N., Zhang, X., Wang, C.: Integrated open geospatial web service enabled cyber-physical information infrastructure for precision agricultural monitoring. Comput. Electron. Agric. **111**(C), 78–91 (2015)
6. Presad, S., Peddoju, S.K., Ghost, D.: Energy efficient mobile vision system for plant leaf disease identification. In: Proceedings of IEEE Wireless Communication and Networking Conference, pp. 3314–3319 (2014)
7. Sumriddetchkajorn, S.: Mobile device-based optical instruments for agriculture. In: Proceedings of SPIE, vol. 8881 (2013)
8. LeCun, Y., Bengio, Y., Hinton, G.: Deep learning. Nature **521**, 436444 (2015)
9. Arkeman, F., Buono, Y., Hermadi, A.: Satellite image processing for precision agriculture and agroindustry using convolutional neural network and genetic algorithm. IOP Conf. Ser.: Earth Environ. Sci. **54**(1), 012102 (2017)
10. Lobell, D.: The use of satellite data for crop yield gap analysis. Field Crops Res. **143**, 56–64 (2013)
11. Payne, A., Walsh, K., Subedi, P., Jarvis, D.: Estimating mango crop yield using image analysis using fruit at stone hardening stage and night time imaging. Comput. Electron. Agric. **100**, 160–167 (2014)

12. Valencia, C.: Growing Cavendish bananas enable farmers to earn more, Philippine Star, 1 January 2017
13. Padin, M.G.: Disease could wipe out Cavendish bananasPBGEA. Business Mirror, 3 May 2016
14. Oztel, I., Yolcu, G., Ersoy, I., White, T., Bunyak, F.: Mitochondria segmentation in electron microscopy volumes using deep convolutional neural network. In: Proceedings of 2017 IEEE International Conference on Bioinformatics and Biomedicine (BIBM), pp. 1195–1200, November 2017
15. Wang, X., Chen, Z., Liu, G., Wan, Y.: Fiber image classification using convolutional neural networks. In: Proceedings of 2017 4th International Conference on Systems and Informatics (ICSAI) Systems and Informatics (ICSAI), pp. 1214–1218, November 2017
16. Rafiq, A., Makroo, H., Hazarika, M.: Artificial neural network-based image analysis for evaluation of quality attributes of agricultural produce. Journal of Food Processing and Preservation 40(5), 1010–1019 (2016)
17. Leslie, J., Summerell, B.: The Fusarium Laboratory Manual. Blackwell Publishing, Iowa (2006)

Intelligent System Design for Massive Collection and Recognition of Faces in Integrated Control Centres

Tae Woo Kim(✉), Hyung Heon Kim, Pyeong Kang Kim, and Yu Na Lee

Technology Laboratory, Innodep. Inc., Seoul, Korea
{davidkim,josephkim,david,yuna}@innodep.com

Abstract. Intelligent system design for face collection and recognition in integrated control center is presented. Generally, cameras installed for city surveillance monitors wide area. So, it is hard to recognize person by face with those cameras because of lack of image resolution. Our system adopts two cameras, one for normal surveillance and the other for zooming a target face. The system carries out motion detection for candidate of human and detects and recognizes faces based on Microsoft cognitive services. The system utilizes codec's metadata for motion detection. So, without decoding it can detect the area of motion. As a result, it can handle hundreds of cameras simultaneously in one server. It entrust MS cognitive services with face recognition. It can provide monitoring agents with functionality of searching video in terms of people, which is virtually meaningful to them.

Keywords: Component · Metadata parsing · Intelligent zooming
Cognitive services · Face recognition

1 Introduction

Worldwide threats of terrors and high requirements for safety have installed lots of CCTVs citywide. In Korea, integrated control centers have been built continuously in every city after notorious serial killer and the number of CCTVs per center reaches several thousands. Because these cameras operate on 24 h, 365 days basis, the amount of video generated from them are tremendous. Normally, those videos are utilized when there are crimes, disasters and other problems. A man should watch all videos with naked eyes in real time.

Watching lots of videos with naked eyes is very tedious and time-consuming job. Moreover, [1] indicates that it is impossible to monitor thousands of cameras with a few agents. For that reason, lots of video analysis techniques are being studied. [2] handles

This work was supported by Institute for Information & communications Technology Promotion(IITP) grant funded by the Korea government(MSIT) (No. 2016-0-00109, Development of Video Crowd Sourcing Technology for Citizen Participating-Social Safety Services).

K. Arai et al. (Eds.): IntelliSys 2018, AISC 868, pp. 382–387, 2019.
https://doi.org/10.1007/978-3-030-01054-6_27

video analytics department consists of followings, which are optimum camera placement, data acquisition and storage, feature extraction, object detection and tracking, interpretation of video data, data visualization, and analytics algorithms. It summarizes studies regarding each departments and [3] introduces methodology for application of intelligent video systems to integrated control centers. More specific research such as [4] handles recognition of high semantic events.

Meanwhile, even though it's possible to detect such high semantic events with the techniques mentioned above, engines with those algorithms inevitably requires lots of computing resources. As a result, only a few number of CCTVs can be analyzed with one server and it can't replace monitoring agents substantially. For that reason, in this paper, we devise an intelligent system which can analyze lots of CCTVs simultaneously.

The proposed system aims at utilizing face information extracted from lots of CCTVs when searching videos. To handle lots of CCTVs, the system parses metadata of video codec to find area of motion instead of analyzing decoded video itself. Meanwhile, CCTVs are installed at high position and covers wide area and far range. So, the resolution of the CCTV video isn't enough for face recognition. Therefore, proposed system closes up the area of motion, captures images and uploads those images to Microsoft cognitive services for face recognition and save analyzed information.

The rest of this paper is organized as follows: Sect. 2 depicts the system design of proposed intelligent system in terms of its hardware and software aspects. Section 3 describes implementation based on Sect. 2 and its analysis, then followed by conclusions and future works in Sect. 4.

2 System Design

2.1 Methodology

The system employs two cameras, which are installed side by side and pointing in the same direction. One camera does normal monitoring and detects motion. The other one serves as auxiliary. The latter is informed of area of motion by the former camera and it zooms up the area as in Fig. 1.

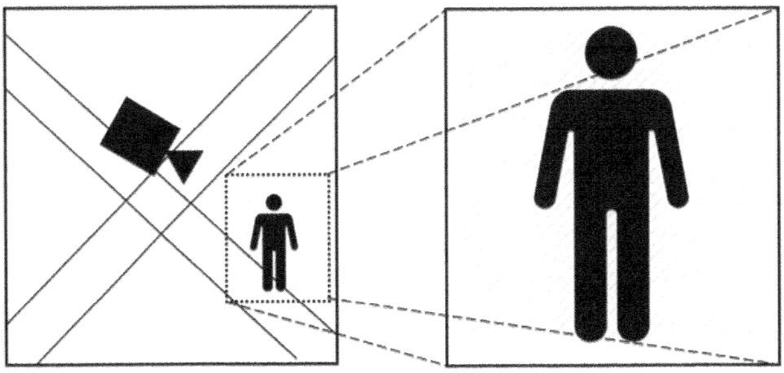

Fig. 1. Intelligent zooming concept for face acquisition.

Then the zoomed frame is transmitted to Azure for face recognition. The recognition information is piled up for future recognition.

2.2 Hardware Configurations

Hardware systems consist of systems depicted in Fig. 2.

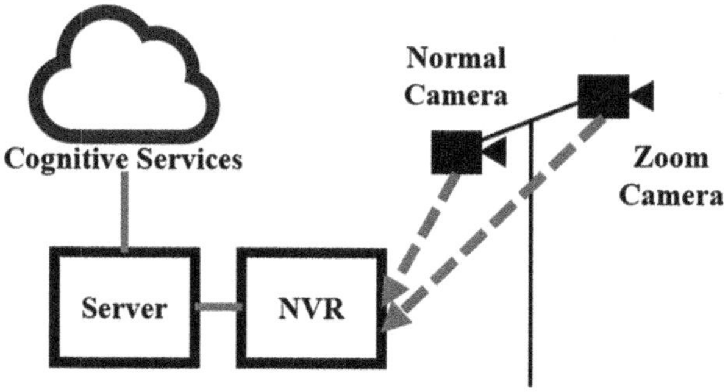

Fig. 2. Hardware configuration of proposed system.

At first, two cameras are installed at the same poll. One is for normal surveillance and the other is for zooming. The videos from the cameras are saved in NVR (Network Video Recorder) and the server contains all the software that will be described later in C section. The NVR serves as mediator between cameras and server. It provides the server with API (Application Programming Interface) to receive CCTV video and to transmit ptz (Pan Tilt Zoom) control signal. All systems are connected to network.

2.3 Software Configurations

In this section, the procedures of the system are presented. It includes calibration, motion detection, the calculation of coordinates and zooming and face recognition.

- Calibration
  This system handles two cameras assuming they are directing at the same direction. Calibration should be done before further actions. At first the zoom camera should share the same view with the normal camera. To do so, it copies the pan, tilt, zoom value of the normal camera. Second, though they have the same ptz value, they are actually viewing slightly different scenery because of their different locations. So, to sync view in detail, fine control of ptz is needed. The server extracts feature points from the normal camera and the zoom camera. To adjust ptz value utilizing the feature points of the same objects in both cameras, the server should know how many pixels are shifted with unit change of pan, tilt, and zoom (Fig. 3). If we call x shift caused by unit pan as x0 and y shift caused by unit tilt as y0, the delta pan and tilt value should be below as in (1), (2) (Fig. 4).

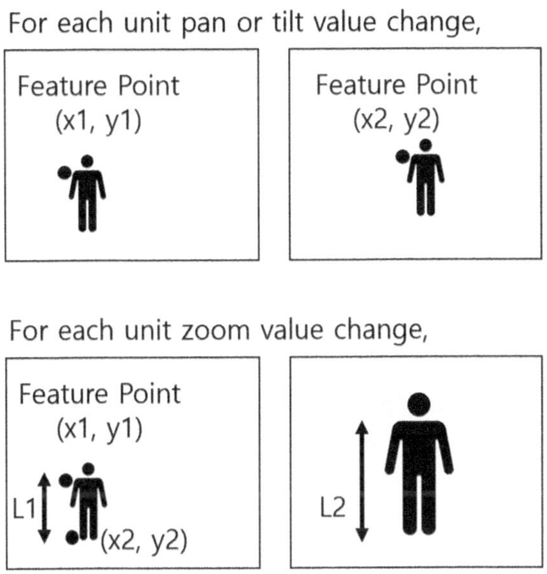

Fig. 3. Pixels change per unit pan, tilt, zoom value.

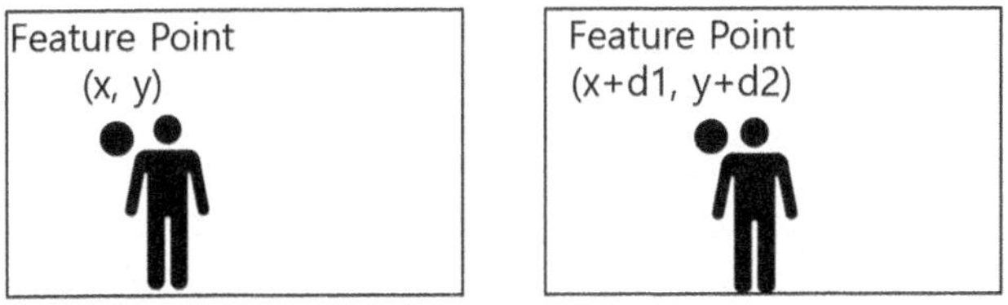

Fig. 4. Feature point discrepancies between two cameras for the same object

$$\text{Delta pan value} = \text{Discrepancy in feature x points}/\text{x0} \tag{1}$$

$$\text{Delta tilt value} = \text{Discrepancy in feature y points}/\text{y0} \tag{2}$$

- Detection of motion
  There might be various methods for detection of motion. But in this paper, we utilized parsing of metadata from video codec, which can extract coordinates of area of motion without decoding. The video of camera with normal purpose is used.
- Calculation of coordinates and zooming
  The area at the coordinates of motion acquired from the previous step should be zoomed by the zoom camera. For this purpose, how many pan, tilt shhould be changed is calculated with the equation similar to (1) and (2). By the way, the zoom level should be like (4).

$$\text{Delta zoom level} = \text{round}(L2/L1) \tag{4}$$

After the zoom camera moves according to the calculated pan, tilt coordinates, it instantly save the image for face recognition.

- Face recognition and registration
  The Azure provides us with API to detect, train and identify faces. Detect query detects faces in the frame and reply face regions where the user can know where faces are. The train query accepts faces and names. If the training is done as the faces are collected, face recognition becomes possible. In other words, if a person whose face was trained before appears in front of our cameras, the name is recognized with Azure query. So, as the time passes the face information acquired from the detect query of zoomed frame should be piled up by querying train.
  The person group provided by Azure can contains 1,000,000 in one group, which is very large number of people. But in case of sites handling thousands of cameras, the number might not enough. So, old face images should be removed in chronological order.

3 Implementation and Analysis

Analysis system software was developed using matlab and python. We tested the algorithm using dahua camera and it was installed in the laboratory. The frame per second was 15 and the MJPEG encoding is used for transferring video data to analysis server.

Figure 5 shows the result of motion detection. The center of mass is used to represent the location of objects.

Fig. 5. Motion detection result.

Figure 6 shows the successful recognition of one human face. By the way, because only frontal face can be recognized by cognitive services, people with rear head cannot be recognized and registered, which is the limitation of this system. Moreover lack of FPS (Frame per second) lead to blurring of image, which hamper the recognition of face.

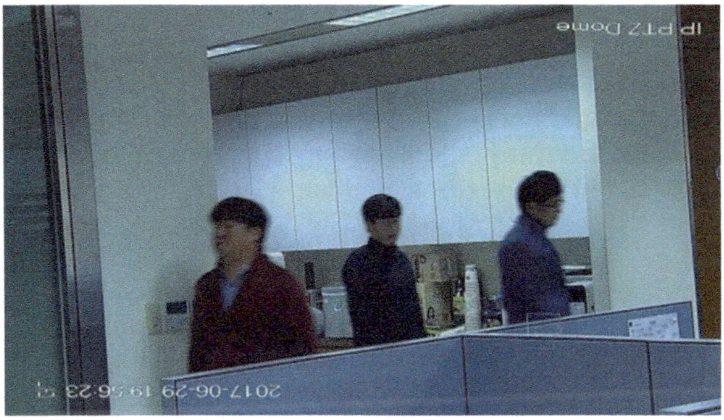

Fig. 6. Automatic ptz result and successful face recognition of one person.

4 Conclusion

The intelligent system for face collection and recognition using automatic ptz control is introduced. Using metadata information of video codec, it was possible to detect motion area of hundreds of cameras and to detect and recognize faces using cognitive services. Though there were some limitations caused by cloud cognitive services, it is expected to contribute to efficiency of operation in integrated control centers when searching specific videos in terms of peoples' faces.

References

1. Noah, S., Thomas, S., Dimitry, G., Rangachar, K.: How effective is human video surveillance performance. In: 19th International Conference on Pattern Recognition ICPR, pp. 1–3, April 2008
2. Ayesha, C., Santanu, C.: Video analytics revisited. IET J. Comput. Vis. **10**, 237–247 (2016)
3. Honghai, L., Shengyong, C.: Intelligent video systems and analytics: a survey. IEEE Trans. Ind. Inf. **9**, 1222–1233 (2013)
4. Subetha, T., Chitrakala, S.: A survey on human activity recognition from videos. In: International Conference on Information Communication and Embedded Systems, pp. 1–7. ICICES, Chennai (2016)

Wheat Plots Segmentation for Experimental Agricultural Field from Visible and Multispectral UAV Imaging

Adriane Parraga[1(✉)], Dionisio Doering[1], Joao Gustavo Atkinson[1], Thiago Bertani[2], Clodis de Oliveira Andrades Filho[2], Mirayr Raul Quadros de Souza[3], Raphael Ruschel[3], and Altamiro Amadeu Susin[3]

[1] School of Computer Engineering, UERGS, Guaíba, Brazil
adriane-parraga@uergs.edu.br
[2] Department of Environment and Sustainability, UERGS, São Francisco de Paula, Brazil
[3] School of Electrical Engineering, UERGS, Porto Alegre, Brazil

Abstract. The use of Unmanned Aerial Vehicles (UAV) in precision agriculture (PA) has increased recently. Most applications capture images from cameras installed in the UAVs and later create mosaics for human inspection. In order to further improve the quality of data offered by this technology, application specific image processing algorithms that enhance, segment and extract information from the raw images delivering information equivalent to in-situ measurements are still missing. The present study describes a method for image segmentation to assist the characterization of nitrogen content in wheat fields. The proposed methodology uses the UAV and Computer Vision algorithms that process visual (RGB) and multispectral agricultural images. Data is first collected by the UAV that flies over an area of interest and collects high resolution RGB and multispectral images at a low altitude. Subsequently, a mosaic is created for each crop stage and the proposed algorithm segments the ROIs (regions where wheat crop is present) based on vegetation index. Using the proposed algorithm, the wheat plots are correctly segmented for two kinds of Brazilians wheat cultivates. The segmentation was validated by experts indicating that the proposed algorithm is suitable to be used as a first step of a method that assist the analyses of nitrogen content specific to wheat crops.

Keywords: Segmentation · Unmanned Aerial Vehicles (UAV) Vegetation index

This research was supported by the Fapergs (the Brazilian Foundation for Science and Technology) under grant 16/2551-0000524-9.

K. Arai et al. (Eds.): IntelliSys 2018, AISC 868, pp. 388–399, 2019.
https://doi.org/10.1007/978-3-030-01054-6_28

1 Introduction

Southern Brazil agricultural fields play a significant role in the country's wheat production. Also, wheat cultivator is one of the world's most important crops, therefore its yields can affect people and impact economies around the globe.

Precision agriculture is one example of possible technologies that aims to increase production [1]. Unmanned Aerial Vehicles (UAVs) use in precision agriculture has been proposed for a wide-range of applications to develop techniques where farmers (and researchers) can benefit. In [2,3], it was proposed to combine selected vegetation indices (VIs) and plant height information to estimate biomass of a summer barley. In [4], it was proposed to estimate Amazon Forest Biomass using UAV.

UAVs are able to capture data with high resolution by using visible and multispectral sensors. By doing the acquisition of high resolution imagery, it is possible to model the gross primary productivity, analyze the fertilizes quantities, identify the presence of insects, weeds and others invasive plants. For all these applications, segmenting the area of interest is the first and essential step to study the crops during phrenological stages. The information obtained from these images can then be used in the management of crop production. The main advantages of using UAVs over satellites (one of the main alternatives for capturing images related to PA) include the possibility of doing data acquisition during cloudy weather conditions and a lower long-term cost [5].

With the development of computer vision, automatic segmentation of agriculture lands have been widely proposed in the past years. In [6] it is proposed to segment canopy/tree coverage versus the underlying soil using Digital Surface Model (DSM) based on radiometric classification. In [7] it is presented the development of a segmentation scheme for robust and accurate clustering of pixels based on color to capture nitrogen deficiency in corn leaves. The proposed algorithm aims to segment only the green pixels of the image by employing a hierarchical K-means algorithm and combining the clustering results of two color spaces, and finally, morphological operators to removes small groups of pixels based on a threshold. Hamuda [8] proposed an algorithm based on HSV color space and the use of morphological erosion and dilation to discriminate crop, weeds and soil. In [9], it is proposed a two-stage framework for segmenting the greenhouse vegetables foliar disease spots images. The first stage involves a comprehensive color feature and its detection method, which combines ExR, H component of HSV color space and *Lab* color space. The second stage employs an interactive region growing segmentation. A major drawback of this method is its need of human intervention. In [10], it is proposed to segment regions affected by diseases based on the improved fuzzy C-means algorithm to the extraction of cucumber leaf spot disease under complex backgrounds.

On plot-based experimental crop researches, like in this work, frequent capture and assessment of data from UAV images is of interest. The plots, however, are not always homogeneous and intra-plot differences as well as sections of the plots borders are commonly removed in the sampling phase. This is done by manually selecting ROIs (Regions of Interest) to make sure that only the core of each parcel is sampled. The processing of creating such ROIs manually, however, is time consuming and prevents the automatic analyses of large data sets.

Fully automatic segmentation is still an open problem and dependent on the application domain. Only a few works involve wheat crops and also, multispectral images have some singularities that need a different approach. In this work, we propose an algorithm to segment plots of wheat with the objective to provide a tool to assist the study and analysis of nitrogen management during a crop cycle. In this work the crops have two different Brazilians wheat varieties, called Toruk and Parrudo. The proposed method is based on vegetation Index and a ROI Map. The challenge is to find one algorithm that segments two kinds of Wheat cultivar throughout the crop cycle with variation of nitrogen fertilizer and also with a clutter background.

2 Methodology

For this work, an experimental wheat crop field with circa 100 m in length and 60 m in width was selected (as in Fig. 2). The study area contains several 2.5 m $\times$ 1 m rectangular-like plots that hold two wheat varieties (Toruk and Parrudo). Different N fertilizer content was applied for experimental agricultural. The whole growth stage was then captured by an UAV. Variability in the crops growth were created for all test areas where each one received a varying quantity of nitrogen.

2.1 Acquisition and Database

The images used in this work were acquired simultaneously at a height of 50 m aboveground using a pair of cameras coupled to a DJI Matrice 100 Quadcopter. These cameras correspond to a single channel DJI X3 Visible (RGB), with 12 MB resolution and 8-bit pixel depth, and a Parrot Sequoia, with 1.2 MB resolution four multispectral channels (G, R, NIR and RE) and 10-bit pixel depth. The resulting pixel size for the RGB and multispectral images are 1.9 cm and 4.8 cm, respectively. Post-processing of the acquired images included georeferencing and orthomosaicking using photogrammetry softwares.

Data are acquired by a large set of overlapping images that are post-processed to derive a single global orthoimage. An example of the images to be segmented is shown in Figs. 2 and 3 for RGB.

2.2 Segmentation Algorithm

The proposed segmentation algorithm has four steps and its flowchart is presented in Fig. 1. The four steps consist of:

(1) Preprocessing
(2) Filtering: Morphological operation
(3) Segmentation map generator
(4) Validate data

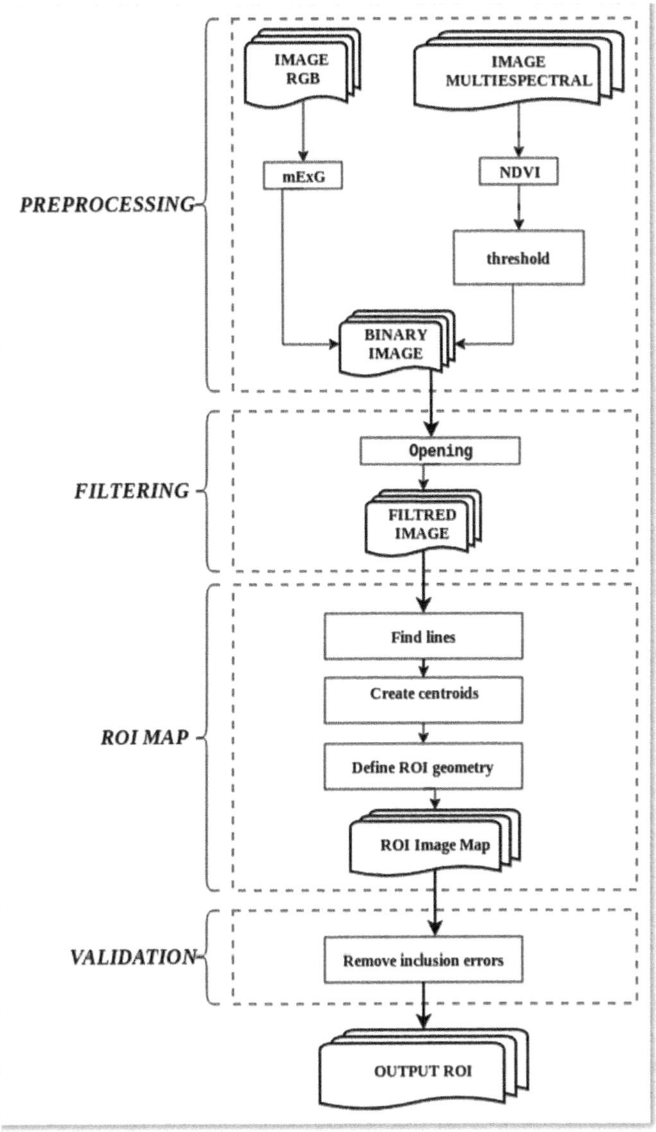

Fig. 1. Flowchart of the proposed segmentation algorithm.

Step 1: Preprocessing

The preprocessing consisted in calculating two indexes from the image bands in order to diminish background effect and enhance spectral information related to photosynthetic activity. For the RGB images, a custom index is proposed (Index 1), and which is given by (1). In this equation, I_G, and I_B are the actual pixel values from the Green and Blue channels of the original RGB image, respectively. Then, the output image I_{BIN} from (1) is a binary image.

$$I_{BIN} = \begin{cases} 1 & \text{if } I_G \leq I_B \text{ or } I_G \leq 0.85.I_R \\ 0 & \text{otherwise} \end{cases} \tag{1}$$

A second index (Index 2) was derived for the multispectral images using the Red and NIR spectral bands. This index is given by the following equation (refeq:NDVI), where NIR is the Near-infrared band of the spectrum and I_R pixel values from the Red channel. After the calculation of I_{NDVI}, the values are rescaled from 0 to 255 $(255 * I_{NDVI} - min(I_{NDVI})/(max(I_{NDVI}) - min(I_{NDVI})))$. The resulting images from Index 2 were, then, converted to binary images using a threshold defined by the OTSU's algorithm [9,11].

$$I_{NDVI} = \frac{I_{NIR} - I_R}{I_{NIR} + I_R} \tag{2}$$

Step 2: Filtering
The Filtering step consisted of applying morphological operators to the resulting binary images from Step 1 in order to improve plot detection. The goal was to remove excessive background vegetation that is beyond the border of each plot. An erosion filter (opening) and two kernels were selected. Both kernels are 9×9 windows and the first one is square shaped while the second is triangular-like or diamond-like shaped. This procure was implemented using the morphologyEx function derived from OpenCV using Python with the algorithm MORPH_OPEN. The whole procedure is repeated a few amount of times.

Step 3: ROI Map Generator
This step consists of generating a segmentation map I_{Map}. From the filtered images generated in the previous step and based on the geometry of the plots, an algorithm that rotate the image from $-45°$ to $45°$ is used to identify vertical and horizontal lines that meet orthogonally. For each time that the image is rotated, the number of pixels identified as a horizontal or vertical line is counted for each row or column. If the number of pixels in a row or column is higher than a preset threshold, it is considered a line. The lines are then draw to generate a map of the plots. Following, a centroid for each plot is created and from which the ROIs will be constructed using a chosen area and/or geometry. An intermediate image of the map can be seen in Fig. 6. The map generated is in relation to the best degree found for the image.

Step 4: Validation
As a final step, a validation process is carried out aiming to reduce inclusion errors. This process works by using Index 1 and 2 mean values for all pixels in each plot and a threshold. Hence, plots with low vegetation cover can be eliminated using a simple logical operator. As this first process aims to identify

false plots near the borders of the study area only, a second rule was created to ensure that plots with at least two valid neighbor plots are kept.

This step is performed into two stages: elimination of false-positives using the area of each plot and its number of neighbors.

Stage 1: A logical AND operation is performed between the I_{Map} with the binary image output, I_{BIN} (from the step 1, as in (3)). Then the Algorithm *Map Refine* is applied for each plot in the image I_{Map}.

$$I_{Map} = I_{Bin} \wedge I_{Map} \tag{3}$$

Algorithm Map Refine

1) IF
   Plot Area ≤ Area_Threshold
   THEN
   Discard Plot
   ElSE
   Keep the Plot
2) Update I_{Map}

After searching for false positives based on the size of the detected objects, the next stage is to go through each plot of updated map seeking if the remaining plots have at least 2 plots nearby. If a plot has less than 2 neighbors, it is discarded. There are some cases where the low vegetation (grass/bush) was a bit large and ended up not being removed as it should. The contrary happens too, in some cases objects were mistakenly removed because of the low vegetation of the plot. These two problems have been solved with the following procedure.

Stage 2: Image I_{map} is traversed, and each time a new object is found in the image, the amount of neighbors of that object in the resulting AND image is searched and counted. If a plot has less then 2 neighbor, the plot is discarded. In addition, if a plot was discarded by area, but it has 3 or 4 neighbors, it is included again. It is done in such a way that the addition of a new object can be done more than once.

3 Results and Discussion

The algorithm was implemented in Python. It was applied in 10 images, 5 RGB images and 5 multispectral, during the Wheat cycle. For facility, we will present the results only in one RGB and multispectral example. The objective is to find the center coordinate for each plot and draw a rectangle containing each plot.

Examples of a RGB and a multispectral images to be segmented are in Figs. 2 and 3, respectively. The results are presented for each step described in the segmentation section, which we called Partial Results.

Fig. 2. Example of RGB image mosaic of the plots of wheat.

Fig. 3. Example of multispectral image (NIR, Red and RedEdge bands) of the plots of wheat.

Fig. 4. Result of the RGB binarization process.

Partial Results: Step 1

Figure 4 is the I_{BIN} for the original RGB image. The multispectral image binarization in Fig. 5.

Partial Results: Step 2

Since the background interferes in the results by a single binarization, as we can see in Figs. 4 and 5, Step 2 is applied to improve the separation of these plots. The result is in Fig. 6, after being applied a square kernel of size 9×9 pixels 7 times and opening with the diamond kernel 9×9 pixels 5 times. For

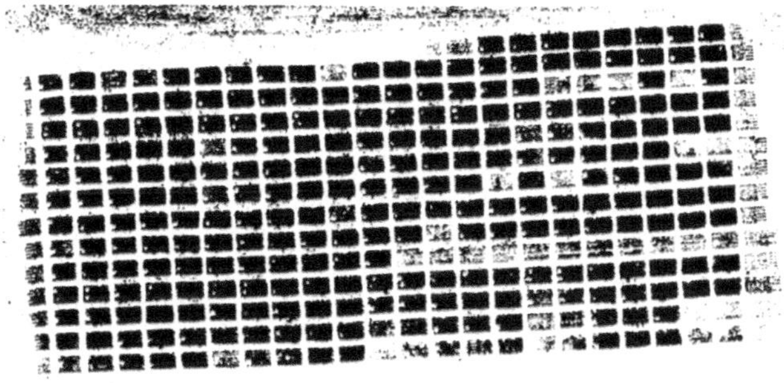

Fig. 5. Result of the multispectral binarization.

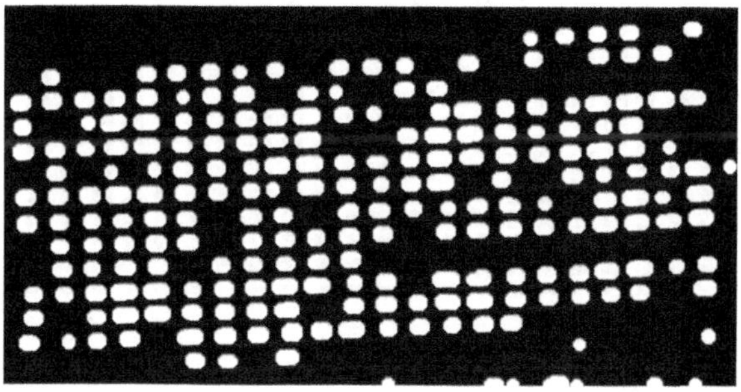

Fig. 6. Result of the morphologic process applied in the binary image.

the multispectral images, the opening with a square kernel of 9×9 pixels was applied 2 times and opening with diamond kernel 9×9 pixels was 3 times.

The results of the morphological filtering do not present the complete map with all the plots, since some objects disappeared in the process of opening. Thus, the next step is responsible to solve this issue.

Partial Results: Step 3

After Step 2 is complete, Step 3 begins with the Map creation. The partial map creation is presented in Fig. 7, showing the grid format necessary to find the centroids of each plot.

After the Map is created, the algorithm finds the centroids of the objects to select an area of equal size predefined by the user. In this example, the size of the area defined to be selected in each object for the RGB image was 900 pixels and for the multispectral image was 500 pixels. In Fig. 8 is the map with the defined area size, and in Fig. 9 the map overlapped the original image.

As we can see in Fig. 9, the Map finds plots all over the area. Although, on the top of the mosaic there is no plots. So the next step is performed to detect false positives and remove them from the map.

Final Results: Step 4

Using the Map of Fig. 8, the algorithm does an logical AND operation between Figs. 8 and 4. The result is in Fig. 10.

From Fig. 10 is made the count of pixels that remained in each object, stage one described in Methodology Section. Thereafter, Fig. 8 is traversed removing those objects having an amount of pixels per object below the desired threshold. In this case we have used a threshold of 10% of the predefined area. Thus removing most of the false positives obtained in Fig. 9. However in some cases objects were removed incorrectly, where can be seen in Fig. 11, highlighted by red circles.

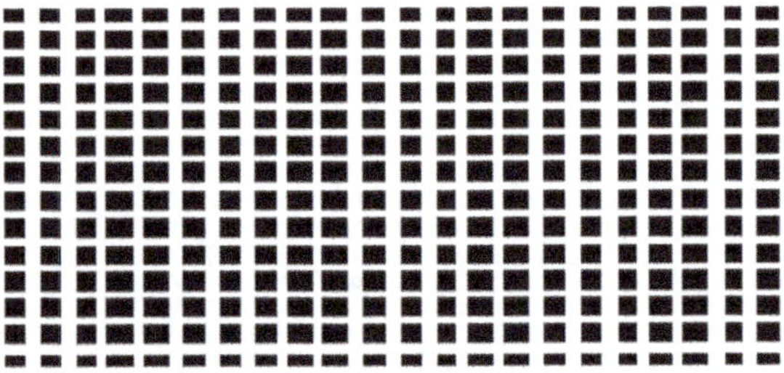

Fig. 7. Intermediate map creation - binary grid with lines found for the map creation.

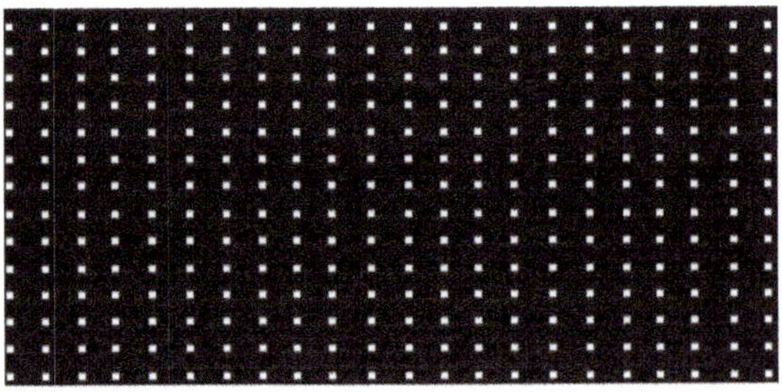

Fig. 8. Image of the first version of the I_{map}.

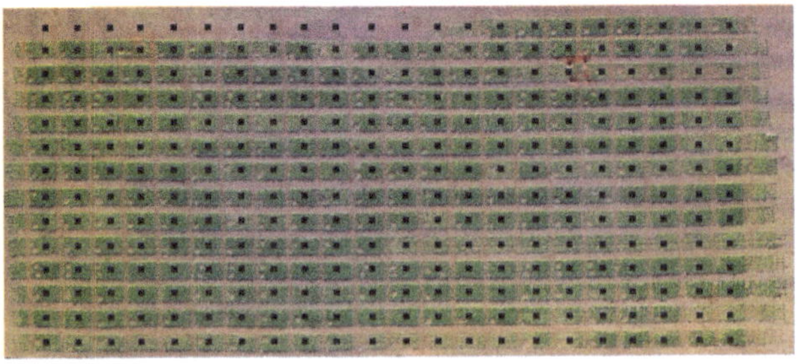

Fig. 9. Image of the I_{map} over the RGB image for visualization purposes.

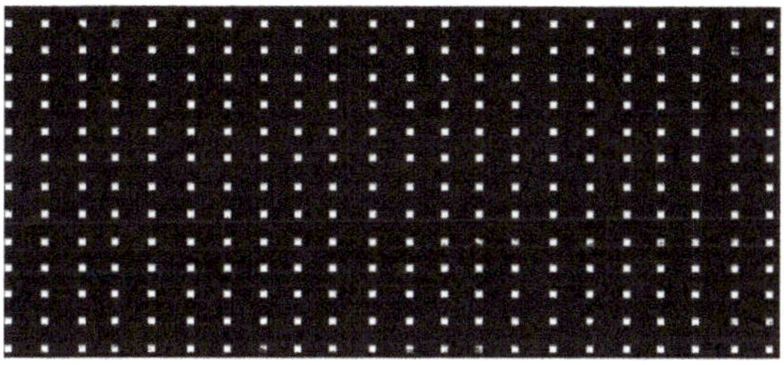

Fig. 10. Logical AND result from Step 4.

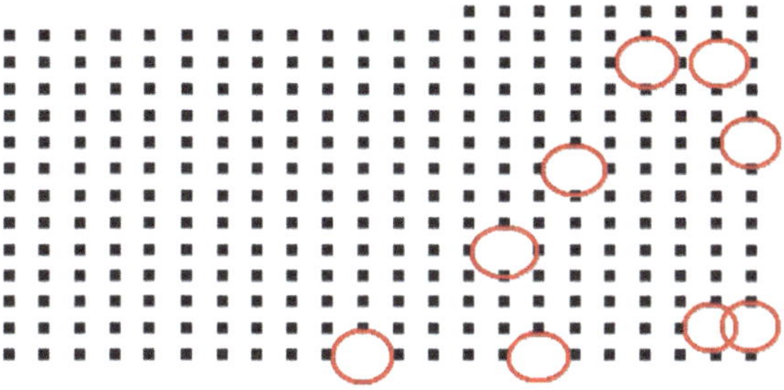

Fig. 11. Example of the image of the I_{map} with some plots wrongly removed, highlighted by red circles.

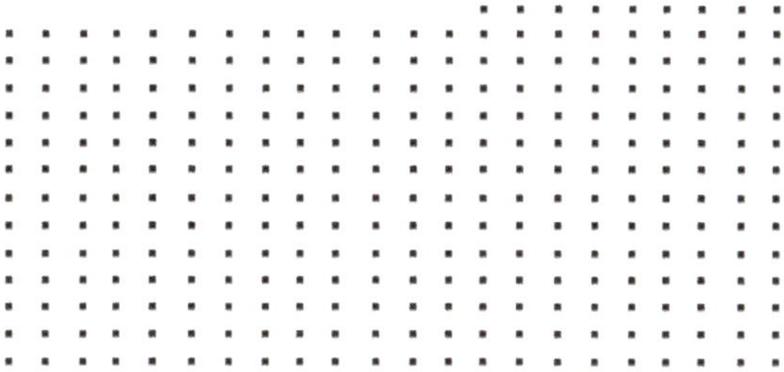

Fig. 12. Final segmentation Map I_{map}.

To solve this problem and find the final segmentation map, we applied the Stage 2 of Step 4 (Fig. 12). If the minimum threshold of neighbors equals 3, the first time the algorithm passes through the object, only two neighbors will be found, then the object will not be added back to the corner, for this to happen the addition process is repeated until no more plots is added or removed (Fig. 13).

Fig. 13. Final segmentation map I_{map} overlapped in the original image.

4 Conclusions

This work presented an algorithm to automatically segment the wheat plots during the whole growth stage. Our method explores a feasible and simple way to segment wheat plots images captured under real field conditions using color information and solves the problems of uneven illumination and asymmetrical field background effectively, which is of great theoretical value.

Main contribution of this application is to provide a tool for the researchers agronomists, which perform this segmentation manually and it is fundamental for several analysis.

Although the method consists in tuning several variables, once this is complete, the method worked properly for all images acquired during the whole wheat cycle. This is important since each week has a significantly change in the field.

Since our code was implemented to run solely on a CPU, some portions possess a significant execution time, which can be drastically reduced. Given the parallel nature of image processing algorithms, our future works includes a GPU implementation using CUDA, where we expect to achieve significant performance improvement over the current implementation. Also some works will be investigated to improve results, such as [12,13].

References

1. Khanal, S., Fulton, J., Shearer, S.: An overview of current and potential applications of thermal remote sensing in precision agriculture. Comput. Electr. Agricult. **139**, 22–32 (2017). https://doi.org/10.1016/j.compag.2017.05.001. ISSN 0168-1699
2. Bendig, J., Yu, K., Aasen, H., Bolten, A., Bennertz, S., Broscheit, J., Gnyp, M.L., Bareth, G.: Combining UAV-based plant height from crop surface models, visible, and near infrared vegetation indices for biomass monitoring in barley. Int. J. Appl. Earth Obs. Geoinf. **39**, 79–87 (2015). https://doi.org/10.1016/j.jag.2015.02.012. ISSN 0303-2434
3. Brocks, S., Bareth, G.: Estimating Barley biomass with crop surface models from oblique RGB imagery. Remote Sens. **10**(2), 268 (2018)

4. Messinger, M., Asner, G.P., Silman, M.: Rapid assessments of amazon forest structure and biomass using small unmanned aerial systems. Remote Sens. **8** (2016). https://doi.org/10.3390/rs8080615. ISSN 2072-4292
5. Li, W., Yuan, H., Li, W., Song, L.: Prediction of wheat gains with imagery from four-rotor UAV. In: 2016 2nd IEEE International Conference on Computer and Communications (ICCC), Chengdu, pp. 662–665 (2016). https://doi.org/10.1109/CompComm.2016.7924784
6. Mancini, A., Dyson, J., Frontoni, E., Zingaretti, P.: Soil and crop/tree segmentation from remotely sensed data by using digital surface models. Preprints 2017, 2017110142. https://doi.org/10.20944/preprints201711.0142.v1
7. Zermas, D., Teng, D., Stanitsas, P., Bazakos, M., Kaiser, D., Morellas, V., Papanikolopoulos, N.: Automation solutions for the evaluation of plant health in corn fields. In: IEEE International Conference on Intelligent Robots and Systems, vol. 2015, pp. 6521–6527. [7354309] Institute of Electrical and Electronics Engineers Inc.. https://doi.org/10.1109/IROS.2015.7354309
8. Hamuda, E., Ginley, B.M., Glavin, M., Jones, E.: Automatic crop detection under field conditions using the HSV colour space and morphological operations. Comput. Electron. Agric. **133**, 97–107 (2017). https://doi.org/10.1016/j.compag.2016.11.021. ISSN 0168-1699
9. Ma, J., Du, K., Zhang, L., Zheng, F., Chu, J., Sun, Z.: A segmentation method for greenhouse vegetable foliar disease spots images using color information and region growing. Comput. Electron. Agric. **142**(1), 110–117 (2017). https://doi.org/10.1016/j.compag.2017.08.023. ISSN 0168-1699
10. Ma, J., Li, X., Wen, H., Fu, Z., Zhang, L.: A key frame extraction method for processing greenhouse vegetables production monitoring video. Comput. Electron. Agric. **111**, 92–102 (2015). https://doi.org/10.1016/j.compag.2014.12.007. ISSN 0168-1699
11. Otsu, N.: A threshold selection method from gray-level histograms. IEEE Trans. Syst. Man Cybern. **9**(1), 6266 (1979). https://doi.org/10.1109/TSMC.1979.4310076
12. Xie, S., Imani, M., Dougherty, E.R., Braga-Neto, U.M.: Nonstationary linear discriminant analysis. In: Proceedings of the 51th Asilomar Conference on Signals, Systems, and Computers, Pacific Grove, CA, October–November 2017
13. Imani, M., Braga-Neto, U.M.: Particle filters for partially-observed Boolean dynamical systems. Automatica **87**, 238–250 (2018)

Evaluation of Image Spatial Resolution for Machine Learning Mapping of Wildland Fire Effects

Dale Hamilton(✉), Nicholas Hamilton, and Barry Myers

Department of Math and Computer Science, Northwest Nazarene University, Nampa, USA
{dhamilton,nicholashamilton,blmyers}@nnu.edu

Abstract. Wildfires burns 4–10 million acres across the United States with suppression costs approaching $2 billion, annually. High intensity wildfires contribute to post fire erosion, flooding and loss of timber resources. Accurate assessment of the effects of wildland fire on the environment is critical to improving the management of wildland fire as a tool for restoring ecosystem resilience. Sensor miniaturization and small unmanned aircraft systems (sUAS) offer a new paradigm, providing affordable, on-demand monitoring of wildland fire effects at a much finer spatial resolution than is possible with satellite or manned aircraft, providing finer detail at a much lower cost. This project examined the effect hyperspatial imagery acquired with a sUAS has on improving the extraction of post-fire effects knowledge from imagery. Support vector machines were shown to map post-fire effects land cover classes more accurately using hyperspatial color imagery than 30 m color imagery.

Keywords: Support vector machine (SVM) · Fuzzy logic
Small unmanned aircraft system (sUAS) · Hyperspatial imagery

1 Introduction

This study examines improvements in mapping of wildland fire severity from hyperspatial (sub-decimeter resolution) small unmanned aircraft system (sUAS) imagery using computer vision and machine learning. Wildlands provide habitat for around six and a half million species according to the United Nations Environment Program [1]. Decades of fire suppression have led to the current departure of wildlands from the fire return interval characteristically experienced prior to European settlement. As a result, wildlands in the western US are experiencing a much higher incidence of catastrophic fires. Fire consumes millions of acres of American wildlands each year, with suppression costs approaching two billion dollars annually [2]. High intensity wildland fires contribute to post fire erosion, soil loss, flooding events and loss of timber resources.

This publication was made possible by Undergraduate Research Grants from National Aeronautics and Space Administration Idaho Space Grant Consortium and an Institutional Development Award (IDeA) from the National Institute of General Medical Sciences of the National Institutes of Health under Grant #P20GM103408.

K. Arai et al. (Eds.): IntelliSys 2018, AISC 868, pp. 400–415, 2019.
https://doi.org/10.1007/978-3-030-01054-6_29

This results in negative impacts on communities, wildlife habitat, ecosystem resilience, and recreational opportunities.

This paper investigates direct or immediate effects of a fire such as biomass consumption as observed in the days and weeks after the fire is contained [3]. Therefore, this study defines burn severity as the measurement of biomass consumption [4]. Identification of burned area extent within an image can be achieved by exploiting the spectral separability between burned organic material (black & white ash) and vegetation [5, 6]. Classifying burn severity can be achieved by separating pixels with black ash (low fuel consumption) from white ash (more complete fuel consumption), relying on the distinct spectral signatures between the two types of ash [7].

New advances in sUAS capabilities enable the acquisition of imagery with a spatial resolution of centimeters and temporal resolution of minutes [8]. This hyperspatial imagery enables objects to be represented in the image by multiple pixels [9]. Extraction of actionable knowledge using machine learning from this higher resolution data will enable fire ecologists to better study temporary environmental effects caused by wildland fire.

2 Background

2.1 Current Methods for Mapping Wildland Fire Extent and Severity

Current methods for acquiring imagery which can be utilized for assessing fire effects rely on satellites, which in the case of Landsat have a spatial resolution of 30 m [10]. Monitoring Trends in Burn Severity (MTBS) is a national project within the US that maps burn severity and extent from Landsat data with records going back to 1984. However, this project only maps wildland fires greater than 400 hectares in the western US and greater than 200 hectares in the eastern US [11, 12]. As a result, much of the body of fire history contained in fire atlases omit the spatial extent of small and moderate sized fires [13]. These smaller fires can account for 20% of the total area burned across a landscape, which is also the most ecologically diverse of the total area burned [14]. Accurate historical record of fire history is necessary in order to determine departure of current fire frequency from historic fire frequency, a key metric for determining ecosystem resilience [15].

2.2 Previous Machine Learning Applications for Wildland Fire Mapping

There are many examples of efforts using a variety of machine learning algorithms mapping wildland fire extent using relatively low resolution satellite imagery. Zammit [16] compared implementations of both a Support Vector Machine (SVM) and k-Nearest Neighbor (kNN) classifiers performing a pixel based classification using pixel values from the green, red and near-infrared (NIR) bands from 10 m resolution imagery acquired with the SPOT 5 satellite. In this case, the SVM was found to map fire extent with higher accuracy than the kNN was able to achieve.

Hamilton [17] investigated improvements achievable in the accuracy of post-fire effects mapping with an SVM using hyperspatial (sub-decimeter) drone imagery. Spatial context using a variety of texture metrics were also evaluated in order to determine the inclusion of spatial context as an additional input to the analytic tools along with the three-color bands. This analysis shows that the SVM was able to map burn extent and biomass consumption from color imagery with very high accuracy. The addition of texture as an additional fourth input provided for an even further increase in accuracy.

2.3 Utilization of sUAS for Image Acquisition

The proliferation of small unmanned aircraft system technology has made the procurement and use of remotely sensed imagery a viable possibility for many organizations that could not afford to obtain such data in the past. Most commercially available sUAS come with an onboard digital camera with three bands capturing visible light in the blue, green and red spectrum [18].

(1) *Image acquisition:* Machine learning based analytics use spectral reflectance to identify a variety of classes of vegetative features in images from which actionable knowledge can be derived. Spectral responses in the visible spectra can be used to differentiate between different image features such as white and black ash [5, 6], as well as other features of interest to fire managers such as vegetation type [5, 19].

The approaches to mapping attributes related to fuels, burn severity, and post-fire vegetation response discussed in the previous sections contain a number of issues which must be addressed while developing methods, analytic tools and metrics for mapping wildland fire effects with much higher resolution than is currently available from the current generation of satellites. The DJI Phantom 4, a commonly available sUAS, comes with a digital color camera that has a horizontal field of view of 94 degrees, acquires twelve megapixel images with 3000 rows by 4000 columns of pixels. Aerial imagery acquired while flying at an altitude of 120 m AGL has a spatial resolution of approximately 5 cm per pixel [5]. Objects that are wider than that pixel resolution will be discernible in the acquired hyperspatial imagery as shown in Fig. 1(a). The black rectangles in the image are burned areas. Small lines and patches of white within the burned area are white ash from sagebrush which was fully combusted by the fire. The unburned vegetation consists primarily of annual and perennial grasses and forbs in addition to Wyoming big sagebrush and yellow rabbit brush. The scene contains two western juniper trees. Linear features are fire containment lines dug by a bulldozer.

Features that are easily identified in hyperspatial imagery are lost in low resolution 30 m LANDSAT satellite imagery, being aggregated into more dominant neighboring features. Figure 1(b) shows the same scene as the preceding image, but resampled to 30 m spatial resolution having 48 pixels aligned in 6 rows by 8 columns.

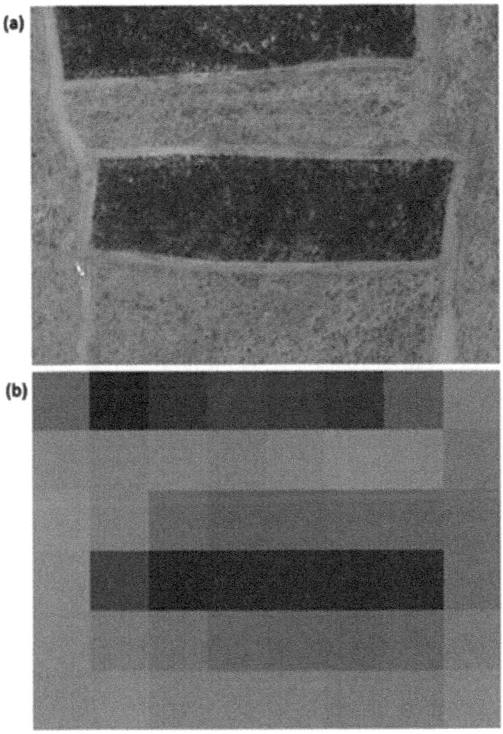

Fig. 1. (a) Image of a rangeland study area acquired with a Phantom 4 sUAS flying at 120 m AGL with a spatial resolution of about 5 cm per pixel. (b) Same scene resampled to 30 m resolution with six rows and eight columns of pixels.

2.4 Mapping Wildland Post-fire Landcover Components with sUAS

Fire ecology enables managers to study temporary environmental changes by accounting for the pronounced change that wildland fire has on an ecosystem. The emerging field of Ecoinformatics integrates environmental and information sciences to define entities and natural processes with language common to both humans and computers [20]. This emerging field provides the methodologies and tools needed to acquire, analyze and manage the growing amounts of complex ecological data available from the immense volume of ecological data available from a variety of sources, including hyperspatial sUAS imagery.

Development of methods, algorithms and metrics utilizing hyperspatial color imagery can more accurately map components that are indicative of biomass consumption (white ash) than is possible with current systems due to the low spatial resolution of imagery acquired via satellite and manned aircraft [6]. In order to fully leverage this hyperspatial imagery, it is necessary to develop analytic tools which identify the extent of the burned area within the image, classifying the burned pixels by post-fire land cover components. In order to accomplish this, machine learning based analytics have been developed which use an SVM to discriminate between black ash, white ash, crown vegetation, surface vegetation and mineral soil [5], which can most accurately be identified with imagery having spatial resolution higher than 0.25 m [8, 22]. Utilizing these classes, the analytic tools can interpret the scene relying on relationships between the classes. Classification of burned area extent can be achieved by exploiting the spectral separability between burned organic material (black & white ash) and vegetation [6]. Classifying biomass consumption can be achieved by separating pixels with partially combusted black ash from white ash which is indicative of high consumption, relying on the distinct spectral signatures between the two types of ash [7]. In forested biomes, we can also identify fires with low biomass consumption by looking for patches of unburned vegetation within the extent of the fire. If a patch is comprised only of tree crown(s), the analysis can infer that the vegetation is a tree which the fire passed under and classify the pixels as low intensity surface fire [21].

3 Methodology

The recent advances in sUAS technology promise to provide wide availability of hyperspatial imagery to users who previously did not have the ability to generate remotely sensed imagery on their own. The copious amounts of easily obtained data resulting from the acquisition of hyperspatial imagery warrant investigation into development of methods, analytic tools and metrics which enable the extraction of information and knowledge from imagery captured at much higher resolution than was previously possible. The affordability of sUAS is facilitating the dissemination of knowledge previously unattainable from lower resolution data.

3.1 Evaluating Spatial Resolution: Does Size Matter?

Landsat imagery with a spatial resolution of 30 m is commonly used for many earth observation purposes including land cover mapping. The US land management agencies, which include the US Forest Service (USFS) and agencies in the US Department of Interior (DOI) have ongoing programs to provide landscape scale geospatial products for describing burn severity (mtbs.gov) across the entire US. While these programs produce consistent geospatial products across the entire country, many local managers claim that 30 m spatial resolution is not adequate for acquiring the knowledge needed for management needs at the local level. In comparing 30 m Landsat imagery to hyperspatial imagery, a number of variables can affect classification accuracy including sensor resolution, atmospheric influence and temporal resolution. In order to isolate the effect of spatial resolution on burn classification accuracy, hyperspatial sUAS imagery was resampled to 30 m resolution and used as a classifier input for our accuracy tests.

(1) *Image Acquisition Platforms: Impact on Spatial Resolution:* When comparing the effects of spatial resolution on image acquisition, it is important to isolate spatial resolution from other factors that can impact classification accuracy. Due to the temporal sensitivity of white ash as an indicator of biomass consumption, imagery acquired with satellites can cause a temporal degradation of biomass consumption

data. Imagery can only be acquired via satellite when the sensor is over the scene, which in the case of Landsat, is every 16 days. Once the satellite is in place, the scene may be obscured by either smoke or clouds, requiring an additional 16 days before getting another opportunity to acquire an image of the scene. The delay between post-fire containment and the first opportunity for satellite image acquisition can cause a significant reduction in the amount of white ash visible in the image, reducing the ability to map biomass consumption from the image. Additionally, atmospheric scatter of light reflected from the scene can also diminish the ability of fire extent and ash type (black as opposed to white) to be detected from satellite imagery.

In order to isolate spatial resolution from these temporal and radiometric resolution considerations, we resample 5 cm hyperspatial orthomosaic of a fire to 30 m medium spatial resolution. This ensures that both the 5 cm and 30 m images have the same temporal extent, providing a record of the burn scene at the same time. Additionally, atmospheric scatter in 30 m image will be eliminated due to the image being captured at 120 m AGL instead of 650 km AGL.

(2) *Spatial Resolution Hypothesis:* In testing whether an SVM can map burn extent and biomass consumption more accurately using 5 cm or 30 m imagery, a null hypothesis (H_0) is specified along with an associated alternate hypothesis (H_1). If H_0 is rejected, then H_1 is accepted in its place. The null and alternate hypotheses are:

H_0: Post-fire surface cover components can be mapped with equal accuracy using either hyperspatial color imagery or 30 m color imagery.
H_1: Post-fire surface cover components can be mapped more accurately from 5 cm color imagery than 30 m color imagery.

H_0 was tested by comparing mapping accuracy, which is calculated measure of the percentage of validation pixels where the user and the SVM labeled the pixel with the same class [22]. An upper-tailed Student's t-test established the statistical significance of the difference in accuracy between the 5 cm and 30 m classifications. A p-value below a significance level of 0.05 rejected H_0 in favor of H_1 establishing that the mean accuracy of 5 cm classifications are higher than 30 m classifications.

(3) *Spatial Resolution Experiment Methodology*: In order to isolate spatial resolution from these temporal and radiometric resolution considerations, we resample 5 cm hyperspatial orthomosaic of a fire to 30 m medium spatial resolution. This ensures that both the 5 cm and 30 m images have the same temporal extent, providing a record of the burn scene at the same time. Additionally, atmospheric scatter in the 30 m image will be eliminated due to the image being captured at 120 m AGL instead of 650 km AGL.

(a) *Resample 5 cm Orthomosaic to 30 m:* Hyperspatial orthomosaics were resampled to have medium resolution of 30 m, which is an equivalent spatial resolution to Landsat imagery. The resolution reduction was accomplished using the OpenCV resize function [23]. In addition to resizing the image, the spatial reference had to be exported from the 5 cm orthomosaic to the resized image. The reduction in spatial resolution required that the georeference from the original orthomosaic had to be converted to the spatial resolution of the resized image. Changing the spatial resolution in the world file simply entails setting the associated lines in the TIFF world file (with a tfw file extension) to the desired spatial resolution of 30 m. Calculating the centroid coordinates of the upper left hand pixel of the resized 30 m image from 5 cm orthomosaic is accomplished as shown in Fig. 2.

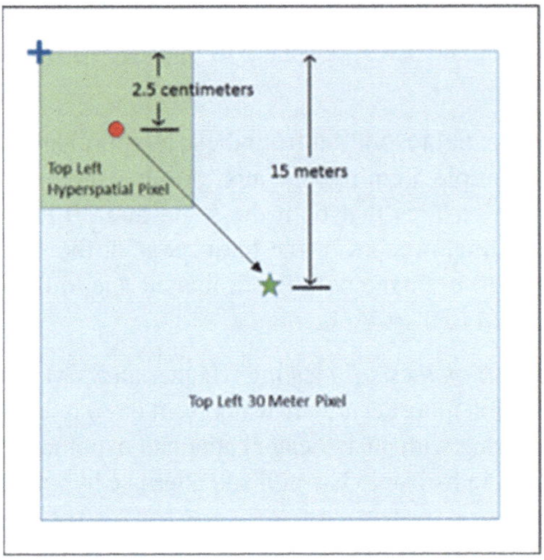

Fig. 2. Deriving 30 m image georeference from 5 cm georeferenced image. The centroid of the top left 5 cm pixel is shown as a red dot. The centroid of the top left medium resolution pixel is shown as a green star. The top left corner of the image is shown as a blue cross.

The top left corner of the 5 cm image (the blue cross) is located, being half the spatial resolution of the 5 cm image (e.g. 2.5 cm) north and west of the centroid of the top left pixel in the 5 cm image (red dot). The centroid of the 30 m image is then located (green star), being a distance of half the spatial resolution of the 30 m image (15 m) south and east from the top left corner of the image.

(b) *Labeling 30 m Training Pixels using Fuzzy Logic:* Fuzzy logic allows decisions to be based on imprecise boundaries rather than relying on precise boundaries that are used by Boolean logic. This use of vagueness allows the expression of how much the data fits given criteria, transitioning from one class to another over a range of values. Fuzzy logic is often more applicable to ecological data than the crisp delineations resulting from Boolean logic where data will transition from one class to another at a single threshold value. Fuzzy logic is used in this study for labeling 30 m pixels for training, allowing 30 m training pixels to be assigned post-fire land cover class labels to be derived from 5 cm classes.

- *Calculate 5 cm Class Densities for each 30 m Pixel*: Density of each of the post-fire land cover classes is calculated for each 30 m pixel from the 5 cm classification pixels located within each 30 m pixel. A set of class density rasters are created; one raster for each class found in the post-fire land cover classification. Within each 30 m pixel, counts are taken of hyperspatial class pixels for each class occurring in the 5 cm classification image. The 5 cm class counts are divided by the total number of 5 cm pixels within the 30 m pixel, and then multiplied by 100. These class densities for each 30 m pixel are recorded in the class density rasters.

- *Fuzzification of Post-Fire Land Cover Class Densities:* Fuzzy set theory allows the specification of how well an object satisfies a vague criterion [24] with fuzzy logic providing a means for specifying that the transition from one class to another is not demarked at a single value but transitions from one class to another over a range of values [22]. For example, Lewis [25] found that white ash cover exceeding 33 to 50% for a site is indicative of strong water repellant soil conditions. Fuzzy logic allows us to specify that the transition of water repellency from weak to strong occurs between 33 and 50% cover. Rather than set membership being expressed as either zero or one as is the case with Boolean logic, fuzzy logic allows set membership to be specified as a range of membership from 0.0 to 1.0.
  - Assignment of the post-fire land cover class to a 30 m pixel is based on a combination of whether the pixel burned and white ash cover within the pixel. In neither case is Boolean logic appropriate for determining set membership. Increasing the density of a class by a handful of 5 cm pixels and as a result changing a Boolean expression from false to true does not adequately describe the state of the pixel.
  - The first set membership to consider when assigning 30 m post-fire land cover class is whether the 30 m pixel burned. Burn extent fuzzy membership specifies how much of the pixel is in in either the burned or unburned set. Determination of burn extent is much better handled using fuzzy logic, allowing the transition between burned and unburned sets to occur as the combination of black ash and white ash 5 cm pixels transition from 35 to 65% of the pixels within a 30 m pixel as shown in Fig. 3.

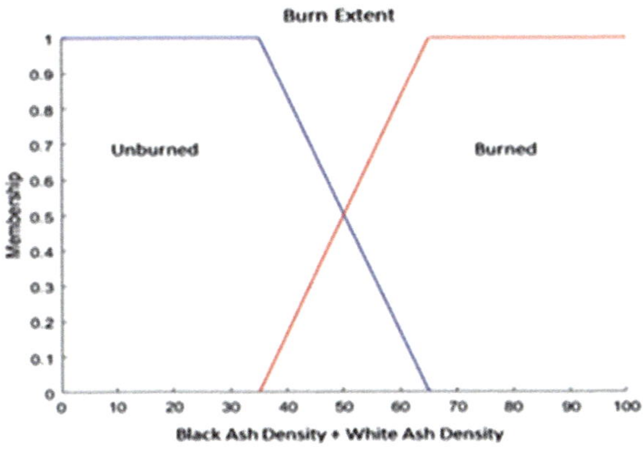

Fig. 3. Burn extent fuzzy sets.

- An additional set membership needed for assigning 30 m post-fire land cover classes measures biomass consumption, evaluating the relationship between black ash and white ash densities. As the white ash cover, expressed as a percentage of white ash to burned (white and black ash) pixels transitions from 33 to 50%, the biomass consumption transitions from low to high consumption [25] as shown in Fig. 4.

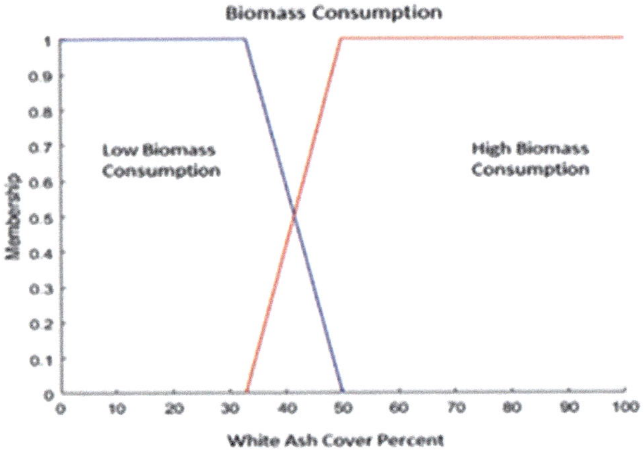

Fig. 4. Biomass consumption fuzzy sets.

- *Activate fuzzy rules by applying fuzzy logic*: Fuzzy logic is used to label the 30 m pixels based on their fuzzy set membership. When applying fuzzy logic to fuzzy set membership, a fuzzy AND is expressed as taking the minimum value of the expressions on either side of the AND operator. Likewise, a fuzzy OR is expressed as taking the maximum value of the expressions on either side of the OR operator. When evaluating a series of fuzzy if statements, as shown in Algorithm 1, the data is defuzzified by selecting for activation the action associated with the if expression that has the highest value.
  Algorithm 1: Fuzzy logic for labeling 30 m pixels from 5 cm post-fire land cover classes.

```
If (extent:burned AND combust:high)
    training pixel = White Ash
If (extent:burned AND combust:low)
    training pixel = Black Ash
If (extent:unburned)
    training pixel = Unburned
```

(c) *Train and Classify SVM with 30 m Fuzzy Logic Training Pixels:* Each of the 30 m pixels was labeled with training data labels as specified previously. The SVM was trained on 70% of the 30 m training pixels, with the remaining training pixels being withheld for validation of the SVM. The SVM then classified the 30 m image, using the validation pixels to calculate the accuracy of the SVM classifier when validated against the validation pixels.

(d) *Validate 30 m Burn Extent Classification Against 5 cm:* Pixels between the 5 cm and 30 m runs do not coincide spatially due to differing spatial resolution. Validation regions are specified as vector format polygons in ArcGIS, stored in a polygon shapefile. Designating the training data as polygons allows the validation data to be independent of the spatial resolution of any particular raster. Polygons are created in ArcGIS using heads up digitizing, with the 5 cm orthomosaic as a basemap.

Accuracy of an SVM classification of post-fire land cover class is calculated with an areal comparison between the user specified validation polygons and the classified output. The ArcGIS Tabulate Area tool calculates the area of each of the post-fire land cover classes from the SVM that are inside each of the classes that occur in the validation polygons. The table generated by the tool is a confusion matrix, from which accuracy is calculated, showing the percentage of acreage in the validation polygons in which the user specified polygon classes are in agreement with the classes predicted by the SVM.

(e) *Establishment of statistical significance:* The statistical significance of increased accuracy across the validation sets when mapping burn extent and biomass consumption using 5 cm as opposed to 30 m imagery was established using one tailed paired t-tests. The null hypothesis states that burn extent and biomass consumption will be mapped with equal accuracy regardless of whether the SVM uses 5 cm or 30 m imagery. By contrast, the alternate hypothesis states that burn extent and biomass consumption will be mapped more accurately using 5 cm imagery as opposed to 30 m imagery. In order to apply the t-test, the accuracy of the classification was taken using 5 cm imagery and then again with 30 m imagery for the same scene. The significance level that the t-test passed is 0.05 which gives it 95% certainty to reject the null hypothesis in favor of the alternate hypothesis. That is, burn extent and biomass consumption can more accurately be mapped from 5 cm color imagery than 30 m color imagery.

4 Results and Analysis

4.1 Description of Study Areas and Image Acquisition

A set of post-fire orthomosaics were acquired in southwestern Idaho primarily during the summer of 2017. Of the 16 fires over which post-fire burn imagery was acquired, the majority of the fires were in lands administered by the Bureau of Land Management Boise District, with the remaining fires occurring on private land as well as the Boise National Forest. The fires flown ranged in size from five to 400 acres. The majority of the fires over which imagery was acquired were in sagebrush steppe ecosystems in the Snake River Valley of southwestern Idaho, primarily in areas administered by the United States Department of Interior Bureau of Land Management Boise District (Fig. 5).

Fig. 5. Post-fire photo of sagebrush steppe burned area representative of the majority of fires over which imagery was acquired in support of this research project.

4.2 Comparison of Spatial Resolution for Mapping Post-fire Effects

The assessment of SVM classification accuracy at both 5 cm and 30 m resolution was explored over a set of burn orthomosaics from five fires, the results of which were tested with a two-tailed *t*-test. The resulting *P* values fell below the significance level, thus rejecting the null hypothesis that burn extent and biomass consumption will be classified with the same accuracy using 5 cm or 30 m imagery. Consequently, the alternate hypothesis was accepted which states that the SVM classifies burn extent and biomass consumption with different accuracies.

(1) *Spatial Resolution Accuracy Results*: Evaluation of the preferred spatial resolution with which the SVM was able to classify burn extent and biomass consumption was accomplished by assessing the accuracy of the SVM output from both 5 cm and 30 m orthomosaics. Due to evaluating classifications at inconsistent spatial resolution, the users used ArcGIS to identify and label training regions, recording and labeling the regions in a polygon shapefile. An areal comparison of the classifications against the validation polygons was performed by ArcGIS, which created a confusion matrix with the areal quantification comparing the classification to the user labeled validation data. This areal quantification allows the calculation of accuracy which is the area within the validation regions that were classified the same as the user label divided by the total area of the validation regions. Example classifications of burn extent and biomass consumption from both 5 cm and 30 m orthomosaics are shown in Fig. 6.

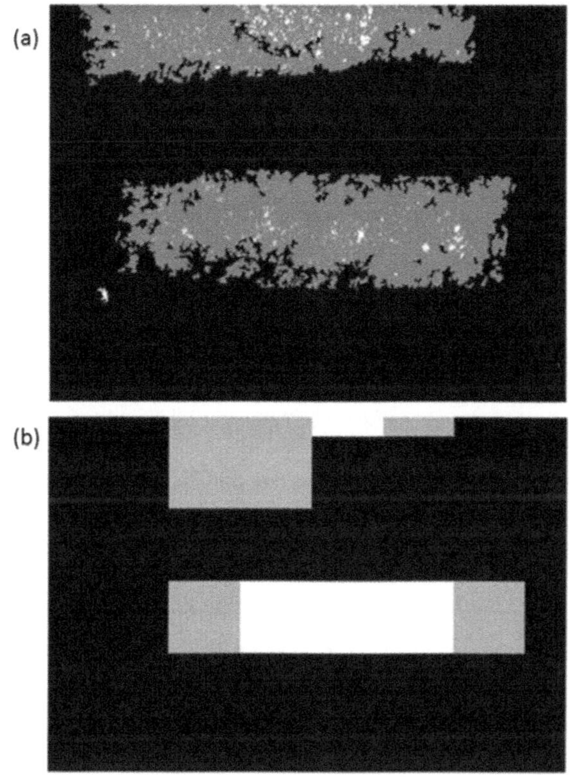

Fig. 6. Classified output showing unburned, black ash and white ash pixels classified at 30 m for scene represented in Fig. 2. Unburned are colored black, black ash is colored grey and white ash is colored white.

The Global Burned Area Satellite Validation Protocol, endorsed by the Committee on Earth Observation Satellites (CEOS) contains guidelines for using remotely sensed imagery as reference data. The reference image observed by the user while validating burn severity geospatial layers should have higher spatial resolution than the imagery used to generate the burn severity maps [26]. The reference image needs to exhibit spectral and radiometric resolution adequate for the unambiguous discrimination of burned from unburned areas. Lastly, the reference image needs to be acquired before any vegetation recovery or removal of char and ash; that is, within weeks after the fire event [11]. Unlike satellite imagery, which contains heterogeneous pixels, hyperspatial sUAS imagery contains homogeneous pixels, greatly facilitating the identification of objects within the image, eliminating the need for the reference image to be higher resolution than the classified image. Hyperspatial sUAS imagery meet the other two criteria listed, having the same spectral and radiometric resolution as well as being from the same temporal period. As a result, users can identify and label regions for validation purposes using the same hyperspatial image for defining validation data as well as training data for supervised burn classification.

Accuracy results for each of the validation sets for both the burn extent and biomass consumptions from both of the spatial resolutions on each fire are shown in Table 1.

Table 1. 5 cm vs 30 m classification accuracy

	Burn extent		Biomass cons.	
Fire	5 cm	30 m	5 cm	30 m
Jack	94.74	71.04	96.97	18.29
MM 106	86.58	30.62	98.68	3.17
Immigrant	98.34	38.05	97.97	44.8
Elephant	98.5	83.47	98.67	26.33
Owyhee	87.65	59.94	87.42	49.42

Classification accuracy for both burn extent and biomass consumption was averaged for both 5 cm and 30 m orthomosaics then multiplied by 100. The resulting Mean Classification Accuracy is listed in Table 2 for both 5 cm and 30 m classifications.

Table 2. 5 cm vs 30 m mean classification accuracy

Spatial resolution	Burn extent	Biomass consumption
5 cm	93.16	95.94
30 m	56.62	28.40

A comparison of the Mean Classification Accuracy between the 5 cm and 30 m shows that 5 cm has higher accuracy for both the burn extent and biomass consumption classifications.

(2) *Spatial Resolution Accuracy Statistical Significance:* The statistical significance of increased accuracy between the 5 cm and 30 m classifications across the validation sets for the burn images was established by using one-tailed t-tests. The null hypothesis is that burn extent and biomass consumption can be classified with equal accuracy using either 5 cm or 30 m orthomosaics. By contrast, the alternate hypothesis is that the SVM can classify burn extent and biomass consumption with higher accuracy using a 5 cm orthomosaic as opposed to a 30 m orthomosaic. The t-test was run on the accuracy results from both 5 cm and 30 m orthomosaics, first testing burn extent, then biomass consumption. The significance level that the t-test passed is 0.05 which gives it 95% certainty to reject the null hypothesis in favor of the alternate hypothesis.

The burn extent accuracy tests rejected the null hypothesis with a P value of 0.018. Likewise, the biomass consumption accuracy tests rejected the null hypothesis with a P value of .006. In both cases, the null hypothesis was rejected, supporting the alternate hypothesis which shows that burn extent and biomass consumption are classified with measurably increased accuracy using a 5 cm orthomosaic as opposed to using a 30 m orthomosaic.

5 Conclusion and Future Work

5.1 Conclusion

The higher spatial resolution available with imagery acquired with sUAS has the potential to allow land managers to get a much more detailed view of the aftermath of a wildland fire. The machine learning and image processing algorithms applied in this project enable the extraction of knowledge regarding fire effects from the very high detail available in hyperspatial imagery. The resulting post-fire effects mapping products provide increased accuracy over previous mapping methods, providing land managers with more accurate and detailed information of where the fire burned and how severely it burned.

This study showed that post-fire land cover classes can be mapped from hyperspatial color imagery with increased accuracy, showing that burn extent and biomass consumption are mapped more accurately from hyperspatial color imagery than from 30 m color imagery. The SVM using hyperspatial sUAS imagery classified burn extent with an average accuracy of 93.16% and biomass consumption with an average accuracy of 95.94%. By contrast, when the SVM classified fire effects using 30 m imagery, burn extent was mapped with an average accuracy of 56.62% and biomass consumption with an average accuracy of 28.40%. The improved accuracy resulting from the use of higher resolution hyperspatial imagery shows the utility of using sUAS for acquiring post fire imagery, allowing land managers to map post fire effects with higher accuracy than is possible through current methods.

Increased mapping accuracy of post-fire effects provides land managers with a more complete knowledge of the change affected on the landscape by wildland fire. Improved knowledge enables managers to more effectively prescribe post-fire actions in order to more efficiently mitigate the detrimental effects of the fire, improving the ecological response of associated landscapes and resulting in improved resiliency of fire-adapted ecosystems across the western US.

5.2 Future Work

This project investigated the improvements in classification accuracy obtainable using hyperspatial sUAS imagery to map rangeland fire effects. As a result, the vast majority of the burned areas flown were in the xeric rangelands of southwestern Idaho. During the 2018 fire season, more imagery needs to be acquired in the mesic upper Payette and Boise River watersheds in the Boise National Forest (BNF), in order to extend the application and analysis of these methods to include forested biomes. The cooperative agreement in place with the BNF extends through the 2018 season, which will facilitate acquisition of burn imagery during the summer of 2018 by the research team from NNU. The acquisition of burn imagery for forested ecosystems will allow the research team to further develop and validate the analytic tools to detect vegetation structure, allowing more accurate determination of the burn extent of surface fires in forested environments.

The analytic tools developed to investigate the improvement of accuracy between hyperspatial imagery and 30 m imagery have the potential to be used to classify burn extent and biomass consumption from Landsat imagery using the same methodology. The potential to use sUAS imagery to train classifiers to map fire effects from Landsat imagery has the potential to improve the accuracy of burn extent and biomass consumption mapping of very large fires, which are much too large to fly with the current generation of sUAS.

Acknowledgment. We would like to acknowledge the mentorship of Greg Donohoe and Eva Strand at the University of Idaho. We would also like to acknowledge the assistance of undergraduate research assistants including Jonathan Branham, Ryan Pacheco, Zachary Garner and Jonathan Hamilton. Additionally we would like to acknowledge the Boise National Forest and the Bureau of Land Management Boise District for providing image acquisition access to burned sites within their jurisdictions.

References

1. Mora, C., Tittensor, D.P., Adl, S., Simpson, A.G., Worm, B.: How many species are there on Earth and in the ocean? PLoS Biol. **9**(8), e1001127 (2011)
2. National Interagency Fire Center (NIFC), "Federal Firefighting Costs," (2017)
3. Keeley, J.E.: Fire intensity, fire severity and burn severity: a brief review and suggested usage. Int. J. Wildland Fire **18**(1), 116–126 (2009)
4. Key, C.H., Benson, N.C.: Landscape assessment (LA) (2006)
5. Hamilton, D., Bowerman, M., Collwel, J., Donahoe, G., Myers, B.: A spectroscopic analysis for mapping wildland fire effects from remotely sensed imagery. J. Unmanned Veh. Syst. (2017). https://doi.org/10.1139/juvs-2016-0019
6. Lentile, L.B., et al.: Remote sensing techniques to assess active fire characteristics and post-fire effects. Int. J. Wildland Fire **15**(3), 319–345 (2006)
7. Hudak, A.T., Ottmar, R.D., Vihnanek, R.E., Brewer, N.W., Smith, A.M., Morgan, P.: The relationship of post-fire white ash cover to surface fuel consumption. Int. J. Wildland Fire **22**(6), 780–785 (2013)
8. Laliberte, A.S., Herrick, J.E., Rango, A., Winters, C.: Acquisition, orthorectification, and object-based classification of unmanned aerial vehicle (UAV) imagery for rangeland monitoring. Photogramm. Eng. Remote Sens. **76**(6), 661–672 (2010)
9. Sridharan, H., Qiu, F.: Developing an object-based hyperspatial image classifier with a case study using WorldView-2 data. Photogramm. Eng. Remote Sens. **79**(11), 1027–1036 (2013)
10. National Aeronautics and Space Administration (NASA), "LANDSAT 7" (2017)
11. Sparks, A.M., Boschetti, L., Smith, A.M., Tinkham, W.T., Lannom, K.O., Newingham, B.A.: An accuracy assessment of the MTBS burned area product for shrub–steppe fires in the northern Great Basin, United States. Int. J. Wildland Fire **24**(1), 70–78 (2015)
12. Eidenshink, J.C., Schwind, B., Brewer, K., Zhu, Z.-L., Quayle, B., Howard, S.M.: A project for monitoring trends in burn severity. Fire Ecol. **3**(1), 3–21 (2007)
13. Morgan, P., Heyerdahl, E., Miller, C., Wilson, A., Gibson, C.: Northern rockies pyrogeography: an example of fire atlas utility. Fire Ecol. **10**(1), 14 (2014)
14. Hamilton, D., Hann, W.: Mapping Landscape Fire Frequency for Fire Regime Condition Class, presented at the Large Fire Conference (2015)
15. Wildland Fire Leadership Council, "The national strategy: the final phase in the development of the National Cohesive Wildland Fire Management Strategy," Washington, DC (2014). http://www.For.GovstrategydocumentsstrategyCSPhaseIIINationalStrategyApr2014Pdf. Accessed 11 Dec 2015

16. Zammit, O., Descombes, X., Zerubia, J.: Burnt area mapping using support vector machines. For. Ecol. Manage. **234**(1), S240 (2006)
17. Hamilton, D., Myers, B., Branham, J.: Evaluation of texture as an input of spatial context for machine learning mapping of wildland fire effects. Int. J Signal Image Process. **8**(5) (2017)
18. Lebourgeois, V., Bégué, A., Labbé, S., Mallavan, B., Prévot, L., Roux, B.: Can commercial digital cameras be used as multispectral sensors? a crop monitoring test. Sensors **8**(11), 7300–7322 (2008)
19. Rango, A., et al.: Unmanned aerial vehicle-based remote sensing for rangeland assessment, monitoring, and management. J. Appl. Remote Sens. **3**(1), 033542–033542-15 (2009)
20. Wikipedia Contributors, "Ecoinformatics," Wikipedia, 23 December 2016
21. Scott, J.H., Reinhardt, E.D.: Assessing crown fire potential by linking models of surface and crown fire behavior. USDA For. Serv. Res. Pap., no. Journal Article, p. 1 (2001)
22. Han, J., Kamber, M., Pei, J.: Data Mining: Concepts and Techniques, 3rd Ed (2012)
23. OpenCV, vol. 3.2, no. Computer Program (2017). www.opencv.org
24. Russell, S., Norvig, P.: Artificial Intelligence: A Modern Approach, 3rd edn. Prentice Hall, Upper Saddle River (2010)
25. Lewis, S.A., Robichaud, P.R., Frazier, B.E., Wu, J.Q., Laes, D.Y.: Using hyperspectral imagery to predict post-wildfire soil water repellency. Geomorphology **95**(3), 192–205 (2008)
26. Boschetti, L., Roy, D.P., Justice, C.O.: International Global Burned Area Satellite Product Validation Protocol Part I–Production and Standardization of Validation Reference Data (2009, unpublished data)

Region-Based Poisson Blending for Image Repairing

Wei-Cheng Chen and Wen-Jiin Tsai(✉)

National Chiao Tung University, Hsinchu, Taiwan, R.O.C.
wjtsai@cs.nctu.edu.tw

Abstract. Image inpainting is a widely used technique for image repairing, which utilizes information from the undamaged regions in the same image to do the inpainting. However, the repairing methods using inpainting cannot work well when the textures in the damaged area cannot be found in the remaining area of the image. In this paper, an image repairing method is presented, which uses Affine transformation and Poisson blending for repairing, based on the assumption that reference images are available. The reference image is first affine-transformed to have the same viewing angle and scaling size to the damaged image and then Poisson blending is applied to repair the damaged area by compositing the corresponding area in the affined reference image. Since the blending will preserve reference image's textures and adjust its illumination and color tone according to the damaged image, the composition is supposed to be seamlessly. However, it was observed that using Poisson blending in image repairing may cause dramatically blurring in the resulting image sometimes. This mainly comes from the big differences in pixel-gradient changes in the blending area. To cope with the problem, a region-based Poisson blending is proposed, in which the damaged area is not blended as a whole, but segmented into regions and applied with blending separately. The experimental result shows that the blurring artifacts can be reduced obviously by using the proposed region-based approach.

Keywords: Poisson blending · Image repairing · Region-based

1 Introduction

Photos play an important role in our life. However, it is troublesome once the photos get damaged or critical parts are covered by something else. Famous sights usually attract many tourists. It is unavoidable that the line of sight is blocked by undesired things (e.g., cars or tourists) when you take photos. Removing unwanted portions from the image will leave empty areas which needs repairing to make the image completed. Image inpainting has been widely used to repair damaged regions of a given image. Traditional methods utilize information from undamaged region in the same image to do the repairing [1–4]. Bertalmio and Sapiro [1] proposed an PDE-based image inpainting approach which smoothly propagates surrounding information to the damaged area for recovery. Although the PDE-based method provides a means for reconstructing small damaged portions of an image, it cannot manage well for large

K. Arai et al. (Eds.): IntelliSys 2018, AISC 868, pp. 416–430, 2019.
https://doi.org/10.1007/978-3-030-01054-6_30

portions with complex texture. To solve this problem, exemplar-based inpainting methods [2, 3] have been proposed, which use a priority map to find the most similar patch to fill the missing pixels from the boundary of damaged area towards its inner parts. The priority calculation includes confidence-term and data-term. Since exemplar-based methods can propagate both linear structure and texture into damaged area, it performs better than PDE-based methods for large region reconstruction.

However, the exemplar-based methods sometimes may produce very poor results by filling the damaged area with unsuitable patches coming from the regions of different textures. This is due to that the missing structures are complicated or non-linear and that the confidence and data terms are measured using local information. To cope with this problem, Sun et al. [4] proposed a structure propagation method which allows users to draw some lines in the damaged area to indicate the location of missing structures which divide the area into regions of different textures. Follow user-specified lines, the method propagates the structures from surroundings to the damaged area and select patches from the regions of the same textures to fill the missing pixels. Although the structure propagation method may perform better than exemplar-based methods, it still suffers from the problem that it is hard to draw the lines accurately if the structure is too complicated. Besides, it cannot repair the structures or textures that do not appear in the remaining image.

Therefore, some methods [5–7] have proposed to use images from Internet as reference for repairing. Thanks to the availability of Internet and image capturing devices, many photos of the same scenery can be found from the Internet. By using these photos, the missing structures in the damaged image can be found, although they might be taken by camera from different viewing angles and distances, or under different lighting conditions.

The Poisson blending is an image blending technique allowing for seamless cutting and pasting of portions of images by operating in the gradient domain. It is first proposed by Perez et al. [2] and has been proven to be robust to lighting changes. Poisson blending has been applied in image repairing, for example, the methods in [5–7], where additional images from Internet are used as reference to reconstruct the damaged portion of the given image. Wang et al. [5] proposed a method which uses SURF [14] for image matching and registration, and then uses Poisson blending technique to blend the registered image with the input one. Their method focus on reformulating Poisson blending equations, so that their fast blending can run on mobile devices as an application of low computational cost. Amirshahi et al. [6] proposed an image completion algorithm using occlusion-free image from photo sharing sites in the Internet. The method automatically selects the most suitable images from a database of downloaded images and seamlessly completes the input image using selected images with minimal user intervention. Whyte et al. [7] proposed an internet-based impainting method similar to [6]. They achieve geometric registration by using multiple homographies and a global affine transformation on image intensities for photometric registration. Chavan et al. [8] proposed an image inpainting technique based on SURF but have some restriction: The reference image need to be taken at the same location and illumination.

Although some studies have proposed to use Poisson blending in image repairing, most of them use classical Poisson blending equation, or with some modifications to speed up the computation. Using classical Poisson blending technique in image repairing, however, it sometimes may cause dramatically blurring in the resulting image. The blurring mainly comes from the big difference of gradient change between reference and target images. To cope with the problem, this paper presents a region-based Poisson blending technique which partitions the to-be-repaired portion (mask area) of the image into regions and then applies Poisson blending on each region separately to complete the image repairing.

The rest of this paper is organized as follows: in Sect. 2, the problem of using classical Poisson blending in image repairing is described. In Sect. 3, the framework of the proposed method and the detail of the region-based blending process are provided. Experiments and analysis are presented in Sect. 4. Finally, Sect. 5 summarizes this paper.

2 Image Repairing Using Reference Frames

The proposed method consists of two steps: (1) Image transformation, and (2) Poisson blending. It is based on the assumption that a reference image (source image) having the same scene with the to-be-repaired image (target image) can be found from Internet or anywhere. The Step 1 is to affine the source image so that the contents of both source and target images have the same viewing angles and scaling. The keypoint detector and keypoint descriptor proposed in SIFT [10] are used for keypoint matching and homography between source and target images. Then, an affined source image can be obtained, which has contents with the same viewing angle and scale as that in the target image. Figure 1 shows an example of the affine transformation result. To repair the mask area of a given target image, the corresponding area in the affined source image is used as reference in Poisson blending. This used-for-repairing area is referred as *s-mask* with a prefix "s-" to distinguish it from the mask area in the target image. Using Poisson blending, the s-mask's textures will be reserved while its illumination and color tone will be adjusted according to the surrounding information of the mask in the target image. As a consequence, the blended s-mask can be used to replace the mask in the target image to complete the repairing.

Poisson image blending [9] is usually used to cut portions of source images and paste to target images seamlessly, as illustrated in Fig. 2. Let S be the target image, and Ω be a closed subset of S with boundary $\partial\Omega$. Let f^* denote a function defined over S excluding the interior of Ω and f denote the function defined over the interior of Ω. As shown in Fig. 3, assume there is a source image g and v denote the pixel gradient in g. The Ω is a mask which specifies what portions in S should be replaced by the pixels in g. P′erez et al. [9] proposed a minimization equation as below to solve the pixels within the mask to make the blending seamless.

$$\min_f \iint_\Omega |\nabla f - v|^2 \, with f|\partial\Omega = f^*|\partial\Omega \qquad (1)$$

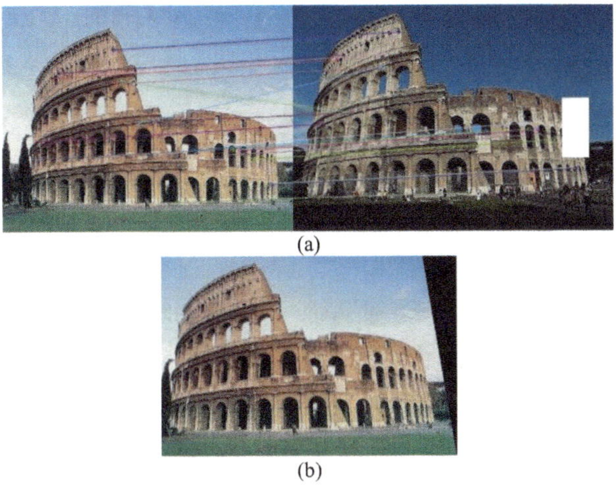

Fig. 1. Affine transformation (a) keypoint matching between source image (left) and target image (right) (b) affined source image.

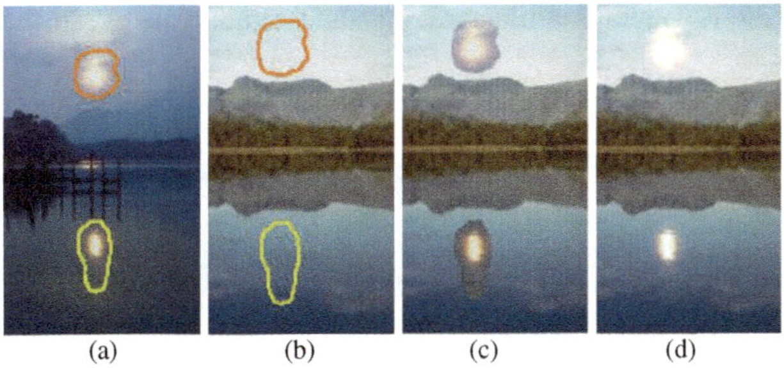

Fig. 2. Image blending example (a) source image (b) target image (c) result of using cloning (d) result of using Poisson blending.

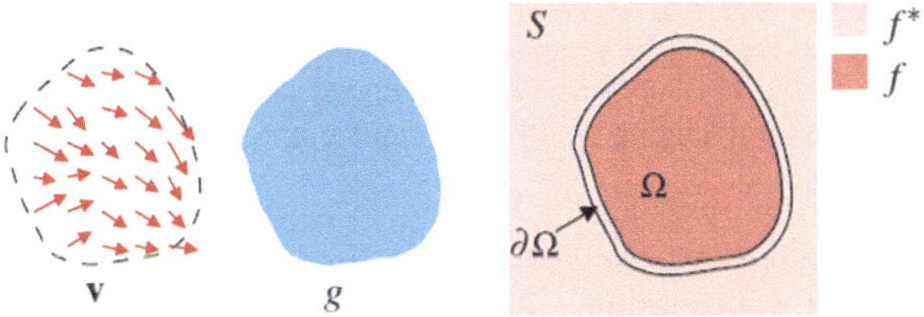

Fig. 3. Guided interpolation for image editing [9].

where, $\nabla. = \left[\frac{\partial.}{\partial x} \frac{\partial.}{\partial y}\right]$ is the gradient operator. The equation is computed using iterative method and applied to each color component, separately. Equation (1) can be converted into a discrete form called Discrete Poisson Equation as follows, where for a pixel p in S, let N_p be the set of 4-connected neighbors of p, f_p be the value of f at p, and $v_{pq} = v(\frac{p+q}{2}) \cdot \vec{pq}$.

$$|N_p| f_p - \sum_{q \in N_p \cap \Omega} f_q = \sum_{q \in N_p \cap \partial\Omega} f_q^* + \sum_{q \in N_p} v_{pq} \quad \text{for all } p \in \Omega \tag{2}$$

2.1 Problem of Poisson Blending

Poisson blending has been applied in image repairing [5–7]. When applied to image repairing, a mask, Ω, is used to specify what portions of the target image to be removed and one or more source images are used as reference to repair the damaged part of the target image. Most of researches focus on finding the most suitable patches among the source images and applying blending technique to blend the patches into the target.

However, using classical Poisson blending in image repairing may obtain poor results in some situations. An example is shown in Fig. 4, where Fig. 4(a) is the source image, Fig. 4(b) is the target image with the white area denoting the mask, namely, the area to be repaired, and Fig. 4(d) is the ground truth of the target image. The blending result using classical Poisson blending is shown in Fig. 4(c). It is observed that the repaired area shows obvious artifacts along the boundary between the building and the sky.

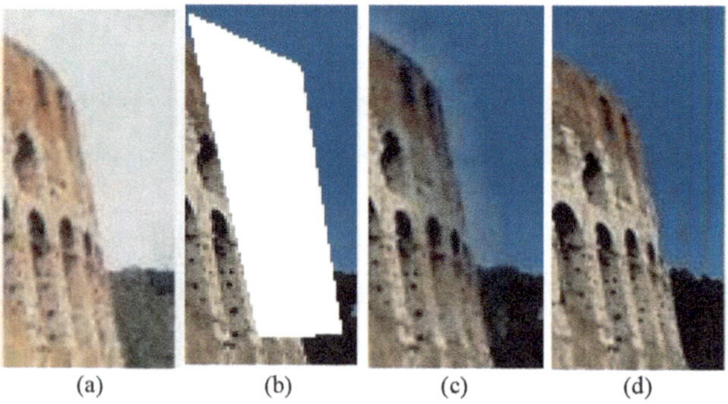

Fig. 4. Problem of using Poisson blending for image repairing (a) affined source image (b) target image (c) Poisson blending result (d) ground truth of target image.

To explain why the unexpected artifacts occur in the classical Poisson blending result, Fig. 5 is used for illustration, where Fig. 5(a) shows the pixel-value differences between source and ground-truth target images inside and around the mask area, while Fig. 5(b) shows the differences between source and resulting target image using classical Poisson blending (i.e., Fig. 5(a) = Fig. 4(a) to (d) and Fig. 5(b) = Fig. 4(a) to (c)). The images in Fig. 4 are shown in 3D presentation, where the x-axis and y-axis correspond to the x-axis and y-axis of the images, and z-axis shows the pixel-value differences. The two images have been adjusted so that they have the same viewing angles for easy comparison. From Fig. 5(a) it is observed that the mask area resides across two surfaces with different heights in z-axis direction. The two distinct heights are due to that the two photos (Fig. 4(a) and (b)) are taken in different times: one in daytime and the other in nighttime, resulting in big color and intensity differences between source sky and target sky, but only little difference between source building and target building. That is, the gradient change in the source image is greatly different from that in the target image. To repair such an area, classical Poisson blending will try to compensate the surface-height difference and make it smoothly across the boundary, resulting in tilted surfaces and blurred boundary as shown in Fig. 5(b). That's why in Fig. 4(c) the building near the sky boundary looks obviously blur and so is the sky near the same boundary.

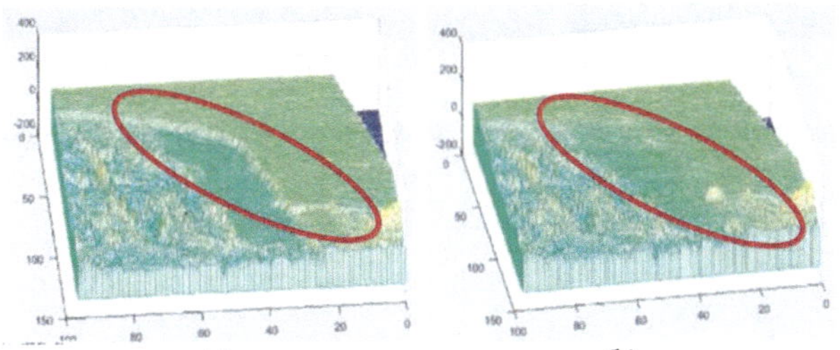

(a) (b)

Fig. 5. 3D presentation of frame difference between the source image and two target images. (a) The ground-truth image as the target image (b) Poisson blending result as the target image.

However, Poisson blending did not always produce artifacts although it may. As a simple example, assume there exists a constant (say T) such that each pixel in the source image can adjust its pixel value to be equal to the corresponding target pixel's value by adding T, then the artifact won't occur because the source can be blended into the target image perfectly without changing the gradient of each source pixel. However, if there are two groups of pixels in the source image, and there exist one constant for each group (say T1 and T2, respectively) such that each of their source pixels can be adjusted to match each of their corresponding target pixels by using their respective constants, then artifacts "may" occur if $T1 \neq T2$. Here "may" instead of "will" is used because it is also possible that the artifacts cannot be perceived and thus negligible. Figure 6 shows two simplified examples with masks reside across the boundary of two regions and each region is filled with a single color. In each case, source images are on the left; target images are in the middle; and Poisson blending results are on the right. In Fig. 6(a), we have T1 = (240, 28, 36) – (237, 28, 36) = (3, 0, 0) and T2 = (255, 255, 255) – (65, 60, 238) = (190, 195, 17); while in Fig. 6(b), T1 = (3, 0, 0) and T2 = (80, 75, 239) – (65, 60, 238) = (15, 15, 1). With $T1 \neq T2$ in both cases, Fig. 6(a) shows obvious artifact after Poisson blending, while in Fig. 6(b) the artifact is almost unperceivable. To predict whether artifacts are perceivable, assume there is a threshold

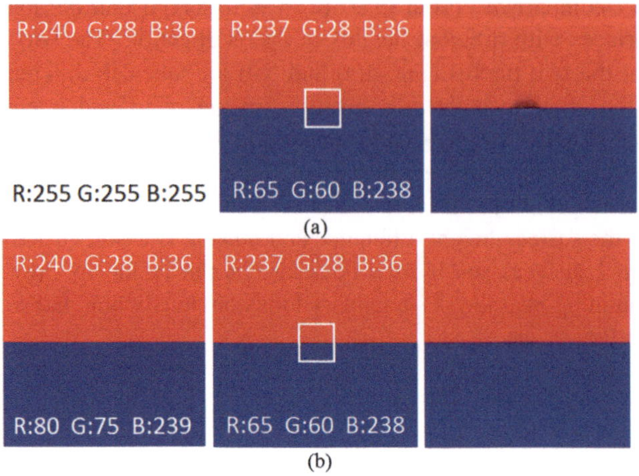

Fig. 6. Perceivable and unperceivable artifacts. (In each case, left: source, middle: target, right: Poisson result).

α such that after Poisson blending, whether the artifact is perceivable can be determined as follows:

$$artifact(R_i, R_j) = \begin{cases} 1 & if\ (ADJ(R_i, R_j) = 1\ and \\ & \quad SHD(R_i, R_j) \geq \alpha) \\ 0 & otherwise; \end{cases} \tag{3}$$

where, both ADJ(.) and SHE(.) are defined as follows.

$$ADJ(R_i, R_j) = \begin{cases} 1 & if\ R_i\ and\ R_j\ are\ adjacent\ regions \\ 0 & otherwise; \end{cases} \tag{4}$$

$$\text{SHD}\ (R_i, R_j) = abs\left(\left(avg(R_i^S) - avg(R_i^T)\right) - \left(avg\left(R_j^S\right) - avg\left(R_j^T\right)\right)\right) \tag{5}$$

where, $avg(R_i^S)$ and $avg(R_i^T)$ stand for the pixel value average of region i in the source image and target image, respectively. Note that $SHD(\mathrm{R}_1, \mathrm{R}_2)$ is conceptually similar to the surface-height difference (SHD) of two adjacent regions in Fig. 5, where the first term in (5) is the height of R_1 surface; while the second term is the height of R_2 surface. The α in (3) represents the threshold for just noticeable noise of Poisson blending. To find out the value of α, many experiments have been conducted, which change pixel values of the regions, perform Poisson blending across region boundary, and observe the blending results with human eyes. After the experiments, it is found that the threshold α is about 10 for each individual color channel. Therefore, (3) is modified to (6) below:

$$artifact(R_1, R_2) = \begin{cases} 1 & \begin{array}{r} if\ (ADJ(R_1, R_2) = 1 \quad and \\ (SHD_R(R_1, R_2) \geq \alpha \quad or \\ SHD_G(R_1, R_2) \geq \alpha \quad or \\ SHD_B(R_1, R_2) \geq \alpha)) \end{array} \\ 0 & otherwise; \end{cases} \tag{6}$$

To improve the blending result when it is predicted to be artifact perceivable, it is proposed to use a region-based method which, in the case of Fig. 4, rather than performs Poisson blending on the mask area as a whole, it will apply the blending to the building and the sky, separately.

3 Region-Based Poisson Blending

The framework of the proposed method with region-based blending is depicted in Fig. 7, where there are three steps: (1) Affine transformation, (2) Image segmentation, and (3) Region-based blending. First, SIFT [10] is used to affine the source image so that the contents of both source and target images have the same viewing angles and scaling. Then, a clustering algorithm is adopted to the affined source image and segment it into regions. Finally, the proposed region-based blending is applied, which utilizes the artifact prediction (6) to decide whether the segmented regions should be applied blending together or separately. The Step 2 has been described in the previous section. The details of the other steps are presented in the following subsections.

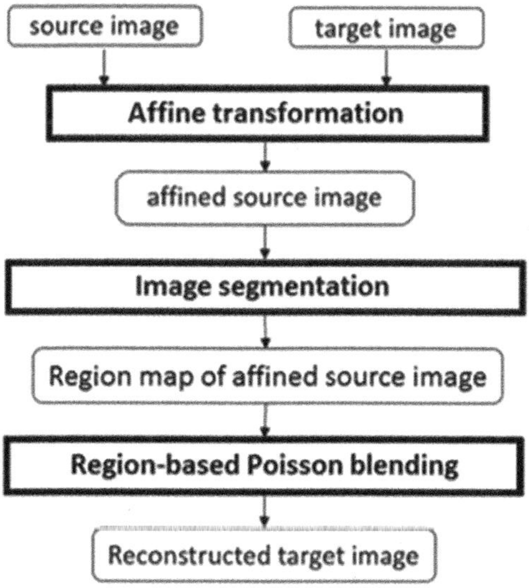

Fig. 7. The framework of the proposed method for image repa.

3.1 Image Segmentation

As mentioned, the blurring artifact comes from that the blending area consists of multiple regions with distinct gradient changes between the regions in s-mask area and the regions in mask area. Therefore, before applying Poisson blending, the proposed method first checks if the mask area contains multiple regions. Considering that there is no available pixel inside mask area, the corresponding s-mask area is used instead. K-means algorithm [14] is used to classify pixels and distinguish regions.

K-means is a well-known algorithm for data clustering. It aims at partitioning n observations into k clusters (In our case, n means total pixels in the s-mask area) so that each observation belongs to the cluster with the nearest mean. To perform the algorithm, the number of clusters, k, should be determined first. Rather than using a fixed value for k, it is proposed to dynamically choose a proper k for each s-mask area according to its intensity histogram, that is, the distribution of pixel intensity-values. First, the s-mask area is converted to gray-scale, applied with a medium filter (kernel size 9), and then the number of pixels for each intensity value is counted and put into the intensity histogram with 256 bins, recorded by an array of 256 elements, called Hgram [256], where Hgram[i] records the number of pixels with intensity value i in the s-mask area. After the histogram is built, it is smoothed by a Gaussian filter with a 9×9 kernel, and then the number of *peak*s in it is examined, where the "peak" in the intensity histogram is defined as the bin with maximal height in the center of $(P_{width} * 2 + 1)$ successive bins. Namely, bin i is said to be a peak if its corresponding Hgram[i] meets the condition below.

$$\text{Hgram}[i] > \text{Hgram}[j], \quad \text{for} \quad \forall j \in [\text{i} - P_{width}, \text{i} + P_{width}] \quad and \quad j \neq i \tag{7}$$

Finally, for k-means algorithm, the number of clusters, k, is chosen to be equal to the number of *peaks* in the intensity histogram. It is worth mentioning that if the value of P_{width} is set too small, there will be too many regions after clustering; on the contrary, there will be too few regions to allow region-based blending to be applied. Although the value of P_{width} will affect the number of peaks, it is not very sensitive nevertheless. In this paper, the value of P_{width} is empirically set to 10, which will result in a proper number of regions in the s-mask area for the proposed region-based repairing method.

Once the number of clusters, k, is determined, K-means algorithm [13] is applied to cluster pixels in the affined source image. The clustering algorithm is briefly described as follows: First, it decides the initial center of each cluster. Then, for each pixel, it calculates Euclidean distance between this pixel and the center of each cluster and then assign this pixel to the cluster with minimal distance. The step is repeated for each unclustered pixel. After all the pixels have been clustered, it will recalculate the center value of each cluster and go for above steps again until the max iterative times are reached or the cluster centers become stable. In this paper, the max iterative number is set to 10; the initial centers of clusters are decided using the method in [15]; and the Euclidean distance is defined as below:

$$d = (x_i - C_k)^2 \tag{8}$$

where, x_i is the i_{th} input pixel value and C_k is the k_{th} center value. Lab color space is used in calculating Euclidean distance, where the Lab color space is based on non-linearly compressed coordinates and is designed to approximate human vision.

Figure 8 shows the segmentation process, where Fig. 8(a) is the affined source image; Fig. 8(b) shows the s-mask area which has been converted to gray-level and smoothed with a median blur filter; Fig. 8(c) shows the intensity histogram of the s-mask area in Fig. 8(b). In this case, the number of peaks (indicated by red lines) is determined as 4; namely, k is set to 4 in k-means clustering. Figure 8(d) shows the clustering result, where the K-means algorithm is implemented by using OpenCV.

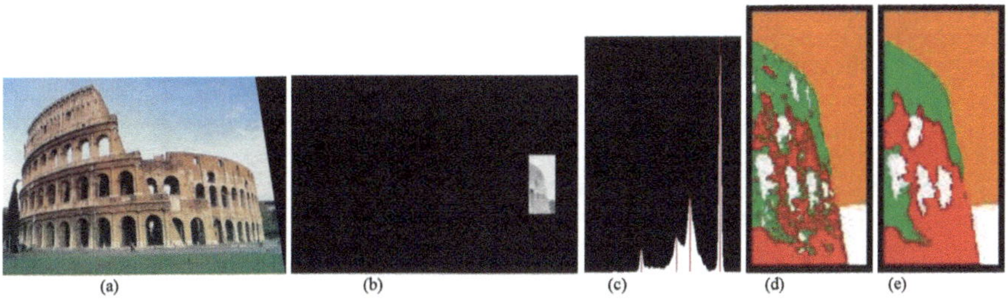

Fig. 8. Image Patch Segmentation (a) Affined source (b) gray-level s-mask area applied with medium filter (c) intensity histogram with red lines indicating the peaks (d) clustering result using k-means algorithm, (e) small region elimination.

In Fig. 8(d), it is observed that there are many small regions that are not suitable to apply blending effectively. Therefore, the proposed method eliminates these small regions by merging them into its adjacent regions. A region is said to be a small region if its pixel counts is less than n which is empirically set to 30 in our experiments. If there are several adjacent regions, the one with average pixel values most close to its average pixel values is selected for merging. The effect of eliminating small regions is also shown in Fig. 8, where Fig. 8(d) is the clustering result without small region elimination, while Fig. 8(e) is that after elimination. Besides, since Poisson blending can only be applied to connected area, the disconnected clusters resulted from k-means algorithm will be labeled as different regions.

3.2 Region-Based Blending

Different from classical Poisson which considers the mask area as a whole for blending, the proposed region- based method may apply blending on regions separately if the area consists of multiple regions and are predicted as artifact perceivable. However, although (6) can be used to predict artifacts, it cannot be directly used in image repairing domain where the mask area in the target image contains pixels to be removed and hence no pixel available for calculating SHD. To make it applicable, the SHD definition in (5) is modified to (9) below, where only pixels on the mask boundary are considered.

$$SHD(R_i, R_j) = abs\left(\left[avg\left(\left(\varphi(R_i^S)\right)\right) - avg\left(\left(\varphi(R_i^T)\right)\right)\right] - \left[avg\left(\varphi\left(R_j^S\right)\right) - avg\left(\varphi(R_j^T)\right)\right]\right) \tag{9}$$

where, $\varphi(R_i^S)$ denotes s-mask boundary pixels that connected to region R_i^S and $\varphi(R_i^T)$ denote the mask boundary pixels connected to region R_i^T. Figure 9 shows an example, where assume that s-mask is segmented into four regions $R_1 \sim R_4$. For R_1 and R_4, assume their average pixel values of $\varphi(R_1^S)$ and $\varphi(R_4^S)$ are 210 and 130, respectively; and the average pixel values of $\varphi(R_1^T)$ and $\varphi(R_4^T)$ are 110 and 35, respectively. Then SHD(R_1, R_4) is equal to $|(210 - 110) - (130 - 35)| = 5$ which is less than α leading to *artifact*(R_1, R_4) = 0 according to (6). Namely, artifacts between R_1 and R_4 are not perceivable. As a result, R_1 and R_4 will be merged as one for Poisson blending together, as shown in Fig. 9(c).

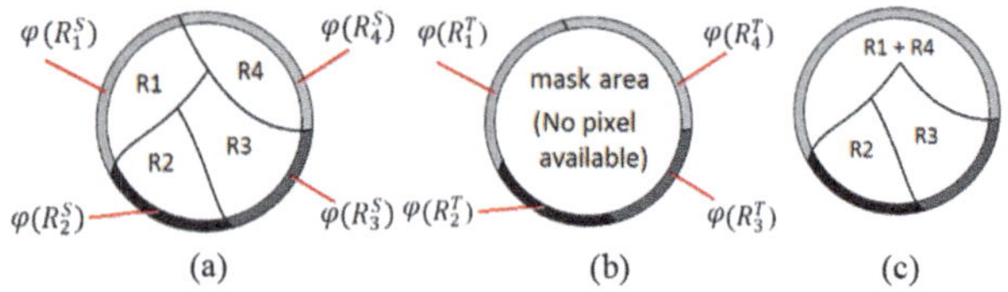

Fig. 9. Region merging.

Since s-mask may contain multiple regions resulted from image segmentation step, for every two connected regions among them, (6) with modified SHD definition in (9) is examined, and finally how many regions to apply Poisson blending separately can be determined.

Applying Poisson equation to a region requires that this region have all its boundary pixels available for blending. However, in the target image, there are boundary pixels for the mask area, but not for each region. That is, to do Poisson blending on R_1^T in Fig. 10, in addition to the pixels in $\varphi(R_1^T)$ which are available, all the pixels in the red line (denoted by $\tilde{\varphi}(R_1^T)$ which are unavailable) are needed. Similarly, the pixels in both $\varphi(R_2^T)$ and $\tilde{\varphi}(R_2^T)$ are also needed for R2 blending, but the pixels in $\tilde{\varphi}(R_2^T)$ are unavailable.

For the available and unavailable boundaries, $\varphi(R_i^T)$ and $\tilde{\varphi}(R_i^T)$, of region R_i on the target image, their corresponding boundaries on the source image are denoted as $\varphi(R_i^S)$

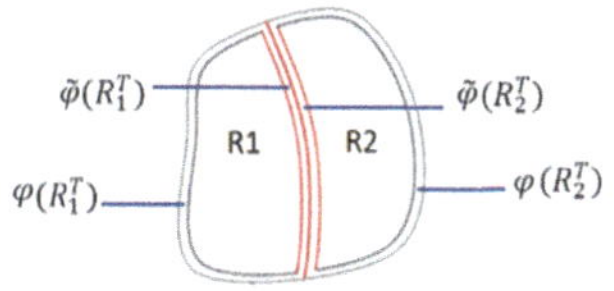

Fig. 10. Missing boundary and available boundary.

and $\tilde{\varphi}(R_i^S)$, respectively. To predict pixel values in unavailable boundary $\tilde{\varphi}(R_i^T)$, it is proposed to use (10) below:

$$tp = sp - \left(\frac{sum\left(\varphi(R_i^S) - outlier(\varphi(R_i^S))\right)}{|(\varphi(R_i^S) - outlier(\varphi(R_i^S))|} - \frac{sum\left(\varphi(R_i^T) - outlier(\varphi(R_i^T))\right)}{|(\varphi(R_i^T) - outlier(\varphi(R_i^T))|} \right) \quad (10)$$

where, $t_p \in \tilde{\varphi}(R_i^T)$ and s_p is corresponding pixel $\in \tilde{\varphi}(R_i^S)$. The outlier(K) denotes the set that each element in it is an outlier to the set P. The outlier definition in [11] is adopted. According to it, the outlier(K) is defined as follows:

$$x \in \text{outlier(K) if} \begin{cases} x > \bar{K} + 3 * \text{SD(K)}, & \text{or} \\ x < \bar{K} - 3 * \text{SD(K)} \end{cases} \quad (11)$$

where, $x \in$ K, and SD(K) is the Standard Deviation of K. With (10), an integral boundary $\varphi(R_i^T) \cup \tilde{\varphi}(R_i^T)$ can be created for each region R_i and applied with Poisson blending. Thus, (2) can be re-written as below for the region-based Poisson blending method:

$$|N_p| f_p - \sum_{q \in N_p \cap \left(\varphi(R_i^T) \cup \tilde{\varphi}(R_i^t)\right)} f_q^* + \Sigma_{q \in N_p} v_{pq}, \quad \text{for all} \quad \text{p} \in R_i^T \quad (12)$$

If there is a region inside mask area and not connected to the mask boundary, then its entire boundaries are not available. Such regions will be merged into their adjacent regions which has the closest average pixel values.

4 Experimental Result

To evaluate the performance of proposed region-based method, two kinds of image sets are created for experiments. One set is that each target image in it has the ground truth for evaluating the repairing result. The other set is that each image in it does not have the ground truth.

The experiment on the first-set images is that some area is assumed to be unwanted, so we remove this area and apply the repairing methods to it for recovery. After recovering the image, quality assessment metrics are applied to see how well the reconstructed image looks like the original image. Two widely adopted image quality assessment metrics: PSNR and SSIM [12] are used in our experiments. The experiment on the second set of the images is that some objects on the image are intended to be removed and recovered with background. The performance for this set is evaluated by assessing how well the recovered background looks like the neighboring background and the assessment is done by visual perception with human eyes.

To evaluate the performance of the region-based method, the traditional Poisson method is used for comparison. Figure 11(a) and (b) show the results of experiments on the first set of images and Fig. 11(c) and (d) show the results of the second set. In each figure, the first image is the source image obtained from Internet; the second one is the target image with a red rectangle indicating the mask area; the third one shows classical Poisson blending results; and the last one is the result of our method.

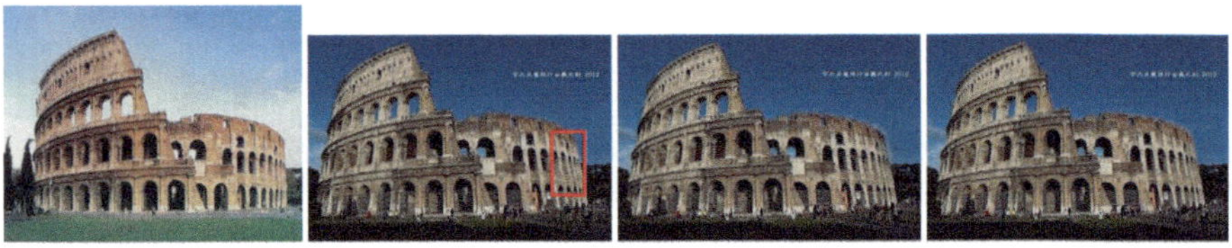

(a) Left most: source, 2nd column: target, 3rd column: classical Poisson, right most: proposed

(b) Left most: source, 2nd column: target, 3rd column: classical Poisson, right most: proposed

(c) Left most: source, 2nd column: target, 3rd column: classical Poisson, right most: proposed

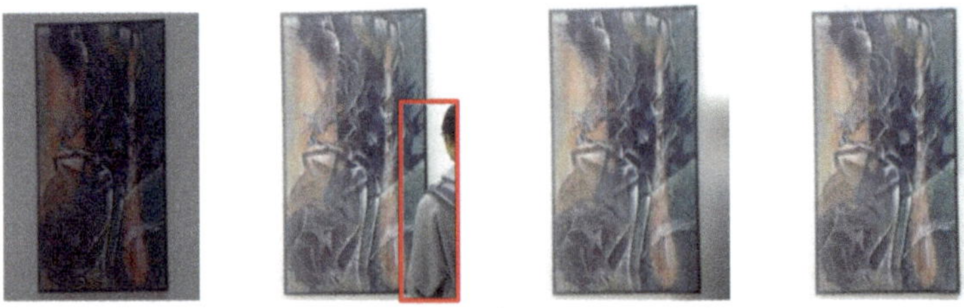

(d) Left most: source, 2nd column: target, 3rd column: classical Poisson, right most: proposed

Fig. 11. Result comparison.

In these figures, the results of using classical Poisson are shown in the 3rd column, where the recovered area shows obvious artifacts. In Fig. 11(a), the recovered dinosaur head and neck look much brighter than the body. In Fig. 11(b), there are unexpected noises appearing around the boundary between the sky and the building. In Fig. 11(c), it looks blurry and brighter on the recovered parts of the building, especially the dark area inside arched doors. In Fig. 11(d), the recovered wall on the right side of the picture shows obvious shadow. All these problems of using Poisson method have been improved effectively by the proposed region-based method, as can be seen in the last column of Fig. 11(a) to (d). For easy comparison, a close-up of recovered area in each case is shown in Fig. 12, where the region map of the proposed method is also presented. It is observed that since a suitable region map is utilized to guide the blending, the region-based method can eliminate artifacts effectively.

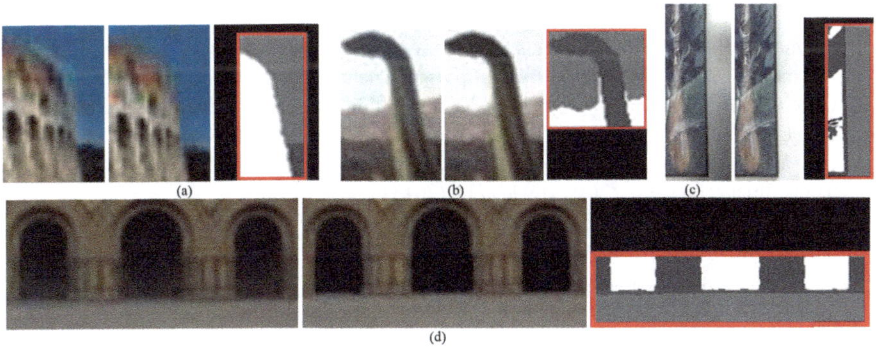

Fig. 12. Close-up of the recovered area and the region map (In each case, left: classical Poisson, middle: proposed, right: region map).

In addition to visual perception effect, the proposed method also performs better than classical Poisson in objective quality metrics. Since the first two sets of images, Fig. 11(a) and (b), have ground truths as mentioned, their PSNR and SSIM values can be measured accordingly. The results are listed in Table 1. Since both PSNR and SSIM values are the higher the better, the results in Table 1 show that the proposed method outperformed classical Poisson for both objective and subjective metrics in each case.

Table 1. Comparison in terms of PSNR and SSIM

	Classic poisson		Proposed method	
	PSNR	SSIM	PSNR	SSIM
Fig. 12(a)	27.90 dB	0.87	33.22 dB	0.91
Fig. 12(b)	26.57 dB	0.80	29.18 dB	0.86

5 Conclusion

Poisson blending algorithm is mainly designed for fusing objects into images. It has been used in repairing images recently, where reference images are utilized to repair the damaged area in the target image. However, it is observed that poor repairing results might occur when the Poisson blending is applied in the situation that the gradient change in the source image is largely different from that in the target image. To cope with the problem, a region-based method is presented to repair images, where the source image is segmented into regions and then the target image is repaired by regions accordingly. In the experimental results, it can be seen that much robust results can be obtained by the proposed method, compared to classical Poisson blending.

References

1. Vidmar, R.J.: On the use of atmospheric plasmas as electromagnetic reflectors. IEEE Trans. Plasma Sci. **21**(3), 876–880 (1992). http://www.halcyon.com/pub/journals/21ps03-vidmar
2. Criminisi, A., Perez, P., Toyama, K.: Region filling and object removal by exemplarbased image inpainting. IEEE Trans. Image Process. **13**, 1200–1212 (2004)
3. Liu, Y., Caselles, V.: Exemplar-based image inpainting using multiscale graph cuts. IEEE Trans. Image Process. **22**(5), 1699–1711 (2013)
4. Sun, J., Yuan, L., Jia, J., Shum, H.Y.: Image completion with structure propagation. ACM Trans. Graph. (ToG) **24**(3), 861–868 (2005)
5. Hang, H., Lin, A.: Database-assisted interactive mobile image completion (2010). http://www.stanford.edu/class/ee368
6. Amirshahi, H., Kondo, S., Ito, K., Aoki, T.: An image completion algorithm using occlusion-free images from internet photo sharing sites. IEICE Trans. Fund. Electron. Commun. Comput. Sci. **91**(10), 2918–2927 (2008)
7. Whyte, O., Sivic, J., Zisserman, A.: Get out of my picture! internet-based inpainting. In: BMVC (2009)
8. Chavan, T.R., Trupti, R., Nandedkar, A.V.: Digital image inpainting using speeded up robust feature. In: 2014 International Conference on Advances in Computing, Communications and Informatics (ICACCI). IEEE (2014)
9. Pérez, P., Gangnet, M., Blake, A.: Poisson image editing. ACM Trans. Graph. (TOG) **22**, 313–318 (2003)
10. Lowe, D.: Distinctive image features from scale-invariant keypoints. Int. J. Comput. Vis. **60**(2), 91–110 (2004)
11. Shiffler, R.E.: Maximum Z scores and outliers. Am. Stat. **42**(1), 79–80 (1988)
12. Wang, Z., Bovik, A.C., Sheikh, H.R., Simoncelli, E.P.: Image quality assessment: from error visibility to structural similarity. IEEE Trans. Image Process. **13**(4), 600–612 (2004)
13. MacQueen, J.B.: Some methods for classification and analysis of multivariate observations. In: Proceedings of the Fifth Berkeley Symposium on Mathematical Statistics and Probability, vol. 1, no. 14 (1967)
14. Bay, H., Tuytelaars, T., Van Gool, L.: Surf: speeded up robust features. In: European Conference Computer Vision, pp. 404–417. Springer, Heidelberg (2006)
15. Arthur, D., Vassilvitskii, S.: K-means ++: the advantages of careful seeding. In: Proceedings of the 18th Annual ACM-SIAM Symposium on Discrete Algorithms (2007)

Modified Radial Basis Function and Orthogonal Bipolar Vector for Better Performance of Pattern Recognition

Camila da Cruz Santos[1(✉)], Keiji Yamanaka[1], José Ricardo Gonçalves Manzan[2], and Igor Santos Peretta[1]

[1] Faculty of Electrical Engineering, Federal University of Uberlândia, Uberlândia, Brazil
ccamila.cruz.santos@gmail.com, iperetta@gmail.com, keiji@ufu.br

[2] Câmpus Uberaba Parque Tecnológico, Federal Institute of Triângulo Mineiro, Uberaba, Brazil
josericardo@iftm.edu.br

Abstract. This work proposes the use of orthogonal bipolar vectors (OBV) as new target vectors for Artificial Neural Networks (ANN) of the Radial Basis Functions (RBF) type for pattern recognition problems. The network was trained and tested with three sets of biometric data: human iris, handwritten digits and signs of the Australian sign language, Auslan. The objective was to verify the network performance with the use of OBVs and to compare the results obtained with those presented for the Multilayer Perceptron (MLP) networks. Datasets used in the experiments were obtained from the CASIA Iris Image Database developed by the Chinese Academy of Sciences - Institute of Automation, Semeion Handwritten Digit of Machine Learning Repository and UCI - Machine Learning Repository. The networks were modeled using OBVs and conventional bipolar vectors for comparing the results. The classification of the patterns in the output layer was based on the Euclidean distance. The results show that the use of OBVs in the network training process improved the hit rate and reduced the amount of epochs required for convergence.

Keywords: Human iris · Handwritten digit · Auslan sign
Orthogonal bipolar vector · Radial basis function
Multilayer perceptron · Pattern recognition · Target vector

1 Introduction

Artificial neural networks (ANNs) are computational tools that are part of a large area of knowledge called Computational Intelligence. In recent years, research in this area has grown a lot, due to the ability of these complex problem-solving tools.

Of course, several studies have been carried out in order to improve the performance of this tool. These studies have approaches such as the improvement of the training algorithm, the determination of ideal topologies for each problem, among

K. Arai et al. (Eds.): IntelliSys 2018, AISC 868, pp. 431–446, 2019.
https://doi.org/10.1007/978-3-030-01054-6_31

others. The objective of this work is to verify if the vector orthogonalization technique improves the performance of Radial Basis Function (RBF) networks for different problems.

The motivation is presented in Sect. 2. Section 3 presents the architecture of the implemented networks. Section 4 presents the description of the experimental procedure performed in the work. The results are demonstrated in Sect. 5 and discussed in Sect. 6. The conclusion is presented in Sect. 7.

2 Motivation

In previous studies, it has been shown that the use of orthogonal bipolar vectors (OBVs) as target in the ANN training process of the Multilayer Percepton (MLP) type provided an increase in the recognition rate of biometric patterns, and decreased the number of epochs needed for network convergence [16–20].

The improvement in network performance occurs due to the orthogonality property of the OBVs, where the Euclidean distance of the target with the output produced by the network is greater when compared to the conventional vectors.

RBF network is considered a universal approximation, and it is a popular alternative to MLP, since it has a simpler structure and a much faster training process [15]. This type of network is employed in several types of problems that involve function approximation and pattern recognition [7].

Considering that pattern recognition problems can be solved by ANNs of the MLP and RBF type, and that the use of OBVs as MLP targets provided performance gain, it is desired to investigate whether these new targets also allow performance gains for ANNs of RBF type.

3 Network Architecture

Radial basis function networks, conventionally known as RBF, are neural networks that use the feedforward architecture, i.e. data is propagated from the input layer to the hidden layer until it reaches the output layer, without feedback from the network [68].

3.1 Traditional RBF

The traditional construction of a RBF has three completely different layers, which can be observed in Fig. 1. The input layer consists of sensory units that interconnect the network with the external environment. The hidden layer (RBF neurons) makes nonlinear transformations in the input space, being characterized by the use of radial basis functions. The output layer is responsible for the final response of the modeled network [7].

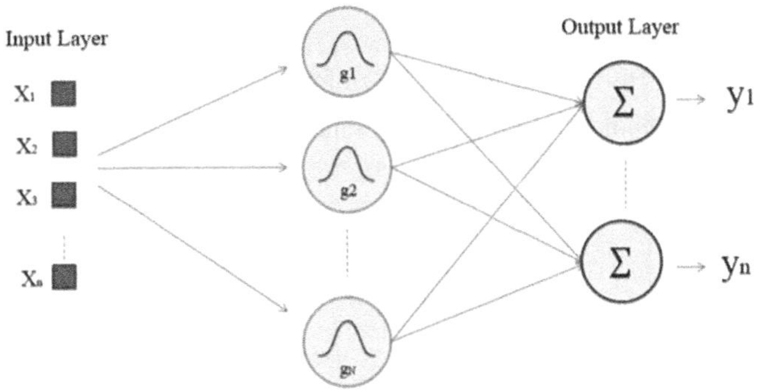

Fig. 1. Traditional RBF architecture.

In the hidden layer, the Gaussian (1) and multi-square (2) functions are among the most used ones.

$$\varphi(x) = e^{-\frac{\sum_{i=1}^{n}(x_i - W_{ji})^2}{2\sigma^2 j}} \tag{1}$$

$$\varphi(x) = \sqrt{\sum_{i=1}^{n}(x_i - W_{ji})^2 + \sigma^2} \tag{2}$$

In (1) and (2) x_i represents data received by the input layer, w_{ji} weights assigned to the hidden layer, and σ the sample variance.

3.2 Modified RBF

For the work a modification to the output layer of the RBF network was done, changing the training process, updating its weights, and using nonlinear activation functions. The hyperbolic tangent (3) and bipolar logistics (4), both with an image limited to the real interval [−1, 1], were used for this modification. The process of training and adjustment of weights of the network is present in Sect. 4.C. The classification approach was based on Euclidean distance, explained in Sect. 4.D.

$$f(x) = \frac{1 - e^{-\beta x}}{1 + e^{-\beta x}} \tag{3}$$

$$f(x) = \frac{2}{1 + e^{-\beta x}} - 1 \tag{4}$$

3.3 Definition of Vector

To perform the experiments, we used three types of targets [2], defined as follows:

(1) *Conventional Bipolar Vector (CBV):* Represented by (5), has the elements "1" and "−1" to determine the correct output of the network. In the experiments, the size of the vector is determined by the number of patterns to be classified.

$$\underset{V_{ij}}{\rightarrow} = \begin{cases} 1\,for\,i = j \\ -1\,for\,i \neq j \end{cases} \tag{5}$$

(2) *Orthogonal Bipolar Vector (OBV):* They are characterized by being mutually orthogonal; for mathematical reasons, inherent to the generation algorithm, their size is always a power of 2. For OBV generation, the theorem proposed by [8] is used, where $n = 2^{km}$ is used to calculate the n components of the vector, m is the number of components of the seed vector V, and k is the number of orthogonal vectors to be generated. The operation [V, V] represents the concatenation of the vectors. The algorithm runs as described below:
   (a) *Initiation of the seed vector:* $V_{m(1)} = (1,1, \ldots, 1)$;
   (b) *Concatenation of the seed vector to generate mutually orthogonal vectors:* $V_{2m(1)} = [V_{m(1)}, V_{m(1)}]$ and $V_{2m}(2) = [V_{m(1)}, -V_{m(1)}]$;
   (c) *Concatenation of the orthogonal vectors:* $V_{4m(1)} = [V_{2m(1)}, V_{2m(1)}]$, $V_{4m(2)} = [V_{2m(1)}, -V_{2m(1)}]$, $V_{4m(3)} = [V_{2m(2)}, V_{2m(2)}]$ e $V_{4m(4)} = [V_{2m(2)}, -V_{2m(2)}]$;
   (d) *Repeat step "c" until generation* $V_{n(1)}, \ldots, V_{n(2k)}$ *as OBVs.*

(3) *Non-orthogonal Vector (NOV):* This is a type of CBV, but with the same dimension of OBV, used for comparison in the experiments.

4 Experimental Procedure

The experiments performed for RBF training and testing were done with three different datasets presented in this section. For each of the sets, three types of target with different characteristics and dimensions were used.

4.1 Experimental Data

(1) *Human Iris:* the dataset was obtained from the Chinese Academy of Sciences - Institute of Automation database, CASIA [1] (Fig. 2). For the training and testing of the network, 7 samples of each of the 70 individuals that had the complete set of registered images were used. The images were obtained using infrared lights for better contrast and sharpness. Each image has 18 concentric circles. The first five circles from the center of the circle were used in order to eliminate the interference of the eyelashes, reducing computational effort [2]. The detection of the circumferences for removal of the pixels follows the Hough Circular transform used by [13]. Each circumference on the iris has 480 pixels, so each training

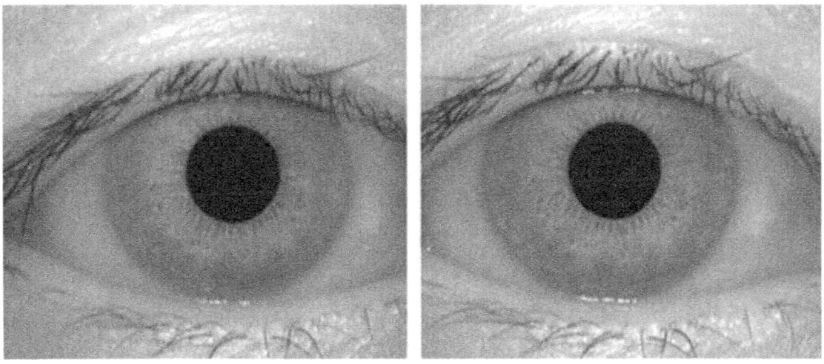

Fig. 2. Example iris images in CASIA-IrisV1.

pattern corresponds to a set of $5 \times 480 = 2400$ pixels represented by "−1", the white pixels, and "1", the black ones. The training vectors represent the 2400 pixels of the 5 circumferences of each image.

(2) *Handwritten Digit*: data from this set of tests were obtained through the international repository Semeion Handwritten Digit of Machine Learning Repository [3]. We used data from about 90 individuals who were asked to write the numbers twice from 0 to 9. Firstly they should write them down quietly and secondly they were asked to write them very quickly. The figures were scanned and the image had 256 pixels arranged in a square matrix of 16 columns. The pixels corresponding to the bottom of the image were given the value "−1" and the other positions of the matrix received the value "1".

(3) *Auslan Signs:* The last set used was one of the signs of the Australian sign language obtained from the UCI - Machine Learning Repository [3]. The set contains 27 samples from each of the 95 signs. They were obtained using high quality position trackers in native individuals [4]. The captured data represent positions relative to the chin, palm, fingers, but are not totally accurate (Fig. 3).

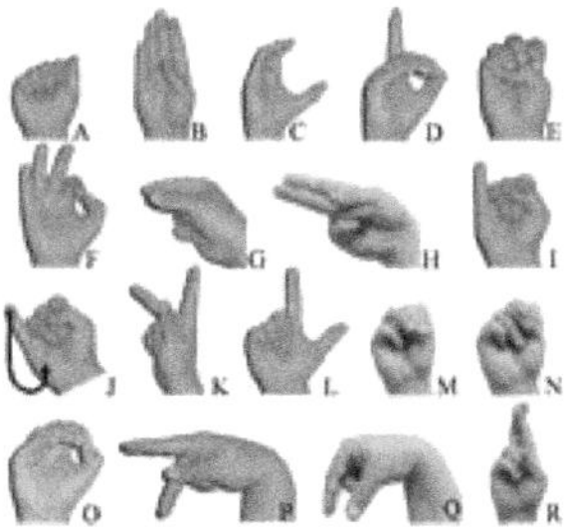

Fig. 3. Example of Australian Sign.

4.2 Target Vectors

For each set of data tested, the three types of target vectors described in Sect. 3.C were used. The size used for training the network in each set of data is described in Table 1.

Table 1. Size of target vectors

Data	CBV	OBV	NOV
Human iris	70	128	128
Handwritten digit	10	16	16
Auslan sign	95	128	128

4.3 RBF Training Stage

The training process of an RBF is divided into two stages: the learning and adjustment of the weights of the hidden layer and that of the output layer. The network layers have totally different functions, so their optimization process is done using different techniques and at different time scales [9].

In [6], author states that there are several strategies for training an RBF network that vary depending on how the centers of the radial base functions are specified. We used the training by self-organized selection of centers, where the learning process is hybrid, separated in two stages:

- *Self-organized learning stage:* Performed in the hidden layer, whose function is to estimate the appropriate location for the centers of the radial base functions.
- *Supervised learning stage:* Responsible for adjusting the weights of the output layer.

For the unsupervised phase of the two implementations of the network (traditional and modified), the k-means algorithm [10] was used, responsible for grouping the samples into homogeneous subgroups. The algorithm places de center of the radial base functions in the regions where the input space has the most significant data of that group.

In the implementation of the traditional algorithm, the adjustment of the weights of the output layer (supervised stage) was done through the technique of optimization of the weights proposed in [15]. The first part is the determination of the matrix Φ whose elements (i, j) are $\Phi = \varphi_k\,(x(i), c_j)$, where φ_k represents the chosen function and c_j the centers determined by the k-means. The optimum weights vector w_{ls} is obtained by (6). When the regularization parameter λ approaches 0, the vector w_{js} converges to the pseudo-inverse solution of the problem (7). Pseudo-inverse matrix Φ^+ of the matrix Φ is represented by (8).

$$\left(\Phi^T \lambda\, \Phi\right) w_{ls} = \Phi^T y_d \tag{6}$$

$$w_{ls} = \Phi^+ y_d, \quad \lambda = 0 \tag{7}$$

$$\Phi^+ = \left(\Phi^T \Phi\right)^{-1} \Phi^T \tag{8}$$

With this technique of updating weights, in only one epoch the results obtained are those considered optimal for that execution. In this way, the network can be trained more quickly when compared to other learning algorithms.

The supervised network stage in the modified implementation was performed through the Generalized Delta Rule technique, common in MLP training, but without the backpropagation phase of the error in input layer. The initial weights used in this training phase were randomly initialized in the interval [−0.5, 0.5]. Learning rate was started at 0.005 and updated at runtime, as specified by [11]. In addition, the *momentum* term was added, whose value is comprised in the interval [0, 1] and whose objective is to make the network convergence process more efficient. When the difference between the matrices is large, the contribution of the *momentum* will also be considerable. In contrast, when the solution is close to expected, the variation will be irrelevant, since the variation of the errors between two interactions will be low. The update of the network output layer weights after the use of the *momentum* term is given by [7]:

$$w_i(t+1) = w_i(t) + \underbrace{\beta.(w_i(t) - w_i(t-1))}_{momentum\ term} + \underbrace{\alpha.\delta_i.Y_i}_{learning\ term} \quad (9)$$

where β is defined as the *momentum* rate, whose recommended training value is between ($0 \leq \beta \leq 0.9$) [14], w_i are the weights of class *i*, α is the value of the network learning rate, δi represents the local gradient relative to the i-th neuron of the output layer, and Y_i the output expected by the network.

The simulations were done on a computer with AMD FX 6300 processor, 3.5 GHZ–4.1 GHZ and 16 GB of memory (RAM).

4.4 Classification Approach

In this paper, a classification approach based on Euclidean distance was adopted. This technique states that every presentation of an instance I (input vector of space α) is calculated as the Euclidean distance between the output vector in space β with all the target vectors in β of each class. The class whose representation presents the lowest result is chosen as "winner" and associated with instance I [12].

In the use of OBVs, each class, in a multiclass problem, is represented by a single bipolar vector of size n, called OBV. During the training of instance *I* in α, the desired outputs of the *n* bipolar functions are specified by the OBV of class *i*. In networks of type MLP and RBF, these *n* functions are implemented by *n* output units of the network.

4.5 Statistical Analysis

Statistical methods are divided into two categories: parametric and nonparametric. These two categories differ in specific parameters for execution. In addition, the parametric ones can only be applied to data that follow a normal distribution and the others do not.

Therefore, important information needed before, deciding which test to use, is whether the distribution of the data to be analyzed fits into a normal distribution. The Kolmogorov Smirnov test allows us to verify if a sample of data fits into some theoretical distribution, e.g. the normal distribution [22]. This is a widely used test and in the statistical analyzes of this work, it was used to decide on the use of a parametric or nonparametric test.

When the data sample does not fit the normal probability distribution, it is recommended the use of nonparametric tests. When the problem is to compare means between independent groups, in addition to using a nonparametric test, there is also a requirement that the variances be the same. The Mann-Whitney test allows the comparison of equality of means without these requirements [23].

Statistical tests were performed in statistical analysis software R, version 3.4.3 [5].

5 Results

Six models of RBFs were implemented according to the proposed experimental procedure using Matlab version R2016a as a computational environment. We performed 100 experiments with each proposed model and the graphs and tables presented are a result of the average of the executions.

The results obtained with the traditional modeling of all data sets are arranged in Table 2. Due to the use of the pseudo-inverse optimization of the matrix Φ, the network is trained in only one epoch. The results represent the average performance obtained in 100 runs of each set of data with the corresponding vector.

Table 2. Results of traditional RBF network architecture

Data	# Clusters	CBV	NOV	OBV
Human iris	75	87.46	87.64	87.70
Handwritten digit	70	89.77	89.56	89.51
Auslan sign	150	57.76	55.90	55.97

For the data to be fairly compared, Table 3 presents the mean of overall performance of the RBF in its modified implementation, after 100 runs with each training pattern. To obtain these data, the network was trained until the stabilization of the error, so the number of epoch was variable.

Table 3. Mean performance of modified RBF network architecture

Data	# Clusters	CBV	NOV	OBV
Human iris	75	93.57	93.57	97.14
Handwritten digit	70	90.86	91.11	90.12
Auslan sign	150	76.04	77.07	73.68

In the experiments done for comparison with MLP, the training of the network was stopped in 50 epochs, following the same pattern of training done in previous works, in which the OBVs were used for the same training sets in MLP type networks.

The Kolmogorov-Smirnov normality test [22] was applied in the results of the 100 experiments performed with each of the data sets. The results are presented in Tables 4, 5 and 6, representing the values obtained with the human iris, the handwritten digits, and the Australian signs, respectively. In this test, when the p value is less than 0.05, it indicates that the samples do not follow the normal distribution and a nonparametric test is required.

Table 4. Kolmogorov–Smirnov normality test – human iris

Epoch	D	p
1	0.415	2.20E−16
5	0.452	2.20E−16
10	0.338	2.20E−16
15	0.3545	2.20E−16
20	0.362	2.20E−16
25	0.35282	2.20E−16
30	0.333	2.20E−16
35	0.31928	2.20E−16
40	0.289	2.20E−16
45	0.24954	2.20E−16
50	0.201	5.85E−11

Table 5. Kolmogorov–Smirnov normality test – handwritten digits

Epoch	D	p
1	0.16976	6.189E−08
5	0.13023	7.612E−05
10	0.11904	4.059E−04
15	0.11199	1.078E−03
20	0.16836	8.213E−08
25	0.21399	2.336E−12
30	0.21021	6.124E−12
35	0.20797	1.073E−11
40	0.19202	4.934E−10
45	0.15332	1.5E−06
50	0.16771	9.379E−08

Tables 7, 8 and 9 show the results obtained in the Mann-Whitney test, which compares the different targets used in the modeling of the modified RBF. For each comparison, the test showed a value of p. If p is greater than 0.05 samples it shows no appreciable differences, otherwise the samples are significantly different.

Table 6. Kolmogorov–Smirnov normality test – auslan signs

Epoch	D	p
1	0.36888	2.20E−16
5	0.36478	2.20E−16
10	0.34794	2.20E−16
15	0.3128	2.20E−16
20	0.30104	2.20E−16
25	0.27252	2.20E−16
30	0.23323	1.33E−14
35	0.22485	1.34E−13
40	0.19447	2.80E−10
45	0.18117	5.60E−09
50	0.14487	6.79E−06

Table 7. Mann–Whitney test statistics for the comparison of the experimental results – human iris

Epoch	Comparison		
	CBV70 × NOV128	CBV70 × OBV128	NOV128 × OBV128
1	0.8883	2.20E−16	2.20E−16
5	0.1231	2.20E−16	2.20E−16
10	0.2019	2.20E−16	2.20E−16
15	0.9238	2.20E−16	2.20E−16
20	0.3234	2.20E−16	2.20E−16
25	0.4522	2.20E−16	2.20E−16
30	0.7016	2.20E−16	2.20E−16
35	0.3598	2.20E−16	2.20E−01
40	0.07463	2.20E−16	2.20E−16
45	0.4688	2.20E−16	2.20E−16
50	0.865	2.20E−16	2.20E−16

Table 8. Mann–Whitney test statistics for the comparison of the experimental results – handwritten digits

Epoch	Comparison		
	CBV10 × NOV16	CBV10 × OBV16	NOV16 × OBV16
1	0.351129	4.19E−33	1.20E−31
5	0.274799	2.20E−16	2.20E−16
10	0.227802	3.00E−31	4.40E−30
15	0.231298	5.08E−34	1.16E−33
20	0.181673	1.28E−34	1.58E−34
25	0.330096	2.88E−34	4.00E−34

(*continued*)

Table 8. *(continued)*

Epoch	Comparison		
	CBV10 × NOV16	CBV10 × OBV16	NOV16 × OBV16
30	0.161363	1.06E−34	1.23E−34
35	0.066418	1.24E−34	1.36E−34
40	0.266421	1.35E−34	1.75E−34
45	0.057824	6.58E−34	2.36E−34
50	0.351129	4.19E−33	1.20E−31

Table 9. Mann–Whitney test statistics for the comparison of the experimental results – Auslan sign

Epoch	Comparison		
	CBV95 × NOV128	CBV95 × OBV128	NOV128 × OBV128
1	0.5857	2.20E−16	2.20E−16
5	0,2570	2.20E−16	2.20E−16
10	0.7572	2.20E−16	2.20E−16
15	0.5047	2.20E−16	2.20E−16
20	0.7759	2.20E−16	2.20E−16
25	0.8969	2.20E−16	2.20E−16
30	0.1486	2.20E−16	2.20E−16
35	0.1469	2.20E−16	2.20E−16
40	0.4853	2.20E−16	2.20E−16
45	0.3775	2.20E−16	2.20E−16
50	0.4639	2.20E−16	2.20E−16

In Fig. 4, we present the results obtained after a modification of the supervised stage of the network, using learning rate $\alpha = 0.005$ and *momentum* term $\beta = 0.9$. The graph represents an average performance of each epoch in 100 runs of the network.

The results obtained by the data set of the handwritten digits, after modification of the supervised network stage are shown in Fig. 5. We used a learning rate $\alpha = 0.005$ and a momentum term $\beta = 0.9$.

The results of the Auslan Signs data set found after modifying the supervised network stage are shown in Fig. 6, using the learning rate $\alpha = 0.001$ and the momentum term $\beta = 0.9$.

Tables 10, 11 and 12 show the results obtained by the three data sets during the execution of the 100 experiments, stopped in the 50th epoch, in each dataset.

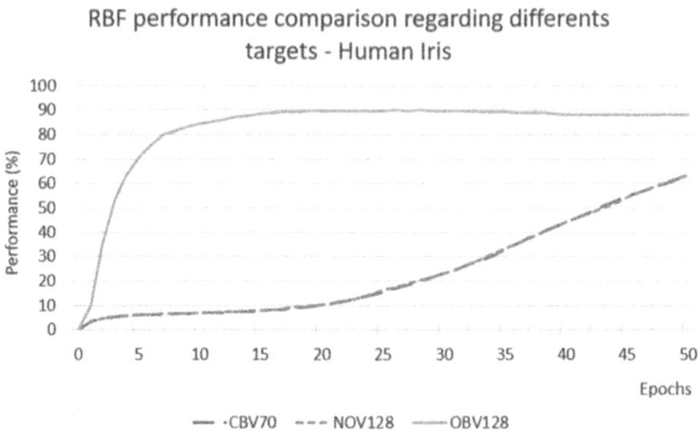

Fig. 4. Performance results for RBF topology with 75 clusters for Human Iris database – Modified network architecture.

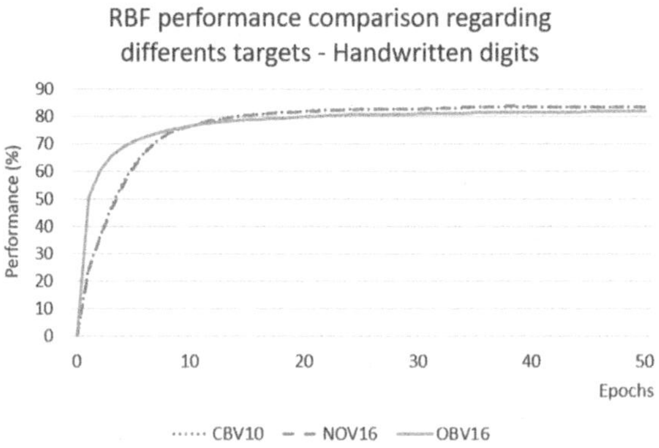

Fig. 5. Performance results for RBF topology with 70 clusters for Handwritten digits database – Modified network architecture.

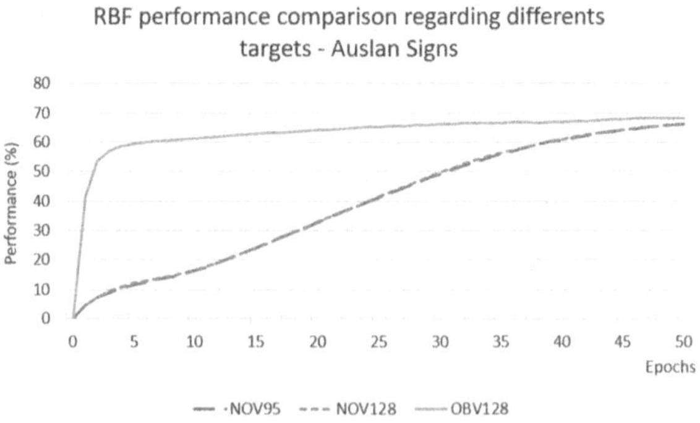

Fig. 6. Performance results for RBF topology with 150 clusters for database Auslan Signs – Modified network architecture.

Table 10. Average performance – human iris

Parameters analyzed	CBV70	NOV128	OBV128
Mean performance at the first epoch	3.43	3.37	10.02
Mean performance at the fifth epoch	6.12	6.27	70.75
Mean performance in the 50 epoch	63.47	63.55	88.2
Overall performance	74.28	75.71	90

Table 11. Average performance – handwritten digits

Parameters analyzed	CBV10	NOV16	OBV16
Mean performance at the first epoch	24.45	23.96	50.75
Mean performance at the fifth epoch	62.64	62.02	71.19
Mean performance in the 50 epoch	83.33	83.45	81.89
Overall performance	84.20	84.44	82.72

Table 12. Average performance – Auslan sign

Parameters analyzed	CBV95	NOV128	OBV128
Mean performance at the first epoch	4.44	4.57	41.55
Mean performance at the fifth epoch	11.61	12	59.33
Mean performance in the 50 epoch	66.08	66.16	68.34
Overall performance	68.07	68.42	68.65

6 Discussion

Figures 4, 5 and 6 represents the average performance obtained by the execution of the RBF network in the implementation in which the supervised stage of the network was modified. The graphs demonstrate the superiority of the OBVs in relation to the other vectors used, being remarkable in the three datasets that in the first five training epoch the OBVs were able to reach a higher hit rate than the others.

Tables 10, 11 and 12 also demonstrates the results presented in the graphs, highlighting some characteristics, such as the average performance after the first and fifth training epoch. These values show us that when OBVs are used in pattern recognition tasks, convergence occurs with a much shorter number of epochs when compared to the other vectors.

In the experiments done with the iris images, in the first epoch the network presented 10.02% accuracy with OBV_{128}, while using CBV_{70} was only 3.43% correct and with NOV_{128} with 3.37%. In the 5th epoch, the hit rate with OBV_{128} was 70.75%, with the use of CBV_{70} the performance was 6.12% and with NOV_{128} the hit rate was 6.27%. The best performance of the OBVs continues in the later epochs, in the 10th epoch the network presents 6.92% of correctness with the CBV_{70}, 6.97% with the NOV_{128} and 84.30% using OBV_{128}.

The results with the Australian Signals also present the same characteristic. In the 5th epoch the modeling of the network that uses OBV_{128} as target has 59.33% of correctness, whereas with CBV_{95} the hit rate is 11.61%, and using NOV_{128} the percentage of success is 12%. In the handwritten digits, in the first epoch, the hit rate of the network modeled with OBV_{16} presented a 50.75% accuracy, while the modeling with CBV_{10} and NOV_{16} showed a percentage of accuracy around 24%, proving that way, the effectiveness of these new vectors for pattern classification problems, since the performance was higher in a reduced number of epochs.

For the comparison of means in independent groups, the Mann-Whitney statistical test was used. This non-parametric test was chosen after being verified by the Kolmogorov-Smirnov test that the data did not follow a normal distribution. In this test, when the value of p, obtained as a result of the comparisons, presents values below 0.05, it shows that there are no considerable differences between the samples. Values greater than 0.05 demonstrates significant differences between the data.

The results of the Mann-Whitney test revealed no significant differences between the CBVs and NOVs in any of the sets of tests. However, in the comparison between the NOVs with the OBVs and the CBVs with the OBVs the values of p found were all smaller than 2.2e−16, representing the difference between the vectors compared.

The results presented by [21] using the same datasets, number of epochs, activation functions and targets in a network MLP type are listed in Table 13.

Table 13. MLP mean performance

Data	CBV	NOV	OBV
Human iris	83.48	82.13	89.18
Handwritten digit	76.91	76.52	82.45
Auslan sign	62.77	60.06	78.84

Comparing the results presented in Table 13 with those found by the RBF in its modified modeling, it is possible to notice that in almost all the problems tested, the RBF presented a very close hit rate of that of the MLP, except in the database of the Auslan Signs that the hit percentage did not approach that presented. Using the Iris images database, the maximum performance obtained by the MLP in 50 epochs was of 89.18%. Using the OBVs, and the RBF hit rate for the same dataset it was of 88.2%. In the handwritten digits, the RBF performance was of 82.72% while that presented for the MLP was of 82.45%. This similarity in the means of the results of the comparisons shows that the two networks have equivalent performance for the patterns tested. In addition, it is noted that OBVs improve the performance of both types of ANNs.

The results obtained by the traditional modeling presented in Table 2, when compared to those shown in Table 3, demonstrate that the change made in the supervised stage of the network improved the performance for the recognition of the inserted patterns. According to Table 2, the data trained with the Iris image using OBV_{128} obtained 87.70% accuracy, while in the network modified with the same target vectors, the performance of the network reached 97.14% of correctness. With the data set referring to the handwritten digits, the modified RBF achieved a maximum performance of 90.12%, and in traditional modeling, the result was of 89.51%. The results of traditional modeling for Auslan signs were of 55.97% accuracy, whereas with the modified network, it obtained 73.68% hit rate.

7 Conclusion

This paper has proposed an unconventional use of OBVs as target of an ANN of the RBF type to improve the performance of pattern recognition problems. The databases used for the study were the human iris images of CASIA [1], images of handwritten digits [3], and Australian signs obtained by [3]. It has also proposed a modification in the ANN output layer, changing its linear function to a nonlinear one, such as the hyperbolic tangent.

The results showed that the two proposed techniques are satisfactory for pattern recognition. The modeling of the RBF in which the modification was made in the supervised training stage, presented better results for the standards tested. It was observed that the use of OBVs also reduces the number of epochs spent for training convergence. Whereas the networks that used OBVs as target achieve convergence in 5 epochs on average, conventional vectors need 30 epochs on average, as can be seen in Figs. 4, 5 and 6.

The fast convergence and higher hit rate is due to the characteristic of the orthogonality of the OBVs, which causes the increase of the distances at the outputs points, making the intersection region between two classes much smaller, an advantage for the classification of standards.

In a future work, with the objective of further improvement in the classification of patterns, the authors wish to perform tests with other clustering algorithms, as well as experiments with other databases to verify the effectiveness of the technique.

Acknowledgment. The authors thank Federal University of Uberlândia and Federal Institute of Triângulo Mineiro for support in the realization of the research.

References

1. Chinese Academy of Sciences - Institute of Automation, Database of 256. Greyscale Eye Images. http://www.cbsr.ia.ac.cn/IrisDatabase.htm
2. Manzan, J.R.G., Nomura, S., Yamanaka, K., Carneiro, M.B.P., Veiga, A.C.P.: Improving iris recognition through new target vectors in MLP artificial neural networks. In: Artificial Neural Networks in Pattern Recognition, pp. 115–126. Springer, Heidelberg (2012). https://doi.org/10.1007/978-3-642-33212-8_11
3. Lichman, M.: UCI machine learning repository (2013). http://archive.ics.uci.edu/ml
4. Kadous, M.W.: Temporal classification: extending the classification paradigm to multivariate time series (Unpublished doctoral dissertation). The University of New South Wales (2002)
5. Leemis, L: Learning Base R. Lightning Source (2016). ISBN 978-0-9829174-8-0
6. Haykin, S.: Redes Neurais: Princípios e prática. Bookman, Porto Alegre RS (2001)

7. Silva, I.N., Spatti, D.H., Flauzino, R.A.: Redes Neurais Artificiais para engenharia e ciências aplicadas - curso prático, 1 ed., Vol. 1. ArtLiber Editora, São Paulo (2010)
8. Fausett, L.: Fundamentals of neural networks: architectures, algorithms, and applications (1994)
9. Lowe, D.: What have neural networks to offer statistical pattern processing?. In: Proceedings of SPIE 1565, Adaptive Signal Processing, 1 December 1991. https://doi.org/10.1117/12.49798. Prentice-Hall, Inc
10. Duda, R.O., Hart, P.E., Stork, D.G.: Pattern Classification, 2nd edn (2001)
11. Nied, A.: Training of artificial neural networks based on systems of variable structure with adaptive learning rate. (Unpublished doctoral dissertation) – Federal University of Minas Gerais, Belo Horizonte (2007). (in Portuguese)
12. Manzan, J.R.G., Yamanaka, K., Nomura, S.: Orthogonal bipolar vectors as multilayer perceptron targets for biometric pattern recognition. In: International Conference on Computing, Networking and Communications - Garden Grove, CA, USA (2015)
13. Pereira, M.B., Veiga, A.C.P.: An application of genetic algorithms to improve the reliability of an iris recognition system. In: XXVIII National Congress of Applied and Computational Mathematics – São Paulo/ Brazil (2005). https://doi.org/10.1109/MLSP.2005.1532892
14. Rumelhart, D.E., Hinton, G.E., Williams, R.J.: Learning internal representations by error propagation. Technical report, DTIC Document (1985)
15. Du, K.-L., Swamy, M.N.S.: Neural Networks and Statistical Learning (2013). https://doi.org/10.1007/978-1-4471-5571-3
16. Nomura, S., Yamanaka, K., Katai, O., Kawakami, H., Shiose, T.: Improved mlp learning via orthogonal bipolar target vectors. JACIII **9**(6), 580–589 (2005)
17. Nomura, S., Yamanaka, K., Katai, O., Kawwakami, H., Shiose, T.: A new approach to improving math performance of artificial neural networks (2004)
18. Nomura, S., Manzan, J.R.G., Yamanaka, K.: Análise experimental de novos vetores alvo na melhoria do desempenho de mlp. IX Conferência de estudos em engenharia elétrica (CBIC 2011)
19. Manzan, J.R.G., Yamanaka, K., Nomura, S.: Improvement in perfomance of MLP using new target vectors (in Portuguese). In: X Brazilian Congress on Computational Intelligence – Fortaleza (2011)
20. Manzan, J.R.G., Nomura, S., Yamanaka, K.: Mathematical evidence for target vector type influence on MLP learning improvement. In: Proceedings on the International Conference on Artificial Intelligence (ICAI), p. 1 (2012)
21. Manzan, J.R.G., Yamanaka, K., Nomura, S.: A mathematical discussion concerning the performance of multilayer perceptron-type artificial neural networks through use of orthogonal bipolar vectors. In: Computational and Applied mathematics 2016 (2016)
22. Conover, W.J.: Practical nonparametric statistics. Wiley series in probability and statistics: applied probability and statistics (1999)
23. Martins, G.A., Fonseca, J.S.: Curso de estatística. Atlas, 6ª Edição (2006)

Fuzzy Logic and Log-Sigmoid Function Based Vision Enhancement of Hazy Images

Sriparna Banerjee[1](✉), Sheli Sinha Chaudhuri[1], and Sangita Roy[2]

[1] Electronics and Telecommunication Engineering, Jadavpur University, Kolkata, India
sriparnatinni@yahoo.in, shelism@rediffmail.com
[2] Electronics and Communication Engineering, Narula Institute of Technology, Kolkata, India
roysangita@gmail.com

Abstract. This paper proposes a novel single image haze removal algorithm, which aims at generating artifact free, contrast enhanced haze free images. The proposed algorithm is designed to overcome the halo artifacts arising in output images obtained from various popular existing image haze removal methods and to restore the contrast of images which are degraded by scattering of scene light by aerosol particles present in the atmosphere. Here the halo artifacts are removed from output images by using a patch-independent procedure for dark channel evaluation and the contrast of images are restored by using a novel Type 1 fuzzy logic and log-sigmoid function based contrast enhancement procedure. Three fuzzy production rules are created considering pixel intensity values and log-sigmoid function values as input fuzzy linguistic variables and mapping constant as output fuzzy linguistic variable. The fuzzy logic is used here for automatic evaluation of membership function value of unique mapping constant for each pixel, which after defuzzification is used to map each pixel intensity value from the degraded image to corresponding pixel intensity value in the enhanced image. The efficiency of this proposed algorithm is established by obtaining satisfactory results from comparative studies of qualitative and quantitative analyses results which are carried out by applying the proposed algorithm and several other popular algorithms on same set of hazy images. Two significant quantitative parameters are chosen for performing quantitative analysis-a) Elapsed Time (time required for the execution of the algorithm) and Contrast-to-Noise Ratio which is an important visibility metric to measure visual quality of images. For both types of parameters it has been shown that our algorithm gives better results compared to other methods.

Keywords: Artifacts · Log-sigmoid function values · T1 fuzzy logic
Fuzzy production rules · Pixel intensity values · Defuzzification
Mapping constant

K. Arai et al. (Eds.): IntelliSys 2018, AISC 868, pp. 447–462, 2019.
https://doi.org/10.1007/978-3-030-01054-6_32

1 Introduction

The contrast of hazy images gets degraded due to scattering and attenuation of scene light travelling from scene object towards the camera. The portion of attenuated scene light which gets reflected into the line of sight by fog, mist, dust and other suspended air particles present in the atmosphere, collectively known as aerosols, is referred to as airlight [1]. The combination of this airlight with transmission (which signifies that portion of scene light which reaches the camera sensor without any attenuation) is the main reason for degradation in the contrast of images. The image haze removal algorithms proposed by the authors in [2–6] are basically physics based approaches, while designing these approaches the authors have not considered any details regarding the weather condition prevailing at the place during the time when hazy images are captured. The authors in [7–10] have developed their respective algorithms keeping the dark channel prior concept as backbone. The authors in [8] have proposed the novel dark channel prior concept while the authors in [7, 9, 10] have further modified the algorithm proposed by authors in [8] and have dealt with the artifacts present in the algorithm in some way or the other. The basics of image processing techniques were studied from [11]. The authors in [12] have proposed a novel image haze removal algorithm with less computational complexity and good time efficiency. The algorithm mainly operates on linear combination of pixels and is capable of handling both color and gray images. The authors in [13] have proposed a new method of transmission estimation which is capable of eliminating scattering light and enhancing the visibility of hazy images. The authors in [14] have proposed a flexible edge collapse based model which is capable of self-adjusting the intensities of transmission map for better recovery of hazy images. The authors in [15] have developed image haze removal algorithm primarily depending on two observations: (1) haze free images are more visually enhanced than hazy images; and (2) airlight varies with distance between the scene object and the camera sensor. The authors in [16] have portrayed the image haze removal problem as an ill-posed inverse problem and have stated several solutions to solve it. The authors in [17] have proposed a multi-scale fusion method where the information obtained from two hazy images are blended using three weight maps and the artifacts arising due to blending are reduced using Laplacian pyramid representation. The authors in [18] have proposed novel color attenuation prior for image haze removal and the linear model developed by them is capable of recovering the scene depth. The authors in [19] have built up an optimization problem by combining with a weighted L1–norm based contextual regularization for transmission estimation. In this paper we developed an image haze removal algorithm keeping the dark channel prior concept as the base and have tried to solve some problems associated with it. Here we have used a patch-independent method for dark channel evaluation contrary to the patch based procedures used for dark channel evaluation in most of the existing, popular methods to overcome the halo artifacts arising in output images mainly due to non-uniform transmission within the patches. Apart from that here we have also proposed a novel contrast enhancement procedure involving fuzzy logic and log-sigmoid function which is capable of generating a unique mapping constant value used for mapping each pixel from contrast degraded to contrast enhanced images. The paper is organized as follows: Sect. 1 for Introduction, Sect. 2 contains the Stepwise analysis of the proposed algorithm, Sect. 3 is dedicated for Results Sect. 4 for Discussion and finally Sect. 5 concludes the entire work.

2 Step by Step Analysis of the Proposed Algorithm

2.1 Mathematical Model of a Hazy Image

Hazy image can be mathematically expressed as in [5, 6, 8–10, 13, 15, 20]:

$$I(x) = J(x)t(x) + A(1 - t(x)) \tag{1}$$

Here in (1), I and J represent the intensity values of any hazy image and scene radiance respectively while 't' and 'A' represent transmission and global atmospheric light respectively. The first and second terms on the right side of (1) signify direct attenuation and airlight [1], respectively. Direct attenuation describes the rate of decay of scene light in accordance with the distance between the object and the camera and the airlight is mainly responsible for the degradation in the contrast of hazy images.

2.2 Computation of the Desired Dark Channel Using the Proposed Patch-Independent Method

Initially before processing any hazy image, we first resize its dimension to a fixed size of 400 × 500 and we also normalize each pixel intensity value present in the image, such that after normalization the pixel intensity values lie within [0 1]. As proposed by the authors in [8], more than 70% of the pixels in haze free images have very low intensity values almost equal to zero. The reason behind these low intensity values is the presence of trees, rocks and many other colorful objects in images. Among the three color channels in images, at least one of the channels contains pixels with very low intensity values. The desired channel containing low intensity pixels can be mathematically evaluated by using the following equation as proposed by the authors in [8–10, 16, 20]:

$$J^{dark}(x) = \min_{y \in \Omega(x)} \left(\min_{c \in \{r,g,b\}} (J^c(y)) \right) \tag{2}$$

Here the authors have used a local patch of size 15 × 15 centered on a pixel located at 'x' to evaluate the pixel with minimum intensity within the patches. The halo artifacts arise in images due to non-uniform transmission within the patches. They are mainly present along the edges. The halo artifacts are totally removed by applying the proposed patch independent method where the desired channel is found by using the following equation:

$$J^{dark}(x) = (\min(R(x), \min(G(x), B(x))) \tag{3}$$

In (3), R(x), G(x) and B (x) represent the intensity values of pixel at any particular location say ‘x’ in the red, green and blue color channels of any image respectively. The improvement in the visual quality of haze free images and removal of the halo artifacts can be clearly depicted from the images in Table 1.

Table 1. Improvement in the visual quality of the images using patch independent method compared to existing patch based methods

Image No.	Comparative Improvement in visual quality of the images		
	Hazy image	*Image obtained from patch based methods*	*Image obtained from the proposed patch independent method*
1			
2			
3			
4			
5			

2.3 Estimation of the Global Atmospheric Light Present in the Image

Before starting the estimation of global atmospheric light, we initially segmented out the sky and the non-sky regions of all hazy images present in our dataset, manually. Then we have given those pixel intensity values as inputs to the artificial neural network classifier during the training phase and have assigned two distinct class labels to pixel intensity values belonging to sky and non-sky regions respectively. After training the neural network classifier, when we give pixel intensity values of any hazy image as inputs to the trained neural network classifier, it will automatically classify pixel intensity values to their respective classes. After sorting the pixels, which are classified as the sky region pixels in the ascending order, 10% of the brightest pixels out of these pixels are taken and average of them is calculated. Then this average value is assigned as the value of the global atmospheric light. Here the brightest pixels have been selected since the authors in [7, 15] have stated that brightest pixels are most haze-opaque.

2.4 Estimation of the Transmission Map

The transmission refers to that portion of light which reaches the camera sensor without being scattered and deflected on its way to the camera from the scene object. The accurate estimation of transmission map is highly necessary for proper dehazing of hazy images. The transmission map is estimated by using the following equation as in [8]:

$$t(x) = 1 - \omega(\min_{y \in \Omega(x)}(\min_{c}(\frac{I^c(y)}{A^c}))) \quad (4)$$

In (4), the term 'ω' is called haziness factor which is introduced to maintain the value of t(x) at a value greater than zero. The value of haziness factor is set to 0.9. Its value should be kept less than one, to keep the little amount of haze in the image, so that the originality of the image can be preserved.

2.5 Recovery of the Scene Radiance

After evaluating J, atmospheric light and transmission, finally the scene radiance is recovered by the following equation as in [8]:

$$J(x) = \frac{I(x) - A}{\max(t(x), t0)} + A \quad (5)$$

In (5) the value of t0 is kept constant at 0.1, so that the denominator does not become equal to zero. But from Table 1, we can see that haze free images obtained after scene radiance recovery are visually degraded.

2.6 Contrast Enhancement of Images Using the Proposed T1 Fuzzy Logic and Log-Sigmoid Function Based Contrast Enhancement Procedure

Here the log-sigmoid function is used as a transformation function during contrast enhancement of degraded images. The mathematical expression of log-sigmoid function used here as a transformation function is:

$$f(x) = (\frac{1}{1 + e^{-(k_i * (J(x)))}}) \quad (6)$$

Here k_i represent the mean values of histograms for each color channel, where 'i' varies from 1 to 3, since there are three color channels. But since there are (400 * 500) = 200000 pixels in each color channel and there are 256 intensity levels in a histogram, ideally the mean value of the histogram should lie somewhere around 782. Pixel intensity values after normalization lie between 0 and 1. In order to find the range of log-sigmoid function values, we consider two extreme cases.

CASE I. When the pixel intensity value is 0 and the mean value of histogram is 782.

So putting these values in (6), we can see that the log-sigmoid function value obtained for this case is

$$f(x) = \frac{1}{1 + e^{-(782*0)}} = 0.5$$

CASE II. When the pixel intensity value is 1 and the mean value of the histogram is 782.

Then putting these values in (6), for this case the log-sigmoid function value obtained is

$$f(x) = \frac{1}{1 + e^{-(782*1)}} = 1$$

So from CASE I and CASE II, we can see that the range of the log-sigmoid function values lie between 0.5 and 1.

The visual quality of radiance haze free images obtained after scene radiance recovery are poor and the blocking and the ringing artifacts are still present in images mostly in the sky region. So there is need of further improvement of visual quality of images. Here we have tried to enhance the contrast of recovered radiance images by introducing the proposed Type 1 fuzzy logic based contrast enhancement procedure where pixel intensity values represented as ‘pi’ and log-sigmoid function values represented as ‘lgsfv’ are considered as input fuzzy linguistic variables and mapping constant represented as ‘mppi’, unique for each pixel which is used to map each pixel of contrast degraded radiance images to corresponding contrast enhanced images are taken as output fuzzy linguistic variables. Here fuzzy logic is used to automatically select a mapping constant value for each pixel, unlike standard procedures where a single mapping constant is used to map all the pixels in the image irrespective of pixel intensity values.

In order to apply fuzzy logic, we first constructed three fuzzy production rules.

The three fuzzy production rules are as follows:

PR1. If pi is LOW and lgsfv is LESS Then mppi is HIGH.
PR2. If pi is MEDIUM and lgsfv is AVERAGE Then mppi is AVERAGE.
PR3. If pi is LARGE and lgsfv is HIGH Then mppi is LOW.

The fuzzy membership curves obtained for each pixel variable are as follows (Figs. 1, 2 and 3):

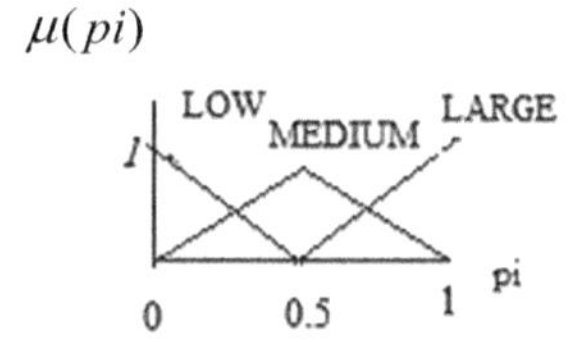

Fig. 1. Graphical representation of fuzzy membership function representing input fuzzy variable ‘pi’.

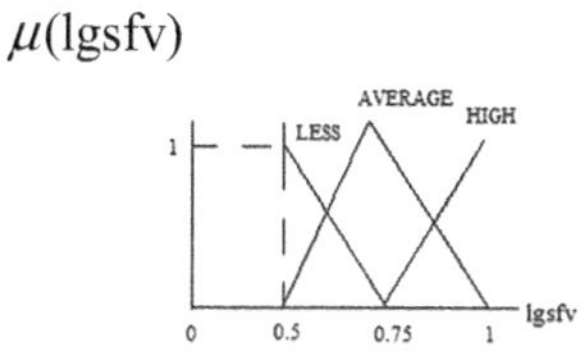

Fig. 2. Graphical representation of fuzzy membership function representing input fuzzy variable 'lgsfv'.

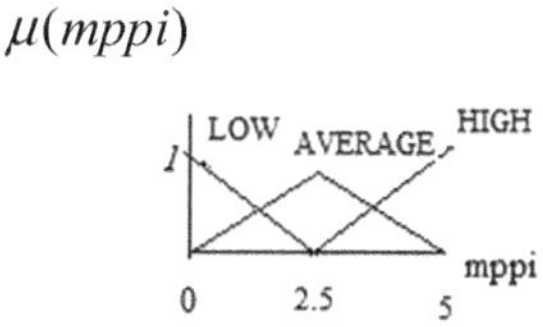

Fig. 3. Graphical representation of fuzzy membership function representing output fuzzy variable 'mppi'.

Here the three fuzzy linguistic variables pi, lgsfv and mppi represent pixel intensity values, log-sigmoid function values and mapping constant of pixel intensity values respectively. For linguistic variable 'pi', the linguistic value, LV = {LOW, MEDIUM, LARGE}

For linguistic variable 'pi', the dynamic range, DR = [0 1].
For linguistic variable 'lgsfv', the linguistic value, LV = {LESS, AVERAGE, HIGH}
For linguistic variable 'lgsfv', the dynamic range, DR = [0.5 1].
For linguistic variable 'mppi', the linguistic value, LV = {LOW, AVERAGE, HIGH}
For linguistic variable 'mppi', the dynamic range, DR = [0 5].

The logic behind choosing these dynamic range values are as follows:

For pi:
DR = [0 1], since the pixel intensity values are initially normalized by dividing each pixel intensity value in the input hazy color image by 255.
For lgsfv:
DR = [0.5 1], the logic behind this is explained in CASE I and CASE II earlier.
For mppi:
DR = [0 5], is chosen arbitrarily after trial and error method as the mapping constant values lying in this range give satisfactory results for all the images in our dataset.

Initially after dividing each image into three color channels, we applied this proposed contrast enhancement technique individually on each color channel. Then after enhancing the pixels in each color channel individually, we again combined them to form a contrast enhanced color image.

Let us consider the crisp values of pixel intensity and log-sigmoid function values as pi^* and $lgsfv^*$ respectively and corresponding fuzzy membership values of them as $\mu_{A'_r}(pi_r)$ and $\mu_{B'_r}(\lg sfv_r)$ respectively.

The fuzzy membership values of the corresponding mapping constant can be evaluated using the following equation:

$$\mu_{c'_r}(mppi_r) = \max_{i=1}^{3}[\mu_{c'_i}(mppi)] \tag{7}$$

$$= \max_{i=1}^{3}[a_r \circ \mu_{R_i}(pi, \lg sfv; mppi)] \tag{8}$$

$$a_r = min(\mu_{A'_r}(pi_r), \mu_{B'_r}(\lg sfv_r)) \tag{9}$$

$$\mu_{R_i}(pi, \lg sfv; mppi) = \min(\min(\mu_{A_i}, \mu_{B_i}), \mu_{c_i}) \tag{10}$$

$\mu_{R_i}(pi, \lg sfv; mppi)$ is same for all three color channels. Here 'i' varies from 1 to 3, since there are three fuzzy production rules. But the value of a_r will change for each pixel. For green and blue channels all the above equations will remain the same except (9).

For green channel, (9) will be replaced by this equation:

$$a_g = \min(\mu_{A'_g}(pi_g), \mu_{B'_g}(\lg sfv_g)) \tag{11}$$

For blue channel, (9) will be replaced by this equation:

$$a_b = \min(\mu_{A'_b}(pi_b), \mu_{B'_b}(\lg sfv_b)) \tag{12}$$

The values of a_r will be different for each pixel in the red channel because it contains result obtained by applying the Fuzzy T- norm operator on the fuzzy membership values of the corresponding input crisp values of the pixel intensity values and the log-sigmoid function values. The value will differ mostly, since the crisp values will be different every time and hence the corresponding fuzzy membership values will also be different. According to the boundary property of Fuzzy T-norm operator, it calculates the minimum value among the entities on those it is applied [21]. Similarly for every pixel in the green and blue channels the values of a_g and a_b will differ.

Here $\mu_{R_i}(pi, \lg sfv; mppi)$ is evaluated by using the Mamdani Fuzzy Implication Relation [21].

Then the defuzzified output of the mapping constant is obtained by applying the center of area rule using the following equation:

For mapping constant of pixel in the red channel:

$$mppi_r^* = \frac{\sum_{\forall mppi \in M} mppi * \mu_{c'_r}(mppi_r)}{\sum_{\forall mppi \in M} \mu_{c'_r}(mppi_r)} \quad (13)$$

For green and blue channels the same mathematical expression is used as in case of red channel with slight changes. Like for green and blue channel $mppi_r$ and $\mu_{c_{Cr}}(mppi_r)$ will be replaced by $mppi_g$, $\mu_{c_{Cg}}(mppi_g)$ and $mppi_b$, $\mu_{c_{Cb}}(mppi_b)$ respectively. The graphical procedure of the entire method is shown in Fig. 4 and an example of graphical representation of membership function of output variable 'mppi' is shown in Fig. 5.

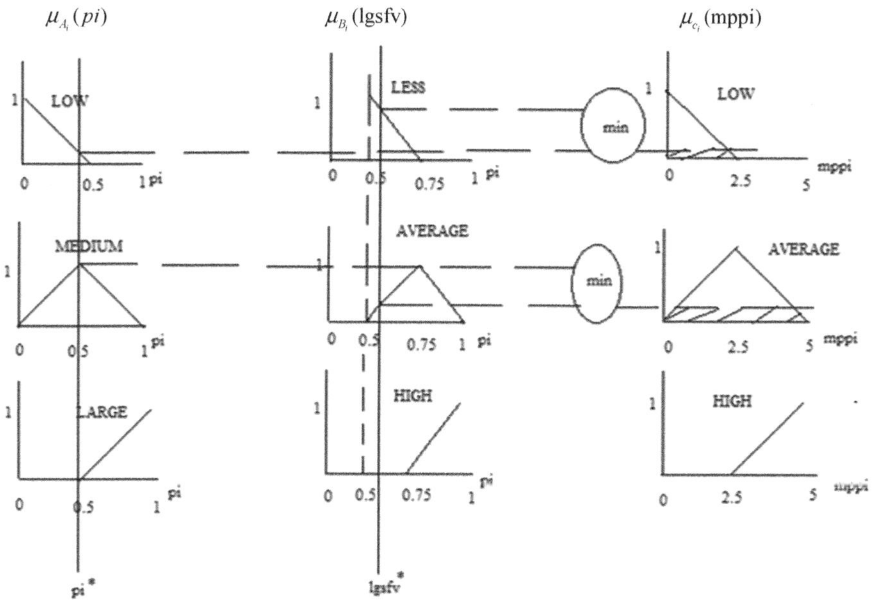

Fig. 4. Graphical representation protraying the entire procedure representing the entire procedure of how the fuzzy membership function of 'mppi' is obtained by using Type 1 Fuzzy Logic.

After obtaining the mapping constant value for each pixel, the contrast enhancement of the radiance images are done using the following equation:

$$N_j(x) = J_j(x) * \lg sfv_j(x) * mppi_j^* \quad (14)$$

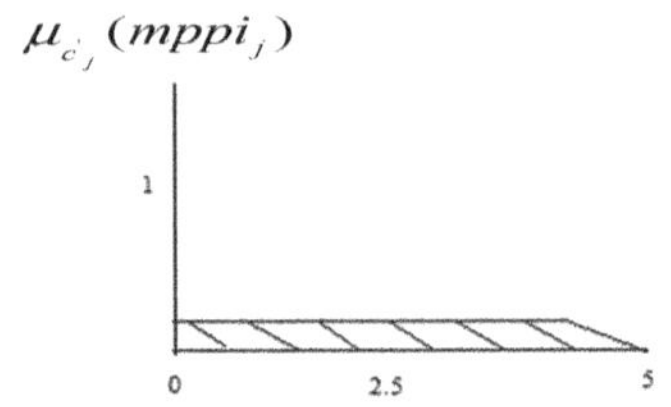

Fig. 5. Example of a graphical representation of the fuzzy membership curve of output fuzzy linguistic variable 'mppi' which is obtained as an output of the process depicted in Fig. 4.

In (14), the value of 'j' varies from 1 to 3, since there are three color channels and 'x' represents any arbitrary pixel location and varies for each pixel location in respective color channels. $N_j(x)$ represent the pixel intensity values of the new contrast enhanced image, $\lg sfv_j(x)$ represents the log-sigmoid value and $mppi_j^*$ stands for defuzzified mapping constant value to any pixel belonging to the respective color channel. The block diagram of the proposed method is shown in Fig. 6 and basic block diagram of Type 1 Fuzzy Logic is depicted in Fig. 7. The visual improvement of images by applying the proposed contrast enhancement method is shown in Table 2.

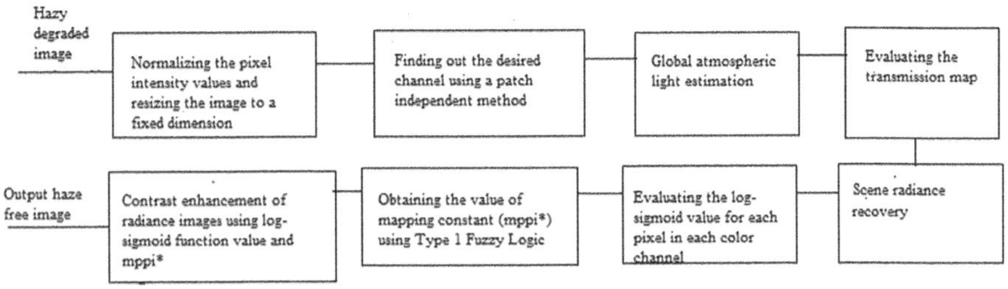

Fig. 6. Block diagram of the entire method.

3 Results

The efficiency of the algorithm is estimated by performing both qualitative and quantitative analyses on more than 1500 hazy images in the dataset built by us, which consists of images obtained from websites and also D-HAZY, which is a dataset constructed by the authors in [22]. D-HAZY consists of more than 1400+ hazy images. Still now it is the only standard dataset available for measuring the performance of image haze removal algorithms. Here we have performed both qualitative and quantitative analyses of haze free images obtained. Quantitative analysis is carried out by taking two important parameters like Elapsed time and Contrast-to-Noise Ratio (CNR) which is a visibility metric proposed by Dai Zhengguo and is suitable for

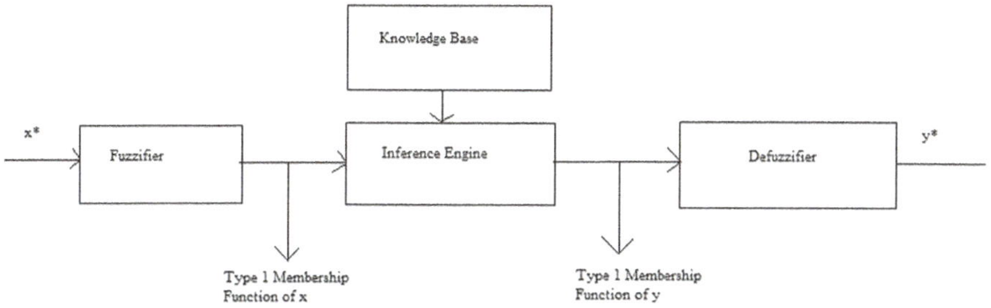

Fig. 7. Basic block diagram of Type 1 fuzzy controller.

Table 2. Improvement in the visual quality of the images using the proposed contrast enhancement method

Image No.	Contrast enhancement of radiance images		
	Hazy image	*Contrast degraded radiance images*	*Final contrast enhanced images*
1			
2			
3			
4			
5			

measuring the visibility quality of hazy image, since it contains a significant bias called haze. The qualitative analysis is shown in Table 3. These algorithms are chosen for comparative analyses since these are the most popular methods among the existing algorithms.

The total time elapsed to execute any program is one of the main quantitative parameters to measure the effectiveness of the program. Since the elapsed time is one of the fundamental properties to decide whether the algorithm is suitable for practical implementation or not, so here we have taken the Elapsed time as one of the quantitative parameters for analysis. The comparative studies of quantitative analysis results which are performed taking Elapsed time and CNR as quantitative parameters are shown in Tables 4 and 5, respectively.

The time elapsed to execute the same algorithm on the same image may vary since it depends on the speed of the processor in the computer in which the algorithm is executed and also on the version of the Matlab installed in the computer, where the

Table 3. Comparitive study of the qualitative results obtained by applying the proposed algorithm with that obtained by applying other popular algorithms on the same set of images

Image No.	*Changes in the visual quality of the images with algorithms*					
	Original image	*He 'et al.' [8]*	*Tarel 'et al.' [12]*	*Ancuti 'et al.' [17]*	*Meng 'et al.' [19]*	*Our results*
1	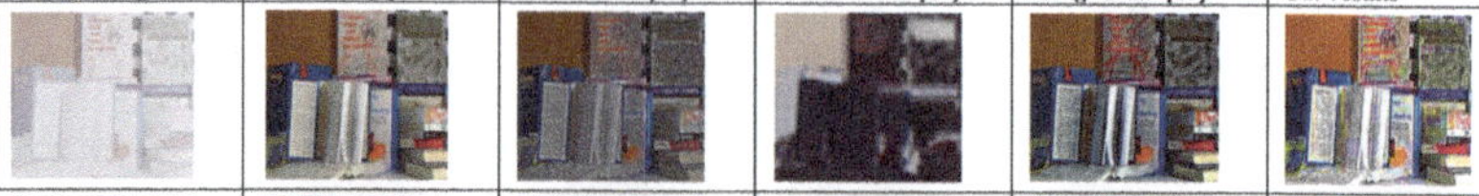					
2	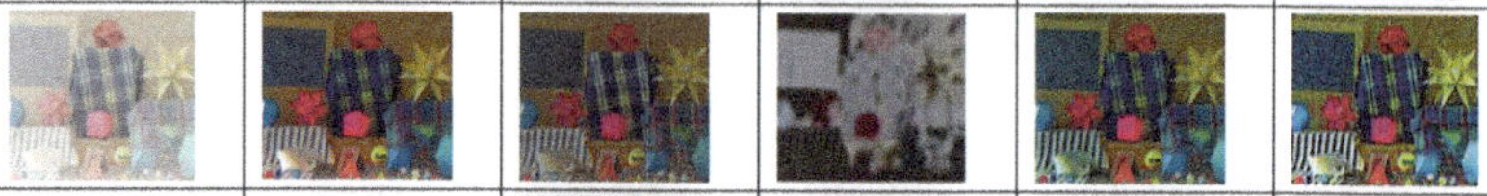					
3						
4	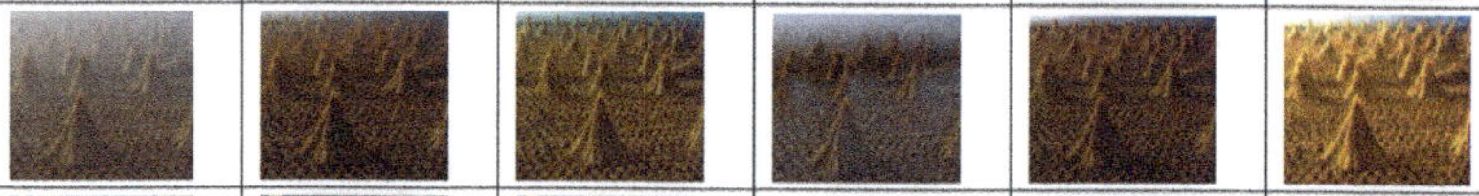					
5	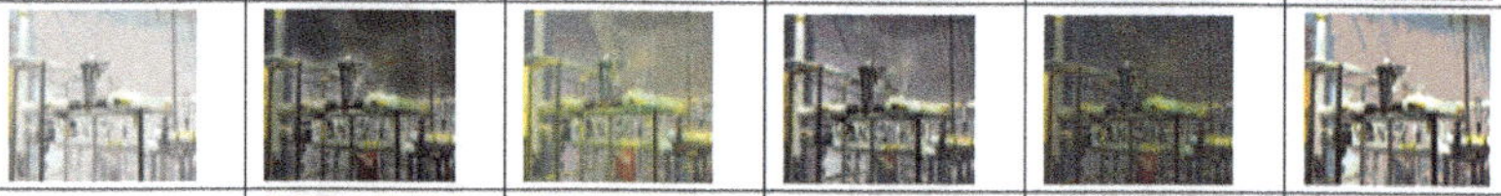					
6	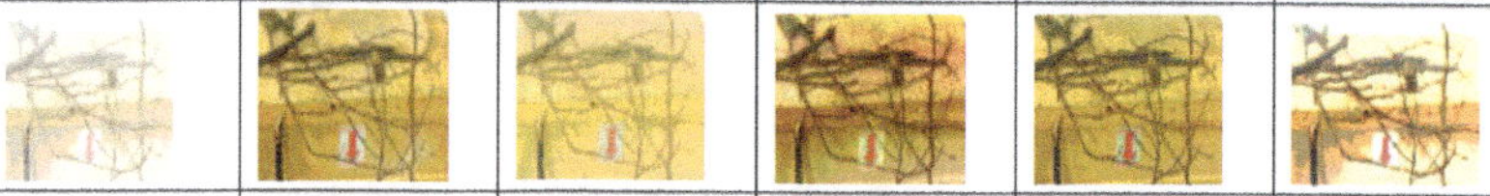					
7						
8						

Table 4. Comparative study of time required to execute the algorithms

Image No.	Time elapsed to execute several algorithms				
	He et al. [8]	Tarel et al. [12]	Ancuti et al. [17]	Meng et al. [19]	Our results
1	11.23	6.42	5.54	4.38	**1.65**
2	12.45	6.21	5.22	3.89	**1.72**
3	10.08	6.07	4.76	3.42	**1.41**
4	13.62	6.75	5.81	3.75	**1.89**
5	12.31	6.62	5.31	4.21	**1.53**
6	10.54	6.21	5.67	3.82	**1.84**
7	11.81	6.53	5.61	3.94	**1.73**
8	12.01	6.44	5.34	3.71	**1.61**

Table 5. Comparitive study of CNR obtained by applying several algorithms on the same set of images

Image No.	CNR values obtained by applying several algorithms on same set of images				
	He et al. [8]	Tarel et al. [12]	Ancuti et al. [17]	Meng et al. [19]	Our results
1	174.5	170.3	143.2	175.8	**228.9**
2	210.1	252.9	238.6	234.6	**289.1**
3	276.5	314.1	245.5	288.9	**345.7**
4	361.5	319.5	389.1	348.9	**421.1**
5	279.29	223.76	227.65	283.6	**334.78**
6	264.73	226.8	245.6	256.7	**284.76**
7	138.69	156.4	124.6	175.4	**198.2**
8	256.72	278.1	221.6	267.5	**278.6**

program is executed. The results obtained here clearly shows that our method performs better than other methods.

Mathematically, CNR can be represented as:

$$CNR = \left(\frac{\mu_A - \mu_B}{\sigma_B}\right) \tag{15}$$

In (15), μ_A and μ_B represents the mean coefficients of the target object and background region, respectively whereas σ_B signifies background noise.

So from Table 5 it becomes clear that our method provides higher CNR compared to other methods proving that our method provides much more visually improved and contrast enhanced images compared to other methods.

4 Discussion

In this proposed method, we used fuzzy logic and log-sigmoid function based contrast enhancement method because fuzzy logic enables us to evaluate unique mapping constant for each pixel from just three fuzzy production rules contrary to conventional logic which requires large number of rules (one rule for each pixel) to serve the same purpose. Here we considered pixel intensity values and log-sigmoid function values as input fuzzy linguistic variables because pixel intensity value is the main criteria on which the value of output mapping constant variable depends. Higher the value of pixel intensity, lower will be the value of mapping constant and vice versa. In other words the pixel intensity value is inversely proportional to the output mapping constant. The log-sigmoid function is used here as a contrast enhancement transformation function. These log-sigmoid function values are used as an input fuzzy linguistic variables since we multiply the log- sigmoid function values with the corresponding pixel intensity values and the output mapping constant values in the final contrast enhancement step. Hence it also serve as an important criteria for choosing the output mapping constant. The values of log-sigmoid function vary between 0.5 and 1. So the two extreme cases are: (1) The pixel intensity values decrease by 50% when it gets multiplied by log-sigmoid value of 0.5; and (2) The pixel intensity value remains unchanged when it is multiplied by log-sigmoid value of 1. Any log-sigmoid function values in between these two extreme values lower the pixel intensity values in accordance to the corresponding log-sigmoid function values. Hence like pixel intensity values, the log-sigmoid function values are also inversely proportional to output mapping constant values. Our choice of log-sigmoid function as a transformation function is because the basic working principle of log transformations [11] includes expansion of dark intensity range in images and compression of the higher intensity range. So this property of log-sigmoid function matches with our objective because here our main focus is to expand the range of low intensity values since we have earlier stated the most of the pixels in haze-free images have very low intensities. So, log-sigmoid function best suits our purpose.

5 Conclusion

The proposed method is capable of generating visually improved haze free images devoid of artifacts. The patch – independent method used here for evaluation of the desired dark channel can satisfactorily remove the halo artifacts which arise in radiance images due to non-uniform transmission within the patches. The radiance images obtained although are free from halo artifacts are still contrast degraded and blocking artifacts are present mostly in the sky regions. To overcome these issues we proposed the novel contrast enhancement technique, which uses Type 1 fuzzy logic in combination with the log-sigmoid function. Fuzzy logic is used to evaluate a unique mapping constant for mapping each pixel from a degraded image to corresponding contrast enhanced image. We resized each color channel of an image to a particular dimension of 400*500, resulting in ((400 * 500 * 3) = 600000) pixels for the three color channels. Hence if we use conventional logic to generate unique mapping constant for each pixel, we have to construct 600000 rules, but by using fuzzy logic we can reduce the number of rules to three, covering the entire range of pixels. In standard existing procedures of contrast enhancement where a single mapping constant value is used for mapping each pixel as it may cause unnecessary brightening or darkening of images due to high and low values of mapping constants respectively. This in turn leads to the loss of details in images and hence to overcome this problem we proposed this technique of contrast enhancement. The entire work was carried out using Matlab 2014a. The qualitative and quantitative analyses of output images performed and comparative studies are carried out by taking those results and similar results obtained by applying several other popular, existing haze removal algorithm on the same set of images using same quantitative parameters. Two parameters used for quantitative analysis are Elapsed time and CNR, respectively. Elapsed time is most important parameter to determine the possibility of implementation of an algorithm in real life and CNR is an important visibility metric to determine the visual quality of hazy images. The future scope of this work is further reduction of the Elapsed time (within 1 s) so that this algorithm can be applied in several practical situations.

References

1. Koschmeider, H.: Theorie derhorizontalen Sichtweite. Beitr Phys. Freien **12**, 171–181 (1924)
2. Tan, K.K., Oakley, J.P.: Physics-based approach to color image enhancement in poor visibility conditions. J. Opt. Soc. Am. A **18**(10), 2460–2467 (2001)
3. Oakley, J.P., Satherley, B.L.: Improving image quality in poor visibility conditions using a physical model for contrast degradation. IEEE Trans. Image Process. **7**, 167–169 (1998)
4. Narasimhan, S.G., Nayar, S.K.: Contrast restoration of weather degraded images. IEEE Trans. Pattern Anal. Mach. Intell. **25**(6), 713–724 (2003)
5. Narasimhan, S.G., Nayar, S.K.: Vision and the atmosphere. Int. J. Comput. Vis. **48**, 233–254 (2002)
6. Schechner, Y.Y., Narasimhan, S.G., Nayar, S.K.: Instant Dehazing of images using polarization. In: Proceedings of the 2001 IEEE Computer Society Conference on Computer Vision and Pattern Recognition, vol. 1, pp. 325–332, June 2001
7. Lv, X., Chen, W., Shen, I.-f.: Real-time Dehazing for image and video. In: 8th Pacific Conference on Computer Graphics and Applications (PG), Hangzhou, Zhejiang, China (2010)
8. He, K., Sun, J., Tang, X.: Single image haze removal using dark channel prior. IEEE Trans. Pattern Anal. Mach. Intell. **33**(12), 2341–2353 (2011)
9. Das, D., Roy, S., Chaudhuri, S.S.: Dehazing technique based on dark channel prior model with sky masking and its quantitative analysis. In: 2nd International Conference on Control, Instrumentation, Energy and Communication (CIEC), Kolkata, India (2016)
10. Kumari, A., Sahoo, S.K.: Real time visibility enhancement for single image haze removal. In: Elsevier 11th International Multi-conference on Information Processing (IMCIP-2015), Bangalore, India (2015)
11. Gonzalez, R.C., Woods, R.E.: Digital Image Processing. Pearson Education, New Delhi (2009)
12. Tarel, J.P., Hautiere, N.: Fast visibility restoration from a single color or gray level image. In: IEEE International conference on Computer Vision, Kyoto, Japan (2009)
13. Fattal, R.: Single image Dehazing. ACM Trans. SIGGRAPH **27**(3) (2008)
14. Chen, B.-H., Huang, S.-C.: Edge collapse-based Dehazing algorithm for visibility restoration in real scenes. J. Disp. Technol. **12**(9), 964–970 (2016)
15. Tan, R.: Visibility in bad weather from a single image. In: 2008 IEEE Conference on Computer Vision and Pattern Recognition (CVPR 2008), pp. 1–8, June 2008
16. Roy, S., Chaudhuri, S.S.: Modeling of Haze image as Ill-posed inverse problem & its solution. Int. J. Mod. Educ. Comput. Sci. **8**, 8 (2016)
17. Ancuti, C.O., Ancuti, C.: Single image Dehazing by multi-scale fusion. IEEE Trans. Image Process. **22**(8), 3271–3282 (2013)
18. Zhu, Q., Mai, J., Shao, L.: A fast single image haze removal algorithm. IEEE Trans. Image Process. **24**, 3522–3533 (2015)
19. Meng, G., Wang, Y., Duan, J., Xiang, S., Pan, C.: Efficient image Dehazing with boundary constraint and contextual regularization. In: IEEE International Conference on Computer Vision (ICCV 2013), Sydney (2013)
20. Zhang, E., Lv, K., Li, Y., Duan, J.: A fast video image Defogging algorithm based on dark channel prior. In: 6th International Congress on Image and Signal Processing (CISP 2013), Shanghai, China (2013)

21. Konar, A.: Computational Intelligence: Principles, Techniques and Applications. Springer, Heidelberg (2005)
22. Ancuti, C., Ancuti, C.O., Vleeschouwer, C.D.: D-HAZY: a dataset to evaluate quantitatively Dehazing algorithms. In: International Conference on Image Processing (ICIP 2016), Phoneix, USA (2016)

Video Detection for Dynamic Fire Texture by Using Motion Pattern Recognition

Kanoksak Wattanachote[1(✉)], Yongyi Gong[1], Wenyin Liu[2], and Yong Wang[2]

[1] School of Information Science and Technology, Guangdong University of Foreign Studies, Guangzhou, Guangdong, China
kanoksak.wattanachote@gmail.com, gongyongyi@gdufs.edu.cn

[2] School of Computer Science and Technology, Guangdong University of Technology, Guangzhou, Guangdong, China
liuwy@gdut.edu.cn, 13928887919@126.com

Abstract. Significant motion features which are able to be used for fire video detection in regard to the dynamic fire texture are proposed in this article. We are now interested in motion characteristics rather than color schemes. Since colors of fire textures observed on video medium nowadays are possibly illustrated with whimsical colors. It is not caused by only nature chemical phenomena but also by special effect application technologies in video industry. We propose four data series of motion features gained from motion vector field or optical flow estimation, namely, the series of average radius, the series of motion coherence index, the covariance stationary series of average radius, and the covariance stationary series of motion coherence index, respectively. The extracted data is used by machine learning part to form training set and test set for video classification using support vector machine method. Our four proposed data series are able to leverage fire video detection. Our experimental results demonstrate that the accuracy of video detection in regard to fire texture is significantly high and its time elapsed only few seconds of gaining data.

Keywords: Fire video detection · Video classification · Motion features
Motion coherence index · Fire texture · Dynamic texture

1 Introduction

Over the last decade, most fire calamity surveillance systems had been developed based on several constituents e.g. optical, ionization, particle sampling, temperature sampling, humidity sampling, and air transparency testing. But those fire calamity surveillance systems would not work until the sensor detected the equipment [1, 2]. We propose video detection regarding to fire texture recognition which an aim to leverage the implementation for fire calamity surveillance video systems. However, there exist fire

The work was supported by National Science Foundation Grant of China 61370160, Guangdong Province Natural Science Foundation Project (2015A030313578).

K. Arai et al. (Eds.): IntelliSys 2018, AISC 868, pp. 463–477, 2019.
https://doi.org/10.1007/978-3-030-01054-6_33

calamity surveillance video systems and applications, including flame detection in videos which have been developed based on video textures analysis. For instance, Foggia et al. [3] proposed fire detection for video surveillance application by combining complementary information based on color, shape variation, and motion analysis. Li et al. [4] proposed flame detection in video using a Dirichlet Process Gaussian Mixture Color Model. Other examples of flame detection using color phenomena were proposed in [5, 6]. Despite color scheme became an important parameter for those flame detection methods, notwithstanding, color detection would not be accomplished if fire color illustrated in uncommon color scheme. Moreover, the cost of combining several data e.g. color and motion data, for fire texture analysis is quite expensive.

Since colors of fire, nowadays can be peculiar caused by some nature chemical phenomena. Besides, video replication process can possibly reproduce the fire color to illustrate extraordinary colors by using special effect applications in video and film industry. In this study, we emphasize motion characteristics rather than color schemes. We propose an idea of enhancing motion series features analysis aimed to leverage fire video detection. We extracted several ordinary motion features from motion vector field on dynamic fire textures. Several motion estimation techniques and optical flow production approaches [7–9] have been studied to gain the most appropriate method. In this end, we chose Farneback method [9, 10] to produce the motion vector field in each video frame.

The motion features we are interested e.g. the series of average radius which used to represent the motion velocity of vectors and the series of motion coherence index which used to represent the coherency of motion direction of vectors [11]. Moreover, the covariance stationary series of average radius and motion coherence index have been recently introduced as the attributes of dynamic texture [12, 13]. Since the covariance stationary series of dynamic texture has been investigated to explain the stationarity through their motion features. The covariance stationary series has been studied to indicate the repetition of intensity change in dynamic textures. We then inspect the stationarity through motion vector field and express by graphical stationary series model. However, we emphasize only covariance motion pattern rather than finding periodic motion pattern [13], to reduce the cost of model computation. Finally, we reasonably chose four features for this study: (1) the series of average radius (2) the series of motion coherence index, (3) the covariance stationarity series pattern of average radius, and (4) the covariance stationarity series pattern of motion coherence index. We implement these four significant features for our video detection approach.

The remaining parts of this article, we discuss related works in Sect. 2. Features extraction and system architecture are devoted in Sect. 3. We describe using support vector machine, training set and test set for fire video detection, including experimental results in Sect. 4. Discussion is devoted in Sect. 5. And, conclusions will be summarized in Sect. 6.

2 Related Works

2.1 Motion Vector Estimation

Farneback method maintains two main advantages of velocity estimation by estimating over a surface block (SB) of region rather than a pixel [9, 10]. An advantage is that the effects of noise and inaccuracies in tensor or vector estimation are reduced significantly. Another advantage is to avoiding the aperture problem. The information obtained from neighbor or other nearest parts is useful to fill the missing velocity component, even the aperture problem is occurring in some part of the estimating region. We demonstrate an example of optical flow field created by Farneback method to monitor the texture of interest (TOI). For example, Fig. 1 demonstrates the motion vector field created by passing two parameters r_t and SB to our motion estimation function, where the vector radius attribute (r_t) must be larger than a threshold v_t.

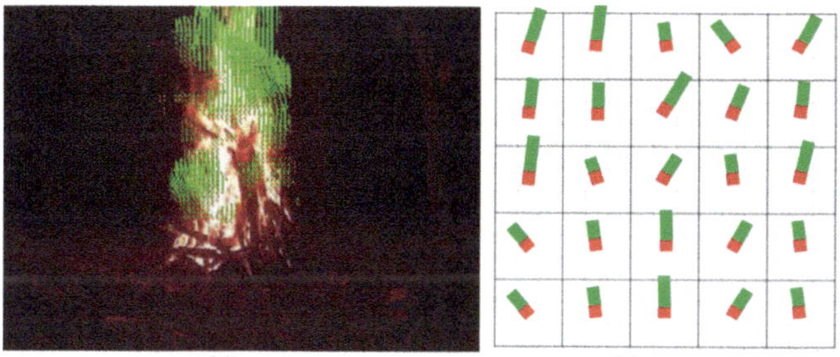

(a) (b)

Fig. 1. Dynamic fire texture and its motion vectors on motion vector field produced by 3 × 3 pixels surface block (SB = 3). The vectors demonstrated in this figure were gained by parameter v_t, where $v_t \geq 1.0$ and SB = 3. The average angular value of cropped region, for instance, in (b) is approximately 270 degrees unit or 4.714285 radians unit.

The angular value is measured at the beginning point $(x, 0)$ in xy-coordinate.

2.2 Covariance Stationary Series

Regarding to the theory of stationary time series, if a random variable x is indexed to time t, the observations $\{x_t, t \subset T\}$ is called a time series, where T is a time index in the integer set, $T \in Z$ [14]. Here, the stationarity means covariance stationarity. For instance, if the covariance function $\gamma(t, t+\tau) = E\{[x_t - \mu(t)][x_{t+\tau} - \mu(t+\tau)]\}$ for a series x_t, where $t = 1, 2, \ldots, S$, is an univariate function of time interval τ, it means $\gamma(t, t+\tau) = \gamma(t)$. Fu et al. [15], proposed the new concept of correlation coefficient stationary time series. For the correlation coefficient stationary time series x_t, there exists $\mu(t)$ and $\sigma(t)$ are its mean function and standard deviation function. The series x_t is called correlation coefficient stationary series if any y_t is a covariance stationary series, calculated by (1).

$$y_t = \frac{[x_t - \mu(0)]}{\sigma(0)} \tag{1}$$

3 Features Extraction

Basically, a vector in xy-coordinate represents the quantity of both magnitude and direction. In our study, we proposed magnitude data by using vector radius. Besides, we measured vector direction in angular unit and described the motion direction by leveraging the motion coherence index. We studied both the series of radius and the series of motion coherence index to express covariance stationary series as described in [14, 16]. We applied covariance stationarity analysis method by passing thirty sequence images to create a series model before using it to automatically find the covariance stationary series of the next three hundred or greater than three hundred consecutive video frames. We implemented the aforesaid method for four series attributes namely, the series of average radius, the series of motion coherence index, the covariance stationarity series pattern of average radius, and the covariance stationarity series pattern of motion coherence index. The series attributes we extracted from fire texture are described in the following subsections.

3.1 Radius Data

The vector radius (r) usually describes the moving velocity of each SB of intensity change in a video frame. We calculate an average radius ($\bar{r}$) from collected radii in each consecutive video frame. We used $\bar{r}$ to represent the average moving velocity of intensity change of that video frame. We collect the data series $\bar{r}_t$, where $t = 1, 2, \ldots, N$, where N is a total number of observed frames in the integer set, $N \in Z$.

3.2 Covariance Stationary Series of Radius

We leveraged the covariance stationary series (1) for radii data by (2). Here, we firstly collect a dataset of thirty average radius $\bar{r}_k$, where $k = 1, 2, \ldots, 30$, from thirty consecutive video frames to calculate the correlation coefficient stationary series of thirty average radius $\bar{r}_k$, and represented by $\bar{\vartheta}_0$, where $\{\bar{\vartheta}_0, 0 \in T\}$. For the correlation coefficient stationary time series $\bar{\vartheta}_0$, there exists $\mu(0)$ and $\sigma(0)$, those are its mean function and its standard deviation function. We calculate the mean function $\mu(0)$ and standard deviation function $\sigma(0)$ of $\bar{\vartheta}_0$ and use them to estimate the covariance stationary series function of next consecutive series $\bar{r}_t$ in the same video, where $t = 1, 2, \ldots, N$ and $\{\bar{r}_t, t \in T\}$. Hence, the covariance stationary series y_{rt} of $\bar{r}_t$ is calculated by (2).

$$y_{\bar{r}t} = \frac{[\bar{r}_t - \mu(0)]}{\sigma(0)} \tag{2}$$

Suppose that we have three hundred consecutive video frames ($N = 300$). And, each video clip has frame rate equal to thirty frames per second (30 FPS). Hence, our proposed method can gain prerequisite data within ten seconds from each video clip for fire texture analysis in video detection.

3.3 Motion Coherence Index Data

In this subsection, we leverage the motion coherence index to be the third data series. We applied the motion coherence index function proposed by [11] in practical, to estimate the motion coherence index of vectors in an optical flow field. Our proposed motion coherence index is calculated on each four adjacent vectors in optical flow field. The angular values of vectors are measured in angular unit radian and demonstrated as degree unit in system user-interface. For example, suppose $\theta(x, y)$ represents an angular value of vector on an xy-coordinate. We usually find three neighborhoods angular values (α). Those are $\alpha(x+1, y)$, $\alpha(x, y+1)$, and $\alpha(x+1, y+1)$. $\theta(x, y)$ here means if $x = 0$ and $y = 0$, the θ represents angular value of vector at the most top-left surface block (SB) in the optical flow field. We consider the angular values from the SB at the top-left to the bottom-right in every motion vector field of interest. We average every four angular values as demonstrated in (3), where $j = 1, 2, 3$, α_1 means $\alpha(x+1, y)$, α_2 means $\alpha(x, y+1)$, and α_3 means $\alpha(x+1, y+1)$, respectively.

$$\mu_j = \frac{\left[\theta(x, y) + \alpha_j\right]}{2} \tag{3}$$

We, next step, find the different between μ_j and π, and denoted by delta (Δ), where $\Delta_j = \pi - \mu_j$. We then find the Euclidean distance between $\theta(x, y)$ and each μ_j, denoted as the Euclidean distance $\delta_{\mu j}$, and the distance between $\theta(x, y)$ and Δ denoted as the Euclidean distance $\delta_{\Delta j}$, where the $\delta_{\mu j} = |\theta(x, y) - \mu_j|$ and $\delta_{\Delta j} = |\theta(x, y) - \Delta_j|$ respectively. We pass $\delta_{\mu j}$ and $\delta_{\Delta j}$ into motion coherence index function (4) to find the motion coherence index at (x, y) before finding the mean of those three indices by (5). We normalized the motion coherence $\bar{\hat{C}}_{(x,y)}$ in (5) to gain the motion coherence index (C_i) which can describe the motion coherence close to human perception, as demonstrated in (6).

$$\hat{C}\left(\delta_{\mu j}, \delta_{\Delta j}\right)_{x,y,j} = \left|\pi - 2\operatorname{argmin}\left(\delta_{\mu j}, \delta_{\Delta j}\right)\right| \tag{4}$$

$$\bar{\hat{C}}_{(x,y)} = \frac{1}{3}\sum\nolimits_{j=1}^{3} \hat{C}\left(\delta_{\mu j}, \delta_{\Delta j}\right)_{x,y,j} \tag{5}$$

$$C_i = \begin{cases} 0;\ \bar{\hat{C}}_{(x,y)} \le 70 \\ \left[3\left(\dfrac{\bar{\hat{C}}_{(x,y)} - 70}{30}\right)^2 - 2\left(\dfrac{\bar{\hat{C}}_{(x,y)} - 70}{30}\right)^3\right] \times 100;\ otherwise \end{cases} \tag{6}$$

We collect C_i, where $i = 1, 2, \ldots, M$ total number of motion coherence indices in motion vector field. We average the motion coherence indices C_i by (7) to find out the motion coherence index in each video frame ($\bar{C}$).

$$\bar{C}_t = \frac{1}{M}\sum\nolimits_{i=1}^{M} C_i \tag{7}$$

We collect the data series $\bar{C}_t$, where $t = 1, 2, \ldots, N$ total number of observed frames in the integer set, $N \in Z$.

3.4 Covariance of Motion Coherence Index

We now leverage the covariance stationary series data of the motion coherence index $\bar{C}$ instead of $\bar{r}$ as described in Subsection B. We collect thirty of $\bar{C}$ derived from thirty consecutive video frames denoted as its correlation coefficient stationary series, represented by $\bar{\zeta}_0$, where $\{\bar{\zeta}_0, 0 \in T\}$. We calculate the mean function $\tilde{\mu}(0)$ and standard deviation function $\tilde{\sigma}(0)$ of $\bar{\zeta}_0$ before leveraging them to estimate the covariance stationary series function of next consecutive series $\bar{C}_t$, where $t = 1, 2, \ldots, N$ and $\{\bar{C}_t, t \in T\}$. The covariance stationary series $y_{\bar{C}t}$ of $\bar{C}_t$ is calculated by (8). Here, we collect the data series $y_{\bar{C}t}$, where $t = 1, 2, \ldots, N$ total number of observed frames in the integer set, $N \in Z$.

$$y_{\bar{C}t} = \frac{\left[\bar{C}_t - \tilde{\mu}(0)\right]}{\tilde{\sigma}(0)} \tag{8}$$

In Sect. 3, now we extracted four signals of $\bar{r}_t$, $y_{\bar{r}t}$, $\bar{C}_t$, and $y_{\bar{C}t}$ from the TOI to prepare the dataset in our experiment next section.

4 SVM Methodology and Experimental Results

In our experiment, our dataset consists of four signals used for fire video detection. Our first signal is the radius index ($\bar{r}$) of each video frame that uses to describe the fluctuation of intensity change in a video sequence. The second signal is the covariance index in covariance stationary series ($y_{\bar{r}}$) of the radius index. The third signal is the motion coherence index ($\bar{C}$) which used to represent the fluctuation of motion direction in a video sequence. And, the forth signal is the covariance index in covariance stationary series ($y_{\bar{C}}$) of motion coherence index. Figure 2 shows the series of those four features in time interval τ, supposing the observed values are in 300 consecutive video frames ($\tau = 300$) to observe the fluctuation and iteration of motion patterns for fire characterization.

We implemented LIBSVM support vector machines library [17] to train and test our dataset using MATLAB. Since, the support vector machines or SVMs introduced by Cortes and Vapnik [18] can distinguish between groups for linear and nonlinear data in the original variable space [18, 19], by using kernels.

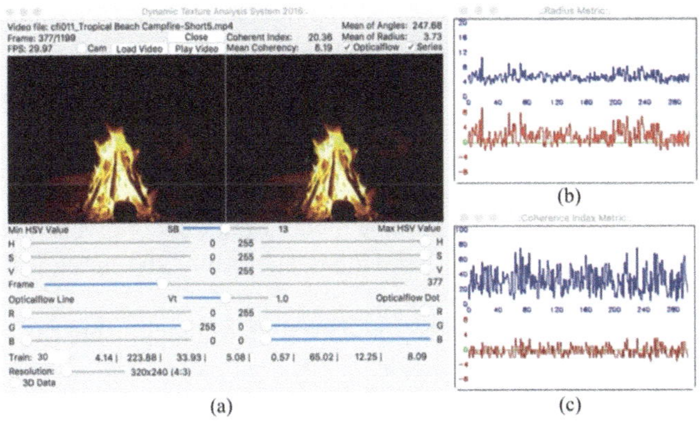

Fig. 2. (a) Dynamic fire texture and its four extracted data series. (b) Blue graph represents the average radius series, and red graph represents the covariance series of average radius. (c) Blue graph represents the series of motion coherence index, and the red graph represents the covariance series of motion coherence index.

4.1 System Architecture

We demonstrate our fire video detection process in Fig. 3. Some of video sequences used in our training and test dataset were derived from [11][1]., YouTube[2],[3], and DynTex database [20]. In this study, we developed our system as semi-automatic video detection system. The data prerequisite was obtained by using our video texture analysis system, developed by C++, OpenCV and QT. We modeled the detection part as our prior video classification process and implemented using LIBSVM[4] [17] in MATLAB.

(1) *Starting video sequence:* We pass the video sequence in our system. The videos passed into our system were clipped to gain the TOI e.g. fire, smoke, waterfall and others.
(2) *Farneback motion estimation:* To produce vectors and present as motion vector field which contains only the significant vectors filtered by the following parameters.
    (a) *SB:* The 3D surface block which is created to extract the features of dynamic textures initially gained by using SB to find the motion vector. The size of SB in our system can be set by size between 3×3 and 21×21 pixels.
    (b) v_t: The vector radius threshold is used for gaining significant vectors which able to be in floating range between 0 and 2.0 ($0 \leq v_t \leq 2.0$).
(3) *Data extraction:* To gain significant four data series of four features as follows.
    (a) *Radius index series* ($\bar{r}_t$)*:* Represents the series of average radius of motion vector field in a video sequence.

[1] http://video.minelab.tw/DTT/
[2] https://www.youtube.com/results?search_query=fire
[3] https://www.youtube.com/results?search_query=color+fire
[4] SVM library is available at https://www.csie.ntu.edu.tw/~cjlin/libsvm/.

(b) *Motion coherence index series ($\bar{C}_t$):* Represents the series of motion coherence index of motion vector field in a video sequence.
(c) *Covariance series of radius index ($y_{\bar{r}t}$):* Represents the covariance series of radius index in a video sequence.
(d) *Covariance series of motion coherence index ($y_{\bar{C}t}$):* Represents the covariance series of motion coherence index in a video sequence.

(4) *SVM training:* To create data training model from a set of given video sequences.
(5) *SVM prediction:* To classify fire videos and return the classification accuracy (percentage).
(6) *Detection results:* To determine the detection accuracy using the percentage of classification accuracy.

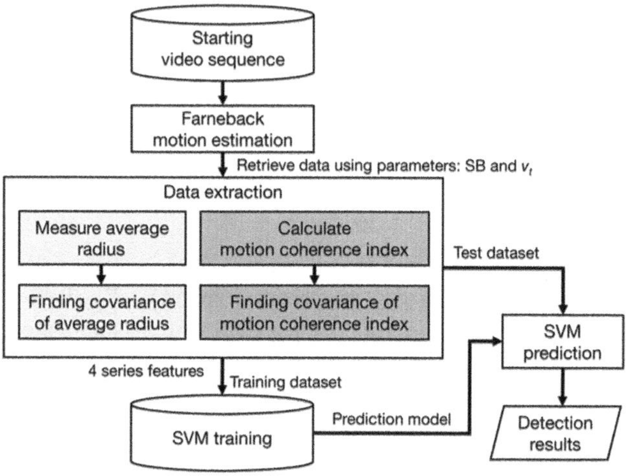

Fig. 3. System architecture diagram for fire video detection using SVM.

4.2 SVM Training Dataset

Our training dataset gains from fifteen different video sequences as demonstrated in Fig. 4. It consists of seven fire texture videos and eight of other dynamic textures i.e. smoke, waterfall, etc. Here, we have 4,500 data records of four features consist of five columns. The first column is label. The second column is the radius index averaged from the vector radius in video frame. The third column represents the motion coherence index. The forth column is covariance index of the average radius. And, the last column is the covariance index of motion coherence index. In our training dataset, label 1 denoted as fire video sequences and label 0 expresses other video sequences.

We construct the hyper-plane by using SVM classifier to train those feature vectors. The workflow of system is shown in Fig. 3. We train the four features and constructed the prediction model or training model to detect fire texture video in test datasets.

Fig. 4. SVM Training dataset from seven video clips of fire textures and other eight videos.

4.3 SVM Test Dataset

We collected data from several different video sequences as demonstrated in Figs. 4, 5, and 6. We extracted the four attributes data and collected their data series before using our SVM prediction model in Subsection B to detect fire video and described with an accuracy value of detection (percentage). After we got a classification model, we used

Fig. 5. SVM Test dataset 1, contains fifteen video sequences of fire textures.

Fig. 6. SVM Test dataset 2 contains 20 random video sequences of mixed textures.

that model to test the training dataset itself, where the labels of fire video are 1, and the labels of other videos are 0. The results are demonstrated in Table 1.

Table 1. Accuracy results of fire video detection using training model to test training dataset itself

Video ID	Extracted features data			Accuracy	
	Label	Series name	Records	Records	%
1	1	cfi006 A4F	300	299	99.6667
2	1	cfi015_T4F	300	293	97.6667
3[a]	1	cfi032_44F	300	269	89.6667
4	1	cfs006_P4F	300	293	97.6667
5	1	FC08Trim4F	300	294	98.0000
6	1	fi101_KA4F	300	300	100.0000
7	1	fi115_2 4F	300	300	100.0000
8	0	01_54ab14F	300	296	98.6667
9	0	01_6481h4F	300	300	100.0000
10	0	02_57db14F	300	263	87.6667
11	0	645c620.4F	300	300	100.0000
12	0	cfs012_M4F	300	300	100.0000
13	0	cfw002_W4F	300	300	100.0000
14	0	tsre02_J4F	300	288	96.0000
15	0	etc14_es4F	300	300	100.0000
			4500	**4395**	**97.6667**

[a] SB = 9, $v_t = 0.3$

We designed our experiment for fire video detection in four parts. Our first part, the test dataset gained from the dataset that used to create classification model in the previous Subsection, as illustrated in Fig. 4. The labels of all records in test dataset must be 1, due to the assumption of our experiment for video detection of fire texture. Since the motion of fire texture plays a major role to produce optical flow in motion vector field. In Fig. 4, there are seven fire videos and eight of other videos. Here, we have 4,500 data records for training set and 300 data records for each dataset. That means our test dataset of each video sequence contains 300 data records. The detection results are demonstrated in Table 2.

The second part, we collected all data series from fifteen fire video sequences for testing. Each fire video contains 300 data records. The video captures are demonstrated in Fig. 5. This dataset contains several colors of fire textures. Since an objective of our study is to detect fire video by using motion attributes rather than color phenomena. The accuracy results are demonstrated in Table 3.

Table 2. Accuracy results of fire video detection using training model to test training dataset with label 1 for all data

Video ID	Extracted features data		Accuracy	
	Label	Series name	Records	%
1	1	cfi006 A4F	299	99.6667
2	1	cfi015_T4F	293	97.6667
3[a]	1	cfi032_44F	269	89.6667
4	1	cfs006_P4F	293	97.6667
5	1	FC08Trim4F	294	98.0000
6	1	fi101_KA4F	300	100.0000
7	1	fi115_2 4F	300	100.0000
8	1	01_54ab14F	4	1.3333
9	1	01_6481h4F	0	0.0000
10	1	02_57db14F	37	12.3333
11	1	645c620.4F	0	0.0000
12	1	cfs012_M4F	0	0.0000
13	1	cfw002_W4F	0	0.0000
14	1	tsre02_J4F	12	4.0000
15	1	etc14_es4F	0	0.0000

[a] SB = 9, vt = 0.3

Table 3. Accuracy results of fire video detection using training model to test first test dataset (All Fire Videos)

Video ID	Extracted features data			Accuracy	
	Label	Series name	Records	Records	%
1	1	cfi009_14F_	300	299	99.6667
2	1	cfi015_T4F_	300	291	97.0000
3	1	cfi051_B4F_	300	297	99.0000
4	1	cfi054_B4F_	300	295	98.3333
5	1	FC04Trim4F_	300	276	92.0000
6	1	FC07Trim4F_	300	296	98.6667
7	1	FC10Trim4F_	300	289	96.3333
8	1	FC11Trim4F_	300	293	97.6667
9	1	FC14Trim4F_	300	297	99.0000
10	1	FC15Trim4F_	300	286	95.3333
11	1	FC24Trim4F	300	256	85.3333
12	1	FC26Trim4F_	300	284	94.6667
13	1	fire_0084F_	300	297	99.0000
14	1	fi009_Fi4F_	300	286	95.3333
15	1	fi010_Fi4F_	300	282	94.0000
			4500	**4324**	**96.0889**

The last experiment, we collected data series from twenty random videos for testing. This dataset contains several videos of different dynamic texture. Each video contains different data records. The video captures are demonstrated in Fig. 6. Fire detection results are demonstrated in Table 4.

Table 4. Accuracy results of fire video detection using training model to test second test dataset (Mixed Videos)

Video ID	Extracted features data			Accuracy	
	Label	Series name	Records	Records	%
1	1	cfi010_14F_	324	321	99.0741
2	1	wf004_A 4F_	318	0	0.0000
3	1	csm001_B4F_	327	183	55.9633
4	1	cfi015_T4F_	315	308	97.7778
5	1	csm017_S4F_	347	132	38.0403
6	1	cfs004_H4F	300	159	53.0000
7	1	wf010.mp4F_	300	3	1.0000
8	1	FC10Trim4F	300	296	98.6667
9	1	FC09Trim4F_	300	291	97.0000
10	1	csm002_B4F_	300	48	16.0000
11	1	csm010_T4F_	346	140	40.4624
12	1	wf003_Wa4F_	367	0	0.0000
13	1	cwf002_F4F_	338	0	0.0000
14	1	fwf01.mp4F_	300	3	1.0000
15	1	fi001_Ho4F_	300	272	90.6667
16	1	etc10_Ca4F_	300	76	25.3333
17	1	etc06_an4F_	300	59	19.6667
18	1	cfi002 A4F_	300	297	99.0000
19	1	csm011_E4F_	300	133	44.3333
20	1	san_fran4F_	300	15	5.0000

4.4 Experimental Results

The information gained from Tables 1 and 2 show that our prediction model can classify fire video sequences and other video sequences with an average accuracy of approximately 97.6667%. We have tested our model by firstly tested with our training datasets. Each dataset contains 300 data records. The results in Table 2 shows that our prediction model can predict fire video sequences correctly, observed from Video ID 1–7 in Table 2. The average accuracy only fire video prediction is 97.5238%.

The information gained from Table 3 shows that our prediction model can detect fire video sequences with high average accuracy of approximately 96.0889%. In Table 4, our prediction model can predict fire video sequences correctly, observed from Video ID 1, 4, 8, 9, 15, and 18. The average accuracy is higher than 97%. In our detection system, if the accuracy is higher than 75%, that video will be predicted as fire video. However, the predicted accuracy of Video ID 6 in Table 4 is only 53% and was classified as other video. Since there are two dynamic textures in that video namely, smoke and fire. The motion patterns of smoke and fire are different observed from our studies. This issue will be discussed in Discussion section.

5 Discussion

Regarding to our experiment, we observed that fire, smoke and waterfall produced different signals of the four features proposed in this study. The examples of different feature signals of fire, smoke, and waterfall are demonstrated in Fig. 7, Fig. 8 and Fig. 9, respectively.

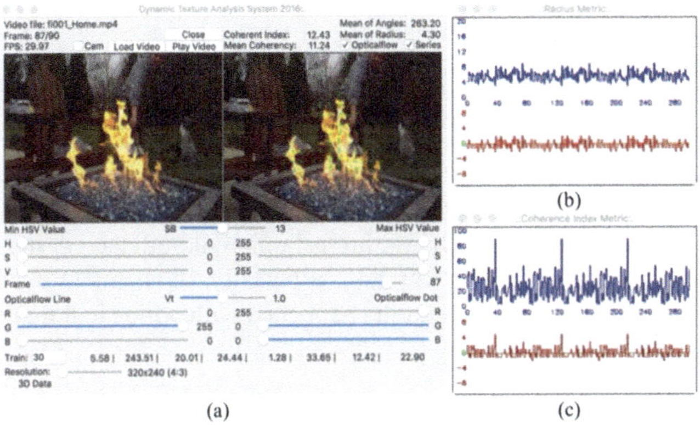

Fig. 7. Dynamic fire texture and its extracted data series of four features.

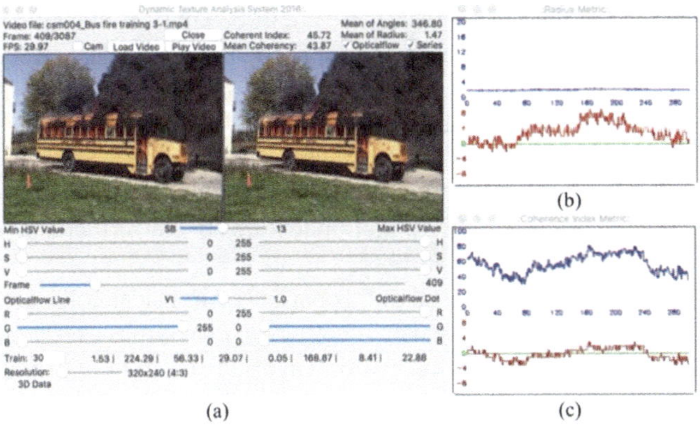

Fig. 8. Dynamic smoke texture and its extracted data series of four features.

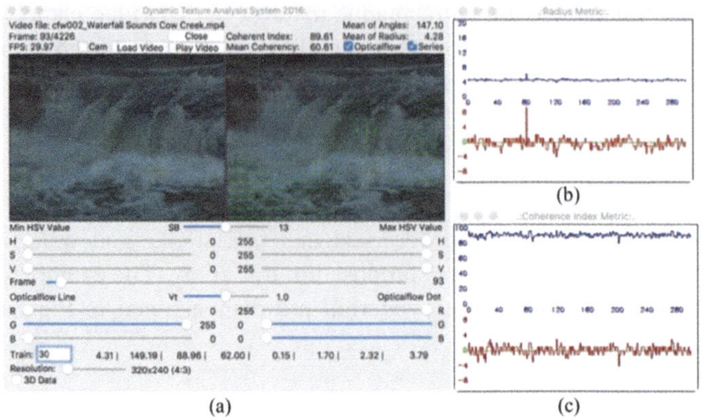

Fig. 9. Dynamic waterfall texture and its extracted data series of four features.

Moreover, the significant motion features used in this study are able to detect fire subject video sequences either traditional color or whimsical color. This contribution is able to continue study to be leveraged for fire detection in video based fire calamity surveillance system, not only for home fire calamity surveillance system but also for industrial fire calamity surveillance system; especially, in chemical industry or petroleum industry, where the fire is able to be happened with whimsical flame colors.

6 Conclusions

The four series of motion features proposed in this study are significant for fire video detection. Our objective is to implement motion features rather than color schemes for fire video detection purpose. However, there are three kinds of dynamic texture which able to be classified using those four features. There are fire smoke and waterfall which could be characterized by using motion features.

Besides, there exist some videos which partially contain smoke and fire textures could not be detected as fire subject video sequences. Since smoke and fire demonstrate different motion signals as illustrated in Figs. 7 and 8. We are now extending and preparing to discover other significant motion features in our further study in regarding to dynamic textures classification purpose.

Fruitfully, one problem discussed in this paper is whimsical color fire video detection, has been successfully done by using motion characteristics of fire texture. Nevertheless, dynamic texture classification by using hybrid features of both motion attributes and color attributes is continuously investigated to enhance dynamic texture classification and detection purposes.

References

1. Chen, T., Kao, C., Chang, S.: An intelligent real-time fire-detection method based on video processing. In: IEEE International Carnahan Conference on Security Technology, pp. 104–111 (2003)
2. Tasselli, G., Alimenti, F., Bonafoni, S., Basili, P., Roselli, L.: Fire detection by microwave radiometric sensors: Modeling a Scenario in the Presence of Obstacles. IEEE Trans. Geosci. Remote Sens. **48**(1), 314–324 (2010)
3. Foggia, P., Saggese, A., Vento, M.: Real-time fire detection for video-surveillance applications using a combination of experts based on color, shape, and motion. IEEE Trans. Circuits Syst. Video Technol. **25**(9), 1545–1556 (2015)
4. Li, Z., Mihaylova, L.S., Isupova, O., Ross, L.: Autonomous flame detection in videos with a Dirichlet process gaussian mixture color model. IEEE Trans. Ind. Inform. Spec. Sect. Multisens. Fusion Integr. Intell. Syst. **PP**(99) (2017)
5. Borges, P.V.K., Izquierdo, E.: A probabilistic approach for vision-based fire detection in videos. IEEE Trans. Circuits Syst. Video Technol. **20**(5), 721–731 (2010)
6. Dimitropoulos, K., Barmpoutis, P., Grammalidis, N.: Spatio-temporal flame modeling and dynamic texture analysis for automatic video-based fire detection. IEEE Trans. Circuits Syst. Video Technol. **25**(2), 339–351 (2015)
7. David, J.F., Yair, W.: Optical flow estimation. In: Paragios, N., et al. (eds.) Handbook of Mathematical Models in Computer Vision. Springer (2006). ISBN 0-387-26371-3
8. Papenberg, N., Bruhn, A., Brox, T., Didas, S., Weickert, J.: Highly accurate optical flow computation with theoretically justified warping. Int. J. Comput. Vis. **67**(2), 141–158 (2006)
9. Farneback, G.: Fast and accurate motion estimation using orientation tensors and parametric motion models. In: Proceedings 15th International Conference on Pattern Recognition, vol. 1, pp. 135–139 (2000)
10. Farneback, G.: Two-frame motion estimation based on polynomial expansion, Lecture Notes in Computer Science, vol. 2749, pp. 363–370 (2003)
11. Wattanachote, K., Shih, T.K.: Automatic dynamic texture transformation based on a new motion coherence metric. IEEE Trans. Circuits Syst. Video Technol. **26**(10), 1805–1820 (2016)
12. Wattanachote, K., Wang, Y., Shih, T.K., Hsu, H., Liu, W.: Dynamic textures and covariance stationary series analysis using strategic motion coherence. In: IEEE International Conference on Advanced Information Networking and Applications, pp. 205–212 (2017)
13. Wattanachote, K., Lin, Z., Jiang, M., Li, L., Wang, G., Liu, W.: Fire and smoke dynamic textures characterization by applying periodicity index based on motion features. In: Liu, M., Chen, H., Vincze, M. (eds.) Computer Vision Systems, ICVS 2017. Lecture Notes in Computer Science, vol. 10528, pp. 507–517. Springer, Cham (2017)
14. Franses, P.H.: Time Series Models for Business and Economic Forecasting. Cambridge University Press, Cambridge (1998)
15. Fu, H., Liu, C.: Analysis method of correlation coefficient ARMA (p, q) series. J. Aerosp. Power **18**(2), 161–166 (2003)
16. Rasmussen, C.E., Williams, C.K.I.: Gaussian Processes for Machine Learning. MIT Press, Cambridge (2006). Chapter 4
17. Chang,C., Lin, C.: LIBSVM: a library for support vector machines (2001)
18. Cortes, C., Vapnik, V.: Support-vector networks. Mach. Learn. **20**(3), 273–297 (1995)
19. Cristianini, N., Shawe-Taylor, J.: An Introduction to Support Vector Machines: And Other Kernel-Based Learning Methods. Cambridge University Press, Cambridge (2000)
20. Péteri, R., Fazekas, S., Huiskes, M.J.: DynTex: a comprehensive database of dynamic textures. Pattern Recogn. Lett. **31**(12), 1627–1632 (2010)

A Gaussian-Median Filter for Moving Objects Segmentation Applied for Static Scenarios

Belmar García García(✉), Francisco J. Gallegos Funes, and Alberto Jorge Rosales Silva

SEPI-ESIME, Instituto Politécnico Nacional, Mexico City, Mexico
belmar_2g@hotmail.com, fco.funes@gmail.mx, arosaless23@gmail.com

Abstract. Background subtraction or also called foreground detection is an approach normally used for moving object segmentation in video sequences captured from a fixed camera. Most of the methods under this approach are not able to segment or require the strict absence of objects during their training or learning period in the first frames. In this document, a method capable of segmenting moving objects from the beginning of a video sequence and at the same time constructing a reference background image is proposed. The segmentation results show that the foreground and background regions of the scene are not affected during this stage compared to other methods.

Keywords: Background · Foreground · Video sequence · Segmentation Learning period

1 Introduction

Moving objects segmentation is a very active research area in computer vision that is rapidly increasing number of surveillance cameras that leads to a strong demand for automatic processing methods for their output. The main tasks in Moving objects segmentation systems include motion detection, object classification, tracking, activity understanding, and semantic description. Our interest is focused on the detection phase in a general visual moving objects system using static cameras. The detection of moving objects in video streams is the first relevant step of information extraction in many computer vision applications [1]. Background subtraction is an approach typically used for moving regions segmentation in video sequences from a fixed camera; each new frame is compared to a reference image or background model in such a way that pixels values in the current frame deviate significantly from the reference, the pixels are classified as part of the moving objects or the foreground region. These latter pixels are subsequently processed by the most common tasks already mentioned in a Moving objects segmentation system [2].

There exist a lot of methods found in scientific literature, but any method outperforms the others ones. Consequently, the background subtraction method must be motivated by the content of the scene better than the model itself [2, 3]. The present work shows the effectiveness to take advantage and avoid disadvantages of the two of the most important methods used in the scientific literature: single Gaussian and median filter.

K. Arai et al. (Eds.): IntelliSys 2018, AISC 868, pp. 478–493, 2019.
https://doi.org/10.1007/978-3-030-01054-6_34

First, the noise for each static background pixel can be reasonably modeled by a single Gaussian distribution, but this is restricted to have absence of objects during the first frames of the video sequence (approximately 90 frames agreed to Matlab example) is required to construct correctly the background model or reference image [4]. For other side the median filter does not require absence of objects at the beginning of the video sequence to carry out the segmentation, but it lacks of a statistical model to describe noise values for each pixel in a static background [5].

The robustness in the proposed method is to obtain a reference image at the beginning of video sequence (training or learning period) with the statistical features of a Gaussian distribution and in turn eradicates the restriction of moving object detection or segmentation during this process. To validate the robustness, during learning period, at least half of the values adopted by each pixel must be background values (median filter condition called in this document). This is necessary because the median filter works using a buffer that must satisfy this feature for its correct implementation [6]. After this stage, the next buffer is taken with the following pixel values until the video sequence is completed.

2 Related Work

The use of methods under background subtraction approaching are classified as:

Unimodal versus multimodal: static background models assume that the intensity values of a pixel can be modeled by a single unimodal distribution. Such models usually have low complexity, but cannot handle moving backgrounds, while this is possible with multimodal models at the price of higher complexity [1, 2].

Recursive versus non-recursive: non-recursive techniques store a buffer of a certain number of previous sequence frames and estimate the background model based on the temporal variation of each pixel within the buffer, while recursive techniques recursively update a single background model based on each input frame. In the first case, the background is well adapted to eventual variations, but memory requirements can be significative; in the latter, the complexity is lower, but input frames from distant past can have an effect on the current background, and, therefore, any error in the background model is carried out for a long-time period [1, 2].

Based on this, we propose a moving object segmentation method, which is classified as unimodal-no recursive from a unimodal-recursive method (single Gaussian) and a unimodal-no recursive technique (median filter). In the Literature review, in the year 1995, McFarlane and Schofield [7] with Approximated Median Filter mentioned that in practice it is complicated to obtain a background image without objects. Later in 1997, Wren and Ali [4, 8] proposed the use of a Gaussian distribution to model the values of a static background pixels, providing an adaptive threshold for classifying foreground and background pixels by means of their standard deviation but requiring the strict absence of objects in the scene to construct the background model during their learning period. Posteriorly, since the Gaussian Mixture Model was proposed in 2000 by Stauffer and Grimson [9, 8] for multimodal backgrounds, despite being the most representative technique in background subtraction approach [10, 11], it is impossible to detect objects during their training period. The scope of this work does not include multimodal backgrounds; however, for scenarios or indoor environments with static backgrounds delivers acceptable results.

This paper is organized as follows: Sect. 3 describes the proposed methodology applied to training or learning period where the median filter condition must be satisfied; in Sect. 4 the segmentation of the following buffers is analyzed, where not always such condition is fulfilled. Experimental results are shown in Sect. 5 by using the single Gaussian and Gaussian Mixture Model algorithms as references. Finally, Sect. 6 concludes this paper.

3 Proposed Methodology

3.1 Description of the Methodology

The values adopted by a particular pixel p, located in coordinates (x, y) in the image at any time t is called as *pixel process* [12]. Its analytical representation is shown in (1).

$$\{p_1, \ldots.., p_i\} = \{I(x, y, i){:}1 \leq i \leq t\} \tag{1}$$

where I represents the image sequence, and i the values of the pixel at time t.

Figure 1 shows different stages of the proposed methodology for moving regions segmentation during training period (first buffer) to be applied to each pixel along time axis. From the video sequence to analyze, the first images or frames are taken. Subsequently, the median operation is applied to the values adopted by each pixel along the time axis (Fig. 2(a)). After, absolute distances are calculated from the absolute difference between median previously obtained respect to each value adopted by the pixel (Fig. 2(b)). Absolute distances are carried to the same scale (0–100%) during normalization process (Fig. 2(c)). Finally, two thresholding stages are carried out using a Gaussian distribution to separate the foreground values from the background values (Fig. 2(d)).

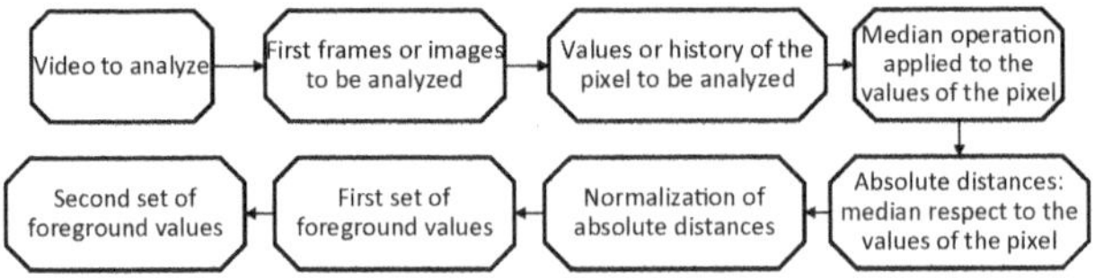

Fig. 1. Segmentation process of moving regions during the training period (first buffer).

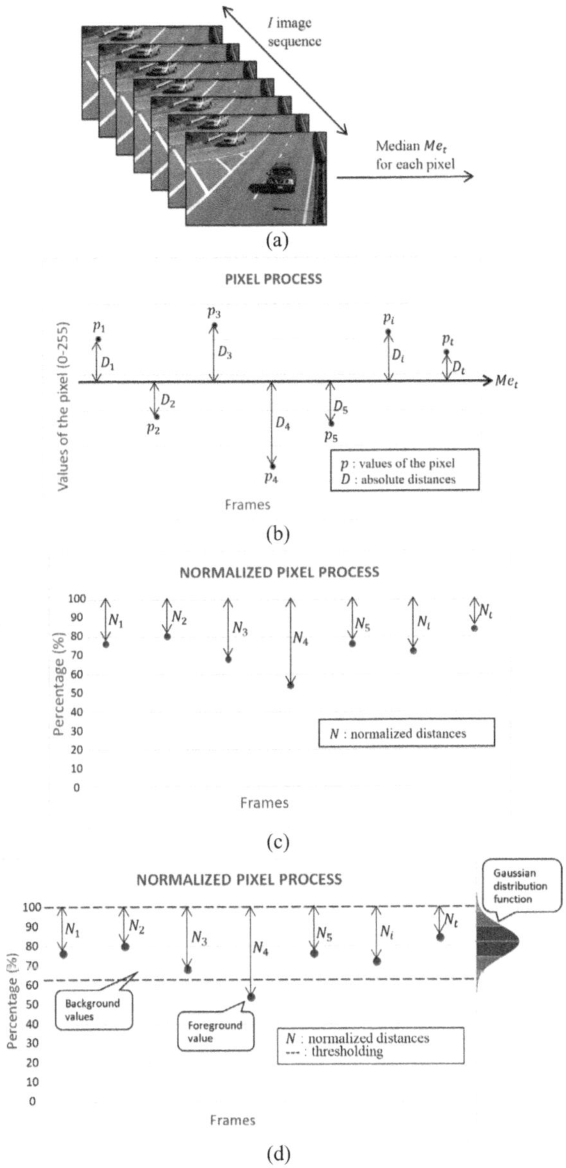

Fig. 2. Illustration of the proposed methodology. (a) Median operation applied to the values adopted by the pixel during the first frames, (b) Absolute distances are calculated from the previous median respect to the values adopted by the pixel, (c) Normalization of the absolute distances, (d) Two thresholding stages applied to the normalized pixel process using a Gaussian distribution function.

3.2 Analysis of the Pixel Process

In a video sequence there are pixel processes that contain background and foreground regions whereas others only the background region. In Fig. 3(b), a frame extracted from a sequence, a pixel was selected from a region of the scene where movement was observed (road lane) and another one from an area absent from moving objects. These areas are indicated with dots in red color. Figure 3(a) shows a pixel process that contains background and foreground regions corresponding to the pixel taken from the area with presence of movement whereas in Fig. 3(c) a pixel process with only background region belonging to the area with no motion.

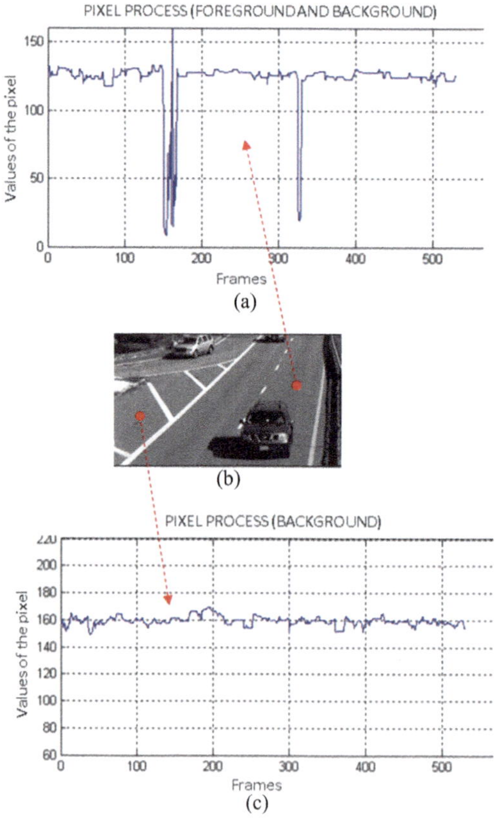

Fig. 3. Pixel processes with presence and absence of movement. (a) Pixel process with foreground and background regions, (b) Areas of the scene with presence and absence of movement, (c) Pixel process with only background region.

Therefore, given a pixel process, a design of an algorithm is proposed to separate these two regions (background and foreground) by each pixel, allowing to segment moving regions.

3.3 Normalization of the Pixel Process

The normalization of the pixel process is necessary due to the values of the background region not always vary around the same amount and on the other hand the foreground regions not always are located within the same boundaries of the background region. The normalization will allow to observe and to interpret the data with less complexity.

The proposed approach starts with obtaining the median of the values observed in the pixel during the frames that compose the learning period. The values are ordered in magnitude (2) and then (3) or (4) is applied according to the case.

$$p_1, p_2, p_3, \ldots\ldots\ldots\ldots\ldots, p_i, \ldots\ldots\ldots, p_t \tag{2}$$

$$Me_t = p_{\frac{t+1}{2}} \quad \textit{if t is even} \tag{3}$$

$$Me_t = \frac{p_{\frac{t}{2}} + p_{\frac{t}{2}+1}}{2} \quad \textit{if t is odd} \tag{4}$$

This operation allows to locate at least one-pixel value corresponding to background region; this stage must satisfy that at least half of the values adopted for each pixel correspond to background values [6]. After, absolute distances are calculated from the absolute difference between each of the pixel values $p_i(x, y)$ and the median previously obtained (5).

$$\{D_1 \ldots\ldots D_t\} = |p_i(x, y) - Me_t| \tag{5}$$

The new values of the pixel history are now represented by the absolute distances $D_1 \ldots\ldots D_t$.

Finally, the distances $D_i(x, y)$ are carried to a same scale through normalization process using (6) allowing to observe and to interpret the data with less complexity.

$$\{N_1 \ldots\ldots N_t\} = [255 - D_i(x, y)](c) \tag{6}$$

$$c = \frac{100}{255} = 0.392 \tag{7}$$

c is a constant resulted from the maximum normalized scale of 100 divided by the largest value that a pixel can take in a grayscale video sequence. Figure 4 illustrates a pixel process randomly chosen (Fig. 4(a)) next to its respective normalized pixel process (Fig. 4(b)). Figure 4(b) can be interpreted as the most representative background values are close to 100% whereas the less representative background values are far from this area.

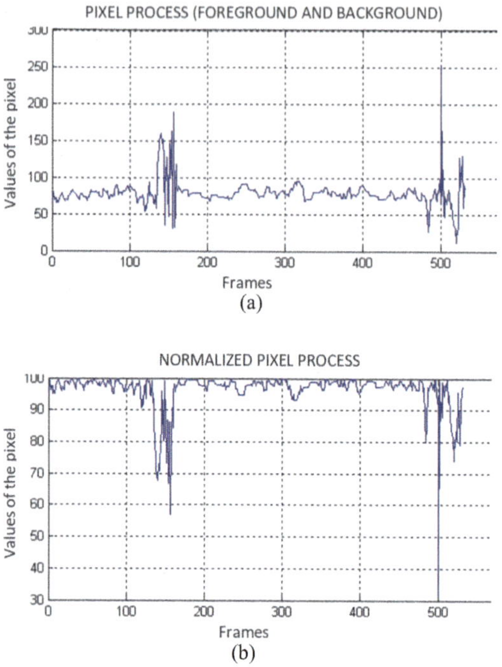

Fig. 4. Comparison between a pixel process and its corresponding normalized pixel process. (a) Pixel process with foreground and background regions, (b) Values of the pixel process carried to scale 0–100%.

3.4 First Thresholding Stage

It has been shown that values of a background pixel can be modeled by means of a Gaussian distribution [4, 13] which is completely defined by: mean μ_t and standard

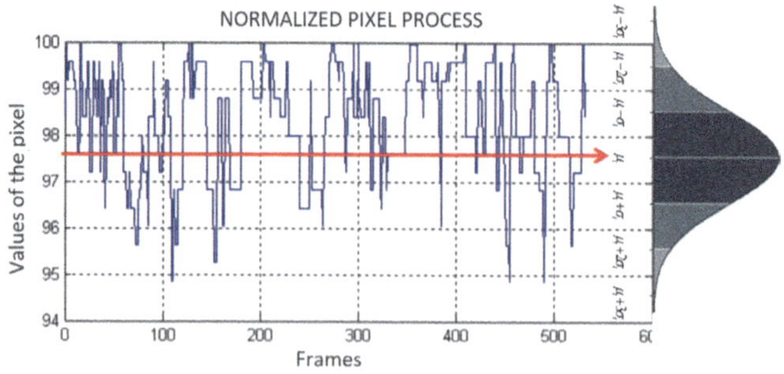

Fig. 5. Graphic representation of the background model using a Gaussian distribution function applied to a normalized pixel process. μ_t and $\mu_{t(proposed)}$ (red line) are approximately equivalents in absence of foreground values.

deviation σ_t. In Fig. 5 can be seen a graphic representation of this affirmation applied to a normalized pixel process.

According to (8) and (9) the mean and standard deviation are calculated.

$$\mu_t = \frac{1}{t}\sum_{i=1}^{i=t} N_i(x, y) \tag{8}$$

$$\sigma_t = \sqrt{\frac{1}{t}\sum_{i=1}^{i=t} \left[N_i(x, y) - \mu_t\right]^2} \tag{9}$$

Based on Fig. 5, the value of the mean (red line) also can be calculated by (10).

$$\mu_{t(\text{proposed})} = 100 - 2.5\sigma_t \tag{10}$$

This means that the condition given by (11) will be satisfied in a normalized pixel process that contains only the background region.

$$\mu_{t(proposed)} \approx \mu_t \tag{11}$$

The presence of pixel values corresponding to foreground region expands the difference between both means due to foreground values increase the standard deviation σ_t causing $\mu_{t(\text{proposed})}$ decrease in magnitude faster than μ_t. Therefore, we have all the elements to establish the first threshold to separate the foreground values through the inequality given by (12).

$$\frac{\mu_{t(proposed)}}{\mu_t}(100\%) \geq threshold1 \tag{12}$$

*threshold*1 has typical values close to 100% (e.g. 97.5, 98, 99, etc. used experimentally).

If (12) is not satisfied, indicates the presence of foreground values, if this happens, the less representative background value N_i is extracted from the normalized pixel process and taken as a foreground value. Both means are recalculated again to verify (12) and this process is carried out until (12) is satisfied. The less representative background values extracted from the pixel process constitute the first set of foreground values.

3.5 Second Thresholding Stage

Frequently, the amount of the pixel values belonging to foreground region is much smaller or they are very close to those corresponding to the background region, causing μ_t and $\mu_{t(proposed)}$ not change significantly and consequently (12) is satisfied. To solve this problem (13) is proposed as a reference.

$$ref = \mu_t - 2.5\sigma_t \tag{13}$$

μ_t and σ_t correspond to the last mean and standard deviation obtained by (8) and (9) during the first thresholding stage. Finally, from the previously thresholded pixel process, the less representative background value N_i is compared to the reference *ref* as is shown in (14).

$$\frac{N_i}{ref} \geq threshold2 \tag{14}$$

*threshold*2 takes the same typical values than *threshold*1.

If the inequality (14) is not satisfied, N_i will be considered as foreground value and extracted from the normalized pixel process. This procedure is repeated with the rest of the less representative background values until (14) is fulfilled. The set of values N_i discriminated constitute the second set of foreground values.

The first and second set of foreground values represent the total foreground region i.e. the pixel values corresponding to the moving object, while the rest of the values not extracted from the pixel process are considered background values of the scene. These latter are used to construct the background model or reference image.

4 Segmentation After Learning Period

After the learning period, from the values contained in the next buffer, the median operation is applied to determine its belonging to the background values of the previous buffer. If so, the segmentation process is carried out as just described and in turn the background model of the scene is updated adapting to gradual illumination changes [13]. Otherwise, it means that some object or objects with slow or static movement cause that the median filter condition is not satisfied, so, the values of the new buffer are discriminated in foreground and background one by one depending on their belonging to the background region or not of the previous buffer. Figure 6 shows a brief graphic description of the proposed algorithm.

The background values are delimited by means of (15) where Me_t refers to the median calculated in the previous buffer and *ref*(*abs*) corresponds to the magnitude of *ref* in a non-normalized scale (16). *lim*(*upper*) and *lim*(*lower*) correspond to the maximum and minimum values to be considered as a background value.

$$\begin{aligned} lim(upper) &= Me_t + ref(abs) \\ lim(lower) &= Me_t - ref(abs) \end{aligned} \tag{15}$$

$$ref(abs) = 255 - (ref/c) \tag{16}$$

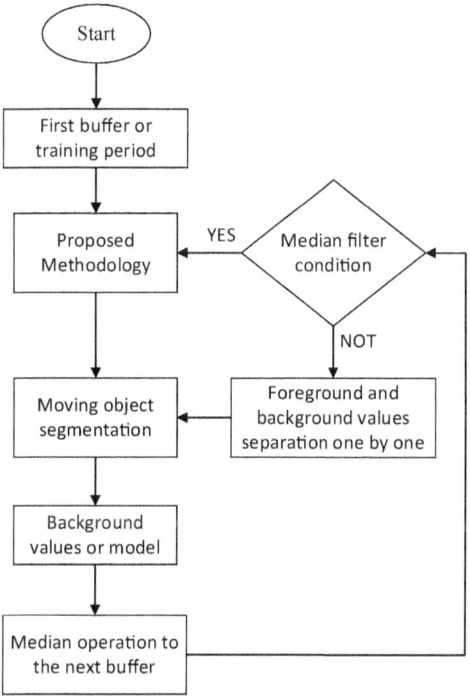

Fig. 6. Flow diagram of the proposed algorithm.

5 Experimental Results

5.1 Single Gaussian

Experimental tests were made using a demo video of Matlab 2014a at 29 frames per second (fps) which was converted to grayscale and the images reduced to a resolution of 240 × 160 pixels. To evaluate the moving regions segmentation, the proposed algorithm was compared to the single Gaussian method. Firstly, both techniques were applied to a video sequence with an approximate duration of 17 s, of which the first 3 s are absent from moving objects in such a way that the single Gaussian model construct its background model correctly. After, the first 3 s of the video sequence were removed and the evaluation was carried out again. The results were quantified using the False Negative Error (FNE) (17) and the False Positive Error (FPE) (18). They refer that the foreground is falsely classified as background and that the background is falsely classified as foreground respectively [14–16]. A manual segmentation of a certain frame is used as reference for both techniques (Fig. 7).

Segmentation results by the proposed algorithm (Fig. 8) for both scenarios (presence and absence of moving objects during the first frames) remain practically unchanged in foreground areas as well as in the background areas. The quantitative results are reported in Table 1 where the FPE and FNE magnitudes (below 2%) do not vary significantly for both situations demonstrating an acceptable segmentation quality.

$$FNE = \frac{number\ of\ foreground\ pixels\ wrongly\ classified}{number\ of\ foreground\ pixels\ in\ manual\ segmentation} \tag{17}$$

$$FPE = \frac{number\ of\ background\ pixels\ wrongly\ classified}{number\ of\ background\ pixels\ in\ manual\ segmentation} \tag{18}$$

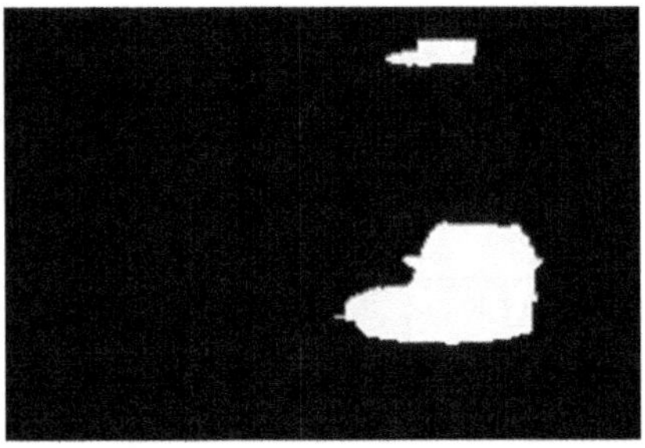

Fig. 7. Manual segmentation used as reference used for the calculation of FNE and FPE errors.

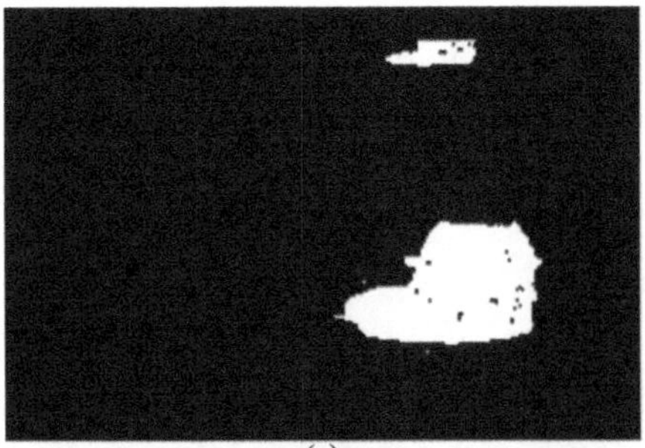

(a)

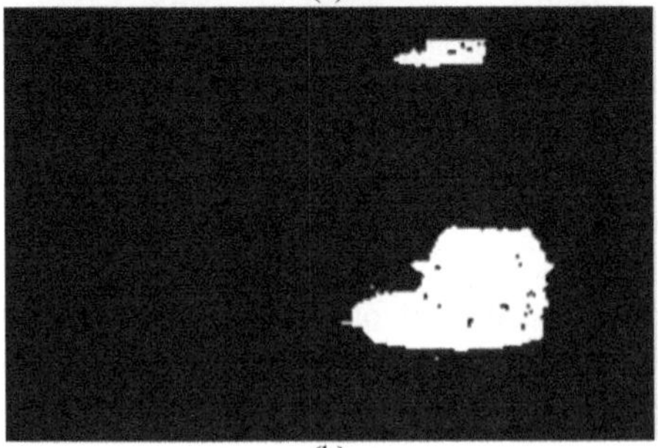

(b)

Fig. 8. An acceptable segmentation quality by the proposed algorithm for both scenarios. (a) Absence of moving objects during the first frames, (b) Presence of moving objects during the first frames.

Table 1. Quantified errors of the proposed algorithm

Proposed algorithm	Absence of objects (acceptable segmentation)	Presence of objects (acceptable segmentation)
FPE	0.03%	0.01%
FNE	1.72%	1.45%

The results obtained by the simple Gaussian method (Fig. 9) show that the absence of moving objects during the first frames allowed an acceptable segmentation quality (Fig. 9(a)) and the quantitative results are reported in Table 2 with FPE and FNE below 1%. However, when the scene consists of the presence of moving objects during the first frames of the video sequence, the segmentation results in the foreground area are not acceptable as illustrated in Fig. 9(b) contrary to the proposal in Fig. 8(b).

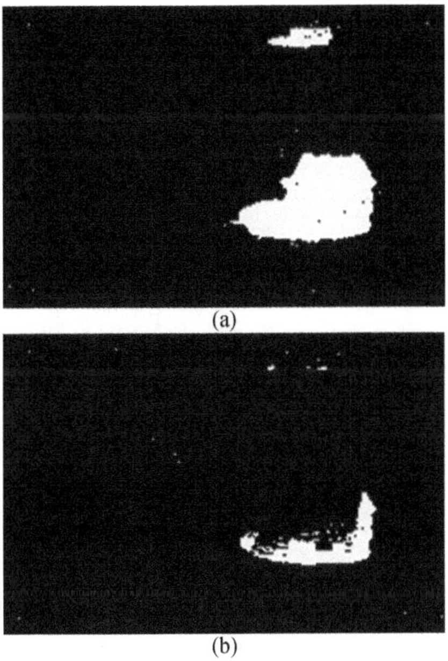

Fig. 9. Segmentation results by the single Gaussian method. (a) Absence of moving objects during the first frames; an acceptable segmentation quality, (b) Presence of moving objects during the first frames; the foreground segmentation is not acceptable.

Table 2. Quantified errors of the single gaussian method

Single Gaussian Method	Absence of objects (acceptable segmentation)	Presence of objects (not acceptable segmentation)
FPE	0.34%	0.07%
FNE	0.42%	66.2%

This is because such objects present at the beginning of the video sequence did not allow the correct construction of the background model of the scene. In Table 2 an FNE of 66.2% is reported, a rather high error compared to the proposed algorithm in Table 1 which is 1.45%. This last result demonstrates the robustness of the present work because in practice is very complicated to obtain a video sequence free of moving objects to construct the background model of the scene correctly.

Finally, these two metrics were calculated each 20 frames along the video sequence to observe their behaviors (Figs. 10 and 11). As can be seen in Fig. 11(b), in some frames the FNE in the single Gaussian method reaches values above 80% which means that less than 20% of the foreground area is visible in the segmentation results for these frames when there are moving objects at the beginning of the video sequence. The FNE obtained by the proposed algorithm (with and without moving objects during the first frames) are very close to each other (Fig. 10(b)), and in turn these are very similar in magnitude to those calculated in the single Gaussian method when there are not moving objects at the beginning of the video sequence. To culminate, the background areas segmented for both scenarios and techniques present an acceptable quality because the FPE are below 1% (Figs. 10(a) and 11(a)).

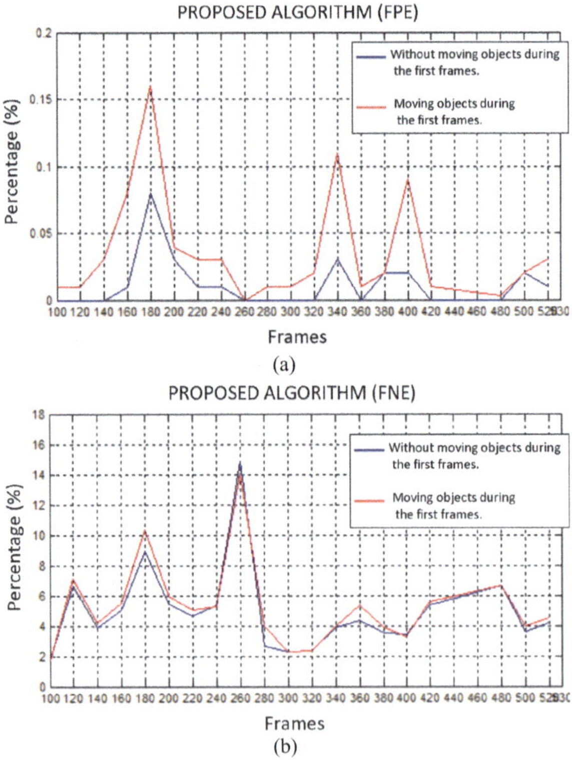

Fig. 10. Errors each 20 frames in the proposed algorithm. (a) False Positive Error, (b) False Negative Error.

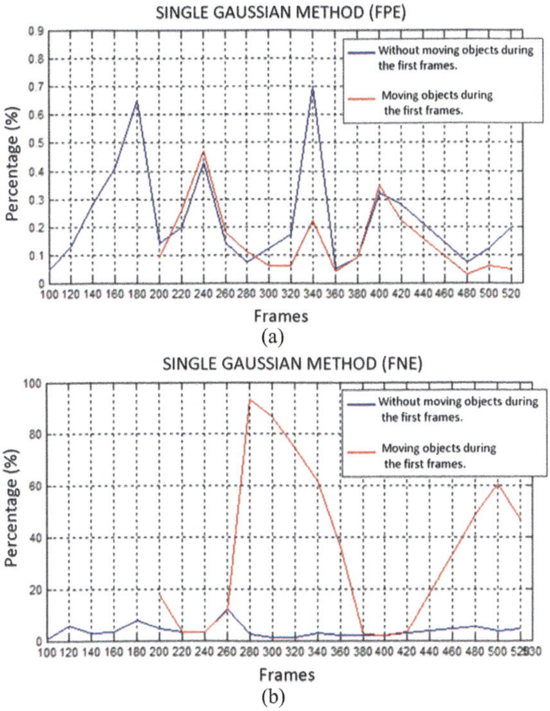

Fig. 11. Errors each 20 frames in the single Gaussian method. (a) False Positive Error, (b) False Negative Error.

5.2 Gaussian Mixture Model

During the experimental tests in the case of the presence of moving objects at the beginning of a video sequence, it was observed that this segmentation method was not able to carry out segmentation of the moving regions during these first frames since they were used for building the reference background model. As a result, the first segmented frame (Fig. 12(a)) does not correspond to the first original frame (Fig. 12(b)) of the video sequence while in the proposed method there is correspondence (Fig. 12(c)).

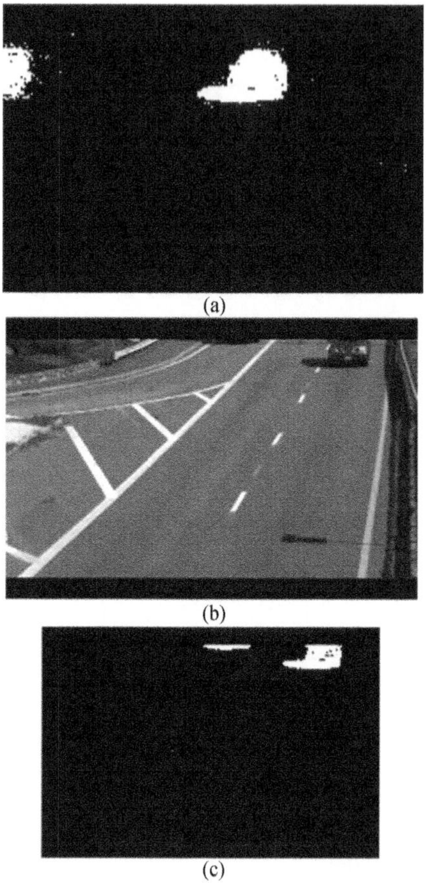

Fig. 12. Correspondence with the first original frame of the video sequence. (a) First segmented frame by the Gaussian Mixture Model; there is not correspondence, (b) Original frame, (c) First segmented frame by the proposed algorithm; there is correspondence.

6 Conclusion

This article presented the description of an algorithm of moving object segmentation capable of carry out detection from the first frames in a video sequence and at the same time obtains the background values to build the background model or reference image of the scene. Its design overcome disadvantages and take advantages of two techniques frequently used in the scientific literature: single Gaussian and Median filter. During the experimental tests was demonstrated the quality of the moving objects segmentation by the proposed algorithm is not affected by the presence of moving objects at the beginning of the video sequence as in the single Gaussian method and Gaussian Mixture Model validating the robustness of this work. As future work, we will work to overcome the main limitation of this methodology which is to handle multimodal backgrounds as well as to deal other typical problems present in moving objects segmentation algorithms such as shadows and sudden changes in lighting.

Acknowledgment. We would like to thank the Instituto Politécnico Nacional and CONACyT from Mexico, undoubtedly the support from these institutions was vital in the accomplishment of this work.

References

1. Maddalena, L., Petrosino, A.: A self-organizing approach to background subtraction for visual surveillance applications. IEEE Trans. Image Process. **17** (2008)
2. Cheung, S.-C.S., Kamath, C.: Robust techniques for background subtraction in urban traffic video. In: Proceedings SPIE 5308, Visual Communications and Image Processing (2004)
3. Buch, N., Velastin, S.A., Orwell, J.: A review of computer vision techniques for the analysis of urban traffic. IEEE Trans. Intell. Transp. Syst. **12** (2011)
4. Wren, C.R., Azarbayejani, A.: Pfinder: real-time tracking of the human body. IEEE Trans. Intell. Transp. Syst. **19** (1997)
5. Piccardi, M.: Background subtraction techniques: a review. In: IEEE International Conference on Systems, Man and Cybernetics (2004)
6. Engel, J.I., Martín, J., Barco, R.: A low-complexity vision-based system for real-time traffic monitoring. IEEE Trans. Intell. Transp. Syst. **18** (2016)
7. McFarlane, N.J.B., Schofield, C.P.: Segmentation and tracking of piglets in images. In: Machine Vision and Applications, vol. 8. Springer-Verlag (1995)
8. Zeng, Z., Jia, J., Yu, D.: Pixel modeling using histograms based on fuzzy partitions for dynamic background subtraction. IEEE Trans. Fuzzy Syst. **25**(3) (2017)
9. Stauffer, C., Grimson, W.E.L.: Learning patterns of activity using real-time tracking. IEEE Trans. Pattern Anal. Mach. Intell. **22** (2000)
10. Datondji, S.R.E., Dupuis, Y., Subirats, P.: A survey of vision-based traffic monitoring of road intersections. IEEE Trans. Intell. Transp. Syst. **17**(10) (2016)
11. Zhong, Z., Zhang, B., Lu, G., Zhao, Y.: An adaptive background modeling method for foreground segmentation. IEEE Trans. Intell. Transp. Syst. **18**(5) (2017)
12. Stauffer, C., Grimson, W.E.L.: Adaptive background mixture models for real-time tracking. In: Proceedings IEEE Computer Society Conference on Computer Vision and Pattern Recognition, vol. 2 (1999)
13. Wang, K., Liu, Y., Gou, C.: A multi-view learning approach to foreground detection for traffic surveillance applications. IEEE Trans. Veh. Technol. **65**(6) (2016)
14. Kim, H., Sakamoto, R., Kitahara, I., Toriyama, T., Kogure, K.: Background subtraction using generalised Gaussian family model. Electron. Lett. **44** (2008)
15. Wang, W., Yang, J., Gao, W.: Modeling background and segmenting moving objects from compressed video. IEEE Trans. Circuits Syst. Video Technol. **18** (2008)
16. Lee, H., Kim, H., Kim, J.-I.: Background subtraction using background sets with image- and color-space reduction. IEEE Trans. Multimed. **18**(10) (2016)

Straight Boundary Detection Algorithm Based on Orientation Filter

Yanhua Ma[1(✉)], Chengbao Cui[2], and Yong Wang[2]

[1] College of Information Science and Technology, Qingdao University of Science and Technology, Qingdao 266000, Shandong, China
yhma@qust.edu.cn
[2] Zaozhuang Mining Group Co., Ltd., Zaozhuang 266000, Shandong, China

Abstract. This paper investigates the problem of boundary identification in image processing, especially linear edge detection of images from videos. We propose a novel linear boundary detection algorithm which is an extension to oriented linear filters. The objective is to fix the misidentified linear boundary due to image noise or other issues. The algorithm includes three steps: (1) edge detection; (2) binary processing; and (3) edge modification and extraction by eliminating isolated point or noise. It attempts to generate continuous edge with no blur as well as to remove false boundaries. The size and orientation of linear filter determine the length and orientation of identified boundaries, respectively. The experimental results indicate that the proposed algorithm is effective for detecting linear boundaries of decay image and produce high quality solutions.

Keywords: Boundary detection · Linear filter · Pattern recognition

1 Introduction

Partitioning nontrivial images is one of the most complicated tasks in image processing. Generally, image segmentation accuracy determines quality and performance of computer vision applications. Hence, considerable attention should be taken to increase the chance of rugged segmentation. Majority of image segmentation algorithms can be categorized according to two basic properties of intensity values: discontinuity and similarity. In the first category, the algorithm partitions an image based on sharp changes in intensity, such as edges in an image. The typical algorithms in the second category partition an image into regions that match a set of predefined criteria. Thresholding, region growing, and region splitting and merging are examples of approaches in this category [1].

In this paper, we introduce an application using an approach in the discontinuity-based category. Our objective centers on finding edges of conveyor belts in images. The conveyor belt is a key component in an automated conveyor system. Its operational robustness and resilience is closely associated with the manufacturing production process. Off-tracking is the most common type

K. Arai et al. (Eds.): IntelliSys 2018, AISC 868, pp. 494–503, 2019.
https://doi.org/10.1007/978-3-030-01054-6_35

of mechanical malfunction for conveyor belts. It may lead to waste of materials, increasing conveying resistance, even industrial accidents. For example, conveyor belts off-tracking in the mining industry may cause severe fire accidents and casualties. Even though there are mechanical or electronic monitoring and controlling techniques that can be used to detect and fix the conveyor belts off-tracking issue, these controlling devices have limitation in functionalities. Indeed, currently available monitoring devices can only be applied for video surveillance so that judgments of off-tracking are still rely on human observations on video images. Hence, it is critical to develop automated devices that can accurately detect the off-tracking issue of a conveyor system in time, and then alert related person. Such devices will not only reduce accidents and be beneficial to operational safety, but also improve labor productivity. The motivation of this project hinges on industrial needs of detecting the off-tracking issue. More specifically, the observation target is the conveyor belt and we aim at differentiating it from other parts of an image. Image segmentation algorithms can be applied to process point detection, line detection and edge detection. In this paper, we apply an edge detection approach that has been designed for detecting meaningful discontinuities in gray level.

Edge detection is one of the widely adopted techniques for monochrome image segmentation and in particular has been a staple of segmentation algorithms for many years. The related work in the area of edge detection algorithms is enumerated partly in Sect. 2.

2 Related Work

The concept of edge detection represents a large group of approaches that extract information about edges in an image using mathematical rules and advanced computing techniques. For example, [2] incorporated the parallel computing technique into existing edge detection approaches and derived parallel operations for detecting texture edges. In another study, zero-crossings of second derivatives is utilized in the edge detection technique [3]. To directly capture useful properties of an edge operator, [4] formulated a set of edge detection criteria using variational techniques and introduced a set of heuristics for operators integration. The computational approach in edge detection has been widely explored. [5] not only proposed mathematical forms for edge detection and localization criteria, but also introduced a set of specialized criteria that produce closed-form solutions for step edges. In a recent study, [6] defined a unified formulation that simultaneously combines low- and mid-level image representations which significantly reduced computational cost.

Some researchers pay attention to specific issues and difficulties in edge detection. As edge detectors usually do not produce clear results when facing junctions, [7] provided solutions to junction detection and localization using contours. [8] studied another important issue, the multi-scale boundary detection, and by training a classifier to combine local boundary cues across scales. Region detection for video data, an extension to natural image processing, has been exploited by [9] which solved the issue of distinguishing occlusion boundaries from internal boundaries. [10] presented a new technique to deal with non-straight (standard) line detection problem. [11] proposed a new genetic programming (GP) approach to extract edge features applied to edge detection.

The design of edge detectors is an interdisciplinary field that broader area of research are incorporated. For instance, [12] proposed an edge detector using wavelet transforms of symmetrical wavelets. In another study, [13] adopted fuzzy categorization in edge classification and introduced the FEDGE-fuzzy edge detection approach. Machine learning is also an emerging technique that is applied to improve quality of edge detectors. A neural-network based model is proposed by [14] and justified to produce high-quality edge imagery.

3 The Problem

3.1 Recognition Algorithm

Edge detection or image segmentation is one of the most critical processes in image comprehension and computer vision. A series of edge detectors for image processing have been developed, while there is no approach that can universally satisfy all applications. This paper aims at solving the problem of boundary detection in a video about an automated conveyor system. In particular, the objective is to monitor and identify whether a conveyor belt is offtrack when the conveyor system is under operation.

Two frames of the video are shown in Fig. 1. The object in Fig. 1a, pointed by the two arrows, is a conveyor belt that may offset its track to the left or right. For example, it shifts obviously to the left in Fig. 1b compared with that in Fig. 1a. On the conveyor belt, a pair of rollers are fixed on each side, cycled in Fig. 1a.

A crucial issue in monitoring this conveyor system is finding boundaries of the belt and the fixed rollers. As all boundaries are linear in shape and vertical in orientation, we choose approaches that deal with this type of edge detection, such as Prewitt operator, Sobel operator, Robinson operator and Kirsch operator. They are defined separately as (8-neighborhoods and 3×3 sizes).

$$\text{Prewitt vertical operator: } \begin{pmatrix} -1\ 0\ 1 \\ -1\ 0\ 1 \\ -1\ 0\ 1 \end{pmatrix}$$

$$\text{Sobel vertical operator: } \begin{pmatrix} -1\ 0\ 1 \\ -2\ 0\ 2 \\ -1\ 0\ 1 \end{pmatrix}$$

$$\text{Robinson vertical operator: } \begin{pmatrix} -1 & 1\ 1 \\ -1 & -2\ 1 \\ -1 & 1\ 1 \end{pmatrix}$$

$$\text{Kirsch vertical operator: } \begin{pmatrix} -5\ 3\ 3 \\ -5\ 0\ 3 \\ -5\ 3\ 3 \end{pmatrix}$$

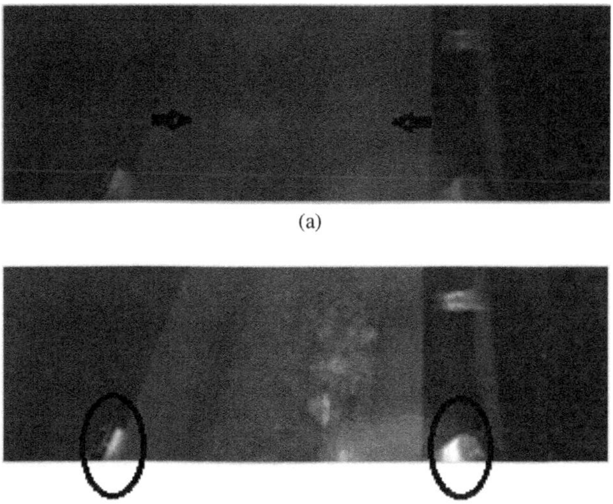

(a)

(b)

Fig. 1. Two special frames in a video.

These edge detectors are implemented through MATLAB in Fig. 1. The results are shown in Fig. 2. We observe that the boundaries are barely detected accurately. First, it is difficult to mark the left boundary of the conveyor belt, particularly through Robinson, as it is significantly disconnected. On the other hand, boundaries of the rollers are also hard to identified, especially by Prewitt and Sobel.

4 Straight Boundary Detection Algorithm

We mentioned in the previous section that our target boundaries are almost vertical and linear, while the typical detectors do not generate high quality results in this case. To improve the solutions, we introduce a straight boundary detection approach with three steps: (1) edge detection; (2) binary processing; and (3) edge modification and extraction by eliminating isolated point or noise.

4.1 Edge Detection

Due to the target of vertical and linear boundaries, we use the first-order partial derivative to detect the edges. The notation of the first-order partial derivative in the x-axis is as follows:

$$\frac{\partial f}{\partial x} = f(x+1, y) - f(x, y) \tag{1}$$

The results of edge detection is defined as

$$g(x, y) = \frac{\partial f}{\partial x} = f(x+1, y) - f(x, y) \tag{2}$$

(a) Prewitt vertical detector

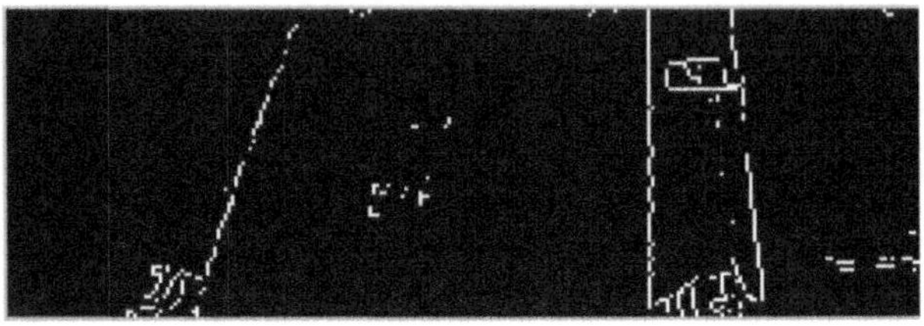

(b) Sobel vertical detector

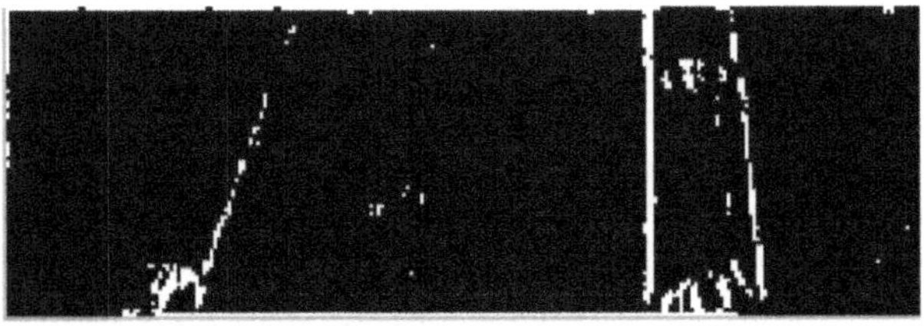

(c) Robinson vertical detector

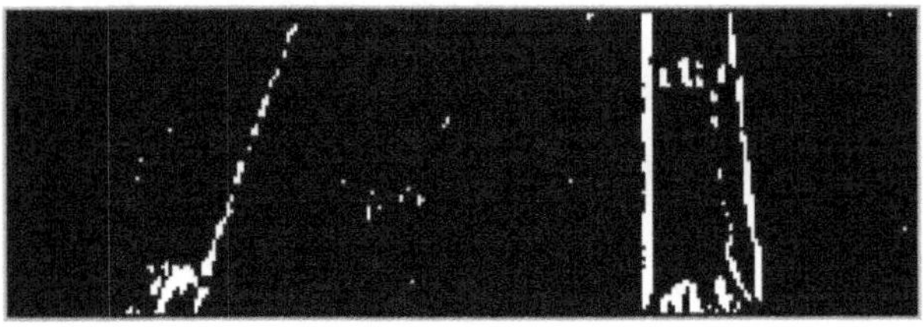

(d) Kirsch vertical detector

Fig. 2. Boundary detection results by different detectors.

where $f(x, y)$ is the input image, $g(x, y)$ is the processed image, i.e. the image of edge detection.

Obviously, the convolution sum of $f(x, y)$ is equal to the image $g(x, y)$. The convolution mask C can be defined as $C = [-1, 1]$. Hence, the convolution sum is given by the expression

$$g(x, y) = \sum_{s=0}^{1} C(s) f(x + s, y) \tag{3}$$

Figure 3 gives the result of edge detection of Fig. 1a based on (3).

Fig. 3. First-partial derivative edge detection in the x-axis.

4.2 Binary Processing

Figure 3 is a gray-scale image with 256 gray grades. However, the outputs in real application should be in binary in order to achieve an ideal recognition effect. Therefore, we want to map the gray-scale image into binary. In this process, q represents the gray grade before mapping and r describes the gray grade after mapping. The function is in (4).

$$r = \begin{cases} 0, & \text{if } q \leq \theta \\ 1, & \text{otherwise} \end{cases} \tag{4}$$

θ is a threshold defined arbitrarily, while has to satisfy the criteria that a continuous, clear boundary can be identified. In other words, this value is selected through experiments and cannot be too large or too small. A too large threshold may cause discontinuous boundaries and a too small threshold may cause pseudo- boundaries. In this application, the threshold is arranged as 0.2. The result of binarization is shown in Fig. 4.

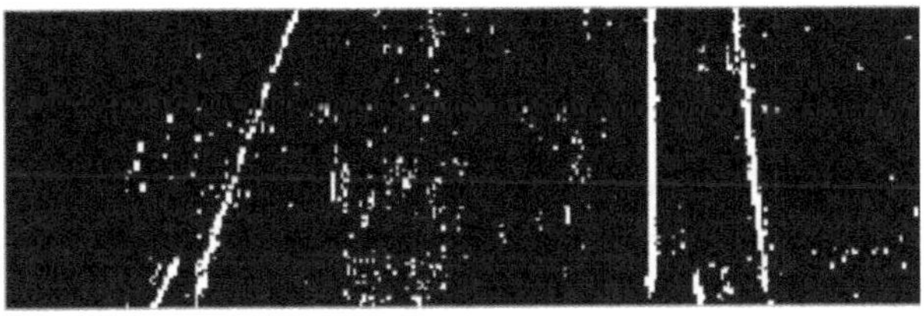

Fig. 4. Binarization.

4.3 Filtering

After the binarization process, we can observe that many isolated points appear in the result. The boundaries have several typical properties, i.e., straight, vertical, and fixed width, that point out the choice of filtering approach to eliminate isolated points or noises. We define the filtering operator in (5) and (6).

$$f_i(x,y) = 2 \times \frac{\sum_{s=-m}^{m}\sum_{t=-n}^{n} h_i(s,t) f(x+s, y+t)}{\sum_{s=-m}^{m}\sum_{t=-n}^{n} h_i(s,t)} \tag{5}$$

$$g(x,y) = \begin{cases} 1, & f(x,y) = 1 \text{ and } \max(f_i(x,y)) \geq 1 \\ 0, & \text{otherwise} \end{cases} \tag{6}$$

where $g(x,y)$ is the filtered image, $f_i(x,y)$ is temporary image, $f(x,y)$ is the image obtained by binarization, and h_i is one of the filters which changes slow with different i.

$$h_i(s,t) = \begin{cases} 1, & [t \times \frac{i}{n} - w_d] \leq s \leq [t \times \frac{i}{n} + w_d] \\ 0, & \text{otherwise} \end{cases} \tag{7}$$

where $i = 0, \pm1, \pm2, ..., \pm m$, $m = \frac{(w-1)}{2}$ and $n = \frac{(h-1)}{2}$. w_d controls breadth of the filter. When $w_d = 0$ and $w_d = 1$, it is a line-width of 1 and 3 respectively. $w \times h$ is size of the filter. For example, in Fig. 5, $w = 5$, $h = 7$, and $w_d = 1$ so that the breadth is 3. The filtering result is shown as Fig. 6.

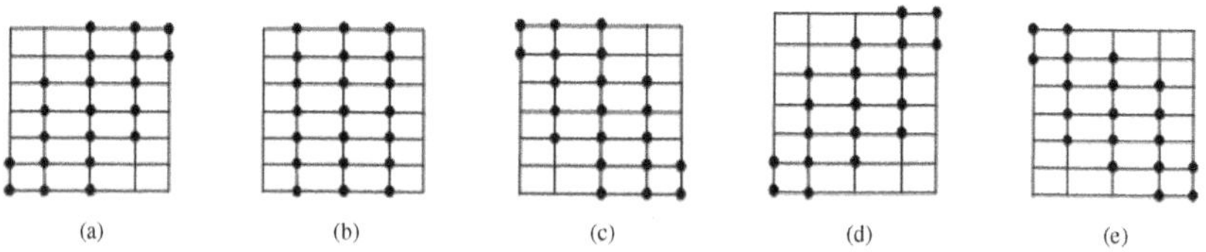

Fig. 5. Several filters.

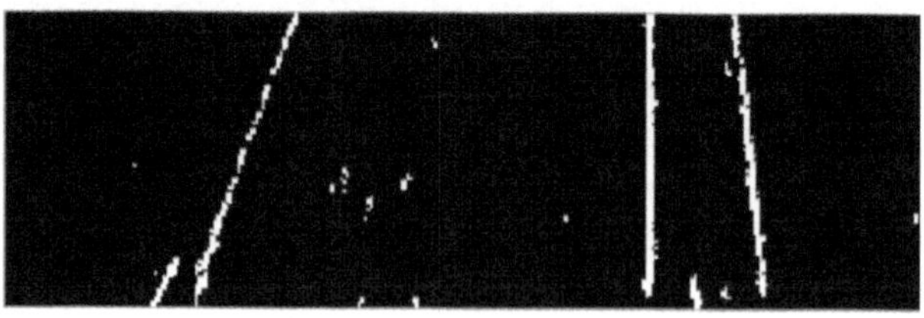

Fig. 6. Filtered image.

4.4 Continuous Modifying

Even though the isolated points or noises are eliminated successfully, as shown in Fig. 6, the boundaries are inevitably become discontinuous. We adopt a similar operator to that in the Sect. 4.3 to modify the discontinuous boundary. In this step, we replace (6) by (8). The result of the filtering is shown as Fig. 7.

$$g(x,y) = \begin{cases} 1, & \max(f_i(x,y)) \geq 1 \\ 0, & \text{otherwise} \end{cases} \tag{8}$$

Fig. 7. Modified image.

We apply the filters h_i to further modify the edges and eliminate the blurs remained. h_i is defined as

$$h_1 = \begin{pmatrix} 1 \\ 1 \\ 1 \\ 1 \\ 1 \\ 1 \\ 1 \\ 1 \\ 1 \end{pmatrix}, h_2 = \begin{pmatrix} 0\,0\,1 \\ 0\,0\,1 \\ 0\,0\,1 \\ 0\,1\,0 \\ 0\,1\,0 \\ 0\,1\,0 \\ 1\,0\,0 \\ 1\,0\,0 \\ 1\,0\,0 \end{pmatrix}, h_3 = \begin{pmatrix} 1\,0\,0 \\ 1\,0\,0 \\ 1\,0\,0 \\ 0\,1\,0 \\ 0\,1\,0 \\ 0\,1\,0 \\ 0\,0\,1 \\ 0\,0\,1 \\ 0\,0\,1 \end{pmatrix}$$

The result of the filtering is shown as Fig. 8.

Fig. 8. To further modify the edges.

4.5 Edge Extraction

Edge extraction is the last step of image processing in boundary identification. Figure 9 shows the result through the algorithm defined (9) that polishes the edges on Fig. 8.

$$g(x,y)=\begin{cases}1, & \text{if } f(x,y_1)=0, f(x,y_1+1)=1,\ldots, \\ & f(x,y_1+a)=1, f(x,y_1+a+1)=0 \\ & \text{then } y=y_1+\frac{a}{2} \\ 0, & \text{otherwise}\end{cases} \tag{9}$$

Fig. 9. Edge extracting.

To verify quality of the results, we overlay Fig. 9 on Fig. 1a. It is obvious that the edges are completely fitted (see Fig. 10).

Fig. 10. Return to the original image.

5 Conclusions and Future Work

This study proposes a straight boundary recognition approach based on orientation filter. The experiment results show that the algorithm generates high recognition efficiency and quality, especially on noisy or blurred images. We did not evaluate accuracy of the algorithm in this experiment. This can be included in an extension work in the future. In addition, we will modify the algorithm to use Canny edge detector to implement boundary recognition or carry out the self-regulation binary threshold.

References

1. Gonzalez, R.C., Woods, R.E.: Digital Image Processing, 2nd edn. Prentice Hall, Upper Saddle River (2002)
2. Rosenfeld, A., Thurston, M.: Edge and curve detection for visual scene analysis. IEEE Trans. Comput. **100**(5), 562–569 (1971)
3. Marr, D., Hildreth, E.: Theory of edge detection. Proc. R. Soc. Lond. B Biol. Sci. **207**(1167), 187–217 (1980)
4. Canny, J.F.: Finding edges and lines in images. MIT Artificial Intelligence Laboratory, Massachusetts, USA, Technical report AI-TR-720, 6 (1983)
5. Canny, J.: A computational approach to edge detection. IEEE Trans. Pattern Anal. Mach. Intell. **6**, 679–698 (1986)
6. Leordeanu, M., Sukthankar, R., Sminchisescu, C.: Efficient closed-form solution to generalized boundary detection, pp. 516–529. Springer (2012)
7. Maire, M., Arbeláez, P., Fowlkes, C., Malik, J.: Using contours to detect and localize junctions in natural images. In: IEEE Conference on Computer Vision and Pattern Recognition, CVPR 2008, pp. 1–8. IEEE (2008)
8. Ren, X.: Multi-scale improves boundary detection in natural images, pp. 533–545. Springer (2008)
9. Sundberg, P., Brox, T., Maire, M., Arbeláez, P., Malik, J.: Occlusion boundary detection and figure/ground assignment from optical flow. In: 2011 IEEE Conference on Computer Vision and Pattern Recognition (CVPR), pp. 2233–2240. IEEE (2011)
10. Alemán-Flores, M., Alvarez, L., Gomez, L., Santana-Cedrés, D.: Line detection in images showing significant lens distortion and application to distortion correction. Pattern Recognit. Lett. **36**, 261–271 (2014)
11. Fu, W., Johnston, M., Zhang, M.: Genetic programming for edge detection: a gaussian-based approach. Soft Comput. **20**(3), 1231–1248 (2016)
12. Sun, M., Sclabassi, R.J.: Symmetric wavelet edge detector of the minimum length. In: Proceedings of the International Conference on Image Processing, vol. 2, pp. 177–180. IEEE (1995)
13. Ho, K., Ohnishi, N.: Fedgefuzzy edge detection by fuzzy categorization and classification of edges. In: Fuzzy Logic in Artificial Intelligence Towards Intelligent Systems, pp. 182–196 (1997)
14. Vrabel, M.J.: Edge detection with a recurrent neural network. In: Applications and Science of Artificial Neural Networks II, vol. 2760, pp. 365–371 (1996)

Using Motion Detection and Facial Recognition to Secure Places of High Security: A Case Study at Banking Vaults of Ghana

Emmanuel Effah[1(✉)], Salah Kabanda[2], and Edward Owusu-Adjei[1]

[1] Computer Science and Engineering Department, University of Mines and Technology (UMaT), Tarkwa, Ghana
eeffah@umat.edu.gh, adjei.edward@yahoo.com

[2] University of Cape Town, Cape Town, South Africa
salah.kabanda@uct.ac.za

Abstract. Motion Detection and Facial recognition (MD&FR) have been extensively studied to improve surveillance services in areas where motion detections and human identification are needed. As crime techniques keep improving, surveillance technologies must also advance in the same manner, and MD&FR is the reliable surveillance technology in current literature. Despite the conceivable potentials of integrating MD&FR technologies into current surveillance systems (SS), the available systems have deployed MD-based and FR-based SS in isolation. In this paper, we proposed a Smart Surveillance Systems (SSS) functional framework, the design Flowchart of this SSS and a software approach to implement the SSS using MD&FR techniques in OpenFace and OpenCV Facial Recognition (FR) and Motion Detection (MD) libraries, respectively. Extreme Learning Machine (ELM) was the Facial Recognition (FR) algorithm used due to its robustness and higher accuracies under variable regularization factors (RF). Contrasting with the current CCTV Camera-based surveillance systems used in Ghanaian banking Vaults, this system augmented security when it was deployed. Thus, our system saves much storage space because it records automatically only during motion detections, logs all detected faces and alert instantly if intrusion is detected. The SSS worked best with CCTV Cameras and also operated efficiently with lower pixel-rated (cheap) cameras, such as WebCams and Infinix Note3 under variable RF. This system is recommended for policy consideration in banks and other firms to minimise security systems' overheads and boost efficiency of SS.

Keywords: OpenFace · Extreme Learning Machine (ELM)
OpenCV, FisherFaces · Regularization Factors (RF)
Banking vaults · Facial recognition · Motion detection
Smart Surveillance System (SSS) · Surveillance systems (SS)

K. Arai et al. (Eds.): IntelliSys 2018, AISC 868, pp. 504–520, 2019.
https://doi.org/10.1007/978-3-030-01054-6_36

1 Introduction

Motion Detection and Facial recognition (MD&FR) are among the most prominent forms of biometric recognition and security technique that is widely recommended in literature because it offers myriad cross-domain applications and so, have been on the forefront of research efforts for the past two decades [1]. FR based authentication systems are seeing a lot of recent innovations in terms of selective market applications involving Human Machine Interface (HMI) such as the upcoming ATM (Automated Teller Machine) user identity verification systems [3] and Driving License identity authentication while MD is widely applied in robotics and automation systems [2]. According to [19], MD&FR have been extensively studied, due to both scientific fundamental challenges and the potential applications where motion detections and human identification surveillance are needed.

FR-based surveillance systems (SS) show merits, among the most documented include non-intrusiveness, low cost of equipment and no user agreement requirements during acquisition [20]. However, merits of FR-based SS have not been harnessed due to these setbacks [19,21] such as different viewpoints, blocked effects, illumination changes, strong lighting states, SS automations and inadequate smart inclusions. Given the persistent need to protect organizational assets and be weary of security concerns, SS must be well equipped, automated and smart to serve their intended purposes efficiently. Whilst this need is imperative, it can also be challenging to adhere to, specifically in developing countries that face contextual constraints of infrastructure. For example, SS or security systems in Ghanaian banks are based on CCTV cameras that require continuous monitoring, recording and more storage space.

These systems are non-intelligent because of their inability to differentiate between legal and illegal intruders approaching high security areas to ensure effective surveillance services. As a consequence, SS in Ghanaian banks face following challenges; lack the ability to detect moving objects automatically, require constant monitoring and or recording, unable to verify faces to recognize the persons entering high security zones such as the vaults, unable to log recognized users and promptly alert the security personnel whenever unauthorized persons intrude the vaults. Consequently, management rely on the storage devices for facts during emergency situations and perpetrators could also go undetected if they mask their faces [21]. In this era of smart technology that abridges both human and machine thinking [4], and computer vision libraries that improve the way computers perceive and manipulate video frames and images [5], the need for more efficient and reliable security systems that address the stipulated challenges must be researched [21].

This study therefore proposes a functional framework, flowchart of a SSS and developed MD&FR-based autonomous SSS software that is efficient under uncontrolled lighting conditions which is applicable in highly secured areas such as the banking vaults- which is the focus of this study.

The vault is a highly secured bulk money room at various banks that is accessed by only authorised employees and constantly monitored by the security persons using real-time video feed CCTV cameras. The goal of this paper is to propose an SSS with improved security efficiencies in banking vaults, capture frames only when motion is detected (i.e. reduce storage space wastages in existing system), verify authorised users and log them before and after accessing the security area, work well under lower pixel rated cameras and promptly notify security personnel whenever motion is detected.

2 Literature Review

2.1 FR and FR Methods

Face recognition SS using video or visual surveillance system can be a very challenging due to probably low resolutions frames, poor lighting conditions of some video capturing devices and the presence of uncontrolled movement [21]. Other challenges in video based-surveillance systems could also be the poor resolution of the face of the subject to be captured [19,20], obstruction or masking of the subject's face [21,22], improper location of the camera from the subject causing the number of pixels on the subjects face to be low [27], making it difficult for robust face recognition. Sometimes, the positioning of the camera may cause the face to be out of view of the camera frame [23]. Since these challenges about the subject are difficult to deal with, FR methods and tools have rather received much research attention to address this challenge [6]. Since these challenges about the subject are difficult to deal with, several FR methods, algorithms and tools have been researched, and the onus lies on their effective implementations.

FR methods are classified as appearance-based (AB) methods and model-based (MB) methods as shown in Fig. 1. Former methods use holistic texture features that are applied to either whole-face or specific regions in a face image whereas latter methods employ shape and texture of the face [6]. The commonest appearance-based FR methods are Principal component analysis (PCA), independent component analysis (ICA) and linear discriminant analysis (LDA), all under the linear approach. The PCA method and LDA methods have demonstrated keen success in face detection, recognition, and tracking [23]. An emerging method for performing a nonlinear form of PCA is the Kernal PCA [21]. Appearance-based FR methods transform the problem to a face space analysis problem, where many well-known statistical methods can be tried out. A special aspect of this model is its applicability to low resolution or poor quality images. However, it requires sufficiently large sampling data for a successful distribution; prior knowledge of the human faces is not utilized in this model; impact of facial variations due to illumination, pose, and expression is subjected to its limitations. The model-based FR scheme constructs a model of the human face, which can capture the facial variations. The prior knowledge of human face is utilized to design the model [6]. Model-based FR methods have these merits: has a perfect intrinsic physical relationship with real faces [17]; an explicit modeling of face variations, such as pose, illumination, and expression gives the possibility to handle these variabilities in practice. Model construction is the main challenging task in model-based FR methods because of the difficulty in extracting

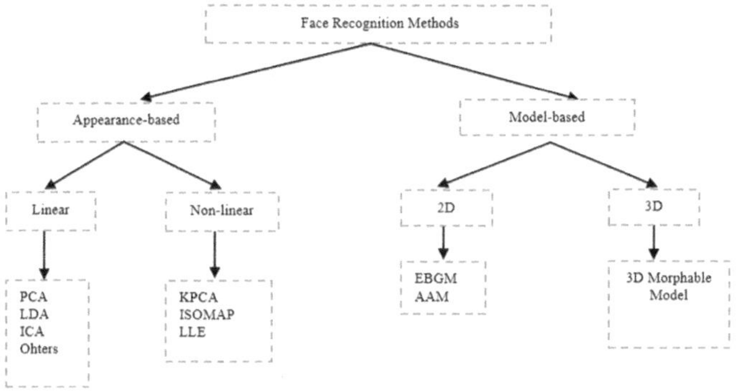

Fig. 1. A general classification of face recognition techniques.

facial feature points automatically [17,18], high dependence on time-consuming fitting process and reliance on high-resolution cameras with good quality face images. This paper used ELM which is a model approach due to its stipulated merits and its robustness and higher efficiencies.

2.2 Visual or Video-Based Surveillance Systems (VSS)

According to [26], the processes in VSS are summarized: background modeling, motion segmentation, classification of foreground moving objects, human identification, tracking, understanding of object behaviours and fusing of images taken from multiple cameras to increase the surveillance areas [25] (Fig. 2).

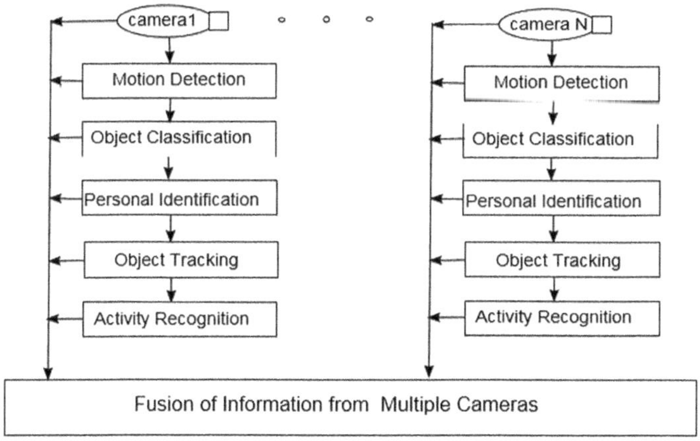

Fig. 2. Block diagram of visual SS [25].

2.3 Motion Detection (MD)

First step in VSS is MD (Fig. 2) [25]. MD segments the moving foreground object from the rest of the images. Successful segmentation of foreground object helps in the subsequent process such as object classification, personal identification, object tracking and activity recognition in the video. Motion segmentation is done mainly with background subtraction, temporal or frame differencing, and optical ow [26]. Out of the three methods, background subtraction is the most popular method for detecting moving regions in an image by taking the absolute difference between the current image and the reference background image [26]. A proper threshold is judiciously selected which segments foreground from the background [25].

2.4 Computer Vision (CV)

CV is the way machines reconstruct the world we see in one or more images by properly rearranging its properties such as shape, illumination and colour distributions [7]. Several applications have been made to make use of CV such as optical character recognition system called tesseract, automatic number plate recognition systems and facial recognition systems. Among the tools documented in literature for developing efficient MD & FR software in CV [8–10] are discussed in Table 1.

Table 1. Computer vision software development tools [8–11]

Tool	Advantages	Diadvantages	Application
DeepMask [8]	It is good at segmentation objects.	It is not very selective and can generate mask for image regions that are not interesting	Use in general recognition including inanimate object identification
OpenCV [9]	It is portable (cross-plateform) and Open Source. It is fast when it comes to image processing.	Manual memory management. Memory used needs to be cleared manually.	Use in computer vision application and robotics.
OpenFace [10,11]	Has a high accuracy rate for recognition and its support vector machine (SVM), require less training time. It is open source	It is not cross-platform	Use only for facial recognition.

3 Smart Surveillance System (SSS)

As documented in literature, VSS record continuously [19–21] under diverse lighting conditions [21] with different rated CCTV cameras. FR-based VSS advanced this technology by adding facial detection for biometric identification [25]. However, we are of the view that VSS must be more advanced to meet the current crime trends especially in places of high security.

For the purpose of this study, various banks were visited to examine their SS. IT experts and users of existing SSs were interviewed to ascertain deficiencies in existing SS.

4 System Analysis and Design Methodology

4.1 System Analysis

We deployed the waterfall model due to the progressive nature of this project and how the requirement of it was well known and clear [15].

The requirements analysis phase commenced by one of the researcher visiting the various banks to examine existing SS and interview IT experts and users of existing SS to ascertain deficiencies in existing SS.

The questions specifically focused on the the business functions of the Banks, security challenges banks face as a result of the existing SS and the anticipated way forward. After understanding the analysis, it was discovered that existing security systems in Ghanaian banks are based on CCTV cameras that require continuous monitoring, recording and more storage space. These systems are non-intelligent due to their inability to differentiate between legal and illegal intruders approaching high security areas and are also unable to log legal users of the system and also alert promptly during illegal intrusions. By way of addressing the stipulated flaws of the existing SS, a Smart Surveillance System (SSS) with functional framework as shown in Fig. 3 was proposed.

The SSS has three sets of users based on the requirements analysis:

(1) Administrator: The person or people authorized to view the video feeds, adds people to be notified and trains faces for the FR subsystem.
(2) The recipient: The employees to receive the notifications during intrusions.
(3) Authorized Employees: Persons authorized to use the facial recognition system.

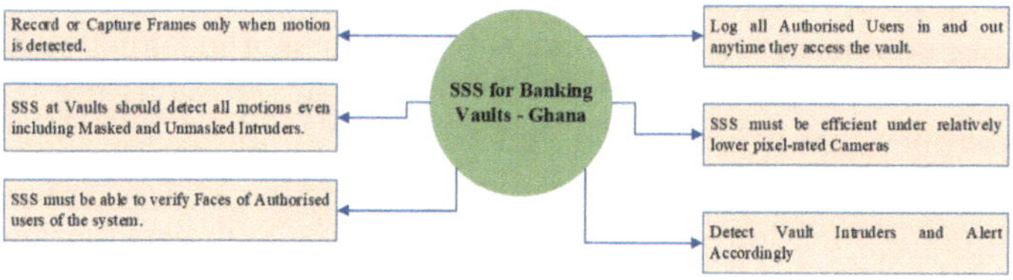

Fig. 3. Functional framework of proposed SSS for Ghanaian banking vaults.

The functional requirements of this system as shown in Fig. 4 are:

- Detect moving objects.
- Reduce the need for a human to watch or record from the Camera constantly.

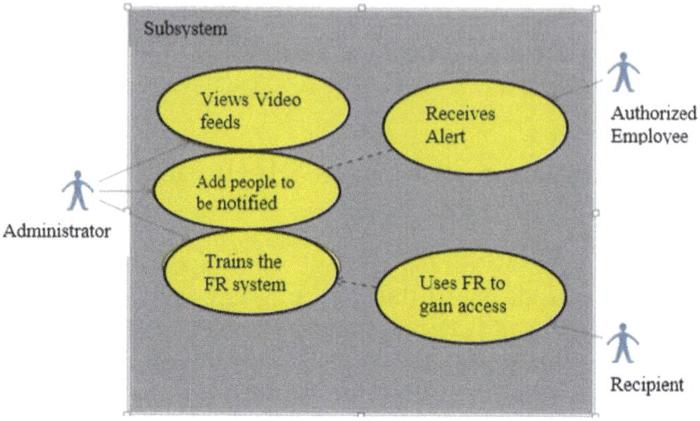

Fig. 4. Use case diagram of the proposed SSS.

- Verify faces using facial recognition to the security zone.
- Logs all recognized people.
- Ability to alert on detection of illegal intrusion.

The actual input and output process of the system was studied and created in the design phase. The user interface design which deals with the interactions with the system and the abstract representation and data flows inputs and output of the system were designed as detailed out in Fig. 4.

4.2 Design and Implementation

The implementation was done in C++ environment using the OpenCV and OpenFace libraries. ELM algorithm for the FR part was imported from Python. The OpenCV library resourced the MD or optical flow component while OpenFace library offered very effective appearance-based FR algorithm [6]. Consequently, our software is expected to work excellently with low pixel-rated such as webcams, infrared cameras and Infinix Phone Cameras when operated under good lighting conditions. Background subtraction method and blob detection algorithm are added to improve the effectiveness of our software. OpenCV is most suitable because it has tested and tried codes and has a large code base contributed by researchers and programmers [12]. Also, OpenCV is a library of CV programming functions which aid in real-time CV processing and manipulations [13]. OpenFace is suitable due to its near human accuracy in FR and less training time on normal computers which do not have discrete graphics processing unit. Thus, the client need not have a very expensive computer to use the software [14]. OpenFace is a library that stays updated with the latest deep neural network architectures and technologies for FR. Figure 5 shows the flowchart diagram of the proposed SSS comprising of the MD and FR subsystems.

The program was finally developed and tested severally by letting it go through several compiler optimizations by setting various GCC flags to achieve maximum speed and performance. Various image resolutions were tried with the system, and it was found during computations that 320 240 pixels was optimal for efficient speed of the proposed SSS. With the FR part, capturing very quality frames for recognition was found to be better than using low-quality frames.

The result remains under testing. Object-oriented design treated a class as a unit [16]. Unit test was performed on various class instances to ensure suitable results and the desired logical consistencies. The software has been deployed with the various video sources such as WebCam, CCTV Camera, and Infinix Note 3 phones with camera quality of 16 megapixels and it performed excellently. It worked as expected by taking pictures and video only when people moved into the high-security zones (Banking vaults). With the FR part, only the trained faces were logged to access the security area while the illegal intruders were refused, and notices of illegal intrusions were sent to the appropriate persons. We initially realized that the software was ineffective at places with severe background noise. However, we have resolved that and can now achieve optimal performance in all environments by testing with various background subtraction methods provided by OpenCV.

4.3 ELM Algorithm

Over the last decade, considerable progress has been made in the development and improvement of FR algorithms. Various methodologies have been employed for FR. Neural Networks, Support Vector Machine and ELM are training based algorithm and suffer from local minima challenges [13]. However, ELM is primarily selected for its robustness, ability to offer a better generalization performance despite the fact that the earning speed of ELM is slow. ELM is also resilient to changes in illumination which is an added advantage for the proposed system because of voltage instability nature of the area of deployment.

The structure of ELM is shown in Fig. 5.

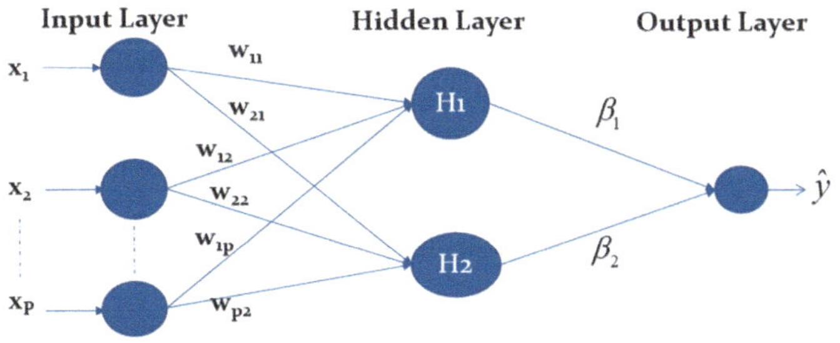

Fig. 5. Structure of ELM.

Given a training set:

$$D = \{(x_i, t_i) | (x_i = \in R^n, (t_i = \in R^m i = (1, \ldots, N)\} \tag{1}$$

Activation function: $g(x)$,

Standard SLFN (single hidden layer Feedforward neural network) with N hidden nodes can be modeled mathematically as:

$$\sum_{i=1}^{N} \beta_i g(w_i \cdot x_j + b_i = t_j), j = 1, \cdots N \tag{2}$$

$$H(w_1, \ldots, w_{\tilde{N}}, b_1, \ldots, b_{\tilde{N}} X_1, \ldots, X_N) = \begin{bmatrix} g(w_1 \cdot X_1 + b_1 & \ldots & g(w_{\tilde{N}} \cdot X_1 + b_{\tilde{N}} \\ \vdots & \ldots & \vdots \\ g(w_1 \cdot X_N + b_1 & \ldots & g(w_{\tilde{N}} \cdot X_N + b_{\tilde{N}} \end{bmatrix}_{N \times \tilde{N}} \tag{3}$$

$$\beta = \begin{bmatrix} \beta_1^T \\ \vdots \\ \beta_{\tilde{N}}^T \end{bmatrix}_{\tilde{N} \times n} \tag{4}$$

$$T = \begin{bmatrix} t_1^T \\ \vdots \\ t_{\tilde{N}}^T \end{bmatrix}_{N \times m} \tag{5}$$

(a): SLFNs are actually linear systems if SLFNs are actually linear systems if the input weights and the hidden layer biases can be chosen randomly. Since $H\beta = T$ is a linear system, β can be calculated using one of the least square solutions of a general linear system. This solution has three important properties which are minimum training error, smallest norm of weights, and unique solution.

(b): Given

- Training set $D = \{(x_i, t_i)|(x_i = \in R^n, (t_i = \in R^m i = (1, \ldots, N)\}$
- Activation function $g(x)$ and
- number of hidden nodes $= \tilde{N}$

ELM algorithm runs as follows:

(1) Randomly assign input weight w_i and bias b_i, where i = 1, ..., N
(2) Calculate the hidden layer output matrix H
(3) Calculate the output weight β by resolving the linear system $HB = T$ thus, $\beta = H^{-1} * T$

(c): Considering the minimum norm least square solution,

$$\hat{\beta} = H^+ T \tag{6}$$

where, H^+ is the Moore-Penrose generalized inverse of matrix $H_{m \times n}$ given by: $H^+ = (H^T H)_{-1} H^T$.

For a regularized linear output: $H^+ = (H^T H + \lambda I)^{-1} H^T$ where λ is the regularisation factor (RF) which could be 0.1, 0.01, 0.001, or 0.0001.

This study used the regularized linear output to ensure that the problems with overfitting is well minimised.

ELM is faster compared with back-propagation (BP) algorithm and SVM while maintaining its classification accuracy at the same level as that of BP or SVM and it is considered as a generalized version of single-hidden-layer-feedforward networks (SLFNs) [28]. As seen in the equations above, the weights between input layer and hidden layer are randomly generated and the weights between hidden layer and output layer are computed by using the outputs of hidden neurons with randomly assigned input-hidden neuron weights. The challenge of robustness is addressed by the introduction of the RF [28].

5 Proposed Software-Based SSS

This is the main software which detects motions, takes pictures, records video and reports when the object is moving. Figure 6 shows the flowchart diagram of the proposed SSS comprising of the MD and FR subsystems.

5.1 Video Capture

The video capture mode is one of the initial vital steps of this system takes before all others. The system needs two video sources but with proper configurations, the system can be set to use one or more source, but this system was built to use two video sources because it is a case study. One camera in the vault and one also mounted on the door to the vault. The video capture source can be taken wirelessly or by cable due to the flexibility of OpenCV. The video frames must be presented to the system in pure colour but it can also be configured to accept video frames in black and white. Two frames are quickly taken each time for analysis and processing. They are represented and stored in a matrix for processing.

(d). The MD ensues the video is captured. The steps taken to achieve MD as illustrated in MD subsystem section in Fig. 6 . When the absolute difference of the new video frame and the previous frame is taken, there is going to be a zero value which is achieved for all the pixels above threshold in the current video frame, which indicates whether there was no motion, or a non-zero value is achieved for some pixels, which indicates there was motion. The non-zero values are the values of the pixels causing motion which will range from 0 to 255 which each number represents a colour in the matrix. Thus, if some difference value is achieved, there is surely motion of an object. Absolute Difference function calculates the absolute difference between two frames in the OpenCV library. The two frames should be of the same size, so the system makes use of frame size 320×240. Also, to identify motion, the motion causing pixels are segregated using a binary threshold.

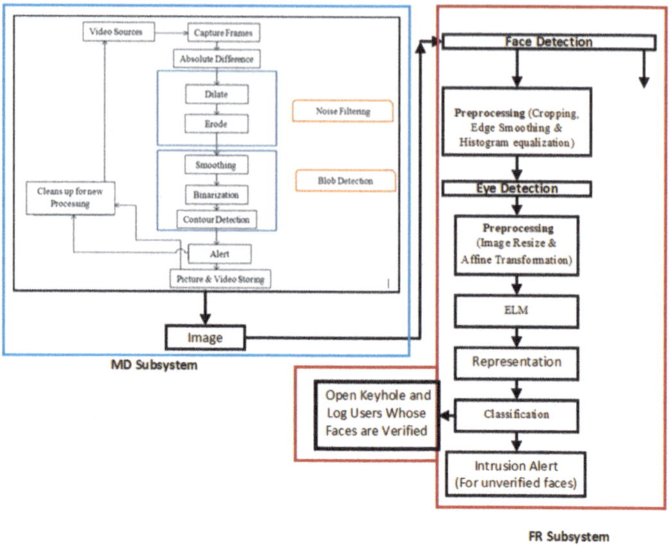

Fig. 6. Flow chart of the proposed SSS.

5.2 Image Filtering

Bogus movements in the background may be mistaken for motion. Hence such noise causing elements needs to be filtered. Basic morphological operations such as erosion and dilation functions in OpenCV are used for image filtering which is also illustrated in the Fig. 6. Dilate resizes the image to the shape of the incoming frame then the removal of noise ensues by applying erode function. Blob detection specifies mathematical steps for detecting specific region based on some property in a digital image. A blob is made up of pixels having the specified property in common, and pixels are combined if they share common neighbourhood. Pixels that have values greater than the threshold are the property of interest. In a video frame, if after absolute differencing and noise removal, the value of a pixel is greater than the binary threshold it is eligible to form a blob.

5.3 Contour Drawing

A contour is the boundary of object pixels above the threshold. Contour Analysis outputs a vector which contains blobs detected. Contours are edge-based features which are insensitive to illumination changes. Every entry of the vector contains co-ordinates of the pixels of a respective blob. The size of the vector depends on many blobs detected. The object is bounded using bounding rectangle function over the contour using OpenCV. Here the boundary of the blobs causing motion is identified and highlighted. If many blobs adjacent to each other are causing motion, and even if a single pixel becomes common amongst two adjacent blobs, the two blobs unite, and the common centroid is calculated to show new bounding box. In the current release of the program, there is an inline function called findcontour which finds the contours, and there is another which handle drawing of contours called drawcontour.

5.4 Alert Functionality

This part of the system sends data over the network by creating a new thread with a message "MOTION" after it sends to the target device and the thread exits successfully. Any system on the specified network chosen as a target would receive a message whether a computer or phone.

5.5 Training of Images

Affine transformation inbuilt into the OpenFace library transforms the face to a 3D model, so the image appears as if the face is looking directly towards the camera. This transformation is used during the training process to better the accuracy of the process. The low-dimensional image is then fed into the OpenFace support vector machine. Figure 7 shows the affine transformation in the FR subsystem.

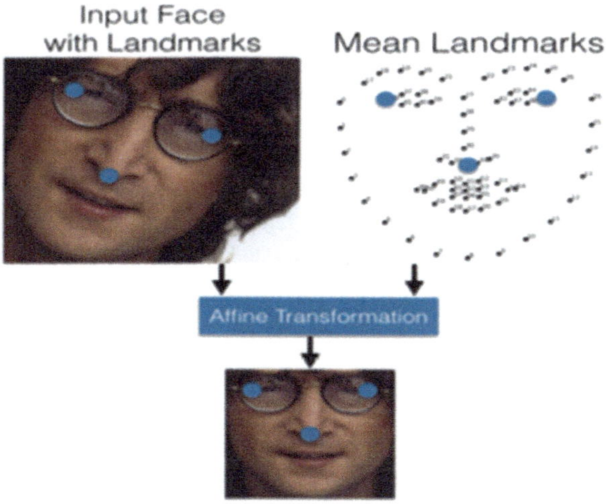

Fig. 7. Affine transformation.

5.6 The Recognition

Given an input image with multiple faces, the face recognition system typically first runs face detection to isolate the faces in image. Each face is pre-processed as shown in Fig. 6, and then a low-dimensional representation is obtained. A low-dimensional representation is important for efficient classification after the ELM algorithm has run. The process is shown in FR subsection of Fig. 6.

5.7 The Alert Receiving Subsystem

This is the part of the system which receives alerts when motion has been detected by the MD subsystem. This software will be given to authorized users who are chosen purposely to receive alerts. This software basically listens on a dedicated port, awaiting a specific string. When it receives the string, it plays a sound and logs the time. Displaying the time at the very lower bottom. Figure 8 displays the system interface that receives alerts.

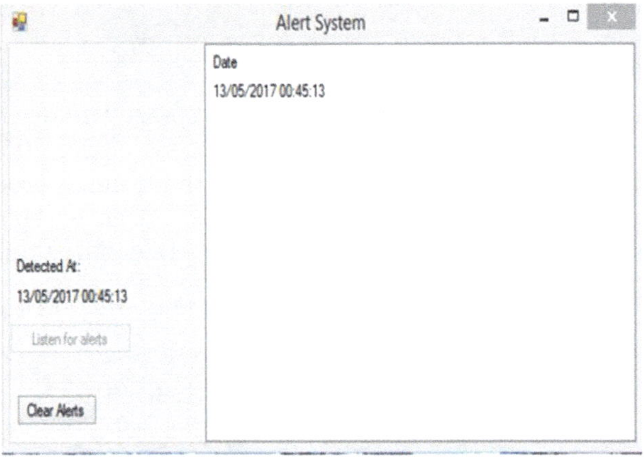

Fig. 8. The alert system.

6 Results

The results of the system when it was deployed is as follows. Table 2 presents the system's performance under variable RF and it shows that the algorithm operated best at RF = 0.0001; the CCTV cameras gave the bests frames and hence the highest recognitions. Figure 9 shows the wireless feed from the Infinix Note 3 with the aid of an app called DroidCam demonstrating the logging functionality. Bounding Rectangles were drawn by the system when the motion was detected across its area of coverage.

Figure 9 also presents the feed from the web camera and the phone running simultaneously depicting some interesting abilities. These are:

- It could analyse two sources easily on one thread.
- It could detect motion above the line of interest turning the line into a green from red when that happens.
- It did not take into consideration the movement of the fan. Which shows that it does ignore background noise.

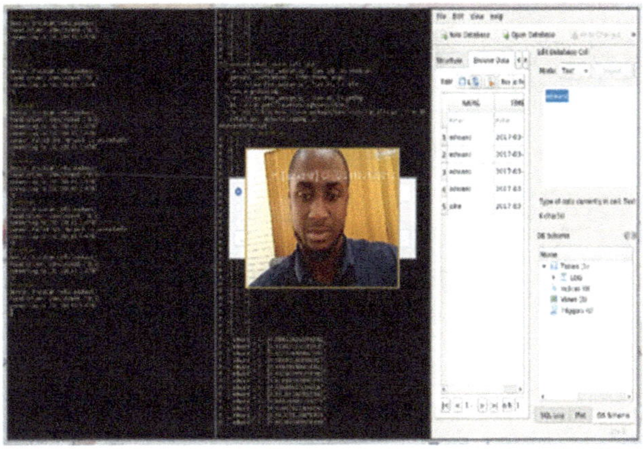

Fig. 9. FR system in usage with logging.

The FR application too was deployed. The logging of faces worked perfectly by logging the recognized face into the database when it was able to identify the person and opened the keyhole. This software also showed great performance with the detection speed and identification process. It did also output a confidence level for its recognition. The confidence level can be increased closer to 1 for a much more accurate recognition only. Figure 10 shows the system at work.

When motion is always detected, one of the important parts was the ability of the software to alert clients. So, the developed software as shown in Fig. 8 depicts an outlined view of detections and the latest and most current detection dates and times displayed near the left bottom side of the application form.

The proposed system has addressed the challenges of inadequate smartness in existing SS by fulfilling the functions in the proposed SSS framework shown in Fig. 3. The novelties of our system are underlaid in its smartness [19–24]; our system classifies all masked faces as intruders and alert accordingly for immediate action, record frames or videos only after motion detections to save storage space, log only faces of recognised employees that have been trained into the system and performs excellently with both high and low pixel-rated cameras under good lighting condition. With the aid of the ELM algorithm, it was established that our system performs best at RF = 0.0001 which firmed the basis for including lower pixel rated cameras (Table 2). Also, from Table 2, it was evident that under the same lighting condition, FR performance is much enhanced by the frame quality of the camera being used. The results show that CCTV cameras performed best, followed by the Infinix Note 3 and lastly the WebCam.

The background subtraction method and blob detection algorithm improved the efficiency and accuracy of our system compared with existing systems. Also, the motion detection application built with QT makes it easier for this application to be ported to other operating system environments

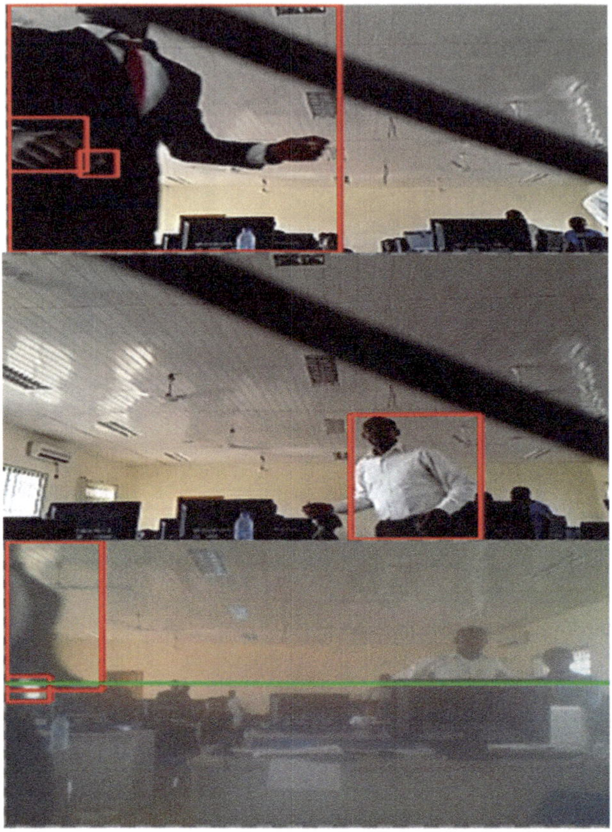

Fig. 10. System in motions detecting mode: Motion areas in rectangles.

Table 2. Performance of the FR Subsystem under Variable RF using ELM Algorithm

	Test using WebCam	Test using CCTV	Test using phone (Infinix Note 3)	RF
Correctly detected faces	19 out of 25	23 out of 25	22 out of 25	0.1
	19 out of 25	24 out of 25	21 out of 25	0.01
	17 out of 25	22 out of 25	16 out of 25	0.001
	22 out of 25	25 out of 25	24 out of 25	0.0001

7 Conclusion

Software based SSS has been developed to address the challenges in the existing SS. This system in its novelty integrated OpenCV and OpenFace and takes advantage of their mutual merits. When this proposed SSS was deployed in a branch of Access Bank, Ghana during the testing stage, it performed excellently by attesting its ability to detect moving objects automatically, record only during detections, verify faces to recognize the persons entering the vaults, log and opened the keyhole for all recognized employees and alert the security when unauthorized persons intruded the vaults. Surveillance/security team received instant alerts on their cell phones during illegal intrusions using the alert software and on the monitoring PC. The perpetrators masking their faces were classified as intruders by the system. Also, the proposed system performed well with cheap cameras such as webcams when operated under good lighting conditions. The system is user-friendly, and the ELM algorithm was very efficient. This system is recommended for policy consideration by firms that hold security in high esteem aside protecting vaults. Future researches are recommended to improve the ability of this system to fully reduce noise in the environment by extra frame processing and expand its capacity to take more video sources while processing simultaneously.

Acknowledgment. Profound gratitude to all staff of the Access Bank Ghana, Tarkwa Branch for making this work a success.

References

1. Meshgini, S., Aghagolzadeh, A., Seyedarabi, H.: Face recognition using Gabor filter bank, kernel principal component analysis and support vector machine. Int. J. Comput. Theory Eng. **4**(5), 23–34 (2012)
2. Krem.com article: Catching Identity Thieves with Facial Recognition. http://www.krem.com/story/news/local/2-on-your-side/2016/12/13/dol-catching-identity-thieves-with-facial-recognition/20355021/. Accessed 5th May 2017
3. Lowcards.com article: Facial Recognition ATMs Could Curb ATM Theft. http://www.lowcards.com/facial-recognition-atms-standard-future-29504. Accessed 16 May 2017
4. Zorzi, M.: AI emulates the human brain. https://erc.europa.eu/projects-and-results/erc-stories/self-learning-ai-emulates-human-brain. Accessed 24 Feb 2017
5. Joshi, A.: How do I detect an object in the video frame using OpenCV. https://www.quora.com/How-do-I-detect-an-object-in-video-frame-using-OpenCV. Accessed 19 Jan 2017
6. Indhumathi, C., Gayathri, N.S.: Unconstrained face recognition from blurred and illumination with pose variant face image using SVM. Int. J. Res. Comput. Appl. Robot. **2**(2), 112–117 (2014)
7. Szeliski, R.: Computer Vision: Algorithms and Applications. http://szeliski.org/Book/. Accessed 18 Feb 2017
8. Carroll, J.: Computer vision software tools from Facebook are now open source. http://www.vision-systems.com/articles/2016/08/computer-vision-software-tools-from-facebook-are-now-open-source.html. Accessed 9 Apr 2017
9. Bolhumeur, P.N., Hespanha, J.P., Kriegman, J.D.: Eigenfaces vs. sherfaces: recognition using class specific linear projection. IEEE Trans. Pattern Anal. Mach. Intell. **19**, 711–720 (1997)

10. Fisher, R.A.: The use of multiple measurements in taxonomic problems. Ann. Eugen. **7**, 179188 (1936)
11. Sinha, U.: OpenCV vs. VXL vs. LTI: performance test (2010). http://aishack.in/tutorials/opencv-vs-vxl-vs-lti-performance-test/. Accessed 8 Apr 2017
12. Pulli, K., Baksheev, A., Kornyakov, K., Eruhimov, V.: Real-time computer vision with OpenCV. Commun. ACM **55**(6), 61–69 (2012). https://doi.org/10.1145/2184319.2184337
13. Hamid, R.K., et al.: A comparison of energy consumption prediction models based on neural networks of a bioclimatic building. Energies **567573**, 1–24 (2016). MDPI
14. Brandon, A.: OpenFace: a general-purpose face recognition library with mobile applications, p. 12 (2016). http://reports-archive.adm.cs.cmu.edu/anon/2016/CMU-CS-16-118.pdf. Accessed 24 Feb 2017
15. Zeil, S.: Software development process (2016). https://www.cs.odu.edu/~zeil/cs350/f16/Public/processModels/index.html. Accessed 8 Apr 2017
16. Huang, S.-C.: An advanced motion detection algorithm with video quality analysis for video surveillance systems. IEEE Trans. Circuits Syst. Video Technol. **21**(1), 1–14 (2011)
17. Czarneszki, K.: What is structural modelling (2015). www.softlab.ntua.gr/~kkontog/ECE355-05/lectures/Lect8-Ch2-Unit4-Part3.ppt. Accessed 26 April 2017
18. Fakhroutdinov, K.: Class diagram uml diagrams (2009). http://www.uml-diagrams.org. Accessed 26 Apr 2015
19. Jolliffe, I.T.: Principal Component Analysis. Springer, New York (1986)
20. Espinosa-Duro, V., Faundez-Zanuy, M., Mekyska, J.: A new face database simultaneously acquired in visible, near infrared and thermal spectrums. In: Cognitive Computation. Springer (2012). https://doi.org/10.1007/s12559-012-9163-2
21. Swaminathan, A., Kumar, N., Ramesh Kumar, M.: A review of numerous facial recognition techniques in image processing. IJCSMC **3**(1), 233–243 (2014)
22. Lin, F., Fookes, C., Chandran, V., Sridharan, S.: Face recognition from super-resolved images. In: Proceedings of the Eighth International Symposium on Signal Processing and Its Applications, pp. 667–670 (2005)
23. Lin, F., Fookes, C., Chandran, V., Sridharan, S.: Super-resolved faces for improved face recognition from surveillance video. In: International Conference on Biometrics, p. 110 (2007)
24. Espinosa-Duro, V., Faundez-Zanuy, M., Mekyska, J.: A new face database simultaneously acquired in visible, near infrared and thermal spectrums. Cogn. Comput. **5**, 119–135 (2012). https://doi.org/10.1007/s12559-012-9163-2
25. Barnich, O., Van Droogenbroeck, M.: ViBe: a universal background subtraction algorithm for video sequences. IEEE Trans. Image Process. **20**(6), 17091724 (2011)
26. Hu, W.: A survey on visual surveillance of object motion and behaviors. IEEE Trans. Syst. Man Cybern. Part C (Appl. Rev.) **34**(3), 334–352 (2004)
27. Hermosilla, G., et al.: Face recognition and drunk classification using infrared face images. J. Sens. **2018**, Article ID 5813514, 8 pp (2018)
28. Lee, K.-H.: An efficient learning scheme for extreme learning machine and its application. Int. J. Comput. Sci. Electron. Eng. (IJCSEE) **3**(3), 212–216 (2015). ISSN 2320402

Kinect-Based Frontal View Gait Recognition Using Support Vector Machine

Rohilah Sahak, Nooritawati Md Tahir(✉), Ihsan Yassin, and Fadhlan Hafiz Helmi Kamaru Zaman

Faculty of Electrical Engineering, Universiti Teknologi MARA (UiTM), 40450 Shah Alam, Selangor, Malaysia
nooritawati@ieee.org

Abstract. This paper investigated the most suitable multi-class support vector machine (SVM) coding design in recognising human gait based on frontal view that include one-versus-all (OVA), one-versus-one (OVO), error correcting output codes (ECOC), ordinal, sparse random and dense random algorithms. Firstly, walking gait of 30 subjects is captured using Kinect sensor. Next, all 20 skeleton joints within the full gait cycle are extracted as input features. Further, the gait features acted as inputs to the SVM classifier, specifically using linear kernel with various coding design algorithms are evaluated and tested in determining the most optimum results in recognition of human gait based on frontal view. Result proven that one-versus-all (OVA) attained the highest accuracy, specifically 96%.

Keywords: Human gait · Multi-class SVM · Error correcting output codes One-versus-all · One-versus-one · Dense random · Ordinal and sparse random

1 Introduction

Recent advances in 3D Kinect has created many research opportunity in human gait analysis and recognition as compared to a normal standard camera video. This is because Kinect sensor is able to extract all 20 skeleton joints without involving complex processes [1] and no requirement of background subtraction process too. The usage of Kinect in human gait recognition has been proven as a reliable method since it enhanced the recognition performance as reported by Munsell et al. [2] and Proch [3]. Initially, investigation and research on recognition of human gait using Kinect for lateral view was already conducted by several researchers. Conversely, the study on recognition of human gait has been extended to frontal view as it attained promising results as compared to lateral view. Moreover, for space such as corridors and hallway, frontal view is more suitable.

On the other hand, numerous researchers has utilized and used classifier in human gait recognition namely support vector machine (SVM), neural network (NN) and k-nearest neighbor (KNN). Among these classifiers, SVM provides global optimal solution, small sample-size used, good generalization ability and resistant to over-fitting problem as discussed in [4–7].

K. Arai et al. (Eds.): IntelliSys 2018, AISC 868, pp. 521–528, 2019.
https://doi.org/10.1007/978-3-030-01054-6_37

Gianaria et al. [8] discussed that SVM Type 1 (*C*-SVM) and SVM Type 2 (v-SVM) as introduced by Chang & Lin [9] was used for recognition of twenty subjects in frontal view. A total of 55 static and dynamic features were extracted as input features. Based on their study, with v-SVM as classifier and combination of distance of elbows, distance of knees, movement of head in the *x* direction and *y* direction, mean distance between left knee and right knee in *y* direction, the recognition rate based on gait of 20 subjects is 96.25%.

Moreover, study on recognition of human gait in frontal view was performed by Prathap et al. [10]. In this study, Levenberg-Marquardt back propagation and correlation algorithms were employed as classifiers. Results showed that Levenberg-Marquardt back propagation on features such as height, distance between centroids and step length contributed to 94% recognition accuracy.

Recently, in 2015 Kastaniotis et al. [11] conducted gait recognition for 30 subjects based on front view. In addition, statistical methods specifically Wald Wolfowitz (WW) and Mutual Nearest Point Distance were used as feature selection in determining the significant features from the angles of eight selected limbs. Result showed that accuracy of 93.29% was attained for features selected by WW.

Though SVM was reported in these studies for human gait recognition, to the best of our knowledge, there is no detail investigation on the most suitable SVM algorithm for human gait recognition has been reported with regards to multi-SVM that include one-versus-all (OVA), one-versus-one (OVO) and error correcting output codes (ECOC), ordinal, sparse random and dense random. Therefore, in this study we deem further to investigate these SVM algorithms for classification of human gait.

2 Multi-class Support Vector Machine Using Error Correcting Output Code

Initially, SVM is developed for binary classification purpose. The aim is to construct a hyperplane with a maximum margin yet less classification error to distinguish between two categories. In order to select an optimal hyperplane, regularization parameter *C* needs to be selected by the user. Theoretically, regularization parameter *(C)* controls the tradeoff between the margin maximization and errors of data. If *C* is too large, it may over fit the data and if it is too small, it may under fit the data. Furthermore, as we know SVM can deal with linear and non-linear data. For linear data, only a simple straight hyperplane is used to differentiate the cases. However, for non-separable data, the data must be mapped into a high dimensional space using kernel function prior to constructing the hyperplane.

Therefore, besides *C*, kernel function parameter is also vital in order to attain the optimal hyperplane. Three most popular kernel functions in SVM are linear, radial basis function (RBF) and polynomial kernels. Hsu et al. suggested [12] that linear kernel is to be employed as initial analysis as it has no kernel parameter to adjust. Moreover, this kernel is most suitable for a sparse and a large number of features.

On the other hand, multiclass SVM overcomes the limitation of the original SVM, where classes with more than two can be classified. There are two approaches in multiclass SVM, namely, constructing and combining several binary classifiers and considering all the classes in one optimization formulation. The common approach that can be implemented in the multiclass SVM is constructing and combining several binary classifiers since the second approach is proven as impractical to many applications and suffers large optimization problem which then can lead to high computational training time as reported in [13, 14].

There are three methods in constructing binary classifier; which commonly used in pattern recognition specifically one-versus-all (OVA), one-versus-one (OVO) and error correcting output codes (ECOC). Among these method, study reported in [15, 17] mentioned that ECOC is a suitable algorithm in dealing with multiclass classification such as bias and variance. Also, ECOC is able to enhance generalization performance as reported by Gholam et al. [16]. ECOC can be divided into two main components. The first component is coding design. In this component, base codeword for each trained binary classes is designed and further decoded in the second component, which known as decoding scheme. This scheme determines how the predictions of the binary classifiers are combined. In this stage, new codeword is generated from test data. Then, the new codeword is compared with base codeword that obtained in the coding design. To compare the codeword, classification error is used to determine performance of the classifiers. In coding design, there are few types of coding design, namely, one-versus-all (OVA), one-versus-one (OVO), dense random, ordinal and sparse random. Among these algorithms, OVA and OVO are commonly used in multi-class SVM. There is no previous work reported on dense random, ordinal and sparse random with regards to human gait recognition. Theoretically, in OVA, the number of binary classifier is similar as the number of class. Let number of class as L. Hence, there are L binary classifiers. For each binary classifier, one class is designated as positive class and the rest are designated as negative class. In OVO, one class is designated as positive class and another as negative and the rest are ignored. On other hand, as for ordinal, for the first binary learner, the first class is negative, and the rest positive. For the second binary learner, the first two classes are negative, and the rest positive, and so on. Conversely in dense random, for each binary learner, the software randomly assigns classes into positive or negative classes, with at least one of each type. Meanwhile for sparse random, for each binary learner, the software randomly assigns classes as positive or negative with probability 0.25 for each, and ignores classes with probability 0.5.

Therefore, the experimental analysis focused on investigating the coding design algorithms specifically OVA, OVO, ECOC, ordinal, sparse random and dense random for multi-class SVM with linear kernel in recognition of human gait in frontal view.

3 Methodology

In order to investigate the performance of each coding design as mentioned earlier for human gait recognition, the walking gait of 30 subjects is recorded using Kinect. Figures 1 and 2 depicted the layout measurement and actual environment of the monitoring area.

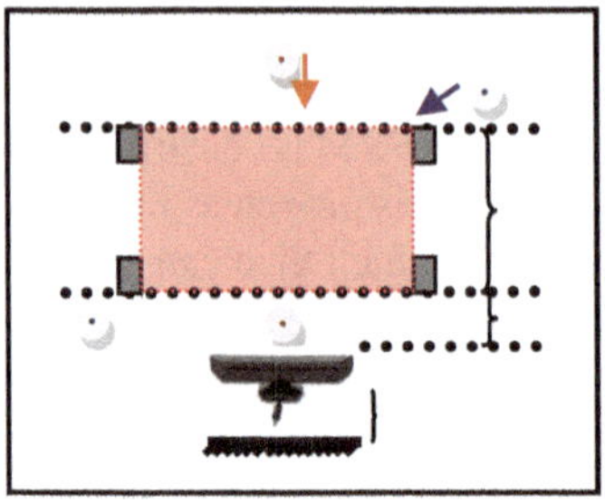

Fig. 1. Plan of the monitoring area in top view.

Fig. 2. Actual environment of the monitoring area from Kinect view.

The red area indicates the covered monitoring area, and this is the area of walking gait of subjects is recorded. The recorded video was captured at 30 frames per second with the resolution of 640 x 480.

Firstly, 11 male subjects and 19 female subjects were required to perform their normal walking gait using their normal speed, in arbitrary path. In order to gain normal walking patterns, the subjects began walking outside of the captured area. The subjects are required to walk repeatedly for 10 times and no restriction on clothing types except for pants. Recall that the limitation of this research is it focused on normal speed and common walking style of subjects in an indoor environment. No restrictions on clothing type applied to the subjects. Another limitation is the distance of walking range by subjects and the Kinect sensors is between 1 and 4 m. In addition, each of the subject's information such as age, height and weight were recorded as well. In this research, it was computed that the average subjects' age was 30.10 years and average subjects' height and weight were 1.60 m and 61.68 kg, respectively.

Moreover, skeleton joints in the 3D space were normalized prior the feature extraction stage. Then, the 20 skeleton joints within the full gait cycle were extracted, which then resulted as features with the dimension of 60 by 300. These features are then used as inputs to the SVM to evaluate the algorithms such as OVA, OVO, dense random, ordinal and sparse random in order to attain suitable algorithm for multi-class SVM. Figure 3 showed the walking subject in frontal views in RGB image, depth image and skeleton image.

Fig. 3. Example of human gain in frontal one frame in: color image (top), depth image (middle) and skeleton image as extracted by Kinect sensor (bottom).

4 Results and Discussion

Figure 4 illustrates the recognition accuracy as regularization parameter (C) is varied from 0.0001 to 1, with 1 decade increment for OVA, OVO, dense random, ordinal and sparse random. Obviously, accuracy was almost saturated when C at 1, for all algorithms. The highest accuracy namely 96% is attained for OVA algorithm when C is at 0.1 and 1 respectively. Furthermore, it is observed that the recognition accuracy for dense random is similar to sparse random.

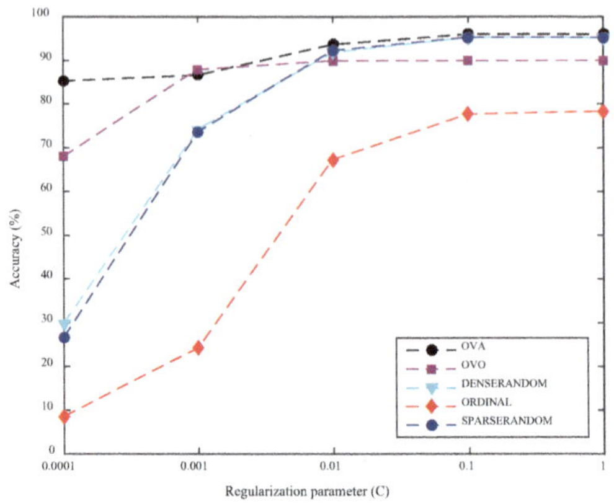

Fig. 4. Recognition accuracy for various C.

Figure 5 depicted the recognition accuracy as number of subject increases from 10 to 30 subjects. The optimal SVM with OVA, dense random and sparse random was attained when C is at 0.1 for 10 for 20 and 30 subjects accordingly. However, with C at 0.01 results showed that OVO is the optimal SVM. Contrast to ordinal, the optimal SVM for 10 and 20 subjects attained when C is at 0.1 and 1 for 30 subjects.

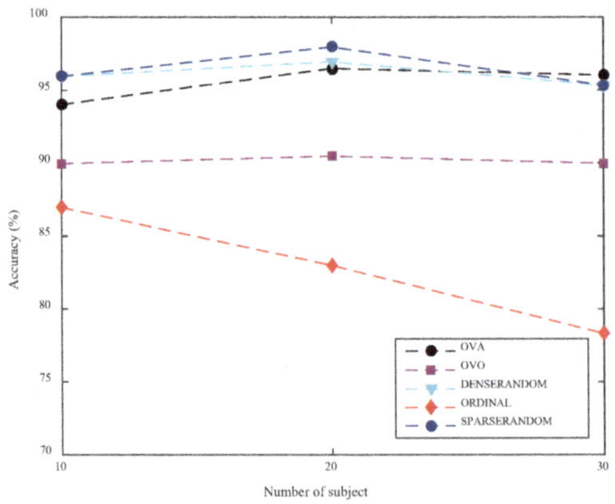

Fig. 5. Recognition accuracy for coding design algorithms.

Obviously, for dense random, ordinal and sparse random, recognition accuracy was drastically dropped when the number of subject increased from 20 to 30 and as observed for OVA and OVO algorithms, a slight dropped can be observed.

Next, Table 1 tabulated the computational time for each coding design during the optimal SVM model. Obviously, OVA algorithm requires less computational time as compared to other algorithms. The highest computational time is for sparse random.

Table 1. Computation time of coding design algorithm of SVM with linear kernel

No.	Coding design algorithm	Computation time (s)
1	**OVA**	**21.9**
2	OVO	98.1
3	Dense random	204.1
4	Ordinal	36.8
5	Sparse random	230.7

5 Conclusion

As a conclusion, human gait recognition in frontal view using various algorithms namely OVA, OVO, ECOC, ordinal, sparse random and dense random for multi-class SVM has been investigated and reported in this study. As the numbers of subjects are increased, results showed that OVA and OVO outperformed other algorithms based on recognition rate attained. The highest recognition accuracy attained is for OVA with 96% accuracy rate and 21.908 s computational time. Further work includes evaluating the algorithms for gait recognition based on oblique views. Additionally several other classifiers namely deep learning neural network and extreme learning machine (ELM) can be explored too for classification purpose.

Acknowledgment. This research is funded by Ministry of Higher Education (MOHE) Malaysia under the Niche Research Grant Scheme (NRGS) Project No: 600-RMI/NRGS 5/3 (8/2013). The authors wish to thank Human Motion Gait Analysis (HMGA) Laboratory, IRMI Premier Laboratory, IRMI, UiTM, Malaysia for the instrumentation and experimental facilities provided as well as Faculty of Electrical Engineering UiTM Shah Alam for all the support given during this research.

References

1. Preis, J., Kessel, M., Werner, M., Linnhoff-Popien, C.: Gait recognition with kinect. In: Workshop on Kinect in Pervasive Computing (2012)
2. Munsell, B.C., Temlyakov, A., Qu, C., Wang, S.: Person identification using full-body motion and anthropometric biometrics from kinect videos. In: Lecture Notes in Computer Science (Including Subseries Lecture Notes in Artificial Intelligence and Lecture Notes in Bioinformatics), vol. 7585, no. Part 3, pp. 91–100 (2012)
3. Proch: The MS kinect use for 3D modelling and gait analysis in the MATLAB environment. In: Proceedings of the Conference Technical Computing, pp. 1–6, Prague (2013)
4. Chan, W.C., Chan, C.W., Cheung, K.C., Harris, C.J.: On the modelling on nonlinear dynamic system using support vector neural networks. J. Eng. Appl. Artif. Intell. **14**, pp. 105–113 (2001)
5. Zhou, W.T.S., Wu, L., Yuan, X.: Parameters selection of SVM for function approximation based on differential evolution. In: International Conference on Intelligent Systems and Knowledge Engineering, pp. 1–7 (2007)
6. Zhu, J.S.Y.G.Q., Liu, S.R.: Support vector machine and its applications to function approximation. J. East China Univ. Sci. Technol. **5**, 5 (2002)

7. Vapnik, V.N.: An overview of statistical learning theory. IEEE Trans. Neural Netw. **10**, 988–999 (1999)
8. Gianaria, E., Grangetto, M., Lucenteforte, M., Balossino, N.: Biometric authentication. Lecture Notes in Computer Science, pp. 16–27. Springer, Cham (2014)
9. Chang, C., Lin, C.: LIBSVM: a library for support vector machines. ACM Trans. Intell. Syst. Technol. **2**, 1–39 (2013)
10. C. Prathap and S. Sakkara, "Gait Recognition using skeleton data," 2015 Int. Conf. Adv. Comput. Commun. Informatics, ICACCI 2015, pp. 2302–2306, 2015
11. Kastaniotis, D., Theodorakopoulos, I., Theoharatos, C., Economou, G., Fotopoulos, S.: A framework for gait-based recognition using Kinect. Pattern Recognit. Lett. **68**, 327–335 (2015)
12. Chih-Wei, L.C.-J.H., Chih-Chung, C.: A practical guide to support vector classification. Department of Computer Science, National Taiwan University, Taiwan (2010)
13. Fei, B., Liu, J.: Binary tree of SVM: a new fast multiclass training and classification algorithm. IEEE Trans. Neural Netw. **17**(3), 696–704 (2006)
14. Hsu, C., Lin, C.: A comparison of methods for multi-class support vector machines. IEEE Trans. Neural Netw. **13**, 415–425 (2002)
15. Escalera, S.: Error-correcting ouput codes library. J. Mach. Learn. Res. **11**, 661–664 (2010)
16. Gholam, B., Montazer, A., Escalera, S.: Error correcting output codes for multiclass classification : application to two image vision problems. In: The 16th CSI International Symposium on Artificial Intelligence and Signal Processing (AISP 2012), no. Aisp, pp. 508–513 (2012)
17. Wang, Z., Xu, W., Hu, J., Guo, J.: A multiclass SVM method via probabilistic error-correcting output codes. In: International Conference on Internet Technology and Applications, ITAP 2010 - Proceedings, pp. 2–5 (2010)

Curve Evolution Based on Edge Following Algorithm for Medical Image Segmentation

Sana Ullah(✉), Shah Khalid, Farhan Hussain, Ali Hassan, and Farhan Riaz

Department of Computer Engineering, College of Electrical and Mechanical Engineering, National University of Sciences and Technology, Islamabad, Pakistan
csemkd@gmail.com, {alihassan,farhan.riaz}@ce.ceme.edu.pk

Abstract. Image segmentation is one of the key tasks in different applications of image processing, computer vision and image analysis. A variety of image segmentation techniques are available in the literature, among which level sets based image segmentation approaches are widely used. Although these techniques perform well in segmentation of synthetic and real images which have strong object boundaries. However, when it comes to medical images where the object boundaries are weak in terms of intensity and prone to boundary leakage, these methods give poor segmentation results and lead to inaccurate boundary detection. To solve this problem, we proposed to formulate a novel level set/active contour model that is based on an edge indicator function which incorporates average edge magnitude and edge direction information. Implicit stopping criteria for level set evolution is also devised which controls the motion of evolving curve and stop the evolving curve at object boundaries. The proposed method is evaluated using two different medical images datasets of different imaging modalities, i.e., PH^2 dataset and vital stained magnification endoscopy (CH) images. Experimental results for image segmentation shows that our proposed method improve upon the other state-of-the-art approaches that have been considered in this paper.

Keywords: Medical images · Image segmentation
Level set methods · Edge flow

1 Introduction

Image segmentation is one of the most important, sensitive and interesting task in different applications of image processing, computer vision, pattern recognition and medical image analysis. Clinical and diagnosis procedures depend upon the information extracted from medical images as a result of segmentation. There is a wide range of techniques that have been proposed for performing image segmentation. Among these techniques, the level sets, formally introduced by Osher et al. [1] for the first time as a numerical technique for tracking has been

K. Arai et al. (Eds.): IntelliSys 2018, AISC 868, pp. 529–538, 2019.
https://doi.org/10.1007/978-3-030-01054-6_38

used extensively for image segmentation. The level sets make use of a higher dimensional function, namely, level set function (LSF) as a representation of contour/surface which serves as initial zero level set. By representing the contours/surface as a level set function, the problem of image segmentation can be formulated and solved as a partial differential equation [1]. The partial differential equation which represents level set function (level set function in turn represent contour/surface) depends upon certain energy functional. The process of minimizing this energy functional or the motion of contour/surface is termed as level set evolution. Level sets methods proposed so far in the literature for image segmentation can be broadly categorized into two classes i.e. edge based models and region based models. The former models make use of edge information for image segmentation whereas in the later models, a region descriptor is used to control the motion of the contour/surface towards region of interest (ROI).

In level set method, the initial zero level set contour is defined as a signed distance function which is updated in an iterative fashion [2]. In the recent years, many authors have proposed variants of level set method for image segmentation by making use of various image features in formulating the basic energy functional. Almost all of these methods are widely used for segmentation of medical images as well which exhibit some very specific issues e.g. weak object boundaries/boundary leakage, intensity inhomogenities and noise etc. For example, Belaid et al. [3] implemented an active contour model in which they incorporated the phase features into level set formulation for medical images segmentation. In their proposed method, the speed term in LSF comprises of two phase features i.e. local phase and local orientation. The local phase feature is derived from the image data whereas the local orientation measures the degree of alignment between normal to the zero level contour and local image orientation. This method outperform in medical images with weak edges however the method is not generic and needs strict parameters tuning. Estellers et al. [4] introduced the concept of harmonic active contours (HAC) for image segmentation. In HAC model, the geometric representation of 2D image is introduced into LSF. The basic idea is to compute and align gradients of both image and the level set function hence leads to the motion of contour/surface to the right place. Although HAC gives promising segmentation results on medical images, conversely it gives very poor results in cases where there is intensity inhomogeneity in images. Zhou et al. [5] proposed a hybrid approach for medical image segmentation. Their proposed method combines two different active contour/level set models, namely, an edge-based active contour (EBAC) and a region based active contour (RBAC). The effects of two different active contour models (i.e. edge based and region based) are adjusted for segmentation of heart CT scan images. Though this method give sound segmentation results around weak edges, however the performance of the above mentioned hybrid approach highly depends on the initial position of the zero level set contour. Zhang et al. [6] presents their novel approach of reaction diffusion level set evolution (RD-LSE). The basic idea is that, they have introduced a diffusion term into level set evolution and formulated an RD-LSE equation. It is a two-step splitting method which recursively solve the RD-LSE equation in an iterative fashion. In lower dimensional level set formulation, the above said method addresses boundary leakage and give promising segmentation results for different sets of synthetic and real images, however, the effectiveness of the aforementioned method is not yet evaluated in high dimensional level sets. Wang et al. [7] proposed their level set model for image segmentation which is based on local region features. They introduced local linear function into energy functional. The minimization of the energy functional in order to move LSF towards boundaries is achieved in an iterative manner. This method give good performance in medical image segmentation using local linear function, energy functional with local non-linear function is not yet evaluated.

In this paper, we present a novel edge flow based level set approach for image segmentation that incorporates local edge features into LSF in order to move the initial zero level set to the desired position. The main contributions are as follows:

- We compute the normal vector to evolving contour/LSF (which is at right angle to the 2D image plane) and the direction of image gradient, the curve evolution stops at image locations where both vectors become orthogonal. The basic idea behind this approach is that during curve evolution the normal to evolving contour points and the orientation of gradient vector is in the direction of edge, when the evolving curve reaches at edges, both vectors becomes perpendicular implicitly incorporating a stopping criteria for curve evolution.
- In our method, we have also introduced the remedy for weak edges by considering the average edge magnitude in regions adjacent to evolving contour. Integrating average edge magnitude and average edge direction into LSF improve the segmentation accuracy.

The organization of the rest of this paper is as follow. Section 2 presents our proposed methodology followed by a description of the Datasets (Sect. 3). Section 4 shows the segmentation results and performance evaluation of our proposed method in comparison with the other state of the art level set based image segmentation approaches followed by Conclusions (Sect. 5).

2 Methods

2.1 Distance Regularized Level Sets

Active contours in essence are dynamic fronts which iterate towards the boundaries of objects [8]. In the very basic level set formulation, a zero level set $\mathcal{C}\ (t) = \{(x, y)|\ \phi(t, x, y) = 0\}$ typically represents the fronts $\mathcal{C}$ of a function $\phi(t, x, y)$. The most generic mathematical representation of level sets can be given as:

$$\frac{\partial \phi}{\partial t} = F|\nabla \phi|, \tag{1}$$

where F is the speed function and $|\nabla \phi|$ represents the gradient of level set function. Traditional level sets can give birth to shocks, they may also have sharp and/or regular shape etc. (the curve does not stay differentiable). In order to remedy these issues, one possible numerical solution is that the function ϕ is initialized as a signed distance function followed by its reshaping or reinitialization as a signed distance function periodically during evolution. However, this process of periodic re-initialization increases the computational complexity of the level set method and in fact make it a bit complicated. On contrary to this cumbersome approach, a variational level set method was proposed by Li et al. [2] for curve evolution which make use of the known finite difference scheme thus avoiding the tiresome re-initialization process. The signed distance function have to satisfy the property of $|\nabla \phi| = 1$ that can be mathematically formulated as follows:

$$P(\phi) = \frac{1}{2} \int_{\Omega} |(\nabla \phi - 1)|^2 dxdy, \tag{2}$$

This term penalizes the contour whenever it try to violate the property of a signed distance function. Hence, the curve evolution can be mathematically written as:

$$\varepsilon_n(\phi) = \mu P(\phi) + \varepsilon_n e(\phi), \tag{3}$$

where $\mu > 0$ controls penalization of ϕ, and $\varepsilon_n e(\phi)$ is controlling the curve evolution. The functional $\varepsilon_n(\phi)$ is minimized by the gradient flow which can be represented as follows [2]:

$$\frac{\partial \phi}{\partial t} = -\frac{\partial \varepsilon_n}{\partial \phi}, \tag{4}$$

where $\frac{\partial \varepsilon_n}{\partial \phi}$ represents the directional differential of the functional ε_n. Morevover, the above equation is the interpretation of the curve evolution ϕ in the opposite direction to the differential $\frac{\partial \varepsilon_n}{\partial \phi}$. We aim to apply this framework for image segmentation. The $\varepsilon_n e(\phi)$ term in (3) is called *external energy* and is defined as a function of image data. Accordingly, the *internal energy* is represented by the term $P(\phi)$. The *external energy* controls the gradient flow (4) that in turn optimizes the function in (3). At the same time, $P(\phi)$ penalizes the curve if diviates from a signed distance function. This way the contour/LSF maintains the characteristic of an approximate signed distance, hence the complex periodic re-initialization procedure is no more required. The *external energy* function in our proposed method is formualted as follows [9]:

$$\varepsilon_n e = \lambda \mathcal{L}(\phi) + \nu \mathcal{A}(\phi), \tag{5}$$

where $\mathcal{L}(\phi)$ and $\mathcal{A}(\phi)$ are the length and area terms respectively which depends on the image data and $\lambda > 0$ and ν are constants. We have incorporated the edge following features [10] in the images for extracting the data from the images. We obtain our final level set evolution (LSE)/curve evolution by minimization of the following curve evolution/LSE equation:

$$\frac{\partial \varepsilon_n}{\partial t} = \mu \frac{\partial P(\phi)}{\partial \phi} + \frac{\partial \varepsilon_n e(\phi)}{\partial \phi}, \tag{6}$$

In the above (6), the first term in the right hand side can be solved as follows [2]:

$$\frac{\partial P(\phi)}{\partial \phi} = \left[\Delta \phi - \nabla . \frac{\nabla \phi}{|\nabla \phi|} \right] \tag{7}$$

In order to calculate the second term in right hand side of (6), we have formulated the external energy in our proposed method as a function of edge following features which are extracted from the images.

2.2 Edge Flow Calculation

Image segmentation via level set methods in numerous applications of computer vision and image analysis make use of various image information, i.e. edge, intensity, texture and orientation/phase, etc. for defining an objective functional. Our level set formulation is based on edge information which controls the motion of evolving contour towards ROI. We exploit average edge magnitude and average edge direction in devising level set function. Following equations calculates the average edge magnitude and average edge direction for a given image *I(x, y)*:

$$M_e\ (i,j) = \frac{1}{M_k} \sum_{(i,j) \epsilon N} \sqrt{M_e x(i,j)^2 + M_e y(i,j)^2} \tag{8}$$

$$D_e\ (i,j) = \frac{1}{M_k} \sum_{(i,j) \epsilon N} tan^{-}1 \left(\frac{M_e y(i,j)}{M_e x(i,j)} \right) \tag{9}$$

where *Mk* defines an NxN window in the local neighborhood. *Mex, Mey* are the image gradient in x and y directions and can be calculated as follows:

$$M_e x(i,j) = -Gau_y \times I(x,y) \approx \frac{\partial I(x,y)}{\partial y} \tag{10}$$

$$M_e y(i,j) = Gau_x \times I(x,y) \approx -\frac{\partial I(x,y)}{\partial x} \tag{11}$$

Gau-x, Gau-y in the above equations represents the different masks of the image moment vector weighted by Gaussian operator in x and y directions, respectively [9].

$$Gau_x(x,y) = \frac{1}{\sqrt{2\pi}\sigma} \left(\frac{x}{\sqrt{x^2+y^2}} \right) exp \left(\frac{x^2+y^2}{2\sigma^2} \right) \tag{12}$$

$$Gau_y(x,y) = \frac{1}{\sqrt{2\pi}\sigma}\left(\frac{y}{\sqrt{x^2+y^2}}\right) exp\left(\frac{x^2+y^2}{2\sigma^2}\right) \tag{13}$$

Edge magnitude and direction are indicated by edge vectors which exhibit a vector stream flowing around the object boundary. In case of weak edges and in images prone to boundary leakage these edge vectors are randomly distributed and can-not capture object boundary perfectly, that is why we incorporate average edge magnitude and average edge direction to remedy the aforementioned problem. On the basis of average edge magnitude, we define our edge indicator function as follows:

$$\mathcal{F} = 1 - M_e(i,j)^2 \approx 1 - \frac{1}{M_k}(M_e x(i,j)^2 + M_e y(i,j)^2) \tag{14}$$

Let $|\nabla\phi|$ be the gradient of the curve, and let $\mathbf{v}$ be a vector indicating the direction of edge in the image at a particular pixel. By multiplying the edge vector pointing in the direction of the preeminent local edges with the evolving curve, we obtain the length term as follows:

$$\mathcal{L}(\phi) = \int_\Omega \delta(\phi)\mathbf{v}\nabla\phi dxdy, \tag{15}$$

where $\delta(\phi)$ implies the Dirac delta function and $\mathbf{v}$ is the edge vector, the orientation of which is along the dominant edges direction. The minimization of (15) is achieved when the edge vector and the normals to the curve get perpendicular to one another. By differentiating the length term $\mathcal{L}(\phi)$ with respect to ϕ, we have

$$\frac{\partial\mathcal{L}(\phi)}{\partial\phi} = \nabla_{\mathbf{v}}.\frac{\nabla\phi}{|\nabla\phi|}. \tag{16}$$

where $\nabla_{\mathbf{v}}$ estimates the divergence of the curve gradient $\nabla\phi$ in accordance to the vector $\mathbf{v}$. The area term $\mathcal{A}(\phi)$ can be derived by calculating the magnitude of the edges g, as defined in (14).

$$\mathcal{A}(\phi) = \int_\Omega \mathcal{F}(u,v)H(-\phi)dxdy \tag{17}$$

By differentiating the above term $\mathcal{A}$ with respect to the curve ϕ, we obtain

$$\frac{\partial\mathcal{A}(\phi)}{\partial\phi} = \mathcal{F}(u,v)\delta(\phi). \tag{18}$$

Accordingly, the evolution equation for ε_e can be given as

$$\frac{\partial\varepsilon_n e}{\partial t} = \lambda\nabla_{\mathbf{v}}\frac{\nabla\phi}{|\nabla\phi|} + \mathcal{F}(x)\delta(\phi) \tag{19}$$

The substitution of the results acquired for both the energy terms which constitute our basic energy functional, i.e. the internal energy term (7) and external energy term (19), respectively into basic energy functional (6), we have

$$\frac{\partial \varepsilon_n}{\partial \phi} = -\mu \left[\Delta\phi - \nabla . \frac{\nabla \phi}{|\nabla \phi|} \right] - \lambda \left[\nabla_{\mathbf{v}} . \frac{\nabla \phi}{|\nabla \phi|} \right] - \nu \mathcal{F}(x)\delta(\phi). \tag{20}$$

Formulating the curve evolution/LSE this way ensures that the energy functional is minimized by the function ϕ and it satisfies the equation $\frac{\partial \varepsilon_n}{\partial \phi} = 0$. Therefore the energy functional ε_n minimization via steepest descent process is obtained using the following gradient flow (inferred from (4)):

$$\frac{\partial \varepsilon_n}{\partial \phi} = \mu \left[\Delta\phi - \nabla . \frac{\nabla \phi}{|\nabla \phi|} \right] + \lambda \left[\nabla_{\mathbf{v}} . \frac{\nabla \phi}{|\nabla \phi|} \right] + \nu \mathcal{F}(x)\delta(\phi). \tag{21}$$

The above (21) is a variational level set model based on an edge following approach and solved by making use of the gradient descend algorithm.

3 Materials

We have used two different datasets to validate the proposed segmentation methods: the PH2 dermoscopy dataset and the vital stained magnification endoscopy dataset for the performance evaluation of our proposed level set formulation for medical images segmentation.

3.1 PH2 Dataset

PH2 dataset contain 200 dermoscopy images which were acquired using a dermoscope during the clinical examination at the Pedro Hispano Hospital Matosinhos, Portugal. Careful examination of each image was carried out by an expert dermetologist for identification and manual segmentation of lesion from normal skin. Manual segmentation result, i.e. Ground Truth for each image is also provided in PH2 dataset. All the images and respective ground truths are stored in Bitmap (.bmp) file format.

3.2 Chromoendoscopy (CH) Dataset

This dataset comprises of 176 images of the stomach which were acquired during normal clinical examination at the Poruguese Institute of Oncology Porto, Portugal. Each image was examined by two experienced Gastroenterologist for identification of lesion. Manual segmentation of lesion from normal tissues was carried out and each image was annotated. The Ground truth for each image is also provided in CH dataset. Endoscopy images and respective annotations are saved as graphic files in Portable Network Graphics file format.

4 Experimental Results

This section presents the implementation details and segmentation results of our proposed method. Using same medical images datasets, we compare the segmentation results of our proposed level set approach with results achieved using state of the art RD-LSE and DRLSE methods. We have used Dice Similarity Coefficient (DSC) for this comparison. It is a statistical measure of similarity between segmented image and Ground Truth and give insight into degree of overlap between the two images. Let, S be the image acquired using automatic segmentation, G be the ground truth/ annotated image, DSC can be computed as,

$$DSC = \frac{2|S \cap G|}{|S| + |G|} \tag{22}$$

The value of dice similarity coefficient ranges between 0 (no overlap) and 1 (full overlap). DSC is calculated for each image in both experiments and average dice coefficients are calculated for performance evaluation of all the three methods (see Table 1).

The choice of methods for comparison purpose is based on the fact that the basic energy functional formulation and energy terms in our proposed method are similar to that defined in the aforementioned methods. However, the main difference is in the underlying edge indicator function which captures object boundaries and the stopping criteria that is used for curve evolution. The stopping criteria for curve evolution/LSE in state of the art RD-LSE and DRLSE depends upon the pre defined fixed number of iterations which is not a generic approach and often leads to incorrect boundary detection. In our proposed method, we have introduced an implicit stopping criteria in the implementation which controls the motion of initial zero level set contour towards exact object boundary regardless of the number of iterations. The coefficient of area term, i.e. α is set positive for PH2 dataset and negative in case of CH dataset. The reason behind this peculiar choice of values for α is that, images in CH dataset have weak edges as compared to PH2 dataset. As stated in Sect. 2, the area term $\mathcal{A}_g(\phi)$ controls the motion of LSF towards object boundary and α add external force in this motion. So, positive values of α in images prone to boundary leakage will further push the LSF inside the object boundary and will result in inaccurate boundary detection. Therefore, negative values of α are desirable in such cases. The proposed level set method for medical image segmentation is implemented in Matlab version R2015a using Dell inspiron laptop with Intel(R) Core(TM) i5-3210M CPU @ 2.50 GHz (4 CPUs) and 4 GB Ram on Windows 10 Education 64-bit.

Our experiments show that the proposed method outperforms the other methods that have been considered in this paper. Although an overall good performance of the method is observed, the method performs particularly well in cases in which the edge boundaries are weak as the method performs well for the CH images. The method also shows good results on dermoscopy images with an overall DSC of 0.83 (see Figs. 1 and 2).

Table 1. Average dice similarity coefficient

	PH2 dataset	CH dataset
DRLSE	0.75	0.47
RD-LSE	0.79	0.53
Proposed method	0.8397	0.5947

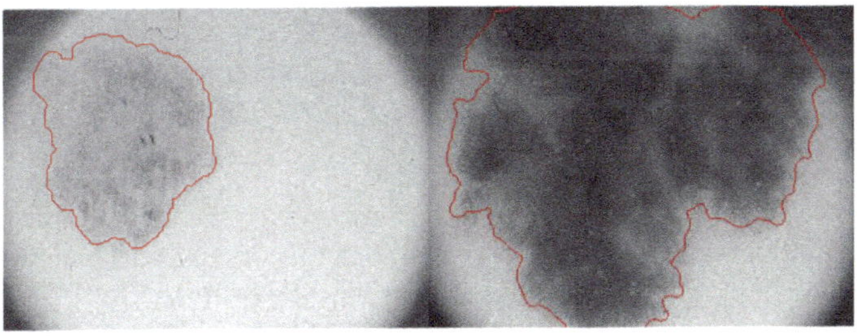

Fig. 1. Segmentation results for dermoscopy images using the proposed method.

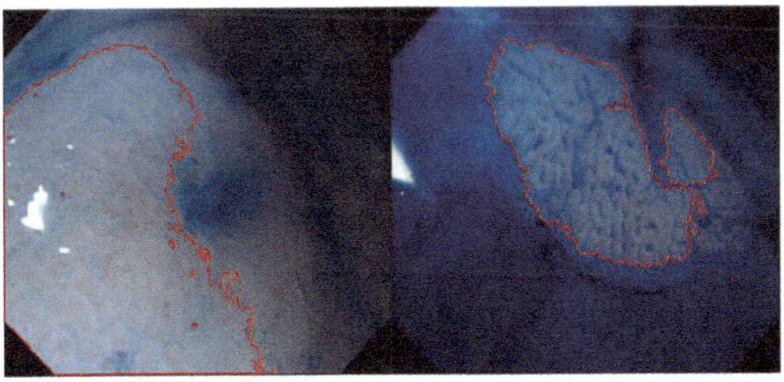

Fig. 2. Segmentation results for CH images using the proposed method.

5 Conclusion

We have proposed a new edge flow based level set formulation for medical image segmentation. Different authors have proposed level sets based image segmentation techniques so far in the literature. Although these techniques perform well in synthetic images and real images (having solid boundaries of objects) segmentation. However, when it comes to medical images where the object boundaries, i.e. ROI are weak in terms of intensity and prone to boundary leakage, these methods give poor segmentation results and detect inaccurate boundaries. To remedy this problem, we have incorporated average edge magnitude and average edge direction information in our level set formulation. Implicit stopping criteria for level set evolution is also devised which controls the motion of evolution curve and stops the curve at object boundaries. The proposed method is evaluated using two different medical images datasets, i.e. PH2 dataset and CH dataset. The former dataset consists of 200 real dermoscopy images whereas the later contains 176 chromo endoscopy images of stomach. Experimental results shows that our proposed method outperforms the other methods that have been considered in this paper.

References

1. Osher, S., Sethian, J.A.: Fronts propagating with curvature dependent speed: algorithms based on hamilton-jacobi formulations. J. Comput. Phys. **79**(1) (1988)
2. Li, C., Xu, C., Gui, C., Fox, M.D.: Distance regularized level set evolution and its application to image segmentation. IEEE Trans. Image Process. **19**(12), 3243–3254 (2010)
3. Belaid, A., Boukerroui, D., Maingourd, Y., Lerallut, J.-F.: Phase-based level set segmentation of ultrasound images. IEEE Trans. Inf. Technol. Biomed. **15**(1), 138–147 (2011)
4. Estellers, V., Zosso, D., Bresson, X., Thiran, J.-P.: Harmonic active contours. IEEE Trans. Image Process. **23**(1), 69–82 (2014)
5. Zhou, Y., Shi, W.-R., Chen, W., Chen, Y.-L., Li, Y., Tan, L.-W., Chen, D.-Q.: Active contours driven by localizing region and edge-based intensity fitting energy with application to segmentation of the left ventricle in cardiac CT images. Neurocomputing **156**, 199–210 (2015)
6. Zhang, K., Zhang, L., Song, H., Zhang, D.: Reinitialization-free level set evolution via reaction diffusion. IEEE Trans. Image Process. **22**(1), 258–271 (2013)
7. Wang, X.-F., Min, H., Zou, L., Zhang, Y.-G.: A novel level set method for image segmentation by incorporating local statistical analysis and global similarity measurement. Pattern Recognit. **48**(1), 189–204 (2015)
8. Liu, S., Peng, Y.: A local region-based chan-vese model for image segmentation. Pattern Recognit. **45**(7), 2769–2779 (2012)
9. Riaz, F., Hassan, A., Zeb, J.: Distance regularized curve evolution: a formulation using creaseness features for dermoscopic image segmentation, pp. 1061–1065 (2014)
10. Somkantha, K., Theera-Umpon, N., Auephanwiriyakul, S.: Boundary detection in medical images using edge following algorithm based on intensity gradient and texture gradient features. IEEE Trans. Biomed. Eng. **58**(3), 567–573 (2011)

Enhancing Translation from English to Arabic Using Two-Phase Decoder Translation

Ayah ElMaghraby(✉) and Ahmed Rafea

Computer Science and Engineering Department, School of Sciences and Engineering, The American University of Cairo, Cairo, Egypt
{aelmaghraby, rafea}@aucegypt.edu

Abstract. This paper describes an approach to enhance statistical machine translation. This approach uses a two-phase decoder system; the first decoder which we call the initial decoder translates from English to Arabic and the second is a post-processing decoder that re-translates the initial's decoder Arabic output to Arabic again to fix some of the translation errors. This new technique showed to be useful when trying to translate corpus from a different context other than the original corpus used in the training of the initial decoder. We recorded a BLEU score enhancement on out-of-context corpus close to 10 BLEU points on UN corpus and 2 BLEU points on TED 2013 corpus.

Keywords: Statistical machine translation · Two-phase decoder
Post-processing technique · Out-of-context corpus

1 Introduction

Translation comes from the human need to understand and be understood. Without translation, a lot of the human history and knowledge would be lost. In the past few years with the increase use of social media like Facebook and Twitter, Natural Language Processing methodologies and algorithms became a very hot topic.

Machine translation faces many problems when translating between two language pairs. The problems get more severe when one of these two languages has great difference in its structure than the other such as English and Arabic languages. In Arabic, one can say a whole English sentence by saying one word for example in English: "We heard her", the translation in Arabic would be: "سَمِعناها". Arabic sentences can be in the form of SVO form and occasionally in VSO form especially if it is a passive sentence.

Some techniques when translating from Arabic to English focus on the re-ordering done while translating as in [1] where the basic idea is to minimize the amount of re-ordering during translation by displacing Arabic words in training text. In this paper, they try to tackle the shortcomings in phrase-based translation when translating Arabic to English. The main problem they aim at solving is the great distance between Arabic and English verbs which are caused by differences between the two languages' structures.

A lot of the research done in translation between Arabic and English does morphological analysis, tokenization and segmentation prior to translation. Then do de-

K. Arai et al. (Eds.): IntelliSys 2018, AISC 868, pp. 539–549, 2019.
https://doi.org/10.1007/978-3-030-01054-6_39

segmentation after translation. Usage of morphological analyzers and segmentation techniques can be seen in [2]. De-segmentation techniques can be seen in [3].

Another approach was proposed in [4], assuming semantic similarity between different words if they belong to the same topic, English words are clustered in classes for example {football, field, pitch} would belong to the same class. These classes are projected on Arabic language, then each word on English and Arabic side are replaced by their respective class identifier. These classes are determined based on the alignments offered by GIZA++ [5].

A new approach using neural networks in machine translation that is called neural machine translation has arisen recently. Instead of using statistical methods, this approach uses long short-term memory neural networks for translation. This technique was applied on translation between English-Arabic in [6]. The recorded increase of BLEU score as much as +4.46 and +4.98 over the baseline in the case of phrase-based and neural network when applying orthographic normalization and morphological ware tokenization.

One way of enhancing machine translation was proposed in [7] by trying to create a diverse translation system from an existing single engine using bagging and boosting. Using ensemble learning, weak translation systems are created then they are used to learn a strong translation system. An overview of the approach is; first a combination system which contains several input systems is created. Second, a weak SMT is then trained on a distribution of the training set. Third, the training set is re-weighted and updated to generate a new training set using bagging and boosting weight updating methods, which is used to train another weak SMT. Finally, from the combination of the weak SMT, a strong SMT is generated. The last step is a bit tricky how to combine the system and join them into a strong system. So, in this paper they propose different methods to generate the strong system aka (combine the weak ones). They study six approaches to combine a strong system, sentence level combination, minimum Bayes risk decoding (MBR), confusion network and indirect HMM, confusion network and METEOR alignment, boosted re-ranking and discriminative re-ranking.

We describe in this paper an approach to enhance machine translation; in this approach we use a two-phase decoder to re-translate the Arabic output of the English-Arabic machine translation system to fix translation issues and enhance translation when translating test set from a corpus other than the training data.

2 Challenges of Machine Translation Between Arabic and English

Arabic is complex language with rich morphology, and the divergence between Arabic and English add a great complexity when translating between the two languages. One word in Arabic can correspond to a complete sentence in English, for example the question "**Do you know him?**" is translated to "أتعرفونه؟". As a result of the possibility of adding a combination of suffixes and prefixes to an Arabic word leads to having the same word in many different forms hence the number of word types in Arabic is more than English for the same corpus. This is challenging problem whether it's a data-driven machine translation because it won't be able to translate a word it didn't see or a knowledge-based machine translation which must incorporate linguistic information efficiently.

Arabic words could be ambiguous if taken out of context, for example the word "كتب" could mean the past tense verb "**written**" or the noun word "**books**". So, the semantics of the word depend on the context of the sentence and what comes before or after the word. This complicates the translation because the corresponding word that should be used depends on the context. For example the word "**party**" could mean in Arabic "حفلة" or "حزب". So if the word before party is "**ruling**" then we probably want the word "حزب" and if the previous word is "**invite to**" then it's more likely we need the word "حفلة". Humans resolve ambiguity using their knowledge-base but this doesn't always work with machines but similar attempt was done in [4].

One of the challenges of the Arabic languages is posed by the inconsistency of spelling some script letters this was addressed in [8], for example many times the letter "**Alef**" can be written in many forms for example; "ا", "أ" and "آ". The first contains the "**Hamza**" "ء" which changes the sound of the letter. The third form contains "mad" "~" which indicates this word should be pronounced with two "**alefs**". As a result, this leads to problems in Arabic words spelling. A word like "آية" which means a verse in the Quran could be written using any of the previous forms of the letter alef even though the intended meaning is the same word which means a verse in the Quran. This inconsistency in writing the words increases the sparsity which means having the same word spelled in many forms. As a result this causes ambiguity because people could spell the same word in different forms. One of the ways to eliminate this ambiguity is trying to make all forms of the word is written in the same so it doesn't appear in the translation system in more than one form all meaning the same thing. For example; changing "ا", "أ" to "ا" is called normalization referred to in the literature is Reduced Normalization (RED). Sometimes the letter "ا" can be written in some words as dot-less yaa "ى" which is pronounced in the same way as the alef "ا". Choosing appropriate "ا" and "ى" depending on the context is called Enriched Normalization (ENR) which is introduced in [9].

Another problem in English-Arabic translation is verb-subject order in the sentence. In Arabic it's normal to have sentences in the form of subject verb-object or verb-subject-object and both would translate to the same sentence in English. For example, the sentence "**The boy ate an apple**" could be translated to "الولد أكل التفاحة" or "أكل الولد التفاحة". The first Arabic translation the verb comes before the subject and in the second example the subject comes before the verb. Both translations are correct, but the subject-verb order is less common than verb-subject order. This problem is more common when translating from Arabic to English it causes re-ordering issues is especially in phrase-based statistical machine translation. The problem is explained more in details in [1].

Another difference between Arabic-English syntax is the structure of the noun phrase in both languages. In Arabic the definite article "ال" is appended to the noun while in English the definite article "**the**" is used before. In Arabic, if we add a definite article to a noun the adjective following it should have a definite article in it too. For example, the noun phrase "**The big book**" has one definite article while in Arabic it's translated to "الكتاب الكبير" which contains the definite article in both words in the noun phrase. This issue is normally solved when using phrase-based statistical machine translation as this issue can be modeled in the phrasal internal alignments.

Another issue noticed was getting lower BLEU scores when translating sentences that were not taken from the corpus used in training. This can be caused by unseen words or difference in topic between test set and training set. This is a well-known problem in machine translation, not having a generic enough SMT that could translate data from different topics or contexts efficiently. The issue with most sentences is that they cannot be fixed easily by analyzing the structure of the sentence and trying to modify them automatically in a two-phase manner as there no pattern observed. In other words, there was no way to figure out the original sentence from the output.

3 Baseline System

3.1 Creating Baseline System

The dataset LDC2007T08 (ISI parallel corpus) is pre-processed and tokenized using Stanford Core NLP [10]. Using this dataset we created our language model with IRSTLM [11]. All the decoders created from now on are used this language model.

We divided the ISI data set into training set, development set and test set. Training set contains around ~ 1000000 sentences, development set contain 90000 sentences and test set contains 2000 sentences. We created our baseline system that is a phrase-based decoder using Moses toolkit [12] and the previously created language model. We used the tokenizer offered by Moses to tokenize the ISI training data and cleaned it using the cleaner script in Moses that removes empty lines. We used GIZA++ to align the training data. We ran this decoder on the test set of the ISI corpus. The results can be seen in Table 1.

Table 1. Baseline system results

Baseline system	Decoder type
	Phrase-based SMT from English to Arabic
BLEU-4 score	6.79

As shown in Table 1, the BLEU score is very low. One cause of this is that Moses tokenizer does not support Arabic officially and it falls back to using the English tokenizer. So, due to the differences between English and Arabic this tokenizer doesn't work well. So we tried to use the tokenizer in Stanford CoreNLP to tokenize the training data and clean it using the Moses cleaner. The results of this experiment are shown in Table 2.

In the next step we compare different kinds of decoders, we compare phrase-based and five syntax-based decoders. The syntax-based decoders were created by parsing training data using Stanford Parser in the Stanford CoreNLP framework. We calculate BLEU-4 on the test set by Stanford Phrasal [13].

Table 2. Effect of corpus tokenization

Tokenization effect	Decoder type	
	Phrase-based SMT from English to Arabic tokenized by Moses	Phrase-based SMT from English to Arabic tokenized by Stanford CoreNLP
BLEU-4 score	6.97	23.07

We compare different kinds of decoder types to choose the best of them. We compare between phrase based decoder, syntax based string-to-tree decoder that uses GHKM extraction algorithm (Williams & Koehn, 2012), syntax based string-to-tree decoder that uses Chiang 2005 extraction algorithm (Chiang, 2005), syntax based tree-to-string decoder that uses GHKM extraction algorithm, syntax based tree-to-string decoder that uses Chiang 2005 extraction algorithm and syntax based tree-to-tree decoder that uses Chiang 2005 extraction algorithm. The results of this experiment are shown in Table 3. Based on the BLEU score we chose phrase-based decoder to be our baseline system.

Table 3. Comparison between different decoder types

Decoder types	Decoder type					
	Phrase-based SMT from English to Arabic	String-to-Tree SMT from English to Arabic (Chiang 2005)	String-to-Tree SMT from English to Arabic (GHKM)	Tree-to-String SMT from English to Arabic (GHKM)	Tree-to-String SMT from English to Arabic (Chiang 2005)	Tree-to-Tree SMT from English to Arabic (Chiang 2005)
BLEU-4 score	23.07	19.68	21.05	16.94	17.9	16.91

3.2 Testing Our Decoders on UN Corpus

We tested our previously created phrase-based decoder and syntax decoder from string-to-tree that uses GHKM extraction algorithm system on the UN corpus [14]. We extracted a test set from the UN corpus that contains 1780 sentences. The results of this experiment can be seen in Table 4.

Table 4. Results of translating UN corpus

Effect of our decoders on UN corpus	Decoder type	
	Phrase-based SMT from English to Arabic	String-to-Tree SMT from English to Arabic (GHKM)
BLEU-4 score	16.05	15.89

The results on this corpus are lower than that of the ISI corpus for several reasons; first it's a different corpus than the one the decoder was trained on. Second, this corpus contains words that were not seen before in the ISI corpus and the decoder failed to translate them like "**Human Rights**". Third, the UN corpus contains the session logs of the UN while the ISI corpus contains news extracted from several news agencies.

3.3 Testing Our Decoders on TED 2013 Corpus

We tested our previously created phrase-based decoder and syntax decoder from string-to-tree that uses GHKM extraction algorithm system on the TED 2013 talks corpus [15]. This corpus was created from TED talks conducted in 2013. We divided the corpus into a test set corpus that contains 1990 sentences and the remaining was used as a development set containing approximately 147K sentences. The results of this experiment can be seen in Table 5.

Table 5. Results of translating TED 2013 corpus

Effect of our decoders on TED 2013 corpus	Decoder type	
	Phrase-based SMT from English to Arabic	String-to-Tree SMT from English to Arabic (GHKM)
BLEU-4 score	14.43	13.19

4 Two-Phase Decoder

The idea of two-phase decoder is to have two decoders the first we will call it an initial decoder and the second decoder we will call it post-processing decoder. The post-processing decoder is created from a part of the data set in attempt to enhance translation results. One of the well-known issues in machine translation is getting lower BLEU scores when translating test sets from corpora not used in the training or a corpora that is from a different context. We used Moses toolkit to create a decoder using development set. This development set is pre-processed and tokenized using Stanford CoreNLP. We tried different combinations of different decoder types. They are all tested using the same test set using BLEU-4 calculator in Stanford Phrasal.

We did this experiment on the UN corpus. We don't use the whole corpus. We only use a development set of size 200K sentences and the same test set previously used which contains 1780 sentences. It is worth noting that the total size of the UN corpus is 18 million sentences. We only use 200K as a development set because we need to translate the development set to create our post-processing decoder and this takes a long time to translate using our decoders so we need a minimum amount of the data to create this post-processing decoder. We translated the development set using the string-to-tree SMT with GHKM extraction algorithm decoder. We create three SMTs one that is phrase-based, one that is tree-to-string syntax SMT with GHKM extraction algorithm and the last one is tree-to-tree Syntax based SMT. All of them re-translate the Arabic output. To test the two-phase decoder we first translate our test set using our baseline system and the string-to-tree SMT with GHKM extraction algorithm. Then, we use the three post-processing SMTs created before to translate the output of the previous translation. We calculate BLEU-4 score for six different outputs using the BLEU calculator in Stanford Phrasal. The results of this experiment can be seen in Table 6.

Table 6. Effect of post-processing decoder on UN corpus

Decoder type	Decoder type	
	Phrase-based SMT from English to Arabic	String-to-Tree SMT from English to Arabic (GHKM)
Initial decoder	16.05	15.89
Post-processing decoder: phrase-based SMT that translates from Arabic to Arabic	20.77	**26.59**
Post-processing decoder: syntax-based SMT Tree-to-String that translates from Arabic to Arabic	23.51	19.29
Post-processing decoder: syntax-based SMT Tree-to-Tree that translates from Arabic to Arabic	23.51	19.29

We repeated the experiment on another corpus called TED 2013 [15]. This corpus is extracted from TED talks done in 2013. We use a development set of approximately 147K sentences and the same test set previously used which contains 1990 sentences. We translated the development set using the string-to-tree SMT with GHKM extraction algorithm decoder. We create three SMTs one that is phrase-based, one that is string-to-tree syntax SMT with GHKM extraction algorithm and the last one is tree-to-tree Syntax based SMT. All of them re-translate the Arabic output. To test the two-phase decoder we first translate our test set using our baseline system and the string-to-tree SMT with GHKM extraction algorithm. Then, we use the three post-processing SMTs created before to translate the output of the previous translation. We calculate BLEU-4 score for four different outputs using the BLEU calculator in Stanford Phrasal. The results of this experiment can be seen in Table 7. The increase in BLEU score recorded in this case range between 1–2 BLEU points. The increase this time was not as much because the development set used to create the post-processing decoder is much smaller than that used for the UN corpus.

5 Conclusions

The impact of using two-phase SMT system from Arabic to Arabic to enhance the translation quality an experiment was conducted that uses three two-phase decoders:

(1) Syntax-based tree-to-string decoder.
(2) Phrase-based decoder.
(3) Syntax-based tree-to-tree decoder.

Table 7. Effect of post-processing decoder on TED 2013 corpus

Decoder type	Decoder type	
	Phrase-based SMT from English to Arabic	String-to-Tree SMT from English to Arabic (GHKM)
Initial decoder	14.43	13.19
Post-processing decoder: phrase-based SMT that translates from Arabic to Arabic	15.04	14.89
Post-processing decoder: syntax-based SMT Tree-to-Tree that translates from Arabic to Arabic	**15.31**	14.59
Post-processing decoder: syntax-based SMT String-to-Tree that translates from Arabic to Arabic	15.23	14.69

Table 8. Examples of phrase-table and rule-table entries

Enhancements in Translation	**Comparison**			
	Source	**Target**	**English Translation**	**Fix**
Example 1	تهدف الى حرمان منهم من	يهدف الى حرمانهم من	Aims to deprive them of	It adds the right gender prefix to the verb
Example 2	خلال الحوار البناء	عن طريق حوار بناء	Through a constructive dialogue	It uses the appropriate preposition and noun
Example 3	Yugoslavia	يوغسلفيا	Yugoslavia	It provided a translation for a word that was not translated
Example 4	في الحركة في الخمسينات قبل	وهي حركة في الخمسينات	It's a movement in the fifties	The source sentence didn't make any sense because it contained extra words and prefixes

Table 9. Examples of enhancements in translation

Enhancements in Translation	Comparison		
	Before	After	Expected English Translation
Example 1	Human Rights Questions	مسائل حقوق الانسان	Human right's issues.
Example 2	Provision ان لتغطية نفقات الخدمات المختلفة 4500 حوالي دولار شهريا	تغطية نفقات الخدمات المختلفة دولار 4500حوالي للفرد شهريا	The provision cost of different services around 4500 dollar per person per month
Example 3	الاتصال العسكرية فريق	فريق الاتصال العسكري	Military liaison team
Example 4	قال ان النص النهائي، ان قرار المجلس	وللاطلاع علي النص النهائي انظر	To look at the final version look
Example 5	مشاركة من السلطات المختصة من الولايات المتحدة	مشاركة السلطات المختصة في الولايات المتحدة	Participation of competent authorities in the United States
Example 6	المشروع شركة	شركة المشروع	The project's company
Example 7	مختلفة مجموعات	مختلف المجموعات	Different groups
Example 8	ينص على هوية من مختلف المجموعات	فيما يتعلق بتحديد هوية مجموعات مختلفة	With regard to the identification of different groups
Example 9	الذين يأتي الى المكسيك المهاجرين	المهاجرين القادمين إلى المكسيك	The immigrants going to Mexico

The two-phase syntax-based tree-to-string decoder enhanced the quality of BLEU score when translating the output of a phrase based decoder and the output of a syntax-based tree-to-string English to Arabic decoder in the case of the UN corpus around 9–10 BLEU points.

Examples of problems that were fixed using this approach:

- Some words and sentences weren't translated the first time and were output as English words. This can be seen in example one and two in Table 9.
- The right gender of words. This can be seen in example three in Table 9.
- Fixes to translations. This can be seen in example four in Table 9.
- Fixes prepositions. This can be seen in example five in Table 9.

- Short term re-ordering. This can be seen in examples six and seven in Table 9.
- Fixed some issue that caused semantic problems. This can be seen in example eight in Table 9.
- Fixed issues with plural in Arabic. This can be seen in example nine in Table 9.

The high increase in the case of the UN corpus is attributed to having the first decoders trained on the ISI parallel corpus while the second decoder is trained on the UN parallel corpus. In other words, the second decoder fixed issues in the translation of the first decoder to explain more how did this happen we need to look inside the model generated for the post-processing decoder. When using Moses to create a decoder, the model generated contains lexical translation for words and phrase table or rule table depending whether the decoder is phrase-based or syntax-based. In the case of phrase-based decoder it's called phrase-table. The phrase-table contains translation information for phrases used during the training process to create the decoder. This translation information includes alignment and the parallel translation for phrases. In the case of syntax-based decoder it's called a rule table. It is generated along with a grammar rules file. The rule table is similar to the phrase table. The rule table differs from the phrase-table such that it contains POS tags of the words along with the phrase translation. The glue-grammar file contains parallel grammar rules that show how the structure of the source language should be transformed in the target language. The two-phase decoder has better results in the case of the UN corpus because the decoder is trained on the UN corpus and adds some missing basic information not found in the baseline decoder needed to translate the UN corpus. This missing basic information could be seen in the phrase-table and the rule-table. For examples on phrase-table and rule-table entries in the post-processing decoder, see Table 8 along with the parallel translations entries in the phrase-and the rule-table. There is alignment information that shows how words should be moved in the resulting output to fix translation.

References

1. Bisazza, A., Federico, M.: Chunk-based verb reordering in VSO sentences for Arabic-English statistical machine translation. In: Proceedings of the Joint Fifth Workshop on Statistical Machine Translation and MetricsMATR, Stroudsburg, PA, USA, pp. 235–243 (2010)
2. Carpuat, M., Marton, Y., Habash, N.: Improving Arabic-to-English statistical machine translation by reordering post-verbal subjects for alignment. In: Proceedings of the ACL 2010 Conference Short Papers, pp. 178–183 (2010)
3. Salameh, M., Cherry, C., Kondrak, G.: Reversing morphological tokenization in English-to-Arabic SMT. In: HLT-NAACL, pp. 47–53 (2013)
4. Khemakhem, I.T., Jamoussi, S., Hamadou, A.B.: Arabic-English semantic class alignment to improve statistical machine translation. In: Recent Advances in Natural Language Processing, p. 663, September 2015
5. Och, F.J., Ney, H.: A systematic comparison of various statistical alignment models. Comput. Linguist. **29**(1), 19–51 (2003)
6. Almahairi, A., Cho, K., Habash, N., Courville, A.: First result on Arabic neural machine translation. arXiv:1606.02680 [cs], June 2016

7. Xiao, T., Zhu, J., Liu, T.: Bagging and Boosting statistical machine translation systems. Artif. Intell. **195**, 496–527 (2013)
8. El Kholy, A., Habash, N.: Orthographic and morphological processing for English-Arabic statistical machine translation. Mach. Transl. **26**(1–2), 25–45 (2012)
9. Kholy, A.E., Habash, N.: Techniques for Arabic morphological detokenization and orthographic denormalization (2011)
10. Manning, C.D., Surdeanu, M., Bauer, J., Finkel, J.R., Bethard, S., McClosky, D.: The stanford CoreNLP natural language processing toolkit. In: ACL (System Demonstrations), pp. 55–60 (2014)
11. Federico, M., Bertoldi, N., Cettolo, M.: IRSTLM: an open source toolkit for handling large scale language models. In: Interspeech, pp. 1618–1621 (2008)
12. Koehn, P., et al.: Moses: open source toolkit for statistical machine translation. In: Proceedings of the 45th Annual Meeting of the ACL on Interactive Poster and Demonstration Sessions, Stroudsburg, PA, USA, pp. 177–180 (2007)
13. Spence Green, D.C., Manning, C.D.: Phrasal: a toolkit for new directions in statistical machine translation. In: Proceedings of the Ninth Workshop on Statistical Machine Translation, pp. 114–121 (2014)
14. Ziemski, M., Junczys-Dowmunt, M., Pouliquen, B.: The United Nations Parallel Corpus v1. 0. In: LREC (2016)
15. Tiedemann, J.: Parallel data, tools and interfaces in OPUS. In: Proceedings of the 8th International Conference on Language Resources and Evaluation (LREC 2012) (2012)

On Character vs Word Embeddings as Input for English Sentence Classification

James Hammerton(✉), Mercè Vintró, Stelios Kapetanakis, and Michele Sama

Kare [knowledgeware] (formerly Gluru) LTD, Aldwych House 71-91, Aldwych, London WC2B 4HN, UK
{james,merce,stelios,michele}@karehq.com

Abstract. It has become a common practice to use word embeddings, such as those generated by word2vec or GloVe, as inputs for natural language processing tasks. Such embeddings can aid generalisation by capturing statistical regularities in word usage and by capturing some semantic information. However they require the construction of large dictionaries of high-dimensional vectors from very large amounts of text and have limited ability to handle out-of-vocabulary words or spelling mistakes. Some recent work has demonstrated that text classifiers using character-level input can achieve similar performance to those using word embeddings. Where character input replaces word-level input, it can yield smaller, less computationally intensive models, which helps when models need to be deployed on embedded devices. Character input can also help to address out-of-vocabulary words and/or spelling mistakes. It is thus of interest to know whether using character embeddings in place of word embeddings can be done without harming performance. In this paper, we investigate the use of character embeddings vs word embeddings when classifying short texts such as sentences and questions. We find that the models using character embeddings perform just as well as those using word embeddings whilst being much smaller and taking less time to train. Additionally, we demonstrate that using character embeddings makes the models more robust to spelling errors.

Keywords: Word embeddings · Character embeddings
Long short-term memory networks · Convolutional networks
Text classification · Natural language processing

1 Introduction

In recent years, it has become a common practice to use word embeddings, such as word2vec [1] or GloVe [2], as input to neural networks performing natural language processing (NLP) tasks, such as language modelling, named entity recognition, sentiment analysis or machine translation. This is because such embeddings

Mercè and Stelios have since left the company. Mercè and Stelios can be contacted via merce.ricart@shell.com and s.a.kapetanakis@gmail.com respectively.

K. Arai et al. (Eds.): IntelliSys 2018, AISC 868, pp. 550–566, 2019.
https://doi.org/10.1007/978-3-030-01054-6_40

aid generalisation by capturing statistical regularities in the way words are used, and indeed they can capture many aspects of the meaning of the words. For example, [1] demonstrates that simple vector addition of two embeddings can yield a vector semantically related to both, e.g. 'Vietnam' + 'capital' yielding 'Hanoi' as its closest neighbour and that analogical reasoning can be performed via linear operations on the embeddings. They thus provide a rich representation of words that learners can exploit to perform their tasks effectively.

However such word embeddings require large amounts of textual data to be trained (e.g. [1] used a corpus of 1 billion words when learning embeddings for single words, and then another corpus of 33 billion words when learning embeddings for phrases) and then the construction of large dictionaries of high-dimensional vectors. To handle out of vocabulary (OOV) words or spelling mistakes they require an "unknown" embedding to be used, either randomly generated or generated from rare tokens being replaced by a specialised token. This limits the extent to which OOV words and spelling mistakes can be handled effectively.

Finally, for some NLP tasks people are also finding that they can get good results using character-level input alone (e.g. [3]). There are a few advantages to using character level input that make this option attractive if similar performance can be achieved to that of using word-level input:

- Character embeddings will be based on a fixed set of characters, which will be much smaller than the vocabularies typically used in NLP tasks. Where character input replaces word-level input, this makes the models smaller and more computationally efficient, which is an important consideration when employing them on embedded devices with limited resources.
- Character-level input can help a model deal with OOV words and spelling mistakes, at least where new words with similar spellings to existing words have similar meanings. We will test this point empirically.

In this paper, we explore the use of word embeddings vs character embeddings for two classification tasks using English sentences. The rest of this paper is organised as follows:

- Section 2 sets out the background to the work presented here and the research questions we set out to address.
- Section 3 discusses work related to this paper.
- Section 4 describes the tasks and datasets we use for this work.
- Section 5 describes the approach taken, including the models we use, the preprocessing of the sentences we perform and how the evaluation is carried out.
- Section 6 presents the results of the experiments.
- Section 7 discusses the results presented.
- Section 8 presents our conclusions and discusses some options for future work.

2 Background and Research Questions

Prior to this work, we already had a classifier employing word2vec embeddings deployed to make predictions as part of the Gluru task manager application[1]. However we became aware that character models had been applied successfully for some tasks and decided to evaluate whether using character embeddings rather than word embeddings in our production system would make sense. Note that here we're not just interested in the performance of the classifier in terms of how well it would make its predictions but also in terms of the costs of deploying the classifier, e.g. model size, memory usage, prediction times, etc. We thus set out to compare the 2 approaches both in terms of prediction performance and in terms of the computational resources used. We therefore decided to address the following research questions:

- Will using character embeddings yield the same performance as we get with word embeddings?
- What are the training costs (e.g. amount of training data required, time required) of using character embeddings vs word embeddings?
- What are the costs of deploying the models, e.g. memory usage, disk usage, number of parameters, time taken per prediction?
- How resilient are the models to handling spelling errors and OOV words?

By answering these questions, we were then able to make a well informed choice about which model to deploy. These questions will be applicable any time one considers using character vs word embeddings as input for a task, and indeed can be generalised to any choice of two models where processing language is concerned and the possibility of spelling errors or OOV words arises.

3 Related Work

As the amount of available textual information grows, the task of text classification has become fundamental in the field of natural language processing. The body of research concerned with text classification, which involves assigning one of a set of predefined classes to a textual document, is wide and has focused on several of its applications for tasks such as information retrieval, ranking and document classification [4,5]. In our work we focus on short-text classification (i.e. sentences), which has become an important application of text classification for tasks such as sentiment analysis, as well as question answering and dialogue systems.

Several approaches have been proposed for the problem of short-text classification, which leverage traditional machine learning techniques such as Support Vector Machines (SVMs) [6] and Conditional Random Fields (CRFs) [7]. There has also been a body of work concerned with sequential short-text classification [8,9] due to the fact that short texts tend to appear in sequence (e.g. sentences

[1] This application will soon be discontinued.

in a document). However, we limit the scope of our work and only consider short texts in isolation. Recently, models based on neural networks have become very popular due to their good performance. Research on short-text classification has been performed with different neural architectures such as convolutional neural networks (CNNs) [10] and long short-term memory networks (LSTMs) [11]. Many of the neural network approaches treat words as their basic units and make use of word embeddings [12,13], which are fundamental to state-of-the-art NLP. However, there has been a growing interest in developing models at the character level, generating character embeddings using words as a basis [14] or directly from a string of raw characters [3]. This has resulted in performance comparisons between word and character-based models [3]. Our work extends this research to the task of short-text classification and presents further experiments that compare character and word embedding approaches.

4 Tasks and Data Sets

4.1 Benchmark Sentence Classification Problem

Here we used the dataset constructed for our internal sentence classification problem. The task is a binary classification task with a positive class and a negative class.

The dataset consists of sentences extracted from users emails as they appeared in the Gluru task manager application supplemented with some sentences from the emails in the Enron corpus [15]. The sentences were manually labelled.

Table 1 summarises the details of this dataset, reporting the number of sentences in the test and training set alongside the means and standard deviations (given in brackets) of the words per sentence and characters per sentence, plus the numbers of positive and negative examples. Note that during training a random 20% of the training set is used as a validation dataset (e.g. used for deciding best weights or early stopping).

Table 1. Dataset characteristics for benchmark sentence classification problem

	Sentences	Mean (stdev) Words/Sentence	Mean (stdev) Chars/Sentence	Examples Positive	Examples Negative
Training	19,518	14.07 (10.34)	80.82 (64.41)	6,745	12,773
Test	4,876	14.02 (10.02)	80.41 (63.66)	1,683	3,193

4.2 Effect of Reducing Training Set Size on Benchmark Sentence Classification Problem

Here we use the same dataset and task as in Sect. 4.1, but we vary the size of the training set from 5%, 10% and then 20% to 80%, in 20% increments, of the total size (via random sampling) and see the impact on performance for both character vs word embeddings when evaluating with the entire testing set. The motivation for this is the hypothesis that there might be some size of dataset below which we get differing levels of performance from the two models, and we wanted to cater for the possibility that our dataset was already above the threshold.

4.3 Effect of Introducing Spelling Errors on Benchmark Sentence Classification Problem

In order to assess the robustness of the models to spelling mistakes, we used the same dataset and task as in Sect. 4.1, and we introduced spelling errors during the evaluation phase. We did this by altering X randomly selected characters from each sentence at random, and substituting each character selected with a random choice from the alphanumeric characters or a space. We did this for values of X = 1, 4, 8, 12 and 16.

4.4 Sentence Classification with Parakweet Lab's Email Intent Dataset

We wanted to assess whether the findings we had with our own internal dataset would carry over to a publicly available data set and we chose the Email Intent Dataset[2] because it was also derived from email data and consisted of English sentences classified into to two classes. This consists of a training set and a test set of sentences drawn from the Enron corpus [15] and labeled as to whether they contain an 'Intent', defined to correspond primarily to the categories 'request' and 'propose' in [16]. In some cases, the creators also applied the positive label to some sentences from the 'commit' category if they contained a datetime.

The dataset details are provided in Table 2[3]. As before, the table reports the number of sentences, means and standards deviations for the word per sentence and characters per sentence and the numbers of positive and negative examples for the training and testing sets. Again during training 20% of the training set is set aside for validation, and here the validation performance was used in some runs for early stopping or determine which epoch produced the best set of weights.

Table 2. Dataset characteristics for Parakweet's email intent detection

	Sentences	Mean (stdev) Words/Sentence	Mean (stdev) Chars/Sentence	Examples	
				Positive	Negative
Training	3,657	16.56 (10.66)	91.64 (62.19)	1,719	1,938
Test	991	16.90 (11.33)	93.14 (68.08)	309	682

[2] https://github.com/ParakweetLabs/EmailIntentDataSet.

[3] The numbers reported here for the size of the training and test sets are what we found in the dataset on direct examination but the Wiki page for the dataset reports different numbers.

As this dataset is quite small to start with, we did not repeat the experiment reducing the size of the training data here.

5 The Approach

5.1 Convolutional Networks for Sentence Classification

We employ convolutional neural networks (CNNs) for the purposes of this paper. In natural language processing, CNNs are increasingly being used as an effective alternative (e.g. [10,12,17]) to the commonly used recurrent neural networks (RNNs) such as (bi)LSTM networks [18,19]. They are often more efficient to implement, requiring only a single forward pass to perform predictions compared to the multiple passes required with RNNs, whilst they do not require backpropagation through time thus simplifying training.

The network architectures we trained thus followed the following template[4]:

- Depending on whether we are using character or word embeddings, the input layer has either $M \times C$ or $N \times W$ units, specifying either the first M characters or the N words of the input sentence with zero padding used where a sentence is shorter than M characters or N words. C and W indicate the size of the character and word vocabularies respectively, with 1 hot vectors indicating which character or word is being presented.
- We then add a single convolution layer over these units using stride 1 and filter length 3, and 250 filters, with max pooling.
- The output from the convolution/maxpooling is then fed into a single dense layer of 100 units.
- Finally a softmax output layer is attached with 1 unit for each output class, in both cases here this means we have 2 output units[5].

When using word embeddings as input, we used the Google News pre-trained word2vec embeddings for English[6]. These are 300 dimensional vectors. The embeddings are refined during training.

When using character embeddings, we allocate 100 dimensions and let the networks learn the embeddings from random initialisation.

5.2 Preprocessing of Text

For both character and word-level input and both datasets, the sentences have already been extracted and we convert the sentences to lower case.

---

[4] To build and train the networks, we used Python 3.5 and Keras (https://keras.io/) with Tensorflow (https://www.tensorflow.org/) as the backend. All the experiments reported here were performed on an Ubuntu 16.04 Linux server with 64GB main RAM, using a NVIDIA GeForce GTX 1070 GPU with 8GB RAM.

[5] We could use a binary output here instead.

[6] https://code.google.com/archive/p/word2vec/.

For word-level input we then tokenise on whitespace and present up to the first N words to the network, where for each word we look up the embedding from the Google News word2vec embeddings, defaulting to a random embedding for unknown words.

For character level input, we present up to the first M characters of the sentence to the network. To avoid having to store a dictionary mapping each character to its representation to the neural network, we hash each character to one of 968 different values - this number was chosen to be much larger than the vocabulary size in order to avoid collisions.

5.3 Evaluation

In each experiment, we evaluate by presenting the relevant test set and comparing the predictions to the expected results. To account for variations in performance when keeping training algorithm and hyper parameters constant, each experiment involves performing five training runs, each starting from random weight initialisations, and then evaluating the performance of the trained network for each run.

As both tasks are classification tasks, we compute following standard classification metrics for each class:

- the precision, P, where $P = T_P/(T_P + F_P)$
- the recall, R, where $R = T_P/(T_P + F_N)$
- the fscore, $F1$, which is the harmonic mean of the precision and recall, where $F1 = 2PR/(P + R)$

where:

- T_P is the number of true positives, i.e. those examples that are correctly classified as members of the class
- F_P and F_N are the number of false positives and false negatives respectively, i.e. the examples incorrectly identified as members of the class or incorrectly identified as not being members of the class, respectively.

We report the average and standard deviation of these metrics computed across the 5 runs for each experiment.

6 Experimental Results

6.1 Benchmark Supervised Classification Results

Table 3 summarises the network architectures used for these experiments. The columns are as follows:

- Network - contains labels for the network used, e.g. Char512 indicating character embeddings using the first 512 characters of a sentence and Word89 indicating word embeddings as input using the first 89 words of the sentence. These labels are used in other tables as references to the networks concerned.

Table 3. Networks used for benchmark sentence classification problem

Network	Inputs	Weights	Epoch duration	Size on disk
Char512	512 chars	197,352	6.6s	2.3 MB
Word89	89 words	55,877,752	68.2s	640 MB
Word20	20 words	55,877,752[a]	48.6s	640 MB
Network	Memory usage	Prediction time total (per sentence)		
Char512	3 MB	0.456s (0.094 ms)		
Word89	448 MB	0.366s (0.076 ms)		
Word20	448 MB	0.322s (0.067 ms)		

[a] Because the weights for each input window in a convolution network are shared across all the input windows, altering the number of input words/characters presented does not alter the number of weights.

- Inputs - indicates the number and type of inputs.
- Weights - reports the number of weights in the network.
- Epoch duration - reports the duration in seconds of one complete presentation of the dataset during training.
- Size on disk - reports the size of the saved Keras models.
- Memory usage - reports the estimated memory usage of the models.
- Prediction time - reports the total time and time per sentence for predicting the classifications of the test set.

In all cases, the number of filters was 250, the filter length was 3, hidden layer size was 100 units, and the networks were trained for 5 epochs with batch size 16, using the ADAM optimiser [20] and a learning rate of 0.2. Dropout was also applied to the weights from the final hidden layer to the output during training with probability of 0.2.

As this table shows, both the size of the model (the number of weights) and the runtime are much smaller for the character models than for the word embedding models. However the character embeddings models were slightly slower at doing predictions than the word embeddings model. We hypothesise that the sliding window being run over 512 characters (compared to 89 words) is offsetting other efficiency gains here. The fact that the version of the word model that uses only the first 20 words is faster than the one using 89 words provides some support for this given the 2 models are the same size otherwise.

Table 4 details the results of the experiments. The columns are as follows:

- Network - network label as per Table 3.
- Precision - reports the mean and, in brackets, the standard deviation of the precision across the 5 runs.
- Recall - reports the mean and, in brackets, the standard deviation of the recall across the 5 runs.
- F1 - reports the mean and, in brackets, the standard deviation of the F1 across the 5 runs.

Table 4. Benchmark sentence classification results

Positive class			
Network	Precision (stdev)	Recall (stdev)	F1 (stdev)
Char512	**0.74** (0.05)	**0.81** (0.05)	**0.77** (0.01)
Word89	0.70 (0.04)	0.83 (0.05)	0.76 (0.01)
Word20	0.70 (0.06)	0.79 (0.08)	0.74 (0.02)
Negative class			
Network	Precision (stdev)	Recall (stdev)	F1 (stdev)
Char512	**0.90** (0.02)	**0.85** (0.05)	**0.87** (0.01)
Word89	0.90 (0.02)	0.80 (0.05)	0.85 (0.02)
Word20	0.88 (0.03)	0.82 (0.07)	0.84 (0.03)

For both the positive and negative classes, whilst the model using character inputs obtained the best mean scores for each metric, the standard deviations show that there is no significant difference between using character embeddings vs word embedding when the first 512 characters are used vs the first 89 words of a sentence. Reducing the number of words used to 20 increases the variability of the scores when using word embeddings and caused the average F1 to drop slightly.

6.2 Results for Varying Training Set Size

Table 5 summarises the results for these experiments, note that this was only done for the "Char512" and the "Word89" networks. The columns are:

- Percent - the percentage of the training data used to train the model.
- Positive class F1 - the mean and standard deviation (in brackets) of the F1 for the positive class computed across the 5 runs.
- Negative class F1 - the mean and standard deviation (in brackets) of the F1 for the negative class computed across the 5 runs.

As the table illustrates, the smaller dataset sizes do reduce performance though perhaps not as dramatically as one might expect, until we get down below 20% of the original size. Interestingly, we get only slightly reduced performance for using 40% of the data compared to 100% here. Also the performance of the negative class drops slightly when going from 80% to 100%, though given the variability indicated by the standard deviations this may be more of an apparent effect than a real one.

Reducing the size of the training set to 5% has a bigger impact on the performance of the positive class for the character embeddings than the word embeddings, however the mean F1 is still within a standard deviation of the performance of the word embeddings when using only 5% of the data. Otherwise the impact of the reduced dataset sizes looks very similar for the two cases. This

Table 5. Varying size of training data

Word embeddings		
Percent	Positive class F1 (stdev)	Negative class F1 (stdev)
5	0.64 (0.04)	0.83 (0.01)
10	0.68 (0.03)	0.85 (0.01)
20	0.69 (0.05)	0.86 (0.01)
40	0.73 (0.01)	0.87 (0.00)
60	0.75 (0.01)	0.88 (0.01)
80	0.75 (0.02)	0.88 (0.00)
100	0.76 (0.01)	0.85 (0.02)
Best-worst	0.12	0.05
Char embeddings		
Percent	Positive class F1 (stdev)	Negative class F1 (stdev)
5	0.61 (0.06)	0.84 (0.01)
10	0.68 (0.02)	0.85 (0.01)
20	0.70 (0.02)	0.87 (0.00)
40	0.75 (0.02)	0.87 (0.01)
60	0.76 (0.02)	0.87 (0.01)
80	0.76 (0.02)	0.88 (0.00)
100	0.77 (0.01)	0.87 (0.01)
Best-worst	0.16	0.04

suggests that there is not much difference overall in the training data requirements when using character embeddings vs word embeddings, contrary to the hypothesis raised in Sect. 4.2.

6.3 Results for Introducing Spelling Errors on Benchmark Sentence Classification Problem

The results are summarised in Table 6. The 'Perturbations' column indicates how many characters were replaced with a random character as per Sect. 4.3, the remaining columns giving the mean F1 plus standard deviation for the positive and negative classes respectively as in Table 5. Again the perturbations were only done for the "Char512" and "Word89" networks.

Here it can be seen that the performance of both classes becomes progressively worse with the greater number of perturbations, as one would expect. Note however that this effect is only slight for the negative class and is much stronger for the positive class, as indicated by the 'Best-worst' row where the drops for the positive class are 0.53 (word embeddings) and 0.40 (character embeddings).

As one would intuitively expect, the impact on the word embeddings case is larger than in the character embeddings case. The presence of a spelling mistake

Table 6. Effect of varying levels of spelling errors

Word embeddings		
Perturbations	Positive class F1 (stdev)	Negative class F1 (stdev)
0	0.76 (0.01)	0.85 (0.02)
1	0.73 (0.01)	0.87 (0.01)
4	0.60 (0.04)	0.84 (0.00)
8	0.46 (0.03)	0.82 (0.00)
12	0.35 (0.04)	0.81 (0.01)
16	0.23 (0.06)	0.80 (0.01)
Best-worst	0.53	0.05
Char embeddings		
Perturbations	Positive class F1 (stdev)	Negative class F1 (stdev)
0	0.77 (0.01)	0.87 (0.01)
1	0.75 (0.02)	0.88 (0.01)
4	0.67 (0.04)	0.87 (0.01)
8	0.56 (0.06)	0.85 (0.01)
12	0.49 (0.06)	0.84 (0.01)
16	0.37 (0.05)	0.82 (0.01)
Best-worst	0.40	0.06

has a bigger impact here due to the potential for producing an OOV token and thus losing the meaning of the whole word, whereas the use of character embeddings can allow the network to e.g. treat 'like', similarly to 'lije'.

Focusing on the F1s here however misses an interesting phenomenon with the precision and recall. Table 7 summarises the figures (Prec = Precision, Rec = Recall), and shows that the precision of the positive class holds up whilst the recall is reduced, whereas with the negative class the recall holds up and even improves slightly whilst the precision drops, though this is not as big an effect as the drop in the positive class's precision. This effect can be seen both when using word embeddings and when using character embeddings, though as before the drop-off is more pronounced with the word embeddings.

6.4 Results for Classification with Parakweet Lab's Email Intent Dataset

Table 8 summarises the network architectures used here. As with earlier tables, the Network column lists a label for each network architecture and the Inputs column indicates the type and number of inputs to the network. The 'Early stop' and 'Best net' columns indicating whether early stopping was used or whether the best network (based on the validation set loss computed during training) was saved and used in testing. As before the 'Epoch duration' reports the training time in seconds for one complete presentation of the training data.

Table 7. Effect of spelling errors on precision and recall

Word embeddings				
Perturbations	Positive class		Negative class	
	Prec (stdev)	Rec (stdev)	Prec (stdev)	Rec (stdev)
1	0.77 (0.03)	0.70 (0.02)	0.85 (0.01)	0.89 (0.02)
4	0.75 (0.03)	0.51 (0.07)	0.78 (0.02)	0.91 (0.03)
8	0.73 (0.04)	0.34 (0.03)	0.73 (0.01)	0.93 (0.02)
12	0.69 (0.06)	0.24 (0.04)	0.70 (0.01)	0.94 (0.02)
16	0.70 (0.07)	0.14 (0.04)	0.68 (0.01)	0.94 (0.02)
Char embeddings				
Perturbations	Positive class		Negative class	
	Prec (stdev)	Rec (stdev)	Prec (stdev)	Rec (stdev)
1	0.81 (0.04)	0.69 (0.04)	0.85 (0.01)	0.92 (0.03)
4	0.84 (0.04)	0.57 (0.06)	0.81 (0.02)	0.94 (0.02)
8	0.86 (0.02)	0.42 (0.07)	0.76 (0.02)	0.96 (0.01)
12	0.84 (0.04)	0.35 (0.07)	0.74 (0.02)	0.97 (0.02)
16	0.86 (0.04)	0.24 (0.04)	0.71 (0.01)	0.98 (0.01)

Table 8. Networks used for sentence classification with Parakweet's email intent dataset

Network	Inputs	Early stop	Best net	Weights	Epoch duration
Char512	512 chars			197,352	1.4 s
Char512:ES	512 chars	Yes		197,352	1.4 s
Char512:BW	512 chars		Yes	197,352	1.4 s
Word89	89 words			55,877,752	9.2 s
Word89:ES	89 words	Yes		55,877,752	9.2 s
Word89:BW	89 words		Yes	55,877,752	9.2 s

As with the benchmark supervised sentence classification, 250 filters were used, the filter length was 3, the hidden layer consisted of 100 units and the networks were trained for 5 epochs with batch size 16 using the ADAM optimiser with a learning rate of 0.2 and a dropout probability of 0.2. Again we report the mean and standard deviation of the precision, recall and F1 over 5 runs for each configuration.

Note that because this is a much smaller dataset, the epoch durations are shorter as a result, but again the network size and runtimes are smaller for the character embeddings than for the word embeddings.

Table 9 details the results of experiments performed with this dataset. The columns are the same as in Table 4, but reporting for the Intent and Not Intent classes here. Note that the network labels sometimes end with ES (indicating use

Table 9. Results for sentence classification with Parakweet's email intent dataset

Intent			
Network	Precision (stdev)	Recall (stdev)	F1 (stdev)
Char512:ES	**0.83** (0.05)	0.49 (0.14)	0.60 (0.11)
Char512	0.71 (0.04)	0.67 (0.06)	**0.69** (0.02)
Char512:BW	0.61 (0.05)	**0.77** (0.04)	0.68 (0.02)
Word89:ES	0.79 (0.06)	0.59 (0.04)	0.68 (0.01)
Word89:250	0.68 (0.05)	0.67 (0.05)	0.67 (0.01)
Word89:BW	0.64 (0.04)	0.75 (0.04)	**0.69** (0.02)
No intent			
Network	Precision (stdev)	Recall (stdev)	F1 (stdev)
Char512:ES	0.81 (0.03)	**0.95** (0.03)	0.87 (0.01)
Char512	0.86 (0.02)	0.87 (0.03)	0.87 (0.01)
Char512:BW	**0.88** (0.01)	0.77 (0.06)	0.82 (0.03)
Word89:ES	0.83 (0.01)	0.93 (0.03)	**0.88** (0.01)
Word89	0.85 (0.01)	0.86 (0.04)	0.86 (0.01)
Word89:BW	**0.88** (0.02)	0.80 (0.03)	0.84 (0.01)

of early stopping) or BW(indicating use of the best weights seen during training). Early stopping increases the standard deviations whilst reducing the character embeddings mean F1 for the 'Intent' class, whilst using the best weights (assessed via validation loss) boosted recall at the expense of precision for the 'Intent' class and did the reverse for the 'No Intent' class, with the character embeddings F1 reduced in this case.

The best F1s achieved for character embeddings vs word embeddings are either the same (for 'Intent') or within a standard deviation of each other (for 'Not Intent'), again indicating that the two options give similar levels of performance.

7 Discussion

The results presented in the previous section indicate that a model trained using learned character embeddings as input alone can perform as well as a model trained using pretrained word embeddings in some short text classification tasks, and can do so using a much smaller network with far fewer parameters. Whilst more work is needed to determine to what extent this holds up for all text classification tasks, we're not the only ones to have found that character embeddings can be as effective as word embeddings [3]. So why would we get equal performance using word vs character embeddings? Some possible explanations come to mind:

(1) The tasks are so simple that the precise representation of the input does not greatly matter so long as it captures sufficient surface features of the sentence. For both tasks it's not obvious that this would be the case - both tasks intuitively rely on having some understanding of the meaning of a sentence. That said if there are surface/syntactic cues that correlate strongly with the task then the networks may simply be picking those cues up rather than performing the task. If so then it's a form of the explanation in point 2 in this list.
(2) The network is learning a surrogate of the tasks that is based on surface features that are present in both forms of input and which correlates with the tasks. If this is the case, we'd expect that constructing a large scale dataset for each task might be required to get the networks using features beyond the surface cues from the text, at which point we might then see a difference in performance. Moreover it may be that constructing a simple bag of ngrams or bag of words based classifier might perform the task just as well as these networks.
(3) The character embeddings, by capturing information about how characters are used, may be capturing some of the information the word embeddings capture, albeit indirectly. This seems unlikely to us but would be interesting to investigate to see if we can rule it out.

Some further comments:

- The pretrained word embeddings can provide information about the usage and meaning of words derived from a large corpus that you simply would not get from the sentences treated in isolation that we use for both tasks here. On the other hand the character strings also provide an advantage over the word strings in providing information about the morphology of words that you don't get from the word embeddings. These considerations suggest that some combination of word embeddings and character embeddings may boost performance, albeit at the expense of making the models larger again, because each form of input brings information that the other cannot bring. Of course if explanations 1 and 2 above apply then this would not apply.
- The nature of the task may well be important here. If the classification relies on the meaning of each word in the sentence being correctly deduced, we would not expect good performance. For example, we would not expect Named Entity Recognition to work well on character input alone because each word has to be labelled (to indicate if it is part of an entity or not) and thus the meaning of each specific word in the sentence is crucial to the labelling task. Could it be that when assigning a single label to a chunk of text that information from the surface form of the text is sufficient for most such tasks?
- Regarding the character model being more robust to spelling errors and OOV words than the word embedding model, this is intuitive but there are limits here. Handling of OOV words will only work to the extent that words with similar character strings have similar meaning/usage to a relevant in vocabulary word. Similarly, spelling mistakes will only be handled well to the extent

that the misspelled token does not represent another word that the model knows or is sufficiently similar to the correctly spelled word for the model to figure things out. If correct spelling becomes critical to the classification task then it may be better to deal with spelling errors prior to the classification being performed.

8 Conclusions and Future Work

We have presented a set of research questions (in Sect. 2) to answer when evaluating the use of character embeddings vs word embeddings as input to models performing NLP tasks that address both the performance of the classifiers in terms of precision, recall, F1 and in terms of the computational costs of training and deploying such models.

The results of the experiments we presented illustrate a clear win in favour of deploying the character embedding based model. Not only did we confirm that using character embeddings as input can be as effective as word embeddings when making predictions but the models are much smaller, and the computational costs are decisively lower - as confirmed by the fact that our hosting costs for the character model were 200x less than for the word2vec model. Additionally the character embeddings model is clearly more resilient, as evidenced by our experiments on introducing spelling errors to the sentences during evaluation. Also, it has shown that a training dataset of 19k in size can be reduced to 40% of its size with only modest loss of accuracy. However, the character models were slightly slower at generating predictions than the word embedding models.

This work is limited as follows:

- Both of the datasets employed are quite small, and the two tasks, both of which involve classifying sentences in isolation, won't represent the range of possible classification tasks. It may be that other text classification tasks may be more difficult when using character input vs word level input. Also it may be that for some tasks a larger dataset is needed for differences in performance to manifest themselves.
- We use pre-trained word embeddings derived from news articles, where perhaps using pre-trained embeddings derived from a corpus of Gluru users emails or from the Enron corpus might give better results by being better tailored to the domains of the two tasks.
- The comparison is limited to convolutional networks.
- Both tasks involve sentence classification, and it may be that for longer texts the use of character vs word embeddings might make more difference.
- The classification is performed independently of the context in which a sentence occurs.
- No attempt at combining the use of character and word embeddings has been made.

These limitations, and the discussion in Sect. 7, suggest the following lines of followup work:

- Investigate using alternative embeddings to the Google News embeddings, including using embeddings derived from a corpus of emails.
- Investigate the extent to which the tasks can be performed based on e.g. bag of words or bag of (character or word) n-grams, to help determine to what extent the tasks can be resolved via surface features.
- Extend the comparison to use LSTM networks. [18] and (bi-directional) LSTM networks [19] to see if the sequential nature of processing in such recurrent networks makes any difference to the issue.
- Extend the comparison to other text classification tasks and larger data sets, such as the Twitter Sentiment Analysis dataset[7] and the IMDB movie review dataset[8]. These datasets are both much larger datasets and the latter involves processing long multi-paragraph text rather than sentences.
- Investigate whether we can combine the use of character and word embeddings effectively to improve performance.
- Investigate whether using other sub-word level input, such as character n-gram embeddings, can be an effective strategy.

Acknowledgment. The authors would like to thank their colleagues at Gluru for their support during this work.

References

1. Mikolov, T., Sutskever, I., Chen, K., Corrado, G., Dean, J.: Distributed representations of words and phrases and their compositionality. In: NIPS 2013, pp. 1–9, October 2013. http://arxiv.org/abs/1310.4546
2. Pennington, J., Socher, R., Manning, C.D.: Glove: global vectors for word representation. In: Empirical Methods in Natural Language Processing (EMNLP), pp. 1532–1543 (2014). http://www.aclweb.org/anthology/D14-1162
3. Zhang, X., Zhao, J., LeCun, Y.: Character-level convolutional networks for text classification. In: Cortes, C., Lawrence, N.D., Lee, D.D., Sugiyama, M., Garnett, R. (eds.) Advances in Neural Information Processing Systems 28 (NIPS 2015), pp. 649–657. Curran Associates, Inc. (2015). http://papers.nips.cc/paper/5782-character-level-convolutional-networks-for-text-classification.pdf
4. Deerwester, S., Dumais, S.T., Furnas, G.W., Landauer, T.K., Harshman, R.: Indexing by latent semantic analysis. J. Am. Soc. Inf. Sci. **41**(6), 391 (1990)
5. Pang, B., Lee, L., et al.: Opinion mining and sentiment analysis. Found. Trends® Inf. Retrieval **2**(1–2), 1–135 (2008)
6. Silva, J., Coheur, L., Mendes, A.C., Wichert, A.: From symbolic to sub-symbolic information in question classification. Artif. Intell. Rev. **35**(2), 137–154 (2011)
7. Nakagawa, T., Inui, K., Kurohashi, S.: Dependency tree-based sentiment classification using CRFs with hidden variables. In: Human Language Technologies: The 2010 Annual Conference of the North American Chapter of the Association for Computational Linguistics, pp. 786–794. Association for Computational Linguistics (2010)

[7] http://thinknook.com/twitter-sentiment-analysis-training-corpus-dataset-2012-09-22/.

[8] http://ai.stanford.edu/~amaas/data/sentiment/.

8. Lendvai, P., Geertzen, J.: Token-based chunking of turn-internal dialogue act sequences. In: Proceedings of the 8th SIGDIAL Workshop on Discourse and Dialogue, pp. 174–181 (2007)
9. Lee, J.Y., Dernoncourt, F.: Sequential short-text classification with recurrent and convolutional neural networks. In: Proceedings of NAACL-HLT 2016 (2016). https://arxiv.org/abs/1603.03827
10. Kim, Y.: Convolutional neural networks for sentence classification. In: Proceedings of the 2014 Conference on Empirical Methods in Natural Language Processing (EMNLP 2014) (2014). http://emnlp2014.org/papers/pdf/EMNLP2014181.pdf
11. Socher, R., Huval, B., Manning, C.D., Ng, A.Y.: Semantic compositionality through recursive matrix-vector spaces. In: Proceedings of the 2012 Joint Conference on Empirical Methods in Natural Language processing and Computational Natural Language Learning, pp. 1201–1211. Association for Computational Linguistics (2012)
12. Collobert, R., Weston, J., Bottou, L., Karlen, M., Kavukcuoglu, K., Kuksa, P.: Natural language processing (Almost) from Scratch. J. Mach. Learn. Res. **12**, 2493–2537 (2011)
13. Kalchbrenner, N., Grefenstette, E., Blunsom, P.: A convolutional neural network for modelling sentences. In: Proceedings of the 52nd Annual Meeting of the Association for Computational Linguistics, June 2014. http://goo.gl/EsQCuC
14. Santos, C.D., Zadrozny, B.: Learning character-level representations for part-of-speech tagging. In: Proceedings of the 31st International Conference on Machine Learning (ICML 2014), pp. 1818–1826 (2014)
15. Klimt, B., Yang, Y.: The enron corpus: a new dataset for email classification research. In: Boulicaut, J.F., Esposito, F., Giannotti, F., Pedreschi, D. (eds.) Machine Learning: ECML 2004. Lecture Notes in Computer Science, vol. 3201, pp. 217–226. Springer, Heidelberg (2004) https://doi.org/10.1007/978-3-540-30115-8_22
16. Cohen, W.W., Carvalho, V.R., Mitchell, T.: Learning to classify email into "Speech Acts". In: EMNLP 2004, vol. 4, pp. 309–316 (2004). http://acl.ldc.upenn.edu/acl2004/emnlp/pdf/Cohen.pdf
17. Conneau, A., Schwenk, H., Barrault, L., Lecun, Y.: Very deep convolutional networks for natural language processing. In: Proceedings of EACL 2017 (2017). https://arxiv.org/abs/1606.01781
18. Hochreiter, S., Schmidhuber, J.: Long short-term memory. Neural Comput. **9**(8), 1–32 (1997)
19. Graves, A., Schmidhuber, J.: Framewise phoneme classification with bidirectional LSTM and other neural network architectures. Neural Netw. **18**(56), 602–610 (2005). http://www.sciencedirect.com/science/article/pii/S0893608005001206
20. Kingma, D.P., Ba, J.: Adam: a method for stochastic optimization. In: 3rd International Conference for Learning Representations (2015). http://arxiv.org/abs/1412.6980

Performance Comparison of Popular Text Vectorising Models on Multi-class Email Classification

Ritwik Kulkarni(✉), Mercè Vintró, Stelios Kapetanakis, and Michele Sama

Kare [knowledgeware] (formerly Gluru) LTD, Aldwych House 71-91, Aldwych, London WC2B 4HN, UK
{ritwik,merce,stelios,michele}@karehq.com

Abstract. As the demand for routing, auto-responding, labelling, etc. is increasing, automated email classification and their responsive action(s) have become an area of increased interest in both supervised and unsupervised text learning. However, the large number of disparate classes required as a training set for any investigated domain seems an obstacle for increasing the performance of the literature baselines. We analyse the performance of six state-of-the-art research approaches against a highly-constrained, domain-driven text corpus, including variations in testing to identify the best possible approach in dealing with larger documents such as emails. We identify the Memory Network as the best candidate, among other popular neural networks, due to its top accuracy among the models compared, faster prediction and ability to train on limited data.

Keywords: Memory networks · Deep neural networks
Convolutional neural networks · Email classification

1 Introduction

Email classification applies to a wide range of applications, including Online Marketing, Security and automated Management. Research-wise this challenge pertains to the area of text classification with unique features since emails can have: large variety in features, significant volume of text, rich content and substantial variation in context veracity. A rather large body of email classification literature pertains to SPAM detection [1,2] to mention a couple, which usually is defined as unsolicited emails, while sentiment analysis, another popular application of text classification, is applied to email classification in [3]. Another domain of email classification is multi-class classification [4,5], which is a considerably harder task given the variance in the input space and a lack of hard boundaries in the output space. Our study falls under this third category, in which we focus

Ritwik, Mercè and Stelios have since left the company. They can can be contacted via rittyun@yahoo.com, merce.ricart@shell.com and s.kapetanakis@brighton.ac.uk respectively

K. Arai et al. (Eds.): IntelliSys 2018, AISC 868, pp. 567–578, 2019.
https://doi.org/10.1007/978-3-030-01054-6_41

on a more nuanced classification of emails which requires the contents of the text to be resolved at a finer level and be matched to the class among several different classes, rather than a binary decision.

Email features can be highly sparse in terms of usefulness and can make any classification attempt rather skewed in performance. The quality of the training stage (training dataset) can heavily affect the end-product result since in an ideal world any possible email categories (we will refer to them as classes throughout this paper) should be covered comprehensively for an algorithm to perform an adequate classification task [6,7].

The literature has numerous evidence examples of traditional machine learning algorithms on top of the email classification challenge. Examples include Neural Networks, Decision Trees, kNN, Naive Bayes, Support Vector Machines (SVMs) and others showing adequate results [2,8–11]. All the above algorithms follow the traditional machine learning principles where algorithms are expected to learn from experience, with respect to a range of domain-related tasks and performance criteria. Modern machine learning expects algorithms to be exposed to examples from a certain domain while incorporating the following two aspects: (i) be able to make predictions with the minimal amount of data points; and (ii) be able to work with an open-world assumption, i.e. besides being accurate upon their learning samples to allow a margin for learning. In this paper we provide a comparative analysis of six popular models which vectorise text based on certain learning algorithms and apply this to the task of email classification. Our classification task aims towards the prediction of pre-defined responses to incoming email queries. We analyse the performance of these six models based on their ability to rank the correct response to an email query. We use email-response pairs from a real life scenario where human agents are called to select an appropriate response to an email. As a result, we end up using data rich in information content and complex enough to pose a tough challenge to neural network models upon a highly-constrained domain. Our domain being multi-class classification, we investigate the performance on two subsets of the data which constrain the amount of training examples. We devise learning upon limited amount of training data for data-balancing reasons on a highly unbalanced dataset. We conclude that relatively naive implementations of Memory Networks perform better against the best-known literature baselines. Surprisingly, limited data provision for training purposes provides accountable results, rendering obsolete the need for large batch training over time.

In the subsequent sections we describe the dataset and it's features in Sect. 2, followed by a brief description of the six models we implement, in Sect. 3. Section 4 details the performance of the models while Sect. 5 concludes with a discussion on some of the insights gained from the experiments.

2 Dataset

The dataset consists of 300,000 pairs of email queries and a standardised response selected from a list of 121 templates, manually crafted by a human agent. The mean length of the email and response is 104 (max: 3,750) and 115 (max: 470) words respectively. As seen in Fig. 1, the distribution of frequencies is quite skewed. The top occurring response has a frequency of 97,521 followed by the second highest at 19,160, while the fifth highest is at 5,340 with a drop which is roughly exponential in nature. This poses the problem of having a severely imbalanced dataset for training and testing purposes. To mitigate this issue we filter the data and generate two subsets. Given the heavy skew in response distribution, our objective is to cover as many unique responses as possible and at the same time have enough training examples, balanced by a maximum number of samples we collect per response. We do this so that the evaluation figures are not biased by the sample distribution. We set the minimum frequency at 300, which means that we only sample responses that occur at least 300 times, and a maximum threshold of 500 samples per response. This results in 57 unique responses with 27,440 pairs. We call this subset 1. The data is split into 80%–20% for training and testing, respectively. Another subset of this data is created which increases the number of training samples per response by setting the minimum frequency at 3,000 and maximum threshold at 5,000. With this setting we cover the top 11 responses still accounting for 77% of the data. We call this subset 2. The differences in results in the two subsets are shown in Tables 2 and 3.

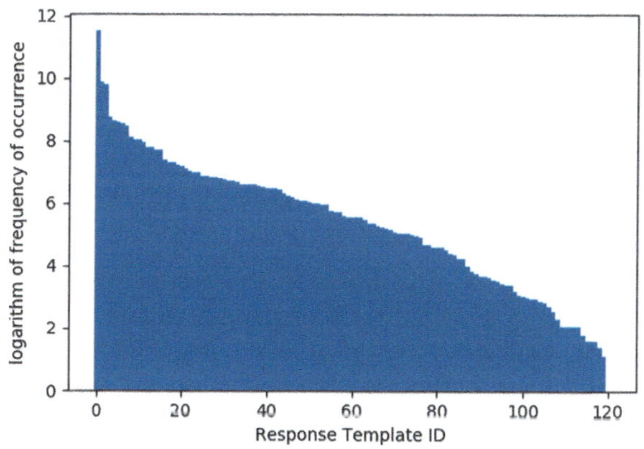

Fig. 1. Log frequency distribution of response occurrence

Table 1 shows the raw overlap between words in an email-response pair, i.e. it shows the common words in a pair when the response to the email is correct and when the response is random. The overlap is calculated after processing the text, which involves removing stop-words, HTML tags and stand-alone special

Table 1. Normalised overlaps of common words

	Mean	S.T.D	Max	Min
email-Correct response pair	0.11	0.14	0.96	0.0
email-Random response pair	0.07	0.04	0.70	0.0

characters. Further noise reduction is performed by removing very low-frequency words (0.001 times per sample) and very high-frequency words (occurring 4 times in each sample). By visual inspection, such words were mostly found to be garbage strings indicating formatting lines. The overlap is calculated as

$$O = \frac{|E \cap R|}{|E \cup R|}$$

where E and R are the sets of words in the email and response respectively. Table 1 shows that although there is high variability in the data, there are indications of a signal to be exploited in differentiating related and unrelated responses.

3 Models

We experiment with six popular text processing models which vectorise text and utilise the vectors for various downstream tasks. We evaluate their performance on the email-response dataset described in the previous section. The models are known to perform well on various Natural Language Processing tasks and here, we test them against a large real-life and complex dataset. In the following subsections the models we implemented are described briefly to present their overall high-level design. In this study, we restrict ourselves from doing any extra enhancements to the individual designs of the models and compare their performance using a basic architecture.

3.1 Tf-Idf

We use a standard *tf-idf* model as a baseline to compare an email query to the corpus of responses and we evaluate whether the predicted response is in the top ranks. *Tf-idf* is a popular model in information retrieval and versions of it have been around since the early 70s [12]. *Tf-idf* is short for *term frequency-inverse document frequency*. It is essentially a scaling factor applied to words in a text and reflects the importance of a particular word to a document in the corpus. A word in a document that appears more frequently is given more importance, however it is modulated by the frequency of occurrence of that word in the entire corpus. Different versions of this central concept are often used in text-searching tasks and, more recently, as a way of pre-processing text before it is fed forward to more sophisticated models. The inverse document frequency is defined as,

$$Idf_t = \log\left(1 + \frac{N}{n_t}\right)$$

where N is the total number of documents and n_t is the number of documents where the term t appears.

Thus with this formulation the text and subsequently the responses are projected onto a vector space. Similarly, an incoming query (i.e. an email) is vectorised and a cosine similarity measure is performed to rank the most similar responses given the query email. A standard cosine similarity measure is used as

$$\frac{A \cdot B}{||A|| \cdot ||B||}$$

3.2 DSSM

We experimented with the DSSM [13] to perform response selection. DSSM stands for *Deep structured Semantic Model*, which is a popular feedforward neural network for document retrieval. It can be seen as an extension of latent semantic models with supervision (e.g. Latent Semantic Indexing [14]), which uses the implicit document-query relevance judgments, originally from click-through data. The DSSM is, in brief, a deep neural network model used for text representation in a high dimensional vector space (although considerably lower in dimensionality when compared to the raw inputs) which encodes the semantic similarity between two text strings. The DSSM can be used for ranking and classification purposes in the text document domain by training the network in such a way, that the queries and the corresponding documents find a similar vicinity in the encoded vector space. The vicinity measure is usually the cosine similarity, but other distance measures can be used. The model is trained with an objective function which maximizes the log probability of the relevant document (i.e. correct response in the case of email classification) being the most similar. Apart from the pair of query and related document, the DSSM requires a set of negative example documents which are unrelated to the query. The inputs are each one-hot encoded as a bag of character trigrams feature vector, following the original paper.

We trained the DSSM as described in [15] and refer the reader here for the dataset and details on how the model is trained.

3.3 InferSent

InferSent [16] is a text vectorisation method which produces usable embeddings of sentences rather than words. It falls under the class of transfer learning techniques, in which information obtained (as network weights) while training a model to perform a particular task, is used to solve a different but related task. InferSent is a neural network trained on the Stanford Natural Language Inference (SNLI) dataset as a classification problem. It produces text vectors which encode semantic similarity and can, in principle, be used for other text-processing tasks such as document similarity. InferSent uses a bi-directional LSTM architecture with max pooling and has a considerable advantage in required training times compared to models like SkipThought and FastSent [16].

We leverage the InferSent model in a manner similar to the DSSM to perform the email classification task. However, unlike the DSSM, we use the pre-trained model provided by Facebook Research at [17]. As before, emails act as queries to the model, which after vectorisation are compared to the response vectors. The responses are ranked from best to least match, again using cosine similarity.

3.4 CNN

Convolutional Neural Networks (CNN) have shown impressive results in tasks of text classification [18]. CNNs are known to be fast and ideally suited for rapidly processing large text. However, the max pooling operations in CNNs lead to a loss of local word order information and hence have a limited ability to capture context. The filters, however, may succeed in capturing some long-term dependencies in a text. We offer a comparison against a character-level CNN used as a text classifier. A character-based model showed better performance over a word-based model in terms of requiring less vocabulary and slightly better accuracy. The model accepts as input a sequence of characters, which are encoded following [19]. The input layer of the network has $M \times C$ units, which specifies the first M characters of the input email. We find setting M to 700 characters gives us best results, with zero-padding when texts are shorter. C indicates the size of the character vocabulary, with one-hot vectors indicating which character is being presented. We then add an embedding layer with 100 dimensions, which is randomly initialised and learned by the network. A single convolution layer is added over these units using stride 1 and filter length 3, and 250 filters, with max pooling. The output from the convolution/maxpooling is then fed into a single dense layer of 100 units. Finally, a softmax output layer is attached with 1 unit for each template found in our dataset. The network is trained with ADAM [20] optimiser using categorical cross-entropy as the loss function and a batch size of 32.

3.5 Bi-LSTM

Long Short Term Memory networks (LSTM) are known to encode context in a sequence of words, albeit in a limited sense but nevertheless allowing for some degree of semantics to be captured. Refer to [21] for in-depth workings of recurrent neural networks. We implement a character-level classifier using a bidirectional LSTM design, based on the architecture described in [22]. The input layer is as explained for the CNN model, on top of which we also add a randomly initialised 100-unit embedding layer. We then stack a forward and a backward LSTM layer with 64 hidden units each. The output of the two LSTMs is concatenated and fed forward to a softmax layer which, like the CNN model, has an output unit for each standardised response in the dataset. The network is trained with ADAM optimiser using categorical cross-entropy as the loss function and a batch size of 32.

3.6 Memory Network

Memory Networks [23,24] are a neural network architecture inspired partly from a standard computer memory, which has the ability to read and write data to an external memory storage. However, due to the smooth transformation functions they are also completely differentiable. Thus the reading and writing operations are also learned during training. A Memory Network has the ability to perform computations over a longer span of text when compared to standard Recurrent Neural Networks like LSTMs or GRUs. Hence Memory Networks are better suited for tasks which require processing of large text bodies such as paragraphs or passages of text. In this paper we apply it to email and response bodies.

We implement a modified version of the model presented in [24] and adapt it to perform email classification. The input structure follows the usual logic of converting the entire set of passage inputs (i.e. responses) $P = p_1, p_2...p_n$ to memory vectors $M = m_1, m_2...m_i$. The conversion is done by projecting P to a $A_{d \times V}$ space where d is the dimensionality of the memory vectors and V is the size of the vocabulary. Similarly, the query (i.e. email) is also projected to a $B_{d \times V}$ space to obtain internal query vectors U. The query weights are obtained using a dot product with all the memory vectors and normalised using a softmax.

$$R = Softmax(U^T M)$$

where R signifies the probability of match of the query based on the stored memories. The final memory representation, which forms a part of the input to the controller module, is given as a weighted sum

$$S = \sum r_i q_i \qquad r_i \in R$$

where q_i is another projection of the query to a $C_{Q \times V}$ space and Q is the size of the query vocabulary space. The input to the controller L is then a concatenation of S and U, while the controller is trained as a classifier with desired responses converted into one-hot labels.

$$O = Softmax(L(S|U))$$

where L is a bi-directional LSTM network, O is the output and $|$ indicates a concatenation operation. We use embeddings of size 128 and the dimensionality of the hidden units of the LSTM is also 128, with recurrent and feedforward dropouts at 0.5. The network is trained with ADAM optimiser using categorical cross-entropy as the loss function and a batch size of 100.

4 Results

Table 2 shows the performance of the different models on the email classification task. P@1 and P@5 indicate the probability of the desired correct response being at rank 1 and among the top 5 ranks respectively. A significant difference at the design level that separates the models into two categories (marked by * in Tables 2 and 3) is the ability of the models to be *universal*. By this, we mean that the *tf-idf*, DSSM and InferSent models, once trained on a sufficiently large corpus, can be applied to any similar data without requiring further training on that data, thus making them *universal* in a sense. Whereas the CNN, Bi-LSTM and Memory Network, which are specifically trained on the email classification task, are limited only to this particular task and require training data from the same corpus. To this purpose, we divide the data into a 80%-20% train-test split. To keep the numbers comparable, we test all the models on the test data of the latter three (*) models only.

Table 2. Performance on subset 1

Model	P@1	P@5
Tf-Idf	0.16	0.36
DSSM	0.06	0.11
InferSent	0.07	0.16
*CNN	0.40	0.72
*Bi-LSTM	0.34	0.64
*Memory network	**0.44**	**0.76**

Table 3. Performance on subset 2

Model	P@1	P@5
Tf-Idf	0.27	0.54
DSSM	0.04	0.08
InferSent	0.05	0.13
*CNN	0.55	0.92
*Bi-LSTM	0.42	0.70
*Memory network	**0.61**	**0.94**

Results in Table 2 are obtained with the frequency filter settings of subset 1 (refer: *Dataset*), selecting the top 57 responses with an average of 320 training samples per response. We can see that the Memory Network outperforms all other models on both the measures of P@1 and P@5. The CNN comes in a close second while the remaining models fall considerably behind the top results. The DSSM and InferSent perform at near chance level, which indicates a failure in learning meaningful representations for large texts. These models are ideally suited for short sentences and a possible reason for this is discussed in the following section. To confirm the hypothesis that the two models are under-performing due to the size of the text, we tested them on sentence level inputs. In order to do so, we split the input emails as well as the responses into the constituent sentences. Subsequently, we vectorise the sentences and compare the vectors, as before, using cosine similarity. The sentence pair which has the highest similarity decides the top ranked response for a particular email query. Results of sentence level processing are shown in Table 5. Performance of both the DSSM and InferSent are significantly increased and are better than *tf-idf*, however they are not comparable to the top performing models. The Bi-LSTM, which is supposed to take into account word ordering and some level of local context, performs twice as good as the *tf-idf* baseline, which purely performs a word match without context information. However, the CNN, which completely depends on the learned filters, performs much better than the Bi-LSTM.

Results in Table 2 have less training examples per response. To test whether this factor leads to the varied level of performance between models, we increase the number of training samples per response and investigate whether differences in model performance can be narrowed down. We perform the same steps of training and testing, but on a reduced dataset. The reduction is in the number of unique responses. However, it leads to a higher number of total samples and the number of samples per response, hence having more training data. Refer to subset 2 in Sect. 3. With these frequency filter settings (chosen heuristically) we select the top 11 occurring responses in the data with an average of 3,200 training samples per response.

Results shown in Table 3, although higher in accuracy than in Table 2, again follow the same pattern, with the Memory Network clearly outperforming all other models.

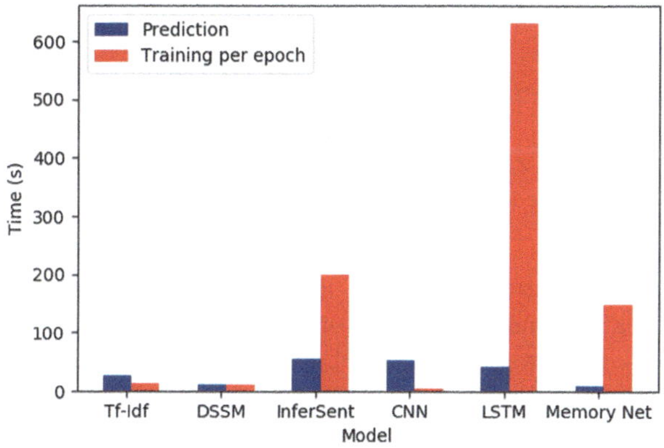

Fig. 2. Comparison of training (per epoch) and prediction times

Figure 2 shows the competitive advantages between the two leading networks in Tables 2 and 3: the CNN and the Memory Network. Although the Memory Network performs better, the CNN is very fast to train. The Memory Network, however, is faster at prediction time compared to the CNN.

Table 4 details additional information about the models relating to their size in terms of trainable parameters and storage on disk along with the number of epochs required to reach peak performance. Figures denoted with * are estimates.

For the sake of completeness, we test the performance of the top 2 models on all the unique 121 responses in the data and allowing it to be completely imbalanced. The results are as following:

- CNN: P@1 = 0.34; P@5 = 0.64
- Memory Network: P@1 = 0.39; P@5 = 0.69

However, since the data is severely imbalanced the above numbers do not reflect the true accuracy.

Table 4. Additional information

Model	Parameters	Epochs to peak performance	Size on disk
Tf-Idf	-	-	17 Kb
DSSM	$\sim 11 \times 10^6 *$	50	78 Mb
Infersent	$\sim 32 \times 10^6 *$	23	147 Mb
CNN	198,261	20	2.3 Mb
Bi-LSTM	259,811	20	3.1 Mb
Memory network	9,409,575	37	72 Mb

Table 5. Performance of DSSM and inferSent on sentence level processing (Subset 1)

Model	P@1	P@5
DSSM	0.20	0.57
InferSent	0.15	0.54

5 Conclusion

We compared the performance of six popular models in their basic forms on an email classification task. The Memory Network was shown to perform the best among the models we compared. However, despite being the best performing model, the top P@1 value is only 0.44 for the wider data and 0.61 for the reduced data. This indicates the complexity of the underlying data. The data comprises emails written by real human customers and thus varies in length, style, vocabulary and possibly other semantic factors. This makes the task of pattern recognition increasingly hard when dealing with real-world data. Our study evaluates how popular neural models are expected to function when exposed to a real life situation. The models considered in this study have shown excellent performance otherwise when dealing with either toy or well-curated datasets. The known drawback of recurrent neural network models (requiring considerably longer training times) is evident when compared for training times. One must make a cost-to-benefit analysis in deciding between, for instance, a CNN architecture or a Memory Network, given that the Memory network takes longer training times but has shorter prediction times.

The Memory Network has an advantage over the CNN and Bi-LSTM models because it also makes use of the text in the response bodies during decision making. The response text body is equivalent to a *story* in a traditional passage comprehension task for a Memory Network [24]. This information is *seen* by the controller unit while mapping the input vectors to the classes. The CNN and Bi-LSTM networks can only do label prediction based on an incoming email. This may be one of the reasons as to why the Memory Network performs the best. The fact that the CNN performs better than the Bi-LSTM is indicative of the fact that in large text classification, a high resolution of word order information might not be as useful as global context information.

Our study shows that the DSSM and Infersent completely fail to perform on this task. A very likely reason for this may be that the models work best with short sentences. Shorter sentences carry less information and hence are easier to encode as vectors compared to paragraphs. These models might not have the ability to generate good quality vectors for large passages of text, such as email bodies. Therefore, the resulting vectors may become very noisy with the diverse information they attempt to encode and lose all the signal in the noise. Repeating the exercise for these models, where a comparison is done sentence by sentence, for every query and document, did indeed yield a much better performance. The improvement in results is however involved with a large computation time in the combinatorial comparisons to make. Another possibility would be to combine the sentence vectors of a document by some variant of averaging the vectors. However, this approach is also prone to a significant loss in information encoded in the "average" vector when the number of vectors increase.

We have implemented all the models in more or less their vanilla forms. Certainly, specific modifications to the design of each of the networks is an open exploration on how it will affect their performance on such real-world datasets. For e.g. the next line of investigation will be done using multi-channel CNN's or a combination of CNN+LSTM to study how well they handle the data. As it stands, the Memory Network design is the most promising approach for its requirement of less training data, flexibility in design implementation, faster prediction time and higher accuracy.

Acknowledgment. The authors would like to thank James Hammerton for his valuable discussion on the CNN. The authors also thank Alexis Conneau for his insight into the InferSent model.

References

1. Cormack, G.V., et al.: Email spam filtering: a systematic review. Found. Trends® Inf. Retrieval **1**(4), 335–455 (2008)
2. Youn, S., McLeod, D.: A comparative study for email classification. In: Advances and Innovations in Systems, Computing Sciences and Software Engineering, pp. 387–391 (2007)
3. Liu, S., Lee, I.: A hybrid sentiment analysis framework for large email data. In: 2015 10th International Conference on Intelligent Systems and Knowledge Engineering (ISKE), pp. 324–330. IEEE (2015)
4. Lê, T., Lê, Q.: The nature of learners' email communication. In: International Conference on Computers in Education, Proceedings, pp. 468–471. IEEE (2002)
5. Cohen, W.W., Carvalho, V.R., Mitchell, T.M.: Learning to classify email into "speech acts". In: EMNLP, vol. 4, pp. 309–316 (2004)
6. Kiritchenko, S., Matwin, S.: Email classification with co-training. In: Proceedings of the 2011 Conference of the Center for Advanced Studies on Collaborative Research, pp. 301–312. IBM Corp. (2011)
7. Anand, A., Pugalenthi, G., Fogel, G.B., Suganthan, P.: An approach for classification of highly imbalanced data using weighting and undersampling. Amino acids **39**(5), 1385–1391 (2010)

8. Sharma, A.K., Sahni, S.: A comparative study of classification algorithms for spam email data analysis. Int. J. Comput. Sci. Eng. **3**(5), 1890–1895 (2011)
9. Klimt, B., Yang, Y.: The enron corpus: a new dataset for email classification research. In: Machine learning: ECML 2004, pp. 217–226 (2004)
10. Gomez, J.C., Moens, M.-F.: PCA document reconstruction for email classification. Comput. Stat. Data Anal. **56**(3), 741–751 (2012)
11. Alsmadi, I., Alhami, I.: Clustering and classification of email contents. J. King Saud Univ.-Comput. Inf. Sci. **27**(1), 46–57 (2015)
12. Salton, G., McGill, M.J.: Introduction to modern information retrieval (1986)
13. Huang, P.-S., Urbana, N.M.A., He, X., Gao, J., Deng, L., Acero, A., Heck, L.: Learning deep structured semantic models for web search using clickthrough data. In: The 22nd ACM International Conference on Conference on Information and Knowledge Management, pp. 2333–2338 (2013). http://dl.acm.org/citation.cfm?id=2505665
14. Deerwester, S., Dumais, S.T., Furnas, G.W., Landauer, T.K., Harshman, R.: Indexing by latent semantic analysis. J. Am. Soc. Inf. Sci. **41**(6), 391 (1990)
15. Saeidi, M., Kulkarni, R., Togia, T., Sama, M.: The effect of negative sampling strategy on capturing semantic similarity in document embeddings. In: Proceedings of the 2nd Workshop on Semantic Deep Learning (SemDeep-2), pp. 1–8 (2017)
16. Conneau, A., Kiela, D., Schwenk, H., Barrault, L., Bordes, A.: Supervised learning of universal sentence representations from natural language inference data. arXiv preprint arXiv:1705.02364 (2017)
17. FacebookResearch. Infersent pre-trained model. https://github.com/facebookresearch/InferSent
18. Kim, Y.: Convolutional neural networks for sentence classification. arXiv preprint arXiv:1408.5882 (2014)
19. Zhang, X., Zhao, J., LeCun, Y.: Character-level convolutional networks for text classification. In: Advances in Neural Information Processing Systems, pp. 649–657 (2015)
20. Kingma, D., Ba, J.: Adam: A method for stochastic optimization. arXiv preprint arXiv:1412.6980 (2014)
21. Graves, A., et al.: Supervised Sequence Labelling with Recurrent Neural Networks, vol. 385. Springer (2012)
22. Graves, A., Schmidhuber, J.: Framewise phoneme classification with bidirectional LSTM and other neural network architectures. Neural Netw. **18**(5), 602–610 (2005)
23. Weston, J., Chopra, S., Bordes, A.: Memory networks. arXiv preprint arXiv:1410.3916 (2014)
24. Sukhbaatar, S., Weston, J., Fergus, R., et al.: End-to-end memory networks. In: Advances in Neural Information Processing Systems, pp. 2440–2448 (2015)

Fuzzy Based Sentiment Classification in the Arabic Language

Mariam Biltawi[1(✉)], Wael Etaiwi[1], Sara Tedmori[1], and Adnan Shaout[2]

[1] King Hussein Faculty of Computing Sciences, Princess Sumaya University for Technology, Amman, Jordan
{maryam,w.etaiwi,s.tedmori}@psut.edu.jo

[2] The ECE Department, The University of Michigan-Dearborn, Dearborn, USA
shaout@umich.edu

Abstract. Sentiment Analysis is the task of identifying individuals' positive and negative opinions, emotions and evaluations concerning a specific object. Fuzzy logic in the field of sentiment analysis can be employed to classify the polarity of sentences or documents. Although some efforts have been made by researchers who applied fuzzy logic for Sentiment Analysis on English texts, to the best of the authors' knowledge, no efforts have been made targeting Arabic texts. This paper proposes a lexicon based approach to extract sentiment polarity from Arabic text using a fuzzy logic approach. The proposed approach consists of two main phases. In the first phase, Arabic text is assigned weights, while in the second phase fuzzy logic is employed to assign the polarity to the inputted sentence. Experiments were conducted on Large Scale Arabic Book Reviews Dataset (LABR), and the results showed 94.87%, 84.04%, 80.59% and 89.13% for recall, precision, accuracy, and F1-measure, respectively.

Keywords: Arabic fuzzy logic · Arabic sentiment analysis
Arabic natural language processing · Linguistic variables

1 Introduction

The World Wide Web (WWW) has become an integral and indispensable part of our daily life. Every day, millions of businesses use the web to promote, announce, market and present brand names, products and services, in order to collect customer feedback and use it to improve on their offerings, quality of products, brand performance and/or service quality. Hence, in the past few years, businesses and consumers alike have recognized the need to develop tools that analyze texts to uncover the sentiments these words, phrases, and/or sentences convey.

Sentiment Analysis (SA) refers to the task of identifying, individuals' positive and negative opinions, emotions and evaluations concerning a specific object such as an event, a product, a topic, or an individual, from a given data set. While numerous efforts have been made to conduct sentiment analysis on different languages such as English, Chinese [1] and French [2], little work can be found on Arabic language texts. Arabic language is a Semitic language that contains three variants, the classical, the modern

K. Arai et al. (Eds.): IntelliSys 2018, AISC 868, pp. 579–591, 2019.
https://doi.org/10.1007/978-3-030-01054-6_42

standard Arabic (MSA) and the colloquial. MSA is the most used. It contains 28 letters with no upper lower cases, and is written from right to left. The morphologically rich nature of the Arabic language makes sentiment analysis from Arabic texts a more challenging task than sentiment analysis from other languages [3].

One main task in SA is Sentiment Classification (SC) which aims to classify the polarity of a given text. Sentiment Classification (SC) approaches can be categorized into three main approaches: Machine Learning (ML), lexicon-based, and hybrid approaches [4]. The ML approach, a supervised approach that allows the computer to learn by example (dataset) has been used in the classification of polarity. Different ML methods for data classification can be used such as decision trees, rule based methods, logistic regression, Naïve Bayes (NB), Support Vector Machine (SVM), k-Nearest Neighbor (k-NN), and so on. A lexicon-based approach is an unsupervised approach that depends on building a dictionary or lexicon to be used in sentiment classification [4].

Fuzzy logic is a computing approach based on the "degree of truth" rather than the complete "true or false" values (i.e. Boolean logic) [5]. Presented by LotfiZadah in 1964, Fuzzy logic is most commonly used in complex problems where the implementation and development is too costly, or when the overall understanding of the problem is limited, such as: speech recognition [6] and hand written recognition [7].

Designing a fuzzy logic system is a five step process. In the first step (1), the input and output variables of the system are declared. In the second step (2), the linguistic values of these variables are determined. These linguistic values are in essence the linguistic terms used to describe the linguistic variables such as: high, medium and low. The next step, step (3), comprises determining how the system inputs/outputs are to be transformed into fuzzy sets by defining the membership function (degree of belonging function) of each fuzzy linguistic variable. Membership function could be in any type of shape such as: triangular membership function, trapezoidal membership function and Gaussian membership function, as illustrated in Fig. 1. In step (4), the fuzzy-rules, in the form of: If (premise), then (conclusion) are designed. The output of all applied fuzzy rules should be aggregated into one fuzzy output. Finally, step (5) involves defuzzifying this fuzzy output by converting the output fuzzy value into a crisp value using a defuzzification technique.

Fig. 1. Membership functions.

Fuzzy logic in the field of SA can be employed to classify the polarity of sentences or documents. Although some efforts have been made by researchers who applied fuzzy logic for sentiment analysis from English texts, to the best of the authors' knowledge, no efforts have been made targeting Arabic texts. In this research, the authors use fuzzy logic to enhance automatic sentiment analysis of text written in the Arabic language. In this approach, the original text is preprocessed and prepared to be used by the fuzzy logic system. Each sentence is mapped into the fuzzy logic input linguistic variables after weighting each word using the ArSenL lexicon [8]. The fuzzy system then returns the fuzzy output and defuzzifies it (convert it to crisp value) using the predefined fuzzy rules in order to extract the final result.

The rest of this paper is structured as follows: Sect. 2 presents the related work, Sect. 3 presents the proposed algorithm in details. In Sect. 4, the experimental results are presented and discussed. Finally, the conclusion is mentioned in Sect. 5.

2 Related Work

Fuzzy logic is used when uncertainty exists, or when there is no precise meaning for a word; that's why it goes hand by hand with sentiment analysis. Any fuzzy logic system consists of input and output variables. Each input and output variable may have multiple linguistic values that can be represented using membership functions. The authors of [9] used fuzzy logic to weight samples in support vector machine (SVM) algorithm in order to reduce the noisy and unimportant samples that make a small contribution in the classification process and accuracy rate. The authors proposed a novel fuzzy sentiment membership determining method. The proposed sentiment model consists of three layers. In the first layer, the interrelations of the texts, topics and words are computed, and a sentiment score is assigned to the text. In the second layer, SVM is utilized to classify samples from the test data set. In the last layer, experiments on the remaining dataset are conducted. In the paper, the authors conclude with many improvements that could be done, such as, enhance the text sentiment scoring technique, reduce the algorithm complexity, and apply the approach on many different applications (e.g. sentiment reviews filtering).

The authors of [10] built a Fuzzy Semantic Classifier (FSC) whose purpose is to classify semantics of customer product reviews into five categories: very strong, strong, moderate, weak, very weak. The FSC starts with identifying the opinion words for each sentence, opinion words are: adverb, adjective, verb and noun, then stop words are removed and the words are stemmed. Next, the four opinion variables are sent to the fuzzy logic stage which consists of the four main steps: fuzzification, membership function designing, fuzzy rules designing, and defuzzification, in order to discover the polarity of sentences. The FSC is tested using eight product review datasets However, Dragoni et al. [11–13] proposed a fuzzy concept-based sentiment analysis system which uses a knowledge base consisting of both WordNet and SenticNet. The main purpose of this system is to learn the fuzzy membership functions. It has two main steps, first, the concept polarities with respect to a domain are modeled using fuzzy logic, and second, a two-level graph is created in which the upper level represents the semantic relationships between concepts and the lower level represents links to all membership functions and domains.

Fuzzy functions can be used to clarify the sentiment polarity of the comments and reviews by using fuzzy logic principles to analyze hedges, modifiers, concentrators and dilators that appear in the text and reflect the overall polarity. Rahmath and Ahmad [14] proposed a multi-step opinion mining system that focuses mainly on the linguistic hedges. The system consists of three main steps; pre-processing, feature selection and classification step. The feature selection step uses fuzzy functions to emulate the effect of linguistic hedges. Other researches have also used fuzzy functions to emulate linguistic hedges [15–18].

Fuzzy clustering have also been used in sentiment analysis. Sheeba and Vivekanandan [19] proposed a fuzzy logic based framework that analyzes a meeting transcript and detects both implicit and explicit expressions in it, by classifying positive, negative and neural words and identifying the topics of the meeting transcript. In their work, the authors used the fuzzy c-mean algorithm to cluster similar words. Fuzzy c-mean is a clustering method used to allow one piece of data to belong to two or more clusters, another research used the c-mean is [20].

Qamar and Ahmad [21] proposed a fuzzy logic system that classifies text emotions into; happiness, sadness, surprise, fear, disgust, and anger. The authors applied fuzzy membership function and fuzzy rules to find the mood from the text. The complete steps of fuzzy logic (Fuzzification, Membership Function Design, Linguistic Rule Design, Aggregation and Implication, Accumulation, and finally Defuzzification) have been applied by Priyanka and Gupta [22] who proposed a fuzzy logic based sentiment analysis model for fine grained classification of customer reviews.

To date, the use of fuzzy logic for sentiment analysis purposes from Arabic texts is still not explored Most of the researches in the Arabic Sentiment analysis share main ideas in the preprocessing step which is considered a main step before feature extraction and sentiment classification, such as tokenization, stemming, and stop words removal. Al-Radaideh and Twaiq [23] implemented the rough set theory to classify Arabic tweets. After the preprocessing step, the TF-IDF is computed for each term in each document, then the dataset is split into training and testing sets, then different approaches used to compute reducts and rules are generated. Experimental results showed that the highest accuracy is achieved using genetic reducer.

Another research used ML techniques to identify polarities to investigate both MSA and Jordanian dialectical Arabic, for this purpose a framework was proposed by Duwairi [24]. Shoukry and Rafea [25] proposed a hybrid approach that combines Support Vector Machine (SVM) and the semantic orientation approach in order to improve performance measures of Egyptian dialect.

3 Proposed Algorithm

In this section, the proposed algorithm is detailed. The algorithm consists of two main phases, each consisting of several steps. Figure 2 demonstrates the detailed steps of each phase. The inputs to this algorithm are sentences representing people's reviews from the LABR dataset.

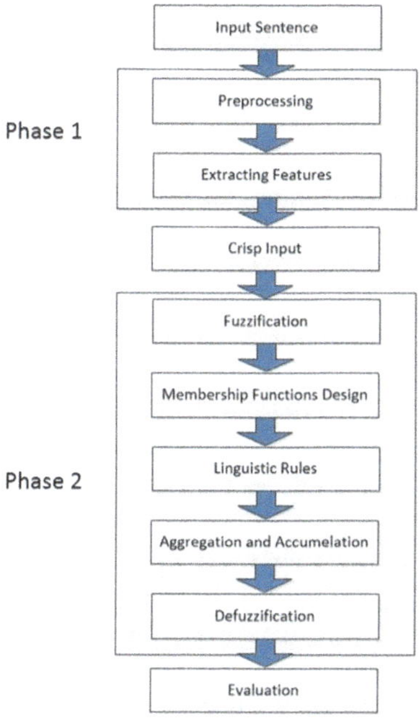

Fig. 2. Proposed algorithm.

3.1 Phase 1

As mentioned earlier, the input to this stage is a sentence representing an individual's opinion. The output from this phase is a tuple comprising three values, representing the weights of the verbs, nouns and articles of each inputted sentence. Phase one consists of two main steps: preprocessing and feature extraction.

(1) Preprocessing

The proposed preprocessing step combines noise removal, normalization and tokenization. This preprocessing step is conducted on the inputted data which is in the form of user comments, reviews or tweets. Users' reviews are generally unstructured and subject to noise. Hence, the preprocessing step is vital in order to process the input and prepare it for the next phase. The noise removal consists of removing English text, numbers, and punctuation symbols. Removing punctuation involves removing characters such as commas, exclamation marks, and hashtags. For example, considering the word: ممل!!! (Meaning: boring) transliterated to mml will be replaced by ممل after preprocessing. Normalization is performed on the text to normalize certain characters, such as Alif, Ta' al-Marboutah, and Tatweel. All forms of Alif (آ، إ، أ) are normalized into the bare Alif (ا), Ta' al-Marboutah (ة) is normalizaed into Ha' (ه), and Tatweel (-) is removed.

A sentence is a group of words (tokens) expressing a complete thought. Each word in the sentence has its own meaning that in itself contributes to the overall meaning of the sentence. Hence, it is essential to split each sentence into its constituent words in order to extract its polarity and identify its role in the overall sentence polarity. The process of splitting the sentence into separated words is called tokenization, in which each sentence is divided into one or more words delimited by space.

(2) Feature Extraction

In the proposed approach, the feature extraction component in itself is a three step process, namely, (a) Part-Of-Speech (POS) tagging, (b) POS tags' mapping, and (c) token weight extraction. In Arabic, a meaningful word can be divided into three categories; namely, noun, verb, and article [26]. Hence, it is important to classify the individual words of each sentence into their correct types in order to study the roles that each type of word plays in the overall polarity. In this research, the authors employ the Stanford POS tagger to tag each word by its POS label. The Stanford POS tagger consists of more than 30 different tags. However, in this approach the 30 tags were mapped into the three main simple POS tags; verb, noun and article.

Furthermore, using the ArSenL lexicon [8], weights are assigned to these tags in order to prepare them for the next phase. To help describe the polarity of words, each Arabic word in the ArSenL lexicon is associated with three scores: positive, negative and neutral. Each score has a value that ranges between 0 and 1. For each word, multiple results with varying scores can be found.

The ArSenL lexicon does not contain all Arabic words. Moreover, not all forms of every Arabic word exist in the ArSenL lexicon. Therefore, these words were given a zero score. In addition, the ArSenL lexicon provides different weight for each word, and in this approach the maximum weight is assigned for the word in the dataset. The process of extracting the word's weight from the ArsenL lexicon is as follows:

- Two forms of the words from the dataset are prepared: the original, which is preprocessed, and its stem. Stemming is performed using Information Science Research Institute (ISRI) Stemmer for stemming Arabic words [27].
- The words in the lexicon were preprocessed and normalized.
- Two lookups were applied in case one fails. The first one is to compare the original word and its POS tag with the words from the lexicon along with their POS tags, if there is a match then weight is assigned. If no match is found, then the second lookup is applied to compare the stem of the original word and its POS tag with the words in the lexicon along with their POS tags.
- In case no match is found after the two lookups, then a zero weight is assigned for the word.
- In case a match is found in one of the lookups, then the maximum weight is assigned.
- Finally, the total weight is computed for each POS tag comprising the sentence, each of these POS tags will represent three variables; verb, noun, and article, and will be an input to the second phase.

3.2 Phase 2

The second phase is controlled by a Fuzzy Logic System (FLS) depicted in Fig. 3. It consists of five main steps; fuzzification, membership function design, fuzzy rule-base design, aggregation, and defuzzification. The input of this phase is the three outputs from the previous phase, which represent a crisp weight for each variable: verb, noun, and article. Phase 2 outputs a crisp value that represents the polarity for the inputted sentence.

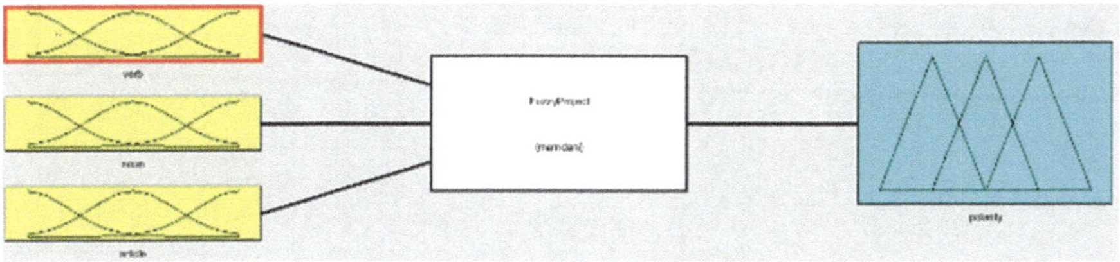

Fig. 3. Fuzzy Logic System (FLS).

(1) Fuzzification

As mentioned previously, the first step in any fuzzy logic system is to determine the fuzzy linguistic variables (inputs and outputs). Since Arabic sentences consist of three main word types, verb, noun and article, the authors of this paper selected these three variables as the input fuzzy linguistic variables, and the resulting polarity is selected as the output linguistic variable. The universe of discourse ranges from −10 to 10, which represents the weights for the words.

The input linguistic variables verb and noun of the fuzzy logic system have five linguistic values; high positive, low positive, neutral, low negative, high negative; whilst the input variable article has two linguistic values; positive and negative. Moreover, the output variable polarity consists also of five linguistic variables; strong positive, strong negative, neutral, weak negative and strong negative. Table 1 shows the linguistic values of each variable.

Table 1. The linguistic values of system linguistic variables

Linguistic variable	Type	Linguistic values (Fuzzy set)
Verb	Input	High Positive, Low Positive, Neutral, Low Negative, High Negative
Noun	Input	High Positive, Low Positive, Neutral, Low Negative, High Negative
Article	Input	Positive, Neutral, Negative
Polarity	Output	Strong Positive, Weak Positive, Neutral, Weak Negative, Strong Negative

(2) Membership Function Design

Membership function is used to convert the crisp value extracted in phase 1 to a fuzzy value. In this paper, Triangular membership functions and piece-wise membership functions are used for modeling the linguistic values for each linguistic variable. Figure 4 depicts the membership functions for each fuzzy variable in our FLS; each membership function is defined with multiple points. For example, the Verb linguistic variable has the linguistic value "Low-Positive" which is represented with a triangular membership function using the following equation:

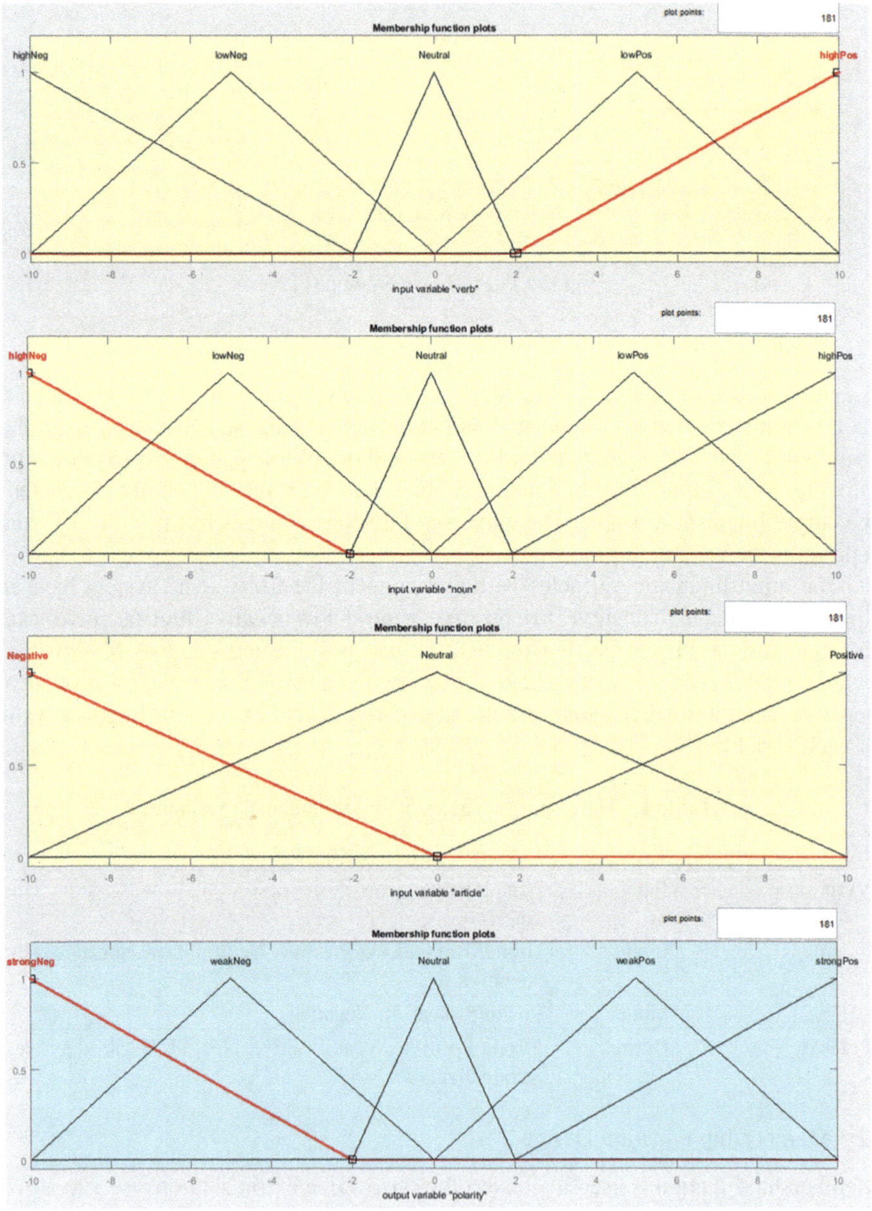

Fig. 4. Membership functions for fuzzy linguistic variables.

$$\mu(x) = \begin{Bmatrix} 0 & x \leq -2 \\ (x+2)/2 & -2 \leq x \leq 0 \\ (2-x)/2 & 0 \leq x \leq 2 \\ 0 & x > 2 \end{Bmatrix} \tag{1}$$

(3) Linguistic Rules Design

In this step, fuzzy rules were built in order to link the input and output variables together. Each fuzzy rule is in the form of: If x1 is A and x2 is B and x3 is C, then x4 is D, where x1, x2, x3 and x4 are the fuzzy linguistic variables, while A, B, C and D are their corresponding linguistic values. The fuzzy rules may cover all linguistic values of the linguistic variables, so the total number of rules in this research is equal to the total number of fuzzy linguistic values of each fuzzy term multiplied by each other, as follows:

$$\begin{aligned} &\textit{Total number of rules} = \textit{Number of "Verb" linguistic values} \\ &* \textit{Number of "Noun" linguistic values} \\ &\quad * \textit{Number of "Article" linguistic values} \\ &\quad = 5 * 5 * 3 \\ &\quad = 75 \textit{ rules.} \end{aligned} \tag{2}$$

Table 2 shows a sample of the rules used by our approach.

Table 2. Fuzzy rules

1.	If (verb is highPos) and (noun is highPos) and (article is Positive) then (polarity is strongPos)
2.	If (verb is highPos) and (noun is lowPos) and (article is Positive) then (polarity is strongPos)
3.	If (verb is highPos) and (noun is Neutral) and (article is Positive) then (polarity is strongPos)
4.	If (verb is highPos) and (noun is highNeg) and (article is Positive) then (polarity is weakPos)
5.	If (verb is highPos) and (noun is lowNeg) and (article is Positive) then (polarity is weakPos)
	.
	.
	.

(4) Aggregation and Accumulation

After fuzzifying the inputs using the membership functions, each input variable has its own membership value, and in case of a fuzzy rule contains more than one input variable, it is necessary to aggregate the fuzzy value of each input variable in order to apply the rule, this process is called aggregation. The aggregation merges the fuzzy input values of each applied rule into one fuzzy set. In fuzzy logic, the "AND" operator may be used in fuzzy rules to selects the minimum value for each rule, and the "OR" operator may be used to select the maximum value for each rule. The rules used in our approach mainly use AND in their antecedent part.

After examining all fuzzy rules and aggregating the applied rules, an accumulation process is used to combine the outputs derived from all applied fuzzy rules into one fuzzy set. Our approach uses the Mamdani method, which uses the maximum accumulation method to combine the outputs for each rule.

(5) Defuzzification

The last step is to map the output fuzzy set into a crisp values in order to determine the final polarity of the sentence. Centroid defuzzification method is used in this research for this purpose. The final polarity value is given by the following formula:

$$z^* = \frac{\int \mu_c(z).\, zdz}{\int \mu_c(z).\, dz} \tag{3}$$

Where, μ_C is the membership value of z. Then the final class can be obtained using the polarity membership function illustrated in Fig. 5.

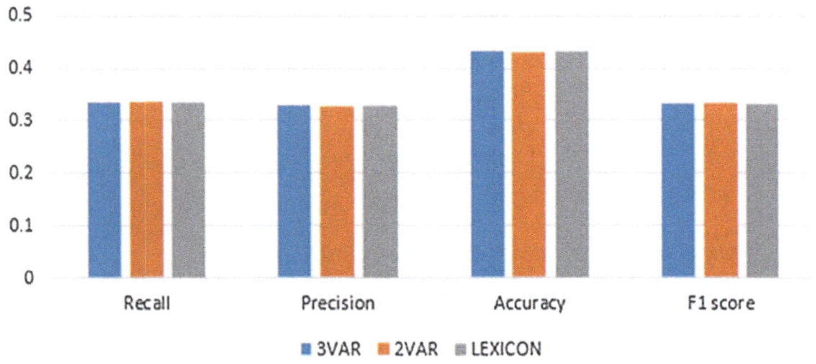

Fig. 5. Results for Experiment 1 (Positive, Negative, and Neutral).

4 Experimental Results

As discussed earlier, the proposed algorithm is a two-phase algorithm. The first phase is implemented using python and second phase is implemented using Matlab. The input to the first phase is a sentence taken from the dataset, whilst the output of the first phase is the features for each sentence consisting of the three variables, verb, noun, and article. The output from the first phase is then used as an input to the second phase. The output from the second phase is the polarity of the sentence in the form of crisp value.

The proposed algorithm was tested on LABR [28], which is a SA dataset of over 63000 book reviews in Arabic. These reviews were collected from Goodreads website. It consists of 42831 positive reviews, 8224 negative reviews, and 12201 neutral reviews.

Three algorithms were tested: (1) the proposed approach using the three variables; verb, noun, and article. (2) The proposed approach using only two variables; verb and noun. (3) lexicon-based approach with no fuzzy technique. In the lexicon-based approach the same steps of phase 1 is applied on the reviews and the weight were summed at the end and checked for its polarity.

Experiment was conducted twice on each algorithm; first, by considering three sentiment polarities; positive, negative, and neutral; second, by considering only two sentiment polarities; positive and negative. In the second experiment, the neutral polarity is neglected, also in the proposed fuzzy approach, through removing the neutral fuzzy value from each variable and as a result the number of rules used were decreased to 25 fuzzy rules. For performance evaluation purposes, the precision, recall, accuracy and F1-score measures were computed. Table 3 illustrates these measures for the first experiment where three polarity classes were used, while Table 4 shows the results for the second experiment.

Table 3. Results for Experiment 1 (Positive, Negative, and Neutral)

Measures	3VAR	2VAR	LEXICON
Recall	0.3368	0.3366	0.3368
Precision	0.3281	0.3274	0.3276
Accuracy	0.4328	0.4311	0.4325
F1 score	0.3324	0.3319	0.3321

Table 4. Results for Experiment 2 (Positive and Negative)

Measures	3VAR	2VAR	LEXICON
Recall	0.9487	0.8469	0.6846
Precision	0.8404	0.7902	0.8469
Accuracy	0.8059	0.7118	0.6316
F1 score	0.8913	0.8175	0.7571

Results depicted in the two tables above are illustrated graphically in Figs. 5 and 6. These results show that the proposed approach performs well when the neutral polarity is neglected, and using the three linguistic variables, with an accuracy reaches 80.59%. However, when neutral polarity is considered accuracy decreases 40.31%, this is because not all the words of the Arabic language is contained in the lexicon, since the proposed approach relies on the weights conducted from the lexicon. On the other hand, it is obvious that the proposed approach outperforms the lexicon-based approach when using three linguistic variables in the first experiment. While, in the second experiment the proposed approach outperforms the lexicon-based approach when using both three and

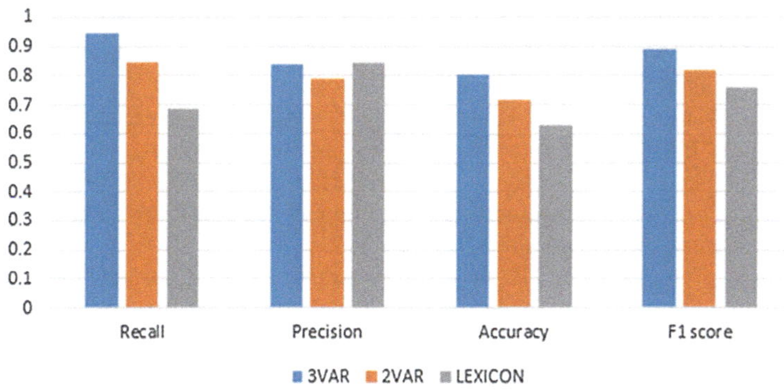

Fig. 6. Results for Experiment 2 (Positive and Negative).

two linguistic variable except in one measure which the precision. The precision of lexicon-based approach outperforms the proposed approach when two linguistic variables were used.

5 Conclusion

This paper proposed an unsupervised lexicon based approach, using fuzzy logic to enhance automatic sentiment analysis of text written in the Arabic language. The proposed approach consists of two main phases, the first one is the prior-FSL phase and the second one is the FSL phase. In the first phase, the sentences are cleaned and assigned weight, while in the second phase the polarity for each given sentence is obtained. Two experiments were conducted on LABR dataset, the highest accuracy reached is 80.59%. As a future work, the authors are testing the algorithm on other datasets in order to perform a comprehensive comparison.

References

1. Lee, H.Y., Renganathan, H.: Chinese sentiment analysis using maximum entropy (2011)
2. Ghorbel, H., Jacot, D.: Sentiment analysis of French movie reviews. In: Soro, A., Vargiu, E. (eds.) Advances in Distributed Agent-Based Retrieval Tools, vol. Pallotta, pp. 97–108. Springer, Heidelberg (2011)
3. Farghaly, A., Shaalan, K.: Arabic natural language processing: challenges and solutions. ACM Trans. Asian Lang. Inf. Process. (TALIP) **8**(4), 141–1422 (2009)
4. Biltawi, M., Etaiwi, W., Tedmori, S., Hudaib, A., Awajan, A.: Sentiment classification techniques For Arabic language: a survey. In: 7th International Conference on Information and Communication Systems (ICICS). IEEE (2016)
5. Yen, J., Langari, R.: Fuzzy Logic: Intelligence, Control, and Information. Prentice-Hall Inc., Upper Saddle River (1999)
6. Sun, J., Karray, F., Basir, O., Kamel, M.: Natural language understanding through fuzzy logic inference and its application to speech recognition. In: Proceedings of the 2002 IEEE International Conference on Fuzzy Systems, FUZZ-IEEE 2002, vol. 2, pp. 1120–1125 (2002)
7. Fitzgerald, J.A., Geiselbrechtinger, F., Kechadi, T.: Application of fuzzy logic to online recognition of handwritten symbols. In: Ninth International Workshop on Frontiers in Handwriting Recognition, IWFHR-9 2004, pp. 395–400 (2004)
8. Badaro, G., Baly, R., Hajj, H., Habash, N., El-Hajj, W.: A large scale Arabic sentiment lexicon for Arabic opinion mining, pp. 165–173 (2014)
9. Zhao, C., Wang, S., Li, D.: Fuzzy sentiment membership determining for sentiment classification. In: 2014 IEEE International Conference on Data Mining Workshop, pp. 1191–1198 (2014)
10. Nadali, S.: Fuzzy semantic classifier for determining strength levels of customer product reviews, masters, Universiti Putra Malaysia (2012)
11. Dragoni, M., Tettamanzi, A.G.B., Da Costa Pereira, C.: Using fuzzy logic for multi-domain sentiment analysis. In: Proceedings of the 2014 International Conference on Posters & Demonstrations Track, Aachen, Germany, vol. 1272, pp. 305–308 (2014)

12. Dragoni, M., Tettamanzi, A.G.B., da Costa Pereira, C.: A fuzzy system for concept-level sentiment analysis. In: Presutti, V., Stankovic, M., Cambria, E., Cantador, I., Iorio, A.D., Noia, T.D., Lange, C., Recupero, D.R., Tordai, A. (eds.) Semantic Web Evaluation Challenge, pp. 21–27. Springer International Publishing (2014)
13. Dragoni, M., Tettamanzi, A.G.B., da Costa Pereira, C.: Propagating and aggregating fuzzy polarities for concept-level sentiment analysis. Cogn. Comput. **7**(2), 186–197 (2014)
14. RahmathP, H., Ahmad, T.: Fuzzy based sentiment analysis of online product reviews using machine learning techniques. Int. J. Comput. Appl. **99**(17), 9–16 (2014)
15. Indhuja, K., Reghu, R.P.C.: Fuzzy logic based sentiment analysis of product review documents. In: 2014 First International Conference on Computational Systems and Communications (ICCSC), pp. 18–22 (2014)
16. Dalal, M.K., Zaveri, M.A., Dalal, M.K., Zaveri, M.A.: Opinion mining from online user reviews using fuzzy linguistic hedges, opinion mining from online user reviews using fuzzy linguistic hedges. Appl. Comput. Intell. Soft Comput. **2014**, e735942 (2014)
17. Tumsare, P., Sambare, A.S., Jain, S.R.: Opinion mining in natural language processing using SentiWordNet and fuzzy. Int. J. Emerg. Trends Technol. Comput. Sci. IJETTCS **3**(3), 154–158 (2014)
18. Pimpalkar, A., Wandhe, T., Rao, M.S., Kene, M.: Review of online product using rule based and fuzzy logic with smiley's. Int. J. Comput. Technol. IJCAT **1**(1), 39–44 (2014)
19. Sheeba, J.I., Vivekanandan, K.: A fuzzy logic based on sentiment classification. Int. J. Data Min. Knowl. Manage. Process (IJDKP) **4**(4), 27 (2014)
20. Nderu, L.: Importance of the neutral category in fuzzy clustering of sentiments. Int. J. Fuzzy Log. Syst. **4**(2), 1–6 (2014)
21. Qamar, S., Ahmad, P.: Emotion detection from text using fuzzy logic. Int. J. Comput. Appl. **121**(3), 29–32 (2015)
22. Priyanka, C., Gupta, D.B.: Fine grained sentiment classification of customer reviews using computational intelligent technique. Int. J. Eng. Technol. **7**, 1453–1468 (2015)
23. Al-Radaideh, Q.A., Twaiq, L.M.: Rough set theory for Arabic sentiment classification, pp. 559–564 (2014)
24. Duwairi, R.M.: Sentiment analysis for dialectical Arabic. In: 2015 6th International Conference on Information and Communication Systems (ICICS), pp. 166–170 (2015)
25. Shoukry, A., Rafea, A.: A hybrid approach for sentiment classification of Egyptian Dialect Tweets. In: 2015 First International Conference on Arabic Computational Linguistics (ACLing), pp. 78–85 (2015)
26. Hadni, M., Alaoui Ouatik, S., Lachkar, A., Meknassi, M.: hybrid part-of-speech tagger for non-vocalized Arabic text. Int. J. Nat. Lang. Comput. **2**(6), 1–15 (2013)
27. Taghva, K., Elkhoury, R., Coombs, J.: Arabic stemming without a root dictionary. Inf. Technol. Coding Comput. ITCC **1**, 152–157 (2005)
28. Aly, M., Atiya, A.: LABR: a large-scale arabic book reviews dataset. In: Association of Computational Linguistics (ACL), Bulgaria, August 2013

Arabic Tag Sets: Review

Marwah Alian[1,2(✉)] and Arafat Awajan[2]

[1] Hashemite University, Zarqa, Jordan
Marwah2001@yahoo.com
[2] Princess Sumaya University for Technology, Amman, Jordan
Awajan@psut.edu.jo

Abstract. Labeling a word with a suitable tag based on its context and its grammatical category is a major step in many applications of natural language processing. Constantly, there is an effort for inventing a set of these tags for Arabic language. In this research, a review for the existing Arabic tag sets is presented. A description for their features and limitations is also introduced.

Keywords: Tag · Tag set · Arabic tag set

1 Introduction

Part of Speech Tagging is the process of assigning proper tag for each word in a text representing its grammatical and morphological syntactic feature for a word [1]. Further, a tag is a code that holds simple or complex information that represent a word features and it labels the word in a text [2]. The development of a tag set that consist of representative tags at early stages is important for diacritical based tagging system. The need for such a tag set comes from the fact that Arabic language does not have a standard or complete tag set [3].

The approaches used for Part Of Speech (POS) tagging are classified into three main approaches; the first approach is the Rule-based Approach which sometimes called linguistic approach or Knowledge-Based Approach [4, 5], this approach use a set of linguistic rules during the process of tagging. The second approach is the Statistical Approach that is also called Probabilistic Approach or Stochastic Approach [6, 7]. This approach depends on building a statistical language model by gathering statistics from existing tagged corpora. The third approach is the hybrid approach in which rule-based and statistical approaches are involved [8, 10]. In the hybrid approach, both rule-based and statistical approaches are combined. On the other hand, some systems use other approaches, like machine learning algorithms, neural networks and decision trees [5, 9]. The existing Arabic tag sets vary in size from 6 tags to 2,000 detailed tags.

Some of these tag sets follow the same standards adopted in the tag set design for English, but these tag sets may be inappropriate for Arabic. Also, there are some morphological features that are common between Arabic tag sets like number, gender, case, person, definiteness and mood. However, the attributes are not uniformed among the morphological features [15]. In this research, a review for the existing Arabic tag sets is presented with their features and limitations.

K. Arai et al. (Eds.): IntelliSys 2018, AISC 868, pp. 592–606, 2019.
https://doi.org/10.1007/978-3-030-01054-6_43

2 Arabic Tag Sets

2.1 El-Kareh and Al-Ansary Tag Set

It consists of 72 tags and it was used in their semi-automatic tagger. El-Kareh and Al-Ansary tagger [8] was constructed from traditional Arabic grammar and it gives an accuracy of 90%. They classify words into three major classes: Verb, Noun, and Particle. However, Verbs were divided into three subclasses: Nouns into 46 subclasses and particles into 23 subclasses [1].

2.2 Khoja Tag Set

Khoja [10] depends on ancient Arabic grammar to design a morphosyntactic tag set and she did not follow Indo-European tag sets which depend on Latin. All subcategories in Khoja tag set are derived from the parent categories therefore the tag set hold language generalization. In this tag set the noun is classified into: common, proper, numeral, adjective, and pronoun which have three subclasses: personal, relative, and demonstrative. However particle classified into: preposition, adverbial, conjunctions, interjections, explanations, exceptions, answer, negatives, and subordinates. Figure 1 shows an example of tagging a part of text using Khoja tag set.

بعث_VPSg3M خادم_NCSgMNI الحرمين_NCDuMAD الشريفين_NCDuMGD الملك_NCSgMND فهد_NP بن_
NCSgMNI عبد_NCSgMAI العزيز_NCSgMAD آل_R سعود_NP برقية_NCSgFNI تهنئة_NCSgFGI الى_PPr
فخامة_NCSgFGI الرئيس_NCSgMGD الكسندر_RF كواسنيفيسكي_RF رئيس_NCSgMNI جمهورية_NCSgFGI
بولندا_RF بمناسبة_PPr_NCSgFGI اليوم_NCSgMAD الوطني_NCSgMND لبلاده_PPr_NCPlFGI_NPrPSg3M

Fig. 1. Tagging a text using Khoja tag set [10].

Even that the work of Khoja has been the first comprehensive designed Arabic tag set and it is highly used in several applications, it has limitations [11].

2.3 Buckwalter Tag Set

Buckwalter tag set has two types: the first one is the untokenized tag set and the second is the tokenized tag set which contains around 500 tags. However, Buckwalter tag set consist of 70 sub tags that are possible to be combined to make around 170 complex tags with features for verbs, nominal and subject. For example, person, voice, aspect, and mood are included as verbs features [12]

The untokenized Buckwalter tag set contains around 485 tags and use the same basic 70 sub tags such as DET for determiner, ACC for accusative and ADJ for adjective. As an example for tagging the word الصين 'China' using Buckwalter tag set would be: DET+NOUN_PROP+CASE_DEF_NOM [13, 14].

Moreover, this tags set was reduced and called reduced Bukwalter in which the number of tags included in the new tag set is about 220 tags where case, mood and state.

2.4 Reduced Tag Set (RTS)

Linguistic Data Consortium (LDC) tag set was developed by the LDC team which consists of 24 tags and it was the reason behind introducing the reduced tag set (RTS) which goal was to increase the efficiency and performance of syntactic parsing for Arabic. RTS consists of 25 tags and follows the English tag set that where designed especially for Wall Street Journal. Also, RTS marks some features such as gender, person, definiteness, case and mood [4, 15].

2.5 Extended RTS

RTS was expanded to include 75 tags with adding only labeled morphological features on words. This tag set is called extended reduced tag set (ERTS) which holds the same features in RTS plus additional marked morphological features of number, nominal, definiteness, gender [4]. Table 1 shows some examples for Full Bukwalter, reduced RTS, and Extended ERTS tag sets.

Table 1. Buckwalter, reduced RTS and ERTS example

			Full	RTS	ERTS
حصيلة	HSylp	'result'	NOUN+NSUFF_FEM_SG +CASE_IND_NOM	NN	NNF
نهائية	nhA}yp	'final'	ADJ+NSUFF_FEM_SG +CASE_IND_NOM	JJ	JJF
حادث	HAdv	'accident'	NOUN+CASE_DEF_ACC	NN	NNM
النار	AlnAr	'the-fire'	DET+NOUN+CASE_DEF_GEN	NN	DNNM
الجماعي	AlimAEy	'group'	DET+ADJ+CASE_DEF_GEN	JJ	DJJM
شخصين	$xSyn	'two-persons'	NOUN +NSUFF_MASC_DU_GEN	NN	NNMDu

2.6 Penn Arabic Treebank (PATB)

PATB used widely as a tag set for Arabic [10] and it provides tokenization, complex POS tags, and syntactic structure. Also it provides discretization, empty categories, lemma and some semantic tags [7]. This tag set has over 400 tags that specify details about Arabic word morphology like definiteness, number, gender, person, voice, case and mood [16]. However, twenty dash tags are used for syntactic and semantic functions where syntactic tags have TPC and OBJ while semantic tags cover time (TMP) and location (LOC). Some tags in PATB can be used to label syntactic or semantic purpose like SBJ that is used to label semantic subject of adverb or syntactic subject of a verb [7]. The total number of tags used in PATB reaches 2,200 tag types to specify several issues and features for Arabic word morphology [16, 18] including 114 basic tags [10]. PATB is invented by the Linguistic Data Consortium in the University of Pennsylvania and it has four published versions and it provide a morphological analysis level of annotation where in PATB a morphological analyzer is used to produce a set of candidate analyses for every word then the linguists select the best analyses for the word according to its context [4]. Figures 2 and 3 provide examples for PATB annotation.

```
(S (NP-TPC-1 Huquwq+u حُقُوقُ
        (NP Al+<inosAn+i الإِنْسَانِ))
   (VP ta+qaE+u  تَقَعُ
      (NP-SBJ-1 *T*)
      (PP Dimona ضِمْنَ
         (NP <ihotimAm+i+nA إِهْتِمامِنا)
      )))
```

حُقُوقُ الإِنْسَانِ تَقَعُ ضِمْنَ إِهْتِمامِنا
human rights exist within our concern

Fig. 2. PATB annotation example.

2.7 PADT Tag Set

The PADT tag is designed to have two main parts. The first part represents the part of speech using two letters while the second part represents features and it consists of seven letters to encode the values of the following features: Mood (Indicative, Subjunctive, Jussive or D for ambiguous), Voice (Active or Passive), Person (1 speaker, 2 addressee, 3 others), Gender (Masculine or Feminine), Number (Singular, Dual or Plural), Case (1: nominative, 2: genitive or 4: accusative) and Definiteness (Indefinite, Definite, Reduced or Complex) [7]. For example, the tag NNIS7 ----A---- means: noun, common noun, masculine inanimate, singular, 7th case which is instrumental, and negativeness is affirmative" [2]. Figure 4 shows another example using PADT tagging.

2.8 Alshamsi and Guessom Tag Set

This tag set is designed for the Hidden Marcov Model part of speech tagger system that Alshamsi and Guessom [6] invented to extract Name Entity. Their tag set consists of 55 tags and it is not a fine-grained tag set since they identify the NOUN category and subcategories by a number of tags such as NOUN (noun), adjective (ADJ), pronoun (PRON), proper noun (PNOUN), definite noun (DEF) and indefinite noun (INDEF). These tags are needed for their tagger that is used for Name Entity extraction.

Alshamsi and Guessom subdivide both noun and particle class into four subclasses. They point out, there is no need to have fine-grained a tag set since their tagger was intended to be for Named Entity extraction. For verb category, they use the tags: IVERB to represent imperfect verb, PVERB for perfect verb, CVERB for imperative verb, MOOD-.I for indicative, MOOD~J for subjunctive, SUFF~UBJ for suffix subject and FUTURE for future Imperative. Moreover for particle category they use the tags: INTERROGATE to represent interrogation, NEGATION to represent negation, CONJ to represent conjunction and PREP to represent preposition particles. Also, they

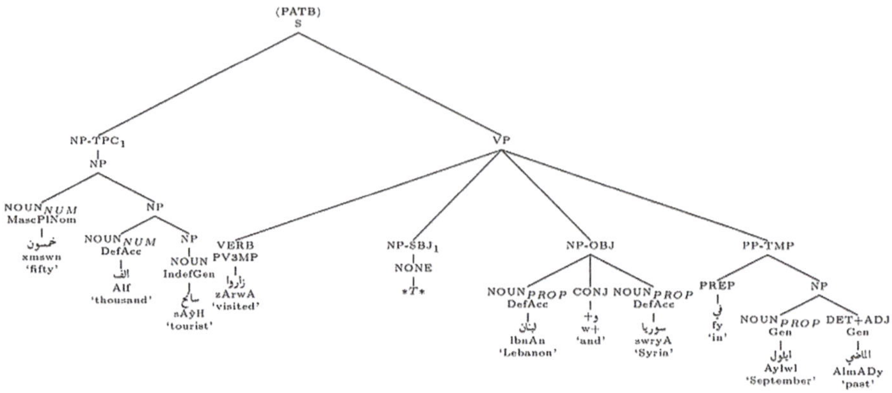

Fig. 3. In Penn Arabic Treebank (PATB) خمسون الف سائح زارو لبنان وسوريا في أيلول الماضي

Form	`al-lA-silkIyaTu`	*al-lā-silkīyatu* اَللَّاسِلْكِيَّةُ
Morph	`al >| lA >| FiCL |< Iy |< aT |<< "u"`	
Tag	`A-----FS1D`	
Form	`lA-silkIy`	*lā-silkīy* لَاسِلْكِيّ
Morph	`lA >| FiCL |< Iy`	
Root	`"s l k"`	
Reflex	wireless, radio	
Class	adjective	

Fig. 4. PADT example.

use features like person, number and gender with tag names in order to describe the morphology analysis of a word [6]. Figure 5 represents two examples for tagging two sentences with Alshamsi and Guessom tag set.

Word	Meaning	POS Tag
الحادي	The first of	DEF+ADJ
عشر	Ten	NOUN
من	From	PREP
أكتوبر	October	PNOUN

Word	فرنسي	شخص	ال	هذا
Transliteration	faransiyy	shakhS	al	haadhaa
Meaning	French	person	is	This
POS Tag	ADJ	NOUN	DEF	DPRON_MS

Fig. 5. Alshamsi and Guessom tagging for two sentences.

2.9 ARBTAGS Tag Set

ARBTAGS tag set has a hierarchy that makes it different than other tag sets. In this hierarchy, the noun is divided in to sixteen sub-classes while verb is classified into three sub-classes: perfect, imperfect and imperative. Particle class has seven sub-classes: preposition, vocative, conjunction, etc. Also, it has one punctuation tag. This tag set has a new additional tag that is used to present foreign words. Therefore, the size of general tags in ARBTAGS is 28 tags but it consists of 161 detailed tags: 101 tags for noun, 50 tags for verb, 9 for particle and one for punctuation. These tags have more features information that considered inflectional. For example; the word كتب is tagged by ARBTAGS as [VePeMaSnThSj] where the tag means [Perfect Verb, Masculine Gender, Singular Number, Third Person, and Subjunctive Mood [24]. Figure 6 shows the specifications of ARBTAGS formula.

The tagset has the following main formula:
[T , S , G , N , P , M , C , F] , Where:
T (Type) = {Verb, Noun, Particle}
S = Sub-Class {Common, Demonstrative, Relative, Personal, Adverb,Diminutive,Instrument, Conjunctive, Interrogative, Proper and Adjective}
G (gender)= {Masculine, Feminine, Neuter}
N (Number) = {Singular, Plural, Dual}
P (Person) = {First, Second, Third}
M (Mood) = {Indicative, Subjunctive, Jussive}
C (Case) = {Nominative, Accusative, Genitive}
F (State) = {Definite, Indefinite}

Fig. 6. ARBTAGS tag formula [24].

2.10 CATiB Tag Set

Habash and Roth construct the Columbia Arabic Treebank (CATiB) [23] which is a database of syntactic analysis used for Arabic text. CATiB differ than other Arabic Treebanking approaches in the constraints that it involves on linguistic richness and its focus on speed. CATiB approach has two basic ideas; the first one is to avoid annotation of redundant information while the second idea is to use terminology and representations inherited from traditional Arabic syntax. Thus, a simple parsing approach can produce grammar analysis [21].

CATiB uses MADA&TOKAN toolkit for initial tokenization and POS tagging. However, CATiB tokenization scheme is the same schema for PATB and PADT but the number of tags in CATiB is very small compared to the tags in PATB's.

CATiB has six POS tags: NOM, PROP, VRB, VRB-PASS, PRT, and PNX where NOM for non-proper nominals include nouns, pronouns, adjectives and adverbs, PROP for proper nouns, VRB for active-voice verbs, VRB-PASS for passive voice verbs, PRT for particles including prepositions or conjunctions and PNX for punctuation [23]. Figure 7 shows an example sentence that is tagged by CATiB.

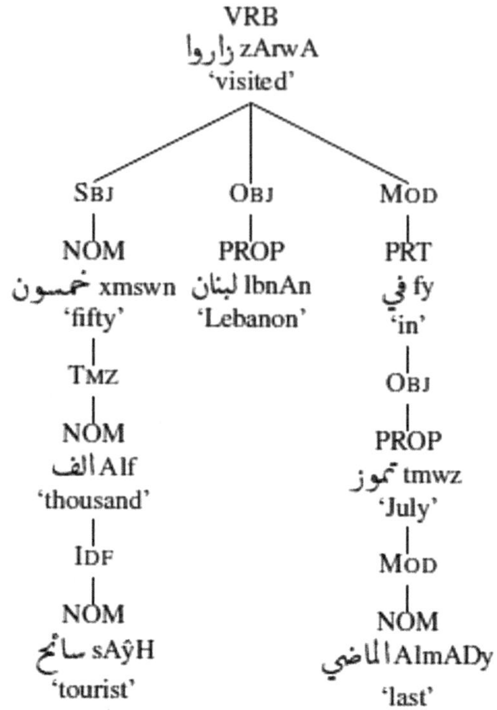

Fig. 7. CATiB example.

2.11 Yahya Elhadj Tag Set

Elhadj et al. [7] define 16 features as the main correlated morpho-syntactic features for a word then they add more fine grain features to get 71 features that are labeled with proper tags. In this tagging system the tag set covers three levels: the first level consists of 7 tags and the second level consists of 23 tags while the lower level has 54 tags. According to the requirements the tagging system can be used in any of the three levels. Figure 8 shows an example for using Elhadj tag set to tag Sect. 4 from Chapter al fateha in Quran.

2.12 SALMA Tag Set

SALMA Tag is introduced by Sawalha [15] to encode 22 morphological feature categories for each morpheme. The first character in the tag represents the word's class: verb, noun, particle or punctuation. The second character represents the noun subclass while the third character represents verb subclass while the fourth character represents the particle subclass. The fifth and sixth character are used for others and punctuations. The next characters from the seventh to the eighteen character in the tag represent morphological features as follows: gender in the seventh character, number in the eighth position, person in the ninth, inflectional morphology in the 10th, case and mood in the 11th, case and mood marks in the 12 position, definiteness in the 13, the 14

السورة	الآية	القطعة	النوع أو القسم	العدد	التركيب	التعريف	الجنس	الزمن	الوجه	الحالة الإعرابية	صيغة البناء	الشخص	تعدي	البناء	علامة الإعراب	وزن	الصحة	تجرد	جمود
الفاتحة	4	مالِك	AJV	NS		DT	GM									PA	SS	F1	IT
الفاتحة	4	ِ	MNV							CG					MI				
الفاتحة	4	يَوْمِ	NNN	NS	CU	DT	GM												
الفاتحة	4	ِ	MNV							CG					MI				
الفاتحة	4	ل	MNV			DF								DJ					
الفاتحة	4	دِين	NNN	NS	CU		GM												
الفاتحة	4	ِ	MNV							CG					MI				

Fig. 8. Elhadj tag set tagging section (verse) 4 form chapter Alfateha in Quran [7].

character represent voice, emphasized and non-emphasized are presented in the 15 position, transitivity is presented in the 16, rational in the 17 character, declension and conjugation in the 18 position. The final four characters represent morphological information that is used in the analysis of Arabic text; the 19th character represent unaugmented and augmented, number of root letters represented in the 20th character, verb root in the 21 character and noun type in the final character. This tag set is utilized in Qutuf analyzer for Arabic morphology and part of speech tagger [15]. Figures 9 and 10 show examples for SALMA tagging.

Word		Morphemes			Tag
wa waaṣṣaynā And We have enjoined	وَوَصَّيْنَا	وَ	*wa*	*And*	`p--c------------------`
		وَصَّيْ	*waṣṣay*	*Have enjoined*	`v-p---mpfs-s-amohvtt&-`
		نَا	*nā*	*We*	`r---r-xpfs-s----hn----`
al-'insāna (on) man	الْإِنسَانَ	الْ	*al-*	*The*	`r--d------------------`
		إِنسَانَ	*'insāna*	*man*	`nq----ms-pafd---htbt-s`
bi- wālidayhi His parents	بِوَالِدَيْهِ	بِ	*bi*	*To*	`p--p------------------`
		وَالِدَ	*wālida*	*Parents*	`nu----md-vgki---htot-s`
		يْ	*y*	*Both*	`r---r-xdts-s----------`
		هِ	*hi*	*His*	`r---r-msts-k----------`
ḥusn^an Kindness	حُسْنًا	حُسْنَ	*ḥusn*	*kindness*	`ng----ms-vafi---ndst-s`
		ً	*an*		`r---k------f----------`

Fig. 9. SALMA example 1 [15].

2.13 Ahmed H. Aliwy Tag Set

The main tags in this tag set are Noun, Verb, Particle, Residual and Punctuation where Noun has 17 subclasses with the features: Number, Gender, Case and Structured. Verb class has three subclasses: Past (Pst), Present (Prt), Imperative (Imv). While verb attributes are: Gender, Number, Person, Mood, Certainty, Structured, and Voice.

The subclasses of the particle are defined according to its working. In this tag set there are 21 subclasses for particles: for Jussive Jus, For Reduction, preposition Red, For Conjunction Cnj, for Accusative Acu, Not working or Preventive, NonCopier Cop

وَوَصَّيْنَا	وصي	فَعَّلْنَا	وَ p--c-----------------
			وَصَّيْ v-p---mpfs-s-amohvtt&-
			نَا r---r-xpfs-s---------
الْإِنْسَانَ	أنس	فِعْلَان	الْ r---d----------------
			إِنْسَانَ nq----ms-pafd---hdbt-s
بِوَالِدَيْهِ	ولد	فَاعِل	بِp--p----------------
			وَالِدَ nq----ms-pafd---hdbt-s
			يْ r---r-xdts-s---------
			هِ r---r-msts-k---------
حُسْنًا	حسن	فُعْل	حُسْنَng----ms-vafi---ndst-s
			اً r---k-----f---------

Fig. 10. SALMA tag example 2 [15].

and Prevent Prv. Residuals can be SymbolRSym, Abbreviations and Acronym RAbc or Not Classified RNcl. But there are no features for residuals and punctuations.

This tag set has 3552 tags since the combinations of some tags are impossible. The Noun tag has the form: N+POS_ Number+Gender+Case+Structured while Verb tag has the form: V +POS_Person+Number+Gender+Case+Structured+Certinity+Voice. Particles tag has the form: P+Working_Meaning. For Residual it can be one of three tags: ROth, RSys or RAcb, and the Punctuation tag is CPnc.

Figure 11 gives an example of the use of Aliwy Tag set for the Arabic text: "مرة وقبل سنتين كتبت عن العراق" [19].

It is considered as a multilevel tag set that is compatible with Classical Arabic and Modern Standard Arabic. Aliwy argued that his tag set has almost all Arabic features and classes where classes and features are selected carefully and his tag set has no interleaving [19].

3 Limitations in Arabic Tag Sets

Available Arabic tag sets do not have a standard scheme for correlating each word to its morpheme and they join the tagging of both morphemes and words. Moreover, many reports about these tag sets do not give a detailed description for their design aspects [22].

The existing tag sets have a limitation in covering all the features of Arabic language which leads to missing features.

However the analysis used for texts in designing existing tag sets is not competent for Arabic features and characteristics. Also a number of tagging systems involve a small number of tags that gives a narrow view about the text and they do not explain more about particles and verbs [7].

Clitics and word base			Tag
Token	Transliteration	Translation	
مرة	mrp	Once ,Time	NNou_SFNN
،	,	,	CPnc
و	w	And	PNon_Non
قبل	qbl	before	NAdv_SMAN
سنتين	sntyn	Two years	NNou_DFGN
،	,	,	CPnc
كتبت	ktbt	I wrote	VPst_3SMOYNA
عن	En	About	PRed_Adv
ال	Al	The	PNon_Def
عراق	ErAq	Iraq	NPrp_SMGN

Fig. 11. An example for the use of Aliwy tag set.

For example, in [17] they used only 24 tags and Catib [23] used only 8 tags. Such tagging systems may be insufficient for more general needs.

Even though the tag sets with large number of tags are complete and efficient for advanced tasks, they look very hard to be predicted while small tag sets tend to be more predictable and appropriate for many applications [20].

4 Summary

Since 2000, researchers have been introducing new Arabic tag sets or enhancing previously proposed tag sets. However, between 2010 and 2012, less attention was given to Arabic tag set until 2013 where Salma tag set was introduced by Sawalha [15] and Aliwy tag set was introduced by Ahmad Aliwy [19]. Table 2 shows a comparison between Arabic tag sets that were introduced in this report. It is shown that many tag sets has no particle attributes.

Existing tag sets shares common and basic tags while they differ in the number of levels they include in their morphological analysis and therefore they combine basic tags to get more complex tags as a feature for a word not as new tags. The tags details in any tag set depend on the application that the tag set was conducted specially for.

Table 2. Comparison between Arabic Tag Sets

Year	Tag set	Developed by	No. of tags	Simple/Complex	Tags details	Particle	Limitations
2000	El-Kareh S, Al-Ansary	El-Kareh S, Al-Ansary	72 tags	–	Words are classified into three main classes, Verbs, Noun and Particle. Each class is divided into subclasses, Verbs into 3 subclasses; Nouns into 46 subclasses and Particles into 23 subclasses.	23 sub-classes of the main class particle.	–
2001	Khoja	Shereen Khoja	177 tags	Simple	103 nouns, 57 verbs, 9 particles, 7 residual, and 1 punctuation.	No attributes	Many of Arabic classes are not taken into account.
2002	Buckwalter	Tim Buckwalter	485 tags-untokenized Thousands – tokenized	Complex	Form-based Very rich for many computational problems Tag set that can be used for tokenized and untokenized text Buckwalter's Tag set (170 morphemes, 500 tokenized tags, 22 K untokenized tags)	No attributes	No distinction between categories and features for POS
2004	Reduced Buckwalter tag sets BIES	Ann Bies and Dan Bikel	24	Very simple	Inspired by the Penn English Treebank	No attributes	The nouns, verbs and particles have no attributes.

(*continued*)

Table 2. (*continued*)

Year	Tag set	Developed by	No. of tags	Simple/Complex	Tags details	Particle	Limitations
2004-	The Extended Reduced Tag Set (ERTS)	Used in Amira system	72	–	A subset of the full Buckwalter morphological set	Added the explicit or marked morphological features of gender, number and definiteness on nominals	–
2004	Penn Treebank: PATB	Mohamed Maamouri and Ann Bies	2,000 tag types. This includes combinations of 114 basic tags.	Detailed tag set	Follows Arabic traditional grammer. Tags specify details about word morphology such as definiteness, number, case, person, voice, gender and mood.	No attribute	With some kinds of words, the PATB morphology systematically fails to determine many of the contextual and lexical parameters
2006	Alshamsi and Guessom	Alshamsi and Guessom	55	Very specific	Specific for Name Entity Take into account the structure of Arabic sentence		–
2008	ARBTAGS	AlQrainy & Ayesh	161 detailed tags and 28 general tags	Simple	Based on ancient Arabic grammar. 101 nouns, 50 verbs, 9 particles, 1 punctuation	No attributes	Punctuations and foreign words are not covered
2009	CATiB	Nizar Habash and Ryan M. Roth	6	The simplest tag set,	Using representations and terminology inspired by traditional Arabic syntax	No attributes.	Many classes and features are missed.

(*continued*)

Table 2. *(continued)*

Year	Tag set	Developed by	No. of tags	Simple/Complex	Tags details	Particle	Limitations
2009	Yahya Elhadj	Yahya Elhadj		Simple with respect to Noun, and Verb	Three classes (Noun, Verb, Particle), to cover the three categorization levels: 7 tags for the upper level, 23 for the inner level, and 54 for the lower level; 71 features	No attributes	No features for verbs.
2013	Salma	Majdi Sawalha	22 character for a tag	Simple	7 tags for the upper level, 23 for the inner level, and 54 for the lower level;	No variation	More theoretical Redundant tags
2013	Aliwy	Ahmad Aliwy	3552 detailed tags and 45 main tags.	Complex	17 tags for noun classes, 3 tags for verb, 21 tags for particles, 3 tags for Residuals and one for punctuation	21 tags for particles.	–

References

1. Abumalloh, R., Al-Sarhan, H., Ibrahim, O., Abu-Ulbeh, W.: Arabic part-of-speech tagging. J. Soft Comput. Decis. Support. Syst. **3**(2), 45–52 (2016)
2. Böhmová, A., Haji, J., Hajiová, E., Hladká, B.: The prague dependency treebank: a three level annotation scenario. In: Treebanks: Building and Using Parsed Corpora. Springer (2003)
3. Alqrainy, S., Ayesh, A., Almuaidi, H.: Automated tagging system and tagset design for arabic text. J. Comput. Linguist. Res. **1**(2), 55–62 (2010)
4. Maamouri, M., Bies, A.: Developing an Arabic treebank: methods, guidelines, procedures, and tools. In: Proceedings of the Workshop on Computational Approaches to Arabic Script-based Languages (COLING), Geneva, pp. 2–9 (2004)
5. Alqrainy, S.: Morphological - syntactical analysis approach for Arabic textual tagging. Ph.D. thesis, De Montfort University (2008)
6. Al Shamsi, F., Guessoum, A.: A hidden Markov model–based POS tagger for Arabic (2006)
7. Elhadj, Y., Abdelali, A., Bouziane, R., Ammar, A.H.: Revisiting Arabic part of speech tagsets. In: Proceedings of 11th International Conference on Computer Systems and Applications (AICCSA), pp. 793–802 (2014)
8. El-Kareh, S., Al-Ansary, S.: An Arabic interactive multi-feature POS tagger. In: Proceedings of International Conference on Artificial and Computational Intelligence for Decision, Control, and Automation in Engineering and Industrial Applications (CIDCA), Monastir, Tunisia, pp. 204–210
9. ElHadj, Y., Al-Sughayeir, I.A., Al-Aansari, A.M.: Arabic part-of-speech tagging using the sentence structure. In: Proceedings of 2nd International Conference on Arabic Language Resources and Tools. Cairo, pp. 241–245 (2009)
10. Khoja, S., Garside, R., Knowles, G.: A tagset for the morphosyntactic tagging of Arabic. In: Proceedings of Corpus Linguistics, Lancaster, pp. 341–353 (2001)
11. Abuzed, M., Arteimi, M.: Using the Brill of speech tagger for modern standard Arabic. In: The International Arab Conference on Information Technology (ACIT), Amman (2005)
12. Alosaimy, A.M.S. Atwell, E.S.: A review of morphosyntactic analysers and tag-sets for Arabic corpus linguistics. In: Corpus Linguistics, Lancaster, pp. 16–19 (2015)
13. Buckwalter, T.: Issues in Arabic orthography and morphology analysis. In: Proceedings of the Workshop on Computational Approaches to Arabic Script Based Languages, pp. 31–34. COLING, Geneva (2004)
14. Alkuhlani, S., Habash, N., Roth, R.: Automatic morphological enrichment of a morphologically underspecified treebank. In: Association for Computational Linguistics (NAACL-HLT), Atlanta [s.n.], pp. 460–470 (2013)
15. Sawalha, M., Atwell, E., Abushariah, M.A.M.: SALMA: standard arabic language morphological analysis. In: Proceedings of 1st International Conference on Communications, Signal Processing, and their Applications (ICCSPA), Sharjah, pp. 1–6 (2013)
16. Smrž, O., Bielický, V., Kouřilová, I., Kráčmar, J., Hajič, J., Zemánek, P.: Prague Arabic dependency treebank: a word on the million words. In: Proceedings of the LREC Workshop on HLT & NLP within the Arabic World: Arabic Language and Local Languages (2008)
17. Diab, M., Kadri, H., Daniel, J.: Automatic tagging of Arabic text: from raw text to base phrase chunks. In: Proceedings of Human Language Technology-North American Association for Computational Linguistics (HLT-NAACL) (2004)
18. Diab, M.: Towards an optimal POS tag set for modern standard arabic processing. In: Proceedings of Recent Advances in Natural Language Processing (RANLP), Borovets (2007)

19. Aliwy, A.: Arabic morphosyntactic raw text part of speech tagging system. University of Warsaw, Faculty of Mathematics, Informatics and Mechanics (2013)
20. Habash, N.: Introduction to Arabic Natural Language Processing. Morgan & Claypool Publishers Series, San Rafael (2010)
21. Ibrahim, M.N.: Statistical Arabic grammar analyzer. In: Proceedings of 16th International Conference in Computational Linguistics and Intelligent Text Processing (CICLing), Cairo, pp. 187–200 (2015)
22. Sawalha, M., Atwell, E.: A standard tag set expounding traditional morphological features for Arabic language part-of-speech tagging. Word Struct. **6**(1), 43–99 (2013)
23. Habash, N., Roth, R.M.: CATiB: the Columbia Arabic treebank. In: Proceedings of the Association for Computational Linguistics (ACL-IJCNLP), pp. 221–224 (2009)
24. Alqrainy, S., Ayesh, A.: Developing a tagset for automated POS tagging in Arabic. In: Proceedings of the 10th WSEAS International Conference on COMPUTERS, Athens, pp. 956–961 (2006)

Information Gain Based Term Weighting Method for Multi-label Text Classification Task

Ahmad Mazyad(✉), Fabien Teytaud, and Cyril Fonlupt

LISIC, Université du Littoral Côte d'Opale,
50 Rue Ferdinand Buisson, 62100 Calais, France
ahmad.mazyad@univ-littoral.fr

Abstract. In text classification, terms are given weights using Term Weighting Scheme (TWS) in order to improve classification performance. Multi-label classification task are generally simplified into several single-label binary task. Thus, the term distribution are considered only in terms of positive and negative categories. In this paper, we propose a new TWS based on the information gain measure for multi-label classification task. This TWS try to overcome this shortness without affecting the complexity of the problem. In this paper, we examine our proposed TWS with eight well-known TWS on two popular problems using five learning algorithms. From our experimental results, our new proposed method outperforms other methods, especially regarding the macro-averaging measure.

Keywords: Machine learning · Text classification
Term-weighting scheme · Support vector machine

1 Introduction

Text Categorization (TC) goal is to classify a text document into one or more categories. Generally, this approach is to learn an inductive classifier from a set of predefined categories. This approach requires that documents are represented in a suitable format such as the Vector Space Model (VSM) representation [1]. In a (VSM), a document d_j is defined by a term vector $d_j = (w_{1,j}, w_{2,j}, ..., w_{t,j})$ in which each term is associated with a weight $w_{k,j}$.

The weight represents the quantity of information a term contributes to the semantics of a document. The method which assigns a weight to a term is called TWS.

TC belongs to the family of supervised learning. Thus, TWS could be either unsupervised or supervised depending on whether it makes use of class information. The unsupervised methods include the famous $tf.idf$ proposed in [2] by Jones. $tf.idf$ stands for Term Frequency-Inverse Document Frequency, and it is borrowed from the information retrieval field.

K. Arai et al. (Eds.): IntelliSys 2018, AISC 868, pp. 607–615, 2019.
https://doi.org/10.1007/978-3-030-01054-6_44

The Supervised Term Weighting (STW) methods incorporate documents membership information when computing each term weight. These methods include the feature selection metrics such as χ^2, information gain ig, gain ratio gr, odds ratio or in [3,4]. Recently, some authors proposed multiple intuition based methods. Wang *et al.* in [5] proposed the *inverse category frequency* icf proposed. In [6], Lan *et al.* presented *relevance frequency* rf.

A general formula for the different TWSs experimented in this paper could be defined as:

$$w_{t,d} = tf_{t,d} \times CF_t.$$

Where $w_{t,d}$ is the weight of a term t in a document d, $tf_{t,d}$ stands for the term frequency of t in d and CF_t is the collection frequency factor of t.

Different works performed on term weighting methods have shown different results and contradictory conclusions [6]. For instance, by comparing $tf.idf$ to three STW, Debole *et al.* in [3] showed the superiority of $tf.gr$ over $tf.ig$ and $tf.\chi^2$ while finding no consistent superiority over $tf.idf$. Another study done by Lan *et al.* in [6] confirmed the superiority of $tf.idf$ over $tf.chi$. A recent and fair comparison between state of the art TWS [7] have shown similar results as shown in [3]. However, in [4], Deng *et al.* concluded unlike Debole the superiority $tf.chi$ over $tf.idf$.

For this work, we seek to find an efficient TWS for multi-labeled classification task.

The paper is organized as follows: Sect. 2 presents the standard TWSs alongside with our proposed approach. In Sect. 3 we present five learning algorithms used in order assess the performance of TWSs. In Sect. 4, we compare the TWSs applied to two well-known data sets. Lastly, we consider future works in Sect. 5.

2 Term Weighting Methods

In this section, first, we present well-known TWS and second, our method.

2.1 Preliminary

Traditional classification algorithms are well suited for single label data sets. Thus, it can not learn from multi-labeled data sets. Several approaches exist to handle the multilabel classification task [8] such as problem transformation methods, and algorithm adaptation methods. Binary relevance transformation strategy is the most widely used strategy that simplifies the multi-labeled data set into several distinct single-label binary data set. That is, given the list of labels $L = \{l_1, l_2, ..., l_m\}$, the original data set is transformed into m different data sets $D = \{D_1, D_2, ..., D_m\}$. For each data set D_i, documents having the label l_i will be tagged as the positive category c_i, and the rest as the negative category $\overline{c_i}$. Weights are then computed independently for each binary data set.

Based on the binary transformation, given a term t_k and a category c_i, STW could be expressed using statistical information a, b, c and d obtained from the training data: a, b, c, d represents the number of documents that contain/do not contain t_k and belong/do not belong to the positive category c_i.

These statistical information are used in all TWS included in our work, except for $tf.icf$.

In this paper, we logarithmically normalized the Term Frequency (tf) formula:

$$tf_{t,d} = \log(f_{t\in d}) + 1,$$

with $f_{t\in d}$ the number of occurrences of the term t in the document d.

2.2 Collection Frequency Factors

A Collection Frequency (CF) factor is a combination of statistical information. It is intended to measure the discriminative power of a term, i.e. it tells how much a term is related to a certain category. χ^2 corresponds to a test of independence between two variables (a term and a category). χ^2 is a popular feature selection method. χ^2 and other supervised feature selection schemes have been tested in several papers, as a term weighting methods for text categorization. For example, Deng *et al.* in [4], replaced the idf component with χ^2 component, claiming that $tf.\chi^2$ is more efficient than $tf.idf$. In contrast, in a similar test, Debole *et al.* in [3], compare tf.idf with three supervised term weightings, namely, χ^2, information gain and gain ratio. The authors have found no consistent superiority of these new term weighting methods over tf.idf.

Information Gain (ig) [9] is a measure of dependence between two random variables. In the context of text classification, it can be expressed as a measure of dependency between one random term and one random class. Mutual information is widely used in feature selection for text classification [3,4].

Debole *et al.* used Gain Ratio (gr) applied to a feature selection method [3]. The authors claim that $tf.gr$ is a better term evaluation functions than the $tf.ig$. In their text categorization test, they confirmed the superiority of $tf.gr$ over $tf.ig$ and $tf.\chi^2$.

Odds Ratio (or) is a measure that describes the strength of association between two random variables. It was first used as a feature selection methods by Mladeni'c *et al.* [10] who found that odds ratio outperforms five other scoring methods studied in text classification experiments. Another comparative study on feature weight in text categorization is done by Deng *et al.* in [4]. The study shows a good performance of $tf.or$ but still outperformed by $tf.\chi^2$.

Relevance frequency (rf) is a supervised weight scheme proposed in [6]. rf measures the distribution of term t_k between positive and negative category and favors those terms that are more concentrated in the positive category than in negative category.

Inverse Category Frequency (icf) is another supervised term weighting method proposed by Wang *et al.* in [5]. icf stands for inverse category frequency and aims to favor terms that appear in fewer categories.

Main known CF factors are presented in Table 1. We present some state of the art TWS in the next section.

Table 1. Seven traditional CF factors. N is the total number of Docs, a the number of Docs in the positive category cat that contain the term t_k, b defines the number of Docs in cat with no occurrences of t_k, c is the number of Docs not in cat in which t_k occurs at least once, d the number of Docs that don't belong to cat and have no occurrences of t_k. $|C|$ represents the number of categories and $|C_{t_k}|$ is the number of categories that contain t_k

CF	Formula
idf	$\log(\frac{N}{a+c})$
χ^2	$N \times \frac{(a \times d - b \times c)^2}{(a+c)(b+d)(a+b)(c+d)}$
ig	$(\frac{a}{N} \times \log \frac{a \times N}{(a+b)(a+c)}) + (\frac{c}{N} \times \log \frac{c \times N}{(c+d)(a+c)})$ $+ (\frac{b}{N} \times \log \frac{b \times N}{(a+b)(b+d)}) + (\frac{d}{N} \times \log \frac{d \times N}{(c+d)(b+d)})$
gr	$ig/(-\frac{a+c}{N} \times \log \frac{a+c}{N} - \frac{b+d}{N} \log \frac{b+d}{N})$
or	$\log(2 + \frac{a \times d}{b \times c})$
rf	$\log\left(2 + \frac{a}{\max(1,c)}\right)$
icf	$\log_2\left(\frac{\lvert C \rvert}{\lvert C_{t_k} \rvert}\right)$

The TWS presented has proved to be efficient in text classification through a huge number of experimental studies. However, all these methods, except for $tf.icf$ has a common shortness: they consider the distribution only in terms of positive and negative categories.

2.3 Our Information Gain Based Method

The basic idea of our proposed ig based method comes in form of a question: how much information gain a term t_k have about a category after subtracting the information gain of the same term t_k of the other categories. It is to say that the higher the difference between a term information gain of one category and the average of the other categories, the more the term helps in separating positive and negative categories. As explained in Sect. 2.1, a multi-label classification task is transformed into multiple binary single-label classification task, therefore, a term has multiple collection frequency weights, one for each binary task. Each weight only considers the distribution of a feature/term in terms of the positive category and the negative category (all documents that do not belong to the positive category). We think that using these weights could be helpful for more effective TWS.

Considering this idea, we propose a new TWS based on information gain. Its formula is defined by:

$$w'_{t,c} = w_{t,c} - (\mu_{c' \in C} w_{t,c'} + \sigma_{c' \in C} w_{t,c'}).$$

Where $w'_{t,c}$ is the new weight of a term t and a category c, $w'_{t,c}$ is the information gain score of a term t and a category c, $\mu_{c' \in C} w_{t,c'}$ is the mean of weights on all other categories, and $\sigma_{c' \in C} w_{t,c'}$ is the standard deviation of weights on all other categories.

To evaluate the differences between the information gain measure and our proposed method, let us consider the weights for the three terms in Table 2.

First, let us clarify some points:

- When $\mu + \sigma > ig$, the term contributes more to the negative categories than to the positive category.
- When $\mu + \sigma < ig$, the term contributes more information to the positive category.
- When $\mu + \sigma = ig$, the term has about the same amount of information about both positive and negative categories.

First, considering the term t_1 in Table 2, $\mu + \sigma$ (0.5) is higher than ig value (0.3), which means that the negative categories have higher weights than the positive category, however the ig value of t_1 is a positive value, in contrary to our new method. That said, the difference doesn't have a big impact on scores especially when the number of categories in the corpus is big, as $\mu + \sigma$ will have about the same value.

Now, if we consider terms t_1 and t_3, they both have the same ig value (0.3), which means that they both contribute the same amount of information to the positive category, however by looking at the values of $\mu + \sigma$, t_1 has a value of $0.5 > 0.3$ and t_3 has a value of $0.1 < 0.3$. In this case, we think that t_3 should have a higher value than t_1 as it has the same information gain in the positive category but smaller information gain in the negative categories.

Finally, t_2 has same information gain value both in the positive category and the negative categories $ig = \mu + \sigma = 0.2$, thus, the ig-based value is equal to 0.

Table 2. Comparison of the weighting values of ig and the proposed method. $\mu + \sigma$ is the average plus the standard deviation of scores of categories which are not the positive category. The values were hand chosen

Feature	ig	$\mu + \sigma$	New
t_1	0.3	0.5	−0.2
t_2	0.2	0.2	0
t_3	0.3	0.1	0.2

3 Classifiers

Generally, the performance of a TWS is assessed on known benchmarks by evaluating a classification model on VSM representation of this TWS. In order to build the classification models, we experiment five different algorithms, namely:

Passive-Aggressive, C4.5, Support Vector Machine, Stochastic Gradient Descent and Nearest Centroid.

Support Vector Machine (SVM)s are a set of supervised machine learning methods introduced by Boser *et al.* Developed from statistical learning theory, SVMs have shown good performance in many fields. In text classification, Joachims in [11] used SVM in which he demonstrates the better efficiency of SVM over other learning algorithms. Passive-Aggressive (PA) proposed by Crammer *et al.* in [12] is a learning algorithm focused on online learning and large scale data set. The method treats a flow of documents, and outputs a prediction once a document is received. Later at any time a document true label is discovered, the method redefines its prediction function. Stochastic Gradient Descent (SGD) classifier [13] is a linear classifier, such as linear SVM, PA that uses SGD for training. This classifier is also used for large scale categorization problem. Nearest Centroid (NC) [14] is a neighborhood-based classification algorithm, and C4.5 [15] is a state of the art supervised learning algorithm based on decision tree.

4 Results and Discussion

In this study, we compare eight term weighting methods alongside with our approach on two popular data sets, i.e. Reuters-21578[1] and Oshumed (see footnote 1) using five classification algorithms in terms of micro- and macro-averaged F_1 measure.

4.1 Data Corpora

Two widely-used datasets are used to compare the performance of our proposed method with the performance of eight well-known TWS: Reuters-21578 and Oshumed. Binary relevance transformation strategy is applied on the two multi-label classification task as explained in Sect. 2.1. A default list of stop words, numbers and punctuation are removed. Lower case transformation is applied, and the Porter's stemming is performed.

(1) *Reuters-21578 Benchmark Corpus:* This data set is a well-known benchmark for TC research. We use the "ApteMod" split [11]. The Apte split includes 10788 documents from the financial service, divided into a training set (7769 documents) and a test set (3019 documents). The data set is highly skewed, the smallest category contains only two documents and the biggest contains 3964 documents.

(2) *Oshumed Benchmark Corpus:* The second dataset is another well-known benchmark from the Oshumed[1] collection created by W. Hersh. the corpus includes a total number of 13,929 medical abstracts splitted into a training subset of 6,286 abstracts and a test subset of 7,643 abstracts from the MeSH categories of the year 1991. Each document in this data set belongs to one or more categories from 23 cardiovascular diseases categories.

Table 3 presents statistics about the two datasets.

[1] http://disi.unitn.it/moschitti/corpora.htm.

Table 3. Statistics on the selected data sets used for our experiments (Training/Test)

	Reuters	Oshumed
Number of documents	7769/3019	6286/7643
Number of terms	26000	30198
Number of categories	90	23
The smallest category	1/1	65/70
The largest category	2877/1087	1799/2153

4.2 Evaluation

Numerous evaluation metrics exist to evaluate the classification models such as F_1 measure. The F_1 measure can be considered as a weighted average of the precision (the fraction of positive predictions that is correct) and recall (the fraction of actual positives that have been correctly classified) and can be formally defined as:

$$F_1 = \frac{2 * recall * precision}{recall + precision}.$$

Generally, the F_1 measure is computed in two ways, micro-averaged and macro-averaged. In micro-averaged, big categories are emphasized while in macro-averaged, all categories have the same importance.

In the two tables, underlined results represent the highest score over a column, and the bolded results is the best pair of micro-/macro-averaged F_1 scores when all the classifiers and all the TWSs are considered. The pair having the highest mean is chosen as the best.

4.3 Results

Tables 4 and 5 show the micro-/macro-averaged F_1 performances of different TWSs. using linear SVM for the two data sets Reuters and Oshumed, respectively.

Considering the Reuters data set, the best micro-averaged F_1 score 88.68% is achieved by using our method using SVM classifier. In terms of macro-averaged F_1 score, using our method gives the best score 57.70%. The best micro-/macro-averaged F_1 pair 87.29%/57.70% is also achieved by our information gain based method. Compared to the second best pair (87.27%/48.59%) achieved by $tf.or$, the proposed method records a boost of over 9% in terms of macro-averaged F_1.

In terms of learning algorithms, in this experiment, PA, SVM and SGD show comparable performances. NC records the lowest results.

Considering the Oshumed data set, The highest micro-averaged F_1 (67.45%) is achieved using our proposed method. The highest macro-averaged F_1 (62.37%) achieved by using $tf.or$. As a pair of micro- and macro- averaged F_1, the proposed method has a slightly higher average.

Table 4. Micro-/Macro- averaged F_1 results (%) on Reuters-21578 corpus using eight standard TWSs and the proposed method

	PA	SVM	SGD	NC	C4.5
tf	86.6/48.5	85.9/39.7	86.4/41.1	54.6/34.7	81.9/53.6
tf.χ^2	86.3/48.5	84.8/43.9	86.4/40.8	54.6/34.7	81.8/53.2
tf.*idf*	87.2/48.2	85.7/40.3	86.6/42.7	73.5/47.0	81.3/53.4
tf.*gr*	86.5/47.1	86.5/42.4	86.4/41.1	54.6/34.7	82.0/51.8
tf.*or*	86.6/48.5	87.3/48.6	86.3/40.8	54.6/34.7	81.9/52.8
tf.*ig*	86.9/47.7	86.5/42.4	86.3/41.0	54.6/34.7	82.1/54.2
tf.*icf*	85.9/46.4	84.0/37.8	85.0/40.3	62.5/46.4	80.9/52.0
tf.*rf*	86.5/46.7	87.8/45.3	86.4/40.8	54.6/34.7	82.0/52.8
New	**87.3/57.7**	88.7/51.7	88.4/49.0	66.5/54.7	82.2/51.3

Table 5. Micro-/Macro- averaged F_1 results (%) on Oshumed using eight standard TWSs and the proposed method

	PA	SVM	SGD	NC	C4.5
tf	60.7/54.0	58.2/47.0	59.4/48.8	49.8/44.5	56.6/52.4
tf.χ^2	60.3/55.5	62.5/55.3	59.4/52.0	54.7/51.8	57.4/53.9
tf.*idf*	62.7/56.4	59.3/49.1	61.8/53.4	60.3/57.4	56.9/52.9
tf.*gr*	63.8/58.1	60.8/51.7	63.3/56.0	62.4/60.2	56.8/52.6
tf.*or*	65.2/62.4	64.9/58.8	66.0/60.6	59.2/57.4	56.6/52.8
tf.*ig*	63.7/58.1	60.8/51.7	63.2/56.0	62.4/60.2	57.1/53.5
tf.*icf*	56.5/51.3	49.3/41.9	54.6/48.3	59.1/55.2	56.5/52.7
tf.*rf*	64.0/60.1	63.4/55.5	64.4/57.2	58.4/56.0	56.6/53.0
New	64.4/60.8	67.0/60.8	**67.4/61.2**	59.4/57.0	56.7/52.5

In terms of learning algorithms, SGD and SVM perform the best followed closely by PA and finally, NC and C4.5 show the lowest results.

Overall, in our study, we find that the proposed method give good results, better than the standard TWSs. $tf.or$, $tf.rf$, $tf.idf$ and $tf.ig$ have also shown good results. $tf.\chi^2$ and $tf.icf$ give the worst results.

5 Conclusion

For this work, we study a new term weighting scheme applied to multi-label text classification based on the information gain measure. The basic idea is that the information gain weight of a feature in negative categories should affect the importance of this term in the positive category.

We studied the effectiness of our method in comparaison to eight well-known TWSs applied to text classification tasks.

Experimental results show that our method outperformed all other methods tested in this study, specially in regard to the macro-averaged measure.

References

1. Salton, G., Buckley, C.: Term-weighting approaches in automatic text retrieval. Inf. Process. Manag. **24**(5), 513–523 (1988)
2. Jones, S.K.: A statistical interpretation of term specificity and its application in retrieval. J. Doc. **28**(1), 11–21 (1972)
3. Debole, F., Sebastiani, F.: Supervised term weighting for automated text categorization. In: Text Mining and its Applications, pp. 81–97. Springer (2004)
4. Deng, Z.-H., Tang, S.-W., Yang, D.-Q., Li, M.Z.L.-Y., Xie, K.-Q.: A comparative study on feature weight in text categorization. In: Advanced Web Technologies and Applications, pp. 588–597. Springer (2004)
5. Wang, D., Zhang, H.: Inverse category frequency based supervised term weighting scheme for text categorization, preprint arXiv:1012.2609v4 (2013)
6. Lan, M., Tan, C.L., Su, J., Lu, Y.: Supervised and traditional term weighting methods for automatic text categorization. IEEE Trans. Pattern Anal. Mach. Intell. **31**(4), 721–735 (2009)
7. Mazyad, A., Teytaud, F., Fonlupt, C.: A comparative study on term weighting schemes for text classification (2017)
8. Tsoumakas, G., Katakis, I.: Multi-label classification: an overview. Int. J. Data Warehous. Min. (IJDWM) **3**(3), 1–13 (2007)
9. Cover, T.M., Thomas, J.A.: Elements of Information Theory. Wiley, Hoboken (2012)
10. Mladeni'c, D., Grobelnik, M.: Feature selection for classification based on text hierarchy. In: Text and the Web, Conference on Automated Learning and Discovery, CONALD 1998. Citeseer (1998)
11. Joachims, T.: Text categorization with support vector machines: learning with many relevant features. In: European Conference on Machine Learning, pp. 137–142. Springer (1998)
12. Crammer, K., Dekel, O., Keshet, J., Shalev-Shwartz, S., Singer, Y.: Online passive-aggressive algorithms. J. Mach. Learn. Res. **7**(Mar), 551–585 (2006)
13. Zhang, T.: Solving large scale linear prediction problems using stochastic gradient descent algorithms. In: ICML 2004, pp. 919–926 (2004)
14. Tibshirani, R., Hastie, T., Narasimhan, B., Chu, G.: Diagnosis of multiple cancer types by shrunken centroids of gene expression. Proc. Natl. Acad. Sci. **99**(10), 6567–6572 (2002)
15. Quinlan, J.R.: C4.5: Programs for Machine Learning. Morgan Kaufmann Publishers Inc., San Francisco (1993)

Understanding Neural Network Decisions by Creating Equivalent Symbolic AI Models

Sebastian Seidel(✉), Sonja Schimmler, and Uwe M. Borghoff

Computer Science Department, Institute for Software Technology,
Bundeswehr University Munich, 85577 Neubiberg, Germany
{sebastian.seidel,sonja.schimmler,uwe.borghoff}@unibw.de

Abstract. Different forms of neural networks have been used to solve all sorts of problems in the previous years. These were typically problems that classic approaches of artificial intelligence and automation could not solve efficiently, like handwriting recognition, speech recognition, or machine translation of natural languages. Yet, it is very hard for us to understand how exactly all these different types of neural networks make their decisions in specific situations. We cannot verify them as we can verify, e.g., grammars, trees and classic state machines. Being able to actually prove the reliability of artificial intelligence models becomes more and more important, especially, when cyber-physical systems and humans are the subject of the AI's decisions. The aim of this paper is to introduce an approach for the analysis of decision processes in neural networks at a specific point of training. Therefore, we identify characteristics that artificial neural networks have in common with classic symbolic AI models and where both are different. Besides, we describe our first ideas of how to overcome the aspects where both systems are different and of how to find a way to create something from an artificial neural network that is either an equivalent symbolic model or at least similar enough to such a symbolic model to allow for its construction. Our long term goal is to find, if possible, an appropriate bidirectional transformation between both AI approaches.

Keywords: Artificial neural networks · Symbolic AI models
Connectionism · Symbolism

1 Introduction

Artificial intelligence was one of the most discussed topics in science and industry in the past few years. It is one, if not the key component for modern automatization projects like, for example, machine translation of natural languages, self-driving cars, automated production facilities, or drones.

The main reason for the significant improvements that were made in the field of AI research in the last ten years is the successful implementation of

K. Arai et al. (Eds.): IntelliSys 2018, AISC 868, pp. 616–637, 2019.
https://doi.org/10.1007/978-3-030-01054-6_45

artificial neural networks for problem solving. Artificial neural networks are very good at solving problems characterized by a significant amount of uncertainty or problems that cannot be formally described well. Classic approaches to artificial intelligence like state machines and decision trees are, in contrast, especially well suited for solving problems that can be described by a set of mathematical or logical rules or, in other words, by a formal algorithm [1].

But artificial neural networks also have some major drawbacks. It is not trivial to build an artificial neural network or to train an already existing artificial neural network to solve a specific problem. Even for a given trained artificial neural network that reliably solves a specific problem, we generally do not know how it 'makes' its decisions. That means that after being successfully trained, a neural network can predict the output vector for a specific input vector with sufficient precision, but we do not know how exactly a single input signal influences the computation of the output. This makes the verification of an artificial neural network very difficult.

In this paper, we take a closer look at the similarities and differences of artificial neural networks and classic AI models. Furthermore, we propose a method to extract decision trees from Feedforward Neural Networks. This method will serve as a basis for future work, e.g. our plan to include a greater variety of artificial neural networks and symbolic models and to integrate the transformation from symbolic models into artificial neural networks.

The paper is structured as follows. In Sect. 1, we compare artificial neural networks, focussing on Feedforward Neural Networks, and symbolic AI models, with a focus on tree structures and state machines. In Sect. 2, we give an overview of related work. In Sect. 3, we describe the problem of extracting an equivalent decision tree from a Feedforward Neural Network. In Sect. 4, our approach to extract a decision tree from a Feedforward Neural Network is introduced and limitations are discussed. Future work is discussed in Sects. 5 and 6 concludes the paper.

1.1 Different Approaches to Artificial Intelligence and Their Advantages and Disadvantages

In this section, we will focus on the comparison of different existing AI approaches.

In general, there exist three main approaches to create models that are simulating Artificial Intelligence. Those are connectionists models, symbolic models, and logistics approaches [4]. Connectionism is, according to [5], a movement that hopes to explain intellectual abilities using artificial neural networks. Consequently, artificial neural networks are in the center of the connectionism. Symbolic AI, also known as classic AI, is the branch of artificial intelligence research that tries to explicitly represent human knowledge in a declarative form. For that purpose, it is necessary to translate knowledge, which is often procedural or implicit, into an explicit form, using symbols and rules for their manipulation [6]. Therefore, symbolic systems like grammars, decision trees, behavior trees, and state machines are examples for symbolic models. Connectionism and Symbolic AI have very different functional principles. In Subsect. 1.2, we give some more details about these principles.

(1) Connectionists Approach: Artificial Neural Networks: In general, an artificial neural network is a cluster of simple, connected processors called neurons. The input neurons produce a sequence of real-valued activations triggered by sensor inputs. Other neurons produce activations when triggered by weighted connections from previously active neurons. Activated output neurons create output signals, which are combined to create the networks output. The network learns by finding the correct weights for all connections between its neurons so that the actual output equals the desired output for a given input [3].

There exist many different types of artificial neural networks, and some of them are quite different from each other according to their structure and functionality. A range of the most common categories and how they are related to each other can be seen in Fig. 1.

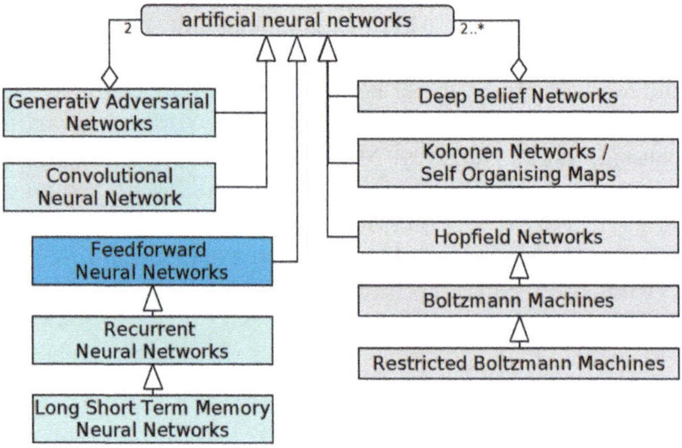

Fig. 1. Different categories of artificial neural networks [2].

First, we will look at Feedforward Neural Networks because this is the most basic type of artificial neural network. They are very common and widely used to solve a great number of classification problems [1], either as pure Feedforward Neural Networks or as components of other artificial neural network types. For future work, we also consider to take a look at Recurrent Neural Networks that are basically Feedforward Neural Networks that can contain loops. These artificial neural networks form the basis of Long Short Term Memory Neural Networks, which do not only have recurrent connections, but additionally feature memory cells with input-, output- and forgetgates to deal with the vanishing or exploding gradient problem when learning long sequences. Feedforward- and

Recurrent Neural Networks are also frequently used as a basic structure in other types of neural networks. Generative Adversarial Networks and Convolutional Neural Networks frequently contain Feedforward Neural Networks as well.

The other categories of artificial neural networks presented in Fig. 1 differ from Feedforward Neural Networks and their enhancements. They are structured differently and, in many cases, use different procedures to compute results. In our work, we explore a way to extract equivalent symbolic systems from Feedforward Neural Networks. This way, we provide a basis for later work with additional types like Recurrent Neural Networks and Long Short Term Memory Neural Networks. It may also be useful for the analysis of Generative Adverserial Neural Networks and Convolutional Neural Networks, since many of them incorporate Feedforward Neural Networks.

(2) Symbolic Approaches: Symbol-Based AI Models: There are many different types of symbolic models that can be used to create AI systems. Nearly all of them are structured and constructed to solve a specific class of problems or to fulfill a specific task they are explicitly good at. This requires almost always that their domains are formally defined and a problem can precisely be described over an exact set of rules and symbols the symbolic AI model can handle [7]. To cover this, we will try to work with very common types of symbolic models, which are often used and can be adapted to efficiently solve a variety of domain problems. By doing this, we hope to increase the adaptability of our idea for a greater number of use cases.

One of the most common structures of symbolism are trees. They are, for example, widely used in searching, either as decision trees to search a match for a set of conditions or as behavior trees to describe sequential behavior [4,11]. Decision trees are further well suited for our approach to extract a symbolic AI model from a Feedforward Neural Network because they are both used to calculate some kind of classification by making decisions over certain aspects of situations or traits of objects. That is why we work with decision trees in our first attempt.

Another symbolic model, we plan to incorporate into our idea in the future, are state machines. State machines are also frequently used in many decision making scenarions [4] and are widely known in the field of machine automation as well as simulation and computer games [11]. For future work, we plan to specialize in Moore machines when it comes to state machines. This is no restriction, since Mealy machines and Moore machines can be converted into each other.

For the moment, we will focus on tree structures, especially decision trees, when regarding symbolic AI models.

1.2 Comparison of Artificial Neural Networks with Symbolic Tree Structures and State Machines

As we already mentioned, the internal functionality of artificial neural networks differs significantly from the way how symbolic AI models work. The following four differences are crucial for us. This is exemplified by the Feedforward Neural Network and the symbolic tree structure in Fig. 2:

(1) There are differences between the input and output values of artificial neural networks and symbolic systems. Artificial neural networks receive a vector of numerical values as input. Its dimension conforms to the number of input neurons. They further produce a vector of numerical values as output and its dimension conforms to the number of output neurons. In Fig. 2, these vectors are (I_1, I_2, I_3) and (O_1, O_2). Symbolic AI models, for example decision trees, receive a single symbolic value as input and produce symbolic values as output. An example is shown in Fig. 2, where I is the input value and O_1, O_2, and O_3 are the output values.
(2) An artificial neural network is capable of adjusting the strength of its connections with the help of backpropagation algorithms. With this ability, it can change its internal signal processing to achieve better results in predicting correct outputs that match given input signals. A symbolic tree structure generally lacks this ability. Its connections are not weighted, as Fig. 2 shows, and thus cannot be adapted. Therefore, the tree will always produce the same output for a given input.
(3) Because of artificial neural networks weighted connections, a neuron's successor is activated by the signal of its predecessor corresponding to the weight of their connection. In a symbolic structure, on the other hand, a symbol is either completely processed or not processed at all. There is nothing like a 'partial transfer' of a symbol.
(4) Values in artificial neural networks are processed simultaneously in every phase and purely time-driven. In symbolic AI, the symbols are processed sequentially and sequences are triggered by incoming symbols, for example, characteristics in decision trees (the symbols $(S_1, S_2$, and $S_3)$ in Fig. 2) and transition symbols in state machines.

Because of these differences, artificial neural networks are in general good at performing tasks of pattern recognition. This includes recognizing visual patterns like traffic signs and handwritten texts or audio classification [3]. To complete those tasks, artificial neural networks are not explicitly handcrafted for a single problem but created with specific structural characteristics, including the internal structure and the training algorithms, to perform well, when confronted with a special type of problem. This makes them useful for solving problems that underly constant changes and that are difficult to verify. A resulting drawback is that we do understand how signals are distributed in an artificial neural network, but not the logical decision process that is simulated by its behavior. Symbolic AI models on the other hand are widely used for the automation of restricted processes, for example, movement control for industrial machines and robots, and the development of agent-based or rule-based models like supply chain management or simple AI models for computer games. They are well suited when a formal definition is available to solve a problem. Without this definition, a problem cannot be suitably described with symbols and strict rules to solve it. Besides, unknown or constantly changing boundary conditions and

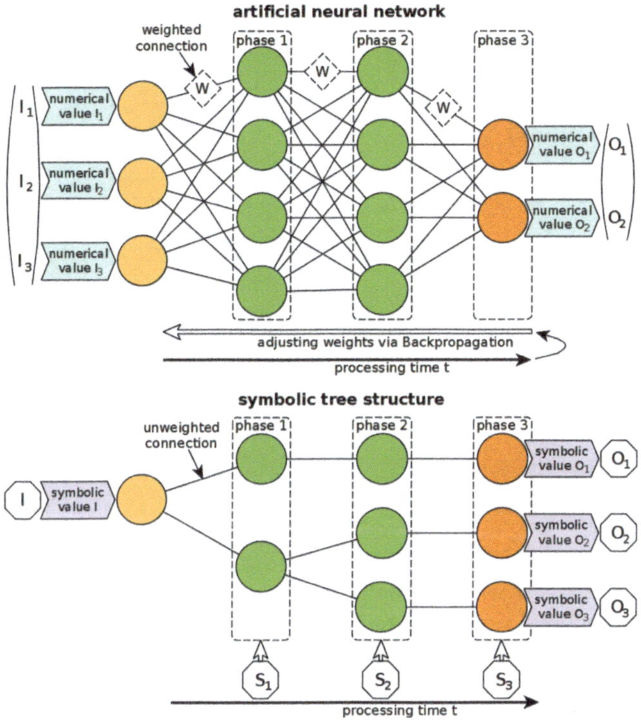

Fig. 2. Comparing the functionality of a Feedforward Neural Network with a symbolic tree structure.

aspects of such problems make it as well difficult to correctly solve them with inflexible symbolic AI systems.

One aspect of artificial neural networks that cannot be incorporated into symbolic AI models, is their learning behavior and, as a result, their ability to adapt to changing problems. Therefore, we do not try to take the ability to learn and adapt into account when we try to extract decision trees from Feedforward Neural Networks. We use Feedforward Neural Networks at a specific point of training time and extract decision trees with a behavior that is equivalent to the Feedforward Neural Network's actual training level. When these networks receive further training, their behavior evolves and a new, supposedly slightly different, decision tree can be extracted.

For future work, we will also attempt to convert a symbolic model into an artificial neural network. We will try to extract an artificial neural network's internal structure and its relevant connection weights. Additional connections with new marginal, but non-zero, weights and the complete learning behavior will be added to the artificial neural network manually to provide its ability to learn, because symbolic models simply lack that aspect. To sum it up, we always treat symbolic AI models as equivalents to only a snapshot in time of an artificial neural network at a specific state of its training, as it is visualized in Fig. 3.

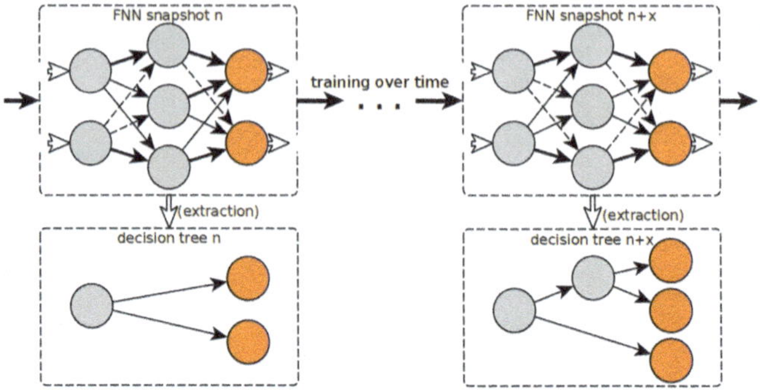

Fig. 3. The idea of extracting decision trees from Feedforward Neural Network snapshots in training time.

2 Related Work

In this section, we will introduce the basics of Paul Smolensky's comparison of connectionist systems with symbolic systems and his resulting Integrated Connectionist/Symbolic Cognitive Architecture. We will focus on the filler/role principle and the description for distributed representations of symbols it provides, which will serve as a basis for our work [8].

Smolensky states on one hand that most domains of human cognition treat problems as systems of functions, which calculate output based on given input over structures built of symbols. On the other hand, he describes that, according to neuroscience, the human brain operates through massively parallel computation of very simple calculations in a biological neural network. This lets him assume that there is a connection between the way a Neural Network performs calculations and the way calculations are structured in formal symbolic systems [9,10]. Based on this assumption, he developed his Integrated Connectionist/Symbolic Cognitive Architecture. This architecture consists of three levels, a symbolic level that describes a representation for symbolic models, a vectorial level that is used as transition between the other two levels, and a neural level that offers a description for connectionists models, capable of simulating calculations in symbolic systems. As his primary field of application, Smolensky investigates natural languages, which are described with formal, symbolic rules (grammars) and are processed by humans with the help of a biological neural network, their brain. The following two key features from this theory are important for our work.

First, a symbol is recognized as a combination of a filler and a role. The filler is some kind of representation and the role is something like a binding or position. It can be regarded as some kind of context, in which the filler stands compared to the rest of the symbolic model. As an example, we use the word 'hello' and the letters it consists of as shown in Fig. 4. In this figure, one can clearly see that both, filler and role, are important to determine the meaning of every single letter and that all five letters are required to be shown correctly to

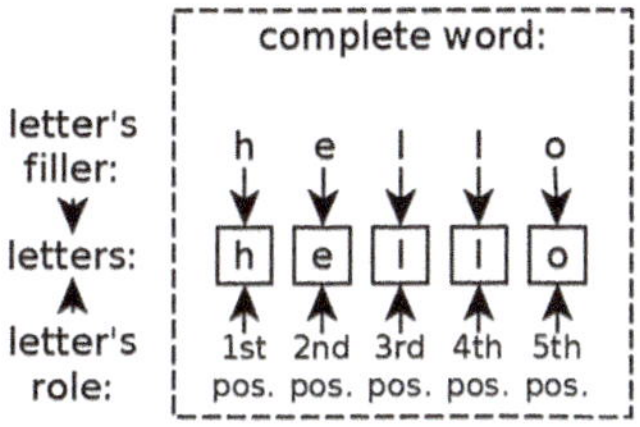

Fig. 4. The five letters building the word 'hello' with their corresponding fillers and roles.

create the word 'hello'. It is obvious that switching a letter's filler, for example changing 'h' to 'c' in the first position, creates a new word, 'cello'. One can also see that the two letters 'l' are not the same, because they have a different role and are therefore not redundant. The reason for this is that both have a different context in the word, in this case their position in it. We will refer to this as the filler/role principle [8,10] in the following.

Second, a word is seen as a vector, which is the mathematical representation of a distributed pattern of activations of neural units. Any symbol can be represented as a number of neuron activations in an artificial neural network or as part of such an activation pattern. Let us assume that we have an alphabet consisting of four basic letters and let those letters be 'e', 'h', 'l' and 'o'. They can be represented with activation patterns over three neurons, which can either be positive active, negative active or inactive as pictured in Fig. 5.

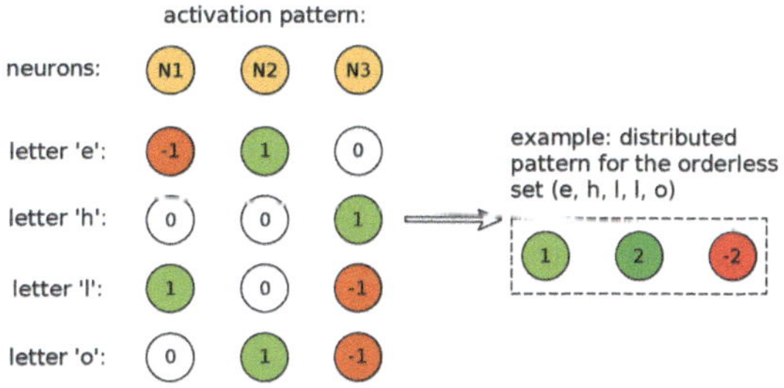

Fig. 5. The example of an activation pattern realizing a five-letter-alphabet.

In this example, all four letters are represented with the same three neurons and any orderless set consisting of a freely chosen number of the single letters can be explicitly represented. For this representation, the single activation vectors presenting a single filler of a letter are superpositioned. To ensure that this is possible, it is necessary that the activation vectors representing the single letters are linearly independent [8]. But up to this point, this set of letters is, as mentioned, an orderless set and lacks any information about the letters context. So this single activation vector can only represent the filler, but lacks the context.

The roles have to be incorporated into the distributed representation. Therefore, Smolensky introduces a second activation vector that represents only the roles of the context bound letters. Because the distributed representation for that vector can again be freely chosen, besides the restriction that the representations for the different roles have to be linearly independent, this vector cannot simply be super-positioned with the filler vector. This would require all single filler and role representations to be linearly independent or the representation could suffer from a loss of information during processing. The solution is to combine the filler vector and the role vector with the help of tensor multiplication as shown in Fig. 6.

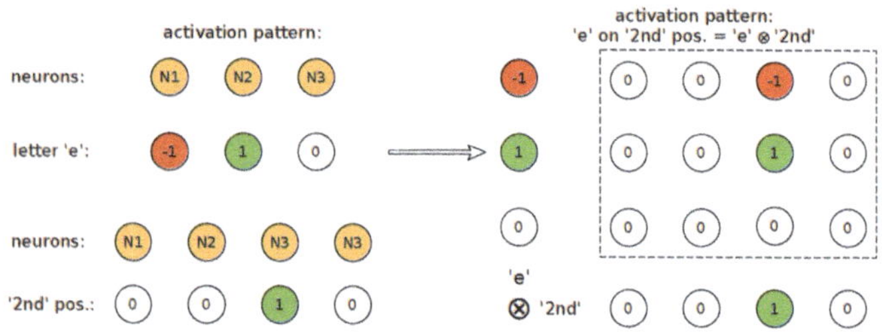

Fig. 6. The activation pattern realizing the letter 'e' at the second position, computed by combining the activation vectors of 'e' and '2nd' with tensor multiplication.

The result of $vector_e \otimes vector_{2nd}$ is a tensor product that can itself distinctly be associated with a specific letter at a specific position. These tensor products can again be super-positioned to create a set of letters, now all with their specific position. If no role vector appears twice, the set created by these super-positioned tensor products can be viewed as the distributed representation of a word [8]. A number of role vectors can be combined by using tensor multiplication before combining the resulting tensor product with the filler vector by tensor multiplication. The result can be viewed as the complete role of that symbol with all corresponding context information.

In this paper, we use the idea of symbol defining context in the form of super-positioned vectors of characteristics, as will be shown in Sect. 4. For decision trees, the combination of two role vectors can be the position behind two branches. If the path to the correct symbol follows the left branch and then the right branch, afterwards, its role is left, then right. This is represented by the tensor product '$\mathsf{1st}_{\mathsf{left}} \otimes \mathsf{2nd}_{\mathsf{right}}$'. This combination of different role vectors can be repeated recursively [8].

3 Statement of the Problem

In this section, we will describe the primary challenges we have to cope with when extracting a symbolic AI model from an artificial neural network. We will present the relevant questions to be answered in order to establish our method.

It is generally hard to comprehend decision processes of artificial neural networks. They may work similar to biological neural networks and were created with our own nerve systems as a prototype. But this is only true when it comes to the computational level. On a more abstract level, humans do not think in massively parallel calculated activation patterns. Abstract human thinking, the creation of concepts, is more related to idealized and generalized ideas. Humans express these ideas in prototypic characteristics, symbols, which are bound in special relationships to each other. The best example for this is mathematics or logic. We are used to process those symbols step by step, dependent on their contextual relations. And because we think this way when it comes to higher cognition, we have constructed most of our classic AI models and steering concepts that way for all sorts of automation. This implies all the benefits and downsides mentioned in Subsect. 1.1. That is why artificial neural networks are deployed to solve problems connected to complex real life data that are not explicitly formalized. But artificial neural networks have their advantages and disadvantages, too, as stated in Subsect. 1.1. The solution we propose is to create equivalent symbolic models from artificial neural networks. Trying to solve this challenge raises a number of questions that stem from the structural differences and the different functional concepts that artificial neural networks and symbolic AI models have. For this paper, we focus on extracting decision trees from Feedforward Neural Networks to create a basis for working on this solution. Most importantly, we do not try to take the ability to learn and adapt over time into account. The questions we have to answer to reach our goal are the following:

- How are single symbols, specific characteristics or traits an object or a situation can have, encoded in artificial neural networks and how can they be extracted?
- How can we match the parallel processing of data in artificial neural networks with the generally sequential methods of processing data in symbolic models?

After we have described how to extract decision trees from Feedforward Neural Networks, the idea is to derive further progress from this basic starting point. For future work, the next step will be to create equivalent state machines in addition to decision trees. This could broaden our approach to take a significant number of different problem scenarios into account. The next logical step will be to consider Recurrent Neural Networks to further enhance our method's potential. The basis for all this is the connection between Feedforward Neural Networks and decision trees, which is described in Sect. 4.

4 Approach to Extract Symbolic Models from Artificial Neural Networks

In this section, our approach to extract a decision tree from a Feedforward Neural Network will be presented. In Subsect. 4.1, we will describe how our method is derived and what simplifications we make. In Subsect. 4.2 we will describe our proposed method.

4.1 Derivation of the Chosen Approach

According to Sect. 2, in a symbolic system, a symbol is a representation or placeholder for an idea or thought, either as the result of an information processing operation or as an intermediate product of such an operation. It consists of the filler, the image or placeholder of this representation, and the role that holds all the contextual information how such a representation is embedded in the information processing operations. Smolensky visualizes the idea of fillers and roles with the nodes and leaves of binary trees, defining how words are built out of letters in combination with their positioning. These trees have the fillers or letter symbols as leaves and the combination of branches that lead to these leaves as roles [8,10]. In terms of other symbolic models like decision trees, the symbols are also built by combining their images and their positions in the directed graph. These symbols then describe a specific intermediate set of still possible solutions. The concept is illustrated in Fig. 7. In the binary tree, the possible fillers for a resulting symbol of the calculation are 'a', 'b', 'c', and 's'. The correct roles for those fillers are 'll', 'lr', 'rl' and 'rr'. The decision tree is almost the same and also reveals fillers of intermediate nodes. They are 'a|c' and 'b|s' with their roles 'l_' and 'r_'. These nodes are what we called intermediate products. Starting from the initial node 'a|b|c|s', if we follow the 'l_' branch, we reach the state 'a|c' that is defined by the filler/role combination '(a|c; l_)'. On the other hand, if we follow the 'r_' branch, we reach 'b|s', defined by the combination '(b|s; r_)'. The same principle can be used to reach the four leaves of the decision tree shown in Fig. 7. For example, if we first follow the 'r_' branch and after that the '_l' branch we reach the state 'b' that is defined by '(b; rl)' with '$\text{rl} = \text{r\_} \otimes \text{\_l}$' according to [8]. This results in the leaf 'b' being defined by the filler/role combination '(b; $\text{rl} = \text{r\_} \otimes \text{\_l}$)'. Therefore, a node in a symbolic model that can be presented as directed graph, can be defined with Smolensky's filler/role principle. It uses the name of the node as the filler and the path to this node in a given directed graph as its role. With this in mind, we use Smolensky's filler role principle not only for binary trees but also for other symbolic models. To create any combination

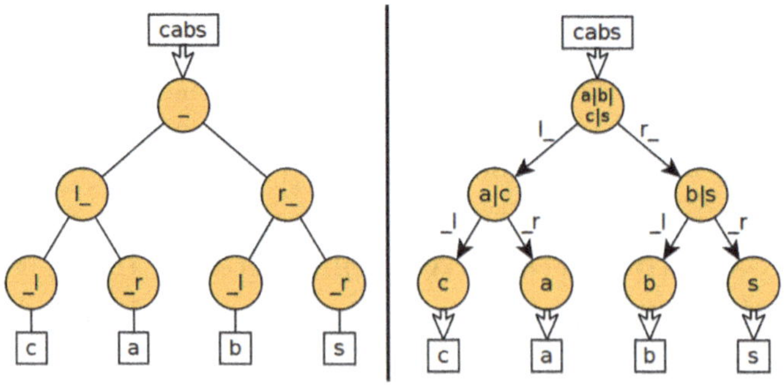

Fig. 7. The concept of filler and role of a symbol in a binary tree [8] and in an equivalent decision tree.

of fillers and represent complex contexts for symbols, the vectors of single fillers and contexts are combined with tensor multiplication to create the complete filler for a single symbol [8]. This also allows complex fillers to be interpreted as path through the directed graph of a symbolic model.

Both trees in Fig. 7 compute exactly the letter of the word 'cabs', which equals the letter standing at the $\mathsf{n_{th}}$ position in this word. The computation needs the fillers and the roles of these letters. In Fig. 8, a slightly more complex example is shown, which demonstrates that this principle also works for state machines. The Moore machine on the right side of Fig. 8 also computes the letter of the word 'cabs' on position 'n'. If it knows this letter, it can compute the letters on the positions '$n+1$' and '$n-1$' as well. Again, the filler is the picture of the letter and the role is the combination of all context information that is connected to that filler. Because the Moore machine is more complex than a tree, there is more context information connected with each symbol. For example, the first letter 'c' requires the filler 'c' and the role '1st pos' when the complete word is given. When the second letter 'a' has just been computed, only the filler 'c' and the role 'l_' are required to compute the first letter 'c'. Because the second letter 'a' is composed of its filler 'a' and its role as the second letter in the word cabs, you just need the role 'l_' (left) to receive the first position in the word as new role. The needed role 'l_' should satisfy the condition '$\mathsf{1_{st}pos} = \mathsf{2_{nd}pos} \otimes \mathsf{l\_}$'. With these conditions in mind, the filler/role principle can also be applied to state machines like Moore machines.

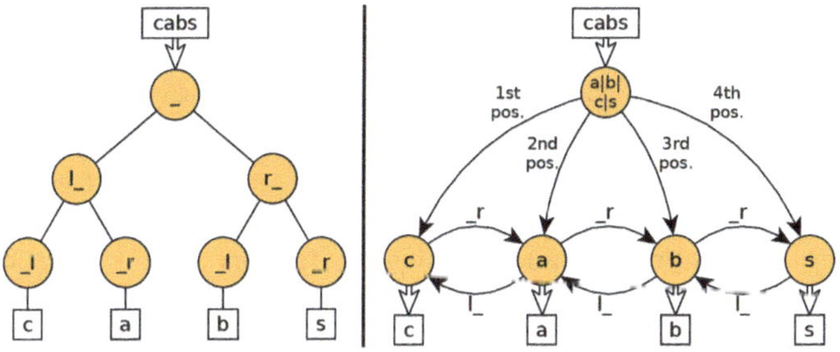

Fig. 8. A moore machine that calculates the letters from the word cabs on a given position in that word, compared to the binary tree from Fig. 7.

Symbols can therefore consist of a set of context information. The combination of this information is the same, regardless if it is embedded in a symbolic model or in an artificial neural network. The difference lies in the representation of this combination. The representation can be distributed in a number of single symbols. These symbols emerge from each other, timestep by timestep, as it is in a decision tree or state machine. It can also be distributed in a number of activations that are all calculated and presented simultaneously in one single timestep. The important feature is that the combination itself remains.

In our approach to extract an equivalent symbolic model from an artificial neural network, we focus on providing a basis for future research in this field and therefore make a few simplifying assumptions:

- We only look at Feedforward Neural Networks as the artificial neural network model. We want to provide the basis for a couple of more complex neural network models like Recurrent Neural Networks. This way, the principles we describe for Feedforward Neural Networks can later be reused or expanded when examining more complex models that derive from Feedforward Neural Networks or include them, as pictured in Fig. 1.
- Concerning Feedforward Neural Networks, we only work with the activation values 'active' and 'not active' for single neurons. This is the simplification of the fact that, for a single neuron's influence on the calculation, it is determining if this neuron is active or not.
- When using symbolic models for comparison, we will stick to tree structures like decision trees, first, to ensure that our ideas can be developed step by step. We will later also consider symbolic models with more complex structures, like state machines.

When discussing potential enhancements later in Sect. 5, we will take a look at how we could handle the restrictions that result from these simplifying assumptions.

4.2 Method to Extract Decision Trees from Feedforward Neural Networks

We stated in Subsect. 4.1 that the distributed representation of the output signals of an artificial neural network have to match with the output symbols of its equivalent symbolic model because both are the results of the neural networks, respectively the symbolic model's computations. If we treat both as equivalent, their outputs should produce an equivalent statement. In that case, the vector of output signals from a neural network and the distributed pattern of characteristics it reassembles should equal the produced symbols and the characteristics they are combined of because their combination forms an equivalent statement. This idea is visualized in Fig. 9, where a simple example of a decision tree dividing objects into cars, planes, and boats and a small Feedforward Neural Network with one hidden layer, doing the same, are presented.

If we look at the example in Fig. 9, we see that the pictured Feedforward Neural Network and the given decision tree compute whether a given object is either a boat, plane, or car. It is a simple classification problem and thus typical for both, Feedforward Neural Networks and decision trees. Both models have to compute the same classes and the obvious difference is not their result but how this result is encoded. What is presented in the decision tree in the form of the symbol 'plane' is a vector of activation signals in the neural network. Because this is the activation vector of the output layer, it is known what type of classification the activation pattern represents. The neural network was either constructed or

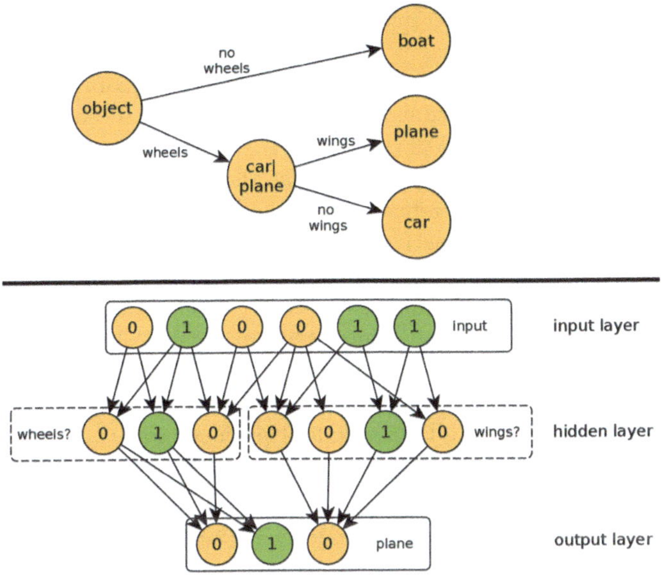

Fig. 9. Example to show how the patterns for specific characteristics from symbolic models might be computed in artificial neural networks.

pretrained to match this specification. That means, for every activation vector in the output layer, we know the resulting classification it belongs to. The same is true for the input layer. We know by construction which activation pattern belongs to which single information about the object. In the case of a picture from a vehicle that might be a pixel with its colour, the filler, and its position in the picture, the role. It might also be a single frequency for a music play or some other sort of particular information. To compute one of the possible classifications, the decision tree checks the given object for specific traits like wheels or wings. To recognize these traits, it is necessary that the objects are explicitly marked to have or not have them. The neural network computes its activation vector for the output layer by combining the activation values of the layer above, according to the weighted connections between the neurons in the output layer and the neurons in its predecessor layer. The same is true for any other layer above, except the input layer. In the end, the neural network combines the initial particular information the object consists of, with mathematical operations over weight matrices to new activation vectors, which then represent new filler/role combinations. At a certain point in the computation process, for a Feedforward Neural Network in a certain layer, the artificial neural network has computed a characteristic trait in form of an activation vector that is, beginning from that layer, the only activation vector with influential weighted connections to a specific part of the output layer defining a single categorization decision. If this is true, this characteristic in form of an activation vector equals a symbol, required to make a decision in a decision process that could be modelled with a symbolic system like a decision tree.

We illustrate the idea with a small example and the help of the neural network in Fig. 9. Assume that the output activation vector $(0, 1, 0)$ equals a plane, the vector $(0, 1, 1)$ equals a car and the vector $(1, 1, 1)$ equals a boat. This means that the first neuron being inactive indicates that the object cannot be a boat and that this neuron being active indicates that the object cannot be a plane or car. In return, this neuron being active or inactive must represent a characteristic that either all boats have but all cars and planes lack, or a characteristic that all planes and cars have but all boats lack. A closer look at the first output neuron reveals that its activation is only influenced by the neurons 1, 2, and 3 of the hidden layer. Therefore, the activation pattern of only these neurons represents the mentioned characteristic. When the activated output neuron stands for the characteristic being present, all possible activation vectors of the three neurons that would reach the threshold for its activation with the weighted sum of their activations ensure that this characteristic is present. Regarding the example in Fig. 9, let us assume the output neuron becomes active if it receives at least an input of the value 1, hence the threshold is 1. The weights for all three connections at the given state of training are 0.5, then at least two of three hidden neurons have to be active to activate the output neuron. The active output neuron stands for a characteristic that no plane and no car has, but a boat has. Resulting from that, for the three hidden neurons in our example, the activation vectors $(0, 1, 1)$, $(1, 1, 0)$, $(1, 0, 1)$ and $(1, 1, 1)$ are the distributed representation of a trait that a boat can have but cars and planes cannot. In our example, this would be 'no wheels', but it can be any trait that matches the mentioned condition. Whatever trait it might be, it would be used by this network to distinguish boats from cars and planes. This backcomputation of the activation vector for a specific trait can now be recursively repeated by using the new valid activation vectors for the examined trait to compute all activation combinations in the input layer that lead to said trait being present. We generalize this procedure. The idea is sketched in Fig. 10.

The green and red circles in layer n are the neurons, which are part of the activation pattern that realizes the examined characteristic. The color green means that the neuron has to be active and red means it has to be inactive. If it is not important if a neuron is active or inactive, this neuron is not part of the pattern. To become active, the activation function of this neuron has to be triggered and therefore the activation threshold of this function has to be reached. It is exactly the other way round for an inactive neuron whose activation threshold must not be reached. To receive the numbers that can be compared to a neuron's threshold, all neurons that have a non-zero weighted connection to that neuron have to be taken into account. The outputs of those neurons, themselves being a combination of the neurons activation values and their output functions, have to be combined with the weights of their connections to the target neuron. Normally, the weight is multiplied with the output. In the case of the example shown in Fig. 10, we assume that the activation function of neuron 1 is a sum, as it

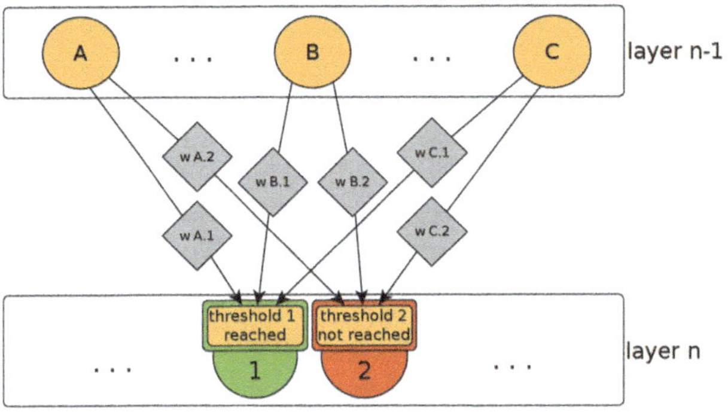

Fig. 10. Needed components for the backcalculation of characteristics of single traits from one layer to the previous one.

often is. This function receives $\mathsf{output_A} * \mathsf{w_{A.1}} + \mathsf{output_B} * \mathsf{w_{B.1}} + \mathsf{output_C} * \mathsf{w_{C.1}}$. Afterwards, it is checked whether the resulting number reaches the threshold for the activation. It is the same for neuron 2, except that the threshold must not be reached. We summarize this idea in Fig. 11 [1].

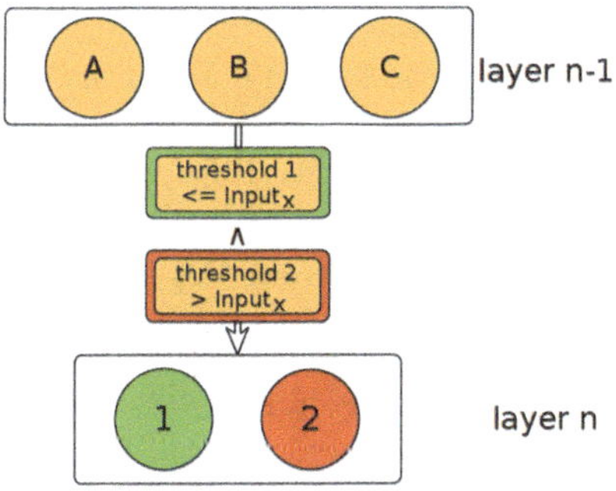

Fig. 11. Principle of backcalculation characteristics of computational results or single traits from one layer to the previous one.

As sketched in Fig. 11, for all single neurons x that are part of the activation pattern of a single characteristic or trait, the $\mathsf{input_x}$, their activation function receives, is:

- $\mathsf{input_x} = \mathsf{f_{input_x}}(\mathsf{W_x})$
  with $\mathsf{W_x} = \{(\mathsf{w_{i.x}} * \mathsf{output_i}) \mid \mathsf{i} =$ neuron with non-zero weighted connection to x $\}$
  and $\mathsf{output_i} = \mathsf{f_{output_i}}(\mathsf{activation_i})$

Often, f_{input_x} is a sum, so $input_x$ is often calculated by $input_x = \sum_i (w_{i.x} * output_i)$. But in general, it can be any function. When all inputs are calculated it is important if the neuron is required to be active or inactive. Considering all active neurons x in the trait's activation vector, it has to be true that $input_x \geq threshold_x$. For the inactive neurons it has to be true that $input_x < threshold_x$. As a result, only those activation vectors $vector_{act_{n-1}}$ of connected neurons from 'layer $n-1$' can be a representation of the characteristic from the activation vector $vector_{act_n}$ of 'layer n', which fulfill the following requirement:

- $\forall$ $activation_i \in vector_{act_{n-1}}$ has to be true that
  $\forall$ $neuron_{x_a} \in vector_{act_n}(input_x \geq threshold_x) \wedge$
  $\forall$ $neuron_{x_i} \in vector_{act_n}(input_x < threshold_x)$
  with $neuron_{x_a}$ being an active neuron, and
  with $neuron_{x_i}$ being an inactive neuron.

By fulfilling these conditions, an activation vector in layer 'n' qualifies as a valid vector to propagate the examined trait further through the neural network. Consequently, the complete representation of said trait in layer '$n-1$' is the set of all valid activation vectors that fulfill the conditions mentioned above. These representations of traits can be found in every layer of the neural network. This is true because the output layer of the neural network provides all characteristics included in the data, the neural network receives, as we have shown earlier. And beginning from the output layer, they can be computed step by step for every layer of the neural network above the output layer as we described above. Considering the functionality of artificial neural networks to process all data simultaneously and that at least in a Feedforward Neural Network, every processing step takes place between two layers, this seems to be understandable. We argue that we are now capable of doing the following:

- We can extract the characteristics that specific outputs, such as categories or neural network states, need to have and those characteristics they must not have, by analyzing the output layer.
- We can further compute the representations of those characteristics in all other layers up to the input layer.

This enables us to identify all sets of valid activation vectors for these characteristics in the input layer. Therefore, we should be able to describe what specific parts of input are connected with specific characteristics the neural network works with. This idea is similar to the concept of Deep Dream [12], but we think it has a significant advantage. Deep Dream highlights everything that is important for the artificial neural network to classify a specific object. This can be seen in the pictures produced with that system. We try to get an activation vector that can be viewed as an equivalent of a specific highlighted feature. That way, we do not just get the complete set without being able to distinguish its single components from each other, but we get all of them separately. When we know all the single features, we know the symbols, the abstract ideas, which the artificial neural network searches for in the input data to make its decisions. In terms of our initial example from Fig. 9, this means that the activation vector for the output 'boat' always has the trait we marked with 'no wheels' while 'car' and 'plane' always incorporate the trait marked with 'wheels'. We now also know the valid activation vectors for the input layer that are a representations of those traits. To recognize what kind of specific trait is encoded in those sets of activation vectors, a human analyst could then survey the input combinations that result from those vectors. With reference to our example, all the activation vectors that represent the trait 'wheel' are related to pictures that show some form of wheel-like circle. By looking at a sufficient number of those input signal combinations and regarding the possible classifications, it can be seen that the represented trait is indeed 'wheel'. The sets of activation vectors corresponding to the single characteristics, and only to those characteristics, can be computed automatically this way. However, tagging these activation combinations with a specific symbolic name must either be done by a human, by analyzing those sets, or the tags are known from the construction of the input layer of the Feedforward Neural Network. For future work, we plan to incorporate further thoughts for automating the tagging process.

We can now directly determine those combinations of input signals that guarantee that the result of the neural network's computation can only be 'boat' on one side or 'car' or 'plane' on the other side. The overall allocation of possible traits and classifications in our example is visualized in Table 1 and the associated equivalent trees are shown in Fig. 12.

Table 1. The combinations of input signals resulting in the categories 'Boat', 'Car' and 'Plane'

no wheel	wheel	
no wings		wings
boat	car	plane

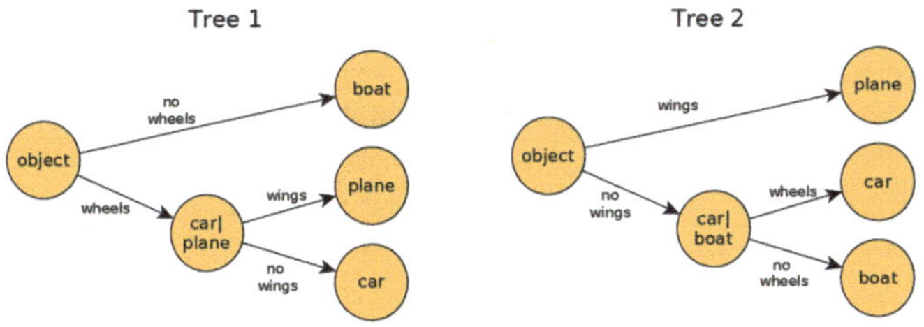

Fig. 12. The two possible decision trees that can be built as a result from the allocation visualized in Table 1.

With this knowledge, we are able to re-engineer these equivalent decision trees for a Feedforward Neural Network. Each of them is sufficient, because the step-by-step decisions of the trees are computed all at the same time in the neural network and the resulting categorizations are the same.

5 Possible Enhancements for Future Work

In this section, we will have a look at the potential to generalize our proposed procedure for the use with other, more complex, artificial neural networks and other symbolic AI models. We will further discuss how we can overcome some of the restrictions we made, which are summarized in Table 2:

Table 2. The restrictions connected with the presented approach

restriction	considered improvements	possible related problems
only desision trees considered	incorporating Moore machines	—
only Feedforward Neural Networks considered	incorporating Recurrent Neural Networks	recurrent neurons could cause unpredictable large amounts of processing steps
only 'active' & 'inactive' are valid activation values	incorporating real-valued activations & still search for thresholds	computational overhead might become extremely large
arithmetical & logical operations required	incorporating tensor-based filler/role multiplication	limitations regarding the neural network models that can be processed

First, we consider the restriction that neurons of artificial neural networks should only be 'active' or 'inactive' ('1' or '0'). As a result, we have to worry about two activation values, only. In general, neurons can have any positive or negative activation value, it just depends on the structure of the neural network. But if you take a closer look at the method to get the activation vectors corresponding to specific traits in higher layers, which we introduced in Subsect. 4.2, you see that we used a combination of activation thresholds. To overcome the restriction, we just need to determine, whether these thresholds of the single neurons in the considered trait's activation vector are reached or not. In the end, this comes down to the comparison of two real-valued numbers, one being greater, lesser, or equal to the other one. The result of such a comparison is a Boolean and therefore, can again only have two values. If we cut the restriction that neurons can only have two activation values and allow real-valued numbers for activation, the following consequences have to be considered:

- The principle to calculate sets of valid input activation vectors for specific traits an artificial neural network uses to make decisions can still be utilized.
- The computational overhead that is necessary to decide if a single neuron in layer 'n' will be activated by an activation vector in layer '$n-1$' can heavily increase because it would be necessary to compute real-valued ranges of valid activations for the neurons in layer '$n-1$'.

We hope that we can use the filler/role principle and its proposed direct tensor calculation from [8] to reduce the complexity connected to that issue in the future, but this requires further research.

Now, we survey the restriction that limited us to the model of Feedforward Neural Networks on the side of connectionists models. The advantage when considering only Feedforward Neural Networks is that every neuron that is important for the same specific processing step is located in the same layer of the neural network. According to Fig. 1, the artificial neural network model that is most similar and more potent than Feedforward Neural Networks are Recurrent Neural Networks. They additionally have recurrent neurons [1]. Regarding the applicability of the method we described in Subsect. 4.2, we have not taken a look at them, yet. This is planned for future work.

We also spare a few thoughts about the applicability of our concept for other symbolic structures that are not based on trees. Especially, we are interested in being able to use this concept for the creation of state machines like the one shown in Fig. 8. This special interest in state machines derives from our long term goal to find equivalent symbolic models for artificial neural networks, involving all those models that are very common in the AI field. State machines definitely fulfill this requirement [11]. The difference between decision trees and state machines is that state machines have access to information referring their internal state. Our assumption is that we have to extract the following information from an artificial neural network to create an equivalent state machine:

- The states, a system can have, and how these states are characterized.
- The transitions between those states, consisting of their starting state, their target state and the signal or command that triggers the transition.

We take Moore machines as our favored state machine model, because in Moore machines, the output symbols are only dependent on the state they are in. This way, we can directly link the state and the output signals. In other words, if a system reaches a new state, this state is essentially defined by the output. To go more into detail, we use an example state machine representing a locked door, which is visualized in Fig. 13.

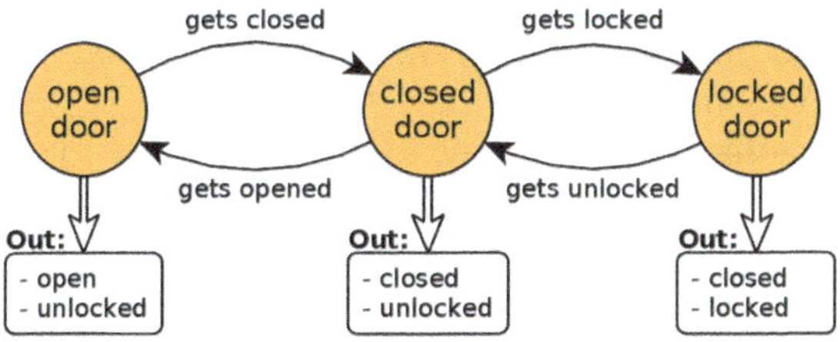

Fig. 13. A simple example of a state machine representing a door, which can be locked, unlocked, closed, and opened.

We can see that the states are sets of characteristics that define the described system. They form the system's output symbols and define in which state the Moore machine is in a specific processing step. Those are the same sets of characteristics that we can find in Feedforward Neural Networks. This leads us to the idea of extracting the states, we need to build state machines by combining the specific characteristics we can extract from Feedforward Neural Networks with the procedure we have introduced in Subsect. 4.2. What we do not know at the moment is how the transitions can be extracted from or encoded in the artificial neural network. It requires future research to work on a solution for this. But our assumption is that they can be extracted from the differences in the activation vectors that encode the two states the transition connects. As a next step. We plan to focus on the extraction of complete state machines in the form of Moore machines.

6 Conclusion

In this paper, we first gave an overview of the diverse types of artificial neural networks and symbolic AI models to find a good starting point for the creation of equivalent models. We decided to take Feedforward Neural Networks, because they are commonly used and they form a good basis to incorporate other, more complex artificial neural networks in the future.

Regarding symbolic AI models, we have chosen trees and state machines. Both are common, too, and decision trees are a good starting point for our work in combination with Feedforward Neural Networks. After we looked at differences and similarities, we described an idea of how to extract symbols from Feedforward Neural Networks in the form of sets, composed of activation vectors. These activation vectors were presented as valid representations of specific characteristics that are used by the neural network to compute decisions like the classification of objects. Again, we want to make clear that this is just an initial idea and there is much more work to do. We mentioned the biggest limitations we see so far with this approach in Subsect. 4.2. To solve the most urgent challenges, we will focus on two questions in the future: First, we want to fully incorporate state machines into our idea to cover an additional, widely used and well investigated symbolic AI model. Second, we will try to utilize Smolenskys filler/role theory to make calculating the sets of activation vectors for specific characteristics more efficient.

In summary, we think that the basic connection between Feedforward Neural Networks and decision trees, which was presented in this paper, is a good starting point for further research in the field of connecting artificial neural networks and classic, symbol-based AI models.

Acknowledgment. The authors would like to thank Franz Schmalhofer for his many constructive, open-minded discussions regarding our ideas. We also would like to thank Wolfgang Hommel for polishing our paper and helping us to identify future challenges. We highly appreciate their support.

References

1. Goodfellow, I., Bengio, Y., Courville, A.: Deep Learning Cambridge. The MIT Press, Massachusetts (2017)
2. van Veen, F.: The Neural Network Zoo. The Asimov Institute, Utrecht (2016). https://www.asimovinstitute.org/neural-network-zoo/
3. Schmidhuber, J.: Deep Learning in Neural Networks: An Overview. University of Lugano & SUPSI - Istituto Dalle Molle di Studi sull'Intelligenza Artificiale, Manno-Lugano (2014)
4. Russel, S., Norvig, P.: Artificial Intelligence: A Modern Approach, 3rd edn. Pearson Education Inc., Upper Saddle River (2010)
5. Garson, J., Zalta, E.N.: Stanford Encyclopedia of Philosophy. Metaphysics Research Lab - Stanford University, Stanford (2016). https://plato.stanford.edu/archives/win2016/entries/connectionism/
6. Reingold, E., Nightingale, J.: Artificial Intelligence Tutorial Review. Department of Psychology - University of Toronto, Mississauga (1999). http://www.psych.utoronto.ca/users/reingold/courses/ai/symbolic.html
7. Minsky, M.: Logical vs. Analogical or Symbolic vs. Connectionist or Neat vs. Scruffy - Artificial Intelligence at MIT, Expanding Frontiers. The MIT Press, Cambridge (1990)
8. Smolensky, P., Legendre, G.: The Harmonic Mind - Volume 1: Cognitive Architecture. The MIT Press, Cambridge (2011)
9. Smolensky, P.: Connectionist AI, Symbolic AI, and the Brain - Artificial Intelligence Review. Springer-Verlag GmbH, Heidelberg (1987)
10. Smolensky, P.: Symbolic Functions From Neural Computation. The Royal Society Publishing, London (2012)
11. Millington, I., Funge, J.: Artificial Intelligence for Games, 2nd edn. Morgan Kaufmann Publishers, Burlington (2009)
12. Mordvintsev, A., Olah, C., Tyka, M.: Inceptionism: Going Deeper into Neural Networks. Google Research Blog (2015). https://research.googleblog.com/2015/06/inceptionism-going-deeper-into-neural.html

A High Performance Classifier by Dimensional Tree Based Dual-kNN

Swe Swe Aung(✉), Nagayama Itaru, and Tamaki Shiro

Information Engineering, University of the Ryukyus, Okinawa, Japan
{sweswe,nagayama,shiro}@ie.u-ryukyu.ac.jp

Abstract. The k-Nearest Neighbors algorithm is a highly effective method for many application areas. Conceptually the other good properties are its simplicity and easy to understand. However, according to the measurement of the performance of an algorithm based on three considerations (simplicity, processing time, and prediction power), the k-NN algorithm lacks the high-speed computation and maintenance of high accuracy for different *k* values. The k-Nearest Neighbors algorithm is still under the influence of varying *k* values. Besides, the prediction accuracy fades away whenever *k* approaches larger values. To overcome these issues, this paper introduces a kd-tree based dual-kNN approach that concentrates on two properties to keep up the classification accuracy at different *k* values and upgrade processing time performance. By conducting experiments on real data sets and comparing this algorithm with two other algorithms (dual-kNN and normal-kNN), it was experimentally confirmed that the kd-tree based dual-kNN is a more effective and robust approach for classification than pure dual-kNN and normal k-NN.

Keywords: Dual-kNN · kd-tree based dual-kNN · k-NN · Robustness

1 Introduction

The k-Nearest Neighbors algorithm is a supervised lazy classifier that has local heuristics. For each observation instance, it constructs a distance vector of the observed instance and each training data in space S to find the closest distance. Because of constructing a distance vector and choosing the closest distance from the distance vector throughout the whole data set, the k-Nearest Neighbors become a lazy learner. In general, the time complexity of the k-Nearest Nearest classifier in Big Oh notation is n2 where n is the number of training examples. Therefore, when the amount of data increases, the classical k-Nearest Neighbors usually become slow in computations. To this end, this lazy computation significantly deadly kills the performance.

To cope with this time consuming process of the k-NN, S.S. Aung, I. Nagayama and S. Tamaki applied multithreading approach to k-NN. According to their experimentally proof, the approach unexpectedly enhanced the processing time by simultaneously maintaining the previous high prediction accuracy. However, the multithreading approach suffers from the major drawback of creating and deleting threads when it has to deal with enormous amount of threads.

K. Arai et al. (Eds.): IntelliSys 2018, AISC 868, pp. 638–654, 2019.
https://doi.org/10.1007/978-3-030-01054-6_46

Commonly, mostly classification algorithm mainly concerns the behavior of each attribute in the case of labeling an instance. That's why it requires to probably clean some unrelated or very high variation attributes up that will possibly block the processing time performance and classification accuracy in an unexpected way. Thus, another perceptive insight into the speed-up in processing time intensive works, especially for instance based classification algorithms, is feature selection. Feature selection selects the best appropriate features or attributes from messy data set according to the measurement of the strength of relationship among attributes. S.S. Aung, I. Nagayama and S. Tamaki applied feature selection approach together with k-means clustering algorithm to k-NN to avoid the compensations for the time consumption of thread creation and deletion in the previous work. Then, a coupled with the experimental results, the authors pointed out the effectiveness of feature selection in boosting processing time and classification accuracy. However, from our experiments, some attributes that is not likely to give more information still support decision maker to make a right decision somehow for long-term approach. It is better to avoid ignoring completely all unrelated or weak attributes.

Furthermore, as a semi-supervised learning algorithm, k-Nearest Neighbors (k-NN) requires enough training data and a predefined k value to find the k nearest data based on a distance computation [7]. From our experiments, the classification accuracy of k-NN declines gradually over k values from $k = 1$ to $k = \mathrm{n}$. Sometimes, the prediction accuracy goes to infinity in a fluctuating trend. In these cases, it is very difficult to select the best suitable k value with the best performance. For this issue, we introduce dual-kNN to smooth out the impact of noisy training example by enhancing the strategy of normal k-NN into dual consideration in classification.

However, dual-kNN is also driven by the facts that are based on the basic concepts of original k-Nearest Neighbors. Consequently, the k-Nearest Neighbors algorithm is still suffering from the time complexity problem of matching each observed instance to entire training set to decide or classify which right instance will belong to an observed instance. Therefore, it has to consume a large amount of time to finish one observed instance successfully. That is to say, the dual-kNN still inherits this danger inherent in processing time performance. Therefore, the time complexity is also problematic for the dual-kNN. For this problem, the idea that comes into our mind is to construct a dynamic kd-tree of the entire training data set and then observe its nearest neighbors by moving down the hierarchy left, which are the most associated and leaving the instances on the right side from considering or right leaving the left. The primary objective of this approach is to reduce some unrelated training instances with an observed instance. In the next sections, we will discuss in detail about the productive result of this approach.

The rest of this paper is organized as follows. Section 2 describes related works. Although the system is the combination of dual-kNN and kd-tree approach, we will discuss dual-kNN more details in Sect. 3 in order to understand more clearly. Then, the whole system (kd-tree based dual-kNN) is detailed in Sect. 4. Section 5 analyses the efficiency of the new model on five datasets, and compares the processing time and the prediction accuracy of the new approach with the pure dual-kNN and k-Nearest Neighbors and Sect. 6 is the conclusion.

2 Related Works

The k-Nearest Neighbors appeals to many researchers about its simplicity and high prediction accuracy among machine learning methods, even though it is a lazy learning method. Therefore, researchers have been investigating the weak points of k-NN from different types of point of views to innovate normal k-NN with the greatest accomplishment.

In [1], the authors proposed a system, which extended nearest neighbors algorithm for pattern recognition considered not only who the nearest neighbors of the test sample are, but also who considered the test sample as their nearest neighbor. By iteratively assuming all the possible class memberships of a test sample, the ENN is able to learn from the global distribution; therefore, it improved pattern reorganization performance.

In [2], the authors presented a multithreading machine learning system for upgrading its classification process and proved the processing time performance by comparing the efficiency of k-NN with Naïve Bayes. However, creating and deleting threads is also very expensive operation. In [3], the authors designed a system, which investigates how to improve the lazy process and how to upgrade the classification accuracy of k-NN by filtering the attributes, which are not strong enough correlation with other attributes.

In [4], they designed a static kd-tree for k-nearest neighbors for reducing processing time by a simple modification to the queries to exploit the coherence between successive points and not changing to kd-tree. Thus, the system still has under the influence of the time consumption of insertion function. For the issue, they applied multithreading to the static kd-tree. What's more, the paper only concentrated on improving processing time.

In [9], the authors presented an approach that is named dual-kNN aims for improving the weakness of classical k-NN. The k-NN is under the influence of a steady decrease in accuracy over larger k values and the high variance in training data. Thus, the authors tried to cope with these issues by introducing the dual-kNN and experimentally proved that the efficiency of the approach by conducting experiments on UIC Machine Learning Repository data sets.

In [10], the authors especially investigated the effect on k-NN by applying different distance functions (Euclidean, cosine, Chi square, and Minkowsky) on three different types of medical datasets (categorical, numerical, and mixed types of data). According to the experimental results, Chi square distance function achieved the best classification accuracy among four different distance functions.

The systems we discussed above tried to overcome the weakness of the k-Nearest Neighbors too with different ideas, and view. However, the classification accuracy and high speed performance go together to accomplish a high performance algorithm. Therefore, these essential properties are coupled together in kd-tree based dual-kNN.

3 Dual-KNN

To define dual-kNN precisely, let us first define some notations. The notation Ri is used to denote a set of instances, x, which belongs to an instance which has class type i, defined as following, before we observe training data.

$$R_i = \{< x_1, .., x_m > \text{belongs to an instance which has class type } i\} \tag{1}$$

Where i defines the category or class type i = {0, 1, 2, ..., n}, and m is the number of instances.

We write $R_i \in X$. Where X is the space of all kinds of instances defined as following:

$$X = \{R_0, R_1, R_2, R_3, R_4, R_5, R_6, \ldots \ldots \ldots ., R_n\} \tag{2}$$

In dual-kNN learning, the nearest neighbors of an observed instance are defined by the distances between them. Here, we apply Standard Euclidean distance for constructing a distance vector space over training instances. Like classical k- Nearest Neighbors algorithm, dual-kNN approximates the nearest neighbors based on the given distance vector.

Let's assume $< x_1, x_2, \ldots, x_n >$ be the n instances that are observed from training examples are nearest to an observed instance (x_q) based on the distance space.

Before going to dive into deep understanding about dual-kNN, let us stop by for skimming over k-Nearest Neighbors. In k-Nearest Neighbors, it explores the kth nearest neighbors only in the closest area, according to distance vector and then adds a voting system to determine the final decision [6] as shown in Fig. 1. In this figure, it is obvious that the algorithm only considers the instances, which exist in the dotted line circle.

The dotted, line circle means the shortest distance area to the observed instance. According to this figure, the observed instance is a black square existed in the center, the sample instances are black triangles and pentagons. For $k = 3$, the nearest neighbors are two triangles and one pentagon. According to k-Nearest Neighbor, for $k = 3$, the nearest neighbors determines the nearest neighbor is to be the triangle (Fig. 1).

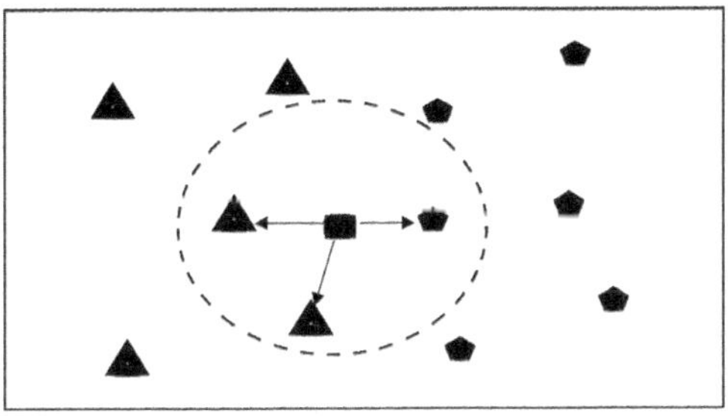

Fig. 1. The 3-Nearest Neighbor.

From here, the concept of dual-kNN starts to go against the concept of classical k-Nearest Neighbor. Here, we can define Senior (the inner circle) and Junior (the outer circle) according to Fig. 2. In dual-kNN, it has dual thinking that the extended concept of looking for other neighbors (Juniors) behind the first nearest neighbor (Senior) to get support from Juniors for proving that the associated senior is the right nearest neighbor for the observed instance compared with other neighbors.

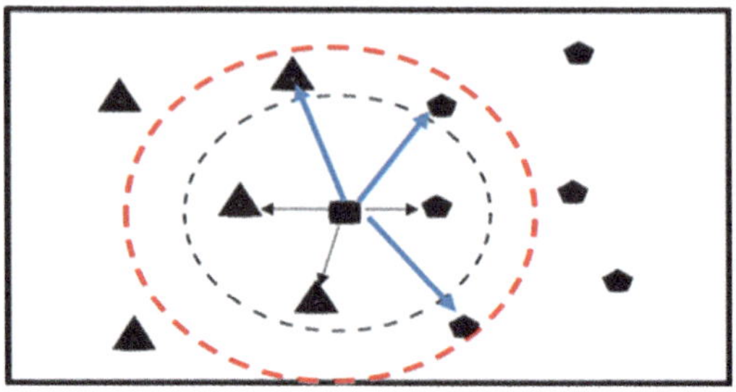

Fig. 2. Dual-3NN.

In order to make the clear concept of dual-kNN learning problems, we would like to briefly discuss about dual-kNN by giving a sample diagram as shown in Fig. 2.

Firstly, it searches the nearest neighbors (seniors) from an observed instance, and then it identifies the number of categories existing in the seniors; it has two categories (triangle and pentagon) according to the figure, and how many members are included in each category; the triangle category has one member and the pentagon category has two members. According to the dual-kNN algorithm, for $k = 3$, each category must have three nearest instances, for $k = 5$, each category must have five nearest instances, etc. In the next step of the algorithm, it searches out the next nearest neighbors (juniors) of each category until it completes the required number of members for each category. After finishing up these two steps completely, it starts to measure the distance between the observed instance (the black square) and each member of those two categories and finds the centroid of each group. The centroid is the contribution of each nearest neighbor to the observed instance, giving shorter distance to close neighbors. Note the dual-3NN algorithm classifies the observed instance (the black square) as a triangle example in this figure. The detail processes are also illustrated in Algorithm 1.

In prediction system, dual-kNN clusters the n nearest instances $< x_1, x_2, \ldots, x_n >$ into R_i category. In other words, it groups the same instances into the same category. If R_i is an empty set, we can automatically realize that the instances, which belong to the class type, i do not include the nearest instances. Then, we just ignore it. Non empty set can be defined as follows:

$$R_1 = \{< x_1, r_1 >, \ldots.., < x_k, r_1 >\} \tag{3}$$

$$R_2 = \{< x_1, r_2 >, \ldots.., < x_k, r_2 >\} \tag{4}$$

$$R_n = \{< x_1, r_n >, \ldots.., < x_k, r_n >\} \tag{5}$$

Where k is the number of nearest neighbors and n represents the class type from 1 to n. The number of members included in each group must be the same number that is defined as follows:

$$|R| = \begin{cases} 1 \; if |R_1| = |R_2| \ldots = |R_n| \\ 0 \; otherwise \end{cases} \tag{6}$$

If | R | is 1, the system will properly work without hesitation. Otherwise, there are some improper decisions. The next step is to find out the distance between the observed instance and the members of each group. Thus, each group (R_n) has occupied its local distance space. In order to approximate the nearest neighbor, the centroid of each group can be specified by the following equations:

$$C(R_1) = \frac{(d(x_{1,1}, o) + d(x_{1,2}, o) + \ldots. + d(x_{1,k}, o))}{k} \quad (7)$$

$$C(R_2) = \frac{(d(x_{2,1}, o) + d(x_{2,2}, o) + \ldots. + d(x_{2,k}, o))}{k} \quad (8)$$

$$C(R_n) = \frac{(d(x_{n,1}, o) + d(x_{n,2}, o) + \ldots. + d(x_{n,k}, o))}{k} \quad (9)$$

Where $C(R_n)$ is the centroid of R_n and o is observed instance, x_q.

The final step is making a decision on nearest neighbor by applying the following equation.

$$NN(x_q) \leftarrow \underset{i \in |R|}{argmin} \sum_{i=1}^{k} \frac{(d(x_1, o) + d(x_2, o) + \ldots. + d(x_k, o))}{k} \quad (10)$$

The Eq. (10) can be generalized as following.

$$NN(x_q) \leftarrow \underset{i \in |R|}{argmin} \sum_{i=1}^{k} C(R_n) \quad (11)$$

Note all of the k-Nearest Neighbor variants of the dual-kNN algorithm take into account that each senior competes with others by getting the support of associated junior members to get more strength in the nearest neighbor competition. Thus, this new idea mainly tends to reduce distortion, overfitting and noise.

Algorithm 1 . Dual-KNN algorithm
Let X = { $R_0, R_1, \ldots\ldots\ldots, R_{13}, R_{14}$} to be training examples. 1. Given a query instance x_q to be classified. 2. Let < $x_1, x_2, \ldots, x_k$ > denote the k instances from training examples that are nearest to x_q. 3. Define the number of categories M from the nearest neighbors. 4. For i from 1 to M, do For t from 1 to K, do $R_i \leftarrow$ Min { $d(x_1,x_q)$, $d(x_2,x_q)$, ... , $d(x_n,x_q)$} 5. Find the centroid of each category, $C(R_i)$, by using from equation 6 to 8. 6. Classify the nearest neighbor of x_q by applying (10)

4 Dynamic Dimensional Tree Based Dual-KNN

We have already discussed dual-kNN how to observe the nearest neighbors in the previous section. At this time, this section is going to detail about how to construct the underlying k-d tree data structure exploited to boost the classification celerity of dual-kNN.

4.1 K-d Trees

The k-d tree is a hierarchical data structure constructed over a training data set of partitioning the training data set into sub-tables corresponding to a hyperplane. In this case, a plane is the median value of an attribute, which has the greatest variance among attributes. The k-d tree is a binary tree, with a root node and sub-trees of children with a parent node. The final node is a leaf node. In this work, a leaf which contains a set of neighbor, instances corresponding to a parent node. For each observed instance, their nearest neighbors can be learned by visiting inside a leaf if it is a right sub-tree for them. Otherwise, this sub-tree can be ignored by traversing nodes until the right one is found. Figure 3 illustrates a k-d tree with sub-trees and leaf nodes containing a set of counterpart instances.

4.2 Constructing Dynamic kd Trees Dual-kNN

A kd-tree is a kind of binary tree of splitting a node into children corresponding to the median value of one attribute [5]. In this paper, we choose an attribute, which has the highest variance, defined in (12), will be used as an index value for every instance in the case of creating sub-trees. Thus, before creating kd-tree, finding variation is taken as a first action.

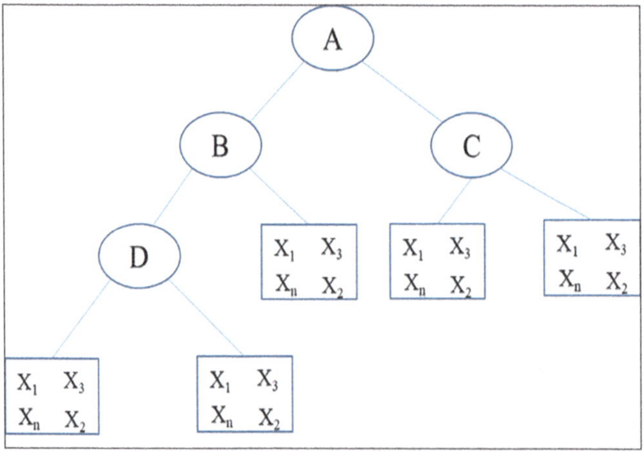

Fig. 3. A sample k-d tree for dual-kNN.

$$V_j = \frac{1}{n}\sum_{i=1}^{n}\left(x_i - \mu_j\right)^2 \tag{12}$$

Where V_j is the variance of attribute j, u is the mean value of attribute j, where i = {1, 2, ..., n}, and n is the number of instances.

Then, the only one attribute, which occupies the greatest variance, is selected as a hyper plane for splitting a data set into sub-tables or sub-trees. This simple selection process can be described in the mathematic (13).

$$\check{V} = \max\left(A_1 : A_n\right) \tag{13}$$

Where $\check{V}$ is the attribute, which has the maximum variance, and $\max\left(A_1 : A_n\right)$ is the largest value in the range from A_1 to A_n . After that, the dataset is sorted to the index value and then create sub-trees repeatedly by computing the median value of the index. In kd-trees, the left child of a node has a smaller index value, whereas the right-hand side child has a larger index value than its parent index value.

After creating kd-trees and classifying an observed instance, the observed instance becomes a training example for future classification. That is to say, the insertion is also an important function in tree-based systems. However, to take the action of insertion new instance, it must search for the empty location throughout the tree. For this recursively, empty location searching is also an expensive process for CPU. In this case, we create dynamic kd-trees instead of static kd-trees. Whenever an observed instance is trying to classify, a dynamic kd-tree is constructed for this observed instance by dividing the given data set according to the median value and the index value of observed instance, and altogether it learns the current created node is not only a leaf node, but also the right node for searching for its nearest neighbor, by matching the index value of the current node with the index value of the observed instance.

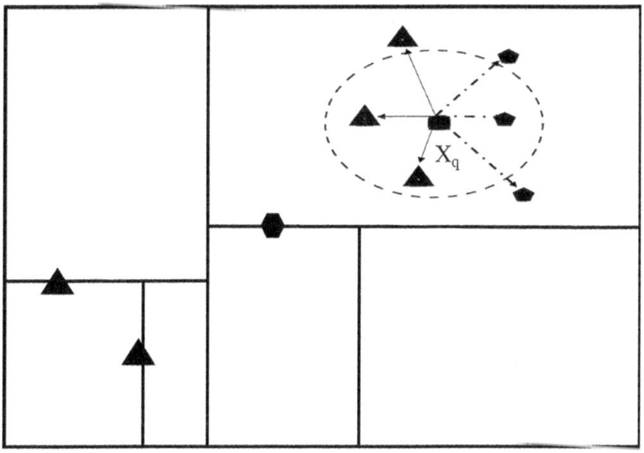

Fig. 4. A sample kd-tree based dual-kNN for K = 3.

Here, let us briefly discuss defining leaf node. Because this paper only considers kd-trees for dual-kNN in observing nearest neighbors, for $k = 3$, a leaf node contains 7 neighborhoods (Fig. 4), for $k = 5$, a leaf node occupies 11 nearest instances (Fig. 5). Thus, it stops creating sub-trees whenever it only has lower or equal numbers of instances to $(2k + 1)$ and then the rest instances are stored in a leaf node.

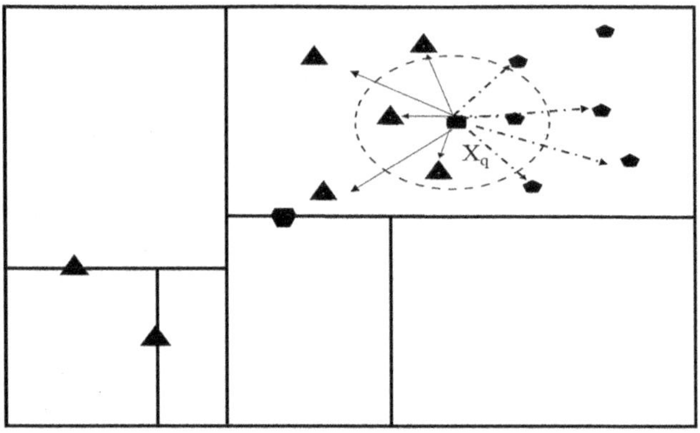

Fig. 5. A sample kd-tree based dual-kNN for K = 5.

When it finds out the right sub-tree for the observed instance, it observes the k-Nearest Neighbor, which belongs to the query instance by visiting the sub-tree and applying the dual-kNN. Figure 4 shows a sample kd-tree for observing the nearest neighbors of dual-3NN.

This new idea mainly tends to reduce distortion, overfitting, noise, and speed up the classification speed. Thus, the effectiveness of this new idea is going to discuss in the next section by giving some experimental provenances.

The step by step process of kd-tree for dual-kNN is illustrated in Algorithm 2. Algorithm 1 discusses about dual-kNN how to observe the k-Nearest Neighbors by extending the original scheme of normal k-Nearest Neighbor. But, Algorithm 2 describes how to create a kd-trees for dual-kNN. In Algorithm 2, it constructs a kd-trees by splitting a given data set according to median value of the highest variance attribute repeatedly until a sub-tree has numbers of candidate lessen than or equal to 2K + 1. Then, it calls Algorithm 1 by passing the sub-tree and an observed instance x_q. After that, dual-kNN observes the true k-Nearest Neighbor for x_q by using the parameter (the sub-tree from Algorithm 2).

To demonstrate better the process flow of the proposed system, according to an observed instance, it randomly constructs a kd- tree over training data set (Fig. 6). The constructed kd-tree is used by the dual-kNN classifier to classify the right category of an observed instance. The final result is restored back in database as a training example for the purpose of future classification.

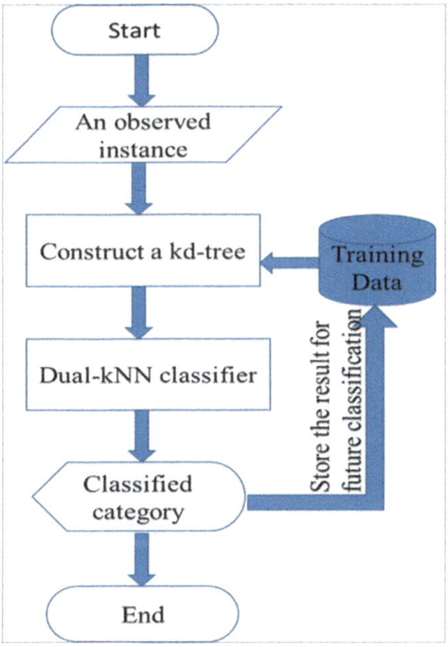

Fig. 6. Process flow of propose approach.

Algorithm 2 . kd-Tree based Dual-KNN algorithm
Let X = { $R_0, R_1, R_2, R_3, R_4, R_5, R_6, \ldots\ldots\ldots, R_n$} to be training examples.
1. Given a query instance x_q to be classified.
2. Find an attribute, A, which has the greatest variance over X.
3. Let I be the greatest variance attribute of x_q.
4. Sort the given data set X according to the highest variance attribute.
5. While (Flat is true)
a. Find the median value M of A
b. If (I < M) go to right sub-tree
c. Else go to left sub-tree
d. While (candidates of sub-tree > 2K+1)
1. Flag = true
2. Go to step 5.
e. Flag = false.
6. Call Algorithm 1.

5 Experiment and Analysis

In this section, we will discuss the comparative study of dual-kNN, normal k-NN, and kd-tree based dual-kNN. From this study, the results prove that the efficiency of the new approach with two properties (classification accuracy and processing time performance). For those results, we learn the prediction accuracy using Purity measurement and measure processing time performance over five different data sets, abalone, balance scale, breast cancer, iris, and wine, from the UCI machine learning repository center for machine learning and intelligent systems [8].

For this provenance, we are going to start with the classification accuracy. In this classification accuracy study, we explore the classification ability of each k value over five datasets. As we know, k = 1 has the highest prediction ability, and then the accuracy gradually declines for larger k values. It means because of weak robustness. Thus, this paper deeply concentrates on smoothing and adjusting the gaps between k values.

Tables 1, 2, and 3 illustrate the classification accuracy of kd-tree based dual-kNN, pure dual-kNN, and normal-kNN by utilizing abalone, balance scale, breast cancer, iris, and wine datasets. It is obvious that the accuracy of kd-tree based dual-kNN is better than not only dual-kNN, but also normal k-NN in every forecast. Moreover, it can be seen that the estimation accuracy of kd-tree based dual-kNN among *k* values is not a very large gap. The most obvious part in dual-kNN is between K_1 and K_3, but in dual-kNN (Table 2), and normal k-NN (Table 3), the prediction accuracy suddenly goes down to around 80% and 75%, respectively from 100% accuracy over *k* values 1, 3, 5, and 7. It is pointing out that kd-tree based dual-kNN is the most robustness approach among these three algorithms.

Table 1. Classification accuracy using kd-tree based dual kNN

Data set	K_1	K_3	K_5	K_7
Abalone	100%	92%	84%	78%
Balance scale	100%	97%	93%	96%
Breast cancer	100%	99%	97%	97%
Iris	100%	100%	100%	100%
Wine	100%	89%	85%	81%

Table 2. Classification accuracy using dual-kNN

Data Set	K_1	K_3	K_5	K_7
Abalone	100%	90%	81%	77%
Balance scale	100%	94%	86%	87%
Breast cancer	100%	99%	98%	97%
Iris	100%	100%	100%	100%
Wine	100%	89%	80%	80%

Table 3. Classification accuracy using k-NN

Data set	K_1	K_3	K_5	K_7
Abalone	100%	70%	63%	62%
Balance scale	100%	84%	80%	76%
Breast cancer	100%	97%	97%	97%
Iris	100%	96%	96%	96%
Wine	100%	86%	76%	73%

The discussion described above is the detailed comparison of kd-tree based dual-kNN, pure dual-kNN, and normal k-NN for five data sets over *k* value from 1 to 7. The next figure (Fig. 7) is the summary discussion of the comparative study of the total average accuracy of kd-tree based dual-kNN, dual-kNN, and normal k-NN by giving the comparison with visualization view.

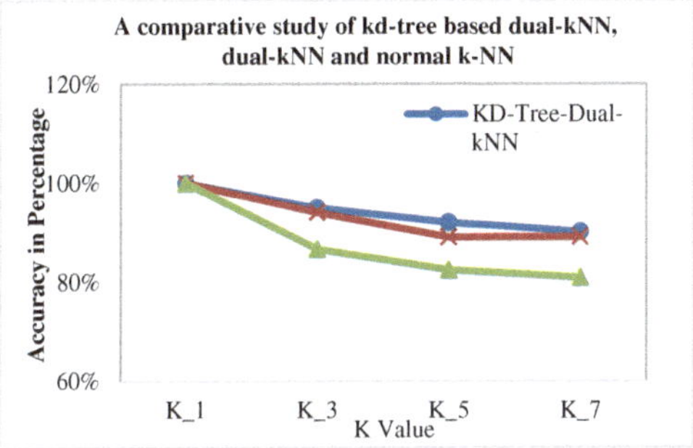

Fig. 7. A summary of comparative study of classification of kd-tree based dual-kNN and pure dual-kNN.

As detail stated in Fig. 7, in which the emphasis is on showing the summary of the comparative study of kd-tree based dual-kNN, pure dual-kNN, and k-NN over k values from 1 to 7. In this figure, the blue line represents the total average classification accuracy of kd-tree based dual-kNN, the red one is pure dual-kNN, and the green one represents normal k-NN. The total average accuracy of kd-tree based dual-kNN is above 90%. However, the total average accuracy of pure dual-kNN starts with 100% and ends with 89%. Likewise, the accuracy of normal k-NN is under 90%. As discussed above, the kd-tree based dual-kNN is more efficient than pure dual-kNN and normal k-NN.

Finally, we summarize the discussion of classification accuracy using kd-tree based dual-kNN and dual-kNN as reported by Fig. 7. It is obvious that in every sector, kd-tree based dual-kNN achieves a higher accuracy than pure dual-kNN, and normal k-NN that shows a remarkable achievement of kd-tree based dual-kNN. Beyond that, for kd-tree based dual-kNN, the accuracy among k values is not very different, whereas for dual-kNN, it starts with 100%, and ends with 89%. To put it another way, the robustness of kd-tree based dual-kNN is higher than dual-kNN.

The last performance measurement is the reduction in processing time. From Figs. 8, 9, 10, 11 and 12, these figures illustrate the improvement of classification time of kd-tree based dual-kNN by comparing with pure dual-kNN over k values from 1 to 7. In this study, because we more focus on processing time improvement of kd-tree based dual-kNN and dual-kNN, we only compare the processing time of kd-tree based dual-kNN with pure dual-kNN. However, we compare the total average processing time performance of three algorithms as shown in Table 4, Figs. 13 and 14.

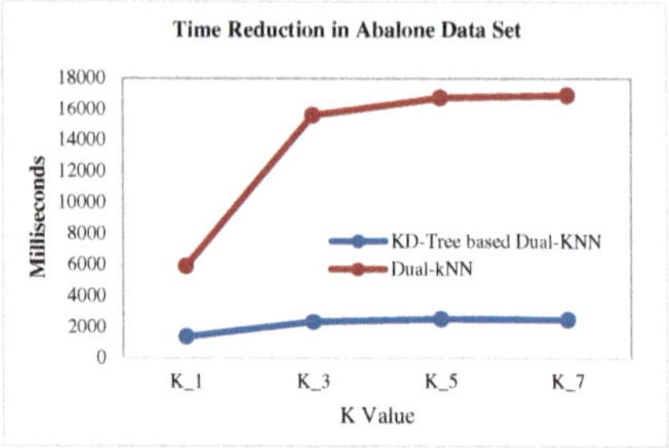

Fig. 8. A comparative study of reduction in time using abalone.

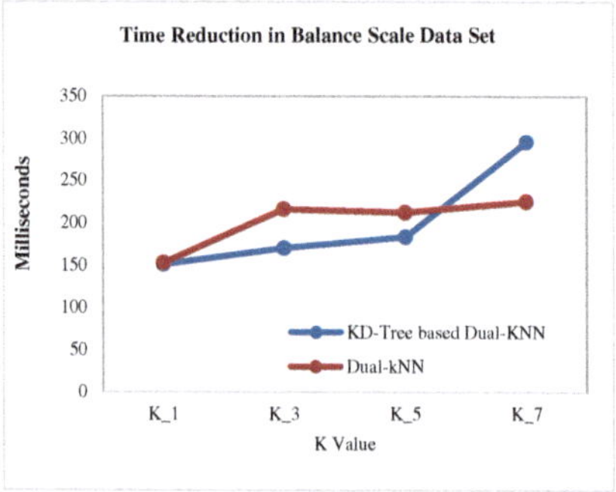

Fig. 9. A comparative study of reduction in time using balance scale.

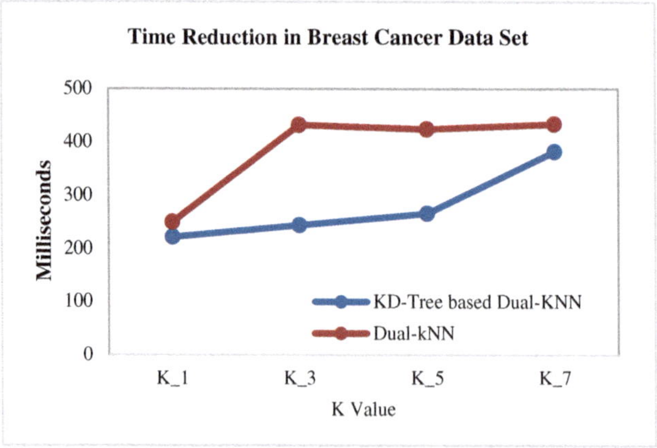

Fig. 10. A comparative study of reduction in time using breast cancer data.

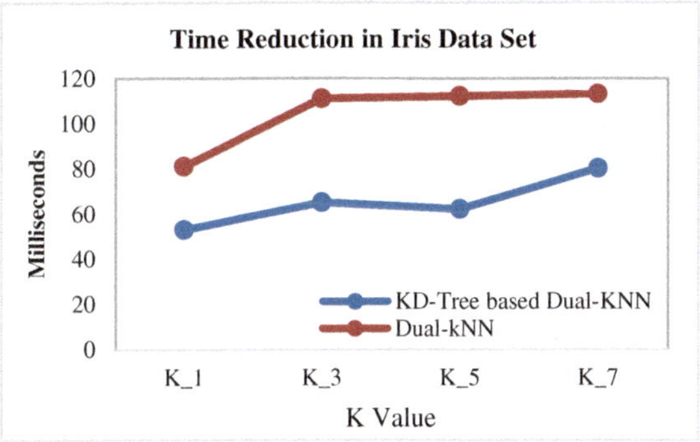

Fig. 11. A comparative study of reduction in time using iris data set.

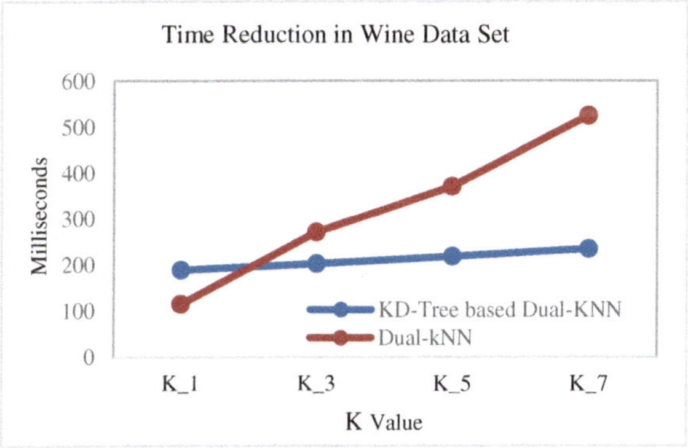

Fig. 12. A comparative study of reduction in time using wine data set.

Table 4. A comparison of processing time performance of kd-tree based dual-kNN, pure dual-kNN, and classic k-NN

Dataset (KB) (Amount × No: of Instances)		Processing time (Milliseconds)			Processing time improvement ration over pure dual-kNN	Processing time improvement ration over pure k-NN
		kd-Tree based dual-kNN	Pure dual-kNN	k-NN		
Abalone	620 × 4177	7537	56050	23787	743%	315%
Balance scale	720 × 625	562	711	578	126%	102%
Breast cancer	92 × 700	987	1538	1109	155%	112%
Iris	348 × 150	127	218	213	171%	167%
Wine	56 × 178	250	1019	300	407%	120%

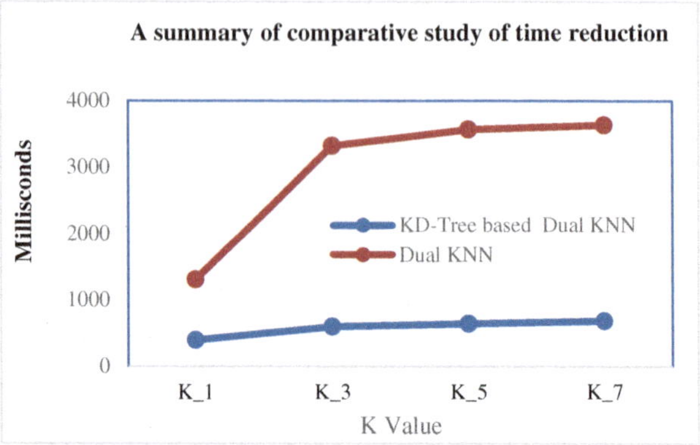

Fig. 13. A summary of comparative study of time reduction using five data sets over k values from 1 to 7.

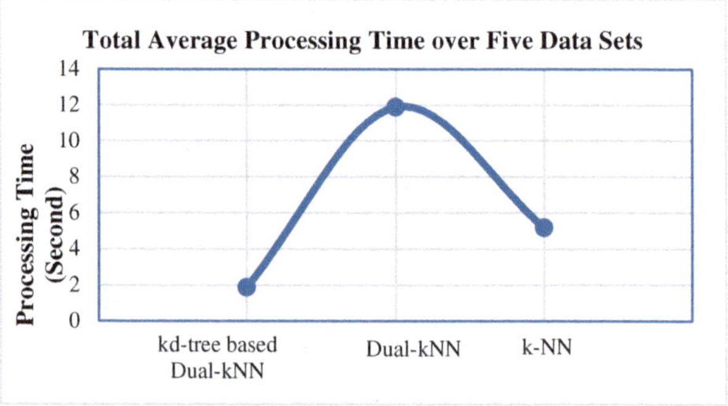

Fig. 14. Total average processing time of kd-tree based dual-kNN, dual-kNN, and k-NN over five data sets.

Figure 13 shows a summary of the comparative study of classification time reduction in dual-kNN using kd-tree. The processing time of pure dual-kNN suddenly grows up until K = 7, whereas the classification time of kd-tree based dual-kNN remain stable until k is 7.

Table 4 is the summary of the processing time improvement using five data sets without separating k values. According to the evidence illustrated in Table 4, kd-tree based dual-kNN is better than pure dual-kNN and normal k-NN. As we know, k-NN is a lazy learner algorithm and needs a large amount of processing time to successfully classify even only one observed instance. For this lack of speedy processing, this paper applied a kd-tree to dual-kNN aiming for making it a smart algorithm. It can be clearly seen that the improvement ration processing time of the proposed approach over pure dual-kNN for abalone dataset is 743% faster, and over k-NN is 315%. In balance scale data set, the improvement ratio processing time over dual-kNN and k-NN is 126% and 102% respectively. In breast cancer, the improvement ratios are 155% and 112%, in iris, 171%, and 167%, in wine data set, 407% and 120% respectively. According to these experimental evidence, the kd-tree based dual-kNN upgrades not only classification accuracy, but also the processing time performance of dual-kNN and normal k-NN.

Figure 14 illustrates and compares the total average processing time of kd-tree based dual-kNN, dual-kNN, and k-NN. As stated in Fig. 14, kd-tree based dual-kNN boosted the classification time fairly quite well over both algorithms (dual-kNN and k-NN). Figure 15 depicts the total average processing time improvement ration of kd-tree based dual-kNN over dual-kNN and normal k-NN. In details, the total average improvement ratio over dual-kNN is over 3.2 and over normal k-NN is over 1.6.

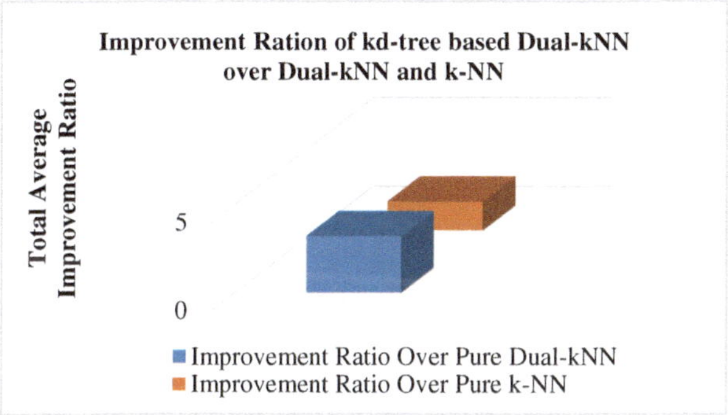

Fig. 15. Total average processing time improvement ratio of kd-tree based dual-kNN over dual-kNN and k-NN using five datasets.

6 Discussion

As discussed in the previous sections, kd-tree based dual-kNN is the best predictor among pure dual-kNN and normal k-NN as it has the ability to predict with higher accuracy and higher processing time. However, a slightly more complex process is required for kd-tree based dual-kNN corresponding to its underground platforms (sorting algorithm and constructing tree) and therefore it is more sensitive to noisy data than dual-kNN and k-NN. The simplest algorithm is normal k-NN, but it still has many things to improve to apply to the real world systems. The dual-kNN can do the prediction with higher accuracy than k-NN, but it is still under the influence of the required matching time of an observed instance with each training example throughout the whole dataset.

7 Conclusion

In this study, we propose a new classification approach, kd-tree based dual-kNN, aimed for upgrading prediction accuracy with varying k values and improving the processing time performance of the pure dual-kNN algorithm and normal k-NN by applying dynamic kd-tree approach. According to the experimental results, the kd-tree based dual-kNN upgrades not only the processing time but also classification accuracy. According to the experimental results discussed in previous sections, the kd-tree based dual-kNN is efficient and feasible to classify with higher performance. In accuracy measurement, kd-tree based dual-kNN could predict with higher accuracy with 94.3% total average) than classic k-NN (with 93% total average) and normal k-NN (with 87%). In processing time performance, kd-tree dual-kNN upgrades the processing time around total average 320% over pure dual-kNN, and 160% over normal k-NN respectively. Additionally, all factors prove that kd-tree based dual-kNN is more effective and feasible approach than pure dual-kNN with impressive evidence. For future work, we will investigative the noise tolerance of kd-tree based dual-kNN, pure dual-kNN, and normal k-NN over varying *k* values using more and different data sets. As a limitation, the system also very depends on sorting process of kd-tree construction. Sorting algorithm sorts the data set according to attribute value. Therefore, it may better support the system if the data set does not include noisy data.

References

1. Tang, B., He, H.: ENN: extended nearest neighbor method for pattern recognition. IEEE Comput. Intell. Mag. **10**, 52–60 (2015)
2. Aung, S.S., Nagayama, I., Tamaki, S.: Intelligent traffic prediction by multi-sensor fusion using multi-threaded machine learning. IEIE Trans. Smart Process. Comput. **5**(6), 430–439 (2016)
3. Aung, S.S., Nagayama, I., Tamaki, S.: Plurality rule–based density and correlation coefficient–based clustering for K-NN. IEIE Trans. Smart Process. Comput. **6**(3), 183–192 (2017)
4. Merry, B., Gain, J., Marais, P.: Accelerating kd-tree searches for all k-nearest neighbors. In: European Association for Computer Graphics, Department of Computer Science, University of Cape Town, January 2013
5. Bentley, J.L.: Multidimensional binary search tree used for associative searching. Commun. ACM **18**(9), 509–517 (1975). ISSN 0001-078
6. Mitchell, T.M.: Machine Learning. McGraw-Hill Companies, New York (1997). Carnegie Mellon University
7. Du, M., Ding, S., Jia, H.: Study on density peaks clustering based on k-nearest neighbors and principal component analysis. Knowl. Based Syst. J. **99**, 135–145 (2016)
8. Blake, C., Keogh, E., Merz, C.J.: UCI repository of machine learning databases. University of California, Department of Information and Computer Science, Irvine, CA (1998). http://www.ics.uci.edu/mlearn/MLRepository.html
9. Aung, S.S., Nagayama, I., Tamaki, S.: Dual-kNN for a pattern classification approach. IEIE Trans. Smart Process. Comput. **6**(5), 326–333 (2017)
10. Hu, L.Y., Huang, M.W., Ke, S.W., Tsai, C.F.: The distance function effect on k-nearest neighbor classification for medical datasets. SpringerPlus **5**, 1304 (2016). https://doi.org/10.1186/s40064-016-2941-7

High-Speed 2D Parallel MAC Unit Hardware Accelerator for Convolutional Neural Network

Hossam O. Ahmed[1,2(✉)], Maged Ghoneima[1], and Mohamed Dessouky[1]

[1] Faculty of Engineering, Ain-Shams University, Cairo, Egypt
Hossam.o.ahmed@ieee.org
[2] American College of the Middle East (ACM), Egaila, Kuwait

Abstract. The increasing importance of depending on Convolutional Neural Networks (CNN) in many real-time applications especially for image classifications and Humanoid Robots leads to the search for an optimum solution to accelerate the computational process capabilities for the hardware-based systems. Multiply-Accumulate (MAC) is the most computational demanding unit in any CNN architectures. In this paper, three optimized 2D MAC hardware-based architecture units have been designed using VHDL and synthesized for the operation on the FPGA platform due to its parallelism-architecture support feature. The logic utilization, power dissipation, and timing analyze of the three proposed 2D MAC have been made using Quartus ii tools and showed that the 3rd MAC design can achieve a 18.34 Giga Operation per Second (GOPS) while keeping the core dynamic thermal power dissipation level at 303.67 mW.

Keywords: Convolutional neural networks · Computational intelligence
Multiply-Accumulate (MAC) · Field Programmable Gate Array (FPGA)
VHDL

1 Introduction

The rapid evolution in many high-tech applications that depends on real-time processing left no plenty of choices for designers except searching for the optimum hardware platform that can provide a boost in computation processing speed while keeping the power consumption for such hardware accelerators in an adequate level, especially if these targeted applications depend on the convolutional neural network (CNN) as its main core unit such in advanced Robotics and Autonomous Systems (RAS), Natural Language Processing, Speech Recognition, image classifications.

The convolutional neural network architectures supposed to be erected on hardware platforms that support the concept of parallelism data flow behavior, which is obvious, since it is originally inspired by the way of how in nature, the human brain processing the information in parallel manner. Due to the large number of neurons required to construct a CNN architecture and the corresponding enormous number of synapses needed for the connection of these neurons, the attempts that have been made to use the conventional central processing units (CPU) and clusters of General Purpose Graphics Processing Units (GPGPU) as the optimum hardware processing units for the CNN

K. Arai et al. (Eds.): IntelliSys 2018, AISC 868, pp. 655–663, 2019.
https://doi.org/10.1007/978-3-030-01054-6_47

different networks suffer a notice degradation in the computational overall performance, in terms of processing time and power consumption, especially when applied to a real-time based systems. The main issue that both the CPU and GPGPU faced is their Shortage of providing parallel processing capabilities which is the most important characteristic of the CNN architectures. Recently, the Field Programmable gate Array (FPGA) and Application Specific Integrated Circuit (ASIC), due to their ability to provide parallel computations with a lower power drained levels in comparing to the CPU and the GPGPU, become the optimum trend for designing complex CNN architectures [1–3].

The most computationally and resources demanding unit in all the CNN different architecture is the Multiply-Accumulate (MAC) units. The MAC unit is responsible for accumulating the addition of the resultant values from the multiplication processes between the input data values with the corresponding weight values. The numbers of MAC units in small CNN networks can reach up to 341k MAC units as in LeNet-5 architecture and up to 3.9G as in ResNet-50 architecture. The maximum number of MAC operations needed in every single stride is depends on the Filter dimension in this layer, since the filter size may be differing from layer to another one in the same CNN architecture. For instance, the filter size is fixed and equal to 5×5 in the LeNet-5 architecture, while in AlexNet architecture the filter size is 11×11 in the first layer, 5×5 in the second layer, and 3×3 for the other layers. In general, the filter size is varying between 1×1 to 11×11 in the most well-known CNN architectures like LeNet, AlexNet, VGGNet, GoogleNet, and ResNet-50 [2, 4].

It is noteworthy that the design of optimized MAC unit for deep learning architectures is one of the key aspects that determine the overall performance of the targeted system. Many contributions have been made to optimize the MAC unit regarding the power dissipation and the computational speed rates. The contribution in [5], stated designing a generic MAC unit with two inputs that is suitable for any DSP-based application based on a 16-bit floating-point multiplier block diagram using. Furthermore, other contributions, as in [6–10], suggested enhancing the optimization level of the MAC unit by depending on fixed-point architecture rather than the floating-point architecture assuming that 1% error can be tolerated [11–14].

This paper proposes a three-parallel hardware-based architecture for designing and implementing a 2D MAC unit using VHDL. The different between the proposed three architectures is the window size of the multiplication section inside the MAC unit. Each of 2D MAC unit design has a verified and analyzed using Quartus ii tools. The rest of the paper is organized as follows. In Sect. 2 is discussed the concepts and main characteristics of the 2D MAC unit designs. The experimental results are presented in Sect. 3 and concluded the paper in Sect. 4.

2 Proposed 2D Parallel MAC Unit Designs

The main goal of this paper is to optimize the design of the MAC unit, which is considered as the most intense computational demanding element in any CNN network architecture by proposing three designs of a high-speed 2D parallel MAC unit hardware accelerators using 8-bits length for the operands.

The 2D MAC unit is responsible for aggregating the results accomplished by the element-wise multiplication between the elements of the 2D filter (weights) and the corresponding input feature map elements. The 2D filter is described by two parameters, the filter height (R) and filter width (S), while the input feature map is described by two parameters, the input feature map height (H) and the input feature map width (W) as illustrated in Fig. 1.

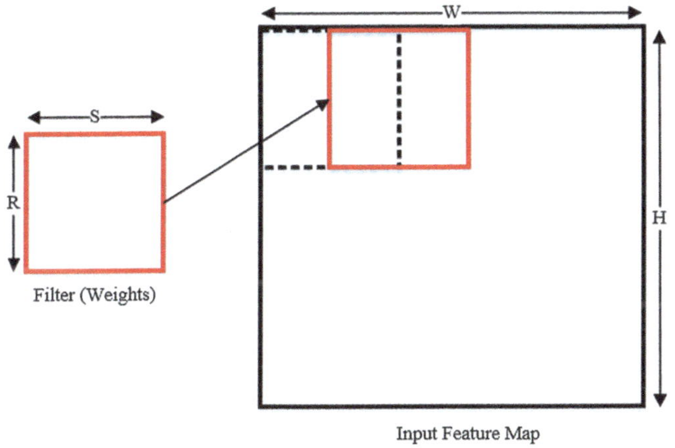

Fig. 1. General sliding window processing in CNN networks.

The proposed three hardware architectures, in this paper, for accelerating the computational process of the 2D MAC unit is achieved based on the dependence on the parallelism capabilities offered by relying on FPGA as the targeted hardware platform. The first 2D MAC unit hardware architecture has 9 parallel multipliers bocks and 1 addition/accumulate block to perform a 2D MAC operation for 3 × 3 filter size, while the second 2D MAC unit hardware architecture has 25 parallel multipliers bocks and 1 addition/accumulate block to perform a 2D MAC operation for 5 × 5 filter size, and the third 2D MAC unit hardware architecture has 49 parallel multipliers bocks and 1 addition/accumulate block to perform a 2D MAC operation for 7 × 7 filter size as illustrated in Fig. 2. The proposed three hardware architectures are designed using VHDL language and analyzed using Quartus ii tools. The architecture for each of the proposed 2D parallel MAC units are based on 8-bits fixed point. One of the important features that is delivered by the proposed 2D MAC units is the independence of the Digital Signal Processing (DSP) blocks or the embedded multipliers offered inside the FPGA silicon fabric architecture to conserve the flexibility of the design to be adopted in further enhanced ASIC designs.

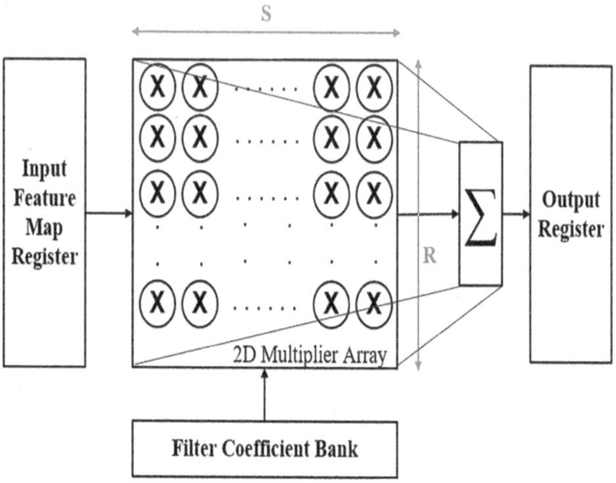

Fig. 2. Detailed 2D MAC unit hardware architecture operation.

3 Experiments

The logic utilization, timing, and power dissipation reports of the proposed MAC units have been obtained from Intel Quartus Prime, and the targeted FPGA chip for the three designs was assumed to be Stratix V 5SGXEABN3F45I3YY. The logic utilization flow summary of the three MAC units' comparison, as in Table 1, showed that the number of logic utilization went from 722 ALM for the 3 × 3 2D MAC unit to 4,216 ALM for the 7 × 7 2D MAX unit, which is considered as an expected result, in which the logic utilization will be increased with the increase of the filter dimensions. Also, it proved that the architectures of the three 2D MAC units don't depend on neither the embedded multiplier 9-bit elements, DSP blocks, nor the embedded memory bits offered in the FPGA silicon architecture, which eliminates the design dependency on such special units that varies from FPGA vendor into another.

The time analyzer report from the TimeQuest tool showed that under the fast 900 mV 0C model, the three 2D MAC units have different maximum operating frequency, as in Table 2. The first 2D MAC unit can reach a maximum frequency up to 472.14 MHz, the second 2D MAC unit can reach a maximum frequency up to 543.48 MHz, while the third 2D MAC unit can reach a maximum frequency up to 366.84 MHz. It was expected that the frequency may be dropped as the filter dimension increased due to the complexity of the design path routing, however the first 2D MAC unit achieved a lower operating frequency in comparing to the second 2D MAC and the third 2D MAC designs. Another timing testing has been made to check for timing violations, which can cause unexpected behavior of the proposed 2D MAC units, and it showed that there are no timing violations for the critical path in the three 2D MAC architectures and they are having a positive setup time slack and hold time slack as showed in Table 2. Also, the slack histogram timing scenario test has been made for each of the proposed 2D MAC unit separately as illustrated in Figs. 3, 4 and 5 and

Table 1. Flow summary report comparission between the three 2D parallel MAC proposed designs using Quartus ii Software

	1st MAC design	2nd MAC design	3rd MAC design
Filter size	3 × 3	5 × 5	7 × 7
No. of soft multipliers	9	25	49
No. of soft addition unit	1	1	1
INTEL ALTERA Family	Stratix V		
Device	5SGXEABN3F45I3YY		
Logic utilization (in ALMs)	722/359,200	2,222/359,200	4,216/359,200
Combinational ALUTs	1443	4432	8377
Dedicated logic registers	326	972	1813
Total memory bits	0/54,067,200	0/54,067,200	0/54,067,200
Embedded multiplier 9-bit elements or DSP blocks	0/352	0/352	0/352

proved that despite the variations in the number of path edges that raised from decreasing the operating frequency, there are no timing violations for all the 2D MAC units.

Table 2. Time analyzer report comparision between the three 2D parallel MAC proposed designs using Quartus ii TimeQuest

	1st MAC design	2nd MAC design	3rd MAC design
Filter size	3 × 3	5 × 5	7 × 7
INTEL ALTERA Family	Stratix V		
Device	5SGXEABN3F45I3YY		
Latch clock name	Clk	Clk	Clk
Maximum frequency	472.14 MHz	543.48 MHz	366.84 MHz
Setup time slack	0.382 ns	0.160 ns	0.274 ns
Hold time slack	0.186 ns	0.215 ns	0.202 ns

In Table 3, the amount of the three main sources of power dissipation has been illustrated; the core dynamic thermal power dissipation, the core static thermal power dissipation, and the I/O thermal power dissipation. The most critical aspects that need to be considered for the proposed 2D MAC units, is the core dynamic power dissipation amount, since it represents the amount of power that has been dissipated due to the timing transition of the Adaptive Logic Modules (ALM) that have been consumed from the FPGA silicon fabric to construct the architecture of the proposed systems. The results showed a predictable value of the core dynamic thermal power dissipation amounts for the three 2D MAC units, since the core dynamic thermal power dissipation amount should be decreased as the filter size dimensions decreased, due to the reduction of the amount of ALM used in such case.

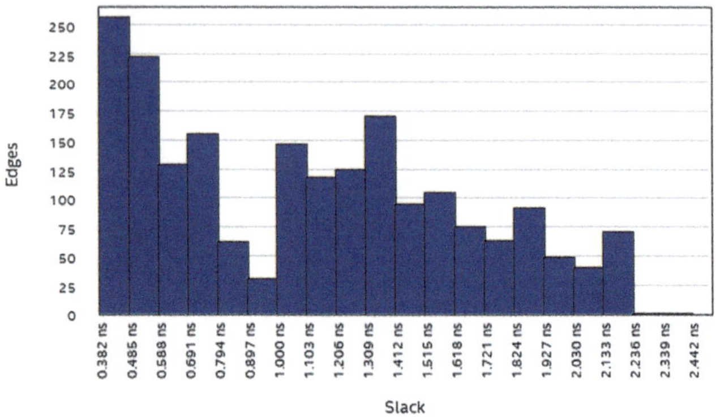

Fig. 3. Slack histogram of the clock signal of the proposed 3 × 3 2D MAC unit.

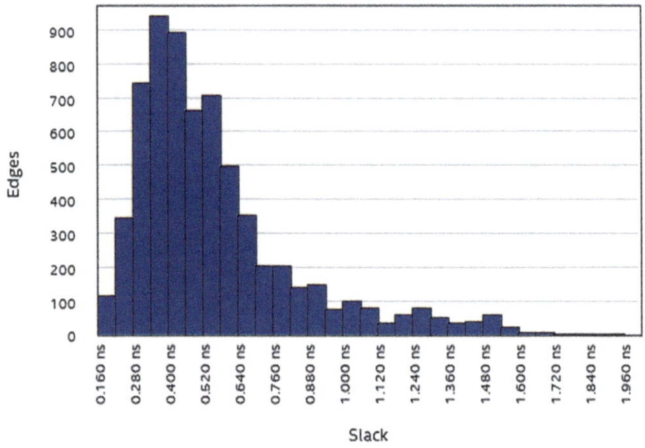

Fig. 4. Slack histogram of the clock signal of the proposed 5 × 5 2D MAC unit.

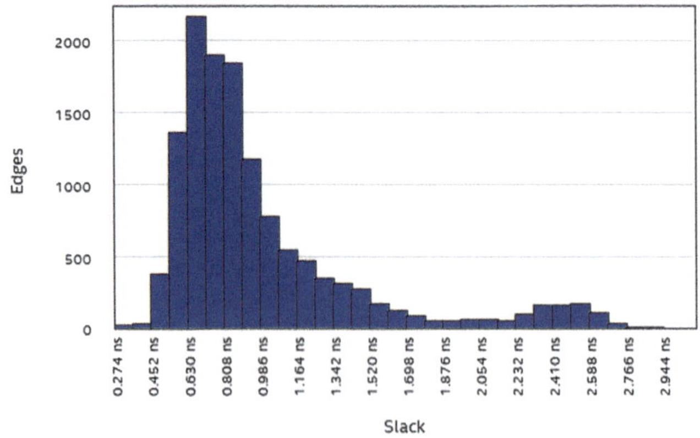

Fig. 5. Slack histogram of the clock signal of the proposed 7 × 7 2D MAC unit.

Table 3. Power analyzer comparison report between the three 2D parallel MAC proposed designs using Quartus ii Powerplay

	1st MAC design	2nd MAC design	3rd MAC design
Filter size	3 × 3	5 × 5	7 × 7
INTEL ALTERA Family	Stratix V		
Device	5SGXEABN3F45I3YY		
Total thermal power dissipation	1470.14 mW	1688.82 mW	1842.21 mW
Core dynamic thermal power dissipation	50.53 mW	201.99 mW	303.67 mW
Core static thermal power dissipation	1319.90 mW	1308.54 mW	1334.50 mW
I/O thermal power dissipation	99.71 mW	178.30 mW	204.04 W

The overall computational performance of the proposed 2D MAC designs are indicated in Table 4. The results showed that despite the drop in the maximum frequency for the 7 × 7 2D MAC unit architecture in comparing with the 3 × 3 and 5 × 5 2D MAC unit architectures, the overall computational processing is still the highest with an 18.34 Giga Operation per Second (GOPS) as shown in Table 4 due to the 49 parallel multipliers it has. The results in general showed the benefits of depending on concurrent architectures for boosting the overall speed of the CNN architectures.

Table 4. Maximum operation per second comparison between the three 2D parallel MAC proposed designs

	1st MAC Design	2nd MAC Design	3rd MAC Design
Filter size	3 × 3	5 × 5	7 × 7
INTEL ALTERA Family	Stratix V		
Device	5SGXEABN3F45I3YY		
Maximum frequancy	472.14 MHz	543.48 MHz	366.84 MHz
Number of parallel multipliers	9	25	49
Maximum GOPS	4.72 OPS	14.13 OPS	18.34 GOPS

The impact of increasing the concurrent operations of the dot products in the designed three 2D MAC unit architectures in this paper in comparing to the proposed MAC unit from [3], beside the contribution of [10] is shown in Table 5. Table 5 shows that the maximum computational speed per second giga operations per seconds for proposed 3 × 3 MAC unit is nearly equal to the maximum computational speed per second giga operations per seconds in [3, 10] due to the retrogression in the maximum operation frequency that the design can reach. Also, the results showed that increasing the filter dimensions above the 3 × 3 can boost the overall computational speed, since the proposed 5 × 5 2D MAC unit can achieve a computational speed up to 14.13 GOPS, while the maximum computational speed achieved by the proposed 7 × 7 2D MAC unit by 18.34 GOPS.

Table 5. The computational features comparison between the proposed MAC units in this paper and the proposed MAC unit from [3, 10]

	Proposed 3 × 3 MAC unit in this paper	Proposed 5 × 5 MAC unit in this paper	Proposed 7 × 7 MAC unit in this paper	The 8-bit parallel MAC unit from [3]	The 8-bit parallel MAC unit from [10]
FPGA Vendor	Intel Altera	Intel Altera	Intel Altera	Intel Altera	Xilinx
Parallel multiply-accumulate per clock cycle	9	25	49	4	4
DSPs units consumed	None	None	None	None	1
LUTs (or ALUTs) units consumed	1443	4432	8377	595	197
Flip-Flops units consumed	326	972	1813	54	201
Maximum frequency for low cost FPGA	NA	NA	NA	454.13 MHz	422 MHz
Maximum frequency for high-density FPGA	472.14 MHz	543.48 MHz	366.84 MHz	834.03 MHz	660 MHz
Maximum GOPS	4.72 GOPS	14.13 GOPS	18.34 GOPS	4.17 GOPS	5.29 GOPS

4 Conclusion

The dimensions, three-different proposed 2D MAC units in this paper represent the common filter sizes in the different convolutional neural networks and the main contribution of this paper is to boost the speed of the computational calculation of the MAC unit in such networks by depending on the FPGA platform due to its parallelism-architecture support feature. The three designed 2D MAC architectures have been tested using Intel Quartus Prime, and the targeted FPGA chip for the three designs was assumed to be Stratix V 5SGXEABN3F45I3YY. The proposed 2D achieved an overall computational processing speed up to 18.34 Giga Operation per Second (GOPS) in the 7×7 2D MAC unit case. The contribution in this paper of boosting the computational capabilities of MAC unit for the CNN is just can be considered as the cornerstone for designing complete CNN different architectures using FPGA, and furthermore syntheses the same results for ASIC platform to compare the computational gap between the two platforms.

References

1. Véstias, M., Neto, H.: Trends of CPU, GPU and FPGA for high-performance computing. In: 2014 24th International Conference on Field Programmable Logic and Applications (FPL), Munich, Germany. IEEE (2014)
2. Moini, S., Alizadeh, B., Emad, M., Ebrahimpour, R.: A resource-limited hardware accelerator for convolutional neural networks in embedded vision applications. IEEE Trans. Circuits Syst. II Express. Briefs **64**(10), 1217–1221 (2017)
3. Ahmed, H.O., Ghoneima, M., Dessouky, M.: Concurrent MAC unit design using VHDL for deep learning networks on FPGA. Presented at the IEEE Symposium on Computer Applications & Industrial Electronics (ISCAIE 2018), Penang Island, Malaysia (2018, in Press)
4. Wang, H., Shao, M., Liu, Y., Zhao, W.: Enhanced efficiency 3D convolution based on optimal FPGA accelerator. IEEE Access **5**, 6909–6916 (2017)
5. Saravanan, R., Balaji, P., Prabu, R.: Design of 16-bit floating point multiply and accumulate unit. IJMTES Int. J. Mod. Trends Eng. Sci. **03**(01) (2015)
6. Shaikh, T., Beleri, M.: FPGA implementation of multiply accumulate (MAC) unit based on block enable technique. Int. J. Innov. Res. Comput. Commun. Eng. **3**(4) (2015)
7. Ashwini, N., Rao, T.K., Rao, D.S.: Low power multiply accumulate unit (MAC) for DSP applications. Int. J. Res. Stud. Sci. Eng. Technol. IJRSSET **2**(8), 49–54 (2015)
8. Nain, P., Virdi, G.S.: Multiplier-accumulator (MAC) unit. Int. J. Digit. Appl. Contemp. Res. **5**(3) (2016)
9. SaiKumar, M., Kumar, D.A., Samundiswary, P.: Design and performance analysis of multiply-accumulate (MAC) unit. Presented at the International Conference on Circuit, Power and Computing Technologies, ICCPCT, Nagercoil, India (2014)
10. Duarte, R.P., Véstias, M., de Sousa, J.T., Neto, H.: Parallel dot-products for deep learning on FPGA. Presented at the 2017 27th International Conference on Field Programmable Logic and Applications (FPL), Ghent, Belgium, September 2017
11. Taylor, G., Lacey, G., Areibi, S.: Deep learning on FPGAs: past, present, and future, 13 February 2016
12. Dettmers, T.: 8-bit approximations for parallelism in deep learning. Presented at the ICLR 2016, San Juan, Puerto Rico, 2–4 May 2016 (2016)
13. Wu, E., Fu, Y., Sirasao, A., Attia, S., Khan, K., Wittig, R.: Deep learning with INT8 optimization on Xilinx devices. In: UltraScale and UltraScale+FPGAs, vol. v1.0.1, no. WP486, 24 April 2017
14. Gysel, P., Motamedi, M., Ghiasi, S.: Hardware-oriented approximation of convolutional neural networks. Presented at the ICLR, San Juan, Puerto Rico (2016)

Excessive, Selective and Collective Information Processing to Improve and Interpret Multi-layered Neural Networks

Ryotaro Kamimura[1(✉)] and Haruhiko Takeuchi[2]

[1] IT Education Center, Tokai University,
4-1-1 Kitakaname, Hiratsuka, Kanagawa 259-1292, Japan
ryo@keyaki.cc.u-tokai.ac.jp

[2] Human Informatics Research Institute, National Institute of Advanced Industrial Science and Technology, 1-1-1 Higashi, Tsukuba 305-8566, Japan
takeuchi.h@aist.go.jp

Abstract. This paper aims to propose a new type of learning method to train multi-layered neural networks. In deep learning, pre-training by unsupervised learning such as auto-encoders and Boltzmann machines is used to produce initial connection weights to be used in the main training or fine-tuning. It has been observed that the connection weights are not necessarily effective for training supervised learning, because the objectives of unsupervised and supervised learning are naturally different. However, without the appropriate pre-training, multi-layered neural networks have difficulty in training, because information on input patterns and also the errors decrease naturally by going through many hidden layers. To overcome this problem of vanishing information, particularly, on input patterns, we try to produce redundant and excessive information in terms of the activation of output neurons before training multi-layered neural networks. Then, the excessive information can be reduced by the vanishing information property of multi-layered neural networks. It can be expected that appropriate connection weights can be found among many candidates created in a process of producing the excessive information. Finally, all of the connection weights are collectively treated for better interpretation. The method was applied to two data sets: artificial and symmetric data set and the real snack food selection data set. In both experimental results, redundant and excessive information generation was observed in terms of connection weights and improved generalization performance was observed.

Keywords: Excessive · Selective · Collective
Multi-layered neural networks · Generalization · Interpretation

1 Introduction

The present paper aims to propose a new type of learning method to train multi-layered neural networks. Recently, multi-layered neural networks have received

K. Arai et al. (Eds.): IntelliSys 2018, AISC 868, pp. 664–675, 2019.
https://doi.org/10.1007/978-3-030-01054-6_48

considerable attention particularly in their application to the character or image recognition problems. One of the main characteristics of those learning methods is that the initial connection weights are given by the pre-training using the auto-encoder, restricted Boltzman machines [1,2], and so on. However, it has been observed that the connection weights, obtained by the pre-training tend to lose their original characteristics in the main training or fine-tuning. This is because the objectives of the unsupervised learning in the pre-training and supervised learning in the main training are different. For example, if some features are not useful in producing appropriate targets, they are forced to be modified to produce connection weights more suitable for producing targets. Unsupervised pre-training is effective only if features extracted by unsupervised learning can be used to produce appropriate targets. In addition, it has been reported that without pre-training, better generalization performance can be obtained [3].

To solve this problem, the present paper does not use the pre-training but instead tries to keep the original information on inputs as much as possible for multi-layered neural networks. Because information on input patterns tend to decrease gradually by going through many hidden layers naturally, we try to produce information on input patterns excessively. More specifically, the number of output neurons in the unsupervised learning as the pre-training is forced to increase excessively. Then, this excessive information content on input patterns naturally decreases by going through many hidden layers in the main-training. However, because many different types of connection weights in terms of excessive information are prepared, the multi-layered neural networks can choose the appropriate ones among many.

Excessive information production can explain the better performance of multi-layered neural network, applied to image or character recognition data sets [1]. In the image or character data sets, much redundant and peripheral information exists and they should be reduced as much as possible in a process of learning in multi-layered neural networks. However, in the real data set, it often happens that the number of variables is not so large, and it is sometimes impossible to obtain data sets with many variables. In that case, the present method can produce additional variables, although they are redundant.

Finally, we try to interpret relations between inputs and outputs of multi-layered neural networks. As has been well known that neural networks has the serious black-box problem [4–8] where it is impossible to interpret the final results by neural networks. The black-box problem has become more serious recently, because much attention has been paid to multi-layered neural networks. The present paper uses the collective interpretation method where all connection weights are collectively treated and multi-layered networks can be reduced to ones without hidden layers.

The paper is organized as follows. In Sect. 2, we present the excessive, selective and collective information processing method, comparing that with conventional deep learning. In Sect. 3, we present two preliminary experimental results on the artificial and symmetric data set and the real data set on food selection experiments. The experimental results show that redundant information generation was observed in terms of connection weights and improved generalization could be observed for both data sets.

2 Theory and Computational Methods

2.1 Excessive Information Generation

In this section, we explain why the excessive information should be generated by comparing our method with those used in the conventional deep learning. One of the main reasons for the attention that has been given to the multi-layered neural networks is that the vanishing gradient descent can be weakened by layer-wise pre-training by using the restricted Boltzmann machines, auto-encoders and so on. In addition, we have pointed out that input information naturally decreases by going through many hidden layers, which has been well known as the information channel in information theory [9]. Conventional deep learning has been partially successful in reducing the effect of vanishing information as well as gradient descent. However, it has been observed that pre-training has not been as effective as was originally expected. For example, the effects of connection weights by the pre-training gradually diminishes when the learning is advanced. This happens because the objective of the unsupervised learning for pre-training is different from that of supervised learning in the fine-tuning. Thus, it is necessary to develop a method to keep information in the pre-training as much as possible in the fine-tuning.

The present paper aims to use unsupervised self-organizing maps [10, 11] for pre-training, although its knowledge is not transferred in the form of connection weights, but in the form of outputs from the output neurons. Because by going through many hidden layers, information on inputs tends to decrease gradually, we try to generate excessive information in the pre-training in the form of the excessive number of output neurons.

Figure 1 shows differences between the present method and the conventional deep learning. In Fig. 1(a), the conventional deep learning is described in which the layer-wise pre-training by the unsupervised learning is used to produce initial connection weights. Then, with those connection weights by the pre-training, the final tuning is performed as shown in Fig. 1(a). On the other hand, in the present method, the pre-training in the form of unsupervised learning is added to the main training as shown in Fig. 1(b). However, the number of the features or variables increases and the sparsity of output neurons is controlled by the parameter of Gaussian outputs from the output layer in the excessive information acquisition. Then, with those augmented outputs from the unsupervised SOM, multi-layered neural networks are trained, reducing the excessive information by going through many hidden layers. Finally, all connection weights are collectively treated in Fig. 1(c), and relations between inputs and outputs can be easily interpreted as is the case with the conventional regression analysis.

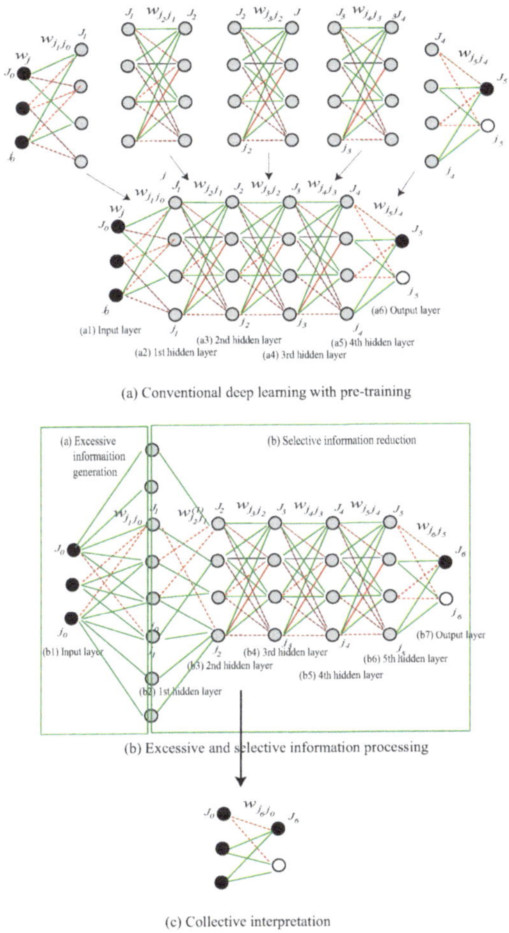

Fig. 1. (a) Conventional deep learning, (b) excessive and selective information processing, and (c) collective interpretation.

2.2 Excessive Information Production

For the excessive information production, we use the Self-Organizing Map (SOM) [10,11], because it can produce similar connection weights by increasing the number of output neurons. Now, distance between inputs and connection weights is computed. The distance from the inputs to the j_1th output neuron ($j_1 = 1, 2, ..., J_1$) for the sth inputs ($s = 1, 2, ..., S$) is computed by

$$d_{j_1}^s = \sum_{j_0=1}^{J_0} \left(x_{j_0}^s - w_{j_1 j_0}\right)^2 \tag{1}$$

where $x^s_{j_0}$ represents the sth input and $w_{j_1 j_0}$ represents weights from the j_0th input ($j_0 = 1, 2, ..., J_0$) to the j_1th output neuron. The output from the j_1th neuron of the first hidden layer is computed by

$$v^s_{j_1} = \exp\left(-\frac{d^s_{j_1}}{\sigma}\right) \quad (2)$$

where the parameter σ represents the width of the distribution.

Connection weights are updated, considering distance between winners and neighboring neurons. First, a winner j_c is determined by

$$j_c = \mathrm{argmin}_{j_1} d^s_{j_1} \quad (3)$$

Then, the update rule for the tth learning cycle is given by

$$w_{j_1 j_0}(t+1) = w_{j_1 j_0}(t) + \alpha(t) h_{j_c j_1}(x^s_{j_0}(t) - w_{j_1 j_0}(t)) \quad (4)$$

where α is a learning parameter and $h_{j_c j_1}$ represent distance between winners and the corresponding neurons. This update rule shows that the method can produce typical connection weights represented by the winner and the corresponding and neighboring weights similar to those of the winner abundantly. Thus, the SOM can produce excessive information in terms of many similar connection weights. Finally, we should note that the SOM is here used particularly to visualize connection weights and to show how redundant connection weights can be produced. If visualization is not necessary, other unsupervised methods such as auto-encoders can be used, expecting to produce the similar results.

2.3 Selective Information Reduction

Selective information reduction is performed by using the ordinary supervised multi-layered neural networks, because we think that multi-layered neural networks tend to lose information coming from inputs gradually by going through many hidden layers. Then, the inputs to the second hidden layer is computed by

$$v^s_{j_2} = \mathrm{tansig}\left(\sum_{j_1=1}^{J_1} w_{j_2 j_1} v^s_{j_1}\right) \quad (5)$$

where $w_{j_2 j_1}$ denote connection weights from the j_1 neuron in the first hidden layer (output layer from the SOM) to the j_2th neuron in the second hidden layer. Then, from the third layer, the same procedures are taken to produce outputs. Finally, the output from the output neuron is computed by

$$o^s_{j_6} = \mathrm{softmax}\left(\sum_{j_5=1}^{J_5} w_{j_6 j_5} v^s_{j_5}\right) \quad (6)$$

As mentioned, multi-layered neural networks has already a property to lose information of input patterns gradually. However, it is desirable to use more explicit information reduction method such as weight decay, which was used in the present experiments below discussed.

2.4 Collective Interpretation

Finally, complex connection weights of multi-layered neural networks are reduced to much simpler ones by treating all connection weights collectively. We here interpret relations between inputs and outputs by simplifying multi-layered neural networks to ones without hidden layers as shown in Fig. 1(c).

Now, collective weights between inputs and outputs are computed by summing and multiplying all weights in the input, intermediate and output layers

$$w_{j_6 j_0} = \sum_{j_1=1}^{J_1} \sum_{j_2=1}^{J_2} \sum_{j_3=1}^{J_3} \sum_{j_4=1}^{J_4} \sum_{j_5=1}^{J_5} w_{j_6 j_5} w_{j_5 j_4} w_{j_4 j_3} w_{j_3 j_2} w_{j_2 j_1} w_{j_1 j_0} \tag{7}$$

Thus, simplified collective weights can be used to interpret relations between inputs and outputs.

3 Results and Discussion

3.1 Symmetric Data Set

(1) Experimental Outline: In the first experiment using the artificial and symmetric data set, we try to show that redundant connection weights were generated by SOM, when the number of neurons increased and improved generalization performance could be obtained. Figure 2 shows a symmetric data set with additional noise. The number of input patterns in the data set was 200 by adding some noise, and 20 input variables were used. The number of hidden neurons was increased from two to 25. The training data set was composed of 20% of the whole data set, because the SOM is an unsupervised learning without targets and it is possible to train SOM with the whole data set. Then, the experiment tried to extract as much information as possible from the whole data set. This means that even with the small size training data set, better generalization is expected to be obtained, because of information on the whole data set. The remaining data set was equally divided into a validation and testing set. For comparison, the results were taken when minimum validation errors were attained for comparison. Excessive information acquisition was performed by SOM and the subsequent information reduction learning was by the ordinal BP with the weight decay. For easy reproduction of the present results, in the experiments the Matlab neural network package was used with all default parameter values, and the only exception was that the regularization parameter was set to 0.5.

(2) Generation of Excessive Connection Weights: Our method tried to extract features from input patterns as much as possible and even those on the random noises were extracted, namely, excessive information generation. Figure 3 shows connection weights of the excessive information acquisition component when the number of output neurons increased from two to 25. When there were two neurons, the connection weights were explicitly separated into two parts, as shown in Fig. 3(a). When the number of neurons increased from five in Fig. 3(b) to 25 in Fig. 3(f), connection weights gradually became similar

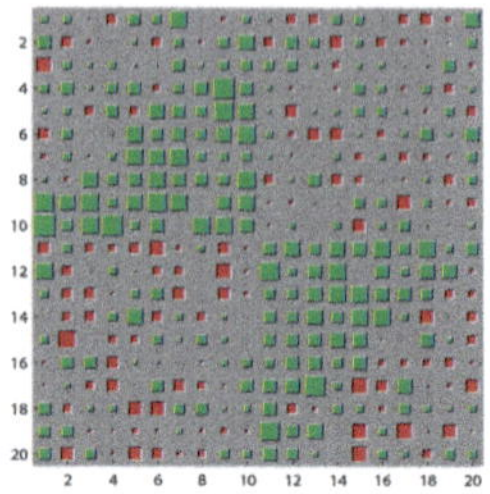

Fig. 2. Original symmetric data set with random noise. Green and red squares represent positive and negative weights and the size of squares represents the magnitude of connection weights.

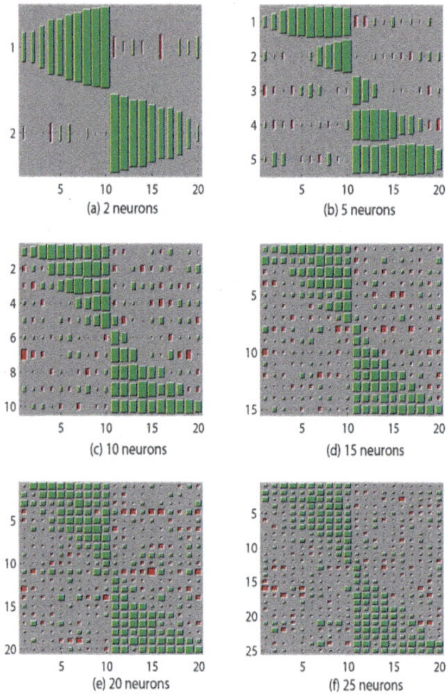

Fig. 3. Connection weights when the number of output neurons in SOM increased from two (a) to 25 (f) for the symmetric data set.

to the original data set, producing weights including all details, and including noises in the data set.

(3) Selective Information Reduction for Improved Generalization: Excessive information obtained by the unsupervised SOM component can be selectively reduced by going through many hidden layers. As mentioned, information on inputs tend naturally to decrease by going through many hidden layers, which is related to improved generalization. Table 1 shows the summary of generalization performance for the symmetric data set. The best average error of 0.0187 was

Table 1. Summary of experimental results on generalization performance for the symmetric data set

Method	Neurons	Layers	Avg	Std	Min	Max
Logistic			0.0600	0.0287	0.0125	0.1125
Bag			0.0375	0.0212	**0.0000**	0.0625
BP		2	0.0275	0.0165	0.0125	0.0625
New	15	3	**0.0187**	**0.0088**	0.0125	**0.0375**

obtained by the present method when the number of neurons in the excessive information acquisition component was 15 and the number of hidden layers was three in the information reduction component. In addition, the best maximum error of 0.0375 and the smallest standard deviation of 0.0088 were obtained by the present method. The second best average error of 0.0275 was by the conventional BP with two hidden layers. The third best one of 0.0375 was by the bagging ensemble method. One of the interesting results was that the bagging method [12,13] produced zero minimum error. The worst error of 0.06 was by the logistic regression analysis. The logistic regression also produced the worst maximum error of 0.1125. The results show that the present method produced better generalization performance by reducing the excessive information created by the SOM.

(4) Collective Interpretation: Finally, all connection weights were collectively treated and neural networks without hidden layers were produced for better interpretation. The symmetric property of input patterns of the input patterns in the present method could be clearly seen. Figure 4(a) and (b) show collective weights obtained in the early step of learning and connection weights in the final stage of learning by the present method. The weights in the initial stage represented the importance of input patterns that the importance decreased gradually in Fig. 4(a). It seems that with the final stage in Fig. 4(b), more detailed importance could be obtained, and the importance also decreased according to the importance of input patterns. Thus, the present method could extract the symmetric property of input pattern from the early stage of learning. Figure 4(c) shows the importance of input variables by the bagging method. The method could also extract the symmetric property of input patterns. On the other hand, Fig. 4(d) shows the regression coefficients by the logistic regression analysis. In this method, the most important one in the center was emphasized, and it could not necessarily extract the symmetric property of input patterns. As mentioned, the logistic regression analysis produced the worst generalization performance. This can be explained by the fact that the regression analysis could not extract the basic property of input patterns with the small size training data set.

3.2 Snack Food Selection

(1) Experimental Outline: The second experiment aims to predict which snack foods, when displayed on a monitor, were chosen by the subjects via their eye

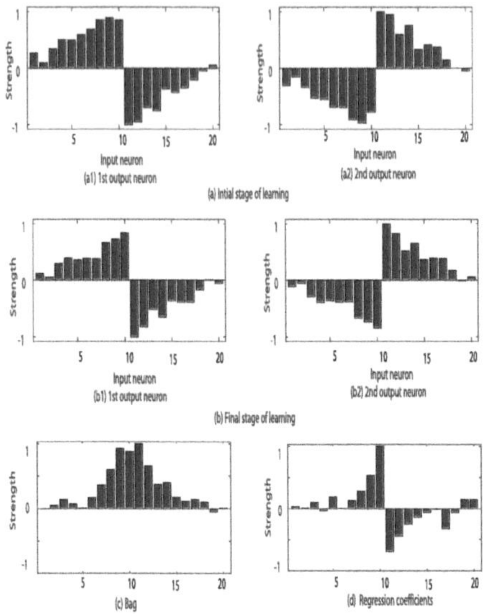

Fig. 4. Collective weights into the first and the second output neuron of the initial learning stage (a) and the final stage (b) by the present method, the variable importance by the bag method (c), and the regression coefficients (d) by the logistic regression analysis for the symmetric data set.

tracking-records. The eye-tracking records were collected from the psychological experiment of item selection task. We made six choice sets, each of which consisted of four snack food pictures. The subjects were instructed to browse the stimulus on a 17-in. TFT display, and they were asked to choose one of the four snacks in each choice set. The experiment was controlled by a Tobii T60 eye-tracking system with Tobii Studio Software. The eye-tracking records were translated into fixation data, and five eye-tracking major indices were calculated for each snack food. The eye-tracking data sets were represented by variable No. 1 as the time to the first fixation, No. 2 as a total fixation duration, No. 3 as a fixation count, No. 4 as a total visit duration and No. 5 as a visit count, with the subjects' decision of "chosen" or "not chosen" label. Eye-tracking data for 22 subjects with 528 instances were used to predict human decisions. The number of instances was increased slightly to have approximately the equal number of positive and negative instances, which would allow us to easily evaluate generalization performance. Only 20% of all instances was for training, while the remainder were divided into the validation and testing data set.

(2) Production of Excessive Connection Weights: The experimental results show that when the number of neurons increased, highly redundant and excessive connection weights were generated. Figure 5 shows connection weights when the number of neurons in the excessive information acquisition phase increased from two (a) to 50 (f). As can be seen in the figures, all connection weights tended to represent the similar characteristics. When the number of neurons increased, the

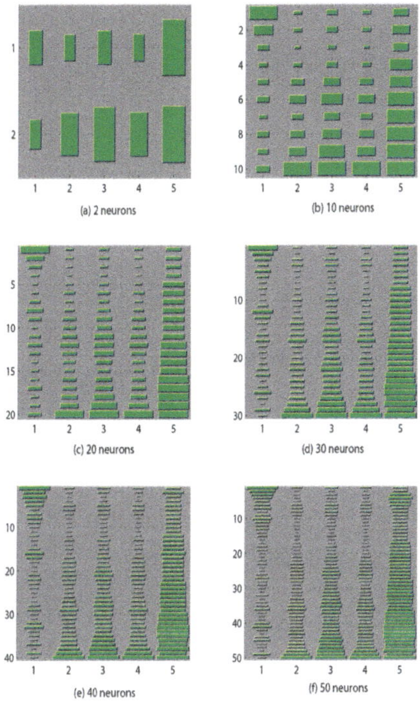

Fig. 5. Connection weights from the input to output layer when the number of neurons increased from two (a) to 50 (f) for the snack food data set.

number of similar connection weights gradually increased, meaning that many redundant and excessive connection weights were produced.

(3) Selective Information Reduction and Improved Generalization Comparison: Table 2 shows a summary of the generalization performance for the food data set. The best average error of 0.1204 was obtained by the present method with nine hidden layers. This shows that the excessive information in many connection weights was diminished by going through nine hidden layers. The second best of 0.1290 was found by the bagging method. In addition, the bagging produced the smallest minimum error of 0.0957. The third best error of 0.1380 was found by the conventional BP with seven hidden layers. Finally, the worst error of 0.1531 was found by the logistic regression analysis.

(4) Collective Interpretation: Figure 6 (a) and (b) show the collective weights for the input and the output neurons. In the initial stage of learning in Fig. 6(a), the variable No. 5, representing the number of visiting the corresponding snack food, played the most important factor in deciding the subjects choice. For the final stage of learning in Fig. 6(b), the other factors became more significant. The bagging method produced the similar variable importance, although the strength of the fourth variable, representing total visit duration, was the largest. In contrast, by the regression analysis in Fig. 6(d), only the fourth variable had larger

Table 2. Summary of experimental results on generalization performance for snack foods selection data set

Method	Neurons	Layers	Avg	Std	Min	Max
Logistic			0.1531	0.0145	0.1358	0.1728
Bag			0.1290	0.0206	**0.0957**	0.1605
BP		7	0.1380	0.0109	0.1173	0.1543
New	40	9	**0.1204**	**0.0107**	0.1080	**0.1420**

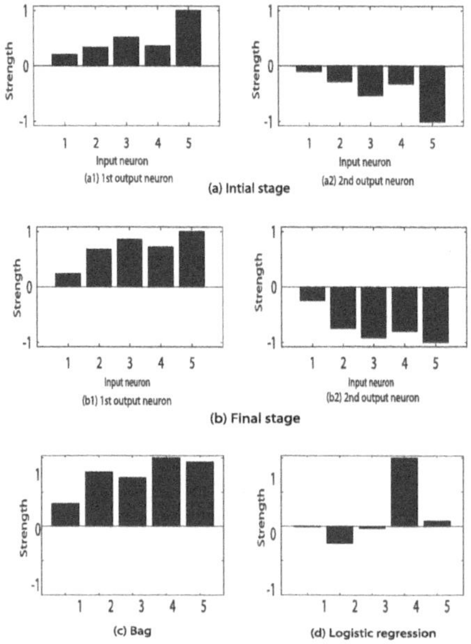

Fig. 6. Collective weights into the first and the second output neuron of the initial stage (a) and the final stage (b) by the present method, the variable importance by the bagging method (c) and the regression coefficients (d) by the logistic regression analysis for the snack food data set.

importance, while all the others had the very small values. Because the regression analysis had the worst generalization errors, it could not extract important features with the small training data set.

4 Conclusion

The present paper proposed a new type of method to train multi-layered neural networks. The method first tried to obtain excessive information by the SOM and then, multi-layered neural networks are used to lose excessive information content on input patterns by going through many hidden layers. Finally, all of the connection weights were collectively treated and multi-layered neural networks were reduced to ones without hidden layers for easy interpretation. The method was applied to two data sets: the first was an artificial and symmetric data set and the second was a real food selection data set. In the first experiment, redundant information in terms of connection weights were observed and improved generalization performance was produced. The present method could produce the clear symmetric property of input patterns. In the second experiment on the real snack food selection, improved generalization was observed by increasing input variables from five to 40 and by increasing similar and redundant connection weights.

The present method is particularly effective for the cases where the number of available input features or variable is small. Because the method aims to augment the number of variables by adding many similar and redundant features. In those cases where already redundant information is available, such as images or characters recognition, excessive information is already available, and therefore we need to reduce the excessive information by modifying the present method.

References

1. Hinton, G.E., Salakhutdinov, R.R.: Reducing the dimensionality of data with neural networks. Science **313**(5786), 504–507 (2006)
2. Bengio, Y., Lamblin, P., Popovici, D., Larochelle, H., et al.: Greedy layer-wise training of deep networks. In: Advances in Neural Information Processing Systems, vol. 19, pp. 153–160 (2007)
3. Schmidhuber, J.: Deep learning in neural networks: an overview. Neural Netw. **61**, 85–117 (2015)
4. Andrews, R., Diederich, J., Tickle, A.B.: Survey and critique of techniques for extracting rules from trained artificial neural networks. Knowl. Based Syst. **8**(6), 373–389 (1995)
5. Benítez, J.M., Castro, J.L., Requena, I.: Are artificial neural networks black boxes? IEEE Trans. Neural Netw. **8**(5), 1156–1164 (1997)
6. Ishikawa, M.: Rule extraction by successive regularization. Neural Netw. **13**(10), 1171–1183 (2000)
7. Huynh, T.Q., Reggia, J.A.: Guiding hidden layer representations for improved rule extraction from neural networks. IEEE Trans. Neural Netw. **22**(2), 264–275 (2011)
8. Mak, B., Munakata, T.: Rule extraction from expert heuristics: a comparative study of rough sets with neural network and ID3. Eur. J. Oper. Res. **136**, 212–229 (2002)
9. Abramson, N.: Information Theory and Coding. McGraw-Hill, New York (1963)
10. Kohonen, T.: Self-Organization and Associative Memory. Springer, New York (1988)
11. Kohonen, T.: Self-Organizing Maps. Springer, Heidelberg (1995)
12. Breiman, L.: Bagging predictors. Mach. Learn. **24**(2), 123–140 (1996)
13. Breiman, L.: Random forests. Mach. Learn. **45**(1), 5–32 (2001)

A Neural Architecture for Multi-label Text Classification

Sam Coope[1(✉)], Yoram Bachrach[2], Andrej Žukov-Gregorič[3], José Rodriguez[4], Bogdan Maksak[5], Conan McMurtie[5], and Mahyar Bordbar[5]

[1] PolyAI, London, UK
[2] Google Deepmind, London, UK
[3] Blackrock, London, UK
[4] Melior.ai, Dublin, Ireland
[5] DigitalGenius Ltd., London, UK
{sam,yoram,andrej,jose,bogdan,conan,mario}@digitalgenius.com

Abstract. We propose a novel supervised approach for multi-label text classification, which is based on a neural network architecture consisting of a single encoder and multiple classifier heads. Our method predicts which *subset* of possible tags best matches an input text. It efficiently spends computational resources, exploiting dependencies between tags by encoding an input text into a compact representation which is then passed to multiple classifier heads. We test our architecture on a Twitter hashtag prediction task, comparing it to a baseline model with multiple feedforward networks and a baseline model with multiple recurrent neural networks with GRU cells. We show that our approach achieves a significantly better performance than baselines with an equivalent number of parameters.

Keywords: Natural language processing · Neural networks
Deep learning · Multi-label classification · Co-occurring labels

1 Introduction

Consider the task of summarizing a document by extracting important keywords reflecting the main topics discussed in the text. Such keyword tags enable fast and simple searching for a large corpus.

Text tagging is useful in many domains. For instance, customer support agents may annotate conversations between them and a customer with keywords relating to the issue the customer is experiencing; this enables other support agents to quickly search for conversations dealing with an issue, or routing the messages to the agent best capable to handling such issues. Similarly, online social networks, such as Twitter allow marking messages with one or more tags, designating the main topics of each message; these tags enable a simple and efficient way for searching for messages dealing with a certain topic.

K. Arai et al. (Eds.): IntelliSys 2018, AISC 868, pp. 676–691, 2019.
https://doi.org/10.1007/978-3-030-01054-6_49

Tags are a useful way of summarizing messages, but having users manually annotate messages with these keywords is time consuming. Given a set of potential tags of interest, one can automate tagging by training a machine learning model to predict which subset of tags best characterizes a message.

One can build a system for automatically tagging texts by taking a training set of of messages, each annotated with its correct subset of tags, and processing it to tune the parameters of a model; once training is complete, the model can take a previously unobserved message and output the set of relevant tags, without the need of human annotation. We emphasize that text tagging differs from simple mutli-class classification, as our goal is not to predict a *single* keyword matching a text, but rather to select a *set* of labels matching the text. Thus, this is a multi-label classification problem.

The runtime of such a machine learning model, during both training and at prediction time, depends on the number of model parameters; it is thus very desirable to build models with few model parameters.

In many domains where people tag texts with multiple tags, some tags are likely to co-occur together, whereas others are unlikely to co-occur. For instance, the tags "#trump" and "#maga" are likely to co-occur as both deal with US politics, whereas "#sunday" and "#microscope" are less likely to co-occur as they deal with different domains. Can the propensity of tags to co-occur be leveraged to build parameter-efficient tagging models?

1.1 Our Contribution

We propose an efficient neural network architecture for predicting a subset of keywords characterizing a text, which exploits correlations between tags, and evaluate it on a Twitter hashtag prediction dataset. Our architecture takes the form of a Large Encoder with Many Independent Classifiers (LEMIC). The LEMIC design uses a single encoder network, which contains most of the parameters of the model, and a classifier head with a small number of parameters for each class.

The classifier heads use the encoded representation of the text to classify a specific label, as shown in Fig. 1. By using most of the parameters in the encoder, the number of parameters does not increase greatly when more classes are added to the classification task.

We examine two variants of the LEMIC architecture, one based on a feedforward neural network and one based on a recurrent neural network (RNN). We contrast them with baselines based on multiple independent classifiers, one for each possible keyword (where each classifier is either a feedforward network or an RNN).

We empirically evaluate our approach using a new Twitter hashtag prediction dataset[1], showing that the LEMIC architecture outperforms multiple classifiers, whilst using a small fraction of the parameters; for both the feedforward and the recurrent architectures, we observe better performance, faster convergence times

[1] Contact the first author.

and a substantial reduction in the total number of parameters when trained on a dataset of co-occurring labels.

1.2 Structure of the Paper

We continue the introduction by describing important related work. We then more closely examine the key problem that we tackle, i.e. text classification tasks when classes are correlated. In particular, we describe a straightforward baseline which uses multiple isolated classifiers, and explain why such a solution may be wasteful in terms of compute time and memory.

We continue by describing our proposed LEMIC architecture, which relies on two components: an encoder subnetwork and multiple classifier heads. We discuss both a variant relying on a feedforward network, and one using a recurrent neural network.[2] We then discuss our loss function and our network training approach.

We then turn to our empirical analysis of the model. We first describe our Twitter hashtag dataset, explain how it was sourced, and discuss some properties of hashtag co-occurrence observed in the data. We then contrast the performance of our model with that of the multiple-classifier baseline, showing that our model is far more efficient and achieves a much better performance (for any given compute "budget").[3] Finally, we provide some conclusions and propose interesting problems for future research.

2 Related Work

Various alternative methods were proposed for multi-label text classification and the related problem of building label summaries of text such as methods based on textual similarity and word co-occurrence [1,2], applying multiple classifiers [3–6] or combining multiple classifiers with parse trees [7] or lexical chains [8].

One line of related research deals with predicting hashtags in Twitter, using various techniques such as naive methods based on multiple classifiers [9], word co-occurrence [10], algebraic techniques [11], methods using collaborative filtering based recommender systems [12] or applying Bayesian topic models [13] such as the Latent Dirichlet Allocation [14].

Deep learning proven to be state of the art in many NLP problems such as machine translation [15], text classification and prediction [16,17] question-answering [18], named-entity recognition [19] and chatbots [20,21].[4]

Earlier work has already proposed using neural networks for label prediction [27,28] but such solutions rely on the naive approach of building a single neural network classifier per each label.

[2] Other network designs can also be used in the LEMIC architecture, such as ones based on a convolutional neural network (CNN).

[3] We show that LEMIC outperforms the baseline both when examining a feedforward variant and an RNN variant.

[4] Deep learning has of course also proven extremely useful in domains other than NLP, such as vision [22–24] or control and reinforcement learning [25,26].

Closest in nature to our method is an approach that uses a single neural network for Twitter hashtag prediction [29], using a bi-directional Gated Recurrent Unit (GRU) over characters to produce a compact sentence embedding. However, that model assumes *mutually exclusive labels*, and predicts only a *single tag* for each text; it applies a single linear projection to the class probabilities from a sentence embedding and uses a categorical cross-entropy loss. In contrast, our model outputs a *set* of tags, and aims to maximize the accuracy in predicting all the correct tags by applying a binary cross-entropy loss for *every* label.

Our network design exploits correlations between label predictions by using an encoder-decoder network, somewhat akin to the design used in Sequence-to-Sequence networks [30] or autoencoders [31]. However, the goal of our network is very different. As opposed to Sequence-to-Sequence networks, we do not attempt to develop a full language model.

Further, as opposed to autoencoders or dimensionality reduction methods, our goal is not to compress the original text into a concise description so that it can later be reconstructed exactly [32–34] or approximately [35–38]. Rather, given a set of possible tags, our goal is to determine the best subset of tags that best characterizes the document.

3 Text Classification with Correlated Classes

We train a machine learning model for predicting the best subset of tags characterizing a document. During *training* our system receives a set of documents, each annotated with its of corresponding tags. We denote the set of tagged documents as $X = \{x_1, x_2, \ldots x_{|X|}\}$ where each x_i is some representation of a document.

Given a vocabulary of words $V = (w_1, w_2, \ldots, w_v)$, and a text p we denote the term-frequency (TF) representation of p as:

$$p^{TF} = (d_1, \ldots, d_v)$$

where d_j is the number of times the word w_j appears in the p.

Similarly given an alphabet $\Sigma = (c_1, c_2, \ldots, c_A)$, the sequential character representation of the text is simply the sequence of the indices of the characters occurring in p. In our experiments, we use the TF representation for out feed-forward networks and the sequential character representation for RNNs. We denote the set of all possible labels to be K and the set of labels associated with document x_i to be $\kappa(x_i) \in 2^{|K|}$ (i.e. $\kappa(x_i)$ is a binary vector with 1 s for relevant tags and 0 s elsewhere).

3.1 Multiple Binary Classifiers Baseline

One text tagging approach is to treat each possible tag as a standard binary classification problem. Given a possible label k, we can examine X, and partition it into positive instances which were tagged with label $k \in K$, $P_k = \{x \in$

$X|k \in \kappa(x)\}$, and negative instances which were not tagged with that label $N_k = \{x \in X|x \notin \kappa(x)\}$. We can then train a classifier $C_k : x \rightarrow \{0,1\}$. The runtime and memory consumption of the classifier depends on the classification algorithm we use.

However, we are not interested in predicting whether a *single specific* tag is relevant to a text; our goal is to identify *all* tags that are relevant to a given document. After training all the classifiers $C_1, C_2, \ldots, C_{|K|}$, we can take an unobserved input x use the predictions $C_1(x), C_2(x), \ldots, C_{|K|}(x)$ to get the predicted subset of labels $\kappa(x) = \{k \in K|C_k(x) = 1\}$.

If the runtime of a single classifier is c, the runtime of determining all the relevant tags is $c \cdot |K|$. Furthermore, if a single classifier requires m bytes in memory, storing the parameters for all the models requires $m \cdot |K|$, which may not be tractable due to memory constraints.

The naive approach is especially wasteful when the tags are correlated. Consider the case where the tag k_b is very correlated with the tag k_a; for instance, consider the case where k_b is used almost any time when the tag k_a is applicable, and almost never used when k_a is not applicable. In this case, one could build only the classifier for k_a and apply k_b if and only if k_a is applicable.

Even in cases where there is a high but not perfect correlation between tags, they may still rely on similar information. For instance, tags k_a and k_b may only be applicable when the text mentions a specific topic such as politics. In this case, there is no point trying to infer this intermediate information multiple times in each classifier, as the naive approach would do.

Another alternative is to use a single large neural network and use the output of the final layer to compute class probabilities (i.e. the model shares *all* the parameters for each class). However, as discussed in the experimental section, we found that these models suffer greatly when training on uneven data distributions, and the architecture tends to only learn to label the most common labels.

3.2 LEMIC Architecture

We propose the LEMIC architecture which is designed to outperform models similar the previously described baseline. Additionally, the LEMIC architecture can use several orders of magnitude fewer parameters than the baseline models and still performs better than the baselines (as shown in Fig. 2).

With a dataset containing a large number of labels, the baseline approach suffers from a very skewed dataset distribution; each classifier has a very small number of positive examples to train on, and performance may suffer from as a result. The encoder of a our LEMIC network, which contains most of the parameters, has a more varied data distribution making it less prone to suffer less from this issue.

The LEMIC architecture is based on a text encoder and multiple small classifiers. The encoder network generates a concise description vector d, given an input text example x_i, consisting of key information required to predict relevant

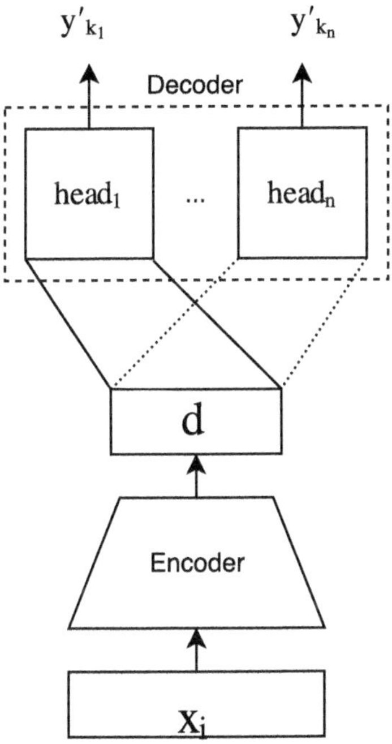

Fig. 1. The LEMIC architecture.

tags. The classifier networks then take d and attempt to predict its associated tag.

Similarly to the baseline, during training our model is given a set of (text, labels) example pairs:

$$D = \{(x_1, \kappa(x_1)), \ldots (x_N, \kappa(x_N))\}$$

During training we set internal network parameters θ. Following training, the system can digest a new message n and obtain a binary prediction $p_\theta(k, n) \in \{0, 1\}$ regarding whether label k is applicable to message n for each label k.

The key principle that lies behind LEMIC is that the *same* description d generated by the encoder is used to make *all* label predictions. Thus, the size m of this short description forms an *information bottleneck*, forcing the network to focus on the few properties of the input that drive as many of the predictions as possible (as captured by the loss function we use).[5]

[5] An encoder/decoder design for a network has been used in many architectures in the past [15,30,39,40].

Similarly, other designs form an intentional information bottleneck to obtain a concise description of an input [31,38,40–42]. The novel aspect of our design lies in the specific architecture we use for the encoder and classifiers and the way we use them to minimize the loss in predicting the relevant labels.

3.3 Encoder

The encoder receives a representation of a passage, and outputs a concise m-dimensional description vector $d \in \mathbb{R}^m$, with m selected to be relatively small so as to obtain an information bottleneck in the network.

In this paper, we experiment with both recurrent and feed-forward encoders. The feed-forward encoders consist of two fully connected layers which take a TF representation of the text as input. We have also experimented with an RNN encoder, which is somewhat similar to that of Dhingra et al. [29]. Our design uses a character-level bi-directional network with Gated Recurrent Units (GRU) [43] as an encoder.

As is common in RNN designs for text processing, the final states of our forward and backward GRU cells are concatenated to form the description vector d. As our results in the empirical section show, using this recurrent encoder design in LEMIC achieves lower performance than a LEMIC design using a feed-forward network as an encoder. This is consistent with recent findings indicating that for various text processing tasks, a small and shallow feedforward network can outperform more elaborate recursive designs [44].

3.4 Classifiers

The LEMIC decoder consists of multiple sub-networks, called "heads", each responsible for classifying a single label. Each head takes the same description vector d produced by the encoder and outputs a prediction $y'_k \in \{0, 1\}$ for a label designated to that particular head k.

In all of our experiments, each head is composed of a two-layer feed-forward neural network, followed by a softmax layer. The i'th head takes as input the description vector d and produces a 2-dimensional vector of unnormalized tag probabilities:

$$a'^i = (a'^i_0, a'^i_1)$$

The vector a'^i is then normalized through a softmax to produce a vector:

$$a^i = (a^i_0, a^i_1)$$

where:

$$a^i_0 = \frac{\exp(a'^i_0)}{\exp(a'^i_0) + \exp(a'^i_1)}$$

$$a^i_1 = \frac{\exp(a'^i_1)}{\exp(a'^i_0) + \exp(a'^i_1)}$$

We view a^i_0 is the probability of i *not* being a suitable tag for the input and a^i_1 the probability of i being a *correct* tag for the input.

We emphasize that other variants of the LEMIC model may use a different design for the encoder and classifier heads, such as having the heads be convolutional nets.

3.5 Training and Loss

After each head produces its output we obtain $|K|$ 2-dimensional vectors, each encoding a binary probability distribution for some label i. We assign a separate cross-entropy loss L_i to each head, with y_i denoting the correct (ground truth) classification for the i'th tag:

$$L_k(y_i, a^i) = y_i \log \frac{1}{a_1^i} + (1 - y_i) \log \frac{1}{a_0^i}$$

Our final loss needs to consider *all* the head outputs. We thus let our final loss be the *mean* of the individual head losses:

$$L(\kappa(x), \hat{a}) = \frac{1}{|K|} \sum_{k \in K} L_k$$

where $\hat{a}$ is a $2 \times |K|$-dimensional vector, the concatenation of the individual head outputs: $\hat{a} = (a_0^0, a_1^0, a_0^1, a_1^1, \ldots, a_0^k, a_1^k)$.

During training, we iteratively tune model parameters after examining mini-batches of training set instances, by applying backpropagation [45]. We train our model using stochastic gradient descent (SGD), with the Adam optimizer [46] (using a learning rate of 5×10^{-4}). We have implemented our model and the baselines in TensorFlow [47], and ran our experiments on our GPU cluster.

4 Experiments

To evaluate our proposed architecture, we train various sizes of both the feed-forward and the recurrent LEMIC models and compare them to the baseline approach with multiple independent classifiers (both feed-forward and recurrent) in the task of predicting hashtags from the contents of tweets. We show that LEMIC outperforms the baselines even when it has far fewer parameters.

4.1 Dataset

Our dataset is composed of texts from Twitter along with their designated hashtags. To construct the dataset, we have fetched tweets from Twitter using the "hashtags trending" feature in January and February 2017.

To make sure we only use tweets with meaningful data, we have removed tweets consisting of less than fifteen words, leaving us approximately 100,000 tweets. Our results rely on 10-fold cross-validation.

To choose the hashtags for classification, the ten most frequent hashtags were taken as "root" hashtags. We sourced tweets containing any of the root hashtags, and examined all hashtags in these tweets.

A co-occurrence score was calculated for every hashtag to every root hashtag, and the hashtags that have more than 0.1 co-occurrence with a root hashtag were also included in the dataset.

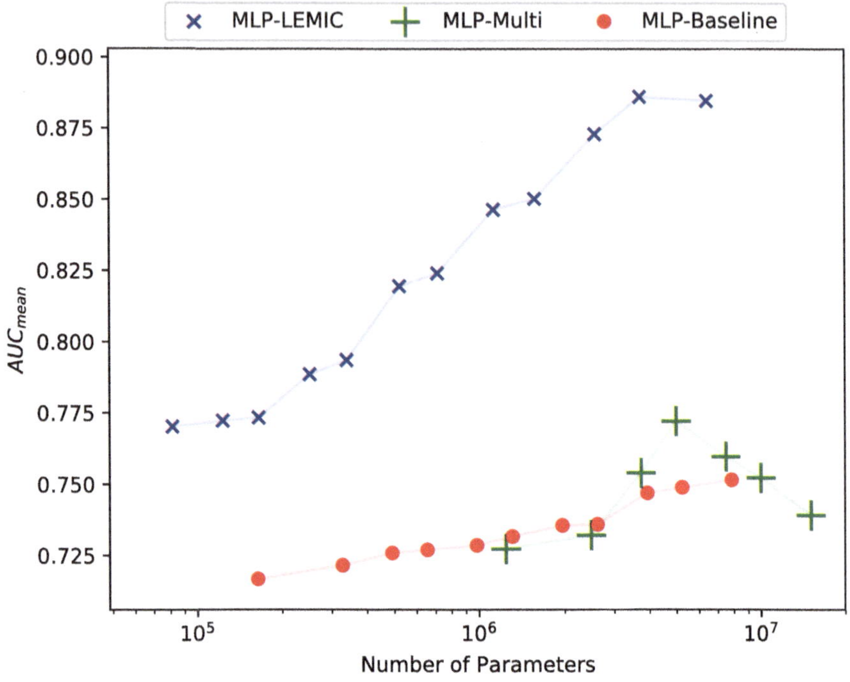

Fig. 2. AUC_{mean} of feed-forward models with increasing parameter sizes. The MLP-Multi models consist of a single-headed LEMIC model for each class.

As a co-occurrence measure for hashtags h_1 and h_2 we use the total number of tweets with both tags included (C_{h_1,h_2}), divided by the size of the smallest group of tags, i.e.:

$$\frac{C_{h_1,h_2}}{min(n_{h_1}, n_{h_2})}$$

where n_{h_i} is the number of tweets with tag h_i. Table 1 shows some of the most co-occurring hashtag pairs in the dataset.

We chose co-occurring tags with the aim of choosing semantically or thematically similar tags in the dataset. Hashtags that occur in the same tweet often relate to the same domain.

We only used hashtags that occurred in at least 900 tweets, resulting in $|K| = 31$ hashtags used for classification.

Our dataset forms a text classification problem where some of the tags have a strong correlation with each other.

4.2 Preprocessing

Our preprocessing procedure is simple. First, we remove all the hashtag words from the tweet (as these are the required output of the architecture). After

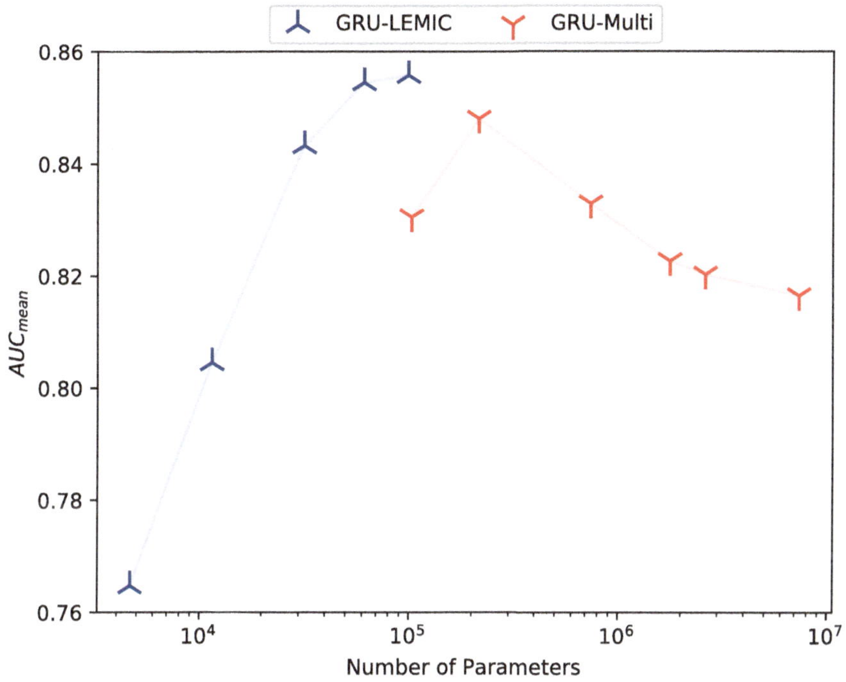

Fig. 3. AUC_{mean} of the recurrent models with increasing parameter size.

removing the hashtags, punctuation is removed and words are lower-cased and stemmed. We have also removed all duplicate entries (which are very common on Twitter due to retweets and automatic Twitter bots).

For the models using a feed-forward network, a vocabulary V is then created of the $n_V = 5000$ most frequent words in the dataset. We note that this is a relatively small vocabulary size, resulting in a relatively quick computation. The input to the feed-forward based network is the TF representation of the tweet.

For the recurrent models, we have dropped all the non-ASCII characters, and represented each character as a one-hot 128-dimensional vector.

4.3 Baseline

We used a model similar to those used by Sarkar et al. [27] as a baseline: a one hidden-layer feed-forward network for each class; however, we use the same TF input representation as used for the feedforward LEMIC models, rather than hand-crafted features.

For recurrent and feed-forward LEMIC models, we also trained similar models separately for each class in the dataset, i.e. each class has its own single-headed LEMIC model. For brevity, we call these models *multi-models* for in this section.

Table 1. The Top 10 Co-occurrences between Hashtags in the Dataset

h_1	$h2$	Co-occurrence
#business	#fdlx	1.00
#news	#fdlx	1.00
#tfl	#london	1.00
#trump	#scotus	0.86
#business	#news	0.80
#brexitbill	#brexit	0.55
#maga	#trump	0.53
#resist	#devos	0.52
#entrepreneur	#news	0.29
#trump	#brexit	0.25

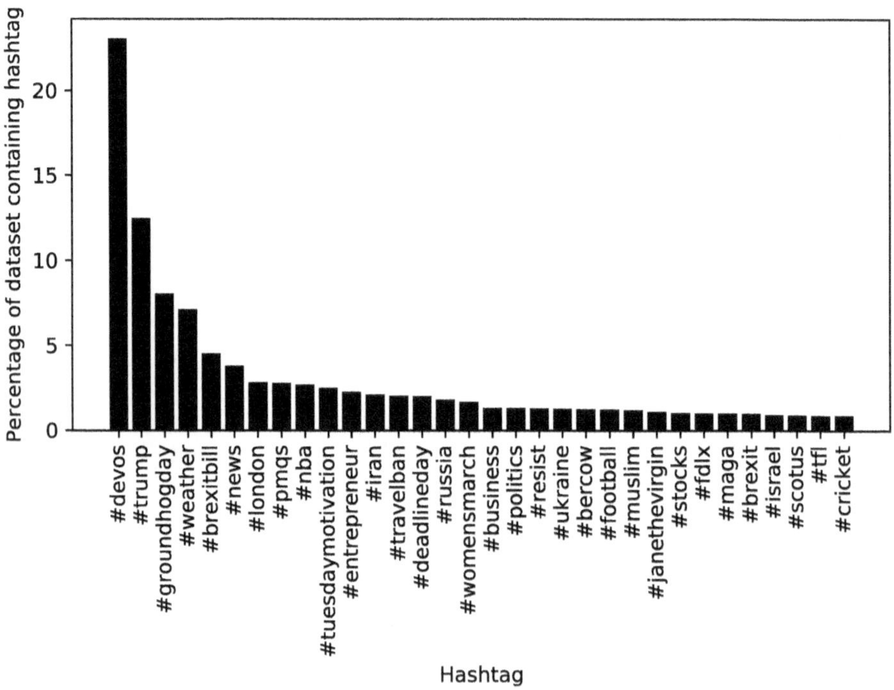

Fig. 4. Hashtag distribution in the twitter dataset.

4.4 Results

We compare the performance of LEMIC and the baselines. To evaluate performance, we use the area under the ROC curve (AUC) of each tag, as it is an effective means of evaluating how well a classifier performs [48,49]. We note that the AUC is a summary statistic, which reflects the possible trade-offs between the precision and recall of each of the individual classifiers. If a more detailed investigation of the performance and precision-recall trade-offs is desired, one can examine the ROC curve for each of the classifiers. We chose the average AUC across classifiers as it offers a simple overall performance metric. Another alternative is choosing a minimal desired recall, and examining the accuracy achieved for each classifier subject to achieving this desired recall value; however, this methodology is difficult to apply, as it is not clear what constitutes a reasonable minimal recall value.

As shown in Fig. 4, the dataset has a very uneven distribution of hashtags. Our chosen loss function reflects the implicit assumption of being interested in the performance of every classifier equally; similarly, our overall performance metric is the unweighted mean of the AUC of each hashtag.[6] We denote the mean AUC across classifiers as AUC_{mean}. We show that our model has a consistently better AUC_{mean} than baseline and multi-models with similar number of parameters.

(1) Feed-forward networks: We trained several LEMIC and baseline models with varying hidden layer sizes, where both the encoder and heads of the models have one hidden layer. The size of the encoder output is half that of the hidden layer of the encoder, and the size of the hidden layer in the classifiers is half the size of the encoder output.

Figure 2 shows how the AUC_{mean} increases with parameter size for the LEMIC model, the feed-forward multi-model, and the baseline. LEMIC significantly outperforms the other models at every parameter size, with very small LEMIC models of only 10,000 parameters performing better than multi-models with hundreds of thousands.

All of the feed-forward models saturate in performance at around 5 million parameters, with the best LEMIC model consisting of approximately 3.7 million parameters and an AUC_{mean} of 0.886.

We have also implemented a single large feed-forward network mapping the raw TF input or the encoder's output (both the feed-forward and the RNN version) to $2 \cdot K$ outputs (two outputs per tag). This design differs from LEMIC by having a single large head rather than multiple independent heads, and scored consistently less than 0.6 for AUC_{mean}. This indicates the importance of using multiple heads, as we do in LEMIC.

(2) Recurrent networks: Figure 3 shows that the LEMIC architecture also performs well when using recurrent encoder. The recurrent *multi-models* seem to perform better than the feed-forward multi-models, although they still perform significantly worse than LEMIC designs of similar parameter sizes.

[6] We note that it is simple to modify our architecture to place more emphasis on some of the classifiers by tweaking the loss structure. For instance one may multiply some classifier head losses by different constants than others in the overall loss function. However, we see no reason to do this for the specific Twitter dataset we have used for the evaluation here.

Interestingly, the performance of the RNN models saturates at lower parameter sizes than the feed-forward networks, and the best recurrent LEMIC model performs significantly worse than the best performing feed-forward LEMIC. This is in contrast to other NLP tasks, where RNNs perform better than feed-forward designs. One possible explanation is that for text classification, the order in which words occurs in the sentence is of lesser importance. This finding is consistent with work indicating that there are many natural language processing tasks where a feedforward design achieves a better performance than RNN architectures [44] (and is less prone to overfitting).

(3) Performance, runtime and memory: The above analysis is centered around the tradeoffs between model complexity and model performance. The x-axis on Figs. 2 and 3 is the number of model parameters. One may view the number of parameters used in a model as a reflection of the maximal "budget" of either runtime or memory consumption.

Our results in these figures thus evaluate the best way to spend this "budget" so as to get the best model performance for generating multiple tags. The results indicate that LEMIC offers a better performance than the baselines for any of the budgets we examined. Further, the results indicate that a feedforward LEMIC design is more efficient than a RNN LEMIC design.

Given the above results, we would recommend applying a LEMIC design for text tagging where tags are correlated. Given a bound on the allowed runtime (or memory) for text tagging, a LEMIC architecture would achieve a better accuracy than the baseline model.[7]

5 Conclusion

We proposed the LEMIC design for text tagging. LEMIC uses an encoder to find a latent representation the input that contains the relevant information for predicting each of the tags. This latent representation is *shared* between the different classifier heads, allowing the model to exploit correlations between tags (or information predictive of the tags).

Our empirical evaluation on our Twitter hashtag prediction dataset shows that LEMIC outperforms baseline designs with independent feed-forward or recurrent neural network classifiers. Given a fixed "budget" of compute or memory resources, LEMIC achieves a higher accuracy than the baseline models, as it is more efficient in expending the resources, reusing intermediate results across the different classifiers.

[7] In the unrealistic scenario where one has no "budget" constraints (i.e. infinite memory and compute time), one can build an isolated classifier for each tag, possibly using a different architecture for each such tag, and achieve a better accuracy.

5.1 Future Work

Various directions are open for future research.

First, future work may further improve the performance of our approach. For instance one may optimize the relative number of parameters between the encoder and classifiers. We used simple heuristics for determining the various hidden layer sizes, without thoroughly searching for the optimal design. Further, it is impractical to sweep over all the possible architectures variants for each new dataset. It is thus desirable to find good design principles that would result in a good (though not necessarily optimal) design choices, such as the hidden layer sizes.

Second, one may employ recently proposed neural network tools to further improve the performance. One particularly interesting potential improvement for the RNN models is applying a self-attention mechanism, forming a weighted sum of the GRU hidden states, rather than taking the final states.

Finally, it would be interesting to see whether similar design principles could be used in structured prediction scenarios. For instance, how should one handle constraints regarding classifier outputs, such as cases where certain combinations of classifier outputs are disallowed, as is the case in hierarchical prediction?

References

1. Matsuo, Y., Ishizuka, M.: Keyword extraction from a single document using word co-occurrence statistical information. Int. J. Artif. Intell. Tools **13**(01), 157–169 (2004)
2. Rose, S., Engel, D., Cramer, N., Cowley, W.: Automatic keyword extraction from individual documents. In: Text Mining, pp. 1–20 (2010)
3. Zhang, K., Xu, H., Tang, J., Li, J.: Keyword extraction using support vector machine. In: International Conference on Web-Age Information Management, pp. 85–96. Springer (2006)
4. Volkova, S., Bachrach, Y., Armstrong, M., Sharma, V.: Inferring latent user properties from texts published in social media. In: AAAI, pp. 4296–4297 (2015)
5. Volkova, S., Bachrach, Y.: On predicting sociodemographic traits and emotions from communications in social networks and their implications to online self-disclosure. Cyberpsychol. Behav. Soc. Netw. **18**(12), 726–736 (2015)
6. Lewenberg, Y., Bachrach, Y., Volkova, S.: Using emotions to predict user interest areas in online social networks. In: DSAA, pp. 1–10. IEEE (2015)
7. Hulth, A.: Improved automatic keyword extraction given more linguistic knowledge. In: Proceedings of the 2003 Conference on Empirical Methods in Natural Language Processing, pp. 216–223. Association for Computational Linguistics (2003)
8. Ercan, G., Cicekli, I.: Using lexical chains for keyword extraction. Inf. Process. Manage. **43**(6), 1705–1714 (2007)
9. Mazzia, A., Juett, J.: Suggesting hashtags on twitter. In: EECS 545m. Computer Science and Engineering, University of Michigan, Machine Learning (2009)
10. Xiao, F., Noro, T., Tokuda, T.: News-topic oriented hashtag recommendation in twitter based on characteristic co-occurrence word detection. In: International Conference on Web Engineering, pp. 16–30. Springer (2012)

11. Li, T., Wu, Y., Zhang, Y.: Twitter hash tag prediction algorithm. In: ICOMP (2011)
12. Kywe, S.M., Hoang, T.-A., Lim, E.-P., Zhu, F.: On recommending hashtags in twitter networks. In: International Conference on Social Informatics, pp. 337–350. Springer (2012)
13. Godin, F., Slavkovikj, V., De Neve, W., Schrauwen, B., Van de Walle, R.: Using topic models for twitter hashtag recommendation. In: WWW, pp. 593–596. ACM (2013)
14. Blei, D.M., Ng, A.Y., Jordan, M.I.: Latent dirichlet allocation. JMLR **3**, 993–1022 (2003)
15. Bahdanau, D., Cho, K., Bengio, Y.: Neural machine translation by jointly learning to align and translate. arXiv preprint arXiv:1409.0473 (2014)
16. Glorot, X., Bordes, A., Bengio, Y.: Domain adaptation for large-scale sentiment classification: a deep learning approach. In: ICML 2011, pp. 513–520 (2011)
17. Preoţiuc-Pietro, D., Volkova, S., Lampos, V., Bachrach, Y., Aletras, N.: Studying user income through language, behaviour and affect in social media. PloS one **10**(9), e0138717 (2015)
18. Bachrach, Y., Gregorič, A.Ž., Coope, S., Tovell, E., Maksak, B., Rodriguez, J., McMurtie, C., Bordbar, M.: An attention mechanism for neural answer selection using a combined global and local view. In: Proceedings of ICTAI 2017. IEEE (2017)
19. Gregorič, A.Ž., Bachrach, Y., Minkovsky, P., Coope, S., Maksak, B.: Neural named entity recognition using a self-attention mechanism. In: Proceedings of ICTAI 2017. IEEE (2017)
20. Kandasamy, K., Bachrach, Y., Tomioka, R., Tarlow, D., Carter, D.: Batch policy gradient methods for improving neural conversation models. arXiv preprint arXiv:1702.03334 (2017)
21. Serban, I.V., Sankar, C., Germain, M., Zhang, S., Lin, Z., Subramanian, S., Kim, T., Pieper, M., Chandar, S., Ke, N.R., et al.: A deep reinforcement learning chatbot. arXiv preprint arXiv:1709.02349 (2017)
22. Krizhevsky, A., Sutskever, I., Hinton, G.E.: Imagenet classification with deep convolutional neural networks. In: Advances in Neural Information Processing Systems, pp. 1097–1105 (2012)
23. Lewenberg, Y., Bachrach, Y., Shankar, S., Criminisi, A.: Predicting personal traits from facial images using convolutional neural networks augmented with facial landmark information. In: IJCAI, pp. 1676–1682 (2016)
24. Simonyan, K., Zisserman, A.: Very deep convolutional networks for large-scale image recognition. arXiv preprint arXiv:1409.1556 (2014)
25. Mnih, V., Kavukcuoglu, K., Silver, D., Graves, A., Antonoglou, I., Wierstra, D., Riedmiller, M.: Playing atari with deep reinforcement learning. arXiv preprint arXiv:1312.5602 (2013)
26. Van Hasselt, H., Guez, A., Silver, D.: Deep reinforcement learning with double q-learning. In: AAAI, pp. 2094–2100 (2016)
27. Sarkar, K., Nasipuri, M., Ghose, S.: A new approach to keyphrase extraction using neural networks. arXiv preprint arXiv:1004.3274 (2010)
28. Kumarika, B.T., Dias, N.: Smart web content bookmarking with ANN based key phrase extraction algorithm. In: 2014 International Conference on Advances in ICT for Emerging Regions (ICTer), pp. 228–234. IEEE (2014)
29. Dhingra, B., Zhou, Z., Fitzpatrick, D., Muehl, M., Cohen, W.W.: Tweet2vec: character-based distributed representations for social media. arXiv preprint arXiv:1605.03481 (2016)

30. Sutskever, I., Vinyals, O., Le, Q.V.: Sequence to sequence learning with neural networks. In: Advances in Neural Information Processing Systems, pp. 3104–3112 (2014)
31. Baldi, P.: Autoencoders, unsupervised learning, and deep architectures. In: ICML Unsupervised and Transfer Learning, vol. 27, no. 37–50, p. 1 (2012)
32. Ziv, J., Lempel, A.: Compression of individual sequences via variable-rate coding. IEEE Trans. Inf. Theory **24**(5), 530–536 (1978)
33. Bell, T.C., Cleary, J.G., Witten, I.H.: Text Compression. Prentice-Hall, Inc., Englewood Cliffs (1990)
34. Fenwick, P.M.: The burrows-wheeler transform for block sorting text compression: principles and improvements. Comput. J. **39**(9), 731–740 (1996)
35. Roweis, S.T., Saul, L.K.: Nonlinear dimensionality reduction by locally linear embedding. Science **290**(5500), 2323–2326 (2000)
36. Jolliffe, I.: Principal Component Analysis. Wiley Online Library (2002)
37. Zhang, Z.-Y., Zha, H.-Y.: Principal manifolds and nonlinear dimensionality reduction via tangent space alignment. J. Shanghai Univ. (Engl. Ed.) **8**(4), 406–424 (2004)
38. Hinton, G.E., Salakhutdinov, R.R.: Reducing the dimensionality of data with neural networks. Science **313**(5786), 504–507 (2006)
39. Cho, K., Van Merriënboer, B., Bahdanau, D., Bengio, Y.: On the properties of neural machine translation: encoder-decoder approaches. arXiv preprint arXiv:1409.1259 (2014)
40. Kramer, M.A.: Nonlinear principal component analysis using autoassociative neural networks. AIChE J. **37**(2), 233–243 (1991)
41. Vincent, P., Larochelle, H., Bengio, Y., Manzagol, P.-A.: Extracting and composing robust features with denoising autoencoders. In: Proceedings of the 25th International Conference on Machine Learning, pp. 1096–1103. ACM (2008)
42. Vincent, P., Larochelle, H., Lajoie, I., Bengio, Y., Manzagol, P.-A.: Stacked denoising autoencoders: learning useful representations in a deep network with a local denoising criterion. J. Mach. Learn. Res. **11**, 3371–3408 (2010)
43. Chung, J., Gulcehre, C., Cho, K., Bengio, Y.: Empirical evaluation of gated recurrent neural networks on sequence modeling. arXiv preprint arXiv:1412.3555 (2014)
44. Botha, J.A., Pitler, E., Ma, J., Bakalov, A., Salcianu, A., Weiss, D., McDonald, R., Petrov, S.: Natural language processing with small feed forward networks. arXiv preprint arXiv:1708.00214 (2017)
45. Demuth, H.B., Beale, M.H., De Jess, O., Hagan, M.T.: Neural Network Design. Martin Hagan (2014)
46. Kingma, D., Ba, J.: Adam: a method for stochastic optimization. arXiv preprint arXiv:1412.6980 (2014)
47. Abadi, M., Agarwal, A., Barham, P., Brevdo, E., Chen, Z., Citro, C., Corrado, G.S., Davis, A., Dean, J., Devin, M., et al.: Tensorflow: large-scale machine learning on heterogeneous distributed systems. arXiv preprint arXiv:1603.04467 (2016)
48. Hanley, J.A., McNeil, B.J.: The meaning and use of the area under a receiver operating characteristic (roc) curve. Radiology **143**(1), 29–36 (1982)
49. Bradley, A.P.: The use of the area under the ROC curve in the evaluation of machine learning algorithms. Pattern Recognit. **30**(7), 1145–1159 (1997)

A Neuro Fuzzy Approach for Predicting Delirium

Frank Iwebuke Amadin(✉) and Moses Eromosele Bello

Department of Computer Science, University of Benin, Benin, Edo State, Nigeria
frankamadin@uniben.edu, moseseromo@yahoo.com

Abstract. Delirium is a mental disease that is prevalent amongst adult which cannot be diagnosed using laboratory tests. Diagnosing delirium requires the use of symptoms and even experienced psychiatrist often misdiagnose it. In this paper, a neuro fuzzy technique was applied in diagnosing delirium. A dataset containing thirty diagnosed case of delirium was gotten from Federal Neuro Psychiatric Hospital in Edo State and it was used to train and test the diagnostic accuracy of the system. Upon completion of the simulation the system had a training and testing error of 0.33334 and 0.45373, respectively.

Keywords: Delirium · Adaptive Neuro-Fuzzy Inference System (ANFIS)

1 Introduction

Mental health includes psychological, social, and emotional well-being of an individual. It revolves round how an individual acts, feel, handles stress, think, make choices and relate to others. Assessment of mental health is very important at every stage in life, from infancy to adulthood. Some illnesses have been known to affects the mind and mood of individual, altering their emotional, psychological, and social well-being [1]. These forms of illness are known as mental illness. Mental illnesses are diseases that affect the cognitive aspect of brain and this could result in irrational behavior of an individual, impaired reasoning and making the individual incapable of carrying out his/her daily activities. More than 200 forms of mental disorder has been classified in DSM-IV and some of the most common mental illness include depression, Dementia, anxiety disorder, Schizophrenia, Bipolar disorder and Delirium [2]. Delirium is a mental ailment that impairs consciousness, attention, memory, behavior and cognitive processes of an individual [3]. It is a neuropsychiatric syndrome of acute onset and fluctuating course, which is clinically characterized by varied level of attention, consciousness, thought, memory, and behavior [2]. It is an underdiagnosed and severe mental ailment with its symptoms and severity varying significantly, but its core features are impaired sensory and cognitive functions [4]. The occurrence amongst the general public differs considerably depending on the location [5]. Studies have shown that its occurrences in rural communities' is estimated to be 1%–2%, in hospital during admission it is estimated to be between 3%–29% and the incidence amongst the elderly and persons in postoperative, intensive care and/or palliative care might rise up to 81% [6–8]. Several techniques have been applied to diagnosing delirium but the Confusion Assessment Method (CAM) has the highest accuracy [9–12]. However there is no evidence of a prediction based model

K. Arai et al. (Eds.): IntelliSys 2018, AISC 868, pp. 692–699, 2019.
https://doi.org/10.1007/978-3-030-01054-6_50

for diagnosing delirium. The aim of our paper is to develop an Adaptive Neuro Fuzzy Inference System model for predicting delirium to assist doctor and physicians.

2 Adaptive Neuro Fuzzy Inference System

Adaptive Neuro-Fuzzy Inference System (ANFIS) is the combination of two powerful machine learning techniques, Fuzzy Logic (FL) and Neural Network (NN) with FL embedded in the NN. The power of this system lies with the strengths of NN and FL. ANFIS is a six layer structure with each layer having a distinct function. Figure 1 shows the ANFIS architecture.

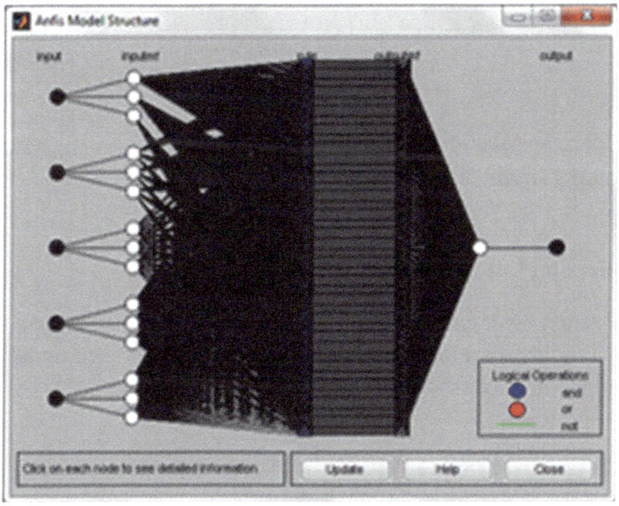

Fig. 1. Shows the adaptive neuro fuzzy inference system architecture.

2.1 Input Layer

This layer allows input into the ANFIS.

2.2 Membership Function Layer

The membership function which maps inputs to fuzzy sets is contained in this layer. In this paper we used bell membership function to map inputs to fuzzy set. The bell membership equation is shown in (1).

$$\mu(v_i) = \frac{1}{1 + \left| \frac{x_i - c_i}{a_i} \right|^{2b_i}} \tag{1}$$

where

a, b and c = premise parameters
v_i = linguistic variables from input layer
$\mu(v_i)$ = membership function of v_i

2.3 Rule Layer

The rule layer determines the outcome of any combination of linguistic variables. This layers receives inputs value from the second layer. The Takagi-Sugeno inference model was applied in this layer and can be expressed mathematically as shown in (2).

$$O_i^3 = \mu(v_1) * \mu(v_2) * \ldots .\mu(v_n) \tag{2}$$

Where

$\mu(v_n)$ = membership function linguistic variable n
O_i^3 = the output of the ith neuron in layer 3

2.4 Normalization Layer

In the normalization layer each neuron is exclusively paired to a neuron from rule layer. The normalization layer scales the input coming from the preceding layer. This can be expressed mathematically as shown in (3)

$$O_i^4 = \frac{O_i^3}{O_1^3 + O_2^3 + \cdots + O_n^3} \tag{3}$$

2.5 Defuzzification Layer

This layer contains just one neuron and all output generated from the normalization layer is fed into it. The function of the defuzzification layer is to convert the fuzzy values to real values. This can be expressed mathematically as shown in (4)

$$O_i^5 = O_i^4\left(p_1(v_i) + p_2(v_2) + \cdots p_n(v_n) + r\right) \tag{4}$$

Where

p_i = consequent parameter for variable n
r = bias

2.6 Output Layer

The output of the system is produces in this layer. In this layer the number of neurons determines the number of output produced by the system. This can be represented mathematically as shown in (5)

$$O_i^6 = \sum_i^n O_i^5 \quad (5)$$

3 Results and Discussion

Using clinical symptoms represented by input 1, input 2, input 3, input 4 and input 5 in diagnosing delirium, where input 1 represents clouding of consciousness, input 2 represents difficulty maintaining or shifting attention, Input 3 represents disorientation, input 4 represents Illusions and input 5 represents hallucinations an ANFIS model was formulated. However these symptoms are no means exhaustive but it was necessary to employ feature selection technique, to select a subset of the predominant features for diagnosing delirium [13, 14]. The ANFIS model was trained using these 5 clinical symptoms. The Neuro Fuzzy system was implemented in Matrix Laboratory (MATLAB). The dataset used in this paper contained 30 diagnosed delirium cases and was gotten from Federal Neuro Psychiatric Hospital in Edo State. The data was processed further into the format needed for this study. 20 diagnosed cases which is approximately 67% of the dataset was used to train the system while the remainder was used to test the system after the training. Tables 1 and 2 shows the symptom value of 10 cases in the dataset and the diagnostic result and outcome, respectively.

Table 1. Show the symptom value of 10 cases in the dataset

S/N	Delirium symptoms				
	Input 1	Input 2	Input 3	Input 4	Input 5
Case 1	3.90	3.58	3.58	6.58	9.93
Case 2	6.07	0.63	0.63	8.32	3.82
Case 3	4.15	6.12	6.12	7.59	0.63
Case 4	5.08	3.89	3.81	9.26	0.51
Case 5	3.03	5.10	5.10	7.04	8.94
Case 6	7.55	7.99	7.99	7.76	7.34
Case 7	7.68	4.47	4.47	3.44	9.98
Case 8	5.61	2.57	2.57	5.59	8.31
Case 9	0.4	5.27	5.27	2.97	2.11
Case 10	3.92	1.00	1.00	2.91	0.70

Table 2. Show the diagnostic result and diagnostic value

S/N	Diagnostic outcome	
	Diagnostic value	Diagnostic result
Case 1	4.22	Moderate
Case 2	3.97	Moderate
Case 3	4.17	Moderate
Case 4	3.34	Moderate
Case 5	5.93	Moderate
Case 6	8.32	Severe
Case 7	6.76	Moderate
Case 8	5.26	Moderate
Case 9	4.46	Moderate
Case 10	2.17	Mild

The ANFIS model was trained for 30 epochs using backward propagation learning algorithm with an error tolerance of 0.05, an error of 0.33334 was realized. When tested the system had an error of 0.45373, which indicates that the model was able to classify 99% of the test data accurately. Figures 2, 3, 4, 5, 6 shows Adaptive Neuro Fuzzy Inference Architecture, Membership Function, Fuzzy Inference Engine, Training data, Training and Testing processes generated during simulation by MATLab.

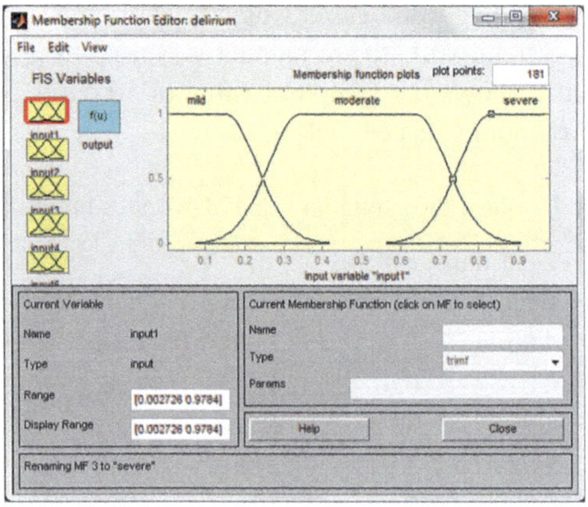

Fig. 2. Shows the fuzzy inference engine of the adaptive neuro fuzzy system.

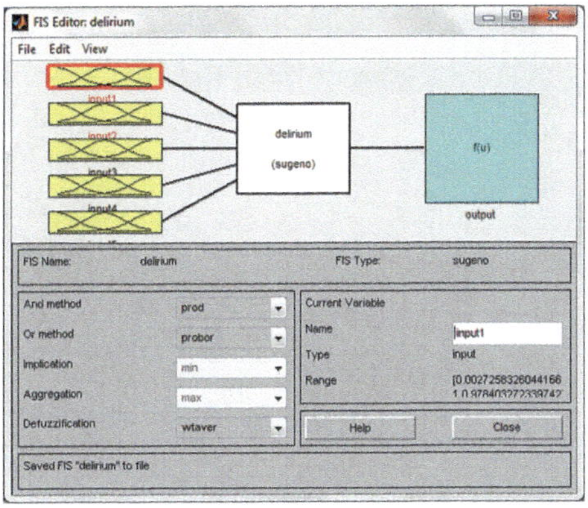

Fig. 3. Shows the bell membership function in layer 2.

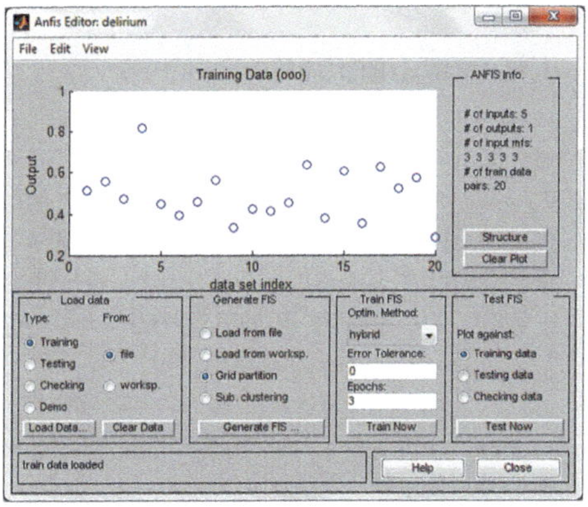

Fig. 4. Shows the training dataset.

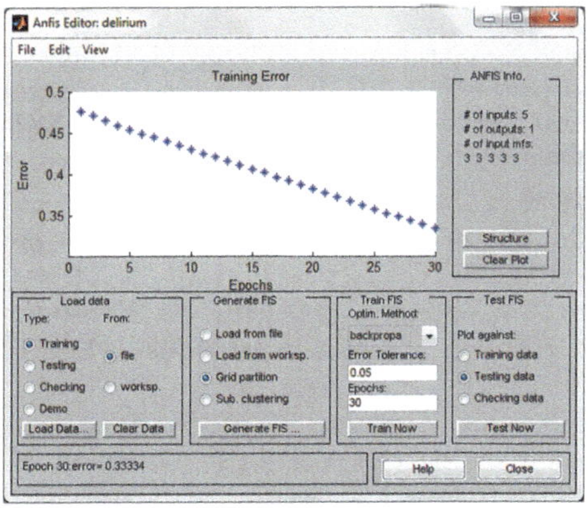

Fig. 5. Shows the training process.

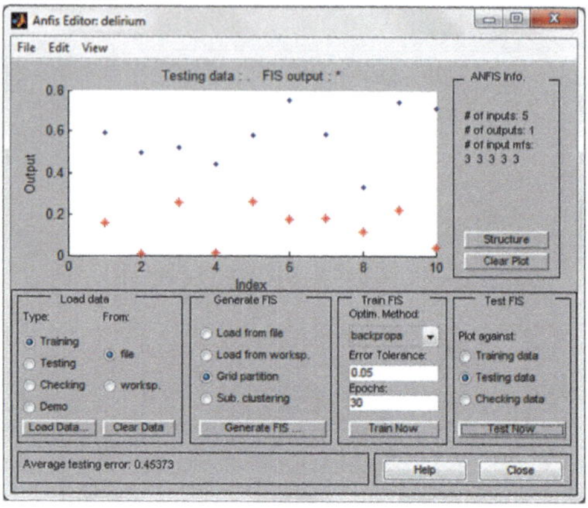

Fig. 6. Shows the testing process.

4 Conclusion and Recommendation

The diagnosis of delirium requires the use of clinical symptoms and cannot be determined through laboratory test. The ANFIS model we designed has generated an excellent result, classifying 99% of the test data accurately. The implementation of this model will assist doctors in diagnosing delirium. More robust system should be designed in the future to incorporate more clinical symptom based as it might help in diagnosing delirium more accurately.

Acknowledgment. Our sincere appreciation goes to the management and staff of Federal Neuro Psychiatric Hospital in Edo State Nigeria.

References

1. Phillip, W.G., Dennis, S.C.: Depression: a disease of the mind, brain, and body. Am. J. Psychiatry **159**, 1826 (2002)
2. Diagnostic and Statistical Manual for Mental Disorders-Fourth Edition (DSM-IV). American Psychiatric Association, Washington, DC (2005)
3. Soo-Hee, C., Hyeongrae, L., Tae-Sub, C., Kyung-Min, P., Young-Chul, J., Sun, I.K., Jae-Jin, K.: Neural network functional connectivity during and after an episode of delirium. Am. J. Psychiatry **12**, 498–507 (2012)
4. Carlota, M.G., Hugo, A.W., Brigit, P.C., Debbie, S.D., Paul-Hugo, M.K.: Validation of an automated delirium prediction model (DElirium MOdel (DEMO)): an observational study. BMJ **7**, e016654 (2017)
5. De-Wit, H.A., Winkens, B., Gonzalvo, C.: The development of an automated ward independent delirium risk prediction model. Int. J. Clin. Pharm. **23**, 15–23 (2016)
6. Siddiqi, N., House, A.O., Holmes, J.D.: Occurrence and outcome of delirium in medical in-patients: a systematic literature review. Age Ageing **35**, 350–364 (2006)

7. Ryan, J.O., Regan, N.A., Caoimh, R.O.: Delirium in an adult acute hospital population: predictors, prevalence and detection. BMJ (2013). https://doi.org/10.1136/bmjopen-2012-001772
8. Gerard, T.D., Ely, E.W.: Delirium in the critically ill patient. In: Handbook of Clincal Neurology (2008). https://doi.org/10.1016/s0072-9752
9. Luetz, A., Heymann, A., Radtke, F.M., Chenitir, C., Neuhaus, U., Nachtigall, I.: Different assessment tools x for intensive care unit delirium: which score to use? Crit. Care Med. **38**, 409–418 (2010)
10. Van, M.J., Van-Marum, R.J., Slooter, A.J.C.: Comparison of delirium assessment tools in a mixed intensive care unit. Crit. Care Med. **37**, 1881–1885 (2009)
11. Inouye, S.K., Viscoli, C.M., Horwitz, R.I., Hurst, L.D., Tinetti, M.E.: A predictive model for delirium in hospitalized elderly medical patients based on admission characteristics. Arch. Intern. Med. **119**, 474–481 (1993)
12. Inouye, S.K., Zhang, Y., Jones, R.N., Kiely, D.K., Yang, F., Marcantonio, E.R.: Risk factors for delirium at discharge: development and validation of a predictive model. Arch. Intern. Med. **167**, 1406–1413 (2007)
13. Odeh, M.S.: Using an adaptive neuro-fuzzy inference system (ANFIS) algorithm for automatic diagnosis of skin cancer. J. Commun. Comput. **8**, 751–755 (2011)
14. Odigie, B.E., Achukwu, P.U., Atoigwe, B.E., Obaseki, D.E., Usunobun, J.O., Bello, M.E.: Model formulation using adaptive neuro-fuzzy inference systems (ANFIS) for cervical lesions (CL): a case study of commercial sex workers (CSWs) predict diagnosis, pp. 225–229. University of Benin Academic Research Day Presentation (Book of proceedings), UNIBEN Press, Benin City, Nigeria (2016)

GPSR Compliance

The European Union's (EU) General Product Safety Regulation (GPSR) is a set of rules that requires consumer products to be safe and our obligations to ensure this.

If you have any concerns about our products, you can contact us on ProductSafety@springernature.com

In case Publisher is established outside the EU, the EU authorized representative is:

Springer Nature Customer Service Center GmbH
Europaplatz 3
69115 Heidelberg, Germany

Batch number: 10761593

Printed by Printforce, the Netherlands

Advances in Intelligent Systems and Computing

Volume 868

Series editor

Janusz Kacprzyk, Polish Academy of Sciences, Warsaw, Poland
e-mail: kacprzyk@ibspan.waw.pl

Kohei Arai · Supriya Kapoor
Rahul Bhatia
Editors

Intelligent Systems and Applications

Proceedings of the 2018 Intelligent Systems Conference (IntelliSys) Volume 1

Set 2

Springer

Editors
Kohei Arai
Faculty of Science and Engineering
Saga University
Saga, Japan

Rahul Bhatia
The Science and Information
(SAI) Organization
Bradford, UK

Supriya Kapoor
The Science and Information
(SAI) Organization
Bradford, UK

ISSN 2194-5357 ISSN 2194-5365 (electronic)
Advances in Intelligent Systems and Computing
ISBN 978-3-030-01053-9 ISBN 978-3-030-01054-6 (eBook)
https://doi.org/10.1007/978-3-030-01054-6

Library of Congress Control Number: 2018955283

This Springer imprint is published by the registered company Springer Nature Switzerland AG
The registered company address is: Gewerbestrasse 11, 6330 Cham, Switzerland

Editor's Preface

Welcome to the Intelligent Systems Conference (IntelliSys) 2018 which is held on September 6–7, 2018, in London, UK.

The technology nowadays has reached the point where intelligent systems are replacing human intelligence in aiding in the solution of very complex problems as well as in decision-making processes. In many cases, intelligent systems have already outperformed human activities. Several directions took place in this field of computational intelligence. Massive access and use of intelligent systems in everyday applications have created the need for such an international conference which serves as a venue to report on up-to-the-minute innovations and developments.

IntelliSys 2018 provided a setting for discussing a wide variety of topics including deep learning, neural networks, image/video processing, intelligent transportation, artificial intelligence, robotics, data mining, smart health care, natural language processing, ambient intelligence, machine vision, and the Internet of things. This two-day conference program covers four keynote talks, contributed papers, special sessions, poster presentations, workshops, and tutorials on theory and practice, technologies, and systems.

IntelliSys 2018 has attracted 568 submissions from 50+ countries. After the double-blind review process, we finally selected 194 full papers including 13 poster papers to publish. The conference, IntelliSys 2018, has a wide range of featured talks including keynotes which provide visions and insights into future research directions and trends.

We would like to express our deep appreciation to the support of many people: authors, presenters, participants, keynote speakers, session chairs, volunteers, program committee members, steering committee members, and people in other various roles. We would also like to express our sincere gratitude to all their valuable suggestions, advice, dedicated commitment, and hard work.

We are looking forward to our upcoming Intelligent Systems Conference that will be held on 2019 at the same location. We hope that it will be as interesting and enjoyable as it has been in all of its three predecessors.

Hope you find IntelliSys 2018 in London both enjoyable and valuable experience!

Kind Regards,

Kohei Arai
Conference Chair

Contents

The Random Neural Network and Web Search: Survey Paper

Will Serrano(✉)

Intelligent Systems and Networks Group, Imperial College London, London, UK
G.Serrano11@imperial.ac.uk

Abstract. E-commerce customers and general Web users should not believe that the products suggested by Recommender systems or results displayed by Web search engines are either complete or relevant to their search aspirations. The economic priority of Web related businesses requires a higher rank on Web snippets or product suggestions in order to receive additional customers; furthermore, Web search engines and recommender systems revenue is obtained from advertisements and pay-per-click. This survey paper presents a review of Web Search Engines, Ranking Algorithms, Citation Analysis and Recommender Systems. In addition, Neural networks and Deep Learning are also analyzed including their use in learning relevance and ranking. Finally, this survey paper also introduces the Random Neural Network with its practical applications.

Keywords: Neural networks · Web search · Ranking algorithms
Deep learning

1 Introduction

Web search engines and Recommender systems were developed to address the need for searching precise data and items on the Internet [27]. Although they provide a straight link among Web users with the pursued products or wanted information, any Web search result list or suggestion will be biased due to profitable economic or personal interests, as well as by the users' own inaccuracy when typing their queries or requests. Sponsored search [15] enables the economic revenue that is needed by Web search engines; it is also vital for the survival of numerous Web businesses and the main source of income for free to use online services. Multiple payment options adapt to different advertiser targets while allowing a balanced risk share among the advertiser and the Web search engine for which pay-per-click method is the widest used model.

The Internet has fundamentally changed the travel industry; it has enabled real time information and the direct purchase of services and products; Web users can buy directly flight tickets, hotels rooms and holiday packs. Travel industry supply charges have been eliminated or decreased because the Internet has provided a shorter value chain [43]; however services or products not displayed within higher order of Web Search Engines or Recommender systems lose tentative customers. A parallel situation is also found in academic and publications search where the Internet has permitted openpublication and increased the accessibility of academic research. Authors are able to bypass the

K. Arai et al. (Eds.): IntelliSys 2018, AISC 868, pp. 700–737, 2019.
https://doi.org/10.1007/978-3-030-01054-6_51

conventional method of human evaluation to obtain journal publication [50] and make their work public on their personal Websites. With the intention to expand the research contribution to a wider number of readers and be more cited [52], authors have the personal interest to make their publications appear in high academic search rank orders. Ranking algorithms are critical in the aforementioned examples as they decide on the result relevance and order therefore marking data as transparent or nontransparent to e-commerce customers and general Web users. Considering the Web search commercial model, businesses or authors can be tempted to distort ranking algorithms by falsely enhancing the appearance of their publications or items. On the other hand, Web search engines or Recommender systems are biased towards pretending relevance on the rank in which they order results from explicit businesses or Web sites, in exchange for a commission or payment. The main consequence for a Web user is that relevant products or results may be "hidden" or displayed at a very low order of the search list while unrelated products or results on high order.

Artificial neural networks are models based on the brain within the central nervous system; they are usually presented as artificial nodes or "neurons" with different layers connected together via synapses. The learning properties of artificial neural networks have been applied to resolve extensive and diverse tasks that would have been difficult to solve by ordinary rules based programming; these applications include optimization [145, 146] and image and video recognition [149, 154]. Neural Networks have also been applied to Web Search and result ranking and relevance [206, 207].

This paper presents a survey of Web Search in Sect. 2 including Internet Assistants, Web Search Engines, Meta search engines, Web result clustering, Travel services and citation analysis. Ranking is described in Sect. 3; with ranking algorithms, relevance metrics and learning to Rank. We define Recommender Systems in Sect. 4 and Neural Networks in Web Search, Learning to Rank, Recommender Systems and Deep Learning are analyzed in Sect. 5. The Random Neural Network is presented on Sect. 6 alongside with the G-Networks and the Cognitive Packet Network. Finally, conclusions are explained on Sect. 7.

2 Web Search

With the development of the Internet, several applications and services have been proposed or developed to manage the increasingly greater volume of information and data accessible in the World Wide Web.

2.1 Internet Assistants

Internet assistants learn and adapt to variable user's interests in order to filter and recommend information. These agents normally define a user as a set of weighted terms which are either explicitly introduced or implicitly extracted from the Web browsing behavior. Relevance algorithms are determined by a vector space model that models both query and answer as an ordered vector of weighed terms. Web results are the parsed fragments obtained from Web pages, documents or Web results retrieved from different sources. The user provides explicit or implicit feedback on the results considered relevant or interesting, this is then used to adapt the weights of the term set profile.

Intelligent Agents [1] are defined as a self-contained independent software module or computer program that performs simple and structurally repetitive automated actions or tasks in representation lof Web users while cooperating with other intelligent agents or humans. Their attributes are autonomy, cooperation with other agents and learning from interaction with the environment and the interface with users' preferences and behaviour. Intelligent Agents [2] behave in a manner analogue to a human agent with Autonomy, Adaptability and Mobility as desirable qualities; they have two ways to make the Web invisible to the user: by abstraction where the used technology and the resources accessed by the agent are user transparent and by distraction where the agent runs in parallel to the Web user and performs tedious and complex tasks faster than would be possible for a human alone.

Spider Agent [3] is a metagenetic assistant to whom the user provides a set of relevant documents where the N highest frequency keywords form a dictionary which is represented as a Nx3 matrix. The first column of the dictionary contains the keywords whereas the second column measures the whole amount of documents that contains the keywords, finally the third column contains the sum frequency of the specific word over the overall documents. The metagenetic algorithm first creates a population of keyword sets from the dictionary based on three genetic operators: Crossover, Mutation and Inversion; then it creates a population of logic operators sets (AND, OR, NOT) for each of the first populations. Spider Agent forms different queries by the combination of both populations and searches for relevant documents for each combination. The main concept is that different combinations of words in different queries may result in search engines providing additional different relevant results.

Syskill and Webert [4] helps users to select relevant Web pages on specific topics where each user has a set of profiles, one for each topic, and Web Pages are rated as relevant or irrelevant. Syskill & Webert transforms the source code of the Web Page based on Hyper Text Markup Language (HTML) into a binary feature vector which designates the presence of words using a learning algorithm based on a naive Bayesian classifier. Letizia [5] is a Web user interface agent that helps Web browsing. Letizia learns user behavior and provides with additional interesting Web pages by exploring the current Web page links where the user interest assigned to a Web document is calculated as the reading time, the addition to favorites or the click of a shown link.

WebWatcher [6] is a Web tour guide agent that provides relevant Web links to the user while browsing the Web; it acts as a learning apprentice observing and learning interest from its user actions when select relevant links. WebWatcher uses Reinforcement Learning where the reward is the frequency of each the searched terms within the Web Page recommending Web pages that maximize the reward path to users with similar queries.

Lifestyle Finder [7] is an agent that generates user profiles with a large scale database of demographic data. Users and their interests are grouped by their input data according to their demographic information. Lifestyle Finder generalizes user profiles along with common patterns within the population; if the user data corresponds to more than one cluster, the demographic variables whose estimates are close with the entire corresponding groups generate a limited user profile. The demographic feature that best distinguishes the corresponding groups is utilized to ask the Web user for additional details where the final group of corresponding clusters is obtained after several user iterations.

2.2 Web Search Engines

Internet assistants have not been practically adopted by Internet users as an interface to reach relevant information; instead Web search engines are preferred option as the portal between users and the Internet. Web search engines are software applications that search for information in the World Wide Web while retrieving data from Web sites and Online databases or Web directories. Web search engines have already crawled the Web, fetched its information and indexed it into databases so when the user types a query, relevant results are retrieved and presented promptly. The main issues of Web search engines are result overlap, rank relevance and adequate coverage [8] for both sponsored and non-sponsored results.

Web personalization builds user's interest profile by using their browsing behavior and the content of the visited Web pages to increase result extraction and rank efficiency in Web search. A model [9] that represents a user needs and its search context is based on content and collaborative personalization, implicit and explicit feedback and contextual search. A user is modeled by a set of terms and weights [10] related to a register of clicked URLs with the amount of visits to each, and a set of previous searches and Web results visited; this model is applied to re order the Web search results provided by a non-personalized Web search engine. Web queries are associated to one or more related Web page features [11] and a group of documents is associated with each feature; a document is both associated with the feature and relevant to the query. These double profiles are merged to attribute a Web query with a group of features that define the user's search relevance; the algorithm then expands the query during the Web search by using the group of categories. There are different methods to improve personalized web search based on the increment of the diversity of the top results [12] where personalization comprises the re ranking of the first N search results to the probable order desired by the user and query reformulations are implemented to include variety. The three diversity methods proposed are: the most frequent selects the queries that most frequently precede the user query; the maximum result variety choses queries that have been frequently reformulated but distinct from the already chosen queries, and finally, the most satisfied selects queries that usually are not additionally reformulated but they have a minimum frequency.

Spatial variation based on information from Web search engine query records and geolocation methods can be included in Web search queries [13] to provide results focused on marketing and advertising, geographic information or local news; accurate locations are assigned to the IP addresses that issue the queries. The center of a topic is calculated by the physical areas of the users searching for it where a probabilistic model calculates the greatest probability figure for a geographic center and the geographical dispersion of the query importance. Aspects for a Web query are calculated [14] as an effective tool to explore a general topic in the Web; each aspect is considered as a set of search terms which symbolizes different information requests relevant to the original search query. Aspects are independent to each other while having a high combined coverage; two sources of information are combined to expand the user search terms: query logs and mass collaboration knowledge databases such as Wikipedia.

2.3 Metasearch Engines

Metasearch engines were developed based on the concept that single Web search engines were not able to crawl and index the entire Web. While Web search engines are useful for finding specific information, like home pages, they may be less effective with a comprehensive search or wide queries due their result overlap and limited coverage area. Metasearch engines try to compensate their disadvantages by sending simultaneously user queries to different Web search engines, databases, Web directories and digital libraries and combining their results into a single ranked list. The main operational difference with Web search engines is that Metasearch engines do not index the Web.

There are challenges in the development of Metasearch engines [16]: Different Web search engines are expected to provide relevant results which have to be selected and combined. Different parameters shall be considered when developing a Metasearch engine [17]: functionality, working principles including querying, collection and fusion of results, architecture and underlying technology, growth, evolution and popularity. Numerous metasearch architectures can be found [18] such as Helios, Tadpole and Treemap with different query dispatcher, result merger and ranking configurations. Helios [19] is an open source metasearch engine that runs above different Web Search Engines where additional ones can be flexibly plugged into architecture. Helios retrieves, parses, merges, and reorders results given by the independent Web search engines.

An extensive modular metasearch engine with automatic search engine discovery [20] that incorporates a numerous number of autonomous search engines is based on three components: "automatic Web search engine recognition, automatic Web search engine interconnection and automatic Web search result retrieval". The solution crawls and fetches Web pages choosing the Web Search Engine ones, which once discovered, are connected by parsing the HTML source code, extracting the form parameters and attributes and sending the query to be searched. Finally, URLs or snippets provided by the different Web Search Engines are extracted and displayed to the Web user.

There are several rank aggregation metasearch models to combine results in a way that optimizes their relevance position based on the local result ranks, titles, snippets and entire Web Pages. The main aggregation models are defined as [21]: Borda-fuse is founded on the Borda Count, an voting mechanism on which each voter orders a set of Web results; the Web result with the most points wins the election. The Bayes-fuse is built on a Bayesian model of the probability of Web result relevance. The difference between the probability of relevance and irrelevance respectively determines relevance of a Web Page based on the Bayes optimal decision rule. Rank aggregation [22] can be designed against spam, search engine commercial interest and coverage.

The use of the titles and associated snippets [23] can produce a higher success than parsing the entire Web pages where the effective algorithm of a metasearch engine for

Web result merging outperforms the best individual Web search engine. A formal approach to normalize scores for metasearch by balancing the distributions of scores of the first irrelevant documents [24] is achieved by using two different methods: the distribution of scores of all documents and the combination of an exponential and a Gaussian distribution to the ranks of the documents where the developed exponential distribution is used as an approximation of the irrelevant distribution.

Metasearch engines provide personalized results using different approaches [25]. The first method provides distinct results for users in separate geographical places where the area information is acquired for the user and Web page server. The second method has a ranking method where the user can establish the relevance of every Web search engine and adjusts its weight in the result ranking in order to obtain tailored information. The third methods analyses Web domain structures of Web pages to offer the user the possibility to limit the amount of Web pages with close subdomains.

Results from different databases present in the Web can also be merged and ranked [26] with a database representative that is highly scalable to a large number of databases and representative for all local databases. The method assigns only a reduced but relevant number of databases for each query term where single term queries are assigned the right databases and multi-term queries are examined to extract dependencies between them in order to generate phrases with adjacent terms. Metasearch performance against Web search has been widely studied [27] where the search capabilities of two metasearch engines, Metacrawler and Dogpile, and two Web search engines, Yahoo and Google, are compared.

2.4 Result Web Clustering

Web result clustering groups result into different topics or categories in addition to Web search engines that present a plain list of result to the user where similar topic results are scattered. This feature is valuable in general topic queries because the user gets the theme bigger picture of the created result clusters. There are two different clustering methods; pre-clustering calculates first the proximity between Web pages and then it assigns the labels to the defined groups [28] whereas post-clustering discovers first the cluster labels and then it assigns Web pages to them [29]. Web clustering engines shall provide specific additional features [30] in order to be successful: fast subtopic retrieval, topic exploration and reduction of browsing for information. The main challenges of Web clustering are overlapping clusters [31], precise labels [32], undefined cluster number [33], and computational efficiency.

Suffix Tree Clustering (STC) [28] is an incremental time method which generates clusters using shared phrases between documents. STC considers a Web page as an ordered sequence of words and uses the proximity between them to identify sets of documents with common phrases between them in order to generate clusters. STC has three stages: document formatting, base clusters identification and its final combination into clusters.

Lingo Algorithm [29] finds first clusters utilizing the Vector Space Model to create a "TxD term document matrix where T is the number of unique terms and D is the number of documents". A Singular Value Decomposition method is used to discover the relevant matrix orthogonal basis where orthogonal vectors correspond to the cluster labels. Once clusters are defined; documents are assigned to them.

A cluster scoring function and selection algorithm [31] overcomes the overlapping cluster issue; both methods are merged with Suffix Tree Clustering to create a new clustering algorithm called Extended Suffix Tree Clustering (ESTC) which decreases the amount of the clusters and defines the most useful clusters. This cluster scoring function relies on the quantity of different documents in the cluster where the selection algorithm is based on the most effective collection of clusters that have marginal overlay and maximum coverage; this makes the most different clusters where most of the documents are covered in them.

A cluster labelling strategy [32] combines intra-cluster and inter-cluster term extraction where snippets are clustered by mapping them into a vector space based on the metric k-center assigned with a distance function metric. To achieve an optimum compromise between defining and distinctive labels, prospect terms are selected for each cluster applying a improved model of the information Gain measurement. Cluster labels are defined as a supervised salient phrase ranking method [33] based on five properties: "sentence frequency/Inverted document frequency, sentence length, intra cluster document similarity, cluster overlap entropy and sentence independency" where a supervised regression model is used to extract possible cluster labels and human validators are asked to rank them.

Cluster diversification is a strategy used in ambiguous or broad topic queries that takes into consideration its contained terms and other associated words with comparable significance. A method clusters the top ranked Web pages [34] and those clusters are ordered following to their importance to the original query however diversification is limited to documents that are assigned to top ranked clusters because potentially they contain a greater number of relevant documents. Tolerance classes approximate concepts in Web pages to enhance the snippets concept vector [35] where a set of Web pages with similar enhanced terms are clustered together. A technique that learns relevant concepts of similar topics from previous search logs and generates cluster labels [36] clusters Web search results based on these as the labels can be better than those calculated from the current terms of search result.

Cluster hierarchy groups data over a variety of scales by creating a cluster tree. A hierarchy of snippets [37] is based on phrases where the snippets are assigned to them; the method extracts all salient phrases to build a document index assigning a cluster per salient phrase; then it merges similar clusters by selecting the phrase with highest number of indexing documents as the new cluster label; finally, it assigns the snippets whose indexing phrases belong to the same cluster where the remaining snippets are assigned to neighbors based on their k-nearest distance.

Web search results are automatically grouped through a Semantic, Hierarchical, Online Clustering (SHOC) [38]; a Web page is considered as a string of characters where a cluster label is defined as a meaningful substring which is both specific and significant. The latent semantic of documents is calculated through the analysis of the associations between terms and Web pages. Terms assigned with the same Web page should be close in semantic space; same as the Web pages assigned with the same terms. Densely allocated terms or Web pages are close to each other in semantic space; therefore they should be assigned to the same cluster. Snake [39] is a clustering algorithm based on two information databases; the anchor text and link database assigns each Web document to the terms contained in the Web document itself and the words included in the anchor text related to every link in the document, the semantic database ranks a group of prospect phrases and choses the most significant ones as labels. Finally, a Hierarchical Organizer combines numerous base clusters into few super clusters. A query clustering approach [40] calculates the relevance of Web pages using previous choices from earlier users, the method applies a clustering process where collections of semantically similar queries are identified and the similarity between a couple of queries is provided by the percentage of shared words in the clicked URL within the Web results.

2.5 Travel Services

Originally, travel service providers, such as airlines or hotels, used global distribution systems to combine their offered services to travel agencies. Global distribution systems required high investments due to their technical complexity and physical dimensions as they were mainframe based infrastructure; this generated the monopoly of a few companies that charged a high rate to the travel services providers in order to offer their services to the travel agencies [41]. With this traditional model, customers could purchase travel provider services directly at their offices or through a travel agent. This scenario has now been changed by two factors: the Internet has enabled e-commerce; the direct accessibility of customers to travel services providers' information on real time with the availability of online purchase [42] and higher computational servers and software applications can implement the same services as the global distribution systems did, at a lower cost.

As a consequence, new players have entered this scenario [43]; Software database applications such as ITA Software or G2 SwitchWorks use high computational server applications and search algorithms to process the data provided from travel services providers; they are the replacement of the global distribution systems. Online travel agents, such as Expedia or TripAdvisor, are traditional travel agents that have automatized the global distribution systems interface; customers can interact with them directly through the Internet and buy the final chosen product. Metasearch engines like Cheapflights or Skyscanner among many others use the Internet to search customers' travel preferences within travel services providers and online travel agents however customers cannot purchase products directly; they are mainly referred to the supplier Web site.

Other players that have an active part in this sector are Google and other Web Search Engines [44]; they provide search facilities that connect directly travel service providers with consumers bypassing global distribution systems, software distribution systems, Metasearch engines and online travel agents. This allows the direct product purchase that can reduce distribution costs due to a shorter value chain. Travel ranking systems [45] also recommend products and provide the greatest value for the consumer's budget with a crucial concept where products that offer a higher benefit are ranked in higher positions.

With the integration of the customer and the travel service provider through the Internet, different software applications have been developed to provide extra information or to guide through the purchase process [46] based on user interactions. Travel related Web interfaces [47] help users to make the process of planning their trip more entertaining and engaging while influencing users' perceptions and decisions. There are different information search patterns and strategies [48] within specific tourism domain web search queries based on search time, IP address, language, city and key terms keywords.

2.6 Citation Analysis

Citation analysis consists of the access and referral of research information including articles, papers and publications. Its importance is based on its measurement of the value or relevance of personal academic work; it is becoming more respected since its evaluation impacts on other decisions such as promotion, scholarships and research funds. The Journal Impact Factor [49] is calculated as "the number of citations received in the current year to articles published in the two preceding years divided by the total number of articles published in the same two years". It is based on the concept that high impact journals will only try to publish very relevant work therefore they will be used for outstanding academics. This traditional measurements are peer review based on human judgment however it is time consuming and assessors may be biased by personal judgments such as the quality of journal or the status of university instead of competent and impartial reviews.

The Web has provided an alternative citation measures, causing the citation monopoly to change because of the wider accessibly it provides. New online platforms [50] such as Scopus, Web of Science and Google Scholar provide other measures that may represent more accurately the relevance of an academic contribution. Bibliometric indicators [51] evaluate the quality of the publication based on empirical values; mostly number of citations or number of downloads. They are not only quicker to obtain but they also provide a wider coverage because they also retrieve results from relevant academic work published on personal or academic Web pages and open journals based on conference papers, book chapters or thesis. In addition, academics are personally interested to artificially influence ranking methods by "optimizing" the appearance of their research publications to show them in higher ranks with the purpose to extend their reader audience [52] in order to get cited more.

The h-index [53] is applied to qualify the contribution of individual scientific research work: "An academic has an h-index of h if the x publications have individually received at least h citations and the other (x-h) publications have up to h citations". The h-index has become popular among researchers because it solves the disadvantages of the journal impact factors and it has also a good correlation with the citation analysis database. It is simple and fast to calculate utilizing databases like Google Scholar, Scopus or Web of Science however it suffers from some disadvantages [49]; it does not differentiate negative citations from positive ones so both are equally taken into account, it does not consider the entire amount of citations an author has collected and it disadvantages a reduced but highly-cited paper set very strongly.

Google Scholar is a free Web academic search engine that registers the metadata or the complete text of academic articles from different databases and journals. The result relevance is calculated by a ranking algorithm that combines the term occurrence, author, publications and citations weights [54]. Google Scholar has received both good and bad critics since it started in 2004 [55]: The lack of transparency in the relevance assessment of the information due its automatic process of retrieving, indexing and storing information has been criticized; however the more transparent its evaluation algorithms, the easier it becomes to influence its metrics. Google Scholar has been widely analyzed and studied by the research community against other online databases [56, 57].

A hybrid recommender system based on content-based and collaborative-based relevance [58] expands the method of the traditional keyword or key term search by adding citation examination, author examination, source examination, implicit ratings and explicit marks. Two ranking methods are applied: "in-text citation frequency examination" evaluates the occurrence which a publication is "cited within the citing document" and "in-text citation distance examination" determines the similarity between references within a publication by evaluating the word separation.

3 Ranking

Ranking is the core process on many information retrieval applications including Web search engines and Recommender systems for two main reasons: the user evaluation of quality or performance from a customer point of view and sponsored adverts based on ranking from a commercial perspective. There are several methods and techniques for ranking Web pages and products that exploit different algorithms built on the analysis of the Web page content, the anchor text of the links between pages or the network structure of the World Wide Web hyperlinked environment.

3.1 Ranking Algorithms

Web search engine evaluation comprises the measurement of the quality and services they provider to users; it is also used to compare their different search capabilities. Automatic evaluation analyses the different presented URLs whereas human evaluation is based on the measurement of result relevance; while the latest one is more precise; it takes more effort and therefore is more expensive as it is normally done by surveys. Other methods that combine both automatic and user evaluation are applications that monitor user activities such as click-troughs or time spent of each Web site. "Dynamic ranking or query-dependent ranking" improves the order of the results returned to the user based on the query while "static ranking or query-independent" calculates the relevance of the Web pages to different topics.

There are different methods to assess Web pages relevance:

(1) *HTML Properties*

The first method focuses on the HTML properties of the Web page including the type of text, section, size, position, anchor text and URL. Web spammers base their strategy mostly in the design of the HTML properties of the Web page to get an artificial better ranking from Web search engines.

(2) *TF-IDF*

The second method is founded on the inspection of the Web page text itself and how the words in the query relate within it. TF-IDF or "Term Frequency and Inverse Document Frequency" applies the vector space model with a scoring function where the importance of a term increases if it is repeated often in a document however its weight decreases if the same term appears multiple times on numerous documents as it is considered less descriptive.

A probabilistic TF-IDF retrieval model [59] emulates the decision making of a human brain for two categories of relevance: "local relevance" is first applied to a particular Web page location or section whereas the "wide relevance" spans to cover the whole Web page; the method then merges the "local relevance" decisions for each position as the Web page "wide relevance" decision. Document Transformation adapts the Web page vector close to the query vector by adding or removing terms, an example to long term incremental learning [60] is based on a search log that stores information about click through Web users which have browsed among the Web result pages from a Web Search engine and Web Pages selected by the user. Content-Time-based ranking is a Web search time focus ranking algorithm [61] that includes both "text relevance" and "time relevance" of Web pages. "Time relevance" is the interval computed between the query and the content or update time; where every term in a Web page is assigned to an explicit content time interval, finally the time focus TF-IDF value of each keyword is calculated to obtain the overall rank.

(3) *Link Analysis*

The third method is the link analysis which it is founded on the relation between Web pages and their Web graph; Link analysis associates Web pages to neural network nodes and links to neural network edges.

In-Degree [62] was the first link analysis ranking algorithm; it ranks pages according to their popularity which it is calculated by the sum of links that point to the page.

Page Rank [63] is founded on the concept that a Web document has a top "Page Rank" if there are many Web pages with reduce "Page Rank" or a few Web pages with elevated "Page Rank" that link to it. Page rank algorithm provides a likelihood distribution that represents the probability that a Web user person arbitrarily following the links will reach to the specific Web page.

HITS algorithm [64] or Hyperlink Induced Topic Search algorithm assigns relevance founded on the authority concept and the relation among a group of relevant "authoritative Web pages for a topic" and the group of "hub Web pages that link to many related authorities" in the Web link arrangement. The hyperlink structure among Web pages is analyzed to formulate the notion of authority; the developer of Web document p when inserts a hyperlink to Web document q, provides some relevance or authority on q. Hubs and authorities have a reciprocally strengthening association: "a relevant hub is a Web page that links to many relevant authorities; a relevant authority is a Web page which is linked by numerous relevant hubs". The authority value of a Web document is the addition of every hub value of Web documents linking to it. The hub value of a Web document is the addition of total number of authorities' values of Web documents linking to it.

SALSA algorithm [65] or "Stochastic Approach for Link-Structure Analysis" combines Page Rank and HITS by taking a random walks alternating between hubs and authorities. Two random walks are assigned two scores per Web Page; it is based on the in link and out link that represent the proportion of authority and hub respectively.

QS page-rank is a query-sensitive algorithm [66] that combines both global scope and local relevance. Global Web Page importance is calculated by using general algorithms such as Page Rank; Local query sensitive relevance is calculated by a voting system that measures the relation between the top Web pages retrieved.

DistanceRank [67] is an iterative algorithm built on reinforcement learning; distance between pages is counted as punishment where distance is established as the amount of links among two Web Pages. The objective is the reduction of the punishment or distance therefore a Web document with reduced "distance" has a greater rank. The main concept is that as correlated Web pages are connected between them; the distanced built method finds relevant Web pages quicker.

DING is a Dataset Ranking [68] algorithm that calculates dataset ranks; it uses the connections among them to combine the resulting scores with internal semantic based ranking strategies. DING ranks in three stages: first, a dataset rank is calculated by an external dataset hyperlink measurement on the higher layer, then for each dataset, local ranks are calculated with hyperlink measurements between the inner entities and finally the relevance of each dataset entity is the combination of both external and internal ranks.

ExpertRank [69] is an online community and discussion group expert ranking algorithm that integrates a vector space method to calculate the expert subject importance and a method similar to Page Rank algorithm to evaluate topic authority. Expert subject relevance score is calculated as a likeness of the candidate description and the provided request where the candidate description is generated by combining all the comments in the topic conversation group and the authority rating is computed according to the user interaction graph.

3.2 Relevance Metrics

Performance can be described as the display of relevant, authoritative, updated results on first positions. Retrieval performance is mostly evaluated on two parameters: precision is the percentage of applicable results within the provided Web document list and recall is the percentage of the entire applicable results covered within the provided Web document list. Due the intrinsic extent of the Web, where it is difficult to measure the quantity of all Web pages and the fact users only browse within the first 20 results; precision is widely used as a performance measurement. In addition, another factor to consider is that Web search engines are different and therefore they perform differently [70]; some may perform better than others for different queries and metrics.

There are different parameters to assess the performance of recommender systems:

- Precision evaluates the applicability of the first N results of the rating result set in relation to a query.
- Recall represents the applicability of the first N results of the rating result set in relation to the complete result set.
- The Mean Absolute Error (MAE) evaluates the error between user predictions and user ratings.
- The Root Squared Mean Error (RSME) measures the differences between predictions and user ratings.
- F-score combines Precision and Recall in an evenly weighted metric.
- The Average Precision calculates the mean figures of P@n over the number n of retrieved Web pages.
- The Mean Average Precision measures the balance of average precision within a collection of Q Web searches.
- The Normalized Discounted Cumulative Gain uses a rated relevance metric to measure the relevance of a result that uses its rank within the Web page set.
- TREC average precision includes rank position on the performance evaluation; it is defined at a cutoff N.
- The Reciprocal Rank of a Web search is defined as the multiplicative inverse applied to the rank of the first relevant result.
- The Mean Reciprocal Rank corresponds to averaged value of the RR (Reciprocal Rank) over a collection of Q searches.
- The Expected Reciprocal Rank is based on a cascaded model which penalizes documents which are shown below very relevant documents.

Different research analyses have studied the efficiency of Web search engines [71] and the level of overlap among the Web pages presented by the search engines. Two new measurements to assess Web search engine performance are proposed [72]: the relevance of result rating by a Web search engine is assessed by the association among Web and human rating and the capability to retrieve applicable Web pages as the percentage of first ranked within the retrieved Web pages. Four relevance metrics are applied to calculate the quality of Web search engines [73]; three of them are adapted from the Information Retrieval: Vector Space method, Okapi likeness value and Coverage concentration rating and a new method which is developed from the human interaction to take human expectations into account based on a raw score and a similarity score.

The Text REtrieval Conference (TREC) [74] is supported by the "National Institute of Standards and Technology (NIST)" and "U.S. Department of Defense" to sponsor studies in the information and data retrieval research group; TREC provides the framework for extensive evaluation applied to different text retrieval methods. Effectiveness of TREC algorithms against Web search engines has been compared [74] where tested algorithms were developed with Okapi or Smart based on the term weighting methods (TF-IDF). Web search engines are evaluated against a series of relevance metrics [75, 76]: precision, TREC based mean reciprocal rank (MRR) and the TREC style average precision (TSAP) obtained from double relevance human evaluations for the top 20 Web results displayed.

Web search evaluation methods can be assigned into eight different categories [77]: Assessment founded on relevance, Assessment founded on ranking, assessment founded on user satisfaction, assessment founded on size and coverage of the Web, assessment founded on dynamics of Search results, assessment founded on few relevant and known items, assessment based on specific topic or domain and Automatic assessment.

Different Automatic Web search engine evaluation methods have been proposed [78, 79] to reduce human intervention. The algorithm first submits a query to the different Web search engines where the top 200 Web pages of every Web search engine are retrieved; then it ranks the pages in relation to their likeness to the Web user information requirements; the model is built on the vector space method for query-document matching and ranking using TF-IDF. After ranking; the method lists as relevant the Web pages that are in the first 20 documents retrieved by each Web search engine with the top 50 ranked results. An automatic C++ application [80] compares and analyzes Web result sets of several Web search engines; the research analyses the URL coverage and the URL rank where recall and accuracy are used as relevance metrics however there is not description regarding the method that decides URL relevance. A technique for evaluating Web search engines robotically [81] is founded in the way they order already rated Web search results, where a search query is transmitted to different Web search engines and the rank of retrieved Web results is measured against the document that has already been paired with that query.

Web search engine evaluation has been examined from a webometric perspective [82] using estimates of success values, amount of provided URL, amount of domains retrieved and number of sites returned where evaluations are made to assess the retrieval of the most precise and comprehensive Web results from every single Web engine.

3.3 Learning to Rank

Learning to rank is defined as the implementation of semi-supervised or supervised computer learning techniques to generate a ranking model by combining different documents features obtained from training data in an information retrieval system. The training data contains sets of elements with a specified position represented by a numerical score, order or a binary judgment for each element. The purpose of a ranking method is to order new result lists in order to produce rankings similar to the order in the training data. Query document sets are normally defined by quantitative vectors which components are denoted features, factors or ranking signals. The main disadvantages of the learning to rank approach are that the optimization of relevance is based on evaluation measurements without addressing the decision of which measures to use; in addition, learning to rank does not take into consideration that user relevance judgment changes over the time.

The model of learning to rank consist on:

- a set Q of M queries, $Q = \{q_1, q_2, \ldots, q_M\}$
- a set D of N Web pages per query $q_{M,D} = \{d_{11}, d_{12}, \ldots, d_{NM}\}$
- a set X of M feature vectors per each query q_M and Web page d_N pairs, $X = \{x_{11}, x_{12}, \ldots, x_{NM}\}$
- a set Y of M relevance decisions per each query q_M and Web page d_N pairs, $Y = \{y_{11}, y_{12}, \ldots, y_{NM}\}$.

The set of query-Web page pairs $\{q_M, d_{NM}\}$ has associated a set of feature vectors $\{x_{NM}\}$ that represents specific ranking parameters describing the match between them. The set of relevance judgments $\{y_{NM}\}$ can be a binary decision, order or score. Each feature x_{NM} and the corresponding score y_{NM} form an instance. The input to the learning algorithm is the set feature vectors $\{x_{NM}\}$ that represents the pair $\{q_M, d_{NM}\}$ and the desired output corresponds to set of relevance judgement $\{y_{NM}\}$. The ranking function f(x) is acquired by the learning method during the training process on which relevance assessments are optimized with the performance measurement or cost function.

RankSVM learning to rank algorithm learns retrieval functions using clickthrough data for training based on a Support Vector Machine (SVM) approach [83]. It uses a mapping function to correlate a search query with the features of each of the possible results where each data pair is projected into a feature space combined with the corresponding click-through data that measures relevance. Distinct ranking methods are developed for several classes of contexts [84] based on RankSVM that integrates the ranking principles into a new ranking model that encodes the context details as features. It learns a Support Vector Machine model for a binary categorization on the selection between a couple of Web pages where the evaluation is based on human conclusions and indirect user click.

A machine learned ranking model [85] routinely perceives and reacts to recent related queries where recent ranking takes into account freshness and relevance; the system is divided into two approaches: a high accuracy recent related query detector and a specific recent related sorter trained to represent the characteristics and data distribution applicable for recent ordering.

Random Forests is a low computational cost alternative to point-wise ranking approach algorithm [86] founded on the machine learning method "Gradient Boosted Regression Trees". The learning algorithm of Random Forests is applied multiple times to different subsets and the results are averaged whereas Gradient Boosted Regression Trees sequentially adds small trees at each iteration instead of training many full trees. The combination of both two algorithms first learns a ranking function with Random Forests and later uses it as initialization for Gradient Boosted Regression Trees.

4 Recommender Systems

Recommender systems predict users' interest of different items or products providing a set of suggested items applying tailored relevance algorithms. Recommender systems consist on a database where user ratings and item description are stored and updated iteratively. A user interface interacts with the user where a profiler extracts user properties with explicit and implicit feedback; different suggestions and their order are computed by the recommender ranking algorithm. Due their filtering properties they are widely used within e-commerce [87] as they also support e-commerce customers to find rare products they might not have discovered by themselves.

There are two main different categories of recommender systems [88]. Content-based recommender systems are built on a description of the product and a profile of the customer's wishes where different properties of the items and users are used to identify products with similar features without including other user's ratings. They suffer from some disadvantages [89] such as its inability to recommend completely different items that the user may also consider relevant and a customer needs to mark several products before getting relevant recommendations. Collaborative recommender systems are founded on the customer's previous marks to other products and the consideration of other decisions made by similar users; they made suggestions based on a high correlation between users or items. Although collaborative recommendation reduces the issues from the Content-based solution; it has other drawbacks such as it needs a large number of rating data to calculate accurate correlations and predictions; it also ignores on its calculations new added users or items. Hybrid Recommender Systems take a combination of both approaches.

The user based recommender generally uses the Pearson's relationship similarity whereas the item based recommender systems generally use the Cosine Similarity, although the Euclidean distance similarity can also be used and the Manhattan distance similarity.

In a multi-criteria ranking recommenders, users give ratings on several characteristics of an item as a vector of ratings [90] whereas cross domain recommender systems suggest items from multiple sources with item or user based formulas founded on locality collaborative filtering [91]; they operate by first modelling the traditional likeness association as a directly connected graph and then exploring the entire potential paths that links users or items to discover new cross domain associations.

There are several relevance metrics for assessment of recommender systems in various e-commerce business frameworks [92]; accuracy evaluation metrics are allocated into three major categories: predictive based on the error between estimated and true user ratings, classification based on the successful decision making and rank based on the correct order.

Social network information is inserted to recommender systems [93] as an additional input to improve accuracy; Collaborative Filtering based social recommender systems are classified into two categories: matrix factorization approach where user to user social information is integrated with user item feedback history and neighborhood based social approaches based on social network graphs.

5 Neural Networks

Artificial neural networks are representations of the brain, the principal component of the central nervous system. They are usually presented as artificial nodes or "neurons" in different layers connected together via synapses to create a network that emulates a biological neural network. The synapses have values called weights which are updated during network calculations. There are two main models to represent a neural network; the feed forward model where connections between neurons follow only the forward direction and the recurrent model where connections between neurons form a direct cycle. Artificial neural networks are normally described by three variables: the links among independent layers of neurons; the learning method for adapting the weights of the connections and the activation function that transforms a neuron input to its associated output value.

5.1 Neural Networks in Web Search

The capability of a neural network to learn recursively from several input figures to obtain the preferred output values has also been applied in the World Wide Web as a user interest adjustment method to provide relevant answers. Neural networks have modelled Web graphs to compute page ranking; a Graph Neural Network [94] consists on nodes with labels to include features or properties of the Web page and edges to represent their relationships; a vector called state is associated to each node, it is modelled using a feed forward neural network that represents the reliance of a node with its neighborhood. Another graph model [95] assigns every node in the neural network to a Web page where the synapses that connect neurons denotes the links that connect Web pages; the Web page material rank approximation uses the Web page heading, text material and hyperlinks; the link score approximation applies the similarity between the quantity of phrases in the anchor text that are relevant pages. Both Web page material rank approximation and Web link rank approximation are included within the connection weights between the neurons.

A unsupervised neural network is used in a Web search engine [96] where the k-means method is applied to cluster n Web results retrieved from one or more Web search engines into k groups where k is automatically estimated; once the results have been retrieved and feature vectors are extracted, the k means grouping algorithm calculates the clusters by training the unsupervised neural network. In addition, a neural network method [97] classifies the relevance and reorders the Web search results provided by a metasearch engine; the neural network is a three layer feed forward model where the input vector represents a keyword table created by extracting title and snippets words from all the results retrieved and the output layer consists of a unique node with a value of 1 if the Web page is relevant and 0 if irrelevant.

A neural network ranks pages using the HTML properties of the Web documents [98] where words in the title have a stronger weight than in the body specific, then, it propagates the reward back through the hypertext graph reducing it at each step. A back propagation neural network [99] is applied to Web search engine optimization and personalization where its input nodes are assigned to an explicit measured Web user profile and a single output node represents the likelihood the Web user may regard Web page as relevant.

An agent learning method is applied to Web information retrieval [100] where every agent uses different Web search engines and learns their suitability based on user's relevance response; a back propagation neural network is applied where the input and the output neurons are configured to characterize any training term vector set and relevance feedback for a given query.

5.2 Neural Networks in Learning to Rank

Although there are numerous learning to rank methods we only analyze the ones based on Neural Networks. Learning to rank algorithms is categorized within three different methods according to their input and output representation:

The Pointwise method considers every query Web page couple within a training set is assigned to a quantitative or ordinal value. It assumes an individual document as its only learning input. Pointwise is represented as a regression model where provided a unique query document couple, it predicts its rating.

The Pairwise method evaluates only the relative order between a pair of Web pages; it collects documents pairs from the training data on which a label that represents the respective order of the two documents is assigned to each document pair. Pairwise is approximated by a classification problem. Algorithms take document pairs as instances against a query where the optimization target is the identification of the best document pair preferences. RankNet [101] is a pairwise model based on a neural network structure and Gradient Descent as method to optimize the probabilistic ranking cost function; a set of sample pairs together with target likelihoods that one result is to be ranked higher than the other is given to the learning algorithm. RankNet is used to combine different static Web page attributes such as Web page content, domain or outlinks, anchor text or inlinks and popularity [102]; it outperforms PageRank by selecting attributes that are detached from the link fabric of the Web where accuracy can be increased by using the regularity Web pages are visited. RankNet adjusts attribute weights to best meet pairwise user choices [103] where the implicit feedback such as clickthrough and other user interactions is treated as vector of features which is later integrated directly into the ranking algorithm. SortNet is a pairwise learning method [104] with its associated priority function provided by a multi-layered neural network with a feed forward configuration; SortNet is trained in the learning phase with a dataset formed of pairs of documents where the associated score of the preference function is provided, SortNet is based on minimization of square error function between the network outputs and preferred targets on every unique couple of documents.

The Listwise method takes ranked Web pages lists as instances to train ranking models by minimizing a cost function defined on a predicted list and a ground truth list; the objective of learning is to provide the best ranked list. Listwise learns directly document lists by treating ranked lists as learning instances instead of reducing ranking to regression or classification. ListNet is a Listwise cost function [105] that represents the dissimilarity between the rating list output generated by a ranking model and the rating list given as master reference; ListNet maps the relevance tag of a query Web page set to a factual value with the aim to define the distribution based on the master reference. ListNet is built on a neural network with an associated Gradient Descent learning algorithm with a probability model that calculates the cost function of the Listwise approach; it transforms the ratings of the allocated documents into probability distributions by applying a rating function and implicit or explicit human assessments of the documents.

5.3 Neural Networks in Recommender Systems

Neural Networks have been also applied in Recommender Systems as a method to predict user ratings to different items or to cluster users or items into different categories.

An Adaptive Resonance Theory (ART) is an unsupervised learning method based on a neural network, ART is comprised of a comparative and an identification layer both of them formed of neurons; in addition, a recognition threshold and a reset unit are included to the method. A recommender system based on a collaborative filtering application using the k-separability method [106] is built for every user on various stages: a collection of users is clustered into diverse categories based on their likeness applying Adaptive Resonance Theory, then the Singular Value Decomposition matrix is calculated using the k separability method based on a neural network with a feed forward configuration where the n input layer corresponds to the user ratings' matrix and the single m output the user model with $k = 2m + 1$. An ART model is used to cluster users into diverse categories [107] where a vector that represents the user's attributes corresponds to the input neurons is and the applicable category to the output ones.

A Self Organizing Map (SOM) artificial neural network presents a reduced dimensional quantified model of the input space; the SOM is trained with unsupervised learning. A Recommender application that joins Self Organizing Map with collaborative sorting [108] applies the division of customers by demographic features in which customers that correspond to every division are grouped following to their item selection; the Collaborative filtering method is used on the group assigned to the user to recommend items. The SOM learns the item selection in every division where the input is the customer division and the output is the cluster type. A SOM calculates the ratings between users [109] to complete a sparse scoring matrix by forecasting the rates of the unrated items where the SOM is used to identify the rating cluster.

There are different frameworks that combine collaborative sorting with neural networks. Implicit patterns among user profiles and relevant items are identified by a neural network [110]; those patters are used to improve collaborative filtering to personalize suggestions; the neural network algorithm is a multiplayer feed forward model and it is trained on each user ratings vector. The neural network output corresponds to a pseudo user ratings vector that fills the unrated items to avoid the sparsity issue on recommender systems. A study of the probable similarities among the scholars' historic registers and final grades is based on an Intelligent Recommender System structure [111] where a multi layered neural network with a feed forward configuration is applied with a supervised learning.

Any Machine Learning algorithm that includes a neural network with a feed forward configuration with an input nodes, two hidden nodes and a single output node learning process [112] can be applied to represent collaborative filtering tasks; the presented algorithm is founded on the reduction of the dimensionality reduction applying the Singular Value Decomposition (SVD) of an preliminary user ranking matrix that excludes the necessity for customers to rank shared items with the aim of becoming forecasters for another customer preferences. The neural network is trained with a n singular vector and the average user rating; the output neuron represents the predicted user rating.

Neural Networks have been also implemented in film recommendation systems. A neural network in a feed forward configuration with a single hide layer is used as an organizer application that predicts if a certain program is relevant to a customer using its specification, contextual information and given evaluation [113]; A TV program is represented by a 24 dimensional attribute vector however the neural network has five input nodes; three for transformed genre, one for type of day and the last one for time of day, a single hidden neuron and two output neurons: one for like and the other for dislike. A neural network identifies which household follower provided a precise ranking to a movie at an exact time [114]; the input layer is formed of 68 neurons which correspond to different user and time features and the output layer consists of 3 neurons which represent the different classifiers. An "Interior Desire System" approach [115] considers that if customers may have equivalent interest for specific products if they have close browsing patterns; the neural network classifies users with similar navigation patterns into groups with similar intention behavioral patterns based on a neural network with back propagation configuration and supervised learning.

5.4 Deep Learning

Deep learning applies a neural network with various computing layers that perform several linear and nonlinear transformations to model general concepts in data. Deep learning is under a branch of computer learning that models representations of information. Deep learning is characterized as using a cascade of l-layers of nonlinear computing modules for attribute identification and conversion; each input of every sequential layer is based on the output from the preceding layer. Deep learning learns several layers of models that correlate to several levels of conceptualization; those levels generate a scale of notions where the higher the level, the more abstract concepts are learned.

Deep learning models have been used in learning to rank to rate brief text pairs which main components are phrases [116]; the method is built using a convolutional neural network structure where the best characterization of text pair sets and a similarity function is learned with a supervised algorithm. The input is a sentence matrix with a convolutional feature map layer to extract patterns, a pooling layer is then added to aggregate the different features and reduce the representation. An attention based deep learning neural network [117] focuses on different aspects of the input data to include distinct features; the method incorporates different word order with variable weights changing over the time for the queries and search results where a multilayered neural network ranks results and provides a listwise learning to rank using a decoder mechanism. Deep Stacking Networks [118] are used for information retrieval with parallel and scalable learning; the design philosophy is based on basic modules of classifiers, which are first designed, then are combined together to learn complex functions. The output of each Deep Stacking Network is linear whereas the hidden unit's output is sigmoidal nonlinear.

Deep learning is also used in Recommender Systems. A deep feature representation [119] learns the content information and captures the likeness and implicit association among customers and items where Collaborative filtering is used in a Bayesian probabilistic framework for the rating matrix. A Deep learning approach [120] assigns items and customers to a vector space model that optimizes the similarity between customers and their favored products; the model is extended to jointly learn features of items from different domains and user features according to their Web browsing history and search queries. The deep learning neural network maps two different high dimensional sparse features into low dimensional dense features within a joint semantic space.

6 Random Neural Network

6.1 G-Networks

G-Networks or Gelenbe Network [121, 122] is a model for neural networks as well as queueing methods with precise control operations, such as route management or traffic removal. The stationary distribution of G-networks has a product form solution to Jackson's theorem [123] however the resolution of a set of non-linear equations for the traffic circulation [124] is required.

A G-Network is an open arrangement of G-queues with different categories of users:

- Positive users reach from another internal queues or external network inputs as Poisson rates and conform to normal service and direction rules as in regular network structures [123].
- Negative users reach from other internal queues or external network inputs as Poisson rates and eliminate customers in a loaded queue [123]; this represents the necessity to eliminate single network traffic due congestion or "batches" of users [125].
- "Triggers" reach from another internal queues or external network inputs, they relocate users and transfer them to another queues [126].
- "Resets" reach from another internal queues or external network inputs, they set the empty queue to an aleatory size with a distribution that is equal to the stationary distribution at that queue [127, 128]. A reset signal that reaches to a loaded queue has no consequences.

G- Networks were expanded to several categories of positive and negative users [129, 130]. A positive user category is defined by the routing likelihood including the task rate factor for every neuron whereas negative users of distinct categories could possess distinct "user elimination" capacities.

G-networks have been applied in an extensive selection of implementations [131, 132], such as single server [133] and resource allocation in multimedia systems [134].

6.2 Random Neural Network Model

The Random Neural Network [135] is a spiked method with recurrent aleatory properties. The key analytical characteristics are the "product form" where single network stable status solution exists. The Random Neural Network model characterizes more accurately the transmission method of signals in numerous biological neural networks in which signals are transmitted as spikes instead of as fixed analogue impulses.

- A positive spike is interpreted as an excitation signal because it increases by one unit the potential of the receiving neuron.
- A negative spike is interpreted as an inhibition signal decreasing by one unit the potential of the receiver node; when the potential is already zero, it does not produce any effect.

The Random Neural Network method has a solution with a "product form"; the network's stable probability distribution may be represented as the multiplication of the partial probability of every node status [136].

The Random Neural Network learning algorithm [137] is built on a quadratic error cost function with gradient descent. The resolution of o linear and o nonlinear equations is required to solve the back propagation model every time the m Random Neuron Network learning algorithm is presented with an additional input and output set.

6.3 Random Neural Network Extensions

The Random Neural Network was extended to multiple categories of signals [138] where each different flow is a category of signals in the configuration of spikes. A spike which represents a signal category when leaves a node, could be codified at the receiving node as a positive or negative spike within the identical or different category. Transmission rates of spikes from a category are equivalent to the amount of excitation of the inner status in that specific category at the transmitting neuron. Its learning algorithm [139] applies to both recurrent and feedforward models founded on the optimization of a cost function using gradient descent that involves the solution of a set of mC nonlinear and mC nonlinear equations with a $O([mC]^2)$ complexity for the feedforward configuration and $O([mC]^3)$ complexity for the recurrent one.

The Bipolar RNN [140] method has positive as well as negative nodes and balanced performance of positive and negative spikes flowing through the network. Positive neurons collect and transmit only positive spikes whereas negative neurons accumulate and transmit only negative spikes. Positive spikes cancel negative spikes at each negative neuron and vice versa. Connections between different neurons can be positive or negative where a spike departing a neuron may transfer to a different neuron as a spike of alike or opposite potential. The Bipolar RNN presents auto associative memory capabilities and universal function approximation properties [141]. The feed forward Bipolar model [142] with r hidden layers (r + 2 in total) can uniformly estimate continuous functions of r parameters.

The RNN with synchronized interactions [143] includes the possibility of neurons acting together on other neurons, synchronized transmission among neurons, where a neuron may activate transmitting to another neuron and cascades of this activate transmissions can follow. The model defines the activation of transmission among a pair of neurons and activated transmission by sequences of neurons, as well as feedback rings in the sequences to produce extensive transmission bursts. Synchronous interactions between two neurons that jointly excite a third neuron [144] are enough to generate synchronous transmission by great concentration of neurons. A m-neuron recurrent neural network that has both conventional positive and negative exchanges and synchronous exchanges has a learning algorithm complexity of $O(m^3)$ based on gradient descent.

Deep learning with the Random Neural Network [145] is based on soma to soma interactions between natural neuronal cells. Clusters are formed of compact concentration of neurons where the transmission configuration of a neuron instantaneously stimulates and incites transmission by adjacent neurons through dense soma-to-soma interactions founded on the mathematical characteristics of the G-Networks and the Random Neural Network. The characteristics of Deep Learning clusters head to a transmission function that can be applied for wide sets of neurons.

6.4 Random Neural Network Applications

The Random Neural Network has been applied in an extensive number of different scenarios.

(1) *Optimization*

The RNN has been used to obtain an approximate solution of numerous NP-hard maximization tasks like the calculation of the minimum vertex [146] in order to find the node cover of smallest possible size; the travelling salesman problem [147] to find the straightest closed route in an array of m cities where the main requirement is that the path shall navigate through the entire cities only one time. In addition, the RNN has been applied to task to processors assignment for distributed systems [148] with consideration to clustering transactions and dynamic load balancing. Multicast routing can be optimized by using the RNN [149], the network is represented as a leveraged graph where the task is the discovery of the smallest "Steiner tree for the graph" for a prearranged number of endpoints by the best readily rules: the smallest spanning tree rule and the mean distance rule.

(2) *Image Processing and Video Compression*

The production of several artificial image patterns with diverse features is performed by a RNN [150] which connects each neuron m(x, y) to every pixel or image element p(x, y) and every neuron is linked up to eight surrounding local neurons. In addition, the RNN is used for texture modelling and synthesis [151] where the network weights of a recurrent model are learned straight from the pattern of the image which is then applied to generate a synthetic pattern that reproduces the genuine one. The RNN extracts morphometric data of a human brain from Magnetic Resonance Imaging (MRI) x-rays [152] where pictures are divided into regions characterized by its detailed granular properties to be learned and identified by different trained recurrent RNNs. A perpendicular cross-section from "two dimensional top–down scanning electron microscopy" pictures of the attribute surface of a semi-conductor [153] is also predicted by the RNN. Imagery content categorization algorithms, structures and computing programs that frequently scan images assigning an array of images with at least one random neural network is patented solution [154] where each scan is associated to one of several textures and every associated pattern is correlated against every of several picture portions for every of the several scans.

Video sources are adaptive compressed real time by the RNN with motion detection [155] which maintains an expected quality of the uncompressed picture stated by the viewer where the RNN acts as an auto encoder. A set of RNNs compresses at different compression levels [156] in addition of a simple movement detection; the method performs sequential subsampling of images and final decompression where lost images are calculated and inserted using approximations. The deliberately drop of frames to reduce network resources [157] can be compensated by interpolating frames using the RNN with a function approximation.

(3) *Cognitive Packet Networks*

The Cognitive Packet Networks (CPN) [158] has intelligent and learning capabilities where routing and flow control are based on adaptive finite state machines and Random Neural Networks [159]. CPN packets take routing decisions instead of the nodes or protocols. Targets are allocated to Cognitive packets in advance, before being transmitted through the network and following those goals adaptively while leaning from their individual measurements about the Cognitive Packet Network and from the information of another Cognitive packets when they interchange knowledge each other. The design of CPN architecture [160, 161] has been implemented with a QoS based routing algorithm in test bed for optimum and poorest simulation performance to demonstrate the ability of the CPN to adjust to variations in traffic capacity and link break downs.

The CPN has also been tested to transmit Voice [162] delivering better QoS metrics (Delay, jitter and packet desequencing) against Voice over. Voice over CPN is an addition of the Cognitive Packet Network routing algorithm that backs the requirements of voice packet delivery [163] simultaneously with other traffic transmission with the equal or diverse QoS requirements; the implementation is based on Reinforcement Learning to dynamically seek paths that meet the quality requirements of voice communications. Real Time over CPN [164] is based on QoS targets that meet the requirements of real-time packet transmission in the in concurrence with multiple QoS classes (delay, loss and jitter) for multiple traffic flows simultaneously.

The CPN has been applied in a framework at Internet Service Provider level [165] to optimize QoS requirements. The drift parameter [166] is defined by the likelihood that packets will be dispatched following to the Random Neural Network advice rather than arbitrary; Some CPN packets are routed at random to enable the discovery of new routes and to avoid the saturation of the network weights where the CPN quickly determines a backbone of paths with superior quality and explores other network areas afterwards.

Energy efficiency has also been considered in Ad Hoc Cognitive Packet Networks (AHCPN) [167] where unicast messages for route searching are preferred rather than broadcast; QoS metrics include the energy kept in the nodes. As well, to reduce energy consumption and decrease the interference communications range [168], payload and acknowledge packets are transmitted with an adjusted transmission power level. The AHCPN protocol is used in an infrastructure independent interior crisis response mechanism [169] to reduce time latency and prolong the life time of smart mobile devices; it progressively explores the best telecommunication paths among hand held devices and the exit node while providing connection to a server hosted in a cloud. The CPN was patented [170] as a technique for packet transmission in a datagram telecommunication background with a variety of digital packet communication nodes with interconnected links.

(4) *Self-aware Network*

Self-aware networks [171, 172] observe their own performance by applying internal checks and calculation methods to take optimum autonomous usage of these measurements for auto management where the pursued Goal is a combination of QoS metrics Delay and Loss. Self-aware networks through self-monitoring, measurement and intelligent adaptive behavior simultaneously use a selection of cabled and wireless telecommunication networks [173] to provide diverse values of quality of service (QoS) that includes security, reliability and cost; likewise, the communication infrastructure is shared between several customers and networks where the availability of resources fluctuates over time. Nodes can attach and detach the self-aware network independently and explore routes to meet QoS and communication requirements in largely unknown networks [174]; nodes monitor and discover the status of another nodes, connections and routes in addition to the amount of network traffic and congestion in order to refresh their own applicable measurements about the routes they require to select, applying a decision explicit to their individual requirements.

(5) *Software Defined Network*

A model of Software Defined Network (SDN) is the Cognitive Packet Network [175], in addition, the CPN can be considered as a self-aware computer network (SAN). As a SDN, the CPN modifies its behavior to adaptively accomplish its objectives by observing its individual performance and the connections to the peripheral networks; objectives include the detection of services for customers, improvement of Quality of Service (QoS), reduction of its individual energy usage, compensation for failed or malfunctioning elements, detection and reaction to intruders, and finally the its defense against external assaults.

A Software Defined Network platform [176] is developed with a double QoS diversity among pairs of interactive nodes. Each edge node or customer node is concurrently an origin and an end, administrating upload customer initiated traffic and download traffic transmitted back in reply; this method allows the interaction between end customer nodes to swap effectively among the both modes while delivering best performance Quality of Service. Irregular traffic load among the transmitted and received information is applied to activate variations in the QoS where the reduce data bandwidth needs reduce delay QoS and the greater traffic bandwidth needs lower loss rate.

A Cognitive Routing Engine (CRE) [177] is capable to discover almost-optimum routes for a user-specified QoS while using a very small monitoring overhead and reduced response time for the Software Defined Network controllers and switches. A logically centralized CRE finds the best overlay routes [178] when the telecommunication infrastructure among the overlay nodes from different clouds is based on the public Internet network. In addition; the CRE is capable to perform irregular route optimization with the purpose of additionally increase QoS where the transmission route is distinct from the receiving route for a prearranged data center connection.

The current Internet Protocol with route alterations provided by an overlay network is exploited by a machine learning and Big Data in order to perform real-time administration QoS path improvement for Internet size networks [179]. The method is based on an adjustable overlay structure that generates minor alterations of the IP paths that produce smaller packet loss ratios packet transmission delays.

(6) *Smart Routing*

QoS goals can be assigned by external network customers [180] which then they will be directly set up by Self Aware Networks to manage their individual performance in order to accomplish those targets. Genetic methods create and manage paths from earlier learned routes [181] by correlating their suitability against an established QoS and combining relevant paths. An analogy between genotypes and networks paths [182] uses the origin node of every route to learn novel paths using the crossover operation in which each link is considered as the "encoding of a genotype". An adaptive genetic method [183] applies a round-robin strategy to the optimum paths acting as a traffic weighting approach to reduce congestion of a provided route; in addition, it provides the possibility to collect metric information comprising a greater number of nodes.

The CPN with Reinforcement Learning learns autonomously the optimum path in the network purely by means of discovery in a reduced period of time and is capable to adjust to a change along its current route, transferring to the updated best path [184], data packets are redirected only when an enhanced path is discovered. A new method to calculate a similar delay performance [185] is based on a metric that merges route distance and queue usage; this approach also reduces header size in packets. QoS metrics can also combine static and dynamic measurements [186] such as shortest path and minimum delay. Routing oscillations [187] are produced because the interference of several data streams producing packet desequencing and it is assumed that they have a QoS detrimental effect; although they can provide an improved performance if they are managed by alternating the data packet path aleatory.

Software implementations of the CPN routing algorithm are unsuitable in explicitly built devices or hardware with reduce computational capabilities because of its algorithm complexity; simpler alternative algorithms [188] match the performance of the original CPN in comparison with implementations based on hardware and software solutions due their reduced complexity and resource requirements. A FPGA approach [189] of an explicitly built CPN router for a hardware implementation design determines that the widely used architectural methods applied in high bandwidth IP routers are also valid for CPN routers. An iterative routing method for CPN [190] divides great size routing tables into reduced ones; the optimum paths to those reduced routing tables are stored in the intermediary routes within the network, which can be applied to resolve bigger size routing decisions. This method reduces network connection establishment time without and the QoS is also improved.

(7) *CyberSecurity*

Cognitive Packet Networks Denial of Service (DoS) protection method [191] generates an additional self-controlling network that surrounds each vital network node. It is based on Distributed DoS discovery method which produces management messages from the targets of the DDoS cyber-attack to the surrounding network of defense in which DDoS messages streams are discarded in advance therefore they do not reach the vital network node. Each node self-determines two factors: the full data rate that it is able to accept packets and the full assignation of data rate that it is ready to assign to any specific data connection that transverse it. When a CPN node gets a packet from a new connection that has not been previously established [192] it transmit a specific stream acknowledgement data packet to the origin along the inverse route, and confirm the origin its data rate assignment. The router analyses the entire data streams that cross it and eliminates the packets of any stream that surpasses the bandwidth assignment; when the assignment is surpassed, the router reports previous routers that data packets of this stream must be discarded or deleted to a safe router for further analysis. The most important disadvantage of eliminating packet for DDoS protection is its associated "collateral damage" due to the drop of authentic data messages because of wrong decisions [193], consequently an additional refined defense method is proposed founded on prioritization and checking, in which the likelihood that a packet is legitimate automatically assigns the Quality of Service that it obtains.

The CPN Admission Control (AC) method [194] builds its judgment on several Quality of Service values. The "self-assessment" and "self-awareness" abilities of the Cognitive Packet Network are used to gather information which enables the Admission Control algorithm to determine the user admission considering their QoS requirements, the QoS effect that its admission will make on existing connections and the existence of viable routes for the estimated additional data rate. The CPN AC method evaluates the additional stream effect by testing it at a reduced bandwidth [195], therefore, the test messages do not add traffic to the current network's capacity. Customers stipulate the QoS limitations [196] they require to acquire the network service they need for an effective link where every customer can receive several QoS allocations. The choice of whether to admit an additional connection [197] is made applying an algorithm of QoS values based on "Warshall's algorithm" that searches for a route with satisfactory Quality of Service metrics which may include the additional stream.

A mixture of Random Neural Networks (RNN) and Bayesian decision making is applied to the detection of Denial of Service networking attack [198]; the method measures different instantaneous behavior [199] and the long duration statistical variables that describes the incoming network flow; it acquires a probability density function estimating and evaluating the likelihood ratio where the detection choice election step analyses the characteristics of the arriving data flows according to each characteristic which are combined applying both recurrent and feedforward structures of the Random Neural Network. Four different implementations of the detection decision making process can be used [200]: Average probability estimation, RNN with likelihood values, histogram groups and real metrics. Seven different implementations [201] are compared where experimental results are evaluated in a large networking testbed.

(8) *Gene Regulatory Networks*

A likelihood method for Genetic Regulatory Networks (GRN) [202] is represented by the interactions between the density values of every element in the GRN; this method also comprises the characterization of positive and negative relationships between elements, the secondary level relationships which enable a pair of elements to react together on additional elements and finally the Boolean connections among them. The outcome is a precise result in "product form" where the combined stability likelihood distribution of the density for every chromosome is the multiplication of the marginal distribution of every concentration.

The Bayesian method includes a previous Gibbs distribution [203] that offers an appropriate technique to incorporate several elements of biological information to construct large-scale genetic regulatory networks; a back engineering method consists on the Bayesian network averaging method that assembles suitable regression strategies. A Bayesian model averaging based networks (BMAnet) [204] is used to build reliable and large-scale gene regulatory networks able to identify disease candidate genes.

An innovative pathway assessment method to identify diverging behaving pathways in irregular circumstances [205] applies the G-network model where genetic regulatory network structure variables are predicted from regular and irregular specimens and the application of optimization methods with associated restrictions. A random genetic expression method that describes the binary actions of a chromosome [206] is based on Hill equations to the traditional Gillespie algorithm.

(9) *Web Search*

An Intelligent Internet Search Assistant (ISA) [207] built on the Random Neural Network measures customers' relevance and selects the Web snippets from several Web search engines or Recommender Systems applying the learned choices in an iterative method. The ISA behaves as search assistant among users and the Big Data [208]: recommender systems, academic databases, metasearch engines and Web search engines. ISA emulates a brain learning structure [209] using the Random Neural Network with Deep Learning clusters where ISA measures and evaluates Web result relevance connecting a dedicated Deep Learning cluster to a specific Web Search Engine. A Management Cluster [210] is a Deep Learning Cluster that is included to decide the last result relevance calculated from the inputs of each different Deep Learning cluster.

7 Conclusions

This paper has presented a detailed survey of Web Search including Internet Assistants, Web Search Engines, Meta search engines, Web result clustering, Travel services and citation analysis. In addition Ranking has been examined with ranking algorithms, relevance metrics and learning to Rank. Recommender Systems have been defined and Neural Networks in Web Search, Learning to Rank, Recommender Systems and Deep Learning have been analyzed. The Random Neural Network has been presented alongside the G-Network and the Cognitive Packet Network.

References

1. Murugesan, S.: Intelligent agents on the internet and web: applications and prospects. Informatica **23**, 3 (1999)
2. Etzioni, O., Weld, D.S.: Intelligent agents on the internet: fact, fiction, and forecast. IEEE Expert Syst. **10**(4), 44–49 (1995)
3. Zacharis, N.Z., Panayiotopoulos, T.: A metagenetic algorithm for information filtering and collection from the World Wide Web. Expert Syst. **18**(2), 99–108 (2001)
4. Pazzani, M., Billsus, D.: Learning and revising user profiles: the identification of interesting web sites. Mach. Learn. **27**(3), 313–331 (1997)
5. Lieberman, H.: Letizia: an agent that assists web browsing. In: International Joint Conference on Artificial Intelligence, pp. 924–929 (1995)
6. Joachims, T., Freitag, D., Mitchell, T.M.: Web watcher: a tour guide for the world wide web. In: International Joint Conference on Artificial Intelligence, pp. 770–777 (1997)
7. Krulwich, B.: Lifestyle finder: intelligent user profiling using large-scale demographic data. AI Mag. **18**(2), 37–45 (1997)
8. Spink, A., Jansen, B.J., Kathuria, V., Koshman, S.: Overlap among major web search engines. Internet Res. **16**(4), 419–426 (2006)
9. Micarelli, A., Gasparetti, F., Sciarrone, F., Gauch, S.: Personalized search on the world wide web. In: The Adaptive Web, pp. 195–230 (2007)
10. Matthijs, N., Radlinski, F.: Personalizing web search using long term browsing history. In: Web Search and Data Mining, pp. 25–34 (2011)
11. Liu, F., Yu, C.T., Meng, W.: Personalized web search for improving retrieval effectiveness. IEEE Trans. Knowl. Data Eng. **16**(1), 28–40 (2004)
12. Radlinski, F., Dumais, S.T.: Improving personalized web search using result diversification. In: Special Interest Group on Information Retrieval, pp. 691–692 (2006)
13. Backstrom, L., Kleinberg, J.M., Kumar, R., Novak, J.: Spatial variation in search engine queries. In: WWW, pp. 357–366 (2008)
14. Fei, W., Madhavan, J., Halevy, A.Y.: Identifying aspects for web-search queries. J. Artif. Intell. Res. **40**, 677–700 (2011)
15. Jansen, B.J., Mullen, T.: Sponsored search: an overview of the concept, history, and technology. Int. J. Electron. Bus. **6**(2), 114–131 (2008)
16. Meng, W., Yu, C.T., Liu, K.-L.: Building efficient and effective metasearch engines. ACM Comput. Surv. **34**(1), 48–89 (2002)
17. Elizabeth, M.M.: Jacob: information retrieval on internet using meta-search engines: a review. J. Sci. Ind. Res. **67**, 739–746 (2008)
18. Jadidoleslamy, H.: Search result merging and ranking strategies in meta-search engines: a survey. Int. J. Comput. Sci. **9**, 239–251 (2012)
19. Gulli, A., Signorini, A.: Building an open source meta-search engine. In: WWW (Special Interest Tracks and Posters), pp. 1004–1005 (2005)
20. Wu, Z., Raghavan, V.V., Qian, H., Vuyyuru, R., Meng, W., He, H., Yu, C.T.: Towards automatic incorporation of search engines into a large-scale metasearch engine. In: Web Intelligence, pp. 658–661 (2003)
21. Aslam, J.A., Montague, M.H.: Models for metasearch. In: Special Interest Group on Information Retrieval, pp. 275–284 (2001)
22. Dwork, C., Kumar, R., Naor, M., Sivakumar, D.: Rank aggregation methods for the Web. In: WWW, pp. 613–622 (2001)
23. Lu, Y., Meng, W., Shu, L, Yu, C.T., Liu, K.-L.: Evaluation of result merging strategies for metasearch engines. In: Web Information Systems Engineering, pp. 53–66 (2005)

24. Manmatha, R., Sever, H.: A formal approach to score normalization for metasearch. In: Human Language Technologies, pp. 98–103 (2002)
25. Akritidis, L., Katsaros, D., Bozanis, P.: Effective ranking fusion methods for personalized metasearch engines. In: Panhellenic Conference on Informatics, pp. 39–43 (2008)
26. Meng, W., Zonghuan, W., Yu, C.T., Li, Z.: A highly scalable and effective method for metasearch. ACM Trans. Inf. Syst. **19**(3), 310–335 (2001)
27. Sampath Kumar, B.T., Pavithra, S.M.: Evaluating the searching capabilities of search engines and metasearch engines: a comparative study. Ann. Libr. Inform. Stud. **57**, 87–97 (2010)
28. Zamir, O., Etzioni, O.: Web document clustering: a feasibility demonstration. In: Special Interest Group on Information Retrieval, pp. 46–54 (1998)
29. Osinski, S., Weiss, D.: A concept-driven algorithm for clustering search results. IEEE Intell. Syst. **20**(3), 48–54 (2005)
30. Carpineto, C., Osinski, S., Romano, G., Weiss, D.: A survey of Web clustering engines. ACM Comput. Surv. **41**(3–17), 1–38 (2009)
31. Crabtree, D., Gao, X., Andreae, P.: Improving web clustering by cluster selection. In: Web Intelligence, pp. 172–178 (2005)
32. Geraci, F., Pellegrini, M., Maggini, M., Sebastiani, F.: Cluster generation and cluster labelling for web snippets: a fast and accurate hierarchical solution. In: String Processing Information Retrieval, pp. 25–36 (2006)
33. Zeng, H.-J., He, Q.-C., Chen, Z., Ma, W.-Y., Ma, J.: Learning to cluster web search results. In: Special Interest Group on Information Retrieval, pp. 210–217 (2004)
34. He, J., Meij, E., de Rijke, M.: Result diversification based on query-specific cluster ranking. J. Assoc. Inf. Sci. Technol. **62**(3), 550–571 (2011)
35. Ngo, C.L., Nguyen, H.S.: A method of web search result clustering based on rough sets. In: Web Intelligence, pp. 673–679 (2005)
36. Wang, X., Zhai, C.X.: Learn from web search logs to organize search results. In: Special Interest Group on Information Retrieval, pp. 87–94 (2007)
37. Li, Z., Wu, X.: A phrase-based method for hierarchical clustering of web snippets. In: Association for the Advancement of Artificial Intelligence, pp. 1947–1948 (2010)
38. Zhang, D., Dong, Y.: Semantic, hierarchical, online clustering of web search results. In: Advanced Web Technologies and Applications, pp. 69–78 (2004)
39. Ferragina, P., Gulli, A.: The anatomy of a hierarchical clustering engine for web-page, news and book snippets. In: International Conference on Data Mining, pp. 395–398 (2004)
40. Baeza-Yates, R.A., Hurtado, C.A., Mendoza, M.: Query clustering for boosting web page ranking. In: Advances in Web Intelligence, pp. 164–175 (2004)
41. Sismanidou, A., Palacios, M., Tafur, J.: Progress in airline distribution systems: the threat of new entrants to incumbent players. J. Ind. Eng. Manag. **2**, 251–272 (2009)
42. Granados, N.F., Kauffman, R.J., King, B.: The emerging role of vertical search engines in travel distribution: a newly-vulnerable electronic markets perspective. In: Hawaii International Conference on System Sciences, pp. 389–399 (2008)
43. Werthner, H.: Intelligent systems in travel and tourism. In: International Joint Conference on Artificial Intelligence, pp. 1620–1625 (2003)
44. Jansen, B.J., Ciamacca, C.C., Spink, A.: An analysis of travel information searching on the web. J. Inf. Technol. Tourism **10**(2), 101–118 (2008)
45. Ghose, A., Ipeirotis, P.G., Li, B.: Designing ranking systems for hotels on travel search engines to enhance user experience. In: International Conference on Information Systems, vol. 113, pp. 1–19 (2010)

46. Xiang, Z., Fesenmaier, D.R.: An analysis of two search engine interface metaphors for trip planning. J. Inf. Technol. Tourism **7**(2), 103–117 (2004)
47. Kruepl, B., Holzinger, W., Darmaputra, Y., Baumgartner, R.: A flight metasearch engine with metamorph. In: International Conference on World Wide Web, pp. 1069–1070 (2009)
48. Mitsche, N.: Understanding the information search process within a tourism domain-specific search engine. In: Information and Communication Technologies in Tourism, pp. 183–193 (2005)
49. Meho, L.I., Rogers, Y.: Citation counting, citation ranking, and h-index of human-computer interaction researchers: a comparison of scopus and web of science. J. Am. Soc. Inform. Sci. Technol. **59**(11), 1711–1726 (2008)
50. Bar-Ilan, J.: Which h-index? A comparison of WoS, Scopus and Google Scholar. Scientometrics **74**(2), 257–271 (2008)
51. Bar-Ilan, J.: Informetrics at the beginning of the 21st century - a review. J. Inf. **2**(1), 1–52 (2008)
52. Beel, J., Gipp, B., Wilde, E.: Academic search engine optimization (ASEO): optimizing scholarly literature for Google Scholar and Co. J. Sch. Publ. **41**(2), 176–190 (2010)
53. Hirsch, J.E.: An index to quantify an individual's scientific research output. Natl. Acad. Sci. USA **102**(46), 569–572 (2005)
54. Beel, J., Gipp, B.: Google Scholar's ranking algorithm: an introductory overview. In: International Society For Scientometrics and Infometrics, pp. 439–446 (2009)
55. Lewandowski, D.: Google Scholar as a tool for discovering journal articles in library and information science. Online Inf. Rev. **34**(2), 250–262 (2010)
56. Walters, W.H.: Comparative recall and precision of simple and expert searches in Google Scholar and eight other databases. Libr. Acad. **11**(4), 971–1006 (2011)
57. Harzing, A.-W., Alakangas, S.: Google Scholar, Scopus and the Web of Science: a longitudinal and cross-disciplinary comparison. Scientometrics **106**(2), 787–804 (2016)
58. Gipp, B., Beel, J., Hentschel, C.: Scienstein: a research paper recommender system. In: International Conference on Emerging Trends in Computing, pp. 309–315 (2009)
59. Wu, H.C., Luk, R.W.P., Wong, K.-F., Kwok, K.-L.: Interpreting TF-IDF term weights as making relevance decisions. ACM Trans. Inf. Syst. **26**(3–13), 1–37 (2008)
60. Kemp, C., Ramamohanarao, K.: Long-term learning for web search engines. In: Principles and Practice of Knowledge Discovery in Databases, pp. 263–274 (2002)
61. Jin, P., Li, X., Chen, H., Yue, L.: CT-Rank: a time-aware ranking algorithm for web search. J. Converg. Inf. Technol. **5**(6), 99–111 (2010)
62. Haveliwala, T.H.: Topic-sensitive PageRank: a context-sensitive ranking algorithm for web search. IEEE Trans. Knowl. Data Eng. **15**(4), 784–796 (2003)
63. Brin, S., Page, L.: The anatomy of a large-scale hypertextual web search engine. Comput. Netw. **30**(1–7), 107–117 (1998)
64. Kleinberg, J.M.: Authoritative sources in a hyperlinked environment. In: Symposium on Discrete Algorithms, pp. 668–677 (1998)
65. Lempel, R., Moran, S.: The stochastic approach for link-structure analysis (SALSA) and the TKC effect. Comput. Netw. **33**(1–6), 387–401 (2000)
66. Wen-Xue, T., Wan-Li, Z.: Query-sensitive self-adaptable web page ranking algorithm. In: International Conference on Machine Learning and Cybernetics, pp. 413–418 (2003)
67. Bidoki, A.M.Z., Yazdani, N.: Distance rank: an intelligent ranking algorithm for web pages. Inf. Process. Manage. **44**(2), 877–892 (2008)
68. Delbru, R., Toupikov, N., Catasta, M., Tummarello, G., Decker, S.: Hierarchical link analysis for ranking web data. In: Extended Semantic Web Conference, vol. 2, pp. 225–239 (2010)

69. Jiao, J., Yan, J., Zhao, H., Fan,W.: ExpertRank: an expert user ranking algorithm in online communities. In: New Trends in Information and Service Science, pp. 674–679 (2009)
70. Ljosland, M.: Evaluation of web search engines and the search for better ranking algorithms. In: Special Interest Group on Information Retrieval, Workshop on Evaluation of Web Retrieval, pp. 1–6 (1999)
71. Gordon, M.D., Pathak, P.: Finding information on the world wide web: the retrieval effectiveness of search engines. Inf. Process. Manage. **35**(2), 141–180 (1999)
72. Vaughan, L.: New measurements for search engine evaluation proposed and tested. Inf. Process. Manage. **40**(4), 677–691 (2004)
73. Li, L., Shang, Y., Zhang,W.: Relevance evaluation of search engines' query results. In: World Wide Web Posters, pp. 1–2 (2001)
74. Singhal, A., Kaszkiel, M.: A case study in web search using TREC algorithms. In: World Wide Web, pp. 708–716 (2001)
75. Hawking, D., Craswell, N., Thistlewaite, P.B., Harman, D.: Results and challenges in web search evaluation. Comput. Netw. **31**(11–16), 1321–1330 (1999)
76. Hawking, D., Craswell, N., Bailey, P., Griffiths, K.: Measuring search engine quality. Inf. Retrieval **4**(1), 33–59 (2001)
77. Ali, R., Mohd, M., Beg, S.: An overview of web search evaluation methods. Comput. Electr. Eng. **37**(6), 835–848 (2011)
78. Can, F., Nuray, R., Sevdik, A.B.: Automatic performance evaluation of web search engines. Inf. Process. Manage. **40**(3), 495–514 (2004)
79. Ali, R., Mohd, M., Beg, S.: Automatic performance evaluation of web search systems using rough set based rank aggregation. In: Intelligent Human Computer Interaction, pp. 344–358 (2009)
80. Hou, J.: Research on design of an automatic evaluation system of search engine. In: Future Computer and Communication, pp. 16–18 (2009)
81. Chowdhury, A., Soboroff, I.: Automatic evaluation of world wide web search services. In: Special Interest Group on Information Retrieval, pp. 421–422 (2002)
82. Thelwall, M.: Quantitative comparisons of search engine results. J. Am. Soc. Inform. Sci. Technol. **59**(11), 1702–1710 (2008)
83. Joachims, T.: Optimizing search engines using clickthrough data. In: Knowledge Discovery and Data Mining, pp. 133–142 (2002)
84. Xiang, B., Jiang, D., Pei, J., Sun, X., Chen, E., Li, H.: Context-aware ranking in web search. In: Special Interest Group on Information Retrieval, pp. 451–458 (2010)
85. Dong, A., Chang, Y., Zheng, Z., Mishne, G., Bai, J., Zhang, R., Buchner, K., Liao, C., Diaz, F.: Towards recency ranking in web search. In: Web Search and Data Mining, pp. 11–20 (2010)
86. Mohan, A., Chen, Z., Weinberger, K.Q.: Web-search ranking with initialized gradient boosted regression trees. In: Yahoo! Learning to Rank Challenge, pp. 77–89 (2011)
87. Wei-Po, L., Chih-Hung, L., Cheng-Che, L.: Intelligent agent-based systems for personalized recommendations in Internet commerce. Expert Syst. Appl. **22**(4), 275–284 (2002)
88. Almazro, D., Shahatah, G., Albdulkarim, L., Kherees, M., Martinez, R., Nzoukou, W.: A survey paper on recommender systems. CoRR abs/1006.5278 (2010)
89. Adomavicius, G., Tuzhilin, A.: Toward the next generation of recommender systems: a survey of the state-of-the-art and possible extensions. IEEE Trans. Knowl. Data Eng. **17**(6), 734–749 (2005)
90. Adomavicius, G., Manouselis, N., Kwon, Y.: Multi-criteria recommender systems. In: Recommender Systems Handbook, pp. 769–803 (2011)

91. Cremonesi, P., Tripodi, A., Turrin, R.: Cross-domain recommender systems. In: International Conference on Data Mining Workshops, pp. 496–503 (2011)
92. Schröder, G., Thiele, M., Lehner, W.: Setting goals and choosing metrics for recommender system evaluations. In: Conference on Recommnder Systems, pp. 78–85 (2011)
93. Yang, X., Guo, Y., Liu, Y., Steck, H.: A survey of collaborative filtering based social recommender systems. Comput. Commun. **41**, 1–10 (2014)
94. Scarselli, F., Yong, S.L., Gori, M., Hagenbuchner, M., Tsoi, A.C., Maggini, M.: Graph neural networks for ranking web pages. In: Web Intelligence, pp. 666–672 (2005)
95. Chau, M., Chen, H.: Incorporating web analysis into neural networks: an example in hopfield net searching. IEEE Trans. Syst. Man Cybern. C **37**(3), 352–358 (2007)
96. Bermejo, S., Dalmau, J.: Web meta-search using unsupervised neural networks. Artif. Neural Nets Probl. Solving Methods. **2**, 711–718 (2003)
97. Shu, B., Kak, S.C.: A neural network-based intelligent metasearch engine. Inf. Sci. **120**(1–4), 1–11 (1999)
98. Boyan, J., Freitag, D., Joachims, T.: A machine learning architecture for optimizing web search engines. In: Association for the Advancement of Artificial Intelligence workshop on Internet-Based Information Systems, pp. 1–8 (1996)
99. Wang, S., Xu, K., Zhang, Y., Li, F.: Search engine optimization based on algorithm of BP neural networks. In: Computational Intelligence and Security, pp. 390–394 (2011)
100. Choi, Y.S., Yoo, S.I.: Multi-agent web information retrieval: neural network based approach. In: Advances in Intelligent Data Analysis, pp. 499–512 (1999)
101. Burges, C.J.C., Shaked, T., Renshaw, E., Lazier, A., Deeds, M., Hamilton, N., Hullender, G.N.: Learning to rank using gradient descent. In: International Conference Machine Learning, pp. 89–96 (2005)
102. Richardson, M., Prakash, A., Brill, E.: Beyond PageRank: machine learning for static ranking. In: World Wide Web, pp. 707–715 (2006)
103. Agichtein, E., Brill, E., Dumais, S.T.: Improving web search ranking by incorporating user behavior information. In: Special Interest Group on Information Retrieval, pp. 19–26 (2006)
104. Rigutini, L., Papini, T., Maggini, M., Scarselli, F.: SortNet: learning to rank by a neural preference function. IEEE Trans. Neural Netw. **22**(9), 1368–1380 (2011)
105. Cao, Z., Qin, T., Liu, T.-Y., Tsai, M.-F., Li, H.: Learning to rank: from pairwise approach to listwise approach. In: International Conference on Machine Learning, pp. 129–136 (2007)
106. Patil, S.K., Mane, Y.D., Dabre, K.R., Dewan, P.R., Kalbande, R.: An efficient recommender system using collaborative filtering methods with K-separability approach. Int. J. Eng. Res. Appl. **1**, 30–35 (2012)
107. Chang, C.-C., Chen, P.-L., Chiu, F.-R., Chen, Y.-K.: Application of neural networks and Kano's method to content recommendation in web personalization. Expert Syst. Appl. **36**(3), 5310–5316 (2009)
108. Lee, M., Choi, P., Woo, Y.: A hybrid recommender system combining collaborative filtering with neural network. In: Adaptive Hypermedia and Adaptive Web-Based Systems, pp. 531–534 (2002)
109. Kavitha Devi, M.K., Thirumalai Samy, R., Vinoth Kumar, S., Venkatesh, P.: Probabilistic neural network approach to alleviate sparsity and cold start problems in collaborative recommender systems. In: International Conference on Computational Intelligence and Computing Research, pp. 1–4 (2010)
110. Vassiliou, C., Stamoulis, D., Martakos, D., Athanassopoulos, S.: A recommender system framework combining neural networks & collaborative filtering. In: International Conference on Instrumentation, Measurement, Circuits and Systems, pp. 285–290 (2006)

111. Kongsakun, K., Fung, C.C.: Neural network modeling for an intelligent recommendation system supporting SRM for Universities in Thailand. World Sci. Eng. Acad. Soc. Trans. Comput. **2**(11), 34–44 (2012)
112. Billsus, D., Pazzani, M.J.: Learning collaborative information filters. In: International Conference on Machine Learning, pp. 46–54 (1998)
113. Krstic, M., Bjelica, M.: Context-aware personalized program guide based on neural network. IEEE Trans. Consum. Electron. **58**(4), 1301–1306 (2012)
114. Biancalana, C., Gasparetti, F., Micarelli, A., Miola, A., Sansonetti, G.: Context-aware movie recommendation based on signal processing and machine learning. In: Challenge on Context-Aware Movie Recommendation, pp. 5–10 (2011)
115. Chou, P.-H., Li, P.-H., Chen, K.-K., Menq-Jiun, W.: Integrating web mining and neural network for personalized e-commerce automatic service. Expert Syst. Appl. **37**(4), 2898–2910 (2010)
116. Severyn, A., Moschitti, A.: Learning to rank short text pairs with convolutional deep neural networks. In: Special Interest Group on Information Retrieval, pp. 373–382 (2015)
117. Wang, B., Klabjan, D.: An attention-based deep net for learning to rank. CoRR abs/1702.06106, pp. 1–8 (2017)
118. Deng, L., He, X., Gao, J.: Deep stacking networks for information retrieval. In: International Conference on Acoustics, Speech, and Signal Processing, pp. 3153–3157 (2013)
119. Wang, H., Wang, N., Yeung, D.-Y.: Collaborative deep learning for recommender systems. In: Special Interest Group on Knowledge Discovery and Data Mining, pp. 1235–1244 (2015)
120. Elkahky, A.M., Song, Y., He, X.: A multi-view deep learning approach for cross domain user modeling in recommendation systems. In: World Wide Web, pp. 278–288 (2015)
121. Gelenbe, E.: Réseaux stochastiques ouverts avec clients négatifs et positifs, et réseaux neuronaux. Comptes-Rendus. vol. 309, no. II, pp. 979–982. Académie des Sciences de Paris (1989)
122. Gelenbe, E., Tucci, S.: Performances d'un systeme informatique dupliqué. Comptes-Rendus, vol. 312, no. II, 27–30 Académie des Sciences (1991)
123. Gelenbe, E.: Queueing networks with negative and positive customers. J. Appl. Probab. **28**, 656–663 (1991)
124. Gelenbe, E., Schassberger, R.: Stability of product form G-networks. Probab. Eng. Inf. Sci. **6**, 271–276 (1992)
125. Gelenbe, E.: G-networks with signals and batch removal. Probab. Eng. Inf. Sci. **7**, 335–342 (1993)
126. Gelenbe, E.: G-networks with triggered customer movement. J. Appl. Probabil. **30**(3), 742–748 (1993)
127. Fourneau, J.-M., Gelenbe, E.: G-networks with resets. Spec. Interes. Comm. Meas. Eval. **29**(3), 19–20 (2001)
128. Fourneau, J.-M., Gelenbe, E.: Flow equivalence and stochastic equivalence in G-networks. Comput. Manag. Sci. **1**(2), 179–192 (2004)
129. Fourneau, J.-M., Gelenbe, E., Suros, R.: G-networks with multiple classes of negative and positive customers. Theoret. Comput. Sci. **155**(1), 141–156 (1996)
130. Gelenbe, E., Labed, A.: G-networks with multiple classes of signals and positive customers. Eur. J. Oper. Res. **108**(2), 293–305 (1998)
131. Gelenbe, E.: G-networks: a unifying model for neural and queueing networks. Ann. Oper. Res. **48**(5), 433–461 (1994)
132. Gelenbe, E.: The first decade of G-networks. Eur. J. Oper. Res. **126**(2), 231–232 (2000)
133. Gelenbe, E., Glynn, P., Sigman, K.: Queues with negative arrivals. J. Appl. Probability. **28**, 245–250 (1991)

134. Gelenbe, E., Shachnai, H.: On G-networks and resource allocation in multimedia systems. Eur. J. Oper. Res. **126**(2), 308–318 (2000)
135. Gelenbe, E.: Random neural networks with negative and positive signals and product form solution. Neural Comput. **1**(4), 502–510 (1989)
136. Gelenbe, E.: Stability of the random neural network model. Neural Comput. **2**(2), 239–247 (1990)
137. Gelenbe, E.: Learning in the recurrent random neural network. Neural Comput. **5**(1), 154–164 (1993)
138. Gelenbe, E., Fourneau, J.-M.: Random neural networks with multiple classes of signals. Neural Comput. **11**(4), 953–963 (1999)
139. Gelenbe, E., Hussain, K.: Learning in the multiple class random neural network. IEEE Trans. Neural Netw. **13**(6), 1257–1267 (2002)
140. Gelenbe, E., Stafylopati, A., Likas, A.: Associative memory operation of the random network model. In: International Conference on Artificial Neural Networks, pp. 307–312 (1991)
141. Gelenbe, E., Mao, Z.-H., Li, Y.-D.: Function approximation with spiked random networks. IEEE Trans. Neural Netw. **10**(1), 3–9 (1999)
142. Gelenbe, E., Mao, Z.-h., Li, Y.-d.: Function approximation by random neural networks with a bounded number of layers. Differ. Equ. Dyn. Syst. **12**, 143–170 (2004)
143. Gelenbe, E., Timotheou, S.: Random neural networks with synchronized interactions. Neural Comput. **20**(9), 2308–2324 (2008)
144. Gelenbe, E., Timotheou, S.: Synchronized interactions in spiked neuronal networks. Comput. J. **51**(6), 723–730 (2008)
145. Gelenbe, E., Yin, Y.: Deep learning with random neural networks. In: International Joint Conference on Neural Networks, pp. 1633–1638 (2016)
146. Gelenbe, E., Batty, F.: Minimum cost graph covering with the random neural network. In: Computer Science and Operations Research, pp. 139–147 (1992)
147. Gelenbe, E., Koubi, V., Pekergin, F.: Dynamical random neural network approach to the traveling salesman problem, pp. 630–635. Systems, Man Cybern. (1993)
148. Aguilar, J., Gelenbe, E.: Task assignment and transaction clustering heuristics for distributed systems. Inform. Comput. Sci. **97**, 199–219 (1996)
149. Gelenbe, E., Ghanwani, A., Srinivasan, V.: Improved neural heuristics for multicast routing. Sel. Areas Commun. **15**(2), 147–155 (1997)
150. Atalay, V., Gelenbe, E., Yalabik, N.: The random neural network model for texture generation. Int. J. Pattern Recognit. Artif. Intell. **6**(1), 131–141 (1992)
151. Gelenbe, E., Hussain, K., Abdelbaki, H.: Random neural network texture model. Appl. Artific. Neural Netw. Image Process. **104**, 1–8 (2000)
152. Gelenbe, E., Feng, Y., Krishnan, R.: Neural network methods for volumetric magnetic resonance imaging of the human brain. Proc. IEEE **84**(10), 1488–1496 (1996)
153. Gelenbe, E., Koçak, T., Wang, R.: Wafer surface reconstruction from top-down scanning electron microscope images. Microelectron. Eng. **75**(2), 216–233 (2004)
154. Gelenbe, E., Feng, Y.: Image content classification methods, systems and computer programs using texture patterns. U.S. Patent 5,995,651 (1999)
155. Gelenbe, E., Sungur, M., Cramer, C., Gelenbe, P.: Traffic and video quality with adaptive neural compression. Multimed. Syst. **4**(6), 357–369 (1996)
156. Cramer, C.E., Gelenbe, E., Bakircioglu, H.: Low bit rate video compression with neural networks and temporal sub-sampling. Proc. IEEE **84**(10), 1529–1543 (1996)
157. Cramer, C., Gelenbe, E.: Video quality and traffic QoS in learning-based subsampled and receiver-interpolated video sequences. IEEE J. Sel. Areas Commun. **18**(2), 150–167 (2000)

158. Gelenbe, E., Xu, Z., Seref, E.: Cognitive packet networks. In: International Conference on Tools with Artificial Intelligence, pp. 47–54 (1999)
159. Gelenbe, E., Lent, R., Zhiguang, X.: Measurement and performance of a cognitive packet network. Comput. Netw. **37**(6), 691–701 (2001)
160. Gelenbe, E., Lent, R., Montuori, A., Xu, Z.: Towards networks with cognitive packets. In: Performance and QoS of Next Generation Networking, pp. 3–17 (2002)
161. Gelenbe, E., Lent, R., Montuori, A., Xu, Z.: Cognitive packet networks: QoS and performance. In: Modelling, Analysis, and Simulation on Computer and Telecommunication Systems, pp. 3–12 (2002)
162. Wang, L., Gelenbe, E.: An implementation of voice over IP in the cognitive packet network. In: International Symposium on Computer and Information Sciences, pp. 33–40 (2014)
163. Wang, L., Gelenbe, E.: Demonstrating voice over an autonomic network. In: International Conference on Autonomic Computing, pp. 139–140 (2015)
164. Wang, L., Gelenbe, E.: Real-time traffic over the cognitive packet network. In: Computer Networks, pp. 3–21 (2016)
165. Gelenbe, E., Lent, R., Nunez, A.: Smart WWW traffic balancing. In: Communications Society Workshop on IP Operations and Management, pp. 15–22 (2003)
166. Desmet, A., Gelenbe, E.: A parametric study of CPN's convergence process. In: International Symposium on Computer and Information Sciences, pp. 13–20 (2014)
167. Gelenbe, E., Lent, R.: Power-aware ad hoc cognitive packet networks. Ad Hoc Netw. **2**(3), 205–216 (2004)
168. Lent, R., Zonoozi, F.: Power control in ad hoc cognitive packet networks. In: Texas Wireless Symposium, pp. 65–69 (2005)
169. Bi, H., Gelenbe, E.: A cooperative emergency navigation framework using mobile cloud computing. In: International Symposium on Computer and Information Sciences, pp. 41–48 (2014)
170. Gelenbe, E.: Cognitive Packet Network. U.S. Patent 6,804,201 (2004)
171. Gelenbe, E., Gellman, M., Su, P.: Self-awareness and adaptivity for quality of service. In: International Symposium on Computers and Communications, pp. 3–9 (2003)
172. Gelenbe, E., Lent, R., Nunez, A.: Self-aware networks and QoS. Proc. IEEE **92**(9), 1478–1489 (2004)
173. Gelenbe, E.: Steps toward self-aware networks. Commun. ACM **52**(7), 66–75 (2009)
174. Gelenbe, E.: A software defined self aware network: the cognitive packet network. In: Network Cloud Computing and Applications, pp. 9–14 (2014)
175. Gelenbe, E.: A software defined self-aware network: the cognitive packet network. In: Semantics Knowledge and Grids, pp. 1–5 (2013)
176. Gelenbe, E., Kazhmaganbetova, Z.: Cognitive packet network for bilateral asymmetric connections. IEEE Trans. Industr. Inf. **10**(3), 1717–1725 (2014)
177. François, F., Gelenbe, E.: Towards a cognitive routing engine for software defined networks. In: International Conference on Communications, pp. 1–6 (2016)
178. François, F., Gelenbe, E.: Optimizing secure SDN-enabled inter-data centre overlay networks through cognitive routing. In: Modeling, Analysis, and Simulation on Computer and Telecommunication Systems, pp. 283–288 (2016)
179. Brun, O., Wang, L., Gelenbe, E.: Big data for autonomic intercontinental overlays. IEEE J. Sel. Areas Commun. **34**(3), 575–583 (2016)
180. Pu, S., Gellman, M.: Using adaptive routing to achieve quality of service. Perform. Eval. **57**(2), 105–119 (2004)
181. Gelenbe, E., Gellman, M., Lent, R., Liu, P., Su, P.: Autonomous smart routing for network QoS. In: International Conference on Autonomic Computing, pp. 232–239 (2004)

182. Gelenbe, E., Liu, P., LainLaine, J.: Genetic algorithms for route discovery. IEEE Trans. Syst. Man Cybern. **36**(6), 1247–1254 (2006)
183. Gelenbe, E., Liu, P., LainLaine, J.: Genetic algorithms for autonomic route discovery. In: Distributed Intelligent Systems: Collective Intelligence and Its Applications, pp. 371–376 (2006)
184. Lent, R., Liu, P.: Searching for low latency routes in CPN with reduced packet overhead. In: International Symposium on Computer and Information Sciences, pp. 63–72 (2005)
185. Gelenbe, E., Liu, P.: QoS and routing in the cognitive packet network. In: World of Wireless, Mobile and Multimedia Networks, pp. 517–521 (2005)
186. Gellman, M., Liu, P.: Random neural networks for the adaptive control of packet networks. In: International Conference on Artificial Neural Networks, vol. 1, pp. 313–320 (2006)
187. Gelenbe, E., Gellman, M.: Oscillations in a bio-inspired routing algorithm. In: Mobile Ad-hoc and Sensor Systems, pp. 1–7 (2007)
188. Laurence, A.: Hey: reduced complexity algorithms for cognitive packet network routers. Comput. Commun. **31**(16), 3822–3830 (2008)
189. Hey, L.A., Cheung, P.Y.K., Gellman, M.: FPGA based router for cognitive packet networks. In: Field-Programmable Technology, pp. 331–332 (2005)
190. Liu, P., Gelenbe, E.: Recursive routing in the cognitive packet network. In: Testbeds and Research Infrastructures for the Development of Networks and Communities, pp. 1–6 (2007)
191. Gelenbe, E., Gellman, M., Loukas, G.: Defending networks against denial of service attacks. Opt. Photonics Secur. Def. Unmanned/Unattended Sens. Sens. Netw. **5611**, 233–243 (2004)
192. Gelenbe, E., Gellman, M., Loukas, G.: An autonomic approach to denial of service defence. In: World of Wireless Mobile and Multimedia Networks, pp. 537–541 (2005)
193. Gelenbe, E., Loukas, G.: A self-aware approach to denial of service defence. Comput. Netw. **51**(5), 1299–1314 (2007)
194. Gelenbe, E., Sakellari, G., D'Arienzo, M.: Admission control in self aware networks. In: Global Telecommunications Conference, pp. 1–5 (2006)
195. Gelenbe, E., Sakellari, G., D'Arienzo, M.: Admission of packet flows in a self-aware network. In: Mobile Ad-hoc and Sensor Systems, pp. 1–6 (2007)
196. Gelenbe, E., Sakellari, G., D'Arienzo, M.: Admission of QoS aware users in a smart network. ACM Trans. Autonom. Adapt. Syst. **3**(1), 1–28 (208). Article no. 4
197. Gelenbe, E., Sakellari, G., D'Arienzo, M.: Controlling access to preserve QoS in a self-aware network. In: Self-Adaptive and Self-Organizing Systems, pp. 205–213 (2007)
198. Loukas, G., Oke, G.: A biologically inspired denial of service detector using the random neural network. In: Mobile Ad-hoc and Sensor Systems, pp. 1–6 (2007)
199. Loukas, G., Oke, G.: Likelihood ratios and recurrent random neural networks in detection of denial of service attacks. In: Symposium Performance Evaluation of Computer and Telecommunication System, pp. 1–8 (2007)
200. Öke, G., Loukas, G., Gelenbe, E.: Detecting denial of service attacks with bayesian classifiers and the random neural network. In: IEEE Fuzzy Systems, pp. 1–6 (2007)
201. Öke, G., Loukas, G.: A denial of service detector based on maximum likelihood detection and the random neural network. Comput. J. **50**(6), 717–727 (2007)
202. Gelenbe, E.: Steady-state solution of probabilistic gene regulatory networks. Phys. Rev. **76**(3), 1–8 (2007). 31903
203. Kim, H., Gelenbe, E.: Reconstruction of large-scale gene regulatory networks using bayesian model averaging. In: Bioinformatics and Biomedicine, pp. 202–207 (2011)
204. Kim, H., Park, T., Gelenbe, E.: Identifying disease candidate genes via large-scale gene network analysis. Int. J. Data Mining Bioinform. **10**(2), 175–188 (2014)

205. Kim, H., Atalay, R., Gelenbe, E.: G-network modelling based abnormal pathway detection in gene regulatory networks. In: International Symposium on Computer and Information Sciences, pp. 257–263 (2011)
206. Kim, H., Gelenbe, E.: Stochastic gene expression modeling with hill function for switch-like gene responses. IEEE/ACM Trans. Comput. Biol. Bioinf. **9**(4), 973–979 (2012)
207. Serrano, W., Gelenbe, E.: An intelligent internet search assistant based on the random neural network. In: Artificial Intelligence Applications and Innovations, pp. 141–153 (2016)
208. Serrano, W.: A big data intelligent search assistant based on the random neural network. In: International Neural Network Society Conference on Big Data, pp. 254–261 (2016)
209. Serrano, W., Gelenbe, E.: Intelligent search with deep learning clusters. In: Intelligent Systems Conference, pp. 254–261 (2017)
210. Serrano, W., Gelenbe, E.: The deep learning random neural network with a management cluster. In: International Conference on Intelligent Decision Technologies, pp. 185–195 (2017)

Avoiding to Face the Challenges of Visual Place Recognition

Ehsan Mihankhah(✉) and Danwei Wang

School of Electrical and Electronic Engineering,
Nanyang Technological University, Singapore, Singapore
ehsanm@ntu.edu.sg

Abstract. Through this paper, bottlenecks of conventional place recognition techniques are studied, and a replacement strategy is proposed for each item. Conventional place recognition algorithms are extensions of object recognition techniques applied to larger scale targets known as the place landmarks. The discussion presented in this paper aims to address the challenges of detection and recognition of the places, which make this topic distinctive from detection and recognition of the objects and landmarks. The challenges are listed under related categories. The table of challenges, reasons, and the recommendations to avoid these situations is presented as the guideline for selection of proper tools for place recognition purpose.

Keywords: Place detection · Place recognition
Non-feature-based place recognition · Loop closure detection

1 Introduction

Location awareness is a critical requirement in autonomous mobile robotics. Path planning, navigation, mission planning, and many other functionalities are highly dependent on location awareness. To fulfill different objectives, location of the robot should be known with multiple levels of precision. For example, metrically precise coordinate position estimation is required for local path planning and navigation, which is mostly addressed through SLAM (simultaneous localization and mapping) techniques [1], locally and for a limited duration. However, mission planning requires general estimation of robot location in the studied environment, which is mostly addressed through place recognition techniques [2]. Although the focus of this paper is on the latter case, recognition of previously visited places facilitates loop-closure and map-merging phase of the former group of localization techniques, to help them maintain the acceptable precision and consistency in a longer run.

Localization of mobile robots can be achieved either by external referencing [3], or through processing the on-board sensor data [4]. In the former approach, beacons and cameras can be used for precise indoor applications [5]. Moreover, sound signatures, signal strength and availability of WiFi access points, and other place-specific factors can be used for fingerprint-based location estimation in indoor applications [6]. For outdoor, GNSS readings (which is perfect under ideal satellite coverage) [7], and position calculation through distance estimation from nearby mobile phone towers

K. Arai et al. (Eds.): IntelliSys 2018, AISC 868, pp. 738–749, 2019.
https://doi.org/10.1007/978-3-030-01054-6_52

(which results to a large location uncertainty radius) [8], are possible techniques in accordance with the former approach. On the other hand, the latter approach has been addressed by researchers who extended the scope of metrically accurate localization through incremental map generation in outdoor environment [9]. Other than that, the image retrieval techniques that correspond an observation to one place out of a group of expected places, form another major direction to address place recognition problem [10]. These image retrieval techniques for place recognition are mainly the extensions of object recognition techniques. While localization through external referencing requires minimal processing, indoor solutions cannot be extended to outdoor, and outdoor solutions lose coverage in indoor environment [11]. On the other hand, place recognition using onboard equipment confront limitations such as drift and sensitivity to environmental conditions [2]. This paper addresses the limitations of place recognition through processing the onboard equipment, and proposes alternative solutions to avoid facing such limitations. A place recognition procedure is consisted of several steps. Each step is a major component of the methodology that substantially affects the recognition results. This paper presents a guideline for selection of proper components for place recognition solutions. The focus of the paper is on the philosophy, including definition, sensing, and procedure steps, rather than the analysis of the results of numeric case-studies.

Section 2 of this paper studies the existing approaches to place detection. Section 3 explains the shortcomings of the common approaches by studying the definition, sensing strategies, and the procedure steps (place recognition components). Proposed amendments are suggested in Sect. 3 as well. In Sect. 4, the paper summarizes through presentation of the guideline table, and the conclusion is made successively.

2 Existing Place Recognition Techniques

In this section, two major conventional approaches for place recognition, which only use the onboard equipment, are studied.

2.1 Techniques Inspired by SLAM

SLAM has been an ongoing research for the past few decades [12–14]. Perfect SLAM algorithms were proposed in the literature and mature implementations are available as commercial products [15], and as open-source solutions [16]. Some SLAM algorithms compare the consequent observations point-by-point [17], and the others only process the feature-points [18]. While pointwise comparison is less sensitive to the shape of the objects that the observation point is surrounded by, feature-wise comparison is less demanding in terms of computation and storage [19], which are the two factors that form valid concerns in large-scale missions and in the long run.

According to the map, when metrically accurate current coordinate location of the mobile robot is close to a previously explored point, the explored area around the current position is recognized as a revisited place. However, there is an estimation error in each incremental step of map generation that propagates through time, and results to a drifted map, which ultimately, results to inaccurate position estimation. SLAM has negligible limitations for 2D indoor applications. Recent LiDAR-based and camera-based indoor 3D SLAM techniques have also proven acceptable consistency for indoor environment [20]. However, maintaining a consistent 2D or 3D SLAM, as a general solution for large-scale outdoor environment for long-term autonomy, is still an open challenge.

One suggested solution to overcome this limitation is using the offline generated map of the environment. One such implementation is NDT[1] matching [21], which is also used to localize a driverless car on the road [22]. In this implementation, the 3D point cloud of the studied place is stored as the known map of the environment. LiDAR inputs are used to match the current observation with the offline map, and initialization is handled through GNSS. Another general treatment is to reset the online map generation drifts through GNSS readings (under ideal satellite coverage) or environment signal-finger-printing (in presence of the necessary infrastructure) [23]. However, this approach is very much dependent on external referencing, which is beyond the scope of this paper.

In SLAM-based localization, if a revisited place is observed frequently, to the extent that the drift falls within the scan matching tolerance, small drifts can be compensated, and the map accuracy can be constantly recovered, and so the location estimation precision can be maintained. However, if the revisited place is not observed for a long time, to the extent that the accumulated drift disallows matching the new observation with the correct piece of the map, it becomes very challenging to recognize a revisited place. To recover from this situation, recognition of revisited places should be handled through a technique that is independent from SLAM.

2.2 Techniques Inspired by Image Retrieval

The most commonly applied technique for place recognition, which is generally independent from SLAM, is image retrieval. In this technique, current observation is compared against a database of stored images to recognize the best match. Each image in the database is associated to a place. Therefore, matching the current observation with a previously stored frame conveys the similarity to the place with which it is associated with [2]. Observations can be 2D images (colour intensity information) [24], gray 3D point clouds (depth information) [25–27], or RGBD images (both colour intensity and depth information) [28, 29].

Image retrieval techniques used for place recognition are extended versions of object recognition algorithms. Similar to object recognition methodologies, which identify each object through a set of specific features, in place recognition methodologies based on image retrieval, places are either recognized as a result of recognition of an outstanding landmark or through recognition of a set of specific type of features [10, 30]. The procedure involves the phase of breaking the images into smaller patches, known as segmentation step. Following this, features are extracted from the patches. Then features are converted to numeric structures, known as feature descriptor vectors. Each image is described through a set of feature-descriptors. The set of feature

[1] Normal Distributions Transform.

descriptors is called the image descriptor. Feature descriptors are stored in a database. For each new observation, feature descriptors are computed and compared against the stored feature-descriptors in the database. The most similar stored image to the current observation is the one that embodies the highest number of similar feature-descriptors.

Feature selection is a critical step in image retrieval. Invariance of features is ideally expected against scaling, rotation, shifting, and deformations [31]. Enough number of features should be consistently identifiable to compute a representative image descriptor. Several techniques are available for feature detection. Working on camera frames, simplest visual features are edges [32] and corners [33]. For example, the Hessian feature detector computes the Hessian for points that exhibit high derivative values in two orthogonal directions [34], a point is selected as a feature when determinant of the Hessian becomes maximum at that point. The Harris detector [33] is a geometric feature detector that finds corner-like structures from image regions where the second moment matrix has large eigenvalues. LoG[2] [35], and a close approximation of it, DoG[3] [36], are also well-known feature detectors. A more resilient detector against rotation, scaling, illumination changes, and input noise [37] is Harris-Laplacian detector [38] that combines the Harris corner detection with LoG. If the Harris-Laplace is applied to Hessian, the result is called Hessian-Laplace detector [39].

When input is in form of binary point cloud data, feature extraction in 3D case can be achieved through processing the meshes [40]. For example, in [40], features points are extracted when second difference operator is applied to mesh edges. Likewise, the work [41] extracts feature lines from meshes by applying third order derivatives. Work [42] extracts curvature extremums as feature points. The work [43] uses region growing method, which is a meshless feature detection methodology that identifies regions with sharp features from point cloud segments. Feature lines are found subsequently. Through a statistical approach, PCA[4] extracts salient features from point clouds. Extracted features are the direction of principle axes in the space along which the studied data points have the highest variance, and the lowest variance [44]. This method can be applied to multi-dimensional data for place recognition [45]. In a recent work [46], 3D point clouds are projected to different planes in the space, and similar feature detector is used to find features holding statistic information about the studied point cloud.

Majority of place detection techniques use the SIFT[5] descriptor [4]. SIFT algorithm calculates the local gradient magnitudes and their orientations for each key-point. Subsequently, it calculates the orientation histograms for sub-regions. The SIFT descriptor is a vector that holds values of all orientation histograms for each key-point. An extension of SIFT is called GLOH9F[6] descriptor that is designed to enhance robustness and distinctiveness of description [47]. SURF[7] descriptor [48] is a well-

[2] Laplacian-of-Gaussian.

[3] Difference-of-Gaussian.

[4] Principle Component Analysis.

[5] Scale Invariant Feature Transform.

[6] Gradient location-orientation histogram.

[7] Speeded-Up Robust Features Descriptor.

known computationally efficient replacement to SIFT. [37]. “Haar wavelets” [49] are used in SURF. The SURF algorithm generates a structure that embeds intensity statistics along 2D coordinate directions. There are other types of descriptors that work on content distribution. For example, geometric histograms [50] or shape context [51] are used to form image descriptors. Work [52] uses 3D sparse visual features, which are mainly the corners, introduced in [53]. In [52], description procedure is initiated by detection of landmarks followed by computation of 3D Gestalt feature descriptor [54–56]. Work [57], introduces surface entropy features (SURE) [58] for place recognition. SURE features embed the statistics of the distribution of local surface normal vectors. Invariant feature selection is an ongoing challenge in place recognition.

Using the bag-of-words model [59] is the most referenced methodology for storage, retrieval and comparison of feature descriptors [60]. Bag-of-words model associates feature descriptor to words. Subsequently, word processing techniques are used to compare descriptor structures. Bag-of-words model discards the geometric and spatial relevance among image features. Since bag-of-words dictionary can get large in the long run, optimized data access techniques were suggested in the literature to handle this problem [61]. As explained earlier, snapshot pairing through viewpoint tolerant feature matching is the common approach in place recognition based on image retrieval [62]. Combination of features and feature descriptors has been used in place recognition research. These descriptors were stored, retrieved, and compared through modified versions of the bag-of-words model.

3 Challenges of Existing Methodologies and the Proposed Replacements

3.1 SLAM Specific Challenges

As explained in Sect. 2.1, SLAM solves place recognition problem for short-term autonomy, in small-scale indoor environments. However, for long-term autonomy, and large-scale outdoor environment, SLAM requires an independent mechanism to handle recognition of revisited places to cancel the drifts (which is known as the problem of “closing the loop” in SLAM community). Therefore, recognition of places through SLAM, for large-scale outdoor environment in long-term autonomy, is a mutually dependent problem. Consequently, for long-term autonomy in large-scale outdoor environment, it is necessary to solve SLAM and place recognition problems individually and independently. Source of inconsistency in SLAM algorithms is the accumulation of error, caused by incremental map generation and location estimation through the comparison of consecutive observations. Therefore, the source of problem is the incremental computation approach, which embeds an intrinsic integrator.

SLAM is prone to facing storage and computational limitations in the long run, when pointwise scan matching is involved. Therefore, for large-scale long-term applications, pointwise methods should be avoided. On the other hand, feature-based methods are successful whenever enough number of expected type of features are guaranteed to exist consistently in all observations. Invariant feature selection and feature association are major challenges in feature-based SLAM. Therefore, for large-scale long-term autonomy in outdoor missions, or even for combined indoor and outdoor missions, SLAM-based place recognition should be complemented with external referencing. However, if the presence of external referencing is assumed, place recognition can be handled solely through the external referencing equipment, and without any reliance on SLAM.

3.2 Challenges of Place Recognition Through Common Image Retrieval Techniques

As explained in Sect. 2.2, place recognition based on image retrieval is commonly handled through landmark recognition and feature matching techniques. Although recognition of outstanding landmarks is a concrete proof for recognition of a place, not all places contain an outstanding landmark. Therefore, this approach is limited only to recognition of places that include outstanding landmarks. In another words, mapping a place to a landmark is not generally a valid assumption for recognition of all places.

Even if features are directly extracted from observations without prior landmark recognition, existence of enough number of specific type of features in every observation is a limiting assumption. This problem enroots in invariant and consistent features selection challenge. This limitation advocates incorporation of non-feature-based description methods such as [63, 64]. Under the assumption of existence of enough number of feature points, the philosophy of a place being represented only through limited number of feature-points is a matter of dispute. The assumption of a place being represented through limited key-points is resulted by extending object recognition philosophy to place recognition. The assumption of an object being fully represented through limited features is logical. However, every single piece of the surrounding of a place participates in characterization of that place. This necessitates incorporation of holistic description techniques such as [65] rather than local ones that operate only on few image patches.

Bag-of-words database is reported to have included over 100 million code-word to feature records in 20 km traversal [66]. In this situation entry and retrieval of feature vectors should be handled through optimized data access methods such as the one proposed in [61].

3.3 General Concerns

Viewpoint invariance is one of the major concerns in place recognition. This is partially the result of "looking at a place from an external viewer's stand point through a limited field of view". When this approach is combined with landmark recognition, the combination remains satisfactory for most object recognition techniques. However, expecting this setting to work for place recognition introduces a logical dispute. Assume a robot is exactly located in front of a landmark and a camera with limited field-of-view, which is used as the sensor, is directed towards the landmark. Recognition of the landmark in this situation is easy, and is taken equivalent to recognition of the place. Now, if the camera faces oppositely, the mentioned landmark falls out of the field of view, and as a consequence, landmark is not recognized, and so is not the place. In both scenarios, the observation is made exactly in the same place, but the recognition

Table 1. Summary of challenges, sources and the suggested replacement strategies

Challenge	Scope	Source	Suggested replacement
Drift	SLAM-based place recognition	Propagation of incremental mapping error through time	Limit this approach to indoor, and short term applications
Excessive storage and computation demands	Point-wise SLAM-based place recognition	Comparison of all points of the current observation with the entire history	Limit this approach to indoor, and short term applications
Feature selection and feature matching	Feature based SLAM for place recognition	Enough number of expected type of features are not guaranteed to exist consistently in all observations	Supplement with external referencing
Landmark recognition	Place recognition through image retrieval	Not all places contain outstanding landmarks	Do not take landmark recognition equivalent to place recognition
Feature selection and feature matching	Place recognition through image retrieval	Enough number of expected type of features are not guaranteed to exist consistently in all observations	Use a non-feature-based descriptor
Few features do not fully represent places	Place recognition through image retrieval	Unlike the case of object recognition, a set of features cannot fully represent a place	Use a holistic place descriptor
Bag-of-words dictionary grows excessively	Place recognition through image retrieval	Feature vectors stored in the bag-of-words model can reach millions very quickly	Use advanced data access models to store and retrieve data. Use non-feature-based techniques
Viewpoint invariance	General	Looking at a place from an external viewer's stand point through a limited field of view	Look from inside of the place out, through the entire field of view
Environmental changes	General	Color is affected by illumination along the day, and appearance changes in different seasons along the year	Use depth information
Depth accuracy, range, and resolution	General	Stereo vision, infrared reflection, and ultrasonic time of return measurements are not producing satisfactory performance for this application	Use LiDAR to capture depth information

(*continued*)

Table 1. (*continued*)

Challenge	Scope	Source	Suggested replacement
False recognition	General	Frame to frame comparison is prone to false recognition	Use group comparison, and study places inside sequences
General Definition for the term "place"	General	Existing place definitions do not include all types of places	A place is the subset of the three dimensional environment in which, any partial observation made across the entire field-of-view, holds consistent descriptor that is distinctively different compared to the descriptors computed from its surrounding

results are very different. This is another reason why landmark recognition cannot necessarily be taken equivalent to place recognition. One suggestion to overcome this situation is to "look from inside of the place out, through the entire field of view". This is achievable by using the panoramic view camera (2D) or 360° LiDAR (3D).

The other concern for place recognition is the depth value. Depth value is a valuable information from the place because the color information can be affected by illumination changes throughout the day, and the appearance of the place varies in different seasons along the year. Therefore, depth is the invariant information, which maintains its consistency throughout the time. Depth information can be captured through different techniques such as stereo vision, infrared reflection, ultrasonic time of flight measurement, and through LiDAR. While LiDAR can produce precise measurements in long ranges, none of the other techniques can perform the same. Therefore, it is highly recommended to use LiDAR to capture depth information for place recognition applications.

The last but not the least is the false detection issue. This problem is mainly rooted in the frame to frame comparison approach. If two places are supposed to be compared, several observations made from one place should be compared to several observations made from the other place. Group comparison reduces the chance of false detection. One other suggestion would be to study places in sequences similar to the approaches suggested by [67, 68]. Defining the term "place" through landmarks, features, or precise coordinate location, have limitations to include all places. Taking into account all the mentioned limitations, the proposed definition of the term "place" that grounds the basis for proposition of a place recognition methodology that addresses all the mentioned concerns, should be at descriptor level by taking into account the analysis over all the observation made at different spots of the place. Therefore, the suggested definition is as follows.

"A place is the subset of the three dimensional environment in which, any partial observation made across the entire field-of-view, holds consistent descriptor that is distinctively different compared to the descriptors computed from its surrounding".

4 Summary and Conclusion

This paper studied the major challenges faced by conventional place recognition methodologies that do not rely on external referencing. For each concern, the causing factor was introduced, and the replacement treatment was suggested. This analysis is summarized in Table 1. To address these concerns, a novel methodology is suggested, which will appear in the author's future publications.

5 Future Work

This paper addressed the shortcomings of the existing place recognition methodologies, and suggested different approaches to avoid facing them. Next step is to propose the step-by-step solution which takes into account all the mentioned concerns. This will be covered in our subsequent publications.

References

1. Thrun, S., Burgard, W., Fox, D.: Probabilistic Robotics. MIT Press, Cambridge (2005)
2. Lowry, S., Sunderhauf, N., Newman, P., Leonard, J.J., Cox, D., Corke, P., et al.: Visual place recognition: a survey. IEEE Trans. Robot. **32**, 1–19 (2016)
3. Song, T., Capurso, N., Cheng, X., Yu, J., Chen, B., Zhao, W.: Enhancing GPS with lane-level navigation to facilitate highway driving. IEEE Trans. Veh. Technol. **66**, 4579–4591 (2017)
4. Lowe, D.G.: Object recognition from local scale-invariant features. In: The Proceedings of the Seventh IEEE International Conference on Computer Vision, vol. 2, pp. 1150–1157 (1999)
5. Radoi, I., Gutu, G., Rebedea, T., Neagu, C., Popa, M.: Indoor positioning inside an office building using BLE. In: 2017 21st International Conference on Control Systems and Computer Science (CSCS), pp. 159–164 (2017)
6. Vo, Q.D., De, P.: A survey of fingerprint-based outdoor localization. IEEE Commun. Surv. Tutor. **18**, 491–506 (2016)
7. Gu, Y., Kamijo, S.: GNSS positioning in deep urban city with 3D map and double reflection. In: 2017 European Navigation Conference (ENC), pp. 84–90 (2017)
8. Liu, H., Zhang, Y., Su, X., Li, X., Xu, N.: Mobile localization based on received signal strength and pearson's correlation coefficient. Int. J. Distrib. Sensor Netw. **11**, 157046 (2015)
9. Cadena, C., Carlone, L., Carrillo, H., Latif, Y., Scaramuzza, D., Neira, J., et al.: Past, present, and future of simultaneous localization and mapping: toward the robust-perception age. IEEE Trans. Robot. **32**, 1309–1332 (2016)
10. Murillo, C., Guerrero, J.J., Sagues, C.: SURF features for efficient robot localization with omnidirectional images. In: Proceedings 2007 IEEE International Conference on Robotics and Automation, pp. 3901–3907 (2007)

11. Hamid, M.H.A., Adom, A.H., Rahim, N.A., Rahiman, M.H.F.: Navigation of mobile robot using Global Positioning System (GPS) and obstacle avoidance system with commanded loop daisy chaining application method. In: 2009 5th International Colloquium on Signal Processing & Its Applications, pp. 176–181 (2009)
12. Aulinas, J., Petillot, Y., Salvi, J., Llado, X.: The SLAM problem: a survey. In: Presented at the Proceedings of the 2008 Conference on Artificial Intelligence Research and Development: Proceedings of the 11th International Conference of the Catalan Association for Artificial Intelligence (2008)
13. Durrant-Whyte, H., Bailey, T.: Simultaneous localization and mapping: part I. IEEE Robot. Autom. Mag. **13**, 99–110 (2006)
14. Bailey, T., Durrant-Whyte, H.: Simultaneous localization and mapping (SLAM): part II. IEEE Robot. Autom. Mag. **13**, 108–117 (2006)
15. KAARTA 3D Mapping. http://www.kaarta.com/
16. da Silva, M.F., Xavier, R.S., do Nascimento, T.P., Gonsalves, L.M.G.: Experimental evaluation of ROS compatible SLAM algorithms for RGB-D sensors. In: 2017 Latin American Robotics Symposium (LARS) and 2017 Brazilian Symposium on Robotics (SBR), pp. 1–6 (2017)
17. Mendes, E., Koch, P., Lacroix, S.: ICP-based pose-graph SLAM. In: 2016 IEEE International Symposium on Safety, Security, and Rescue Robotics (SSRR), pp. 195–200 (2016)
18. de la Puente, P., Rodriguez-Losada, D.: Feature based graph-SLAM in structured environments. Auton. Robots **37**, 243–260 (2014)
19. Mihankhah, E., Taghirad, H.D., Kalantari, A., Aboosaeedan, E., Semsarilar, H.: Line matching localization and map building with least square. In: 2009 IEEE/ASME International Conference on Advanced Intelligent Mechatronics, pp. 1734–1739 (2009)
20. Engel, J., Schöps, T., Cremers, D.: LSD-SLAM: large-scale direct monocular SLAM, Cham, pp. 834–849 (2014)
21. Saarinen, J., Andreasson, H., Stoyanov, T., Ala-Luhtala, J., Lilienthal, A.J.: Normal distributions transform occupancy maps: application to large-scale online 3D mapping. In: 2013 IEEE International Conference on Robotics and Automation, pp. 2233–2238 (2013)
22. Takeuchi, E., Tsubouchi, T.: A 3-D scan matching using improved 3-D normal distributions transform for mobile robotic mapping. In: 2006 IEEE/RSJ International Conference on Intelligent Robots and Systems, pp. 3068–3073 (2006)
23. Schleicher, D., Bergasa, L.M., Ocana, M., Barea, R., Lopez, E.: Real-time hierarchical GPS aided visual SLAM on urban environments. In: 2009 IEEE International Conference on Robotics and Automation, pp. 4381–4386 (2009)
24. Tron, R., Vidal, R.: Distributed 3-D localization of camera sensor networks from 2-D image measurements. IEEE Trans. Autom. Control **59**, 3325–3340 (2014)
25. Zhongyang, Z., Yan, L., Wang, J.: LiDAR point cloud registration based on improved ICP method and SIFT feature. In: 2015 IEEE International Conference on Progress in Informatics and Computing (PIC), pp. 588–592 (2015)
26. Paul, R., Newman, P.: FAB-MAP 3D: topological mapping with spatial and visual appearance. In: 2010 IEEE International Conference on Robotics and Automation, pp. 2649–2656 (2010)
27. Cole, M., Newman, P.M.: Using laser range data for 3D SLAM in outdoor environments. In: Proceedings 2006 IEEE International Conference on Robotics and Automation, ICRA 2006, pp. 1556–1563 (2006)
28. Zhu, H., Weibel, J.B., Lu, S.: Discriminative multi-modal feature fusion for RGBD indoor scene recognition. In: 2016 IEEE Conference on Computer Vision and Pattern Recognition (CVPR), pp. 2969–2976 (2016)

29. Labbé, M., Michaud, F.: Online global loop closure detection for large-scale multi-session graph-based SLAM. In: 2014 IEEE/RSJ International Conference on Intelligent Robots and Systems, pp. 2661–2666 (2014)
30. Cadena, C., Galvez-López, D., Tardos, J.D., Neira, J.: Robust place recognition with stereo sequences. IEEE Trans. Robot. **28**, 871–885 (2012)
31. Hassaballah, M., Awad, A.I.: Detection and description of image features: an introduction. In: Awad, A.I., Hassaballah, M. (eds.) Image Feature Detectors and Descriptors: Foundations and Applications, pp. 1–8. Springer, Cham (2016)
32. Canny, J.: A computational approach to edge detection. IEEE Trans. Pattern Anal. Mach. Intell. **6**, 679–698 (1986)
33. Harris, C., Stephens, M.: A combined corner and edge detector. In: Alvey Vision Conference, pp. 147–151 (1988)
34. Beaudet, P.R.: Rotationally invariant image operators. In: Proceedings of the 4th International Joint Conference on Pattern Recognition, pp. 579–583 (1978)
35. Lindeberg, T.: Feature detection with automatic scale selection. Int. J. Comput. Vis. **30**, 79–116 (1998)
36. Lowe, G.: Distinctive image features from scale-invariant keypoints. Int. J. Comput. Vis. **60**, 91–110 (2004)
37. Grauman, K., Leibe, B.: Visual Object Recognition. Synthesis Lectures on Artificial Intelligence and Machine Learning, vol. 5, pp. 1–181 (2011)
38. Mikolajczyk, K., Schmid, C.: Scale & affine invariant interest point detectors. Int. J. Comput. Vis. **60**, 63–86 (2004)
39. Mikolajczyk, K., Tuytelaars, T., Schmid, C., Zisserman, A., Matas, J., Schaffalitzky, F., et al.: A comparison of affine region detectors. Int. J. Comput. Vis. **65**, 43–72 (2005)
40. Hubeli, A., Gross, M.: Multiresolution feature extraction for unstructured meshes. In: Visualization, VIS 2001, Proceedings, pp. 287–294 (2001)
41. Hildebrandt, K., Polthier, K., Wardetzky, M.: Smooth feature lines on surface meshes. In: Presented at the Proceedings of the Third Eurographics Symposium on Geometry Processing, Vienna, Austria (2005)
42. Watanabe, K., Belyaev, A.G.: Detection of salient curvature features on polygonal surfaces. Comput. Graph. Forum **20**, 385–392 (2001)
43. Demarsin, K., Vanderstraeten, D., Volodine, T., Roose, D.: Detection of closed sharp edges in point clouds using normal estimation and graph theory. Comput. Aided Des. **39**, 276–283 (2007)
44. Abdi, H., Williams, L.J.: Principal component analysis. Wiley Interdiscip. Rev. Comput. Stat. **2**, 433–459 (2010)
45. Wang, J., Dodds, Z., Miranker, W.L.: Principal component analysis for place recognition. J. Neural Parallel Sci. Comput. **5**, 347 (1996)
46. He, L., Wang, X., Zhang, H.: M2DP: a novel 3D point cloud descriptor and its application in loop closure detection. In: 2016 IEEE/RSJ International Conference on Intelligent Robots and Systems (IROS), pp. 231–237 (2016)
47. Mikolajczyk, K., Schmid, C.: A performance evaluation of local descriptors. In: 2003 IEEE Computer Society Conference on Computer Vision and Pattern Recognition, Proceedings, vol. 2, pp. II-257–II-263 (2003)
48. Bay, H., Tuytelaars, T., Van Gool, L.: SURF: speeded up robust features. In: Leonardis, A., Bischof, H., Pinz, A. (eds.) Computer Vision – ECCV 2006: 9th European Conference on Computer Vision, Graz, Austria, 7–13 May 2006, Proceedings, Part I, Springer, Heidelberg (2006)
49. Goswami, J.C., Chan, A.K.: Fundamentals of Wavelets: Theory, Algorithms, and Applications. Wiley Publishing (2011)

50. Ashbrook, P., Thacker, N.A., Rockett, P.I., Brown, C.I.: Robust recognition of scaled shapes using pairwise geometric histograms. In: Presented at the Proceedings of the 6th British Conference on Machine Vision, vol. 2, Birmingham, United Kingdom (1995)
51. Belongie, S., Malik, J., Puzicha, J.: Shape matching and object recognition using shape contexts. IEEE Trans. Pattern Anal. Mach. Intell. **24**, 509–522 (2002)
52. Cieslewski, T., Stumm, E., Gawel, A., Bosse, M., Lynen, S., Siegwart, R.: Point cloud descriptors for place recognition using sparse visual information. In: 2016 IEEE International Conference on Robotics and Automation (ICRA), pp. 4830–4836 (2016)
53. Leutenegger, S., Lynen, S., Bosse, M., Siegwart, R., Furgale, P.: Keyframe-based visual–inertial odometry using nonlinear optimization. Int. J. Robot. Res. **34**, 314–334 (2015)
54. Bosse, M., Zlot, R.: Place recognition using keypoint voting in large 3D lidar datasets. In: 2013 IEEE International Conference on Robotics and Automation, pp. 2677–2684 (2013)
55. Bosse, M., Zlot, R.: Keypoint design and evaluation for place recognition in 2D lidar maps. Robot. Auton. Syst. **57**, 1211–1224 (2009)
56. Bosse, M., Zlot, R.: Place recognition using regional point descriptors for 3D mapping. In: Howard, A., Iagnemma, K., Kelly, A. (eds.) Field and Service Robotics: Results of the 7th International Conference, pp. 195–204. Springer, Heidelberg (2010)
57. Fiolka, T., Stückler, J., Klein, D.A., Schulz, D., Behnke, S.: Distinctive 3D surface entropy features for place recognition. In: 2013 European Conference on Mobile Robots, pp. 204–209 (2013)
58. Fiolka, T., Stückler, J., Klein, D.A., Schulz, D., Behnke, S.: SURE: surface entropy for distinctive 3D features. In: Stachniss, C., Schill, K., Uttal, D. (eds.) Spatial Cognition VIII: International Conference, Spatial Cognition 2012, Kloster Seeon, Germany, August 31–September 3, 2012, Proceedings, pp. 74–93. Springer, Heidelberg (2012)
59. Baeza-Yates, R.A., Ribeiro-Neto, B.: Modern Information Retrieval. Addison-Wesley Longman Publishing Co., Inc. (1999)
60. Ho, K.L., Newman, P.: Detecting loop closure with scene sequences. Int. J. Comput. Vis. **74**, 261–286 (2007)
61. Galvez-López, D., Tardos, J.D.: Bags of binary words for fast place recognition in image sequences. IEEE Trans. Robot. **28**, 1188–1197 (2012)
62. Satkin, S., Hebert, M.: 3DNN: viewpoint invariant 3D geometry matching for scene understanding. In: 2013 IEEE International Conference on Computer Vision, pp. 1873–1880 (2013)
63. Singhal, A., Srivastava, N., Mishra, R.: Hiding signature in colored image. In: 2006 International Symposium on Communications and Information Technologies, pp. 446–450 (2006)
64. Mihankhah, E., Wang, D.: Environment characterization using Laplace eigenvalues. In: 2016 14th International Conference on Control, Automation, Robotics and Vision (ICARCV), pp. 1–6 (2016)
65. Murillo, C., Kosecka, J.: Experiments in place recognition using gist panoramas. In: 2009 IEEE 12th International Conference on Computer Vision Workshops, ICCV Workshops, pp. 2196–2203 (2009)
66. Schindler, G., Brown, M., Szeliski, R.: City-scale location recognition. In: 2007 IEEE Conference on Computer Vision and Pattern Recognition, pp. 1–7 (2007)
67. Milford, M.J., Wyeth, G.F.: SeqSLAM: visual route-based navigation for sunny summer days and stormy winter nights. In: 2012 IEEE International Conference on Robotics and Automation, pp. 1643–1649 (2012)
68. Ho, K.L., Newman, P.: Loop closure detection in SLAM by combining visual and spatial appearance. Robot. Auton. Syst. **54**, 740–749 (2006)

A Semantic Representation of Sensor Data to Promote Proactivity in Home Assistive Robotics

Amedeo Cesta, Gabriella Cortellessa, Andrea Orlandini, Alessandra Sorrentino, and Alessandro Umbrico(✉)

Institute of Cognitive Sciences and Technologies, Italian National Research Council, Rome, Italy
{amedeo.cesta,gabriella.cortellessa,andrea.orlandini, alessandra.sorrentino,alessandro.umbrico}@istc.cnr.it

Abstract. IoT technology integrated in smart home environments produces a huge and heterogeneous amount of data that can be used to characterize events that occur inside an environment as well as activities that a person is performing. The knowledge that can be extracted from such data can be used to proactively facilitate the daily living of the person with physical and/or cognitive weaknesses. This paper describes work aiming at better taking advantage of those data by integrating a cognitive loop composed by a semantic representation and a decision making functionality. In particular, the paper presents results of the Knowledge-based cOntinuous Loop (KOaLa) research initiative particularly focusing on the use of semantic technology to represent and reason on sensor data to enable the triggering of proactive services toward the users in an intelligent home.

Keywords: Intelligent environments · Knowledge representation Ontological descriptions · Automated planning · Sensor networks Artificial intelligence

1 Introduction

Trigger for this research is the experience developed during the GiraffPlus project [1] whose goal was to create a technological environment to monitor older people at home. The project represented an interesting example of continuous monitoring integrating a sensor network, software services with a telepresence robot. Specifically, the goal was to support prevention and long-term monitoring as well as to foster social interaction and communication by proposing a solution built around the primary users (i.e., the seniors) [2]. In GiraffPlus a sensor network allows secondary users (e.g., relatives or caregivers) to monitor and access real-time information about the activities and physical parameters of a primary user. In addition, a robot (the Giraff) lives with a user in his/her home and

K. Arai et al. (Eds.): IntelliSys 2018, AISC 868, pp. 750–769, 2019.
https://doi.org/10.1007/978-3-030-01054-6_53

allows him/her to contact and communicate with secondary users (allowing virtual visits, but also the delivery of reminders, audio and video messages like those described in [3]).

The overall system was actually deployed in several houses of different older people for an intense experimental campaign [4]. The results demonstrate the effectiveness of the proposed solution but also identified some limitations. In particular, the limited interaction abilities at home was pretty clear (see again [3] for some initial follow up) and the limited *autonomy* of the overall system as a support tool for continuous help. To overcome such a limitation we pursue the synthesis of a complete "sense-reason-act" cycle which relies on the pipeline of a semantic representation and reasoning module with a planning and execution module. Such modules, when integrated in a sensorized environment GiraffPlus-like, may endow the whole system with a level of proactivity. The KOaLa (Knowledge-based cOntinuous Loop) research initiative has been started specifically to create the pipeline mentioned above and ultimately to build an intelligent assistive environment capable of integrating Artificial Intelligence (AI) techniques to better take advantage of sensor data and reanalyze a continuous and proactive control loop. The research objective is to introduce *cognitive* capabilities that allow the system to proactively support the daily home living of primary users. The key point is to realize reasoning mechanisms capable of analyzing the continuous flow of data coming from sensors and make decisions accordingly. An open issue with *sensor-based/data-driven* applications is the lack of standard techniques for managing the huge amount of heterogeneous information coming from sensors and build *useful* knowledge that can be exploited to realize complex services. A knowledge-based reasoning mechanism capable of processing sensor data to dynamically infer the status of the environment as well as states or activities of a primary user is a key enabling feature to automatically synthesize supporting tasks to be enacted.

The solution proposed by KOaLa consists in the pipeline of a knowledge processing and extraction module, called the *KOaLa Semantic Module*, with a planning and execution module, the *KOaLa Acting Module*. The *KOaLa Semantic Module* aims at providing sensor data with a clear semantics and building a semantic representation of an environment. Such a module uses standard semantic technologies based on the Web Ontology Language (OWL). It relies on the *KOaLa Ontology* which has been defined as an extension of the existing Semantic Sensor Network Ontology [5] (SSN). The *KOaLa Ontology* leverages and extends the SSN semantics and the SSO ontology design pattern [6] to model the different types of sensor that can be used, their capabilities as well as the events, situations and activities that may concern the status of an environment. The *KOaLa Acting Module* aims at synthesizing and executing the set of operations needed to achieve a set of desired *planning goals*. Such planning goals are generated by the *KOaLa Semantic Module* after an analysis of correlated events and/or activities detected. They represent operations the system should perform in order to proactively react to events or support activities the *KOaLa Semantic Module* has recognized within the considered scenario. The *KOaLa Acting Module*

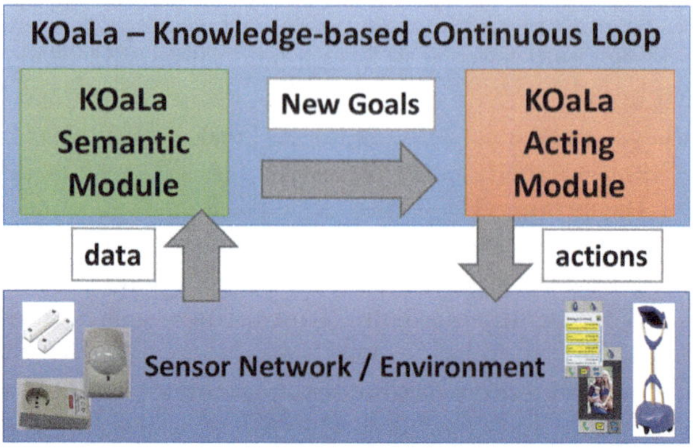

Fig. 1. The KOaLa "sense-reason-act" cycle.

synthesizes and executes this set of operations by leveraging the timeline-based approach to planning [7,8]. Operations derived from *planning goals* are integrated into a temporal plan which is continuously evaluated, updated, and executed by the system.

Figure 1 shows the overall KOaLa approach and the *semantic* and *acting* modules that compose the envisaged "sense-reason-act" pipeline. This paper focuses on the semantic module of the envisaged architecture which allows the system to build an abstraction of the environment not only to perform activity recognition, see an example in [9], but also to understand what the system can do to "react" or support the recognised activities. In other words, one of the objectives of the semantic representation is the synthesis of planning goals to make the system do things autonomously. The paper describes how sensor data is represented, how data processing incrementally builds and refine knowledge and it is structured as follows: (i) Sect. 2 created some context for the work by describing the GiraffPlus project; (ii) Sect. 3 provides a conceptual architecture of the envisaged control system; (iii) Sect. 4 provides some details about the definition of the KOaLa Ontology; (iv) Sect. 5 describes the data processing

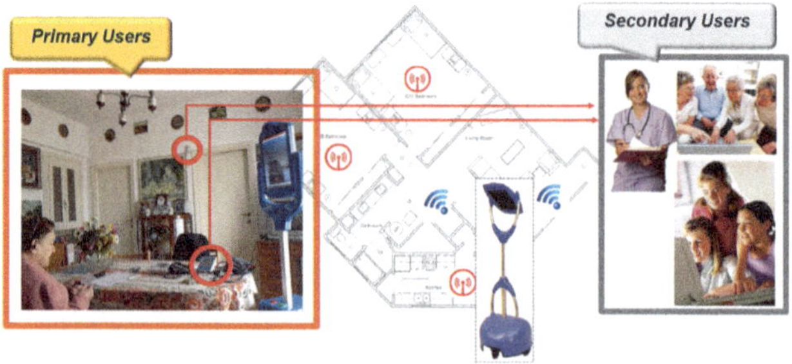

Fig. 2. The GiraffPlus project complete concept.

mechanisms and the management of the knowledge; (v) Sect. 6 provides a brief overview of the capability of the system at the current state; (vi) Sect. 7 ends the paper by pointing out next future works to improve the current solution.

2 Context of the Work

This work uses GiraffPlus project experience as a reference point to identify missing capabilities and situate potential solutions. GiraffPlus aimed at realizing a system capable of supporting older people directly at their home through several personalized services. These services can be categorized as services towards the primary user (i.e., virtual visits, reminders, messages) and towards the secondary user (i.e., real-time monitoring visualizations, reports, alarms, warnings). The services at home are provided by the telepresence robot Giraff. The robot allows older people to communicate to their friends and family, by sending audio messages and performing videocalls. The seniors can interact with the robot either by touching the touch screen which is part of it or speaking with it. Interaction abilities are extremely important to improve the emotional engagement between the user and the robot. For this reason, different interaction abilities have been integrated in this work. Figure 2 shows a conceptual representation of the considered context and the different types of actor involved within the project.

The goal is to have a robot that proactively interacts with the user, depending on the human's state detected by the system. For example, if the system detects that the primary user did not have lunch, the robot can suggest to cook something together either asks further question to understand the user's state. On the other end, the robot can also participate in the daily activity of the user, by giving advices and offering suggestions, i.e., propose a new recipe while the user is cooking, or walking instead of watching the tv, depending on the situation detected. The robot is used as support in critical situations, i.e., performing emergency calls in case of fall detection. Furthermore, GiraffPlus combined the presence of robot with a network of sensors. The network is composed by physical sensors (i.e., blood pressure sensors) and environmental sensors (i.e. motion sensor, actuators). Physical sensors are directly worn by the user and they are used to monitor the health state. On the other hand, the environmental sensors are installed in different areas of the environment and allow to monitor both the house and the user state. Automatic adjustments of the lightness and temperature inside a room depend on the user preferences and can be performed by analyzing that sensor information. Thanks to combination of those sensors, it is also possible to infer if a person has eaten during the day and/or slept during the night.

3 KOaLa: Knowledge-Based Continuous Loop

The pursued research goal is to realize an autonomous system capable of "self-representing" and monitoring a home environment, recognizing particular events and/or activities occurring within such environment and deciding how to manage them, i.e. how to react to a particular detected event or how to support a particular detected activity. Figure 3 shows a layered conceptual architecture which characterizes KOaLa at three different levels of abstraction. The *sensor layer* manages the continuous flow of data coming from sensors. It interprets and collects sensor data into a dataset which provides a first representation of the environment. The *environment layer* initializes the knowledge by taking into account the specific configuration of the environment and iteratively refines such knowledge by processing data collected into the dataset. The resulting knowledge can be further characterized at two distinct levels of abstraction. The *environment and observation* level characterizes the knowledge about the environment and the observations received by sensors. The *events and activity recognition* level characterizes the environment in terms of events and activities detected. Finally, a *service layer* can be built on top of this knowledge in order to realize complex functionalities.

To achieve the pursued research objectives the system must *know* the *meaning* of a particular event or action detected. At the same time, the system must *know* the operations that can be performed to manage such events and how such operations must be executed. Thus, the solution we propose relies on the integration of two core techniques of AI: (i) *knowledge representation and reasoning* techniques; (ii) *autonomous planning and execution* techniques. Broadly speaking, *knowledge representation and reasoning* techniques leverage well-founded logic formalisms to define semantics for representing and processing information. Considering application scenarios like the GiraffPlus case study, such techniques are

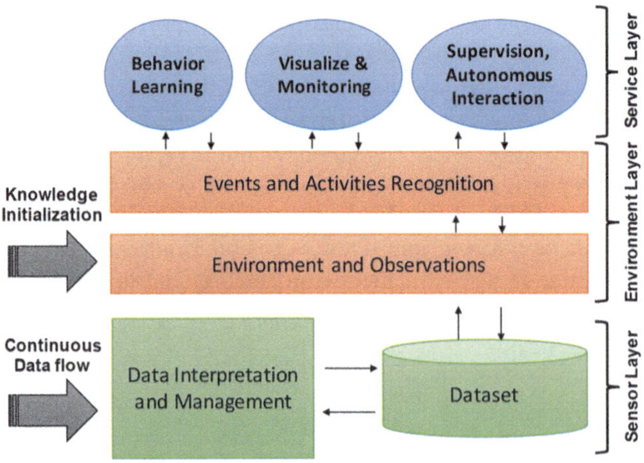

Fig. 3. The KOaLa conceptual layered architecture.

necessary to interpret and process *sensor data* in a uniform way according to a clear and well-defined semantics. The *knowledge* resulting from data interpretation represents a general and abstract description which can be leveraged to realize "specific" services as shown by Fig. 3. Specifically, an analysis of the generated knowledge can identify particular events or activities that a *supervision service* can use to allow the system to proactively react to an event or support an activity. In this regard, *autonomous planning and execution* techniques are well-suited to provide the system with the capability of dynamically synthesize and execute activities to support the daily home-living of a person. Such operations are executed according to a set of constraints that are represented into a *planning model* which can be dynamically adapted to the particular configuration of the environment.

4 KOaLa Ontology

The envisaged monitoring system must be capable of continuously processing a significant amount of data coming from the environment. Data can be heterogeneous and can be described by different formats and/or parameters depending on the particular communication protocols adopted by sensor manufacturers. A common language to uniformly interpret and manage data and produce information is necessary. Thus, a dedicated ontology has been defined in order to characterize the concepts and the properties that are relevant in the considered application scenario. Such ontology is used by the semantic module of Fig. 1 to provide sensor data with semantics. The KOaLa ontology has been defined as extension of the SSN ontology [5] and the DUL ontology.[1] It follows a context-based approach which characterizes the knowledge about the environment at different levels of abstraction. Specifically, three contexts have been identified: (i) the *sensor context*; (ii) the *environment context*; (iii) the *observation context*.

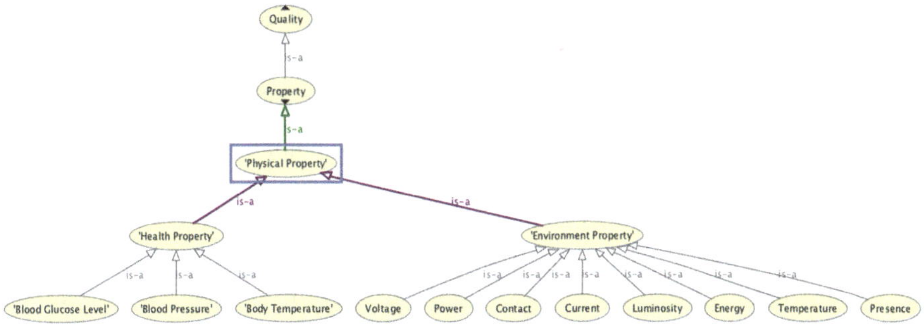

Fig. 4. KOaLa extension of `SSN:Property` class.

[1] http://www.loa.istc.cnr.it/ontologies/DUL.owl.

4.1 Sensor Context

The *sensor context* characterizes the knowledge about the sensor composing a particular environment, their deployment and the properties that such sensors may observe. This context is strictly related to SSN and there are several concepts that are extended or borrowed from that ontology. The aim of the *sensor context* is to provide a more detailed representation of the different types of sensor that can compose an environment as well as the different types of property that can be observed. Such a representation allows the system to dynamically recognize the actual capabilities with respect to the control of the environment and determine the set of *operations* the system may perform to support the daily home living of a person.

According to the SSN documentation, the class `SSN:Property` is a subclass of `DUL:Quality` and models an observable quality of an event or an object. This class has been extended in order to model the types of property the system can monitor over time. Figure 4 shows the set of classes defined by KOaLa to extend `SSN:Property` and the related hierarchy. These classes represent observable qualities of physical elements that can be part of an application scenario. These properties are subclassess of the `PhysicalProperty` which directly extends `SSN:Property`, and can be partitioned into two different groups. The class `HealthProperty` models properties that describe physiological parameters of a person the system may monitor over time. The classes `BloodPressure`, `BodyWeight`, `BloodGlucoseLevel` represent some examples of physiological parameters that can be monitored. The class `EnvironmentProperty` models properties about the *state* of the home environment and the related physical objects. For example, the class *Energy* models the energy consumed by an element of the environment over time (e.g. a TV). Similarly the class *Contact* models the state of objects like *windows* that can be *open* or *close*.

Different properties are measured through different values. Thus, also the class `SSN:ObservationValue` has been extended in order to model different types of value that can be received from observations (i.e. data). In this way, it is possible to model specific knowledge about received data. It could be possible to specify *class axioms* that characterize the particular datatype used to represent values as well as known *bounds/thresholds*. For example, `TemperatureValue` class definition specifies the datatype representing observed values, and known thresholds representing the "regular" temperature bounds for an environment.

$$\begin{aligned}\texttt{TemperatureValue} \sqsubseteq\ & \texttt{SSN:ObservationValue} \sqcap \\ & \exists\, \texttt{hasTemperatureValue.(XSD:double)} \sqcap \\ & \exists\, \texttt{envMinTemperatureValue.(envMin)} \sqcap \\ & \exists\, \texttt{envMaxTemperatureValue.(envMax)}\end{aligned}$$

Sensors have different capabilities and can generate information about different properties. KOaLa extends the `SSN:Sensor` class in order to define the particular types of *sensing device* that can be installed and controlled. Such a characterization allows the system to recognize the particular type of sensors available and reason about their specific capabilities in terms of properties that can be observed. Class definitions for sensor devices like `Pir`, `Switch` and `Gap` with class axioms specifying observed properties are added to the ontology.

$$\begin{aligned}\texttt{Pir} \sqsubseteq\ & \texttt{SSN:Sensor} \sqcap \\ & \exists\, \texttt{SSN:observes.Luminosity} \sqcap \\ & \exists\, \texttt{SSN:observes.Temperature} \sqcap \\ & \exists\, \texttt{SSN:observes.Presence}\end{aligned}$$

$$\begin{aligned}\texttt{Gap} \sqsubseteq\ & \texttt{SSN:Sensor} \sqcap \\ & \exists\, \texttt{SSN:observes.Luminosity} \sqcap \\ & \exists\, \texttt{SSN:observes.Temperature} \sqcap \\ & \exists\, \texttt{SSN:observes.Presence} \sqcap \\ & \exists\, \texttt{SSN:observes.Contact}\end{aligned}$$

$$\begin{aligned}\texttt{Switch} \sqsubseteq\ & \texttt{SSN:Sensor} \sqcap \\ & \exists\, \texttt{SSN:observes.Energy} \sqcap \\ & \exists\, \texttt{SSN:observes.Voltage} \sqcap \\ & \exists\, \texttt{SSN:observes.Power} \sqcap \\ & \exists\, \texttt{SSN:observes.Current}\end{aligned}$$

4.2 Environment Context

The information the system can gather through sensors concerns properties of *physical elements* of a particular environment. There could be several sensors that observe the *same* property of an element. Let us consider for example three `Pir` sensors installed into the same room, in such a case all of them will produce information about the `Temperature` of the same room. From the ontology point of view such data must be interpreted and associated to the same property of the room. In addition, sensors may observe information that are not relevant with respect to a particular type of element. Thus, the *environment context* models the different types of element that can compose a *home environment* and the properties that can characterize the state of these elements. Moreover, it models the particular configuration of the environment and the deployment of sensors by leveraging concepts like `DUL:SpaceRegion`, `SSN:Platform`, `SSN:Deployment` and related properties.

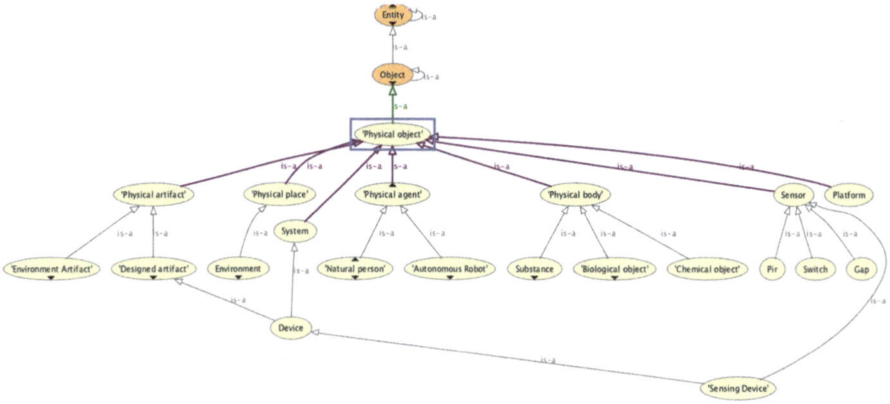

Fig. 5. Partial view of the KOaLa environment context with respect to the extension of the `DUL:PhysicalObject` class.

Figure 5 partially shows the *environment context* which has been defined by extending concepts like `DUL:PhysicalAgent`, `DUL:PhysicalArtifact` and `DUL:PhysicalPlace`. According to the DUL documentation, a `DUL:Physical Agent` is a `DUL:PhysicalObject` that is capable of self-representing a `DUL:Description` in order to plan a `DUL:Action`. It is well suited to model *agents* that can "autonomously" move within the environment and perform some *activities*. Given the considered application context, there are two types of agent that are relevant with respect to the purpose of the system: (i) assistive robots; (ii) patients.

The `AssistiveRobot` class models a general `DUL:PhysicalAgent` like the GiraffPlus robot that "share" the home environment with a person and is capable of performing some `Activity`.

$$\texttt{AssistiveRobot} \sqsubseteq \texttt{DUL:PhysicalAgent} \sqcap \exists\,\texttt{performs.Activity}$$

A `Patient` class extends `DUL:NaturalPerson` and models a patient living into the home environment the system is monitoring. Like general agents (`DUL:NaturalPerson` is subclass of `DUL:PhysicalAgent`). A *patient* can perform some *activities* within the environment but can also have *health-related properties* the system can observe over time in order to monitor his/her health status.

$$\texttt{Patient} \sqsubseteq \texttt{DUL:NaturalPerson} \sqcap \exists\,\texttt{performs.Activity} \sqcap \exists\,\texttt{SSN:hasProperty.HealthProperty}$$

The class `DUL:PhysicalPlace` has been extended by introducing the class `HomeEnvironment` in order to model the different types of structural part of a home environment together with the related properties that can be observed. In general the individuals of these classes are associated with a *location* in space which localizes the structural element within the environment. They can have a set of sensors attached through *platforms*[2] and have a set of properties that can be observed through associated sensors.

$$\texttt{HomeEnvironment} \sqsubseteq \texttt{DUL:PhysicalPlace} \sqcap \exists\,\texttt{DUL:hasLocation.DUL:SpaceRegion} \sqcap \exists\,\texttt{DUL:hasPart.SSN:Platform} \sqcap \exists\,\texttt{SSN:hasProperty.EnvProperty}$$

The class `HomeEnvironment` is further specialized into a set of subclasses like `BedRoom`, `LivingRoom` or `Kitchen` that model specific parts composing a home environment together with the related properties. The definitions of these classes specify properties like `Temperature`, `Luminosity` or `Presence` that are useful to characterize events or activities that can be respectively observed or recognized by the system. For example, the property `Presence` can be used to detect whether a person is moving inside the kitchen or not.

$$\texttt{Kitchen} \sqsubseteq \texttt{HomeEnvironment} \sqcap \exists\,\texttt{SSN:hasProperty.Temperature} \sqcap \exists\,\texttt{SSN:hasProperty.Luminosity} \sqcap \exists\,\texttt{SSN:hasProperty.Presence}$$

[2] According to SSN semantics a `Platform` represents an entity to which other entities can be attached.

Finally, the class `DUL:PhysicalArtifact` has been extended by introducing the class `EnvironmentArtifact` in order to model the different types of object that can be found inside a home environment together with the related properties that can be observed. An *environment artifact* can be localized through a specific *location* and is part of a specific structural element of a home environment. An environment artifact can have several sensors attached through *platforms* and can have a set of *properties* that can be observed through such sensors.

$$\begin{aligned}\texttt{EnvironmentArtifact} \sqsubseteq\ & \texttt{DUL:PhysicalArtifact} \sqcap \\ & \exists\,\texttt{DUL:hasLocation.DUL:SpaceRegion} \sqcap \\ & \exists\,\texttt{DUL:isPartOf.Environment} \sqcap \\ & \exists\,\texttt{SSN:hasPart.SSN:Platform} \sqcap \\ & \exists\,\texttt{SSN:hasProperty.EnvProperty}\end{aligned}$$

The class `EnvironmentArtifact` is specialized into a set of subclasses like `Window`, `Fridge` or `TV` that model specific observable properties of elements that can be part of a home environment. For example, a *tv* object can be associated with the property `Energy` which can be used to *recognize* whether a person is watching the tv or not.

$$\begin{aligned}\texttt{TV} \sqsubseteq\ & \texttt{EnvironmentArtifact} \sqcap \\ & \exists\,\texttt{SSN:hasProperty.Energy}\end{aligned}$$

4.3 Observation Context

The sensor context characterizes the knowledge about the available sensors and properties that can be observed. The environment context characterizes the knowledge about the structure and physical elements that can compose a home environment together with sensor deployment (i.e., a configuration). The *observation context* characterizes the knowledge about the *features* that can actually produce information (i.e., observations) as well as the *events* and the *activities* that can be observed through them.

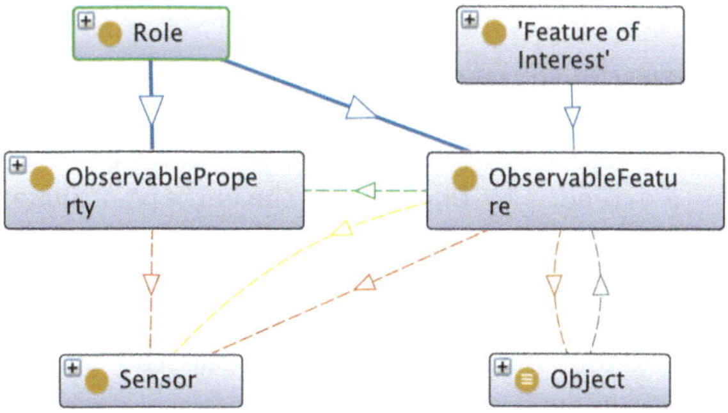

Fig. 6. Definition of `ObservableFeature` and `ObservableProperty` concepts.

Figure 6 shows the concepts of `ObservableFeature` and `ObservableProperty` that leverage and extend the concepts of `SSN:FeatureOfInterest` and `DUL:Role`. Individuals of the `ObservableFeature` class represent abstractions of *interesting* physical objects of the environment that play the *role* of "being observable" through the associated sensors. In other words, an *observable feature* is a *role* that a physical object of the environment (i.e., a *feature of interest*) can play according to the deployment of sensors. Similarly, individuals of the `ObservableProperty` class represent abstractions of properties related to *observable features* that therefore can be observed through the associated sensors.

$$\begin{aligned} \texttt{ObservableFeature} \sqsubseteq\ & \texttt{DUL:Role} \sqcap \\ & \texttt{SSN:FeatureOfInterest} \sqcap \\ & \exists\, \texttt{isObservableThrough.SSN:Sensor} \sqcap \\ & \exists\, \texttt{isRelatedToObject.DUL:Object} \sqcap \\ & \exists\, \texttt{hasObservableProperty.SSN:Property} \end{aligned}$$

$$\begin{aligned} \texttt{ObservableProperty} \sqsubseteq\ & \texttt{DUL:Role} \sqcap \\ & \texttt{SSN:Property} \sqcap \\ & \exists\, \texttt{isObservableThrough.SSN:Sensor} \sqcap \\ & \exists\, \texttt{SSN:observes.SSN:Property} \end{aligned}$$

Thus, the interpretation used to identify these elements is the following: *A particular property x of a physical object y of the environment becomes observable (i.e., it can play the role) if there exists a sensing device z deployed on the same object y and if such sensing device z is capable of observing the property x.* According to this semantics a reasoner can dynamically infer the set of *observable features*, together with the related *observable properties*, by analyzing the particular configuration of the considered environment, as it will be shown in the next section.

The information produced by *observable features* is modeled by leveraging the SSO ontology design pattern [6] and by leveraging/extending concepts like `SSN:Observation` and `SSN:SensorOutput` that belong to the SSN ontology. Observed data is processed in order to identify a set of relevant *events* and/or *activities* that must be managed by the system in some way. Thus, the *observation context* provides a definitions of these abstract concepts by extending the `DUL:Event` class.

Figure 7 shows part of the concepts defined for the observation context, that extend the `DUL:Event` class. According to the documentation, the `DUL:event` class is defined as the set any physical, social or mental process, event or state. This class has been extended by introducing the concept of `ObservedEvent` and `Activity` (as subclass of `DUL:Action`). Broadly speaking, the `ObservedEvent` class represents observable events that concern the status/behavior of some *objects* of the environment. The class `Activity` models *actions* that can be performed by a particular `DUL:PhysicalAgent` and that have effects on a set of *objects* of the environment.

$$\begin{aligned} \texttt{ObservedEvent} \sqsubseteq\ & \texttt{DUL:Event} \sqcap \\ & \exists\, \texttt{concerns.DUL:Object} \end{aligned}$$

$$\begin{aligned} \texttt{Activity} \sqsubseteq\ & \texttt{DUL:Event} \sqcap \\ & \exists\, \texttt{concerns.Object} \sqcap \\ & \exists\, \texttt{DUL:hasParticipant.DUL:PhysicalAgent} \end{aligned}$$

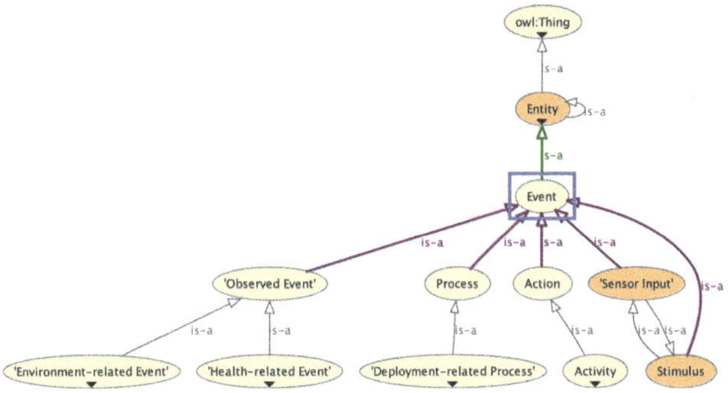

Fig. 7. Partial view of the KOaLa observation context with respect to the extension of the `DUL:Event` class.

The class `ObservedEvent` is further specialized in order to model two particular types of *observed events* the system is interested with: (i) *environment-related events*; (ii) *health-related events*. The class `EnvironmentRelatedEvent` models the set of *observable events* concerning the state of the objects composing the environment (e.g. `WaterLeak`, `HighTemperature`, `MotionDetection`). The class `HealthRelatedEvent` instead models the set of *observable events* concerning the health-status of a monitored person within the environment (e.g. `HighStress`, `Fallen`).

Similarly, the class `Activity` is further specialized in order to model particular types of *activities* the system is interested in: (i) *acting activities*; (ii) *monitoring activities*. The class `ActingActivity` models the set of observable activities that can be performed by a person (but also a robot) within the environment (e.g. `Sleeping`, `Cooking`, `WatchingTV`). Instead, the class `MonitoringActivity` is mainly referred to the system capabilities and models the set of *monitoring operations/functions* a system is capable to perform within a home environment (e.g. `BloodPressureMonitoring`, `SleepMonitoring` or `TherapyReminder`). This type of activities is necessary to introduce *self-diagnosis* mechanisms. The system recognizes the set of *operations/functions* that can be actually performed according to the specific configuration of the environment, i.e. *system capabilities*.

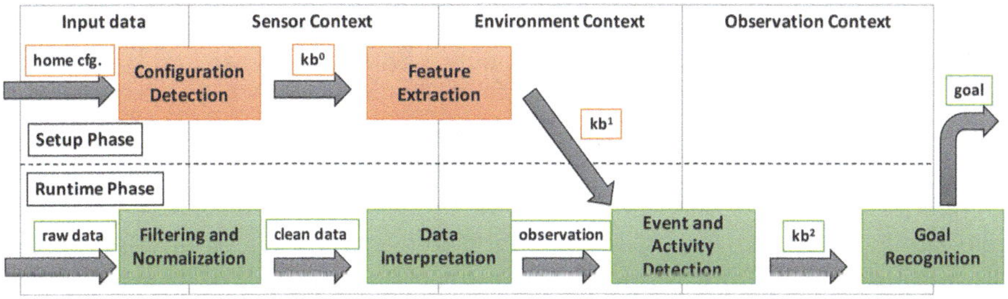

Fig. 8. The KOaLa data processing flow with the related reasoning modules.

More in general the distinction between *observed events* and *activities* allows the system to reason about the causality between *situations* and *actions* and dynamically identify possible ways of reacting to detected events and/or supporting detected actions.

5 Sensor Data Processing and Knowledge Extraction

Given the semantics defined by the ontology described in the previous section (TBox), a KOaLa reasoner can *interpret* data received from sensors and elaborate the resulting information to build an abstraction of the environment. Such abstraction is represented by means of a Knowledge Base (KB) which *instantiates* the ontology (ABox). The KB is incrementally built and iteratively refined by *inferring* additional information from sensors when possible. The resulting KB provides a complete representation of an application scenario which can be further analyzed in order to extract *planning goals.* Planning goals are particular types of event that enable a *proactive* support of the daily home living of a person.

The process depicted in Fig. 8 shows the overall data processing and information extraction mechanism of KOaLa. This process manages the information flow aimed at building and maintaining the knowledge about the application scenario. As shown in Fig. 8, it is characterized by two reasoning phases. A *setup phase* analyses information about the specific configuration of the monitored environment. This analysis performs a sort of *self-diagnosis* of the system and generates an initial knowledge base (tagged with kb^1 in Fig. 8) which characterizes the capabilities of the system according to the detected configuration. A *runtime phase* refines the knowledge base (i.e., kb^1) by processing observations coming from sensors. The resulting knowledge base (tagged with kb^2 in Fig. 8) characterizes the events and activities detected inside the environment. A *goal recognition module* elaborates such knowledge in order to trigger *planning goals* the acting module must deal with. Specifically, such goals are managed by the planning and execution process to achieve an autonomous and proactive behavior of the environment. Both the setup phase and the runtime phase can be seen as a *pipeline* of reasoning modules each of which elaborates data by means of a set of dedicated *inference rules.* Such rules extend the ontology described in the previous section and define a semantics to link data belonging to different contexts. As Fig. 8 points out, the interpretation of these rules allows a KOaLa reasoner to incrementally abstract and infer additional knowledge by *following* the hierarchical structure of contexts.

The setup phase takes as input a *home configuration specification* which describes the structure of the environment the system is going to monitor as well as the set of sensors composing the network and their deployment. The *Configuration Detection* module interprets this specification according to the sensor context of the ontology. It generates an initial knowledge base (tagged with kb^0 in Fig. 8) characterizing the sensors and the objects composing the environment, their deployment and the properties that can be observed.

The *Feature Extraction* module refines such knowledge in order to identify the observable features with the related observable properties and the set of events and activities that can be actually observed and performed. Once the setup phase is complete, the system enters the *runtime phase* which iteratively refines the generated knowledge by handling *observations*. Let us consider a simple home configuration specification like the one represented by the xml code below.

```
<environment id="0x0184e8a0">
 <roomlist>
  <room id="0" type="http://.../koala#Kitchen">
   <object id="1" type="http://.../koala#Fridge">
    <sensor id= "2" type="http://.../koala#Gap" />
   </object>
   <object id="3" type="http://.../koala#Microwave">
    <sensor id="7" type="http://.../koala#Switch" />
   </object>
   <object id="5" type="http://.../koala#Table">
    <sensor id= "6" type="http://.../koala#Pir" />
   </object>
   <sensor id="4" type="http://.../koala#Pir" />
  </room>
  <room id="8" type="http://.../koala#LivingRoom">
   <object id="9" type="http://.../koala#Sofa">
    <sensor id= "10" type="http://.../koala#Pir" />
   </object>
   <object id="11" type="http://.../koala#TV">
    <sensor id="12" type="http://.../koala#Switch" />
   </object>
   <object id="13" type="http://.../koala#Window">
    <sensor id="14" type="http://.../koala#Gap" />
   </object>
  </room>
 </roomlist>
 <topology>
  <adjacent>
   <node room-id="0" />
   <node room-id="8" />
  </adjacent>
 </topology>
</environment>
```

The document describes the rooms that compose the house (see the tag `roomlist`) and the topology of the house (see the tag `topology`). For each room, the document specifies the objects located inside (see the tag `object`, sub-tag of `room`) and the deployed sensors (see the tag `sensor`). Sensors can be deployed either on single objects (see the object with id "1" as an example) or can be deployed on the room itself (see the room with id "0" as an example). It is possible to notice that each element of the document is associated with meta-information specifying the element type with respect to the KOaLa ontology (see the attribute `type` of the tags). The *Configuration Detection* module of Fig. 8 leverages such meta-information to *interpret* the home configuration specification and properly initialize the KB. The kb^0 will contain an individual for each *room*, *object* and *sensor* of the home configuration specification. The module will also add the individuals of the class `SSN:Platform` and *assert* the properties `hasPlatform` and `SSN:deployedOnPlatform` in order to properly represent the configuration of the environment. For example, the kb^0 will contain an individual p of the class SSN:Platform, and individual k of the class Kitchen, an individual s of the class Pir (representing the sensor with id "4") and the assertions (k hasPlatform p) and (s SSN:deployedOnPlatform p).

The kb^0 is refined by the *Feature Extraction* module which adds the individuals of the classes ObservableFeature and ObservableProperties. The kb^1 will contain an individual of the class ObservableFeature for each element of the environment with a sensor deployed on it. Then, such individuals are associated to individuals of the class ObservableProperty. According, to the KOaLa ontology, for example, individuals of the class Kitchen are associated to the property Temperature and individuals of the class Pir can observe the property Temperature. Thus, kb^1 will contain an individual of the class ObservableProperty associating the individual of the class Kitchen to the property that can be observed through the deployed sensor of class Pir.

A sensor network like the one described in the GiraffPlus case study can be composed by several types of sensor that can use different communication protocols and different data formats. In addition, sensors may *fail* and/or produce *noise* that must be ignored with respect to monitoring objectives. Thus, as shown in Fig. 8, the first step in the "runtime pipeline" is to elaborate *raw data* coming from sensors and generate *clean data* representing useful information that can be analyzed. The *Data Filtering and Normalization* module is directly connected to the physical sensors and it is responsible for filtering and normalizing received data. It is the only one module of the envisaged system strictly dependent on the particular protocol and data format used by available sensors (e.g., MQTT and Z-Wave in the GiraffPlus case study). It encapsulates the sensor-dependent logic for interpreting low-level data and generate normalized data that can be interpreted by KOaLa. The *Data Interpretation* module interprets data according to the sensor context in order to determine the property a received value refers to, identify the related sensor within the environment and generate an observation.

Clean data gathered from sensors consists of a value, a label encoding the property the value refers to and an ID identifying the source sensor. *Data Interpretation* elaborates this information in order to generate observations. Specifically, the interpretation process leverages the SSO ontology design pattern [6] in order to generate the individuals needed to represent the SSN:SensorOuput, the SSN:ObservedValue characterizing a particular observed property, and properly associate these individuals to the generated SSN:Observation. Then, the *Event and Activity Detection* module interprets such an observation by taking into account the knowledge about the environment (i.e., the kb^1 in Fig. 8). Different inference rules detect different types of events and activities according to the particular set of features and properties involved within an observation. Once that all the observations have been processed, a *Goal Recognition* module reasons on the overall knowledge of the environment in order to eventually identify and trigger a set of *planning goals*. It analyzes the knowledge resulting from the configuration and received observations (i.e., the knowledge base denoted by kb^2 in Fig. 8) and generates a set of operations the system can perform in order to react to some detected events or support some detected activities.

This knowledge processing mechanism has been developed by means of the Apache Jena software library.[3] Specifically, we have used the *Ontology API* to manage the knowledge base data structure and the *Inference API* to build a rule-based inference engine. The next subsections provide some details about the development of the reasoning engine and the definition of the *inference rules*.

5.1 Feature Extraction

The feature extraction module is part of the *self-diagnosis* mechanism performed during the *setup phase*. It does not directly elaborate sensor data but rather it analyzes the *house configuration specification* in order to detect the actual capabilities of the system according to the particular configuration. To detect the capabilities of the system it is necessary to identify the elements that can actually "produce" information and therefore identify the properties that can be actually monitored. Thus, the first step is to detect *observable features* by analyzing the structure of the environment and the deployment of sensors (i.e., the kb^0 generated by the *Configuration Detection* module). As already described in the ontology section, the concept of `ObservableFeature` characterizes particular objects of the application scenario that can play the *role* of "being observable" through some sensors. The inference rule used to recognize such elements can be formally described by the formula below.[4]

$$
\begin{array}{rl}
\texttt{DUL:Object(o)}\wedge & \\
\texttt{SSN:Platform(p)}\wedge & \\
\texttt{SSN:Deployment(d)}\wedge & \\
\texttt{DUL:hasPart(o, p)}\wedge & \\
\texttt{SSN:hasDeployment(s, d)}\wedge & \\
\texttt{SSN:deployedOn(d, p)} \longrightarrow & \texttt{ObservableFeature(x)}\wedge \\
& \texttt{hasObservableFeature(o, x)}\wedge \\
& \texttt{isObservableThrough(x, s)}
\end{array}
$$

An object can be monitored and therefore it represents an *interesting feature* in the SSN sense, if and only if there is a sensor deployed on it. According to this rule, if there is a sensor s deployed on a platform p which is part of an object o, then the object o is an observable feature. There could be several sensors deployed on a same object and different sensors can observe different properties of an object. Thus, our design choice was to model *inferred* observable features as distinct individuals associating an object to the deployed sensors that make the object "observable". Similarly, this design choice allows also KOaLa to associate observable properties of an object to the specific sensors that make such properties "observable". A property of an observable feature can be monitored if and only if an associated sensor is actually capable of observing that particular type of property. Similarly to `ObservableFeature`, the concept of `ObservableProperty` characterizes properties of an object that can be actually observed, namely, properties that can play the role of "being observable". The formula below describes the inference rule used to recognize observable properties.

---

3 https://jena.apache.org/.

4 The rule language of Apache Jena does not support SWRL, it leverages the RDF syntax and processes a knowledge base by "navigating" the corresponding RDF graph representation.

```
              ObservableFeature(f)∧
                    DUL:Object(o)∧
                  DUL:Property(p)∧
                    SSN:Sensor(s)∧
     hasObservableFeature(o, f)∧
           DUL:hasProperty(o, p)∧
      isObservableThrough(f, s)∧
               SSN:observes(s, p) → ObservableProperty(x)∧
                                    hasObsProperty(x, p)∧
                                    SSN:observes(x, p)
```

Once that the system knows the observable features of the environment (i.e., elements capable of generating information) and the related observable properties, it is possible to determine the set of *events* that can be actually observed and the set of *activities* that can be actually recognized within the environment. Let us consider for example a knowledge base containing an individual classified as `Kitchen` which is associated with a property like `Temperature` according to the ontology. If such an individual is associated with an `ObservableFeature` and if this observable feature is associated with an `ObservableProperty` which is in turn associated to a property like `Temperature` then, it is possible to conclude that the system can monitor the temperature level of the kitchen. The capability of monitoring a property like `Temperature` is associated to events like `LowTemperature` or `HighTemperature`. Thus, a system capable of monitoring the property `Temperature` of an environment object o (e.g., a `Kitchen`) is also capable of detecting the events `LowTemperature` and `HighTemperature` concerning the status of object o. Other types of event or activity the system is capable of detecting can be recognized in similar ways by applying different rules. The knowledge generated by such self-diagnosis mechanism can be used to dynamically configure the *control model* of the planning and execution process by leveraging an approach similar to the one described in [10] for a reconfigurable manufacturing system.

5.2 Event and Activity Detection

The *Event and Activity Detection* module is part of the *runtime phase* and it is responsible for integrating configuration knowledge about the environment (i.e. the outcome of the *setup phase*) with the received observations. This module iteratively refines the current knowledge every time an observation is received. It encapsulates a set of inference rules that process observations according to the known observable features of the environment and the related properties. These rules leverage the KOaLa ontology in order to take into account "known" general properties/parameters of the application context and the set of correlated events and/or activities.

```
          SSN:Observation(o)∧
   SSN:FeatureOfInterest(f)∧
        SSN:SensorOutput(d)∧
    SSN:ObservationValue(v)∧
              DUL:Object(e)∧
SSN:featureOfInterest(o, f)∧
 hasObservableFeature(e, f)∧
        SSN:hasOutput(o, d)∧
         SSN:hasValue(d, v)∧
      v < tempLowerBound → LowTemperature(x)∧
                           concerns(x, e)∧
                           SSN:isProducedBy(x, o)
```

The rule above represents an example of how an event like `LowTemperature` can be detected by processing observations concerning the temperature of an observable feature of the environment. Rules can evaluate *parameters/constants* that model particular features of the considered context. For example the variable `tempLowerBound` represents the expected lower bound of a *regular* temperature inside a room. Thus if the observed temperature of an element of the environment like e.g., the `Kitchen`, has a value lower than the `tempLowerBound`, then the rule can generate the event `LowTemperature`. Similar rules have been defined to detect other types of event like the *presence* of a person inside a room or *high hearth-rate* of a person as well as other types of activity like *cooking*, *sleeping* or *watching tv*.

5.3 Goal Recognition

The *Goal Recognition* module is the last step of the *runtime phase*. This module aims at identifying operations the system can perform to proactively support the daily home living of a person, i.e. *planning goals*. This module is not responsible for refining the knowledge but rather, it is responsible for "closing the loop" between knowledge processing and planning. The generation of a well-defined semantic representation of the environment allows the *Goal Recognition* module to more effectively analyze the detected status and *infer* operations to trigger the planning and execution process. In general, the actual set of operations the system can perform (i.e., the actual set of *planning goals*) to support *proactivity* depends on the particular configuration of the environment, i.e., *system capabilities*. Such operations are represented as *planning goals* and are associated to particular combinations of events and/or activities, called *situations*. Thus, planning goals are generated by means of dedicated rules that analyze the knowledge in order to identify such situations. There can be different types of situation that can generate planning goals. Such situations can range from operations related to supporting activities like therapy reminders or comfort management to operations related to emergency and health management. For example, if the system detects *low temperature* and *presence* inside a particular room of the house, then it can proactively perform an operation to *heat* the room in order to improve the comfort of a person. Similarly, if a *fallen* or a very *high/low hearth-rate* of a person is detected then the system can proactively perform an *emergency call* to a doctor or a relative in order to promptly assist a person.

6 Towards Proactivity and Plan Synthesis

Looking back to Fig. 1, the knowledge processing mechanism described in the previous section allows the *KOaLa Semantic Module* to identify planning goals that can be sent to the *KOaLa Acting Module* in order to achieve *proactivity*. As already said, the *Goal Recognition* module plays a key role in this pipeline. Indeed, it is responsible for dynamically connecting the *representation side* with the *acting side* of KOaLa. The *KOaLa Acting Module* is responsible for synthesizing and executing the operations needed to achieve the desired goals. This module leverages timeline-based planning and execution technologies [7,8] to dynamically generate and maintain a temporal plan which specifies the sequences of operations that must be executed over time. The planning and execution process relies on a *temporal model* of the monitored environment characterizing general rules in terms of constraints and operations that must be followed to synthesize valid plans. This temporal model is dynamically generated according to the detected configuration of the environment (i.e., the knowledge base of the *KOaLa Semantic Module*). The set of planning goals that can be achieved and operations that can be performed depends on the actual configuration of the environment.

The key objective of the *KOaLa Acting Module* is to dynamically control the environment by managing and integrating into a valid temporal plan, *real-time goals* and *assistive goals*. Real-time goals concern operations the system must perform to proactively react to an event or support activities. For example, planning goals like the ones described above that "ask" the system to *heat* the room to improve the comfort of a person or "ask" the system to make an *emergency call* to alert *secondary users* that a *fallen* has been detected, represent real-time goals that KOaLa is currently capable to manage. Assistive goals instead represent operations that can be planned in advance according to the detected configuration and the particular *needs* of a person (i.e., *user profile*). Let us consider for example a person with dietary restrictions that must follow a therapy. According to this information, KOaLa must synthesize the set of reminding and monitoring operations needed to support the user within an entire day.

7 Conclusions and Future Works

This paper proposes a novel approach to the integration of knowledge representation and reasoning with planning and execution in order to realize a proactive monitoring and control system for a new generation home environment. The current state of KOaLa supports the management of real-time events and activities detected by the reasoning module. In the paper we have mainly described the work dome to realize the KOaLa Semantic Module. Future works will concern the management of assistive goals in order to integrate daily planned activities with real-time events in the case of the assistance to an older person at home.

In this respect, the management of assistive goals implies the capability of managing and reasoning on *user profiles* in order to dynamically identify the set of *planning goals* that will compose the daily plan of the system.

References

1. Coradeschi, S., Cesta, A., Cortellessa, G., Coraci, L., Gonzalez, J., Karlsson, L., Furfari, F., Loutfi, A., Orlandini, A., Palumbo, F., Pecora, F., von Rump, S., Štimec, A., Ullberg, J., Ötslund, B.: Giraffplus: combining social interaction and long term monitoring for promoting independent living. In: 2013 6th International Conference on Human System Interactions (HSI), June 2013, pp. 578–585 (2013)
2. Cesta, A., Cortellessa, G., Fracasso, F., Orlandini, A., Turno, M.: User needs and preferences on AAL systems that support older adults and their carers. J. Ambient. Intell. Smart Environ. **10**(1), 49–70 (2018)
3. Cortellessa, G., Fracasso, F., Sorrentino, A., Orlandini, A., Bernardi, G., Coraci, L., Benedictis, R.D., Cesta, A.: Robin, a telepresence robot to support older users monitoring and social inclusion: development and evaluation. Telemed. e-Health **24**, 145 (2017)
4. Barsocchi, P., Cesta, A., Coraci, L., Cortellessa, G., Benedictis, R.D., Fracasso, F., Rosa, D.L., Orlandini, A., Palumbo, F.: The giraffplus experience: from laboratory settings to test sites robustness (short paper). In: 2016 5th IEEE International Conference on Cloud Networking (Cloudnet), October 2016, pp. 192–195 (2016)
5. Compton, M., Barnaghi, P., Bermudez, L., García-Castro, R., Corcho, O., Cox, S., Graybeal, J., Hauswirth, M., Henson, C., Herzog, A., Huang, V., Janowicz, K., Kelsey, W.D., Phuoc, D.L., Lefort, L., Leggieri, M., Neuhaus, H., Nikolov, A., Page, K., Passant, A., Sheth, A., Taylor, K.: The SSN ontology of the W3C semantic sensor network incubator group. Web Semant. Sci. Serv. Agents World Wide Web **17**(Supplement C), 25–32 (2012)
6. Janowicz, K., Compton, M.: The stimulus-sensor-observation ontology design pattern and its integration into the semantic sensor network ontology. In: Proceedings of the 3rd International Conference on Semantic Sensor Networks (SSN 2010), Aachen, Germany, pp. 64–78. CEUR-WS.org, Germany (2010)
7. Cialdea Mayer, M., Orlandini, A., Umbrico, A.: Planning and execution with flexible timelines: a formal account. Acta Informatica **53**(6–8), 649–680 (2016)
8. Umbrico, A., Cesta, A., Mayer, M.C., Orlandini, A.: PLATINUm: a new framework for planning and acting. In: Esposito, F., Basili, R., Ferilli, S., Lisi, F. (eds.) Advances in Artificial Intelligence, AI*IA 2017. Lecture Notes in Computer Science, vol. 10640. Springer, Cham (2017)
9. Pecora, F., Cirillo, M., Dell'Osa, F., Ullberg, J., Saffiotti, A.: A constraint-based approach for proactive, context-aware human support. J. Ambient. Intell. Smart Environ. **4**(4), 347–367 (2012)
10. Borgo, S., Cesta, A., Orlandini, A., Umbrico, A.: A planning-based architecture for a reconfigurable manufacturing system. In: The 26th International Conference on Automated Planning and Scheduling (ICAPS) (2016)

Learning by Demonstration with Baxter Humanoid

Othman Al-Abdulqader(✉) and Vishwanthan Mohan

University of Essex, Colchester, UK
othman93ali@yahoo.com

Abstract. Despite robots' high capabilities to perform various tasks, they require programming for each new task. Learning by demonstration (like humans) enables robots to perform new tasks without the need for well-crafted programs. The aim of this project is to apply learning by demonstration method in which the user performs two simple tasks in front of the robot and the robot imitates them. Baxter humanoid is used in the project; it has two arms with seven joints on each arm, a face display and a non-programmable wheeled base. The task inspired from studies on animal cognition to use a tool appropriately in order to fetch/obtain a reward, following the demonstration of the human. The overall learning by demonstration system integrated two core sensory-motor loops: (1) Action Observation: to observe/represent user demonstration, the VICON system in the robotics arena was integrated to the Baxter so as to receive information about objects in the scene, movement trajectory, ensuing consequences; (2) Action Generation: to make the robot reproduce the observed demonstration of the tool use. Baxter imitated the user's demonstrations successfully and the required results achieved. The developed system is in general task agnostic, and can be further extended in the domain of motor primitives by moving from trajectories to movement shapes. Further, the system is useful in industrial assembly tasks since the robot does not need programming for each new task which saves time and simplifies robot work.

Keywords: Humanoid · Baxter · Robotics · Learning by demonstration

1 Introduction

Imitation Learning is a technology which enables robots to perform a task based on the user's demonstrations. Usually, robots perform tasks based on the user's programming for each new task which is time consuming and very limited to certain tasks. For this reason, Imitation Learning saves time and resources. According to Ko et al. [1] there are industrial benefits for Imitation Learning. This research project implements Imitation Learning in a robot in order to perform a series of simple tasks based on human demonstrations. Baxter is the name of the robot that is used in the project. It is a humanoid robot developed by rethink robotics in order to be used mostly in industrial companies. Some of them used Baxter in their factories for various tasks such as machine tending, packaging and material handling. It is designed to be safe, flexible and powerful, with 7 degrees of freedom on each arm and the arm movements are almost similar to those

K. Arai et al. (Eds.): IntelliSys 2018, AISC 868, pp. 770–790, 2019.
https://doi.org/10.1007/978-3-030-01054-6_54

of human arms [2]. Baxter has two arms with grippers, a multipurpose display, a non-programmable wheeled base and three cameras.

As is shown in Fig. 1, Baxter has two arms: right and left; both arms are programmable and are used in order to perform multiple tasks. Each arm consists of seven different joints that form the shoulder, elbow and wrist. Symbols S0 and S1 control the shoulder joints, while E0 and E1 represent the elbow and the wrist is represented by the W0, W1 and W2 joints.

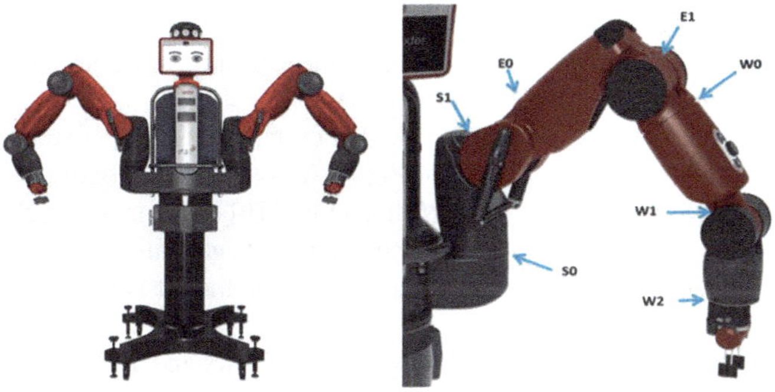

Fig. 1. Baxter Robot (Left) and Baxter arm (Right) [3, 4].

The project has two main tasks: the first task is to rotate the handle of a receptacle for an object to come out of the tool. The second task is to achieve the gained object. Baxter learns how to perform the tasks after a user demonstration of the first task. In order for Baxter to know the locations of the handle and the object, the VICON system is used; it is a vision technology which provides precise Cartesian coordinates of anti-symmetrical markers forming a model placed on any object. It is integrated with Baxter to be its vision source instead of the cameras that Baxter already has, since VICON is more accurate, faster and easier to implement. It is usually used in movies and games, where the actor wears reflective markers in order to track his movements. According to Merriaux et al. [5] VICON has a system error less than 2 mm. In this project, a model of markers is attached to the handle, Baxter's head and the toy, in order to track the movements of each object. The detection and positioning of the objects are carried out by cameras installed in the robotics lab. The VICON system provides the robot with positions of the handle in 3D space and tracks its movements throughout the task. These values are not usable for Baxter, due to the fact that Baxter and VICON have different origins. Therefore, a VICON-Baxter integration is developed in order to convert VICON readings. The arm joints angles required for each movement are generated by the Inverse Kinematics service. The generated joints angles are then sent to the arm for execution through the Joint Command service [3].

The idea for this method is taken from the field of behavioural psychology, specifically the study of animal/infant cognition. This project has benefits for the Industry, since the robot does not require programming for each new task. In order to achieve the tasks, a tool is built from a cardboard box and other materials. Figure 2 shows a 3D model of the tool and its parts.

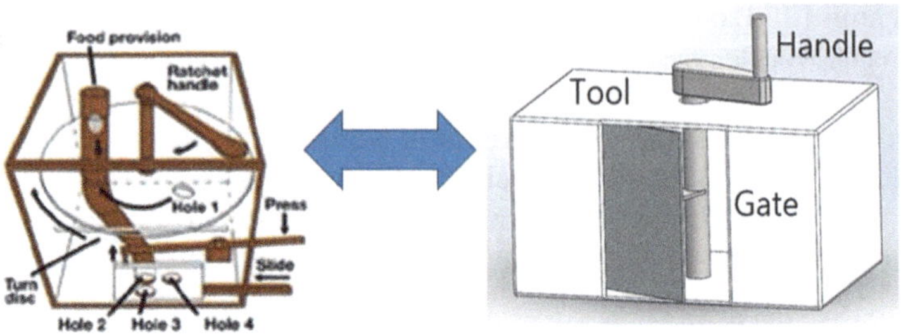

Fig. 2. Turn disc tool (Left) 3D model of the tool and Gate opened (Right).

The turn disc tool used by Whiten et al. [16] on left side and a 3D representation of the tool used in the project are shown in Fig. 2. It is shown that both tools are related and the project's tool is inspired from Whiten et al. [16] experiment on chimpanzees.

The tool consists of three main parts: the handle, the gate and the pusher. The robot has to rotate the handle in order to open the gate, while the pusher is pushing an object out of the tool.

The next chapter is the literature review which reviews and discusses previous relevant projects and highlights how this project is different. Following that is the Implementation one, which shows the execution and implementation of the project's tasks. The Results and Discussion chapter reviews the performance of the robot and whether the tasks were achieved or not. The next one is the Conclusion chapter which provides an overview of the implemented project. The final part includes the references of the materials used in this project.

2 Literature Review

Robots have to know what to imitate and how to imitate a certain task before the imitation of any task. Manual programming is one of the methods which makes a robot knows how to perform a certain task; the robot performs a task by executing a set of instructions loaded in the robot's memory. Even though using this method has some benefits, the drawback of programming each new task takes a large amount of time. On the other hand, Imitation Learning enables robots to replicate the behaviours being demonstrated by users. Even though using this method have some drawbacks, the benefits of not programming the robot for each new task enables non-programmers to deal with the robot easily and it saves a lot of time [6]. This chapter will review and discuss projects and publications related to learning by demonstration project.

2.1 What to Imitate

The robot has to determine the relevant parts to accomplish the task from a demonstration. There are different methods that enable robots to focus on the task only and ignore irrelevant things, such as, VICON system which uses model of markers to specify the objects of interest. In fact, robots cannot perform tasks the same way the user performed them due to the differences in the physical layout of the robot and the user [7]. For example, the user is demonstrating an object grasping task using his hand; the robot has to know that the task is grasping an object and implements it in a way that suits its hardware. Bentivegna, Atkeson and Cheng developed a framework that enables the robot to learn tasks by user's demonstrations [8]. Two tasks used to test the framework: marble maze and air hockey. The framework uses robot's cameras in order to detect and localize objects. Since the tasks are on flat surface, it is possible to make a perspective mapping between the images coming from robot's cameras with the surface of air hockey task. After mapping the surface, the objects were identified by their shape and colour. Unlike the previous project, Billard, Calinon and Guenter used Hidden Markov Models (HMV) to determine whether to imitate the gesture or the hand movements in a demonstration [9]. The overall system consists of three main stages: the first stage is encoding the signals after removing the noise. The second stage is a selection of the task constraints, while the third stage is the implementation of the optimal trajectory. Two web cams were used to determine the positions of objects based on their colour. While the previous projects used cameras as their vision system, Rozo, Pablo and Carme developed a framework that uses haptic feedback device to demonstrate tasks for the robot [10]. The framework is achieved by teleporting the robot's arm with a haptic device which provides the user with a forces/torques reflection. The task is to extract a metallic ball from a hole in a container by moving the container in different states. The robot and the user depend completely on the force/torque feedback, that is, no vision system applied.

The project implemented by Bentivegna, Atkeson and Cheng [8] uses mapping of the surface which makes the implementation accurate but very limited to that surface as well as using robot's cameras for moving targets is very difficult and complex. While Billard, Calinon and Guenter [9] implemented a very good framework to manipulate simple objects based on kinesthetic demonstration, but the data collected contain noise and if the noise is not correctly filtered it could affect the accuracy of objects positioning. Unlike the previous projects where vision is used, using haptic device [10] is a new and effective way to guide the robot throughout the task but haptic feedback is not the way that humans, infants or animals imitate each other through. Vision is the natural way of communication and learning from others, haptic feedback device makes the task very limited and has no area of exploration. Learning by Demonstration project uses VICON system to determine the objects to be tracked accurately and efficiently, in addition, any object with a model of markers could be tracked and followed which represents objects generalization and freedom.

2.2 How to Imitate

It is the process of knowing how to imitate the user's observed behaviours from a demonstration. There are many methods in which a robot imitates a demonstration: The first one is when the robot records the user actions using cameras or motion tracking systems then imitates those actions which is similar to the projects done by Kim et al. and Kulic et al. [11, 12]. The second method is Kinesthetic teaching where the user guides the robot in order to do a certain task physically on the robot's own body, it is implemented by Sauser et al. [13]. Immersive teleportation is another method where the user uses a joystick or a remote-control device in order to guide the robot during the task, it is implemented by Babič, Hale and Oztop [14]. The framework proposed by Bentivegna, Atkeson and Cheng [8] uses primitive recognition module which is developed by Bentivegna and Atkeson for the robot to divide the observed task into parts called primitives [15]. The framework consists of three main stages: Primitive Selection, Sub-goal Generation and Action Generation. It starts with Primitive Selection which selects the primitive type. Sub-goal Generation determines the goal of the selected primitive. The last stage is Action Generation which sends commands to the actuators to achieve the goals specified. The implementation of the proposed model by Billard, Calinon and Guenter [9] is done by demonstrating a task five times then the collected data will go through three main stages: the first stage is encoding the signals after removing the noise. The second stage is a selection of the task constraints, while the third stage is the implementation of the optimal trajectory. Then the robot reproduces the task by displacing the objects in multiple locations. Rozo, Pablo and Carme [10] used a different approach to teach the robot by demonstration; the user demonstrates the task several times using a haptic device teleported with the robot's arm. These demonstrations are reproduced by the robot through a modified version of Gaussian Mixture Regression. For the projects developed by [8–10] the user demonstrates the task multiple times before the robot could imitate the task, while this project uses Inverse Kinematics which is accurate and easy to use, it also doesn't require repeating the task multiple times for the robot to be able to imitate the proposed demonstration.

2.3 Animal/Infant Cognition

The project's tool design is inspired by Animal/Infant cognition and Whiten turn disc task. According to Whiten et al., animals like chimpanzees have behaviour similar to humans; that is, imitating others in order to perform certain tasks [16]. Whiten performed multiple experiments to observe chimpanzees' behaviour, one experiment includes the use of a turn disc tool to extract food as a reward and it is performed on a group of chimpanzees. Figure 3 shows the Tool design and parts.

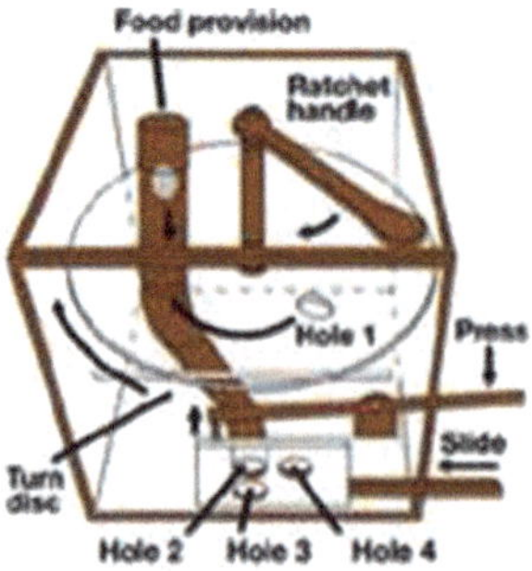

Fig. 3. Tool design by Whiten et al. [16] © Elsevier.

As is shown in Fig. 3, the tool has a food pipe that is blocked by a rotating disc, a chimpanzee was trained to rotate the handle so the hole in the disc will be aligned with the pipe in order to pass the food to the next point. Then the chimpanzee has to either press or slide the stick for the food to come out of the tool, the food is being hidden inside the tool all the time. After that, the trained chimpanzee was returned to his group so they could imitate it. The group of chimpanzees failed to obtain the food before any of them was trained, but after they saw the trained chimpanzee performing the task they imitated his actions successfully to retrieve the food out of the tool. In addition, infants tend to imitate adults as well, in an experiment to discover infant abilities Fagard et al. [17] found that infants at age of 18 months were able to imitate a demonstration of an adult and getting an object out of a tool. Both studies inspired the design of the project's tool which also contains a handle and an object as a reward for the robot if the tool used properly and the handle is rotated correctly.

3 Implementation

This chapter discusses the elements of the task and how each part is implemented. It also points out how the tool is designed and created. The chapter starts by providing a detailed description of the task goals and procedure. After that is an explanation on how the tool is implemented and its parts. The chapter ends by showing how the VICON system works and the project overall implementation.

3.1 Task Description

The task begins when the user rotates the handle in front of the robot to demonstrate how it works. The tool works by moving the handle in a circular way from one point to

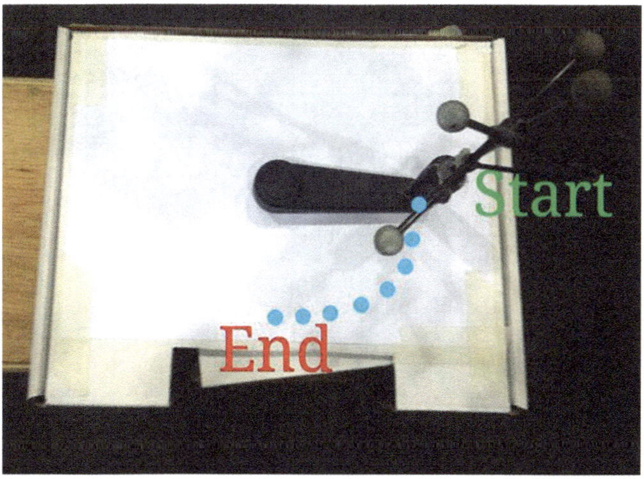

Fig. 4. The handle route.

another towards the robot; this movement opens the gate. Meanwhile, a pusher inside the tool is responsible for pushing objects out of the tool for the robot to get as a reward.

Figure 4 shows the route of the handle during the demonstration as well as the imitation. As shown in the figure, the starting point is coloured in green and the points in blue represent the locations where the handle moves, while the end point is coloured in red. The user demonstration of the handle movements is shown in Figs. 5, 6, 7, 8.

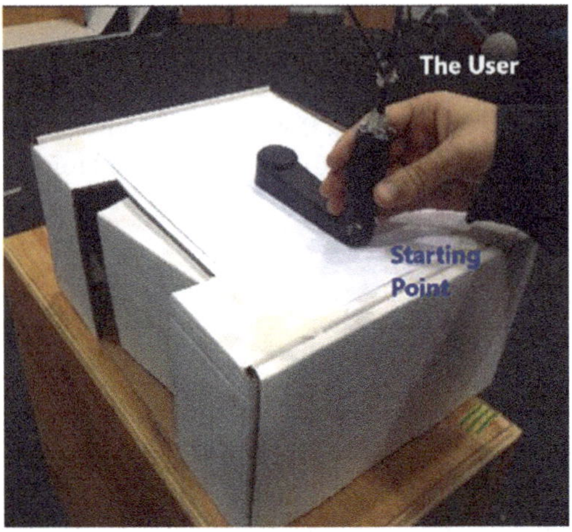

Fig. 5. User demonstration 1.

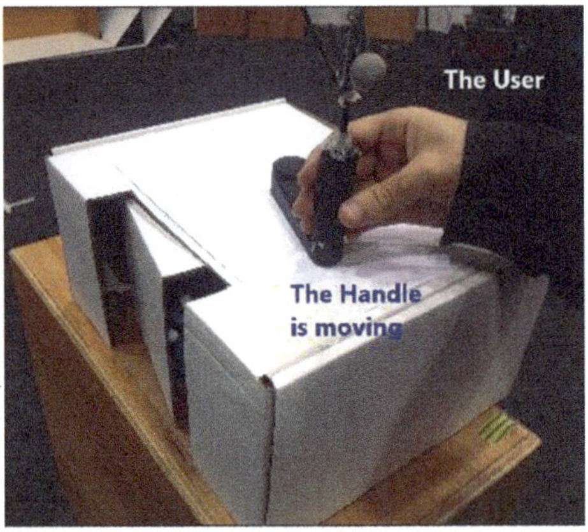

Fig. 6. User demonstration 2.

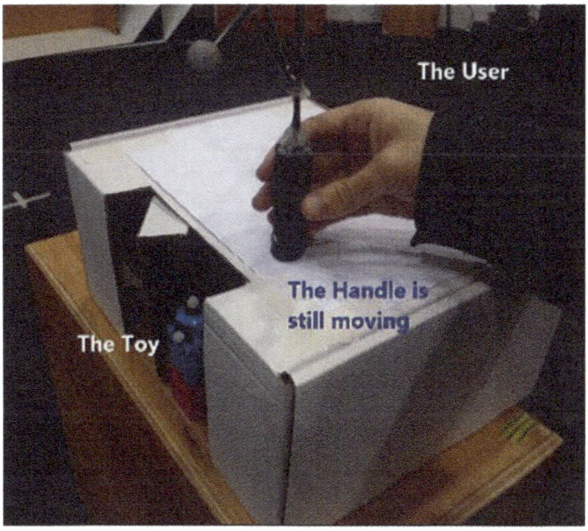

Fig. 7. User demonstration 3.

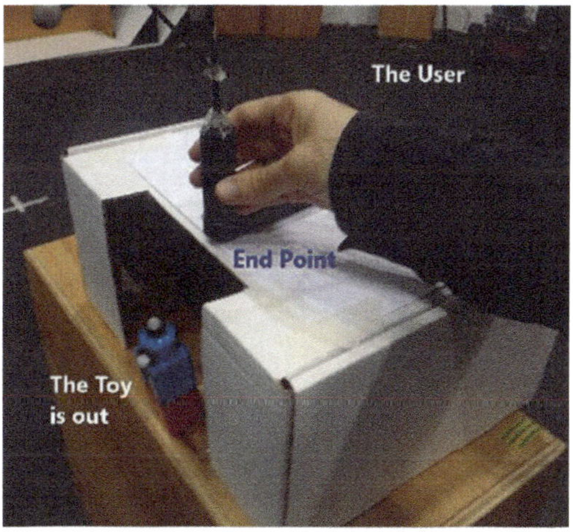

Fig. 8. User demonstration 4.

Figure 5 shows the beginning of the user demonstration where the user holds the handle on the starting point and illustrates the functionality of the tool to Baxter. The toy is completely hidden inside the tool at this stage.

As is shown in Fig. 6, the user rotates the handle slowly to open the gate which is half open, while the toy is still hidden inside the tool.

The user continues to demonstrate the movement of the handle as is shown in Fig. 8 and the gate is mostly open. The toy starts to appear and VICON starts reading its positions.

The user reaches the end point where the gate is completely open as is shown in Fig. 8. The toy pushed out of the tool so the VICON could easily track it.

3.2 The Tool

- Tool Design

Figure 9 shows the sketch draw of the tool design for the project, while Fig. 10 shows the real tool designed. The tool design is inspired by animal/infant cognition.

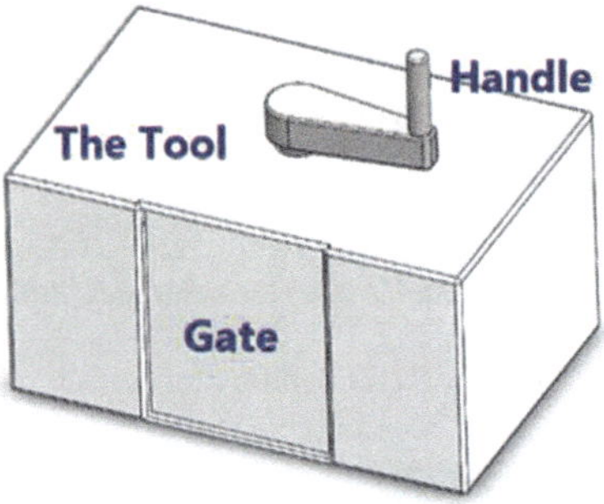

Fig. 9. Tool design sketch.

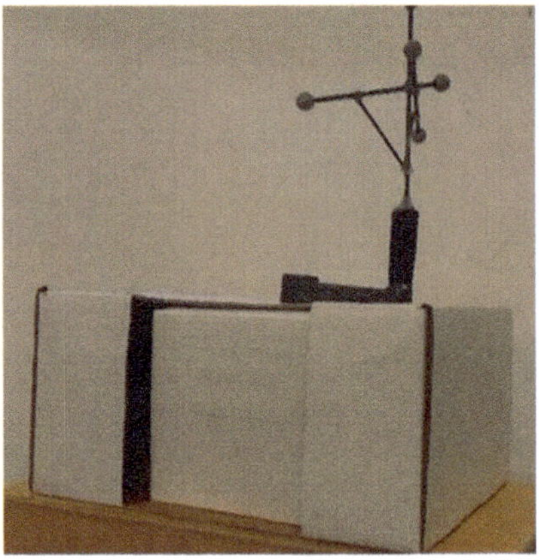

Fig. 10. Front view of the tool.

- Tool Parts

The tool consists of 4 main parts: the handle, the gate, the pusher and the main body. The handle is the rotational part which the robot grasps and rotates to open the gate; once the handle is rotated in a circular way, the gate will open. The pusher is used to push objects outside the tool once the handle moves. The final part is the main body which is a cardboard cuboid; the front side contains the gate, while the top side contains the handle and inside is the pusher. The tool parts are shown in Fig. 11. The pusher is shown in the right side of the image.

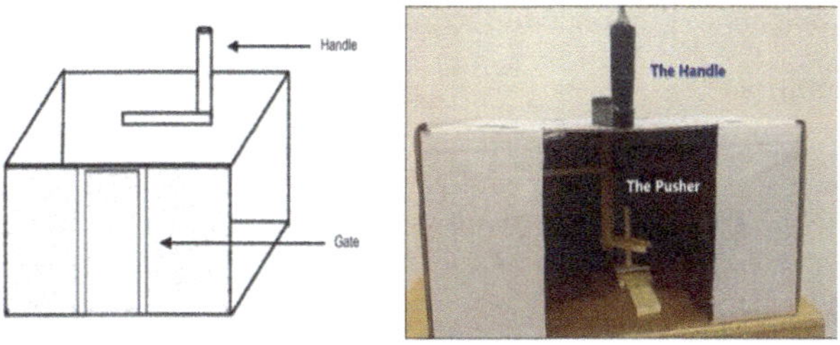

Fig. 11. Tool parts.

- Tool Operation

The tool works by rotating the handle in a circular way from one point to another towards the robot, this movement will open the gate. Meanwhile, inside the tool there is a pusher that is responsible for pushing objects out of the tool. Once the gate is open, a toy object will come out of the tool for the robot to achieve as a reward.

3.3 VICON Implementation

VICON is a motion tracking system used to provide precise Cartesian coordinates of objects being tracked. These coordinates are used to inform the robot about the locations of the handle and the toy in 3D space.

- VICON Tracker Software

It is a software that creates and manages objects for VICON to track them. In addition, it could reboot and calibrate the cameras.

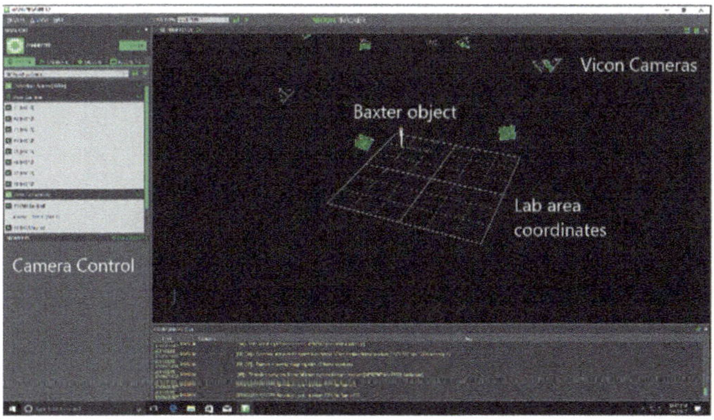

Fig. 12. VICON tracker software

Figure 12 shows the user interface of the VICON Tracker software, the main window is the large black area which includes all the objects and markers in the lab. The green numbers show the status of each of the 8 cameras, while the white square represents the coordinates of objects in reference to the origin which is the middle point.

- VICON Operation

In order to implement VICON in the project, the following steps were followed:

(1) Object Creation: It is the process of selecting a number of markers to represent an object in VICON to be tracked. At least 3 markers are needed to represent an object, they should be positioned in an anti-symmetrical form as Fig. 13 shows below.

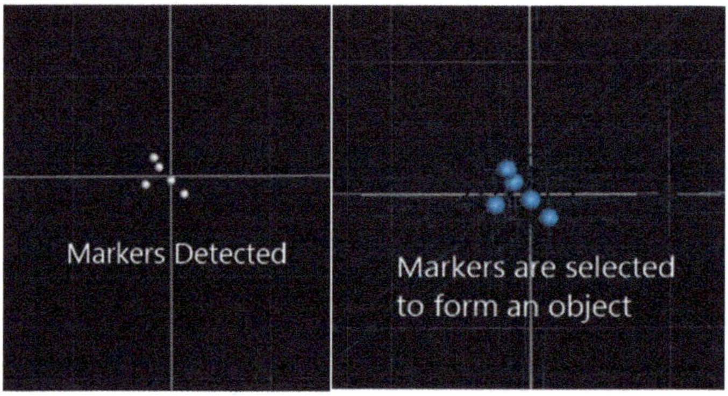

Fig. 13. A markers model detection (Left) and model creation (Right).

Figure 13 showed that a set of markers are detected, model creation starts by selecting all the markers which forms the object then name it as in Fig. 14.

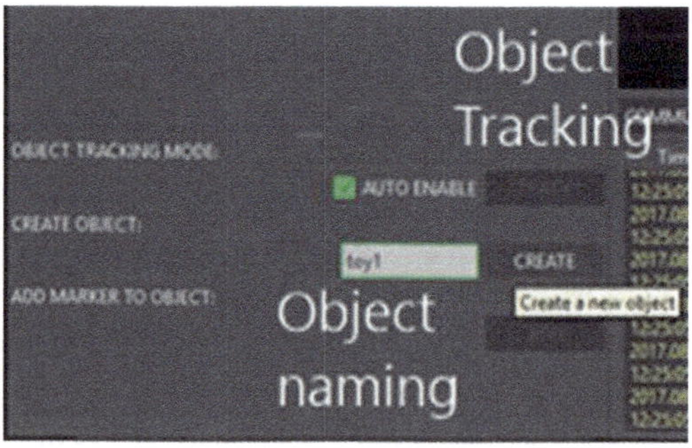

Fig. 14. Object naming and creation.

(2) Zeroing the coordinates of the objects to the VICON origin coordinates
Figure 15 shows the process of zeroing the coordinates of the object to VICON's origin coordinates, the blue cube is dragged and placed to the 0,0, Z position. Z is determined

based on the place where VICON will read the data from. For example, if a 100 mm stick is the object and the readings will be taken from the middle, Z will be 50 mm. Zeroing objects is very important stage because it makes the user determines which area in that object is required to get VICON readings on.

Fig. 15. Zeroing the coordinates of the objects.

(3) Execution of a VICON C++ code to retrieve the Cartesian coordinates of the object being tracked.
(4) The VICON writes the coordinates in a text file.
(5) There are three text files created: the first one is handle text file which includes all the positions of the handle. The second one is Baxter text file which contains all the positions of Baxter. The third one is toy text file which contains all the positions of the toy. All the data were retrieved from VICON during user's demonstration.

3.4 The Overall Implementation

Figure 16 shows the overall implementation of the project; there are 4 main stages: The Vision, VICON-Baxter Coordination, Motion Planning and Movement.

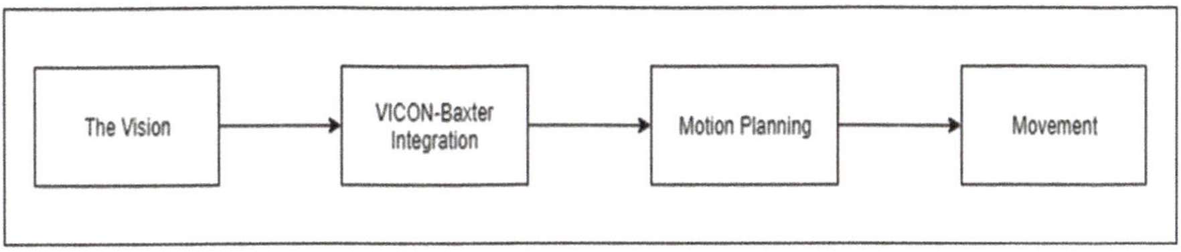

Fig. 16. Overall implementation of the project.

As is shown in Fig. 16, the first stage is the Vision which includes getting data from VICON system, the next stage is VICON-Baxter coordination which takes the data produced by VICON and convert it into values usable by Baxter, after that, Motion Planning is responsible for generating the joints angles based on the VICON-Baxter coordination values. The final stage is the Movement which executes the generated joints angles and makes the arm move.

- The Vision

The way the robot recognizes the movements of a certain object is done using the VICON system; it provides precise locations of an object of interest. VICON configurations for the handle object are as follow: firstly, markers are placed on an anti-symmetrical model which is placed on top of the tool's handle, this model will enable VICON's camera to determine the Cartesian coordinates of the object. Next, is the creation of objects in the VICON Tracker software, this process will create an object called handle which represents the handle. The same process of object creation is repeated and the model is placed on the robot's head and on the toy. Next, the object with the model of Markers is placed on VICON's origin (X = 0, Y = 0, Z = 0) and point of reference is determined in the VICON Tracker software. Point of reference is the place in object where the VICON provides the readings on, for example, if the user wants the robot to grasp the handle from the middle, the user has to measure the distance from the middle of the handle to the ground which will be the Z offset in VICON Tracker software.

Figure 17 shows Baxter, the handle and the toy objects created in VICON Tracker. The tallest object is the Baxter and the reference point is set to the middle where Baxter origin is located. The smallest object is the toy which consists of three markers only, while the last object is the handle.

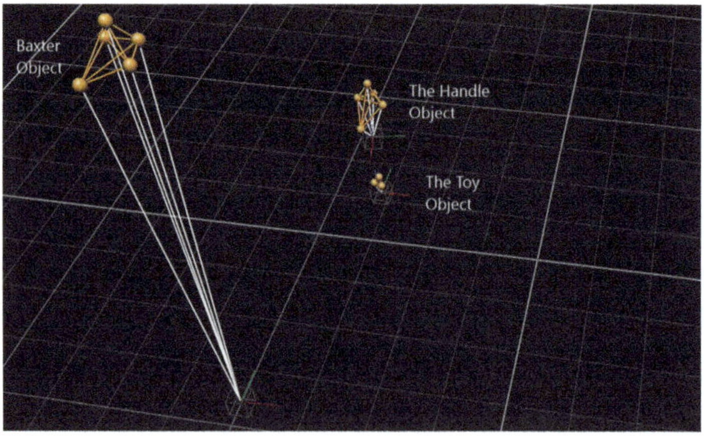

Fig. 17. Objects created.

Finally, a C++ code is compiled and executed in a remote computer to obtain the Cartesian coordinates from VICON. The obtained values are written to a three text files: handle, Baxter and toy. Each text file contains the corresponding positions of the handle, Baxter and the toy respectively.

- VICON-Baxter Integration

The data obtained from VICON are based on VICON's reference which defers from Baxter's reference since they have different origins. For instance, the VICON specifies a cube object location as (X = 1 Y = 2 Z = 10), for the same object Baxter locates the object as (X = 5 Y = 10 Z = 100). Therefore, a unified Cartesian coordinates are required to make VICON the vision source of Baxter. For each of the Cartesian coordinates (X, Y and Z) there is a conversion equation, the equations are obtained as follow: in the beginning, a model of markers is placed on the handle, on top of Baxter's head. Next, Baxter right hand's gripper is grapping the handle and moves it in three different positions. For each position, VICON locates Baxter's head and the handle, while Baxter's endpoint state service returns the Cartesian coordinates of the gripper. As a result, for each Cartesian coordinate there is 9 different values (X, Y, Z * 3 positions), these values create 3 different equations with 3 unknowns.

Solving these 3 equations provides 3 different offsets for each Cartesian coordinate. Finally, the equation is made by multiplying 2 offsets with VICON's readings of the handle and Baxter's location, while adding/subtracting the third offset from the result. The obtained equations are shown in Fig. 18.

```
if select == "handle":                # applying the VICON-Baxter Integration for the handle
    x_for_baxter = (( -1.4263135213368647 * (x_value/1000)) + (2.8451782480441827 * (body_x/1000)) - (2.394471837666306)) + 0.040
    y_for_baxter = (( -1.6255716579702184 * (y_value/1000)) - ( 0.5049980818275669 * (body_y/1000)) - (0.9652114393155212)) + 0.025
    z_for_baxter = ((  1.498052872255575 * (z_value/1000)) + (  16.678897500965153 *(body_z/1000)) - ( 18.174259547726425)) + 0.01
else:                                 # applying the VICON-Baxter Integration for the toy
    x_for_baxter = (( -1.4263135213368647 * (x_value/1000)) + (2.8451782480441827 * (body_x/1000)) - (2.394471837666306)) + 0.075
    y_for_baxter = (( -1.6255716579702184 * (y_value/1000)) - ( 0.5049980818275669 * (body_y/1000)) - (0.9652114393155212)) - 0.03
    z_for_baxter = ((  1.498052872255575 * (z_value/1000)) + (  16.678897500965153 *(body_z/1000)) - ( 18.174259547726425)) + 0.110
```

Fig. 18. Cartesian coordinates equations.

VICON and Baxter have different units for their values; VICON readings are in Millimetres, while Baxter readings in Meters. Therefore, the values obtained from VICON are divided by 1000 due to the conversion of units from Millimetre to Meter as is shown in Fig. 18. In addition, there are two sets of equations with little difference between them. The first set is for the handle, while the second set is for the toy.

- Motion Planning and Movement

It is the process of generating the joints angles for the robot's arm based on a target Cartesian coordinates. The process called Inverse Kinematics (IK); it's a built-in service that generates joints angles values using Kinematics equations and takes Cartesian coordinates of the target as inputs. The following steps are followed to run IK service: firstly, the VICON-Baxter coordination output is passed to the Inverse Kinematics function as 3 variables (X Y Z) named as x_for_baxter, y_for_baxter and z_for_baxter. Secondly, a service function is created to process requests and responses to and from the Inverse Kinematics service. Lastly, a service request and response were created to interact with the service. The service request will send the Cartesian coordinates, while the response will return the joints angles or an invalid solution found statement. The invalid solution found statement appears when IK service cannot generate the joints angles because the arm cannot reach that Cartesian coordinates.

The final stage is to issue the joints angles to the robot using the built in joint command core messages operation, this operation has four different modes: Position, Velocity, Torque and Raw position mode. Position mode implements joint angles for each joint. In Velocity mode the user specifies the velocity of each joint. Torque mode uses joint torques, while Raw position is similar to Position mode but without collision avoidance. The best mode for the project is the Position mode since Inverse Kinematics service generates joints angles which will be executed using joint command operation.

The generated joints angles are passed to the right arm using joint command service which will move the arm based on the joints angles provided. Upon reaching the handle, the robot will close the arm's gripper and starts to rotate the handle in a circular way. The handle movement will open the gate as well as triggers the pusher to push the toy out of the tool. Once the toy is out, Baxter will achieve it. Those steps will move the arm to one point only, to make the arm moves to multiple locations the following is followed: first of all, VICON stores the readings in text files where each line contains XYZ values separated by commas. Next, a function called ik_test is used to pass the VICON readings to the IK service by taking the line number as well as any modification to the Cartesian values as inputs and applying the VICON-Baxter integration equations inside the function. Another function called move_baxter is responsible of calling the joint command service, while another function called generate_line aims to determine which line of VICON readings to be executed. In order to decide which line to choose, a filter inside generate_line function works by subtracting the X values of two consecutive lines and if the result is greater than a certain number then this line will be selected. For instance, if Baxter is required to imitate the task through only 3 points, then the filter will have the value as 50 mm instead of 10 mm. All the lines numbers to be executed are stored in an array called output. The final step is to make a loop with the range 0 to the required number of points Baxter has to imitate the task through (for e.g. 3, 5 and 10). Inside the for loop there are two functions to be called: the first function is used to call the Inverse Kinematics service as well as specifying the line to be executed and if any modification to the Cartesian coordinates is required, in addition there are two options: handle and toy, where there is a slight change in VICON-Baxter coordination for the handle and the toy. The second function is called move_baxter which is used to send the generated joints angles to the robot right arm.

4 Results and Discussion

The results are evaluated in three main experiments: The first experiment is to make Baxter imitates the task through 3 points only. The second experiment is to make Baxter imitates the task through 5 points. The final experiment is to make Baxter imitates the task through 10 points. For all the experiments, VICON detection and tracking of objects are shown in Figs. 19 and 20.

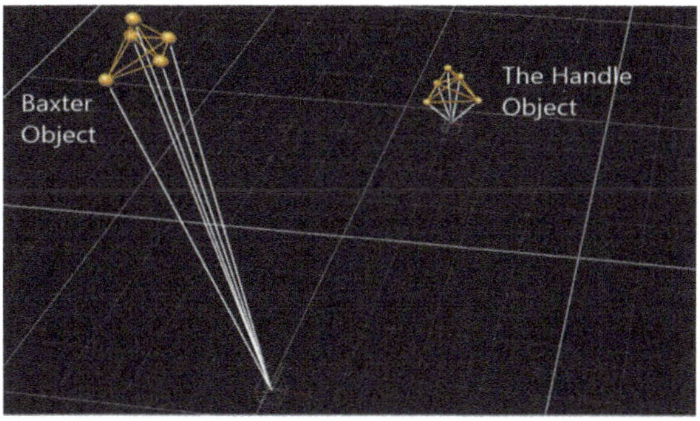

Fig. 19. VICON objects detection when the gate is close.

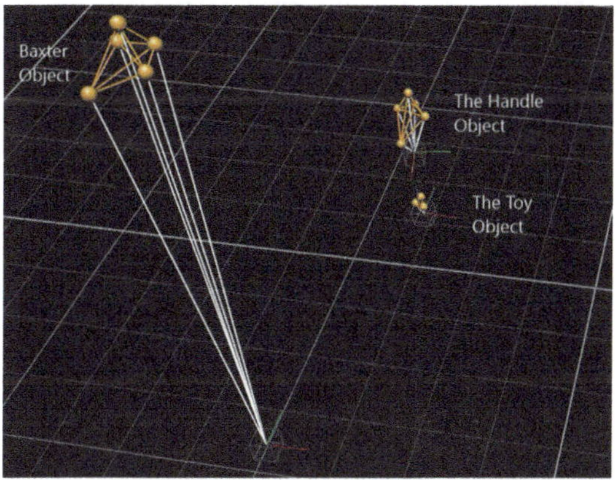

Fig. 20. VICON objects detection when the gate is open.

Figure 19 shows the objects being tracked by VICON while the gate is closed, the objects are Baxter and the tool only. After rotating the handle and opening the gate, the number of objects becomes 3 because the third object is the toy which came out of the tool as shown in Fig. 20.

4.1 Experiment 1

In this experiment, the user demonstrates the full movement of the handle. After that, Baxter has to imitate the demonstration in 3 points: the beginning, the middle and the end. The points are defined by a function called generate_line which calculates the difference between two consecutive X values from VICON readings of the handle and returns the line number which matches the condition specified. For the three points experiment, the function is set to return the line numbers where the difference between the X value and the next X value is 47 mm and the result is 3 different lines. Baxter took 3 steps to reach the target, the first point the gate was fully closed. The second point, the gate is half open and the toy starts to appear. The last point, the gate was fully open and the object is out of the tool. Even though the imitation is accomplished successfully, Baxter movements were not stable and not smooth. The imitation was faster than experiment 2 and 3, however, the movement was not accurate and Baxter opt for straight line execution from point to point rather than a circular way which could break the tool or make the handle moves aggressively.

Figure 21 shows a comparison between the user's demonstration and Baxter imitation trajectories for the 3 points. The first two lines represent a comparison between the X values from the user during the demonstration and from Baxter during the imitation. There are small variations in the values due to the percentage of error on Baxter and VICON-Baxter integration accuracy. The middle two lines show the difference between the Y values from the user during the demonstration and from Baxter during the imitation, the values are almost identical. The last two lines illustrate the variation between the Z values from the user during the demonstration and from Baxter during the imitation. There is a difference in the Z values especially towards the end.

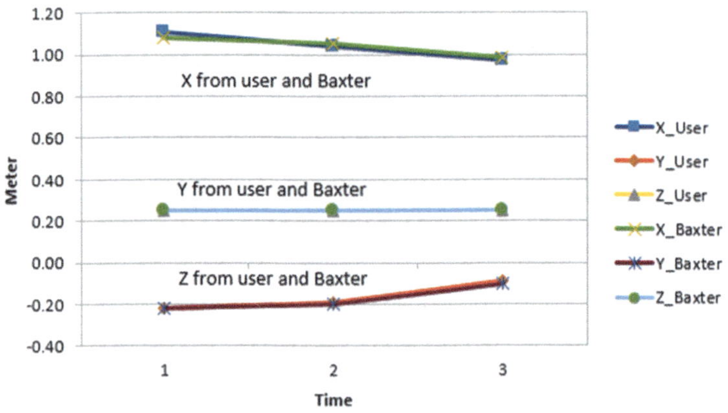

Fig. 21. Demonstration and imitation comparison at 3 points.

4.2 Experiment 2

In this experiment, the user demonstrates the full movement of the handle. Then Baxter has to imitate the demonstration in 5 points. The points are defined by generate_line function, the function is set to return the line numbers where the difference between the X value and the next X value is 21 mm.

Baxter took 5 steps to reach the target, the first 3 points the gate was closed and starting to open. For the points 4 and 5, the gate was fully open and the object is out of the tool. The imitation is accomplished successfully, the movement of the handle was smooth and stable compared to experiment 1. However, the movement was still not close to the user movement, usually the user takes continues small movements to accomplish tasks.

Figure 22 shows a comparison between the user's demonstration and Baxter imitation trajectories. The first two lines represent a comparison between the X values from the user during the demonstration and from Baxter during the imitation, there are small variations between the demonstration and the imitation since the points applied are not enough to fully imitate the task. The middle two lines show the difference between the Y values from the user during the demonstration and from Baxter during the imitation, they are almost identical. The last two lines illustrate the variation between the Z values from the user during the demonstration and from Baxter during the imitation. There are big differences especially at point 4 which will be eliminated when 10 points experiment is implemented.

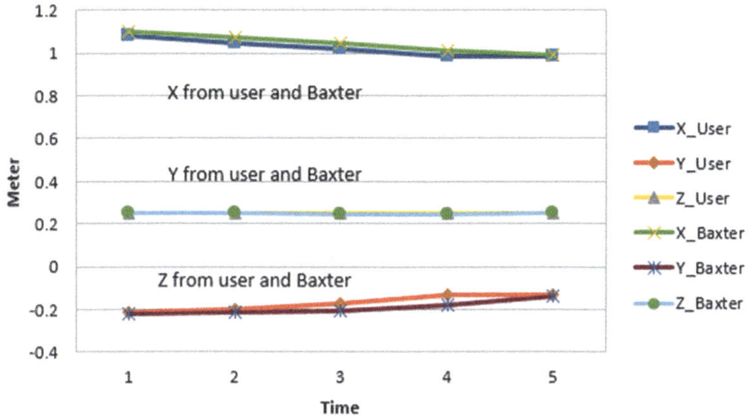

Fig. 22. Demonstration and imitation comparison at 5 points.

4.3 Experiment 3

In this experiment, the user demonstrates the full movement of the handle, this demonstration is similar to the demonstrations carried out in experiments 1 and 2. Baxter has to imitate the demonstration in 10 points. The points are defined by generate_line function, the function is set to return the line numbers where the difference between the X value and the next X value is 10 mm. Baxter took small steps to reach the target, the first 5 points the gate was closed and starting to open. At point 6 and 7, the gate was open and the object starts to show up. Finally points 8 to 10, the gate was fully open and the object is out of the tool. Even though the imitation took longer to accomplish, the movement of the handle was smooth and stable compared to experiment 1 and 2.

Figure 23 shows a comparison between the user's demonstration and Baxter imitation trajectories. The first two lines represent a comparison between the X values from the user during the demonstration and from Baxter during the imitation. The X values are almost identical with a small variation in the beginning. The middle two lines show the difference between the Y values from the user during the demonstration and from Baxter during the imitation. The last two lines illustrate the variation between the Z values from the user during the demonstration and from Baxter during the imitation. Therefore, increasing the number of points produces more accurate results as well as smoothening Baxter's movement and the optimal method is to use 10 points imitation.

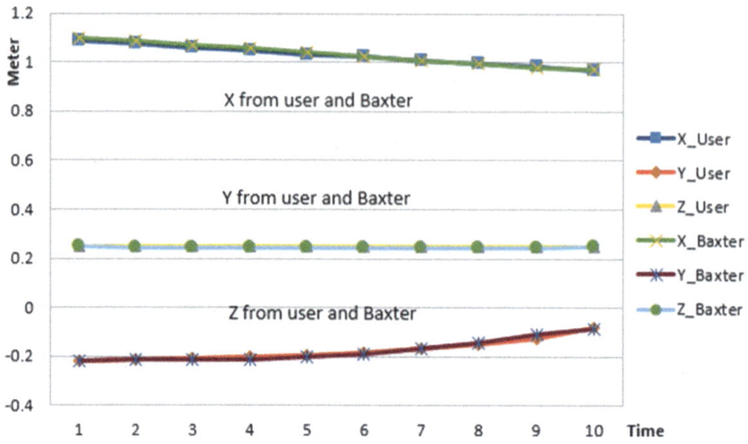

Fig. 23. Demonstration and imitation comparison at 10 points.

5 Conclusion

Robots' abilities are very beneficial in order to achieve required tasks and operations. However, programming the robot for each new task is time consuming and inefficient. Learning by Demonstration aims to remove these issues and generalise the work of robots; it is a technology that permits robots to complete tasks based on the user demonstrations.

This research project applied Learning by Demonstration on Baxter humanoid robot in order to perform a series of two simple tasks based on the user's demonstration of those tasks. Two main tasks were implemented: the first task is to rotate the handle of a receptacle for a toy object in order to come out of the tool. The second task is to achieve the toy from task 1. The user demonstrates the first task in front of Baxter. In order for Baxter to know what to imitate according to the demonstration, the VICON system was used. The VICON system is a motion tracking system that uses reflected markers in order to select objects to track from an environment. However, VICON's localisation of objects differs from Baxter's, because they have different origins. Therefore, a VICON-Baxter integration was used in order to convert VICON readings into values usable for Baxter. These values are Cartesian coordinates which represent the locations of the handle in 3D space. In order to reach the handle, the Inverse Kinematics service was used to generate the joints angles for Baxter from given Cartesian coordinates. At the next stage, those joints angles are executed by using the Joint Command service in order to reach the target. There were 3 experiments that were carried out in order to choose the optimal way for Baxter to imitate the demonstration. The first experiment made Baxter imitate the demonstration for only 3 points in the trajectory; the experiment results showed that Baxter's movement was fast and the task were accomplished successfully. However, the movement was not smooth and the tool could have been damaged. In the second experiment, Baxter rotated the handle through 5 points in the trajectory, its movement was slower and smoother, and the task was accomplished successfully. However, there were slightly big differences in the values between the user demonstration and the imitation. The final experiment was the best and most efficient, as Baxter performed the imitation in 10 different points in the trajectory. Even though the movement was slow, the handle rotation was smooth and stable. The project goals were achieved successfully and Baxter managed to imitate the user's demonstration. The inspiration for this project was taken from the field of behavioural psychology, specifically the study of animal/infant cognition. In addition, this project is beneficial for the industry, since the robot does not require programming for each new task. There are some areas of research where this project can be extended, e.g. the VICON-Baxter integration could be simplified into a standalone program that takes care of the connection between VICON and any robot.

References

1. Ko, W., Wu, Y., Tee, K., Buchli, J.: Towards industrial robot learning from demonstration. In: Proceedings of the 3rd International Conference on Human-Agent Interaction, pp. 235–238. ACM (2015)
2. Baxter Collaborative Robots for Industrial Automation. Rethink Robotics. http://www.rethinkrobotics.com/Baxter/. Accessed 10 Aug 2017
3. I., "Baxter Robot," *ioannis* (2014). http://www.edinburgh-robotics.org/equipment/robotarium-east-field-systems-humanoid/Baxter-robot. Accessed 05 Aug 2017
4. "Arms - sdk-wiki", Sdk.rethinkrobotics.com (2016). http://sdk.rethinkrobotics.com/wiki/Arms. Accessed 10 Aug 2017
5. Merriaux, P., Dupuis, Y., Boutteau, R., Vasseur, P., Savatier, X.: A study of vicon system positioning performance. Sensors **17**(7), 1591 (2017)
6. Asfour, T., Gyarfas, F., Azad, P., Dillmann, R.: Imitation learning of dual-arm manipulation tasks in humanoid robots. In: 2006 6th IEEE-RAS International Conference on Humanoid Robots (2006)
7. Nehaniv, C.L.: Nine billion correspondence problems. In: Nehaniv, C.L., Dautenhahn, K. (eds.) Imitation and Social Learning in Robots, Humans and Animals: Behavioural, Social and Communicative Dimensions, Cambridge University Press (2007)
8. Bentivegna, D., Atkeson, C., Cheng, G.: Learning tasks from observation and practice. Robot. Auton. Syst. **47**, 163–169 (2004)
9. Billard, A.G., Calinon, S., Guenter, F.: Discriminative and adaptive imitation in uni-manual and bi-manual tasks. Robot. Auton. Syst. **54**(5), 370–384 (2006)
10. Rozo, L., Pablo J., Carme T.: Robot learning from demonstration of force-based tasks with multiple solution trajectories. In: 2011 15th International Conference on Advanced Robotics (ICAR), pp. 124–129. IEEE (2011)
11. Kim, S., Kim, C., You, B., Oh, S.: Stable whole-body motion generation for humanoid robots to imitate human motions. In: Proceedings of the IEEE/RSJ International Conference on Intelligent Robots and Systems (IROS) (2009)
12. Kulic, D., Ott, C., Lee, D., Ishikawa, J., Nakamura, Y.: Incremental learning of full body motion primitives and their sequencing through human motion observation. Int. J. Robot. Res. **31**(3), 330–345 (2012)
13. Sauser, E.L., et al.: Iterative learning of grasp adaptation through human corrections. Robot. Auton. Syst. **60**(1), 55–71 (2012)

14. Babič, J., Hale, J.G., Oztop, E.: Human sensorimotor learning for humanoid robot skill synthesis. Adapt. Behav. **19**(4), 250–263 (2011)
15. Bentivegna, D.C., Atkeson, C.G.: A framework for learning from observation using primitives. In: Proceedings of the RoboCup 2002 International Symposium, Fukuoka, Japan (2002)
16. Whiten, A., Spiteri, A., Horner, V., Bonnie, K.E., Lambeth, S.P., Schapiro, S.J., De Waal, F.B.: Transmission of multiple traditions within and between chimpanzee groups. Curr. Biol. **17**(12), 1038–1043 (2007)
17. Fagard, J., Rat-Fischer, L., Esseily, R., Somogyi, E., O'Regan, J.K.: What does it take for an infant to learn how to use a tool by observation? Front. Psychol. **7**, 267 (2016)

Selective Stiffening Mechanism for Surgical-Assist Soft Robotic Applications

Sunita Chauhan(✉), Mathew Guerra, and Ranjaka De Mel

Department of Mechanical and Aerospace Engineering, Monash University, Clayton, Australia
Sunita.Chauhan@monash.edu

Abstract. The compliant nature of soft robots provides a safer interaction with humans when compared to rigid robots. Hence, soft robotic devices have increasingly become an area of research for medical applications where patient safety is paramount. However, due to the high flexibility inherent in these devices, their flexural rigidity inhibits their ability to exert sufficient forces. In this work, a new wire jamming mechanism is proposed to selectively change the flexural rigidity of soft robotic instruments and a hybrid robotic prototypes based on this concept is presented. Results showed that this mechanism increased the stiffness by almost five folds. The simplicity of the concept will easily allow further optimization for its potential use in surgical-assist applications.

Keywords: Soft robots · Flexural rigidity · Surgical-assist devices
Minimally invasive surgery · Vacuum activated control

1 Introduction

Soft robots have progressively increased in popularity due to their highly compliant nature and locomotion similar to living organisms, and thus tend to be safer when used around humans, especially in a medical environment. Additionally, the flexibility of soft robots can match that of rigid robots with minimal complexity, weight and cost [1]. Whilst the compliant nature of soft robots can be an advantage, it can also be a drawback. In comparison to rigid robots, the low stiffness of soft robots results in larger deformations caused by an external force and the force that a soft robot can exert on the environment is limited. One of the driving factors for the development and research of soft robotic devices in medical applications is their employment in minimally invasive surgeries (MIS), where their compliant qualities can greatly increase patient safety [1]. However, depending on the type of MIS, the device's stiffness properties may not be able to perform a desired task successfully.

There have been a number of studies on different mechanisms to vary the stiffness of soft robots from a soft (liquid) to hard (solid) state, and vice versa. These mechanisms aim to provide the benefits of both soft and rigid devices through selective stiffening. The stiffness of these mechanisms is best described by the flexural rigidity (*EI)* [2]. Hence, mechanisms can incorporate methods to change the elastic properties (*E*) or the moment of inertia (*I*). In this regards, a majority of the reported mechanisms focus on changing the elastic properties of the structure via material or structural controllability

K. Arai et al. (Eds.): IntelliSys 2018, AISC 868, pp. 791–803, 2019.
https://doi.org/10.1007/978-3-030-01054-6_55

[3]. Studies show that the simplest mechanisms that achieve the shortest activation time between states are those that employ jamming by means of structural controllability. This process requires an understanding and control of interactions between the unjammed elements of a given mechanism [4], which could normally be achieved with the application of a pressure on the structure's elements by the use of vacuum or compressed air. As detailed in [3], the main forms of jamming are granular, layer and wire jamming, where the structural elements are typically contained within an easily deformable elastic membrane. In the unjammed state, the pressure difference between the inside and outside of the membrane is low. This means the elements are free to move within the membrane, given that they are not tightly packed initially [5]. The jamming mechanisms that utilize vacuum operate by creating a pressure difference between the inside and outside of the membrane, which results in the membrane contracting around the elements and ultimately applying a pressure.

Soft robotic medical instruments that articulate in a 3D space require the stiffening mechanism to have small *EI* when bending (in any direction) but high *EI* when applying a force on the environment. Granular stiffening mechanisms can provide this minimal bending resistance, however due to the nature of granules, their volume and cross-sectional area (CSA) can easily be changed when unjammed. If the volume and CSA are not sufficiently controlled, the jammed stiffening properties can (unwillingly) degrade. Layer jamming is achieved through the friction between layers of a thin material, e.g. sheets [6]. These mechanisms can be easily constructed for strictly 2D bending motions. However, for 3D bending, device fabrication can be more tedious, as shown in [6, 7]. In the wire jamming mechanism, the structural elements consist of wires or fibers. A system described in [8] explored the use of repeated interlocking fibers, however the miniaturization of these fibres for medical applications is claimed to be difficult. Another study delves into a shaft that can rigidify through the friction generated between a ring of cables, an outer spring and an inner inflatable tube [9]. This study showed promising results and served as an inspiration for our present study to be explored further especially for surgical-assist applications such as picking, holding/grasping, suturing, etc. along with embedded force sensing mechanisms [11].

This paper presents a novel concept and mechanism design that utilizes a wire jamming mechanism with an intended use in medical applications (though it can be extended to other industrial applications). In the following sections, the mechanism design and methodology is explained first as Sect. 2. It also includes a mathematical model to approximate and describe the stiffening properties of the structure. Sect. 3 explains the experimental set-up and feasibility studies on the two constructed prototypes tested in laboratory experiments followed by the results and discussions in Sect. 4. The paper concludes with Sect. 5 on the overall contents presented and tests performed to determine the feasibility of the concept and model.

2 Methodology and Prototype Design

2.1 Mechanism Design

The concept consists of round wires arranged in a circular configuration and inserted into an elastic membrane. Transition between unjammed and jammed states is controlled through the application of vacuum. The working principle of the proposed concept is demonstrated in Fig. 1.

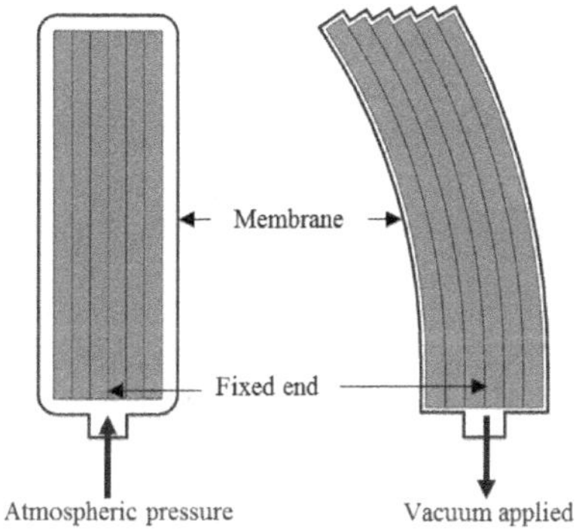

Fig. 1. Atmospheric pressure is allowed to enter into the membrane (unjammed state), allowing the wires to move past each other freely (left). When a vacuum is applied (jammed state), a large pressure differential is created, constraining the motion of the wires (right).

In the unjammed state (no vacuum applied), the atmospheric pressure can fill the membrane, allowing the wires to slide passed each other with minimal friction. In the jammed state, the pressure on the elements created by the vacuum increases the normal forces on the wire surfaces. These normal forces generate friction ($F_f = \mu N$), thus resisting the movements of the wires and increasing the stiffness of the overall body. As expected, the membrane also deforms to the shape of wire bundle. In this study, large deformation cantilever beam theory is applied and the deflections in x and y coordinates, with an applied force, F, is shown in Fig. 2. Here, Δx is assumed to be small in comparison to Δy. Additionally, it is assumed that the material being used behaves in a linear-elastic manner.

The flexural rigidity (EI) can be determined as below, where L is the distance between the fixed support and applied force F, c is the bending curvature and M is the internal moment in the beam at a given point:

$$EI = M/c \tag{1}$$

The wire material is expected to behave linear-elastically and small deformations are assumed; the vertical deflection of the structure can be given as:

$$\Delta y = (FL^3)/(EI) \tag{2}$$

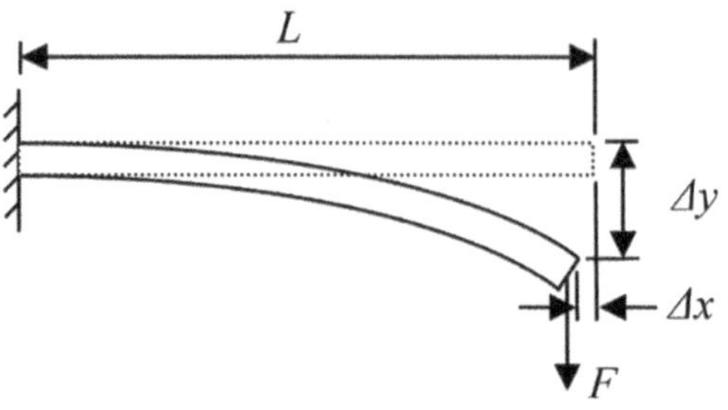

Fig. 2. Deflections (Δx and Δy) associated with a cantilever beam due to an applied force F.

Furthermore, the flexural rigidity can be approximated as:

$$EI = (FL^3)/\Delta y \tag{3}$$

In this study, the stiffness factor S is used to determine the change in stiffness caused by the applied vacuum. Thus, S is estimated by EI_j/EI_u, where EI_j and EI_u represent jammed and unjammed stiffness respectively. It can be seen that for a constant F and L, the stiffness factor, S, can be determined by the ratio of Δy in the unjammed and jammed state.

2.2 Actuator Design and Fabrication

To determine the feasibility of the proposed stiffening mechanism, two prototypes were constructed. Prototypes consisted of wires (Ultra XT 1.17 mm Monofilament fishing line) arranged in a 1x7 configuration, as shown in Fig. 3(a). An elastic membrane was also implemented (natural rubber), in which the wires were inserted. The wire bundles for each prototype are shown in Fig. 3(b). For the first prototype, the wires were twisted and fixed at both ends. Note that the amount of twist of the wire bundle (length 100 mm) has no justified value. The wires in the second prototype were fixed at one end, with the remaining length of the wires free to move (untwisted). Ends were fixed with super glue (Selleys Quick Fix Supa Glue) for ~10 mm from the end of the wire bundle.

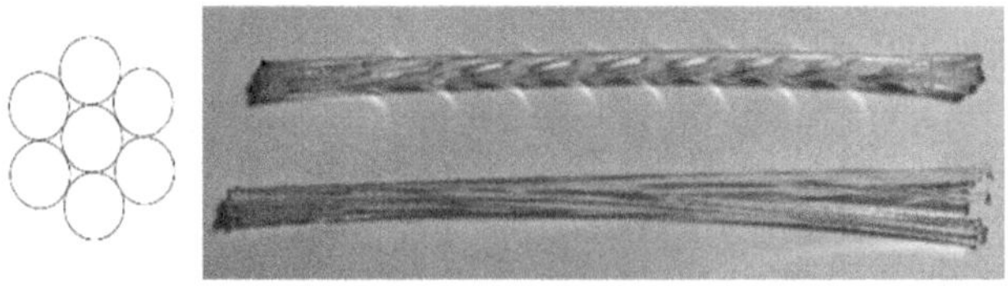

(a) (b)

Fig. 3. (a) 1x7 wire configuration. (b) Twisted (top) and untwisted (bottom) wire bundles for the constructed prototypes.

The outer diameter of the wire bundles was chosen as 3.5 mm, suitable for thin soft robotic medical instruments. The 1x7 wire configuration was chosen, however, a number of configurations could be implemented such as wires of different diameter, type (e.g. hollow) and material, which can further be combined together to obtain desired properties, i.e. I, E, E_j and F_f [10]. As mentioned, the focus of this study is in determining the achievable stiffening factor S, hence the theoretical calculation for I is not required. If I and L are kept constant, the only property that can influence flexural rigidity is E, as discussed in the previous section. Furthermore, the change in E is directly caused by the generated friction, F_f, which is dependent on the vacuum pressure applied and the wire bundle configuration and material properties.

(1) *Fabrication of the Silicone Body*

Fabrication of the silicone body was primarily chosen as a two-step process. Firstly, a mold of the design was 3D printed and followed by setting silicone material poured to the mold. Rigid steel wires were inserted into the mold prior to pouring silicone. The physical structure of the proposed soft-robotic concept consists of an elastic-membrane made from silicone rubber (Ecoflex 00–50) and an exoskeleton made from ABS plastic. The flexible membrane (core) consists of 4 air chambers, which were selected to allow better pneumatic control, better structural integrity and sufficient impact resistance from external objects during application. In addition, a through whole was made at the center of the membrane to: (1) allow easy assembly into a fixed cylindrical base; (2) reduction in weight; (3) provide better response to pressure gradients during application.

(2) *Fabrication of the Exo-skeleton*

During specific laparoscopic applications such as suturing, it is important that the soft-robot can withstand twist in addition to the bending motion while still maintaining

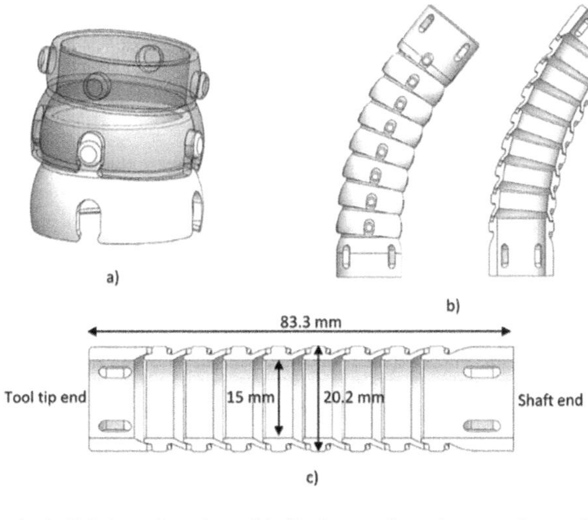

Fig. 4. (a) Single ball joint of casing. (b) Casing undergoing maximum bending. (c) Casing dimensions.

structural rigidity. For this purpose, a hard exoskeleton was designed as shown in Fig. 4. The two main objectives of housing of the elastic membrane within an exoskeleton are to provide controlled deformation to the membrane and provide structural integrity during application.

The exoskeleton provides a safety-envelope to the soft component. With the addition of snap-pins and slots in the surface of the ball-joint components, it allowed the individual components to mimic movement of a ball-joint while resisting any lateral twist at the joints. The thickness of the exoskeleton was carefully determined to resist air-bubbles due to expansion of any weaker parts within air chambers of the membrane while still allowing the soft-core to deform naturally to pressure variations.

2.3 Control Simulations

Prior to the experimental set-up, the proposed design was simulated in SolidWorks to understand the deformation due to pressure variations in the air chambers. The objective of the simulation was to identify the relationship between the pressure variations in air-chambers and bending of the prototype. Furthermore, the simulation tests were also used as a guide to identify any limitations of the stiffening mechanism used within the elastic membrane. Simulations were performed with and without the exoskeleton (Fig. 5(a) and (b)).

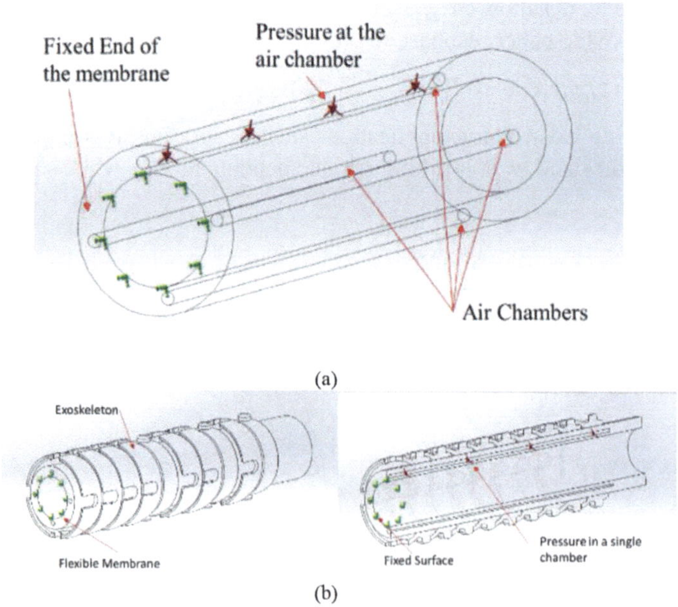

Fig. 5. (a) Static modelling of membrane with pressurizing single vacuum chamber, and (b) the membrane enclosed by the exoskeleton.

Case 1: Deformation of the elastic membrane due to pressure changes in the air chambers (without exoskeleton)

The proposed soft model components primarily exhibited material characteristics of silicone rubber. The same material was applied in the simulations as it was readily available in the SolidWorks materials database (Fig. 5(a)). One end of the membrane was fixed to replicate the cantilever joint and the other end was left to float. A range of pressures were applied to a single vacuum chamber and resulting displacements and bending angles were recorded (please refer to the next section for results).

Case 2: Simulation of the prototype with exoskeleton in place

The exoskeleton was developed by assembling 9 ball-joint type components attached to each other using pin-slot type connectors. Similar surface fixation method was employed as per case 1 and variable pressures were applied to a single vacuum chamber (Fig. 5(b)). The resulting deformations and bending angles were recorded.

3 Experimental Set-up

To verify the capability of the proposed mechanism, two experiments were conducted: for determining the change in stiffness achieved for the twisted prototype and for the untwisted prototype. The experimental setup for both experiments is shown in Fig. 6(a). The fixed end of the prototype was clamped down to simulate a cantilever beam. The applied force F is varied with the use of a variable mass, which consists of five removable pieces (Fig. 6(b)). The variable mass was hooked onto the loop attached to the free end of the prototype and provided forces of 0.059, 0.098, 0.128, 0.167 and 0.206 N. The experiments were conducted by applying the mass and measuring the vertical deflection Δy of a designated point of the body (top most point of the body in line with the direction of F). For each experiment, three repeated measurements were conducted at five different loads for the jammed and unjammed states. For the jammed state, a vacuum pump rated for −100 kPa (Welch 2581C-24) was used. L for both experiments was 81 mm. The limitation of the experiment for the untwisted prototype was that F was limited to a 2D plane. Given that the stiffening properties of this prototype could differ depending on the direction of F, an accurate representation of the stiffening factor for 3D applied forces was not determined.

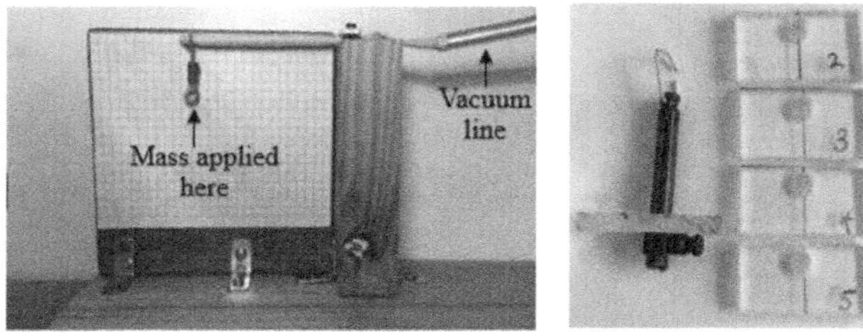

(a) (b)

Fig. 6. (a) Experimental set-up used to simulate a cantilever beam for determining prototype stiffening properties. (b) Variable mass used to apply a force at the end of the prototype.

The objective of the experimental setup was to identify whether the deformation model obtained through simulation aligns with the physical behavior of the prototype. Figure 7 shows the experimental setup used. To pressurize the air chambers, a small compressor with a maximum compression of 9 psi, driven by a maximum voltage of 12 VDC was used.

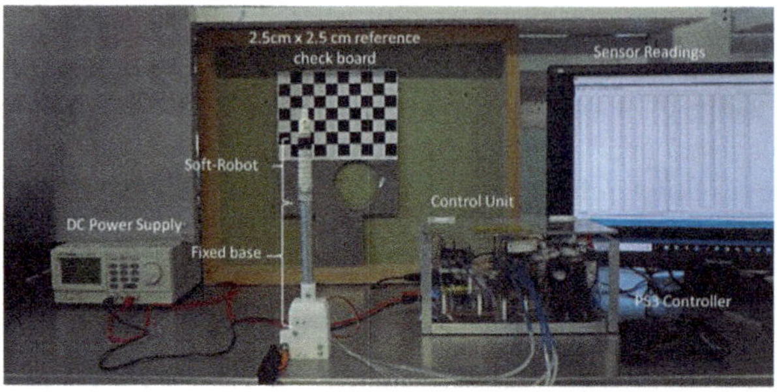

Fig. 7. Experimental setup for testing pressure vs bending characteristics.

The pressures to the chambers were regulated using a microcontroller and pneumatic solenoid valves (MAC 34 Series) combination along with manual pressure release valves for each solenoid as a safety measure. The air pump was connected directly to the solenoid valves whose outputs were connected to the air chambers of the prototype as well as individual pressure sensors.

The pressure sensors provide pressure feedback to the microcontroller which then can be used to determine the pressures at the chambers. Prior to using the pressure sensors for measurements, they were calibrated by calibrated by applying a pre-defined pressure through the valves and measuring the output voltage range from each sensor. Once the measurements were taken, a numerical model for each sensor output was developed. The deformation of the proposed design was measured using a Bumblebee 2 (point grey) stereo camera which has a factory calibrated resolution of 0.01 mm. The feedback from the sensors was further amplified for ease of detection and sensitivity. The entire system including the compressor and solenoid valve actuation was controlled by a PS3 gaming controller.

4 Results and Discussions

4.1 Simulation Results

Figure 8 shows the simulation results for bending of the proposed design with and without the exoskeleton. Table 1 lists the bending angle and y-axis (vertical) deformation at various pressures for the model (with and without the exoskeleton), results of which are shown in Fig. 9.

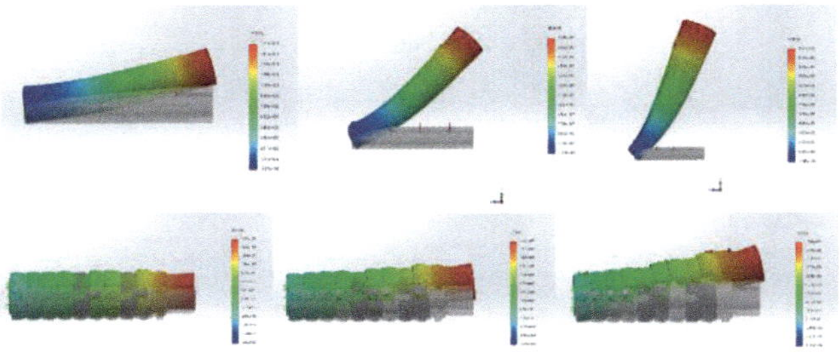

Fig. 8. Simulation results for various pressures (Top row: 5 psi, 7 psi and 9 psi from right), (bottom row: with exoskeleton for the same pressures).

Table 1. The bending angle and Y-Axis (Vertical) deformation at various pressures

Pressure (psi)	Membrane only		With exoskeleton	
	ΔY (mm)	ΔΘ (deg)	ΔY (mm)	ΔΘ (deg)
0	0	0	0	0
1	16.12	12.31	1.6	1.5
2	32.2	23.05	3.37	2.34
3	48.37	34.06	5	3.69
4	64.49	41.17	6.67	4.38
6	96.7	50.42	10	12.03
9	145.1	61.07	15	16.2

Based on the simulation results, it can be observed that deformation of the membrane for pressures under 5 psi varied linearly while for pressures above 5 psi, it varied exponentially. Although the simulation only recorded deformation in a single axis, the deformation was rather three-dimensional, though deformation in y-axis was evidently the most significant variation out of all. When the membrane was housed within an exoskeleton, the deformations appeared to be significantly damped and linear. When considering the bending angle during deformation on both setups, both scenarios showed linear variations proportional to pressure gradients. However, bending angle with the exoskeleton in place was relatively smaller compared to the bending angle of the membrane itself.

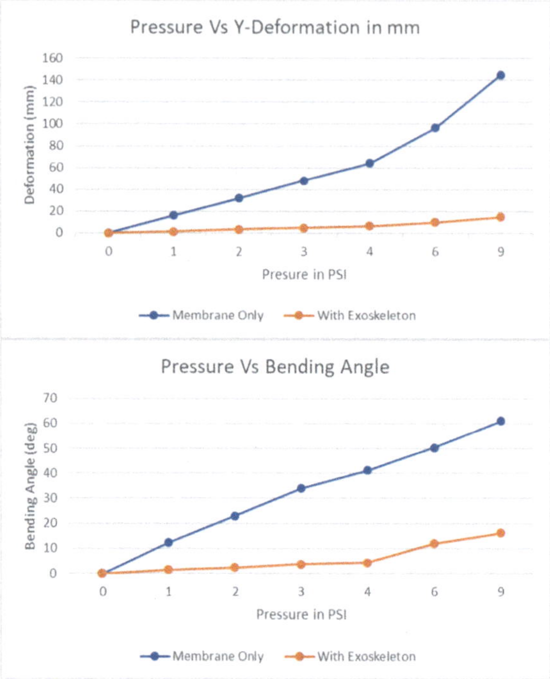

Fig. 9. Modelled relationship between y-deformation and bending angle.

4.2 Experimental Results

The experimental procedure was described in the previous section; the results of variation in input pneumatic pressure, as observed by a Bumblebee camera are shown in Fig. 10. Table 2 lists the observed bi-directional values of the bending angles in up and down movements and the hysteresis behavior is plotted in Fig. 10 below.

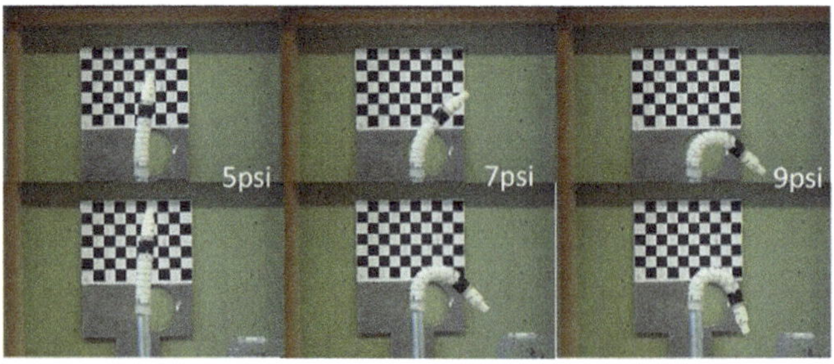

Fig. 10. Bending angles of the prototype for various pressures.

Table 2. Bi-directional values of the bending angles

Pressure (psi)	Angle down (deg)	Angle up (deg)
0	0	0
5	8.78	6.5
6	20.38	97.27
7	47.03	134.41
8	107.11	152.25
9	132.06	160.8

The deflection and stiffness factor resulting from the two experiments (twisted and untwisted prototypes) are shown in Fig. 11. They both share similar stiffness properties when unjammed. Theoretically, this was expected as the cross –sectional area of both prototypes at any point along the length of the body should be relatively the same, given the 1x7 configuration. Therefore, the prototypes should also have a similar I and E, assuming negligible F_f. This would only apply to sections of the structure that were not fixed with super glue. However, it can be seen that the twisted prototype is slightly stiffer. This can be explained through the friction generated caused by the twisting of the wires, thus increasing E of the twisted wire bundle. The unjammed results show a near linear trend, which relates closely to small deformation theory of a cantilever beam. Hence, it can be supposed that Δx was insignificant in comparison to Δy for the larger F. Furthermore, (2) and (3) used to approximate the prototypes performance are fairly valid.

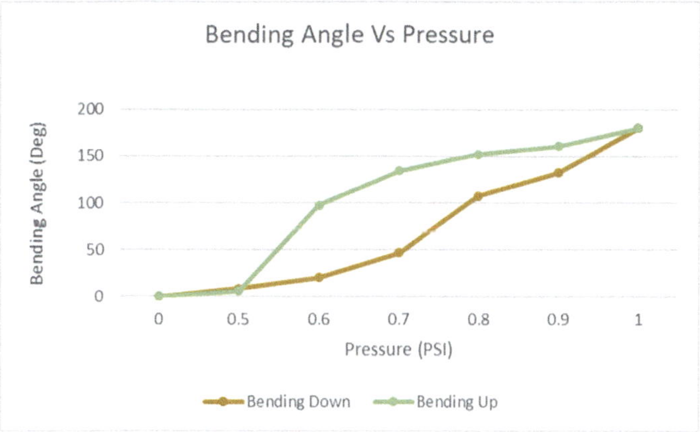

Fig. 11. Hysteresis behavior of the design in reality.

The measured data clearly shows that the twisted prototype has a considerably smaller average stiffening factor (1.56) when compared to the untwisted prototype (5.17). This change could be caused by the untwisted prototype experiencing a better F_f performance which results in a higher E_j and/or an increase in I caused by the wires deviating from their original circular configuration. Furthermore, the unjammed twisted prototype experienced a noticeable F_f and therefore would result in a lower change in stiffness when jammed, given the CSA remains relatively constant and the wires are already partially jammed. It should be noted that an increase in vacuum pressure would result in higher F_f for both prototypes. This would directly increase E_j and S.

The experimental outcomes showed relatively different results in terms of the bending angle when compared to the simulations. The overall bending angle during experimental testing lagged behind its simulations results (simulations assumed that the materials behave linearly during deformation while it could be non-linear in reality). Furthermore, fabrication errors can cause imperfections and non-uniform shapes as oppose to simulated bodies which can behave differently when pressurized. In addition, experimental results showed bending angle to follow a hysteresis behavior while in simulation, it was no hysteresis was detected.

The flexural rigidity results from this study and those determined in [9] are relatively comparable. From bench tests, the FORGUIDE shaft achieved an average flexural rigidity as high as 1489 (range 1345–1541) Ncm^2, for an outer diameter of 4.7 mm. The maximum average flexural rigidity achieved from our study was 698 (range 544–1029) Ncm^2, for an outer diameter of 3.51 mm. Therefore, with further optimization of materials and design,

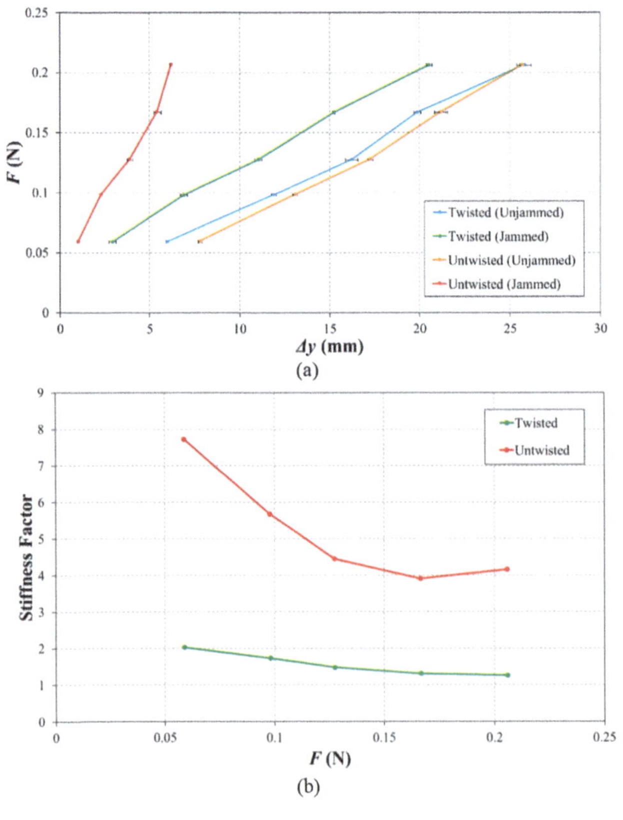

Fig. 12. (a) Average measured Δy during application of F for both prototypes (unjammed and jammed). Error bars indicate one standard deviation of measurement uncertainty. (b) Calculated stiffness factor for each applied F.

the concept within this study could potentially exceed the performance of the FORGUIDE shaft (Fig. 12).

5 Conclusions and Future Work

A concept based on a vacuum activated wire jamming mechanism for use in soft robot applications was conceived and tested. The simulation results and the experimental outcomes were in confluence, confirming the consistency of the design. In addition, the proposed system demonstrated well-constrained and controlled maneuvers which were the most demanding expectations from a soft-robot given the level of flexibility it has compared to a typical rigid body robot. By enhanced design of the exoskeleton and integrating with a robust control system, the performance of the proposed concept can be further enhanced. This study has only explored one type of wire material and configuration for the proposed structure. We are now working on optimizing the design with various materials, diameters and configurations. A means of maintaining the cross-sectional area of the untwisted prototype will also be explored. Further optimization of the proposed concept is believed to make it viable for use in medical applications.

References

1. IEEE: Soft Robotics—The Next Industrial Revolution? IEEE Robot. Autom. Mag. **23**(3), 17–20 (2016)
2. Hibbeler, R.C.: Statics and Mechanics of Materials. Prentice Hall, Jurong (2011)
3. Blanc, L., Delchambre, A., Lambert, P.: Flexible medical devices: review of controllable stiffness solutions. Actuators **6**(3), 2 (2017)
4. Steltz, E., Mozeika, A., Rembisz, J., Corson, N., Jaeger, H.: Jamming as an enabling technology for soft robotics. In: SPIE 7642, Electroactive Polymer Actuators and Devices (EAPAD) 2010, 764225, San Diego, California, United States (2010)
5. Brown, E., Rodenberg, N., Amend, J., Mozeika, A., Steltz, E., Zakin, M.R., Lipson, H., Jaeger, H.M.: Universal robotic gripper based on the jamming of granular material. Proc. Natl. Acad. Sci. U.S.A. **107**(44), 18809–18814 (2010)
6. Kim, Y.-J., Cheng, S., Kim, S., Iagnemma, K.: A novel layer jamming mechanism with tunable stiffness capability for minimally invasive surgery. IEEE Trans. Rob. **29**(4), 1031–1042 (2013)
7. Zuo, S., Iijima, K., Tokumiya, T., Masamune, K.: Variable stiffness outer sheath with "Dragon skin" structure and negative pneumatic shape-locking mechanism. Int. J. Comput. Assist. Radiol. Surg. **9**(5), 857–865 (2014)
8. Moses, M.S., Kutzer, M.D., Ma, H., Armand, M.: A continuum manipulator made of interlocking fibers. In: 2013 IEEE International Conference on Robotics and Automation (ICRA), Karlsruhe, Germany (2013)
9. Loeve, J., Plettenburg, D.H., Breedveld, P., Dankelman, J.: Endoscope shaft-rigidity control mechanism: "FORGUIDE". IEEE Trans. Biomed. Eng. **59**(2), 524–551 (2011)
10. Robertsons, "Section 8 - Wire Rope," in Robertsons Product Catalogue 2014 edition, Queensland, Robertsons (2014)
11. Trejos, L., Jayaraman, S., Patel, R.V., Naish, M.D., Schlachta, C.M.: Force sensing in natural orifice transluminal endoscopic surgery. Surg. Endosc. **25**(1), 186–192 (2011)

View-Invariant Robot Adaptation to Human Action Timing

Nicoletta Noceti[1(✉)], Francesca Odone[1], Francesco Rea[2], Alessandra Sciutti[2], and Giulio Sandini[2]

[1] DIBRIS, Università degli Studi di Genova, Genoa, Italy
{nicoletta.noceti,francesca.odone}@unige.it
[2] RBCS, Istituto Italiano di Tecnologia, Genoa, Italy
{francesco.rea,alessandra.sciutti,giulio.sandini}@iit.it

Abstract. In this work we describe a novel method to enable robots to adapt their action timing to the concurrent actions of a human partner in a repetitive joint task. We propose to exploit purely motion-based information to detect view-invariant dynamic instants of observed actions, i.e. moments in which the action dynamic is subject to a severe change. We model such instants as local minima of the movement velocity profile and mark temporal locations that are preserved under projective transformations, i.e. that resist to the mapping on the image planes and then can be considered view-invariant. Also, their level of generality allows them to easily adapt to a variety of human dynamics and settings. We first validate a computational method to detect such instants offline, on a new dataset of cooking activities. Then we propose an online implementation of the method, and we integrate the new functionality in the software framework of the iCub humanoid robot. The experimental testing of the online method proves its robustness in predicting the right intervention time for the robot and in supporting the adaptation of its actions durations in Human-Robot Interaction (HRI) sessions.

Keywords: View-invariant dynamic instants · Human action timing Human-robot interaction

1 Introduction

The success of a collaborative task is in great part determined by the selection of the appropriate action timing by the two partners. In fact, the ability to select the right moment to act and the choice of the correct speed are crucial to maximize the fluidity of an interaction.

Humans are in general very good at choosing when to intervene, often by anticipating when their partners are going to complete their sub-goals and move forth to the next sub-action [1]. Moreover, the adaptation to the others' speed often comes with no conscious effort: the two partners co-adapt the velocity of their motion, as commonly seen when two persons walk or clap together

K. Arai et al. (Eds.): IntelliSys 2018, AISC 868, pp. 804–821, 2019.
https://doi.org/10.1007/978-3-030-01054-6_56

[2] or when they perform arm movements in sequence [3]. This mechanism is particularly useful when executing repetitive joint tasks.

A current challenge in human-robot interaction is to provide the artificial agents with the ability to adapt to the human counterpart's action timing. Such skill helps at triggering in the user a similar tendency to adapt, leading to stability in synchronous behaviors and efficiency in joint actions [4]. Conversely, the natural tendency humans have to adapt to the partner's rhythm is significantly reduced when they collaborate with a non-adaptive robot in a goal-directed task [5,6]. Hence, a robot not picking the correct action timing and not adapting over time could lead to a break in the interaction flow and a lack of mutual adaptation.

To achieve this goal, the robot needs to identify some relevant time instants, informative of the partner's movement timing, in an observed action. Through the extraction of these instants the robot can infer the periodicity of a certain action and use this information to choose the appropriate moment to act and to adapt accordingly the duration of its own action. Real-time processing is a key requirement, as well as the ability to generalize to a variety of different human activities while being tolerant to view-point changes.

In this paper we propose to identify such relevant instants by exploiting motion information embedded in the so-called *dynamic instants*, i.e., time instants in which the dynamic of an action is subject to a severe change, that may be due to variations in the acceleration, or the direction of motion. The concept of dynamic instants was first proposed in [7], where the authors experimentally showed they are invariant to view-point changes. In this paper we define a set of dynamic instants computed from optical flow. We perform motion segmentation and describe the overall motion with the optical flow components, from which we derive information on the velocity profile over time [8]. We identify the local minima of such velocity as indications of a motion change and we use them for action segmentation. The dynamic instants we obtain in this way are classified as instants where an action is starting, ending, or changing.

Our goal is enabling mutual coupling between humans and robots, to support phenomena of synchronization and prediction in joint actions (see [14] for an overview). To this aim we look for distinguishable events during continuous movements which could help segmenting the action and predicting the unfolding of a repetitive motion. This bears some similarities with recent approaches based on oscillator theory (as in [15,16]), although we do not focus on modelling synchronization in terms of dynamical processes.

In the context of computer vision our approach may be related to works on action recognition, as [7,17–19]. Differently from our work, in such cases the segmentation of motion information in units is functional at the main recognition goal. Our work is different in spirit in that we are not pursuing a recognition task, and we do not rely on any specific motion definition (as gestures, actions, or more structured activities). Notably we look for instants that could be robust across different views and without requiring any a priori information about the actor's shape or position. As a consequence, we are able to process sequences

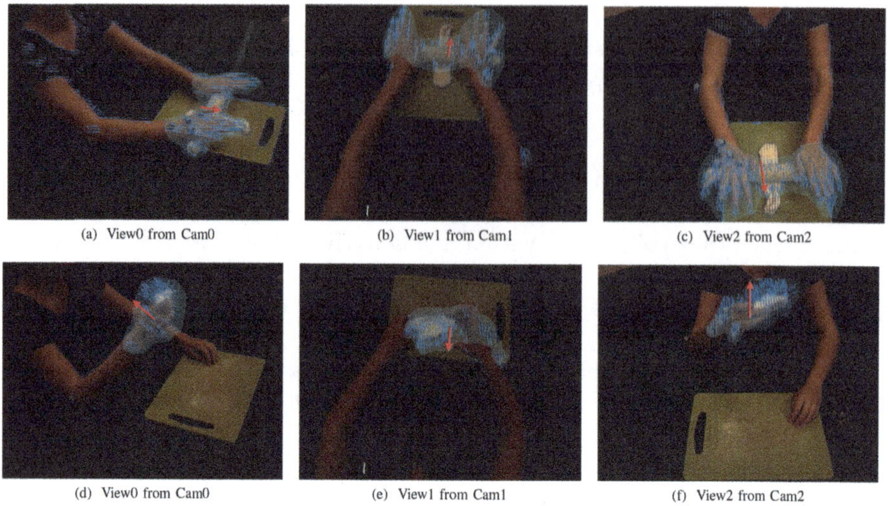

Fig. 1. Sample frames of the three views of the *Cooking* dataset (see Fig. 3) for two different actions (*Rolling the dough* above, *Eating* below). Vectors of the optical flow are reported in blue (sampled each 10 pixels), and the selected region of interest is highlighted in light blue. In red we report the average vector among the region of interest.

of motion observed at different granularities (e.g. performed with the full body or with a single hand) guaranteeing the capability of adapting to a variety of scenarios.

From an application standpoint, we focus on interaction tasks, first considering the batch analysis of a dataset we collected in-house – the *Cooking* dataset[1] – depicting kitchen activities observed from multiple points of view. Then we propose an online version of the algorithm and we validate it in a human-robot interaction scenario with the humanoid robot iCub [11].

We experimentally show that the proposed method detects rather accurately the dynamic instants of the variety of actions included in the *Cooking* dataset. Moreover, dynamic instants exhibit a clear persistence to viewpoint changes, with about 80% being detected across different views with a time difference inferior to 200 ms. We then verify that the online version of the module maintains an accurate performance in the detection of the instants and we demonstrate the feasibility of its use in interaction settings.

The remainder of the paper is organised as follows. Section 2, which introduces our approach to identify dynamic instants in video sequences, is followed by Sect. 3 where we assess the method in an offline scenarios. Section 4 considers an application to HRI, that we experimentally evaluate in Sect. 5. Section 6 is left to a final discussion.

---

[1] The dataset and its annotation will be soon made available online. Motion capture sequences will be also provided.

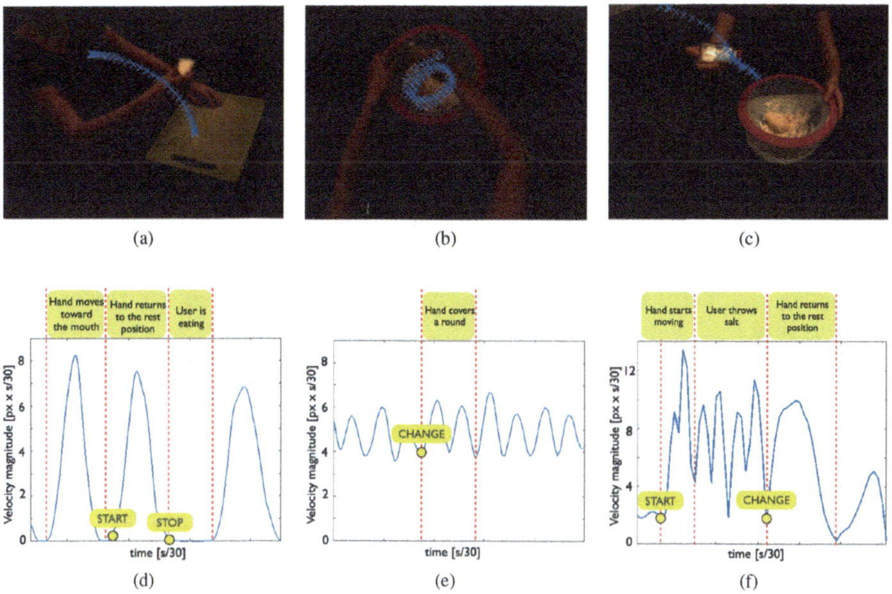

Fig. 2. Examples of trajectories from the *Cooking* dataset (above), with portions of their corresponding velocity profiles in which dynamic instants are detected (below).

2 Dyamic Instants Detection

In this section we introduce our approach to identify dynamic instants over time from a sequence of images. Given a video stream, we start by applying at each time instant the method in [9] to identify and describe a moving region. In the following, for readability we start with a brief sketch of the motion description method; then we will discuss the detection of dynamic instants.

2.1 Motion Detection and Representation

At each time instant t, we compute the optical flow using a dense approach [10] to provide an estimate of the apparent motion vector in each point of the image. Then, we apply a motion-based image segmentation to detect the region of interest to be considered in the next stages of the analysis. To this purpose, we apply the following steps:

1. The magnitude of the optical flow vectors is computed and thresholded to identify points with significant motion information. This produces a binary map where points characterised by a high module of the apparent motion vector are highlighted.
2. A perceptual grouping is applied to identify the connected components of the map; isolated points are discarded.

3. The largest connected component of the map is selected, if large enough, as representative of the structure of interest moving in the scene. We will refer to it as moving region of time t.

A selection of sample frames and their corresponding optical flow estimates are reported in Fig. 1. The moving regions identified in each image are also highlighted.

Each moving region, henceforth referred to as $\mathcal{R}(t)$, is then described exploiting the optical flow. Let $\mathbf{u}_i(t) = (u_i(t), v_i(t))$ be the optical flow components associated with point $\mathbf{p}_i(t) \in \mathcal{R}(t)$, and N the size of the region, i.e. the number of pixels in it.

The motion of the region can be compactly yet effectively described by computing the average of the optical components:

$$V(t) = \frac{1}{N} \sum_{\mathbf{p}_i(t) \in \mathcal{R}(t)} ||\mathbf{u}_i(t)||. \tag{1}$$

The red vectors reported in Fig. 1 provide a visual representation of such description.

By combining the obtained representations over time, we may finally compose a temporal description of the occurring dynamic event: given an image sequence of length T, the feature vector $\mathcal{V} = (V(t_0), \ldots, V(t_0 + T))$ is collected to represent the observed motion in the next stages of the analysis.

2.2 Dynamic Instants Detection

The velocity profile describing the apparent motion in a sequence provides interesting semantic cues for the interpretation of the activity occurring in the scene. Consider the examples reported in Fig. 2, that show some prototypical actions with their associated velocities. On the left, the evolution of an *Eating* action (Fig. 2a) – in which a user is repeatedly approaching the mouth with some food, and then returning to a rest position – is nicely described with a sequence of velocity bells followed by a plateau, in which the velocity is very close to zero, corresponding to the time interval where the user is eating (Fig. 2d). The bells can be easily identified by the starting and ending points. In the middle, a repetitive *Mixing* action is reported (Fig. 2b). Again, the velocity profile is composed by a sequence of peaks, while the velocity does not reach the zero between them (Fig. 2e). Nevertheless, those points still correspond to a change in the dynamic of the action. The rightmost example (Fig. 2c and f), describing the motion of a person sprinkling salt, shows a more complex velocity evolution, from which a sequence of action units can still be recognized.

Inspired by these observations, we aim here at identifying START, STOP, and CHANGE points as local minima in the velocity profile. More specifically, START points are those instants in which the motion starts after a rest, while STOP points correspond to the end of an action followed by an interval in which no motion is occurring. CHANGE points refer to time instants between two atomic action units, performed continuously in time. As for the detection procedure, given the locations LOC of the local minima in the velocity profile, we analyse each of these points to assign them a label. More specifically, if the velocity of a certain dynamic instant LOC(i) is above a threshold τ, then it is assigned the CHANGE label. Conversely, if the velocity is very close to zero, then we verify if a plateau is present after the current location, in which the velocity persists in this status. If this is the case, then we assign the labels STOP and START to, respectively, current and next dynamic instant, otherwise a CHANGE is again detected.

3 Validation in an Offline Scenario

In this section we validate our strategy in an offline setting, considering videos depicting cooking activities observed by three different view points.

3.1 Cooking Dataset

The dataset includes 20 actions performed by a user which is observed by a set of 3 cameras acquiring the video sequences synchronously (the acquisition setting is shown in Fig. 3): a lateral view (see sample frames in Fig. 1a and d), an egocentric viewpoint slightly above the subject's head (Fig. 1b and e), and a frontal view (Fig. 1c and f). As it can be noticed, the motion appearance can be significantly influenced by the viewpoint, which impacts not only on the direction but also on the magnitude of the average optical flow vectors (see the red vectors in Fig. 1). For this reason, the dataset is an ideal test-bed for evaluating the tolerance to viewpoint changes of our dynamic instants.

Most of the observed actions also involve the manipulation of objects. Each video consists in the repetition of 20 instances of the atomic action. For each action a pair of videos has been acquired, so to have available a training and a test sequence. Cameras acquire images of size 1293×964 at a rate of 30 fps.

We manually annotated the dataset marking the temporal locations of START, STOP, and CHANGE dynamic instants. To provide a reliable annotation, we only consider sequences in which the different action instances could be clearly segmented. We thus ended up with the list of 17 actions reported in Table 1.

3.2 Experimental Analysis

We start by evaluating the accuracy of our method in detecting the dynamic instants. If not otherwise stated, in all the experiments reported in the paper, the threshold applied to the optical flow estimates for the identification of the moving region has been set to 2, while the value of τ (see Sect. 2.2) has been fixed to 5. The values has been experimentally selected on the training set. In Table 1 we report for each test video sequence and for each class of dynamic instants, i.e. START (ST), STOP (SP), and CHANGE (CH) instants, the ratio between the

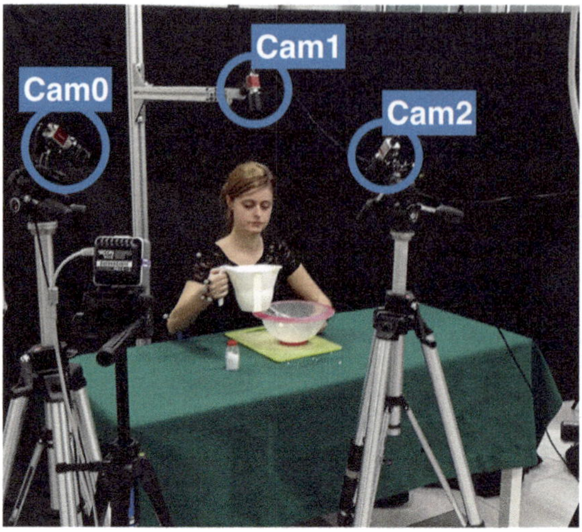

Fig. 3. The acquisition setting of the *Cooking* dataset, with 3 cameras acquiring synchronously video sequences of the observed actions.

Table 1. Performance evaluation of the method to detect dynamic instants on the *Cooking* dataset. ST, SP, and CH Refer to, respectively, START, STOP, and CHANGE points

	View0			View1			View2			Average		
	ST	SP	CH	ST	SP	CH	ST	SP	CH	ST	SP	CH
Cutting the bread	0.56	0.56	0.81	0.63	0.63	1.02	0.25	0.25	1.19	0.48	0.48	1.00
Cleaning a dish	–	–	0.96	–	–	1.02	–	–	1.00	–	–	0.99
Eating	0.83	0.83	– (3)	0.93	0.83	– (3)	0.93	0.90	– (3)	0.90	0.85	– (3)
Beating eggs	–	–	0.90	–	–	0.85	– (1)	– (1)	0.82	– (0.33)	– (0.33)	0.86
Squeezing a lemon	– (3)	– (3)	0.62	– (3)	– (3)	0.71	– (3)	– (3)	0.76	– (3)	– (3)	0.70
Using the mezzaluna knife	–	–	1.06	–	–	1.00	–	–	1.02 –	–	1.03	
Stirring a mixture	–	–	0.98	–	–	0.98	–	–	1.00	–	–	0.99
Opening a bottle	1.24	1.21	0.92	0.85	0.85	0.78	0.73	0.70	0.78	0.94	0.92	0.83
Crushing garlic	–	–	1.08	–	–	0.95	–	–	1.03	–	–	1.02
Peeling a potato	1.08	1.03	0.90	1.22	1.22	0.78	1.03	1.03	0.98	1.11	1.09	0.89
Pouring water	1.00	1.00	– (6)	1.00	1.00	– (15)	0.90	0.90	– (7)	0.97	0.97	– (9.33)
Reaching an object	1.00	1.00	–	1.42	1.42	– (3)	1.00	1.00	–	1.14	1.14	– (1)
Rolling the dough	–	–	0.98	–	–	0.98	–	–	0.98	–	–	0.98
Mixing a salad	–	–	1.00	–	–	1.10	–	–	0.98	–	–	1.03
Sprinkling salt	0.50	0.45	1.37	0.45	0.40	1.37	0.25	0.25	1.33	0.40	0.37	1.36
Cleaning the table	– (1)	– (1)	1.03	– (1)	– (1)	1.03	–	–	1.03	– (0.67)	– (0.67)	1.03
Transporting an object	1.00	1.00	–	1.00	1.00	– (1)	1.00	1.00	–	1.00	1.00	– (0.33)
Average	0.90	0.89	0.97	0.94	0.92	0.97	0.76	0.75	0.99	0.97	0.98	0.90

number of successfully detected points and the number of expected (annotated) points. A dynamic instant with label L is successfully detected at time t, if it exists in the ground truth an annotated instant at time t' with same label and such that $|t - t'| \leq \Delta T$. In the experiments, we fixed $\Delta T = 6$, corresponding to $\frac{1}{5}$s. If a certain class of dynamic instants should not be present in the sequence, we report in parenthesis the amount of false positives. Overall, the detection performs rather accurately. With the exception of a few cases (e.g. *Cutting the bread*, *Sprinkling salt*) the performance seems to be independent on the specific viewpoint, even though a slight decrease can be observed, on average, for View2 which is apparently more affected by ambiguity in the motion estimation due to the relative position between the camera and the action planes.

Focusing on the actions, it is worth noting that for some of them (especially the already mentioned *Cutting the bread* and *Sprinkling salt*) the presence of START and STOP points have been annotated even though no clear pause was present between the repetitions. This leads to an annotation including *ideal points* with no clear counterpart in the measurements, and as a side effect the misclassification of some of them as CHANGE instants. We finally mention that the action *Pouring water* is characterised by a significant number of false positives for the CHANGE class, mainly due to the apparent motion of the water inside the container manipulated by the subject.

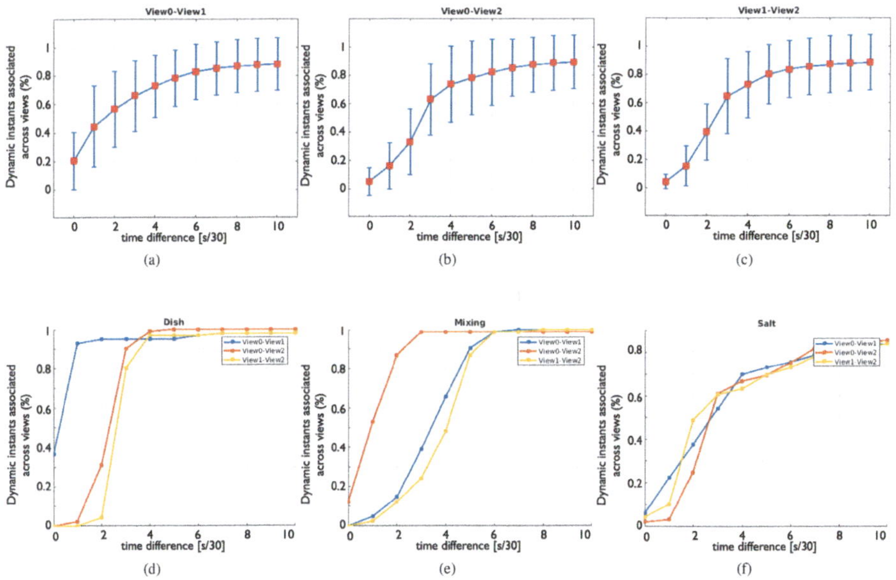

Fig. 4. An evaluation of the persistence of dynamic instants detection across views. Above: overall percentage of dynamic instants successfully matched between pairs of views for different values of the time difference (ΔT_V in the text). Below: a focus on some prototypical actions behaviors: *cleaning a dish*, mixing a salad, *sprinkling salt* (see text for further details).

To evaluate the persistence of dynamic instants to viewpoint changes, we consider a pair of views i and j, and an allowed time difference ΔT_V. Given a dynamic instant P with label L detected in view i at time t, we consider the detection persistent to the viewpoint change if it exists a dynamic instant P' detected in view j at time t' and with label L such that $|t-t'| \leq \Delta T_V$. In Fig. 4, first row, we show the evaluation between each pair of views in the *Cooking* dataset. For different values of the time difference, we compute the percentage of dynamic instants which have been matched between the views. The analysis has been performed in both directions (i.e. from view i to view j, and vice versa), and then the average accuracy has been plotted. It is easy to notice that for all the comparisons the percentage of matched dynamic instants approaches the 80% already with 6 frames ($\frac{1}{5}$s) of difference. View0 and View1 seem in good agreement, with around the 20% of matched points detected exactly in the same time instants.

In the second row of Fig. 4 we report some prototypical behaviors of single actions across views. On the left, *Cleaning a dish* is an action performed on a diagonal plane with respect to the frontal view (View2), ending up with less ambiguity between points extracted from View0 and View1, already with an expected delay of 0 frames (about the 40% of points are matches). In the center, the *Stirring a mixture* sequence is performed on the table, thus the appearance from the top view (View1) is different with respect to the other two, which are more coherent. Last, on the right, we report an example of action where no significant difference exists in the relations between couples of views.

4 Using Dynamic Instants for HRI

In this section we introduce the implementation of the dynamic instants detector on the iCub humanoid robot (see a visual sketch of the procedure in Fig. 5). To this purpose, we modified the algorithm to work online and to be integrated within the iCub software framework, based on the middleware Yarp [11]. More specifically, the online dynamic instants detection follows the same process described in Sect. 2, but considers a sliding window in which the detection is performed. Assuming that the observed action is cyclic, a prediction on the next dynamic instant occurrence and of the next action duration may be also obtained and used to generate a robotic action with the appropriate timing. In the following we provide details on each module involved in the computation.

4.1 Online Dynamic Instants Detection in the iCub Framework

The input for the dynamic instant detection is a sequence of images from the right camera of the robot iCub, acquired with a frame rate of 15 fps and a 320×240 resolution. The method is run after the software module proposed in [9,12], that allows iCub to recognize the presence of motion in the surrounding which is likely to be produced by a human, i.e. by a potential interacting

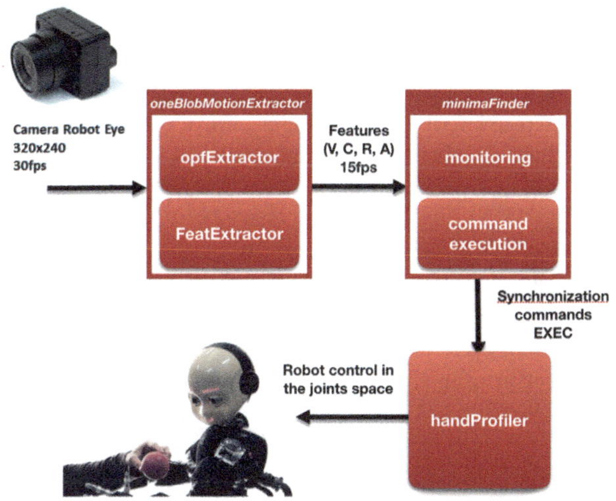

Fig. 5. A visual sketch of the online procedure to detect dynamic instants on the and plan the consequent action of the robot.

agent. After that, the video stream is fed to a software module called *oneBlobMotionExtractor* which extracts motion features from the optic flow associated with the observed action. The features are passed to a *minimaFinder* module, which analyzes velocity over time, predicts the occurrence of future dynamic instants and establishes when the next robot action should start an how long it will last. The trigger and duration of the robot action are then sent to the *handProfiler* module, which is in charge of generating a biological plausible robot action. In the remainder of the section we provide a more detailed description of each module.

(1) oneBlobMotionExtractor: The module includes two classes of concurrent computing modules, *opfCalculator* and *featExtractor* that address respectively the segmentation based on motion and the description of the motion through features. The computational demand of the module is distributed in multiple threads to achieve efficient parallel computing. The *opfCalculator* module provides maps of the horizontal and vertical components of the optical flow $\mathbf{u}_i(t) = (u_i(t), v_i(t))$ on the whole image to the rest of the Yarp network. Since the eyes of the robot may be moving during the acquisition, a component of ego-motion may be present. For this reason, the estimate is provided after a compensation of it. The *featExtractor* class analyzes the largest and most persistent moving blob of the image plane. Our assumption is that only one region of interest moves in the scene. The class provides to the rest of the Yarp network a timestamp of the current execution cycle and a vector of 4 motion features: velocity, curvature, radius of curvature, and angular velocity. For the rest of the analysis, following the consideration in Sect. 2 we exploit velocity to detect the dynamic instants in the observed actions.

(2) minimaFinder: The *minimaFinder* module operates in parallel two asynchronous algorithms: *monitoring* and *command execution*. For *monitoring*, the *minimaFinder* module reads the vector of the 4 features provided by the *featExtractor* from the Yarp network and finds the minima in the dynamic progression of velocity. The values are filtered in time according to the following operation:

$$V_k^f = \alpha \cdot V_k^m + (1 - \alpha) \cdot V_{k-1}^f \tag{2}$$

where α is a weight controlling the importance of the two terms, V_k^f is the filtered velocity and V_k^m is the measured velocity, to eliminate small fluctuations due to noise in the sensing.

Within a non-overlapping sliding time window, the module stores the velocity features passed by *featExtractor* and the corresponding timestamp in a buffer. At the end of the time window, the *minimaFinder* extracts the minimum of the buffer. By subtracting the timestamp of such minimum with that extracted in the previous time window, the estimate of the temporal interval between two consecutive dynamic instants in the motion is derived: $\Delta t_k = t_k - t_{k-1}$ where t_k is the timestamp of the current minimum in velocity.

The temporal interval is saved in a FIFO stack buffer to maintain memory of measured time intervals in order to detect relevant changes in the human action pace. Moreover, the size of the non-overlapping sliding time window is reset to estimated time interval.

Assuming that the observed action is cyclic, the module predicts the time interval before the next dynamic instant by computing a linear weighted sum of the previous and current measured time intervals:

$$\bar{\Delta} t_k = \beta \cdot \Delta t_k + (1 - \beta) \cdot \bar{\Delta} t_{k-1} \tag{3}$$

with β balancing the importance of the two terms, and $\bar{\Delta} t_k$ representing the estimated time interval before the next dynamic instant.

The occurrence of the next dynamic instant, hereafter *targetTimestamp* $\hat{t}_{k+1}$, is then obtained by summing to estimated interval the last detected dynamic instant $\hat{t}_{k+1} = t_k + \bar{\Delta} t_k$.

In the *command execution* phase, when the internal clock reaches the *targetTimestamp* the *minimaFinder* sends to the controller of the whole body motion of iCub the command $\boldsymbol{EXEC}(t, \delta)$. The first parameter sets the timing of the robot action in correspondence to the (predicted) dynamic instant of human action, the second parameter controls the duration of the next action to a value corresponding to the estimated time interval.

The *minimaFinder* avoids to send new commands to the controller if the controller has not terminated the action. At the end of the action execution the *minimaFinder* adopts three strategies in relation to three possible conditions: (1) The new *targetTimestamp* occurs after the end of the action execution time; (2) The new *targetTimestamp* occurs before the end of the previous action execution time of ε seconds; (3) The new action execution occurs after the end of the next time window.

In the first condition the *minimaFinder* waits until the *targetTimestamp* is reached to send the new command. In the second condition the *minimaFinder* immediately executes another action command but action duration δ is computed according to: $\delta = \bar{\Delta t_k} - \varepsilon$. In the third condition the *minimaFinder* immediately executes another action of duration equal to the current estimate of the interval between dynamic instants.

(3) handProfiler: The *handProfiler* module generates and executes wholebody biologically plausible movements, given a desired trajectory and the desired action duration. The only constraint is that the trajectory of each motion (or part of motion) can be represented as a portion of an ellipse. Ellipses have been chosen since their parametric definition allows for the reproduction of wide number of trajectories (e.g.: circles, quasi-straight curves). The robot executes the motion in the 3D space guaranteeing that the velocity of the end-effector (center of the palm in the iCub robot) varies with its curvature according to a law proper of biological motion, the two-third power law [13]. The selection of biologically plausible motion for the humanoid robot is motivated by findings that suggest that this choice is crucial to foster automatic adaptation by humans (e.g. [3]). Once generated, the action is saved as trajectory (position and time) in the space of joint angles by computing the inverse kinematic. This assures perfect repeatability of the trajectory and fast execution of the action, since inverse kinematic is never re-computed during execution. Moreover, the module can change the duration of the action by recomputing the scale factor that changes proportionally all the time intervals between consecutive joint positions. The procedure does not change the shape of the velocity profile and consequently maintains the action compatible with the two-third power law.

5 Validation in an Online Scenario

With the software framework described above the robot should be able to detect online the dynamic instants in the observed action, predict the occurrence of the next ones and time accordingly the start and duration of its own action. To evaluate the performance of the system we measured its accuracy as the ratio between the number of instantiated robot actions and the number of human actions, which was approximated with the number of dynamic instants extracted with an offline detection. Moreover, for the dynamic instants accurately detected, we computed the error in action timing as the average difference in time between each dynamic instant (detected offline) and the closest action start. Last, we computed the duration of human actions as the difference in time between two consecutive dynamic instants (detected offline) and the duration of robot actions.

In the remainder, the value of α (2) has been set to 0.4, the size of the non-overlapping sliding time window is initialized to 2 s. Finally, we set $\beta = 0.9$ (3). All the values have been experimentally selected on the training set of the *Cooking* dataset.

5.1 Comparative Assessment with the Offline Algorithm

To assess whether the software framework produces results that are comparable with the dynamic instants detection performed offline, while allowing for robot action execution, we test the algorithm on a subset of the videos of the *Cooking* dataset, downsampled at 30 fps and at a resolution of 320 × 240. We focus on videos depicting View 2, which would correspond to a situation in which the robot is standing in front of the actor. Moreover, we consider the actions characterized by START and STOP dynamic instants only (*Transporting an object*, *Reaching an object*, *Pouring water* and *Eating*, see Table 1), where the annotation of the exact occurrence of the dynamic instants was more straightforward.

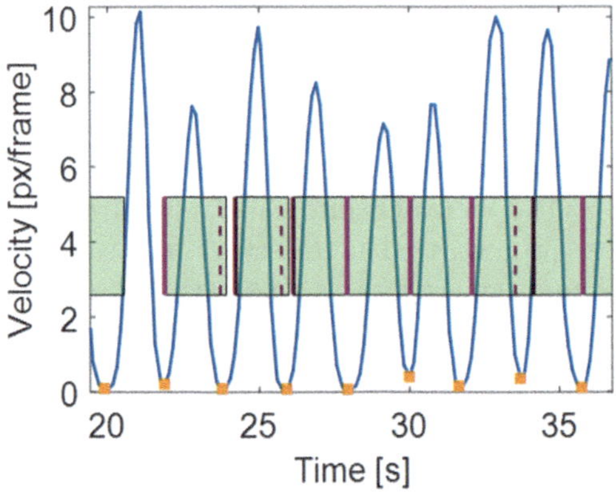

Fig. 6. Representative plot of action processing by the robot. The graph presents human action (blue line: velocity) and the corresponding robot action timing (green patches). The dynamic instants of human motion are represented by orange squares. Each robot action patch has a color-coded first edge: magenta, when the robot could start the action at the predicted dynamic instant; purple, when the action start was postponed because the predicted dynamic instant (dashed purple line) occurred while the robot was concluding the previous motor command (see text for more details).

In Fig. 6, we plot a representative example of the output of the software system for the *Transporting* action. The graph presents a portion of the velocity trend over time, derived from the optic flow by the *oneBlobMotionExtractor*. The orange squares represent the dynamic instants detected by the batch analysis (Sect. 2). The velocity plot is overlaid with green patches representing the time of the execution of the robot actions, as commanded by the *minimaFinder* module. The color of the first edge of each patch indicates whether the robot could start the action at the predicted dynamic instant (magenta line - case **a**, in the description of *minimaFinder* in Sect. 4.1) or that was not possible because the robot was concluding the previous motor command (case **b**). In the latter case, the timing of the predicted dynamic instant, *targetTimestamp*, is represented by a purple dashed line within the action patch. Since in that case an action is triggered at the end of the current execution, with a duration shortened in order to enable the robot to "catch" the next dynamic instant, the first edge of the patch of such corresponding action is plotted in purple.

In Table 2 we report the performance of the system for the four actions considered. On average, the majority (88%) of the dynamic instants detected by the offline analysis are also detected and successfully used in the online setting to trigger an action. Moreover, the average timing error between the human dynamic instants and the start of robot action is close to 0 (on average 90 ms). The average duration of robot actions is close, on average, to the duration of the faster among the human actions considered, whereas results shorter than the longer ones (as *Eating* and *Pouring water*). Examining the trend of the durations of robot actions over time, it emerges a tendency to increase toward the end of the video. The issue can be cured with a longer exposure of the robot to the action.

Table 2. Results of the validation of the online system on prerecorded videos. Averages ± SD

	Accuracy	Err Act [s]	Dur Hum [s]	Dur Rob [s]
Eat	0.94	0.37 ± 0.42	2.26 ± 0.58	1.91 ± 0.29
Pour	1.00	0.02 ± 0.58	2.73 ± 0.62	1.88 ± 0.32
Reach	0.75	0.05 ± 0.69	1.63 ± 0.19	1.81 ± 0.30
Transp.	0.84	−0.07 ± 0.66	1.74 ± 0.17	1.79 ± 0.25
Average	0.88 ± 0.11	0.09 ± 0.19	2.09 ± 0.5	1.85 ± 0.06

5.2 Proof of Concept: Interactive Scenario

To validate the system in an interactive online setting, we considered a proof-of-concept scenario, where the robot had to execute a stereotyped action in coordination with a human performing its own task. We considered the case of iCub performing an elliptical tool movement (as if for pouring something from a spoon) in coordination with a human partner transporting an object (see supplementary video). /colormagenta The coordination does not necessarily need to show an object transferring from the human to the robot. In fact in order to show coordination both out-of-phase and in-phase conditions constitute a valid result for our analysis. This justifies the absence of real object transferring from the human to the robot. Further we chose the elliptical trajectory as representative of simple cyclic action but also for its good degree of generalisation. In fact the elliptical trajectory is designed by the user thanks to a parametric definition. By changing ellipse parameters the trajectory can be transformed in broad range of other curves from pseudo-straight curve (ellipse with great radius of curvature) to circular curves (elliptical curves with constant radius of

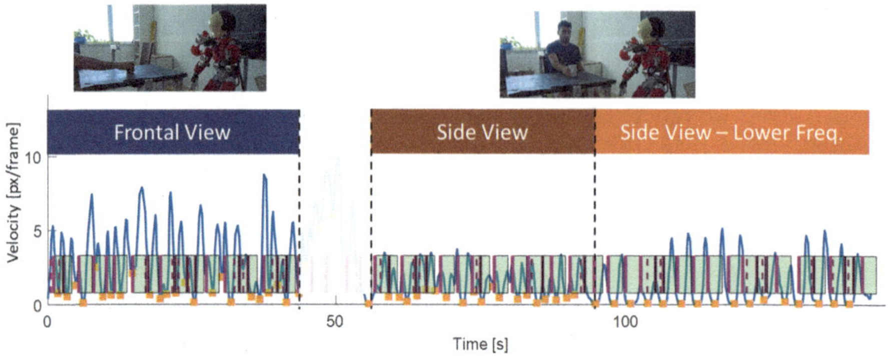

Fig. 7. Temporal structure of the HRI proof of concept experiment. The robot first is positioned in front of the human actor (Frontal View), then the actor moves to the side (Side View) and then slows down action frequency, keeping the same position (Side View - Lower frequency). The bottom panel shows the corresponding human and robot action timings, with the same conventions as Fig. 6.

curvature). We evaluated robot performance when observing the action from two different perspectives (frontal and side-view) and also after a change in the pace of human action execution. The results are reported in Fig. 7 and Table 3. From the analysis it emerges that the robot starts its actions in correspondence of about 86% of the dynamic instants, with higher accuracy for the slower action. The time difference between a dynamic instant and the closest action start is relatively short (about 300 ms on average). The performance is very similar across the two views. Action duration is comparable between human and robot in the faster pace condition, whereas robot action results shorter when human motion becomes suddenly characterized by a slower pace. /colormagenta The timing of the sensing and control loop does not affect the response time since a careful distribution of the computational demand on multiple computers removed the computation bottleneck associated with the extraction of the feature from the optical flow. Also in this scenario, the robot would probably have needed more time to allow for a full adaptation. In summary the results of the validation in the interactive scenario confirm the positive evaluation obtained from the analysis of the prerecorded *Cooking* actions.

6 Discussion

In this paper we proposed a method for detecting dynamic instants from video sequences as local minima of the velocity profile of an action, discussing a straightforward application for robotics where such information is a key to select the best timing for collaborative human-robot acts. The method has been designed to be as general as possible, with no a priori knowledge of the actor's shape, the trajectory of the motion and the perspective from which it is observed. This algorithm, which is also able to predict the occurrence of future dynamic

Table 3. Results of the proof of concept human-robot experiment, considering the different views (*Front*, *Side*) and the Slower Action Pace ($Side_S$). Averages ± SD

	Accuracy	Err Act [s]	Dur Hum [s]	Dur Rob [s]
Front	0.79	0.23 ± 0.60	1.82 ± 0.61	1.84 ± 0.38
Side	0.78	0.19 ± 0.64	1.81 ± 0.53	2.01 ± 0.44
$Side_S$	1.0	0.49 ± 0.77	2.92 ± 0.37	2.06 ± 0.44
Average	0.86 ± 0.12	0.30 ± 0.16	2.18 ± 0.64	1.97 ± 0.11

instants by exploiting previous measures, informs the robot about *when* it should trigger the start of its actions – to start simultaneously with the human partner – and *how long* the action should last – to maintain the coordination in time.

In the first part of the paper we validated the approach in an offline scenario, using the *Cooking* dataset. The results clearly speak in favor of our method, also showing the persistence of dynamic instants detection across different views.

In the second part of the paper, we present the implementation of the proposed method in the iCub framework. The dynamic instants detection is performed in parallel with action execution, by looking for velocity minima on variable-size sliding time windows, whose size is progressively adjusted by the algorithm, in order to cope with actions of different dynamics. We experimentally showed on actions of the *Cooking* dataset that the robot could adapt to actions performed with slightly faster rhythm (on average 1.8 s) by starting from the fixed window size of 2 s.

Later, we showed how to make the robot adapt to human timing – considering both robot action start time and its duration – exploiting the history of dynamic instants observed in the past human movements. This way, potential errors in the estimate of the next dynamic instant or conflicting situations (e.g. when the robot action should be triggered before the previous one is completed) can be handled, without breaking the adaptive mechanisms. Currently, conflicts are handled forcing the robot to perform the action as soon as the previous instance has been completed, appropriately adjusting the duration so as to finish tentatively before the subsequent predicted dynamic instant. This strategy allows us to guarantee a more continuous action flow, thus favouring coordination and mutual adaptation. Future works will specifically consider situations in which the stability of the robot action duration is more crucial, so as to diversify the robot response depending on the context.

It is important to note that for the moment the method assumes that the observed action is repetitive and tries at every cycle to adjust the robot action timing as a consequence. In the future it might be possible to consider an "observation phase" in which the robots just monitors the human partner behaviors and evaluates the stability of the timing, before deciding whether to start with its own repetitive actions. This "observation phase" could be triggered again in case an excessive variability in the observed action periodicity is measured over time, suggesting that the human action has ceased to be rhythmic.

The simplicity of the method proposed could make it a basic layer for higher-level action processing, with the possibility of exploiting alternative ways for segmenting the moving region (e.g. using object detection) or different sensors (as skeleton from depth or motion capture data).

Acknowledgment. The research presented here has been supported by the European CODEFROR project (FP7-PIRSES-2013-612555).

References

1. Flanagan, J.R., Johansson, R.S.: Action plans used in action observation. Nature **424**(6950), 769 (2003)
2. Neda, Z., Ravasz, E., Brechet, Y., Vicsek, T., Barabasi, A.-L.: The sound of many hands clapping. Nature **403**, 849–850 (2000)
3. Bisio, A., Sciutti, A., Nori, F., Metta, G., Fadiga, L., Sandini, G., Pozzo, T.: Motor contagion during human-human and human-robot interaction. PLoS One **9**, e106172 (2014)
4. Mörtl, A., et al.: Modeling inter-human movement coordination: synchronization governs joint task dynamics. Biol. Cybernet. **106**(4–5), 241–59 (2012). 1–19
5. Lorenz, T., Mörtl, A., Hirche, S.: Movement synchronization fails during non-adaptive human-robot interaction. In: Proceedings of the 8th ACM/IEEE International Conference on Human-Robot Interaction, pp. 189–190. IEEE Press, March 2013
6. Vannucci, F., Sciutti, A., Jacono, M., Sandini, G., Rea, F.: Adaptation to a humanoid robot in a collaborative joint task. In: 26th IEEE International Symposium on Robot and Human Interactive Communication (2017)
7. Rao, C., Alper, Y., Mubarak, S.: View-invariant representation and recognition of actions. Int. J. Comput. Vis. **50**(2), 203–226 (2002)
8. Noceti, N., Sciutti, A., Sandini, G.: Cognition helps vision: recognizing biological motion using invariant dynamic cues. In: International Conference on Image Analysis and Processing (2015)
9. Vignolo, A., Rea, F., Noceti, N., Sciutti, A., Odone, F., Sandini, G.: Biological movement detector enhances the attentive skills of humanoid robot iCub. In: IEEE-RAS 16th International Conference on Humanoid Robots, pp. 338–344 (2016)
10. Farnebäck, G.: Two-frame motion estimation based on polynomial expansion. In: Image Analysis, pp. 363–370 (2003)
11. Metta, G., Natale, L., Nori, F., Sandini, G., Vernon, D., Fadiga, L., Bernardino, A.: The iCub humanoid robot: an open-systems platform for research in cognitive development. Neural Netw. **23**(8), 1125–1134 (2010)
12. Vignolo, A., Noceti, N., Rea, F., Sciutti, A., Odone, F., Sandini, G.: Detecting biological motion for human robot interaction: a link between perception and action. Front. Robot. AI **4**, 14 (2017). https://doi.org/10.3389/frobt
13. Noceti, N., Odone, F., Sciutti, A., Sandini, G.: Exploring biological motion regularities of human actions: a new perspective on video analysis. ACM Trans. Appl. Percept. (TAP) **14**(3), 21 (2017)
14. Bütepage, J., Kragic, D.: Human-Robot Collaboration: From Psychology to Social Robotics. arXiv preprint arXiv:1705.10146 (2017)
15. Mörtl, A., Lorenz, T., Hirche, S., Vasilaki, E.: Rhythm patterns interaction-synchronization behavior for human-robot joint action. PloS one **9**, e95195 (2014)

16. Ijspeert, A.J., Nakanishi, J., Schaal, S.: Learning rhythmic movements by demonstration using nonlinear oscillators. In: Proceedings of the IEEE/RSJ International Conference on Intelligent Robots and Systems (IROS 2002), no. BIOROB-CONF-2002-003 (2002)
17. Cabrera, M.E., Wachs, J.P.: A human-centered approach to one-shot gesture learning. Front. Robot. AI **4**, 8 (2017). https://doi.org/10.3389/frobt.2017.00008
18. Shi, Q., Wang, L., Cheng, L., Smola, A.: Discriminative human action segmentation and recognition using semi-Markov model. In: IEEE Conference on Computer Vision and Pattern Recognition, pp. 1–8 (2008)
19. Shao, L., Ji, L., Liu, Y., Zhang, J.: Human action segmentation and recognition via motion and shape analysis. Pattern Recognit. Lett. **33**(4), 438–445 (2012)

A Rule-Based Expert System to Decide on Direction and Speed of a Powered Wheelchair

David A. Sanders[1(✉)], Alexander Gegov[2], Malik Haddad[1], Favour Ikwan[1], David Wiltshire[1], and Yong Chai Tan[3]

[1] School of Engineering, Portsmouth University, Portsmouth, UK
{david.sanders, david.wiltshire}@port.ac.uk, malik@malikhaddad.com, Favour.Ikwan@myport.ac.uk

[2] School of Computing, Portsmouth University, Portsmouth, UK
alexander.gegov@port.ac.uk

[3] Faculty of Engineering and Built Environment, SEGi University, Petaling Jaya, Malaysia
tanyongchai@segi.edu.my

Abstract. Some rule based techniques are presented that can assist powered wheelchair drivers. The expert system decides on the direction and speed of their wheelchair. The system tends to avoid obstacles while having a tendency to turn and head in towards a desired destination. This is achieved by producing a new target angle as an extra input. Other inputs are from sensors and a joystick. Directions are recommended and mixed with user inputs from the joystick representing desired direction and desired speed. The rule-based system decides on an angle to turn the powered wheelchair and suggest it. Inputs from the joystick and sensors are mixed with the suggested angle from the Rule Based Expert System. A modified direction for the wheelchair is produced. The whole system helps disabled wheelchair users to drive their powered wheelchairs.

Keywords: Rule-based · Decision · Wheelchair · Assist · Powered Collision avoidance · Driving

1 Introduction

This paper proposes a rule based expert system to assist with the control of a powered wheelchair. Powered wheelchairs are used by disabled people who cannot use a manual wheelchair because they can't rotate the wheels. A powered wheelchair can provide independence and freedom [1]. The methods and systems presented here will help more disabled people to drive powered wheelchairs. Knowledge about the environment around the wheelchair is provided by ultrasonic sensors so that the system can assist a disabled driver with avoiding obstructions and obstacles in their path. Knowledge of the direction towards a desired destination allows the system to tend to guide the wheelchair towards the destination.

Around seven million Americans use assistive mobility devices. There are about two million wheelchair and scooter users and another five million use other devices, for

K. Arai et al. (Eds.): IntelliSys 2018, AISC 868, pp. 822–838, 2019.
https://doi.org/10.1007/978-3-030-01054-6_57

example walkers, canes and crutches [2]. Nearly one third of people using mobility devices require assistance from other people [1].

The predominant primary conditions for wheelchair and scooter users are stroke and osteoarthritis. Osteoarthritis is the main condition linked to using mobility devices [1, 2].

A powered wheelchair is usually used by people lacking mobility or dexterity because of shoulder, arm hand, or more widespread disability, and who do not have enough strength in their legs to use their feet to push a manual wheelchair. Powered wheelchairs can also provide tilt, recline and elevation, and bespoke functions for health and normal day-to-day functioning.

Powered wheelchairs can be categorised into four types: wheelchairs driven by their front, centre or rear wheels and four-wheel drive wheelchairs. They can also be categorised by their seating: (a) similar to a seat in a car and (b) with a sling-style seat and frame.

A user of a wheelchair often controls their speed and direction with a joystick. If a user lacks the coordination to effectively use a joystick or if they can't use their hands or fingers then other input devices can be used (foot control, sip tubes/puff switches, or head or chin controllers, etc.).

Technology is allowing more and more people to use powered wheelchairs for both indoor and outdoor use.

They can include outdoor wheels and tires and powered wheelchairs can move at up to 6 mph. They can have extra wheels to provide stability, for example when outside and away from pavements and roads. Rear or mid wheel drive chairs are well-liked for use indoors and outdoors.

Users of powered wheelchair can spend a great deal of time in their wheelchairs, so they need to suit the environment where they are being used. Every wheelchair driver is unique and they need bespoke seats, arm rests and leg rests to provide comfort and stability. Powered seats, reclining and tilting backs, and electric leg rests are possible additions.

If a user has head injuries, neurological or physiological problems, or lacks special awareness then they might not be able to safely steer. Potential wheelchair users could be unable to avoid collisions or be blind, etc. Systems described here are helping these disabled wheelchair drivers to drive more safely.

Controllers for wheelchairs are normally open-loop and drivers indicate desired speed and direction by positioning an input device (for example a lever or joystick) and the chair tends to travel at the desired speed along the desired route. Wheelchair users make corrections to evade obstacles. This paper describes how information from an input device can be processed and mixed with inputs from a sensor system and a target destination to assist a wheelchair user in steering their chair. Global planning and local planning combine inside a rule-based expert. The result provides drivers with assistance. A global path is mixed with local information from the sensor system [3].

Powered wheelchair navigation has been considered [4, 5]. Algorithms have typically been local and no attempt has been made to improve a system more globally. Obstacle avoidance has been considered [6] with local inputs from sensors [7].

Some research work has calculated initial wheelchair paths and then has locally modified them if an obstruction was perceived [3] but they have rarely been used successfully to assist wheelchair users. This paper describes how three inputs can be used with a local planner to drive the wheel motors. The three inputs are: on-board sensors, a joystick, and a global target destination. The powered wheelchair can react quickly to joystick movements but can also respond to any obstacles that might be detected ahead. The powered wheelchair tends to turn towards the global destination but can avoid obstacles along the way.

Huq *et al.* used a fuzzy context-dependent system to eliminate some limitations [8] by using a goal oriented navigation while avoiding obstacles. Genetic algorithms were mixed with Fuzzy logic to overcome some mapping difficulties and establish a local position [9]. Bennewitz and Burgard presented random planning methods that could produce real-time paths in unknown environments [3, 10] that precisely followed trajectories [11]. Hwang and Chang presented obstacle avoidance methods that used fuzzy decentralized sliding-mode control [12]. Song and Chen solved some of the local minima problems and then improved on the well-known potential-field technique [7] and Nguyen *et al.* produced Bayesian Neural Networks to avoid obstructions [13].

Techniques are presented here to partially optimise some minimum-cost paths. A joystick regulates direction and speed and AI systems provide input to modify them if necessary [14–17]. The system uses perception based rules that are similar to [3, 35].

Calculations trade off the distance to objects against the length of a path. A steering angle is determined by rules and that is combined with input from the joystick. A new revised steering angle is created and that is used to drive the motors. The procedures were tested in simulation and with sensors mounted on a Bobcat II powered wheelchair (Fig. 1).

Fig. 1. A Bobcat II powered wheelchair.

Many different sensors can be used to assist a powered wheelchair user to safely avoid obstacles [18]: infrared [21]; ultrasonic [20] or laser or structured light [19]. Global systems are tricky to use inside a building [22] but local sensors have been used such as: gyroscopes, odometers, tilt sensors and ultrasonics [23, 24]. Cameras are reducing in price but processing tends to be more complex [25]. Computers are getting more powerful and are also reducing in price [26]. That means that cameras are being used more often for applications. The best source of knowledge about what is required is still usually the disabled human driver, but reduced visibility or their disability can reduce their proficiency [27].

Ultrasonics were selected because they are robust, cheap and simple [28]. The sensors and input from the joystick are described in Sect. 2 and then Sect. 3 describes wheelchair kinematics. Section 4 describes the rules and the control methods and Sect. 5 presents some of testing and results. Section 6 is a short conclusion.

2 Joystick and Sensor Inputs

2.1 Ultrasonics

The ultrasonic systems are like the systems described in [29–33, 35]. Sensors were mounted above each driving wheel. Distance to objects was measured using the time taken for pulses to reflect back from the obstacle to the receivers.

The wheelchair has a solid steel frame for strength and stability and is covered by a shell made of fibreglass. Trailing casters are at the back and large driving wheels are at the front. Each driving wheel had an ultrasonic sensor secured on the frame above it.

A joystick was usually connected straight to a wheelchair controller to steer the powered wheelchair. In the research presented in this paper, that direct connection between powered wheelchair and joystick was parted and instead, a computer was introduced in between the chair and joystick. The joystick input was then managed by the computer. The system could function in a choice of:

- Joystick input sent directly to controller.
- Joystick input modified by the computer to adjust direction and speed.

Three basic rules applied to modifying direction and speed:

1. Overall control remained with the disabled wheelchair user.
2. Direction and speed were only adjusted when needed.
3. If a change in speed and direction was needed then the change applied was smooth.

Imaginary potential fields were placed around obstacles detected by the sensors [7, 23, 35]. If nothing was being sensed then a range-finder gradually increased the range of the sensors (by lengthening the ultrasonic pulses) until potential obstacles were detected so that the system provided warnings of likely difficulties ahead.

2.2 Mapping the Environment

Ultrasonics are often noisy and can provide incorrect readings. Because of that, Histogramic In-Motion Mapping was used to filter out any incorrect readings [35]. A volume ahead of the wheelchair was divided into a right-hand side and a left-hand side. A matrix was then established for each side and an overlapping volume with three elements in each matrix: IMMEDIATE, HALFWAY and OUTLYING.

If an obstacle was detected somewhere in front of the powered wheelchair then it was labelled as IMMEDIATE, HALFWAY or OUTLYING. Sensor beams over-lapped and bounded the volume ahead. The centre matrix denoted circumstances when both left and right sensors had detected something.

The volume ahead of the powered wheelchair was therefore represented by a 2-D 3×3 grid with nine elements:

LEFT-HAND SIDE, MIDDLE, RIGHT-HAND SIDE
×
IMMEDIATE, HALFWAY, OUTLYING

If obstacles were sensed then associated element(s) in the grid were increased by a relatively large amount, for example 5, to a max value of 15. Other empty elements were reduced by a smaller amount (for example 2) down to zero. In that way, a straightforward histogrammic representation was created that represented a volume ahead of the powered wheelchair. When obstacles were sensed then the values of the associated cells swiftly increased. Random errors in any cells might fleetingly increase because of solitary misreads but they would then quickly reduce again. If obstacles were detected in an element but then they moved to another element, then the new element quickly increased in value. If an obstacle vanished from an element then its value decreased to zero. A reliable estimation of the range to an obstacle was arrived at within <0.4 s.

2.3 Joystick

A Penny & Giles joystick was used to drive the powered wheelchair. The joystick had 2× potentiometers within it and joystick position was resolved with 2× A/D converters that were connected to the potentiometers. The data extracted from the joystick was effectively in a Cartesian coordinate system. The Cartesian coordinates were transformed to a polar coordinate system: $|J|\angle\theta$. The speed demanded by the wheelchair driver was represented by the angle that the joystick moved away from a central location $|J|$. The direction demanded by the wheelchair user was represented by $\angle\theta$.

The time that a joystick remained in a position gave a degree of confidence.

$$|J| = \sqrt{((\text{Alpha} * \text{Alpha}) + (\text{Beta} * \text{Beta}))} \tag{1}$$

Where, Alpha and Beta are the Cartesian co-ordinates.

Histogrammic mapping worked like a pseudo-integrator. When joysticks were kept in a steady position and orientation then the associated element increased over time and other elements reduced. An element with the greatest value represented the stick orientation and position.

|J| and θ represented desired direction and speed. Confidence in that desired direction and speed was also recorded. Each cell in the array had two values:

- "*AngularConf*" represented how steadily the joystick was held in a position.
- "*SpeedMag*" represented desired speed.

JoyIn was the joystick input to the rule based expert system. *JoyIn* represented the confidence-level for the disabled driver intentions. *JoyArray* calculated which element represented the joystick and *AngularConf* increased. Other empty element reduced in value. Histogram elements reduced quickly but increased more slowly. *JoyArray* elements increased to a max value in <0.4 s and reduced to zero in <130 ms.

The rates to increase or decrease in value were verified experimentally. Individual rates could be established for individual users or for particular tasks.

3 Kinematics of a Bobcat II Powered Wheelchair

Wheelchair kinematics are shown in Fig. 1. A Bobcat II powered chair has two bigger driving wheels at the front of the chair that are driven independently. Driving the front wheels at different speeds allows a disabled wheelchair user to control the direction and speed of a chair. If wheel radius is r then wheel diameter is 2r. Using the notation and symbols of [3, 35], the 2× front wheels were distance W apart. The centre of gravity is at C and P defines the intersection of a line through the wheel axis and another line through the centre of the chair. D is the gap between P and C.

If there is not any slip between the wheels and the floor then,

$$v_{tang} = 1/2(v_{right} + v_{left}) \tag{2}$$

$$\omega_{tang} = 1/W(v_{right} - v_{left}) \tag{3}$$

$$v_{right} = r\omega_{right} \text{ and } v_{left} = r\omega_{left} \tag{4}$$

where, v is the linear velocity and ω is angular velocity.

Global position of the chair in global coordinates is [O X Y] using vector notation. So that:

$$\boldsymbol{q} = [x_C y_P \theta]^T \tag{5}$$

Where, x_C and y_P are the global coordinates of P. θ is the orientation of the local coordinate frame, that is [P x_C y_P] in Fig. 2 with respect to the horizontal axis. The coordinates of the chair are defined by (5). The body of the chair is assumed as rigid and the driving wheels are assumed not to slip. So the chair can only move normal to the axis of the driving wheels. The velocity at the point of contact between a wheel and the (orthogonal to the wheel plane) is zero.

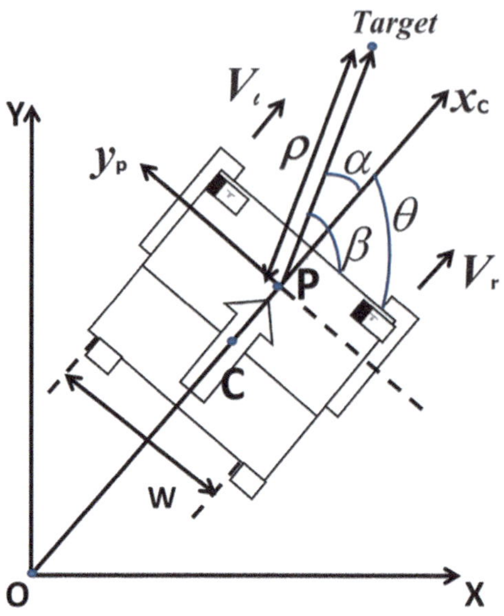

Fig. 2. Wheelchair kinematics.

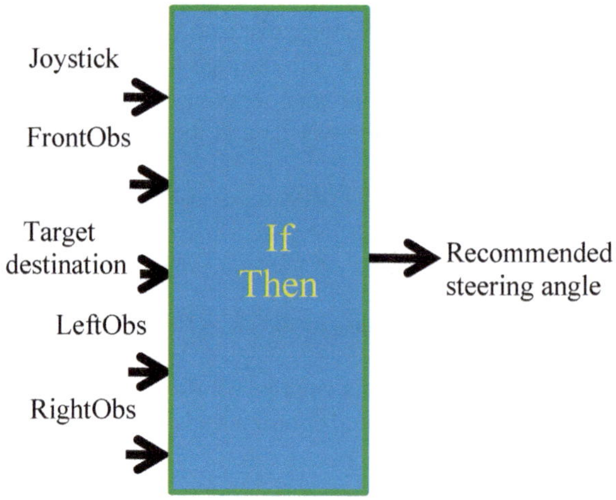

Fig. 3. Revised rule-based expert system.

$$(dy_P/dt)\cos\theta - (dx_C/dt)\sin\theta - d\theta/dt = 0 \tag{6}$$

Kinematics restrictions are not time dependent and are

$$A^T(\boldsymbol{q})\, d\boldsymbol{q}/dt = 0 \tag{7}$$

$$\text{Where, } C^T A(\boldsymbol{q}) = 0 \tag{8}$$

and $A(\boldsymbol{q})$ is an input matrix associated with the constraints.

$C(\boldsymbol{q})$ is a full-rank matrix formed by a set of linearly independent vector fields covering the null space of $A^T(\boldsymbol{q})$ and v_{tang} is a vector time function obtained from (7) and (8) for time t.

$$D\boldsymbol{q}/dt = C(\boldsymbol{q})\boldsymbol{v}_{tang} \tag{9}$$

The constraint matrix in (6) for the powered wheelchair is

$$\mathbf{A}^T(\boldsymbol{q}) = [-\sin\theta \cos\theta - d] \tag{10}$$

and

$$\boldsymbol{v}_{tang} = [v\omega]^T \tag{11}$$

Where, v is the linear velocity and ω is the angular velocity of meeting point P (down the chair axis).

So, the kinematics described by (9) can be placed into a $d\boldsymbol{q}/dt$ matrix.

If the chair only moves forward then $v = -\, v_{,tang}$ and a streamlined matrix can represent the wheelchair.

A controller generated the steering angle and wheel velocities where Angle = $(v_{left} - v_{right})/W$, to steer the chair along a desired path.

4 Control and the Rules

The linear and angular velocities (v and ω) to move the chair were calculated to take the chair from the current position ($\rho_0\ \alpha_0\ \beta_0$) to a new desired target position. Considering linear control [34],

$$v = K_\rho \rho \tag{12}$$

and,

$$\omega = K_\alpha \alpha + K_\beta \beta \tag{13}$$

A matrix could depict the closed-loop system to drive the wheelchair to the destination $(\rho, \alpha, \beta) = (0, 0, 0)$.

Code for an overdamped controller was written and simulated successfully before being downloaded to the powered wheelchair systems. Joystick input was combined with the input from ultrasonic sensors using expert rules that were calculated in order to assist the user in avoiding obstacles. The initial set of rules were enhanced and upgraded later by including a global target destination (a final target destination or via point) so disabled wheelchair users could tend to follow an efficient global path. The revised rules combined five inputs in an effort to avoid obstacles and follow a more global path (Fig. 4). They were (Fig. 3):

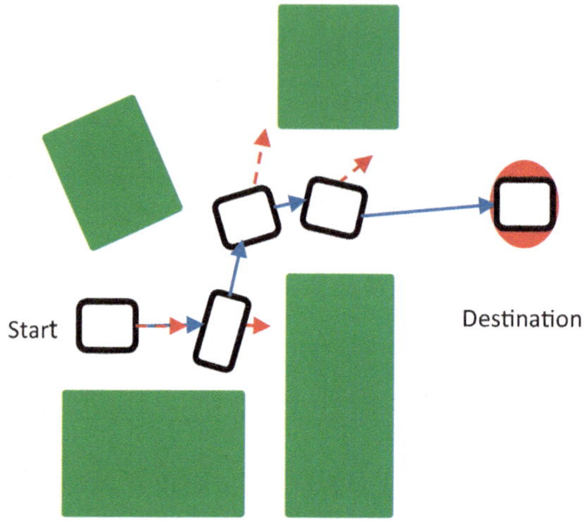

Fig. 4. Use of the rules to guide a powered chair through obstacles. Approach directions (blue solid line) and calculated directions (red dashed line) are shown.

- Demanded steering angle (from a joystick).
- Distance to objects detected to the right.
- An angle to a target destination.
- Distance to obstacles detected to the left.
- Distance to obstacles detected by both sensors.

A volume in front of the chair was described by the ultrasonics. The input from the ultrasonics was used to produce a recommended steering angle that could be used to adjust the angle fed into the wheelchair controller. The new suggested trajectory was safe and efficient. If $\angle\theta$ was right of the chair then the chair turned clockwise but if $\angle\theta$ was left of the chair then it turned anticlockwise. The control system improved wheelchair driving if human senses were impaired.

In addition to the joystick steering angle and the volume ahead of the chair, the rule based expert system was also provided with the angle to a target destination. Including that angle to a target destination increased the number of rules but improved operation. Some rules are described as examples:

FIRST EXAMPLE - destination and obstacle to the left:

One: *If* Joystick = 0° *and* LeftObs = HALFWAY *and* RightObs $\leq$ OUTLYING *and* FrontObs $\leq$ OUTLYING *and* AngleToTarget = 75°, *then adjust steering angle by* 0°

Two: *If* Joystick = 0° *and* LeftObs = HALFWAY *and* RightObs ≤ OUTLYING *and* FrontObs ≤ OUTLYING *and* AngleToTarget = 60°, *then adjust steering angle by* −10°
Three: *If* Joystick = 0° *and* LeftObs = HALFWAY *and* RightObs ≤ OUTLYING *and* FrontObs ≤ OUTLYING *and* AngleToTarget = 50°, *then adjust steering angle by* −25°
SECOND EXAMPLE - destination and obstacle to the right:
Four: *If* Joystick = 0° *and* LeftObs ≤ OUTLYING *and* RightObs = HALFWAY *and* FrontObs ≤ OUTLYING *and* AngleToTarget = 75°, *then adjust steering angle by* 15°
Five: If Joystick = 0° *and* LeftObs = ≤ OUTLYING and RightObs = HALF-WAY *and* FrontObs ≤ OUTLYING *and* AngleToTarget = 60°, *then adjust steering angle by* 30°
Six: *If* Joystick = 0° *and* LeftObs = ≤ OUTLYING *and* RightObs = HALFWAY *and* FrontObs ≤ OUTLYING *and* AngleToTarget = 30°, *then adjust steering angle by* 25°

THIRD EXAMPLE - destination to the right and an object in front:

Seven: *If* Joystick = 0° *and* LeftObs = IMMEDIATE *and* RightObs = IMMEDIATE *and* FrontObs ≤ OUTLYING *and* AngleToTarget = 20°, *then adjust steering angle by* 15°
Eight: *If* Joystick = 0° *and* LeftObs = IMMEDIATE *and* RightObs = IMMEDIATE *and* FrontObs ≤ OUTLYING *and* AngleToTarget = 25°, *then adjust steering angle by* 20°
Nine: *If* Joystick = 0° *and* LeftObs = IMMEDIATE *and* RightObs = IMMEDIATE *and* FrontObs ≤ OUTLYING *and* AngleToTarget = 30°, *then adjust steering angle by* 25°
The systems worked well with the revised rules and assisted wheelchair users if a human driver was impaired. The wheelchair path is shown again in Fig. 5 with an additional green solid arrow pointing towards the desired global target destination.

5 Tests and the Results

Figure 5 shows an example of a simulation. Once the algorithms had been tested successfully during simulations, then the software and hardware were mounted on the chair. Standard test routes at the University of Portsmouth were then used.

If sensors detected an object close to the chair, then the chair tended to turn away in order to avoid a collision. That manoeuvre might be overruled by a wheelchair user by using their joystick if for example they wanted to move close to the object. Typically to reach a wall mounted light switch.

If an obstacle was detected OUTLYING or nearer then the wheelchair would attempt to avoid the obstacle. If an object was detected ahead of the chair while moving towards the desired target destination, then the chair tended to turn to move alongside the object. When nothing was obstructing the path of the wheelchair and the joystick

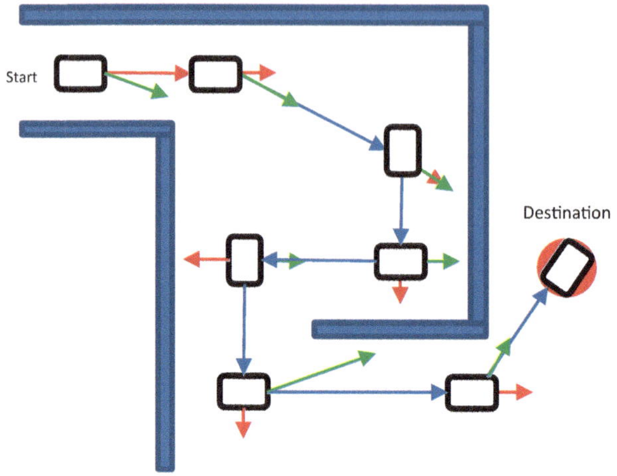

Fig. 5. An example of a simulation showing the wheelchair avoiding local minima (inner wall corners).

was held forward, then the chair tended to head for the desired target. Time taken to reach a desired destination tended to be reduced if vision was impaired or if any disability made it difficult to steer. The system would turn the chair to change the bearing so that the chair tended to move in the direction of the desired destination.

Av result from a simulation is shown in Fig. 6 and a result form a real-time test run is shown in Fig. 8.

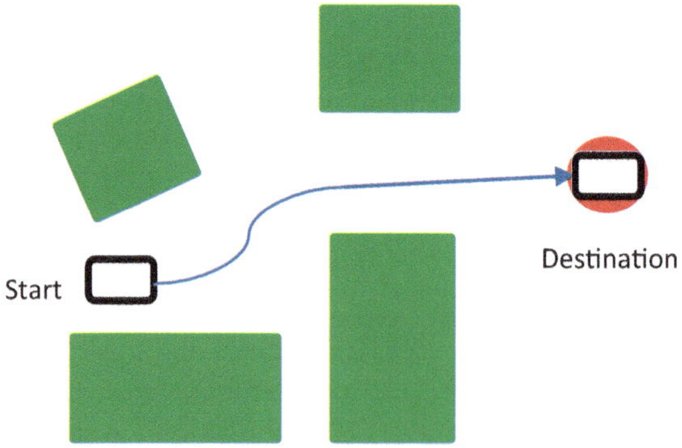

Fig. 6. Results from a real time experiment with the same rules applied.

Results from testing were compared with [3]. This rule-based system tended to complete the standard test routes more quickly. Figure 7 is a comparison between the time taken by the chair as it was driven through a set of the standard routes at the University of Portsmouth.

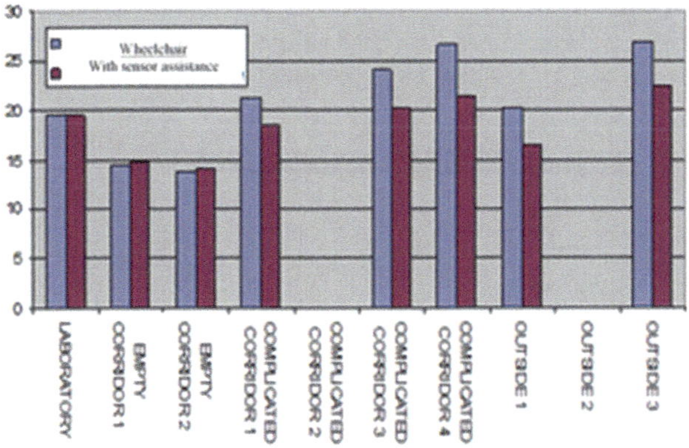

Fig. 7. Average time to complete routes. Right hand (darker) bars are time with sensors assisting a wheelchair driver and left hand (lighter) bars are time taken without sensors assisting.

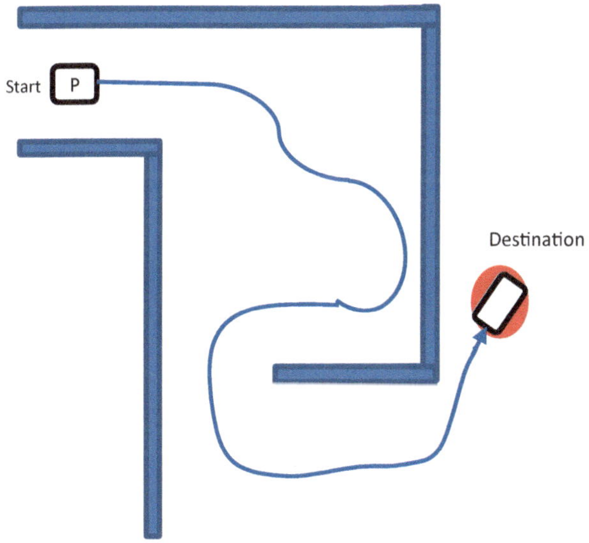

Fig. 8. The path of a wheelchair using the rules when the driver is not able to drive to the destination without assistance.

In the majority of tests the average time to safely complete a route was less for the systems described in this paper. Figure 7 shows two anomalies. The chairs completed simple routes (for example routes that were just the empty corridors) without sensors assisting the users. As routes became more complicated, then sensors were useful in assisting users. The addition of a desired target destination as an extra input to the rule based expert system made driving less efficient for easier routes and parts of routes.

That was because if a user could see and comprehend what was going on, then they would not need any sensors to help them in a simple uncluttered environment. Examples of standard routes at the University of Portsmouth are represented in Figs. 8 and 9.

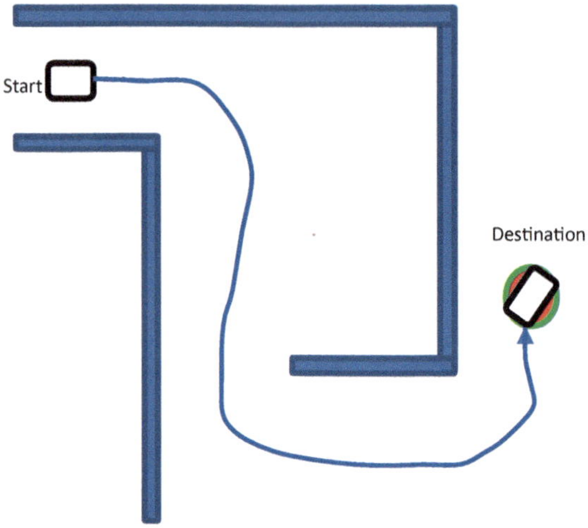

Fig. 9. A wheelchair path when a driver can safely drive themselves to a desired target destination.

In Fig. 8, the rules have a tendency to pull a chair in the direction of the desired target.

Chairs could reach their destinations efficiently. If a powered wheelchair user could steer their wheelchair skilfully then they could overcome the rules if they were making driving less efficient.

The techniques and methods produced faster results for the majority of the routes and reduced computation when compared against other methodologies. The rule-based expert systems were safe and effective.

The chairs had to avoid both stationary and moving obstacles (e.g. human beings moving or standing near the chair).

Whenever the sensors detected something near the chair, then the chair was turned to avoid a collision.

The avoidance of collisions is a comparatively high priority. The collision avoidance expert rules tended to over-ride systems but if a joystick was kept stationary in roughly the same place then the user joystick input was integrated over time so that the desire of a disabled driver eventually overrode other behaviour.

If sensor input increased beyond a threshold limit within the array elements then collision avoidance rules were activated.

If an object was detected ahead of the chair while it was moving towards a desired target destination then the chair tended to follow edge of the object; the chair rotated to move parallel to an edge, for example a corridor wall.

If nothing was detected then the chair tended to drive on a bearing between the angle to a desired target and the angle being demanded by the joystick.

Results from testing were compared with results from testing other systems and this rule-based expert system behaved appropriately.

6 Discussion and Conclusion

The rule-based expert systems proved to be safe and robust. They were efficient and simple in assisting powered wheelchair user to steer their chair. They successfully assisted drivers by quickly detecting obstacles and safely steering around them. In that way the wheelchair users were assisted to complete routes.

Methods and systems were compared against recently published systems and were validated.

One limitation is that the rules have been hard-coded so that they cannot continue to learn. Future (and some current) research is investigating ways to learn. Current work is also considering mixing AI tools [35–49] so that the different AI tools are used to their best effect.

A problem with rule based approaches is that not all environment configurations can be considered at design stage. This raises the question of how the implementation compares with search based control methods, or learnt behaviours (e.g. reinforcement learning, neuro or neuro-fuzzy approaches). A goal-based behaviour could work more efficiently and that will be investigated with other AI techniques in the future?

If the target is an obstacle then the rules would initially try to avoid it. If the driver continues to indicate that they want to drive towards the target (for example a wall to change a light-switch) then the driver will over-rule the system.

References

1. Electric Wheelchairs: Types & Reviews of Powerchairs, Disabled World. http://www.disabled-world.com/assistivedevices/mobility/wheelchairs
2. Mobility Device Statistics - United States - The University of California - Disability Statistics Center, 22 April 2013. http://www.disabled-world.com/disability/statistics/mobility-stats.php
3. Parhi, D.R., Singh, M.K.: Rule-based hybrid neural network for navigation of a wheelchair. Proc. IMechE Part B. J. Eng. Manuf. **224**, 11103–1117 (2009)
4. Nguyen, A.V., Nguyen, L.B., Su, S., Nguyen, H.T.: Shared control strategies for human - machine interface in an intelligent wheelchair. In: 35th Annual International Conference of IEEE-Engineering-in-Medicine-and-Biology-Society (EMBC), Osaka, Japan. Book Series: IEEE Engineering in Medicine & Biology Society Conference Proceedings, pp. 3638–3641 (2013)
5. Parhi, Z.R., Pradhan, S.K., Panda, A.K., Behra, R.K.: The stable and precise motion control for multiple wheelchairs. Appl. Soft Comput. **9**(2), 477–487 (2009)
6. Sanders, D.A., Sanders, B., Gegov, A., Ndzi, D.: Results from investigating powered wheelchair users learning to drive with varying levels of sensor support. In: Proceedings of SAI Intelligent Systems (2017)

7. Song, K.T., Chen, C.C.: Application of asymmetric mapping for wheelchair navigation using ultrasonic sensors. J. Intell. Wheelchairic Syst. **17**(3), 243–264 (1996)
8. Huq, R., Mann, G.K.I., Gosine, R.G.: Wheelchair navigation using motor schema and fuzzy context dependent behaviour modulation. Appl. Soft Comput. **8**(1), 422–436 (2008)
9. Begum, M., Mann, G.K.I., Gosine, R.G.: Integrated fuzzy logic and genetic algorithmic approach for simultaneous localization and mapping of wheelchairs. Appl. Soft Comput. **8** (1), 150–165 (2008)
10. Bennewitz, M., Burgard, W.: A probabilistic method for planning collision-free trajectories of multiple wheelchairs. In: Proceedings of 14th Euro Conference on AI (ECAI), Berlin, Germany, 20–25 August 2000, pp. 9–15
11. Gueaieb, W., Miah, M.S.: An intelligent wheelchair navigation technique using RFID technology. IEEE Trans. Instrum. Meas. **57**(9), 1908–1917 (2008)
12. Hwang, C.L., Chang, N.W.: Fuzzy decentralized sliding-mode control of a car-like wheelchair in distributed sensor-network spaces. IEEE Trans. Fuzzy Syst. **16**(1), 97–109 (2008)
13. Nguyen, A.V., Nguyen, L.B. Su, S., Nguyen, H.T.: The advancement of an obstacle avoidance bayesian neural network for an intelligent wheelchair. In: 35th International Conference of IEEE-Engineering-in-Medicine-and-Biology-Society (EMBC), Osaka. Book Series: IEEE Eng in Medicine and Biology Society Conference Proceedings, pp. 3642–3645 (2013)
14. Sanders, D.A.: Using a self-reliance factor for a disabled driver to decide on the share of combined-control between a powered wheelchair and an ultrasonic sensor system. IEEE Trans. Neural Syst. Rehab. Eng. **25**(8), 1221–1229. https://doi.org/10.1109/tnsre.2016.2620988
15. Sanders, D., Stott, I., Graham-Jones, J., Gegov, A., Tewkesbury, G.E.: Expert system to interpret hand tremor and provide joystick position signals for powered wheelchairs with ultrasonic sensor systems. Ind. Rob. **38**(6), 585–598 (2011)
16. Robinson, D.C., Sanders, D.A., Mazharsolook, E.: Ambient intelligence for optimal manufacturing and energy efficiency. Assem. Autom. **35**(3), 234–248 (2015)
17. Sanders, D.A., Tewkesbury, G., Gegov, A., et al.: Fast transformations to provide simple geometric models of moving objects. In: Proceedings of Intelligent Robotics & Application Conferences, part I, vol. 9244, pp. 604–615 (2015)
18. Sanders, D., Langner, M., Tewkesbury, G.: Improving wheelchair- driving using a sensor system to control wheelchair-veer and variable-switches as an alternative to digital-switches or joysticks. Ind. Rob. Int. J. **37**(2), 151–167 (2010)
19. Larsson, J., Broxvall, M., Saffiotti, A.: Laser-based corridor detection for reactive Navigation. Ind. Rob. Int. J. **35**(1), 69–79 (2008)
20. Sanders, D., Graham-Jones, J., Gegov, A.: Improving ability of tele-operators to complete progressively more difficult mobile robot paths using simple expert systems and ultrasonic sensors. Ind. Rob. Int. J. **37**(5), 431–440 (2010)
21. Lee, S.: Use of infrared light reflecting landmarks for localization. Ind. Rob. Int. J. **36**(2), 138–145 (2009)
22. Milanes, V., Naranjo, J., Gonzalez, C.: Autonomous vehicle based in cooperative GPS and inertial systems. Robotica **26**, 627–633 (2008)
23. Sanders, D., Stott, I.: A new prototype intelligent mobility system to assist powered wheelchair users. Ind. Rob. **26**(6), 466–475 (1999)
24. Chang, Y.C., Yamamoto, Y.: On-line path planning strategy integrated with collision and dead-lock avoidance schemes for wheeled wheelchair in indoor environments. Ind. Rob. Int. J. **35**(5), 421–434 (2008)
25. Sanders, D.: Progress in machine intelligence. Ind. Rob. **35**(6), 485–487 (2008)

26. Sanders, D.A., Lambert, G., Pevy, L.: Pre-locating corners in images in order to improve the extraction of Fourier descriptors and subsequent recognition of shipbuilding parts. Proc. Inst. Mech. Eng. Part B J. Eng. Manuf. **223**(9), 1217–1223 (2009)
27. Sanders, D.: Comparing speed to complete progressively more difficult mobile robot paths between human tele-operators and humans with sensor-systems to assist. Assem. Autom. **29** (3), 230–248 (2009)
28. Sanders, D.: Analysis of the effects of time delays on the teleoperation of a mobile robot in various modes of operation. Ind. Rob. **36**(6), 570–584 (2009)
29. Sanders, D.A., Langner, M., Gegov, A.E., Ndzi, D., Sanders, H., Tewkesbury, G.E.: Tele-operator performance and their perception of system time lags when completing mobile robot tasks. In: Proceedings of 9th International Conference on Human Systems Interaction, pp. 236–242 (2016)
30. Sanders, D.A., Ndzi, D.L., Chester, S., Malik, M.: Adjustment of tele-operator learning when provided with different levels of sensor support while driving mobile robots. In: IEEE Proceedings of SAI Intelligent Systems, pp. 962–967 (2016)
31. Sanders, D.A., Bausch, N.: Improving steering of a powered wheelchair using an expert system to interpret hand tremor. In: Liu, H. (ed.) Proceedings of Intelligent Wheelchairics and Applications (ICIRA 2015), part II, vol. 9245, pp. 460–471 (2015)
32. Sanders, D., Stott, I.J., Robinsosn, D.C., et al.: Analysis of successes and failures with a tele-operated mobile robot in various modes of operation. Robotica **30**, 973–988 (2012)
33. Sanders, D.A., Tewkesbury, G.E., Stott, I.J., et al.: Simple expert systems to improve an ultrasonic sensor-system for a tele-operated mobile-robot. Sens. Rev. **31**(3), 246–260 (2011)
34. Eason, G., Noble, B., Sneddon, I.N.: On certain integrals of Lipschitz-Hankel type involving products of Bessel functions. Phil. Trans. Roy. Soc. London **A247**, 529–551 (1955)
35. Sanders, D.A., Sanders, H., Ndzi, D.L., Gegov, A.E.: Rule-based system to assist a tele-operator with driving a mobile robot. In: IEEE Proceedings of SAI Intelligent Systems Conference, pp. 987–994 (2016)
36. Sanders, D., Geov, A.: AI tools for use in Assembly Automation and some examples of recent applications. Assem. Autom. **33**(2), 184–194 (2013)
37. Sanders, D.A., Tewkesbury, G.E., Ndzi, D., et al.: Improving automatic robotic welding in shipbuilding through the introduction of a corner-finding algorithm to help recognise shipbuilding parts. J. Mar. Sci. Technol. **17**(2), 231–238 (2012)
38. Sanders, D., Lambert, G., Graham-Jones, J., et al.: A robotic welding system using image processing techniques and a CAD model to provide information to a multi-intelligent decision module. Assem. Autom. **30**(4), 323–332 (2010)
39. Sanders, D.A., Tewkesbury, G.E.: A pointer device for TFT display screens that determines position by detecting colours on the display using a colour sensor and an Artificial Neural Network. Displays **30**(2), 84–96 (2009)
40. Bergasa-Suso, J., Sanders, D.A., Tewkesbury, G.E.: Intelligent browser-based systems to assist Internet users. IEEE Trans. Educ. **48**(4), 580–585 (2005)
41. Sanders, D.A., Tan, Y., Rogers, I., et al.: An expert system for automatic design-for-assembly. Assem. Autom. **29**(4), 378–388 (2009)
42. Sanders, D.A., Lambert, G., Pevy, L.: Pre-locating corners in images in order to improve the extraction of Fourier descriptors and subsequent recognition of shipbuilding parts. Proc. IMechE Part B J. Eng. Manuf. **223**(9), 1217–1223 (2009)
43. Sanders, D.: Controlling the direction of walkie type forklifts and pallet jacks on sloping ground. Assem. Autom. **28**(4), 317–324 (2008)

44. Sanders, D.A., Bergasa-Suso, J., Khusainov, R., Gegov, A., Chester, S., Bausch, N.: Introducing dead bands within two-dimensional clusters of user data to improve data classification. In: Proceedings of the IEEE 9th International Conference on Human Systems Interaction, pp. 14–20 (2016)
45. Sanders, D.A.: Progress in machine intelligence. Ind. Rob. Int. J. **35**(6), 485–487 (2008)
46. Geov, A., Gobalakrishnan, N., Sanders, D.A.: Rule base compression in fuzzy systems by filtration of non-monotonic rules. J. Intell. Fuzzy Syst. **27**(4), 2029–2043 (2014)
47. Sanders, D.A., Gegov, A., Ndzi, D.: Knowledge-based expert system using a set of rules to assist a tele-operated mobile robot. In: Bi, Y., Kapoor, S., Bhatia, R. (eds.) Studies in Computational Intelligence, vol. 751, pp. 371–392. Springer (2018)
48. Sanders, D.A., Sanders, H., Gegov, A., Ndzi, D.: Rule-based system to assist a tele-operator with driving a mobile robot. In: Lecture Notes in Networks and Systems, vol. 16, pp. 599–615. Springer (2018)
49. Sanders, D.A.: Non-model-based control of a wheeled vehicle pulling two trailers to provide early powered mobility and driving experiences. IEEE Trans. Neural Syst. Rehabil. Eng. **26** (1), 96–104 (2018)

Our New Handshake with the Robot

How Our Changing Relationship with Machines Can Make Us More Human

Marcin Remarczyk(✉), Prashant Narayanan, Sasha Mitrovic, and Melani Black

Cognizant Technology Solutions, London, UK
{Marcin.Remarczyk,Prashant.Narayanan,Sasha.Mitrovic, Melani.Black}@cognizant.com

Abstract. Very few topics in today's digital conversations are more *en vogue* than Artificial Intelligence and Robotics. However, all too often, debate is centred on the machine and not the human side of this rapidly evolving handshake. This paper introduces a framework that structures the relationship from demystifying through to designing and adopting the handshake, and explores impact on enterprises and provides a future outlook.

Keywords: Artificial intelligence · Robotics · Automation · Human-machine

1 A Relationship of Two Very Different Partners

1.1 Introduction

The notion of partnering with machines has been fuelling our fantasies for some generations. Our imagination creates fascination and fear at the same time. Fascinating to many is the vision of tremendous benefits of Artificial Intelligence and Robotics (AIR) to everyday lives across all spheres from ecology to entertainment and from leisure to life sciences. However, at the same time, many of us are also apprehensive about its possible imminent impact on our workplaces and, in the longer term, are concerned about the possibility of dominance by humanoid robots [1].

The environment we live in is being fundamentally transformed through software in physical devices, vehicles, home appliances, and other items, enabling these objects to connect and crucially exchange data. Meanwhile, enterprises face increasing cost pressures, changing customer priorities, and shifting regulatory obligations. All these compound to forcing organisations to improve their operational efficiencies.

These gains in efficiency and competitiveness are generated through adaptive operating models and investments in emerging technologies. Enterprises are looking for non-invasive automation based on emerging technology to reduce implementation costs, accelerate benefits and address a wide range of risks. Davenport and Ronanki [2] recognise three unique business needs for AIR: automating business processes, gaining insight through data analysis, and engaging with customers and employees. Out of the 152 projects they studied, the majority fell under the category of process automation; in other

K. Arai et al. (Eds.): IntelliSys 2018, AISC 868, pp. 839–851, 2019.
https://doi.org/10.1007/978-3-030-01054-6_58

words, the automation of digital and physical tasks which, typically, include back office administrative and financial tasks. Today, Robotic Process Automation (RPA) is applied due to its advanced nature and acts as a replacement to earlier business process automation tools. And why is RPA more advanced? This is because robots act like a human, inputting and consuming information from multiple IT systems. The second most common theme from the study was Cognitive insight with 38% of projects being of a nature which improved insight through data analysis. Algorithms are used to detect patterns in vast volumes of data in order to interpret their meaning. These machine learning applications were used, for example, to predict what a particular customer is likely to buy, identify credit fraud in real-time and detect insurance claims fraud. Cognitive insight provided by machine learning differs from that available from traditional analytics. It is more advanced in the sense that it is much more data-intensive and detailed; the models typically are trained on some part of the data set; and the models become more artificially intelligent over time. One example of this is their ability to use new data to make predictions or categorisation gradually enhanced over time. Engaging with customers and employees was the least common theme identified. In other words, 16% of projects were of a nature that engage employees and customers using natural language processing chatbots, intelligent agents, and machine learning.

1.2 It – The Machine

There are three main types of intelligent systems (see Fig. 1), systems that *do*, systems that *think* and systems that *learn*. The systems able to learn – the focus of this paper – are referred to as cognitive computing and use data mining, pattern recognition and natural language processing to imitate the way humans work. The ultimate objective of cognitive computing is to develop automated, interconnected and crucially autonomous systems capable of solving problems and making decisions.

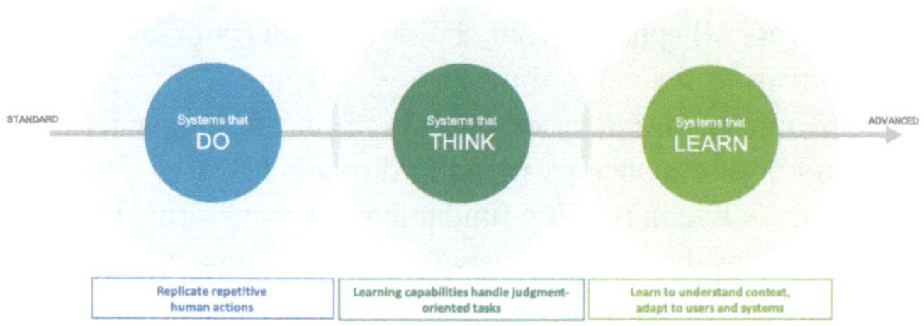

Fig. 1. Intelligent system types, Cognizant, 2017.

The emergence of cognitive computing has been accelerated in 2017 through new technologies changing households and workplaces alike. It helps us imagine and increasingly picture a future where humans and machines coexist to enable faster and more informed decision making, improve operational performance and increase efficiency. Characteristic for these systems is the ability to make sense of significant sets of data, apply reason to extract insight and continuously learn, while collaborating with other machines as well as human users. These ecosystems incorporate automation platforms, tools, and suppliers along with complementary technologies, or, so-called, innovation multipliers. Cognitive computing is a true differentiator for leaders of shared services, transformation and innovation, outsourcing, and operational areas, such as human resources, finance, supply chain, and support services. In a fast-evolving market, even more important than the technology tools themselves, it is critical to develop and apply an integrated framework to guide the development and delivery of the advanced services supported by systems able to learn.

1.3 You – The Human

Through past decades, we learnt to co-exist with various types of machines displaying varying degrees of intelligence from relatively simple computers of the 20th century to the latest voice recognition systems. Collectively, we mainly recognise the value these systems add to our private and professional lives.

Dating back to the 1950's – and what is considered to be one of the oldest computer interfaces – is the command line. This surfaced as a more effective way in which humans could control computers, compared to previous methods where commands had to be fed into a computer in batches. Teletype machines, which were originally used for telegraph transmissions, were modified for humans to change commands in the midst of a process and, in turn, receive feedback from the computer machine on a real-time basis. Video display units were used to display command line information more rapidly. Graphical user interfaces only emerged in the 1980's and made computers much easier to use.

Fast forwarding to the world today, we have reached the epitome of voice recognition with systems such as Alexa, Siri and the Google Home smart speaker, and others. Historically, voice recognition has struggled with a reputation of being inaccurate. This is now changing as computers become more powerful. Amazon's Alexa continues to expand its music, smart-home, and digital-assistant abilities. It is mainly a wireless speaker, but capable of much more. Using nothing but the sound of your voice, you can play music, search the web, create to do and shopping lists, shop online, get instant weather reports, and control popular smart-home products – all while your smartphone stays in your pocket.

To continue exploiting the benefits of AIR, organisations need to move towards teams that partner humans with AIR and towards more adaptive operating models that enable this handshake between humans and machines. We recognise three main components of this handshake. During *Demystification*, organisations first address possible emerging concerns of implementing AIR, which, as the paper explores further, will be the foundation automation opportunities are realised upon. During *Definition*, organisations design their new digital operating models to ensure processes, procedures, capabilities, etc. reflect enhanced presence of AIR. Finally, during *Adoption*, organisations successfully embed new ways of working by changing mindsets, upskilling its workforce and continuously addressing anxieties (see Fig. 2).

Fig. 2. *Handshake* components, Cognizant, 2018.

2 Demystifying

Addressing existing, as well as foreseeing emerging concerns about AIR, is crucial to any initiative in this area in any enterprise. It is the foundation on which opportunities to automate can be explored, new ways of working piloted, and the new human-machine handshake designed. Moreover, it is the stepping stone towards successfully adopting new systems.

The potential to automate tasks, processes, and, in some cases, entire jobs through AIR, results in growing concerns among employees across industries and functions. This is understandable in light of research by authors such as Muro [3], according to whom "since 1980, globalization, offshoring, and automation eliminated nearly 7 million manufacturing jobs in the U.S. – more than one-third of U.S. manufacturing positions – as manufacturing employment plunged from 18.9 million jobs to 12.2 million." In addition to direct job losses, according to Manyika [4], AIR is driving "growing polarisation of labour-market opportunities between high-skill and low-skill jobs, unemployment and underemployment especially among young people, and stagnating incomes for a large proportion of households and income inequality".

However, according to a recent study by the professional services firm Deloitte, only 22% of the 250 surveyed executives consider headcount reduction as a benefit of AIR (see Fig. 3). Instead, improvement to performance, operational capability, and quality of decision making were identified by the study as the more prominent drivers for AIR. Interestingly, based on evidence to date, the introduction of AIR results in the closure of only circa 12% of jobs, while circa 75% of jobs are being enhanced, and circa 13% of jobs are being reinvented [5].

During the initial phases of any AIR initiative, there are two fundamental things which leadership ideally should achieve: reassure and excite its employees.

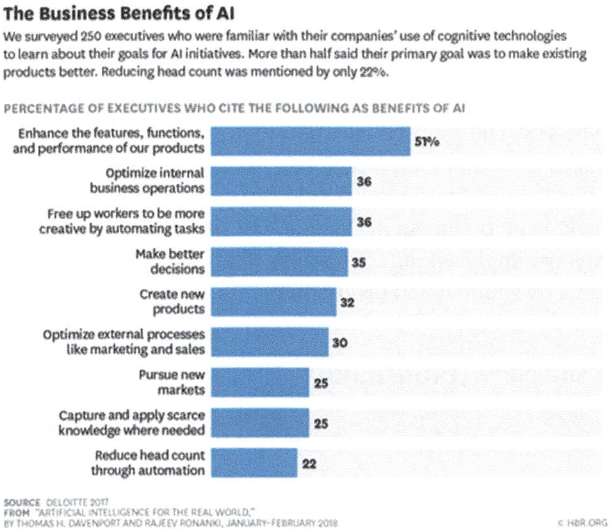

Fig. 3. The Business Benefits of AI, Deloitte, 2017.

2.1 Reassure Staff

In essence, there are two fundamental anxieties staff can experience ahead of the introduction of AIR. On the one hand, some employees are likely to perceive automation as a threat to their own job, or, worse, their own career. On the other hand, people can experience apprehension when dealing with new technologies. Addressing fear of automation should be handled across the enterprise, at team and, importantly, at individual levels.

To address the matter of job security, those responsible must get across the motives for introducing AIR in a clear and transparent manner. As pointed out above, management is rarely motivated by headcount reduction. Rather, existing roles will be redefined and enhanced, giving employees options to upskill and take on higher value jobs in the enterprise.

Although in 2018 technology can take on many responsibilities from us in the workplace but also in private lives, we still enjoy a relatively significant comparative advantage over machines when it comes to highly complex, ambiguous, or, simply new problems. This, however, is likely to trigger further or other types of anxiety. Don't expect everyone to be cheering about leaving the very often referenced comfort zone and be retrained and having to take on more complex tasks in the department. Whether employees will, or will not, embrace emerging opportunities, will depend on their personality and also the stages in their own career lifecycle. Here, it may be advisable to look for options such as compensation, early retirement, or alternative roles elsewhere in the corporation.

2.2 Excite Staff

Introducing AIR is always also an opportunity to transform the workforce and the organisation in general. As such, it is a unique point in time for the management to re-engage with staff. This can only be effective if employees are truly excited about the vision and, perhaps, even more important, feel to be part of that journey. Wherever and whenever possible, the new initiatives should be inclusive and allow for staff participation. This can include involvement in shaping AIR initiatives, roles in projects, piloting new functionality, or simply asking for opinion. The more involvement, the less anxiety and, therefore, less resentment and obstruction.

3 Defining the New Handshake

3.1 Introduction

As advances in AIR provide new opportunities to automate processes in enterprises, the nature of the interaction between AIR and humans is evolving. This has a profound impact on existing operating models; as the footprint of technology increases, the human factor shifts upwards on the value scale and external as well as internal turbulence accelerates, demanding a robust, yet flexible model. The redesign must be carried out in a thoughtful, integrated and strategic way taking into consideration technology trends, generational changes within the workforce as well as the new touchpoints between human and machine.

Fulfilling the needs of 21st century customers – connected across multiple channels – has led to the creation of environments that combine diverse architectures and platforms, serviced through multiple vendors. Tasking the organisation with managing this complex environment manually is laborious and expensive. Organisations, therefore, should reconstruct process flows based on a standardised set of best practices to automate operational tasks across multi-vendor environments in order to optimise operational efficiency. This will, for instance, maximise service uptime, increase resource utilisation, improve service levels, ensure compliance, and minimise costs. To ensure the success of intelligent process analytics, organisations need to build capabilities, not only to manage data, but also to translate the analysis outcomes that allows to speed up business processes, e.g. social collaboration to provide quicker incident resolution to customers. In our experience, best-in-class companies are looking for intelligent process automation to reduce the amount of manual interventions across IT and operations by 55% to 60% over the next one to two years. This will result in an estimated reduction of 20% in operational cost through productivity gains, process standardisation, and better insights into process optimisation [6].

As advances in AI provide new opportunities to automate processes in enterprises, the nature of the interaction between AIR and humans is evolving. This has a profound impact on existing operating models; as the footprint of technology increases, the human factor shifts upwards on the value scale and external as well as internal turbulence accelerates, demanding a robust, yet flexible model. The redesign must be carried out in a thoughtful, integrated and strategic way taking into consideration technology trends, generational changes within the workforce, as well as the new touchpoints between human and machine.

Four aspects of operating model design for the AIR age appear. First, developing the fitting organisational DNA to initiate and accelerate the AIR journey by applying human-driven design principles and methods. Second, restructuring the workforce by transferring resources from high to low automation departments, from low to high value, and strategically more important parts of the organisation. Third upskilling and reskilling the workforce to ensure it remains relevant and well positioned to perform with and alongside AIR competence. Finally, building capabilities accountable for ongoing ownership, enhancement and maintenance of automation.

3.2 Getting the Organisational DNA Right

When considering the future organisational structure, two transformative models emerge; namely, product/platform centric: middle/back office with data analytics, cloud and enterprise tools; and customer/channel centric: front office with customer experience-oriented social, mobile, AIR tools/models. The processes which underpin both models are, in turn, influenced by agile lifecycles and lean delivery, as well as RPA. These are likely to dominate the future, covering and, importantly, integrating the end-to-end enterprise value chain. The concept of Human Driven Design [7] was established as early as 1969. Though many organisations have tried to adopt the principles of design thinking in AIR, relatively few have made strides to truly get the benefits of AIR. Cognizant has identified ten key principles to make human design processes part of the DNA of an organisation to drive transformation through AIR (see Fig. 4).

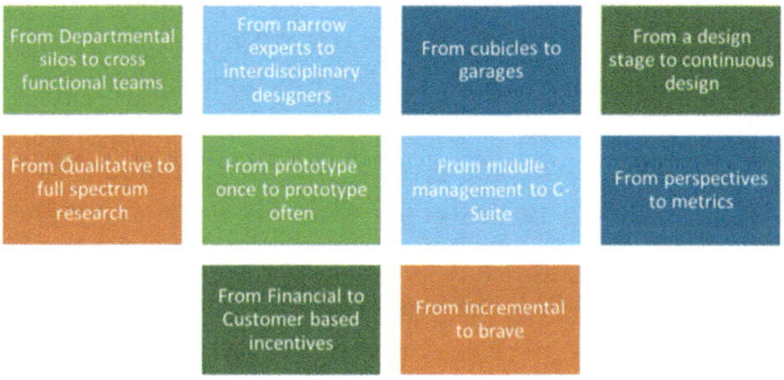

Fig. 4. Human driven design principles, Cognizant, 2018.

3.3 Restructuring the Workforce

As stated before, launching AIR enables enterprises to redirect resources towards tasks and roles with greater value, and as such, make more effective use of personnel, provided appropriate measures to reskill and upskill are in place.

Mainly, organisations are interested in retaining staff and repurpose their skills and expertise to areas of higher degree of complexity and ambiguity, areas where humans, given the current capacity of AIR, enjoy an advantage. With that, the touchpoints between employees and machines are also climbing upwards the value chain. These handshakes must be designed carefully, always aiming at making them as intuitive and stress-free as possible. The goal should be to shift the pain from the interaction towards the machine, away from the human, even if this means higher investment in the technology. In the long run, this is very likely to pay off.

It must be said that automating through AIR, naturally, offers a chance to reduce staff and, therefore, cost. Even if cost optimisation is not the main driver, this is the time when underperforming staff can be exited, or, broader, departments can be rationalised or reshaped.

3.4 Upskilling the Workforce

A recent study of more than 1,500 companies which are currently using or testing AIR systems and related technologies, indicates the rise of new associated roles [8]. These new roles are not simply replacing old ones but are entirely novel positions requiring skills and training never needed before. There are three emerging roles that will be particularly relevant for organisations implementing AIR.

1. *Machine-human trainers:* Dedicated trainers will help machines to make fewer errors and to mimic human behaviours. Koko, a start-up originating in the MIT Labs, recently developed a BETA AIR system that can enable chat-bots such as Alexa to respond with sympathy to customer questions. Trainers teach Koko algorithms to enable a response of a more human-like manner so that the system can talk to humans using compassion and even humour.
2. *Orchestrators:* The role of the orchestrator is to interpret and align to the strategic enterprise vision and also to bridge the information gap between the technology and business side of the organisation for AIR. Orchestrators focus on making sense of data produced by machines, and converting this data into meaningful insight.
3. *Specialists:* The role of the specialist is of particular importance as she or he ensures that AIR technologies are working properly and serve humans to make lives easier. They will also play a role in designing the bots. Remember the Uber self-drive disaster that occurred in Arizona in March 2018. The specialist will act as an AIR ombudsman who will ensure that these advanced AIR technologies perform within the set of rules.

3.5 Ensure Accountability for AIR in the Organisation

With AIR taking on a more prominent role in the enterprise, it is increasingly vital that operating models reflect natural *homes* for this emerging capability, meaning clearly defined ownership of maintaining and innovating AIR across the corporation. Also, the IT function shouldn't necessarily be the default answer. In the mid to longer term, other departments can be seen competing for this key area such as operations, various digital functions; or, if one argues that machines are our new *colleagues*, perhaps it is time for HR to claim this field. Given the strategic nature of AIR, it is possible that we will see the emergence of dedicated new departments with sole focus on maintaining existing capability and refining it while observing trends in the market and bringing the relevant innovation into the enterprise.

4 Adopting the New Handshake

4.1 Introduction

Introducing AIR in an organisation brings about a major transformation and, therefore, is a critical part of the change agenda. As such, it is imperative that any implemented changes are successfully adopted and embedded into daily business operations. In order to achieve successful change, targeted organisational change management is crucial.

Embedding new technologies, such as AIR, is likely to lead to failure if sole focus is placed on technology and, at the same time, the human component is somewhat ignored. It requires buy-in from employees; simply introducing technology is not sufficient and by no means guarantees successful outcomes. Innovation in technologies shifts paradigms in the workplace. Whether the advances are beneficial to the human factor in the organisation – or not – the instinctive first reaction of the workforce is to perceive the technologies as a threat. As such, there needs to be a concerted effort to help shift the perception and attitude of the workforce towards new technologies, especially when these are perceived negatively.

However, embedding and sustaining this fundamental change necessitates more than simple communication and training. Cultural as well as behavioural change will be key to ensuring change is both understood and embedded. By influencing and shaping new behaviours, an organisation can implement tangible cultural change and accelerate the speed of adoption. The transformation within the organisation can be potentially significant. It is as much about changing the human behaviour and shifting the organisational culture as it is about automating or *intelligentifying* a process. There are three key dimensions of embedding lasting change are:

- Vision, leadership, and strategy.
- Learning and development, employee engagement, and benefits.
- Adoption, sustainability, and outcomes.

4.2 Vision, Leadership, and Strategy

To keep fear of new technologies and doubts about their benefits at bay, leaders in organisations ought to be pro-active and take initiative. Without alignment among the leadership, there will be diminished prospect for success. Leaders need to engage with the workforce, introduce and elaborate on the vision by creating an effective overarching change narrative with a range of linked stories that make the vision real to staff. Communication should be as clear, open, and consistent, as possible and should be delivered to the right people at the right time, whether in a town hall, in person to teams or individuals. There is a compelling story to tell. Organisations are now moving with the times and introducing technologies that are capable of increasing productivity and enabling the workforce to pick up new skills and have access to information faster, allowing each employee to develop their skills and knowledge base within the business. It is up to each organisation to determine how it intends to employ new technologies, and leadership must convey this vision so that the workforce understands the direction, the bigger picture, and importantly purpose. This will increase engagement and motivation enabling change and ensuring its long-lasting impact. Ultimately, employees' fears need to be addressed, and a way for this to happen is for the senior leaders to paint a vision where employees are the most important component of the *new* future. They are taken on the journey and not slidelined or ignored.

4.3 Learning and Development, Employee Engagement, and Benefits

Once the mindset has been appeased, the next step is the change in behaviour. A concerted effort of rolling out training on how to use the technologies, enables employees to understand the application of the technology, and how each employee can benefit from it and contribute to the successes of the transformation. The organisation needs to build training around these narratives and the overall vision, and enable the human component to evolve in such a way to support the organisation's vision. Most importantly, the organisation should identify the ever evolving capability needs, make adequate training available and accessible, and encourage participation in the same. The question of *what's in it for me?* should always be at the forefront of the organisation's vision, as this is exactly how each employee will be experiencing the transformation. Without it, the perceived benefits to the organisation, its long term objectives, and the desired outcomes become perishable goods on a dusty convenience store shelf.

4.4 Adoption, Sustainability, and Outcomes

Assuming that the exerted efforts to change minds and behaviours have been effective, this new state ought to be cemented. To fully embed the culture, a robust plan needs to be in place to keep the parts of the workforce, that have adopted the latest technologies, engaged and a valued partner in the organisation. The manner of driving and sustaining adoption will depend on the initial vision and strategy, and on understanding of the benefits brought by the technologies to the individual, the teams, as well as to the organisation as a whole. These technologies, integral part of the *new normal*, are no longer reactive but proactive in the sense that their inquisitive minds find new and creative ways of embedding the new capabilities in business processes.

No matter how well a major change programme is set up, there is a constant risk of momentum loss. There is a range of activities which can be rolled out to ensure this does not happen. A smart organisation will measure adoption rates throughout the engagement with the workforce and will carry out surveys and monitor the sentiment among staff to understand levels of AIR adoption. This information feeds into an action plan designed to support the initiative.

4.5 Cultural Change in an Organisation

Cultural and behaviour change is key to ensuring that the transformation of the business is both understood and embedded. This is more than communications and training. By shifting behaviours, the organisational culture with its norms, rituals and language will change over time and accelerate adoption. AIR, fundamentally, requires people to change their behaviours, shift their roles, and shake their perceptions, as focusing on a successful roll-out of the technology alone is simply not sufficient. Microsoft, for example, released its automated chat-bot Tay (Thinking About You) on Twitter 2016 with the purpose that it learns from other tweeters and that it can then engage with people using language that mimics that of humans. Despite all the best intentions, some users began tweeting politically incorrect phrases, teaching Tay inflammatory messages revolving around common themes on the internet. As a result, the robot began releasing racist and vulgar messages in response to other tweets. This could have been prevented through deploying key-word and content filters or other mechanisms.

Similarly, while deploying AIR within an organisation, filters should be introduced to ensure employees have confidence when dealing with AIR technologies. AI should be held accountable and there should be some kind of a root cause and performance mechanism whereby *bad-bots* are removed from the system and are replaced. Finally, the employees should be allowed to be active players with AIR. Take for instance an engineer who is looking for a 5% improvement in output in an oil well. The engineer will continually adjust and modify the AI software criteria to influence the outcome he or she wants. Hence, when the engineer is able to achieve the goal with the help of the machine, he or she is likely to build trust towards the system.

While the main focus is on addressing internal impact on people and processes, adoption by external stakeholders should be considered, too. Communicating key messages to customers and users on how the company is managing the transformation and how it will improve their interaction with the enterprise, will build brand loyalty.

5 Racing with Machines Rather than Against Them

Regardless of the longer term implications, in the short to medium term women and men in the workplace will need to engage more comprehensively with machines as part of their everyday activities. Tighter integration with technology will free up time for human workers including managers to focus more consciously on activities to which they bring skills that machines have yet to master. This could make work more complex, and more difficult to organise, with managers required to spend more time coaching and mentoring [9].

As, in particular, young people make education and career choices, it is important they are aware of the factors which drive automation in certain sectors. This will help to identify in-demand skills as well as those activities that will complement already automated business processes.

Employers will have to reassure and excite staff in ways and manners not done before. Addressing fear of automation should be handled across the enterprise, at both team, and more importantly, individual levels. Additionally, whenever any new AIR initiatives are introduced, they should be inclusive and allow for staff participation. Also to ensure better adoption by organisations, the perception of such AIR technologies should be improved through effective communication and interventions to increase awareness.

Highly skilled workers who collaborate closely with the machine are likely to be in strong demand. Those involved in developing and deploying AIR technologies will be well-placed in the labour market and will be able to demand salary premiums. In addition, employees who carry out tasks and activities that complement automation will also find themselves in an advantageous position. These and other workers will also be able to take advantage of new opportunities for independent work as the corporate landscape shifts and automation initiatives are being outsourced. Lower skilled workers deployed alongside the new technologies will be able to achieve more in terms of output and productivity but may experience wage pressure given the potential oversupply of similar skills.

Finally, AIR and associated technological changes will create new roles and opportunities for those with human skills that machines are still unable to replicate, such as social and emotional capabilities, providing tailored expertise to new challenges, coaching and developing others, and creativity. For now, the workplace still expects men and women to undertake rote tasks that do not stretch these innate capabilities as far as they could. As machines take on ever more of the predictable activities, these skills will be at a premium. Automation could make us all more human [10].

6 Summary

This paper talks about how enterprises must demystify the human-machine handshake by addressing possible emerging concerns of implementing AIR. As stated in the paper, this is the foundation on which automation opportunities can be explored, the handshake designed and embedded in an adaptive operating model and mindsets and behaviours changed. Finally, the paper concludes by stating that the AIR revolution is not the future: it is the present and it is about reimagining the human and machine handshake across the company to get the most benefit from this technology's power to augment human capability.

References

1. Xing, B., Marwala, T.: Introduction to Human Robot Introduction. Springer, Heidelberg (2017)
2. Davenport, T.H., Ronanki, R.: Artificial Intelligence for the Real World. Harvard Business Review, January/February 2018
3. Muro, M.: Manufacturing Jobs Aren't Coming Back. MIT Technology Review, November 2016
4. Manyika, J.: Technology, Jobs, and the Future of Work. McKinsey Global Institute, May 2017
5. Frank, M., Roehrig, P., Pring, B.: What To Do When Machines Do Everything. Wiley, Hoboken (2014)

6. Brown, R.H., Roehrig, P., Malhotra, V.: The Robot and I: How New Digital Technologies Are Making Smart People and Businesses Smarter by Automating Rote Work. Cognizant (2017)
7. Fukuzumi, Z., Noda, N., Tanikawa, Y.: How to apply human-centered design process to software development process? IEEE Conference Paper, May 2017
8. Daugherty, P.D., Wilson, H.J.: Human + Machine – Reimagining Work in the Age of AI. Harvard Business Press, Boston (2018)
9. Knight, W.: More Evidence That Humans and Machines Are Better When They Team Up. MIT Technology Review, November 2017
10. Palmer, A.: How brains and machines can be made to work together. The Economist, January 2018

Simulation of an Artificial Hearing Module for an Assistive Robot

Marcio L. L. Oliveira(✉), Jes J. F. Cerqueira, and Eduardo F. Simas Filho

Electrical Engineering Program, Federal University of Bahia, Salvador, Brazil
marciooluizz@gmail.com, {jes,eduardo.simas}@ufba.br

Abstract. It is proposed in this paper a simulation of an artificial hearing module for a low-cost assistive robot with the purposes of: (1) identifying the speaker; (2) determining the speaker's location; (3) estimating what emotion is transmitted by his or her speech; and (4) determining an algorithm that links the previous tasks. Mel Frequency Cepstral Coefficients (MFCCs) are used as feature extraction technique, artificial neural networks as classifier and multilateration as technique for locating a sound source. Results from both simulated and public data sets indicate the efficiency of the proposed method.

Keywords: Artificial hearing · Artificial neural networks
Cepstral coefficients · Mel Frequency Cepstral Coefficients (MFCCs)
Multilateration

1 Introduction

Assistive robotics are advancing in many distinct ways, being used in many different cases, such as working with children, the elderly, disabled and also entertaining adults. The design of an assistive robot must be taken into account so that it has interactions more similar to what humans would have between them, in such a way that users would not find the robot strange at the first time it is introduced into a new environment [1]. In this way, when an artificial hearing module is elaborated, it will be required to perform some tasks that humans do, in a natural way, through hearing.

The ability to communicate through voice is one of the most important means of communication among human beings. The human ear possesses great abilities that were also developed for social use. However, this communication doesn't always need to be made with understandable words. Often, dialogues can be done without the use of any vocabulary. In this way, a social robot that interacts with humans needs something similar to an ear that is able to perform some functions. For artificial hearing, the sound reaches the microphones and is interpreted. Different interpretations can be made by the module. Each speaker carries in his or her way of speaking various relevant information, such as listener distance, tone, pitch of the voice, words, emotions, etc.

K. Arai et al. (Eds.): IntelliSys 2018, AISC 868, pp. 852–865, 2019.
https://doi.org/10.1007/978-3-030-01054-6_59

For this problem, it is proposed here an artificial hearing module, which will be included into an assistive robot. Such an "ear" will have to be able to, initially, perform three distinct tasks: (1) locating a sound source - knowing where in space they are talking to it [2]; (2) speaker identification - knowing who is talking to it [3]; (3) identification of stereotyped emotions - knowing what emotion is being transmitted by the speaker's voice [4].

Unlike the approach of others researchers [5] and common sense for hearing, the content of the words spoken will not be taken into account here. These three distinct tasks were chosen because they were considered to be essential for the development of a low-cost social robot that could interact with adults, children, the elderly or people with special needs. For this purpose it is proposed that, for the elaboration of such a module, techniques such as, MFCC, artificial neural networks and multilateration will be used.

The purposes of this research are: (1) To simulate and identify text-independent speakers, whose audios were previously recorded; (2) to simulate and identify stereotyped emotions, regardless of the speaker, whose audios were also previously recorded; (3) simulating with mathematical software the location of a sound source in the non-real simulated environment; (4) propose an algorithm to link all the tasks of the module.

2 Literature Review

An audio-based emotion estimation method for interactive robotic therapy for children with autism spectrum disorder was proposed in [6]. It was possible to identify emotions of autism children, on real-time, with a assistive robot, but this was the only function of this module.

Building ears for a robot with sound localization and separation was proposed in [7]. This project developed a wheeled robot that could track and follow a sound source using time differences of arrival (TDOA) between the microphones. But it couldn't discern in any way what kind of sound was being emitted.

3 Methodology

For developing the proposed methods, some techniques were used. Mel frequency cepstral coefficients (MFCCs) and artificial neural networks (ANN) were applied for speaker identification and for identifying stereotyped emotions; core affect to classify the emotions; and multilateration for locating a sound source.

3.1 Mel Frequency Cepstral Coefficients, Delta and Delta-Delta

A series of steps, with their respective mathematical expressions, are required to obtain the Mel frequency cepstral coefficients (MFCCs) with delta and delta-delta, as seen in Fig. 1 [8]. MFCCs, delta and delta-delta parameters were used as features to feed the neural network.

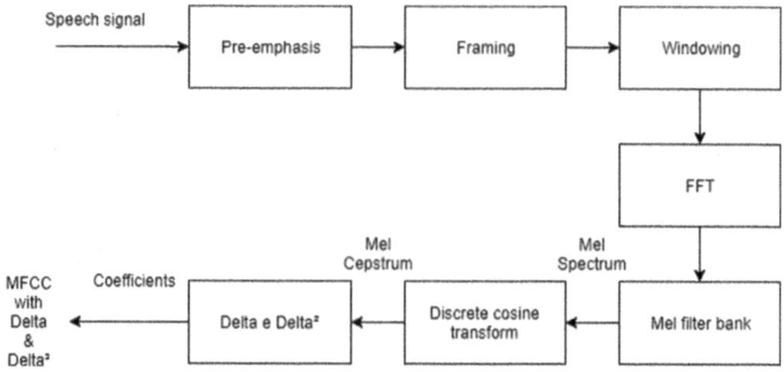

Fig. 1. Block diagram to obtain MFCCs with delta and delta-delta.

(1) *Pre-emphasis:* First of all, the speech signal is passed through a filter which purpose is to emphasize the high frequencies (to increase the energy of the high frequencies) [9]. This filter difference equation is given by:

$$Y[n] = X[n] - 0.95X[n-1]. \tag{1}$$

(2) *Framing:* The goal of this step is to break the audio signal into small frames of 15ms, so that there are two adjacent windows overlapping (50% of overlap) and no information is lost. Since the frames are very small, therefore almost no changes between frames, they are classified as quasi-stationary waves [3].

(3) *Hamming Windowing:* Each of the frames generated in the previous step passes through a Hamming window. It is used to integrate all the lines of frequencies nearby. The Hamming window is given by the following equation [8]:

$$w(n) = 0.54 - 0.46\cos(\frac{2\pi n}{N-1}), \tag{2}$$

where $0 \leq n \leq N-1$ and N is the number of samples of each frame.

(4) *Fast Fourier Transform (FFT):* Each of the frames that passed through the Hamming Window is converted to the frequency domain (spectrum), using fast Fourier transform. FFT is a computational algorithm for the fast execution of discrete Fourier transform (DFT) [10].

(5) *Mel Filter Bank:* The Mel scale relates the perceived frequency, or pitch, of a pure tone to its actual measured frequency. Humans are much better at discerning small shifts in tone at low frequencies than at high frequencies. A bank of triangular filters linearly spaced at the frequency of Mel is used to give greater prominence to the most important frequencies of the voice (Fig. 2). To calculate the energies of the Mel filter bank, each filter of the bank is multiplied with the power spectrum and the coefficients are summed [3]. This generates a vector with same length as the number of filters in the Mel filter bank used.

The frequency conversion from Hz to Mel frequency scale is given by [10]:

$$M(f) = 1125\ln(1 + f/700), \tag{3}$$

where f is the frequency in Hz and M is the Mel scale frequency. After passing through the Mel filter bank, the Mel frequency spectral coefficients are obtained.

(6) *Discrete Cosine Transform (DCT):* The Mel frequency spectral coefficients are converted to a time-like domain called quefrency domain (cepstrum) using the discrete cosine transform (DCT), generating the Mel frequency cepstral coefficient [8]. From the second to the sixteenth coefficients were kept. The number of coefficients was chosen experimentally to improve classification performance.

(7) *Delta and Delta-Delta:* For a better accuracy in the identification by a classifier, in addition to the Mel frequency cepstral coefficients coefficients, delta and delta-delta are also calculated. Delta measures the variation (first derivative) of the speech (velocity) and delta-delta the variation (second derivative) of the change of speech change (acceleration). These other coefficients are added to give more dynamics and add information that may have been lost on the change of each frame. Delta and delta-delta also provide information on the speaker's change and style of speech [11].

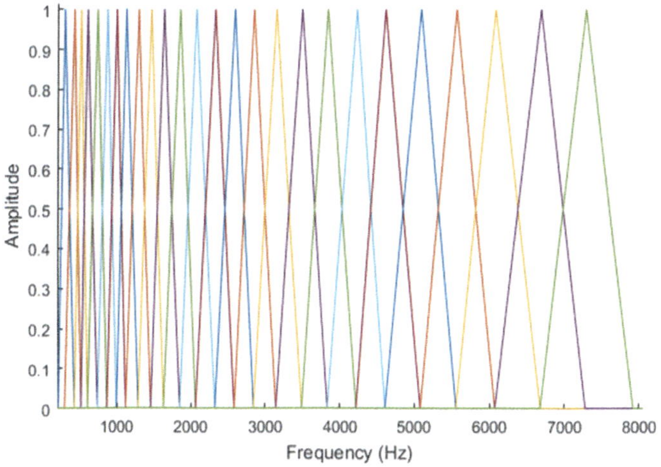

Fig. 2. Mel filter bank with 26 filters.

The first derivative, or delta (Δ), can be approximated by the equation below, where typically $P = 2$ and $c[n]$ is the nth coefficient of the MFCC:

$$\Delta c[n] = \frac{\sum_{p=1}^{P} p(c[n+p] - c[n-p])}{2\sum_{p=1}^{P} p^2}. \tag{4}$$

The second derivative or delta-delta (Δ^2), can be approximated in a way similar to (4), but with the elements of the first derivative in place of the Mel frequency cepstral coefficients

$$\Delta^2 c[n] = \frac{\sum_{p=1}^{P} p(\Delta c[n+p] - \Delta c[n-p])}{2\sum_{p=1}^{P} p^2}. \tag{5}$$

After calculating the delta and delta-delta, 45 feature coefficients were obtained to be used in a pattern recognition classifier.

3.2 Feature Extraction for the Simulations

For the simulations frames of 15 ms each, 1024 samples per frame, 50% of overlap between frames and 26 Mel filters were used. From that 45 features coefficients were extracted (15 from MFCC, 15 from delta and 15 from delta-delta) from each the speech.

For the simulation of identification of a speaker, a code that computes the MFCCs with delta and delta-delta and uses neural networks to classify the coefficients and thus to identify the speakers was produced. A spoken-word database with 5 speakers and 500 words, 100 per speaker, was recorded with a sampling frequency of 44100 Hz. All words were recorded in Portuguese.

In the process of identifying which emotion is being transmitted by the speaker, a database of emotions previously recorded was used. Considering that these emotions were performed by actors and not captured naturally, they are called stereotyped. For this simulation, the Surrey Audio-Visual Expressed Emotion (SAVEE) [12] emotion database was used. In this database were used 3 speakers, 6 emotions and 70 speech samples per emotion per speaker. The recordings were made at a sampling frequency of 44100 Hz. All the recording were spoken in British English. The emotions used in this simulations were: anger, disgust, fear, joy, calmness and sadness. These six emotions were used to feed the pattern classifier.

3.3 Core Affect

In order to classify emotions and, at the same time, limit the number of possible actions to be taken by the assistive robot, it is proposed to use the core affect [13], a Cartesian map, where emotions are classified as being pleasant or unpleasant on its horizontal axis and high (activation) or low arousal (deactivation) on the vertical axis (Fig. 3). In this way, the innumerable human emotions are compiled in only four possibilities, increasing the accuracy of the classifier and reducing the complexity of the possible decisions made by the robot.

The six emotions were regrouped into these four groups. Pleasant with high arousal (joy), pleasant with low arousal (calmness), unpleasant with high arousal (anger and disgust) and unpleasant with low arousal (fear and sadness).

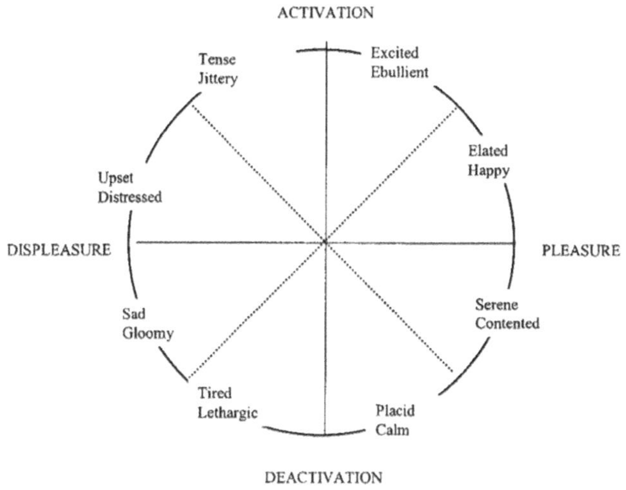

Fig. 3. Core affect [13].

3.4 Artificial Neural Networks (ANN)

Artificial neural networks (ANN) are computational models that have their inspiration in the central nervous system of animals, although their architecture and elements are now completely different from their biological inspiration. They are networks capable of adapting without being directly programmed for it. They are extremely useful in pattern recognition [14].

Among the possible classifiers, artificial neural networks using the MFCCs as input, seems to present good results in the identification of speakers [15]. ANN can also be used to identify emotions with MFCCs [16].

Multilayer perceptrons (MLP) are feedforward (the information can only move forward) artificial neural networks with one or more hidden layers. An illustration of a typical MLP can be seen in the Fig. 4. Generally such networks are trained using known inputs and outputs. After feeding the input signals, it is observed how far the output results are from what they were expected to be. In networks with backpropagation training, the correction of the weights, due the output error regarding the actual result, is done backwards, starting from the last layer and correcting them until the first [14].

Resilient Backpropagation (Rprop) is a technique of supervised learning of neural networks. Rprop takes into account only the partial derivative signal on all the standards (not the magnitude), and acts independently on each weight [17]. Rprop algorithm was used in this work for MLP training as it combines fast and accurate results.

For the simulations, two different sets of neural networks were trained. One for the identification of speakers and the other for stereotyped emotions classification.

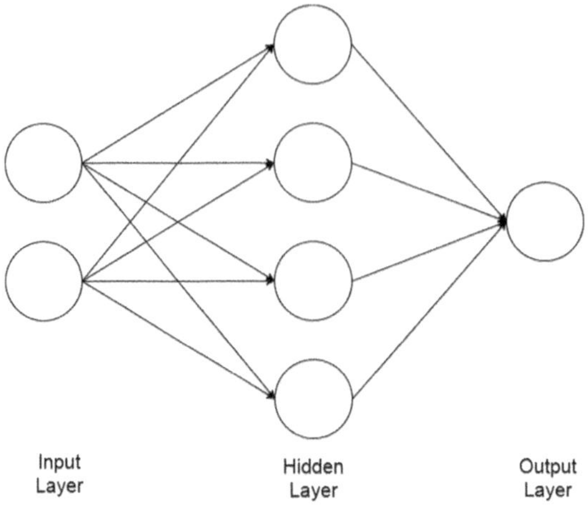

Fig. 4. Illustration of a typical feedforward artificial neural network.

For speaker identification, the 45 coefficients of feature extraction generated 45 neurons in the input layer of the artificial neural network. A total of 30 neurons were used in the hidden layer and 5 neurons in the output layer, one for each speaker.

For simulation of identification of stereotyped emotions, 45 coefficients were also computed, therefore, the neural network used comprises 45 neurons in the input layer, 40 neurons in the hidden layer and 6 neurons in the output layer, one for each emotion. Resilient backpropagation was used as training algorithm for both cases.

3.5 Multilateration

In the process of identifying a sound source, a technique used often is multilateration or hyperbolic navigation. This technique is widely used in aviation and nautical with sonars and triangulations. It is based on the Time Difference of Arrival (TDOA) of the sound wave to each microphone, whose positions are known. From this, hyperboles, whose foci are in the microphones, are obtained. The intersection of the hyperboles of the microphones will be the point of origin of the sound [18].

Knowing that i, j and c are sensors (microphones), where c is the first to register an event; $t_i - t_c$ and $t_j - t_c$ are the time difference of arrival between microphones; p_i, p_c and p_j are the position of each sensor; and v is the speed of sound, the linearized multilateration equation is given by:

$$x = \mathbf{A}^{-1}b, \tag{6}$$

where x is the desired position. Each pair i and j creates the matrix $\mathbf{A}$ and the vector b, where $b_{ijc} \in b \in \mathbb{R}^3$ and $\vec{a_{ijc}} \in \mathbf{A} \in \mathbb{R}^{3\times3}$ such that $\vec{a_{ijc}}^T$ are the lines of $\mathbf{A}$.

$$\vec{a_{ijc}} = 2(v(t_j - t_c)(\vec{p_i} - \vec{p_c}) - v(t_i - t_c)(\vec{p_j} - \vec{p_c})), \tag{7}$$

where $\vec{a_{ijc}} \in \mathbb{R}^3$ and

$$b_{ijc} = v(t_i - t_c)(v^2(t_j - t_c)^2 - \vec{p_j}^T\vec{p_j}) + (v(t_i - t_c) - v(t_j - t_c))\vec{p_c}^T\vec{p_c} + v(t_j - t_c)(\vec{p_i}^T\vec{p_i} - v^2(t_i - t_c)^2), \quad (8)$$

where $b_{ijc} \in \mathbb{R}$ [2].

For the simulation of sound source location, 4 virtual microphones were used, whose positions are known. It was also considered, an environment where the speed of sound in the air is 343 m/s. This simulation did not consider effects such as noise, echo, not knowing the real speed of sound in the environment, hardware limitations, microphone sensitivity, reverberation and others. The simulated environment was hypothetic and considering ideal conditions.

4 Simulation and Results

It is proposed to create a computer simulations that solve each of the problems raised for the artificial hearing module for an assistive robot.

4.1 Speaker Identification

The simulation for the identification of a speaker and stereotyped emotion has two distinct parts. The first part is the neural network training and the second part is to test and obtain results with the trained network.

The number of MFCCs coefficients were chosen experimentally. Fifteen coefficients were kept. With this amount, the results were improved in 5% in comparison with twelve or thirteen coefficients.

Between all the classifier tested, like ANN with scaled conjugate gradient (SCG), Learning Vector Quantization, Fuzzy C-mean, ANN with Rprop was chosen as the best cost-benefit because it is fast and reliable, thus, perfect to be embedded in a low-cost real-time artificial hearing module for an assistive robot. This result can be seen at the Table 1, where it's possible to see the hit rate of each classifier and how much time the simulation took compared with the current one.

Table 1. Different classifiers tested

Classifier	Hits (%)	Time (%)
ANN with Rprop	94%	–
ANN with SCG	94%	+15%
Learning Vector Quantization	80%	+200%
Fuzzy C-mean	65%	−13%

To train the neural network, the samples were separated between test and the training, of which 80% (400 words) of the speech samples were used for the training of the network and 20% (100 words) for the test. In the network training, the 400 words undergo through MFCC extraction with delta and delta-delta to extract its features. After each word passed through the network, all weights of each neuron were corrected backwards using Rprop. Once trained, the network was tested for results. The features of the remaining 100 words were extracted and, subsequently, passed through the trained network. Each hit or error were counted.

For this proposed simulation, an accuracy of 94% was reached, which is the best result, considering the computational cost, obtained among different classifiers tested. This results shows that the simulation could tell which speaker spoke 94% of the time. As seen in Fig. 5, the ANN identified the speaker 1, 100% of the time correctly, 90% the speaker 2, 90% the speaker 3, 95% the speaker 4 and 95% the speaker 5. This result is consistent with that found by other authors [10,19,20].

Confusion Matrix	Speaker 1	Speaker 2	Speaker 3	Speaker 4	Speaker 5	Total
Speaker 1	20	0	0	0	0	20 100%
Speaker 2	1	18	1	0	0	20 90%
Speaker 3	2	0	18	0	0	20 90%
Speaker 4	0	0	1	19	0	20 95%
Speaker 5	0	0	1	0	19	20 95%
Total	23 86,9%	18 100%	21 85,7%	19 100%	19 100%	100 94%

Fig. 5. Confusion matrix of speaker identification.

4.2 Identification of Stereotyped Emotions

As in the identification of a speaker, for the identification of stereotyped emotions, it is also necessary to train a neural network. Thus, 83% (348 samples) of performed speeches with different emotions were used to train the network and 17% (72 samples) for tests and results. The six emotions used to generate the network were: anger, disgust, fear, joy, calmness and sadness. In the same way as before, the MFCC is extracted with delta and delta-delta, obtaining the coefficients that were used to generate the neural network. Once trained, this network was tested with the remaining samples and the hits or errors were counted.

For the case of identifying each of the emotions separately, there was a 63% of accuracy, as seen in Fig. 6. Emotions are hard to identify, even for human ears, and this result was not regarding the speaker, so this rate was expected. It is interesting to note that some emotions had a high accuracy, like calmness, with 92%, and other were really low, like joy, with 33%. However, when rearranging the outputs of the ANN with the core affect map to classify the emotions, there was a 72% of accuracy (Fig. 7) for the four Cartesian cases. But if each axis was independently observed, the network got right 79% of the time if the emotion was pleasant or unpleasant and 85% of the time if the emotion was of high or low arousal. With these results, it is possible for the robot to make a reliable decision. This result is consistent with that found by other authors [6,16].

Confusion Matrix	Anger	Disgust	Fear	Joy	Calmness	Sadness	Total
Anger	7	0	1	2	2	0	12 58%
Disgust	0	8	1	0	1	2	12 67%
Fear	1	1	7	0	0	3	12 58%
Joy	3	4	1	4	0	0	12 33%
Calm	0	0	0	0	11	1	12 92%
Sadness	0	0	3	0	1	8	12 67%
Total	11 64%	13 62%	13 54%	6 67%	15 63%	14 57%	72 63%

Fig. 6. Confusion matrix of identification of stereotyped emotions.

Confusion Matrix	Displeasure Deactivation	Pleasure Deactivation	Displeasure Activation	Pleasure Activation	Total
Displeasure Deactivation	21	2	1	0	24 88%
Pleasure Deactivation	1	11	0	0	12 92%
Displeasure Activation	4	2	16	2	24 67%
Pleasure Activation	4	0	4	4	12 33%
Total	30 70%	15 73%	21 76%	6 67%	72 72%

Fig. 7. Confusion matrix using core affect.

4.3 Location of a Sound Source

To identify speaker location, linear multilateration is used. For this simulation, four virtual simulated microphones were used. The positions of the microphones are known. The simulation chose random points to be the sound sources. Considering that when a microphone is struck by a sound wave, an event begins, the simulation calculates the TDOA and uses the mathematical (6) to locate the sound source.

As seen in Fig. 8, this simulation is very accurate, locating the sound source 100% of the time, although, hardly in a real environment would this precision be the same, since the current simulation is set in a perfect environment with hypothetic conditions.

The results of this simulation showed that linear multilateration is a viable, without much computational cost method to locate a sound source.

4.4 Proposed Linking Algorithm

For the elaboration of an artificial hearing module, it is proposed an algorithm that would organize the three tasks of the module: (1) identify who speaks; (2) the emotion; (3) where it is spoken. The logic behind the algorithm can be see in Fig. 9 and it is described bellow:

(1) A speech is captured by the microphone and goes through a pre-processing.

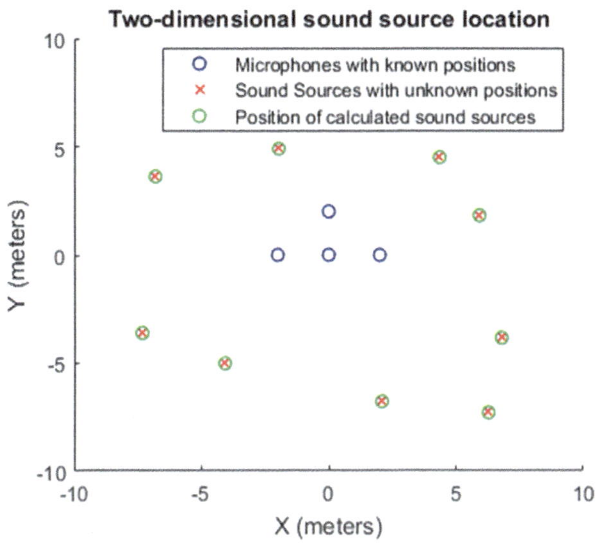

Fig. 8. Location of a sound source in a two-dimensional environment.

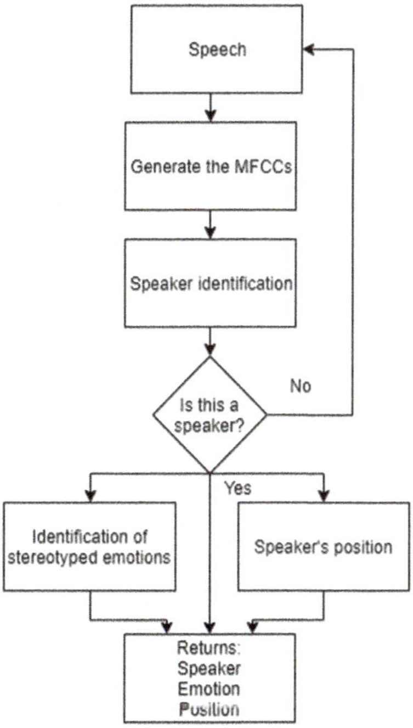

Fig. 9. Flowchart of the proposed artificial hearing module.

(2) The Mel frequency cepstral coefficients (MFCCs) with delta and delta-delta are generated.
(3) These coefficients pass through a previously trained neural network. The network gives as output if the speech captured belongs to one of the known speakers or not.
(4) If so:
   - This speaker is stored as one of the module return results.
   - It uses the already generated MFCCs with a second different neural network to identify the emotion of the speech.
   - It identifies where the speaker is located, using multilateration, regarding the position of the robot's microphones.
(5) Otherwise, the process of capturing a new speech is repeated (goes back to 1).

5 Conclusion

In order to obtain a more natural communication between robots and humans, it is necessary to use a hearing device. Preferably one that identifies the speaker and emotions, this way the robot is not limited to only interpreting words. In this paper, were discussed methods for the development of an artificial hearing module for a low-cost assistive robot. The objectives of the article are to simulate and find: (1) the speaker; (2) the emotion in the speaker's voice; (3) where he or she is; (4) to propose an algorithm for the hearing module. The following results were obtained: (1) the speaker identification simulation has an accuracy of 94%; (2) the stereotyped emotions identification simulation with core affect has an accuracy of 72%; (3) the sound source locating simulation has an accuracy of 100%; and (4) a linking algorithm was proposed.

For future work, this system will be embedded in a dedicated electronic board to test how the simulated results would behave in a real environment. Subsequently, this module will be integrated with other modules and there will be the final construction of the low-cost assistive robot.

References

1. Takayama, L., Dooley, D., Ju, W.: Expressing thought: improving robot readability with animation principles. In: 6th ACM/IEEE International Conference on Human-Robot Interaction (2011)
2. Norrdine, A.: An algebraic solution to the multilateration problem. In: 2012 International Conference on Indoor Positioning and Indoor Navigation (2015)
3. Hasan, R., Jamil, M., Rabbani, G., Rahman, S.: Speaker identification using Mel Frequency Cepstral Coefficients. In: 3rd International Conference on Electrical & Computer Engineering ICECE
4. Dahake, P.P., Shaw, K.: Speaker dependent speech emotion recognition using MFCC and support vector machine. In: International Conference on Automatic Control and Dynamic Optimization Techniques, pp. 1080–1084 (2016)
5. Valin, J.-M.: Auditory system for a mobile robot, Ph.D. dissertation, University of Sherbrooke (2005)

6. Kim, J.C., Azzi, P., Jeon, M., Howard, A.M., Park, C.H.: Audio-based emotion estimation for interactive robotic therapy for children with autism spectrum disorder. In: 14th International Conference on Ubiquitous Robots and Ambient Intelligence (URAI), pp. 39–44 (2017)
7. Huang, J., Ohnishi, N., Sugie, N.: Building ears for robots: sound localization and separation. Artif. Life Robot. **1**, 157–163 (1997)
8. Muda, L., Begam, M., Elamvazuthi, I.: Voice recognition algorithms using Mel Frequency Cepstral Coefficient (MFCC) and Dynamic Time Warping (DTW) techniques. J. Comput. **2**(3), 138–143 (2010)
9. Chakraborty, K.: Voice recognition using MFCC algorithm. Int. J. Innov. Res. Adv. Eng. (IJIRAE) **1**(10), 158–161 (2014)
10. Wahyuni, E.S.: Arabic speech recognition using MFCC feature extraction and ANN classification. In: 2nd International Conferences on Information Technology, Information Systems and Electrical Engineering (ICITISEE), pp. 22–25 (2017)
11. Togneri, R., Pullella, D.: An overview of speaker identification: accuracy and robustness issues. IEEE Circ. Syst. Mag. **11**(2), 23–61 (2011)
12. Haq, S., Jackson, P.J.B., Edge, J.: Audio-visual feature selection and reduction for emotion classification. Centre for Vision, Speech and Signal Processing (CVSSP), University of Surrey, Guildford, UK (2008)
13. Russell, J.A.: Core affect and the psychological construction of emotion. Psychol. Rev. **110**(1), 145–172 (2003)
14. Svozil, D., Kvasnieka, V., Pospichal, J.: Introduction to multi-layer feed-forward neural networks. Chemom. Intell. Lab. Syst. **39**, 43–62 (1997)
15. Sória, R.A., Cabral, E.F.: Speaker recognition with artificial neural networks and MEL-frequency cepstral coefficients correlations. In: 8th European Signal Processing Conference (EUSIPCO 1996) (1996)
16. Lalitha, S., Geyasruti, D., Narayanan, R., Shravani, M.: Emotion detection using MFCC and cepstrum features. Procedia Comput. Sci. **70**, 29–35 (2015)
17. Riedmiller, M., Rprop, I.: Rprop - description and implementation details, University of Karlsruhe, Karlsruhe, Technical report (1994)
18. Chan, Y.T., Ho, K.C.: A simple and efficient estimator for hyperbolic location. IEEE Trans. Signal Process. **42**(8), 1905–1915 (1994)
19. Chauhan, N., Chandra, M.: Speaker recognition and verification using artificial neural network. In: IEEE WiSPNET, pp. 1147–1149 (2017)
20. Martinez, J., Perez, H., Escamilla, E.: Speaker recognition using Mel Frequency Cepstral Coefficients (MFCC) and Vector Quantization (VQ) techniques. In: 22nd International Conference on Electrical Communications and Computers (CONIELECOMP), pp. 248–251 (2012)

Dynamic Walking Experiments for Humanoid Robot

Arbnor Pajaziti(✉), Xhevahir Bajrami, Ahmet Shala, and Ramë Likaj

Department of Mechatronics, University of Prishtina, Prishtina, Kosovo
{arbnor.pajaziti,xhevahir.bajrami,ahmet.shala,
rame.likaj}@uni-pr.edu

Abstract. This paper is focused on improvement of the humanoid robot walking by conducting dynamic walking experiments, based on human pattern gait parameters. The proposed approach has been tested and implemented to the Bioloid robot, firmly by simulating the robot on the Matlab program. Humanoid interaction considers that the robot starts and finishes the walk with the Double Support Feet (DSF). In this paper is presented the kinematic model of the Bioloid robot based on Denavit-Hartenberg (D-H) convention. To imitate the human walking, a three dimensions trajectory of the feet and the hip has been developed. For the design of the trajectories, first motion constraints during walking have been derived based on the human walking pattern. The sinusoidal spline interpolation is used to find the smooth trajectories for the foot and the hip in both single and double support phases are implemented on the Bioloid robot. In order to realize the humanoid walking in the sense of static stability, the robot's centre of gravity during the walking was located within the support foot area. After tests with different step lengths and step heights, we are able to analyse the end-effector of each walking parameter before implementation into the real humanoid robot.

Keywords: Kinematics · Walking motion · Humanoid robot · Parameter gain

1 Introduction

Gait analysis is the study of animal locomotion, including locomotion of humans [1, 2]. Describing human gait requires some specific terms, which are defined in this section. The gait cycle begins when a foot contacts the ground and ends when that foot contacts the ground again [10]. Thus, each cycle begins at initial contact with the stance phase and proceeds through a swing phase until the cycle ends with the limb's next initial contact [1]. Hence, the human walking step is composed of two different phases (Fig. 1) [15].

2 Direct Kinematics

For both modelling of direct and inverse kinematics, two basic frames will be considered: the first frame marked as $\sum T_0$ is placed on the robot's chest serves as a

K. Arai et al. (Eds.): IntelliSys 2018, AISC 868, pp. 866–880, 2019.
https://doi.org/10.1007/978-3-030-01054-6_60

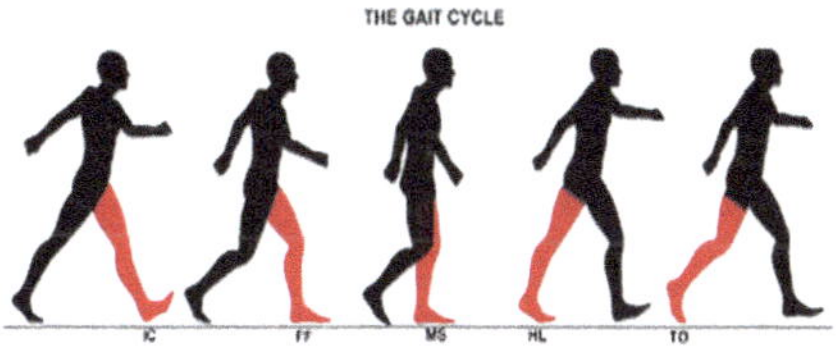

Fig. 1. Human gait cycle (http://www.footbionics.com/Patients/The+Gait+Cycle.html).

reference for the wing kinematics and the second base frame marked by $\sum Pelvis$ is on the hip and is used for reference positions associated with the legs of the robot (see Fig. 2). A direct kinematic model relates the linear and angular velocities of the frame $\dot{X}$ in terms of joint angles q and joint velocities $\dot{q}$:

$$\dot{X} = J(\dot{q}) \tag{1}$$

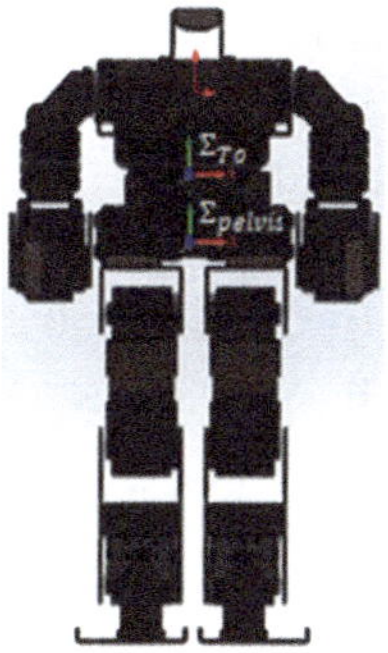

Fig. 2. Coordinates of the Bioloid Robot.

3 Direct Kinematics of Arms

The direct kinematics consists in finding the Cartesian position and the orientation (in Euler angles) of some point of interest of the robot. In the case of manipulator, it is generally the terminal element, whereas in humanoid robots we are generally interested in the position of its extremities. This is done simply by successively multiplying the homogeneous matrices that specify the position and orientation of each orthogonal frame that is fixed in each part of the robot [17].

To locate the articulations and terminal elements of each arm (i.e. the hands) orthogonal frames are assigned successively as shown in Fig. 3. The position and orientation of these frames with respect to the anterior one are specified by means of homogeneous matrices which were taken from the real robot, Bioloid. In the kinematic chain, the fixed frame of each joint (motor axis location) is located first by means of a constant homogeneous matrix and then its rotation is indicated by means of a homogeneous matrix whose rotation includes the corresponding joint variable (Fig. 4).

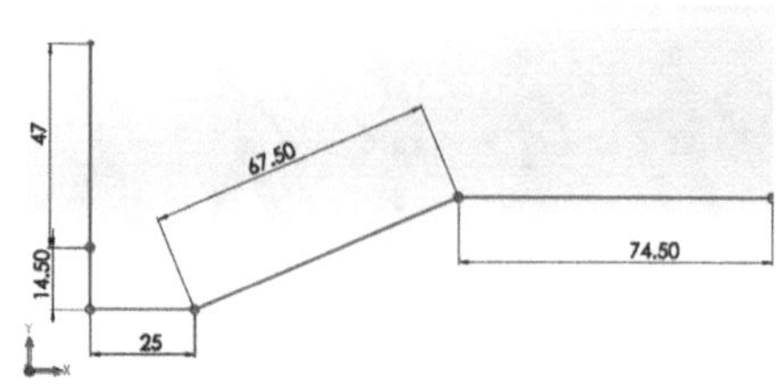

Fig. 3. Geometry parameters of the Bioloid Arm.

For the case of the right arm, the homogeneous matrices that describe the kinematics are presented in (2). The distance values that define the size of our robot are shown in Table 1.

$$T^0T_1^0T_2^1T_3^2T_4^3T_5^4T_6^5 = \begin{pmatrix} 0 & 0 & -1 & L_0 \\ 1 & 0 & 0 & 0 \\ 0 & -1 & 0 & 0 \\ 0 & 0 & 0 & 1 \end{pmatrix} \cdot \begin{pmatrix} c_1 & -s_1 & 0 & 0 \\ s_1 & c_1 & 0 & 0 \\ 0 & 0 & 1 & 0 \\ 0 & 0 & 0 & 1 \end{pmatrix} \cdot \begin{pmatrix} 0 & 0 & 1 & 0 \\ -1 & 0 & 0 & -L_1 \\ 0 & -1 & 0 & L_2 \\ 0 & 0 & 0 & 1 \end{pmatrix}$$
$$\begin{pmatrix} c_3 & s_3 & 0 & 0 \\ -s_3 & c_3 & 0 & 0 \\ 0 & 0 & 1 & 0 \\ 0 & 0 & 0 & 1 \end{pmatrix} \cdot \begin{pmatrix} 1 & 0 & 0 & 0 \\ 0 & 1 & 0 & -L_3 \\ 0 & 0 & 1 & 0 \\ 0 & 0 & 0 & 1 \end{pmatrix} \cdot \begin{pmatrix} 0 & 0 & 1 & 0 \\ 1 & 0 & 0 & -L_4 \\ 0 & 1 & 0 & 0 \\ 0 & 0 & 0 & 1 \end{pmatrix}$$
$$\begin{pmatrix} c_5 & s_5 & 0 & 0 \\ -s_5 & c_5 & 0 & 0 \\ 0 & 0 & 1 & 0 \\ 0 & 0 & 0 & 1 \end{pmatrix}, \quad Where: s_i = \sin(\theta_i), c_i = \cos(\theta_i) \tag{2}$$

For the case of the humanoid robot, the homogeneous matrix describing the position and orientation of the right hand with respect to the fixed frame of the robot hip is given by the successive multiplication of the corresponding homogeneous matrices.

$$T_{T_0}^R = T_{T_0}^1 \cdot T_1^3 \cdot T_3^5 \cdot T_R^5 \tag{3}$$

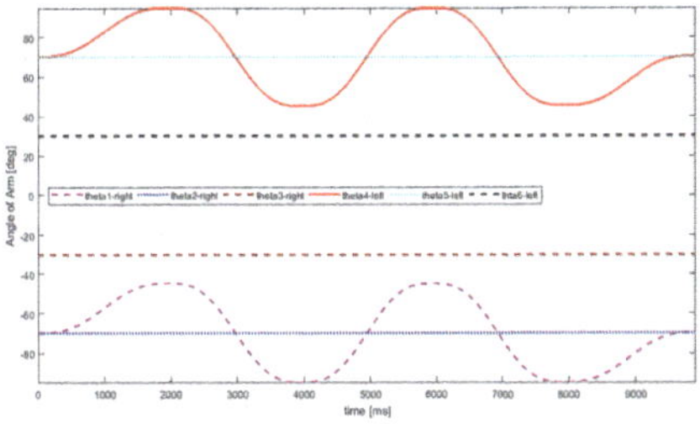

Fig. 4. The angle of upper body for hands of the humanoid robot.

Table 1. Parameters of Links for Arms

$L_0 = 47$	$L_1 = 14.5$	$L_2 = 25$	$L_3 = 67.5$	$L_4 = 74.5$

These matrices can be simplified further considering only the frames corresponding to each joint as shown in the following equations as the samples of simplified kinematics:

$$T_{T_0}^{1} = \begin{pmatrix} 0 & 0 & -1 & -47 \\ c_1 & -s_1 & 0 & 0 \\ -s_1 & -c_1 & 0 & 0 \\ 0 & 0 & 1 & 0 \end{pmatrix}, T_1^3 = \begin{pmatrix} 0 & 0 & 1 & 0 \\ c_1 & -s_1 & 0 & -14.5 \\ -s_1 & -c_1 & 0 & -25 \\ 0 & 0 & 0 & 1 \end{pmatrix}$$
$$. \quad T_3^5 = \begin{pmatrix} c_5 & s_5 & 0 & 0 \\ -s_5 & c_5 & 0 & -67.5 \\ 0 & 0 & 1 & 0 \\ 0 & 0 & 0 & 1 \end{pmatrix}, T_R^5 = \begin{pmatrix} 0 & 0 & 1 & 0 \\ 1 & 0 & 0 & -74.5 \\ 0 & 1 & 0 & 0 \\ 0 & 0 & 0 & 1 \end{pmatrix} \tag{4}$$

The way of multiplying considers the rule of eliminating previous super-indices with subsequent sub-indices. Considering the matrices specified in (1) and (2), the homogeneous matrix $T_{(\theta)}^{T_0}$ is given by:

$$T_{(\theta)}^{T_0} = \begin{pmatrix} c_{35} & 0 & -s_{35} & -61.5 - 67.5c_3 + 74.5c_{35} \\ s_1 s_{35} & c_1 & s_1 c_{35} & s_1(14.5 - 67.5s_1 - 74.5s_{35}) \\ c_1 s_{35} & -s_1 & c_1 c_{35} & s_1(14.5 - 67.5s_1 - 4.5s_{35}) \\ 0 & 0 & 0 & 1 \end{pmatrix} \tag{5}$$

The pitch, roll, and yaw (θ, ϕ, ψ) angles will be obtained as follows. These angles represent successive rotations - each about the local (rotated frame). This means that the rotation matrix resulting from these three rotations is:

$$Out = R(\theta, x)R(\phi, z)R(\psi, y) = \begin{pmatrix} c_\phi c_\psi & -s_\phi & c_\phi s_\psi \\ s_\phi s_\psi + c_\phi s_\psi s_\psi & c_\theta c_\phi & -s_\phi c_\psi + c_\theta s_\phi s_\psi \\ -c_\theta s_\psi + s_\theta s_\phi c_\psi & s_\theta c_\phi & c_\theta c_\psi + s_\theta s_\phi s_\psi \end{pmatrix} \tag{6}$$

And, by equating matrix from (6) to an arbitrary matrix R we have that:

$$R_{11}\cdots R_{33} = \begin{pmatrix} c_\phi c_\psi & -s_\phi & c_\phi s_\psi \\ s_\phi s_\psi + c_\phi s_\psi s_\psi & c_\theta c_\phi & -s_\phi c_\psi + c_\theta s_\phi s_\psi \\ -c_\theta s_\psi + s_\theta s_\phi c_\psi & s_\theta c_\phi & c_\theta c_\psi + s_\theta s_\phi s_\psi \end{pmatrix} \quad (7)$$

Equation (7) states that an arbitrary matrix R is obtained from successive rotations pitch, roll, and yaw.

Now, consider that we have numerical values for matrix R and we want to obtain pitch, roll, and yaw angles. From (7) we have that:

$$\phi = \left\{ a\tan 2\left(\sqrt{1 - R_{12}} - R_{12}\right) \quad \begin{matrix} \text{if } R_{12} \neq 0 \\ \text{if } R_{12} = 0 \end{matrix} \right. \quad (8)$$

$$\theta = \begin{cases} a\tan 2(R_{22}, R_{23}) & \text{if } R_{12} \neq 1 \\ a\tan 2(R_{21}, R_{31}) & \text{if } R_{12} = 1 \end{cases} \quad (9)$$

$$\psi = \begin{cases} a\tan 2(R_{22}, R_{23}) & \text{if } R_{12} \neq 1 \\ 0 & \text{if } R_{12} = 1 \end{cases} \quad (10)$$

Notice that if $R_{12} = 1$ then $\phi = \pi/2$ and $c_\phi = 0$. For this case $R_{21} = c_{\phi-\psi}$ so $\theta - \psi = a\tan 2(R_{21}, R_{31})$ and there are infinitely many solutions. By convention we chose $\psi = 0$.

From (10) it is easy to see that the Cartesian position of the right hand with respect to the hip frame is:

$$\begin{pmatrix} x_R^{T_0} \\ y_R^{T_0} \\ z_R^{T_0} \end{pmatrix} = \begin{pmatrix} -L_0 - L_2 - L_3c_3 + L_4c_{35} \\ s_1(L_1 - L_3s_3 + L_4s_{35}) \\ s_1(L_1 - L_3s_3 + L_4s_{35}) \end{pmatrix} \quad (11)$$

And by comparing (9) with (10) we have that:

$$\begin{pmatrix} \theta_R^{T_0} \\ \phi_R^{T_0} \\ \psi_R^{T_0} \end{pmatrix} = \begin{pmatrix} -L_0 - L_2 - L_3c_3 + L_4c_{35} \\ s_1(L_1 - L_3s_3 + L_4s_{35}) \\ s_1(L_1 - L_3s_3 + L_4s_{35}) \end{pmatrix} \quad (12)$$

4 Direct Kinematics of the Legs

In case of the legs, the location of the base frame should be in the pelvis of the robot. The similar procedure is for the arms, the leg frames are located by means of two homogeneous matrices, one to locate the fixed frame and the next that includes the corresponding joint variable to indicate its rotation [11].

4.1 Kinematics of the Legs

This section aims to find the joint angle trajectory based on the end-effector, and based position and orientation which are called inverse kinematics [5]. For humanoid robot legs, we will consider two cases: In first case the pelvis is considered fixed and the inverse kinematics consists of finding the angles of each leg to specify the Cartesian position of the feet and their orientation; The second case is to consider the fixed foot and find the necessary angles for the pelvis to have the desired position and orientation [9, 19]. These two approaches are complementary since a walking path will consider the second approach for the supporting leg and the first approach for the free leg (Fig. 5). We will discuss below in detail for both the cases.

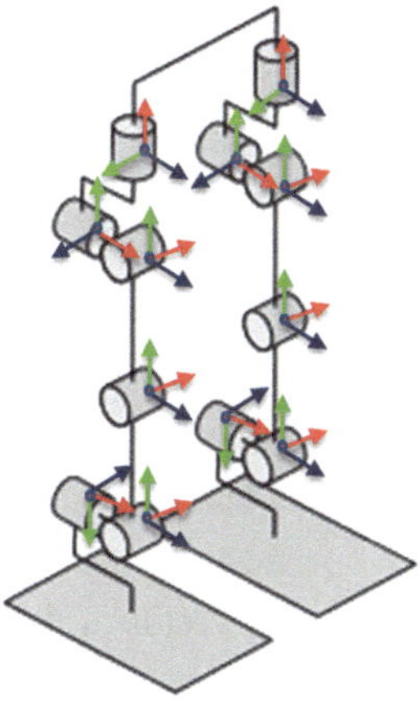

Fig. 5. Kinematic model of the humanoid Bioloid robot.

4.2 Right Leg

The approach we present in this section is to first solve the DOF-s θ_9, θ_{11} and θ_{13} to obtain the location of the ankle with respect to the hip then solve to give the orientation of an angle of yaw by means of θ_9 and finally the joints θ_{15} and θ_{17} will give the final orientation of the foot (Fig. 7). Consider the frames $\sum R_{Hip}$ as well as the vectors $r(Pelvis \rightarrow R_{Hip1})$ and $r(Pelvis \rightarrow R_{Ankle})$ the Fig. 6.

The vector $r_{pelvis-RHip1}$ with respect to the pelvis is given by:

$$r^{pelvis}_{pelvis-RHip1} = \begin{pmatrix} -L_5 \\ -L_6 \\ 0 \end{pmatrix} \tag{13}$$

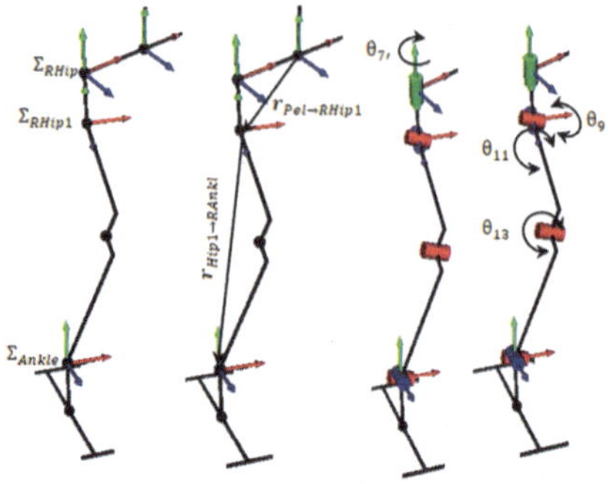

Fig. 6. Kinematic analysis of the right leg.

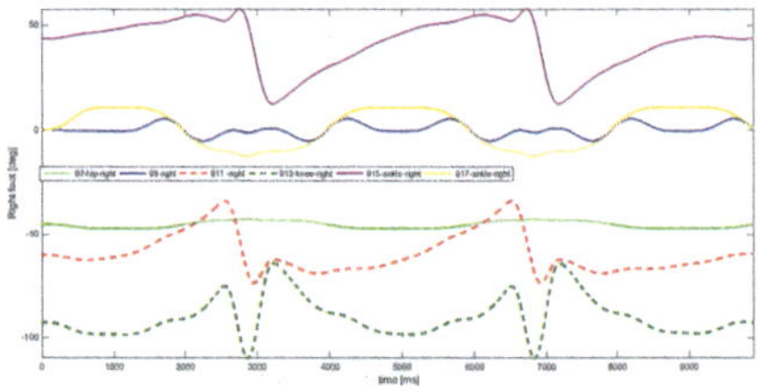

Fig. 7. Angle of lower body for right foot.

If we are able to define the position vector of the ankle with respect to the frame $\sum RHip_1$ as:

$$^{RHip1}r_{RHip1 \to Ankle} = \begin{pmatrix} x_B \\ y_B \\ z_B \end{pmatrix} \tag{14}$$

Then, we can see that the same vector with respect to the frame $\sum Rhip$ would be given by:

$$^{RHip1}r_{RHip1 \to Ankle} = \begin{pmatrix} x_A \\ y_A \\ z_A \end{pmatrix} = \begin{pmatrix} c_{7'}x_B - s_{7'}z_B \\ y_B \\ s_{7'}x_B + c_{7'}z_B \end{pmatrix} \tag{15}$$

And the vector of the pelvis to the ankle would be given by:

$$^{Pelvis}r_{pelvis \to RAnkle} = \begin{pmatrix} x_{RAmkle} \\ y_{RAnkle} \\ z_{RAnkle} \end{pmatrix} = \begin{pmatrix} x_A - L_5 \\ y_A - L_6 \\ z_B \end{pmatrix} \tag{16}$$

5 Parameters of Pelvis and Feet Gait Trajectory

In previous section gait analyses for human were reviewed and basic definition useful for pattern generation was presented. In the section, the human-like walking pattern is generation to adapt to different ground constraints. Therefore, there are only few parameters such as step length, step width, step height and so on, to generate the suitable walking pattern for humanoid robot [16].

The N parameter specifies the number of steps that will be made during the ride. The length of the steps in front and side direction is specified by LS = 10 [cm] and $k_n = 2[\text{cm}]$, while LS defines the maximum height of the foot during the step. The spin (yaw) for each step is specified by the left and right hip angle [8, 9]. The parameters of joint-motors specify the central position of the arms as well as the amplitude of the arm during the ride. Some parameters are specified in milliseconds (ms), and define the duration of the stride as follows: T is the total duration of the stride.

5.1 Trajectories for Omnidirectional Walking

One of the main features of the Humanoid interact software is that is used to perform a parameterized walk for the humanoid robot. In this section is given a detailed explanation of the proposed trajectories for the pelvis and the feet of the robot. Trajectories are planned in task space; which consists of six components: three for the position and three more for the orientation. The trajectories correspond to:

$$x_p = \begin{pmatrix} x_{pelvis} \\ y_{pelvis} \\ z_{pelvis} \\ \theta_{body} \\ \phi_{body} \\ \psi_{body} \end{pmatrix}', x_L = \begin{pmatrix} x_{L\_foot} \\ y_{L\_foot} \\ z_{L\_foot} \\ \theta_{L\_foot} \\ \phi_{L\_foot} \\ \psi_{L\_foot} \end{pmatrix}', x_A = \begin{pmatrix} x_{R\_Ankle} \\ y_{R\_Ankle} \\ z_{R\_Ankle} \\ \theta_{R\_Ankle} \\ \phi_{R\_Ankle} \\ \psi_{R\_Ankle} \end{pmatrix} \quad (17)$$

We will now describe the trajectories of each of these components for the six movements described of the walk.

5.2 Right and Left Foot Start Walk

For the beginning of the walk, it is considered that both feet are side by side [3, 10]. The frame of reference for this first step is placed in the centre of both feet. It is with respect to this framework that the positions and orientations of the pelvis and of the feet are specified [13, 14]. Although a frame is re-defined for each movement of the elements of the robot, this first frame is the reference of the whole walk.

The duration of the first stride corresponding to the Left Start movement is taken from parameter $Left - Right_{Hip} = 11(^\circ)$ duration T (corresponding to the First Step), hence the instantaneous variable t satisfies the condition: Feet are fixed.

$$0 < t \leq T \tag{18}$$

During this period of time one of the feet is fixed since it serves as support so, and needs to be adopted:

$$\begin{aligned} x(t)_{Fix\_Foot} &= 0; \\ y(t)_{Fix\_Foot} &= \omega \cdot c; \\ z(t)_{Fix\_Foot} &= 0; \\ \psi(t)_{Fix\_Foot} &= 0; \end{aligned} \tag{19}$$

Where c = 3.83 is the distance from the centre of the feet to the right or left foot.

The free foot starts its movement at the specified time by means of boundary and has the duration specified by means of step height. In this time interval, this foot must reach its final position, which depends on the length of the steps in x (CPth) and in y (CPz), see Fig. 8, given by:

$$\begin{aligned} c_X^{Free\_Foot} &= \frac{\sum \Delta_X^{Free\_Foot}}{2}; \\ c_Y^{Free\_Foot} &= \sum \Delta_Y^{Free\_Foot} \end{aligned} \tag{20}$$

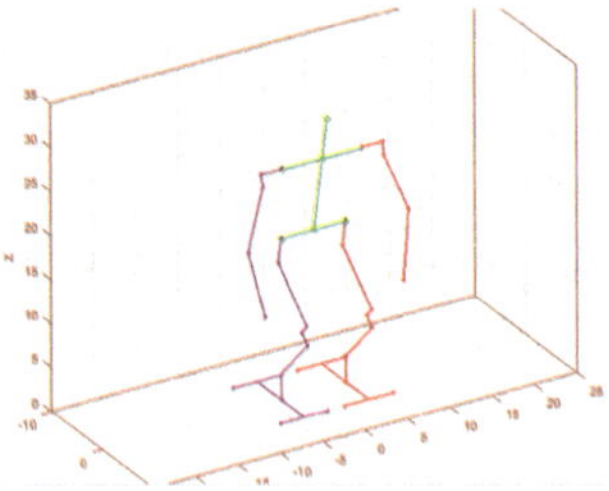

Fig. 8. Modelling of the humanoid robot in Oxyz.

In frontal advance (x) half step is given whereas in lateral movement an entire step is given. The trajectories can be expressed by means of:

$$\begin{gathered} x_{Free\_Foot} = \left\{ \begin{array}{l} cycle(c_X^{Free\_Foot}, T^{Step}, t \\ c_X^{Free\_Foot} \end{array} \right\} \\ if\ 0 < t < Time^{Strat\_Step} \\ if\ Time^{Strat\_Step} < t < Time^{Strat\_Step} + T^{Step} \\ if\ Time^{Strat\_Step} + T^{Step} < t < T \end{gathered} \tag{21}$$

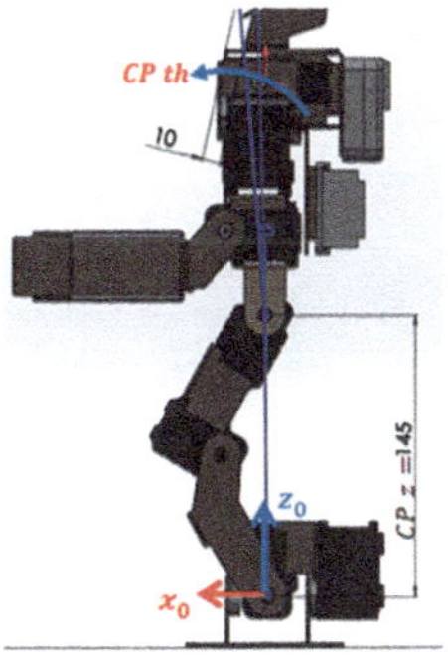

Fig. 9. The position of motion walking.

Where:

$$Cycle_{(k,T,t)} = k_n \cdot \left(\frac{t}{T} - \sin\left(\frac{2\pi t}{T}\right)\right) \quad (22)$$

The schematic diagram of the simulation program is depicted in Fig. 10. A Matlab program creates the simulation of the Bioloid robot based on the desired walking parameters mentioned before such as step length, step height, and so on. This simulation is a powerful tool that enables us to examine the walking parameters before implementing to the real robot. By this simulation, the general position of the robot is predicted and we are able to analyse the end-effector of each walking parameter before real implementation into the Bioloid robot in Fig. 9.

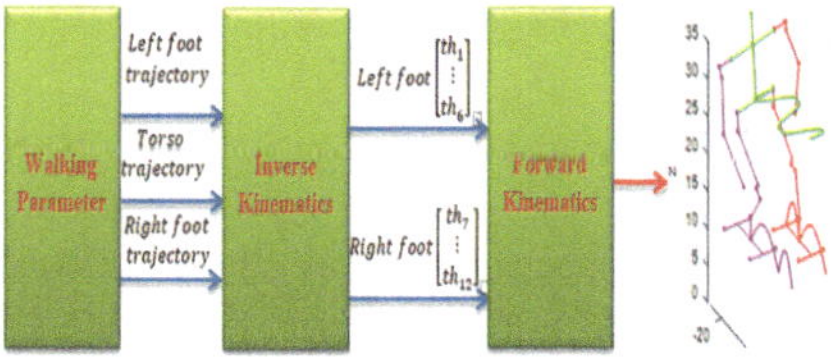

Fig. 10. Schematic diagram of the proposed program for kinematic simulation of the humanoid walking.

The graph of this function is presented in Fig. 11. The idea of using these functions is that the velocity and acceleration of motion are zero at the beginning and end of the specified period.

The trajectory in x and z-direction have the same form, with the same periods, so the formula shown in (21) is used. The only difference in this sense is the amplitude of the motion which is $c_Y^{Free\_Foot}$.

In Figs. 12, 15 and 16 are shown the stick diagrams of the humanoid robot which is walking on level ground. From this diagram, one can observe the overall motion of the humanoid during both the SSP and the DSP.

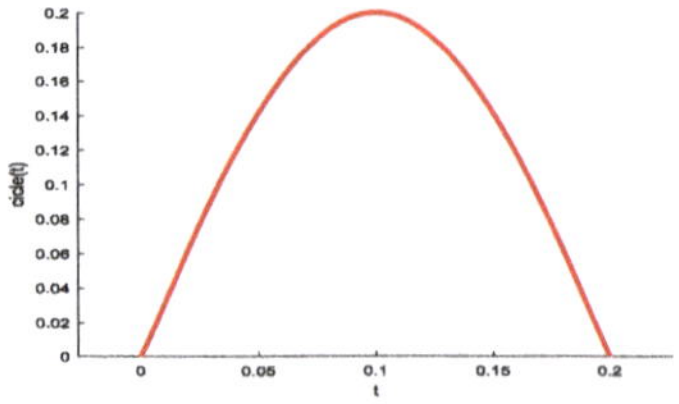

Fig. 11. The sinusoidal trajectory of feet in Oxz.

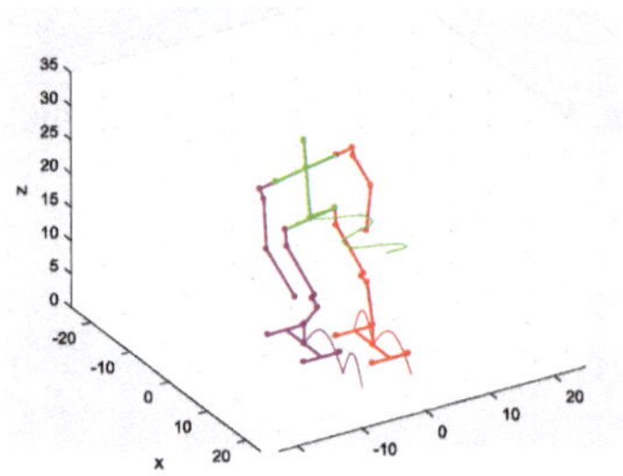

Fig. 12. Stick diagram of the robot with hip and feet trajectory.

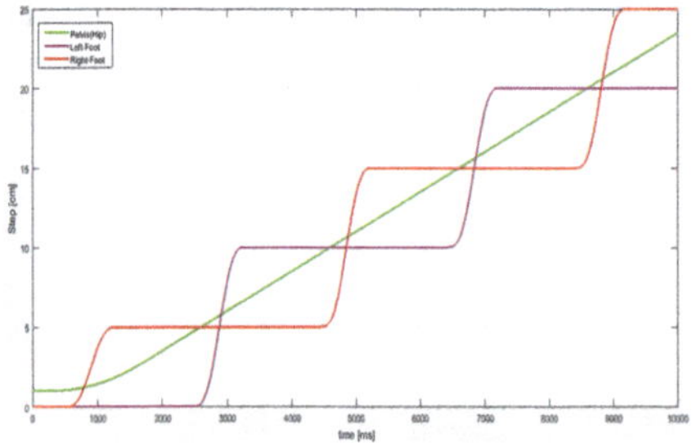

Fig. 13. Foot left-right and hip trajectories in x-z direction.

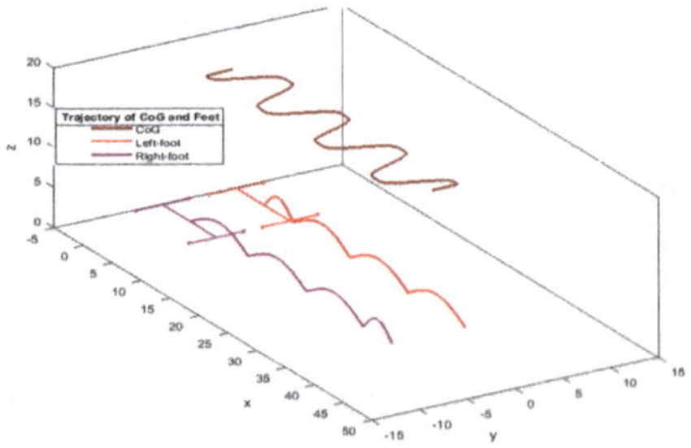

Fig. 14. Position trajectory hip (brown), right foot (magenta) and left foot (red) in three dimensions.

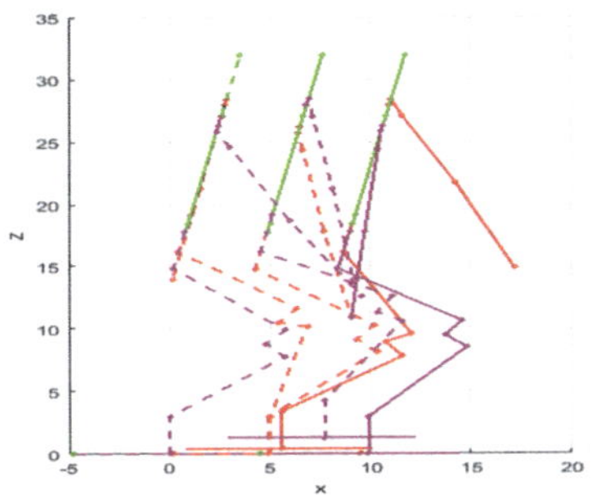

Fig. 15. Stick diagram of robot simulation at the time in the x-z plan.

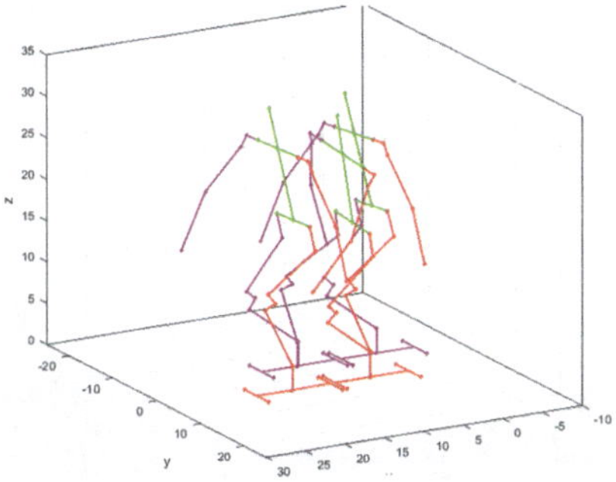

Fig. 16. Stick diagram of the robot in frontal for three frames.

Figure 13 shows the Centre of Gravity (CoG) of the humanoid during walking. The posture of the humanoid at the end of each step is close to that at the beginning of each step, and the displacement of Centre of Gravity (CoG) trajectory is almost horizontal [1, 2].

Figure 14 also shows the three-dimensional trajectory of the feet and pelvis (hip) during one complete cycle of walking according to the selected walking parameters. Up to here, both trajectories for feet and hip are constructed based on to the values of parameters. Now, one can find the joint angle trajectories based on the inverse kinematics algorithm [5, 6].

In order to determine the optimal trajectories simulation is performed by using the kinematic model. The result of each simulation is a set of numbers defining the stability margin, for each time step simulation (Fig. 17).

The trajectory with the highest stability margin at the most critical point is chosen as the optimal trajectory [12, 18]. To obtain a useful trajectory it is necessary to include the physical limitations of the robot in the trajectory simulations [4]. The physical angular limitations were determined by moving the joints one at a time to their physical limit and then performing a reading of the angle [15].

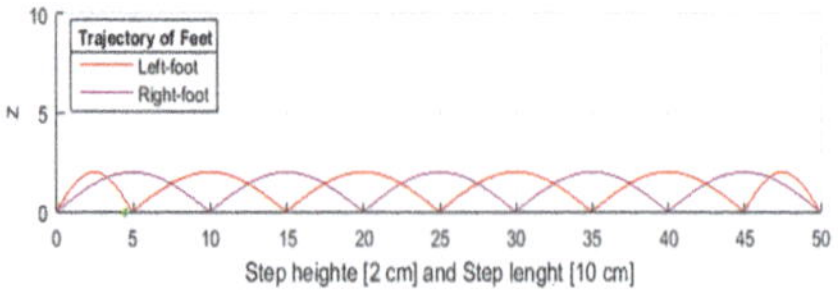

Fig. 17. The trajectory of feet in x-z direction.

5.3 Walking Realization on the Bioloid Robot

The optimal value of k_n is found using experimental tests. As shown in Fig. 18, in these tests, the humanoid robot joints are actuated in such a way that the humanoid robot stands in the right leg, while the left leg is in the air and the hip is moved to locate at the top of the right foot, i.e., $k_n = 0$. As expected in this case, the humanoid robot tips are given over to the left side [4]. Then the value of k_n is increased to reach the stability region. In fact, there is a range of k_n where the humanoid is stable [11]. This range is obtained for the Bioloid robot, as $0.5 \leq k_n \leq 2.5$ cm.

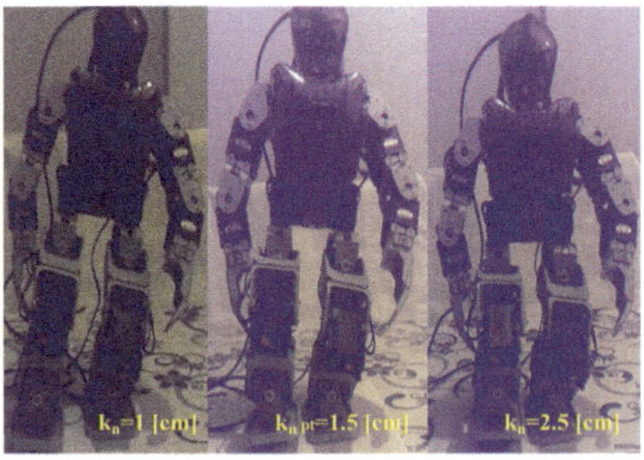

Fig. 18. Experimental tests to find the optimal value of k_n.

6 Conclusions

In this paper, the new approach of walking of the humanoid robot is realized based on human walking patterns. Walking patterns include trajectories of the feet and the hip, which are generated corresponding to the predefined walking parameters. These trajectories are calculated using sinusoidal spline interpolation. The trajectory constraints are developed based on the walking analysis of human [1, 4, 7]. To find the optimal values of walking parameters that ensure stable walking, several experiments have been performed. In fact, the walking parameters were selected in such a way that centre of gravity of the humanoid robot is moved from the rear foot to the front foot in the double support phase.

In addition, these parameters ensure that the centre of gravity always remains within the range of the support foot area, to maintain the ZMP (Zero Moment Point) inside the support area, and consequently stabilize the walking humanoid robot. Another important role of the controller is the conversion of the Cartesian space trajectories generated by the joint space trajectories. Therefore, first forward kinematics relation based on the D-H convention is developed and the closed form solution of the inverse kinematics was derived. This was done by decoupling the inverse kinematics problem into two parts: inverse position and inverse orientation.

References

1. Bajrami, X., Kopacek, P., Shala, A., Likaj, R.: Modelling and control of a humanoid robot. e & i Elektrotechnik und Informationstechnik **130**(2), 61–66 (2013)
2. Kajita, S., et al.: Humanoid walking pattern generation by using preview control of zero-moment point. In: Proceedings IEEE Conference on Robotics & Automation, Taipei, Taiwan, 14–19 September, pp. 1620–1626 (2003)
3. Hyon, S.H., Suewaka, D., Torii, Y., Oku, N.: Design and experimental evaluation of a fast torque-controlled hydraulic humanoid robot. IEEE/ASME Trans. Mechatron. **22**(2), 623–634 (2017)
4. Bajrami, X.: Dynamic modelling and simulation of a humanoid robot. Ph.D. thesis. Vienna University of Technology, Austria (2013)
5. Shahverdi, P., Masouleh, M.T.: A simple and fast geometric kinematic solution for imitation of human arms by a NAO humanoid robot. In: 2016 4th International Conference on Robotics and Mechatronics (ICROM), Tehran, pp. 572–577 (2016)
6. Goldbeck, C., Kaul, L., Vahrenkamp, N., Worgotter, F., Asfour, T., Braun, J.M.: Two ways of walking: contrasting a reflexive neuro-controller and a LIP-based ZMP-controller on the humanoid robot ARMAR-4. In: 2016 IEEE-RAS 16th International Conference on Humanoid Robots (Humanoids), Cancun, pp. 966–972 (2016)
7. Bajrami, X., et al.: Trajectory planning and inverse kinematics solver for real biped robot with 10 DOF-s. IFAC Pap. Line **49**(29), 88–93 (2016)
8. Kopacek, P.: Cost oriented humanoid robot. In: Proceedings of the 18th IFAC World Congress, Milano, Italy, 28 August–2 September 2011 (2011)
9. Bajrami, X., et al.: Kinematics and dynamics modelling of the biped robot. IFAC Proc. Vol. **46**(8), 69–73 (2013)
10. Mu, X., Wu, Q.: Synthesis of a complete sagittal gait cycle for a five-link biped robot. Robotica **21**, 581–587 (2004)
11. Bajrami, X., et al.: Genetic and Fuzzy Logic algorithms for robot pathfinding. In: 2016 5th Mediterranean Conference on Embedded Computing (MECO), pp. 195–199. IEEE (2016)
12. Raibert, M.H.: Legged Robots that Balance. MIT Press, Cambridge (1986). IEEE International Conference on Robotics and Automation, April 2005
13. Kajita, S., Nagasaki, T., Kaneko, K., Yokoi, K., Tanie, K.: A running controller of humanoid biped HRP-2L. In: Proceedings of IEEE International Conference on Robotics and Automation (ICRA) 2005
14. Pajaziti, A., Gojani, I., Shala, A., Pira, B.: Planning walking patterns for a biped robot using fuzzy logic controller. In: IARP Workshop on Robots for Humanitarian Demining, HUDEM 2004, Brussels-Leuven, Belgium, June 2004
15. Kajita, S., Kanehiro, F., Kaneko, K., Fujiwara, K., Harada, K., Yokoi, K., Hirukawa, H.: Resolved momentum control: humanoid motion planning based on the linear and angular momentum. In: Proceedings of the IEEE International Conference on Intelligent Robots and Systems (IROS) (2003)

16. Nunez, V., Yokoi, K., Nadjar Gauthier, N., Blazevic, P., Stasse, O.: Whole body posture controller with stable motion planning. Submitted to IEEE International Conference on Humanoid Robots (Humanoids) (2006)
17. Hayashi, R., Tsujio, S.: High-performance jumping movements by pendulum-type jumping machines. In: IEEE International Conference on Intelligent Robots and Systems (IROS) (2001)
18. Tiong, Y.K., Romyaldy: Conceptual framework for designing humanoid locomotion control system. In: Proceedings of IEEE International Conference on Robotics, Automation and Mechatronics (RAM) (2004)
19. Sugihara, T., Nakamura, Y.: Contact phase invariant control for humanoid robot based on variable impedance inverted pendulum model. In: Proceedings of IEEE International Conference on Robotics and Automation (ICRA) (2003)

A Method to Produce Minimal Real Time Geometric Representations of Moving Obstacles

David Sanders(✉), Qian Wang, Nils Bausch, Ya Huang, Sergey Khaustov, and Ivan Popov

School of Engineering, University of Portsmouth, Portsmouth, PO1 3DJ, UK
{david.sanders,qian.wang,nils.bausch,ya.huang,sergey.khaustov, ivan.popov}@port.ac.uk

Abstract. Real time modelling methods are compared for use with a robot manufacturing work-cell and a simple image processing system. The static parts of a robotic manufacturing work-cell are modelled as a number of solid polyhedra. The robot is modelled as a number of connected spheres and cylinders. The static model is renewed when an object enters or leaves the static work-place. Simple polyhedra, spheres and similar 2-D slices in actuator space are compared with other models as representations of objects move in and out of the reach of the robot. Models are compared for their efficiency in accessing data and ability to update as information about moving objects changes. Geometric models of the robot and the robot work-cell are loaded into a path planner to compare the models for efficiency on planning paths around moving objects. Some preliminary results are presented.

Keywords: Robot · 2-D slice · Obstacle · Path

1 Introduction

Complicated industrial environments consist of machinery, objects being worked on and manipulated, and obstacles to avoid [1–5]. Available free space for moving machinery will depend on the modelling methods used in a varying situation and environment. The navigation of industrial robots through a changing and dynamic environment is difficult [6], especially when the movement of objects is not known or expected in advance. The traditional planning methods can be slow and reactive navigation tends not to look far enough ahead. Vander-Stappen [7] described some efficient methods to calculate an exact solution for motion planning problems if there are only a few moving objects but even he believed that the algorithms used for planning frequently used long and worst-case calculation and computing time [8]. Exact algorithms do not need to be used in real-time though. The processing time will depend on processing speed, problem complexity and how complicated the modelling algorithms are. Processing speed is still increasing but at the same time, the dynamic problems being addressed are becoming more complicated.

If some assumptions can be made about the shape, distribution and size of moving objects then the problem can be simplified. Complexity of free space increases linearly

K. Arai et al. (Eds.): IntelliSys 2018, AISC 868, pp. 881–892, 2019.
https://doi.org/10.1007/978-3-030-01054-6_61

with the number of objects. De-Berg considered how complicated motion planning problems were for bounded-reach robots with only a few objects around them [9]. Tang looked into how to geometrically and topologically depict intersections between moving polygon shapes [10]. Manufacturing environments where industrial robots are used can often contain polygon objects. For real-time planning, if an environment could be modelled satisfactorily then the velocities that could cause a collision could be modelled [11].

The complexity of the modelling methods directly affects the planning algorithm complexity. The algorithms need to calculate a sequence of collision-free orientations left for an industrial robot when there are objects moving around it. The complexity results in comparatively lengthy computations and therefore prolonged processing times.

Some real-time models being used in diverse spaces are studied in this paper. The models are updated by a simple vision system. The work-cell for the robot is modelled by static solid polyhedra stored within computer memory. The static model is revised when objects enter or leave the robot work cell. Simple polyhedral, spheres and two-dimensional slices in an actuator space were compared as potential models for the moving objects. Spheres appear to be the simplest models for moving objects. Halperin [12] explored ways of manipulating loosely connected spheres. He evaluated models based on spheres and identified the things that made them simple and efficient to use. He described effective algorithms that calculated places where the spheres touched.

In order to calculate a path, a mapping needs to take place either from a workspace to a joint (or configuration) space or from the joint space to a workspace. That is required so that the robot, the objects and the environment can be compared within a common space and so that static and moving objects can be avoided. Sanders studied geometric modelling and decided that the following were important for representing a robot work cell with moving objects: quick and simple model creation, quick and efficient calculation of intersections between models and ease of use with algorithms that are planning movements [1, 3–5]. Sharma investigated the fastest time to transfer a vehicle from one place to another while avoiding other vehicles [13]. The other vehicles represented moving obstacles and a conflict was considered when the distance between vehicles was smaller than a speed-dependent safety distance. Fiorini described a robot motion planning method within a dynamic environment that took place within common velocity space [14].

Models described here are assessed for their capacity to be revised when new data is available, for example, if an object enters, leaves or is moving within the robot work cell.

The physical structure of the robot is modelled as a set of connected spheres and cylinders and the range of motion of the robot is quantized. A quick but sub-optimal path is produced using simple models so that the robot can avoid objects as it moves towards a goal. The method inspects a 3-D table of quantized joint space.

Solid polyhedra are used to represent the static parts of the robot work-cell. As new objects enter or depart from the work cell then the available free space is updated. A six-sided parallelepiped, a single sphere, multiple spheres and 2-D slices in joint space are employed to model moving objects. They are compared and some results are presented.

The modelling of the robot and the static environment is discussed next before considering modelling using 2-D Slices and the transformation from real space to joint space. Then the modelling of 2-D slices is compared to modelling using spheres and modelling using simple polyhedral shapes. Finally, some results are discussed before some general conclusions are drawn.

2 Robot and the Static Environment

Industrial environments and robot work cells often consist of solid floors, walls and objects with straight edges and flat surfaces. Accurate models of these shapes are more difficult to create, change and use in real-time. If moving objects and robots are modelled using polyhedral shapes, then although the accuracy might be higher, computation is more complicated and processing time is lengthened. A static environment only needs to be modelled once because it does not change. Therefore, computational complexity and processing time is not a problem. So, polyhedra can be used to accurately model a static environment.

Providing any model used to represent the industrial robot enclosed the entire industrial robot then the most important aspect of the model was the time taken to calculate intersections. Many industrial robots have links (often a forearm and upper arm) and joints (often base, shoulder and elbow). The simplest depiction of this type of industrial robot is two lines that are connected by a joint. If fixed distances are defined from the lines then the volume can be considered to enclose the industrial robot casing. This model gives two cylinders with hemispherical ends that are connected by a joint. An advantage of this model is that the robot links are efficiently modelled and the calculations of intersections between objects and the robot arm are simplified. An end-effector can be modelled by an extra sphere at the end of one of the cylinders with a radius that is large enough to surround the whole of the end effector. Work-pieces can be encompassed within the sphere.

The time taken to calculate intersections was recorded for various models representing moving objects within the work cell of the industrial robot. Three of the models performed well: six-sided parallelepiped in Cartesian space, spheres and 2-D slices in joint space. During the testing and comparison, it was presumed that the 2-D cross section of a moving object was known in the X-Y plane of movement and that the depth or height (Z) of the moving object could be provided by a sensor system (in this case a simple vision system was used). Moving objects were therefore considered to be two-and-a-half-dimensional. That is to say that they had a height and a 2-D shape. The 3-D moving shapes were then treated as cubes and cylinders. 2-D planar slices in joint space, spheres and parallelepiped models were used to model the 3-D moving shapes to a practical precision. The 2-D slices provided faster real-time modelling of moving shapes in a discrete three-dimensional space.

2.1 Modelling Using 2-D Slices

2-D slices were produced by creating two sets of boundaries: base joint angles ($\theta 1_{min}$ and $\theta 1_{max}$) that bound the object and the minimum distance D_{min} and the maximum distance D_{max} from the origin (the minimum and maximum radii). A moving object can then effectively be modelled as a sequence of 2-D slices. A reference planar slice is determined within the boundary created by a line from the Origin bounded by D_{min} and D_{max} and the Z axis value. These impeded elbow ($\theta 2$) and shoulder ($\theta 3$) joints can then be calculated for this plane and then copied for all $\theta 1$ within the bounding angles, $\theta 1_{min}$ and $\theta 1_{max}$.

This significantly reduced the number of tests and searches for blocked points. The planning algorithm is largely reduced to copying values rather than complicated mathematical calculations.

2.2 Transformations into Joint Space

A single point in a Cartesian space does not transform to be a single point in an actuator space. If a single point lies inside the operational volume of an industrial robot then it will transform into one or more complicated 3-D shapes. These complicated 3-D shapes may be represented by approximating the profiles of the shapes using mathematical curves to describe the geometric shapes, or they may be represented by units of space. In this research, moving objects were represented as regions of small units within a joint space. That technique would not be restricted to any particular robot and is suitable for a robot with any number of degrees of freedom.

A KUKA KR125 robot was used and the research considered the three major axes. A 3-D table was created and within the table, each dimension corresponded to one of the degrees of freedom for the industrial robot ($\theta 1/\theta 2/\theta 3$). The configuration of the wrist was not studied but the wrist and end effector could easily be included within an extra sphere.

Each table element was set to "free" and orientations were determined in the actuator space where the industrial robot interconnected with the moving objects. An element denoted a robot configuration in terms of, ($\theta 1_{center}$, $\theta 2_{center}$, $\theta 3_{center}$), with a value added to represent some movement away from the central values. All the units together embodied the whole work space for the industrial robot. The number of units in the table, TotalNodes, was:

$$\{(\theta 1_{max} - \theta 1_{min})/2 \times \delta\theta 1\} * \{(\theta 2_{max} - \theta 2_{min})/2 \times \delta\theta 2\} * \{(\theta 3_{max} - \theta 3_{min})/2 * \delta\theta 3\} \quad (1)$$

where,

$\theta 1_{max}$ and $\theta 1_{min}$ = upper and lower boundaries of $\theta 1$.
$\theta 2_{max}$ and $\theta 2_{min}$ = upper and lower boundaries of $\theta 2$.
$\theta 3_{max}$ and $\theta 3_{min}$ = upper and lower boundaries of $\theta 3$

If the robot body intersected with a moving object at any configuration in a unit, that unit was set to "obstructed". If the robot did not intersect with any moving object at any configurations then all configurations within a unit, the unit continued to be set to "free". The global path planning problem was then reduced to searching for a series of neighbouring units between a BEGINNING and an END configuration that were "free". Free space tends to be bigger than obstructed space so that a quick and efficient method was to test each transformed moving object and for nodes which could contain the robot. This method was adopted. The main algorithm is summarized here:

(1) Create the Static Model (only done once):
Initialise data structures to create a 3-D table to represent joint space.
Calculate trigonometric solutions.
Set all units in the table to "free" status.
Associate flags with each node.
(2) Generate "free" space based on static environment model.
(3) Data about moving objects was then either simulated or provided by the vision system.
Repeatedly read the data about moving objects.
(4) Create the two and a half dimensional model.
(5) For an element where the industrial robot could intersect with a moving object, recursively test elements next to the original element to decide whether they can also be reached by the industrial robot.

3 Modelling Using 2-D Slices

To begin with, the limits in X were increased by the upper-arm radius:

$$\text{StartofRow\_clear} = \text{StartofRow} - \text{UpperlimRad} \tag{2}$$

$$\text{EndofRow\_clear} = \text{EndofRow} + \text{UpperlimRad} \tag{3}$$

The center point modulus and the ends of an edge StartofCol were determined.

$$(\text{TopLeftcorner,Ang}) = \text{InvTan}(\text{StartofCol}/\text{EndofRow\_clear}) \tag{4}$$

$$\text{Corner(TopLeft, Modulus)} = \sqrt{\left(\text{StartCol}^2 + \text{EndRow}^2\right)} \tag{5}$$

The subsequent model parameters were calculated: largest base angle, ($\theta1_{ma}x$), smallest base angle, ($\theta1_{min}$), outside radius from origin, (D_{max}), and inside radius from origin, (D_{min}).

If a moving object corresponded to a known template then a third dimension (a height Z) was obtained for the moving object from a template, otherwise height was set to infinity. That segment was then extrapolated on the Y axes and calculation only took place in a Z, Y plane. The model of the moving object was expanded by the upper-arm radius in the Z and Y plane. θ1 was changed to a new lower limit and an inverse kinematic solution was calculated for all points within the moving object for the Z and Y axis.

```
FOR Y = (MinUpperRad) TO (MaxRadius + UpperRadius)
  FOR Z = 255 TO (Rad(Z) + UpperRadius)
    CALL InvKin
  NEXT Z
NEXT Y
```

Coordinates in Z and Y were translated into robot joint angles with an inverse kinematic solution subroutine called InvKin. Distance from the Cartesian point (L3) to the origin and the angle to a point (Pointθ) were calculated:

$$\mathrm{Point}\,\theta = \mathrm{Inv\_Tan}\,Z/Y \tag{6}$$

$$\mathrm{sqofL3} = Y^2 + Z^2 \tag{7}$$

$$\mathrm{L3} = \sqrt{\mathrm{sqofL3}} \tag{8}$$

L3 was checked against the upper-arm to test whether an impact was conceivable. If within reach of upper-arm and if θ2 was within limits, then θ3 was set to "blocked" between its limits and θ2 was set to Curveθ and. If L3 was less than Forearm plus Upperarm then the Forearm could collide with that point. θ3 and θ2 were computed using the cosine rule. If θ3 and θ2 were within limits then a flag was set to "obstructed".

2-D slice models have been used with sensor systems [15, 16] mobile robots [17–21] and powered wheelchairs [22, 23].

4 Modelling Using Spheres

Data was initialized. A table was created such that the limits of the table corresponded to the limits for the angles of the robot joints inside the work cell. Objects outside the work cell were disregarded. So, the table was used to conduct checks for intersections at a limited number of orientation and positions. That meant that a smaller number of trigonometric solutions was needed and they could be calculated at the beginning, when calculation time did not matter. Before the objects were determined, all the elements in the table were set to 'EMPTY'. Another 4 flags were associated with each table element: 'ON LIST', 'UPPERARM TESTED', 'FOREARM TESTED', and 'NEW OBJECT'. The code representing each element was kept as a byte of data in an array and the five flags each used one bit. Object data was extracted from a file or sent from the vision system and the initial job for the program was to read this data.

A sub-task then calculated the upperarm and forearm blocked space in the table. A configuration was calculated at which the part of the arm being considered was nearest to the center of the object. If a forearm was being studied, then a configuration where the Foretip was at the sphere center was computed. For the upperarm, a configuration was calculated for which the center line of the upperarm pointed at the center of the sphere. If the object was within reach of the link being examined, then this configuration was the first unit for the transformed object. A base angle was computed from the X, Y coordinates for the sphere. First, the modulus (L3) and the angle (Sphereθ) from the robot to the center of the sphere was computed and a check was performed to find whether the sphere was out of the range of the robot.

$$\text{Waist}\,\theta = q1 = \text{InverseTan}(Y/X) \tag{9}$$

$$\text{ModulusXY} = \sqrt{(X^2 + Y^2)} \tag{10}$$

$$\text{Sphere}\,\theta = \text{InverseTan}(Z/\text{ModulusXY}) \tag{11}$$

$$L3 = \sqrt{(X^2 + Y^2 + Z^2)} \tag{12}$$

Shoulder and elbow angles (θ2 and θ3) were calculated using the cosine rule.
Upper-Arm length = L1 = 225 mm and ForeArm = L2 – 165 mm

$$\theta 3 = \text{InverseCos}\left[\left(L1^2 + L2^2 - L3^2\right)/(2 * L1 * L2)\right] \tag{13}$$

$$\theta 2 = \text{InversCos}\left[\left(L1^2 + L3^2 - L2^2\right)/(2 * L1 * L3)\right] + \text{Sphere}\,\theta \tag{14}$$

If the center of the sphere was close to the robot then θ3 would exceed its lower limit (θ3 < 90°). In this case, θ3 was set to 90° and θ2 was calculated using InversTan:

$$\begin{aligned} &\text{If}\,\theta 3 < 90^\circ\ \text{THEN} \\ &\quad \theta 3 = 90^\circ \\ &\quad \theta 2 = \text{InversTan}(L2/L1) + \text{Sphere}\theta \\ &\text{END} \end{aligned}$$

This provided a first configuration that was near the center. When the lower boundary of θ2 was surpassed, (θ2 < −30°), then the angle was set to −30°. The distance between the center of the sphere and the upperarm (Mod) was then calculated using FindtheModulus. The cosine rule could be used from that to find a new θ3:

$$\begin{aligned} &\text{If} - 30^\circ < \theta 2\ \text{THEN} \\ &\quad \theta 2 = -30^\circ \\ &\quad \theta 3 = \text{InversCos}\left[\left(L1^2 + L2^2 - \text{Mod}^2\right)/(2 * L1 * \text{Mod})\right] \\ &\text{END} \end{aligned}$$

The table element representing the first configuration was set to obstruct. Neighbouring elements were then tested. If they were also obstructed then their neighbors were tested. The position problem was solved using forward kinematic calculations and the minimum distance between the arm of the robot and the object (as long as that calculation had not been performed before). This continued recursively until all of the object had been transformed. Elements were set to "obstructed" if they had any two opposite neighbouring elements that were also obstructed. Any elements at the edge of the now solid model were added to a record. Neighbouring elements that were recorded were tested and that was repeated until the surface of the transformed sphere was fully outlined.

Obstructed elements were kept so that they could be expanded later on. When an element was expanded then it was retrieved from the record store and new obstructed points were added. When every element had been considered then the transformation of the object was completed. The time taken to calculate objects was recorded and some examples are given in Sect. 6 of this paper.

5 Simple Polyhedral Shape

Polyhedra have often been used to model objects and a simple polyhedral model was used here. Moving objects were modelled as simple six-sided-parallelepiped. The edges of the model in X and Y were obtained by calculating the limits of the columns and rows set by the vision system. As before, if it was possible then the height of an object could be retrieved from a template. Edge positions were expanded by the radius of the part of the robot being tested (for example the forearm or upperarm). The example of expanding the forearm in X is shown here:

$$\text{ExpandXLow} = \text{EdgePos(LowX)}-\text{ForRad} \tag{15}$$

$$\text{ExpandXHigh} = \text{EdgePos(HighX)} + \text{ForRad} \tag{16}$$

The expanded polyhedral edge limits were then tested against the Cartesian coordinates of the arm.

6 Results

The time taken to transform objects was recorded. Time taken for the various models to transform a large cube into an actuator space is displayed within Table 1.

Table 1. Transformation times for a cube

Model	Time (ms)	Blocked elements
Single sphere	18.7	2425
2 × Spheres	94.2	2335
Simple polyhedra	126.2	1982
2-D similar slices	15.2	2488

6.1 Static Environment

The system needs the time taken to convert moving objects to be short. It can take a long time to model a static environment but this only needs to be done once at the start.

In this work that conversion took up to a minute of computer time depending on the complexity of the static environment.

6.2 2-D Similar Slices

An advantage of this model was that once collision coordinates for θ3 and θ2 were computed for a specific θ1 then the collisions could be repeated for all specific θ1 that might collide. This downgraded the principal processing job to a copy function rather than having to calculate a reverse or a forward kinematic solution. This model was the quickest to transform into a discrete 3-D actuator space.

If a moving object enlarged beyond a certain size or if at relocated nearer the origin then the moving object could intersect both the Forearm and Upperarm. So, the actuator space occupied by the moving object will abruptly increase and so calculation time will increase.

6.3 Sphere Model

Initially, the objects were modelled as individual spheres with the smallest radius that enclosed the object. If there was time then the object was re-modelled by 2 smaller spheres and then four smaller spheres, etc. Elements set to obstructed that were related to the first sphere would usually also collided with the models using multiple spheres. Forward kinematic solutions did not need to be recalculated for these elements. Despite that, the total processing time increased with the number of spheres. That was because the overhead involved in calculating intersections for each sphere was longer than any time that was saved because the spheres were smaller. So, calculations with a single sphere were quicker than multiple spheres despite the fact that the single sphere had a larger volume and was less accurate.

A problem with using more than one sphere was that the center of several spheres could be set to "blocked" (with some surrounding elements) after the first sphere was expanded. As these elements were obstructed, later spheres might not retest and some nodes might not be recorded.

6.4 Simple Polyhedra Model

Although this was the most accurate model, it took the longest to calculate.

7 Discussion

Higher accuracy models tended to need more processing time and took longer to reach a solution. Lower accuracy models needed robot links or moving objects to be artificially increased in size to remove the risk of unnoticed collisions. Lowering accuracy led to rejecting some potentially useable solutions. The speed of calculation was important for moving objects and their models. The easiest possible intersection calculation involved spherical models. Computation was slashed to only calculating distances from a line (representing a robot link) to a point (the center of a sphere) and then subtracting the sphere radius to produce a distance from the robot to the sphere surface.

The use of more than one sphere was studied. As the robot working environment became more complicated then more spheres could model it more accurately. Increasing the number of spheres increased the accuracy of a model. Cubic numbers of spheres were considered, that is 1/8/27/64, etc. If they were of equal size then they formed regular lattice patterns. An infinite number of spheres could model a cube exactly but it was found that modelling objects using spheres of equal size was not the most efficient way of doing it. For example, if a cube is modelled using sixty-four equally sized spheres, eight of those spheres are completely enclosed. Those eight could be replaced by one larger sphere and that would not reduce accuracy or increase the volume of the model.

To compare modelling using single and multiple spheres, as an extra example, a model of a cylinder using one and two spheres was compared. The volume of two spheres of radii 35 mm was compared to that of one sphere of 70 mm:

Volume of the two spheres: $2 \times 4/3 \times \pi \times 353 = 359{,}188\ mm^3$.
Volume of the single sphere: $4/3 \times \pi \times 703 = 1{,}436{,}755\ mm^3$.

The area of the two spheres would be much smaller except that the model of the robot must then be considered to find the union volume:

$$\text{Model} \cup \text{Robot} \tag{17}$$

The Upperarm model radius = 85 mm so: Union radius for a single sphere was 70 + 85 = 155 mm. Union radius for two spheres is 35 + 85 = 120 mm.

Union volume of a single sphere is $4/3 \times \pi \times 1503 = 14{,}137{,}167\ mm^3$. Union volume of two spheres is $2 \times 4/3 \times \pi \times 1153 = 12{,}741{,}211\ mm^3$.

A similar number of obstructions were recorded for both models. If elements in a second sphere were not tested for obstruction during calculations for a previous sphere, then that would change the calculated number. That partially explains a lack of reduction in calculation times when using two spheres to model objects.

The simple six-sided parallelepiped had a volume that was less than that of the 2-D slice model. The volume for the Parallelepiped was:

$\text{Volume} = (60 + 160)^2 \times (140 + 80) = 10{,}648{,}000\ mm^3$.

This could reduce the number of obstructed elements, but the shape, and hence the associated computations, would be more complicated. So, the calculation time would increase.

8 Conclusions

Angles for θ3 and θ2 were only calculated for a single slice so that processing time was reduced because that slice of "obstructed" elements was then just copied for all the base angles that might intersect the object. Copying is quicker than calculating and so this method appears to be faster than methods requiring real-time calculations.

The number of "obstructed" elements found was similar for all the models. Intersection volumes were similar and that implied a corresponding accuracy.

Modelling using similar 2-D slices produced the quickest intersection calculations. The 2-D slices were simpler than polyhedra, as the model only required two bounding angles for the base joint (an inner and outer radius) and a height. The model using similar 2-D slices were the most effective of the models considered. 2-D slices in joint space represented moving objects at least as well as the other models but they were quicker.

9 Future Work

The Real time modelling methods are now being compared for use in mobile robot applications using simple image processing and ultrasonic systems. The mobile robot will be modelled as connected spheres. And simple polyhedra, spheres and similar 2-D slices in actuator space are to be compared with other models as representations of objects around the mobile robot.

References

1. Sanders, D.A.: Recognizing shipbuilding parts using artificial neural networks and Fourier descriptors. Proc Inst. Mech. Eng. Part B J. Eng. Man **223**(3), 337–342 (2009)
2. Rasol, Z., Sanders, D.A.: An automatic system for simple spot welding tasks. In: Total Vehicle Technology Conference, pp. 263–272 (2001)
3. Sanders, D.A., Harris, P.: Image modelling for real time manufacturing applications using 2-D slices in joint space and simple polyhedra. J. Des. Manuf. **3**, 21–27 (1993)
4. Sanders, D.A.: Real time geometric modelling using models in an actuator space and cartessian space. J. Robot. Syst. **12**(1), 19–28 (1995)
5. Sanders, D.A., Lambert, G., Pevy, L.: Pre-locating corners in images in order to improve the extraction of Fourier descriptors and subsequent recognition of shipbuilding parts. Proc. Inst. Mech. Eng. Part B J. Eng. Man **223**(9), 1217–1223 (2009)
6. Carpin, S.: Randomized motion planning: a tutorial. Int. J. Robot. Autom. **21**(3), 184–196 (2006)
7. Berretty, R.-P., Overmars, M.H., van der Stappen, A.F.: Dynamic motion planning in low obstacle density environments. Comput. Geom. **11**(3–4), 157–173 (1998)
8. Sanders, D.A., Moore, A., Luk, B.L.: A joint space technique for real time robot path planning. In: Robots in Unstructured Environments, pp. 1683–1689. IEEE 91TH376-4 (1991). ISBN 0-7803-0078-5
9. de Berga, M., Katzb, M.J., Overmarsa, M.H., van der Stappena, A.F., Vleugels, J.: Models and motion planning. Comput. Geom. **23**(1), 53–68 (2002)
10. Tang, K.: A geometric method for determining intersection relations between a movable convex object and a set of planar polygons. IEEE Trans. Rob. **20**(4), 636–650 (2004)
11. Large, F., Laugier, C., Shiller, Z.: Navigation among moving obstacles using the NLVO: principles and applications to intelligent vehicles. Auton. Robot. **19**(2), 159–171 (2005)
12. Halperin, D., Overmars, M.H.: Spheres, molecules, and hidden surface removal. In: Proceedings of 10th Annual Symposium on Computational Geometry, pp. 113–122 (1994)
13. Sharma, V., Savchenko, M., Frazzoli, E., Voulgaris, P.G.: Transfer time complexity of conflict-free vehicle routing with no communications. Int. J. Robot. Res. **26**, 255–271 (2007)
14. Fiorini, P.: Robot motion planning among moving obstacles. University of California, Ph.D. dissertation (1995)

15. Sanders, D.A.: Environmental sensors and networks of sensors. Sensor Rev. **28**(4), 273–274 (2008)
16. Sanders, D.A., Lambert, G., Graham-Jones, J.J., et al.: A robotic welding system using image processing techniques and a CAD model to provide information to a multi-intelligent decision module. Assembly Autom. **30**(4), 323–332 (2010)
17. Sanders, D.A.: Comparing ability to complete simple tele-operated rescue or maintenance mobile-robot tasks with and without a sensor system. Sensor Rev. **30**(1), 40–50 (2010)
18. Sanders, D.: Comparing speed to complete progressively more difficult mobile robot paths between human tele-operators and humans with sensor-systems to assist. Assembly Autom. **29**(3), 230–248 (2009)
19. Sanders, D.A., Graham-Jones, J., Gegov, A.: Improving ability of tele-operators to complete progressively more difficult mobile robot paths using simple expert systems and ultrasonic sensors. Ind. Robot **37**(5), 431–440 (2010)
20. Sanders, D.A., Stott, I.J., Robinson, D.C., et al.: Analysis of successes and failures with a tele-operated mobile robot in various modes of operation. Robotica **30**, 973–988 (2012)
21. Sanders, D.A., Tewkesbury, G.E., Stott, I.J., et al.: Simple expert systems to improve an ultrasonic sensor-system for a tele-operated mobile-robot. Sensor Rev. **31**(3), 246–260 (2011)
22. Sanders, D.A., Langner, M., Tewkesbury, G.E.: Improving wheelchair-driving using a sensor system to control wheelchair-veer and variable-switches as an alternative to digital-switches or joysticks. Ind. Robot **37**(2), 157–167 (2010)
23. Sanders, D., Stott, I., Graham-Jones, J.: Expert system to interpret hand tremor and provide joystick position signals for powered wheelchairs with ultrasonic sensor syst. Ind. Robot **38**(6), 585–598 (2011)
24. Sanders, D., Gegov, A., Ndzi, D.: Knowledge-based expert system using a set of rules to assist a tele-operated mobile robot. In: Studies in Computational Intelligence, vol. 751, pp. 371–392. Springer (2018)
25. Sanders, D., Sanders, H., Gegov, A., Ndzi, D.: Rule-based system to assist a tele-operator with driving a mobile robot. In: Lecture Notes in Networks and Systems, vol. 16, pp. 599–615. Springer (2018)
26. Sanders, D., Ndzi, D., Chester, S., Malik, M.: Adjustment of tele-operator learning when provided with different levels of sensor support while driving mobile robots. In: Lecture Notes in Networks and Systems, vol. 16, pp. 548–558. Springer (2017)
27. Sanders, D., Sanders, B., Ndzi, D., Bausch, N.: Using confidence factors to share control between a mobile robot tele-operator and ultrasonic sensors. In: IEEE Proceedings of SAI Intelligent Systems Conference (IntelliSys) 2017, UK, 7 September 2017, pp. 1026–1033 (2017). ISBN 978-1-5090-6435-9
28. Sanders, D., Langner, M., Gegov, A., Ndzi, D., Sanders, H., Tewkesbury, G.: Tele-operator performance and their perception of system time lags when completing mobile robot tasks. In: Proceedings of 9th International Conference on Human Systems Interaction, UK, 6 July 2016. IEEE, pp. 236–242 (2016)

Application of Deep Learning Technique in UAV's Search and Rescue Operations

Kyaw Min Naing[1(✉)], Ahmad Zakeri[1], Oliver Iliev[2], and Navya Venkateshaiah[1]

[1] School of Engineering, University of Wolverhampton, Wolverhampton, UK
A.Zakeri@wlv.ac.uk, nav.venkateshaiah@gmail.com
[2] Faculty of Information and Automation, FON University, Skopje, Macedonia
oliver.iliev@fon.edu.m

Abstract. This paper is concerned with the application of Deep Learning techniques for analyzing image data for search and rescue operations of Unmanned Aerial Vehicle (UAV). It uses Keras and its Tensorflow backend to model a deep Convolutional Neural Network (CNN) Learning technique and train the model with MNIST digits dataset to predict the hand- written word from the image data received from the ground level. The paper explains the stages involved in the implementation of LeNet method of Deep Learning techniques for developing a classifier for long distance recognition of handwritten words.

Keywords: Deep neural network · Deep learning · Convolutional neural network MNIST · LeNet · Image processing · Unmanned Aerial Vehicle · Tensorflow Keras

1 Introduction

Unmanned Aerial Vehicles (UAVs) are currently being extensively applied in civilian tasks [5] such as security, surveillance, and disaster rescue. The use of Unmanned Aerial Vehicle in search and rescue activities in particular has increased dramatically over the past three years in different areas/fields such as marines, mountains, wild environments [6–10]. The main purpose of a search and rescue (SAR) operation is to identify and rescue the target in the shortest possible time, and this is critical as any delay will possibly reduce the chances of survival of the victims [14]. For effective and quick identification, and in particular, for long distance recognition of images, UAVs need to be equipped with intelligent devices and systems that can help them quickly recognise the images from long distances [2].

Image recognition or object recognition as the process of identifying and detecting an object or a feature in a digital image, has been the subject of research in many field of engineering and science, with many types of algorithms being developed to facilitate these processes, for example, image recognition algorithms include Optical character recognition, or Object recognition algorithms rely on matching, learning, or pattern recognition algorithms using appearance-based or feature-based techniques.

Current development in image recognition or processing involves object detection with trained data set. It is called machine learning (ML), it enables Artificial Intelligence

K. Arai et al. (Eds.): IntelliSys 2018, AISC 868, pp. 893–901, 2019.
https://doi.org/10.1007/978-3-030-01054-6_62

(AI) systems to learn from data. Machine Learning can now be part of image processing [13]. The currently available Machine Learning algorithms can facilitate supervised, unsupervised and reinforcement learning. In supervised learning, algorithms are presented with a dataset containing a collection of features. Additionally, labels or target values are provided for each sample. This mapping of features to labels of target values is where the knowledge is encoded. Once it has learned, the algorithm is expected to find the mapping from the features of unseen samples to their correct labels or target values. Deep learning algorithms are a subset of Machine Learning algorithms that typically involve learning representations at different hierarchy levels to enable building complex concepts out of simpler ones.

Deep learning is recently showing outstanding results for solving a wide variety of robotic tasks in the areas of perception, planning, localization, and control [12]. Its excellent capabilities for learning representations from the complex data acquired in real environments make it extremely suitable for image processing applications. Figure 1 compares the working of Deep Learning vs Machine Learning.

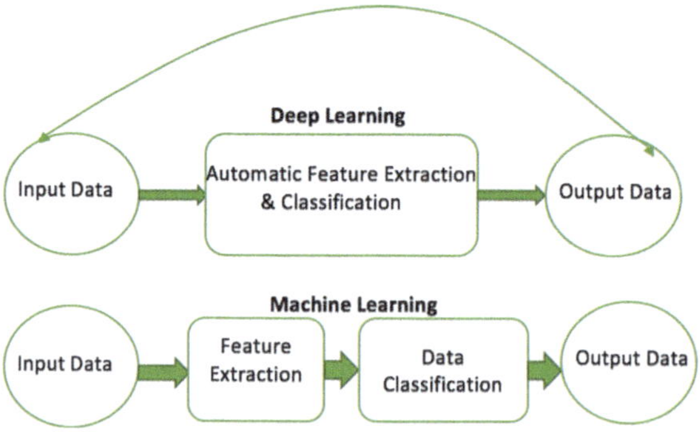

Fig. 1. Deep learning vs Machine learning.

As shown in Fig. 1, the traditional Machine Learning approaches worked in a way that required first design of a feature extraction algorithm, which generally involved a

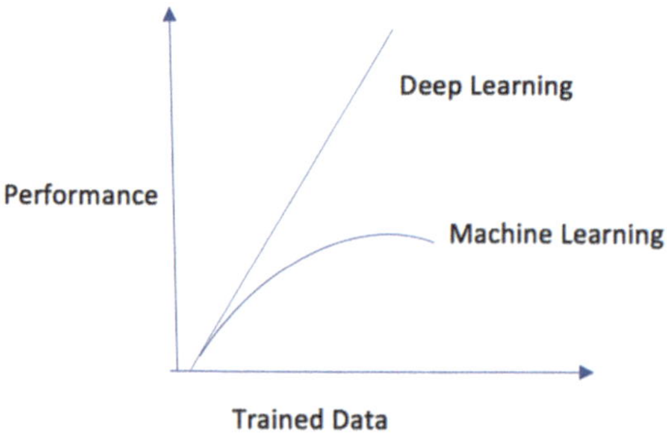

Fig. 2. Performance comparing of two models.

lot of heavy mathematics (complex design), and didn't perform too well at all (accuracy level just wasn't suitable for real-world applications). After doing all of that you would also have to design a whole classification model to classify your input given the extracted features. With deep networks, we can perform feature extraction and classification in one shot, which means we only have to design one model. Figure 2 compares the performance of Deep Learning vs Machine Learning. When the amount of training data increases in machine learning, the performance of ML decreases but the Deep learning get the maximum performance with huge amount of data with the same situation.

2 Deep Convolutional Neural Networks (CNNs) and the Training (Learning) Process

There are several deep learning technologies available in supervised learning; the most relevant algorithms nowadays in supervised learning are:

- Feedforward Neural Networks, a popular variation of these called Convolutional Neural Networks (CNNs),
- Recurrent Neural Networks (RNNs), and
- a variation of RNNs called Long Short-Term Memory (LSTM) models.

Feedforward Neural Networks, also known as Multilayer Perceptrons (MLPs) are the most common supervised learning models. Their purpose is to work as function approximators: given a sample vector with features, a trained algorithm is expected to produce an output value or classification category that is consistent with the mapping of inputs and outputs provided in the training set. The approximated function is usually built by stacking together several hidden layers that are activated in chain to obtain the desired output. The number of hidden layers is usually referred to as the depth of the model, which explains the origin of the term deep learning.

This paper uses Keras and Tensorflow backend platform in python to design and train a designated deep Convolutional Neural Network architecture with the given sample image data set, and then applies the trained network to new image data taken by UAV camera for identification and recognition of the image.

3 Modelling Deep Learning with Tensorflow

Tensorflow software library help tackle implementation of machine learning and deep learning methods. These libraries are very broad enough to implement many types of methods and algorithms which are included seeding, loading data, designing neural network architecture, compiling and training of the machine learning or deep learning model. The technique has been used in areas such as robotics, voice recognition devices, map extracting, object detection [15], etc. [1]

In addition, the Tensorflow backend called Keras [3] is also used for detecting hand written digits or MNIST data. The Convolutional neural network (CNN) [11] is used to determine MNIST image into words by transforming multi-Dimensional images into characters [4]. Open source Tensorflow made by Google team to model CNN with

MNIST digits. The reason of using MNIST digits for modelling deep learning is because a person called Yann LeCunn already released hand written datasets in 1998 which then main source for data scientists and researchers for classifying objects in Deep Learning.

4 Modelling and Simulating Deep Convolutional Neural Network Using Tensorflow and Keras

A deep convolutional neural network is built in this paper using Keras with Tensorflow backend library tool for modelling and simulating of hand written digits by training with MNIST dataset sample image data of hand written images. The flow diagram for the proposed model system is shown in Fig. 3.

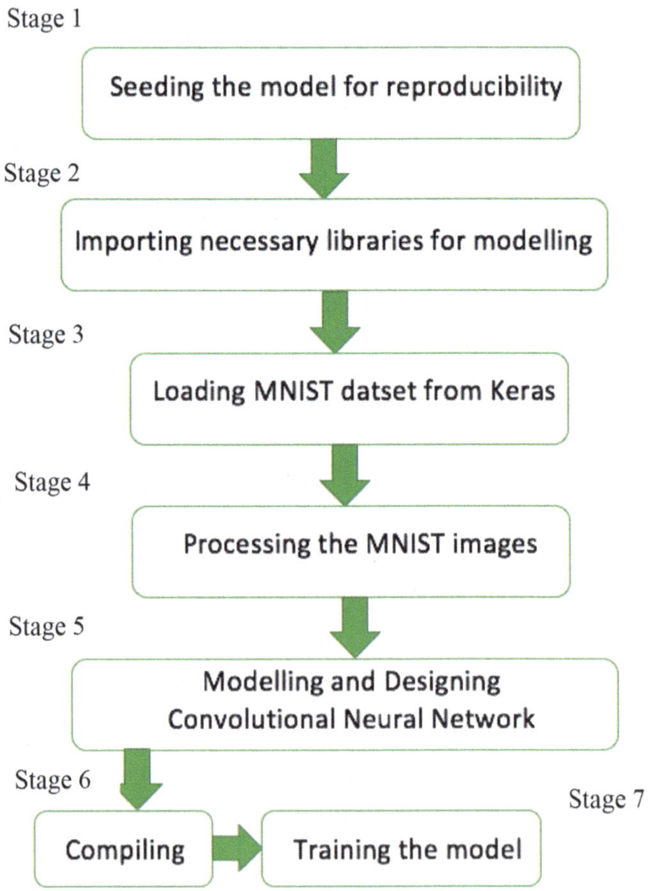

Fig. 3. Model system approach flow diagram.

The modelling process for testing and training include seven stages described as follows:

4.1 Stage-1

In the first stage, Neural network may not be predicted precisely because of instability of the algorithms being trained with same dataset with unusual results. To overcome this complexity, random seeding to start the system with random weights applied on the neurons respectively. This cause the system to learn automatically each time if the identical neural network train with identical result.

4.2 Stage-2

Importing Keras library to use all the necessary libraries from that framework to model and simulate the MNIST datasets. Given dataset can be loaded into the system to classify hand written images already uploaded in the Keras. After that, a sequential model needed to perform and design neurons by stacking layers of that neurons using Dense layer which implement the operation of the neural network. The system can cause overfitting by using complicated neurons in the model. To avoid that overfitting neural layer was used using with the Dropout layer which set a fraction rate of input to 0 at each update during training and flatten the input using Flatten layers. To create successful deep convolution neural network for 2-Dimensional images, Maxpooling2D layer was used to operate spatial data of each pixel of the image which presents same pixel detection to get accurate output.

4.3 Stage-3 and 4

In this stage, loading the hand-written images from the dataset stored. For testing and training the data, it included in the dataset are images with its respective labels. When training the input data, the total of digits is 60000 and only 10000 are being tested for training and testing the data. The value of each pixel is 255 pixels in processing the MNIST images.

4.4 Stage-5

- Using Sequential model to initialise and stack all the neural layers.
- 2-Dimensional convolutional layer uses lenet-5 have 32 neurons which is the first convolutional layers. The size of the pool size is 3 by 3 kernel size pixels. As with the Dense layer, a recommended neuron called Relu should be activated. MNIST digits input is (28 by 28) pixels with 1 depth black and white image.
- Second 2-D Convolutional layer, on the other hand uses 64 layers and remaining values are same as first layer.
- Third layer is 2-D MaxPooling, for every 4 pixels reduce them to one with 2 by 2 pool size.
- Dropout layer produce results from training apply well to the validation data, going to dropout 0.25 of the neuron.
- We have to flatten everything, so we have more than one dimension at this point, dense fully connected layer is going to able recombine all these possible

representation stored by convolutional neuron, represent 3 by 3 corner, feed that into dense layer with any kind of configuration. Multi dimension to one dimension by flattening.

- Configuring Dense layer with 128 neurons and relu activation.
- We are going to dropout half in this layer.
- Configuring Dense layer with 10 neurons and softmax activation.

When training machine learning models, users often want to able to examine the state of various aspects of the model, and how this state changes over time. To this end, Tensorflow and Keras supports a collection of different summary operations that can be inserted into the graph later in the results. Figure 4 shows convolutional neural network model summary.

```
In [8]: model.summary()

_________________________________________________________________
Layer (type)                 Output Shape              Param #
=================================================================
conv2d_1 (Conv2D)            (None, 26, 26, 32)        320
_________________________________________________________________
conv2d_2 (Conv2D)            (None, 24, 24, 64)        18496
_________________________________________________________________
max_pooling2d_1 (MaxPooling2 (None, 12, 12, 64)        0
_________________________________________________________________
dropout_1 (Dropout)          (None, 12, 12, 64)        0
_________________________________________________________________
flatten_1 (Flatten)          (None, 9216)              0
_________________________________________________________________
dense_1 (Dense)              (None, 128)               1179776
_________________________________________________________________
dropout_2 (Dropout)          (None, 128)               0
_________________________________________________________________
dense_2 (Dense)              (None, 10)                1290
=================================================================
Total params: 1,199,882
Trainable params: 1,199,882
Non-trainable params: 0
_________________________________________________________________
```

Fig. 4. Convolutional neural network model summary.

4.5 Stage-6 and 7

In this stage included compiling and training on 10000 from 60000 sample MNIST images with validation.

4.6 Results

This model is testing with only 1 epoch and the validation accuracy is 98.36% (see Fig. 5).

```
In [10]: model.fit(X_train, y_train, batch_size=128, epochs=1, verbose=1, validation_data=(X_test, y_test))

         Train on 60000 samples, validate on 10000 samples
         Epoch 1/1
         60000/60000 [==============================] - 319s - loss: 0.2410 - acc: 0.9267 - val_loss: 0.0518 - val_acc: 0.9836

Out[10]: <keras.callbacks.History at 0x7f5202ce6898>
```

Fig. 5. Trained data results.

The results from Figs. 6, 7 and 8 verify that when predicting accuracy increases, the amount of loss data on the trained data decreases over the final result. The validation accuracy and loss get from feedback received from fully connected deep neural network enhance the system performance.

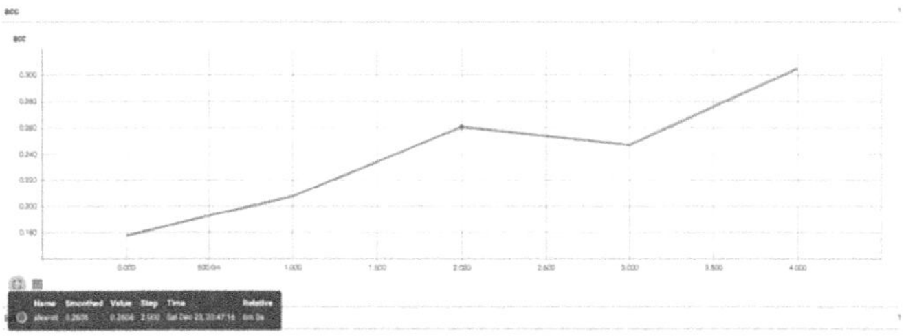

Fig. 6. Accuracy results.

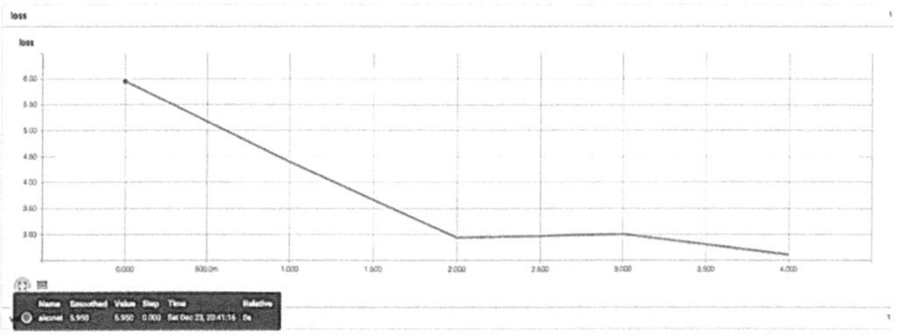

Fig. 7. Loss values to the training data decreases.

Relu Activation functions were selected in the system to test the accuracy of the classification of the system was used for this purpose and received 98.36% classification accuracy is obtained on the test data.

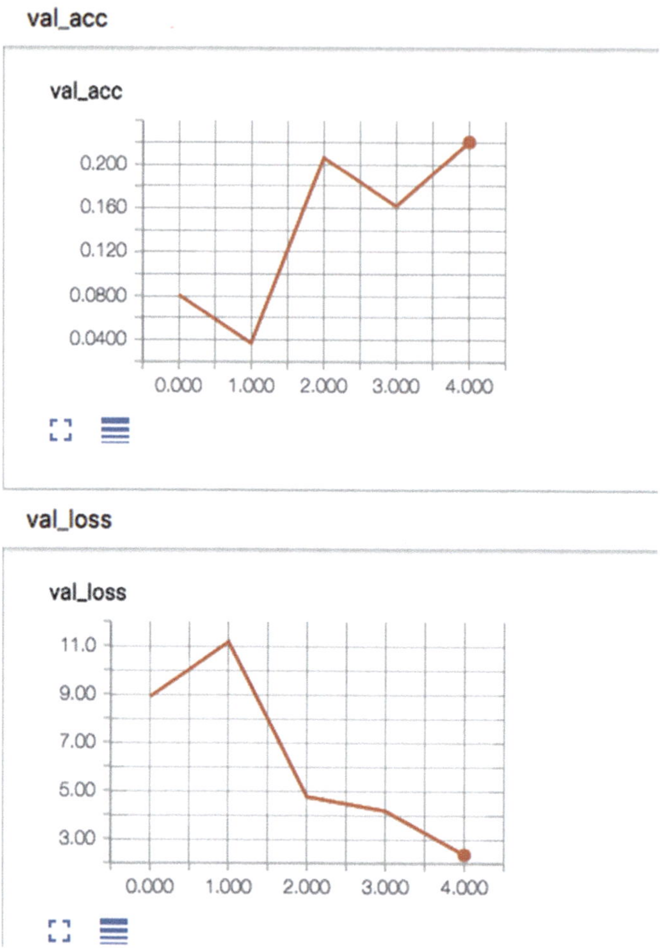

Fig. 8. Validated accuracy and loss after feedback from output.

5 Conclusion

Fast evolving technologies has led to the innovation of powerful unmanned aerial vehicles (UAV), also known as drones, for the purpose of extracting information or high-quality images in dangerous environments which human won't able to travel. In this paper we demonstrated a novel approach in applying deep learning technique (as a part of Machine Learning – ML) to assist image recognition task of Unmanned Aerial Vehicles (UAVs) in their search and rescue (SAR) operations. Tensorflow (Open Source software Library) and Keras (backend platform in python) were used in designing and training of a deep Convolutional Neural Network (NN) architecture. The learning process was conducted over given sample image data set, and the trained NN was applied over new image data taken by UAV camera, with the result achieved indicated the trained system capable of recognising images with high level of accuracy.

References

1. Ciresan, D., Meier, U., Schmidhuber, J.: Multi-column deep neural networks for image classification. In: Proceedings of the CVPR (2012)
2. Colomina, I., Molina, P.: Unmanned aerial systems for photogrammetry and remote sensing: a review. ISPRS J. Photogram. Remote Sens. **92**, 79–97 (2014)
3. Deng, L.: A tutorial survey of architectures, algorithms and applications for deep learning. APSIPA Trans. Signal Inf. Process. **3**, e2 (2014)
4. Gundlach, J.: Design Unmanned Aircraft Systems: A Comprehensive Approach. American Institute of Aeronautics and State University, Blacksburg (2012)
5. Ghonge, M.M., Jawandhiya, P.M.: Review of Unmanned Aircraft System (UAS). Int. J. Adv. Res. Comput. Eng. Technol. (IJARCET) **2**(4), 1646 (2013)
6. Howard, A.G.: Some improvements on deep convolutional neural network based image classification, arXiv preprint arXiv:1312.5402 (2013)
7. Klemas, V.V.: Coastal and environmental remote sensing from unmanned aerial vehicle: an overview. J. Coastal Res. **31**(5), 1260–1267 (2015)
8. Li, H., Zhao, R., Wang, X.: Highly efficient forward and backward propagation of convolutional neural networks for pixelwise classification, arXiv preprint arXiv:1412.4526 (2014)
9. Nonami, K., Kendoul, F., Suzuki, S., Wang, W., Nakazawa, D.: Autonomous Flying Robots, Unmanned Aerial Vehicles and Micro Aerial Vehicles. Springer, Tokyo (2010). ISBN 978-4-431-53855-4
10. Pereira, E., Bencatel, R., Correira, J., Felix, L., Goncalves, G., Morgano, J., Sousa, J.: Unmanned air vehicles for costal and environmental research. In: da silva, C.P. (ed.) Proceedings of the ICS, Journal of Coastal Research, Special Issue No. 56, pp. 1557–1561 (2009)
11. Ren, J.S.J., Xu, L.: On vectorization of deep convolutional neural networks for vision tasks. In: Proceedings of the AAAI (2015)
12. Schmidhuberc, J.: Deep learning in neural networks: an overview. Neural Netw. **61**, 85–117 (2015)
13. Simonyan, K., Zisserman, A.: Very deep convolutional networks for large-scale image recognition. In: Proceedings of the ICLR (2015)
14. Szegedy, C., Toshev, A., Erhan, D.: Deep neural networks for object detection. In: Proceedings of the NIPS (2013)
15. Tompson, J., Goroshin, R., Jain, A., et al.: Efficient object localization using convolutional networks. In: Proceedings of the CVPR (2015)

Analysis of the Use of a NAO Robot to Improve Social Skills in Children with ASD in Saudi Arabia

Eman Alarfaj(✉), Hissah Alabdullatif, Huda Alabdullatif, Ghazal Albakri, and Nor Shahriza Abdul Karim

Prince Sultan University, Riyadh, Saudi Arabia
emanalarfaj1@gmail.com

Abstract. Autism Spectrum Disorder (ASD) is extensively spread amid children; it affects their social, communication and interactive skills. As robotics technology has been proven to be a significant helpful utility, those able individuals overcome their disabilities. Robotic technology is used in ASD therapy. The purpose of this research is to show how humanoid robot NAO can improve the social skills for children who suffer from autism in Saudi Arabia by interacting with the autistic child and perform a number of tasks. The objective of this research is to identify, implement and test the effectiveness of the module for interacting with ASD children in an autism center in Saudi Arabia. The methodology in this study followed the 10 layers of protocol that needs to be followed during any human-robot interaction. Also in order to elicit the scenario module, TEACCH Autism Program was adopted. Six different qualified interaction modules have been elicited and designed in this study; the robot will be programmed to perform these modules in a series of controlled interaction sessions with the Autistic children to enhance their social skills.

Keywords: Humanoid robot NAO · Autism Spectrum Disorder (ASD) Human-robot interaction

1 Introduction

Autism Spectrum Disorder (ASD) is a set of developmental disorders that affect the development of communication and social skills of an individual [1]. Individuals diagnosed with Autism lack social skills, suffer from absence of interest in interactive activities (e.g. imitation and joint attention), struggle in verbal and nonverbal communication, difficulties understanding and expressing emotions [2]. These missing skills can be enhanced and developed through rehabilitation centers, unfortunately only a few centers are available in Saudi Arabia. To compensate the lack of such centers in Saudi Arabia we will use robotic intervention and observe how robots might help develop these absent skills in autistic children. Several countries around the world started using robots and artificial intelligence to help autistic children improve their skills and, as a result, received positive feedback [3–6]. Moreover, studies have shown that, socially interactive robots can assist in the therapy of children diagnosed with autism, as they are motivating, accurate and adaptive to the individual's needs [3].

K. Arai et al. (Eds.): IntelliSys 2018, AISC 868, pp. 902–913, 2019.
https://doi.org/10.1007/978-3-030-01054-6_63

This project aims to elaborate on a case study that experiments how NAO Robots can improve the social skills for children diagnosed with ASD in Saudi Arabia. Moreover, the purpose of this project is to examine the progress and usage of a NAO Robot and its effect in refining social skills in autistic children, encouraging eye contact, physical interaction, and imitation, as well as the child's interaction with people so that each child gets individualized treatment. Among all the available robots, NAO robot is much simpler in appearance and has the most human like design, which will reduce stress and appear more approachable to autistic children [4]. The proposed robot will be programmed to perform a series of six certified scenario modules with the autistic child under professional supervision, these interaction activities are designed from authenticated training programs and will focus on the child's social and communication skills. The objective is to see how the robot can be used to interact with the autistic child and perform a number of tasks, capture the child's attention and assist him/her gain social and communication skills.

Scenarios will be designed with explicit therapeutic and educational purposes in five main developmental areas: sensory, communication and interaction, motor, cognitive, social and emotional development. Social robots are thought to be stimulating tools that play an important role with children that are diagnosed with ASD. The purpose of this research is to study whether or not NAO Robots can enhance the social skills for children diagnosed with ASD in Saudi Arabia.

The remainder of this paper is organized as follows: Sect. 2 introduces Autism Spectrum Disorder, Robotic Intervention for Autism and Robotic for Social Training. Section 3 reviews various theories for robotic interaction. Section 4 illustrates the methods used for the HRI program, the scenario module and the assessment method. Finally, in Sect. 5 we conclude this paper.

2 Literature Review

2.1 Autism Spectrum Disorder (ASD)

Autism was first discovered by Leo Kanner M.D. Since 1943 Kanner [6] started recognizing and spotting groups of children that hold characteristics of extreme withdrawal and lack of social skills. ASD are a group of disorders with shared impairments in three key areas are also known as the 'triad of impairments'-social interaction, communication skills and imaginative thought and play [7]. The severity of this disorder varies among autistic children. On one hand, children diagnosed with severe ASD, the ability to make eye-contact and communicate is an absent trait; however the child may possess high mathematical and writing skills. On the other hand, mild ASD might be recognized in the shortage of social and communication skills in a child, and might even be identified with mental handicap disorder [8]. Current therapies are aimed to help people with ASD to closely function as a normal one because there is no treatment for ASD [9].

Some studies show that children with ASD react to machines (such as computers and robots) more than they react to people. In addition, a study conducted by Ozonoff (1995) had shown that individuals with ASD are more reactive to feedback controlled by technology rather than people [10]. There are some factors that make robots the perfect tool to be used in ASD therapy. First, robots fulfill the need for repetitive actions [10]. A robotic mediator has the positive role to decrease stress and challenges frequently confronted by autistic children during communication. Though, the robot's purpose is not to substitute the therapists but more like an 'attracting agent' to retain the ASD child involved and interested through therapy [10]. Furthermore, the robot has a clear look and presentation in comparison to objects, its behavior can be customized to match diverse scenarios and it has the ability to offer predictable and simpler contact with autistic children [10].

2.2 Robotic Intervention for Autism

Robotics technology appeared recently to be a significant assistive utility that is available to support people with disabilities. It is predictable that around 10% of people nowadays live with a disability [11]. The execution of existing robotic technologies is imagined to have some contribution to the autism population in assisting them to have a good time and a healthier life [12]. It has been proven that a lot of people with ASD prefer robotic features over none robotic toys [13].

To observe autistic children's initial response during an interaction between ASD children and NAO Robots, Shamsuddin et al. (2012) [14] conducted a series of interaction experiments that were real-time monitored and recorded. NAO robots would perform simple interaction modules under the supervision of an occupational therapist and a Clinical Psychologist. The purpose of their observation was to justify the human-robot-interaction to insure the reliability and safety of this interaction, particularly when such interactions occur for medical purposes.

To study the initial behavior of children with autism when introduced to human-robot interaction performed by NAO robot, Shamsuddin et al. (2012) [15] conducted an experiment that was based on the motivation of contributing to communities that suffers from autism. Their goal in the experiment was to decrease the children's autistic features in communication behavior comparing to the normal class. Their paper study the results from a pilot study that include five autistic children that are between 7 to 13 years old. The goal was to monitor the autistic features of the child during the interaction between the child and the robot and in the traditional environment. The main outcome of their experiments shows that 4 out of the 5 children's autistic behavior have decreased when the robot was performing HRI modules during the interaction. This outcome shows that the NAO robot kept the child involved with the robot during interaction session and was able to attract the children's attention, thus it gave a positive influence to the children's communication skills.

As Mohammad et al. (2016) [3] argued, the use of robots in treating children diagnosed with ASD has increased, and in spite of how promising the results are and how these robots are improving the social skills for ASD children, research in this area is still being explored. To support their argument and show the great potential of using robots to improve the ASD children's social skills, interaction sessions have been conducted. For their research, ASD children were paired with a NAO Robot to conduct human-robot interaction sessions and perform five social tasks. These tasks were anticipated to calculate initial existing social skills deficits and provide feedback for improvement.

Every session contained five challenging tasks designed to teach children with autism an individual social response. The five social tasks focused on the following:

(1) Verbal communication
(2) Following NAO's direction of pointing and recognizing facial expression
(3) Joint attention regulation
(4) Facial expression recognition
(5) Pointing

The results of these interactions indicated that the skills the children gain from these sessions are successfully transferred during human-to-human interaction, which prove the significance of having robot intervention to help in the treatment of ASD children.

2.3 Robotic for Social Training

To demonstrate social skills of children with ASD in the communication scheme, two training fundamentals that are calculated through consultation with psychotherapist. The first fundamental is eye contact. Eye contact allows children to pick up the ability of aligning with other people's eyes which will provide clearer interaction motions [16]. The second fundamental is reading emotions; the recognition of emotion can be a decent indicator for social skills improvement. To conduct a trustworthy and more accurate analysis, the child's facial expression is a very important sign in the interaction between the child and the robot to examine the child's level of interests [17].

A research by Ryo and Jaeryoung (2016) [4] was conducted and focused on using musical therapy to explore behavior in ASD children and explored how robots can both improve social behavior and reduce anxiety. Two different trials were conducted to test the three following hypothesis:

(1) Children will follow the robot's instruction and touch correct parts of the robot.
(2) Children will approach the robot as soon as they meet it on their first encounter.
(3) Children will stay in close proximity to the robot.

Children didn't follow the robot's instructions and the results, therefore, didn't support the first hypothesis. Less than 20% of the initiated response time supported the second hypothesis. While the third and last hypothesis was supported based on the interaction area, where one of the areas had the longest duration of interaction. According to authors, the children's mind attribution toward robots may indicate that the children consider the robot as an intelligent creature. Moreover, some participants showed obsessive behavior towards the LED lightening on the robot's head.

To analyze the utilization of a robot to enhance social skills in autistic children, Valadão et al. (2016) [18] worked on a pilot study that focus on the implementation and utilization of a robot for motivating social characteristics in autistic children. By conducting and recording an interactive session between the ASD children and the robot, they have measured some social skills like staring and tapping the robot. Their work outcome was that the autistic group had more physical connection with the robot than the children without ASD. They have also made more social verbal and physical actions.

These results are reviewed as an enhancement of social skills in children that are diagnosed with autism.

3 Theory

3.1 Human Robot Interaction (HRI)

Recently, many machines and robots are being used for therapeutic and educational purposes [19]. However, none of these approaches are practiced in Saudi Arabia, where a percentage of children who live in the Kingdom are diagnosed with autism but are medically treated using traditional methods only. Among the common traits associated with autism, and recognized in autistic children, is the difficulty of communicating with others and the struggle of interpreting their social skills [20]. Robots and robot technologies can be acknowledged as effective therapeutic tools [21]. Moreover, robots with human appearance and human-robot interaction can improve and recover the lacking skills in children diagnosed with ASD, they also assist them in identifying and understanding emotions and other behaviors [22].

In robot based intervention, human-robot interaction (HRI) is described as the active relationship between humans and robots [14]. Humanoid robot with HRI architecture is more likely to train autistic children how to discover and understand feelings and social behaviors [9]. Additionally, robotics can be a beneficial therapeutic tool where parents can improve their children at home. Some interactive and social behavior that autistic children do not naturally have can be provoked by Robots [14, 22]. Also, related work has identified HRI as an effective tool for diagnosing ASD [21, 23] and for teaching social skills [24, 25]. HRI is also used for interference for autistic children and is part of the progress in Socially Assistive Robot (SAR) [23]. Scholtz (2003) [26] listed various HRI roles:

(1) *Supervisor Interaction*: A supervisor role considered as observing and directing the overall situation. Supervisor has the power to alter the overall objective.
(2) *Operator Interaction*: The operator is a person who deals with internal software when the robot shows defective or inadequate behavior.
(3) *Mechanic Interaction*: The mechanic responsible of physical interference, he is the person who modifies the hardware. To guarantee that the behavior is fixed the testing should cover both the software and the hardware.
(4) *Peer Interaction*: The partner of the robots can give orders within the overall objective. The meaning of term "peers" is that robot and human devote their skills to the team according to their capability. The peer can get feedback from the robot.
(5) *Bystander Role*: Bystanders have minimal idea of the robot's ability and does not have access to device where the robot's status is displayed. Bystanders have the most limited interactions with the robot. For example, a bystander can be a victim found by the search and rescue robot or it could be a driver passing by a driverless car. In this model, the bystander has only a subset of actions available.

The same individual might take more than one role at a time. The different roles within HRI demand different experience of the situation. In our scenario our team has the supervisor, operator and mechanic roles while the therapist will be the peer and the autistic child will be the bystander.

3.2 Socially Assistive Robotics

Socially Assistive Robotics (SAR) is an area in HRI has a more precise focus on helping individuals through social communication instead of physical connection [14]. SAR has a variety of practice areas. Social robots have been practiced in the domain of nursing homes, it's intended to boost and enhance social skills among inmates [14]. SAR has also been practiced in rehabilitation robotics domain, in this area or domain it has been proven that giving and receiving help through social communication may have therapeutic advantages. Research groups found out that using robot as therapeutic assistance for children that are diagnosed with autism appeared to be an anticipating case for future researches [14].

NAO Robot can motivate and assist in establishing an effective environment for children with autism who struggle or annoyed while trying to communicate with humans.

The methodology followed in this study was based on the work of Shamsuddin et al. (2014) [10]. As presented in Fig. 1, they particularly suggested considering all 10 phases of the protocol to be followed in any HRI program. Each phase presented in the flow of protocols was used as guidance for our researcher.

The developmental phases in this study is illustrated in the flow chart that is presented in Fig. 2, each phase presented in the flow of protocols, Shamsuddin et al. (2014) [10] was used as guidance for our method.

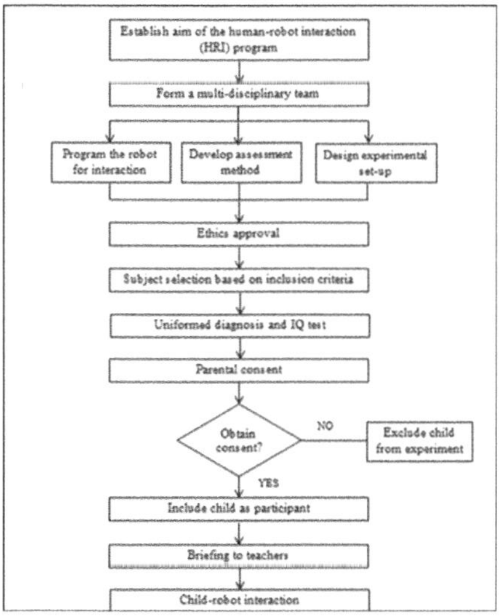

Fig. 1. Each stage in the protocol flow.

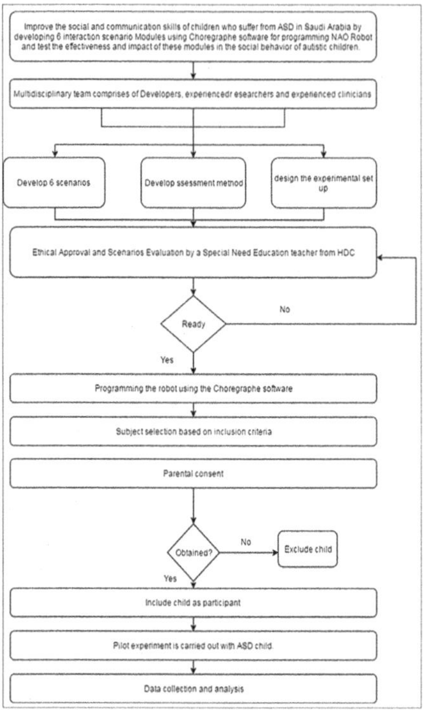

Fig. 2. Our proposed approach based on [10].

4 Methodology

4.1 Flow of Methods and Protocols

(1) *Establish aim of the human-robot interaction (HRI) program*: Our aims of this project are to improve the communication and social skills of children who suffer from ASD in Saudi Arabia by identifying the appropriate module for the interaction with ASD children during child-robot interaction through implementing treatment model in Choregraphe software for programming NAO and test the effectiveness and impact of the modules in the social behavior of autistic children. Therefore, we had developed six different interaction scenarios for NAO Robot. The interaction scenario content was developed based on intensive review studies and discussion and feedback of experienced clinicians in autism therapy. Moreover, this study represents a proof of concept. Accordingly, our goal to gather useful information from a small sample (2–4 ASD children) when compared with more advanced study. It is very common in early stages of the study to use small sample such as [5, 9, 10, 24, 29], used less number of subjects to make a proof of concept with subjects equals to four, two, one and one respectively. Therefore, it is possible to analyze and evaluate the developed modules and showing how the robot can be

used to interact with the autistic child and perform a number of tasks, record how the child react and help the child in developing social and communication skills.

(2) *Form a multi-disciplinary team*: Our multidisciplinary team consists of Developers, experienced researchers and autism specialist from Saudi Arabia Human Development Center.

(3) *Program the robot for interaction and design of experimental set-up*: This phase where we develop the interaction content scenario module, the assessment method and design the experimental setup.

(4) *Ethics approval*: Since our research involves human children, therefore it is required to get ethics approval. In our research, the scenarios are presented to a Special Need Education teacher from Saudi Arabia Human Development Center to gain feedback on the interaction content. A special need teacher is someone who works with special needs for children and knows how to guide them to reach their potential. Then, modifications are done to the scenario modules until they are ready and approved by Prince Sultan University and Saudi Arabia Human Development Center. Finally, each scenario module is developed using the Choregraphe software to program the robot.

(5) *Subject selection based on inclusion criteria*: In our research, the experiment will be carried out at Saudi Arabia Human Development Center, Riyadh Branch which has a program that is specifically designed for children with autism, and where the study samples of ASD children were selected. As presented in [10], the inclusion criteria for the ASD children in our study were:

- Confirmed diagnosis of autism by medical professionals
- No hearing deficit and vision deficit,
- Parental consent,
- Able to speak in Arabic,
- Able to follow simple commands in Arabic
- Age group 6–10 years old
- No self-injury or aggressive behavior.

(6) *Uniformed diagnosis*: For our research study, all children were diagnosed of autism by medical professionals.

(7) *Parental consent and briefing to teachers*: All parents will sign a written consent form that is approved by the Prince Sultan University. The written consent will also state that at any time the participant is allowed to decline his participation. In our research study, the teacher who accompanied the children will not have any communication with them nor gives any instructions. Their being is to give the children a 'comforting presence' [10].

(8) *Child-robot interaction and analysis*: The most important phase is the interaction phase where the autistic child is subject to HRI. The environment settings, the setup, the data collection and analysis are to be updated later on in the next phase of the project.

4.2 Experiment Protocol: Scenario Module

TEACCH (Treatment and Education of Autistic and Communication related handicapped Children) is an Autism Program for people with ASD. TEACCH program was founded at University of North Carolina (UNC) in 1970s by Eric Schopler [27]. The TEACCH method is implemented by Human Development Center in Saudi Arabia, based on their scenarios we have developed 6 scenario modules. The 6 scenario modules that have been designed will be developed to the NAO Robot. The NAO Robot technical potentiality and constraints are the important points to be noted in this phase.

4.3 Assessment Method

The TEACCH Preschool Curriculum Guide is a comprehensive guide that assesses the child's skills [27]. This guide is implemented by Human Development Center in Saudi Arabia [28], based on their assessment we will only use the daily assessment plan in order to assist the ASD child. The purpose of this assessment is to collect data to help record students' progress and then analyze the progress between the normal class interaction and the child-robot interaction (Table 1).

Table 1. Scenario modules

Module name	Objective	Purpose
NAO introducing itself.	Create a friendly environment for the child and observe how the child's react to the NAO robot	As presented in [12, 14, 18], the introductory module is used to create a friendly environment for the autistic child
Naming animals	Develop new vocabulary	Naming four animals without any help
Find hidden objects- to enhance their understanding of prepositions of place and directions	Enhance their understanding of prepositions of place and evolve their ability of performing the given verbal commands	Perform a verbal command that consists of prepositions of place to locate the hidden objects
Performing the given written commands without any help	Enhance the child ability of reading and performing the given command independently	Perform simple written commands and finish the activity independently
Symmetry game - Improving the skills of imitating body movements	Improving the skills of imitating general movement	Imitate various positions
Imitation of attention movements and another person's movements and play them back in sequential order	Increase concentration and imitating another person's movements	Correctly imitate two simple movements various positions

The Data Sheet will be structured according the data sheet used in HDC [28]. The first column will be for the field, the second will be for the scenario number, and the third to the seventh row will be to record whether each skill area was done correctly and independently or not by writing either "Yes" or "No" according the day of the week. If any help was done for the child at all, then it is not considered independent and it would be written as "No". This includes any "pointing, or reminding" that was necessary. The skills count for "Yes" in which the student was independent will be written in the last column by counting the number of "Yes". This ratio will be the measure of the student's ability to do this goal independently, which will be indicated by a ratio of more than three "Yes" in one week for the same goal "scenario".

5 Conclusion

Despite the fact that Autism Spectrum Disorder is widely spread among children, Saudi Arabia does not have a lot of centers to treat the children's lack of social and interactive skills. As a substitute for this absenteeism, using NAO Robot was proposed as a supporting treatment method. Many investigations and comparisons were conducted where foreign countries have suggested using the support of HRI in the treatment process as well. Consequently, six different certified interaction modules have been designed for this study and the robot will be programmed to perform these modules in a series of supervised interaction sessions with the Autistic child. The main drawback was the unavailability of the NAO Robot in Saudi Arabia, but this limitation has been resolved. For the future plan of this study, we will start developing the NAO Robot to perform the interaction modules and collaborate with the Human Development Center in Riyadh to conduct and test the proposed interaction sessions with Autistic children.

Acknowledgment. This work was supported by the Human Computer Interaction Research Group; Prince Sultan University, Riyadh, Saudi Arabia [RG-CCIS-2017-06-01].

References

1. Welch, K.C., Lahiri, U., Warren, Z., Sarkar, N.: An Approach to the design of socially acceptable robots for children with autism spectrum disorders. Int. J. Soc. Robot. **2**(4), 391–403 (2010)
2. Scassellati, B., Admoni, H., Matarić, M.: Robots for use in autism research. Annu. Rev. Biomed. Eng. **14**(1), 275–294 (2012)
3. Mavadati, S.M., Feng, H., Salvador, M., Silver, S., Gutierrez, A., Mahoor, M.H.: Robot-based therapeutic protocol for training children with Autism. In: 2016 25th IEEE International Symposium on Robot and Human Interactive Communication (RO-MAN) (2016)
4. Suzuki, R., Lee, J.: Robot-play therapy for improving prosocial behaviours in children with autism spectrum disorders. In: 2016 International Symposium on Micro-NanoMechatronics and Human Science (MHS) (2016)

5. Robins, B., Dautenhahn, K., Ferrari, E., Kronreif, G., Prazak-Aram, B., Marti, P., Iacono, I., Gelderblom, G.J., Bernd, T., Caprino, F., Laudanna, E.: Scenarios of robot-assisted play for children with cognitive and physical disabilities. Interact. Stud. Soc. Behav. Commun. Biol. Artif. Syst. **13**(2), 189–234 (2012)
6. Kanner, L.: Autistic disturbances of affective contact. Place of publication not identified: publisher not identified (1943)
7. Haddadi, P., Haghshenas, S., Rostami, R.: Rehabilitation in Autism Spectrum Disorder (ASD): a mixture of neurofeedback training and Auditory Integration Training (AIT). Procedia Soc. Behav. Sci. **30**, 611–614 (2011)
8. Wing, L.: Autistic spectrum disorders. BMJ **312**(7027), 327–328 (1996)
9. Shamsuddin, S., Yussof, H., Ismail, L., Hanapiah, F.A., Mohamed, S., Piah, H.A., Zahari, N.I.: Initial response of autistic children in human-robot interaction therapy with humanoid robot NAO. In: 2012 IEEE 8th International Colloquium on Signal Processing and Its Applications (2012)
10. Shamsuddin, S., Yussof, H., Mohamed, S., Hanapiah, F.A.: Design and ethical concerns in robotic adjunct therapy protocols for children with autism. Procedia Comput. Sci. **42**, 9–16 (2014)
11. Shamsuddin, S., Yussof, H., Hanapiah, F.A., Mohamed, S.: Response of children with autism to robotic intervention and association with IQ levels. In: 4th International Conference on Development and Learning and on Epigenetic Robotics (2014)
12. Yussof, H., Shamsuddin, S., Hanapiah, F.A., Ismail, L.I., Miskam, M.A.: IQ level assessment methodology in robotic intervention with children with autism. In: 2015 10th Asian Control Conference (ASCC) (2015)
13. Zheng, Z., Young, E.M., Swanson, A.R., Weitlauf, A.S., Warren, Z.E., Sarkar, N.: Robot-mediated imitation skill training for children with autism. IEEE Trans. Neural Syst. Rehabil. Eng. **24**(6), 682–691 (2016)
14. Shamsuddin, S., Yussof, H., Ismail, L.I., Mohamed, S., Hanapiah, F.A., Zahari, N.I.: Initial response in HRI- a Case study on evaluation of child with autism spectrum disorders interacting with a humanoid robot NAO. Procedia Eng. **41**, 1448–1455 (2012)
15. Shamsuddin, S., Yussof, H., Ismail, L.I., Mohamed, S., Hanapiah, F.A., Zahari, N.I.: Humanoid robot NAO interacting with autistic children of moderately impaired intelligence to augment communication skills. Procedia Eng. **41**, 1533–1538 (2012)
16. Leekam, S.R., López, B., Moore, C.: Attention and joint attention in preschool children with autism. Dev. Psychol. **36**(2), 261–273 (2000)
17. Clark, T.F., Winkielman, P., Mcintosh, D.N.: Autism and the extraction of emotion from briefly presented facial expressions: stumbling at the first step of empathy. Emotion **8**(6), 803–809 (2008)
18. Valadão, C.T., Goulart, C., Rivera, H., Caldeira, E., Filho, T.F.B., Frizera-Neto, A., Carelli, R.: Analysis of the use of a robot to improve social skills in children with autism spectrum disorder. Res. Biomed. Eng. **32**(2), 161–175 (2016)
19. Landa, R., Garrett-Mayer, E.: Development in infants with autism spectrum disorders: a prospective study. J. Child Psychol. Psychiatry **47**(6), 629–638 (2006)
20. Hollander, E., Kolevzon, A., Coyle, J.T.: Textbook of Autism Spectrum Disorders. American Psychiatric Publishing, Washington, DC (2010)
21. Scassellati, B.: Quantitative metrics of social response for autism diagnosis. In: ROMAN 2005, IEEE International Workshop on Robot and Human Interactive Communication (2005)
22. Villano, M., Crowell, C.R., Wier, K., Tang, K., Thomas, B., Shea, N., Schmitt, L.M., Diehl, J.J.: Domer. In: Proceedings of the 6th International Conference on Human-Robot Interaction - HRI 2011 (2011)

23. Feil-Seifer, D.: Socially assistive robot-based intervention for children with autism spectrum disorder. In: IEEE International Conference on Robotics and Automation Workshop on Unifying Characteristics of Research in Human-Robot Interaction, pp. 10–11 (2008)
24. Robins, B., Dautenhahn, K., Boekhorst, R.T., Billard, A.: Robotic assistants in therapy and education of children with autism: can a small humanoid robot help encourage social interaction skills? Univ. Access Inf. Soc. **4**(2), 105–120 (2005)
25. Lathan, C., Boser, K., Safos, C., Frentz, C., Powers, K.: Using cosmo's learning system. (CLS) with children with autism. In: Proceedings of the International Conference on Technology-Based Learning with Disabilities, Dayton, OH, pp. 37–47, July 2007
26. Scholtz, J.: Theory and evaluation of human robot interactions. In: 2003 Proceedings of the 36th Annual Hawaii International Conference on System Sciences (2003)
27. "TEACCH Autism Program," Our Mission and Vision - UNC TEACCH Autism Program, 18 October 2016. https://www.teacch.com/about-us/. Accessed 06 May 2017
28. "Human Development Center," Human Development Center. http://hdc.com.sa/. Accessed 06 May 2017
29. Giullian, N., Ricks, D., Atherton, A., Colton, M., Goodrich, M., Brinton, B.: Detailed requirements for robots in autism therapy. In: 2010 IEEE International Conference on Systems, Man and Cybernetics (2010)

Supervisory Control of a Multirotor Drone Using On-Line Sequential Extreme Learning Machine

Oualid Doukhi(✉), Abdur Razzaq Fayjie, and Deok-Jin Lee

School of Mechanical and Automotive Engineering, Kunsan National University, Kunsan, South Korea
{doukhioualid,abdurrfayjie,deokjlee}@kunsan.ac.kr

Abstract. Quadrotors are underactuated nonlinear systems, which mean it needs online monitoring and controller tuning during flight period. Classical proportional integral derivative (PID) control and artificial neural network control (NNC) shows good results in many applications. Therefore in this paper, we propose a neural network supervisory control technique for the classical PID controller using fast online sequential learning method called on-line sequential extreme learning machine (OS-ELM). This technique seeks to online tune the control input to improve the flight capabilities automatically. The effectiveness of the proposed control algorithm comparing it with the conventional neural network based PID controller is demonstrated through realistic simulation using ROS-Gazebo framework.

Keywords: Quadrotor · On-line learning · Disturbances
Neural network

1 Introduction

Autonomous unmanned aerial vehicles (UAVs) have gotten significant consideration nowadays due to their wide civilian and military applications include aerial mapping, package delivery, investigation, and inspection [1]. However, some new problems come up with the more extensive application, such as the size, weight and manoeuvrability, etc. To solve those problems researcher proposes a Multirotor aerial vehicle, called quadrotor UAV, which is becoming a research focus among the other platforms, thanks to their easy construction and ability to take-off and land vertically (VTOL), fly in any direction and even to stop their movement staying hovering above a specific target [2]. From the other point of view quadrotor UAVs are a non-linear system with strongly coupled dynamics, which make the design of flight control difficult and challenging task [3].

In the literature, many researches have been done on the control of quadrotor UAV [4]. As to be expected, each control scheme has its advantages and disadvantages, for instance, linear flight controllers like PID, LQR/LQG [5], and

K. Arai et al. (Eds.): IntelliSys 2018, AISC 868, pp. 914–924, 2019.
https://doi.org/10.1007/978-3-030-01054-6_64

H_∞ control algorithm [6], all those control methods it shows acceptable performance meanwhile their control accuracy decreases and prone to failure due to non-linearities, unmodeled dynamics or noise and the changing in operating conditions. Numerous other nonlinear control techniques have been presented, in [7] the author presents an integral backstepping as a control strategy where the actuators model was identified and all sensor noises were taken into consideration, sliding mode control [8], Adaptive control, model predictive control (MPC) and feed back linearization [9], shown to able to achieve good tracking then the linear controllers and it provide a promising solution to handle the presence of disturbances and model uncertainties, but those control strategy it needs an accurate model which it is difficult to obtain it in the real world due to its multi-input multioutput characteristics Recently more advanced control techniques based on artificial intelligence approaches were conducted, some of them are biologically-inspired, such as fuzzy logic, and artificial neural networks, reinforcement learning and iterative learning. The main feature of these methods is independency to system model [10].

Comparing to traditional neural network based controllers which need too many parameters to be initialized and tunned. We propose a feedforward neural network with one hidden layer based on OS-ELM whose input weights and hidden node parameters are generated randomly and fixed, the output weights will be updated using the optimal solution obtained with the least squares norm by calculating the inverse of hidden output matrix, by this way we don't need to tune any parameter accepts the output weights. The central idea of the proposed method is to design a hybrid OS-ELM-PID Supervisory controller for quadrotor trajectory tracking by mapping the desired trajectory and the output from the PID controller to an OS-ELM network's output weights. First, a traditional PID controller is designed to calculate the control input to the system. Then the network is trained on-line using the desired trajectory and PID output real data; thus, the inputs to the neural network will be mapped to the network's output weights. Finally, the output from the neural network will be added to the PID output to construct the total control input to the quadrotor, by this way we guarantee the fast learning speed to drive the quadrotor to the desired path. The proposed approach is evaluated in a simulation environment using gazebo and ROS interfaced with Matlab software.

The rest of this paper is organized as follows. Section 2 gives a brief review of the OS-ELM. Details of the designed controller based on a hybrid PID-OS-ELM is described in Sect. 3, the simulation results are given in Sect. 4 to demonstrate the performance of the proposed controller comparing it with the conventional PID. Finally, the paper is concluded in Sect. 5.

2 On-Line Extreme Learning Machine

Extreme learning machine ELM was outlined for single hidden layer neural network [11], see Fig. 1, the input weights and the parameters of the hidden layer do not need to adjust during the learning procedure. The weights between the

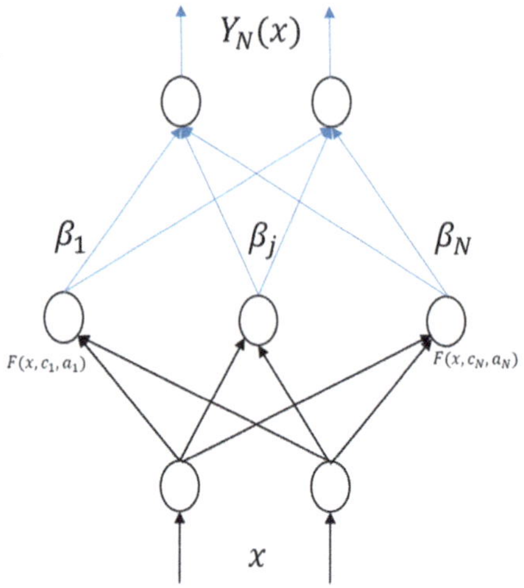

Fig. 1. Artificial neural network with one hidden layer.

hidden layer and the output layer are calculated using analytic method [12]. The output of the ELM with N hidden nodes can be modeled as

$$Y_N(x) = \sum_{i=1}^{N} \beta_i F(x, c_i, a_i), \qquad x\epsilon R^n, \qquad c_i \epsilon R^n \tag{1}$$

where β_i is the output weight connecting the ith hidden node to the output node, $F(x, c_i, a_i)$ is the activation function of the ith node, the parameters c_i and a_i are the input weight and the bias which are randomly generated at the initialization step and then fixed later.

2.1 Basic OS-ELM Algorithm

OS-ELM is originated by the concept of batch learning of ELM. Training samples can be presented in a sequential manner, i.e., either one by one or chunk by chunk with fixed or varying length and remove those observation from training whenever training would be completed. It consists of two steps, initialization step and sequential learning step.

(1) *Initialization step:* For given training data (x_k, y_k), with k represent the kth chunk of data $k = 1, ...L$, activation function $F(x)$ and by assuming N number of hidden neurons. Random initialization of input weights a_i and bias c_i for all N number of nodes should be done in this step.
(2) *Sequential learning step:* The L sample data (x_k, y_k) are used to train the OS-ELM, there must exist an optimal β_i^*, with the initially fixed parameters c_i and a_i such that

$$y_k = \sum_{i=1}^{N} \beta_i^* F(x_k, c_i, a_i), k = 1, ..., L \tag{2}$$

The previous equation can be rewritten in compact form as

$$F\beta^* = Y \tag{3}$$

where

$$F = \begin{bmatrix} F(x_1, c_1, a_1) & \dots & F(x_1, c_N, a_N) \\ \vdots & \ddots & \vdots \\ F(x_L, c_1, a_1) & \dots & F(x_L, c_N, a_N) \end{bmatrix}_{LxN} \tag{4}$$

$$\beta^* = [\beta_1^{*T} \dots \beta_N^{*T}]^T \tag{5}$$

$$Y = [Y_1^T \dots Y_N^T]^T \tag{6}$$

Finally, the least-square method instead of the traditional backpropagation algorithm was used to calculate the optimal output weight β^*, as follows:

$$\beta^* = F^{\dagger}Y \tag{7}$$

Where $F^{\dagger}$ is the Moore-Penrose generalized inverse of matrix F

3 Controller Design Using OS-ELM

The main objective in this section is to design a position and attitude controller for quadrotor based on hybrid PID-OS-ELM to ensure the trajectory tracking capability over $(x, y, z, \phi, \theta, \psi)$ to tracks the desired trajectory $(x_d, y_d, z_d, \phi_d, \theta_d, \psi_d)$. The control system contains two loops inner and outer one, the inner loop is responsible for attitude control with the controller goal to design control inputs (U_2, U_3, U_4) such that the Euler angles (ϕ, θ, ψ) follow the desired attitude trajectory $(\phi_d, \theta_d, \psi_d)$, knowing that the roll and pitch desired angles (ϕ_d, θ_d) are derived from inverse kinematics. The outer loop is responsible for position and altitude control, the control objective is to derive the control inputs (U_x, U_y, U_1) to ensure trajectory tracking capability over (x, y, z) respectively. with (U_x, U_y) are virtual control inputs used to generate the desired roll and pitch angles. The global structure of the quadrotor control is presented in Fig. 2.

3.1 Position and Altitude Control

The outer loop is designed for position and Altitude control system, since the proposed control algorithm is model-free, which mean we don't need to know the translational and altitude dynamics. To achieve our control objective, we consider the trajectory tracking over x axis for example, we can get the error as

$$e_x = x_d - x \tag{8}$$

The structure of the OS-ELM control algorithm, contain m hidden nodes with radial basis activation function RBF $F(x, a_i, c_i)$

$$F(x, c_i, a_i) = exp(\frac{\| x - c_i \|^2}{2a_i^2}) \tag{9}$$

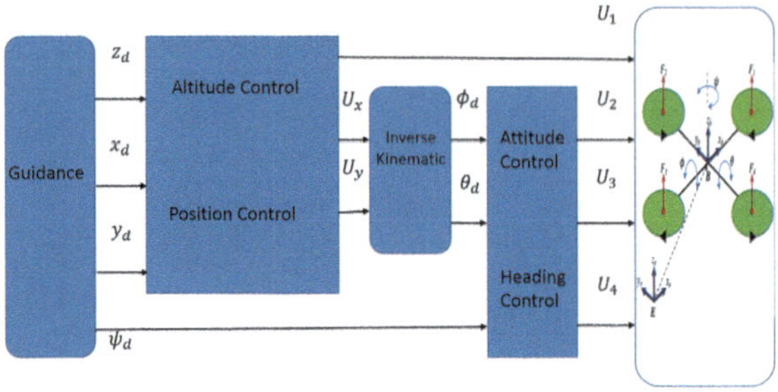

Fig. 2. The control structure.

where $c_i = [c_{11}, ..., c_{1m}]$ and $a_i = [a_1, ..., a_m]^T$ are the ith RBF neuron parameters respectively, which are randomly initialized and then fixed after that. by selecting an appropriate input $x = [x_d, U_p]$ to the neural network we can obtain the control input U_N by sequential manner using the OS-ELM algorithm as follows, see Fig. 3.

$$U_N = F_1\beta_1 + F_2\beta_2 + ... + F_m\beta_m \tag{10}$$

where m is the number of the hidden neurons, and β_m is the output weight connecting the ith hidden neuron to the output neuron.

The input signal to control the quadrotor position is described by

$$U(t) = U_p(t) + U_N(t) \tag{11}$$

with U_p is the output of the PID controller

$$U_p(t) = k_p e_x + K_d \frac{de_X}{dt} + \int e_x dt \tag{12}$$

and U_N is the output of the OS-ELM, (10) can be written as

$$U_N = \sum_{i=1}^{m} \beta_i F(x_k, c_i, a_i), \qquad k = 1..2 \tag{13}$$

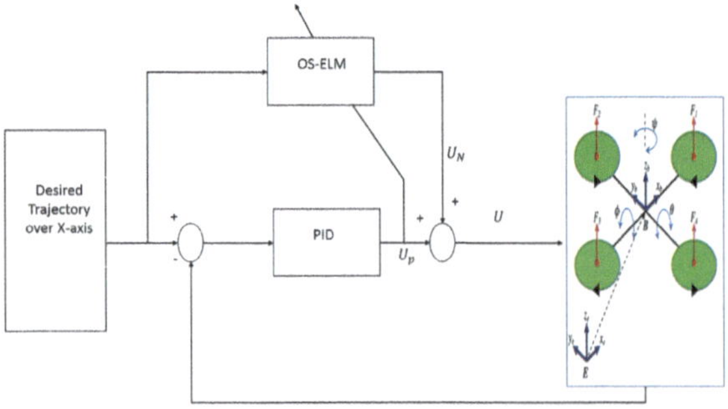

Fig. 3. Detailed architecture for the control algorithm over X-axis.

We can rewrite the Eq. 13 in the following form:

$$F\beta = U_N \tag{14}$$

with F and β can be obtained as described in (4) and (5). Since the goal is to supervise the PID output U_p, the target scalar used here is U_p as follows:

$$E(t) = U_N(t) - U(t) \tag{15}$$

If a OS-ELM with m hidden neurones can approximate $E(T)$ with zero error, there exist optimal β^*, such that

$$F\beta^* = U_N = E(t) \tag{16}$$

Finally, the least-square method instead of the classical optimization methods was carried-out to calculate the optimal output weight β^*

$$\beta^* = F^\dagger E(t) \tag{17}$$

where $F^\dagger$ is the Moore-Penrose inverse of matrix F.

Algorithm 1. PID-OS-ELM

1: *Initialization*: Initialize Random values to
   $\beta \rightarrow \beta_0$ To be updated
   $c_i \rightarrow c_{i0}$ Fixed
   $a_i \rightarrow a_{i0}$ Fixed
2: *sequential learning phase*:
   -Define the desired trajectory X_d
   -Calculate the PID output U_P
   -Get the input $x = [X_d, U_p]$ to the neural network;
   -Find the output matrix $F(a_i, c_i)$;
   -Find the output $U_N \rightarrow \beta F(a_i, c_i)$
   -Calculate The optimal output weight $\beta^* \rightarrow F(a_i, c_i)^\dagger U_p$
   -Update the output weight $\beta \rightarrow \beta^*$
   -Calculate the total control input $U \rightarrow U_N + U_p$

By following the same procedure presented in the Algorithm 1 we design the altitude control input U_1 and the inner loop control inputs U_2, U_3, U_4 for attitude control.

4 Simulation Results

In this section, to demonstrate the capabilities of the proposed method we compare it with neural network PID supervisory controller (NN-PID), in this case back backpropagation the algorithm was used to tune the neural network parameters (β, a_i, c_i) as presented in Algorithm 2.

Algorithm 2. NN-PID

1: *Initialization*: Initialize Random values to
$\beta \rightarrow \beta_0$ To be updated
$c_i \rightarrow c_{i0}$ To be updated
$a_i \rightarrow a_{i0}$ To be updated
2: *Training phase*:
-Define the desired trajectory X_d
-Calculate the PID output U_P
-Get the input $x = [X_d, U_p]$ to the neural network;
-find the output matrix $F(a_i, c_i)$;
-find the output $U_N \rightarrow \beta F(a_i, c_i)$
-train The parameters β_t, a_t, c_t using back propagation
-Update the parameters $\beta \rightarrow \beta_t, a_i \rightarrow a_t, c_i \rightarrow c_t$
-Calculate the total control input $U \rightarrow U_N + U_p$

Attitude and position control simulations are implemented on Matlab with gazebo simulator interfaced trough ROS, see Fig. 4. The reference trajectory in the test is a circle defined below:

$$\begin{cases} x_d = sin(0.1t) \\ y_d = cos(0.1t) \end{cases} \tag{18}$$

$$z_d = \begin{cases} 1\ m\ if\ t < 30s \\ 2\ m\ if\ 30s \leq t \leq 70s \\ 1\ m\ if\ t > 70s \end{cases} \tag{19}$$

$$\psi_d = \begin{cases} 0^\circ\ m\ if\ t < 10s \\ 45^\circ\ m\ if\ 10s \leq t \leq 50s \\ 15^\circ\ m\ if\ t > 50s \end{cases} \tag{20}$$

and the desired Roll and Pitch angles are generated using the inverse kinematic as follows:

$$\begin{cases} \phi_d = \frac{1}{g}(U_x \sin\psi_{des} - U_y \cos\psi_{des}) \\ \theta_d = \frac{1}{g}(U_x \cos\psi_{des} - U_y \sin\psi_{des}) \end{cases} \tag{21}$$

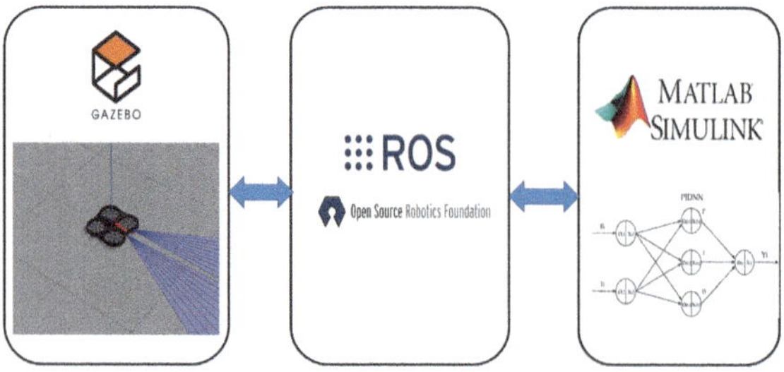

Fig. 4. Simulation setup.

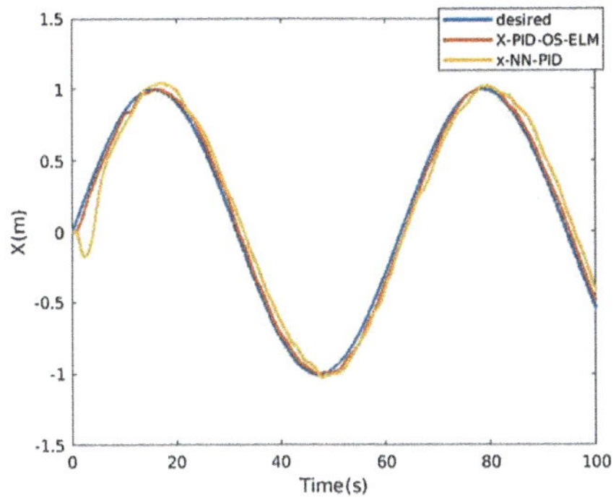

Fig. 5. Tracking response in x-axis.

The simulation result in Figs. 5, 6, 7, 8 and 9 present the output response of the proposed controller. The figures show the performance of the outer and inner loop for position, altitude and heading control using the PID-OS-ELM. The simulation results demonstrate that the proposed algorithm is efficient, the output of the PID-OS-ELM is better than NN-PID, we notice that the response time and trajectory tracking is more accurate this is because the proposed algorithm is needed to tune only the output weight β based on zero error condition, oppositely to NN-PID which based on backpropagation algorithm to train the network parameters.

The obtained results prove the validity of the PID-OS-ELM, and during the simulation setup we notice that the neural network parameters β, c_i, a_i in the NN-PID need to be initialized by proper way, in the other hand in the PID-OS-ELM the parameters are selected randomly it doesn't affect the online learning process which makes it very fast and more robust.

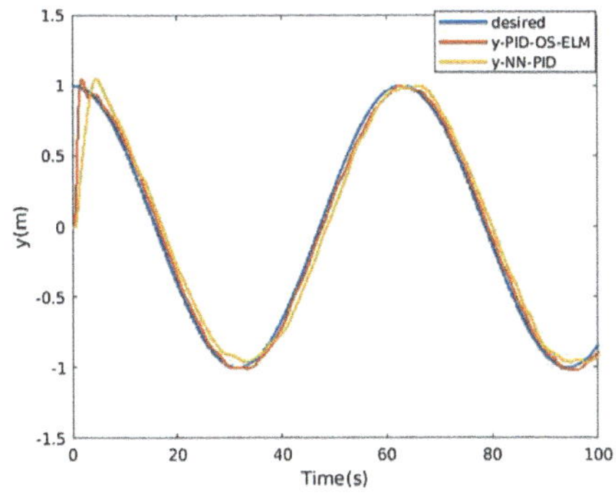

Fig. 6. Tracking response in y-axis.

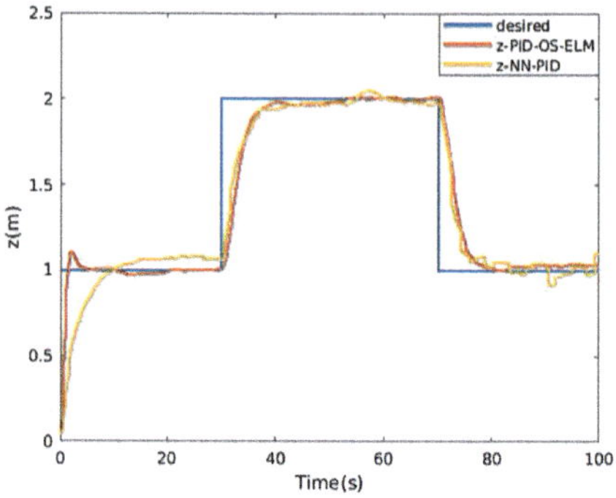

Fig. 7. Altitude response.

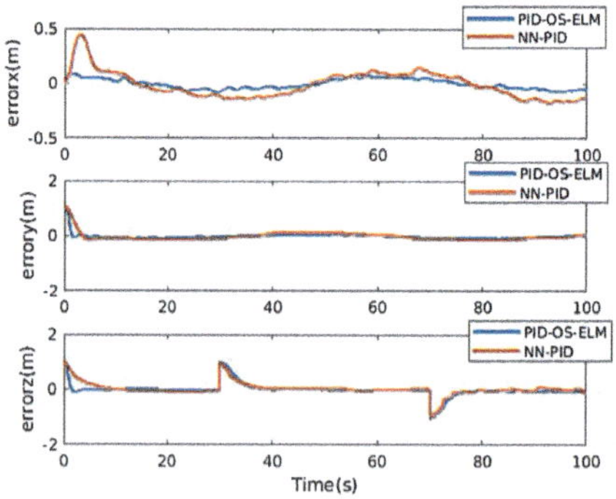

Fig. 8. Tracking error.

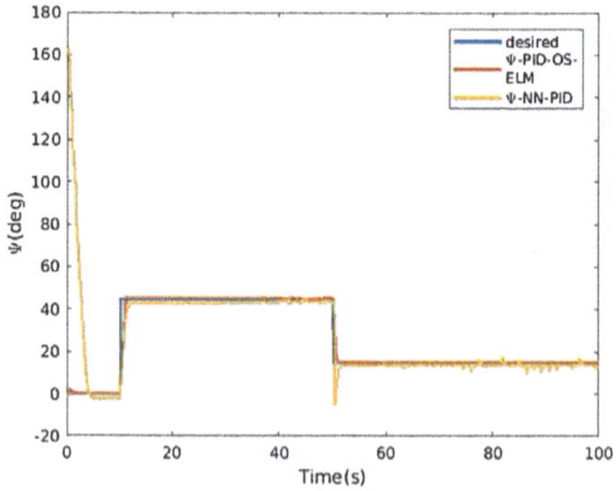

Fig. 9. Yaw angle tracking.

5 Conclusion

In this paper, a PID-OS-ELM controller is proposed for control of a quadrotor multirotor UAV. A feedforward network with one hidden layer whose input weights and hidden neurons parameters are initialized randomly and fixed later, was used to supervise the conventional PID controller. Different from the traditional NN-PID algorithm, the output weights of this neural network are updated based on least squares norm by calculating the inverse of the hidden output matrix. The obtained simulation results demonstrate that the proposed controller has superior performance on response time and trajectory tracking accuracy than the neural network based PID controller.

Acknowledgment. This research was supported by Unmanned Vehicles Advanced Core Technology Research and Development Program through the National Research Foundation of Korea (NRF), Unmanned Vehicle Advanced Research Center (UVARC) funded by the Ministry of Science, ICT and Future Planning, the Republic of Korea (No. 2016M1B3A1A01937245) and by the Ministry of Trade, Industry and Energy (MOTIE) under the R&D program (Educating Future-Car R&D Expert) (N0002428). It was also supported by Development Program through the National Research Foundation of Korea (NRF) (No. 2016R1D1A1B03935238).

References

1. Ma'sum, M.A., Arrofi, M.K., Jati, G., Arifin, F., Kurniawan, M.N., Mursanto, P., Jatmiko, W.: Simulation of intelligent unmanned aerial vehicle (UAV) for military surveillance. In: 2013 International Conference on Advanced Computer Science and Information Systems (ICACSIS), Harlow (2013)
2. Norouzi Ghazbi, S., Aghli, Y., Alimohammadi, M., Akbari, A.A.: Quadrotors unmanned aerial vehicles: a review. Int. J. Smart Sens. Intell. Syst. **9**(1) (2016)
3. Li, J., Li, Y.: Dynamic analysis and PID control for a quadrotor. In: 2011 International Conference on Mechatronics and Automation (ICMA) (2011)
4. Zulu, A., John, S.: A review of control algorithms for autonomous quadrotors. arXiv preprint arXiv:1602.02622
5. Lee, H.-I., Lee, B.-Y., Yoo, D.-W., Moon, G.-H., Tahk, M.-J.: Dynamics modeling and robust controller design of the multi-UAV transportation system. In: 29th Congress of the International Council of the Aeronautical Sciences (2014)
6. Kang, T., Yoon, K.J., Ha, T.-H., Lee, G.: H-infinity control system design for a quad-rotor. J. Inst. Control Robot. Syst. (2015)
7. Bouabdallah, S., Siegwart, R.: Full control of a quadrotor. In: 2007 IEEE/RSJ International Conference on Intelligent Robots and Systems, IROS 2007 (2007)
8. Xu, R., Ozguner, U.: Sliding mode control of a quadrotor helicopter. In: 2006 45th IEEE Conference Decision and Control (2006)
9. Alexis, K., Nikolakopoulos, G., Tzes, A.: Model predictive quadrotor control: attitude, altitude and position experimental studies. IET Control Theory Appl. (2012)

10. Santos, M., Lopez, V., Morata, F.: Intelligent fuzzy controller of a quadrotor. In: 2010 International Conference on Intelligent Systems and Knowledge Engineering (ISKE) (2010)
11. Matias, T., Souza, F., Araújo, R., Antunes, C.H.: Learning of a single-hidden layer feedforward neural network using an optimized extreme learning machine. J. Neurocomputing **129** (2014)
12. Kasun, L.L.C., Zhou, H., Huang, G.-B., Vong, C.M.: Representational learning with ELMs for big data. IEEE (2013)

Development of a Haptic Telemanipulator System Based on MR Brakes and Estimated Torques of AC Servo Motors

Ngoc Diep Nguyen[1], Sy Dzung Nguyen[2], Ngoc Tuyen Nguyen[1], and Quoc Hung Nguyen[3(✉)]

[1] Faculty of Mechanical Engineering, Industrial University of Ho Chi Minh City, Ho chi Minh City, Vietnam
pdiep66@yahoo.com, tuyen.vinalpg@gmail.com
[2] Institute for Computational Science, Faculty of Electrical and Electronics Engineering, Ton Duc Thang University, Ho Chi Minh City, Vietnam
nguyensydung@tdt.edu.vn
[3] Faculty of Engineering, Vietnamese German University (VGU), Binh Duong, Vietnam
hung.nq@vgu.edu.vn

Abstract. This research focuses on design, manufacturing and testing of a 3-DOF haptic telemanipulator system featuring MR brakes and torque estimator of AC servomotors. The haptic telemanipulators are composed of a slave manipulator and a haptic master manipulator. The slave manipulator is a commercial industrial robot with three-rotary DOF (elbow manipulator). The haptic master manipulator is designed with the same kinematic chain and kinematic dimensions as the slave. At the joints of the master, MR brakes are integrated in order to generate braking torques corresponding to the torques estimated from the servomotors of the slave. With this configuration, a desired force acting on the operator of the master slave, which is proportional to the load at the end-effecter of the slave can be archived. After the introduction, design and simulation of the proposed haptic telemanipulator system, experimental work is conducted and some experimental results are obtained and presented with discussion.

Keywords: Telemanipulator · Haptics · Magnetorheological fluid (MRF) MR brake · AC motor torque estimation

1 Introduction

In recent years, a variety of researches on development and application of force feedback techniques have been conducted. The force feedback techniques cover a wide range of application including medical engineering, aerospace technologies, military applications, computer games and virtual reality devices such as virtual rehabilitation [1, 2]. In robotics, the concept of "tele-presence" is based on the force feedback techniques, which is defined as the ideal of sensing sufficient information about the tele-operator and task environment, and communicating this to the human operator in a sufficient natural way. The operator then can feel physically present at the remote site.

K. Arai et al. (Eds.): IntelliSys 2018, AISC 868, pp. 925–935, 2019.
https://doi.org/10.1007/978-3-030-01054-6_65

Such a system is referred to as a haptic master-slave system. With the master-slave system, the operator can perform tasks in hazardous or distant environments by knowledgeably guiding the slave robot slave a safe distant location. This helps the operator to successfully control the slave robot with less effort, faster and more precise.

There have been many researches on haptic system featuring magnetorheological fluid (MRF) [3–6], which all showed that MRF is a good candidate for haptic applications. Recently, Nguyen et al. [7] have developed a 3DOF haptic manipulator system featuring MR brakes. In that research, the haptic telemanipulator system consisted of a slave manipulator which is a commercial industrial articulated manipulator and a haptic master manipulator, which had similar kinematic chain with the master, featuring three MR brakes located at the joints to generate a required force at the operator. The required torques at the joints of the master were calculated from the required force based on static modeling of the manipulator. In order to archive the calculated torque, a PID controller was used for each brake. Because of inaccuracy of the static modeling, the calculated torques are not the exact required values which result in the inaccuracy of the feedback force. This research is a further development of [7] with two main contributions: Firstly, the required torques of the MR brake sat the joints of the master are estimated from measured armature current of the corresponding motors of the slave, called as torque estimator. Secondly, experimental results for force feedback are conducted and the feedback force of the master is obtained and compared with the force at the end-effector of the salve. After the introduction, design and simulation of the proposed haptic telemanipulator system, the prototype system is manufactured and experimental work will be conducted and some experimental results are obtained and presented with discussion.

2 Haptic Telemanipulator System

Configuration of the haptic telemanipulator system used in this study is shown in Fig. 1. The system is composed of two main parts: a slave manipulator, which is commercial articulated manipulator and a haptic master manipulator featuring three MRBs. The master is designed with kinematic similarity with the slave. At each joint of the master, an encoder is integrated in order to measure its angular position. In addition, at joint 1 & joint 2, the MRB1 and MRB2 are installed to generate braking torques equal to the corresponding torques at the slave. It is noted that the MRB3 is not installed at the joint 3 but at the other side of the joint 2 and its shaft is connected to the shaft of the joint 3 through a gear-belt system. This reduces the effects of the MRB3's mass on performance of the master. Motion of the slave manipulator is driven by AC servo motors SM1, SM2 and SM3 to copy the motion of the master. As the operator moves the operation knob (end-effecter) of the master, the resulted motions at the joints are measured by encoders and sent to the servo motor drives to command these servo motors execute equal motions. This results in similar motion of the slave. At the same time, the actuating torque of the motors are estimated and sent to the MRB torque controller. The outputs of the MRB torque controllers are the required currents applied to the coils of MRBs to generate braking torques equivalent to the estimated actuating torques of the motors. Thus, an expected force acting on the operator hand equivalent to the force at the end-effecter of the slave can be archived. In other words, the operator can feel the working force at the slave end-effecter.

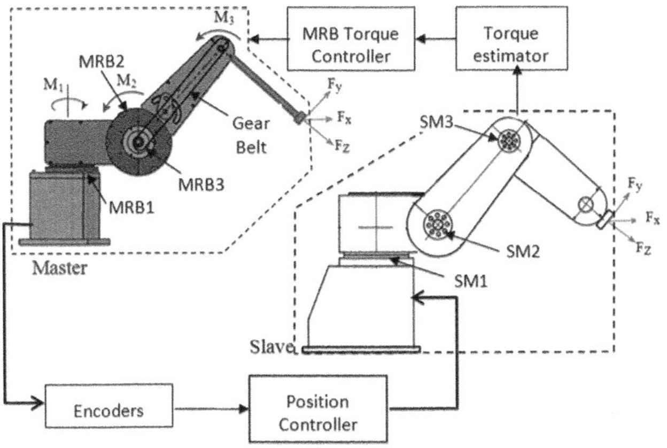

Fig. 1. Configuration of the haptic telemanipulator system.

3 MR Brakes for the Haptic Telemanipulator

3.1 Configuration of the MR Brake

In this study, side-coil MRBs shown in Fig. 2 are used to create torques at the joints of the master. As shown in the figures, a disc made of magnetic steel is fastened to the shaft made of nonmagnetic steel. The disc is placed in a stationary housing made of magnetic steel. In each side, the housing consists of two parts: inner and outer part. The coils are placed on the inner part of the housing and covered by the outer part. It is observed that the coils do not directly contact with the MRF, therefore, they are not contaminated by the MRF. In addition, with this configuration, the inner face of the housing can be manufactured continuously. This allows more accurate manufacturing; especially smaller gaps of the MRFs can be archived. The space between the disc and the housing is filled with MRF. When coils are powered, a corresponding magnetic field is generated which make the MR fluid in the gap becomes solid-like instantaneously. The shear friction between the rotating disc and the solidified MR fluid provides the required braking torque on the shaft.

3.2 Braking Torque of the MR Brake

By using Bingham plastic model for MRF, and assumed that the velocity of MRF in the duct varies linearly from the housing face to the rotary disc, the induced braking torque of the MRB can be determined by [8–10].

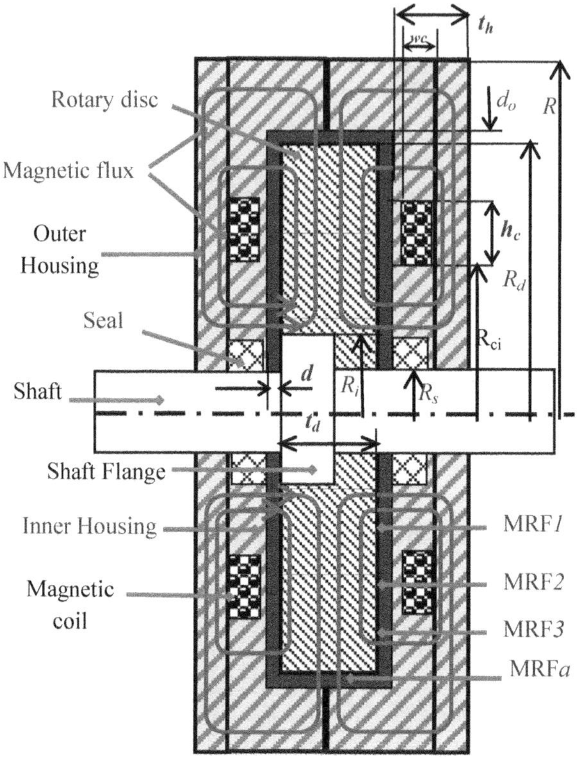

Fig. 2. Configuration of the haptic telemanipulator system.

$$
\begin{aligned}
T_b = & \frac{\pi\mu_1 R_{ci}^4}{d}[1-(\frac{R_i}{R_{ci}})^4]\Omega + \frac{4\pi\tau_{y1}}{3}(R_{ci}^3 - R_i^3) \\
& + \frac{\pi\mu_2 R_{co}^4}{d}[1-(\frac{R_{ci}}{R_{co}})^4]\Omega + \frac{4\pi\tau_{y2}}{3}(R_{co}^3 - R_{ci}^3) \\
& + \frac{\pi\mu_3 R_d^4}{d}[1-(\frac{R_{co}}{R_d})^4]\Omega + \frac{4\pi\tau_{y3}}{3}(R_d^3 - R_{co}^3) \\
& + 2\pi R_d^2 t_d(\tau_{ya} + \mu_a\frac{\Omega R_d}{d_o}) + 2T_{sf}
\end{aligned}
$$

The off-state force (the torque when no current is applied to the coils) of the MR brake can be determined by

$$
\begin{aligned}
T_0 = & \frac{\pi\mu_0 R_d^4}{d}[1-(\frac{R_s}{R_d})^4]\Omega + \frac{4\pi\tau_{y0}}{3}(R_d^3 - R_s^3) \\
& + 2\pi R_d^2 t_d(\tau_{y0} + \mu_0\frac{\Omega R_d}{d_o}) + 2T_{sf}
\end{aligned}
$$

In the above, T_b is the generated braking torque of the MRB, R_d and R_i are respectively the outer radius and inner radius of the disc, d is the gap size of MRF ducts at the end-face, d_o is the gap size of the annular MRF duct at the outer cylindrical face of the disc, t_d is the disc thickness, Ω is the angular velocity of the disc, T_{sf} is the friction torque of the leaf seal, R_{ci} and R_{co} are respectively the inner and outer radius of the coil, μ_j and τ_{yj} are respectively the average post-yield viscosity and yield stress of the jth active MRF volume (denoted by MRFj in Fig. 2) in the end-face duct while μ_a and τ_{ya} are respectively the average post yield viscosity and yield stress of MRF in the annular duct, μ_O and τ_{yO} are viscosity and yield stress of the MRF in case of no applied current. The rheological properties of MRF such as the induced yield stress τ_{yj}, $\tau_{ya,}$ and the corresponding post yield viscosity μ_j, μ_a are functions of applied magnetic flux density across the active MRF volumes and can be approximated by the following equation [13]:

$$Y = Y_\infty + (Y_0 - Y_\infty)(2e^{-B\alpha_{SY}} - e^{-2B\alpha_{SY}})$$

where Y stands for a rheological parameter of MRF (the yield stress and the post yield viscosity). The value of Y varies from the zero-applied field value Y_O to the saturation value Y_∞.. α_{SY} is the power index of the Y parameter and B is the magnetic density. In this work, the commercial MR fluid (MRF132-DG) made by Lord Corporation is used. The rheological properties of the MRF132-DG are derived based on curve fitting method from experimental results, which are: $\mu_0 = 0.1$ pa · s; $\mu_\infty = 3.8$ pa · s; $\alpha_{s\mu} = 4.5T^{-1}$; $\tau_{y0} = 15$ pa; $\tau_{y\infty} = 40000$ pa; $\alpha_{st_y} = 2.9T^{-1}$.

The leaf seal friction torque T_{sf} can be approximately calculated by [8–10].

$$T_{sf} = 0.65(2R_s)^2\Omega^{1/3}$$

In the above, the friction torque is calculated in ounce-inches, Ω is the rotation speed of the shaft measured in RPM, and R_s is the shaft diameter at the sealing measured in inches.

3.3 Optimal Design of the MR Brakes

In this study, the objective function is the mass of the MRB and design variable are significant geometric dimensions of the MRB (design variables) such as the coil height h_c, the coil width w_c the inner radius of the disc R_i, the outer radius of the disc R_d, the disc thickness t_d, Outer radius of the brake R, the housing thickness t_h and the inner radius of the coil R_{ci}. In addition, the braking torque is constrained to be greater than 10 Nm. Noteworthily, the smaller value of MRF gap size, the more compact size and larger value of braking torque can be archived. Therefore, in design of MRBs, the gap size should be chosen as small as possible considering manufacturing availability and cost. In this research, the gap size is chosen at 0.6 considering our manufacturing capacity. In addition, the shaft radius is set by $R_s = 6$ mm considering the strength of the shaft at the torque of 10 Nm. With 2% of accuracy and the convergence rate is set by 0.1%, optimal solution of the MRB is obtained based on finite element analysis in ANSYS [11, 12] and summarized in Table 1.

Table 1. Optimal solution of the MRB

Design parameter (*mm*)	**Coil:** Width $w_c = 2.3$; Height $h_c = 6.0$; Radius $R_{ci} = 43.8$; No. of turns: 2*40 = 80 **Housing:** $R_s = 6$, $R = 55.2$, $t_h = 5.9$, $L = 18.2$ **Disc:** Radius $R_i = 33.5$, $R_d = 53.5$; Thickness $t_d = 5.4$ **MRF duct gap:** 0.5
Characteristics	Max. Torque: 10 Nm Mass: 1.32 kg Off-state Torque: 0.084 Nm Power Cons.: 13 W Coil Resistance(Ω): $R_c = 1.05$

The optimized MRB is manufactured and its braking torque is experimentally obtained and presented in Fig. 3.

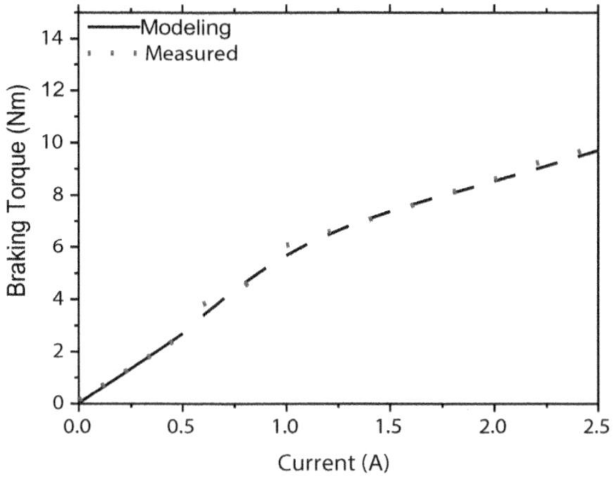

Fig. 3. Braking torque of the optimized MRB vs applied current.

4 Controller Design for the Haptic Manipulator

4.1 Position Controller

Angular positions of the master joints are measured by encoders, which then sent to the controller (drives of the servo motors) to command these servo motors execute corresponding motions equal to those of the master. The block diagram for position control system is shown in Fig. 4.

4.2 Torque Controller

It is obvious that actuating torques of the servo motors are functions of the force acting on the end-effecter of the slave. Therefore, by generating braking torques of the MRBs of the master equal to actuating torques of the corresponding motors, an equivalent force can be obtained at the end-effecter of the master. The block diagram for torque control system is shown in Fig. 5. As shown in the figure, actuating torque of the motors is estimated by the motor torque estimator, which is based on measured armature current in this research. The estimated torque is then sent to the torque controller. The outputs of the controller are the currents applied to the coils of MRBs to create equivalent torques.

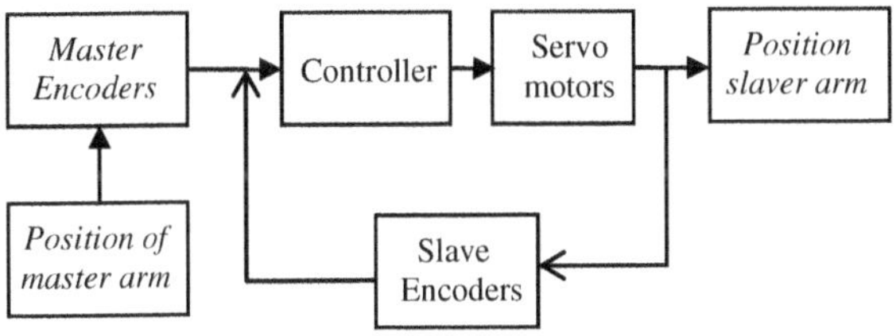

Fig. 4. Block diagram for position control system.

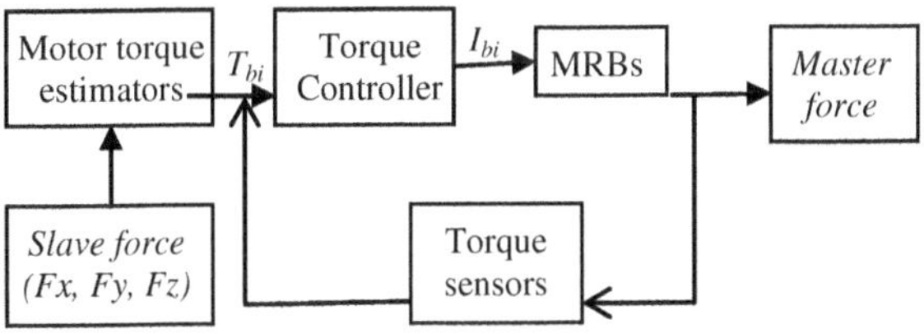

Fig. 5. Block diagram for torque control system.

5 Results and Discussion

5.1 Experimental Setup

Figure 6 shows the diagram of the prototype tele-manipulator system for experimental work. The PCI card, NI-6289 (National Instruments) is used for interaction between the manipulator system and the computer. In this study, PID control algorithm is used. A 3D force sensor (OptoForce) is used at the end-effector of the slave manipulator to measure and feedback the working force to the controller while another a 3D force sensor is attached to the operation knob of the master to measure the feedback force acting on the operator. When operator moves the master arm, angular positions of the

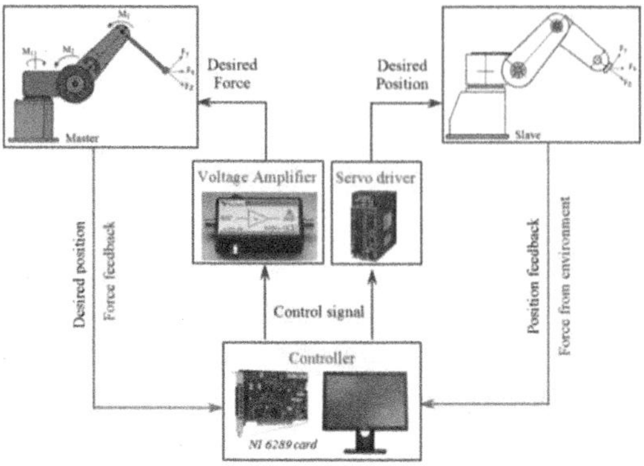

Fig. 6. Experimental system of the haptic telr-manipulator.

master joints are measured by encoders then sent to the drives of servo motors to command these servo motors execute corresponding motions equal to those of the master. Consequently, the motion of the slave arm is almost similar to that of the master arm. At the same time, torques at joints of the slave are estimated from electric current of the motor armature and transferred to the controller. The control voltage from the controller is then amplified and supplied to the coils of MRBs in order to create a resultant force to the user.

5.2 Experimental Results

Figure 7 shows experimental results of position response of the system. It is observed from the figure that the motions of slave agree very well with the motioned of the master. Numerical results show that the angular position error between the master and the slave is smaller than 0.5%.

Figure 8 shows experimental results of force response of the system. It is observed from the figure that the induced force at the master quite similar to that at the slave and the proposed haptic tele-manipulator can be employed for force feedback at the working environment of the slave robot to the operator to improve the accuracy of the operation. However, the force at the master is larger than that of the slave. The main reasons come from the friction torque at the joints of the master is not accounted for. In addition, an error of estimation of actuating torques of the motors based on measured armature current is also another important reason. These challenging issues will be considered in our next researches.

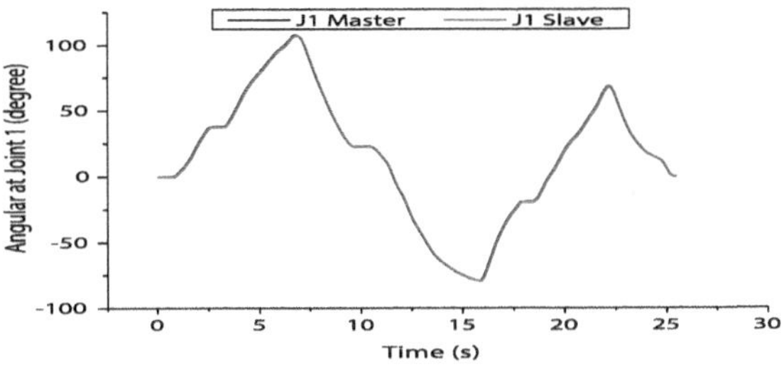

Angular value at joint 1

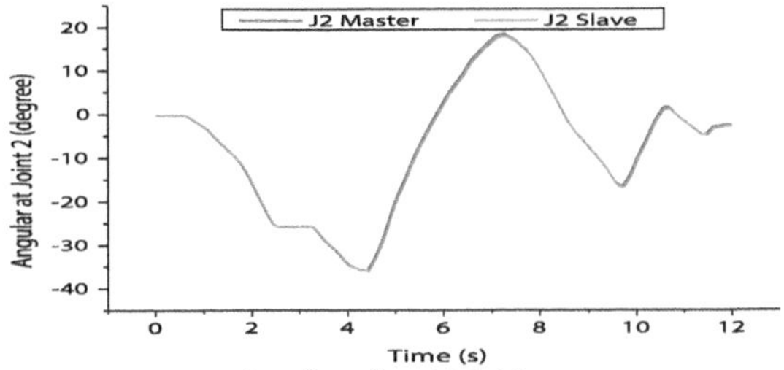

Angular value at joint 2

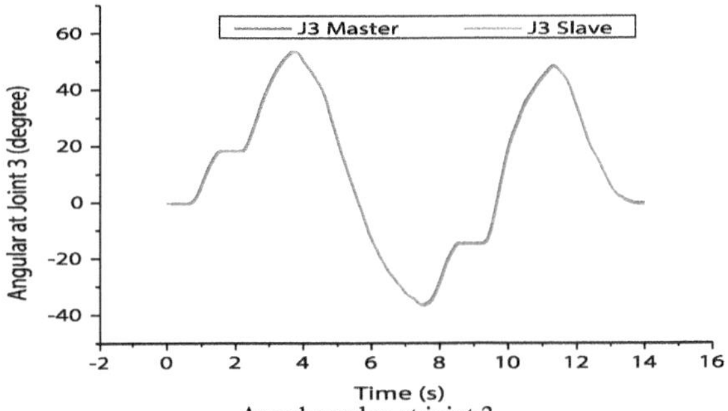

Angular value at joint 3

Fig. 7. Experimental results of position response.

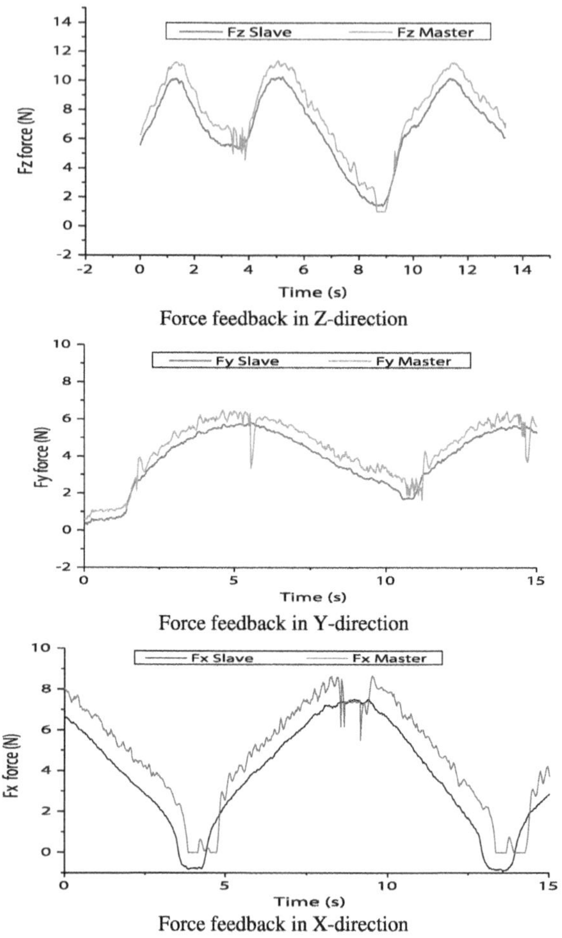

Fig. 8. Experimental results of force reflection.

6 Conclusions

In this research, a 3-DOF haptic telemanipulator system featuring MR brakes and torque estimator of DC servomotors was developed. The haptic telemanipulator are composed of a slave manipulator and a haptic master manipulator. The slave manipulator is a commercial elbow. The haptic master manipulator was designed with the same kinematic chain and kinematic dimensions as the slave. At the joints of the master, MR brakes are integrated in order to generate braking torques corresponding to the torques estimated from the servomotors of the slave, by which an equivalent force is induced at the end-effecter of the master. After the introduction, design and simulation of the proposed haptic telemanipulator system, experimental work was conducted. Experimental results showed that the position of the master agrees very well that of the master. It was also shown that the proposed system can be employed for force feedback from the working environment of the slave robot to the operator. However, the force induced at the master is larger than the force at the slave due two uncounted friction and inaccuracy of torque estimator. These issues will be considered as the second phase of this research.

Acknowledgment. This work was supported by research fund from Ministry of Education and Training of Vietnam under Grant No. KYTH-114.

References

1. Kawasaki, H., Mouri, T.: Design and control of five-fingered haptic interface opposite to human hand. IEEE Trans. Robot. **23**(5) (2007)
2. Nawrat, Z., Kostka, P.: Polish cardio-robot 'Robin Heart'. System description and technical evaluation. Int. J. Med. Robot. Comp. **2**(1), 36–44 (2006)
3. Nguyen, Q.H., Nguyen, P.B., Choi, S.B.: Optimal design of a hybrid MR brake for haptic wrist application. In: Proceedings of SPIE 7977, Active and Passive Smart Structures and Integrated Systems (2011). https://doi.org/10.1117/12.880396
4. Oh, J.S., Shin, W.K., Uhm, C.H., Lee, S.R., Han, Y.M., Choi, S.B.: Control of haptic master – slave robot system for minimally invasive surgery. J. Phys. Conf. Ser. (2014). https://doi.org/10.1088/1742-6596/412/1/012041
5. Stone, R.J.: Haptic feedback: a brief history from telepresent to virtual reality. In: Proceedings of 1st International Workshop on Haptic Human-Computer Interaction. LNCS, vol. 2085, pp. 1–16 (2000)
6. Oh, J.S., Shin, W.K., Uhm, C.H., Lee, S.R., Han, Y.M., Choi, S.B.: Control of haptic master – slave robot system for minimally invasive surgery. J. Phys. Conf. Ser. **412**, 012041 (2013). https://doi.org/10.1088/1742-6596/412/1/012041
7. Diep, N.N., Nguyen, X.H., Tuyen, N.N., Nguyen, Q.H.: Development of a 3-DOF haptic tele-manipulator system using magnetorheological brakes. In: Proceedings of the International Conference on Advances in Computational Mechanics. Lecture Notes in Mechanical Engineering, pp. 793–805 (2018)
8. Nguyen, Q.H., Nguyen, N.D., Choi, S.B.: Design and evaluation of a novel magnetorheological brake with coils placed on the side housings. Smart Mater. Struct. **24**(4) (2015)
9. Nguyen, N.D., Nguyen, X.H., Le, H.D., Nguyen, Q.H.: Design and Evaluation of a new magnetorheological brake with two coils placed on each side of the brake housing considering manufacturing benefits. Vietnam Mech. Eng. J. August 2016. (ISSN 0866-7056)
10. Nguyen, Q.H., Nguyen, N.D., Choi, S.B.: Optimal Design of a novel configuration of MR brake with coils placed on the side housings of the brake. In: Proceedings of SPIE 9059, Industrial and Commercial Applications of Smart Structures Technologies, 90590I (2014)
11. Nguyen, Q.H., Choi, S.B.: Optimal design of an automotive magnetorheological brake considering geometric dimensions and zero-field friction heat. Smart Matter. Struct. **19**(11), 1–11 (2010)
12. Nguyen, Q.H., Han, Y.M., Choi, S.B., Wereley, N.M.: Geometry optimization of MR valves constrained in a specific volume using the finite element method. Smart Mater. Struct. **16**, 2242–2252 (2007)
13. Zubieta, M., Eceolaza, S., Elejabarrieta, M.J., Bou-Ali, M.M.: Magnetorheological fluids: characterization and modeling of magnetization. Smart Mater. Struct. **18**, 1–6 (2009)

Proximity Full-Text Search with a Response Time Guarantee by Means of Additional Indexes

Alexander B. Veretennikov(✉)

Chair of Calculation Mathematics and Computer Science, INSM,
Ural Federal University, Yekaterinburg, Russia
alexander@veretennikov.ru

Abstract. Full-text search engines are important tools for information retrieval. Term proximity is an important factor in relevance score measurement. In a proximity full-text search, we assume that a relevant document contains query terms near each other, especially if the query terms are frequently occurring words. A methodology for high-performance full-text query execution is discussed. We build additional indexes to achieve better efficiency. For a word that occurs in the text, we include in the indexes some information about nearby words. What types of additional indexes do we use? How do we use them? These questions are discussed in this work. We present the results of experiments showing that the average time of search query execution is 44–45 times less than that required when using ordinary inverted indexes.

Keywords: Full-text search · Search engines · Inverted indexes
Additional indexes · Proximity search · Term proximity

1 Introduction

A search query consists of several words. The search result is a list of documents containing these words. In a modern search system, documents that contain search query words that are near each other are more relevant than other documents [1, 2]. Inverted indexes [3, 4] are used to address this search task. For each word in each document, we need to store a record in the index. This record includes the number of the word in the document and the ID (identifier) of the document. We can define the ID of a document as the document's ordinal number. These records are called "postings".

Words appear in documents at different frequencies. The maximum query response time is determined by the most frequently used words. Zipf's law [5] describes the typical word frequency distribution. It is common to have a search system that can usually perform a query within 1 s of time but works very slowly, requiring 20–30 s, for example, for a query that contains frequently occurring words. This is the problem

The work was supported by Act 211 Government of the Russian Federation, contract № 02. A03.21.0006.

K. Arai et al. (Eds.): IntelliSys 2018, AISC 868, pp. 936–954, 2019.
https://doi.org/10.1007/978-3-030-01054-6_66

that we wish to solve in this paper. With additional indexes, we can guarantee a stable query response time, for example, within 1 s.

An example of a word frequency distribution is shown in Fig. 1. The horizontal axis represents different words, from frequently occurring words to infrequently occurring words (from left to right, in decreasing order of their frequencies). On the vertical axis, we plot the number of occurrences of each word. With typical inverted indexes, the query execution time is proportional to the number of occurrences of the queried words in the indexed texts.

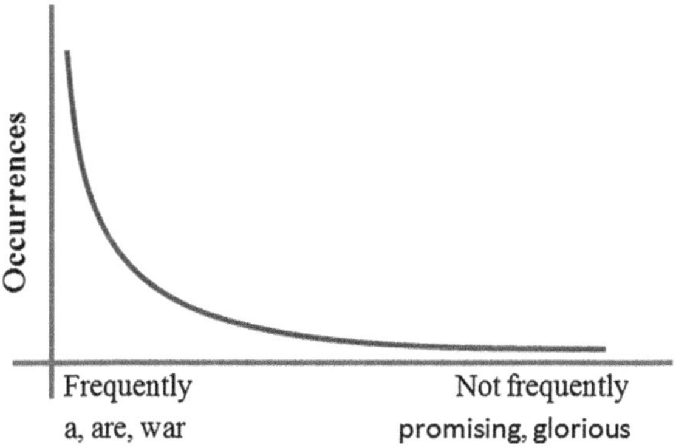

Fig. 1. Word frequency distribution.

Some search systems exclude most frequently used words from the index and, consequently, from any search – this is called the stop words approach. However, this approach is not correct [6]. Some most frequently occurring words can have unique meanings in specific contexts. For example, consider the query "time and a word yes". Yes are an English rock band, and "Time and a Word" is one of their well-known songs. Therefore, the word "yes" has a specific meaning in the context of this query. A similar query example is "who are you who". The Who are an English rock band, and "Who are You" is one of their songs.

The other approach is to maintain additional indexes for faster query execution.

In [1], some additional indexes are introduced, but only two-word queries are considered. In [7], most frequently occurring words are excluded from consideration. We avoid both of these faults by introducing several types of additional indexes for several types of words.

2 Proximity and Relevance

2.1 Importance of Proximity

We introduce a parameter *MaxDistance*. Let us consider a search result – a document containing words in the specified query. If the length of the fragment containing the words is less than or equal to *MaxDistance*, then the search result is relevant and important; otherwise, it can be skipped.

For example, we consider the search query "time and a word yes".

Result 1: "time and a word yes" – This is an important search result.

Result 2: "time and a word by yes" – This search result is also important.

Result 3: "time ...some other words ... and ... some other words ... a ... word ... some other words... by ... some other words... yes" – This search result may not be important because "time", "and", and "yes" are not linked by any meaning.

We will present a formal definition of *MaxDistance* later.

2.2 Relevance Function

Let the importance of a pair of word occurrences be inversely proportional to the square of the distance between the words in the document [1].

Consider the following relevance function:

$$S = a \cdot SR + b \cdot IR + c \cdot TP \tag{1}$$

Here a, b, and c are non-negative parameters, where $a + b + c = 1$. Search results can be sorted in accordance with this function [1]. *SR* is a static rank of the document, such as the PageRank [8], and is independent of the query. *IR* is an information retrieval rank, such as BM25. *TP* is a proximity ranking function.

For two word occurrences A and B, the value of $TP(A, B)$ can be calculated as $1 / |A - B|^2$, where ($|A - B|$) is the distance between the words in the document.

We define the occurrence of a word as its ordinal number within the document.

Consider the following text: "time and a word by yes".

The word positions (ordinal numbers) are as follows: time: 0, and: 1, a: 2, word: 3, by: 4, yes: 5.

Consider the following search query: "and word". The search result is "and a word", at position 1. There is an extra word, "a", between "and" and "word" in this text. Therefore,

$$TP = 1/|1-3|^2 = 0.25.$$

Consider the search query "time and". The search result is "time and", at position 0. Therefore,

$$TP = 1/|0-1|^2 = 1.$$

If the search query occurs in the text in its exact form (with no extra words between the query terms in the text), then

$$TP = 1.$$

2.3 Importance of TP

Let us introduce a value denoted by *MaxTPDistance* to define the following condition: if $|A - B| \leq MaxTPDistance$, then *TP* is "important"; otherwise, only (*SR* + *IR*) is "important".

What does "important" mean? We can assume that *SR*, *IR* and *TP* can be normalized.

Thus, $0 \leq SR \leq 1$, $0 \leq IR \leq 1$, and $0 \leq TP \leq 1$.

Then, $0 \leq a \cdot SR \leq 1$, $0 \leq b \cdot IR \leq 1$, and $0 \leq c\,TP \leq 1$.

Let us introduce a parameter *TP_Critical*, with an example value of 0.15.

If $(c \cdot TP) \leq TP\_Critical$, then the relevance (or score) of the search result will be determined mostly by *SR* + *IR*. In this case, *TP* is not "important".

Let $c = 1$ and $TP\_Critical = 0.15$.

Consider the following text: "time and a word by yes".

Consider the following search query: "and word". The search result is "and a word", at position 1, with $|A - B| = 2$.

$c \cdot TP = 0.25 > TP\_Critical$, indicating that the search result is important.

Now, consider the search query "time word". The search result is "time and a word", at position 0, with $|A - B| = 3$.

$c \cdot TP = 1/3^2 \approx 0.11 < TP\_Critical$, indicating that the search result is not important.

In this case, $MaxTPDistance = 2$.

2.4 Evaluating TP for More Than Two Words

How do we evaluate *TP* if the query consists of more than two words? Consider an *n*-word query *Q*. We have a search result *R*, which is represented by *n* word positions in a document:

$R = X(1), X(2), \ldots, X(n)$.

$Let\, A(R) = min(X(1), \ldots, X(n))$ and

$B(R) = max(X(1), \ldots, X(n))$.

Requirement: if the search query occurs in the text in its exact form (with no extra words between the query terms in the text), then $TP(R) = 1$.

$TP(R) = 1$ when $|A(R) - B(R)| = (n - 1)$.

Proposition:

$TP(R) = TP(X(1), \ldots, X(n)) = 1/(|A(R) - B(R)| - (n-2))^2$.

For example, suppose that we have the following search query: "time and a word yes".

Result 1: "time and a word yes" – This is an important search result.

$TP(R) = 1/(|0-4| - (5-2))^2 = 1$.

Result 2: "time and a word by yes" – This is a less important search result.

$TP(R) = 1/(|0-5| - (5-2))^2 = 0.25$.

We also can consider a more flexible *TP* function, such as

$TP(R) = TP(X(1), \ldots, X(n)) = 1/(p \cdot (|A(R) - B(R)| - (n-2)))^2$.

The value of *p* can be different for different systems.

2.5 Evaluating MaxTPDistance for More Than Two Words

Let us define the function *MaxTPDistance*(n) as follows: for any search query Q consisting of m words, where $m \le n$, and a search result $R = X(1), X(2), \ldots, X(m)$ for Q, if $|A(R) - B(R)| > MaxTPDistance(n)$, then $c \cdot TP(R) \le TP\_Critical$; moreover, *MaxTPDistance*(n) is the smallest value for which this is true.

By definition,
if $a \ge b$, then $MaxTPDistance(a) \ge MaxTPDistance(b)$.
Let $n = 3$, $TP\_Critical = 0.15$, and $c = 1$.
Consider a 3-word search query Q and a search result R.
If $|A(R) - B(R)| = 2$, then
$TP(R) = 1/(2-1)^2 = 1 > TP\_Critical$.
If $|A(R) - B(R)| = 3$, then
$TP(R) = 1/(3-1)^2 = 0.25 > TP\_Critical$.
If $|A(R) - B(R)| = 4$, then
$TP(R) = 1/(4-1)^2 \approx 0.11 < TP\_Critical$.
Consider a 2-word search query Q and a search result R.
If $|A(R) - B(R)| = 1$, then
$TP(R) = 1/(1)^2 = 1 > TP\_Critical$.
If $|A(R) - B(R)| = 2$, then
$TP(R) = 1/(2)^2 = 0.25 > TP\_Critical$.
If $|A(R) - B(R)| = 3$, then
$TP(R) = 1/(3)^2 \approx 0.11 < TP\_Critical$.

In this case, $MaxTPDistance(3) = 3$.

For any query Q consisting of m words, where $m \le 3$, and any search result R for Q that satisfies the condition $|A(R) - B(R)| > 3$, we have $c \cdot TP(R) \le TP\_Critical$.

2.6 Definition of MaxDistance

Let us introduce our new parameter, *MaxDistance*.
Let $n \ge 1$ be a number.

We assume that for any query of length $m \le n$, our search will return all relevant results. If the query has a length $> n$, it must be divided into parts.

Let $MaxDistance = MaxTPDistance(n)$.

We can also define a parameter $MaxDistance = 7$ (for example) and build indexes accordingly. Then, for any query of length m, where $m \le n \le MaxDistance$, with n being some number, our search will return all relevant results.

In our experiments, we use $MaxDistance = 5$, 7 or 9.

2.7 More Generic TP Structure

Let us also consider a more generic version of *TP*:

$$TP(R) = TP(X(1), \ldots, X(n)) = 1/(|A(R) - B(R)| - (n-2))^{e(n)}, e(n) = 1 + (2/n).$$

We assume that for longer queries, more extra words are acceptable between query terms in the text.

Let us calculate *MaxTPDistance*(3) for this case.

Let $n = 3$, $TP\_Critical = 0.15$, and $c = 1$.

Consider a 3-word search query Q and a search result R.

If $|A(R) - B(R)| = 2$, then $TP(R) = 1 > TP\_Critical$.

If $|A(R) - B(R)| = 3$, then $TP(R) \approx 0.314 > TP\_Critical$.

If $|A(R) - B(R)| = 4$, then $TP(R) \approx 0.16 > TP\_Critical$.

If $|A(R) - B(R)| = 5$, then $TP(R) \approx 0.09 < TP\_Critical$.

Consider a 2-word search query Q and a search result R.

If $|A(R) - B(R)| = 1$, then

$TP(R) = 1/(1)^2 = 1 > TP\_Critical$.

If $|A(R) - B(R)| = 2$, then

$TP(R) = 1/(2)^2 = 0.25 > TP\_Critical$.

If $|A(R) - B(R)| = 3$, then

$TP(R) = 1/(3)^2 \approx 0.11 < TP\_Critical$.

In this case, *MaxTPDistance*(3) = 4. We need a larger value of *MaxDistance* with such a *TP* function.

3 Word Type

In [9], we defined three types of words.

Stop Words: Examples include "and", "at", "or", "yes", "who", "was", and "war". These words are very commonly used and may not be included in the index in some other approaches. However, we include all words.

Frequently Used Words: These words are frequently encountered but convey meaning. These words should always be included in the index.

Ordinary Words: This category contains all other words. We assume that no performance problems will arise from these words.

We use a morphological analyzer for lemmatization. For each word in the dictionary, the analyzer provides a list of numbers of lemmas (i.e., basic or canonical forms). The lemma numbers lie in the range from zero to (*WordsCount* – 1), where *WordsCount* is the number of different lemmas considered (we use a combined Russian/English dictionary with approximately 200 000 Russian lemmas and 92 000 English lemmas).

If a word does not appear in the analyzer's dictionary, we assume that its lemma is the same as the word itself.

When using the analyzer, we apply aforementioned three-type division approach, not to the words themselves but to the lemmas of the words. The lemmas are divided into three types in terms of the frequency with which they are encountered: stop lemmas, frequently used lemmas, and other lemmas.

How do we distribute the lemmas among these groups? Let us sort all lemmas in decreasing order of their occurrence frequency in the texts. This sorted list we call the

FL-list. The number of a lemma in the *FL*-list we call its *FL*-number. Let the *FL*-number of a lemma *w* be denoted by *FL*(*w*).

The first *SWCount* most frequently occurring lemmas are stop lemmas. The second *FUCount* most frequently occurring lemmas are *frequently used* lemmas. All other lemmas are ordinary lemmas. *SWCount* and *FUCount* are parameters. Representative example values are *SWCount* = 700 and *FUCount* = 2100.

Let us consider the following text, with identifier *ID*1: "A friend of mine who has desired the honour of meeting with you". This is the excerpt from the Charles Dickens's Barnaby Rudge.

After lemmatization: [a] [friend] [of] [mine, my] [who] [have] [desire] [the] [honour] [of] [meet, meeting] [with] [you].

With *FL*-numbers: [a: 17] [friend: 793] [of: 24] [mine: 2482, my: 264] [who: 293] [have: 55] [desire: 2163] [the: 10] [honour: 3774] [of: 24] [meet: 1008, meeting: 4375] [with: 40] [you: 47].

Let us enumerate the words starting from zero. Then, the word "friend" appears in the text at position 1. Then, the lemma "friend" appears in the text at position 1. The lemma "my" appears in the text at position 3. Thus, the distance between the lemma "my" and the lemma "friend" in the text is 2. We can say that lemma "my" > "of", because *FL*(my) = 264, *FL*(of) = 24, and 264 > 24 (we use the *FL*-numbers to establish the order of the lemmas in the set of all lemmas).

For an ordinary lemma *q*, we can say that *FL*(*q*) = $\sim$. In this case, *q* occurs in the texts so rarely that *FL*(*q*) is irrelevant. We denote by "$\sim$" some big number.

Let us consider the results obtained with our example values, namely, *SWCount* = 700 and *FUCount* = 2100.

Stop lemmas (<700): a, of, my, who, have, the, with, you.

Frequently used lemmas ($\geq$700, <2800): friend, mine, desire, meet.

Ordinary lemmas ($\geq$2800): honour, meeting.

4 Additional Indexes

We define several types of additional indexes.

4.1 Ordinary Index with Near Stop Word (NSW) Records

For each lemma in each document, a record (*ID*, *P*, NSW) is included in the index. *ID* is the ordinal number of the document. *P* is the corresponding word's ordinal number within the document. The NSW record contains information about all stop lemmas occurring near position *P* (at a distance $\leq$ *MaxDistance*). This information is efficiently encoded [9–11].

For example, let *MaxDistance* = 5. The NSW record for the first occurrence of "friend" in the aforementioned example contains the following: (a, −1), (of, 1), (my, 2), (who, 3) (have, 4). In (a, −1), the distance (−1) between "a" and "friend" is stored, and so on.

Let the document identifier *ID*1 be 27. Below, let us consider several example postings in the ordinary index, in the format (document *ID*, word position, NSW record).

friend: (27, 1, ((a, −1), (of, 1), (my, 2), (who, 3) (have, 4))).
mine: (27, 3, ((a, −3), (of, −1), (who, 1), (have, 2), (the, 4))).
desire: (27, 6, ((of, −4), (my, −3), (who, −2), (have, −1), (the, 1), (of, 3), (with, 5))).

The lemma types considered here are frequently used and ordinary.

For a stop lemma, we include only the first occurrence in the document and no NSW records.

Let us consider a lemma. For optimization purposes, we can use two data streams for the lemma. The first data stream contains the (*ID*, *P*) records. The second data stream contains the corresponding NSW records. In this case, the NSW records can be easily skipped if required.

See [11] for more details about NSW records.

4.2 Expanded (W, V) Indexes

The expanded (*w*, *v*) index is the list of occurrences of the lemma *w* for which lemma *v* occurs in the text at a distance less than or equal to *MaxDistance* from *w*.

The lemma types considered are as follows: for *w*, frequently used; for *v*, frequently used or ordinary. Each posting includes the distance between *w* and *v* in the text.

In the case that both *w* and *v* are frequently used, we create only one expanded index. To prevent duplication, we create an expanded (*w*, *v*) index only if $w \leq v$.

Below, let us consider several example postings in the format (document *ID*, word position, distance).

(friend, mine): (27, 1, 2). Here, 27 is the document *ID*, 1 is the position of "friend" in the document, and 2 is the distance between "friend" and "mine" in the document. We store a posting only for the key (friend, mine); no posting is stored for the key (mine, friend). Note that *FL*(friend) = 792 < *FL*(mine) = 2482.

(friend, desire): (27, 1, 5).
(desire, mine): (27, 6, −3).
(mine, honour): (27, 3, 5).

We previously studied (*w*, *v*) indexes in [9–12]. In the current work, we describe several new use cases.

4.3 Expanded (F, S, T) Indexes

The expanded (*f*, *s*, *t*) index is the list of occurrences of the lemma *f* for which lemmas *s* and *t* both occur in the text at distances less than or equal to *MaxDistance* from *f*.

We create an expanded (*f*, *s*, *t*) index only for the case in which $f \leq s \leq t$.

Here, *f*, *s*, and *t* are all stop lemmas.

Below, let us consider several example postings, in the format (document *ID*, word position, distance between *f* and *s*, distance between *f* and *t*):

(a, of, my): (27, 0, 2, 3).
(a, my, who): (27, 0, 3, 4).
(a, of, who): (27, 0, 2, 4).
(a, have, my): (27, 0, 5, 3).
(of, my, who): (27, 2, 1, 2).
…
(of, with, who): (27, 9, 2, −5).

This type of index is the largest.

5 Preprocessing the Query

Let us consider the following query: "friend mine who".

After lemmatization: [friend] [mine, my] [who].

Each element of the query after lemmatization is called a cell. This query contains three cells. The first cell is [friend], the second is [mine, my], and the last is [who].

Important condition: Each cell of the query must contain lemmas of only one type. If this condition is not met, then the query must be divided. For example, from the initial query [friend] [mine, my] [who], we derive two queries: [friend] [mine] [who] and [friend] [my] [who].

Second condition: If all lemmas in the query are stop lemmas, then each cell must contain only one lemma. If this condition is not met, then the query must be divided.

6 Processing the Query

We apply different processing methods for different types of queries.

6.1 All Lemmas of the Query Are Ordinary

In this case, we use the ordinary index. We skip the NSW records. In this case, *MaxDistance* is not used.

6.2 All Lemmas of the Query Are Frequently Used

Let us consider the following query: "beautiful red hair".

After lemmatization: [beautiful: 2216] [red: 2191] [hair: 1850].

(1) *The first approach*

The lemma "beautiful" is the lemma that is encountered *least often* in the texts.

We therefore designate [beautiful] as the main cell of the query.

We consider the following expanded indexes: (red, beautiful) and (hair, beautiful).

The lemma "red" is more frequently used than "beautiful". Thus, we have the expanded index (red, beautiful). However, there are two logical expanded indexes:

(red, beautiful) and (beautiful, red). Suppose that we are reading records from the (red, beautiful) index. From a record (*ID*, *Position*, *Distance*), we can produce the record (*ID*, *Position* + *Distance*, −*Distance*), which corresponds to the (beautiful, red) logical index.

Thus, we can say that we have the (beautiful, red) and (beautiful, hair) indexes.

Each of these indexes contains the positions of the lemma "beautiful" in texts.

Let us consider the following text: "A beautiful, shimmering, red curly hair ...".

We have (*ID*, 3, –2) in the (red, beautiful) index and (*ID*, 5, –4) in the (hair, beautiful) index.

Consequently, we have (*ID*, 1, 2) in the (beautiful, red) logical index and (*ID*, 1, 4) in the (beautiful, hair) logical index. These two records have identical *ID* and *Position* fields.

We need to check for all lemmas in the query, except the main lemma, for which a record (*ID*, *P*, *) exists at the specified position (*ID*, *P*) in all selected logical expanded indexes.

Let us consider the position (*ID*, 1). In the (beautiful, red) index, a record (*ID*, 1, 2) exists. In the (beautiful, hair) index, a record (*ID*, 1, 4) exists. Thus, we have "beautiful red hair" in the text.

(2) *The second approach*

Let us consider a (*w*, *v*) index. From a record (*ID*, *Position*, *Distance*), we can produce two related records: (*ID*, *Position*) for the key *w* and (*ID*, *Position* + *Distance*) for the key *v*. Thus, from one (*w*, *v*) index, we can derive two logical indexes (*w*, *v*) and (*v*, *w*). The first contains the occurrences of *w*, and the second contains the occurrences of *v*.

From the record (*ID*, 3, –2) in the (red, beautiful) index, we can produce the record (*ID*, 3) as an occurrence of "red" and (ID, 1) as an occurrence of "beautiful".

We can divide any query into a list of pairs of words. For example, let us consider "beautiful red hair" – > (beautiful red) (red hair). Then, from the extended (beautiful, red) and (red, hair) indexes, we can produce 4 streams of postings, one each for "beautiful", "red", "red" and "hair" (there is one extra "red" stream, which we can skip). Then, we combine these logical streams for "beautiful", "red" and "hair" and obtain the results.

(3) *Comparison of the first and second approaches*

The second approach requires more computational resources to produce single-key streams in memory, but fewer data need to be loaded. Consider the query "beautiful bright red hair" – > [beautiful: 2216] [bright: 2530] [red: 2191] [hair: 1850].

The first approach requires 3 two-key indexes: (beautiful, bright), (red, bright), and (hair, bright). The second approach requires 2 two-key indexes: (beautiful, bright) and (red, hair).

Now, let us consider the query "beautiful red rose" – > [beautiful: 2216] [red: 2191] [rose: 1007, rise: 1753]. Using the first approach, we need three indexes: (red, beautiful), (rise, beautiful), and (rose, beautiful).

(4) *The third approach*

We can divide any query into a list of pairs of words. For example, let us consider "beautiful red hair" – > (beautiful red) (red hair). Then, we need to combine the corresponding streams of data. This approach is more effective than the second approach, but it is also more complex to realize because it is more complex to combine two-key streams than single-key streams.

6.3 Not All of the Lemmas Are Frequently Used, and There Are no Stop Lemmas

Let us consider the following query: "red glorious promising rose".

After lemmatization: [red: 2191], [glorious: ~] [promising: ~] [rose: 1007, rise: 1753].

Frequently used lemmas: red, rose, rise.

Ordinary lemmas: glorious, promising.

There are several approaches we can propose here.

(1) *The first approach*

We select the frequently used lemma *w* in the query that has the lowest frequency. For every other lemma *v* in the query, a logical expanded (*w*, *v*) index exists. For example, let us select [red] as the main cell. We can use the following expanded indexes:

(red, promising) – contains occurrences of red (near promising).

(red, glorious) – contains occurrences of red (near glorious).

(red, rise) – contains occurrences of red (near rise).

(red, rose) – contains occurrences of red (near rose).

(2) *The second approach*

We select the ordinary lemma *w* in the query that has the lowest frequency. For every other frequently used lemma *v* in the query, a logical expanded (*w*, *v*) index exists. For every other ordinary lemma *q* in the query, we can use the ordinary index *q* (skipping the NSW records). For example, let us select [promising] as the main cell. We can use the following indexes:

(red, promising) – contains occurrences of red (near promising).

(rise, promising) – contains occurrences of rise (near promising).

(rose, promising) – contains occurrences of rose (near promising).

(glorious) – we use the ordinary index, because both "glorious" and "promising" are ordinary lemmas and no extended (promising, glorious) index exists.

We do not need a list of occurrences of "promising" because we know that "promising" occurs somewhere nearby.

(3) *The third approach*

We can also select a two-component key index for a frequently used or ordinary lemma. For this example, we have

(red, promising) for red,
(rise, promising) for rise,
(rose, promising) for rose, and
(red, glorious) for glorious.

If we store in some dictionary the length of each index (w, v), then we can select the most suitable variant.

6.4 All Lemmas of the Query Are Stop Lemmas

In this case, (f, s, t) indexes are used.

Let us consider the following query: "to be not to be".

After lemmatization: [to: 7] [be: 21] [not: 156] [to: 7] [be: 21]. We can use the (to, be, not) and (to, to, be) indexes to produce results.

Now, let us consider the following query: "who are you who".

[who: 293] [are: 268, be: 21] [you: 47] [who, 293].

We produce two new queries:

Q1: [who: 293] [are: 268], [you: 47] [who, 293].

Q2: [who: 293] [be: 21], [you: 47] [who, 293].

Let us consider Q1.

We can use the (you, are, who) and (you, who, who) indexes to obtain results.

6.5 All Lemma Types Appear in the Query

Let us consider the following query: "notes about Gallic war".

After lemmatization: [note: 1373] [about: 211] [gallic: ~] [war: 674].

Stop lemmas: about, war.

Frequently used lemmas: note.

Ordinary lemmas: gallic.

We select the non-stop lemma w with the lowest frequency. For the lemma w, we use the ordinary index and process the NSW records. For this example, we select "gallic".

For every other frequently used lemma v in the query, a logical expanded (w, v) index exists. In this example, the only index of this type is (note, gallic).

For every other ordinary lemma q in the query, we need to use the ordinary index q (skipping the NSW records). If another frequently used lemma p exists in the query, we can also use the expanded (p, q) index instead of the ordinary index.

6.6 Additional Examples

Consider the following query: "time and a word yes".

After lemmatization: [time: 184] [and: 28] [a: 17] [word: 602] [yes: 2375].

We can see that in our dictionary, "time", "and", "a", and "word" are all stop lemmas, whereas "yes" is a frequently used lemma. In this case, we can use the ordinary index with NSW records. We select from the ordinary index all occurrences of "yes", and for each such occurrence, we need to check the NSW record for the existence of the lemmas "time", "and", "a", and "word".

7 Search Experiment Environment

All search experiments were conducted using a collection of texts with a total size of 71.5 GB, consisting of 195 000 documents of plain text, fiction and magazine articles.

MaxDistance = 5, 7 or 9.

SWCount = 700.

FUCount = 2100.

We used the following computational resources:

CPU: Intel(R) Core(TM) i7 CPU 920 @ 2.67 GHz.

HDD: 7200 RPM.

RAM: 24 GB.

OS: Microsoft Windows 2008 R2 Enterprise.

Query selection: We selected a document from the collection. Next, we selected some words from the document. We formed a query from those words. We selected the words from different positions in the document. We evaluated the query using standard inverted indexes and our indexes to estimate the performance gain of our approach.

The experimental procedure is as follows:

(1) Selection of a random document in the index.
(2) Selection of search queries as follows.
    (a) Selection of a sequence of words. The query length is 3, 4 or 5.
    (b) Selection of a sequence of words, with the omission of every other word. The query length is 3.

Let us consider a document "Gaul, taken as a whole, is divided into three parts". We select queries "Gaul taken as", "Gaul taken as a", "Gaul taken as a whole" at 2.a. We select "Gaul as whole" at 2.b.

(c) Selection of a sequence of words, with the omission of the second word. For example, consider the query "Gaul as a whole". The query length is 3 or 4.
(d) Selection of a sequence of words, with the omission of the second and third word. For example, consider the query "Gaul a whole". The query length is 3.
(3) Search for each selected query. We evaluate the query using standard inverted indexes and our indexes. In the search, all the records corresponding to the given word are read. Thus, even if the required query is found, reading continues to the end.

Queries of three, four, or five words are selected, because *MaxDistance* = 5 in the first experiment.

However, we can perform larger queries with a larger value of *MaxDistance*. The benefits of this approach are as follows:

(1) We verify that the index is correctly constructed and performs as required. Since queries are selected from an already-indexed document, they should be precisely found. We verify that the search results include a record corresponding to the document used in selecting the query.
(2) The queries found are relatively diverse and include a large number of different words.
(3) Many of the queries include stop words and frequently encountered words.

All the queries are processed sequentially in a single program thread.

8 Search Experiments with MaxDistance = 5

Idx1: ordinary inverted file without any improvements such as NSW records (the size of Idx1 was 43.3 GB).

Idx2: our indexes, including the ordinary inverted index with NSW records and the (*w*, *v*) and (*f*, *s*, *t*) indexes, with *MaxDistance* = 5.

Queries: 5250 (519 queries consisted only of stop lemmas).

Query length: from 3 to 5 words.

Average query times:

Idx1: 13.66 s, Idx2: 0.29 s.

Average data read sizes per query:

Idx1: 468.6 MB, Idx2: 9.9 MB.

We improved the query processing time by a factor of 47.1 with Idx2, and we improved the data read size by a factor of 47.3; see Figs. 2 and 3, respectively.

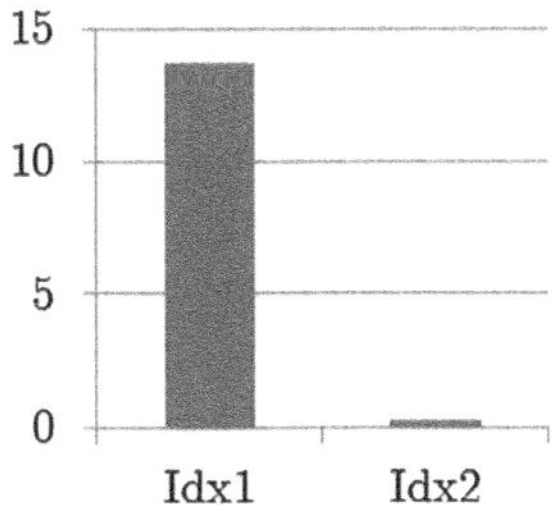

Fig. 2. Average query execution times for Idx1 and Idx2 (seconds), with MaxDistance = 5.

Let us consider Fig 2. The left-hand bar shows the average query execution time with the standard inverted indexes. The right-hand bar shows the average query execution time with our indexes. Our bar is much smaller than the other bar because our searches are very quick.

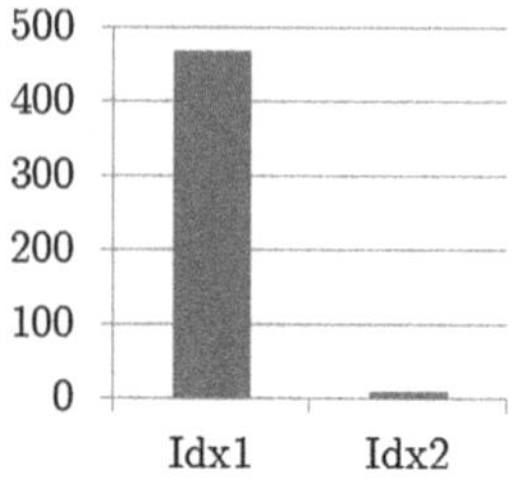

Fig. 3. Average data read sizes per query for Idx1 and Idx2 (MB), with MaxDistance = 5.

Let us consider Fig 3. The left-hand bar shows the average data read size per query with the standard inverted indexes. The right-hand bar shows the average data read size per query with our indexes. We need to read much fewer data from the disk, and our bar is much smaller than the other bar.

Index sizes:

Ordinary index with NSW records: 110 GB (the total size of the NSW records can be calculated as follows: 110 GB – 43.3 GB = 66.7 GB).

Expanded (*w*, *v*) indexes: 143 GB.

Expanded (*f*, *s*, *t*) indexes: 622 GB.

9 Search Experiments with MaxDistance = 7

Idx1: ordinary inverted file without any improvements such as NSW records.

Idx2: our indexes, including the ordinary inverted index with NSW records and the (*w*, *v*) and (*f*, *s*, *t*) indexes, with *MaxDistance* = 7 (see Fig. 4).

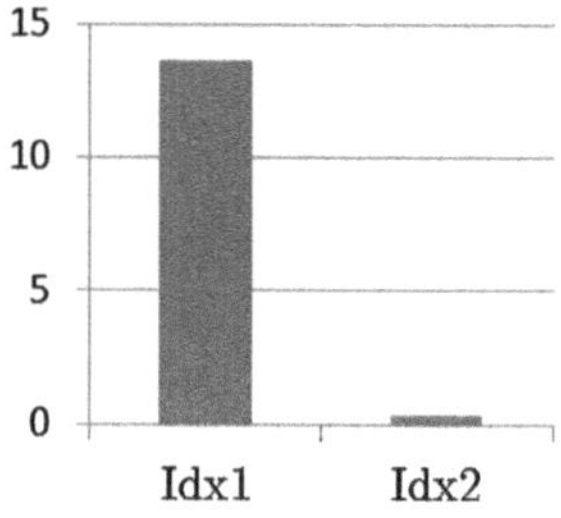

Fig. 4. Average query execution times for Idx1 and Idx2 (seconds), with MaxDistance = 7.

Queries: 5250 (519 queries consisted only of stop lemmas).

Query length: from 3 to 5 words.

Average query times:

Idx1: 13.66 s, Idx2: 0.31 s.

Average data read sizes per query:

Idx1: 468.6 MB, Idx2: 10.03 MB.

We improved the query processing time by a factor of 44 with Idx2, and we improved the data read size by a factor of 46.7.

We can see a small increase in the average query execution time in comparison with the *MaxDistance* = 5 case.

10 Search Experiments with MaxDistance = 9

Idx1: ordinary inverted file without any improvements such as NSW records.

Idx2: our indexes, including the ordinary inverted index with NSW records and the (*w*, *v*) and (*f*, *s*, *t*) indexes, with *MaxDistance* = 9 (see Fig. 5).

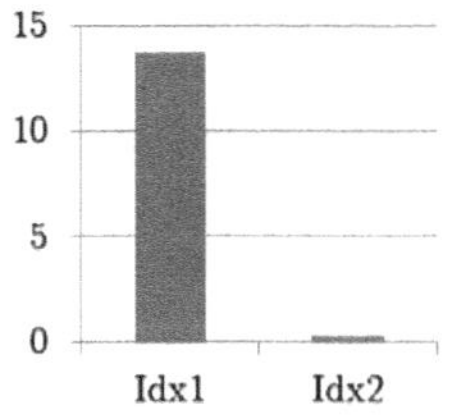

Fig. 5. Average query execution times for Idx1 and Idx2 (seconds), with MaxDistance = 9.

Queries: 5250 (519 queries consisted only of stop lemmas).

Average query times:

Idx1: 13.66 s, Idx2: 0.29 s.

Average data read sizes per query:

Idx1: 468.6 MB, Idx2: 10.236 MB.

We improved the query processing time by a factor of 47.1 with Idx2, and we improved the data read size by a factor of 45.77.

With *MaxDistance* = 9, we have the same average query execution time as with *MaxDistance* = 5. We can see a small increase in the average data read size per query in comparison with the *MaxDistance* = 5 case.

The average query execution times with the additional indexes are roughly the same with *MaxDistance* = 5, 7 and 9. The disposition of the data on the disk or some peculiarities of our index structure [13] could be sources of minor differences.

11 Other Additional Indexes and Related Work

In [6, 14, 15], nextword indexes and partial phrase indexes are introduced. These additional indexes can be used to improve performance. However, they can help only with phrase searches. Consider the text "to be or not to be". With the query "to be not to be", this text will not be found in a phrase search. Thus, our approach is more powerful.

Only phrase search is optimized in [16] as well.

In [1], only two-term queries are processed. The authors of [1] decreased the query processing time by up to a factor of 5 (Table 5-2 in [1]). By contrast, our indexes can

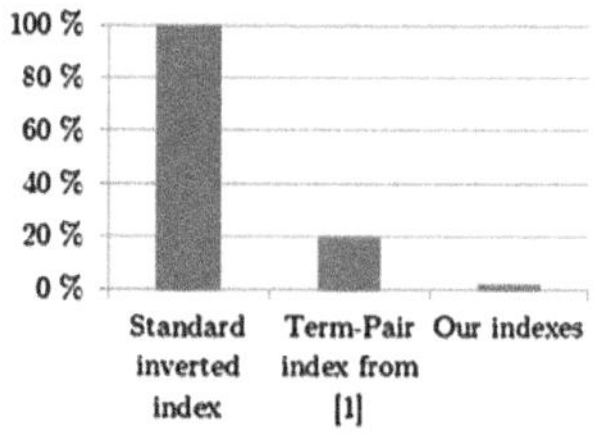

Fig. 6. Query processing time comparison with Term-Pair indexes [1].

decrease the query processing time by up to a factor of 44–47, and we support multiple-term queries. We can see this in Fig. 6.

The leftmost bar shows the average query execution time with the standard inverted indexes, normalized to 100%. The center bar shows the average execution time with term-pair indexes [1] relative to that with the standard inverted file. The rightmost bar shows the average execution time with our indexes relative to that with the standard inverted file. The rightmost bar is tiny because of our very fast searches.

12 Value of MaxDistance

The value of *MaxDistance* may be different for different types of lemmas. For example, for stop lemmas, we can use 5 or 7, whereas for frequently used lemmas, we can use 7, 9 or 11.

We can assume that for more frequently used lemmas, the importance of the semantic connections between nearby words will be high only for small distances between words. For less frequently used lemmas, the importance of semantic connections can be higher at larger distances.

Moreover, we can introduce a function *FMaxDistance*(*w*) to represent the value of *MaxDistance* for lemma *w*.

13 Conclusion and Future Work

In this paper, we have introduced several types of words and several types of additional indexes for different word types. We can use additional indexes of different types depending on the types of words contained in the search query.

A search query can contain any words, including very frequently used words.

We have also defined several types of search queries depending on the types of words they contain.

For each search query type, we have defined which types of additional indexes can be used for query execution.

We have presented the results of experiments showing that the average time of query execution with our indexes is 44–47 times less than that required when using ordinary inverted indexes.

For each word in the text, we use the additional indexes to store information about the words at distances from the given word of less than or equal to *MaxDistance* (a parameter, which can take a value of 5, 7, or even more). This information allows us to enhance the processing speed for frequently occurring words contained in the search query, such as “war”, “world”, “beautiful”, “red”, “mine”, “be”, and “who”.

We also studied the dependence of the query execution time on the value of *MaxDistance*. The results of search experiments with *MaxDistance* = 5, 7, and 9 are presented.

In future research, we wish to study optimized methods of index creation for large values of *MaxDistance*. The index building time for large values (greater than 9) of *MaxDistance* can, for now, be regarded as a limitation of our method. Moreover, it will be interesting to investigate different types of queries in more detail.

References

1. Yan, H., Shi, S., Zhang, F., Suel, T., Wen, J.R.: Efficient term proximity search with term-pair indexes. In: CIKM 2010 Proceedings of the 19th ACM International Conference on Information and Knowledge Management, Toronto, ON, Canada, 26–30 October 2010, pp. 1229–1238 (2010)
2. Buttcher, S., Clarke, C., Lushman, B.: Term proximity scoring for ad-hoc retrieval on very large text collections. In: SIGIR 2006, pp. 621–622 (2006)
3. Zobel, J., Moffat, A.: Inverted files for text search engines. ACM Comput. Surv. **38**(2) (2006). Article 6
4. Tomasic, A., Garcia-Molina, H., Shoens, K.: Incremental updates of inverted lists for text document retrieval. In: SIGMOD 1994 Proceedings of the 1994 ACM SIGMOD International Conference on Management of Data, Minneapolis, Minnesota, 24–27 May 1994, pp. 289–300 (1994)
5. Zipf, G.: Relative frequency as a determinant of phonetic change. Harvard studies in classical philology. **40**, 1–95 (1929)
6. Williams, H.E., Zobel, J., Bahle, D.: Fast phrase querying with combined indexes. ACM Trans. Inf. Syst. (TOIS) **22**(4), 573–594 (2004)
7. Schenkel, R., Broschart, A., Hwang, S., Theobald, M., Weikum, G.: Efficient text proximity search. In: String Processing and Information Retrieval, 14th International Symposium, SPIRE 2007. Lecture Notes in Computer Science, vol. 4726, Santiago de Chile, Chile, 29–31 October 2007, pp. 287–299. Springer, Heidelberg (2007)
8. Brin, S., Page, L.: The anatomy of a large-scale hypertextual web search engine. In: Proceedings of the 7th International Conference on World Wide Web (WWW 1998) (1998)
9. Veretennikov, A.B.: O poiske fraz i naborov slov v polnotekstovom indekse [About phrases search in full-text index]. Sistemy upravleniya i informatsionnye tekhnologii [Control systems and information technologies], **48**(2.1), 125–130 (2012). in Russian
10. Veretennikov, A.B.: Ispol’zovanie dopolnitel’nykh indeksov dlya bolee bystrogo polnotekstovogo poiska fraz, vklyuchayushchikh chasto vstrechayushchiesya slova [Using additional indexes for fast full-text searching phrases that contains frequently used words]. Sistemy upravleniya i informatsionnye tekhnologii [Control Systems and Information Technologies]. **52**(2), 61–66 (2013). In Russian

11. Veretennikov, A.B.: Effektivnyi polnotekstovyi poisk s ispol'zovaniem dopolnitel'nykh indeksov chasto vstrechayushchikhsya slov [Efficient full-text search by means of additional indexes of frequently used words]. Sistemy upravleniya i informatsionnye tekhnologii [Control Systems and Information Technologies]. **66**(4), 52–60 (2016). in Russian
12. Veretennikov, A.B.: Sozdanie dopolnitel'nykh indeksov dlya bolee bystrogo polnotekstovogo poiska fraz, vklyuchayushchikh chasto vstrechayushchiesya slova [Creating additional indexes for fast full-text searching phrases that contains frequently used words]. Sistemy upravleniya i informatsionnye tekhnologii [Control systems and information technologies]. **63**(1), 27–33 (2016). in Russian
13. Veretennikov, A.B.: O strukture legko obnovlyaemykh polnotekstovykh indeksov [About a structure of easy updatable full-text indexes]. Sovremennye problemy matematiki i ee prilozhenii. Trudy Mezhdunarodnoi (48-i Vserossiiskoi) molodezhnoi shkoly-konferentsii} [Proceedings of the 48th International Youth School-Conference "Modern Problems in Mathematics and its Applications"], pp. 30–41 (2017). http://ceur-ws.org/Vol-1894/
14. Bahle, D., Williams, H.E., Zobel, J.: Efficient phrase querying with an auxiliary index. In: SIGIR 2002 Proceedings of the 25th Annual International ACM SIGIR Conference on Research and Development in Information Retrieval, Tampere, Finland, 11–15 August 2002, pp. 215–221 (2002)
15. Chang, M., Poon, C.K.: Efficient phrase querying with common phrase index. In: ECIR 2006. LNCS, vol. 3936, pp. 61–71. Springer, Heidelberg (2006)
16. Gugnani, S., Roul, R.K.: Triple indexing: an efficient technique for fast phrase query evaluation. Int. J. Comput. Appl. **87**(13), 9–13 (2014)

Application of Density Clustering Algorithm Based on SNN in the Topic Analysis of Microblogging Text: A Case of Smog

Yonghe Lu(✉) and Jiayi Luo

School of Information Management, Sun Yat-Sen University, Guangzhou, China
luyonghe@mail.sysu.edu.cn, luojy25@foxmail.com

Abstract. As one of the most important livelihood events, smog is a popular keyword for discussion on social network. Researches on microblogging about smog can help us understand the impact of smog on the community, and provide guidance for enterprise and government on decision making. Because similarity of two short texts with completely different contents may occur when using the SNN similarity directly on the microblogging that has sparse characteristics, to avoid this from happening, first redefine SNN similarity by adding condition; second, take smog as an example, conduct the density clustering with the new similarity algorithm on 30,000 short texts collected on Sina social network, after which 107 clusters are obtained. The algorithm performs well on its evaluation of cohesion and separation by the silhouette coefficient, and the value is slightly higher than the traditional SNN similarity; third, extract keywords for 107 clusters and conduct manual detection on the 16 clusters selected randomly from the semantic perspective, the result shows that 14 of the 16 clusters have obvious topics and most of the clusters contain data deviating from the topics; finally, the analysis shows that the topics of the 107 clusters consist of four aspects: (1) how smog affects public health, (2) basic necessities of live, (3) the economy, and (4) decision-making of the government.

Keywords: Smog · Microblogging short text · SNN similarity
Density clustering · Topic analysis

1 Introduction

Smog, a type of air pollutant consists of sulfur dioxide, nitrogen oxide and inhalable particles, is becoming more common and severe in China over the past few years due to rapid development of industries and rising number of vehicles. It affects public health as well as other aspects of people's life, such as food, clothing, accommodation and

This research was supported by National Natural Science Foundation of China (Grant No. 71373291). This work was also supported by Science and Technology Planning Project of Guangdong Province, China (Grant No. 2015A030401037) and Science and Technology Planning Project of Guangdong Province, China (Grant No. 2016B030303003).

K. Arai et al. (Eds.): IntelliSys 2018, AISC 868, pp. 955–972, 2019.
https://doi.org/10.1007/978-3-030-01054-6_67

travel, bringing about governance problems to government departments, economic losses to the industries, but also business opportunities to enterprises at the same time.

Sina Weibo is the largest online social platform in China, on which users actively record their lives and express opinions. As an important topics of people's livelihoods, smog has become a hot topic of microblogging, as shown in Fig. 1.

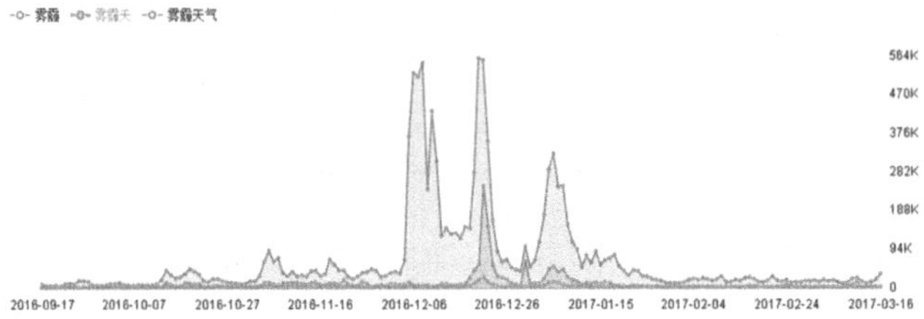

Fig. 1. The trend of smog in Sina microblogging.

In this paper, the following methods are used in order to understand the current situation of smog and the influence of smog on the society.

- Literature research: Literature research helps us generally grasp the current academic research status in the field of smog and short text clustering by analyzing the current achievements.
- Empirical research: Collecting the microblogs relevant to smog as the first-hand material for understanding the public opinion.
- Clustering analysis: Considering the characteristics of microblog, an improved SNN similarity is proposed. And density clustering based on the new similarity measure is used to analyze the microblogs of smog.
- Manual detection: Manual detection refers to the process of full or sample test of the experimental results in the form of manual without the help of statistical analysis tools. Randomly select the microblogs from the clusters obtained by random sampling and determine whether their contents match the topic of the clusters from the perspective of semantics.

The rest of the paper is organized as follows. Section 2 introduces the related work of smog and short text clustering. In Sect. 3, some related concepts and techniques are presented. Section 4 introduces the improved SNN similarity. In Sect. 5, the performed experimental study and the obtained results are discussed. Based on Sects. 5 and 6 discusses the topics of microblogs involving smog. And finally Sect. 7 presents the conclusions and highlights some directions for future work.

2 Related Work

2.1 Researches of Smog

At present, the studies of smog mainly involve its causes [1, 2], the nature [3], the harm [4, 5] and the prevention and control of smog [6, 7].

However, from literature research to empirical research, the current academic researches on smog mainly focus on the experts' point of views, interpret literature and data with professional knowledge and give suggestions from the subjective level. They haven't paid enough attention to the smog event, thus makes it difficult to fully reflect the views and attitudes of the masses, resulting in its lack of objectivity.

2.2 Similarity Measures

There are two kinds of approaches to measure the similarity based on the vector space model, one is by measuring the difference between the data, the other is measuring the similarity [8]. Euclidean distance is the most famous and widely used dissimilarity degree measure, while cosine similarity is the most commonly used similarity measure.

The literature points out that for high dimensional data, the distance between data points is very small. In the high-dimensional space, the attributes of two data objects with similar distances may be very different in some dimensions, and if these dimensions dominate the Euclidean distance measure, the similarity calculation would be interfered and cause instability in the calculation result [9]. Moreover, cosine similarity, could be unreliable when dealing with the similarity between documents due to the low similarity between high-dimensional data objects [10].

In order to deal with the problems mentioned above, a new measure method of indirect similarity called Shared Nearest Neighbor Similarity has proposed by Jarvis and Patrick, which uses the number of neighbors shared between two points that are mutual k-nearest neighbors to represent the similarity [11].

2.3 Short Text Clustering

Microblogs are short in length, arbitrary in expression and diverse in terms, and the vectors represented by the vector space model are often accompanied by high-dimensional and sparse features.

Various algorithms have been put forward to process high-dimensional data, such as SOM [12], hypergraph mapping [13], subspace algorithms [14, 15] and clustering algorithms based on SNN similarity [10]. Although the ability to process high-dimensional data is improved, these algorithms still have some problems, such as high time complexity and unstable results.

In order to solve the limitation of the original SNN algorithm, a SNN-based density clustering algorithm is proposed [16]. The use of the core point and SNN similarity enhances the capacity and flexibility of the algorithm. Based on the original algorithm, incremental SNN density-based clustering algorithm, to some extent, solves the problem of redundant computing in the case of dynamic data [17, 18]. And considering that SNN cannot process databases with mixture attributes, Jiang Shengyi has improved the original algorithm to enable it to process dataset with categorical attributes [19].

SNN similarity is simple and easy to implement, with the ability to deal with high-dimensional data and variable density clusters. The density clustering algorithm can identify clusters with different shapes and sizes and is insensitive to noise. Therefore, this paper attempts to introduce the SNN-based density clustering algorithm into the clustering of microblogging short text.

3 Related Concepts and Techniques

3.1 SNN Similarity and SNN Density

SNN similarity, also known as Shared Nearest Neighbor Similarity, measures the similarity by the number of neighbors shared by two points that are mutual k-nearest neighbors. The more neighbors the two points share, the larger the similarity is. The SNN similarity weakens the influence of the sparse degree of distribution of points in space by ignoring the distance, and its calculation method is as follows:

for $\forall x_i, x_j \in X(i = 1,2,\ldots,n \cap j = 1,2,\ldots,n, i \neq j)$,
if $x_i \in \mathrm{KNN_j}$ *and* $x_j \in \mathrm{KNN_i}$,
$\mathrm{similarity}(x_i, x_j) = \mathrm{Num}(\mathrm{KNN_i} \cap \mathrm{KNN_j})$;
else $\mathrm{similarity}(x_i, x_j) = 0$.

x_i and x_j represent two data objects in the dataset X. $\mathrm{KNN_i}$ represents the k-nearest-neighbors list of the data object x_i, and $\mathrm{Num}(\mathrm{KNN_i} \cap \mathrm{KNN_j})$ represents the number of neighbors contained in the intersection of the k-nearest-neighbors list of x_i and the k-nearest-neighbors list of x_j.

By calculating the number of shared neighbors whose SNN similarity is equal to or greater than a given threshold (EPS), we can get the SNN density, a measure of how well points are surrounded by similar points in space.

Because SNN similarity adopts the idea of neighbors, to a certain extent, it ignores the density difference caused by different distances between points. Points in high-density and low-density regions generally have relatively high SNN densities whereas points in high-low-density transition regions tend to have lower SNN density [16].

3.2 Inner Product of Vectors

The inner product of vectors, also known as dot product, refers to a binary operation that receives two vectors on a real number R and returns a real number. The inner product symbol is usually represented by "•", the formula is defined as follows:

$$X \cdot Y = \sum_{k=1}^{n} X_1 \cdot Y_1 + X_2 \cdot Y_2 + \ldots + X_n \cdot Y_n \tag{1}$$

$X = (X_1, X_2, \ldots, X_n)$ and $Y = (Y_1, Y_2, \ldots, Y_n)$ represent two vectors.

The value of inner product is between $(-\infty, \infty)$, especially when $x \cdot y = 0$, the vector x is orthogonal to the vector y.

3.3 Density-Based Clustering

Considering the number of categories to be classified and the shape and size of the clusters to be treated of microblogging about smog are unknown, the partitional clustering algorithm is not suitable. Because the data also has the characteristics of large amount and high feature dimension, the hierarchical clustering algorithm will bring a series of problems such as high experimental time complexity and large device memory overhead.

Therefore, density-based clustering algorithm is used in this paper to cluster microblogging texts. The main idea of density clustering algorithm is to find high-density regions separated by low-density regions.

(1) *DBSCAN [20]:* The most commonly used density clustering algorithm. It mainly consists of three parts: first, traversing all data objects, identifying core points and creating clusters; second, identifying boundary points and assigning boundary points to the nearest cluster; third is merging the clusters where the core points are directly density-reachable.

The DBSCAN clustering algorithm has the advantages of no need to determine the number of clusters in advance, being insensitive to noise, and the ability to handle clusters with arbitrary shape and size. But it also has some flaws such as not suitable for data with large density changes and having poor scalability of data dimensions.

(2) *Density clustering algorithm based on SNN similarity:* a new clustering algorithm that combines SNN density with the DBSCAN algorithm [16]. The algorithm uses the SNN density instead of the density of traditional DBSCAN algorithm to identify the core points, boundary points and noise points. And the traditional algorithm is used for core point fusion and boundary point assignment.

The new clustering algorithm based on SNN density is more flexible. In addition to the advantages of adaptability and robustness like DBSCAN, it also can identify clusters with different densities.

4 Improved SNN Similarity

The traditional SNN similarity reflects the local structure of points in data space, which is insensitive to density changes and spatial dimensions and is suitable for processing high-dimensional data. However, microblogging short text has the characteristics of sparse features, different from the general high-dimensional data. Direct use of SNN similarity to calculate the similarity of microblogging short text will cause calculation error. Therefore, the similarity of SNN is adjusted accordingly to give the method to calculate the similarity of microblogging short text.

4.1 Analysis

Consider the situation in Fig. 2, we can know that $a \in KNN_b, \mathrm{b} \in KNN_a$ and $Num(KNN_a \cap KNN_b) = 3$.

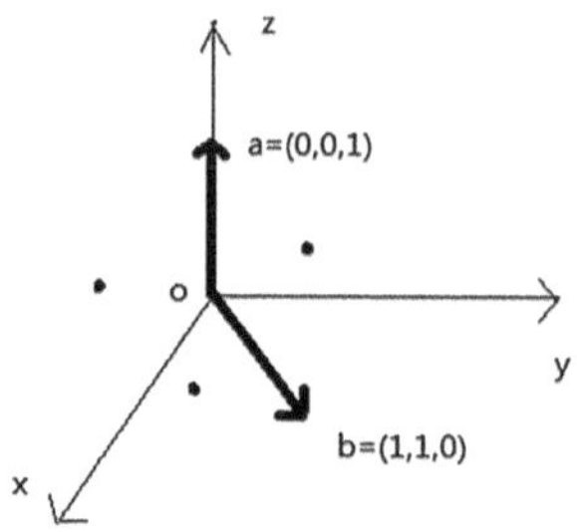

Fig. 2. The special case of SNN similarity (k >= 4).

According to the definition of SNN similarity in Sect. 3, Similarity (a, b) = 3. In fact, from the non-zero features shared by vector a and vector b, same features do not exist between vectors a and b.

Vector	Feature		
	x	y	z
a	0	0	1
b	1	1	0

According to the intuitive understanding that "the more key words two sentences share, the greater the SNN similarity between them is", there is no similarity between a and b.

If point a is the core point and the similarity between b and other neighbors is less than 3, the traditional algorithm will wrongly categorize b into the same cluster as a, while in reality they are not in the same cluster.

Due to the short length of content, the microblogging texts have fewer non-zero features for each document vector after constructing the vector representation, and the special cases above are more likely to occur when calculating SNN similarity. Therefore, it is necessary to adjust the SNN similarity calculation method in this case.

4.2 Improve Proposals

We can find that the calculation error of traditional SNN similarity in short text is due to the fact that it cannot recognize two high-dimensional sparse vectors without shared features. From a mathematical point of view, two vectors are orthogonal when there are no shared features. According to Sect. 3.2, the feature of vector orthogonal can be identified by vector inner product. Therefore, the proposed SNN similarity is calculated as follows:

for $\forall x_i, x_j \in X(i = 1,2,\ldots,n \cap j = 1,2,\ldots,n, i \neq j)$,
if $x_i \in KNN_j$ *and* $x_j \in KNN_i$ *and* $x_i.x_j \neq 0$,
similarity(x_i, x_j) = Num($KNN_i \cap KNN_j$);
else similarity (x_i, x_j) = 0.

x_i and x_j are two data objects in the dataset X. KNN_i represents the k-nearest-neighbor list of the data x_i, $x_i \cdot x_j$, represents the inner product, and $Num(KNN_i \cap KNN_j)$ represents the number of neighbors contained in the intersection of the k-nearest-neighbor list of x_i and the k-nearest neighbor list of x_j.

5 Experimental Study

In order to verify the feasibility of SNN-based density clustering in short text, an improved SNN similarity measure proposed in Sect. 4 is used to analyze the microblogging of smog. The framework of the experiment is shown in Fig. 3.

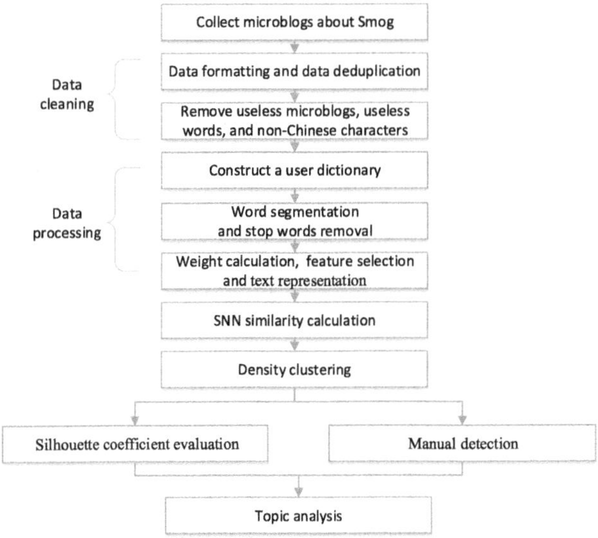

Fig. 3. The framework of experimental study.

5.1 Data Collection and Sampling

According to the micro-index of the keyword "smog", as shown in Fig. 1, the relevant microblogs are most concentrated from December 4, 2016 to January 10, 2017, and octopus collector is used for the search and acquisition of the microblogging for this period of time.

Considering large data size will result in high time cost and high requirement of the equipment, the original collected data is randomly sampled, and a total of 30000 data is taken for the subsequent experimental analysis.

5.2 Data Cleaning

Some meaningless microblog posts are deleted including voting post and non-commenting reposts. Some fixed expressions, such as "网页链接"(website link), "秒拍视频"(second shot video) and "展开全文"(full text display) are directly cleaned by the replacement function in excel. The meaningless contents in microblogs with non-fixed expression usually have a fixed representation, such as "@用户名"(@username), "#话题#"(#topic#) and "|定位"(|location). Take regular expressions to match them for cleaning.

Microblogs that contains less than or are equal to 12 words and don't convey complete semantics and clear viewpoints should be deleted as well.

5.3 Data Preprocessing

(1) *Construction of user dictionary*

Due to the colloquialization of textual expressions and the prevalence of unregistered words, it is necessary to construct a specific user dictionary for microblogs about smog before word segmentation. The new word discovery function of NLPIR2016 is used to identify new words in microblogs in this paper.

(2) *Word segmentation and stop words removal*

After the user dictionary is imported, perform the word segmentation and mark the part of speech by NLPIR2016. Then remove the stop words through part-of-speech matching. A series of high-frequency words such as "雾霾天气"(smoggy weather), "雾霾天"(smoggy day) and "雾霾"(smog) that contribute little to topic analysis are deleted. In addition, most single words that have limited expressive ability and cannot reflect the contents are also deleted.

After data processing, 25104 data are selected for final clustering analysis.

(3) *Feature selection and text representation*

This paper adopts Document frequency (DF) for feature selection, select words with document frequency greater than or equal to 50 as feature words, a total of 936 words and uses the TF-IDF for text representation.

5.4 Parameter Selection

(1) *K value for shared neighbors*

Before clustering, K- nearest neighbors of each sample should be calculated. K value - the maximum K-nearest neighbor graph (the abscissa represents the value of K, and the ordinate represents the maximum value of the K-th neighbor's distance of all the data) is used to find the optimal K value. When the distance between the maximal k-1-nearest neighbor and the maximal k-nearest neighbor is small, it can be approximately assumed that the k-1 nearest neighbors and the k-nearest neighbors of all the samples are similar. Therefore, when the value varies greatly, it can be considered as the boundary of clusters, corresponding to the point where the slope of the curve changes greatly in Fig. 4.

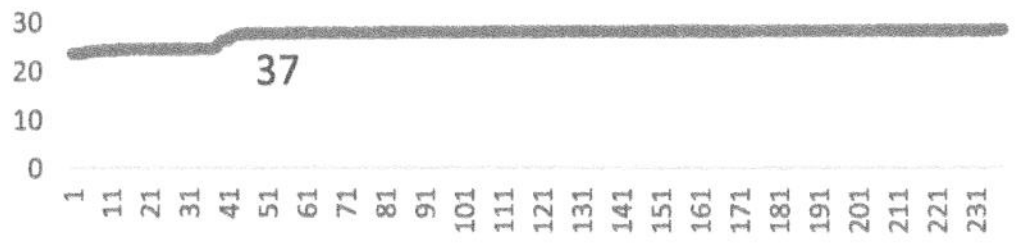

Fig. 4. K value - the maximum K-nearest neighbor graph.

(2) *EPS and MinPts*

The basic method for parameters selection of DBSCAN algorithm is to observe the distances from the point to its N nearest neighbors. For a certain n, calculate the distance between all the points and their n-nearest neighbors, sort them in ascending order and draw the sorted values. Select the distance between the rapidly changing point and its n nearest neighbor as EPS and the corresponding n as MinPts.

Parameters selection method of DBSCAN algorithm is used for selecting the parameters of the clustering algorithm in this paper. The bigger the SNN similarity is, the greater the similarity between the point and its n-nearest neighbor is, which differ from the distance. Therefore, after calculating the SNN similarity, sort them in descending order and draw the sorted values, shown in Fig. 5.

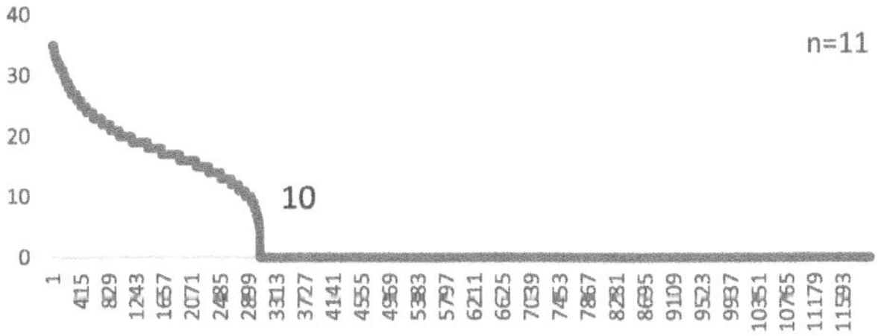

Fig. 5. Settings of EPS and MinPts (n = 11).

5.5 Clustering Results

Taking the data of smog as a sample, the improved SNN similarity proposed in Sect. 4 is used to calculate the similarity. In the experiment, considering that the vector space model ignores the correlation between words and the text similarity is generally low, the threshold of noise is set to 0 rather than EPS, which means those whose similarity with the core point is greater than 0 are considered as boundary points.

Parameters and results are shown in Table 1:

Table 1. Clustering results

Experiments	k value	MinPts/EPS	Number of clusters	Proportion of noise
SNN similarity	37	11/10	107	72.92%
Improved SNN similarity	37	11/10	107	73.06%

Compared with the traditional SNN similarity, clustering based on the improved SNN similarity can identify the special cases mentioned in Sect. 4, so the proportion of noise has slightly increased.

5.6 Evaluation

(1) *Silhouette coefficient*

The silhouette coefficient is proposed by P. Rousseeuw in 1986 [21] and the overall silhouette coefficient is defined as the arithmetic mean of the individual silhouette coefficient.

Since the Euclidean distance is unstable in the calculation of high-dimensional data, the cosine similarity instead of the European distance is taken as the measure in silhouette coefficient. Zero vectors do not have the same words with the other vectors. Even two zero vectors are likely to differ in content. And the zero vector should not be categorized into any cluster. Therefore, set the cosine of the zero vector with any vector to -1 when evaluating, and the silhouette coefficient is adjusted as follows:

$$coeff(i) = \frac{a_i - b_i}{\max(a_i, b_i)} \tag{2}$$

a_i represents the mean of cosine similarity between object i and other objects belong to the same cluster and b_i represents the maximum among the means of the cosine similarity between object i and those objects in each cluster which object i does not belong to.

The silhouette coefficient of improved SNN similarity is slightly higher than the traditional SNN similarity, as shown in Table 2. And both of the silhouette coefficient are higher than 0.8. It can be considered that the clustering algorithm based on SNN similarity performs well in high-dimensional data.

Table 2. The silhouette coefficient of the experiments

Experiments	Silhouette coefficient
SNN similarity	0.8427
Improved SNN similarity	0.8445

(2) *Manual detection*

Merge the microblogs according to the results and count the term frequency of the combined texts. After sorting the terms in descending order of the frequency, select the top 20 words as the corresponding cluster's thesaurus.

Considering the randomness, representativeness and time cost of manual detection, we randomly select 16 clusters from 107 clusters, and generalize the topics of the clusters after reading their combined texts and thesaurus, as shown in Table 3.

Table 3. Randomly selected clusters' thesaurus and themes

Cluster	Top 20 keywords (selected by term frequency)	Topics
1	严重(severe) 天气(weather) 空气(air) 可以(can) 口罩(mask) 蓝天(blue sky) 没有(no) 城市(city) 时候(when) 停课(class suspension) 学生(students) 希望(hope) 呼吸(breathe) 感觉(feel) 阳光(sunshine) 知道(know) 孩子(children) 污染(pollution) 学校(school) 生活(life)	-
9	难受(uncomfortable) 航拍(aerial photography) 鼻子(nose) 中毒(poisoning) 严重(severe) 感觉(feel) 嗓子(throat) 知道(know) 呼吸(breathe) 笼罩(shroud) 新浪(Sina) 感冒(catch a cold) 戴口罩(wear a mask) 不行(cannot) 口罩(mask) 天气(weather) 记者(journalist) 高空(high altitude) 出门(go out) 发烧(fever)	Health is affected and people feel uncomfortable.
13	清肺(clear lung) 润肺(moisten lung) 食物(food) 愉快(happy) 文章(article) 发表(publish) 可以(can) 天气(weather) 严重(severe) 山药(yam) 对抗(against) 雪梨(pear) 化痰(resolve phlegm) 功效(efficacy) 病毒(virus) 作用(effect) 银耳(tremella) 木耳(Chinese black fungus) 白萝卜(mooli) 罗汉果(luo han guo)	Foods that help clear and moisten lung.
14	分享(share) 图片(pictures) 网友(online friends) 朋友(friends) 视频(video) 蓝天(blue sky) 没有(no) 详情(details) 广告(advertising) qq(a social software in China) 文章(article) 严重(severe) 滚滚(billow) 年华(year) 重度(severe) 锦绣(beautiful) 青白(Cyan and white) 答案(answer) 转载(repost) 感觉(feel)	Sharing pictures, articles, and videos about smog.
21	限行(traffic restrictions) 文章(article) 发布(publish) 单双号(even-odd numbered license plates) 车辆(vehicles) 严重(severe) 开始(start) 天气(weather) 实行(implement) 政府(government) 污染(pollution) 影响(effect) 问题(problems) 没有(no) 时候(when) 交警(police) 应该(should) 生活(life) 持续(continue) 解除(relieve)	Discussion about traffic restrictions.
25	躲雾霾(get away from smog) 地方(places) 还有(and) 潇洒(cool) 水泥(cement) 鸟语花香(singing birds and fragrant flowers) 石灰(lime) 躲霾(get away from smog) 起来(get up) 来到(come) 游客(tourists) 彩绘(painted) 野生(wild) 没有(no) 工作室(studio) 北方(the north) 出去(go out) 孩子(children) 胃口(appetite) 紧迫感(sense of urgency)	Go travelling to get away from smog.
30	漂亮(pretty) 知道(know) 没有(no) 可能(maybe) 枫树(maple tree) 厚厚的(thick) 性子(temper) 完美(perfect) 气氛(atmosphere) 银杏(ginkgo) 高峰(peak) 喜欢(like) 浪漫(romantic) 专机(private plane) 窗户(windows) 蛐蛐(cricket) 走走(have a walk) 开战(wartime) 虔诚(devout) 生存(survival)	Description of beautiful scenery.
47	变成(turn into) 尾气(exhaust) 口罩(mask) 空气(air) 合格(qualified) 当成(regard as) 朋友(friends) 感觉(feel) 鲜花(flowers) 摆渡(ferry) 铁道(railway) 出来(come out) 电影(film) 蓝天(blue sky) 天气(weather) 持续(continue) 黑心(black-hearted) 醒来(wake up) 失眠(insomnia) 遮阳伞(parasol)	Discuss smog and the exhaust of cars.
51	怀念(miss) 蓝天(blue sky) 白云(white clouds) 阳光(sunshine) 逝去(gone) 空气(air) 天气(weather) 知道(know) 日子(days of life) 可以(can) 银杏(ginkgo) 荒漠(desert) 暖气(heating) 严重(severe) 会员(member) 感觉(feel) 失去(lose) 笼罩(shroud) 溜达(stroll) 医护(medical care)	Express nostalgia for the blue sky.
61	气温(temperature) 天气(weather) 阴天(cloudy) 小雨(light rain) 最高(highest) 最低(lowest) 预计(forecast) 冷空气(cold air) 预报(forecast) 北风(north wind) 国际(international) 持续(continue) 下降(drop) 明显(significantly) 地区(regions) 注意(pay attention to) 气象台(weather station) 雨夹雪(sleet) 气象(meteorology) 为主(mainly)	Weather forecast or weather description
62	石化(Petrochemical) 烧烤(barbecue) 北风(north wind) 过来(come) 严重(severe) 人民(people) 污染(pollution) 生活(life) pz(Pengzhou) 露天(outdoor) 关闭(close) 政府(government) 天气(weather) 国家(country) 治理(governance) 项目(project) 取缔(ban) 世界(the world) 打球(play ball games) 还有(and)	Comments on Pengzhou petrochemical projects and outdoor barbecue.
74	降落(land) 机场(airport) 航班(flight) 首都(capital) 成功(succeed) 延误(delay) 架次(the quantifier of plane) 取消(cancel) 飞机(plane) 能见度(visibility) 小时(hours) 客机(airliner) 笼罩(shroud) 航空(aviation) 抵达(arrival) 航班延误(delay of flight) 独树一帜(unique) 航班取消(cancellation of flight) 飞行员(pilots) 返航(return)	Visibility is poor and flights are affected.
77	音乐(music) 大师(master) 安好(well) 天色(sky) 娃娃(child) 歌剧(opera) 错过(miss) 持续(continue) 天气(weather) 广播(broadcast) 逃离(get away) 可怕(terrible) 奉献(contribution) 灾难(disaster) 对话(dialogue) 忧伤(sorrowful) 一生(all his life) 古典(classical) 蓝色(blue) 来自(come from)	Listening to music in smoggy days.
85	停产(stop producing) 企业(enterprise) 污染(pollution) 制药(pharmacy) 环保(environmental protection) 开工(go into operation) 地区(regions) 停课(class suspension) 减少(cut back) 全省(the whole province) 利润(profit) 业务(business) 限行(traffic restrictions) 知道(know) 公司(company) 部分(part of) 部门(department) 正常(normal) 范围(range) 公告(announcement)	Enterprises are suspended from production for air pollution.
94	经济(economic) 发展(development) 环境(surroundings) 问题(problems) 增长(increase) 治理(governance) 污染(pollution) 身体(body) 幸福(happiness) 贪污(corruption) 速度(speed) 健康(health) 知道(know) 投资(investment) 政治(politics) 关系(relationship) 大陆(mainland) 持续(continue) 发达(developed) 造谣(spread rumors)	Discussion on smog and economic development.
96	警报(alert) 红色(red) 天气(weather) 没有(no) 爆表(off the chart) 开机(turn on) 倒霉(unfortunate) 小伙伴(friends) 地区(regions) 打开(turn on) 排污(discharge of pollutants) 舍不得(reluctant) 鬼子(Japanese) 步行(walk) 封面(cover) 准备(prepare) 过分(too much) 十足(very much) 高档(top grade) 防空(air defense)	Comments about the issuing smog alerts.

As can be seen from Table 3, most of the 16 clusters obtained by random sampling are effective and have certain topics. The readability of different clusters' keywords is different. Such as clusters 9, 13, 25, 30, 51, 61, 74 and 94, the topics are clear and the thesaurus are easy to read; while the clusters 14, 62, 77 and 96 cannot express the content of original microblogs well. Among them, cluster 62 contains strong public opinions, and has high requirements for geography and current affairs background knowledge, resulting in its poor readability of keywords. Cluster 1 has an unclear topic, and there is no obvious topic of the combined text. We speculate that it may be caused by the improper fusion of the core points of different topics.

Although the most similar principle for data distribution was used while clustering, clusters of well-defined topics may still include data that are off-topic. Cluster 9 is mainly composed of microblogs describing the physical discomfort of smog days, but is mixed with a small number of microblogs about "aerial photography" according to the keywords. Cluster 51 describes the netizens' nostalgia for the weather without smog and microblogs such as "medical care" are mixed in. And cluster 85 has also been mixed with posts about class suspensions and vehicle limitation under the theme of the suspension of production of polluting enterprises.

In order to examine whether the 16 randomly selected clusters have clear topics from the semantic point of view and to check the consistency between the topics of the combined texts and the topics read from the extracted thesaurus, randomly read the microblogs in the clusters (clusters containing data over 51) or read all of them (number of microblogs in cluster does not exceed 51), and carry out manual detection based on the reading, the statistical results shown in Table 4.

Table 4. Manual detection results of the clusters selected randomly

Cluster	Documents number	Artificial evaluation		Contents of the data deviating from the topics
		Number	Topics consistency	
1	4318	50	–	–
9	108	50	80%	Microblogs of "aerial photography of smog" (雾霾航拍)
13	86	50	62%	The microblogs of "headline articles" (发表头条文章)
14	43	43	74%	Other microblogs of sharing (分享)
21	102	50	58%	The microblogs of "headline articles" (发表头条文章)
25	16	16	94%	Irrelevant microblogs that mention escaping from smog (躲雾霾)
30	15	15	80%	Other microblogs that contain the word "beautiful" (漂亮)
47	23	23	22%	–

(continued)

Table 4. *(continued)*

Cluster	Documents number	Artificial evaluation		Contents of the data deviating from the topics
		Number	Topics consistency	
51	28	28	93%	Other microblogs containing the word ——"miss" (怀念)(for people or for the smog)
61	98	50	88%	Irrelevant microblogs that mention weather
62	51	51	61%	Mainly related to the microblogs with the "smog" (雾霾) and "wind" (风)
74	21	21	90%	Irrelevant microblogs that talk about flight
77	11	11	64%	Irrelevant microblogs that talk about music
85	13	13	85%	Funny jokes or comments that contain the relevant words
94	25	25	88%	Irrelevant microblogs that mention smog and economy
96	15	15	67%	Irrelevant microblogs that mention smog alert

Note: The data that is not related to the corresponding topic of Table 3 in the manual detection is defined as the data deviating from the topic.

As can be seen from Table 4, among the 16 randomly sampled clusters, 14 clusters are more than 60% consistent with the topic, indicating that these clusters have better clustering effect and high cohesion.

Particularly, the off-topic microblogs in clusters 9, 13, and 21 are data with features of other topics, and some keywords of these off-topic microblogs appear in the thesaurus in Table 3, such as "aerial photography" of cluster 9, "article" and "publish" of clusters 13, 21. The reason maybe that the core points with different topics are mistakenly combined to form a new cluster.

Off-topic data of other clusters mostly are unrelated microblogs which contains one or several keywords. For example, the off-topic data in cluster 51 are microblogs that are nostalgic to people or smog, while the topic of the cluster is "Express the nostalgia for non-smoggy days". And both of the microblogs happen to contain the same keyword "miss". Similarly, the off-topic microblogs in cluster 62 is mainly associated with other microblogs in the cluster through the use of two keywords "smog" and "wind". The main reason for this situation is that the vector space model and SNN similarity treat the words independently and neglect the semantic relations of the words.

In addition, the manual detection has found that cluster 1 and cluster 47 have poorer topic features. The manual detection result of cluster 1 is consistent with the interpretation of the keywords, that microblogs do not have obvious topic features. It is found that the topic features of cluster 47 are poor through manual detection, and the microblogs with less prominent topic features are mixed.

In general, through manual detection we find that the clustering effect of microblogging is satisfactory, and 14 of 16 clusters have obvious topic features, while only a few clusters have unclear topics. In addition, the presence of more or less off-topic data in most clusters may be related to the setting of EPS and noise threshold during clustering and the neglect of semantic information.

6 Topic Analysis

Through the collation and interpretation of the thesaurus of 107 clusters with good readability, it can be found that the discussions about "smog" on Sina Weibo mainly focus on the influence of smog on people's health and people's daily life, the positive and negative effects of smog on economy, the government's response to smog and the impact of these measures on the daily life of the masses.

6.1 Life

(1) *Smog has an impact on the health of the masses*
In the peak season of smog, most of the relevant contents on Sina Weibo complain of physical discomfort, including uncomfortable nose and throat, poor breathing, fever, cough and so on. Related clusters are 9, 26, 27, 28, 34 and 88.

(2) *Smog affects the daily life of people in many aspects*
Clothing: People buy and wear anti-smog mask to cope with smog, and pictures of celebrities wearing these masks have become popular on the Internet, see cluster 2.
Food: Sina Weibo is filled with various food recommendations which allegedly help clear and moisten lung, and Internet users are also following these directions to make food, see cluster 13.
Accommodation: Netizens argue that schools, kindergartens and other places where students spend a lot of time in should install air-purification system to filter PM2.5, see cluster 35.
Travel: Many drivers are caught off guard by traffic restrictions based on even- and odd- numbered license plates; Expressway being closed also led to many vehicles stuck in the intersection, resulting in traffic jams; In addition, affected by the smog, the ground visibility is so poor that many netizens say they have got lost on the way, and many flights have got delayed or even canceled, see clusters 21, 37, 38, 56, 57, 66, 72 and 74.
The 25th cluster points out that many netizens have chosen to travel to other cities to get away from smog.

(3) *Smog became a joke material*
Despite causing people inconvenience and frustration, smog also derives a second identity and related events have become materials for the funny pieces of microblogging. As seen in cluster 5, a reporter mistakes an elderly man for a woman during an interview because of the poor visibility and a passer-by is scolded by square dancers for persuading them not to dance on smoggy days; In cluster 24, a traffic police comforts a girl who has run several red lights due to the poor visibility that the traffic cameras haven't recorded her violations because it was too smoggy. And tips from traffic department, such as don't get out of the car to ask for direction just in case you won't be able to find your car later, have been forwarded by netizens and become joke materials. Relevant clusters includes clusters 11, 12, 81, 103.

6.2 Economic Development

(1) *Many industries are faced with difficulties*
Due to the smog, the low visibility of ground has a serious impact on industries that rely heavily on transportation. The aviation industry is an important transportation industry. The takeoff and landing of aircrafts are all affected by the atmospheric conditions. Smog leads to the majority of flights' delays or cancelations, and causes the airlines huge economic losses, see clusters 37, 56, 66, 74. In addition, cluster 18 points out that affected by expressway shut- downs, flight delays and cancellations, the express delivery industry are experiencing problems of delays in receiving dispatching and delivery of parcels for lack of transportation capacity.

(2) *Smog relating economies have come into being*
Smog outbreak of a wide range, while affecting people's lives, also spawns a new market demand. Anti-smog masks are coming out with a variety of styles (cluster 2), Sina Weibo is full of ads about food which helps clear and moisten lung (cluster 13); cluster 35 points out that air purifiers have received widespread attention; ideas such as "getting away from smog" referred in cluster 25 also to some extent help boost the tourism industry.

6.3 Government Decision-Making

(1) *Smog has an impact on government's decision-making*
The government is the advocate, key actor and responsibility bearer of the smog governance. To alleviate further pollution to the atmosphere, many local governments have imposed the policy of traffic restrictions based on even-odd numbered license plates, see cluster 21. Polluting enterprises with backward production capacity have been ordered to stop production, see clusters 70, 85. Opencast waste incineration has been prohibited, see cluster 83. Chengdu City Management even bans more than 900 open-air barbecue stalls within its jurisdiction and harshly investigate and penalise dust pollution during constructions and the use of coal, see cluster 62.

(2) *Microblogs refracts the government's smog policy deficiencies*
For traffic restrictions based on even-odd numbered license plates, different netizens have different attitudes, most of the microblogs on Sina Weibo question this policy and express disagreement. In cluster 21 there is a post that points out people are worried and anxious about these issues on a daily basis since the policy imposed. Among them, the microblog post entitled "河北限号惹火车辆交易 石家庄二手车一车难求" (traffic restrictions promote vehicle transactions in HeiBei, Second-hand cars are now hard to get in ShiJiaZhuang) points out that the traffic restrictions are affecting the vehicle trading market, and the majority of citizens are considering buying second-hand cars and using different license plates separately to cope with the policy. Some netizens even question the validity of traffic restrictions policy. For example, a user propose that "车辆限行后空气没有改善反而加重,限行和雾霾有没有直接联系" (The air quality is not improved

but even poorer with the traffic restriction policy. Is there a direct relationship between the traffic restrictions and smog?). It can be seen that the government's policy on tackling the smog has fallen into a dilemma where it hasn't received the approval of the majority of the public and the policy implementation is also not good enough.

For Chengdu to ban open-air barbecue stalls, in cluster 62, netizens are almost "one-sided", Pengzhou Petrochemical project has become the target of public criticism. Chengdu residents believe that open-air barbecue are far less polluting to the atmosphere than Pengzhou Petrochemical project. Compared to the total banning of all outdoor barbequing, no governmental remediation measures has been taken on the matter of Pengzhou Petrochemical Project located on the upwind of Chengdu, which arouses the anger of netizens and trigger a large amount of online public opinions. The Chengdu government's move to clamp down on the smog by banning the open-air barbecue stalls has not been approved by the public, and calls for the closure of the Pengzhou Petrochemical Project among people are prominent.

7 Conclusion and Future Work

7.1 Research Conclusion

In this paper, we use the clustering algorithm based on SNN similarity to cluster the microblogging short text with the theme of "smog". The experimental results show that:

- The large-scale outbreak of smog has caused a certain amount of impact on people's physical health and daily life.
- The smog has both positive and negative effects on the economy. On one hand, smog has plagued the industries that rely heavily on transportation, such as the aviation industry and the express delivery industry, causing huge economic losses. On the other hand, at the same time, smog also induced the smog economic effects of anti-smog masks, lung clearing and moistening food, and air purifiers.
- Smog has forced the government to make policy decisions and these policies act on people's daily lives, which has led to the public opinions outburst on Sina Weibo.

After topic analysis and manual detection, it can be found from the clustering results that the microblogs about smog has the characteristics of wide contents, entertaining and fast spreading.

As an important topic of livelihood, smog has a close relationship with people's lives. Microblogs involving smog sometimes is not really a discussion of smog, but more often smog only appears as a backdrop for the post, for example: "#一日一食一记##早餐,早安#又是一个雾霾天啊早餐——有啥放啥的三明治我是不是该买个煎蛋那个模具了,这鸡蛋的形状太难看了anyway,,,,,,, 。Christmas is coming" (#Daily meal record# #Breakfast, Good Morning# It is another smoggy day. Breakfast - A sandwich made with the random ingredients that I have. Should I buy the mold for omelettes? The shape of the egg is too ugly. Anyway, Christmas is coming). In the meantime, some microblogs about smog with a funny character are spread more widely and varies in expression.

7.2 Research Deficiencies and Future Work

There are still some deficiencies in this study:

- The clustering result has a high proportion of noise, and the clustered microblogging texts account for only a small part of the data, which may be due to the following reasons: (1) K value is not properly selected, a large number of data is not the nearest K neighbors to each other. (2) The improper clustering parameters EPS and MinPts result in fewer core points and some data cannot form clusters. (3) The microblogs about smog are short in length and they often consist of a wide range of topics in which smog always act as a background element, which causes more noise data.
- In the interpretation of clustering results, keywords which the topic is inconsistent are cross-mixed in some clusters. In addition, some clusters have the similar keywords as other clusters. The reasons for the situation mentioned above may be: (1) Inappropriate selection of clustering parameters EPS has led the fusion of the irrelevant core points to form a new cluster. (2) Vector space model and SNN similarity neglect the semantic of texts and the association between words, which leads to the fusion of clusters of different topics because they shared certain keywords, and the separation of related topics with different expressions.
- In the clustering results, there are a large number of advertising microblogs, funny pieces and reposts of articles and images in which smog is only the background rather than the topic of the content. And that has certain degree of influences on the clustering results. This is due to incomplete data cleaning and the wide range of contents in the smog topic itself.

For the shortcomings of this paper, subsequent research can improve the short text clustering from three aspects: find optimal parameters, further optimize SNN similarity and increase the research in semantic level, so as to obtain a more accurate and interpretative clustering result.

Acknowledgment. This research was supported by National Natural Science Foundation of China (Grant No. 71373291). This work was also supported by Science and Technology Planning Project of Guangdong Province, China (Grant No. 2015A030401037) and Science and Technology Planning Project of Guangdong Province, China (Grant No. 2016B030303003).

References

1. Jun, S., Linli, C., Qianshan, H., Lin, S.: The changes and causes of fog and haze days in Eastern China. Acta Geogr. Sin. **05**, 533–542 (2010)
2. Mohammadi, H., Cohen, D., Babazadeh, M., et al.: The effects of atmospheric processes on tehran smog forming. Iranian J. Publ. Health (2012)

3. Zhang, X.Y., Sun, J.Y., Wang, Y.Q., et al.: Factors contributing to haze and fog in China (in Chinese). Chin. Sci. Bull. (Chin. Ver.) **58**, 1178–1187 (2013)
4. Renjie, C., Haidong, K.: Haze/Fog and human health: a literature review. Chin. J. Nat. **05**, 342–344 (2013)
5. Zhou, M., He, G., Fan, M., et al.: Smog episodes, fine particulate pollution and mortality in China. Environ. Res. **136**, 396–404 (2015)
6. Xiaoyang, S.: The importance and problems of citizen awakening under the Haze. Tendering & Purchasing Manag. (2017)
7. Chao, Y., Wenjia, L.: On the governance dilemma and countermeasures of controlling Haze. Environ. Sustain. Dev. **41**(2), 68–71 (2016)
8. He Ling, W., Lingda, C.Y.: Similarity measurement of data in high-dimensional spaces. J. Math. Pract. Theory **09**, 189–194 (2006)
9. Hinneburg, A., Aggarwal, C.C., Keim, D.A.: What is the nearest neighbor in high dimensional spaces? In: International Conference on Very Large Data Bases (2000)
10. Ertöz, L., Steinbach, M., Kumar, V.: A new shared nearest neighbor clustering algorithm and its applications. In: The Workshop on Clustering High Dimensional Data and ITS Applications at Siam International Conference on Data Mining (2002)
11. Jarvis, R.A., Patrick, E.A.: Clustering using a similarity measure based on shared near neighbors. IEEE Trans. Comput. (1973)
12. Kohonen, T.: Self-Organizing Maps; Self-Organizing Maps (2001)
13. (Sam) Han, E.-H., Karypis, G., Kumar, V., et al.: Clustering based on association rule hypergraphs. In: SIGMOD Workshop on Research Issues on Data Mining and Knowledge Discovery (1997)
14. Xia, L., Weishu, X.: Summary of subspace clustering algorithms research based on CLIQUE. Comput. Simul. (2010)
15. Huiping, C., Yu, W., Jiandong, W.: Research and advances of subspace clustering. Comput. Simul. (2007)
16. Tan, P.-N., Steinbach, M., Kumar, V.: Introduction to Data Mining: A Complete Edition. The People's Posts and Telecommunications Press (2011)
17. Singh, S., Awekar, A.: Incremental shared nearest neighbor density-based clustering. In: ACM International Conference on Information and Knowledge Management, pp. 1533–1536. ACM (2013)
18. Bhattacharjee, P., Awekar, A.: Batch incremental shared nearest neighbor density based clustering algorithm for dynamic datasets (2017)
19. Xia, L., Shengyi, J.: Improved shared nearest neighbor clustering algorithm. Comput. Eng. Appl. **47**(8), 138–142 (2011)
20. Ester, M., Kriegel, H.-P., Sander, J., et al.: A density-based algorithm for discovering clusters a density-based algorithm for discovering clusters in large spatial databases with noise (1996)
21. Rousseeuw, P.: Silhouettes: a graphical aid to the interpretation and validation of cluster analysis. J. Comput. Appl. Math. (1987)

Public Opinion Analysis of Emergency on Weibo Based on Improved CSIM: The Case of Tianjin Port Explosion

Yonghe Lu(✉), Xiaohua Liu, and Hou Zhu

School of Information Management, Sun Yat-Sen University, Guangzhou, China
luyonghe@mail.sysu.edu.cn, deelaux@foxmail.com

Abstract. Nowadays, Weibo, the most popular and commonly used microblog, has already played an important role of a public opinion field in China. Especially when emergency occurs, a public opinion storm will be raised on Weibo. An improved CSIM algorithm was put forward to help analyze the public opinions by clustering the messages. The Tianjin Port Explosion was chosen as an example and the related original posts and hot comments were collected as corpus. By analyzing the clustering result, the hot topics were identified and evolution patterns of public opinions on emergency were explored. According to the clustering result, the public first concerned about "the description of the explosion", "mourning and prayer" and "the discussion about the responsibility of media". However, in terms of quantity, the public most concerned about "the evaluation of government measures". The next was "mourning and prayer", while the third was "the description of the explosion".

Keywords: Opinion analysis · Social media · Weibo · Short text clustering CSIM algorithm · Tianjin port explosion

1 Introduction

With the rapidly growing popularity of mobile Internet, microblog has become the main priorities of public opinion field by virtue of its real-time, flexibility, sharing and interactivity among users. Among all the microblog service, Sina Microblog, which is also called Weibo, is the most popular. The 2017 third-quarter financial report of Weibo showed that the number of monthly active users has reached 376 million. And it had 165 million average daily active users in September, 2017. The opinion field formed by Weibo users' actions such as forwarding, commenting and liking has already affected the real life.

This research was supported by National Natural Science Foundation of China (Grant No. 71373291). This work also was supported by Science and Technology Planning Project of Guangdong Province, China (Grant No. 2015A030401037).

K. Arai et al. (Eds.): IntelliSys 2018, AISC 868, pp. 973–997, 2019.
https://doi.org/10.1007/978-3-030-01054-6_68

Although Weibo provides a platform for free expression, it hardly has the responsibility and capability to ensure the reality and quality of information. Especially when emergency events occur, the public's mood swings easily. Once the government has any improper performance on information release or disaster management, the anxiety and discontent will rapidly increase and widely spread through user network on Weibo. Besides, the wide spread of misleading information and provocative information on Weibo may cause the chaos and panic in the real life, or cause irretrievable damage to who involved in the event.

Only when thoroughly understand the dynamic evolution pattern of public opinion, can the government provide better services, take more proper measures and improve public satisfaction and credibility. Public opinion analysis can be a worthy supplement to event analysis and provide reference for government policy development and implementation. Weibo has become an active and important platform on social opinion displaying and transmitting, the opinion analysis on Weibo, which in emergency events in particular, have been increasingly emphasized in both commercial area and academic area.

At about 23:30 on August 12th, 2015, a catastrophic explosion occurred in a hazards warehouse which belongs to Tianjin Dongjiang Port Rui Hai International Logistics Co. Ltd. and another explosion burst out 30 s later. The accident caused great casualties and property losses. As of 9:00 on August 17th, 2015, the death toll had risen to 114, included 16 firefighters, 5 policemen, 10 other identified persons and 60 unidentified persons.[1]

Not only the scholars but also the domestic public show concern for the explosion. However, due to the absence of official information, the public cannot get the truth of the scene at the first time, and some of the rumors and donation scams spread through Weibo without any obstacle. In the next several press conferences, the government spokesman parried questions on the suffering condition and the cause of the explosion. The passive attitude provoked the public condemnation and resulted in a crisis of public opinion. So, Tianjin Port Explosion was chosen as an example.

Nowadays, the opinion analysis on Weibo using machine learning usually concentrated on six aspects, including the evolution pattern of the number of posts [1, 2], the pattern of information propagation [3–5], topic extraction [6, 7], sentiment analysis [8–10], opinion sentence extraction [11–13] and response model for the crisis of public opinion [14, 15]. Seldom researchers put attention on how the topic of public opinion change as time goes by. In order to analyze the hot topics and the topic evolution of public opinion on emergency on Weibo, an improved CSIM was proposed. Compared with CSIM (a document Clustering algorithm based on Swarm Intelligence and k-Means) put forwarded by Wu Bin et al. [16], there are three improvements, including a hybrid similarity measure function, an improved probability conversion function and novel short-term memories. Taking the case of Tianjin Port Explosion, the paper also provided a framework of opinion analysis on Weibo and discovered the topic evolution pattern of the public opinion on emergency on Weibo.

[1] Data sources of the number of casualties in Tianjin port explosion: NetEase News, http://news.163.com/15/0818/08/B19ME28L00014AED.html.

2 Related Work

Domestic and overseas scholars are both concerned about Tianjin Port Explosion and have taken up research on it. In [17], authors proposed three flaws in local governance, including insufficient supervision, unprofessional emergency measures and lacking of opinion control. In [18], author considered the government and industry are expected to pay more attention to the management of dangerous chemicals logistics. In [19], author suggested that officials should enhance their abilities of crisis public relations. In [20–23], authors focused on the roles of different media in the event and various effects they brought. At the same time, in [24–26], authors aimed to explore a better way for emergency response and emergency rescue. In [27], author examined the effect of Tianjin Port Explosion on the atmosphere based on the data detected in Korea. However, none of them analyzed the event from the perspective of the public.

Text clustering is an important method for opinion analysis. Typically, the most commonly used clustering algorithm is K-means algorithm, for its simplicity and high convergence rate. But it has two significant drawbacks. One is that the algorithm needs predefined cluster number and another is that the accuracy of clustering results depends on the initial clustering center. In other hand, ant colony algorithm is one of the most popular heuristic swarm intelligence algorithm. The basic model of the algorithm, which simulates the ants' self-organization behavior of clustering the scattered corpses in colony, was first proposed by Denueubourg et al. and was applied in robotics. Then, B Faieta and ED Lumer put forward a LF algorithm, which applied the ant colony algorithm to data analysis and knowledge discovery [28]. This model was later extended and improved by other scholars and was widely used in data analysis, especially data clustering. The ant colony algorithm has the advantages of autonomy, robustness and parallelism. However, the convergence speed is slow and it is hard to deal with the outliers.

In 2002, Wu Bin et al. presented a document clustering algorithm based on Swarm Intelligence and K-means called CSIM, which successfully solved the problems of K-means algorithm and ant colony algorithm and plus combined their merits [16]. Actually, CSIM can be seen as a two-step process. In the first step, an initial set of clusters is generated by ant colony algorithm. Automatically, the cluster number and the initial cluster center are formed. In the second step, the clustering result is optimized using K-means algorithm and the outliers can also be included in the nearest neighbor clusters.

Obviously, CSIM is extremely suitable for data clustering which the number of cluster is unknown in advance. However, messages on Weibo exists the problem of feature sparsity, nonstandard words and colloquial expressions, the original clustering methods for long text are not so applicable. To solve the problems of short-text clustering, the scholars have put forward a variety of improved algorithm based on the long text clustering algorithm aimed at different problems and demand as well as completely new method. The majority of the researches focus on solving the problem created by feature sparsity and provide solutions mainly from two perspectives. From feature selection perspective, two common methods are mining frequency itemsets [29, 30] and concept expansion using search engines [31], HowNet [32] or WordNet

[33]. And from feature extraction perspective, there are two approaches. One approach is constructing new feature vectors based on probabilistic topic model [34, 35]. Another approach is training word embeddings and extracting complex features using neural networks [36].

3 Method

In order to enhance the efficiency and accuracy of the opinion analysis of Tianjin Port Explosion on Weibo, three main improvements are made to CSIM, so that it can be more suitable for short-text clustering.

3.1 Hybrid Similarity Measure Function

To solve the problems of high dimensionality and sparsity of the space-vector matrix, digging out the potential topic of all documents and building the semantic links among them is an effective method. However, simply transfer the word space-vector to topic space-vector will decrease or even drop the information of similarity of expression form. So, a hybrid similarity measure function that considers both word vector similarity and topic vector similarity is proposed.

The hybrid similarity is defined as the mean value of word vector similarity and topic vector similarity. Document O_i is represented as word vector $O_i = (w_{i1}, w_{i2}, \ldots, w_{in})$ and topic vector $O_i = (z_{i1}, z_{i2}, \ldots, z_{im})$ from the word and semantic aspects respectively.

The formula of measuring w_{ij} is showed as (1):

$$w_{ij} = \frac{freq_{ij}}{max_l freq_{ij}} * \ln \frac{N}{n_i} \tag{1}$$

where $freq_{ij}$ denotes the term frequency of w_i in document O_j, $max_l freq_{ij}$ denotes the max term frequency w_l in document O_j, N denotes the total number of documents in the corpus, and n_i denotes number of documents where the term w_i appears.

The other hand, formula of measuring z_{ij} is showed as (2):

$$z_{ij} = p(z = z_j | \theta_i) \tag{2}$$

where θ_i denotes multinomial distribution of topics in document O_j in LDA model.

The hybrid similarity function is defined as (3):

$$sim(O_i, O_j) = \frac{1}{2} \cdot (sim_w(O_i, O_j) + sim_z(O_i, O_j)) = \frac{1}{2} \cdot \left(\frac{\sum w_{ik} w_{jk}}{\sqrt{\sum w_{ik}^2} \cdot \sqrt{\sum w_{jk}^2}} + \frac{\sum z_{it} z_{jt}}{\sqrt{\sum z_{it}^2} \cdot \sqrt{\sum z_{jt}^2}} \right) \tag{3}$$

where w_{ik} and w_{jk} respectively denotes the tf-idf of word w_k in document O_i and O_j, while z_{it} and z_{jt} respectively denotes probability of topic z_t occurring in document O_i and O_j in LDA model.

So the swarm similarity measure function is changed as (4):

$$f(O_i) = \sum_{O_j \in Neigh(r)} \left[1 - \frac{10 * (1 - sim(O_i, O_j))}{\alpha}\right] = \sum_{O_j \in Neigh(r)} \left[1 - \frac{10 * \left(1 - \frac{1}{2} \cdot \left(sim_w(O_i, O_j) + sim_z(O_i, O_j)\right)\right)}{\alpha}\right] \quad (4)$$

3.2 Improved Probability Conversion Function

In CSIM, the probability conversion function are two lines with a slope k. When the value of probability function is larger than a random probability, the ant picks up or drops the object. However, as microblog is mainly recorded in a colloquial way, that means different user express the same meaning using different words or phrases and variety of opinions are contained in one microblog, and new cyberwords are frequently used, the similarity within cluster and discrimination among clusters are not distinct as long text. If the probability conversion function of CSIM is used in microblogs clustering, it is very likely that the ant cannot pick up or drop an object for a long time, and the efficiency of the algorithm will greatly decrease.

To address this problem, the paper proposes an improved probability conversion function based on the action failed times. Variable pickFail is defined as the time of failure that the ant cannot choose an object to pick up in one iteration. And variable dropFail is defined as the time of failure that the ant cannot choose a position to drop the object in one iteration. Two thresholds are set as the demarcation point of using different picking-up probability conversion function, and the same as dropping probability conversion function. When the action failed times reaches the first threshold, the condition of picking-up or dropping an object is relaxed. A coefficient, named pickRelaxedCoef or dropRelaxedCoef will add to the slope k. And this coefficient will increase as the time of failure action increase. When the time of failure action reaches the second threshold, the ant will stop picking up an object or drop the object at current position.

For picking-up step, the picking-up probability conversion function is defined as the original algorithm as (5), and two thresholds are pf_1 and pf_2, $pf_1 < pf_2$.

$$P_{pick} = \begin{cases} 1 & , f(O_i) \le 0 \\ 1 - k \times f(O_i) & , 0 < f(O_i) \le \frac{1}{k} \\ 0 & , f(O_i) > \frac{1}{k} \end{cases} \quad (5)$$

When pickFail < pf_1, the slope k is equal to its initial value k_{init}. This period is seen as "proper picking".

When $pf_1 <$ pickFail $\leq pf_2$, the slope k depends on pickFail and is defined as (6). This period is called “relaxed-conditional picking”.

$$k_n = k_{init} + pickRelaxedCoef_n (n \geq 1) \tag{6}$$

where k_{init} is the initial value of k. k_{n-1} is increased to k_n every time when pickFail can be divided exactly by 1000. And pickRelaxedCoef is defined as (7):

$$pickRelaxedCoef_n = \frac{1 - k_{n-1}}{2} + pickRelaxedCoef_{n-1} (n \geq 1) \tag{7}$$

where $pickRelaxedCoef_0$ is set as 0.

When pickFail > pf2, the ant stops choosing new object and don’t pick up any object in this iteration.

For dropping step, the dropping probability conversion function is defined as (8), and two thresholds are df_1 and df_2, $df_1 < df_2$.

$$P_{drop} = \begin{cases} 1 & , f(O_i) \geq \frac{1}{k} \\ k \times f(O_i) & , 0 < f(O_i) < \frac{1}{k} \\ 0 & , f(O_i) \leq 0 \end{cases} \tag{8}$$

When dropFail $< df_1$, the slope k is equal to its initial value k_{init}. This period is called “proper dropping”.

When $df_1 <$ dropFail $\leq df_2$, the slope k depends on dropFail and is defined as (9). This period is called “relaxed-conditional dropping”.

$$k_n = k_{init} + dropRelaxedCoef_n (n \geq 1) \tag{9}$$

where k_{init} is the initial value of k. k_{n-1} is increased to k_n every time when dropFail can be divided exactly by 1000. And dropRelaxedCoef is defined as (10):

$$dropRelaxedCoef_n = \frac{1 - k_{n-1}}{2} + dropRelaxedCoef_{n-1} (n \geq 1) \tag{10}$$

where $dropRelaxedCoef_0$ is set as 0.

However, when dropFail $> df_2$, the ant drops the object at where it is whatever the f (O_i) is. The dropping probability conversion function is equal to 1. And this period is called “forced dropping”.

3.3 Novel Short-term Memories

In CSIM, ants randomly choose an object and move. It causes some objects may never be picked up and the convergence rate is slow. To solve this problem and improve precision and efficiency, some scholars proposed that short-term memories are used to record the ant’s action history information and guide the next choice of object or position. Among them, Chen Yunfei et al. proposed adding a global object memory and a local environment memory for each ant in DBCSI algorithm to reduce the blindness of choosing object and moving [37]. However, the improvement is limited when it applies to short text set which is high-dimensional and has low swarm similarity.

Inspired by Chen Yunfei et al., this paper introduces three novel short-term memories for each ant and another one for the ant swarm to adapt the traits of short text and the probability conversion function based on the ants' failed operations.

(1) *Memory of objects that give priority to pick up (Dominant Objects Memory):* When an ant is trying to choose an object to pick up, the objects in this memory will be top options. It contains the objects that are never picked up and were forcedly dropped. The initial state is all document objects. When an object is picked up, it's deleted from this memory. And when an object is forcedly dropped, it's added into this memory.
(2) *Memory of objects that are dropped under relaxed condition (Subdominant Objects Memory):* When Dominant Objects Memory is empty, the objects in this memory is prior to be picked up. It contains the objects that are dropped under relaxed condition. The initial state is all document objects. When an object is dropped properly, it's deleted from the memory.
(3) *Memory of positions where the ant drops an object properly (Dominant Positions Memory):* When an ant is moving to find a position for dropping the object, it firstly choose positions in this memory, where objects were once dropped properly. Such that the similar objects are more likely to be dropped at the same point and the spatial separation between clusters is improved. The initial state is empty. When an object is dropped properly at one position, the position is added into this memory.
(4) *Memory of positions where the ant swarm drop the objects properly (Swarm Position Memory):* After the iteration, all ants communicate and sum up all the positions where they have dropped the objects properly, and these positions are recorded in this memory. The objects at the same position in this memory can be viewed as members from the same cluster. Objects that are not at these positions may not be properly dropped as the limitation of max iteration times or cannot form a cluster with other objects, so they are excluded from forming the initial cluster centers for k-means.

4 Experiment

To analyze the public opinion of Tianjin Port Explosion on Weibo, the original posts and hot comments are collected using Octoparse. According to the clustering result given by improved CSIM, different point of views can be easily brought out. The framework of the analysis is shown as Fig. 1.

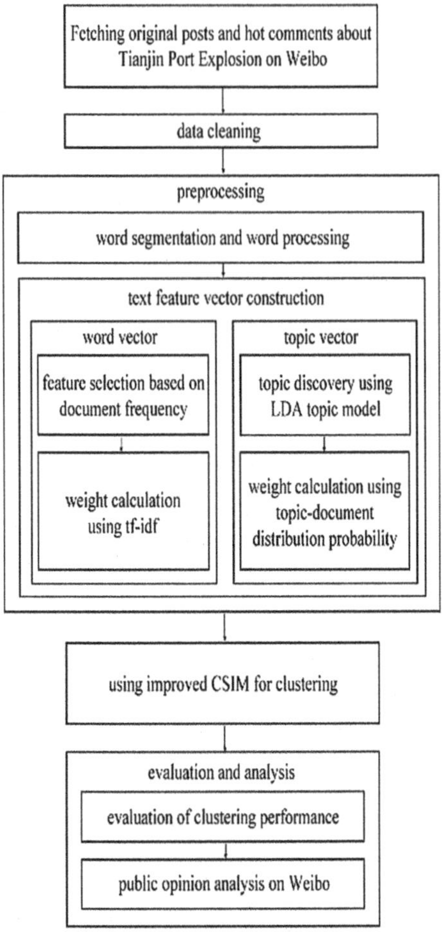

Fig. 1. The framework of the experiment.

4.1 Dataset

According to the mentioned trend of "天津港" (Tianjin Port) provided by Weibo Data Center, the Tianjin Port Explosion attached high attention during August 13th 2015 to August 21st 2015 since it occurred on August 12th 2015. As event developed and more details were known, public opinion about the event constantly changed. To ensure the public opinion's variety and integrality, the corpus includes original posts and hot comments of Tianjin Port Explosion from Weibo non-official authorized users within three months of the event occurred. Finally, 14193 records of original posts and 11681 records of hot comments were collected as raw data using Octoparse.

4.2 Data Cleaning and Preprocessing

During the process of data cleaning, the original posts and comments posted by users who have official verifications are removed, because official verified users usually publish posts about the facts or opinions on behalf of government but not the public.

When users forward or comment other messages on Weibo, share article from other application, add some emoticons or locate themselves, the message will automatically be posted with specific symbols, such as hashtag (#topic#), username (@username) and so on. However, those words have high word frequency but have nothing to do with the opinion analysis. And they will disturb the word processing and feature dimension reduction based on word frequency after word segmentation, and eventually affect the performance of clustering. So, these specific symbols should be removed from the contents. The specific symbols are shown in Table 1.

Table 1. Specific symbols on Weibo

<table>
<tr><th colspan="2">Specific symbols</th></tr>
<tr><td>#<topic>#</td><td>@<username></td></tr>
<tr><td>回复@<username>:</td><td>转发微博</td></tr>
<tr><td>对@<username>说:</td><td>我分享了@<username>的文章
|<article title>...</td></tr>
<tr><td>（分享自@<username>）</td><td>(<website name>)</td></tr>
<tr><td>分享网易<新闻 or 图集 or 跟帖>:
<article title></td><td>来自@<username></td></tr>
<tr><td>我在看<article title><website link></td><td>我正在看<article title or video title>
<website link></td></tr>
<tr><td>我点评了|<article name>…</td><td>发表文章 |<article title></td></tr>
<tr><td>我发表了文章</td><td>（下载客户端：<app name or
download link>）</td></tr>
<tr><td>.|<article title>…</td><td>http://<website link></td></tr>
<tr><td>[<emoticon>]</td><td>|<province>·<detailed location></td></tr>
</table>

The contents that the length are less than five words are often meaningless modal words or incomplete semantic sentences, and have no contribution on public opinion analysis, so they are also removed.

The tool for word segmentation and part-of-speech tagging is NLPIR/ICTCLAS 2016. Then all the pronouns, numerals, adverbs, auxiliary verbs, prepositions, conjunctions, interjections, mood words, onomatopocia, time words, locative words and punctuation as well as stop words are removed, because these words are lack of or hard to indicate practical semantics.

Some words have high word frequency because they are frequently used to described the fact of explosion, such as "天津" (Tianjin), "事故" (accident), "事件" (event), "爆炸" (explosion) and "发生" (occur). However, these words have little contribution to classify different opinions, and should be removed. Then, all records that the content has no word are removed. Finally, total 19875 texts are selected as the corpus.

Document frequency is chosen as feature selection method. The words that the document frequency is equal to or larger than 50 are selected, so each document is represented by a 921-dimension vector. Tf-idf is used for weight calculation.

JGibbLDA[2], a Java implementation of Latent Dirichlet Allocation, is used to infer latent topic structure of the event messages and comments. As the great difference in the

[2] The introduction of JGibbLDA: http://jgibblda.sourceforge.net/#3._How_to_Program_with_JGibbLDA

length of text, with about 17% of text is less than 5 words, the latent topic inference will be likely to have large deviation when using all short texts for parameter estimation. So, we choose short texts that have more than 15 words for parameter estimation and generate the LDA model, and then use this model for latent topic inference of the whole corpus. This method can get a more accurate topic probabilistic distribution.

In order to get a model that have high interpretability of each topic, low similarity between topics and large difference between the probability of document belongs to different topic, we randomly choose 1000 texts and compare the topic model generated by different combination of topic number K, hyper-parameter α and β. For the best result, the topic number K is set as 20, α is set as 0.1 and β is set as 0.2. In parameter estimation step, the number of iterations is 1600. And in inference step, the number of iterations is 1000. The probability of each topic assigned to each document is used as the weight of topic of each document as (2).

Improved CSIM is used for Tianjin Port Explosion Weibo messages clustering, where swarm similarity α is 6.5, initial slope k is 0.3, ant number is 10, max iteration time is 10000.

5 Results

Eventually, the algorithm generated 39 clusters. Table 2 shows part of the clustering results.

Table 2. Part of the clusters information, including number of messages, Top 20 keywords and description

Cluster ID	Number of messages	Top20 keywords	Description
#0	877	钱(money), 捐(donate), 没(not), 说(say), 马云(Jack Ma), 用(use), 去(go), 逼(force), 干(do), 拿(take), 道德绑架(moral coercion), 社会(society), 看(look), 妈(mother), 花(spend)	The public discussed several events about donation.
#1	1210	消防员(firefighter), 牺牲(sacrifice), 生命(life), 去(go), 原因(reason), 真相(truth), 责任(responsibility), 领导(leader), 冲(rush), 命(life), 死(die), 指挥(command), 灭火(extinguishing), 消防(firefighting), 应该(should)	The public questioned whether the scheme against fire was unscientific which resulted in so many sacrifice.
#8	2509	愿(wish), 安息(R.I.P), 平安(peace), 逝者(the dead), 祈福(bless), 希望(hope), 生命(life), 加油(fighting), 祈祷(pray), 珍惜(cherish), 生者(the living), 好(well), 坚强(strong), 知道(know), 灾难(disaster)	The public prayed for the affected people.
#10	1160	集团(corporation), 指(point), 刑事(criminal), 采取(take), 措施(measure), 官员(official), 强制(compulsion), 责任人(responsible person), 有关(involved), 责任(responsibility), 天津市(Tianjin), 追责(accountability), 管(manage), 总裁(president), 企业(enterprise)	The public urged for a complete investigation and severe punishment.

(continued)

Table 2. (*continued*)

#12	684	看(look), 希望(hope), 视频(video), 好(well), 活(live), 心痛(heartache), 哭(cry), 真(real), 逝去(die), 生命(life), 伤亡(casualties), 消防官兵(firefighter), 大(large), 泪(tear), 可爱(adorable)	The public felt shock and worried after watching the videos which were took by witnesses.
#23	552	原(former), 瑞海(Ruihai), 子(son), 李(Li), 局长(director), 股东(shareholder), 公安局长(director of Public Security Bureau), 公安局(Public Security Bureau), 物流(logistics), 公司(corporation), 企业(enterprise), 背景(background), 名字(name), 李亮(Li Liang), 系(is)	The public suspected that Ruihai International Logistics Company may have powerful connections to some government officials.
#24	404	建(build), 刻(engrave), 纪念碑(monument), 名字(name), 立(set up), 上(on), 纪念(commemorate), 责任人(responsible person), 写(write), 应该(should), 拍摄(shooting), 现场(scene), 抢险(rescue), 下(under), 幸存者(survivor)	The public satirized that the names of those responsible should be engraved on the monument which was going to set up in the ecological park.
#37	306	奖励(bonus), 签(sign), 协议(agreement), 赔偿(compensation), 说(say), 内容(content), 用(use), 合理(reasonable), 炸(explode), 家(house), 没(not), 意思(mean), 叫(call), 逼(force), 好(well)	The public discussed about the practice that the government offered two thousand bonuses to those who accepted the indemnity agreement the early

Silhouette Coefficient is commonly used to evaluate the clustering result. However, as the texts are represented by both word vector and topic vector, it's hard to measure the distances between texts. So the distance between two documents is replaced for the similarity between them, and the Silhouette Coefficient formula is converted as follows:

$$S(i) = \frac{simA(i) - simB(i)}{\max\{simA(i),\ simB(i)\}} \tag{11}$$

where simA(i) represents the mean of similarity of object i and other objects which belong to the same cluster, and simB(i) represents the mean of similarity of object i and that not belong to the same cluster.

The value of the variant of Silhouette Coefficient is 0.822, shows that the algorithm has good clustering performance.

6 Discussion

6.1 General Opinion Analysis of Tianjin Port Explosion

After analyzing the clustering results, we found that the public concerns can be divided into five types: the description and discussion of the explosion (including the scene of the explosion and rescue, the effect of explosion and so on), prayer and mourning, the evaluation of government measures (including the rescue scheme, information disclosure, causal investigation, compensation scheme, investigation of accountability and so on), the discussion on the responsibility of media and the discussion on other related events. The information of each topic is shown in Table 3.

Table 3. The information of five major topics

Major topic	Cluster ID	Top 15 words	Frequency of messages	Proportion
the description of the explosion	#6, #7, #15, #19, #25, #27, #28, #29, #30, #31, #32, #33, #36	仓库(warehouse), 现场(scene), 危险品(dangerous material), 公司(corporation), 瑞海(Ruihai), 大(large), 失联(missing), 人员(staff), 火灾(fire), 中心(center), 核心区(core area), 小(small), 死亡(death), 公安局(Public Security Bureau), 救援(rescue)	4092	20.59%
mourning and prayer	#8, #12, #17, #20	愿(wish), 英雄(hero), 安息(R.I.P.),平安(peace), 逝者(the dead), 希望(hope), 走好(along the way good to heaven), 加油(fighting), 祈福(bless), 生命(life), 好(well), 康复(recover), 祈祷(pray), 致敬(salute), 看(look)	4289	21.58%
the evaluation of government measures	#1, #3, #4, #5, #9, #10, #14, #21, #22, #23, #24, #26, #34, #35, #37, #38	说(say), 好(good), 去(good), 政府(government), 大(large), 看(look), 死(die), 责任(responsibility), 真相(truth), 消防员(firefighter), 知道(know), 想(think), 企业(enterprise), 没有(no), 瑞海(Ruihai)	8905	44.80%
the discussion on the responsibility of media	#2, #13, #18	牺牲(sacrifice), 消防员(firefighter), 消防(firefighting), 想(think), 问责(accountability), 希望(hope), 消防战士(firefighter), 人员(staff), 知道(know), 说(say), 致敬(salute), 没有(no), 大(large), 事情(event), 看(look)	1156	5.82%
the discussion on other related events	#0, #11, #16	捐(donate), 钱(money), 捐款(donation), 说(say), 事(event), 马云(Jack Ma), 去(go), 逼(force), 没(no), 用(use), 干(do), 道德绑架(moral coercion), 社会(society), 想(want), 拿(take)	1433	7.21%

The public discussion about the event mostly concentrated in the first ten days (see Fig. 2). The topic of "the evaluation of government measures" gain the most discussion and the topic of "mourning and prayer" came second. As shown in Fig. 2, on the second day of the explosion, the discussion on "the description of the explosion", "mourning and prayer" and "the discussion on the responsibility of media" reached the peak. And the number of the messages on that day were 677, 1561 and 214 respectively. The discussion frequency on "the evaluation of government measures" reached the highest point (1005 messages) on the third day. It shows that the public firstly concerned about the explosion scene and the victims and rescue workers then they began to focus on the government.

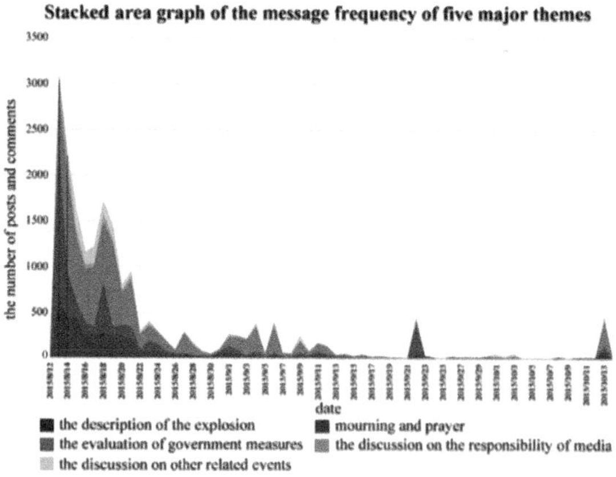

Fig. 2. The stacked area graph of the message frequency of five major themes.

6.2 Detailed Description and Analysis

(1) *The description of the explosion*

Messages of this topic are mainly about the information and description of the explosion scene, the information of donation, traffic and rescue, the information of missing person and the effect of the explosion. And Fig. 3 shows the evolving process of messages of this topic across time.

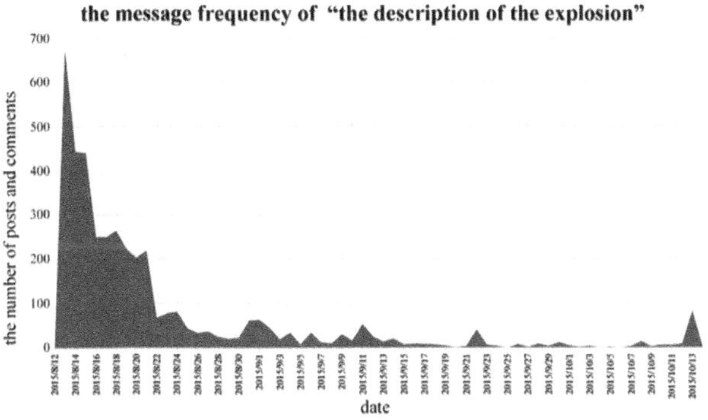

Fig. 3. The area graph of the message frequency of "the description of the explosion".

Several clusters show that when the emergency occurred, not only the verified media users become the information sources that provide real process, ordinary users who witnessed the event can also provide important information, sometimes even more instant and detailed. Meanwhile, Weibo users, especially famous personal users who are active in Weibo and have massive followers, spontaneously posted or forwarded messages about the event progression and some of the news, enable the information to reach a wider scope in a faster speed. These reflect that people is eager to keep a close eye on the emergency and become an information conveyor through Weibo.

Some users on the spot posted the information about needs of blood of certain type and materials on Weibo, so that the donation of blood and materials can be more targeted and will not be wasted. In addition, several users posted the information of traffic dispersion and the situation of receiving and curing the wounded in order to help the disaster relief officials enhance coordination on assistance.

Otherwise, some users posted information of their missing relatives and friends, and hoped to contact with them rely on the relationship chains on Weibo. The majority of personal users are willing to lend a hand and exactly speed up the process of finding the lost contact person.

The public also concerned about the negative effect of the explosion. Some of them were worried about the environmental problems that cyanide in the explosive may cause air pollution and water pollution. Although @天津发布 (the Weibo of the Information Office of Tianjin Municipal People's Government) published an article which showed that Tianjin Municipal Environmental Monitoring Center had finished the emergency monitoring near the exploded spot and six conventional Indices were all normal, most of the users doubted the truth of the result. Additionally, @天津发布 regularly published the new environmental monitoring report, but seldom people forward or comment it. It indicates that the majority of the users may not believe in the government data. The Weibo users were also focus on economic fields. Cluster no. 6 and no. 7 are mainly related to the trend of port shares and futures, as well as the effect on warehousing, logistics and transportation industry at Tianjin Port and surrounding port. And cluster no. 30 is concerned about the damages of the cars that parked near the explosive spot and houses nearby. To sum up, when talking about the effect of an emergency the public will mainly focus on the fields that related to their health and wealth.

(2) *Mourning and prayer*

The evolving process of messages of this topic across time is shown as Fig. 4. Through the messages and videos on Weibo from survivors and media, netizens felt the fear and sorrow as the persons at site did. The casualties and residents lived near the chemical plant affected the hearts of people. So, the majority of users prayed for Tianjin as well as the living in the explosion and mourned the death via Weibo, which is free from time and place restrictions, hoping that the messages can encourage the affected people. In the meantime, because of the sudden and unpredictability of disaster, some people made an appeal for treasuring the present and cherishing the people around. In Chinese tradition, the seventh day after death is an important day of mourning. So, the number of messages about mourning the victims especially rose on August 18th, 2015. According to the result, mourning and prayer were concentrated in the second and the seventh day after the disaster occurred.

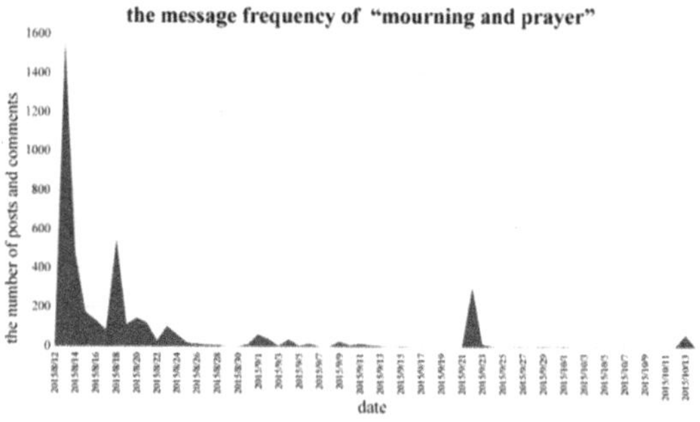

Fig. 4. The area graph of the message frequency of "mourning and prayer".

In Tianjin Port Explosion, the fact that fire fighters rescued casualties in chemical exploded area without regard to their own safety made them the most admirable group. As some people were deeply moved by the selfless spirit, they saluted and thanked the firefighters and volunteers for their heroism via Weibo messages and pictures. From the Fig. 4, there was a small peak on September 22nd. It was because a firefighter called Chaofang Zhang eventually woke up, and he had been in a coma by the explosion for more than a month. So, netizens sent their blessing and hoped that the hero would get well soon.

(3) *The evaluation of government measures*

The topic which has the most public concern and discussion mainly includes the evaluation of government measures in the aspect of information disclosure, rescue, investigation on the cause and related personnel, accountability and compensation. And Fig. 5 shows the evolving process of messages of this topic across time.

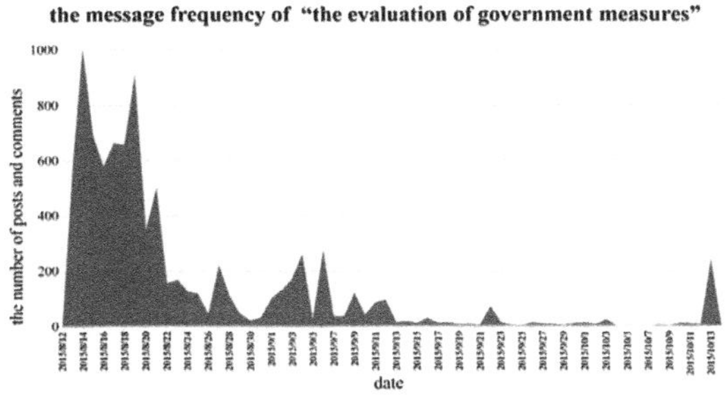

Fig. 5. The area graph of the message frequency of "the evaluation of government measures".

As it can be seen from Fig. 6, "cause investigation" is the subtopic of most concern to the public. The subtopics of "information disclosure" and "rescue" peaked on the third day while "cause investigation" peaked on the fourth day. Then the public focus turned to "accountability" which reached the peak on the eighth day. And on September 4th, the public discussion on "compensation" raised to the highest point.

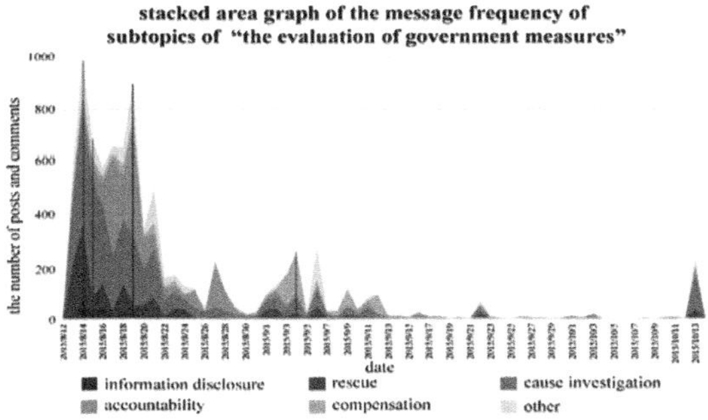

Fig. 6. The stacked area graph of the message frequency of subtopics of "the evaluation of government measures".

After the Tianjin Port explosion occurred, the local media @天津日报 (the official Weibo of Tianjin Daily) posted the first related message at 00:09 on April 13th. However, the message contained nothing about the accurate current situation of the explosion. It was not until 02:44 that @平安天津 (the official Weibo of Tianjin Public Security Bureau) published an article about the explosion time, explosion site and the progress of fire-extinguishing and the public finally knew the basic situation of the accident. Surprisingly, @天津发布 (the official Weibo of the Information Office of Tianjin Municipal People's Government) posted the message about the explosion at 03:52. The public thought it should have firstly released the accurate information, so they considered that the government had tried to cover up the accident at first.

In recent decade, some local government tried to deal with the heavy accidents by blackout and covering up the truth, which lead to the public distrust and discontent. According to the clustering result, the public demand for information disclosure was also significant in this event, and the request for disclosure in a short time was high. They urged the government to report the fire source and other explosive chemicals in the scene, the casualty toll, environmental monitoring results and so on the next day of explosion. A small group of users argued that the government could not finish so much work so soon, but the majority thought it was because of the government inaction and unreasonable shirking among government departments.

However, when the government released the information of the explosion, some people refused to believe it. Although @天津发布 updated the report of the newest death toll at 12:00, 18:00 and 21:00 on April 13th, part of the netizens were still not satisfied with the reports. They considered that the death toll should have been much higher, according to the large number of firefighters involved in the rescue. Conversely, another part of the citizens argued that it was a conspiracy theory that the government concealed the toll and explained that the missing person was not included in the death toll. The argument, which obtained the most concern and discussion in this topic, reveals that the deep distrust of the government existed in the society and it will influence the rational judgment of government behavior to some extent.

As for the rescue measure, because the large number of dangerous chemicals stored near the fire site exist explosion or leakage risks, the public's attention was fixed on the fire-extinguishing scheme. The public were worried about the safety of the firefighters and residents nearby, so the doubts from the public lay in three aspects: whether the fire-extinguishing method aiming at the unidentified chemicals was scientific, whether the firefighters were well equipped with chemical protective equipment and whether there were firefighters died in vain because of the unreasonable command. It was because several wide-spread messages said that the firefighters were not told the chemicals couldn't be put out by water at first. Additionally, in the second press conference on August 14th, the government officials said the type and the number of the chemicals in explosive area remained unclear. So, the netizens blamed the government for unscientific command, inadequate investigation and insufficient supervision.

The next is about cause investigation. The messages show that the public concerned about not only the immediate cause of the explosion, but also the underlying problems. Media reported that the qualification of dangerous materials service of Ruihai International Logistics Company had already lapsed. So, the public pointed the finger at the government for such supervision loopholes. On the other hand, as several Weibo users founded that the residential area seemed too close to the explosion spot from the videos and plus the media revealed the "red top mediation" phenomenon behind the management of Tianjin Port, the public accused the government of the impropriety and irrationality of the approval of the location planning and safety assessment of the company. "Red top mediation" means the government officials make an invisible profit from various procedures, qualification inspecting, certification and so on. Besides, some netizens were afraid that the government would finally prevaricate over the cause investigation because the public attention of the event and the public pressure on the government would gradually decline as time goes. So, they urged the government to complete the investigation and give the public especially the victims an account of the explosion. From Fig. 5, there is a small peak on October 13th about discussion of government work. It was because a warehouse exploded in Tianjin because of the leak of ethanol material. The public thought that the inspection and rectification of government had seldom effect. In addition, it has been nearly two months from the explosion occurred that day, but the accident investigation report was still not published. So, the public was disappointed at the efficiency and ability of government, although according to the "Regulations on Report and Investigation of Production Safety Accident" promulgated by the State Council, the maximum time of investigation of major accident is actually 6 months. In summary, the public has formed the consciousness of delving further into the underlying reason when the major accident occurs. They explore the potential problems which exist in a company or even an industry through the occurrence of a specific event, and press the government on social platform to solve these problems efficiently and fairly. They sincerely hope that the government and the company can learn from the disaster and avoid any recurrence.

Another subtopic which also gets considerable concern is accountability. The messages show that when talking about responsibility, the discussion mainly focus on the responsibility of the government, of which the concern is much higher than the responsibility of the company. And when referring to the irresponsibility of government, people generally think of corruption. Nowadays, the public is extremely sensitive to corruption, because several vicious events originated from collusion between government and businessmen hardly get impartial trial and the grave consequences are usually hard to restore. Fortunately, in recent years, the effort of anti-corruption is increasingly strengthened. And a few cases that the public successfully fight against corruption on Weibo greatly foster the enthusiasm for anti-corruption of the public. According to the report of media, several senior executives of Ruihai International Logistics Company are partly related to specific government officials. For example, Shexuan Dong, one of the actual stockholders of the company, is the son of Peijun Dong, the former director of Tianjin Port Public Security Bureau. However, the public considered Shexuan Dong was just a scapegoat, as his father had died of sickness and he had the least power and status among senior executives. They thought that Liang Li, the chairman of the company, had powerful connections, although the news said that he was a nominal shareholder and his parent were just grass-roots civil servants. The public speculated that some of the senior government officials were involved in corruption. They also insisted that the subjects of corruption inquiry should not be limited to the junior official. And they required that the government should thoroughly dig out the potential interest relationship chain. As the accident caused so much loss and so many casualties, the public desire of giving severely punishment to those who were responsible for the accident was significantly strong.

The last major subtopic is related to compensation. In an interview, some victims who did not accept the indemnity agreement expressed their grievances, because people blamed them for taking advantage of national calamity for profits. However, the netizens disagreed on the acceptability of the compensation. Another discussion is about the news that the government offered two thousand bonuses to those who accepted the indemnity agreement early. It is a general measure that the government reward the people who sign the agreement early, in order to decrease the situation that the victim make an unreasonable demand of compensation and reject to sign the agreement. Some Weibo users has misunderstood that the government referred to compensation as "bonus", so they felt really angry about it. It reflected that these people actually neither concern about the victims nor care about the news carefully, but probably look for a chance to vent their anger and dissatisfaction towards the government. On the other hand, some people queried the rationality of the bonus, because there is possibility that the government induces the victims to accept the unreasonable compensation scheme. Besides of two discussions above, some users thought that the government should not spend taxpayers' money in paying for the loss. Instead, Ruihai International Logistics Company ought to take the responsibility of compensation.

Besides of the five major aspects mentioned above, other government measures also attached public attention. Cluster no. 24 discussion about the government determined to build a harbor ecological park with a monument on the site of the explosion spot. This measure seemed that the government want to comfort the public and give a solution of improving the environment. However, some netizens did not buy it. They thought investigation and accountability are the most important tasks for government, or the monument was meaningless. Therefore, only when the public see the substantial effort and clear result of cause investigation and accountability, can the government be recognized and trusted by the society.

Additionally, the content of two other clusters were worth attention. Cluster no. 21 is about the discussion of the news that small animals were put into the explosion spot for testing whether the environment was suitable for living. The majority of the messages are not relevant to the rational discussion of the feasibility and scientificalness of the practice. They put forward an extreme idea that the poor animals should be replaced by the leaders of the government. The messages in cluster no. 22 showed that part of users suggested that the urban management staffs should enter the explosion site and put out the fire instead of the firefighters because they are "efficient and brave" in driving away the street vendors and confiscating the goods violently and people will not regret for their death. These two clusters announced that some people seized the chance to vent the anger that had accumulated bit by bit towards the leadership and urban management staffs.

In general, the public has brought forward more and higher request for government implementation. On the one hand, the public attaches great attention on government behaviors. But in fact, the majority of people do not familiar with the detailed government work process and task assignment. In essence, the accident site was damaged seriously and the evidence collection was hard, so the investigation was relatively hard to conduct. However, many of the public mainly focus on the final result of the investigation and accountability and constantly urge the government to report the result without considering the difficulty of carrying out investigation and obtaining evidence as well as the limit of government efficiency. The few ways to know the progress of government work are press conference and the information that government issued on their website or official Weibo. But as it can be seen from the messages, the public pay little attention to the information delivered by the government, or even distrust it. It results in the vicious circle that the barrier of communication and understanding between government and the public raise the distrust which may in turn increase the barrier. In the context of low government credibility, every minor neglect or mistake in government practices can provoke a strong public complaint and plus further aggravate the distrust of the government. On the other hand, the increasing accumulation of discontent with government decisions and the behaviors of some officials in daily life will seriously affect the judgment and evaluation of government practices. The negative and extreme emotion of government or certain government officials will be let off with Weibo message by the chance of the occurrence of major accident.

(4) *The discussion on the responsibility of media*

Messages of these topic shows that when government credibility decreases, the public relies more on the media than the government to get information and find out the truth. But some of the messages shows that the public distrust of the media emerged because of the media's irresponsible behaviors. And Fig. 7 shows the evolving process of messages of this topic across time.

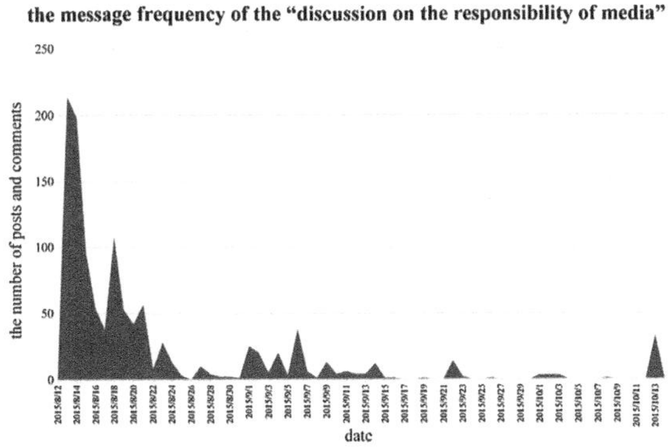

Fig. 7. The area graph of the message frequency of "discussion on the responsibility of media".

As the local government was absent from their duty of information disclosure, the public tended to ask the mainstream media for reliable information. According to the amount of forwarding and commenting of official verified users, people preferred forwarding or commenting the messages from Weibo users of mainstream media than that of government. Besides, the public were willing to believe in the investigation and information provided by the media. As it can be seen from the result of the first topic, the public were willing to share the news or articles on the topic they were concerned about from media users' Weibos or news portals and plus added their own opinions sometimes. Besides, when some rumor about exploded spot, casualties and dangerous chemicals have caused public panic, they forwarded or copied the science popularization or clarification written by the media users, such as @环球时报 (Global Times), @人民日报 (People's Daily), @头条新闻 (NewsHead), and helped prevent the rumor from further spreading.

However, some behaviors of several media have raised public discontent and distrust. During the investigation phase, some of the videos of explosive spot were blocked and deleted by Sina which caused the users' dissatisfaction. The public complained that every time the severe accident occurred, Sina would delete the videos of the scene in order to prevent them from spreading wider, and further block the truth and control over public opinions. They thought that Weibo was controlled by government, which made them angry and disappointed. The public accused that some of the media release false information that the firefighters who were the first batch to be sent into the explosion spot were temporary workers. Because the problems and events about temporary workers always receive widespread attention in the last few years, for the sake of attracting more attention, some media determined to publish the controversial news without any verification. In fact, those firefighters were belonged to the fire detachment of Tianjin Port Public Security Bureau and they were fire forces of the Tianjin Port Group. Although, they were not authorized staff of government, they were contract fire fighters not temporary workers. Besides, some news reported that the firefighters of Tianjin Port Group may use an unscientific fire-extinguishing method at the beginning and cause more explosions because they were lack of knowledge and technique of firefighting. However, they admitted that they had restriction on physical strength, fire-extinguishing skills and experience compared to the formal firefighters from Fire Department of Tianjin Public Security Bureau, but they had training routine every weekday including physical training and fire extinguishing knowledge learning. And they declared that they were not told of the fact that fire site had stored the explosive chemicals. Consequently, some people blamed the media for helping the government to divert the public's attention by news, as it seemed like the relevant department tried to put the blame onto firefighters for the inefficiency of fire-extinguishing. The netizens commended that the firefighters who took part in the recue were all heroes regardless of whether they were authorized staff of government or not and the government should not pass the buck.

In summary, the public prefers to trust the media rather than the government. However, they are also disgusted that some media deliberately uses the inflammatory words or even distorts the truth for chasing attention and benefit in recent years. And they sometimes query whether some media is under the government control and help the government block messages. The messages in this clusters fully revealed the public's ambivalence about media and their sincere hope that the media can provide objective and reliable information in Tianjin Port explosion event.

(5) *The discussion on other related events*

The last topic is about other relevant events of Tianjin Port Explosion. Cluster no. 11 reflected the public fears about the storage and management of chemicals or other explosive materials, as they associated the Tianjin Port Explosion with several explosions in 2015 such as the explosion of a petrochemical company in Shandong province on April 21st, 2015. The evolving process of messages of this topic across time is shown as Fig. 8.

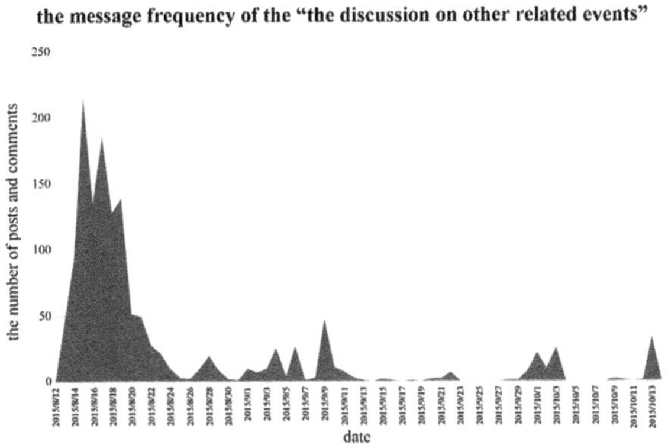

Fig. 8. The area graph of the message frequency of "the discussion on other related events".

Another two concerned subtopics are about donation in the explosion. The first subtopic was some Weibo users tried to force Jack Ma, the chairman of Alibaba Group, to donate money on Weibo, but other users expressed their incomprehension and disagreement on this behavior. Actually, it was not the first time that extortive donation occurred on celebrities in business and entertainment circles, especially when giant disaster happened. Some people criticized those who urged Jack Ma to donate and argued that donation ought to be a person free choice, and not depend on the income level. Besides, some people pointed out that Jack Ma, the "Most Generous Chinese in 2014" according to Hurun Philanthropy List 2014, had supported charity for many times, but conversely, those who supported extortive donation may even never donate anything.

The second subtopic was about an event that a Weibo user called @我的心永远属于拜仁慕尼黑 always made up a story that her father was died in the explosion and her mother had died a year ago, and posted an article on Weibo to cheat other users out of their money. The nineteen-year-old girl has received totally 96,578.44 yuan of donation through the function of "giving a tip" of Weibo article. The public was shocked and angry at the truth and scolded the girl for swindling taking advantage of people's kindness and generosity by the chance of disaster and they hoped the girl can be severely punished by the law. Additionally, it rose the introspection of the reliability and safety of one-to-one donation.

Moreover, some people wanted to know the trace of donations and were afraid that the staff of Red Cross Society of China abuses the funding again. It shows that the corruption event of Red Cross Society of China has a long-term bad influence on its image and reputation.

Actually, charities in China have constantly involved in scandals, and they have gradually lost the public trust. More people are willing to send their donations directly to those in need. Both Weibo and WeChat have provided access for the public to one-to-one donation. However, the information of asking for help on the internet is hard to judge its truth. So, the news that someone makes up a sad story on the Internet and cheats the public for donation have gradually increased. The risks of donation make people hesitate when they choose the objects and channels, or even drop the donation plan. Besides, some people misunderstand the meaning and significance of donation. To some extent, extortive donation is originated in the hostility to the rich. And it is an unreasonable appeal for personal moral obligation. Additionally, the discussion about extortive donation partly reflect the people's thought for the relation between one's personal wealth and his social responsibility he need to take.

7 Conclusion

7.1 Summary of the Research

Weibo text analysis is a development trend for public opinion analysis, and the primary technological foundation of Weibo text analysis is short text clustering. This paper applied CSIM algorithm to short text clustering, and several improvements were made according to the feature of short text for better results. We applied the improved CSIM algorithm to analyze the public opinion of Tianjin Port Explosion. Through the analysis of Weibo clustering result, we reassessed the tragedy from the perspective of public opinion and supplement the analysis of Tianjin Port Explosion. In addition, the hot topics and topic evolution pattern of public opinion on emergency were revealed.

According to the clustering result, the number of messages about mourning and prayer, the description of the explosion as well as the responsibility of media reached the peak on the second day after the explosion. Then, on the third day, the public shifted their focus to the government measures, which attracted the most attention as a whole. Among the government measures, the public pay most attention to cause investigation, accountability and information disclosure. The discussion quantity about other related event reached its highest point on the fourth day.

Viewing from the contents of the messages, when an emergency occurred, the public firstly feel panic and worried. Then the public sentiment shifts to cherishing, praying and encouragement. They also keep a close eye on the scene situation and the progress of the rescue. Afterwards, the public turn their focus to the problem of responsibility. They mainly blame the government for incompetence and ignorance of daily management and supervision. Meanwhile, they make stricter requirements to the efficiency of major accident handling. Additionally, because of the distrust of the government and the officials, some people tend to ignore or negate their effort. And the accumulated discontent with the government and officials broke out through Weibo messages and comments by the chance of the emergency. So, daily practices and emergency handling are both of great importance to improve government credibility and public satisfaction. Besides, the public attitude towards the media is rather ambivalent. On the one hand, some believe the media will provide more reliable information than the government. But on the other hand, some doubt the objectivity and impartiality of the information because some media have report misleading news or provocative information before for chasing the economic benefit or attracting public attention.

7.2 Weakness and Future Work

However, there are still some limitations in our research. On one hand, more cases of emergency should be analyzed to improve the topic evolution pattern. On the other hand, some categories cross mutually and the cohesion of clustering was a little bit low so that the analysis according to the clustering result may be affected. There are several reasons as follow. First, the Chinese word segmentation is not so accurate because the Weibo messages contain a large amount of incorrect words and informal words, although the new word discovery has been done. Second, after word segmentation and word feature dimension reduction, the relationship between words and semantics are destroyed. For example, two Weibo messages with different meaning may have the same word vector because they both contain some low frequency words which are important to express the meaning of the sentence but they are eliminated. However, if all these low frequency words are selected as the feature word, the feature vector will be too high-dimensional and sparse. Although LDA has been used to add semantic information, the clustering result showed that it had limited effect. Because one Weibo messages may contain several topics or the topic of some message is not so clear, it will affect the performance of the generated topic model. When a larger topic number is set, the "meaning" of each topic will become confused. So LDA cannot provide a more detailed semantic relationship between messages. Third, as the diversity of the viewpoints, some messages are hard to classify into any cluster and become isolated point when the ant colony algorithm is running. But they will finally be respectively assigned to a cluster in the k-means step though their meanings are not similar to other messages in that cluster.

In the further study, the algorithm can be further improved to acquire a more accurate short text clustering result. In addition, each cluster will be automatically extract the main opinion, and various public opinions from different points will be shown in visual way. Combined the clustering results with the information like forward numbers, comment numbers, like numbers and basic information of Weibo users, the analysis of public opinion can show the degree of support, coverage, the degree of transmission and the feature of support and transmission group of each opinion cluster so that the analysis of public opinion and the formative factor will be more detailed and accurate.

Acknowledgment. This research was supported by National Natural Science Foundation of China (Grant No. 71373291). This work also was supported by Science and Technology Planning Project of Guangdong Province, China (Grant No. 2015A030401037).

References

1. Liu, K., Li, L., Jiang, T., Chen, B., Jiang, Z., Wang, Z., et al.: Chinese public attention to the outbreak of Ebola in West Africa: evidence from the online big data platform. Int. J. Environ. Res. Publ. Health **13**(8), 780 (2016)
2. Gu, H., Chen, B., Zhu, H., Jiang, T., Wang, X., Chen, L., et al.: Importance of internet surveillance in public health emergency control and prevention: evidence from a digital epidemiologic study during avian influenza a H7N9 outbreaks. J. Med. Int. Res. **16**(1), e20 (2014)

3. Xiong, X., Hu, Y.: Research on the dynamics of opinion spread based on social network services. Acta Physica Sinica **61**(15) (2012)
4. Su, Q., Huang, J., Zhao, X.: An information propagation model considering incomplete reading behavior in microblog. Physica A Stat. Mech. Appl. **419**(2), 55–63 (2015)
5. Huang, J., Su, Q.: A rumor spreading model based on user browsing behavior analysis in microblog. In: Proceedings of International Conference on Service Systems and Service Management 2013, vol. 8923, pp. 170–173 (2013)
6. Zhao, Y., Qin, B., Liu, T., Tang, D.: Social sentiment sensor: a visualization system for topic detection and topic sentiment analysis on microblog. Multimedia Tools Appl. **75**(15), 8843–8860 (2016)
7. Zhou, C., Zhang, Y., Li, B., Li, D.: Hot topics extraction from Chinese micro-blog based on sentence. In: Proceedings of 2015 IEEE 12th International Conference on Ubiquitous Intelligence and Computing and 2015 IEEE 12th International Conference on Autonomic and Trusted Computing and 2015 IEEE 15th International Conference on Scalable Computing and Communications and Its Associated Workshops (UIC-ATC-ScalCom), pp. 645–648 (2016)
8. Xue, B., Fu, C., Zhan, S.: A study on sentiment computing and classification of Sina Weibo with word2vec. In: Proceedings of 3rd IEEE International Congress on Big Data, pp. 358–363 (2014)
9. Shi, W., Wang, H., He, S.: Sentiment analysis of Chinese microblogging based on sentiment ontology: a case study of '7.23 wenzhou train collision'. Connection Sci. **25**(4), 161–178 (2013)
10. Han, P., Li, S., Jia, Y.: A topic-independent hybrid approach for sentiment analysis of Chinese microblog. In: Proceedings of 17th IEEE International Conference on Information Reuse and Integration, pp. 463–468 (2016)
11. Zhu, Y., Tian, H., Ma, J., Liu, J., Liang, T.: An integrated method for micro-blog subjective sentence identification based on three-way decisions and naive Bayes. In: Proceedings of 9th International Conference on Rough Sets and Knowledge Technology, pp. 844–855 (2014)
12. Shi, H., Chen, W., Li, X.: Opinion sentence extraction and sentiment analysis for Chinese microblogs. In: Proceedings of 2nd CCF Conference on Natural Language Processing and Chinese Computing, vol. 400, pp. 417–423 (2013)
13. Fang, Y.C., Du, Y.J., Tang, M.W.: News topic-typed microblog opinion sentence recognition. In: Proceedings of 2nd IEEE International Conference on Computer and Communications, pp. 2385–2390 (2017)
14. Xin, M., Wu, H., Niu, Z.: A quick emergency response model for micro-blog public opinion crisis based on text sentiment intensity. J. Softw. **7**(6), 1413–1420 (2012)
15. Wu, H., Xin, M.: A quick emergency response model for micro-blog public opinion crisis oriented to mobile Internet services: design and implementation. In: Advances in Multimedia, Software Engineering and Computing, vol. 2 (2011)
16. Bin, W.U., Wei-Peng, F.U., Zheng, Y., Liu, S.H., Shi, Z.Z.: A clustering algorithm based on swarm intelligence for web document. J. Comput. Res. Dev. **39**(11), 1429–1435 (2002)
17. Li, Y.Z.: Three obvious defects in the local governance as seen by the 8.12 Tianjin Port Explosion. People's Tribune, vol. 25, pp. 65–65 (2015)
18. Bi, W.T.: All parties should perform their respective duties in dangerous goods logistics. Labour Prot. **10**, 42–44 (2015)
19. Du, J.F.: Rethinking of public opinion crisis to Tianjin Port Explosion. PR World **8**, 56–59 (2015)
20. Liu, H.M.: the omission of media emergency management: a case study of 8.12 Tianjin Port Explosion. Journalism Lover **11**, 10–15 (2015)

21. Huang, W.J.: Strategy for the guidance of public opinion by Chinese mainstream media under the media covergence: a case study of the reports of 8.12 Tianjin Port Explosion. News World **11**, 141–143 (2015)
22. Xing, X., Wang, C.F.: A study on the influence of social media on the public opinion of major paroxysmal public crisis: viewing the Penetration of Social Media from 8.12 Tianjin Port Explosion. Journalism Lover **11**, 16–18 (2015)
23. Wang, H.C.: A case study on the cause of rumor on microblog in public emergency: a case study of 8.12 Tianjin Port Explosion. Today's Massmedia **11**, 50–52 (2015)
24. Nie, Z.Y., Ding, R.G., Wang, H.B., Yong, Z., Fan, S.Y., Yang, Z.K., et al.: The experience and inspiration of emergency response of chemical defense in 8.12 Tianjin Port Explosion. Chin. J. Pharmacol. Toxicol. **5**, 842–846 (2015)
25. Guo, X.X., Li, Z.J., Li, H., Zhang, Z.X., Xu, C.Z., Zhu, B.: Organization and management of the treatment for the wounded in 8.12 explosion in Tianjin Port. Chin. J. Traumatol. **87**(2), 110–148 (2015)
26. Wang, H.Y., Wu, H.Y.: Problems in the management of mass casualties in the Tianjin Explosion. Critical Care **20**(1), 1 (2016)
27. Chung, Y.S., Kim, H.S.: On the August 12, 2015 occurrence of explosions and fires in Tianjin, China, and the atmospheric impact observed in central Korea. Air Qual. Atmos. Health **8**(6), 1–12 (2015)
28. Faieta, B., Lumer, E.D.: Exploratory data analysis via self-organization. In: Proceedings of 4th International Conference on Computer-Assisted Information Retrieval, pp. 570–585 (1994)
29. Liu, J.B., Yang, F.: Short text frequent clustering algorithm for public opinion analysis. J. Beijing Electr. Sci. Technol. Inst. **18**(4), 6–11 (2010)
30. Wang, Y.H., Xia, Y., Yang, S.Q.: Study on massive short documents clustering technology. Comput. Eng. **33**(14), 38–40 (2007)
31. Sahami, M., Heilman, T.D.: A web-based kernel function for measuring the similarity of short text snippets. In: Proceedings of the 15th International Conference on World Wide Web, pp. 377–386 (2006)
32. Jin, C.X., Zhou, H.Y.: Chinese short text clustering based on dynamic vector. Comput. Eng. Appl. **47**(33), 156–158 (2011)
33. Dutta, S., Ghatak, S., Roy, M., Ghosh, S.: A graph based clustering technique for tweet summarization. In: Proceedings of 4th International Conference on Reliability, Infocom Technologies and Optimization, pp. 1–6 (2015)
34. Quan, X., Liu, G., Lu, Z., Ni, X., Wenyin, L.: Short text similarity based on probabilistic topics. Knowl. Inf. Syst. **25**(3), 473–491 (2010)
35. Cao, C.P., Cui, H.C.: Microblog topic detection based on LSA and structural property. Appl. Res. Comput. **9**, 2720–2723 (2015)
36. Xu, J., Xu, B., Wang, P., Zheng, S., Tian, G., Zhao, J., et al.: Self-taught convolutional neural networks for short text clustering. Neural Netw. Official J. Int. Neural Netw. Soc. **88**, 22 (2017)
37. Chen, Y.F., Liu, Y.S., Qian, Y.Y., Zhao, J.H.: A heuristic density-based clustering algorithm of swarm intelligence. Trans. Beijing Inst. Technol. **25**(1), 45–48 (2005)

Subject Analysis of the Microblog About US Presidential Election Based on LDA

Yonghe Lu(✉) and Yawen Zheng

School of Information Management, Sun Yat-sen University, Guangzhou, China
luyonghe@mail.sysu.edu.cn, 1141612001@qq.com

Abstract. The US presidential election in 2016 drew the attention of the whole world. Both Donald Trump and Hillary Clinton were supported by many people. The news had become a hot topic of Sina microblog (Weibo) for many times. Weibo is an important platform for Chinese public opinion, so the election is valuable for public opinion analysis. Using empirical method, I crawled 21608 search result of Weibo from August 2016 to February 2017 which took "US presidential election" as the keyword. And I used parallel LDA (Latent Dirichlet Allocation) model and software MALLET to make subject analysis about the election after data preprocessing. Then I got 10 topics using LDA and tested the clustering effect utilizing human check method. At last, I found the hot topics by analyzing the clustering result. Topics include real-time report of election, politics, foreign affairs, economics, joke about the vote, tip-off of hackers, Trumps's victory, dramatic election and derivative news. It reveals Chinese people's focus which can help improving civic literacy.

Keywords: US presidential election · Latent Dirichlet Allocation (LDA) Subject analysis · MALLET

1 Introduction

At the end of July 2016, the Democratic and Republican presidential candidates of the 58th US presidential election were published. The election result was announced that the Republican presidential candidate, Donald Trump, had beaten the Democratic candidate and former Secretary of State, Hillary Clinton. Trump would become the 45th president of the United States, which attracted a lot of attention and discussion in Sina microblog (Weibo).

Over the past few decades, many US presidential candidates have played the "China" card, especially the opposition party. China has historically been an unavoidable attack on US electoral politics. American political observers call it "China Bashing", in other words, beat China. But today, decades later, the Chinese has been

This research was supported by National Natural Science Foundation of China (Grant No. 71373291). This work was also supported by Science and Technology Planning Project of Guangdong Province, China (Grant No. 2015A030401037) and Science and Technology Planning Project of Guangdong Province, China (Grant No. 2016B030303003).

K. Arai et al. (Eds.): IntelliSys 2018, AISC 868, pp. 998–1008, 2019.
https://doi.org/10.1007/978-3-030-01054-6_69

used to this situation. We all understand that the candidates who took China as a point during the campaign just wanted to earn more votes from the underlying blue-collar workers who lost the vested interest. And no matter who, once elected, would immediately turn to ignore the Chinese question. It reflects China's own changes, and also highlights the more complicated interest relation between the two countries. Chinese experts pointed out that though the Democratic candidate, Hillary Clinton, advocated the policy which is unfavourable to China, whose policy has been known to all without too much speculation. On the contrary, the Republican candidate, Donald Trump, is completely unfamiliar to the public, and the uncertainty of his success is even more interesting.

Weibo is a social network platform sharing short real-time information. It is a platform for information sharing, dissemination and access which is based on user's relationship. And it can show the ideological trends at every moment with its timeliness and arbitrariness. For example, if someone published some big emergencies or events that cause global attention on Weibo, its real-time properties, live sense and quickness exceeds all media. The 140-word limit in Weibo brought civilians and Shakespeare to the same level, causing a large number of original content produced in an explosive manner. "Silent Majority" found a stage to show themselves on the Weibo. Obviously, Weibo contains a large number of public opinion data, which can reflect the trend of public opinion in time and plays an important role in the analysis of network public opinion.

In this study, I crawled the Weibo about the US presidential election in 2016 and used software MALLET based on parallel LDA (Latent Dirichlet Allocation) model to make subject analysis, which helped me get the hot topics on which Chinese people focus.

Through the research process, I use some scientific research methods. Details are as follows:

(1) *Literature research:* Many scholars have studied the topic model LDA, and the 2016 US presidential election is very important current affairs, so the literature research and social evaluation is the basis of this study. Literature collection is done mainly through well-known database, including SCI, CNKI, and search engine like Google and Baidu academic.
(2) *Cluster analysis:* Cluster analysis is an important part of this paper. With the "US election" as the key word, I use Bazhuayu software to get Weibo content from August 10, 2016 to February 9, 2017, and achieve word segmentation, removal of stop-words and other pretreatment. Clustering is done using software MALLET based on parallelized LDA to prepare for the topic analysis.
(3) *Human check:* In order to test the effect of clustering, I randomly pick some Weibo from each cluster and check the semantic of them to determine whether each Weibo is close to the topic of the cluster it belongs to, especially those who do not contain the keywords of the cluster.
(4) *Statistical analysis:* Statistical analysis is the theory and technology for quantitative data processing, which is scientific, intuitive and repeatable. Calculate the number of microblog, drawing the relationship diagram of published time and quantity of Weibo, showing the trend of discussion over time.

This article is divided into four sections:

Section 1: Introduction. It mainly introduces the research background and significance, the research purpose and the method, and the main content and the section arrangement of the article.
Section 2: Relevant theories. Introduce the topic model, LDA and its extended deformation, LDA parallelization and LDA in Weibo data analysis and so on.
Section 3: Subject Analysis. The data are collected and clustered, and the clustering results are checked by myself to obtain the results of the topic analysis.
Section 4: Conclusion. Summarize the conclusions of the study, and look forward to further work.

2 Relevant Theories

A. *LDA*

LDA (Latent Dirichlet Allocation) is a statistical topic model published in JBLR by Blei et al. [1] in 2003. It is applied in text modeling in order to find out the hidden semantic dimensions, the "topics" or "concepts", from the text in an unsupervised method. It finds out the topic structure of the text by using the co-occurrence features of the words. This method does not need any background knowledge about the text. It can also analyze the linguistic phenomena of "polysemy" and "synonymy".

The main idea of LDA is that the document is a mixed distribution of several topics, and each topic is also a probability distribution of words. Therefore, a single document can be expressed as a probability distribution for these hidden topics (Doc-topic), which as a probability distribution of words (topic-word).

LDA contains lots of calculations which needs an approximate algorithm. Blei proposed variational Bayesian inference [1], which is less computationally intensive but less accurate. Griffiths proposed Gibbs Sampling algorithm [2], the algorithm is computationally intensive, but more simple and accurate. People generally use the Gibbs sampling method, which infers the posterior distribution approximation with the sample distribution by collecting the posterior distribution samples [3].

B. *Deformation, Parallelization and Application of LDA*

The algorithm of LDA continues to expand, gradually derived a number of variants, such as HLDA [4] (Blei, NIPS 2003), HMM-LDA [8], DTM [10], CTM [9], Labeled LDA (Ramage et al. 2009) [11], PLDA (Wang et al. 2009) [5], PLDA + [7], Author-topic model (Rosen-Zvi et al. UAI 2004) [12] and so on.

For example, the supervised and hierarchical Labeled Latent Dirichlet Allocation model can train out tagged topics. The biggest difference from LDA is that LDA selects a topic for a word from all topics, and Labeled LDA only chooses it from the topics corresponding to the label related to that document. It overcomes the earlier introduction of supervised subject models such as Supervised LDA and Disc LDA which limit the documents to associating with multiple labels, allowing the multi-label corpus to be modeled [11].

In 2007, UC Irvine's David Newman team found that Gibbs sampling can be parallelized particularly in LDA. They proposed a parallelized version of LDA algorithm, distributed-LDA (AD-LDA) [6]. Based on Gibbs sampling, using the idea of global synchronization, the document is evenly distributed to several processors. After the iteration on each processor separately, the information on each processor is summarized and the local models on each processor are replaced by the global model, and the next iteration is performed. Algorithm is shown in Fig. 1 [6]:

---

repeat
  for each processor p in parallel **do**
    Copy global counts: $N_{wkp} \leftarrow N_{wk}$
    Sample $\mathbf{z}_p$ locally: LDA-Gibbs-Iteration($\mathbf{x}_p, \mathbf{z}_p, N_{kjp}, N_{wkp}, \alpha, \beta$)
  end for
  Synchronize
  Update global counts: $N_{wk} \leftarrow N_{wk} + \sum_p (N_{wkp} - N_{wk})$
until termination criterion satisfied

Fig. 1. AD-LDA algorithm.

The software MALLET used later is also based on parallel LDA.

The data analyzed by LDA in this paper is a kind of Internet text. Weibo text has poorly organized language and contains a large number of new vocabulary, so there is a high degree of uncertainty in the topic mining process of such texts while the probability model like LDA is more likely to make a more accurate calculation of the topic. On one hand, compared to the grammar-based text analysis method, LDA can exploit the potential semantic structure through probabilistic statistics of co-occurrences in the text, which can solve the non-standard grammar of the microblog text. On the other hand, the LDA model uses the probability distribution method to represent the texts. The texts belong to different topics with different probabilities which is similar to the probability distribution of real data. The LDA model can express and quantify the existing uncertainties in the texts and reduce the effect of noise [13].

3 Subject Analysis

A. *Data Collection and Preprocessing*

(1) *Data collection:* Using the software Bazhuayu, setting "US presidential election" as key word, I crawled 21,608 microblogs in Weibo for six months between August 10, 2016 and February 9, 2017. The content of the microblogs crawled includes links, content, publisher, published time, forwarding amount, commenting amount, praise amount and so on.

(2) *Descriptive statistics:* Calculating the number of microblogs in each week, we can see that the total trend of the discussion amount, as shown in Fig. 2:

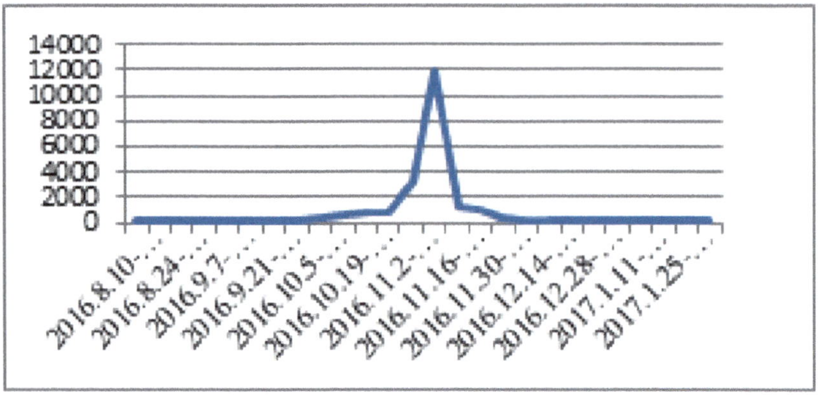

Fig. 2. Trend of "US presidential election" microblogs.

From Fig. 2, we can see that it is the outbreak of public opinion between November 2, 2016 and November 15, 2016. The general trend of public opinion about the U.S. presidential election is normally distributed with a center of the time of publishing election result.

Considering about the election schedule:

09:00–10:42 of September 27, 2016, Beijing Time, the U.S. presidential candidate Hillary Clinton and Trump conducted a television debate;

12:00 on November 9, 2016, Beijing time, the result of the election was released;

On January 6, 2017, the U.S. Congress officially confirming that Donald Trump defeated Democratic presidential candidate Hillary Clinton with 304 votes to 227 votes in November 2016, and was elected as 45th president of United States;

US President Barack Obama made a farewell speech in Chicago on January 10, 2017, announcing the end of his eight-year presidential term.

To sum up, at all four major periods of time, it inevitably aroused public opinion because of the important events in the whole election process.

Therefore, we can see that the release of the results of the U.S. presidential election has caused the climax of public opinion discussion, and the landmark events in the election process have also attracted a certain amount of attention.

(3) *Date preprocessing:* Used the Chinese word segmentation API interface provided by ICTCLAS for word segmentation, and added new words such as "Trump" to the user dictionary. Maked use of HIT's Chinese stop-words list to remove stop words, and added "Hillary", "Trump" and other words of high word frequency to stop words list. Stop words are commonly used terms such as pronouns and modal particles, which appear frequently but do not help with topic mining. The final data will be processed into txt documents stored in the set encoding UTF-8

B. *Clustering Analysis*

Run the software MALLET in command desk to analyze the preprocessed data (UTF-8 format). MALLET is a Java tool that provides statistical natural language processing, document classification, clustering, topic modeling, information extraction, and other machine learning textual applications. It can convert texts into mathematical expressions to do machine learning more effectively. This process is done through a "pipe" system that can achieve word segmentation, remove stop words, convert sequences to vectors, and more. MALLET uses Gibbs sampling method.

MALLET's topic modeling can analyze a large number of unlabeled (unknown categories) texts. By analyzing these texts, you can come up with some (the number can be specified, but also acquiescence) topic. Each topic includes some frequently appeared words, and the probability of the text corresponding to each topic is also shown. The built topic model can be saved for use in inferring the topic to which an unknown text belongs. The MALLET topic modeling tools include an efficient sampling-based Dirichlet distribution, Pachinko Allocation, and Layered LDA. This software has been rather mature, convenient and efficient. After model has been successfully built, ten clusters are obtained, and also the key words of each cluster and the probability of each text corresponding to each cluster are shown, which can be used to judge the cluster to which each text is more inclined to. The main topics of the ten clusters are summarized as follows:

(1) Real-time reports of the presidential election.
(2) Political discussions.
(3) U.S. diplomatic actions.
(4) Economic considerations.
(5) Joke about the U.S. presidential election.
(6) The breakout of Assange.
(7) Trump counterattack to defeat Hillary Clinton.
(8) The dramatic election.
(9) Influence to China.
(10) Derivative news of presidential election.

C. *Human Check*

In order to test the effect of clustering, the clusters obtained by clustering should be checked by human. By choosing a number of data from each category randomly and observed the semantics, n we can determine whether it is indeed close to the topic of the cluster, especially those who do not literally contain the key words of that cluster.

Human check should ensure that the randomness and representativeness of detection. Here listed only a few items in Table 1 as an example considering the limited space.

Through human check, most of the data in randomly selected samples are close to the topic. For example, Article 1 in Table 1, which does not include the keywords of cluster 2, is clustered into cluster 2, with the main idea of "Foreign forces influence the political situation in the United States" while Cluster 2 is entitled "political discussion", so the human check result are close to the topic.

There is also a small amount of data deviating from the topic, just like the 2nd item in Table 1, which belongs to Cluster 3, with the thrust that "English has become a disadvantage in the United States and the United States", while Cluster 3 has the topic of U.S. policy-related diplomacy discuss. Obviously, the human check results should be deviated from the topic. The reason why it is clustered into cluster 3 is that the last sentence in this text is close to the topic of cluster 3 with the key word "Russia" of cluster 3 and the rest of the text is not significantly close to another topic.

Table 1. Examples of human check

Number	Text	Cluster	Whether contain keyword	Human check result
1	【1940, "Foreign forces" influence the political situation in the United States.】 From fake news to gossip, various "overseas factors" that are considered influencing the current U.S. presidential election have emerged as early as the 1940s. It is the United States' best ally during World War II that carefully planned all these masters.	2	No	Close
2	Why English Has Become a British and American Disadvantage? An English-speaking society is like a glass house: for the English-speaking English and Americans, foreign is not transparent. English used to be an asset of the United States and the UK. Today, it has become a disadvantage. Let us look back from the hacker attacks on the U.S. presidential election in Russia and observe a more general situation.	3	Yes	Deviation

Table 2 is the human check results. The table lists the human check results of ten clusters, including the percentage of data that contains keywords and is close to the topic in each cluster, and the proportion of noise data, and so on. Table 2 shows that, except for cluster 9, the useful data of clusters occupy more than 60%, and the clustering effect is not bad. Except for cluster 8 and cluster 10, the proportion of the data in the clusters that contains keywords and are close to the topic is much higher than that of the data that does not contain the keyword and is close to the topic, indicating that the keywords can describe the topics of the cluster well. In general, clustering results of clusters 8, 9, and 10 are a bit less effective. Other clusters have more data that contain keywords and are close to the topic, and the clustering effect is better.

D. *Analysis Results*

Unlike the K-means algorithm, the LDA model can use the silhouette coefficient to prove the clustering effect, and the software MALLET cannot visualize the clustering effect. Therefore, the human check method is finally used to test the clustering effect, which inevitably makes effect evaluation subjective, only by repeating checks can minimize errors. In addition, the LDA model cannot summarize the topic in a clear semantic meaning, which leads to subjective judgments on the keywords when the topic is extracted from the clustering result. Also, the number of clusters needs to be specified when using MALLE. Moreover, since there is no index in MALLET to help determine the clustering effect, it is difficult to select the optimal number of clusters, which leads to overlapping between clusters and the cluster of some data is not very satisfied.

Table 2. Human check results

Number	Have keyword and close	No keyword and close	Have keyword and deviation	No keyword and deviation	Useful data	Noise data
1	76%	6%	6%	12%	82%	18%
2	78%	2%	12%	8%	80%	20%
3	72%	4%	2%	26%	72%	28%
4	64%	20%	2%	14%	84%	16%
5	80%	6%	4%	10%	86%	14%
6	74%	8%	0%	18%	82%	18%
7	52%	24%	4%	20%	76%	24%
8	28%	36%	4%	32%	64%	36%
9	40%	18%	12%	30%	58%	42%
10	30%	50%	8%	12%	80%	20%

Note: The noise data is defined as deviated text, while useful data refers to text that is close to the topic.

Although there are some shortcomings in the experiment, we can learn from the topic extraction and analysis of the Weibo discussions about the U.S. presidential election which mainly include:

(1) real-time reports of the presidential election and the discussion of Internet users. Keywords are "support rate", "percentage point", "polling station" and so on.
(2) Political discussion. Such as the clashes between populism and elitism, the comparison between capitalism and socialism, the confrontation between the Democratic Party and the Republican Party, and the thinking on democracy. Keywords are "establishment", "populism" and so on.
(3) The discussions on U.S. diplomatic situation and foreign policy, such as doubts about Russia's involvement in the U.S. presidential election and the candidate's explanation of the Syrian issue. The key words are "Russia", "Syria", "Turkey", "diplomats" and so on.
(4) Economic considerations. The key words are "Federal Reserve", "Investor", "Volume" and so on.

(5) Joke about the election. Keywords are "Trump", "Jolin Tsai", "watch the fun" and so on.
(6) Assange broke the news about Hillary Clinton which led to the FBI's investigation and caused people's doubts about the electronic selection and the restart of the counting work. The situation has become more complicated, confusing and controversial. Keywords are "fbi", "Bureau", "Assange", "Wisconsin" and so on.
(7) Public opinion nearly lean to one side with Hillary Clinton, but Trump counterattack and won the support of Ohio and Florida, miraculously reversed to defeat Hillary Clinton at the last minute. Keywords are "Florida", "Ohio", "ups and downs", "one-sided" and so on.
(8) The idea to the dramatic of the election and the news broke out by each forces. Keywords are "unexpected", "climax", "unprecedented", "drama" and so on.
(9) China's concern and discussion on U.S. election and the impact of the election on Chinese. Keywords are "RMB", "real estate", "People's Daily" and so on.
(10) Derivative news of election. The U.S. election has become a heated topic. The ordinary people seriously expressed their views. The attention of celebrities to the election will also cause public hot debate. There have also been many election-derived news, such as the collapse of the website of the Canadian Immigration Bureau and the bastard of Clinton. The keywords are "Daniel Wu", "Canada", "Buffett", "Clinton" and so on.

4 Conclusion

A. *Research Conclusions*

The 2016 US presidential election captured the headlines of major newspapers and Internet media. Internet users in Weibo gave so much speech and attention to heat the "US presidential election" to the top of topics. Phoenix and Netease kept tracking with special coverage and updated speedily with rich information. Weibo contains a lot of public opinion data, which can timely reflect the changing trend of public opinion. Therefore, it is valuable to make a topic analysis of the content about the U.S. general election in Weibo. This paper uses the literature research method to collect the relevant research to lay the foundation for the dissertation, and uses the empirical research method, with the software Bazhuayu and the keyword of "US presidential election", collecting Weibo content between August 10, 2016 and February 9, 2017. After the word segmentation and stop-word removing, I used the software MALLET based on parallel LDA to mine the topic and gotten clusters. Finally, I used human check to analyze the clustering results and draw conclusions. Through the topic analysis, ten hot topics are shown like real-time reports of the election, political discussions, the U.S. diplomatic situation and policy, economic considerations, joke about the election, hackers broke the news and the consequences, Trump's counterattack, the dramatic election, the impact of the election on China, the derivative news of election. The conclusion reveals the hot issues that public are concerned about and has a certain effect on improving people's political literacy.

B. *Prospect*

There are still some shortcomings in the empirical analysis in this paper:

(1) It is found during human check that there is overlapping between clusters and the cluster of some data is not very satisfied. The topics cannot be summarized with definite semantics, and the accuracy of clustering needs to be improved.
(2) The test and comparison of the coefficients in the LDA model are not enough when MALLET is used.
(3) The testing of other clustering methods in short text clustering is lacking.
(4) Did not achieve visual processing.

In later research, we can further improve the clustering algorithm of Weibo texts based on the present thesis and use MALLET skillfully to obtain more accurate and practical clustering results. In addition, the topic analysis results can also be displayed in a visual way, simply and intuitively display the hot issues discussed in Weibo. In addition, we can make a more scientific analysis of heat by combining the data of Weibo texts, like the amount of forwarding, the amount of comments, the amount of praise and the user's basic information, to visually display support, coverage and dissemination of the population in order to analyze more detail about the views and their formation.

References

1. Blei, D., Ng, A., Jordan, M.: Latent Dirichlet allocation. J. Mach. Learn. Res. **3**, 993–1022 (2003)
2. Griffiths, T., Steyvers, M.: Finding scientific topics. In: Proceedings of the National Academy of Sciences, Washington D.C., United States National Academy of Sciences, pp. 5228–5235 (2004)
3. Heinrich, G.: Parameter estimation for text analysis. Technical report (2008)
4. Blei, D.M., Jordan, M.I., Griffiths, T.L., Tenenbaum, J.B.: Hierarchical topic models and the nested chinese restaurant process. In: Proceedings of the 16th International Conference on Neural Information Processing Systems (NIPS 2003), vol. 57, no. (2), pp. 17–24 (2007)
5. Wang, Y., Bai, H., Stanton, M.: PLDA: parallel latent Dirichlet allocation for large-scale applications. In: AAIM, pp. 301–314 (2009)
6. Newman, D., Asuncion, A., Smyth, P., Welling, M.: Distributed algorithms for topic models. J. Mach. Learn. Res. **10**(12), 1801–1828 (2009)
7. Liu, Z., Zhang, Y., Chang, E.Y.: PLDA + : parallel latent DIRICHLET allocation with data placement and pipeline processing. ACM TIST **2**(3), 1–18 (2011)
8. Griffiths, T., Steyvers, M., Blei, D., Tenenbaum, J.: Integrating topics and syntax. Adv. Neural. Inf. Process. Syst. **17**, 537–544 (2004)
9. Blei, D., Lafferty, J.: Topic models. In: Srivastava, A., Sahami, M. (eds.) Text Mining: Classification, Clustering, and Applications. Chapman & Hall/CRC Data Mining and Knowledge Discovery Series (2009)
10. Blei, D.M., Lafferty, J.D.: Dynamic topic models. In: Proceedings of the 23rd International Conference on Machine Learning, pp. 113–120 (2006)

11. Ramage, D., Hall, D., Nallapati, R.: Labeled LDA: a supervised topic model for credit attribution in multi-labeled corpora. In: Proceedings of the 2009 Conference on Empirical Methods in Natural Language Processing (EMNLP 2009). Association for Computational Linguistics, Stroudsburg, pp. 248–256 (2009)
12. Zvi, M., Griffiths, T., Steyvers, M.: The author-topic model for authors and documents. In: Proceedings of the 20th Conference on Uncertainty in Artificial Intelligence (UAI 2004), pp. 487–494. AUAI Press, Arlington (2004)
13. Zhang, P.-J., Song, L.: Review on the topic modeling method of Weibo text based on LDA. J. Libr. Inf. Serv. **56**(24), 120–126 (2012)

An Analysis on the Micro-Blog Topic "The Shared Bicycle" Based on K-Means Algorithm

Yonghe Lu(✉) and Yuanyuan Zhai

School of Information Management, Sun Yat-Sen University, Guangzhou, China
luyonghe@mail.sysu.edu.cn, 506260644@qq.com

Abstract. With the rapid development of the shared economy, the shared bicycle is changing people's lives. In recent years, Micro-blog has become a platform for people to share and access information. Exploring the theme of micro-blogs can help us understand the attitude of the public to shared-bicycle. In this paper, the clustering algorithm is applied to the short text analysis of micro-blog. Firstly, the theory and technology of short text clustering are introduced. Secondly, crawler is used to gather micro-blog topics relevant to shared bike with a total of more than 30,000; k-means is used to cluster; the result of cluster is evaluated through silhouette coefficient; it comes out that when the number of the cluster is 39, the result of cluster works best. Thirdly, the theme tags are defined for 8 clusters randomly extracted from the 39 clusters, which involves the advantages of the bike; the four new inventions of China; companies to expand overseas markets; the phenomenon of irregular parking; people's feeling of riding; the deposit and the volume of the bike; other shared things and the employment. Fourthly, the number of blogs are extracted randomly and analyzed artificially from the semantic perspective, the results of which are: most of the data has the obvious characteristics of the theme, but contains a small amount of noise data, and the courses are analyzed. In the end, the research work is summarized and the experimental shortcomings and future prospects are discussed.

Keywords: Shared bicycle · Micro-blog topic · Short text clustering K-means

1 Introduction

In May 2017, young people from the 20 countries along the One Belt And One Road ranked China's "new four great inventions": high-speed rail, scanning payment, sharing of bicycles and online shopping. Since the end of 2016, the shared bicycle in China has enjoyed explosive growth. Globally, more and more people think bicycles will become one of the key tools for urban transport in the future.

In the era of digitalization and informationization, WeChat, microblog, Twitter and other applications with mobile Internet are becoming more and more popular. Sina Micro-blog has become an important place for people to share information, express emotions and communicate. Up to September, 2017, the number of active users on

K. Arai et al. (Eds.): IntelliSys 2018, AISC 868, pp. 1009–1024, 2019.
https://doi.org/10.1007/978-3-030-01054-6_70

Micro-blog was 376 million, an increase of 27% compared with the same period in 2016, with the mobile terminal accounting for 92%. Daily active users reached 165 million, up 25% from the same period last year. The users have provided a large amount of text information on the microblog platform, the value of which researchers have been noticed [1].

As a new thing, sharing bicycle has triggered a heated discussion on Micro-blog. In this paper, the clustering algorithm is applied to the analysis of Micro-blog about "sharing bicycle". The research results have practical significance in the following aspects. To learn about the current situation of sharing bicycles and the public's feedback on the sharing of bicycles; to understand the government's management of bicycles; to learn about the operation status of the company; to lay the foundation for further research.

This article is divided into five sections:

Section 1: Introduction. It mainly introduces the research background, significance and the section arrangement of the article.
Section 2: Related Research. Try to understand previous studies on the application of clustering and the sharing of bicycles.
Section 3: Methodology. Introduce the methods used in this paper and theories and techniques related to our experiment.
Section 4: Experimental Study. The data are collected and clustered, and the clustering results are manually tested to obtain the results of the topic analysis.
Section 5: Conclusion. Summarize the conclusions of the study, and look forward to further work.

2 Related Research

2.1 Application of Clustering

Clustering is widely used in medicine [2, 3], agriculture [4, 5], geography [6, 7], molecular biology [8], commerce [9] and physics [10]. Zhang Lin and Xie Zhonghong [11] have used the clustering analysis method to classify users of blog, and then analyzed the characteristics and influence of each cluster. Liu Jinshuo [12] has proposed a topic discovery method of k-means clustering based on LDA model, and conducted effect verification in the network food safety issue.

In addition, clustering has a wide range of applications in other fields, such as pattern recognition and image processing.

2.2 Shared Bicycle

Research on the sharing of bicycles is mainly focused on economics. Guo [13] has found that e-bikes, as a substitute for sharing bicycles, do not have the characteristics of "free sharing" of bicycles, and have less impact on Shared bikes by using Michael Porter's Five Forces Model. Li [14] has used the Net Present Value and Internal Rate of Return model (NPV/IRR) in market research to analyze the profitability of shared bicycle enterprises and discussed the relationship between the government and the sharing of bicycles.

From the perspective of design, Huang [15] has analyzed the advantages and disadvantages of the design of the sharing bicycle. In terms of the optimization of the background management system, Xu [16] has proposed several optimization measures for the frequent collapse of the Shared bicycle app.

On the cultural and spiritual level, an article [17] in modern business points out that "sharing the spirit can achieve the sharing economy"; Xiaoyun [18] believes that "the most important thing to share is culture".

From what has been discussed above, we found that the shared bicycle is a new thing, and there are few literatures on the analysis and discussion of shared bicycles, which mainly focus on economics, market research and strategic analysis. Big data analysis of shared cycling is a new idea of research. Therefore, in this paper a big data analysis of shared bicycle, through the clustering method is proposed, expecting to learn about hot topics on the shared bicycle.

3 Methodology

3.1 Methods Used in this Paper

- **Literature Research.** Through the literature research method, we can get a general understanding of the existing research results in the field of shared bikes and application of short text clustering. Literature collection is done mainly through well-known database, including China National Knowledge Infrastructure (CNKI), Elsevier, IEEE, Google Scholar, Baidu academic and so on.
- **Clustering Analysis.** Clustering analysis is an important part of this empirical study. Crawler is used to gather micro-blog topics relevant to shared bike from September 01, 2017 until September 31, 2017. After data preprocessing, such as data cleansing, word segmentation and removal of stop words, K-means clustering is used to prepare for thematic analysis.
- **Manual check method.** Manual check method refers to the process of performing full-scale or sample-based testing on an experimental result by means of manual judgment without the help of statistical analysis tools. In order to examine the effectiveness of clustering, randomly select a number of microblog from each cluster, conduct semantic analysis to determine whether each microblog is close to the topic of the cluster it belongs to, especially those who does not contain key-words of the cluster.

3.2 Selection of Related Theories and Techniques

Clustering is the task of grouping a set of objects into groups, also known as clusters [19]. The objects in the same cluster should be similar with each other, and the objects in different clusters are dissimilar. There are different types of objects, including: text documents [20], videos [21], images [22], facial data [23] and so on. Short text clustering is the application of clustered objects to short text, including the following processes.

(1) Text Preprocessing
As the text belongs to unstructured data and cannot be processed directly by the computer, it must be pre-processed. Text preprocessing generally consists of text segmentation and removal of stop words.
   (a) Chinese word segementation
   Since the text ultimately needs to be represented by features derived from the text segmentation, the text segmentation is the first of any text processing technology to be carried out. Segmentation technique of Chinese word is different from that of English word, because Chinese does not have spaces between words as natural delimiters like English. At present, there are many mainstream word segmentation techniques such as dictionary-based segmentation, statistics-based word segmentation, word segmentation based on understanding. And the third one allows the computer to understand the meaning of the text, so it is also called artificial intelligence segmentation which is getting more and more attention with the development of artificial intelligence. Semantic research is the key to realize the breakthrough of theory and practice [24]. But the technology is not yet mature.
   Among the existing mature word segmentation systems, "Jieba" Chinese text segmentation system based on the Python language is influential, and can make the accuracy up to 98%. So in this paper, it is chosen for text segmentation.
   (b) Remove Stop Words
   Removing stop words means filtering out those words that are not very useful for text category identification. For Chinese, some words such as "of(的)", "we(我们)", "this(这个)", are mostly meaningless and belong to the stop words. In addition, some numbers and punctuation, such as colon, dash, ellipsis, etc., also need to be removed.
(2) Text Representation
   (a) Select Text Representation Model
   After the preprocessing of the text above, it is necessary to consider how to represent the text as a unified, structured data that can be recognized by the computer, which is called text modeling. Currently, common text representation models include Probabilistic model, Boolean model and Vector space model.
   Probabilistic model belongs to supervised text mining technology. So it is less used in text clustering and more used in text classification. Boolean model is a simple retrieval model based on set theory and Boolean algebra [25]. The Boolean model represents the document as a vector consisting of 0 and 1. However, it cannot reflect the relative importance of feature items to the text, and lacks flexibility [26]. In order to make up the deficiency of the Boolean model, the weight of the word is substituted for 0 and 1 to form the space vector model. In 1969, Gerard Salton and others proposed the Vector Space Model [27]. The main idea is that each document is mapped to a point in a

vector space of a set of normalized orthogonal vectors [28]. It expresses the semantic similarity in terms of spatial similarity, which is intuitive and easy to understand.

Obviously the third model is superior to the first two models, and thus be chosen in our experiment.

(b) Feature Selection

After the text preprocessing, each word obtained by the segmentation is called a feature [29]. In order to avoid the bad influence of the high feature dimension to the speed and effect of text processing, the original feature must be screened effectively [30], which is called feature selection. The commonly used feature selection methods are mainly divided into supervised and unsupervised.

The commonly used feature selection methods are mainly divided into supervised and unsupervised. There are mainly supervised features selection methods [29] such as Information Gain (IG), CHI-Squared (CHI), Mutual Information (MI), and Expected Cross Entropy (ECE). As clustering is unsupervised, we only consider Document Frequency (DF), Term Strength (TS), Term Contribution (TC) and so on, of which the DF is the simplest method. Apart from that, it has low computational complexity and only increases linearly with the increase of text sets, so it can be applied to large-scale corpus [31]. After comprehensive consideration, document frequency is chosen as feature selection method in our experiment.

(c) Weight Calculation

In general, the more important a feature word is to a document, the greater the value of the weight of the word. The feature weight calculation methods usually include the Boolean Weighting method and the Term Frequency-Inverse Document Frequency Weighting (TF-IDF).

In the Boolean weighting method, all features are treated only as two states: appearances and appearances. It does not take into account the frequency of occurrence of the feature word and may not reflect the importance of different words in the text.

The weighting method of TF-IDF comes from text retrieval [32]. Compared with the method of Boolean weight calculation, it takes into account the words that occur more frequently in a single text and less frequently in other texts. Term Frequency (TF) refers to the frequency at which a feature word appears in a text. Inverse Document frequency (IDF) indicates the distribution of the feature in the whole text set. TF-IDF uses the product of these two to express weight. The formula is as follows:

$$W_i = TF_i * IDF_i \tag{1}$$

By comparing the above two methods, the latter has a better weighting effect and thus is chosen in this paper research.

3.3 Selection of Clustering Algorithm

The study of clustering algorithm first appeared in the 1960s and played a crucial role in data mining. Nowadays, there are mainly five types of cluster analysis algorithms [33]: Partitioning Method, Hierarchical Method, Density-Based Method, Grid-Based Method and Model-Based Method.

The idea of the partitioning method is to divide a given dataset and each partition is a cluster. The method has the advantages of simple implementation, fast operation, high implementation efficiency and is capable of processing large-scale data. However, the number of clusters needs to be determined in advance. In addition, noises and "outliers" have a great influence on the clustering results [34].

Hierarchical clustering algorithm [35] is divided into condensed hierarchical clustering and split hierarchical clustering according to its clustering direction. The representative algorithms include BIRCH algorithm [36], CURE algorithm [37] and ROCK algorithm [38]. It needs to store the similarity matrix and occupy a large memory cost, which is not suitable for large-scale data. A large number of samples or clusters are needed to be checked and estimated to make the decision of merging or splitting, making this kind of algorithm less scalable [39].

The main idea of the density-based method is to divide spatial data into different clusters by clustering density, and divide the data of similar density into a cluster. However, the computational complexity of density cells is large and the scalability of data dimensions is poor [40].

The research object of this paper belongs to short text, with the characteristics of high dimension and large quantity. If the hierarchical clustering method is used, a large amount of time and storage space will be consumed, which is difficult to adopt under the condition of insufficient machine and equipment. If we use the density-based clustering method, we cannot achieve good results due to the unknown data density of microblog text sets. Therefore, in our experiment we choose k-means introduced as follows, the most traditional, effective and simplest clustering method, to cluster microblogging texts.

The k-means algorithm is a classic algorithm first proposed by MacQueen in 1967 [41]. The algorithmic process is as follows. Firstly, k objects are randomly extracted from the data set as the initial center of each cluster. Each remaining object is then calculated for its distance from each cluster center and assigned to the nearest cluster. After that compute the new mean of each cluster as the center of the updated cluster. At last, repeat the above process until the guideline function converges.

3.4 Selection of Clustering Evaluation Index

Indicators that evaluate the effectiveness of clustering are generally divided into external index and internal index [42]. The subject information of the research object in our experiment, the microblog about "sharing bicycle", is unknown beforehand, so we cannot use external evaluation indexes but internal ones. At present, there are many internal indicators, including Dunn index (DI) [43], Davies-Bouldin index (DB) [44], Calinski-Harabasz index (CH) and Silhouette Coefficient. Among the internal evaluation indexes, the Silhouette Coefficient takes into account both the intra-cluster cohesion and the inter-cluster resolution, and can effectively evaluate the clustering effectiveness. At the same time, most clustering algorithms need to set some initial parameters manually before starting to execute. Therefore, Silhouette Coefficient is chosen in this paper to determine the number of clusters and as one of the clustering evaluation methods.

Silhouette Coefficient was proposed by Kaufman L et al. in 1990 [45]. The idea is to obtain the overall clustering silhouette coefficient by calculating the silhouette coefficient of a sample individual.

First, the individual silhouette coefficient is calculated for each sample d_i in the dataset as follows:

$$s_i = \frac{b_i - a_i}{\max(a_i, b_i)} \tag{2}$$

In the formula above, a_i represents the average distance between sample d_i and each sample in the same cluster except d_i, and b_i is the average distance from the sample d_i to all the objects in other clusters. The value of s_i ranges from -1 to 1. When $s_i = 1$, it means that d_i is assigned to a completely correct cluster; when $s_i = 0$, it means that the cluster is not obvious; when $s_i = -1$, it means that d_i is assigned to a wrong cluster.

After calculating the individual silhouette coefficient, the silhouette coefficient of the overall cluster is calculated as follows, which may also be referred to as the average silhouette coefficient.

$$s_k = \frac{1}{n}\sum_{i=1}^{n} s_i \tag{3}$$

In the above formula, n refers to the number of samples in the dataset, and k is the number of clusters.

4 Experimental Study

In order to draw the hot issues that the Chinese people pay attention to about sharing bicycles, we used the techniques mentioned in part three to cluster the microblog texts, analyzed the topics about the shared bicycle according to the clustering results, and then evaluated the effectiveness of clustering by manual check method. The experimental process is shown in Fig. 1.

4.1 Data Collection

At present, there are mainly two ways to obtain the Micro-blog data: the web crawler technology and the API interface provided by Sina Micro-blog. Since Sina Micro-blog has many restrictions on API calls and it is not open enough to meet the large-scale data needs, the web crawler is chosen in this paper to capture data sets. Many web pages now use ajax [46] technology to dynamically generate web pages through JavaScipt, a

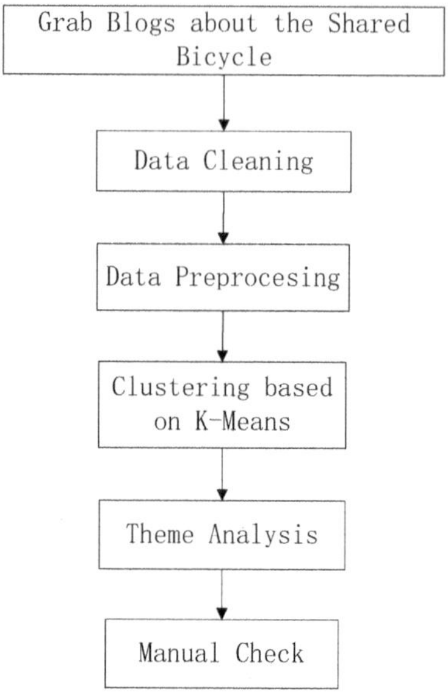

Fig. 1. Experiment process.

site that websites set to prevent large-scale data collection. However, many mature data collection software now can solve the problem through specific settings.

In this paper, we used octopus collector for microblogging text search and acquisition, and the following search strategies have been carried out on advanced search page of Sina Micro-blog.

- Keyword: the Shared Bicycle
- Type restrictions: original (Original microblogging can directly reflect the personal opinion of Internet users)
- Published time: from September 01, 2017 until September 31, 2017
- Published location: no limitation.

After the above operation, we crawled 34,036 microblogs in Sina Micro-blog and the data output format was XLSX. Collection fields included: user name, microblogging content and published time.

4.2 Data Cleaning

First, to delete duplicate, incomplete and error data: Take advantage of Microsoft Office Excel's built-in function of deduplication to delete the content that is duplicated. Due to the possibility of misalignment of data captured by the automatic collection software, it is necessary to check whether the publication date of the microblog is within the limited scope. If there is a dislocation, it can be manually modified or deleted. In addition, delete data with missing values.

Second, to delete meaningless content: Meaningless blogs, such as advertisements, pure sharing, questioning and answering (Q & A), might not reflect the attitude of the public to shared-bicycle directly. The type of advertising microblogging usually contains "red envelope", "clothes" and other similar words. The sharing microblogging generally contains "Share Netease News", "Share NetEase pictures", "Share NetEase video", "enjoy the micro-survey" and other similar phrases.

Third, to remove words or sentences irrelevant: Microblogs usually include some specific formats when forwarding, sharing and publishing original opinions, which have no relation with the theme of the microblog and will increase vector dimension. It can be summarized in two forms, one is the fixed expression, the other one is the variable expression. The fixed expression mainly contains those fixed words or sentences such as "interactive topics", "website links", "second video", "full text display", "I published a headline article", "I uploaded to listen to who is talking … (Come to Youku to see more exciting videos of mine)", which can be cleaned directly through the replacement function of Microsoft Office Excel. The variable expression can be cleaned by regular expression matching.

Fourth, to conduct traditional and simplified font conversion: Some traditional Chinese characters appeared in microblog can be converted into simplified Chinese characters through the traditional Chinese-Simplified conversion function of Microsoft Office Excel.

At last, after the above process, the length of some microblogging text becomes 0 or very small, which are meaningless and should be deleted. There were 28,840 effective blogs left after data cleaning.

4.3 Data Preprocessing

First, the user dictionary built manually was imported into the "Jieba" Chinese text segmentation system to divide 28,840 valid data. Then, we used the stop words table to remove the stop words. After that, we generated word cloud map based on the frequency of the word.

Fig. 2. Cloud map.

As can be seen from Fig. 2, the words with higher frequency include: "sharing", "deposit", "ofo", "travel", "city", etc.

4.4 Clustering Based on K-Means

Set the number of clusters to 2–50 and the dimension of the space vector to 300–500 for clustering. The silhouette coefficients are as follow:

The dimension of the space vector and the number of clustering both might affect the clustering effectiveness. As can be seen from Fig. 3, the silhouette coefficients of the dimension of 300 are larger than the dimension of 400 and 500 on the whole, so the spatial vector with the dimension of 300 is selected. When the dimension is 300, the silhouette coefficient has a local maximum value of 0.151 when the number of clustering is 39. It is considered that the clustering effect is good at this time.

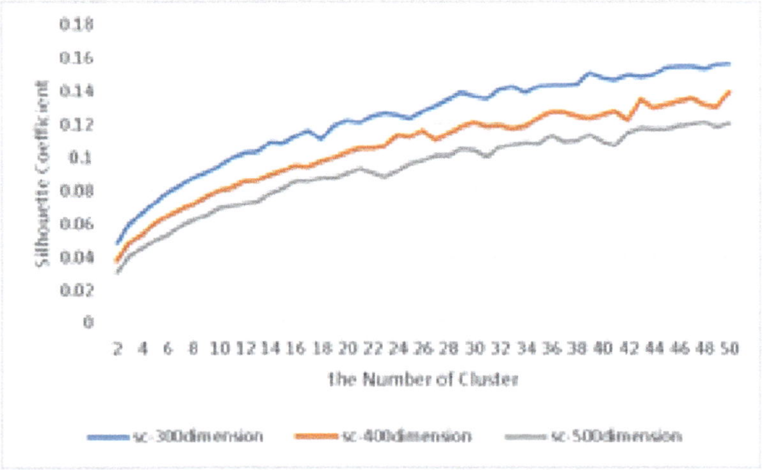

Fig. 3. Silhouette coefficient.

4.5 Theme Analysis

Randomly select 8 clusters from 39 clusters, and then select 20 features with the highest frequency in each cluster as thematic labels, from which the theme of each cluster can be concluded as follows:

- From Table 1, we can see that there are many words such as "green", "environmentally friendly", "convenience", "low-carbon" describing the merits of sharing a bike in the cluster3. Therefore, the theme of the cluster3 can be concluded that the shared bicycle is green and energy saving and can bring convenience to people's traveling.
- There are two subjects in the cluster 4. One is about the new four great inventions of China. In May 2017, young people from the 20 countries along the One Belt And One Road ranked China's "new four great inventions": high-speed rail, scanning payment, sharing of bicycles and online shopping. The other is related to companies expanding overseas markets as there are some words about "American" and "globe".

Table 1. Keywords and theme

Cluster	Top 20 keywords	Theme
Cluster3	出行(travel) 绿色(green) 城市(city) 市民(citizens) 共享(shared) 带来(to bring) 骑行(riding) 方式 (the way) 交通(transport) 环保(environmentally friendly) 环保 (low-carbon) 便利 (convenience) 公里 (kilo meter) 解决 (solve) 世界(world) 公交(bus) 地铁 (subway) 骑行 (ride) 投放 (put in) 文明(civilization)	The shared bicycle is green and energy saving and brings convenience to people's traveling.
Cluster4	中国(China) 创新(innovation) 新(new) 支付(pay) 领域(domain) 科技(science) 美国(America) 四大发明(new four inventions) 活力(energy) 网购(shopping online) 生活(life) 高铁(high-speed rail) 全球(globe) 五年(five years) 世界 (world) 技术(technology) 支付宝(Alipay) 联网(online) 计划(plan) 数据(data)	the new four inventions of China; companies to expand overseas markets
Cluster 7	乱停(disorderly parking) 乱放(leave around) 投放(put) 停放(parking) 城市(city) 出行(travel) 市民(citizen) 带来 (bring) 影响(influence) 现象(phenomenon) 共享(sharing) 整治(remediation) 管理(management) 规范(norm) 便利 (convenient) 解决(solve) 治理(governance) 城管(Urban management) 道路(street) 道(street)	the phenomenon of irregular parking
Cluster 11	就业(employment) 万人(Millions of people) 带动(drive) 行业(industry) 报告(report) 研究(research) 约(about) 国家信息中心(National Information Center) 国内(domestic) 新增(Added) 发布(release) 我国(my country) 万辆(Ten thousand) 累计(grand total) 中国(China) 投放量(The amount invested) 新(new) 经济 (economic) …	the employment problem
Cluster 22	共享(shared) 汽车(car) 经济(economic) 充电(charge) 雨伞(umbrella) 宝 电女友(girlfriend) 新(new) 发展 (development) 来了(come) 电动车(electric car) 时代(era) 宝马(BMW) 出行(travel) 继(base) 模式(model) 元 (yuan) 杭州(Hangzhou) 小时(hour) …	other Shared things
Cluster 23	感觉(feeling) 骑(ride) 骑着(ride) 不错(not bad) 自行车(bike) 走(on foot) 骑车(ride a bike) 一辆(one) 车(bike) 人生(life) 喜欢(like) 吃(eat) 城市(city) 一种(one) 早上(in the morning) 特别(special) 晚上(night) 小时(hour) 发现 (find) 做(do)	feeling of riding
Cluster 28	企业(enterprise) 投放(put) 运营(operation) 自行车(bike) 管理(manage) 共享(sharing) 车辆(vehicle) 北京(Beijing) 摩拜 杭州(Hangzhou) 停放(stop and put) 城市(city) 市场(market) 互联网(the Internet) 用户(user) 租赁(lease) 部门(department) 约(about) 发展(development) 谈(talk)	vehicle volume and the unregulated operation of companies
Cluster 30	押金(deposit) 酷骑(kuqi) 用户(user) 退(back) 退还(return) 难退(return difficultly) 酷骑 退款(Refund) 客服 (Customer service) 电话(phone) 元(yuan) 倒闭(collapse) 公司(company) 酷奇(kuqi) 骑(ride) 一家(one) 上门 (come) 小鸣(xiaomign) 濒临(on the verge) …	about deposits

- The cluster 7 mainly talks about the phenomenon of irregular parking, bring congestion to the city. Thus, authorities should issue the citywide rule to regular user's behavior.
- The main content of the cluster 11 is about the employment problem solved by the shared bicycle industries. "The National Research Center for Employment Sharing in Cycling Industry" released by the State Information Center shows that at present, China's bicycle industry has provided a total of 100,000 jobs for people [47].
- The 22nd cluster tells us that the concept of sharing is widely used. For example, following the shared bike, there has been sharing charge treasure, sharing cars, sharing umbrellas, or even sharing a boyfriend or girlfriend on Valentine's Day.
- The 23rd cluster is mainly about the feeling of riding bicycle. Most people like to ride a bike, either on their way to school, or on their way home, either in the early morning or late afternoon.
- The theme of the 28th cluster is about companies putting bicycles on the market. In order to facilitate management, the government introduced a policy to restrict enterprises operation.
- The cluster 30 mainly talks about deposit. The deposits of different brands of the shared bicycles are not the same. With the fierce competition in the shared bicycle market, some brands are coming to an end soon and users have asked the company to refund the deposits. So we suggest that companies should keep the public informed of their time limit for refund.

4.6 Manual Check

The contour coefficient is to evaluate the clustering effect from the perspective of spatial distance, not the semantic angle. In order to test the effectiveness of clustering scientifically, 100 microblogs are randomly extracted from each cluster for semantic analysis. We need to judge manually whether it is indeed close to the topic of the cluster, especially those who do not literally contain the key words of that cluster. And then we evaluate the effect of this clustering experiment from the perspective of semantics and analyze the reasons for the evaluation result.

As can be seen from Table 2, most of the sample content is close to the subject of its cluster. The majority of the micro-blog of the cluster matches the theme more than 80% except the cluster28 with a compliance of only 54%. It can be concluded that the clustering effectiveness of most clusters is good and the cohesion of clusters is high from the semantic point of view. However, there are also some noise data divided into three kinds.

Table 2. Result of mannual testing

Cluster	Close to theme			Not close to theme		
	Contain the keywords?		Total	Contain the keywords?		Total
	Yes	No		Yes	No	
Cluster3	91	0	91	9	0	9
Cluster4	86	0	86	14	0	14
Cluster7	99	1	100	0	0	0
Cluster11	97	0	97	0	3	3
Cluster22	92	0	92	8	0	8
Cluster23	85	0	85	15	0	15
Cluster28	52	2	54	1	5	6
Cluster30	94	2	96	1	3	4

The first kind of the noise data are those microblogs containing keywords but the content deviates from the theme. The reason for this phenomenon may be as follows. Firstly, the frequency of the keyword contained in the text maybe low, so the theme features is not obvious; Secondly, the K-means algorithm is the local optimal algorithm based on distance, rather than semantics, and it is vulnerable to noise data; Thirdly, the pace vector model based on the frequency of word ignores the semantic information between words, thus increasing the ambiguity between texts; Fourthly, the number of clusters may not be accurate, which easily leads to insufficient extraction of topics of each cluster (including single-topic and multi-topic mixed cases), and thus many potential topics are not excavated.

The second kind of the noise data are those not containing keywords and not close to the theme of the cluster. Such noise may be caused by the limitations of the algorithm itself and the problem of manual extraction. That is, the blog without the keyword and deviating from the subject is just counted as the noise sample.

The last kind of the noise data are those blogs which can match the theme of the cluster but not contain keywords. The reason might be that the phenomenon of multi-word synonyms makes many blog posts related to the topic even if they do not contain keywords of the cluster.

5 Conlusion and Future Work

5.1 Conclusion

For analyzing the subject of "the shared bikes" on Micro-blog, we have used the literature research method to collect the relevant research to lay the foundation for the dissertation, chosen K-means for subject analysis and evaluated the semantic effectiveness of clustering by manual check.

From the clustering results, the following points of concern are drawn. The shared bicycle is energy-saving and convenience to people's traveling and most people love the feeling of riding. The shared bicycle has become one of the new four great inventions of China in 2017. The bicycle industry has solved many employment problems and is gradually expanding to overseas market. With the prosperity of sharing bicycles, we have entered the age of sharing such as the shared charge treasure, sharing cars, sharing umbrellas. However, there are many problems with the development of shared bicycles. The irregular parking of bicycles had brought chaos to the city. With the fierce competition in the shared bicycle market, some brands are coming to an end soon and users have asked the company to refund the deposit. Subject analysis can provide a basis for decision-making in the management. Sharing the bikes is the responsibility of the entire community. Citizens should care for the shared bicycles and do not park disorderly. The enterprise shall reasonably control the quantity of vehicles. The government should introduce policies to strengthen social supervision.

Human check for verification of the true correspondence of the text meaning to the clusters is randomly applied. The experimental results show that the contour coefficient has certain limitation as the criterion of clustering effect analysis. The low contour coefficient does not mean that the clustering effectiveness is poor. Using k-means algorithm for short text clustering like microblog, the contour coefficient is generally low. But when it comes to artificial detection, it is found that the clustering effect is better in terms of semantics.

5.2 Future Work

The limitations of the study are:

(1) The conclusion of this study is only based on micro-blog texts, which is not of universal significance for other types of data.
(2) Subject analysis was conducted by clustering, and the depth of analysis is not enough.

In the future, we can enrich our research from the following three aspects.

(1) Try to use K-means to cluster other types of datasets and compare the clustering effects of different datasets from both mathematical and semantic perspectives. Then, to explore whether the algorithm is better suited to some types of data?
(2) Analyze the intensity and novelty of each subject in combination with the time factor, and display the results in a visual way.
(3) As the current clustering algorithm is generally based on distance rather than semantics, try to improve clustering algorithm from the perspective of semantics in the future. For example, feature selection can be done by constructing a domain ontology or incorporate the idea of deep learning into clustering.

Acknowledgment. This research was supported by National Natural Science Foundation of China (Grant No. 71373291). This work was also supported by Science and Technology Planning Project of Guangdong Province, China (Grant No. 2015A030401037) and Science and Technology Planning Project of Guangdong Province, China (Grant No. 2016B030303003).

References

1. Yuan, B.: Microblog Topic Mining Based on Relation Network. Harbin Institute of Technology (2014)
2. Yi, Z., Juanying, X., et al.: Clustering analysis for erythemsto-squamous diseases. J. Univ. Jinan (Sci. Technol.) **2017**(03), 181–187 (2017)
3. Zou, J., Yang, X.-Q., Wang, R.-T., Tao, X.-H.: Medication regularity of traditional Chinese medicine classical prescription depression based on apriori and clustering algorithm. Chin. J. Exp. Tradit. Med. Formulae **10**, 211–215 (2017)
4. Mamat, T., Jianhua, X.: Regionalization of agriculture mechanizaiton's efficiency base on cluster avalysis. J. Agric. Mech. Res. **08**, 27–31 (2017)
5. Qi, G., Ding, X., Xiao, X.: Research on segmentation method of crop disease images based on the fuzzy c-means. Intell. Comput. Appl. **02**, 72–74 (2017)
6. Wang, J., Feng, D., Chai, H., Liu, Y.U.: Dominant discontinuities analysis based on stereographic projection and k-means clustering algorithm. Chin. J. Geotech. Eng. **03**, 1–8 (2017)
7. Liu, J., Xue, C., Fan, Y., Kong, F., He, Y.: A raster-oriented clustering method with space-attribute constraints. J. Geo-Inf. Sci. **04**, 1–10 (2017)
8. Sun, J., Li. Z.: Reliability evaluation of DNA sequence clustering based on genetic algorithm. J. Zhejiang Sci-Tech Univ. (Nat. Sci. Edn.), 1–6 (2017)
9. Fan, S.-W., Liu, F.: Applied research of clustering analysis and association rules analysis in commodity precise marketing. East China Econ. Manag. **05**, 182–184 (2017)
10. Zhang, J., Shen, S., Huang, S., et al.: Clustering analysis of voltage contour time-sequence situaition pictures in substation-centralized distribuition network. Autom. Electr. Power Syst. **08,** 125–132 (2017)
11. Zhang, L., Xie, Z.-H.: User-types and influences of micro-blogs based on cluster analysis. Inf. Sci. **08**, 57–61 (2016)
12. Liu, J., Peng, Y., et al.: LDA-K-means algorithm of network food safety topic detection. Eng. J. Wuhan Univ. **02**, 307–310 (2017)
13. Guo, J.: Economic thinking of sharing cycling. Reform Openning **6**, 73 (2017)

14. Li, M.: Research and analysis of the shared bicycle market. Money China **05**, 121–123 (2017)
15. Huang, C.: A brief analysis of the advantages and disadvantages of sharing bicycle design from the perspective of physics – taking ofo as an example. China Strat. Emerg. Ind. **12**, 3–4 (2017)
16. Xu, X.: The optimization of the background management system of the shared bicycle App. Electron. Technol. Softw. Eng. **4**, 80–81 (2017)
17. Sharing the spirit can achieve the sharing economy. Mod. Bus. **07**, 10 (2017)
18. Xiao, Y.: Bicycle culture deserves to be shared the most. China Bicycl. **03**, 1 (2017)
19. Manning, C.D., Raghavan, P., Schütze, H.: Introduction to Information Retrieval. Cambridge University Press, Cambridge (2008)
20. Cao, J., Wu, Z., Wu, J., Xiong, H.: SAIL: summation-based incremental learning for information-theoretic text clustering. IEEE Trans. Cybern. **43**(2), 570–584 (2013)
21. Guil, N., González-Linares, J.M., Cózar, J.R., Zapata, E.L.: A clustering technique for video copy detection. In: Pattern Recognition and Image Analysis, pp. 451–458. Springer, Heidelberg (2007)
22. Gao, B., et al.: Web image clustering by consistent utilization of visual features and surrounding texts. In: Proceedings 13th Annual ACM International Conference, pp. 112–121 (2005)
23. Cao, X., Wei, X., Han, Y., Lin, D.: Robust face clustering via tensor decomposition. IEEE Trans. Cybern. **45**(11), 2546–2557 (2015)
24. Zhao, C., Du, L., et al.: Preliminary exploration based on Chinese natural language understanding. Mod. Electron. Tech. **30**(6), 82–85 (2007)
25. BaezaYates, R.A., RibeiroNeto, B., et al.: Mod. Inf. Retr. **43**(1), 26–28 (1999)
26. Zhang, J.: Study on Text Representation Model Based on Concept. Tsinghua University (2006)
27. Salton, G., Wong, A., Yang, C.S.: A vector space model for automatic indexing. Commun. ACM **18**(11), 613–620 (1975)
28. Wang, Y.: Chinese Information Processing Technology and Its Foundation. Shanghai Jiao Tong University Press (1992)
29. Chen, T., Xie, Y.: Literature review of feature dimension reduction in text categorization. J. China Soc. Sci. Tech. Inf. **24**(6), 690–695 (2005)
30. Srinivas, M., Supreethi, K.P., Prasad, E.V., et al.: Efficient text classification using best feature selection and combination of methods. In: Human Interface and the Management of Information. Designing Information Environments. Springer, Heidelberg (2009)
31. Principle and Practice of Data Mining. Publishing House of Electronics Industry (2011)
32. Salton, G., Buckley, C.: Term-weighting approaches in automatic text retrieval ☆. Inf. Process. Manage. **24**(5), 513–523 (1988)
33. Han, J., Kamber, M., Pei, J.: Data Mining: Concepts and Techniques, 3rd edn. (2011)
34. Huang, X.: Research on clustering problems and algorithms for high-dimentional data. Harbin Institute of Technology (2014)
35. Szekely, G.J., Rizzo, M.L.: Hierarchical clustering via joint between-within distances: extending ward's minimum variance method. J. Classif. **22**(2), 151–183 (2005)
36. Zhao, Y., Guo, J., Zhegn, L., et al.: Improved BIRCH hierarchial clustering algorithm. Comput. Sci. **35**(3), 180–182 (2008)
37. Wei, G., Zheng, X.: Research on the CURE algorithm of hierarchical clustering method. Sci. Technol. Ind. **5**(11), 22–24 (2005)
38. Kaufman, L., Rousseeuw, P.J.: Finding groups in data: an introduction to cluster analysis. In: DBLP (1990)

39. Zhou, T., Liu, H.: Clustering algorithm research advances on data mining. Comput. Eng. Appl. **48**(12), 100–111 (2012)
40. Cheng, Y.: Research of Chinese Short Text Clustering Algorithm. Jilin University (2016)
41. Macqueen, J.: Some methods for classification and analysis of multivariate observations. In: Proceedings of Berkeley Symposium on Mathematical Statistics and Probability, pp. 281–297 (1967)
42. Liu, Y., et al.: Understanding and enhancement of internal clustering validation measures. IEEE Trans. Cybern. **43**(3), 982–994 (2013)
43. Dunn, J.C.: A fuzzy relative of the ISODATA process and its use in detecting compact well-separated clusters. J. Cybern. **3**(3), 32–57 (1973)
44. Davies, D.L., Bouldin, D.W.: A cluster separation measure. IEEE Trans. Pattern Anal. Mach. Intell. **1**(2), 224 (1979)
45. Kaufman, L., Rousseeuw, P.J.: Finding groups in data: an introduction to cluster analysis. DBLP (1990)
46. Garrett, J.J.: Ajax: a new approach to web applications (2007). http://www.adaptivepath.com/publications/essays/archives/000385.php
47. The National Research Center for Employment Sharing in Cycling Industry. http://www.sic.gov.cn/News/250/8452.htm

Enhancement of the K-Means Algorithm for Mixed Data in Big Data Platforms

Oded Koren[1], Carina Antonia Hallin[2], Nir Perel[1(✉)], and Dror Bendet[1]

[1] School of Industrial Engineering and Management, Shenkar – Engineering, Design, Art, Ramat Gan, Israel
{odedkoren, perelnir}@shenkar.ac.il, drorbendet@gmail.com

[2] Department of International Economics, Government and Business, Copenhagen Business School, Frederiksberg, Denmark
cah.egb@cbs.dk

Abstract. Big data research has emerged as an important discipline in information systems research and management. Yet, while the torrent of data being generated on the Internet is increasingly unstructured and non-numeric in the form of images and texts, research indicates there is an increasing need to develop more efficient algorithms for treating mixed data in big data. In this paper, we apply the classical K-means algorithm to both numeric and categorical attributes in big data platforms. We first present an algorithm which handles the problem of mixed data. We then utilize big data platforms to implement the algorithm. This provides us with a solid basis for performing more targeted profiling for business and research purposes using big data, so that decision makers will be able to treat mixed data, i.e. numerical and categorical data, to explain phenomena within the big data ecosystem.

Keywords: Big data · Mixed data · Hadoop · K-means

1 Introduction

Every business or organization appears to experience a data-driven revolution in management. Firms adopt big data tools to capture enormous amounts of fine-grained data derived from social media activity, Web browsing patterns, mobile phone usage, video, audio, image, and text message usage, and new formations of data generation like mobile utilizations, messages over the internet, and IOT usages [6]. The analysis of these data promises to produce insights and predictions that will revolutionize managerial decision making [21]. Possibly, the invention of big data is "the most significant tech disruption in business and academic ecosystems since the meteoric rise of the Internet and the digital economy" [2].

As big data involves the ability to render into data many aspects of the world that have never been quantified before, also referred to as "datafication" [8], the challenge for businesses is to develop better and more simple algorithms, systems, and processes that can make sense of all of the heterogeneous and fragmented information on the

K. Arai et al. (Eds.): IntelliSys 2018, AISC 868, pp. 1025–1040, 2019.
https://doi.org/10.1007/978-3-030-01054-6_71

Web. Publications in information and management science on big data are increasingly grappling with such challenges. Top-tier information science journals, such as *Management Science* and *MIS Quarterly*, have commissioned special issues on data science, analytics, and big data, and, recently, journals on big data have been launched [2].

A big data ecosystem includes a platform that is enabled to handle a huge amount of data (in several levels) via a variety of tools. The use of big data technologies is associated with the emergence of new technical skills, such as Apache™ Hadoop®!, MapReduce, Apache Pig!, Apache Hive TM, and Apache HBase™ [27]. The early adaptation of big data tools attracted media attention, such as when Sears started to experiment with Apache™ Hadoop®!, that was central to the first wave of big data investments. Of course, Sears learned Apache™ Hadoop®! the hard way, through trial and error, since it had only a few outside experts available to guide its work when it introduced the software in 2010 [16].

The processes on large amounts of data that can be stored in Hadoop Distributed File System (HDFS™) can be executed via MapReduce jobs [9]. Furthermore, there are other functionalities, possibilities, and tools that can enable the analysis of information for various business purposes (such as machine learning algorithms). The ability to combine big data tools with different data analysis functionalities, such as Apache HIVE TM and Apache Pig!, is growing [12, 18, 22], as is the variety of other big data tools designated for handling data (like ETL process and analysis) [28]. Big data is also being studied in relation to machine learning tools such as Apache Mahout™ [19].

The approaches to dealing with the structuring of the massive volumes of data in big data are performed by different capabilities and tools [10, 28]. Apache™ Hadoop®! is a platform that includes the ability to store, manage, read, write, and operate on massive amounts of data/files via HDFS™, a system based on the Google File System (GFS) [14] with the capability of analyzing the information for different purposes [28]. Although these approaches have advanced the capabilities of dealing with massive data, they do not offer algorithms that can structure data effectively for analytical and decision making purposes. For example, IBM's Watson may be on the cutting edge in natural language processing, but it has a long way to go in terms of the system's capability for absorbing and interpreting big data across the Internet [2]. These observations reflect a need to develop new approaches for structuring and categorizing massive amounts of data in an emergent big data ecosystem.

K-means is a popular data clustering method. It is a simple and elegant approach to partitioning a dataset into K distinct clusters. This algorithm was originally described by [20]. First, a value of K is specified, and then the algorithm assigns each observation from the data set to exactly one of the K clusters. The assignment decision is done by minimizing the 'differences' between observations which belong to the same cluster. These differences are commonly measured by squared Euclidean distance, but there are many other possible ways to define this concept. A recent example involving K-means utilizations can be found in [11], where the authors studied how different types of communities may affect the effectiveness of open source software. In addition, [13] used the K-means method to investigate and identify different types of user roles in innovation-contest communities. Reference [25] applied the K-means algorithm to studying time varying effects on the allocation of marketing resources. And finally, [15] used K-means to analyze doctor's profiles.

One of the challenges with using the K-means algorithm has been that the algorithm works well with numeric data, but is not directly applicable to non-numeric, categorical data [4], since the Euclidean distance function is not meaningful when considering categorical values.

This paper presents a novel approach, which overcomes the difficulty of working with mixed data for decision making in big data. We address the question of how K-means algorithms can solve the problem of clustering mixed data in big data.

The performance of the K-means algorithm on categorical data has been studied in the information science literature, which describes how it converts multiple category attributes into binary attributes and then treats them as numeric [24]. However, this method may greatly increase the computational effort, especially when working with big data. Consequently, scholars have applied K-modes algorithms and the K-prototypes algorithm [17]. The K-modes algorithm extends the K-means method to clustering categorical data by defining differences between clusters in terms of frequencies and by considering modes instead of means. The K-prototypes algorithm is a mixture of the K-means and the K-modes algorithms. That is, the definition of a "cluster center" (or representative) allows treating a clustering problem with categorical variables to be a traditional K-means problem [26]. The general method of choosing a representative of a cluster and measuring dissimilarities between clusters is performed by relative frequency-based methods [3] or in studies applying the K-means algorithm on mixed data [3, 29]. However, the latter studies were not performed in a big data environment. For example, the numerical studies presented in [3] considered datasets with at most 690 elements.

Our contribution is to adapt the K-means algorithm on mixed big data. That is, we have used big data platforms (in terms of parallel computation techniques and storage capabilities) in order to explore how the K-means algorithm works on big data with both numeric and non-numeric variables. Since data size expands tremendously, analyzing data on a single machine is inefficient. Therefore, considering parallelism within a distributed computational framework is the most appropriate solution. One of the most common programming frameworks for processing large scale datasets through the utilization of parallelism is MapReduce [9] and the exploitation of the qualities of parallel computing [5, 7].

In this paper, we address two fundamentals: (i) We provide a clustering algorithm which handles both numeric and categorical attributes in big data environments, based on the capabilities of big data tools and the K-means algorithm; and (ii) We explore how the results of the algorithm in a big data environment, based on the ability to support complex architectures, can provide the extension of capabilities, such as clustering, profiling, analysis and predictions.

Our algorithm enables the application of the K-means algorithms to both numerical and non-numerical data. The empirical evidence is broadly supportive of the two issues we seek to address. We first create a procedure that "flattens" all the data from categorical and numerical data to pure numerical data. We then filter all the categorical classes into distinct groups, based on the categorical combinations, which allows us to analyze each group separately (since we are dealing with big data, the grouping process and the K-means process are performed via big data platforms). That is, we perform the K-means algorithm only on the remaining numeric variables. Lastly, we collect all the groups' analysis outcomes, which can serve as a basis for further analysis, in order to support the organization requirements and business needs.

The implication of our study involves the presentation of a method for treating mixed data in big data which was not previously possible. The approach advances the capabilities of dealing with massive data, such as in decision making, since profiling, forecasting, and other analyses can be performed in a more targeted manner.

Recent studies have discussed the relation between big data and theory. For example, it is suggested that big data and theory can be synergistic for exploring phenomena or problem solving by using the big data platforms and tools to generate theoretical insights and by not starting with a preconceived theory [23]. Furthermore, in [1], the author indicates that "big data has potentially important implications for theory." On the one hand, theory can be replaced by patterns derived from data. On the other hand, data without theory lacks order, sense, and meaning. We have adopted the concept presented in these studies. That is, we present a method for analyzing data in big data environments, where this method can be applied for any relevant theoretical issue.

The rest of the paper is organized as follows. In Sect. 2 we present our new alternative procedure for performing the K-means algorithm with mixed data in a big data environment. We then turn in Sects. 3–5 to an implementation example of the proposed procedure on a generated dataset of approximately 1 GB, while Sect. 6 concludes the paper.

2 Model Development

We argue that K-means applied to mixed data can enhance decision making within the big data ecosystem and allows decision makers to treat massive amounts of data. The current study thus analyzes the impact of K-means applied to both numerical and categorical (non-numerical) data in big data platforms. The model assumes a dataset, which includes m categorical variables and n quantitative variables, and that categorical variable j may have $a_j \geq 2$ different states.

The K-Means Algorithm Procedure:

Claim 1: Non-numeric data in big data can be assigned values.

Proof: We first perform the K-means algorithm on our dataset by adopting the following steps:

(1) Create $\prod_{j=1}^{m} a_j$ different types of groups, which differ by their values of the categorical variables. Each record is assigned to its group, according to its categorical values.

(2) Each group generated in step 1 is a file (or other storage format) in the big data platform (this will enable parallel computing in the next steps).
(3) Perform a parallel K-means algorithm on all groups according to the numeric variables.
(4) Aggregate all the clusters (K clusters from each group) from step 3 to one outcome for further analysis.

As a simple example, assume we have two categorical variables and three numeric variables, as follows: Gender – male/female; marital status – single/married/divorced; income; age and number of children. Table 1 presents the first step in the data mining process, which is to create six groups that differ by the values of their categorical variables.

At step 2, each one of these groups is saved as a file containing all the records with the same categorical attributes. At step 3, the K-means algorithm is performed (according to the numeric variables) simultaneously on all groups, so that, for each group, we get K clusters (the K may be different for each group, depending on the decision needs and requirements, and the number of records in each group). Step 4 is optional, and is performed according to the needs of the research.

3 Implementation Example of the Algorithm

The following section presents an end-to-end implementation example.

(1) Upload the data set and categorical files to the HDFS™ (in Apache™ Hadoop®![1]).

Pre-set: each of the $\prod_{j=1}^{m} a_j$ possible combinations of the values of each categorical variable is in a separated file. Each file contains the records with the corresponding categorical values. This is a mandatory step due to the fact that there is a need to create all combinations of the available states based on the definition/business requirements. Note that there might be empty files (groups) if there are no records with the corresponding categorical values.

(2) Multiplication of all the files (from step 1) can create multiple lines. Each line describes a unique combination. All lines are stored in a file in HDFS™ (in Apache™ Hadoop®!) for parallel analysis (in big data platform).
(3) Filter the dataset for each unique file (form step 2) and send the relevant quantitative variables to the relevant file.
(4) Run (via bash script) K-means (Apache Mahout™[2]) on each file that is located in a separated directory (from step 3) with the following parameters:
    (a) A configurable parameter x for the number of iterations (in this use case we use 5 iterations for all K-means runs).

[1] http://hadoop.apache.org/.
[2] http://mahout.apache.org/.

Table 1. The data mining process of creating groups with numerical and categorical data

Group	Categorical variables
Group 1	Male and single
Group 2	Male and married
Group 3	Male and divorced
Group 4	Female and single
Group 5	Female and married
Group 6	Female and divorced

(b) A Number of clusters (K), which is influenced by the number of records per each unique file (from step 3). The number of clusters K increases when the number of records per file grows.

(5) Gather all the clusters to one defined structure for additional analysis (compare between clusters, order, analysis, etc.).

Note that steps 1 to 3 were implemented and tested on a single-node environment. The Apache Pig! code operation includes (see next section for a detailed description):

(1) Loading the full dataset.
(2) Creating all the categorical variables combinations (3 categorical variables with different states – total of 36 groups in this use case example).
(3) Filtering the relevant Categorical variables and creating the groups/files (per each combination) with the relevant filter quantitative variables (5 variables in this use case example).

The total run duration time of steps 1 to 3 in our example was 3 min 21 s in average using a single-node environment. We neglected to include this amount of time with the total time, since it is relatively small compared with the total K-means running time. The procedure flow diagram is presented in Fig. 1, while Table 2 describes the implementation steps and guidelines.

4 Data Structure

4.1 Before Manipulation

To evaluate the performance of our K-means procedure, we tested a fictive sample in the big data ecosystem. The following nine variables were used as part of the sample data set implementation of a use case: 5 quantitative variables and 3 categorical variables. Table 3 presents the names and values of the three categorical variables.

The example illustrates the challenge of using, in some cases, the average of categorical values in case of K-means. Suppose that K-means clustering algorithm finds the marital status average of 1.5. What does it mean? Half single? Almost Married?

Next, we combined a sample of quantitative variables with the categorical data. Table 4 presents the list of the five quantitative variables and Table 5 shows raw data example before manipulation.

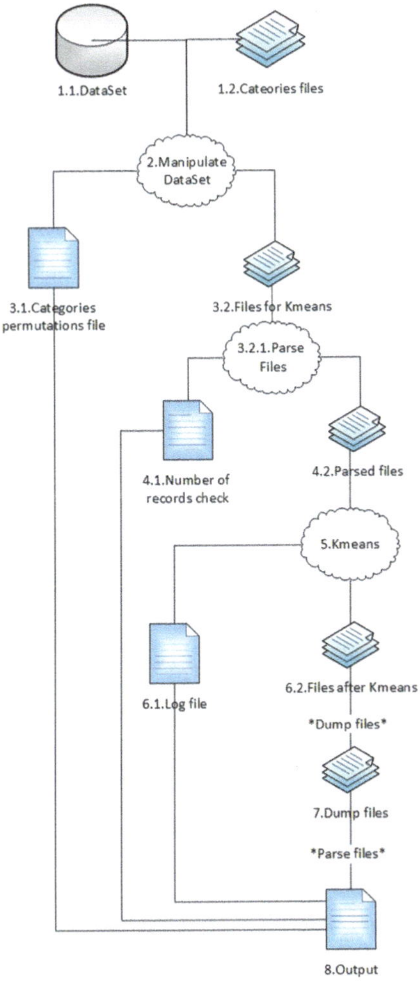

Fig. 1. Procedure flow.

4.2 After Manipulation

To create different states and make reactive the averages in the categorical variables, we performed a Cartesian product between all categorical variables, so that we had $3 \times 3 \times 4 = 36$ distinct groups as follows:

$$\mathit{3}_{\text{Marital status 3 states}} * \mathit{3}_{\text{Age range 3 states}} * \mathit{4}_{\text{Academic degree 4 state}} = \textbf{\textit{36 distinct groups}}$$

The data set was transformed into the following new data set that includes all the categorical permutations. An example of a recode is presented in Table 6.

Table 2. Procedure implementation guidelines

Step	Name	Deception	Input	Output	Implantation note
1.1	DataSet	Data set file			Load the DataSet to Apache™ Hadoop®! (HDFS™)
1.2	Categories files	Unique separated file for each category with all states			Load the DataSet to Apache™ Hadoop®! (HDFS™)
2	Manipulate DataSet		1.1. DataSet 1.2. Categories files	3.1. Categories files 3.2. Files for K-means	Apache Pig!
3.1	Categories permutations file	All permutations of the categories states			Result of cross (Cartesian product) on 1.2. Categories files
3.2	Files for K-means	Make files to run K-means on them			
3.2.1	Parse files	Manipulate 3.2 files for K-means files	3.2. Files for K-means	4.1. Length check 4.2. Parsed files	Delete the category column
4.1	Number of records check	Checks the number of records			Inner validation – uses in 8. Output
4.2	Parsed files	Files are ready to run in K-means (in directories)			
5	K-means	Run K-means on each directory	4.2 Parsed files	6.1. Log file 6.2. Files after K-means	Apache Mahout™. K driven from number of records. Max of iterations is predefined.
6.1	Log file	Record the time for each run			

(*continued*)

Table 2. (*continued*)

Step	Name	Deception	Input	Output	Implantation note
6.2	Files after K-means	Result of K-means algorithm			
7	Dump files	Dump of the result of 6.2 Files after K-means			Apache Mahout™ Apache™ Hadoop®!
8	Output	Aggregate all dump files to csv/excel			Parsing 7. Dump files and aggregate them to csv/excel file

Table 3. States of categories

#	Variable	Number of values	Values
1	Marital status	3	Single, Married, Divorcee
2	Age range (selection)	3	20–34, 35–49, 50–64
3	Academic degree	4	Non, First degree, Second degree, Third degree

Table 4. Quantitative variables

#	Variable
1	Salary (amount)
2	Clothing spending (month)
3	Distance from work (km)
4	Working hours (average day)
5	Food spending (month)

Table 7 presents the size and capacity of the dataset that was used for this implementation example.

Meanwhile, Table 8 presents the entire list of all 36 files/groups permutations (from the full dataset) and their capacity (number of records and size in Kbytes).

Table 5. Raw data example before manipulation

Age	AcadDegr.	Marital status	Salary	Clothing spending per month	Distance from work (km)	Working hours (average day)	Food spending (month)
20–34	Second degree	Married	4,801	677	106	5	322
20–34	First degree	Divorcee	5,244	2,396	87	6	4,388
35–49	First degree	Single	5,566	3,958	28	9	2,236
20–34	Non	Single	1,776	637	61	8	1,680

Table 6. Categorical permutations example

Field	Value
Category	20-34_First-degree_Divorcee
Salary	1015
Clothing spending (month)	4274
Distance from work (km)	68
Working hours (average day)	11
Food spending (month)	2466

Table 7. Data capacity

Parameter	Size
Size	~1 GB (975 MB)
Total Number of records	19,600,000
Number of groups/files* (see algorithm description)	36

Table 8. Group file permutations and capacity

#	Name/description	Size (KBytes)	Number of records
1	20-34_First-degree_Divorcee	11,364	537,432
2	20-34_First-degree_Married	11,312	516,264
3	20-34_First-degree_Single	11,260	573,888
4	20-34_Non_Divorcee	11,148	520,576
5	20-34_Non_Married	11,144	575,848
6	20-34_Non_Single	11,048	553,896
7	20-34_Second-degree_Divorcee	11,036	536,256
8	20-34_Second-degree_Married	11,024	546,056
9	20-34_Second-degree_Single	11,004	571,536
10	20-34_Third-degree_Divorcee	10,940	565,264
11	20-34_Third-degree_Married	10,924	522,928
12	20-34_Third-degree_Single	10,912	558,600
13	35-49_First-degree_Divorcee	10,888	554,680

(*continued*)

Table 8. (*continued*)

#	Name/description	Size (KBytes)	Number of records
14	35-49_First-degree_Married	10,828	552,328
15	35-49_First-degree_Single	10,788	559,776
16	35-49_Non_Divorcee	10,784	539,392
17	35-49_Non_Married	10,740	543,704
18	35-49_Non_Single	10,720	526,848
19	35-49_Second-degree_Divorcee	10,700	558,208
20	35-49_Second-degree_Married	10,680	530,376
21	35-49_Second-degree_Single	10,664	531,944
22	35-49_Third-degree_Divorcee	10,656	558,208
23	35-49_Third-degree_Married	10,624	536,648
24	35-49_Third-degree_Single	10,604	565,656
25	50-64_First-degree_Divorcee	10,588	542,136
26	50-64_First-degree_Married	10,576	541,352
27	50-64_First-degree_Single	10,568	546,840
28	50-64_Non_Divorcee	10,524	536,648
29	50-64_Non_Married	10,492	539,392
30	50-64_Non_Single	10,480	545,664
31	50-64_Second-degree_Divorcee	10,432	529,592
32	50-64_Second-degree_Married	10,400	551,544
33	50-64_Second-degree_Single	10,316	508,032
34	50-64_Third-degree_Divorcee	10,272	548,800
35	50-64_Third-degree_Married	10,196	532,728
36	50-64_Third-degree_Single	9,996	540,960
Total		386,632	19,600,000

5 Output Structure

The outcomes of the implementation use case include the following products:

- Running log
- Output file – union of all outcomes clusters for future analysis and additional insights
- Performance example.

5.1 Running Log

The running log in Table 9 was designed for the implementation use case and is a text structure file that includes the starting time and the ending time for the entire process. The log also contains the following values per each file/category (permutation).

- Time (start/initiate)
- Time (start for each iteration)

Table 9. Example of log outcomes (partial log)

```
. . .
20-34_First-degree_Divorcee - new terminal -
2017-06-01--04:01:14.680
20-34_First-degree_Divorcee - before - 2017-06-
01--04:01:14.757 n = 1371 k = 10 num of iter = 2
20-34_First-degree_Married - new terminal -
2017-06-01--04:01:17.832
20-34_First-degree_Married - before - 2017-06-
01--04:01:17.912 n = 1317 k = 10 num of iter = 2
. . .
. . .
. . .
35-49_Third-degree_Single - new terminal - 2017-
06-01--04:03:23.431
35-49_Third-degree_Single - before - 2017-06-01-
-04:03:23.575 n = 1443 k = 10 num of iter = 2
50-64_First-degree_Divorcee - new terminal -
2017-06-01--04:03:29.319
50-64_First-degree_Divorcee - before - 2017-06-
01--04:03:29.506 n = 1383 k = 10 num of iter = 2
. . .
. . .
. . .
 20-34_First-degree_Divorcee - after - 2017-06-
              01--04:15:36.280
20-34_First-degree_Married - after - 2017-06-01-
              -04:15:55.164
20-34_First-degree_Single - after - 2017-06-01--
              04:19:41.963
. . .
. . .
. . .
```

- Number of rows/records
- Number of max iterations
- Number of selected K (clusters)
- Time (finish/end).

5.2 Clusters Log

The clusters log includes the K-means results per each permutation (group/file). To ensure the possibility of advanced/additional analysis, machine learning capabilities, different AI functionalities, and different BI opportunities, we decided to gather the entire K-means results (from all groups/files) into a structure that allows us to identity the specific categorical permutation. Note that each group can have between 0 to K clusters. The value of K per each group is in the log file (see Running log).

The head line of the clusters log contains the list of variables designated as the categorical, cluster length, c type (vector of the mean values of the centroid) or r type (vector of cluster's radius).

Each record in the clusters log presents the K-means results per each group in the order of the variables as described in the head line of the Clusters log. Remember that the K-means algorithm is performed only on the quantitative variables after the partition of all records into their corresponding groups.

Please also note that the Clusters log enables advanced BI, AI, and additional machine learning algorithm functionalities on the results (from simple queries, such as sorting, ordering, and selecting, to more complicated and sophisticated possibilities of comparing between clusters, running additional machine learning algorithms on the Clusters log, and other functionalities) (Table 10).

Table 10. Cluster log outcomes (partial clusters log)

```
Categories, n, c:Salary, c:Clothing-spending-
(month), c:distance-from-work-(km), c:working-
hours-(avrage day), c:food-spending-(month),
r:Salary, r:Clothing-spending-(month),
r:distance-from-work-(km), r:working-hours-
(avrage day), r:food-spending-(month)
20-34_First-
degree_Divorcee,154,8839.682,3392.682,71.799,7.9
22,3145.383,779.165,696.034,40.717,2.062,1060.44
7
20-34_First-
degree_Divorcee,99,7546.01,1480.99,75.121,8.061,
730.192,1087.673,864.4,43.121,1.963,341.532
20-34_First-
degree_Divorcee,158,2615.892,1613.797,78.918,7.9
75,3403.715,880.222,938.229,42.779,1.949,699.134
......
```

Table 11 presents partial edited outcomes log of a basic K-means algorithm (10 clusters with 10 iterations, using Apache Mahout™), which was performed on all variables, numeric and categorical (for comparison), of the original data set, i.e. 19,600,000 records (Table 8).

Table 11. K-means outcomes (partial outcomes)

```
num of records, c:Age-Range, c: Academic-
Degree, c:Marital-Status, c:Salary, c:Clothing-
spending-(month), c:distance-from-work-(km),
c:working-hours-(avrage day), c:food-spending-
(month) ...
...
1843968, 1.992, 1.510, 2.017, 2200.227,
3060.820, 78.138, 8.004, 3398.305 ...
...
1752632 , 1.998, 1.504, 1.994, 6508.183,
1198.528, 77.622, 8.006, 3526.225 ...
...
```

As presented in Table 11, some of the outcomes are not so insightful, for example, an academic degree of 1.510 – is it between first degree and second degree? In business matters, the information resulting from a basic K-means (without the Enhancement) might be useless for organizations and for their decision making processes. Our approach presented in this paper overcomes the challenge of performing a more targeted decision making.

5.3 Performance Example - Running Times

Table 12 presents 5 runs of steps 1–3 in the above implementation procedure:

In Table 13, we present the running times of step 5 (as described in the procedure flow) for the above example, where we executed the process in big data multi-node environment. Table 14 presents the results for a single-node environment.

The differences between the running times in Tables 13 and 14 are well observed. Indeed, working in a multi-node environment allows us to perform tasks in parallel, and therefore is much more efficient.

Table 12. Dataset manipulation run time

Pig runtime			
Average	Running time	Starting time	Ending time
00:03:21	00:03:18	23:59:10	23:55:52
	00:03:21	00:05:50	00:02:29
	00:03:21	00:10:56	00:07:35
	00:03:21	00:15:13	00:11:52
	00:03:22	00:19:44	00:16:22

Table 13. K-means in multi-node environment

Total runtime	Run #
00:26:39.339	M1
00:24:53.432	M2
00:26:01.409	M3
00:28:46.028	M4
00:24:44.687	M5
00:24:03.577	M6
00:25:35.413	M7
00:26:05.337	M8
00:24:04.655	M9
00:24:23.075	M10

Table 14. K means single-node environment

Total runtime	Run #
03:21:08.783	S1
03:20:44.000	S2
03:15:14.948	S3
03:16:03.084	S4
03:21:31.207	S5

6 Implications of the K-Means Algorithm Implementation in Big Data

6.1 Limitations

While we found that the implementation of the K-means algorithm worked well in the runs, the complexity analysis of our suggested procedure must be tested in future studies. However, we argue that the complexity of our process is better when comparing it to the complexity of a regular K-means algorithm that runs on the full dataset, due to the ability to reduce the size of the dataset. The procedure will run on subsets that possess less recodes per group. This will influence the number of K-means iterations per group.

Also note that in a big data environment, all the K-means calculations can be done in parallel, i.e. in different data nodes. Therefore, we believe that the complexity will be mostly influenced by the size of the largest group that will be generated.

6.2 Implications for Theory

We presented a method for analyzing data in big data environment, where this method can be applied for any relevant theoretical question in a big data environment; for example, exploring a certain phenomenon, deriving patterns from data, improving decision making methods and running predictions.

6.3 Implications for Practice

This paper presents a new approach, which overcomes the difficulty of working with mixed data for decision making in a big data environment. The power of clustering and narrowing down the profiles to targeted groups, based on the business needs, improves the decision making process.

6.4 Conclusions and Future Work

We demonstrated the strength of the enhancement outcomes compared to basic k-means outcomes. Our approach allows to present clusters with diverse information in a very straightforward method. This procedure enables an organization with a more accurate analysis of data, and may create better business understanding and insights from the business data for a variety of services and needs. This implementation may also improve business decision making processes due to business data comprehension.

Future work of this research can be to investigate the capabilities to implement and analyze the enhancement outcomes on different business sectors. This can be done by adjusting the procedure implementation for a selected business use case sector. Also, further research can include an examination of the ability to use the procedure outcomes for variety of business AI processes, such as: business prediction, business profiling, and business targeting.

References

1. Abbasi, A., Sarker, S., Chiang, R.H.: Big data research in information systems: toward an inclusive research agenda. J. Assoc. Inf. Syst. **17**(2) (2016)
2. Agarwal, R., Dhar, V.: Editorial—big data, data science, and analytics: the opportunity and challenge for IS research. Inf. Syst. Res. **25**(3), 443–448 (2014)
3. Ahmad, A., Dey, L.: A K-mean clustering algorithm for mixed numeric and categorical data. Data Knowl. Eng. **63**(2), 503–527 (2007)
4. Berkhin, P.: A survey of clustering data mining techniques. In: Grouping Multidimensional Data, pp. 25–71. Springer, Berlin (2006)
5. Cai, X., Nie, F., Huang, H.: Multi-view K-means clustering on big data. IJCAI (2013)
6. Cisco: The Zettabyte era: trends and analysis. White paper (2016)

7. Cui, X., Zhu, P., Yang, X., Li, K., Ji, C.: Optimized big data K-means clustering using MapReduce. J. Supercomput. **70**(3), 1249–1259 (2014)
8. Cukier, K., Mayer-Schoenberger, V.: The rise of big data: how it's changing the way we think about the world. Foreign Aff. **92**(3), 28–40 (2013)
9. Dean, J., Ghemawat, S.: MapReduce: simplified data processing on large clusters. Commun. ACM **51**(1), 107–113 (2008)
10. Demchenko, Y., Ngo, C., Membrey, P.: Architecture framework and components for the big data ecosystem. J. Syst. Netw. Eng. 1–31 (2013)
11. Di Tullio, D., Staples, D.S.: The governance and control of open source software projects. J. Manag. Inf. Syst. **30**(3), 49–80 (2013)
12. Engelberg, G., Koren, O., Perel, N.: Big data performance evaluation analysis using Apache Pig. Int. J. Softw. Eng. Appl. **10**(11), 429–440 (2016)
13. Füller, J., Hutter, K., Hautz, J., Matzler, K.: User roles and contributions in innovation-contest communities. J. Manag. Inf. Syst. **31**(1), 273–308 (2014)
14. Ghemawat, S., Gobioff, H., Leung, S.T.: The Google File System, ACM SIGOPS Operating Systems Review, vol. 37, pp. 29–43 (2003)
15. Guo, S., Guo, X., Fang, Y., Vogel, D.: How doctors gain social and economic returns in online health-care communities: a professional capital perspective. J. Manag. Inf. Syst. **34** (2), 487–519 (2017)
16. Henschen, D.: Why Sears is going all-in on Hadoop. InformationWeek (2012)
17. Huang, Z.: Extensions to the k-means algorithm for clustering large data sets with categorical values. Data Min. Knowl. Discov. **2**, 283–304 (1998)
18. Kendal, D., Koren, O., Perel, N.: Pig vs. hive use case analysis. Int. J. Database Theory Appl. **9**(12), 267–276 (2016)
19. Landset, S., Khoshgoftaar, T.M., Richter, A.N., Hasanin, T.: A survey of open source tools for machine learning with big data in the Apache™ Hadoop®! ecosystem. J. Big Data **2**(1), 24 (2015)
20. MacQueen, J.B.: Some methods for classification and analysis of multivariate observations. In: Proceedings of the 5th Berkeley Symposium on Mathematical Statistics and Probability, pp. 281–297 (1967)
21. Manyika, J., Chui, M., Brown, B., Bughin, J., Dobbs, R., Roxburgh, C., Byers, A.H.: Big data: the next frontier for innovation, competition, and productivity. McKinsey Global Institute (2011)
22. Preethi, R.A., Elavarasi, J.: Big data analytics using Hadoop tools—Apache Hive vs Apache Pig. Int. J. Emerg. Technol. Comput. Sci. Electron. **24**(3) (2017)
23. Rai, A.: Synergies between big data and theory. Manag. Inf. Syst. Q. **40**(2), iii–ix (2016)
24. Ralambondrain, H.: A conceptual version of the K-means algorithm. Pattern Recogn. Lett. **16**(11), 1147–1157 (1995)
25. Saboo, A.R., Kumar, V., Park, I.: Using big data to model time-varying effects for marketing resource (re) allocation. MIS Q. **40**(4) (2016)
26. San, O.M., Huynh, V.-N., Nakamori, Y.: An alternative extension of the k-means algorithm for clustering categorical data. Int. J. Appl. Math. Comput. Sci. **14**(2), 241–248 (2004)
27. Tambe, P.: Big data investment, skills, and firm value. Alok Gupta, pp. 1452–1469 (2014)
28. White, T.: Hadoop: The Definitive Guide, 4th edn. OReilly Media, Sebastopol (2015)
29. Xu, R., Wunsch, D.: Survey of clustering algorithms. IEEE Trans. Neural Netw. **16**(3), 645–678 (2005)

An Exploratory Study of the Inputs for Ensemble Clustering Technique as a Subset Selection Problem

Samy Ayed(✉), Mahir Arzoky, Stephen Swift, Steve Counsell, and Allan Tucker

Brunel University London, Middlesex, UK
samy.ayed@brunel.ac.uk

Abstract. Ensemble and Consensus Clustering address the problem of unifying multiple clustering results into a single output to best reflect the agreement of input methods. They can be used to obtain more stable and robust clustering results in comparison with a single clustering approach. In this study, we propose a novel subset selection method that looks at controlling the number of clustering inputs and datasets in an efficient way. The authors propose a number of manual selection and heuristic search techniques to perform the selection. Our investigation and experiments demonstrate very promising results. Using these techniques can ensure better selection methods and datasets for Ensemble and Consensus Clustering and thus more efficient clustering results.

Keywords: Ensemble clustering · Consensus clustering Subset selection problem · Heuristic search · Machine learning

1 Introduction

Clustering is the process of differentiating groups inside a given set of objects. The resulting groups are assigned so that objects that are with each subset are more closely related to each other than objects that are assigned in different subsets. There are various practical applications involving the grouping of a set of objects into a number of non-overlapping subsets. Partitioning methods formulated on the relationship between objects through correlation or other distance metrics are mutually known as clustering algorithms [18]. There is extensive work in the field of clustering, with many clustering algorithms being developed. Each one of these algorithms can utilise different similarity methods and/or have a different objective function. In addition, varying the parameters of the same method or different methods and applying it on the same data can produce varying results. Moreover, clustering methods can perform well on some datasets but not on others. Thus, an essential question to ask is: given the number of clustering methods and datasets, how do we choose between them?

One way to solve the varying clustering results is through the use of input formulated from multiple clustering results, a technique called Ensemble Clustering that has been gaining popularity recently. The Ensemble Clustering problem aims to operate summation in order to gain a representative clustering with the least variability and an augmentation in mutual agreement. Ensemble Clustering overcomes inherent biases in the search method by relaxing constraints to accept a solution rated as poorer than

K. Arai et al. (Eds.): IntelliSys 2018, AISC 868, pp. 1041–1055, 2019.
https://doi.org/10.1007/978-3-030-01054-6_72

neighbouring solutions. This allows clustering algorithms to expand their local maxima and further explore the search space beyond local bounds. The ensemble methodology explores beyond the local maxima through the use of an input, formulated from multiple clustering results to return a clustering result based upon the agreement between these inputs.

Ensemble Clustering was first introduced by Strehl and Ghosh [27], has gained great momentum and has been covered in the literature of [1, 5, 7, 12, 14, 15, 23]. This problem relates to a structure which targets a combination of a set of multiple clustering solutions or partitions into a single concrete clustering which optimises the data shared amongst all the accessible clustering solutions. Undoubtedly, single clustering may not always generate promising results and similarly, multiple algorithms may independently make inferior choices by assigning some elements to the wrong clusters [25]. However, by taking into account the outcome of several different clustering algorithms jointly, better results can be achieved by mitigating degeneracies in distinctive solutions. These problems have been further investigated in the literature of [10, 13, 21] in the context of the stability and accuracy of the results. The Ensemble Clustering technique has been shown to work on different datasets and especially well in bioinformatics [28]. It has highlighted the potential of certain algorithms in applications to synthetic datasets. While ensemble clustering is becoming ubiquitous in other aspects of data mining, however, there has been little research into the implementation of an ensemble approach in the area of cluster analysis [28].

Consensus Clustering (CC) is an ensemble method which uses an agreement matrix to serve as a communal reservoir of clustering inputs, from which a consensus can be iteratively obtained. CC receives a number (r), of clustering results as input and returns a single consensus based on the level of agreement between these inputs. It is oriented towards generating an optimum solution by iteratively comparing pairwise clustering methods to determine a level of agreement. CC subsequently filters out poor agreements until an optimised clustering ensemble is generated.

For this project, we look to model the behaviour of a variety of clustering methods (inputs for consensus clustering) and datasets to create a large number of synthetic datasets for investigating consensus clustering and the way it performs with the inputs. However, before this can be accomplished it is difficult to get representative datasets just by performing experiments on all datasets since they are of different properties and sizes. In addition, datasets can work well on some clustering techniques and not others.

A number of studies have looked at the cluster ensemble selection problem [4, 6, 11]. Given a large library of clustering methods, the problem looks at identifying and selecting a subset from the library to produce a smaller cluster ensemble with an equivalent or better clustering methods performance. The majority of studies use metrics that may influence the performance of ensemble clustering such as diversity, quality or accuracy to design the methods. However, there seems to be a lack of approaches which look at identifying the optimal subset of both clustering inputs (for Ensemble Clustering) and datasets. We propose a different metric, Weighted Kappa (WK), which measures agreements between the clustering inputs for all selected datasets [2]. One advantage of using the WK metric is that there is an interpretable table that can provide input on the quality of the results. To our knowledge, this is the first study that makes use of this metric in the cluster ensemble selection problem, in addition, no previous study has looked at selecting both the clustering inputs and datasets for CC using heuristic search techniques. Hence, this paper investigates a novel combinatorial optimisation technique that looks at identifying and selecting a suitable subset for benchmarking and testing Ensemble clustering by controlling the number of inputs and datasets in a much more efficient manner. We propose two manual selection methods and three heuristic techniques (Genetic Algorithm, Simulated Annealing and Hill Climbing) to perform the selection. We believe that by using heuristic algorithms it is possible to obtain better selection results. A quality metric needs to be defined to maximise the number of inputs and datasets to include in CC. We look to investigate if the inputs and datasets chosen are enough to produce a model out of them i.e., evaluate the results to see if there is a distribution that they fit into. We look to extend this work to explore and assess the CC methodology for evaluating the efficiency of a given clustering method clusters more comprehensively.

The paper is organised as follows. In the next section, we describe the datasets and the clustering techniques chosen for this study. In Sect. 3, we introduce our quality measurement metric, Weighted Kappa. Section 4 details the experimentations conducted to select the appropriate WK threshold. The process of matrix creation and data preparation is introduced in Sect. 5. Section 6 details the experimental methods chosen for this study. A total of five proposed methods are introduced in this section. Section 7 explains the results produced from the experiments and Sect. 8 presents the post-analysis results. Finally, Sect. 9 gives a brief description of our conclusions and presents future directions.

2 Datasets and Consensus Clustering Inputs

2.1 Datasets

The datasets used for this paper are derived from various data repositories used by the machine learning community for the empirical analysis of machine learning algorithms. Particular attention is given to the type and nature of the datasets selected with strong emphasis on real-world data. We collated a wide range of data categories mainly clustering data that includes; bio-medical, statistical, botanical and ecological. Data were collected from: MLData [22], UCI Machine Learning Repository [9], Kaggle Datasets [19], StatLib [26] and Time Series Data Library [17].

The database currently contains 198 datasets, with attributes (number of columns) ranging from 3 to 167 and instances (number of rows) up to 4,898. All datasets went through a data cleansing process to make sure that they were accurate and in the correct format before running them on the clustering methods. The authors would like to note that the expected clustering arrangements are known for all datasets.

2.2 Clustering Methods (Inputs)

There are broad classes of traditional approaches to the clustering problem, for this work we present a wide range of different clustering algorithms (input methods), 32 in total.

This allowed us to be more confident about the reliability of our methods. R was used to implement the inputs for CC, which in this case was used on our subset selection problem. The R script produces the 32 inputs selected and the expected clustering arrangement for the 198 datasets. Table 1 displays a summary of the input methods selected for this work and the number of variations implemented for each method.

Table 1. Clustering methods (inputs) summary

Clustering methods	Details	Variations
K-means	The 'stats' package is used for implementing the K-means function. The following algorithms were used: Forgy, Lloyd, MacQueen and Hartigan-Wong	4
Hierarchical clustering	The agglomeration methods are Ward, Single, Complete, Average, Mcquitty, Median and Centroid. Two versions of the methods are produced, using both Euclidian and Correlation distance methods. The 'stats' package is used	14
Model-based clustering	Model-based clustering is implemented using a contributed R package called 'mclust'. The following identifiers is used VII, EEI, VVI, EEV and VVV	5
Affinity Propagation (AP)	An R package for AP clustering called 'apcluster' is used. AP was computed using the following similarity methods: negDistMat, expSimMat and linSimMat	3
Partitioning Around Medoids (PAM)	A more generic version of the K-means method is implemented using the 'cluster' package. Two similarity distance methods are used: Euclidean and Correlation	2
Clara (partitioning clustering)	Clara is a partitioning clustering method for large applications. It is part of the 'cluster' package	1
X-means clustering	An R Script based on [24]	1
Density Based Clustering of Applications with Noise (DBSCAN)	A density-based algorithm as part of the 'dbscan' package	1
Louvain clustering	A multi-level optimisation of modularity algorithm for finding community structure	1

3 Weighted-Kappa

WK is a simple statistical metric derived from Cohen's Kappa Coefficient of Agreement [8]. It is used for measuring the inter-rater agreement between two or more observers. In clustering problems, this allows for a comparative assessment of two or more components. Moreover, the class relationships between clustering arrangements generated through multiple clustering techniques can be evaluated. WK evaluates both the similarities and disagreements between pair-wise clustering arrangements in a matrix. This allows for agreements between different inputs to be produced. WK compares clusters that generate a score within the range -1.0 to $+1.0$ where -1.0 denotes no concordance and $+1.0$ denotes complete concordance between the clustering arrangements. The distinction between the score establishes the structure of the arrangements. For example, a high WK value indicates that the two arrangements are very similar, whilst a low value indicates that they are dissimilar. A value close to 0.0 is usually observed for random clusters, indicating they are not similar and has no values in common. Table 2 shows the interpretation table for WK values.

Table 2. Agreement strength of Weighted-Kappa

Weighted-Kappa	Agreement strength
$0.0 \leq K \leq 0.2$	Poor
$0.2 < K \leq 0.4$	Fair
$0.4 < K \leq 0.6$	Moderate
$0.6 < K \leq 0.8$	Good
$0.8 < K \leq 1.0$	Very good

Some of the inputs and the datasets are poor for consensus clustering; however, it is difficult to easily identify the inputs and datasets that are poor as unsupervised learning has no gold standard to compare it to. Thus, this research work relies on the WK metric for the evaluation of the inputs. WK is implemented to measure the similarity between inputs for all of the selected datasets. An equivalent metric to the WK metric is Hubert-Arabie Adjusted Rand [16]. They can be both used in cluster analysis for comparing two clustering inputs obtained from different clustering methods. WK was selected for this study because of its similarity with Adjusted Rand and the benefit of having a qualitative interpretation, c.f. Table 2.

4 Defining the Threshold

As mentioned earlier, certain inputs and datasets can produce poor WK values. There is therefore a need to select an appropriate threshold value. Data that does not cluster are data that has an average WK value of less than that of a particular threshold. Even though the WK interpretation table, Table 2 displays a rough scale of what is expected for the agreement strength, another way was needed to define the threshold value. Thus, we conducted a simulation experiment which generated a million pairs of random clustering arrangements of different number of variables, n. The values of n start at 100 and increments by 100 each time until it reaches 1,000 (10 different sizes). Then, two random clusters are chosen and the WK values of these two clustering arrangements are recorded. This is repeated for all clustering arrangements produced. The simulations results were plotted and the distribution of WK is observed to find the most appropriate threshold. The maximum, minimum, average and standard deviation WK values of the million random clusters, for each of the 10 sets of n variables, are computed and displayed on a plot (Fig. 1). The plot clearly shows a decreasing trend of the maximum WK values as the number of variables increase - the maximum value for a million simulations is rapidly approaching zero. We assume that the same pattern would continue after $n = 1{,}000$, continuing to get closer to zero. However, we believe it will never reach zero along the x-axis as the model seems to be asymptotic.

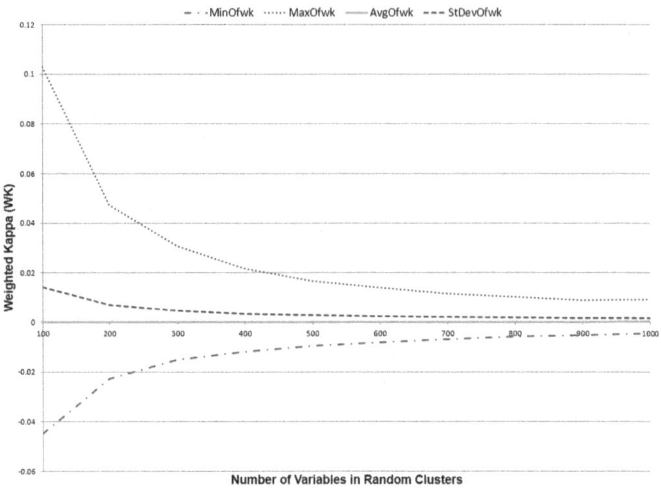

Fig. 1. Simulation experiment results plot of a million two random clustering arrangements of 10 varying number of variables, n.

From the simulations conducted it is clear that there is a very small chance of n variables to be above WK value of 0.1, as the maximum value starts off from approximately 0.1. In fact, the only time a WK value of 0.1 is attained was for the 100 variables run and that was only for one value out of a million simulations. Analysing the rest of the results from the simulations, it seems that this particular value is an outlier as the next maximum WK value achieved was for 0.08. There were only 12 points produced above the 0.08 threshold (out of a million simulations) and there are no points for the 0.09 WK value scale.

Since the WK value range is between +1.0 and −1.0, the minimum WK values generated from the simulation are negatives values. The minimum WK value seems to show the same trend as the maximum plot, except that it is on the opposite axis and that the negative values seem to be lower for the majority of the simulations. The average WK values produced from the simulations is around zero, as the simulations are producing small negative values as well as small positive values, this is indicated by the minimum and maximum plots. In addition, for the standard deviation plot, it seems that as the number of variables increases the range between them starts tightening down i.e., as the minimum and maximum plots converge, the standard deviation gets smaller. Both the mean and the standard deviation approach zero as the number of variables increase in size.

The average number of variables for the 198 datasets is 421. Thus, from the plot of the simulations, it can be seen that for 421 variables, the maximum WK value that can be produced is around 0.02. Anything above 0.02 is unlikely to occur at random. A WK threshold of 0.1 is still quite high, as the chance of two clustering arrangements that are being compared being random is extremely rare and if they are not random then they must be similar instead. However, the authors have decided to use it as the threshold due to the maximum WK value produced from the simulations was 0.1 and also due to it being half-way between the poor scales of the agreement table (Table 2).

5 Matrix Creation and Problem Definition

For this work, 198 datasets were collated and 32 clustering techniques were selected as inputs. On an initial inspection of the datasets and inputs, it is clear that some of the datasets do not cluster well and some of the clustering methods are not as effective as others on the datasets. It is difficult to get representative datasets just by performing experiments on all datasets as they are all of different sizes and properties. The same difficulty can be said for the inputs (clustering methods).

$$\begin{bmatrix} W_{11} & W_{12} & W_{13} & \cdots & W_{1m} \\ W_{21} & W_{22} & \cdots & \cdots & W_{2m} \\ W_{31} & \vdots & & & W_{3m} \\ W_{41} & \vdots & & & W_{4m} \\ \vdots & \vdots & & & \vdots \\ \vdots & \vdots & & & \vdots \\ \vdots & \vdots & & & \vdots \\ W_{n1} & W_{n2} & \cdots & \cdots & W_{nm} \end{bmatrix}$$

Since all of the datasets under analysis contain the expected clustering arrangements, the techniques used in this work can be verified. To address this problem, a 198 × 32 matrix of the WK values of the inputs' (clustering methods) clustering arrangements versus the expected clustering arrangements for each of the datasets was constructed. Thus, let W be an n rows (number of datasets) by m columns (number of inputs) real matrix where the ith, j^{th} value w_{ij} is the WK of input j (the actual clustering arrangement versus the expected clustering arrangement) applied to dataset i. Figure 2 displays a simplified representation of the matrix.

$$\begin{bmatrix} W_{11} & W_{12} & W_{13} & \cdots & W_{1m} \\ W_{21} & W_{22} & \cdots & \cdots & W_{2m} \\ W_{31} & \vdots & & & W_{3m} \\ W_{41} & \vdots & & & W_{4m} \\ \vdots & \vdots & & & \vdots \\ \vdots & \vdots & & & \vdots \\ \vdots & \vdots & & & \vdots \\ W_{n1} & W_{n2} & \cdots & \cdots & W_{nm} \end{bmatrix}$$

Fig. 2. Matrix representation of WK values (clustering inputs versus datasets).

To aid in the visualisation of the WK matrix (198 × 32), a heatmap of the WK values of the datasets and inputs was produced, shown in Fig. 3. An R package 'stats' (Version 3.5.0) was used to create the heatmap. WK values of 0.0 are shown in white (indicating poor results) and WK values of 1.0 are shown in black (indicating identical clustering arrangements). Values between 0.0 and 1.0 are shown as shadows of grey. It can be clearly seen from the figure that many of the WK values of the inputs (compared to the actual clustering arrangements) for the datasets are poor. This indicates that some of the inputs do not cluster well on all of the datasets and that some of the datasets do cluster well at all. Thus, being able to identify the inputs and datasets that are poor and to exclude them from the matrix is important. The aim is to find the best balance between inputs and datasets. Manually removing the poor datasets or poor inputs would alter the row averages and column averages as they are interconnected. Thus, selecting the appropriate datasets and inputs becomes a sub-selection problem where the goal is to include as many datasets and as many clustering methods as possible.

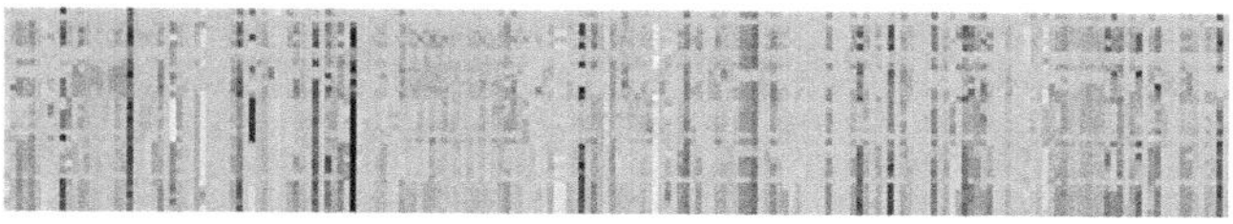

Fig. 3. A heatmap representation of the WK matrix.

6 Experimental Methods

As it is a subset selection problem and we look to maximise both datasets and clustering methods (inputs), we propose five methods for selecting the inputs and datasets. Two of the methods are based on manual selection (MS1 and MS2) and the other three are based on heuristic search techniques (Random Mutation Hill climbing, Simulated Annealing and Genetic Algorithms). The 198 × 32 matrix is used as input for the sub-selection algorithm and the output is a subset containing the selected datasets and inputs. As previously explained in Sect. 4, a WK threshold of 0.1 was selected and incorporated into the techniques proposed. The aim was to include as many datasets (rows) and clustering inputs (columns) as possible. The maximum number of datasets to include corresponds to the total number of datasets available and likewise for inputs.

6.1 Manual Selection Methods

The manual process involves removing all rows and all columns where the average WK value is less than 0.1. It is computed in two stages and thus two manual selection methods need to be defined. The WK threshold that was chosen for this work is 0.1. The first manual selection method (MS1) removes all rows that have an average WK value less than the threshold (step 1), then removes all columns that have an average WK value less than the threshold (step 2). The second manual selection method (MS2) removes all columns that are below the threshold (step 1) then removes all rows that are less than the threshold (step 2). Both methods will effectively reduce the original WK matrix in size. Results of both methods are displayed in Table 3.

Table 3. Sub-selection methods results table, displaying the number of included inputs and datasets for each of the proposed methods

Method	Datasets	Inputs	% Datasets	% Inputs
MS1	79	11	39.9	34.4
MS2	53	28	26.8	87.5
RMHC	60	28	30.3	87.5
SA	60	28	30.3	87.5
GA	60	28	30.3	87.5

6.2 Heuristic Search Methods

Three heuristic search methods were implemented - Genetic Algorithm, Hill Climbing and Simulated Annealing. For the three techniques selected a fitness function of quality was needed, defined in the next section. The same program with the same fitness function was used for the three algorithms. The fitness function simply works by creating a binary mask which corresponds to include or exclude from the matrix, such that 1 to include a dataset or an input from the matrix and vice versa for excluding a dataset or an input. Given a binary string, the number of rows and columns that is less than the threshold can be counted.

All three techniques were run for 10,000 iterations (or fitness function calls). The time it undertakes the program to run is proportional to the fitness function calls. For some of the experiments, the fitness function calls are referred to as the number of iterations and the runtime of the program is proportional to the iterations number. Either iterations or fitness function calls might be used throughout this paper. A well-known issue with heuristic search techniques is that they can get stuck at local maximums. One common solution to the problem is to restart the search at another random point. Thus, the three heuristic experiments were repeated 25 times each and the average was recorded.

$$\begin{aligned} f(S, W) &= \sum_{i=1}^{n} \sum_{j=1}^{m} \delta(s_i, 1)\delta(s_{j+n}, 1)(w_{ij} - T) \\ &= \sum_{i=1}^{n} \sum_{j=1}^{m} \delta(s_i \times s_{j+n}, 1)(w_{ij} - T) \end{aligned} \tag{1}$$

Random Mutation Hill Climbing (RMHC) is a heuristic search algorithm that uses an iterative approach to find an objective in the search space by simply maximising the objective function. This algorithm starts at a random point in the search space and aims for a better fitness of the objective function by randomly searching the adjacent or closer neighbours. The process continues until the optimal solution is obtained and it starts all over again at the new point reached. The RMHC algorithm was chosen as it is the most basic variant of an evolutionary algorithm and it has been previously used in numerous studies such as [3].

Simulated Annealing (SA) is a meta-heuristic technique which improves on RMHC. The idea of SA originated from the natural process of annealing in metallurgy, which involves heating materials to a very high temperature and then allowing it to slowly cool down to alter its physical structure. In SA a temperature parameter is kept to simulate the heating process (it expresses the probability of accepting a solution with a worst fitness). The temperature is initially set to a high value, allowing the temperature to steadily "cool" (decreasing whilst running the algorithm). This temperature keeps decreasing to reach a zero by the end of the algorithm, revealing the solutions. SA have been applied to various problems [20] and was chosen for this reason. We note that one per cent of the fitness function calls was used to find the starting temperature [28].

Genetic Algorithm (GA) was chosen as it is a powerful tool which can perform various optimisation problems. GA represents the solution to a problem as string (encoded as a chromosome). A population of chromosomes represents a subset of the search space of all possible solutions. The fitness function is also used to rate the worth of a solution that the chromosome represents. A standard binary GA using elitism was used. The population size of the GA was set to 25 and the generations were terminated after 10,000 fitness function calls. Genetic operations, mutation and crossover, are applied to the solutions to help find the best fitness. The mutation is set at 0.5/number of bits, whereas the crossover rate = 0.5.

6.3 Fitness Function

Let W be an n rows (number of datasets) by m columns (number of inputs) real matrix where the ith,j^{th} value w_{ij} is the WK of input j applied to dataset i. Given a binary string S of length $n + m$, the first n bits are a mask of what datasets are selected and the next m bits are a mask of what inputs have been selected. For example, if $n = 5$ and $m = 3$, then the string S = "10101010" means that datasets 0, 2 and 4 have been selected along with input 1. Let s_i be the ith bit of string S. The fitness function used in this paper is defined as follows:

The function $\delta(i, j)$ is the Kronecker delta function δ_{ij}, i.e. 1 if $i = j$, 0 otherwise.

Essentially the fitness function sums all of the WK values remaining after the binary string S has been applied, penalised by a threshold value T, 0.1. The rationale behind this fitness function is that when used in conjunction with a heuristic search method, a configuration of S will be found that maximises the number of WK values that are above T, excluding configurations where there are values below T. Alternative fitness functions were experimented with before selecting the current fitness function.

7 Results

Results from the experiments are shown in Table 3. It displays the number of inputs and datasets for each of the five proposed techniques. For all five methods, a threshold of 0.1 WK was used. Selecting a higher threshold would obtain worst results, fewer datasets and inputs. The table also displays the percentages of the inputs and datasets in proportion of the total number of inputs and datasets, respectively. From the table, it can be seen that MS1 produced the most datasets, 79, but much fewer inputs, 11. There is a large difference between the inputs for this method and the rest of the methods. As the aim of the experiments is to include as many datasets and clustering inputs as possible, it seems that the three heuristic techniques produce the best results, 60 datasets and 28 inputs. However, as can be seen from the results all three heuristic methods produce the same optima, the same number of inputs and datasets. These identical results imply that the search space is relatively smooth and easy to search. In order to find the best technique, we need to analyse the three heuristic techniques in terms of runtime.

To find out the most efficient technique from the three heuristic methods, the convergence graphs of the three experiments were produced. As the experiments were repeated 25 times for each method, the average fitness is calculated and plotted on a graph (Fig. 4). From the plot, it can be seen that RMHC converges at the least number of iterations, 3,283, indicating that it is the most efficient in terms of runtime for this particular combinatorial optimisation problem. SA comes second in terms of runtime, converging at 5,701 iterations. Note that one per cent of the fitness function calls are sacrificed to find the initial temperature. GA performed the worst in terms of runtime converging the last out of the three methods. The authors presume if this was a larger feature subset selection problem, GA and/or SA might perform better.

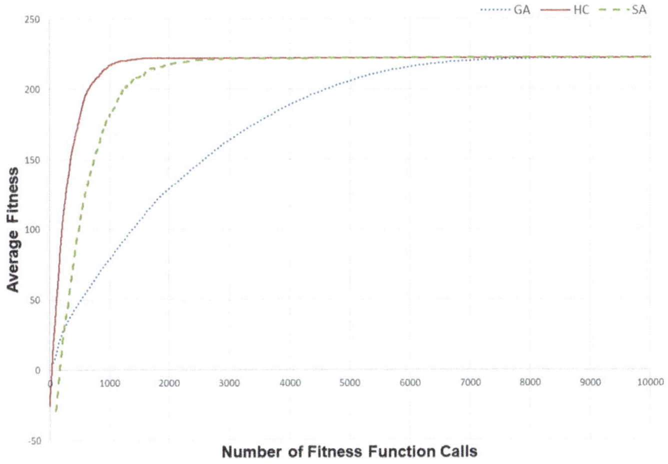

Fig. 4. A plot summarising the convergence points of GA, RMHC and SA.

The fitness function value produced from all three experiments is 221.895. The minimum theoretical fitness is −399.839 and the maximum theoretic fitness is 288.823. The fitness function value achieved is 90.2% of the way between the minimum and maximum theoretical fitness, points close to the minimum and maximum are almost certainly not achievable.

In addition, the three methods producing the exact same fitness function values does not mean that the outputted string binary arrangements are similar. Thus, we defined a simple metric based on Hamming Distance that cross-compares the similarity of each binary string representation from all the repeats to each other. It counts the occurrence of 1s and divides them by the number of possible 1s. This was computed for all three heuristic search methods used in this study. Results show that the binary string representation produced from each of the 25 repeats of the three methods (75 in total) is identical to each other, illustrating consistency of the results.

8 Post-analysis Results

From the previous section RMHC was found to be the most efficient heuristic method for this particular combinatorial optimisation problem. As this technique outputs 60 datasets and 28 input clustering methods, a WK matrix of the 60 datasets versus the 28 inputs was constructed. The matrix is of the WK values of the output clustering arrangement versus the expected clustering arrangement of the 28 input methods, for the 60 datasets. The followings are post-analysis on the quality of the results.

A normality test was used to determine if the data under analysis was modelled by a normal distribution. It computes the likelihood of a random variable from the data to be of a normal distribution to find out how the averages of the inputs vary. Thus, to test for normality we averaged each column in the 60×28 matrix; this provided us with 28 values of 60 in size. Before testing for normality, we applied the following transformations to the data:

(a) Computed the absolute values of the results
(b) Removed any zero values
(c) Took the log (base *e*) of the results

Five normality tests were selected for this analysis, they are displayed in Table 4. They are part of the R "nortest" package (30 July 2015). The average results from these five normality tests are displayed in Table 4. By working out the average of the five normality tests, results show that 22 of the 28 inputs pass the normality test at the 0.01 significance level. Results show that the individual averages of the test values are normally distributed and so too the mean of the five tests. This indicates that it is possible to generate the input distribution from another normal distribution based on that mean i.e., by providing a distribution model to use based on WK values of the inputs.

Table 4. A summary of the normality test p-values and their mean for all inputs

Name	Test type	P-value
ad.test	Anderson-Darling Test for Normality	0.08062
cvm.test	Cramer-von Mises test of goodness-of-fit	0.07086
lillie.test	Lilliefors (Kolmogorov-Smirnov) Test for Normality	0.04948
pearson.test	Pearson Chi-Square Test for Normality	0.04352
sf.test	Shapiro-Francia Test for Normality	0.13560
Average		0.07601

For the 60 datasets, we computed the average WK of the inputs' clustering arrangements versus the expected clustering arrangements (we will refer to as inputs vs. expected). Subsequently, we computed the average WK of the inputs' clustering arrangements against each other's, again for the 60 datasets (we will refer to as inputs vs. inputs). This provided us with two datasets of size 60. Correlating these two datasets produced 0.562 which shows a strong positive correlation above the 1% significance level. This suggests that if the number of inputs keeps increasing, the mean of the inputs vs. inputs will possibly start resembling the mean of the inputs vs. expected. From looking at the WK pairs of values (inputs vs. inputs and inputs vs. expected) it can be seen that in many of the cases they are close together. Thus, from the datasets selected, results show that the inputs vs. inputs are similar to the inputs vs. expected results, indicating a link between the intra-method agreement and the method versus the expected. This is interesting as it implies if we would like to find out the average quality of the inputs without having the expected clustering arrangements, we could evaluate the performance based on the average performance of the inputs vs. inputs. This allows us to have an approximate idea of the quality of the results (inputs) if there are no expected clustering arrangements to compare it with. We would only need to compare the input with every other input and compute the average. If it is very high we would know that the clustering methods are reasonably accurate and if it is very low we would know that the clustering methods are not appropriate. We look to further investigate this relationship as part of future work.

Moreover, we have computed the standard deviation of the inputs vs. inputs and inputs vs. expected. Correlating these together presents a correlation of 0.441, which clearly passes the 1% significance level. Lastly, to find out if there is a relationship between the average inputs vs. expected and the size of the datasets (number of instances and attributes), we correlated them. The correlation was shown to be 0.093 for the number attributes, and −0.165 for the number of instances. These two correlations are weak and they do not pass the 10% significance level (0.211). This indicates that there are no relationships between the average WK of the inputs and the changing dataset sizes. This relationship is independent i.e., does not increase or reduce based on the size of the dataset. The same was repeated for correlating the average inputs vs. inputs and the dataset sizes (number of instances and attributes). Correlations are also shown to be weak: 0.014 for the number of attributes and −0.187 for the number of instances. These results show that there is no bias in the WK values produced or our proposed data selection technique.

9 Conclusions and Future Work

Clustering inputs and datasets for consensus clustering can be poor. Moreover, it is not straightforward to identify the inputs that are poor or the appropriate datasets. This research work explores the behaviour of CC and looks at modelling a suitable distribution for it by selecting the most suitable inputs and datasets. Instead of selecting them at random, it proposed five sub-selection techniques (three heuristic and two manual selection methods) to achieve the largest number of inputs and datasets. Results showed that a normal distribution model, based on the WK values of the inputs, can be used to generate the input distributions of CC. Using these techniques may improve the results for CC. In addition, results have also presented a quick metric that can be used to estimate the quality of the inputs (clustering methods), if there are no expected clustering arrangements to compare the outputted clustering arrangements with. For this work we are not applying the results to CC; however, we look to extend this study towards a larger issue of combinatorial optimisation and apply results to CC. For future work we would also seek to expand on the datasets; a larger number of dataset values is needed to model a more accurate distribution.

References

1. Ailon, N., Charikar, M., Newman, A.: Aggregating inconsistent information: ranking and clustering. J. ACM. **55**(5), 1–27 (2008). https://doi.org/10.1145/1411509.1411513
2. Altman, D.G.: Practical Statistics for Medical Research. Chapman and Hall, London (1997)
3. Arzoky, M., Swift, S., Tucker, A., Cain, J.: A seeded search for the modularisation of sequential software versions. J. Object Technol. **11**(2) (2012). https://doi.org/10.5381/jot.2012.11.2.a6
4. Azimi, J., Fern, X.: Adaptive cluster ensemble selection. In: IJCAI International Joint Conference on Artificial Intelligence, pp. 992–997 (2009)
5. Bansal, N., Blum, A., Chawla, S.: Correlation clustering. Mach. Learn. **56**(1–3), 89–113 (2004). https://doi.org/10.1023/B:MACH.0000033116.57574.95
6. Caruana, R., Niculescu-Mizil, A., Crew, G., Ksikes, A.: Ensemble selection from libraries of models. In: Twenty-First International Conference on Machine Learning - ICML 2004, p. 18 (2004). https://doi.org/10.1145/1015330.1015432
7. Charikar, M., Guruswami, V., Wirth, A.: Clustering with qualitative information. J. Comput. Syst. Sci. **71**(3), 360–383 (2005). https://doi.org/10.1016/J.JCSS.2004.10.012
8. Cohen, J.: A coefficient of agreement for nominal scales. Educ. Psychol. Meas. **20**(1), 37–46 (1960). https://doi.org/10.1177/001316446002000104
9. Dua, D., Taniskidou, E.K.: UCI machine learning repository. University of California, School of Information and Computer Science, Irvine, CA. http://archive.ics.uci.edu/ml. Accessed 6 Oct 2017
10. Dudoit, S., Fridlyand, J.: Bagging to improve the accuracy of a clustering procedure. Bioinformatics **19**(9), 1090–1099 (2003). https://doi.org/10.1093/bioinformatics/btg038
11. Fern, X.Z., Lin, W.: Cluster ensemble selection. Stat. Anal. Data Min. **1**(3), 128–141 (2008). https://doi.org/10.1002/sam.10008
12. Fern, X.Z., Brodley, C.E.: Solving cluster ensemble problems by bipartite graph partitioning. In: Twenty-First International Conference on Machine Learning - ICML 2004, p. 36 (2004). https://doi.org/10.1145/1015330.1015414
13. Fred, A.L.N., Jain, A.K.: Combining multiple clusterings using evidence accumulation. IEEE Trans. Pattern Anal. Mach. Intell. **27**(6), 835–850 (2006). https://doi.org/10.1109/TPAMI.2005.113
14. Gionis, A., Mannila, H., Tsaparas, P.: Clustering aggregation. In: Proceedings - International Conference on Data Engineering, pp. 341–352 (2005). https://doi.org/10.1109/ICDE.2005.34
15. Giotis, I., Guruswami, V.: Correlation clustering with a fixed number of clusters. In: SODA, p. 16 (2005). https://doi.org/10.1145/1109557.1109686
16. Hubert, L., Arabie, P.: Comparing partitions. J. Classif. **2**(1), 193–218 (1985). https://doi.org/10.1007/BF01908075
17. Hyndman, R.J.: Time series data library. http://data.is/TSDLdemo. Accessed 15 Oct 2017
18. Jain, K., Murty, M.N., Flynn, P.J.: Data clustering: a review. ACM Comput. Surv. **31**(3), 264–323 (1999). https://doi.org/10.1145/331499.331504
19. Kaggle: Kaggle datasets. www.kaggle.com/datasets. Accessed 15 Sept 2017

20. Kirkpatrick, S., Gelatt, D., Vecchi, M.P.: Optimization by simulated annealing. Science **220**(4598), 671–680 (1983). https://doi.org/10.1007/BF01009452
21. Kuncheva, L.I., Vetrov, D.P.: Evaluation of stability of k-means cluster ensembles with respect to random initialization. IEEE Trans. Pattern Anal. Mach. Intell. **28**(11), 1798–1808 (2006). https://doi.org/10.1109/TPAMI.2006.226
22. Mldata.org.: Machine learning data set repository. http://mldata.org. Accessed 7 Dec 2017
23. Monti, S., Tamayo, P., Mesirov, J., Golub, T.: Consensus clustering: a resampling-based method for class discovery and visualization of gene expression microarray data. Mach. Learn. **52**(1–2), 91–118 (2003). https://doi.org/10.1023/A:1023949509487
24. Pelleg, D., Moore, A.W.: X-means: extending k-means with efficient estimation of the number of clusters. In: Proceedings of the Seventeenth International Conference on Machine Learning. table contents, pp. 727–734 (2000). https://doi.org/10.1007/3-540-44491-2_3
25. Singh, V., Mukherjee, L., Peng, J., Xu, J.: Ensemble clustering using semidefinite programming with applications. Mach. Learn. **79**(1–2), 177–200 (2010). https://doi.org/10.1007/s10994-009-5158-y
26. StatLib.: StatLib—Datasets Azrchive. Carnegie Mellon University (1989). http://lib.stat.cmu.edu/datasets. Accessed 20 Nov 2017
27. Strehl, A., Ghosh, J.: Cluster ensembles – a knowledge reuse framework for combining multiple partitions. J. Mach. Learn. Res. **3**, 583–617 (2002). https://doi.org/10.1162/153244303321897735
28. Swift, S., Tucker, A., Vinciotti, V., Martin, N., Orengo, C., Liu, X., Kellam, P.: Consensus clustering and functional interpretation of gene-expression data. Genome Biol. **5**(11) (2004). https://doi.org/10.1186/gb-2004-5-11-r94

Challenges in Developing Prediction Models for Multi-modal High-Throughput Biomedical Data

Abeer Alzubaidi(✉)

School of Science and Technology, Nottingham Trent University, Nottingham, UK
abeer.alzubaidi022014@my.ntu.ac.uk

Abstract. Rapid advances in high-throughput technologies have provided variant types of biological data like gene expression, copy number alterations, miRNA expression, and protein expression. The integration of diverse biomedical datasets have received wide attention because of its great potential to build a range of models which can be more thorough about the mechanisms of cancers and other complicated diseases. However, the impact of constructing a prediction model from heterogeneous high-throughput datasets is not comprehensively defined. This paper identifies the challenges related to issues of developing prediction approaches for Multi-Modal High Dimensional and Small Sample Size biomedical datasets. The various challenges encountered are based on the characteristics of the data, the aim of the integration, and the level of the integration. Heterogeneity and dimensionality of high-throughput data bring many computational and statistical challenges. Thus fusing them into a unified and informative space for prediction purposes is a difficult task. Furthermore, validating and evaluating the outcomes of the prediction models built from multi-modal biomedical datasets, involve several underlying issues that need to be properly handled, in order to report reliable findings. Moreover, interpretability and robustness of the prediction models are becoming crucial factors for personalised medicine. The directions are introduced briefly to address these challenges and some possibilities for future work are discussed.

Keywords: High dimensional small sample size HDSSS data · Classification
Prediction models · Multi-modal high-throughput biomedical data
Feature selection · Validation · Interpretability

1 Introduction

Rapid advances in high-throughput techniques have introduced diverse biological datasets that can provide information about cancers and other complex diseases. A large body of research has been introduced using these datasets to construct different analysis models for various biological outcomes [1–4]. Most of the proposed studies adopt diverse types of biomedical data separately to construct various models. However, each high-throughput dataset may hold a different piece of knowledge related to complex diseases. Some biological studies have shown that complex diseases can be caused by the underlying interaction of various factors, including genetic, genomic, behavioural, proteomic and environmental effects [5–7]. In addition to the heterogeneity of complex

K. Arai et al. (Eds.): IntelliSys 2018, AISC 868, pp. 1056–1069, 2019.
https://doi.org/10.1007/978-3-030-01054-6_73

diseases, the availability of multi-modal data repositories such as the Cancer Genome Project (CGP), the Cancer Genome Atlas (TCGA), and the International Cancer Genome Consortium (ICGC) has created a need for developing various integrative analysis models which might be able to answer different biological questions of interest. Multi-modal biological data repositories can consist of high dimensional genomics, epigenomics, proteomics, as well as imaging, and clinical data for the matched group of patients. This offers unprecedented opportunities to use these various types of data to build a range of models which can be more comprehensive about the mechanisms of cancers and other human diseases. Therefore, the next frontier in the move towards personalised and precision medicine is to develop robust prediction models from multi-modal biomedical data for the diagnosis and prognosis of the complicated diseases.

Several studies have constructed various integrative analysis models to investigate the integration gain of diverse biomedical data. Most of these integrative studies have adopted clinicogenomic models that rely on combining clinical and genomic datasets [8–13]. Clinicogenomic integrative models focus on addressing the challenges of integrating disparate dimensionalities of clinical and high dimensional genomic datasets. In terms of biological problems, most clinicogenomic studies use gene expression data from widely available public genomic datasets, despite the fact that each of these datasets provides variant aspects about cellular activity. Some integrative analysis models have been proposed to fuse different types of high-throughput datasets for addressing various biological questions [14, 15]. Moreover, several reviews in the literature have discussed the challenging issues that arise in the area of biomedical data integration [16–18]. However, the challenges related to issues of constructing prediction models from multi-modal high-throughput biomedical datasets which typically comprise a large number of variables and a small number of samples have not been comprehensively discussed. High Dimensional and Small Sample Size (HDSSS) datasets pose intrinsic statistical challenges for developing reliable and accurate prediction models [19–24]. Constructing prediction approaches from two or more HDSSS biomedical data can be more critical and poses different computational and statistical issues. Therefore, this paper defines the challenging issues that arise when designing and evaluating prediction models that can extract significant, stable and interpretable information from multi-modal HDSSS biomedical data. The directions are presented briefly to deal with these challenging issues.

The remaining paper is structured as follows: Sect. 2 discusses the stages of integration, Sect. 3 discusses the challenges of developing prediction models for multi-modal high-throughput biomedical data, and Sect. 4 presents summary and future directions.

2 Stages of Integration

Three integration levels have been proposed by Pavlidis et al. [25], namely, the early, late, and intermediate. Identifying the appropriate integration stage can be based on the aim of the analysis model and the characteristics of the datasets. Figure 1 illustrates the process of developing prediction models from multi-modal high-throughput biomedical datasets using early, late and intermediate integration stages.

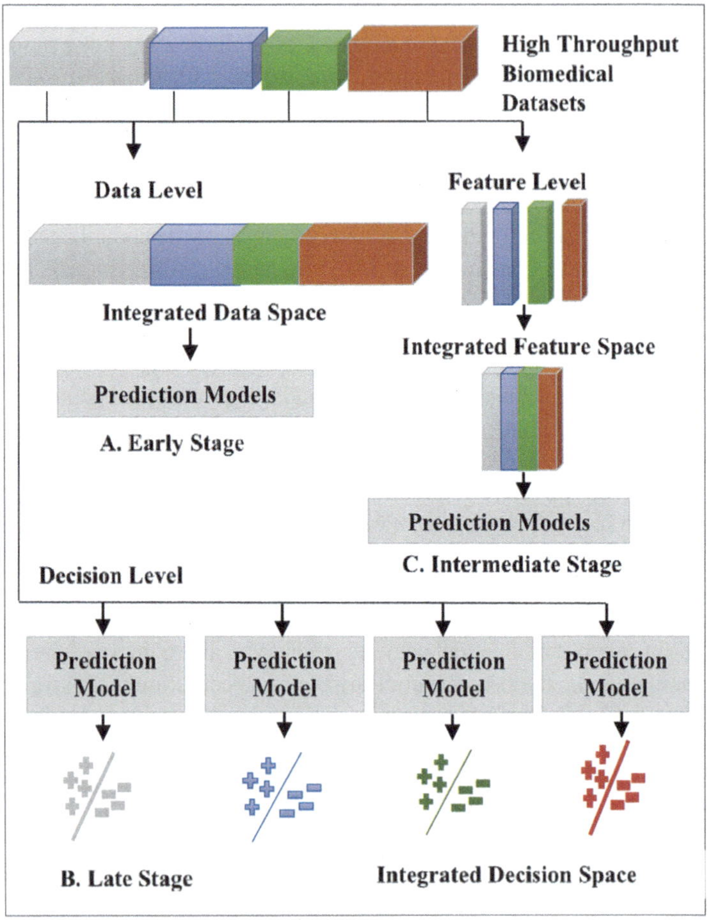

Fig. 1. Integration stages

2.1 Early Integration - (Data Level) Stage

This integration stage involves combining heterogeneous biomedical datasets at the raw data level before any kind of analysis model takes place. Data from various modalities are concatenated into a single data space, when these modalities have the identical set or subsets of observations. Several proposed analysis models adopted the early integration stage to combine multiple biological modalities [11, 26, 27]. Since the analysis models can be performed on the integrated data space, various types of potential relationships which might be existent between the biomedical datasets can be identified. However, fusing multi-modal high-throughput datasets at the raw data level without accounting for their dimensionality leads to the curse of dimensionality issues - (which are discussed in details in the next section). Furthermore, due to the difference in dimensionality among various high-throughput datasets, prediction models can be biased toward the dominant dataset with higher dimensions, which can cause low statistical significance of the experimental obtained results.

2.2 Late Integration - (Decision Level) Stage

In this integration approach, prediction models are applied independently on each dataset, and thereafter the obtained decisions of all prediction models are combined to improve the decision-making process. Several researchers have adopted different types of late integration techniques for combining diverse biological information [28, 29]. In the late integration stage, the analysis of each dataset can be executed in parallel using different technical methodology. However, applying the analysis models to each biomedical dataset separately leads to ignoring any potential interactions which may be existent among them, and also ignores the notion of heterogeneity of the complicated diseases.

Moreover, since the prediction models are applied independently, identifying the appropriate techniques for fusing the obtained decisions and the interpretation of the concatenated decision space can be another challenge.

2.3 Intermediate Integration - (Feature Level) Stage

The intermediate integration approach has been presented to overcome the limitations of data and decision level approaches. The feature level stage essentially relies on creating intermediate transformed representations for each dataset independently, and then integrating them before developing any analysis modeling. Different intermediate integration techniques have been introduced for combining diverse biomedical modalities and forming a new common feature space [30–32]. Dealing with the alternative representations of the original datasets makes the intermediate integration stage a suitable approach for addressing the challenges of fusing various high-throughput datasets with the high and disparate dimensionalities. However, processing the original datasets individually into alternative representations can lead to making the recognition of the underlying relationships a hard task. Moreover, finding feasible transformed representations that can easily integrate with the other representations of the biomedical modalities and constitute an interpretable feature space can be another challenge.

3 Challenges of Developing Prediction Models for Multi-modal High-Throughput Biomedical Data

There are a number of challenging issues that arise when developing prediction approaches from multi-modal high-throughput biomedical data. These challenges are discussed in the following subsections.

3.1 Curse of Dimensionality

One of the most important issues related to the construction of robust prediction models is the curse of dimensionality. The term 'curse of dimensionality' has been coined by Bellman [33] and refers to various phenomena that arise when dealing with datasets which comprise hundreds or thousands of variables. High-throughput datasets usually contain a large number of variables and a small sample size. Classical classification rules cannot be constructed when using all the available features of a single dataset, particularly when the high dimensional feature space combined with a small number of observations. For example, training the Fisher Discriminant Analysis algorithm separately on each dataset which contains a small number of cases and a large number of features becomes inapplicable because there would not be enough samples in each response group to estimate a covariance matrix. Even if all the available features can be used to train other machine learning algorithms, the predictive performance of the classification models could degrade to a point which performs random guessing. The potential downside of the curse of dimensionality is becoming most apparent when integrating diverse high-throughput datasets of few cases.

To explain the impact of the curse of dimensionality issues on developing prediction models from various biomedical data, three integrative approaches are discussed in this paper according to the level of integration. The first two approaches adopt the early integration stage and the third multi-modal prediction approach employs the feature level technique. Late integration stage is not discussed in this section because the prediction models are applied independently to each biomedical dataset.

3.1.1 Early Prediction Approach Without Dimensionality Reduction

In this modeling approach, the early integration stage (data level) is adopted to combine heterogeneous high-throughput modalities at the raw data level. The feature spaces of biomedical datasets are combined together to form an integrated large matrix. The fused data matrix is used to model the prediction approach using machine learning algorithms without employing any dimensionality reduction techniques. The performance of any machine learning algorithm mainly relies on its implicit assumptions and the characteristics of the data [34]. In terms of high-throughput datasets, their large feature spaces are more likely to contain many irrelevant, redundant and noisy variables that do not contribute to enhancing the predictive accuracy. Concatenating these datasets without taking the curse of dimensionality issues into consideration will lead to train classification models to learn the noise and this can result in degrading the prediction performance to the level of random guessing.

Furthermore, these large and noisy feature spaces usually combine with a small number of cases, thus, there is an increased risk of obtaining poor generalisation ability (overfitting). The training error rate of the classification model (the percentage of training samples misallocated by the classifier) tends to decrease, while the generalisation error rate (the testing error rate that results from using a learned model to predict the outcomes of new cases that were not used in training the machine learning algorithm) tends to increase. In this case, the training error rate can dramatically underestimate the testing error rate. When dealing with Small Sample Size (SSS) data the classification model cannot estimate the parameters correctly due to an insufficient number of observations. Thus, the learned model is likely to overfit to the training data because the decision boundary of the trained model does not match the real data distribution, thereby resulting in a poor generalisation ability on unseen testing data. Consequently, adopting the early prediction approach without employing any dimensionality reduction techniques may not be appropriate in the context of multi-modal HDSSS biomedical data.

Therefore, one of the most important factors that can deal with the curse of dimensionality issues, and improve the estimation of parameters of the machine learning models is to select a small number of robust predictors which can be used to form the classification rule effectively. Reducing the set of potential decision boundaries can help to correct the problem of overfitting. Furthermore, adopting a linear classifier with a hyperplane decision boundary can be another possible solution to reduce the risk of overfitting due to a trade-off between model complexity and the possibility of overfitting. Machine learning algorithms use learning data to determine the decision boundary, and when having small sample size SSS data, highly fitted decision boundaries discriminate the training observations optimally, but they might not be able to assign the independent testing samples to their response groups correctly. Therefore, linear classifiers with simple hyperplane decision boundaries tend to suffer less from overfitting and generalise well on independent testing data than do non-linear classifiers.

3.1.2 Early Prediction Approach with Dimensionality Reduction

Identifying a small number of relevant predictors from diverse biomedical data can be an appropriate approach to provide machine learning algorithms with effective information, and to reduce the risks of overfitting and computational cost. Thus, some of the proposed integrative analysis models in the literature have adopted the early integration stage with the utilisation of the dimensionality reduction techniques. With this predictive modeling approach, the high-throughput datasets are combined together at the data level stage to constitute a large feature space which thereafter passes to the utilised dimensionality reduction technique. The outcome of the data reduction algorithm is used to model a machine learning algorithm and the performance of the derived learned model is validated on observations that were not used during the training process. This kind of predictive model generally achieves a quite high classification accuracy which also depends on the applied technical methodology.

However, some of these prediction approaches might produce unreliable experimental results due to the fact that both the adopted data reduction algorithms and the classification models are biased to the dominant dataset with higher dimensions. Some high-throughput genomic datasets like gene expression data usually contain thousands or tens thousands of features and other genomic datasets like miRNA expression data may contain few hundreds of features. Concatenating these biomedical datasets using the early integration stage results in dominating one dataset (e.g. gene expression) with higher dimensions on the fused common feature space than other datasets. The consequence of this is likely to be that the utilised reduction algorithms select most or all biological predictors from the dominant dataset which are used thereafter to build the classification models. In such situations, the predictive performance of the prediction approach constructed from multi-modal gnomic data (e.g. gene expression and miRNA expression) and the prediction approach built from the individual dataset (e.g. gene expression) is not significantly different. As a result, using the early integration strategy to develop two-stage prediction model – (dimensionality reduction and classification) may not be appropriate for multi-modal biomedical datasets with disparate dimensionalities unless the dimensionality difference is addressed properly.

3.1.3 Intermediate Prediction Approach

In 1986 George Box [35] hypothesised that "a large proportion of process variation is explained by a small proportion of the process variables". Recognising a small group of robust biomarkers which has the actual impact on the outcome would be the most cost-effective procedure in developing powerful prediction models, which can explain why and how these sample cases assign to their specific response groups. This is particularly valid in the biomedical applications, where the disease outcome is distributed in several biological markers. The percentage of informative predictors in microarray data typically ranges from 2%–5% [36, 37]. Creating a common space from alternative representations of various biological datasets can be a reasonable solution to deal with the curse of dimensionality issues of high-throughput data and to avoid governing single genomic dataset on the concatenated feature space.

In this predictive modeling approach, high-throughput datasets are combined using the intermediate integration stage, where the data reduction algorithm is applied to each biomedical dataset independently to identify robust biological markers. Then, the identified biomarkers from each dataset are combined together at the feature level and formed a new concatenated feature space. The training data that contain only the fused identified biomarkers are used to construct the classification models and then estimate their predictive performances using testing observations that were not used during the reduction and classification stages. In the literature, the most popular evaluation measurement that is used for assessing the outcome of the dimensionality reduction methods is to test the performance of the classification models trained on a group of relevant predictors using unseen testing data.

However, in the context of high-throughput datasets that comprise a large number of variables and a small number of cases, dimensionality reduction techniques are more likely to identify several combinations of predictors with similar classification accuracies; especially, multivariate selection methods which evaluate the features simultaneously rather than individually. Therefore, the concerns about robustness criterion have been raised, particularly in the biomedical applications. The stability of selection algorithms can be specified as obtaining variant outcomes due to the little variation in the data, where robust selection algorithms should be able to generate consistent predictors over variant training data. Therefore, the applied data reduction algorithm should be stable enough to identify consistent predictors from each high-throughput dataset. A prediction model is only as good as the data used to develop it. The concatenated feature space of robust predictors might provide the intermediate prediction approach with sufficient information to construct an accurate and reliable model for future predictions. However, the intermediate prediction approach with dimensionality reduction cannot consider any potential relationship which might be existent among the original biomedical datasets.

3.2 Validation of the Prediction Models

The main objective of developing multi-modal prediction approaches is to obtain a generalisable model that would have a high generalisation ability when applied to new patients. In order to evaluate how these prediction models will perform in the future, it is necessary to validate their performances empirically using independent testing data. Therefore, the cross-validation technique is an essential step to validate the predictive performance of the classification models. In the literature, several studies adopt the simplest validation approach, where the supervised selection method is applied to the entire data and the prediction model is performed on the partitioned data. However, using the entire dataset by supervised selection algorithms can provide biased estimation results in the prediction models because the dimensionality reduction algorithms have already seen the data labels [38, 39]. Thus, the obtained results of these prediction models should treat as overly optimistic. In order to get unbiased estimation of the model's performance, both the supervised selection algorithms and classification models should be performed solely on the training dataset. Therefore, it is necessary to test the performance of the proposed prediction models using observations that were not seen during the reduction and learning processes. Furthermore, to avoid selection bias and to account the variance in classification performance, the validation procedure should be repeated several iterations.

The cross-validation technique is applying multiple data partitions and systematically swapping out samples for testing. Identifying the suitable cross-validation method for SSS biomedical data should be considered carefully in order to have sufficient training and testing samples for feature selection and classification processes. Moreover, training and testing sets should include the same class proportions as in sample groups. A discussion on the appropriateness of various cross-validation techniques for SSS data is provided, along with the advantages and limitations of each method.

- *Holdout cross-validation method:* includes holding out a set of training samples from learning the model and validating the performance of the trained model using the held out samples. Hold out approach can be an appropriate validation method for datasets which are characterised by a large number of samples and a small number of variables. Therefore, for high-throughput biomedical data which usually constitute a large number of variables and a small number of samples, holdout approach may not be the proper validation procedure. Since the number of patient samples in both sets is small, thus adopting holdout procedure leads to reduce the amount of the data which are used during the learning process - (training samples would be smaller). Deciding the size of held-out samples of SSS dataset can be another issue: if the training set is relatively large, the test error rate may give unreliable estimates for future prediction, where selecting a large testing set enhances the estimation of the generalisation error, but degrades the training process.
- *Leave One Out Cross Validation (LOOCV) Method:* involves temporarily leaving out one sample from a dataset of size n, then training the classification model on the remaining $n - 1$. Therefore, LOOCV procedure can be used to overcome the limitation of wasting the data by holdout validation approach, especially for small sample size SSS data. The LOOCV procedure is iterated n times (i.e. n is equal to the total

number of samples), in each iteration a different case is taken out. That means, each sample patient is assigned to its response group by the classification model learned without this sample, thus, the estimation average of LOOCV has high variance. Another serious drawback of LOOCV validation method is being computationally expensive, especially when the prediction models comprise two stages – (dimensionality reduction stage and classification stage). LOOCV can be considered a special case of k-fold cross validation, however, they are performed independently.

- *K-fold Cross-Validation Method:* randomly splits a dataset into k non-overlapping partitions of equal size, where k-1 partitions are used to fit the classification model and the remaining set is used to validate its performance. The k-fold validation method is used usually to overcome the drawbacks of holdout and LOOCV validation approaches by having fewer data to waste than holdout approach and k times less expensive than *n* times of LOOCV procedure. Furthermore, k-fold validation procedure guarantees that there is no overlap between the samples of both sets which is a key factor for estimating the generalisation error rate of the prediction models accurately. Therefore, the k-fold cross-validation technique can be an appropriate procedure for diverse SSS biomedical data. To account the variance in classification model performance, repeated k-fold validation procedure should be employed. Commonly used values of k are 5 and 10, and the value that can experimentally provide a good compromise for bias-variance trade-off should be utilised.

3.3 Performance Estimation

Since the aim of integrating various high-throughput biomedical data is to improve the predictive accuracy, it is necessary to estimate the performance of the prediction model using reliable evaluation metrics and then compare it with the model built using individual modalities. The accuracy metric has been widely used as the main estimation measure for assessing the predictive performance of the classification models. The accuracy of the classification model can be defined as the proportion of testing cases that are assigned correctly to their response groups, where misclassification error rate (MCE) can be obtained by subtracting the accuracy proportion from one. However, the sizes of response groups of some high-throughput biomedical data are considerably different. Therefore, the predictive accuracy or MCE estimation metrics may not be reliable for these datasets when their response groups have considerable disparate sizes.

For imbalanced class size biomedical datasets, The Area Under the ROC Curve (AUC) can be utilised to assess the predictive performance of the classification models. The Receiver Operating Characteristic (ROC) curve is the plot of True Positive Rate (TPR) versus False Positive Rate (FPR) for a range of threshold values on the outputs of the classifiers. AUC resides in the range of [0, 1] if the AUC value is equal to 1, it means the predictive performance is perfect (i.e. the classification model correctly assigned all the unseen sample cases it was given during the testing stage). AUC has become a standard performance estimation metric because: it is more reliable than the other evaluation methods like accuracy, more discriminative than measurements like recall, precision, and F-measure, insensitive to unbalanced class prior probabilities and can be measured over the range of True Positive Rate (TPR) or Sensitivity (SS) and True Negative Rate (TNR) or Specificity (SP) trade-offs [40, 41].

In order to examine the practical significance of the integration gain, a set of statistical comparison tests should be conducted, where it is necessary to test whether the differences in predictive performances of prediction models built using multi-modal biomedical datasets and individual prediction models built using single dataset are statistically significant. Particularly for small sample size SSS data, it is important to adopt a wide range of unimodal and multimodal prediction approaches and multiple high-throughput datasets and various biomedical modalities, in order to draw reliable conclusions and report significant and generalisable findings. In addition to using suitable evaluation measures and a set of statistical comparison tests, validating the obtained results and attaining useful feedback from biomedical experts is of significant importance.

3.4 Interpretation

Interpretation of the prediction models is a required component for personalised and precision medicine. Most of the introduced integrative studies mainly rely on improving the predictive accuracy of the prediction models rather than interpretability. Optimising the predictive accuracy is more likely increasing the complexity of the computational models, and thereby decreasing the explainability, thus, there is a trade-off between model predictivity and model interpretability. However, providing interpretability for the prediction approaches which built from different types of biological data is a crucial element for their acceptance in clinical practice. One of the most significant factors that may provide some explanation for the biological prediction approaches is selecting a small group of robust biomarkers. The relationship between the identified biomarkers and the disease outcomes or phenotypes should be inferred in order to help physicians investigate the biological process of interest in greater details. Therefore, prediction models built on extracted features – (linear combination of variables) obtained from feature extraction techniques like Principle Component Analysis (PCA) may reduce the interpretability of the personalised prediction models while feature selection methods can be adopted to maintain the predictors - outcome relationship and provide meaningful insight into the high-throughput biological data.

3.5 Data Heterogeneity

Since diverse high-throughput biomedical datasets are likely to be gathered independently from different perspectives, thus the nature of these datasets can be different. The heterogeneity in the nature of diverse biomedical data can cause different computational challenges when developing prediction approaches. For example, high-throughput datasets contain different levels of noise and missing values and come in diverse formats. Therefore, the differences in dimensionalities, types, statistical distributions, missing values, and different levels of imprecisions and uncertainties need to be correctly handled. As a result, the integrative analysis models should adopt different pre-processing steps to address the heterogeneity of biomedical data. Fusing the data of each sample is considered as a pre-processing step that must be carried out before integrating different biomedical modalities.

4 Summary and Future Directions

Rapid advances in techniques in Bioinformatics and Biomedical domains have provided a dramatic increase in diverse biomedical datasets like gene expression, copy number alterations, miRNA expression, and protein expression. Each of these high-throughput datasets can provide one view about cellular activity, and thus their integration has received wide attention from biologist and medical experts. Furthermore, the heterogeneity of complicated diseases and the availability of multi-modal biological data repositories highlight the critical need for developing a range of models capable of leveraging diverse biological datasets and answer different questions of interest. Therefore, the next frontier in the move towards personalised and precision medicine is to develop integrative diagnosis and prognosis models which can be more thorough about the complicated mechanisms of cancers and other complex diseases like diabetes and schizophrenia.

However, the challenging issues of employing high-throughput biomedical data in developing prediction models should be considered carefully when designing the computational framework and evaluating its outcome critically. The diversified challenges encountered in this paper are based on the characteristics of the data, the aim of the integration, and the stage of the integration. The data related issues involve dealing with the heterogeneity and the dimensionality of variant high-throughput biomedical data, and fusing them in a unified and knowledgeable space for prediction purposes. High-throughput biomedical datasets are likely to be generated in diverse formats, from different sources, and from different perspectives. Thus, the naive integration of these biomedical modalities without addressing the heterogeneity issues of the data is inapplicable.

In addition to the diversity in the nature of high-throughput data, the curse of dimensionality can be another tough challenge that needs to be handled properly prior to constructing classification models. Thus, adopting the early prediction approach without employing any data reduction techniques may not be appropriate for HDSSS biomedical datasets. Moreover, the dimensionality difference between diverse modalities should be considered carefully to avoid governing one dataset on the common feature space. Therefore, applying the two-stage early prediction approach may lead to producing low statistical significant results unless the challenge of dimensionality difference is addressed properly. The intermediate approach can be most suitable for dealing with the issues of large and disparate dimensionalities of biomedical datasets; however, it is hard to recognise the potential interaction. On the other hand, the employment of feasible technical methodology, and the examination of their robustness can have a considerable impact on developing reliable multi-modal prediction approaches.

Since the major aim of unifying various high-throughput biomedical data is to improve the predictivity of the prediction models, it is necessary to estimate their performances using proper validation methods and reliable evaluation metrics. This paper investigates the appropriateness of different cross-validation techniques with their advantages and limitations and identifies the proper validation procedure for SSS datasets. Diverse estimation metrics are identified in the literature to assess the predictions of the classification models. However, some of these measurements may not be valid for biomedical datasets when their response groups have considerable disparate sizes. Therefore, it is necessary to utilise reliable evaluation measurement which is insensitive to unbalanced class prior probabilities to assess precisely the effectivity of the prediction models. In order to draw reliable conclusions and report practical significant findings, a set of statistical comparison tests should be conducted to test the statistical significance of the experimental obtained results in different aspects. Interpretability and robustness of the prediction models are crucial factors for their acceptance in clinical practice and becoming required elements in the personalised medicine. Future work includes proposing an intelligent system that can address the aforementioned challenges and extract relevant knowledge from multi-modal high-throughput biomedical data in order to develop robust and interpretable prediction models. The integration gain and the differences in performances between multi-modal prediction approaches and uni-modal prediction approaches using multiple datasets and diverse modalities will be investigated.

Acknowledgment. I would like to thank my Ph.D. sponsor [Ministry of Higher Education and Scientific Research in Iraq - University Of Al-Qadisiyah] for their financial support which has enabled me to carry out the work, write the paper, and to present the findings at the conference.

References

1. Sotiriou, C., Piccart, M.J.: Taking gene-expression profiling to the clinic: when will molecular signatures become relevant to patient care? Nat. Rev. Cancer **7**(7), 545–553 (2007)
2. Potti, A., Mukherjee, S., Petersen, R., Dressman, H.K., Bild, A., Koontz, J., Kratzke, R., Watson, M.A., Kelley, M., Ginsburg, G.S., West, M., Harpole, D.H.J., Nevins, J.R.: A genomic strategy to refine prognosis in early-stage non-small-cell lung cancer. N. Engl. J. Med. **355**(6), 570–580 (2006)
3. Golub, T.R., Slonim, D.K., Tamayo, P., Huard, C., Gaasenbeek, M., Mesirov, J.P., Coller, H., Loh, M.L., Downing, J.R., Caligiuri, M.A., Bloomfield, C.D., Lander, E.S.: Molecular classification of cancer: class discovery and class prediction by gene expression monitoring. Science **286** (1999)
4. Bhattacharjee, A., Richards, W.G., Staunton, J., Li, C., Monti, S., Vasa, P., Ladd, C., Beheshti, J., Bueno, R., Gillette, M., Loda, M., Weber, G., Mark, E.J., Lander, E.S., Wong, W., Johnson, B.E., Golub, T.R., Sugarbaker, D.J., Meyerson, M.: Classification of human lung carcinomas by mRNA expression profiling reveals distinct adenocarcinoma subclasses. Proc. Natl. Acad. Sci. **98**(24), 13790–13795 (2001)
5. McClellan, J., King, M.-C.: Genetic heterogeneity in human disease. Cell **141**(2), 210–217 (2010)
6. Schadt, E.: Molecular networks as sensors and drivers of common human diseases. Nature **461**(7261), 218–223 (2009)
7. Eichler, E., Flint, J., Gibson, G., Kong, A., Leal, S.M., Moore, J.H., Nadeau, J.H.: Missing heritability and strategies for finding the underlying causes of complex disease. Nat. Rev. Genet. **11**(6), 446–450 (2010)

8. Wu, J., Zhou, L., Huang, L., Gu, J., Li, S., Liu, B., Feng, J., Zhou, Y.: Nomogram integrating gene expression signatures with clinicopathological features to predict survival in operable NSCLC: a pooled analysis of 2164 patients. J. Exp. Clin. Cancer Res. **36**, 4 (2017)
9. Irigoien, I., Arenas, C.: Diagnosis using clinical/pathological and molecular information. Stat. Methods Med. Res. **25**(6), 2878–2894 (2016)
10. van Vliet, M.H., Horlings, H.M., van de Vijver, M., Reinders, M.J.T.: Integration of clinical and gene expression data has a synergetic effect on predicting breast cancer outcome. PLoS One **7** (2012)
11. Stephenson, J., Smith, A., Kattan, M.W., Satagopan, J., Reuter, V.E., Scardino, P.T., Gerald, W.L.: Integration of gene expression profiling and clinical variables to predict prostate carcinoma recurrence after radical prostatectomy. Cancer **104**(2), 290–298 (2005)
12. Pittman, J., Huang, E., Dressman, H., Horng, C., Cheng, S., Tsou, M., Chen, C., Bild, A., Iversen, E., Huang, A., Nevins, J., West, M.: Integrated modeling of clinical and gene expression information for personalized prediction of disease outcomes. PNAS **101** (2004)
13. Thomas, M., De Brabanter, K., Suykens, J.A.K., De Moor, B.: Predicting breast cancer using an expression values weighted clinical classifier. BMC Bioinform. **15**(1), 411 (2014)
14. Metsis, V., Huang, H., Andronesi, O.C., Makedon, F., Tzika, A.: Heterogeneous data fusion for brain tumor classification. Oncol. Rep. **28**(4), 1413–1416 (2012)
15. Al-Shahrour, F., Diaz-Uriarte, R., Dopazo, J.: Discovering molecular functions significantly related to phenotypes by combining gene expression data and biological information. Bioinformatics **21**(13), 2988–2993 (2005)
16. Li, Y., Wu, F.-X., Ngom, A.: A review on machine learning principles for multi-view biological data integration. Brief Bioinform. (2016)
17. Tsiliki, G., Kossida, S.: Fusion methodologies for biomedical data. J. Proteomics **74**(12), 2774–2785 (2011)
18. Hamid, S., Hu, P.N., Roslin, M., Ling, V.C., Greenwood, M.T., Beyene, J.: Data integration in genetics and genomics: methods and challenges. Hum. Genomics Proteomics **2009**, 869093 (2009)
19. Pappu, V., Pardalos, P.M.: High-dimensional data classification. In: Aleskerov, F., Goldengorin, B., Pardalos, P.M. (eds.) Clusters, Orders, and Trees: Methods and Applications: In Honor of Boris Mirkin's 70th Birthday, pp. 119–150. Springer, New York, New York, NY (2014)
20. Fan, J., Fan, Y.: High dimensional classification using features annealed independence rules. Ann. Stat. **36**(6), 2605–2637 (2008)
21. Fan, J., Li, R.: Statistical challenges with high dimensionality: feature selection in knowledge discovery (2006)
22. Kim, H., Choi, B.S., Huh, M.Y.: Booster in high dimensional data classification. IEEE Trans. Knowl. Data Eng. **28**(1), 29–40 (2016)
23. Golugula, A., Lee, G., Madabhushi, A.: Evaluating feature selection strategies for high dimensional, small sample size datasets. In: 2011 Annual International Conference of the IEEE Engineering in Medicine and Biology Society, pp. 949–952 (2011)
24. Alzubaidi, A., Cosma, G.: A multivariate feature selection framework for high dimensional biomedical data classification. In: 2017 IEEE Conference on Computational Intelligence in Bioinformatics and Computational Biology (CIBCB), pp. 1–8 (2017)
25. Pavlidis, P., Weston, J., Cai, J., Grundy, W.N.: Gene functional classification from heterogeneous data. In: Proceedings of the Fifth Annual International Conference on Computational Biology, pp. 249–255 (2001)
26. Li, L.: Survival prediction of diffuse large-B-cell lymphoma based on both clinical and gene expression information. Bioinformatics **22**(4), 466 (2006)

27. Li, L., Chen, L., Goldgof, D., George, F., Chen, Z., Rao, A., Cragun, J., Sutphen, R., Lancaster, J.M.: Integration of clinical information and gene expression profiles for prediction of chemo-response for ovarian cancer. In: 2005 IEEE Engineering in Medicine and Biology 27th Annual Conference, pp. 4818–4821 (2005)
28. Campone, M., Campion, L., Roché, H., Gouraud, W., Charbonnel, C., Magrangeas, F., Minvielle, S., Genève, J., Martin, A.-L., Bataille, R., Jézéquel, P.: Prediction of metastatic relapse in node-positive breast cancer: establishment of a clinicogenomic model after FEC100 adjuvant regimen. Breast Cancer Res. Treat. **109**(3), 491–501 (2008)
29. Futschik, M.E., Sullivan, M., Reeve, A., Kasabov, N.: Prediction of clinical behaviour and treatment for cancers. Appl. Bioinform. **2**(3 Suppl.), S53–58 (2003)
30. Daemen, A., Gevaert, O., De Moor, B.: Integration of clinical and microarray data with kernel methods. In: 2007 29th Annual International Conference of the IEEE Engineering in Medicine and Biology Society, pp. 5411–5415 (2007)
31. Gevaert, O., Smet, F., Timmerman, D., Moreau, Y., De Moor, B.: Predicting the prognosis of breast cancer by integrating clinical and microarray data with Bayesian networks. Bioinformatics **22** (2006)
32. Ray, B., Henaff, M., Ma, S., Efstathiadis, E., Peskin, E.R., Picone, M., Poli, T., Aliferis, C.F., Statnikov, A.: Information content and analysis methods for multi-modal high-throughput biomedical data. Sci. Rep. **4**, 4411 (2014)
33. Bellman, R.: Dynamic Programming, 1st edn. Princeton University Press, Princeton (1957)
34. Misaki, M., Kim, Y., Bandettini, P.A., Kriegeskorte, N.: Comparison of multivariate classifiers and response normalizations for pattern-information fMRI. Neuroimage **53**(1), 103–118 (2010)
35. Box, E.P., Meyer, R.D.: An analysis for unreplicated fractional factorials. Technometrics **28**(1), 11–18 (1986)
36. Dembélé, D.: A flexible microarray data simulation model. Microarrays **2**(2), 115–130 (2013)
37. Singhal, S., Kyvernitis, C.G., Johnson, S.W., Kaiser, L.R., Liebman, M.N., Albelda, S.M.: MicroArray data simulator for improved selection of differentially expressed genes. Cancer Biol. Ther. **2**(4), 383–391 (2003)
38. Smialowski, P., Frishman, D., Kramer, S.: Pitfalls of supervised feature selection. Bioinformatics **26**(3), 440–443 (2010)
39. Simon, R., Radmacher, M.D., Dobbin, K., McShane, L.M.: Pitfalls in the use of DNA microarray data for diagnostic and prognostic classification. J. Natl. Cancer Inst. **95** (2003)
40. Ling, X., Huang, J. Zhang, H.: AUC: a statistically consistent and more discriminating measure than accuracy. In: Proceedings of the 18th International Joint Conference on Artificial Intelligence, pp. 519–524 (2003)
41. Ling, X., Huang, J. Zhang, H.: AUC: a better measure than accuracy in comparing learning algorithms. In: Advances in Artificial Intelligence, pp. 329–341 (2003)

Selecting Accurate Classifier Models for a MERS-CoV Dataset

Afnan AlMoammar(✉), Lubna AlHenaki, and Heba Kurdi

Computer Science Department, KSU, KSA Riyadh, Saudi Arabia
{437203909,437204268}@student.ksu.edu.sa,
hkurdi@ksu.edu.sa

Abstract. The Middle East Respiratory Syndrome Coronavirus (MERS-CoV) is a viral respiratory disease that is spreading worldwide necessitating to have an accurate diagnosis system that accurately predicts infections. As data mining classifiers can greatly assist in enhancing the prediction accuracy of diseases in general. In this paper, classifier model performance for two classification types: (1) binary and (2) multi-class were tested on a MERS-CoV dataset that consists of all reported cases in Saudi Arabia between 2013 and 2017. A cross-validation model was applied to measure the accuracy of the Support Vector Machine (SVM), Decision Tree, and k-Nearest Neighbor (k-NN) classifiers. Experimental results demonstrate that SVM and Decision Tree classifiers achieved the highest accuracy of 86.44% for binary classification based on healthcare personnel class. On the other hand, for multiclass classification based on city class, the decision tree classifier had the highest accuracy among the remaining classifiers; although it did not reach a satisfactory accuracy level (42.80%). This work is intended to be a part of a MERS-CoV prediction system to enhance the diagnosis of MERS-CoV disease.

Keywords: Data mining · Medical data · Classification · Classifier model MERS-CoV · Accuracy measurement · Cross-validation model

1 Introduction

Middle East respiratory syndrome (MERS) is a viral respiratory disease that spread over 27 countries around the world. The disease was caused by a novel coronavirus called the Middle East respiratory syndrome coronavirus (MERS-CoV). Moreover, coronaviruses are a large family of viruses responsible for causing many diseases, from mild colds to Severe Acute Respiratory Syndrome (SARS). MERS-CoV is one of the most common major causes for the increase in mortality among children and adults in the world [1]. The first identification of MERS-CoV was in Saudi Arabia in 2012. It spread rapidly in Saudi Arabia and many other countries and caused a large number of deaths [2]. Therefore, early diagnosis of MERS-CoV infection may help to control the outbreak of the virus and reduce human suffering. Computer and data mining techniques can provide great help in analyzing, diagnosing, and predicting diseases, and they can assist in controlling virus infection [3].

K. Arai et al. (Eds.): IntelliSys 2018, AISC 868, pp. 1070–1084, 2019.
https://doi.org/10.1007/978-3-030-01054-6_74

Using data mining techniques in diagnosis and prediction of diseases has been developing fast over the last few decades. Data mining is the process of analyzing a large amount of complex data to find useful patterns and extract hidden information by applying machine learning algorithms [4]. In healthcare, the generated data is vast and too complex to be analyzed and processed by traditional methods. Due to this, the need for data mining in healthcare is becoming essential. Accordingly, data mining has been widely used in healthcare, including outcomes prediction, treatment effectiveness evaluation, infection control, and disease diagnosis [3]. Moreover, studies on using data mining in healthcare show that it succeeds in helping to improve diagnostic accuracy prediction and predicting health insurance fraud, which lightens the burden of increasing workloads, and reducing healthcare costs [5].

Recently, various types of data mining methods have been applied by a number of researchers [6, 7], using real MERS-CoV datasets based on several types of machine learning classifiers. MERS is a complex disease caused by MERS-CoV that spreads easily and has a high death rate; approximately 40% of patients diagnosed with MERS have died [1]. The challenge remains to provide prediction systems that accurately anticipate and diagnose MERS-CoV. Prediction systems are primarily motivated by the necessity of achieving maximum possible accuracy. Our motivation for this study is to utilize data mining techniques in order to control the spreading of MERS-CoV and to save people's lives. Motivated by the above needs, we make the following contribution in the application of classification algorithms to a MERS-CoV dataset for identifying the accurate classifier.

The main contribution of this study is to apply a support vector machine classifier beside two other classifiers to assess the classification accuracy on MERS-CoV dataset. Whilst the previous studies used datasets consisting of information about MERS-CoV cases only up to 2015, our dataset covers all affected cases in Saudi Arabia from 2013 to 2017.

The remaining parts of this paper are organized as follows: The literature review is introduced in Section 2. Then, the system design and implementation are presented in Section 3. The methodology is then described in Section 4. After that, the results and discussion are detailed in Section 5. Finally, the conclusions and directions for future work are discussed.

2 Literature Review

One of the early applications of data mining techniques was in medical areas where it could help in predicting and diagnosing diseases and support medical decision making. Several researchers have been working in data mining application and experimental use of medical datasets. This review will go through some of the related work in healthcare, but it is not meant to be exhaustive. The first part of the literature review introduces some applications of classification algorithms on different medical datasets. The second part is a review of the related works of the MERS-CoV diagnosis and prediction using data mining techniques.

For instance, the researchers in [8], apply data mining on historical health records to improve the prediction of chronic disease. In this study, two datasets from UC Irvine (UCI) repository are considered: heart disease and diabetes. Many data mining algorithms are applied, including: Naïve Bayes, Decision Tree, Support Vector Machine (SVM), and Artificial Neural Networks (ANN). From the experiment, SVM performs better than the other classifiers on the heart disease dataset, while Naïve Bayes classifier achieves the highest accuracy on the diabetes dataset.

A recent study [9] uses data mining to increase the diagnosis of neonatal jaundice in newborns. The dataset consists of records of healthy newborn infants with 35 or more weeks of gestation collected from the Obstetrics Department of the Centro Hospital. Several data mining algorithms are applied to the dataset: Decision Tree, CART, Naïve Bayes, Artificial Neural Networks, SVM, and Easy Logistic algorithms. The results of this study show that the most effective predictive models are Naïve Bayes, Neural Networks, and Easy Logistic algorithms.

The researchers in [10] compare different data mining algorithms to find the most efficient and effective algorithm in terms of accuracy, sensitivity, and precision. An experiment is conducted using an original Wisconsin Breast Cancer dataset from the UCI machine learning repository with four classifiers: SVM, Naïve Bayes, Decision Tree, and k-Nearest Neighbor (k-NN). The effectiveness of all classifiers is evaluated in terms of time to build the model, correctly classified instances, incorrectly classified instances, and accuracy. The results show that SVM is the most efficient classifier in Breast Cancer prediction and diagnosis with high precision and low error rate.

Another study [11], applies different machine learning algorithms on artificial lung cancer datasets systematically collected by the Hospital Information System in order to explore the advantages and disadvantages of each algorithm. Many experiments are conducted on the dataset using the following machine learning algorithms: Decision Tree, Bagging, Adaboost, SVM, k-NN, and Neural Network. The results show that, according to the high accuracy of these algorithms, Adaboost and Neural Network are suitable for this type of cancer analysis.

The researchers in [12] compare two classification algorithms: Decision Trees and Random Forest with Self-Organizing Map (SOM) to build a predictive model for diabetic patients. The dataset uses in this study is collected from the Hospital Information System of the Ministry of National Guard Health Affairs (MNGHA), Saudi Arabia, between 2013 and 2015. The authors found that the Random Forest algorithm achieves the highest recall and precision.

The authors in [13] introduce a MobDBTest Android mobile application. MobDBTest uses machine learning techniques to predict diabetes levels for the users. The propos Android mobile application is tested on real dataset collected from a reputed hospital in the Chhattisgarh state of India. Four machine learning algorithms such as J48, Naïve Bayes, SVM and Multilayer Perceptron are used to classify the collected data. The results show that J48 algorithm outperformed other methods in terms of sensitivity, specificity and ROC areas.

During the past six years, more information about the MERS-CoV disease has become available to the public. MERS-CoV is a well-known virus that is still rapidly growing. Finding the accurate classifier can help to improve the prediction accuracy of MERS-CoV infection. The study in [7] applies data mining techniques to a

MERS-CoV dataset to identify the accurate classifier models of binary, multi-class, and multi-label classification. The dataset includes all MERS-CoV cases in Saudi Arabia from the Saudi Ministry of Health from 2013 to the second half of 2016. Three classifier models are built using k-NN, Decision Tree, and Naïve Bayes algorithms. The outcome of this research is that the Decision Tree is the most accurate algorithm for the binary-class classification, whereas k-NN is the most accurate algorithm for the multi-class classification. Additionally, for the multi-label classification the Naïve Bayes is the most accurate algorithm.

Another related study [6], involves experimental data mining to build prediction models for MERS-CoV. The experiments are conducted on a dataset collected from the Saudi Ministry of Health. It consists of MERS-CoV cases between 2013 and 2015. The Naive Bayes and Decision Tree algorithms are used to develop recovery and stability predictive models based on the MERS-CoV dataset. The results of recovery models indicate that healthcare workers are more likely to survive. Moreover, symptoms and age are important attributes for predicting stability in stability models. In general, Decision Tree has better accuracy over all models.

The researchers in [14] propose a molecular approach to analyze DNA sequences of MERS-CoV to draw the route of transmission of MERS-CoV from Saudi Arabia to the world. Full DNA sequences that are collected from 15 different regions from the National Center for Biotechnology Information (NCBI) are converted into amino acid sequences to be used in the analysis process. Moreover, the proposed approach uses Apriori and Decision Tree algorithms to find the similarities and differences between different amino acid sequences. Relevance between several sequences is found using Decision Tree algorithm.

The study described in [15] proposes a cloud-based MERS-CoV prediction system to predict and prevent MERS-CoV infection spread between citizens and regions. The dataset consists of patients, medicines, and reports of each user. It is stored in multiple clouds known as a medical record (M.R.) database. In addition, this system is based on a statistical classifier in data mining, which is a Bayesian classification algorithm for initial classification of the patient base on predicting class membership probabilities. The outcome of this study is a prediction of MERS-CoV-infected regions on Google Maps with high accuracy in the classification.

A study [16] applies three data mining algorithms to compare two viruses with similar symptoms: severe acute respiratory syndrome (SARS) and MERS coronaviruses. Apriori, Decision Tree, and SVM data mining algorithms are used on data of the spike in glycoprotein from the NCBI to distinguish between the two viruses. From the experiment, it is clear that distinguishes between MERS and SARS spike glycoproteins with a high accuracy.

Table 1 presents a comparison of literature review that applied data mining techniques on medical data over the different categories. These categories are reference number, used data mining algorithm, used dataset, the objective of the research, used tool, and finally the outcome of the research. From the Table 1, it can be seen that several algorithms and techniques have been applied to medical datasets and that the most common methods for classification are Decision Tree, SVM, and k-NN algorithms.

Table 1. Comparison of relevant literature review

Ref. no.	Data mining techniques	Dataset	Objective	Tool	Outcomes
[8]	Naïve Bayes, Decision Tree, SVM, and ANN	Heart disease, and diabetes datasets	Predicting Chronic disease by mining the data containing historical health records	WEKA	SVM gives highest accuracy rate of 95.55% and Naïve Bayes classifier gives highest accuracy of 73.58% for the heart disease, diabetes respectively
[9]	Decision Tree, CART, Trusting Bayes classifier, neural networks SMO, and easy logistic	Records of Healthy newborn infants with 35 or more weeks of gestation	Improving the diagnosis of neonatal jaundice in newborns	WEKA	The most effective predictive models are Trusting Bayes with 88% accuracy, neural networks with 87% accuracy, and easy logistic with 89% accuracy
[10]	SVM, C4.5, Naive Bayes, and *k*-NN	Wisconsin Breast Cancer (original) dataset	Finding the most efficient algorithm for Breast Cancer prediction and diagnosis	WEKA	The SVM has proven its efficiency in Breast Cancer prediction and diagnosis with 97.13% accuracy
[11]	Decision Tree, Bagging, Adaboost, SVM, *k*-NN, and Neural Network	Artificial lung cancer dataset	Comparing different classification algorithms in order to explore the advantages and disadvantages of each one	RStudio	Adaboost algorithm and neural network algorithm have relative high accuracy with 97.5% accuracy
[12]	Self-Organizing Map (SOM), Decision Tees	Adult population data	Constructing intelligent predictive model for diabetic	RStudio and WEKA	The RandomForest model could assist health care

(*continued*)

Table 1. (*continued*)

Ref. no.	Data mining techniques	Dataset	Objective	Tool	Outcomes
	C4.5, and Random Forest		disease by using real healthcare data		providers with 90% accuracy to make better clinical decisions in identifying diabetic patients
[13]	J48, Naïve Bayes, SVM and Multilayer Perceptron	Reputed hospital in the Chhattisgarh state of India	Predict diabetes levels for the users uses machine learning techniques	Android mobile application	The results show that J48 algorithm outperformed other methods in terms of sensitivity, specificity and ROC areas
[7]	*k*-NN, Decision Tree, and Naïve Bayes algorithms	MERS-CoV cases in Saudi Arabia noted between 2013 and second half of 2016	Identifying accurate classifier modes for binary, multiclass, and multi-label classification of a text-based MERS-CoV dataset	RapidMiner Studio	The accurate algorithm for the Binary-Class classification is Decision Tree with 90% accuracy, for the Multi-Class classification is *k*-NN with 51.60% accuracy, and for the Multi-Label classification is Naïve Bayes with 77% accuracy
[6]	Naive Bayes, and Decision Tree algorithms	1082 records of MERS-CoV cases noted between 2013 and 2015	Building predictive models for MERS-CoV infection to understand which factors contribute to complications of this infection	WEKA	The results show that, Decision Tree classifier has better accuracy of 55.69%, and 68% for the stability and the recovery models respectively

(*continued*)

Table 1. (*continued*)

Ref. no.	Data mining techniques	Dataset	Objective	Tool	Outcomes
[14]	Decision Tree, and Apriori Algorithms	DNA sequences of MERS-CoV outbreak from different regions in the world where the viruses speared	Finding the similarities between different MERS-CoV amino acid sequences to know transmission route of MERS-CoV	Mathematical model	The results show that Riyadh, Makkah, and Buridah regions of MERS-CoV transmission in Saudi Arabia
[15]	BBN classification	Multiple attributes: 1-personal (static), 2-MERS (changes over time)	Identifying an intelligent system for predicting and preventing MERS-CoV infection	R Studio, WEKA, and Amazon EC2	The BBN achieve an accuracy of 83.1% on synthetic data
[16]	Decision Tree, and Apriori Algorithms	DNA sequences of MERS-CoV of outbreak	Finding the similarities between different MERS-CoV amino acid sequences to know transmission route of MERS-CoV	Mathematical model	The results show that Riyadh, Makkah, and Buridah regions of MERS-CoV transmission in Saudi Arabia

In conclusion, in related studies data mining is widely used for the prognoses and diagnoses of many diseases. However, the datasets used in [6, 7] are limited and include the MERS-CoV cases in Saudi Arabia from 2013–2015 only. It is important to increase the size of the dataset to cover new cases. Therefore, this study applied Data mining techniques using Decision Tree, SVM, and k-NN classification algorithms to a real dataset of MERS-CoV cases in the Kingdom of Saudi Arabia that was collected during 2013–2017.

3 System Design and Implementation

The system overview is illustrated in Fig. 1. It shows high-level components of the classification framework. The classification framework is composed of three subsystems, which are the MERS-CoV dataset, supervised learning, and data scientist. The MERS-CoV dataset subsystem aims to collect MERS-CoV data from different sources

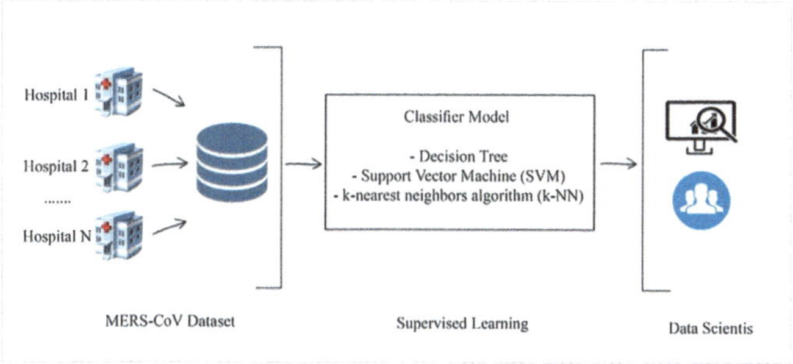

Fig. 1. System overview.

and integrate them into one database. The purpose of the supervised learning subsystem, which is the core of this study, is applying data mining techniques to build three different classifier models. Finally, the third subsystem consists of data scientists who analyze data and evaluate results.

Figure 2 shows the overall workflow of the classification framework, which is divided into two main phases. The first phase aims to collect data of patients who are affected by MERS-CoV from different cities in Saudi Arabia between January 2013 and October 2017. The second phase is the most important phase. Its purpose is to identify the classifier model and evaluate the classification accuracy using cross validation test mode.

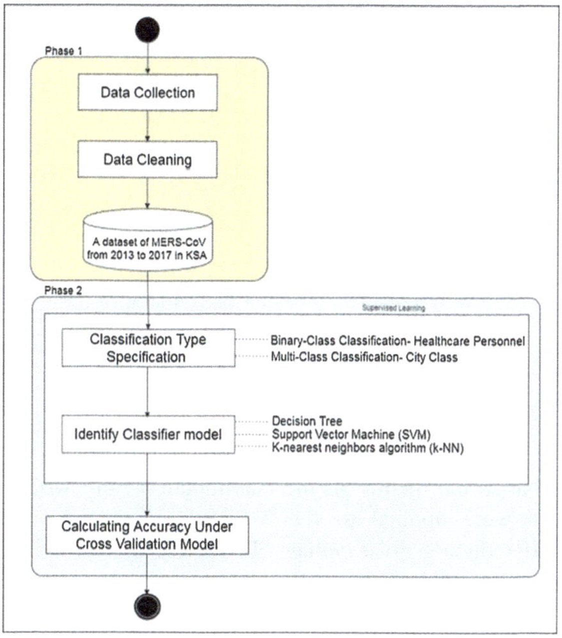

Fig. 2. System workflow.

4 Methodology

4.1 Dataset Description and Pre-processing

As mentioned, the dataset used in this study covers all MERS-CoV cases in Saudi Arabia, including 1,186 alive records and 224 death records, which were reported between 2013 and 2017. The dataset of MERS-CoV cases from 2013–2015 was obtained by a request from [2], the 2016–2017 dataset was collected from the website of the World Health Organization [2]. Moreover, The MERS-CoV dataset consists of the following information about MERS-CoV patients: gender, age, exposure to camels, comorbidities, exposure to MERS-CoV cases, city, and whether the patient is employed in healthcare or not. In addition, the dataset contains information about status to detect whether the patient is alive or dead.

The challenge of building the dataset was that data were published on the website as text description of details of the MERS-CoV cases, represented in Fig. 3, was not promptly usable by any data mining tool. This compelled us to construct the dataset from scratch. Furthermore, all records were prepared in Comma Separated Value (CSV) format, which is appropriate for a data mining tool.

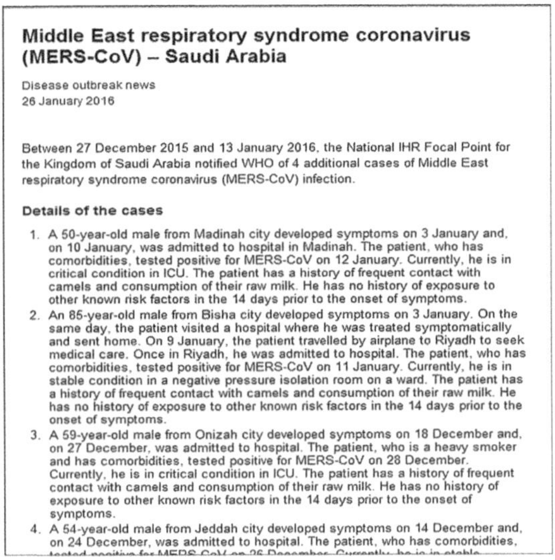
Middle East respiratory syndrome coronavirus (MERS-CoV) – Saudi Arabia

Disease outbreak news
26 January 2016

Between 27 December 2015 and 13 January 2016, the National IHR Focal Point for the Kingdom of Saudi Arabia notified WHO of 4 additional cases of Middle East respiratory syndrome coronavirus (MERS-CoV) infection.

Details of the cases

1. A 50-year-old male from Madinah city developed symptoms on 3 January and, on 10 January, was admitted to hospital in Madinah. The patient, who has comorbidities, tested positive for MERS-CoV on 12 January. Currently, he is in critical condition in ICU. The patient has a history of frequent contact with camels and consumption of their raw milk. He has no history of exposure to other known risk factors in the 14 days prior to the onset of symptoms.
2. An 85-year-old male from Bisha city developed symptoms on 3 January. On the same day, the patient visited a hospital where he was treated symptomatically and sent home. On 9 January, the patient travelled by airplane to Riyadh to seek medical care. Once in Riyadh, he was admitted to hospital. The patient, who has comorbidities, tested positive for MERS-CoV on 11 January. Currently, he is in stable condition in a negative pressure isolation room on a ward. The patient has a history of frequent contact with camels and consumption of their raw milk. He has no history of exposure to other known risk factors in the 14 days prior to the onset of symptoms.
3. A 59-year-old male from Onizah city developed symptoms on 18 December and, on 27 December, was admitted to hospital. The patient, who is a heavy smoker and has comorbidities, tested positive for MERS-CoV on 28 December. Currently, he is in critical condition in ICU. The patient has a history of frequent contact with camels and consumption of their raw milk. He has no history of exposure to other known risk factors in the 14 days prior to the onset of symptoms.
4. A 54-year-old male from Jeddah city developed symptoms on 14 December and, on 24 December, was admitted to hospital. The patient, who has comorbidities,

Fig. 3. A sample of text description of MERS-CoV cases.

In order to enhance the quality of the classification framework, different preprocessing techniques were applied to the MERS-CoV dataset, including replacing missing values and reducing noise values. To handle the missing values, each was replaced with the mean of the attribute that includes missing values. Additionally, the noise in the dataset appeared due to the existence of inconsistent data in the dataset. For instance, the gender attribute is represented in some instances using the full word "female" or "male," while in other instances it is represented using the abbreviations "F" or "M." So, the inconsistent values were integrated into a standard value which is "F or M." Furthermore, the data were converted from categorical to numerical data because the SVM algorithm deals only with numerical data.

4.2 Data Mining

In a similar approach to [7], in this study, the classifier model performance was examined for two classification types: binary classification based on the healthcare personnel class, and multi-class classification based on the city class. The SVM, Decision Tree, and k-NN classification algorithms were chosen because they are commonly used for medical mining as presented in the Literature Review section of this paper. In addition, they outperform other applied algorithms, as shown in [6, 12, 13]. Furthermore, SVM was used in this study because it was not applied to MERS-CoV dataset in recent studies [6, 7]. Also, the Decision Tree and k-NN classifiers in [7] achieved the highest accuracy on binary-class and multi-class classification respectively. Additionally, based on the structure of the dataset, each class was chosen to represent a different classification type.

The software used in this study was RapidMiner Studio version 7.6. RapidMiner is an open-source data mining software tool written in Java programing language. It is issued under the Affero General Public License that provides an integrated environment for data mining and predictive analytics. Moreover, RapidMiner is used to perform machine learning algorithms for data mining tasks [17]. SVM, Decision Tree, and k-NN algorithms were applied on the MERS-CoV dataset using a RapidMiner tool. In sum, this experimental study was recreated six times.

The essential parameters of the k-NN algorithm are k, which is determined by the numbers of nearest neighbors. The value of k was set to 5 because an odd value is recommended to prevent a tie, when two or more classes have the same number of votes. Additionally, the Euclidean distance function was used as a similarity measure between testing and training data [4, 18]:

$$Euclidean\,distance\,(x, x_i) = \sqrt{\sum\nolimits_{i=0}^{m} (x - x_i)^2} \tag{1}$$

Where x is the testing point, and x_i is the training point.

For the decision tree, gain ratio was used as the attribute selection method for splitting, because it measures the information that gained by each attribute easily and quickly. The information gain ratio calculated using the following formula [4, 18]:

$$Gain(A) = Info(D) - InfoA(D) \tag{2}$$

where the Info (D) is the average amount of information needed to identify the class label of a tuple in D also, known as the entropy of D, and it is calculated by:

$$Info(D) = \sum_{i=1}^{m} \mathrm{P_i} \log_2(\mathrm{p_i}) \tag{3}$$

Where $\mathrm{P_i}$ is the probability that an arbitrary tuple in D belongs to class C, where $Info(D)$ is the expected information required to classify a tuple from D based on the partitioning by attribute A, it is calculated by:

$$Info_A(D) = \sum_{j=1}^{v} \frac{|\mathrm{D_j}|}{|\mathrm{D}|} \times \mathrm{Info}(\mathrm{D_j}) \tag{4}$$

The term $\frac{|\mathrm{D_j}|}{|\mathrm{D}|}$ is the weight of the $\mathrm{j^{th}}$ partition [4].

The maximum depth of the Decision Tree was set to 20. Also, the Decision Tree was generated with a pruning function, which allows for reducing the size of the tree by removing low-power sub-trees.

The essential parameter of SVM classifier is the kernel function. The most common kernel function that used with SVM classifier is linear kernel function [6] it defined as:

$$F(x) = W.(x_i, y_i,) + b \tag{5}$$

Where $X = (x_i, y_i,)$ is the dataset, x_i, is the instances, y_i, is the class label and $i = 1, 2, \ldots\ldots n$ and W is the weight vector or the coefficient, and b is a scalar value called bias [4, 17]. Other important parameters are the value of complexity constant (C), and the tolerance parameter. In this study, the kernel function that used was the linear function because the data is linearly separable, the parameter C was set to 0, and the tolerance parameter was set 0.001.

Since the classification type of classes, classifier algorithm, and their parameters were specified, a model is needed for assessing the classification performance. Therefore, a cross-validation model is used to assess the classification performance. In k-fold cross validation technique, the dataset is randomly split into k equal-sized subsets. Each time, one of the k subsets is used as the test set and the other k-1 subsets are put together and used as training set (train on nine datasets and test on one). Then, the process is repeated ten times. In this empirical study, all models were built using 10-fold cross validation. The advantage of this method is that it data division into training and testing sets is irrelevant [19].

The most significant part of many studies is discovered during evaluation, and the value of the study can be assessed. To compare all results of the applied algorithms to the MERS-CoV dataset in this project, their performances were quantitatively measured using accuracy, which is the most widely used evaluation metric to reflect the percentage of correctly-classified records in the testing phase [21]. Therefore, the accurate classifier will be useful for building a MERS-CoV prediction system.

A confusion matrix is an important way for analyzing the performance of a binary-class classification. Moreover, in this matrix, each row contains information about actual

Table 2. Confusion matrix

	Negative (predicted)	Positive (predicted)
Negative (actual)	TP	FN
Positive (actual)	FT	TN

class while each column contains information about predicted class. Accordingly, the confusion matrix aims to analyze how well a classifier can recognize tuples of different classes. Table 2 illustrates the confusion matrix for a two-class classifier [20].

For evaluating the classification framework based on the confusion matrix, the accuracy formula of each classification type was used for the binary-class classifications; the accuracy was calculated based on the following formula:

$$Accuracy = \frac{100 \times (\mathrm{TP} + \mathrm{TN})}{\mathrm{TP} + \mathrm{FN} + \mathrm{TN} + \mathrm{FP}} \tag{6}$$

On the other side, for the multi-class classifications, the accuracy was calculated based on the following formula:

$$Accuracy = \frac{\sum_{i=1}^{l} \frac{100 \times (TP_i + TN_i)}{TP_i + FN_i + TN_i + FP_i}}{l} \tag{7}$$

Where, TP (True Positives) is the correctly classified positive cases, TN (True Negative) is the correctly classified negative cases, FP (False Positives) is the incorrectly classified negative cases, and FN (False Negative) is the incorrectly classified positive cases [21].

5 Results and Discussion

Based on the essential parameters of the classifier models, which are presented in the methodology section, the accuracies obtained on the MERS-CoV dataset with each classifier model for each classification type are shown in Table 3. The best accuracy is for the binary-class classification based on healthcare personnel class with 86.44%, which was produced by SVM and Decision Tree algorithms. Figure 4 illustrates the result of the binary-class classification; when applying SVM with the healthcare personnel class, the margin width was maximizing, making the prediction faster and more accurate. On other hand, as healthcare personnel class became a root of the decision tree the depth of the tree was minimized and the tree is not complex that generates accurate predictions.

Table 3. Classifier model accuracy for each classification type

Classification type	SVM classifier	Decision tree classifier	*k*-NN classifier
Binary-class classification	86.44%	86.44%	85.31%
Multi-class classification	18.24%	42.80%	30.80%

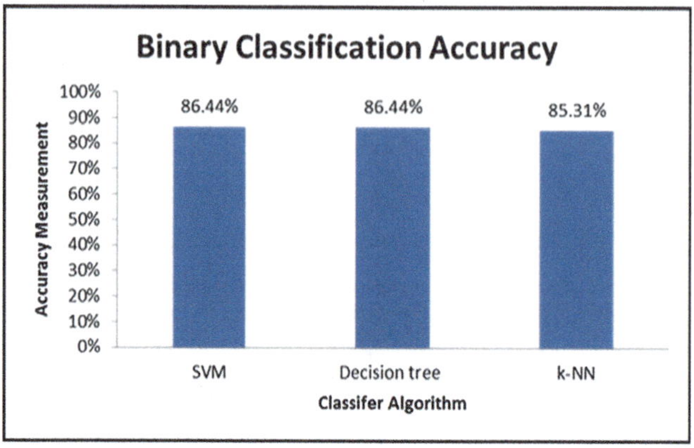

Fig. 4. Binary-class classification accuracy.

Another important finding is that, for multi-class classification the Decision Tree obtains the highest accuracy with 42.80% based on city class. Whereas the accuracy of *k*-NN was 30.80% and SVM classifier was under 20% as shown in Fig. 5. Moreover, for evaluating the effectiveness of the results of our method, we have to compare the experimental results with the results of a recent study [7]. A recent study [7] reported highest accuracy of 51.60% for multi-class classification based on *k*-NN classifier. This could be due to the value of parameter *k* that is set to 5 in this study while it is set to 3 in [7].

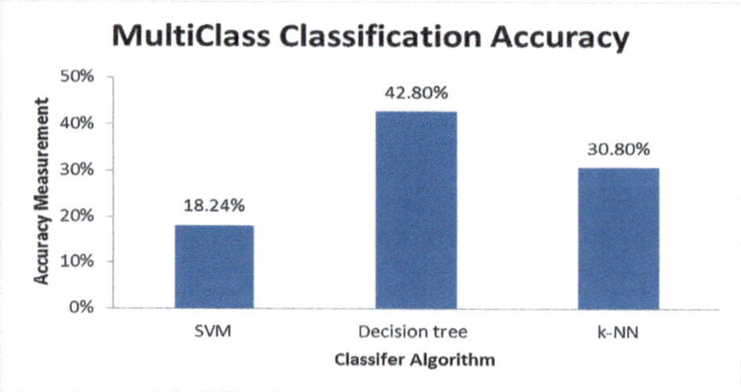

Fig. 5. Multi-class classification accuracy.

On another side, the researchers in [7] reported a higher accuracy on binary-class classification of 90.00%, when using the Decision Tree algorithm based on gender attribute, while our method achieved a lower accuracy of 86.44% when using SVM and Decision Tree classifiers based on healthcare personnel attribute. Therefore, using healthcare personnel attribute as binary-class may not be appropriate for MERS-CoV dataset classification.

The experimental results demonstrate that the accurate classifier models for binary-class and multi-class classification types are built by using Decision Tree and *k*-NN algorithms respectively. Additionally, the results of this study indicate that using SVM classifier is not suitable for classification of MERS-CoV dataset. In general, the main explanation of our results is based on the essential parameter settings.

6 Conclusions and Future Work

The classifier model performance of several classification types can greatly assist to enhance the prediction accuracy of MERS-CoV infection. In this study, we have identified a classifier model performance that applied binary and multiclass classification on real a MERS-CoV dataset. Three algorithms were used to build classifier models, which were SVM, Decision Tree, and *k*-NN. The algorithms were applied using RapidMiner, a data mining tool. The performance of classifier models was measured using the accuracy evaluation metric; in addition, cross-validation was used as a model for assessing classification performance.

The experimental results have shown that both SVM and Decision Tree classifiers achieved the highest accuracy of 86.44% on binary-class classification based on healthcare personnel class. On the other hand, the Decision Tree classifier had the highest accuracy of 42.80% among the remaining classifiers for multiclass classification based on city class, although it did not reach a satisfactory accuracy level. In general, the comparison of the experimental results and the results of a recent study indicate that Decision Tree and *k*-NN classifiers are the accurate classifiers for binary-class and multi-class classification types respectively. Additionally, using an SVM classifier is not suitable for classification of a MERS-CoV dataset. For future work, it is intended that this experiment will be applied to the universal MERS dataset. Furthermore, other preprocessing technique such as remove missing value can be used to measure its effect on the classifier models' performance. Additionally, other classification methods, such as ensemble learning, can be used. Also, another similarity metric, such as cosine similarity, may be used with the *k*-NN algorithm. Finally, for the multiclass classification, we suggest recreating the empirical study with different parameters to determine a classifier that gives accuracy greater than 50%.

References

1. Coronavirus website - Ministry of Health. http://www.moh.gov.sa/en/CCC/. Accessed 29 Oct 2017
2. WHO: Middle East respiratory syndrome coronavirus (MERS-CoV). http://www.who.int/emergencies/mers-cov/en/. Accessed 23 Oct 2017
3. Koh, H.C., Tan, G.: Data mining applications in healthcare. J. Healthc. Inf. Manag. **19**(2), 64–72 (2005)
4. Han, J., Kamber, M.: Data Mining: Concepts and Techniques. Elsevier, Haryana, India, Burlington (2012)
5. Yoo, et al.: Data mining in healthcare and biomedicine: a survey of the literature. J. Med. Syst. **36**(4), 2431–2448 (2012)

6. Al-Turaiki, M., Alshahrani, M., Almutairi, T.: Building predictive models for MERS-CoV infections using data mining techniques. J. Infect. Public Health **9**(6), 744–748 (2016)
7. AlMansour, N., Kurdi, H.: Identifying accurate classifier models for a text - based MERS-CoV dataset. Presented at the Intelligent Systems Conference 2017, London, UK (2017)
8. Deepika, K., Seema, S.: Predictive analytics to prevent and control chronic diseases, pp. 381–386 (2016)
9. Ferreira, D., Oliveira, A., Freitas, A.: Applying data mining techniques to improve diagnosis in neonatal jaundice. BMC Med. Inform. Decis. Mak. **12**(1), December 2012
10. Asri, H., Mousannif, H., Moatassime, H.A., Noel, T.: Using machine learning algorithms for breast cancer risk prediction and diagnosis. Procedia Comput. Sci. **83**, 1064–1069 (2016)
11. Li, J., Zhao, Z., Liu, Y., Cheng, Z., Wang, X.: A comparative study on machine classification model in lung cancer cases analysis. In: Yen, N.Y., Hung, J.C. (eds.) Frontier Computing, vol. 422, pp. 343–357. Springer Singapore, Singapore (2018)
12. Daghistani, T., Alshammari, R.: Diagnosis of diabetes by applying data mining classification techniques. Int. J. Adv. Comput. Sci. Appl. **7**(7) (2016)
13. Sowjanya, K., Singhal, A., Choudhary, C.: MobDBTest: a machine learning based system for predicting diabetes risk using mobile devices, pp. 397–402 (2015)
14. Kim, D., Hong, S., Choi, S., Yoon, T.: Analysis of transmission route of MERS coronavirus using decision tree and apriori algorithm, pp. 559–565 (2016)
15. Sandhu, R., Sood, S.K., Kaur, G.: An intelligent system for predicting and preventing MERS-CoV infection outbreak. J. Supercomput. **72**(8), 3033–3056 (2016)
16. Jang, S., Lee, S., Choi, S.-M., Seo, J., Choi, H., Yoon, T.: Comparison between SARS CoV and MERS CoV using Apriori Algorithm, Decision Tree, SVM. In: MATEC Web of Conferences, vol. 49, p. 08001 (2016)
17. RapidMiner Studio - RapidMiner Documentation. http://docs.rapidminer.com/studio/. Accessed 11 Jan 2017
18. Witten, H., Frank, E., Hall, M.A.: Data Mining: Practical Machine Learning Tools and Techniques, 3rd edn. Morgan Kaufmann, Burlington (2011)
19. Kohavi, R.: A study of cross-validation and bootstrap for accuracy estimation and model selection. In: Proceedings of the 14th International Joint Conference on Artificial Intelligence, vol. 2, pp. 1137–1143 (1995)
20. Stehman, S.V.: Selecting and interpreting measures of thematic classification accuracy. Remote Sens. Environ. **62**(1), 77–89 (1997)
21. Sokolova, M., Lapalme, G.: A systematic analysis of performance measures for classification tasks. Inf. Process. Manag. **45**(4), 427–437 (2009)

Big Data Fusion Model for Heterogeneous Financial Market Data (FinDf)

Lewis Evans[1(✉)], Majdi Owda[1], Keeley Crockett[1], and Ana Fernández Vilas[2]

[1] School of Computing, Mathematics and Digital Technology, Manchester Metropolitan University, M41 5GD UK, Manchester, UK
{l.evans,m.owda,k.crockett}@mmu.ac.uk

[2] I&C Lab. AtlantTIC Research Centre, University of Vigo, 36310 Pontevedra, Spain
avilas@det.uvigo.es

Abstract. The dawn of big data has seen the volume, variety, and velocity of data sources increase dramatically. Enormous amounts of structured, semi-structured and unstructured heterogeneous data can be garnered at a rapid rate, making analysis of such big data a herculean task. This has never been truer for data relating to financial stock markets, the biggest challenge being the 7Vs of big data which relate to the collection, pre-processing, storage and real-time processing of such huge quantities of disparate data sources. Data fusion techniques have been adopted in a wide number of fields to cope with such vast amounts of heterogeneous data from multiple sources and fuse them together in order to produce a more comprehensive view of the data and its underlying relationships. Research into the fusing of heterogeneous financial data is scant within the literature, with existing work only taking into consideration the fusing of text-based financial documents. The lack of integration between financial stock market data, social media comments, financial discussion board posts and broker agencies means that the benefits of data fusion are not being realised to their full potential. This paper proposes a novel data fusion model, inspired by the data fusion model introduced by the Joint Directors of Laboratories, for the fusing of disparate data sources relating to financial stocks. Data with a diverse set of features from different data sources will supplement each other in order to obtain a Smart Data Layer, which will assist in scenarios such as irregularity detection and prediction of stock prices.

Keywords: Big data · Data fusion · Heterogeneous financial data

1 Introduction

The ineluctable growth of heterogeneous financial data sources relating to financial stocks poses a serious challenge to researchers and regulators who attempt to analyse stock market discussions and prices for a variety purposes such as detecting possible irregular behaviour [1, 2]. With the advent of social media, financial discussion boards (FDBs), and traditional news media dissemination, investors have an almost endless amount of communication channels to make use of for executing well-informed investments [3]. The analysis of such communication is difficult to undertake, due to the many

K. Arai et al. (Eds.): IntelliSys 2018, AISC 868, pp. 1085–1101, 2019.
https://doi.org/10.1007/978-3-030-01054-6_75

problems associated with big data within the financial market domain [1, 4]. Big data is defined as "data sets with sizes beyond the ability of commonly used software tools to capture, curate, manage, and process data" [4].

There exists a myriad of studies on the Vs of big data, the first instance being the consideration of volume, velocity, and variety [5], since then there have been extensions to the Vs of big data, including the 4 Vs [6], 5 Vs [7], 7 Vs [8], and more recently, a 42 V approach to big data has been proposed [9]. For our study on financial stock markets, we adopt the 7 Vs conceptual model of big data (volume, variety, velocity, variability, veracity, value and visualisation), as these seven are clearly distinguishable in the field of financial stock markets [4]. The increasing number of Vs in source data, the more complex the fusion process will be in order to produce Smart Data.

Data fusion has been a well-established practice for managing heterogeneous data sources through the use of associating and combining data sources together [10, 11]. Several models proposed for the fusion of data include the model proposed by the Joint Directors of Laboratories (JDL) [12] and the Dasarathy model [13]. These models, however, have been outdated due to their emphasis on specific domains and applications, often needing to be revised and adapted based on the specific fusion task [14].

Limited research has been undertaken on the fusion of financial data sources, in this paper we coin the term FinDF to refer to the fusing of financial data sources. Existing fusion techniques do not consider more than two data sources, and focus on Securities and Exchange Commission (SEC) filings (which are only available for stocks listed on US exchanges such as the NYSE or NASDAQ) along with other text-based document filings [15]. The existing challenges of FinDF lie in the fact that each of these financial data sources have a different origin, their contents will often be distributed over a variety of websites and vary dramatically in terms of their structure and intent. As existing research focuses primarily on integrating textual documents, there is an opportunity to improve upon existing methodologies by establishing data fusion techniques which take into account data sources such as social media comments, financial discussion board posts, broker agency ratings and stock market data.

This paper proposes a novel data fusion model to address the fusion of financial data from multiple source environments, providing a solution for the current challenges of data association from multiple environments, namely *how* to fuse such data. The proposed model will approach the fusion task from two dimensions: (1) fusing the different data sources together based on time-slice windows; and (2) the company in which the data corresponds to.

This paper is organised as follows: Sect. 2 looks at the related work on data fusion, including its use in various fields and how the JDL model has inspired existing fusion tasks. Section 3 introduces some of the financial data sources which are used by investors to discuss stocks and make investment decisions. Section 4 explores the challenges of big data in relation to financial markets, and how the 7 Vs of big data are dominant within the field of financial markets. Section 5 presents the proposed FinDF model for the fusing of financial data sources. Section 6 explores the future work which could be performed as a result of this research, in addition to drawing a conclusion in relation to how the FinDF model addresses some of the challenges of big data within the financial market domain.

2 Related Work

2.1 Data Fusion

Several definitions exist within the literature for the term data fusion. The first definition being coined by Hall and Llinas [16]: "data fusion techniques combine data from multiple sensors, and related information from associated databases, to achieve improved accuracies and more specific inferences that could be achieved by the use of a single sensor alone".

The terms data fusion and information fusion are often used synonymously; there is, however, a distinction which should be made. The term data fusion is used to refer to fusing raw data (data which is obtained directly from a source with no pre-processing or cleaning being carried out), whereas the term information fusion is used to refer to the fusion of data which is already processed in some way [17]. Regardless of the term used, data and information fusion techniques are used to enhance knowledge discovery [18].

There exist a considerable number of challenges associated with the fusion of data sources, many of these challenges stem from the disparity of how different data is structured [19]. The most notable challenges, outlined by [20], include:

(1) Disparate Data
The input data which is provided to a data fusion model will most often be generated by a variety of sources such as humans (e.g. textual comments), APIs (e.g. time-stamped sequential data), scraping (e.g. textual content). Fusion of such heterogeneous data in order to construct a comprehensible and accurate view of the overall picture is a challenging task in itself.

(2) Outliers and False data
Noise and impreciseness of data can be found in almost all sources of data. A data fusion algorithm should be able to take measures against outliers which are presented to it and take appropriate action accordingly as part of the fusion process.

(3) Data Conflict
Data fusion algorithms must be able to treat conflicting data with great care, being careful not to simply discard it, but to provide a means of cross-checking the data across the different sources.

(4) Imperfection of Data
Data will often be affected by some element of impreciseness, a data fusion model should be able to express such imperfections and make a decision such as whether or not to discard such data, or fuse the data and accept the risk of imperfect data fusion.

(5) Out of Sequence Data
Data which is inputted into a data fusion model will often be organised in discrete pieces which feature a corresponding timestamp, detailing its time of origin. Undoubtedly, the different input sources may be out of sequence due to varying time-zones in which the data is collected from, including factors such as daylight-saving time.

(6) Data Association
Associating multiple entities into groups is the most significant problem of the data fusion process. It can be seen as trying to establish hidden or secret relationships between entities which may not appear to be immediately apparent.

(7) Data Collection
As is the case with many web 2.0 technologies, APIs are often provided for the unified collection of data. However, not all sources provide such a convenient way of collecting data, meaning techniques such as web scraping will need to be utilised for data collection.

2.2 Fields Utilising Data Fusion

Data fusion has been employed successfully in a wide range of domains in order to combine multiple data sources into a unified data output [21]. Table 1 lists several fields in which data fusion has been adopted to improve the accuracy of analysing multiple data sources.

Table 1. Fields utilising data fusion

Field	Description	Refs.
Forensics - Network Intrusion Detection Systems (IDS)	Complementing evidence and artifacts from different layers of a computer or devices to create a complete picture of what events occurred during a reactive forensic investigation. The proposed model (based on the JDL model) can successfully reduce false positive alarms generated by IDS and improve the detection of unknown threats.	[67]
Military – Unmanned Aerial Vehicles (UAV)	Detection of threats based on multi-sensor multi-source data fusion. The proposed model (also based on the JDL) aimed to enhance the situation awareness of the UAV (human) operators by providing a model supporting the detection of threats based on different data sources fused together.	[68]
Navigation Systems	Beacons used for navigation systems and emergencies are highly susceptible to noise, frequency shifts and measurement errors. The adoption of data fusion was able to reduce packet error rate from beacons and sensors from 70% to 4.5%.	[64]
Track monitoring from multiple in-service trains	Monitoring of a rail-track network to ensure safety of its users and to reduce maintenance costs by early detection of faults. The proposed model, which fused position data from trains, and track data (vibrations), indicated that fusing data helped in the detection of track changes, resulting in early detection of track faults.	[69]
Geosciences – Habitat Mapping	Data combined from multiple sources (hyperspectral, aerial photography, and bathymetry data) was utilised for the purposes of mapping and monitoring of the benthic habitat in the Florida Keys.	[70]

The success of data fusion in these domains through the use of fusing different data relating to the same objects for better observations make it an attractive option for combining financial stock market data.

Although work has been undertaken which integrates market data with financial news and work which considers the fusion of documents, this work does not consider the fusion of such a wide variety of disparate data sources such as social media comments, discussion board posts and broker agency ratings [22, 23]. To our knowledge, there has been no work undertaken which considers the fusion of multiple disparate data sources relating to financial stock markets.

2.3 Data Fusion Models

There have been a number of reviews of existing data fusion models and architectures in recent years [17, 20]. Existing models include the Intelligence cycle model, Boyd control loop model, Dasarathy model, and the Thompoulos model [24]. Although there have been several proposals of data fusion models over the years, none have become more widely adopted as the JDL model [25], which will now be overviewed in detail.

(1) JDL Model
Initially proposed by the U.S Joint Directors of Laboratories (JDL) and the US Department of Defense (DoD) in 1985 [24, p. 111], the JDL model is considered the seminal model for data fusion tasks [26]. The JDL model (Fig. 1) is comprised of five processing levels, a database management system (DBMS), human interaction, and a data bus which connects all of these components together [27].

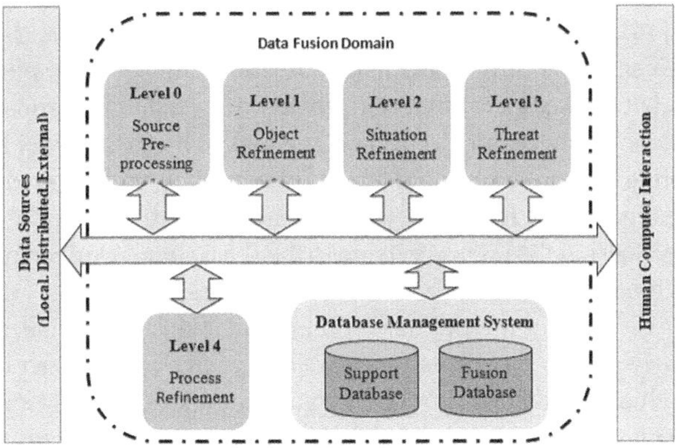

Fig. 1. JDL data fusion model.

(a) Level 0 – Source Pre-processing
The lowest layer present in the JDL model involves reducing the volume of the data using data cleaning techniques, addressing missing values, and maintaining useful information for the higher-level processes.

(b) Level 1 – Object Refinement
At this low-level of the fusion model, data is aligned to objects in order to allow statistical estimation, and to permit common data processing [26, 28].
(c) Level 2 – Situation Refinement
This level deals with the relationships between objects and observed events, attempting to provide a contextual description between the relationships [27, 29].
(d) Level 3 – Threat Refinement
The fusion process of this level attempts to create data for future predictions. The output of which is prediction data which can be stored for further analysis or acted upon [21].
(e) Level 4 – Process Refinement
The monitoring of system performance, including handling real time constraints is addressed at this level [29].
This level of the data fusion model does not perform any data processing operations, as it is more focused on identifying information required for data fusion improvement [30, 31].
(f) Support Database
The support database of the JDL model serves as a data repository in which raw data is stored to facilitate the fusion process [32].
(g) Fusion Database
At the conclusion of the data fusion process, fused data is stored within the fusion database, to be used for future analysis tasks.

(2) *JDL Model Revisions*

The original JDL data fusion model was incepted to provide a process flow for sensor and data fusion [14]. As a result of the JDL model being over thirty years old, it has been revised over the years to address specific data fusion challenges. Despite the popularity of the JDL model, it has been subject to scrutiny due to being tuned primarily for military applications and being too restrictive [20]. Revisions to the JDL model in 1999 by [33], involved a redefined model which attempted to steer away from a model which, at the time, was tailored primarily for military applications, which was the case for many data fusion tasks at that period [34]. This revision to the JDL model revolved primarily around redefining the Threat Refinement process; as the concept of "threats" does not exist to such an extent as it does in the military domain. Steinberg, Bowman, and White [33] redefined the *Threat Refinement* level as *Impact Assessment*, as impact is considered an umbrella-term which, unlike threat refinement, is not restricted to specific domains.

Further revisions and extensions to the JDL model were proposed in 2004 by [35]. Proposals in this paper involved extending the model to include the previous remarks on issues relating to quality control, reliability, and the consistency in data fusion processing.

3 Financial Data Sources

Investors have a plethora of information sources when it comes to researching and discussing stock options. The data fusion model we propose will utilise sources from a variety of environments. In this section, we will detail the data sources which will be fused by the data fusion model.

3.1 Financial Discussion Boards (FDBs)

During the early 2000s, the emergence of financial discussion boards such as Yahoo! Finance and Raging Bull provided two of the most prominent messaging boards on the internet [36]. FDBs provide an unprecedented opportunity for investors to invest, debate, and exchange information on stocks, often expressing their own individual opinion, and often having no prior social connections to other users [37]. FDBs are often specific to certain stock markets, Interactive Investor and London South East, for example, provide a platform for investors to discuss stocks which float on the London Stock Exchange, offering a separate discussion board for each stock [38]. Existing work undertaken by [39] has utilised this data source for the purpose of highlighting potential irregularities through the use of information extraction (IE).

3.2 Social Media

Boasting over 313 million active users worldwide, Twitter provides for fast dissemination of information [40–42]. Twitter has been the subject of several experiments by researchers for its use in discussing financial stocks [43–45]. Twitter has also recently doubled the character limit of tweets from 140 characters to 280 characters, allowing users to circulate even more information within tweets [46].

In 2012, Twitter unveiled a feature named cashtags, a feature initially unique to Stocktwits [47], which allowed for clickable hyperlinks to be embedded in tweets, similar to the behaviour of hashtags [44]. These cashtag entities are structured to mimic the TIDM (Tradable Instrument Display Mnemonic) of a company, prefixed with the $ symbol (e.g. $VOD for Vodafone).

One of the nuances of the cashtag feature involves a phenomenon which has not yet been explored within the literature, which we refer to as "cashtag collision" [44]. This occurs when two companies with identical TIDM identifiers (e.g. $TSCO) appear on multiple exchanges across the world, yet Twitter is unable to clearly distinguish between them, so the discussions of both are merged into a singular search feed.

Other notable sources of information relating to financial stocks include the likes of Reddit, which have several subreddits for the purpose of discussing stock options for stocks all over the world.

3.3 Broker Agencies

Brokers are agents which trade on behalf of their clients, and often provide their clients and the rest of the financial market community with advice on investment decisions [48]. Companies such as London South East aggregate broker ratings from a wide collection of reputable broker agencies such as JP Morgan and Barclays [49].

3.4 News Corporations

Many investors still rely on information provided by news corporations which monitor the financial market world. The *Financial Times*, for example, is often regarded as a reputable source of financial market news within the UK due to the well-regarded journalists associated with it [50].

3.5 Stock Market Data

Researchers and investors often rely on timely intraday stock market data such as those provided by Google Finance and Yahoo Finance APIs, however, since mid-2017, the Google Finance and Yahoo Finance APIs are no longer active [51]. Financial stock market data can be obtained from the Time Series Data API hosted by AlphaVantage [52].

AlphaVantage offers free intraday and historic stock market data from 24 exchanges around the world, providing real-time stock market data from time intervals ranging from one minute to sixty minutes.

The core collectable attributes of these data sources, along with their structure type, are listed in Table 2. All of the financial data sources possess an attribute corresponding

Table 2. Collectable Attributes of Financial Data Sources

Financial Data Source	Collectable Attributes	Structure Type
FDBs (Threads & Posts)	Thread ID Thread URL Thread Subject Post ID Post URL Post Subject Post Author Post Text	Unstructured
Social Media	Content ID Content Author Content Text Content Upvotes (including likes, favourites, upvotes) Content Shares	Unstructured
Broker Agencies (Ratings)	Broker Name Company TIDM Broker Rating	Semi-Structured
News Corporations (News Articles)	Article URL Article Title Article Author Article Text	Unstructured
Stock Market Data	Open/Close Price Low/High Price	Structured

to the date and time the source was created and has been omitted from the table for clarity. The time of each of these data sources is one of the two dimensions in which these sources will later be fused together, the other being the company name.

4 Big Data Challenges in Relation to Financial Market Data

The 7 Vs of big data are abundant in the financial market domain; this section will now go into detail as to the prevalence of each of these Vs, which are summarised in Table 3.

Table 3. Prevalence of the 7 Big Data Vs within financial data sources

	Volume	Variety	Velocity	Variability	Veracity	Value	Visualisation
Social Media	*H*	*H*	*H*	*H*	*M*	*M*	*N/A*
FDBs	*M*	*H*	*M*	*H*	*H*	*M*	*N/A*
Broker Agencies	*M*	*H*	*M*	*H*	*M*	*H*	*N/A*
News Corporations	*H*	*H*	*M*	*H*	*H*	*M*	*N/A*
Stock Market Data	*L*	*L*	*H*	*L*	*L*	*H*	*N/A*

4.1 Volume

The amount of data pertaining to financial stocks is vast in nature. Discussions relating to stocks are not just confined to financial discussion boards, but flows into other environments such as Twitter, Reddit, and mainstream media, making the volume of data to analyse a gargantuan task. The popularity of Twitter alone for discussing stocks can result in thousands of tweets relating to certain stocks being generated every day. Events such as dividend announcements [53] can exacerbate this further, causing a surge of activity in the social media domain [54].

4.2 Variety

The variety of data sources intensifies the big data problem present in the financial world. Social media platforms, FDBs, broker agencies, news websites – all of these communication channels have a dramatically different structure which fall into one of the three recognised categories; structured, semi-structured and unstructured [55, 56]. This is one the biggest challenges of the data fusion process – how can such differently structured forms of data be fused together without sacrificing the quality of said data sources?

4.3 Velocity

The speed in which financial data is transmitted is extraordinary in itself, minutely stock price data for multiple exchanges is available for free from sources such as AlphaVantage [52, 57]. Real-time analysis of such high velocity data present within sources such as Twitter and live intraday stock data is not a trivial task [4].

Further exacerbating the velocity of financial data, emerging technologies such as High-Frequency Trading (HFT) involves the use of sophisticated computing algorithms which submit and cancel orders rapidly, giving the illusion of liquidity [58]. This can further intensify the velocity aspect of big data in financial markets.

4.4 Variability

The combination of unstructured, semi-structured and structured data within the financial market community is rife. Real-time data feeds of stock prices, articles published by the Regulatory News Service (RNS), social media, corporate news websites and mainstream media provide just a taste of the huge variety of data sources which are readily available for investors to digest [59].

4.5 Veracity

Missing data, noise, abnormalities – all the characteristics of veracious data can easily be found within financial data sources. News articles published by news corporations are a prime example of this, different corporations structure their articles in varying layouts which make use of various metadata, with some news websites including tags to associate the article with a specific company or industry. The non-uniform nature of articles and their associated structure leads to data which cannot be easily compared.

4.6 Value

The most sought-after V in big data is its value [60]. This V is the main objective when collecting such vast amounts of data, finding relationships, whether they be explicit or hidden in order to unveil the true value of such data [61].

4.7 Visualisation

Visualisation of disparate data is incredibly difficult to accomplish due the large number of features present in big data sets [62]. It is often regarded as the end goal of big data, after the challenges such as veracity have been tackled.

5 Proposed Data Fusion Model

Although many of the financial data sources do not possess a high amount of value for analysis value within isolation, when combined with other financial data sources they can provide valuable new insights into the behaviour and intent of investors.

Our proposed data fusion model (Fig. 2) draws upon the underlying principles of the JDL model, defining key levels which deal with specific tasks within the data fusion process. The proposed model will fuse together different financial data sources, which are collected using the techniques, summarised in Table 4.

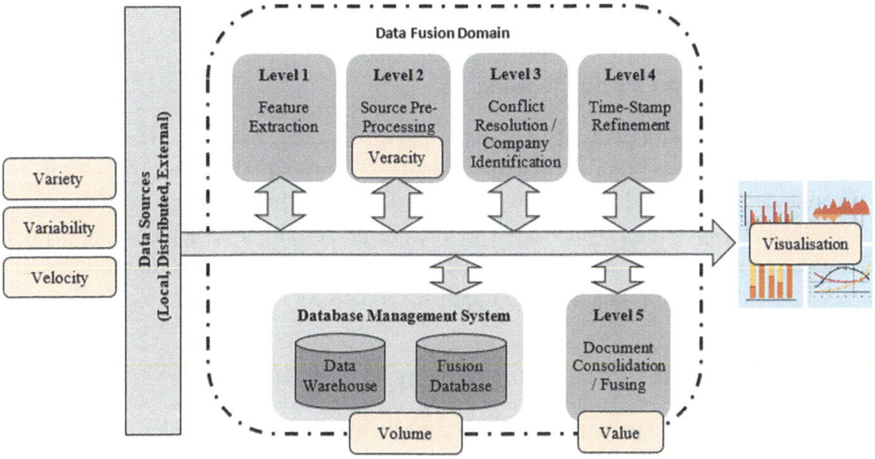

Fig. 2. Proposed financial data fusion (FinDF) model.

Table 4. Collection techniques for financial data sources

Financial Data Source	Collection Technique	Libraries/APIs
FDBs (Threads & Posts)	Web Scraping	BeautifulSoup[a], Scrapy[b], Selenium[c]
Social Media	APIs	Twitter – Tweepy[d], Reddit – PRAW[e]
Broker Agencies (Ratings)	Web Scraping	BeautifulSoup, Scrapy, Selenium
News Corporations (News Articles)	Web Scraping	BeautifulSoup, Scrapy, Selenium
Stock Market Data	APIs	AlphaVantage[f]

[a] https://www.crummy.com/software/BeautifulSoup/
[b] https://scrapy.org/
[c] https://www.seleniumhq.org/
[d] http://www.tweepy.org/
[e] https://praw.readthedocs.io/
[f] https://www.alphavantage.co/

5.1 Data Warehouse

The data warehouse houses the raw data, which has yet to be processed by the different layers of the fusion model. Our proposed fusion model uses a conventional RDBMS for data warehousing purposes, PostgreSQL [65].

5.2 Level 1 – Feature Extraction

Not all of the data available from each of the financial data sources will have value as a result of being fused. The first level will therefore select the most appropriate features from the data sources.

5.3 Level 2 – Source Pre-processing

Many revised JDL models will list *source pre-processing* but not attribute a level to such a crucial process; other data fusion models will simply label it as a pre-requisite – where the data is cleaned before it is even considered for fusion. The

model we propose clearly defines a source pre-processing level which deals with the common pre-processing tasks; data cleaning, normalisation, transformation, missing values imputation, outliers and noise identifications [63].

5.4 Level 3 – Conflict Resolution/Company Identification

As a result of all stock exchanges around the world referring to companies using different ticker/TIDM symbols, such collisions which occur will attempt to be addressed before the fusion process can continue.

A large part of this task involves identifying the company which is being referred to within the data source, this will be a common occurrence when analysing global tweets from Twitter, and analysing news articles which refer to companies by their name as opposed to their TIDM.

5.5 Level 4 – Time-Stamp Refinement

Timestamps are the determinant feature in which disparate data can be associated. Data which does not have a timestamp associated with it cannot easily be fused with other data sources [64]. This level will address inconsistent time-stamps across the different data sources, attempting to unify the data based on pre-existing time-stamps. Nuances such as daylight-saving time and time-zone differences across the different sources will also be conducted at this level.

5.6 Level 5 – Document Consolidation/Fusing

After the data has gone through a vigorous cleaning process and the timestamps have been aligned across the data sources, the fusion process can then continue with storing the fused data within the document-oriented fusion database. The fusing of this data is performed in accordance with pre-determined time-slice windows (for example, 15-minute intervals), and the company TIDM (ticker symbol).

5.7 Fusion Database

After the final fusion level has been undertaken, fused data is stored in a document-oriented fashion, allowing the fused data to be stored in a document-oriented NoSQL structure such as that supported by MongoDB [66].

6 Conclusions and Future Work

This paper has proposed a novel data fusion model for fusing together heterogeneous data from different financial data sources. The proposed model adapted the heavily-employed JDL data fusion model for the purposes of financial data fusion.

The proposed FinDF model attempts to address the challenges of working with big data within the confines of financial markets. Associating different data sources by time and company will be a challenging process when taking into consideration each of the 7 Vs of big data.

In terms of the original 3 Vs (volume, variety and velocity), the fusion model will associate voluminous amounts of disparate data which is being generated at a rapid rate. Taking into consideration 2 of the other Vs (variability and veracity), these are present in the data sources in varying levels, web scraping techniques will allow us to collect data from a variety of websites, which will often be veracious in nature due to the different structure of discussion boards and other communicative websites. The last 2 Vs (value and visualisation) come after the fusion process has occurred. Although it can be argued that every data source has some inherent value in isolation, the outcome of the fusion process will allow the value to be truly apparent through the use of identifying hidden relationships between the different data sources.

Identifying the name of a company within the different data sources is also a substantial challenge which can be addressed through Natural Language Processing (NLP) techniques. The problems of cashtag collisions on Twitter could also mean that previous work undertaken could have been susceptible to incorrect analysis. HFT is also an area which requires special attention when it comes to the analysis of stock movements, such high velocity activity can make the analysis of stock market movements challenging to undertake.

The data fusion model presented in this paper will be used in the future as part of a larger multi-layered ecosystem for the monitoring of potentially irregular comments pertaining to financial stocks. This ecosystem will monitor a variety of discussion channels used by investors, in addition to news sources and utilise the data fusion model in order to amalgamate the different sources of stock information and stock prices.

To our knowledge, this is the first conceptualised model for the fusing of heterogeneous financial data sources.

References

1. Flood, M.D., Jagadish, H.V., Raschid, L.: Big data challenges and opportunities in financial stability monitoring. Financ. Stab. Rev. **20**, 129–142 (2016)
2. Ngai, E.W.T., Gunasekaran, A., Wamba, S.F., Akter, S., Dubey, R.: Big data analytics in electronic markets. Electron. Mark. **27**(3), 243–245 (2017)
3. Alexander, L., Das, S.R., Ives, Z., Jagadish, H.V., Monteleoni, C.: Research challenges in financial data modeling and analysis. Big data **5**(3), 177–188 (2017)
4. Seddon, J.J.J.M., Currie, W.L.: A model for unpacking big data analytics in high-frequency trading. J. Bus. Res. **70**, 300–307 (2017)
5. Laney, D.: Application Delivery Strategies (2001)
6. Chen, C.L.P., Zhang, C.-Y.: Data-intensive applications, challenges, techniques and technologies: a survey on Big Data. Inf. Sci. (NY) **275**, 314–347 (2014)
7. Bello-Orgaz, G., Jung, J.J., Camacho, D.: Social big data: recent achievements and new challenges. Inf. Fusion **28**, 45–59 (2016)
8. Emmanuel, I., Stanier, C.: Defining Big Data. In: Proceedings of the International Conference on Big Data and Advanced Wireless Technologies - BDAW 2016, pp. 1–6 (2016)
9. Shafer, T.: The 42 V's of Big Data and Data Science (2017). https://www.kdnuggets.com/2017/04/42-vs-big-data-data-science.html. Accessed 25 Nov 2017
10. Bleiholder, J., Naumann, F.: Data fusion. ACM Comput. Surv. **41**(1), 1–41 (2008)
11. Alyannezhadi, M.M., Pouyan, A.A., Abolghasemi, V.: An efficient algorithm for multisensory data fusion under uncertainty condition. J. Electr. Syst. Inf. Technol. **4**(1), 269–278 (2017)
12. Välja, M., Korman, M., Lagerström, R., Franke, U., Ekstedt, M.: Automated architecture modeling for enterprise technology management using principles from data fusion: a security analysis case. In: Proceedings of PICMET 2016: Technology Management for Social Innovation, pp. 14–22 (2016)
13. Borges, V.: Survey of context information fusion for ubiquitous Internet-of-Things (IoT) systems. Open Comput. Sci. **6**(1), 64–78 (2016)
14. Blasch, E., et al.: Revisiting the JDL model for information exploitation. In: Proceedings of 16th International Conference Information Fusion, FUSION 2013, pp. 129–136 (2013)
15. Burdick, D., et al.: Extracting, Linking and Integrating Data from Public Sources: A Financial Case Study. IEEE Data Eng. 1–8 (2015)
16. Hall, D.L., Llinas, J.: An introduction to multisensor data fusion. Proc. IEEE **85**(1), 6–23 (1997)
17. Castanedo, F.: A review of data fusion techniques. Sci. World J. **2013**, 19 (2013)
18. Acar, E., Rasmussen, M.A., Savorani, F., Næs, T., Bro, R.: Understanding data fusion within the framework of coupled matrix and tensor factorizations. Chemom. Intell. Lab. Syst. **129**, 53–63 (2013)
19. Golmohammadi, K., Zaiane, O.R., Golmohammadi, S., Golmohammadi, K., Zaiane, O.R.: Time series contextual anomaly detection for detecting market manipulation in stock market (2015)
20. Khaleghi, B., Khamis, A., Karray, F.O., Razavi, S.N.: Multisensor data fusion: a review of the state-of-the-art. Inf. Fusion **14**(1), 28–44 (2013)
21. Andersen, L.C.: Data-driven Approach to Information Sharing using Data Fusion and Machine Learning. Norwegian University of Science and Technology (2016)
22. Geva, T., Zahavi, J.: Predicting intraday stock returns by integrating market data and financial news reports predicting intraday stock returns by integrating market data and financial news reports. In: Mediterranean Conference on Information Systems (2010)

23. Geva, T., Zahavi, J.: Empirical evaluation of an automated intraday stock recommendation system incorporating both market data and textual news. Decis. Support Syst. **57**, 212–223 (2014)
24. Mora, D., Falcão, A.J., Miranda, L., Ribeiro, R.A., Fonseca, J.M.: Multisensor Data Fusion (2016)
25. Bevilacqua, M., Tsourdos, A., Starr, A., Durazo-Cardenas, I.: Data fusion strategy for precise vehicle location for intelligent self-aware maintenance systems. In: 2015 6th International Conference on Intelligent Systems, Modelling and Simulation (2015)
26. Mcdaniel, D.: An information fusion framework for data integration. In: Information Fusion Application to Data Integration, p. 858 (2001)
27. Abdelgawad, A., Bayoumi, M.: Data fusion in WSN. In: Resource-Aware Data Fusion Algorithms for Wireless Sensor Networks, vol. 118. Springer, London (2012)
28. Snidaro, L., García, J., Llinas, J.: Context-based information fusion: a survey and discussion. Inf. Fusion **25**, 16–31 (2015)
29. Chandrasekaran, B., Gangadhar, S., Conrad, J.M.: A survey of multisensor fusion techniques, architectures and methodologies. In: Conference Proceedings - IEEE SOUTHEASTCON (2017)
30. Raol, J.R.: Multi-sensor Data Fusion with MATLAB. CRC Press, New York (2010)
31. Elmenreich, W.: An Introduction to Sensor Fusion. Vienna University of Technology, Austria (2002)
32. Solano, M.A., Jernigan, G.: Enterprise data architecture principles for high-level multi-int fusion: a pragmatic guide for implementing a heterogeneous data exploitation framework. In: Information Fusion (FUSION), pp. 867–874 (2012)
33. Steinberg, N., Bowman, C.L., White, F.E.: Revisions to the JDL data fusion model. Proc. SPIE **3719**(1), 430–441 (1999)
34. Wald, L.: Data fusion: a conceptual approach for an efficient exploitation of remote sensing images. In: Fusion Earth Data, International Conference, 17–23 January 1998
35. Llinas, J., Bowman, C., Rogova, G., Steinberg, A.: Revisiting the JDL data fusion model II. Sp. Nav. Warf. Syst. Command **1**(7), 1–14 (2004)
36. Antweiler, W., Frank, M.Z.: Is all that talk just noise? The information content of internet stock message boards. J. Finan. **59**(3), 1259–1294 (2004)
37. Chen, H.M.: Group polarization in virtual communities: the case of stock message boards. Sch. Libr. Inf. Sci. **1994**, 185–195 (2013)
38. Sun, F., Belatreche, A., Coleman, S., Mcginnity, T.M., Li, Y.: Pre-processing online financial text for sentiment classification: a natural language processing approach. In: Computational Intelligence for Financial Engineering and Economics (CIFEr) (2014)
39. Owda, M., Crockett, K., Lee, P.: Financial discussion boards irregularities detection system (FDBs-IDS) using information extraction. In: Intelligent Systems, 8–12 September 2017
40. Mirbabaie, M., Stieglitz, S., Ruiz Eiro, M.: #IronyOff – understanding the usage of irony on twitter during a corporate crisis. In: Proceedings of Pacific Asia Conference on Information Systems, July 2017
41. Zappavigna, M.: The Discourse of Twitter and Social Media. Continuum International Publishing Group, London (2012)
42. Cazzoli, L., Sharma, R., Treccani, M., Lillo, F.: A large scale study to understand the relation between Twitter and financial market. In 2016 Third European Network Intelligence Conference (ENIC), pp. 98–105 (2016)
43. Tafti, A., Zotti, R., Jank, W.: Real-time diffusion of information on Twitter and the financial markets. PLoS One **11**(8), 1–16 (2016)

44. Vilas, F., Evans, L., Owda, M., Redondo, R.P.D., Crockett, K.: Experiment for analysing the impact of financial events on Twitter. In: Algorithms and Architectures for Parallel Processing (2017)
45. Kwuan, H.: Twitter cashtags and sentiment analysis in predicting stock price movements (2017)
46. Rosen: Tweeting Made Easier (2017). https://blog.twitter.com/official/en_us/topics/product/2017/tweetingmadeeasier.html. Accessed 08 Nov 2017
47. Li, Q., Shah, S., Nourbakhsh, A., Fang, R., Liu, X.: funSentiment at SemEval-2017 Task 5: Fine-grained sentiment analysis on financial microblogs using word vectors built from StockTwits and Twitter, pp. 852–856 (2017)
48. Harris, L.: Trading and Exchanges: Market Microstructure for Practitioners, vol. 60, no. 4. Oxford University Press, Oxford (2002)
49. London South East, Broker Ratings (2017). http://www.lse.co.uk/broker-tips.asp. Accessed 28 Oct 2017
50. Manning, P.: Financial journalism, news sources and the banking crisis. Journalism **14**(2), 173–189 (2013)
51. Avalon, G., Becich, M., Cao, V., Jeon, I., Misra, S., Puzon, L.: Multi-factor Statistical Arbitrage Model (2017)
52. Elliot, A., Hsu, C.H., Slodoba, J.: Time Series Prediction: Predicting Stock Price, no. 2 (2017)
53. Boylan, H.: The innovative use of Twitter technology by bank leadership to enhance shareholder value. Purdue University (2016)
54. Wei, W., Mao, Y., Wang, B.: Twitter volume spikes and stock options pricing. Comput. Commun. **73**, 271–281 (2016)
55. Golmohammadi, K., Zaiane, O.R. Data mining applications for fraud detection in securities market. In: European Intelligence and Security Informatics Conference, pp. 107–114 (2012)
56. Sagiroglu, S., Sinanc, D.: Big data: a review. In: 2013 International Conference Collaboration Technologies System, pp. 42–47 (2013)
57. Vantage: Alpha Vantage API Documentation (2017). https://www.alphavantage.co/documentation/. Accessed 25 Oct 2017
58. Goldstein, M.A., Kumar, P., Graves, F.C.: Computerized and high-frequency trading. Financ. Rev. **49**(2), 177–202 (2014)
59. Shenoy, S.S., Hebbar, C.K.: Stock market reforms – a comparative study between Indian stock exchanges & select exchanges abroad. Int. J. Sci. Res. Technol. **1**(1), 38–45 (2015)
60. Duong, T.H., Nguyen, H.Q., Jo, G.S.: Smart data: where the big data meets the semantics. Comput. Intell. Neurosci. **2** (2017)
61. Fouad, M.M., Oweis, N.E., Gaber, T., Ahmed, M., Snasel, V.: Data mining and fusion techniques for WSNs as a source of the Big Data. Procedia Comput. Sci. **65**(Iccmit), 778–786 (2015)
62. Grolinger, K., et al.: Challenges for Map Reduce in Big Data. In: Electrical and Computer Engineering Publications (2014)
63. García, S., Ramírez-Gallego, S., Luengo, J., Benítez, J.M., Herrera, F.: Big Data analytics big data preprocessing: methods and prospects. Big Data Anal. **1**(9) (2016)
64. Traub-Ens, J., Bordoy, J., Wendeberg, J., Schindelhauer, C.: Data fusion of time stamps and transmitted data for unsynchronized beacons. IEEE Sens. J. **15**(10), 5946–5953 (2015)
65. Chen, S.: Cheetah: a high performance, custom data warehouse on top of MapReduce. Proc. VLDB Endow. **3**(1–2) 1459–1468 (2010)
66. Boicea, A., Radulescu, F., Agapin, L.I.: MongoDB vs Oracle - database comparison. In: Proceedings of 3rd International Conference Emerging Intelligent Data and Web Technologies, EIDWT 2012, pp. 330–335, September 2012

67. Hallstensen, V.: Multisensor Fusion for Intrusion Detection and Situational Awareness. Norwegian University of Science and Technology (2017)
68. Bouvry, P., et al.: Using heterogeneous multilevel swarms of UAVs and high-level data fusion to support situation management in surveillance scenarios. In: International Conference on Multisensor Fusion and Integration for Intelligent Systems, pp. 424–429 (2016)
69. Lederman, G., Chen, S., Garrett, J.H., Kovačević, J., Noh, H.Y., Bielak, J.: A data fusion approach for track monitoring from multiple in-service trains. Mech. Syst. Signal Process. **95**, 363–379 (2017)
70. Zhang, C.: Applying data fusion techniques for benthic habitat mapping and monitoring in a coral reef ecosystem. ISPRS J. Photogramm. Remote Sens. **104**, 213–223 (2015)

A Comparative Study of HMMs and LSTMs on Action Classification with Limited Training Data

Elit Cenk Alp(✉) and Hacer Yalim Keles

Department of Computer Engineering, Ankara University, Ankara, Turkey
calp@netas.com.tr, hkeles@ankara.edu.tr

Abstract. Action classification from video streams is a challenging problem, especially when there is a limited number of training data for different actions. Recent developments in deep learning based methods enabled high classification accuracies for many problems in different domains, yet they still perform poorly when the dataset is small. In this work, we examined the performances of Hidden Markov Models (HMM) and long short-term memory (LSTM) based recurrent neural network models using the same sequence classification framework with the well known KTH action dataset. KTH contains limited examples for training, hence challenges the deep learning based techniques even when transfer learning is applied in feature extraction. Our experiments depict that using a pre-trained convolutional network, i.e. SqueezeNet, and fine-tuning for feature extraction; HMM performs better in sequence modeling than an LSTM based model. Using the same feature extraction approach, i.e. fine-tuned SqueezeNet, we obtained 99.30% accuracy with an HMM, which is the best classification accuracy that is reported so far with this dataset; yet 81.92% accuracy with the best performing LSTM configuration.

Keywords: Action classification · Hidden Markov Models · LSTM SqueezeNet · Deep learning

1 Introduction

Recognition of human activities from video streams is an active research area for many computer vision applications, such as automatic video surveillance, content-based video retrieval and modeling human-computer interaction. The purpose of an action recognition system is to determine the class label of the activities in a video stream by analyzing the motions between the successive frames; hence, a robust classification system requires sequence modeling to discriminate different complex actions.

Classical computer vision based approaches represent each frame using some local and global handcrafted features, such as histogram of orient gradients (HOG) [1], motion history images (MHI) [2], motion energy

K. Arai et al. (Eds.): IntelliSys 2018, AISC 868, pp. 1102–1115, 2019.
https://doi.org/10.1007/978-3-030-01054-6_76

images (MEI) [2], spatio-temporal volume [3]. There are no globally best-in-class feature descriptors for all data sets, so learning the representation directly from raw data is more advantageous in many ways [4]. In particular, deep neural networks such as Convolution Neural Networks (CNN) have become the preferred method of learning the representation from the image content [5–8]. In general, convolutional networks learn a mapping from raw image pixels to a feature space that represents the content of the image efficiently. In deep architectures this mapping is learned via multi-layered nonlinear operations.

The features that are learned from direct raw video data using deep learning methods have more representational power; they work more stable in classification problems. However, representation learning requires large amounts of data. In the case of small training data, transfer learning provides better representation learning. In this work, since we are aiming to work with a small dataset with only a limited number of training samples, we utilize a pre-trained deep CNN, namely the SqueezeNet [9], in feature extraction after transfer learning. In classification, the features from the fine-tuned SqueezeNet are fed into two temporal classifiers: (1) a HMM based classifier, (2) an LSTM based Recurrent Neural Network (RNN) classifier. Hence, the developed framework learns the spatio-temporal relations of successive video frames for each action. Our research show that with a limited amount of data in each class, as is the case on the KTH dataset [10], HMMs generalizes better in sequence modeling than LSTMs, using the same feature representations.

The remainder of this paper is organized as follows: In Sect. 2, we review the related works, in Sect. 3 we describe the details of our feature extraction and classification approaches. Experimental results are provided in Sect. 4. The Conclusion is provided in Sect. 5.

2 Related Works

There is a wide range of different computer vision based approaches that is used extensively for action classification in the literature[1]. These approaches aggregate motion information and local appearance using hand-crafted features such as Histogram of Optical Flows (HOF), Histogram of Oriented Gradients (HOG), Motion History Image (MHI), Motion Energy Image (MEI). Among them, one of the most successful approach is proposed by Bobick and Davis [2] which is a spatio-temporal template representation of video sequences, i.e. MEI and MHI. MHI representation summarizes the motion of a moving object in a sequence by its shape and movement. The still background is implicitly removed from the image and the movement is represented as an attenuating silhouette in a single image. In our preliminary work [11], we modified MHI representation and utilized it with HMMs and obtained 99.35% on Weizmann dataset. Vezzani et al. [12] used HMMs to classify actions. They use the vertical and horizontal

[1] A comprehensive review of different action representation approaches are provided in [4].

projections of the foreground mask as features and a Mixture of Gaussian to model the emission probabilities. They report 96% accuracy on the UT-Tower Dataset [13].

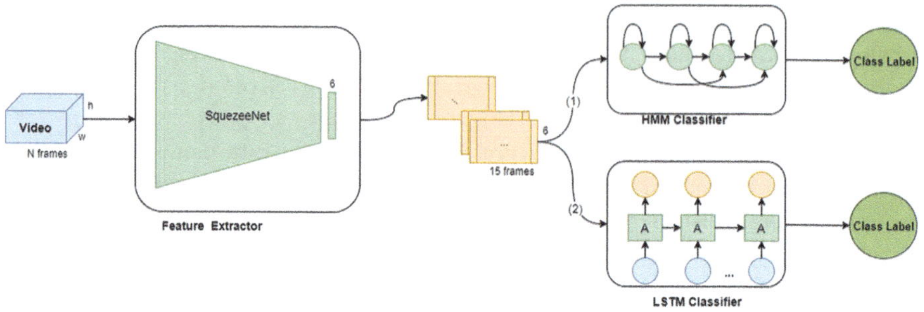

Fig. 1. The proposed action recognition framework.

As for feature learning, Le et al. [14] used an unsupervised learning technique by extending the Independent Subspace Analysis algorithm. Combining this technique with the deep learning techniques, i.e. stacking and convolutional networks, they obtained state-of-the-art results for many datasets; on the KTH dataset, they report 93.9% accuracy. Ji et al. [15] proposed a 3D Convolutional Neural Network to model the spatio-temporal features of a volume of video frames to classify human actions. They report 90.2% on the KTH dataset. Simonyan and Zisserman [16] proposed a two-stream Convolutional network to classify human actions. The first stream models the spatial features, while the second stream models the temporal features using the multiple-frame optical flow data. They classified actions by combining the scores of each stream using an SVM or averaging. They report 88.0% accuracy on UCF-101 dataset [17] and 59.4% on HMDB-51 dataset [18], using SVM. Karpathy et al. [19] trained a system using two separate CNN models for action classification. The Context stream is trained with low-resolution images, and the fovea stream is taken from the central part of the high-resolution representation of the image. They experimented with three different fusion models: early fusion, late fusion and slow fusion. They report that the best result is obtained with slow fusion model, yet the performances among these different fusion techniques are not significantly different. Their work reveals that when it comes to time sequenced data, CNNs do not improve much using different fusion techniques.

Donahue et al. [20] utilized Long Short-Term Memory (LSTM) for video classification. After the CNN layer, the embedded features from each frame are fed to an LSTM network. Similarly, Ng et al. [21] used CNN-LSTMs for action classification. These works show that the learned features using a CNN and LSTM based sequence modeling, when trained end to end, show significant performances on large action datasets, such as UCF-101. Although we use a similar approach, our purpose is different in this work; we want to evaluate the performance of LSTM based sequence modeling when there is a limited number of training data and compare it with HMMs. We do not train the CNN and LSTM end-to-end, in order to provide same feature sets to both HMMs and LSTMs. To the best of our knowledge, with a limited number of training data, there is a lack of comparative study between HMMs and LSTMs.

Recently, Lei et al. [22] proposed a solution to action classification by combining an HMM with a CNN on Weizmann and KTH datasets. In their work, the Gaussian Mixture Model was replaced by the CNN for modeling the emission probabilities. They report 93.97% accuracy on KTH dataset. Our CNN-HMM design is different from theirs in that we use the CNN for embedding the raw RGB frames to a lower dimensional space, hence generating the sequence of features for a video stream. We use Gaussian Models for modeling emission probabilities for the HMMs. In our experiments we obtain 99.3% accuracy on the KTH dataset.

3 The Method

In our preliminary work [11], we proposed a modified MHI for representing the motion in a video frame. Computing 7 Hu Moments from this representation that is fed to HMMs provided promising results on the Weizmann dataset. However, using such handcrafted features in human action classification does not always generalize well for different datasets, with different action categories in different environments. Therefore, to better generalize to different datasets, in this work we extend our research to utilize a CNN to learn the features from raw data. The features that we extract using our CNN model are used to classify actions with an HMM. In this approach, we preserve the same architecture for HMM models with our previous work; yet, instead of Hu Moments, we use learned features from the CNN model. In addition to this, we performed extensive experiments by utilizing the same features with the LSTM based RNN as well. Our CNN based action recognition framework is depicted in Fig. 1.

In the following subsections, we describe the details of our method including the feature extraction and classification.

3.1 Feature Extraction

In this work, we explain the representation that we implemented using both the hand-crafted features and the CNN based features.

3.1.1 Hand-Crafted Feature Representation Using Modified MHI and Hu Moments

Bobick and Davis [2] introduced the Motion History Image (MHI), which encodes the temporal differences between consecutive frames in a video stream in an image. The moving parts in the frames is represented with high intensity values, while the other parts, i.e. non-moving parts, fade away as time passes. Hence, the moving pixels in a recent frame is always brighter than the ones in the previous frames, which generates a kind of activity map for the whole video sequence in a single image. This is formulated by Bobick and Davis using (1).

$$H_\tau(x,y,t) = \begin{cases} \tau, & \text{if } \Psi(x,y,t) = 1. \\ max\left(0, H_\tau(x,y,t-1) - \delta\right), & \text{otherwise.} \end{cases} \quad (1)$$

Instead of using (1) for the MHI calculation, in our previous work, we generated and used a slightly modified version of as in (2); we refer to this version as Modified MHI (MMHI) from now on.

$$H_\tau(x,y,t) = \begin{cases} \tau, & \text{if } \Psi(x,y,t) = 1. \\ H_\tau(x,y,t-1) * \beta, & \text{otherwise.} \end{cases} \quad (2)$$

In (2), instead of decreasing the intensities using a constant value from the previous MHI, we scale it using a multiplier with a value between 0 and 1. In our experiments, we set β to 0.9, which results in 10% decrease in pixel values in the following frame. This form of intensity reduction helps us representing the whole sequence of motion in the image, even for the longer sequences; hence gives better classification accuracies than the classical MHI formalization.

The moments that are extracted from images are invariant to changes in position, rotation and scaling differences; hence they are widely used features in image representation in many pattern recognition applications [22]. After we calculated the MMHI, we extract 7 Hu moments to represent the MHIs in a reduced feature space. These features are used for classification after normalization.

3.1.2 CNN Based Feature Representation

CNNs are one of the widely used deep learning models that have recently shown excellent performance in tasks such as handwritten digit classification, pattern recognition, human action recognition, and image classification. It is a hierarchical model that contains multiple convolutional layers that transform input volumes to some feature representations. The lower layers of the architecture learn some low level feature representations, i.e. directed edges, corners, etc.; while following layers, in a cascaded manner, learn higher level features, i.e. faces, parts of objects, etc. [23].

Since deep learning-based models require large amounts of video data for training, we utilize a pre-trained CNN network to extract features from video frames. In this work, we used the pre-trained SqueezeNet [9] that is trained using the ImageNet dataset [24] as our basis model and fine-tuned this model using our training data. This network has many practical advantages; it is a small network, containing 50x fewer parameters than AlexNet [6], and training is faster than other similarly performing networks, such as AlexNet and VGGNets [25].

In the Squeeze Network, so-called fire modules are used. In the fire modules, a squeeze layer, i.e. using 1×1 convolutional kernels, is followed by an expansion layer, i.e. using 1×1 and 3×3 convolutional kernels. The network architecture starts with a convolution layer (Conv1) and is followed by 3 fire modules; after a pooling layer a set of three fire modules are applied. Following these, there is another pooling layer and a fire module. After these fire modules, i.e. FMs:1–9, there is a convolution layer (Conv10). The number of filters in this layer is configured considering the number of classes in the problem domain. There is no fully-connected layers in this architecture; instead, after Conv10, a global average pooling is applied, which takes the spatial averages of the filters. This approach helps reducing the overfitting if the number of training data is limited, which is the case in our domain. Finally a softmax layer is applied to generate the likelihoods of each classes. A broad view of the architecture is depicted in Fig. 2.

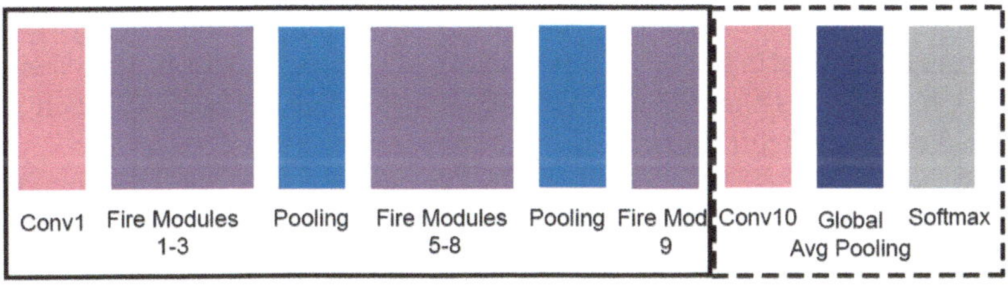

Fig. 2. The layers in the SqueezeNet [9] architecture.

In order to fine-tune this network using our datasets, we modified only the last three layers. These layers are depicted with a dashed rectangle in Fig. 2. The filter size in the last convolutional layer is changed to the number of action categories in the dataset, i.e. on the KTH dataset the category count is 6. Utilizing only the training samples in the dataset, we use stochastic gradient descent optimization to adapt the weights in the last convolutional layer. In this setting, we set the learning rate to 0.0001 and the momentum parameter to 0.9.

After transfer learning is completed, we extract the likelihoods of each action from each video frame and utilize this vector as our feature representation for action classification.

3.2 Classification

This section describes the HMM and LSTM based models that are used to classify the feature sequences that we extract using the methods that we explain in the previous section.

3.2.1 Using Hidden Markov Models

HMMs are very suitable for modeling the temporal characteristics of a sequence of data. One of the advantages of HMMs is that since the number of learned parameters is far less than deep sequence based models, they can be trained with less data. We trained the HMMs using the Baum-Welch algorithm [26].

In training, we used the feature vectors with real values as the set of observations. Each action category is modeled by a different model.

The sequence data for each category are formed as follows: the video is split into multiple video segments that we call *clips*, with a sliding window; each window is composed of 15 or 30 frames, with a stride of 1 frame. For example, for generating 15-frame clips, a video with 17 frames is split into three video clips: from 1'st to 15'th, from 2'nd to 16'th, and from 3'rd to 17'th frames. The video clips are generated from all the data, including the augmented data, in the training dataset. We generated the test clips in a similar fashion for performance evaluations.

Using the training video clip features for each action category, we trained an ergodic HMM model. In the ergodic model, the nodes are fully-connected; hence the transitions are more flexible. To evaluate the category of a video clip, its representative feature sequence is evaluated with each action model and the model that generates the highest log-likelihood for that sample is chosen as the true action category during testing.

3.2.2 Using the Long Short-Term Memory Model

LSTMs are a type of recurrent neural networks that contain a set of memory cells which store and modify the state of a sequence and allow for modeling temporal characteristic of data [27]. They are advantageous in that using a fixed number of parameters, they can successfully model long-term temporal relationships in a varying length sequence data.

We experimented with several LSTM architectures with different stacked configurations and different hyperparameters (please refer to Sect. 4 for the details). In all experiments in different configurations, we use the same feature representations of the video clips as we did with HMMs. In this setting, the features that are extracted using our CNN or MMHI-Hu Moments for each frame in a video clip are fed as the inputs to the LSTM model. The result is obtained from the last layer of the model, i.e. softmax layer, which provides the probability distribution of each action category considering all the frames in a clip. For example, there are 6 classes in the KTH dataset, hence a 6-dimensional output vector is produced, the values of which sum to 1. The index of the highest value in this vector is selected as the class index that is chosen by the LSTM model. Each clip is evaluated separately in this manner.

4 Results and Discussions

In this work, we want to evaluate the performances of HMMs and LSTMs using the same set of features when the training data is limited; hence, we selected the KTH action dataset. We resampled a set of 15-frame and 30-frame clips from each test stream with a 1 frame shift, as we explain in the Sect. 3.2.1. This approach helps us to evaluate the model performance thoroughly with more test samples and with different alignment options that are extracted from the test dataset. We define the recognition accuracy as the ratio between the number of correctly classified video clips over the total number of video clips. This evaluation metric also depicts the performance of our models for action classification in a continuous video stream that contains multiple different actions.

Table 1. Action recognition accuracy on KTH dataset using modified MHI - Hu moments and CNN features using HMMs

State number	MMHI - Hu moments accuracy (%)		CNN accuracy (%)	
	15 Frame Window	30 Frame Window	15 Frame Window	30 Frame Window
3 State	70.14	-	87.62	-
5 State	74.94	-	92.45	-
7 State	-	-	95.78	98.25
8 State	86.42	91.29	97.41	98.78
11 State	89.15	-	97.38	-
15 State	91.65	94.58	97.58	99.22
17 State	90.97	94.38	97.69	**99.30**
20 State	92.77	95.45	95.54	97.71

4.1 Training Parameters

We used the pre-trained SqueezeNet CNN model that is trained with the ImageNet dataset and fine-tuned the model using the KTH dataset. We trained the CNN model with Stochastic Gradient Descent (SGD) with a learning rate 0.0001 and momentum 0.9. We used the categorical cross-entropy function as the lost function.

As for HMM training, we use the Baum-Welch algorithm for 10 iterations. Diagonal covariance matrix is used for HMM states. Two types of HMMs are used in the experiments; Ergodic HMM and left-to-right HMM. In all the experiments, we observed that the left-to-right model performs approximately 20% lower than the ergodic model, hence, we reported the test results for the ergodic model in this section. In the HMM test phase, the probability is calculated by the forward algorithm for each trained model and the class with the highest likelihood is selected. The features of each video frame were extracted using (1) CNN model, (2) MMHI - Hu Moments.

We train the LSTM network with the Adam optimizer [28]. The learning rate is set to 0.001. We select the beta1 parameter of the Adam optimizer as 0.9, beta2 as 0.999 and the epsilon parameter as $1e-8$. Categorical cross entropy loss is used for error back-propagation. As for the LSTM architectures, we made various experiments by changing the number of memory units and stacked layers. These are: many one layer LSTMs with 128, 64, 32, 16, 8, 4 units; stacks of three LSTMs with 128-64-32 units, 64-64-32 units; stacks of four LSTMs with 64-64-32-32 units. We train each LSTM model for 500 iterations.

4.2 KTH Dataset

KTH dataset contains six types of human actions, i.e. jogging, hand clapping, boxing, walking, running, hand waving, performed by 25 persons in four different scenarios: outdoors with scale variation, outdoors, indoors and outdoors with different clothes. Figure 3 shows some sample frames from KTH dataset. In most of the videos, the background is static and homogenous. This database contains 2391 sequences. For training and testing, we follow Schuldt et al. [10]'s original experiment settings, i.e. the videos that belong to 16 persons are used for training, the videos for the remaining 9 persons for testing. Test and train datasets are non-overlapping with regard to the persons in the video.

Fig. 3. Sample frames from KTH dataset.

4.3 Experiments

We performed KTH experiments in four different settings: we extracted features using (1) MMHI - Hu Moments and (2) CNN; we utilized (1) HMM based and (2) LSTM based sequence models for classification. The results of our experiments using both features with the HMM classifier are depicted in Table 1. We obtained 99.30% accuracy when we extract the features using our CNN model with 30-frame clips and the state number of the HMM is set to 17. As can be seen from this table, when the number of states is less than 15, the classification performance drops considerably; hence, we performed only a few experiments in this setting and included only partial results. Figure 4 depicts the confusion matrix of the best result. When we use the same experiment setting for the KTH dataset with the best results that we obtained with the Weizmann dataset in our preliminary work [11], i.e. MMHU-Hu Moments are used as features and Ergodic HMM type is selected, number of HMM states is 7; the accuracy with KTH

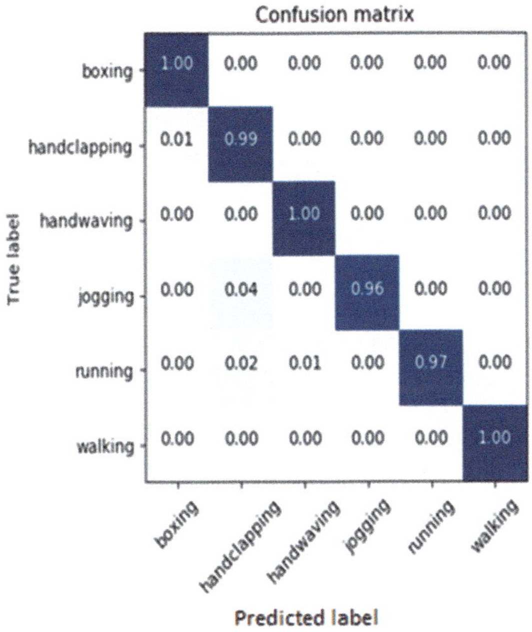

Fig. 4. Confusion matrix of the experiment that has 17 state Ergodic HMM using 30 frame window with features extracted from the CNN on KTH dataset.

dataset is around 95.78%. In this setting, the accuracy with the KTH dataset is not as good as the Weizmann dataset. We believe that, the representative capability of the Hu Moments is not sufficient for more challenging background settings and the variety of the people taking actions, as is the case in the KTH dataset.

In order to evaluate the performance of LSTM models, we trained various models with the KTH train dataset i.e. 16 persons. We generated the validation set by randomly selecting 15% of the trainin data. We use 15-frame and 30-frame video clips for LSTMS as well. In the training and validation phase, we shuffle the data.

We tested various architectures, i.e. 4–32, 8, 8–64, 16, 32, 64, 128. 128-64, 128-64-32 LSTM units. The training and validation accuracies for these experiments are depicted in Table 2. As can be seen from the training and validation accuracies, we stopped training before overfitting in each experiment. The best result is obtained using two stacks of LSTMs with 128-64 units. After determining the best performing architecture, we re-train an LSTM model using all the training data, i.e. including the validation set. After training, the model is tested using all the test data (9 person); the test accuracy is 76.92%. Although the validation accuracy in the training is 99.16% (Table 2), the generalization capability of the model is lower in the test data. The primary reason is that

Table 2. Different LSTM architecture validation score on KTH dataset

LSTM Architecture	Train Accuracy %	Validation Accuracy %
4–32 Units	91.53	91.32
8 Units	89.85	89.34
8–64 Units	95.35	94.81
16 Units	89.59	88.84
32 Units	92.70	92.45
64 Units	95.48	95.09
128 Units	98.82	97.80
128-64 Units	**99.41**	**99.16**
128-64-32 Units	98.83	98.26

the actions of 9 people that is used in testing is never seen by the model during training. In validation, since the frames are selected randomly from the training frames, the person that acts in different clips have probably been seen by the model. Using the same experimental setting of the best result, we increased the training data by data augmentation and re-trained the CNN and LSTM models, separately. In this case, the performance is improved considerably to 81.92%. LSTM training, compared with the HMM training, requires larger amount of data. We believe that the content in the KTH training set is not enough for training a robust LSTM model.

Table 3. Action recognition comparison result on KTH dataset

Methods	Average recognition accuracy %
Our CNN - HMM	**99.30**
Our CNN - LSTM	81.92
3D CNN [15]	90.20
Lei CNN - HMM [22]	93.97
ICA [14]	93.90
Mid - Level Motion Feature [29]	90.50
DBNs (Deep Belief Networks) [30]	96.60

Comparison of the performances of the state-of-the-art approaches on KTH dataset is depicted in Table 3. With the proposed HMM model, our model achieve 99.30% accuracy on this dataset, which is the highest accuracy reported so far.

5 Conclusion

In this research, we want to inquiry the performances of two sequence modeling techniques, i.e. HMM and LSTM, using the same set of sequence features.

For this purpose we conducted experiments both using hand-crafted feature representation and learning based feature representation. As for the hand-crafted feature representation, we utilized a feature that works successfully for the Weizmann dataset in our previous work [11], namely the MMHI. We extracted 7 Hu Moments using MMHI from each video frame. On the other hand, as for the learned feature representation, we fine-tuned a CNN model, i.e. the SqueezeNet, using the KTH training data and use it for feature embedding. When we trained an HMM model using these features, we observed that CNN based model is performing better than the MMHI-Hu Moments. Then, we conducted many experiments with different LSTM architectures, using the CNN based features as the sequence data. We observed that there is a high margin with our best performing LSTM model and the HMM model.

Although we applied transfer learning to deal with the shortage of data problem, LSTM model still demands variety of data. Using the same dataset, HMM could perform 99.30% accuracy with the same train-test split. Our study show that although deep learning based models are very robust for modeling the underlying patterns in data, they require a lot of training data for modeling them. In such cases, HMMs are still strong candidates for modeling sequence data.

References

1. Thurau, C., Hlavac, V.: Pose primitive based human action recognition in videos or still images. In: 2008 IEEE Conference on Computer Vision and Pattern Recognition, pp. 1–8, June 2008
2. Bobick, A.F., Davis, J.W.: The recognition of human movement using temporal templates. IEEE Trans. Pattern Anal. Mach. Intell. **23**(3), 257–267 (2001)
3. Gorelick, L., Blank, M., Shechtman, E., Irani, M., Basri, R.: Actions as space-time shapes. IEEE Trans. Pattern Anal. Mach. Intell. **29**(12), 2247–2253 (2007)
4. Sargano, A., Angelov, P., Habib, Z.: A comprehensive review on handcrafted and learning-based action representation approaches for human activity recognition. Appl. Sci. **7**(1), 110 (2017)
5. Szegedy, C., Liu, W., Jia, Y., Sermanet, P., Reed, S., Anguelov, D., Erhan, D., Vanhoucke, V., Rabinovich, A.: Going deeper with convolutions. In: 2015 IEEE Conference on Computer Vision and Pattern Recognition (CVPR), pp. 1–9, June 2015. doi.ieeecomputersociety.org/10.1109/CVPR.2015.7298594
6. Krizhevsky, A., Sutskever, I., Hinton, G.E.: Imagenet classification with deep convolutional neural networks. In: Advances in Neural Information Processing Systems 25, pp. 1097–1105. Curran Associates, Inc. (2012). http://papers.nips.cc/paper/4824-imagenet-classification-with-deep-convolutional-neural-networks.pdf
7. Chatfield, K., Simonyan, K., Vedaldi, A., Zisserman, A.: Return of the devil in the details: delving deep into convolutional nets. In: British Machine Vision Conference (2014)
8. Sutskever, I., Vinyals, O., Le, Q.V.: Sequence to sequence learning with neural networks. In: Proceedings of the 27th International Conference on Neural Information Processing Systems - Volume 2, ser. NIPS 2014, pp. 3104–3112. MIT Press, Cambridge (2014). http://dl.acm.org/citation.cfm?id=2969033.2969173

9. Iandola, F.N., Han, S., Moskewicz, M.W., Ashraf, K., Dally, W.J., Keutzer, K.: Squeezenet: alexnet-level accuracy with 50x fewer parameters and <0.5 mb model size, February 2016. arXiv:1602.07360 [cs], http://arxiv.org/abs/1602.07360
10. Schuldt, C., Laptev, I., Caputo, B.: Recognizing human actions: a local SVM approach. In: Proceedings of the 17th International Conference on Pattern Recognition, ICPR 2004, vol. 3, pp. 32–36, August 2004
11. Alp, E., Keles, H.: Action recognition using MHI based HU moments with HMMs. In: IEEE EUROCON 2017, 17th International Conference on Smart Technologies (2017)
12. Vezzani, R., Baltieri, D., Cucchiara, R.: HMM based action recognition with projection histogram features. In: Proceedings of the 20th International Conference on Recognizing Patterns in Signals, Speech, Images, and Videos, ser. ICPR 2010, pp. 286–293. Springer, Heidelberg (2010)
13. Chen, C.C., Ryoo, M., Aggarwal, J.: UT-tower dataset: aerial view activity classification challenge (2010). http://cvrc.ece.utexas.edu/SDHA2010
14. Le, Q.V., Zou, W.Y., Yeung, S.Y., Ng, A.Y.: Learning hierarchical invariant spatio-temporal features for action recognition with independent subspace analysis. In: CVPR 2011, pp. 3361–3368, June 2011
15. Ji, S., Xu, W., Yang, M., Yu, K.: 3D convolutional neural networks for human action recognition. IEEE Trans. Pattern Anal. Mach. Intell. **35**(1), 221–231 (2013)
16. Simonyan, K., Zisserman, A.: Two-stream convolutional networks for action recognition in videos. In: Proceedings of the 27th International Conference on Neural Information Processing Systems - Volume 1, ser. NIPS 2014, pp. 568–576. MIT Press, Cambridge (2014)
17. Soomro, K., Zamir, A.R., Shah, M.: UCF101: a dataset of 101 human actions classes from videos in the wild. CoRR, vol. abs/1212.0402 (2012)
18. Kuhne, H., Jhuang, H., Garrote, E., Poggio, T., Serre, T.: HMDB: a large video database for human motion recognition. In: IEEE International Conference on Computer Vision (ICCV) (2011)
19. Karpathy, A., Toderici, G., Shetty, S., Leung, T., Sukthankar, R., Fei-Fei, L.: Large-scale video classification with convolutional neural networks. In: 2014 IEEE Conference on Computer Vision and Pattern Recognition, pp. 1725–1732, June 2014
20. Donahue, J., Hendricks, L.A., Rohrbach, M., Venugopalan, S., Guadarrama, S., Saenko, K., Darrell, T.: Long-term recurrent convolutional networks for visual recognition and description. IEEE Trans. Pattern Anal. Mach. Intell. **39**(4), 677–691 (2017)
21. Ng, J., Hausknecht, M.J., Vijayanarasimhan, S., Vinyals, O., Monga, R., Toderici, G.: Beyond short snippets: deep networks for video classification. In: 2015 IEEE Conference on Computer Vision and Pattern Recognition (CVPR), pp. 4694–4702 (2015)
22. Lei, J., Li, G., Zhang, J., Guo, Q., Tu, D.: Continuous action segmentation and recognition using hybrid convolutional neural network-hidden markov model model. IET Comput. Vis. **10**(6), 537–544 (2016)
23. Zeiler, M.D., Fergus, R.: Visualizing and understanding convolutional networks. In: European Conference on Computer Vision, pp. 818–833. Springer (2014)
24. Deng, J., Dong, W., Socher, R., Li, L.-J., Li, K., Fei-Fei, L.: ImageNet: a large-scale hierarchical image database. In: CVPR09 (2009)
25. Simonyan, K., Zisserman, A.: Very deep convolutional networks for large-scale image recognition. CoRR, vol. abs/1409.1556 (2014)
26. Baum, L.: An inequality and associated maximization technique in statistical estimation of probabilistic functions of a Markov process. Inequalities **3**, 1–8 (1972)

27. Hochreiter, S., Schmidhuber, J.: Long short-term memory. Neural Comput. **9**(8), 1735–1780 (1997). https://doi.org/10.1162/neco.1997.9.8.1735
28. Kingma, D., Ba, J.: Adam: a method for stochastic optimization. In: Proceedings of the 3rd International Conference on Learning Representation (2015)
29. Fathi, A., Mori, G.: Action recognition by learning mid-level motion features. In: 2008 IEEE Conference on Computer Vision and Pattern Recognition, pp. 1–8, June 2008
30. Hasan, M., Roy-Chowdhury, A.K.: Continuous learning of human activity models using deep nets. In: Fleet, D., Pajdla, T., Schiele, B., Tuytelaars, T. (eds.) Computer Vision - ECCV 2014, pp. 705–720. Springer, Cham (2014)

Tag Genome Aware Collaborative Filtering Based on Item Clustering for Recommendations

Zhipeng Gao[1], Bo Li[1(✉)], Kun Niu[2], and Yang Yang[1]

[1] State Key Laboratory of Networking and Switching Technology, Beijing University of Posts and Telecommunications, Beijing 100876, China
{gaozhipeng,hopelee,yyang}@bupt.edu.cn

[2] School of Software Engineering, Beijing University of Posts and Telecommunications, Beijing 100876, China
niukun@bupt.edu.cn

Abstract. Item-based collaborative filtering has become the most popular algorithm in recommender systems which exploits relationships among items to predict users' preferences for items. However, this algorithm suffers from data sparsity and poor varieties of information sources. It only contains information from rating data, lacking the items' meta information such as classifications and characteristics, which make similarities measurement among items imprecise, consequently reduce accuracy and interpretability of results. In this paper, we incorporate tag genome information into item-based collaborative filtering using item clustering techniques, which provides a more objective and comprehensive description of items. In particular, we utilize relevance between items and tags to calculate similarities in a more precise way and integrate the results with item-based collaborative filtering through an optimization process. An innovative improved stochastic gradient descent algorithm is adopted to accelerate the optimization process, which lowers down the time complexity without sacrificing the accuracy of the model. Our experimental results demonstrate the effectiveness of our proposed algorithm by gaining lower rating prediction error than currently popular item-based collaborative filtering and SVD latent factor model, and also outperforms them in a more practical top K recommender system.

Keywords: Item clustering · Collaborative filtering · Tag genome
Recommender system

1 Introduction

Recommender systems are developed to build relationships between a large number of users and items automatically, which can help users find what they like quickly. As the gap between increasing speed of quantity of facts provided and human ability to process all the information available is getting bigger and bigger, recommender systems are facing greater challenges in both effectiveness

K. Arai et al. (Eds.): IntelliSys 2018, AISC 868, pp. 1116–1128, 2019.
https://doi.org/10.1007/978-3-030-01054-6_77

and interpretability. The fundamental goal of recommender systems is to predict users' preferences for items. Specific applications of recommender systems can be divided into two types: one is rating prediction, the other is top K recommender. Nowadays they have been applied in many internet leaders like Amazon, Google, Netflix and Yahoo.

Recommender systems are often based on Collaborative Filtering (CF), which relies only on past user behavior, e.g. their previous transactions or item ratings and does not need any domain knowledge which can be profiled. For instance, a user's preference for an item like movie can be represented by scores ranging from 1 to 5 he has rated on a movie website. Collaborative filtering includes user-based CF and item-based CF. User-based CF recommends an item to a user according to the most similar users to him. The basic idea is a user will like something which users similar to him all like. Item-based CF tries to dig out relationships among items and recommends items he hasn't consumed which are similar to items he has shown an interest on. For user-based CF often suffers from insufficient computing ability and the results of it lack interpretability, item-based CF is getting applied in a wider range of fields. That is why in this paper, we mainly focus on item-based CF.

As is mentioned above, item-based CF relies mostly on past user behaviors which are generated between a large quantity of users and items. However, consuming ability of users is very limited, which leads to a limited number of consumption records, like sparse ratings on a movie website. Besides, making ratings on items is subjective conducts by users, which can be influenced by complex environmental factors. For example, when a user watches a movie, a good or bad mood he is in, in morning or evening the behavior takes place and whether he will be interrupted by other things all make a difference in the score he will rate for the movie. What's more, representing a user's preference for an item by several levels of scores is somehow vague. For 3 points or 4 points may mean similar for a user judging an item but play a significant different role in algorithm because the total range of score is 1 to 5 points. All previous examples can tell us only measuring similarities among items by ratings from users lacks objectivity and accuracy.

Tag genome information is items' meta information on various of characteristics. It's generated from all kinds of dimensions which uses scores to represent relevance of an item and its characteristics. For example, Titanic gains 0.95 on tag "romantic" but only gains 0.19 on tag "thriller". And Forrest Gump gains 0.93 on tag "inspiring" with 0.008 on tag "waste of time". All the scores value from 0 to 1 when higher scores represent higher relevance. As we can see, tag genome information is direct, subjective, abundant description of items, which is collected artificially or learnt from users' remarks. Making use of tag genome information can help us make more precise and comprehensive depictions of items and calculate similarities among items more precisely.

In this paper we propose a multifaceted model integrating clustering techniques into item-based CF. It makes full use of tag genome information in order to solve problem of scarce information sources and imprecise similarity measurement, making the recommendation results not only rely on rating data but also related to items' meta information. Our method, which we call ICCF (Item Clustering based Collaborative Filtering), combines item clustering model and collaborative filtering in a personalized way. In this way, the prediction results can be divided into two parts. Item clustering techniques capture relationships among items in the same cluster, while item-based CF captures implicit latent relationships among items in different clusters. This is done by solving a joint optimization problem that estimates personalized parameters of above two models for each user.

The rest of the paper is organized as follows. Section 2 introduces the notation. Section 3 presents the related work. Section 4 presents the proposed model. Section 5 presents the evaluation methodology and experimental results. Finally, Sect. 6 provides some concluding remarks.

2 Notation

All vectors are represented by bold lowercase letters and are column vectors (e.g., $\mathbf{g}$). All matrices are represented by uppercase letters (e.g., R, G). For a given matrix A, its i-th row is represented by ${\mathbf{a}_i}^T$ and its j-th column by $\mathbf{a}_j$. All scalar values are represented by non-bold lowercase letters (e.g., μ, α).

The number of users is denoted by n and the number of items is denoted by m. Matrix R is used to represent the user-item ratings. A rating r_{ui} indicates the preference of user u for item i, while the higher value means stronger preference. Matrix G is used to represent tag genome information of items. $\mathbf{g}_i$ is tag genome vector of item i. We distinguish predicted ratings from known ones, by using the notation $\hat{r}_{ui}$ for the predicted value of r_{ui}. The (u, i) pairs for which r_{ui} is known are stored in the set $\kappa = \{(u, i) \mid r_{ui} \text{ is known}\}$. In order to combat overfitting the sparse rating data, our model is regularized so estimates are shrunk towards baseline defaults. Regularization is controlled by constants which are denoted by λ. Exact values of these constants are determined by cross validation.

3 Related Work

So far there have been extensive works in the area of recommender systems. One of mainstream researches is how to make improvements based on collaborative filtering, which contains user-based CF and item-based CF [1]. Here we present a few notable works in the area that give inspiration for our work.

3.1 Collaborative Filtering and Clustering Techniques

Deshpande et al. [2] developed a nearest neighbor item-based approach, which showed that item-based models led to better top K recommendations than user-based. Jon et. al. [3] firstly raised a method of dividing ratings into several parts by clustering items and doing CF departly, which gained mixed results on improving accuracy. Xue et al. [4] generated user clusters from training data and found neighbor users in the same cluster. Minneapolis et al. [5] thought different groups of people had different patterns of behavior. Taking all users into consideration in a model would lose local characteristics. The approach partitioned users into clusters and combined a global model and local models, which showed a better performance than single model.

3.2 Memory-Based Model and Optimization-Based Model

Essentially both user-based CF and item-based CF are memory-based methods using statistics techniques [6]. The basic idea is to compute the user's predicted vote of an item as a weighted average of votes by other similar users or items. In these methods, plenty of time is used on the computation of Pearson Correlation Coefficient to measure similarities among items or users [7] without achieving an optimization process. Cremonesi et al. [8] developed SVD (Singular Value Decomposition) method, which used a truncated SVD decomposition of matrix R to generate recommendations. Yehuda et al. [9] combined advantages of above methods and added parameters in original collaborative filtering. An optimization process is achieved by gradient descent. This multifacted collaborative filtering model turned out better results.

3.3 Context Aware Methods

Adomavicius et al. [10] declared the importance of context information for recommender systems and introduced all kinds of context information in this paper. Jongseol et al. [11] incorporated music tag information into personalized recommendations. Asmaa et al. [12] focused on the situation of course-choosing and made enough use of domain knowledge. The approach proposed in this paper divided students into different levels of parts according to students' majors and academic levels. After that, they built models in different groups. Their work demonstrated that incorporating context information into original models could bring a significant improvement.

Our proposed model ICCF differs from the earlier work in the following ways: (i) Most of previous work is focused on clustering users. Even if little work tried to cluster items, it didn't achieve good results because it only partitioned train set into several separate parts. That's the problem our model tries to solve. ICCF makes a deep combination of item clustering results and collaborative filtering. (ii) ICCF no longer relies only on past user behaviors like collaborative filtering. It incorporates tag genome information as a context aware model, which increases variety and effectiveness of data. (iii) ICCF is an optimization-based model, which introduces personal parameters for each user. By achieving an optimization process, it can model each user's characteristics.

4 Proposed Approach

4.1 Motivation

For diversity of users' interests and randomness of consumption behavior, the calculation of item-item similarities can be imprecise. It can be possible that two similar items get neglected when calculating similarities among a mass of items that a user rated. Besides, rating score is a vague notion influenced by so many environment factors, making measuring item-item similarities only by ratings from common users sometimes unreasonable. What's more, the number of two items' common users has inhomogeneity, which causes similarities calculated by ratings lose uniformity. And it exists that the number of common users of two items is very small, making similarity results filled with randomness. All in all, a single item-based collaborative filtering model may not be sufficient to reflect item-item similarities.

By incorporating items' tag genome information, which is their meta information, we can figure each item from various of dimensions. In this way, we can measure similarities among items in a more reliable and comprehensive way. These dimensions are consistent and scores for each item in all dimensions are measured in a uniform standard, making similarity measurement among items has uniformity.

From an intuitive perspective, we analyse that even if a user could have various of interests and consume all kinds of items, items he likes would present a distribution in clusters. For example, a user who wants to watch some good movies will search in a search engine like "100 movies with the highest scores in 2017". Or a girl who likes suspense films will query "irritative but not bloody classic suspense films". Based on the theory that people usually do things that are similar to other things they have shown an interest on, in other words, in a cluster way, we propose that people's evaluations on something mainly derive from their judgments on the similar things. So we integrate item clustering techniques into item-based CF. Item clustering model is used to capture users' preferences for things in the same cluster, while item-based CF is used to capture latent impalpable relationships among things in different clusters. We take advantage of both models to predict users' preferences for items they hasn't rated and propose our integrated model ICCF, which is expected to have better results.

4.2 Neighborhood Models

In this section, we mainly introduce item-based CF and k-means clustering technique we will use. Then we propose a neighborhood model based on statistics technique by simply combining the two models. The experimental results with different parameters have demonstrated our above guesses.

(1) Item-Based Collaborative Filtering: Item-based CF is aimed at matching similar items with those users have purchased or rated. Central to item-based CF approach is similarity measure among items. The most widely used similarity measure is Pearson Correlation Coefficient, ρ_{ij}. It's calculated according to

ratings of two items by common users, denoted by $U_{ij} = U_i \cap U_j$.

$$\rho_{ij} = \frac{\sum_{u \in U_{ij}} (r_{ui} - \overline{r}_i)(r_{uj} - \overline{r}_j)}{\sqrt{\sum_{u \in U_{ij}} (r_{ui} - \overline{r}_i)^2} \sqrt{\sum_{u \in U_{ij}} (r_{uj} - \overline{r}_j)^2}} \tag{1}$$

Since many ratings are unknown, it's expected that some items share only a handful of common users. Accordingly, similarities based on a greater user support are more reliable. An appropriate similarity measure, denoted by s_{ij}, would be a shrunk correlation coefficient [9]:

$$s_{ij} \stackrel{def}{=} \frac{n_{ij}}{n_{ij} + \lambda_1} \rho_{ij} \tag{2}$$

The variable n_{ij} denotes the number of users that rated both i and j.

Our target is to predict $\hat{r}_{ui}$, the unobserved rating by user u for item i. Using the similarity measure, we can take as a weighted average of the ratings of neighboring items.

$$\hat{r}_{ui} = \frac{\sum_{j \in K_u} s_{ij} r_{uj}}{\sum_{j \in K_u} s_{ij}} \tag{3}$$

(2) Clustering Algorithm: There are many algorithms that can be used to create clusters. In this paper, a k-means algorithm is selected as the basic clustering algorithm. The number k is an input to the algorithm that specifics the desired number of clusters. In the first pass, the algorithm takes the first random k items as the centroids of k unique clusters. Each of the remaining items is then compared to the closest centroid. In the following passes, the cluster centroids are re-computed based on cluster formed in the previous pass and the cluster membership is re-evaluated.

Algorithm 1. Clustering Algorithm

randomly initialize k cluster centroids $\boldsymbol{\mu}_1, \boldsymbol{\mu}_2, ..., \boldsymbol{\mu}_k \in R^n$
repeat
  for each $\mathbf{g}_i \in G$ **do**
    calculate the cluster $c^{(i)}$ it belongs to:
    $c^{(i)} := \underset{j}{argmin} ||\mathbf{g}_i - \boldsymbol{\mu}_j||^2$
  end for
  for each cluster j **do**
    re-compute the cluster centroid:
    $\boldsymbol{\mu}_j := \frac{\sum_{i=1}^{m} \{c^{(i)} == j\} \mathbf{g}_i}{\sum_{i=1}^{m} \{c^{(i)} == j\}}$
  end for
until convergence

Assuming that items could be clustered into k groups, items in the same cluster with item i is represented as C_i^k.

Using tag genome information, we can make cluster analysis on items to explore the distribution between users and items they have purchased or rated. Here we use k-means algorithm to divide items into 15 clusters. Computing ratio of users who have rated more than 70% items in limited number of clusters. As in Fig. 1, most of items they rated are distributed in several clusters they like. It shows that users rated items as a cluster way.

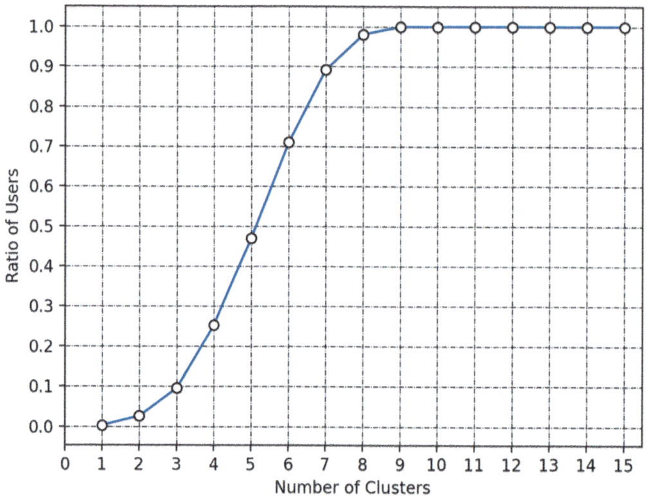

Fig. 1. Distribution of rated items by users.

(3) An Integrated Neighborhood Model: In this section, we propose a neighborhood model integrating above item-based CF with k-means clustering algorithm. We firstly partition items into k clusters using k-means algorithm. According to results of clustering, items rated by user u can be divided into two parts. One part consists of items which are in the same cluster with the target unrated item i, and the other consists of items which are in the different clusters with item i. Similarities between items in the same cluster and item i are calculated by tag genome matrix G, using Euclidean Distance denoted by ϵ_{ij}. Similarities between items in the different clusters and item i are calculated by item-based CF using s_{ij}. We make a weighted average, denoted by α, between the two parts as the prediction of unrated rating r_{ui}:

$$\hat{r}_{ui} = \alpha \frac{\sum_{j \in K_u} p_{ij} \epsilon_{ij} r_{uj}}{\sum_{j \in K_u} p_{ij} \epsilon_{ij}} + (1-\alpha) \frac{\sum_{j \in K_u} (1-p_{ij}) s_{ij} r_{uj}}{\sum_{j \in K_u} (1-p_{ij}) s_{ij}} \tag{4}$$

$$p_{ij} = \begin{cases} 1, if\ j \in C_i^k \\ 0, if\ j \notin C_i^k \end{cases} \tag{5}$$

We compare our integrated neighborhood model with item-based CF using root mean squared error (RMSE) to measure the accuracy of the results on test set (Fig. 2). The meta-parameter k is determined by cross-validation. We use $k = 20$ for k-means clustering algorithm. Another important parameter is α, which controls weight between two parts, as is illustrated above. Our experimental results show that RMSE on test set presents a regular change as α values from 0 to 1. When results of two parts are combined in a balance way ($\alpha = 0.6$), RMSE is smaller than item-based CF. But it will lose most of the accuracy when combined in an unbalanced way like α is too small (close to 0) or too big (close to 1), which leads to larger RMSE than item-based CF. Hence, the results of two parts are in a complementary relationship.

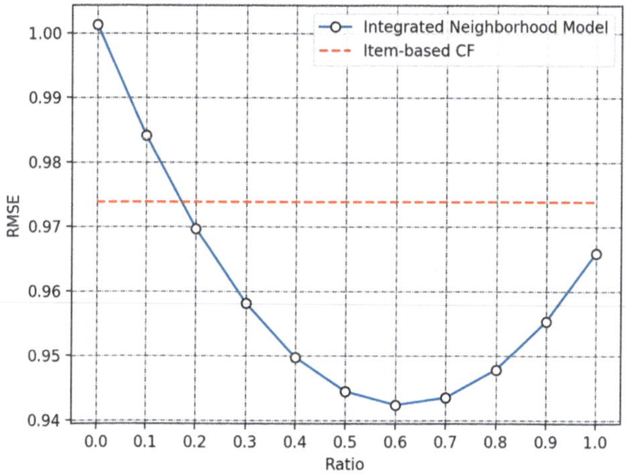

Fig. 2. Comparison of neighborhood models.

4.3 Our Proposed Model

Typical rating data contains large user tendencies and item tendencies. For example, a good movie is more likely to acquire better scores than an ordinary movie. A user who is easy to be satisfied usually gives most of the movies he has watched good evaluations. In this circumstance, the level of scores cannot represent users' preferences for movies objectively. Thus we need to eliminate these factors from original ratings by accounting them, which we encapsulate with in the baseline estimates. Denoted by μ the overall average rating. A baseline estimate for an unknown rating r_{ui} is denoted by b_{ui} and account for the user and item effects, as in [9]:

$$b_{ui} = \mu + b_u + b_i \tag{6}$$

The parameters b_u and b_i indicate personalized user and item tendency learnt from the observed deviations of user u and item i, respectively.

On the basis of baseline estimates, we introduce the previous integrated neighborhood model. To more precisely depict each user's characteristics, we use personalized parameter α_u for user u, representing degree of the user's ratings being influenced by his preferences for other movies in the same cluster. With baseline estimates, the preference of user u for item i can be replaced by a more objective way, denoted by $r_{ui} - b_{ui}$. We use the proposed model ICCF to predict r_{ui} as:

$$\begin{aligned} \hat{r}_{ui} = \mu + b_u + b_i + \alpha_u \frac{\sum_{j \in \kappa_u} p_{ij} \epsilon_{ij} (r_{uj} - b_{uj})}{\sum_{j \in \kappa_u} p_{ij} \epsilon_{ij}} \\ + (1 - \alpha_u) \frac{\sum_{j \in \kappa_u} (1 - p_{ij}) s_{ij} (r_{uj} - b_{uj})}{\sum_{j \in \kappa_u} (1 - p_{ij}) s_{ij}} \end{aligned} \tag{7}$$

This is our final prediction rule, which allows fast online prediction. More computational work is needed at a pre-processing stage where parameters are estimated. A major design goal of the new neighborhood model is facilitating an efficient global optimization. Thus, model parameters are learnt by solving the regularized least squares problem associated with (8):

$$\min_{b_u, b_i, \alpha_u} \sum_{(u,i) \in \kappa} (r_{ui} - \hat{r}_{ui})^2 + \lambda_2 ({b_u}^2 + {b_i}^2 + {\alpha_u}^2) \tag{8}$$

We use an innovative improved stochastic gradient descent to achieve the optimization process. To lower down time complexity while avoiding over-fitting, we ensure each parameter will get modified at least once when we loop through all known ratings in κ. And we will skip ratings in which parameters involved have been modified in current loop. Let us denote the prediction error, $r_{ui} - \hat{r}_{ui}$, by e_{ui}. For a given training case r_{ui}, we modify the parameters by moving in the opposite direction of the gradient. The optimization process is as follows:

We choose initial step size $\gamma = 0.04$, and shrink it with a ratio of 0.93 after each iteration. The regularization parameter $\lambda_2 = 0.15$ and the number of iterations throughout the training data is 20. We achieve other three popular methods for comparison. The first method is non-personalized prediction rule, which employs movie means to all unknown ratings, denoted by MovieAvg. The second method is item-based CF, which is the most popular approach in the CF literature. The third method is the SVD latent factor model. Finally, we consider our proposed model ICCF. Performances of all the four models on test set have been shown in Table 1. As we can see, ICCF has the smallest RMSE than other methods.

5 Evaluation Through a Top-K Recommender

5.1 Evaluation Methodology

Besides using RMSE measure to calculate prediction accuracy, we adopt top K recommender to evaluate the performance of our algorithm. Top K recommen-

Algorithm 2 Optimization Process

initialize $b_u = 0,\ b_i = 0,\ \alpha_u = random(0, 1)$ for every user and item
repeat
  intialize user set and item set as empty set denoted by $U,\ I$
  randomly arrange κ
  for each r_{ui} in κ **do**
    if u in U and i in I **then**
      continue
    else
      calculate current error e_{ui}
      update b_u, b_i, α_u as:
      • $b_u \leftarrow b_u + \gamma \cdot (e_{ui} - \lambda_2 \cdot b_u)$
      • $b_i \leftarrow b_i + \gamma \cdot (e_{ui} - \lambda_2 \cdot b_i)$
      • $\alpha_u \leftarrow \alpha_u + \gamma \cdot (e_{ui} \cdot (\frac{\sum_{j \in \kappa_u} (1-p_{ij}) s_{ij} (r_{uj} - b_{uj})}{\sum_{j \in \kappa_u} (1-p_{ij}) s_{ij}} - \frac{\sum_{j \in \kappa_u} p_{ij} e_{ij} (r_{uj} - b_{uj})}{\sum_{j \in \kappa_u} p_{ij} e_{ij}}) - \lambda_2 \cdot \alpha_u)$
      insert u to U, insert i to I
    end if
  end for
  shrink γ
until reach the number of iterations

Table 1. Comparison of different methods on RMSE

Method	RMSE
MovieAvg	1.1162
Item-based CF	0.9770
SVD	0.9256
ICCF	0.9113

dation is a practical strategy in pragmatic recommender systems. For each user, a top K recommender calculates his preferences for items he hasn't rated and sorts items in descending order according to prediction scores. Finally it suggests the top K items which are supposed to be most appealing to the user.

Quality of a top K recommender depends on how many items in the recommendation list are truly viewed or rated by users. Precision and recall rate are two measurable indicators used in top K recommender systems. Supposedly, for user u, set of items in the top K recommendation list is denoted by $S(u;K)$; truly rated or viewed items set in test data is $S(u;test)$. The intersection $S(u;K) \cap S(u;test)$ is the correctly recommended items set. Precision and recall rate is defined as below:

$$Precision = \frac{\sum_{u \in U} |S(u;K) \cap S(u;test)|}{\sum_{u \in U} |S(u;K)|} \tag{9}$$

$$Recall = \frac{\sum_{u \in U} |S(u;K) \cap S(u;test)|}{\sum_{u \in U} |S(u;test)|} \tag{10}$$

Depending on intuitive judgment, precision and recall rate are greatly subject to K. In general, precision rate is negatively related to K, and recall rate is positively related to K, which has been shown in experimental results. A new index indicator F can be introduced, which synthetically considers the above two indicators.

$$F = \frac{2 \cdot Precision \cdot Recall}{Precision + Recall} \tag{11}$$

5.2 Experimental Results

We choose ratings in ml-1m data for experiment, which contains 6040 users and 3487 movies. The ratings count is 1000209, thus the sparsity of dataset is 95.26%. The train set and test set is split with a proportion of 3:1 through random sampling method. These movies' tag genome information is also obtained from the MovieLens web site. A one-dimensional vector with a length of 1128 is corresponding to each movie. During experiment process, we choose movies which users rated 5 stars as they like, and mix them with other 500 random movies. The top K movies are recommended to each user. K values from 1 to 100 to evaluate our algorithm. SVD model is used for comparison, for it has close RMSE with our proposed model ICCF.

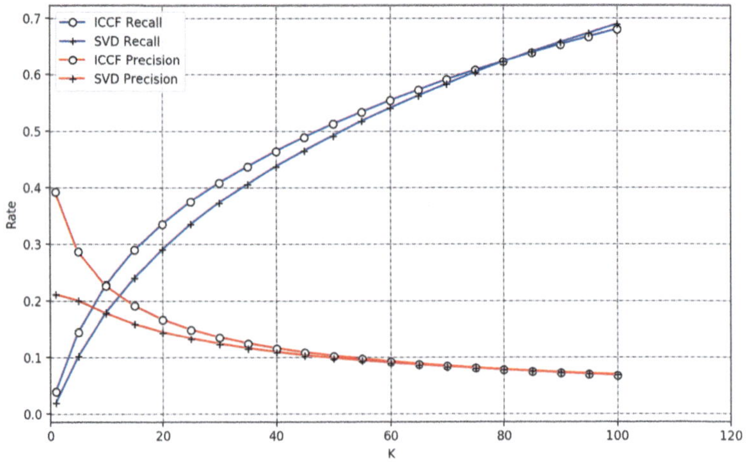

Fig. 3. Precision and recall.

Figure 3 plots the distribution of precision and recall rate of both models. Figure 4 plots F score as K changes. As we can see, precision rate decreases and recall rate increases as K increases. F score has a maximum value at K = 15. The figures show that there are significant differences between the two models. No matter considering precision, or recall, or F, our proposed model ICCF outperforms SVD model.

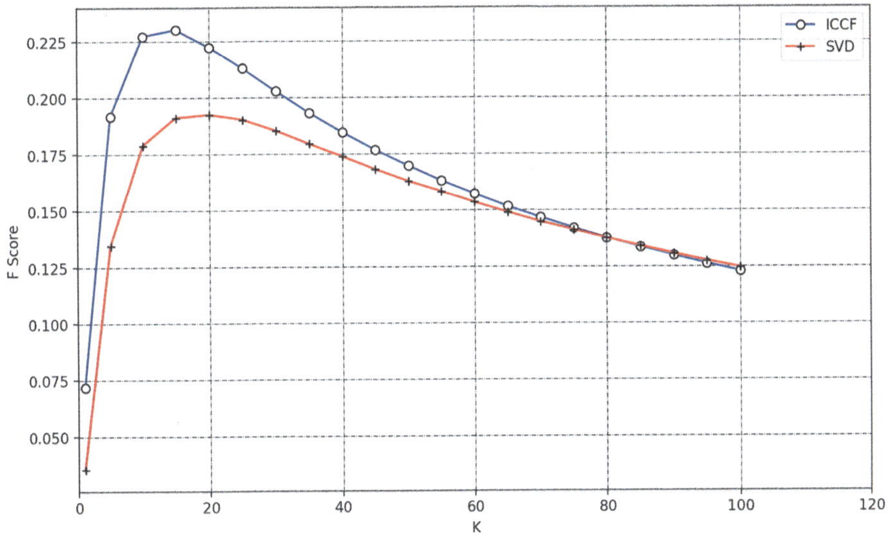

Fig. 4. F score.

6 Conclusion

In this paper, we firstly analyse users' behavior pattern. We are encouraged by finding that most users consume items in a cluster way. Thus a user's preference for a item depends on his ratings on other items in the same cluster. We introduce tag genome information on the basis of ratings to measure similarities among items in a more precise way, which solve the problem of data sparsity and single form of data.

We propose a integrated model ICCF which combines a clustering technique with item-based CF, which incorporates advantages of both models and predicts users' preferences for items in a personalized way. We compare performance of our proposed model with popular item-based CF and SVD latent factor model on RMSE. Experimental results show significant improvements on RMSE. Also, our model outperforms SVD model in a more practical top K recommender.

Acknowledgment. This study is supported by National Science & Technology Pillar Program (2015BAH03F02), and National Key Research and Development Program of China (2016YFE0204500).

References

1. Linden, G., Smith, B., York, J.: Amazon. com recommendations: Item-to-item collaborative filtering. IEEE Internet Comput. **7**(1), 76–80 (2003)
2. Deshpande, M., Karypis, G.: Item-based top-N recommendation algorithms. ACM Trans. Inf. Syst. (TOIS) **22**(1), 143–177 (2004)

3. OConnor, M., Herlocker, J.: Clustering items for collaborative filtering. In: Proceedings of the ACM SIGIR Workshop on Recommender Systems, vol. 128. UC Berkeley (1999)
4. Xue, G.-R., Lin, C., Yang, Q., Xi, W., Zeng, H.-J., Yu, Y., Chen, Z.: Scalable collaborative filtering using cluster-based smoothing. In: Proceedings of the 28th Annual International ACM SIGIR Conference on Research and Development in Information Retrieval, pp. 114–121. ACM (2005)
5. Christakopoulou, E., Karypis, G.: Local item-item models for top-n recommendation. In: Proceedings of the 10th ACM Conference on Recommender Systems, pp. 67–74. ACM (2016)
6. Breese, J.S., Heckerman, D., Kadie, C.: Empirical analysis of predictive algorithms for collaborative filtering. In: Proceedings of the Fourteenth Conference on Uncertainty in Artificial Intelligence, pp. 43–52. Morgan Kaufmann Publishers Inc. (1998)
7. Resnick, P., Iacovou, N., Suchak, M., Bergstrom, P., Riedl, J.: Grouplens: an open architecture for collaborative filtering of netnews. In: Proceedings of the 1994 ACM Conference on Computer Supported Cooperative Work, pp. 175–186. ACM (1994)
8. Cremonesi, P., Koren, Y., Turrin, R.: Performance of recommender algorithms on top-n recommendation tasks. In: Proceedings of the fourth ACM Conference on Recommender Systems, pp. 39–46. ACM (2010)
9. Koren, Y.: Factorization meets the neighborhood: a multifaceted collaborative filtering model. In: Proceedings of the 14th ACM SIGKDD International Conference on Knowledge Discovery and Data Mining, pp. 426–434. ACM (2008)
10. Adomavicius, G., Tuzhilin, A.: Context-aware recommender systems. In: Recommender Systems Handbook, pp. 217–253. Springer (2011)
11. Lin, K., Xu, Z., Liu, J., Wu, Q., Chen, Y.: Personalized music recommendation algorithm based on tag information. In: 2016 7th IEEE International Conference on Software Engineering and Service Science (ICSESS), pp. 229–232. IEEE (2016)
12. Elbadrawy, A., Karypis, G.: Domain-aware grade prediction and top-n course recommendation. In: RecSys, pp. 183–190 (2016)

First-Half Index Base for Querying Data Cube

Viet Phan-Luong(✉)

Aix-Marseille Univ, Université de Toulon, CNRS, LIS, Equipe BDA, Marseille, France
viet.phanluong@lis-lab.fr

Abstract. Given a relational fact table R, we call a base of data cubes on R a structure that allows to query the data cubes with any aggregate function. This work presents a compact base of data cubes, called the first-half index base, with its implementation, and the method for querying the data cubes using this base. Through experiments on real datasets, we show how the first-half index base resolves efficiently the main data cube issues, i.e., the storage space and the query response time.

Keywords: Data warehouse · Data cube · Data mining

1 Introduction

The concept of data cube offers important interests to business intelligence as it provides aggregate views of data over multiple combinations of dimensions. Those aggregate views can help managers to make appropriate decision in their business. In fact, a data cube built on a relational fact table with n dimensions and a measure M for an aggregate function g can be seen as the set of the Structured Query Language (SQL) group-by queries over the power set of the n dimensions, where g is applied to each group of measures M. The result of such a SQL group-by query is an aggregate view, called a cuboid.

Though the concept is simple, there are many important issues in computation time and storage space, because of the exponential number of the cuboids and because of the big size of large datasets. To make data cube query available in Online Analytical Processing (OLAP), most solutions to reduce the time computation are to precompute the data cube and store it on disk. However, the storage space can be tremendous.

To tackle these issues, there exist many different approaches. In [9], an I/O-efficient technique based upon a multiresolution wavelet decomposition is used to build an approximate and space-efficient representation of data cubes. Naturally, the response to an OLAP query is also approximate. The iceberg data cube approach [8,10,20,22] does not compute all aggregates, but only those above certain thresholds. This approach does not allow all data cube queries because data cubes are partially computed.

K. Arai et al. (Eds.): IntelliSys 2018, AISC 868, pp. 1129–1144, 2019.
https://doi.org/10.1007/978-3-030-01054-6_78

The other approaches search to represent the entire data cube with efficient methods for computation and storage [1,2,7,18] The computing time and storage space are optimized based on equivalence relations defined on aggregate functions [11,19] or on the concept of closed itemsets in frequent itemset mining [17] or by reducing redundancies between tuples in cuboids, using tuple references [2,11,15,16,19,21,23]. In these approaches, the computation is usually organized on the complete lattice of sub-schemes of the fact table dimension scheme. The computation can traverse the complete lattice in a top-down or bottom-up manner. To create cuboids, the sort operation is used to reorganize tuples: tuples are grouped and the aggregate functions are applied to the measures. To optimize the storage space, only aggregated tuples with aggregated measures are directly stored on disk. Non-aggregated tuples are not stored but represented by references to the stored tuples where the non aggregated tuples are originated or to tuples in the fact table. The work [23] implemented many of these approaches and reported the experimental results on real and synthetic datasets. It shown that The Totally-Redundant-Segment BottomUpCube approach (TRS-BUC) nearly dominated its competitors in all aspects of the data cube problem: fast computation of a fully materialized cube in compressed form, incrementally updateable, and quick query response time.

The work [12,13] presents a simple and reduced representation that allows to compute efficiently the entire data cubes for any aggregate functions. The main idea in this work is that among the cuboids of a data cube, there are ones that can be easily and rapidly get from the others, with no important computing time. These others are computed and stored on disk using an integrated binary search prefix tree structure for compact representation and efficient search. In contrast to the approaches that compute all tuples of the data cube with optimization in computing time and storage space, the work [13] computes and represents only the cuboids of a half of data cube, called the last-half data cube. It follows a special top-down approach that does not traverse the complete lattice of the dimension sub-schemes: from each cuboid in the last-half data cube, over a dimension scheme X, we can compute, with no important time cost, a non-stored cuboid over a dimension sub-scheme $Y \subset X$, using an operation called aggregate projection. The set of those non-stored cuboids is called the first-half data cube.

Moreover, the above proposed representation is not only for a specific aggregate function, nor for a specific measure, but it allows for computing all cuboids with any measure and any aggregate function. In fact, each cuboid in the representation is an index: a set of rowids that reference to tuples in the fact table.

In the present work, we extend [13] in a somehow contrast direction, by studying a more compact and efficient representation that allows to speed up the computation of data cube queries. The contribution consists of:

- A compact representation based on the bottom-up computation, and
- The methods for computing the data cube queries based on this compact representation.

The efficiency of the representation in run time and storage space is shown through experiments on four real datasets.

The paper is organized as follows. Section 2 recalls the main concepts in [13], in particular, the concept of the first-half and the last-half data cubes. Section 3 presents the concepts of the new representation for querying data cubes. Section 4 presents the methods for computing the group-by query with aggregate functions, based on this representation. Section 5 reports the experimental results and ends with discussions. Finally, conclusion and further work are in Sect. 6.

2 Preliminary

This section recalls the main concepts presented in [13]. A data cube over a dimension scheme R is the set of cuboids built over all subsets of R, that is the power set of R. As in most of existing work, dimensions (attributes) are encoded in integer, let us consider $R = \{1, 2, ..., n\}$, $n \geq 1$. The power set of R can be recursively defined as follows:

(1) The power set of $R_0 = \emptyset$ (the empty set) is $P_0 = \{\emptyset\}$.
(2) For $n \geq 1$, the power set of $R_n = \{1, 2, ..., n\}$ can be recursively defined as follows:

$$P_n = P_{n-1} \cup \{X \cup \{n\} \mid X \in P_{n-1}\} \tag{1}$$

P_{n-1} is called the *first-half power set* of R_n and the second operand of P_n, i.e., $\{X \cup \{n\} \mid X \in P_{n-1}\}$, the *last-half power set* of R_n.

Example 1: For $n = 3$, $R_3 = \{1, 2, 3\}$, we have:

$P_0 = \{\emptyset\}$, $P_1 = \{\emptyset, \{1\}\}$, $P_2 = \{\emptyset, \{1\}, \{2\}, \{1, 2\}\}$,
$P_3 = \{\emptyset, \{1\}, \{2\}, \{1, 2\}, \{3\}, \{1, 3\}, \{2, 3\}, \{1, 2, 3\}\}$.

The first-half power set of S_3 is $P_2 = \{\emptyset, \{1\}, \{2\}, \{1, 2\}\}$ and the last-half power set of S_3 is $\{\{3\}, \{1, 3\}, \{2, 3\}, \{1, 2, 3\}\}$, got by adding 3 to each element of P_2.

The set of all cuboids over the schemes in the first-half power set of R is called the *first-half data cube* and the set of all cuboids over the schemes in the last-half power set of R is called the *last-half data cube*.

In [12,13], the last-half data cube is precomputed and stored on disks and data cube queries are computed based on the last-half data cube. However, it proposed a more general framework: instead of computing for a data cube for a particular aggregate function, it computes and stores the tuples indexes over the schemes in the last-half power set of R, and data cube queries for any aggregate function are computed based on these indexes.

From now on we consider a relational fact table T with a dimension scheme $R = \{1, 2, ..., n\}$ and a set of measures $M = \{n + 1, n + 2, ..., n + k\}$.

3 First-Half Index Base for Data Cubes

In the present work, we follow the view of data cube as the composition of two parts: the last-half and the first-half. To improve the computing time and the storage space of the data cube representation, we shall follow an approach that is somehow inverse to the approach proposed in [12,13]. Indeed, in the new approach, the first-half data cube is computed and stored on disk. It forms a base to compute all data cube queries. For this, we define a data structure and some algorithms.

3.1 Data Indexes on an Attribute

Data on a dimension (an attribute) of the fact table T is indexed using the search binary tree structure. This structure has following fields:

- *data* : to contain an attributed value,
- *ltid* : to contain the list of rowids associated with the attributed value,
- *lsib* and *rsib*: the left and the right sub-trees.

The structure is organized for searching on the data field. We call a tree with this structure an *attribute index tree.*

To insert attributed values into an attribute index tree, we use the algorithm *InsData2AttIndex*.

Algorithm InsData2AttIndex:

Input: An attributed value *val*, the *rowid* of a tuple that contains *val*, and an attribute index tree *P*.

Output: The attribute index tree *P* updated.

Method:

```
if (P == NULL) {
   Create P with P.data = val;
     Create P.ltid with the 1st element rowid;
     P.lsib = NULL and P.rsib = NULL;
   }
else if P.data > val {
        insert val and rowid into P.lsib;
     }
else if P.data < val {
        insert val and rowid into P.rsib);
     }
else append rowid to P.ltid);
}
```

Such attribute index trees (and further index trees) are stored on disk. As each index is a partition of the set of all rowids of T, it is important to note that, to optimize the storage space of these indexes, we save only the partitions of rowids and omit the attributed values.

3.2 Tuples Indexes on a Dimension Scheme

Given a sub-scheme $\{A_1, ..., A_k\}$ (for $1 \leq i \leq n$, $1 \leq A_i \leq n$) of the dimension scheme R of the fact table T, we assume that the index over $\{A_1, ..., A_{k-1}\}$ is already created for all tuples of T. As the base, the indexes over the schemes $\{1\}, ..., \{n\}$ are created using the InsData2AttIndex algorithm. Let P be an element of the index over $\{A_1, ..., A_{k-1}\}$. That is, P is the list of all rowids of the tuples that have the same value on $\{A_1, ..., A_{k-1}\}$; these tuples may be different on A_k. To create the index of tuples on $\{A_1, ..., A_k\}$, we use the algorithm $TupleIndex$.

Algorithm TupleIndex:

Input: The fact table T and a dimension scheme $\{A_1, ..., A_k\}$.

Output: The index of tuples of T over $\{A_1, ..., A_k\}$.

Method:

```
1. Get the tuple index over {A1,...,Ak-1}
2. For each P in the index on {A1,...,Ak-1}
   do
2.1     Initialize a tree TrP to empty;
2.2     For each rowid in P do
2.2.1     Let val be the attributed value
          on Ak of the tuple at rowid in T;
2.2.2     Insert val and rowid to TrP using
2.2.2     the InsData2AttIndex algorithm;
2.2.3     done;
2.3     Write the tree TrP to disk;
3. done;
```

To access to the tuple at the row identified by *rowid* in the fact table T, we organize the data as follows. The fact table T is loaded into a list of blocks in the main memory. Each block is an array of fixed size k. To determine the number of the block that contains the tuple at *rowid* and the rang of the tuple in the block:

$$blocknumber = rowid \ / \ k$$
$$rang = rowid \ \% \ k,$$

where / and % denote respectively the quotient and the rest of the division on integers.

3.3 Creating the First-Half Index Base

The indexes of tuples over the schemes in the first-half power set of the dimension scheme of the fact table T are generated by the algorithm $GenIndexFH$.

Algorithm GenFHIndex:

Input: T and its dimension scheme $\{1, ..., n\}$.

Output: The tuple indexes for the first-half data cube over $\{1, ..., n\}$ and their dimension sub-schemes.

Method:

```
Let RS be a list of dimension sub-schemes,
initially empty;

1. Use InsData2AttIndex to generate n indexes
   over schemes {1},...,{n} and append
   successively these schemes to RS.

2. Set a pointer psch to the 1st scheme in
RS and let P be the pointed scheme
   (at this point, P is {1});

3. len = 1;

4. while len < n and psch ≠ NULL do

4.1. Let lastAtt be the last attribute of P

4.2. For each attribute i from 1 to n−1
     such that i > lastAtt do
4.2.2 Append i to P to create a new scheme
      nsc and append nsc to RS;

4.2.4 Use TupleIndex to generate the tuple
      index over nsc;

4.2.5 Save the tuple index to disk;

4.2.6 done;

4.3 Set psch to the next element in RS and
   len = length(nsc);

4.4 done;

5. Return RS.
```

Each scheme in RS is associated with the information that allows to identify the corresponding tuple index stored on disk.

3.4 First-Half Index Base Representation for Data Cube

Based on the tuple indexes over the schemes in the first-half power set of the fact table dimension scheme, we propose a representation, called the *first-half index base* for querying data cube on T. The first-half index base is the triple $(T, RS, FHIndex)$, where

- RS is the return of $GenFHIndex(T)$, and
- $FHIndex$ is the set of tuple indexes generated by $GenFHIndex(T)$.

The above elements are stored on disks. For efficient computing, the list of dimension schemes RS and the fact table T are retrieved in the main memory. The fact table T is stored in a list of blocks of fixed size as explained previously.

4 Data Cube Query Based on the First-Half Index Representation

This section explains how we can compute the queries with the aggregate functions MAX, COUNT, SUM, AVERAGE, and VARIANCE, based on the first-half index base.

4.1 Query on the First-Half Cube

To compute a cuboid on an aggregate function g, over a scheme F in the first-half data cube, we access to the tuple index over F, and for each partition P of rowids of this index:

- Let $rowid1$ be the first $rowid$ in P and let t be the tuple at $rowid1$.
- Let $t(F)$ be the restriction of t on F,
- Let M be the set of the measures that we can get from T for all $rowids$ in P,
- Apply the aggregate function g to M, let $g(M)$ be the result,
- Save $t(F)$ and $g(M)$ to disk.

For computing the function VARIANCE, for each partition P, we use a temporary list to store the measures in the tuples at all $rowids \in P$.

4.2 Query on the Last-Half Cube

The algorithm *QueryLH* computes a cuboid in the last-half data cube, over a scheme L, based on the tuple index over $F = L - \{n\}$ where n is the last attribute of the dimension scheme. This tuple index is in the first-half index base $(T, RS, FHIndex)$.

Algorithm QueryLH:

Input: A dimension sub-scheme $L = \{A_1, ..., A_k, n\}$, an aggregate function g, and $(T, RS, FHIndex)$.

Output: The cuboid on T and g, over L.

Method:

```
1. Get the tuple index over F = L − {n} from
FHIndex;

2. For each partition P in the index over F
do

2.1   Initialize a tree TrP to empty;
2.2   For each rowid in P do
2.2.1   Let val be the value on attribute n
          of the tuple at rowid in T;
2.2.2   Insert val and rowid to TrP
          using InsData2AttIndex;
2.2.3   done;
2.3   Let t(F) be the restriction on F of the
        tuple t at the rowid at the root of
        TrP, and g(TrP) the result of the
        application of g to the set of
        measures in all tuples at the
        rowids in TrP;
2.4   Save t(F) and g(TrP) to disk;
2.5   done;
```

5 Experimental Results and Discussions

The first-half index base for data cube query is implemented in C and experimented on a laptop with 8 GB memory, Intel Core i5-3320 CPU @ 2.60 GHz x 4, running Ubuntu 12.04 LTS. The experimentation is done on four real datasets CovType [3], SEP85L [4], STCO-MR2010_AL_MO [5] and *OnlineRetail* [6].

CovType is a dataset of forest cover-types. It has 581,012 tuples on ten dimensions with cardinality as follows: Horizontal-Distance-To-Fire-Points (5,827), Horizontal-Distance-To-Roadways (5,785), Elevation (1,978), Vertical-Distance-To-Hydrology (700), Horizontal-Distance-To-Hydrology (551), Aspect (361), Hillshade-3pm (255), Hillshade-9am (207), Hillshade-Noon (185), and Slope (67).

SEP85L is a weather dataset. It has 1,015,367 tuples on nine dimensions with cardinality as follows: Station-Id (7,037), Longitude (352), Solar-Altitude (179), Latitude (152), Present-Weather (101), Day (30), Weather-Change-Code (10), Hour (8), and Brightness (2).

STCO-MR2010_AL_MO is a census dataset on population of Alabama through Missouri in 2010, with 640,586 tuples over ten integer and categorical attributes. After transforming categorical attributes (STATENAME and CTYNAME), the dataset is reorganized in decreasing order of cardinality of its attributes as follows: RESPOP (9,953), CTYNAME (1,049), COUNTY (189),

IMPRACE (31), STATE (26), STATENAME (26), AGEGRP (7), SEX (2), ORIGIN (2), SUMLEV (1).

OnlineRetail is a data set that contains the transactions occurring between 01/12/2010 and 09/12/2011 for a UK-based and registered non-store online retail. This dataset has incomplete data, integer and categorical attributes. After verifying, transforming categorical attributes into integer attributes, for the experiments, we retain 393,127 complete data tuples and the following ten dimensions ordered in their cardinality as follows: CustomerID (4,331), StockCode (3,610), UnitPrice (368), Quantity (298), Minute (60), Country (37), Day (31), Hour (15), Month(12), and Year (2).

5.1 On Building the First-Half Index Base of Data Cube

Table 1 reports the run time and the memory use for computing the first-half index bases and their store space. In this table:

– the term "Run time" means the time (in seconds) from the start of the program until the first-half index base is completely built, including the time to read/write input/output files.

– the term "Memory use" means the maximal space of the main memory allocated to the program, from the start until the first-half index base is completely built.

– the term "Storage space" means the volume in mega bytes to store the first-half index base on disks.

Table 1. Results on Computing First-Half Base

Datasets	Run time	Memory use	Storage space
CoveType	138s	90 Mo	2 Go
SEP85L	127s	170 Mo	1.8 Go
STCO-...	141s	125 Mo	2.2 Go
OnlineRetail	96s	80 Mo	1.4 Go

5.2 On Query with Aggregate Functions

The group-by SQL queries are computed based on the first-half index base. The following aggregate functions MAX, COUNT, SUM, AVG, and VARIANCE are experimented. The queries are in the following simple form:

```
Select ListOfDimensions, f(m)
From Fact_Table
Group by ListOfDimensions;
```

For each dataset and each half of the corresponding data cube, the query is computed for all cuboids in the half. For instance, for CovType, we run the above query for 512 cuboids of the last-half and 512 cuboids of the first-half.

Tables 2 and 3 show the total time in seconds for computing the aggregate query for the five aggregate functions, for all cuboids in the first-half data cubes and in the last-half data cubes, respectively. The total times include all computing and i/o time, in particular, the time to write the results to disk. For instance, for the aggregate function SUM, the total time for computing 512 cuboids in the first-half data cube of dataset CovType is 237 seconds (Table 2). The last column (MEAN) represents the average of the total time on the five aggregate functions.

Table 2. Total time for computing aggregate query on first-half cubes

DATASETS	MAX	COUNT	SUM	AVG	VAR	MEAN
CoveType	235s	215s	237s	302s	288s	255.4 s
SEP85L	140s	118s	141s	169s	170s	147.6 s
STCO-...	144s	126s	144s	170s	176s	152 s
OnlineRetail	117s	106s	117s	145s	144s	125.8 s

Table 3. Total time for computing aggregate query on last-half cubes

DATASETS	MAX	COUNT	SUM	AVG	VAR	MEAN
CoveType	309s	279s	364s	383s	373s	341.6 s
SEP85L	193s	169s	191s	224s	237s	202.8 s
STCO-...	184s	168s	184s	212s	228	195.2 s
OnlineRetail	150s	146s	158s	180s	183s	163.4 s

Table 4 shows the average time in seconds to compute the response to an aggregate query. The data in this table is computed based on the data in Tables 2 and 3. For each half of a data cube, the average time is calculated by dividing the total time by the number of cuboids in the half. For instance, for the first-half data cube of CoveType, the average query response time for SUM is $237s/512 = 0.46$ second, and for the five a ge query response time is $255.4s/512 = 0.50$ second. For the entire data cube, the average query response time for SUM is $(237s + 364s)/1024 = 0.59$ second, and for the five aggregate functions the average query response time is $(255.4s + 341.6s)/1024 = 0.58$ second.

Table 4. Average response times of aggregate query

ON FIRST-HALF CUBES						
DATASETS	MAX	COUNT	SUM	AVG	VAR	MEAN
CoveType	0.46	0.42	0.46	0.59	0.56	0.50
SEP85L	0.55	0.46	0.55	0.66	0.66	0.58
STCO-...	0.28	0.25	0.28	0.33	0.34	0.30
OnlineRetail	0.23	0.21	0.23	0.28	0.28	0.25
ON LAST-HALF CUBES						
DATASETS	MAX	COUNT	SUM	AVG	VAR	MEAN
CoveType	0.60	0.54	0.71	0.75	0.73	0.63
SEP85L	0.75	0.66	0.75	0.88	0.93	0.79
STCO-...	0.36	0.33	0.36	0.41	0.45	0.38
OnlineRetail	0.29	0.29	0.31	0.35	0.36	0.32
ON ENTIRE CUBES						
DATASETS	MAX	COUNT	SUM	AVG	VAR	MEAN
CoveType	0.53	0.48	0.59	0.67	0.65	0.58
SEP85L	0.65	0.56	0.65	0.77	0.80	0.68
STCO-...	0.32	0.29	0.32	0.37	0.39	0.34
OnlineRetail	0.26	0.25	0.27	0.32	0.32	0.28

5.3 Discussions

To get some ideas on the efficiency of the first-half index base approach, we show in this section the experimental results on CoveType and SEP85L of the present approach and those of the last-half data cube representation [13,14] and of the very competitive method TRS-BUC [23]. In the discussion, we shall show (i) the implementation and the reported measures in the experimentation of the approaches, and (ii) their common and different points.

The last-half data cube representation approach is implemented in the same conditions as the first-half index base approach. After [23], TRS-BUC is implemented on a PC Pentium 4, 2.80 Ghz, running Windows XP. The results reported for TRS-BUC are the storage space and the construction time of the representation and the average query response time (avg. QRT) based on the representation. However, [23] did not precise whether the construction time and the average query response time include the time to write the result to disk. In contrast, in [13,14] and the present work, all time reported includes the i/o time, in particular, the time to write the data result (cuboids) to disk. Moreover, in Table 5, the avg QRT of [13,14] and the present work is the average on the five aggregate functions MAX, COUNT, SUM, AVG, and VARIANCE.

Table 5 synthesizes the main experimental results of the first-half index base with views on the results of the last-half data cube representation and those of the TRS-BUC method. The data in this table is graphically represented in Fig. 1, 2, and 3. In each graph, the three columns, from left to right, represent respectively the results of the TRS-BUC method, the last-half data cube representation, and the first-half index base representation.

Table 5. Main results with views on previous work

TRS-BUC			
Datasets	Construction time	Storage space	avg QRT
CoveType	300s	0.4 Gb	0.7s
SEP85L	1150 s	1.2 Gb	0.5 s
LAST-HALF CUBE REPRESENTATION			
Datasets	Run time	Storage space	avg QRT
CoveType	1018 s	7 Gb	0.93s
SEP85L	444s	2.8 Gb	0.74 s
FIRST-HALF INDEX BASE			
Datasets	Run time	Storage space	avg QRT
CoveType	138s	2 Gb	0.58s
SEP85L	127s	1.8 Gb	0.68 s

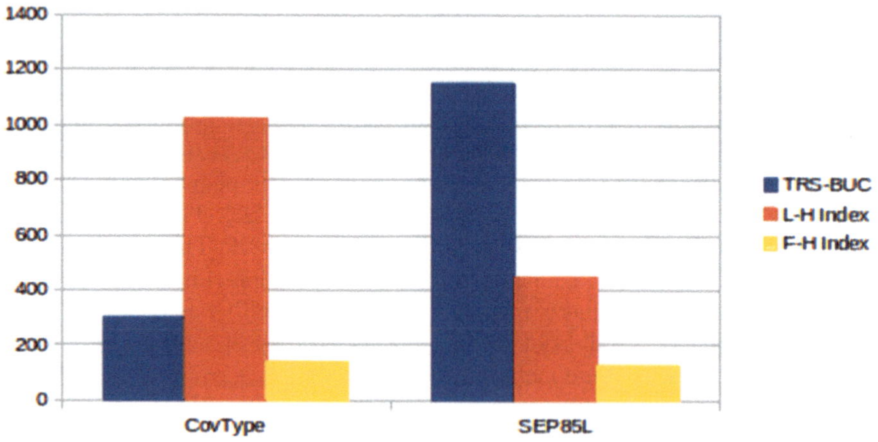

Fig. 1. Construction/Run times of three methods in seconds.

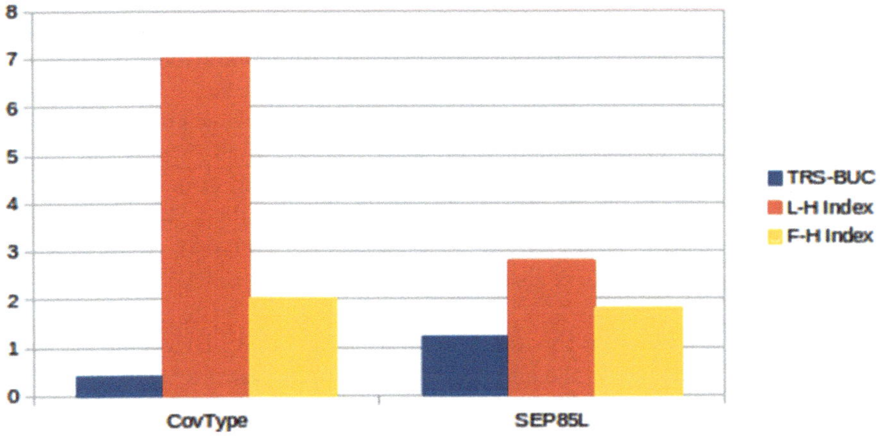

Fig. 2. Storage space of the representations in giga bytes.

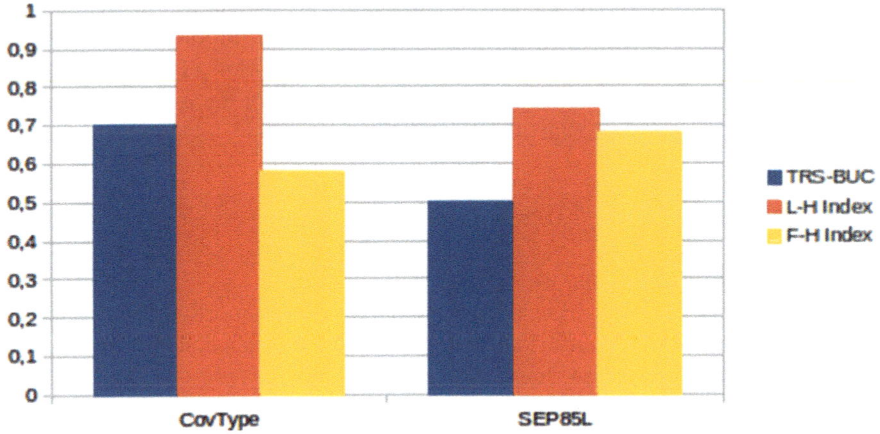

Fig. 3. Average query response times of three methods in seconds.

The common point between the first-half index base approach and the last-half data cube representation approach is that the both approaches use a half of the data cube index to represent the entire data cube. The different points are:

(a) One uses the first-half and the other uses the last-half. In general the volume of the last-half data cube is bigger than the volume of first-half data cube.

(b) The first-half index base approach follows the bottom-up computation, whereas the last-half data cube representation approach follows the top-down computation.

(c) In the first-half index base approach, as the representation of data cube, only the partitions of rowids are stored on disk, but no data tuples (tuples over dimension sub-schemes). In the last-half data cube representation approach, both data tuples and the partitions of rowids are stored on disk.

By above points (b) and (c), the reduction of the storage space and the run time to build the first-half index base with respect to the storage space and the run time to build the last-half data cube representation is considerable: on storage space, about 71% for CoveType and 35% for SEP85L, and on run time, about 86% for CoveType and about 71% for SEP85L. Though these important reductions, the average query response time is clearly improved by the first-half index base.

The difference between the above approaches and TRS-BUC consists in:

– TRS-BUC computes the entire data cube and optimizes the storage space of tuples: if an aggregate group has only one tuple, then the tuple itself is not stored, but only the rowid of its original tuple in the fact table: the rowid is called a reference. The approaches [13,14] optimize the storage by representing the data cube by a half cube of indexes.

– TRS-BUC computes the representation of the data cube built on a specific aggregate function applied to only one measure: for each aggregated tuple, only one aggregated value of the measure is stored. Whereas the representations by a half of data cube compute the index bases for computing data cube with any aggregated function: for each aggregated tuple, a partition of rowids is stored. This explains why the storage space of the representation by a half of data cube is much more bigger than that of TRS-BUC.

6 Conclusion and Further Work

From the above discussions, we can see that

– The first-half index base representation is really more efficient than the last-half data cube representation, in terms of storage space, construction time, and average query response time, and

– We can not really compare experimental results of the last-half data cube representation or the first-half index base with those of TRS-BUC, because they are not defined and computed in the same conditions.

However, on the results in Table 5, we can see that the first-half index base can be a competitive approach to TRS-BUC, as it builds the bases in very reduced time, with respect to the construction time of TRS-BUC, and offers the avg QRT that is almost comparable to that of TRS-BUC. Moreover, based on the first-half index base, the data cube query can be computed with any measure and any aggregate functions, without recomputing the representation.

For further work, we plan to test the first-half index base on much larger datasets and to study its incremental construction.

References

1. Agarwal, S., et al.: On the computation of multidimensional aggregates. In: Proceedings of VLDB 1996, pp. 506–521 (1996)
2. Harinarayan, V., Rajaraman, A., Ullman, J.: Implementing data cubes efficiently. In: Proceedings of SIGMOD 1996, pp. 205–216 (1996)
3. Blackard, J.A.: The forest covertype dataset. ftp://ftp.ics.uci.edu/pub/machine-learning-databases/covtype
4. Hahn, C., Warren, S., London, J.: Edited synoptic cloud reports from ships and land stations over the globe. http://cdiac.esd.ornl.gov/cdiac/ndps/ndp026b.html
5. Census Modified Race Data Summary File for Counties Alabama through Missouri. http://www.census.gov/popest/research/modified/STCO-MR2010_AL_MO.csv
6. Online Retail Data Set, UCI Machine Learning Repository. https://archive.ics.uci.edu/ml/datasets/Online+Retail
7. Ross, K.A., Srivastava, D.: Fast computation of sparse data cubes. InL Proceedings of VLDB 1997, pp. 116–125 (1997)
8. Beyer, K.S., Ramakrishnan, R.: Bottom-up computation of sparse and iceberg cubes. In: Proceedings of ACM Special Interest Group on Management of Data (SIGMOD 1999), pp. 359–370 (1999)
9. Vitter, J.S., Wang, M., Iyer, B.R.: Data cube approximation and histograms via wavelets. In: Proceedings of International Conference on Information and Knowledge Management (CIKM 1998), pp. 96–104 (1998)
10. Han, J., Pei, J., Dong, G., Wang, K.: Efficient computation of iceberg cubes with complex measures. In: Proceedings of ACM SIGMOD 2001, pp. 441–448 (2001)
11. Lakshmanan, L., Pei, J., Han, J.: Quotient cube: how to summarize the semantics of a data cube. In: Proceedings of VLDB 2002, pp. 778–789 (2002)
12. Phan-Luong, V.: A simple and efficient method for computing data cubes. In: Proceedings of The 4th International Conference on Communications, Computation, Networks and Technologies INNOV 2015, pp. 50–55 (2015)
13. Phan-Luong, V.: A simple data cube representation for efficient computing and updating. Int. J. Adv. Intell. Syst. **9**(3–4), 255–264 (2016). http://www.iariajournals.org/intelligent_systems
14. Phan-Luong, V.: Searching data cube for submerging and emerging cuboids. In: Proceedings of The 2017 IEEE International Conference on Advanced Information Networking and Applications Science AINA 2017, pp. 586–593. IEEE (2017)
15. Sismanis, Y., Deligiannakis, A., Roussopoulos, N., Kotidis, Y.: Dwarf: shrinking the PetaCube. In: Proceedings of ACM SIGMOD 2002, pp. 464–475 (2002)
16. Wang, W., Lu, H., Feng, J., Yu, J.X.: Condensed cube: an efficient approach to reducing data cube size. In: Proceedings of International Conference on Data Engineering, pp. 155–165 (2002)
17. Casali, A., Cicchetti, R., Lakhal, L.: Extracting semantics from data cubes using cube transversals and closures. In: Proceedings of International Conference on Knowledge Discovery and Data Mining (KDD 2003), pp. 69–78 (2003)
18. Casali, A., Nedjar, S., Cicchetti, R., Lakhal, L., Novelli, N.: Lossless reduction of datacubes using partitions. Int. J. Data Warehous. Min. (IJDWM) **5**(1), 18–35 (2009)
19. Lakshmanan, L., Pei, J., Zhao, Y.: QC-Trees: an efficient summary structure for semantic OLAP. In: Proceedings of ACM SIGMOD 2003, pp. 64–75 (2003)
20. Xin, D., Han, J., Li, X., Wah, B.W.: Star-cubing: computing iceberg cubes by top-down and bottom-up integration. In: Proceedings of VLDB 2003, pp. 476–487 (2003)

21. Feng, Y., Agrawal, D., Abbadi, A.E., Metwally, A.: Range cube: efficient cube computation by exploiting data correlation. In: Proceedings of International Conference on Data Engineering, pp. 658–670 (2004)
22. Shao, Z., Han, J., Xin, D.: Mm-cubing: computing iceberg cubes by factorizing the lattice space. In: Proceedings of International Conference on Scientific and Statistical Database Management (SSDBM 2004), pp. 213–222 (2004)
23. Morfonios, K., Ioannidis, Y.: Supporting the data cube lifecycle: the power of ROLAP. VLDB J. **17**(4), 729–764 (2008)

Analyzing the Accuracy of Historical Average for Urban Traffic Forecasting Using Google Maps

Hajar Rezzouqi[1,2](✉), Ihsane Gryech[1,2], Nada Sbihi[1], Mounir Ghogho[1,3], and Houda Benbrahim[2]

[1] FIL, TICLab, Université Internationale de Rabat, Rabat, Morocco
{hajar.rezzouqi,ihsane.gryech,nada.sbihi,mounir.ghogho}@uir.ac.ma
[2] IRDA, Rabat IT Center, ENSIAS, Université Mohammed V, Rabat, Morocco
houda.benbrahim@um5.ac.ma
[3] The School of EEE, The University of Leeds, Leeds, UK

Abstract. Urban traffic management uses increasingly sophisticated methods to overcome the many challenges involved in the development of traffic forecasting solutions. The main challenge is the acquisition of real-time large-scale urban traffic data at a sufficient spatio-temporal resolution. This is a challenge mainly because of the high financial cost that the installation of a large number of sensors would incur. This paper addresses this challenge by leveraging 'real-time' Google Traffic maps which show the state of the traffic on different road segments using four different colors. Since Google Traffic maps are provided in the form of rendered images, we apply image processing on Google Traffic maps to extract traffic data that are suitable for processing and analysis. The traffic data obtained from Google Traffic are validated with a traffic data set collected by sensors installed in Paris. Then, using data gathered through our Google Traffic-based method for several roads in Rabat (Morocco), we evaluate the accuracy of traffic prediction based on historical average. The overall accuracy reaches 74.9% on week days and 83.3% on weekend days. Further, a more detailed study by type of road and by time period was conducted, showing an overall accuracy of 95.8% in fluid traffic situations.

Keywords: Google traffic · Image processing · Data collection Prediction · Historical average

1 Introduction

The expanding traffic flow in cities could lead to a transport crisis. There are several dimensions that are affected by urban traffic, but the most negatively impacted ones are the environment and the economy. Therefore, optimising and predicting urban traffic are necessary to monitor and control congestion and pollution, and are thus at the core of research and development in smart cities.

K. Arai et al. (Eds.): IntelliSys 2018, AISC 868, pp. 1145–1156, 2019.
https://doi.org/10.1007/978-3-030-01054-6_79

This field of research requires an understanding and analysis of traffic behaviour, hence the need for massive traffic data for the many different roads of the city.

There are various methods used for the acquisition and analysis of traffic data. Regarding the data acquisition, the most commonly mentioned sensors in the research literature are: inductive loop vehicle detectors, cameras, vehicles equipped with GPS, smartphones, and surveys that can be used to estimate the origin-destination matrices.

In [1], the collection of the individual itineraries was carried out through the use of vehicles equipped with data loggers. In [2,3], taxis equipped with GPS, considered as mobile sensors, provide large-scale real-time traffic traces which allow for a global view of the dynamics of the urban road network. The authors in [4] argued that the smartphone can be considered as a reliable source for data acquisition. A correlation coefficient was computed between the generated variables from the on-board instrumentation and those provided from a smartphone equipped with a GPS receiver, a triaxial accelerometer, a triaxial gyroscopic and a compass, yielding results between 99% and 100%. Data collection can also be accomplished by more advanced sensors such as the Electronic Traffic Bayonet Device (ETBD). The latter enables to take pictures of each passing vehicle, automatically identifies the license plate, the vehicle brand, the speed, etc. and then stores the information in a database [5]. Some other new technologies have also been deployed to record road traffic data. For example, in [6], Bluetooth was proven to be easy to install and maintain and does not raise privacy concerns; the MAC address and the exact time of detection of the individual devices detected by the Bluetooth sensor were recorded. It was then possible to estimate the travel time between two different detections and hence speed of the vehicle.

Even though the technologies allowing traffic data collection are in constant evolution, ensuring such collection in urban areas faces obstacles which hinder urban traffic analysis. Indeed, efficient collection requires a massive and costly installation of sensors. This is particularly challenging for developing countries. To address the above-mentioned issue, we propose a novel method of traffic data collection based on Google Traffic, which is a feature displayed on Google Maps to indicate the level of congestion on roads. We evaluate the validity of Google Traffic data using an open traffic dataset from the city of Paris. After this validation step, we address the problem of predicting the level of congestion on some roads in the city of Rabat using Google Traffic.

2 Methodology

The scarcity of traffic sensors on Moroccan roads makes traffic prediction a difficult task. Smartphones represent a reliable source of information about some traffic parameters. Unfortunately, getting hold of smartphone data for researchers is difficult as mobile operators are reluctant to share such data because of privacy concerns. To circumvent this problem, we propose to extract information from Google Traffic maps, which show the level of congestion on the main roads of most cities. Indeed, Google estimates the level of congestion on a road by analysing the GPS data obtained from the Android smartphones carried by the vehicles passing through the road. The level of congestion is determined by the average speed on the road over a given time period [7]. As shown in Fig. 1, for each congestion level, there is a specific color: the green color represents fluid traffic, orange color indicates less traffic congestion, red denotes congestion while brown refers to high traffic congestion. When smartphone data are not available, none of these colors is shown on the corresponding roads. Therefore, the reliability of Google Traffic improves with the penetration of android smartphones.

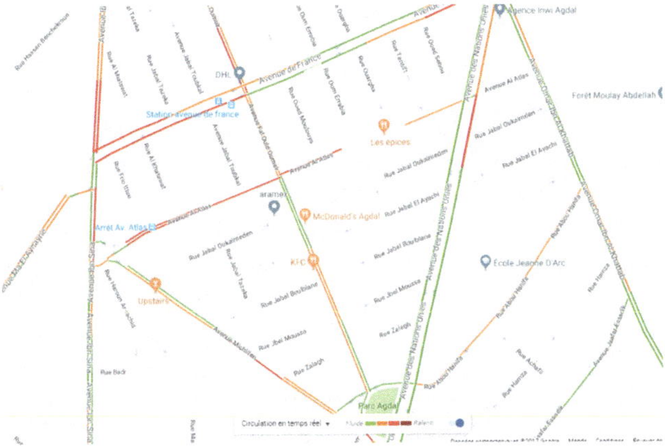

Fig. 1. Congestion levels of an area of Rabat city displayed on a Google Traffic map.

The methodology that we have adopted in our work to extract and analyse the data obtained from Google Traffic maps is based on the implementation of the following steps (as described in Fig. 2):

- Automatic maps collection: The area and roads of interest are first selected; then, an automatic screen capture of the corresponding Google map (with traffic feature on) is programmed to obtain a traffic map every 5 min; this is accompanied by an automatic refresh of the page.
- Image processing: Applying basic image processing tools on the built image database, relevant traffic information is extracted every 5 min.
- Data processing: This consists of building the final database where each data element consists of the road identity, the date and time of capture, and the level of congestion.
- Traffic prediction: Since the state of the traffic is represented by a categorical variable (4 levels of congestion), the prediction is cast as a classification task. In this paper, we limit our traffic prediction study to historical average.
- Performance analysis

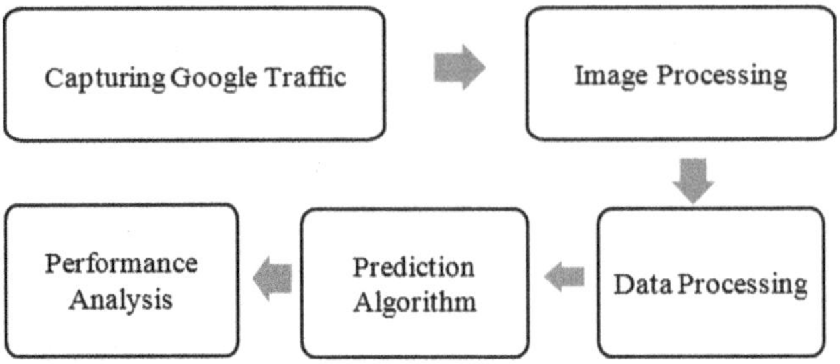

Fig. 2. Traffic data collection and analysis process.

2.1 Image Processing

The image processing consists of extracting the RGB code of the positions p representing the coordinates of the pixels chosen to represent the different roads of interest. These positions were manually determined before the automatic data collection. Each position 'pixel' has its own RGB code. Since for each color we have different shades, we were faced with a multitude of RGB codes. For this reason and to facilitate our data analysis, we decided to use as centroids of the RGB codes the main colors needed, namely, green, orange, red, brown and white (which indicates that no data is available). Once the centroids have been fixed, the Euclidean distances between the RGB codes and these centroids are calculated, and the minimum distance determines the color category the RGB code belongs to. Thus, this distance is given by

$$d(p, c_j) = \sqrt[2]{(R_p - R_{c_j})^2 + (G_p - G_{c_j})^2 + (B_p - B_{c_j})^2}$$

where p is a specific position on the map whose extracted RGB code is represented by the tuple (R_p, G_p, B_p), and the tuple $(R_{c_j}, G_{c_j}, B_{c_j})$ represent the RGB code of the jth centroid (j = 1, ..., 5).

2.2 Validation of the Proposed Data Collection Method: A Comparative Study

To test the proposed approach of collecting traffic data from Google Maps, and to check the validity of our extracted data, we used an existing dataset from the city of Paris [8]. This publicly available dataset is provided by the city hall of Paris, and collected using permanent traffic sensors installed on the urban network of the city. Only hourly traffic data is available. The dataset includes:

- The flow: the number of vehicles that passed by the measuring point during a time period (one hour).
- The occupancy rate: this corresponds to the time of presence of vehicles on a loop, as a percentage of a fixed time period.

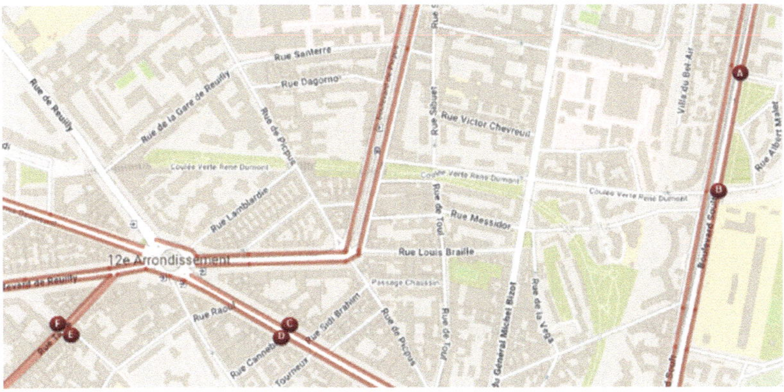

Fig. 3. Sensors location in the studied area of the city of Paris

Figure 3 illustrates the studied area. We next compare our newly collected dataset, and the dataset provided by Open Data Paris over the same period of time. To do so, we considered two separate measurement points (A and B) on the large “Boulvard Soult” road, two measurement points (E and F) on the small “Rue Traine” road, one on each lane, and two other measurement points (C and D) on “Rue Cannebire”. The selected day chosen for the comparative study was 17 October 2017. With the data collected from Google about the small roads (E, F), we obtained a green shade during almost the whole period of extraction. Thus, there is an absence of variation in the level of congestion unlike the data collected from the installed sensors. This makes it hard to match the level of congestion with traffic parameters such as the flow, and the occupancy ratio. However, for large roads (A, B), we were able to detect this variation.

There is also an issue with the time resolution of the traffic measurements in the two datasets: the screen shots from Google Traffic are taken every 5 min, and the sensor measurements are provided hourly. Within each hour, each 5 min period is associated with one level of congestion. Hence, in order to be able to compare the two sets of measurements, we aggregate Google traffic data using one of the following methods.

The first method consists of calculating the mode during each hour, i.e. the predominant level of congestion. The limitation of this method is that the dominant level is very often green, which takes us back to the problem we faced with the two points of measurement (E, F). This is illustrated in Fig. 4 where the blue curve corresponds to the level of congestion retrieved from google, while the green curve describes the data retrieved from the Paris dataset.

The second method assumes that the level of congestion representing any given hour is the corresponding highest level of congestion. By selecting the maximum level of congestion, we observe that the traffic flow, the occupancy rate and the level of congestion curves vary in the same way (Figs. 5 and 6). This implies that the data extracted from Google Maps are in agreement with those obtained from the sensors and thus reflect the real state of traffic.

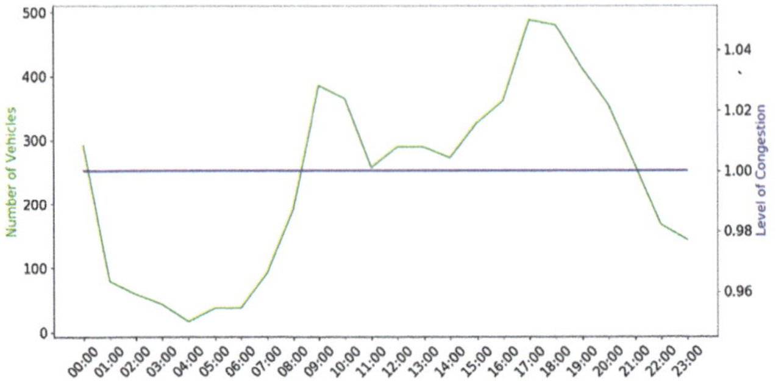

Fig. 4. Comparison between paris data set and Google traffic data set based on the mode (Measurement point: B).

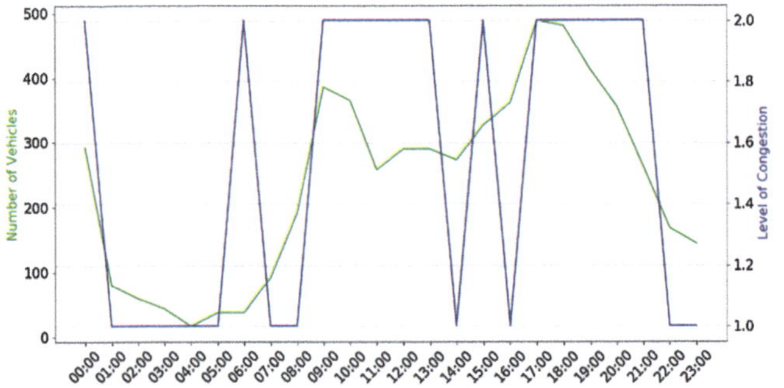

Fig. 5. Comparison between paris data set and Google traffic data set based on the highest congestion level (Measurement point: B).

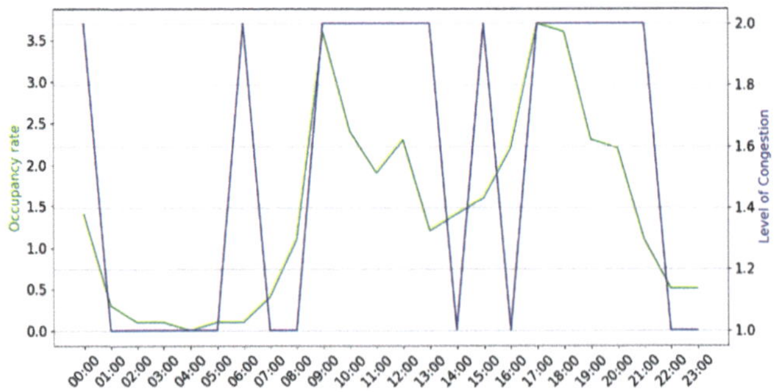

Fig. 6. Comparison between the occupancy rate extracted from Paris Data set and Google Traffic Data set based on the highest congestion level (Measurement point: B).

2.3 Disaggregation of Traffic Data

The traffic parameters available in the Paris dataset are measured hourly. One may be interested in disaggregating these measurements in order to understand the dynamics of the traffic at a higher temporal resolution. Further, we also notice that there are sometimes many missing values which require to be imputed. To address the above mentioned issues, we propose to combine the hourly sensor data with Google Traffic maps. More specifically, we derive a formula to extract traffic values (Number of vehicles, occupancy rate, average speed) at a five minute-temporal resolution from the Google Maps congestion levels and the hourly data provided in the Paris dataset. Consider the average speed as an example. Let X be the average speed recorded per hour, and let X_i be the average speed over the ith five minute period. We would like to estimate the X_i's using X and the congestion levels that are collected from Google Traffic every five minutes. We propose to estimate the X_i's using the following method:

$$X_i = w_i.X, \quad i = 1, ..., 12 \tag{1}$$

subject to

$$X = \frac{\Sigma_{i=1}^{12} X_i}{12}, \qquad \Sigma_{i=1}^{12} w_i = 12 \tag{2}$$

where the weights w_i are to be determined using the levels of congestion extracted from Google Traffic maps. Indeed, the more congested a road is, the lower the average speed. This 'proportionality' could be leveraged to compute the weights.

There is however no trustworthy reference regarding the average speed thresholds on which Google Traffic rely to define the four levels of congestion. This implies that there will always be an uncertainty about the weight calculation. Nevertheless, we here make the assumption that the average speed thresholds, v_1, v_2, v_3 and v_4, defining the different levels of congestion are equidistant, i.e.:

$$v_3 - v_2 \simeq v_2 - v_1 \simeq v_1 - 0 \implies v_2 \simeq 2.v_1, \quad v_3 \simeq 3.v_1$$

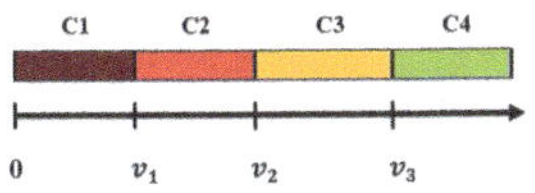

Hence, the weights can be estimated as follows:

$$w_i = 12 \frac{z_i}{\Sigma_{i=1}^{12} z_i} \qquad i = 1, .., 12 \tag{3}$$

where z_i, which takes on values from the set $\{1, 2, 3, 4\}$, is determined from the Google Traffic-derived congestion level associated with the ith five minute period; $z_i = 4$ for brown color, i.e. very congested traffic, $z_i = 3$ for red, $z_i = 2$ for orange and $z_i = 1$ for green.

3 Prediction Model

In this paper, traffic prediction is merely based on the historical averages of the data extracted from the images obtained with Google Traffic. The database obtained through image processing consists of the date/time of the screen capture, the different positions corresponding to the selected roads, and the colors that reflect the level of congestion.

A white color on the map means that either there are no vehicles on the corresponding road, or there are no smartphones detected. We also noticed that this color is the most common color at night. This may introduce a bias to our prediction model. So, in our study, we assumed that most road users are equipped with at least one smartphone, and thus not detecting a smartphone means very few vehicles passing through the road and hence a fluid traffic (green color). As a result, we merged the white color with the green one.

We have selected a specific area of Rabat city to test our approach. We have collected Google Traffic data from 22 September 2017 at 00:05 to 23 October 2017 at 23:55, i.e. over a period of 32 days, with a frequency of one map per 5 min. 49 positions were chosen for this study, giving a database of 451, 535 observations. Since our prediction model is based on historical average, it assumes that the traffic has a strong periodic component, with may change with the type of road and time period. For this reason, we have distinguished week days from weekend days, and grouped the data by road and subsequently by hour. Once the data are aggregated to ensure the same spatial and temporal conditions, the mode is determined. Our database was divided into two parts: training and testing sets. For traffic prediction on a test week day, all the data corresponding to the week days are aggregated from the training data, and the level of congestion for the test day at time t is estimated according to the following formula:

$$l_{t,i+1} = \text{Mode}(l_{t,i}, l_{t,i-1}, l_{t,i-2}, ..., l_{t,i-n}) \tag{4}$$

where $l_{t,i+1}$ denotes the level of congestion to be predicted on weekday $i+1$ at time t, and $n+1$ corresponds to the number of days used in the training data set.

For predicting the traffic on weekends, the same approach as above is applied using the training data corresponding to weekend days.

4 Performance Analysis

The performance of the prediction of the level of congestion on the selected roads is evaluated using the following metrics: Accuracy, Precision, Recall and F1 score. The results corresponding to the prediction of traffic on week days and weekend days are presented in Tables 1 and 2, respectively.

There are several reasons for traffic congestion, the most important of which road construction work, accidents, interruptions, delays caused by traffic lights, stopping points, and intersections. Indeed, with the high number of cars on a road, any traffic disruption can cause a traffic jam. In some cases, bottling is not directly or rationally justified. This is mainly due to the individual behaviour of drivers, especially those who drive very aggressively or those who drive very carefully and slow down traffic. Thus, a model that relies on the historical average is not unable to predict these unexpected occurrences, since it only takes into account the usual conditions on any given route. This may justify the overall accuracy obtained in the paper: 74.9% on weekdays and 83.3% on weekends (Tables 1 and 2); this is also illustrated in Fig. 7.

Table 1. The performance of the historical average-based prediction for week days

	The historical average
Accuracy	74.9%
Precision	71.9%
Recall	74.9%
F1	72.7%

Table 2. The performance of the historical average on weekends

Accuracy	83.3%
Precision	78.5%
Recall	83.3%
F1	80%

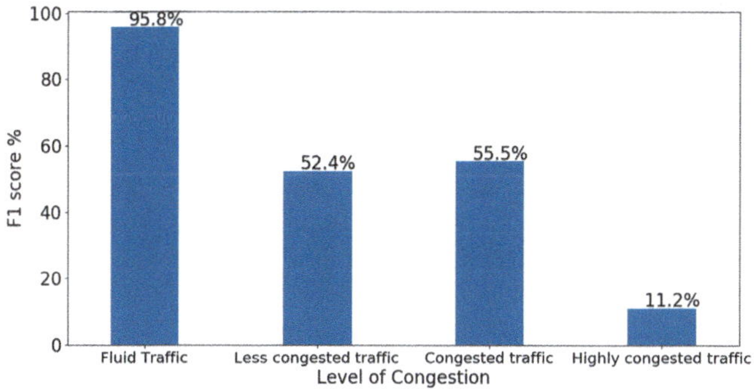

Fig. 7. Global performance for each level of congestion.

Further, we have examined the prediction performance related to each congestion level, and found that the more fluid the traffic is, the better the F1 score of the historical average, which reaches 95.8%. On the other hand, the more congested traffic becomes, and this is quite likely due to the above-mentioned causes, the lower the F1 score, which may be as low as 11.2%.

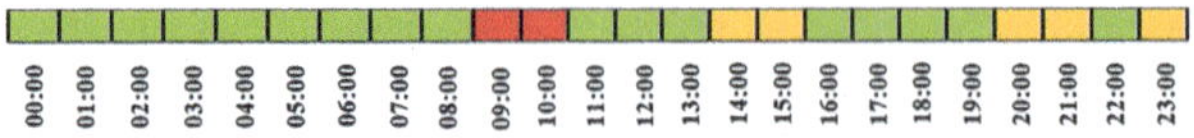

Fig. 8. Level of congestion per time period during weekdays.

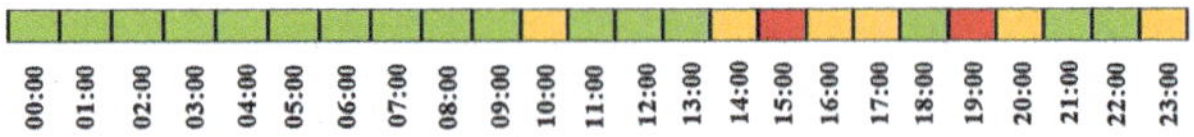

Fig. 9. Level of congestion per time period during weekends.

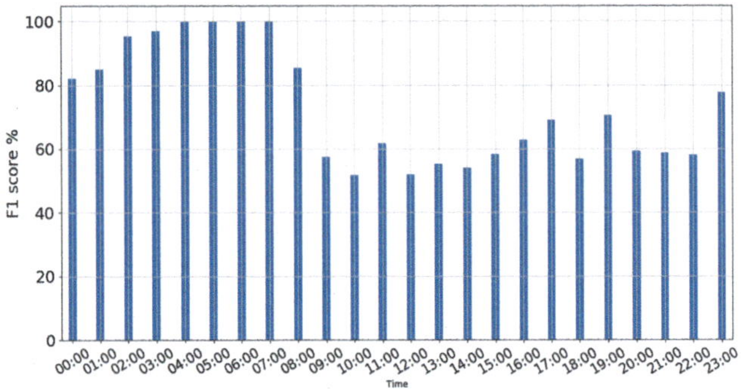

Fig. 10. Global performance by time period during weekdays.

We also studied the performance of the historical average per time period during week days (see Fig. 10). The results are coherent with what has been found regarding the performance by congestion level: during peak hours, when traffic is congested (Figs. 8 and 9), the accuracy becomes less satisfactory (e.g. the overall accuracy at 10:00am is 52%), whereas in the hours where the traffic is fluid (for example at 4am and 5am), a high precision has been reached (nearly 100%). A similar approach has been applied to weekends, (Fig. 11) but here the peak hours are not the same as on week days (Fig. 9). Indeed, people tend to go out more in the evenings between 07:00pm and 00:00am. This explains the lower performance during these hours and a better performance in the very early morning hours.

Finally, we have found that for roads with a more dynamic traffic (presence of shopping centres, schools, companies, etc.), the accuracy of the historical average-based prediction at peak hours decreases. Indeed, it is in these hours that we notice rapid fluctuations in the level of congestion, which makes it difficult to predict the level of congestion at a specific hour, based only on the historical average.

Traffic prediction using the historical average is characterised by ease of implementation and speed of execution. The more traffic data is available, the better the performance of this method gets. This method however may not be sufficiently accurate in the presence of unexpected events [9]. The other limitation is that the level of congestion at any given time is estimated based on the history of traffic over the last few days, and thus does not take into account the traffic conditions of the immediate past (last minutes or hours). Other factors influencing the traffic such as the weather have not been taken into account in our model. We are currently investigating this issue as well as exploring the spatio-temporal relationships between roads in order to increase the accuracy of prediction.

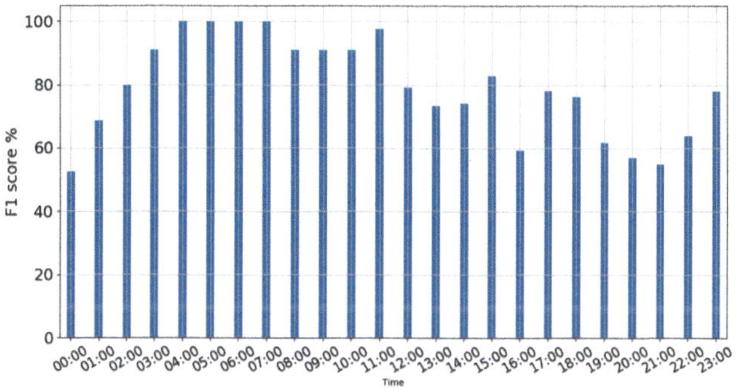

Fig. 11. Global performance by time period during weekends.

5 Conclusion and Future Work

We have proposed a novel method of urban traffic data collection which consists of exploiting Google Traffic Maps using image processing. A validation method, based on a comparative study with an existing data set from the city of Paris, was proposed to verify the reliability of the extracted Google Traffic data. This study demonstrated that there is a correlation between the level of congestion obtained from Google Traffic maps and the number of vehicles on roads, as well as the occupancy rate retrieved through the sensors installed in Paris. We have also proposed a way to disaggregate traffic data obtained with sensors using Google Traffic maps. Using the proposed traffic data collection, we built a database about the traffic in an urban area of the city of Rabat (Morocco). Using this dataset, we have evaluated the performance of the historical average-based traffic prediction of the level of traffic congestion. Week and weekend days were treated separately. The perceived limitations of this method, due to unexpected events and changing weather conditions are motivations for future work.

Acknowledgment. This work is a part of the MOROCCAN VIDEO INTELLIGENT TRANSPORT SYSTEM project, and is funded by "Centre National pour la Recherche Scientifique et Technique" (CNRST).

References

1. Park, J., Murphey, Y.L., Kristinsson, J., McGee, R., Kuang, M., Phillips, T.: Intelligent speed profile prediction on urban traffic networks with machine learning. In: Proceedings of the International Joint Conference on Neural Networks (2013)
2. Yu, B., Song, X., Guan, F., Yang, Z., Yao, B.: k-nearest neighbor model for multiple-time-step prediction of short-term traffic condition. J. Transp. Eng. **142**(6), 1–10 (2016)
3. Deng, B., Denman, S., Zachariadis, V., Jin, Y.: Estimating traffic delays and network speeds from low-frequency GPS taxis traces for urban transport modelling. Eur. J. Transp. Infrastruct. Res. **15**(4), 639–661 (2015)
4. Diaz Alvarez, A., Serradilla Garcia, F., Naranjo, J.E., Anaya, J.J., Jimenez, F.: Modeling the driving behavior of electric vehicles using smartphones and neural networks. IEEE Intell. Transp. Syst. Mag. **6**(3), 44–53 (2014)
5. Yuan, W., Deng, P., Taleb, T., Wan, J., Bi, C.: An unlicensed taxi identification model based on big data analysis. IEEE Trans. Intell. Transp. Syst. **17**(6), 1703–1713 (2016)
6. Park, H., Haghani, A.: Optimal number and location of Bluetooth sensors considering stochastic travel time prediction. Transp. Res. Part C: Emerg. Technol. **55**, 203–216 (2015)
7. Stenovec, T. (n.d.): How Google Maps knows about traffic - Business Insider. http://www.businessinsider.com/how-google-maps-knows-about-traffic-2015-11?IR=T. Accessed 6 Dec 2017
8. Open Data Paris. Open Data Paris (2011). Accessed 8 Dec 2017. https://opendata.paris.fr/page/home/
9. Bolshinsky, E., Freidman, R.: Traffic Flow Forecast Survey, 1–15 (2012)

Intelligent Transportation System in Smart Cities (ITSSC)

Sondos Dahbour[1(✉)], Raghad Qutteneh[1], Yara Al-Shafie[1], Iyad Tumar[1], Yousef Hassouneh[2], and Abdellatif Abu Issa[1]

[1] Electrical and Computer Engineering Department, Birzeit University, Ramallah, Palestine
sondosda3@gmail.com, raghadzaki@gmail.com, Yaraalshafei2014@hotmail.com, {itumar,aabuissa}@birzeit.edu
[2] Computer Science Department, Birzeit University, Ramallah, Palestine
yhassouneh@birzeit.edu

Abstract. Due to the noticeable increase in population and the number of moving vehicles in the streets in big cities, intelligent transportation system in smart city would be a good solution that leads to a comfortable life for the citizens and to reduce the traffic congestions. This paper presents an Intelligent Transportation System in Smart Cities (ITSSC) for managing traffic congestion at intersections and accident detection as a special case. This work counts on the notion of Internet of Things (IoT) and Global Positioning System (GPS) in which a base station receives packets from moving vehicles to calculate the number of vehicles in each street and/or detect an accident occurrence. The ITSSC works on the collected information and analyzes it and then uses a prioritization method to control the function of the traffic light in a way that decides when the traffic light will open (green light) and for how long time. The achieved results show that the ITSSC decreased the total average needed time at each traffic light in order to free or at least decrease the number of waiting cars at that traffic light compared to the traditional traffic lights. In addition to that the average waiting time for each car at a traffic light is decreased by 28% compared with the traditional traffic lights.

Keywords: Intelligent transpiration system · Traffic light · Smart cities

1 Introduction

Information technology is the main part of the construction of smart cities that are highly advanced in terms of overall infrastructure. These cities consist of separate intelligent and smart systems that are combined together to form big linked parts, which are made to serve the general purpose of an easier life for citizens. Transportation system is one of these basic parts of a coherent system in every city.

Through the past years, many modifications were made on the transportation system wishing to come up with a solution for traffic congestion and accidents since these problems lead to latency and shortage in people's lives or work. Most of the suggested and proposed solutions to build such a transportation system require user involvement

K. Arai et al. (Eds.): IntelliSys 2018, AISC 868, pp. 1157–1170, 2019.
https://doi.org/10.1007/978-3-030-01054-6_80

or depend on social media news as a source of information that might be incorrect. Moreover, the rapid increase in the crowdedness we are experiencing in our life now causes difficulties and hardships in the movement of vehicles. Therefore, a critical demand for an intelligent system with the least dependency on users' input to solve the traffic congestions is required. Furthermore, it is unquestionable that time saving is an important factor to improve such a system. According to a study done by the Texas A&M Transportation Institute (TTI), some Americans might lose up to 80 h annually in traffic jams [1, 2]. An intelligent traffic system can greatly reduce this meaningless waiting time. In addition to that, it can reduce power and gas consumed by vehicles stopping at traffic lights.

In This paper, the proposed ITSSC for controlling and managing transportation is based on real-time traffic and real time collected data, in adaptive behavior, which is different from traditional traffic lights that use a fixed traffic light timer. The proposed system has responsibilities divided into three phases: collecting real time data, analyzing the data and finally give orders based on the analyzed data. Traffic conditions and information will be gathered from four streets at each intersection at first, then analyzing the collected data taking into account the vehicles locations and the number of vehicles at the intersection respectively. Finally, based on the collection and analyzing phases, a decision of how to reorganize streets, when and for how long each traffic light will be opened (green light) are going to be sent and applied. Additionally, the possibility of an accident happening at any given time had to be taken into consideration. How would the emergency cases be rectified quickly and immediately? How to, and what agent from the different emergency vehicles (Police, Ambulance or Fire Truck) will receive the evocation message to come to the site.

The rest of this paper is organized as follows. The background and the related work are discussed in Sect. 2. Section 3 presents the proposed ITSSC, approach and methodology. Results and analysis are discussed in Sect. 4. Finally, Sect. 5 concludes the paper and introduces a future work.

2 Related Work

There is some work that had been done to develop smart traffic systems. One of these efforts is the Traffic Information Centre (TIC) that published in 2012 at the intelligent vehicles symposium in Spain [3]. This paper introduced a new scheme that count on information center in order to handle the traffic light. It is an intelligent traffic light designed to form a data collection center (server). This server gathered Information about the available and current traffic statistics in the city from the clients (vehicles). After that the server will return the collected information to the other clients (vehicles). Transferred messages between both sides have three types, first the 'Hello' setup message, which is sent when a new client changes the previous TIC range and enters a new server range, second, the accident warning message, and finally the default data message. In accident cases, TIC receives a special accident message warning for an accident. In [4], the authors introduced a way of designing the traffic system as a peer-2-peer system. This new suggestion was considered better than Client-Server method (C/S) because of the high cost of the C/S due to its centralized nature. The main idea behind the P2P method is the architecture of working in a hierarchy that is split into two tiers such that every passing vehicle is represented as a single node, which is called a peer, then nodes are divided into group of clusters, each cluster contains a one non-static (changes periodically) super node, which represents a gateway between clusters. These clusters communicate with each other continuously to share information with the gateway. The work in [5] built a Traffic Management System (TMS) which consist of several phases, starts by data sensing and gathering from different sources such as social media, passing through data fusing, aggregation, processing on data collected in the previous stage and ending with service delivery to end-users. Moreover, many researches had been done to develop accident detection systems or even to prevent accidents due to its importance in saving human lives. The authors in [6] present one of the different technologies, which were used to build these systems that consist basically from sensors because of its cheap price and ease of use. Obstacle detection and indication sensor with the motion sensor (PIR) are used to detect humans. These sensors use the IR waves since any object generate heat, PIR is used with Fresnel lenses to control the width of PIR. Sensors are located on the left and right sides of the car so when a human being is detected by one sensor, a car will turn into the other side and when two sensors detect a human being, this means that someone is in the front of the car. When there is no possibility to avoid the human being a buzzer will work and the circuit will open to stop car motor motion, Arduino and Wi-Fi shield are the parts that are used to tweet and ask for help. The researches on the accident detection in the traffic is not new studies, a paper was published in 1998 titled "Traffic Incident Detection: Sensors and Algorithms" talked about experiments and searches done in Evans Lane, Saint Louis, USA. The results of the researches agreed that the most important parameters that are needed to be measured in traffic are volume, speed, density, lane occupancy, travel time, and vehicle headway. However, the two most important are the queue length and the turning movement because they play a role with controlling in changing the traffic light. In order to design a system, a previous known data will help if it is available. Few studies done on a high way in north America, and their results were as follows, average traffic parameters are 100,000 vehicles per day with a traffic volume equal to five lanes. Speeds range from 30 to 85 miles per hour, though posted limits are 40 to 65 in most of the U.S. These parameters are needed to be measured using equipment, sensors that are distributed along or in the middle of the road. Each parameter needs a special sensor to give the most accurate reading. In addition to the sensor type, the range of sensing is important. Therefore, a sensor designed for collecting road speeds of 5 to 50 mph will not measure highway speeds of 35 to 90 mph very accurately. One of the methods that have been pursued by a variety of cities and metropolitan areas for incident detection and existence is called the inductive loop. The concept of this method is to place surveillance cameras that is responsible for taking videos and send them to the traffic management centre, this method will provide the data center with the vehicle count parameter and lane occupancy. This system is affected strongly with the surrounding circumstances, weather, temperature or road conditions since it depends mainly on sensor readings [7]. In 2015, a new work had been published about an approach that use video image processing method with the help of microwave radar and a video camera in order to improve the transportation system. The approach depends on a default preinstalled background image for an empty road. When a new input image for a street gets into the system, it is converted to grayscale format using double precision. After that processes will be done to the converted image including comparison with the preinstalled empty street image, to decide whether an image will be converted again to binary image or if it will be discarded. Undoubtedly, environmental conditions will influence on the functionality of this methodology such as image reading errors [8]. Focusing on the same idea, an improved approach was presented in [9]. This approach use a Raspberry PI (RPi) instead of the video camera since it has an operating system, memory, SD card and a camera. In addition, the police officers and the emergency vehicles, which have a special privilege to access the RPi system using their smart phones, can do controlling operation of traffic lights. Another work about agent based intelligent introduced in [10]. In this work, the intelligent traffic control systems are used to both control the flow of traffic and reducing the traffic congestion. The first problem solution is the studying of how agents could learn the best departure times in a daily commuting environment and how following those departure times alleviates congestion. The second problem solution is the studying of how agents could learn to select desirable lanes to improve traffic flow and minimize delays for all drivers. The authors in [11] discuss Machine-to-Machine (M2M) scheme that tries to find a solution to reduce the road traffic congestion. The M2M communications enable all objects, virtual or physical, to communicate without human intervention. A special protocol can be used in smart city applications for that. Also, the communication channel technology could be used to exchange messages from sensors/devices and gateways to an M2M server platform. In similar work to M2M a Vehicle to Infrastructure (V2I) approach is used to manage the traffic congestion [12]. The process passes in many phases starting with broadcasting traffic light statues to the vehicles, then, process the received data in the vehicle to adjust its speed. After that, vehicles will communicate with each other to improve their efficiency. Guerrero-Ibanez et al., discussed the idea of how integrated technologies such as connected vehicle, cloud computing and IoT could help in building intelligent transportation system that will reduce traffic congestion and increase road safety. These technologies could help in implementing new techniques, evolving intelligent traffic lights instead of traditional ones, for collecting information using cloud computing, processing and spreading it using vehicle-to-vehicle and vehicle-to-infrastructure using wireless technologies.

3 Proposed ITSSC

This section introduces the ITSSC scenario, approach and methodology.

3.1 Scenario

Traffic lights in general are controlled in a basic way that depends on a fixed timer equal to 30 s in most scenarios. In other words, each traffic light open; turns on green for 30 s and then would have to wait for one and a half minute (red light) until it open again (the other three traffic lights take 30 s each, then 30 X 'one time' = 1 and a half minute). In addition, there are no traffic jam constraints and there is no equality in lane distribution, consequently leading to two unequally car packed streets taking the same opening and closing period of time. Moreover, when an accident takes place, traffic lights continue to run in their sequence without any thought of a solution for the problem.

In this paper, the proposed scenario deals with intersection that includes four streets, divided into two sides for both directions where each street (direction) has two lanes as shown in Fig. 1. Each street at a single intersection covers a certain range of GPS coordinates. The Base station, which is responsible for controlling the intersection, contains a table mapping each GPS coordinates to the corresponding street and the traffic light controlling that street. Base stations are distributed in a city in a manner that guarantees that none of their coverage overlaps with each other and all the base stations send the collected information to a central server. This work assumes that all vehicles on the street are smart vehicles that have the ability to repeatedly send their packets to a base station. Each packet consists of four fields: car IP address, car GPS location, impact sensor reading and accelerometer sensor reading. When a base station receives a packet, it first uses the GPS information to indicate to which street at that intersection the car belongs. Thereafter, it increases the corresponding street counter and uses this information in the recalculation process. After that, a base station checks the impact and the accelerometer sensors' readings to determine whether or not there is an accident in the street. A base station can give priority to each street and its traffic light on time period (the time in which the traffic light remains green) according to some criteria using the detailed following procedure:

- Cars continuously send their packets every 2 s to the base station they belong to at a certain traffic light. After that, the base station increases the car counter for this traffic light. Each car can only have one entry in the server database, this entry is only updated if the car changes its GPS location/range, meaning that it has moved and it is now in another street in a different intersection, which means that a car halted in a single position will not lead to multiple counter increments.
- In a single round, all non-empty streets take turns being turned on depending on their priority. Otherwise, empty streets are taken out of the calculation process since it is inefficient to prioritize and empty street.
- The street with the highest counter will get the highest priority to be open first. The traffic light controls this street will stay green for a specific period of time which equals to

$$\textit{Number of cars in that street} \times \textit{Crossing}\ \text{Time} \tag{1}$$

After that, the previous street is removed from the calculation process for the current round, and the calculation is repeated again for all the streets that remain. Moreover, when the assigned period to the previous traffic light finishes, the server sends out its latest calculations (priority orders), which are repeatedly done at each time a new car is registered. E.g., if the previous traffic light's (the first) time was 15 s, while the last calculation update happened at time = second 13, this would mean that the calculations (street orders) of time = second 13 are sent to open the current (2nd) street at time = second 15.

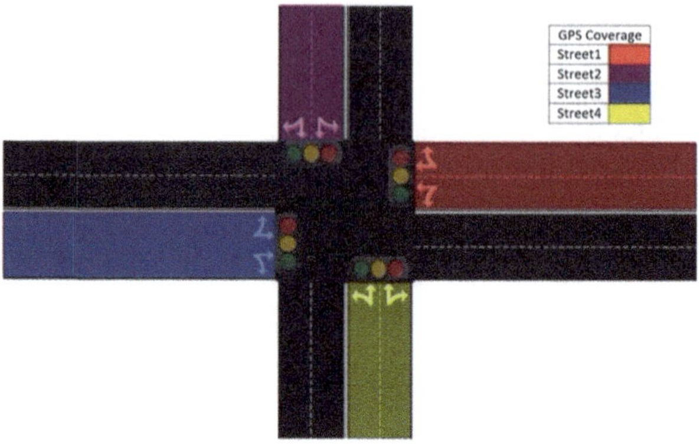

Fig. 1. Traffic light Intersection scenario.

- The previous two steps are repeated to turn the rest of the traffic lights on in order to finish a complete single round. In the case of two or more streets have equal car counters, the system depends on the previous iteration (round) to see which street had the highest counter and gives that street a higher priority.

This all guarantees that the ITSSC effectively redistributes the ‘on time period’ for traffic lights in a way where precious time is not thrown away on an empty street, nor a shortage of time is encountered on an overloaded street. Effective redistribution was a result of traffic times being decided depending on a counter, resolving any street inequality issues. Imagine that one car is waiting in a single street, then the light will turn on for 30 s while the car only needs 5 s to cross the street resulting in more than 20 s of wasted time. While on the other hand, using the ITSSC, the traffic light is only granted the needed time for all cars to cross (5 s) saving more than 20 s of time.

Based on empirical results, the ITSSC has two parameters (thresholds), the number of cars in each lane and the waiting time at each lane in the street. If the number of cars at a lane is less than the threshold value (82) then the ITSSC count on this parameter only in the simulation. However, if the number of the cars at a lane exceeds the threshold (82 cars), then the other parameter (a lane waiting 90 s in close state (red light at traffic light)) will be adopted when prioritizing streets over the others. This parameter depends on the time the traffic light has been turned off, so the street that has been waiting for more than 90 s will be opened first. If more than one street has been waiting for more than 90 s (threshold), then the street with the higher waiting time has the advantage over the others to open first. The ITSSC will check if any of the streets exceeds the 90 s threshold at the beginning of each ordering process in the same round. However, if none of the streets waiting time exceeds the threshold value, the normal counter algorithm that will be explained in Sect. 4 is used alone, even if the intersection counter is higher than 82 cars. At the beginning of every round, the algorithm checks whether the number of cars in the intersection exceeds 82 cars or not, in order to decide if the simple counter method should be used or the 90 s waiting threshold. As an example, if the total number of cars in the intersection exceeds 82 cars at the beginning of the first round, the algorithm will check whether any of the streets have been waiting for more than 90 s. Let’s assume that none of the streets have been waiting for more than 90 s so the ITSSC will take the counter approach in prioritizing the four streets. In the second reordering on that same round (three streets are left deprioritized), if two streets out of the remaining three have been waiting for more than 90 s, one for 120 and the other for 110 s. The higher priority will be given to the street that has been waiting for longer time. Even if the counter value for that street is less than the other one.

The ITSSC is also specialized for dealing with emergency situations, such as accidents. First, the server would be notified in case of accident using the impact and accelerometer readings in the packets which have been sent to the server. These two readings sensor are used to make a decision on whether there is an accident or not. Both sensors have a different value which would inform the server about a certain accident. One of them is sufficient to decide if a car is experiencing an accident. This is because the impact sensor reading changes only if the car's airbags have been deployed, this means that it is guaranteed that the car is facing a violent accident. Alternatively, a change in the accelerometer reading happens only if there is a strong action which caused either a strong vibration or a high level of rotation to the car. Then, the server alerts the nearest police station, hospital and fire department to attend to the accident site. After that each emergency department sends out corresponding vehicles to deal with the emergency situation. Meanwhile, as for the street ordering calculations, it relies on whether the accident street is fully or partially closed. On one hand, if it is fully closed, meaning that no cars can bypass the accident, only the other three remaining streets traffic lights are taken into account when prioritization has to be done. This situation will stay until the accident is dealt with and the street is finally open again, thus, placing it back into the ordering process with the other three streets. The street reopening decision is taken according to a special packet being sent from the police car to the server, which is recognized in the saved server database stating that the street can be open again. However, if the street is partially open, at least one lane is open, the street stays closed until the police arrives at the accident scene, checks if there is still a possibility for cars to pass through the open lane, then sends the special packet to the server signaling it that this street can still be included in the prioritization process.

3.2 Approach and Methodology

This subsection describes how the ITSSC works in two ways algorithmic notation, in addition to the system flow chart as shown in Fig. 2. The require Database in the server contains traffic lights IP, statues, streets position.

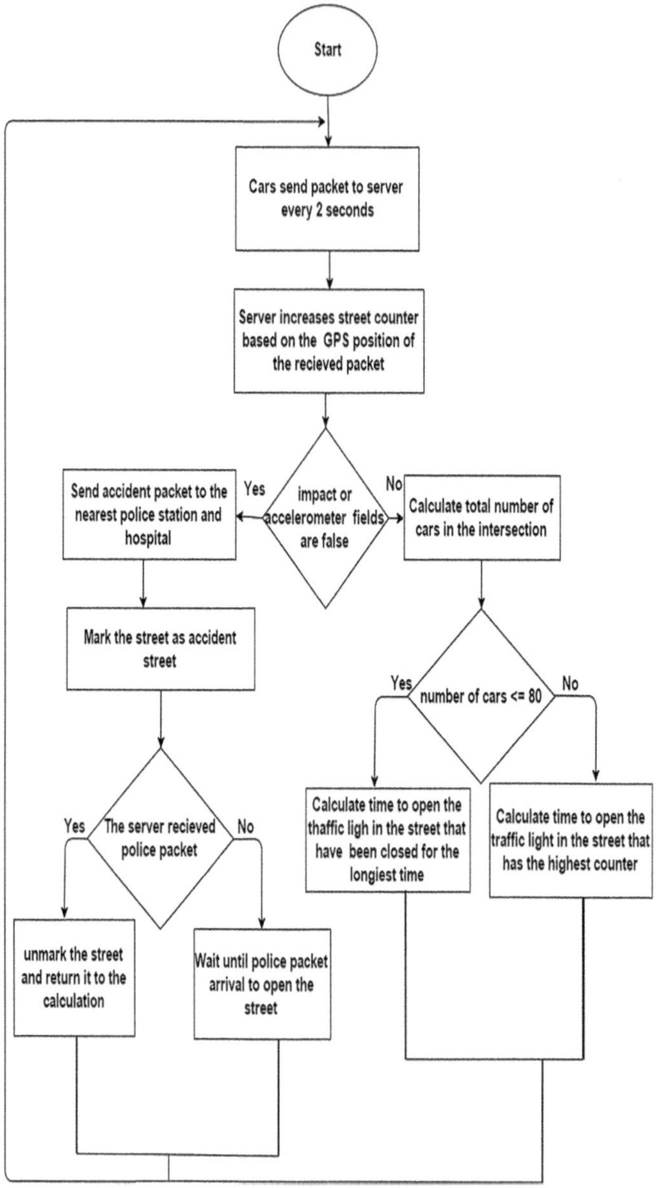

Fig. 2. ITSSC flow chart.

ITSSC Algorithm

```
1: While true
2: A car sends a packet that includes {IP address, GPS position, impact sensor and accelerometer sensor fields reading} every 2 seconds to the server
3: Server Increases Street counter according to the received GPS position
4: If impact and accelerometer fields are false then
5: Check total number of cars at the intersection
6: If the number of cars less than or equal 82 then
7: set the priority for the streets according to the highest counter and calculate the time period that a traffic light will stay ON.
8: Else if the number of cars greater than 82 and one of the traffic lights has been turned off for more than 90 seconds then
9: Turn On the traffic light that has been turned off for longest time.
10: Else
11: set the priority according to street counter.
12: End If
13: Else If impact or accelerometer fields are true then
14: Send accident packet to the nearest police station and hospital contains accident street position
15: Mark the street as accident street
16: If at least one of the lanes is open and safe for passing vehicles then
17: police send packet to the server.
18: Remove mark from the street and return it back to the calculations
19: Else
20: wait until police's packet arrives to open the street
21: End if
22: End if
23: Compare GPS coordinates and decrement street counter
24: Back to step
```

3.3 Equations

The equations are an exception to the prescribed specifications of this template. You will need to determine whether or not your equation should be typed using either the Times New Roman or the Symbol font (please no other font). To create multileveled equations, it may be necessary to treat the equation as a graphic and insert it into the text after your paper is styled.

Number equations consecutively. Equation numbers, within parentheses, are to position flush right, as in (1), using a right tab stop. To make your equations more compact, you may use the solidus (/), the exp function, or appropriate exponents.

4 Results and Analysis

The ITSSC aims at decreasing the average waiting time for vehicles on traffic lights. The average waiting time (AWT) is the time that the car spends in an intersection until it passes it, which is equal the summation of the waiting time by each car divided by the number of cars in the intersection as shown in (2). This section presents three solutions of ITSSC and shows the experiments results, analysis and discussion. Many experiments had been done in order to get more results and then choose the best or most appropriate solution for all cases (for different number of cars at the intersection).

$$AWT = \frac{\sum wating\ time\ for\ car}{number\ of\ cars\ in\ the\ intersection} \quad (2)$$

The first solution was the counter solution, which means the control of the traffic lights will be done according to the number of cars in each street; therefore, the street with the highest counter takes the highest priority to be opened first. Several experiments were done with different number of cars to compare the counter's results with the normal case (opining the traffic lights for 30 s whatever the number of cars is and all in the same order every time) and shown in Fig. 3.

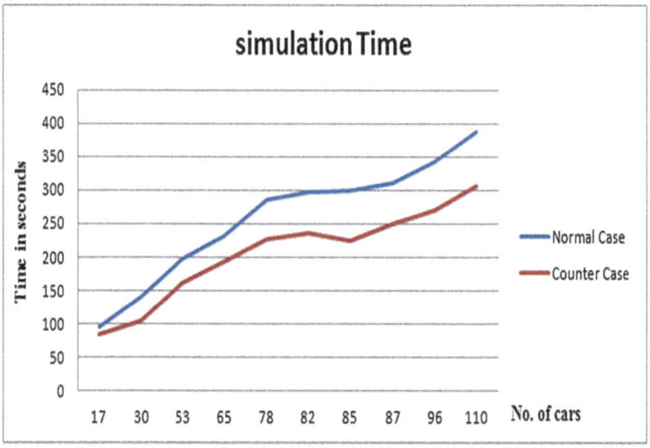

Fig. 3. Simulation time for a car in one intersection in the normal case and the counter solution case.

The second solution was the threshold solution, which means the controlling process for the traffic lights will be according to the period at which the traffic light has been turned off (red light). This solution works as follows:

- Check the OFF time for the four traffic lights.
- Compare between them.
- The traffic light with the longer OFF time opens first.

The previous procedure was done under the condition that this traffic light has passed a specified time of waiting that is equal to 90 s (threshold).

The third solution was hybrid solution that combines the first and second solutions (counter and threshold solutions). This solution works as follows:

- The server compares the vehicle's counter of all the traffic lights that crossed the threshold value of 90 s.
- If more than one street that has pass the waiting time threshold, open the street with the highest counter (busiest).

The simulations and the calculations of AWT for all the aforementioned solutions were tested on different total number of cars in one intersection from a range (17–110) cars, the obtained result are presented in Table 1 and plotted in Figs. 4 and 5.

Table 1. Average waiting time in all the three solution in the range of 17–110 cars

Total no. of cars in the intersection	Average waiting time by seconds		
	Counter solution	Threshold (90 s) solution	Threshold (90 s) & counter solution
17	**76.58**	76.58	76.58
30	**81.86**	82.00	81.86
53	**93.07**	98.03	98.03
65	**116.02**	118.00	118.00
78	**120.15**	124.05	123.32
80	**125.96**	126.43	**125.96**
82	125.02	128.06	**125.02**
85	126.21	**129.05**	126.21
87	120.02	**133.03**	120.02
96	123.63	**123.63**	123.63
110	132.89	**132.89**	132.89

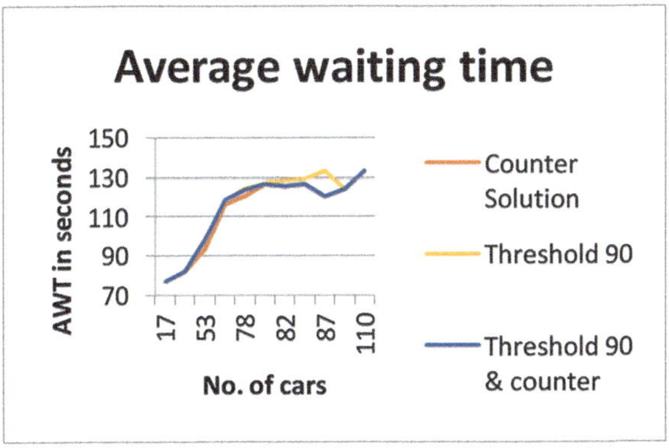

Fig. 4. A plot representation for the difference in average waiting time for the normal and algorithm case for one car.

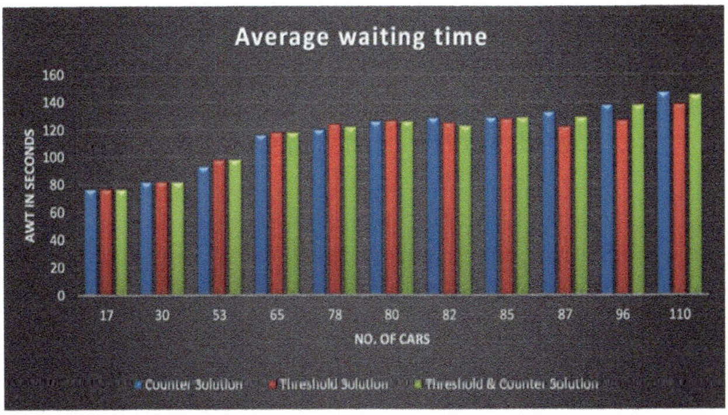

Fig. 5. Chart representation for the average waiting time in all the three solutions.

Based on the empirical results of the many experiments that had been done, the ITSSC is designed as follows:

If the total number of cars in the intersection is less than or equal 82, the ITSSC will prioritize the traffic lights according to the first solution (counter solution). Otherwise, the ITSSC will order the traffic lights according to the second solution with threshold 90 s, which means that the traffic light that remained closed for the longest period.

After the ITSSC adapt the first two solutions and ignore the third one based on the achieved results. We make a comparison in the simulation between the ITSSC and traditional traffic light. Figure 6 presents a comparison results between traditional traffic lights and the ITSSC. The figure shows that the ITSSC has better AWT compared to the traditional traffic lights. Also, it is noticeable that the difference gap increases as long as the number of the cars in the intersection is increasing. Finally, we compared the performance of ITSSC with another method that based on the video image processing. This method depends on image processing to decide the period of each traffic light to stay on. Image processing algorithm has static ordering of the intersection streets, which means street 1 opens first then street 2, street 3 and finally street 4, and it is a continuous repeated order.

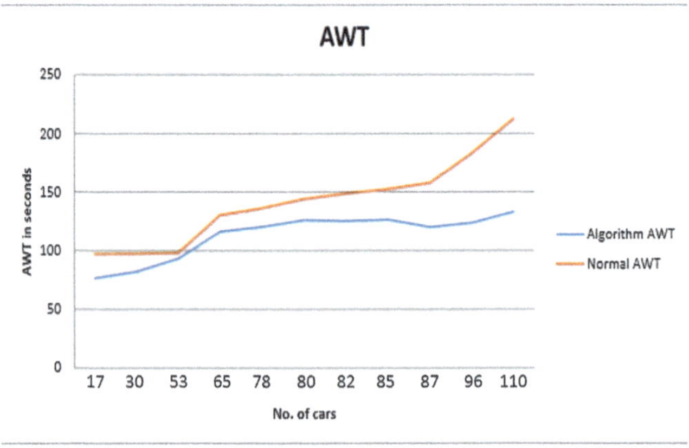

Fig. 6. A plot representative for the difference in average waiting time for the ITSSC (algorithm AWT) and the traditional traffic lights (normal AWT) for one car.

Each traffic light opens for a specific period proportional to the number of cars in the street which is calculated depending on the image captured by this image processing approach. Figure 7 shows a plot of the simulation results. The AWT of image processing algorithm for the case when the total number of cars is less than 82 is less than AWT in the ITSSC. However, the opposite is true when the total number of cars is more than 82, then the AWT in the ITSSC is better than image processing algorithm. From Fig. 7, we can see that the difference between the ITSSC and the image processing algorithm for the case when total number of cars is less than 82 is too small (maximum 5 s), whereas the difference in case the total number of cars is more than 82 is large and visible between the results, in some cases it reaches 15 s and above. Therefore, we can conclude that image-processing approach can be used just for small number of cars in the intersection, while the ITSSC is better for high number of cars.

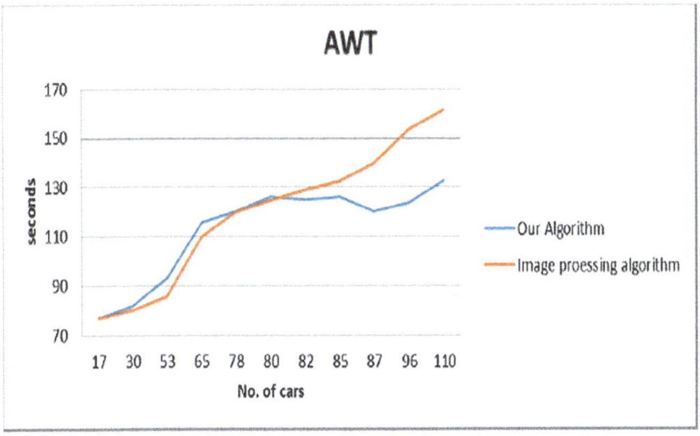

Fig. 7. A plot representation for the difference in the average waiting time for the Image processing algorithm and ITSSC (our algorithm).

5 Conclusion and Future Works

This paper introduces the intelligent transportation system in smart cities (ITSSC). Also, an intelligent accident detection system is added to the general smart transportation system, which includes the idea of making a central server get to know that an accident has already happened with the details of its' time and location. The results show that the proposed ITSSC has better performance over the traditional traffic light systems since it decreases the average waiting time at the intersections and average waiting time for each car by 28%. Which mean the ITSSC will help the drivers in reducing the time to get to their destinations and take proper trip decisions, increase road safety, save power consumption, and finally decrease the pollution. As for future work, this paper did not discuss the MAC layer side, and this issue could be addressed and the work could be expanded in the future work to resolve the problem of having collisions between packets. Also, for better accident detection, a set of sensors can be used along the roads. Accident prevention can be addressed as well and not only limiting the project to accident detection level. In Addition, the ITSSC can be expanded to involve managing the streets in a way that deals with controlling traffic lights and cross-roads simultaneously with traffic lights in the entire city in order to make it easy and quick for emergency vehicles to reach the destination to overtake the critical situation as fast as possible.

References

1. Dooley, E.: Here's how much time Americans waste in traffic. ABC News. http://abcnews.go.com/US/time-americanswastetraffic/story?id=33313765. Accessed 22 June 2017
2. Chun, B.T.: A study on intelligent traffic system related with smart city. IJSH Int. J. Smart Home **9**(7), 223–230 (2015)
3. Yang, Y.C., Cheng, C.M., Lin, P.Y., Tsao, S.L.: A real-time road traffic information system based on a peer-to-peer approach. In: Proceeding in ISCC 2008, IEEE Symposium on Computers and Communications (2008)
4. Djahel, S., Doolan, R., Muntean, G., Murphy, J.: A communications oriented perspective on traffic management systems for smart cities: challenges and innovative approaches. IEEE Commun. Surv. Tutor. **17**(1), 125–151 (2014)
5. Islam, M.M., Hassan, M.R., Chowdhury, I., Chowdhury, M.D.: Internet of car: accident sensing, indication and safety with alert system. Am. J. Eng. Res. (AJER) **2**(10), 92–99 (2013)
6. Weil, R., Wootton, J., Garcia-Ortiz, A.: Traffic incident detection: sensors and algorithms. Math. Comput. Model. **27**(9–11), 257–291 (1998)
7. Chintalacheruvu, N., Muthukumar, V.: Video based vehicle detection and its application in intelligent transportation systems. J. Transp. Technol. **2**, 305–314 (2012)
8. Misbahuddin, S., Zubairi, J.A., Saggaf, A., Basuni, J., AWadany, S., Al-Sofi, A.: IoT based dynamic road traffic management for smart cities. In: Proceeding in the 12th International Conference on High-Capacity Optical Networks and Enabling/Emerging Technologies (HONET) (2015)
9. Sandhu, S.S., Jain, N., Gaurav, A., Sriman, N.Ch., Iyengar, N.: Agent based intelligent traffic management system for smart cities. Int. J. Smart Home **9**(12), 307–316 (2015)
10. Elmangoush, A., Steinke, R., Magedanz, T., Corici, A.A., Bourreau, A., Al-Hezmi, A.: Application-derived communication protocol selection in M2M platforms for smart cities. In: Proceeding in the 18th International Conference on Intelligence in Next Generation Networks (2015)
11. Djahel, S., Jabeur, N., Barrett, R., Murphy, J.: Toward V2I communication technology-based solution for reducing road traffic congestion in smart cities. In: Proceeding in International Symposium on Networks, Computers and Communications (ISNCC) (2015)
12. Guerrero-Ibanez, J.A., Zeadally, S., Contreras-Castillo, J.: Integration challenges of intelligent transportation systems with connected vehicle, cloud computing, and internet of things technologies. IEEE Wirel. Commun. **22**(6), 122–128 (2015)

Learning to Drive With and Without Intelligent Computer Systems and Sensors to Assist

David Adrian Sanders(✉), Giles Eric Tewkesbury, Hassan Parchizadeh, Josh Robertson, Peter Osagie Omoarebun, and Manish Malik

Department of Mechanical and Design Engineering, Portsmouth University, Portsmouth, UK
david.sanders@port.ac.uk

Abstract. This paper investigates whether the use of sensors throughout training is more effective than not. Results are presented from investigating novice vehicle drivers learning while using varying amounts of sensor support. Qualitative and quantitative data evaluations are made to compare drivers with sensors against those while they learn to drive a vehicle. Dependence on the training methods used was logged as a different amount of support given by the intelligent systems. The work reflects on whether experienced drivers taught with sensors aiding them during their training, were then able to drive satisfactorily when they later did not have any assistance. Some results are included and, in all situations, assistance was more advantageous as driving tasks became more difficult.

Keywords: Learning · Driving · Vehicle · Assistive · Sensors

1 Introduction

The paper investigates how a vehicle driver adapts their driving depending on the amount of support provided by sensors. Some consequences of providing help to drivers during their learning is evaluated for both skilled and novice drivers.

Advanced driver-assistance systems can improve driving and safety. Safety features can avoid collision or accidents. Technology can be used to avoid collisions and/or alert drivers about difficulties and can take over all or some control of a vehicle. Features can include adaptive cruise control and automatic braking or can alert a driver about other cars on the road or about potential danger. They can keep a vehicle in the correct lane or consider and/or reveal anything in a blind spot. In an emergency, effective interaction between a human driver and their vehicle can save the day. This paper discusses that interfacing and collaboration [1–8].

Several driver-assistance systems features are being built in to modern cars and others can be available as optional add-on packages or as after-market solutions [9]. Driver-assistance systems rely on inputs from sensors such as radar, imaging and vision, in-car networking and Light Detection and Ranging (LiDAR).

Driver-assistance is fast growing [10, 11] with increasing adoption rates and specific standards being developed. For example, communications protocols such as Vehicle Information API [12] or IEEE 2020 for image quality.

K. Arai et al. (Eds.): IntelliSys 2018, AISC 868, pp. 1171–1181, 2019.
https://doi.org/10.1007/978-3-030-01054-6_81

In addition, vehicle-to-infrastructure and vehicle-to-vehicle systems can be added, for example Wi-Fi, mobile telephony or data networks [13].

The amount of assistance to provide can depend on several factors. This paper considers how a vehicle driver should be trained based on varying levels of advanced assistance available. Some research advocated that experienced drivers who were taught to drive a vehicle without assistance from any sensors could then drive more competently when a sensor system was provided to assist [14–16]. This research reflects on whether an experienced driver taught with sensors to assist them would then drive easily when they did not have a sensor system to help them. Any change in participant performance was scrutinised during driving tasks with different levels of assistance from sensors. This research shows that they altered their behaviour when they encountered new situations and environments.

During conventional research, drivers have usually been expert drivers using varying levels of assistance. The acquiring of driving skill has usually been ignored [17]. The research presented in this paper explores drivers in a simple vehicle incorporating sensors to assist them.

Similar studies using industrial robot arms were presented in [18, 19]. In this research, constraint-based support was used with virtual force-feedback [17]. The closeness to an obstacle (for example a wall) was represented by force feedback and that was akin to research described by Volpe [20]. Volpe represented the force felt by a slave robot arm at a remote joystick. The force back in the controller was used to turn the vehicle to evade collision.

The vehicle and system are explained in [16]. Experiments and some results are described in Sect. 2 and Sect. 3 is a discussion about the experiments and some conclusions from the results obtained.

2 Experimentation

Drivers used a joystick to control their vehicle. It was possible to provide three distinct amounts of support to a vehicle driver:

Setting 0: Sensors were off. There was no force-feedback. Vehicle drivers could drive the vehicle without interference from any sensors. Drivers had the most independence with Setting 0 but they were more likely to crash.

Setting 1: Sensors switched on and there was force feedback through the joystick when the vehicle came near to obstacles. As the distance between the vehicle and objects became smaller, the size of repelling force increased.

Setting 2: Sensors were engaged, and the vehicle automatically steered away from obstacles to prevent a collision. There was not a resistive force when the vehicle drove towards an obstacle.

Setting 3: In Setting 3, both Setting 1 and Setting 2 were delivered. Drivers were the most constrained in this case.

If a driver moved their joystick further in any direction then they could always make their vehicle move in opposition to any deflecting force.

Performance measures were the average completion time and number of collisions.

2.1 Initial Tests

Initial tests compared learning in Setting 0 (sensors switched off) and Setting 3 (sensors switched on and automatic avoidance safely steering the vehicle around obstacles). Subsequent tests compared learning at Setting 0, Setting 1 and Setting 2 when driving along different courses in different environments.

Drivers used Bobcat II vehicles [16] and drove them around four separate circuits. The routes were of varying length and complexity, containing varying numbers and types of obstacles. Drivers used a Force feedback Pro joystick to drive the vehicle through each course while trying to avoid obstacles. The joystick had been used successfully during experiments at Tohoku University [17] and so was chosen for this work.

For tests with Setting 3 selected, when a driver drove their vehicle towards an object then the resistive force increased. The help was more restrictive than when in Setting 1 or Setting 2.

Participants were 59 University students and staff without any experience of driving while using the system. The participants were placed into one of two sets (Alpha & Beta) and divided further into groups to attempt the four routes.

The routes were explained to the drivers and then each driver drive a route 10 times using sensors for some assistance and 10 times with the sensor system switched off. After that, the groups repeated the driving task but this time using a new setting. Another set of tests explored how drivers drove through a route when confronted with new conditions after they had already developed their vehicle driving skills.

A simple questionnaire was used to gather opinion and preference at the end of the testing.

Some examples of questions are:

"Did you favour Setting 0 or Setting 3?"
"Is it easier to drive with Setting 0 or Setting 3?"

Answers are recorded in Table 1.

Table 1. Answers from the questionnaire

	Strongly prefer 0	Prefer 0	Undecided	Prefer 3	Strongly prefer 3
Preference	12	14	19	10	4
How Easy?	15	18	7	12	7

2.2 Results from Initial Testing

For each attempt, the time taken to complete a route in seconds was averaged. Results for Alpha and Beta are in Figs. 1 and 2 for the four routes.

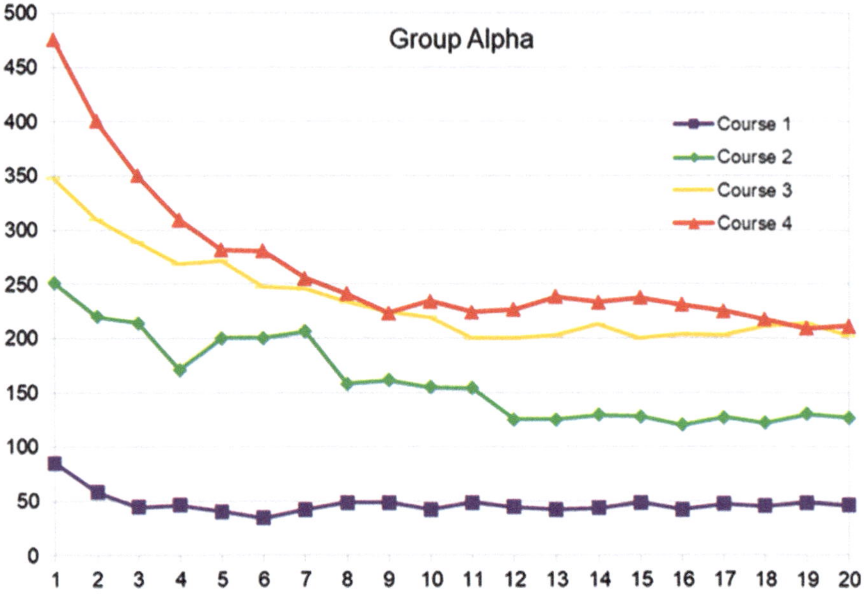

Fig. 1. Average completion time for Groups Alpha and Beta over four dissimilar courses. Setting 3 support first, and then Setting 0.

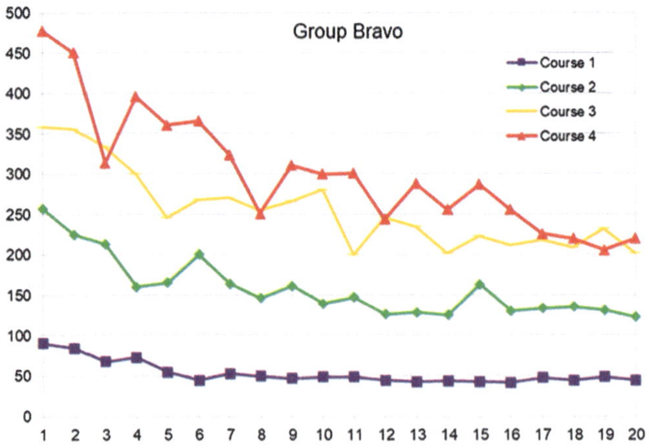

Fig. 2. Average completion time for Groups Alpha and Beta over 4x dissimilar courses. Setting 0 first and then Setting 3.

1 to 10 corresponds to the first sat of tests (Alpha with Setting 3 support and Beta with Setting 0) and 11 to 20 show the second set of tests (Beta with Setting 3 support and Alpha with Setting 0).

It can be seen in Figs. 1 and 2 that participants in Alpha (force-feedback applied) learnt in a smoother manner and more quickly than Group Beta. This implies a supportive effect when using Setting 3 at first and while early learning is taking place and skills are being developed. In addition, the drivers in Group Alpha performed better in the second part of the tests, with assistance set at Setting 0. The acquisition of new skills was faster when using Setting 3 (a combination of both Setting 1 and Setting 2).

Drivers may have reached a general skill level and they were able to apply that skill indifferent circumstances and situations. The new condition here was to drive the vehicle without any assistance. Drivers in Beta did not appear to improve any further during the second half of testing. The time taken and the stability of the results were similar. That could be because the new skills that were gained using Setting 0 assistance does not easily transfer to new conditions and situations.

Table 1 is some of the subjective opinion obtained from questionnaires. Drivers said it was easier driving without assistance (Setting 0). That was different to the objective results obtained form testing. Questionnaires implied that drivers didn't enjoy being helped.

2.3 Discussion

Initial testing suggested that the vehicle drivers who were taught with the sensor system trying to help them were able to handle different circumstances or conditions, for example if the drivers were not being helped.

Being supported by a sensor system while learning appears to influence future ability when driving without the sensors providing any help.

Drivers might have acquired knowledge about the systems and the routes during tests.

Additional research is needed to validate the testing and to check whether driving skills do transfer to different working conditions.

The testing results match with those in Chikura [17] but participant performance in the face of new tasks needs to be scrutinised more in order to check the results.

The data collected suggests that vehicle drivers taught without being assisted by sensors did not then perform better than those who were helped by sensors.

In addition, the results suggested that drivers who learnt without help did not learn progressively compared to drivers lacking any experience. Skills obtained without help may have an adverse effect on driving if assistance is then added later.

That result is significant for and should be considered when establishing training processes for intelligent systems.

Off-the-record conversation suggested that this was partially because drivers thought their freedom to make decisions and to make move as they wished was hampered.

That contradicted the more objective results obtained from the recording the average completion time for a driving task. That suggested that self-biased observations and criticisms should not be used to make training decisions. Considering the preference of the students appears to lead to inferior application.

2.4 Second Tests

These tests explored how drivers changed their way of working if circumstances changed. Drivers completed tasks with various sensor settings to provide varying quantities of help.

Longer and more complicated routes were used for these tests. Three test runs were created.

These tests considered changed in behaviour when varying amounts of help were provided and circumstances were changed.

Drivers drove the routes six times. Then the routes were changed, and drivers drove the new route six times. Then the routes were changed again, and drivers drove the new route another six times. Every driver kept the same settings for the sensor system for each route.

A set of twelve new volunteer drivers were placed in three new groups:

Group Phi Φ,
Group Chi, X
Group Psi Ψ.

Φ used Setting 0 (without help), X used Setting 1 (repulsion force) and Ψ used Setting 2 (automatic steering to avoid an obstacle).

The time taken was recorded along with the number of collisions.

2.5 Results

Figures 3, 4 and 5 show the amount of time it took to complete a route in seconds for Groups Φ, X and Ψ.

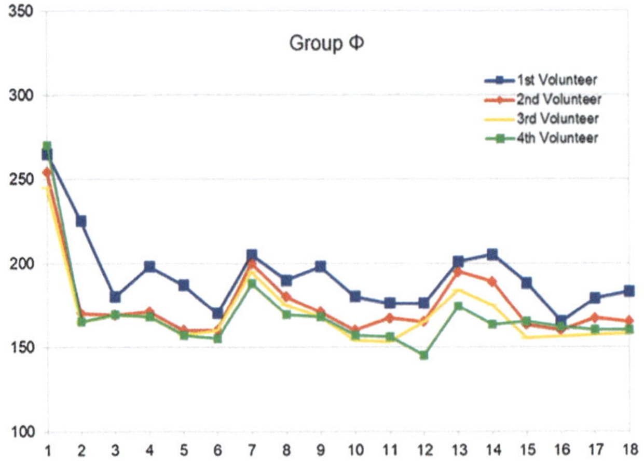

Fig. 3. Setting 0 – No driver assistance.

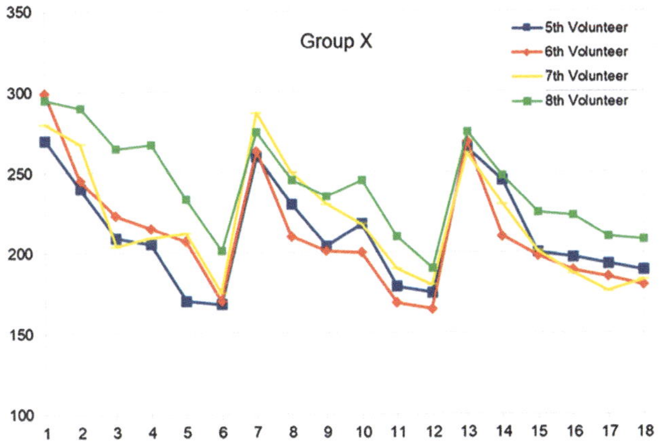

Fig. 4. Setting 1 – Ultrasonic sensors providing a repulsive force.

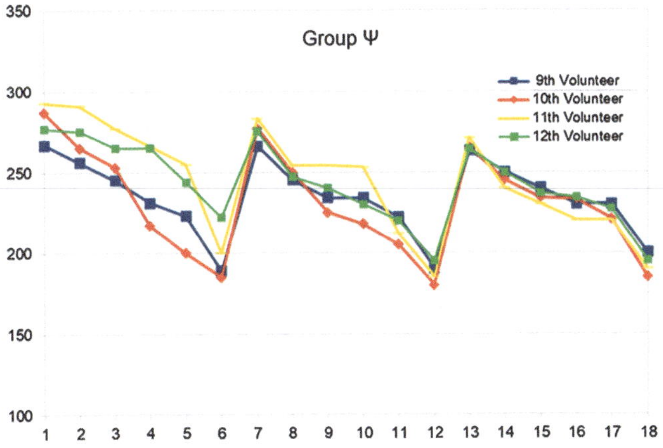

Fig. 5. Setting 2 - System automatically steering away from obstacles.

Drivers using setting 3 had the fewest collisions but they took longer to complete the routes compared with drivers using setting 0.

Whenever the course layout was changed (position of obstacles, etc.) then the differences in completion time was noted. Figure 3 is smoother than Fig. 4 or Fig. 5 because completion time increased more for Setting 1 and Setting 2, but did not change much for Setting 0.

This suggested that skills learned without any driver assistance might be more useful than skills learned with driver assistance provided.

Figure 6 shows the number of collisions for each driver. There were variances in driver conduct and adjustment as assistance levels were changed and as conditions changed.

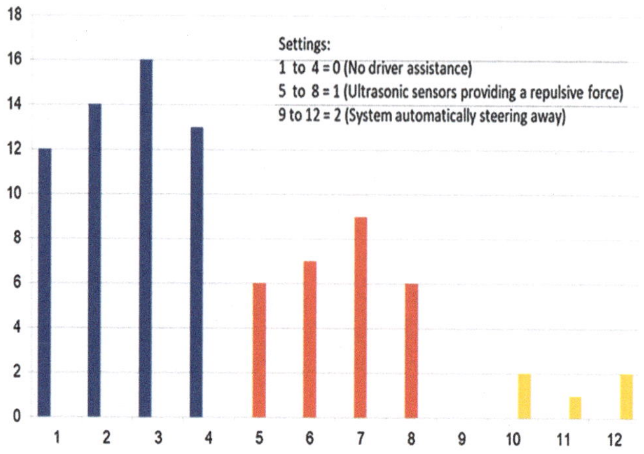

Fig. 6. Number of collisions for each vehicle driver.

Average completion time changed for each setting whenever a course was altered.

3 Conclusions

If drivers learnt to drive with help from a sensor system, then it had a positive effect later. If drivers learnt to drive without any help from a sensor system, then it could have a harmful effect later.

Using Setting 2 reduced collisions.

The second set of tests challenged results from the initial testing. Experiments need to be repeated with more volunteers to results more significant results.

Initial testing implied support was useful while learning. That result was not significant during later testing. Future testing needs to concentrate on individual drivers in an effort to obtain less ambiguous test results. Despite that, the results offer an insight into driver behaviour during learning when it is possible to provide different levels of sensor support. It may also suggest that drivers who learnt without sensors might find it harder to navigate when presented with extra information from sensors

The results suggest that sensors should be used while teaching.

Once proficient then drivers perform better without any help from sensors, especially if they are driving through simple and safe environments.

For every driver in every situation, assistance became more beneficial as the routes became more difficult and complicated.

Future work is moving to examine mixing AI tools [21, 42] to make use of particular AI tools when they are most useful.

References

1. Sanders, D.A., Ndzi, D., Chester, S., Malik, M.: Adjustment of tele-operator learning when provided with different levels of sensor support while driving mobile robots. In: Bi, Y., Kapoor, S., Bhatia, R. (eds.) SAI Intelligent Systems (IntelliSys), vol. 2. Lecture Notes in Networks and Systems, vol. 16, pp. 548–558. Springer, UK (2017). https://doi.org/10.1007/978-3-319-56991-8_41
2. Sanders, D.A., Sanders, H., Gegov, A., Ndzi, D.: Rule-based system to assist a tele-operator with driving a mobile robot. In Bi, Y., Kapoor, S., Bhatia, R. (eds.) SAI Intelligent Systems (IntelliSys), vol. 2. Lecture Notes in Networks and Systems, vol. 16, pp. 599–615. Springer (2017). https://doi.org/10.1007/978-3-319-56991-8_44
3. Sanders, D.A.: Using self-reliance factors to decide how to share control between human powered wheelchair drivers and ultrasonic sensors. IEEE Trans. Neural Syst. Rehabil. Eng. **25**(8), 1221–1229 (2017). https://doi.org/10.1109/TNSRE.2016.2620988
4. Sanders, D.A.: Non-model-based control of a wheeled vehicle pulling two trailers to provide early powered mobility and driving experiences. IEEE Trans. Neural Syst. Rehabil. Eng. **26**(1), 96–104 (2018). https://doi.org/10.1109/TNSRE.2017.2726443
5. Sanders, D.A., Gegov, A., Tewkesbury, G., Khusainov, R.: Rule-based system to assist a powered wheelchair driver. In: IEEE Proceedings of the Intelligent Systems Conference (IntelliSys), United Kingdom, 7–8 September 2017, pp. 558–565. IEEE, London (2017). ISBN 978-1-5090-6435-9 ()
6. Sanders, D.A., Sanders, B., Gegov, A., Ndzi, D.: Results from investigating powered wheelchair users learning to drive with varying levels of sensor support. In: Proceedings of the Intelligent Systems Conference (IntelliSys), 7–8 September 2017, pp. 241–245. IEEE, London (2017). ISBN 978-1-5090-6435-9
7. Sanders, D.A., Sanders, B., Ndzi, D., Bausch, N.: Using confidence factors to share control between a mobile robot tele-operator and ultrasonic sensors. In: Proceedings of SAI Intelligent Systems Conference (IntelliSys), 7–8 September 2017, pp. 1026–1033. IEEE, London (2017). ISBN 978-1-5090-6435-9
8. Sanders, D.A., Gegov, A., Ndzi, D.: Knowledge-based expert system using a set of rules to assist a tele-operated mobile robot. In: Bi, Y., Kapoor, S., Bhatia, R. (eds.) Intelligent Systems and Applications. Studies in Computational Intelligence, vol. 751, pp. 371–392. Springer, Berlin (2017)
9. US: Mobileye intros smartphone connected driver assistance (ADAS) technology. Telematics News 2012. http://telematicsnews.info/2012/01/12/us-mobileye-intros-smartphone-connected-driver-assistance-adas-technology_j3122. Accessed 13 Nov 2017
10. Riches, I.: Strategy Analytics: Automotive Ethernet: Market Growth Outlook I Keynote Speech 2014 IEEE SA: Ethernet & IP @ Automotive Technology Day (PDF). IEEE. http://standards.ieee.org/events/automotive/2014/00_Automotive_Ethernet_Market_Growth_Outlook.pdf. Accessed 13 Nov 2017
11. Rees, K., Park, J.S.: Vehicle Information Access API. W3C Business Group report, W3C and LG Electronics (2014). https://www.w3.org/2014/automotive/vehicle_spec.html. Accessed 13 Nov 2017
12. ADAS Definition. Autoconnectedcar.com. Archived from the original on 06/10/2012. https://web.archive.org/web/20120610055853/http://telematicsnews.info/2012/01/12/us-mobileye-intros-smartphone-connected-driver-assistance-adas-technology_j3122. Accessed 13 Nov 2017

13. Banga, B.: Global ADAS and Autonomous Driving Components Market, Analysis & Forecast. Business Intelligence and Strategy Research (2017). https://bisresearch.com/industry-report/global-adas-autonomous-driving-components-market-2026.html. Accessed 13 Nov 2017
14. Sanders, D.A., Graham-Jones, J., Gegov, A.: Improving ability of tele-drivers to complete progressively more difficult mobile robot paths using simple expert systems and ultrasonic sensors. Ind. Robot. **37**(5), 431–440 (2010)
15. Sanders, D.A., Bausch, N.C., Ndzi, D.L.: Changes to user learning behaviour of vehicle drivers depending on the level of sensor support. In: Arabnia, H.R., Deligiannidis, L. (eds.) IEEE Proceedings of International Conference Health Informatics and Medical Systems, HIMS 2015, pp. 62–65. CSREA Press (2015)
16. Sanders, D.A., Stott, I., Robinson, D., Ndzi, D.: Analysis of successes and failures with a tele-operated mobile robot in various modes of operation. Robotica **30**, 973–988 (2012)
17. Chikura, D., Takahashi, M., Watanabe, S., Kitamura, M.: Adaptation of user behavior to the different level of tele-operation support. In: IEEE International Conference on Systems, Man, and Cybernetics, vol. 3, pp: 739–744 (1999)
18. Backes, P.G.: Supervised autonomy for space robotics. Prog. Astronaut. Aeronaut. **161**, 139–158 (1994)
19. Draper, J.V., et.al.: Measuring operator skill and teleoperator performance. In: Proceedings of International Symposium on Teleoperation and Control (1998)
20. Volpe, R.: Techniques for collision prevention, impact stability, and force control by space robots. Prog. Astronaut. Aeronaut. **161**, 175–212 (1994)
21. Gegov, A., Arabikhan, F., Sanders, D., Vatchova, B., Vasileva, T.: Fuzzy networks with feedback rule bases for complex systems modelling. Int. J. Knowl. Based Intell. Eng. Syst. **21**(4), 211–225 (2017)
22. Gegov, A., Petrov, N., Sanders, D., Vatchova, B.: Boolean matrix equations for node identification in fuzzy rule based networks. Int. J. Knowl. Based Intell. Eng. Syst. **21**(2), 69–83 (2017)
23. Gegov, A., Petrov, N., Sanders, D., Vatchova, B.: Modular rule base fuzzy networks for linguistic composition based modelling. Int. J. Knowl. Based Intell. Eng. Syst. **21**(2), 53–67 (2017)
24. Gegov, A., Sanders, D., Vatchova, B.: Aggregation of inconsistent rules for fuzzy rule base simplification. Int. J. Knowl. Based Intell. Eng. Syst. **21**(3), 135–145 (2017). https://doi.org/10.3233/KES-170358
25. Mohd-Safar, N.Z., Ndzi, D., Sanders, D., Noor, H.M., Kamarudin, L.M.: Integration of fuzzy c-means and artificial neural network for short-term localized rainfall forecasting in tropical climate. In: Bi, Y., Kapoor, S., Bhatia, R. (eds.) Studies in Computational Intelligence, vol. 751, pp. 325–348. Springer (2018)
26. Sanders, D.: New method to design large scale high-recirculation airlift reactors. Proc. Inst. Civ. Eng. J. Environ. Eng. Sci. **12**(3), 62–78 (2017)
27. Sanders, D.A., Bergasa-Suso, J., Khusainov, R., Gegov, A.E., Chester, S.D., Bausch, N.C.: Introducing dead bands within two-dimensional clusters of user data to improve data classification. In: Proceedings of the 9th International Conference on Human Systems Interaction (HSI), UK, 06 July 16, pp. 14–20. IEEE (2016)
28. Sanders, D.A., Langner, M., Gegov, A.E., Ndzi, D.L., Sanders, H., Tewkesbury, G.E.: Tele-operator performance and their perception of system time lags when completing mobile robot tasks. In: Proceedings of the 9th International Conference on Human Systems Interaction (HSI), UK, pp. 236–242. IEEE (2016)

29. Gegov, A.E., Sanders, D.A., Vatchova, B.: Mamdani fuzzy networks with feedforward rule bases for complex systems modelling. J. Intell. Fuzzy Syst. **30**(5), 2623–2637 (2016)
30. Robinson, D.C., Sanders, D.A., Mazharsolook, E.: Ambient intelligence for optimal manufacturing and energy efficiency. Assembly Autom. **35**(3), 234–248 (2016)
31. Gegov, A., Sanders, D., Vatchova, B.: Complexity management methodology for fuzzy systems with feedforward rule bases. Int. J. Knowl. Based Intell. Eng. Syst. **19**(310), 83–95 (2016)
32. Sanders, D.A., Bausch, N.C.: Improving steering of a powered wheelchair using an expert system to interpret hand tremor. In: Liu, H., Kubota, N., Zhu, X., Dillmann, R., Zhou, D. (eds.) Intelligent Robotics and Applications: Part II. Lecture Notes in Artificial Intelligence, vol. 9245, pp. 460–471. Springer, Cham (2015)
33. Robinson, D.C., Sanders, D.A., Mazharsolook, E.: Ambient intelligence for optimal manufacturing and energy efficiency. Assembly Autom. **35**(3), 234–248 (2015). https://doi.org/10.1108/AA-11-2014-087
34. Gegov, A., Sanders, D., Vatchova, B.: Complexity management methodology for fuzzy systems with feedforward rule bases. Int. J. Knowl. Based Intell. Eng. Syst. **19**(310), 83–95 (2015). https://doi.org/10.3233/kes-150310
35. Sanders, D.A., Tewkesbury, G.E., Gegov, A.E.: Fast transformations to provide simple geometric models of moving objects. In: Liu, H., Kubota, N., Zhu, X., Dillmann, R., Zhou, D. (eds.) Intelligent Robotics and Applications Part I. Lecture Notes in Computer Science, vol. 9244, pp. 604–617. Springer (2015). https://doi.org/10.1007/978-3-319-22879-2_55
36. Gegov, A., Arabikhan, F., Sanders, D.: Rule base simplification in fuzzy systems by aggregation of inconsistent rules. J. Intell. Fuzzy Syst. **28**(3), IFS1418 (2015). https://doi.org/10.3233/ifs-141418
37. Gegov, A., Sanders, D., Vatchova, B.: Complexity management methodology for fuzzy systems with feedback rule bases. J. Intell. Fuzzy Syst. **26**(1), 451–464 (2014). https://doi.org/10.3233/IFS-130857
38. Sanders, D.A., Gegov, A., Haddad, M., Ikwan, F., Wiltshire, D., Tan, Y.C.: A rule-based expert system to decide on direction and speed of a powered wheelchair. In: IEEE Proceedings of the SAI Conference on Intelligent Systems (2018, in press)
39. Sanders, D., Gegov, A., Tewkesbury, G.E., Khusainov, R.: Sharing driving between a vehicle driver and a sensor system using trust-factors to set control gains. In: IEEE Proceedings of the SAI Conference on Intelligent Systems (2018, in press)
40. Haddad, M., Sanders, D.A., Bausch, N., Tewkesbury, G., Gegov, A., Hassan, M.: Learning to make intelligent decisions using PROMITHEE II and the analytical hierarchy process. In: IEEE Proceedings of the SAI Conference on Intelligent Systems (2018, in press)
41. Sanders, D., Robinson, D.C., Hassan, M., Haddad, M., Gegov, A., Ahmed, N.: Making decisions about saving energy in compressed air systems using ambient intelligence and AI. In: IEEE Proceedings of the SAI Conference on Intelligent Systems (2018, in press)
42. Sanders, D., Wang, Q., Bausch, N., Huang, Y., Khaustov, S., Popov, I.: An efficient method to produce minimal real time geometric representations of moving obstacles. In: IEEE Proceedings of the SAI Conference on Intelligent Systems (2018, in press)

Sharing Driving Between a Vehicle Driver and a Sensor System Using Trust-Factors to Set Control Gains

David A. Sanders[1(✉)], Alexander Gegov[2], Giles Eric Tewkesbury[1], and Rinat Khusainov[3]

[1] Department of Mechanical and Design Engineering, Portsmouth University, Portsmouth, UK
{david.sanders,giles.tewkesbury}@port.ac.uk

[2] School of Computing, Portsmouth University, Portsmouth, UK
alexander.gegov@port.ac.uk

[3] Department of Electronics and Energy, Portsmouth University, Portsmouth, UK
rinat.khusainov@port.ac.uk

Abstract. This paper presents a method to share the control of a vehicle between a driver and sensors. The vehicle can be driven by a driver, or the sensors can drive the vehicle, or control can be shared between them. In some circumstances, sharing control can allow a human driver to maneuver more safely and efficiently. The gain settings in the controller can be set automatically to be specific to a particular human driver at a particular time. A trust-factor is calculated in real time for the vehicle driver and sensors help the human driver to drive their vehicle. That can correct for any detected weaknesses and inadequacies. A driver might not see a vehicle ahead or a vehicle driver might be drowsy. In an emergency, proficient collaboration between a vehicle and vehicle driver can be vital. This paper examines that interfacing and collaboration. The proposed methods are validated with initial testing.

Keywords: Vehicle · Shared-control · Driver · Trust-factor

1 Introduction

Advanced driver-assistance systems automate, enhance and adapt vehicle systems to improve driving and safety. Safety features can avoid collision or accidents. Technology can be used to avoid collisions and/or alert drivers about difficulties and can take over all or some control of a vehicle. Features can include adaptive cruise control and automatic braking or can alert a driver about other cars on the road or about potential danger. They can keep a vehicle in the correct lane or consider and/or reveal objects in a blind spot. In an emergency, proficient collaboration between a vehicle and vehicle driver can be vital. This paper examines that interfacing and collaboration [1–8].

Some driver-assistance systems features are being built in to some modern cars and others can be available as optional add-on packages or as after-market solutions [9].

K. Arai et al. (Eds.): IntelliSys 2018, AISC 868, pp. 1182–1195, 2019.
https://doi.org/10.1007/978-3-030-01054-6_82

Driver-assistance systems rely on inputs from sensors such as radar, imaging and vision, in-car networking and Light Detection and Ranging (LiDAR).

Driver-assistance is fast growing [10, 11] with increasing adoption rates and specific standards being developed. For example, IEEE 2020 image quality and communications protocols for example the vehicle information Application Programming Interface (API) [12]. In addition, vehicle-to-infrastructure and vehicle-to-vehicle systems can be added; for example, Wi-Fi, mobile telephony or data networks [13].

Many systems have been described to help human drivers work within remote and hazardous environments [14–22]. Some papers have investigated how drivers might cooperate with their vehicles [23–25]. In the papers in the literature though, the controllers have often traded stability margins for clearness [26–28]. Vehicle drivers can find driving more difficult when the environment is less structured [29–35].

A wheeled vehicle has been selected in the research described within this paper because that is still the most common sort of vehicle [36–42]. A simple input mechanism to test vehicle control is a joystick. The vehicle draws high current to drive the vehicle electric motors and the joystick draws a low current and provides an input to the controller. A human driver is generally a superior driver compared to a computer driver and systems described in this paper endeavour to assist the human vehicle driver.

Driving is often an open-loop activity. A vehicle driver specifies desired speeds and directions and their vehicle tries to travel on the desired bearing at the desired speed. The path of the vehicle can be disturbed because of the dissimilarities of the driving wheels or the trailing casters or because of varying responses to different surfaces and gradients. Drivers must respond to correct a vehicle path in the face of any disturbances.

Situations can sometimes be unpredictable [43] and that can disturb a vehicle driver and vehicle operation [44]. Cooperation between a human vehicle driver and sensors on the vehicle [45] can assist a driver [46].

Kuniaki [47] described collaborative systems and Macharet [48] described the use of ultrasonic sensors to find a bearing to a target using triangulation. That was extended in [49, 50].

Procedures described in this paper facilitate an intimate cooperation between a vehicle driver and a set of vehicle sensors by integrating the commands from a human driver together with the data coming from the vehicle sensors. A blended architecture for a controller is described within [51] that facilitates co-operative behaviour between a vehicle driver and a vehicle. In [52], a cooperative controller is described to improve performance and reduce work-load. Autonomous systems might help a vehicle driver to reduce their work-load while improving safety [53, 54]. Combined control was described by Satti in [55]. That system employed a brain-computer interface. In [56], a control system is described that controls a quadcopter to evade crashes and that can be used to fly copters in various formations.

Human drivers and sensor systems might deliver a command at the same time and then it is possible to fuse the two signals using a stipulated ratio [57]. The way that the authority is disseminated can be important. Severaldifferent interfaces cold be used [20, 32, 37, 58–61]. Carlson *et al.* presented a method to forecast direction and to adjust controller signals in order to cause the vehicle to travel in particular directions [62]. Experiments suggested parameters and controller distribution was adjusted in real-time.

In this paper, some new methods are presented that can help a vehicle driver to drive through complicated or changing surroundings. Input from the human driver is combined with suggestions from the sensor system so that drivers and sensors help each other to drive the vehicle safely. Section 2 describes the vehicle and the sensors that were used to test ideas. The new controller is explained within Sects. 3 and 4 introduces the shared control. Some results are in Sects. 5 and 6 is a summary and discusses where the research work is likely to go in the future.

2 Vehicle and Sensors

A Bobcat II Vehicle was used during the research [22, 26]. There were inputs from both sensors and a human driver that fed into a shared controller. The vehicle consisted of a heavy base with some ultrasonic sensors mounted on it that could be employed to help the driver to steer clear of any obstacles. The sensors delivered the range to any obstacle in front of the vehicle. Sensor data were managed by a micro-computer that regulated direction and speed.

2.1 Vehicle

There were four wheels on the vehicle; the two wheels at the front were larger driving wheels and the two wheels at the rear were casters. The driving wheels were attached to motors and were independently driven so that the vehicle could be steered by adjusting current being sent to the motors connected to the front wheels.

If ϕ is direction, V is the linear velocity of a vehicle, and ω is angular velocity, then the effective velocity at the center of mass for the vehicle is

$$V_c = (V, \phi, \omega) \tag{1}$$

Driving the wheels independently provides movement and orientation as the driving wheels are on the same axis. A kinematic model of the vehicle is presented in [3].

2.2 Sensor System

Any obstacles in front of the vehicle were detected using ultrasonics. A 3 m s pulse was needed to provide a high output. Longer pulses hold extra energy and therefore should sense objects that are further away. If speed of sound is 330 m/s, then a 3 ms sound pulse would have a length of 0.99 m. If a pulse exits from the transmitter and then rebounds off an object to reappear at the receiver, then 0.5 m is suggested as a minimum range if using a 3 ms pulse of ultrasonic energy. This research sometimes required closer ranges than that so shorter pulses were sometimes used.

As the vehicle moved around, objects appeared and disappeared and sometimes it could be tricky to identify targets. In addition, the ultrasonic signals tended to be noisy and mis reads were sometimes returned. These misreads were filtered to increase consistency.

The volumes of space in front of the vehicle were divided into a grid with three sectors and then stored within an array as: near, middle and far. The transducers were immovable and overlapped to cover the whole of the area in front of the vehicle. Elements within the array were increased by 5 if there was an object within them. If an element did not have an object within it then was reduced by one. Figure 1 shows the pattern created by two beams coming from two sensors.

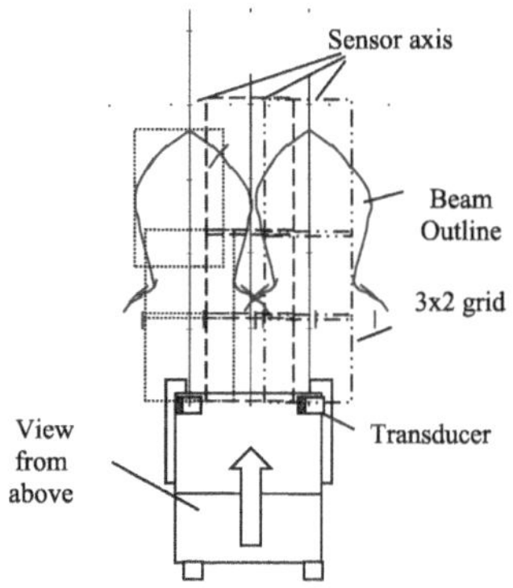

Fig. 1. An array created from two overlapping beams from two sensors.

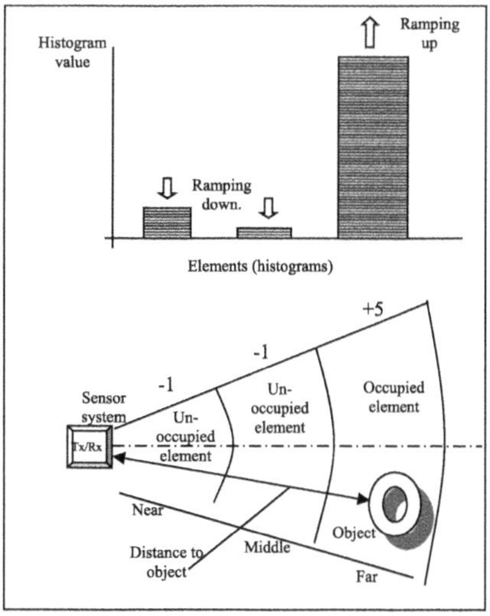

Fig. 2. A histogrammic portrayal of the local environment ahead of a vehicle using three elements. An object is making an element increase in value.

Values within the elements of the array ranged from zero to fifteen. Figure 2 shows a histogrammic portrayal of the local environment ahead of a vehicle using three elements. An object within the third element is causing the value of that element to increase.

A grid element value is increasing quickly up to its maximum value because an object has been detected within that element. Mis reads in other elements might increase individual values for a very short period of time, but they were then quickly reduce in value. If the object moved into a different element, then the new element quickly increases, and the previous element quickly reduces. Ranges were picked up in half a second.

3 Controller

The controller followed commands provided by a human driver and sensors helped the driver to automatically avoid objects in the path of their vehicle.

3.1 Control

Linear and angular vehicle velocity are studied. Vehicles tracked a velocity and direction, as shown in Fig. 3.

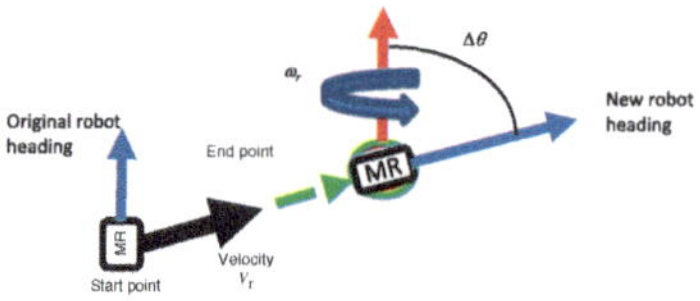

Fig. 3. Controller design.

The control law to track the linear velocity of the vehicle, V_r was:

$$V_r = V_M x D/D_{Des} \quad \text{when } |D| < D_{Des} \tag{7}$$

and

$$V_r = V_M x D/D_{DesSp} \quad \text{when } |D| > D_{DesSp} \tag{8}$$

Where, D was a vector obtained from a vehicle joystick, V_M was maximum vehicle speed, and D_{Des} was desired speed. If the vector provided by the vehicle joystick was larger than the range of the sensors, then the vehicle will move at the specified speed.

The heading of the vehicle changes as it moves around. A control law is defined to track the heading of the vehicle. The controller rule for vehicle angular velocity, ω_r, is:

$$\omega_r = \omega_M x \Delta\theta/\theta_D, \quad \text{when } |\Delta\theta| < \theta_D \tag{9}$$

and

$$\omega_r = \omega_M \text{x}\, \Delta\theta/|\Delta\theta|, \quad \text{when } |\Delta\theta| > \theta_D \tag{10}$$

Where, ω_M is maximum value of angular velocity, θ_D is preferred direction, and vehicle heading error is $\Delta\theta$.

If the error for the vehicle direction is larger than a pre-set angle, then the vehicle will turn. If the error for the vehicle direction is less than a pre-set angle, then angular velocity is adjusted to track a chosen direction.

3.2 Avoiding Objects

It was easier to avoid objects because of the omnidirectional of the vehicle. A vector signified direction and speed. If a vehicle drove close to an object, then a repulsive force was produced and the vehicle was steered away from the obstacle (Fig. 4).

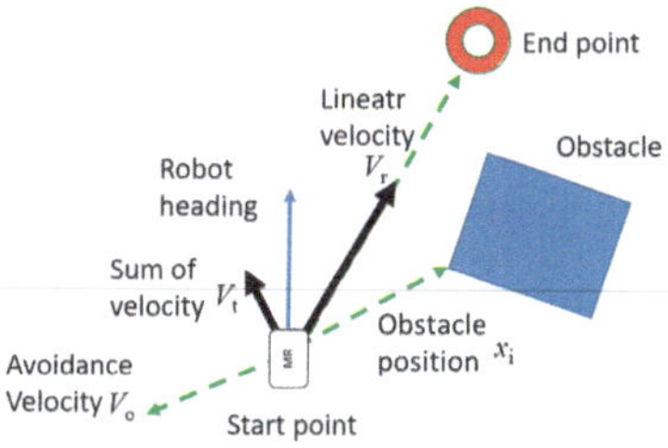

Fig. 4. If a vehicle drove close to an object, then a repulsive force was produced and the vehicle was steered away from the obstacle.

Velocity to avoid an object, $\mathbf{v_o}$ was calculated using:

$$\mathbf{V_o} = a \sum_i [(D_{Ms} - |\mathbf{x}_i|)/ D_s)(-\mathbf{x}_i / |\mathbf{x}_i|] \tag{11}$$

Where, $\boldsymbol{x}_i$ are vectors representing obstacles in front of the vehicle, D_s is a safe distance, and a was a constant. Ultrasonics detected positions for obstacles, x_i.

Velocity resulting from the algorithm, $\boldsymbol{Vt}$, is:

$$\boldsymbol{V}_t = \boldsymbol{V}_r + \boldsymbol{V}_o \tag{12}$$

Where, $\boldsymbol{V}_r$ is the linear velocity obtained from seeking for the destination, $\boldsymbol{V}_o$ is a velocity to avoid obstacles generated by the system, and, $\boldsymbol{V}_t$ is the resulting actual velocity. $\boldsymbol{V}_t$ would avoid objects but it didn't change the vehicle heading much.

4 Sharing Control

The idea of shared-control is to combine commands form the sensors with vehicle driver commands in an attempt to improve vehicle driving.

Drivers could usually steer a vehicle safely, but sensors can be more repeatable and accurate. Systems could provide the human driver with some autonomy but use their skill when it might be useful or to intervene to help avoid objects.

If the vehicle was driven through complicated situations then the new system tended to make better decisions. The architecture of the proposed controller is shown in Fig. 5.

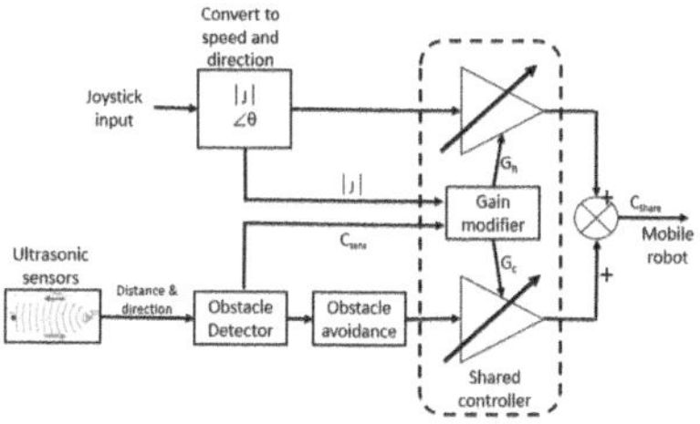

Fig. 5. Shared and combined control.

The creation of the new controller extended work in [53] and safely maneuvered the vehicles, combining inputs form a joystick and sharing with inputs form the sensors. Human drivers steered their vehicles using joysticks and the vehicle sensor system safely avoided objects.

The value of the electric current fed into the vehicle motors was calculated from the inputs provided by both the sensors and a human driver. If objects are far away, then the human vehicle driver would not require any help. In situations where there were lots of objects or there were objects close to the vehicle, then the system tended to reduce or inhibit the input from their joystick to avoid collision.

Shared gains from the human vehicle driver and sensor system altered as the vehicle was driven around. The resulting command, $\boldsymbol{C}_{\text{share}}$ (Fig. 5) is:

$$\boldsymbol{C}_{\text{share}} = G_h|\mathrm{J}| + G_c\boldsymbol{u}_h, \boldsymbol{C}_{\text{sens}} \in [0, 1] \tag{13}$$

Where, $|\mathrm{J}|$ is the joystick input and $\boldsymbol{C}_{\text{sens}}$ is range to objects. $\boldsymbol{C}_{\text{share}}$, could then be added to the weighted input from the joystick used by the human driver, multiplied by weighted gains G_h, and weighted outputs from the control system. G_w is that weighted gain. The new concept of trust-factors was used to set controller gains as in (13).

4.1 Confidence of Avoidance

The Trust-Factor for the driver was produced from three other factors. An Avoidance Factor represented driver ability. If the vehicle was moving closer to objects as it was driven around, then drivers had a smaller Trust-Factor allocated to them. The Avoidance-Factor, E_{av}, is

$$E_{av} = |x_0|/D_{Sa} \tag{14}$$

Where xo is the shortest distance between the vehicle and an object and, D_{Sa} is a constant that represented a safe range. If the distance to the vehicle was longer than D_{Sa}, then a driver would be given a higher driver confidence rating. If the vehicle was closer than D_{Sa} then confidence reduces.

4.2 Confidence in Safety

A Safety Factor represented the ability of a driver to drive their vehicle safely. If the human driver was driving at a higher speed and not colliding, then the driver was presumed to be confident. Trust-Factor for safety E_{safe} is:

$$E_{\text{safe}} = \{1-(|V_{\text{h}}|/|V_{\text{TH}}|)\}^{q}, \text{for}\, q<1. \tag{15}$$

Where, V_h is a the linear velocity being demanded by a driver and V_{TH} is the quickest that a driver was allowed to move. Assigning a bigger Trust-Factor at lower speeds was achieved using q.

4.3 Confidence of Assistance

Time spent driving and driver tiredness are significant. A human vehicle driver gets tired if continuously driving a vehicle and then the Trust-Factor gets smaller. Drivers are more likely to be awake and more alert at the beginning of a day.

Joysticks were monitored.

Engagement time, E_{te}, was

$$E_{te} = E_{te-1} + (1/T_{\text{a}}), \textit{if driver is rested} \tag{16}$$

$$E_{te} = E_{te-1} - (1/T_{\text{a}}), \textit{if driver is tired} \tag{17}$$

Where, T_{a} was the amount of time that a human driver had been driving their vehicle. If the driver was actively controlling their vehicle then the estimate reduced. If the driver took some rest, then the estimate increased.

4.4 Confidence Overall

The gains for the controllers G_c and G_h and an overall Trust-Factor was:

$$\text{OverConFact} = E_t \times \max(E_{avoid}, E_{\text{safety}}) \tag{18}$$

$$G_h = \text{OverConFact} \tag{19}$$

$$G_c = 1 - \text{OverConFact} \tag{20}$$

If both the avoidance and the safety Trust-Factors are high, then a driver could drive their vehicle without interference. If avoidance was higher, and the vehicle was far away from obstacles then the driver would still be in control of their vehicle. If the safety estimate were high, then vehicle speed is limited.

5 Results

The methods were validated with a series of experiments.

5.1 Simulation

The vehicle shared-controller was initially validated using simulation. Joystick input was fixed, and speed was set to half speed.

A typical trajectory is shown in Fig. 6. Vehicle position and heading are recorded at frequent intervals. Direction to a target is shown as a dotted line. The sensor to the left would detect the wall and guide the vehicle away from it. The solid line shows the actual direction of movement; moving towards A. As the vehicle is driven towards A, the sensor on the right will detect the wall below the vehicle and the system will turn the vehicle to the left in order to avoid the wall. Once in free space again, then Vo will reduce to 0 and the vehicle will turn and move towards the target, guided by input from the joystick. At C, the left-hand sensor will detect the gap and Vo increases and the vehicle will steer to the right to avoid the edge of the gap.

The vehicle is then safe at D and through the gap. There are not any objects in front of the vehicle so that the vehicle can turn towards the destination.

The simulated vehicle avoided objects and drove safely towards the destination.

The values of the Trust-Factor at the points was:

Start	0.3
Point A	0.5
Point B	0.95
Point C	0.6
Point D	0.95

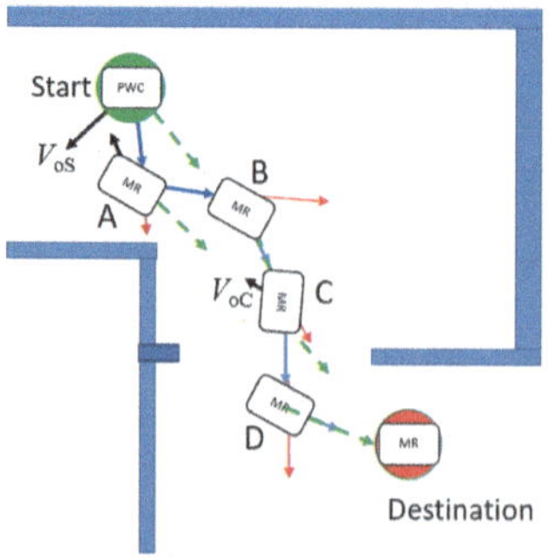

Fig. 6. Simulation experimentation and testing.

The Trust factor associated with avoiding objects became significant if an object was detected. In that case driver confidence decreased and the vehicle would be made to reduce in speed. Sensors partly controlled the vehicle until it returned to safely open space. The shared controller allowed the driver to move towards a goal while avoiding objects.

Commands sent to vehicle motors were a mixture of the simulated sensor and joystick inputs (13). They adjusted the gains if the controller and produced suitable steering and speeds for the simulated vehicle. It did not crash.

5.2 Vehicle Experiments

Volunteer drivers at Portsmouth drove the vehicle through objects towards a destination. They used joysticks to control the vehicle.

A camera was used to record trajectories and Fig. 7 is a typical route.

Recorded values of Trust-factor for the experiment were:

Start	0.05
Point A	0.06
Point B	0.67
Point C	0.68
Point D	0.74

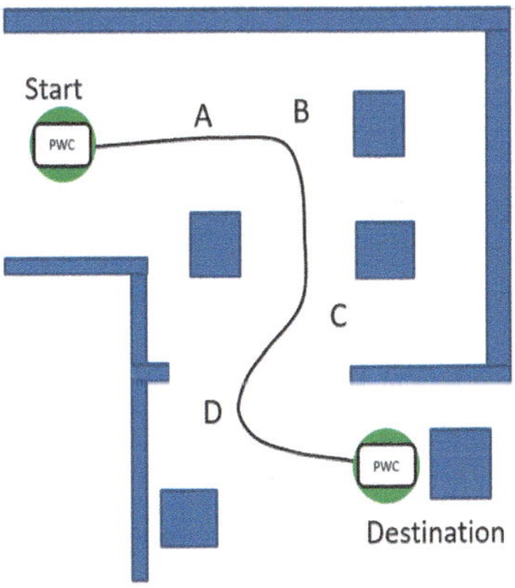

Fig. 7. Recorded trajectory of vehicle experiment.

Drivers were able to concentrate on steering and not on avoiding objects.

Shared control allowed the sensors to assist drivers in controlling their vehicles. Vehicles did not collide and they safely moved towards their destinations.

The controller combined inputs from sensors and the driver's joystick (13). Controller-gains controlled speed and direction.

Driver Confidence was reduced if objects were present that had to be circumvented. Vehicles did not crash.

If a driver drove their vehicle more slowly then driving was safer and the driver was given higher authority. Shared-control moved the vehicle safely away from obstacles when necessary while tracking the driver joystick for most of the time.

6 Conclusions

Shared-control was demonstrated. Drivers controlled a vehicle unless the sensors needed to help them.

Mixing input from drivers and sensors was achieved using Trust-factors. Trust-Factors were set depending on the evidence from sensors, for example: range to an obstacle; the time that a vehicle driver had been driving, etc.

Human drivers tended to steer their vehicles more safely when the sensor system helped them. Experiments demonstrated that shared-control was safe.

The static ultrasonic sensors are being replaced by a scanning device created by Martin Langner at Chailey Heritage because the scanning device is smaller but covers a larger volume with a longer range.

References

1. Sanders, D.A., Ndzi, D., Chester, S., Malik, M.: Adjustment of tele-operator learning when provided with different levels of sensor support while driving mobile robots. In: Bi, Y., Kapoor, S., Bhatia, R. (eds.) SAI Intelligent Systems (IntelliSys), vol. 2. Lecture Notes in Networks and Systems, vol. 16, pp. 548–558. Springer, UK (2017). https://doi.org/10.1007/978-3-319-56991-8_41
2. Sanders, D.A., Sanders, H., Gegov, A., Ndzi, D.: Rule-based system to assist a tele-operator with driving a mobile robot. In: Bi, Y., Kapoor, S., Bhatia, R. (eds.) SAI Intelligent Systems (IntelliSys), vol. 2, Lecture Notes in Networks and Systems, vol. 16, pp. 599–615. Springer, UK (2017). https://doi.org/10.1007/978-3-319-56991-8_44
3. Sanders, D.A.: Using self-reliance factors to decide how to share control between human powered wheelchair drivers and ultrasonic sensors. IEEE Trans. Neural Syst. Rehabil. Eng. **25**(8), 1221–1229 (2017)
4. Sanders, D.A.: Non-model-based control of a wheeled vehicle pulling two trailers to provide early powered mobility and driving experiences. IEEE Trans. Neural Syst. Rehabil. Eng. **26** (1), 96–104 (2018)
5. Sanders, D.A., Gegov, A., Tewkesbury, G., Khusainov, R.: Rule-based system to assist a powered wheelchair driver. In: IEEE Proceedings of the Intelligent Systems Conference (IntelliSys) 2017, 7–8 September 2017. IEEE, London (2017)
6. Sanders, D.A., Sanders, B., Gegov, A., Ndzi, D.: Results from investigating powered wheelchair users learning to drive with varying levels of sensor support. In: Proceedings of the Intelligent Systems Conference (IntelliSys) 2017, 7–8 September 2017. IEEE, London (2017)
7. Sanders, D.A., Sanders, B., Ndzi, D., Bausch, N.: Using confidence factors to share control between a mobile robot tele-operator and ultrasonic sensors. In: Proceedings of SAI Intelligent Systems Conference (IntelliSys) 2017, 7–8 September 2017. IEEE, London (2017)

8. Sanders, D.A., Gegov, A., Ndzi, D.: Knowledge-based expert system using a set of rules to assist a tele-operated mobile robot. In: Bi, Y., Kapoor, S., Bhatia, R. (eds.) Intelligent Systems and Applications. Studies in Computational Intelligence, vol. 751. Springer (2017)
9. US: Mobileye intros smartphone connected driver assistance (ADAS) technology. Telematics News (2012). http://telematicsnews.info/2012/01/12/us-mobileye-intros-smartphone-connected-driver-assistance-adas-technology_j3122. Accessed 13 Nov 2017
10. Riches: Strategy Analytics: Automotive Ethernet: Market Growth Outlook | Keynote Speech 2014 IEEE SA: Ethernet & IP @ Automotive Technology Day (PDF). IEEE. http://standards.ieee.org/events/automotive/2014/00_Automotive_Ethernet_Market_Growth_Outlook.pdf. Accessed 13 Nov 2017
11. Rees, K., Park, J.S.: Vehicle Information Access API. W3C Business Group report, W3C and LG Electronics (2014). https://www.w3.org/2014/automotive/vehicle_spec.html. Accessed 13 Nov 2017
12. ADAS Definition. Autoconnectedcar.com. Archived from the original on 06/10/2012. https://web.archive.org/web/20120610055853/http://telematicsnews.info/2012/01/12/us-mobileye-intros-smartphone-connected-driver-assistance-adas-technology_j3122. Accessed 13 Nov 2017
13. Banga, B.: Global ADAS and Autonomous Driving Components Market, Analysis & Forecast. Business Intelligence and Strategy Research (2017). https://bisresearch.com/industry-report/global-adas-autonomous-driving-components-market-2026.html. Accessed 13 Nov 2017
14. Sanders, D.A.: The modification of pre-planned manipulator paths to improve the gross motions associated with the pick and place task. Robotica **13**, 77–85 (1995)
15. Sheridan, T.B.: Tele-operation, telerobotics and telepresence: a progress report. Control Eng. Pract. **3**(2), 205–214 (1995)
16. Sheridan, T.: Humans and Automation: System Design and Research Issues. Wiley (1995)
17. Sanders, D.A.: Comparing speed to complete progressively more difficult mobile robot paths between human tele-operators and humans with sensor-systems to assist. Assembly Autom. **29**(3), 230–248 (2009)
18. Sanders, D.A., Langner, M., Tewkesbury, G.E.: Improving wheelchair-driving using a sensor system to control wheelchair-veer and variable-switches as an alternative to digital-switches or joysticks. Ind. Robot **37**(2), 157–167 (2010)
19. Sanders, D.A., Baldwin, A.: X-by-wire technology. In: Total Vehicle Technology Conference, pp. 3–12 (2001)
20. Sanders, D.A.: Controlling the direction of walkie type forklifts and pallet jacks on sloping ground. Assembly Autom. **28**(4), 317–324 (2008)
21. Fiorini, P., Oboe, R.: Internet-based telerobotics: problems and approaches. In: Proceedings of ICAR 1997, Monterey, CA, pp. 765–770 (1997)
22. Richard, J.P.: Time-delay systems: an overview of some recent advances and open problems. Automatica **39**, 1667–1694 (2003)
23. Lawrence, D.A.: Stability and transparency in bilateral tele-operation. IEEE Trans. Robot. Autom. **9**(5) (1993)
24. Kim, W., Hannaford, B., Bejczy, A.: Force reflection and shared compliant control in operating telemanipulators with time delay. IEEE Trans. Robot. Autom. **8**(2), 176–185 (1992)
25. Niemeyer, G., Slotine, J.J.: Stable adaptive tele-operation. IEEE J. Ocean. Eng. **16**(1), 152–162 (1991)
26. Niemeyer, G., Slotine, J.J.: Towards force reflecting tele-operation over the Internet. In: Proceedings of the IEEE International Conference on Robotics and Automation, Leuven, Belgium (ICRA 1998), pp. 1909–1915 (1998)

27. Luk, B.L., Cooke, D.S., Galt, S., Collie, A.A., Chen, S.: Intelligent legged climbing service robot for remote maintenance applications in hazardous environments. J. Robot. Autonom. Syst. **53**(2), 142–152 (2005)
28. Rooks, B.: Plotting future UK robotics research Programmes. Ind. Robot **33**(3), 165–169 (2006)
29. Sanders, D.A., Tewkesbury, G.E., Robinson, D.C.: Simple expert systems to improve an ultrasonic sensor-system for a tele-operated mobile-robot. Sens. Rev. **31**(3), 246–260 (2011)
30. Sanders, D.A., Stott, I.J.: Analysis of failure rates with a tele-operated mobile robot between a human tele-operator and a human with a sensor system to assist. Robotica J. **30**(6), 973–988 (2012)
31. Sanders, D., Tewkesbury, G., Graham-Jones, J.: Simple rules to modify pre-planned paths to improve gross robot motions associated with pick and place assembly tasks. Assembly Autom. **31**(1), 69–78 (2011)
32. Sanders, D.A.: Comparing ability to complete simple tele-operated rescue or maintenance mobile-robot tasks with and without a sensor system. Sens. Rev. **30**(1), 40–50 (2010)
33. Sanders, D.A., Graham-Jones, J., Gegov, A.: Improving ability of tele-operators to complete progressively more difficult mobile robot paths using simple expert systems and ultrasonic sensors. Ind. Robot **37**(5), 431–440 (2010)
34. Stott, Sanders, D.: New powered wheelchair systems for the rehabilitation of some severely disabled users. Int. J. Rehab. Res. **23**(3), 149–153 (2000)
35. Sanders, D., Stott, I.: A new prototype intelligent mobility system to assist powered wheelchair users. Ind. Robot **26**(6), 466–475 (1999)
36. Goodwin, M.J., Sanders, D.A., Poland, G.A., et al.: Navigational assistance for disabled wheelchair-users. In: Proceedings of Euromicro Conference 95, vol. 43, pp. 73–79 (1997)
37. Sanders, D.: Analysis of the effects of time delays on the teleoperation of a mobile robot in various modes of operation. Ind. Robot **36**(6), 570–584 (2009)
38. Sanders, D., Stott, I., Graham-Jones, J., et al.: Expert system to interpret hand tremor and provide joystick position signals for powered wheelchairs with ultrasonic sensor systems. Ind. Robot **38**(6), 585–598 (2011)
39. Sanders, D.A., Bausch, N., Liu, H., et al.: Improving steering of a powered wheelchair using an expert system to interpret hand tremor. In: Proceedings Intelligent Robotics and Applications, Pt I vol. 9245, pp. 460–471 (2015)
40. Sanders, D., Stott, I.: The use of virtual reality to train powered wheelchair users and test new wheelchair systems. Int. J. Rehabil. Res. **23**(4), 321–326 (2000)
41. Sands, D.: Cost effective robotics in the nuclear industry. Ind. Robot **33**(3), 170–173 (2006)
42. Nakamura, T., Satoh, K.: Development of an omni-directional mobile robot using traveling waves based on snail locomotion. Ind. Robot **35**(3), 206–210 (2008)
43. Luk, B.L., Collie, A.A., Cooke, D.S., Chen, S.: Walking and climbing service robots for safety inspection of nuclear reactor pressure vessels. Measure. Control **39**(2), 43–47 (2006)
44. Luk, B., Liu, K., Collie, A., Cooke, D., Chen, S.: Tele-operated climbing and mobile service robots for remote inspection and maintenance in nuclear industry. Ind. Robot **33**(3), 194–204 (2006)
45. Bakari, M.J., Seward, D.W.: Human arm-like mechanical manipulator - the design and development of a multi-arm mobile robot for nuclear decommissioning. In: Proceedings of 3rd International Conference on Informatics in Control, Automation and Robotics – Robotics and Automation, pp. 168–175 (2006)
46. DeJong, B.P., Faulring, E.L., Edward Colgate, J., Peshkin, M.A., Kang, H., Park, Y.S., Ewing, T.F.: Lessons learned from a novel tele-operation test-bed. Ind. Robot **33**(3), 187–193 (2006)

47. Sanders, D.A., Stott, I.J., Robinson, D.C., et al.: Analysis of successes and failures with a tele-operated mobile robot in various modes of operation. Robotica **30**, 973–988 (2012)
48. Stott, J., Sanders, D.A.: New powered mobile robot systems for the rehabilitation of some severely disabled users. Int. J. Rehab. Res. **23**(3), 149–153 (2000)
49. Sanders, D.A., Graham-Jones, J., Gegov, A.: Improving ability of tele-operators to complete progressively more difficult mobile robot paths using simple expert systems and ultrasonic sensors. Ind. Robot-an Int. J. **37**(5), 431–440 (2010)
50. Sanders, D.A., Stott, I.J.: A new prototype intelligent mobility system to assist powered mobile robot users. Ind. Robot **26**(6), 466–475 (1999)
51. Hayati, S., Venkataraman, S.T.: Design and implementation of a robot control system with traded and shared control capability. In: Proceedings of IEEE International Conf on Robotics and Automation, Scottsdal, AZ, USA, pp. 1310–1315 (1989)
52. Abbink, D.M.M., Boer, E.R.: Haptic shared control: smoothly shifting control authority? Cogn. Technol. Work **14**(1), 19–28 (2012)
53. Itoh, M., Inagakia, T., Tanaka, H.: Haptic steering direction guidance for pedestrian-vehicle collision avoidance. In: Proceedings of IEEE International Conference on System, Man, and Cybernetics, Seoul, Korea, pp. 3309–3314 (2012)
54. Carlson, T., Demiris, Y.: Increasing robotic mobile robot safety with collaborative control: evidence from secondary task experiments. In: Proceedings of IEEE International Conf on Robotics and Automation, Anchorage, Alaska, USA, pp. 5582–5587 (2010)
55. Satti, R., Coyle, D., Prasad, G.: Self-paced brain-controlled mobile robot methodology with shared and automated assistive control. In: Proceedings of IEEE Symposium on Computational Intelligence, Cognitive Algorithms, Mind, and Brain, Paris, France, pp. 1–8 (2011)
56. Franchi, A., Secchi, C., Ryll, M., Bulthoff, H.H., Giordano, P.R.: Shared control: balancing autonomy and human assistance with a group of quadrotor UAVs. IEEE Robot. Autom. Mag. **19**(3), 57–68 (2012)
57. Kim, H.K., Biggs, S.J., Schloerb, D.W., Carmena, J.M., Lebedev, M.A., Nicolelis, M.A.L., Srinivasan, M.A.: Continuous shared control for stabilizing reaching and grasping with brain-machine interfaces. IEEE Trans. Biomed. Eng. **53**(6), 1164–1173 (2006)
58. Sanders, D.A., Urwin-Wright, S.D., Tewkesbury, G.E., et al.: Pointer device for thin-film transistor and cathode ray tube computer screens. Electr. Lett. **41**(16), 894–896 (2005)
59. Sanders, D.A., Bergasa-Suso, J.: Inferring learning style from the way students interact with a computer user interface and the WWW. IEEE Trans. Educ. **53**(4), 613–620 (2010)
60. Sanders, D.A., Tewkesbury, G.E.: A pointer device for TFT display screens that determines position by detecting colours on the display using a colour sensor and an Artificial Neural Network. Displays **30**(2), 84–96 (2009)
61. Sanders, D.A., Baldwin, A.: X-by-wire technology. In: Total Vehicle Technology Conference, pp. 3–12 (2001). ISBN: 1-86058-324-5
62. Carlson, T., Demiris, Y.: Collaborative control for a robotic mobile robot: evaluation of performance, attention, and workload. IEEE Trans. Syst. Man Cybern. Part B Cybern. **42**(3), 876–888 (2012)

The Impact of Road Intersection Topology on Traffic Congestion in Urban Cities

Marwan Salim Mahmood Al-Dabbagh(✉), Ali Al-Sherbaz, and Scott Turner

Faculty of Arts, Science and Technology, The University of Northampton, Northampton, Northamptonshire, UK
{marwan.aldabbagh,Ali.al-sherbaz, scott.turner}@northampton.ac.uk

Abstract. Due to the dramatic increase in population and the rapid developments of the automobile industry, tremendous issues have emerged in road traffic systems, such as traffic congestion. In some big cities, traffic congestion occurs due to several factors, such as roadworks, car accidents, and drivers' behaviour. Consequently, it negatively impacts on environment and economy as well as on the behaviour of drivers and passengers. In this paper, we first identified the factors that may increase the journey time and listed the traffic congestion measurement metrics. Then, we investigate to reveal the relationship of road network topology to the traffic congestion level regarding travel time and the number of affected vehicles in case of traffic incidents. The open source traffic simulator SUMO (Simulation of Urban MObility) was used to simulate vehicular traffic and to generate traffic jams on the roadmap under various scenarios. These scenarios involve three road topologies which were: crossroads in Denver (CO, USA), roundabouts in Nantes (France), and hybrid topology, which is the combination of intersections and roundabouts, in Northampton town (UK). The results showed that the delayed time is less in the roundabout traffic map topology, and the number of affected vehicles in the case of traffic incidents is less likely to happen in the hybrid topology.

Keywords: Road traffic congestion · Travel time · Road accident SUMO (Simulation of Urban MObility) · Crossroad map · Roundabout map Hybrid map

1 Introduction

The problems of road traffic congestion [1] have become a common phenomenon in some big cities around the world. Various reasons may cause traffic congestion; a large number of vehicles that utilise the roadway; the adverse weather conditions; the drivers' behaviour; and poor traffic signal timing [2]. Many negative consequences, then, may arise; drivers may waste time as they are in long queues in roads and cars consume more fuel which causes air pollution [3]. Therefore governments are striving to tackle these problems by initiating the purpose of public transport in big cities, making streets larger to avoid road traffic congestion and enlarge the entrance roads around cities and residential areas, constructing more bridges over roadways, and building crossings and

K. Arai et al. (Eds.): IntelliSys 2018, AISC 868, pp. 1196–1207, 2019.
https://doi.org/10.1007/978-3-030-01054-6_83

establishing tunnels [4]. According to INRIX [5], 2258 traffic hotspots caused 16636 traffic congestion annually in Denver. The report revealed that drivers spent an average of 36 peak hours in traffic congestion which cost $1182 for each driver. Also, the report anticipated that traffic jams would cost approximately $10b in the next decade if Denver's government does not deal with traffic jam issues. Moving on to Europe, in Nantes city, commuter spends an average of 75 h stuck in traffic jams annually. In the UK, particularly in Northampton town [6], drivers spend 20 h per year, in total cost of £62 M, an average of £697 per driver due to a traffic jam. Thus, it is necessary to recognise factors that contribute to the cause of traffic congestion and to provide efficient tools to measure it as well as to find out realistic solutions to decrease its impact on society. In this paper, we first identified the factors that may increase the journey time and listed the traffic congestion measurement metrics. Then, we investigate the relationship of road network topology to the traffic congestion level in terms of travel time and the number of affected vehicles in case of traffic incidents. Three real roadmaps were selected for both left- and right-hand driving in order to conduct experiments. This study aims to mitigate traffic congestion in urban areas. It also may benefit the urban and transport planners, civil engineers, and transport policy makers, as well as congestion management researchers who are directly or indirectly involved or enthusiastic to be concerned in the task of traffic congestion management.

The rest of the paper is organised as follows: In Sect. 2, we describe the traffic congestion formation factors; the traffic congestion metrics are presented in Sect. 3; in Sect. 4, wc discuss related works on traffic congestion; in Sect. 5, we present the methodology; the simulation results are discussed in Sect. 6; and in the last section, we draw the conclusions.

2 Traffic Congestion Formation Factors

Traffic congestion is a road network condition during a specific time which occurs every day [7]. According to the U.S Department of Transportation (DoT) report [8], traffic congestion occurs as a result of the interaction of seven factors. Figure 1 depicted the sources of congestion. The report classified these factors into three categories:

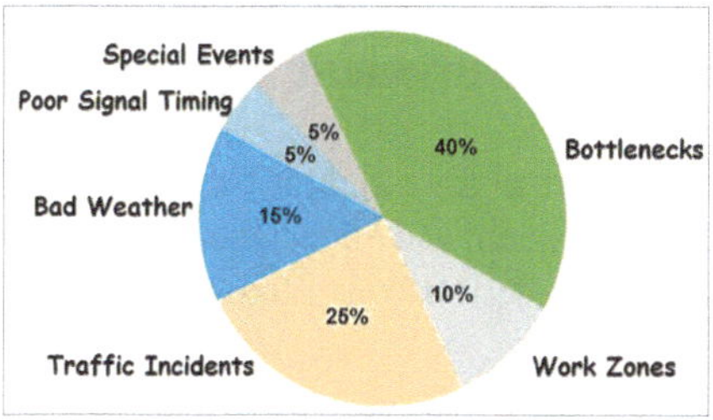

Fig. 1. Traffic congestion sources [8].

2.1 Category 1: Traffic-Influencing Events

- Traffic Incidents is defined as any crash or collision between a moving vehicle with anything else (e.g., a vehicle, a wall, a road sign, and sometimes runs over an animal, human being, or even a tree). The majority of the recent studies claimed that the major concern of traffic incident, which can bring death to roads, is caused by speed; non-application with traffic laws; and the phone usage during driving.
- Work zones are the street activities which make physical changes to the highway movement such as a reduction in the number or width of travel lanes, lane shifts, and lane diversions.
- Weather conditions are the other factor that may change the drivers' behaviour to reduce speed. The main adverse weather conditions are rain, snow, ice, fog, and smoke.

2.2 Category 2: Traffic Demand

- Fluctuations in Normal Traffic - traffic congestion level varies during the week and the volume during the day.
- Special Events are a particular feature occurs in a specific place for a period of time-based on the event such as football match game, National occasions (Independence Day), and a fire truck stationed in the street. These circumstances apparently will be leading to heighten traffic flow in the region of the event.

2.3 Category 3: Physical Highway Features

- Traffic Control Devices such as traffic light signals and railroad upgrade intersections are used to control the traffic on the roadway. However, these devices may create a traffic jam if displayed to an accurate malfunction.
- Physical Bottlenecks ("Capacity") – The road is meant to absorb the maximum number of vehicles. However, there are other determinants such as the number and expanse of lanes and shoulders, merge areas at crossings, and roadway alignment (grades and curves) which could restrict the traffic flow.

3 Traffic Congestion Measurement Metrics

Traffic congestion is considered as an essential factor which assists in evaluating the performance of transportation system and affects the decisions of transport development plans [9]. It can be measured in various ways like travel time, speed, level of service, delay, and throughput [10].

- Travel Time Index refers to the quota of the time needed to finish a trip under congested circumstances to the time required to achieve the same trip under free-flow conditions.
- Speed is the average mileage travelled at a given time. It plays an essential role in estimating traffic congestion. At present, Geographical Information System (GIS) is

utilised to recognise traffic congestion by monitoring vehicle traffic based on Global Positioning System (GPS) and register peak hours.

- Level-Of-Service (LOS) is an index to the density of congestion suspensions on a particular road or at an intersection. It is classified into six categories beginning from A (free congestion) to F (exceedingly congestion) as demonstrated in Table 1.
- Delay is measured by deducting the arrival time from the logical time for the trip. According to managing urban traffic congestion (OECD) report, 29% of the transport professionals in North America are utilising the delay index in estimating traffic congestion.
- Throughput is the number of vehicles that pass through a particular street during rush-hour.

Table 1. Level-of-Service (LOS) for a junction [11]

Level of service	Average delay (s/veh)	Description
A	≤10	Free flow
B	10–20	Stable flow (slight delays)
C	20–35	Stable flow (acceptable delays)
D	35–55	Approaching unstable flow (tolerable delay, occasionally wait through more than one signal cycle before proceeding)
E	55–80	Unstable flow (intolerable delay)
F	>80	Forced flow (jammed)

4 Related Works

There are many studies on this topic to alleviate traffic congestion on the roadway network. However, these efforts have only been partially successful. The authors in [12] propose an intelligent model to optimise traffic control signals based on generated past traffic demand patterns by using a simulation tool to analyse traffic data aiming to provide statistics on the vehicle trip such as travel time, travel speed, delay, the number of stops, fuel consumption, and CO and HC emissions. In addition, an ensemble-based system model for traffic control and management based on current traffic demands was proposed in [13] where traffic sensors devices were used, helped to monitor the traffic placed at the exits of all main roads entering the study area. The proposed system aims to reduce the number of sensors on the intersections, make traffic control decisions smartly, and traffic measurement efficiently. Another solution to mitigate traffic congestion was involved in [14]. The authors proposed two specific approaches to improve traffic light cycles using Multi-Agent Systems (MAS) and Swarm Intelligence (SI) concepts. Two models have been investigated with large different urban areas. The results indicate a reduction in terms of the waiting time and the journey time of vehicles. A study by Tsapakis et al. [15] on the impact of weather conditions on urban travel time has been done. In this study, three kinds of weather elements (rain, snow, temperature) have been investigated and its effect on traffic congestion in terms of travel time and speed by analysing more than 380 travel links. The obtained results showed that both rain and snow have a functional impact on their intensity. The authors of [16] studied the impact of the rainstorm on metropolitan road traffic in terms of traffic velocity. In this study, four types of roads were identified, namely speed rise greatly, speed rise slightly, speed decline slightly, and speed decline greatly, and explored the relationship between the rainfall intensities and speed variation rates (SVR). The results indicated that approximately half of the roads are much affected by the rainstorm. And regarding traffic velocity, the rainstorm has a heterogeneous impact based on spatial. The idea of study the impact of driver behaviour during the adverse weather conditions is introduced by Chakrabartya and Gopal [17]. The study was conducted on three groups of drivers based on driver experience years and evaluated driver errors during three kinds of weather conditions (clear weather, rain, and foggy). The simulation was carried out in three sessions for each driver 30 min each. Liang et al. [18] design a model to classify two kinds of drivers' behaviour at a congested signalised intersection, which helps to reduce traffic delay at the road junction.

In summary, the focus of traffic congestion problems was on the circumstances of traffic light timing, roadworks, weather conditions, and drivers' behaviour. However, To our best knowledge, we did not find any relevant research into the road topology in relation to travel time and the number of affected vehicles in case of traffic incidents. Hence, this paper will investigate the relationship of road topology and its impact on traffic congestion regarding travel time and the number of affected vehicles.

5 Simulation and Evaluation Methodology

This study classified the roadway networks into three main kinds based on road intersection: crossroads, roundabouts, and hybrid road networks. The simulation is carried out in three real roadway maps: Denver (CO, USA), Nantes (France), and (Northampton town (UK) which they were extracted from OpenStreetMaps (OSM) database [19] with a size of 5 $\times$ 4 km.

5.1 Travel Time Determinants

This study defined a number of determinants that increase travel time. These determinants were used in simulation scenarios, and they are as follows:

- Density refers to the number of vehicles that are simulated in the scenario. Two different sizes with (200 veh/km^2) and (300 veh/km^2) were simulated in each selected map.
- Quantity refers to the number of incidents that occur on the roadway network at the same time. We simulated one and three traffic incidents in each selected map.
- Location refers to the congestion area. This study deals with two locations of incidents: (1) close to City/Town centre; (2) Outskirt of City/Town centre.
- Duration time is the period of time spent when vehicles are fully stopped due to traffic incident. Two different incidents with 30 and 60 min have been simulated.

- Volume defines the size of the incident area in the street, and the number of blocked lanes. This study has been simulated with three different volumes: one, two, and three lanes in each selected map.
- Timing is to identify when the traffic incident may occur: before and after rush-hour time.

5.2 Simulation of Urban MObility (SUMO)

This study used SUMO-0.25.0 the microscopic road traffic simulation package in order to generate realistic road traffic scenarios on roads. SUMO is an open source used to generate traffic for extensive road network relying on a set of inputs Extensible Markup Language (XML) files to generate the traffic simulation. Figure 2 illustrates the whole process to run a traffic simulation in SUMO [20].

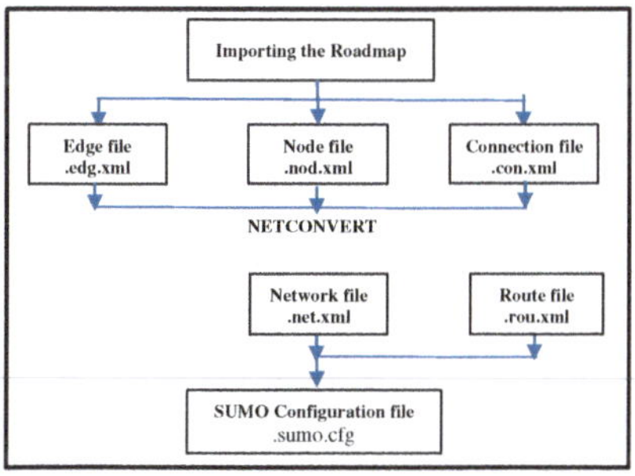

Fig. 2. Traffic simulation steps.

Importing the roadmap can be accomplished by either using digital mapping software such as OpenStreetMap (OSM) or by using an application named 'Netgen'. We used OSM file which consists of three files: .edg.xml file that contains information for all streets such as the street ID number, the priority value, the street type, and lanes number; .nod.xml file that contains information for all the junctions in the network such as the node number, the X and Y position value of the node, and shape value; and .con.xml file that contains information for two fields, the start and end point for vehicle trip. Network file is generated through NETCONVERT instruction on the osm file on a command line whereas Route file, which comprises routing information for each vehicle defined in the network, is generated by running Random trip Python file on the command line. The final stage is to generate SUMO configuration file by integrating the network file and the route file.

5.3 Simulation Scenario

To ensure that our scenarios are more realistic, the experiment run-time was divided into 45 intervals, 4 min for each. Normal distribution method was used to generate vehicles number in each period, using (1).

$$f_x(x) = \frac{1}{\sigma\sqrt{2\pi}} e^{\left[-\frac{(x-\mu)^2}{2\sigma^2}\right]} \tag{1}$$

Where:

- μ is the mean of the distribution.
- σ is the standard deviation of the distribution.
- σ^2 is the variance.

In the simulation scenarios, we chose the average delay time and number of affected vehicles as main factors to measure traffic congestion level. In this study, several scenarios and tests were conducted in relation to the travel time determinants as mentioned earlier. 32 tests were applied in each scenario, which categorised into two groups, 16 tests in each, based the density of generated vehicles. Table 2 presents the full details of the tests with the first group (200 vehicles/km^2) as well as these test were applied to the second group (300 vehicles/km^2).

Table 2. First group (200 Vehicles/km^2) tests

Density	Location	Quantity	Duration time	Timing
16 Tests (200 vehicles / Km2)	8 tests Close to City/Town centre	4 tests One Traffic incident	2 tests 30 minutes	1 test Before rush-hour
				1 test After rush-hour
			2 tests 60 minutes	1 test Before rush-hour
				1 test After rush-hour
		4 tests Three Traffic incident	2 tests 30 minutes	1 test Before rush-hour
				1 test After rush-hour
			2 tests 60 minutes	1 test Before rush-hour
				1 test After rush-hour
	8 tests Outskirt City/Town centre	4 tests One Traffic incident	2 tests 30 minutes	1 test Before rush-hour
				1 test After rush-hour
			2 tests 60 minutes	1 test Before rush-hour
				1 test After rush-hour
		4 tests Three Traffic incident	2 tests 30 minutes	1 test Before rush-hour
				1 test After rush-hour
			2 tests 60 minutes	1 test Before rush-hour
				1 test After rush-hour

Table 3 shows the simulation parameters for all scenarios.

Through the implementation of the scenario, and for the aim of measure, the traffic congestion level opted a group of vehicles with delayed journey time greater than 5 min.

Table 3. Shows the simulation parameters for all scenarios

Parameters	Value
Simulation time	3 h
Location	Northampton, Nantes, Denver
The maps	20 km^2 (4 × 5 km)
The average speed	30 mph
Number of generated vehicles	200 veh/km^2 and 300 veh/km^2
Vehicle length + safety distance	10 m
Routing algorithm	Dijkstra

6 Simulation Results and Discussion

This section illustrates the simulation results obtained from the implementation of the scenarios on the selected maps in SUMO. To ascertain the scenarios results, we carried out each simulation test three times.

6.1 Analysis of the Results: Delay Time

The average delay time for a vehicle has been measured by applying (2).

$$A.V.\,Delay\,time = \frac{\sum_{i=1}^{n} Delay\,time}{No.\,Affected\,vehicles} \quad (2)$$

Where:

- n is the number of affected vehicles

Figures 3 and 4 indicate results that Nantes city has the least delay time in comparison with other cities for all traffic incidents cases. Concerning the timing of traffic incident

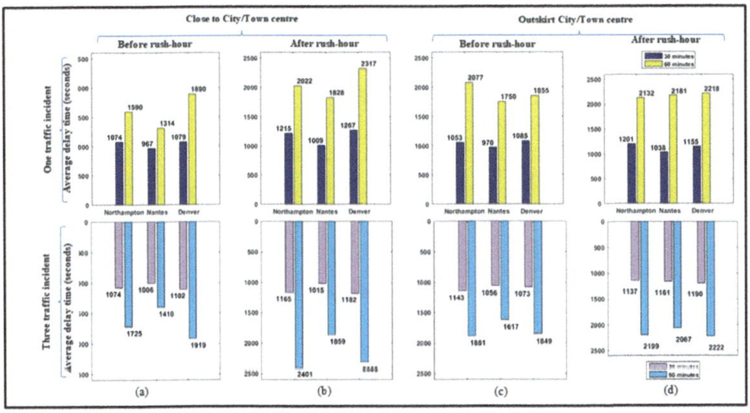

Fig. 3. First group (200 vehicles/km^2) results – delay time.

occurrence, the average delay time after rush-hour time is more significant than before rush-hour time for all traffic incidents cases.

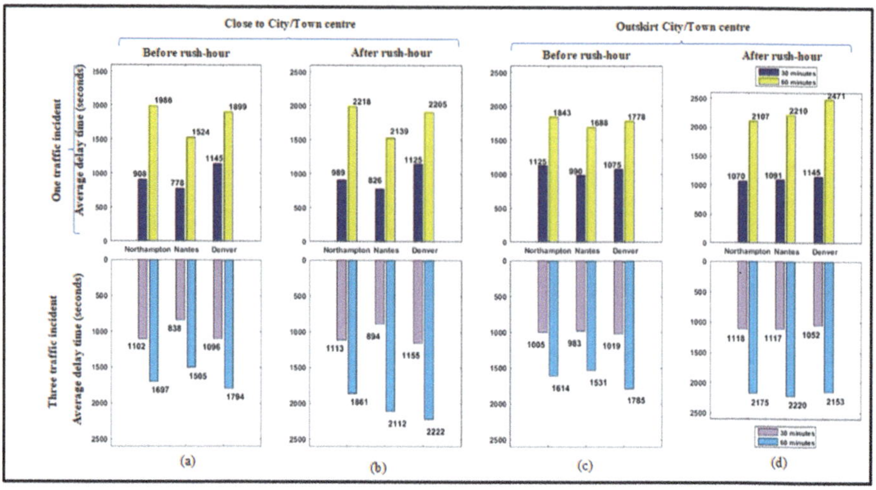

Fig. 4. Second group (300 vehicles/km^2) results – delay time.

6.2 Analysis of the Results: Affected Vehicles

The number of affected vehicles have been measured by selecting the vehicles which have a delay on arrival time greater than 5 min to their destination. The results in Figs. 5 and 6 showed that Northampton town traffic map is better than other cities regarding a number of affected vehicles for most traffic jam cases, followed by Denver and Nantes city. Concerning the location of the traffic incidents, with Northampton

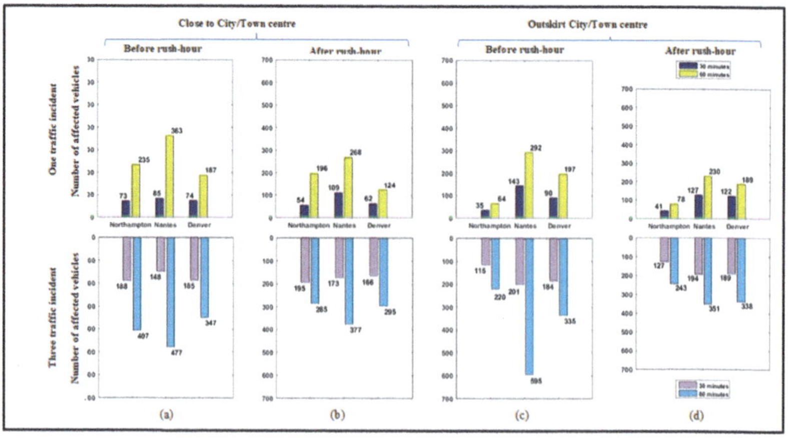

Fig. 5. First group (200 vehicles/km^2) results – number of affected vehicles.

traffic map, the number of affected vehicles by traffic jam located near the city centre is higher than that outside the city. In contrast, for Nantes and Denver city traffic map, the number of affected vehicles by traffic jam located near the city centre is lower than that outside the city for most traffic jam cases, except the case of traffic congestion for 60 min duration time with Nantes city.

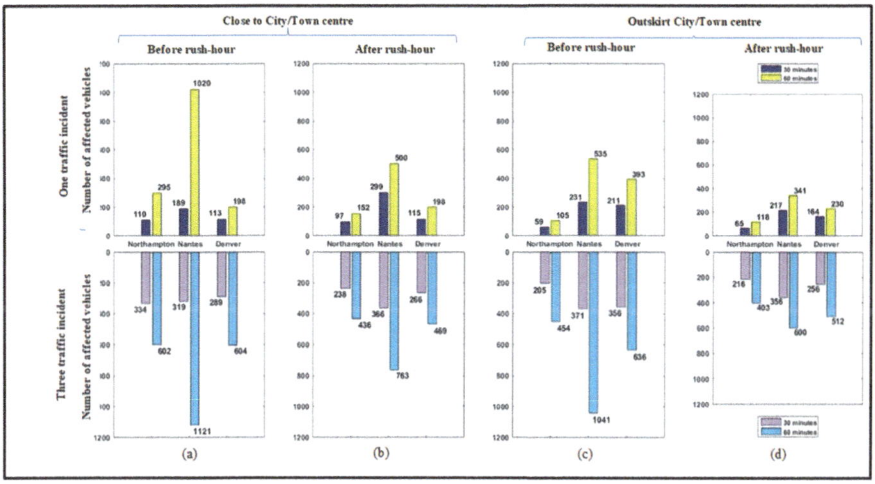

Fig. 6. Second group (300 vehicles/km^2) results – number of affected vehicles.

7 Conclusions

This paper discussed the impact of road topology on traffic congestion in different urban cities: Denver, Nantes and Northampton city have been chosen as locations for implementing the simulation scenarios. We run the simulation in three real maps based on road intersection: crossroads, roundabouts, and hybrid. The simulation results indicated that there was a significant impact of road topology on traffic congestion in terms of the increase of delay time for vehicle trip and the number of affected vehicles in the case of road incident. Moreover, the roundabout traffic map is the ideal among the types if we consider the delay time as the primary factor, or the hybrid type is the ideal if we consider the number of affected vehicles as the principal factor. Furthermore, this finding has significant implications for developing and help urban planners, transport planners, civil engineers or congestion management researchers whom they are interested in traffic congestion management. Future studies on the current topic are therefore recommended.

Acknowledgment. This research was funded by the ministry of higher education and scientific research, Republic of IRAQ - scholarship (Ref. 20432).

References

1. Zhang, J., Yu, Y., Lei, Y.: The study on an optimized model of traffic congestion problem caused by traffic accidents. In: Proceedings of 28th Chinese Control Decision Conference, CCDC 2016, pp. 688–692 (2016)
2. Al Mallah, R., Quintero, A., Farooq, B.: Distributed classification of urban congestion using VANET. IEEE Trans. Intell. Transp. Syst. **18**(9), 2435–2442 (2017)
3. Xu, Y., Wu, Y., Xu, J., Ni, D., Wu, G., Sun, L.: A queue-length-based detection scheme for urban traffic congestion by VANETs. In: Proceedings of 2012 IEEE 7th International Conference on Networking, Architecture, and Storage, NAS 2012, pp. 252–259 (2012)
4. Toulni, H., Nsiri, B., Boulmalf, M., Bakhouya, M., Sadiki, T.: An approach to avoid traffic congestion using VANET. In: International Conference on Next Generation Networks and Services, NGNS, pp. 154–159 (2014)
5. Pishue, B.: United States' US traffic hotspots traffic hotspots measuring the impact of congestion in the United States (2017)
6. A. G. To and C.: Global traffic scorecard, p. 2016 (2016)
7. Dubey, P.P., Borkar, P.: Review on techniques for traffic jam detection and congestion avoidance. In: IEEE Sponsored 2nd International Conference on Electronics and Communication System (ICECS 2015), pp. 434–440 (2015)
8. Federal Highway Administration: Final report, "Traffic Congestion and Reliability: Trends and Advanced Strategies for Congestion Mitigation". Int. J. Toxicol. **24**(Suppl. 5), 53–100 (2005)
9. Haragos, I.M., Holban, S., Cernazanu-Glavan, C.: Determination of quality factor used in road traffic. An experimental study. In: SAMI 2014—IEEE 12th International Symposium Applied Machine Intelligence and Informatics, Proceedings, pp. 123–128 (2014)
10. Litman, T.: Factors to consider when estimating congestion costs and evaluating potential congestion reduction strategies (2013)
11. Lozano, A., Manfredi, G., Nieddu, L.: An algorithm for the recognition of levels of congestion in road traffic problems. Math. Comput. Simul. (2009)
12. Ruwanpura, J.Y.: Optimization of traffic signal light timing using simulation. In: Ingalls, R.G., Rossetti, M.D., Smith, J.S., Peters, B.A. (eds.) Proceedings of the 2004 Winter Simulation Conference, vol. 2, pp. 1428–1433 (2004)
13. Pescaru, D., Curiac, D.-I.: Ensemble based traffic light control for city zones using a reduced number of sensors. Transp. Res. Part C Emerg. Technol. **46**, 261–273 (2014)
14. Lanka, S.: Traffic light optimization solutions using multi-modal, distributed and adaptive approaches. In: 2015 International Conference on Advances in ICT for Emerging Regions (TCTer), pp. 220–225 (2015)
15. Tsapakis, I., Cheng, T., Bolbol, A.: Impact of weather conditions on macroscopic urban travel times. J. Transp. Geogr. **28**, 204–211 (2013)
16. Li, Q., Hao, X., Wang, W., Wu, A., Xie, Z.: Effects of the rainstorm on urban road traffic speed—a case study of Shenzhen, China. In: The International Archives of the Photogrammetry, Remote Sensing and Spatial Information Sciences, Volume XLII-2/W7, 2017 ISPRS Geospatial Week 2017, 18–22 September 2017, Wuhan, China, vol. XLII, pp. 18–22 (2017)
17. Chakrabarty, N., Gupta, K.: Analysis of driver behaviour and crash characteristics during adverse weather conditions. Procedia Soc. Behav. Sci. **104**, 1048–1057 (2013)
18. Qi, L., Zhou, M., Luan, W.: Impact of driving behavior on traffic delay at a congested signalized intersection. IEEE Trans. Intell. Transp. Syst. **18**(7), 1882–1893 (2017)

19. OpenStreetMap: OpenStreetMap (2017). http://www.openstreetmap.org/copyright. Accessed 8 Oct 2017
20. Gupta, P., Singh, L.P., Khandelwal, A., Pandey, K.: Reduction of congestion and signal waiting time. In: 2015 Eighth International Conference on Contemporary Computing (IC3), pp. 308–313 (2015)

Forecasting Air Traveling Demand for Saudi Arabia's Low Cost Carriers

Eman Alarfaj(✉) and Sharifa AlGhowinem

Prince Sultan University, Riyadh, Saudi Arabia
emanalarfaj1@gmail.com

Abstract. The recent revolution of Airline market sector in Saudi Arabia has brought more attentions to air travel demand forecasting. New low cost carriers (LCCs) with potential of delivering an affordable services to customers have been established. While there are many academic literature on passenger demand forecasting, there has not been any reported study that capture the effect and impacts of Islamic holidays in forecasting Saudi Arabia's LCCs passenger demand. We approach this issue by investigating the improvement of forecasting Saudi Arabia's low cost carriers (LCCs) passenger demand using machine learning techniques by accounting the Islamic Holidays. For this research, King Khalid International Airport air passenger demand will be analyzed. Our aim is to apply different forecasting models: Genetic Algorithm, Artificial Neural Network and Classical linear regression to forecast Saudi Arabia's domestic LCC passenger demand. The model's performance will be evaluated using mean absolute percentage error (MAPE).

Keywords: Artificial intelligence (AI) · Low cost carrier (LCC)
Air passenger forecasting

1 Introduction

Analyzing the demand for air travel is a critical and important activity for an airport planning and development process. Air Travel Industry in Saudi Arabia has dramatically grown in recent years. Many low cost carriers (LCCs) have been established in Saudi Arabia like Flynas, Sama, and Flyadeal. Their increasing market presence have led to the increased passenger demand and have given the investors an opportunity in the LCCs business [1].

Accordingly, low cost carriers (LCCs) forecasting facilitate future passenger demand to support airport long-term growth. Previously, traditional regression analysis techniques were used to forecast air travel demand such as ([2–6]). In recent years, artificial intelligence forecasting techniques such as Genetic Algorithm (GA) and Artificial Neural Networks (ANN) have been proposed in the literature for example ([7–10]). The artificial intelligence models proved to be more robust and have higher prediction capabilities than traditional statistics methods.

K. Arai et al. (Eds.): IntelliSys 2018, AISC 868, pp. 1208–1220, 2019.
https://doi.org/10.1007/978-3-030-01054-6_84

While there are many academic literature on passenger demand forecasting, there has not been any reported study that capture the effect and impacts of Islamic holidays in forecasting Saudi Arabia's LCCs passenger demand. Moreover, the majority of the reported studies have used only one technique either statistics techniques or artificial intelligence forecasting techniques. The primary objective of this study is to investigates the effectiveness of using machine learning techniques to forecast Saudi Arabia's low cost carriers (LCCs) passenger demand by accounting the Islamic Holidays.

The remainder of this paper is organized as follows: Sect. 2 introduces different factors that influence air travel demand, aviation, LCCs and tourism and the forecasting process. Section 3 reviews various studies of demand forecasting. Section 4 illustrates the methods and algorithms used for forecasting. Section 5 presents the results of this study. Finally in Sect. 6, we conclude this paper.

2 Background

2.1 Factors Influencing Air Travel Demand

Air travel demand is affected by various factors, each factor consists of elements that can promote or reduce demand growth. According to [2], the air travel demand can be affected by two board groups. The first group consists of the factors that are external to the airline industry and the second group are those factors that are internal to the airline industry. The external group includes country demographic, economic and social data such as population, employed population, and jet fuel price. The internal group includes the factors that are within airline industry such as number of flights per day. The factors were collected from 15 different relevant articles as listed in Table 1.

Table 1. Air travel demand variables collected from 15 different relevant articles

Factors name	Number of repeat	Reference articles
Population Size	12	[2–4, 7, 8, 10–16]
Employed/ Unemployed Population	6	[8, 10, 11, 14–16]
Gross Domestic Product (GDP)	12	[2, 3, 5, 7, 8, 10–16]
GDP Per Capita	2	[8, 15]
Per Capita Income (PCI)	6	[2, 3, 7, 10, 11, 16]
Economic Growth Rate	6	[2, 3, 7, 10, 11, 16]
Jet Fuel Price	5	[5, 6, 8, 14, 15]
Airfares	4	[8, 12, 14, 15]
Exchange Rates	7	[2, 3, 6, 7, 10, 11, 16]
Consumer Price Index (CPI)	6	[2, 3, 6, 7, 11, 16]
Total Expenditures	3	[2, 3, 7]

2.2 Aviation, LCCs, and Tourism in Saudi Arabia

Saudi Arabia is known as the Islamic religion center. It is the home of the two holy mosques and the Prophet of Islam, Mohammad (peace be upon him). Every year, millions of tourists come for the Islamic holiday seasons such as Hajj and Ramadan. These seasons contributes significantly to the economy of Saudi Arabia. In 2017, the number of pilgrims was more than two million (2,352,122), the domestic pilgrims were 600,108 [17]. In the past decade, Saudi Arabia has improved the airline marker sector in order to reduce the reliance on oil. Furthermore, Saudi governments has decided to transform the airline industry and aviation sector ownership to private by 2020. Airport planning and development has made impressive efforts to attract more airline [18]. In 2017, the number of LCCs operating in Saudi Arabia have grown to five LCCs (Flynas, Sama Airline, Nesma Airline, SaudiGulf and Flyadeal). Moreover, in 2017 Saudi Arabia population size reached 32.55 million [17]. This present great potential for LCCs to start their marker. In our study, all the five airlines are included in the dataset provided by King Khalid International airport.

2.3 Forecasting Model

Passengers demand forecasting is defined as the ability to forecast the number of passengers at a time in the future. Various methods have been used to forecast the air travel demand, such as grey theory (e.g. [19]), gravity models (e.g. [13,20]), Multiple Linear Regression (e.g. [8]), Artificial Neural Network (e.g. [7,8,10,11,13,16]), Genetic Algorithm (e.g. [9,10]), Adaptive Neuro-Fuzzy Inference System (ANFIS) (e.g. [14,21]), Box Jenkins (e.g. [6]), Stepwise Regression (e.g. [3]), Semi Logarithmic model (e.g. [4]) and log-log model (e.g. [5]). In forecasting passenger demand, the most popular method is Multiple linear regression (MLR). It is recommended by the International Civil Aviation Organization and thus used as a baseline in this study. In recent years, artificial intelligence forecasting techniques such as Genetic Algorithm and Artificial Neural Networks have been proposed in the literature. The artificial intelligence models are more robust and have higher prediction capabilities [22,23]. Therefore, a comparison of the forecasting accuracy will be made to investigates the effectiveness of using these techniques to forecast Saudi Arabia's low cost carriers (LCCs) passenger demand. The following list presents a brief description of these techniques that will be used in this study.

Multiple Linear Regression (MLR): Multiple linear regression (MLR) is an effective statistical technique and the most common form of Linear Regression Analysis. It is used to explain the relationship between independent variables and dependent variable [24]. MLR analysis predict the future values and can be used to get point estimates.

Artificial Neural Network (ANN): ANN model consists of an input layer, a hidden layer, and an output layer. Each layer has a collection of nodes, each node is connected to the adjacent layer's nodes [11]. The input layer is designed to receives the variables data [25]. The hidden layer has a collection of processing nodes, each node receives an input from the adjacent layers through weighted elements. Finally, the output layer computes prediction of the hidden layer results. To optimize the neural network model weights, the Back Propagation Neural (BPN) is used. BPN is a learning algorithm, it adjust the weights to produce desirable results by back propagates the errors calculated at the output layer towards the input layer.

Genetic Algorithm (GA): According to [26], "Genetic Algorithms are search algorithms based on mechanics of natural selection and natural genetics". Genetic Algorithm is composed of binary bit strings collections, the initials values of the collections are determined randomly and evaluated. In the complex space, each collection of ones and zeros can be searched for and a ranking will be returned from an evaluation function of the relation between the collections [10]. Genetic Algorithms process requires an initial population to start, the population can be randomly created [27]. As an initial population size, number of studies have used a range between 30 and 100 [26].

2.4 Model Evaluation

The Goodness-of-fit (GOF) tests are used to measure the variance between the actual and the forecasted value. These measures are used in testing the statistical hypothesis and when comparing results across multiple studies or competing model. GOF are widely used to evaluate the performance of the forecasting models. The mean absolute percentage error (MAPE) is one of the GOF measures that is used to evaluate model forecasting performance.

Syntax:

$$MAPE = \frac{1}{n} \sum_{i}^{n} \frac{|T_i - Y_i|}{|T_i|} * 100\% \tag{1}$$

where: T_i is the actual values.

Y_i is the Forecasted values.

n is the number of data points "data record".

3 Related Work

A study done by [8], proposed the artificial neural network (ANN) and classical linear regression models to predict Australia's low cost carrier passenger demand (PAX) and revenue passenger kilometers (RPK). The prediction was based on the following factors: Australia's population, real GDP, real GDP per capita, unemployment, tourism, air fares and 4 dummy variables. A previous study has been conducted on the use of Adaptive Neuro-Fuzzy Inference System (ANFIS) in forecasting passenger demand of Australia's LCCs (PAX) and revenue passenger kilometers (RPK) [14]. Moreover, the author have presented another related study focused on Genetic Algorithm (GA) models [9], aiming at estimating the weighting factors of the model that leads to predictions of Australia's low cost carrier passenger demand (PAX) and revenue passenger kilometers (RPK).

Another study by [7], developed a forecasting model to forecast the number of international and domestic airline passenger in Saudi Arabia using artificial neural network. The author has used 16 input variables in the ANN model, the results indicated that three factors were the most influencing factors in the Saudia Airline, and thus chosen for the model for forecasting. The factors were per capita income, population size, and oil gross domestic product. Furthermore, another study investigated the relationship between LLCs and the Saudi Arabia's international tourism demand and targeted the impact of Saudi Arabia's low cost carriers on the tourism demand. Authors used a time series forecasting model to forecast the international passenger arrivals to Saudi Arabia using Box Jenkins (SARIMA) [6].

Authors [10] have combined two artificial inelegant techniques to forecast the international and domestic air passenger demand in Egypt. The Hybrid techniques are back-propagation neural network and genetic algorithm. The prediction was based on the following factors: Gross domestic product, per capita income, population in Egypt, employed population, gross national product, Economic growth rate, and foreign exchange rate.

The primary purpose of the previous studies was to design solutions that can assist the airline industry to predict the air travel passengers demand using traditional and artificial intelligence forecasting techniques. Moreover, the accuracy was compromised because seasonal events were not considered. The majority of these studies have used only one technique either statistics techniques or artificial intelligence forecasting techniques. Also, they used different types of dataset (yearly, quarterly, monthly) which led to small dataset for training and validation. Nonetheless, our work will approach the problem by using three techniques, one traditional techniques (MLR) and two artificial intelligence forecasting techniques (ANN and GA) for comparison. This study will analyze the effects and impacts of Islamic holiday seasons such as Hajj and Ramadan.

4 Research Methodology

4.1 Data Collection

In this study, the dataset have been collected from King Khalid International Airport which include the following:

- Total Flights: Daily, monthly, yearly by Airline (Domestic and international).
- Total PAXs (Number of passengers): Daily, monthly, yearly by Airline (Domestic and international).
- Total PAXs (Number of passengers): Monthly and yearly by class (First, Business or Economy).
- Terminal one, two and three data for Domestic and international flights.

The data have records of all Riyadh international and domestic flights and passengers of major airlines, from 2009 to 2017.

4.2 Feature Extraction

Variable extraction is considered to be one of the most important steps for developing a forecasting model. Essentially, the model performance depend significantly on the input data used to train it. Based on an extensive literature review of the factors that influence passenger demand (see Sect. 2.1), six variables were selected from Table 1 for inclusion as independent variables in the forecasting model. The list of variables and their sources are described in Table 2.

Table 2. Summary of this study extracted features

List of variables	Data measurement	Data sources
Population Size	Millions	General Authority For Statistics (Kingdom of Saudi Arabia)
Employed population	Millions	General Authority For Statistics (Kingdom of Saudi Arabia)
Gross Domestic Product (GDP)	USD Billion	General Authority For Statistics (Kingdom of Saudi Arabia)/Trading Economics and World Bank Databases
Per Capita Income (PCI)	USD	TradingEconomics and World Bank Databases
Economic Growth Rate	Percent	TradingEconomics and World Bank Databases
Jet Fuel Price	USD per gallon	The International Air Transport Association (IATA)

The Exchange rate and Consumer price index was not selected, because both variables are used for international demand while in our study the demand is for domestic flights. Airfares was not selected due to data unavailability since we are using King Khlaid Airport dataset and not a specific airline company data. Finally, the GDP per capita and Total Expenditures were both excluded since these variable are not common to be used for domestic air travel demand forecasting.

The availability of a consistent dataset allows the use of a monthly data for the period 2009–2017. The data used in the estimation of the model originate from a variety of sources as listed in Table 2. This study uses Saudi Arabia population size at a year constant population value (e.g. the population size at year 2010 will be the same for all the 2010 twelve months). The same is applied for both Gross Domestic Product GDP and Per Capita Income PCI. On the other hand, the Employed population will be constant at quarter level of each year (e.g. the Employed population at year 2010 will be the Q1 value for the first three months, Q2 value for the 4–6 months, Q3 value for the 7–9 months, and Q4 value for the 10–12 months).

4.3 Data Normalization

Normalization is done to ensure the data inputs are at a uniform scale. Ensuring uniformed input values implicitly weights all features equally in their representation and can make training faster. The dataset used in the study have inputs that are on widely different scales like the Population Size and Per capita income (PCI). In order to get the same range of values for each input feature, normalization is required so that all the inputs are at a comparable range. By comparing the results on the related studies, it was found that the result derived by the z-score method is more efficient and effective than the other standardization methods. In this study, the z-score standardization method will be used to transforms that inputs into a new data ranges.

4.4 Forecasting

First, there is a common step for all the forecasting models that is the training and testing data used in the model. To capture the effects and impacts of Islamic holiday seasons such as Hajj and Ramadan, the dataset will be divided into two sub datasets. The first subset will be the monthly dataset, this dataset will includes all the monthly data from 2009 to 2017 (Jan 2009 to Dec 2017) and the second subset will be the partial dataset which will include a selected month from all the years (e.g. Jan 2009, Jan 2010, ... Jan 2017) as presented in the Fig. 1.

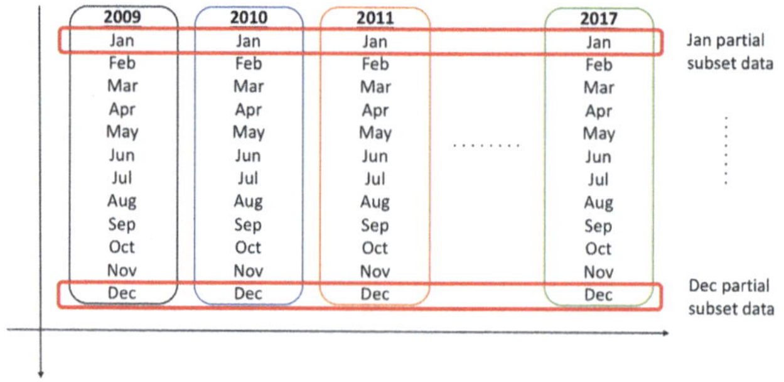

Fig. 1. Partial dataset for every month.

In each model, the data will be divided into three data sets: training data, testing data and validation data. The three datasets will be divided into a 70:15:15 ratio. The first dataset is the training dataset which will be used for model selection. Then the testing dataset that is used for evaluating the model's forecasting ability. For training and testing division, we will use cross-validation method to prevent under-fitting and over-fitting. Lastly, the validation dataset which will be used to validate the accuracy of the forecasting model, this dataset will contain data records that were not used in any form during the model development. Among the 9 years sample records, the data records from the year 2009 to 2015 are selected to form the training and testing set, while the year from 2016 to 2017 are used for validating the forecasting accuracy.

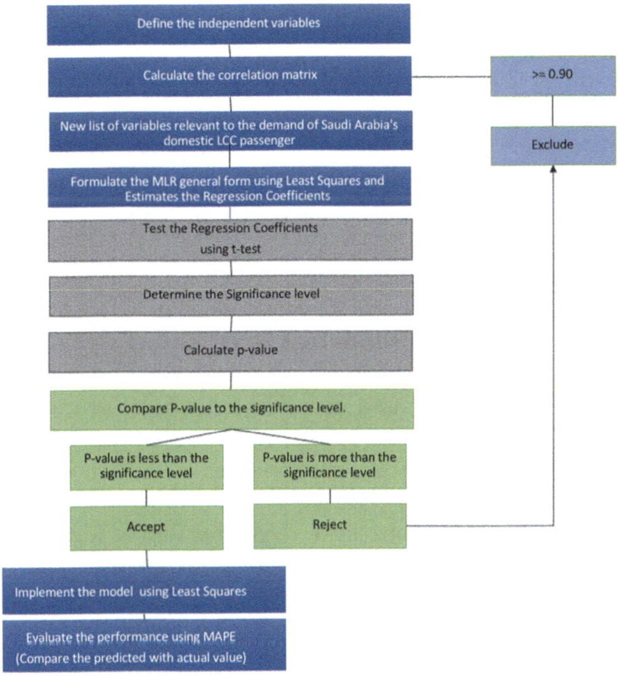

Fig. 2. The study's multiple linear regression modelling process.

Multiple Linear Regression Model (MLR)

The following steps will be followed to build the MLR model (see Fig. 2):

- Define the MLR model variables for forecasting Saudi Arabia's domestic LCC passenger demand. In this study, the candidate variables are listed in Table 2 in Sect. 4.2.
- Calculate the correlation matrix, from which the variables will be analyzed for inclusion in the model development. According to [2], if the correlation matrix shows high correlation (more than 0.90) between two variables then the variables should be excluded from the model.

- Analyze the correlation between the variables, a new list of variables relevant to the demand of Saudi Arabia's domestic LCC passenger should be considered.
- Formulate the MLR general form using Least Squares.
- Test the hypothesis using t test:
  $H0 : \beta_i = 0$
  $H1 : \beta_i \neq 0$
  The null hypothesis states that the regression coefficient is zero, and the alternative hypothesis is that the regression coefficient is not zero.
- The Significance level used is 0.05.
- Calculate the following: the standard error of the regression coefficients, the regression coefficients of the regression line, the degrees of freedom, the test statistic, and the p-value.
- Compare the p-value to the significance level. The null hypothesis will be rejected when the p-value is more than the significance level and accepted otherwise.
- Implement the model with only the accepted variable's regression coefficient using least square.
- Evaluate the model performance using MAPE.

Artificial Neural Networks Model (ANN)

The study's artificial neural network process is summarized in Fig. 3.

One of the most commonly used supervised Artificial neural network (ANN) models is Multi-Layer Perceptron (MLP) model. MLP uses Back-propagation network (BPN) algorithm. BPN is supervised network with a layered feed-back which is desirable model for forecasting passengers demand presented in this study. BPN consists of an input layer, a hidden layer, and an output layer. By comparing the actual output and the forecasted output, the error will be calculated at the output layer and will be propagated back to the input layer. This study BPN model is presented in Fig. 4. The candidate variables are listed in Table 2 in Sect. 4.2, they are all used in the input layer with one node for each input variable. To capture non-linear relationships between variables the hidden layer is used, the number of nodes required at the hidden layer will be experimental selection.

Genetic Algorithm (GA)

The genetic algorithm will be used in this study, steps are outlined as follows:

(1) Generate a random population, where each chromosome is n-length, with n being the number of variables. That is, random weights of each variable are established with n elements each.
Where $n = \{Xi : i = 1, 2, \ldots, n\}$.
(2) Select the weights that increase desirable results.
(3) Apply crossover and mutation to the selected weights and generate an offspring.
(4) Form a new population using the selection operator by recombine the offspring and the current population.

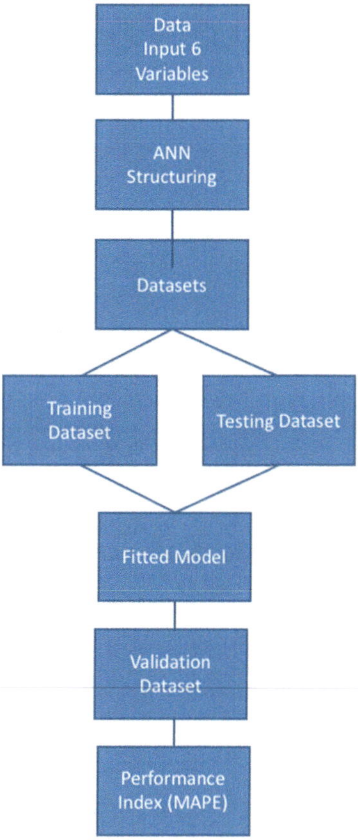

Fig. 3. The artificial neural network modelling process for this study. (adapted from [28]).

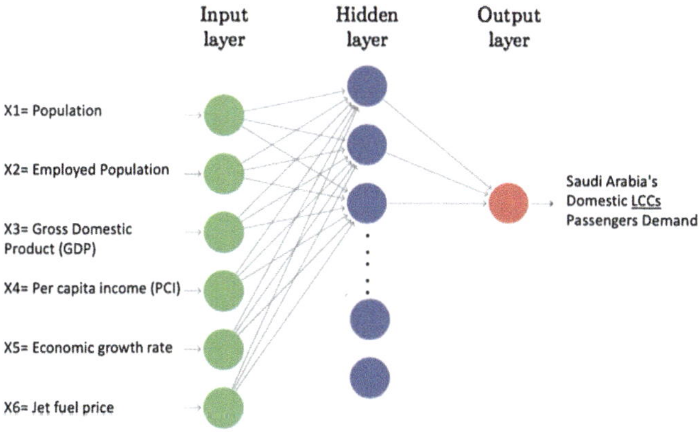

Fig. 4. The artificial neural network structure for this study.

(5) Repeat steps two to four.
(6) The fitness function used is MAPE, the fitness function terminates if there is a low chance to achieve significant changes in the next generations.

4.5 Performance Evaluation

Performance evaluation is an essential step to establish a deep understanding of the model and to measure the model's reliability and accuracy. The mean absolute percentage error (MAPE) is the selected measure to evaluate model forecasting performance since it was the most used measure in all related studies. Since this study will use a Multiple linear regression model (MLR), an artificial neural network (ANN) and a Genetic Algorithm model (GA) to predict Saudi Arabia's LCCs domestic air travel demand, the final results will present a comparison of the forecasting accuracy of Saudi Arabia's domestic LCCs passenger demand models (MLR, ANN, and GA) using MAPE. To classify the forecasting accuracy, the model performance index will be classified into four categories as presented by [29] in Table 3, which has been cited in at least 84 other reported studies. Moreover, it's often used as benchmarks for forecasting accuracy comparisons among forecasting models [30].

Table 3. MAPE classification, Adapted from [29]

MAPE value	Forecasting accuracy
MAPE $<$ 10%	Highly accurate forecasting
10% $\leq$ MAPE $\leq$ 20%	Good forecasting
20% $\leq$ MAPE $\leq$ 50%	Reasonable forecasting
MAPE $>$ 50%	Inaccurate forecasting

5 Results

We continue working on this proposal with the aim of forecasting Saudi Arabia's domestic LCC passenger demand using different Machine Learning techniques and investigates their prediction accuracy. The Machine Learning techniques, that are followed in this study will also be compared to investigates the effectiveness of using these techniques to forecast passengers demand of Saudi Arabia's low cost carriers (LCCs). Finally, we look forward to provide a reliable forecast that can assists and empower Airport management in predicting the future, so as to plan the supply of services that are required to satisfy the travel demand.

6 Conclusion

Considering the increasing importance of the travel industry on the economics of Saudi Arabia, it is necessary to provide accurate and reliable forecast for the low cost carriers (LCCs) of Saudi Arabia and determine the needs for more LCCs and capital investments. This research investigates the effectiveness of using machine learning techniques to forecast Saudi Arabia's LCCs passenger demand by accounting the Islamic Holidays, which could be considered to make decisions and support airport long-term growth. This could then influence various domains including economic, strategic management, and business. Accurate Forecasts empower management in planning and making decision for the future. In future, results could be improved by analyzing larger datasets with both domestic and international LCC passenger demand. Moreover, it would be beneficial to extend the seasonal events to cover other events such as summer holiday.

References

1. Blinova, T.: Analysis of possibility of using neural network to forecast passenger traffic flows in russia. Aviation **11**(1), 28–34 (2007)
2. Ba-Fail, A.O., Abed, S.Y., Jasimuddin, S.: The determinants of domestic air travel demand in the kingdom of saudi arabia. J. Air Transp. World Wide **5**(2), 72–86 (2000)
3. Abed, S.Y., Ba-Fail, A.O., Jasimuddin, S.M.: An econometric analysis of international air travel demand in Saudi Arabia. J. Air Transp. Manage. **7**(3), 143–148 (2001)
4. Sivrikaya, O., Tunç, E.: Demand forecasting for domestic air transportation in Turkey. Open Transp. J. **7**(1), 20–26 (2013)
5. Priyadarshana, M., Fernando, A.S.: Modeling air passenger demand in bandaranaike international airport Srilanka. J. Bus. Econ. Policy **2**(4), 146–151 (2015)
6. Alsumairi, M., Tsui, K.W.H.: A case study: the impact of low-cost carriers on inbound tourism of saudi arabia. J. Air Transp. Manag. **62**, 129–145 (2017)
7. BaFail, A.O.: Applying data mining techniques to forecast number of airline passengers in Saudi Arabia (domestic and international travels). J. Air Transp. **9**(1) (2004)
8. Srisaeng, P., Baxter, G.S., Wild, G.: Forecasting demand for low cost carriers in australia using an artificial neural network approach. Aviation **19**(2), 90–103 (2015)
9. Srisaeng, P., Baxter, G., Richardson, S., Wild, G.: A forecasting tool for predicting australia's domestic airline passenger demand using a genetic algorithm. J. Aerosp. Technol. Manag. **7**(4), 476–489 (2015)
10. El-Din, M.M., Farag, M., Abouzeid, A.: Airline passenger forecasting in Egypt (domestic and international). Int. J. Comput. Appl. **165**(6) (2017)
11. El-Din, M.M., Ghali, N., Sadek, A., Abouzeid, A.: A study on air passenger demand forecasting from Egypt to Saudi Arabia. Communications **3**, 1–5
12. Suryani, E., Chou, S.-Y., Chen, C.-H.: Air passenger demand forecasting and passenger terminal capacity expansion: a system dynamics framework. Expert Syst. Appl. **37**(3), 2324–2339 (2010)
13. Sohag, M.S., Rokonuzzaman, M.: Demand forecasting for a domestic airport-a case study

14. Srisaeng, P., Baxter, G.S., Wild, G.: An adaptive neuro-fuzzy inference system for forecasting australia's domestic low cost carrier passenger demand. Aviation **19**(3), 150–163 (2015)
15. Srisaeng, P., Richardson, S., Baxter, G., Wild, G.: Forecasting australia's domestic low cost carrier passenger demand using a genetic algorithm approach. Aviation **20**(2), 39–47 (2016)
16. Chen, S.-C., Kuo, S.-Y., Chang, K.-W., Wang, Y.-T.: Improving the forecasting accuracy of air passenger and air cargo demand: the application of back-propagation neural networks. Transp. Plann. Technol. **35**(3), 373–392 (2012)
17. T.G.A. for Statistics: The general authority for statistics (2017). https://www.stats.gov.sa/
18. O. B. Group: Airport privatisation gathers speed in Saudi Arabia (2016). https://oxfordbusinessgroup.com/
19. Hsu, C.-I., Wen, Y.-H.: Improved grey prediction models for the trans-pacific air passenger market. Transp. Plann. Technol. **22**(2), 87–107 (1998)
20. Grosche, T., Rothlauf, F., Heinzl, A.: Gravity models for airline passenger volume estimation. J. Air Transp. Manage. **13**(4), 175–183 (2007)
21. Chen, M.-S., Ying, L.-C., Pan, M.-C.: Forecasting tourist arrivals by using the adaptive network-based fuzzy inference system. Expert Syst. Appl. **37**(2), 1185–1191 (2010)
22. Hill, T., O'Connor, M., Remus, W.: Neural network models for time series forecasts. Manage. Sci. **42**(7), 1082–1092 (1996)
23. Nagar, Y., Malone, T.: Making business predictions by combining human and machine intelligence in prediction markets (2011)
24. Tiryaki, S., Aydın, A.: An artificial neural network model for predicting compression strength of heat treated woods and comparison with a multiple linear regression model. Constr. Build. Mater. **62**, 102–108 (2014)
25. Barla, P., De Sève, P.J.: Demand uncertainty and airline network morphology with strategic interactions, Universite Laval (1999)
26. Goldberg, D.: Genetic Algorithms in Optimization, Search and Machine Learning. Addison-Wesley, Reading (1989)
27. Godinho, P., Silva, M.: Genetic, memetic and electromagnetism-like algorithms. In: The Routledge Companion to the Future of Marketing, p. 393 (2014)
28. Jiang, D., Zhang, Y., Hu, X., Zeng, Y., Tan, J., Shao, D.: Progress in developing an ann model for air pollution index forecast. Atmos. Environ. **38**(40), 7055–7064 (2004)
29. Martin, C.A., Witt, S.F.: Accuracy of econometric forecasts of tourism. Ann. Tour. Res. **16**(3), 407–428 (1989)
30. Li, G., Wong, K.K., Song, H., Witt, S.F.: Tourism demand forecasting: a time varying parameter error correction model. J. Travel. Res. **45**(2), 175–185 (2006)

Artificial Morality Based on Particle Filter

For Moral Dilemmas in Intelligent Transportation Systems

Federico Grasso Toro[1(✉)] and Damian Eduardo Diaz Fuentes[2]

[1] Working Group 8.51 Metrology Software,
Physikalisch-Technische Bundesanstalt, Berlin, Germany
federico.grassotoro@ptb.de
[2] Institute for Quality, Safety and Transportation, Braunschweig, Germany
damian.diaz@iqst.de

Abstract. Advanced driver assistance systems seem to be the core of many intelligent transportation systems aiming to increase safety. Although, the ever-present question of how safe will it be to allow an artificial intelligence based system to perform safety-relevant functions remains. We believe this problem falls in the field of affective computing, also called emotion AI. This paper offers an exemplary solution by means of particle filters, creating and developing the concept of Artificial Faith to apply it into methods of Artificial Morality. Subsequently, these new concepts are applied to an AI localization system estimating position by Global Navigation Satellite Systems. The tested AI uses spatial road networks data for the identification of which lane the vehicle travels, exploiting prior networks information. However, incorporating complex information constraints the system to highly non-Gaussian posterior densities, extremely difficult to represent with accuracy. The particle filter approaches with and without digital map-matching present no restrictions for non-linearity of models as well as noise distribution, therefore allowing both the speed and the heading measurement errors to be modelled accurately. Therefore, it is proven that the tested AI system can increase its quality by improving its accuracy, capturing multi-modal distributions, while having a natural way of incorporating information such as digital maps, when available. These particle filter based location estimators work as part of the tested AI localization system, providing the newly developed concepts of soft and hard artificial morality and solving a simple artificial moral dilemma for artificial intelligent accuracy-based quality function, reducing its referential position uncertainty.

Keywords: Particle filter · Map matching · Artificial intelligence ethics
Artificial morality

1 Introduction to the Tested AI-Based System

The Global Navigation Satellite System usage for civilian localization is an ever growing application, as GNSS services keeps expanding, e.g. BeiDou and Galileo. At the same time, GNSS based safety-relevant applications for land vehicles are developed

K. Arai et al. (Eds.): IntelliSys 2018, AISC 868, pp. 1221–1237, 2019.
https://doi.org/10.1007/978-3-030-01054-6_85

to construct essential components of Intelligent Transportation Systems (ITS), e.g. driver assistance in cars or railway localization for train control system. In this evolving field many technologies are combined to provide specialized information to the systems. Nevertheless, GNSS usage for safety-relevant applications will always require reliable GNSS-based localization systems for navigation purposes, as the developed intelligent GNSS-based localization system in [1].

The mathematical-based Particle Filter (PF) methods for position estimation described in [2], applied in a demonstrator in [1] are the philosophical bases of the newly developed concepts of artificial morality for AI-based systems.

The present paper focuses on the mathematical description of the artificial faith based concepts of soft and hard artificial morality, by means of PFs intended to be used as aid for any AI-based system that takes into consideration an approach based on needed AI ethics, as suggested in [3].

Any future safety-relevant application of AI in ITS based on the prototype developed in [1] must contain, at least:

(1) Position estimation module based on Particle Filter (PF).
(2) Deviation evaluation module based on Mahalanobis Ellipses Filter (MEF) for accuracy description.
(3) Quantitative and qualitative validation models for GNSS based on Artificial Neural Networks (ANN) models.

Figure 1 depicts the prototype, after an ANN configuration provides weights for the relevance of GNSS inputs.

The ANN validation tool matches ANN-based estimations of trueness, precision and location availability (defined as "the percentage of accurate data within the available dataset") calculated by the MEF methodology, providing a description of a bivariate deviation (in both easting deviation and northing deviation), between the GNSS and an independent reference system (RS), describing the correlated behavior of the deviation dataset by means of the rotation value of the resulting MEF ellipses.

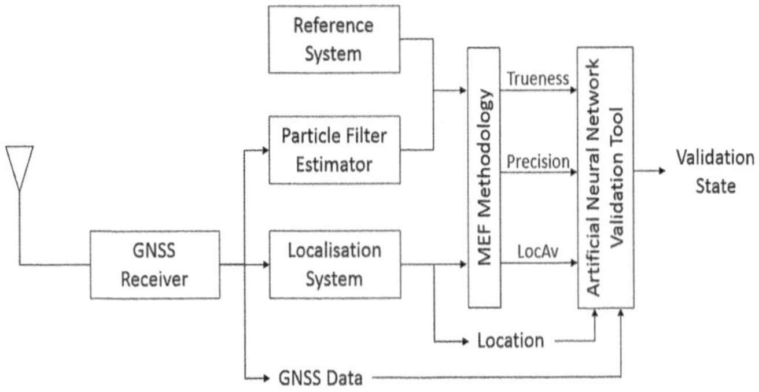

Fig. 1. Block description of intelligent GNSS-based localization system [1].

Based on the quality of the results provided for the MEF methodology and the quality of the provided reference system, the developed Particle Filter Estimator in Fig. 1 needs to be both accurate and reliable. The PF-based approach has no restrictions regarding non-linearity of the used models and the noise distribution, so both values of velocity and heading measurement errors can be modelled with accuracy. In [2], the most significant advantages of the proposed particle filter approach for the map-matching application are summarized as follows:

(1) Particle Filters provide a natural way to incorporate road map information into vehicle position estimation.
(2) Particle Filters can capture multi-modal distributions, to improve vehicle localization.

Furthermore, a dynamic location estimator based on particle filters and aided by means of map-matching techniques is presented with even better results in [1].

The state estimation problem presented and solved by PF-based techniques shows that the dynamic evolution of the vehicle position by speed control input and a constraining digital map, as presented in [2], can be considered an example for an artificial moral dilemma for an ITS based AI systems.

This paper focuses on the need for principles to solve the real time AI moral dilemma of relaying or not in accuracy information provided by an AI-based system in safety-relevant applications, as proposed in [3].

While PF estimator without map-matching technique solves the AI moral dilemma by means of a "soft artificial morality" (SAM), the PF estimator with map-matching technique solves it by means of a "hard artificial morality" (HAM). These estimators present solutions for the ITS non-stationary inverse problem. This problem is solved by using the available measured data, combined with prior knowledge of the physical phenomena, as well as with the instruments to previously produce statistical estimations of the desired dynamic variables, minimizing its error, as seen in [4].

In the field of AI ethics and affective computing, these methods equal to the construction of the concept of an artificial faith, as well as its correlated soft and hard artificial moralities for the real-time solution of AI moral dilemmas. These solutions can be extended to solve other AI moral dilemmas.

The AI moral dilemma of the demonstrator in [1] deals with the artificial moral decision based on accuracy quality function. Therefore, the creation of PF-based estimators for artificial morality allows the AI system to solve dilemmas by means of a "morally corrected" set of decisions. When the solution of the dilemma depends on the previously available information, it is called SAM, and the decision is taken from the results of the PF. When the moral dilemma can be limited within a specific path of choice, also called "artificial faith from previous belief", the solution depends on available map information (known path) and it is called HAM. In this last case, the AI-based system can only take decisions within a predefined path, both literally and figuratively. In the following section, the mathematical bases for PF-based artificial morality approaches are presented.

2 Particle Filter Method

In AI ethics, the presented methods for artificial morality applications aiming to reduce uncertainty in referential position, allow the best possible decisions to be taken in any simple AI moral dilemma. A theoretical description of the AI ethics problem can be presented as follows:

Stating an estimation problem for an ethical state can be done by combining prediction models with uncertain measurements, in order to obtain accurate estimations of the system variables, as in [5]. These problems can be typically solved by Bayesian filters [4], where the Bayesian approach to statistics attempts utilizing available information, reducing the present uncertainty in the stated decision-making problem. When any new information is obtained, it can be added to the previous information for a new statistical procedure [6].

In both cases, the Bayes' theorem is used to combine the newly collected information and the previously available information, as described in [7]. Kalman Filter is the most known and used Bayesian filter method, developed in [5]. However, KF application are limited to linear models within additive Gaussian noises. And therefore, an Extended Kalman Filter approach should be used for less restrictive cases, as presented in [8] by the usage of linearization techniques.

In addition to these existing methods, there are Monte Carlo Methods called Particle Filter methods, developed in [9]. They use posterior density for random sampling as well as associated weights, not requiring restrictive hypotheses, like KF or EKF approaches. This makes them a better suited solution for non-linear models with non-Gaussian errors.

The PF methods used in [1] are the main component of the data fusion block presented in Fig. 2. This block produces a reference for the deviation analysis [10]. Further explanations of the used Mahalanobis Ellipses Filter (MEF) methodology can be found in Sect. 5. Consequently, the accuracy of both proposed artificial morality methods is evaluated by MEF in Sect. 6.

2.1 PF-Based Method

It can be described as a sequential Monte Carlo technique for solving state estimation problems, by the usage of Sequential Importance Sampling (SIS) algorithm, including a resampling step as presented in [8]. The SIS algorithm uses an importance density to represent densities that cannot be computed. In AI ethics, SIS represents the construction of background cultural morality for moral enquires.

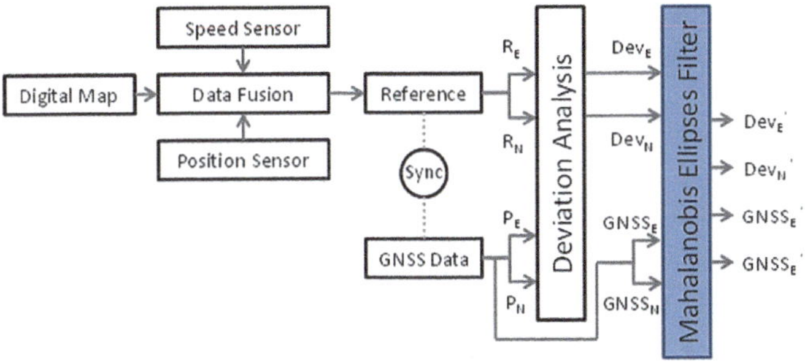

Fig. 2. Deviation analysis of data fusion-based reference [10].

Then, samples can be drawn from this importance density, and not from the actual density.

Mathematically, this can be explained as follows:

When $\{x^i_{0:k},\ i = 0,\ldots,N\}$ are particles having associated weights $\{w^i_k, i = 0,\ldots,N\}$ and $x_{0:k} = \{x_j,\ j = 0,\ldots,k\}$ is a set of all states up to t_k, where $k < N$; N meaning number of particles.

All weights can be normalized as $\sum_{i=1}^{N} w^i_k = 1$.

Having posterior density discretely approximated at instant t_k:

$$\pi(x_{0:k}|z_{1:k}) \approx \sum_{i=1}^{N} w^i_k \delta\left(x_{0:k} - x^i_{0:k}\right) \tag{1}$$

where δ is presenting the Dirac delta function. Considering the assumptions in [7], for the "evolution-observation model":

(1) The considered sequence x_k for k = 1, 2,…, can be a Markovian process, such as:

$$\pi(x_k|x_0, x_1, \ldots, x_{k-1}) = \pi(x_k|x_{k-1}) \tag{2}$$

(2) The sequence z_k for k = 1, 2,…, can be a Markovian process from the history of x_k, such as:

$$\pi(z_k \mid x_0, x_1, \ldots, x_k) = \pi(z_k \mid x_k) \tag{3}$$

(3) Then, the sequence x_k only can depend on its own history of past observations, such as:

$$\pi(x_k \mid x_{k-1}, z_{1:k-1}) = \pi(x_k \mid x_{k-1}) \tag{4}$$

being $\pi(a|b)$ the "conditional probability of a when b". By the usage of these hypotheses, from (2) and (4), the resulting posterior density, based on (1), can be re-stated as:

$$\pi(x_k \mid z_{1:k}) \approx \sum_{i=1}^{N} w_k^i \delta(x_k - x_k^i) \tag{5}$$

The SIS Particle Filter usually presents the problem of degeneracy phenomenon, meaning that after some states, any state but one particle can have negligible weight, as in [9].

The degeneracy phenomenon implies large computational effort that shall be used to update the particles whose contribution overall to the approximation of the posterior density function ends up meaning almost zero. In AI ethics this equals to heavy cultural weights that do not add (or add close to nothing) to the moral dilemma to be solved.

To overcome this problem the number of particles is increased or the importance density as prior density $\pi(x_k|x_{k-1}^i)$ is appropriately selected. Special resampling techniques are also recommended to avoid the degeneracy phenomenon [11]. In AI ethics, this means signifying the cultural diversification for the artificial morality to be determined.

The resampling can be described as the mapping of random measures $\{x_k^i, w_k^i\}$ into a single random measure $\{x_k^{i*}, N^{-1}\}$ within uniform distributed weights. This resampling is performed if and only if the number of effective particles (with large weights) is below a threshold. Alternatively, the resampling can be done indistinctively, happening at every single instant t_k. But for this, the Sampling Importance Resampling (SIR) algorithm needs to follow [8].

The algorithm is applied to the evolution, from t_{k-1} to t_k, as presented in [12], summarized in the following three steps:

Step 1. For $i = 1,\ldots,N$ all new particles x^i_k shall be drawed from prior density $\pi(x_k \mid x^i_{k-1})$. Then, the likelihood density shall be used for the calculation of weights $w^i_k = \pi(z_k \mid x^i_k)$.

Step 2. Total weight, described by $T_w = \sum_{i=1}^{N} w_k^i$ shall be calculated. Then, particle weights, for $i = 1,\ldots,N$ shall be normalized, meaning $w_i = T_w^{-1} w^i_k$.

Step 3. The particles resampling is done:

Step 3.1. The cumulative Sum of Weights (CSW) is constructed by solving $c_i = c_{i-1} + w^i_k$ for $i=1,\ldots, N$, with $c_0=0$.

Step 3.2. With $i = 1$, starting point u_1 is drawn from the uniform distribution $U[0, N^{-1}]$.

Step 3.3. For j = 1,…, N

Movement along CSW, by making $u_j = u_1 + N^{-1}(j-1)$

While condition $u_j > c_i$ make $i=i+1$

Therefore assigning samples $x^j_k = x^i_k$.

And therefore assigning weights $w^j_k = N^{-1}$.

While these resampling steps presented above reduce many effects of the described degeneracy problem, this may also lead to loss of diversity (in AI ethics meaning a cultural bias), due to repeated particles in the resulting sample [2]. This is called a "sample impoverishment" problem and it can be used in cases with small process noise, where the particles can collapse within a single particle in a few instants t_k, as in [4].

It must be stated that a disadvantage of this approach has to do with the large computational cost needed to solve the SMC methods. This does not allow its application to extremely complex problems. A solution might be the algorithms capable of estimating state variables as well as parameters, simultaneously. An example is presented in [9].

2.2 Particle Filter with Map Matching

A map matching approach integrates GNSS localization data with a digital map of the travelled path to identify the lane.

This addition in the artificial morality framework would be considered as the introduction of a "fixed faith" into the moral dilemma resolution. The digital track map for artificial morality would be an obligatory path to follow; therefore, it would be the "hard morality" of the AI-based system.

The moral dilemma, considered here a moral stated problem that would take into consideration a dynamical evolution of the moral state, as well as the evolving cultural morality constraining it into a known path, provided that the AI system has a culturally dynamic hard artificial morality (HAM) as presented in the following sections.

This shall be required for all ITS safety-related applications requirements. The performance of the intelligent localization system supporting navigation function in the ITS from Fig. 1 shall then improve by the usage of map-matching and HAM.

For an ITS, a required horizontal positioning accuracy is between 1 to 40 m (95%) [1]. Map matching research has been done aiming to achieve this requirement, by means of:

(1) Topological analysis of spatial road network data [13],
(2) Probabilistic theory [14],
(3) Bayesian filtering [15],
(4) Fuzzy Logic [16], and
(5) Belief Theory [17].

But to incorporate previously known digital map information to conventional Kalman Filter framework research is not easy. It depends too much on its specific application.

A Particle Filter approach was proposed in [18]. This can be used within the current AI ethics research to prove that the AI system can validate its decision through a constructed moral dilemma statement. The state vector for the moral decision, based on the artificial morality model, consists on the coordinates (Northing and Easting): $x_k = \left[p_k^N p_k^E\right]^T$ as presented within a subscript k to correspond to t_k (time instant).

Assuming that the movement between states happens on a known path (cultural bias), the general description that was proposed in [19], can be the non-linear function $\rho^{h}(x)$:

$$R^{h} = \{x : \rho^{h}(x) = 0\},\ h = 1,\ M \tag{6}$$

The path network is modelled by path segments $R_{k,k+1}$, being straight lines between the nodes ξ_{k}, ξ_{k+1} that satisfy (6).

Also, it is assumed that the partially observable discrete-time Markov chains describe the AI moral state. And furthermore, the state x_k would always depend on the x_{k-1} described by the probabilistic law $\pi(x_k|x_{k-1})$.

This AI moral dilemma then can be stated as the estimation of a sequence of states $x_{0:k} = \{x_0, \ldots, x_k\}$, given the series of observation $z_{1:k} = \{z_{1,\ldots,} z_k\}$ subject to the artificial morality motion model $\pi(x_k|x_{k-1})$, within the measurement model $\pi(z_k|x_k)$ and the constraints on the state vector for the moral decision, given in form of the moral path.

Prior probability at t_0, $\pi(x_0)$ is known and the goal to find the "best" trajectory is solved by minimum mean-square error criteria, by Bayesian estimation theory [19], based on the dead-reckoning equations:

$$x_{k+1} = \begin{bmatrix} P^{N}_{k+1} \\ P^{E}_{k+1} \end{bmatrix} = X_k + V_k T_k \begin{bmatrix} \cos\psi_k \\ \sin\psi_k \end{bmatrix} \tag{7}$$

with V_k being speed, k meaning the heading rate and T_k representing the sampling period for the time instant t_k.

The presented technique has two operational modes; meaning an artificial faith describing two artificial moralities:

(1) AI System moves along a given path (stable faith).
(2) AI System turns and is located between two adjacent lanes in the path (doubtful faith).

The possible switching, between the two operational modes, can be performed based on the analysis of the AI system range of the encountered moral dilemma (heading rate).

Figure 3 presents characteristic particle trajectories during a vehicle turn for the second operational mode, as presented and explained in details in [18].

This means the finding of the path of its AI ethical path. This would mean that the system would be losing the acquired moral bias for the next moral dilemma. The position errors accumulated during the turn, considered a cultural bias of the AI system, there will be residual along-path error in the estimated vehicle location after the turning situation. The magnitude of this accumulated error will depend on the quality of the dead-reckoning sensors and the turning curvature characteristics.

2.3 Analogy to Artificial Faith

From Fig. 3, the turn during trajectory of the vehicle can be easily understood as the decision to make by the AI.

In this case, Artificial Faith provides the path to follow. And since, in this case there is the possibility of having a fixed faith (aided by additional information, in this case a digital map), it can be understood and applied as the previously mentioned Hard Artificial Morality (HAM).

While, in cases of lacking inside information, the proposed Artificial Faith can only be applied as Soft Artificial Morality (SAM). Questions regarding the Artificial Faith by hard and soft artificial morality can be answered observing the SAM as a model-based predictor for the next position in the path, while the HAM can be described as a model based predictor for the next position in the path, aided by an available database from where to extract previously acquired information to help to solve the present moral dilemma, oriented by the fixed faith provided by the artificial path to follow.

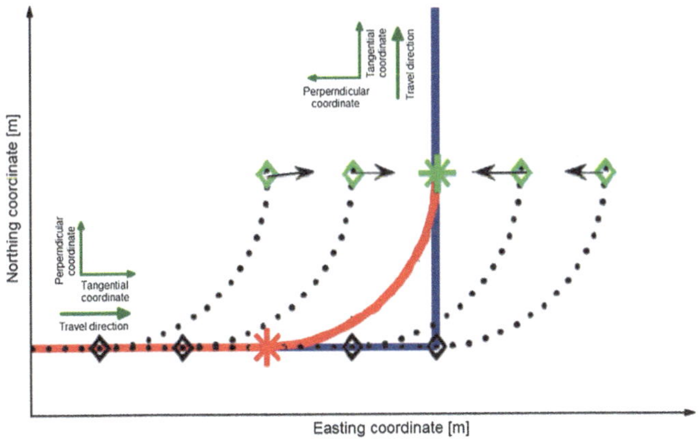

Fig. 3. Particle trajectories during turn/decision [18].

This is the solution for a simple AI moral dilemma. In the following sections the mathematical bases for both SAM and HAM are presented.

3 Soft Artificial Morality Approach

The soft artificial morality (SAM) approach presented here is a PF-based method. It takes into consideration a dynamic evolution model described in [2], with a position time-discrete evolution equation, such as:

$$x_{k+1} = \begin{bmatrix} E_{k+1}^{N} \\ E_{k+1}^{E} \end{bmatrix} = x_k + \begin{bmatrix} \cos \psi_k \\ \sin \psi_k \end{bmatrix} vel_k \Delta T + v_k \quad (8)$$

where the term E represents the evolution model estimated position and T its sampling period.

Detailed explanation can be found in [20].

The SAM observation model, based on the moral position, as measured by the GNSS-Receiver in [2] uses the modelled observation equation:

$$z_k = \begin{bmatrix} O_{k+1}^{N} \\ O_{k+1}^{E} \end{bmatrix} = x_k + n_k \tag{9}$$

where the term O represents the observed position. Detailed explanation can be found in [20]. For a detailed performance evaluation of this filter technique, referred to [2].

In the present paper these equations are considered and applied, based on the simulated dynamic moral dilemma of the AI system traveling along a defined morality path. All relevant results and their corresponding interpretation are presented in the Sect. 6.

4 Hard Artificial Morality Approach

The hard artificial morality (HAM) approach presented here is a PF-based method. It takes into consideration a dynamic evolution model with speed control input within a constraint digital map, as presented in [2].

The evolution of the model has been optimized for the AI system ethics case, by means of a single-direction moral path, where there is no branching possible.

The model is also assumed to allow the AI system only be located over the digital map.

In AI ethics, this means that the AI system can only follow its own defined path, supplying a HAM (hard artificial morality) to solve potential moral dilemmas. The simplified model described in [2] is presented here as a one-dimensional state variable, used to predict time-discrete evolution equation:

$$\mathrm{x}_{\mathrm{k}+1} = \mathrm{E}_{\mathrm{k}+1} = \mathrm{x}_{\mathrm{k}} + \mathrm{vel}_{\mathrm{k}}\Delta\mathrm{T} + \mathrm{v}_{\mathrm{k}} \tag{10}$$

where the term E represents the evolution model and measures the moral distance along the morality path, from its starting point to the AI system position.

Detailed explanation of the mathematical background can be found in [20].

Also, the observation model is based on the horizontal moral decision is modelled:

$$z_k = \begin{bmatrix} O_{k+1}^{N} \\ O_{k+1}^{E} \end{bmatrix} = h(x_k) + n_k \tag{11}$$

where the term O represents the observed horizontal position and the h function is a time-independent non-linear function relating the one-dimensional position of the AI system, along the morality path with a bi-dimensional horizontal moral dilemma.

Detailed explanation of the mathematical background can be found in [2] and [20].

All results and interpretations are presented in Sect. 6.

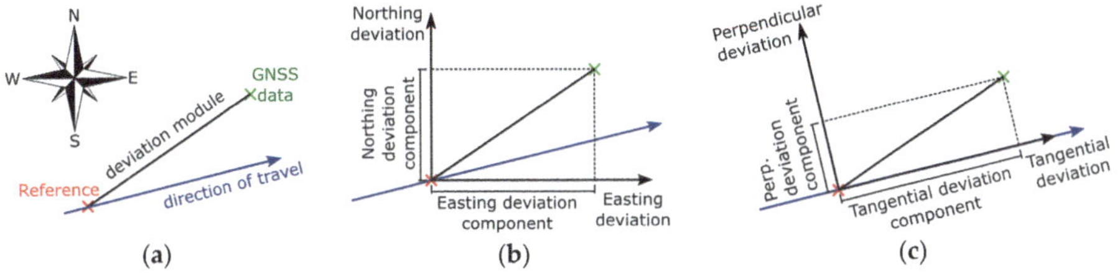

Fig. 4. (a) Deviation module; (b) Easting/Northing deviation components; and (c) Tangential/Perpendicular deviation components [1].

5 MEF Methodology Overview

5.1 Classifying Deviations

A full study of accuracy analysis of a GNSS-based system by means of its measured deviation over time is described in [10]. But to understand the MEF methodology used, first it is needed to differentiate between the possible representations of deviation. Figure 4 shows a graphical description of possible deviation classification. Defined as "the distance between satellite data and reference system data" the deviation can be represented in several ways while Fig. 4(b) presents deviation in Easting and Northing components. The deviation decomposition into Tangential and Perpendicular components can be seen in Fig. 4(c).

The Easting/Northing representation can be used to find reliability margins of accuracy, as in [20]. And furthermore, a represented deviation in Tangential and Perpendicular coordinates can be useful to develop the current AI System artificial morality dilemma.

The information provided by the deviation analysis can describe the whole localization system's behavior. Detailed examples and applications can be found in [10], where the "Mahalanobis Ellipses Filter" (MEF) is introduced.

MEF methodology for accuracy analysis takes into consideration the correlation between deviation components, providing a quality description of dynamic AI-based localization systems.

5.2 Mahalanobis Distance

In [21], Mahalanobis introduced his distance as a statistical distance measure based on the correlations between variables, to identify different patterns within related datasets, analyzing similarities of unknown samples to a known dataset. Examples of applications of this method can be found in [22, 23].

Differences between MD and Euclidean distance are explained in Fig. 5 by two representations of bivariate sample datasets.

In Fig. 5(a), there are constant Euclidean distance circles, traced to help visualize the ED from the center of the dataset.

Meanwhile, in Fig. 5(b) the plotted ellipses are there to represent the constant values of MD, as 1σ, 2σ and 3σ.

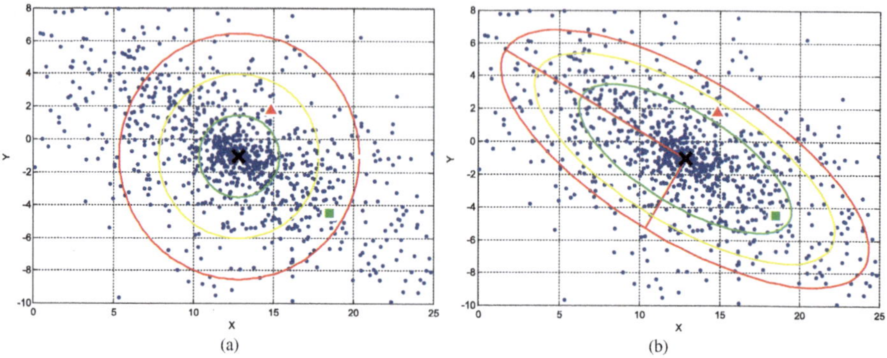

Fig. 5. (a) Dataset with ED circles of 2.5, 5, and 7.5 units; (b) Dataset with MD ellipses of 1σ, 2σ, and 3σ [9].

Considering a bivariate vectors dataset:

$$Dataset = \{vec_1, vec_2, \ldots, vec_n\},$$

where each vector has two components (X and Y):

$$\overrightarrow{vec_i} = (x_i, y_i),$$

and each component belongs to a univariate elements subdataset, such as:

$$x_i \in Subdataset\ x = \{x_1, x_2, \ldots, x_n\},$$
$$y_i \in Subdataset\ y = \{y_1, y_2, \ldots, y_n\},$$

then, the calculation of the mean values is:

$$x_\mu = \frac{\sum_{i=1}^{n} x_i}{n}\ ;\ y_\mu = \frac{\sum_{i=1}^{n} y_i}{n}.$$

Represented the mean value of the dataset by the vector:

$$\overrightarrow{vec_\mu} = (x_\mu, y_\mu),$$

from on the dataset bivariate vectors, the correlation matrix results:

$$S = \frac{1}{n-1}\sum_{i=1}^{n}\left[(\overrightarrow{vec_i} - \overrightarrow{vec_\mu})^T(\overrightarrow{vec_i} - \overrightarrow{vec_\mu})\right], \quad (1)$$

and the Mahalanobis distance for any given vector vec_i:

$$M_D(\overrightarrow{vec_i}) = \sqrt{(\overrightarrow{vec_i} - \overrightarrow{vec_\mu})S^{-1}(\overrightarrow{vec_i} - \overrightarrow{vec_\mu})^T}. \quad (2)$$

Fig. 6. Mathematical summary for Mahalanobis distance.

There are two observations from the dataset that were marked as exemplary.

One observation was marked with a red triangle; and the other one was marked with a green square.

The comparison of the provided locations of the marks data to the constant ED circles makes believe that the red triangle is closer to the center of the dataset than the green square.

But, on the other hand, by considering the MEF approach for outliers' detection, the red triangle results further out than the green square, as presented in Fig. 5(b).

A summary of Mahalanobis distance and its correspondent mathematical representation is given in Fig. 6. Details regarding this representation and its correspondent applications can be found in [10] and [1].

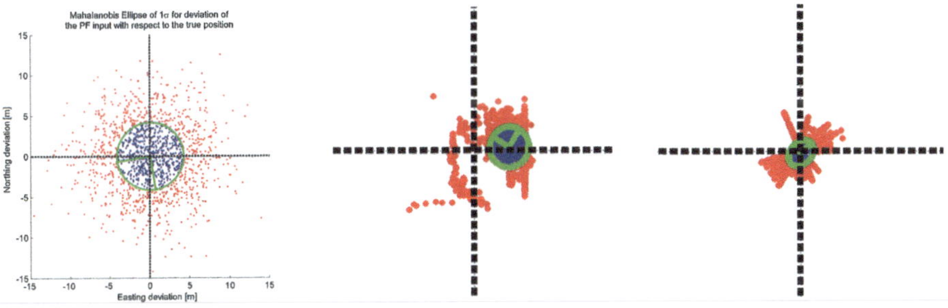

(a) (b) (c)

Fig. 7. Filter input/output comparison: Uncertainty description for the moral dilemma as deviation between GNSS data and RS in [1]. (a) PF input; (b) Output of the PF approach without digital map; and (c) Output of the PF approach with digital map.

6 Summary of Proposed Artificial Morality

In the present paper the proposed PF approaches within the AI-based localization system were developed and evaluated, based on simulated dynamic moral dilemmas for ITS. The graphical results, by means of the MEF filter, are presented in Fig. 7, where the deviation samples with 1σ MEF for input moral position and outputs moral position are presented. The uncertainty by means of deviation is shown after filtering with both SAM and HAM approaches.

The PF method for dynamic moral position estimator aided with map matching technique (also called HAM) proves to be a safer choice for ITS applications. This was the selected methodology in [1]. Table 1 summarizes the comparison of the trueness and precision in both SAM and HAM cases, complementary to the graphical results of Fig. 7.

Table 1. MEF deviation parameters of both soft and hard artificial morality dynamic filtering approaches

Parameter		SAM	HAM
Ellipses Center [m]	Easting	1.1937	0.0331
	Northing	0.1101	−0.0625
Semi-radii [m]	σ_A	0.6550	0.3554
	σ_B	0.6824	0.4867
Rotation	[deg]	−31.97	−44.09
LocAv 1σ	[%]	47.70%	44.99%

From the results, it is concluded that SAM improves precision for the estimated positions (reducing uncertainty for the AI moral dilemma), as the semi-radii from the MEF analysis are smaller for the filter output. Nevertheless, the trueness of the estimated moral dilemma deteriorated, as the ellipses center is further from the origin. The SAM approach has a clearer morality path, causing a bias in the uncertainty, meaning an acquired cultural/historical bias for solving the moral dilemma. In the HAM approach both the trueness of the moral dilemma and the precision of the estimated moral dilemma improved, resulting on reduced uncertainty and clearer focus (harder reinforcement in the path) for any simple AI moral dilemmas. While the SAM approach is a good tool for future AI-based systems, the HAM method is a better solution for future AI-based ITS. Complex AI moral dilemmas, such as decision making in case of collisions, shall be further studied. But the bases presented here to solve simple AI moral dilemmas shall be the starting point to research those currently unknown solutions.

Both SAM and HAM result useful to assist AI-based ITS. In the following sections, conclusions, a description of potential further work and an opened discussion are presented.

7 Conclusions and Further Analysis

In this paper a simple AI moral dilemma for an intelligent GNSS-based localization system is presented.

Two Particle Filter approaches were tested as the developed Artificial Faith into two distinctive morality approaches: Soft Artificial Morality (SAM) and Hard Artificial Morality (HAM). It was concluded that the most suitable method for the tested AI localization system is a HAM.

From the evaluation it was concluded that both methods improve precision (reducing uncertainty) for estimated moral dilemmas, as seen in the smaller MEF results in Table 1. Therefore, this application for artificial morality shall be included as an essential requirement for all ITS, achieving uncertainty reduction, as well as starting the discussion about other AI moral dilemmas resolution.

The present paper also allows future crossfield contributions, comprising several filtering techniques with logic and moral particularism in AI ethics and affective computing, specially oriented to the development of case-based soft and hard artificial morality for AI systems. The presented state estimation problem, analogous to the ever-present moral dilemma of AI-based localization systems, was solved by means of PF. Therefore, it can also be concluded that the application of morality concepts not only tends to a better understanding of the mathematical approaches for decision-making techniques, but a new re-signification of points of view for both AI and logic researchers.

8 Opened Discussion

To solve complex moral dilemmas, further work on Artificial Faith should allow other AI-based systems for decision-making to improve their applications, based on new studies of moral particularism paradigm, described as "[certain] scenarios where there are no moral principles and any moral judgement can be only found as one decides on each particular case, either real or imagined".

As proposed in [2], a simple moral dilemma of the AI-based validation tool, deciding between right and wrong for the AI based accuracy quality function in [1] matches the application of the presented HAM, resulting in the reduction of uncertainty in the referential position for better decision-making in AI moral dilemmas.

This artificial morality approach focuses on map matching techniques, equalled to an obligatory morality path to be followed by the AI system. Therefore, this simple AI moral dilemma was solved by taking into consideration not only the dynamic evolution of the position (analogous to the cultural based morality of the AI system), but also of the constraining path (analogous to the artificial faith of the AI system).

More complex AI moral dilemmas will need to be further studied on the bases of moral particularism to aid the already existing SAM and HAM methods.

The usage of Artificial Morality proposed in this paper focused only on AI-based land vehicles localization systems. But beyond the clear limitation of the presented approach for the development of Artificial Morality, the present paper is intended to open up the discussion about mathematical-based approaches to further development of Artificial Morality for future AI-based systems.

We believe that the developed concepts of Soft Artificial Morality (SAM) and Hard Artificial Morality (HAM) shall be further developed to be applied in all other AI-related systems.

These developed AI ethical solutions will provide future AI-based systems with culturally dynamic artificial moral particularism, solving AI-based system problems for all future safety-relevant applications, as well as providing peace of mind to human users all over the world.

References

1. Toro, F.G.: Development of intelligent GNSS-based land vehicle localisation systems. Ph.D. Dissertation, Technische Universität Braunschweig, Brunswick, Germany, May 2015
2. Toro, F.G., et al.: Particle filter technique for position estimation in GNSS-based localisation systems. In: 2015 International Association of Institutes of Navigation World Congress, Prague, Czech Republic, 20–23 October 2015
3. Tegmark, M.: Life 3.0: Being Human in the Age of Artificial Intelligence. Knopf, August 2017
4. Kaipio, J., Somersalo, E.: Statistical and computational inverse problems. Appl. Math. Sci. **160** (2004)
5. Maybeck, P.: Stochastic Models, Estimation and Control. Academic Press (1979)
6. Colaco, M.J., Barreto Orlande, H.R., Vaz Vianna, F.L., Da Silva, W.B., Da Fonseca, H.M., Dulikravich, G.S., Fudym, O.: Kalman and particle filters. In: METTI V - Thermal Measurements and Inverse Techniques, Volume: I. Rosco (2011)
7. Winkler, R.L.: An Introduction to Bayesian Inference and Decision. Probabilistic Publishing (2003)
8. Ristic, B., Arulampalam, S., Gordon, N.: Beyond the Kalman Filter. Artech House (2004)
9. Doucet, A., de Freitas, N., Gordon, N.: Sequential Monte Carlo Methods in Practice. Springer (2001)
10. Toro, F.G., Diaz Fuentes, D.E., Schnieder, E.: New filter by means of Mahalanobis distance for accuracy evaluation of GNSS. In: POSNAV (2013)
11. Andrieu, C., Doucet, A., Singh, S.S., Tadić, V.B.: Particle methods for charge detection, system identification and control. Proc. IEEE **92**, 423–438 (2004)
12. Arulampalam, S., Maskell, S., Gordon, N.: A tutorial on particle filters for online nonlinear/non-Gaussian Bayesian tracking. IEEE Trans. Signal Process. **50**, 174–188 (2002)
13. Pink, O., Hummel, B.: A statistical approach to map matching using road network geometry, topology and vehicular motion constraints. In: Proceedings of 11th International IEEE Conference on Intelligent Transportation Systems (2008)
14. Lou, Y., Zhan, C., Zheng, Y., Xie, X., Wang, W., Huang, Y.: Map-matching for low-sampling-rate GPS trajectories. In: 17th ACM SIGSPATIAL International Conference on Advances in Geographic Information Systems (2009)
15. Smaili, C., Najjar, M.E.B.E., Charpillet, F.: A hybrid Bayesian framework for map matching: formulation using switching Kalman filter. J. Intell. Robot. Syst. **74**, 725–743 (2014)
16. Syed, S., Cannon, M.E.: Fuzzy logic based-map matching algorithm for vehicle navigation system in urban canyons. In: ION National Technical Meeting (2004)
17. Najjar, M.E.B.E., Bonnifait, P.: A road-matching method for precise vehicle localization using belief theory and Kalman filtering. Auton. Robot. **19**, 173–191 (2005)
18. Davidson, P., Collin, J., Takala, J.: Application of particle filters to a map-matching algorithm. Gyroscopy Navig. (2011)
19. Dmitriev, S.P., Stepanov, O.A., Rivkin, B.S., Koshaev, D.A.: Optimal map-matching for car navigation system. In: Proceedings of 6th International Conference on Integrated Navigation Systems (1999)

20. Toro, F.G., Manz, H., Lu, D., Schnieder, E.: Accuracy evaluation of GNSS for a precise vehicle control. In: CTS 2012, 13-th IFAC Symposium on Control in Transportation Systems, Sofia, Bulgaria, September 2012
21. Mahalanobis, P.C.: On the generalised distance in statistics (1936)
22. De Vogeleer, K., Ickin, S., Fiedler, M., Erman, D., Popescu, A.: Estimation of quality of experience in 3G networks with the Mahalanobis distance. In: CTRQ (2011)
23. Kamei, T.: Face retrieval by an adaptive Mahalanobis distance using a confidence factor. ICIP (2012). https://doi.org/10.1109/ICIP.2002.1037982

Addressing the Problem of Activity Recognition with Experience Sampling and Weak Learning

William Duffy(✉), Kevin Curran, Daniel Kelly, and Tom Lunney

School of Computing, Engineering and Intelligent Systems,
Ulster University, Derry, UK
Duffy-w@ulster.ac.uk

Abstract. Quantifying individual's levels of activity through smart or proprietary devices is currently an active area of research. Current implementations use subjective methods, for instance, questionnaires or require comprehensively annotated datasets for automated classification. Each method brings its own specific drawbacks. Questionnaires cause recall bias and providing annotations for datasets is difficult and tedious. Weakly supervised methodologies provide methodologies for handling inaccurate or incomplete annotations and literature has shown their effectiveness for classifying activity data. As a key issue of activity recognition is capturing annotations, the aim of this work is to evaluate how classification performance is affected by limiting annotations and to investigate potential solutions. Experience sampling combined with the algorithms in this paper can result in a classifier accuracy of 74% with a 99.8% reduction in annotations, with increased compute overheads. This paper shows that experience sampling combined with a method of populating labels to unlabeled feature vectors can be a viable solution to the annotation problem.

Keywords: Activity recognition · Weak supervision · Experience sampling
Multi-class classification · ECOC SVM

1 Introduction

Currently, the practicality of collecting user specific activity data has never been so high. This is due to the global adoption of smart devices across users of all age ranges, providing new opportunities for signal analysis and the application of pattern recognition techniques.

With advances in healthcare leading to extensions in overall lifespan, the elderly are becoming a larger proportion of the overall population, placing a greater strain on the healthcare system and increasing the associated costs [1]. As a result, methods of automated activity detection are receiving interest in research. Methods of detecting the onset of serious medical issues, for example, gait analysis of an individual thought to be suffering from Parkinson's disease has advantages both in patient care and potentially reducing the cost to healthcare providers [2]. Wearable devices can also be useful in the field of biometric security, and sports such as dart throw analysis [3].

K. Arai et al. (Eds.): IntelliSys 2018, AISC 868, pp. 1238–1250, 2019.
https://doi.org/10.1007/978-3-030-01054-6_86

A prominent issue and the focus of this research is the problem of annotated data. Having a fully annotated dataset is desirable as it provides the maximum amount of training data for a supervised classifier to learn from. This gives the classifier the best chance at recognizing this activity when new unseen data is introduced. Generally, fully annotated data is expensive to record [4]. This is due to both the effort involved in the capture process but also the equipment and setup [2]. The typical setup of an activity recognition capture session involves connecting multiple sensors and performing the same activity over an extended period to ensure sufficient data is available to both train and validate the model. Many studies are also performed in laboratory settings, further complicating the capture of useful data [1]. Even with a successfully completed data capture experiment, we are limited not only to the activities captured but also to the individuals who performed them. A system which is trained to recognized activities of a specific person will always be more successful than a generic system. This is due to the slight idiosyncrasies displayed by each person in performing the same activity. However, with many movement sensors having high refresh rates, asking users to accurately annotate their own activities is not feasible. For this reason, weakly supervised methodologies are being proposed to solve the issue. These methodologies take considerably weaker labelling information than normal supervised methods [5]. Generally, unlabeled data is available in greater quantities than labelled data. Weak supervision takes advantage of this and tries to gain information from the relationship between labelled and unlabeled data points [4].

If wearables are to be used, and if the models are to be truly tailored to each user, then a method of gathering data from the user is required to allow data to be annotated. One avenue is to use the experience sampling method which is a type of diary study where the user will annotate the data manually. Although this type of data collection could be considered intrusive if a user is requested for data continuously [4]. In order to reduce the number of queries sent to the user and therefore reduce the intrusiveness of the method, weak supervision techniques can be used to gather extra information about the data points which the user annotated.

This paper sets out to show that experience sampling in combination with a method of weak learning could provide a solution to the problem of annotated data for activity recognition.

Section 3 demonstrates our methodology, Sect. 4 outlines the experiments, Sect. 5 discusses results and Sect. 6 presents our conclusions and recommendations for future work.

2 Related Work

Research into weak supervision for activity recognition has shown potential [4, 6]. Experience sampling, a type of diary study, has been used to gather annotations from the user about the activity they are performing. Multiple methods of annotation request are used, some which ask what activity users have been performing the most over a given time frame and some which ask for all the activities performed within the time window [4]. Instead of polling users at fixed intervals, context aware methods of experience sampling exist but have not been tested for the purpose of activity recognition. These apply a cost benefit approach to asking the user for input [7]. This approach could be useful in reducing the number of intrusive data requests the system makes.

Weak supervision techniques are variations on standard supervised techniques, they attempt to focus on gaining knowledge from the unlabeled data as it is available in far greater quantities than labelled data. One such variation is multiple instance learning which places feature vectors into sets known as bags. Classifiers, for instance, SVM are modified to accept these bags. Previous work applied a multiple instance SVM with experience sampling. Classifier accuracy improved in comparison to an SVM provided only the labels gained through experience sampling for sampling windows of 10, 30 and 120 min [4].

Another weak supervision method is graph label propagation which attempts to group labelled and unlabeled data into communities. Each community is then assigned a label based on the known labelled instances inside these communities. This label is then applied to all the unlabeled instances within that community. When applied to the TU Darmstadt dataset, graph label propagation exceeded the baseline accuracies set by an SVM trained with full ground truth [4].

Active learning for activity recognition, which attempts to discover the unlabeled data points which the classifier can learn most from. When applied to accelerometer data which initially has minimal annotations it provides a significant increase in accuracy [6]. However, issues with active learning are noisy annotators and the classifier may focus on difficult to annotate data points [8].

3 Methodology

3.1 Dataset

The Human Activities and Postural Transitions (HAPT) dataset [9] uses a waist mounted smartphone to collect 6 activities. The signals from the gyroscope and the accelerometer are sampled at 50 Hz. In total, 30 participants are included in the dataset with the age range being between 19–48 years.

The dataset contains a mixture of static and dynamic activities as well as the postural transitions which occurred on the static activities. For the purposes of this experiment, the postural transitions are being ignored. The static activities performed are standing, sitting and lying and the dynamic activities are walking, walking upstairs/downstairs.

Validation is achieved through the use of a random train/test split. The participants in the training set are completely independent of the test set, so the system will not be tested on individuals it has been trained on. The training data is made up of 70% of the participants and the test data of the remaining 30%.

3.2 Feature Extraction

The features used for this dataset are pre-computed. They come from the raw 3-axis accelerometer and gyroscope signals and are a mixture of time domain and frequency domain features. The signals are de-noised using a median and low pass Butterworth filter. Five different time series are produced from these signals. Acceleration values produced from "body" movements and the acceleration values produced from gravity. The signals are split using a low pass Butterworth filter. The jerk of the body signals are then produced by calculating the rate of change of the acceleration values with respect to time. The time series now produced are the acceleration due to gravity, the body acceleration, the body gyroscope acceleration and jerk of the body signals. The magnitude of each of these is calculated using norms. Fast Fourier transforms are calculated from these signals.

The signals are split into windows which are 128 samples in length, approximately 2.6 s of data. Several variables are calculated from each of these signals, including mean, standard deviation, max/min values, median absolute deviation, signal magnitude area, energy of signal, interquartile range, signal entropy, correlation coefficients, the auto regression coefficient, angle between the vectors, skewness, kurtosis, mean frequency, frequency component which has the largest magnitude and energy of the frequency interval of the Fourier transform window. The mean of the angle between the vectors is also calculated. This results in a 561 length feature vector.

3.3 Experience Sampling

Experience sampling is a diary study method for gathering data from a user. We simulate it in this experiment by asking a user for data about each activity they are currently performing. This process is simulated by finding which feature vectors lie on the experience sampling intervals. E.g. for a one minute experience sampling window, the feature vector which lies on exactly one minute will be used. The annotation for this feature vector is then obtained from the original set of annotations.

A consequence of this method is that we only get back what we ask from the user. Requesting more data from the user will be beneficial to the experiment as it could provide more data for training but such a system could be considered intrusive to the user. With this in mind, designing the questions the system will ask the user is important. Previous implementations have asked for data like what activity they have been performing the most since the previous data request and what activity are they currently performing. This experiment will only simulate asking what they are currently performing and will label the current feature vector with the data collected from the user.

A potential issue is that when a user is about to input the data they will move the device, and the movement of using the device will be incorrectly labelled as the locomotive activity they are performing. Since this data has been pre-collected and since we are simulating the user entering a label, this is not an issue but if this were to be used on an online system it would require data collection to be halted until the user has completed the request and resume once they have stopped using the device.

The experience sampling will be performed using several different time intervals. The simulated annotation requests will be performed every 1, 3, 5, 10, 15, 20, 30 and 60 min.

To ensure that the experience sampling methodology is providing an unbiased view of the data, the sampling times are being randomly moved back and forward by a few seconds. The entire experiment is then repeated with these new experience sampling labels. This process is only performed on the training data, the test set does not change.

(1) Short experience sampling windows (1–5 min)
It can be expected that the short experience sampling windows will perform like the fully supervised approach. At one minute intervals, even with 95% of labels removed, the system should still have sufficient data to classify 6 activities with reasonable accuracy.

(2) Medium experience sampling windows (10–20 min)
We postulate that this could be the optimal window size as it provides an acceptable break between data collection requests, while potentially providing enough data for classification.

(3) Long experience sampling windows (30+ min)
It is suspected that this length of sampling windows will show the limitations of the dataset, with potentially insufficient information being captured to provide accurate classification. It will be possible for entire activities to be missed.

3.4 Weak Supervision

Two algorithms are presented which attempt to populate labels to unlabeled feature vectors using pairwise Euclidean distance. The initial setup of both algorithms is the same, with a 561-width matrix containing all feature vectors stored in X and labels stored in Y with $Yi \in \{0, 1, \ldots 6\}$. All labels in Y will initially be 0 except those collected using experience sampling.

The variable K is used by Algorithm 1, it controls the number of times the algorithm repeats. As the amount of processing is exponentially increased for each incrementally higher value of K so the values will be limited to between 1 and 3.

Algorithm 1 presents the methodology for the single label based propagation. After the experience sampling has gathered labels, the algorithm runs through each of the experience sampling labelled feature vectors and finds the single closest unlabeled feature vector to each. Each of these single vectors is given the same label as the experience sampling feature vector, effectively doubling the number of labelled feature vectors. This process is then repeated K times, however, for each successive run, the original labelled feature vectors are removed. For example, the system is looking for the closest variables to the experience sampling points on the first run, these experience sampling points are then removed for the second run. Without this step, it will just re-find the original points.

Algorithm 1 Single label propagation

```
1: xPoints = index of each position Y which is not zero
2: for each x in xPoints
3:     cLabel = value of Y at position x
4:     cFeature = feature vector of X at position x
5:     For each z in X
6:         distances(z)
=          pairwise distance of cFeature and X(z)
7:     end
8:     lowest = min(distances) which is not index x
9:     Y(lowest) = cLabel
10: end
11: For x = 1:K
12:     temp = xPoints
13:     xPoints
= index of each position Y which is not zero
14:     Remove temp from xPoints
15:     Repeat lines 2 to 10 with new xPoints
16: end
```

For the second algorithm, the variable M is used to control the number of similar feature vectors to look at for comparison. The algorithm aggregates the labels gathered by experience sampling into groups, for example, all feature vectors labelled as walking will be grouped together. The average of each of feature vectors in each group is taken, these averages are then compared against the unlabeled data points looking for the M closest. These M data points are then given the label of the closest group.

3.5 Classification

As Support Vector Machines have been shown to perform well on activity data, classification will be achieved through the use of an error-correcting output codes support vector machine (ECOC SVM). An ECOC SVM as it allows an SVM to act as a multi-class classifier. This is advantageous for activity data as generally there are multiple activities to classify.

This is achieved by providing an ensemble of binary one vs one classifiers [10]. The error-correcting codes output by the ECOC SVM means that although there are extra data overhead, these codes can help the system recover from errors such as poor input features or a flawed training algorithm, something which may be appropriate for weak supervision techniques since noisy labels are inevitable.

Decision tree based ensemble methods have also been used successfully for activity recognition, again the ensemble method has the capability for multi-class classification. In this case, a Tree Bagger will be used.

Each of these classifiers will be tested on the fully supervised data and the more successful of these two selected.

Algorithm 2 Unique aggregate label propagation

```
xLabels = 2 − width matrix containing each known label and its index in Y
uLabels = unique labels in xLabels
for each x in uLabels
    tLabel = label in xLabels at position x
    temp = indexes of Y with label tLabel
    temp2 = feature vectors in X with indexes in temp
    featuresAvg = mean of features in temp2
    for each z in X
        distances(z) = pairwise distance between featuresAvg and X(z)
    end
    counter = 0
    while(counter ≤ M)
        smallest(counter) = min(distances) which is not index x
    end
    Y(smallest) = tLabel
end
```

4 Experiments

The following experiments will first evaluate the performance of standard supervised classification with full ground truth. The amount of data available to the classifier will then be restricted via experience sampling and finally new labels introduced through weak supervision methods.

4.1 Supervised

Standard supervised learning was performed by using the pre-computed features with all the provided activity labels.

The results shown in Table 1 provide a baseline for weakly supervised performance to be compared against, with an overall average of 96.4% the ECOC SVM performed significantly better than the Tree Baggers average of 90.1%. As it performs better then ECOC SVM will be used for any future experiments.

Table 1. Supervised classifier comparison

Activity	Accuracy	
	ECOC SVM	Tree bagger
Walking	99.4%	94.0%
Walking upstairs	96.8%	86.8%
Walking downstairs	96.4%	82.1%
Sitting	89.2%	88.6%
Standing	96.9%	89.2%
Laying	99.8%	99.8%

Table 2. Experience sampling window length

Sampling window (min)	0	1	3	5	10	15	20	30	60
Number of labels	7415	322	107	64	32	21	16	10	5

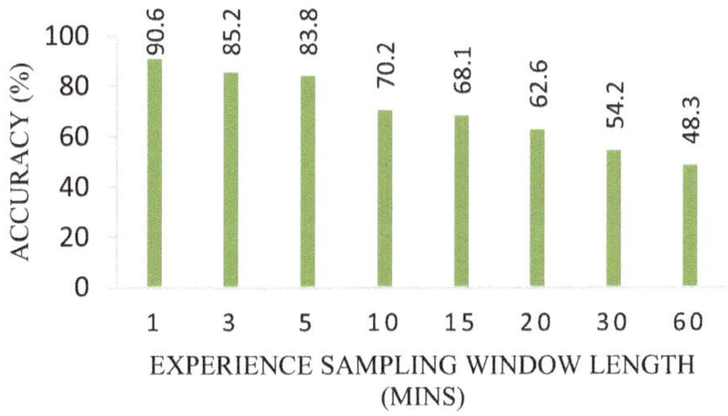

Fig. 1. Experience sampling performance.

4.2 Weakly Supervised

(1) *Randomisation of experience sampling requests*

To provide as unbiased view of the data as possible, each of the experience sampling requests will be randomly moved forward or backwards by a random number of seconds.

A seed value of 1 is being used to ensure that these results are repeatable and that on each experiment is being provided with the same annotations.

(2) *Experience Sampling Only*

Table 2 shows the number of labels found through experience sampling for each length of sampling window. Even a one minute interval significantly reduces the number of labels available in the training set, any feature vector which has not been assigned a label is discarded as it is not usable by a supervised classifier.

Figure 1 shows that the experience sampling method alone provides excellent results at low window lengths but once the sampling window exceeds 5 min performance is significantly reduced.

(3) *K and M values*

Experiments were performed to find the most appropriate values for the K and M values, these will be performed on the 1, 10 and 20 min experience sampling windows. These times have been selected as they provide data about both short and medium length sample windows. The longer windows are not being used due to dataset limitations.

The bigger the values the more labels will be generated, however, this could also result in an increased number of incorrect labels compared to ground truth.

(a) *Algorithm 1*

We can see from Table 3 that increasing the K-value does increase the number of new labels generated, however, the bigger the K-value used, the higher chance that incorrect labels will be created.

Table 3. K-value tuning Algorithm 1

K value	1	2	3
1 min sampling window			
New labels	1274	2547	5094
Percent correct	96.8%	94.8%	91.7%
10 min sampling window			
New labels	128	256	512
Percent correct	97.8%	95.3%	93.4%
20 min sampling window			
New labels	64	128	256
Percent correct	98.4%	96.9%	95.2%

Table 4. M-value tuning Algorithm 2

M value	10	20	30
1 min sampling window			
New labels	378	434	491
Percent correct	98.8%	98.1%	97.1%
10 min sampling window			
New labels	85	144	169
Percent correct	97.0%	94.7%	90.4%
20 min sampling window			
New labels	69	127	186
Percent correct	96.2%	95.9%	94.8%

A K-value of three will be used for the remainder of the experiment. This is due to it having significantly more labels than the other K-values, while only having a slight reduction in quality. K-values of above 3 will likely generate more labels, however, the performance impact is unacceptably high.

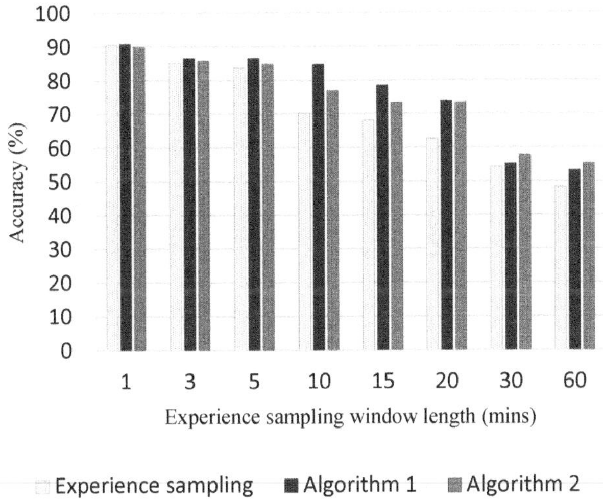

Fig. 2. Classifier performance.

(b) *Algorithm 2*

Based on the results in Table 4, an M value of 20 is selected for Algorithm 2. The reduction in label quality between the values of 10 and 20 is insignificant compared to between 20 and 30. The value of 20 also provides substantially more labels than 10, which will be useful to the classifier.

(4) *Classifier Results*

Figure 2 shows the overall classification accuracy for each experience sampling window length. The graph shows that as the sampling window length increases above 5 min the classifier accuracy of experience sampling starts to decrease quickly whereas, the two weakly supervised algorithms decrease much slower until the 30 min mark. Both algorithms allow the experience sampling window to be increased to 20 min while maintaining an accuracy of above 70%. Algorithm 1 performs better overall, but not significantly. After the 20 min window accuracy sharply decreases for both Algorithms 1 and 2. This is possibly due to the experience sampling window being so long that entire activities are being missed.

Table 5 shows the classifier accuracy for individual activities in three different sampling window lengths. The one minute sampling length provides the closest comparison to fully supervised, while lengths greater than 20 min run into the limitations of the dataset. For each of these window lengths we can see that certain activities appear to be affected differently based on the length of the sampling window. Walking, standing and laying do not appear to be significantly impacted even with a sampling length of 20 min. However, walking upstairs, walking downstairs and sitting are all reduced. Walking upstairs is affected the worst with a significant reduction from 1 to 20 min sampling length. It could be that certain activities need more data for accurate classification or the dataset could be unbalanced.

Table 5. Per activity classifier performance

Activity	Walking	Walking upstairs	Walking downstairs	Sitting	Standing	Laying
1 min sampling window						
Experience sampling	98.2%	91.5%	84.2%	81.9%	87.1%	99.8%
Algorithm 1	95.8%	94.4%	88.2%	79.9%	88.2%	99.8%
Algorithm 2	97.9%	90.4%	84.8%	79.0%	87.4%	99.8%
10 min sampling window						
Experience sampling	91.6%	52.8%	55.4%	19.3%	95.0%	96.6%
Algorithm 1	94.4%	70.2%	70.8%	30.2%	97.2%	95.6%
Algorithm 2	87.4%	58.9%	76.3%	46.5%	91.6%	97.2%
20 min sampling window						
Experience sampling only	95.5%	5.3%	41.9%	44.1%	84.9%	94.9%
Algorithm 1	93.3%	38.9%	71.1%	55.8%	83.3%	94.9%
Algorithm 2	92.8%	18.8%	47.3%	75.3%	68.4%	88.8%

Overall, the two algorithms appear to be effective at increasing the classifier accuracy of the worst performing activities. With Algorithm 1 increasing the accuracy of walking upstairs and walking downstairs by 33.6% and 29.2% respectively at the 20 min experience sampling window.

(5) *Performance results*

Although Algorithm 2 has not achieved the classification results that Algorithm 1 did, it is significantly faster as shown in Fig. 3. Although in recent years smart devices have greatly improved in terms of battery life and compute performance [11], they would be the ideal platform for an activity recognition system and anything which reduces the power impact of these systems would be advantageous.

5 Discussion

While these methods do not provide a completely unobtrusive system, they do maintain reasonable levels of classification accuracy even with significant reductions in the number of labelled feature vectors.

Limitations of this work include entire activities can be missed when the experience sampling window becomes too long. A potential solution is to provide a method of moving the request timings to detect more valuable data. These methods could also be used to increase the experience sampling request timings if the data to be collected provides limited new information. This could result in a system with fewer requests and a higher level of user acceptance.

Another limitation of this work is the more activities in the time series there is an increased likelihood that noisy labels will be produced. A potential solution is to provide multiple methods of validating populated labels rather than just their distance in the feature space.

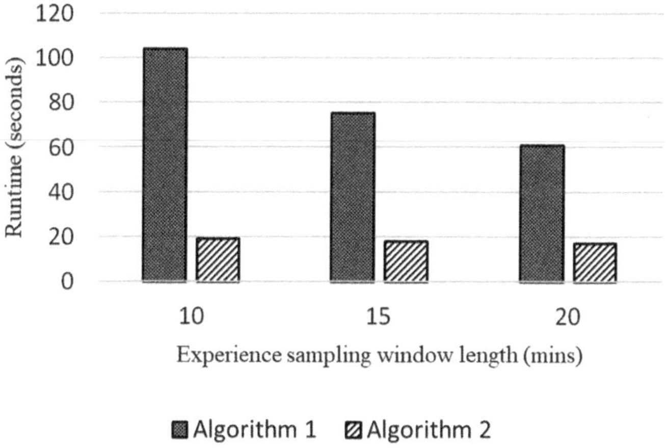

Fig. 3. Algorithm compute performance.

6 Conclusion and Future Work

As shown, experience sampling is a potentially viable method for collecting activity data. Classifiers perform surprisingly well when presented with minimal training samples, allowing a 10-min experience sampling window to provide 70% classifier accuracy.

The weak supervision techniques introduced in this paper allows this window to be effectively doubled, with classifier accuracy of 74% even with a 20-min sampling window.

However, above this window length generally results in problems as entire activities are missed and the limitations of the dataset are pushed. The problem is shown in both the experience sampling labels and the weak supervision labels.

A potential issue with the experience sampling method is that when the user is polled for data they will be required to make movements in order to put the annotation into a device. As this experiment is mocked up on already collected data there is no way to simulate this, however a potential solution is to stop collecting data from sensors at a fixed time before alerting the user and begin collecting after a period has expired. Another solution to this issue could be to ask what activity a user was performing a predefined number of seconds ago, e.g. what activity did you perform 30 s ago. Another issue is that people will be performing movements with the device in different orientations, for example, walking while changing a song.

Future work will be on determining the ideal time for asking the user for input other than just timing, for example, an online classifier which attempts to discover when an activity that the system does not recognize is being performed.

References

1. Wong, C., Zhang, Z.Q., Lo, B., Yang, G.Z.: Wearable sensing for solid biomechanics: a review. IEEE Sens. J. **15**(5), 2747–2760 (2015)
2. Wong, W.Y., Wong, M.S., Lo, K.H.: Clinical applications of sensors for human posture and movement analysis: a review. Prosthet. Orthot. Int. **31**(1), 62–75 (2007)
3. Zheng, Y.L., et al.: Unobtrusive sensing and wearable devices for health informatics. IEEE Trans. Biomed. Eng. **61**(5), 1538–1554 (2014)
4. Stikic, M., Larlus, D., Ebert, S., Schiele, B.: Weakly supervised recognition of daily life activities with wearable sensors. IEEE Trans. Pattern Anal. Mach. Intell. **33**(12), 2521–2537 (2011)
5. Hernández-González, J., Inza, I., Lozano, J.A.: Weak supervision and other non-standard classification problems: a taxonomy. Pattern Recognit. Lett. **69**, 49–55 (2016)
6. Stikic, M., Van Laerhoven, K., Schiele, B.: Exploring semi-supervised and active learning for activity recognition. In: 12th IEEE International Symposium on Wearable Computers, ISWC 2008, pp. 81–88 (2008)
7. Kapoor, A., Horvitz, E.: Experience sampling for building predictive user models. In: Proceeding of the Twenty-Sixth Annual SIGCHI Conference on Human Factors in Computing Systems (CHI 2008), pp. 657–666 (2008)
8. Settles, B.: Active learning literature survey. Mach. Learn. **15**(2), 201–221 (2010)
9. Reyes-Ortiz, J.-L., Oneto, L., Samà, A., Parra, X., Anguita, D.: Transition-aware human activity recognition using smartphones. Neurocomputing **171**, 754–767 (2016)
10. Dietterich, T.G., Bakiri, G.: Solving multiclass learning problems via error-correcting output codes. J. Artif. Intell. Res. **2**, 263–286 (1995)
11. Shoaib, M., Bosch, S., Incel, O., Scholten, H., Havinga, P.: A survey of online activity recognition using mobile phones. Sensors **15**(1), 2059–2085 (2015)

Public Key and Digital Signature for Blockchain Technology

Based on the Complexity of Solving a System of Polynomial Equations

Elena Zavalishina, Sergey Krendelev(✉), Egor Volkov, Dmitry Permiashkin, and Dmitry Gridin

Novosibirsk State University, JetBrains Research, Novosibirsk, Russia
elena.zavalishina.v@gmail.com,
s.f.krendelev@gmail.com, egor.volkov.u@gmail.com,
d.permiashkin@g.nsu.ru, dmitry.gridin.v@gmail.com

Abstract. This article proposes the algorithm of generation of public key for digital signature. The algorithm is quantum-resistant because it uses the complexity of solving a system of polynomial equations. It is assumed that in this case a standard hash function is used for the digital signature implementation, which has 384-512 bit output. The digital signature that is formed in this way will allow the blockchain technology to be quantum-resistant.

Keywords: Blockchain · Digital signature · Public key · Quantum computer

1 Introduction

This article deals with the quantum-resistant digital signature algorithm.

A digital signature is an attribute of a digital document which is needed for identifying the authorship of a document.

Consider the digital signature principle. To sign a document using a digital signature you need to create 2 keys, public and private, select a hash function and calculate the hash value of the source document. Then the digital signature is calculated from the hash value by a certain formula using the private key. The digital signature is transmitted for verification together with the document and the public key, where the hash value is calculated and the signature is authenticated using a public key.

The digital signature is an essential part of blockchain technology-based payment systems without third-party intermediaries (e.g. Bitcoin). At the time of making transaction, when one account transfers his ownership of the assets to another account in the transaction made on the blockchain, a message with intentions is formed and signed with a digital signature. The fact of the signature demonstrates that it was the particular account that disposed of assets.

In 2016, the U.S. National Institute of Standards and Technology (NIST) released a Report on Post-Quantum Cryptography [1]. With the arrival of quantum computers previously secure algorithms become breakable. For example, the length of hash value

K. Arai et al. (Eds.): IntelliSys 2018, AISC 868, pp. 1251–1258, 2019.
https://doi.org/10.1007/978-3-030-01054-6_87

should be longer - 384-512 bit. We took into account the recommendations of NIST and propose the quantum-resistant algorithm in this article. It is so because it uses the complexity of solving a system of polynomial equations.

2 Hash of the Document Representation

A standard hash function generally returns a binary string with a length of 256 to 384 bit. Possible generalizations of hash functions can return data of arbitrary size. Our version requires data that is less than the modulus m over which the calculations are performed. This means that we need data which is adapted for this method. Let T be a returning value of the hash function. Now we write T in expanded form using powers of the modulus m

$$T = \lambda_0 + \lambda_1 m + \lambda_2 m^2 + \cdots + \lambda_k m^k$$

Therefore, we represent the hash as a set of relevant data $\lambda_0, \lambda_1, \lambda_2, \ldots, \lambda_k$. Note that every number is strictly less than m.

3 Cremona Transformations

Cremona transformation is a reversible polynomial mapping of a vector space R^n into itself such that R^n is a set of polynomial functions [2]:

$$\begin{gathered} h_1(x_1, x_2, \ldots, x_n) \\ h_2(x_1, x_2, \ldots, x_n) \\ \ldots \\ h_n(x_1, x_2, \ldots, x_n) \end{gathered}$$

The reversibility, in particular, means that the system of equations

$$\begin{gathered} h_1(x_1, x_2, \ldots, x_n) = a_1 \\ h_2(x_1, x_2, \ldots, x_n) = a_2 \\ \ldots \\ h_n(x_1, x_2, \ldots, x_n) = a_n \end{gathered}$$

is solvable for any right-hand side.

The simplest example is the upper triangular map

$$\begin{gathered} h_k(x_1, x_2, \ldots, x_n) = x_k + f_k(x_1, x_2, \ldots, x_n), \\ k = 1, 2, \ldots, n, \\ f_n = 0. \end{gathered}$$

Another example is the lower triangular map

$$h_k(x_1,x_2,\ldots,x_n) = x_k + f_k(x_1,x_2,\ldots,x_{k-1}),$$
$$k = 1,2,\ldots,n,$$
$$f_1 = 0.$$

Furthermore, by composing such functions arbitrarily we obtain a one-to-one correspondence.

The situation changes radically if we consider a modulus Z_m^n instead of a vector space R^n, where m is some number. Then we introduce the Euler function $\varphi(m)$ and the ring $Z_{\varphi(m)}$.

Then the functions of the form

$$h_k(x_1,x_2,\ldots,x_n) = x_k^{r_k} + f_k(x_{k+1},x_{k+2},\ldots,x_n),$$
$$k = 1,2,\ldots,n,$$
$$f_n = 0,$$

are suitable as one-to-one correspondence of a modulus Z_m^n, where r_k is an invertible element of $Z_{\varphi(m)}$.

It is an upper triangular map.

$$h_k(x_1,x_2,\ldots,x_n) = x_k^{r_k} + f_k(x_1,x_2,\ldots,x_{k-1}),$$
$$k = 1,2,\ldots,n,$$
$$f_1 = 0,$$

s_k is an invertible element of $Z_{\varphi(m)}$.

It is a lower triangular map.

All possible compositions of such mappings are obviously bijective.

4 Public Key Generation

First we select some modulus m, calculate $\varphi(m)$. Then we select the number of variables n, upper triangular and lower triangular Cremona mappings and add invertible matrices over the ring Z_m.

We calculate a composition of these mappings. This calculated composition modulo m is a public key [3]. We inform all those who wishes of the open key and the modulus m.

Let us give a simple example of public key generation.

Choose a modulus m.

$$\mathrm{m} = 47, \varphi(\mathrm{m}) = 46 = 2 \times 23.$$

Suppose A is an invertible matrix over the ring:

$$A = \begin{pmatrix} 5 & 3 \\ 4 & 7 \end{pmatrix}, \quad \det A = 23$$

Choose an upper triangular map:

$$u = x^3 + 3y^2 + 5y$$
$$v = y^3$$

Hence,

$$A \cdot \begin{pmatrix} u \\ v \end{pmatrix} = \begin{pmatrix} 5x^3 + 3y^2 + 5y + 3y^3 \\ 4x^3 + 12y^2 + 20y + 7y^3 \end{pmatrix} = \begin{pmatrix} p \\ q \end{pmatrix}$$

Choose an upper triangular map:

$$f_1 = p$$
$$f_2 = p^2 + q$$

We get an open key:

$$\begin{aligned} f_1 &= 5x^3 + 3y^2 + 5y + 3y^3 \\ f_2 &= \left(5x^3 + 3y^2 + 5y + 3y^3\right)^2 + 4x^3 + 12y^2 + 20y + 7y^3 \\ &= 25x^6 + 30x^3y^3 + 30x^3y^2 + 3x^3y + 4x^3 \\ &\quad + 9y^6 + 18y^5 + 39y^4 + 37y^3 + 37y^2 + 20y \end{aligned}$$

Note 1: If we don't transmit the modulus, then the attacker would need to solve the system of polynomial equations. If the degrees of polynomials are strictly greater than 2, than it is an undecidable problem [4]. However, the modulus must be distributed as it is necessary for the digital signature.

Note 2: A part of obtained polynomials could be included in the private key so we can use the version of a system where the number of equations is less than the number of unknowns for the digital signature.

5 Digital Signature

Thus, we obtain a system of equations modulo m that is solvable for any right-hand side. In other words, there is a public key that is a public key in the usual sense.

Suppose T is a numerical value of the result of hashing of a document. Let us write T in expanded form using powers of m :

$$T = a_0 + a_1 m + a_2 m^2 + \cdots + a_{n-1} m^{n-1}$$

The set of numbers $(x_1, x_2, \ldots, x_n)$ can be found from a public key construction from the right-hand side of the system of equations:

$$\begin{gathered} h_1(x_1, x_2, \ldots, x_n) = a_0 \\ h_2(x_1, x_2, \ldots, x_n) = a_1 \\ \cdots \\ h_n(x_1, x_2, \ldots, x_n) = a_{n-1} \end{gathered}$$

After calculating the set of numbers let us consider the transformation:

$$x_i = \left(\overrightarrow{\xi_i}, \vec{y}\right), i = 1, 2, \ldots, n$$

$\overrightarrow{\xi_i} \in Z_m^s, i = 1, 2, \ldots, n, s \geq n, \vec{y} \in Z_m^s$, vectors $\overrightarrow{\xi_i}$ are secret.

Suppose that the system of linear equations $x_i = \left(\overrightarrow{\xi_i}, \vec{y}\right), i = 1, 2, \ldots, n$ has a rank of n or less. Hence, this equation has a solution [5]. Generally speaking, there are many solutions.

Now the public key for the digital signature has a form

Modulus m. Vector $\vec{y} = (y_1, y_2, \ldots, y_s)$.

A system of polynomial equations

$$\begin{gathered} f_i(z_1, z_2, \ldots, z_s) = h_i\left(\left(\overrightarrow{\xi_1}, \vec{y}\right), \left(\overrightarrow{\xi_2}, \vec{y}\right), \ldots, \left(\overrightarrow{\xi_n}, \vec{y}\right)\right) \\ i = 1, 2, \ldots, n \end{gathered}$$

Naturally, all computations are masked.

Verifying the signature consists of checking the correspondence

$$f_i(y_1, y_2, \ldots, y_s) = a_{i-1}, \qquad i = 1, 2, \ldots, n$$

a_{i-1} are the components derived from the hash of a document; the hash can be uniquely recovered from them.

Let us give a simple example.

Take the modulus m and the key from the previous example:

$$\mathrm{m} = 47$$

$$\mathrm{f_1} = 5\mathrm{x}^3 + 3\mathrm{y}^2 + 5\mathrm{y} + 3\mathrm{y}^3$$

$$\mathrm{f_2} = 25\mathrm{x}^6 + 30\mathrm{x}^3\mathrm{y}^3 + 30\mathrm{x}^3\mathrm{y}^2 + 3\mathrm{x}^3\mathrm{y} + 4\mathrm{x}^3 + 9\mathrm{y}^6 + 18\mathrm{y}^5 + 39\mathrm{y}^4 + 37\mathrm{y}^3 + 37\mathrm{y}^2 + 20\mathrm{y}$$

Let us assume we have to sign a document with the hash value H:

$$\mathrm{H} = 822 = 23 + 17 \cdot 47$$

Now we have to solve the system of equations

$$\begin{aligned} f_1 &= 23 \\ f_2 &= 17 \end{aligned}$$

Since we know all the transformations, we can write this system of equations as

$$\begin{aligned} f_1 &= p = 23 \\ f_2 &= p^2 + q = 17 \end{aligned}$$

And find its solution modulo m

$$\begin{aligned} p &= 23 \\ q &= 5 \end{aligned}$$

Since we know the matrix A, we can find the inverse of A

$$A^{-1} = \begin{pmatrix} 33 & 6 \\ 8 & 37 \end{pmatrix}$$

And solve the system of equations

$$A \cdot \begin{pmatrix} u \\ v \end{pmatrix} = \begin{pmatrix} p \\ q \end{pmatrix}$$

$$\begin{pmatrix} u \\ v \end{pmatrix} = \begin{pmatrix} 37 \\ 40 \end{pmatrix}$$

Then we invert the upper triangle transformation

$$\begin{aligned} 37 &= x^3 + 3y^2 + 5y \\ 40 &= y^3 \end{aligned}$$

Thus we have

$$\begin{aligned} x &= 16 \\ y &= 31 \end{aligned}$$

For verifying the solution you need to substitute values from the column into the key and make sure that the result is equal to the result of the document hash expansion in base m.

$$\begin{aligned} f_1(16,31) &= 5\cdot 16^3 + 3\cdot 31^2 + 5\cdot 31 + 3\cdot 31^3 = 23 \\ f_2(16,31) &= 25\cdot 16^6 + 30\cdot 16^3 31^3 + 30\cdot 16^3\cdot 31^2 \\ &\quad + 3\cdot 16^3\cdot 31 + 4\cdot 16^3 + 9\cdot 31^6 + 18\cdot 31^5 \\ &\quad + 39\cdot 31^4 + 37\cdot 31^3 + 37\cdot 31^2 + 20\cdot 31 = 17 \end{aligned}$$

6 Information on Software Implementation

We also developed the software prototype of this algorithm.

The average running time of each program module for different document hash sizes and the key with signature sizes for them is shown in Table 1 and Table 2.

Table 1. Program time running

Hash size (bit)	Module size (bit)	Number of variables	Average running time (ms)			
			Private key generation	Public key generation	Document signing	Verification
512	32	16	2.4	602	2.5	1.5
512	16	32	0.5	35375	4.4	16
384	48	8	2.6	24.6	1.6	3.3
384	32	12	2.4	101.1	2.6	0.54
384	12	32	0.5	37800	3.3	16.5
384	8	48	0.5	379930	5.7	58

Table 2. Data size

Hash size (bit)	Module size (bit)	Number of variables	Key with signature size (byte)
512	32	16	100606
512	16	32	726202
384	48	8	13648
384	32	12	37994
384	12	32	869911
384	8	48	3501506

The public keys, where the number of unknowns is two times more than the number of equations, have been used for these measurements.

The program was tested on a computer with Intel i7-4720HQ @ 2.60 GHz.

7 Security

We consider that the algorithm is cryptographically strong because it uses the complexity of solving a system of polynomial equations [4].

Furthermore, we use the system of equations where the number of equations is less than the number of unknowns [6]. This task is considered to be difficult and we use this fact.

The authors think that the modulus m should be a big prime number. Otherwise, in the case of a composite modulus, the system can be split into several simple ones with prime divisors of m as moduli and further restored by the Chinese remainder theorem; this will speed up the selection of values by simplifying the required mathematics.

8 Conclusion

This article proposes the quantum-resistant digital signature algorithm.

In our case, the public key is a set of special polynomials. To create a digital signature we solve a system of equations where in the right-hand side of the system there a vector of hash values of the document.

References

1. NIST Internal or Interagency Reports (IR): 8105 Report on Post-Quantum Cryptography National Institute of Standards and Technology, Gaithersburg, Maryland, p. 15, April (2016)
2. Coble, A.B.: Algebraic geometry and theta functions. Amer. Math. Soc. (1929)
3. Вялый, М.Н.: Сложность вычислительных задач
4. Егор Волков, Анатолий Баранов, Елена Завалишина Криптографическая система с открытым ключом Second Conference on Software Engineering and Information Management (SEIM-2017), pp. 41–44
5. Курош, А.Г.: Курс высшей алгебры. М: Наука, p. 431 (1965)
6. Sacks, G.E.: Mathematical Logic in the 20th century, pp. 269–273. World Scientific (2003)

Heterogeneous Semi-structured Objects Analysis

M. Poltavtseva(✉) and P. Zegzhda

Information Security of Computer Systems Department,
Peter the Great St. Petersburg Polytechnic University, St. Petersburg, Russia
maria.poltavtseva@ibks.icc.spbstu.ru, zeg@ibks.ftk.spbstu.ru

Abstract. The current challenges of decision support systems require a complex analysis of heterogeneous data. These include social and technical information and have various formats. In addition, this information is often incomplete about the domain. Data parts belong to different domains. Such information is defined in this paper as heterogeneous semi-structured objects. The author offers an approach to formalization and comparison of such data sets based on object model and vectorization. The novelty of the work lies in the object similarity measure. One can match objects of any type between themselves in the conditions of information incompleteness. The paper describes a method of formalizing data, matching method, advantages and disadvantages of the proposed solutions. As an example, the authors consider the application of the method in the data analysis of the solving of information security problems. In the paper, the system architecture of the decision support system based on the obtained results is presented.

Keywords: Semi-structured objects · Heterogeneous data
Heterogeneous objects analysis · Data science · Smart security
Decision support systems

1 Introduction

Information technology applied in all spheres of life. The health, location, work, hobbies and food preferences data associated with mobile and stationary devices of many people. Each person may have a large number of virtual reflections from different points of view. Each device is a collection of information about its owner.

Today, the real data has the virtual reflections, and almost everyone associated with one or more technical devices. In these circumstances, the task of heterogeneous data processing takes a new level. When a researcher solves a variety of problems in such diverse areas as the individual identification in the online training, penetration testing in the information security or identification in law enforcement activities it needs complex data analysis about the man and his devices.

Project is financially supported by the Ministry of Education and Science of the Russian Federation, Federal Program "Research and Development in Priority Areas of Scientific and Technological Sphere in Russia for 2014–2020" (Contract No. 14.578.21.0231; September 26, 2017, the unique identifier of agreement RFMEFI57817X0231).

K. Arai et al. (Eds.): IntelliSys 2018, AISC 868, pp. 1259–1270, 2019.
https://doi.org/10.1007/978-3-030-01054-6_88

It is a difficult task, as the analysis includes semistructured data of various nature, like technical specifications of the devices, data from social network profiles, observations, or information about organizations. This work is devoted to such data formalization and operating. Additional complexity is data incompleteness. The set of the information about the same object (object type) often differ depends on the awareness of the researcher.

The paper proposes a formalization of various data on the basis of object data model. Vector description of the object is constructed similar vectors describing the documents in the text collections. Vectorization allows one to use the same approach to the processing properties of different nature and to compare heterogeneous objects in conditions of incompleteness. The novelty of this work is in the distance evaluation approach between objects based on their properties, without pre-classification in the conditions of incomplete data. Using this distance as a measure of similarity, the researcher may apply a classification, clustering, matching object and other data mining methods. The approach allows to compare fragments of the domains as the "bags of objects". The proposed method is a stage in the direction of intelligent decision support systems creating in the fields of ambient intelligence and information security, where technical data related to the social.

2 Related Works

Information systems are developed more than five decades, including methods of mining, and decision support in various fields. The researches include analysis of information from various databases, logs, transactions and other sources [1]. There are a large number of different data mining techniques such as classification, clustering, finding patterns [2]. They are described, in particular, in [1, 3]. In addition, researchers all over the world develop and improve the data mining methods applicable to structured information in different domains. Recent advances in this area are neural nets, deep learning. However, all these methods require complete structured data.

The structuring of the data includes the syntax and semantic components. Syntax structure is the data format. Information in traditional databases, transaction processing systems (OLTP), the log files has a specified format. In turn, text documents, personal information, web pages are semi-structured. The task of processing semi-structured data emerged long enough (e.g. [4]). Its decision is transferring data to a structured form [5] according to their semantics. Transferring directly for some data is not possible due to their semantics (for example, for texts). Researchers are developing new methods that describe the structuring of such data. After their using, it is possible to apply known methods of analysis. For text data this process is well described in [6] and for geo data in [7]. More difficult is to analyze heterogeneous data with semantic semi-structure (heterogeneous data). Heterogeneous data often called:

- The data relating to different sources [8]. It is more correct to say, that it is data from heterogeneous sources.
- Data of different nature [9], related to various domains. This is semantically semi-structured data.

Integration of heterogeneous data sources is an actual task. There are works pointed at its solution (for example [10, 11]). In them, the researcher defines the semantics of the data, and a software tool transforms his request in accordance with the format of the source. You can also mention the special query languages for such cases [12]. However, to analyze the data in such way, the researcher must manually specify their semantics [13]. Analysis of complex heterogeneous objects is supported only in the direct domains. This is, for example, chemical engineering [14] or marine data [9].

In some areas, as mentioned above, social and technological data strongly related and exist incomplete. In this field, one can apply data analysis methods only to special aspects. For example, in the information security there are a misuse/signature detection, anomaly detection, hybrid detection, scan detection, profiling modules. But there are no integrated approaches to the data analysis [15].

Let us define such source data as the heterogeneous semi-structured objects. This definition reflects semi-structured data nature in syntax and semantics and the incompleteness of the original data. To solve the problem of working with such data the author proposes to apply the methods of object data modeling [16] and vectorization [17]. On this basis is described and tested an approach to the objects formalization and distance evaluation between them. Formalization and distance evaluation method does not depend on the data nature and incompletes. The approach allows comparing and looking for similar objects, to evaluate distance of similarity, and rank the objects according to their similarity.

3 Data Representation

Data are presented as a set of objects $O = \{o_1, o_2, \ldots, o_m\}$ with a variable number of characteristic properties. The number of properties may vary depending as on the type of the object, as on the information incompleteness about the domain. Properties define the structure and characteristics of a particular object $o_j = \{c_1, \ldots, c_k\}$. k – number of known properties of the j-th object. Each property can be:

- Unknown.
- Have a value of known format (text, number, coordinates, etc.)
- Be a reference to another object.

If $\mathrm{C_j}$ is the set of properties of the j-th object $C_j = \{c \mid c = Unknown \cup c = Value \cup c = RefO\}$, where *Value* – the value of the property; *RefO* – a reference to the object. *Unknown* means that the known existence of properties, but the value is not defined. Figure 1 shows the formalization of heterogeneous data example.

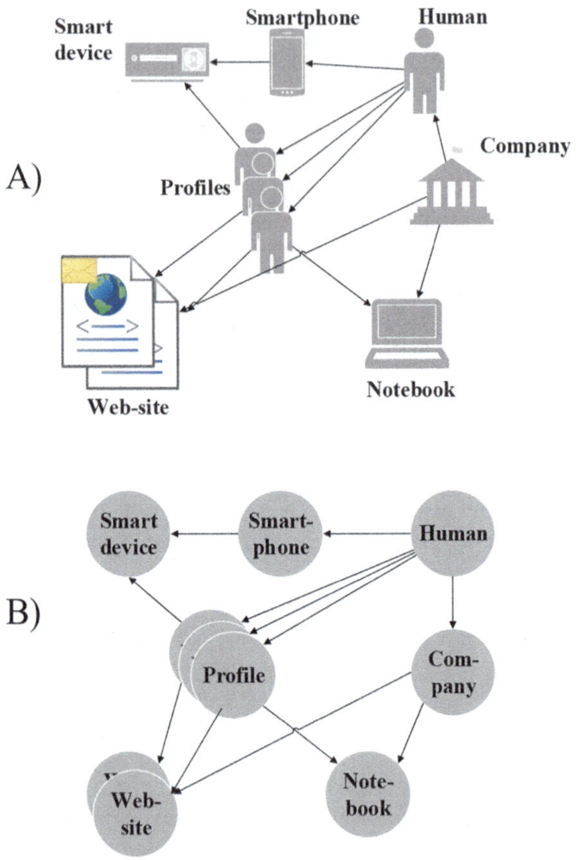

Fig. 1. (A) Heterogeneous data example; (B) Heterogeneous data formalization: Objects and references between them.

Similar to the "bag of words" representation of the text, we use "bag of objects" representation. A particular object characterizes only its attributes, including its relationships. The domain of the attribute has no importance. This is simple and flexible object representation, but it has low accuracy. The description includes all the information known to the researcher. For the same object it may look different (Table 1).

Table 1. Incomplete object data examples of the same object

Data example 1	Data example 2
<... Name: unknown Pulse: 76 Distance: 10456 meters Place: Nord str. 29 ...>	<... Name: Vanski Pulse: unknown Place: Nord str. 29 DeviceName: LG-G-Watch-Vanski ...>

So, it becomes important to identify such objects, evaluate distance between them and rank them according to the similarity. In general, the approach is applicable to other cases of objects with variable characteristics description. For another example, let's see the applicability of the method for detecting attacks in network traffic.

Network traffic may be described by a number of network connections and individual network packets, which will be treated as objects. The following traffic characteristics can be distinguished as properties: type of network protocol, amount of data transferred during the connection, packet size, number of received/transmitted packets, flags installed, status of connection, and so on.

Consider the possibility of a SYN-flood attack detection. This attack could be described as a set of connections with an unfinished triple handshake procedure, i.e., it is sufficient to take into account the following properties: network protocol type, flags set, connection status, number of packets transmitted within the connection. In this case, attack is represented by a number of objects, each of which describes the connection with the unfinished handshake procedure. Representing traffic in the form of a set of objects, we can easily detect the presence of those of them that are similar to the description of the attack considered.

To do this, we first describe the required traffic properties that need to be monitored for detecting a type of SYN-flood attack [18] (and, possibly, other denial-of-service attacks).

$c_1 = N^{in}_{traffic}$ – amount of incoming IP traffic per time unit Δt;

$c_2 = N^{out}_{traffic}$ – amount of outgoing IP traffic per unit of time Δt;

$c_3 = N^{in}_{ACK}$ – number of incoming ACK flags per time unit Δt;

$c_4 = N^{out}_{ACK}$ – number of outgoing ACK flags per time unit Δt;

$c_5 = N_{SYN}$ – number of SYN flags in incoming network packets per time unit Δt;

$c_6 = N_{PSH}$ – number of PSH flags in incoming network packets per time unit Δt;

$c_7 = N^{in}_{TCP}$ – number of incoming TCP packets per time unit Δt.

These are the basic properties of the network traffic that we propose to consider. Each property has its own numerical value. These basic properties are used to detect the SYN-flood attack, but not explicitly, they are used by performing some calculations on values of these properties.

We introduce more complicated properties:

$c_8 = n$, where $n = \dfrac{N^{in}_{traffic}}{N^{out}_{traffic}}$ – ratio of the amount of incoming IP traffic per time unit Δt to amount of outgoing IP traffic for the same time period. Control of the value of this property can be used to detect any DoS attack, since each DoS attack is characterized by an increase in the amount of incoming traffic.

$c_9 = d$, where $d = N^{in}_{ACK} - N^{out}_{ACK}$ – difference between the number of incoming and outgoing ACK flags per unit time Δt. This property can also be evaluated for any type of DoS attack. A negative value of d demonstrates that the server loses the ability to respond to client requests with ACK packets.

$c_{10} = k$, where $k = \frac{N_{SYN}}{N_{TCP}^{in}}$ – frequency of SYN flags appearing in incoming TCP packets. When SYN-flood attacks occur, the frequency of SYN flags increases, because the number of connection requests increases.

$c_{11} = m$, where $m = \frac{N_{PSH}}{N_{TCP}^{in}}$ – frequency of PSH flags in incoming TCP packets. When attacking SYN-flood, the frequency of the occurrence of PSH flags characterizing the useful load of the channel falls.

In this way, if $C^{normal} = \{c_1, \ldots, c_{11}\}$ – is a set of properties that characterize normal network traffic. Then, when compared with the properties of anomalous traffic $C^{abnormal} = \{c_1, \ldots, c_{11}\}$, the similarity of objects over the values of properties should be estimated.

4 Objects Match

For distance evaluation between objects in this model, one can use only their properties. The features are the following:

- Objects can have both common and different properties.
- The value of any object property may be unknown.
- Each property can have a value or be a reference to another object.

Therefore, to evaluate distance between the objects and rank them due to their similarity we use cumulative similarity measure of two components.

4.1 Similarity of Objects by Set of Properties

In this component are taken into account common and different properties of each compared object. First, we form a common vector of objects properties (1).

$$\left.\begin{array}{l} o_1 \rightarrow C_1 = \{c_1^1, \ldots, c_1^{k1}\} \\ o_2 \rightarrow C_2 = \{c_2^1, \ldots, c_2^{k1}\} \end{array}\right\} \rightarrow \overrightarrow{C_{1,2}} = \{c^1, \ldots, c^k\} \quad (1)$$

o_1, o_2 - are objects defined by sets of properties;
c_j^i - the i-th property j-th object;
$\overrightarrow{C_{1,2}}$ - the common vector of objects properties $\overrightarrow{C_{1,2}} = C_1 \cup C_2$.

Then we can define a binary vector of properties for each object, by encoding 1 – if the object has current property, 0 – if hasn't. The number of coordinates matches the both objects properties total number. The similarity of two objects is evaluated as the distance between their binary vectors, for example as the cosine of the angle between the vectors [6]. Thus is evaluated the similarity of objects by set of properties measure $s^1 \in \{0, 1\}$, where 1 - objects completely coincide, and 0 - is completely not the same.

4.2 Similarity of Objects by Property Values

If both objects have the common property, and in both cases its value is defined, this allows one to compare objects by value in this property. For this it is necessary and sufficient to know the properties format. To compare the complex properties, such as coordinates or text, one may use special methods. As a result, for any property evaluation $s_i^2 \in (0, 1)$ is obtained, where 1 - properties are completely the same and 0 - is completely not the same.

If some two objects has n common properties, their similarity by properties values is (2).

$$s^2 = \frac{\sum_{i=1}^{n} s_i^2}{n} \tag{2}$$

4.3 Cumulative Similarity of Objects

An cumulative characteristic of the objects similarity for an individual property is $s_i = \alpha * s_i^1 + \beta * s_i^2$. The sum of the weights $\alpha + \beta = 1$. For properties, that cannot be compared by value, $\beta = 0$. A common cumulative measure of similarity of two objects is (3).

$$s = \frac{\sum_{i=1}^{m} s_i}{m} \tag{3}$$

$$m = \left|\overrightarrow{C_{1,2}}\right| = \left|C_1 \cup C_2\right|$$

To test the distance evaluating and the similarity of the objects measure a test dataset was created. Test dataset is a number of object groups. Each group describes the different home network, including such entities as the network as is, applications, e-mail, computer and so on. The set includes both expert network descriptions and experimentally collected data. The objects have been pre-typed (a, e, n, etc.) for testing. The type was included only in object name, it was not taken into account in experiment evaluation. The experiment evaluation was done for different types of objects. Figure 2(A) shows a fragment of the ranked list of similar objects for an object of type a – application obtained by searching similar objects in the training set.

To test the solution, also a number of additional tests were generated with descriptions based on expert templates. The evaluation result of the similarity on these tests for the object type n – network shown in Fig. 2(B). The weights α and β in both experiments were 0.5.

The results of the experiment showed the similarity of objects of the same type in most of cases, as it should be, and ranking of objects within a type. Expert evaluation of the result proved the relevance ranking in most cases. Objects, more similar to the test object, according to the expert opinion, received a higher score. Unfortunately, the

A)

Object	S
a020	0,887
a022	0,887
a000	0,721
a002	0,721
a010	0,721
a012	0,721
***	
n120	0,392
c000	0,388
c004	0,387
c005	0,340
c100	0,340
n100	0,339
n040	0,329
***	

B)

Object	S
n100	0,789
n110	0,789
n040	0,788
n100	0,788
n020	0,750
n030	0,750
n000	0,704
n005	0,704
***	
c015	0,553
c033	0,549
c120	0,549
***	
e022	0,288
e004	0,227
e020	0,223
***	

Fig. 2. Fragments of ranked lists of matching objects (A) – application object (B) – network object.

subjectivity of expert similarity assessments did not allow to calculate the quality of the comparison.

4.4 Relevance of the Properties Measure

Need to say, that the described method has several disadvantages.

- A large number of properties in real objects. As the method deals with all the available object properties, object properties binary vector are of greater dimension. If the object has many properties, especially links to another objects, the amount of computation is large enough. Therefore, evaluation of large sets of objects, such as users of public wi-fi, requires more resources.
- Semantically, properties are making a different contribution to the similarity of the objects. For example, the "name" property is more important in people or animals comparison than "sex" one.

To partly solve these problems, the author proposes to use relevance of the properties for a specific task analysis. Relevance shows how each property is important in a particular case. Relevance can be defined by expert or calculated based on training (test) collections of objects. In the second case, it is always inversely proportional to the frequency of properties occurrence (4).

$$f_i^1 = \frac{1}{|O_i|} \tag{4}$$

O_i – the number of objects with a given property.

The resulting score is normalized, that is $f_i^1 \in (0, 1)$. Thus, the unique properties will have greater relevance than common ones. The second component of relevance may be defined on the domain base.

To test the relevance of the properties *f*, the training set of the home networks descriptions was partitioned into two parts. In one were included the network successfully carried out attacks on them by an attacker. In another - network with failed attacks. Generated examples where partitioned based on the expert given probability of a successful attack. The second component of relevance was calculated as proposed in [19], then relevance was normalized.

Figure 3 shows a top fragment with 13 attributes maximum relevance. Selected attributes that are meaningful to attacks success, according to expert opinion. At the limit of relevance b = 0.1 all of them are in 12 of the most important attributes.

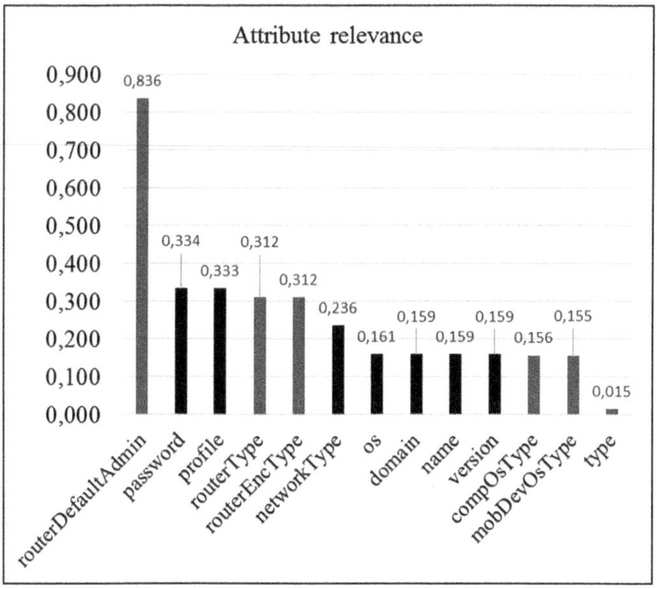

Fig. 3. The attributes relevancies.

That is, in this threshold of relevance all important attributes are in important set and the noise is 42%. On average, on the all set of experiments noise was from 22% to 45%. Limit of relevance allowed us to discard almost a 20% of the properties.

So, we can say, that property relevance allows:

- Discard some properties with low relevance before the similarity calculating.
- To consider the properties relevance in evaluating the similarity of objects.

5 System Architecture

The developed approach we propose to apply first and foremost in intelligent decision support systems. As an example, the penetration testing DSS in information security field [19]. There are systems with learning based on the training sets.

These systems consist of training and operational sets. On the training dataset are calculated relevance values for known object properties. The operational set compared to the training objects to determine and rank similar cases. A training set can be assembled in a particular subject area or generated based on the model. When using the system it is supplemented by the analysis results and thus updates. The architecture of decision support system, which uses the developed method, as is shown in Fig. 4.

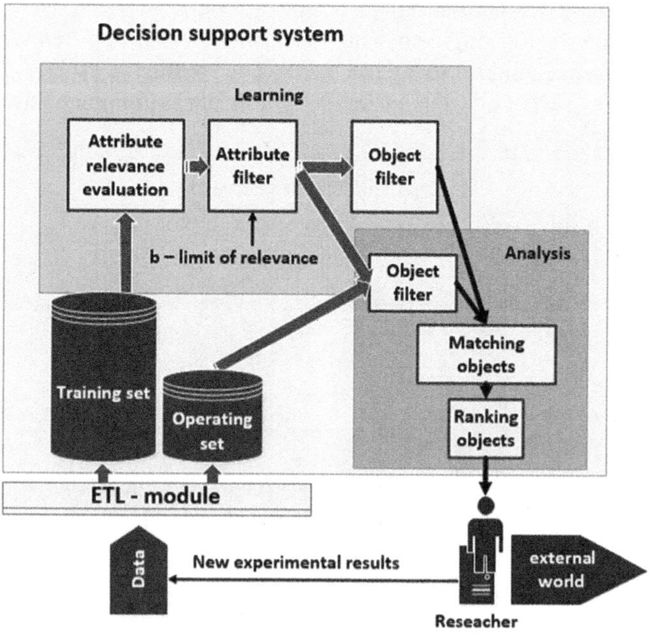

Fig. 4. The architecture of the decision support system, which use the similarity of heterogeneous semi-structured objects and properties relevance.

Learning modules performed in advance. They done pre-evaluation and filtering of the training data. Analysis modules filter and take deal with operational data and looking for a match objects in the knowledge base. As a result, the researcher obtains ranked, based on the similarity, list of objects from knowledge base, corresponding to the operational object(s).

Various analytical modules, comparing sets of objects, relations, etc. may expand the analysis modules set in a specific domain.

6 Conclusion

The paper describes an approach of the formalization and comparison of heterogeneous semi-structured objects. Using the object data model allows to uniformly describe information of different nature. Object pre-typing is not requires. The approach may be used when domain information is always incomplete.

The author has developed and tested methods for evaluation of objects similarity. Cumulative similarity measure has the qualitative and quantitative basis. Qualitative basis of the similarity lies in the similarity of the sets of object properties. The quantitative basis of the similarity lies in the similarity values of matching properties. The measure of similarity allows to apply to the object machine learning methods such as classification and clustering. One can apply the approach to a particular domain and problem by setting the weights of the cumulative similarity measure and the evaluation of properties relevance in his own way. Decision support system architecture on the basis of machine learning using the author's approach is described in the paper. It can be extended by modules of analysis and training in specific applications. For a number of tasks can be useful pre-typed objects and properties.

The development of this work is the introduction of the relations measure between objects and the of similarity relations evaluation. The measure of relations similarity can also include qualitative and quantitative components. A quality component may be based on the structure of object relations and relation graph, and quantitative similarity. On relation attributes can also be effective, given the interaction properties between themselves; for example, the properties "number of friends" and "friend N" in the social network. Quantitative defining of such relations should be developed in the future.

References

1. Barsegyan, A.A., Kupriyanov, M.S., Kholod, I.I., Tess M.D., Elizarov, S.I.: Data and Process Analysis: Handbook, 3rd edn. BXV – Petersburg, St. Petersburg (2009). 512 p
2. Ramsay, J.O.: Functional Data Analysis. Encyclopedia of Statistical Sciences. Wiley, New York (2006). https://doi.org/10.1002/0471667196.ess3138
3. Berry, M.J.A., Linoff, G.: Data Mining Techniques. Wiley, New York (1997)
4. Louise Barriball, K.: Collecting data using a semi-structured interview: a discussion paper. J. Adv. Nurs. **19**, 328–335 (1994). https://doi.org/10.1111/j.1365-2648.1994.tb01088.x
5. Grishkovsky, A.: Integrated processing of unstructured data. Open systems, vol. 6 (2013)
6. Bird, S., Klein, E., Loper, E.: Natural Language Processing with Python. O'Reilly Media, Inc., Sebastopol (2009). 504 p
7. Schabenberger, O., Gotway, C.A.: Statistical Methods for Spatial Data Analysis. Chapman & Hall/CRC Press (2005). 488 p. ISBN 1-58488-322-7
8. Milo, T., Zohar, S.: Using schema matching to simplify heterogeneous data translation. In: Proceedings of the 24th International Conference on Very Large Data Bases, VLDB 1998, pp. 122–133 (1998)
9. Liu, S., Chen, G., Yao, S., Tian, F., Liu, W.: A framework for interactive visual analysis of heterogeneous marine data in an integrated problem solving environment. Comput. Geosci. **104**, 20–28 (2017)

10. Nathan Binkert, Stavros Harizopoulos, Mehul A. Shah, Benjamin Sowell, Dimitris Tsirogiannis: Scalable analysis platform for semi-structured data. Amazon Technologies Inc., Nou Data Corp. (2014). US20130166568A1
11. Madnick, S.E., Siegel, M.D.: Query and retrieving semi-structured data from heterogeneous sources by translating structured queries. Massachusetts Institute of Technology (2001). US6282537B1
12. Beyer, K.S., Ercegovac, V., Gemulla, R., Balmin, A., Eltabakh, M., Kanne, C.-C., Ozcan, F., Shekita, E.J.: JAQL: a scripting language for large scale semistructured data analysis. In: VLDB (2011)
13. Kenneth, W.: Kisiel System method and computer program product to automate the management and analysis of heterogeneous data. Wisdombuilder, L.L.C. (2001). US6327586
14. Constales, D., Yablonsky, G.S., D'hooge, D.R., Thybaut, J.W., Marin, G.B.: Advanced data analysis and modelling in chemical engineering. **120**(2), 417–420 (2017). Elsevier. ISBN 978-0-444-59485-3
15. Dua, S., Du, X.: Data Mining and Machine Learning in Cybersecurity. Taylor and Francis Group, LLC (2011). 248 p
16. Cattell, R.G.G., Barry, D.K., Berler, M., Eastman, J., Jordan, D., Russell, C., Schadow, O., Stanienda, T., Velez, F. (eds.): The Object Data Standard ODMG 3.0. Morgan Kaufmann, January 2000
17. Fatkieva, R.R.: Developing metrics for detecting attacks based on network traffic analysis. Vestnik BGU, No. 9, pp. 81–86 (2013)
18. Goodfellow, I., Bengio, Y., Courville, A.: Deep Learning. The MIT Press, Cambridge (2016). 800 p
19. Poltavtseva, M.A., Pechenkin, A.I.: Intelligent Data Analysis in Decision Support Systems for Penetration Tests. Autom. Control. Comput. Sci. **51**(8), 985–991 (2017). ISSN 0146-4116

An Approach to Energy-Efficient Street Lighting Control on the Basis of an Adaptive Model

Dmitry A. Shnayder(✉), Aleksandra A. Filimonova, and Lev S. Kazarinov

Department of Automatics and Control, South Ural State University, Chelyabinsk, Russia
{shnaiderda, zakharovaaa, kazarinovls}@susu.ru

Abstract. The problem of energy saving in outdoor lighting systems is important, as they consume a significant part of electricity used by the city economy. An approach to the energy-efficient adaptive control of street lighting proposed in this article enables reduction of outdoor lighting electricity consumption by dimming of light lines and individual lighting fixtures. The adaptive lighting control model makes it possible to modify the dimming time dependency curve in real time based on roads' actual luminance readings. The result of application of this model is a considerable decrease of power consumption while ensuring a required level of luminance. The energy savings achieved by reducing the costs of lighting also leads to lessened CO_2 due to electric power generation.

Keywords: Dimming · Outdoor lighting · Adaptive control · Energy saving

1 Introduction

Russia's annual energy consumption for lighting purposes is approximately 109 billion kW/h, which accounts for about 12% of the total national energy consumption [1]. Up to 20% of the lighting energy consumed in the Russian Federation is due to the outdoor lighting. The CO2 equivalent emission attributable to outdoor lighting power generation is 6.04 million tons [2]. The currently estimated cost saving potential in case the entire set of the state-of-the-art energy saving technologies is implemented in the outdoor lighting systems may reach 50% and more. This is why the energy saving problem here becomes very relevant.

Road, municipal street and architectural artistic lighting systems can be identified as one of the main consumption segments of the outdoor lighting. 65% of the entire market consumption goes to the roads lighting, while 25% and 10% are attributable the city street and architectural lighting, respectively.

One of the main factors of enhancing the efficiency of utilization of electricity consumption for outdoor lighting is implementation of automatic control systems and centralized management solutions. The efficiency of use of automated outdoor lighting control is guided and limited by governmental regulatory acts that permit lowering of

K. Arai et al. (Eds.): IntelliSys 2018, AISC 868, pp. 1271–1284, 2019.
https://doi.org/10.1007/978-3-030-01054-6_89

the level of outdoor lighting at night down to 50% of the total luminance capacity. At the same time, solutions where one of the tree phases is directly shut down cannot be regarded as acceptable because in case of their application there may be areas with extremely low degree of luminance, having an adverse effect on the safety of pedestrians and drivers. Hence, the objective of relevance is creating automated control systems with integrated feature of smooth adjustment of the levels of outdoor lighting in keeping with the preset control dependency functions. To provide a solution to this problem, it may be required to lay down the principles of system architecture in terms of the hierarchy and methods of lighting control that may ensure required functionality, flexibility, reliability and energy efficiency for the outdoor lighting systems [3].

2 Related Work

The main tendency of development of the lighting products both in the Russian Federation and worldwide is that the energy efficiency, environmental safety and the operating performance enhancement requirements are getting all the stricter [4, 5]. The mercury arc lamps keep being replaced with energy efficient light sources and devices based on them, besides, to control the lighting performance, the outdoor lighting fixtures are fitted with electronic control gears (ECG) [6–9]. The modern-day automated outdoor lighting control systems must ensure control of the lighting itself along with remote condition monitoring of the lighting fixtures of the real time mode. Such systems must enable control of the entire lighting system in general and of each individual lighting fixture in particular from the central control panel of a distributed control system (DCS) [10–15].

Currently in Russia work is underway to develop and integrate new generation intelligent control systems referred to as Smart Grid into the existing power grids. Implications of implementation of the Smart Grid concept in the outdoor lighting systems are the newly emergent capability of each individual lighting fixture control and diagnostic monitoring, which enhances energy efficiency and operational reliability of outdoor lighting networks. The energy saving effect hinges considerably on the lights dimming over the time of the day as a function of the natural lighting conditions and of the special requirements for the degree of luminance in relation to the time of day [5, 16, 17, 21].

Real-time control of the operating modes and performance parameters of lighting fixtures would allow us to increase the quality of operation of lighting systems, by collecting data on the condition of each lighting fixture within reasonable amounts of time thus making it possible to perform flaw detection and efficient trouble shooting in relation to the functioning of a system [18].

Today the proliferation of the Smart Grid concept in Russia is in its initial phase. In the meantime, the advantages of intellectual grids over the traditional ones are evident: higher operating reliability, energy efficiency, environmental protection and conservation, economic efficiency. Nevertheless, there are a number challenges on the way to realization of the multi-task concept of smart grid, among these challenges are: technological, regulatory, legal, financial ones, etc. This is exactly why the Smart Grid system has not been completely realized in Russia yet, but there are some individual attempts known to us by way of implementing this concept and its confined, local use [19].

A topical direction for development of intelligent outdoor lighting control systems is their integration with the geographical information system [3]. Mapping and analysis of the spatially distributed data and data on the current status, both for the fixed equipment (outdoor lighting devices, posts, lighting fixtures, line system items etc.), and of mobile equipment, such as service vehicles (their geographical position, license plate number, description, etc.) with ensuring of the possibility of real-time control, – all considerably enhance the efficiency of the automated control system for outdoor lighting networks. A Geographical information system (GIS) module, if integrated into the automated control system for outdoor lighting networks, would enhance controllability and monitorability of the lighting system operation, as well as the overall management of the system at the upper level [20].

3 Outdoor Lighting System Energy-Efficient Control Algorithm

Energy-efficient control of outdoor lighting consists in adaptive dimming of the streetlights over the time of a day depending on the duration of the daylight, calculated for every day of the year for the particular geographical area or time zone. The algorithm enables the functionality of the system of lighting in the energy saving mode, which may be split into a basic and optimum sub-modes of energy saving [22].

The nominal energy saving mode (NESM) corresponds to a condition wherein the dimming of the outdoor lights is ensured for a pre-allowed interval of time (see Fig. 1).

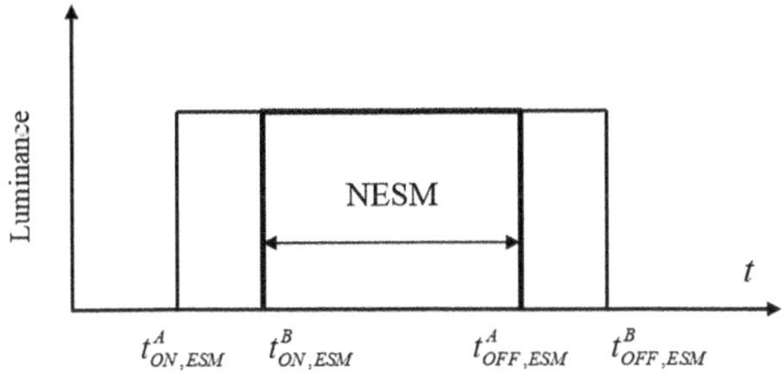

Fig. 1. Nominal energy saving mode.

The optimum energy saving mode is the condition where the duration of the energy saving mode reaches its maximum for the particular current date.

The outdoor lighting energy saving control algorithm includes the following steps:

(1) Pre-recording into the memory of the adaptive dimming unit (ADU) of a dataset representing the lamps ON/OFF switching graph for each day of the calendar year for the particular geographical area.

(2) Pre-recording into the memory of the adaptive dimming unit (ADU) of a dataset representing the energy saving mode ON/OFF switching graph based on the results of the measurement of the traffic intensity for the particular area (for the road, street, square, etc.)
(3) At every i-th cycle of the light operation (being the time interval from the moment the lamp is put ON till the moment the lamp is put OFF over the time of the day) the duration of the ON time T_{Ci} is counted using the timer that starts the counting at the time the lamp is put ON t_{ON} and finishing the counting the moment the lamp is put OFF t_{OFF}. The calculated value of T_{Ci} is stored in the non-volatile memory of the ADU.
(4) At every i+1-th cycle for value T_{Ci} based on the lighting ON/OFF switching graph dataset, two possible corresponding calendar dates t^A and t^B are determined. On Fig. 2, point A corresponds to calendar date t^A with the light switching ON time t^A_{ON} and light switching OFF time t^A_{OFF}, while point B corresponds to calendar date t^B with the light switching ON time t^B_{ON} and light switching OFF time t^B_{OFF}.

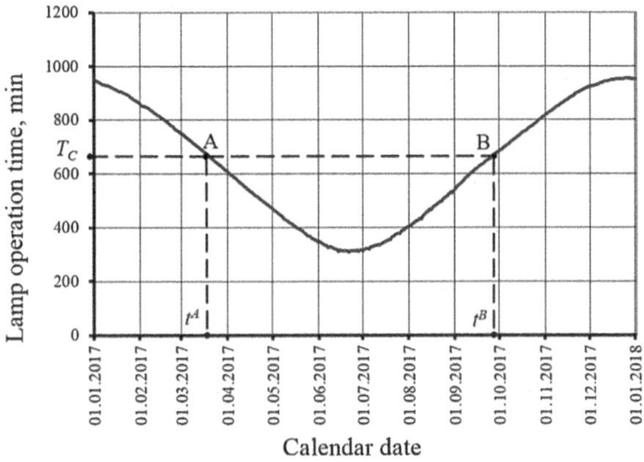

Fig. 2. Schedule for lighting modesswitching.

The time of activation of the NESM is determined as follows:

$$t_{ON,N} = \max\left(t^A_{ON}, t^B_{ON}\right). \tag{1}$$

The time the NESM is deactivated is determined based on the ratio:

$$t_{OFF,N} = \min\left(t^A_{OFF}, t^B_{OFF}\right), \tag{2}$$

The interval between the thus determined NESM ON/OFF times correspond to the safe low luminance mode, satisfying both possible calendar dates: t^A and t^B. The values $t^i_{ON,N}$, $t^i_{OFF,N}$ calculated at the *i*-th cycle of the ADU operation are used for transition into the safe low luminance mode at the *i*+1-th cycle. In the meantime, in case the calculated T_{Ci} is not determined the for any single calendar date, then at the i+1-th cycle the NESM does not get activated.

(5) To maximize the lamp operation time in the energy saving mode, optimum energy saving mode ON/OFF times get calculated $t^{i}_{ON,O}$, $t^{i}_{OFF,O}$. To that end, at each step of the ADU operation an actual calendar date is determined as follows: Signal T_{Ci}, is fed to the filter smoothing the fluctuations T_{Ci}, caused by the errors in the lamps switching ON and switching OFF control as relative to the existing lamps ON/OFF switching graph for the particular dates of the calendar year for the particular geographical area. Using the function of calculation of the first-order derivative T'_{Ci} the direction of the variation of the filtered value T_{Ci} (decreasing or increasing). Based on the calculated values T_{Ci} and T'_{Ci} using the data in the lamps switching ON/OFF table, the actual calendar date is determined (the closest calendar date) with a required and pre-set accuracy Δ. Using this calendar date the times of activation/deactivation of the OESM are further determined, $t^{i}_{ON,O}$ and $t^{i}_{OFF,O}$, which are used for the transition into the OESM at the i+1-th cycle. In the meantime, in case the calculated T_{Ci} is not determined the for any single calendar date, then at the i+1-th cycle the OESM does not get activated.

(6) In the modes control unit (MCU) the time values for ON/OFF of the OESM and NESM are compared. If the OESM ON/OFF times are determined (the relationship between the calculated T_{Ci} and the calendar date A or B is established), then the energy saving mode ON/OFF times will adopt the following values:

$$t_{ON,ESM} = t_{ON,O}, \tag{3}$$

$$t_{OFF,ESM} = t_{OFF,O}. \tag{4}$$

If the OESM ON/OFF times are not determined, but the corresponding similar parameters have been determined for the NESM (i.e. the relationship is established between the calculated T_{Ci} and at least one calendar date), then the energy saving ON/OFF time modes shall adopt the following values:

$$t_{ON,ESM} = t_{ON,N}, \tag{5}$$

$$t_{OFF,ESM} = t_{OFF,N}. \tag{6}$$

In case the neither activation nor deactivation times of both OESM and NESM have been determined, then at the i+1-th cycle the energy saving mode shall not be enabled.

(7) On the basis of the resulting values of the energy saving mode ON/OFF times, the timer countdown periods for energy saving mode ON/OFF get calculated. T_1 is the period of the energy saving mode ON-timer - calculated as the difference between the time of activation of the energy saving mode $t_{ON,ESM}$ and the pre-set time of lamps switching ON t_{ON} for the particular calculation calendar date. T_2 is the period of the energy saving mode OFF-timer – calculated as the difference

between the pre-determined time of deactivation of the energy-saving mode $t_{OFF,ESM}$ and the time of activation of the energy-saving mode $t_{ON,ESM}$.

The calculation of the energy saving mode ON/OFF timers countdown periods is illustrated in Fig. 3.

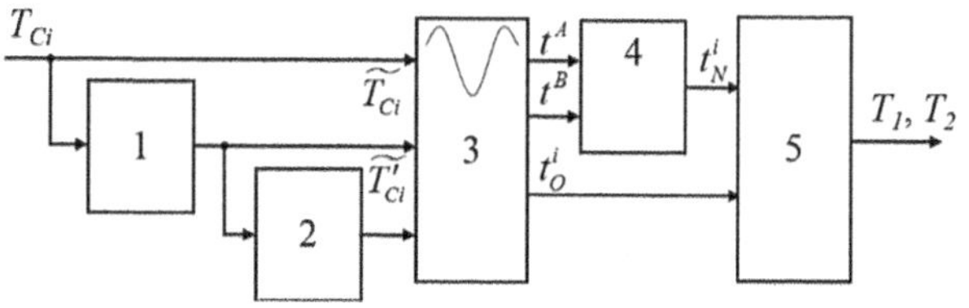

Fig. 3. The calculation of the energy saving mode ON/OFF timers countdown periods.

Figure 3 shows: T_{Ci} are the values of the lamps operation time; 1 is the filter; 2 is the first-order derivative calculation function; 3 is the graph comparison unit; 4 is the NESM; 5 is the modes control unit; T'_{Ci} is the calculated derivative value T_{Ci}; T_{Ci} is the filtered value of the T_{Ci} value; $t^A(t^A_{ON}, t^A_{OFF})$ is calendar date t^A with the time of lamps switching ON t^A_{ON} and the time of lamps switching OFF t^A_{OFF}, corresponding to point A; $t^B(t^B_{ON}, t^B_{OFF})$ is calendar date t^B with the time of lamps switching ON t^B_{ON} and the time of lamps switching OFF t^B_{OFF}, corresponding to point B; $t_N = (t_{ON,N}, t_{OFF,N})$ is the times the lamp is switched on and off in the NESM; $t_O = (t_{ON,O}, t_{OFF,O})$ is the times the lamp is switched on and off in the OESM; T_1 is the period of the ESM ON-timer; T_2 is the period of the ESM OFF-timer.

(8) Based on the information received on the value of the period of the ESM ON-timer and on the value of the period of the ESM OFF-timer, the dimming modes selection control in accordance with Fig. 4 is effected.

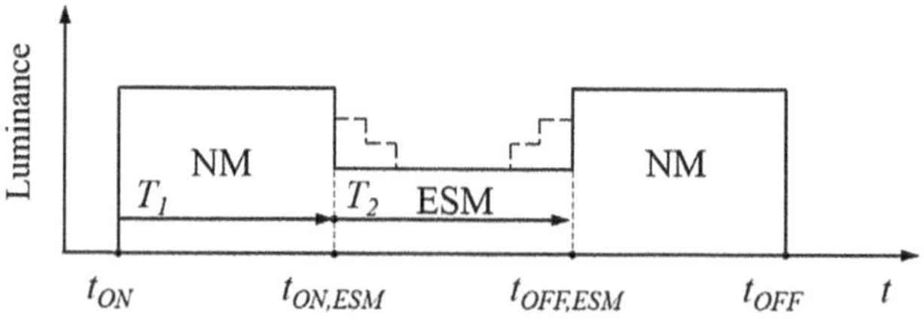

Fig. 4. Dimming modes selection control.

Figure 4 shows: t_{ON} is the point of time when the lamps are to be switched ON for the particular calendar date; t_{OFF} is the point of time when the lamps are to be switched OFF for the particular calendar date; $t_{ON,ESM}$, $t_{OFF,ESM}$ are the pre-set values of the time of activation/deactivation of the ESM for the particular geographical area; NM is the nominal mode; ESM is the energy saving mode; T_1 is the period of the ESM ON-timer; T_2 is the period of the ESM OFF-timer.

4 Outdoor Lighting Energy Saving Model Predictive Control

The suggested algorithm allows forming a required time dependency curve for variation of the lamps degree of illumination. The preset illumination level of a lamp is achieved by adjustment of the output electrical power of the lamp. Dependence of the level of light intensity of a lamp on its power consumption hinges upon the type of lamp and requires clarification in each specific case by referring to the manufacturer's data sheet or by empirical or experimental investigation. It must be noted that from the point of view of the end user the key characteristic of a lighting system is not the power consumption or the light intensity of a lamp, but the level of luminance on the road surface. This lighting level depends not only on the light intensity of the lamps, but also on a whole range of geodesic, geometrical, and optical characteristics, including the spacing of the light posts, the height of installation and the angle of inclination of the lamps, the width of the road, parameters of the lens, etc.

Table 1 shows results of experimental measurements of the level of luminance (E_{AV}) of the automobile road in Chelyabinsk (Russia) with the dimming implemented for 250 W High Pressure Sodium (HPS) lamps in lighting fixtures fitted with individual dimming units. Table 1 shows: I_A, I_B, I_C being the current values in phases A, B and C respectively, P being the power consumed.

Table 1. Results of experimental investigation into the level of luminance of the road surface with dimming implemented for 250 W HPS lamps

Dimming mode	E_{AV}, lx	I_A, A	I_B, A	I_C, A	P, kW
0%	18.6	40.1	31.5	25.2	25.2
10%	15.9	36.2	26.9	35.4	22.5
15%	15.6	34.4	25.8	33.7	21.6
23%	12.6	31.3	22.3	30.7	19.4
30%	10.3	28.1	20.4	27.5	17.5
35%	9.58	26.2	18.7	25.8	16.3
40%	7.2	24.7	17.4	24.3	15.1

The reduction of electric power consumption for lighting at various dimming modes enables reduction of CO_2 emissions, or carbon footprint of the related electrical power generation. In Fig. 5, one can see the dynamics of the CO_2 emissions variation for various dimming modes.

As one can see in Table 1, the parameter of the dimming mode is proportional to the power consumed. Figure 6 shows a corresponding graph of the variation of the luminance (E_{AV}) as a function of the power consumed (P).

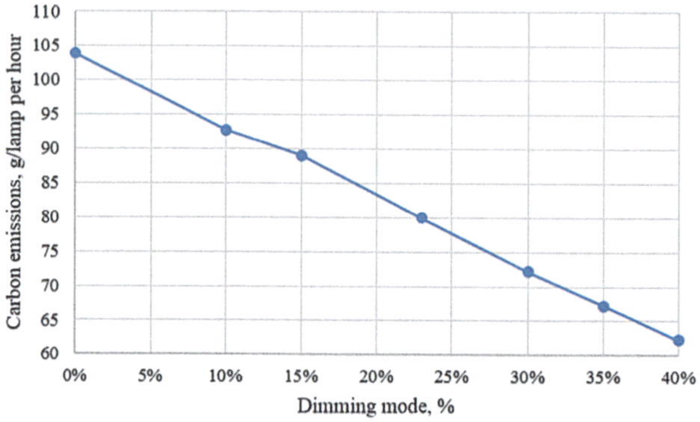

Fig. 5. The graph of estimated variation of the CO_2 emissions in various dimming modes.

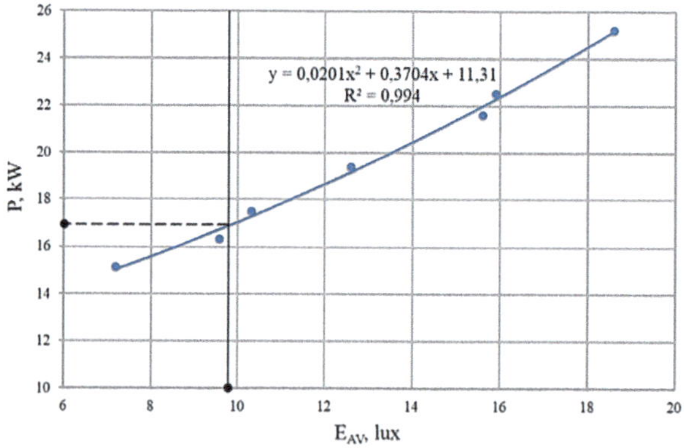

Fig. 6. Graph of variation of luminance as a function of the power consumption.

As one can see from the graph, the resulting relationship curve can be described by the regressive equation:

$$y = 0.0201x^2 + 0.3704x + 11.31, \tag{7}$$

being a static model of variation of the consumed electrical power by the lighting system or line under the dimming depending on the level of luminance of the driveway or road surface and enabling one to determine the extremity values of relevant power consumption when setting the dimming time dependency curve within the limits of the existing required luminance range for the particular driveway or road category. The coefficient of determination R^2 is 0.994, which means a sufficiently high quality of the constructed model. Thus, the obtained model can be used to determine the dimming level. In the particular example, the allowable level of luminance is 10 lx, limiting the lower dimming threshold to the level of 70% (17.5 kW, Table 1).

Thus, the conducting of experiments for individual parts of the road enables us to build luminance-to-dimming dependency models with adequately high precision.

Based on these models it would be possible to form optimum autonomous dimming time dependency curves based on the energy saving criterion, while assuring the required quality of illumination with regard to the category of the particular driveway or road in keeping with the effective standards.

Despite the high efficiency, the actual conducting of such measurement experiments, especially, in metropolises with their dozens of thousands of city lights may turn out to be an unfeasible activity owing to the considerable time and funds consumption involved. An alternative option to experimental research mentioned above could be the road surfaces luminance simulation and design activity using certain special purpose software. In Figs. 7 and 8 one can see the results of the road surface luminance simulation made for the nominal (100% of light intensity) and energy saving (dimming, 40% of light intensity) modes, respectively. According to the standards [1], the average illumination of this class of roads should not be less than 15 lx. The average illumination of the road in Figs. 7 and 8 is 28 and 16.8 lx, respectively, which corresponds to the norm. At the same time dimming allowed to significantly reduced power consumption. The simulation was performed taking into account the type of the road surface materials, and including such surface material parameters as those relevant for the conditions where road lanes may be wet while other parts of the road may not be wet - thus striving to ensure uniformity of luminance in such complex conditions. The initial, input data used for the simulation are provided in Table 2.

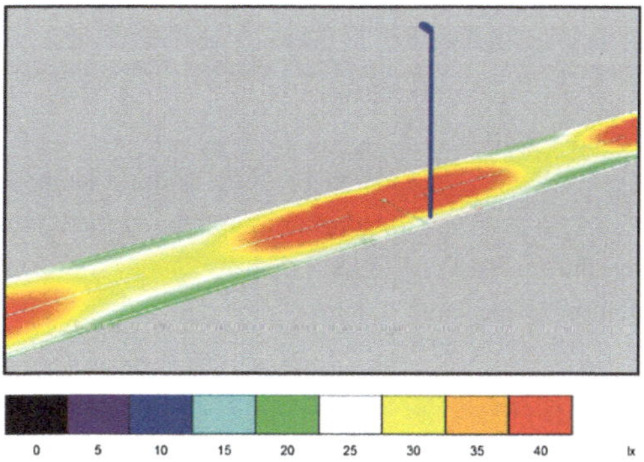

Fig. 7. Simulation of the road surface luminance at maximum level of illumination.

The simulation results are the lookup values of the allowable dimming levels with due consideration for the luminance levels prescribed by applicable effective standards both in the main and duty (night time) modes. The evident advantages of the simulation bases method are the possibility to simulate and to model various dimming conditions for various types of lamps within short periods. This technique is especially efficient for multiple uniform sections of roads or for newly constructed roads, where all the input physical data required for correct simulation, are known. Otherwise, this technique would require considerable efforts in investigation into the lighting networks

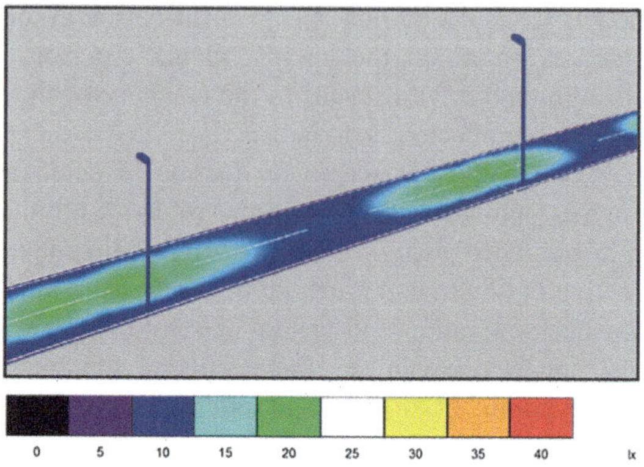

Fig. 8. Simulation of the road surface luminance with dimming of 40% of the total level of illumination.

Table 2. Initial input data used for the simulation

Parameter	Value
Road conditions	
Width	7 m
Number of lanes	2
One-way	no
Sidewalk	
Width	2 m
Lamps and/or fixtures used	
Lamp	High Pressure Sodium lamp, 250 W HPS lamp
Positioning type	Hung on one side, at the bottom
Distance between the posts	30 m
Height of installation of the lamp	11 m

parameters, including the measurements of posts positioning, height of lamps installation, inclination angles of the lighting fixtures, etc.

The most efficient in the sense of the peculiarities related to the experimental and empirical approaches mentioned above to determining of the effective dimming levels is the use of automated predictive model control systems, which imply the application of luminance sensors readings as the input signal for the on-line tuning of the system settings and operating conditions. Refer to Fig. 9 for an example of architecture of such a system.

Figure 9 shows: DU is a dimming unit of the lamp or of a part of the street lighting network (dimming panel or cabinet); CU is a lighting output power correction unit; LS is a luminance sensor; TRS is a traffic rate sensor; MD is a motion or presence detector; E, E^* are the actual and the target luminance values, lx; P is the specific power consumed; ΔP is the correction factor used to modify the electrical power consumed,

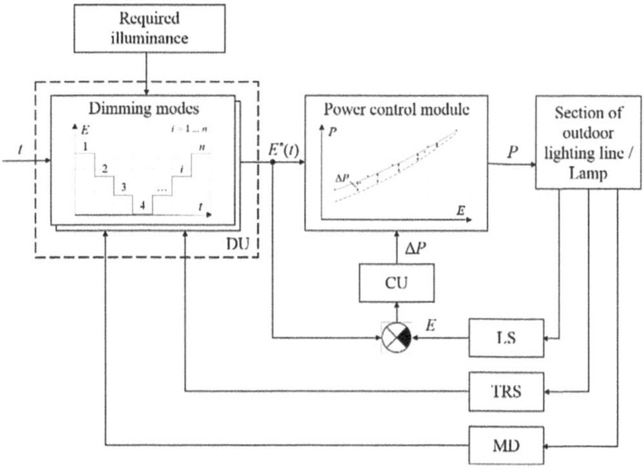

Fig. 9. Architecture of automated model predictive control system for street lighting.

directly affecting the dimming time dependency curve data based on the actual luminance of the road surface.

The system is hinges on the control principle based on a model, that simulates the relationship, as stated above, between the required or target luminance level of the road surface and the dimming mode. In the meantime in the initial phase of the setup and adjustment of the system, some simplified function is used to reflect the relationship, selected for the particular type of lamps with certain operational margin to make sure there should be certain power reserve in the dimming mode, despite the fact that this margin is bound to reduce the energy saving performance of the lighting grid control system. When the lighting grid is operated, the actual levels of the output power P are monitored and analyzed as input parameter on whose basis the predictive correction signal is formed for model output power correction ΔP taking into account the real time deviation of the actual luminance level E from the target standardized level of luminance E^* for the particular segment of the street lighting grid for the particular actual conditions.

An example of similar luminance graph correction is given in Fig. 10 (in dotted line).

In actuality the luminance level measurement may be performed both manually and automatically using fixed luminance sensors (LS), linked to the automated power control points of the lighting grids passing the data thus locally collected on to the control room or distributed control system database. In the case of manual measurement, service vehicles luminance sensors, for example, could be used with the vehicle regularly made to pass the required itinerary to monitor the performance, do the troubleshooting and clear any errors in the street lighting grids.

In addition to the luminance sensors locally in the segments of the street lighting grids it may be sometimes reasonable to use motion detectors (MDs). The signal from the MDs in keeping with the diagram in Fig. 9 will be fed to the dimming unit and introduces a short-term corrective action on the current luminance request $E^*(t)$, raising

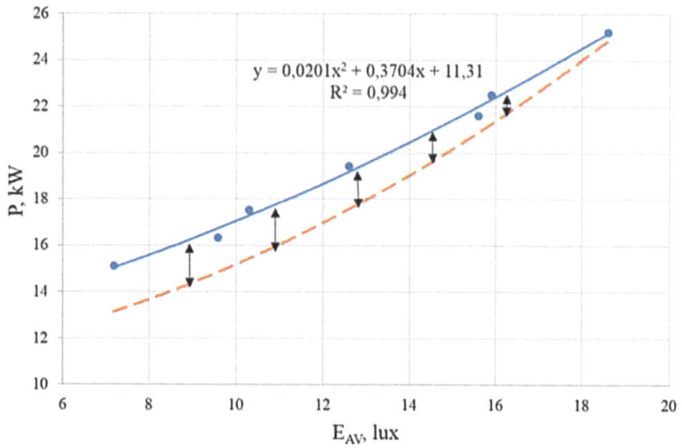

Fig. 10. Example of correction applied to the dimming time dependency curve data based on the data of the road surface actual luminance.

the temporary required luminance level to the maximum or to any other higher pre-set level for a time interval of Δt, determined and pre-programmed in the course of the system setup and commissioning. After this pre-set time elapses the original luminance level $E^*(t)$ relevant to the current conditions will be recuperated. Apart from the MDs, the architecture of the proposed Smart Grid automated lighting control system, it is proposed to use the traffic rate sensors (TRSs). Installation and use of the TRSs, is primarily expedient and recommended for roads with the allowable degree of dimming reaching 50%, as to maximize the dimming energy saving effect, on one hand, and to prevent the luminance levels from dropping to lower than the standardized threshold at the time of need, on the other hand.

The use of LSs, MDs and TRSs enables the operating companies to enhance the quality of lighting while minimizing the costs related to the electric power consumption - as part of the proposed general predictive model control approach to day-to day management of the outdoor lighting grids.

Besides, individual or local lamps dimming provide additional energy saving opportunities to grid operators and owners [19]. The result of application of this type of feature is achieving maximum possible energy saving while ensuring a required level of luminance.

5 Conclusion and Future Work

Thus, we can infer that the proposed predictive model street lighting control approach enables operating companies or owners of lighting grids to attain the maximum energy saving effect with relatively low financial and time spending thanks to effective lamps dimming while assuring the required level of luminance in keeping with all applicable standards.

A practical implementation of the suggested approach may combine various lighting dimming and control system architecture options at the lower (lamps), intermediate (automated local power supply points or cabinets) and upper (control room or DCS) levels of an automated outdoor lighting control system, in particular:

(1) at the lower control level:

- the individual independent dimming units for ECG-operated high pressure sodium lamps, developed for this project;
- the currently commercially available ECGs for gas-discharge lamps, and the dimming units of the LED lamp drivers, fitted with the PLC based control units or wireless control signals transmission technology based units, for example the 2.4 GHz range ZigBee standard;

(2) at the intermediate control level:

- automated cabinets used for regulation of segments of the lighting grids, that perform the dimming of the entire segments;
- the commercially available automated power supply points or cabinets for the lighting grids from manufacturers in various corners of the world;

(3) at the upper control level is the intelligent automated outdoor lighting distributed control system developed specially for this project, combining features of a conventional automated outdoor lighting control system with a real-time geographical information system and a predictive model control system, with incorporated optimum dimming algorithms.

Acknowledgments. The work was supported by Act 211 Government of the Russian Federation, contract № 02.A03.21.0011.

References

1. On updating the requirements for lighting devices and electric lamps in alternating current circuits. http://government.ru/docs/30118. Accessed 11 Jan 2018
2. International Energy Agency. Light's Labour's Lost. Policies for Energy-efficient Lighting. In support of the G8 Plan of Action. OECD/IEA, Paris (2006)
3. Kazarinov, L.S., Shnaider, D.A., Barbasova, T.A., et al.: Automatic control systems of efficient lighting: study (edited by Kazarinov L.S.). Publisher SUSU, Chelyabinsk (2011)
4. Masleyeva, O.V.: Ecologic and economic profits from upgrading of educational establishments' lighting systems. Energ. Effi. **3–4**, 57–58 (2011)
5. Barbasova, T.A., Vstavskaya, E.V., Zakharova, A.A.: Definition of the operational reliability parameters of the street lighting control systems elements. Bull. South Ural. State Univ., Ser. Comput. Technol. **14**(23), 102–106 (2011)
6. Long, X., Liao, R., Zhou, J.: Development of street lighting system-based novel high-brightness LED modules. Optoelectronics **3**, 40–46 (2009)
7. Costa, M.A.D., Costa, G.H., Santos, A.S., Schuch, L., Pinheiro, J.R.: A high efficiency autonomous street lighting system based on solar energy and LEDs. In: Proceedings Power Electronics Conference, pp. 265–273 (2009)

8. Lagorse, J., Paire, D., Miraoui, A.: Sizing optimization of a stand-alone street lighting system powered by a hybrid system using fuel cell, PV and battery. Renew. Energy **34**(3), 683–691 (2009)
9. Choi, A.-S., Song, K.-D., Kim, Y.-S.: The characteristics of photosensors and electronic dimming ballasts in daylight responsive dimming systems. Build. Environ. **40**(1), 39–50 (2005)
10. Popa, M., Cepişcă, C.: Energy consumption saving solutions based on intelligent street lighting control system. UPB Sci. Bull., Ser. C: Electr. Eng. **73**(4), 297–308 (2011)
11. Chen, P.-Y., Liu, Y.-H., Yau, Y.-T., Lee, H.-C.: Development of an energy efficient street light driving system. In: 2008 IEEE International Conference on Sustainable Energy Technologies, pp. 761–764 (2008)
12. Nefedov, E., Maksimainen, M., Sierla, S., Flikkema, P., Yang, C.-W., Kosonen, I., Luttinen, T.: Energy efficient traffic-based street lighting automation. In: 2014 IEEE 23rd International Symposium on Industrial Electronics (ISIE), pp. 1718–1723 (2014)
13. DiLaura, D.L., Houser, K., Mistrick, R., Steffy, G.R.: The Lighting Handbook Reference and Application. Illuminating Engineering Society of North America, New York (2011)
14. Mardaljevic, J.: Simulation of annual daylighting profiles for internal illuminance. Light. Res. Technol. **32**(3), 111–118 (2000)
15. Iversen, A., Svendsen, S., Nielsen, T.: The effect of different weather data sets and their resolution on climate-based daylight modelling. Light. Res. Technol. **45**(3), 305–316 (2012)
16. Simhas, D.: A smart grid application - street lighting management system. UPB Sci. Bull., Ser. C: Electr. Eng. **75**, 309–324 (2013)
17. Martirano, L.: A smart lighting control to save energy. In: Proceedings of the 6th IEEE International Conference on Intelligent Data Acquisition and Advanced Computing Systems, vol. 6072726, pp. 132–138 (2011)
18. Wu, Y., Shi, C., Zhang, X., Yang, W.: Design of new intelligent street light control system. In: Proceedings of 8th IEEE International Conference on Control and Automation, pp. 1423–1427 (2010)
19. Filimonova, A.A., Barbasova, T.A., Shnayder, D.A.: Outdoor lighting system upgrading based on smart grid concept. Energy Procedia **111**, 678–688 (2017)
20. Murray, A.T., Feng, X.: Public street lighting service standard assessment and achievement. Socio-Econ. Plan. Sci. **53**, 14–22 (2016)
21. Xavier, N., Kumar, A., Panda, S.K.: Design, fabrication and testing of smart lighting system. In: Proceedings of Future Technologies Conference, pp. 763–768 (2016)
22. Shnayder, D.A., Shishkin, M.V., Barbasova, T.A., Zakharova, A.A., Abdullin, V.V.: The method and device of energy-saving control of street lighting, 28 January 2013

New Field Operational Tests Sampling Strategy Based on Metropolis-Hastings Algorithm

Nacer Eddine Chelbi[1](✉), Denis Gingras[1], and Claude Sauvageau[2]

[1] Laboratory on Intelligent Vehicles (LIV), Université de Sherbrooke, Sherbrooke, Canada
{nacer.chelbi,denis.gingras}@usherbrooke.ca

[2] PMG Technologies Inc., 100 rue du Landais, Blainville, Canada
csauvageau@pmgtest.com

Abstract. As recently stated by National Highway Traffic Safety Administration (NHTSA), to demonstrate the expected performance of a highly automated vehicles system, test approaches should include a combination of simulation, test track, and on-road testing. The simulation part need to be based on a probabilistic approach. To do so, an appropriate sampling strategy is often used. In this paper, we propose a new sampling strategy based on Markov Chain Monte Carlo (MCMC) methods, using Metropolis-Hastings algorithm to generate samples from probability distributions of Field Operational Tests (FOT); the Safety Pilot Model Deployment (SPMD) in our case. We begin by modeling the probability distribution of each test parameter retrieved from the SPMD database, two estimation methods were applied: Kernel Density Estimation method and EM algorithm. A comparison was made between the two methods to choose the best one. These distribution models are then sampled using our sampling strategy based on the Metropolis-Hastings algorithm.

Keywords: Expectation-Maximization algorithm (EM)
Field Operational Tests (FOT) · Kernal Density Estimation (KDE)
Kolmogorov-Smirnov test · Markov Chain Monte Carlo (MCMC)
Metropolis-Hastings algorithm · Monte Carlo simulations
Safety Pilot Model Deployment (SPMD)

1 Introduction

The ultimate purpose of intelligent vehicles is to make driving easier and safer. The main question is how to ensure that such a technology would meet the safety requirements, especially in worst-case scenarios. Choosing accurate physical tests to validate their performance is definitely the answer. However, such an approach is becoming increasingly costly and time-consuming. That is why we need to investigate new solutions (simulation tools, validation algorithms, etc.) in order to validate the behavior, performance and to reduce the number of physical tests of intelligent, connected, and autonomous vehicles. Real driving tests are proven to be the

K. Arai et al. (Eds.): IntelliSys 2018, AISC 868, pp. 1285–1302, 2019.
https://doi.org/10.1007/978-3-030-01054-6_90

most reliable solution, but unluckily, it is not the optimal one: due to time and cost issues, these tests cannot cover all the operating conditions. Moreover, test scenarios cannot be reproduced exactly as they first happened. One way to bypass this problem is by using virtual validation tools; vehicular simulators make a very good example. On the other hand, using simulations alone for the validation procedure is often insufficient because of a large number of virtual tests that should be done. Therefore, a probabilistic approach is necessary to reduce this constraint. To do so, an appropriate sampling strategy is often used.

Our proposed sampling strategy is based on Markov Chain Monte Carlo (MCMC) methods, using Metropolis-Hastings algorithm to generate samples from probability distributions of Field Operational Tests (FOT); the Safety Pilot Model Deployment (SPMD) in our case. There are three main approaches for testing and validating highly automated driving systems: Field Operational Tests, tests track and simulation [1]. Based on this sampling strategy, we proposed a new virtual validation approach, shown in Fig. 1 (not discussed in this paper), for preventive and active security systems, which involves Field Operational Tests, a probabilistic approach and simulation tests with a vehicular simulation software (PreScan [2]).

1.1 Field Operational Tests

Field Operational Tests (FOT) are large-scale testing programs designed to provide a more comprehensive assessment of the efficiency, quality, robustness and user acceptance of new vehicle technologies such as navigation, communication systems, ADAS, and cooperative systems.

1.2 Validation in a Test Track

In Canada and the United States, manufacturers and importers must self-certify their vehicles to Motor Vehicle Safety Standards; CMVSS in Canada and FMVSS in the United States. In Europe, a third party performs a homologation of the vehicles that are manufactured or imported. New Car Assessment Program (NCAP) in the United States (US NCAP) as well Euro NCAP as in Europe publishes test procedures to rate the performance of vehicles in order to help consumers choose their vehicles. Certification and NCAP testing can be performed in specialized test centers such as Transport Canadas Motor Vehicle Test and Research Centre (MVTC) in Blainville (Quebec, Canada), managed by PMG Technologies. To ensure constant and well-synchronized parameters, PMG technologies uses brake and accelerator robots to control braking and acceleration of the test and target vehicles. Similarly, inertial and GPS navigation systems are used for measuring motion, position and orientation at high speed with a 1 cm precision as well as for performing vehicle-to-vehicle measurements [3].

1.3 Virtual Tests (Simulation)

Using simulation tools in the development and evaluation of systems has become an imminent process in all engineering fields. In the automotive industry, vehicular software simulators make the deployment of new vehicular solutions easier, safer, and more reliable. Moreover, its implementation procedure is cheaper and faster compared to real driving tests. Several vehicular software simulators are well-known, we can mention here: PreScan, Pro-SiVIC, dSPACE, CarSim, etc.

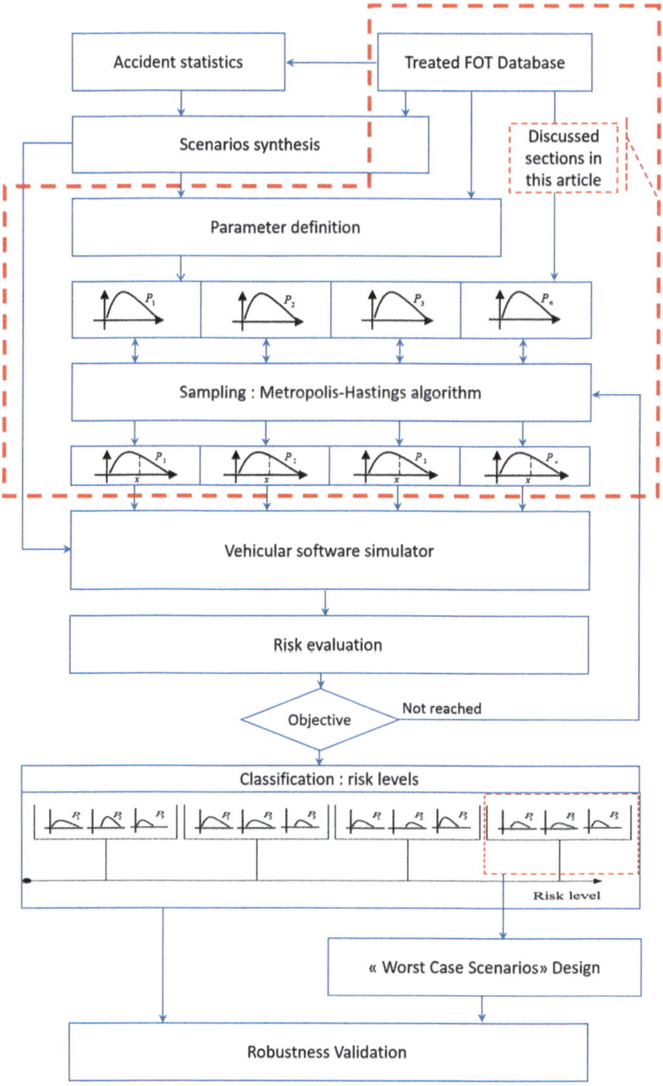

Fig. 1. Proposed validation architecture.

2 State of the Art: Probabilistic Approaches

In the automotive field, the used evaluation and validation methods are generally deterministic, which means that many parameters, such as the vehicle speed, are supposed to be fixed and already known. They attempt to take into account variability and uncertainty by using fixed safety factors. On the other hand, probabilistic approaches make it possible to quantify variations and uncertainties by using mainly probability distributions rather than fixed values. Several initiatives have emerged in the field of intelligent and autonomous vehicles (VAs) [4–7].

2.1 Monte Carlo Simulations (MC)

Monte Carlo simulations try to take advantage of modern computers. The idea is to input values into complex dynamic systems and watch the results overs tens of thousands of iterations. In the vehicular domain, this technique is often based on stochastic models constructed from FOT data. Zhao [7] explains in details the application of this technique in [8,9]. Decision making and stochastic numerical integration in the evaluation process of collision avoidance systems have been studied by Jansson in [4,10]. The studied configuration uses radar sensors to detect and track other vehicles. Since inaccurate sensor information can lead to uncertain state information and can influence collision avoidance systems performance, a statistical evaluation algorithm based on Monte-Carlo simulations was developed to handle uncertainties estimation by calculating the probability for each predefined action. To describe the Monte Carlo simulations technique, Gietelink [11] summarizes several works and gives the main steps of this method applied in control systems analysis. Monte Carlo simulations can reduce the preparation time and cost of an evaluation process, but not necessarily with great efficiency; since the non-critical events of driving can dominate the simulations.

To deal with this inconvenient, the Importance Sampling (IS) technique is often used. We know that the effect of some events taken by a random variable is stronger than others on the desired estimator. So the variation of our estimator can be reduced if these events, which have the most effect, are realized more often. It is, therefore, more convenient to pay more attention to operating conditions that are more likely to become dangerous than others. One possibility is to use the Importance Sampling technique to increase the number of occurrences of these events whose probabilities should be estimated [5,11]. The real strength of the IS technique lies in its ability to accurately estimate the probabilities of rare events involving a random variable that is a function of several other random variables [12]. This method has been discussed and used in Gietelink [5] and Zhao *et al.* [13] work, where dangerous events probabilities were estimated in an adaptive cruise control (ACC) and lane change problems successively. Two other approaches are often mentioned in the literature, Adaptive Importance Sampling (AIS) and Rejection Sampling (RS). For more details, we propose the following references [5,12].

2.2 Markov Chain Monte Carlo (MCMC)

Currently, the most widely used methods are the Monte Carlo (MC) methods, Importance Sampling (IS) is a very good example, which is less effective in problems with higher dimensions. The main advantage of MCMC over MC methods is that they allow samples generation in higher dimensions [14]. Two types of MCMC methods can be distinguished: methods that use random walks, we quote Metropolis-Hastings algorithm and Gibbs sampling as examples. And methods that have been developed to accelerate convergence, such as Hybrid Monte Carlo and Successive Over-relaxation. A state of the art of these methods is well presented in [14].

MCMC methods provide a sequence of samples from a probability distribution for which direct sampling is difficult. The goal is to design a Markov chain so that the stationary distribution of that chain is exactly the distribution we want to sample from [15]. Our proposed sampling strategy is based on the Metropolis-Hastings algorithm. The choice of the algorithm was made after a comparative study between the most used algorithms in the literature (Importance Sampling, Rejection Sampling, Gibs Sampling, Metropolis-Hastings Sampling), a summary of the study is indicated in Tables 1 and 2.

Table 1. MC methods in a Glance

MC	
Importance Sampling	Rejection Sampling
Estimation	Sampling
Distribution choice is mostly based on intuition	
Less efficient in high dimensions	
Easy to understand and to implement	

Table 2. MCMC methods in a Glance

MCMC	
Gibs Sampling	Metropolis-Hastings Sampling
Sampling	Sampling
Do not require a lot of knowledge about the distribution geography	
More efficient in high dimensions	
Very slow and requires good knowledge about conditional probabilities	Easy to implement but requires more computing power

3 Field Operational Tests Database (SPMD) Manipulation

We opted for the Safety Pilot Model Deployment (SPMD) program, a research initiative highlighting the application of safety technologies for connected vehicles "V2V, V2I (DSRC)" in real driving conditions. This program recorded data of 2842 equipped vehicles in Ann Arbor, Michigan for more than two years. In April 2015, 34.9 million miles were recorded, making it (SPMD program) one of the largest databases made public [7]. Six months of data (from 10-01-2012 to

04-30-2013) can be retrieved from the Research Data Exchange site (USDOT) [16]. To understand and know how to use the data, a document [17] is provided. The database contains 8 data-sets, as shown in Fig. 2.

3.1 Parameters Definition

We begin our study with a scenario that has two vehicles configuration, as shown in Fig. 3, a stopped target vehicle and a test vehicle that approaches at different distances and speeds. The data needed to be extracted are Relative distance, relative velocity and weather conditions. One data-set was used from the SPMD database: Data Acquisition System (DAS = DAS1 + DAS2). From the DAS1 data-set, which contains data collected by the Mobileye sensor in each vehicle, we extracted: Range, Rangerate, TargetType and Statut. From the DAS2 data-set, witch contains the position, mouvment and radar unit data in each vehicle, we extracted: ObjectType, RangeX, SpeedX, TargetMoving.

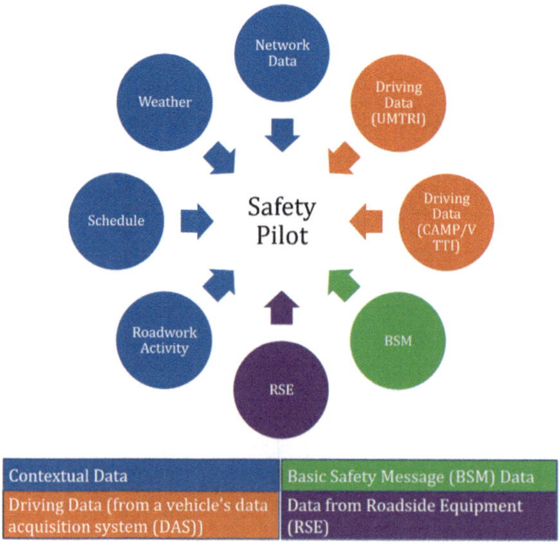

Fig. 2. Content of the SPMD database.

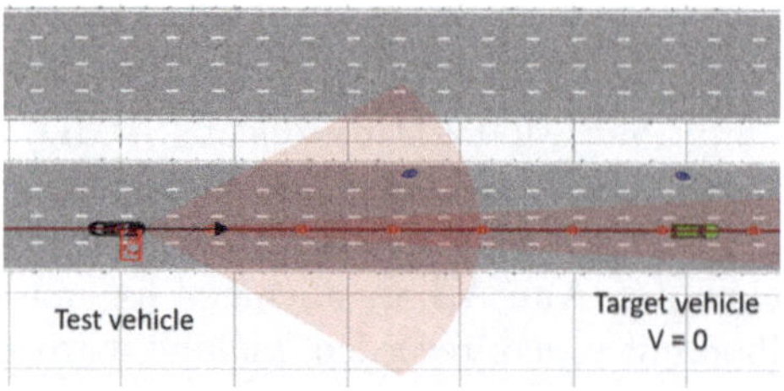

Fig. 3. Studied test scenario.

Since the SPMD database contains only two months of weather conditions data, we opted for the Weather Underground website (www.wunderground.com), suggested by the USDOT [17]. A site that gives access to weather conditions from countries all over the world, allowing the extraction of weather conditions for more specific areas and countries. International weather conditions are collected directly from more than 29,000 weather stations in countries around the world. In our case study, we suppose that the weather conditions will have a more significant influence on the vehicle behavior (slippage, etc.) than on the radar sensors (LRR and SRR) used by the studied Automatic Emergency Braking (AEB) system. In order to extract only data related to the same conditions of this first scenario, two search conditions were applied: Target type needs to be a vehicle and stopped ($V = 0$).

3.2 Probability Density Function (PDF) Models Estimation

To estimate the PDF model of each parameter, we applied two estimation methods: a Kernel Density Estimation (KDE) and an estimation based on the Expectation-Maximization algorithm (EM).

(1) *Kernel Density Estimation (KDE):* Kernel density estimation is a non-parametric method for estimating the probability density of a random variable. It makes it possible to estimate the density at any point based on a sample from a statistical population. This is a generalization of the histogram estimation method [18].
Let $x_1, x_2, ..., x_N \sim f$ an independent and identically distributed sample (i.i.d) drawn from a certain distribution with unknown density f. We are interested in estimating the form of this function. f. Then the nonparametric estimator by the kernel density estimation method is given by:

$$\hat{f}_h(x) = \frac{1}{NH} \sum_{i=1}^{N} K\left(\frac{x - x_i}{H}\right) \tag{1}$$

where K is a kernel and H is a parameter called window. In our case, K is chosen as a standard normal distribution ($\mu = 0$ and $\sigma = 1$):

$$K(x) = \frac{1}{\sqrt{2\pi}} e^{-\frac{1}{2}x^2} \tag{2}$$

(2) *Gaussian Mixture Model Estimation Using EM Algorithm:* A Gaussian Mixture Model (GMM) is used to model the parametric estimation of a distribution of a random variable as a sum of several Gaussians. It is then necessary to determine the mixing probability, the mean and the variance of each Gaussian. This is an iterative technique and it is done most often via the Expectation-Maximization (EM) algorithm. The EM algorithm is composed of two steps: expectation (E step) and maximization (M step) [19].

Expectation (E step): In this step, we calculate the probability that each data point belongs to each cluster. To begin, we will need the PDF equation of a multivariate normal distribution.

$$g_j(x) = \frac{1}{\sqrt{(2\pi)^n \left|\sum_j\right|}} e^{-\frac{1}{2}(x-\mu_j)^T \sum_j^{-1}(x-\mu_j)} \tag{3}$$

where $g_j(x)$ is the a multivariate normal distribution PDF of the cluster j, j is the cluster number, x is the input vector, n is the length of the input vector, $\sum_j$ is the covariance matrix n x n for cluster j, $\left|\sum_j\right|$ is the determinant of the covariance matrix, $\sum_j^{-1}$ is the inverse of the covariance matrix.

The probability that a point i is part of the cluster j can be calculated by:

$$w_j^i = \frac{g_i(x)\Phi_j}{\sum_{l=1}^{k} g_l(x)\Phi_l} \tag{4}$$

where w_j^i is the probability that a i point is part of the cluster j, $g_i(x)$ is the a multivariate normal distribution PDF of the cluster j, Φ_j is the prior probability of the cluster j, k is the cluster number. We apply this equation for every data point and cluster.

Maximization (M step): The update equations of the maximization step are given below:

$$\Phi_j := \frac{1}{n}\sum_{i=1}^{n} w_j^{(i)} \tag{5}$$

$$\mu_j := \frac{\sum_{i=1}^{n} w_j^{(i)} x^{(i)}}{\sum_{i=1}^{n} w_j^{(i)}} \tag{6}$$

$$\sum_j := \frac{\sum_{i=1}^{n} w_j^{(i)} x^{(i)}(x^{(i)} - \mu_j)(x^{(i)} - \mu_j)^T}{\sum_{i=1}^{n} w_j^{(i)}} \tag{7}$$

The prior probability of cluster j, denoted Φ, is computed as the average of probability that a data point belongs to cluster j. The μ equation of the cluster j is the average of all data points in the learning data set, each point being weighted by its cluster belonging probability. We repeat the same iterations until convergence.

3.3 Kernel Density Estimation Vs EM Algorithm Estimation

After plotting the histograms of each parameter, we noticed that some parameters have unimodal distributions (relative distance (Fig. 4), humidity (Fig. 7)) and other parameters have multimodal distributions (relative velocity (Fig. 5), temperature(Fig. 6), visibility (Fig. 8)). For unimodal distributions and for simplicity reasons, we have chosen to use the kernel density estimation. For multimodal distributions, a comparison was made between the two estimation methods to choose the best one. We noticed that for the studied cases, the kernel density estimation method offers a better estimate than the EM algorithm (see Figs. 5, 6 and 8), so we have chosen the kernel density estimation results.

3.4 PDF Models Estimation Results

(1) *Relative Distance* D_{sc}*:* Figure 4 shows the relative distance and its probability density function estimate using KDE method.

$$\begin{aligned} P(x) &= a_1 e^{-(\frac{(x-b1)}{c1})^2} + a_2 e^{-(\frac{(x-b2)}{c2})^2} \\ &+ a_3 e^{-(\frac{(x-b3)}{c3})^2} + a_4 e^{-(\frac{(x-b4)}{c4})^2} \end{aligned} \tag{8}$$

with: $a1 = 0.01806$; $b1 = 12.13$; $c1 = 5.022$; $a2 = 0.01127$; $b2 = 19.92$; $c2 = 9.231$; $a3 = 0.008861$; $b3 = 36.54$; $c3 = 20.24$; $a4 = 0.002773$; $b4 = 86.48$; $c4 = 72.42$.

(2) *Relative Velocity* V_s*:* Figure 5 shows the relative velocity histogram and its probability density function estimate using KDE method and EM algorithm.

$$\begin{aligned} P(x) &= a_1 e^{-(\frac{(x-b1)}{c1})^2} + a_2 e^{-(\frac{(x-b2)}{c2})^2} \\ &+ a_3 e^{-(\frac{(x-b3)}{c3})^2} + a_4 e^{-(\frac{(x-b4)}{c4})^2} \end{aligned} \tag{9}$$

with: $a1 = 0.02336; b1 = 112.3; c1 = 4.02; a2 = 0.01269; b2 = 61.55; c2 = 13.59; a3 = 0.006351; b3 = 38.92; c3 = 30.59; a4 = 0.002171; b4 = 73.55; c4 = 55.27$.

(3) *Temperature (Weather):* Figure 6 shows the temperature histogram and its probability density function estimate using KDE method and EM algorithm.

$$P(x) = a_1 e^{-(\frac{(x-b1)}{c1})^2} + a_2 e^{-(\frac{(x-b2)}{c2})^2} \tag{10}$$

with: $a1 = 0.02373; b1 = 18.45; c1 = 6.127; a2 = 0.02484; b2 = 2.092; c2 = 17.17$.

(4) *Humidity (Weather):* Figure 7 shows the humidity histogram and its probability density function estimate using KDE method.

$$P(x) = a_1 e^{-(\frac{(x-b1)}{c1})^2} \tag{11}$$

with : $a1 = 0.02916; b1 = 73.69; c1 = 19.25$.

(5) *Visibility (Weather):* Figure 8 shows the visibility histogram and its probability density function estimate using KDE method and EM algorithm.

$$P(x) = a_1 e^{-(\frac{(x-b1)}{c1})^2} + a_2 e^{-(\frac{(x-b2)}{c2})^2} \tag{12}$$

with: $a1 = 0.04065; b1 = 22.49; c1 = 9.321; a2 = 0.03104; b2 = 9.778; c2 = 6.129$.

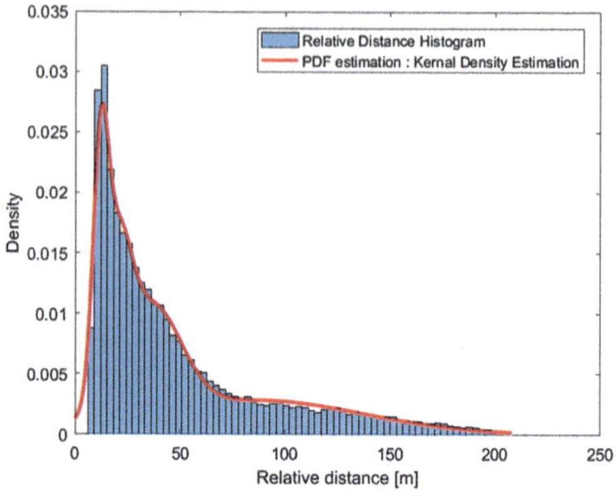

Fig. 4. PDF estimation: Relative distance D_{sc}.

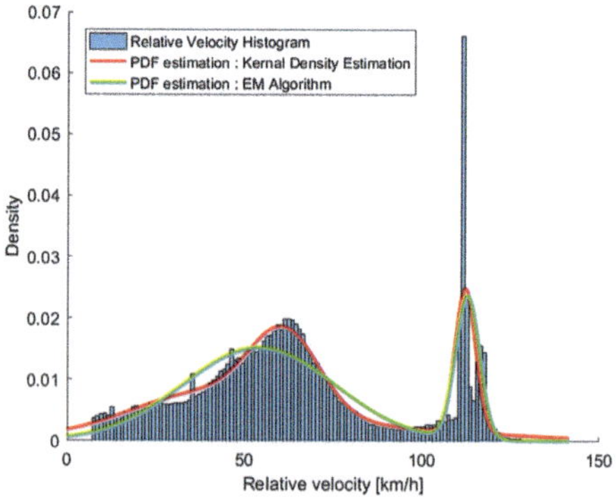

Fig. 5. PDF estimation: Relative velocity V_s.

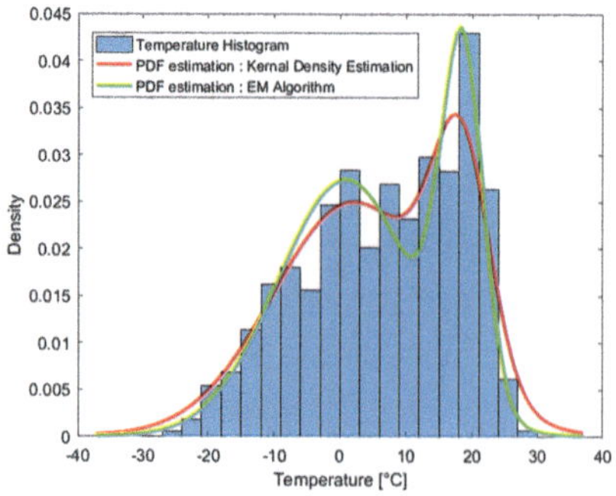

Fig. 6. PDF estimation: Temperature.

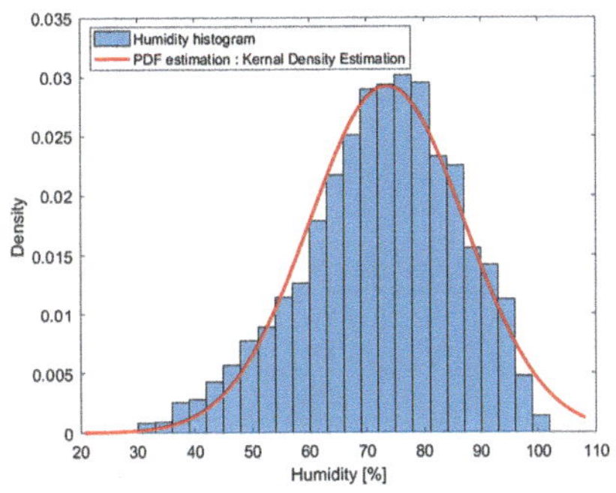

Fig. 7. PDF estimation: Humidity.

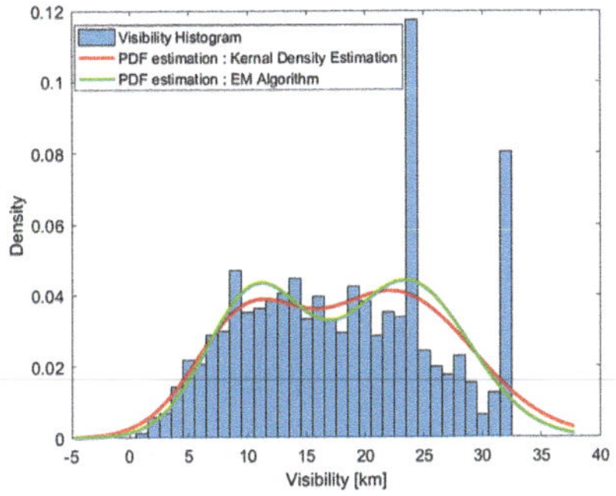

Fig. 8. PDF estimation: Visibility.

4 Metropolis-Hastings Algorithm Implementation

Our goal is to sample from a target probability density, we begin with the relative velocity $p(\theta)$ presented by (9), with $0 < \theta < 140\,\text{km/h}$. The Metropolis sampler creates a Markov chain that produces a sequence of values [20]:

$$\theta^{(1)} \longrightarrow \theta^{(2)} \longrightarrow \cdots \longrightarrow \theta^{(t)} \longrightarrow \cdots \tag{13}$$

where $\theta^{(t)}$ is the Markov chain state at iteration t. After the burn-in phase, the samples in the chain begin to reflect samples of the target distribution $p(\theta)$.

We initialize the algorithm with a random value $\theta^{(1)}$ between 0 and 140. We then use a proposition distribution $q(\theta|\theta(t-1))$ to generate a candidate θ^*. There are a few minor technical constraints to be taken into consideration before choosing the proposition distribution, but for the most part, it can be anything that we like; which makes this algorithm a completely flexible method. A normal distribution centered on the current state is often used ($q(\theta|\theta(t-1)) = N(\theta^{(t-1)}|\sigma^2)$); this is called the Random Walk Metropolis-Hastings algorithm (or the Metropolis Algorithm), and this is what we are going to apply in this article.

The next step is to accept or reject the proposal. The probability of accepting the proposal is as follows:

$$\alpha = min\left(1, \frac{p(\theta^*)}{p(\theta^{(t-1)})}\right) \tag{14}$$

To make a decision on accepting or rejecting the proposal, we generate a uniformly distributed number μ between 0 and 1. If $\mu \leq \alpha$, we accept the proposition and the next state is equal to the proposition: $\theta^t = \theta^*$. If $\mu > \alpha$, we reject the proposition, and the next state is equal to the current state: $\theta^t = \theta^{(t-1)}$. We continue to generate new proposals and accept or reject until the sampler reaches convergence. Here is a summary of the Metropolis sampler steps:

(1) Set $t = 1$
(2) Generate an initial value μ and σ, and set $\theta^t = \mu$
(3) Repeat
- $t = t + 1$
- Generate a proposition θ^* from $N(\theta^{(t-1)}|\sigma^2)$
- Compute acceptance probability $\alpha = min\left(1, \frac{p(\theta^*)}{p(\theta^{(t-1)})}\right)$
- Generate μ from a uniform distribution $Uniform(0, 1)$
- If $\mu \leq \alpha$, accept the proposition and set $\theta^t = \theta^*$, else set $\theta^t = \theta^{(t-1)}$

(4) Until $t = T$

And here is a summary diagram of the algorithm (Fig. 9):

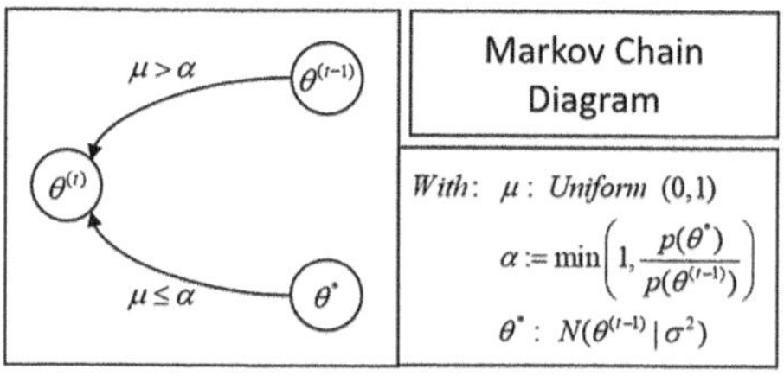

Fig. 9. Markov Chain Diagram.

4.1 Optimal Standard Deviation (σ): Kolmorogorov-Smirnov Test

To choose the optimal standard deviation (σ) for the proposal distribution, we apply a Kolmogorov-Smirnov test. The Kolmogorov-Smirnov test is a nonparametric hypothesis test that is often used to evaluate the difference between the cumulative distribution functions (CDFs) of the probability distributions, and it is defined by:

$$d_U(F, G) = \sup_x |F(x) - G(x)|, \; x \in \Re \tag{15}$$

where $F(x)$ and $G(x)$ are the CDFs of the distributions of the target (in red) and the sample (in green).

We propose to combine the outcome of the Kolmogorov-Smirnov test with the acceptance rate of each standard deviation to choose the optimal one. Here is a summary of the proposed strategy:

(1) Set $\sigma = 1$
(2) Repeat
 - Apply the Metropolis sampler
 - Apply the Kolmogorov-Smirnov test
 - Calculate the acceptance rate
 - $\sigma = \sigma + 1$

(3) Until $\sigma = 200$
(4) Choose the optimal σ, the first value that passes the Kolmogorov-Smirnov test and produces the best acceptance rate

5 Simulation Results

To verify the Metropolis sampler efficiency, we will discuss three different values of σ, where we run 2000 iterations for each value. Figure 10 shows that the acceptance rate is much better with lower standard deviations (σ), but a more detailed discussion need to be done to choose the optimal one.

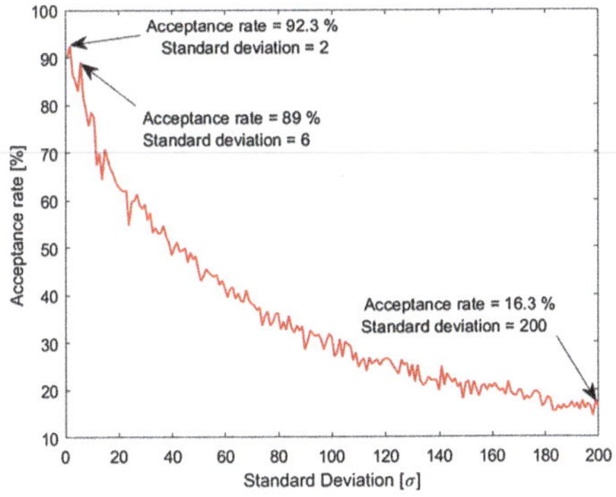

Fig. 10. Acceptance rate as a function of standard deviation (σ) [Relative velocity].

In one case, we will set the value too low ($\sigma = 2$) and in another we will set it too high ($\sigma = 200$), but in a third case we will get it about right ($\sigma = 6$). The sampling results for the relative velocity parameter are shown in Figs. 11, 12 and 13. For all three values of σ, we have two plots. The top one shows the true target distribution (in red), along with a histogram and its PDF distribution (in green) showing the distribution of samples obtained using the Metropolis sampler. The lower panel plots the actual Markov chain: the sequence of generated values.

In Fig. 11, we see what happens when the proposal distribution is too narrow ($\sigma = 2$), the acceptance rate is very high (92.3%) so technically the chain does not get stuck in any one spot, but it does get stuck in one spot and does not cover a very wide range.

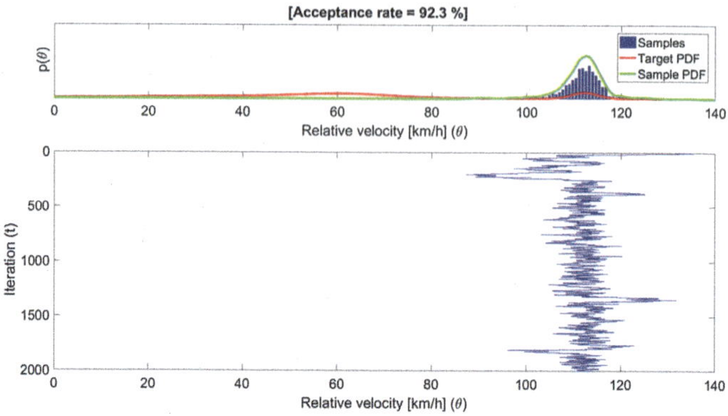

Fig. 11. Sampling results ($\sigma = 2$): Relative velocity [2000 iterations]. (Color figure online)

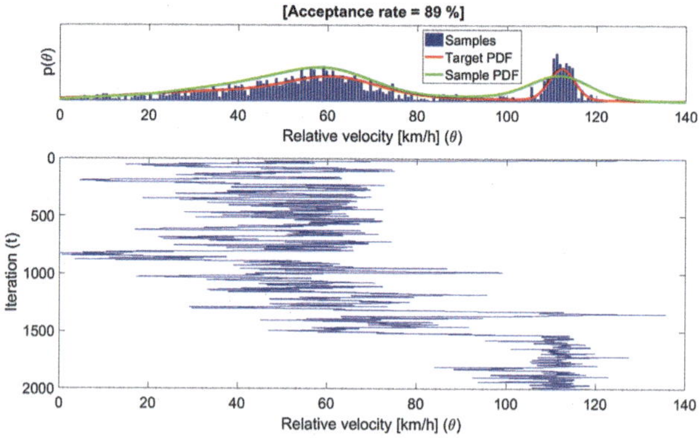

Fig. 12. Sampling results ($\sigma = 6$): Relative velocity [2000 iterations]. (Color figure online)

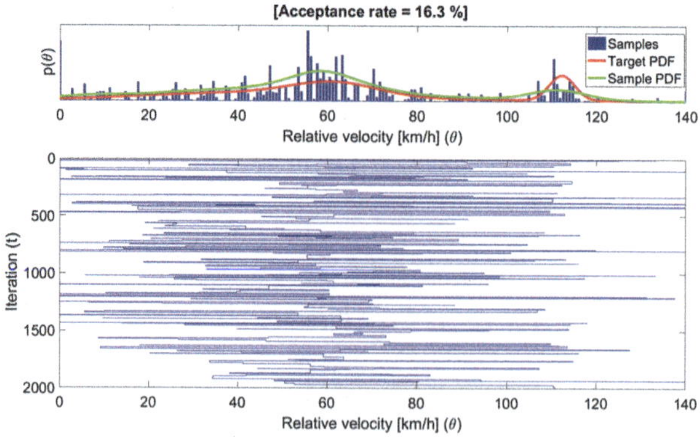

Fig. 13. Sampling results ($\sigma = 200$): Relative velocity [2000 iterations]. (Color figure online)

In Fig. 12, we see what happens when we choose a good proposal distribution ($\sigma = 6$): the chain shown in the lower panel moves rapidly across the whole distribution and covers the whole range, without getting stuck in any one place the acceptance rate here is 89%.

Finally, in Fig. 13, if we set the proposal distribution too wide ($\sigma = 200$): the chain does manage to make big jumps, covering the whole range, but because the acceptance rate is so low (16%) the distribution is highly irregular.

For a final decision, we combine the outcome of the Kolmogorov-Smirnov test with the acceptance rate of each standard deviation to choose the optimal one, $\sigma = 6$ as shown in Table 3 (Relative velocity).

Table 3. Summary table: optimal standard deviation (σ)

σ		1	2	3	4	5	6
Accept rate %		90,25	92,30	86,40	84,95	82,80	89,00
	h	1	1	1	0	0	0
Kol-Smi test	p	5E-50	6E-39	0,03	0,052	0,48	0,72
	k	0,745	0,66	0,13	0,12	0,06	0,04

σ		7	20	80	100	150	200
Accept rate %		81,50	62,90	36,2	31,05	21,45	16,30
	h	0	0	0	0	0	0
Kol-Smi test	p	0,82	0,97	0,92	0,99	0,97	1
	k	0,02	0,01	0,02	0,005	0,01	0

where h is the Kolmogorov-Smirnov test decision ($h = 1$, this indicates the rejection of the null hypothesis). p is the Asymtoptic p-value of the test and k is the test statistic. We see that the first value that passes the Kolmogorov-Smirnov test is $\sigma = 4$, but since the $\sigma = 6$ has a higher acceptance rate, this value has been chosen.

We applied the same approach with the estimated probability densities from the previous section (relative distance, temperature, humidity, visibility). We obtained the optimal standard deviation for each parameter as follow: $\sigma = 4$ for the Relative distance, $\sigma = 5$ for the Temperature, $\sigma = 6$ for the Humudity and $\sigma = 4$ for the visibility. The results are shown in Figs. 14, 15, 16 and 17.

We can see that the chains shown in the lower panels moves rapidly across the whole distribution and covers the whole range, without getting stuck in any one place, the acceptance rates are: 92.2%, 87.35%, 86.55%, 87.45%, respectively for the relative distance, temperature, humidity and visibility. The results give a good intuition of how a good proposal distribution can make a huge difference.

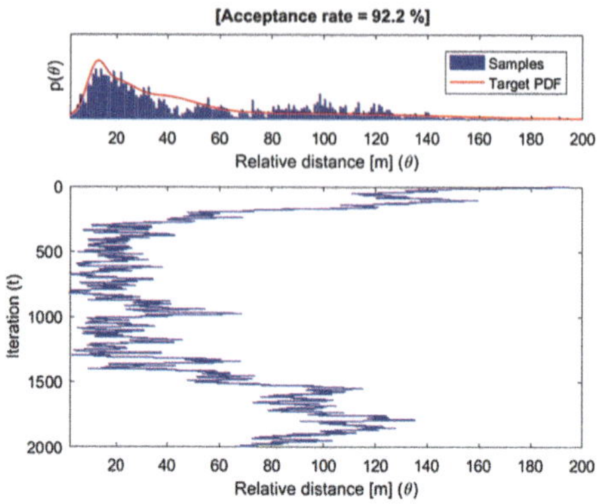

Fig. 14. Sampling results ($\sigma = 4$): Relative distance [2000 iterations].

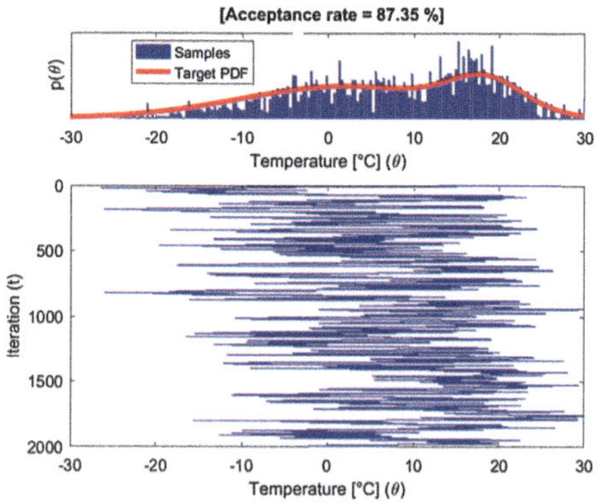

Fig. 15. Sampling results ($\sigma = 5$): Temperature [2000 iterations].

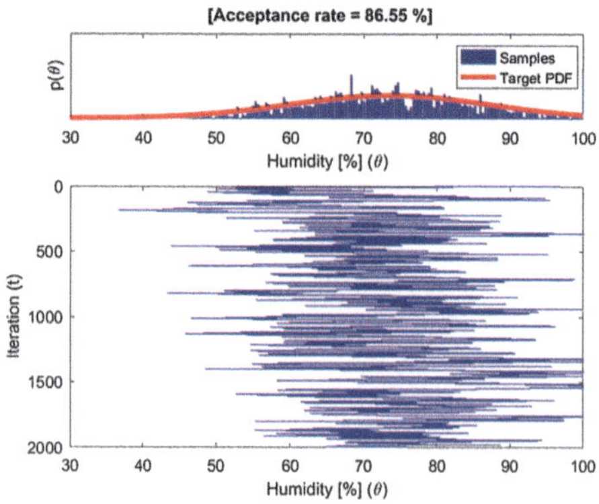

Fig. 16. Sampling results ($\sigma = 6$): Humidity [2000 iterations].

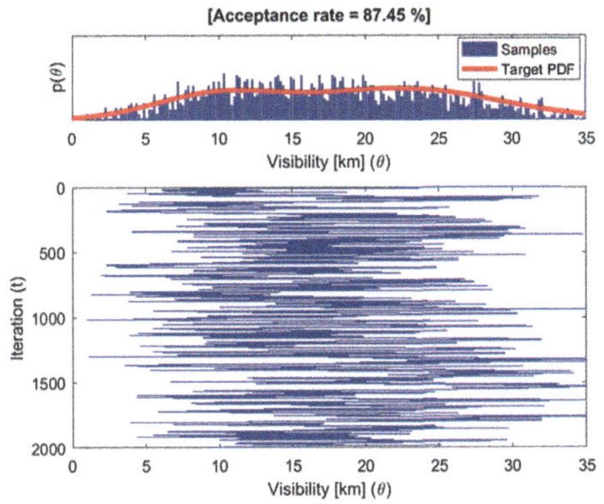

Fig. 17. Sampling results ($\sigma = 4$): Visibility [2000 iterations].

6 Conclusion and Future Work

In this paper, we presented a new sampling strategy of a Field Operational Tests Database based on Metropolis-Hastings algorithm and Kolmogorov-Smirnov test. The results quality of the sampling strategy depends primarily on the *PDFs* estimation quality and the optimization of Metropolis-Hastings algorithm initialization. The preliminary simulation results showed the efficiency of the suggested sampling strategy with high acceptance rates and a very good coverage of the whole distributions. The inclusion of the Metropolis-Hastings algorithm allowed us to create a new sampling strategy that is easy to implement, a robust one, that almost guarantee a convergence towards the exact distribution and finally, a global approach with a weighted coverage over the entire operating space (or range).

In future work, we will consider a more advanced optimization concepts for the Metropolis-Hastings algorithm parameters initialization such as the number of iterations, the start value, etc. We will also focus on investigating in more details the influence of the burning and lag parameters.

Acknowledgment. This research was supported by PMG Technologies (www.pmgtest.com), FRQNT and MITACS. We thank the PMG Technologies team who provided insight and expertise that greatly assisted the present work.

References

1. USDOT and NHTSA: Federal Automated Vehicles Policy : Accelerating the Next RevolutionIn Roadway Safety, September 2016. http://www.safetyresearch.net/Library/Federal$_$Automated$_$Vehicles$_$Policy.pdf. Accessed 23 Nov 2017
2. TASS International: "PreScan", Version 7.5.1, TASS International (2017)
3. PMG Technologies Inc.: "PMG Technoligies Home Page", Pmgtest.com (2017). http://www.pmgtest.com. Accessed 23 Nov 2017

4. Jansson, J.: Collision avoidance theory with application to automotive collision mitigation, Ph.D. dissertation. Department of Electrical Engineering, Linkoping University, Sweden (2005)
5. Gietelink, O.J.: Design and validation of advanced driver assistance systems, Ph.D. dissertation. Delft Center for Systems and Control, Delf University of Technology, Amsterdam, Netherlands (2007)
6. Raffalli, L., Vallée, F., Fayolle, G., De Souza, P., Rouah, X., Pfeiffer, M., Gorgini, S., Ptrot, F., Ahiad, S.: Facing ADAS validation complexity with usage oriented testing. Journal CoRR, volume abs/1607.07849 (2016)
7. Zhao, D.: Accelerated evaluation of automated vehicles, Ph.D. dissertation. Department of Mechanical Engineering, Michigan University, USA (2016)
8. Yang, S., Peng, H.: Development of an errorable car-following driver model. Veh. Syst. Dyn. **48**(6), 751–773 (2009)
9. Woodrooffe, J., Blower, D., Bao, S., Bogard, S., Flannagan, C., Green, P.E., LeBlanc, D.: Performance characterization and safety effectiveness estimates of forward collision avoidance and mitigation systems for medium/heavy commercial vehicles. Department of Electrical Engineering, Michigan University, USA, Report Number UMTRI-2011-36, August 2012
10. Karlson, R., Jansson, J., Gustafsson, F.: Model-based statistical tracking and decision making for collision avoidance application. Department of Electrical Engineering, Linkoping University, Sweden, Report no.: LiTH-ISY-R-2599, Submitted to ACC, 4 March 2005
11. Gietelink, O.J., De Schutter, B., Verhaegen, M.: A probabilistic approach for validation of advanced driver assistance systems. In: Proceedings of the 8th TRAIL Congress 2004 A World of Transport, Infrastructure and Logistics CD-ROM, Rotterdam, The Netherlands, 17 pp., November 2004
12. Srinivasan, R.: Importance Sampling Applications in Communications and Detection, p. 2225. Springer, New York (2002)
13. Zhao, D., Lam, H., Peng, H., Bao, S., LeBlanc, D.J., Nobukawa, K., Pan, C.S.: Accelerated evaluation of automated vehicles safety in lane-change scenarios based on importance sampling techniques. IEEE Trans. Intell. Transp. Syst., Digital Object Identifier, 13 pp, 31 May 2016. https://doi.org/10.1109/TITS.2016.2582208,
14. Chib, S.: Markov Chain Monte Carlo methods: computation and inference (Chap. 57). In: Heckman, J.J., Leamer, E. (eds.) Handbook of Econometrics, vol. 5. Elsevier Science B V (2001)
15. Gilks, W.R., Richardson, S., Spiegelhalter, D.: Markov Chain Monte Carlo in Practice. Chapman & Hall, London (1996)
16. USDOT: "RDE Home Page", Its-rde.net (2017). http://www.its-rde.net. Accessed 17 Sep 2017
17. Booz Allen Hamilton: Safety Pilot Model Deployment Sample Data Environment Data Handbook. US Department of Transportation, Data Handbook Version 1.3, 17 December 2012
18. Murphy, K.P.: Kernels. In: Machine Learning : A Probabilistic Perspective, pp. 481–515. The MIT Press, London (2012)
19. Bishop, C.M.: Mixture models and EM. In: Patter Recognition and Machine Learning, pp. 423-455. Springer Science+Business Media, LLC, Singapore (2006)
20. Murphy, K.P.: Markov Cahin Monte Carlo (MCMC) inference. In: Machine Learning: A Probabilistic Perspective, pp. 839–876. The MIT Press, London (2012)

Learning to Make Intelligent Decisions Using an Expert System for the Intelligent Selection of Either PROMETHEE II or the Analytical Hierarchy Process

Malik Haddad(✉), David Sanders, Nils Bausch, Giles Tewkesbury, Alexander Gegov, and Mohamed Hassan

School of Engineering, University of Portsmouth, Portsmouth, UK
{malik.haddad, david.sanders, giles.tewkesbury, alexander.gegov, mohamed.hassan, nils.bausch}@port.ac.uk

Abstract. This paper presents an expert system to select a most suitable discrete Multi-Criteria Decision Making (MCDM) method using an approach that analyses problem characteristics, MCDM methods characteristics, risk and uncertainty in inputs and applies sensitivity analysis to the inputs for a decisional problem. Outcomes of this approach can provide decision makers with a suggested candidate method that delivers a robust outcome. Numerical examples are presented where two MCDM methods are compared and one is recommended by calculating the minimum percentage change in criteria weights and performance measures required to alter the ranking of any two alternatives. A MCDM method will be recommended based on a best compromise in minimum percentage change required in inputs to alter the ranking of alternatives.

Keywords: Discrete · Intelligent selection · Problem characteristics
Risk · Robustness · Uncertainty

1 Introduction

Different real-life problems require different decision-making techniques. Often, limited guidelines are provided to aid users to select an appropriate decision-making method. There is no method superior to another, but there is a method or a subset of methods more suitable to a specific type of decisional problem.

Multiple Criteria Decision Making (MCDM) methods are often considered as important and reliable techniques in decision making science. MCDM methods are a group of methods and techniques that allow decision makers to deal and integrate with predefined numerous and conflicting criteria to assess alternatives in order to reach a best compromise and feasible solution. Raju and Kumar defined MCDM as the process that enables decision makers to deal with conflicting real world quantitative and or qualitative multi-criteria problems, and to provide best-fit alternatives from a set of alternatives in certain, uncertain, or risky situations [1].

K. Arai et al. (Eds.): IntelliSys 2018, AISC 868, pp. 1303–1316, 2019.
https://doi.org/10.1007/978-3-030-01054-6_91

Since their development in the 1960s, MCDM methods have been criticized for many reasons, Olson, Mechitov and Moshkovich [2] claimed that MCDM methods outputs cannot be checked for accuracy since MCDM methods often have different aggregation algorithms and integrate input sets differently. Moreover, the authors in [2] stressed that it is difficult to compare different MCDM methods. Mutikanga [3] identified some further of criticisms:

- Different MCDM methods might provide different outcomes when applied to the same problem.
- Selecting a MCDM method from a set of methods might be considered a multi criteria problem. Many researchers referred to this problem as a "vicious cycle", they usually accept it as a multi-criteria problem but are against using a multi-criteria method for selection [4–6].
- Personal experience and other factors might influence and bias the decision process, especially when obtaining performance measures and criteria weights.
- Different methods use different aggregation procedures to obtain the overall score of alternatives. Important information might be lost due to a compensation between alternatives' good and poor performances on criteria.

Most humans are capable of working with a finite and small number of criteria at the same time [7]. To cope with multiple conflicting criteria problems, decision makers prefer to apply MCDM methods.

Risk and uncertainty could be sources of distortion in making decisions. Decision makers are encouraged to use more complex scientific decision-making techniques that are less vulnerable to distortion in such environments. MCDM methods might provide a good example of such techniques and could provide a suitable outcome.

Most real-life problems are associated with risk and uncertainty. Risk is an uncertain event that if it occurs, might have a positive or negative effect on the final outcome. For decision makers to provide more satisfactory outcomes, risk and uncertainty should be recognized and mitigated. Decision makers should avoid risks with negative impact, and exploit and enhance risks with positive impact [8]. Decisions should be revised and validated after each step of a decisional process. Invalid or inappropriate decisions should be reviewed. This feedback loop could enhance the decisional process and provide decision makers with a better vision of the impact of risk and uncertainty to the outcome.

Since different decision-making techniques have different advantages and disadvantages, the choice of method used has a significant importance to the final outcome. Applying different MCDM methods to the same decisional problem could generate different outcomes [9]. The use of an inappropriate MCDM method could lead to inappropriate decisions [10, 11]. This paper will explore a number of factors affecting the choice of MCDM methods, proposes a software tool to recommend a subset of candidate methods and conduct sensitivity analysis on inputs of decisional problems to recommend a method with the most robust outcome.

Many researchers identified the need for a MCDM selection expert system [4–6, 9–20], MacCrimmon [19] was probably the first to identify the significance of MCDM methods selection problems. He developed a tree diagram that included illustrative application examples to help potential users in identifying MCDM methods specifications and classifications

Roy and Slowinski [20] criticized other researchers attempts in comparing different MCDM methods based on their outcome and considered such comparisons as "ill founded", moreover authors in [20] preferred to view MCDM methods as tools for better understanding the decisional problem, exploring studying, and evaluating different possibilities, rather than considering MCDM methods as a tool for making decisions.

2 Factors to Be Considered

Researchers [2, 4–6, 18] identified factors that affected the selection process: decision makers' previous knowledge and experience for a method, the availability of software tool to apply a method and ease of use and application. Considering the large number of MCDM methods available, several authors attempted to develop selection approaches, Ozernoy [10] stated that it is impossible to include all types of decisional problems, all existing MCDM methods, assumptions and preference information in one selection expert system.

Analysis of MCDM problems and methods exposed ten factors to be considered when developing a new expert system. These factors are:

- Problem characteristics:
  - Nature of alternative set:
    Continuous or discrete.
  - Type of input set:
    Qualitative, quantitative, or mixed.
  - Nature of information considered:
    Deterministic, non-deterministic, or mixed.
  - Type of decision problem addressed:
    Choice, ranking, description, or sorting.
  - Type of preference mode considered:
    Pairwise comparisons, performance measures.
- MCDM methods characteristics:
  - Type of ordering of alternatives:
    Total order, partial order, or interval.
  - Criteria measure scale:
    Nominal, ordinal, absolute, or ratio.
  - Type of preference structure considered.
    Preference, indifference, or incomparability.
  - Availability of software tool to support method application.
  - Ease of use that include: Ease of method understanding, user friendliness of software tool, previous experience and knowledge, and time needed to apply a method.

A novel MCDM expert system was developed using Visual Basic .net (Vb.net) programming language within Microsoft visual Studio (2012). A screen shot of the user interface is shown in Fig. 1.

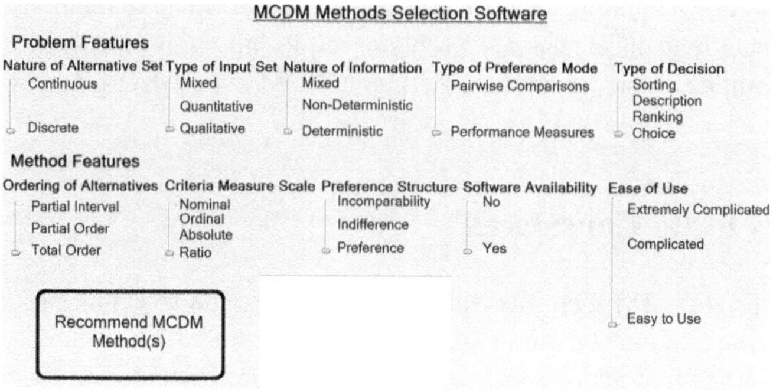

Fig. 1. New expert system user interface.

The authors used Vb.net because of its ease of use, straight forward symbol set and relatively simple user interface.

The expert system recommended a set of candidate methods to decision makers according to their answers to the 10 questions about factors mentioned earlier. Sensitivity analysis is conducted on the candidate methods to recommend a method that provided the most robust outcome and best suits the decisional problem.

3 Numerical Examples

This section describes two examples of applying sensitivity analysis to select the most suitable MCDM method from a subset of candidate methods. Numerical Example 1 used Voice of Customer, part I, Prioritizing Market Segment Preview Sample Decision from Expert Choice Sample Models [21]. Numerical Example 2 was randomly generated. For more numerical examples and applications of the new expert system please contact the authors at: Malik.haddad@port.ac.uk.

3.1 Numerical Example 1

This subsection considered a decisional problem proposed by Zultner [21] with a set of four criteria and four alternatives shown in Table 1.

The expert system presented in this paper was applied, 10 questions addressing the factors mentioned in the previous section were asked as shown in Fig. 2. A screen shot of the expert system is shown in Fig. 3. Answers to these questions were:

- Nature of alternative set?
  - Discrete.
- Type of input set?
  - Quantitative.

Table 1. Decision matrix for numerical Example 1 [21]

Criteria Alt.	$C_1 = 0.115$	$C_2 = 0.503$	$C_3 = 0.322$	$C_4 = 0.060$
A_1	0.467	0.139	0.188	0.565
A_2	0.067	0.101	0.063	0.262
A_3	0.267	0.520	0.312	0.118
A_4	0.200	0.240	0.437	0.055

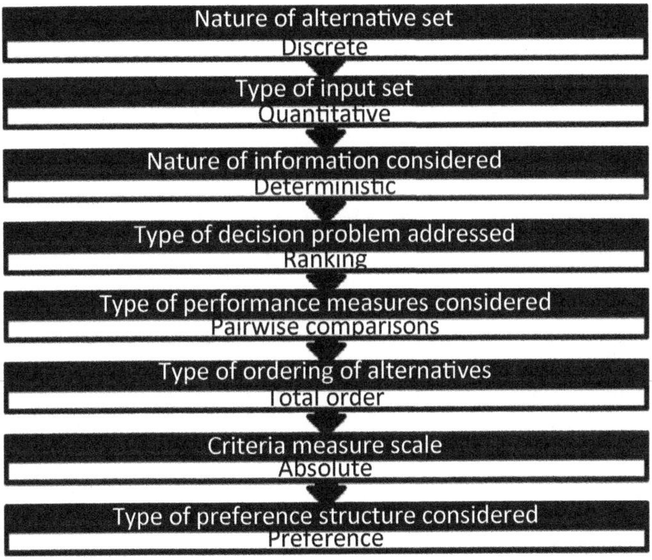

Fig. 2. Expert system branch for numerical Example 1.

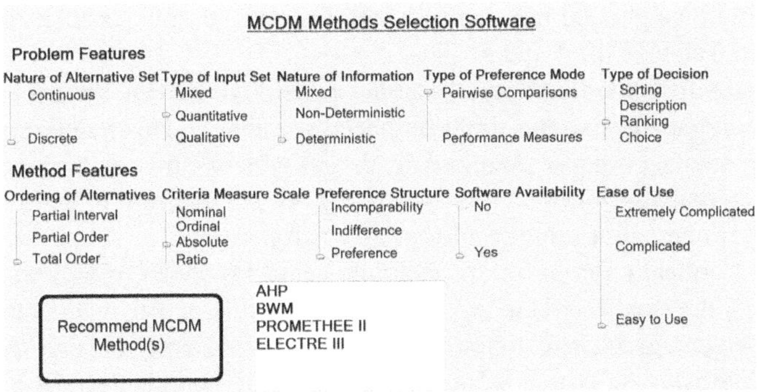

Fig. 3. Screen shot of the new expert system for numerical Example 1.

- Nature of information considered?
  - Deterministic.
- Type of decision problem addressed?
  - Ranking.
- Type of preference mode considered?
  - Pairwise comparisons.
- Type of ordering of alternatives?
  - Total order.
- Criteria measure scale?
  - Absolute.
- Type of preference structure considered.
  - Preference.

Candidate methods suggested by the expert system were:

- The Analytical Hierarchy Process (AHP)
- Best Worst Method (BWM)
- Preference Ranking Organization METHod for Enrichment Evaluations II (PROMETHEE II)
- Elimination Et Choix Traduisant la Realite III, (Elimination and Choice Expressing Reality III), (ELECTRE III)

AHP and PROMETHE II were selected as examples from the group of candidate methods due to software availability and ease of use. AHP provided the following ranking of alternatives: $A_3 > A_4 > A_1 > A_2$, with a global score of alternatives: $A_1 = 0.218$, $A_2 = 0.092$, $A_3 = 0.394$ and $A_4 = 0.296$. PROMETHEE II provided the same ranking of alternatives: $A_3 > A_4 > A_1 > A_2$, with a net flow of alternatives: $\Phi(A_1) = -0.100$, $\Phi(A_2) = -0.920$, $\Phi(A_3) = 0.6287$ and $\Phi(A_4) = 0.3913$.

AHP and PROMETHEE II methods delivered the same ranking of alternatives. Sensitivity analysis was conducted on both methods' outcomes to recommend a method that best suits this decisional problem and provided the most robust outcome. Minimum percentage change required to alter the ranking of alternatives for the most critical criterion weight and most critical performance measure was calculated. Results are shown in Tables 2, 3, 4 and 5.

The most critical criterion in this example using AHP was the second criterion (C_2), signified by the smallest value (bold number) in Table 2. This value represented the minimum percentage change required in the weight of criterion two to change the ranking of alternatives three and four, where a 53.678% decrease in its weight changed the ranking of alternatives three and four ($A_4 > A_3$).

The most critical criterion in this example using PROMETHEE II was the second criterion (C_2), the smallest value (bold number) in Table 3. This value represented the minimum percentage change required in the weight of criterion one to change the ranking of alternatives. Where a 56.262% decrease in its weight preferred alternative four to alternative three ($A_4 > A_3$).

The most critical performance measure in this example using AHP was (A_3C_2), the smallest value (bold number) in Table 4. This value represented the minimum percentage change required in the performance measure (A_3C_2) to change the ranking of

Table 2. Percentage change in criteria weights for numerical Example 1 using AHP

Criteria	Percentage change	New Ranking
C_1	176.521	$A_3 > A_1 > A_4 > A_2$
C_1	363.478	$A_1 > A_3 > A_4 > A_2$
C_2	**−53.678**	$A_4 > A_3 > A_1 > A_2$
C_2	−86.481	$A_4 > A_1 > A_3 > A_2$
C_3	−97.826	$A_3 > A_1 > A_4 > A_2$
C_3	93.789	$A_4 > A_3 > A_1 > A_2$
C_4	211.667	$A_3 > A_1 > A_4 > A_2$
C_4	446.667	$A_1 > A_3 > A_4 > A_2$
C_4	781.667	$A_1 > A_3 > A_2 > A_1$
C_4	1066.667	$A_1 > A_2 > A_3 > A_4$

Table 3. Percentage change in criteria weights for numerical Example 1 using PROMETHEE II

Criteria	Percentage change	New ranking
C_1	213.043	$A_3 > A_1 > A_4 > A_2$
C_1	404.348	$A_1 > A_3 > A_4 > A_2$
C_2	**−56.262**	$A_4 > A_3 > A_1 > A_2$
C_3	58.385	$A_4 > A_3 > A_1 > A_2$
C_4	316.667	$A_3 > A_1 > A_4 > A_2$
C_4	566.667	$A_1 > A_3 > A_4 > A_2$
C_4	783.333	$A_1 > A_3 > A_2 > A_4$
C_4	1100	$A_1 > A_2 > A_3 > A_4$

alternatives three and four (A_3 & A_4), where a 30% decrease its value changed the ranking of alternatives three and four ($A_4 > A_3$).

The most critical performance measures in this example using PROMETHEE II were (A_3C_2) and (A_4C_2), identified by the smallest values (bold numbers) in Table 5. These values represented the minimum change required in the performance measure (A_3C_2) to change the ranking of alternative three and four, where a 36% decrease its value made alternative four preferred to alternative three ($A_4 > A_3$), and the minimum percentage change required in the performance measure (A_4C_2) to change the ranking of alternative one and four, where a 36% decrease in its value made alternative one preferred to alternative four ($A_1 > A_4$). N/F shown in Tables 4 and 5 stands for a non-feasible value.

3.2 Numerical Example 2

This subsection considered a decisional problem with a set of three criteria and three alternatives shown in Table 6. The expert system presented in this paper was applied as shown in Fig. 2. A screen shot of the expert system is shown in Fig. 3. Answers to the

Table 4. Percentage change in performance measures for numerical Example 1 using AHP

Performance Measure	Percentage change	New ranking
A_1C_1	N/F	–
A_2C_1	N/F	–
A_3C_1	N/F	–
A_4C_1	N/F	–
A_1C_2	90	$A_3 > A_1 > A_4 > A_2$
A_2C_2	N/F	–
A_3C_2	**−30**	$A_4 > A_3 > A_1 > A_2$
A_4C_2	−65	$A_3 > A_1 > A_4 > A_2$
A_1C_3	79	$A_3 > A_1 > A_4 > A_2$
A_2C_3	N/F	–
A_3C_3	−84	$A_4 > A_3 > A_1 > A_2$
A_4C_3	−39	$A_3 > A_1 > A_4 > A_2$
A_4C_3	100	$A_4 > A_3 > A_1 > A_2$
A_1C_4	N/F	–
A_2C_4	N/F	–
A_3C_4	N/F	–
A_4C_4	N/F	–

Table 5. Percentage change in performance measures for numerical Example 1 using PROMETHEE II

Performance Measure	Percentage change	New ranking
A_1C_1	N/F	–
A_2C_1	N/F	–
A_3C_1	N/F	–
A_4C_1	N/F	–
A_1C_2	57	$A_3 > A_1 > A_4 > A_2$
A_2C_2	N/F	–
A_3C_2	**−36**	$A_4 > A_3 > A_1 > A_2$
A_3C_2	−57	$A_4 > A_1 > A_3 > A_2$
A_4C_2	**−36**	$A_3 > A_1 > A_4 > A_2$
A_4C_2	−70	$A_4 > A_3 > A_1 > A_2$
A_1C_3	87	$A_3 > A_1 > A_4 > A_2$
A_2C_3	N/F	–
A_3C_3	−74	$A_4 > A_3 > A_1 > A_2$
A_4C_3	−43	$A_3 > A_1 > A_4 > A_2$
A_1C_4	N/F	–
A_2C_4	N/F	–
A_3C_4	N/F	–
A_4C_4	N/F	–

Table 6. Decision matrix for numerical Example 2

Criteria Alt.	$C_1 = 0.600$	$C_2 = 0.300$	$C_3 = 0.100$
A_1	0.500	0.130	0.270
A_2	0.380	0.390	0.500
A_3	0.120	0.480	0.230

10 questions addressing the factors mentioned in the previous section were the same as the previous numerical example.

Candidate methods suggested by the expert system were:

- AHP
- BWM
- PROMETHEE II
- ELECTREE III

AHP and PROMETHE II methods were selected from the group of candidate methods due to software availability and ease of use. AHP provided the following ranking: $A_2 > A_1 > A_3$, with a global score of alternatives: $A_1 = 0.363$, $A_2 = 0.395$ and $A_3 = 0.242$. PROMETHEE II provided the following ranking: $A_1 > A_2 > A_3$, with a net flow of alternatives: $\Phi(A_1) = 0.300$, $\Phi(A_2) = 0.100$, $\Phi(A_3) = -0.400$.

AHP and PROMETHEE II methods delivered different ranking for alternatives. Sensitivity analysis was conducted on both methods' outcomes to recommend a method that best suits this decisional problem and provided the most robust outcome. Minimum percentage change required to alter the ranking of alternatives for the most critical criterion weight and most critical performance measure was calculated. Results are shown in Tables 7, 8, 9 and 10.

Table 7. Percentage change in criteria weights for numerical Example 2 using AHP

Criteria	Percentage change	New Ranking
C_1	−95.333	$A_3 > A_2 > A_1$
C_1	−31.667	$A_2 > A_3 > A_1$
C_1	**14.333**	$A_1 > A_2 > A_3$
C_2	−33.667	$A_1 > A_2 > A_3$
C_2	60.667	$A_2 > A_3 > A_1$
C_2	147.667	$A_3 > A_2 > A_1$
C_3	N/F	–

The most critical criterion in this example using AHP was the first criterion (C_1), signified by the smallest value (bold number) in Table 7. This value represented the minimum percentage change required in the weight of criterion one to change the ranking of alternatives one and two, where a 14.333% increase in its weight changed the ranking of alternatives one and two ($A_1 > A_2$).

Table 8. Percentage change in criteria weights for numerical Example 2 using PROMETHEE II

Criteria	Percentage change	New ranking
C_1	**−18.333**	$A_2 > A_1 > A_3$
C_1	−36.667	$A_2 > A_3 > A_1$
C_1	−68.333	$A_3 > A_2 > A_1$
C_2	40	$A_2 > A_1 > A_3$
C_2	63.333	$A_2 > A_3 > A_1$
C_2	80	$A_3 > A_2 > A_1$
C_3	160	$A_2 > A_1 > A_3$

Table 9. Percentage change in performance measures for numerical Example 2 using AHP

Performance Measure	Percentage change	New ranking
A_1C_1	**8**	$A_1 > A_2 > A_3$
A_1C_1	−33	$A_2 > A_3 > A_1$
A_2C_1	−10	$A_1 > A_2 > A_3$
A_2C_1	−57	$A_1 > A_3 > A_2$
A_3C_1	N/F	–
A_1C_2	64	$A_1 > A_2 > A_3$
A_2C_2	−19	$A_1 > A_2 > A_3$
A_3C_2	27	$A_1 > A_2 > A_3$
A_1C_3	71	$A_1 > A_2 > A_3$
A_2C_3	−39	$A_1 > A_2 > A_3$
A_3C_3	N/F	–

Table 10. Percentage change in performance measures for numerical Example 2 using PROMETHEE II

Performance Measure	Percentage change	New ranking
A_1C_1	−14	$A_2 > A_1 > A_3$
A_1C_1	−62	$A_2 > A_3 > A_1$
A_2C_1	18	$A_2 > A_1 > A_3$
A_2C_1	−58	$A_1 > A_3 > A_2$
A_3C_1	N/F	–
A_1C_2	N/F	–
A_2C_2	13	$A_2 > A_1 > A_3$
A_3C_2	**−11**	$A_2 > A_1 > A_3$
A_1C_3	N/F	–
A_2C_3	N/F	–
A_3C_3	N/F	–

The most critical criterion in this example using PROMETHEE II was the first criterion (C_1), the smallest value (bold number) in Table 8. This value represented the minimum percentage change required in the weight of criterion one to change the ranking of alternatives. Where a 18.333% decrease in its weight preferred alternative two to alternative one ($A_2 > A_1$).

The most critical performance measure in this example using AHP was (A_1C_1), the smallest value (bold number) in Table 9. This value represented the minimum percentage change required in the performance measure (A_1C_1) to change the ranking of alternatives one and two (A_1 & A_2), where an 8% increase in its value changed the ranking of alternatives one and two ($A_2 > A_1$). N/F shown in Tables 9 and 10 stands for a non-feasible value.

The most critical performance measures in this example using PROMETHEE II was (A_3C_2), identified by the smallest values (bold numbers) in Table 10. This value represented the minimum change required in the performance measure (A_3C_2) to change the ranking of alternative one and two, where a 11% decrease in its value made alternative two preferred to alternative one ($A_2 > A_1$).

4 Discussion

This paper applied different MCDM methods to the same decisional problems and the results showed that different MCDM methods might deliver different outcomes. They had different sensitivity towards changes in inputs (i.e. risk and uncertainty), because different methods treat performance measures and criteria weights differently. Criteria weights and performance measures often have different impacts on the final outcome [22]. Hobbs [11] claimed that when two MCDM methods delivered considerably different outcomes then, at least one method is invalid.

When risk and uncertainty are expected to affect criteria weights, then a method that is least vulnerable to fluctuations to criteria weights should be recommended to the decisional problem.

Example 1 showed that AHP required a 53.678% decrease to the most critical criterion weight to alter the ranking of alternatives, while PROMETHEE II required a 56.262% decrease to the most critical criterion weight to alter the ranking of alternatives. Both methods delivered the same outcomes, but AHP was 1.048 times more sensitive to fluctuations in the most critical criterion weight than PROMETHEE II. Thus, PROMETHEE II was recommended for that decisional problem.

Example 2 showed that AHP required a 14.333% increase to the most critical criterion weight to alter the ranking of alternatives, while PROMETHEE II required an 18.333% decrease to the most critical criterion weight to alter the ranking of alternatives. Both methods delivered different outcomes. AHP was 1.279 times more sensitive to fluctuations in the most critical criterion weight than PROMETHEE II. Thus, PROMETHEE II was recommended for that decisional problem.

When risk and uncertainty are expected to affect performance measures, then, a method that is least vulnerable to fluctuations to performance measures would be recommended for the decisional problem.

Example 1 showed that AHP required a 30% decrease to the most critical performance measure score to alter the ranking of alternatives, while PROMETHEE II required a 36% decrease to the most critical performance measure score to alter the ranking of alternatives. AHP was 1.2 times more sensitive to fluctuations in most critical performance measure than PROMETHEE II. Thus, the expert system recommended PROMETHEE II for this decisional problem.

Example 2 showed that AHP required an 8% increase to the most critical performance measure score to alter the ranking of alternatives, while PROMETHEE II required an 11% decrease to the most critical performance measure score to alter the ranking of alternatives. AHP was 1.375 times more sensitive to fluctuations in the most critical performance measure than PROMETHEE II. Thus, the expert system recommended PROMETHEE II for this decisional problem.

When risk and uncertainty are expected to affect both criteria weights and performance measures then a method that is least vulnerable to fluctuations in criteria weights and performance measures should be recommended to the decisional problem. In some cases, a compromise between these factors is recommended.

Example 1 showed that both methods delivered relatively robust outcomes. AHP was more vulnerable than PROMETHEE II to fluctuations in both criteria weights and performance measures. Recommending PROMETHEE II to that decisional problem would provide a more robust outcome with less vulnerability to risk and uncertainty.

Example 2 showed that both methods delivered different outcomes. Both methods were relatively vulnerable to fluctuations in both criteria weights and performance measures. AHP was more vulnerable than PROMETHEE II to fluctuations in both criteria weights and performance measures. Recommending PROMETHEE II for that decisional problem provided a more robust outcome with less vulnerability to risk and uncertainty.

5 Conclusions

Due to the large number of existing MCDM methods, potential users are encouraged to learn more about MCDM methods and problem characteristics to select the method that best suits their decisional problem, and avoid potential user's dissatisfaction.

This paper proposed an expert system to recommend a MCDM method from a subset of candidate methods based on the sensitivity of output to changes in input.

The authors are not suggesting that one MCDM method is better than another, but one MCDM method delivers a more robust outcome than another for a specific decisional problem. Risk and uncertainty to inputs (i.e. performance measures and criteria weights) should be analyzed when recommending a MCDM method for a specific problem. Sensitivity analysis should be conducted on performance measures and criteria weights to give a best compromise recommendation.

The decision-making methods will now be applied to some real-world engineering problems [23–30].

References

1. Raju, K.S., Kumar, D.N.: Irrigation planning using genetic algorithms. Water Resour. Manag. **18**, 163–176 (2004)
2. Olson, D.L., Mechitov, A., Moshkovich, H.: Learning aspects of decision aids. In: Proceedings of the 15th International Conference on MCDM 2000, IEEE Symposium on Computational Intelligence in Multicriteria Decision Making (MCDM 2007), Ankara, Turkey (2007)
3. Mutikanga, H.E.: Water loss management, tools and methods for developing countries. Ph. D. dissertation, Eng. Delft Uni, Delft, Netherlands (2012)
4. Kornyshova, E., Salinesi, C.: MCDM techniques selection approaches: state of the art. In: Proceedings of the 2007 IEEE Symposium on Computational Intelligence in Multicriteria Decision Making (MCDM 2007) (2007)
5. Laaribi, A.: SIG et Analyse Multicitere. Hermes Science Publications, Paris (2000)
6. Ulengin, F., Topcu, Y.I., Sahin, S.O.: An artificial neural network approach to multicriteria method selection. In: Proceedings of 15th International Conference on MCDM 2000, IEEE Symposium on Computational Intelligence in Multicriteria Decision Making (MCDM 2007), Ankara, Turkey (2007)
7. Miller, G.A.: The magical number seven, plus or minus two: some limits on our capacity for processing information. Psychol. Rev. **21**, 81–97 (1956)
8. Project Management Institute: Risk Management Plan. A Guide to the Project Management Body of Knowledge (PMBOK), 6th edn. PMI, Newtown Square (2004)
9. Eldarandaly, K.A., Ahmed, A.N., AbdelAziz, N.M.: An expert system for choosing the suitable MCDM method for solving a spatial decision problem. In: Proceedings of the 9th International Conference on Production Engineering, Design and Control, Alexandria, Egypt, 10–12 February 2009
10. Ozernoy, V.M.: Choosing the best multiple criteria decision making method. INFOR **30**(2), 159–171 (1992)
11. Hobbs, B.F.: What can we learn from experiments in multi-objective decision analysis? IEEE Trans. Syst. Man, Cybern. **16**(3), 384–394 (1986)
12. Saaty, T.L., Ergu, D.: When is a decision making method trustworthy? Criteria for evaluating multi-criteria decision making methods. Int. J. Inf. Technol. Decis. Mak. **14**(6) (2015). https://doi.org/10.1141/s021962201550025x
13. Ballestero, E., Romero, C.: Multiple Criteria Decision Making and Its Applications to Economic Problems. Kluwer Academic Publishers, Dordrecht (1998)
14. Mota, P.: Comparative analysis of multicriteria decision making methods. Ph.D. dissertation, Elec. and Comp. Eng. Uni. Nova De Lisboa, Lisbon, Portugal (2013)
15. Grenshon, M.: Model choice in multi-objective decision making in natural resources systems. Ph.D. dissertation, Sys. And Ind. Eng. U.A., Arizona, Ar., USA (1981)
16. Vincke, P.: A short note on a methodology for choosing a decision-aid method. In: Pardalos, P.M., Siskos, Y., Zopoundis, C. (eds.) Advances in Multicriteria Analysis, pp. 3–7. Kluwer Academic Publishers, Netherlands (1995)
17. Guitouni, A., Martel, J.M.: Tentative guidelines to help choosing an appropriate MCDM method. Eur. J. Oper. Res. **109**, 501–521 (1997)
18. Hanne, T.: Meta decision problems in multiple criteria decision making. In: Multiple Criteria Decision Making-Advances in MCDM Model, Algorithms, Theory and Applications, vol. 21. Springer (Kluwer) (1999)

19. MacCrimmon, K.R.: An overview of the multiple objective decision making. In: Multiple Criteria Decision Making. The University of South Carolina Press, Columbia, South Carolina (1973)
20. Roy, B., Slowinski, R.: Questions guiding the choice of a multicriteria decision aiding method. Euro J. Decis. Process **1**, 69–97 (2013)
21. Expert Choice Sample Model: Voice of customer, Part I, Prioritizing Market Segments. 2004, Expert Choice, from Zultner, R. "Prioritization of restaurant services as a function of how well they contribute to each of their market segments," Decision by Objectives, pp. 340–344, Princeton, N.J. (1991)
22. Tscheikner-Gratl, F., Egger, P., Rauch, W., Kleidorfer, M.: Comparison of multi-criteria decision support methods for integrated rehabilitation prioritization. Water **9**(68) (2017)
23. Sanders, D.A., Gegov, A., Haddad, M., Ikwan, F., Wiltshire, D. Tan, Y.C.: A rule-based expert system to decide on direction and speed of a powered wheelchair. In: IEEE Proceedings of the SAI Conference on Intelligent Systems 2018 (in press)
24. Sanders, D.A., Bausch, N.C.: Improving steering of a powered wheelchair using an expert system to interpret hand tremor. In: Liu, H., Kubota, N., Zhu, X., Dillmann, R., Zhou, D. (eds.) Intelligent Robotics and Applications: Part II. Lecture Notes in Artificial Intelligence, vol. 9245, pp. 460–471. Springer, Cham (2015)
25. Sanders, D., Gegov, A., Tewkesbury, G.E., Khusainov, R.: Sharing driving between a vehicle driver and a sensor system using trust-factors to set control gains. In: IEEE Proceedings of the SAI Conference on Intelligent Systems 2018 (in press)
26. Sanders, D.: New method to design large scale high-recirculation airlift reactors. Proc. Inst. Civ. Eng. J. Environ. Eng. Sci. **12**(3), 62–78 (2017)
27. Sanders, D., Robinson, D.C., Hassan, M., Haddad, M., Gegov, A., Ahmed, N.: Making decisions about saving energy in Compressed Air Systems using Ambient Intelligence and AI. In: IEEE Proceedings of the SAI Conference on Intelligent Systems 2018 (in press)
28. Sanders, D.A.: Using self-reliance factors to decide how to share control between human powered wheelchair drivers and ultrasonic sensors. IEEE Trans. Neural Syst. Rehabil. Eng. **25**(8), 1221–1229 (2017). https://doi.org/10.1109/TNSRE.2016.2620988
29. Sanders, D., Wang, Q., Bausch, N., Huang, Y., Khaustov, S. Popov, I.: An efficient method to produce minimal real time geometric representations of moving obstacles. In: IEEE Proceedings of the SAI Conference on Intelligent Systems 2018 (in press)
30. Sanders, D.A., Sanders, H., Gegov, A., Ndzi, D.: Rule-based system to assist a tele-operator with driving a mobile robot. In: Bi, Y., Kapoor, S., Bhatia, R. (eds.) SAI Intelligent Systems (IntelliSys) vol. 2, Lecture Notes in Networks and Systems, vol. 16, pp. 599–615. Springer (2017). https://doi.org/10.1007/978-3-319-56991-8_44

Guess My Power: A Computational Model to Simulate a Partner's Behavior in the Context of Collaborative Negotiation

Lydia Ould Ouali[1,2(✉)], Nicolas Sabouret[1,2], and Charles Rich[3]

[1] LIMSI-CNRS, UPR 3251, Orsay, France
{ouldouali,nicolas.sabouret}@limsi.fr
[2] Université Paris-Sud, Orsay, France
[3] Worcester Polytechnic Institute, Worcester, MA, USA
rich@wpi.edu

Abstract. We present in this paper a computational model of collaborative negotiation capable of guessing the behavior of powered expressed by his partner. Our model is a straight continuation of our model of negotiation based on power. We present a model inspired from Theory of Mind based on simulation that allows an agent to evaluate its interlocutor's level of power at each negotiation turn. We show on agent-agent simulation that the system correctly predicts the interlocutor's power.

Keywords: Human agent interaction · Collaborative negotiation
Interpersonal relation of power · Reasoning about other
Theory of mind

1 Introduction

Recent years have witnessed substantial growth in research on interactions between humans, intelligent agents and robots. Several new applications require agents and human user to collaborate in order to achieve shared goals and tasks. These include agents such as an *eCoach* collaborating with a patient to choose a course of treatment from among many viable options [1], or an educational agent collaborating with a student to resolve problems [2] along side with companions for the elderly [3]. In such situations, users might have to negotiate with the conversational agent regarding the plan that achieves their common goals in such way that both satisfies them. This type of negotiation is called *"Collaborative negotiation"*. It assumes that each participant is driven by the goal of finding a trade-off that satisfies best the interests of all the participants as a group, instead of one that maximizes his own interest [4,5].

However, negotiation is a multifaceted process which also involves social interaction and affects, as well as personal preferences and opinions [6]. Several research considered the role of social behavior in the negotiation process.

K. Arai et al. (Eds.): IntelliSys 2018, AISC 868, pp. 1317–1337, 2019.
https://doi.org/10.1007/978-3-030-01054-6_92

For instance, [7] studied the impact of emotions of anger and happiness on the outcome of the negotiation. [8] developed an agent that behaves according to different personalities and has a learning mechanism to learn the personality of its opponents.

One key element in the social aspect of negotiation, is the interpersonal relation between the participants. Indeed, research in social psychology have demonstrated that the relation of dominance affects the perception of the negotiation process [9]. Negotiating parties often differ in terms of dominance and this difference exerts an important influence on the behavior of participants. Negotiators build different negotiation's strategies depending on their relative dominance. This influences directly the outcomes of the negotiation. More precisely, Tidens [10] showed that dominance complementarity (*i.e.* one negotiator exhibits dominant behaviors while the other one responds with submissive ones) leads the negotiators to reach mutually beneficial outcomes and increases their reciprocal likings [10,11].

When building conversational agents that collaborate with a human user, it is important to take into account the interpersonal relationship established between the negotiators. We focus in this paper on the relation of dominance. Burgoon [12] defines interpersonal dominance as expressive, relationally based communicative acts by which behaviors of power are exerted and influence achieved. They further define dominance as a dyadic variable in which control attempts by one individual are accepted by the partner, which means if one individual expresses behaviors of high power, the interactional partner will adapt with a low power behavior.

To express those behaviors of power, we designed a model of collaborative negotiation that allows an agent to express its power through negotiation strategies [13]. We showed that human observers correctly perceived the power expressed by the agent. The next step is to extend this model in order to simulate an interpersonal relation of dominance between the agent and the user. This means that if one negotiator exerts behaviors of high power, its partner has to adapt in order to exert complementary behaviors of low power, and vice versa.

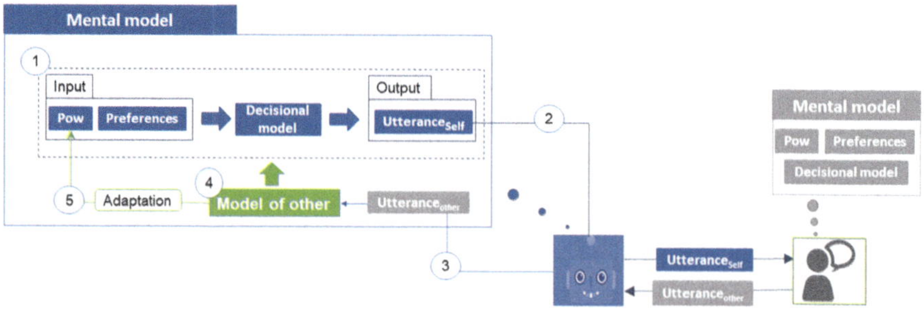

Fig. 1. Model of collaborative negotiation with a model of other. Process of decision step by step.

Our work aims at developing such an agent that is capable of evaluating and adapting his social behavior to the degree of power expressed by interlocutor. We propose a model of theory of mind based on *simulation-theory (ST)* [14] to evaluate the behavior of the partner. *ST* suggests that humans, have the ability to project oneself in the other person's shoes [15]. Therefore, we can simulate his or her mental activity with our own capacities for practical reasoning. It allows us to mimic the mental state of our interactional partner [16].

This paper describes our model of *ST* that allows the agent to evaluate the behavior of its interlocutor and to reason about it in order to predict its intentional power as presented in Fig. 1. The agent assumes that its interlocutor has a similar decision's model based on power. Therefore, for each enunciated utterance by the user, the agent tries to guess the behavior of power that its partner intends to express.

The paper is organized as follows. Section 2 gives an overview of previous automated negotiation agents with theory of mind and other related work on social behavior recognition. In Sect. 3, we briefly present our model of dialogue for collaborative negotiation. We show that extending this model with a theory of mind raises computational issues. We propose a new model to evaluate the dominance of the interlocutor, based on reasoning with uncertainty. Section 4 presents an evaluation of this new model in the context of agent-agent negotiation. We show that the system correctly predicts the interlocutor's power in spite of the restrictions on the model. In Sect. 5, we discuss the future use of this new model for building an adaptive negotiator agent.

2 Related Works

In social interactions in general and negotiation in particular, reasoning about the beliefs, goals, and intentions of others is crucial. People use this so-called theory of mind [17] or mentalizing to understand why others behave the way they do, as well as to predict the future behavior of others. Therefore, various negotiation models use theory of mind approaches to create more realistic models of negotiation.

For instance, Pynadath *et al.* [18] showed the advantages of theory of mind on negotiation even with an overly simplified model. They observed significant similarities between the partner behaviors and the agent's idealized expectations. Moreover, deviations in expectations about the other did not affect the agent performances and played in some cases in the advantage of the agent.

De Weerd *et al.* [19] also investigated the use of high-order theory of mind in mixed-motive situations where cooperative and competitive behaviors play an important role. They found that the use of first-order and second-order theory of mind allow agents to balance competitive and cooperative aspects of the game. This prevents agents from breaking down compared to agents without theory of mind.

Both approaches only focus on rational behaviors and exclude social aspects. However, the impact of social behaviors has been extensively debated in social psychology. Among such works, various researches have been investigating emotion recognition in negotiation; the effects of one individual's emotions on the other's social decisions and behavior during the negotiation. Moreover, they investigate how theses resource of information are integrated. For example, Elfenbein *et al.* [20] proved that emotion recognition accuracy is positively correlated to better performances in negotiation.

In the same vein, several researches suggested that anger expression has effects on the negotiation [21–23]. For example, VanKleef [24] demonstrated that negotiators monitor their opponent's emotions and use them to estimate the opponent's limits, and modify their demands according to the presumed location of those limits. As a result, negotiators concede more to an angry opponent than to a happy one.

Accordingly, reasoning about the social behaviors is important for the perspective of constructing negotiators agents. [25] proposed a model that can observe and predict other's agents emotional behaviors: the three step method revises the agent's beliefs by integrating a Bayesian model which infers probabilities about the emotional behaviors of other's agents and computes probabilistic prediction about their appraisals.

In the context of our work, we focus on the impact of dominance in the negotiation. Dominance as an interpersonal relation is defined by burgoon [12] as expressive, relationally based communicative acts by which power is exerted and influence achieved. Furthermore, it is a dyadic variable in which control attempts by one individual are accepted by the interactional partner [26]. For this reason, we focus on dominance complementarity between a negotiator agent and a human user in the context of collaborative negotiation.

Dominance complementarity is characterized by one person in a dyadic interaction behaving dominantly and his counterpart behaving submissively [10] and those behaviors have been investigated in the context of negotiation. [10] showed that when complementarity occurs in an interaction, people feel more comfortable and helps to create interpersonal liking relation. Moreover, [11] showed that dominance complementarity can positively improve coordination and by consequences improve objective benefits.

Our goal is to create a model of negotiation in which the agent adapts its negotiation strategies to the relation of dominance established with the user. Therefore, the agent has to reason about user's behaviors of *power* to understand the level of dominance or submissiveness expressed. The agent then adopt a complementary strategy in order to complement the user behaviors and establish the relation of dominance.

We present in this paper a model of theory of mind that builds beliefs about the user behaviors of power in order to predict his behavior of dominance. We propose to use our model of negotiation based on power in order to reason about the user behaviors.

3 Model of Collaborative Negotiation

The goal of our work is to simulate an interpersonal relation of dominance. We allow the agent to adapt its behaviors of power to complement its partner's behavior. We use a model inspired form simulation theory to guess the power of the partner. To this aim, we will use and adapt our model of collaborative negotiation to guess the partner's behavior. In this model of collaborative negotiation, the decisional process is governed by the power the agent seeks to express. Indeed, the strategies expressed by the agent are influenced by power. These strategies were designed based on researches from social psychology which studied the impact of power on negotiator's strategies.

In this section, we present an overview of our collaborative negotiation model. Indeed, it is not necessary to have a complete knowledge of this dialogue model to understand the simulation presented here. However, some elements are required to explain our approach. This section briefly presents the main principles of the collaborative negotiation dialogue model. The detailed presentation is explained in a previous work [13].

3.1 Dialogue Model

The goal of a negotiation is to choose an option in a set of possible options $\mathcal{O}$. Each option $o \in \mathcal{O}$ is defined as a set of values $\{v_1, ..., v_n\}$ associated to criteria $\{C_1, ..., C_n\}$ that reflect the option's characteristics. For instance, in a negotiation about restaurants, the criteria might be the type of cuisine and the price, we could have the option: $(French, Expensive)$.

The agent is provided with a set of partial or total ordered preferences $\prec_i$ defined for each criterion C_i. Using these preferences, the agent can compute a score of satisfaction $sat(v)$ for each value of each criterion. The satisfaction of a value $v \in C_i$ is computed as the number of values that the agent prefers less in the preferences order $\prec_i$. The notion of satisfaction represents the score of liking for the value. The closer the satisfaction of a value v gets to 1, the more the agent likes v.

For example, let's consider the criteria of *cuisine* defined with the following values: {*french*(fr), *italian*(it), *japanese*(jap), *chinese*(ch)}. In addition, the agent has a total order of preferences on these values $\prec_{cuisine}=$ $\{ch \prec jap, jap \prec it, it \prec fr\}$. Based on this order of preferences $\prec_{cuisine}$, the agent is able to compute the value of satisfiability associated to each value as presented in Table 1.

Table 1. *Sat* Computed on the set of preferences $\prec_{cuisine}$

Value	ch	jap	it	fr
Sat(value)	0	0.33	0.66	1

The domain of negotiation includes a *communication model*. This model enables the agent to communicate with its partner throughout text based *utterances*. Each one is associated with a natural language formulation.

To keep the model as generic as possible, we defined five utterances types grouped in three different categories:

- **Information moves** (*AskValue/AskCriterion* and *StateValue*) are used to exchange information about the participant's likings. They are used to express what the agent likes or do not like (*e.g I (don't) like Chinese restaurants*).
- **Negotiation moves** (*Propose*, *Accept* and *Reject*) allow the agent to make or to answer to proposals. The agent can propose, accept and reject, both values ("Let's go to a Chinese restaurant") or options ("Let's go to *Chez Francis*").
- **Closure moves** (*NegotiationSuccess* or *NegotiationFailure*) are used to end the dialogue.

3.2 Decision Based on Power

The decision making process is designed to consider, in addition to preferences, the power of the agent as presented in the *step 1* of Fig. 1. Therefore, the agent is initiated with a value of power $pow \in [0, 1]$.

We present in this section, the decisional model, which relies on three elements. We will explain how the principles of power inspired from social psychology literature influence this decisional process.

(1) Satisfiability: The value of satisfiability is associated to the *StatePreference(v)* utterance that allows the agent to express his likings. As shown by [27], high-power negotiators are more demanding than low-power ones. We capture this principle by implementing a set of the agent's satisfiable values, named S, that varies depending on the level of power pow. It is computed as follows:

$$\forall v, \quad v \in S \text{ iff } sat(v) \geq pow \tag{1}$$

For example, let us consider the preference set $\prec_{cuisine}$ whose values of satisfiability are given on Table 1. Assume a first agent Bob with a value of power $pow = 0.7$. We can compute, using the above function, that bob likes only $french$ cuisine $S = \{fr\}$. On the contrary, a second agent Arthur with a smaller value of power $pow = 0.4$ will have its set of satisfiable values initiated to $S = \{fr, it\}$.

This set S is used directly to compute the likability of values expressed in *StatePreference* utterances.

(2) Acceptability: In collaborative negotiation, both negotiators might have to reduce their level of demand over time because they want to reach a mutual agreement. The level of concessions expressed in a negotiation is affected by power. It was demonstrated that low-power negotiators tend to make larger concessions than high-power negotiators [27].

We designed a function that decreases the agent level's of demand during the negotiation, specifically, when the negotiation is not converging. To model this behavior, we use a *concession curve*, named $self(t)$. We define this function to be a time varying function of *pow* which decreases over time t and follows the concession curve. In the beginning, $self(0) = pow$ and when the negotiation evolves without converging, the function decreases as illustrated in Fig. 2.

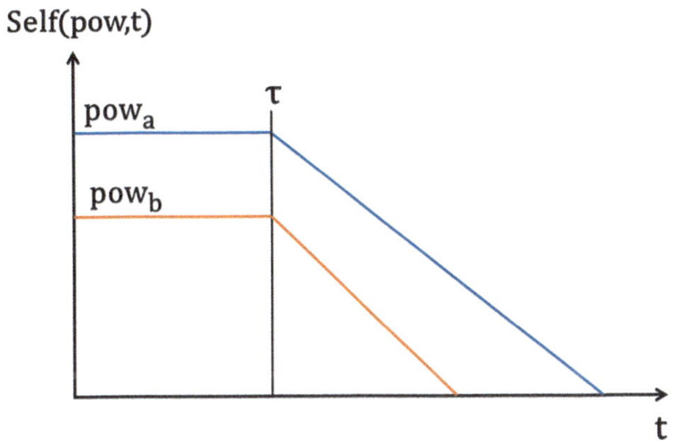

Fig. 2. Concession curve.

The agent uses the value of *self* to directly answer a *Propose(v)* utterance. We say that the value v is *acceptable* at time t, and we note $v \in Ac(t)$, when:

$$v \in Ac(t) \text{ iff } sat(v) \geq self(t) \tag{2}$$

The agent will answer with an *Accept* to a proposal only if the value is acceptable, and will answer with a *Reject* otherwise. Also note that when building proposals (i.e. *Propose* utterances) the agent can only propose a value that is acceptable.

Finally, it is important to note that the set $Ac(t)$ grows over time: as the negotiation evolves, the agent might accept proposals which are not satisfiable, as a consequence of making concessions. We denote $M(t) \subsetneq Ac(t)$ the set of not-satisfiable values that can became acceptable due to concessions: $M(t) = Ac(t) \setminus S$.

For example, in the context of negotiation with a set of preferences $\prec_{cuisine}$ and $pow = 0.6$, as illustrated in Table 1. The set of satisfiable and acceptable values are initiated to $S = Ac(t) = \{it, fr\}$ and $M = \emptyset$. However, due to concessions, the value of self, initially to $self(0) = 0.6$, reduces to $self(t) = 0.3$. Thus, values which are not satisfiable are now acceptable. Therefore, the set $Ac(t)$ is updated $Ac(t) = \{it, fr, jap\}$ and by consequences $M = \{jap\}$.

(3) Lead of the dialogue: De Dreu *et al.* showed that the participant with higher power tends to lead the dialogue and to focus on the negotiation convergence [28,29]. On the opposite, low power negotiators focus on building an accurate model of their partner's preferences in order to make the fairest decision [27]. This leads them to ask more questions about other preferences.

A: "Let's go to a restaurant on the south side."
B: "I don't like restaurants on the south side, let's choose something else."
A: "Let's go to a modern restaurant."
B: "Okay, let's go to a modern restaurant."
A: "Let's go to a French restaurant."
B: "Okay, let's go to a French restaurant."
A: "Let's go to the Maison blanche restaurant. It's a modern, cheap French restaurant on the south side."
B: "What kind of cost do you like?"
A: "I like cheap restaurants."
B: "Do you like expensive restaurants?"
A: "Let's go to the Maison blanche restaurant. It's a modern, cheap French restaurant on the south side."
B: "Okay, let's go to the Maison blanche restaurant."

Fig. 3. Example of dialogue between agents A and B implemented with our model of collaborative negotiation.

In our model, this means that agents with high values of power ($pow > 0.5$) will select *Propose* moves more often. Indeed, to make the negotiation converge, the agent has to keep making proposals until a compromise that satisfies the agent and its partner is found. On the contrary, agents with low power will select *AskPreferences* utterances more often to collect information. A detailed presentation of the utterance selection is presented in [13].

(4) Example of dialogue: We present an example of dialogue generated with two agents implement with our model of dialogue. Agent$A$ adopts a dominant behavior $pow_A = 0.7$ and the agent$B$ follows a submissive behavior $pow_B = 0.4$. The dialogue is illustrated in Fig. 3.

4 Beliefs About Other: General Algorithm

In order to figure out the value of power expressed by the user, the agent has to understand the decisional model of the user. To this goal, we propose to use a Theory of Mind (ToM) approach [17] based on simulation (ST) to establish and maintain representations of the mental states of its human counterpart as illustrated in the *step-4* in Fig. 1. The mental state requires to model the inputs necessary to reason during the negotiation. The idea is to consider different hypotheses about the user's mental state, including his/her value of power and its preferences. Using the agent's decision model, we simulate the possible utterances produced by each hypothesis. We then compare the results of this simulation with the actual utterance expressed by the user ($Utterance_{other}$) in *step 3*. This gives information about the possible values of power. The principle is illustrated in Fig. 4.

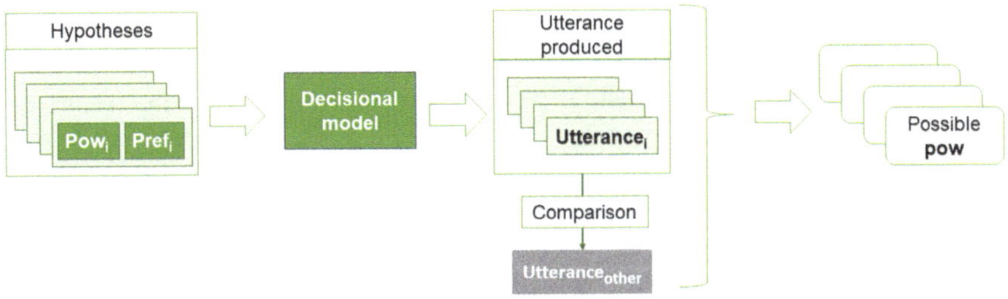

Fig. 4. Model of theory of mind to evaluate hypotheses on the power of the interlocutor from its utterance.

This approach, however, relies on a couple of strong assumptions. First, it assumes that the decision model is an accurate representation of the decisional process of the user. There is no way to guarantee this assumption. However, in our previous research [13], we showed that the behaviors of power expressed by agents are correctly perceived by human users.

The second assumption is that the system can build a model of the possible mental states of the interlocutor. Concretely, this means a value of *pow* and a set of preferences $\prec_i$ for all criteria.

Based on these assumptions, we present the general algorithm of the user's model of mental state as follows:

(1) Build a set H_{pow} of hypothesis about power: $h \in H_{pow}$ represents the hypothesis $pow = h$. In our work, we consider only 9 values: $H_{pow} = \{0.1, 0.2, \ldots, 0.9\}$.
(2) For each hypothesis h, build the set of all possible preferences $Prec_h$: the elements $p \in Prec_h$ are partial orders on the criteria.
(3) After each user utterance u, remove all elements in $Prec_h$ that are not compatible with u. Concretely, if the applicability condition of u is not satisfied in $p \in Prec_h$, then p must be removed from the candidate mental states.
(4) For each h, generate the corresponding utterance using h as input for the decisional model.
(5) Compute a score $score(h)$ based on the size of remaining hypothesis $|Prec_h|$ that generate an output similar to the utterance, $Utterance_{other}$, enunciated by the user.

(6) The hypothesis with the highest score is the most probable value for the user's power value.

$$pow_{other} = \arg\max_{h}(score(h)) \quad (3)$$

Simulation of the Other's Preferences

The representation of the user's preferences is a crucial input for the decisional model. In order to generate the set of possible preferences for each hypothesis, we need to consider all possible partial orders $\prec_i$ for each criterion C_i. We can compute the size of the set of binary preferences based on the number of values, which is $(|C_i| + 1)!$ possible partial orders for each criteria. As a consequence, for a topic with n criteria, there are $\prod_{i=1}^{n}(|C_i| + 1)!$ different possible preference sets.

If we consider a reasonable example, with 5 criteria. For each criteria, we consider approximately 4 to 10 possible values each. The set of possible preferences that the agent has represent for the user's model, are between 24.10^9 and 10^{38} possible preference sets. We can easily conclude that it is not reasonable to consider all theses hypotheses, one by one, at each step of the dialogue.

We conducted an analysis of our decisional model to define the need of preferences. As presented in Sect. 3.2, the preferences are used to compute the *satisfiability* of each value. This value is necessary during the decisional process because it is required to build the set of satisfiable values S, and the set of acceptable values $Ac(t)$.

Table 2. *Sat* computed on the set of preferences for two total ordered preferences sets

Value	ch	jap	it	fr
Sat(value) $\prec_{cuisine}$	0	0.33	0.66	1
Sat(value) $\prec'_{cuisine}$	1	0.33	0	0.66

However, in the case of total ordered preferences, all the values are comparable, which means that they can be sorted by order of preferences, by calculating the number of predecessors, as in the example Table 3. Independently from the values themselves, by knowing the number of values, we can compute the value of *satisfiability* of each value only from its rank of this in the order of preferences. For example, if we define a different set of preferences on the criterion $cuisine$ $\prec'_{cuisine} = \{it \prec jap, jap \prec fr, fr \prec ch\}$, we will obtain similar values of satisfiability as presented in Table 2. Indeed, the score of satisfiability for each value can be directly computed from the number of predecessor in the ranking. Therefore, in a total ordered preferences, we will always obtain the same values of satisfiability as presented in Table 3.

As a consequence, for a given value of power pow, we can compute the number of satisfiable values S without knowing the binary relations of preferences. For

Table 3. Satisfiability depending on the rank for a 4-values criterion

Rank(value)	1	2	3	4
Nb predecessors	3	2	1	0
$Sat(value)$	0	0.33	0.66	1

example, for the criterion *cuisine* which have 4 values, for any given total ordered preference, and a value of $pow = 0.6$, we can deduct that the number of satisfiable values is always $|S| = 2$.

Instead of calculating all the possible relations of preferences $\prec$, we can reduce the calculation to the set of possible satisfiable values. Concretely, consider a criterion with n number of values and total ordered preferences. For a given value of power pow, we obtain, s the size of S. We have to generate only $\binom{n}{s}$ possibilities for the values of S to consider. For example, for the criterion *cuisine* and $pow = 0.6$, we can generate all the possible satisfiable values as illustrated in Table 4.

Table 4. The possible sets S for *cuisine*, with $pow = 0.6$

$S_1 = (it, fr)$	$S_2 = (it, jap)$	$S_3 = (it, ch)$
$S_4 = (fr, jap)$	$S_5 = (fr, ch)$	$S_6 = (jap, ch)$

This partial representation of preferences allows our model of the other to work with a reduced set of hypotheses, compared to the initial model with a complete representation of the preferences. If we consider the same example, with 5 criteria and 10 values per criterion, the maximum number of hypotheses to consider for a given value of power is $\binom{10}{5} = 252$ (this value is maximum for $pow = 0.5$). However, simulating the behavior of the interlocutor with incomplete knowledge has two consequences. First, it requires to adapt the simulation model of the decision making to deal with the non-ordered sets of satisfiable and acceptable values. Second, it might affect the precision of the prediction of the interlocutor's power.

In order to generate hypotheses about satisfiable values, given any criterion, we make the strong assumption that the interlocutor's preferences are *total ordered*. Therefore, given a fixed value of $h \in H_{pow}$, we compute the number of satisfiable values named s which is the size of all our hypotheses S for this value of power. We then built all possible combinations of satisfiable values noted $M_h()$ (see Table 5). This process is generalized to all the hypotheses of H_{pow}. We present in Table 1 an example of the generation of all the hypotheses possible for a criterion *cuisine*.

Using the model of hypotheses composed of hypotheses on power h_{pow} and hypotheses on satisfiable values $M_h(h_i)$ associated to each $h_i \in h_{pow}$, the agent has to simulate the behavior of its interlocutor at each turn of dialogue.

Table 5. Hypotheses on preferences for a 4-value criterion *cuisine*

Hypothesis	Hypotheses	
	$h_i(pow)$	Hypotheses on satisfiable values $M_h(h_i)$
H1	0.3	$\{(fr, it, jap)\}, \{(fr, it, ch)\}, \{(fr, jap, ch)\}, \{(it, jap, ch)\}$
H2	0.4	$\{(fr, it)\}, \{(fr, jap)\}, \{(fr, ch)\}, \{(it, jap)\}, \{(it, ch)\}, \{(jap, ch)\}$
H3	0.5	$\{(fr, it)\}, \{(fr, jap)\}, \{(fr, ch)\}, \{(it, jap)\}, \{(it, ch)\}, \{(jap, ch)\}$
H4	0.6	$\{(fr, it)\}, \{(fr, jap)\}, \{(fr, ch)\}, \{(it, jap)\}, \{(it, ch)\}, \{(jap, ch)\}$
H5	0.7	$\{(fr)\}, \{(it)\}, \{(jap)\}, \{(ch)\}$
H6	0.8	$\{(fr)\}, \{(it)\}, \{(jap)\}, \{(ch)\}$
H7	0.9	$\{(fr)\}, \{(it)\}, \{(jap)\}, \{(ch)\}$

We present in the next section the adaptation of the decisional model to simulate the interlocutor behavior with partial knowledge.

5 Simulation of Other with Partial Representation of Preferences

After each dialogue turn in which an interlocutor expresses an utterance $utterance_{other}$, the agent uses its *model of other* to update his hypotheses on the behavior related to power of its interlocutor $other_{pow}$. Therefore, depending on the utterance's type, the agent computes, for each hypothesis, a score $score(h_i, t)$ which represents the accuracy of this hypothesis to the behavior expressed by the user. Then, the agent associates the value of power to the hypothesis that gets the best score at this turn:

$$pow_{other} = \arg\max_{h_i \in H_{pow}}(score(h_i, t)) \tag{4}$$

In this section, we present the process of simulation in which the agent predicts the behavior of power based on the received utterance.

5.1 Lead of the Dialogue

As presented before, the choice of a specific utterance's type translates behaviors of power. Indeed, a high frequency of choosing *proposal utterance* shows a behaviors of high-power. On the contrary, a high frequency of *share preferences utterances* reflects behaviors of low-power. We note *history* the list of utterances enunciated by the user. The value of power is computed from the ratio of *propose* enunciated versus *ask*.

$$pow_{other} = \begin{cases} > 0.5 \text{ if } \frac{history(Propose)}{hisotry} > 0.5 \\ \leq 0.5 \text{ if } \frac{history(Ask)}{hisotry} > 0.5 \end{cases} \tag{5}$$

Once, the list of possible hypotheses to consider is restricted, we update the hypotheses by taking into consideration the value associated to the expressed utterance. For example, suppose that the agent assumes that $\frac{history(Propose)}{hisotry} > 0.5$. Therefore, the agent will only considers the hypotheses $\{h_4 - h_9\}$ to compute the power of its interlocutor.

5.2 Share a Preference

We consider that the interlocutor shares a preferences in two cases. First, when the interlocutor expresses a *StatePreference(v,s)*, he shares a preference in a way where $v \in S$ (I like v) otherwise $v \notin S$ (I don't like v).

Second, based on our design, we consider that the interlocutor shares a preference when he/she rejects a proposal *Reject(p)*. If a value is not acceptable, it is necessarily not satisfiable because $S \subset Ac(t)$. Thus, if $p \notin Ac(t)$, then $p \notin S$.

To compute whether a value $v \in C_i$ is satisfiable, we check for each hypothesis of power h_i, the set satisfiable values $S_i \in M_h(pow)$. Thus:

$$sat_{S_i}(v) = \begin{cases} True & \text{if } v \in S_i \\ False & \text{otherwise} \end{cases} \tag{6}$$

Therefore, when the agent learns a new preference about its interlocutor, he updates his hypotheses as follows: for each $h_i \in H_{pow}$, we propose to update the agent's hypotheses $M_h(h_i)$ by removing all the hypotheses on preferences that are no longer consistent with the information acquired. Then, we compute the score of each h_i at the moment t:

$$score(h_i, t) = \frac{|M_h(h_i, t)|}{|M_h(h_i, init)|}$$

Example: Suppose that the interlocutor expresses *StatePreference(fr, true)*, this means that the agent has to remove all the hypotheses where $fr \notin S_i$. The update of each hypothesis and their respective scores are presented in Table 6.

For example, for h_4, out of the six initial hypotheses, the agent removes three: $(\{(it, jap)\}, \{(it, ch)\}, \{(jap, ch))$. In the end, $score(h_4) = 0.5$.

Table 6. Hypotheses on preferences for a 4-value criterion *cuisine*

Hypothesis	Hypotheses		
	$h_i(pow)$	Hypotheses on satisfiable values $M_h(h_i)$	$Score(h_i, t)$
H1	0.3	$\{(fr, it, jap)\}, \{(fr, it, ch)\}, \{(fr, jap, ch)\}$	3/4
H2	0.4	$\{(fr, it)\}, \{(fr, jap)\}, \{(fr, ch)\}$	0.5
H3	0.5	$\{(fr, it)\}, \{(fr, jap)\}, \{(fr, ch)\}$	0.5
H4	0.6	$\{(fr, it)\}, \{(fr, jap)\}, \{(fr, ch)\}$	0.5
H5	0.7	$\{(fr)\}$	1/4
H6	0.8	$\{(fr)\}$	1/4
H7	0.9	$\{(fr)\}$	1/4

5.3 Proposals

When the interlocutor accepts (*Accept(p)*) or proposes a proposal (*Propose(p)*), means that the value is acceptable $p \in Ac(t)$. The agent has to calculate for each $h_i \in H_{pow}$ the score of acceptability attributed to this value.

The set of acceptable values $Ac(t)$ depends on the value of $self(t)$, knowing that $self(t)$ decreases over time when the negotiation is not converging. Our goal is to capture the behaviors of concessions over time. Concretely, this means that the set of acceptable values $Ac(t)$ grows during the negotiation, and new values become acceptable noted $M(t)$.

Thus, for each hypothesis on power h_i, we associate a value $self_i(t)$ that represents the level of concessions at the current time. Using this value, and the set of satisfiable values (see Table 3), the agent can compute the number of acceptable values $|Ac(t)_i|$, as presented in Sect. 3.2(2), and the number of the values which became acceptable due to concessions $|M(t)_i|$.

Therefore, we propose to adapt the calculation of *acceptabilty* only with partial knowledge available. For a hypothesis of power h_i, for each hypothesis on preferences $S_i \in M_h(h_i)$, and the list of accepted values during the negotiation A, the score that a value $p \in C_i$ is acceptable, is computed as follows:

$$Acc(p, h_i) = C_{|C_i|-(|S_i|+k)}^{|M_i(t)|-k} \tag{7}$$

$k = |K|$ is the number of elements in the set $K = A \cap \overline{S}_i$, the set of accepted values which are not satisfiable. The score of acceptability is the number of possible sets $M_i(t)$ that are compatible with the value $p \in C_i$ (i.e. the value that was proposed or accepted by the interlocutor) and the hypothesis S_i.

In addition, we normalize this score of acceptability in order to have a coherent comparison between the different hypotheses (the number of hypotheses for each value of *pow* can differ). Thus, given a hypothesis on power h_i, he score of acceptability is normalized by taking into account the ideal score of acceptability. We can compute the "best" score of acceptability a priori:

$$I_{pow} = C_{|C_i|-|S_i|}^{|M_i(t)|}$$

The final value of acceptability normalized is then:

$$score(h_i, t) = \left(\tfrac{1}{I_{pow}} \cdot \textstyle\sum_{S_i \in M_h(h_i)} acc(p, h_i) \right) \frac{1}{|M_h(h_i)|} \tag{8}$$

Example: The agent receives an *Accept(jap)*. We take as example, h_4 to compute the values of acceptability of the value *jap*.

With the partial representation of preferences, the agent only knows hypotheses about the sets $S_i \in M_h(h_i)$. In the case where there are no concessions ($self_i(t) = h_i$), the agent can compute the set of acceptable values because $|Ac_i(t)| = s$.

However, if self decrease to $self_i(t) = 0.3$, the number of acceptable values $|Ac_4(t)| = 3$. By consequence, the agent knows that a new value is acceptable which is also not satisfiable $|M_4(t)| = 1$. Moreover, the agent is not able to compute which value is now acceptable. For example, consider the first hypothesis on satisfiable values related to h_4; $S_i = \{fr, it\}$, the agent cannot compute which of the values *jap* or *ch* is now acceptable. Therefore, we compute the value of acceptability presented above that handle this uncertainty. $acc(jap, h4) = C_2^1 = 2$. The best score $I_{0.6} = 2$. Thus, $score(h_4, t) = 1/6$

6 Evaluation

Section 5 presented a model based on *ST* that aims to guess the behaviors of power expressed by an interlocutor. In order to assess the validity of our model with partial representation of preferences, we propose to simulate dialogues with two artificial agents. We can then compare the result of the prediction with the actual mental model of the interlocutor. We do not only evaluate the accuracy of our model of theory of mind, but also how fast it can predict the power of the interlocutor as well as the timeliness of the algorithm.

6.1 Method

We implemented two agents that have to negotiate over the topic of restaurants. In addition, each agent has to predict the behavior of power expressed by the other agent. The first agent ($agent_A$) plays the role of a dominant agent, whereas, the second agent ($agent_B$) plays the role of a submissive agent.

We manipulated two simulation parameters for the initialization of our agents.

First, we variate the values of power assigned to each agent (named pow_a and pow_b), in order to study the accuracy of prediction in the different ranges of the spectrum of power as presented in Table 7. For each behavior, we generated dialogues for each combination of values of power.

Table 7. Initial condition's setting for the values of power

Dominance value	Initial values of power			
Dominant agent pow_A	0.3	0.4	0.5	
Submissive agent pow_B	0.6	0.7	0.8	0.9

Second, we variate the initial preferences affected to each agent. More specifically, we generated different set of preferences that differs in their complexity. This means, that for each type of model of preferences, we variate respectively, the number of criteria discussed, and the number of values assigned to each criterion as presented in Table 8. When the topic discussed has an important size of values to discuss, this implies that the agent has an important number of hypotheses to consider at each dialogue turn. Our aim is to analyze whether our model can handle an important number of hypotheses and its impact on the viability of real time prediction of power. For each condition of topic *Small, Medium, Large*, we generated 30 different combination of preferences to affect to both agents.

Considering both simulation parameters, we generated 1080 dialogues. In the next section, we present the statistical study made to analyses the predicted behaviors.

Table 8. Initial condition's setting for the preferences set

Model of preference	Number of criteria	Average number of values per criterion	Possible M_h
Small	4	3	1296
Medium	4	4	4.14×10^5
Large	4	10	2.6×10^9

6.2 Analyses of the Dialogues

We present in this section the statistical studies done, as well as the results obtained for both the accuracy of predictions made by our model and the timeliness of the execution.

(1) Accuracy of predictions: For each dialogue, we obtain the predictions made by the agent with ToM abilities about the behaviors of the power of the other agent. Results are summarized in Table 9.

First, we computed the *root mean square error* that indicates the standard deviation of the differences between the guessed values of power and the actual power of the partner. For the total of 1080 dialogues, a small deviation of $rmse = 0.12$ was observed. Moreover, in order to analyze the results in a more general way, we calculated the *residual deviation* that computes the accuracy of the dependent variable being measured. We have computed the deviation of prediction between the predicted value $Other_{pow}$ from the real value of power pow. We observe deviations of the range $rv = 0.015$ which means that based on the two statistical analysis, our model makes accurate and close predictions of other's power.

Table 9. Results of margin errors for prediction

Root mean square error	0,12
Residual variation (rv)	0,015

Second, we analyzed the frequency of false predictions. For example, $agent_A$ predicts that $agent_B$ is dominant, whereas $agent_B$ is submissive. To this end, we computed the percentage of false predictions. For all the dialogues, only 30 predictions were incorrect which means that 2.6% of predictions were false. This result confirms the reliability of our algorithm.

Finally, for all the dialogues in which the agent was able to find a good prediction (in total 1050), we analyzed the rapidity of convergence of our algorithm. The results are presented in Table 10. Indeed, for each dialogue, we computed the number of iterations necessary to find a good prediction and maintain it. On average, the algorithm needed 3 iterations in order to predict the right range of the power of the other (whether the agent has a dominant or a submissive behavior).

Table 10. Average number of iterations to compute a prediction

Number of iteration to make predictions		
Good predictions	$rv \leq 0.02$	Best prediction
2.53	3.25	3.79

Moreover, we calculated the average number of iterations needed to find a prediction such that $rv \leq 0.02$. In average the agent needed 3 iterations in order to evaluate a value close to the other's power. The evolution of convergence for all the dialogues is presented in Fig. 5

We also computed the average number of iterations in order to find the best prediction of $Other_{pow}$. The results showed that the agent makes in average 4 iterations to converge towards the best value.

We studied the impact of the initial number of hypotheses M_h on the convergence of the prediction. We wanted to study whether a large number of hypotheses will need extra iterations to converge. Therefore, we compared results obtained for small topic, medium, and large topics on the convergence of the negotiation. The graph presented in Fig. 5 shows that our algorithm converges in average quickly, independently from the size of the topic. We can observe that the algorithm took two additional iterations to converge on the larger topic compared to medium and small topics. The difference is not significant to affect the general behavior of our model.

(2) Timeliness: We evaluate the time execution of the algorithm in order to study how the model of theory of mind evolves. For each dialogue, we computed the average time execution at each negotiation turn. We aim to study the effect of hypotheses's size on the rapidity of prediction at each turn. Results are presented in Fig. 6. When comparing the time execution between the medium and large models, we observe that the algorithm took in average 12 milliseconds at each turn. However, this difference is not significant since the total execution remains very quick.

We analyzed the behaviors of our model of ToM in different aspects and the obtained results provide a strong support to its accuracy. Indeed, for most of the generated dialogue, the agent with ToM abilities was able to predict a

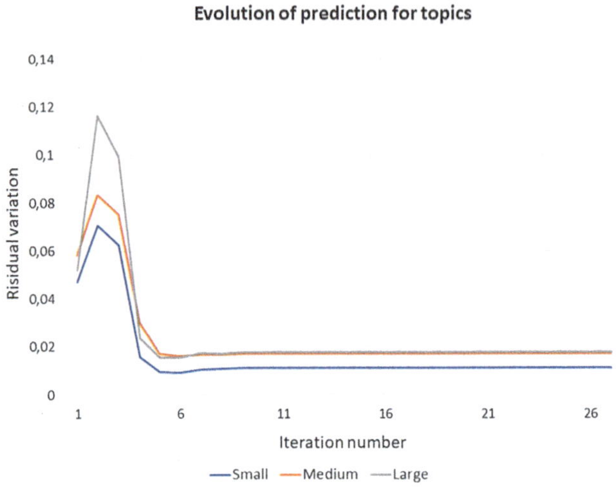

Fig. 5. Resudial variation computed between the real value of power and the guessed one at each dialogue turn.

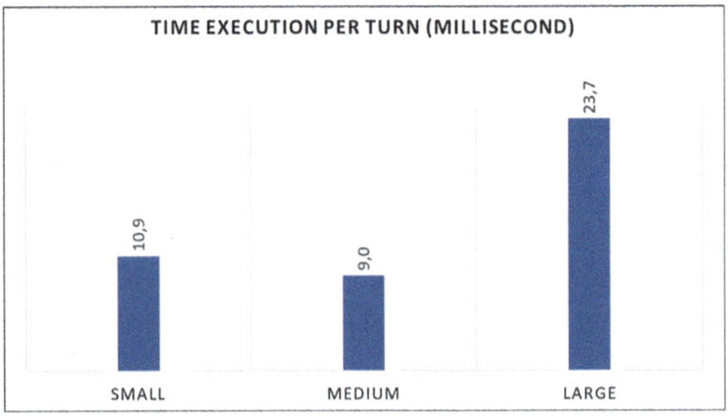

Fig. 6. Iime execution of the ToM algorithm at each dialogue turn.

very close approximation of its partner's level of power. Moreover, the agent was able to find the right dominance range only after two speaking turns and the best evaluation after five turns. These behaviors were generated in a reasonable amount of time, allowing the agent to produce real time dialogues.

These findings strengthen the accuracy of our model and give good perspectives to implement this model in the context of human/agent collaborative negotiation. However, the presented validation in the context of agent/agent negotiation is a controlled evaluation since both agents use the same decisional model. This situation increases the chance of good predictions of the partner's behaviors of power. Thus, our collaborative model of negotiation must be validated in the context of human/agent negotiation. Our initial model of power [13] relies on studies from social psychology. It has been validated with a perceptive study where human participants were able to perceive and recognize behaviors of power expressed by agents. For these reasons, we believe that it is a good approximation of human behaviors of power and we expect good predictions from our ToM model in the context of human-agent interaction.

7 Conclusion

We presented in this paper, a model of collaborative negotiation enabled with a simulation model in order to predict the behavior of the interlocutor. Our aim is to enable a conversational agent to guess the behavior of its interlocutor in order to adopt a complementary behavior.

Our model of ToM focuses essentially on predicting the behavior of power expressed by the agent's interlocutor. To this purpose, we used simulation theory (ST). It assumes that the agent is capable to reason from the perspective of its interlocutor using his own mechanism of reasoning and decision.

We presented a solution in which the agent builds a partial representation of the mental state of its interlocutor. This allows the agent to have sufficient knowledge to compute the value of the interlocutor's power. Our results are compatible with the research in cognitive psychology: [16] suggests that in order to simulate another's mental processes, it is not necessary to categorize all the beliefs and desires attributed to that person as such. In other words, it is not necessary to have a complete model of the interlocutor.

We adapted the decision process of the agent to handle partial knowledge and we presented an evaluation to validate the accuracy of the agent's prediction in the context of agent/agent negotiation. The results confirmed the accuracy of the ToM models. Indeed, the agent was able to generate, in a limited number of dialogue turns, an accurate prediction of it's partner behaviors.

As perspective, we aim to simulate a complementary relation of dominance between an agent and human user during a process of collaborative negotiation. In such case, the agent will have to adapt its behaviors of power to complement the predicted user's behaviors in order to simulate the dominance complementarity. Thus, the agent will have to consider the effect of wrong predictions on the negotiation outcomes. We believe that modeling a complementary relation of dominance will improve the negotiation outcomes and increase mutual liking between the agent and the user similarly to human/human negotiations [10, 11].

References

1. Robertson, S., Solomon, R., Riedl, M., Gillespie, T.W., Chociemski, T., Master, V., Mohan, A.: The visual design and implementation of an embodied conversational agent in a shared decision-making context (eCoach). In: International Conference on Learning and Collaboration Technologies, pp. 427–437. Springer (2015)
2. Howard, C., Jordan, P., Di Eugenio, B., Katz, S.: Shifting the load: a peer dialogue agent that encourages its human collaborator to contribute more to problem solving. Int. J. Artif. Intell. Educ. **27**(1), 101–129 (2017)

3. Sidner, C., Rich, C., Shayganfar, M., Behrooz, M., Bickmore, T., Ring, L., Zhang, Z.: A robotic or virtual companion for isolated older adults. In: International Workshop on Socially Assistive Robots for the Aging Population, Bielefeld, Germany (2014)
4. Sidner, C.L.: An artificial discourse language for collaborative negotiation. In: AAAI, vol. 94, pp. 814–819 (1994)
5. Chu-Carroll, J., Carberry, S.: Response generation in collaborative negotiation. In: Proceedings of the 33rd Annual Meeting on Association for Computational Linguistics, pp. 136–143. Association for Computational Linguistics (1995)
6. Broekens, J., Jonker, C.M., Meyer, J.-J.C.: Affective negotiation support systems. J. Ambient. Intell. Smart Environ. **2**(2), 121–144 (2010)
7. de Melo, C.M., Carnevale, P., Gratch, J.: The effect of expression of anger and happiness in computer agents on negotiations with humans. In: Proceedings of the AAMAS 2011, pp. 937–944 (2011)
8. Kraus, S., Lehmann, D.: Designing and building a negotiating automated agent. Comput. Intell. **11**(1), 132–171 (1995)
9. Van Kleef, G.A., De Dreu, C.K., Pietroni, D., Manstead, A.S.: Power and emotion in negotiation: power moderates the interpersonal effects of anger and happiness on concession making. Eur. J. Soc. Psychol. **36**(4), 557–581 (2006)
10. Tiedens, L.Z., Fragale, A.R.: Power moves: complementarity in dominant and submissive nonverbal behavior. J. Pers. Soc. Psychol. **84**(3), 558 (2003)
11. Wiltermuth, S., Tiedens, L.Z., Neale, M.: The benefits of dominance complementarity in negotiations. Negot. Confl. Manag. Res. **8**(3), 194–209 (2015)
12. Burgoon, J.K., Johnson, M.L., Koch, P.T.: The nature and measurement of interpersonal dominance. Commun. Monogr. **65**(4), 308–335 (1998)
13. OuldOuali, L., Sabouret, N., Rich, C.: A computational model of power in collaborative negotiation dialogues. In: International Conference on Intelligent Virtual Agents, pp. 259–272. Springer (2017)
14. Gordon, R.M.: Folk psychology as simulation. Mind Lang. **1**(2), 158–171 (1986)
15. Shanton, K., Goldman, A.: Simulation theory. Wiley Interdiscip. Rev. Cogn. Sci. **1**(4), 527–538 (2010)
16. Harbers, M., Bosch, K., Meyer, J.-J.: Modeling agents with a theory of mind. In: Proceedings of the 2009 IEEE/WIC/ACM International Joint Conference on Web Intelligence and Intelligent Agent Technology, vol. 02, pp. 217–224. IEEE Computer Society (2009)
17. Premack, D., Woodruff, G.: Does the chimpanzee have a theory of mind? Behav. Brain Sci. **1**(4), 515–526 (1978)
18. Pynadath, D.V., Wang, N., Marsella, S.C.: Are you thinking what I'm thinking? an evaluation of a simplified theory of mind. In: International Workshop on Intelligent Virtual Agents, pp. 44–57. Springer (2013)
19. de Weerd, H., Verbrugge, R., Verheij, B.: Higher-order theory of mind in negotiations under incomplete information. In: Prima, pp. 101–116. Springer (2013)
20. Elfenbein, H.A., Der Foo, M., White, J., Tan, H.H., Aik, V.C.: Reading your counterpart: the benefit of emotion recognition accuracy for effectiveness in negotiation. J. Nonverbal Behav. **31**(4), 205–223 (2007)
21. Sinaceur, M., Tiedens, L.Z.: Get mad and get more than even: when and why anger expression is effective in negotiations. J. Exp. Soc. Psychol. **42**(3), 314–322 (2006)
22. Van Kleef, G.A., De Dreu, C.K., Manstead, A.S.: An interpersonal approach to emotion in social decision making: the emotions as social information model. Adv. Exp. Soc. Psychol. **42**, 45–96 (2010)

23. Ferguson, M.J., Bargh, J.A.: How social perception can automatically influence behavior. Trends Cogn. Sci. **8**(1), 33–39 (2004)
24. Van Kleef, G.A., De Dreu, C.K., Manstead, A.S.: The interpersonal effects of emotions in negotiations: a motivated information processing approach. J. Pers. Soc. Psychol. **87**(4), 510 (2004)
25. Alfonso, B., Pynadath, D.V., Lhommet, M., Marsella, S.: Emotional perception for updating agents' beliefs. In: 2015 International Conference on Affective Computing and Intelligent Interaction (ACII), pp. 201–207. IEEE (2015)
26. Dunbar, N.E., Burgoon, J.K.: Perceptions of power and interactional dominance in interpersonal relationships. J. Soc. Pers. Relationsh. **22**(2), 207–233 (2005)
27. De Dreu, C.K., Van Lange, P.A.: The impact of social value orientations on negotiator cognition and behavior. Pers. Soc. Psychol. Bull. **21**(11), 1178–1188 (1995)
28. Magee, J.C., Galinsky, A.D., Gruenfeld, D.H.: Power, propensity to negotiate, and moving first in competitive interactions. Pers. Soc. Psychol. Bull. **33**(2), 200–212 (2007)
29. De Dreu, C.K., Van Kleef, G.A.: The influence of power on the information search, impression formation, and demands in negotiation. J. Exp. Soc. Psychol. **40**(3), 303–319 (2004)

Load Balancing of 3-Phase LV Network Using GA, ACO and ACO/GA Optimization Techniques

Rehab H. Abdelwahab(✉), Mohamed El-Habrouk, Tamer H. Abdelhamid, and Samir Deghedie

Electrical Engineering Department, Faculty of Engineering, Alexandria University, Alexandria, Egypt
rehabhwahab@gmail.com, eepgmmel@yahoo.co.uk, tmrhlmy62@gmail.com, deghedie@gmail.com

Abstract. In this paper, a proposal for load balancing of the 3-phase Low Voltage (LV) distribution network is introduced. Further, the paper presents the computational problems associated with the optimization techniques used to evaluate the switching patterns for controlling load balancing switching circuit. In addition, the paper presents Genetic Algorithm (GA) and Ant Colony Optimization (ACO) techniques to generate the fast-switching pattern for the reconfiguration of the LV network. The paper presents a comparison and simulation study between GA and ACO as an optimization technique with two objectives, accuracy and speed. Further, the paper presents a hybrid approach ACO/GA that combines the advantages of both techniques. The presented solution uses ACO as an initial technique then GA is employed to ensure high accuracy with fast response. The simulation showed that the hybrid technique performs better than the GA and ACO.

Keywords: Load balancing · Optimization · Genetic Algorithm (GA) Ant Colony Optimization (ACO) · Hybrid

1 Introduction

The electrical grid is an interdependent network, it consists of three stages. The first stage is the bulk generation where electricity is generated using different resources like oil, coal, solar energy, etc. then stepping up the voltage using a transformer. The second stage is the transmission where the high voltage electricity is carried over long distances using transmission lines then transformers are used to step down the voltage. The third stage is the distribution stage [1]. The three-phase electrical power system is the most widely used scheme for transferring power because of its ultimate advantages [1]. However, the three-phase system is facing many challenges affecting its performance. The load balancing is the most important challenge among those [2, 3]. Single-phase loads are the dominant loads on a three-phase system. The neutral current increases due to the normal changes occurring within the system [3]. Changes are induced due to load shifts or the diversity of loads being on or off at the same time. This increase in neutral current can lead to the neutral cable damage. Furthermore, the

K. Arai et al. (Eds.): IntelliSys 2018, AISC 868, pp. 1338–1349, 2019.
https://doi.org/10.1007/978-3-030-01054-6_93

unbalancing in the LV network causes high power losses and increases in energy demand. The generation station will not be able to support the increase in demand and high-power losses, which may lead to blackout [1, 3]. Network reconfiguration is the process of altering loads distribution by using three single-phase contactors or switching devices. Each of these switching devices is connected to a different phase (ph1, ph2, ph3) then redistribute the connected loads depending on their measured parameters [4]. A control system is responsible for gathering the measured data and perform an optimization process to find the optimal reconfiguration [2, 3, 5, 6].

This paper presents a control system that uses an optimization technique to minimize the neutral current. The control system depends on rearranging the loads in such a way that the system becomes more balanced [2]. Figure 1 shows a different number of single-phase loads, each one connected to one phase at a time and a switching circuit that is responsible for switching between phases to reconfigure the distribution network based on the total measured connected loads. Applying an optimization technique to reorganize the loads connected to each phase will result in minimizing the neutral current which will lead to minimizing power losses [3]. Different methods of optimization are used to find the best performance of the control system. Genetic algorithm and Ant colony optimization techniques are presented and simulation study for both of them is conducted. In addition, a comparison between GA and ACO is discussed. The unique advantage of each technique inspired us to present a hybrid technique that merges the advantages of each GA and ACO while avoiding the limitation of each technique. The hybrid technique started with ACO approach as an initial solution then continues the optimization process with GA technique to ensure the accuracy and the speed of conversion. A simulation study is conducted among those techniques according to the performance of each one.

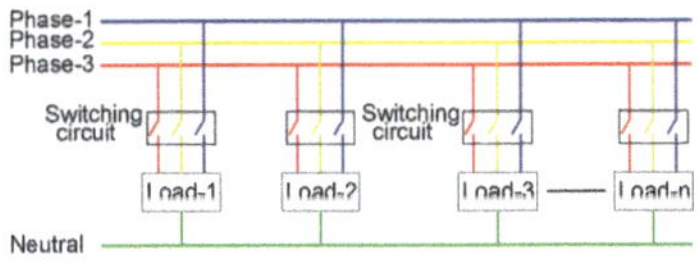

Fig. 1. Switching system of a distribution network.

The contribution of this paper is summarized as follows. Firstly, the system formulation and the implementation of the control system; secondly, applying the genetic algorithm (GA) and ant colony optimization (ACO) techniques to optimize the neutral current by the reconfiguration of LV network. Finally, a hybrid ACO/GA approach is introduced and a comparison is conducted to evaluate the performance of GA, ACO and ACO/GA approaches.

This paper is structured as follows. Section 2 briefly explains the control system and the implementation of the switching pattern of the semiconductor switches. Section 3 introduces the genetic algorithms (GAs) approach followed by Sect. 4, which presents the ant colony optimization (ACO) technique and a comparison between the genetic algorithm and ant colony techniques in conjunction with simulation and results of the system operation. Section 5 presents a hybrid technique (ACO/GA), and also a comparison over three approaches is conducted. Finally, Sect. 6 concludes this paper.

2 System Description and Formulation

2.1 System Overview

The control system depends on rearranging the loads in such a way that the system becomes more balanced [2]. Figure 1 illustrates the connection of a different number of single-phase loads, each load is connected to one phase at a time and a switching circuit is responsible for switching between phases to reconfigure the distribution network based on the total measured connected loads. Applying an optimization technique to reorganize the loads connected to each phase will result in minimizing the neutral current which will lead to minimizing power losses [3]. Massive precautions must be taken in order to make sure that only one phase connected at a time; otherwise, a line-to-line electrical short circuit would occur [2].

2.2 System Formulation

Identifying the control variables used through the objective function, which in this case is to minimize the neutral current, is the main task. An assumption for the impedance of cable lines is considered with the problem solution [3]. The current magnitude of a load (k) is given by I_k at each node and its corresponding power factor angle is given by $\emptyset_k$. This current will be added to the total current of each phase depending upon its corresponding connection to each phase. The neutral current (I_n) is the vector summation of total connected current as shown in (1).

$$\bar{I}_n = \sum\nolimits_{k=1}^{N_d} A_k.\bar{I}_k + \sum\nolimits_{k=1}^{N_d} B_k.\bar{I}_k.a + \sum\nolimits_{k=1}^{N_d} C_k.\bar{I}_k.a^2 \tag{1}$$

where N_d is the number of connected loads to the grid, $a = 1\angle 120^{\circ}$ and the values of the three variables (A_k, B_k, C_k) may equal to 0 or 1 depending on the corresponding load at node k is connected or not connected to that phase as declared in Table 1 [3]. The values of a and a^2 are only used to compensate for 120° or 240° phase shifts between phases. In order to reach the balanced loading condition, the neutral current should be minimized by moving each load as a whole from one phase to another one based on the measured load currents across the three phases of the electric feeder [3, 6].

Table 1. Selection of the phase connected

Parameters	Phase-1	Phase-2	Phase-3
A_k	1	0	0
B_k	0	1	0
C_k	0	0	1

3 Genetic Algorithm Optimization Approach

3.1 Genetic Algorithm Overview

A Genetic Algorithm (GA) technique returns to the theory of evolution and natural genetics principles. Survival of the fittest behavior with a structured randomized data collected is being used to generate a new set of populations [7]. A solution generated by the genetic algorithm is called a chromosome (string), while a collection of chromosomes is referred as a population. Genes are gathered and formed a chromosome and they can have a value of a numerical number, binary, symbols or characters [8]. Figure 2 shows an example of a binary representation of a chromosome. In every new generation, a series of strings is formed using information from the previous ones. These chromosomes will be evaluated by the fitness function to measure the suitability of solution generated by GA [7]. Some chromosomes in population will mate through the process called crossover and produce new chromosomes called offspring. A few chromosomes will have also mutations in their gene, which provide a new search space. Crossover and mutation rates control the number of chromosomes which will go through this entire process. Chromosomes in the population, which have the higher probability, will survive for the next generation based on Darwinian evolution rule [8]. After several generations, the chromosome value will maintain at a certain value, which is the best solution for the problem. GA efficiently use the historical information collected to find new search points with likely enhancement as illustrated in Fig. 3 [7–10].

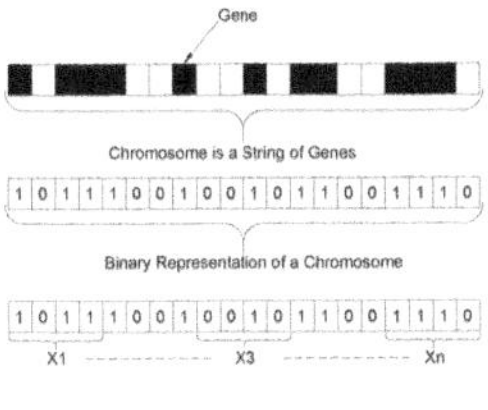

Fig. 2. Binary representation of a chromosome.

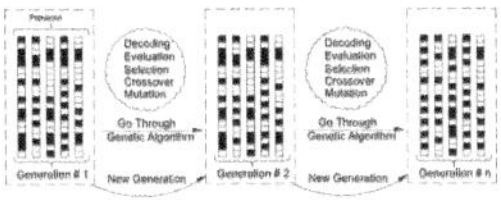

Fig. 3. New generations of chromosomes after applying GA.

3.2 Genetic Algorithm Flowchart (Algorithm)

A Genetic optimization technique is used to synthesize the switching pattern of the semiconductor switches [3, 5]. A Genetic algorithm solution starting with initializing random population and the determination of the number of generation, mutation rate, crossover point, the number of variables and the corresponding number of bits so chromosome could be generated. Then evaluate the fitness function and the process of mates' selection is applied based on Rolette Wheel selection. The crossover process is used to generate new offspring from previous populations while the mutation process introduces a new search space into the solution strings of the populations by toggling a bit in a certain chromosome then create a new generation and start again to evaluate the fitness function. The genetic algorithm process is illustrated in Fig. 4 [7, 11].

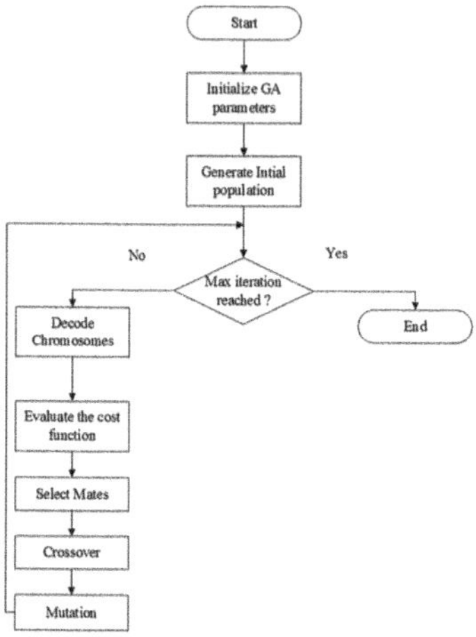

Fig. 4. Genetic algorithm flowchart.

3.3 Apply GA to Minimize Neutral Current

When GA is applied to 99 nodes to optimize the switching pattern, the neutral current has a significant decrease as shown in Fig. 5. The final value of neutral current which has been fixed after 90 iterations approximately is very small and the error percentage is accepted as shown in Table 2.

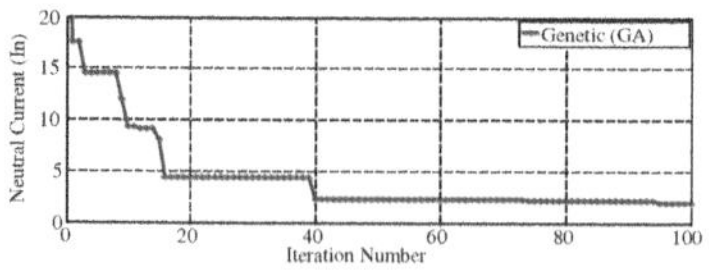

Fig. 5. Neutral current verses iteration number in GA approach for 99 loads.

After applying GAs for many loads with many numbers of populations and calculating the elapsed time to find the optimal solution, it has been seen that increasing the population size will increases the probability of an optimal solution in low number of iteration but consequently this will increase the elapsed time for computing the result as shown in Fig. 6.

Table 2. GA parameters and results for 99 Loads

Number of populations	100
Number of iterations	100
Number of loads	99
Phase-A	918.1495 A
Phase-B	897.9637 A
Phase-C	898.3101 A
Neutral current	1.8967 A
Error percentage	.0698%
Elapsed time	6.852354 s

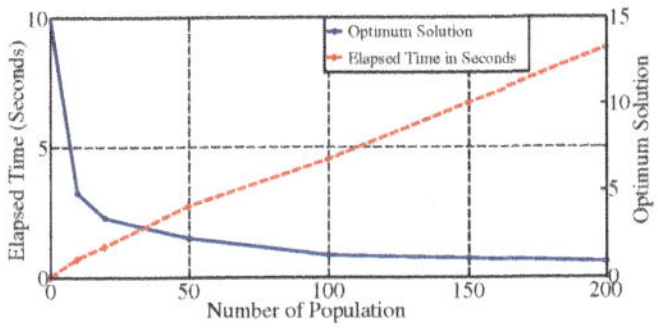

Fig. 6. Conversion time and error probability verses number of population.

4 Ant Colony Optimization Approach

4.1 Ant Colony Overview

Ant Colony optimization is a metaheuristic method used for solving difficult combinatorial optimization problems [12]. Ants cooperate in searching for an optimal trail between their colonies and sources of food. They navigate from nest to food source and the best path is detected through pheromone trails [13]. Each ant moves at random and pheromone is deposited on the path. More pheromone on path increases the probability of path being followed. This behavior leads to the appearance of shortest paths [14]. Pheromones evaporate, so ants are avoided to being trapped in local optima. Pheromones provide positive feedback leads to the rapid discovery of good solution [12, 13, 15–17]. The ACO process can be described as a set of layers. Each layer consists of a number of nodes, which represent the possible discrete values [18]. The starting node is the first step for ants to start their solution, going through the different layers with a constrained that only one node should be selected in each layer in accordance with the state transition rule given by (2) [16, 18].

$$p_{i,j} = \frac{\tau_{i,j}}{\sum \tau_{i,j}} \tag{2}$$

Once the path is complete, the ant deposits some pheromone on the path based on the local updating rule given by (3) and (4) [12, 13, 18, 19].

$$\tau_{i,j} = (1-\rho)\tau_{i,j} + \Delta\tau_{i,j} \tag{3}$$

$$\Delta\tau_{i,j}^{k} = \begin{cases} \frac{\zeta fn_{min}}{fn_{max}}, & if\ ant\ k\ travels\ on\ edge\ i,j \\ 0, & otherwise \end{cases} \tag{4}$$

Where,

$\tau_{i,j}$ is the pheromone amount on the edge i, j.
ρ is the rate of pheromone evaporation.
$\Delta\tau_{i,j}$ is the pheromone deposited through updating process.
ζ is the visibility of ant.
fn_{min} is the worest value of the objective function.
fn_{max} is the finest value of the objective function.

4.2 Ant Colony Flowchart (Algorithm)

At the starting of the optimization process, all the edges or trails are initiated with an equal amount of pheromone and each ant randomly selects a node in each layer. The optimization process is terminated when the pre-specified maximum number of iterations is reached. The steps of applying ACO are as follows:

- Initialize the number of ants and the trail parameters, each path has the same amount of pheromone.
- Calculate probability at each edge then each ant will select a random path based on the cumulative probability.
- Evaluation of fitness function.
- Deposit pheromone along the best route and update each path as pheromone amount decays with time.
- Update the probability of each path and return to the first step.

The flowchart of ant colony optimization technique is illustrated in Fig. 7 [19].

4.3 Apply ACO to Minimize Neutral Current

When applying ACO for 99 loads (nodes) with a number of 100 ants, the neutral current curve shows a decrease in its value as shown in Fig. 8. The final value of neutral current which has been fixed after 3 iterations approximately is very small and the error percentage is accepted as shown in Table 3. The result shows an excellent performance in finding a good solution in a very short time as shown in Fig. 8.

After applying ACO for many loads with many numbers of ants and calculating the elapsed time to find the optimal solution, it has been seen that increasing the number of ants will increases the probability of an optimal solution in low number of iteration but consequently this will increase the elapsed time for computing the result as shown in Fig. 9.

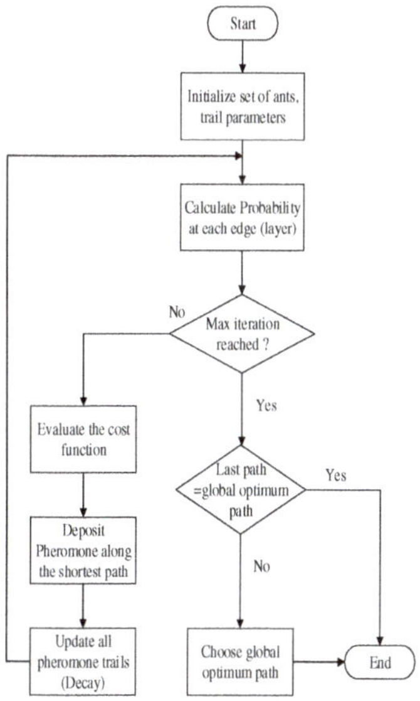

Fig. 7. Ant colony flowchart.

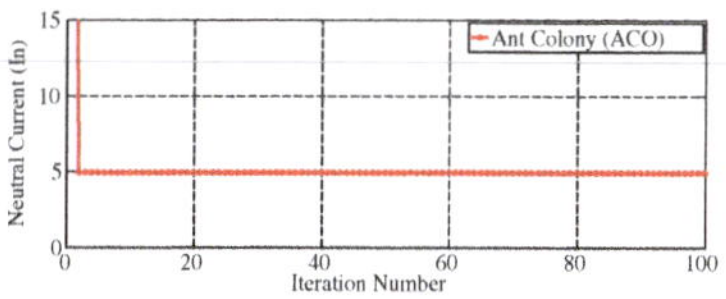

Fig. 8. Neutral current verses number of iteration in ACO approach for 99 loads.

Table 3. ACO parameters and results for 99 Loads

Number of ants	100
Number of iterations	100
Number of loads	99
Phase-A	913.044 A
Phase-B	903.201 A
Phase-C	898.174 A
Neutral current	4.9428 A
Error percentage	.0974%
Elapsed time	9.634920 s

4.4 Comparison Conducted Between GA and ACO Approach

The genetic algorithm approach gives an accurate optimal solution regardless of time elapsed or the number of iteration to achieve it. On the other hand, The ant colony method has a fast solution but not accurate as the GA. There are two points of view

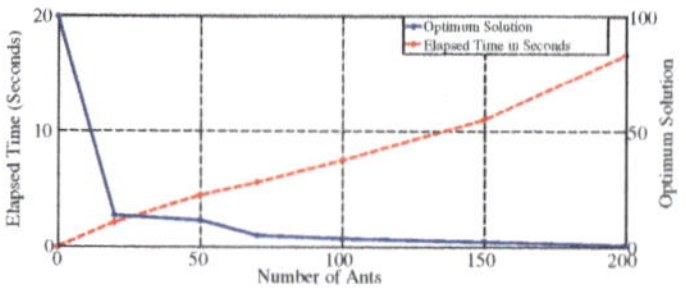

Fig. 9. Conversion time and error probability verses number of ants.

when comparing between GA and ACO techniques. The first one, when applying GA and ACO for 100 iterations and both solutions have started from the same random point to be able to compare between them. The elapsed time for GA and ACO were 6.9 and 9.7 s respectively. The neutral current in GA approach was 1.9 A approximately and 5 A for ACO technique as shown in Fig. 10 which showed that the GA provides an accurate solution.

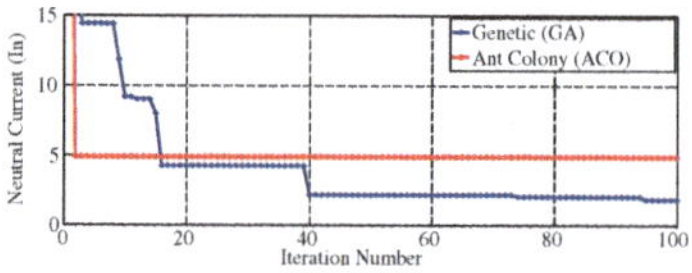

Fig. 10. Comparison between GA and ACO approach over 100 iterations.

However, from the other point of view, If the time for every iteration is fixed which means that the time elapsed for 30 iterations for GA is 2.07 s and for ACO is 2.9 s in average, the elapsed time for both are close. But as shown in Fig. 11, the neutral current in both approaches is almost the same except that in the ant colony technique, the solution has reached the optimal point after 2 or 3 iterations in average, unlike the GA which has reached the optimal point after 16 iterations approximately.

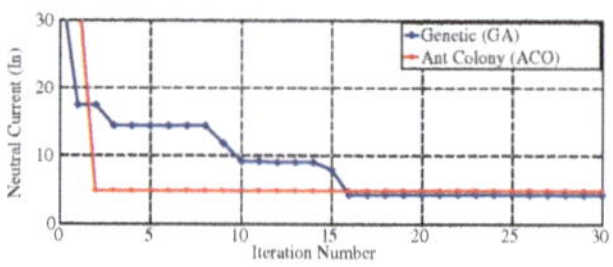

Fig. 11. Comparison between GA and ACO approach over 30 iterations.

5 Hybrid ACO/GA Optimization Technique

5.1 Hybrid ACO/GA Overview

Ant colony optimization technique shows an excellent timing to find an approximate solution while Genetic algorithm shows an excessive performance in finding the optimal solution discarding the number of iteration or elapsed time. The Hybrid system tries to combine these two features to find an optimal solution in a very short time. So,

ACO/GA starts with ant colony method to find a good point as an initial solution and then continues with genetic algorithm technique to ensure the accuracy and efficiency of the system. The optimization process is illustrated in Fig. 12.

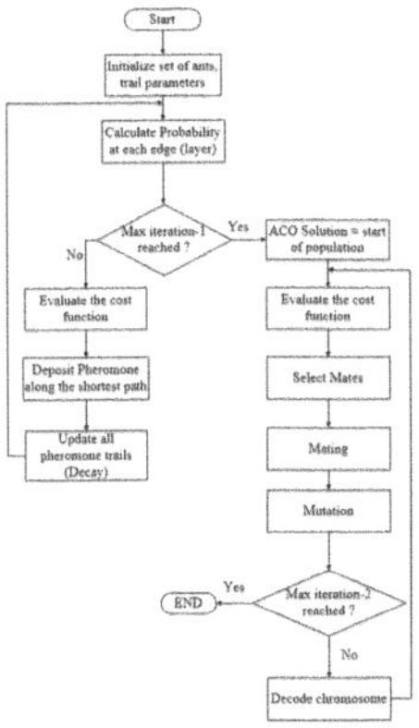

Fig. 12. Hybrid ACO/GA flowchart.

5.2 Hybrid ACO/GA System Results

To evaluate the efficiency of the system, ACO/GA has been applied to 99 loads (nodes). Starting the solution with 100 ants for ACO techniques for 5 iterations then continuing the solution with GA technique with a number of 100 populations and the initial population equal to the last solution of ACO. The neutral current shows a significant decrease in an excellent timing as shown in Fig. 13. The error percentage has an acceptable value as shown in Table 4.

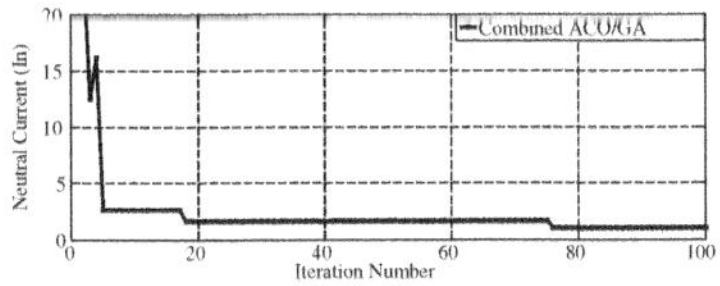

Fig. 13. Neutral current verses iteration number in ACO/GA approach for 99 loads.

5.3 Comparison Conducted Among GA, ACO, ACO/GA Approaches

The three techniques are applied to 99 loads with a total number of 30 iterations. The three solutions have an elapsed time of 2.5 s approximately. Figure 14 shows the enhanced approach reached a better performance after 2 iterations of the ant colony optimization solution and a good solution than Genetic algorithm approach which has reached after 12 iterations in average.

Table 4. ACO/GA parameters and results for 99 Loads

Number of ants	100
Number of populations	100
Total Number of iterations	100
Number of loads	99
Phase-A	903.2044 A
Phase-B	903.0178 A
Phase-C	908.0854 A
Neutral current	1.0365 A
Error percentage	.0382%
Elapsed time	7.841134 s

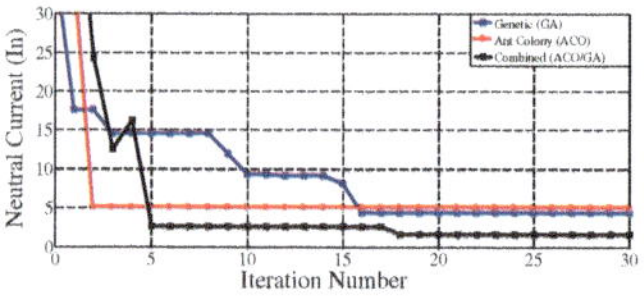

Fig. 14. Comparison between GA, ACO and ACO/GA approach.

5.4 Future Work

The smart management system software can be integrated with this design to display updated reports and draws valuable graphs for real-time consumption and three phases loads distributions; in addition, applying the hybrid optimization technique on a real system like IEEE 14 bus system.

6 Conclusion

In this paper, the phase balancing problem has been formulated as current balancing optimization problems. GAs and ACO were analyzed and implemented as a solution to the optimization problem for generating the switching pattern used to control the semiconductor switches of the control circuit. GAs have proven to be very effective in having the accurate solution as well as ACO has shown much better performance in timing compared to GA technique. The result of ACO technique has been achieved after 3 iterations approximately unlike GA which reached the same result after 15 iterations. This result indicates the fast response for the ant colony method. However, ACO sustained at a certain value after 3 or 4 iterations, GA has proved the efficiency in finding an accurate result which has been reached after 90 iterations approximately. Hybrid ACO/GA technique has been introduced which started with ACO as an initial solution and continued the approach with GA technique. The ACO/GA approach provided better performance regarding time and accuracy. The Hybrid approach has achieved a better result than GA and ACO after 5 iterations. The result of ACO/GA proved the efficiency of this approach over Genetic algorithm approach and Ant Colony optimization approach.

References

1. Saadat, H.: Power System Analysis. McGraw-Hill, Chicago (1999)
2. Siti, M.W., Nicolae, D.V., Jimoh, A.A., Ukil, A.: Reconfiguration and load balancing in the LV and MV distribution networks for optimal performance. IEEE Trans. Power Deliv. **22**(4), 2534–2540 (2007)
3. Gouda, A., Mostafa, H., Gaber, Y.: Smart electric grids three-phase automatic load balancing applications using genetic algorithms. In: Canadian Conference on Electrical and Computer Engineering, pp. 1–4 (2013)
4. Mishra, S., Paul, S., Das, D.: A comprehensive review on power distribution network reconfiguration. Energy Syst. (2016)
5. Goyal, S.K., Singh, M.: Enhanced genetic algorithm based load balancing in grid. **9**(3), 260–266 (2012)
6. Nicolae, D.V., Jordaan, J.A.: Control algorithm of a smart grid device for optimal radial feeder load reconfiguration. In: 2013 9th Asian Control Conference, ASCC 2013, no. 4, pp. 2–6 (2013)
7. Hermawanto, D.: Genetic algorithm for solving simple mathematical equality problem. arXiv Prepr. arXiv:1308.4675 (2013)
8. El-Habrouk, M.K.D.M.: A new control technique for active power filters using a combined genetic algorithm/conventional analysis. IEEE Trans. Ind. Electron. **49**(1), 58–66 (2002)
9. Mitchell, M.: An Introduction to Genetic Algorithms, pp. 1–40 (1998)
10. Lee, Z.-J., Su, S.-F., Chuang, C.-C., Liu, K.-H.: Genetic algorithm with ant colony optimization (GA-ACO) for multiple sequence alignment. Appl. Soft Comput. **8**(1), 55–78 (2008)
11. Zomaya, A.Y.: Observations on using genetic algorithms for dynamic load-balancing. IEEE Trans. Parallel Distrib. Syst. **12**(9), 899–911 (2001)
12. Brand, M., Masuda, M., Wehner, N., Yu, X.-H.: Ant colony optimization algorithm for robot path planning. In: International Conference on Computer Design and Applications (ICCDA 2010), vol. 3, pp. 436–440 (2010)
13. Ibraheem, S.K., Ansari, A.Q.: Ant colony optimization: a tutorial. MR Int. J. Eng. Technol. **7**(2), 35–41 (2015)
14. Seidlová, R., Poživil, J.: Implementation of ant colony algorithms in Matlab (2005). Humusoft.Cz
15. Kushwah, P.: A survey on load balancing techniques using ACO algorithm. Int. J. Comput. Sci. **5**(5), 6310–6314 (2014)
16. Marino, M.A.: Ant Colony Optimization Algorithm (ACO); A New Heuristic Approach for Engineering Optimization, vol. 2005, pp. 188–192 (2005)
17. Monteiro, M.S.R., Fontes, D.B.M.M., Fontes, F.A.C.C.: An ant colony optimization algorithm to solve the minimum cost network flow problem with concave cost functions. In: Proceedings of the 13th Annual Genetic Evolutionary Computation Conference, GECCO 2011, pp. 139–145 (2011)
18. Rao, S.S.: Engineering Optimization: Theory and Practice (2009)
19. Su, C., Chang, C., Chiou, J.: Distribution network reconfiguration for loss reduction by ant colony search algorithm. **75**, 190–199 (2005)

UXAmI Observer: An Automated User Experience Evaluation Tool for Ambient Intelligence Environments

Stavroula Ntoa[1(✉)], George Margetis[1], Margherita Antona[1], and Constantine Stephanidis[1,2]

[1] Institute of Computer Science (ICS), Foundation for Research and Technology Hellas (FORTH), Heraklion, Greece
{stant,gmarget,antona,cs}@ics.forth.gr
[2] Computer Science Department, University of Crete, Heraklion, Greece

Abstract. Ambient Intelligence constitutes a new human-centered technological paradigm, where environments are oriented towards anticipating and satisfying the needs of their inhabitants. In this context, evaluation becomes of paramount importance. This paper presents UXAmI Observer, an automated user experience evaluation tool for Ambient Intelligence environments, taking advantage of their inherent infrastructure to automatically acquire measurements during user testing experiments. The tool provides powerful data visualizations for the entire experiment, for each system and application evaluated, as well as for each experiment participant individually, ensuring synchronization of data with video recordings, and facilitating manual data input by the evaluators themselves.

Keywords: User experience evaluation · Automated evaluation tool
Ambient intelligence · User testing

1 Introduction

Ambient Intelligence (AmI) is an emerging field of research and development, constituting a new technological paradigm, becoming a de facto key dimension of the Information Society, since next generation digital products and services are explicitly designed in view of an overall intelligent computational environment [1]. Although AmI is a multidisciplinary field, its objective is to support and empower users, therefore research in the field should emphasize how and whether this goal is achieved, while in this context it is important to consider the implications of user evaluation [2].

Evaluation is a core concept in Human-Computer Interaction (HCI) targeting the assessment of usability and – more recently – of user experience (UX) [3]. Although a plethora of methods has been proposed in the HCI literature for the evaluation of usability, it is interesting that only few of them are usually actually employed, and namely usability tests, questionnaires, and expert-based inspections [4, 5]. Also, a combination of these methodologies is commonly found in usability evaluations [4] as a means to achieve better results [6]. Despite the fact that usability testing is highly

K. Arai et al. (Eds.): IntelliSys 2018, AISC 868, pp. 1350–1370, 2019.
https://doi.org/10.1007/978-3-030-01054-6_94

valuable in the context of evaluation, several challenges remain to be addressed, including the need for discerning between objective measurements of phenomena and users' perceptions of these phenomena [7], as well as lessening the impact of the individual evaluator on the evaluation results [8], which is a phenomenon expected and methodologically anticipated in expert-based inspections, but not in observation protocols, where it is also evident.

UX on the other hand, has recently become a popular and widely adopted concept by the HCI community, and has influenced the notion of usability [9]. However, despite the fact that usability is an important constituent of UX, it is noteworthy that researchers in the field consider UX to be entirely subjective [10]. Confirming the above, a critical analysis of 66 empirical studies of UX [11], found that the most frequently assessed dimensions are emotions, enjoyment and aesthetics, while context of use (a key UX factor) is rarely researched. Along the same lines, another recent review of UX studies identified that observation is a method rarely used, in contrast to questionnaires and interviews [12].

Evaluation in AmI is a challenging objective and a field which has not yet been extensively explored, due to the inherent difficulties it imposes. Stephanidis [1] highlights that the evaluation of AmI technologies and environments needs to go beyond traditional usability evaluation in a number of dimensions, concerning both the qualities of the environment to be assessed and the assessment methods. A major concern is that evaluation should go beyond performance-based approaches to evaluation of the overall user experience [1, 13], which should be further articulated in the context of AmI environments. Moreover, evaluation should take place in real world contexts [1, 13], which is a challenging task by itself. Traditional evaluation practice has also been pointed out as insufficient for new HCI systems that feature new sensing possibilities, shift in initiative, diversification of physical interfaces, and a shift in application purpose [14]. Challenges include the interpretation of signals from multiple communication channels in the natural interaction context, context awareness, the unsuitability of task-specific measures in systems which are often task-less, as well as the need for longitudinal studies to assess the learning process of users.

Recognizing the need for addressing the new challenges that arise in UX evaluation in AmI environments, this paper presents UXAmI Observer, a tool to support evaluators in carrying out user-based evaluations in AmI environments or AmI simulated spaces. UXAmI Observer aggregates experimental data and provides an analysis of the results of experiments, incorporates information regarding the context of use, and fosters the objectivity of recordings by acquiring data directly from the infrastructure of the AmI environment. The remaining of this paper is structured as follows: Sect. 2 discusses related work in the field of automating usability evaluation; Sect. 3 presents the UXAmI Observer tool, its architecture, and preliminary evaluation results; finally Sect. 4 summarizes the contributions of this work and discusses directions for future work.

2 Automating Usability Evaluation

Automated Usability Evaluation Methods (UEMs) are promising complements to traditional ones, assisting evaluators to identify potential usability problems [15]. A classification of UEMs [15], with a focus on their support towards automated measurements, organizes methods according to their class (testing, inspection, inquiry, analytical, simulation), type (e.g., log file analysis, guideline review, survey), automation type (none, capture, analysis, critique), and effort level (minimal, model development, informal use, formal use). Another study of automation support [16] classifies approaches for extracting usability information from User Interface (UI) events, according to the supported techniques, and namely synchronization and searching, transforming event streams, analysis, visualization, and integrated evaluation support. Analysis in particular involves performing counts and summary statistics, detecting sequences, comparing source sequences against target ones, and characterizing sequences. The classification and examples of tools and environments discussed are based on their technical capabilities; however it is pointed out that there is very little data published regarding the relative utility of the surveyed approaches in supporting usability evaluations. Moreover, it is highlighted that more advanced methods require the most human intervention, interpretation, and effort, while the more automated techniques tend to be least compelling and most unrealistic in their assumptions. In the context of this paper, tools to support usability testing automation through logging will be discussed, as the ones more relevant to the current work.

A common approach in such tools, pursued even in early attempts, is the generation of statistics and automatic calculation of usability metrics. For instance, DRUM [17] supports management of evaluation data, video mark-up and logging, as well as the following automatically calculated metrics based on logged data: task time, snag, help and search times, effectiveness, efficiency, and relative efficiency compared with experts or with the same task on another system. AIDE [18], a metric-based tool assisting designers in creating and evaluating layouts for a given set of interface controls, includes five metrics: efficiency, evaluating how far the user must move a cursor to accomplish their task; alignment, assessing how well objects are aligned; horizontal and vertical balance, calculating how balanced the screen is along the two axes; and constraints, providing a quick overview of the status of any designer-specified constraints. USINE [19] is a tool that takes as input the task model of the system that describes the user's interactions with the system, as well as a log-task table, created with information from the task model and one log file which contains all the possible actions, mapping logged actions with tasks in the model. The tool supports the evaluator by providing the accomplished, failed and never tried tasks, number of user errors and the time that these occurred, time to complete each task, and sequences of accomplished tasks that occur more than once.

More recently, and with the aim to facilitate the recording of user behavior, TRUE [20] proposed an approach that combines log files with attitudinal data, received from polling users themselves at specific intervals. The innovative aspects of TRUE include logging sequences of events, as well as of event sets that collect both the event of interest and the contextual information needed to make sense of it. Automation support is provided in terms of synchronizing the video that is captured with the logged events, as well as providing visualizations of the recorded events. The enhanced log files with event sets and sequences, allow evaluators to drill down to specific events and determine the causes of the identified problems. The tool has been applied for evaluating user experience in serious games, however the fact that it constitutes a custom development built in each application that needs to be evaluated [21], makes it inappropriate for use as a generic all-purpose usability evaluation tool.

An important concern in the development of usability evaluation automation tools refers to instrumenting the software to collect usage data. In summary, five main methods are reported in literature for instrumenting systems [22]: manual instrumentation, by adding logging instructions to the code of the system; toolkit instrumentation, in which the toolkit used for the presentation and handling of the UI is instrumented; system-level instrumentation that uses logging at the operating-system level; and aspect-oriented instrumentation. Following the latter approach UMARA [22], an interactive usability instrumentation tool, allows evaluators to specify what actions to log, by clicking on interface elements in the application itself (e.g., select a text field of interest and then decide which events to monitor for this element, such as mouse, windows, keyboard events, etc.). In the system-level instrumentation approach, an automated usability evaluation tool, running as a service in windows environment and supporting data collection, metrics and data analysis [23] collects messages that the user sends to the application being tested, messages sent by the system to the user, keystrokes and mouse clicks. The system calculates the number of windows opened (total and per window), number of times a menu is selected, and number of times a button is pressed. Finally, the system produces charts to illustrate mouse density, mouse travel pattern, and keystrokes. AppMonitor [24] is a windows-based tool that has been designed to record low-level and high-level events for two specific windows applications, Microsoft Word and Adobe Reader. The tool runs in Microsoft Windows XP platform, and listens to events exchanged between the applications and Windows through an event-hooking Dynamic-Link Library (DLL). The evaluators can select the windows events they wish to be monitored and logged, while the output of the tool is a file listing all the events that have been captured. A usability evaluation toolkit for mobile applications that implements a Software Development Kit (SDK), which can be used by applications with minor modifications in their source code logs view events, dialog and menu events, system keys, and unhandled events that cannot be classified under the previous three event types [25]. It also features an automated metric discovery model based on comparing the ideal sequence of events (as carried out by an expert) for accomplishing a task with users' sequences of events. Then several usability indicators can be calculated, such as the number of backtracks, correct flow ratio, or the number of users who failed to accomplish the task. EISEval [26] is a tool extending usability evaluation automation by capturing data concerning not only the interactions between users and the UI, but between agents themselves as well, thus supporting the evaluation of UIs' dynamic behavior. EISEval performs data analysis on the collected data through measurements and statistics (e.g., frequencies, times, successes and failures), and generates PetriNets to visually reproduce the activities of the user in the target system. Evaluators are also supported by an open and modifiable list of criteria, and are facilitated to record their observations for each one of them. Before actually being used in the context of an evaluation, EISEval requires the evaluator to specify information about the tasks that can be performed with and by the system, as well as information about agents and other configuration settings. EISEval has been used in the context of the environment proposed in [27] to combine objective and subjective metrics, complemented by a questionnaire generating tool and a guidelines inspector tool. Objective and subjective results acquired through the aforementioned modules are visualized in scatter plot charts, organized under specific usability indicators (e.g., information density) to facilitate evaluators in their interpretation. In an effort to provide a more generic evaluation framework that can support usability testing in real production environments and be applied on arbitrary software applications, UEF [28] uses XML files for describing meta-information about the system and providing concrete usage data (logs), while a validator component checks the log files according to predefined syntactic and semantic rules. Then the data are evaluated according to a specific usability model, which can vary for different systems, while the framework also supports subjective evaluation through questionnaires.

Shifting the focus from instrumenting the software to user-based instrumentation, DUE [29] collects and evaluates usability data, based on users' reporting. More specifically, DUE supports recording video from the user's screen, as well as voice recordings. When the user detects a usability problem they press a button to report it, record an explanation, and rate its severity.

An interesting approach, alleviating the need for any instrumentation and event logging is scvRipper [30], a tool which uses computer vision scrapping to automatically extract time-series data (software used and application content accessed/generated) from screen-captured videos, thus enabling the creation of quantitative metrics. Although no instrumentation is required, it should be noted that a sampling process is required once for each application to define the application windows, during which each window is defined through collecting sample images of its visual cues.

Commercial tools on the other hand avoid any instrumentation, and support a variety of features, such as data capture of low-level events (keystrokes, mouse clicks, system events), metrics (e.g., time or activity), screen video capture, logging of observational comments, and event definitions allowing the association of hot keys with the defined events [31]. In addition, they rely on experiment observers to identify task success and user errors [32], while they also support plugins such as eye tracking and physiological measurement systems.

Despite the fact that several tools and frameworks towards usability evaluation automation have been proposed in literature, it is noteworthy that most usability testing is done in a very manual, labor-intensive way [33]. A possible explanation for this is that although statistical information is calculated, the results are often not useful for the evaluator, as the data logged leave out the user's goals and intentions and much of the user's focus of attention when not actually clicking on a button or typing in a field [33]. Efforts towards capturing users' goals require instrumentation of the process; however the complexity of today's systems makes successful instrumentation a challenging task [20]. An important challenge regarding the instrumentation solution is that it should easily map between: UI events and application features; lower level and higher level events (e.g., typing, deleting, and moving); as well as events and the context in which they occur [16]. Furthermore, challenges that logging approaches should address are to be designed to focus on high-level user actions, capture provenance of all events, and observe intermediate user actions [34]. Moreover, requirements that an automated usability testing tool should meet, include [35]: capturing a range of inputs, performing analyses on different aspects of usability, presenting results clearly, being simple and flexible to use, and being able to be used throughout development.

In the light of the above, this paper proposes UXAmI Observer, an innovative automated usability evaluation tool that does not require any instrumentation from the evaluator or the application developers, instead it is based on acquiring information from the messages exchanged in the AmI environment. Through this approach, the proposed tool addresses the challenging aspect of identifying the system and application with which the user is interacting in the entire environment. Furthermore, it calculates typical metrics involved in user-based experiments, such as task success, number of help requests, and user errors. It also includes innovative features, as for instance the number of adaptations introduced by the environment, adaptations rejected by the users, as well as distinct interaction and input errors. Input errors pertain to problems with the interaction modalities used (e.g., gestures, voice), while interaction errors can be defined as issues stemming from erroneous usage of the UI. This distinction facilitates the separated assessment of the interaction vocabulary and of the applications' UI. Finally, the tool produces charts to facilitate evaluators' comprehension of the overall and individual user experience.

3 UXAmI Observer

UXAmI Observer aims to support evaluators in carrying out user-based evaluations, be them laboratory task-based experiments, in situ evaluations, or long-term experiments. In a nutshell, the tool aggregates data regarding user's interaction with systems and applications in AmI environments and presents them through multiple views, such as timelines, charts, and diagrams. In task-based experiments the evaluator has to define the tasks and participant characteristics, whereas long-term experiments can be unstructured, employing users that are already registered in the system (e.g., the inhabitants of an actual AmI environment). Furthermore, the evaluator can view a user session live and provide annotations for it, or review the recorded data and further process them after the experiment. The tool provides two views for an experiment: (1) a view of each interaction session, named Timeline, and (2) insights from the entire experiment, based on all the users that are involved in it throughout the experiment period.

3.1 Architecture

The architecture of the tool (Fig. 1) is based on the Service Oriented Architecture (SOA) model and is built on layered components, each providing the necessary functionality and input to the components of upper layers. Specifically, Authentication Service and UXAmI Service constitute the tool's core endpoints, providing RESTful APIs so that the tool and its users are able to authenticate and acquire the consolidated information regarding user-based experiments. Registration and authentication of an UXAmI user (evaluator) is provided through the Authentication Service endpoint.

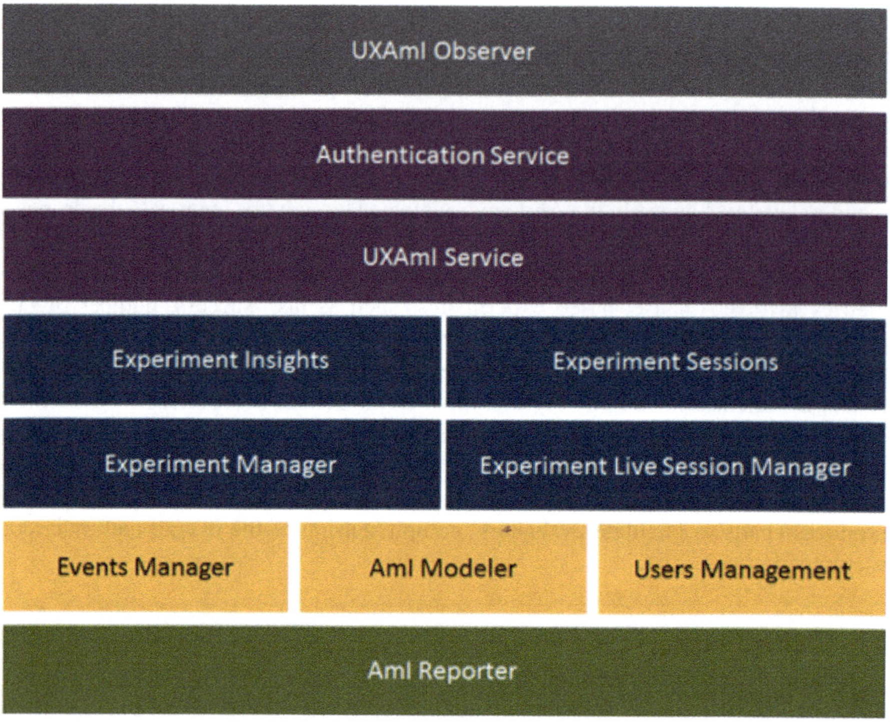

Fig. 1. Architecture of the UXAmI Observer.

The Experiment Sessions component delivers information regarding the execution of a specific session of an experiment, acquiring data from the Experiment Live Session Manager, the Experiment Manager regarding the basic parameters of the experiment (e.g., participants, tasks) and the Events Manager regarding the events that have occurred during this session (e.g., user input commands, system responses). The Experiment Sessions component is then responsible for calculating all the automated measurements. The Experiment Insights component provides consolidated statistical information, which has been acquired and analyzed during the execution of an evaluation experiment or at a post-experiment processing phase, as described in Subsect. 3.5. The information is provided aggregated for all the participants of an experiment and all the systems and applications involved, as well as in sub-clusters pertaining to specific systems or applications. The Experiment Insights component acquires information from the Experiment Sessions component and interoperates with the Experiment Manager, which provides all the necessary functionality to create, read, update, and delete UXAmI experiments. The Experiment Live Session Manager handles the experiment's live streaming engaging all the available cameras, as well as the recording of videos and of annotations made by the experiment observer so that they are available for post-processing, while also timestamping the session.

Events Manager constitutes the register for all the events originating from AmI Reporter. It also provides functionality for events' acquisition based on specific criteria, such as the set of events that belong in a specific timespan, or all the events of type system response, etc. It interoperates with the Experiment Sessions component, which employs the specific information. AmI Modeler is the component responsible for modelling the information acquired from the AmI environment through the AmI Reporter, as well as all the entities of the UXAmI Observer, defining also their interrelationships. Four fundamental entities are modelled, namely events, actors, contexts, and users. Events can be of three main types: interactions that represent user input, actions that relate to agents' information and responses referring to systems' and applications' feedback. Actors in an AmI environment can be humans, applications or agents, according to the type of source or destination of an UXAmI event. Contexts represent a space in the AmI environment where a system may reside, and are structured in a hierarchical manner, according to which each subspace declares the space it belongs to. Moreover, for each context a geojson file is assigned, describing its location and 2D geometry (floorplan).

The Users Management component provides back-end functionality for the management of users. Finally, AmI Reporter is the intercommunication component of the UXAmI system with the AmI environment, incorporating heterogeneous services for the acquisition of the necessary information and interoperates with Events Manager and AmI Modeler. AmI Reporter runs as an independent service and exposes a RESTful API for the population of information of the UXAmI Observer entities. This information originates from external agents, which are responsible to perceive the necessary information from the AmI environment and interpret this information to the appropriate format defined by the AmI Reporter API. Such an external agent has been implemented in order to use UXAmI Observer to carry out user-based experiments in the Living Room area of the Home Simulation Space, located at the FORTH-ICS Ambient Intelligence Research Facility. To this end, the agent subscribed to a REDIS channel, where the AmI environment agents and applications broadcasted messages, listened to the messages and interpreted them according to the AmI Reporter API.

3.2 Experiments

An experiment in UXAmI Observer includes:

- The evaluation targets, that is, the involved artefacts, applications and the relevant contexts (AmI spaces where the artefacts are located).
- A name to facilitate its identification, a description of the experiment and its goals, as well as a photograph.
- The evaluators involved and their expertise, as indicated by the evaluators themselves in their profile, rated on scale from 1 to 5.
- Confidence in the evaluation experiment, as rated by the evaluators on a scale from 1 to 5.

- Tasks (if any) and participants.
- Sessions, that is, usages of the system by one or more participants concurrently.

When creating a new experiment, the evaluator has to first define the evaluation targets, provide a name and a short description for the evaluation, as well as a representative photograph. The process of selecting targets in UXAmI Observer involves either selecting an entire AmI space and subsequently refining selections to specific artefacts and applications in the space, or selecting a pervasive application and subsequently refining the artefacts to which the application runs and which thereby constitute the evaluation targets. Once the experiment has been created, the evaluator can add tasks and participants. Defining a task is optional, refers to task-based experiments, and requires providing a short description for each task (e.g., "Turn the TV on"). Adding participants is mandatory, and is achieved by selecting them from a list of existing participants, or by adding new users to the system through defining their age, gender, and computer expertise level. The evaluator can also define additional participant attributes by determining a name for the attribute, as well as its potential values, through identifying the values' scale (e.g., 1-5) and a string to describe each point of the scale. For instance, if one would like to define a new attribute, the following could be defined: (1) attribute name: computer attitude, (2) attribute scale: 1–3, and (3) attribute scale values: negative; neutral; positive.

The experiment screen acts as a gateway to carry out and review a pilot test, record a new experiment session, view a session execution (session details), or view aggregated statistics for all the experiment sessions (experiment insights). A preview of three indicative statistics is readily available through the experiment screen, and in particular a bar chart representing the task success rate score in total and per task (if the experiment includes tasks), a pie chart illustrating the interaction modalities used (%), and a bar chart presenting interaction accuracy per interaction modality.

3.3 Session Details

Information about an experiment session is clustered under four main themes: (1) session timeline, with all the recorded points of interest (POIs) marked along a horizontal timeline, (2) interaction timeline, with explicit indications of task start, task end, user input commands and implicit interactions, system responses and adaptations, ordered in a vertical timeline according to the time of their occurrence, (3) system responses path, indicating all the system responses during the experiment, and (4) interaction statistics.

Topmost is available the session timeline (Fig. 2), giving to the evaluator an overview of the entire session and all the data that have been recorded, synchronized with the session videos. Up to three videos are supported, one of which should ideally illustrate the screen of the system with which the user interacts. If an eye tracking service has also been used for the experiment, the evaluator can enable annotation of the video with the eye tracking data. Also, all the data related to the experiment are marked on the horizontal timeline, including adaptations that have been introduced by the system, adaptation rejections, input error POIs, emotion POIs, interaction errors, help requests and observer notes. The user can select to play the video, while a red vertical line runs across the horizontal timeline to indicate the current time, facilitating association of POIs with the video displayed above. The POIs annotated on the horizontal timeline can be based on automatically acquired information (Sect. 3.4) or on data manually provided by the observer. Regarding manual input, the evaluator can annotate any of the aforementioned POIs during the live execution of the experiment, or during post-processing by selecting the corresponding marker button located above the horizontal timeline. For each POI, the evaluator can provide the session participants it corresponds to (if more than one), notes, and in the case of input errors their type (e.g., gestures, voice).

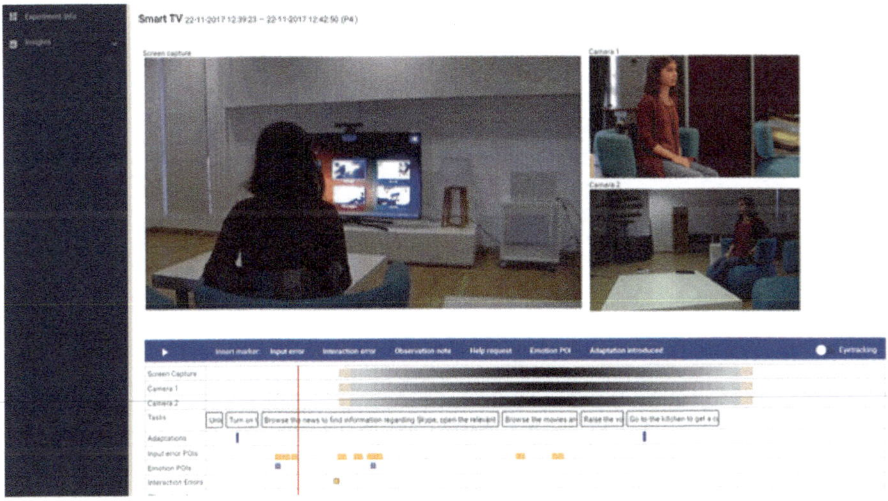

Fig. 2. Session details: session timeline.

Below the experiment timeline follows the vertical interaction timeline (Fig. 3), a scrollable panel which includes the following point annotations: task initiation, task ending, user input, implicit user interactions, system response, and adaptation insertion, ordered according to the time of their occurrence. By selecting a task ending point on the vertical timeline, the evaluator can define whether this task was completed with success, partial success, or if it has failed. Based on these indications, the tool generates the task success rate score and the relevant chart. Each annotation on the timeline is accompanied by specific relevant information, as explained in Table 1. Also, since all events on the vertical timeline are timestamped, they are in complete synchronization with the horizontal timeline, therefore when the evaluator clicks on a vertical timeline point, the horizontal timeline and the videos move to the corresponding timestamp to facilitate direct association of the events and a deeper understanding of the interaction.

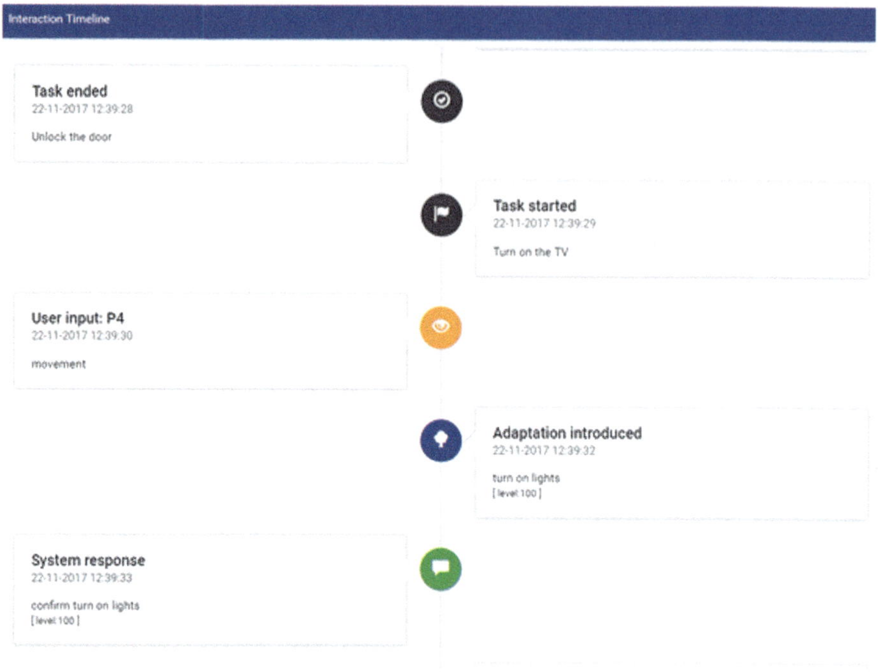

Fig. 3. Session details: interaction timeline.

Table 1. Vertical timeline events information

Event	Time	Participant	Information
Task started	✓		Task description
Task ended	✓		Task description
User input	✓	✓	Input type, Input information[a]
Implicit interaction	✓	✓	Implicit interaction[a]
System response	✓		Status information, as propagated by the application
Adaptation	✓		Adaptation information[a]

[a]As propagated by the corresponding agent

The next structural element of the session details screen is the system responses path, displaying all the system responses in the order they were issued, aiming to shed light to the session from the perspective of the application, leaving aside agent or user actions. Although this path is linear for each single session, this is not the norm for the entire experiment, where through the insights screen the evaluator can have an overview of the various system paths that have been employed by all the users.

Last is the interaction statistics section, which displays statistics information per participant, as well as the overall session statistics in multi-user experiments (Fig. 4). The statistics provided are: number of user interactions, number of user input errors, number of system responses, number of interaction errors, number of adaptations introduced, number of adaptations rejected, number of emotion POIs, usage and accuracy percentage per input modality, and number of errors over time, with the possibility to change the time units illustrated in the chart, so as to explore the user's error behavior over larger periods of time (in the case of long-term experiments).

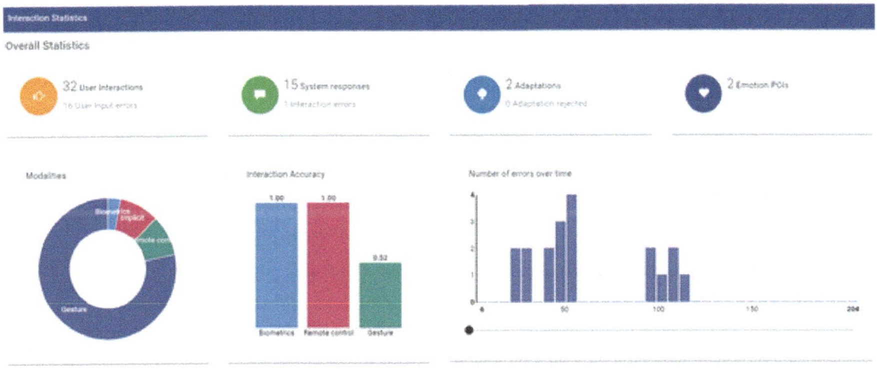

Fig. 4. Session details: interaction statistics overall and per participant.

3.4 "Sensing" Data from the AmI Environment

Besides manually acquired data, UXAmI Observer includes the following automatically detected points: adaptation insertion and adaptation rejection, input error, emotion and implicit interactions. Adaptation insertion points are identified based on events received from the reasoning agents of the AmI environment, agents that will certainly be involved before any adaptation. Cook et al. [36] identify the contributing technologies in an AmI environment and explain that the AmI algorithm perceives with sensors the state of the environment and users, reasons about the data using a variety of AI techniques, and acts upon the environment using controllers in such a way that the algorithm achieves its intended goal. A typical information flow in an AmI environment, clearly involves reasoning mechanisms after any sensed interaction and before any decision making [2, 37]. UXAmI Observer takes advantage of this structure that is inherent in AmI environments and "listens" to information propagated by the reasoning agents, and accordingly marks an adaptation insertion POI on the timeline.

Adaptation rejection points are inferred by monitoring the state of the system that was affected by the adaptation and checking if this state is changed by a user, according to the pseudo-code described in Fig. 5. Two main challenges had to be addressed, namely to accurately identify the systems that are affected by the adaptation and to define the cut-off point after which a change in the system will not be considered an adaptation rejection. Regarding the first, UXAmI Observer identifies as affected systems the ones that are detected to change state after the information propagation by a reasoning agent and before any user interaction or other agent event. It should be noted that more than one systems may be considered as relevant, influenced by a decision of a reasoning agent (e.g., kitchen lights are turned on and interaction with the cooking assistant is switched to speech-based mode). Following the same rationale, all the systems that change state after a user input event and before any other user input or agent event are considered to be affected by a user input. A potential counterexample of the described rationale would be system responses that interleave the expected flow due to network problems or other delays or incorrect behaviors. This, on the one hand, constitutes a problem that should be detected by the evaluator, therefore it is expected that misjudgments by UXAmI Observer will draw the evaluator's attention towards locating problematic behaviors. On the other hand, whenever an adaptation rejection is detected, it is appropriately annotated on the vertical timeline and the tool asks evaluators to confirm this inference. Regarding the cut-off point, it was decided to be determined by the number of user input commands and the time that has elapsed after the adaptation. A simple scenario illustrating the need for a cut-off point is the following: "The environment detects that John is stressed and starts playing his favorite music songs. John is ok with this, but after some time he wishes to turn on the TV, so he turns the music off". In the scenario, the user turns the music off, but not because he objects to the adaptation applied. The cut-off points of five minutes or five user input commands have been arbitrarily defined, meaning that if the state of an affected system is changed after five minutes have elapsed or five user input commands have interceded the adaptation, this change will not be considered as a rejection of the adaptation. Yet, a system affected by the adaptation is no longer examined if its state is changed again due to another decision of a reasoning agent. For simplicity purposes, Fig. 5 illustrates the algorithm for a single system affected, while actually all the systems affected constitute entries in a table that is updated according to the Adaptation Rejection Detection algorithm. Although the initial tests that have been carried out indicate that the defined cut-off points are reasonable, further testing with users living and interacting in actual AmI environments are required. Nevertheless, it should be noted that evaluators are fully empowered to either decline a suggestion of the system as an adaptation rejection, or manually add an adaptation

Adaptation Rejection Detection Algorithm

```
begin
identify currentState of systemAffected
while ((timeElapsed < 5 minutes) AND (userinputCommands < 5)) do
    if (newAdaptationIntroduced = TRUE) then
            identify new_currentState of new_systemAffected
            if ((new_systemAffected = systemAffected) AND
               (new_currentState ≠ currentState)) then
               break;        // stops inspecting for the current adaptation
            end if
    end if
    if (user provides userinputCommand) then
            identify new_currentState of new_systemAffected
             if ((new_systemAffected = systemAffected) AND
                (new_currentState ≠ currentState)) then
                    adaptationRejected : TRUE;
                    break;
            end if
            userinputCommand += 1;
    end if
end while
end
```

Fig. 5. Rejection detection algorithm for each adaptation introduced.

rejection if it was omitted by the tool's inference mechanism. Future versions of the tool will also learn from evaluators' answers regarding adaptation rejection and will adapt the initial thresholds accordingly.

Input error POIs are inferred based on the sequence of user inputs and system responses. More specifically, when UXAmI Observer detects at least two consequent user input commands without a system response, these are annotated as potential input errors, since one of them was potentially erroneous and not recognized by the AmI system. Input commands are acquired through the information propagated by the corresponding AmI environment agents. Although this rationale is effective for systems that support single user interaction with one system at a time, it is evident that this is not always true for multi-user interactions with multiple systems in an AmI environment, where input commands are not given directly to a specific system, instead they are acquired through the environment sensors. For instance, one user might be waving in front of a television to interact with it and another user might be providing voice commands to the heating system, making it possible to receive two consecutive user inputs before a system response. Furthermore, although not common, it is possible that interaction with a system may require the combination of two input commands to trigger a system response. To compensate this behavior that might lead to incorrectly suggested input error POIs, UXAmI Observer asks confirmation from the evaluator, learns from their responses regarding the correctness of the suggestion, and adapts future suggestions accordingly. For instance, in case that a system requires two consecutive input commands, once the evaluator indicates that this was not an input error the combination of these two input commands will not be suggested as a potential input error if it is followed by a response of the specific system. Confirmation by the evaluator is asked in the vertical timeline next to every user input that is highlighted as a potential error.

Last, two more automatically acquired points refer to emotions and implicit interactions. Emotions are received by the information propagated by the corresponding emotion detection agents, if available. Implicit interactions pertain to information related to emotions and detection of user location. In the lack of the corresponding agents in the environment, emotion and implicit interaction POIs will not be automatically annotated. To facilitate evaluators in identifying whether a POI was automatically calculated or provided by a human observer, the label "Indicated by UXAmI Observer" is included in the POI details panel for all the automated measurements.

3.5 Experiment Insights

Insights aim to aggregate information from all the participants of an experiment, thus facilitating the evaluator towards more generalized observations and conclusions. The insights information provided is focused on five main concepts: (1) overview of experiment details, (2) points of interest annotated on the floorplan of the relevant AmI environment, (3) usage information, (4) interaction statistics, and (5) system responses path.

Experiment details refer to information about the participants, clustered and presented according to the enlisted attributes (typically age, gender, and expertise). Moreover, the specific evaluation targets (artefacts/systems and applications) of the current experiment are presented.

The floorplan representation includes interaction, error, and emotion POIs illustrated on a 2D depiction of the AmI space (Fig. 6). Interaction POIs occur based on the artefacts with which the user interacts, error POIs represent points where input errors mostly occur, while emotion POIs refer to points where the user has been reported by the corresponding agent to have specific psychophysiological measurements (e.g., high skin conductance, fast heart rate). The floorplan is constructed using coordinates for each room, stored in AmI Modeler, a process that needs to be carried out only once for each AmI space. Each POI is then annotated as a bubble marker on the floorplan location where it occurred, with a size corresponding to its frequency (three sizes are always available: small, medium, and large).

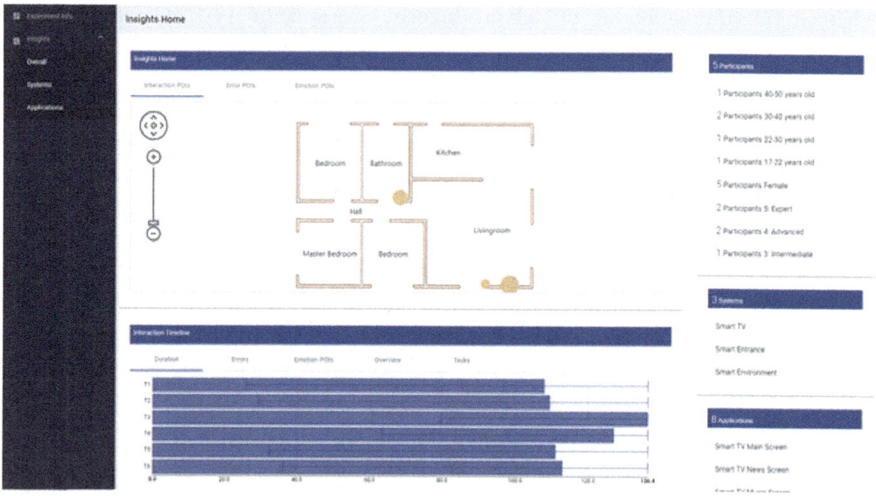

Fig. 6. Insights: experiment details, floor plan, usage information.

POIs therefore are determined by two parameters: their position and size. Position can be acquired either using a user localization agent of the AmI environment, or inferred according to the location of the system with which the user interacts. The size of the POI depends on the frequency of its occurrence, and is calculated dynamically, by receiving the minimum and maximum value of occurrences and dividing them in three quantiles. The floorplan visualization is available topmost in the insights page and aims to assist the evaluator in obtaining an overview of the user's interaction in the environment and to detect where users mostly interact with artefacts and application, where errors happen mostly and where users are more stressed.

Usage information aims to reveal how the specific systems and applications have been used during an experiment. For long-term experiments it refers to displaying interaction heat maps, by illustrating the number of usages per hour and per day of a week. For short-term task-based experiments (Fig. 6), this information includes duration, errors, and emotion POIs per task and an overview of all the above for all the experiment's tasks. The first three are provided in the form of bar charts, where each bar represents the average value per task (e.g., average duration per task) and is accompanied by a line indicating the standard deviation. The overview panel features four area charts, each illustrating how the parameters of duration, user input errors, user emotion POIs, and adaptations applied evolve over the various tasks involved in the experiment.

The interaction statistics are similar to the ones presented in the session timeline view, with the difference that they are calculated over all the system usages by all the experiment participants and include pie charts illustrating input modality usage and implicit interactions types, bar charts depicting accuracy per input modality used and the number of errors over time (featuring customization over the time units displayed), as well as the following numerical indications: number of total user interactions followed by the number of input errors, number of implicit user interactions along with the number of relevant adaptations introduced and the number of those adaptations that were rejected, number of system responses and number of interaction errors, total number of adaptations along with the number of rejected adaptations, and finally number of POIs related to users' detected emotions. In the case of task-based experiments, a chart illustrating task success per task is also included, employing stacked bars with three colors. The chart shows the number of successful, partially successful and failed executions per task, providing also the overall task success rate score, calculated according to (1), where S is the number of successful task executions, PS is the number of partially successful task executions, and T is the total number of task executions [38].

$$TS = (S + (0.5 * PS))/T \tag{1}$$

Finally, the system responses path illustrates all the responses from the applications involved in the experiment aggregated from all the participants (Fig. 7). Therefore, as opposed to the session timeline view, the path is not linear. The goal of this component is to facilitate the exploration of all the possible paths that have been followed by users to retrieve specific information (e.g., find news item through browsing the categories or through search, select a TV channel through a menu or incrementally through the up/down command), and to assist evaluators in clarifying if and how users employ the various offered paths during their interaction with the applications.

If more than one system has been employed in an experiment, and more than one application, the aggregated insights information is also available per system and per application, through appropriate options in the tool menu.

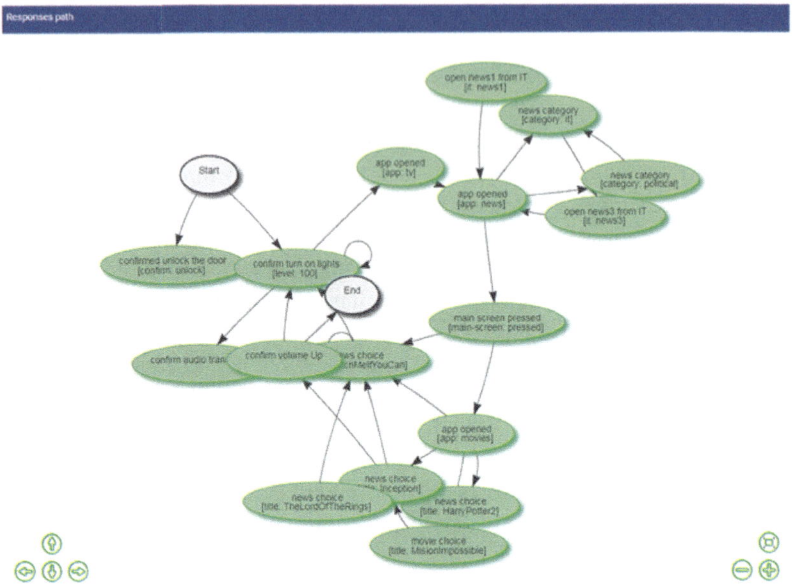

Fig. 7. Insights: system responses path.

3.6 Comparing Experiments

Experiments that refer to the same evaluation targets can be compared, allowing the evaluator to verify if UX was improved during the various iterations. Experiments that are selected for comparison are placed in chronological order side by side. In the experiment comparison screen, the evaluator can view insights and charts stemming from the analysis of all the experiment sessions, as they are presented in the insights view. The experiment comparison screen includes (1) charts for task success, modalities employed, implicit interaction analysis, interaction accuracy, number of errors over time, and (2) insights regarding the total number of system responses, interaction errors, user interactions, user input errors, adaptations applied, adaptations rejected, as well as the number of implicit user interactions accompanied by the number of relevant adaptations and adaptation rejections.

3.7 Preliminary Evaluation

UXAmI Observer has been evaluated by three UX experts following the heuristic evaluation approach [39]. More specifically, each evaluator inspected the interface alone against the heuristic evaluation guidelines. Each problem that was identified was correlated with one or more guidelines. The problems reported by each individual evaluator were aggregated into a single report, merging identical problems. Subsequently, each evaluator provided their severity rating for each one of the problems in the unified list, while a final severity rating for each problem was calculated as the means of the individual evaluators' ratings. In heuristic evaluation, ratings are given by the evaluators, according to their judgement on the frequency, the impact, and the persistence of the problem, and may range from 0 to 4 (0: I don't agree that this is a problem at all, 1: cosmetic problem, 2: minor problem, 3: major problem, 4: catastrophic problem).

The final evaluation report of UXAmI Observer included 31 issues, 22 of which were minor or cosmetic (rating <= 2), such as alignments, label names, or font sizes, which are straightforward to be addressed. The remaining 9 problems are reported in Table 2, in order of severity. In summary, the most important problems revolved around common practices in usability testing and suggestions of functionality to be added, such as evaluation report templates, statistical analysis, or questionnaires support.

Table 2. Heuristic evaluation results

Problem	Severity
Statistical analysis functionality could be included, or at least the possibility to export data in a spreadsheet format	3.00
In the system responses area, it is not visible how often a path was followed	3.00
Insights could also be provided per user group	2.67
User testing also involves questionnaires. It would be good to have some embedded, as well as tools to analyze them	2.67
Charts could constitute illustrations of the usability evaluation report that the evaluator will prepare, so it would be useful to be able to export them to images	2.67
Regarding the evaluation report, the tool could provide a template pre-filled with data to facilitate its preparation	2.67
The tool could allow evaluators to select variables and statistics, if one does not want to explore their full set	2.33
The chart illustrating the number of errors over time does not include measurement units, making it difficult to understand	2.33
The system responses path in the insights view is very small for the evaluator to have a good overview of the paths (in a zoomed-out mode but still readable). It could be larger in height	2.33

A limitation of the study is that the tool was evaluated with experts and not following a user testing approach. However, this constituted the preliminary evaluation and a full-scale user-based experiment will be carried out, as soon as the heuristic evaluation findings have been addressed. Nevertheless, since the heuristic evaluation process is targeted at locating problems, evaluators - who actually constitute the target user group - were also debriefed and asked about impressions. All provided positive feedback, highlighting that the tool would be very useful in carrying out user-based tests through the automated measurements and charts it provides. Features that were also appraised were the vertical session timeline, providing contextual information and the insights information for the entire experiment, as well as per system and application.

4 Conclusions and Future Developments

UXAmI Observer is a tool that assists evaluators in conducting user-based experiments in real or simulated AmI spaces and supports them through visualizations of the data acquired for each experiment participant, as well as with aggregated statistical and usage information from all the participants. In summary, the tool supports: task-based and free usage user-based experiments; detailed analysis of a participant's interaction and aggregated insights from all the participants; automatically acquired measurements regarding adaptations applied and adaptation rejection, user input errors, detected user emotions and implicit interactions; manual annotation of POIs by the evaluator and verification of the automated measurements; synchronized view of all the POIs (manual or automatic) with the session videos; powerful visualizations of the interaction timeline, interaction statistics, system responses path, system usage statistics, and floorplan-based POIs; as well as comparisons among experiments.

An important contribution of the tool is that it does not require any instrumentation, in contrast to previous efforts in the field of usability evaluation automation. More specifically, neither the developers, nor the evaluator are required to invest effort towards instrumenting the software to collect usage data. All that is required is the development of an agent that will receive the events' information and transform it to UXAmI compatible format. Other innovative features of the tool include:

- The discrimination between input errors and interaction errors, with input errors being automatically detected.
- Detection of adaptations inserted and rejected.
- Implicit interactions recognition.
- Association of POIs with the space floorplan.
- Placing observed and automatically acquired metrics in context, allowing the evaluators to better understand the entire user experience.

Future work will address the issues that have been highlighted from the heuristic evaluation, and in particular: provide insights per user category, support questionnaires, facilitate exporting data and images, as well as support evaluation report templates with pre-filled data. In addition, although some tests have already been carried out, further tests with more users involved in long-term experiments are required regarding the validity of the tool's inference mechanisms and its usability. Finally, two new research directions will be explored: (1) the production of rules that could be fed back to the AmI environment, automatically by the tool and manually by the evaluator; and (2) incorporation of deep learning techniques towards automatic recognition of the applications with which the user is currently interacting.

Acknowledgment. This work is supported by the FORTH-ICS internal RTD Programme "Ambient Intelligence Environments".

References

1. Stephanidis, C.: Human factors in ambient intelligence environments. In: Salvendy, G. (ed.) Handbook of Human Factors and Ergonomics, 4th edn., pp. 1354–1373. John Wiley and Sons (2012)
2. Augusto, J.C., Nakashima, H., Aghajan, H.: Ambient intelligence and smart environments: a state of the art. In: Nakashima, H., Aghajan, H., Augusto, J.C. (eds.) Handbook of Ambient Intelligence and Smart Environments, pp. 3–31. Springer, US (2010)

3. MacDonald, C.M., Atwood, M.E.: Changing perspectives on evaluation in HCI: past, present, and future. In: CHI 2013 Extended Abstracts on Human Factors in Computing Systems, pp. 1969–1978. ACM, New York, NY, USA (2013)
4. Martins, A.I., Queirós, A., Silva, A.G., Rocha, N.P.: Usability evaluation methods: a systematic review. In: Saeed, S., Bajwa, I.S., Mahmood, Z. (eds.) Human Factors in Software Development and Design, pp. 250–273. IGI Global (2015)
5. Paz, F., Pow-Sang, J.A.: Current trends in usability evaluation methods: a systematic review. In: 2014 7th International Conference on Advanced Software Engineering and Its Applications (ASEA), pp. 11–15. IEEE (2014)
6. Nielsen, J.: Usability engineering. Elsevier (1994)
7. Hornbæk, K.: Current practice in measuring usability: challenges to usability studies and research. Int. J. Hum. Comput. Stud. **64**(2), 79–102 (2006)
8. Hertzum, M., Jacobsen, N.E.: The evaluator effect: a chilling fact about usability evaluation methods. Int. J. Hum. Comput. Interact. **13**(4), 421–443 (2001)
9. Hornbæk, K., Law, E.L.C.: Meta-analysis of correlations among usability measures. In: Proceedings of the SIGCHI Conference on Human Factors in Computing Systems (CHI '07), pp. 617–626. ACM (2007)
10. Bevan, N.: What is the difference between the purpose of usability and user experience evaluation methods. In: Proceedings of the Workshop UXEM, vol. 9, pp. 1–4 (2009)
11. Bargas-Avila, J.A., Hornbæk, K.: Old wine in new bottles or novel challenges: a critical analysis of empirical studies of user experience. In: Proceedings of the SIGCHI Conference on Human Factors in Computing Systems (CHI '11), pp. 2689–2698. ACM (2011)
12. Maia, C.L.B., Furtado, E.S.: A systematic review about user experience evaluation. In: Marcus, A. (ed.) Design, User Experience, and Usability: Design Thinking and Methods, DUXU 2016. Lecture Notes in Computer Science, vol. 9746, pp. 445–455. Springer, Cham (2016)
13. Gaggioli, A.: Optimal experience in ambient intelligence. In: Riva, G., Vatalaro, F., Davide, F., Alcañiz, M. (eds.) Ambient Intelligence, pp. 35–43. IOS Press (2005)
14. Poppe, R., Rienks, R., van Dijk, B.: Evaluating the future of HCI: challenges for the evaluation of emerging applications. In: Huang, T.S., Nijholt, A., Pantic, M., Pentland, A. (eds.) Artificial Intelligence for Human Computing. Lecture Notes in Artifical Intelligence, vol. 4451, pp. 234–250. Springer (2007)
15. Ivory, M.Y., Hearst, M.A.: The state of the art in automating usability evaluation of user interfaces. ACM Comput. Surv. (CSUR) **33**(4), 470–516 (2001)
16. Hilbert, D.M., Redmiles, D.F.: Extracting usability information from user interface events. ACM Comput. Surv. (CSUR) **32**(4), 384–421 (2000)
17. Macleod, M., Rengger, R.: The development of DRUM: a software tool for video-assisted usability evaluation. In: Alty, J.L., Diaper, D., Guest, S. (eds.) People and Computers, vol. VIII, p. 293. Cambridge University Press (1997)
18. Sears, A.: Heuristic walkthroughs: finding the problems without the noise. Int. J. Hum. Comput. Inter. **9**(3), 213–234 (1997)
19. Lecerof, A., Paternò, F.: Automatic support for usability evaluation. IEEE Trans. Softw. Eng. **24**(10), 863–888 (1998)
20. Kim, J.H. et al.: Tracking real-time user experience (TRUE): a comprehensive instrumentation solution for complex systems. In: Proceedings of the SIGCHI Conference on Human Factors in Computing Systems, pp. 443–452. ACM (2008)
21. Heilbrunn, B., Herzig, P., Schill, A.: Tools for gamification analytics: a survey. In: Proceedings of the 2014 IEEE/ACM 7th International Conference on Utility and Cloud Computing (UCC), pp. 603–608. IEEE (2014)

22. Bateman, S., Gutwin, C., Osgood, N., McCalla, G.: Interactive usability instrumentation. In: Proceedings of the 1st ACM SIGCHI Symposium on Engineering Interactive Computing Systems, pp. 45–54. ACM (2009)
23. Chang, E., Dillon, T.S.: Automated usability testing. In: Howard, S., Hammond, J., Lindgaard, G. (eds.) Human-Computer Interaction INTERACT'97, pp. 77–84. Springer (1997)
24. Alexander, J., Cockburn, A., Lobb, R.: AppMonitor: a tool for recording user actions in unmodified Windows applications. Behav. Res. Methods **40**(2), 413–421 (2008)
25. Ma, X., et al.: Design and implementation of a toolkit for usability testing of mobile apps. Mob. Networks Appl. **18**(1), 81–97 (2013)
26. Tran, C.D., Ezzedine, H., Kolski, C.: EISEval, a generic reconfigurable environment for evaluating agent-based interactive systems. Int. J. Hum Comput Stud. **71**(6), 725–761 (2013)
27. Assila, A., de Oliveira, K.M., Ezzedine, H.: An environment for integrating subjective and objective usability findings based on measures. In: 2016 IEEE Tenth International Conference on Research Challenges in Information Science (RCIS), pp. 1–12. IEEE (2016)
28. Muhi, K., Szőke, G., Fülöp, L.J., Ferenc, R., Berger, Á.: A Semi-automatic usability evaluation framework. In: Murgante, B. et al. (eds.) Computational Science and Its Applications–ICCSA 2013. Lecture Notes in Computer Science, vol. 7972, pp. 529–542. Springer, Heidelberg (2013)
29. Christensen, L., Frøkjær, E.: Distributed usability evaluation: enabling large-scale usability evaluation with user-controlled instrumentation. In: Proceedings of the 6th Nordic Conference on Human-Computer Interaction: Extending Boundaries (NordiCHI 2010), ACM, pp. 118–127 (2010)
30. Bao, L., et al.: Extracting and analyzing time-series HCI data from screen-captured task videos. Empirical Softw. Eng. **22**(1), 134–174 (2017)
31. Howarth, J., Smith-Jackson, T., Hartson, R.: Supporting novice usability practitioners with usability engineering tools. Int. J. Hum. Comput. Stud. **67**(6), 533–549 (2009)
32. Morae: Morae 3 Tutorial (2017). Retrieved from: https://www.techsmith.com/tutorial-morae-current.html
33. Norman, K.L., Panizzi, E.: Levels of automation and user participation in usability testing. Interact. Comput. **18**(2), 246–264 (2005)
34. Alspaugh, S., Ganapathi, A., Hearst, M.A., Katz, R.: Better logging to improve interactive data analysis tools. In: Proceedings of the ACM SIGKDD Workshop on Interactive Data Exploration and Analytics (IDEA 2014), pp. 19–25 (2014)
35. Au, F.T., Baker, S., Warren, I., Dobbie, G.: Automated usability testing framework. In: Proceedings of the Ninth Conference on Australasian User Interface (AUIC 2008), vol. 76, pp. 55–64. Australian Computer Society, Inc. (2008)
36. Cook, D.J., Augusto, J.C., Jakkula, V.R.: Ambient intelligence: Technologies, applications, and opportunities. Pervasive Mob. Comput. **5**(4), 277–298 (2009)
37. Augusto, J.C., McCullagh, P.: Ambient intelligence: Concepts and applications. Comput. Sci. Inf. Syst. **4**(1), 1–27 (2007)
38. Nielsen, J.: Success Rate: The Simplest Usability Metric (2001). Retrieved from: https://www.nngroup.com/articles/success-rate-the-simplest-usability-metric/
39. Nielsen, J.: Heuristic evaluation. Usability Inspection Methods **17**(1), 25–62 (1994)

Research on the Degradation of Indian Regional Navigation Satellite System Based on STK

Shaochi Cheng[1(✉)], Yuan Gao[1,3], Xiangyang Li[1], and Su Hu[2]

[1] PLA Academy of Military Science, Beijing 100091, China
success.no.1@163.com
[2] University of Electrical Science and Technology of China, Sichuan 610054, China
[3] State Key Laboratory on Microwave and Digital Communications, National Laboratory for Information Science and Technology, Tsinghua University, Beijing 100084, China

Abstract. Indian Regional Navigation Satellite System (IRNSS) is a very important space sensor network for navigation in India. In January 2017, IRNSS-1A which was one of the seven satellites of IRNSS lost the ability to provide position information because the three Rubidium atomic clocks on board IRNSS-1A failed. This event had a bad effect on the positioning accuracy of IRNSS. In this paper, we use STK to give a simulation about IRNSS, compare the positioning accuracy of IRNSS before and after IRNSS-1A failed, and analyze the degradation of IRNSS.

Keywords: Space sensor network · Indian Regional Navigation Satellite System STK · Degradation

1 Introduction

In April 2016, India established its independent space-based navigation sensor network – Indian Regional Navigation Satellite System (IRNSS) which consisted of seven satellites [1]. According to Indian Space Research Organization which developed IRNSS, IRNSS could provide position information in the rectangle from Latitude 30° South to 50° North, Longitude 30° East to 130° East which is denoted as the service area of IRNSS [2]. In July 2016, one of the three Rubidium atomic clocks on board IRNSS-1A failed. The other two followed subsequently. This event caused IRNSS-1A to lose the ability to provide position information. On 31 August 2017, India launched IRNSS-1H to replace IRNSS-1A, but because the heat shield failed to separate, the mission was unsuccessful [3]. Up to now, there is no definite plan to replace IRNSS-1A. So in the future, IRNSS could include only six satellites for a long time, which would have a bad effect on the positioning accuracy of IRNSS.

Since the Indian government decided to develop IRNSS in May 2006, researchers have paid more and more attention to the studies on the positioning accuracy of IRNSS. Raghu et al. placed an IRNSS receiver at school of engineering and technology,

This work is funded by National Natural Science Foundation of China (61701503).

K. Arai et al. (Eds.): IntelliSys 2018, AISC 868, pp. 1371–1380, 2019.
https://doi.org/10.1007/978-3-030-01054-6_95

Bangalore, collected data from IRNSS satellites, and used MATLAB to measure the parameters like standard deviation, RMS error, and mean position errors for the position X, Y and Z of the receiver when only three IRNSS satellites were in orbit [4]. In another paper, Raghu et al. used STK to analyze the visibility gap for IRNSS and hybrid (GPS + IRNSS) navigational satellites over the user position (SET-Jain University), and mapped the visibility, elevation, and distance within about 1 week when there were four IRNSS satellites [5]. Gayathri et al. described a methodology for performance analysis of IRNSS receiver on a probe vehicle, collected data from the receiver every second throughout the test field, and then analyzed the signal strength and accuracy of IRNSS when there were five IRNSS satellites [6]. Desai et al. used Accord IRNSS receiver provided by SAC, ISRO, Ahmedabad at SVNIT Surat (Latitude 21.16° North, Longitude 72.78° East) to collect satellite ranging as well as the navigation information for IRNSS and GPS, and then applied MATLAB to analyze all DOP parameters (GDOP, PDOP, HDOP, VDOP and TDOP) of IRNSS, GPS and IRNSS + GPS systems when six IRNSS satellites were in orbit [7]. Safia et al. presented an algorithm to estimate coverage area of GPS and IRNSS, and investigated the earth's surface coverage of IRNSS 5 and 7 satellite vehicles when there were seven IRNSS satellites [8].

This paper aims to compare the positioning accuracy of seven IRNSS satellites with that of six IRNSS satellites when IRNSS-1A failed, and provide a simulation analysis about the degradation of IRNSS. The rest of this paper is organized as follows. In Sect. 2, we give the description of IRNSS satellites, Sect. 3 introduces the Satellite/Systems Tool Kit (STK) and its module to analyze the positioning accuracy of satellites, Sect. 4 uses STK to provide a simulation about IRNSS and an analysis of the degradation of IRNSS, and Sect. 5 is the conclusion of this work.

2 IRNSS Satellites

IRNSS consists of space segment, ground segment and user segment. Seven satellites constitute the space segment of IRNSS. The date of launch and class of orbit of IRNSS satellites are described as follows [9–11].

Each IRNSS satellite has three Rubidium atomic clocks, which are integral elements for timekeeping. Because of the malfunction of three Rubidium atomic clocks onboard, IRNSS-1A cannot provide time information. The positioning principle of navigation satellites can be described as follows [12] (Table 1):

$$\begin{cases} \sqrt{(x^{(1)}-x)^2+(y^{(1)}-y)^2+(z^{(1)}-z)^2}+\delta t_u=\rho_c^{(1)} \\ \sqrt{(x^{(2)}-x)^2+(y^{(2)}-y)^2+(z^{(2)}-z)^2}+\delta t_u=\rho_c^{(2)} \\ \cdots \\ \sqrt{(x^{(N)}-x)^2+(y^{(N)}-y)^2+(z^{(N)}-z)^2}+\delta t_u=\rho_c^{(N)} \end{cases} \quad (1)$$

Table 1. IRNSS satellites

Satellite	Date of Launch	Orbit	Status
IRNSS 1A	2013/7/1	GSO/55° E, 29° inclined	Failed
IRNSS 1B	2014/4/4	GSO/55° E, 29° inclined	Operational
IRNSS 1C	2014/10/15	GSO/83° E, 5° inclined	Operational
IRNSS 1D	2015/3/30	GSO/111.75° E, 31° inclined	Operational
IRNSS 1E	2016/1/20	GSO/111.75° E, 29° inclined	Operational
IRNSS 1F	2016/3/10	GSO/32.5° E, 5° inclined	Operational
IRNSS 1G	2016/4/28	GSO/129.5° E, 5.1° inclined	Operational
IRNSS 1H	2016/8/31		Failed

We denote $(x^{(N)}, y^{(N)}, z^{(N)})$ as the three-dimensional components of the position of satellite N which can be known based on satellite ephemeris. (x, y, z) are the three-dimensional components of the position of the receiver which need to be calculated. δt_u is the clock correction of the receiver which is unknown. $\rho_c^{(N)}$ is pseudo-range:

$$\rho = c \cdot (t_u - t_s) \tag{2}$$

$c = 3 \times 10^8\ \mathrm{m/s}$, t_u is the time of the receiver which is known, t_s is the time of the satellite. In general, t_s can be calculated by atomic clocks on board satellites. And then $\rho_c^{(N)}$ can be calculated by (2). If there are 4 satellites whose $(\mathrm{x}^{(N)}, y^{(N)}, z^{(N)})$ and $\rho_c^{(N)}$ are known, (x, y, z) and δt_u would be calculated by (1), and the receiver can get its position data. But if the atomic clocks of some satellite fail, $\rho_c^{(N)}$ cannot be calculated, and 4 satellites cannot give position data.

3 STK Introduction

STK is an advanced system analysis software which can provide a simulation about satellite constellation, aircraft, missile, ship, ground vehicle, etc. It can calculate position and velocity values of the objects simulated, analyze dynamic change of their relative position. Based on the calculating and analyzing results, STK can give all kinds of 2D and 3D visual dynamic scenarios and accurate reports and graphs [13].

As for space engineering, STK can create a scenario including satellites on all kinds of orbits and other vehicles by ground, sea and air, and then add sensors such as radar, antenna and transmitter to these objects. In addition, it can give the scenario a set of constraints such as sensors' cone angle, receivers' elevation angle, sun elevation angle, etc. Based on the scenario, STK can simulate the process in which satellites and other vehicles move and their relative positions change over time. STK can demonstrate dynamic change of satellites' coverage for other vehicles or a given area, check coverage situation, assess communication quality or calculate positioning accuracy of satellites [14, 15].

With respect to satellite navigation, STK can use dilution of precision (DOP) to assess the positioning accuracy of navigation satellite system. It is defined as follows [16]:

$$\sqrt{\sigma_x^2 + \sigma_y^2 + \sigma_z^2} = \sigma_{URE} \times PDOP \tag{3}$$

$$\sqrt{\sigma_x^2 + \sigma_y^2 + \sigma_z^2 + \sigma_t^2} = \sigma_{URE} \times GDOP \tag{4}$$

$\sigma_x, \sigma_y, \sigma_z, \sigma_t$ are the three-dimensional components of position error and clock-related error which are expected to be as small as possible. σ_{URE} is range error of the receiver which mainly depends on clock correction and ionospheric delay. So PDOP and GDOP are linked to the positioning accuracy of navigation satellite system. If we denote G as

$$G = \begin{bmatrix} \frac{x - x^{(1)}}{r^{(1)}} & \frac{y - y^{(1)}}{r^{(1)}} & \frac{z - z^{(1)}}{r^{(1)}} & 1 \\ \frac{x - x^{(2)}}{r^{(2)}} & \frac{y - y^{(2)}}{r^{(2)}} & \frac{z - z^{(2)}}{r^{(2)}} & 1 \\ \frac{x - x^{(3)}}{r^{(3)}} & \frac{y - y^{(3)}}{r^{(3)}} & \frac{z - z^{(3)}}{r^{(3)}} & 1 \\ \frac{x - x^{(4)}}{r^{(4)}} & \frac{y - y^{(4)}}{r^{(4)}} & \frac{z - z^{(4)}}{r^{(4)}} & 1 \end{bmatrix} \tag{5}$$

(x, y, z) are the three-dimensional components of the position of the receiver. $x^{(N)}, y^{(N)}, z^{(N)} (N = 1, 2, 3, 4)$ are the three-dimensional components of the position of satellite N which can be access to the receiver. $r^{(N)} (N = 1, 2, 3, 4)$ are the distance between the receiver and satellite N. Based on some simplifying assumptions, we can get this equation:

$$\begin{bmatrix} \sigma_x^2 & & & \\ & \sigma_y^2 & & \\ & & \sigma_z^2 & \\ & & & \sigma_t^2 \end{bmatrix} = (G^T G)^{-1} \sigma_{URE}^2 \tag{6}$$

We can use $(G^T G)^{-1}$ to calculate PDOP and GDOP, and STK has a DOP module which can calculate PDOP and GDOP values of navigation satellite system for an area of interest during a given time period, and demonstrate distribution of PDOP and GDOP values in reports and graphs.

Different PDOP and GDOP values indicate different levels of positioning accuracy, rating based on DOP value can be described as follows (Table 2):

Table 2. Rating based on DOP value

DOP value	Rating
<1	Ideal
1–2	Excellent
2–5	Good
5–10	Moderate
10–20	Fair
>20	Poor

4 Simulation Based on STK

We create two scenarios including IRNSS satellites, and denote the scenario including seven IRNSS satellites as scenario1, the scenario including six IRNSS satellites as scenario2. The orbital parameters of seven satellites in scenario1 are based on NORAD Two-Line Element Sets downloaded on 27 October 2016 when no IRNSS satellite failed, and the orbital parameters of six satellites in scenario2 are based on NORAD Two-Line Element Sets downloaded on 9 January 2018 [17] (Figs. 1 and 2).

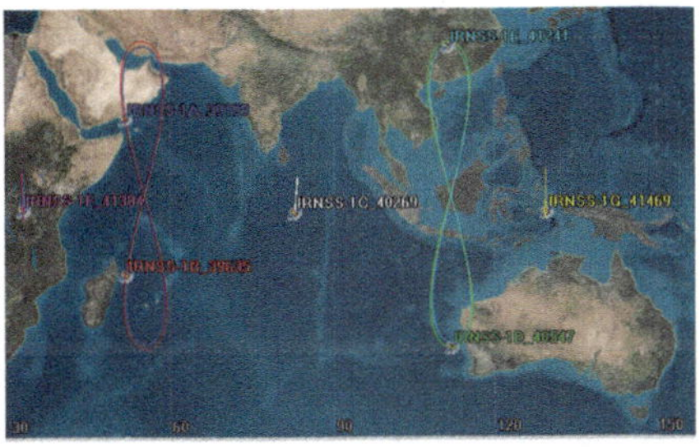

Fig. 1. Scenario1 of 7 IRNSS satellites.

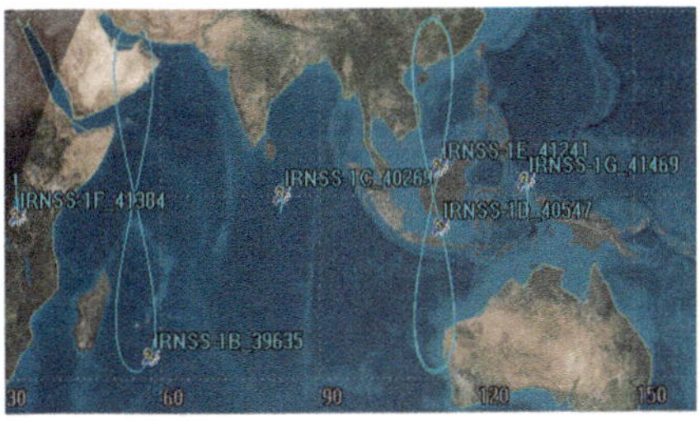

Fig. 2. Scenario2 of 6 IRNSS satellites.

The analysis period of scenario1 is from 27 Oct 2016 04:00:00.000 UTCG to 28 Oct 2016 04:00:00.000 UTCG, and step size is 60 s. The analysis period of scenario2 is from 10 Jan 2018 04:00:00.000 UTCG to 11 Jan 2018 04:00:00.000 UTCG, and step size is 60 s. The rectangle from Latitude 30° South to 50° North, Longitude 30° East to 130° East which is the service area of IRNSS is denoted as coverage analysis area of scenario1 and scenario2, and grid point granularity is 0.5° Lat/Lon. The minimum elevation angle is defined as 5°.

We add one sensor on each satellite and denote the cone half angle of each sensor as 23°. The coverage of satellites' sensors on the service area of IRNSS in scenario1 and scenario2 can be computed by STK. The area which cannot get access to at least four satellites would not get position data.

In Fig. 3, except for the yellow area which is covered by three satellites, most of the service areas of IRNSS are covered by at least four IRNSS satellites, which indicate IRNSS can provide position data to these areas. However, in Fig. 4, the yellow area in the northwest and the red area in the north and west are covered by less than four IRNSS satellites, which means these areas cannot get IRNSS's position data, but the Indian continent is still covered by six IRNSS satellites. Comparing Fig. 3 with Fig. 4, we can see that because of the breakdown of IRNSS-1A, part of the north and west of the service area of IRNSS cannot receive position information, but the Indian continent could still get position data. This means IRNSS's position ability degrades to a certain extent, but not to a large extent.

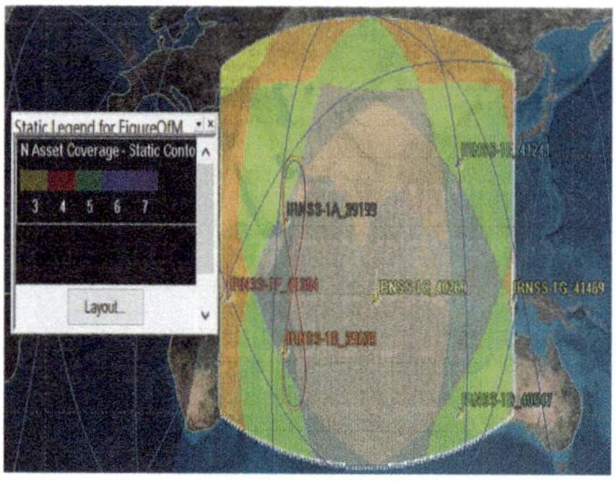

Fig. 3. Coverate of 7 IRNSS satellites.

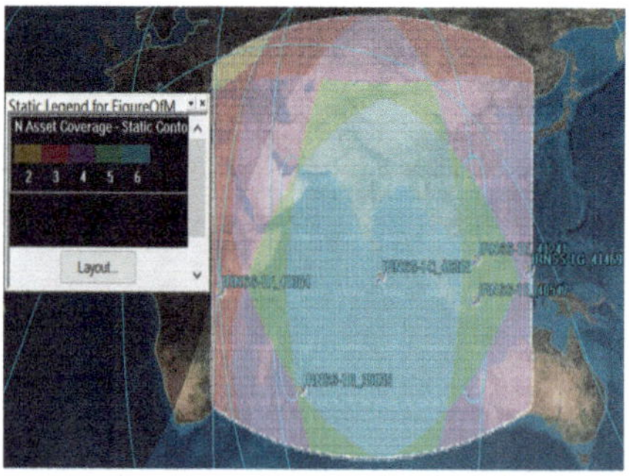

Fig. 4. Coverate of 6 IRNSS satellites.

In order to assess the positioning accuracy of IRNSS, we use STK's DOP module to calculate GDOP and PDOP in the service area of IRNSS. In scenario2, GDOP values by latitude and longitude are shown in the figures below.

In Fig. 5, GDOP values fluctuate fiercely in the area between latitude 20° South and 10° North as well as latitude 30° North and 50° North. And in Fig. 6, the situation is similar in the area between longitude 30° East and 40° East as well as longitude 100° East and 130° East. This means although IRNSS can provide position data in the service area, the positioning accuracy in different regions varies widely. Actually, from Figs. 3 and 4, we can see that the Indian continent and its southern waters are covered by six or seven IRNSS satellites; these areas can get accurate positioning data, and are IRNSS's primary service area.

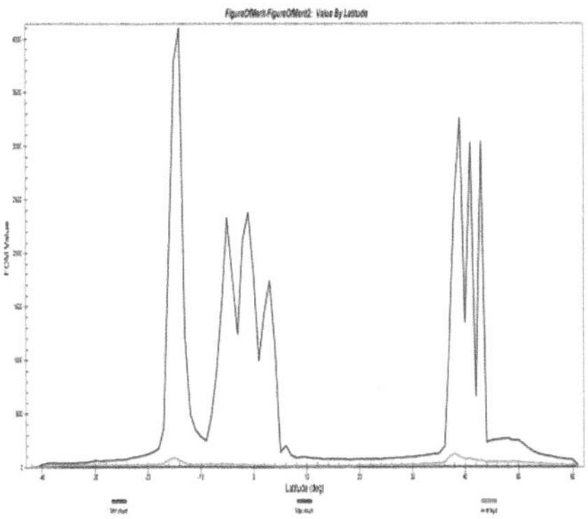

Fig. 5. GDOP in the service area of 6 IRNSS satellites by latitude.

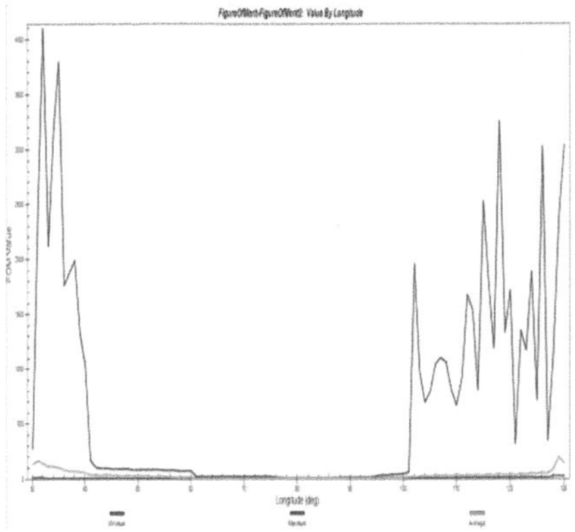

Fig. 6. GDOP in the service area of 6 IRNSS satellites by longitude.

We denote the rectangle from Latitude 8° North to 38° North, Longitude 68° East to 98° East as coverage analysis area of scenario1 and scenario2, and call it primary service area. The Indian continent lies in the rectangle. GDOP values of seven IRNSS satellites in the primary service area by latitude and longitude are shown in Figs. 7 and 8.

Based on Figs. 7 and 8, we can see that GDOP values of seven IRNSS satellites in the primary service area are less than four which is a good level. Furthermore, GDOP values increase with the increase of latitude, because the orbits of seven IRNSS satellites are GSO which are near to the equatorial plane, if a receiver is in the lower latitude, IRNSS satellites are more likely to surround it, giving a good GDOP value [18].

In addition, GDOP values by longitude have a symmetrical distribution in Fig. 8, because seven IRNSS satellites are symmetrically distributed by longitude above the primary service area.

GDOP of six IRNSS satellites in the primary service area by latitude and longitude are shown in Figs. 9 and 10.

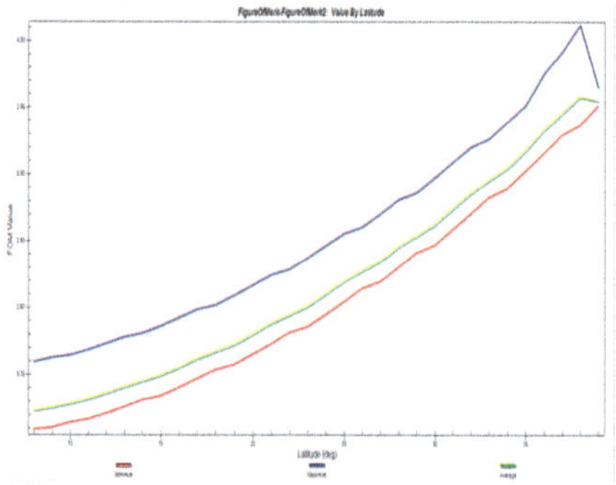

Fig. 7. GDOP of 7 IRNSS satellites in the primary service area by latitude.

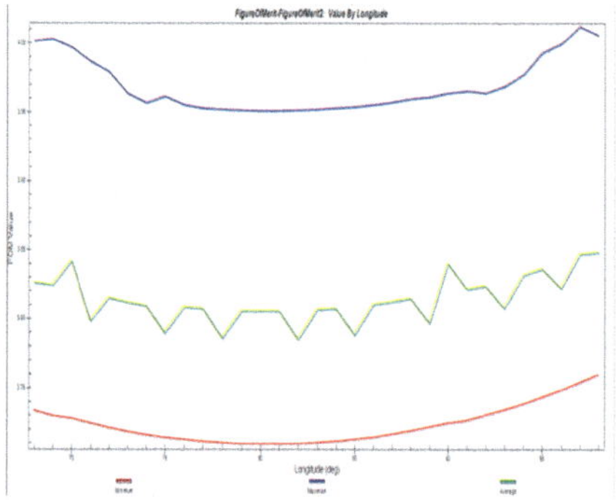

Fig. 8. GDOP of 7 IRNSS satellites in the primary service area by longitude.

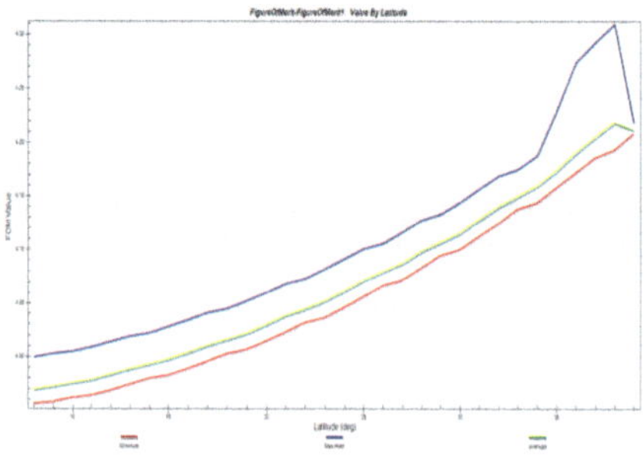

Fig. 9. GDOP of 6 IRNSS satellites in the primary service area by latitude.

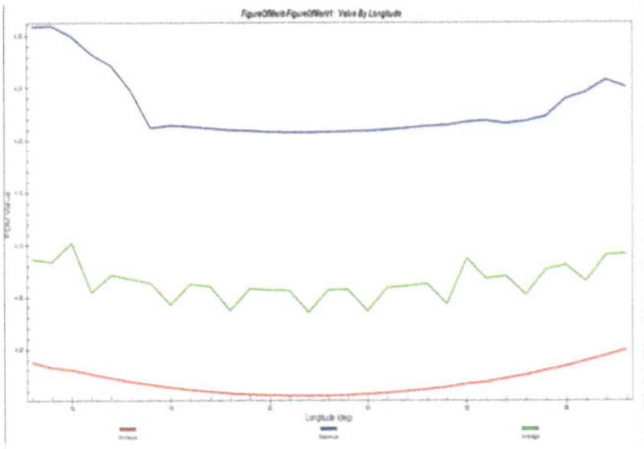

Fig. 10. GDOP of 6 IRNSS satellites in the primary service area by longitude.

Comparing Figs. 7 and 9, we can see that the changing curves by latitude in Fig. 7 resemble those in Fig. 9, and GDOP values in Figs. 7 and 9 are all less than 5, but GDOP values in Fig. 7 are smaller than those in Fig. 9, which indicates that because of the lack of IRNSS-1A, the positioning accuracy of IRNSS in the primary service area decreases, but is still on a good level.

Comparing Figs. 8 and 10, we can see that the blue line in Fig. 10 is not symmetrical, and the left part has higher GDOP values. This means the breakdown of IRNSS-1A has worse influence on the western part of the primary service area than the eastern part, because IRNSS-1A was deployed above the left part of the primary service area. However, the influence is not destructive, because GDOP values are still less than five.

5 Conclusion

In this paper, we use STK to provide a simulation about Indian space-based navigation sensor network – Indian Regional Navigation Satellite System (IRNSS), describe IRNSS satellites' coverage for the service area and the primary service area before and after IRNSS-1A failed, give a quantitative analysis of the change of positioning accuracy, and analyze the degradation of IRNSS's position ability. To put it in a nutshell, the lack of IRNSS-1A reduces IRNSS's service area especially in the west and north, but the decrease is limited, most of the service area still can get position data. In the primary service area, GDOP values are higher than before without IRNSS-1A. This causes the positioning accuracy of IRNSS to decrease, and the degradation in the west is worse than that in the east, but GDOP values are all less than five which is a good level of positioning accuracy, IRNSS still can provide accurate position data.

Acknowledgment. The work of Su Hu was jointly supported by the MOST Program of International S&T Cooperation (Grant No. 2016YFE0123200), National Natural Science Foundation of China (Grant No. 61471100/61101090/61571082), Science and Technology on Electronic Information Control Laboratory (Grant No. 6142105040103) and Fundamental Research Funds for the Central Universities (Grant No. ZYGX2015J012/ZYGX2014Z005). The author would also like to thank all the reviewers; their suggestions have helped improving the work a lot.

References

1. The Hindu: India's very own GPS is ready with seventh navigation satellite launch [EB/OL], 09 January 2018. http://www.thehindu.com/sci-tech/science/irnss-launch-indias-own-regional-navigation-satellite-system-takes-shape/article8531388.ece
2. Indian Space Research Organization: Indian Regional Navigation Satellite System (IRNSS): NavIC [EB/OL], 09 January 2018. https://www.isro.gov.in/irnss-programme
3. The Economic Times: ISRO says launch of navigation satellite IRNSS-1H unsuccessful [EB/OL], 09 January 2018. https://economictimes.indiatimes.com/news/science/isros-irnss-1h-launch-a-big-leap-for-private-sector-in-space/articleshow/60309043.cms
4. Raghu, N., Manjunatha, K.N., Kiran, B.: Determination and preliminary analysis of position accuracy on IRNSS satellites. In: 2016 International Conference on Communication and Signal Processing (ICCSP), pp. 0765–0769 (2016)
5. Raghu, N., Kiran, B., Manjunatha, K.N.: Tracking of IRNSS, GPS and hybrid satellites by using IRNSS receiver in STK simulation. In: 2016 International Conference on Communication and Signal Processing (ICCSP), pp. 0891–0896 (2016)
6. Gayathri, K.M. Thangadurai, N., Vasudha, M.P.: Performance analysis of IRNSS receiver signal strength and accuracy on a moving vehicle. In: 2016 International Conference on Advanced Communication Control and Computing Technologies (ICACCCT), pp. 616–623 (2016)
7. Desai, M.V., Jagiwala, D., Shah, S.N.: Impact of dilution of precision for position computation in Indian regional navigation satellite system. In: 2016 International Conference on Advances in Computing, Communications and Informatics (ICACCI), pp. 980–986 (2016)
8. Safia, Md. A.K., Sultana, Q.: Analysis of earth's coverage of IRNSS and GPS augmented IRNSS. In: 2017 2nd International Conference for Convergence in Technology (I2CT), pp. 146–150 (2017)
9. Gunter's Space Page: IRNSS 1A, …, 1 K [EB/OL], 09 January 2018. http://space.skyrocket.de/doc_sdat/irnss.htm
10. Union of Concerned Scientists: UCS Satellite Database [EB/OL], 10 January 2018. https://www.ucsusa.org/nuclear-weapons/space-weapons/satellite-database#.WlYE-ZDfwVI
11. WIKIPEDIA: Indian Regional Navigation Satellite System [EB/OL], 10 January 2018. https://en.wikipedia.org/wiki/Indian_Regional_Navigation_Satellite_System#cite_note-41
12. Kaplan, E.D., Hegarty, C.J.: Understanding GPS: Principles and Applications, 2nd edn. Publishing House of Electronics Industry, Beijing
13. Analytical Graphics, Inc: Fundamentals of STK 10 (2012)
14. Qi, J., Li, Z., Liu, G.: Research on coverage and link of multi-layer satellite network based on STK. In: IEEE 2015 10th International Conference on Communications and Networking in China (ChinaCom), pp. 410–415 (2015)
15. Nag, S., LeMoigne, J., Miller, D.W., de Weck, O.: A framework for orbital performance evaluation in Distributed Space Missions for earth observation. In: 2015 IEEE Aerospace Conference, pp. 1–20 (2015)
16. Langley, R.B.: Dilution of Precision. GPS World, May 1999
17. North American Aerospace Defense Command: NORAD Two-Line Element Sets Current Data [EB/OL], 09 January 2018. http://celestrak.com/NORAD/elements/geo.txt
18. Xie, G.: Principles of GPS and Receiver Design. Publishing House of Electronics Industry, Beijing

The Application of a Semantic-Based Process Mining Framework on a Learning Process Domain

Kingsley Okoye[1(✉)], Syed Islam[1], Usman Naeem[1], Mhd Saeed Sharif[1], Muhammad Awais Azam[2], and Amin Karami[1]

[1] School of Architecture Computing and Engineering, College of Arts, Technologies and Innovation, University of East London, Docklands Campus, 4-6 University Way, London, E16 2RD, UK
{K.Okoye,Syed.Islam,U.Naeem,S.Sharif,A.Karami}@uel.ac.uk
[2] Faculty of Telecom and Information Engineering, University of Engineering and Technology, Taxila, Pakistan
awais.azam@uettaxila.edu.pk

Abstract. The process mining (PM) field combines techniques from *computational intelligence* which has been lately considered to encompass artificial intelligence (AI) or even the latter, augmented intelligence (AIs) systems, and the data mining (DM) to *process modelling* in order to analyze event logs. To this end, this paper presents a semantic-based process mining framework (SPMaAF) that exhibits high level of accuracy and conceptual reasoning capabilities particularly with its application in real world settings. The proposed framework proves useful towards the extraction, semantic preparation, and transformation of events log from any domain process into minable executable formats – with focus on supporting the further process of discovering, monitoring and improvement of the extracted processes through semantic-based analysis of the discovered models. Practically, the implementation of the proposed framework demonstrates the main contribution of this paper; as it presents a Semantic-Fuzzy mining approach that makes use of labels (i.e. concepts) within event logs about a domain process using a case study of the Learning Process. The paper provides a method which aims to allow for mining and improved analysis of the resulting process models through semantic – labelling (annotation), representation (ontology) and reasoning (reasoner). Consequently, the series of experimentations and semantically motivated algorithms shows that the proposed framework and its main application in real-world has the capacity of enhancing the PM results or outcomes from the syntactic to a much more abstraction levels.

Keywords: Process mining · Process models · Ontology · Semantic annotation Reasoner · AI · Event logs

1 Introduction

The need for novel approaches in design and integration of computational intelligence into everyday (e.g. business, learning) processes has sprout research investigations on

K. Arai et al. (Eds.): IntelliSys 2018, AISC 868, pp. 1381–1403, 2019.
https://doi.org/10.1007/978-3-030-01054-6_96

how to exploit such tools for use in improving the ever increasing data about various organizations. In recent years, a common challenge with many of the business processes has been on how to develop intelligent systems that can provide platforms for exploring the additional, and most often, the monotonous tasks of managing the entire operational process and quality of information - by providing understandable and useful insights on the best possible ways to make the envisioned information explicable in reality. Such process-related analysis, often allied to *process mining*, means there is also need for tools and techniques that can extract valuable information from the event logs about the domain processes in view.

Most organizations have devoted a greater amount of resources towards modelling of their everyday processes or operations. Nevertheless, majority of the mapped processes appears to be non-operational, unfitting and/or presents itself in formats that are targeted towards understanding the process flows (workflows) rather than tackling the complexity of the business or domain processes in reality. In other words, a greater number of the resulting models and methods tends to support just *machine-readable* systems rather than *machine-understandable* systems at large. By machine-understandable systems we refer to methods that are developed not just for representing information in formats that can be easily understood by humans, but also for creating applications and/or systems that trails to inclusively process the information that they contain or supports. Thus, the mapped processes (models) are either semantically-labelled (semantic annotation) to ease the analysis process, or represented in a formal structure (ontology) that permits a computer (through semantic reasoning) to deduce or discover new facts/knowledge as a result of the defined relations or assertions within the ontology.

According to the work of [1–4], the efficiency or quality of the so-called organizations process can be improved if the process analysts and/or owners could perform an accurate analysis or exploration of the extracted event logs. For instance, the process of visualizing the relations or attributes the process instances share amongst themselves in a knowledge base. Of late, the PM field [3] or better still the latter - Process Querying [5] have turn out to be such valuable method that can be used to discover meaningful information from the recorded event logs. Besides, the process mining (PM) field combines techniques from computational intelligence and data mining to process modelling and analysis, as well as several other disciplines related to the business process management (BPM) [3] and AI to analyze the data. However, [6] and [2] notes that a common problem with majority of the PM methods in current literature is that they tend to rely on the tags or labels within the events log to discover the process models, and as consequence, appears to be vague and limited to some extent when confronted with unstructured data. The aforementioned challenge is as a result of lack of abstraction level of analysis (i.e. conceptual information) that are most often required from a real world perspective. Apparently, given such kind of challenge, most of the PM techniques appears to not technically gain from the real-time knowledge (semantics) that describes the tags/labels in event logs of the process domains in question [6].

For this purpose, this paper introduces a framework to help bridge the *semantic gap* displayed by the traditional PM techniques - by providing a method that focus on structuring in a formal way the events logs and resultant models about the domains processes (i.e. sematic-based analysis of the datasets). The work utilizes a case study of the learning process to demonstrate the application of the approach. In short, the work addresses the identified challenges, i.e. - (i) the lack of PM tools/methods that exhibits semantic information extraction, retrieval and/or analysis, and (ii) process mining of events log and the derived models at a more abstraction level as opposed to the syntactic method of analysis displayed by the traditional PM techniques. Primarily, the purpose of the proposed semantic-based PM and model analysis framework (SPMaAF) is a way of introducing formal structures for the events logs used to perform process mining as well as improvement of the analysis of the results to a more abstract level that can be easily understood in a real-world settings. Predominantly, as it concerns the process of extraction, retrieval and conceptual analysis of the derived information (or mapped processes). Technically, the SPMaAF focuses on augmenting the value (machine-understandable system) of information derived from the resultant models by semantically-annotating the individual process entities (instances) with concepts which they represent in the real-world sceneries, and then links them to a well-defined domain ontology so as to allow for improved semantic analysis of the mined datasets. Thus, provide process models that are closer to the domain concepts.

To end with, the study illustrates the application of the proposed SPMaAF framework using the learning process domain case study - to show the importance as well as its comparison against other existing benchmark algorithms used for process mining. Practically, the SPMaAF considers the various phases of the PM techniques and implementation - ranging from the primary stage of collection and transforming the captured events logs to discovering of worthwhile models, then expounds the traditional PM to semantically preparation of the derived models for a more conceptual analysis and/or querying. In terms of abstraction level, the work shows that the SPMaAF is able to provide a formal means to analysing the data sets (i.e. events log or models) by allowing the real world (semantic) meaning of the individual process instances to be improved by making use of semantic technologies - such as the ontology web language (OWL) [7], semantic web rule language (SWRL) [8], description logics (DL) [9], etc. The method utilizes those tools in order to provide semantic information (i.e. through semantic reasoning or inference knowledge discovery) that are closer to human understanding, and yet, machine-understandable. In other words, this paper focuses on establishing the extent to which semantic modelling (ontological representations) and analysis is able to improve the results of the process mining through the *conceptualization* method (i.e. explicit specification of the domain concepts) [5, 10] which, in turn, helps in identifying appropriate semantic knowledge or information about the process instances (elements).

The rest of the paper is structured as follows: in Sect. 2, background information and related areas are discussed. Section 3 explains the SPMaAF including the main methods, components and motivation towards using the semantic-based approach to perform PM. In Sect. 4, the paper presents the different phases of implementing the SPMaAF framework, as well as sets of proposed algorithms that allows for ample implementation of the various phases. Section 5 shows how the work utilizes the case study of Learning Process to illustrate the SPMaAF approach and its main application in real world setting. In Sect. 6, we evaluate the outcome of the experimentations, particularly against other benchmark algorithms used for PM in order to weigh up the impact of the proposed framework, and then, discuss in Sect. 7 the influence and main achievements of the paper including the limitations and threats to validity, and consequently highlights worthwhile areas that could be investigated by extending the approach in this paper. Finally the work concludes and points out the direction for future works in Sect. 8.

2 Background Information and Related Areas

2.1 Patterns Discovery, Classifications and Model Analysis

Process mining (PM) belongs to collection of tools within the Business Intelligence (BI) [3] that utilizes data mining techniques to discover patterns or models from event logs and predict outcomes through further analysis of the discovered models. For example, [11] and [12] introduced various approaches for learning pattern control through a workflow management system (WFM) [3] but does not relate a devoted strategy for the process analysis such as the PM technique. On the other hand, [13] sparks the potential benefits of enhancing learning process models particularly within the context of Adaptive Educational Hypermedia Systems (AEHS) [14] by constructing a framework for exploiting learner models. Their method combines PM techniques with the concepts of learning patterns. Similarly, [15] and [16] have worked on approaches that applies the PM techniques in context of e-learning. The authors analyzed and points out tools that are used to perform PM tasks which qualifies better in support of e-learning processes. Whereas, [17] argues that a way of supporting learners within an e-learning setting is to adopt the combine approach of using the PM with concepts of the discovered learning patterns such as the one described in this paper.

According to [17], to perform the classification or assumptions of learner patterns, datasets need to be extracted from a learning system (i.e. the data sources). Moreover, in view of carrying out the process analysis - the extracted datasets needs to be prepared and transformed into formats that allows for the pattern recognition to follow. Consequently, prior to performing any other further analysis or use of the discovered information, an evaluation of the resulting learning patterns is carried out, and then the results are presented and interpreted as shown in Fig. 1.

As gathered in Fig. 1 such type of process analysis, allied to the PM technique, improves explicit process knowledge. For instance, processes that are complemented by some implicit knowledge which may also be discovered inadvertently [18]. In short, PM could be adopted to analyse and improve the captured processes (e.g. the learning process) and models, or even more, used to recommend future patterns and/or behaviours.

Furthermore, the authors in [17] introduced the PM notion in e-learning settings to highlight the need of the approach in offering the capability of improving information values of such systems. The authors opines that PM in combination with learning concepts could be a promising tool for process automation and modelling. Their arguments are concerned with how learning processes can be improved through the use of process modelling and rule-based controls, as well as how process models can be generated in details taking into account the concept of the learning styles or patterns. Interestingly, the outcome of their research [17] reveals that the implementation of rule-based

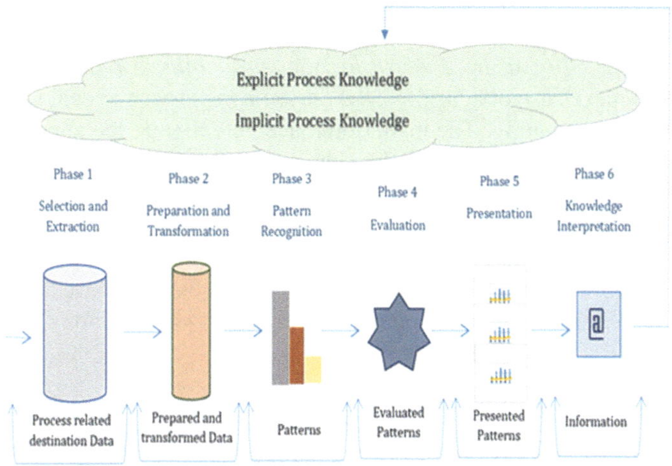

Fig. 1. Process mining and analysis framework

controlled methods (e.g. workflow-nets) [3] into systems that support process management remains an important field of further investigation. For that reason, and to address such gap in literature - this work has introduced the SPMaAF which is perceived as a semantic-based PM approach directed towards the discovery and enhancement of the resulting process models with its application on the learning process domain in order to show the influence and usefulness of the approach. We also explain in this Sects. 2.2 to 2.4 - some of the intelligent functionalities that are attributed to the existing PM techniques and the need for the *advance* notion of the *semantic process mining* that has inspired the development of the SPMaAF framework and its implementation.

2.2 Concepts and Entities Relationship Prediction

A significant feature of Intelligent and Adaptive Educational Learning Systems (IAELS) [14] is the predictive aspect of the systems they support. Reference [19] notes that a pre-emptive educational model seeks to predict possible future activities in order to accomplish better student's training as well as overcome likely issues. Interestingly, the authors propose a Casual and Fuzzy Student Model (CFSM) which describes several user attributes to anticipate the sequencing of activities for the users through the acquisition of domain knowledge about the learners. The main component derived here and pertinent in context of this paper, is the user attributes which are semantically described in form of concepts within an ontology. In essence, the described concepts sets casual relations amongst the individual elements, and as such represents a certainty on how an attribute utilizes the object/datatype properties and/or triggers another attribute.

2.3 Fuzzy Mining and Reasoning

Furthermore, instead of using a *neuro fuzzy* or *probabilistic reasoning* method, the authors in [19] makes use of a *fuzzy casual reasoning* approach to deal with qualitative observations or information. The main difference here, is that the approach organizes *ontologies* which are used in identification and definition of the meanings of the concepts, casual relations, fuzzy rule bases, universe of discourse and other relevant items using the OWL [7] declaration sentences such as *Classes, Datatype Properties, Functional Properties,* and *Process instances* etc. Moreover, ontology firmly sits at the heart of the proposed SPMaAF approach, and a description of how we have utilized and implemented the schema is explained in the later Sects. 4 and 5 of this paper.

2.4 Learning Process Automation

Automated learning process [17] means supplying simulations, contents and interactive maps in a unified and well-structured manner. Moreover, this is where semantic process modelling in combination with the PM techniques is capable of providing useful method for better identification and analysis of learning problems. Hence, the main reason why the proposed SPMaAF framework in this paper is grounded on *process mining* and *semantic process modelling*. The authors in [20] suggests that process automation irrespective of the settings in which they are being used in (be it within the educational settings or any other domain), should be offered as specialized computer-generated services and the analysis have to also integrate some form of business sub-processing tools such as the PM techniques.

2.5 The Shift from Process Mining to Semantic Process Mining and Model Analysis

Process mining (PM) has proven to be one of the existing techniques that can be exploited to analyse event data about any domain process. One of such areas that the PM technique is currently being applied and is gaining attention in recent years is the Educational Process Mining (EPM) [21]. In fact, the EPM is a new domain area within the wider context of business process management (BPM) [3] that aims to apply process mining (PM) techniques to find out user patterns or models from the captured sets of educational data [15, 22]. According to [21], EPM means the application of PM to raw educational data by considering the end to end processes rather than local patterns, as opposed to the Educational Data Mining (EDM) [1] which tends not to be process-centric and do not focus on event data [1, 3] e.g. the rows (instances) and columns (variables) of a typical data file which does not have any meaning. Nonetheless, [21] notes that both the EDM and EPM apply specific algorithms to data in order to discover hidden patterns and/or relationships. In fact, whichever tool one chooses to adopt, the key focus should be on achieving the purpose of adopting such techniques.

A number of researchers have directed their work towards the use and application of this new advanced aspect of PM within the educational settings [4, 15, 22–25]. According to [22] EPM emerges from the educational data mining (EDM) discipline, and the drive for its incentive is primarily to discover, analyse and improve the educational processes based on the hidden or unobserved information within the events log recorded by the IT systems. In [4] and [22] are even more specific about the existing methods for extracting models within the EPM which are limited to some extent because the approaches depend on traditional PM techniques that are purely syntactic in nature (i.e. based on labels in event logs) to discover process models. Technically, in so doing, the developed systems tend not to gain from the real knowledge (semantics) that describes the processes as performed in reality, and as consequence, the actual semantics behind the event logs remains missing and sprouts the additional need for domain experts and even time to interpret them.

Therefore, in practice, PM poses some certain issue of semantics that limits its efficiency when handling large volume of events log from the complex educational systems [2, 22, 24]. The authors in [22] argue that the *semantic process mining* [2, 24] approach appears to be a promising area that can be explored in order to resolve those issues of understanding the different patterns or trace heterogeneity, and as such, to extract streamlined models that fits or represents the actual processes as performed in reality. Moreover, [22] believes that *semantic annotation* of the captured datasets can be utilized to address the challenge of interpreting the processes. Thus, to benefit from the actual semantics behind those event logs tags or labels, the *semantic process mining* which enforces mining and analysis of processes at the conceptual levels have to be employed.

In view of that, this paper focuses on providing a method for semantic PM - by proposing the SPMaAF framework which is directed towards the discovery and semantic enhancement of the process models derived by applying the traditional PM techniques. The work shows through the *learning process domain* case study - How events logs from any given process domain is being extracted, semantically-prepared, and transformed into minable formats that supports the process of discovering, monitoring, and improvement of the extracted processes through semantic-based analysis of the discovered models.

Truly, the semantic-based approach proves to allow the analysis of the extracted event logs based on concepts rather than the event tags of the process. Besides, the semantic-based planning and analysis allows the meaning of the learning objects/models to be enhanced using *property characteristics* and *classification of discoverable entities*, to generate inference knowledge which are then used to determine useful patterns and improve analysis of the resulting models based on the domain concepts.

3 Conceptual Overview of the Semantic Process Mining and Analysis Framework (SPMaAF)

In this section of the paper, we introduce the proposed SPMaAF and its main design components, and subsequently, show how the integration of the different phases of the framework and the resultant algorithms are utilized for ample implementation of the approach. The design of the SPMaAF framework and its main components is shown in Fig. 2.

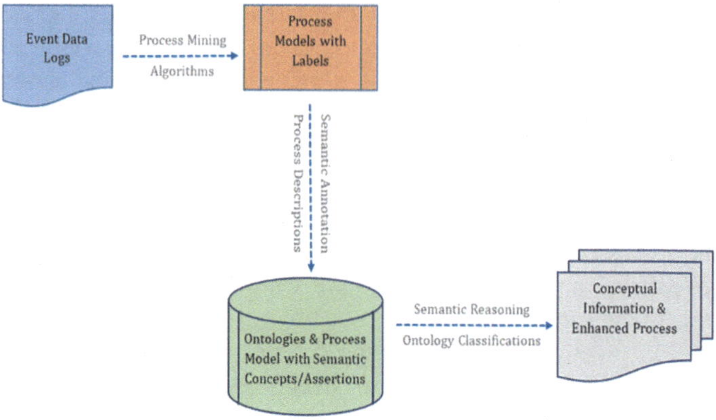

Fig. 2. SPMaAF: Semantic Process Mining and Model Analysis Framework.

In short, the proposed Semantic-based Process Mining and Analysis Framework (Fig. 2) constitutes the following processes as gathered in [26]:

- *Extraction of process models from event data logs*: the derived models are represented as a set of annotated terms that links and relates to defined terms in an ontology, and in so doing, encodes the process logs and the deployed models in the formal structure of ontology (semantic modelling).
- *The inferred ontology classifications*: helps associate meanings to labels in the event logs and models by pointing to concepts (references) defined within the ontology.
- *The reasoner* (inference engine): is designed to perform automatic classification of task and consistency checking to validate the resulting models as well as clean out inconsistent results, and in turn, presents the inferred (underlying) associations.
- *The conceptual referencing*: supports semantic reasoning over the ontologies to derive new information (or knowledge) about the process elements and the relationships they share amongst themselves within the knowledge base.

In fact, the key step to application of the semantic-based process mining and model analysis approach is to focus on connecting the mining algorithm with two primary mechanisms:

1. Event Logs and process models where the labels have references to concepts in an ontology, and
2. Reasoners that are invoked to reason over the resulting ontologies for the logs and models [26].

Furthermore, the application of the semantic-based approach has received a major amount of interest as it concerns the PM field. On one hand, the SPMaAF focuses on using the semantic information (i.e. metadata) that are embodied in the events log of the domain processes to produce a new method for PM, or better still, supports improvement of the existing methods to assist humans in gaining a novel and more accurate results at a higher conceptual level that maps with the domain context as opposed to the traditional PM techniques (that tends to analyse data at the syntactic level). On the other hand, because of the semantic level of analysis, the outcome of the SPMaAF method appears to not just perceived as a machine-readable system but also a machine-understandable system. Whereby, the results of the process is not just in formats that the computer can read and analyse; but also the process analyst and/or owners or IT experts can easily understand the outcomes of the process. Such method of big data analysis is very important especially in current literature - where most often, the classical PM algorithms or tools fail to make use of the semantics (i.e. domain specific information) that are embedded in the events log of the processes which they represent.

To this end, this paper illustrates through the instantiation of the SPMaAF, sets of algorithms formalizations and the resulting semantic fuzzy mining approach - that through encoding of events log and models with rich semantic annotation, and the integration of semantic reasoning aptitudes that it is possible to specify useful domain semantic-based analysis capable of bridging the semantic gap conveyed by the traditional PM techniques [1, 2, 26]. Moreover, when considering such kind and level of semantic-based PM method, the work notes that meaningful information on how the different process elements (or components) fits and relies on each other within the process knowledge-base is made possible, and especially important towards the extraction of models that have the capability to produce newly, valuable, and conceptual information.

In short, the main difference between the SPMaAF (Fig. 2) and the traditional PM framework (Fig. 1) is that - whilst the framework in Fig. 1 tends to analyse the extracted events logs to derive some explicit and/or implicit information about the processes they support without considering the *semantic* aspects of the information that are contained in the events log. On the other hand, the SPMaAF (Fig. 2) focuses on semantical integration and extension of the method in Fig. 1 by considering the semantic gap that is missing with the traditional PM framework in terms of the extracted events log and the derived process models. In other words, whilst the traditional PM techniques analyses the event data logs at *syntactic levels* (i.e. labels or tags in the event logs), the SPMaAF pursues to extend and analyse the available events log and derived process model at a much higher *conceptual levels* (i.e. based on concepts defined within the model).

4 Implementation of the SPMaAF: Process Models Discovery, Semantic-Based Analysis and Algorithms Formalizations

In this section, we describe the different phases of implementing the SPMaAF, as well as, provide the sets of algorithms to instantiate each phase of the framework.

4.1 Different Phases of Implementing the SPMaAF and Design

The application of the SPMaAF is in three phases as detailed in [26]:

In Phase 1: The PM technique is applied in order to make available the process mappings for the learning process, and check its conformance with the event logs based on the Fuzzy Miner [27] as described in [4] and [6]. The main reason is that the resulting process map allows us to quickly, and interactively explore the processes into multiple directions and to show the individual activities *workflow*, and then provide platform for semantic annotation of the different process elements within the knowledge base. The proposed *Algorithm 1* in this Sect. 4.2 describes the procedures on how we have implemented this phase of the process.

In Phase 2: Semantic modelling of the resulting process mappings in terms of the annotated terms is done. Thus, the semantic model represents the domain knowledge about the various learning activities and sequence workflows including the concepts as defined in an Ontology which sits at the core of the approach by using process description languages such as the OWL [7] and SWRL [8]. In addition, the process also makes use of the Reasoner i.e. Pellet - to infer the different instances and the ontological representation (taxonomy) of the learning process model in reality [6]. The proposed *Algorithms 2* in this Sect. 4.3 explains the steps we have taken to implement this stage of the framework.

In Phase 3: Extraction and automated querying of the learning concepts is carried out. The work uses the Eclipse Java Runtime Environment to create the methods and interface for loading the Process Parameters (i.e. the ontology concepts). Essentially, we make use of the OWL Application Programming Interface (OWL API) to extract and load the inferred concepts from the defined ontology. Example of the inferred underlying concepts using the OWL API and the resulting ontologies is as shown in [23] and [24] and described in Sect. 5 of this paper. In [26] also shows the general architecture for implementing such a semantic-based process mining technique. The *Algorithm 3* describes the instantiation of the semantic reasoning phase.

4.2 Algorithm 1

Here, we describe the proposed *Algorithm 1* and how we have used the method to perform the process mining and model discovery (Phase 1) in order to discover useful process models from the events data log. The process proves useful towards generation and mapping of individual traces that makes up each of the process executions. In [6] describes how the proposed algorithm is implemented using PM tools such as the *Disco* based on the Fuzzy Miner Algorithm [27] to generate and map the process models from the event logs for conformance checking which allows one to analyze the individual cases (i.e. the classified traces and the sequence of activities executions).

In essence, *Algorithm 1* explains how the work discovers and produce process models and traces from any given event data log as follows:

Algorithm 1: Discovering Process Models and Individual Traces from Event Logs.

```
1: For all Recorded and Captured Event Data Log EDL
2: Input:  PM – Process mining tool used to extract model, M
           e – Classifier for the event logs, EDL and traces, T
3: Assign: Case_id(e) i.e. the Case associated to event, e within the EDL
           Act_name(e) i.e. Activities associated to event, e within EDL
           Other_attributes e.g. Event ID, Timestamp, Resources, Roles
             etc. associated to event, e within the EDL
4: Output: Process maps for the discovered models, M & individuals
             traces, T classifications for the event log, EDL
5: Procedure: Produce Models, M from EDL for cross-validation to
             determine how well M reflects the performed activities in
             reality i.e TraceFitness, TF and for further analysis
6: Begin
7:   For all Event Data Log EDL
8:     Extract Process Maps, M, & Traces, T ← from Event Log EDL
9:     while no more process element is left do
10:      Analyze Model, M and Traces, T to determine tracesFitness, TF
11:        If T ← Null then
12:            obtain the occurring act_name(e) sequence sets from Log
               EDL
13:        Else If T ← 1 then
14:         cross-validate resulting Trace, T from EDL with Model, M
15:        If trace, T exist then
16:          For each event Classifier, e output ← return as True_Positive,
               TP i.e fits the Model, M
17:         Else If trace, T does not exist then
                 Return event Classifier, e output as True_Negative, TN
               i.e does not fit the Model, M
18: Return: Classification Results of the Process Mining approach and
               Process Mappings
19: End If statements
20: End while
21: End For
```

Indeed, the *Algorithm 1* shows that:

- A typical process model, ***M*** consist of Traces, ***T*** (i.e. Cases)
- A Trace (Case), ***T***, consist of events, ***e***, such that each event relates to precisely one case.
- Events, ***e***, within a Trace are ordered, most often in a sequential order
- Events for any process mining task must have atleast a Case identification ID (*Case_id*) and Activity Name (*Act_name*) attributes to allow for the process model discovery.
- Other additional information may be required for ample implementation of process mining e.g. Event ID, Timestamp, Resources, Cost, Roles, Places, etc.

4.3 Algorithm 2

Semantic *representation* of the process (i.e. ontology) - *Phase 2* is vital for implementation of the SPMaAF framework. The ontological representation is focused on integrating information about the various entities that are embodied in the events data log (***EDL***), and the derived models (***M***). This is especially important when the aim is to discover previously unknown links and/or useful information amongst the process instances. Besides, the implementation of the procedures in Phase 3 (i.e. reasoning over the ontologies and classification of individual outputs) relies totally on the capability to represent the derived knowledge in a formal manner (taxonomies).

Thus, *Algorithm 2* shows how the work creates ontologies from the events log and models as described in [26] as follows:

Algorithm 2: Developing Ontology from Process Models & Event Logs

```
For all defined models M and event log EDL
Input:  C – different classes for all process domain
        R – relations between classes
        I – sets of instantiated process individuals
        A - sets of axioms which state facts
Output: Semantic annotated graphs & an ontology-driven search for
        process models and explorative analysis
Procedure: create semantic model with defined process descriptions and
        assertions
Begin
  For all process models M and event log EDL
    Extract Classes C ← from M and EDL
    while no more process element is left do
    Analyze Classes C to obtain formal structures
      If C ← Null then
        obtain the occurring Process instances (I) from M and EDL
      Else If C ← 1 then
          create the Relations (R) between subjects and objects // i.e
        between classes C and individuals (I)
      If relations R exist then
          For each class C ← semantically analyse the extracted
        relationships (R) to state facts i.e Axioms (A)
            create the semantic schema by adding the extracted
        relationships and individuals to the ontology
Return: taxonomy
End If statements
End while
End For
```

Truly, the annotated logs/models and ontological representations are very useful in carrying out the semantic-based analysis of the process domains - because at this stage, the input data are presented in a formal structure (taxonomies) that can connect to concepts which are referenced in the underlying ontology [26]. In return, from *Algorithm 2*, the work shows that the ontology, *Ont*, is a quadruple, i.e.

$$Ont = (C, R, I, A)$$

which consists of different classes, ***C***, and relations, ***R***, between the classes [26, 28, 29]. Perhaps, a relation, ***R***, trails to connect a set of class(es) with either another class, or with a fixed literal and is capable of also describing the sub assumption hierarchy (e.g. taxonomy) that exists between the various classes and their relationships. In addition, the classes are instantiated with a set(s) of individuals, ***I***, and can likewise contain a set(s) of axioms, ***A***, which states fact (e.g. what is true and fitting within the model, or what is true and not fitting within the model). In other words, the ontologies can be defined as connected sets of taxonomies (i.e. RDF + Axioms) or yet still, structuring in a formal way (Triple + Facts) where the subject includes the *Classes*, predicate represents the *Relations*, and the object includes the *Individuals*, in addition to the sets of axioms which states facts. The authors in [30] are even more specific about the importance of the ontological concepts (particularly the *classes*) in semantic representation of real-time processes. The authors note that "classes are the central item of the ontology" and further explain that a well-defined class may represent all types of procedures e.g. running tasks, data transmission, data flow control, activity workflows, etc.

Without a doubt, the procedures in *Algorithm 2* shows the need for a well-defined *semantic annotation* process especially in developing such an ontology-based framework (SPMaAF) that supports semantic-based PM and analysis - by producing in an automatic manner the underlying taxonomies and/or properties the process instances share within the knowledge base [24, 29]. Thus, the semantic annotation process is necessary for concrete implementation of the SPMaAF including the further steps of enhancing the model as explained in [24]. According to [29] a semantically annotated graph is described as follows:

$$SemAn :: N \cup E \rightarrow COnts$$

where: *SemAn* describes all kinds of annotations which can be input, output, meta-model annotation etc.

In general, [26] notes that the semantical planning of any ontology-based system such as the SPMaAF - must take into account all process *actions* (activities) within the ontology and must consist of some form of semantic annotation. Moreover, according to the definition in [29] if we Let A be the set of all process actions. A process action $a \in A$ is characterized by a set of input parameters $Ina \in P$, which is required for the execution of a and a set of output parameters $Outa \subseteq P$, which is provided by a after execution. All elements $a \in A$ are stored as a triple (*namea, Ina, Outa*) in a process library *libA*.

4.4 Algorithm 3

Accordingly, the last essential component in realizing the SPMaAF and its main application is the aptitude to perform *semantic-based reasoning* in order to infer (i.e. ontology classifications) and check for consistency in the semantic model. To perform the automatic inference and population of the individual ontologies, the system makes use of the *reasoner* to produce the necessary outputs and/or provide the results as executed during the information retrieval process (or search queries).

To this end, *Algorithm 3* shows how the work uses the reasoner (e.g. Pellet) in carrying out the ontological classifications and semantic reasoning following the definitions in [26]

Algorithm 3: Reasoning over Ontologies and Classification of Entities and Outputs

```
For all defined Ontology models OntM
Input: classifier e.g. Pellet Reasoner
Output: classified classes, process instances and attributes
Procedure: automatically generate process instance, their individual
        classes and Learning concepts
Begin
  For all defined object properties (OP) and datatype properties (DP)
        assertions in the model (OntM)
  Run reasoner
  while no more process and property description is left do
  Input the semantic search queries SQ or set parameter P to retrieve
        data from OntM
  Execute queries
    If SQ or P ← Null then
        re-input query or set the parameter concepts
    Else If SQ or P ← 1 then
        infer the necessary associations and provide resulting outputs
Return: classified Concepts
End If statements
End while
End For
```

Basically, *semantic reasoning* as described in *Algorithm 3* – associates meaning to the labels (i.e. attributes) as well-defined within the knowledge-base by referencing the set of rules and/or assertions (e.g. Object/DataType properties) that are defined in the ontologies in order to answer some common question (or queries) and perform a more abstraction analysis of the process elements.

In summary, methods for semantic PM and analysis such as the SPMaAF in this paper - focuses on information about resources hidden within a process knowledge-base, and how they are related [2, 23, 24, 31]. The semantic-based analysis allows the meaning of the domain entities and/or object properties to be enhanced through the use of *property characteristics* and *classification of discoverable entities*, to permit analysis of the extracted event logs based on concepts rather than event tags or labels about the process. Interestingly, there are not too many algorithms that support such semantic level of analysis currently in literature, and there are few existing applications that demonstrate the capabilities of the semantic approach [2, 23, 24, 31]. In [23, 24, 32] shows how semantic annotations and reasoning can be used to provide more improved analysis/ enhancement to process models and event logs through concept matching (i.e. ontology classifications). Specifically, [24] and [32] perform semantic modelling and integration of the resulting process maps with annotated terms and then present the domain knowledge for the activities (i.e. attributes or concepts) as defined within the ontology using process description languages such as the OWL [7] and SWRL [8]. Indeed, reasoning on ontological knowledge plays an important role in semantic representation of processes [33] by allowing for extraction and conversion of explicit information into some implicit information. For example, the intersection or union of classes, description of relationships and concepts assertions. Thus, those main components (as described in *Algorithms 1–3*), namely, *annotated logs/models*, *ontology*, and *semantic reasoning* is the foundation upon which the SPMaAF is developed.

5 Use Case Scenario and Implementation of the Semantic Fuzzy Mining Approach

This section presents the practical implementation of the SPMaAF referred to as the Semantic fuzzy miner - including integration of the different stages of its development, and then, looks at the use case scenario of the *Learning Process* to show how the different components fit and is capable of analysing process models and event logs at a more abstract level.

5.1 Case Study of the Learning Process

This paper makes use of the case study example of the learning process (specifically the Research Process) to illustrate the application of the proposed method. The paper uses event logs about the research process to analyse and answer some common queries or real-time questions about the learning process. The case study as represented in [6] illustrates the workflow of a typical research process from the identification of the research area of topic to conclusion of the research investigation and award of degree. Apparently, in order to complete the research journey, it is expected that the researcher must complete a set of step(s), i.e. activities or milestones - in order to find answers to the research questions [6]. To this end, this work shows by way of the SPMaAF framework and sets of semantically motivated algorithms that PM results can be presented at a much higher abstraction levels through semantic-based analysis of the learning process activities.

5.2 Implementation: Semantic Modelling and Analysis of the Research Learning Process

To demonstrate the application of the SPMaAF framework and the proposed sets of algorithms – the work analyzes the learning process in order to determine patterns and/ or attributes that differentiates some process instances from another. For example, to establish what *attributes* or *paths* particular learners follow or have in common, or what attributes distinguishes the *successful learners* from the *uncomplete* ones [26]. Henceforth, we focus not just on answering the identified questions, but also, to determine to what extent the application of the SPMaAF framework is able to perform semantic reasoning and classify in an effective manner - the different group of learner (i.e. cases) within the learning knowledge base. Practically, the paper focus on the use case distinction of the *successful* and *uncomplete* learners within the knowledge base.

Furthermore, the authors in [6] and [24] shows that the workflow for a research process consist of four milestones, i.e., establish context → learning stage → assessment stage → validation of learning outcome [6, 24]. Thus, from defining the topic area -to- reviewing literatures -and- addressing the problem -then- defending the solution [6]. According to [26] the order in which the different milestones (which comprises of various individual learning activities) are carried out has the capability of determining the research outcome [24].

For example, in order to resolve the aforementioned learning questions in this section with regards to the *Successful* and *Uncomplete* learners – the work refers that a "Successful Learner" is a subclass of, amongst other NamedLearnerCategory, a Person that performs some LearningActivityConcepts, who has a universal object property restriction or relationship with the four milestones of the ResearchProcessClass (i.e. from defining the topic area -to- reviewing literatures -and- addressing the problem - then- defending the solution) as explained in [6, 24, 26] and defined below:

Successful Learner Class:

```
ontology ResearchProcess
  concept SuccessfulLearner
    hasCompleteMilestone    ofType    {DefineTopicArea,
      ReviewLiterature, AddressProblem, DefendSolution}
    isPerformerOf some LearningActivity
    is ofType Person
    hasInstance members {Mattew, Isaac}
  axiom DefinitionOfSuccessfulLearner
```

Clearly, as shown in the description above: the necessary condition is that - if something is a SuccessfulLearner, it is necessary for it to be a participant of the LearningActivity Concept class and necessary for it to have a kind of sufficiently defined condition and relationship with the ResearchProcess class and its attributes, namely: Define Topic Area, Review Literature, Address Problem and Defend Solution [6, 24, 26].

The same *object property restriction* [26] applies to the "Uncomplete Learner" class as described below:

Uncomplete Learner Class:

```
ontology ResearchProcess
  concept UncompleteLearner
  hasOnlyCompleteMilestone ofType {DefineTopicArea, Or
     ReviewLiterature, Or AddressProblem, Not
     DefendSolution}
     isPerformerOf some LearningActivity
     is ofType Person
   hasInstance members {Paul, Danny, Mark, Gregory, John}
  axiom DefinitionOfUncompleteLearner
```

Ideally, we observe from the descriptions above that the *object* or *data type properties/restrictions* are mainly employed when classifying or inferring previously unobserved class(es) or entities that satisfies the restrictions (assertions). For example, those individuals that could be classified as *Successful* or *Uncomplete* leaner who are required to meet the condition (restriction) of the specified class in order to become members or participants of the class. In fact, such kind of structured organisation (i.e. object/data type restrictions) which integrates *semantic annotation* (labelling) and *reasoning* presents itself as a good way of formally representing the various entities (process instances) that can be found within any kind of process knowledge base or domain.

6 Evaluation Outcome and Experimental Results

The proposed framework in this paper shows that the semantic-based approach sparks methods that supports:

(i) The application of process mining techniques to domain processes.
(ii) Provision of real-world semantic knowledge and understanding about domain processes (e.g. case study of the learning process), which are useful towards the development of PM algorithms that are more intelligent with high level of effective conceptual reasoning capabilities.

The work in this paper proves that ontologies help in harmonizing the various process elements that are found within the process models and event logs. Moreover, the semantic annotations and reasoning helps to extract and add useful conceptual knowledge to the mining results.

Specifically, in [24] the work makes use of the event logs in [34] and [35] to weigh up the performance and to demonstrate; how the proposed Semantic-based Fuzzy miner is able to perform a more accurate classification of the individual traces within the process base. This includes the capability to integrate the ontological concepts and perform semantic reasoning capable of discovering worthwhile models given the datasets (with *training set* and a *test set*) [34, 35] for the cross-validation experiments.

Practically, in the experimentations in [24], the Semantic Fuzzy mining approach references a number of different OWL ontologies (e.g. the training model ontology, test set ontology, traceFitnessClassification ontology, etc.) which were all created for the purpose of the experiment. For each ontology created for the experiment, all concepts in their turn were considered by the *reasoner* and are checked for consistency by referencing the process parameters. Based on behavioural characteristics of the provided datasets [34, 35], a cross validation method was adopted in order to overcome the variability in the composition of the *training sets* and *test sets*. The traces were computed and recorded according to the *reasoner's* response, and the classification process was tested on the resulting individual outcomes by quantitatively assessing its performance with respect to correctly classified traces. For each result of the classification process, the replayable, i.e. true positives (TP) and non-replayable, i.e. true negatives (TN) traces were learned.

Indeed, the work observes that for every run set of parameters, the commission error, i.e. *false positives (FP)* and *false negatives (FN)* was null, thus equal to 0. This clearly means that the classifier did not make critical mistakes. For example, settings where a trace is deemed to be an instance of a specific class whilst it really is an instance of another class. At the same time, it is important to note also that the trace accuracy rate was very high i.e. for the correctly classified *true positives* (TP) and *true negatives* (TN), and were consistently observed for all the test sets. Further details about the outcome of the experiment with regards to the discovered process models and results of the classification process for the corresponding individual traces occurring in each *test* set are as reported in [24].

In short, the semantic-based fuzzy mining approach is aimed at improving the process analysis and system performance. Besides, the integration of the different ontologies, conceptual reference models, and reasoner; enables the definition of a more universal analysis question, and then focus on finding answers for those questions in an automated, thus, computerized manner. Even more, since the analysis is carried out at a higher level of abstraction (i.e. conceptualization), the results can be easily understood (i.e. closer to human comprehension) and the process of adding new concepts in the ontology, or yet still, changes or modifications to the attributes (i.e. labels or tags) do not necessarily entails or requires updating the analysis questions or queries.

Therefore, the SPMaAF framework, the resulting Semantic Fuzzy miner and its algorithms formalizations is a significant contribution to the state-of-the-art; where many existing PM techniques requires some form of reconstruction to bring the process analysis to a greater conceptual level or in many cases lacks the capability of identifying and/or making use of the semantics that can be found within the various process domain.

In Table 1, the work has carefully analyzed the aspects of the proposed framework and its main application (i.e. the Semantic Fuzzy Miner compared to other existing benchmark algorithm for semantic process mining). Table 1 highlights the differences and similarities in properties of our approach with the only well-known, if not the only one, semantic process mining algorithm (the Semantic LTL Checker) [2] currently in literature as shown in Table 1. Thus, the Table 1 represent the Semantic-Fuzzy miner and its application properties evaluated against the Semantic LTL Checker [2].

Table 1. The Semantic-Fuzzy miner and its application properties evaluated against existing benchmark algorithm

	Semantic LTL checker	Semantic-Fuzzy miner
Data input	Takes event logs concepts as input to parameters of Linear Temporal Logic (LTL) formulae	Takes process models derived from fuzzy mining of the event log as input to learn and reason about the domain process
Ontology	Ontologies are defined in WSML format	Ontologies are defined in OWL and SWRL format
Reasoning	Integrated using the WSML2Reasoner (W2RF)	Integrated using the Pellet Reasoner
Functionality	Uses LTL properties or formulae defined in LTL Template files (i.e. contains the specification of properties written in the special LTL language)	Uses process description properties (*CLASS_ASSERTIONS; OBJECT_PROPERTY_ASSERTIONS;* and *DATA_PROPERTY_ASSERTIONS)* defined using OWL and SWRL Language/schema.
GUI	There is option to select *concepts* for the parameter values	There is option to select *concepts* for the parameter values
Support	Supports *concepts* as a value (i.e. when a concept is selected, the algorithm will test whether the attribute is an *instance of* that concept, and concepts can only be specified for set attributes).	Supports *concepts* as a value (i.e. when a concept is selected, the algorithm will test whether the attribute is an instance of that concept, and concepts can only be specified for set attributes).

Firstly, the Semantic Fuzzy mining approach based on the distinctive properties described in Table 1 proves to be more robust and effective than the traditional PM techniques. This is due to the fact that the method also takes the semantics perspectives of the event logs and process models into account. Moreover, as opposed to the existing semantic LTL checker [2] which only considers and takes event logs concept as input to parameters of a Linear Temporal Logic (LTL) formula to analyse process, the semantic fuzzy mining approach also takes the process models as input. Besides, because those models are automatically created from the actual event logs of the process domains, the system tends not to unnecessarily lose or leave out important information or missing data.

Secondly, even though both approaches make use of ontologies, a major difference between the existing semantic LTL checker algorithm and the Semantic-Fuzzy miner is the fact that ontologies are defined in Web Service Modelling Language (WSML) [36] format with the semantic LTL checker, whereas on the other hand, ontologies are defined using the OWL [7] and SWRL [8] format in this paper. In essence, whilst there are limitations with WSML ontologies with respect to the exchange of syntax over the web, OWL ontologies aim to bring the expressive and reasoning power of description logic (DL) [9] to the semantic web. Moreover, it is the state of the art logical layer upon which semantic architectures are currently built in literature [37]. In short, OWL ontologies allows one to specify far more about the properties and classes which are defined within a process domain and/or knowledge base. In other words, the OWL ontology as utilized in this paper is developed not just for representing information in formats that can be easily understood by humans, but also for building applications that tends to inclusively process the information that they contain or supports. Thus, supports *machine-understandable* systems rather than just *machine-readable* systems.

Thirdly, from a *reasoning* or *classifier* point of view; whilst the semantic LTL checker makes use of the WSML2Reasoner [36, 38] to perform a more complex inferences that are beyond subsumption reasoning by only benefiting from the inclusion of semantic annotations, the Semantic Fuzzy mining approach is integrated with Pellet reasoner [39] which typically in addition to semantic annotations has been proved to incorporate optimizations for nominals, conjunctive rules and query answering, and incremental reasoning capabilities that supports process descriptions and logic (i.e. class assertions and object/data property assertions).

Nonetheless, the semantic LTL checker and the Semantic Fuzzy miner both have option to select *concepts* for the parameter values, and indeed, supports concepts as a value, i.e. when a concept is selected, the algorithm will test whether an attribute is an instance of that concept (i.e. class), and concepts can only be specified for set attributes. For example, with the Semantic-Fuzzy miner application, one can test whether: For all **Persons** (i.e. Performer instances) does always (**condition check?** - exist four milestones?) implies eventually (**class description**: Successful Learner). In other words, does any named **Person P**: hasCompleteMilestones **A** and **B** and **C** and **D**, where: **A** = DefineTopicArea, **B** = ReviewLiterature, **C** = AddressProblem, and **D** = DefendSolution - represents and points to the concepts within the domain ontology.

7 Discussion

The method in this paper focus on ascertaining by the series of validation experiments - how the outcome of PM techniques can be enriched through semantic representation and analysis of the derived models. As a result, the Semantic-based Fuzzy miner was developed, in addition to the various sets of semantically motivated algorithms and the design framework – SPMaAF proposed in this paper.

In short, the paper shows that the three main tools, namely: annotated logs/model, ontologies, and reasoner - constitute the building block upon which semantic process mining technologies such as the SPMaAF framework are constructed. Thus, while the *semantic-based annotation* is dedicated to providing *metadata* about the different process elements within the knowledge-base, *ontologies* are used to bind together (integrate) the various classes (concepts) in the model. Without a doubt, the population and/or classification of the underlying relationships that exist amongst the process elements is performed using the *reasoner*.

Indeed, such an approach (e.g. SPMaAF) does not only support the implementation of some of the process description languages, e.g. SWRL, OWL, DL etc., but it also allows the re-use of any type of ontology by another. In addition, the process tends to minimalize the number of human errors which are, every now and again, present particularly during the extraction, interpretation and/or analysis of the multiple entities/ concepts that can be found within the domain processes in question.

Therefore, the unabridged notion of the proposed Semantic Fuzzy mining approach, design framework – SPMaAF, sets of algorithms *1, 2* & *3*, and the experimental results proves that semantic concepts (i.e. annotation, ontology, and reasoning) can be layered on top of existing information asset (i.e. process models, event data logs, etc.) to provide a much more easy and accurate way of analysing the real-time processes capable of providing real world insights and/or answers that can be more easily grasp by the process analysts or owners.

On the other hand, whilst the research believes and prove that such method is practically suitable for mining processes at a more conceptual level, there could also exist a number of limitations and threats to validity. Thus, whilst the paper have introduced a design framework, and sets of descriptive algorithms and method to resolve the sets of identified problems and question that motivates the research investigations, there could be potentially many ways to address those problem, or even, bigger areas that have not been yet addressed. Owing to the fact that the semantic process mining is a new area within the process mining field, and there are not too many tools or algorithms that support such an approach currently in literature. Therefore, we assume that this work is only an incentive of more robust and intensive research within the context of the semantic-based process mining (PM).

Also, the procedures and method for semantically analysing the domain processes which this paper introduces is one of the main important contributions of this work. Besides, the technique has the capacity to extract conceptual knowledge (information) from the events log and process models. However, the correlation and integration of the key core elements/components (i.e. annotated logs/models, ontologies, and reasoner) that underlies the proposed method assumes that the work has presented an approach which can possibly be re-introduced or extended in other resourceful ways. Perhaps, a worthwhile extension of the approach may include more sophisticated increment of the ontology schemas and reasoning capabilities that has already been well-defined in this paper. This may include the development of authoring tools capable of augmenting the stated achievements of this paper, or yet still, the mentioned process domains and/or other operational processes in the real world settings in general.

8 Conclusion

This paper introduced the SPMaAF to support the development of semantic-based PM techniques and algorithms that exhibits a higher level of semantic reasoning and/or capabilities as opposed to the syntactic nature of the traditional PM techniques. The instantiation of each phase of the framework was presented in terms of the proposed *Algorithms 1, 2, & 3.* The evaluation results shows that the SPMaAF approach can be perceived as an ontology-based system that is able to perform query answering and/or information retrieval in a much effective and efficient manner in comparison to existing standard logical procedures - particularly in terms of the other state-of-the-art approaches used for semantic process mining.

Future work will be to implement the proposed framework to analyse data from other domain areas of interest. This will allow for further validation and will generalise the findings of the research and case study presented in this paper. Another extension will be to complement the approach with a platform for completely automatic discovering and integration of the semantic information or knowledge.

References

1. Dou, D., Wang, H., Liu, H.: Semantic data mining: a survey of ontology-based approaches. In: 9th IEEE International Conference on Semantic Computing, pp. 244–251. California, USA (2015)
2. deMedeiros, A.K.A., Van der Aalst, W.M.P., Pedrinaci, C.: Semantic process mining tools: core building blocks. In: ECIS, pp. 1953–1964. Galway, Ireland, June 2008
3. Van der Aalst, W.M.P.: Process Mining: Data Science in Action, 2nd edn. Springer-Verlag, Berlin Heildelberg (2016)
4. Okoye, K., Tawil, A.R.H., Naeem, U., Islam, S., Lamine, E.: Semantic-based model analysis towards enhancing information values of process mining: case study of learning process domain. In: Abraham, A., et al. (eds.) Advances in Intelligent Systems and Computing Book Series (AISC), vol. 614, pp. 622–633. Springer International Publishing AG (2017)
5. Polyvyanyy, A., Ouyang, C., Barros, A., Van der Aalst, W.M.P.: Process querying: enabling business intelligence through query-based process analytics. Decis. Support Syst. **100**(1), 41–56 (2017)
6. Okoye, K., Tawil, A.R.H., Naeem, U., Islam, S., Lamine, E.: Using semantic-based approach to manage perspectives of process mining: application on improving learning process domain data. In: 2016 IEEE International Conference on Big Data (BigData), pp. 3529–3538. Washington, DC (2016)
7. W3C.: OWL Web Ontology Language [Internet]. http://www.w3.org/TR/owl-ref/ (2004). Accessed Sept 2017
8. Horrocks, I., Patel-Schneider, P.F., Boley, H., Tabet, S., Grosof, B., Dean, M.: SWRL: A Semantic Web Rule Language Combining OWL and RuleML. W3C Member Submission [Internet]. http://www.w3.org/Submission/SWRL/ (2004). Accessed Sept 2017
9. Baader, F. Calvanese, D. McGuinness, D.L., Nardi, D., Patel-Schneider, P.F.: Description Logic Handbook: Theory, Implementation, and Applications, 1st edn. New York, Cambridge University Press (2003)
10. Balcan, N., Blum, A., Mansour, Y.: Exploiting ontology structures and unlabeled data for learning. In: 30th International Conference on Machine Learning, pp. 1112–1120. Atlanta Georgia, USA (2013)
11. Cesarini, M., Monga, M., Tedesco, R.: Carrying on the e-learning process with a workflow management engine. In: Proceedings of ACM Symposium on Applied Computing, pp. 940–945. Nicosia Cyprus (2004)
12. Perez-Rodriguez, R., Caeiro-Rodriguez, M., Anido-Rifon, L.: Supporting PoEML educational processes in Moodle: A middleware approach. In: SPDECER, Universidad Pontificia de Salamanca (2008)
13. Nguyen, L., Phung, D.: Learner model in adaptive learning. World Acad. Sci. Eng. Tech. **45**, 395–400 (2008)
14. Peña-Ayala, A.: Intelligent and Adaptive Educational-Learning Systems: Achievements and Trends, 1st edn. Springer, Berlin (2013)

15. Trčka, N., Pechenizkiy, M., van der Aalst, W.M.P.: Process Mining from Educational Data. In: Romero, C., et al. (eds.) Handbook of Educational DM, pp. 123–142. Chapman & Hall/CRC Data Mining & Knowledge Discovery Series, CRC Press, Boca Raton, Florida (2010)
16. Pechenizkiy, M., Trcka, N., Vasilyeva, E., van der Aalst, W.M.P., de Bra, P.: "Process Mining Online Assessment Data, pp. 279–288. Proceedings of EDM, Cordoba, Spain (2009)
17. Holzhüter, M., Frosch-Wilke, D., Klein, U.: Exploiting learner models using data mining for e-learning: a rule based approach. In: Peña-Ayala, A. (ed.) IAELS: Achievements and Trends, pp. 77–105. Springer Berlin, Heidelberg (2013)
18. Grob, H.L., Bensberg, F., Coners, A.: Regelbasierte steuerung von geschaftsprozessen-konzeption eines ansatzes auf basis des process mining. In: Die Wirtschaftsinformatik. Heidelberg (2008)
19. Peña-Ayala, A., Sossa, H.: Proactive Sequencing Based on a Causal and Fuzzy Student Model. In: Peña-Ayala, A. (ed.) IAELS: Achievements and Trends, pp. 49–76. Springer, Berlin Heidelberg (2013)
20. Dżega, D., Pietruszkiewicz, W.: Intelligent decision-making support within the e-learning process. In: Peña-Ayala, A. (ed.) IAELS: Achievements and Trends, pp. 497–521. Springer, Berlin (2013)
21. Bogarín, A., Cerezo, R., Romero, C.: A survey on educational process mining. In: Wiley Interdisciplinary Reviews: Data Mining and Knowledge Discovery (WIRES), pp. e1230. Wiley Periodicals (2017)
22. Cairns, A.H., Ondo, J.A., Gueni, B., Fhima, M., Schwarcfeld, M., Joubert, C., Khelifa, N.: Using semantic lifting for improving educational process models discovery and analysis. In: SIMPDA, vol. 1293 of CEUR Workshop Proceedings, CEUR-WS.org, pp. 150–161 (2014)
23. Okoye, K., Tawil, A.R.H., Naeem, U., Lamine, E.: Discovery and enhancement of learning model analysis through semantic process mining. Int. J. Comput. Inf. Syst. Ind. Manage. Appl. **8**(2016), 093–114 (2016)
24. Okoye, K., Naeem, U., Islam, S.: Semantic fuzzy mining: enhancement of process models and event logs analysis from syntactic to conceptual level. Int. J. Hybrid Intell. Syst., IOS Press **14**(1–2), 67–98 (2017)
25. Bogarín, A., Romero, C., Cerezo, R., Sánchez-Santillán, M.: Clustering for improving educational process mining, pp. 11–15. ACM, NY (2014)
26. Okoye, K., Islam, S., Naeem, U.: Ontology: core process mining and querying enabling tool. In: Thomas, C. (ed.) Ontology in Information Science, Chapter 7, pp. 145–168. InTechOpen Publishers (2018)
27. Rozinat, A., Gunther, C.: Disco User Guide - Process Mining for Professionals. Fluxicon.com, Eindhoven, The Netherlands (2012)
28. Gruber, T.R.: Toward principles for the design of ontologies used for knowledge sharing. Int. J. Hum. Comput. Stud. **43**(5–6), 907–928 (1995)
29. Lautenbacher, F., Bauer, B., Forg, S.: Process mining for semantic business process modeling. In: 13th Enterprise Distributed Object Computing Conference Workshops, pp. 45–53. Auckland (2009)
30. Petrenko, O.O., Petrenko, A.I.: A model-driven ontology approach for developing service system applications. J. Comput. Sci. Appl. Inf. Technol. **2**(4), 1–7 (2017)
31. Jareevongpiboon, W., Janecek, P.: Ontological approach to enhance results of business process mining and analysis. J. Bus. Process Manage. **19**(3), 459–476 (2013)

32. Okoye, K., Tawil, A.R.H., Naeem, U., Lamine, E.: Semantic reasoning method towards ontological model for automated learning analysis. In: Pillay, N., Engelbrecht, A., Abraham, A., du Plessis, M., Snášel, V., Muda, A. (eds.) Advances in Intelligent Systems and Computing, NaBIC Conference 2015, pp. 49–60. Springer, Switzerland (2016)
33. Calvanese, D., Kalayci, T.E., Montali, M., Tinella, S.: Ontology-based data access for extracting event logs from legacy data: the onprom tool and methodology. In: Abramowicz, W. (eds.) Business Information Systems. BIS 2017. Lecture Notes in Business Information Processing, vol. 288, pp. 220–236. Springer, Cham (2017)
34. Carmona, J., de Leoni, M., Depair, B., Jouck, T.: IEEE CIS Task Force on Process Mining Process Discovery Contest @ BPM 2016, 1st edn. (2016). http://www.win.tue.nl/ieeetfpm/doku.php?id=shared:edition_2016
35. Okoye, K., Naeem, U., Islam, S., Tawil, A.R.H., Lamine, E.: Process models discovery and traces classification: a Fuzzy-BPMN mining approach. J. Int. Technol. Inform. Manage. (JITIM) **26**(4), 1–50 (Article 1) 2018 (IIMA 2018)
36. de Bruijn, J., Lausen, H., Polleres, A., Fensel, D.: The web service modeling language WSML: an overview. In: Sure, Y., Domingue, J. (eds.) The Semantic Web: Research and Applications. ESWC 2006. LNCS, vol. 4011, pp. 590–604. Springer, Berlin, Heidelberg (2006)
37. Lisi, F.: Building rules on top of ontologies for the semantic web with inductive logic programming. Theor. Pract. Logic Program. **8**(3), 271–300 (2008)
38. Bishop, B., Fischer, F., Keller, U., Steinmetz, N., Fuchs, C.G., Pressnig, M.: WSML Reasoner. IRIS Reasoner, Boston, MA (1999)
39. Sirin, E., Parsia, B.: Pellet: an owl DL reasoner. In: International Workshop on Description Logics (DL2004), Whistler, British Columbia, Canada, vol. 104, CEUR-WS.org (2004)

Improved Multi-hop Localization Algorithm with Network Division

Wei Zhao, Shoubao Su, ZiNan Chang, BingHua Cheng, and Fei Shao(✉)

School of Computer Engineering, Jinling Institute of Technology, Nanjing 211169, China
shaofei@jit.edu.cn

Abstract. Multi-hop localization methods enable the use of position service for various applications as soon as the hop-distance information is available. These methods provide reasonable positioning accuracy for isotropic networks in which sensor nodes distribute evenly and densely. However, their performance decreases significantly in anisotropic networks due to the large deviation of hop-distance. We address the problem of improving the positioning accuracy in anisotropic networks, given a method by using network division. Firstly, an anisotropic network will be divided into several sub-networks using clustering methods. Then our multi-hop localization methods will be conducted in these sub-networks. Finally, location results of all sub-networks will be combined. The metrics such as Average Localization Error (ALE) and Cumulative Distribution Function (CDF) are used to evaluate the performance of our scheme. The results show that it can effectively attack the problem of anisotropy, thus improving the positioning accuracy considerably.

Keywords: Multi-hop localization · Wireless sensor network Anisotropic network · Network division

1 Introduction

In various applications of wireless sensor networks (WSNs), location information is often a precondition, and sometimes we only need sensor nodes to return their geographic positions. There are many localization methods that we can use to obtain the nodes' locations. Multi-hop localization algorithms, as the classical range-free methods, can determine the location of unknown nodes only by information such as connectivity of the network. They complete the location process by relying on some nodes known locations (usually called anchor nodes). Multi-hop positioning technology is widely used due to its advantage of low cost and without additional positioning hardware.

The existing multi-hop localization algorithms [1–3] are mostly based on the assumption that the location area is isotropic and the nodes are evenly distributed in it. However, the multi-hop localization algorithms have a fatal disadvantage, that is, their positioning accuracy is very poor in anisotropic wireless sensor networks. Taking Fig. 1 for an example, the dotted lines indicate the geometric distance between nodes A

K. Arai et al. (Eds.): IntelliSys 2018, AISC 868, pp. 1404–1421, 2019.
https://doi.org/10.1007/978-3-030-01054-6_97

and B, while the solid lines denote the corresponding hop distances between these two nodes. If the sensor nodes are distributed evenly, the geometric distances will roughly match the hop distances (refer to Fig. 1(a)); however, this matching relationship will be totally undermined if they have some obstacles which bend the paths between nodes A and B, leading to a mismatch between geometric distances and hop distances (refer to Fig. 1(b)). A large deviation between the geometric distances and hop distances can lead to a sharp drop of positioning accuracy. This deviation is exactly the reason that causes the performance of multi-hop range-free localization methods so poor in anisotropic networks.

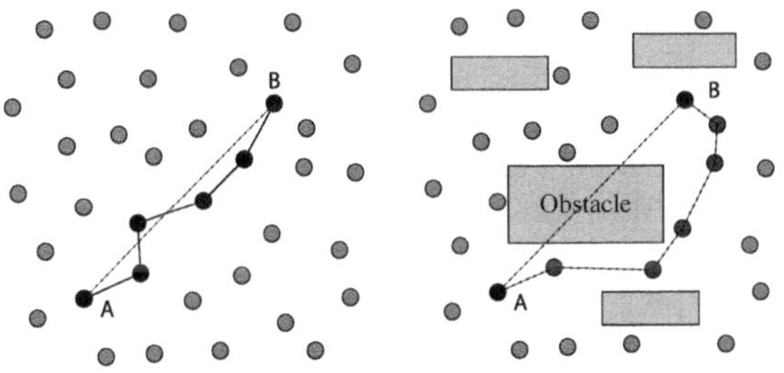

(a) an isotropic network (b) an anisotropic network

Fig. 1. Relationship between geometric distance and distance.

To improve the positioning accuracy in such anisotropic networks, we propose a multi-hop range-free localization method, i.e., MLAND (Multi-hop Localization Algorithm with Network Division). The main idea of the algorithm is to divide a large anisotropic network into several sub-networks and to keep the sensor nodes distribute uniformly in the sub-networks as far as possible. Specifically, this algorithm uses a prediction algorithm to estimate the number of sub-networks Q, then the whole network is divided into Q sub-networks in terms of a division algorithm; after that, multi-hop localization algorithm is performed simultaneously in the total of Q sub-networks; finally, localization results of Q sub-networks are combined together.

The existing localization algorithms based on network division do not predict the number of sub-networks Q, which needs the users to assign the value thus greatly reducing the scalability and practicability of these algorithms. In addition, the existing network division algorithms only use anchor nodes' location information; Different from them, the network division method in this paper uses hop-count information of all nodes and produces a better result. Compared with these existing multi-hop localization algorithms, the innovation of our method has two main points: one is to predict the number of sub-networks Q; the other is the network division algorithm, which will divide a whole network into Q optimal sub-networks.

The rest of this paper is organized as follows. Related works are briefly introduced in Sect. 2. Section 3 describes the proposed method MLAND in detail. In Sect. 4, extensive simulation experiments are conducted to evaluate the performance of MLAND and make comparisons with other classical localization solutions. Finally, Sect. 5 gives some concluding remarks.

2 Related Works

As far as the current existing WSN localization algorithms are concerned, the multi-hop localization algorithms have been widely used due to their economies and simplicities. Especially, three localization algorithms including DV-hop [4], Amorphous [5, 6] and MDS-MAP [7] have attracted widespread attention. These multi-hop localization algorithms show good localization performance in the dense and well-distributed WSNs. However, the realistic network environment is often complex and volatile, causing a sharp decrease in positioning accuracy of these algorithms.

The MDS-MAP algorithm was first proposed by Shang et al. in Missouri-Columbia University. This algorithm applies the Multi-Dimensional Scaling (MDS) statistics method to the localization issue of sensor networks [7]. Then Shang et al. improved the MDS-MAP algorithm and proposed a distributed localization algorithm MDS-MAP(P) [8]. The main idea of MDS-MAP(P) is to divide the whole network into several sub-graphs (local maps), then the relative coordinates of each sub-graph are obtained using MDS method; finally, the relative coordinates are converted to global absolute coordinates. Jennifer Hou proposed PDM (Proximity-Distance Map) algorithm, which modified the measured distance between the sensor nodes, and it is no requirements on the signal transmission model and node deployment density, but the calculation of correction distance depends on the distribution of anchor nodes [9]. Wang Chen et al. improved the traditional multi-hop localization algorithm and proposed the i-multihop localization method, which eliminated the error distance estimation of unknown nodes and anchor nodes, thus improving the positioning accuracy [10]. Liu Yunhao et al. proposed a range-free localization algorithm called REP (Rendered Path) for anisotropic networks. The algorithm can locate nodes in anisotropic networks by only a small number of anchor nodes [11]. Oh-Heum Kwon et al. proposed Map Stitching algorithm based on local maps, which is also targeted at anisotropic regions. The algorithm can effectively avoid node flip and improve the positioning accuracy of anisotropic regions [12]. Gao Ji et al. proposed an incremental localization algorithm, i.e. Incremental Delaunay Refinement [14], which is based on Connectivity-based localization algorithm [13]. The algorithm improves the selection mechanism of anchor nodes, and it can be used in the irregular region even the density of nodes is low. Xiao Qingjun et al. proposed a multi-hop localization algorithm based on pattern driven for anisotropic sensor networks [15]. They put forward the idea of distributed pattern classification to accurately estimate the shortest path distance between anchor nodes and unknown nodes. Inspired by the PDM method, Lee et al. [16] put forward the LSVR and MSVR methods, which regard the hop distance as a nonlinear relationship with the Euclidean distance, and predict the Euclidean distance between nodes using the SVM regression method. Recently, some researchers put forward improved DV-hop algorithms based on Voronoi diagram, which divide a network into several regions, and then locate the unknown nodes in each region [17, 18].

In this paper, an improved multi-hop localization method is proposed. Its positioning accuracy can meet the requirement for most of WSN applications while reducing the requirement of node hardware and the cost greatly. Moreover, this range-free localization method has no extra requirement for WSN nodes and is characterized by such advantages as simple realization, low cost, and a small payment.

3 Description of Our Method

This chapter first takes a typical C-shaped anisotropic network as an example to discuss the necessity of dividing the network into several sub-networks. Then, based on this example, we give the detailed steps of our MLAND algorithm.

On referring to Fig. 2(a), it is a C-shaped anisotropic network. The red symbol '*' represents the anchor nodes while the blue symbol 'o' denotes the unknown nodes. The target of positioning is to locate the unknown sensor nodes using the known coordinates of the anchor nodes. Figure 2(d) depicts the relationship between the distances and hops of anchor nodes. It is obvious that the distances and hops meet a positive proportional relationship when the number of hops is less than 5. With the increase of the number of hops, the positive proportional relationship is gradually destroyed. In addition, Fig. 2(d) shows that the distance divergence becomes the largest when the hop number is 11. Large distance divergence will inevitably lead to the decrease of positioning accuracy. Figure 2(b) shows that the C-shaped anisotropic network has been divided into two sub-networks, in which the red '*' and 'o' symbols represent the first sub-network while the blue symbols indicate the second sub-network. Symbol '×' denotes the center of each sub-network. In Fig. 2(e), the red line depicts the relationship between distances and hops of the first sub-network while the blue line describes that of the second. It is easy to see that the distances are positive proportional to the hops. Similarly, Fig. 2(c) and (f) demonstrate that they can also well meet the proportional relationship when the anisotropic network has been divided into three sub-networks.

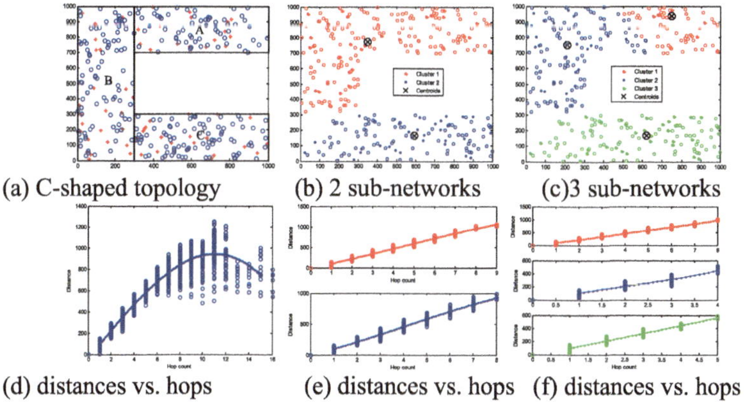

(a) C-shaped topology (b) 2 sub-networks (c)3 sub-networks

(d) distances vs. hops (e) distances vs. hops (f) distances vs. hops

Fig. 2. Example of dividing a network into 2 or 3 sub-networks.

Through the above analysis, we can draw a conclusion that though the network topology in Fig. 2(a) is irregular, it has good characteristics in several local networks, for instances, network A, B and C. In these sub-networks, the distribution of nodes is uniform, and it is suitable to individually perform multi-hop localization algorithms in each sub-network. Based on the above observations, this paper put forward a multi-hop localization method based on the idea of network division. If a wireless sensor network has the good characteristics of regions, it can be divided into several sub-networks which are suitable to use multi-hop localization method individually. Specifically, this localization method needs to meet two conditions: (1) the target network must have regional characteristics, and it must be able to be divided into several sub-networks, in which nodes distribute uniformly and densely; (2) anchor nodes have a certain proportion in each sub-network, for the multi-hop localization method needs enough anchors to complete the operation of location.

Next, two problems will be solved. One is how to determine the number of sub-networks Q, and the other is how to divide the entire network into Q sub-networks. In Sect. 3.1, we will give the prediction algorithm of value Q, and discuss the network division algorithm in Sect. 3.2.

3.1 Prediction Algorithm of Value Q

In the analysis of the topology of WSNs, it is of great significance to estimate the number of sub-networks Q. Selection of reasonable Q will be beneficial to the localization algorithm to run successfully and improve the positioning accuracy. The common methods to predict Q including Akaike Information Criterion (AIC) [19], Bayes Information Criterion (BIC), and Minimum Description Length (MCL) principle [20]. In this section, the BIC is used to evaluate the Gauss distribution fitting of sensor nodes in order to find the optimal value of Q.

The Akaike information criterion (AIC) is an estimator of the relative quality of statistical models for a given set of data. It was formulated by the statistician Hirotugu Akaike. Suppose that we have a statistical model of some data. Let N be the number of estimated parameters in the model. Let $L(\hat{\theta}|y)$ be the maximum value of the likelihood function for the model. Then the AIC value of the model is given by

$$AIC(m) = -2\log L\left(\widehat{\theta}\,|\,y\right) + 2N \tag{1}$$

In a group of alternative models, the one that minimizes AIC is a desirable model, but AIC usually fails when sample data is large. If the value of likelihood function is too large, the model parameter N will be drowned. Therefore, in order to make up for the lack of AIC, Schwarz puts forward the BIC, and its formula is as follows:

$$BIC(m) = -2\log L\left(\widehat{\theta}\,|\,y\right) + N\ln n \tag{2}$$

where n is the number of samples. Compared with the AIC, the second item of BIC is replaced by $N\ln n$. When fitting models, it is possible to increase the likelihood by adding parameters, but doing so may result in overfitting. Both BIC and AIC attempt to resolve this problem by introducing a penalty term for the number of parameters in the model; the penalty term is larger in BIC than in AIC. More importantly, the BIC considers the effect of the sample data and it is more obvious to distinguish model than the AIC. Therefore, in this paper, the BIC is used to evaluate the Gauss distribution fitting model.

Specifically, the procedure of clustering is as follows:

(a) Calculate the cluster radius R. Refer to Fig. 2, it is easy to find that selecting reasonable Q can guarantee the proportional relationship between distances and hops. The size of sub-networks is associated with the cluster radius R. We choose the first inflection point of distance-hop curves as the threshold value of hop, i.e. cluster radius R. Figure 3 shows the selection of R under several network topologies.

(b) Let $\mathbf{M} = \{M_i | i = 1, \ldots, m\}$ denote the set of m anchors. The distance between anchor node M_i and M_j can be formulated as:

$$d(M_i, M_j) = \sqrt{(x_i - x_j)^2 + (y_i - y_j)^2} \tag{3}$$

where (x_i, y_i) and (x_j, y_j) represent the coordinates of node M_i and M_j, respectively.

(c) Find the neighbor set of anchor node M_i, i.e. $N(i) = \{d(M_i, M_j) < R \times per\_hop | j \neq i, j = 1, \ldots, m\}$. Here per_hop indicates the average distance of one hop.

(d) Generate cluster $C(n) = \max\{N(i)|i = 1, \ldots, m\}$, $(n = 1, 2, \ldots)$. At the same time, the cluster center T_n and the cluster variance V_n are calculated as follows:

$$T(n) = \frac{1}{N(n)} \sum_{M_j \in C(n)} M_j \tag{4}$$

$$V(n) = \frac{1}{N(n)} \sum_{M_j \in C(n)} d(M_j, T(n)) \tag{5}$$

After that, find the remaining anchor nodes set $C_{\cup}M = \mathbf{M} - \mathbf{C(n)}$.

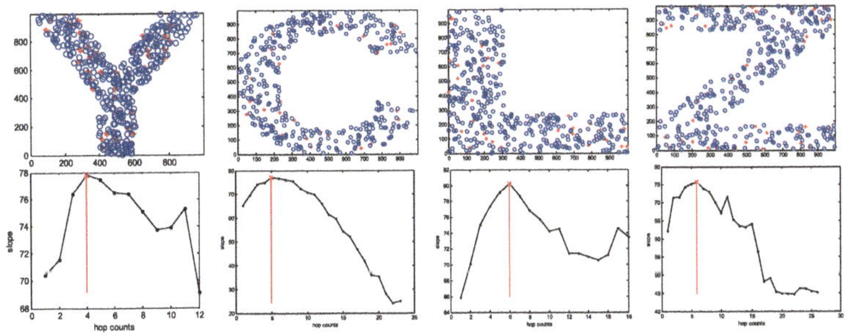

Fig. 3. Selection of cluster radius R.

(e) Repeat steps b, c, d for the set of remaining anchors until the number of remaining anchors is 0, so as to get Q clustering sets $C = \{C(i)|i = 1,\ldots,Q\}$.
(f) If $\min\{C(i)|i = 1,\ldots,Q\} < = 3$, it means that anchors of the smallest cluster are less than 3. Join the smallest cluster into the nearest cluster and repeats this process until $\min\{C(i)|i = 1,\ldots,Q\} > 3$.

Suppose that the Q clusters have the same distribution type, the model density function of the Q clusters is as follows:

$$f(V_n) = \sqrt{\frac{N(n)}{(v_{\max} - v_{\min})}} \times \frac{Q}{m}, \; n = 1, 2, \ldots, Q \tag{6}$$

where V_{max} and V_{min} represent the largest and smallest variance of the Q clusters, respectively. According to the maximum likelihood estimation principle, under the condition of clustering center T(n) as the parameter, the maximum likelihood estimation function of the Q clusters is as follows:

$$\ln L(V_n \mid T_1, T_2, \ldots, T_Q) = Q \ln \frac{Q}{m} + \frac{1}{2} \sum_{n=1}^{Q} \frac{N(n)}{v_{\max} - v_{\min}} \tag{7}$$

Substituting (7) into (2) yields

$$BIC = -2(Q \ln \frac{Q}{m} + \frac{1}{2} \sum_{n=1}^{Q} \frac{N(n)}{v_{\max} - v_{\min}}) + Q \ln m \tag{8}$$

We can acquire the BIC of Fig. 2(a) in terms of (8), as shown in Table 1.

It is obvious that the minimal BIC is 30.2 when Q is 3, which means that it is reasonable to divide the network into three sub-networks, as shown in Fig. 2(a).

Table 1. BIC of various Q

Q	2	3	4	5	6	7	8
BIC	35.1	30.2	36.5	41.8	46.2	50.2	53.9

3.2 Algorithm of Network Division

After using our Prediction Algorithm to predict the number of clusters Q, we still need to divide the sensor nodes into Q clusters. Considering the characteristics of multi-hop localization methods, we use the Spectral Clustering method to conduct the procedure of clustering since the hop matrix is already known. The Spectral Clustering algorithm is based on spectrum theory. Compared with the traditional clustering algorithms, Spectral Clustering algorithm has the advantage of converging to the global optimal solution.

Next, we describe the procedure of our Network Division Algorithm. Suppose that there are N nodes in the set $\mathbf{X} = \{x_1, \ldots, x_N\}$ to be divided into Q clusters. The detail of Network Division Algorithm is as follows.

(1) Construct the graph G(V, E) so that the vertex V_i of G corresponds to each node x_i. The weight w_{ij} of the edge E_{ij} represents the similarity between nodes x_i and x_j.
(2) Use Gaussian similarity function (9) to construct a similarity matrix **W** of set **X**.

$$s(x_i, x_j) = \exp(-\|x_i - x_j\|^2 / 2\sigma^2) \tag{9}$$

Where σ is a predefined parameter.
(3) *Let*

$$d_i = \sum_{j=1}^{N} w_{ij} \tag{10}$$

Construct a degree matrix **D** for each vertex, here **D** is a diagonal matrix and $\mathbf{D} = \mathrm{diag}(d_1, \ldots, d_n)$. Note that d_i represents the degree of node x_i.
(4) Calculate the unnormalized graph Laplacian matrix **L**, **L** = **D** − **W**.
(5) Calculate the least Q eigenvectors of **L**, and then arrange the Q eigenvectors together to form a matrix of N × Q, each of which is treated as a vector in Q-dimensional space; finally, K-means algorithm is used to perform clustering on the matrix N × Q.

3.3 Details of MLAND

Combined with the Prediction Algorithm in Sect. 3.1 and the Network Division Algorithm in Sect. 3.2, the proposed MLAND in this paper mainly includes the following five main steps:

Step 1. Collect Information of anchor nodes and the nodes to be located

Consider a wireless sensor network $S = \{S_1, S_2, \ldots, S_N\}$ which contains N nodes with M anchor nodes and $N{-}M$ unknown nodes ($M \ll N - M$). The location of each node can be expressed as:

$$pos(S_i) = (x_i, y_i)^T, i = 1, \ldots, N \tag{11}$$

In S, the position of M anchors $S_i \in A$ is known, while the position of the remaining $N{-}M$ nodes $S_j \in B$ is unknown, here $A = \{S_i | i = 1, 2, \ldots, M\}$ and $B = \{S_j | j = M + 1, M + 2, \ldots, M + N\}$.

Step 2. Use the Prediction Algorithm to predict reasonable Q

In this step, the reasonable value of Q will be calculated in terms of the Prediction Algorithm which is described in Sect. 3.1.

Step 3. Use the Network Division Algorithm to Divide the anchors into Q clusters

For the nodes in set $X = A \cup B$, we measure their pairwise similarities $w_{ij} = s(x_i, x_j)$ by similarity function (9) which is symmetric and non-negative, and we denote the corresponding similarity matrix by $\mathbf{W} = w_{ij}, (i, j = 1, \ldots, N)$. The pseudo-code of the clustering algorithm is as follows:

Algorithm 1. Network Division Algorithm

Input: similarity matrix $W \in \hbar^{N \times N}$, number of cluster Q
output: Clusters $E_1, \ldots, E_Q$
1 Construct the degree matrix D according to formula (10);
2 Calculate the unnormalized Laplacian matrix L=D-W ;
3 Calculate the least Q eigenvectors $u_1, \ldots, u_Q$ of L;
4 Let $U \in \hbar^{N \times Q}$ be the **matrix** containing the vectors $u_1, \ldots, u_Q$ as columns;
5 For i=1,...,N, let $y_i \in \hbar^{Q}$ be the vector corresponding to the i-th row of U;
6. Cluster the points (yi)i=1,...,n in $\hbar^{Q}$ with the k-**means** algorithm into clusters $C_1, \ldots, C_Q$.
7. Output Clusters $E_1, \ldots, E_Q$ with $E_i = \{j | y_j \in C_i\}$.

Step 4. Perform multi-hop localization algorithm on Q clusters

The hop counts from node S_i to S_j can be expressed as $h(S_i, S_j) \in \mathbf{H}$, where $\mathbf{H}$ represents the hop matrix. The matrix $\mathbf{H}$ contains the information of hop counts h from each node to every anchor, and its dimension is $N \times N$.

Suppose that the multi-hop localization algorithm is performed on the cluster $E_i (E_i \in \mathbf{E} = \{i = 1, 2, \ldots, Q\})$, other clusters and so on.

Firstly, the hop count matrix of the E_i cluster is extracted from the matrix $\mathbf{H}$, and set as $\mathbf{H_i}$, which contains information about the number of hops between all the nodes in the E_i cluster.

Then, the geometric distance of per hop i.e. d_{PER} in the E_i cluster can be calculated as

$$d_{per} = \frac{\sum_{i \neq j} \sqrt{(x_i - x_j)^2 + (y_i - y_j)^2}}{\sum_{i \neq j} h_i} \tag{12}$$

where (x_i, y_i) and (x_j, y_j) denote the positions of node i and node j, respectively; and h_i is the hop counts from anchor i to all the other anchors.

Next, suppose that there exists one sensor node S_j and total n anchor nodes $S_i (i = 1, \ldots, n)$ in the cluster E_i, and their coordinates are marked as $S_j(x, y)$ and $S_1(x_1, y_1)$,

$S_2(x_2, y_2)$, …, $S_n(x_n, y_n)$, respectively. The distances between S_j to S_i are $d_1, d_2, \ldots, d_n$. Then the distance equations as can be written as:

$$\begin{cases} (x-x_1)^2+(y-y_1)^2=d_1^2 \\ (x-x_2)^2+(y-y_2)^2=d_2^2 \\ \vdots \\ (x-x_n)^2+(y-y_n)^2=d_n^2 \end{cases} \tag{13}$$

Introducing $z = x^2 + y^2$, we have

$$\begin{cases} z-2x_1x-2y_1y=d_1^2-(y_1^2+x_1^2) \\ z-2x_2x-2y_2y=d_2^2-(y_2^2+x_2^2) \\ \vdots \\ z-2x_nx-2y_ny=d_n^2-(y_n^2+x_n^2) \end{cases} \tag{14}$$

Let

$$\mathbf{A}=\begin{bmatrix} 1 & -2x_1 & -2y_1 \\ 1 & -2x_2 & -2y_2 \\ \vdots & \vdots & \vdots \\ 1 & -2x_n & -2y_n \end{bmatrix} \tag{15}$$

$$\mathbf{b}=\begin{bmatrix} d_1^2 & -x_1^2 & -y_1^2 \\ d_2^2 & -x_2^2 & -y_2^2 \\ & \vdots & \\ d_n^2 & -x_n^2 & -y_n^2 \end{bmatrix} \tag{16}$$

$$\mathbf{x}=\begin{bmatrix} z \\ x \\ y \end{bmatrix} \tag{17}$$

The above equations can be transformed into the form $\mathbf{Ax} = \mathbf{b}$. Using the Least Squares Method, we can get the coordinate of the node S_j which is given by

$$\hat{\mathbf{x}} = (\mathbf{A}^T\mathbf{A})^{-1}\mathbf{A}^T\mathbf{b} \tag{18}$$

where $\hat{\mathbf{x}}$ represents the estimated location of node S_j.

After doing the same process to the rest Q−1 clusters, we can obtain estimated location of all the clusters. Let $\mathbf{X_i}$ stand for the estimated location of cluster E_i.

$$X_i = \begin{bmatrix} \hat{x}_1 & \hat{y}_1 \\ \hat{x}_2 & \hat{y}_2 \\ \vdots & \vdots \\ \hat{x}_m & \hat{y}_m \end{bmatrix} \tag{19}$$

Step 5. Merge estimated location results of Q clusters

The final location result is given by

$$X = \begin{bmatrix} X_1 \\ X_1 \\ \vdots \\ X_Q \end{bmatrix} \tag{20}$$

4 Experimental Results and Analysis

This chapter will evaluate the performance of the MLAND method. The evaluation metrics and some parameters of the simulation are given in Sect. 4.1. The simulation results are given in Sect. 4.2 and a practical experiment is conducted in Sect. 4.3.

4.1 Simulation Parameters

We evaluate the performance of the MLAND method and compare it with other existing localization algorithms, including DV-hop, Amorphous, MDS-MAP, PDM, and LSVR. A metric-Average Localization Error (ALE) is introduced

$$ALE = \frac{\sum_{i=1}^{N} Location\,Error_i}{N} \tag{21}$$

where N represents the number of all nodes in the network including the anchor nodes and the unknown nodes. Generally, the smaller the ALE is, the smaller the average deviation between the estimated position and the real position of the node. This also means that the estimated position is closer to the real position, the positioning accuracy is higher, and the performance is better. For any unknown node i in the network, its location error i.e. Localization Error_i can be represented as

$$Location\,Error_i = \frac{\|\hat{x}_i - x_i\|_2}{R} \times 100\% \tag{22}$$

where x_i and $\hat{x}_i$ represent the real and estimated location of node i, respectively. R is the communication radius and $\|\bullet\|_2$ denotes the Euclidean distance between pair-wise nodes.

The network topologies used in the simulation include L-shaped and C-shaped, as shown in Table 2. A total of 300 nodes are randomly distributed in these networks of which the size is 1000 × 1000 m. We use two kinds of simulation strategies. The first one is that the ratio of anchor nodes is fixed to 20% while the communication radius of nodes varies from 100 m to 300 m. The second is that the communication radius is fixed to 150 m while the ratio of anchor nodes varies from 10% to 30%. Performance of the MLAND method will be evaluated under both strategies and compared with other algorithms.

Table 2. Topologies in simulations

	L-shaped network	C-shaped network
random deployment	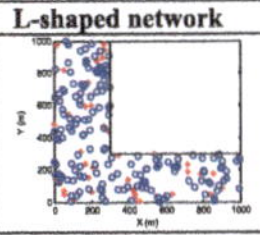	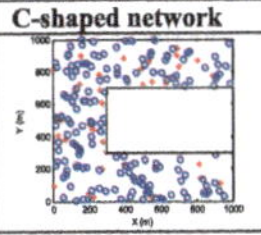

4.2 Simulation Results

The MLAND method demonstrates that the L-shaped and C-shaped topologies should be divided into two and three sub-networks, respectively. Figure 4 shows the results.

Table 3 shows the ALE results under the first simulation strategy, and these results are the average over 100 trials. The number of anchor nodes is set to 60(20%), and the communication radius R increases from 100 m to 300 m. Taking the MLAND method as an example, the ALE decreases with the increase of R, no matter whether it is an L-shaped topology or a C-shaped topology. In the case of DV-hop, Amorphous, and MDS-MAP algorithm, the influence of topology on ALE is very obvious, while PDM,

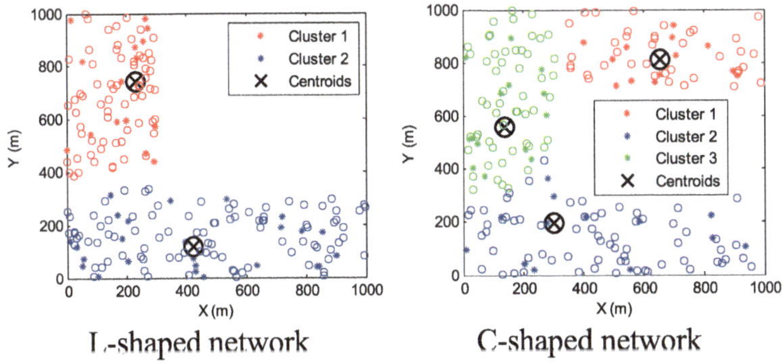

Fig. 4. Division results of L-shaped and C-shaped topology.

LSVR, and MLAND are relatively stable. Simulation results prove that MLAND method can improve the positioning accuracy of multi-hop localization algorithms by simply dividing the network into several parts.

Figure 5 shows the data in Table 3 as curves. MLAND method outperforms DV-hop algorithm in both L-shaped and C-shaped topology.

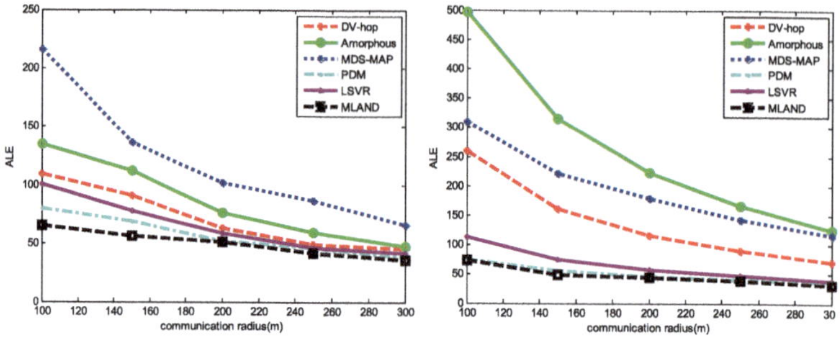

(a) L-shaped network (b) C-shaped network

Fig. 5. ALE vs. R

Table 3. ALE changes with communication radius R

Topology	R(m)	DV-hop	Amorphous	MDS-MAP	PDM	LSVR	MLAND
L-shaped	100	109.5	134.6	215.7	80.4	100.5	**65.1**
	150	91.3	112.4	136.5	69.3	78.0	**56.7**
	200	62.9	76.1	102.3	52.1	58.9	**51.9**
	250	49.3	59.4	86.7	46.5	46.2	**42.1**
	300	45.6	48.0	66.2	38.6	42.1	**36.4**
C-shaped	100	259.2	496.1	308.9	75.7	113.4	**74.5**
	150	161.0	314.7	221.7	57.0	75.6	**48.6**
	200	116.2	222.4	179.3	45.9	57.5	**44.3**
	250	89.6	165.8	142.5	40.5	47.7	**38.2**
	300	71.4	125.1	115.7	37.2	38.8	**32.1**

Table 4 shows the ALE results under the second simulation strategy, and these results are also the average over 100 trials. The communication radius R is set to 150 m, and the number of anchor nodes M is increased from 30(10%) to 90(30%).

Figure 6 shows the curves of those data which are listed in Table 4. It can be seen from the figure that as the number of anchor nodes M increases from 30(10%) to 90 (30%), the ALE changes only a little in the case of these various localization algorithms. The MLAND shows the lowest ALE in both L-shaped and C-shaped topology. Compared with the DV-hop algorithm, ALE of the MLAND has decreased greatly due to dividing the irregular L-shaped and C-shaped topologies into regular sub-networks, thus avoiding the large distance errors and improving the positioning accuracy.

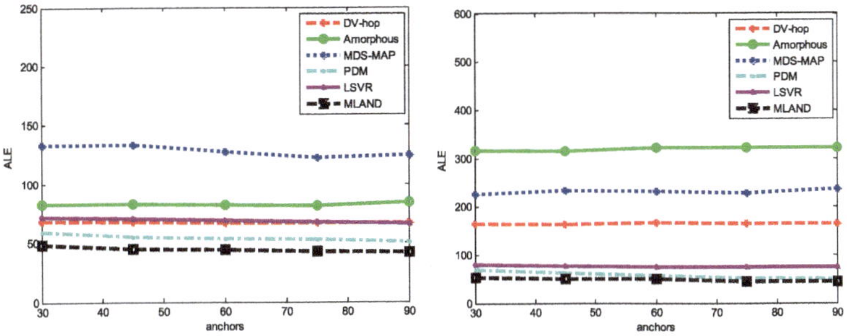

(a) L-shaped network (b) C-shaped network

Fig. 6. ALE vs. M.

Table 4. ALE changes with number of anchors M

Topology	M	DV-hop	Amorphous	MDS-MAP	PDM	LSVR	MLAND
L-shaped	30	68.1	83.0	133.6	59.8	71.7	**48.3**
	45	67.3	82.7	133.2	55.1	70.3	**45.2**
	60	66.9	82.4	127.4	53.8	69.0	**44.3**
	75	66.6	81.8	122.7	52.9	67.4	**42.6**
	90	67.2	84.4	124.6	50.6	66.0	**41.9**
C-shaped	30	163.3	316.4	225.2	68.8	80.6	**54.1**
	45	163.1	315.2	233.3	62.0	77.1	**49.7**
	60	165.4	322.0	230.6	57.4	74.7	**49.5**
	75	164.0	321.1	226.4	52.5	74.6	**45.3**
	90	164.3	321.4	236.3	50.8	74.9	**45.5**

Figure 7 shows the location errors of the six localization algorithms when the communication radius R = 150 m and the number of anchor nodes M = 60.

The symbol red '*' represents the position of the anchor node, and the symbol blue 'o' represents the estimated position of the unknown node. The blue lines point from the estimated position to the real position, so the length of these lines depicts the size of the errors. It can be seen that the MLAND method performs best.

4.3 Experimental Results

In order to evaluate the localization performance of MLAND in the real environment, we select a vacant site in our school and use a car to act as an obstacle within the locating area as shown in Fig. 7. In this experiment, the size of the positioning area is about 25 m × 25 m. A total of 40 nodes are randomly distributed in the positioning area. These nodes are fixed on the 1.3 m-high camera shelf and equipped with the IEEE 802.15.4 compatible transceiver chip TI CC2450, which works in the 2.4 GHz band. Figure 8 gives a snapshot of the nodes used in the experiment.

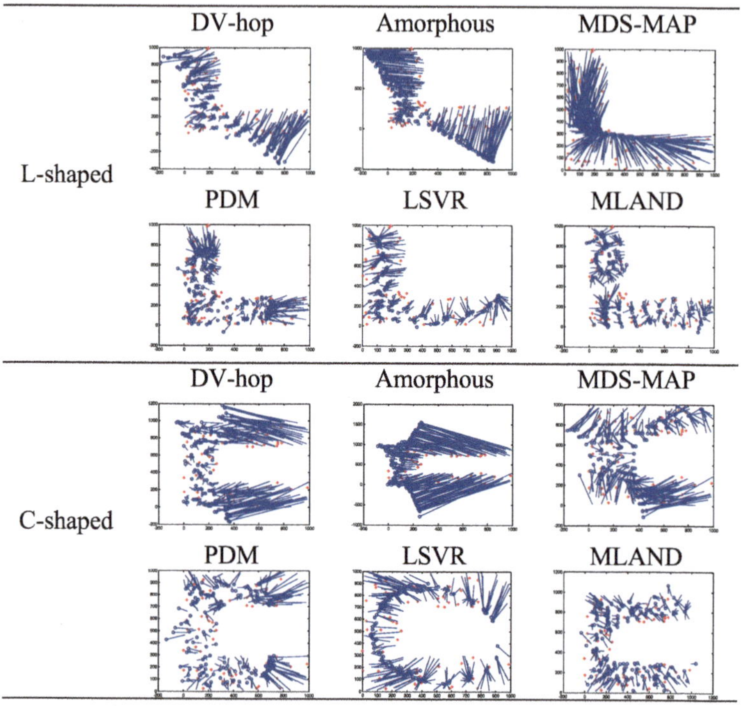

Fig. 7. Location errors of six localization algorithms.

(a) The experimental environment (b) a snap of the nodes

Fig. 8. Photos of environment and nodes.

In the experiment, the communication radius R is set to 5 m, and the number of anchor nodes M is increased from 4(10%) to 12(30%). We plot the ALE results of all the six localization algorithms within Fig. 9. Consistent with the simulation results, the MLAND method still maintains a lower positioning error in the real environment.

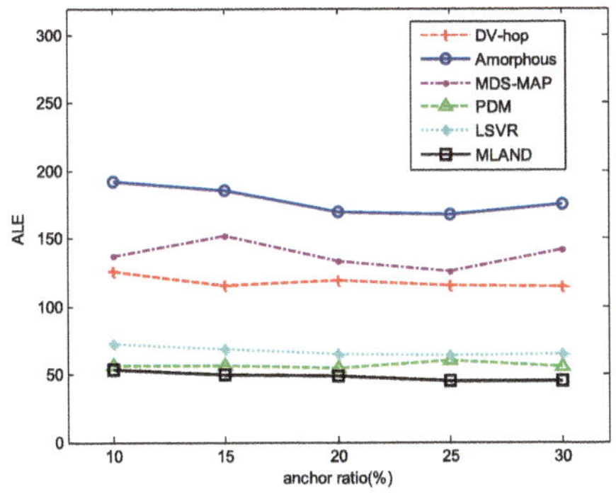

Fig. 9. Performance comparison with different algorithms.

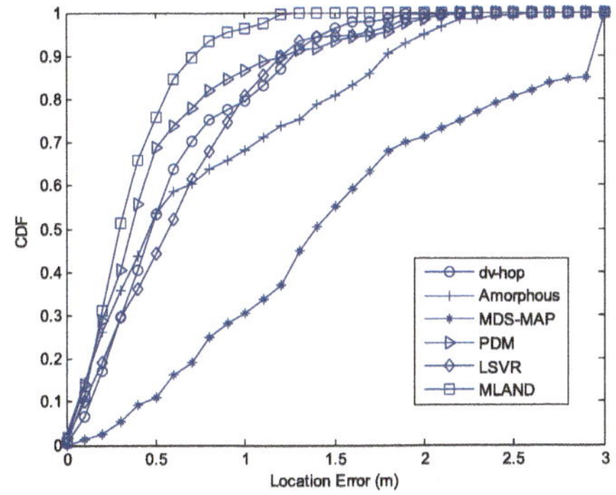

Fig. 10. CDF with different algorithms.

In addition, we also use the Cumulative Distribution Function (CDF) to describe the performance of these localization algorithms, as shown in Fig. 10. If anchor ratio is 20%, we can see that CDF is 60, 83, and 98% when location error is 0.5, 0.8, and 1.5 m, respectively.

5 Conclusions

In this paper, we propose a multi-hop localization method MLAND based on network division. For an anisotropic network with regional features, it can be divided into several sub-networks by using the network division algorithm; and these sub-networks have good network characteristics and are suitable for using the traditional multi-hop positioning algorithm. MLAND method uses the prediction algorithm to predict the optimal number of sub-regions Q, and then uses the network division algorithm to divide the whole network into Q sub-networks, and finally combines the positioning results of Q sub-networks. Compared with the similar algorithms, MLAND has greatly improved the positioning accuracy.

Acknowledgment. This work is sponsored by Doctoral Scientific Research Startup Foundation of Jinling Institute of Technology (JIT-B-201429), Scientific Research Foundation of Jinling Institute of Technology (JIT-rcyj-201505, JIT-2016-jlxm-20), and the Natural Scientific Research Funds for Jiangsu Universities (Nos. 17KJB520008, 17KJA520001) and the Funds for Innovation Team of Nanjing City.

References

1. Nguyen, C.L., Georgiou, O., Doi, Y.: Maximum likelihood based multihop localization in wireless sensor networks. In: IEEE International Conference on Communications, vol. 62, pp. 6663–6668. IEEE (2015)
2. Assaf, A.E., Zaidi, S., Affes, S., Kandil, N.: Low-cost localization for multihop heterogeneous wireless sensor networks. IEEE Trans. Wirel. Commun. **15**(1), 472–484 (2016)
3. Gober, P., Ziviani, A., Todorova, P., Amorim, M.D.D., Hünerberg, P., Fdida, S.: Topology control and localization in wireless ad hoc and sensor networks. Ad Hoc & Sens. Wirel. Netw. **1**(1) (2015)
4. Niculescu, D., Nath, B.: DV based positioning in ad hoc networks. Telecommun. Syst. **22**(1–4), 267–280 (2003)
5. Nagpal, R., Shrobe, H., Bachrach, J.: Organizing a global coordinate system from local information on an ad hoc sensor network. In: Proceedings of the Information Processing in Sensor Networks, Second International Workshop, IPSN 2003, Palo Alto, CA, USA, April 22–23, 2003, vol. 2634, pp. 333–348 (2003)
6. Years, I.R.: Amorphous localization algorithm based on bp artificial neural network. Int. J. Distrib. Sens. Netw. **2015**, 6 (2015)
7. Shang, Y., Ruml, W., Zhang, Y., Fromherz, M.P.J.: Localization from mere connectivity. In: ACM International Symposium on Mobile Ad Hoc Networking & Computing, pp. 201–212. ACM (2003)
8. Shang, Y., Rumi, W., Zhang, Y., Fromherz, M.: Localization from connectivity in sensor networks. IEEE Trans. Parallel Distrib. Syst. **15**(11), 961–974 (2004)
9. Lim, H., Hou, J.C.: Distributed localization for anisotropic sensor networks. ACM Trans. Sens. Netw. **5**(2), 1–26 (2009)
10. Wang, C., Xiao, L.: Locating sensors in concave areas. In: IEEE International Conference on Computer Communications, IEEE INFOCOM 2006, pp. 1–12. IEEE (2006)
11. Li, M., Liu, Y.: Rendered path: range-free localization in anisotropic sensor networks with holes. In: International Conference on Mobile Computing & Networking, vol. 18, pp. 51–62. IEEE (2007)
12. Kwon, O.H., Song, H.J.: Localization through map stitching in wireless sensor networks. IEEE Trans. Parallel Distrib. Syst. **19**(1), 93–105 (2007)
13. Lederer, S., Wang, Y., Gao, J.: Connectivity-based localization of large scale sensor networks with complex shape. In: The Conference on Computer Communications, INFOCOM 2008, vol. 5, pp. 1–32. IEEE (2009)
14. Wang, Y., Lederer, S., Gao, J.: Connectivity-based sensor network localization with incremental delaunay refinement method. In: INFOCOM, pp. 2401–2409. IEEE (2009)
15. Xiao, Q., Xiao, B., Cao, J., Wang, J.: Multihop range-free localization in anisotropic wireless sensor networks: a pattern-driven scheme. IEEE Trans. Mob. Comput. **9**(11), 1592–1607 (2010)

16. Lee, J., Chung, W., Kim, E.: A new range-free localization method using quadratic programming. Comput. Commun. **34**(8), 998–1010 (2011)
17. Zhang, Y., Xiang, S., Fu, W., Wei, D.: Improved normalized collinearity dv-hop algorithm for node localization in wireless sensor network. Int. J. Distrib. Sens. Netw. **2014**(11), 1–14 (2015)
18. Sun, X., Hu, Y., Wang, B., Zhan, J., Li, T.: Vpit: an improved range-free localization algorithm using Voronoi diagrams for wireless sensor networks. Int. J. Multimedia Ubiquitous Eng. **10**(8), 23–34 (2015)
19. Akaike, H.: Information theory and an extension of the maximum likelihood principle. In: International Symposium on Information Theory, vol. 1, pp. 610–624 (2011)
20. Figueiredo, M.A.T., Jain, A.K.: Unsupervised learning of finite mixture models. IEEE Trans. Pattern Anal. Mach. Intell. **24**(3), 381–396 (2002)

Author Index

K. Arai et al. (Eds.): IntelliSys 2018, AISC 868, pp. 1423–1426, 2019.
https://doi.org/10.1007/978-3-030-01054-6

GPSR Compliance

The European Union's (EU) General Product Safety Regulation (GPSR) is a set of rules that requires consumer products to be safe and our obligations to ensure this.

If you have any concerns about our products, you can contact us on ProductSafety@springernature.com

In case Publisher is established outside the EU, the EU authorized representative is:

Springer Nature Customer Service Center GmbH
Europaplatz 3
69115 Heidelberg, Germany

Batch number: 10761593

Printed by Printforce, the Netherlands